CHAPTER 3 POLYNOMIAL AND RATIONAL FUNCTIONS

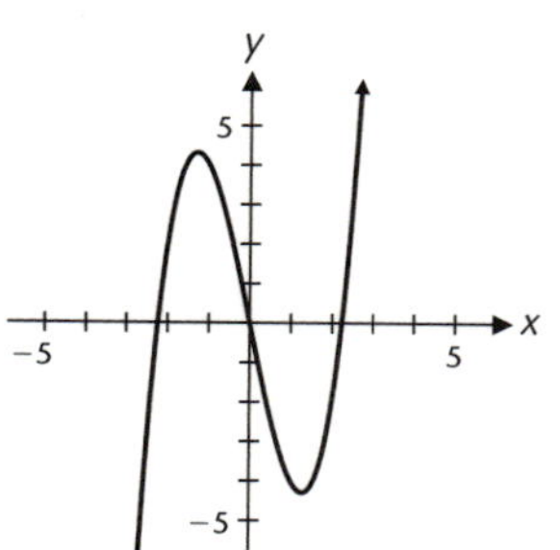

Third-degree polynomial
$g(x) = x^3 - 5x$

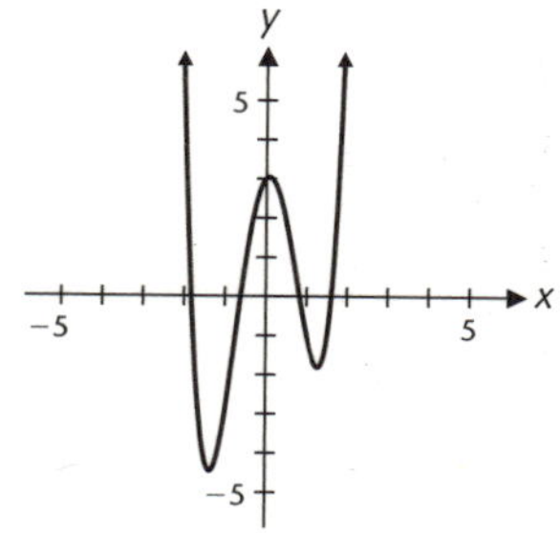

Fourth-degree polynomial
$G(x) = 2x^4 - 7x^2 + x + 3$

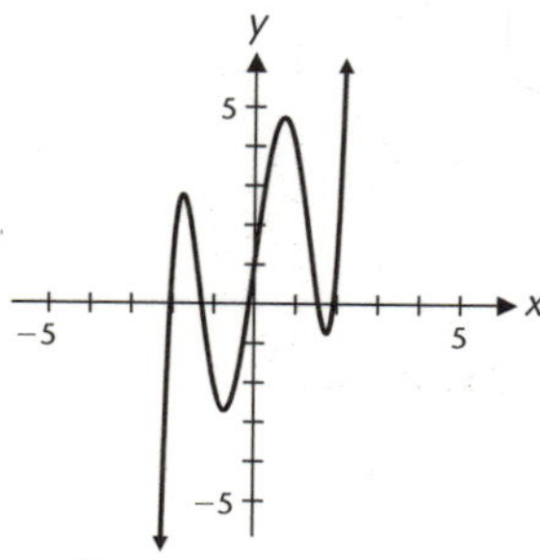

Fifth-degree polynomial
$h(x) = x^5 - 6x^3 + 8x + 1$

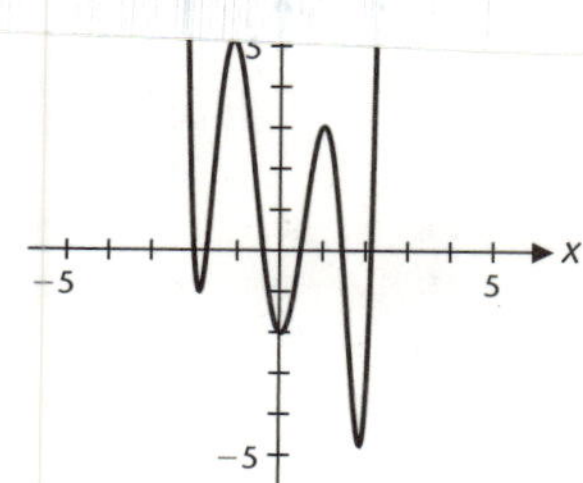

Sixth-degree polynomial
$H(x) = x^6 - 7x^4 + 12x^2 - x - 2$

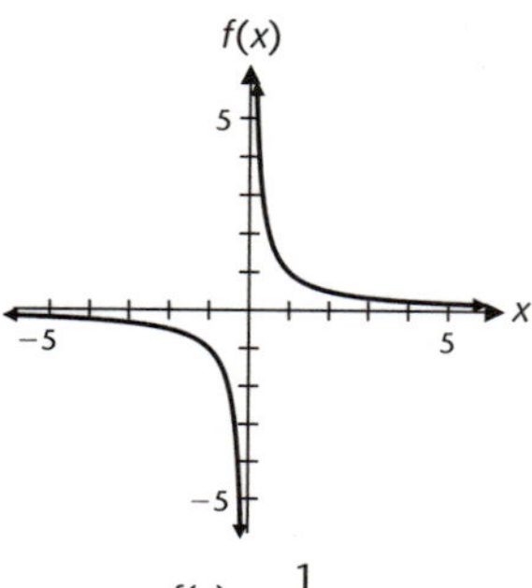

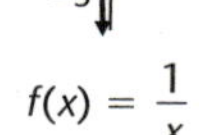

$f(x) = \dfrac{1}{x}$

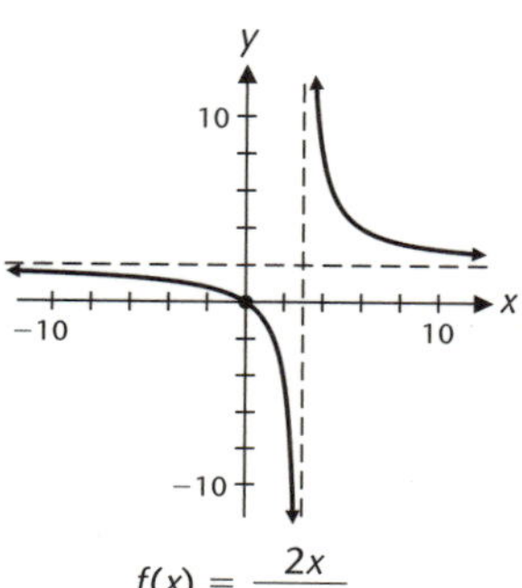

$f(x) = \dfrac{2x}{x - 3}$

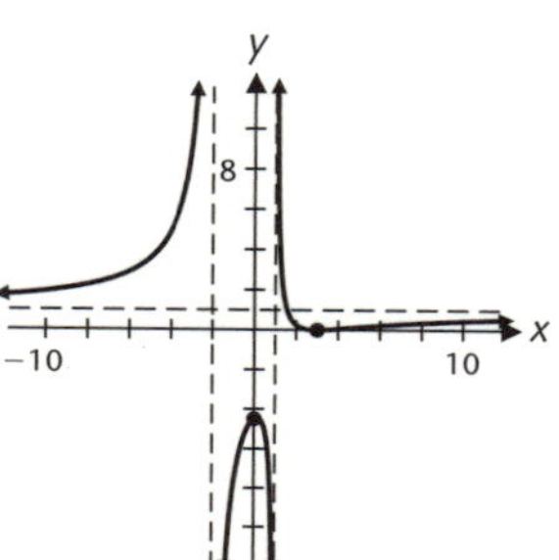

$f(x) = \dfrac{x^2 - 6x + 9}{x^2 + x - 2}$

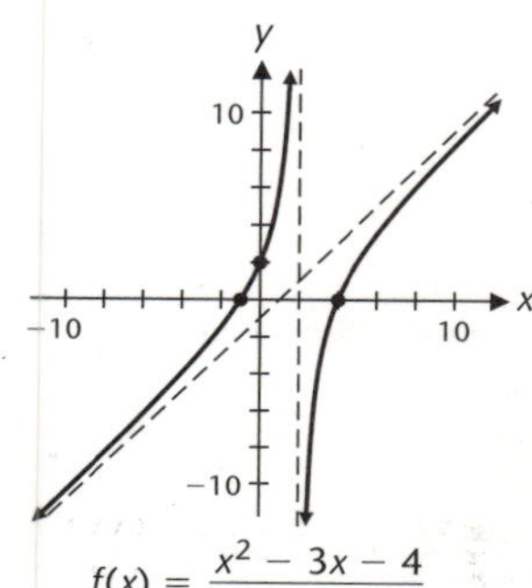

$f(x) = \dfrac{x^2 - 3x - 4}{x - 2}$

CHAPTER 4 EXPONENTIAL AND LOGARITHMIC FUNCTIONS

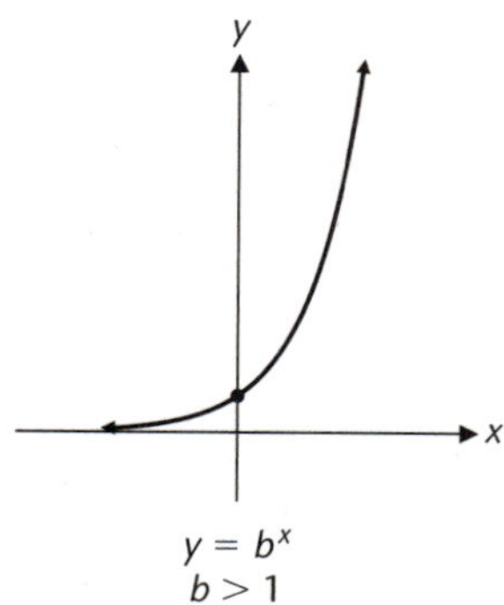

$y = b^x$
$b > 1$

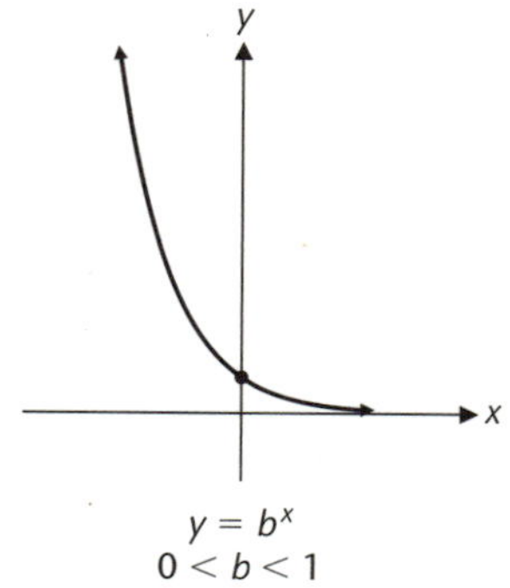

$y = b^x$
$0 < b < 1$

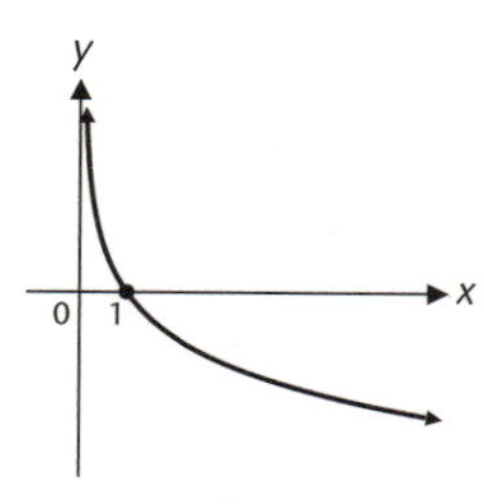

$y = \log_b x$
$0 < b < 1$

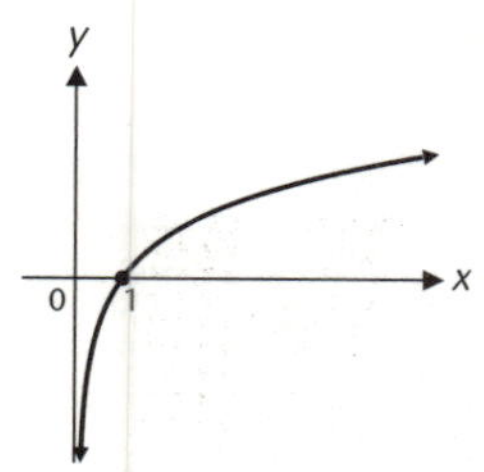

$y = \log_b x$
$b > 1$

CHAPTER 5 TRIGONOMETRIC FUNCTIONS

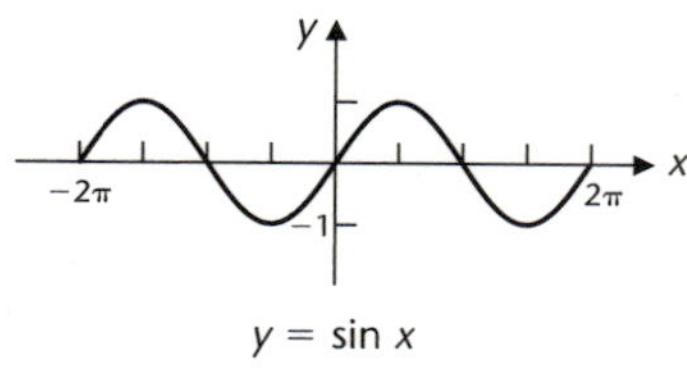

$y = \sin x$

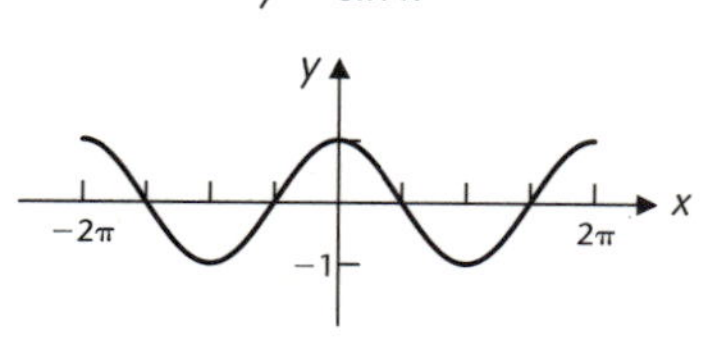

$y = \cos x$

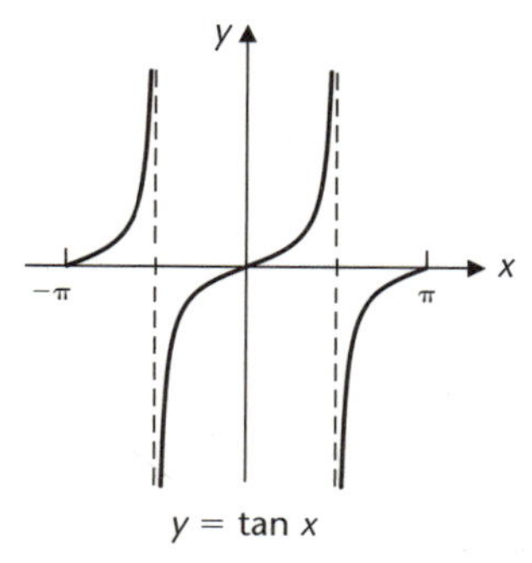

$y = \tan x$

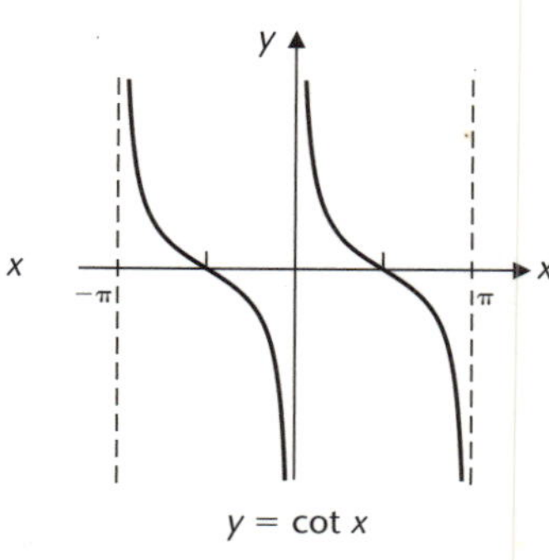

$y = \cot x$

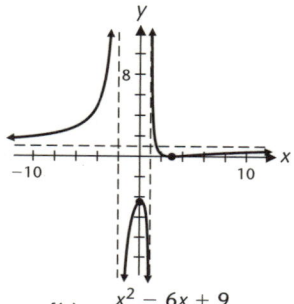

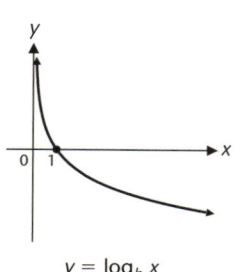

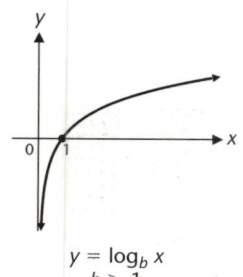

Inequalities and Intervals (A.1)

$a < b$	a is less than b
$a \le b$	a is less than or equal to b
$a > b$	a is greater than b
$a \ge b$	a is greater than or equal to b
(a, b)	Open interval; $\{x \mid a < x < b\}$
$(a, b]$	Half-open interval; $\{x \mid a < x \le b\}$
$[a, b)$	Half-open interval; $\{x \mid a \le x < b\}$
$[a, b]$	Closed interval; $\{x \mid a \le x \le b\}$

Quadratic Formula (2.5)

If $ax^2 + bx + c = 0, a \neq 0$, then

$$x = \frac{-b \pm \sqrt{b^2 - 4ac}}{2a}$$

Rectangular Coordinates (A.2, A.3, 2.1)

(x_1, y_1)	Coordinates of point P_1
$d = \sqrt{(x_2 - x_1)^2 + (y_2 - y_1)^2}$	Distance between $P_1(x_1, y_1)$ and $P_2(x_2, y_2)$
$\left(\dfrac{x_1 + x_2}{2}, \dfrac{y_1 + y_2}{2}\right)$	Midpoint of line joining P_1 and P_2
$m = \dfrac{y_2 - y_1}{x_2 - x_1}, \quad x_1 \neq x_2$	Slope of line through P_1 and P_2

Arithmetic Sequence (10.3)

$a_1, a_2, \ldots, a_n, \ldots$

$a_n - a_{n-1} = d$	Common difference
$a_n = a_1 + (n - 1)d$	nth-term formula
$S_n = a_1 + \cdots + a_n = \dfrac{n}{2}[2a_1 + (n - 1)d]$	Sum of n terms
$S_n = \dfrac{n}{2}(a_1 + a_n)$	

Geometric Sequence (10.3)

$a_1, a_2, \ldots, a_n, \ldots$

$\dfrac{a_n}{a_{n-1}} = r$	Common ratio
$a_n = a_1 r^{n-1}$	nth-term formula
$S_n = a_1 + \cdots + a_n = \dfrac{a_1 - a_1 r^n}{1 - r}, \quad r \neq 1$	Sum of n terms
$S_n = \dfrac{a_1 - r a_n}{1 - r}, \quad r \neq 1$	
$S_\infty = a_1 + a_2 + \cdots = \dfrac{a_1}{1 - r}, \quad \lvert r \rvert < 1$	Sum of infinitely many terms

Permutations and Combinations (10.4, 11.6)

$n! = n(n - 1) \cdots 2 \cdot 1, \quad n \in N$ $\quad$ n factorial

$0! = 1$

For $0 \le r \le n$,

$P_{n,r} = \dfrac{n!}{(n - r)!}$	Permutation
$C_{n,r} = \dbinom{n}{r} = \dfrac{n!}{r!(n - r)!}$	Combination
$(a + b)^n = \displaystyle\sum_{k=0}^{n} \binom{n}{k} a^{n-k} b^k, \quad n \ge 1$	Binomial formula

Circle (A.3)

$(x - h)^2 + (y - k)^2 = r^2$	Center at (h, k); radius r
$x^2 + y^2 = r^2$	Center at $(0, 0)$; radius r

Parabola (2.3, 11.1)

$y^2 = 4ax,$	$a > 0$, opens right; $a < 0$, opens left Focus: $(a, 0)$; Directrix: $x = -a$; Axis: x axis
$x^2 = 4ay,$	$a > 0$, opens up; $a < 0$, opens down Focus: $(0, a)$; Directrix: $y = -a$; Axis: y axis

Ellipse (11.2)

$\dfrac{x^2}{a^2} + \dfrac{y^2}{b^2} = 1,$	$a > b > 0$ Foci: $F'(-c, 0)$, $F(c, 0)$; $c^2 = a^2 - b^2$
$\dfrac{x^2}{b^2} + \dfrac{y^2}{a^2} = 1,$	$a > b > 0$ Foci: $F'(0, -c)$, $F(0, c)$; $c^2 = a^2 - b^2$

Hyperbola (11.3)

$\dfrac{x^2}{a^2} - \dfrac{y^2}{b^2} = 1$	Foci: $F'(-c, 0)$, $F(c, 0)$; $c^2 = a^2 + b^2$
$\dfrac{y^2}{a^2} - \dfrac{x^2}{b^2} = 1$	Foci: $F'(0, -c)$, $F(0, c)$; $c^2 = a^2 + b^2$

Translation Formulas (11.4)

$x = x' + h, y = y' + k; \quad x' = x - h, y' = y - k$

New origin (h, k)

Trigonometric Functions (5.2)

Let x be a real number and let (a, b) be the coordinates of the circular point $W(x)$ that lies on the terminal side of the angle with radian measure x. Then:

$\sin x = b \qquad \csc x = \frac{1}{b} \quad b \neq 0$

$\cos x = a \qquad \sec x = \frac{1}{a} \quad a \neq 0$

$\tan x = \frac{b}{a} \quad a \neq 0 \qquad \cot x = \frac{a}{b} \quad b \neq 0$

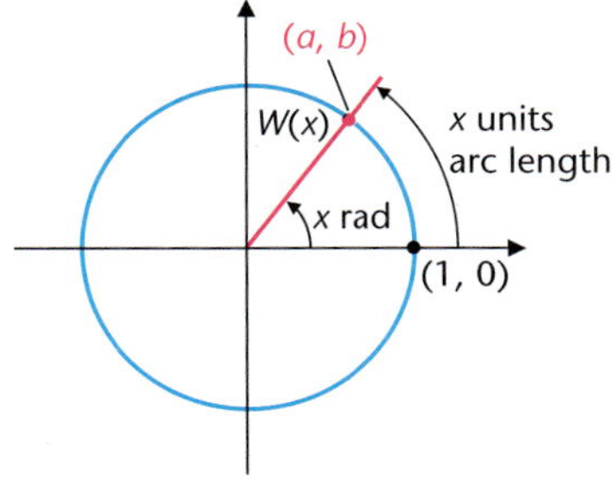

Amplitude, Period, and Phase Shift (5.5)

$y = A \sin (Bx + C) \qquad y = A \cos (Bx + C)$

For $B > 0$,

Amplitude $= |A| \qquad$ Period $= \frac{2\pi}{B} \qquad$ Phase shift $= -\frac{C}{B}$

$y = A \tan (Bx + C) \qquad y = A \cot (Bx + C)$

For $B > 0$,

Period $= \frac{\pi}{B} \qquad$ Phase shift $= -\frac{C}{B}$

Inverse Trigonometric Functions (5.6)

$y = \sin^{-1} x$ means $x = \sin y$, where $-\pi/2 \le y \le \pi/2$ and $-1 \le x \le 1$

$y = \cos^{-1} x$ means $x = \cos y$, where $0 \le y \le \pi$ and $-1 \le x \le 1$

$y = \tan^{-1} x$ means $x = \tan y$, where $-\pi/2 < y < \pi/2$ and x is any real number

Significant Digits (5.3, 7.1, 7.2)

Angle to Nearest	Significant Digits for Side Measure
1°	2
10′ or 0.1°	3
1′ or 0.01°	4
10″ or 0.001°	5

Law of Sines (7.1)

$$\frac{\sin \alpha}{a} = \frac{\sin \beta}{b} = \frac{\sin \gamma}{c}$$

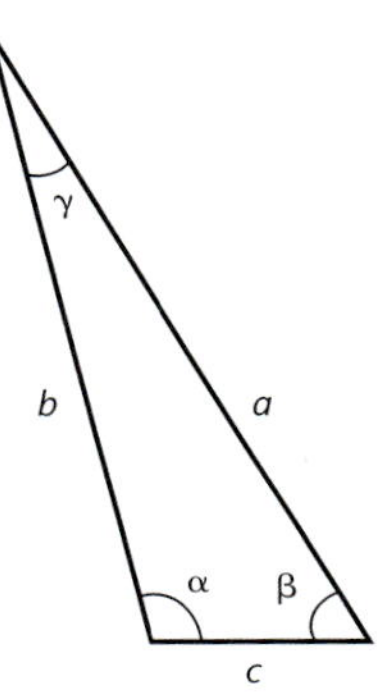

Law of Cosines (7.2)

$$a^2 = b^2 + c^2 - 2bc \cos \alpha$$
$$b^2 = a^2 + c^2 - 2ac \cos \beta$$
$$c^2 = a^2 + b^2 - 2ab \cos \gamma$$

If $\gamma = 90°$, then

$c^2 = a^2 + b^2$ Pythagorean theorem

Polar Coordinates (7.5)

$$r^2 = x^2 + y^2$$
$$x = r \cos \theta$$
$$y = r \sin \theta$$
$$\tan \theta = y/x$$

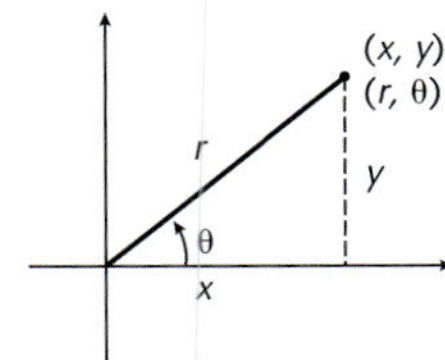

Polar Form of a Complex Number (7.6)

$$x + iy = r[\cos (\theta + 2k\pi) + i \sin (\theta + 2k\pi)]$$
$$= re^{i(\theta + 2k\pi)},\ k \text{ an integer}$$

De Moivre's Theorem (7.7)

nth power of z:

$$z^n = (x + iy)^n = (re^{i\theta})^n = r^n e^{in\theta}$$

nth roots of z:

$$r^{1/n} e^{(\theta/n + k \cdot 360°/n)i},\ k = 0, 1, \ldots, n - 1$$

Trigonometric Identities (5.2, 6.1–6.4)

Reciprocal Identities (5.2, 6.1)

$$\csc x = \frac{1}{\sin x} \qquad \sec x = \frac{1}{\cos x} \qquad \cot x = \frac{1}{\tan x}$$

Quotient Identities (6.1)

$$\tan x = \frac{\sin x}{\cos x} \qquad \cot x = \frac{\cos x}{\sin x}$$

Identities for Negatives (6.1)

$$\sin(-x) = -\sin x \qquad \cos(-x) = \cos x$$
$$\tan(-x) = -\tan x$$

Pythagorean Identities (6.1)

$$\sin^2 x + \cos^2 x = 1 \qquad \tan^2 x + 1 = \sec^2 x$$
$$1 + \cot^2 x = \csc^2 x$$

Sum Identities (6.2)

$$\sin(x + y) = \sin x \cos y + \cos x \sin y$$
$$\cos(x + y) = \cos x \cos y - \sin x \sin y$$
$$\tan(x + y) = \frac{\tan x + \tan y}{1 - \tan x \tan y}$$

Difference Identities (6.2)

$$\sin(x - y) = \sin x \cos y - \cos x \sin y$$
$$\cos(x - y) = \cos x \cos y + \sin x \sin y$$
$$\tan(x - y) = \frac{\tan x - \tan y}{1 + \tan x \tan y}$$

Cofunction Identities (6.2)

$$\sin\left(\frac{\pi}{2} - y\right) = \cos y \qquad \tan\left(\frac{\pi}{2} - y\right) = \cot y$$
$$\sin(90° - \theta) = \cos\theta \qquad \tan(90° - \theta) = \cot\theta$$
$$\cos\left(\frac{\pi}{2} - y\right) = \sin y$$
$$\cos(90° - \theta) = \sin\theta$$

Double-Angle Identities (6.3)

$$\sin 2x = 2\sin x \cos x$$
$$\cos 2x = \begin{cases} \cos^2 x - \sin^2 x \\ 1 - 2\sin^2 x \\ 2\cos^2 x - 1 \end{cases}$$
$$\tan 2x = \frac{2\tan x}{1 - \tan^2 x} = \frac{2\cot x}{\cot^2 x - 1} = \frac{2}{\cot x - \tan x}$$

Half-Angle Identities (6.3)

$$\sin\frac{x}{2} = \pm\sqrt{\frac{1 - \cos x}{2}}$$
$$\cos\frac{x}{2} = \pm\sqrt{\frac{1 + \cos x}{2}}$$

Signs are determined by the quadrant in which $x/2$ lies

$$\tan\frac{x}{2} = \frac{1 - \cos x}{\sin x} = \frac{\sin x}{1 + \cos x} = \pm\sqrt{\frac{1 - \cos x}{1 + \cos x}}$$

Product-Sum Identities (6.4)

$$\sin x \cos y = \tfrac{1}{2}[\sin(x + y) + \sin(x - y)]$$
$$\cos x \sin y = \tfrac{1}{2}[\sin(x + y) - \sin(x - y)]$$
$$\sin x \sin y = \tfrac{1}{2}[\cos(x - y) - \cos(x + y)]$$
$$\cos x \cos y = \tfrac{1}{2}[\cos(x + y) + \cos(x - y)]$$

Sum-Product Identities (6.4)

$$\sin x + \sin y = 2\sin\frac{x + y}{2}\cos\frac{x - y}{2}$$
$$\sin x - \sin y = 2\cos\frac{x + y}{2}\sin\frac{x - y}{2}$$
$$\cos x + \cos y = 2\cos\frac{x + y}{2}\cos\frac{x - y}{2}$$
$$\cos x - \cos y = -2\sin\frac{x + y}{2}\sin\frac{x - y}{2}$$

PRECALCULUS

Barnett, Ziegler, and Byleen's Precalculus Series

College Algebra, Seventh Edition
This book is the same as *College Algebra with Trigonometry* without the three chapters on trigonometry.
ISBN 0-07-236868-3

College Algebra with Trigonometry, Seventh Edition
This book is the same as *College Algebra* with three chapters of trigonometry added. Comparing *College Algebra with Trigonometry* with *Precalculus, College Algebra with Trigonometry* has more intermediate algebra review and starts trigonometry with angles and right triangles.
ISBN 0-07-236869-1

Precalculus: Functions and Graphs, Fifth Edition
This book differs from *College Algebra with Trigonometry* in that *Precalculus* starts at a higher level, placing intermediate algebra review in the appendix; and it starts trigonometry with the unit circle and circular functions.
ISBN 0-07-236871-3

College Algebra: Graphs and Models, Second Edition
This book is the same as *Precalculus: Graphs and Models* without the three chapters on trigonometry. This text assumes the use of a graphing utility.
ISBN 0-07-242428-1

Precalculus: Graphs and Models, Second Edition
This book is the same as *College Algebra: Graphs and Models* with three additional chapters on trigonometry. The trigonometric functions are introduced by a unit circle approach. This text assumes the use of a graphing utility.
ISBN 0-07-242430-3

College Algebra with Trigonometry: Graphs and Models
This book is the same as *Precalculus: Graphs and Models* except that the trigonometric functions are introduced by right triangle trigonometry. This text assumes the use of a graphing utility.
ISBN 0-07-291699-0

PRECALCULUS

GRAPHS AND MODELS

RAYMOND A. BARNETT
Merritt College

MICHAEL R. ZIEGLER
Marquette University

KARL E. BYLEEN
Marquette University

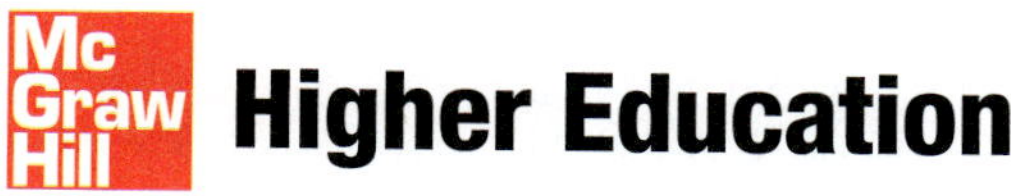

Boston Burr Ridge, IL Dubuque, IA Madison, WI New York San Francisco St. Louis
Bangkok Bogotá Caracas Kuala Lumpur Lisbon London Madrid Mexico City
Milan Montreal New Delhi Santiago Seoul Singapore Sydney Taipei Toronto

The McGraw-Hill Companies

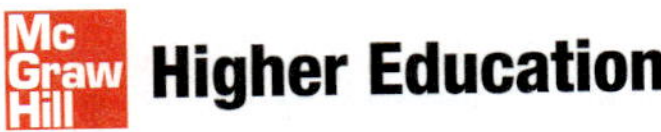

PRECALCULUS: GRAPHS AND MODELS

Published by McGraw-Hill, a business unit of The McGraw-Hill Companies, Inc., 1221 Avenue of the Americas, New York, NY 10020.

This book is printed on acid-free paper.

3 4 5 6 7 8 9 0 VNH/VNH 0 9 8 7 6

ISBN 0–07–242430–3

ISBN 978–0–07–242430–0

Publisher: *William K. Barter*

Executive editor: *Robert E. Ross*

Director of development: *David Dietz*

Executive marketing manager: *Marianne C. P. Rutter*

Lead project manager: *Susan J. Brusch*

Production supervisor: *Kara Kudronowicz*

Media project manager: *Sandra M. Schnee*

Senior media technology producer: *Jeff Huettman*

Design manager: *K. Wayne Harms*

Cover/interior designer: *Rosa & Wesly, Inc.*

Lead photo research coordinator: *Carrie K. Burger*

Supplement producer: *Brenda A. Ernzen*

Compositor: *The GTS Companies/York, PA Campus*

Typeface: *10.5/12 Times Roman*

Printer: *Von Hoffmann Corporation*

Cover image: *Frank Stella, Eskimo Curlew, 1976. Portland Art Museum, Portland, Oregon. Funds from Mr. and Mrs. Howard Vollum. © 2004 Frank Stella/Artists Rights Society (ARS), New York*

Photo credits:

Chapter Opening Photos: 1: © SS11/PhotoDisc; 2: © Vol. 102/Corbis; 3: © Vol. 8/PhotoDisc; 4: © Vol. 8/PhotoDisc; 5: © Vol. 6/PhotoDisc; 6: © Vol. 3/PhotoDisc; 7: © Vol. 19/PhotoDisc; 8: © Vol. 4/PhotoDisc; 9: © Vol. 101/Corbis; 10: © Vol. 7/PhotoDisc; 11: © Vol. 19/PhotoDisc

Library of Congress Cataloging-in-Publication Data

Barnett, Raymond A.

Precalculus : graphs and models / Raymond A. Barnett, Michael R. Ziegler, Karl E. Byleen.—1st ed.

p. cm.

Includes index.

ISBN 0–07–242430–3 (acid-free paper)

1. Functions. 2. Functions—Graphic methods. I. Ziegler, Michael R. II. Byleen, Karl. III. Title.

QA331.3.B394 2005 2003059296

515—dc22 CIP

www.mhhe.com

About the Authors

Raymond A. Barnett, a native of and educated in California, received his B.A. in mathematical statistics from the University of California at Berkeley and his M.A. in mathematics from the University of Southern California. He has been a member of the Merritt College Mathematics Department and was chairman of the department for four years. Associated with four different publishers, Raymond Barnett has authored or co-authored 18 textbooks in mathematics, most of which are still in use. In addition to international English editions, a number of the books have been translated into Spanish. Co-authors include Michael Ziegler, Marquette University; Thomas Kearns, Northern Kentucky University; Charles Burke, City College of San Francisco; John Fujii, Merritt College; and Karl Byleen, Marquette University.

Michael R. Ziegler received his B.S. from Shippensburg State College and his M.S. and Ph.D. from the University of Delaware. After completing postdoctoral work at the University of Kentucky, he was appointed to the faculty of Marquette University where he currently holds the rank of Professor in the Department of Mathematics, Statistics, and Computer Science. Dr. Ziegler has published more than a dozen research articles in complex analysis and has co-authored more than a dozen undergraduate mathematics textbooks with Raymond Barnett and Karl Byleen.

Karl E. Byleen received his B.S., M.A., and Ph.D. degrees in mathematics from the University of Nebraska. He is currently an Associate Professor in the Department of Mathematics, Statistics, and Computer Science of Marquette University. He has published a dozen research articles on the algebraic theory of semigroups and co-authored more than a dozen undergraduate mathematics textbooks with Raymond Barnett and Michael Ziegler.

Contents

4 Exponential and Logarithmic Functions 317

5 Trigonometric Functions 385

6 Trigonometric Identities and Conditional Equations 473

7 Additional Topics in Trigonometry 529

Preface

A Legacy of Success

We take great satisfaction from the fact that more than 100,000 students have learned college algebra or precalculus from a Barnett Series textbook. Ray Barnett is one of the masters of college textbook writing. His central approach is proven and remains effective for today's students.

The Barnett Series **maximizes student comprehension** by emphasizing computational skills, ideas, and problem solving rather than mathematical theory. Completely **worked examples** are used to introduce concepts and to demonstrate problem-solving techniques. These examples are then followed by a similar **matched problem** that the student can work. Answers to the matched problems are located at the end of each section for easy reference. This **active involvement** in the learning process helps students develop a thorough understanding of concepts and processes.

Precalculus: Graphs and Models is the second, retitled edition of *Precalculus: A Graphing Approach.* The word *models* was added to better describe the teaching approach. More than simply a reworking of our widely used *Precalculus: Functions and Graphs*, Fifth Edition, this textbook presents an approach grounded in use of technology and employing data analysis (regression) and modeling techniques.

A Central Theme

In the Barnett Series the **function concept** serves as a unifying theme. A brief look at the table of contents will reveal this emphasis. A major objective of this book is the development of a **library of elementary functions,** including their important properties and uses. Employing this library as a basic working tool, students will proceed through this book with greater confidence and understanding.

Balance by Design

Although technology is employed throughout, we strive to balance **algebraic skill development** with **use of technology as an aid to learning and problem solving.** Additionally, a major objective of the book is to encourage students to investigate mathematical principles and processes graphically and numerically, as well as algebraically. In this way, students gain a broader, deeper, and more useful understanding.

Side-by-Side Presentation

Many solved examples in the book provide **graphical solutions side-by-side** with algebraic solutions. By seeing the same answer result from their symbol manipulations *and* from graphical approaches, students gain insight into the power of algebra and make important conceptual and visual connections. See, for instance, Example 6 on page 27, Example 4 on page 139, Example 5 on page 369, and Example 3 on page 662. Complementing this graphical perspective, we take a **numerical approach** to key concepts when it aids understanding—as in our discussion of asymptotes on page 292.

Data Analysis (Regression) and Modeling

Another important purpose of *Precalculus: Graphs and Models* is to give students significant experience in **data analysis and modeling** of real-world problems. Our accessible presentations of regression and curve fitting, while omitting the theoretical background, give students a feel for the relationship between data that arise in actual situations and the functions they are studying: linear, quadratic, polynomial, exponential, logarithmic, and trigonometric. Enough modeling-oriented applications are included to convince even the most skeptical student that mathematics really is useful—see the Applications Index on the inside back cover, and examples on pages 141 to 142, 188 to 189, and 339 to 340.

Included within most exercise sets—at the ends of sections and chapters—are sets of **Applications** problems and **Modeling and Data Analysis** problems. Furthermore, the **Group Activity** after each chapter usually employs modeling and/or starts with data. See the annotated sample pages in the Features walk-through following this preface for information on these features.

Technology Use

The generic term *graphing utility* is used to refer to any of the various graphing calculators or computer programs that might be available to a student using this book. We assume that each student has easy access to a graphing utility that can perform the following operations:

- Simultaneously display multiple graphs in a user-selected viewing window
- Explore graphs using trace and zoom
- Approximate roots and intersection points
- Approximate maxima and minima
- Plot data sets and find associated regression equations
- Perform basic matrix operations, including row reduction and inversion

Most popular graphing calculators perform all of these operations. The majority of the graphing utility images in this book are "screen dumps" from a Texas Instruments TI-83 Plus graphing calculator. Students not using that TI calculator should be able to produce similar results on any calculator or software meeting the requirements listed. The proper use of such utilities is covered in Section 1.1.

Explore/Discuss Boxes

Interspersed throughout each section, **Explore/Discuss boxes** foster conceptual understanding by asking students to think about a relationship or process before a result is stated. Verbalization of mathematical concepts, results, and processes is strongly encouraged in these explanations and activities.

Review Material

Because the backgrounds and preparation of students in precalculus vary widely, we have provided extensive and flexible review material both within Appendix A and on the Online Learning Center that supports this book. Our **Basic Algebra Review** provides explanations, examples, and exercises in Adobe Acrobat (PDF) format. Additional algorithmically generated practice problems and quizzes and audiovisual instruction are also available for online study. Topics covered include properties of numbers, polynomial operations and factoring, operations on rational expressions, integer and rational exponents, and radicals.

Appendix A of this text reviews basics of linear equations, the Cartesian coordinate system, and analytic geometry formulas: these topics, which are essential for a course emphasizing graphing, can be covered systematically before commencing with the rest of the book or reviewed as needed.

A Final Thought

We consider every student enrolling in a precalculus course to be a potential mathematics major. We appreciate that many students take this course simply to fulfill a requirement. However, while we have these students' attention, we hope to excite an interest in the larger, wonderful world of mathematics. We hope this text assists instructors in their efforts to do the same.

Changes for the Second Edition

Based on abundant advice from users of the first edition, and from numerous reviewers of the book and manuscript, we have made numerous improvements to the pedagogy, organization, and coverage of *Precalculus: Graphs and Models* for this second edition.

Greater Emphasis on Graphing Technology Since we completed the first edition, we have continued to construct (often with the help of reviewers) new and better ways of using the power of graphing utilities to enhance learning. Many new graphing-based teaching approaches, examples, and exercises appear in this edition.

More Modeling Applications Translating real-world problems into the language of mathematics—especially, moving from a data set to a function approximation—is an important skill for any student, whether or not they will move on to a calculus course. This central theme of the text receives greater attention in the second edition, as evidenced by the many new Mathematical Modeling and Modeling and Data Analysis subsections. Many sections and chapters in the second edition end with Modeling and Data Analysis problem sets.

Expanded Coverage of Regression Graphing calculators give students the power to analyze numerical data to find a function that provides a useful model for a given data set. New material on curve fitting appears in a number of chapters; see, for example, the material on logarithmic regression beginning on page 362. Many exercise sets in the second edition include a group of Data Analysis and Regression problems, and most of the Modeling and Data Analysis exercises use regression as well.

Side-by-Side Solutions: Algebraic and Graphical As noted earlier, many solved examples in the second edition provide graphical solutions next to the algebraic solutions. See the Features walk-through following this preface.

Chapter 1 This chapter now includes function operations and inverses, better reflecting the primacy of functions in this course. The Cartesian coordinate system is now covered in an appendix (it was formerly Section 1.1).

Chapter 2 Sections 2.2 and 2.5 now place a greater emphasis on modeling, of linear and quadratic equations, respectively. Linear and quadratic inequalities are covered in a new Section 2.7 (and here, too, the emphasis is on modeling).

Chapter 3 Sections 3.2 and 3.3, which cover zeros of polynomials, have been reorganized. We now treat approximate real zeros first, in Section 3.2, applying the same techniques to solve polynomial inequalities. Rational and complex zeros of polynomials appear in Section 3.3. Rational inequalities are added to Section 3.4.

Chapter 4 Sections 4.1 and 4.2 of the first edition are now covered in Chapter 1. The new Sections 4.2 and 4.4 place a greater emphasis on modeling, as their new titles—"Exponential Models" and "Logarithmic Models"—imply.

Chapter 5 The first of three chapters on trigonometry, have been reorganized and streamlined. The nine sections of the first edition are now covered in six sections. We give an early introduction to the graphs of the trigonometric functions in Section 5.2 and present a thorough analysis of properties and graphs of trigonometric functions in Section 5.4.

Chapters 8 and 9 Because many colleges include determinants and Cramer's rule in their precalculus syllabus, we have added these topics to the second edition. This change necessitated the formation of a new Chapter 9, "Matrices and Determinants," which includes the new topics combined with the material on matrix operations from Chapter 8 of the first edition. Note that the material on determinants and Cramer's rule at the end of Chapter 9 is easily skipped without loss of continuity.

Acknowledgments

In addition to the authors, many others are involved in the development of a manuscript leading to successful publication of a textbook. We wish to thank the following people personally:

James Allis, Elon University
Timothy Beaver, Isothermal Community College
Dave Bergenzer, Utah State University
Susan Bradley, Angelina College
Robert Brown, Montgomery College
Linda Buchanan, Howard College
Wayne Clark, Parkland College
Joe DiCostanzo, Johnson County Community College
Andrew Engelward, Harvard University
Russell Euler, Northwest Missouri State University
Mark Farris, Midwestern State University
Bill Frederick, Indiana University—Purdue University Fort Wayne
David George, Ohio State University

Raghu Gompa, Indiana University Kokomo
Geoffrey Goodson, Towson University
Angie Grant, University of Memphis
Sally Haas, Angelina College
Richard Hill, Michigan State University
Cheryl Kane, University of Nebraska Lincoln
Edward Koslowska, Southwest Texas Junior College
Richard Kubelka, San Jose State University
Janice Lyon, Tallahassee Community College
Roger Marty, Cleveland State University
Laurie McManus, St. Louis Community College
Denise Meeks, Pima Community College
Tom Metzger, University of Pittsburgh
Barbara Sausen, Fresno City College
Ingrid Scott, Montgomery College
Mehrdad Simkani, University of Michigan—Flint
Dave Sobecki, Miami University
Joe Stickles, University of Evansville
Badri Varma, University of Wisconsin—Fox Valley
James Vicich, Scottsdale Community College
Bruce Washington, St. Louis Community College
Steven Winters, University of Wisconsin—Oshkosh
Janet Wyatt, Longview Community College

We also wish to thank the following people:

Hossein Hamedani and Caroline Woods for providing a careful and thorough check of all the mathematical calculations in the textbook, the student's solution manual, and the instructor's answer manual (a tedious but extremely important job).

Dave Sobecki for developing the supplemental manuals that are so important to the success of a text.

Jeanne Wallace for accurately and efficiently producing most of the manuals that supplement the text.

Katherine Aiken for carefully editing and correcting the manuscript.

Susan Brusch for guiding the book smoothly through all publication details.

All the people at McGraw-Hill who contributed their efforts to the production of this book. Producing this new edition with the help of all these extremely competent people has been a most satisfying experience.

—Michael Ziegler and Karl Byleen

Supplements for the Instructor

Instructor's Testing and Resource CD-ROM This cross-platform CD-ROM provides a wealth of resources for the instructor. Supplements featured on this CD-ROM include a computerized test bank using Brownstone Diploma® testing software to quickly create customized exams. This user-friendly program allows instructors to search for questions by topic, format, or difficulty level; edit existing questions or add new ones; and scramble questions and answer keys for multiple versions of the same test.

Instructor's Solutions Manual Prepared by Dave Sobecki of Miami University, this supplement provides detailed solutions to all the exercises in the text. The methods used to solve the problems in the manual are the same as those used to solve the examples in the textbook.

MathZone McGraw-Hill's MathZone is a powerful new online system for homework, quizzing, and testing. The MathZone environment enables assignment of free-response and multiple-choice exercises over the Internet. Exercises are generated algorithmically, to allow for extended practice. Using MathZone, instructors can select algorithmically generated exercises to create multiple versions of assignments and quizzes, with results reported automatically to an online grade book.

ALEKS ALEKS® (Assessment and LEarning in Knowledge Spaces) is an artificial intelligence–based system for individualized math learning, available over the World Wide Web. ALEKS® delivers precise, qualitative diagnostic assessments of students' math knowledge, guides them in the selection of appropriate new study material, and records their progress toward mastery of curricular goals in a robust classroom management system. For more information on ALEKS, see page xxviii.

PageOut PageOut is McGraw-Hill's unique point-and-click course website tool, which enables instructors to create a full-featured, professional-quality course website without knowing HTML coding. With PageOut instructors can post their course syllabus, assign McGraw-Hill Online Learning Center content, add links to important off-site resources, and maintain student results in the online grade book. Instructors can also send class announcements, copy their course site to share with colleagues, and upload original files.

Supplements for the Student

Student's Solutions Manual Prepared by Dave Sobecki of Miami University, the Student's Solutions Manual provides complete worked-out solutions to all the odd-numbered section exercises from the text. The procedures followed in the solutions in the manual match exactly those shown in worked examples in the text.

Online Learning Center Web-based interactive learning is available for students at the Online Learning Center, located at www.mhhe.com/barnett. Student resources are located in the Student Center, and include algorithmically generated practice exams and quizzes, audiovisual tutorials, and web links.

Video Series The video series is composed of a set of videocassettes. An on-screen instructor introduces topics and works through examples using the methods presented in the text. The video series is also available on video CD-ROMs.

NetTutor NetTutor is a revolutionary system that enables students to interact with a live tutor over the World Wide Web. Students can receive instruction from live tutors using NetTutor's web-based, graphical chat capabilities. They can also submit questions and receive answers, browse previously answered questions, and view previous live chat sessions.

Features

Chapter Opener ▶

Each chapter opens with an outline showing the section titles of the chapter, and a chapter introduction that sets the chapter in the context of the entire course. The outline and introduction help students to get their bearings, and instructors to plan lectures and homework. The **Preparing for this chapter** box directs students to review background material from previous courses or chapters required to learn the chapter's content.

CHAPTER 3

Polynomial and Rational Functions

OUTLINE

3.1 Polynomial Functions and Models
3.2 Real Zeros and Polynomial Inequalities
3.3 Complex Zeros and Rational Zeros of Polynomials
3.4 Rational Functions and Inequalities
Chapter 3 REVIEW
Chapter 3 GROUP ACTIVITY: Interpolating Polynomials

RECALL THAT THE ZEROS OF A FUNCTION f ARE THE SOLUTIONS OR ROOTS of the equation $f(x) = 0$, if any exist. There are formulas that give the exact values of the zeros, real or imaginary, of any linear or quadratic function (Table 1).

TABLE 1 Zeros of Linear and Quadratic Functions

Function	Linear	Quadratic
Form	$f(x) = ax + b,\ a \neq 0$	$f(x) = ax^2 + bx + c,\ a \neq 0$
Equation	$ax + b = 0$	$ax^2 + bx + c = 0$
Zeros/Roots	$x = -\frac{b}{a}$	$x = \frac{-b \pm \sqrt{b^2 - 4ac}}{2a}$

Linear and quadratic functions are also called first- and second-degree *polynomial functions*, respectively. Thus, Table 1 contains formulas for the zeros of any first- or second-degree polynomial function. What about higher-degree polynomial functions such as

$p(x) = 4x^3 - 2x^2 + 3x + 5$ Third degree (cubic)
$q(x) = -2x^4 + 5x^2 - 6$ Fourth degree (quartic)
$r(x) = x^5 - x^4 + x^3 - 10$ Fifth degree (quintic)

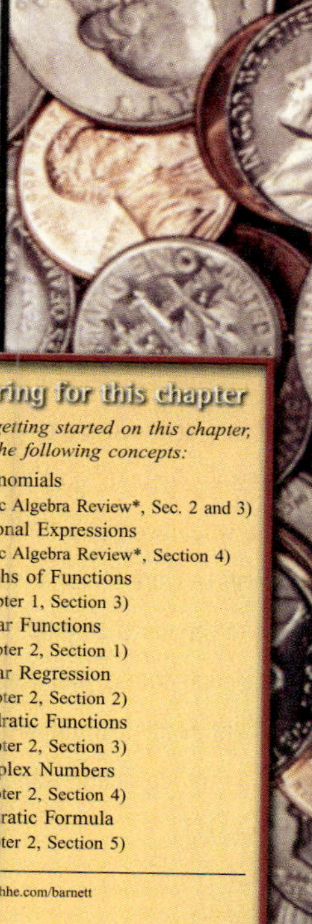

Preparing for this chapter

Before getting started on this chapter, review the following concepts:

- Polynomials (Basic Algebra Review*, Sec. 2 and 3)
- Rational Expressions (Basic Algebra Review*, Section 4)
- Graphs of Functions (Chapter 1, Section 3)
- Linear Functions (Chapter 2, Section 1)
- Linear Regression (Chapter 2, Section 2)
- Quadratic Functions (Chapter 2, Section 3)
- Complex Numbers (Chapter 2, Section 4)
- Quadratic Formula (Chapter 2, Section 5)

*At www.mhhe.com/barnett

EXAMPLE 1 **Solving an Equation**

Solve $5x - 8 = 2x + 1$.

SOLUTION

Algebraic Solution

We use the familiar properties of equality (see Appendix A, Section A.1) to transform the given equation into an equivalent equation with an obvious solution.

$5x - 8 = 2x + 1$	Original equation
$5x - 8 - 2x = 2x + 1 - 2x$	Subtract $2x$ from both sides.
$3x - 8 = 1$	Combine like terms.
$3x - 8 + 8 = 1 + 8$	Add 8 to both sides.
$3x = 9$	Combine like terms.
$\frac{3x}{3} = \frac{9}{3}$	Divide both sides by 3.
$x = 3$	Simplify.

It follows from the properties of equality that $x = 3$ is also the solution set of all the preceding equations in our solution, including the original equation.

Graphical Solution

Enter each side of the equation in the equation editor of a graphing utility (Fig. 1) and use the intersect command (Fig. 2).

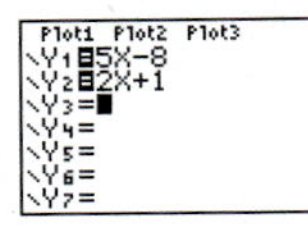

FIGURE 1

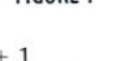

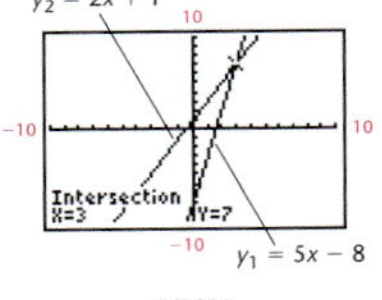

FIGURE 2

Thus, $x = 3$ is the solution to the original equation.

◄ Side-by-Side Solutions: Algebraic and Graphical

Many solved examples in the book provide graphical solutions side-by-side with algebraic solutions. By seeing the same answer result from their symbol manipulations *and* from graphical approaches, students gain insight into the power of algebra and make important conceptual and visual connections.

Examples and Matched Problems ►

Integrated throughout the text, completely worked examples and practice problems are used to introduce concepts and demonstrate problem-solving techniques—algebraic, graphical, and numerical. Each **Example** is followed by a similar **Matched Problem** for the student to work through while reading the material. Answers to the matched problems are located at the end of each section, for easy reference. This active involvement in the learning process helps students develop a thorough understanding of algebraic concepts and processes.

328 | 4 EXPONENTIAL AND LOGARITHMIC FUNCTIONS

EXAMPLE 6 **Continuous Compound Interest**

If \$100 is invested at an annual rate of 8% compounded continuously, what amount, to the nearest cent, will be in the account after 2 years?

SOLUTION

Algebraic Solution

Use the continuous compound interest formula to find A when $P = \$100$, $r = 0.08$, and $t = 2$:

$$A = Pe^{rt}$$
$$= \$100e^{(0.08)(2)} \quad \text{8\% is equivalent to } r = 0.08.$$
$$= \$117.35$$

Compare this result with the values calculated in Table 2.

Graphical Solution

Graphing

$$A = 100e^{0.08x}$$

and using trace (Fig. 9) shows $A = \$117.35$.

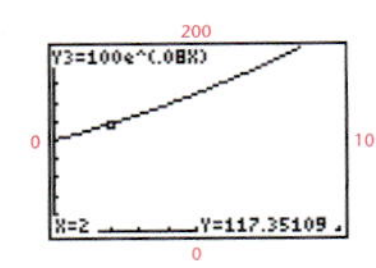

FIGURE 9

MATCHED 6 PROBLEM

What amount will an account have after 5 years if \$100 is invested at an annual rate of 12% compounded annually? Quarterly? Continuously? Compute answers to the nearest cent.

ANSWERS MATCHED PROBLEMS

1. $y = \frac{1}{2}(4^{-x})$

x	y
−3	32.00
−2	8.00
−1	2.00
0	0.50
1	0.13
2	0.03
3	0.01

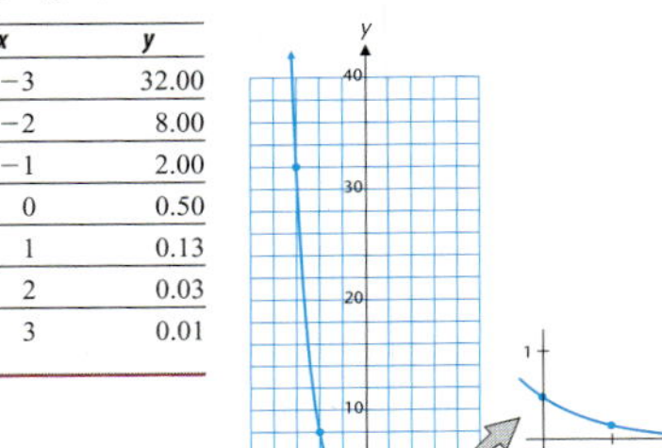

2. $x = -\frac{1}{3}$
3. y intercept: −3; x intercept: 1.83; increasing for all x; horizontal asymptote: $y = -5$
4. \$2,707.04
5. After 23 quarters
6. Annually: \$176.23; quarterly: \$180.61; continuously: \$182.21

Vertical and Horizontal Shifts

If a new function is formed by performing an operation on a given function, then the graph of the new function is called a **transformation** of the graph of the original function. For example, if we add a constant k to $f(x)$, then the graph of $y = f(x)$ is transformed into the graph of $y = f(x) + k$.

FIGURE 2

TABLE 1

	x	$f(x)$
A	-4	0
B	-2	3
C	0	0
D	2	-3
E	4	0

EXPLORE/DISCUSS 1

The following activities refer to the graph of f shown in Figure 2 and the corresponding points on the graph shown in Table 1.

(A) Use the points in Table 1 to construct a similar table and then sketch a graph for each of the following functions: $y = f(x) + 2$, $y = f(x) - 3$. Describe the relationship between the graph of $y = f(x)$ and the graph of $y = f(x) + k$ for k any real number.

(B) Use the points in Table 1 to construct a similar table and then sketch a graph for each of the following functions: $y = f(x + 2)$, $y = f(x - 3)$. [*Hint:* Choose values of x so that $x + 2$ or $x - 3$ is in Table 1.] Describe the relationship between the graph of $y = f(x)$ and the graph of $y = f(x + h)$ for h any real number.

EXAMPLE 1 **Vertical and Horizontal Shifts**

(A) How are the graphs of $y = x^2 + 2$ and $y = x^2 - 3$ related to the graph of $y = x^2$? Confirm your answer by graphing all three functions simultaneously in the same viewing window.

◄ Exploration and Discussion

Interspersed at appropriate places in every section, **Explore/Discuss** boxes encourage students to think critically about mathematics and to explore key concepts in more detail. Verbalization of mathematical concepts, results, and processes is encouraged in these Explore/Discuss boxes, in some matched problems, and in specified problems in almost every exercise set. Explore/Discuss material can be used in class or as an out-of-class activity.

Balanced Exercise Sets ►

Precalculus: Graphs and Models, second edition, contains more than 5,500 problems. Each Exercise set is designed so that an average or below-average student will experience success and a very capable student will be challenged. Exercise sets are found at the end of each section in the text, and are divided into A (routine, easy mechanics), B (more difficult mechanics), and C (difficult mechanics and some theory) levels of difficulty so that students at all levels can be challenged. Problem numbers that appear in blue indicate problems that require students to apply their reasoning and writing skills to the solution of the problem.

EXERCISE 2.4

In Problems 1–26, perform the indicated operations and write each answer in standard form.

1. $(2 + 4i) + (5 + i)$ **2.** $(3 + i) + (4 + 2i)$

3. $(-2 + 6i) + (7 - 3i)$ **4.** $(6 - 2i) + (8 - 3i)$

5. $(6 + 7i) - (4 + 3i)$ **6.** $(9 + 8i) - (5 + 6i)$

7. $(3 + 5i) - (-2 - 4i)$ **8.** $(8 - 4i) - (11 - 2i)$

9. $(4 - 5i) + 2i$ **10.** $6 + (3 - 4i)$

11. $(4i)(6i)$ **12.** $(3i)(8i)$

13. $-3i(2 - 4i)$ **14.** $-2i(5 - 3i)$

15. $(3 + 3i)(2 - 3i)$ **16.** $(-2 - 3i)(3 - 5i)$

17. $(2 - 3i)(7 - 6i)$ **18.** $(3 + 2i)(2 - i)$

19. $(7 + 4i)(7 - 4i)$ **20.** $(5 + 3i)(5 - 3i)$

21. $\dfrac{1}{2 + i}$ **22.** $\dfrac{1}{3 - i}$ **23.** $\dfrac{3 + i}{2 - 3i}$

24. $\dfrac{2 - i}{3 + 2i}$ **25.** $\dfrac{13 + i}{2 - i}$ **26.** $\dfrac{15 - 3i}{2 - 3i}$

In Problems 27–34, evaluate and express results in standard form.

27. $\sqrt{2}\sqrt{8}$ **28.** $\sqrt{3}\sqrt{12}$

29. $\sqrt{2}\sqrt{-8}$ **30.** $\sqrt{-3}\sqrt{12}$

31. $\sqrt{-2}\sqrt{8}$ **32.** $\sqrt{3}\sqrt{-12}$

33. $\sqrt{-2}\sqrt{-8}$ **34.** $\sqrt{-3}\sqrt{-12}$

In Problems 35–44, convert imaginary numbers to standard form, perform the indicated operations, and express answers in standard form.

35. $(2 - \sqrt{-4}) + (5 - \sqrt{-9})$

36. $(3 - \sqrt{-4}) + (-8 + \sqrt{-25})$

37. $(9 - \sqrt{-9}) - (12 - \sqrt{-25})$

38. $(-2 - \sqrt{-36}) - (4 + \sqrt{-49})$

39. $(3 - \sqrt{-4})(-2 + \sqrt{-49})$

40. $(2 - \sqrt{-1})(5 + \sqrt{-9})$

41. $\dfrac{5 - \sqrt{-4}}{7}$ **42.** $\dfrac{6 - \sqrt{-64}}{2}$

43. $\dfrac{1}{2 - \sqrt{-9}}$ **44.** $\dfrac{1}{3 - \sqrt{-16}}$

Write Problems 45–50 in standard form.

45. $\dfrac{2}{5i}$ **46.** $\dfrac{1}{3i}$

47. $\dfrac{1 + 3i}{2i}$ **48.** $\dfrac{2 - i}{3i}$

49. $(2 - 3i)^2 - 2(2 - 3i) + 9$

50. $(2 - i)^2 + 3(2 - i) - 5$

51. Let $f(x) = x^2 - 2x + 2$.

(A) Show that the conjugate complex numbers $1 + i$ and $1 - i$ are both zeros of f.

(B) Does f have any real zeros? Any x intercepts? Explain.

52. Let $g(x) = -x^2 + 4x - 5$.

(A) Show that the conjugate complex numbers $2 + i$ and $2 - i$ are both zeros of g.

(B) Does g have any real zeros? Any x intercepts? Explain.

53. Simplify: i^{18}, i^{32}, and i^{67}.

54. Simplify: i^{21}, i^{43}, and i^{52}.

In Problems 55–58, solve for x and y.

55. $(2x - 1) + (3y + 2)i = 5 - 4i$

56. $3x + (y - 2)i = (5 - 2x) + (3y - 8)i$

APPLICATIONS

57. Geometry. The diagonal of a rectangle is 10 inches and the area is 45 square inches. Find the dimensions of the rectangle, correct to one decimal place.

58. Geometry. The hypotenuse of a right triangle is 12 inches and the area is 24 square inches. Find the dimensions of the triangle, correct to one decimal place.

59. Physics–Well Depth. If the splash of a stone dropped into a well is heard 14 seconds after the stone is released, how deep (to the nearest foot) is the well?

60. Physics–Well Depth. If the splash of a stone dropped into a well is heard 2 seconds after the stone is released, how deep (to the nearest foot) is the well?

61. Manufacturing. A lumber mill cuts rectangular beams from circular logs that are 16 inches in diameter (see the figure).

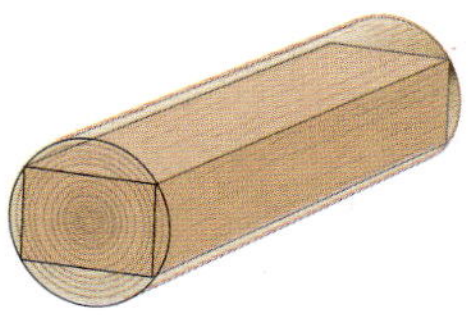

(A) Find a model for the cross-sectional area of the beam. Use the width of the beam as the independent variable.

(B) If the cross-sectional area of the beam is 120 square inches, find the dimensions correct to one decimal place.

(C) Find the dimensions of the beam that has the largest cross-sectional area and find this area. Round answers to one decimal place.

62. Design. A food-processing company packages an assortment of their products in circular metal tins 12 inches in diameter. Four identically sized rectangular boxes are used to divide the tin into eight compartments (see the figure).

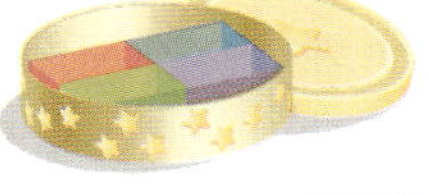

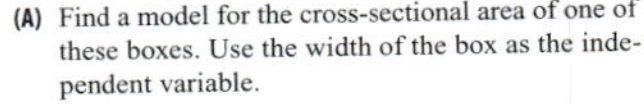

(A) Find a model for the cross-sectional area of one of these boxes. Use the width of the box as the independent variable.

(B) If the cross-sectional area of the box is 15 square inches, find the dimensions correct to one decimal place.

(C) Find the dimensions of the box that has the largest cross-sectional area and find this area. Round answers to one decimal place.

★ **63. Construction.** A water trough is constructed by bending a 4- by 6-foot rectangular sheet of metal down the middle and attaching triangular ends (see the figure). If the volume of the trough is 9 cubic feet, find the width correct to two decimal places.

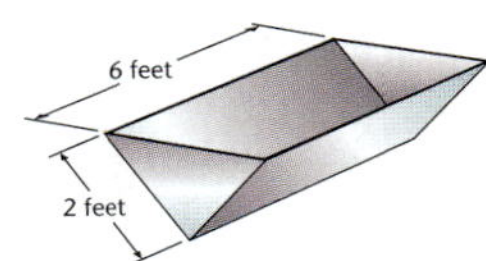

★ **64. Design.** A paper drinking cup in the shape of a right circular cone is constructed from 125 square centimeters of paper (see the figure). If the height of the cone is 10 centimeters, find the radius correct to two decimal places.

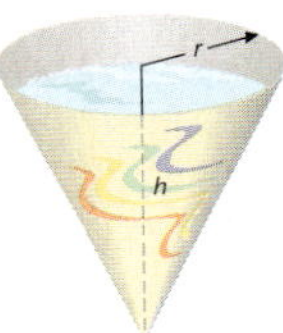

Lateral surface area:
$S = \pi r\sqrt{r^2 + h^2}$

◄ Applications

One of the primary objectives of this book is to give the student substantial experience in modeling and solving real-world problems. More than 700 application exercises help convince even the most skeptical student that mathematics is relevant to everyday life. The most difficult application problems are marked with two stars (★★), the moderately difficult application problems with one star (★), and the easier application problems are not marked. An Applications Index is included following the table of contents to help locate particular applications.

Group Activities ►

A **Group Activity** is located at the end of each chapter and involves many of the concepts discussed in that chapter. These activities strongly encourage the verbalization of mathematical concepts, results, and processes. All of these special activities are highlighted to emphasize their importance.

CHAPTER 1 GROUP ACTIVITY

Mathematical Modeling: Choosing a Long Distance Calling Plan

The number of companies offering residential long distance telephone service has grown rapidly in recent years. The plans they offer vary greatly and it can be difficult to select the plan that is best for you. Here are five typical plans:

Plan 1: A flat fee of \$50 per month for unlimited calls.

Plan 2: A \$30 per month fee for a total of 30 hours of calls and an additional charge of \$0.01 per minute for all minutes over 30 hours.

Plan 3: A \$5 per month fee and a charge of \$0.04 per minute for all calls.

Plan 4: A \$2 per month fee and a charge of \$0.045 per minute for all calls; the fee is waived if the charge for calls is \$20 or more.

Plan 5: A charge of \$0.05 per minute for all calls; there are no additional fees.

(A) Construct a mathematical model for each plan that gives the total monthly cost in terms of the total number of minutes of calls placed in a month. Graph each model on a graphing utility. You may find Boolean expressions like $(x > a)$ helpful in entering your model in a graphing utility (see Example 4 in Section 1.6).

(B) Compare plans 1 and 2. Determine how many minutes per month would make plan 1 cheaper and how many would make plan 2 cheaper.

(C) Repeat part (B) for plans 1 and 3; plans 1 and 4; plans 1 and 5.

(D) Repeat part (B) for plans 2 and 3; plans 2 and 4; plans 2 and 5.

(E) Repeat part (B) for plans 3 and 4; plans 3 and 5.

(F) Repeat part (B) for plans 4 and 5.

(G) Is there one plan that is always better than all the others? Based on your personal calling history, which plan would you choose and why?

◀ Interpretation of Graphs

A course implementing a true "graphing approach" to algebra equips students with the skills to read and interpret graphs. Many exercises in this text ask students to make determinations about equations or functions based on graphs. These exercises are marked with an icon .

EXERCISE 4.1

A

1. Match each equation with the graph of f, g, m, or n in the figure.

(A) $y = (0.2)^x$ **(B)** $y = 2^x$

(C) $y = (\frac{1}{3})^x$ **(D)** $y = 4^x$

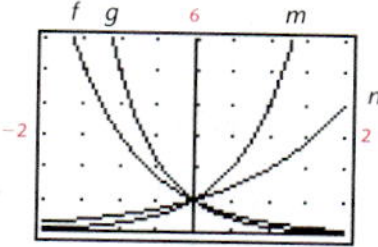

2. Match each equation with the graph of f, g, m, or n in the figure.

(A) $y = e^{-1.2x}$ **(B)** $y = e^{0.7x}$

(C) $y = e^{-0.4x}$ **(D)** $y = e^{1.3x}$

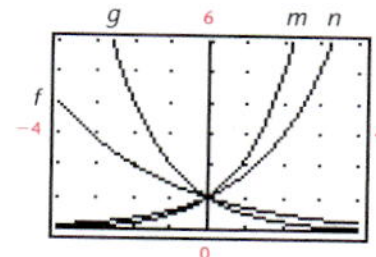

In Problems 3–10, compute answers to four significant digits.

3. $5^{\sqrt{3}}$ **4.** $3^{-\sqrt{2}}$

5. $e^2 + e^{-2}$ **6.** $e - e^{-1}$

7. $\sqrt{e}$ **8.** $e^{\sqrt{2}}$

9. $\dfrac{2^\pi + 2^{-\pi}}{2}$ **10.** $\dfrac{3^\pi - 3^{-\pi}}{2}$

In Problems 11–18, simplify.

11. $10^{3x-1}10^{4-x}$ **12.** $(4^{3x})^{2y}$ **13.** $\dfrac{3^x}{3^{1-x}}$

14. $\dfrac{5^{x-3}}{5^{x-4}}$ **15.** $\left(\dfrac{4^x}{5^y}\right)^{3z}$ **16.** $(2^x3^y)^z$

17. $\dfrac{e^{5x}}{e^{2x+1}}$ **18.** $\dfrac{e^{4-3x}}{e^{2-5x}}$

19. (A) Explain what is wrong with the following reasoning about the expression $[1 + (1/x)]^x$: As x gets large, $1 + (1/x)$ approaches 1 because $1/x$ approaches 0, and 1 raised to any power is 1, so $[1 + 1/x]^x$ approaches 1.

(B) Which number does $[1 + (1/x)]^x$ approach as x approaches ∞?

B

Before graphing the functions in Problems 21–30, classify each function as increasing or decreasing, find the x and y intercepts, and identify any asymptotes. Round any approximate values to two decimal places. Examine the graph to check your answers.

21. $y = 3^x$ **22.** $y = 5^x$

23. $y = (\frac{1}{3})^x = 3^{-x}$ **24.** $y = (\frac{1}{5})^x = 5^{-x}$

25. $g(x) = -3^{-x}$ **26.** $f(x) = -5^x$

EXAMPLE 8 Evaluating and Simplifying a Difference Quotient

For $f(x) = x^2 + 4x + 5$, find and simplify:*

(A) $f(2)$ (B) $f(2 + h)$ (C) $\dfrac{f(2+h) - f(2)}{h}$

(D) $f(x + h)$ (E) $\dfrac{f(x+h) - f(x)}{h}$

SOLUTIONS

(A) $f(2) = 2^2 + 4(2) + 5 = 17$

(B) To find $f(2 + h)$, replace x with $2 + h$ everywhere it occurs in the equation that defines f and simplify:

$$f(2+h) = (2+h)^2 + 4(2+h) + 5 = 4 + 4h + h^2 + 8 + 4h + 5 = h^2 + 8h + 17$$

(C) Using parts (A) and (B), we have

$$\frac{f(2+h) - f(2)}{h} = \frac{h^2 + 8h + 17 - 17}{h} = \frac{h^2 + 8h}{h} = \frac{h(h+8)}{h} = h + 8$$

(D) To find $f(x + h)$, we replace x with $x + h$ everywhere it appears in the equation that defines f and simplify:

$$f(x+h) = (x+h)^2 + 4(x+h) + 5 = x^2 + 2xh + h^2 + 4x + 4h + 5$$

(E) Using the result of part (D), we get

$$\frac{f(x+h) - f(x)}{h} = \frac{x^2 + 2xh + h^2 + 4x + 4h + 5 - (x^2 + 4x + 5)}{h} = \frac{x^2 + 2xh + h^2 + 4x + 4h + 5 - x^2 - 4x - 5}{h} = \frac{2xh + h^2 + 4h}{h} = \frac{h(2x + h + 4)}{h} = 2x + h + 4$$

MATCHED 8 PROBLEM

Repeat Example 8 for $f(x) = x^2 + 3x + 7$.

*The symbol denotes problems that are related to calculus.

Foundation for Calculus ▶

Because many students will use this book to prepare for a calculus course, examples and exercises that are especially pertinent to calculus are marked with an icon .

Technology ►

The generic term *graphing utility* is used to refer to any of the various graphing calculators or computer software packages that might be available to students using this book. The use of technology is integrated throughout the text for visualization, investigation, and verification. The majority of the graphing utility images in this book are "screen dumps" from a Texas Instruments TI-83 Plus graphing calculator. Students not using that TI calculator should be able to produce similar results on any standard graphing calculator or computer algebra system.

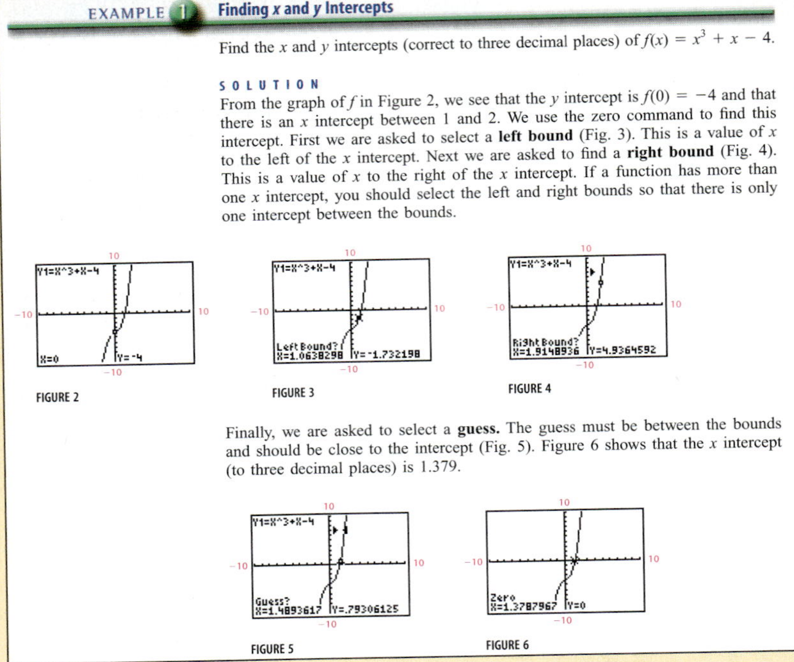

EXAMPLE 1 **Finding *x* and *y* Intercepts**

Find the x and y intercepts (correct to three decimal places) of $f(x) = x^3 + x - 4$.

SOLUTION

From the graph of f in Figure 2, we see that the y intercept is $f(0) = -4$ and that there is an x intercept between 1 and 2. We use the zero command to find this intercept. First we are asked to select a **left bound** (Fig. 3). This is a value of x to the left of the x intercept. Next we are asked to find a **right bound** (Fig. 4). This is a value of x to the right of the x intercept. If a function has more than one x intercept, you should select the left and right bounds so that there is only one intercept between the bounds.

FIGURE 2

FIGURE 3

FIGURE 4

Finally, we are asked to select a **guess.** The guess must be between the bounds and should be close to the intercept (Fig. 5). Figure 6 shows that the x intercept (to three decimal places) is 1.379.

FIGURE 5

FIGURE 6

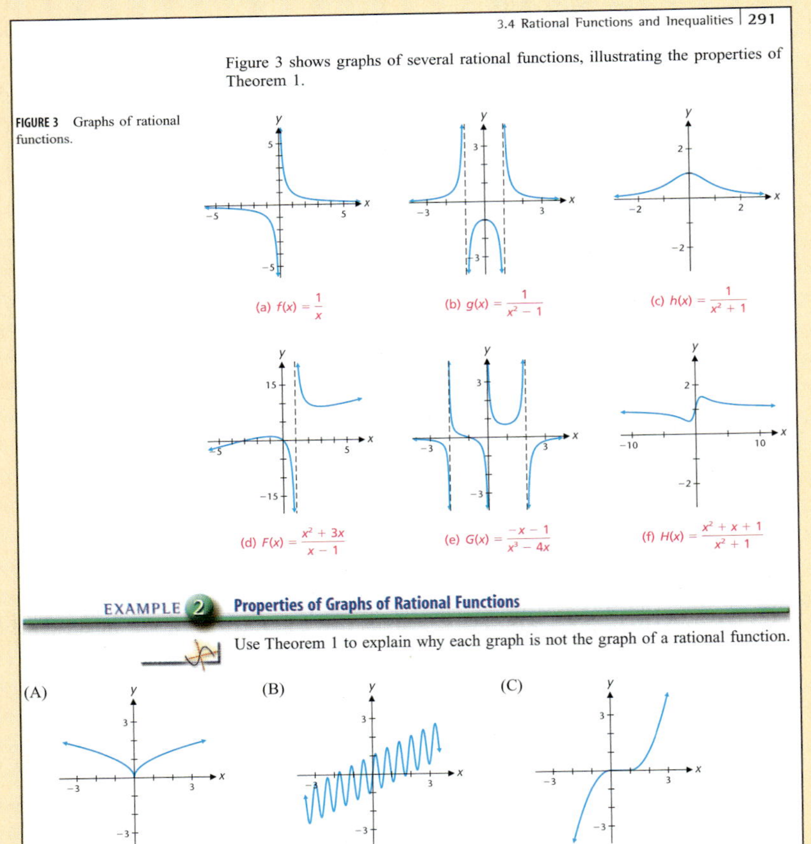

Figure 3 shows graphs of several rational functions, illustrating the properties of Theorem 1.

FIGURE 3 Graphs of rational functions.

(a) $f(x) = \frac{1}{x}$ (b) $g(x) = \frac{1}{x^2 - 1}$ (c) $h(x) = \frac{1}{x^2 + 1}$

(d) $F(x) = \frac{x^2 + 3x}{x - 1}$ (e) $G(x) = \frac{-x - 1}{x^3 - 4x}$ (f) $H(x) = \frac{x^2 + x + 1}{x^2 + 1}$

EXAMPLE 2 **Properties of Graphs of Rational Functions**

Use Theorem 1 to explain why each graph is not the graph of a rational function.

(A) (B) (C)

◄ Graphs and Illustrations

All graphs in this text are computer generated to ensure mathematical accuracy. Graphing utility screens displayed in the text are actual output from a graphing calculator.

Student Aids ▶

Annotation of examples and developments, in small colored type, is found throughout the text to help students through critical stages. Think Boxes are dashed boxes used to enclose steps that are usually performed mentally.

EXAMPLE 2 Solving an Equation

Solve $\frac{7}{2x} - 3 = \frac{8}{3} - \frac{15}{x}$.

SOLUTION

Algebraic Solution

Note that 0 must be excluded from the permissible values of x because division by 0 is not permitted. To clear the fractions, we multiply both sides of the equation by $3(2x) = 6x$, the least common denominator (LCD) of all fractions in the equation.

$$\frac{7}{2x} - 3 = \frac{8}{3} - \frac{15}{x} \qquad x \neq 0$$

$$6x\left(\frac{7}{2x} - 3\right) = 6x\left(\frac{8}{3} - \frac{15}{x}\right) \qquad \text{Multiply by } 6x\text{, the LCD.}$$

$$6x \cdot \frac{7}{2x} - 6x \cdot 3 = 6x \cdot \frac{8}{3} - 6x \cdot \frac{15}{x}$$

$$21 - 18x = 16x - 90 \qquad \text{The equation is now free of fractions.}$$

$$21 - 18x - 16x = 16x - 90 - 16x \qquad \text{Subtract } 16x \text{ from both sides.}$$

$$21 - 34x = -90 \qquad \text{Combine like terms.}$$

$$21 - 34x - 21 = -90 - 21 \qquad \text{Subtract 21 from both sides.}$$

$$-34x = -111 \qquad \text{Combine like terms.}$$

Graphical Solution

Enter $y_1 = \frac{7}{2x} - 3$ and $y_2 = \frac{8}{3} - \frac{15}{x}$ (Fig. 3) in the equation editor of a graphing utility. Note the use of parentheses in Figure 3 to be certain that $\frac{7}{2x}$ is evaluated correctly. Now use the intersect command (Fig. 4).

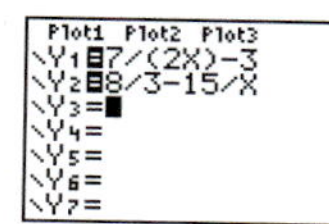

FIGURE 3

Complex Numbers and Radicals

Recall that we say that a is a square root of b if $a^2 = b$. If x is a positive real number, then x has two square roots, the principal square root, denoted by $\sqrt{x}$, and its negative, $-\sqrt{x}$. If x is a negative real number, then x still has two square roots, but now these square roots are imaginary numbers.

DEFINITION 4

Principal Square Root of a Negative Real Number

The **principal square root of a negative real number,** denoted by $\sqrt{-a}$, where a is positive, is defined by

$\sqrt{-a} = i\sqrt{a} \qquad \sqrt{-3} = i\sqrt{3} \qquad \sqrt{-9} = i\sqrt{9} = 3i$

The other square root of $-a$, $a > 0$, is $-\sqrt{-a} = -i\sqrt{a}$.

Note in Definition 4 that we wrote $i\sqrt{a}$ and $i\sqrt{3}$ in place of the standard forms $\sqrt{a}i$ and $\sqrt{3}i$. We follow this convention whenever it appears that i might accidentally slip under a radical sign ($\sqrt{ai} \neq \sqrt{a}i$, but $\sqrt{a}i = i\sqrt{a}$). Definition 4 is motivated by the fact that

$$(i\sqrt{a})^2 = i^2a = -a$$

◀ **Screen Boxes** are used to highlight important definitions, theorems, results, and step-by-step processes.

Caution Boxes appear throughout the text to indicate where student errors often occur. ▶

182 | 2 MODELING WITH LINEAR AND QUADRATIC FUNCTIONS

CAUTION

1. One side of an equation must be 0 before the zero property can be applied. Thus

$$x^2 - 6x + 5 = -4$$
$$(x - 1)(x - 5) = -4$$

does not imply that $x - 1 = -4$ or $x - 5 = -4$. See Example 1, part B, for the correct solution of this equation.

2. The equations

$$2x^2 = 3x \qquad \text{and} \qquad 2x = 3$$

are not equivalent. The first has solution set $\{0, \frac{3}{2}\}$, whereas the second has solution set $\{\frac{3}{2}\}$. The root $x = 0$ is lost when each member of the first equation is divided by the variable x. See Example 1, part C, for the correct solution of this equation.

Do not divide both members of an equation by an expression containing the variable for which you are solving. You may be dividing by 0.

MATCHED 3 PROBLEM

Describe the graph of $f(x) = 2e^{x/2} - 5$, including x and y intercepts, increasing and decreasing properties, and horizontal asymptotes. Round any approximate values to two decimal places.

Compound Interest

The fee paid to use another's money is called **interest.** It is usually computed as a percentage, called the **interest rate,** of the principal over a given time. If, at the end of a payment period, the interest due is reinvested at the same rate, then the interest earned as well as the principal will earn interest during the next payment period. Interest paid on interest reinvested is called **compound interest.**

Suppose you deposit \$1,000 in a savings and loan that pays 8% compounded semiannually. How much will the savings and loan owe you at the end of 2 years? Compounded semiannually means that interest is paid to your account at the end of each 6-month period, and the interest will in turn earn interest. The **interest rate per period** is the annual rate, 8% = 0.08, divided by the number of compounding periods per year, 2. If we let A_1, A_2, A_3, and A_4 represent the new amounts due at the end of the first, second, third, and fourth periods, respectively, then

$$A_1 = \$1{,}000 + \$1{,}000\left(\frac{0.08}{2}\right)$$
$$= \$1{,}000(1 + 0.04)$$
$$A_2 = A_1(1 + 0.04)$$
$$= [\$1{,}000(1 + 0.04)](1 + 0.04$$
$$= \$1{,}000(1 + 0.04)^2$$
$$A_3 = A_2(1 + 0.04)$$
$$= [\$1{,}000(1 + 0.04)^2](1 + 0.0$$
$$= \$1{,}000(1 + 0.04)^3$$
$$A_4 = A_3(1 + 0.04)$$
$$= [\$1{,}000(1 + 0.04)^3](1 + 0.0$$
$$= \$1{,}000(1 + 0.04)^4$$

What do you think the savings an
If you guessed

$$A = \$1{,}000(1 + 0.04)^{12}$$

you have observed a pattern that is ge
est formula:

◀ **Boldface Type** is used to introduce new terms and highlight important comments.

APPLICATIONS

75. Computer Science. Let $f(x) = 10[\![0.5 + x/10]\!]$. Evaluate f at 4, −4, 6, −6, 24, 25, 247, −243, −245, and −246. What operation does this function perform?

76. Computer Science. Let $f(x) = 100[\![0.5 + x/100]\!]$. Evaluate f at 40, −40, 60, −60, 740, 750, 7,551, −601, −649, and −651. What operation does this function perform?

★ **77. Computer Science.** Use the greatest integer function to define a function f that rounds real numbers to the nearest hundredth.

★ **78. Computer Science.** Use the greatest integer function to define a function f that rounds real numbers to the nearest thousandth.

79. Revenue. The revenue (in dollars) from the sale of x car seats for infants is given by

$$R(x) = 60x - 0.035x^2 \qquad 0 \le x \le 1{,}700$$

Find the number of car seats that must be sold to maximize the revenue. What is the maximum revenue (to the nearest dollar)?

80. Profit. The profit (in dollars) from the sale of x car seats for infants is given by

$$P(x) = 38x - 0.035x^2 - 4{,}000 \qquad 0 \le x \le 1{,}700$$

Find the number of car seats that must be sold to maximize the profit. What is the maximum profit (to the nearest dollar)?

★ **81. Manufacturing.** A box is to be made out of a piece of cardboard that measures 18 by 24 inches. Squares, x inches on a side, will be cut from each corner and then the ends and sides will be folded up (see the figure).

Find the size of the cutout squares that will make the maximum volume. What is the maximum volume? Round answers to two decimal places.

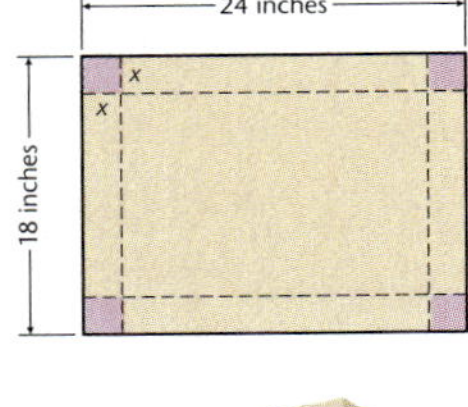

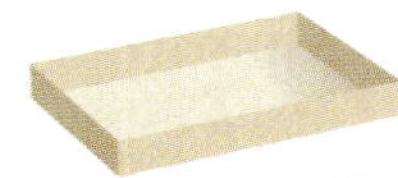

★ **82. Manufacturing.** A box with a hinged lid is to be made out of a piece of cardboard that measures 20 by 40 inches. Six squares, x inches on a side, will be cut from each corner and the middle of the sides, and then the ends and sides will be folded up to form the box and its lid (see the figure).

Find the size of the cutout squares that will make the maximum volume. What is the maximum volume? Round answers to two decimal places.

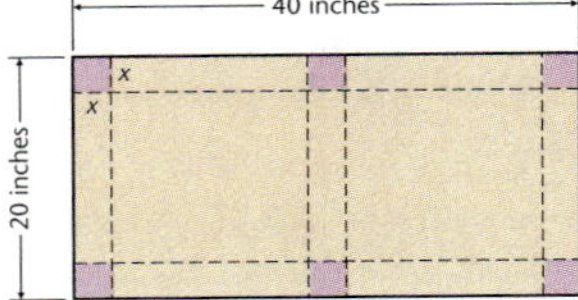

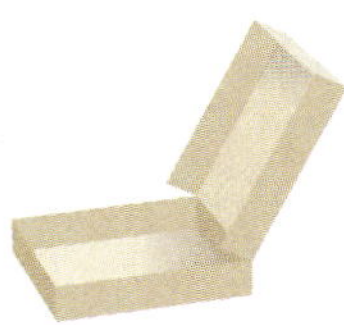

83. Construction. A freshwater pipe is to be run from a source on the edge of a lake to a small resort community on an island 8 miles offshore, as indicated in the figure. It costs \$10,000 per mile to lay the pipe on land and \$16,000 per mile to lay the pipe in the lake. The total cost $C(x)$ in thousands of dollars of laying the pipe is given by

$$C(x) = 10(20 - x) + 16\sqrt{x^2 + 64} \qquad 0 \le x \le 20$$

Find the length (to two decimal places) of the land portion of the pipe that will make the production costs minimum. Find the minimum cost to the nearest thousand dollars.

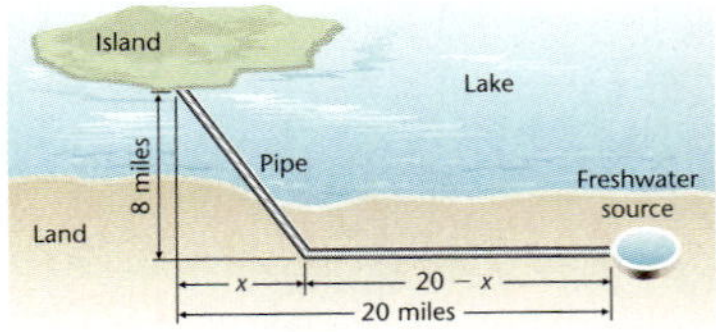

Functional Use of Four Colors improves the clarity of many illustrations, graphs, and developments, and guides students through certain critical steps. ▶

Chapter Review sections are provided at the end of each chapter and include a thorough review of all the important terms and symbols. This recap is followed by a comprehensive set of review exercises. ▶

CHAPTER 10 REVIEW

10.1 Sequences and Series

A **sequence** is a function with the domain a set of successive integers. The symbol a_n, called the ***n*th term,** or **general term,** represents the range value associated with the domain value n. Unless specified otherwise, the domain is understood to be the set of natural numbers. A **finite sequence** has a finite domain, and an **infinite sequence** has an infinite domain. A **recursion formula** defines each term of a sequence in terms of one or more of the preceding terms. For example, the **Fibonacci sequence** is defined by $a_n = a_{n-1} + a_{n-2}$ for $n \geq 3$, where $a_1 = a_2 = 1$. If $a_1, a_2, \ldots, a_n, \ldots$ is a sequence, then the expression $a_1 + a_2 + \cdots + a_n + \cdots$ is called a **series.** A finite sequence produces a **finite series,** and an infinite sequence produces an **infinite series.** Series can be represented using the summation notation:

$$\sum_{k=m}^{n} a_k = a_m + a_{m+1} + \cdots + a_n$$

where k is called the **summing index.** If the terms in the series are alternately positive and negative, the series is called an **alternating series.**

10.2 Mathematical Induction

A wide variety of statements can be proven using the **principle of mathematical induction:** Let P_n be a statement associated with each positive integer n and suppose the following conditions are satisfied:

1. P_1 is true.
2. For any positive integer k, if P_k is true, then P_{k+1} is also true.

Then the statement P_n is true for all positive integers n.

To use mathematical induction to prove statements involving laws of exponents, it is convenient to state a **recursive definition of a^n:**

$$a^1 = a \quad \text{and} \quad a^{n+1} = a^n a \quad \text{for any integer } n > 1$$

To deal with conjectures that may be true only for $n \geq m$, where m is a positive integer, we use the **extended principle of mathematical induction:** Let m be a positive integer, let P_n be a statement associated with each integer $n \geq m$, and suppose the following conditions are satisfied:

1. P_m is true.
2. For any integer $k \geq m$, if P_k is true, then P_{k+1} is also true.

Then the statement P_n is true for all integers $n \geq m$.

10.3 Arithmetic and Geometric Sequences

A sequence is called an **arithmetic sequence,** or **arithmetic progression,** if there exists a constant d, called the **common difference,** such that

$$a_n - a_{n-1} = d \quad \text{or} \quad a_n = a_{n-1} + d \quad \text{for every } n > 1$$

The following formulas are useful when working with arithmetic sequences and their corresponding series:

$a_n = a_1 + (n-1)d$	***n*th-Term Formula**
$S_n = \frac{n}{2}[2a_1 + (n-1)d]$	**Sum Formula—First Form**
$S_n = \frac{n}{2}(a_1 + a_n)$	**Sum Formula—Second Form**

A sequence is called a **geometric sequence,** or a **geometric progression,** if there exists a nonzero constant r, called the **common ratio,** such that

$$\frac{a_n}{a_{n-1}} = r \quad \text{or} \quad a_n = ra_{n-1} \quad \text{for every } n > 1$$

The following formulas are useful when working with geometric sequences and their corresponding series:

$a_n = a_1 r^{n-1}$		***n*th-Term Formula**
$S_n = \frac{a_1 - a_1 r^n}{1 - r}$	$r \neq 1$	**Sum Formula—First Form**
$S_n = \frac{a_1 - ra_n}{1 - r}$	$r \neq 1$	**Sum Formula—Second Form**
$S_\infty = \frac{a_1}{1 - r}$	$\lvert r \rvert < 1$	**Sum of an Infinite Geometric Series**

10.4 Multiplication Principle, Permutations, and Combinations

Given a sequence of operations, **tree diagrams** are often used to list all the possible combined outcomes. To count the number of combined outcomes without actually listing them, we use the **multiplication principle:**

(D) How much profit will the distributor make if all these sets are sold? If there is more than one way to use all the available locomotives and cars, which one will produce the largest profit?

CUMULATIVE REVIEW EXERCISES — CHAPTERS 8 9

Work through all the problems in this cumulative review and check answers in the back of the book. Answers to all review problems are there, and following each answer is a number in italics indicating the section in which that type of problem is discussed. Where weaknesses show up, review appropriate sections in the text.

A

1. Solve using substitution or elimination by addition:
$$\begin{aligned} 3x - 5y &= 11 \\ 2x + 3y &= 1 \end{aligned}$$

2. Solve by graphing: $\begin{aligned} 2x - y &= -4 \\ 3x + y &= -1 \end{aligned}$

3. Solve by substitution or elimination by addition:
$$\begin{aligned} -6x + 3y &= 2 \\ 2x - y &= 1 \end{aligned}$$

4. Solve by graphing: $\begin{aligned} 3x + 5y &\leq 15 \\ x, y &\geq 0 \end{aligned}$

◀ **A Cumulative Review Exercise** set is provided after every second or third chapter, for additional reinforcement.

ALEKS ▶

What Is ALEKS®?

ALEKS is . . .

- A comprehensive course management system. It tells you exactly what your students know and don't know.
- Artificial intelligence. It totally individualizes assessment and learning.
- Customizable. Click on or off each course topic.
- Web based. Use a standard browser for easy Internet access.
- Inexpensive. There are no setup fees or site license fees.

ALEKS (**A**ssessment and **LE**arning in **K**nowledge **S**paces) is an artificial intelligence-based system for individualized math learning available via the World Wide Web.

http://www.highedmath.aleks.com

ALEKS delivers precise, qualitative diagnostic assessments of students' math knowledge, guides them in the selection of appropriate new study material, and records their progress toward mastery of curricular goals in a robust course management system.

ALEKS interacts with the student much as a skilled human tutor would, moving between explanation and practice as needed, correcting and analyzing errors, defining terms and changing topics on request. By sophisticated modeling of a student's "knowledge state" for a given subject matter, ALEKS can focus clearly on what the student is most ready to learn next, thereby building a learning momentum that fuels success.

The ALEKS system was developed with a multimillion-dollar grant from the National Science Foundation. It has little to do with what is commonly thought of as educational software. The theory behind ALEKS is a specialized field of mathematical cognitive science called "Knowledge Spaces."

Knowledge Space theory, which concerns itself with the mathematical dynamics of knowledge acquisition, has been under development by researchers in cognitive science since the early 1980s. The Chairman and founder of ALEKS Corporation, Jean-Claude Falmagne, is an internationally recognized leader in scientific work in this field.

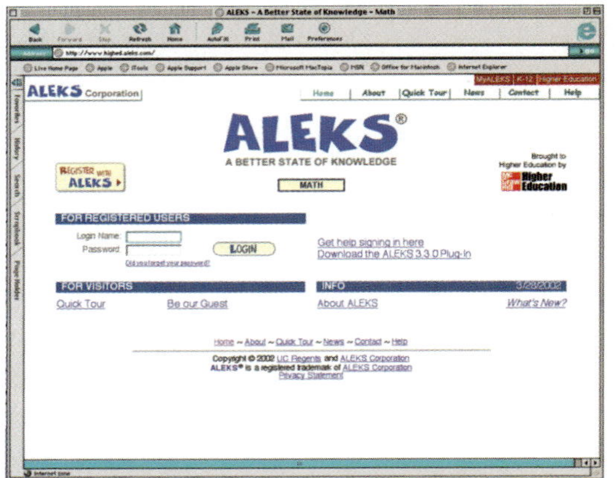

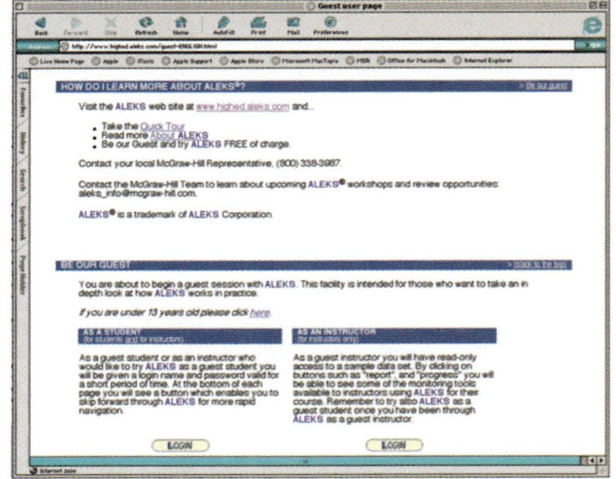

Please visit http:/www.highedmath.aleks.com for a trial run.

To the Student

You are now a student of mathematics. By signing up for this course, whether you did so because you had to or because you wanted to, you are about to begin a course of study that requires—and teaches—abstract mathematical thought. The study of algebra is not merely an exercise in matching equations to examples. It is not only the study of how mathematical ideas can be better translated into more common language. It is not simply the study of how these ideas are applicable to everyday life.

The study of algebra and trigonometry forms the basis for developing new thought patterns. It involves a different way of using your mind. That in itself has value. A successful study of algebra can enhance your reasoning skills, problem-solving abilities, geometric visualization, and creative thought processes. As we develop the concepts and ideas in this text, we strive to help you to think about mathematics in a way that mathematicians, engineers, physical and life scientists, economists and business managers, and other professionals find useful in their professions.

Mathematics is not a spectator sport. Just as you cannot learn to swim by watching someone swim, you cannot learn mathematics by simply reading worked examples. You must work the problems yourself—lots of them. No mathematics text should be read without a pencil and paper at hand. To learn mathematics, you must *do* mathematics.

As you study the text, refer to the "Preparing for this chapter" box on the opening page. Review the topics covered either in Appendix A or in the Basic Algebra Review on the text's Online Learning Center at www.mhhe.com/barnett. This review will remind you of the prerequisite material you studied in previous courses, and help you to reinforce the skills learned in those courses. At the end of each chapter, you will find a Chapter Review, and at the end of every second or third chapter, Cumulative Review Exercises. These can assist you in preparing for tests. Then we recommend the following five-step approach:

1. Read the section from beginning to end, making note of the theorems, the "Cautions," and "Remarks"; do not get discouraged if you cannot understand everything the first time through
2. Work through the examples provided
3. Work the Matched Problems (the solutions to each are included at the end of each section)
4. Review the main ideas of the section
5. Work the assigned (and, when you can, the unassigned) exercises

We assume you will have access to a graphing calculator or other utility. We have included many screen shots from the popular Texas Instruments TI-83 Plus. You can, however, use any standard graphing calculator to complete this course. If you are unsure of which calculator to use, ask your instructor. As you read this text, reproduce each screen shot on your calculator. This will develop the skills necessary to complete the matched problems and exercise sets. For further insight into using a graphing utility, access the McGraw-Hill Graphing Calculator Workshop on the Online Learning Center.

The following supplemental materials can help you to succeed in the course.

Online Learning Center If you purchased a new copy of this text, you have access to the OLC that accompanies it. Check it out!
This site offers many features including:

1. A Basic Algebra Review
2. Practice exercises, sample quizzes, and tests
3. Audio/visual tutorials
4. McGraw-Hill Graphing Calculator Workshop
5. A variety of additional resources

ALEKS If your instructor has not adopted ALEKS® for this course, consider subscribing to this individualized online tutorial. ALEKS® is an artificial intelligence–based tutorial that isolates what you know and then teaches what you need to learn to pass this course. Students who enroll in and complete the course in ALEKS® demonstrate a 15–20 percent increase in test scores. It is worth your time and money to assure your success. Learn more at www.highedstudent.aleks.com.

NetTutor Online support is also available in the form of access to a tutor via the World Wide Web. Receive instruction from live tutors using NetTutor's web-based, graphical chat capabilities during normal study hours. You can also submit questions and receive answers, browse previously answered questions, and view previous live chat sessions, 24 hours a day. Access NetTutor via the Online Learning Center.

Student's Solutions Manual We also encourage you to inquire whether your instructor has made the Student's Solutions Manual available in your college bookstore. Within this supplement, you will find the solutions to all the odd-numbered problems in this text.

Finally, whether your initial plan is simply to make it through this course alive or whether you are planning on further study in mathematics, we urge you to keep an open mind. You might find, as so many students have discovered before you, that once you master algebra, further study in mathematics becomes more attractive than you anticipate. Learn from your instructor, learn from each other, and learn from this text. Remember: you can do it!

PRECALCULUS

CHAPTER 1

Functions, Graphs, and Models

OUTLINE

THE FUNCTION CONCEPT IS ONE OF THE MOST IMPORTANT ideas in mathematics. The study of either the theory or the applications of mathematics beyond the most elementary level requires a firm understanding of functions and their graphs. In the first section of this chapter we discuss the techniques involved in using an electronic graphing device such as a graphing calculator or a computer. In the remaining sections, we introduce the important concept of a function, discuss basic properties of functions and their graphs, and examine specific types of functions. Much of the remainder of this book is concerned with applying the ideas introduced in this chapter to a variety of different types of functions, as is evidenced by the chapter titles following this chapter. Efforts made to understand and use the function concept correctly from the beginning will be rewarded many times in this course and in most future courses that involve mathematics.

Preparing for this chapter

Before getting started on this chapter, review the following concepts:

- Set Notation (Basic Algebra Review*, Section 1)
- Polynomials (Basic Algebra Review*, Sec. 2 and 3)
- Rational Expressions (Basic Algebra Review*, Section 4)
- Square Root Radicals (Basic Algebra Review*, Section 7)
- Interval Notation (Appendix A, Section A.1)
- Cartesian Coordinate System (Appendix A, Section A.2)
- Distance Formula (Appendix A, Section A.3)
- Pythagorean Theorem (Appendix C)

*At www.mhhe.com/barnett

SECTION 1.1 Using Graphing Utilities

Graphing Utilities • Screen Coordinates • The Trace, Zoom, and Intersect Commands • Mathematical Modeling

The use of technology to aid in drawing and analyzing graphs is revolutionizing mathematics education and is the reason for this book. Your ability to interpret mathematical concepts and to discover patterns of behavior will be greatly increased as you become proficient with an electronic graphing device. In this section we introduce some of the basic features of electronic graphing devices. Additional features will be introduced as the need arises. If you have already used an electronic graphing device in a previous course, you can use this section to quickly review basic concepts. If you need to refresh your memory about a particular feature, consult Graphing Utility Features in the index to locate the textbook discussion of that feature.

Graphing Utilities

We now turn to the use of electronic graphing devices to graph equations. We will refer to any electronic device capable of displaying graphs as a **graphing utility.** The two most common graphing utilities are handheld graphing calculators and computers with appropriate software. You should have such a device as you proceed through this book.

We will discuss graphing utilities only in general terms. Refer to the manual or to the graphing utility supplement accompanying this text for specific details relative to your own graphing utility.

An image on the screen of a graphing utility is made up of darkened rectangles called **pixels** (Fig. 1). The pixel rectangles are the same size, and don't change in size during any application. Graphing utilities use pixel-by-pixel plotting to produce graphs.

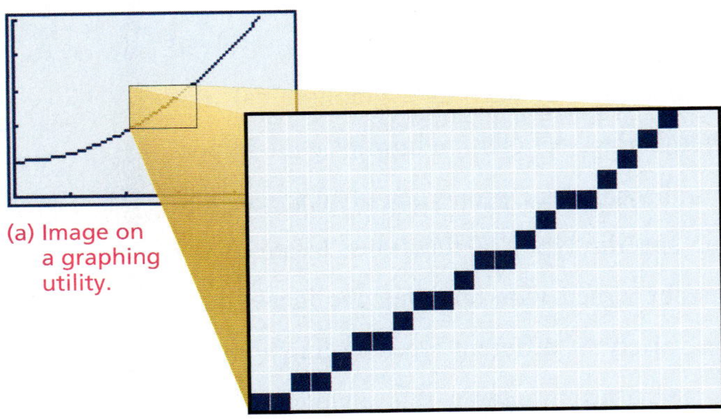

(a) Image on a graphing utility.
(b) Magnification to show pixels.

FIGURE 1 Pixel-by-pixel plotting on a graphing utility.

The accuracy of the graph depends on the resolution of the graphing utility. Most graphing utilities have screen resolutions of between 50 and 75 pixels per inch, which results in fairly rough but very useful graphs. Some computer

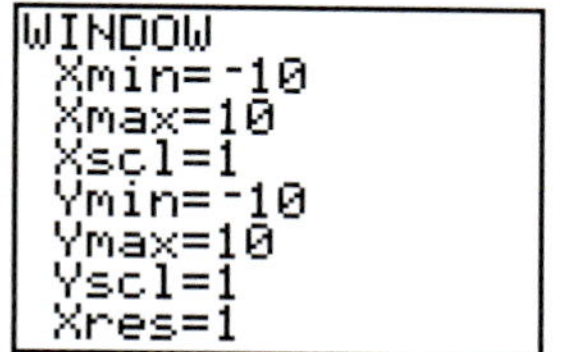

(a) Standard window variable values

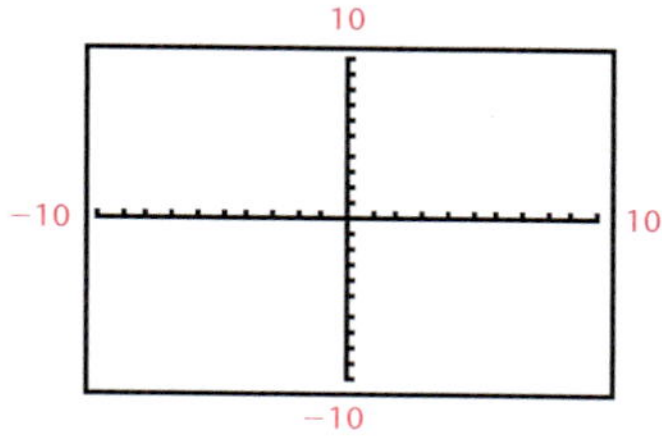

(b) Standard viewing window

FIGURE 2 A standard viewing window and its dimensions.

systems can print very high quality graphs with resolutions greater than 1,000 pixels per inch.

Most graphing utility screens are rectangular. The graphing screen on a graphing utility represents a portion of the plane in the rectangular coordinate system. But this representation is an approximation, because pixels are not really points, as is clearly shown in Figure 1. Points are geometric objects without dimensions, whereas a pixel has dimensions. The coordinates of a pixel are usually taken at the center of the pixel and represent all the infinitely many geometric points within the pixel. This does not cause much of a problem, as we will see.

The portion of a rectangular coordinate system displayed on the graphing screen is called a **viewing window** and is determined by assigning values to six **window variables:** the lower limit, upper limit, and scale for the x axis and the lower limit, upper limit, and scale for the y axis. Figure 2(a) illustrates the names and values of **standard window variables,** and Figure 2(b) shows the resulting **standard viewing window.**

The names **Xmin, Xmax, Xscl, Ymin, Ymax,** and **Yscl** will be used for the six window variables. Xscl and Yscl determine the distance between tick marks on the x and y axes, respectively. **Xres** is a seventh window variable on some graphing utilities that controls the screen resolution; we will always leave this variable set to the default value 1. The window variables may be displayed slightly differently by your graphing utility. In this book, when a viewing window of a graphing utility is pictured in a figure, the values of Xmin, Xmax, Ymin, and Ymax are indicated by labels to make the graph easier to read [see Fig. 2(b)]. These labels are always centered on the sides of the viewing window, irrespective of the location of the axes.

REMARK We think it is important that actual output from existing graphing utilities be used in this book. The majority of the graphing utility images in this book are screen dumps from a Texas Instruments TI-83 graphing calculator. Occasionally we use screen dumps from a TI-86 graphing calculator, which has a wider screen. You may not always be able to produce an exact replica of a text figure on your graphing utility, but the differences will be minor and should cause no difficulties.

We now turn to the use of a graphing utility to graph equations that can be written in the form

$$y = (\text{some expression in } x) \tag{1}$$

Graphing an equation of the type shown in equation (1) using a graphing utility is a simple three-step process:

Graphing Equations Using a Graphing Utility

Step 1. Enter the equation.

Step 2. Enter values for the window variables. (A rule of thumb for choosing Xscl and Yscl, unless there are reasons to the contrary, is to choose each about one-tenth the corresponding variable range.)

Step 3. Press the graph command.

The following example illustrates this procedure for graphing the equation $y = x^2 - 4$. (See Example 1 of Appendix A, Section A.2 for a hand-drawn sketch of this equation.)

EXAMPLE 1 Graphing an Equation with a Graphing Utility

Use a graphing utility to graph $y = x^2 - 4$ for $-5 \le x \le 5$ and $-5 \le y \le 15$.

SOLUTION

Press the Y= key to display the equation editor and enter the equation [Fig. 3(a)]. Press WINDOW to display the window variables and enter the values for these variables [Fig. 3(b)]. Press GRAPH to obtain the graph in Figure 3(c). (The form of the screens in Figure 3 may differ slightly, depending on the graphing utility used.)

FIGURE 3 Graphing is a three-step process.

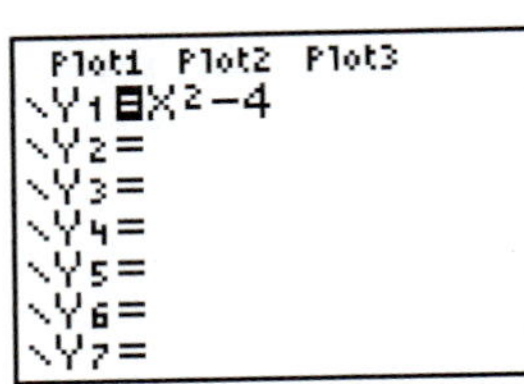

(a) Enter equation.

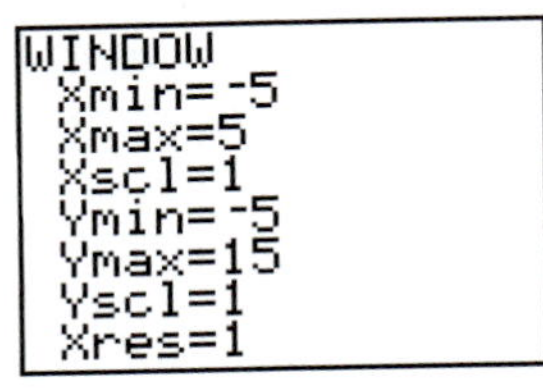

(b) Enter window variables.

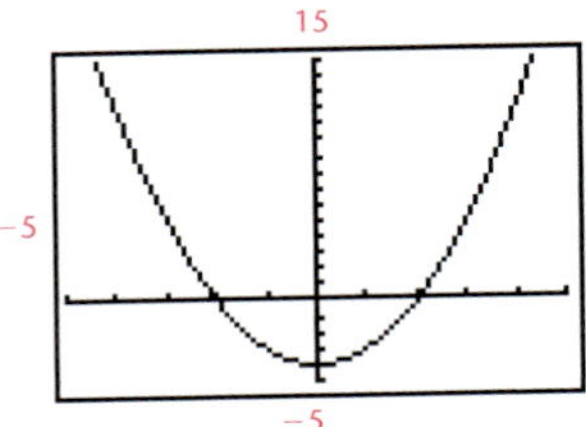

(c) Press the graph command.

MATCHED 1 PROBLEM*

Use a graphing utility to graph $y = 8 - x^2$ for $-5 \le x \le 5$ and $-10 \le y \le 10$.

REMARK For Example 1, we displayed a viewing window for each step in the graphing procedure. Generally, we will show only the final results, as illustrated in Figure 3(c).

The next example illustrates how a graphing utility can be used as an aid to sketching the graph of an equation by hand. The example illustrates the use of algebraic, numeric, and graphic approaches, which add considerably to the understanding of a problem.

EXAMPLE 2 Using a Graphing Utility as an Aid to Hand Graphing—Net Cash Flow

The net cash flow y in millions of dollars of a small high-tech company from 1991–1999 is given approximately by the following equation

$$y = 0.4x^3 - 2x + 1 \qquad -4 \le x \le 4 \tag{2}$$

where x represents the number of years before or after 1995, when the board of directors appointed a new CEO.

(A) Construct a table of values for equation (2) for each year from 1991 to 1999, inclusive. Compute y to one decimal place.

(B) Obtain a graph of equation (2) in the viewing window of your graphing utility. Plot the table values from part A by hand on graph paper, then join these points with a smooth curve using the graph in the viewing window as an aid.

*Answers to matched problems in a given section are found near the end of the section, before the exercise set.

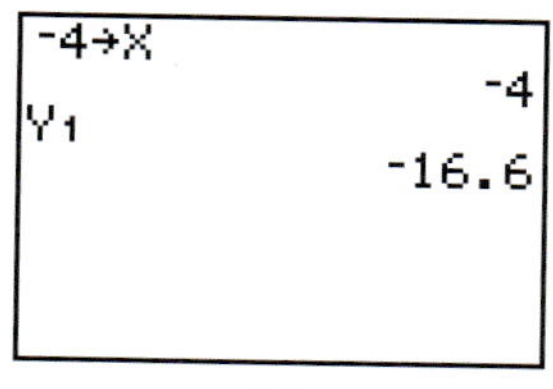

(a)

X	Y1
-4.0	-16.6
-3.0	-3.8
-2.0	1.8
-1.0	2.6
0.0	1.0
1.0	-.6
2.0	.2

Y1■.4X^3-2X+1

(b)

FIGURE 4

```
WINDOW
 Xmin=-5
 Xmax=5
 Xscl=1
 Ymin=-20
 Ymax=20
 Yscl=5
 Xres=1
```

FIGURE 5

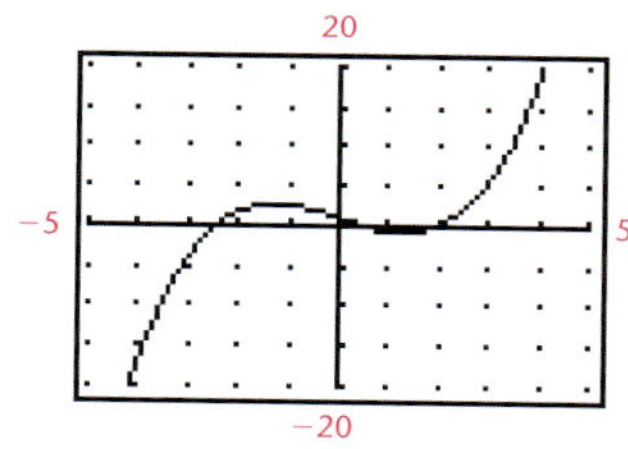

(a) Graphing utility graph

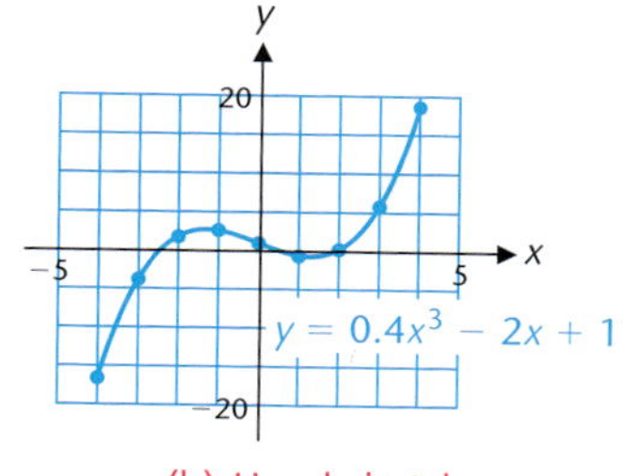

(b) Hand sketch

FIGURE 6 Net cash flow.

SOLUTIONS

(A) After entering the given equation as y_1, we can find the value of y for a given value of x by storing the value of x in the variable X and simply displaying y_1, as shown in Figure 4(a). To speed up this process, many graphing utilities can compute an entire table of values directly, as shown in Figure 4(b). We organize these results in Table 1.

TABLE 1 Net Cash Flow

Year	1991	1992	1993	1994	1995	1996	1997	1998	1999
x	–4	–3	–2	–1	0	1	2	3	4
y (million \$)	–16.6	–3.8	1.8	2.6	1	–0.6	0.2	5.8	18.6

(B) To create a graph of equation (2) in the viewing window of a graphing utility, we select values for the viewing window variables that cover a little more than the values shown in Table 1, as shown in Figure 5. We add a grid to the viewing window to obtain the graphing utility graph shown in Figure 6(a). The corresponding hand sketch is shown in Figure 6(b).

REMARK Table 1 gives us specific detail and the equation with its graph gives us an overview. Each viewpoint has its specific use.

MATCHED PROBLEM 2

Given the equation $y = 1 + 1.9x - 0.2x^3$, complete a table of values for the integers from -4 to 4, plot these points by hand, and then hand sketch the graph of the equation with the aid of a graphing utility.

EXPLORE/DISCUSS 1

The choice of the viewing window has a pronounced effect on the shape of a graph. Graph $y = -x^3 + 2x$ in each of the following viewing windows:

(A) $-1 \le x \le 1,\ -1 \le y \le 1$

(B) $-10 \le x \le 10,\ -10 \le y \le 10$

(C) $-100 \le x \le 100,\ -100 \le y \le 100$

Which window gives the best view of the graph of this equation, and why?

Screen Coordinates

We now take a closer look at *screen coordinates* of pixels. Earlier we indicated that the coordinates of the center point of a pixel are usually used as the **screen coordinates of the pixel,** and these coordinates represent all points within the pixel. As expected, screen coordinates of pixels change as you change values of window variables.

To find screen coordinates of various pixels, move a *cursor* around the viewing window and observe the coordinates displayed on the screen. A **cursor** is a special symbol, such as a plus (+) or times (×) sign, that locates one pixel on the screen at a time. As the cursor is moved around the screen, it moves from pixel to pixel. To see this, set the window variables in your graphing utility so that $-5 \leq x \leq 5$ and $-5 \leq y \leq 5$, and activate a grid for the screen. Move the cursor as close as you can to the point (2, 2) and observe what happens. Figure 7 shows the screen coordinates of the four pixels that are closest to (2, 2). The coordinates displayed on your screen may vary slightly from these, depending on the graphing utility used.

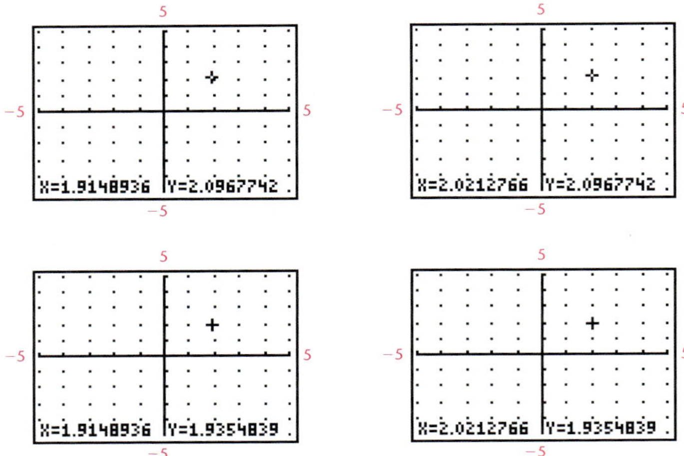

FIGURE 7 Screen coordinates of pixels near (2, 2).

Any of the four pixels in Figure 7 can be used to approximate the point (2, 2), but it is not possible to find a pixel in this viewing window whose screen coordinates are exactly (2, 2). It is instructive to repeat this exercise with different window variables, say, $-7 \leq x \leq 7$ and $-7 \leq y \leq 7$.

The Trace, Zoom, and Intersect Commands

Analyzing a graph of an equation frequently involves finding coordinates of points on the graph. Using the **TRACE** command on a graphing utility is one way to accomplish this. The trace feature places a cursor directly on the graph and only permits movement left and right along the graph. The coordinates displayed during the tracing movement are coordinates of points that satisfy the equation. In most cases, these coordinates are not the same as the pixel screen coordinates displayed using the unrestricted cursor movement discussed earlier.

EXPLORE/DISCUSS 2

Graph the equation $y = x$ in a standard viewing window.

(A) Without selecting the TRACE command, move the cursor to a point on the screen that appears to lie on the graph of $y = x$ and is as close to (5, 5) as possible. Record these coordinates. Do these coordinates satisfy the equation $y = x$?

(B) Now select the TRACE command and move the cursor along the graph of $y = x$ to a point that has the same x coordinate found in part A. Is the y coordinate of this point the same as you found in part A? Do the coordinates of the point using trace satisfy the equation $y = x$?

(C) Explain the difference in using trace along a curve and trying to use unrestricted movement of a cursor along a curve.

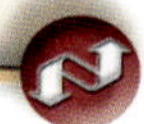

Most graphing utilities have a **ZOOM** command. In general, **zooming in** on a graph reduces the window variables and magnifies the portion of the graph visible in the viewing window [Fig. 8(a)]. **Zooming out** enlarges the window variables so that more of the graph is visible in the viewing window [Fig. 8(b)].

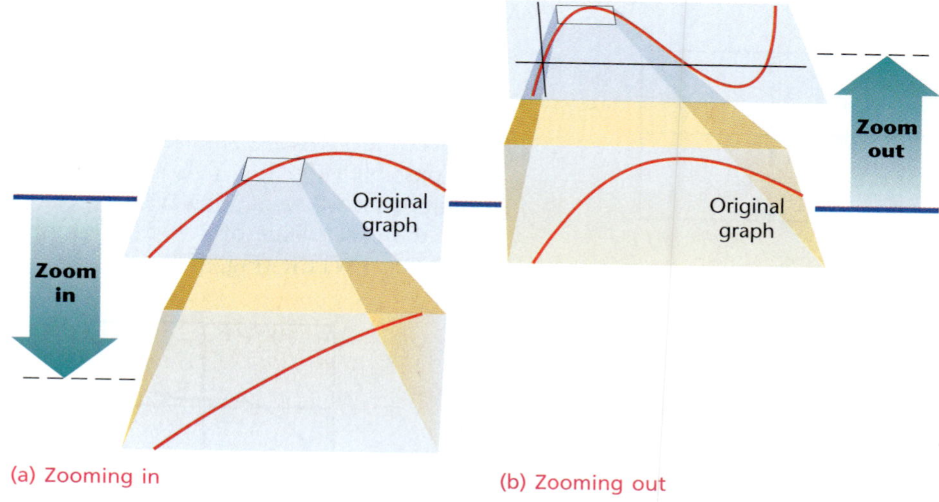

FIGURE 8 The zoom operation.

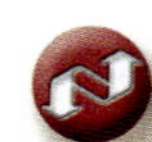

EXPLORE/DISCUSS 3

Figure 9 shows the ZOOM menu on a TI-83.* We want to explore the effects of some of these options on the graph of $y_1 = x$. Enter this equation in the equation editor and select ZStandard from the ZOOM menu. What are the window variables? In each of the following, position the cursor at the origin and select the indicated zoom option. Observe the changes in the window variables and examine the coordinates displayed by tracing along the curve.

(A) ZSquare (B) ZDecimal (C) ZInteger (D) ZoomFit

```
ZOOM MEMORY
1:ZBox
2:Zoom In
3:Zoom Out
4:ZDecimal
5:ZSquare
6:ZStandard
7:ZTrig
8:ZInteger
9:ZoomStat
0:ZoomFit
```

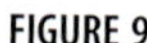

FIGURE 9 The ZOOM menu on a TI-83.

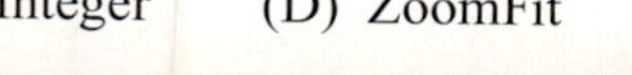

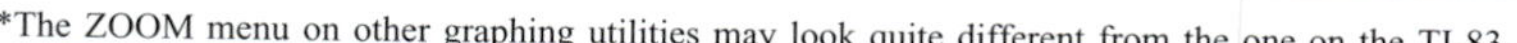

*The ZOOM menu on other graphing utilities may look quite different from the one on the TI-83.

Another command found on most graphing utilities is **intersect*** or **isect.** This command enables the user to find the point(s) where two curves intersect without using trace or zoom.

The use of trace, zoom, and intersect is best illustrated by examples.

EXAMPLE 3 Using Trace, Zoom, and Intersect

Let $y = 0.01x^3 + 1$

(A) Use the TRACE command to find y when $x = 5$.

(B) Use the TRACE and ZOOM commands to find x when $y = 5$.

(C) Use the intersect command to find x when $y = 5$.

Round answers to two decimal places.

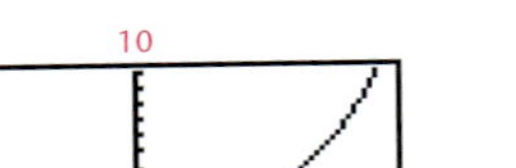

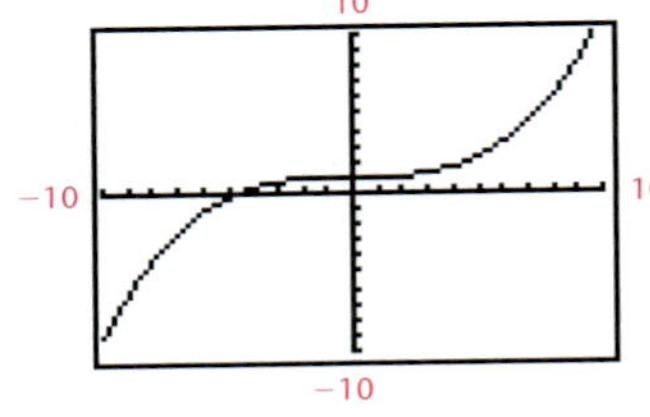

FIGURE 10

SOLUTION

(A) Enter $y_1 = 0.01x^3 + 1$ in a graphing utility (Fig. 10). In Example 2, we discussed two ways to find the value of y_1 for a given value of x. Now we want to discuss a third way, the TRACE command. Graph y_1 in the standard viewing window (Fig. 11).

Select the TRACE command and move the cursor as close to $x = 5$ as possible (Fig. 12). This shows that $y = 2.33$ when $x = 5.11$, not quite what we want. However, we can direct the trace command to use the exact value of $x = 5$ by simply entering 5 (Fig. 13) and pressing ENTER (Fig. 14).

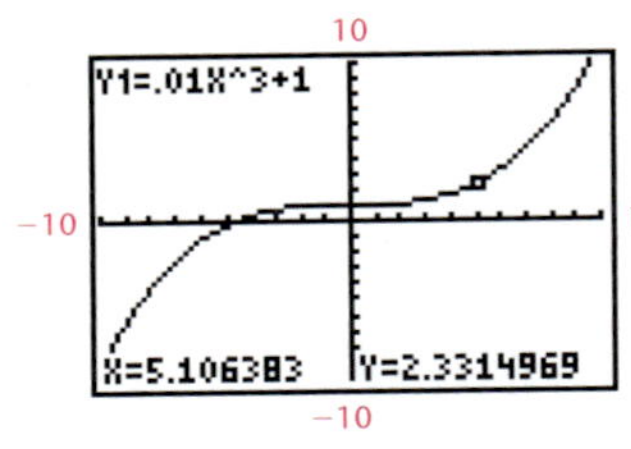

FIGURE 11

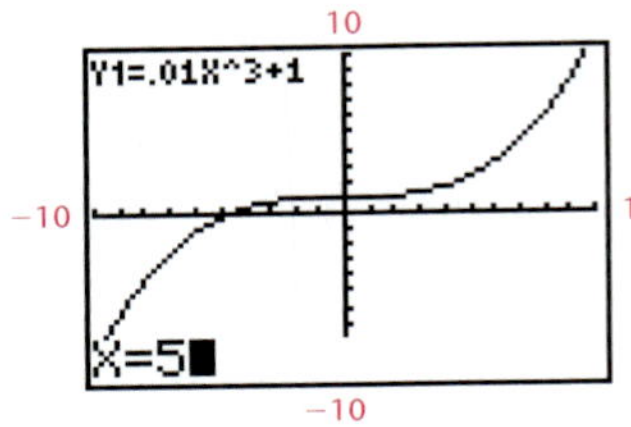

FIGURE 12

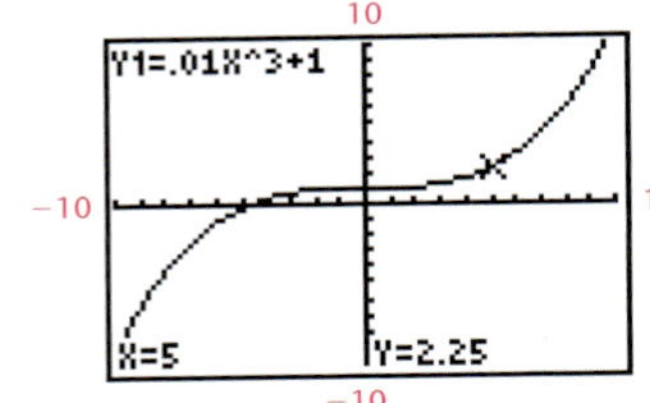

FIGURE 13

FIGURE 14

From Figure 14 we see that $y = 2.25$ when $x = 5$.

(B) Select the TRACE command and move the cursor as close to $y = 5$ as possible (Fig. 15). This shows that $x = 7.45$ when $y = 5.13$, again not exactly what we want. We cannot direct the TRACE command to use the exact value $y = 5$. Instead we press the ZOOM command and select Zoom In to obtain more accuracy (Fig. 16). Then we select the TRACE command and move the cursor as close to $y = 5$ as possible (Fig. 17).

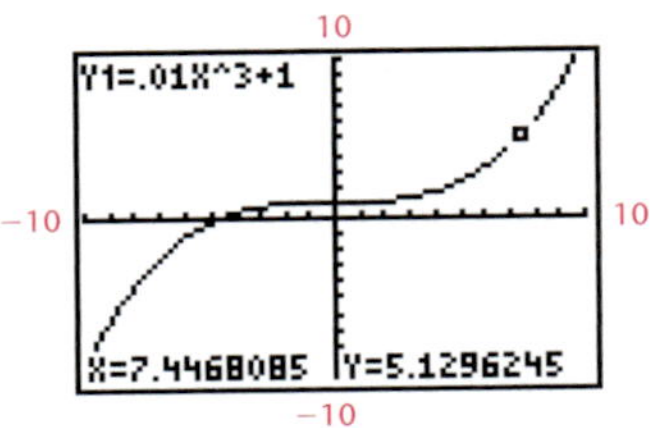

FIGURE 15

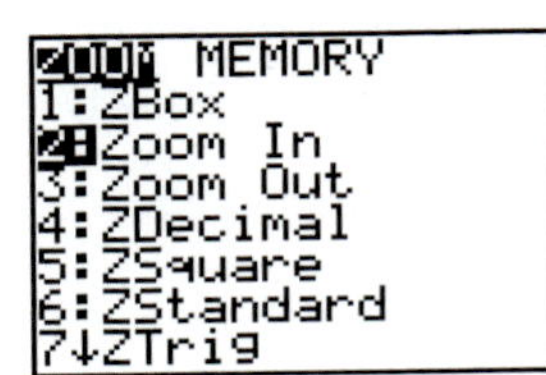

FIGURE 16

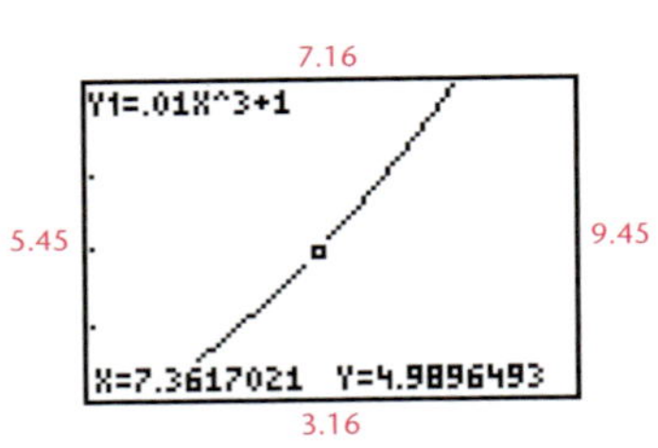

FIGURE 17

*On the TI-83, intersect is found on the CALC (2nd-TRACE) menu.

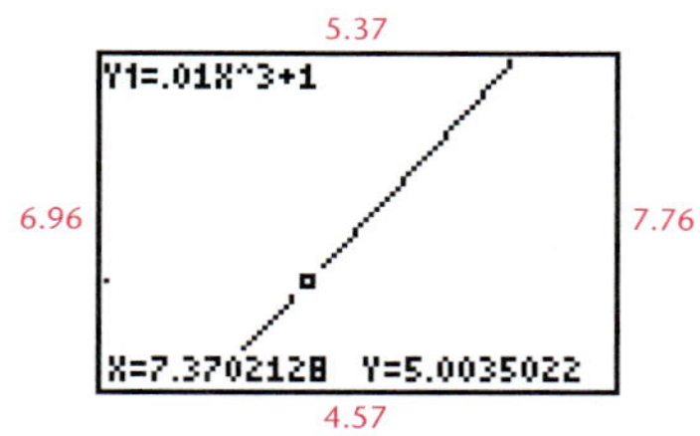

FIGURE 18

Now we see that $x = 7.36$ when $y = 4.99$. This is an improvement, but we can do better. Repeating the Zoom In command and tracing along the curve (Fig. 18), we see that $x = 7.37$ when $y = 5.00$.

(C) Enter $y_2 = 5$ in the graphing utility and graph y_1 and y_2 in the standard viewing window (Fig. 19). Now there are two curves displayed on the graph. The horizontal line is the graph of $y_2 = 5$ and the other curve is the familiar graph of y_1. The coordinates of the intersection point of the two curves must satisfy both equations. Clearly, the y coordinate of this intersection point is 5. The x coordinate is the value we are seeking. We use the intersect command to find the coordinates of the intersection point in Figure 19. When we select the intersect command, we are asked to make three choices: the first curve, the second curve, and a guess. When the desired equation is displayed at the top of the screen, press ENTER to select it (Figs. 20 and 21). (If there are more than two curves, use the up and down arrows to display the desired equation, then press ENTER.)

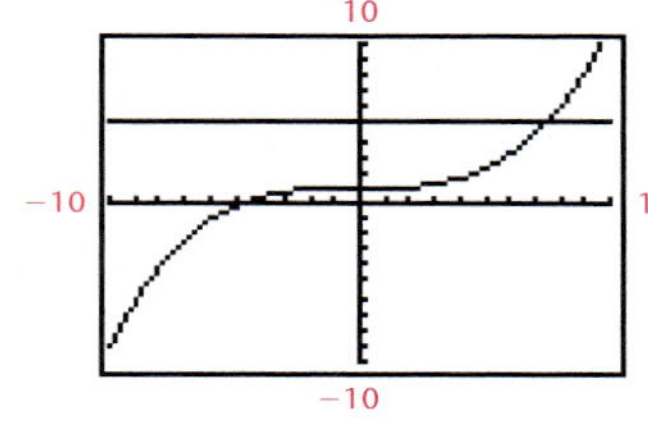

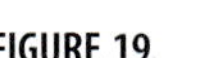
FIGURE 19

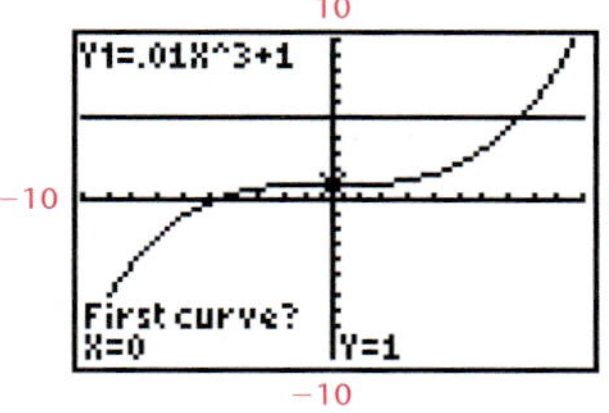

FIGURE 20

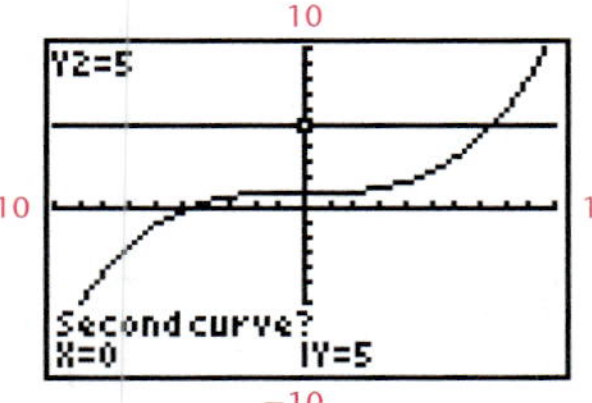

FIGURE 21

To enter a guess, move the cursor close to the intersection point (Fig. 22) and press ENTER (Fig. 23). (We will see a more important use of entering a guess in Example 4.) Examining Figure 23, we see that $x = 7.37$ when $y = 5$.

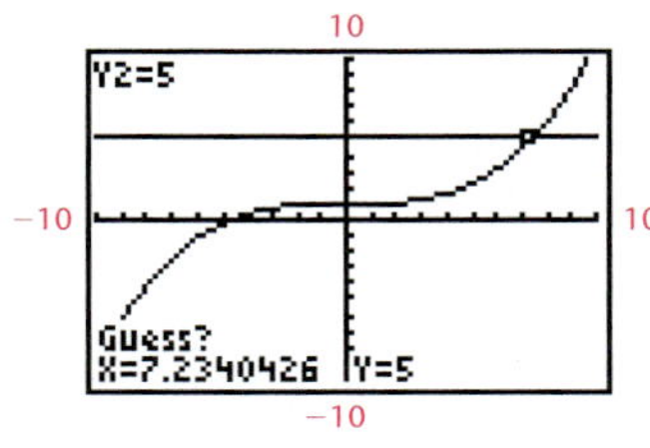

FIGURE 22

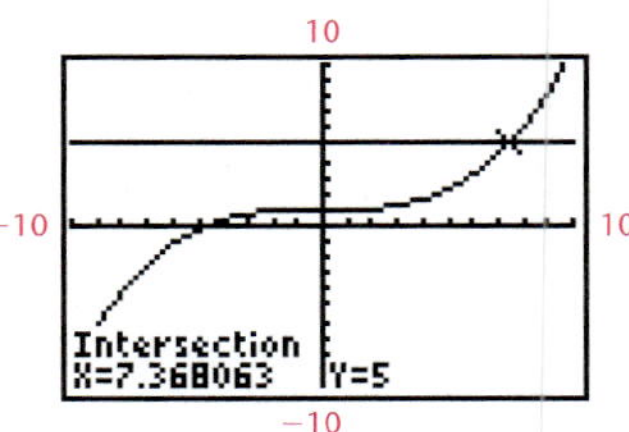

FIGURE 23

MATCHED 3 PROBLEM

Repeat Example 3 for $y = 1 - 0.02x^3$.

EXAMPLE  Solving an Equation with Multiple Solutions

Solve the equation

$$x^3 + 20x^2 + 60x = 200$$

Round answers to two decimal places.

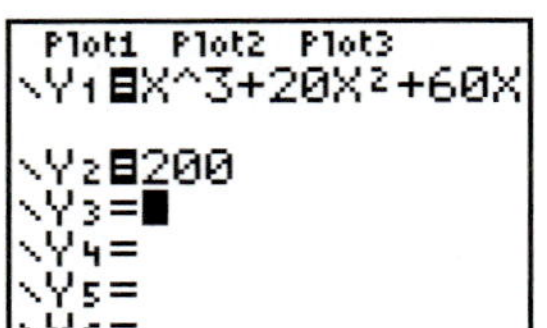

FIGURE 24

SOLUTION

We are going to solve this equation by graphing both sides in the same viewing window and finding the intersection points. First we enter $y_1 = x^3 + 20x^2 + 60x$ and $y_2 = 200$ in the graphing utility (Fig. 24) and graph in the standard viewing window (Fig. 25).

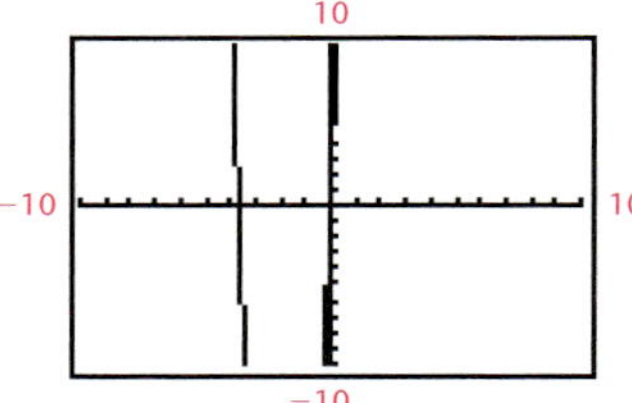

FIGURE 25

Many students always start with the standard viewing window, but in this case it is a poor choice. Because we are seeking the values of x that make the left side of the equation equal to 200, we must choose a value for Ymax that is larger than 200. Changing Ymax to 300 and Yscl to 30 produces a new graph (Fig. 26). The two curves intersect twice in this window. The x coordinates of these points are the solutions we are seeking. But first, could there be other solutions that are not visible in this window? To find out, we must investigate the behavior outside this window. A table is the most convenient way to do this (Figs. 27 and 28).

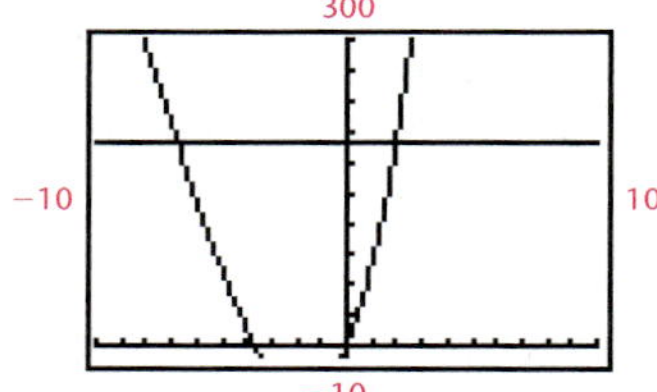

FIGURE 26

X	Y1	Y2
11	4411	200
12	5328	200
13	6357	200
14	7504	200
15	8775	200
16	10176	200
17	11713	200

Y1 = X^3+20X²+60X

FIGURE 27

X	Y1	Y2
-17	-153	200
-16	64	200
-15	225	200
-14	336	200
-13	403	200
-12	432	200
-11	429	200

Y1 = X^3+20X²+60X

FIGURE 28

Examining Figure 27, we see that the values of y_1 are getting very large. It is unlikely that there will be additional solutions to the equation $y_1 = 200$ for larger values of x. Examining Figure 28, we see that there are values of y_1 that are less than 200 so it is likely that there will be additional solutions for more negative values of x. Based on the table values in Figure 28, we make the following changes in the window variables: Xmin $= -20$, Xscl $= 5$, Ymax $= 500$, Ymin $= -200$, Yscl $= 50$. This produces the graph in Figure 29. Examining the values of y_1 for values of x to the left of this window (Fig. 30), we conclude that there are no other intersection points.

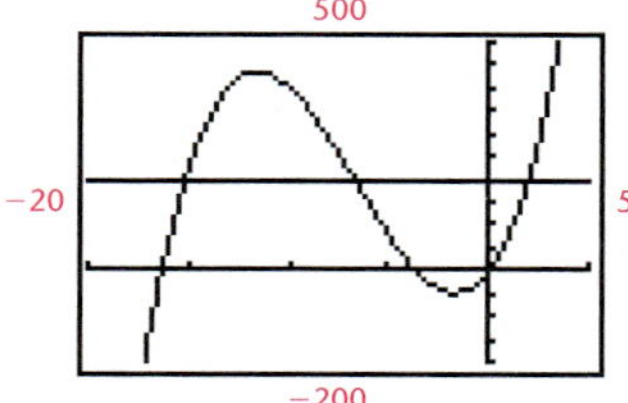

FIGURE 29

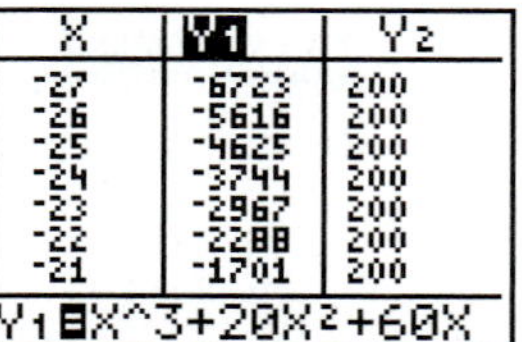

X	Y1	Y2
-27	-6723	200
-26	-5616	200
-25	-4625	200
-24	-3744	200
-23	-2967	200
-22	-2288	200
-21	-1701	200

Y1 = X^3+20X²+60X

FIGURE 30

Now we use the intersect command to find the x coordinates of the three intersection points in Figure 29. We select the two equations as before (Figs. 31 and 32). To specify which of the three intersection points we want to locate, we make a guess that is close to the desired point. We first select the leftmost intersection point by moving the cursor to that point (Fig. 33) and pressing ENTER.

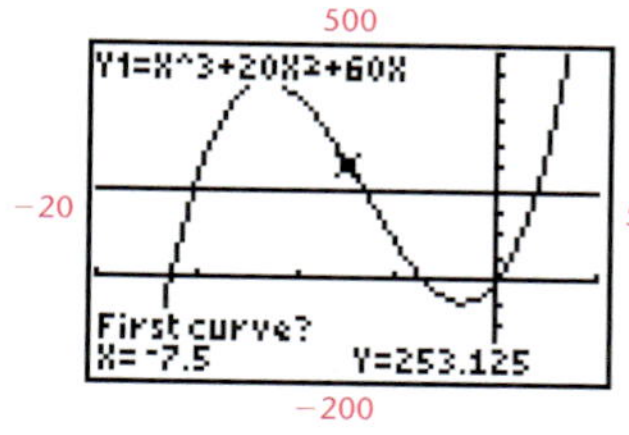

FIGURE 31

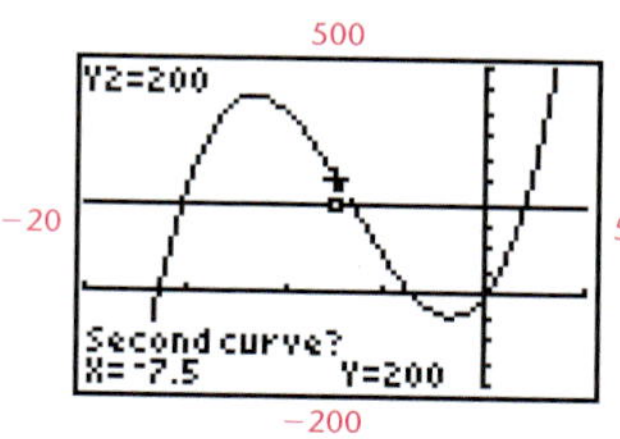

FIGURE 32

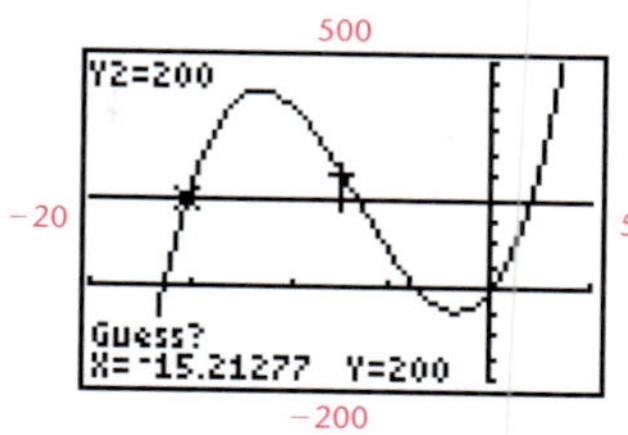

FIGURE 33

The coordinates of the leftmost intersection point are displayed at the bottom of the screen (Fig. 34). To find the other two intersection points, we repeat the entire process. When we get to the screen that asks for a guess, we place the cursor near the point we are looking for and press ENTER (Figs. 35 and 36).

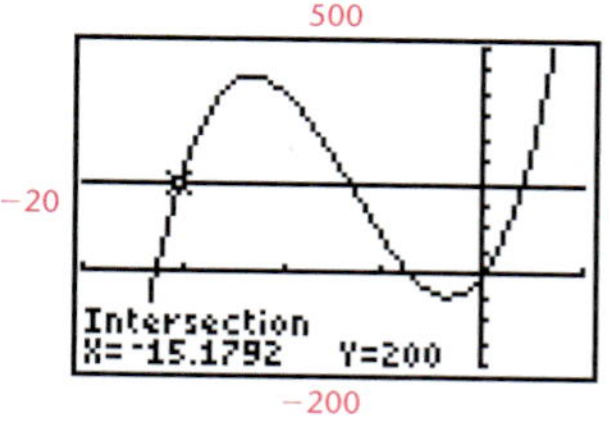

FIGURE 34

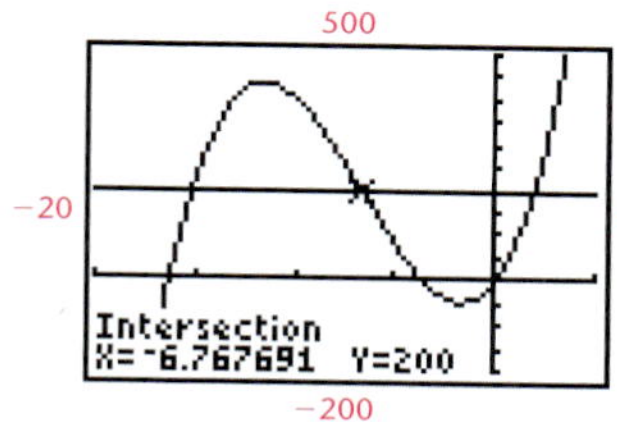

FIGURE 35

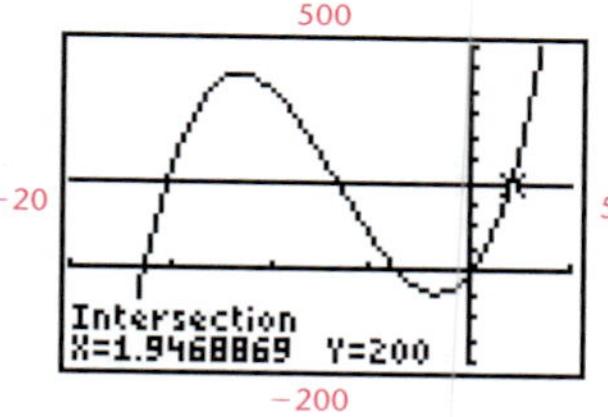

FIGURE 36

Thus, we see that the solutions to $x^3 + 20x^2 + 60x = 200$ are $x = -15.18$, $x = -6.77$, and $x = 1.95$.

MATCHED 4 PROBLEM

Solve $x^3 + 10x^2 - 100x = 100$. Round answers to two decimal places.

In the solution to Example 4, we had to rely on examining tables and our intuition to conclude that the two curves intersected only three times. One of the major objectives of this course is to broaden our knowledge base so that we can be more definitive in our reasoning. For example, in Chapter 3 we will show that any equation like the one in Example 4 can have no more than three solutions.

Examples 1 through 4 dealt with a variety of methods for finding the value of y that corresponds to a given value of x and the value(s) of x that correspond to a given value of y. These methods are summarized in the following box.

Finding Solutions to an Equation

Assume the equation is entered in a graphing utility as $y_1 =$ (expression in x).

To find solutions (x, y) given $x = a$, use any of the following methods:

Method 1. Store a in X on the graphing utility and display y_1 on the home screen.

Method 2. Set Tblstart to a and display the table.

Method 3. Graph y_1, select the trace command, and enter a.

To find solutions (x, y) given $y = b$, use either of the following methods:

Method 1. Graph y_1 and use trace and zoom.

Method 2. Graph y_1 and $y_2 = b$ and use intersect.

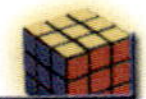

Mathematical Modeling

Now that we have developed the ability to solve equations on a graphing utility, there are many applications that we can investigate. The next example develops a mathematical model for manufacturing a box.

EXAMPLE 5 Manufacturing

Rectangular open boxes are to be manufactured from 11- by 17-inch sheets of cardboard by cutting x- by x-inch squares out of the corners and folding up the sides, as shown in Figure 37.

(A) Write an equation for the volume y of the resulting box in terms of the length x of the sides of the squares that are cut out. Indicate appropriate restrictions on x.

(B) Graph the equation for appropriate values of x. Adjust the window variables for y to include the entire graph of interest.

(C) Find the smallest square that can be cut out to produce a box with a volume of 150 cubic inches.

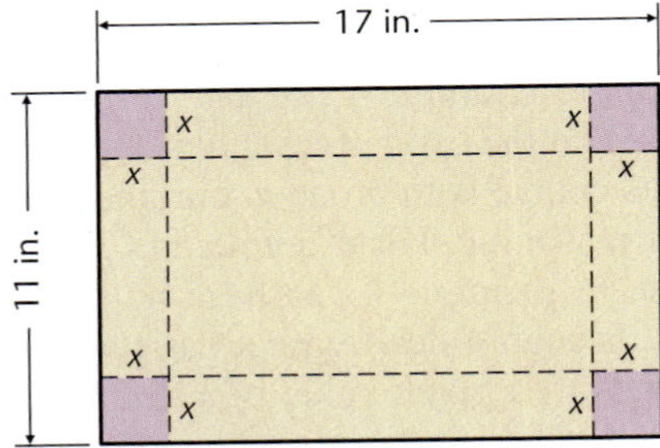

FIGURE 37 Template for boxes.

SOLUTION

(A) The box is shown in Figure 38(b) with the sides folded up and dimensions added. From this figure we can write the equation of the volume in terms of x and establish the restrictions on x. No dimension can be negative; that is,

$$x \geq 0 \qquad \begin{aligned} 11 - 2x &\geq 0 \\ 11 &\geq 2x \\ 5.5 &\geq x \end{aligned} \qquad \begin{aligned} 17 - 2x &\geq 0 \\ 17 &\geq 2x \\ 8.5 &\geq x \end{aligned}$$

Because x must satisfy all three of the above inequalities, we conclude that $0 \leq x \leq 5.5$. (See Fig. 39. Inequalities and intervals are reviewed in Appendix A, Section A.1.) Thus, the volume of the box is given by:

$$y = x(17 - 2x)(11 - 2x) \qquad 0 \leq x \leq 5.5 \tag{3}$$

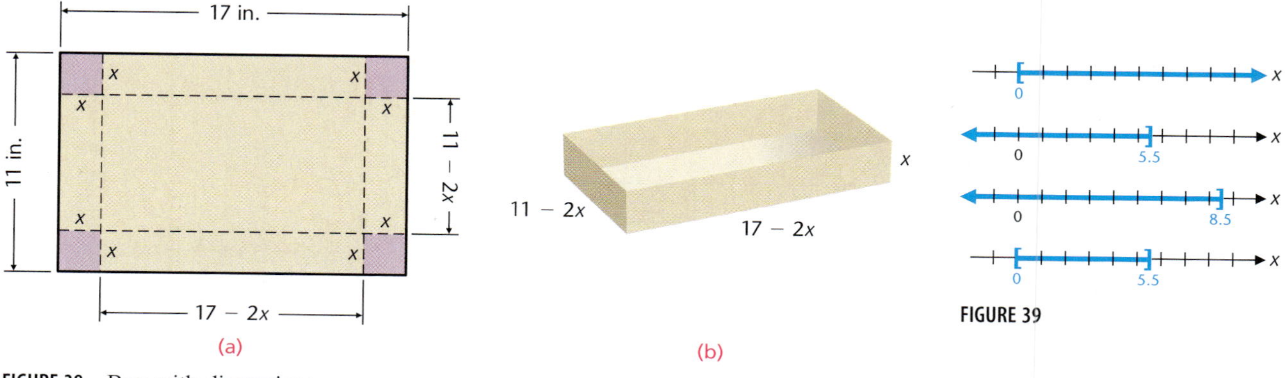

FIGURE 38 Box with dimensions added.

FIGURE 39

(B) Entering this equation in a graphing utility (it does not need to be multiplied out) and evaluating it for several integers between 0 and 5 (Fig. 40), it appears that a good choice for the window dimensions for y is $0 \leq y \leq 200$. This choice can easily be changed if there is too much space above the graph or if part of the graph we are interested in is out of the viewing window. Figure 41 shows the graph of equation (3) in the viewing window selected above.

X	Y1	
0	0	
1	135	
2	182	
3	165	
4	108	
5	35	
6	-30	

Y1■X(17-2X)(11-...

FIGURE 40

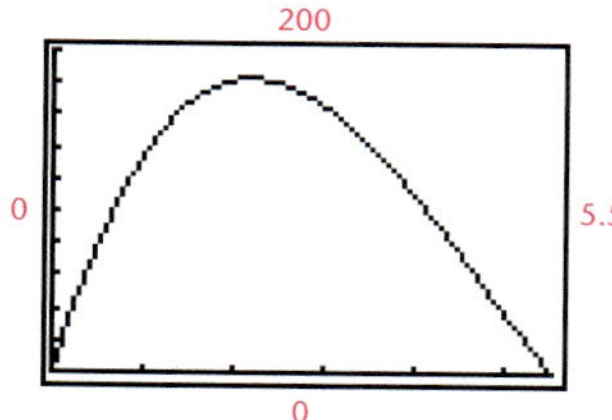

FIGURE 41

(C) We want to find the smallest value of x for which $y = 150$. That is, we want to solve the equation

$$x(17 - 2x)(11 - 2x) = 150$$

We choose to solve this equation with the intersect command. Entering $y_2 = 150$ in the graphing utility and pressing GRAPH produces the two curves shown in Figure 42. The curves intersect twice. Because we are looking for the smallest value of x that satisfies the equation, we want the intersection point on the left (Fig. 43).

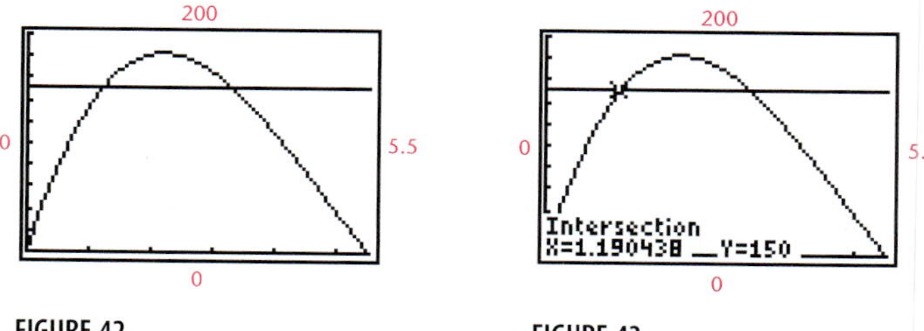

FIGURE 42

FIGURE 43

From Figure 43 we see that the volume of the box is 150 cubic inches when x is 1.19 inches.

MATCHED 5 PROBLEM

Refer to Example 5. Approximate to two decimal places the size of the largest square that can be cut out to produce a volume of 150 cubic inches.

ANSWERS MATCHED PROBLEMS

1.

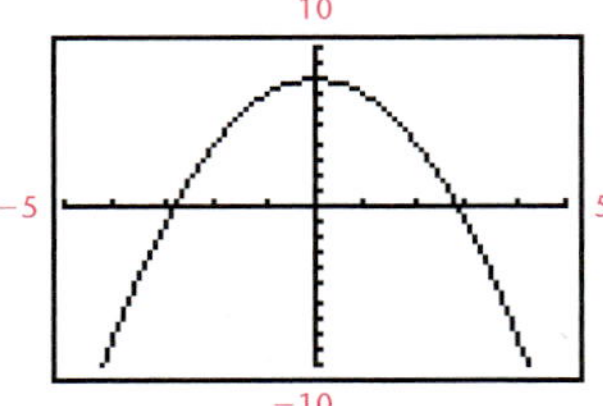

2.

x	−4	−3	−2	−1	0	1	2	3	4
y	6.2	0.7	−1.2	−0.7	1	2.7	3.2	1.3	−4.2

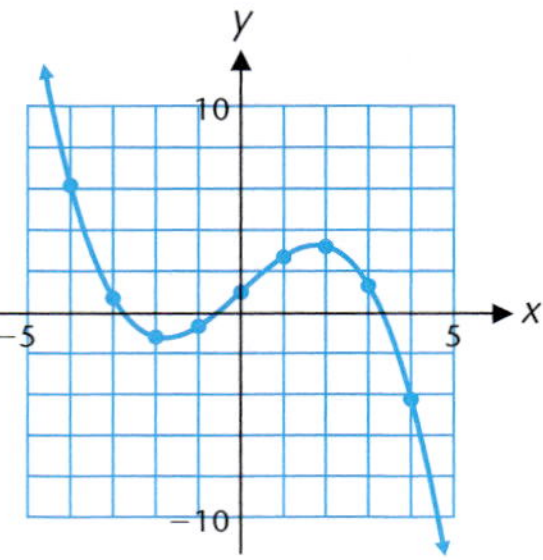

3. (A) −1.5 (B) −5.85 (C) −5.85 **4.** −15.90, −0.92, 6.82 **5.** 3.32 in.

EXERCISE 1.1

A

In Problems 1–6, determine if the indicated point lies in the viewing window defined by

$$\text{Xmin} = -7, \text{Xmax} = 9, \text{Ymin} = -4, \text{Ymax} = 11$$

1. (0, 0) **2.** (0, 10) **3.** (10, 0)

4. (−3, −5) **5.** (−5, −3) **6.** (−8, 12)

7. Consider the points in the following table:

x	3	6	−7	−4	0
y	−9	−4	14	0	2

(A) Find the smallest rectangle in a Cartesian coordinate system that will contain all the points in the table. State your answer in terms of the window variables Xmin, Xmax, Ymin, and Ymax.

(B) Enter the window variables you determined in part A and display the corresponding viewing window. Can you use the cursor to display the coordinates of the points in the table on the graphing utility screen? Discuss the differences between the rectangle in the plane and the pixels displayed on the screen.

8. Repeat Problem 7 for the following table.

x	−4	0	−2	7	4
y	2	−4	0	−2	3

In Problems 9–14, graph each equation in a standard viewing window.

9. $y = -x$

10. $y = 0.5x$

11. $y = 9 - 0.4x^2$

12. $y = 0.3x^2 - 4$

13. $y = 2\sqrt{x + 5}$

14. $y = -2\sqrt{x + 5}$

B

For each equation in Problems 15–20, use the table on a graphing utility to construct a table of values over the indicated interval, computing y values to the nearest tenth of a unit. Plot these points on graph paper, then with the aid of a graph on a graphing utility, complete the hand sketch of the graph.

15. $y = 4 + 4x - x^2$, $-2 \le x \le 6$ (Use even integers for the table.)

16. $y = 2x^2 + 12x + 5$, $-7 \le x \le 1$ (Use odd integers for the table.)

17. $y = 2\sqrt{2x + 10}$, $-5 \le x \le 5$ (Use odd integers for the table.)

18. $y = \sqrt{8 - 2x}$, $-4 \le x \le 4$ (Use even integers for the table.)

19. $y = 0.5x(4 - x)(x + 2)$, $-3 \le x \le 5$ (Use odd integers for the table.)

20. $y = 0.5x(x + 3.5)(2.8 - x)$, $-4 \le x \le 4$ (Use even integers for the table.)

In Problems 21–24, graph the equation in a standard viewing window. Approximate to two decimal places the x coordinates of the points in this window that are on the graph of the equation and have the indicated y coordinates.

21. $y = 4 - 3\sqrt[3]{x + 4}$

(A) $(x, 8)$ **(B)** $(x, -1)$

22. $y = 3 + 4\sqrt[3]{x - 4}$

(A) $(x, 8)$ **(B)** $(x, -6)$

23. $y = 3 + x + 0.1x^3$

(A) $(x, 4)$ **(B)** $(x, -7)$

24. $y = 2 - 0.5x - 0.1x^3$

(A) $(x, 7)$ **(B)** $(x, -5)$

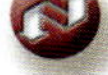

The graphs of each pair of equations in Problems 25–34 intersect in exactly two points. Find a viewing window that clearly shows both points of intersection (there are many windows that will do this). Then use intersect to find the coordinates of each intersection point to two decimal places.

25. $y = x^2 - 10x$, $y = 12 - 5x$

26. $y = x^2 + 15x$, $y = 15 + 10x$

27. $y = 15x - x^2$, $y = 10 - 4x$

28. $y = -x^2 - 20x$, $y = 10x + 15$

29. $y = 0.4x^2 - 5x + 10$, $y = 5 + 9x - 0.3x^2$

30. $y = 0.2x^2 + 7x - 15$, $y = 9 - 7x - 0.1x^2$

31. $y = \sqrt{9x + 20}$, $y = 0.2x + 10$

32. $y = 20 - \sqrt{5x + 5}$, $y = 14 - 0.1x$

33. $y = \sqrt{x + 10}$, $y = 0.1x^2 - 5x - 10$

34. $y = \sqrt{5 - x}$, $y = 0.1x^2 + 5x$

35. **(A)** Sketch the graph of $x^2 + y^2 = 9$ by hand and identify the curve.*

(B) Graph $y_1 = \sqrt{9 - x^2}$ and $y_2 = -\sqrt{9 - x^2}$ in the standard viewing window of a graphing utility. How does this graph compare to the graph you drew in part (A)?

(C) Apply each of the following ZOOM options to the graph in part (B) and determine which options produce a curve that looks like the curve you drew in part (A):

ZDecimal, ZSquare, ZoomFit

36. **(A)** Sketch the graph of $x^2 + y^2 = 4$ by hand and identify the curve.

(B) Graph $y_1 = \sqrt{4 - x^2}$ and $y_2 = -\sqrt{4 - x^2}$ in the standard viewing window of a graphing utility. How does this graph compare to the graph you drew in part (A)?

(C) Apply each of the following ZOOM options to the graph in part (B) and determine which options produce a curve that looks like the curve you drew in part (A):

ZDecimal, ZSquare, ZoomFit

In Problems 37–40, use the intersection command on a graphing utility to solve each equation for the indicated values of b. Round answers to two decimal places.

37. $0.1x^3 - x^2 - 5x + 100 = b$

(A) $b = 25$ **(B)** $b = 75$ **(C)** $b = 125$

38. $0.1x^3 + x^2 - 7x - 100 = b$

(A) $b = -125$ **(B)** $b = -75$ **(C)** $b = 75$

39. $0.01x^4 + 4x + 50 = b$

(A) $b = 25$ **(B)** $b = 75$

40. $0.01x^4 - 3x + 50 = b$

(A) $b = 25$ **(B)** $b = 75$

*Graphs of equations of this form are reviewed in Appendix A, Section A.3.

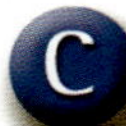

In Problems 41–44, use the intersection command on a graphing utility to solve each equation for the indicated values of b. Round answers to two decimal places.

41. $1{,}200x - x^3 = b$

(A) $b = 12{,}000$ **(B)** $b = 16{,}000$ **(C)** $b = 20{,}000$

42. $x^3 - 300x = b$

(A) $b = 1{,}000$ **(B)** $b = 2{,}000$ **(C)** $b = 3{,}000$

43. $x^4 + 4{,}000x = b$

(A) $b = 30{,}000$ **(B)** $b = -30{,}000$ **(C)** $b = -40{,}000$

44. $800x^2 - x^4 = b$

(A) $b = 100{,}000$ **(B)** $b = 160{,}000$ **(C)** $b = 200{,}000$

45. The point $(\sqrt{2}, 2)$ is on the graph of $y = x^2$. Use trace and zoom to approximate $\sqrt{2}$ to four decimal places. Compare your result with the direct calculator evaluation of $\sqrt{2}$.

46. The point $(\sqrt[3]{4}, 4)$ is on the graph of $y = x^3$. Use trace and zoom to approximate $\sqrt[3]{4}$ to four decimal places. Compare your result with the direct calculator evaluation of $\sqrt[3]{4}$.

47. In a few sentences, discuss the difference between the mathematical coordinates of a point and the screen coordinates of a pixel.

48. In a few sentences, discuss the difference between the coordinates displayed during unrestricted cursor movement and those displayed during the trace procedure.

APPLICATIONS

49. Manufacturing. An oil tank in the shape of a right circular cylinder* has a volume of 40,000 cubic feet. If regulations for such tanks require that the radius plus the height must be 50 feet, find the radius and the height to two decimal places.

50. Manufacturing. A drinking container in the shape of a right circular cone* has a volume of 50 cubic inches. If the radius plus the height of the cone is 8 inches, find the radius and the height to two decimal places.

51. Manufacturing. A rectangular open-top box is to be constructed out of an 8.5-inch by 11-inch sheet of thin cardboard by cutting x-inch squares out of each corner and bending the sides up, as in Figures 37 and 38 in Example 5. What size squares to two decimal places should be cut out to produce a box with a volume of 55 cubic inches? Give the dimensions to two decimal places of all possible boxes with the given volume. Check your answers.

52. Manufacturing. A rectangular open-top box is to be constructed out of a 9-inch by 12-inch sheet of thin cardboard by cutting x-inch squares out of each corner and bending the sides up as shown in Figures 37 and 38 in Example 5. What size squares to two decimal places should be cut out to produce a box with a volume of 72 cubic inches? Give the dimensions to two decimal places of all possible boxes with the given volume. Check your answers.

53. Manufacturing. A box with a lid is to be cut out of a 12-inch by 24-inch sheet of thin cardboard by cutting out six x-inch squares and folding as indicated in the figure. What are the dimensions to two decimal places of all possible boxes that will have a volume of 100 cubic inches? Check your answers.

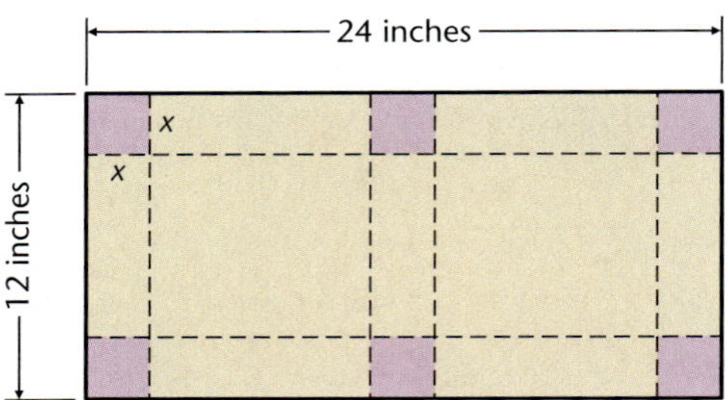

54. Manufacturing. A box with a lid is to be cut out of a 10-inch by 20-inch sheet of thin cardboard by cutting out six x-inch squares and folding as indicated in the figure. What are the dimensions to two decimal places of all possible boxes that will have a volume of 75 cubic inches? Check your answers. (Refer to the figure for Problem 53.)

*Geometric formulas can be found in Appendix C.

55. **Price and Demand.** A nationwide office supply company sells high-grade paper for laser printers. The price per case y (in dollars) and the weekly demand x for this paper are related approximately by the equation

$$y = 100 - 0.6\sqrt{x} \qquad 5{,}000 \le x \le 20{,}000$$

(A) Complete the following table. Approximate each value of x to the nearest hundred cases.

x			
y	20	25	30

(B) Does the demand increase or decrease if the price is increased from \$25 to \$30? By how much?

(C) Does the demand increase or decrease if the price is decreased from \$25 to \$20? By how much?

56. **Price and Demand.** Refer to the relationship between price and demand given in Problem 55.

(A) Complete the following table. Approximate each value of x to the nearest hundred cases.

x			
y	35	40	45

(B) Does the demand increase or decrease if the price is increased from \$40 to \$45? By how much?

(C) Does the demand increase or decrease if the price is decreased from \$40 to \$35? By how much?

57. **Price and Revenue.** Refer to Problem 55. The revenue from the sale of x cases of paper at \$$y$ per case is given by the product $R = xy$.

(A) Use the results from Problem 55 to complete the following table of revenues.

y	20	25	30
R			

(B) Does the revenue increase or decrease if the price is increased from \$25 to \$30? By how much?

(C) Does the revenue increase or decrease if the price is decreased from \$25 to \$20? By how much?

(D) If the current price of paper is \$25 per case and the company wants to increase revenue, should it raise the price \$5, lower the price \$5, or leave the price unchanged?

58. **Price and Revenue.** Refer to Problem 56. The revenue from the sale of x cases of paper at \$$y$ per case is given by the product $R = xy$.

(A) Use the results from Problem 56 to complete the following table of revenues.

y	35	40	45
R			

(B) Does the revenue increase or decrease if the price is increased from \$40 to \$45? By how much?

(C) Does the revenue increase or decrease if the price is decreased from \$40 to \$35? By how much?

(D) If the current price of paper is \$40 per case and the company wants to increase revenue, should it raise the price \$5, lower the price \$5, or leave the price unchanged?

SECTION 1.2 Functions

Definition of Function • Functions Defined by Equations • Function Notation • Modeling and Data Analysis • A Brief History of the Function Concept

The idea of correspondence plays a central role in the formulation of the function concept. You have already had experiences with correspondences in everyday life. For example:

To each person there corresponds an age.

To each item in a store there corresponds a price.

To each automobile there corresponds a license number.

To each circle there corresponds an area.

To each number there corresponds its cube.

One of the most important aspects of any science (managerial, life, social, physical, computer, etc.) is the establishment of correspondences among various types of phenomena. Once a correspondence is known, predictions can be made. A chemist can use a gas law to predict the pressure of an enclosed gas, given its temperature. An engineer can use a formula to predict the deflections of a beam subject to different loads. A computer scientist can use formulas to compare the efficiency of algorithms for sorting data stored in a computer. An economist would like to be able to predict interest rates, given the rate of change of the money supply. And so on.

Definition of Function

What do all the preceding examples have in common? Each describes the matching of elements from one set with elements in a second set. Let's consider the correspondences between the sets A and B listed in Tables 1–3.

TABLE 1

A	B
–2	–5
–1	–3
0	0
1	3
2	5

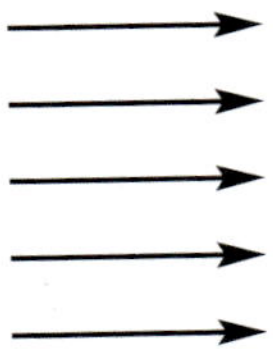

TABLE 2

A	B
–2	–5
–1	–3
0	0
1	3
2	5

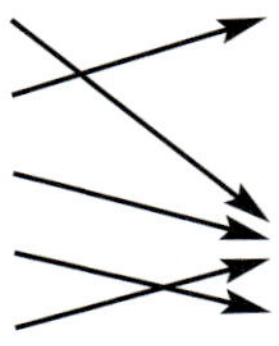

TABLE 3

A	B
–2	–5
–1	–3
0	0
1	3
2	5

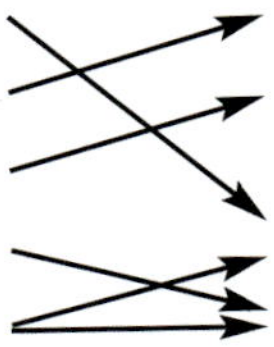

Tables 1 and 2 specify functions, but Table 3 does not. Why not? The definition of the concept of a *function* will explain.

DEFINITION 1

Definition of Function

A **function** is a correspondence between two sets of elements such that to each element in the first set there corresponds *one and only one* element in the second set.

The first set is called the **domain** and the set of all corresponding elements in the second set is called the **range.**

Table 1 specifies a function with domain $A = \{-2, -1, 0, 1, 2\}$ and range $B = \{-5, -3, 0, 3, 5\}$ because to each domain value in A there corresponds exactly one element in B. Table 2 also specifies a function with domain A and range $C = \{-5, 3, 5\}$. Notice that the range value 3 corresponds to three different domain values, -2, 0, and 2. This does not violate the conditions in Definition 1. On the other hand, Table 3 does not specify a function, because two different range values, 3 and 5, correspond to the domain value 2.

REMARK Some graphing utilities use the term *range* to refer to the window variables. In this book, *range* will always refer to the range of a function.

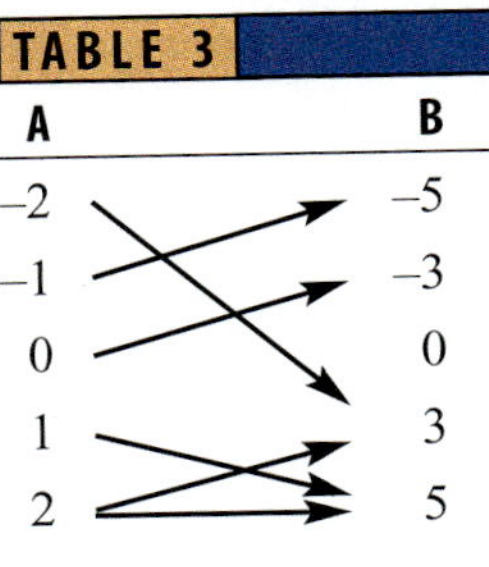

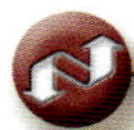

EXPLORE/DISCUSS 1

Consider the set of students enrolled in a college and the set of faculty members of that college. Define a correspondence between the two sets by saying that a student corresponds to a faculty member if the student is currently enrolled in a course taught by the faculty member. Is this correspondence a function? Discuss.

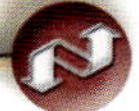

Every function can be specified by using ordered pairs of elements, where the first component represents a domain element and the second element represents the corresponding range element. Thus, the functions specified in Tables 1 and 2 can be written as follows:

$$F = \{(-2, -5), (-1, -3), (0, 0), (1, 3), (2, 5)\}$$
$$G = \{(-2, 3), (-1, -5), (0, 3), (1, 5), (2, 3)\}$$

In both cases, notice that no two ordered pairs have the same first component and different second components. On the other hand, if we list the set H of ordered pairs determined by Table 3, we have

$$H = \{(-2, 3), (-1, -5), (0, -3), (1, 5), (2, 3), (2, 5)\}$$

Notice that the last two ordered pairs, (2, 3) and (2, 5), have the same first component and different second components. Thus, set H does not specify a function.

This ordered pair approach suggests an alternative but equivalent way of defining the concept of function that provides additional insight into this concept.

DEFINITION 2
Set Form of the Definition of Function

A **function** is a set of ordered pairs with the property that no two ordered pairs have the same first component and different second components.

The set of all first components in a function is called the **domain** of the function, and the set of all second components is called the **range.**

EXAMPLE 1 Functions Specified as Sets of Ordered Pairs

Determine whether each set specifies a function. If it does, then state the domain and range.

(A) $S = \{(1, 4), (2, 3), (3, 2), (4, 3), (5, 4)\}$

(B) $T = \{(1, 4), (2, 3), (3, 2), (2, 4), (1, 5)\}$

SOLUTIONS

(A) Because all the ordered pairs in S have distinct first components, this set specifies a function. The domain and range are

$$\text{Domain} = \{1, 2, 3, 4, 5\} \qquad \text{Set of first components}$$
$$\text{Range} = \{2, 3, 4\} \qquad \text{Set of second components}$$

(B) Because there are ordered pairs in T with the same first component [for example, (1, 4) and (1, 5)], this set does not specify a function.

MATCHED PROBLEM 1

Determine whether each set specifies a function. If it does, then state the domain and range.

(A) $S = \{(-2, 1), (-1, 2), (0, 0), (-1, 1), (-2, 2)\}$

(B) $T = \{(-2, 1), (-1, 2), (0, 0), (1, 2), (2, 1)\}$

Functions Defined by Equations

We can specify or define a particular function in various ways: (1) by a verbal description; (2) by a table; (3) by a set of ordered pairs; or, if the domain and range are sets of numbers, (4) by an equation, or (5) by a graph. If the domain of a function is a large or infinite set, it may be impractical or impossible to actually list all of the ordered pairs that belong to the function, or to display the function in a table. Such a function can often be defined by a verbal description of the "rule of correspondence" (for example, "to each real number corresponds its square"), that clearly specifies the element of the range that corresponds to each element of the domain. When the domain and range are sets of numbers, the algebraic and graphical analogs of the verbal description are the equation and graph, respectively. We will find it valuable to be able to view a particular function from multiple perspectives—algebraic (in terms of an equation), graphical (in terms of a graph), and numeric (in terms of a table).

Both versions of the definition of function are quite general, with no restriction on the type of elements that make up the domain or range, and no restriction on the numbers of elements in the domain or range. We are primarily interested, however, in functions with real number domains and ranges. In this text, unless otherwise indicated, **the domain and range of a function will be sets of real numbers.** For such a function we often use an equation in two variables to specify both the rule of correspondence and the set of ordered pairs.

Consider the equation

$$y = x^2 + 2x \qquad x \text{ any real number} \tag{1}$$

This equation assigns to each domain value x exactly one range value y. For example,

If $x = 4$, then $y = (4)^2 + 2(4) = 24$

If $x = -\frac{1}{3}$, then $y = (-\frac{1}{3})^2 + 2(-\frac{1}{3}) = -\frac{5}{9}$

Thus, we can view equation (1) as a function with rule of correspondence

$$y = x^2 + 2x \qquad x \text{ corresponds to } x^2 + 2x$$

or, equivalently, as a function with set of ordered pairs

$$\{(x, y) \mid y = x^2 + 2x, x \text{ a real number}\}$$

The variable x is called an *independent variable,* indicating that values can be assigned "independently" to x from the domain. The variable y is called a *dependent variable,* indicating that the value of y "depends" on the value assigned to x and on the given equation. In general, any variable used as a placeholder for domain values is called an **independent variable;** any variable used as a placeholder for range values is called a **dependent variable.**

Which equations can be used to define functions?

FUNCTIONS DEFINED BY EQUATIONS

In an equation in two variables, if to each value of the independent variable there corresponds exactly one value of the dependent variable, then the equation defines a function.

If there is any value of the independent variable to which there corresponds more than one value of the dependent variable, then the equation does not define a function.

Notice that we have used the phrase "an equation defines a function" rather than "an equation is a function." This is a somewhat technical distinction, but it is employed consistently in mathematical literature and we will adhere to it in this text.

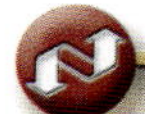

EXPLORE/DISCUSS 2

(A) Graph $y = x^2 - 4$ for $-5 \le x \le 5$ and $-5 \le y \le 5$ and trace along this graph. Discuss the relationship between the coordinates displayed while tracing and the function defined by this equation.

(B) The graph of the equation $x^2 + y^2 = 16$ is a circle. Because most graphing utilities will accept only equations that have been solved for y, we must graph both of the equations $y_1 = \sqrt{16 - x^2}$ and $y_2 = -\sqrt{16 - x^2}$ to produce a graph of the circle. Graph these equations for $-5 \le x \le 5$ and $-5 \le y \le 5$. Then try different values for Xmin and Xmax until the graph looks more like a circle. Use the trace feature to find two points on this circle with the same x coordinate and different y coordinates.

(C) Is it possible to graph a single equation of the form $y =$ (expression in x) on your graphing utility and obtain a graph that is not the graph of a function? Explain your answer.

REMARK If we want the graph of a circle to actually appear to be circular, we must choose window variables so that a unit length on the x axis is the same number of pixels as a unit length on the y axis. Such a window is often referred to as a **squared viewing window.** Most graphing utilities have an option under the zoom menu that does this automatically.

Not all equations determine functions. One way to determine if an equation does determine a function is to examine its graph. The graphs of the equations

$$y = x^2 - 4 \qquad \text{and} \qquad x^2 + y^2 = 16$$

are shown in Figure 1.

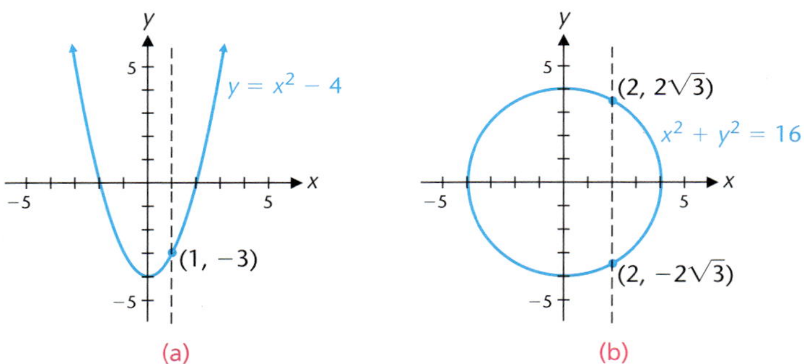

FIGURE 1 Graphs of equations and the vertical line test.

The graph in Figure 1(a) is a parabola and the graph in Figure 1(b) is a circle.* Each vertical line intersects the parabola in exactly one point. This shows that to each value of the independent variable x there corresponds exactly one value of the dependent variable y. For example, to the x value 1 there corresponds only the y value -3 [Fig. 1(a)]. Thus, the equation $y = x^2 - 4$ defines a function. On the other hand, there are vertical lines that intersect the circle in Figure 1(b) in two points. For example, the vertical line through $x = 2$ intersects the circle in the points $(2, 2\sqrt{3})$ and $(2, -2\sqrt{3})$ [Fig. 1(b)]. Thus, to the x value 2 there correspond two y values, $2\sqrt{3}$ and $-2\sqrt{3}$. Consequently, the equation $x^2 + y^2 = 16$ does not define a function. These observations are generalized in Theorem 1.

THEOREM 1
Vertical Line Test for a Function

An equation defines a function if each vertical line in the rectangular coordinate system passes through at most one point on the graph of the equation.

If any vertical line passes through two or more points on the graph of an equation, then the equation does not define a function.

Refer to Figure 1. Because the expression $x^2 - 4$ represents a real number for all real values of x, the function defined by the equation $y = x^2 - 4$ is defined for all real numbers. Thus, its domain is the set of all real numbers, often denoted by the letter R or the interval† $(-\infty, \infty)$. On the other hand, the expression $\sqrt{16 - x^2}$

*Parabolas and circles are discussed extensively later in the book.

†See Appendix A, Section A.1, for a discussion of interval notation and inequalities.

represents a real number only if $16 - x^2 \geq 0$. This inequality is equivalent to $x^2 \leq 16$ or $-4 \leq x \leq 4$. Thus, the domain of the function $y = \sqrt{16 - x^2}$ is $\{x | -4 \leq x \leq 4\}$ or $[-4, 4]$. Unless stated to the contrary, we will adhere to the following convention regarding domains and ranges for functions defined by equations.

AGREEMENT ON DOMAINS AND RANGES

If a function is defined by an equation and the domain is not indicated, then we assume that the domain is the set of all real number replacements of the independent variable that produce *real values* for the dependent variable. The range is the set of all values of the dependent variable corresponding to these domain values.

EXAMPLE 2 Finding the Domain of a Function

Find the domain of the function defined by the equation $y = 4 + \sqrt{x}$, assuming x is the independent variable.

SOLUTION

For $\sqrt{x}$ to be real, x must be greater than or equal to 0. Thus,

$$\text{Domain: } \{x \mid x \geq 0\} \quad \text{or} \quad [0, \infty)$$

Note that in many cases we will dispense with set notation and simply write $x \geq 0$ instead of $\{x \mid x \geq 0\}$.

MATCHED PROBLEM 2

Find the domain of the function defined by the equation $y = 3 + \sqrt{-x}$, assuming x is the independent variable.

Function Notation

We will use letters to name functions and to provide a very important and convenient notation for defining functions. For example, if f is the name of the function defined by the equation $y = 2x + 1$, then instead of the more formal representations

$f{:}y = 2x + 1$ Rule of correspondence

or

$f{:}\{(x, y) \mid y = 2x + 1\}$ Set of ordered pairs

we simply write

$f(x) = 2x + 1$ Function notation

The symbol $f(x)$ is read "f of x," "f at x," or "the value of f at x" and represents the number in the range of the function f that is paired with the domain value x. Thus, $f(3)$ is the range value for the function f associated with the domain value 3.

We find this range value by replacing x with 3 wherever x occurs in the function definition

$$f(x) = 2x + 1$$

and evaluating the right side,

$$\begin{aligned} f(3) &= 2 \cdot 3 + 1 \\ &= 6 + 1 \\ &= 7 \end{aligned}$$

The statement $f(3) = 7$ indicates in a concise way that the function f assigns the range value 7 to the domain value 3 or, equivalently, that the ordered pair (3, 7) belongs to f.

The symbol $f\!:\!x \rightarrow f(x)$, read "f maps x into $f(x)$," is also used to denote the relationship between the domain value x and the range value $f(x)$ (Fig. 2). Whenever we write $y = f(x)$, we assume that x is an independent variable and that y and $f(x)$ both represent the dependent variable.

FIGURE 2 Function notation.

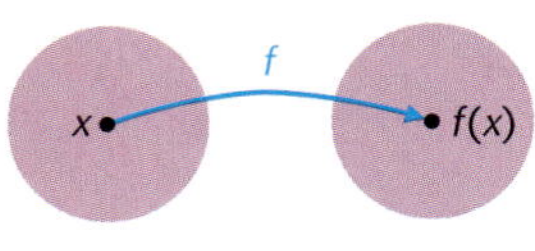

The function f "maps" the domain value x into the range value $f(x)$.

Letters other than f and x can be used to represent functions and independent variables. For example,

$$g(t) = t^2 - 3t + 7$$

defines g as a function of the independent variable t. To find $g(-2)$, we replace t by -2 wherever t occurs in

$$g(t) = t^2 - 3t + 7$$

and evaluate the right side:

$$\begin{aligned} g(-2) &= (-2)^2 - 3(-2) + 7 \\ &= 4 + 6 + 7 \\ &= 17 \end{aligned}$$

Thus, the function g assigns the range value 17 to the domain value -2; the ordered pair $(-2, 17)$ belongs to g.

It is important to understand and remember the definition of the symbol $f(x)$:

DEFINITION 3

The Symbol $f(x)$

The symbol $\mathbf{f(x)}$, read "f of x," represents the real number in the range of the function f corresponding to the domain value x. The symbol $f(x)$ is also called the **value of the function f at x.** Symbolically, $f\!:\!x \rightarrow f(x)$. The ordered pair $(x, f(x))$ belongs to the function f. If x is a real number that is not in the domain of f, then f is **not defined** at x and $f(x)$ **does not exist.**

CAUTION

Do not confuse function notation with multiplication:

$f(x)$ is the value of the function f at x

$2(x) = 2x$ is algebraic multiplication

EXAMPLE 3 Evaluating Functions

For

$$f(x) = \frac{15}{x - 3} \qquad g(x) = 16 + 3x - x^2 \qquad h(x) = \frac{6}{\sqrt{x} - 1}$$

find:

(A) $f(6)$ (B) $g(-7)$ (C) $h(-2)$ (D) $f(0) + g(4) - h(16)$

SOLUTIONS

(A) $f(6) = \frac{15}{6 - 3} = \frac{15}{3} = 5$

(B) $g(-7) = 16 + 3(-7) - (-7)^2 = 16 - 21 - 49 = -54$

(C) $h(-2) = \frac{6}{\sqrt{-2} - 1}$

But $\sqrt{-2}$ is not a real number. Because we have agreed to restrict the domain of a function to values of x that produce real values for the function, -2 is not in the domain of h and $h(-2)$ is not defined.

(D) $f(0) + g(4) - h(16)$

$$= \frac{15}{0 - 3} + [16 + 3(4) - 4^2] - \frac{6}{\sqrt{16} - 1}$$

$$= \frac{15}{-3} + 12 - \frac{6}{3}$$

$$= -5 + 12 - 2 = 5$$

MATCHED PROBLEM 3

Use the functions in Example 3 to find

(A) $f(-2)$ (B) $g(6)$ (C) $h(-8)$ (D) $\frac{f(8)}{h(9)}$

EXAMPLE 4 Finding the Domain of a Function

Find the domain of $g(t) = 5 + 2t - 3t^2$.

SOLUTION

Because $5 + 2t - 3t^2$ represents a real number for all replacements of t by real numbers, the domain of g is R, the set of all real numbers. To express this domain in interval notation, we write

Domain of $g = (-\infty, \infty)$

MATCHED 4 PROBLEM

Find the domain of $h(w) = 3w^2 + 2w - 9$.

The reasoning used in Example 4 can be applied to any polynomial: **The domain of any polynomial is *R*, the set of real numbers.**

EXAMPLE 5 Finding the Domain of a Function

Find the domain of $F(w) = \dfrac{5}{w^2 - 9}$.

SOLUTION

Because division by 0 is not defined, we must exclude all values of w that would make the denominator 0. Factoring the denominator, we can write

$$F(w) = \frac{5}{w^2 - 9} = \frac{5}{(w - 3)(w + 3)} \qquad a^2 - b^2 = (a - b)(a + b)$$

Thus, we must exclude $w = 3$ and $w = -3$ from the domain of F. That is,

$$\begin{aligned} \text{Domain of } F &= \{w \mid w \neq 3, w \neq -3\} && \text{Set notation} \\ &= (-\infty, -3) \cup (-3, 3) \cup (3, \infty) && \text{Interval notation} \end{aligned}$$

We often simplify this by writing

$$F(w) = \frac{5}{w^2 - 9} \qquad w \neq 3, w \neq -3$$

MATCHED 5 PROBLEM

Find the domain of $G(w) = \dfrac{5}{w^2 - 1}$.

EXAMPLE 6 Finding the Domain of a Function

Find the domain of $s(x) = \sqrt{3 - x}$.

SOLUTIONS

Algebraic Solution

Because $\sqrt{3 - x}$ is not a real number when $3 - x$ is a negative real number, we must restrict the domain of s to the real numbers x for which

$$3 - x \geq 0$$
$$3 \geq x$$

Thus, we have

$$\text{Domain of } s = \{x \mid x \leq 3\} = (-\infty, 3]$$

or, more informally,

$$s(x) = \sqrt{3 - x} \qquad x \leq 3$$

Graphical Solution

Entering $y_1 = \sqrt{3 - x}$ in the equation editor and graphing in a standard viewing window produces Figure 3.

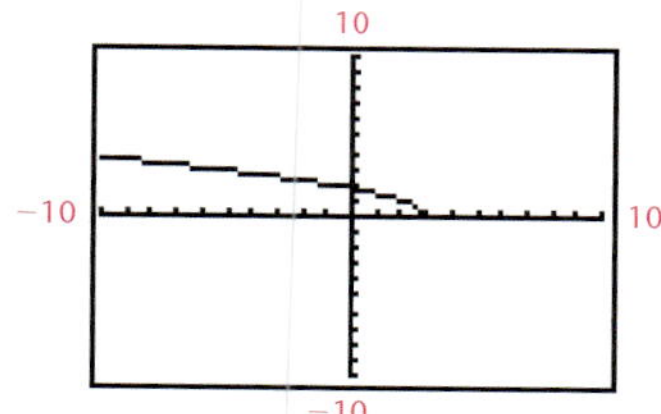

FIGURE 3

Using a table (Fig. 4), we see that evaluating s at a number greater than 3 produces an error message, whereas evaluating s for large negative values produces no errors.

X	Y1
3	0
3.0001	ERROR
-1000	31.67
-1E6	1000
-1E9	31623
-1E12	1E6
-1E15	3.16E7

Y1 ■ √(3−X)

FIGURE 4

Thus, we conclude that the domain of s is $(-\infty, 3]$.

MATCHED 6 PROBLEM

Find the domain of $r(t) = \sqrt{t - 5}$.

EXAMPLE 7 Finding the Domain of a Function

Find the domain of $f(x) = \dfrac{2}{2 - \sqrt{x}}$.

SOLUTIONS

Algebraic Solution

Because $\sqrt{x}$ is not a real number for negative real numbers x, x must be a nonnegative real number. Because division by 0 is not defined, we must exclude any values of x that make the denominator 0. Thus, we solve

$$\begin{aligned} 2 - \sqrt{x} &= 0 \\ 2 &= \sqrt{x} \\ 4 &= x \end{aligned}$$

and conclude that the domain of f is all nonnegative real numbers except 4. This can be written as

$$\begin{aligned} \text{Domain of } f &= \{x \mid x \geq 0, x \neq 4\} \\ &= [0, 4) \cup (4, \infty) \end{aligned}$$

Graphical Solution

Figure 5 shows the graph of $y = 2/(2 - \sqrt{x})$ in a standard viewing window.

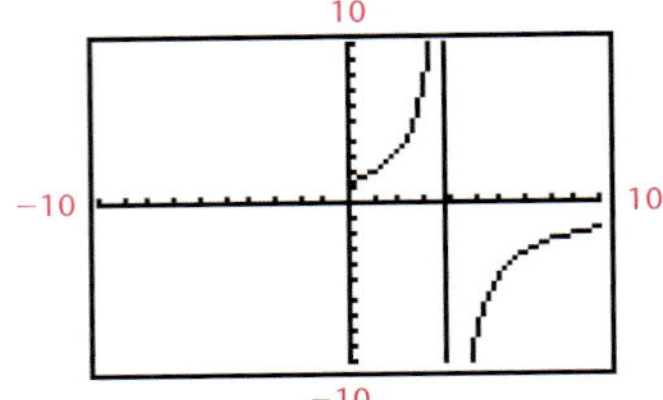

FIGURE 5

The curve starts at $x = 0$, indicating that f is not defined for $x < 0$. Evaluating f at a negative number confirms this (Fig. 6). The vertical line on the graph indicates some strange behavior at $x = 4$. Evaluating f at $x = 4$ (Fig. 6) shows that f is not defined at $x = 4$. Evaluating f at large positive numbers produces no errors.

X	Y1
-1E-4	ERROR
0	1
4	ERROR
100	-.25
1000	-.0675
100000	-.0064
1E7	-6E-4

Y1 = 2/(2−√(X))

FIGURE 6

Thus, we conclude that

$$\begin{aligned} \text{Domain of } f &= \{x \mid x \geq 0, x \neq 4\} \\ &= [0, 4) \cup (4, \infty) \end{aligned}$$

MATCHED PROBLEM 7

Find the domain of $g(x) = \dfrac{1}{\sqrt{x} - 1}$.

Refer to Figure 5. What caused the vertical line in this graph? Explore/Discuss 3 will help you find out.

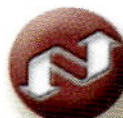

EXPLORE/DISCUSS 3

Graph $y = 2/(2 - \sqrt{x})$ in a standard viewing window (see Fig. 5).

(A) Press TRACE and move the cursor as close to $x = 4$ on the left side of $x = 4$ as possible. What is the y coordinate of this point?

(B) Now move the cursor as close to $x = 4$ on the right side of 4 as possible. What is the y coordinate of this point?

(C) Change Ymax to a value greater than the y coordinate in (A) and Ymin to a value less than the y coordinate in (B).

(D) Redraw the graph in the window from part (C) and discuss the result.

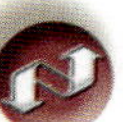

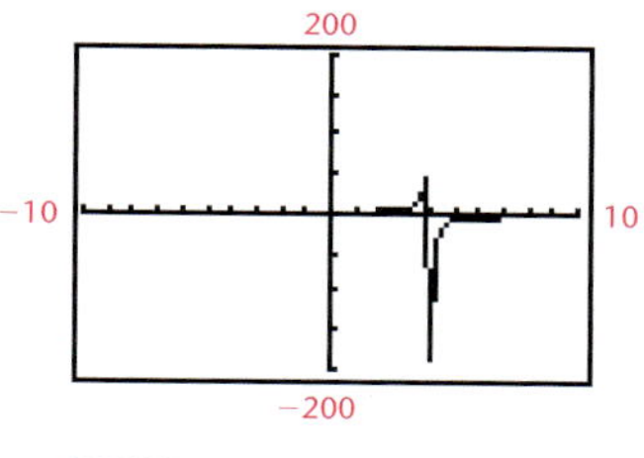

FIGURE 7

Your explorations should have produced a graph similar to Figure 7. The nearly vertical line is produced by connecting the last point on the left of $x = 4$ with the first point on the right of $x = 4$.

In addition to evaluating functions at specific numbers, it is important to be able to evaluate functions at expressions that involve one or more variables. For example, the **difference quotient**

$$\frac{f(x + h) - f(x)}{h} \qquad x \text{ and } x + h \text{ in the domain of } f,\ h \neq 0$$

is studied extensively in a calculus course.

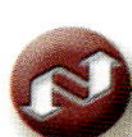

EXPLORE/DISCUSS 4

Let x and h be any real numbers.

(A) If $f(x) = 3x + 2$, which of the following is correct?

(i) $f(x + h) = 3x + 2 + h$

(ii) $f(x + h) = 3x + 3h + 2$

(iii) $f(x + h) = 3x + 3h + 4$

(B) If $f(x) = x^2$, which of the following is correct?

(i) $f(x + h) = x^2 + h^2$

(ii) $f(x + h) = x^2 + h$

(iii) $f(x + h) = x^2 + 2xh + h^2$

(C) If $f(x) = x^2 + 3x + 2$, write a verbal description of the operations that must be performed to evaluate $f(x + h)$.

EXAMPLE 8 Evaluating and Simplifying a Difference Quotient

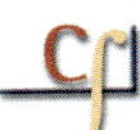

* For $f(x) = x^2 + 4x + 5$, find and simplify:

(A) $f(2)$ (B) $f(2 + h)$ (C) $\dfrac{f(2 + h) - f(2)}{h}$

(D) $f(x + h)$ (E) $\dfrac{f(x + h) - f(x)}{h}$

SOLUTIONS

(A) $f(2) = 2^2 + 4(2) + 5 = 17$

(B) To find $f(2 + h)$, replace x with $2 + h$ everywhere it occurs in the equation that defines f and simplify:

$$\begin{aligned} f(2 + h) &= (2 + h)^2 + 4(2 + h) + 5 \\ &= 4 + 4h + h^2 + 8 + 4h + 5 \\ &= h^2 + 8h + 17 \end{aligned}$$

(C) Using parts (A) and (B), we have

$$\begin{aligned} \frac{f(2 + h) - f(2)}{h} &= \frac{h^2 + 8h + 17 - 17}{h} \\ &= \frac{h^2 + 8h}{h} = \boxed{\frac{h(h + 8)}{h}} = h + 8 \end{aligned}$$

(D) To find $f(x + h)$, we replace x with $x + h$ everywhere it appears in the equation that defines f and simplify:

$$\begin{aligned} f(x + h) &= (x + h)^2 + 4(x + h) + 5 \\ &= x^2 + 2xh + h^2 + 4x + 4h + 5 \end{aligned}$$

(E) Using the result of part (D), we get

$$\begin{aligned} \frac{f(x + h) - f(x)}{h} &= \frac{x^2 + 2xh + h^2 + 4x + 4h + 5 - (x^2 + 4x + 5)}{h} \\ &= \frac{x^2 + 2xh + h^2 + 4x + 4h + 5 - x^2 - 4x - 5}{h} \\ &= \frac{2xh + h^2 + 4h}{h} \boxed{= \frac{h(2x + h + 4)}{h}} = 2x + h + 4 \end{aligned}$$

MATCHED 8 PROBLEM

Repeat Example 8 for $f(x) = x^2 + 3x + 7$.

---

*The symbol [calculus symbol] denotes problems that are related to calculus.

CAUTION

1. If f is a function, then the symbol $f(x + h)$ represents the value of f at the number $x + h$ and must be evaluated by replacing the independent variable in the equation that defines f with the expression $x + h$, as we did in Example 8. Do not confuse this notation with the familiar algebraic notation for multiplication:

$$f(x + h) \neq fx + fh \qquad f(x + h) \text{ is function notation.}$$
$$4(x + h) = 4x + 4h \qquad 4(x + h) \text{ is algebraic multiplication notation.}$$

2. There is another common incorrect interpretation of the symbol $f(x + h)$. If f is an arbitrary function, then

$$f(x + h) \neq f(x) + f(h)$$

It is possible to find some particular functions for which $f(x + h) = f(x) + f(h)$ is a true statement, but in general these two expressions are not equal.

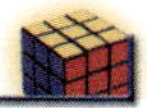

Modeling and Data Analysis

The next example explores the relationship between the *algebraic* definition of a function, the *numeric* values of the function, and a *graphical* representation of the function. The interplay between the algebraic, numeric, and graphical aspects of a function is one of the central themes of this book. In this example, we also see how a function can be used to describe data from the real world, a process that is generally referred to as *mathematical modeling*.

EXAMPLE 9 Consumer Debt

Revolving-credit debt (in billions of dollars) in the United States over a 20-year period is given in Table 4. A financial analyst used statistical techniques to produce a mathematical model for this data:

$$f(x) = 1.2x^2 + 7.5x + 58$$

where $x = 0$ corresponds to 1980.

TABLE 4 Revolving-Credit Debt

Year	Total Debt (Billions)
1980	\$58.5
1985	\$128.9
1990	\$234.8
1995	\$443.2
2000	\$663.8

Source: Federal Reserve System.

Compare the data and the model both numerically and graphically. Use the modeling function f to estimate the debt to the nearest tenth of a billion in 2001 and in 2005.

SOLUTION

Most graphing utilities have the ability to manipulate a list of numbers, such as the total debt in Table 4. The relevant commands are usually found by pressing STAT* (Fig. 8). Then select EDIT and enter the data. Enter the years as L1 and the debt as L2 (Fig. 9). Unlike the TABLE command, which computes a y value for each x value you enter, the EDIT command does not assume any correspondence between the numbers in two different lists. It is your responsibility to make sure that each pair on the same line in Figure 9 corresponds to a line in Table 4.

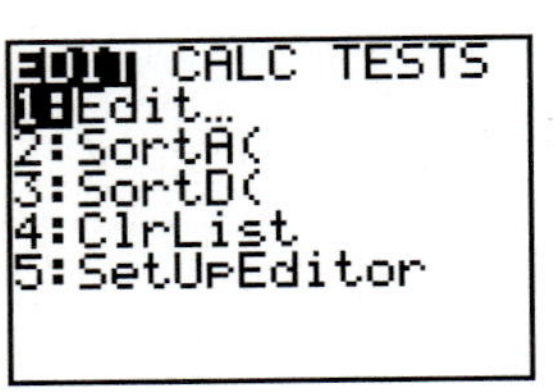

FIGURE 8

L1	L2	L3
0	58.5	
5	128.9	
10	234.8	
15	443.2	
20	663.8	

L3(1)=

FIGURE 9

A graph of a finite data set is called a **scatter plot.** To form a scatter plot for the data in Table 4, first press STAT PLOT (Fig. 10) and select Plot1 (Fig. 11). The Plot1 screen contains a number of options, some of which you select by placing the cursor over the option and pressing ENTER. First, select ON to activate the plot. Then select the type of plot you want. The darkened choice in Figure 11 produces a scatter plot. Next use the 2nd key to enter L1 for the Xlist and L2 for the Ylist. Finally, select the mark you want to use for the plot. Before graphing the data, we must enter values for the window variables that will produce a window that contains the points in the scatter plot. Examining the data in Figure 9, we see that Xmin $= -5$, Xmax $= 30$, Xscl $= 5$, Ymin $= 0$, Ymax $= 700$, and Yscl $= 100$ should provide a viewing window that contains all of these points. (We chose Xmin $= -5$ to clearly show the point at $x = 0$.) Pressing GRAPH displays the scatter plot (Fig. 12).

STAT PLOTS
1:Plot1...Off L1 L2
2:Plot2...Off L1 L2
3:Plot3...Off L1 L2
4↓PlotsOff

FIGURE 10

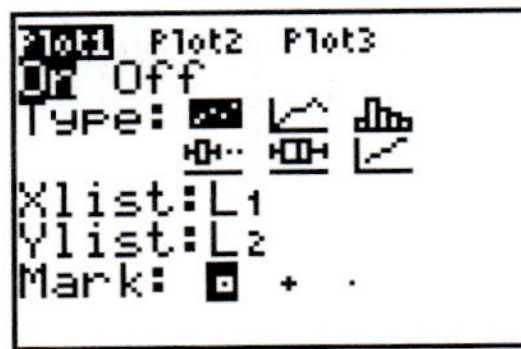
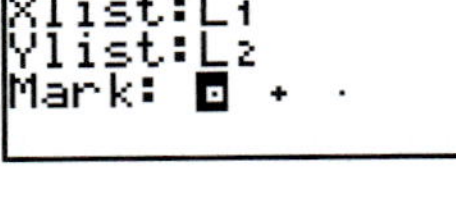

FIGURE 11

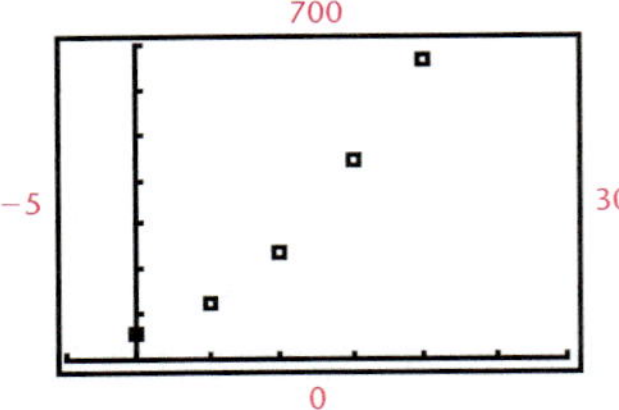

FIGURE 12

Now we enter the modeling function in the equation editor (Fig. 13) and use the TABLE command to evaluate the function (Fig. 14). To compare the data and the function numerically, we enter the data from Figure 9 and the values from Figure 14 in Table 5. To compare them graphically, we press GRAPH to graph both the model and the scatter plot (Fig. 15).

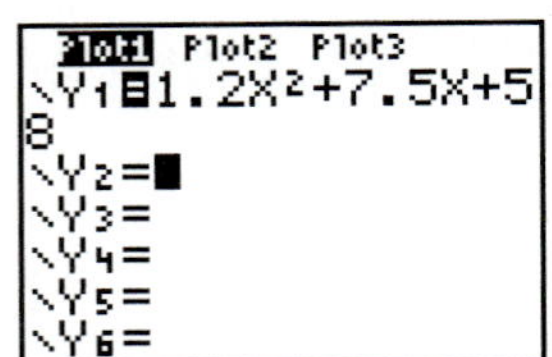

FIGURE 13

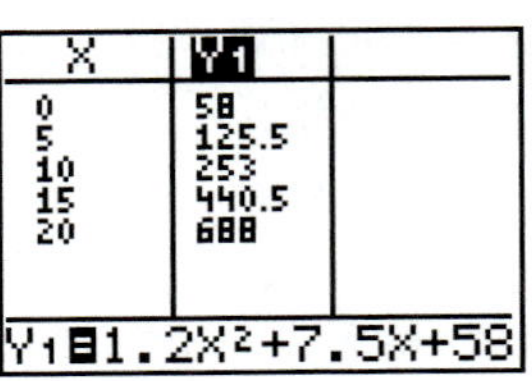

FIGURE 14

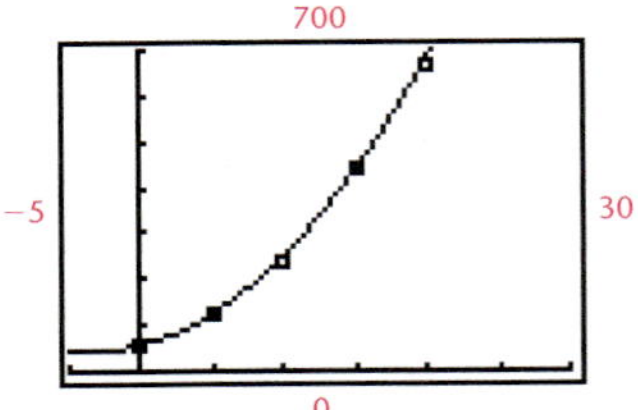

FIGURE 15

*We used a TI-83 to produce the screen dumps in this section. If you are using a different graphing utility, consult your manual for the appropriate commands.

TABLE 5

x	0	5	10	15	20
Debt	58.5	128.9	234.8	443.2	663.8
$f(x)$	58	125.5	253	440.5	688

To estimate the debt in 2001 and 2005, we evaluate $f(x)$ at 21 and at 25:

$$f(21) = 744.7 \qquad f(25) = 995.5$$

Thus, the revolving-credit debt should be $744.7 billion in 2001 and $995.5 billion in 2005.

MATCHED 9 PROBLEM

Credit union debt (in billions of dollars) in the United States is given in Table 6. Repeat Example 9 using these data and the modeling function

$$y = f(x) = 0.5x^2 + 5.5x + 48$$

TABLE 6 Credit Union Debt

Year	Total Debt (Billions)
1970	$48.7
1975	$82.9
1980	$147.0
1985	$245.1
1990	$347.1

Source: Federal Reserve System.

REMARKS

1. Modeling functions like the function f in Example 9 provide reasonable and useful representations of the given data, but they do not always correctly predict future behavior. For example, the model in Example 9 indicates that the revolving-credit debt in 2001 should be about $744.7 billion. But the actual debt for 2001 turned out to be $689.5 billion, which differs from the predicted value by more than $55 billion. Proper use of mathematical models requires both an understanding of the techniques used to develop the model and frequent reevaluation, modification, and interpretation of the results produced by the model.
2. In Chapter 2 we will discuss methods for finding a function f that models a given set of data. It turns out that this is easy to do with a graphing utility.

A Brief History of the Function Concept

The history of the use of functions in mathematics illustrates the tendency of mathematicians to extend and generalize a concept. The word *function* appears to have been first used by Leibniz in 1694 to stand for any quantity associated with a curve. By 1718, Johann Bernoulli considered a function any expression made up of constants and a variable. Later in the same century, Euler came to regard a function as any equation made up of constants and variables. Euler made extensive use of the extremely important notation $f(x)$, although its origin is generally attributed to Clairaut (1734).

The form of the definition of function that had been used until well into the twentieth century (many texts still contain this definition) was formulated by Dirichlet (1805–1859). He stated that, if two variables x and y are so related that for each value of x there corresponds exactly one value of y, then y is said to be a (single-valued) function of x. He called x, the variable to which values are assigned at will, the independent variable, and y, the variable whose values depend on the values assigned to x, the dependent variable. He called the values assumed by x the domain of the function, and the corresponding values assumed by y the range of the function.

Now, because set concepts permeate almost all mathematics, we have the more general definition of function presented in this section in terms of sets of ordered pairs of elements.

ANSWERS MATCHED PROBLEMS

1. (A) S does not define a function.
   (B) T defines a function with domain $\{-2, -1, 0, 1, 2\}$ and range $\{0, 1, 2\}$.
2. $x \le 0$
3. (A) -3 (B) -2
   (C) Does not exist (D) 1
4. All real numbers or $(-\infty, \infty)$
5. All real numbers except -1 and 1 or $(-\infty, -1) \cup (-1, 1) \cup (1, \infty)$
6. $t \ge 5$ or $[5, \infty)$
7. $x \ge 0$, $x \ne 1$ or $[0, 1) \cup (1, \infty)$
8. (A) 17 (B) $h^2 + 7h + 17$
   (C) $h + 7$ (D) $x^2 + 2xh + h^2 + 3x + 3h + 7$ (E) $2x + h + 3$
9.

x	0	5	10	15	20
Debt	48.7	82.9	147.0	245.1	347.1
$f(x)$	48	88	153	243	358

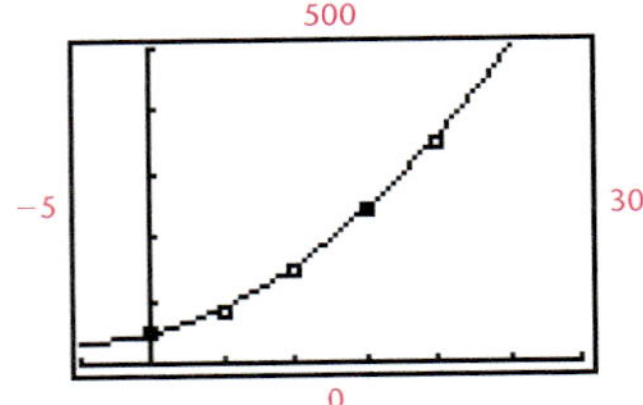

The credit union debt is approximately \$699 billion in 2001 and \$853 billion in 2005.

EXERCISE 1.2

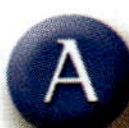

Indicate whether each table in Problems 1–6 defines a function.

1.

Domain	Range
−1	1
0	2
1	3

2.

Domain	Range
2	1
4	3
6	5

3.

Domain	Range
1	3
3	5
	7
5	9

4.

Domain	Range
−1	0
−2	5
−3	8

5.

Domain	Range
−1 → 3	
0 → 3	3
1 → 3	
2 → 3	

6.

Domain	Range
2 → 8	8
3 → 8	
4 → 9	9
5 → 9	

Indicate whether each set in Problems 7–12 defines a function. Find the domain and range of each function.

7. {(2, 4), (3, 6), (4, 8), (5, 10)}

8. {(−1, 4), (0, 3), (1, 2), (2, 1)}

9. {(10, −10), (5, −5), (0, 0), (5, 5) (10, 10)}

10. {(−10, 10), (−5, 5), (0, 0), (5, 5), (10, 10)}

11. {(0, 1), (1, 1), (2, 1), (3, 2), (4, 2), (5, 2)}

12. {(1, 1), (2, 1), (3, 1), (1, 2), (2, 2), (3, 2)}

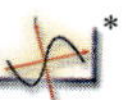*

Indicate whether each graph in Problems 13–18 is the graph of a function.

13.

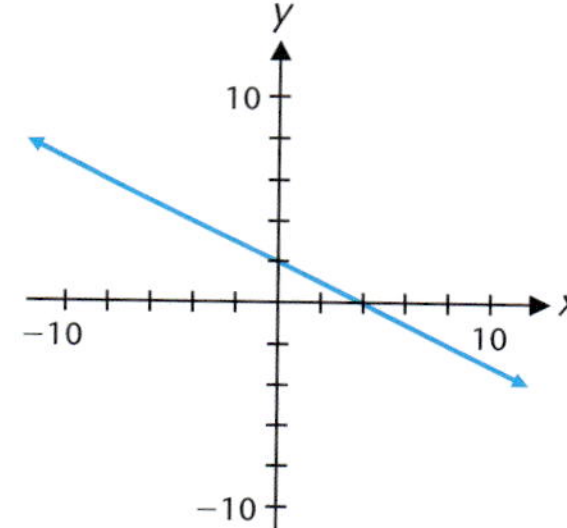

14.

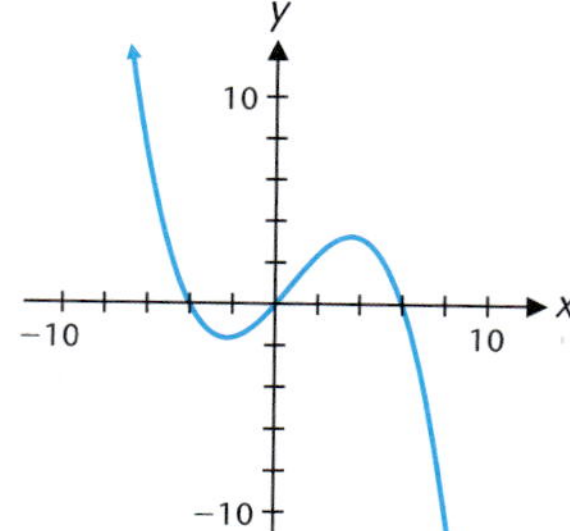

15.

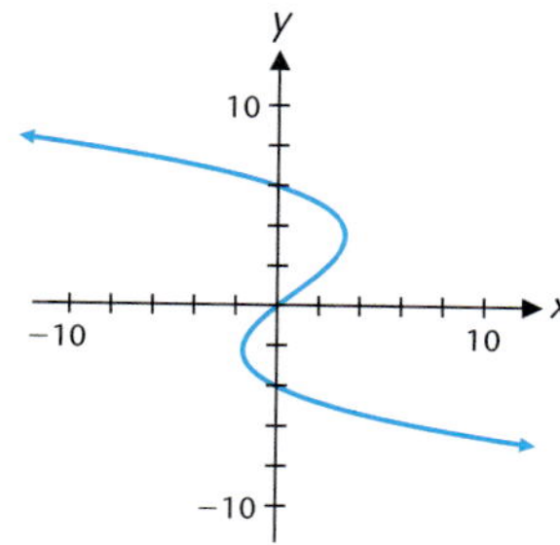

16.

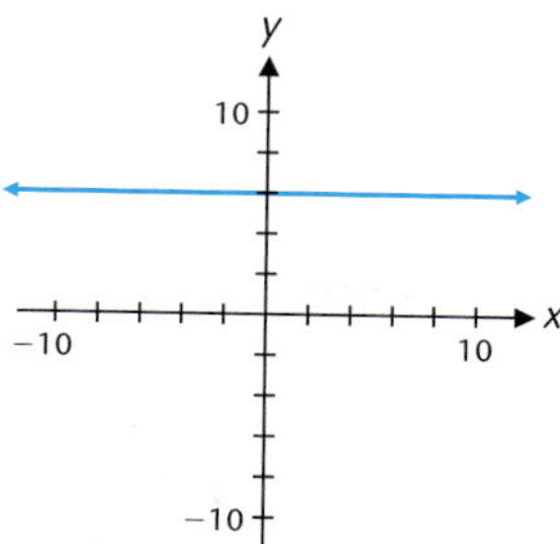

17.

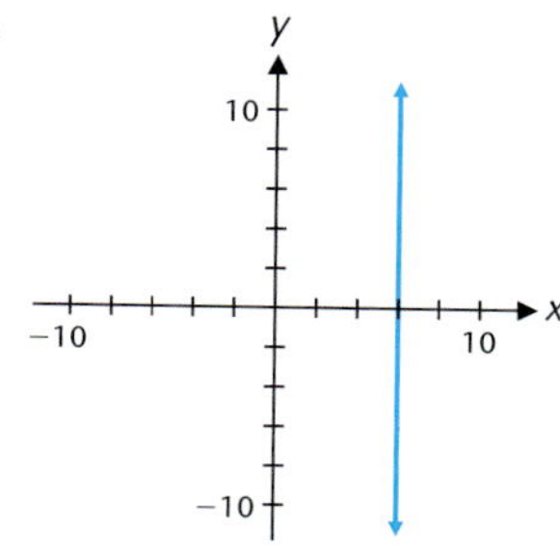

18.

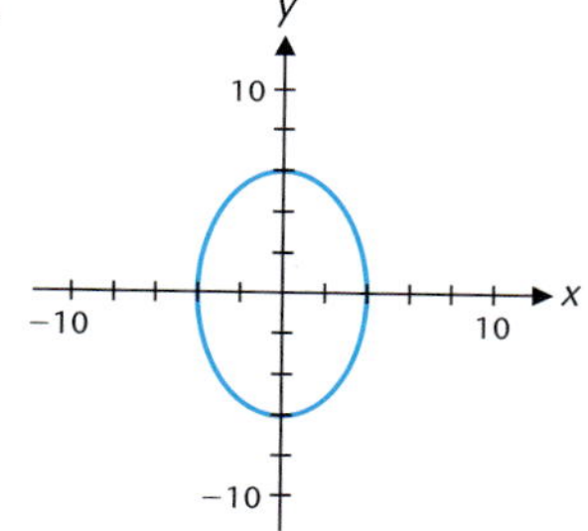

In Problems 19 and 20, which of the indicated correspondences define functions? Explain.

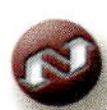

19. Let W be the set of residents of Wisconsin and let R and S be the set of members of the U.S. House of Representatives and the set of members of the U.S. Senate, respectively, elected by the residents of Wisconsin.

(A) A resident corresponds to the congressperson representing the resident's congressional district.

(B) A resident corresponds to the senator representing the resident's state.

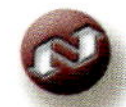

20. Let P be the set of patients in a hospital and let D be the set of doctors and N the set of nurses, respectively, that are on the staff of the hospital.

(A) A patient corresponds to a doctor if that doctor admitted the patient to the hospital.

(B) A patient corresponds to a nurse if that nurse cares for the patient.

*The symbol 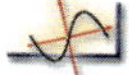 denotes problems that require graphical interpretation.

In Problems 21–24, determine if the indicated equation defines a function. Justify your answer.

21. $x + y = 4$

22. $x^2 + y = 4$

23. $x + y^2 = 4$

24. $x^2 + y^2 = 4$

Problems 25–36 refer to the functions

$f(x) = 3x - 5 \qquad g(t) = 4 - t$

$F(m) = 3m^2 + 2m - 4 \qquad G(u) = u - u^2$

Evaluate as indicated.

25. $f(-1)$

26. $g(6)$

27. $G(-2)$

28. $F(-3)$

29. $F(-1) + f(3)$

30. $G(2) - g(-3)$

31. $2F(-2) - G(-1)$

32. $3G(-2) + 2F(-1)$

33. $\dfrac{f(0) \cdot g(-2)}{F(-3)}$

34. $\dfrac{g(4) \cdot f(2)}{G(1)}$

35. $\dfrac{f(4) - f(2)}{2}$

36. $\dfrac{g(5) - g(3)}{2}$

B

In Problems 37–52, find the domain of the indicated function. Express answers informally and formally using interval notation.

37. $f(x) = 4 - 9x + 3x^2$

38. $g(x) = 1 + 7x - x^2$

39. $h(z) = \dfrac{2}{4 - z}$

40. $k(z) = \dfrac{z}{z - 3}$

41. $g(t) = \sqrt{t - 4}$

42. $h(t) = \sqrt{6 - t}$

43. $k(w) = \sqrt{7 + 3w}$

44. $j(w) = \sqrt{9 + 4w}$

45. $H(u) = \dfrac{u}{u^2 + 4}$

46. $G(u) = \dfrac{u}{u^2 - 4}$

47. $L(v) = \dfrac{v + 2}{v^2 - 16}$

48. $K(v) = \dfrac{v + 8}{v^2 + 16}$

49. $M(x) = \dfrac{\sqrt{x + 4}}{x - 1}$

50. $N(x) = \dfrac{\sqrt{x - 3}}{x + 2}$

51. $s(t) = \dfrac{1}{3 - \sqrt{t}}$

52. $r(t) = \dfrac{1}{\sqrt{t} - 4}$

In Problems 53–56, find a function f that makes all three equations true. [Hint: There may be more than one possible answer, but there is one obvious answer suggested by the pattern illustrated in the equations.]

53. $f(1) = 2(1) - 3$
$f(2) = 2(2) - 3$
$f(3) = 2(3) - 3$

54. $f(1) = 5(1)^2 - 6$
$f(2) = 5(2)^2 - 6$
$f(3) = 5(3)^2 - 6$

55. $f(1) = 4(1)^2 - 2(1) + 9$
$f(2) = 4(2)^2 - 2(2) + 9$
$f(3) = 4(3)^2 - 2(3) + 9$

56. $f(1) = -8 + 5(1) - 2(1)^2$
$f(2) = -8 + 5(2) - 2(2)^2$
$f(3) = -8 + 5(3) - 2(3)^2$

57. If $F(s) = 3s + 15$, find $\dfrac{F(2 + h) - F(2)}{h}$.

58. If $K(r) = 7 - 4r$, find $\dfrac{K(1 + h) - K(1)}{h}$.

59. If $g(x) = 2 - x^2$, find $\dfrac{g(3 + h) - g(3)}{h}$.

60. If $P(m) = 2m^2 + 3$, find $\dfrac{P(2 + h) - P(2)}{h}$.

61. If $L(w) = -2w^2 + 3w - 1$, find $\dfrac{L(-2 + h) - L(-2)}{h}$.

62. If $D(p) = -3p^2 - 4p + 9$, find $\dfrac{D(-1 + h) - D(-1)}{h}$.

The verbal statement "function f multiplies the square root of the domain element by 2 and then subtracts 5" and the algebraic statement $f(x) = 2\sqrt{x} - 5$ define the same function. In Problems 63–66, translate each verbal definition of the function into an algebraic definition.

63. Function g multiplies the domain element by 3 and then adds 1.

64. Function f multiplies the domain element by 7 and then adds the product of 5 and the cube of the domain element.

65. Function F divides the domain element by the sum of 8 and the square root of the domain element.

66. Function G takes the square root of the sum of 4 and the square of the domain element.

In Problems 67–70, translate each algebraic definition of the function into a verbal definition.

67. $f(x) = 2x - 3x^2$

68. $g(x) = 5x^3 - 8x$

69. $F(x) = \sqrt{x^4 + 9}$

70. $G(x) = \dfrac{x}{3x - 6}$

C

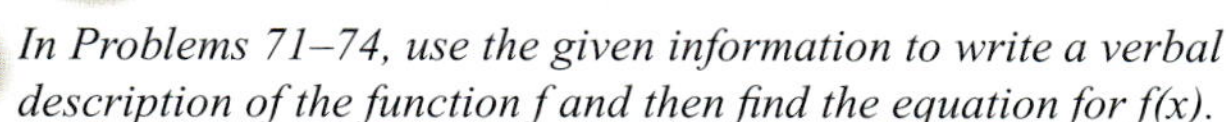

In Problems 71–74, use the given information to write a verbal description of the function f and then find the equation for f(x).

71. $f(x + h) = 2(x + h)^2 - 4(x + h) + 6$

72. $f(x + h) = -7(x + h)^2 + 8(x + h) + 5$

73. $f(x + h) = 4(x + h) - 3\sqrt{x + h} + 9$

74. $f(x + h) = 2\sqrt[3]{x + h} - 6(x + h) - 5$

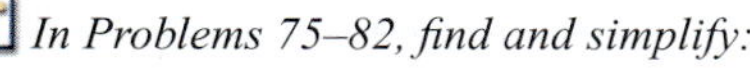

In Problems 75–82, find and simplify:

(A) $\dfrac{f(x + h) - f(x)}{h}$ **(B)** $\dfrac{f(x) - f(a)}{x - a}$

75. $f(x) = 3x - 4$

76. $f(x) = -2x + 5$

77. $f(x) = x^2 - 1$

78. $f(x) = x^2 + x - 1$

79. $f(x) = -3x^2 + 9x - 12$

80. $f(x) = -x^2 - 2x - 4$

81. $f(x) = x^3$

82. $f(x) = x^3 + x$

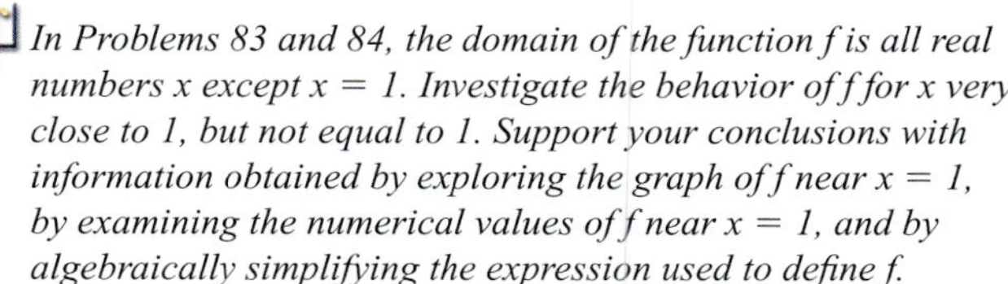

In Problems 83 and 84, the domain of the function f is all real numbers x except x = 1. Investigate the behavior of f for x very close to 1, but not equal to 1. Support your conclusions with information obtained by exploring the graph of f near x = 1, by examining the numerical values of f near x = 1, and by algebraically simplifying the expression used to define f.

83. $f(x) = \dfrac{x^2 - 1}{x - 1}$

84. $f(x) = \dfrac{x^3 - 1}{x - 1}$

APPLICATIONS

85. Physics—Rate. The distance in feet that an object falls in a vacuum is given by $s(t) = 16t^2$, where t is time in seconds.

(A) Find $s(0)$, $s(1)$, $s(2)$, and $s(3)$.

(B) Find and simplify $\dfrac{s(2 + h) - s(2)}{h}$

(C) Evaluate the expression in part (B) for $h = \pm 1, \pm 0.1, \pm 0.01, \pm 0.001$.

(D) What happens in part (C) as h gets closer and closer to 0? Interpret physically.

86. Physics—Rate. An automobile starts from rest and travels along a straight and level road. The distance in feet traveled by the automobile is given by $s(t) = 10t^2$, where t is time in seconds.

(A) Find $s(8)$, $s(9)$, $s(10)$, and $s(11)$.

(B) Find and simplify $\dfrac{s(11 + h) - s(11)}{h}$

(C) Evaluate the expression in part (B) for $h = \pm 1, \pm 0.1, \pm 0.01, \pm 0.001$.

(D) What happens in part (C) as h gets closer and closer to 0? Interpret physically.

87. Boiling Point of Water. At sea level, water boils when it reaches a temperature of 212°F. At higher altitudes, the atmospheric pressure is lower and so is the temperature at which water boils. The boiling point $B(x)$ in degrees Fahrenheit at an altitude of x feet is given approximately by

$$B(x) = 212 - 0.0018x$$

(A) Complete the following table.

x	0	5,000	10,000	15,000	20,000	25,000	30,000
$B(x)$							

(B) Based on the information in the table, write a brief verbal description of the relationship between altitude and the boiling point of water.

88. Air Temperature. As dry air moves upward, it expands and cools. The air temperature $A(x)$ in degrees Celsius at an altitude of x kilometers is given approximately by

$$A(x) = 25 - 9x$$

(A) Complete the following table.

x	0	1	2	3	4	5
$A(x)$						

(B) Based on the information in the table, write a brief verbal description of the relationship between altitude and air temperature.

89. Car Rental. A car rental agency computes daily rental charges for compact cars with the function

$$D(x) = 20 + 0.25x$$

where $D(x)$ is the daily charge in dollars and x is the daily mileage. Translate this algebraic statement into a verbal statement that can be used to explain the daily charges to a customer.

90. Installation Charges. A telephone store computes charges for phone installation with the function

$S(x) = 15 + 0.7x$

where $S(x)$ is the installation charge in dollars and x is the time in minutes spent performing the installation. Translate this algebraic statement into a verbal statement that can be used to explain the installation charges to a customer.

MODELING AND DATA ANALYSIS 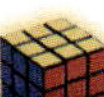

Table 7 contains the average price of admission (in dollars) to a motion picture and the total box office gross (in millions of dollars) for all theaters in the United States.

TABLE 7 Selected Financial Data for the Motion Picture Industry

	1995	1996	1997	1998	1999	2000	2001
Average Price of Admission ($)	$4.35	$4.42	$4.59	$4.69	$5.08	$5.39	$5.66
Box Office Gross ($ in millions)	$5,490	$5,910	$6,360	$6,950	$7,450	$7,660	$8,410

91. Data Analysis. A mathematical model for the average price of admission to a motion picture is

$A(t) = 0.23t + 4.2$

where t represents time in years and $t = 0$ corresponds to 1995.

(A) Compare the model and the data graphically and algebraically.

(B) Estimate (to the nearest cent) the average price of admission in 2002 and 2003.

92. Revenue Analysis. A mathematical model for the total box office gross is given by

$G(t) = 477t + 5{,}460$

where t represents time in years and $t = 0$ corresponds to 1995.

(A) Compare the model and the data graphically and algebraically.

(B) Estimate (to three significant digits) the total box office gross in 2002 and 2003.

Merck & Co., Inc. is the world's largest pharmaceutical company. Problems 93–96 refer to the data in Table 8 taken from the company's 2001 annual report.

TABLE 8 Selected Financial Data for Merck & Co., Inc.

($ in billions)	1997	1998	1999	2000	2001
Sales	$24	$27	$33	$40	$48
R & D Expenses	$1.6	$1.8	$2.1	$2.3	$2.5
Net Income	$4.6	$5.2	$5.9	$6.8	$7.3

93. Sales Analysis. A mathematical model for Merck's sales is given by

$S(t) = 6.1t + 22$

where t is time in years and $t = 0$ corresponds to 1997.

(A) Compare the model and the data graphically and numerically.

(B) Estimate (to two significant digits) Merck's sales in 2002 and in 2004.

(C) Write a brief verbal description of Merck's sales from 1997 to 2001.

94. Income Analysis. A mathematical model for Merck's net income is given by

$I(t) = 0.7t + 4.6$

where t is time in years and $t = 0$ corresponds to 1997.

(A) Compare the model and the data graphically and numerically.

(B) Estimate (to two significant digits) Merck's net income in 2002 and in 2004.

(C) Write a brief verbal description of Merck's net income from 1997 to 2001.

95. Research and Development Analysis. A mathematical model for Merck's sales as a function of research and development (R & D) expenses is given by

$S(r) = 20r^2 - 56r + 63$

where r represents R & D expenditures.

(A) Compare the model and the data graphically and numerically.

(B) Estimate (to two significant digits) Merck's sales if the company spends \$2.7 billion on research and development and if the company spends \$3 billion.

96. Research and Development Analysis. A mathematical model for Merck's net income as a function of R & D expenses is given by

$$I(r) = 0.5r^2 + r + 1.8$$

where r represents R & D expenditures.

(A) Compare the model and the data graphically and numerically.

(B) Estimate (to two significant digits) Merck's net income if the company spends \$2.7 billion on research and development and if the company spends \$3 billion.

SECTION 1.3 Functions: Graphs and Properties

Basic Concepts • Increasing and Decreasing Functions • Local Maxima and Minima • Mathematical Modeling • Piecewise-Defined Functions • The Greatest Integer Function

One of the primary goals of this course is to provide you with a set of mathematical tools that can be used, in conjunction with a graphing utility, to analyze graphs that arise quite naturally in important applications. In this section, we discuss some basic concepts that are commonly used to describe graphs of functions.

Basic Concepts

Each function that has a real number domain and range has a graph—the graph of the ordered pairs of real numbers that constitute the function. When functions are graphed, domain values usually are associated with the horizontal axis and range values with the vertical axis. Thus, the **graph of a function *f*** is the same as the graph of the equation

$$y = f(x)$$

where x is the independent variable and the abscissa of a point on the graph of f. The variables y and $f(x)$ are dependent variables, and either is the ordinate of a point on the graph of f (Fig. 1).

The abscissa or x coordinate of a point where the graph of a function intersects the x axis is called an **x intercept** or **real zero** of the function. The x intercept is also a real solution or **root** of the equation $f(x) = 0$. The ordinate, or y coordinate of a point where the graph of a function crosses the y axis, is called the **y intercept** of the function. The y intercept is given by $f(0)$, provided 0 is in the domain of f. Note that a function can have more than one x intercept but can never have more than one y intercept—a consequence of the vertical line test discussed in Section 1.2.

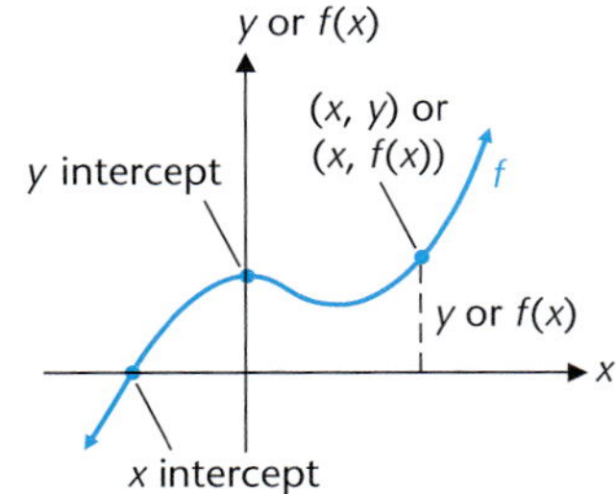

FIGURE 1 Graph of a function.

In Section 1.1, we solved equations of the form $f(x) = c$ on a graphing utility by graphing both sides of the equation and using the intersect command. Most graphing utilities also have a **zero** or **root** command that finds the x intercepts of a function directly from the graph of the function. Example 1 illustrates the use of this command.

EXAMPLE 1 Finding *x* and *y* Intercepts

Find the x and y intercepts (correct to three decimal places) of $f(x) = x^3 + x - 4$.

SOLUTION

From the graph of f in Figure 2, we see that the y intercept is $f(0) = -4$ and that there is an x intercept between 1 and 2. We use the zero command to find this intercept. First we are asked to select a **left bound** (Fig. 3). This is a value of x to the left of the x intercept. Next we are asked to find a **right bound** (Fig. 4). This is a value of x to the right of the x intercept. If a function has more than one x intercept, you should select the left and right bounds so that there is only one intercept between the bounds.

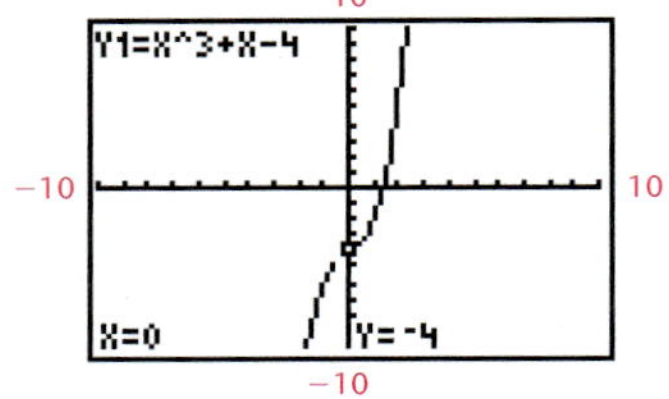

FIGURE 2

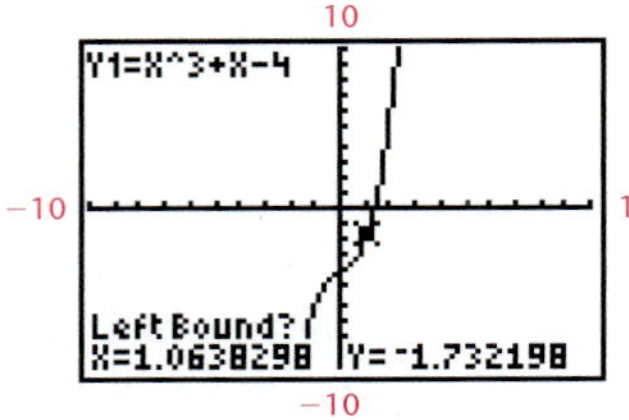

FIGURE 3

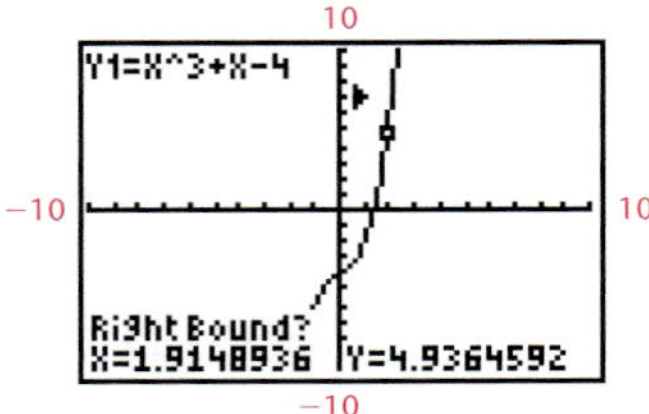

FIGURE 4

Finally, we are asked to select a **guess.** The guess must be between the bounds and should be close to the intercept (Fig. 5). Figure 6 shows that the x intercept (to three decimal places) is 1.379.

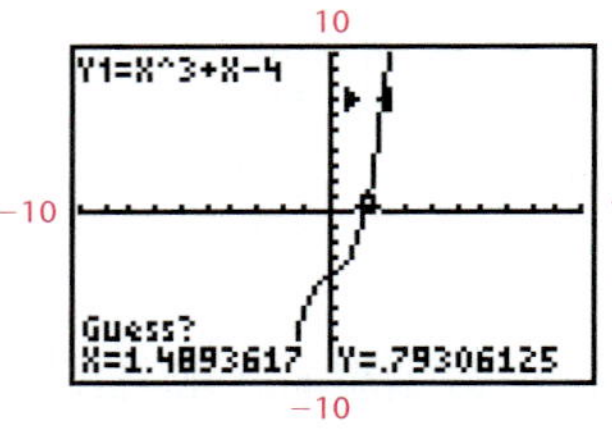

FIGURE 5

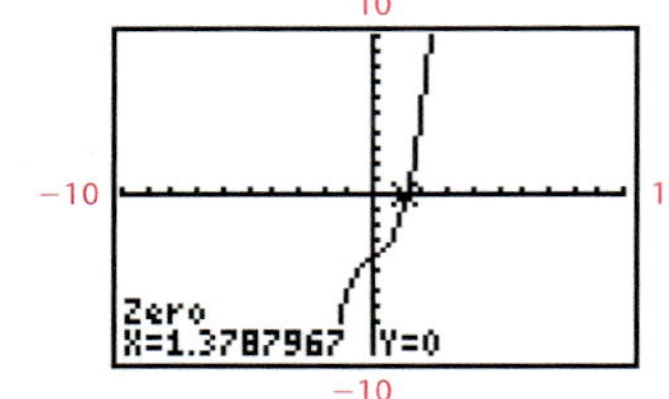

FIGURE 6

MATCHED PROBLEM 1

Find the x and y intercepts (correct to three decimal places) of $f(x) = x^3 + x + 5$.

EXPLORE/DISCUSS 1

Let $f(x) = x^2 - 2x - 5$.

(A) Use the zero command on a graphing utility to find the x intercepts of f.

(B) Find the solutions to the equation $x^2 - 2x - 5 = 0$.

(C) Discuss the differences between the graph of f, the x intercepts, and the solutions to the equation $f(x) = 0$.

The domain of a function is the set of all the x coordinates of points on the graph of the function and the range is the set of all the y coordinates. It is instructive to view the domain and range as subsets of the coordinate axes as in Figure 7. Note the effective use of interval notation* in describing the domain and range of the functions in this figure. In Figure 7(a) a solid dot is used to indicate that a point is on the graph of the function and in Figure 7(b) an open dot to indicate that a point is not on the graph of the function. An open or solid dot at the end of a graph indicates that the graph terminates there, whereas an arrowhead indicates that the graph continues beyond the portion shown with no significant changes in shape [see Fig. 7(b)].

FIGURE 7 Domain and range.

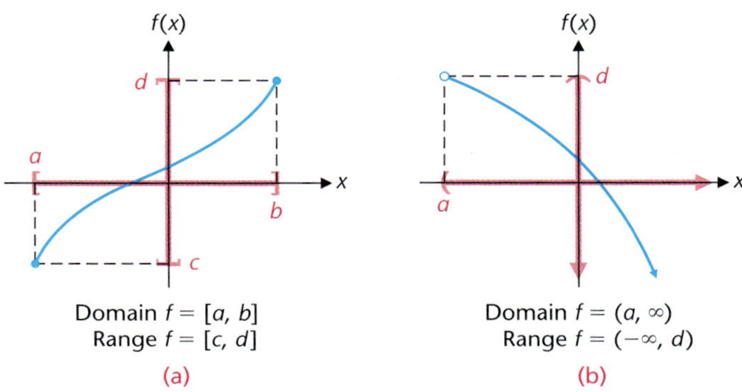

EXAMPLE 2 Finding the Domain and Range from a Graph

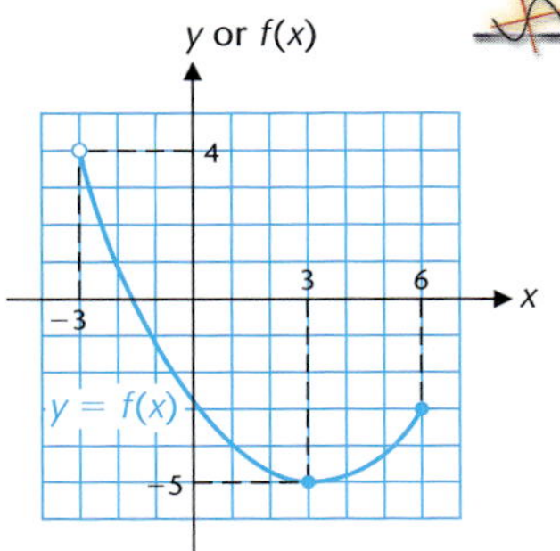

FIGURE 8

Find the domain and range of the function f whose graph is shown in Figure 8.

SOLUTION

The dots at each end of the graph of f indicate that the graph terminates at these points. Thus, the x coordinates of the points on the graph are between -3 and 6. The open dot at $(-3, 4)$ indicates that -3 is not in the domain of f, whereas the closed dot at $(6, -3)$ indicates that 6 is in the domain of f. That is,

Domain: $-3 < x \leq 6$ or $(-3, 6]$

The y coordinates are between -5 and 4 and, as before, the open dot at $(-3, 4)$ indicates that 4 is not in the range of f and the closed dot at $(3, -5)$ indicates that -5 is in the range of f. Thus,

Range: $-5 \leq y < 4$ or $[-5, 4)$

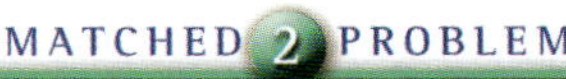

MATCHED 2 PROBLEM

Find the domain and range of the function f given by the graph in Figure 9.

FIGURE 9

*Interval notation is reviewed in Appendix A, Section A.1.

Increasing and Decreasing Functions

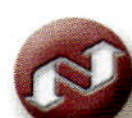

EXPLORE/DISCUSS 2

Graph each function in the standard viewing window, then write a verbal description of the behavior exhibited by the graph as x moves from left to right.

(A) $f(x) = 2 - x$ (B) $f(x) = x^3$

(C) $f(x) = 5$ (D) $f(x) = 9 - x^2$

We now take a look at *increasing* and *decreasing* properties of functions. Intuitively, a function is increasing over an interval I in its domain if its graph rises as the independent variable increases (moves from left to right) over I. A function is decreasing over I if its graph falls as the independent variable increases over I (Fig. 10).

FIGURE 10 Increasing, decreasing, and constant functions.

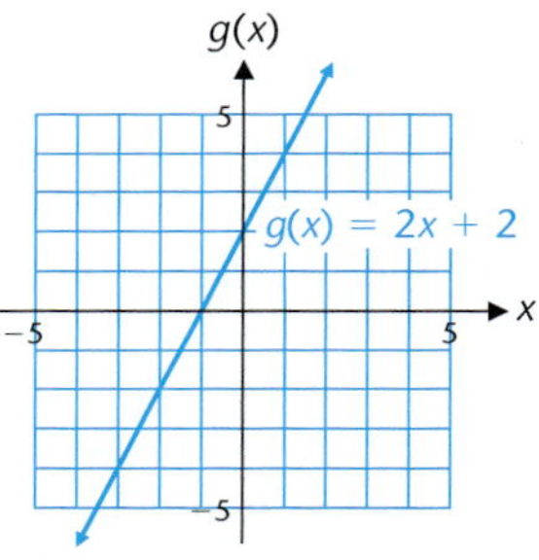

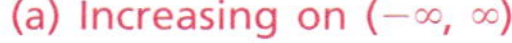

(a) Increasing on $(-\infty, \infty)$

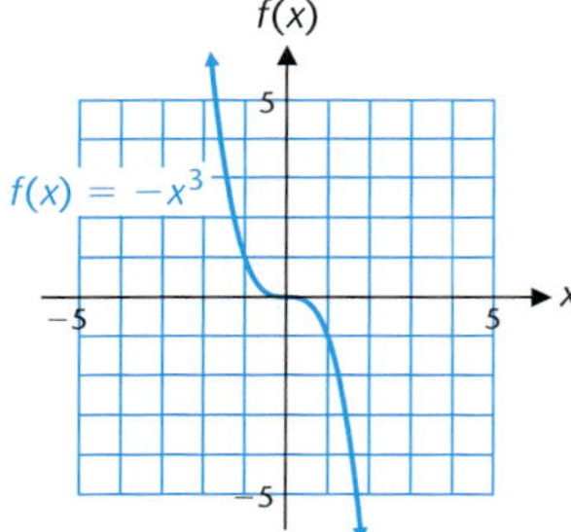

(b) Decreasing on $(-\infty, \infty)$

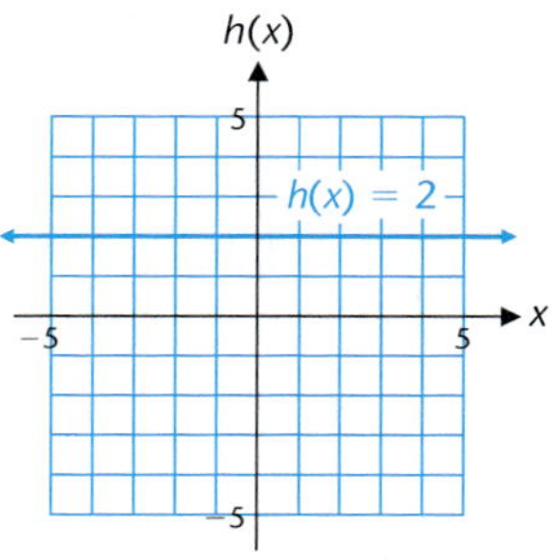

(c) Constant on $(-\infty, \infty)$

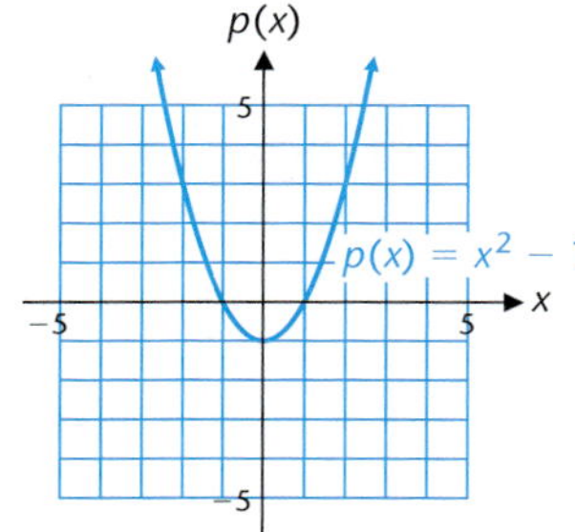

(d) Decreasing on $(-\infty, 0]$
Increasing on $[0, \infty)$

More formally, we define increasing, decreasing, and constant functions as follows:

> **DEFINITION 1**
> **Increasing, Decreasing, and Constant Functions**
>
> Let I be an interval in the domain of function f. Then,
> 1. f is **increasing** on I and the graph of f is **rising** on I if $f(a) < f(b)$ whenever $a < b$ in I.
> 2. f is **decreasing** on I and the graph of f is **falling** on I if $f(a) > f(b)$ whenever $a < b$ in I.
> 3. f is **constant** on I and the graph of f is **horizontal** on I if $f(a) = f(b)$ whenever $a < b$ in I.

Refer to Figure 10(a). As x moves from left to right, the values of g increase and the graph of g rises. In Figure 10(b), as x moves from left to right, the values of f decrease and the graph of f falls.

EXAMPLE 3 Describing a Graph

The graph of

$$f(x) = x^3 - 12x + 4$$

is shown in Figure 11. Use the terms *increasing, decreasing, rising,* and *falling* to write a verbal description of this graph.

SOLUTION

The values of f increase and the graph of f rises as x increases from $-\infty$ to -2. The values of f decrease and the graph of f falls as x increases from -2 to 2. Finally, the values of f increase and the graph of f rises as x increases from 2 to ∞.

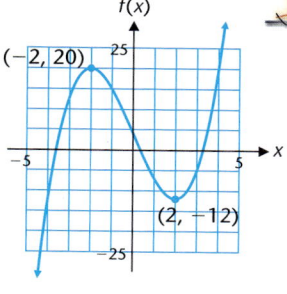

FIGURE 11 $f(x) = x^3 - 12x + 4$.

MATCHED PROBLEM 3

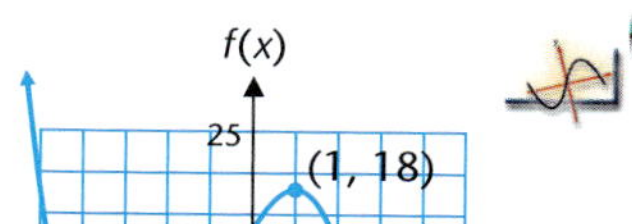

The graph of

$$f(x) = -x^3 - 3x^2 + 9x + 13$$

is shown in Figure 12. Use the terms *increasing, decreasing, rising,* and *falling* to write a verbal description of this graph.

FIGURE 12

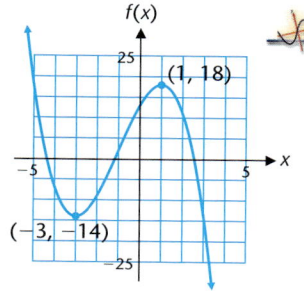

FIGURE 13 $f(x) = x^3 - 12x + 4$.

Local Maxima and Minima

Refer to $f(x) = x^3 - 12x + 4$ in Example 3 (Fig. 13). At the point $(-2, 20)$, the function changes from increasing to decreasing. This implies that the functional value $f(-2) = 20$ is greater than any of the nearby values of the function. At the point $(2, -12)$, the function changes from decreasing to increasing. This implies that the functional value $f(2) = -12$ is less than any nearby values of the function. These concepts are made more formal in the following definition.

DEFINITION 2

Local Maxima and Local Minima

The functional value $f(c)$ is called a **local maximum** if there is an interval (a, b) containing c such that

$$f(x) \leq f(c) \text{ for all } x \text{ in } (a, b)$$

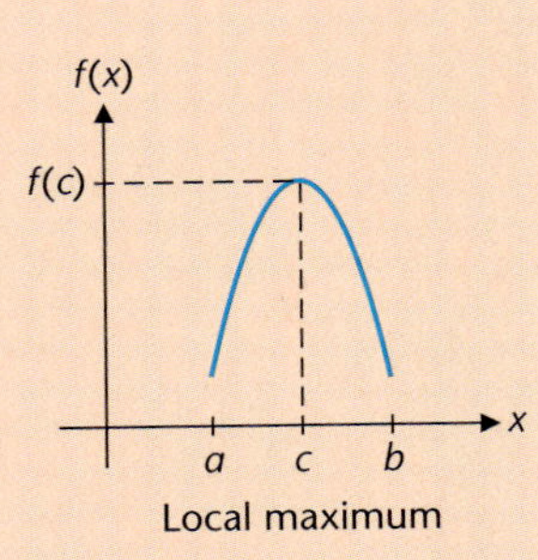

Local maximum

The functional value $f(c)$ is called a **local minimum** if there is an interval (a, b) containing c such that

$$f(x) \geq f(c) \text{ for all } x \text{ in } (a, b)$$

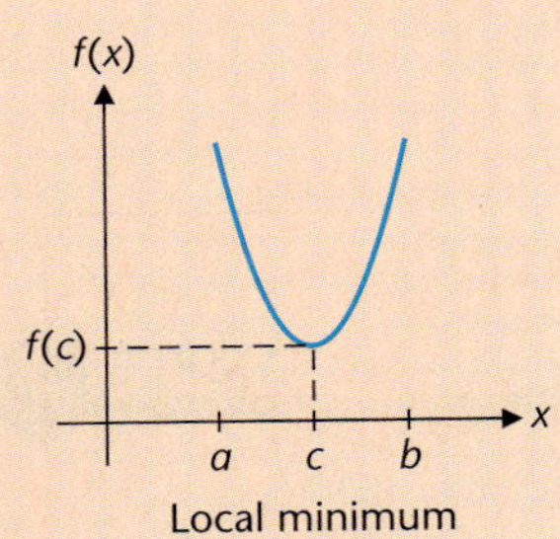

Local minimum

The functional value $f(c)$ is called a **local extremum** if it is either a local maximum or a local minimum.

EXPLORE/DISCUSS 3

Plot the points $A = (0, 0)$, $B = (3, 4)$, $C = (7, 1)$, and $D = (10, 5)$ in a coordinate plane. Draw a curve that satisfies each of the following conditions.

(A) Passes through A and B and is always increasing.

(B) Passes through A, B, and C with a local maximum at $x = 3$.

(C) Passes through A, B, C, and D with a local maximum at $x = 3$ and a local minimum at $x = 7$.

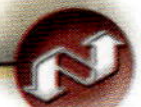

The location of local extrema plays an essential role in the analysis of functions and graphs. Local extrema are also crucial to the solution of many applied problems involving the maximum or minimum values of a function. Most graphing utilities have commands that approximate local maxima and minima. Examples 4 and 5 illustrate the use of these commands.

EXAMPLE 4 **Finding Local Extrema**

Find the domain, any local extrema, and the range of

$$f(x) = x^2 - 40\sqrt{x}$$

Round answers to two decimal places.

SOLUTION

Because $\sqrt{x}$ represents a real number only if $x \geq 0$, the domain of f is $[0, \infty)$. First we must select a viewing window. Because the domain of f is $[0, \infty)$, we choose Xmin = 0. We construct a table of values on a graphing utility (Figs. 14 and 15) to help select the remaining window variables. From Figure 14 we see that Ymin should be less than -64.44. Figure 15 indicates that Xmax = 15 and Ymax greater than 70.081 should produce a reasonable view of the graph. Our choice for the window variables is shown in Figure 16.

X	Y1
0	0
1	-39
2	-52.57
3	-60.28
4	-64
5	-64.44
6	-61.98

Y1 = -40√(X)+X²

FIGURE 14

X	Y1
9	-39
10	-26.49
11	-11.66
12	5.4359
13	24.778
14	46.334
15	70.081

Y1 = -40√(X)+X²

FIGURE 15

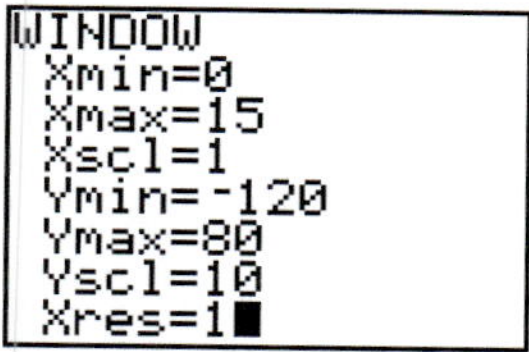

FIGURE 16

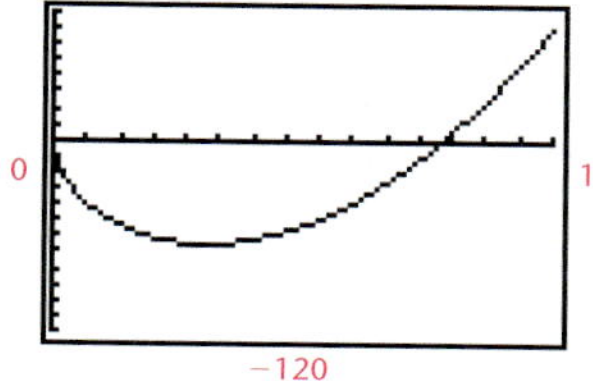

FIGURE 17

The graph of f is shown in Figure 17. Notice that we adjusted Ymin to provide space at the bottom of the screen for the text that the graphing utility will display.

Both the table in Figure 14 and the graph in Figure 17 indicate that f has a local minimum near $x = 5$. After selecting the **minimum** command on our graphing utility, we are asked to select a left bound (Fig. 18), a right bound (Fig. 19), and a guess (Fig. 20).

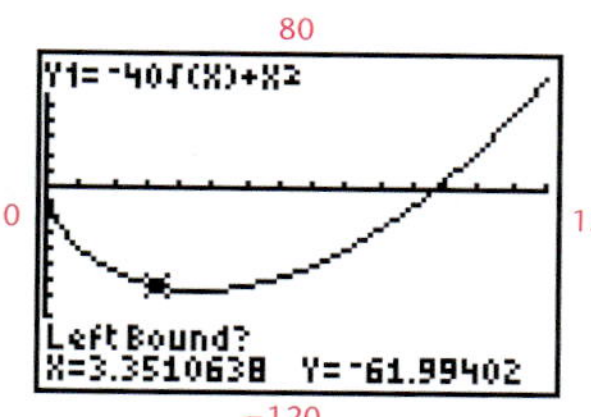

FIGURE 18

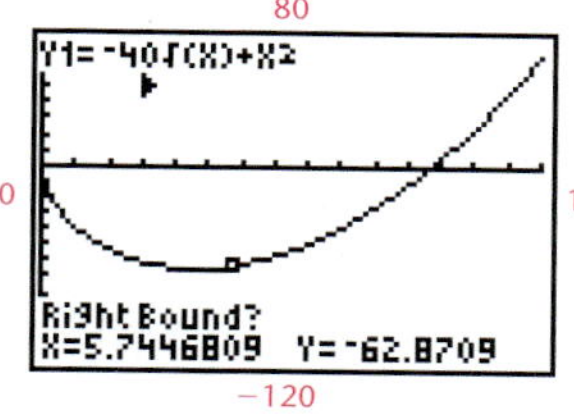

FIGURE 19

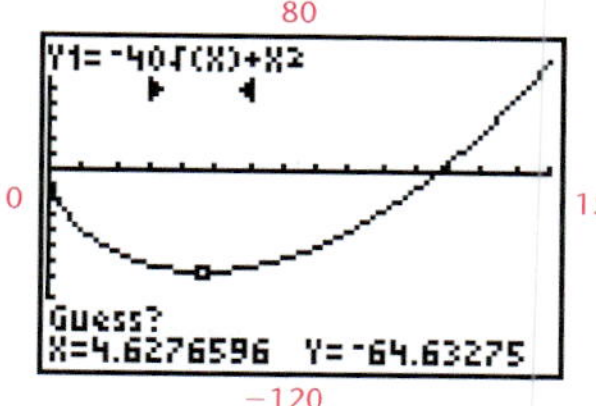

FIGURE 20

The final graph (Fig. 21) shows that, to two decimal places, f has a local minimum value of -64.63 at $x = 4.64$. The curve in Figure 21 suggests that as x increases to the right without bound, the values of $f(x)$ also increase without bound. The graph in Figure 22 and the table in Figure 23 support this suggestion. Thus, we conclude that there are no other local extrema and that the range of f is $[-64.63, \infty)$.

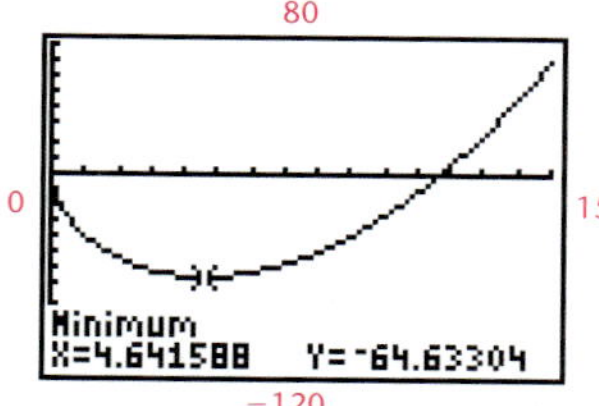

FIGURE 21

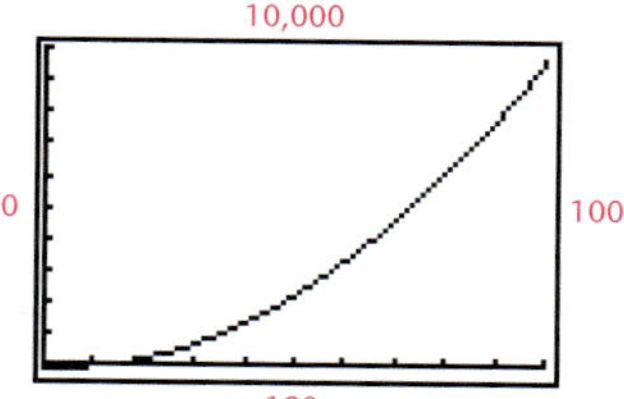

FIGURE 22

X	Y1
40	1347
50	2217.2
60	3290.2
70	4565.3
80	6042.2
90	7720.5
100	9600

Y1 = -40√(X)+X²

FIGURE 23

Summarizing our results, we have

Domain of $f = [0, \infty)$
Range of $f = [-64.63, \infty)$
Local minimum: $f(4.64) = -64.63$

MATCHED 4 PROBLEM

Find the domain, any local extrema, and the range of

$$f(x) = x - 5\sqrt{x}$$

Round answers to two decimal places.

Refer to Example 4. Calculus techniques are required to be certain that f continues to increase as x increases to the right. Without calculus, we have to rely on intuitive arguments involving graphical and numerical techniques, as we did in the solution to Example 4. As we broaden our experience base and become familiar with a larger variety of functions, we will strengthen our intuition. This is one of the major objectives of this book.

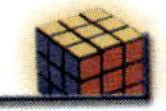

Mathematical Modeling

In Example 5, we use the maximum command to find the maximum value of a revenue function.

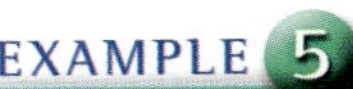

EXAMPLE 5 Maximizing Revenue

The revenue (in dollars) from the sale of x bicycle locks is given by

$$R(x) = 21x - 0.016x^2 \qquad 0 \le x \le 1{,}300$$

Find the number of locks that must be sold to maximize the revenue. What is the maximum revenue, to the nearest dollar?

SOLUTION

We begin by entering the revenue function as Y_1 and constructing a table of values for the revenue (Fig. 24.) From the limits given in the problem, we select Xmin = 0 and Xmax = 1,300. The table in Figure 24 suggests that Ymax = 7,000 is a good choice and, as before, we select Ymin so the text displayed by the graphing utility does not cover any important portions of the graph. We enter the window variables (Fig. 25) and use the maximum command (Fig. 26). (The maximum command works just like the minimum command used in Example 4. The details of selecting the bounds and the initial guess are omitted.)

X	Y1	
300	4860	
400	5840	
500	6500	
600	6840	
700	6860	
800	6560	
900	5940	

Y1=21X-0.016X²

FIGURE 24

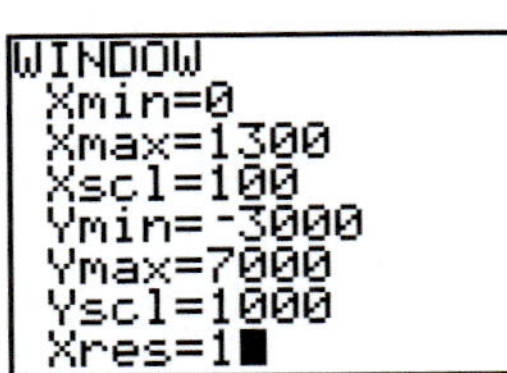

FIGURE 25

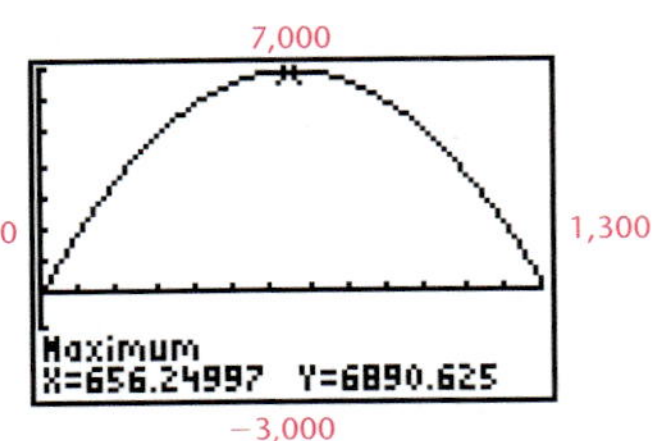

FIGURE 26

X	Y1
656	6890.6
657	6890.6

Y1 = 21X−0.016X²

FIGURE 27

The results in Figure 26 show that, to two decimal places, the maximum revenue is $6,890.63 when $x = 656.25$ locks. But this cannot be the answer to the problem. It is not possible to sell one-fourth of a lock. Examining the values of R at $x = 656$ and $x = 657$ (Fig. 27), we conclude that the maximum revenue, to the nearest dollar, is $6,891 when either 656 locks or 657 locks are sold.

MATCHED 5 PROBLEM

The profit (in dollars) from the sale of x bicycle locks is given by

$$P(x) = 17.5x - 0.016x^2 - 2{,}000 \qquad 0 \le x \le 1{,}300$$

Find the number of locks that must be sold to maximize the profit. What is the maximum profit, to the nearest dollar?

Example 5 illustrates an important step in the mathematical modeling process. Solutions obtained from a model must be interpreted in terms of the original real-world problem. In the case of Example 5, the revenue function R is defined only for integer values of x, $x = 0, 1, 2, \ldots, 1{,}300$. However, for the purposes of mathematical analysis and as an aid in visualizing the behavior of the function R, we assume that the revenue function is defined for all x, $0 \le x \le 1{,}300$. After finding that the maximum value of the revenue function occurs at $x = 656.25$, we must remember to interpret this solution to mean either $x = 656$ or $x = 657$.

Piecewise-Defined Functions

The **absolute value function** is defined by

$$f(x) = |x| = \begin{cases} -x & \text{if } x < 0 \\ x & \text{if } x \ge 0 \end{cases} \qquad \begin{aligned} |-4| &= -(-4) = 4 \\ |3| &= 3 \end{aligned}$$

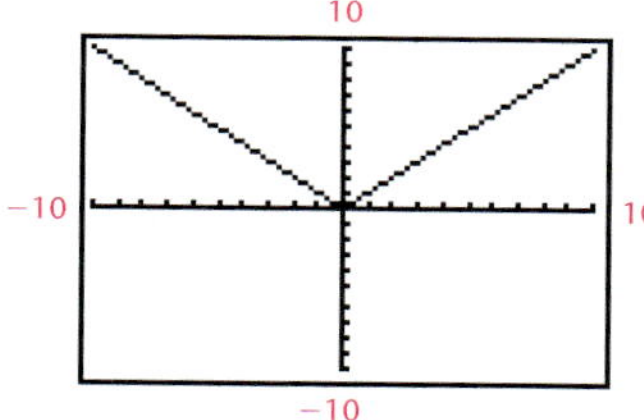

FIGURE 28 Graph of $f(x) = |x| = \text{abs}(x)$.

The graph of $|x|$ is shown in Figure 28. Most graphing utilities use abs or ABS to denote this function, and the graph is produced directly using $y_1 = \text{abs}(x)$.

The absolute value function is defined by different formulas for different parts of its domain. Functions whose definitions involve more than one formula are called **piecewise-defined functions.** Notice that the graph of the absolute value function has a **sharp corner** at (0, 0), a common characteristic of piecewise-defined functions. As Example 6 illustrates, piecewise-defined functions occur naturally in many applications.

EXAMPLE 6 Rental Charges

A car rental agency charges $0.25 per mile if the mileage does not exceed 100. If the total mileage exceeds 100, the agency charges $0.25 per mile for the first 100 miles and $0.15 per mile for any additional mileage.

(A) If x represents the number of miles a rented vehicle is driven, express the mileage charge $C(x)$ as a function of x.

(B) Complete the following table.

x	0	50	100	150	200
$C(x)$					

(C) Sketch the graph of $y = C(x)$ by hand, using a graphing utility as an aid, and indicate the points in the table on the graph with solid dots.

SOLUTIONS

(A) If $0 \le x \le 100$, then

$$C(x) = 0.25x$$

If $x > 100$, then

$$\begin{aligned} C(x) &= \underbrace{0.25(100)}_{\text{Charge for the first 100 miles}} + \underbrace{0.15(x - 100)}_{\text{Charge for the additional mileage}} \\ &= 25 + 0.15x - 15 \\ &= 10 + 0.15x \end{aligned}$$

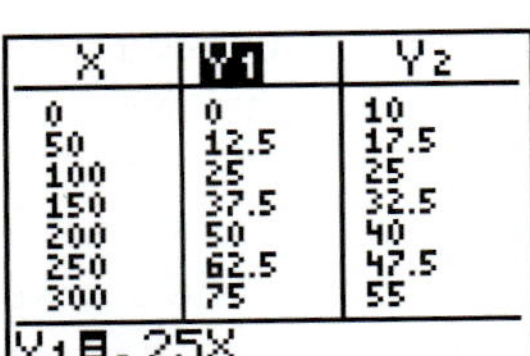

FIGURE 29
$y_1 = 0.25x$, $y_2 = 10 + 0.15x$.

Thus, we see that C is a piecewise-defined function:

$$C(x) = \begin{cases} 0.25x & \text{if } 0 \le x \le 100 \\ 10 + 0.15x & \text{if } x > 100 \end{cases}$$

(B) Piecewise-defined functions are evaluated by first determining which rule applies and then using the appropriate rule to find the value of the function. To begin, we enter both rules in a graphing utility and use the TABLE command (Fig. 29). To complete the table, we use the values of $C(x)$ from the y_1 column if $0 \le x \le 100$, and from the y_2 column if $x > 100$.

x	0	50	100	150	200
$C(x)$	\$0	\$12.50	\$25	\$32.50	\$40

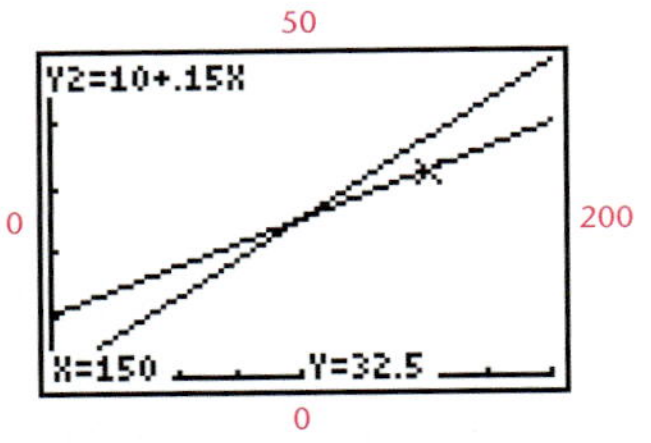

FIGURE 30
$y_1 = 0.25x$, $y_2 = 10 + 0.15x$.

(C) Using a graph of both rules in the same viewing window as an aid (Fig. 30). we sketch the graph of $y = C(x)$ and add the points from the table to produce Figure 31.

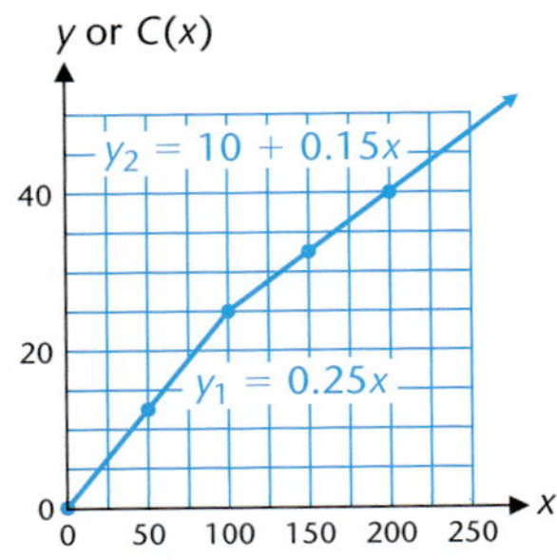

Notice the sharp corner at (100, 25).

FIGURE 31 Hand sketch of the graph of $y = C(x)$.

MATCHED PROBLEM 6

Another car rental agency charges \$0.30 per mile when the total mileage does not exceed 75, and \$0.30 per mile for the first 75 miles plus \$0.20 per mile for the additional mileage when the total mileage exceeds 75.

(A) If x represents the number of miles a rented vehicle is driven, express the mileage charge $C(x)$ as a function of x.

(B) Complete the following table.

x	0	50	75	100	150
$C(x)$					

(C) Sketch the graph of $y = C(x)$ by hand, using a graphing utility as an aid, and indicate the points in the table on the graph with solid dots.

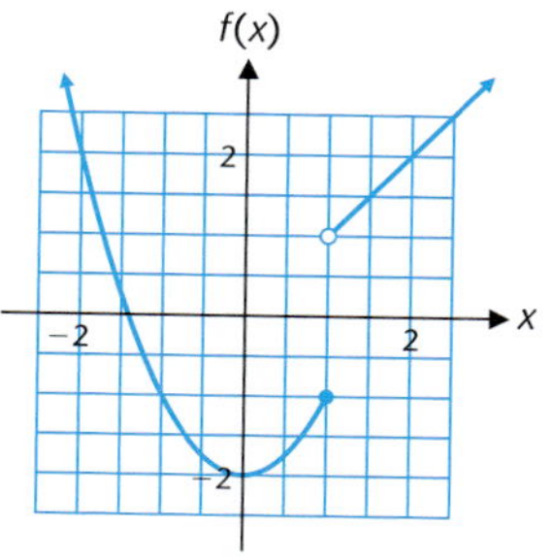

FIGURE 32 Graph of
$$f(x) = \begin{cases} x^2 - 2 & \text{for } x \le 1 \\ x & \text{for } x > 1 \end{cases}$$

Refer to Figures 29 and 31 in the solution to Example 6. Notice that the two formulas in the definition of C produce the same value at $x = 100$ and that the graph of C contains no breaks. Informally, a graph (or portion of a graph) is said to be **continuous** if it contains no breaks or gaps and can be drawn without lifting a pen from the paper. A graph is **discontinuous** at any points where there is a break or a gap. For example, the graph of the function in Figure 32 is discontinuous at $x = 1$. The entire graph cannot be drawn without lifting a pen from the paper. (A formal presentation of continuity can be found in calculus texts.)

The Greatest Integer Function

We conclude Section 1.3 with a discussion of an interesting and useful function called the *greatest integer function.*

The **greatest integer** of a real number x, denoted by $[\![x]\!]$, is the integer n such that $n \le x < n + 1$; that is, $[\![x]\!]$ is the largest integer less than or equal to x (Fig. 33).

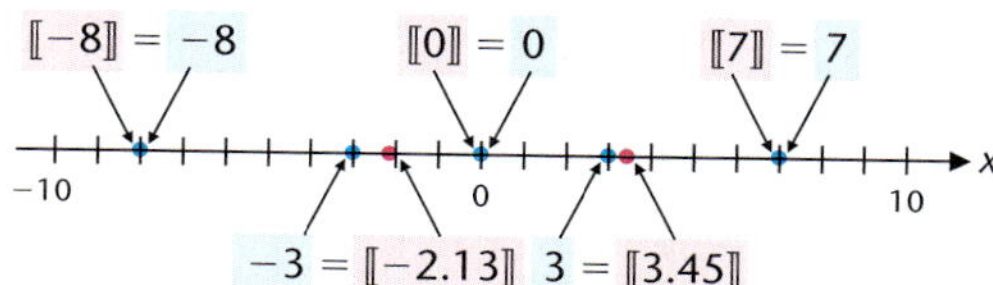

FIGURE 33

The **greatest integer function** f is defined by the equation $f(x) = [\![x]\!]$. A piecewise definition of f for $-2 \le x < 3$ is shown below and a sketch of the graph of f for $-5 \le x \le 5$ is shown in Figure 34. Since the domain of f is all real numbers, the piecewise definition continues indefinitely in both directions, as does the stairstep pattern in the figure. Thus, the range of f is the set of all integers. The greatest integer function is an example of a more general class of functions called **step functions.**

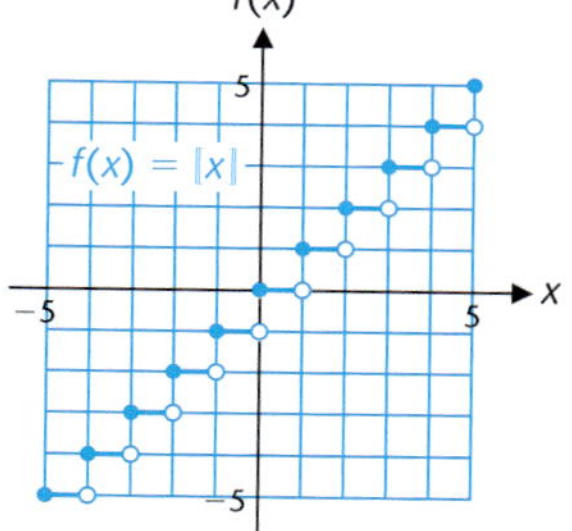

FIGURE 34 Greatest integer function.

$$f(x) = [\![x]\!] = \begin{cases} \vdots & \\ -2 & \text{if } -2 \le x < -1 \\ -1 & \text{if } -1 \le x < 0 \\ 0 & \text{if } 0 \le x < 1 \\ 1 & \text{if } 1 \le x < 2 \\ 2 & \text{if } 2 \le x < 3 \\ \vdots & \end{cases}$$

Notice in Figure 34 that at each integer value of x there is a break in the graph, and between integer values of x there is no break. Thus, the greatest integer function is discontinuous at each integer n and continuous on each interval of the form $[n, n + 1)$.

Most graphing utilities will graph the greatest integer function, usually denoted by int, but these graphs require careful interpretation. Comparing the sketch of $y = [\![x]\!]$ in Figure 34 with the graph of $y = \text{int}(x)$ in Figure 35(a), we see that the graphing utility has connected the endpoints of the horizontal line segments. This gives the appearance that the graph is continuous when it is not. To obtain a correct graph, consult the manual to determine how to change the graphing mode on your graphing utility from **connected mode** to **dot mode** [Fig. 35(b)].

FIGURE 35 Greatest integer function on a graphing utility.

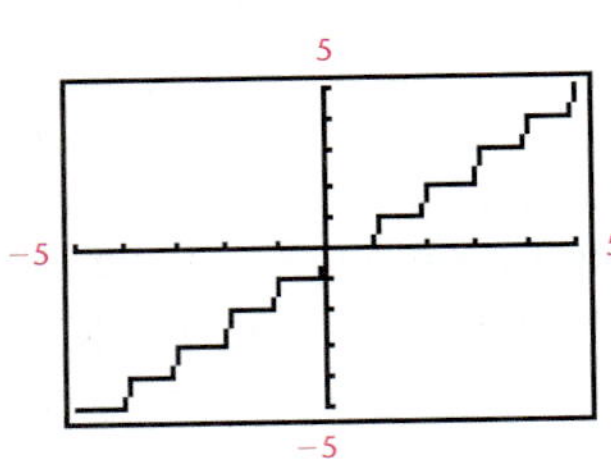

(a) Graph of $y = \text{int}(x)$ in the connected mode.

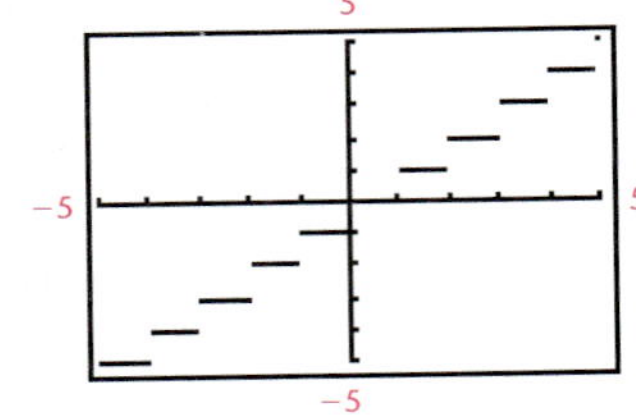

(b) Graph of $y = \text{int}(x)$ in the dot mode.

To avoid misleading graphs, use the dot mode on your graphing utility when graphing a function with discontinuities.

EXAMPLE 7 Computer Science

Let

$$f(x) = \frac{[\![10x + 0.5]\!]}{10}$$

Find

(A) $f(6)$ (B) $f(1.8)$ (C) $f(3.24)$ (D) $f(4.582)$ (E) $f(-2.68)$

What operation does this function perform?

SOLUTIONS

(A) $f(6) = \dfrac{[\![60.5]\!]}{10} = \dfrac{60}{10} = 6$

(B) $f(1.8) = \dfrac{[\![18.5]\!]}{10} = \dfrac{18}{10} = 1.8$

(C) $f(3.24) = \dfrac{[\![32.9]\!]}{10} = \dfrac{32}{10} = 3.2$

(D) $f(4.582) = \dfrac{[\![46.32]\!]}{10} = \dfrac{46}{10} = 4.6$

(E) $f(-2.68) = \dfrac{[\![-26.3]\!]}{10} = \dfrac{-27}{10} = -2.7$

x	f[x]
6	6
1.8	1.8
3.24	3.2
4.582	4.6
−2.68	−2.7

Comparing the values of x and $f(x)$ in the table, we conclude that this function rounds decimal fractions to the nearest tenth.

MATCHED PROBLEM 7

Let $f(x) = [\![x + 0.5]\!]$. Find

(A) $f(6)$ (B) $f(1.8)$ (C) $f(3.24)$ (D) $f(-4.3)$ (E) $f(-2.69)$

What operation does this function perform?

ANSWERS MATCHED PROBLEMS

1. x intercept: -1.516; y intercept: 5
2. Domain: $-4 < x < 5$ or $(-4, 5)$
   Range: $-4 < y \le 3$ or $(-4, 3]$
3. The values of f decrease and the graph of f is falling on $(-\infty, -3)$ and $(1, \infty)$. The values of f increase and the graph of f is rising on $(-3, 1)$.
4. Domain: $[0, \infty]$; range: $[-6.25, \infty)$; local minimum: $f(6.25) = -6.25$
5. The maximum profit of \$2,785 occurs when 547 locks are sold.
6. (A) $C(x) = \begin{cases} 0.3x & \text{if } 0 \le x \le 75 \\ 7.5 + 0.2x & \text{if } x > 75 \end{cases}$

(B)

x	0	50	75	100	150
$C(x)$	\$0	\$15	\$22.50	\$27.50	\$37.50

(C) $C(x)$; $y_2 = 7.5 + 0.2x$; $y_1 = 0.3x$

7. (A) 6 (B) 2 (C) 3 (D) -4 (E) -3; f rounds decimal fractions to the nearest integer.

EXERCISE 1.3

A

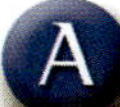

Problems 1–6 refer to functions f, g, h, k, p, and q given by the following graphs. (Assume the graphs continue as indicated beyond the parts shown.)

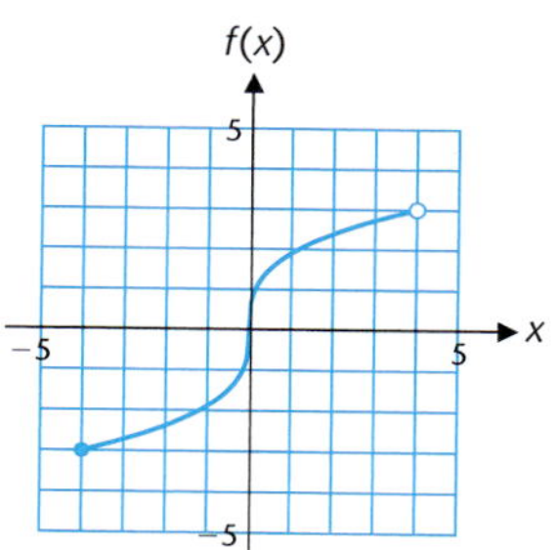

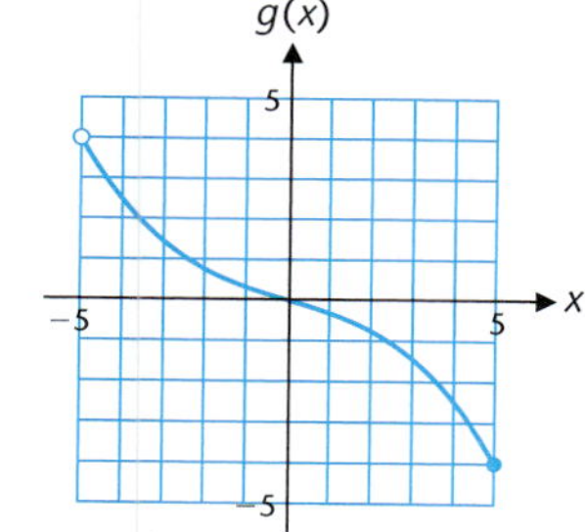

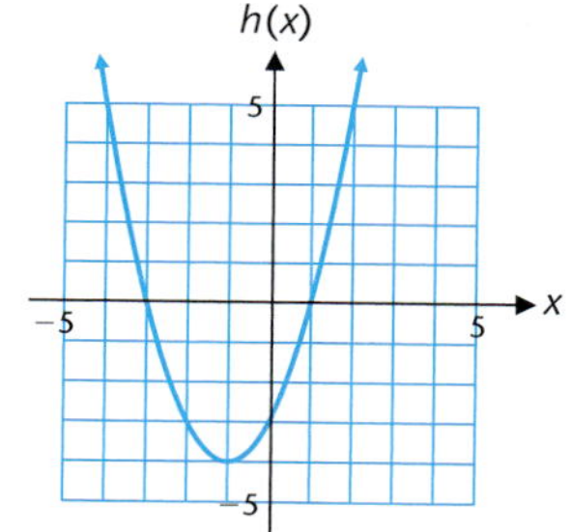

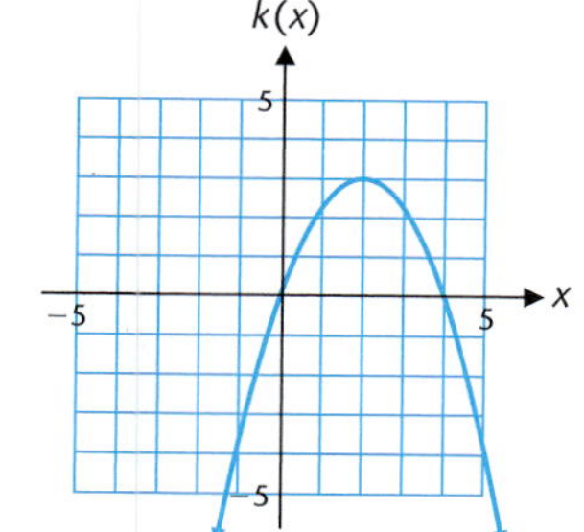

1. For the function f, find
   (A) Domain
   (B) Range
   (C) x intercepts
   (D) y intercept
   (E) Intervals over which f is increasing
   (F) Intervals over which f is decreasing
   (G) Intervals over which f is constant
   (H) Any points of discontinuity
2. Repeat Problem 1 for the function g.
3. Repeat Problem 1 for the function h.
4. Repeat Problem 1 for the function k.
5. Repeat Problem 1 for the function p (see page 52).
6. Repeat Problem 1 for the function q (see page 52).

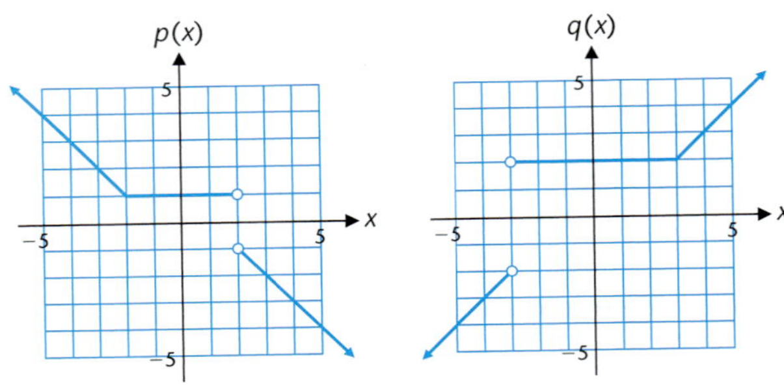

In Problems 7–12, examine the graph of the function to determine the intervals over which the function is increasing, the intervals over which the function is decreasing, and the intervals over which the function is constant. Approximate the endpoints of the intervals to the nearest integer.

7. $f(x) = |x + 2| - 5$

8. $k(x) = |x - 2| - x$

9. $m(x) = |x - 3| - |x + 4|$

10. $q(x) = |x + 2| + |x - 4|$

11. $r(x) = |x + 4| - |x| + |x - 4|$

12. $s(x) = |x| - |x + 5| - |x - 3|$

In Problems 13–18, find the x intercepts, y intercept, and any local extrema. Round answers to three decimal places.

13. $f(x) = x^2 - 5x - 9$

14. $g(x) = -x^2 + 7x + 14$

15. $h(x) = -x^3 + 4x + 25$

16. $k(x) = x^3 + 3x^2 - 15$

17. $m(x) = \sqrt{|x^2 - 12|}$

18. $n(x) = \sqrt{|x^3 - 12|}$

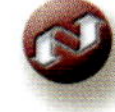

In Problems 19–24, sketch by hand the graph of a continuous function f over the interval [−5, 5] that is consistent with the given information.

19. The function f is increasing on $[-5, -2]$, constant on $[-2, 2]$, and decreasing on $[2, 5]$.

20. The function f is decreasing on $[-5, -2]$, constant on $[-2, 2]$, and increasing on $[2, 5]$.

21. The function f is decreasing on $[-5, -2]$, constant on $[-2, 2]$, and decreasing on $[2, 5]$.

22. The function f is increasing on $[-5, -2]$, constant on $[-2, 2]$, and increasing on $[2, 5]$.

23. The function f is decreasing on $[-5, -2]$, increasing on $[-2, 2]$, and decreasing on $[2, 5]$.

24. The function f is increasing on $[-5, -2]$, decreasing on $[-2, 2]$, and increasing on $[2, 5]$.

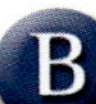

B

In Problems 25–30, sketch the graph of $y = f(x)$ and evaluate $f(-2)$, $f(-1)$, $f(1)$, and $f(2)$.

25. $f(x) = \begin{cases} x + 1 & \text{if } x \le 0 \\ -x + 1 & \text{if } x > 0 \end{cases}$

26. $f(x) = \begin{cases} x & \text{if } x < 1 \\ -x + 2 & \text{if } x \ge 1 \end{cases}$

27. $f(x) = \begin{cases} x + 2 & \text{if } x < -1 \\ x - 2 & \text{if } x \ge -1 \end{cases}$

28. $f(x) = \begin{cases} -x - 1 & \text{if } x \le 2 \\ -x + 5 & \text{if } x > 2 \end{cases}$

29. $f(x) = \begin{cases} x^2 + 1 & \text{if } x < 0 \\ -x^2 - 1 & \text{if } x > 0 \end{cases}$

30. $f(x) = \begin{cases} -x^2 - 2 & \text{if } x < 0 \\ x^2 + 2 & \text{if } x > 0 \end{cases}$

In Problems 31–38, find the domain, range, y intercept, and x intercepts. Round answers to two decimal places.

31. $m(x) = x^3 + 45x^2 - 30$

32. $f(x) = x^3 - 35x^2 + 25$

33. $n(x) = 200 + 200x^2 - x^4$

34. $g(x) = 200 - 40x^3 - x^4$

35. $h(x) = 8\sqrt{x} - x$

36. $s(x) = x - 18\sqrt{x}$

37. $k(x) = x^2 - 50\sqrt{x + 5}$

38. $t(x) = 30\sqrt{20 - x} - x^2$

In Problems 39–44, write a verbal description of the graph of the given function using the terms increasing, decreasing, rising, and falling, and indicate any local maximum and minimum values. Approximate to two decimal places the coordinates of any points used in your description.

39. $f(x) = x^3 - 12x^2 + 3x - 10$

40. $h(x) = x^3 + 15x^2 + 5x + 15$

41. $m(x) = 24|x| - x^2$

42. $n(x) = 15\sqrt{|x|} - x$

43. $g(x) = |x^2 - 5x - 300|$

44. $k(x) = |x^2 - 4x - 480|$

In Problems 45–50, write a verbal description of the graph of the given function using increasing and decreasing terminology, and indicating any local maximum and minimum values. Approximate to two decimal places the coordinates of any points used in your description.

45. $f(x) = x^2 + 4.3x - 32$

46. $g(x) = -x^2 + 6.9x + 25$

47. $h(x) = x^3 - x^2 - 74x + 60$

48. $k(x) = -x^3 + x^2 + 82x - 25$

49. $p(x) = |x^2 - x - 18|$

50. $q(x) = |x^2 - 2x - 30|$

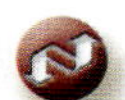

In Problems 51–56, sketch by hand the graph of a function f that is continuous on the interval [−5, 5], except as noted, and is consistent with the given information.

51. The function f is increasing on $[-5, 0)$, discontinuous at $x = 0$, increasing on $(0, 5]$, $f(-2) = 0$, and $f(2) = 0$.

52. The function f is decreasing on $[-5, 0)$, discontinuous at $x = 0$, decreasing on $(0, 5]$, $f(-3) = 0$, and $f(3) = 0$.

53. The function f is discontinuous at $x = 0$, $f(-3) = -2$ is a local maximum, and $f(2) = 3$ is a local minimum.

54. The function f is discontinuous at $x = 0$, $f(-3) = 2$ is a local minimum, and $f(2) = -3$ is a local maximum.

55. The function f is discontinuous at $x = -2$ and $x = 2$, $f(-3) = -2$ and $f(3) = -2$ are local maxima, and $f(0) = 0$ is a local minimum.

56. The function f is discontinuous at $x = -2$ and $x = 2$, $f(-3) = 2$ and $f(3) = 2$ are local minima, and $f(0) = 0$ is a local maximum.

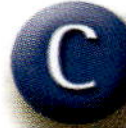

In Problems 57–62, graph y = f(x) in a standard viewing window. Assuming that the graph continues as indicated beyond the part shown in this viewing window, find the domain, range, and any points of discontinuity. (Use the dot mode on your graphing utility.)

57. $f(x) = \dfrac{|5x - 10|}{x - 2}$

58. $f(x) = \dfrac{4x + 12}{|x + 3|}$

59. $f(x) = x + \dfrac{|4x - 4|}{x - 1}$

60. $f(x) = x + \dfrac{|2x + 2|}{x + 1}$

61. $f(x) = |x| - \dfrac{|9 - 3x|}{x - 3}$

62. $f(x) = |x| + \dfrac{|2x + 4|}{x + 2}$

In Problems 63–68, write a piecewise definition of f and sketch by hand the graph of f, using a graphing utility as an aid. Include sufficient intervals to clearly illustrate both the definition and the graph. Find the domain, range, and any points of discontinuity.

63. $f(x) = [\![x/2]\!]$

64. $f(x) = [\![x/3]\!]$

65. $f(x) = [\![3x]\!]$

66. $f(x) = [\![2x]\!]$

67. $f(x) = x - [\![x]\!]$

68. $f(x) = [\![x]\!] - x$

69. The function f is continuous and increasing on the interval $[1, 9]$ with $f(1) = -5$ and $f(9) = 4$.

(A) Sketch a graph of f that is consistent with the given information.

(B) How many times does your graph cross the x axis? Could the graph cross more times? Fewer times? Support your conclusions with additional sketches and/or verbal arguments.

70. Repeat Problem 69 if the function does not have to be continuous.

71. The function f is continuous on the interval $[-5, 5]$ with $f(-5) = -4$, $f(1) = 3$, and $f(5) = -2$.

(A) Sketch a graph of f that is consistent with the given information.

(B) How many times does your graph cross the x axis? Could the graph cross more times? Fewer times? Support your conclusions with additional sketches and/or verbal arguments.

72. Repeat Problem 71 if f is continuous on $[-8, 8]$ with $f(-8) = -6$, $f(-4) = 3$, $f(3) = -2$, and $f(8) = 5$.

73. The function f is continuous on $[0, 10]$, $f(5) = -5$ is a local minimum, and f has no other local extrema on this interval.

(A) Sketch a graph of f that is consistent with the given information.

(B) How many times does your graph cross the x axis? Could the graph cross more times? Fewer times? Support your conclusions with additional sketches and/or verbal arguments.

74. Repeat Problem 73 if $f(5) = 1$ and all other information is unchanged.

APPLICATIONS

75. Computer Science. Let $f(x) = 10[\![0.5 + x/10]\!]$. Evaluate f at 4, −4, 6, −6, 24, 25, 247, −243, −245, and −246. What operation does this function perform?

76. Computer Science. Let $f(x) = 100[\![0.5 + x/100]\!]$. Evaluate f at 40, −40, 60, −60, 740, 750, 7,551, −601, −649, and −651. What operation does this function perform?

★ **77. Computer Science.** Use the greatest integer function to define a function f that rounds real numbers to the nearest hundredth.

★ **78. Computer Science.** Use the greatest integer function to define a function f that rounds real numbers to the nearest thousandth.

79. Revenue. The revenue (in dollars) from the sale of x car seats for infants is given by

$$R(x) = 60x - 0.035x^2 \qquad 0 \le x \le 1{,}700$$

Find the number of car seats that must be sold to maximize the revenue. What is the maximum revenue (to the nearest dollar)?

80. Profit. The profit (in dollars) from the sale of x car seats for infants is given by

$$P(x) = 38x - 0.035x^2 - 4{,}000 \qquad 0 \le x \le 1{,}700$$

Find the number of car seats that must be sold to maximize the profit. What is the maximum profit (to the nearest dollar)?

★ **81. Manufacturing.** A box is to be made out of a piece of cardboard that measures 18 by 24 inches. Squares, x inches on a side, will be cut from each corner and then the ends and sides will be folded up (see the figure).

Find the size of the cutout squares that will make the maximum volume. What is the maximum volume? Round answers to two decimal places.

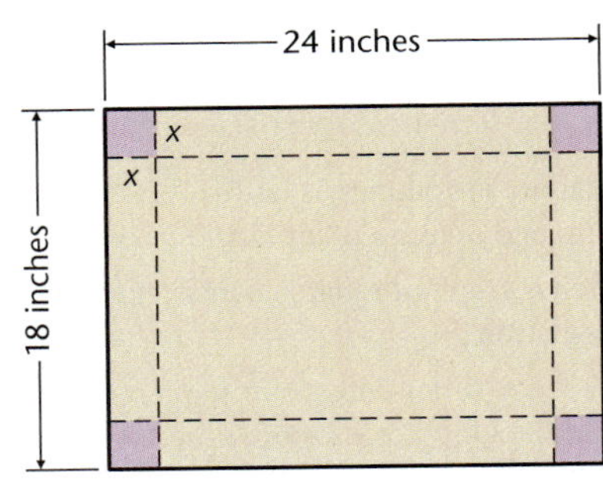

★ **82. Manufacturing.** A box with a hinged lid is to be made out of a piece of cardboard that measures 20 by 40 inches. Six squares, x inches on a side, will be cut from each corner and the middle of the sides, and then the ends and sides will be folded up to form the box and its lid (see the figure).

Find the size of the cutout squares that will make the maximum volume. What is the maximum volume? Round answers to two decimal places.

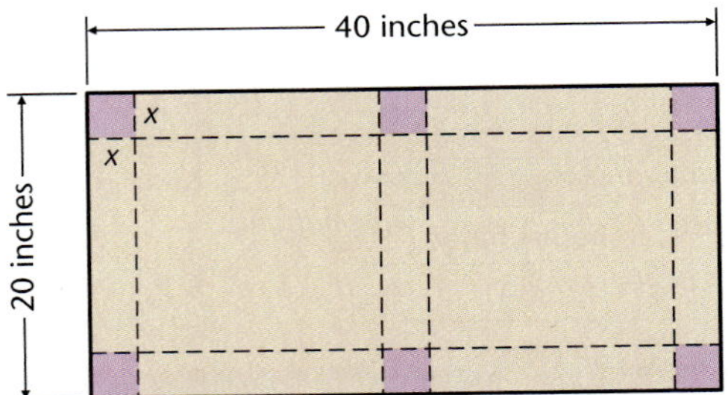

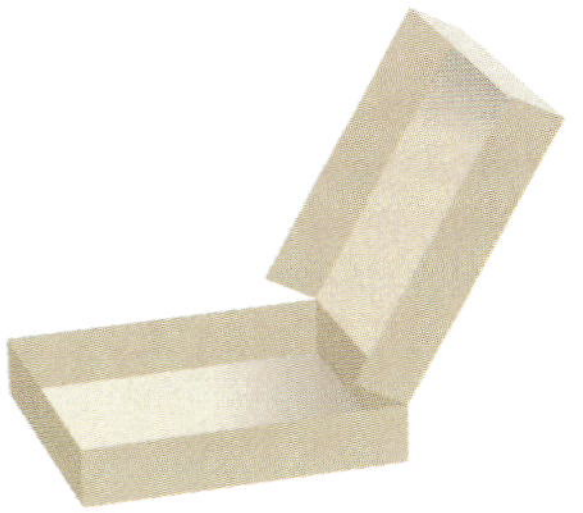

83. Construction. A freshwater pipe is to be run from a source on the edge of a lake to a small resort community on an island 8 miles offshore, as indicated in the figure. It costs $10,000 per mile to lay the pipe on land and $16,000 per mile to lay the pipe in the lake. The total cost $C(x)$ in thousands of dollars of laying the pipe is given by

$$C(x) = 10(20 - x) + 16\sqrt{x^2 + 64} \qquad 0 \le x \le 20$$

Find the length (to two decimal places) of the land portion of the pipe that will make the production costs minimum. Find the minimum cost to the nearest thousand dollars.

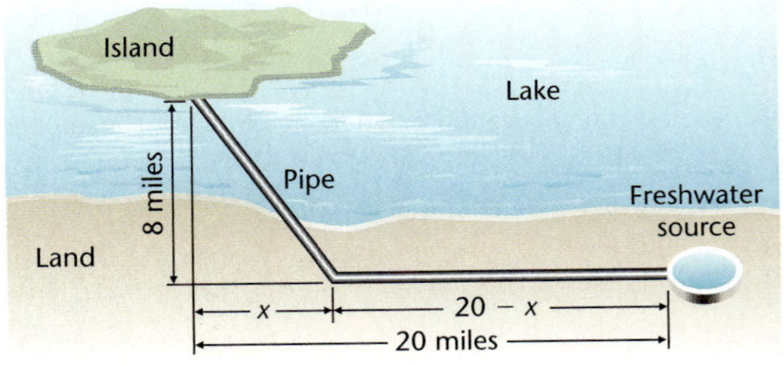

84. Transportation. The construction company laying the freshwater pipe in Problem 83 uses an amphibious vehicle to travel down the beach and then out to the island. The vehicle travels at 30 miles per hour on land and 7.5 miles per hour in water. The total time $T(x)$ in minutes for a trip from the freshwater source to the island is given by

$$T(x) = 2(20 - x) + 8\sqrt{x^2 + 64} \quad 0 \le x \le 20$$

Find (to two decimal places) the length of the land portion of the trip that will make the time minimum. Find the minimum time to the nearest minute.

MODELING AND DATA ANALYSIS 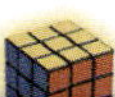

Table 1 contains daily automobile rental rates from a New Jersey firm.

TABLE 1

Vehicle Type	Daily Charge	Included Miles	Mileage Charge*
Compact	\$32.00	100/Day	\$0.16/mile
Midsize	\$41.00	200/Day	\$0.18/mile

*Mileage charge does not apply to included miles.

Source: www.gogelauto.com

85. Automobile Rental. Use the data in Table 1 to construct a piecewise-defined model for the daily rental charge for a compact automobile that is driven x miles.

86. Automobile Rental. Use the data in Table 1 to construct a piecewise-defined model for the daily rental charge for a midsize automobile that is driven x miles.

87. Delivery Charges. A nationwide package delivery service charges \$15 for overnight delivery of packages weighing 1 pound or less. Each additional pound (or fraction thereof) costs an additional \$3. Let $C(x)$ be the charge for overnight delivery of a package weighing x pounds.

(A) Write a piecewise definition of C for $0 < x \le 6$ and sketch the graph of C by hand.

(B) Can the function f defined by $f(x) = 15 + 3[\![x]\!]$ be used to compute the delivery charges for all x, $0 < x \le 6$? Justify your answer.

88. Telephone Charges. Calls to 900 numbers are charged to the caller. A 900 number hot line for tips and hints for video games charges \$4 for the first minute of the call and \$2 for each additional minute (or fraction thereof). Let $C(x)$ be the charge for a call lasting x minutes.

(A) Write a piecewise definition of C for $0 < x \le 6$ and sketch the graph of C by hand.

(B) Can the function f defined by $f(x) = 4 + 2[\![x]\!]$ be used to compute the charges for all x, $0 < x \le 6$? Justify your answer.

★ **89. Sales Commissions.** An appliance salesperson receives a base salary of \$200 a week and a commission of 4% on all sales over \$3,000 during the week. In addition, if the weekly sales are \$8,000 or more, the salesperson receives a \$100 bonus. If x represents weekly sales (in dollars), express the weekly earnings $E(x)$ as a function of x, and sketch its graph. Identify any points of discontinuity. Find $E(5{,}750)$ and $E(9{,}200)$.

★ **90. Service Charges.** On weekends and holidays, an emergency plumbing repair service charges \$2.00 per minute for the first 30 minutes of a service call and \$1.00 per minute for each additional minute. If x represents the duration of a service call in minutes, express the total service charge $S(x)$ as a function of x, and sketch its graph. Identify any points of discontinuity. Find $S(25)$ and $S(45)$.

Table 2 contains the income tax schedule for the state of Minnesota in a recent year.

TABLE 2

Status	Taxable Income Over	But Not Over	Tax Is	Of the Amount Over
Single	$ 0	$ 18,120	..... 5.35%	$ 0
	18,120	59,500	$ 969 + 7.05%	18,120
	59,500	.....	3,887 + 7.85%	59,500
Married	0	26,480	..... 5.35%	0
	26,480	105,200	1,417 + 7.05%	26,480
	105,200	.....	6,966 + 7.85%	105,200

Source: www.taxsites.com/state.html

★ **91. State Income Tax.** Use the schedule in Table 2 to construct a piecewise-defined model for the taxes due for a single taxpayer with a taxable income of x dollars. Find the tax on the following incomes: \$10,000, \$30,000, \$100,000.

★ **92. State Income Tax.** Use the schedule in Table 2 to construct a piecewise-defined model for the taxes due for a married taxpayer with a taxable income of x dollars. Find the tax on the following incomes: \$20,000, \$60,000, \$200,000.

93. Tire Mileage. An automobile tire manufacturer collected the data in the table relating tire pressure x, in pounds per square inch (lb/in.^{2}), and mileage in thousands of miles.

x	28	30	32	34	36
Mileage	45	52	55	51	47

A mathematical model for these data is given by

$f(x) = -0.518x^2 + 33.3x - 481$

(A) Compare the model and the data graphically and numerically.

(B) Find (to two decimal places) the mileage for a tire pressure of 31 lb/in.^{2} and for 35 lb/in.^{2}.

(C) Write a brief description of the relationship between tire pressure and mileage, using the terms *increasing, decreasing, local maximum,* and *local minimum* where appropriate.

94. Automobile Production. The table lists General Motors's total U.S. vehicle production in millions of units from 1989 to 1993.

Year	89	90	91	92	93
Production	4.7	4.1	3.5	3.7	5.0

A mathematical model for GM's production data is given by

$f(x) = 0.33x^2 - 1.3x + 4.8$

where $x = 0$ corresponds to 1989.

(A) Compare the model and the data graphically and numerically.

(B) Estimate to two decimal places the production in 1994 and in 1995.

(C) Write a brief verbal description of GM's production from 1989 to 1993, using *increasing, decreasing, local maximum,* and *local minimum* terminology where appropriate.

SECTION 1.4 Functions: Graphs and Transformations

A Beginning Library of Elementary Functions • Vertical and Horizontal Shifts • Expansions and Contractions • Reflections in the *x* and *y* Axes • Even and Odd Functions

The functions

$$g(x) = x^2 + 2 \qquad h(x) = (x + 2)^2 \qquad k(x) = 2x^2$$

can be expressed in terms of the function $f(x) = x^2$ as follows:

$$g(x) = f(x) + 2 \qquad h(x) = f(x + 2) \qquad k(x) = 2f(x)$$

In this section we will see that the graphs of functions g, h, and k are closely related to the graph of function f. Insight gained by understanding these relationships will help us analyze and interpret the graphs of many different functions.

A Beginning Library of Elementary Functions

As we progress through this book, we will encounter a number of basic functions that we will want to add to our library of elementary functions. Figure 1 shows six basic functions that you will encounter frequently. You should know the definition, domain, and range of each of these functions, and be able to recognize their graphs. You should graph each basic function in Figure 1 on your graphing utility.

FIGURE 1 Some basic functions and their graphs.
[*Note*: Letters used to designate these functions may vary from context to context; R is the set of all real numbers.]

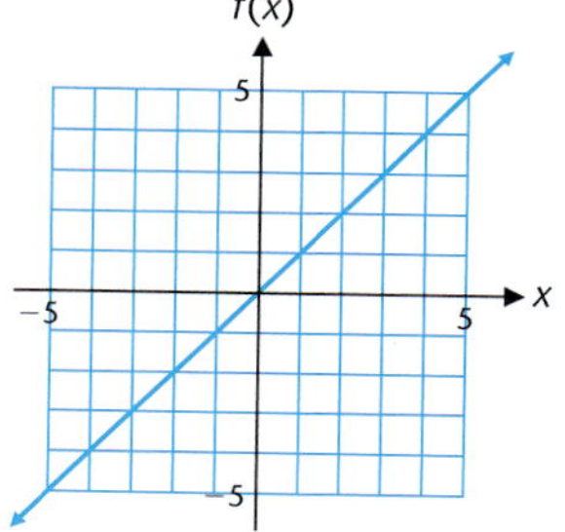

(a) Identity function
$f(x) = x$
Domain: R
Range: R

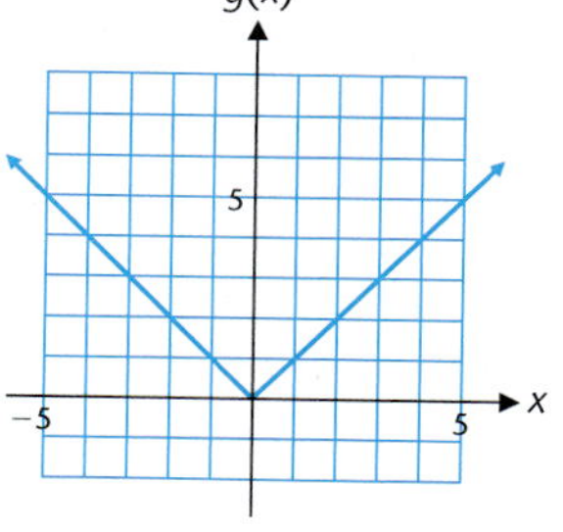

(b) Absolute value function
$g(x) = |x|$
Domain: R
Range: $[0, \infty)$

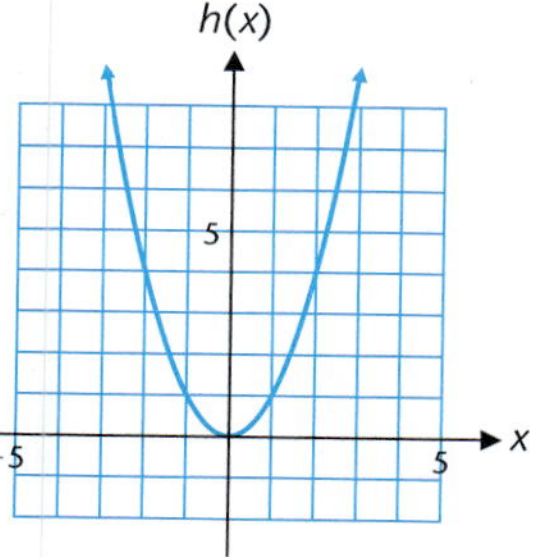

(c) Square function
$h(x) = x^2$
Domain: R
Range: $[0, \infty)$

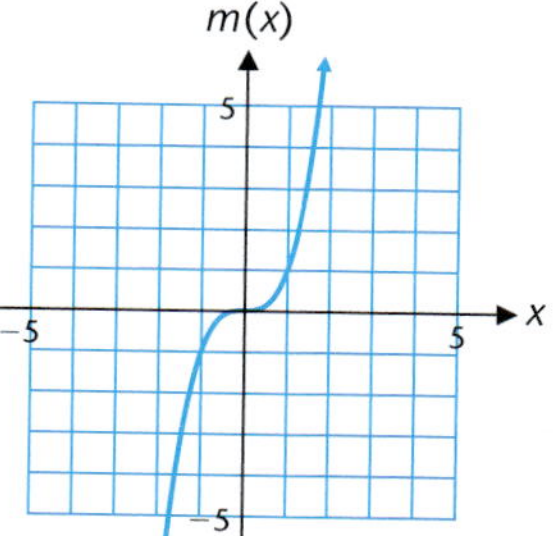

(d) Cube function
$m(x) = x^3$
Domain: R
Range: R

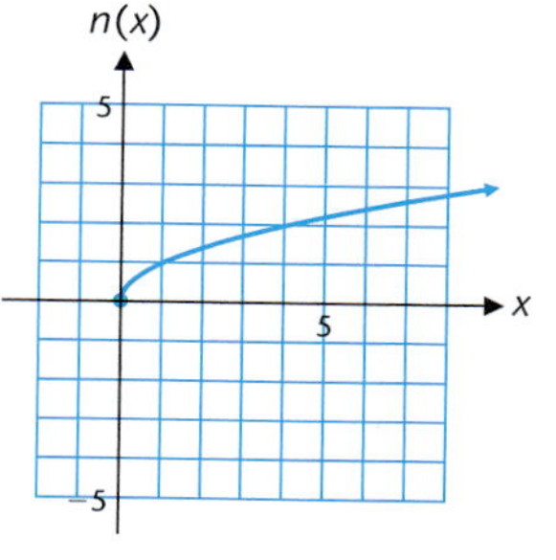

(e) Square root function
$n(x) = \sqrt{x}$
Domain: $[0, \infty)$
Range: $[0, \infty)$

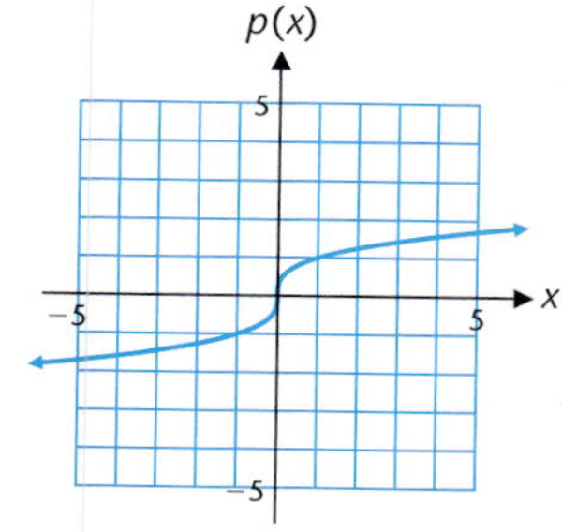

(f) Cube root function
$p(x) = \sqrt[3]{x}$
Domain: R
Range: R

Most graphing utilities allow you to define a number of functions, usually denoted by $y_1, y_2, y_3, \ldots$. You can graph all of these functions simultaneously, or you can select certain functions for graphing and suppress the graphs of the others. Consult your manual to determine how many functions can be stored in your graphing utility at one time and how to select particular functions for graphing. Many of our investigations in this section will involve graphing two or more functions at the same time.

Vertical and Horizontal Shifts

If a new function is formed by performing an operation on a given function, then the graph of the new function is called a **transformation** of the graph of the original function. For example, if we add a constant k to $f(x)$, then the graph of $y = f(x)$ is transformed into the graph of $y = f(x) + k$.

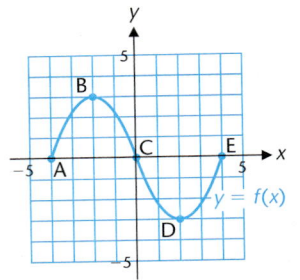

FIGURE 2

TABLE 1

	x	$f(x)$
A	−4	0
B	−2	3
C	0	0
D	2	−3
E	4	0

EXPLORE/DISCUSS 1

The following activities refer to the graph of f shown in Figure 2 and the corresponding points on the graph shown in Table 1.

(A) Use the points in Table 1 to construct a similar table and then sketch a graph for each of the following functions: $y = f(x) + 2$, $y = f(x) - 3$. Describe the relationship between the graph of $y = f(x)$ and the graph of $y = f(x) + k$ for k any real number.

(B) Use the points in Table 1 to construct a similar table and then sketch a graph for each of the following functions: $y = f(x + 2)$, $y = f(x - 3)$. [*Hint:* Choose values of x so that $x + 2$ or $x - 3$ is in Table 1.] Describe the relationship between the graph of $y = f(x)$ and the graph of $y = f(x + h)$ for h any real number.

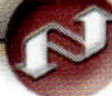

EXAMPLE 1 Vertical and Horizontal Shifts

(A) How are the graphs of $y = x^2 + 2$ and $y = x^2 - 3$ related to the graph of $y = x^2$? Confirm your answer by graphing all three functions simultaneously in the same viewing window.

(B) How are the graphs of $y = (x + 2)^2$ and $y = (x - 3)^2$ related to the graph of $y = x^2$? Confirm your answer by graphing all three functions simultaneously in the same viewing window.

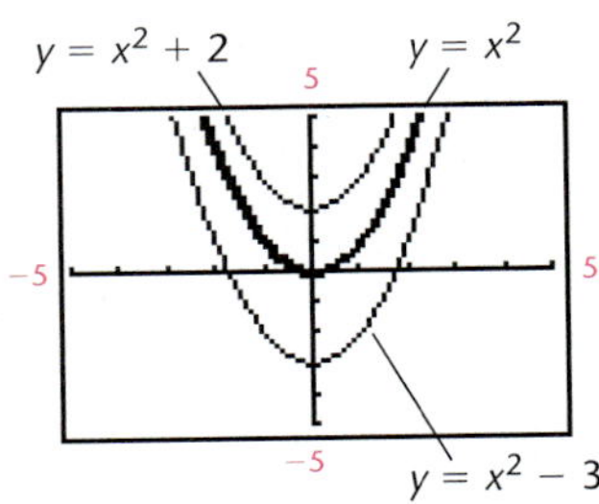

FIGURE 3 Vertical shifts.

SOLUTIONS

(A) The graph of $y = x^2 + 2$ is the same as the graph of $y = x^2$ shifted upward two units, and the graph of $y = x^2 - 3$ is the same as the graph of $y = x^2$ shifted downward three units. Figure 3 confirms these conclusions. (It appears that the graph of $y = f(x) + k$ is the graph of $y = f(x)$ shifted up if k is positive and down if k is negative.)

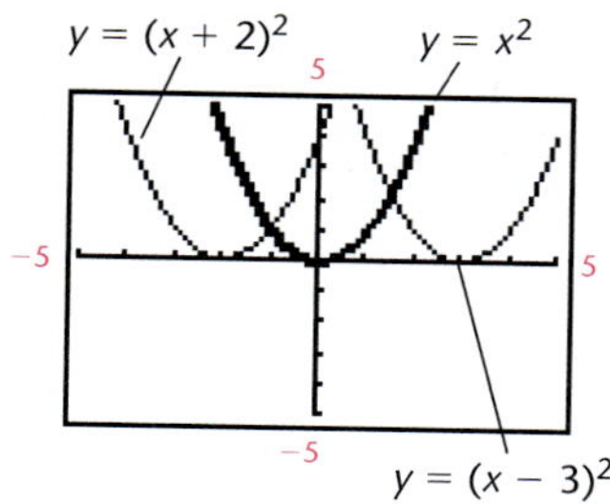

FIGURE 4 Horizontal shifts.

(B) The graph of $y = (x + 2)^2$ is the same as the graph of $y = x^2$ shifted to the left two units, and the graph of $y = (x - 3)^2$ is the same as the graph of $y = x^2$ shifted to the right three units. Figure 4 confirms these conclusions. [It appears that the graph of $y = f(x + h)$ is the graph of $y = f(x)$ shifted right if h is negative and left if h is positive—the opposite of what you might expect.]

MATCHED 1 PROBLEM

(A) How are the graphs of $y = \sqrt{x} + 3$ and $y = \sqrt{x} - 1$ related to the graph of $y = \sqrt{x}$? Confirm your answer by graphing all three functions simultaneously in the same viewing window.

(B) How are the graphs of $y = \sqrt{x + 3}$ and $y = \sqrt{x - 1}$ related to the graph of $y = \sqrt{x}$? Confirm your answer by graphing all three functions simultaneously in the same viewing window.

Comparing the graph of $y = f(x) + k$ with the graph of $y = f(x)$, we see that the graph of $y = f(x) + k$ can be obtained from the graph of $y = f(x)$ by **vertically translating** (shifting) the graph of the latter upward k units if k is positive and downward $|k|$ units if k is negative. Comparing the graph of $y = f(x + h)$ with the graph of $y = f(x)$, we see that the graph of $y = f(x + h)$ can be obtained from the graph of $y = f(x)$ by **horizontally translating** (shifting) the graph of the latter h units to the left if h is positive and $|h|$ units to the right if h is negative.

EXAMPLE 2 Vertical and Horizontal Translations [Shifts]

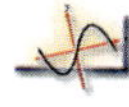

The graphs in Figure 5 are either horizontal or vertical shifts of the graph of $f(x) = |x|$. Write appropriate equations for functions G, H, M, and N in terms of f.

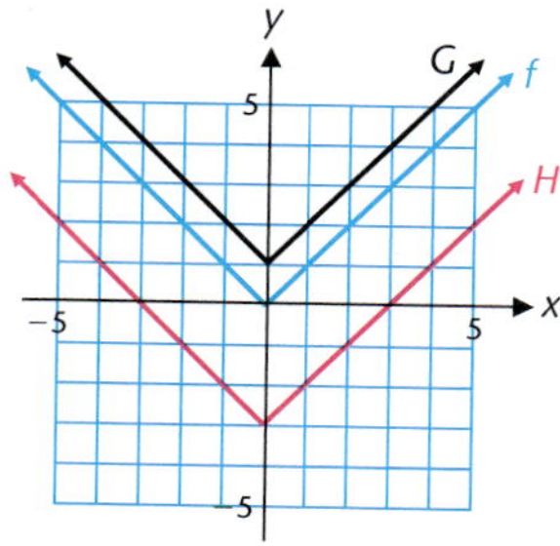

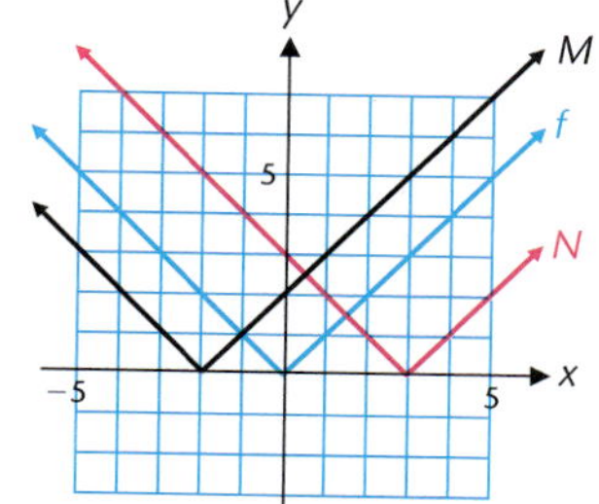

FIGURE 5 Vertical and horizontal shifts.

SOLUTION

Functions H and G are vertical shifts given by

$$H(x) = |x| - 3 \qquad G(x) = |x| + 1$$

Functions M and N are horizontal shifts given by

$$M(x) = |x + 2| \qquad N(x) = |x - 3|$$

MATCHED 2 PROBLEM

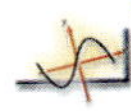

The graphs in Figure 6 are either horizontal or vertical shifts of the graph of $f(x) = x^3$. Write appropriate equations for functions H, G, M, and N in terms of f.

FIGURE 6 Vertical and horizontal shifts.

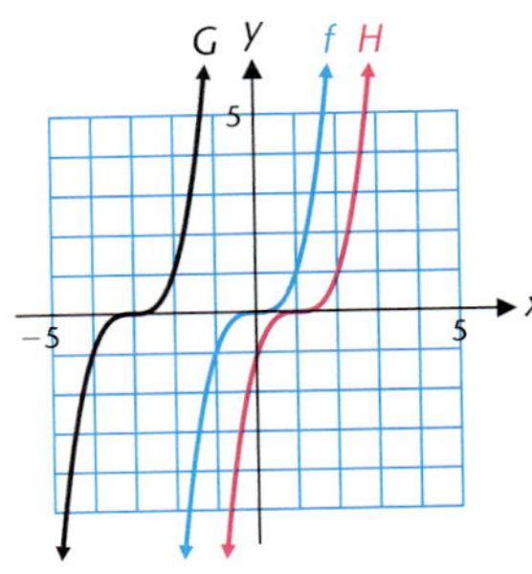

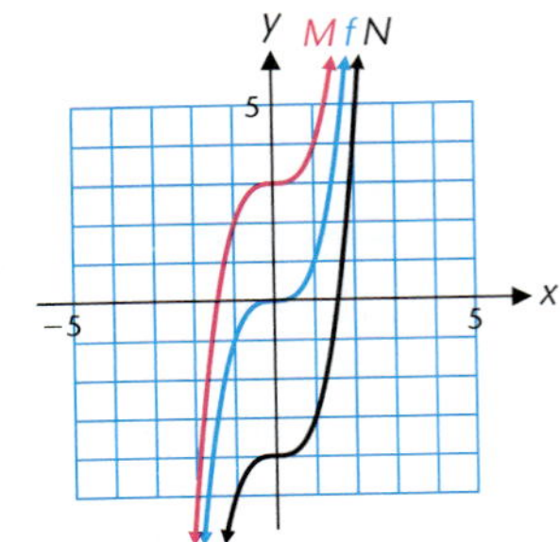

Expansions and Contractions

We now investigate how the graph of $y = f(x)$ is related to the graph of $y = Af(x)$ and to the graph of $y = f(Ax)$ for different positive real numbers A.

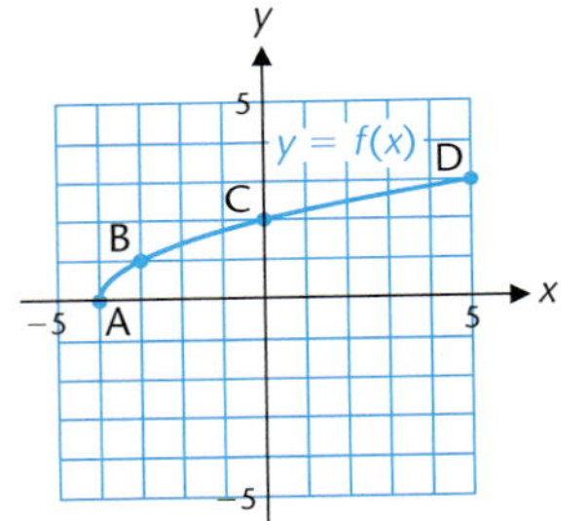

FIGURE 7

TABLE 2	x	$f(x)$
A	−4	0
B	−3	1
C	0	2
D	5	3

EXPLORE/DISCUSS 2

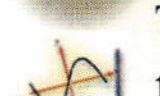

The following activities refer to the graph of f shown in Figure 7 and the corresponding points on the graph shown in Table 2.

(A) Construct a similar table and then sketch a graph for each of the following functions: $y = 2f(x)$, $y = \frac{1}{2}f(x)$. Describe the relationship between the graph of $y = f(x)$ and the graph of $y = Af(x)$ for A any positive real number.

(B) Construct a similar table and then sketch a graph for each of the following functions: $y = f(2x)$, $y = f(\frac{1}{2}x)$. [*Hint:* Select the x values for your table so that the multiples of x are in Table 2.] Describe the relationship between the graph of $y = f(x)$ and the graph of $y = f(Ax)$ for A any positive real number.

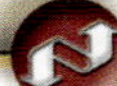

Comparing the graph of $y = Af(x)$ with the graph of $y = f(x)$, we see that the graph of $y = Af(x)$ can be obtained from the graph of $y = f(x)$ by multiplying each ordinate value (y coordinate) of the latter by A. The result is a **vertical expansion** of the graph of $y = f(x)$ if $A > 1$ and a **vertical contraction** of the graph of $y = f(x)$ if $0 < A < 1$. Likewise, comparing the graph of $y = f(Ax)$ with the graph of $y = f(x)$, we see that the graph of $y = f(Ax)$ can be obtained from the graph of $y = f(x)$ by multiplying each abscissa value (x coordinate) of the latter by $\frac{1}{A}$. The result is a **horizontal expansion** of the graph of $y = f(x)$ if $0 < A < 1$ and a **horizontal contraction** of the graph of $y = f(x)$ if $A > 1$.

EXAMPLE 3 Expansions and Contractions

(A) How are the graphs of $y = 2\sqrt[3]{x}$ and $y = 0.5\sqrt[3]{x}$ related to the graph of $y = \sqrt[3]{x}$? Confirm your answer by graphing all three functions simultaneously in the same viewing window.

(B) How are the graphs of $y = \sqrt[3]{2x}$ and $y = \sqrt[3]{0.5x}$ related to the graph of $y = \sqrt[3]{x}$? Confirm your answer by graphing all three functions simultaneously in the same viewing window.

SOLUTIONS

(A) The graph of $y = 2\sqrt[3]{x}$ is a vertical expansion of the graph of $y = \sqrt[3]{x}$ by a factor of 2, and the graph of $y = 0.5\sqrt[3]{x}$ is a vertical contraction of the graph of $y = \sqrt[3]{x}$ by a factor of 0.5. Figure 8 confirms this conclusion.

(B) The graph of $y = \sqrt[3]{2x}$ is a horizontal contraction of the graph of $y = \sqrt[3]{x}$ by a factor of $\frac{1}{2}$, and the graph of $y = \sqrt[3]{0.5x}$ is a horizontal expansion of the graph of $y = \sqrt[3]{x}$ by a factor of 2. Figure 9 confirms this conclusion.

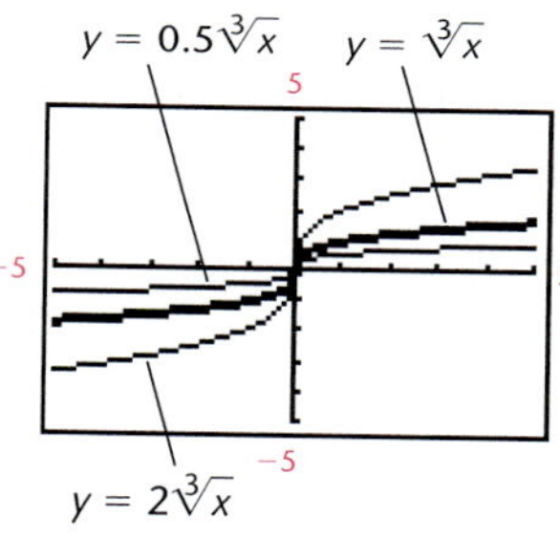

FIGURE 8 Vertical expansion and contraction

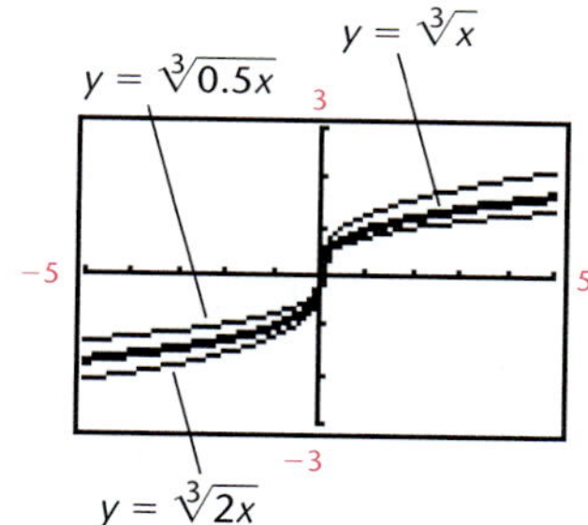

FIGURE 9 Horizontal expansion and contraction.

MATCHED 3 PROBLEM

(A) How are the graphs of $y = 2x^3$ and $y = 0.5x^3$ related to the graph of $y = x^3$? Confirm your answer by graphing all three functions simultaneously in the same viewing window.

(B) How are the graphs of $y = (2x)^3$ and $y = (0.5x)^3$ related to the graph of $y = x^3$? Confirm your answer by graphing all three functions simultaneously in the same viewing window.

Refer to Example 3. If $f(x) = \sqrt[3]{x}$, then we can write

$$f(Ax) = \sqrt[3]{Ax} = \sqrt[3]{A}\sqrt[3]{x} = Bf(x)$$

where $B = \sqrt[3]{A}$. Thus, for certain functions, a horizontal contraction or expansion can also be interpreted as a vertical contraction or expansion.

Reflections in the *x* and *y* Axes

We now investigate how the graphs of $y = -f(x)$ and $y = f(-x)$ are related to the graph of $y = f(x)$.

EXPLORE/DISCUSS 3

The following activities refer to the graph of f shown in Figure 10 and the corresponding points on the graph shown in Table 3.

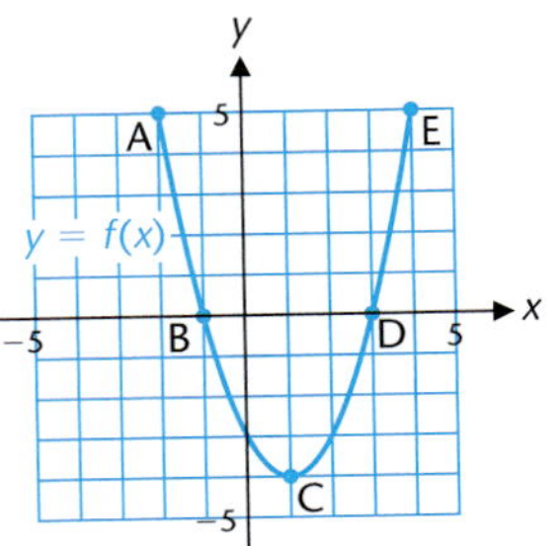

FIGURE 10

TABLE 3

	x	$f(x)$
A	-2	5
B	-1	0
C	1	-4
D	3	0
E	4	5

(A) Construct a similar table and then sketch a graph for $y = -f(x)$. Describe the relationship between the graph of $y = f(x)$ and the graph of $y = -f(x)$.

(B) Construct a similar table and then sketch a graph for $y = f(-x)$. [*Hint:* Choose x values so that $-x$ is in Table 3.] Describe the relationship between the graph of $y = f(x)$ and the graph of $y = f(-x)$.

(C) Construct a similar table and then sketch a graph for $y = -f(-x)$. [*Hint:* Choose x values so that $-x$ is in Table 3.] Describe the relationship between the graph of $y = f(x)$ and the graph of $y = -f(-x)$.

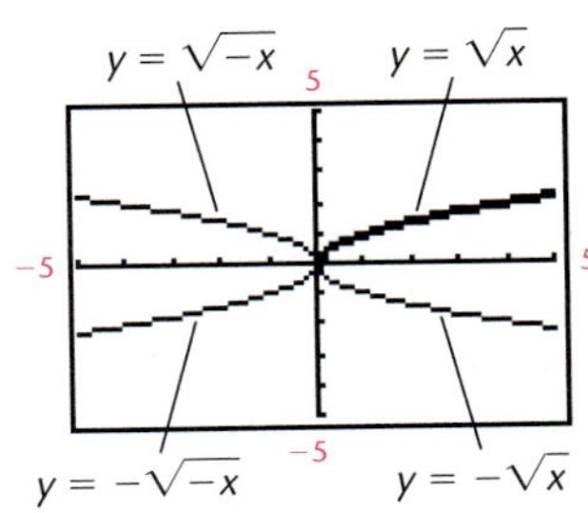

FIGURE 11 Reflections of the graph of $y = \sqrt{x}$

The graph of $y = f(-x)$ can be obtained from the graph of $y = f(x)$ by changing the sign of each abscissa (x coordinate). This has the effect of reflecting the graph of $y = f(x)$ in the y axis. The graph of $y = -f(x)$ can be obtained from the graph of $y = f(x)$ by changing the sign of each ordinate (y coordinate). This has the effect of reflecting the graph of $y = f(x)$ in the x axis. And finally, the graph of $y = -f(-x)$ can be obtained from the graph of $y = f(x)$ by changing the sign of each ordinate (y coordinate) and of each abscissa (x coordinate). This is referred to as *reflecting the graph of* $y = f(x)$ *in the origin* and is equivalent to reflecting in one coordinate axis and then in the other coordinate axis. Figure 11 illustrates these reflections for $f(x) = \sqrt{x}$.

The various transformations considered above are summarized in the box for easy reference.

FIGURE 12 Graph transformations.

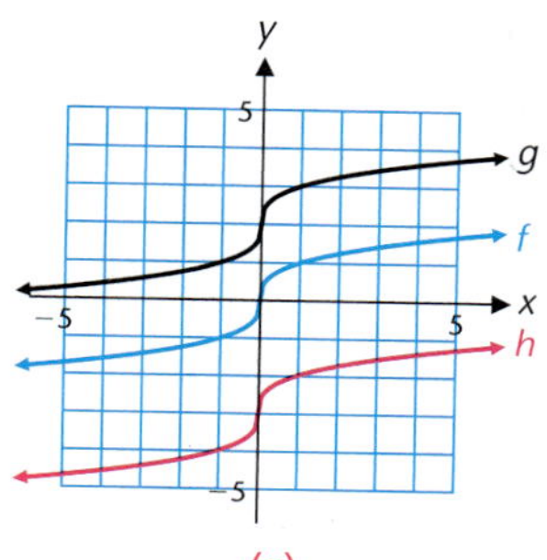

(a)
Vertical translation
$g(x) = f(x) + 2$
$h(x) = f(x) - 3$

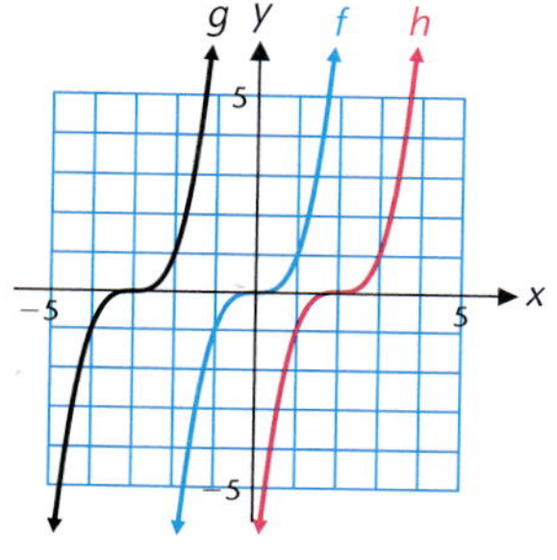

(b)
Horizontal translation
$g(x) = f(x + 3)$
$h(x) = f(x - 2)$

Graph Transformations (Summary)

Vertical Translation [Fig. 12(a)]:

$y = f(x) + k$ $\begin{cases} k > 0 & \text{Shift graph of } y = f(x) \text{ up } k \text{ units} \\ k < 0 & \text{Shift graph of } y = f(x) \text{ down } |k| \text{ units} \end{cases}$

Horizontal Translation [Fig. 12(b)]:

$y = f(x + h)$ $\begin{cases} h > 0 & \text{Shift graph of } y = f(x) \text{ left } h \text{ units} \\ h < 0 & \text{Shift graph of } y = f(x) \text{ right } |h| \text{ units} \end{cases}$

Vertical Expansion and Contraction [Fig. 12(c)]:

$y = Af(x)$ $\begin{cases} A > 1 & \text{Vertically expand the graph of } y = f(x) \text{ by multiplying each } y \text{ value by } A \\ 0 < A < 1 & \text{Vertically contract the graph of } y = f(x) \text{ by multiplying each } y \text{ value by } A \end{cases}$

Horizontal Expansion and Contraction [Fig. 12(d)]:

$y = f(Ax)$ $\begin{cases} A > 1 & \text{Horizontally contract the graph of } y = f(x) \text{ by multiplying each } x \text{ value by } \frac{1}{A} \\ 0 < A < 1 & \text{Horizontally expand the graph of } y = f(x) \text{ by multiplying each } x \text{ value by } \frac{1}{A} \end{cases}$

Reflection [Fig. 12(e)]:

$y = -f(x)$ Reflect the graph of $y = f(x)$ in the x axis
$y = f(-x)$ Reflect the graph of $y = f(x)$ in the y axis
$y = -f(-x)$ Reflect the graph of $y = f(x)$ in the origin

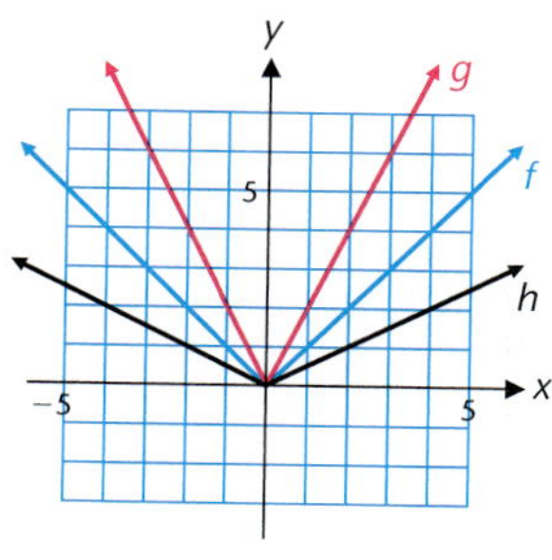

(c)
Vertical expansion and contraction
$g(x) = 2f(x)$
$h(x) = 0.5f(x)$

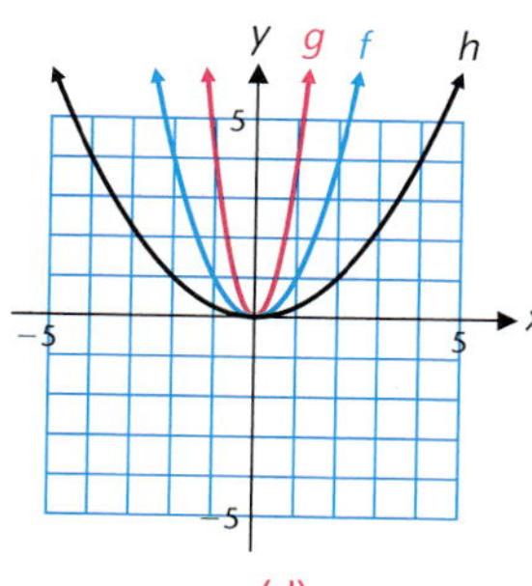

(d)
Horizontal expansion and contraction
$g(x) = f(2x)$
$h(x) = f(0.5x)$

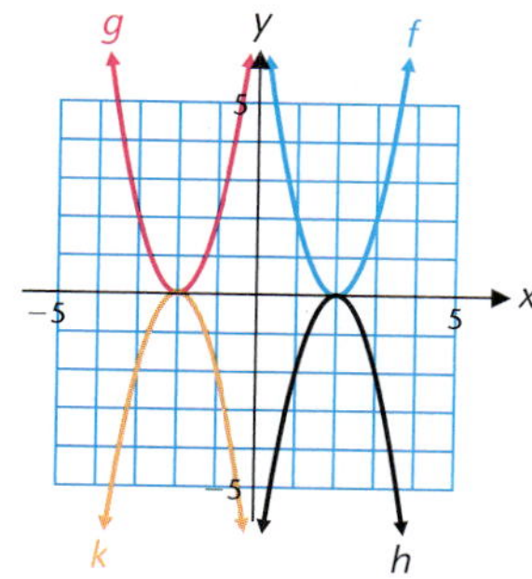

(e)
Reflection
$g(x) = f(-x)$
$h(x) = -f(x)$
$k(x) = -f(-x)$

EXPLORE/DISCUSS 4

Use a graphing utility to explore the graph of $y = A(x + h)^2 + k$ for various values of the constants A, h, and k. Discuss how the graph of $y = A(x + h)^2 + k$ is related to the graph of $y = x^2$.

EXAMPLE 4 Combining Graph Transformations

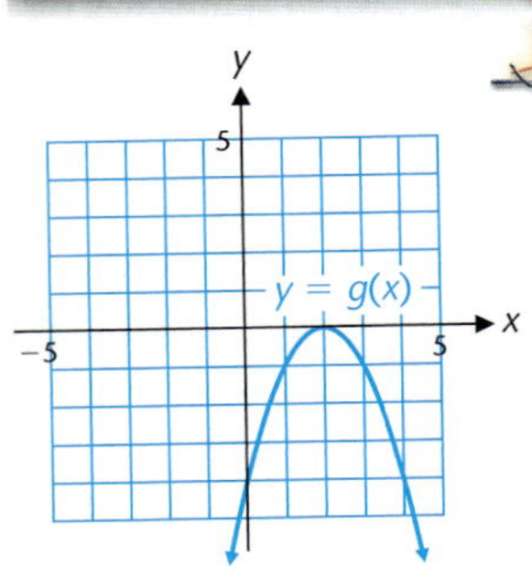

FIGURE 13

The graph of $y = g(x)$ in Figure 13 is a transformation of the graph of $y = x^2$. Find an equation for the function g.

SOLUTION

To transform the graph of $y = x^2$ [Fig. 14(a)] into the graph of $y = g(x)$, we first reflect the graph of $y = x^2$ in the x axis [Fig. 14(b)], then shift it to the right two units [Fig. 14(c)]. Thus, an equation for the function g is

$$g(x) = -(x - 2)^2$$

FIGURE 14

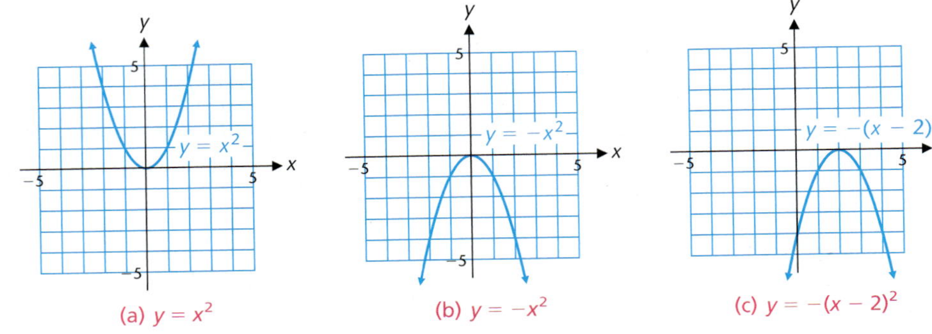

(a) $y = x^2$ (b) $y = -x^2$ (c) $y = -(x - 2)^2$

MATCHED PROBLEM 4

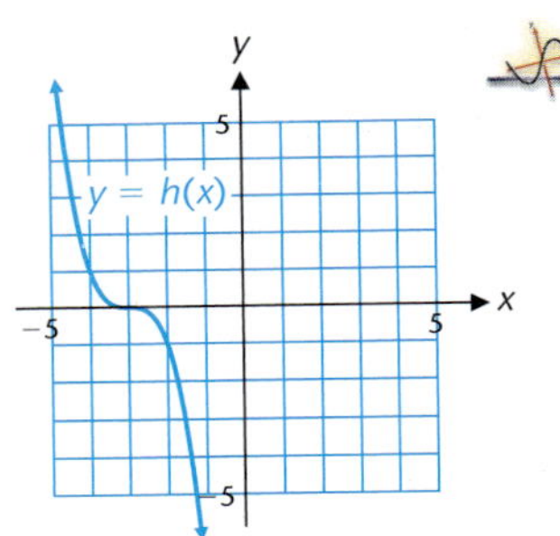

FIGURE 15

The graph of $y = h(x)$ in Figure 15 is a transformation of the graph of $y = x^3$. Find an equation for the function h.

Even and Odd Functions

Certain transformations leave the graphs of some functions unchanged. For example, reflecting the graph of $y = x^2$ in the y axis does not change the graph. Functions with this property are called *even functions*. Similarly, reflecting the graph of $y = x^3$ in the x axis and then in the y axis does not change the graph. Functions with this property are called *odd functions*. More formally, we have the following definitions.

EVEN AND ODD FUNCTIONS

If $f(x) = f(-x)$ for all x in the domain of f, then f is an **even function.**
If $f(-x) = -f(x)$ for all x in the domain of f, then f is an **odd function.**

The graph of an even function is said to be **symmetric with respect to the y axis** and the graph of an odd function is said to be **symmetric with respect to the origin** (Fig. 16).

FIGURE 16 Even and odd functions.

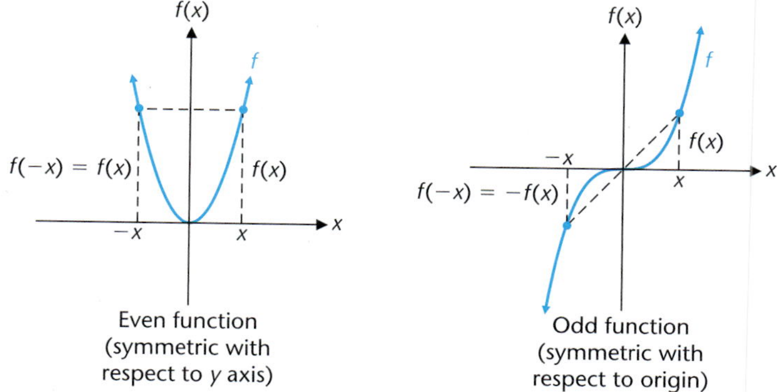

Refer to the graphs of the basic functions in Figure 1. These graphs show that the square and absolute value functions are even functions, and the identity, cube, and cube root functions are odd functions. Notice in Figure 1(e) that the square root function is not symmetric with respect to the y axis or the origin. Thus, the square root function is neither even nor odd.

EXAMPLE 5 Testing for Even and Odd Functions

Determine whether the functions f, g, and h are even, odd, or neither.

(A) $f(x) = x^4 + 1$ (B) $g(x) = x^3 + 1$ (C) $h(x) = x^5 + x$

SOLUTIONS

(A) Algebraic Solution

$$\begin{aligned} f(x) &= x^4 + 1 \\ f(-x) &= (-x)^4 + 1 \\ &= [(-1)x]^4 + 1 \\ &= (-1)^4x^4 + 1 \\ &= x^4 + 1 \\ &= f(x) \end{aligned}$$

Therefore, f is even.

(A) Graphical Solution

Enter $y_1 = x^4 + 1$ and $y_2 = y_1(-x)$ (Fig. 17), draw the graph (Fig. 18), and use the trace command or a table to see if the graphs are identical.

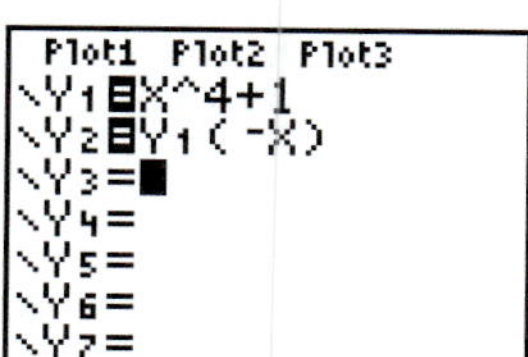

FIGURE 17

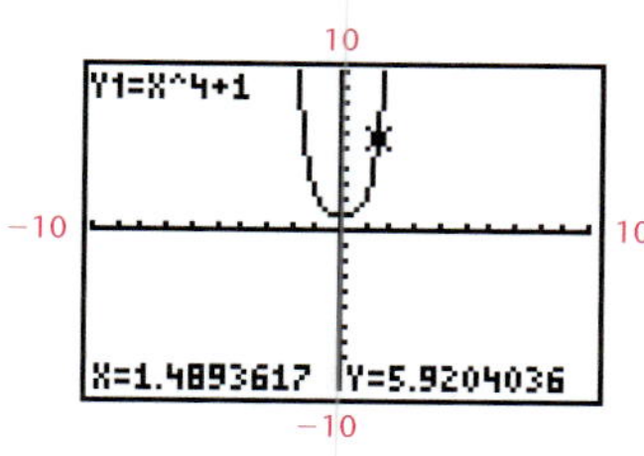

FIGURE 18

(B) Algebraic Solution

$$\begin{aligned} g(x) &= x^3 + 1 \\ g(-x) &= (-x)^3 + 1 \\ &= [(-1)x]^3 + 1 \\ &= (-1)^3x^3 + 1 \\ &= -x^3 + 1 \\ -g(x) &= -(x^3 + 1) \\ &= -x^3 - 1 \end{aligned}$$

Because $g(-x) \neq g(x)$ and $g(-x) \neq -g(x)$, g is neither even nor odd.

(B) Graphical Solution

Enter $y_1 = x^3 + 1$, $y_2 = y_1(-x)$, and $y_3 = -y_1(x)$ (Fig. 19), graph (Fig. 20), and observe that no two of these functions are identical. Thus, g is neither even nor odd.

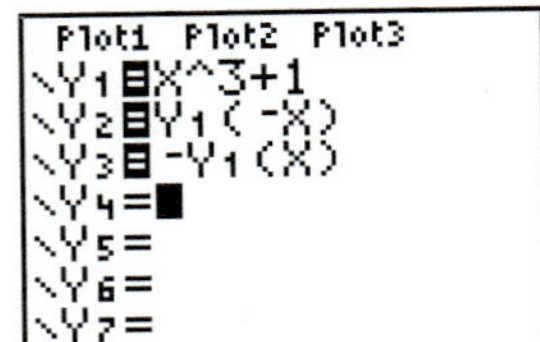

FIGURE 19

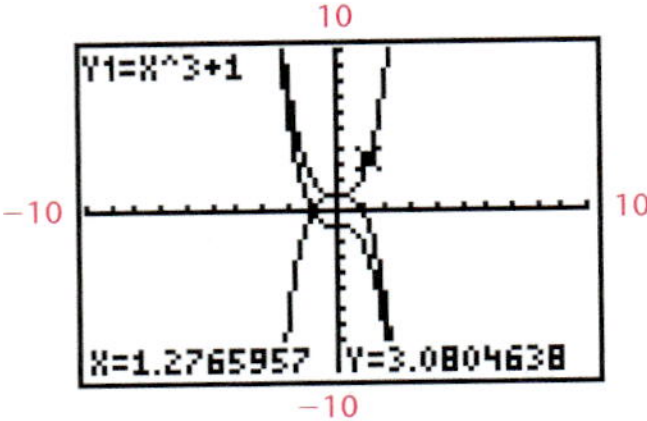

FIGURE 20

(C) Algebraic Solution

$$\begin{aligned} h(x) &= x^5 + x \\ h(-x) &= (-x)^5 + (-x) \\ &= -x^5 - x \\ &= -(x^5 + x) \\ &= -h(x) \end{aligned}$$

Therefore, h is odd.

(C) Graphical Solution

Enter $y_1 = x^5 + x$, $y_2 = y_1(-x)$, and $y_3 = -y_1(x)$ (Fig. 21), graph (Fig. 22), and use the trace command or a table to show that y_2 and y_3 are identical. Thus, h is an odd function.

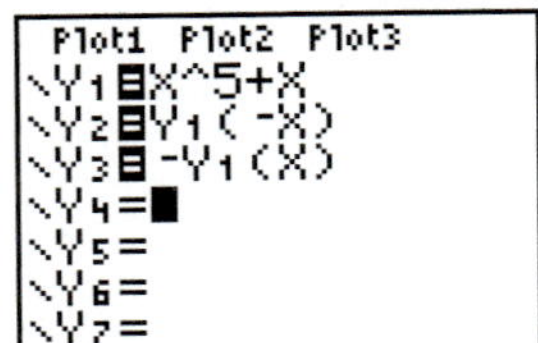

FIGURE 21

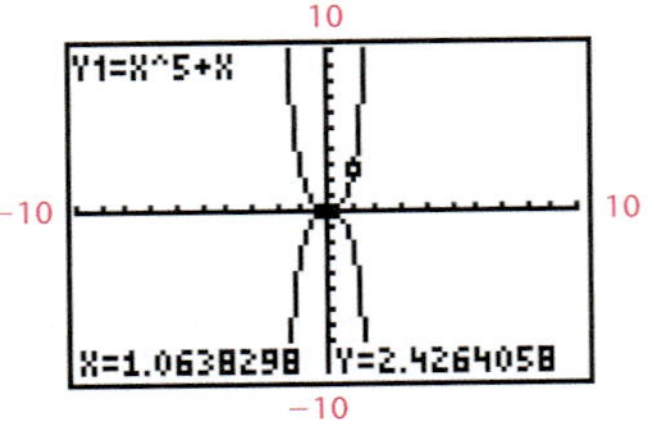

FIGURE 22

MATCHED 5 PROBLEM

Determine whether the functions F, G, and H are even, odd, or neither:
(A) $F(x) = x^3 - 2x$ (B) $G(x) = x^2 + 1$ (C) $H(x) = 2x + 4$

In the solution of Example 5, notice that we used the fact that

$$(-x)^n = \begin{cases} x^n & \text{if } n \text{ is an even integer} \\ -x^n & \text{if } n \text{ is an odd integer} \end{cases}$$

It is this property that motivates the use of the terms *even* and *odd* when describing symmetry properties of the graphs of functions. In addition to being an aid to graphing, certain problems and developments in calculus and more advanced mathematics are simplified if we recognize the presence of either an even or an odd function.

ANSWERS MATCHED PROBLEMS

1. (A) The graph of $y = \sqrt{x} + 3$ is the same as the graph of $y = \sqrt{x}$ shifted upward three units, and the graph of $y = \sqrt{x} - 1$ is the same as the graph of $y = \sqrt{x}$ shifted downward one unit. The figure confirms these conclusions.

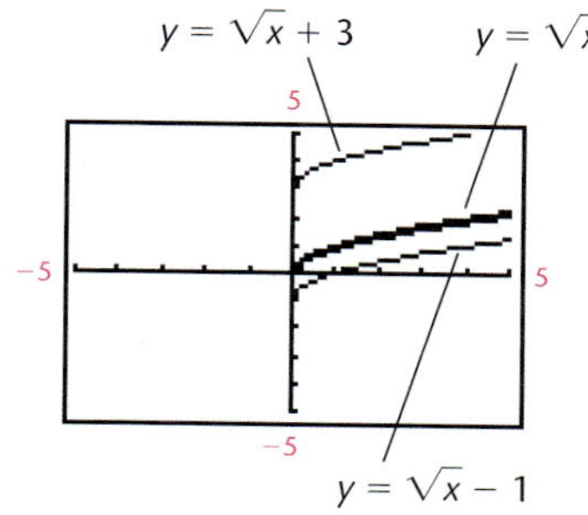

(B) The graph of $y = \sqrt{x + 3}$ is the same as the graph of $y = \sqrt{x}$ shifted to the left three units, and the graph of $y = \sqrt{x - 1}$ is the same as the graph of $y = \sqrt{x}$ shifted to the right one unit. The figure confirms these conclusions.

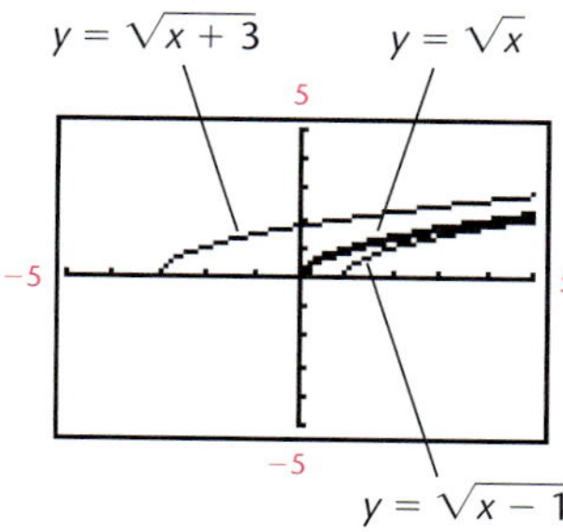

2. $G(x) = (x + 3)^3$, $H(x) = (x - 1)^3$, $M(x) = x^3 + 3$, $N(x) = x^3 - 4$

3. (A) The graph of $y = 2x^3$ is a vertical expansion of the graph of $y = x^3$ by a factor of 2, and the graph of $y = 0.5x^3$ is a vertical contraction of the graph of $y = x^3$ by a factor of 1/2. The figure confirms these conclusions.

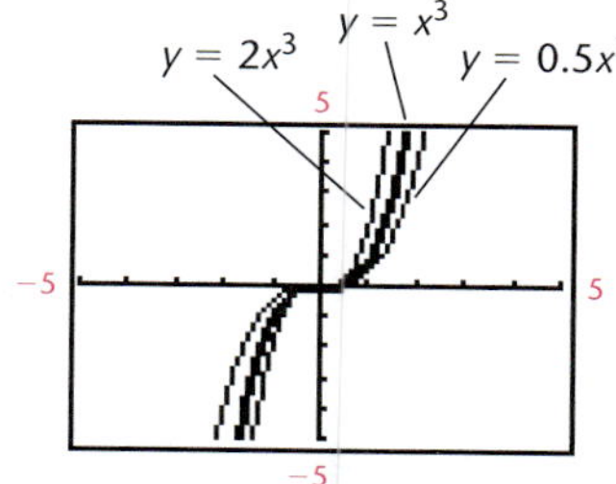

(B) The graph of $y = (2x)^3$ is a horizontal contraction of the graph of $y = x^3$ by a factor of 1/2, and the graph of $y = (0.5x)^3$ is a horizontal expansion of the graph of $y = x^3$ by a factor of 2. The figure confirms these conclusions.

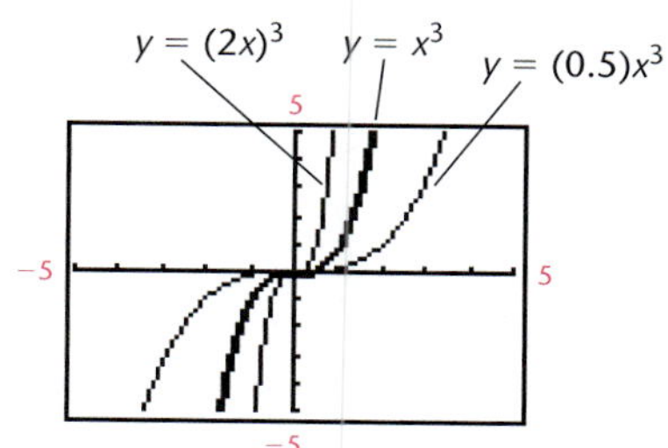

4. The graph of function h is a reflection in the x axis and a horizontal translation of three units to the left of the graph of $y = x^3$. An equation for h is $h(x) = -(x + 3)^3$.

5. (A) Odd (B) Even (C) Neither

EXERCISE 1.4

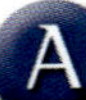

A

Indicate whether each function in Problems 1–10 is even, odd, or neither.

1. $g(x) = x^3 + x$
2. $f(x) = x^5 - x$
3. $m(x) = x^4 + 3x^2$
4. $h(x) = x^4 - x^2$
5. $F(x) = x^5 + 1$
6. $f(x) = x^5 - 3$
7. $G(x) = x^4 + 2$
8. $P(x) = x^4 - 4$
9. $q(x) = x^2 + x - 3$
10. $n(x) = 2x - 3$

Problems 11–26 refer to the functions f and g given by the graphs below (the domain of each function is [−2, 2]). Use the graph of f or g, as required, to graph each given function.

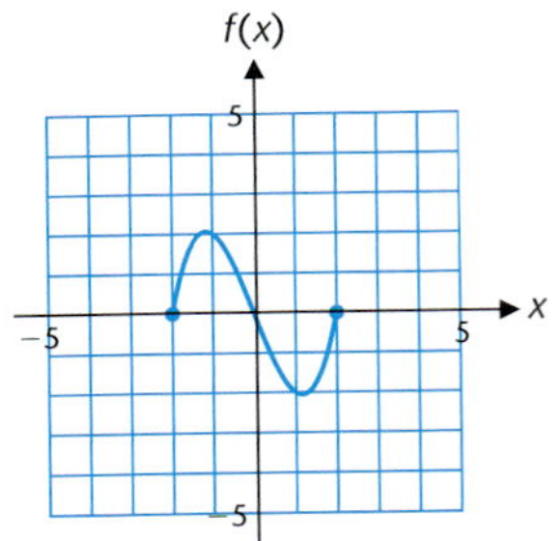

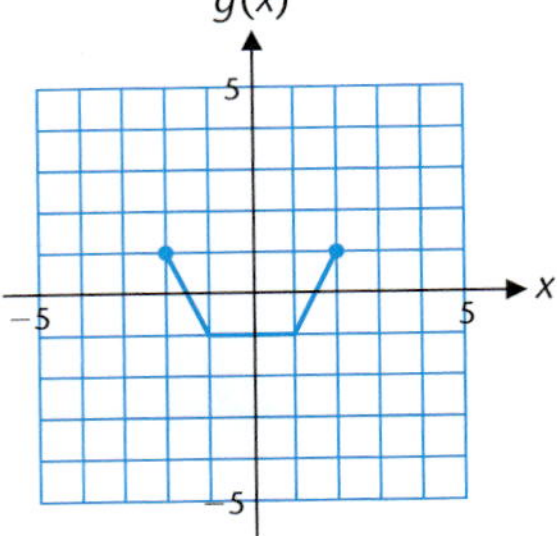

11. $f(x) + 2$
12. $g(x) - 1$
13. $g(x) + 2$
14. $f(x) - 1$
15. $f(x - 2)$
16. $g(x - 1)$
17. $g(x + 2)$
18. $f(x - 1)$
19. $-f(x)$
20. $-g(x)$
21. $2g(x)$
22. $\frac{1}{2}f(x)$
23. $g(2x)$
24. $f(\frac{1}{2}x)$
25. 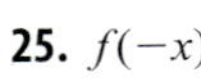 $f(-x)$
26. $-g(-x)$

B

Each graph in Problems 27–34 is the result of applying a transformation to the graph of one of the six basic functions in Figure 1. Identify the basic function, describe the transformation verbally, and find an equation for the given graph. Check by graphing the equation on a graphing utility.

27.

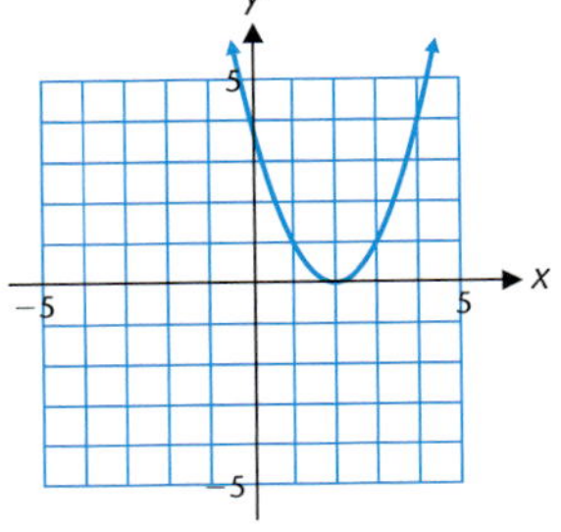

28.

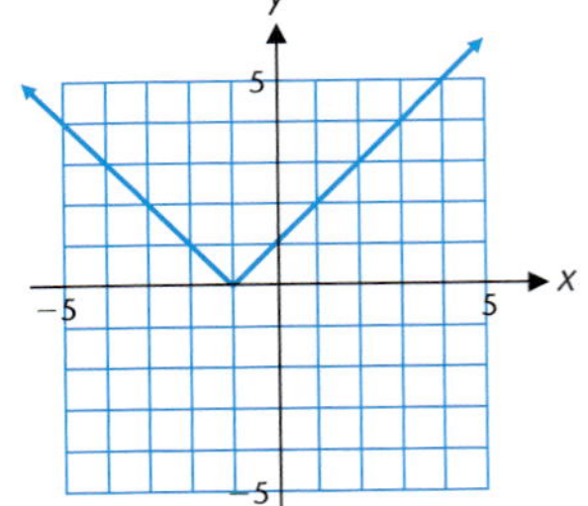

29.

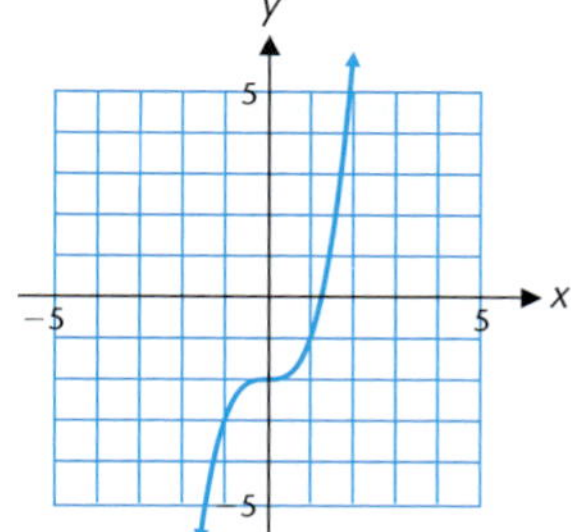

30.

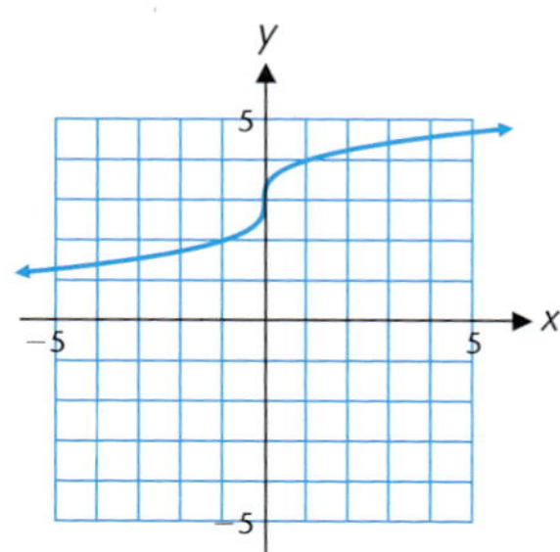

31.

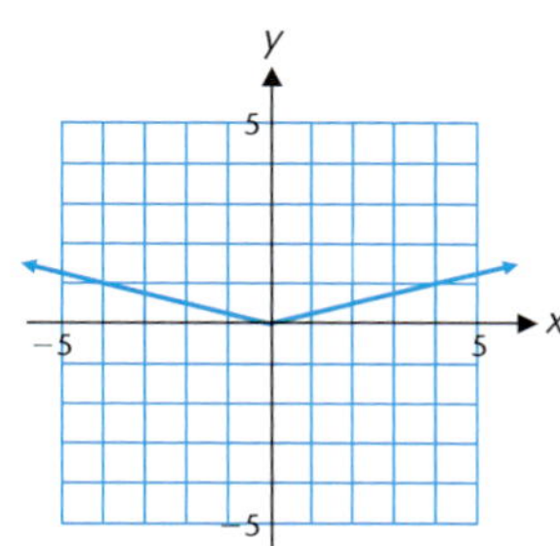

32.

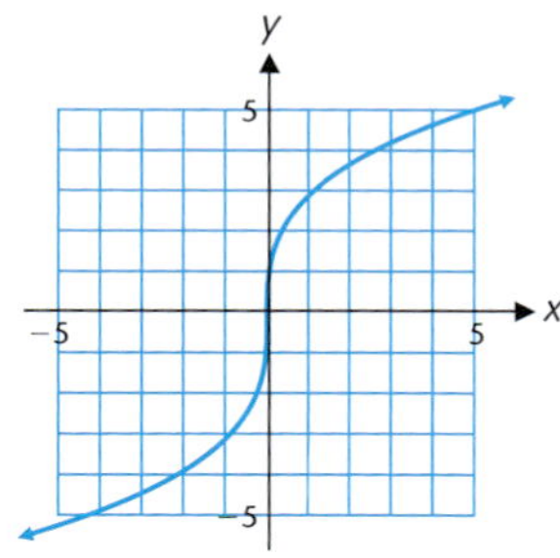

33.

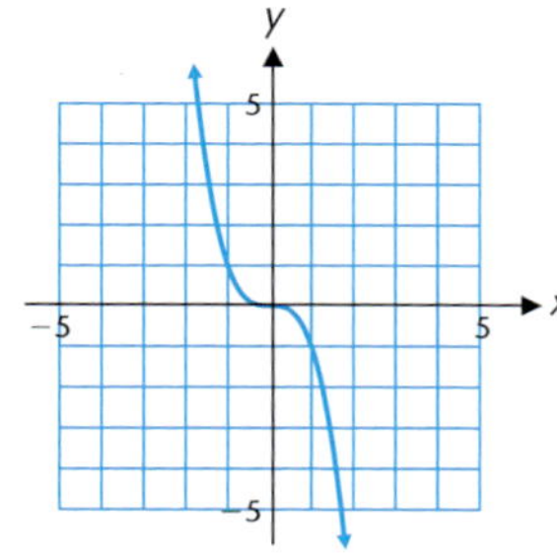

34.

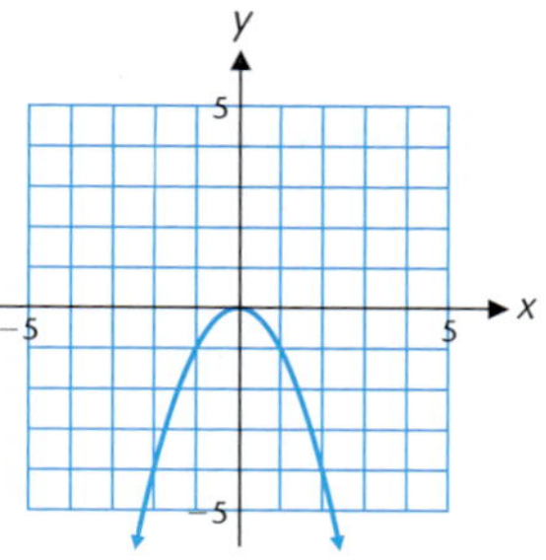

In Problems 35–42, the graph of the function g is formed by applying the indicated sequence of transformations to the given function f. Find an equation for the function g. Check your work by graphing f and g in a standard viewing window.

35. The graph $f(x) = \sqrt[3]{x}$ is shifted four units to the left and five units down.

36. The graph of $f(x) = x^3$ is shifted five units to the right and four units up.

37. The graph of $f(x) = \sqrt{x}$ is shifted six units up, reflected in the x axis, and contracted by a factor of 0.5.

38. The graph of $f(x) = \sqrt{x}$ is shifted two units down, reflected in the x axis, and vertically expanded by a factor of 4.

39. The graph of $f(x) = x^2$ is reflected in the x axis, vertically expanded by a factor of 2, shifted four units to the left, and shifted two units down.

40. The graph of $f(x) = |x|$ is reflected in the x axis, vertically contracted by a factor of 0.5, shifted three units to the right, and shifted four units up.

41. The graph of $f(x) = \sqrt{x}$ is horizontally expanded by a factor of 0.5, reflected in the y axis, and shifted two units to the left.

42. The graph of $f(x) = \sqrt[3]{x}$ is horizontally contracted by a factor of 2, shifted three units up, and reflected in the y axis.

In Problems 43–50, indicate how the graph of each function is related to the graph of one of the six basic functions in Figure 1.

43. $f(x) = (x + 7)^2 + 9$

44. $g(x) = (x - 4)^2 - 6$

45. $h(x) = -|x - 8|$

46. $k(x) = -|x + 5|$

47. $p(x) = 3 - \sqrt{x}$

48. $q(x) = -2 + \sqrt{x + 3}$

49. $r(x) = -4x^2$

50. $s(x) = -0.5|x|$

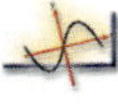

Each graph in Problems 51–58 is the result of applying a sequence of transformations to the graph of one of the six basic functions in Figure 1. Find an equation for the given graph. Check by graphing the equation on a graphing utility.

51.

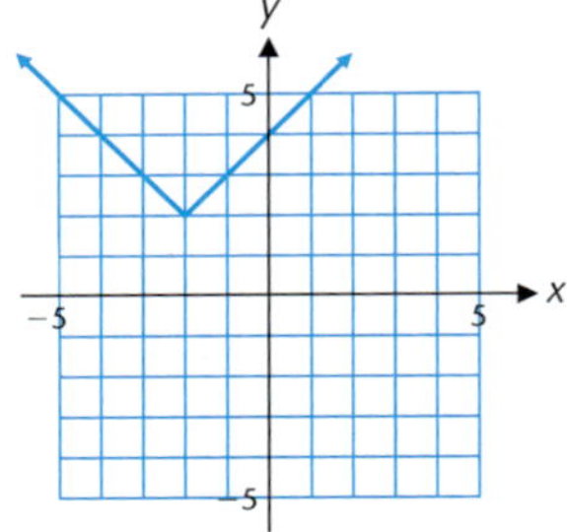

52.

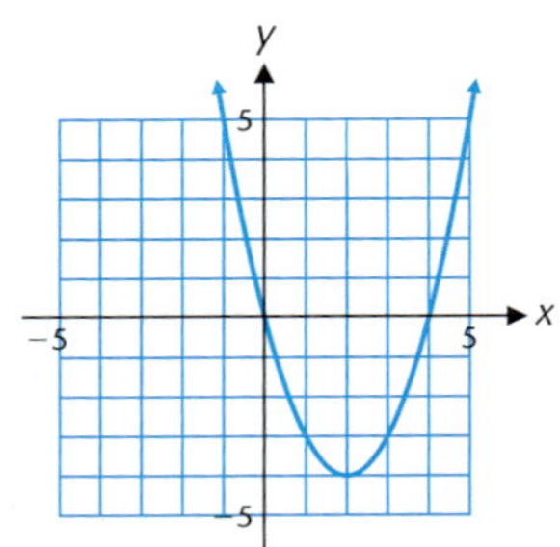

53.

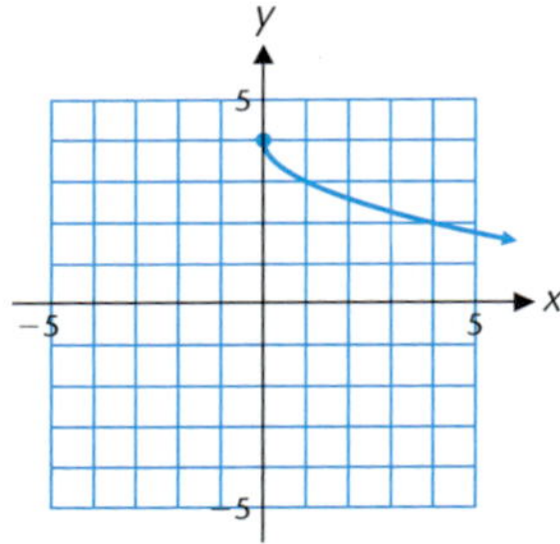

54.

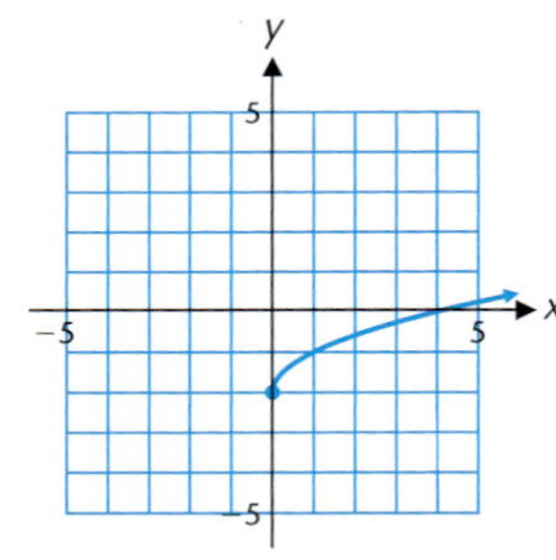

55.

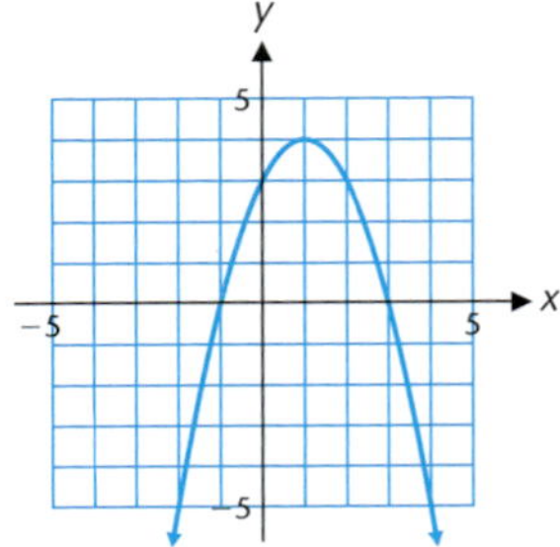

56.

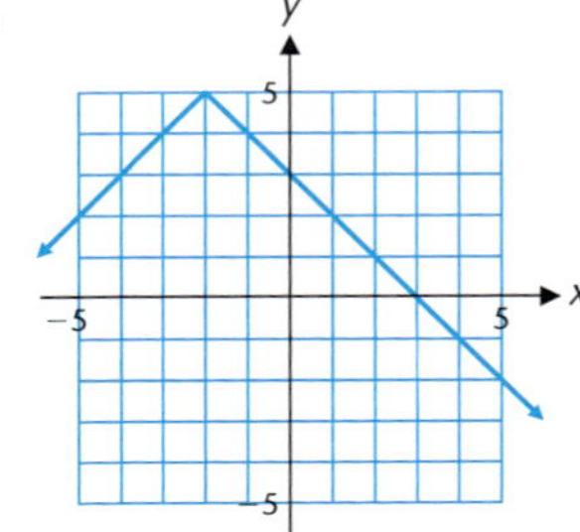

57.

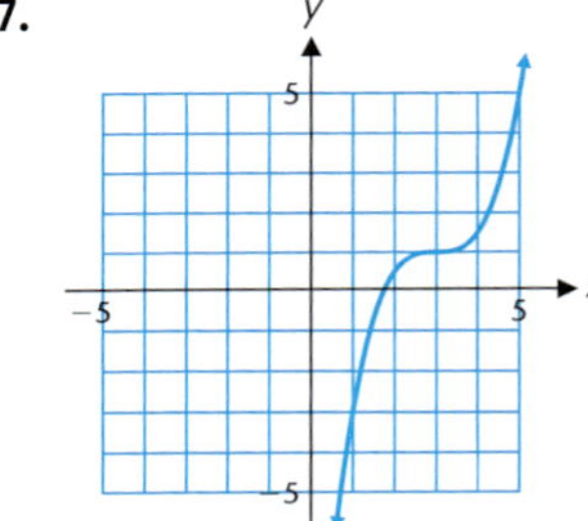

58.

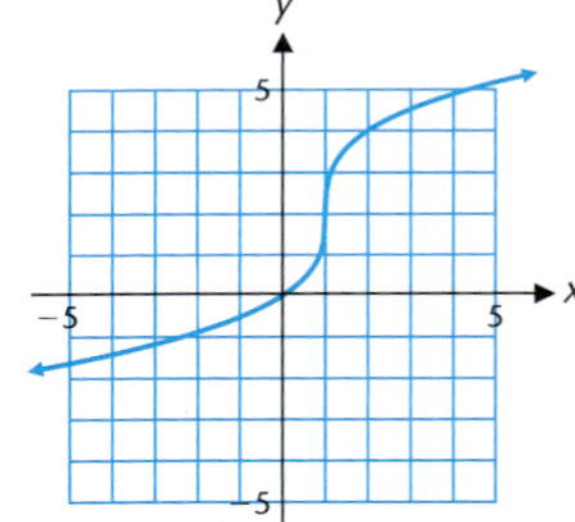

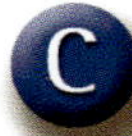

Changing the order in a sequence of transformations may change the final result. Investigate each pair of transformations in Problems 59–64 to determine if reversing their order can produce a different result. Support your conclusions with specific examples and/or mathematical arguments.

59. Vertical shift, horizontal shift

60. Vertical shift, reflection in y axis

61. Vertical shift, reflection in x axis

62. Vertical shift, expansion

63. Horizontal shift, reflection in x axis

64. Horizontal shift, contraction

Problems 65–68 refer to two functions f and g with domain [−5, 5] and partial graphs as shown below.

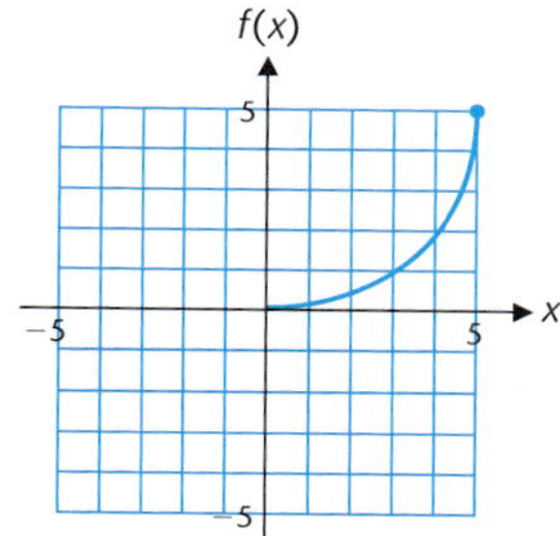

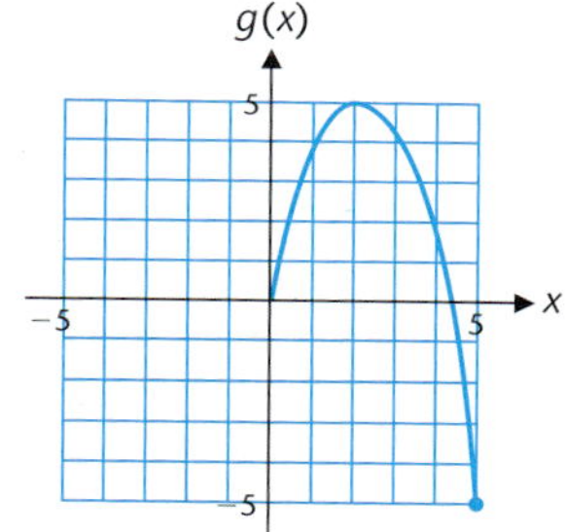

65. Complete the graph of f over the interval $[-5, 0]$, given that f is an even function.

66. Complete the graph of f over the interval $[-5, 0]$, given that f is an odd function.

67. Complete the graph of g over the interval $[-5, 0]$, given that g is an odd function.

68. Complete the graph of g over the interval $[-5, 0]$, given that g is an even function.

69. Let f be any function with the property that $-x$ is in the domain of f whenever x is in the domain of f, and let E and O be the functions defined by

$$E(x) = \tfrac{1}{2}[f(x) + f(-x)]$$

and

$$O(x) = \tfrac{1}{2}[f(x) - f(-x)]$$

(A) Show that E is always even.

(B) Show that O is always odd.

(C) Show that $f(x) = E(x) + O(x)$. What is your conclusion?

70. Let f be any function with the property that $-x$ is in the domain of f whenever x is in the domain of f, and let $g(x) = xf(x)$.

(A) If f is even, is g even, odd, or neither?

(B) If f is odd, is g even, odd, or neither?

In Problems 71–74, graph $f(x)$, $|f(x)|$, and $-|f(x)|$ in a standard viewing window. For purposes of comparison, it will be helpful to graph each function separately and make a hand sketch.

71. $f(x) = 0.2x^2 - 5$

72. $f(x) = 4 - 0.25x^2$

73. $f(x) = 4 - 0.1(x + 2)^3$

74. $f(x) = 0.25(x - 1)^3 - 1$

75. Describe the relationship between the graphs of $f(x)$ and $|f(x)|$ in Problems 71–74.

76. Describe the relationship between the graphs of $f(x)$ and $-|f(x)|$ in Problems 71–74.

APPLICATIONS

77. Production Costs. Total production costs for a product can be broken down into fixed costs, which do not depend on the number of units produced, and variable costs, which do depend on the number of units produced. Thus, the total cost of producing x units of the product can be expressed in the form

$$C(x) = K + f(x)$$

where K is a constant that represents the fixed costs and $f(x)$ is a function that represents the variable costs. Use the graph of the variable-cost function $f(x)$ shown in the figure to graph the total cost function if the fixed costs are $30,000.

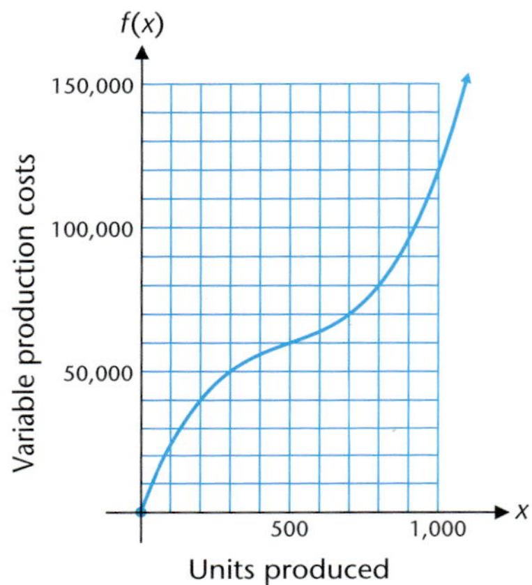

★ **78. Cost Functions.** Refer to the variable-cost function $f(x)$ in Problem 77. Suppose construction of a new production facility results in a 25% decrease in the variable cost at all levels of output. If F is the new variable-cost function, use the graph of f to graph $y = F(x)$.

79. Timber Harvesting. To determine when a forest should be harvested, forest managers often use formulas to estimate the number of board feet a tree will produce. A board foot equals 1 square foot of wood, 1 inch thick. Suppose that the number of board feet y yielded by a tree can be estimated by

$$y = f(x) = C + 0.004(x - 10)^3$$

where x is the diameter of the tree in inches measured at a height of 4 feet above the ground and C is a constant that depends on the species being harvested. Graph $y = f(x)$ for $C = 10$, 15, and 20 simultaneously in the viewing window with Xmin = 10, Xmax = 25, Ymin = 10, and Ymax = 35. Write a brief verbal description of this collection of functions.

80. Safety Research. If a person driving a vehicle slams on the brakes and skids to a stop, the speed v in miles per hour at the time the brakes are applied is given approximately by

$$v = f(x) = C\sqrt{x}$$

where x is the length of the skid marks and C is a constant that depends on the road conditions and the weight of the vehicle. The table lists values of C for a midsize automobile and various road conditions. Graph $v = f(x)$ for the values of C in the table simultaneously in the viewing window with Xmin = 0, Xmax = 100, Ymin = 0, and Ymax = 60. Write a brief verbal description of this collection of functions.

Road Condition	C
Wet (concrete)	3.5
Wet (asphalt)	4
Dry (concrete)	5
Dry (asphalt)	5.5

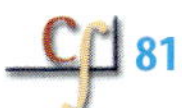

81. Family of Curves. In calculus, solutions to certain types of problems often involve an unspecified constant. For example, consider the equation

$$y = \frac{1}{C}x^2 - C$$

where C is a positive constant. The collection of graphs of this equation for all permissible values of C is called a **family of curves.** Graph the members of this family corresponding to $C = 2$, 3, 4, and 5 simultaneously in a standard viewing window. Write a brief verbal description of this family of functions.

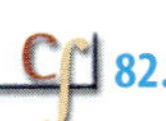

82. Family of Curves. A family of curves is defined by the equation

$$y = 2C - \frac{5}{C}x^2$$

where C is a positive constant. Graph the members of this family corresponding to $C = 1$, 2, 3, and 4 simultaneously in a standard viewing window. Write a brief verbal description of this family of functions.

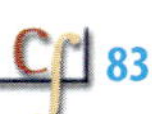

83. Fluid Flow. A cubic tank is 4 feet on a side and is initially full of water. Water flows out an opening in the bottom of the tank at a rate proportional to the square root of the depth (see the figure). Using advanced concepts from mathematics and physics, it can be shown that the volume of the water in the tank t minutes after the water begins to flow is given by

$$V(t) = \frac{64}{C^2}(C - t)^2 \qquad 0 \le t \le C$$

where C is a constant that depends on the size of the opening. Sketch by hand the graphs of $y = V(t)$ for $C = 1, 2, 4$, and 8. Write a brief verbal description of this collection of functions.

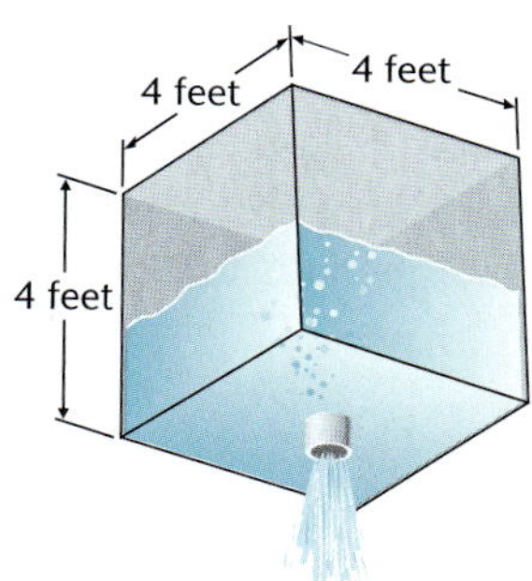

84. **Evaporation.** A water trough with triangular ends is 9 feet long, 4 feet wide, and 2 feet deep (see the figure). Initially, the trough is full of water, but due to evaporation, the volume of the water in the trough decreases at a rate proportional to the square root of the volume. Using advanced concepts from mathematics and physics, it can be shown that the volume after t hours is given by

$$V(t) = \frac{1}{C^2}(t + 6C)^2 \qquad 0 \le t \le 6|C|$$

where C is a constant. Sketch by hand the graphs of $y = V(t)$ for $C = -4, -5$, and -6. Write a brief verbal description of this collection of functions.

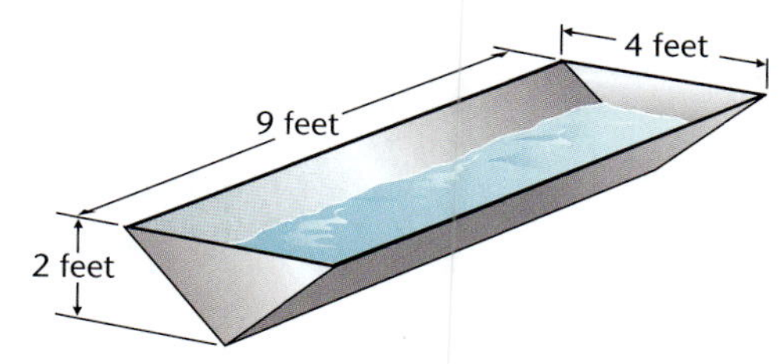

SECTION 1.5 Operations on Functions; Composition

Operation on Functions • Composition • Mathematical Modeling

If two functions f and g are both defined at a real number x, and if $f(x)$ and $g(x)$ are both real numbers, then it is possible to perform real number operations such as addition, subtraction, multiplication, or division with $f(x)$ and $g(x)$. Furthermore, if $g(x)$ is a number in the domain of f, then it is also possible to evaluate f at $g(x)$. In this section we see how operations on the values of functions can be used to define operations on the functions themselves.

Operations on Functions

The functions f and g given by

$$f(x) = 2x + 3 \qquad \text{and} \qquad g(x) = x^2 - 4$$

are defined for all real numbers. Thus, for any real x we can perform the following operations:

$$f(x) + g(x) = 2x + 3 + x^2 - 4 = x^2 + 2x - 1$$
$$f(x) - g(x) = 2x + 3 - (x^2 - 4) = -x^2 + 2x + 7$$
$$f(x)g(x) = (2x + 3)(x^2 - 4) = 2x^3 + 3x^2 - 8x - 12$$

For $x \ne \pm 2$ we can also form the quotient

$$\frac{f(x)}{g(x)} = \frac{2x + 3}{x^2 - 4} \qquad x \ne \pm 2$$

Notice that the result of each operation is a new function. Thus, we have

$$(f + g)(x) = f(x) + g(x) = x^2 + 2x - 1 \qquad \text{Sum}$$
$$(f - g)(x) = f(x) - g(x) = -x^2 + 2x + 7 \qquad \text{Difference}$$
$$(fg)(x) = f(x)g(x) = 2x^3 + 3x^2 - 8x - 12 \qquad \text{Product}$$
$$\left(\frac{f}{g}\right)(x) = \frac{f(x)}{g(x)} = \frac{2x + 3}{x^2 - 4} \qquad x \neq \pm 2 \qquad \text{Quotient}$$

Notice that the sum, difference, and product functions are defined for all values of x, as were f and g, but the domain of the quotient function must be restricted to exclude those values where $g(x) = 0$.

DEFINITION 1
Operations on Functions

The **sum, difference, product,** and **quotient** of the functions f and g are the functions defined by

$$(f + g)(x) = f(x) + g(x) \qquad \text{Sum function}$$
$$(f - g)(x) = f(x) - g(x) \qquad \text{Difference function}$$
$$(fg)(x) = f(x)g(x) \qquad \text{Product function}$$
$$\left(\frac{f}{g}\right)(x) = \frac{f(x)}{g(x)} \qquad g(x) \neq 0 \qquad \text{Quotient function}$$

The **domain** of each function is the intersection of the domains of f and g, with the exception that the values of x where $g(x) = 0$ must be excluded from the domain of the quotient function.

EXPLORE/DISCUSS 1

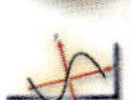

The following activities refer to the graphs of f and g shown in Figure 1 and the corresponding points on the graph shown in Table 1.

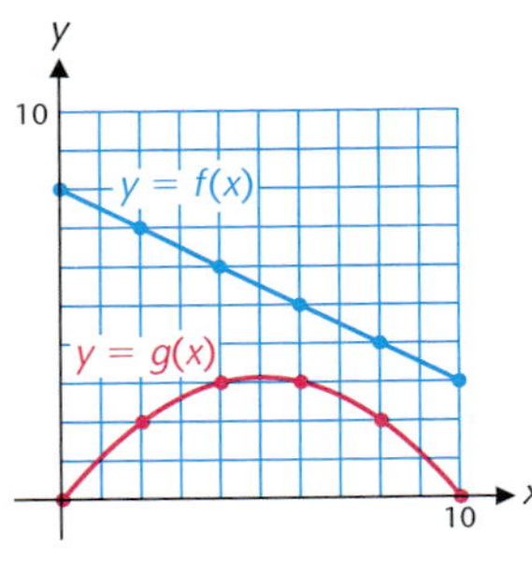

FIGURE 1

TABLE 1

x	f(x)	g(x)
0	8	0
2	7	2
4	6	3
6	5	3
8	4	2
10	3	0

For each of the following functions, construct a table of values, sketch a graph, and state the domain and range.

(A) $(f + g)(x)$ (B) $(f - g)(x)$ (C) $(fg)(x)$ (D) $\left(\frac{f}{g}\right)(x)$

EXAMPLE 1 Finding the Sum of Two Functions

Let $f(x) = \sqrt{4 - x}$ and $g(x) = \sqrt{3 + x}$. Find $f + g$ and find its domain.

SOLUTIONS

Algebraic Solution

$$\begin{aligned}(f + g)(x) &= f(x) + g(x) \\ &= \sqrt{4 - x} + \sqrt{3 + x}\end{aligned}$$

The domains of f and g are

Domain of f: $x \le 4$ or $(-\infty, 4]$
Domain of g: $x \ge -3$ or $[-3, \infty)$

The domain of $f + g$ is the intersection* of these two sets (Fig. 2):

$$(-\infty, 4] \cap [-3, \infty) = [-3, 4]$$

Domain of f

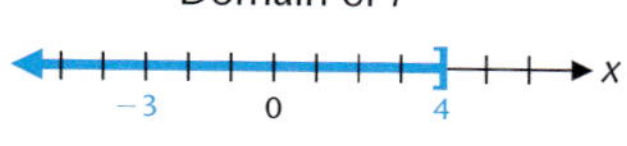

Domain of g

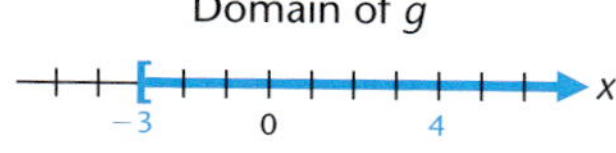

Domain of $f + g$

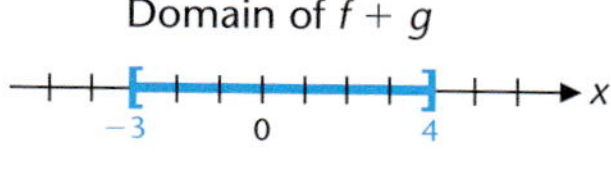

FIGURE 2

Graphical Solution

We enter $y_1 = \sqrt{4 - x}$, $y_2 = \sqrt{3 + x}$, and $y_3 = y_1 + y_2$ in the equation editor of a graphing utility and graph in a standard viewing window (Fig. 3). To get a better look at y_3, we turn off the graphs of y_1 and y_2, and change the viewing window (Fig. 4).

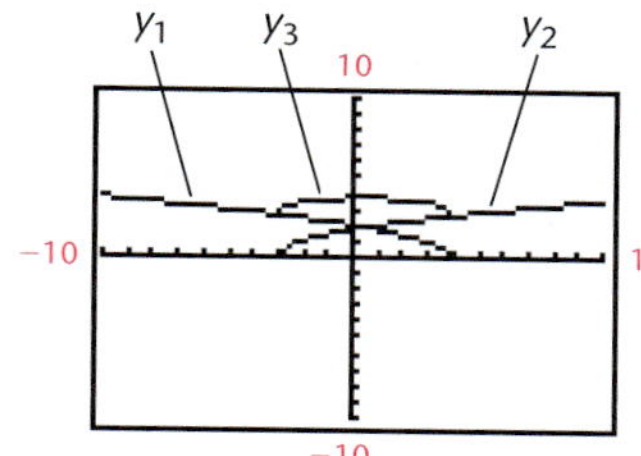

FIGURE 3 Graphs of y_1, y_2, and y_3.

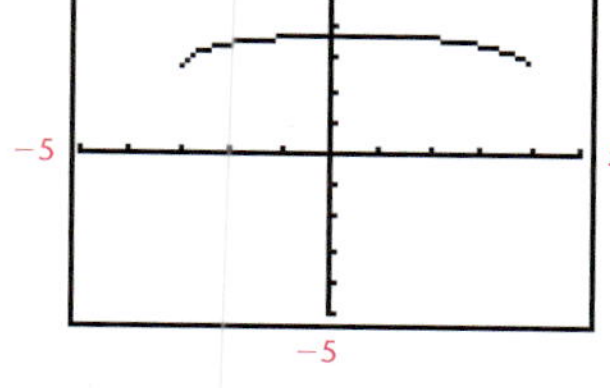

FIGURE 4 Graph of y_3.

Next we press TRACE and enter -3 (Fig. 5). Pressing the left cursor shows that y_3 is not defined for $x < -3$ (Fig. 6).

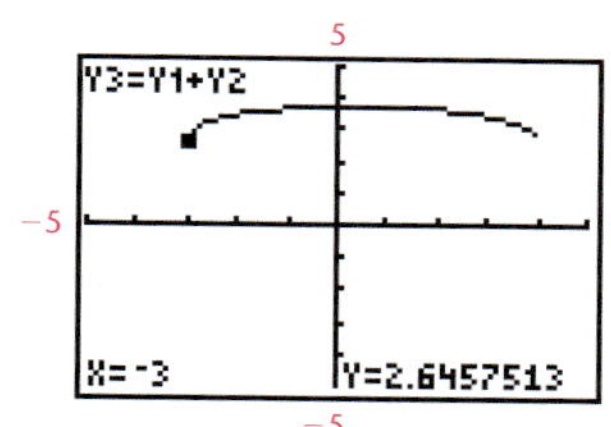

FIGURE 5

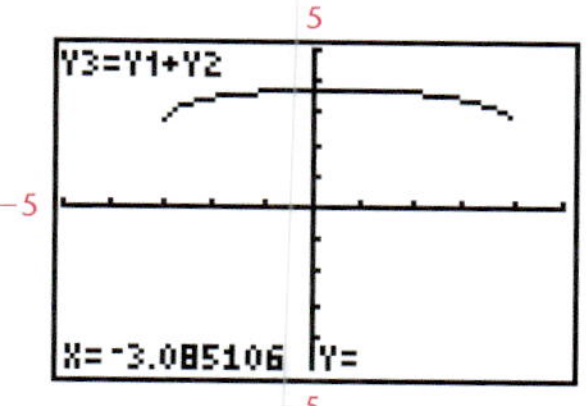

FIGURE 6

Figures 7 and 8 show that y_3 is not defined for $x > 4$. Thus, the domain of $y_3 = f + g$ is $[-3, 4]$.

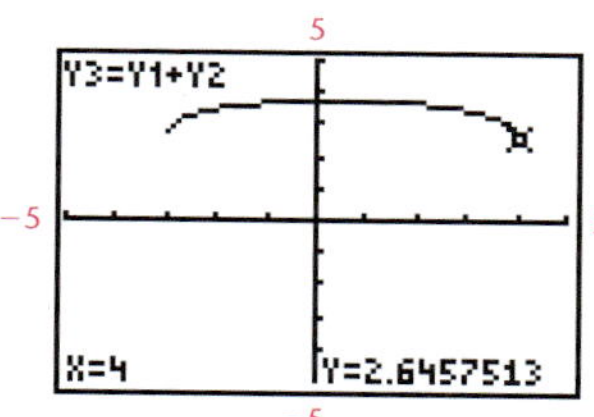

FIGURE 7

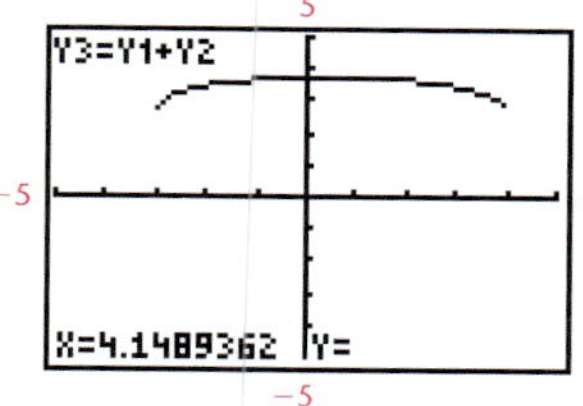

FIGURE 8

*Intersection of intervals is discussed in Appendix A, Section A.1.

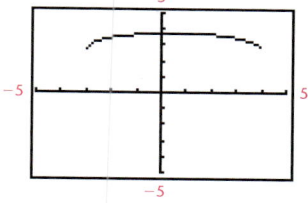

MATCHED 1 PROBLEM

Let $f(x) = \sqrt{x}$ and $g(x) = \sqrt{10 - x}$. Find $f - g$ and find its domain.

EXAMPLE 2 **Finding the Quotient of Two Functions**

Let $f(x) = \dfrac{x}{x - 1}$ and $g(x) = \dfrac{x - 4}{x + 3}$. Find the function $\dfrac{f}{g}$ and find its domain.

SOLUTION

Because division by 0 must be excluded, the domain of f is all x except $x = 1$ and the domain of g is all x except $x = -3$. Now we find f/g.*

$$\left(\frac{f}{g}\right)(x) = \frac{f(x)}{g(x)} = \frac{\dfrac{x}{x - 1}}{\dfrac{x - 4}{x + 3}}$$

$$= \frac{x}{x - 1} \cdot \frac{x + 3}{x - 4}$$

$$= \frac{x(x + 3)}{(x - 1)(x - 4)} \qquad (1)$$

The fraction in equation (1) indicates that 1 and 4 must be excluded from the domain of f/g to avoid division by 0. However, equation (1) does not indicate that -3 must be excluded also. Although the fraction in equation (1) is defined at $x = -3$, -3 is excluded from the domain of g, hence it must be excluded from the domain of f/g. Thus,

Domain of f/g: All real numbers x except -3, 1, and 4

MATCHED 2 PROBLEM

Let $f(x) = \dfrac{1}{x + 2}$ and $g(x) = \dfrac{x - 5}{x}$. Find the function $\dfrac{f}{g}$ and find its domain.

Composition

Consider the functions f and g given by

$$f(x) = \sqrt{x} \qquad \text{and} \qquad g(x) = 4 - 2x$$

We form a new function h by combining f and g as follows:

$$h(x) = f(g(x)) = f(4 - 2x) = \sqrt{4 - 2x}$$

Combining two functions in this manner is called *composition*.

*Operations on fractions are discussed in The Basic Algebra Review, Sec. 4, www.mhhe.com/barnett.

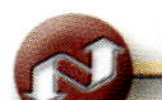

EXPLORE/DISCUSS 2

Refer to functions f, g, and h above. Complete Table 2.

TABLE 2

x	$g(x)$	$h(x) = f(g(x))$
0	$g(0) = 4$	$h(0) = f(g(0)) = f(4) = 2$
1		
2		
3		
4		

```
Plot1 Plot2 Plot3
\Y1=4-2X
\Y2=√(Y1)
\Y3=■
\Y4=
\Y5=
\Y6=
\Y7=
```

FIGURE 9

Now enter g and h in the equation editor of a graphing utility as shown in Figure 9 and check your table. The domain of f is $\{x \mid x \geq 0\}$ and the domain of g is the set of all real numbers. What is the domain of h? Support your conclusion both algebraically and graphically.

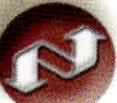

A special function symbol is often used to represent the composition of two functions, which we define in general terms below.

DEFINITION 2
Composition

The composition of function f with function g is denoted by $f \circ g$ and is defined by

$$(f \circ g)(x) = f(g(x))$$

The domain of $f \circ g$ is the set of all real numbers x in the domain of g such that $g(x)$ is in the domain of f.

Thus, the composition of $f(x) = \sqrt{x}$ with $g(x) = 4 - 2x$ is

$$(f \circ g)(x) = f(g(x)) = f(4 - 2x) = \sqrt{4 - 2x}$$

Note that the order of the functions in a composition is important. The composition of g with f is

$$(g \circ f)(x) = g(f(x)) = g(\sqrt{x}) = 4 - 2\sqrt{x}$$

which is not the same as $f \circ g$.

As an immediate consequence of Definition 2, we have (Fig. 10):

The domain of $f \circ g$ is always a subset of the domain of g, and the range of $f \circ g$ is always a subset of the range of f.

FIGURE 10 Composite functions.

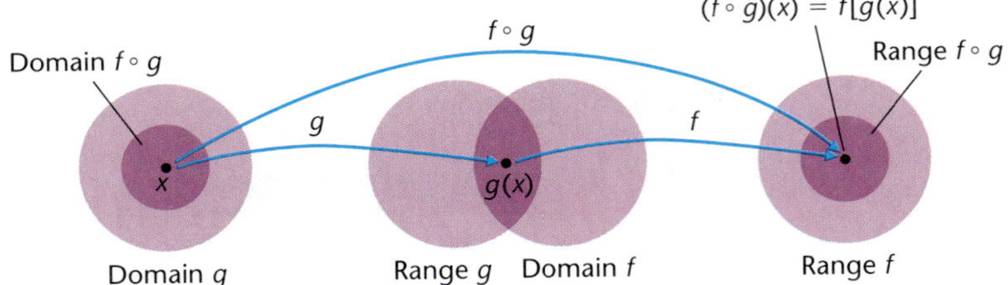

EXAMPLE 3 Finding the Composition of Two Functions

Find $(f \circ g)(x)$ and $(g \circ f)(x)$ and their domains for $f(x) = x^{10}$ and $g(x) = 3x^4 - 1$.

SOLUTION

$$(f \circ g)(x) = f(g(x)) = f(3x^4 - 1) = (3x^4 - 1)^{10}$$
$$(g \circ f)(x) = g(f(x)) = g(x^{10}) = 3(x^{10})^4 - 1 = 3x^{40} - 1$$

The functions f and g are both defined for all real numbers. If x is any real number, then x is in the domain of g, $g(x)$ is in the domain of f, and, consequently, x is in the domain of $f \circ g$. Thus, the domain of $f \circ g$ is the set of all real numbers. Using similar reasoning, the domain of $g \circ f$ also is the set of all real numbers.

MATCHED PROBLEM 3

Find $(f \circ g)(x)$ and $(g \circ f)(x)$ and their domains for $f(x) = 2x + 1$ and $g(x) = (x - 1)/2$.

If two functions are both defined for all real numbers, then so is their composition.

EXPLORE/DISCUSS 3

Verify that if $f(x) = 1/(1 - 2x)$ and $g(x) = 1/x$, then $(f \circ g)(x) = x/(x - 2)$. Because division by 0 is not defined, $f \circ g$ is not defined at $x = 2$. Are there any other values of x where $f \circ g$ is not defined? Explain.

If either function in a composition is not defined for some real numbers, then, as Example 4 illustrates, the domain of the composition may not be what you first think it should be.

EXAMPLE 4 Finding the Composition of Two Functions

Find $(f \circ g)(x)$ for $f(x) = \sqrt{4 - x^2}$ and $g(x) = \sqrt{3 - x}$. Find the domain algebraically and check graphically.

SOLUTION

We begin by stating the domains of f and g, a good practice in any composition problem:

Domain f: $-2 \le x \le 2$ or $[-2, 2]$
Domain g: $x \le 3$ or $(-\infty, 3]$

Next we find the composition:

$$\begin{aligned}(f \circ g)(x) &= f(g(x)) = f(\sqrt{3-x}) \\ &= \sqrt{4 - (\sqrt{3-x})^2} \\ &= \sqrt{4 - (3-x)} \qquad (\sqrt{t})^2 = t \text{ as long as } t \ge 0 \\ &= \sqrt{1+x}\end{aligned}$$

Although $\sqrt{1+x}$ is defined for all $x \ge -1$, we must restrict the domain of $f \circ g$ to those values that also are in the domain of g. Thus,

Domain $f \circ g$: $x \ge -1$ and $x \le 3$ or $[-1, 3]$

To check this, enter $y_1 = \sqrt{3-x}$ and $y_2 = \sqrt{4 - y_1^2}$. This defines y_2 as the composition $f \circ g$. Graph y_2 and use trace or a table to verify that $[-1, 3]$ is the domain of $f \circ g$ (Figs. 11–13).

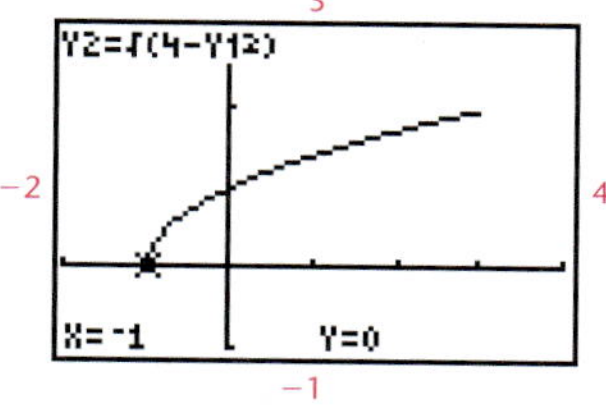

FIGURE 11

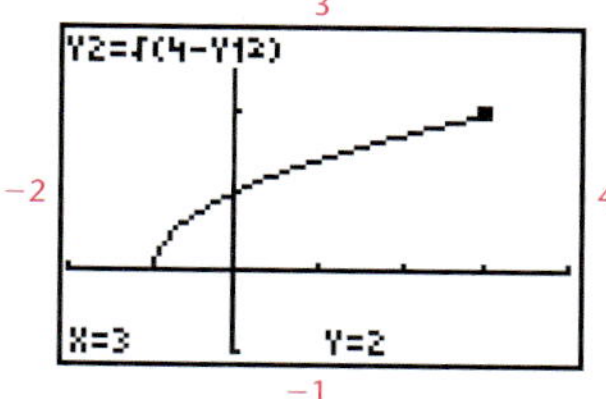

FIGURE 12

X	Y2
-1.001	ERROR
-1	0
0	1
1	1.4142
2	1.7321
3	2
3.001	ERROR

Y2=√(4-Y1²)

FIGURE 13

MATCHED 4 PROBLEM

Find $f \circ g$ for $f(x) = \sqrt{9 - x^2}$ and $g(x) = \sqrt{x - 1}$. Find the domain of $f \circ g$ algebraically and check graphically.

CAUTION

The domain of $f \circ g$ cannot always be determined simply by examining the final form of $(f \circ g)(x)$. Any numbers that are excluded from the domain of g must also be excluded from the domain of $f \circ g$.

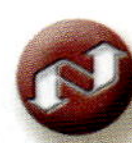

EXPLORE/DISCUSS 4

Here is another way to enter the composition of two functions in a graphing utility. Refer to Example 4. Enter $y_1 = \sqrt{3 - x}$, $y_2 = \sqrt{4 - x^2}$, and $y_3 = y_2(y_1(x))$ in the equation editor of your graphing utility and graph y_3. Does this graph agree with the graph we found in Example 4? Does your graphing utility seem to handle this composition correctly? (Not all do!)

In calculus, it is not only important to be able to find the composition of two functions, but also to recognize when a given function is the composition of simpler functions.

EXAMPLE 5 Recognizing Composition Forms

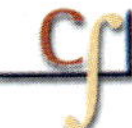

Express h as a composition of two simpler functions for

$$h(x) = \sqrt{2 + 3x^4}$$

SOLUTION

If we let $f(x) = \sqrt{x}$ and $g(x) = 2 + 3x^4$, then

$$h(x) = \sqrt{2 + 3x^4} = f(2 + 3x^4) = f(g(x)) = (f \circ g)(x)$$

and we have expressed h as the composition of f with g.

MATCHED PROBLEM 5

Express h as the composition of two simpler functions for $h(x) = (4x^3 - 7)^4$.

The answers to Example 5 and Matched Problem 5 are not unique. For example, if $f(x) = \sqrt{2 + 3x}$ and $g(x) = x^4$, then

$$f(g(x)) = \sqrt{2 + 3g(x)} = \sqrt{2 + 3x^4} = h(x)$$

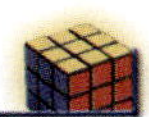

Mathematical Modeling

You will encounter the operations discussed in this section in many different situations. The next example shows how these operations are used to construct a model in economics.

EXAMPLE 6 Modeling Profit

The research department for an electronics firm estimates that the weekly demand for a certain brand of audiocassette player is given by

$$x = f(p) = 20{,}000 - 1{,}000p \qquad 0 \le p \le 20 \qquad \text{Demand function}$$

where x is the number of cassette players retailers are likely to buy per week at p dollars per player. The research department also has determined that the total cost (in dollars) of producing x cassette players per week is given by

$$C(x) = 25{,}000 + 3x \qquad \text{Cost Function}$$

and the total weekly revenue (in dollars) obtained from the sale of these cassette players is given by

$$R(x) = 20x - 0.001x^2 \qquad \text{Revenue Function}$$

Express the firm's weekly profit as a function of the price p and find the price that produces the largest profit.

SOLUTION

The basic economic principle we are using is that profit is revenue minus cost. Thus, the profit function P is the difference of the revenue function R and the cost function C. Because R and C are functions of x, we first express P as a function of x:

$$\begin{aligned} P(x) &= (R - C)(x) \\ &= R(x) - C(x) \\ &= 20x - 0.001x^2 - (25{,}000 + 3x) \\ &= 17x - 0.001x^2 - 25{,}000 \end{aligned}$$

Next we use composition to express P as a function of the price p:

$$\begin{aligned} (P \circ f)(p) &= P(f(p)) \\ &= P(20{,}000 - 1{,}000p) \\ &= 17(20{,}000 - 1{,}000p) - 0.001(20{,}000 - 1{,}000p)^2 - 25{,}000 \\ &= 340{,}000 - 17{,}000p - 400{,}000 + 40{,}000p - 1{,}000p^2 - 25{,}000 \\ &= -85{,}000 + 23{,}000p - 1{,}000p^2 \end{aligned}$$

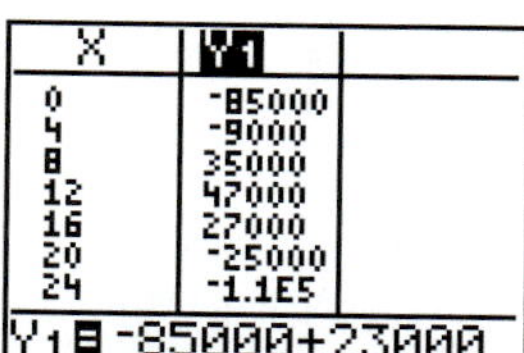

FIGURE 14

Technically, $P \circ f$ and P are different functions, because the first has independent variable p and the second has independent variable x. However, because both functions represent the same quantity, it is customary to use the same symbol to name each function. Thus,

$$P(p) = -85{,}000 + 23{,}000p - 1{,}000p^2$$

expresses the weekly profit P as a function of price p. To find the price that produces the largest profit, we must examine the graph of P. To do this, we change variables from P and p to y_1 and x and define

$$y_1 = -85{,}000 + 23{,}000x - 1{,}000x^2 \qquad 0 \le x \le 20$$

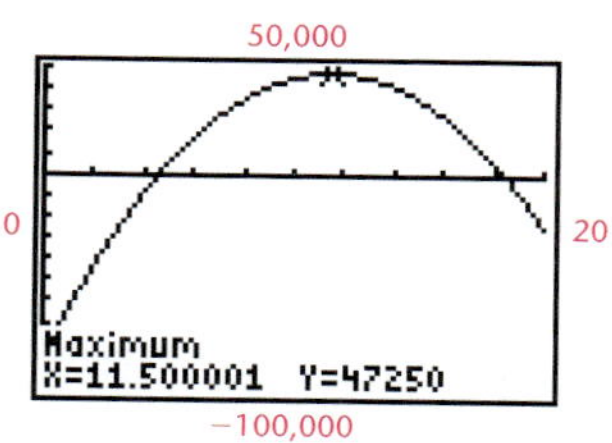

FIGURE 15

The limits for x were given in the statement of the problem. Examining a table (Fig. 14) suggests that reasonable limits on y_1 are $-100{,}000 \le y_1 \le 50{,}000$. Graphing y_1 and using the maximum command (Fig. 15) shows that the largest profit occurs when the price of a cassette player is \$11.50.

MATCHED 6 PROBLEM

Repeat Example 6 for the functions

$$x = f(p) = 10{,}000 - 1{,}000p \qquad 0 \le p \le 10$$

$$C(x) = 10{,}000 + 2x \qquad R(x) = 10x - 0.001x^2$$

ANSWERS MATCHED PROBLEMS

1. $(f - g)(x) = \sqrt{x} - \sqrt{10 - x}$; domain $= [0, 10]$
2. $\left(\frac{f}{g}\right)(x) = \frac{x}{(x + 2)(x - 5)}$; domain = all real numbers x except -2, 0, and 5
3. $(f \circ g)(x) = x$, domain $= (-\infty, \infty)$; $(g \circ f)(x) = x$, domain $= (-\infty, \infty)$
4. $(f \circ g)(x) = \sqrt{10 - x}$; domain: $x \ge 1$ and $x \le 10$ or $[1, 10]$
5. $h(x) = (f \circ g)(x)$ where $f(x) = x^4$ and $g(x) = 4x^3 - 7$
6. $P(p) = -30{,}000 + 12{,}000p - 1{,}000p^2$
The largest profit occurs when the price is \$6.

EXERCISE 1.5

A

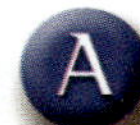

Problems 1–12 refer to functions f and g whose graphs are shown below.

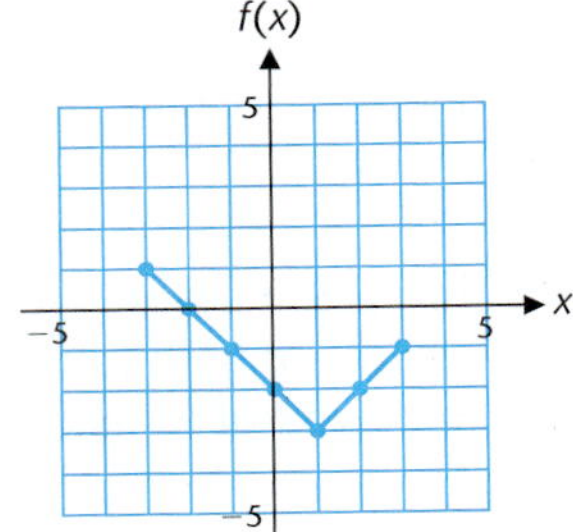

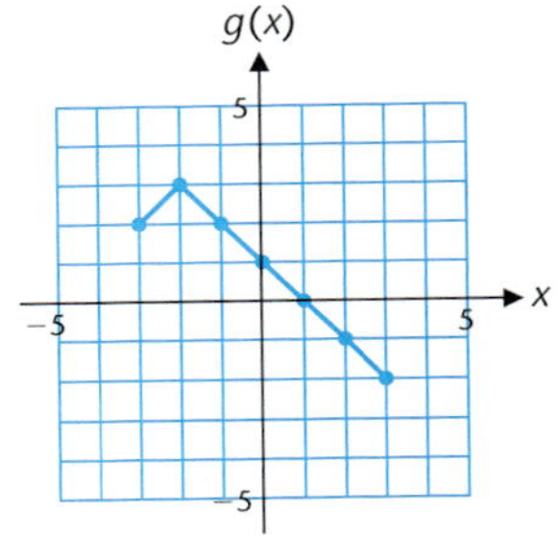

In Problems 1–4 use the graphs of f and g to construct a table of values and sketch the graph of the indicated function.

1. $(f + g)(x)$
2. $(g - f)(x)$
3. $(fg)(x)$
4. $(f - g)(x)$

In Problems 5–12, use the graphs of f and g to find each of the following:

5. $(f \circ g)(-1)$
6. $(f \circ g)(2)$
7. $(g \circ f)(-2)$
8. $(g \circ f)(3)$
9. $f(g(1))$
10. $f(g(0))$
11. $g(f(2))$
12. $g(f(-3))$

In Problems 13–18, for the indicated functions f and g, find the functions f + g, f − g, fg, and f/g, and find their domains.

13. $f(x) = 4x$; $g(x) = x + 1$
14. $f(x) = 3x$; $g(x) = x - 2$
15. $f(x) = 2x^2$; $g(x) = x^2 + 1$
16. $f(x) = 3x$; $g(x) = x^2 + 4$
17. $f(x) = 3x + 5$; $g(x) = x^2 - 1$
18. $f(x) = 2x - 7$; $g(x) = 9 - x^2$

In Problems 19–24, for the indicated functions f and g, find the functions f ∘ g, and g ∘ f, and find their domains.

19. $f(x) = x^3$; $g(x) = x^2 - x + 1$
20. $f(x) = x^2$; $g(x) = x^3 + 2x + 4$
21. $f(x) = |x + 1|$; $g(x) = 2x + 3$
22. $f(x) = |x - 4|$; $g(x) = 3x + 2$
23. $f(x) = x^{1/3}$; $g(x) = 2x^3 + 4$
24. $f(x) = x^{2/3}$; $g(x) = 8 - x^3$

B

In Problems 25–28, find $f \circ g$ and $g \circ f$. Graph f, g, $f \circ g$, and $g \circ f$ in a squared viewing window and describe any apparent symmetry between these graphs.

25. $f(x) = \frac{1}{2}x + 1$; $g(x) = 2x - 2$

26. $f(x) = 3x + 2$; $g(x) = \frac{1}{3}x - \frac{2}{3}$

27. $f(x) = -\frac{2}{3}x - \frac{5}{3}$; $g(x) = -\frac{3}{2}x - \frac{5}{2}$

28. $f(x) = -2x + 3$; $g(x) = -\frac{1}{2}x + \frac{3}{2}$

In Problems 29–34, for the indicated functions f and g, find the functions $f + g$, $f - g$, fg, and f/g, and find their domains.

29. $f(x) = \sqrt{2 - x}$; $g(x) = \sqrt{x + 3}$

30. $f(x) = \sqrt{x + 4}$; $g(x) = \sqrt{3 - x}$

31. $f(x) = \sqrt{x} + 2$; $g(x) = \sqrt{x} - 4$

32. $f(x) = 1 - \sqrt{x}$; $g(x) = 2 - \sqrt{x}$

33. $f(x) = \sqrt{x^2 + x - 6}$; $g(x) = \sqrt{7 + 6x - x^2}$

34. $f(x) = \sqrt{8 + 2x - x^2}$; $g(x) = \sqrt{x^2 - 7x + 10}$

In Problems 35–40, for the indicated functions f and g, find the functions $f \circ g$ and $g \circ f$, and find their domains.

35. $f(x) = \sqrt{x}$; $g(x) = x - 4$

36. $f(x) = \sqrt{x}$; $g(x) = 2x + 5$

37. $f(x) = x + 2$; $g(x) = \dfrac{1}{x}$

38. $f(x) = x - 3$; $g(x) = \dfrac{1}{x^2}$

39. $f(x) = |x|$; $g(x) = \dfrac{1}{x - 1}$

40. $f(x) = |x - 1|$; $g(x) = \dfrac{1}{x}$

Use the graphs of functions f and g shown below to match each function in Problems 41–46 with one of graphs (a)–(f).

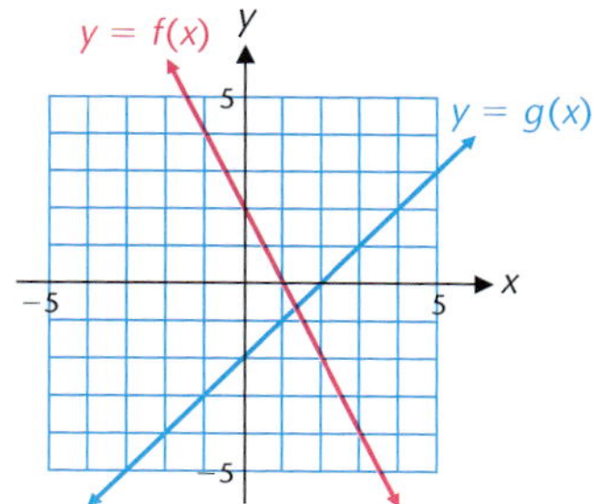

41. $(f + g)(x)$

42. $(f - g)(x)$

43. $(g - f)(x)$

44. $(fg)(x)$

45. $\left(\dfrac{f}{g}\right)(x)$

46. $\left(\dfrac{g}{f}\right)(x)$

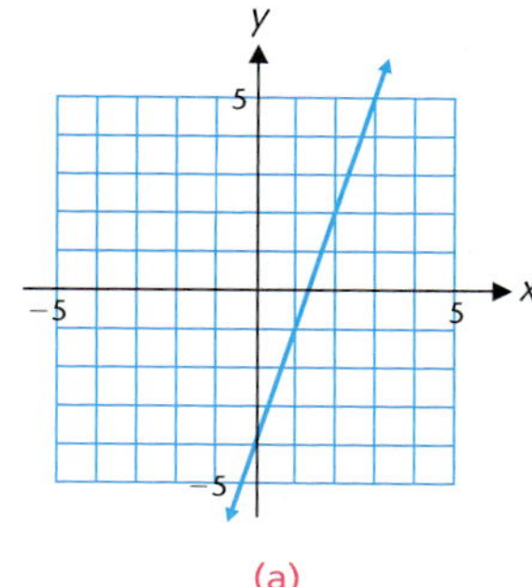

(a)

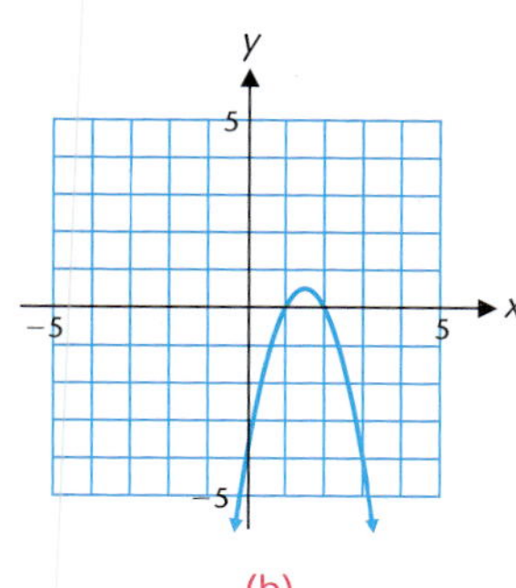

(b)

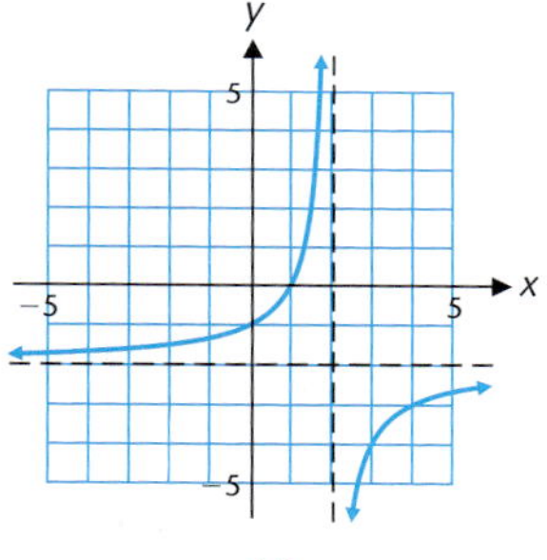

(c)

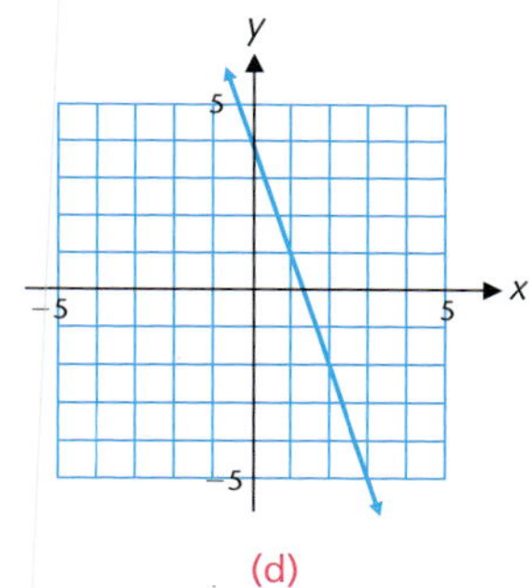

(d)

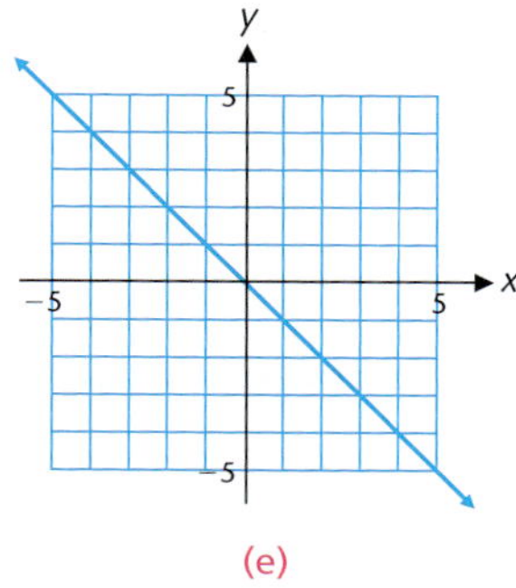

(e)

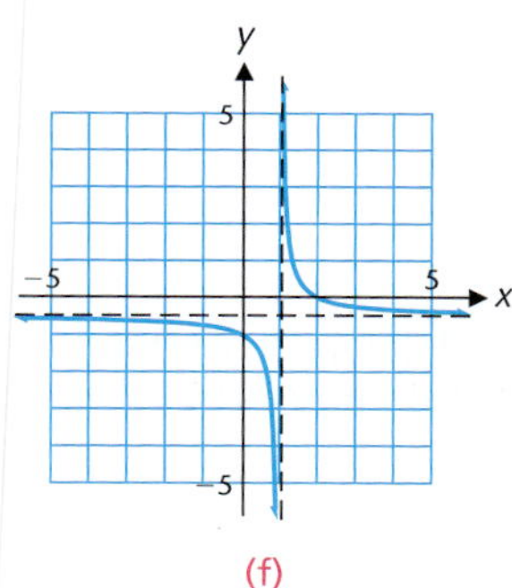

(f)

In Problems 47–54, express h as a composition of two simpler functions f and g of the form $f(x) = x^n$ and $g(x) = ax + b$, where n is a rational number and a and b are integers.

47. $h(x) = (2x - 7)^4$

48. $h(x) = (3 - 5x)^7$

49. $h(x) = \sqrt{4 + 2x}$

50. $h(x) = \sqrt{3x - 11}$

51. $h(x) = 3x^7 - 5$

52. $h(x) = 5x^6 + 3$

53. $h(x) = \frac{4}{\sqrt{x}} + 3$

54. $h(x) = -\frac{2}{\sqrt{x}} + 1$

55. Are the functions fg and gf identical? Justify your answer.

56. Are the functions $f \circ g$ and $g \circ f$ identical? Justify your answer.

57. Is there a function g that satisfies $f \circ g = g \circ f = f$ for all functions f? If so, what is it?

58. Is there a function g that satisfies $fg = gf = f$ for all functions f? If so, what is it?

C

In Problems 59–62, for the indicated functions f and g, find the functions $f + g$, $f - g$, fg, and f/g, and find their domains.

59. $f(x) = x + \frac{1}{x}$; $g(x) = x - \frac{1}{x}$

60. $f(x) = x - 1$; $g(x) = x - \frac{6}{x - 1}$

61. $f(x) = 1 - \frac{x}{|x|}$; $g(x) = 1 + \frac{x}{|x|}$

62. $f(x) = x + |x|$; $g(x) = x - |x|$

In Problems 63–68, for the indicated functions f and g, find the functions $f \circ g$ and $g \circ f$, and find their domains.

63. $f(x) = \sqrt{4 - x}$; $g(x) = x^2$

64. $f(x) = \sqrt{x - 1}$; $g(x) = x^2$

65. $f(x) = \frac{x + 5}{x}$; $g(x) = \frac{x}{x - 2}$

66. $f(x) = \frac{x}{x - 1}$; $g(x) = \frac{2x - 4}{x}$

67. $f(x) = \sqrt{25 - x^2}$; $g(x) = \sqrt{9 + x^2}$

68. $f(x) = \sqrt{x^2 - 9}$; $g(x) = \sqrt{x^2 + 25}$

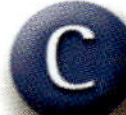

In Problems 69–74, enter the given expression for $(f \circ g)(x)$ exactly as it is written and graph on a graphing utility for $-10 \le x \le 10$. Then simplify the expression, enter the result, and graph in a new viewing window, again for $-10 \le x \le 10$. Find the domain of $f \circ g$. Which is the correct graph of $f \circ g$?

69. $f(x) = \sqrt{5 - x^2}$; $g(x) = \sqrt{3 - x}$;
$(f \circ g)(x) = \sqrt{5 - (\sqrt{3 - x})^2}$

70. $f(x) = \sqrt{6 - x^2}$; $g(x) = \sqrt{x - 1}$;
$(f \circ g)(x) = \sqrt{6 - (\sqrt{x - 1})^2}$

71. $f(x) = \sqrt{x^2 + 5}$; $g(x) = \sqrt{x^2 - 4}$;
$(f \circ g)(x) = \sqrt{(\sqrt{x^2 - 4})^2 + 5}$

72. $f(x) = \sqrt{x^2 + 5}$; $g(x) = \sqrt{4 - x^2}$;
$(f \circ g)(x) = \sqrt{(\sqrt{4 - x^2})^2 + 5}$

73. $f(x) = \sqrt{x^2 + 7}$; $g(x) = \sqrt{9 - x^2}$;
$(f \circ g)(x) = \sqrt{(\sqrt{9 - x^2})^2 + 7}$

74. $f(x) = \sqrt{x^2 + 7}$; $g(x) = \sqrt{x^2 - 9}$;
$(f \circ g)(x) = \sqrt{(\sqrt{x^2 - 9})^2 + 7}$

APPLICATIONS

75. Market Research. The demand x and the price p (in dollars) for a certain product are related by

$$x = f(p) = 4{,}000 - 200p \qquad 0 \le p \le 20$$

The revenue (in dollars) from the sale of x units is given by

$$R(x) = 20x - \frac{1}{200}x^2$$

and the cost (in dollars) of producing x units is given by

$$C(x) = 2x + 8{,}000$$

Express the profit as a function of the price p and find the price that produces the largest profit.

76. Market Research. The demand x and the price p (in dollars) for a certain product are related by

$$x = f(p) = 5{,}000 - 100p \qquad 0 \le p \le 50$$

The revenue (in dollars) from the sale of x units and the cost (in dollars) of producing x units are given, respectively, by

$$R(x) = 50x - \frac{1}{100}x^2 \quad \text{and} \quad C(x) = 20x + 40{,}000$$

Express the profit as a function of the price p and find the price that produces the largest profit.

77. **Pollution.** An oil tanker aground on a reef is leaking oil that forms a circular oil slick about 0.1 foot thick (see the figure). The radius of the slick (in feet) t minutes after the leak first occurred is given by

$$r(t) = 0.4t^{1/3}$$

Express the volume of the oil slick as a function of t.

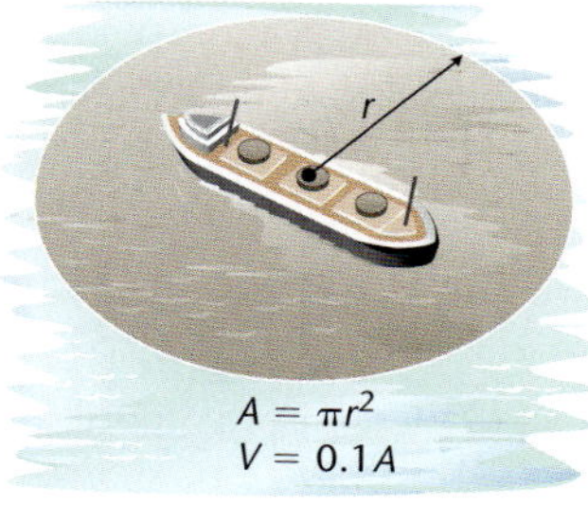

78. **Weather Balloon.** A weather balloon is rising vertically. An observer is standing on the ground 100 meters from the point where the weather balloon was released.

(A) Express the distance d between the balloon and the observer as a function of the balloon's distance h above the ground.

(B) If the balloon's distance above ground after t seconds is given by $h = 5t$, express the distance d between the balloon and the observer as a function of t.

★ 79. **Fluid Flow.** A conical paper cup with diameter 4 inches and height 4 inches is initially full of water. A small hole is made in the bottom of the cup and the water begins to flow out of the cup. Let h and r be the height and radius, respectively, of the water in the cup t minutes after the water begins to flow.

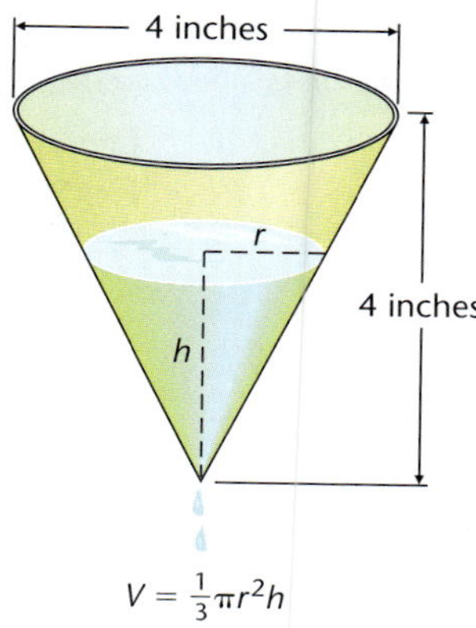

(A) Express r as a function of h.

(B) Express the volume V as a function of h.

(C) If the height of the water after t minutes is given by

$$h(t) = 0.5\sqrt{t}$$

express V as a function of t.

★ 80. **Evaporation.** A water trough with triangular ends is 6 feet long, 4 feet wide, and 2 feet deep. Initially, the trough is full of water, but due to evaporation, the volume of the water is decreasing. Let h and w be the height and width, respectively, of the water in the tank t hours after it began to evaporate.

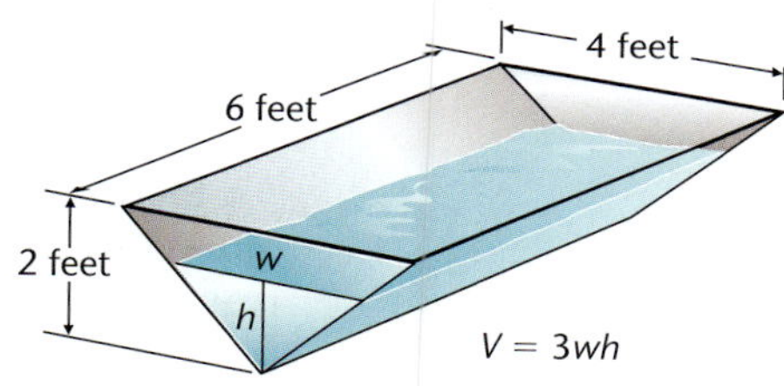

(A) Express w as a function of h.

(B) Express V as a function of h.

(C) If the height of the water after t hours is given by

$$h(t) = 2 - 0.2\sqrt{t}$$

express V as a function of t.

SECTION 1.6 Inverse Functions

One-to-One Functions • Inverse Functions • Mathematical Modeling • Graphing Inverse Functions

Many important mathematical relationships can be expressed in terms of functions. For example,

$C = \pi d = f(d)$ — The circumference of a circle is a function of the diameter d.

$V = s^3 = g(s)$ — The volume of a cube is a function of the edge s.

$d = 1{,}000 - 100p = h(p)$ — The demand for a product is a function of the price p.

$F = \frac{9}{5}C + 32$ — Temperature measured in °F is a function of temperature in °C.

In many cases, we are interested in *reversing* the correspondence determined by a function. Thus,

$d = \frac{C}{\pi} = m(C)$ — The diameter of a circle is a function of the circumference C.

$s = \sqrt[3]{V} = n(V)$ — The edge of a cube is a function of the volume V.

$p = 10 - \frac{1}{100}d = r(d)$ — The price of a product is a function of the demand d.

$C = \frac{5}{9}(F - 32)$ — Temperature measured in °C is a function of temperature in °F.

As these examples illustrate, reversing the relationship between two quantities often produces a new function. This new function is called the *inverse* of the original function. Later in this text we will see that many important functions (for example, logarithmic functions) are actually defined as the inverses of other functions.

In this section, we develop techniques for determining whether the inverse function exists, some general properties of inverse functions, and methods for finding the rule of correspondence that defines the inverse function. A review of Section 1.2 will prove very helpful at this point.

One-to-One Functions

Recall the set form of the definition of function:

A function is a set of ordered pairs with the property that no two ordered pairs have the same first component and different second components.

However, it is possible that two ordered pairs in a function could have the same second component and different first components. If this does not happen, then we call the function a *one-to-one function*. It turns out that one-to-one functions are the only functions that have inverse functions.

DEFINITION 1
One-To-One Function

A function is **one-to-one** if no two ordered pairs in the function have the same second component and different first components.

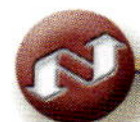

EXPLORE/DISCUSS 1

Given the following sets of ordered pairs:

$f = \{(0, 1), (0, 2), (1, 1), (1, 2)\}$
$g = \{(0, 1), (1, 1), (2, 2), (3, 2)\}$
$h = \{(0, 1), (1, 2), (2, 3), (3, 0)\}$

(A) Which of these sets represent functions?

(B) Which of the functions are one-to-one functions?

(C) For each set that is a function, form a new set by reversing each ordered pair in the set. Which of these new sets represent functions?

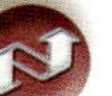

EXAMPLE 1 Determining Whether a Function Is One-To-One

Determine whether f is a one-to-one function for

(A) $f(x) = x^2$ (B) $f(x) = 2x - 1$

SOLUTIONS

(A) To show that a function is not one-to-one, all we have to do is find two different ordered pairs in the function with the same second component and different first components. Because

$$f(2) = 2^2 = 4 \qquad \text{and} \qquad f(-2) = (-2)^2 = 4$$

the ordered pairs $(2, 4)$ and $(-2, 4)$ both belong to f and f is not one-to-one.

(B) To show that a function is one-to-one, we have to show that no two ordered pairs have the same second component and different first components. To do this, we assume there are two ordered pairs $(a, f(a))$ and $(b, f(b))$ in f with the same second components and then show that the first components must also be the same. That is, we show that $f(a) = f(b)$ implies $a = b$. We proceed as follows:

$$f(a) = f(b)$$ Assume second components are equal.

$$2a - 1 = 2b - 1$$ Evaluate $f(a)$ and $f(b)$.

$$2a = 2b$$ Simplify.

$$a = b$$ Conclusion: f is one-to-one.

Thus, by Definition 1, f is a one-to-one function.

MATCHED 1 PROBLEM

Determine whether f is a one-to-one function for

(A) $f(x) = 4 - x^2$ (B) $f(x) = 4 - 2x$

The methods used in the solution of Example 1 can be stated as a theorem.

THEOREM 1
One-to-One Functions

1. If $f(a) = f(b)$ for at least one pair of domain values a and b, $a \neq b$, then f is not one-to-one.
2. If the assumption $f(a) = f(b)$ always implies that the domain values a and b are equal, then f is one-to-one.

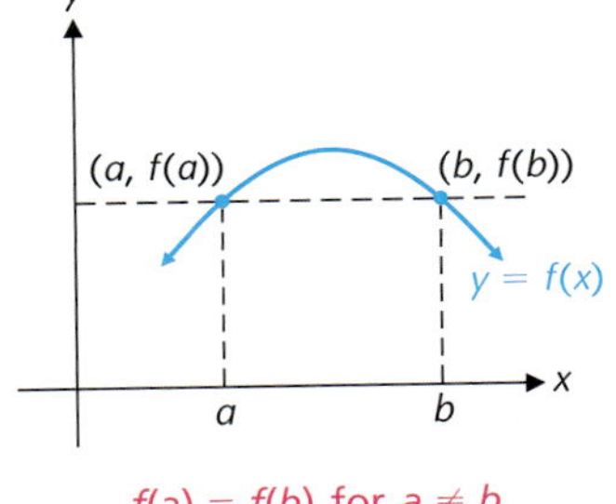

$f(a) = f(b)$ for $a \neq b$
f is not one-to-one
(a)

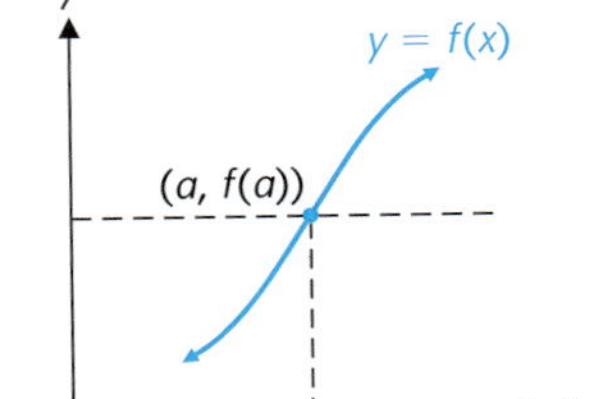

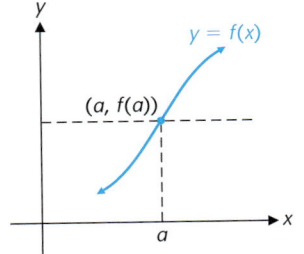

Only one point has ordinate $f(a)$; f is one-to-one
(b)

FIGURE 1 Intersections of graphs and horizontal lines.

Applying Theorem 1 is not always easy—try testing $f(x) = x^3 + 2x + 3$, for example. However, if we are given the graph of a function, then there is a simple graphical procedure for determining if the function is one-to-one. If a horizontal line intersects the graph of a function in more than one point, then the function is not one-to-one, as shown in Figure 1(a). However, if each horizontal line intersects the graph in one point, or not at all, then the function is one-to-one, as shown in Figure 1(b). These observations form the basis for the *horizontal line test.*

THEOREM 2
Horizontal Line Test

A function is one-to-one if and only if each horizontal line intersects the graph of the function in at most one point.

The graphs of the functions considered in Example 1 are shown in Figure 2. Applying the horizontal line test to each graph confirms the results we obtained in Example 1.

A function that is increasing throughout its domain or decreasing throughout its domain will always pass the horizontal line test [Figs. 3(a) and 3(b)]. Thus, we have the following theorem.

FIGURE 2 Applying the horizontal line test.

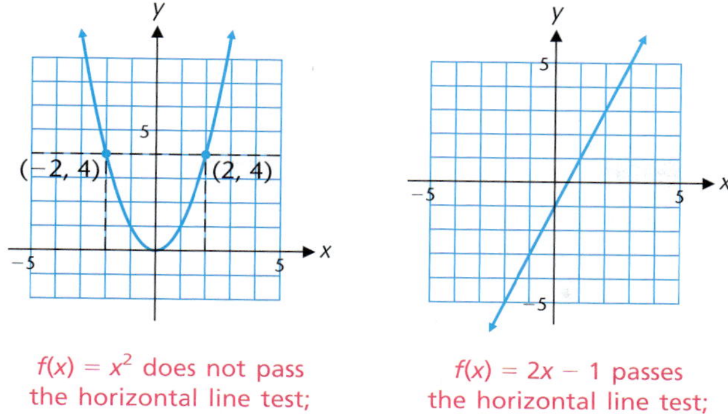

$f(x) = x^2$ does not pass the horizontal line test; f is not one-to-one
(a)

$f(x) = 2x - 1$ passes the horizontal line test; f is one-to-one
(b)

THEOREM 3
Increasing and Decreasing Functions

If a function f is increasing throughout its domain or decreasing throughout its domain, then f is a one-to-one function.

FIGURE 3 Increasing, decreasing, and one-to-one functions.

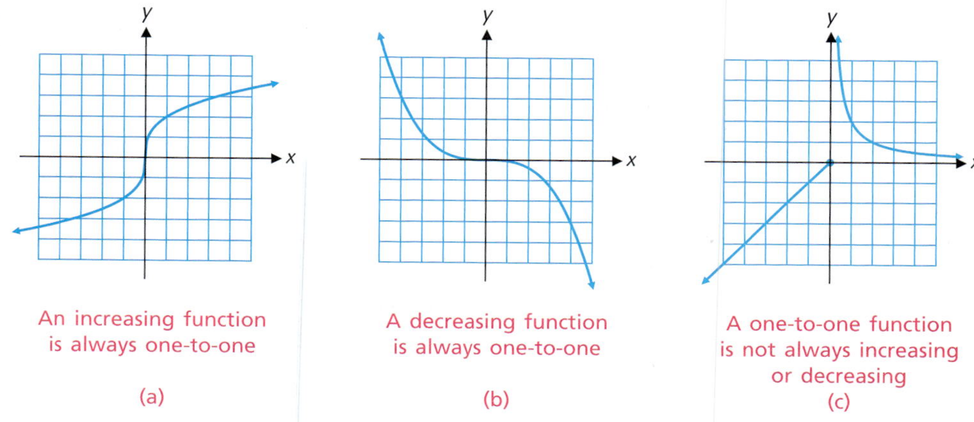

An increasing function is always one-to-one
(a)

A decreasing function is always one-to-one
(b)

A one-to-one function is not always increasing or decreasing
(c)

The converse of Theorem 3 is false. To see this, consider the function graphed in Figure 3(c). This function is increasing on $(-\infty, 0]$ and decreasing on $(0, \infty)$, yet the graph passes the horizontal line test. Thus, this is a one-to-one function that is neither an increasing function nor a decreasing function.

Inverse Functions

Now we want to see how we can form a new function by reversing the correspondence determined by a given function. Let g be the function defined as follows:

$$g = \{(-3, 9), (0, 0), (3, 9)\} \qquad g \text{ is not one-to-one.}$$

Notice that g is not one-to-one because the domain elements -3 and 3 both correspond to the range element 9. We can reverse the correspondence determined by function g simply by reversing the components in each ordered pair in g, producing the following set:

$$G = \{(9, -3), (0, 0), (9, 3)\}$$ G is not a function.

But the result is not a function because the domain element 9 corresponds to two different range elements, -3 and 3. On the other hand, if we reverse the ordered pairs in the function

$$f = \{(1, 2), (2, 4), (3, 9)\}$$ f is one-to-one.

we obtain

$$F = \{(2, 1), (4, 2), (9, 3)\}$$ F is a function.

This time f is a one-to-one function, and the set F turns out to be a function also. This new function F, formed by reversing all the ordered pairs in f, is called the *inverse* of f and is usually denoted by f^{-1}. Thus,

$$f^{-1} = \{(2, 1)\ (4, 2), (9, 3)\}$$ The inverse of f

Notice that f^{-1} is also a one-to-one function and that the following relationships hold:

$$\text{Domain of } f^{-1} = \{2, 4, 9\} = \text{Range of } f$$

$$\text{Range of } f^{-1} = \{1, 2, 3\} = \text{Domain of } f$$

Thus, reversing all the ordered pairs in a one-to-one function forms a new one-to-one function and reverses the domain and range in the process. We are now ready to present a formal definition of the inverse of a function.

DEFINITION 2

Inverse of a Function

If f is a one-to-one function, then the **inverse** of f, denoted f^{-1}, is the function formed by reversing all the ordered pairs in f. Thus,

$$f^{-1} = \{(y, x) \mid (x, y) \text{ is in } f\}$$

If f is not one-to-one, then f **does not have an inverse** and f^{-1} **does not exist.**

CAUTION

Do not confuse inverse notation and reciprocal notation:

$$2^{-1} = \frac{1}{2}$$ Reciprocal notation for numbers

$$(f(x))^{-1} = \frac{1}{f(x)}$$ Reciprocal notation for functions

$$f^{-1}(x) \neq \frac{1}{f(x)}$$ Inverse notation is not reciprocal notation.

The following properties of inverse functions follow directly from the definition.

THEOREM 4
Properties of Inverse Functions

If f^{-1} exists, then

1. f^{-1} is a one-to-one function.
2. Domain of f^{-1} = range of f
3. Range of f^{-1} = domain of f

Finding the inverse of a function defined by a finite set of ordered pairs is easy; just reverse each ordered pair. But how do we find the inverse of a function defined by an equation? Consider the one-to-one function f defined by

$$f(x) = 2x - 1$$

To find f^{-1}, we let $y = f(x)$ and solve for x:

$$y = 2x - 1$$
$$y + 1 = 2x$$
$$\tfrac{1}{2}y + \tfrac{1}{2} = x$$

Because the ordered pair (x, y) is in f if and only if the reversed ordered pair (y, x) is in f^{-1}, this last equation defines f^{-1}:

$$x = f^{-1}(y) = \tfrac{1}{2}y + \tfrac{1}{2} \qquad (1)$$

Something interesting happens if we form the composition of f and f^{-1} in either of the two possible orders.

$$(f^{-1} \circ f)(x) = f^{-1}(f(x)) = f^{-1}(2x - 1) = \tfrac{1}{2}(2x - 1) + \tfrac{1}{2} = x - \tfrac{1}{2} + \tfrac{1}{2} = x$$

and

$$(f \circ f^{-1})(y) = f(f^{-1}(y)) = f(\tfrac{1}{2}y + \tfrac{1}{2}) = 2(\tfrac{1}{2}y + \tfrac{1}{2}) - 1 = y + 1 - 1 = y$$

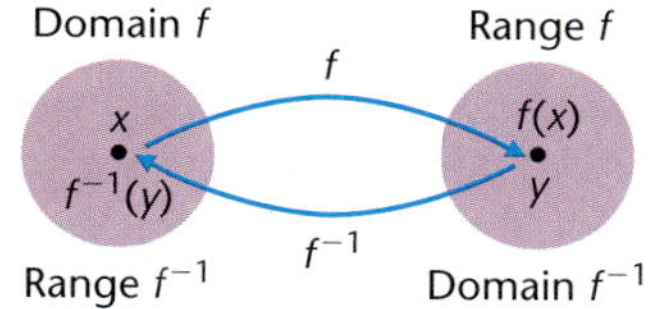

FIGURE 4 Composition of f and f^{-1}.

These compositions indicate that if f maps x into y, then f^{-1} maps y back into x and if f^{-1} maps y into x, then f maps x back into y. This is interpreted schematically in Figure 4.

Finally, we note that we usually use x to represent the independent variable and y the dependent variable in an equation that defines a function. It is customary to do this for inverse functions also. Thus, interchanging the variables x and y in equation (1), we can state that the inverse of

$$y = f(x) = 2x - 1$$

is

$$y = f^{-1}(x) = \tfrac{1}{2}x + \tfrac{1}{2}$$

In general, we have the following result:

THEOREM 5

Relationship Between f and f^{-1}

If f^{-1} exists, then

1. $x = f^{-1}(y)$ if and only if $y = f(x)$.
2. $f^{-1}(f(x)) = x$ for all x in the domain of f.
3. $f(f^{-1}(y)) = y$ for all y in the domain of f^{-1} or, if x and y have been interchanged, $f(f^{-1}(x)) = x$ for all x in the domain of f^{-1}.

If f and g are one-to-one functions satisfying

$$f(g(x)) = x \qquad \text{for all } x \text{ in the domain of } g$$

$$g(f(x)) = x \qquad \text{for all } x \text{ in the domain of } f$$

then it can be shown that $g = f^{-1}$ and $f = g^{-1}$. Thus, the inverse function is the only function that satisfies both these compositions. We can use this fact to check that we have found the inverse correctly.

EXPLORE/DISCUSS 2

Find $f(g(x))$ and $g(f(x))$ for

$$f(x) = (x - 1)^3 + 2 \qquad \text{and} \qquad g(x) = (x - 2)^{1/3} + 1$$

How are f and g related?

The procedure for finding the inverse of a function defined by an equation is given in the next box. This procedure can be applied whenever it is possible to solve $y = f(x)$ for x in terms of y.

Finding the Inverse of a Function f

Step 1. Find the domain of f and verify that f is one-to-one. If f is not one-to-one, then stop, because f^{-1} does not exist.

Step 2. Solve the equation $y = f(x)$ for x. The result is an equation of the form $x = f^{-1}(y)$.

Step 3. (Optional) Interchange x and y in the equation found in step 2. This expresses f^{-1} as a function of x.

Step 4. Find the domain of f^{-1}. Remember, the domain of f^{-1} must be the same as the range of f.

Check your work by verifying that

$$f^{-1}(f(x)) = x \qquad \text{for all } x \text{ in the domain of } f$$

and

$$f(f^{-1}(x)) = x \qquad \text{for all } x \text{ in the domain of } f^{-1}$$

EXAMPLE 2 Finding the Inverse of a Function

Find f^{-1} for $f(x) = \sqrt{x - 1}$.

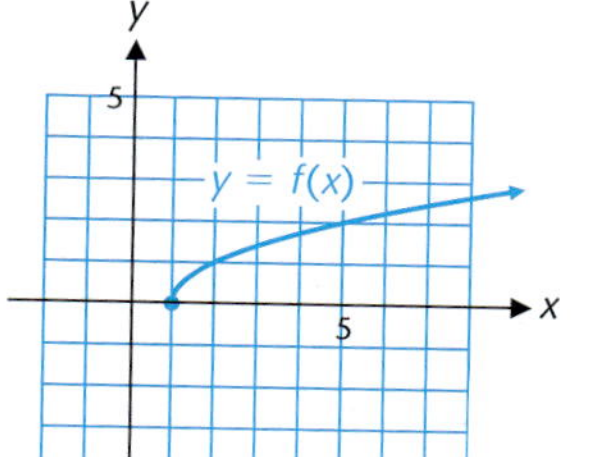

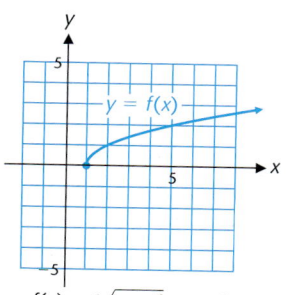

$f(x) = \sqrt{x - 1},\ x \geq 1$

FIGURE 5

SOLUTION

Step 1. Find the domain of f and verify that f is one-to-one. The domain of f is $[1, \infty)$. The graph of f in Figure 5 shows that f is one-to-one, hence f^{-1} exists.

Step 2. Solve the equation $y = f(x)$ for x.

$$y = \sqrt{x - 1}$$
$$y^2 = x - 1$$
$$x = y^2 + 1$$

Thus,

$$x = f^{-1}(y) = y^2 + 1$$

Step 3. Interchange x and y.

$$y = f^{-1}(x) = x^2 + 1$$

Step 4. Find the domain of f^{-1}. The equation $f^{-1}(x) = x^2 + 1$ is defined for all values of x, but this does not tell us what the domain of f^{-1} is. Remember, the domain of f^{-1} must equal the range of f. From the graph of f, we see that the range of f is $[0, \infty)$. Thus, the domain of f^{-1} is also $[0, \infty)$. That is,

$$f^{-1}(x) = x^2 + 1 \qquad x \geq 0$$

CHECK

For x in $[1, \infty)$, the domain of f, we have

$$\begin{aligned} f^{-1}(f(x)) &= f^{-1}(\sqrt{x - 1}) \\ &= (\sqrt{x - 1})^2 + 1 \\ &= x - 1 + 1 \\ &\stackrel{\checkmark}{=} x \end{aligned}$$

For x in $[0, \infty)$, the domain of f^{-1}, we have

$$\begin{aligned} f(f^{-1}(x)) &= f(x^2 + 1) \\ &= \sqrt{(x^2 + 1) - 1} \\ &= \sqrt{x^2} \\ &= |x| \\ &\stackrel{\checkmark}{=} x \end{aligned}$$

$\sqrt{x^2} = |x|$ for any real number x.
$|x| = x$ for $x \geq 0$.

MATCHED 2 PROBLEM

Find f^{-1} for $f(x) = \sqrt{x + 2}$.

EXPLORE/DISCUSS 3

Most basic arithmetic operations can be reversed by performing a second operation: subtraction reverses addition, division reverses multiplication, squaring reverses taking the square root, and so on. Viewing a function as a sequence of reversible operations gives insight into the inverse function concept. For example, the function $f(x) = 2x - 1$ can be described verbally as a function that multiplies each domain element by 2 and then subtracts 1. Reversing this sequence describes a function g that adds 1 to each domain element and then divides by 2, or $g(x) = (x + 1)/2$, which is the inverse of the function f. For each of the following functions, write a verbal description of the function, reverse your description, and write the resulting algebraic equation. Verify that the result is the inverse of the original function.

(A) $f(x) = 3x + 5$ (B) $f(x) = \sqrt{x - 1}$ (C) $f(x) = \dfrac{1}{x + 1}$

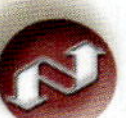

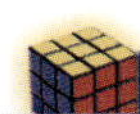

Mathematical Modeling

Example 3 shows how an inverse function is used in constructing a revenue model. See Example 6 in Section 1.5.

EXAMPLE 3 **Modeling Revenue**

The research department for an electronics firm estimates that the weekly demand for a certain brand of audiocassette player is given by

$$x = f(p) = 20{,}000 - 1{,}000p \qquad 0 \le p \le 20 \qquad \text{Demand function}$$

where x is the number of cassette players retailers are likely to buy per week at p dollars per player. Express the revenue as a function of the demand x.

SOLUTION

If x cassette players are sold at p dollars each, the total revenue is

$$\begin{aligned}\text{Revenue} &= (\text{Number of players})(\text{price of each player})\\ &= xp\end{aligned}$$

To express the revenue as a function of the demand x, we must express the price in terms of x. That is, we must find the inverse of the demand function.

$$x = 20{,}000 - 1{,}000p$$

$$x - 20{,}000 = -1{,}000p \qquad \text{Multiply both sides by } \frac{-1}{1{,}000} = -0.001$$

$$-0.001(x - 20{,}000) = p$$

$$-0.001x + 20 = p$$

Thus, the inverse of the demand function is

$$p = f^{-1}(x) = 20 - 0.001x$$

and the revenue is given by

$$R = xp$$

$$R(x) = x(20 - 0.001x)$$

$$= 20x - 0.001x^2$$

MATCHED 3 PROBLEM

Repeat Example 3 for the demand function

$$x = f(p) = 10{,}000 - 1{,}000p \qquad 0 \le p \le 10$$

The demand function in Example 3 was defined with independent variable p and dependent variable x. When we found the inverse function, we did not rewrite it with independent variable p. Because p represents price and x represents number of players, to interchange these variables would be confusing. In most applications, the variables have specific meaning and should not be interchanged as part of the inverse process.

Graphing Inverse Functions

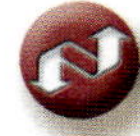

EXPLORE/DISCUSS 4

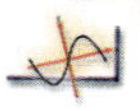

The following activities refer to the graph of f in Figure 6 and Tables 1 and 2.

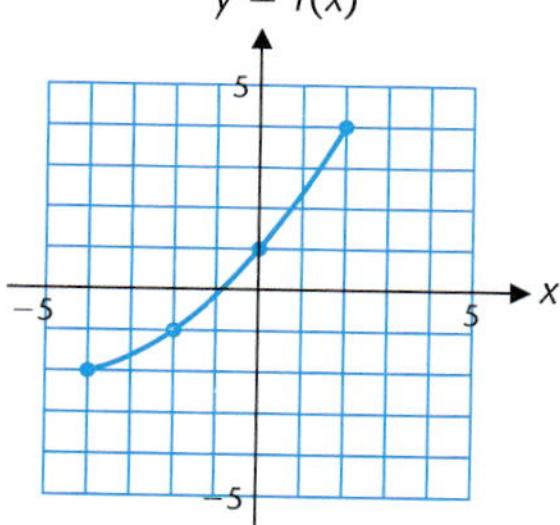

FIGURE 6

TABLE 1

x	$f(x)$
−4	
−2	
0	
2	

TABLE 2

x	$f^{-1}(x)$

(A) Complete the second column in Table 1.

(B) Reverse the ordered pairs in Table 1 and list the results in Table 2.

(C) Add the points in Table 2 to Figure 6 (or a copy of the figure) and sketch the graph of f^{-1}.

(D) Discuss any symmetry you observe between the graphs of f and f^{-1}.

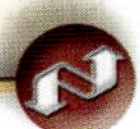

There is an important relationship between the graph of any function and its inverse that is based on the following observation: In a rectangular coordinate system, the points (a, b) and (b, a) are symmetrical with respect to the line $y = x$ [Fig. 7 (a)]. Theorem 6 is an immediate consequence of this observation.

FIGURE 7 Symmetry with respect to the line $y = x$.

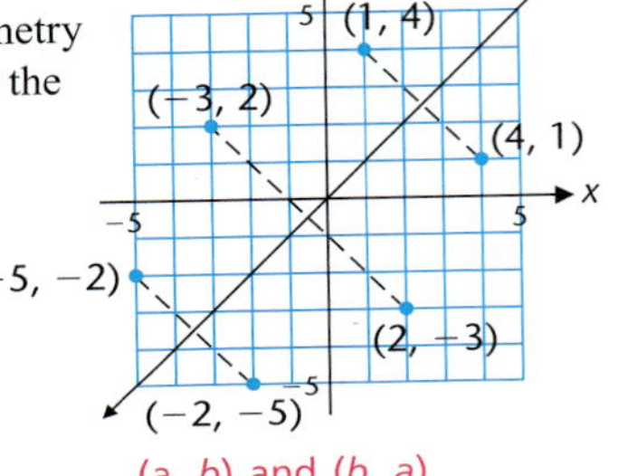

(a, b) and (b, a) are symmetric with respect to the line y = x

(a)

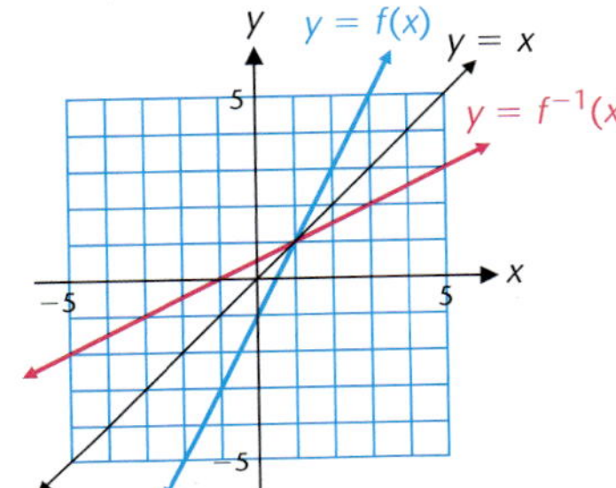

$f(x) = 2x - 1$
$f^{-1}(x) = \frac{1}{2}x + \frac{1}{2}$

(b)

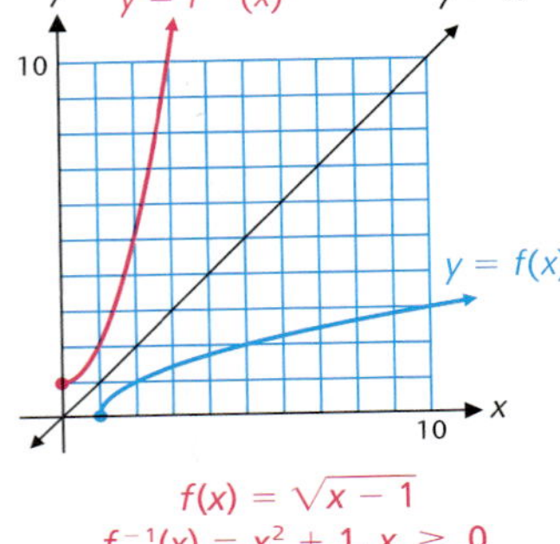

$f(x) = \sqrt{x - 1}$
$f^{-1}(x) = x^2 + 1, x \geq 0$

(c)

THEOREM 6

Symmetry Property for the Graphs of f and f^{-1}

The graphs of $y = f(x)$ and $y = f^{-1}(x)$ are symmetrical with respect to the line $y = x$.

Knowledge of this symmetry property allows us to graph f^{-1} if the graph of f is known, and vice versa. Figures 7(b) and 7(c) illustrate this property for the two inverse functions we found earlier.

If a function is not one-to-one, we usually can restrict the domain of the function to produce a new function that is one-to-one. Then we can find an inverse for the restricted function. Suppose we start with $f(x) = x^2 - 4$. Because f is not one-to-one, f^{-1} does not exist [Fig. 8(a)]. But there are many ways the domain of f can be restricted to obtain a one-to-one function. Figures 8(b) and 8(c) illustrate two such restrictions.

FIGURE 8 Restricting the domain of a function.

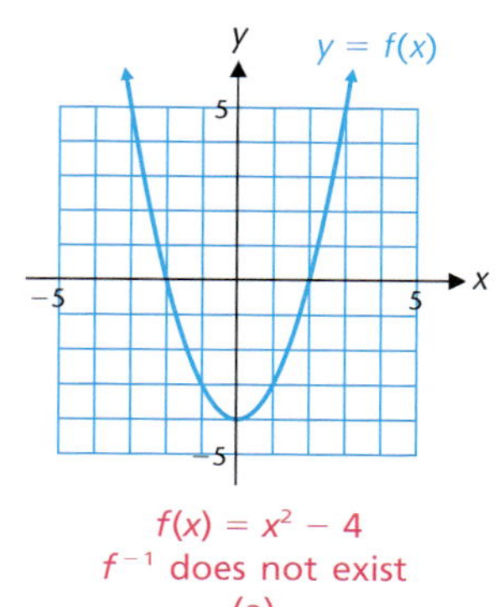

$f(x) = x^2 - 4$
f^{-1} does not exist
(a)

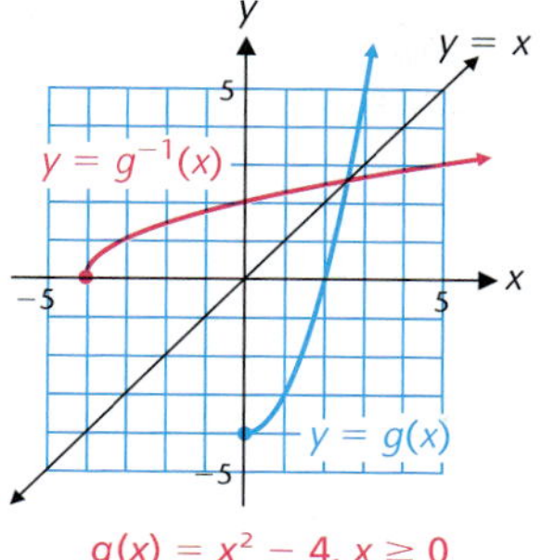

$g(x) = x^2 - 4, x \geq 0$
$g^{-1}(x) = \sqrt{x + 4}, x \geq -4$
(b)

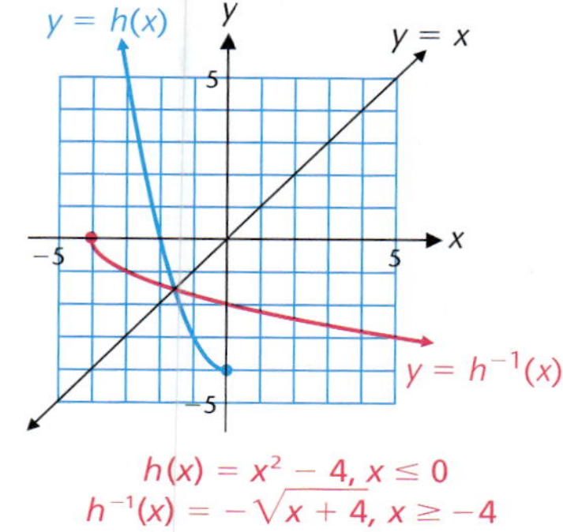

$h(x) = x^2 - 4, x \leq 0$
$h^{-1}(x) = -\sqrt{x + 4}, x \geq -4$
(c)

EXPLORE/DISCUSS 5

To graph the function

$$g(x) = 4x - x^2, \qquad x \geq 0$$

on a graphing utility, enter

$$y_1 = (4x - x^2)/(x \geq 0)$$

(A) The Boolean expression $(x \geq 0)$ is assigned the value 1 if the inequality is true and 0 if it is false. How does this result in restricting the graph of $4x - x^2$ to just those values of x satisfying $x \geq 0$?

(B) Use this concept to reproduce Figures 8(b) and 8(c) on a graphing utility.

(C) Do your graphs appear to be symmetrical with respect to the line $y = x$? What happens if you use a squared window for your graph?

Recall from Theorem 3 that increasing and decreasing functions are always one-to-one. This provides the basis for a convenient and popular method of restricting the domain of a function:

If the domain of a function *f* is restricted to an interval on the *x* axis over which *f* is increasing (or decreasing), then the new function determined by this restriction is one-to-one and has an inverse.

We used this method to form the functions g and h in Figure 8.

EXAMPLE 4 Finding the Inverse of a Function

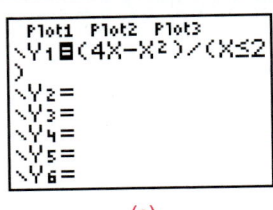

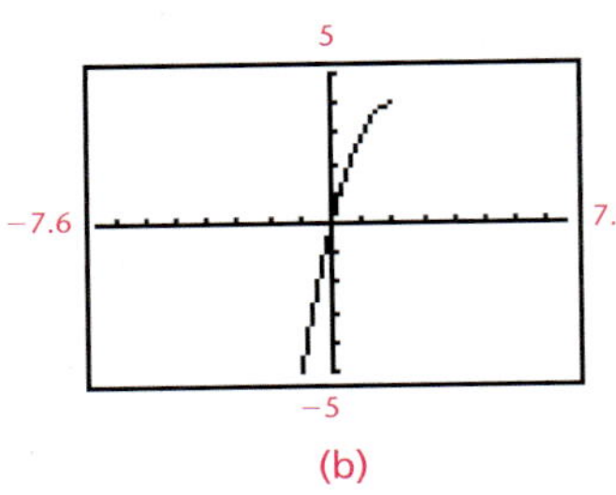

FIGURE 9

Find the inverse of $f(x) = 4x - x^2, x \leq 2$. Graph f, f^{-1}, and $y = x$ in a squared viewing window on a graphing utility and then sketch the graph by hand, adding appropriate labels.

SOLUTION

Step 1. Find the domain of f and verify that f is one-to-one. We are given that the domain of f is $(-\infty, 2]$. Using the Boolean expression $x \leq 2$ [Fig. 9(a)] to restrict the graph to this domain produces the graph of f in Figure 9(b). This graph shows that f is one-to-one.

Step 2. Solve the equation $y = f(x)$ for x.

$$y = 4x - x^2$$

$$x^2 - 4x = -y \qquad \text{Rearrange terms.}$$

$$x^2 - 4x + 4 = -y + 4 \qquad \text{Add 4 to complete the square on the left side.}$$

$$(x - 2)^2 = 4 - y$$

Taking the square root of both sides of this last equation, we obtain two possible solutions:

$$x - 2 = \pm\sqrt{4 - y}$$

The restricted domain of f tells us which solution to use. Because $x \leq 2$ implies $x - 2 \leq 0$, we must choose the negative square root. Thus,

$$x - 2 = -\sqrt{4 - y}$$

$$x = 2 - \sqrt{4 - y}$$

and we have found

$$x = f^{-1}(y) = 2 - \sqrt{4 - y}$$

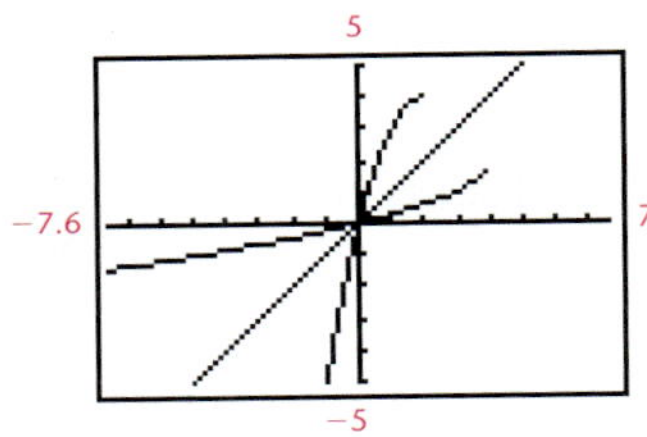

FIGURE 10

Step 3. Interchange x and y.

$$y = f^{-1}(x) = 2 - \sqrt{4 - x}$$

Step 4. Find the domain of f^{-1}. The equation $f^{-1}(x) = 2 - \sqrt{4 - x}$ is defined for $x \leq 4$. From the graph in Figure 9(b), the range of f also is $(-\infty, 4]$. Thus,

$$f^{-1}(x) = 2 - \sqrt{4 - x} \qquad x \leq 4$$

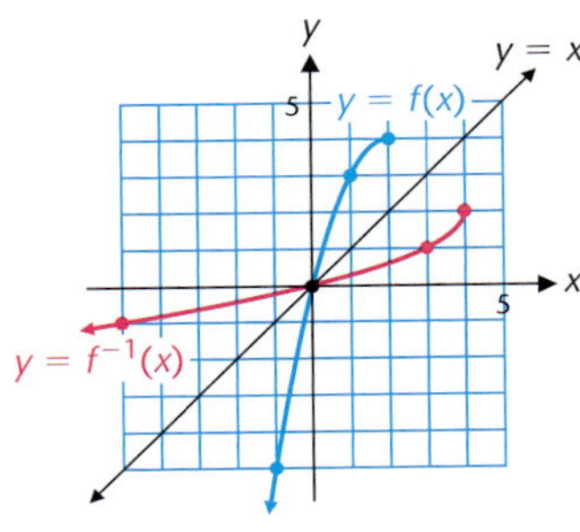

FIGURE 11

The check is left for the reader.

The graphs of f, f^{-1}, and $y = x$ on a graphing utility are shown in Figure 10 and a hand sketch is shown in Figure 11. Note that we plotted several points on the graph of f and their reflections on the graph of f^{-1} to aid in preparing the hand sketch.

MATCHED PROBLEM 4

Find the inverse of $f(x) = 4x - x^2$, $x \geq 2$. Graph f, f^{-1}, and $y = x$ in the same coordinate system.

ANSWERS MATCHED PROBLEMS

1. (A) Not one-to-one
   (B) One-to-one
2. $f^{-1}(x) = x^2 - 2$, $x \geq 0$
3. $R(x) = 10x - 0.001x^2$
4. $f^{-1}(x) = 2 + \sqrt{4 - x}$, $x \leq 4$

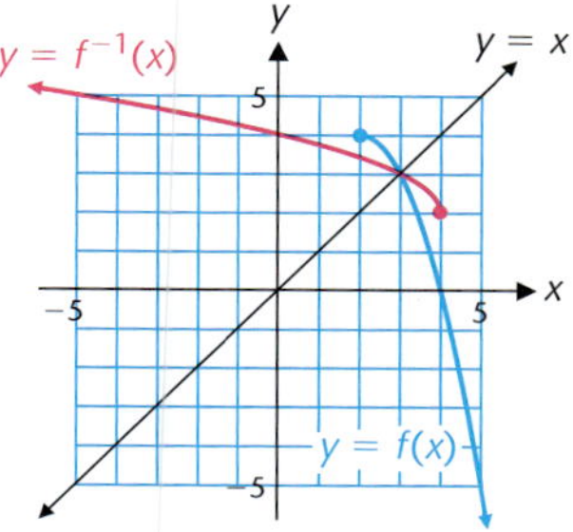

EXERCISE 1.6

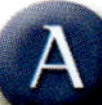

For each set of ordered pairs in Problems 1–6, determine if the set is a function, a one-to-one function, or neither. Reverse all the ordered pairs in each set and determine if this new set is a function, a one-to-one function, or neither.

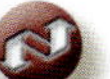

1. $\{(1, 2), (2, 1), (3, 4), (4, 3)\}$
2. $\{(-1, 0), (0, 1), (1, -1), (2, 1)\}$
3. $\{(5, 4), (4, 3), (3, 3), (2, 4)\}$
4. $\{(5, 4), (4, 3), (3, 2), (2, 1)\}$
5. $\{(1, 2), (1, 4), (-3, 2), (-3, 4)\}$
6. $\{(0, 5), (-4, 5), (-4, 2), (0, 2)\}$

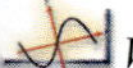

Which of the functions in Problems 7–18 are one-to-one?

7.

Domain	Range
−2 →	−4
−1 →	−2
0 →	0
1 →	1
2 →	5

8.

Domain	Range
−2 →	−3
−1 →	−3
0 →	7
1 →	9
2 →	9

9.

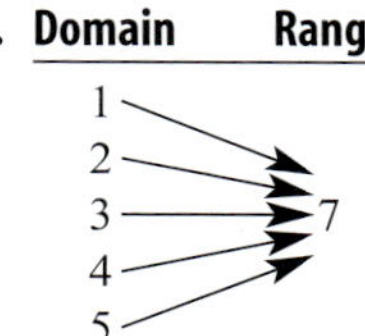

10.

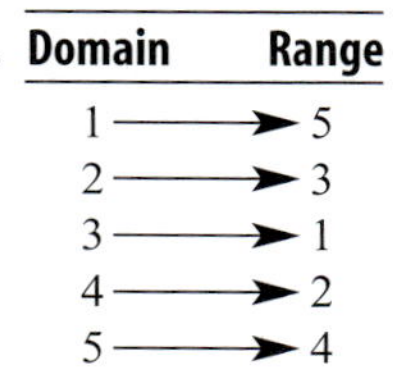

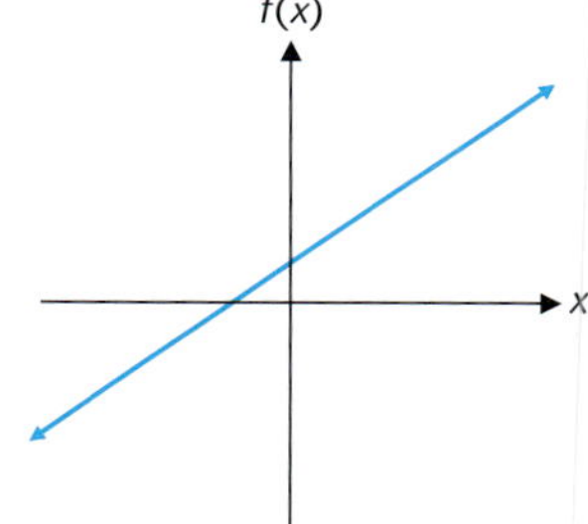

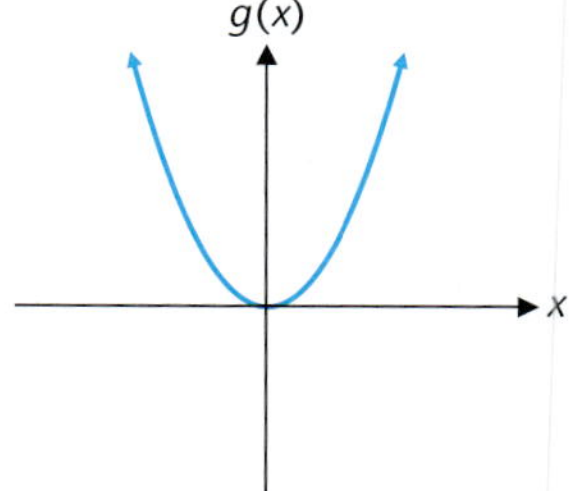

13.

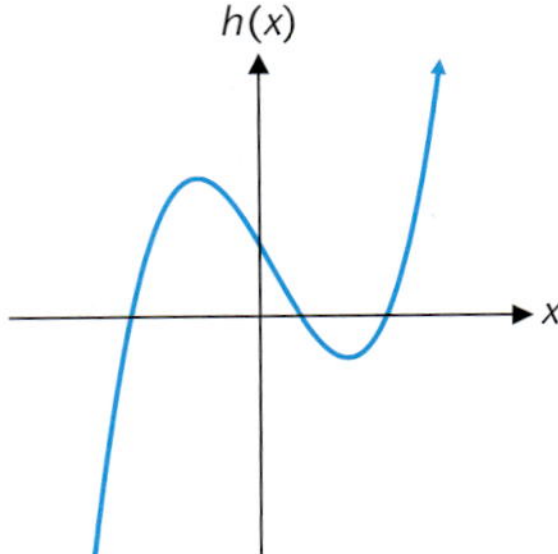

14.

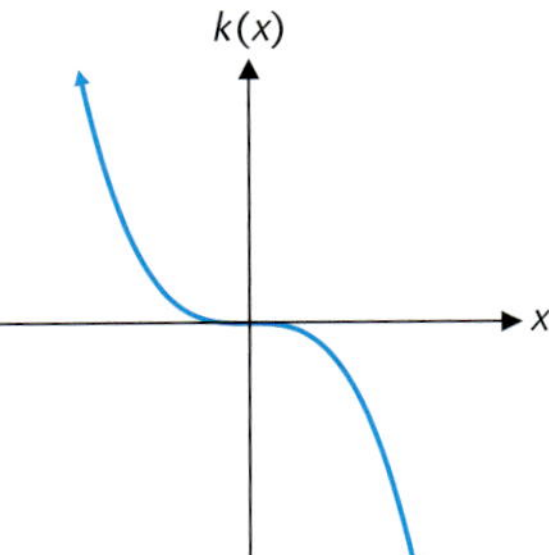

15.

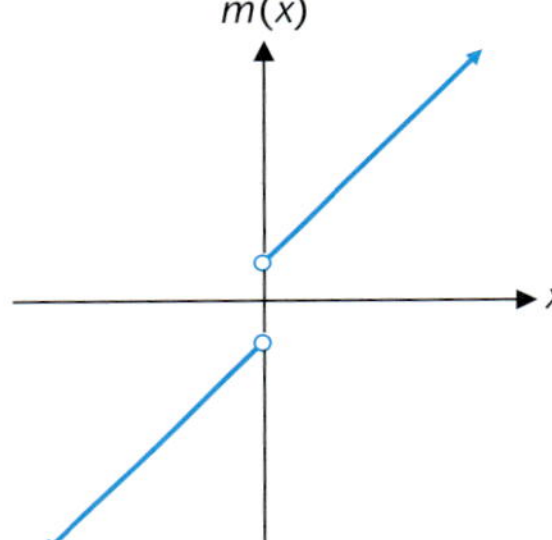

16.

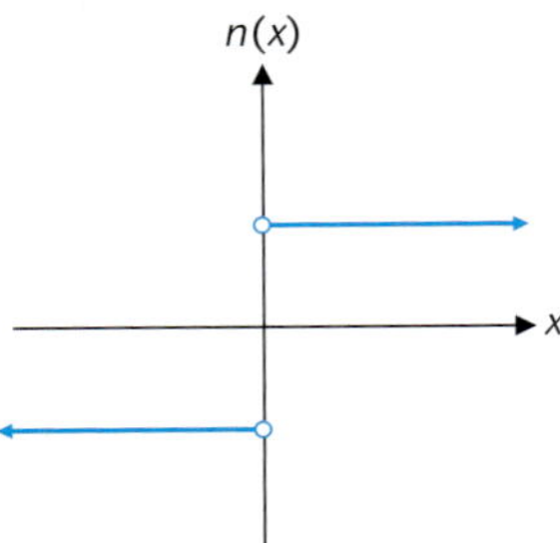

17.

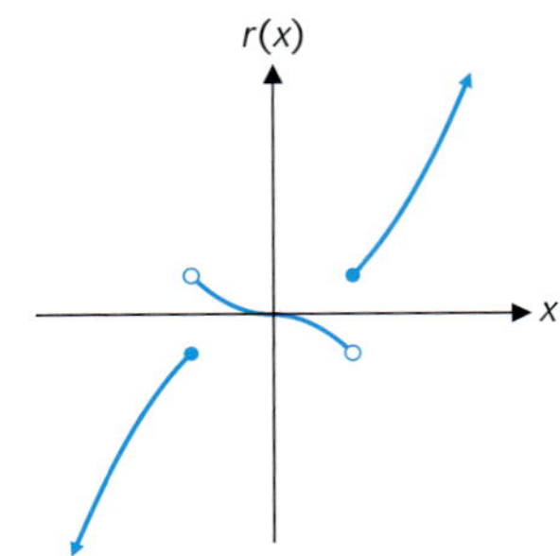

18.

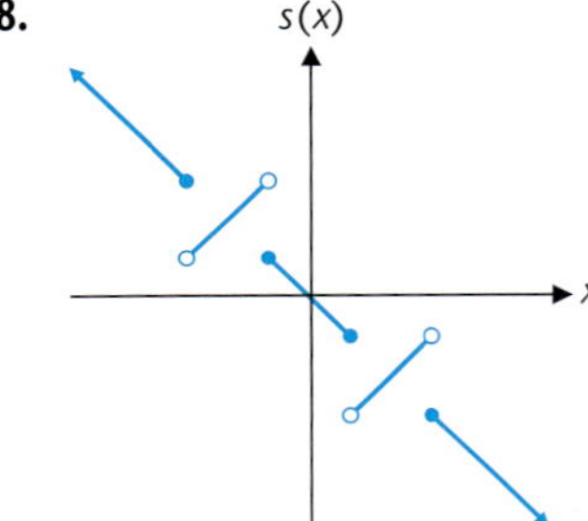

B

In Problems 19–24, use Theorem 1 to determine which functions are one-to-one.

19. $F(x) = \frac{1}{2}x + 2$ **20.** $G(x) = -\frac{1}{3}x + 1$

21. $H(x) = 4 - x^2$ **22.** $K(x) = \sqrt{4 - x}$

23. $M(x) = \sqrt{x + 1}$ **24.** $N(x) = x^2 - 1$

In Problems 25–32, use the horizontal line test (Theorem 2) to determine which functions are one-to-one.

25. $f(x) = \dfrac{x^2 + |x|}{x}$ **26.** $f(x) = \dfrac{x^2 - |x|}{x}$

27. $f(x) = \dfrac{x^3 + |x|}{x}$ **28.** $f(x) = \dfrac{|x|^3 + |x|}{x}$

29. $f(x) = \dfrac{x^2 - 4}{|x - 2|}$ **30.** $f(x) = \dfrac{1 - x^2}{|x + 1|}$

31. $f(x) = \dfrac{x^3 - 9x}{|x^2 - 9|}$ **32.** $f(x) = \dfrac{4x - x^3}{|x^2 - 4|}$

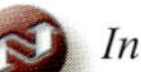

In Problems 33–38, determine if g is the inverse of f.

33. $f(x) = 3x + 5; \quad g(x) = \frac{1}{3}x - \frac{5}{3}$

34. $f(x) = 2x - 4; \quad g(x) = \frac{1}{2}x - 2$

35. $f(x) = 2 - (x + 1)^3; \quad g(x) = \sqrt[3]{3 - x} - 1$

36. $f(x) = (x - 3)^3 + 4; \quad g(x) = \sqrt[3]{x - 4} + 3$

37. $f(x) = \dfrac{2x - 3}{x + 4}$; $g(x) = \dfrac{3 + 4x}{2 - x}$

38. $f(x) = \dfrac{x + 1}{2x - 3}$; $g(x) = \dfrac{3x + 1}{2x + 1}$

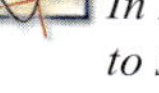

In Problems 39–44, write a verbal description of the given function, reverse your description, and write the resulting algebraic equation. Verify that the result is the inverse of the original function.

39. $h(x) = 3x - 7$

40. $k(x) = 6 - 9x$

41. $m(x) = \sqrt[3]{x + 11}$

42. $n(x) = \sqrt[5]{2 + x}$

43. $s(x) = (3x + 17)^5$

44. $t(x) = (2x + 7)^3$

In Problems 45–48, use the graph of the one-to-one function f to sketch the graph of f^{-1}. State the domain and range of f^{-1}.

45.

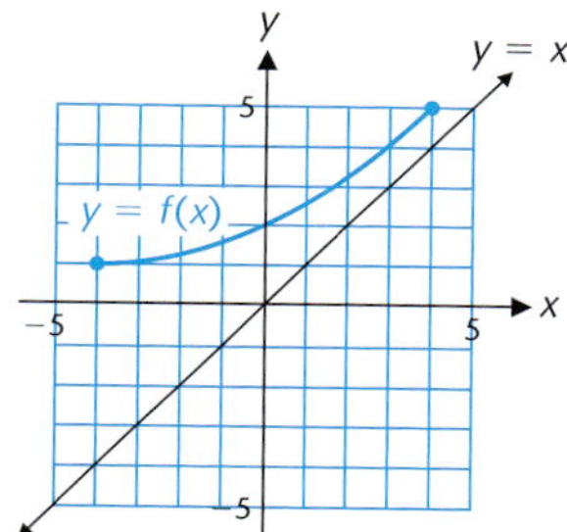

46.

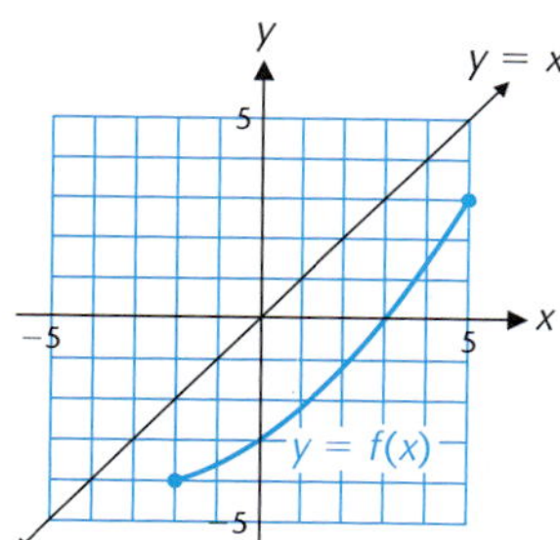

47.

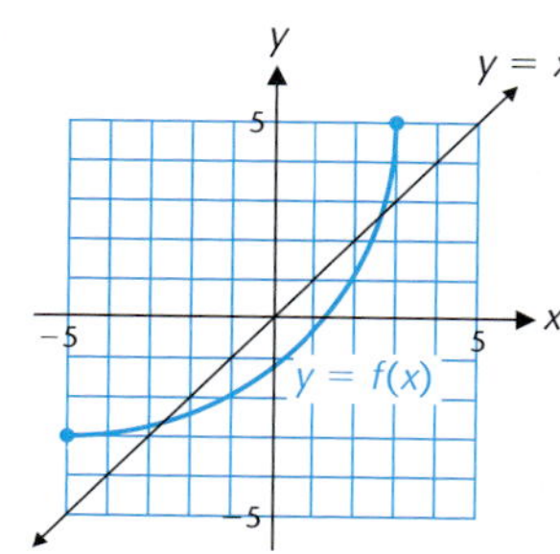

48.

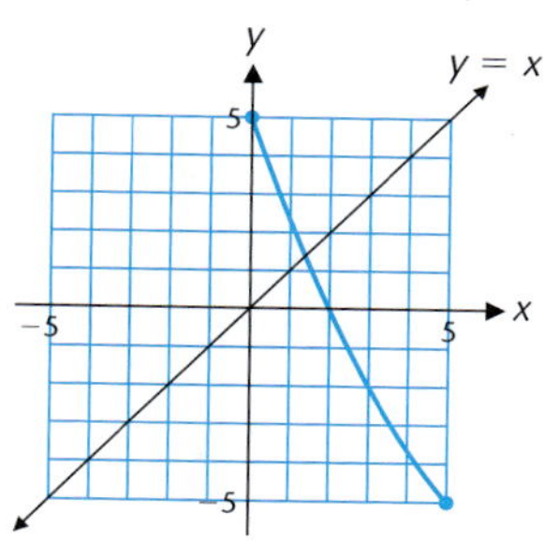

In Problems 49–54, verify that g is the inverse of the one-to-one function f by showing that $g(f(x)) = x$ and $f(g(x)) = x$. Sketch the graphs of f, g, and $y = x$ in the same coordinate system and identify each graph.

49. $f(x) = 3x + 6$; $g(x) = \frac{1}{3}x - 2$

50. $f(x) = -\frac{1}{2}x + 2$; $g(x) = -2x + 4$

51. $f(x) = 4 + x^2, x \geq 0$; $g(x) = \sqrt{x - 4}$

52. $f(x) = \sqrt{x + 2}$; $g(x) = x^2 - 2, x \geq 0$

53. $f(x) = -\sqrt{x - 2}$; $g(x) = x^2 + 2, x \leq 0$

54. $f(x) = 6 - x^2, x \leq 0$; $g(x) = -\sqrt{6 - x}$

The functions in Problems 55–74 are one-to-one. Find f^{-1}.

55. $f(x) = 3x$

56. $f(x) = \frac{1}{2}x$

57. $f(x) = 4x - 3$

58. $f(x) = -\frac{1}{3}x + \frac{5}{3}$

59. $f(x) = \frac{1}{10}x + \frac{3}{5}$

60. $f(x) = -2x - 7$

61. $f(x) = \dfrac{2}{x - 1}$

62. $f(x) = \dfrac{3}{x + 4}$

63. $f(x) = \dfrac{x}{x + 2}$

64. $f(x) = \dfrac{x - 3}{x}$

65. $f(x) = \dfrac{2x + 5}{3x - 4}$

66. $f(x) = \dfrac{5 - 3x}{7 - 4x}$

67. $f(x) = x^3 + 1$

68. $f(x) = x^5 - 2$

69. $f(x) = 4 - \sqrt[5]{x + 2}$

70. $f(x) = \sqrt[3]{x + 3} - 2$

71. $f(x) = \frac{1}{2}\sqrt{16 - x}$

72. $f(x) = -\frac{1}{3}\sqrt{36 - x}$

73. $f(x) = 3 - \sqrt{x - 2}$

74. $f(x) = 4 + \sqrt{5 - x}$

75. How are the x and y intercepts of a function and its inverse related?

76. Does a constant function have an inverse? Explain.

C

The functions in Problems 77–80 are one-to-one. Find f^{-1}.

77. $f(x) = (x - 1)^2 + 2, x \geq 1$

78. $f(x) = 3 - (x - 5)^2,\ x \leq 5$

79. $f(x) = x^2 + 2x - 2, x \leq -1$

80. $f(x) = x^2 + 8x + 7, x \geq -4$

In Problems 81–88, find f^{-1}, find the domain and range of f^{-1}, sketch the graphs of f, f^{-1}, and $y = x$ in the same coordinate system, and identify each graph.

81. $f(x) = -\sqrt{9 - x^2}, 0 \leq x \leq 3$

82. $f(x) = \sqrt{9 - x^2}, 0 \leq x \leq 3$

83. $f(x) = \sqrt{9 - x^2}, -3 \leq x \leq 0$

84. $f(x) = -\sqrt{9 - x^2}, -3 \leq x \leq 0$

85. $f(x) = 1 + \sqrt{1 - x^2}, 0 \leq x \leq 1$

86. $f(x) = 1 - \sqrt{1 - x^2}, 0 \leq x \leq 1$

87. $f(x) = 1 - \sqrt{1 - x^2}, -1 \leq x \leq 0$

88. $f(x) = 1 + \sqrt{1 - x^2}, -1 \leq x \leq 0$

89. Find $f^{-1}(x)$ for $f(x) = ax + b, a \neq 0$.

90. Find $f^{-1}(x)$ for $f(x) = \sqrt{a^2 - x^2}, a > 0, 0 \leq x \leq a$.

91. Refer to Problem 89. For which a and b is f its own inverse?

92. How could you recognize the graph of a function that is its own inverse?

93. Show that the line through the points (a, b) and (b, a), $a \neq b$, is perpendicular to the line $y = x$ (see the figure for Problem 94).

94. Show that the point $((a, + b)/2, (a, + b)/2,)$ bisects the line segment from (a, b), to (b, a), $a \neq b$ (see the figure).

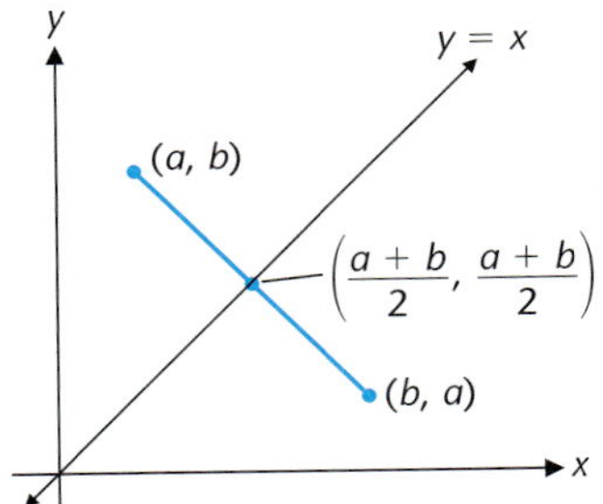

In Problems 95–98, the function f is not one-to-one. Find the inverses of the functions formed by restricting the domain of f as indicated. Check by graphing f, f^{-1}, and the line $y = x$ in a squared viewing window on a graphing utility. [Hint: To restrict the graph of $y = f(x)$ to an interval of the form $a \leq x \leq b$, enter $y = f(x)/((a \leq x)*(x \leq b))$.]

95. $f(x) = (2 - x)^2$:

(A) $x \leq 2$ **(B)** $x \geq 2$

96. $f(x) = (1 + x)^2$:

(A) $x \leq -1$ **(B)** $x \geq -1$

97. $f(x) = \sqrt{4x - x^2}$:

(A) $0 \leq x \leq 2$ **(B)** $2 \leq x \leq 4$

98. $f(x) = \sqrt{6x - x^2}$:

(A) $0 \leq x \leq 3$ **(B)** $3 \leq x \leq 6$

APPLICATIONS

99. Price and Demand. The number q of CD players consumers are willing to buy per week from a retail chain at a price of \$$p$ is given approximately by

$$q = d(p) = \frac{3{,}000}{0.2p + 1} \qquad 10 \leq p \leq 70$$

(A) Find the range of d.

(B) Find $p = d^{-1}(q)$, and find its domain and range.

(C) Should you interchange p and q in part B? Explain.

100. Price and Supply. The number q of CD players a retail chain is willing to supply at a price of \$$p$ is given approximately by

$$q = s(p) = \frac{900p}{p + 20} \qquad 10 \leq p \leq 70$$

(A) Find the range of s.

(B) Find $p = s^{-1}(q)$, and find its domain and range.

(C) Should you interchange p and q in part B? Explain.

101. Revenue. The demand x and the price p (in dollars) for a certain product are related by

$$x = f(p) = 2{,}000 - 40p \qquad 0 \le p \le 50$$

Express the revenue as a function of x.

102. Revenue. The demand x and the price p (in dollars) for a certain product are related by

$$x = f(p) = 3{,}000 - 30p \qquad 0 \le p \le 100$$

Express the revenue as a function of x.

CHAPTER  REVIEW

1.1 Using Graphing Utilities

A **graphing utility** is any electronic device capable of displaying the graph of an equation. The smallest darkened rectangular area that a graphing utility can display is called a **pixel.** The **window variables** for a **standard viewing window** are

$$\text{Xmin} = -10, \text{Xmax} = 10, \text{Xscl} = 1, \text{Ymin} = -10,$$
$$\text{Ymax} = 10, \text{Yscl} = 1$$

Other viewing windows can be defined by assigning different values to these variables. Most graphing utilities will construct a **table** of ordered pairs that satisfy an equation. A **grid** can be added to a graph to aid in reading the graph. A **cursor** is used to locate a single pixel on the screen. The coordinates of the pixel at the cursor location, called **screen coordinates,** approximate the mathematical coordinates of all the points close to the pixel. The **TRACE** command constrains cursor movement to the graph of an equation and displays coordinates of points that satisfy the equation. The **ZOOM** command enlarges or reduces the viewing window. The **intersect** or **isect** command finds the intersection points of two curves.

1.2 Functions

A **function** is a correspondence between two sets of elements such that to each element in the first set, there corresponds one and only one element in the second set. The first set is called the **domain** and the set of all corresponding elements in the second set is called the **range.** Equivalently, a *function* is a **set of ordered pairs** with the property that no two ordered pairs have the same first component and different second components. The *domain* is the set of all first components and the *range* is the set of all second components. An **equation** in two variables **defines a function** if to each value of the **independent variable,** the placeholder for domain values, there corresponds exactly one value of the **dependent variable,** the placeholder for range values. A **vertical line** will intersect the graph of a function in at most one point. Unless otherwise specified, the **domain of a function defined by an equation** is assumed to be the set of all real number replacements for the independent variable that produce real values for the dependent variable. The symbol $f(x)$ represents the real number in the range of the function f that is paired with the domain value x. Equivalently, the ordered pair $(x, f(x))$ belongs to the function f. The **STAT** editor on a graphing utility is used to enter data and the **STAT PLOT** command will produce a **scatter plot** of the data.

1.3 Functions: Graphs and Properties

The **graph of a function** f is the graph of the equation $y = f(x)$. The abscissa of a point where the graph of a function intersects the x axis is called an ***x* intercept** or **real zero** of the function. The x intercept is also a real solution or **root** of the equation $f(x) = 0$. The ordinate of a point where the graph of a function crosses the y axis is called the ***y* intercept** of the function. The y intercept is given by $f(0)$, provided 0 is in the domain of f. Most graphing utilities contain a built-in command, usually called **root** or **zero,** for approximating x intercepts. A solid dot on a graph of a function indicates a point that belongs to the graph and an open dot indicates a point that does not belong to the graph. Dots are also used to indicate that a graph terminates at a point, and arrows are used to indicate that the graph continues with no significant changes.

Let I be an open interval in the domain of a function f. Then,

1. f is **increasing** on I and the graph of f is **rising** on I if $f(a) < f(b)$ whenever $a < b$ in I.
2. f is **decreasing** on I and the graph of f is **falling** on I if $f(a) > f(b)$ whenever $a < b$ in I.
3. f is **constant** on I and the graph of f is **horizontal** on I if $f(a) = f(b)$ whenever $a < b$ in I.

The functional value $f(c)$ is called a **local maximum** if there is an interval (a, b) containing c such that $f(x) \le f(c)$ for all x in (a, b) and a **local minimum** if there is an interval (a, b) containing c such that $f(x) \ge f(c)$ for all x in (a, b). The functional value $f(c)$ is called a **local extremum** if it is either a local maximum or a local minimum. Most graphing utilities have a **maximum** command and a **minimum** command for finding local extrema.

A **piecewise-defined function** is a function whose definition involves more than one formula. Graphs of piecewise-defined functions may have **sharp corners.** The graph of a function is **continuous** if it has no holes or breaks and **discontinuous** at any point where it has a hole or break. Intuitively, the graph of a continuous function can be sketched without lifting a pen from the paper. The **greatest integer** of a real number x, denoted by $[\![x]\!]$, is the largest integer less than or equal to x; that is, $[\![x]\!] = n$, where n is an integer, $n \le x < n + 1$.

The **greatest integer function** f is defined by the equation $f(x) = [\![x]\!]$. Changing the mode on a graphing utility from **connected mode** to **dot mode** makes discontinuities on some graphs more apparent.

1.4 Functions: Graphs and Transformations

The first six basic functions in a library of elementary functions are defined by $f(x) = x$ (identity function), $g(x) = |x|$ (absolute value function), $h(x) = x^2$ (square function), $m(x) = x^3$ (cube function), $n(x) = \sqrt{x}$ (square root function), and $p(x) = \sqrt[3]{x}$ (cube root function) (see Figure 1, Section 1-4). Performing an operation on a function produces a transformation of the graph of the function. The basic transformations are the following:

Vertical Translation:

$y = f(x) + k$ $\begin{cases} k > 0 & \text{Shift graph of } y = f(x) \text{ up } k \text{ units} \\ k < 0 & \text{Shift graph of } y = f(x) \text{ down } |k| \text{ units} \end{cases}$

Horizontal Translation:

$y = f(x + h)$ $\begin{cases} h > 0 & \text{Shift graph of } y = f(x) \text{ left } h \text{ units} \\ h < 0 & \text{Shift graph of } y = f(x) \text{ right } |h| \text{ units} \end{cases}$

Vertical Expansion and Contraction:

$y = Af(x)$ $\begin{cases} A > 1 & \text{Vertically expand the graph of } y = f(x) \text{ by multiplying each } y \text{ value by } A \\ 0 < A < 1 & \text{Vertically contract the graph of } y = f(x) \text{ by multiplying each } y \text{ value by } A \end{cases}$

Horizontal Expansion and Contraction:

$y = f(Ax)$ $\begin{cases} A > 1 & \text{Horizontally contract the graph of } y = f(x) \text{ by multiplying each } x \text{ value by } \dfrac{1}{A} \\ 0 < A < 1 & \text{Horizontally expand the graph of } y = f(x) \text{ by multiplying each } x \text{ value by } \dfrac{1}{A} \end{cases}$

Reflection:

$y = -f(x)$	Reflect the graph of $y = f(x)$ in the x axis
$y = f(-x)$	Reflect the graph of $y = f(x)$ in the y axis
$y = -f(-x)$	Reflect the graph of $y = f(x)$ in the origin

A function f is called an **even function** if $f(x) = f(-x)$ for all x in the domain of f and an **odd function** if $f(-x) = -f(x)$ for all x in the domain of f. The graph of an even function is said to be **symmetrical with respect to the *y* axis** and the graph of an odd function is said to be **symmetrical with respect to the origin.**

1.5 Operations on Functions; Composition

The **sum, difference, product,** and **quotient** of the functions f and g are defined by

$$(f + g)(x) = f(x) + g(x) \qquad (f - g)(x) = f(x) - g(x)$$

$$(fg)(x) = f(x)g(x) \qquad \left(\frac{f}{g}\right)(x) = \frac{f(x)}{g(x)} \qquad g(x) \ne 0$$

The **domain** of each function is the intersection of the domains of f and g, with the exception that values of x where $g(x) = 0$ must be excluded from the domain of f/g.

The **composition** of functions f and g is defined by $(f \circ g)(x) = f(g(x))$. The **domain** of $f \circ g$ is the set of all real numbers x in the domain of g such that $g(x)$ is in the domain of f. The domain of $f \circ g$ is always a subset of the domain of g.

1.6 Inverse Functions

A function is **one-to-one** if no two ordered pairs in the function have the same second component and different first components. A **horizontal line** will intersect the graph of a one-to-one function in at most one point. A function that is increasing (or decreasing) throughout its domain is one-to-one. The **inverse** of the one-to-one function f is the function f^{-1} formed by reversing all the ordered pairs in f.

If f is a one-to-one function, then:

1. f^{-1} is one-to-one.
2. Domain of f^{-1} = Range of f.
3. Range of f^{-1} = Domain of f.
4. $x = f^{-1}(y)$ if and only if $y = f(x)$.
5. $f^{-1}(f(x)) = x$ for all x in the domain of f.
6. $f(f^{-1}(x)) = x$ for all x in the domain of f^{-1}.
7. To find f^{-1}, solve the equation $y = f(x)$ for x. Interchanging x and y at this point is an option.
8. The graphs of $y = f(x)$ and $y = f^{-1}(x)$ are symmetrical with respect to the line $y = x$.

CHAPTER 1 REVIEW EXERCISES

Work through all the problems in this review and check answers in the back of the book. Answers to most review problems are there, and following each answer is a number in italics indicating the section in which that type of problem is discussed. Where weaknesses show up, review appropriate sections in the text.

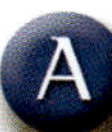

1. Find the smallest viewing window that will contain all the points in the table. State your answer in terms of the window variables.

x	-3	5	-4	0	9
y	2	-6	7	-5	1

2. Indicate whether each set defines a function. Indicate whether any of the functions are one-to-one. Find the domain and range of each function. Find the inverse of any one-to-one functions. Find the domain and range of any inverse functions.

(A) $\{(1, 1), (2, 4), (3, 9)\}$
(B) $\{(1, 1), (1, -1), (2, 2), (2, -2)\}$
(C) $\{(-2, 2), (-1, 2), (0, 2), (1, 2), (2, 2)\}$
(D) $\{(-2, 2), (-1, 3), (0, -1), (1, -2), (2, 1)\}$

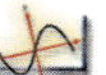
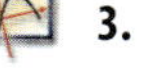

3. Indicate whether each graph specifies a function:

(A)

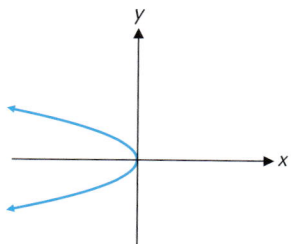

(B)

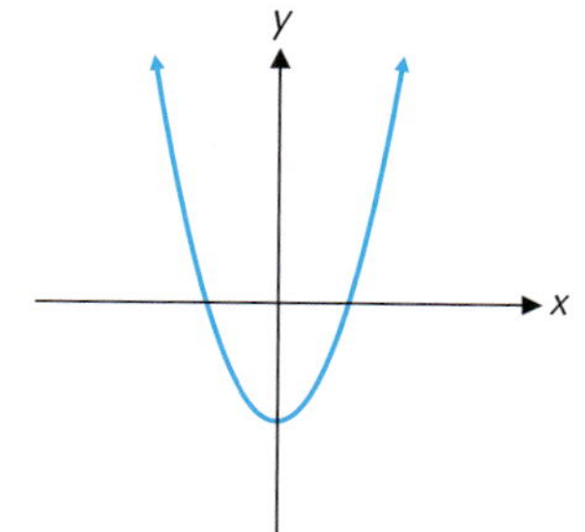

(C)

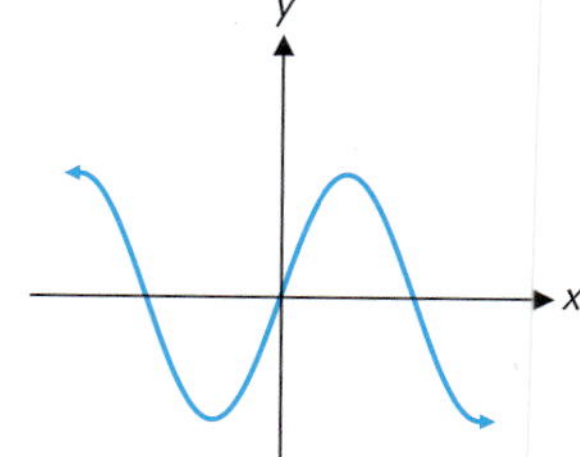

(D)

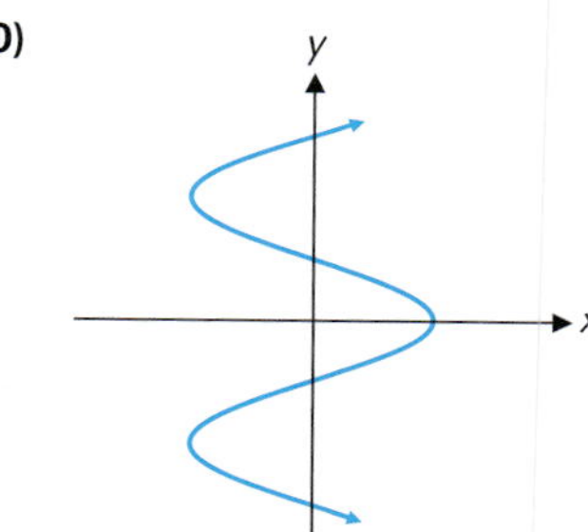

4. For $f(x) = x^2 - 2x$, find:

(A) $f(1)$ **(B)** $f(-4)$ **(C)** $f(2) \cdot f(-1)$ **(D)** $\dfrac{f(0)}{f(3)}$

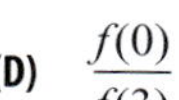

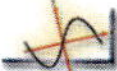

Problems 5–12 refer to the graphs of f and g shown below.

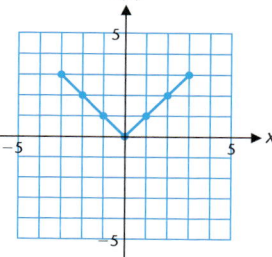

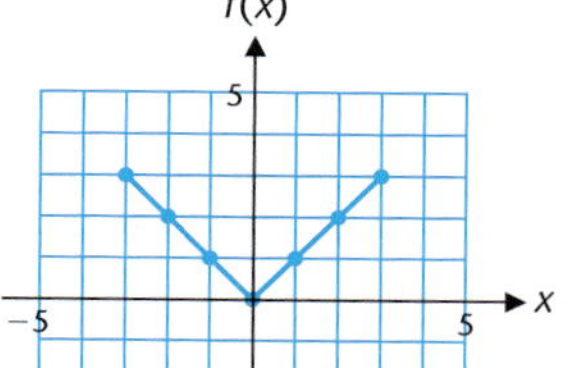

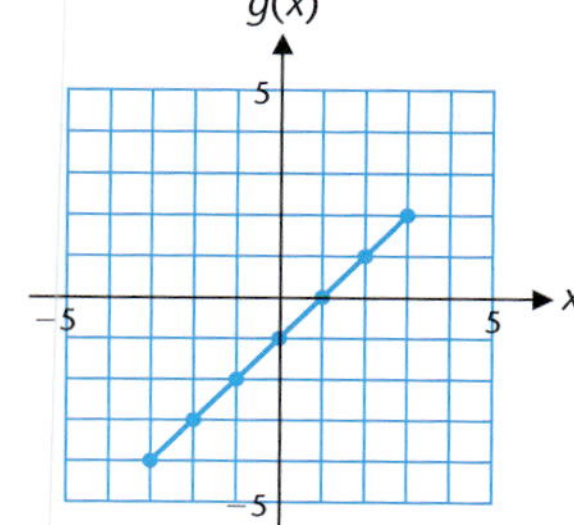

5. Construct a table of values of $(f - g)(x)$ for $x = -3, -2, -1, 0, 1, 2,$ and 3, and sketch the graph of $f - g$.

6. Construct a table of values of $(fg)(x)$ for $x = -3, -2, -1, 0, 1, 2,$ and 3, and sketch the graph of fg.

In Problems 7–10, use the graphs of f and g to find:

7. $(f \circ g)(-1)$

8. $(g \circ f)(-2)$

9. $f[g(1)]$

10. $g[f(-3)]$

11. Is f a one-to-one function?

12. Is g a one-to-one function?

13. Indicate whether each function is even, odd, or neither:
(A) $f(x) = x^5 + 6x$ (B) $g(t) = t^4 + 3t^2$
(C) $h(z) = z^5 + 4z^2$

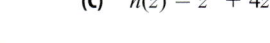

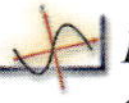

Problems 14–24 refer to the function f given by the following graph.

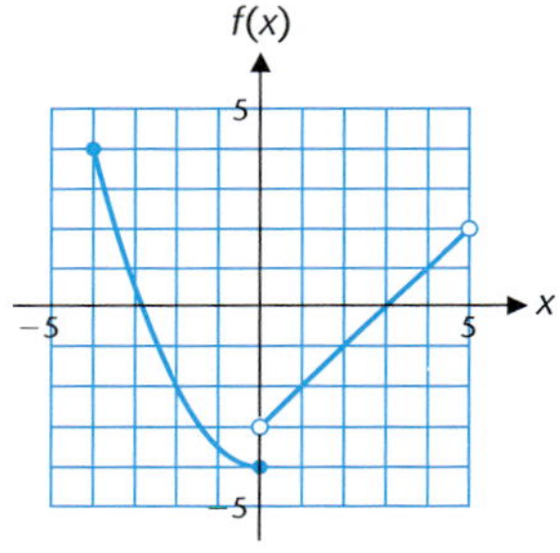

14. Find $f(-4)$, $f(0)$, $f(3)$, and $f(5)$.

15. Find all values of x for which $f(x) = -2$.

16. Find the domain and range of f.

17. Find the intervals over which f is increasing and decreasing.

18. Find any points of discontinuity.

Sketch the graph of each of the following,

19. $f(x) + 1$

20. $f(x + 1)$

21. $-f(x)$

22. $0.5f(x)$

23. $f(2x)$

24. $-f(-x)$

25. Match each equation with a graph of one of the functions f, g, m, or n in the figure. Each graph is a graph of one of the equations and is assumed to continue without bound beyond the viewing window.
(A) $y = (x - 2)^2 - 4$ (B) $y = -(x + 2)^2 + 4$
(C) $y = -(x - 2)^2 + 4$ (D) $y = (x + 2)^2 - 4$

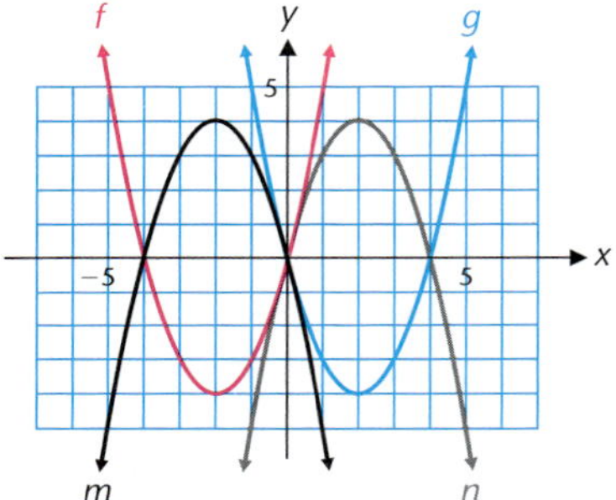

26. Let $f(x) = x^2 - 4$ and $g(x) = x + 3$. Find each of the following functions and find their domains.
(A) f/g (B) g/f (C) $f \circ g$ (D) $g \circ f$

B

Problems 27–33 refer to the function q given by the following graph. (Assume the graph continues as indicated beyond the part shown.)

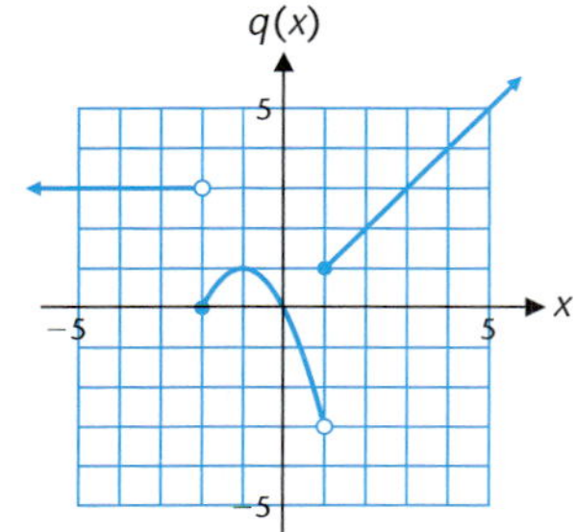

27. Find y to the nearest integer:

(A) $y = q(0)$ **(B)** $y = q(1)$
(C) $y = q(2)$ **(D)** $y = q(-2)$

28. Find x to the nearest integer:

(A) $q(x) = 0$ **(B)** $q(x) = 1$
(C) $q(x) = -3$ **(D)** $q(x) = 3$

29. Find the domain and range of q.

30. Find the intervals over which q is increasing.

31. Find the intervals over which q is decreasing.

32. Find the intervals over which q is constant.

33. Identify any points of discontinuity.

The graphs of each pair of equations in Problems 34 and 35 intersect in exactly two points. Find a viewing window that clearly shows both points of intersection. Use intersect to find the coordinates of each intersection point to two decimal places.

34. $y = x^2 - 20x, y = 4x - 15$

35. $y = \sqrt{10x + 50}, y = 0.3x + 4$

36. Solve the following equation for the indicated values of b. Round answers to two decimal places.

$$0.1x^3 - 2x^2 - 6x + 80 = b$$

(A) $b = 0$ **(B)** $b = 100$
(C) $b = -50$ **(D)** $b = -150$

In Problems 37 and 38, determine if the indicated equation defines a function. Justify your answer.

37. $x + 2y = 10$ **38.** $x + 2y^2 = 10$

39. Find the domain of each of the following functions:

(A) $f(x) = x^2 - 4x + 5$ **(B)** $g(t) = \dfrac{t + 2}{t - 5}$
(C) $h(w) = 2 + 3\sqrt{w}$

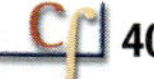

40. If $g(t) = 2t^2 - 3t + 6$, find $\dfrac{g(2 + h) - g(2)}{h}$.

41. The function f multiplies the cube of the domain element by 4 and then subtracts the square root of the domain element. Write an algebraic definition of f.

42. Write a verbal description of the function $f(x) = 3x^2 + 4x - 6$.

In Problems 43 and 44, find the x intercepts, y intercept, local extrema, domain, and range. Round answers to two decimal places.

43. $g(x) = 6\sqrt{x} - x^2$ **44.** $s(x) = x^3 + 27x^2 - 300$

45. Let

$$f(x) = \begin{cases} -x - 5 & \text{for } -4 \le x < 0 \\ 0.2x^2 & \text{for } \quad 0 \le x \le 5 \end{cases}$$

(A) Sketch the graph of $y = f(x)$.
(B) Find the domain and range.
(C) Find any points of discontinuity.
(D) Find the intervals over which f is increasing, decreasing, and constant.

46. Let $f(x) = 0.1x^3 - 6x + 5$. Write a verbal description of the graph of f using increasing and decreasing terminology and indicating any local maximum and minimum values. Approximate to two decimal places the coordinates of any points used in your description.

47. How are the graphs of the following related to the graph of $y = x^2$?

(A) $y = -x^2$ **(B)** $y = x^2 - 3$
(C) $y = (x + 3)^2$ **(D)** $y = (2x)^2$

48. Each of the following graphs is the result of applying one or more transformations to the graph of one of the six basic functions in Figure 1, Section 1.4. Find an equation for the graph. Check by graphing the equation on a graphing utility.

(A)

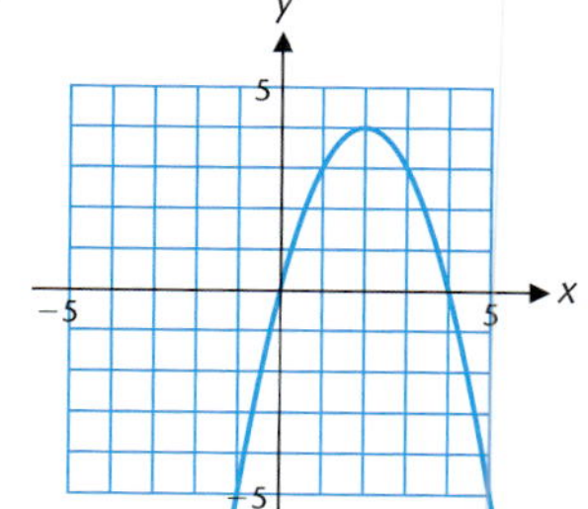

(B)

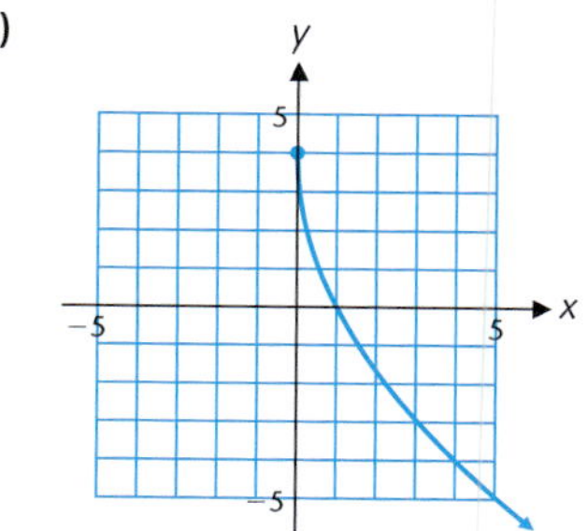

49. The graph of $f(x) = |x|$ is expanded vertically by a factor of 3, reflected in the x axis, shifted four units to the right and eight units up to form the graph of the function g. Find an equation for the function g and graph g.

50. The graph of $m(x) = x^2$ is expanded horizontally by a factor of 2, shifted two units to the left and four units down to form the graph of the function t. Find an equation for the function t and graph t.

51. Is $u(x) = 4x - 8$ the inverse of $v(x) = 0.25x + 2$?

52. Let $k(x) = x^3 + 5$. Write a verbal description of k, reverse your description, and write the resulting algebraic equation. Verify that the result is the inverse of the original function.

53. Find the domain of $f(x) = \dfrac{x}{\sqrt{x} - 3}$.

54. Given $f(x) = \sqrt{x} - 8$ and $g(x) = |x|$,

(A) Find $f \circ g$ and $g \circ f$.

(B) Find the domains of $f \circ g$ and $g \circ f$.

55. Which of the following functions are one-to-one?

(A) $f(x) = x^3$

(B) $g(x) = (x - 2)^2$

(C) $h(x) = 2x - 3$

(D) $F(x) = (x + 3)^2, x \geq -3$

In Problems 56–58, find f^{-1}, find the domain and range of f^{-1}, sketch the graphs of f, f^{-1}, and $y = x$ in the same coordinate system, and identify each graph.

56. $f(x) = 3x - 7$

57. $f(x) = \sqrt{x - 1}$

58. $f(x) = x^2 - 1, x \geq 0$

59. Sketch by hand the graph of a function that is consistent with the given information.

(A) The function f is continuous on $[-5, 5]$, increasing on $[-5, -3]$, decreasing on $[-3, 1]$, constant on $[1, 3]$, and increasing on $[3, 5]$.

(B) The function f is continuous on $[-5, 1)$ and $[1, 5]$, $f(-2) = -1$ is a local maximum, and $f(3) = 2$ is a local minimum.

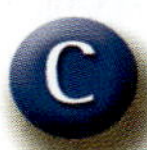

60. Write a verbal description of the function g and then find an equation for $g(t)$.

$$g(t + h) = 2(t + h)^2 - 4(t + h) + 5$$

61. Graph in the standard viewing window:

$$f(x) = 0.1(x - 2)^2 + \frac{|3x - 6|}{x - 2}$$

Assuming the graph continues as indicated beyond the part shown in this viewing window, find the domain, range, and any points of discontinuity. [*Hint:* Use the dot mode on your graphing utility, if it has one.]

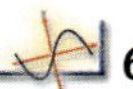

62. A partial graph of the function f is shown in the figure. Complete the graph of f over the interval $[0, 5]$ given that:

(A) f is an even function.

(B) f is an odd function.

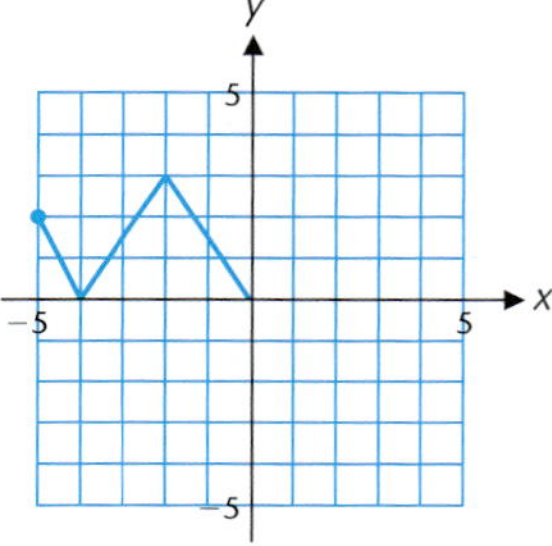

63. For $f(x) = 3x^2 - 5x + 7$, find and simplify:

(A) $\dfrac{f(x + h) - f(x)}{h}$

(B) $\dfrac{f(x) - f(a)}{x - a}$

64. The function f is decreasing on $[-5, 5]$ with $f(-5) = 4$ and $f(5) = -3$.

(A) If f is continuous on $[-5, 5]$, how many times can the graph of f cross the x axis? Support your conclusion with examples and/or verbal arguments.

(B) Repeat part A if the function does not have to be continuous.

65. Let $f(x) = [\![|x|]\!]$.
(A) Write a piecewise definition of f. Include sufficient intervals to clearly illustrate the definition.
(B) Sketch by hand the graph of $y = f(x)$, using a graphing utility as an aid. Include sufficient intervals to clearly illustrate the graph.
(C) Find the range of f.
(D) Find any points of discontinuity.
(E) Indicate whether f is even, odd, or neither.

APPLICATIONS

66. Price and Demand. The price $\$p$ per hot dog at which q hot dogs can be sold during a baseball game is given approximately by

$$p = g(q) = \frac{9}{1 + 0.002q} \qquad 1{,}000 \le q \le 4{,}000$$

(A) Find the range of g.
(B) Find $q = g^{-1}(p)$ and find its domain and range.
(C) Express the revenue as a function of p.
(D) Express the revenue as a function of q.

67. Market Research. If x units of a product are produced each week and sold for a price of $\$p$ per unit, then the weekly demand, revenue, and cost equations are, respectively,

$$x = 500 - 10p$$
$$R(x) = 50x - 0.1x^2$$
$$C(x) = 10x + 1{,}500$$

Express the weekly profit as a function of the price p and find the price that produces the largest profit.

68. Physics: Position of a Moving Object. In flight shooting distance competitions, archers are capable of shooting arrows 600 meters or more. An archer standing on the ground shoots an arrow. After x seconds, the arrow is y meters above the ground as given approximately by

$$y = 55x - 4.88x^2$$

(A) Find the time (to the nearest tenth of a second) the arrow is airborne.
(B) Find the maximum altitude (to the nearest meter) the arrow reaches during its flight.

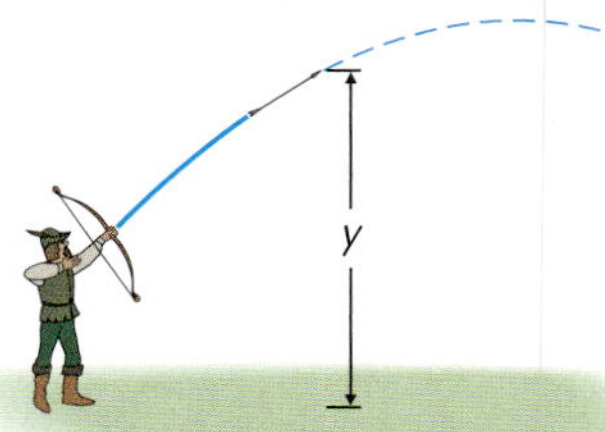

69. Manufacturing. A box with four flaps on each end is to be made out of a piece of cardboard that measures 48 by 72 inches. The width of each flap is x inches and the length of one pair of opposite flaps is $2x$ inches to ensure that the other pair of flaps will meet when folded over to close the box (see the figure). Find the width of the flap (to two decimal places) that will produce a box with maximum volume. What is the maximum volume?

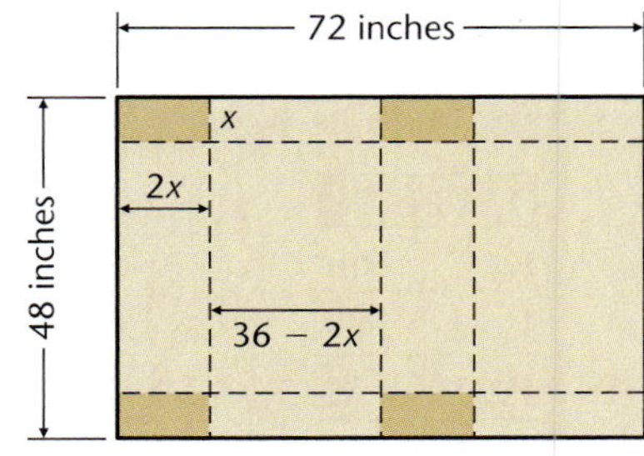

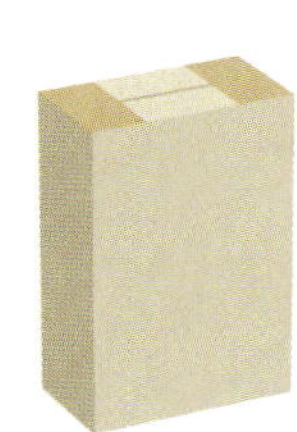

70. Medicine. Proscar is a drug produced by Merck & Co., Inc. to treat symptomatic benign prostate enlargement. One of the long-term effects of the drug is to increase urine flow rate. Results from a 3-year study show that

$$f(x) = 0.00005x^3 - 0.007x^2 + 0.255x$$

is a mathematical model for the average increase in urine flow rate in cubic centimeters per second where x is time taking the drug in months.

(A) Graph this function for $0 \le x \le 36$.

(B) Write a brief verbal description of the graph using increasing, decreasing, local maximum, and local minimum as appropriate. Approximate to two decimal places the coordinates of any points used in your description.

71. Computer Science. In computer programming, it is often necessary to check numbers for certain properties (even, odd, perfect square, etc.). The greatest integer function provides a convenient method for determining some of these properties. Consider the function

$$f(x) = x - ([\![\sqrt{x}]\!])^2$$

(A) Evaluate f for $x = 1, 2, \ldots, 16$.

(B) Find $f(n^2)$, where n is a positive integer.

(C) What property of x does this function determine?

MODELING AND DATA ANALYSIS 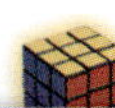

72. Data Analysis. Winning times in the men's Olympic 400-meter freestyle event in minutes for selected years are given in Table 1. A mathematical model for these data is

$$f(x) = -0.021x + 5.57$$

where x is years since 1900.

(A) Compare the model and the data graphically and numerically.

(B) Estimate (to three decimal places) the winning time in 2008.

TABLE 1

Year	Time
1912	5.41
1932	4.81
1952	4.51
1972	4.00
1992	3.75

73. Use the schedule in Table 2 to construct a piecewise-defined model for the taxes due for a single taxpayer in Virginia with a taxable income of x dollars. Find the tax on the following incomes: \$2,000, \$4,000, \$10,000, \$30,000.

TABLE 2 Virginia Tax Rate Schedule

Status	Taxable Income Over	But Not Over	Tax is	Of the Amount Over
Single	\$ 0	\$ 3,000	2%	\$ 0
	\$ 3,000	\$ 5,000	\$ 60 + 3%	\$ 3,000
	\$ 5,000	\$17,000	\$120 + 5%	\$ 5,000
	\$17,000	......	\$720 + 5.75%	\$17,000

Mathematical Modeling: Choosing a Long Distance Calling Plan

The number of companies offering residential long distance telephone service has grown rapidly in recent years. The plans they offer vary greatly and it can be difficult to select the plan that is best for you. Here are five typical plans:

Plan 1: A flat fee of \$50 per month for unlimited calls.

Plan 2: A \$30 per month fee for a total of 30 hours of calls and an additional charge of \$0.01 per minute for all minutes over 30 hours.

Plan 3: A \$5 per month fee and a charge of \$0.04 per minute for all calls.

Plan 4: A \$2 per month fee and a charge of \$0.045 per minute for all calls; the fee is waived if the charge for calls is \$20 or more.

Plan 5: A charge of \$0.05 per minute for all calls; there are no additional fees.

(A) Construct a mathematical model for each plan that gives the total monthly cost in terms of the total number of minutes of calls placed in a month. Graph each model on a graphing utility. You may find Boolean expressions like $(x > a)$ helpful in entering your model in a graphing utility (see Example 4 in Section 1.6).

(B) Compare plans 1 and 2. Determine how many minutes per month would make plan 1 cheaper and how many would make plan 2 cheaper.

(C) Repeat part (B) for plans 1 and 3; plans 1 and 4; plans 1 and 5.

(D) Repeat part (B) for plans 2 and 3; plans 2 and 4; plans 2 and 5.

(E) Repeat part (B) for plans 3 and 4; plans 3 and 5.

(F) Repeat part (B) for plans 4 and 5.

(G) Is there one plan that is always better than all the others? Based on your personal calling history, which plan would you choose and why?

CHAPTER 2

Modeling with Linear and Quadratic Functions

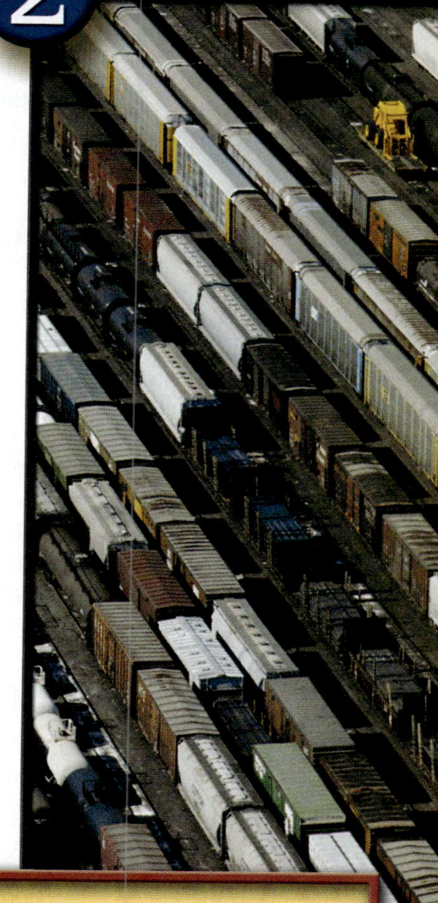

OUTLINE

IN CHAPTER 1 WE INVESTIGATED THE GENERAL CONCEPT OF function using graphs, tables, and algebraic equations. In this and subsequent chapters we investigate particular types of functions in more detail. By the time we finish, we will have a library of elementary functions that form a very important addition to our mathematical toolbox. These elementary functions are used with great frequency in almost any place where mathematics is used: the physical, social, and life sciences; most technical fields; and most mathematical courses beyond this one. Take a few moments to look at the chapter titles in the table of contents and observe how the various types of elementary functions form the structure on which the course is organized.

Preparing for this chapter

Before getting started on this chapter, review the following concepts:

- Properties of Real Numbers (Basic Algebra Review*, Section 1)
- Polynomials (Basic Algebra Review*, Sec. 2 and 3)
- Least Common Denominator (Basic Algebra Review*, Section 4)
- Rational Exponents (Basic Algebra Review*, Section 6)
- Square Root Radicals (Basic Algebra Review*, Section 7)
- Linear Equations and Inequalities (Appendix A, Section A.1)
- Set Operations (Appendix A, Section A.1)
- Cartesian Coordinate System (Appendix A, Section A.2)
- Distance Formula (Appendix A, Section A.3)
- Functions and Graphs (Chapter 1, Sec. 2, 3, and 4)

*At www.mhhe.com/barnett

In this chapter we investigate linear and quadratic functions. As you will see, many significant real-world problems require these functions in their representation and solution. In addition, to find all solutions to quadratic equations, we need to extend the real number system to include complex numbers.

SECTION 2.1 Linear Functions

Constant and Linear Functions • Graph of $Ax + By = C$ • Slope of a Line • Equations of Lines—Special Forms • Parallel and Perpendicular Lines • Mathematical Modeling: Slope as a Rate of Change

The straight line is a fundamental geometric object and an important tool in mathematical modeling. In this section we will add linear functions to our library of elementary functions and explore the relationship between graphs of linear functions and straight lines. We will also determine how to find the equation of a line, given information about the line. And we will see how slope is used to model quantities that have a constant rate of change.

Constant and Linear Functions

One of the elementary functions introduced in Section 1.4 was the identity function $f(x) = x$ (Fig. 1).

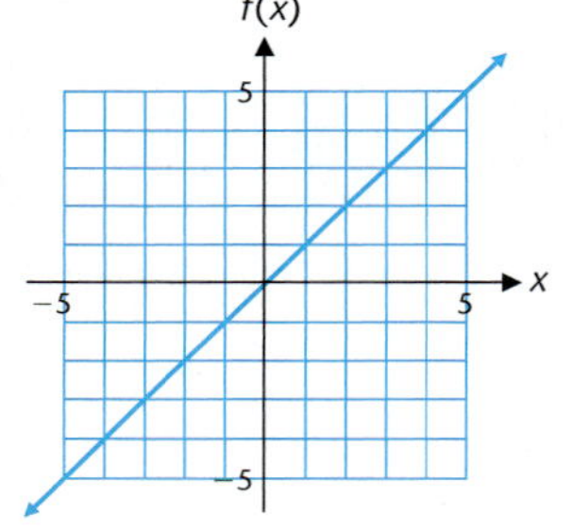

FIGURE 1 Identity function: $f(x) = x$.

EXPLORE/DISCUSS 1

Use the transformations discussed in Section 1.4 to describe verbally the relationship between the graph of $f(x) = x$ and each of the following functions. Graph each function.

(A) $g(x) = 3x + 1$ (B) $h(x) = 0.5x - 2$ (C) $k(x) = -x + 1$

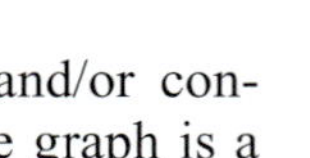

If we apply a sequence of translations, reflections, expansions, and/or contractions to the identity function, the result is always a function whose graph is a straight line. Because of this, functions like g, h, and k in Explore/Discuss 1 are called *linear functions*. In general:

Linear and Constant Functions

A function f is a **linear function** if

$$f(x) = mx + b \qquad m \neq 0$$

where m and b are real numbers. The **domain** is the set of all real numbers and the **range** is the set of all real numbers. If $m = 0$, then f is called a **constant function,**

$$f(x) = b$$

which has the set of all real numbers as its **domain** and the constant b as its **range.**

Figure 2 shows the graphs of two linear functions f and g, and a constant function h.

FIGURE 2 Two linear functions and a constant function.

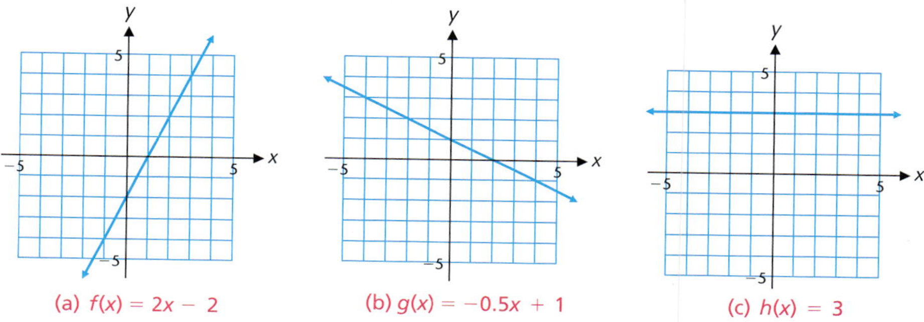

It can be shown (see Problem 90 in Exercise 2.1 for a sketch of a proof) that

The graph of a linear function is a straight line that is neither horizontal nor vertical. The graph of a constant function is a horizontal straight line.

What about vertical lines? Recall from Chapter 1 that the graph of a function cannot contain two points with the same x coordinate and different y coordinates. Because *all* the points on a vertical line have the same x coordinate, the graph of a function can never be a vertical line. Later in this section we will discuss equations of vertical lines, but these equations never define functions.

Recall from Section 1.3 that the y intercept of a function f is $f(0)$, provided $f(0)$ exists, and the x intercepts are the solutions of the equation $f(x) = 0$.

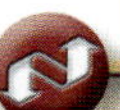

EXPLORE/DISCUSS 2

(A) Is it possible for a linear function to have two x intercepts? No x intercept? If either of your answers is yes, give an example.

(B) Is it possible for a linear function to have two y intercepts? No y intercept? If either of your answers is yes, give an example.

(C) Discuss the possible numbers of x and y intercepts for a constant function.

EXAMPLE 1 **Finding *x* and *y* Intercepts**

Find the x and y intercepts for $f(x) = \frac{2}{3}x - 3$.

SOLUTION

The y intercept is $f(0) = -3$. The x intercept can be found algebraically using standard equation-solving techniques, or graphically using the zero command on a graphing utility.

Algebraic Solution

$$f(x) = 0$$
$$\frac{2}{3}x - 3 = 0$$
$$\frac{2}{3}x = 3$$
$$x = \frac{9}{2} = 4.5$$

Graphical Solution

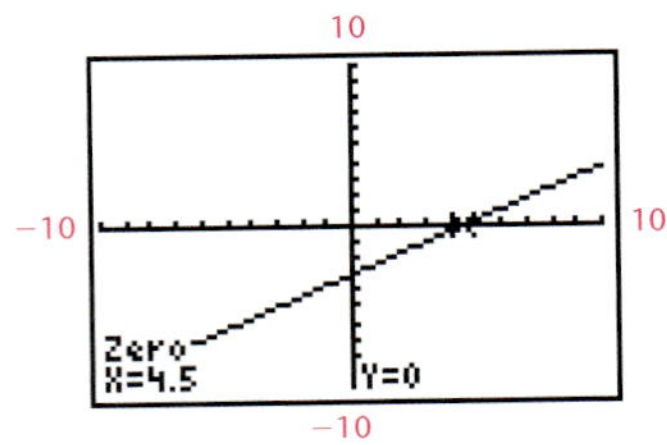

MATCHED 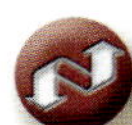 PROBLEM

Find the x and y intercepts of $g(x) = -\frac{4}{3}x + 5$.

Graph of $Ax + By = C$

EXPLORE/DISCUSS 3

Graph each of the following cases of $Ax + By = C$ in the same coordinate system:

1. $3x + 2y = 6$
2. $0x - 3y = 12$
3. $2x + 0y = 10$

Which cases define functions? Explain why or why not.

Graph each case using a graphing utility (check your manual on how to graph vertical lines).

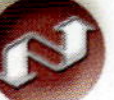

We now investigate graphs of linear equations in two variables:

$$Ax + By = C \tag{1}$$

where A and B are not both zero. Depending on the values of A and B, this equation defines a linear function, a constant function, or no function at all. If $A \neq 0$ and $B \neq 0$, then equation (1) can be written in the form

$$y = -\frac{A}{B}x + \frac{C}{B} \qquad \text{Linear function (slanted line)} \tag{2}$$

which is in the form $f(x) = mx + b$, $m \neq 0$, hence is a linear function. If $A = 0$ and $B \neq 0$, then equation (1) can be written in the form

$$0x + By = C$$

$$y = \frac{C}{B} \quad \text{Constant function (horizontal line)} \tag{3}$$

which is in the form $g(x) = b$, hence is a constant function. If $A \neq 0$ and $B = 0$, then equation (1) can be written in the form

$$Ax + 0y = C$$

$$x = \frac{C}{A} \quad \text{Not a function (vertical line)} \tag{4}$$

We can see that the graph of equation (4) is a vertical line because the equation is satisfied for any value of y as long as x is the constant $\frac{C}{A}$. Hence this form does not define a function.

The following theorem is a generalization of the preceding discussion:

THEOREM 1

Graph of a Linear Equation in Two Variables

The graph of any equation of the form

$$Ax + By = C \quad \text{Standard form} \tag{5}$$

where A, B, and C are real constants (A and B not both 0) is a straight line. Every straight line in a Cartesian coordinate system is the graph of an equation of this type. Vertical and horizontal lines are special cases of equation (5):

Horizontal line with y intercept b: $y = b$

Vertical line with x intercept a: $x = a$

To sketch the graph of an equation of the form

$$Ax + By = C \qquad \text{or} \qquad y = mx + b$$

all that is necessary is to plot any two points from the solution set and use a straightedge to draw a line through these two points. The x and y intercepts are often the easiest points to find.

EXAMPLE 2 Sketching Graphs of Lines

(A) Describe the graphs of $x = -2$ and $y = 3$ verbally. Graph both equations in the same rectangular coordinate system by hand and in the same viewing window on a graphing utility.

(B) Write the equations of the vertical and horizontal lines that pass through the point $(1, -4)$.

(C) Graph the equation $3x - 2y = 6$ by hand and on a graphing utility.

SOLUTIONS

(A) The graph of $x = -2$ is a vertical line with x intercept -2 and the graph of $y = 3$ is a horizontal line with y intercept 3.

Hand-Drawn Solution

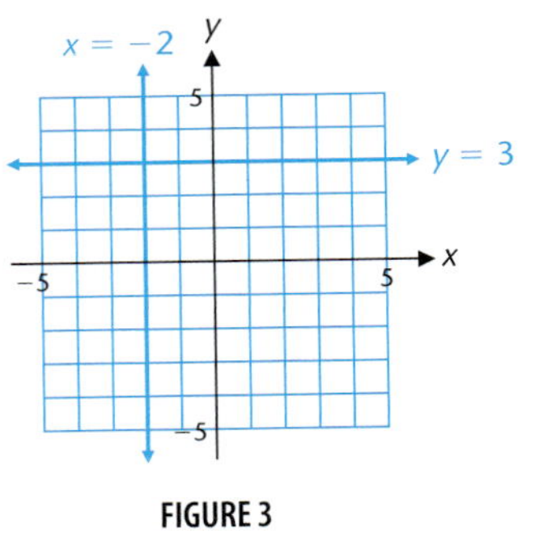

FIGURE 3

Graphing Utility Solution

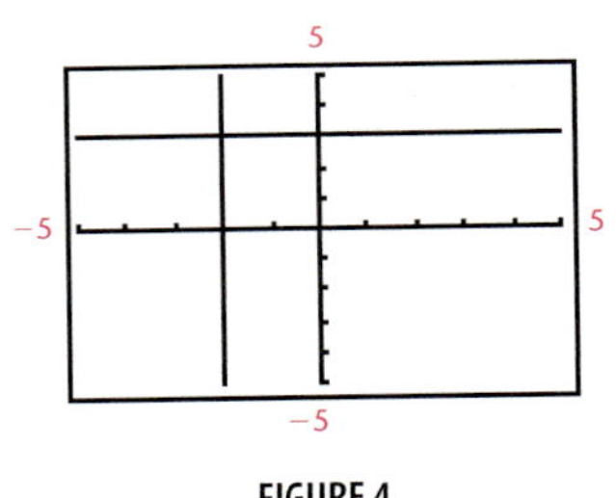

FIGURE 4

(B) Horizontal line through $(1, -4)$: $y = -4$
Vertical line through $(1, -4)$: $x = 1$

(C) Hand-Drawn Solution

Find the x intercept by substituting $y = 0$ and solving for x, and then find the y intercept by substituting $x = 0$ and solving for y. Then draw a line through the intercepts (Fig. 5).

x intercept	y intercept
$3x - 2(0) = 6$	$3(0) - 2y = 6$
$3x = 6$	$-2y = 6$
$x = 2$	$y = -3$

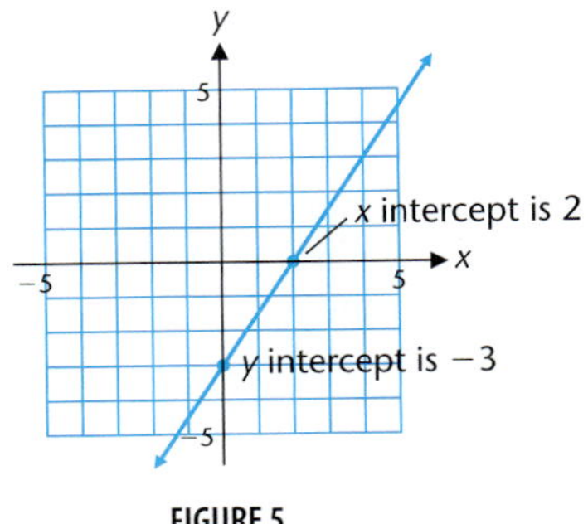

FIGURE 5

(C) Graphing Utility Solution

Solve $3x - 2y = 6$ for y, enter the result in the equation editor, and graph (Fig. 6).

$$
\begin{aligned}
3x - 2y &= 6 \\
-2y &= -3x + 6 \\
y &= 1.5x - 3
\end{aligned}
$$

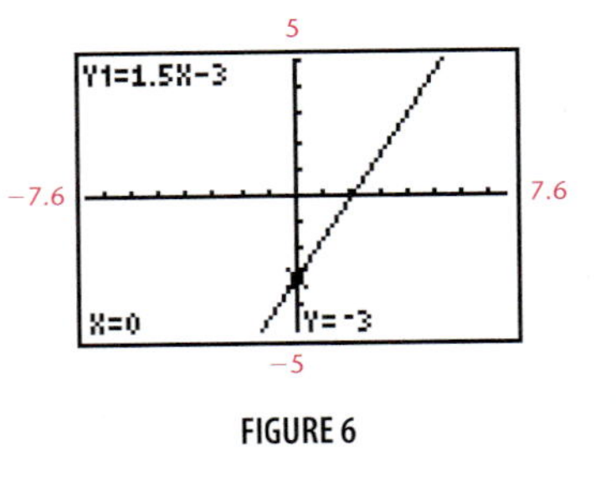

FIGURE 6

Note that we used a *squared viewing window* in Figure 6 to produce units of the same length on both axes. This makes it easier to compare the hand sketch with the graphing utility graph.

MATCHED 2 PROBLEM

(A) Describe the graphs of $x = 4$ and $y = -3$ verbally. Graph both equations in the same rectangular coordinate system by hand and in the same viewing window on a graphing utility.

(B) Write the equations of the vertical and horizontal lines that pass through the point $(-7, 5)$.

(C) Graph the equation $4x + 3y = 12$ by hand and on a graphing utility.

Slope of a Line

If we take two different points $P_1 = (x_1, y_1)$ and $P_2 = (x_2, y_2)$ on a line, then the ratio of the change in y to the change in x as we move from point P_1 to point P_2 is called the **slope** of the line. Roughly speaking, slope is a measure of the "steepness" of a line. Sometimes the change in x is called the **run** and the change in y the **rise.**

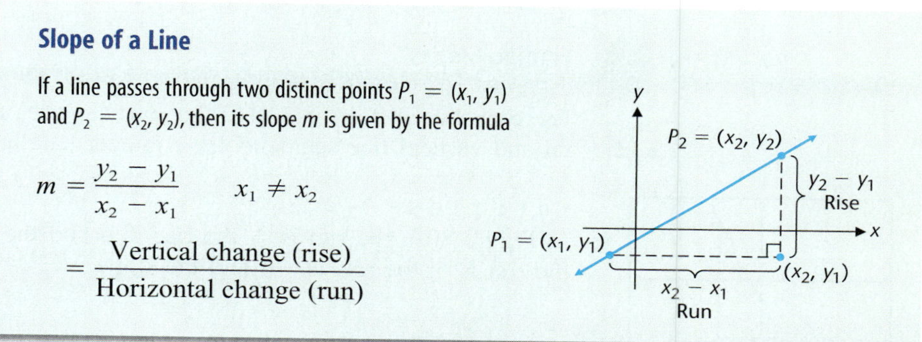

Slope of a Line

If a line passes through two distinct points $P_1 = (x_1, y_1)$ and $P_2 = (x_2, y_2)$, then its slope m is given by the formula

$$m = \frac{y_2 - y_1}{x_2 - x_1} \qquad x_1 \neq x_2$$

$$= \frac{\text{Vertical change (rise)}}{\text{Horizontal change (run)}}$$

For a horizontal line, y doesn't change as x changes; hence, its slope is 0. For a vertical line, x doesn't change as y changes; hence, $x_1 = x_2$, the denominator in the slope formula is 0, and its slope is not defined. In general, the slope of a line may be positive, negative, zero, or not defined. Each case is illustrated geometrically in Table 1.

TABLE 1 Geometric Interpretation of Slope

Line	Slope	Example
Rising as x moves from left to right	Positive	
Falling as x moves from left to right	Negative	
Horizontal	0	
Vertical	Not defined	

In using the formula to find the slope of the line through two points, it doesn't matter which point is labeled P_1 or P_2, because changing the labeling will change the sign in both the numerator and denominator of the slope formula:

$$\frac{y_2 - y_1}{x_2 - x_1} = \frac{y_1 - y_2}{x_1 - x_2}$$

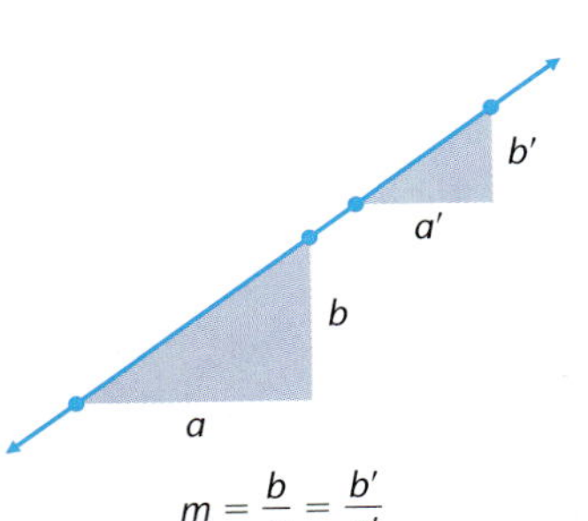

FIGURE 7

For example, the slope of the line through the points (3, 2) and (7, 5) is

$$\frac{5 - 2}{7 - 3} = \frac{3}{4} = \frac{-3}{-4} = \frac{2 - 5}{3 - 7}$$

In addition, it is important to note that the definition of slope doesn't depend on the two points chosen on the line as long as they are distinct. This follows from the fact that the ratios of corresponding sides of similar triangles are equal (Fig. 7).

EXAMPLE 3 Finding Slopes

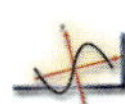
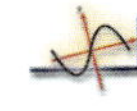

For each line in Figure 8, find the run, the rise, and the slope. (All the horizontal and vertical line segments have integer lengths.)

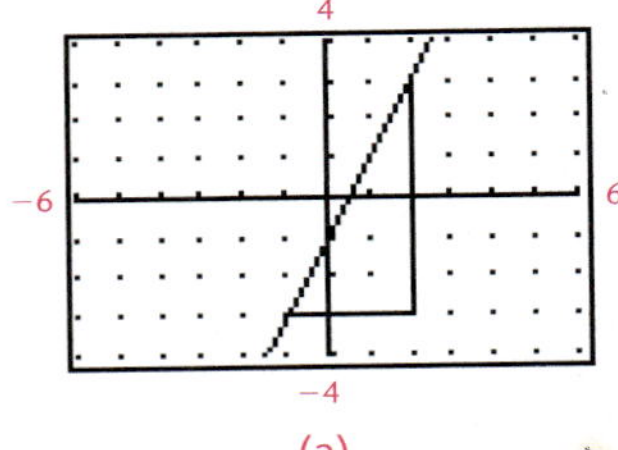

(a)

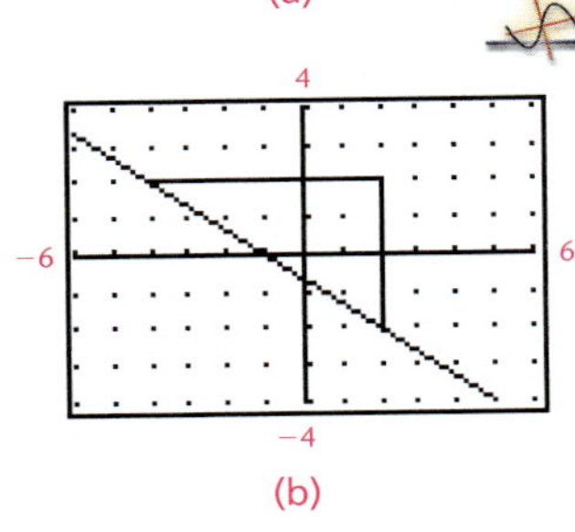

(b)

FIGURE 8

SOLUTION
In Figure 8(a), the run is 3, the rise is 6 and the slope is $\frac{6}{3} = 2$. In Figure 8(b), the run is 6, the rise is -4 and the slope is $\frac{-4}{6} = -\frac{2}{3}$.

MATCHED 3 PROBLEM

For each line in Figure 9, find the run, the rise, and the slope. (All the horizontal and vertical line segments have integer lengths.)

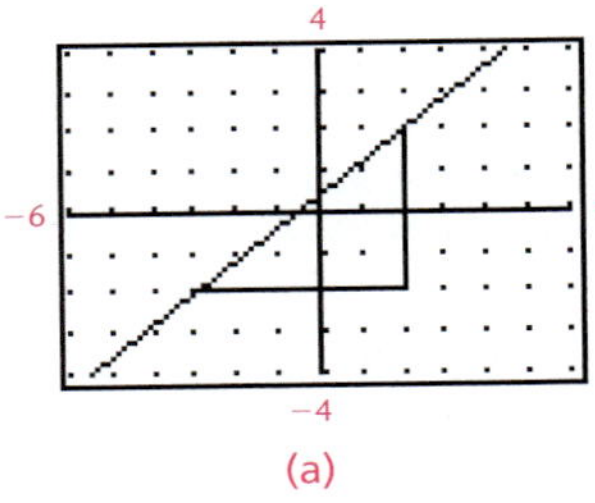

(a)

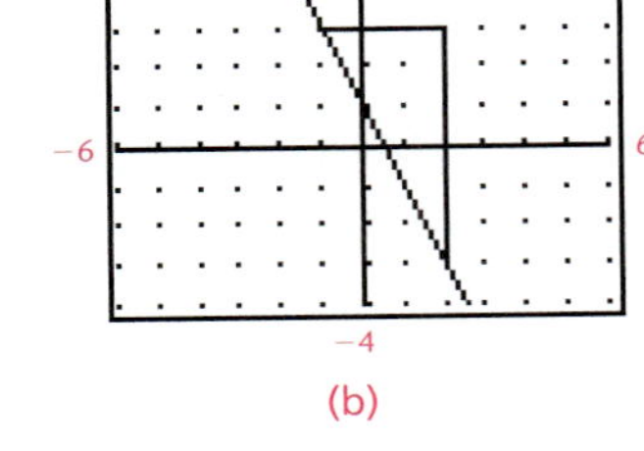

(b)

FIGURE 9

EXAMPLE 4 Finding Slopes

Sketch a line through each pair of points and find the slope of each line.

(A) $(-3, -4), (3, 2)$ (B) $(-2, 3), (1, -3)$
(C) $(-4, 2), (3, 2)$ (D) $(2, 4), (2, -3)$

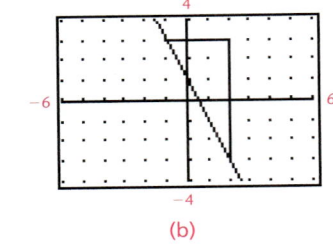

SOLUTIONS

(A)

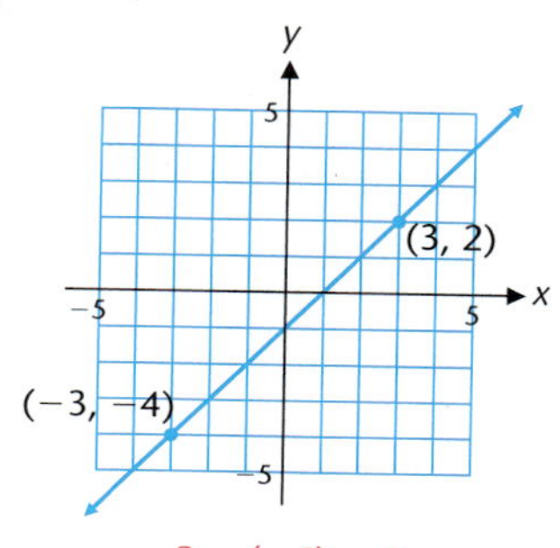

$$m = \frac{2 - (-4)}{3 - (-3)} = \frac{6}{6} = 1$$

(B)

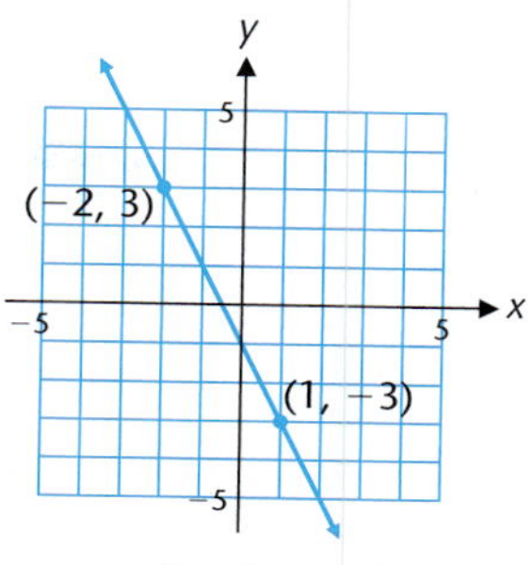

$$m = \frac{-3 - 3}{1 - (-2)} = \frac{-6}{3} = -2$$

(C)

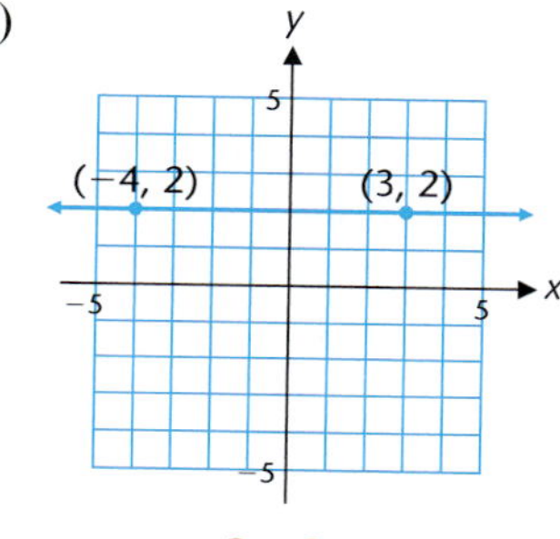

$$m = \frac{2 - 2}{3 - (-4)} = \frac{0}{7} = 0$$

(D)

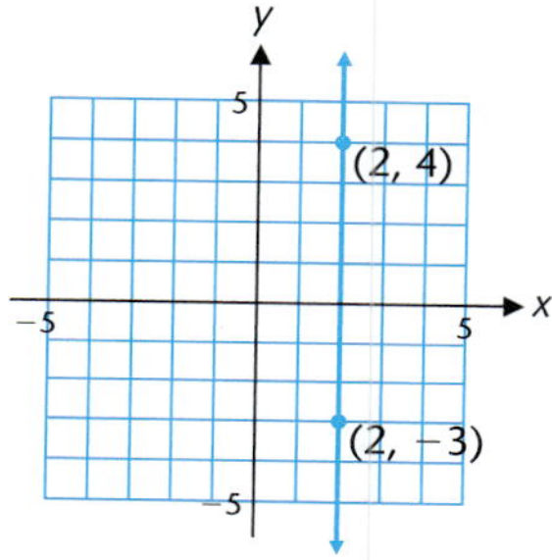

$$m = \frac{-3 - 4}{2 - 2} = \frac{-7}{0}$$

slope is not defined

MATCHED PROBLEM 4

Sketch a line through each pair of points and find the slope of each line.

(A) $(-3, -3), (2, -3)$ (B) $(-2, -1), (1, 2)$

(C) $(0, 4), (2, -4)$ (D) $(-3, 2), (-3, -1)$

The graphs in Example 4 serve to illustrate the following summary:

TABLE 2 Graph Properties of Linear and Constant Functions

Linear Functions		Constant Function
$f(x) = mx + b$, $m > 0$	$f(x) = mx + b$, $m < 0$	$f(x) = b$
Domain $= (-\infty, \infty)$	Domain $= (-\infty, \infty)$	Domain $= (-\infty, \infty)$
Range $= (-\infty, \infty)$	Range $= (-\infty, \infty)$	Range $= \{b\}$
Increasing on $(-\infty, \infty)$	Decreasing on $(-\infty, \infty)$	Constant on $(-\infty, \infty)$

Equations of Lines—Special Forms

Let us start by investigating why $y = mx + b$ is called the *slope–intercept form* for a line.

EXPLORE/DISCUSS 4

(A) Using a graphing utility, graph $y = x + b$ for $b = -5, -3, 0, 3,$ and 5 simultaneously in a standard viewing window. Verbally describe the geometric significance of b.

(B) Using a graphing utility, graph $y = mx - 1$ for $m = -2, -1, 0, 1,$ and 2 simultaneously in a standard viewing window. Verbally describe the geometric significance of m.

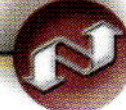

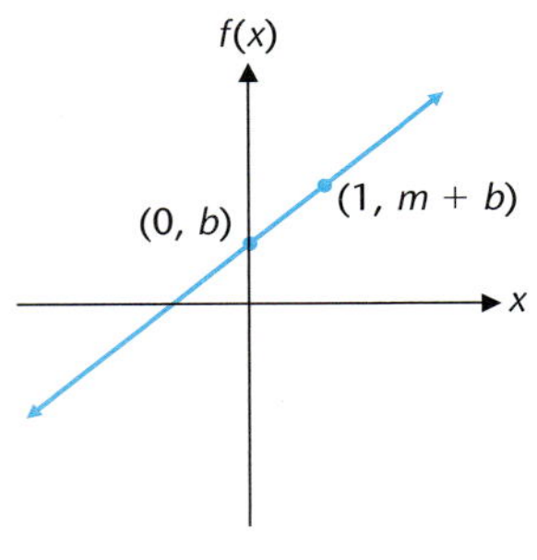

FIGURE 10

As you see, constants m and b in $y = mx + b$ have special geometric significance, which we now explicitly state.

If we let $x = 0$, then $y = b$ and the graph of $y = mx + b$ crosses the y axis at $(0, b)$. Thus, the constant b is the y intercept. For example, the y intercept of the graph of $y = 2x - 7$ is -7.

We have already seen that the point $(0, b)$ is on the graph of $y = mx + b$. If we let $x = 1$, then it follows that the point $(1, m + b)$ is also on the graph (Fig. 10). Because the graph of $y = mx + b$ is a line, we can use these two points to compute the slope:

$$\text{Slope} = \frac{y_2 - y_1}{x_2 - x_1} = \frac{(m + b) - b}{1 - 0} = m \qquad \begin{matrix}(x_1, y_1) = (0, b) \\ (x_2, y_2) = (1, m + b)\end{matrix}$$

Thus, m is the slope of the line with equation $y = mx + b$.

Slope–Intercept Form

The equation

$$y = mx + b \qquad m = \text{slope}, b = y \text{ intercept}$$

is called the **slope–intercept form** of the equation of a line.

The slope and the y intercept in a slope–intercept form are unique. Thus, given the equation $y = 7x - 9$, we can conclude that the slope is 7 and the y intercept is -9.

EXAMPLE 5 Using the Slope–Intercept Form

Graph the line with y intercept -2 and slope $\frac{5}{4}$.

SOLUTION

Hand-Drawn Solution

If we start at the point (0, −2) and move four units to the right (run), then the y coordinate of a point on the line must move up five units (rise) to the point (4, 3). Drawing a line through these two points produces the graph shown in Figure 11.

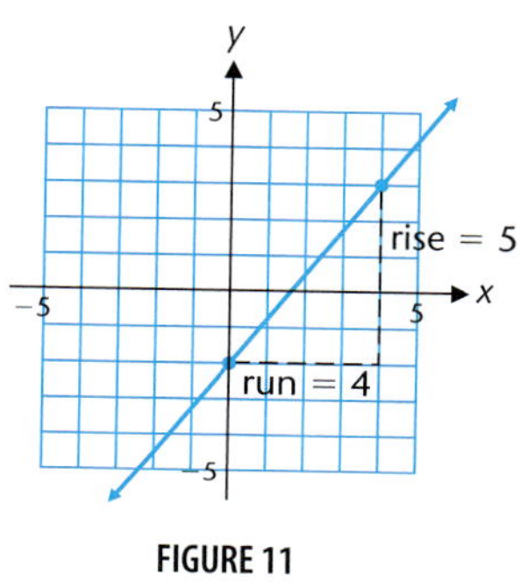

FIGURE 11

Graphing Utility Solution

To graph the line on a graphing utility, we first use the slope–intercept form to find the equation of the line. The equation of a line with y intercept -2 and slope $\frac{5}{4}$ is

$$y = \frac{5}{4}x - 2$$

Graphing this equation on a graphing utility produces the graph in Figure 12.

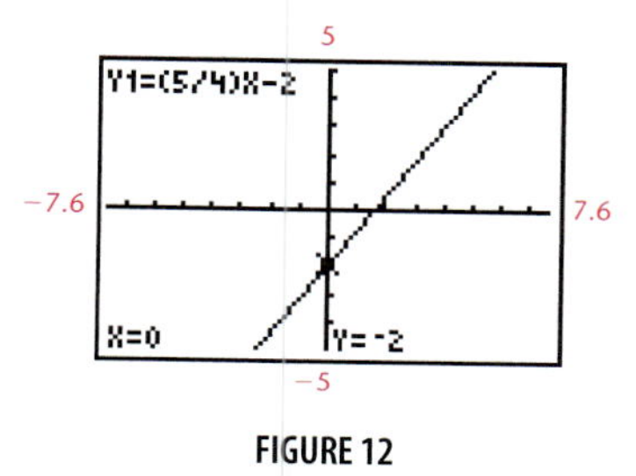

FIGURE 12

MATCHED 5 PROBLEM

Graph the line with y intercept 3 and slope $-\frac{3}{4}$ by hand and on a graphing utility.

Suppose a line has slope m and passes through the point (x_1, y_1). If (x, y) is any other point on the line (Fig. 13), then

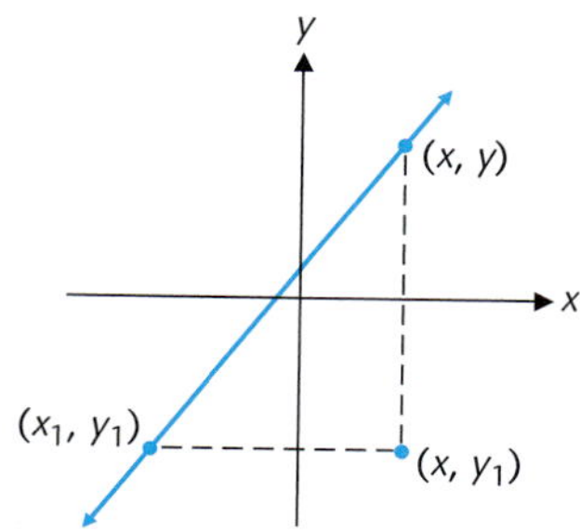

FIGURE 13

$$\frac{y - y_1}{x - x_1} = m$$

that is,

$$y - y_1 = m(x - x_1) \tag{6}$$

Because the point (x_1, y_1) also satisfies equation (6), we can conclude that equation (6) is an equation of a line with slope m that passes through (x_1, y_1).

Point–Slope Form

An equation of a line with slope m that passes through (x_1, y_1) is

$$y - y_1 = m(x - x_1)$$

which is called the **point–slope form** of an equation of a line.

If we are given the coordinates of two points on a line, we can use the given coordinates to find the slope and then use the point–slope form with either of the given points to find the equation of the line.

EXAMPLE 6 Point–Slope Form

(A) Find an equation for the line that has slope $\frac{2}{3}$ and passes through the point $(-2, 1)$. Write the final answer in the form $Ax + By = C$.

(B) Find an equation for the line that passes through the two points $(4, -1)$ and $(-8, 5)$. Write the final answer in the form $y = mx + b$.

SOLUTIONS

(A) If $m = \frac{2}{3}$ and $(x_1, y_1) = (-2, 1)$, then

$$y - y_1 = m(x - x_1)$$

$$y - 1 = \frac{2}{3}[x - (-2)]$$

$$3(y - 1) = 2(x + 2)$$

$$3y - 3 = 2x + 4$$

$$-2x + 3y = 7 \quad \text{or} \quad 2x - 3y = -7$$

(B) First use the slope formula to find the slope of the line:

$$m = \frac{y_2 - y_1}{x_2 - x_1} = \frac{5 - (-1)}{-8 - 4} = \frac{6}{-12} = -\frac{1}{2}$$

Now we choose $(x_1, y_1) = (4, -1)$ and proceed as in part A:

$$y - y_1 = m(x - x_1)$$

$$y - (-1) = -\frac{1}{2}(x - 4)$$

$$y + 1 = -\frac{1}{2}x + 2$$

$$y = -\frac{1}{2}x + 1$$

Verify that choosing $(x_1, y_1) = (-8, 5)$, the other given point, produces the same equation.

MATCHED 6 PROBLEM

(A) Find an equation for the line that has slope $-\frac{2}{5}$ and passes through the point $(3, -2)$. Write the final answer in the form $Ax + By = C$.

(B) Find an equation for the line that passes through the two points $(-3, 1)$ and $(7, -3)$. Write the final answer in the form $y = mx + b$.

The various forms of the equation of a line that we have discussed are summarized in Table 3 for convenient reference. Note that the standard form includes all the other forms as special cases.

TABLE 3 Equations of a Line

Standard form	$Ax + By = C$	A and B not both 0
Slope–intercept form	$y = mx + b$	Slope: m; y intercept: b
Point–slope form	$y - y_1 = m(x - x_1)$	Slope: m; point: (x_1, y_1)
Horizontal line	$y = b$	Slope: 0
Vertical line	$x = a$	Slope: undefined

Parallel and Perpendicular Lines

EXPLORE/DISCUSS 5

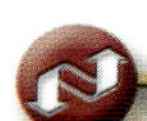

(A) Graph all of the following lines in the same viewing window. Discuss the relationship between these graphs and the slopes of the lines.

$$y = 2x - 5 \qquad y = 2x - 1 \qquad y = 2x + 3$$

(B) Graph each pair of lines in the same *squared* viewing window. Discuss the relationship between each pair of lines and their respective slopes.

$$y = 2x \quad \text{and} \quad y = -0.5x$$

$$y = -3x \quad \text{and} \quad y = \frac{1}{3}x$$

$$y = \frac{4}{5}x \quad \text{and} \quad y = -\frac{5}{4}x$$

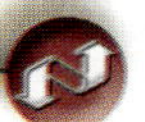

From geometry, we know that two vertical lines are parallel and that a horizontal line and a vertical line are perpendicular to each other. How can we tell when two nonvertical lines are parallel or perpendicular to each other? Theorem 2, which we state without proof, provides a convenient test.

THEOREM 2

Parallel and Perpendicular Lines

Given two nonvertical lines L_1 and L_2, with slopes m_1 and m_2, respectively, then

$$L_1 \parallel L_2 \qquad \text{if and only if } m_1 = m_2$$

$$L_1 \perp L_2 \qquad \text{if and only if } m_1 m_2 = -1$$

The symbols $\parallel$ and $\perp$ mean, respectively, "is parallel to" and "is perpendicular to." In the case of perpendicularity, the condition $m_1m_2 = -1$ can also be written as

$$m_2 = -\frac{1}{m_1} \quad \text{or} \quad m_1 = -\frac{1}{m_2}$$

Thus:

Two nonvertical lines are perpendicular if and only if their slopes are the negative reciprocals of each other.

EXAMPLE 7 Parallel and Perpendicular Lines

Given the line L with equation $3x - 2y = 5$ and the point P with coordinates $(-3, 5)$, find an equation of a line through P that is

(A) Parallel to L (B) Perpendicular to L

SOLUTIONS

First we write the equation for L in the slope–intercept form to find the slope of L:

$$\begin{aligned} 3x - 2y &= 5 \\ -2y &= -3x + 5 \\ y &= \tfrac{3}{2}x - \tfrac{5}{2} \end{aligned}$$

Thus, the slope of L is $\frac{3}{2}$. The slope of a line parallel to L will also be $\frac{3}{2}$, and the slope of a line perpendicular to L will be $-\frac{2}{3}$. We now can find the equations of the two lines in parts A and B using the point–slope form.

(A) Parallel ($m = \frac{3}{2}$):

$$\begin{aligned} y - y_1 &= m(x - x_1) \\ y - 5 &= \tfrac{3}{2}(x + 3) \\ y - 5 &= \tfrac{3}{2}x + \tfrac{9}{2} \\ y &= \tfrac{3}{2}x + \tfrac{19}{2} \end{aligned}$$

(B) Perpendicular ($m = -\frac{2}{3}$):

$$\begin{aligned} y - y_1 &= m(x - x_1) \\ y - 5 &= -\tfrac{2}{3}(x + 3) \\ y - 5 &= -\tfrac{2}{3}x - 2 \\ y &= -\tfrac{2}{3}x + 3 \end{aligned}$$

MATCHED 7 PROBLEM

Given the line L with equation $4x + 2y = 3$ and the point P with coordinates $(2, -3)$, find an equation of a line through P that is

(A) Parallel to L (B) Perpendicular to L

Mathematical Modeling: Slope as a Rate of Change

If (x_1, y_1) and (x_2, y_2) are two distinct points on the graph of $y = mx + b$, then

$$m = \frac{y_2 - y_1}{x_2 - x_1} \qquad \frac{\text{Change in } y}{\text{Change in } x}$$

When viewed as the ratio of the change in y to the change in x, the slope of a linear function is often referred to as the **rate of change** or **average rate of**

change of y with respect to x. This interpretation is used widely in everyday life. For example, if an automobile travels at an average speed of 50 miles per hour for x hours, then the distance traveled is $y = 50x$ miles. If you collect pledges of $25 for each mile you ride in a charity bicycle event, then the total amount the charity will receive if you ride x miles is $y = 25x$ dollars. The next two examples illustrate this important application of linear functions.

EXAMPLE 8 Cost Analysis

A hot dog vendor pays $25 per day to rent a pushcart and $1.25 for the ingredients in one hot dog.

(A) Find the cost of selling x hot dogs in 1 day.

(B) What is the cost of selling 200 hot dogs in 1 day?

(C) If the daily cost is $355, how many hot dogs were sold that day?

SOLUTIONS

(A) The rental charge of $25 is the vendor's **fixed cost**—a cost that is accrued every day and does not depend on the number of hot dogs sold. The cost of ingredients does depend on the number sold. The cost of the ingredients for x hot dogs is $\$1.25x$. This is the vendor's **variable cost**—a cost that depends on the number of hot dogs sold. The total cost for selling x hot dogs is

$$C(x) = 1.25x + 25 \quad \text{Total Cost} = \text{Variable Cost} + \text{Fixed Cost}$$

(B) The cost of selling 200 hot dogs in 1 day is

$$C(200) = 1.25(200) + 25 = \$275$$

(C) The number of hot dogs that can be sold for $355 is the solution of the equation

$$1.25x + 25 = 355$$

Algebraic Solution

$$\begin{aligned} 1.25x + 25 &= 355 \\ 1.25x &= 330 \\ x &= \frac{330}{1.25} \\ &= 264 \text{ hot dogs} \end{aligned}$$

Graphical Solution

Entering $y_1 = 1.25x + 25$ and $y_2 = 355$ in a graphing utility and using the intersect command (Fig. 14) shows that 264 hot dogs can be sold for $355.

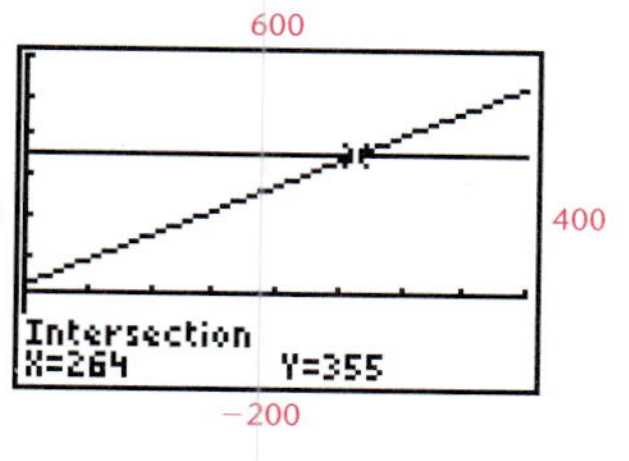

FIGURE 14

MATCHED 8 PROBLEM

It costs a pretzel vendor \$20 per day to rent a cart and \$0.75 for each pretzel.

(A) Find the cost of selling x pretzels in 1 day.

(B) What is the cost of selling 150 pretzels in 1 day?

(C) If the daily cost is \$275, how many pretzels were sold that day?

Refer to Example 8. The vendor's cost increases at the rate of \$1.25 per hot dog. Thus, the rate of change of the cost function $C(x) = 1.25x + 25$ is the slope $m = 1.25$. This constant rate can also be viewed as the cost of selling one additional hot dog. In economics, this quantity is referred to as the **marginal cost.**

EXAMPLE 9 Underwater Pressure

The atmospheric pressure at sea level is 14.7 pounds per square inch. As you descend into the ocean, the pressure increases linearly at a rate of about 0.445 pounds per square inch per foot.

(A) Find the pressure p at a depth of d feet.

(B) If a diver's equipment is rated to be safe up to a pressure of 40 pounds per square inch, how deep (to the nearest foot) is it safe to use this equipment?

SOLUTIONS

(A) Let $p = md + b$. At the surface, $d = 0$ and $p = 14.7$, so $b = 14.7$. The slope m is the given rate of change, $m = 0.445$. Thus, the pressure at a depth of d feet is

$$p = 0.445d + 14.7$$

(B) The safe depth is the solution of the equation

$$0.445d + 14.7 = 40$$

Algebraic Solution

$$\begin{aligned} 0.445d + 14.7 &= 40 \\ 0.445d &= 40 - 14.7 \\ &= 25.3 \\ d &= \frac{25.3}{0.445} \\ &\approx 57 \text{ feet} \end{aligned}$$

Graphical Solution

Entering $y_1 = 0.445x + 14.7$ and $y_2 = 40$ in a graphing utility and using the intersect command (Fig. 15) shows that $p = 40$ when $d \approx 57$ feet.

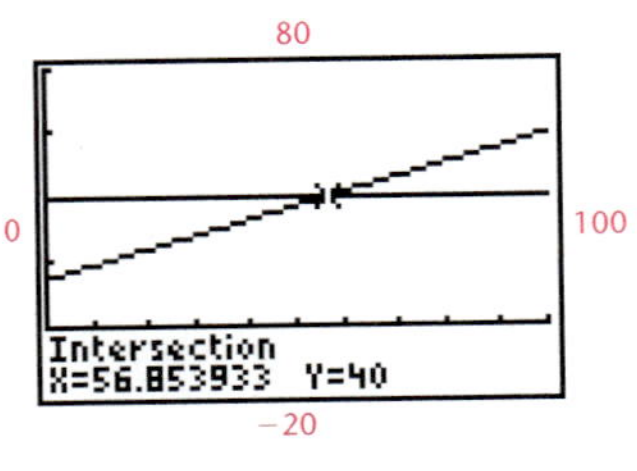

FIGURE 15

MATCHED PROBLEM 9

The rate of change of pressure in freshwater is 0.432 pounds per square inch per foot. Repeat Example 9 for a body of freshwater.

ANSWERS MATCHED PROBLEMS

1. x intercept: $\frac{15}{4} = 3.75$; y intercept: 5
2. (A) The graph of $x = 4$ is a vertical line with x intercept 4. The graph of $y = -3$ is a horizontal line with y intercept -3.

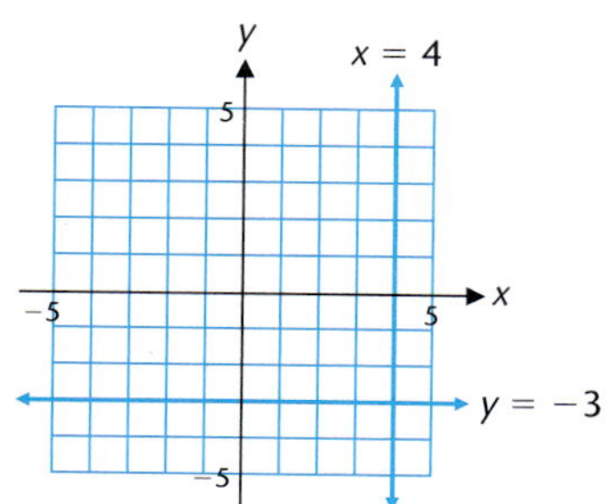

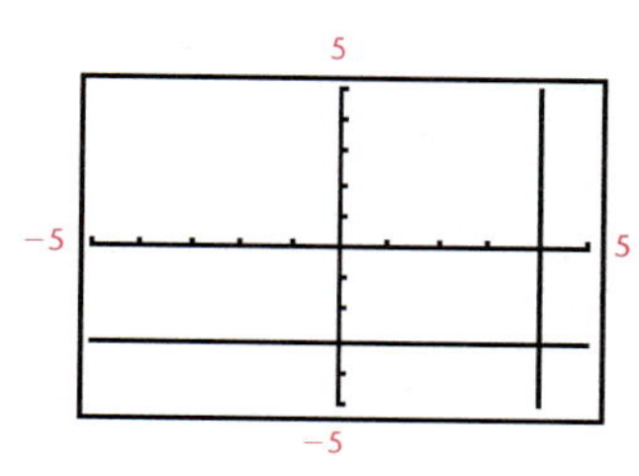

(B) Vertical: $x = -7$; horizontal: $y = 5$

(C)

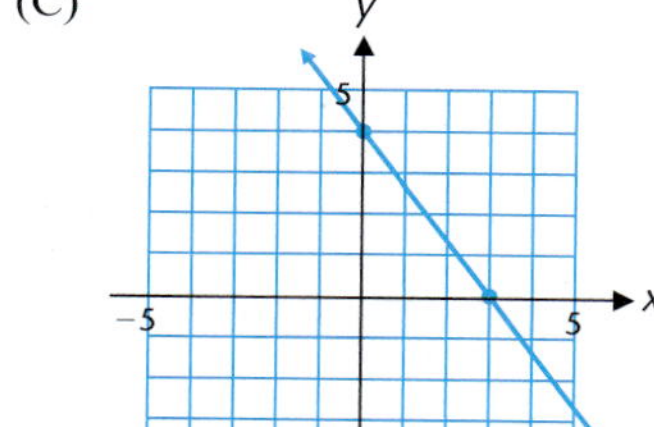

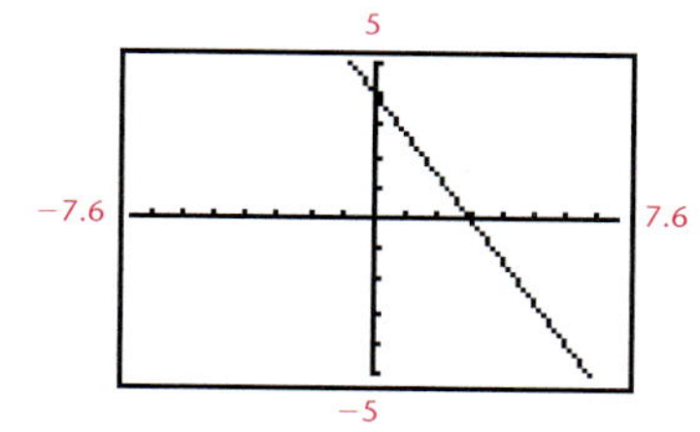

3. (A) Run = 5, rise = 4, slope = $\frac{4}{5} = 0.8$
   (B) Run = 3, rise = -6, slope = $\frac{-6}{3} = -2$
4. (A) $m = 0$ (B) $m = 1$

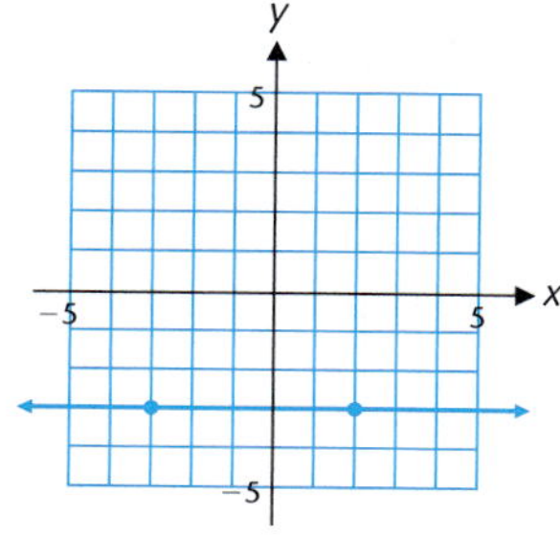

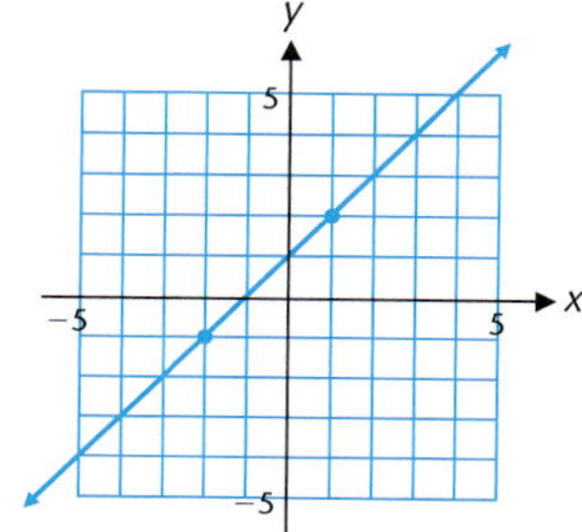

(C) $m = -4$

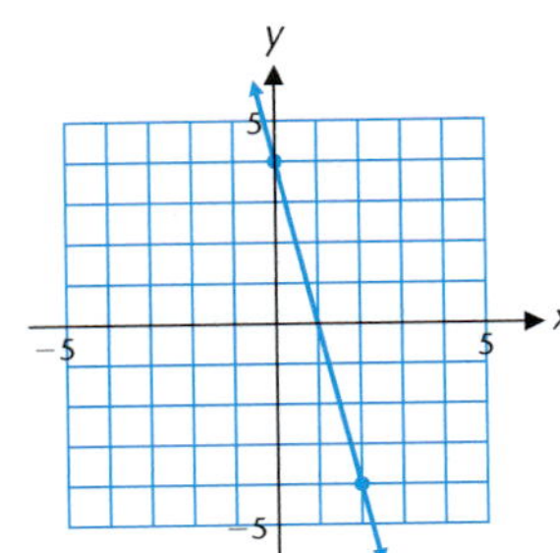

(D) m is not defined

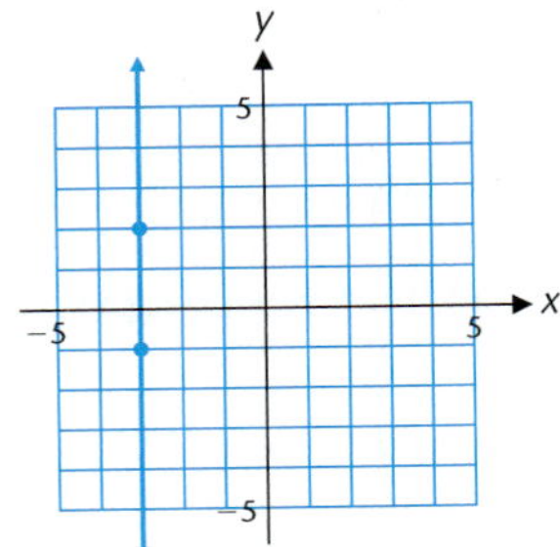

5.

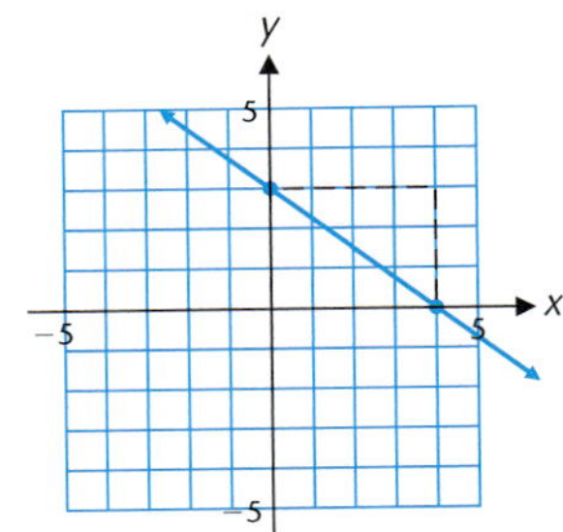

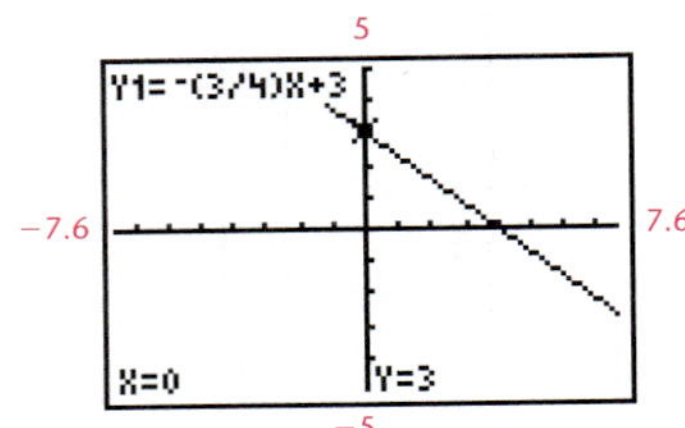

6. (A) $2x + 5y = -4$ (B) $y = -\frac{2}{5}x - \frac{1}{5}$

7. (A) $y = -2x + 1$ (B) $y = \frac{1}{2}x - 4$

8. (A) $C(x) = 0.75x + 20$ (B) \$132.50 (C) 340 pretzels

9. (A) $p = 0.432d + 14.7$ (B) 59 feet

EXERCISE 2.1

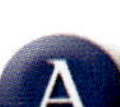

A

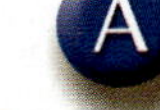

In Problems 1–6, use the graph of each linear function to find the rise, run, and slope. Write the equation of each line in the standard form $Ax + By = C$, $A \geq 0$. (All the horizontal and vertical line segments have integer lengths.)

1.

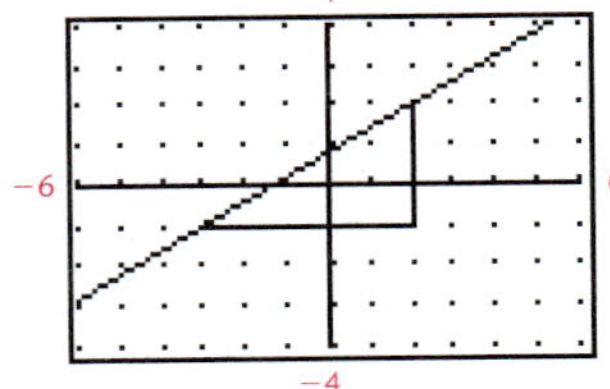

2.

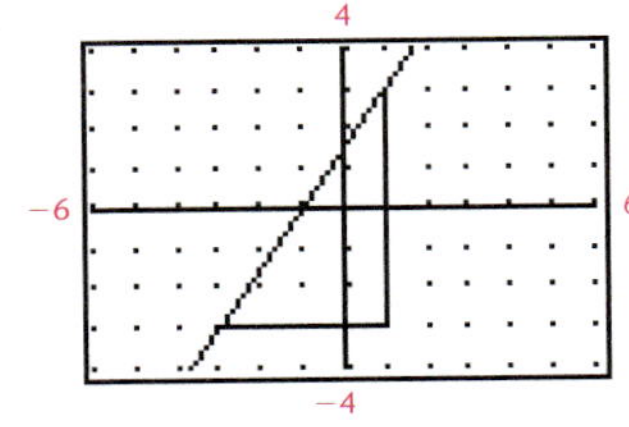

3.

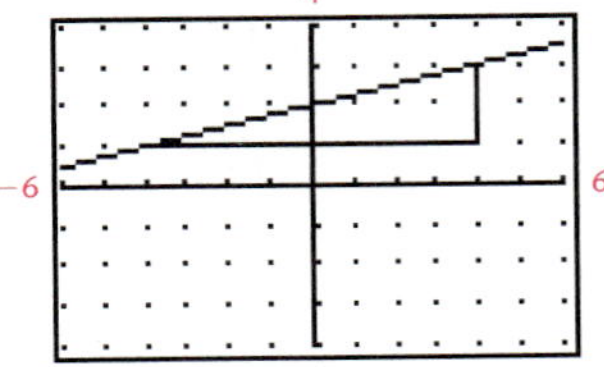

4.

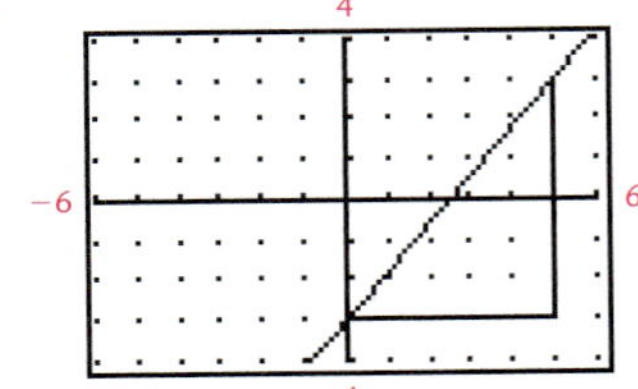

5.

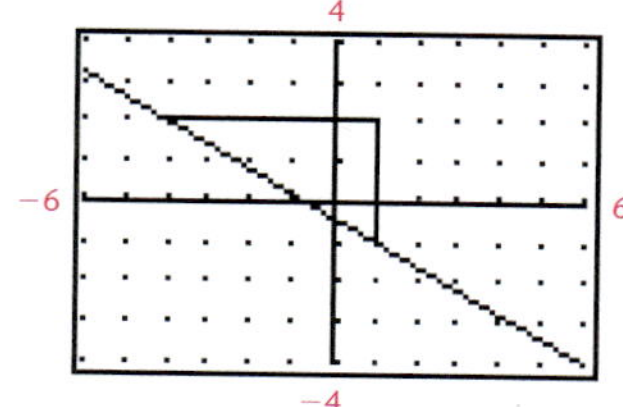

6.

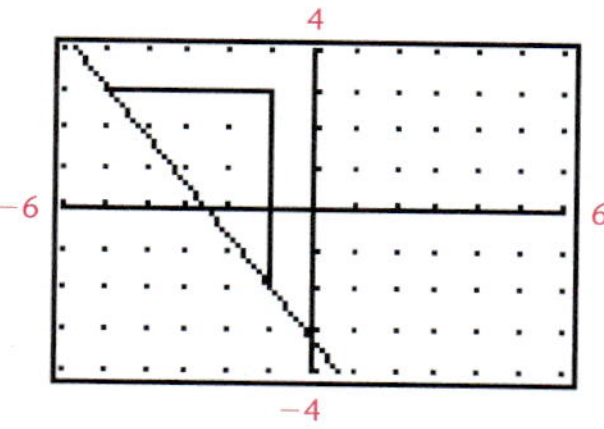

In Problems 7–12, use the graph of each linear function to find the x intercept, y intercept, and slope. Write the slope–intercept form of the equation of each line.

7.

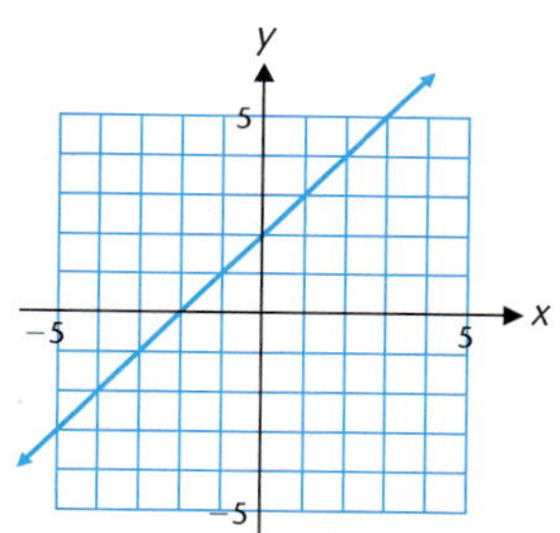

8.

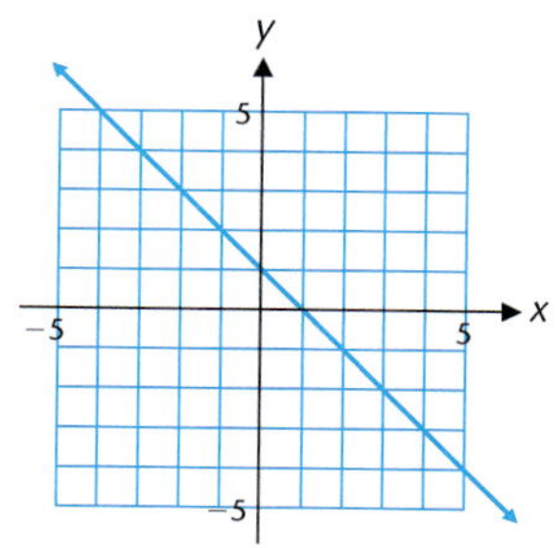

9.

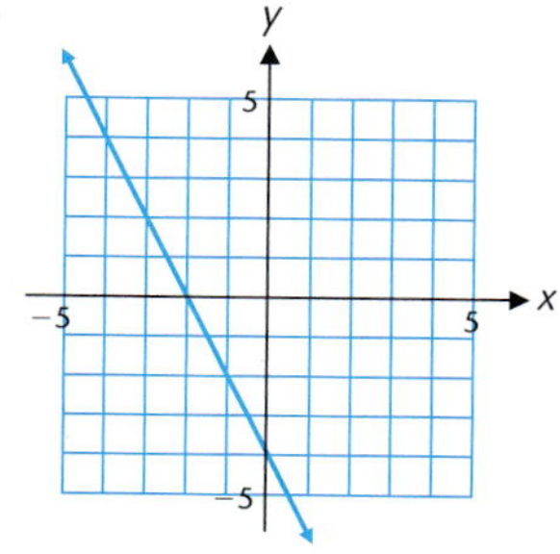

10.

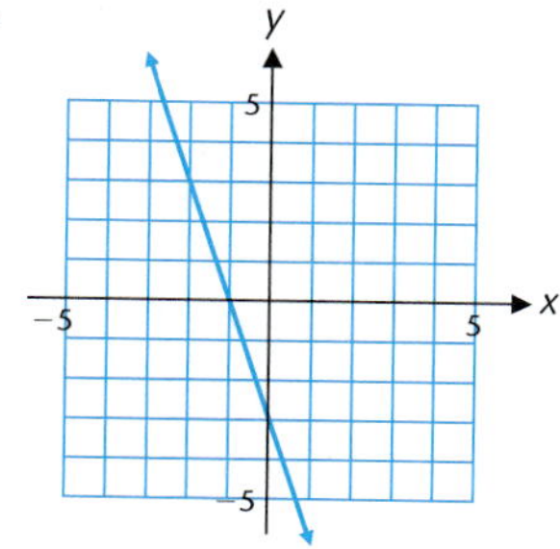

11.

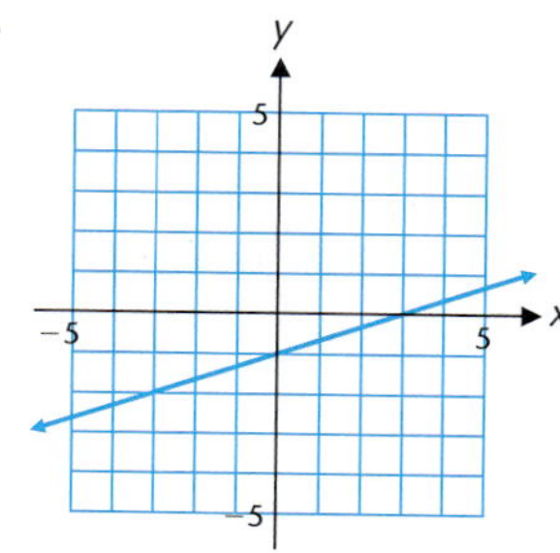

12.

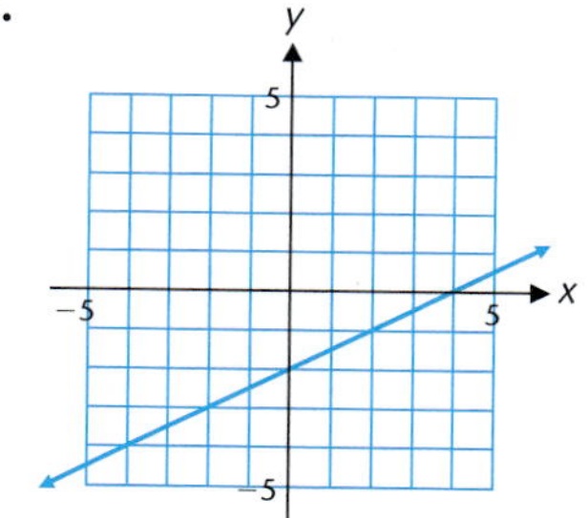

Use the transformations discussed in Section 1.4 to describe verbally the relationship between the graph of $f(x) = x$ and the graphs of the indicated functions in Problems 13–16.

13. $g(x) = 3x - 7$

14. $h(x) = -2x + 9$

15. $k(x) = -\frac{1}{2}x - 4$

16. $m(x) = \frac{2}{3}x + 3$

Which equations in Problems 17–26 define linear functions? Justify your answer.

17. $y = 2x^2$

18. $y = 5 - 3x^3$

19. $y = \dfrac{x - 5}{3}$

20. $y = \dfrac{3 - x}{2}$

21. $y = \frac{2}{3}(x - 7) - \frac{1}{2}(3 - x)$

22. $y = -\frac{1}{5}(2 - 3x) + \frac{2}{7}(x + 8)$

23. $y = \frac{1}{4}(2x + 2) + \frac{1}{2}(4 - x)$

24. $y = \frac{4}{3}(2 - x) + \frac{2}{3}(x + 2)$

25. $y = \dfrac{3}{x - 5}$

26. $y = \dfrac{2}{3 - x}$

In Problems 27–38, find the x intercept, y intercept, and slope, if they exist, and graph each equation.

27. $y = -\frac{3}{5}x + 4$

28. $y = -\frac{3}{2}x + 6$

29. $y = -\frac{3}{4}x$

30. $y = \frac{2}{3}x - 3$

31. $2x - 3y = 15$

32. $4x + 3y = 24$

33. $\frac{y}{8} - \frac{x}{4} = 1$

34. $\frac{y}{6} - \frac{x}{5} = 1$

35. $x = -3$

36. $y = -2$

37. $y = 3.5$

38. $x = 2.5$

In Problems 39–42, write the slope–intercept form of the equation of the line with indicated slope and y intercept.

39. Slope $= 1$; y intercept $= 0$

40. Slope $= -1$; y intercept $= 7$

41. Slope $= -\frac{2}{3}$; y intercept $= -4$

42. Slope $= \frac{5}{3}$; y intercept $= 6$

B

In Problems 43–56, sketch a graph of the line that contains the indicated point(s) and/or has the indicated slope and/or has the indicated intercepts. Then write the equation of the line in the slope–intercept form $y = mx + b$ or in the form $x = c$ and check by graphing the equation on a graphing utility.

43. $(0, 4)$; $m = -3$

44. $(2, 0)$; $m = 2$

45. $(-5, 4)$; $m = -\frac{2}{5}$

46. $(-4, -2)$; $m = \frac{1}{2}$

47. $(1, 6)$; $(5, -2)$

48. $(-3, 4)$; $(6, 1)$

49. $(-4, 8)$; $(2, 0)$

50. $(2, -1)$; $(10, 5)$

51. $(-3, 4)$; $(5, 4)$

52. $(0, -2)$; $(4, -2)$

53. $(4, 6)$; $(4, -3)$

54. $(-3, 1)$; $(-3, -4)$

55. x intercept -4; y intercept 3

56. x intercept -4; y intercept -5

In Problems 57–68, write an equation of the line that contains the indicated point and meets the indicated condition(s). Write the final answer in standard form $Ax + By = C$, $A \geq 0$.

57. $(-3, 4)$; parallel to $y = 3x - 5$

58. $(-4, 0)$; parallel to $y = -2x + 1$

59. $(2, -3)$; perpendicular to $y = -\frac{1}{3}x$

60. $(-2, -4)$; perpendicular to $y = \frac{2}{3}x - 5$

61. $(2, 5)$; parallel to y axis

62. $(7, 3)$; parallel to x axis

63. $(3, -2)$; vertical

64. $(-2, -3)$; horizontal

65. $(5, 0)$; parallel to $3x - 2y = 4$

66. $(3, 5)$; parallel to $3x + 4y = 8$

67. $(0, -4)$; perpendicular to $x + 3y = 9$

68. $(-2, 4)$; perpendicular to $4x + 5y = 0$

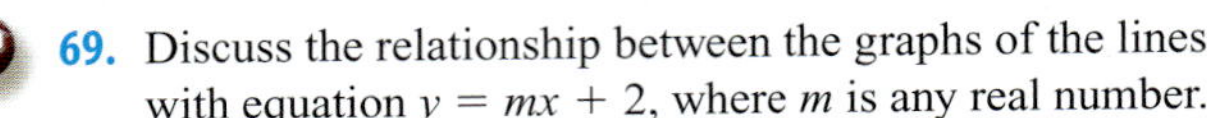

69. Discuss the relationship between the graphs of the lines with equation $y = mx + 2$, where m is any real number.

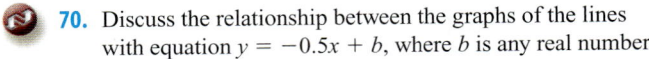

70. Discuss the relationship between the graphs of the lines with equation $y = -0.5x + b$, where b is any real number.

71. **(A)** Find the linear function f whose graph passes through the points $(-1, -3)$ and $(7, 2)$.

(B) Find the linear function g whose graph passes through the points $(-3, -1)$ and $(2, 7)$.

(C) Graph both functions and discuss how they are related.

72. **(A)** Find the linear function f whose graph passes through the points $(-2, -3)$ and $(10, 5)$.

(B) Find the linear function g whose graph passes through the points $(-3, -2)$ and $(5, 10)$.

(C) Graph both functions and discuss how they are related.

Problems 73–78 refer to the quadrilateral with vertices $A(0, 2)$, $B(4, -1)$, $C(1, -5)$, and $D(-3, -2)$.

73. Show that $AB \parallel DC$.

74. Show that $DA \parallel CB$.

75. Show that $AB \perp BC$.

76. Show that $AD \perp DC$.

77. Find an equation of the perpendicular bisector of AD.

78. Find an equation of the perpendicular bisector of AB.

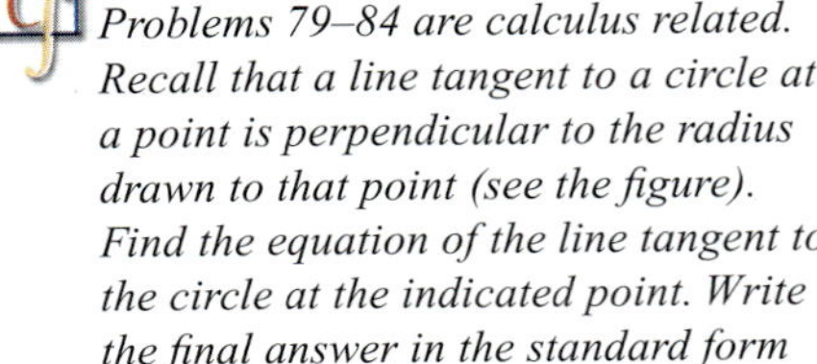

Problems 79–84 are calculus related. Recall that a line tangent to a circle at a point is perpendicular to the radius drawn to that point (see the figure). Find the equation of the line tangent to the circle at the indicated point. Write the final answer in the standard form $Ax + By = C$, $A \geq 0$. Graph the circle and the tangent line on the same coordinate system.

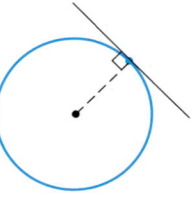

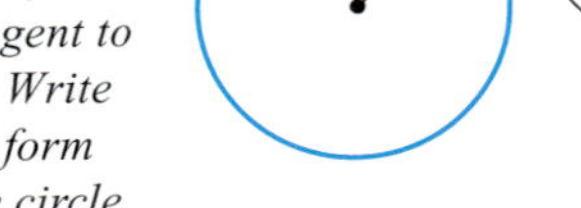

79. $x^2 + y^2 = 25, (3, 4)$

80. $x^2 + y^2 = 100, (-8, 6)$

81. $x^2 + y^2 = 50, (5, -5)$

82. $x^2 + y^2 = 80, (-4, -8)$

83. $(x - 3)^2 + (y + 4)^2 = 169, (8, -16)$

84. $(x + 5)^2 + (y - 9)^2 = 289, (-13, -6)$

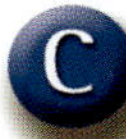

85. **(A)** Graph the following equations in a squared viewing window:

$$3x + 2y = 6 \qquad 3x + 2y = 3$$
$$3x + 2y = -6 \qquad 3x + 2y = -3$$

(B) From your observations in part A, describe the family of lines obtained by varying C in $Ax + By = C$ while holding A and B fixed.

(C) Verify your conclusions in part B with a proof.

86. **(A)** Graph the following two equations in a squared viewing window:

$$3x + 4y = 12 \qquad 4x - 3y = 12$$

(B) Graph the following two equations in a squared viewing window:

$$2x + 3y = 12 \qquad 3x - 2y = 12$$

(C) From your observations in parts A and B, describe the apparent relationship of the graphs of

$$Ax + By = C \text{ and } Bx - Ay = C.$$

(D) Verify your conclusions in part C with a proof.

87. Describe the relationship between the graphs of $f(x) = mx + b$ and $g(x) = |mx + b|$, $m \neq 0$, and illustrate with examples. Is $g(x)$ always, sometimes, or never a linear function?

88. Describe the relationship between the graphs of $f(x) = mx + b$ and $g(x) = m|x| + b$, $m \neq 0$, and illustrate with examples. Is $g(x)$ always, sometimes, or never a linear function?

89. Prove that if a line L has x intercept $(a, 0)$ and y intercept $(0, b)$, then the equation of L can be written in the **intercept form**

$$\frac{x}{a} + \frac{y}{b} = 1 \qquad a, b \neq 0$$

90. Prove that if a line L passes through $P_1 = (x_1, y_1)$ and $P_2 = (x_2, y_2)$, then the equation of L can be written in the **two-point form**

$$(y - y_1)(x_2 - x_1) = (y_2 - y_1)(x - x_1)$$

APPLICATIONS

91. Physics. The two temperature scales Fahrenheit (F) and Celsius (C) are linearly related. It is known that water freezes at 32°F or 0°C and boils at 212°F or 100°C.

(A) Find a linear equation that expresses F in terms of C.

(B) If a European house thermostat is set at 20°C, what is the setting in degrees Fahrenheit? If the outside temperature in Milwaukee is 86°F, what is the temperature in degrees Celsius?

(C) What is the slope of the graph of the linear equation found in part A? Interpret verbally.

92. Physics. Hooke's Law states that the relationship between the stretch s of a spring and the weight w causing the stretch is linear (a principle on which all spring scales are constructed). For a particular spring, a 5-pound weight causes a stretch of 2 inches, whereas with no weight the stretch of the spring is 0.

(A) Find a linear equation that expresses s in terms of w.

(B) What weight will cause a stretch of 3.6 inches?

(C) What is the slope of the graph of the equation? Interpret verbally.

93. Business–Depreciation. A copy machine was purchased by a law firm for $8,000 and is assumed to have a depreciated value of $0 after 5 years. The firm takes straight-line depreciation over the 5-year period.

(A) Find a linear equation that expresses value V in dollars in terms of time t in years.

(B) What is the depreciated value after 3 years?

(C) What is the slope of the graph of the equation found in part A? Interpret verbally.

94. Business–Markup Policy. A clothing store sells a shirt costing $20 for $33 and a jacket costing $60 for $93.

(A) If the markup policy of the store for items costing over $10 is assumed to be linear, write an equation that expresses retail price R in terms of cost C (wholesale price).

(B) What does a store pay for a suit that retails for $240?

(C) What is the slope of the graph of the equation found in part A? Interpret verbally.

95. Cost Analysis. A doughnut shop has a fixed cost of \$124 per day and a variable cost of \$0.12 per doughnut. Find the total daily cost of producing x doughnuts. How many doughnuts can be produced for a total daily cost of \$250?

96. Cost Analysis. A small company manufactures picnic tables. The weekly fixed cost is \$1,200 and the variable cost is \$45 per table. Find the total weekly cost of producing x picnic tables. How many picnic tables can be produced for a total weekly cost of \$4,800?

97. Cost Analysis. A plant can manufacture 80 golf clubs per day for a total daily cost of \$8,147 and 100 golf clubs per day for a total daily cost of \$9,647.

(A) Assuming that the daily cost function is linear, find the total daily cost of producing x golf clubs.

(B) Write a brief verbal interpretation of the slope and y intercept of this cost function.

98. Cost Analysis. A plant can manufacture 50 tennis rackets per day for a total daily cost of \$4,174 and 60 tennis rackets per day for a total daily cost of \$4,634.

(A) Assuming that the daily cost function is linear, find the total daily cost of producing x tennis rackets.

(B) Write a brief verbal interpretation of the slope and y intercept of this cost function.

★ **99. Medicine.** Cardiovascular research has shown that above the 210 cholesterol level, each 1% increase in cholesterol level increases coronary risk 2%. For a particular age group, the coronary risk at a 210 cholesterol level is found to be 0.160 and at a level of 231 the risk is found to be 0.192.

(A) Find a linear equation that expresses risk R in terms of cholesterol level C.

(B) What is the risk for a cholesterol level of 260?

(C) What is the slope of the graph of the equation found in part A? Interpret verbally.

Express all calculated quantities to three significant digits.

★ **100. Demographics.** The average number of persons per household in the United States has been shrinking steadily for as long as statistics have been kept and is approximately linear with respect to time. In 1900, there were about 4.76 persons per household and in 1990, about 2.5.

(A) If N represents the average number of persons per household and t represents the number of years since 1900, write a linear equation that expresses N in terms of t.

(B) What is the predicted household size in the year 2015?

Express all calculated quantities to three significant digits.

101. Flight Conditions. In stable air, the air temperature drops about 5°F for each 1,000-foot rise in altitude.

(A) If the temperature at sea level is 70°F and a commercial pilot reports a temperature of −20°F at 18,000 feet, write a linear equation that expresses temperature T in terms of altitude A (in thousands of feet).

(B) How high is the aircraft if the temperature is 0°F?

(C) What is the slope of the graph of the equation found in part A? Interpret verbally.

★ **102. Flight Navigation.** An airspeed indicator on some aircraft is affected by the changes in atmospheric pressure at different altitudes. A pilot can estimate the true airspeed by observing the indicated airspeed and adding to it about 2% for every 1,000 feet of altitude.

(A) If a pilot maintains a constant reading of 200 miles per hour on the airspeed indicator as the aircraft climbs from sea level to an altitude of 10,000 feet, write a linear equation that expresses true airspeed T (miles per hour) in terms of altitude A (thousands of feet).

(B) What would be the true airspeed of the aircraft at 6,500 feet?

(C) What is the slope of the graph of the equation found in part A? Interpret verbally.

★ **103. Oceanography.** After about 9 hours of a steady wind, the height of waves in the ocean is approximately linearly related to the duration of time the wind has been blowing. During a storm with 50-knot winds, the wave height after 9 hours was found to be 23 feet, and after 24 hours it was 40 feet.

(A) If t is time after the 50-knot wind started to blow and h is the wave height in feet, write a linear equation that expresses height h in terms of time t.

(B) How long will the wind have been blowing for the waves to be 50 feet high?

Express all calculated quantities to three significant digits.

SECTION 2.2 Linear Equations and Models

Solving Linear Equations • Modeling Distance-Rate-Time Problems • Modeling Mixture Problems • Data Analysis and Linear Regression

In this section we will discuss methods for solving equations that involve linear functions. Some problems are best solved using algebraic techniques, whereas others benefit from a graphical approach. Because graphs often give additional insight into relationships, especially in applications, we will usually emphasize graphical techniques over algebraic methods. But you must be certain to master both. There are problems in this section that can only be solved algebraically. And we are going to introduce an important new tool—linear regression—that requires computation best done with a graphing utility.

Solving Linear Equations

EXAMPLE 1 Solving an Equation

Solve $5x - 8 = 2x + 1$.

SOLUTION

Algebraic Solution

We use the familiar properties of equality (see Appendix A, Section A.1) to transform the given equation into an equivalent equation with an obvious solution.

$5x - 8 = 2x + 1$	Original equation
$5x - 8 - 2x = 2x + 1 - 2x$	Subtract $2x$ from both sides.
$3x - 8 = 1$	Combine like terms.
$3x - 8 + 8 = 1 + 8$	Add 8 to both sides.
$3x = 9$	Combine like terms.
$\frac{3x}{3} = \frac{9}{3}$	Divide both sides by 3.
$x = 3$	Simplify.

It follows from the properties of equality that $x = 3$ is also the solution set of all the preceding equations in our solution, including the original equation.

Graphical Solution

Enter each side of the equation in the equation editor of a graphing utility (Fig. 1) and use the intersect command (Fig. 2).

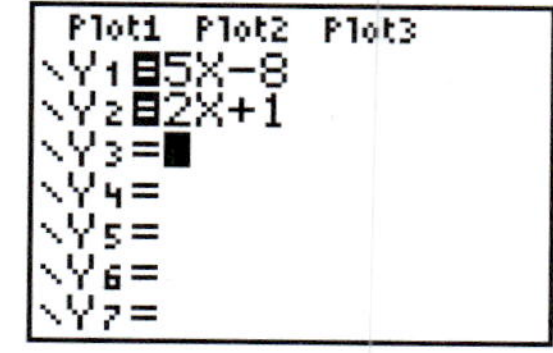

FIGURE 1

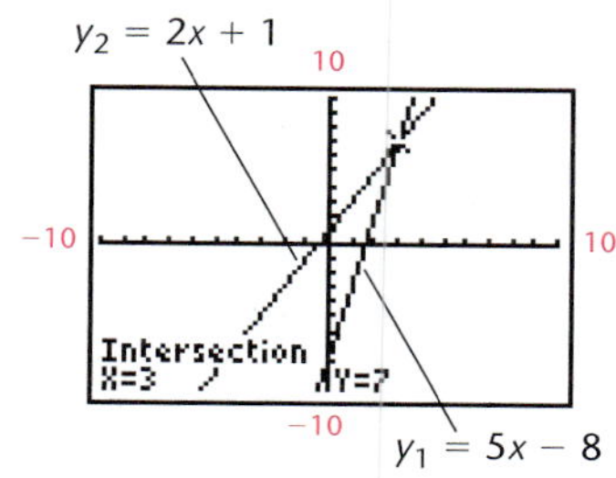

FIGURE 2

Thus, $x = 3$ is the solution to the original equation.

MATCHED PROBLEM 1

Solve $2x + 1 = 4x + 5$.

Refer to the solution of Example 1. Note that we used the informal notation $x = 3$ for the solution set rather than the more formal statement: solution set $= \{3\}$.

EXPLORE/DISCUSS 1

An equation that is true for all values of the variable for which both sides of the equation are defined is called an **identity.** An equation that is true for some values of the variable and false for others is called a **conditional equation.** An equation that is false for all permissible values of the variable is called a **contradiction.** Use algebraic and/or graphical techniques to classify each of the following as an identity, a conditional equation, or a contradiction. Solve any conditional equations.

(A) $2(x - 4) = 2x - 12$ (B) $2(x - 4) = 3x - 12$

(C) $2(x - 4) = 2x - 8$ (D) $\dfrac{2}{x - 1} + 3 = \dfrac{x}{x - 1}$

(E) $\dfrac{1}{x - 1} + 3 = \dfrac{x}{x - 1}$ (F) $\dfrac{1}{x - 1} + 1 = \dfrac{x}{x - 1}$

EXAMPLE 2 Solving an Equation

Solve $\dfrac{7}{2x} - 3 = \dfrac{8}{3} - \dfrac{15}{x}$.

SOLUTION

Algebraic Solution

Note that 0 must be excluded from the permissible values of x because division by 0 is not permitted. To clear the fractions, we multiply both sides of the equation by $3(2x) = 6x$, the least common denominator (LCD) of all fractions in the equation.

$$\frac{7}{2x} - 3 = \frac{8}{3} - \frac{15}{x} \qquad x \neq 0$$

$$6x\left(\frac{7}{2x} - 3\right) = 6x\left(\frac{8}{3} - \frac{15}{x}\right) \qquad \text{Multiply by } 6x\text{, the LCD.}$$

$$6x \cdot \frac{7}{2x} - 6x \cdot 3 = 6x \cdot \frac{8}{3} - 6x \cdot \frac{15}{x}$$

$$21 - 18x = 16x - 90 \qquad \text{The equation is now free of fractions.}$$

$$21 - 18x - 16x = 16x - 90 - 16x \qquad \text{Subtract } 16x \text{ from both sides.}$$

$$21 - 34x = -90 \qquad \text{Combine like terms.}$$

$$21 - 34x - 21 = -90 - 21 \qquad \text{Subtract 21 from both sides.}$$

$$-34x = -111 \qquad \text{Combine like terms.}$$

Graphical Solution

Enter $y_1 = \frac{7}{2x} - 3$ and $y_2 = \frac{8}{3} - \frac{15}{x}$ (Fig. 3) in the equation editor of a graphing utility. Note the use of parentheses in Figure 3 to be certain that $\frac{7}{2x}$ is evaluated correctly. Now use the intersect command (Fig. 4).

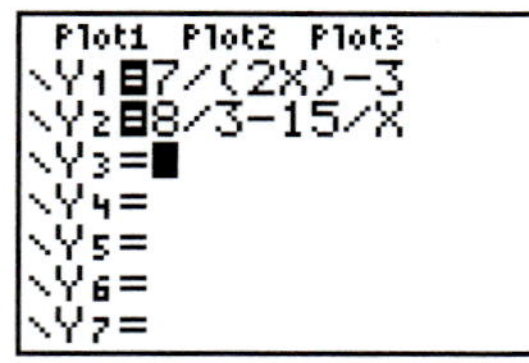

FIGURE 3

$$\frac{-34x}{-34} = \frac{-111}{-34}$$ Divide both sides by -34.

$$x = \frac{111}{34}$$ The solution set is $\left\{\frac{111}{34}\right\}$.

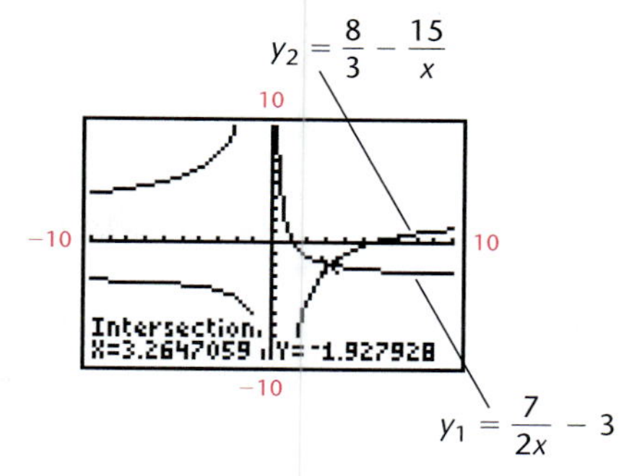

FIGURE 4

Thus, $x = 3.2647059$ is the solution of the original equation. Note that $\frac{111}{34} \approx 3.2647059$ to seven decimal places.

MATCHED 2 PROBLEM

Solve $\frac{7}{3x} + 2 = \frac{1}{x} - \frac{3}{5}$.

We frequently encounter equations involving more than one variable. For example, if L and W are the length and width of a rectangle, respectively, the area of the rectangle is given by (Fig. 5)

$$A = LW$$

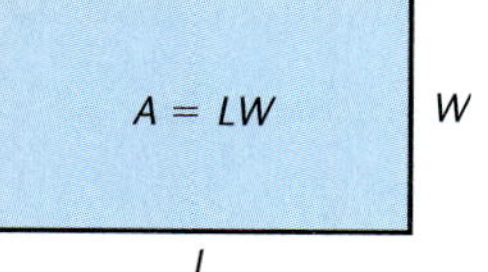

FIGURE 5 Area of a rectangle.

Depending on the situation, we may want to solve this equation for L or W. To solve for W, we simply consider A and L to be constants and W to be a variable. Then the equation $A = LW$ becomes a linear equation in W, which can be solved easily by dividing both sides by L:

$$W = \frac{A}{L} \qquad L \neq 0$$

EXAMPLE 3 Solving an Equation with More than One Variable

Solve for P in terms of the other variables: $A = P + Prt$.

SOLUTION

$A = P + Prt$ Think of A, r, and t as constants.

$A = P(1 + rt)$ Factor to isolate P.

$\frac{A}{1 + rt} = P$ Divide both sides by $1 + rt$.

$P = \frac{A}{1 + rt}$ Restriction: $1 + rt \neq 0$

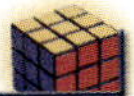
MATCHED 3 PROBLEM

Solve for r in terms of the other variables: $A = P + Prt$.

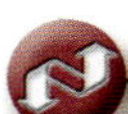

Modeling Distance-Rate-Time Problems

To construct models for word problems we translate verbal statements into mathematical statements. Explore/Discuss 2 will help you review this process.

EXPLORE/DISCUSS 2

Translate each of the following sentences involving two numbers into an equation.

(A) The first number is 10 more than the second number.

(B) The first number is 15 less than the second number.

(C) The first number is half the second number.

(D) The first number is three times the second number.

(E) Ten times the first number is 15 more than the second number.

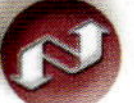

If a quantity Q is changing at a constant rate r with respect to time t, then one of the most important linear models is

$$Q = rt$$

where r and t are expressed in compatible units (miles per hour and hours, gallons per minute and minutes, etc.). The familiar distance equation, $d = rt$, is a special case of this model.

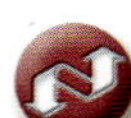

EXPLORE/DISCUSS 3

A bus leaves Milwaukee at 12:00 noon and travels due west on Interstate 94 at a constant rate of 55 miles per hour. A passenger that was left behind leaves Milwaukee in a taxicab at 1:00 P.M. in pursuit of the bus. The taxicab travels at a constant rate of 65 miles per hour. Let t represent time in hours after 12:00 noon.

(A) How far has the bus traveled after t hours?

(B) If $t \geq 1$, how far has the taxicab traveled after t hours?

(C) When will the taxicab catch up with the bus?

EXAMPLE 4 A Distance-Rate-Time Problem

An excursion boat takes 1.5 times as long to go 60 miles up a river than to return. If the boat cruises at 16 miles per hour in still water, what is the rate of the current in the river?

SOLUTION

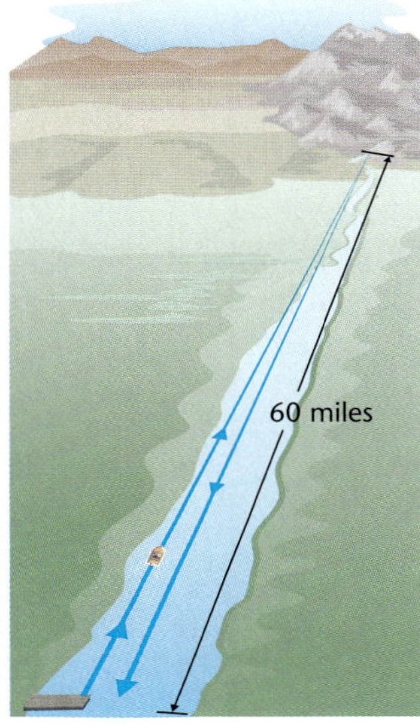

Constructing the Model

Let d_1, r_1, and t_1 represent the distance, rate, and time, respectively, for the trip upstream and d_2, r_2, and t_2 represent the distance, rate, and time, respectively, for the trip downstream. Because both trips cover the same distance, we know that $d_1 = d_2 = 60$. And because the trip upstream takes 1.5 times as long as the trip downstream, we also know that $t_1 = 1.5t_2$. Using the basic model, $d = rt$, we have

$$d_1 = r_1t_1 \qquad d_2 = r_2t_2$$
$$60 = r_1(1.5t_2) \qquad 60 = r_2t_2 \tag{1}$$

We are given that the rate of the boat in still water is 16 miles per hour and we are asked to find the rate of the current. Let

x = Rate of current (in miles per hour)
$r_1 = 16 - x$ = Rate of boat upstream
$r_2 = 16 + x$ = Rate of boat downstream

Substituting for r_1 and r_2 in the distance equations in (1) and solving for t_2, we have

$$60 = (16 - x)(1.5t_2) \qquad 60 = (16 + x)t_2$$
$$\frac{40}{16 - x} = t_2 \qquad \frac{60}{16 + x} = t_2 \qquad x \neq 16,\; x \neq -16$$

The solution to the problem is the x coordinate of the intersection point of the graphs of these equations.

Algebraic Solution

$$\frac{40}{16 - x} = \frac{60}{16 + x}$$ Multiply both sides by $(16 - x)(16 + x)$.

$$40(16 + x) = 60(16 - x)$$
$$640 + 40x = 960 - 60x$$
$$100x = 320$$
$$x = 3.2 \text{ miles per hour}$$

Graphical Solution

Entering $y_1 = \frac{40}{16 - x}$, $y_2 = \frac{60}{16 + x}$ and using the intersect command (Fig. 6) shows that $x = 3.2$ miles per hour is the solution.

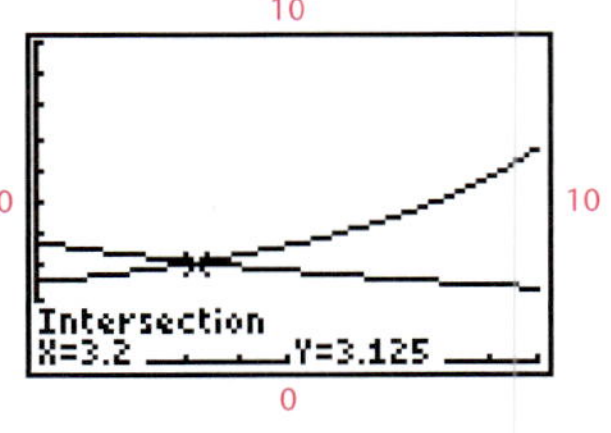

FIGURE 6

MATCHED 4 PROBLEM

A jetliner takes 1.2 times as long to fly from Paris to New York (3,600 miles) as to return. If the jet cruises at 550 miles per hour in still air, what is the average rate of the wind blowing in the direction of Paris from New York?

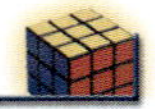

Modeling Mixture Problems

A variety of applications can be classified as mixture problems. Although the problems come from different areas, their mathematical treatment is essentially the same.

EXAMPLE 5 A Mixture Problem

How many liters of a mixture containing 80% alcohol should be added to 5 liters of a 20% solution to yield a 30% solution?

SOLUTION

Let x = amount of 80% solution used.

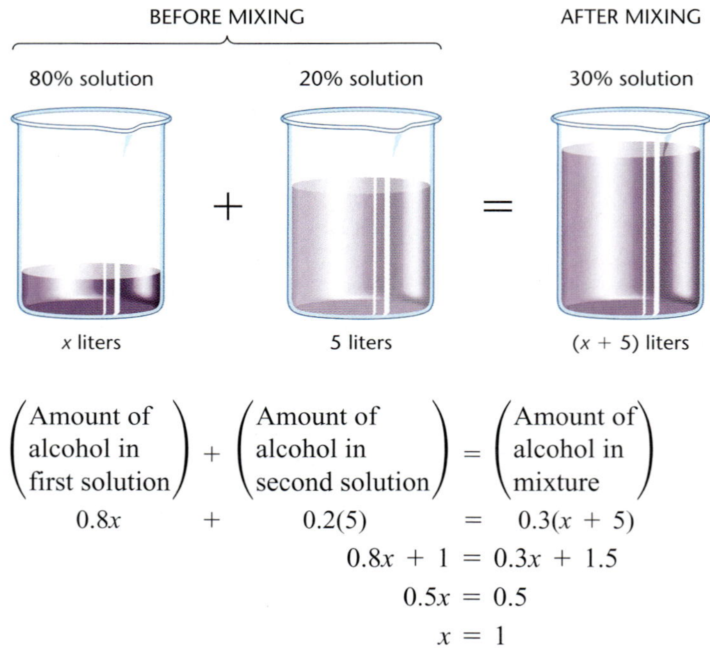

$$\begin{pmatrix}\text{Amount of}\\ \text{alcohol in}\\ \text{first solution}\end{pmatrix} + \begin{pmatrix}\text{Amount of}\\ \text{alcohol in}\\ \text{second solution}\end{pmatrix} = \begin{pmatrix}\text{Amount of}\\ \text{alcohol in}\\ \text{mixture}\end{pmatrix}$$

$$0.8x + 0.2(5) = 0.3(x + 5)$$

$$0.8x + 1 = 0.3x + 1.5$$

$$0.5x = 0.5$$

$$x = 1$$

Add 1 liter of the 80% solution.

CHECK

	Liters of Solution	Liters of Alcohol	Percent Alcohol
First solution	1	0.8(1) = 0.8	80
Second solution	5	0.2(5) = 1	20
Mixture	6	1.8	1.8/6 = 0.3, or 30%

MATCHED 5 PROBLEM

A chemical storeroom has a 90% acid solution and a 40% acid solution. How many centiliters of the 90% solution should be added to 50 centiliters of the 40% solution to yield a 50% solution?

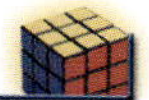

Data Analysis and Linear Regression

In real-world applications numerical data are often encountered in the form of a table. The very powerful mathematical tool *regression analysis* is used to analyze numerical data. In general, **regression analysis** is the process of finding a function that provides a useful model for a set of data points. Graphs of equations are often called **curves** and regression analysis is also referred to as **curve fitting.** In Example 6, we use **linear regression** to construct a mathematical model in the form of a linear function that fits a data set.

EXAMPLE 6 Diamond Prices

Prices for round-cut diamonds taken from an on-line trader are given in Table 1.

(A) Use linear regression on a graphing utility to find a linear model $y = f(x)$ that fits these data, where x is the weight of a diamond (in carats) and y is the associated price of that diamond (in dollars). Round the constants a and b to three significant digits. Compare the model and the data both graphically and numerically.

(B) Use the model to estimate the cost of a 0.85 carat diamond and the cost of a 1.2 carat diamond. Round answers to the nearest dollar.

(C) Use the model to estimate the weight of a diamond that sells for $3,000. Round the answer to two significant digits.

TABLE 1 Round-Cut Diamond Prices

Weight (Carats)	Price
0.5	$1,340
0.6	$1,760
0.7	$2,540
0.8	$3,350
0.9	$4,130
1.0	$4,920

Source: www.tradeshop.com

SOLUTIONS

(A) The first step in fitting a curve to a data set is to enter the data in two lists in a graphing utility, usually by pressing STAT and selecting EDIT (see Fig. 7 on page 142).* We enter the given values of the independent variable x in L_1 and the corresponding values of the dependent variable y

*Remember, we are using a TI-83 to produce the screen images in this book. Other graphing utilities will produce different images.

in L_2 (Fig. 8). Next, we select a viewing window that will show all the data (Fig. 9).

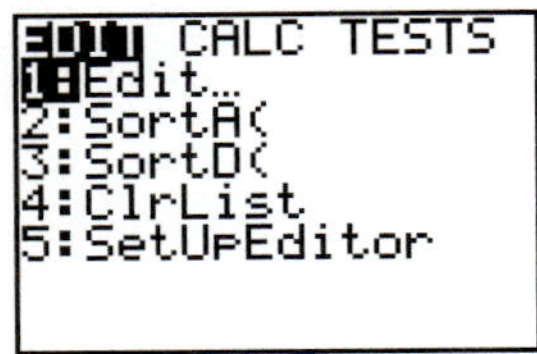

FIGURE 7

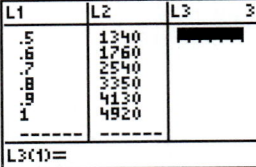

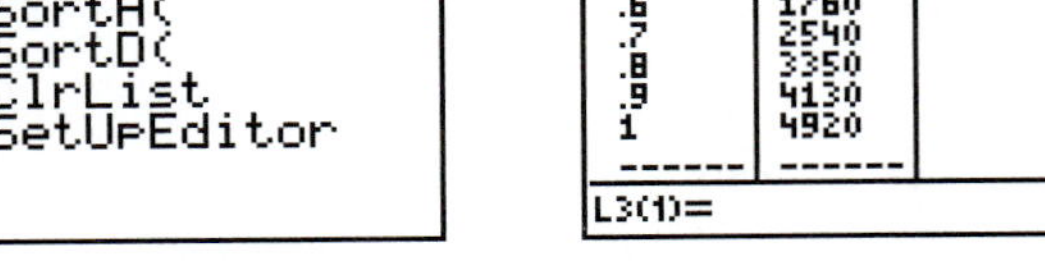

FIGURE 8

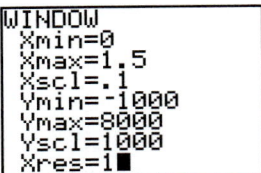

FIGURE 9

To check that all the data will be visible in this window, we need to graph the points in the form (x, y), where x is a number in list L_1 and y is the corresponding number in list L_2. This is called a **scatter diagram.** On most graphing utilities, a scatter diagram can be drawn by first pressing STAT PLOT and selecting the options displayed in

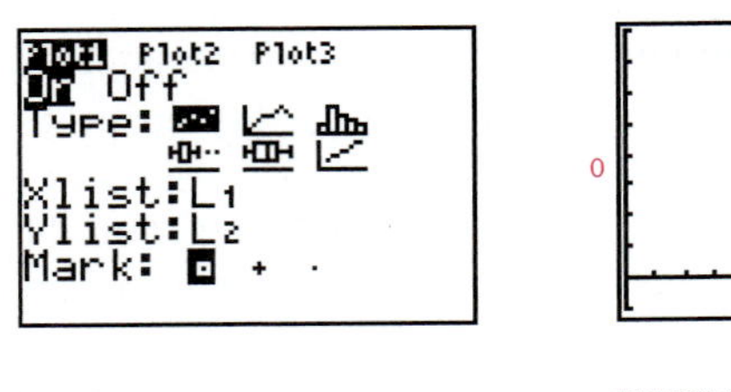

FIGURE 10

FIGURE 11

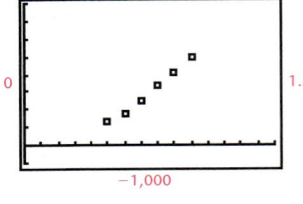

Figure 10. Then press GRAPH to display the scatter diagram (Fig. 11). Now we are ready to fit a curve to the data graphed in Figure 11. First we find the screen on the graphing utility that lists the various regres-

FIGURE 12

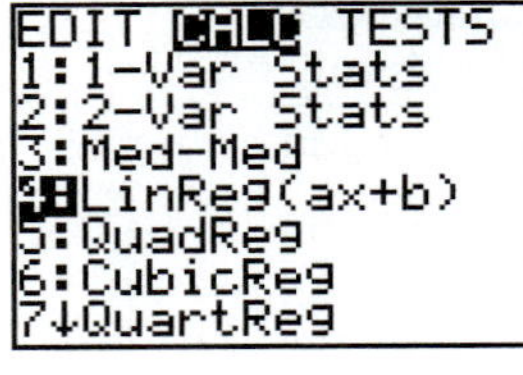

(a)

(b)

sion options, usually by pressing STAT and selecting CALC (Fig. 12). Any of options 4 through C in Figure 12 can be used to fit a curve to a data set. As you progress through this text, you will become familiar with most of the choices in Figure 12. In this example, we are directed to select option 4 (or, equivalently, option 8), linear regression (Fig. 13). Notice that we entered the names of the two lists of data, L_1 and L_2, after the command LinReg($ax + b$) in Figure 13. The order in which we enter these two names is important. The name of the list of independent values must precede the name of the list of dependent values. Press ENTER to obtain the results in Figure 14. The values r^2 and r displayed in Figure 14 are called **diagnostics.** They provide a measure of how well the regression curve fits the data. Values of r close to 1 or -1 indicate a good fit. Values of r close to 0 indicate a poor fit.

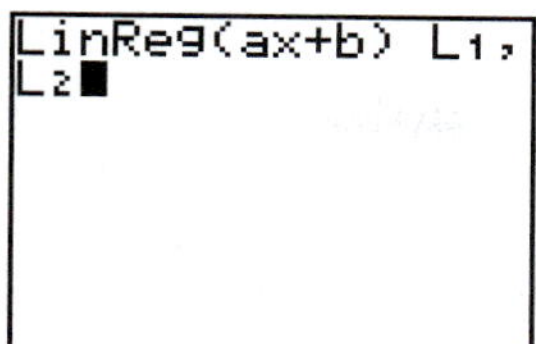

FIGURE 13

LinReg
y=ax+b
a=7377.142857
b=-2526.190476
r²=.9930719921
r=.9965299755

FIGURE 14

After rounding a and b in Figure 14 to three significant digits, the linear regression model for these data is

$$y = f(x) = 7{,}380x - 2{,}530$$

Enter this equation in the equation editor (Fig. 15). The graph of the model and the scatter plot of the data are shown in Figure 16 and a table comparing the data and the corresponding values of the model is shown in Figure 17.*

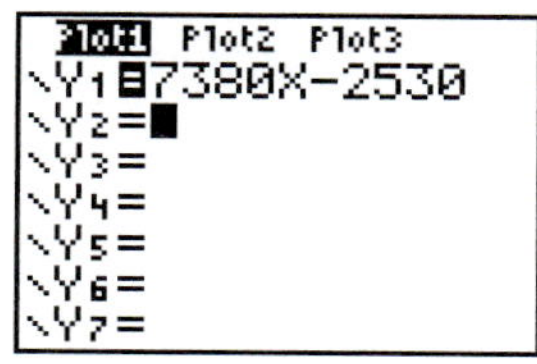

FIGURE 15

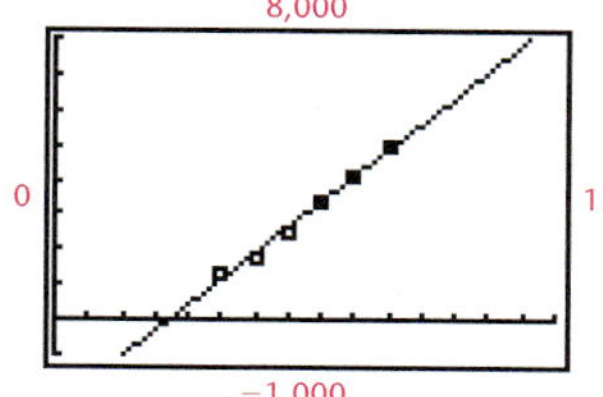

FIGURE 16

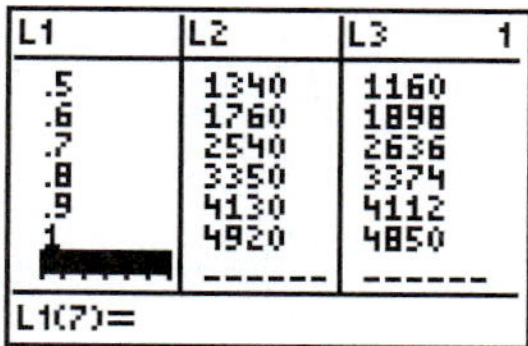

FIGURE 17

Examining Figures 16 and 17, we see that the model does seem to provide a reasonable fit for these data.

(B) Because $x = 0.85$ and $x = 1.2$ are not in Table 1, we use the model to estimate the corresponding prices. From Figure 18 we see that the estimated price of a 0.85-carat diamond is \$3,743. Figure 19 shows that the estimated price of a 1.2-carat diamond is \$6,326. Figure 20 shows how a table can be used in place of Trace to obtain the same results.

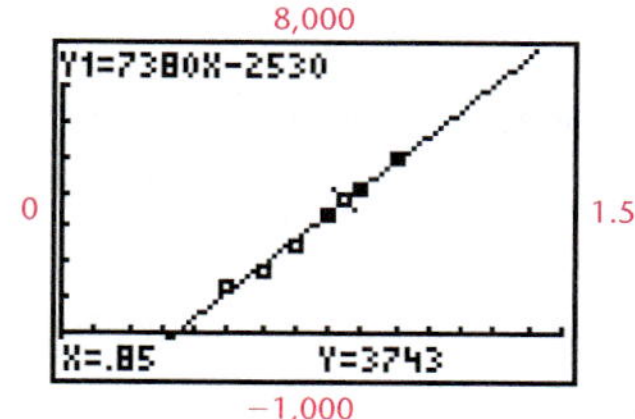

FIGURE 18

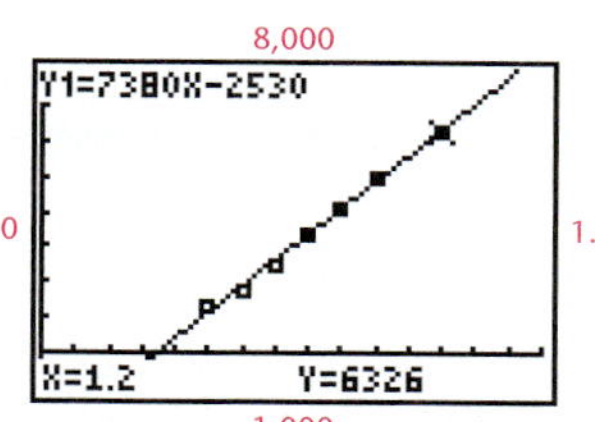

FIGURE 19

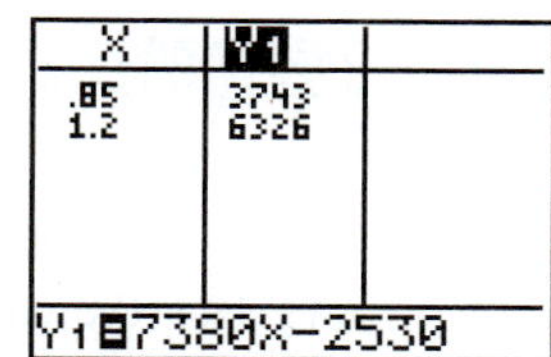

FIGURE 20

*On most graphing utilities, the values of the model displayed as L_3 in Figure 17 can be computed in a single operation by entering $y_1(L_1) \rightarrow L_3$ on the home screen.

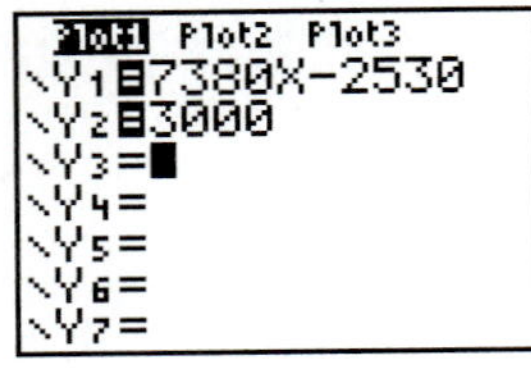

FIGURE 21

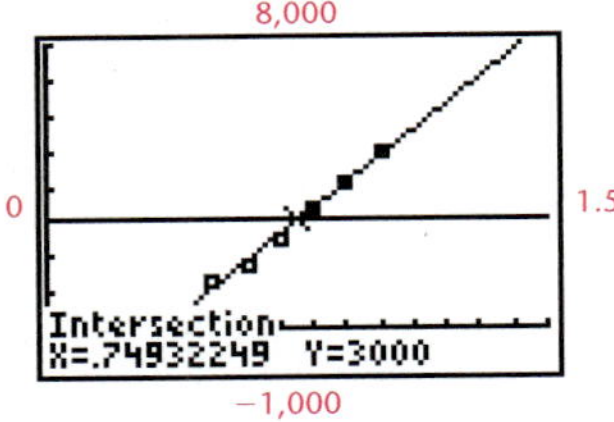

FIGURE 22

(C) This time we are given a value of the dependent variable y and asked to solve for the independent variable.

$$y = 7{,}380x - 2{,}530$$
$$3{,}000 = 7{,}380x - 2{,}530$$

To solve this equation, we add $y_2 = 3{,}000$ to the equation list (Fig. 21) and use the intersect command (Fig. 22).

From Figure 22 we see that $x = 0.75$ (to two significant digits) when $y = 3{,}000$. Thus, a \$3,000 diamond should weigh approximately 0.75 carats.

MATCHED 6 PROBLEM*

Prices for emerald-cut diamonds taken from an on-line trader are given in Table 2. Repeat Example 6 for this data set.

TABLE 2 Emerald-Cut Diamond Prices

Weight (Carats)	Price
0.5	\$1,350
0.6	\$1,740
0.7	\$2,610
0.8	\$3,320
0.9	\$4,150
1.0	\$4,850

Source: www.tradeshop.com

The quantity of a product that consumers are willing to buy during some period depends on its price. Generally, the higher the price, the lower the demand; the lower the price, the greater the demand. Similarly, the quantity of a product that producers are willing to sell during some period also depends on the price. Generally, a producer will be willing to supply more of a product at higher prices and less of a product at lower prices. In Example 7 we use linear regression to analyze supply and demand data and construct linear models.

EXAMPLE 7 Supply and Demand

Table 3 contains supply and demand data for broccoli at various price levels. Express all answers in numbers rounded to three significant digits.

(A) Use the data in Table 3 and linear regression to find a linear supply model $p = f(s)$, where s is the supply (in thousand pounds) and p is the corresponding price of broccoli (in cents).

---

*Be certain to delete the old values in the lists L_1 and L_2 before you work Matched Problem 6. CLEARALL is a short program for TI-83 graphing calculators that will delete all the values in all the lists. This program can be found on the website for this book.

(B) Use the data in Table 3 and linear regression to find a linear demand model $p = g(d)$, where d is the demand (in thousand pounds) and p is the corresponding price of broccoli (in cents).

(C) Graph both functions in the same viewing window and discuss possible interpretations of any intersection points.

TABLE 3 Supply and Demand for Broccoli

Price (Cents)	Supply (Thousand lbs.)	Demand (Thousand lbs.)
76.8	853	1,680
81.5	1,010	1,440
85.2	1,040	1,470
87.5	957	1,280
97.2	1,280	1,040
104	1,620	1,130
105	1,600	1,010

SOLUTION

(A) First, we enter the data from Table 3 in the statistics editor of a graphing utility (Fig. 23). Next we select the linear regression command LinReg($ax + b$) followed by L_2, L_1 to make supply the independent variable and price the dependent variable. This produces the results shown in Figure 24. Thus, the linear model for the supply function is

$$p = f(s) = 0.0344s + 50.0$$

To graph the supply data we use STAT PLOT, setting Xlist to L_2 and Ylist to L_1. Figure 25 shows a graph of the supply data and the supply model.

L1	L2	L3
76.8	853	1680
81.5	1010	1440
85.2	1040	1470
87.5	957	1280
97.2	1280	1040
104	1620	1130
105	1600	1010

L1(1)=76.8

FIGURE 23

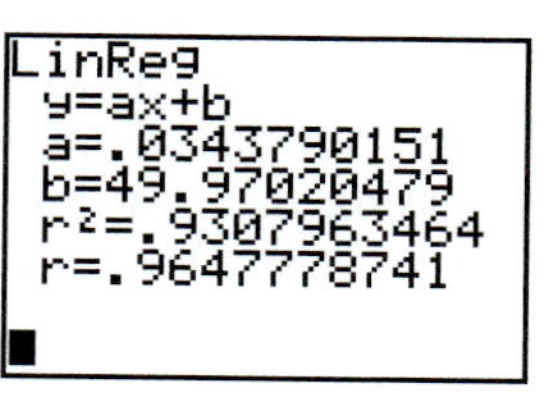

FIGURE 24

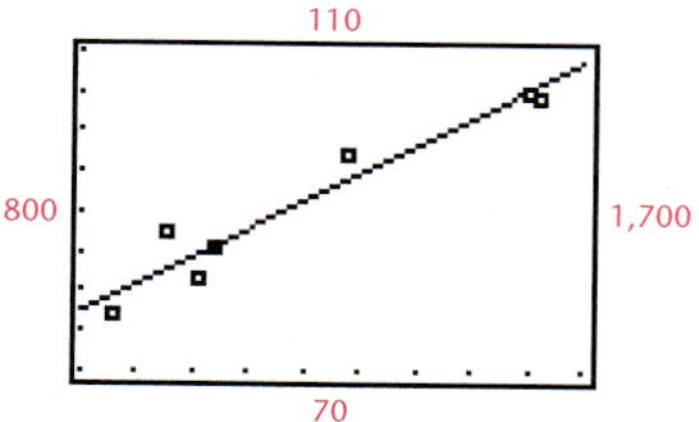

FIGURE 25

(B) This time we use the command LinReg($ax + b$) followed by L_3, L_1 to make demand the independent variable and price the dependent variable. This produces the results shown in Figure 26. Thus, the linear model for the demand function is

$$p = g(d) = -0.0416d + 145.$$

To graph the demand data we use STAT PLOT, setting Xlist to L_3 and Ylist to L_1. Figure 27 shows a graph of the demand data and the demand model.

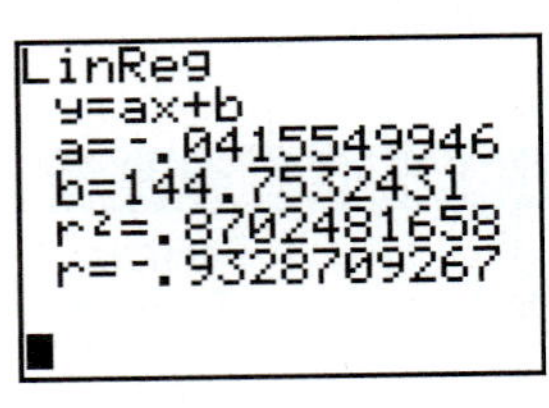

FIGURE 26

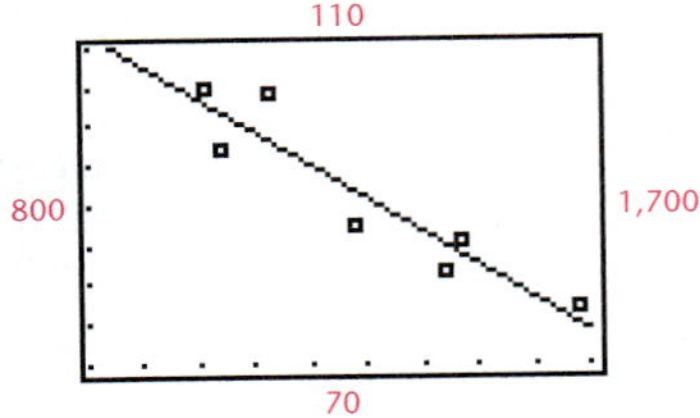

FIGURE 27

(C) We graph both models in the same viewing window and use the intersect command to find the intersection point (Fig. 28). The graphs intersect at $p = 93$ and $s = d = 1{,}250$. This point is called the **equilibrium point,** the value of p is called the **equilibrium price,** and the common value of s and d is called the **equilibrium quantity.** To help understand price fluctuations, suppose the current price of broccoli is 100 cents. We add the constant function $p = 100$ to the graph (Fig. 29).

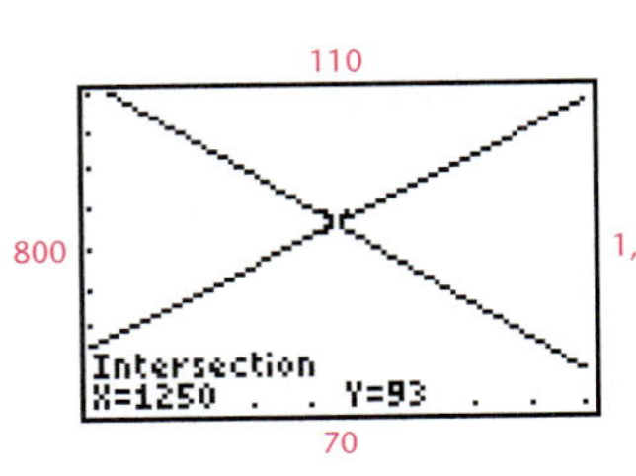

FIGURE 28

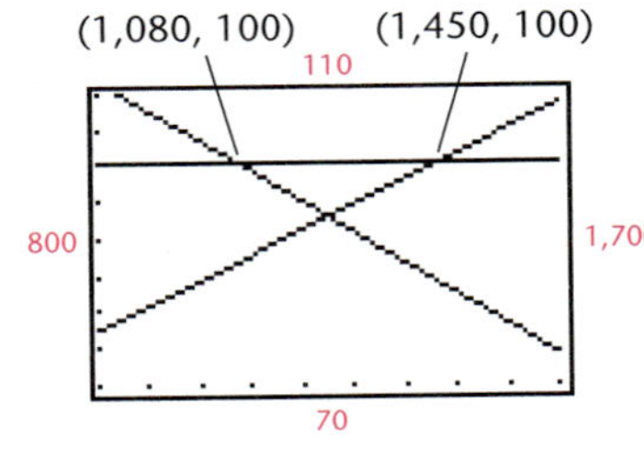

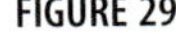

FIGURE 29

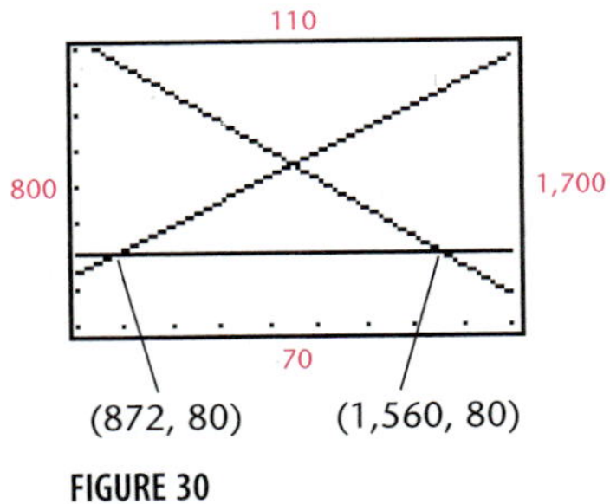

FIGURE 30

Using the intersect command (details omitted), we find that the constant price line intersects the demand curve at 1,080 thousand pounds and the supply curve at 1,450 thousand pounds. Because the supply at a price level of 100 is greater than the demand, the producers will lower their prices. Suppose the price drops to 80 cents per pound. Changing the constant function to $p = 80$ produces the graph in Figure 30. This time the constant price line intersects the demand curve at 1,560 thousand pounds and the supply curve at 872 thousand pounds. Now supply is less than demand and producers will raise their prices. If the producers set the price at $p = 93$ cents, then, as we saw in Figure 28, the supply and demand are equal.

MATCHED 7 PROBLEM

Table 4 contains supply and demand data for cauliflower at various prices. Repeat Example 7 with these data.

TABLE 4 Supply and Demand for Cauliflower

Price (Cents)	Supply (Thousand lbs.)	Demand (Thousand lbs.)
26.5	583	653
27.1	607	629
27.2	596	635
27.4	627	631
27.5	604	638
28.1	661	610
28.6	682	599

ANSWERS MATCHED PROBLEMS

1. $x = -2$
2. $x = -\frac{20}{39} \approx -0.5128205$
3. $r = \frac{A - P}{Pt} \qquad Pt \neq 0$
4. 50 miles per hour
5. 12.5 centiliters
6. (A) $y = 7{,}270x - 2{,}450$
   (B) \$3,730; \$6,270
   (C) 0.75 carats
7. (A) $p = 0.0180s + 16.3$
   (B) $p = -0.0362d + 50.2$
   (C) The price stabilizes at the equilibrium price of 27.6 cents.

EXERCISE 2.2

A

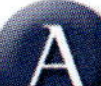

Use the graphs of functions u and v in the figure to solve the equations in Problems 1–4. (Assume the graphs continue as indicated beyond the portions shown here.)

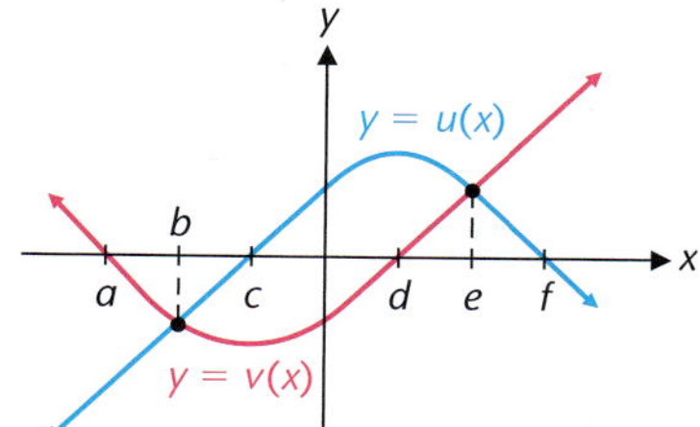

1. $u(x) = 0$
2. $v(x) = 0$
3. $u(x) = v(x)$
4. $u(x) - v(x) = 0$

In Problems 5–18, classify each equation as an identity, a conditional equation, or a contradiction. Solve each conditional equation.

5. $3(x - 2) - 2(x + 1) = x - 8$
6. $4(x - 1) - 2(x + 2) = 2x + 7$
7. $2(x - 1) - 3(2 - x) = 3x - 8$
8. $4(2 - x) + 2(x - 3) = 5x + 2$
9. $5(x + 2) - 3(x - 1) = 2x + 4$
10. $2(x + 1) + 3(2 - x) = 8 - x$
11. $\frac{3}{x - 2} + 1 = \frac{x}{x - 2}$
12. $\frac{1}{x + 3} + 3 = \frac{x}{x + 3}$
13. $\frac{1}{x - 1} + 1 = \frac{x}{x - 1}$
14. $\frac{3}{x - 3} + 4 = \frac{x}{x - 3}$
15. $\frac{-4}{x + 4} + 1 = \frac{x}{x + 2}$
16. $\frac{2}{x - 2} + 3 = \frac{x}{x - 2}$
17. $\frac{1}{x + 1} + 1 = \frac{x}{x + 1}$
18. $\frac{1}{x + 2} + 4 = \frac{x}{x + 2}$

Solve Problems 19–24.

19. $3(x + 2) = 5(x - 6)$
20. $5x + 10(x - 2) = 40$
21. $5 + 4(t - 2) = 2(t + 7) + 1$
22. $5w - (7w - 4) - 2 = 5 - (3w + 2)$
23. $5 - \frac{2x - 1}{4} = \frac{x + 2}{3}$
24. $\frac{x + 3}{4} - \frac{x - 4}{2} = \frac{3}{8}$

B

Solve Problems 25–40.

25. $\frac{7}{t} + 4 = \frac{2}{t}$

26. $\frac{9}{w} - 3 = \frac{2}{w}$

27. $\frac{1}{m} - \frac{1}{9} = \frac{4}{9} - \frac{2}{3m}$

28. $\frac{2}{3x} + \frac{1}{2} = \frac{4}{x} + \frac{4}{3}$

29. $(x - 2)(x + 3) = (x - 4)(x + 5)$

30. $(x - 2)(x - 4) = (x - 1)(x - 5)$

31. $(x + 2)(x - 3) = (x - 4)(x + 5)$

32. $(x - 2)(x + 4) = (x + 3)(x - 5)$

33. $(x - 2)^2 = (x - 1)(x + 2)$

34. $(x + 3)^2 = (x + 2)(x - 4)$

35. $\frac{2x}{x - 3} = 2 + \frac{6}{x - 3}$

36. $\frac{2x}{x + 4} = 2 - \frac{8}{x + 4}$

37. $\frac{2x}{x - 3} = 7 + \frac{4}{x - 3}$

38. $\frac{2x}{x + 4} = 7 - \frac{6}{x + 4}$

39. $\frac{2x}{x - 3} = 7 + \frac{6}{x - 3}$

40. $\frac{2x}{x + 4} = 7 - \frac{8}{x + 4}$

In Problems 41–48, solve for the indicated variable in terms of the other variables.

41. $a_n = a_1 + (n - 1)d$ for d (arithmetic progressions)

42. $F = \frac{9}{5}C + 32$ for C (temperature scale)

43. $\frac{1}{f} = \frac{1}{d_1} + \frac{1}{d_2}$ for f (simple lens formula)

44. $\frac{1}{R} = \frac{1}{R_1} + \frac{1}{R_2}$ for R_1 (electric circuit)

45. $A = 2ab + 2ac + 2bc$ for a (surface area of a rectangular solid)

46. $A = 2ab + 2ac + 2bc$ for c

47. $y = \frac{2x - 3}{3x + 5}$ for x

48. $x = \frac{3y + 2}{y - 3}$ for y

49. Discuss the relationship between the graphs of $y_1 = x$ and $y_2 = \sqrt{x^2}$.

50. Discuss the relationship between the graphs of $y_1 = |x|$ and $y_2 = \sqrt{x^2}$.

Problems 51–58 refer to a rectangle with width W and length L (see the figure). Write a mathematical expression in terms of W and L for each of the verbal statements in Problems 51–58.

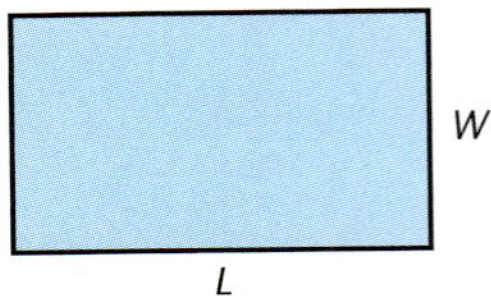

51. The length is twice the width.

52. The width is three times the length.

53. The width is half the length.

54. The length is one-third of the width.

55. The length is three more than the width.

56. The width is five less than the length.

57. The length is four less than the width.

58. The width is ten more than the length.

C

59. Use linear regression to fit a line to each of the following data sets. How are the graphs of the two functions related? How are the two functions related?

(A)

x	y
1	−1
5	1

(B)

x	y
−1	1
1	5

60. Repeat Problem 59 for the following data sets.

(A)

x	y
−1	0
3	5

(B)

x	y
0	−1
5	3

In Problems 61–64, solve for x.

61. $\dfrac{x - \dfrac{1}{x}}{1 + \dfrac{1}{x}} = 1$

62. $\dfrac{x - \dfrac{1}{x}}{x + 1 - \dfrac{2}{x}} = 1$

63. $\dfrac{x + 1 - \dfrac{2}{x}}{1 - \dfrac{1}{x}} = x + 2$

64. $\dfrac{x - 1 - \dfrac{2}{x}}{1 - \dfrac{2}{x}} = x$

65. Find three consecutive integers whose sum is 84.

66. Find four consecutive integers whose sum is 182.

67. Find four consecutive even integers so that the sum of the first three is 2 more than twice the fourth.

68. Find three consecutive even integers so that the first plus twice the second is twice the third.

69. Find the dimensions of a rectangle if the perimeter is 60 inches and the length is twice the width.

70. Find the dimensions of a rectangle if the perimeter is 60 inches and the length is half the width.

APPLICATIONS

71. Sales Commissions. One employee of a computer store is paid a base salary of \$2,150 a month plus an 8% commission on all sales over \$7,000 during the month. How much must the employee sell in 1 month to earn a total of \$3,170 for the month?

72. Sales Commissions. A second employee of the computer store in Problem 71 is paid a base salary of \$1,175 a month plus a 5% commission on all sales during the month.

(A) How much must this employee sell in 1 month to earn a total of \$3,170 for the month?

(B) Determine the sales level at which both employees receive the same monthly income. If employees can select either of these payment methods, how would you advise an employee to make this selection?

73. Wildlife Management. A naturalist for a fish and game department estimated the total number of trout in a certain lake using the popular capture–mark–recapture technique. She netted, marked, and released 200 trout. A week later, allowing for thorough mixing, she again netted 200 trout and found 8 marked ones among them. Assuming that the ratio of marked trout to the total number in the second sample is the same as the ratio of all marked fish in the first sample to the total trout population in the lake, estimate the total number of fish in the lake.

74. Wildlife Management. Repeat Problem 73 with a first (marked) sample of 300 and a second sample of 180 with only 6 marked trout.

75. Chemistry. How many gallons of distilled water must be mixed with 50 gallons of 30% alcohol solution to obtain a 25% solution?

76. Chemistry. How many gallons of hydrochloric acid must be added to 12 gallons of a 30% solution to obtain a 40% solution?

77. Chemistry. A chemist mixes distilled water with a 90% solution of sulfuric acid to produce a 50% solution. If 5 liters of distilled water is used, how much 50% solution is produced?

78. Chemistry. A fuel oil distributor has 120,000 gallons of fuel with 0.9% sulfur content, which exceeds pollution control standards of 0.8% sulfur content. How many gallons of fuel oil with a 0.3% sulfur content must be added to the 120,000 gallons to obtain fuel oil that complies with the pollution control standards?

79. Aeronautics. The cruising speed of an airplane is 150 miles per hour (relative to the ground). You wish to hire the plane for a 3-hour sightseeing trip. You instruct the pilot to fly north as far as he can and still return to the airport at the end of the allotted time.

(A) How far north should the pilot fly if the wind is blowing from the north at 30 miles per hour?

(B) How far north should the pilot fly if there is no wind?

80. Navigation. Suppose you are at a river resort and rent a motor boat for 5 hours starting at 7 A.M. You are told that the boat will travel at 8 miles per hour upstream and 12 miles per hour returning. You decide that you would like to go as far up the river as you can and still be back at noon. At what time should you turn back, and how far from the resort will you be at that time?

81. Earthquakes. An earthquake emits a primary wave and a secondary wave. Near the surface of the Earth the primary wave travels at about 5 miles per second, and the secondary wave travels at about 3 miles per second. From the time lag between the two waves arriving at a given seismic station, it is possible to estimate the distance to the quake. Suppose a station measures a time difference of 12 seconds between the arrival of the two waves. How far is the earthquake from the station? (The *epicenter* can be located by obtaining distance bearings at three or more stations.)

82. Sound Detection. A ship using sound-sensing devices above and below water recorded a surface explosion 39 seconds sooner on its underwater device than on its above-water device. If sound travels in air at about 1,100 feet per second and in water at about 5,000 feet per second, how far away was the explosion?

★**83. Earth Science.** In 1984, the Soviets led the world in drilling the deepest hole in the Earth's crust—more than 12 kilometers deep. They found that below 3 kilometers the temperature T increased 2.5°C for each additional 100 meters of depth.

(A) If the temperature at 3 kilometers is 30°C and x is the depth of the hole in kilometers, write an equation using x that will give the temperature T in the hole at any depth beyond 3 kilometers.

(B) What would the temperature be at 15 kilometers? (The temperature limit for their drilling equipment was about 300°C.)

(C) At what depth (in kilometers) would the temperature reach 280°C?

★**84. Aeronautics.** Because air is not as dense at high altitudes, planes require a higher ground speed to become airborne. A rule of thumb is 3% more ground speed per 1,000 feet of elevation, assuming no wind and no change in air temperature. (Compute numerical answers to three significant digits.)

(A) Let

V_s = Takeoff ground speed at sea level for a particular plane (in miles per hour)

A = Altitude above sea level (in thousands of feet)

V = Takeoff ground speed at altitude A for the same plane (in miles per hour)

Write a formula relating these three quantities.

(B) What takeoff ground speed would be required at Lake Tahoe airport (6,400 feet), if takeoff ground speed at San Francisco airport (sea level) is 120 miles per hour?

(C) If a landing strip at a Colorado Rockies hunting lodge (8,500 feet) requires a takeoff ground speed of 125 miles per hour, what would be the takeoff ground speed in Los Angeles (sea level)?

(D) If the takeoff ground speed at sea level is 135 miles per hour and the takeoff ground speed at a mountain resort is 155 miles per hour, what is the altitude of the mountain resort in thousands of feet?

DATA ANALYSIS AND LINEAR REGRESSION

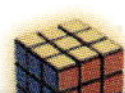

In Problems 85–90, use linear regression to construct linear models of the form $y = f(x) = ax + b$. Round a, b, and your answers to three significant digits.

85. Ticket Prices. Find a linear model $y = f(x)$ for the average ticket price data given in Table 5 where x is years since 1995 and y is average ticket price (in dollars). Use your model to predict the average ticket price in 2005.

TABLE 5 Motion Picture Data

	Average Ticket Price	Box Office Revenue (Billion)
1995	$4.35	$5.49
1996	$4.42	$5.91
1997	$4.59	$6.36
1998	$4.69	$6.95
1999	$5.08	$7.45
2000	$5.39	$7.66
2001	$5.66	$8.41

Source: Motion Picture Association of America

86. Box Office Revenue. Find a linear model $y = f(x)$ for the box office revenue data given in Table 5 where x is years since 1995 and y is box office revenue (in billions of dollars). Use your model to predict the box office revenue in 2005.

87. Olympic Games. Find a linear model $y = f(x)$ for the men's 100-meter freestyle data given in Table 6 where x is years since 1968 and y is winning time (in seconds). Do the same for the women's 100-meter freestyle data. Do these models indicate that the women will eventually catch up with the men? If so, when? Do you think this will actually occur?

TABLE 6 Winning Times in Olympic Swimming Events

	100-Meter Freestyle		200-Meter Backstroke	
	Men (Seconds)	Women (Seconds)	Men (Minutes: Seconds)	Women (Minutes: Seconds)
1968	52.20	60.00	2:09.60	2:24.80
1972	51.22	58.59	2:02.82	2:19.19
1976	49.99	55.65	1:59.19	2:13.43
1980	50.40	54.79	2:01.93	2:11.77
1984	49.80	55.92	2:00.23	2:12.38
1988	48.63	54.93	1:59.37	2:09.29
1992	49.02	54.65	1:58.47	2:07.06
1996	48.74	54.50	1.58.54	2:07.83
2000	48.30	53.83	1:56.76	2:08.16

Source: www.infoplease.com

88. Olympic Games. Find a linear model $y = f(x)$ for the men's 200-meter backstroke data given in Table 6 where x is years since 1968 and y is winning time (in seconds). Do the same for the women's 200-meter backstroke data. Do these models indicate that the women will eventually catch up with the men? If so, when? Do you think this will actually occur?

89. Supply and Demand. Table 7 contains price–supply data and price–demand data for corn. Find a linear model $y = f(x)$ for the price–supply data where x is price (in dollars) and y is supply (in billions of bushels). Do the same for the price–demand data. Find the equilibrium price for corn.

TABLE 7 Supply and Demand for U.S. Corn

Price \$/bu	Supply (Billion bu)	Price \$/bu	Demand (Billion bu)
2.15	6.29	2.07	9.78
2.29	7.27	2.15	9.35
2.36	7.53	2.22	8.47
2.48	7.93	2.34	8.12
2.47	8.12	2.39	7.76
2.55	8.24	2.47	6.98
2.71	9.23	2.59	5.57

Source: www.usda.gov/nass/pubs/histdata.htm

90. Supply and Demand. Table 8 contains price–supply data and price–demand data for soybeans. Find a linear model $y = f(x)$ for the price–supply data where x is supply (in billions of bushels) and y is price (in dollars). Do the same for the price– demand data. Find the equilibrium price for soybeans.

TABLE 8 Supply and Demand for U.S. Soybeans

Price \$/bu	Supply (Billion bu)	Price \$/bu	Demand (Billion bu)
5.15	1.55	4.93	2.60
5.79	1.86	5.48	2.40
5.88	1.94	5.71	2.18
6.07	2.08	6.07	2.05
6.15	2.15	6.40	1.95
6.25	2.27	6.66	1.85
6.65	2.53	7.25	1.67

Source: www.usda.gov/nass/pubs/histdata.htm

SECTION 2.3 Quadratic Functions

Quadratic Functions • Completing the Square • Properties of Quadratic Functions and Their Graphs • Modeling with Quadratic Functions

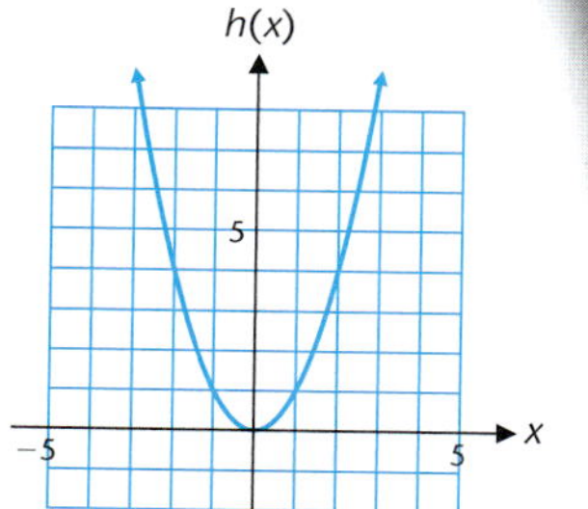

FIGURE 1 Square function $h(x) = x^2$.

Quadratic Functions

The graph of the square function $h(x) = x^2$ is shown in Figure 1. Notice that h is an even function; that is, the graph of h is symmetrical with respect to the y axis. Also, the lowest point on the graph is (0, 0). Let's explore the effect of applying a sequence of basic transformations to the graph of h. (A brief review of Section 1.4 might prove helpful at this point.)

EXPLORE/DISCUSS 1

Indicate how the graph of each function is related to the graph of $h(x) = x^2$. Discuss the symmetry of the graphs and find the highest or lowest point, whichever exists, on each graph.

(A) $f(x) = (x - 3)^2 - 7 = x^2 - 6x + 2$

(B) $g(x) = 0.5(x + 2)^2 + 3 = 0.5x^2 + 2x + 5$

(C) $m(x) = -(x - 4)^2 + 8 = -x^2 + 8x - 8$

(D) $n(x) = -3(x + 1)^2 - 1 = -3x^2 - 6x - 4$

Graphing the functions in Explore/Discuss 1 produces figures similar in shape to the graph of the square function in Figure 1. These figures are called *parabolas*. The functions that produced these parabolas are examples of the important class of *quadratic functions*, which we now define.

Quadratic Functions

If a, b, and c are real numbers with $a \neq 0$, then the function

$$f(x) = ax^2 + bx + c$$

is a **quadratic function** and its graph is a **parabola.***

Because the expression $ax^2 + bx + c$ represents a real number for all real number replacements of x,

the domain of a quadratic function is the set of all real numbers.

We will discuss methods for determining the range of a quadratic function later in this section. Typical graphs of quadratic functions are illustrated in Figure 2.

FIGURE 2 Graphs of quadratic functions.

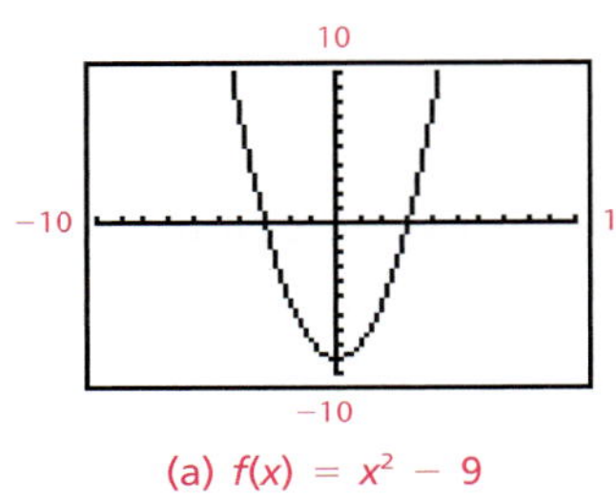

(a) $f(x) = x^2 - 9$

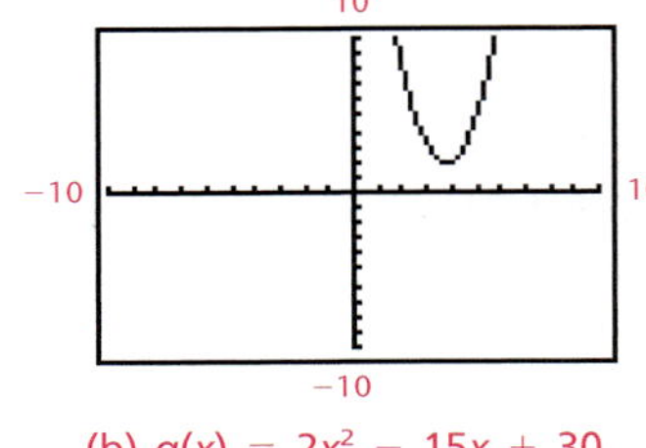

(b) $g(x) = 2x^2 - 15x + 30$

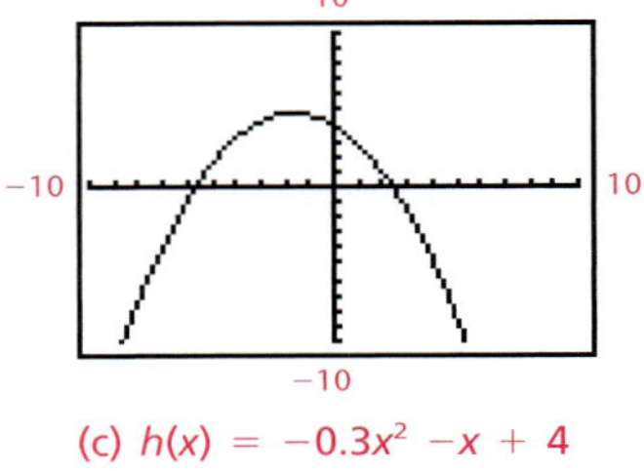

(c) $h(x) = -0.3x^2 - x + 4$

Completing the Square

In Explore/Discuss 1 we wrote each function as two different, but equivalent, expressions. For example,

$$f(x) = (x - 3)^2 - 7 = x^2 - 6x + 2$$

It is easy to verify that these two expressions are equivalent by expanding the first expression. The first expression is more useful than the second for analyzing the graph of f. If we are given only the second expression, how can we determine the first? It turns out that this is a routine process, called *completing the square*,† that is another useful tool to be added to our mathematical toolbox.

---

*A more general definition of a parabola that is independent of any coordinate system is given in Section 8.1.

†See Appendix A, Section A.3, for more examples of completing the square.

EXPLORE/DISCUSS 2

Replace ? in each of the following with a number that makes the equation valid.

(A) $(x + 1)^2 = x^2 + 2x + ?$

(B) $(x + 2)^2 = x^2 + 4x + ?$

(C) $(x + 3)^2 = x^2 + 6x + ?$

(D) $(x + 4)^2 = x^2 + 8x + ?$

Replace ? in each of the following with a number that makes the expression a perfect square of the form $(x + h)^2$.

(E) $x^2 + 10x + ?$ (F) $x^2 + 12x + ?$ (G) $x^2 + bx + ?$

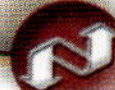

Given the quadratic expression

$$x^2 + bx$$

what must be added to this expression to make it a perfect square? To find out, consider the square of the following expression:

$$(x + m)^2 = x^2 + 2mx + m^2$$

m^2 is the square of one-half the coefficient of x.

We see that the third term on the right side of the equation is the square of one-half the coefficient of x in the second term on the right; that is, m^2 is the square of $\frac{1}{2}(2m)$. This observation leads to the following rule:

Completing the Square

To **complete the square** of the quadratic expression

$$x^2 + bx$$

add the square of one-half the coefficient of x; that is, add

$$\left(\frac{b}{2}\right)^2 \quad \text{or} \quad \frac{b^2}{4}$$

The resulting expression can be factored as a perfect square:

$$x^2 + bx + \left(\frac{b}{2}\right)^2 = \left(x + \frac{b}{2}\right)^2$$

EXAMPLE 1 **Completing the Square**

Complete the square for each of the following:

(A) $x^2 - 6x$ (B) $x^2 + \frac{4}{5}x$ (C) $x^2 - \frac{3}{4}x$

SOLUTION

(A) $x^2 - 6x$

$$x^2 - 6x + 9 = (x - 3)^2 \qquad \text{Add } (-3)^2\text{; that is, 9}$$

(B) $x^2 + \frac{4}{5}x$

$$x^2 + \frac{4}{5}x + \frac{4}{25} = \left(x + \frac{2}{5}\right)^2 \qquad \text{Add } \left(\frac{1}{2} \cdot \frac{4}{5}\right)^2\text{; that is, } \frac{4}{25}$$

(C) $x^2 - \frac{3}{4}x$

$$x^2 - \frac{3}{4}x + \frac{9}{64} = \left(x - \frac{3}{8}\right)^2 \qquad \text{Add } \left(\frac{1}{2} \cdot \frac{-3}{4}\right)^2\text{; that is, } \frac{9}{64}$$

MATCHED 1 PROBLEM

Complete the square for each of the following:

(A) $x^2 - 8x$ (B) $x^2 + \frac{7}{4}x$ (C) $x^2 - \frac{2}{3}x$

It is important to note that the rule for completing the square applies to only quadratic expressions in which the coefficient of x^2 is 1. This causes little trouble, however, as you will see.

Properties of Quadratic Functions and Their Graphs

We now use the process of completing the square to transform the quadratic function

$$f(x) = ax^2 + bx + c$$

into the **vertex form***

$$f(x) = a(x - h)^2 + k$$

Many important features of the graph of a quadratic function can be determined by examining the vertex form. We begin with a specific example and then generalize the results.

Consider the quadratic function given by

$$f(x) = 2x^2 - 8x + 4 \qquad (1)$$

*This terminology is not universally agreed upon. Some call this the *standard form*.

We use completing the square to transform this function into vertex form:

$$\begin{aligned} f(x) &= 2x^2 - 8x + 4 \\ &= 2(x^2 - 4x) + 4 && \text{Factor the coefficient of } x^2 \text{ out of the first two terms.} \\ &= 2(x^2 - 4x + ?) + 4 \\ &= 2(x^2 - 4x + 4 - 4) + 4 && \text{We add 4 to complete the square inside the parentheses.} \\ &= 2(x^2 - 4x + 4) - 2(4) + 4 && \text{But, because of the 2 outside the parentheses,} \\ &= 2(x^2 - 4x + 4) - 8 + 4 && \text{we have actually added 8, so we must subtract 8.} \\ &= 2(x^2 - 4x + 4) - 4 && \text{Factor the perfect square trinomial.} \\ &= 2(x - 2)^2 - 4 && \text{The transformation is complete and can be checked by expanding.} \end{aligned}$$

Thus, the vertex form is

$$f(x) = 2(x - 2)^2 - 4 \tag{2}$$

If $x = 2$, then $2(x - 2)^2 = 0$ and $f(2) = -4$. For any other value of x, the positive number $2(x - 2)^2$ is added to -4, making $f(x)$ larger. Therefore,

$$f(2) = -4$$

is the *minimum value* of $f(x)$ for all x—a very important result! Furthermore, if we choose any two values of x that are equidistant from $x = 2$, we will obtain the same value for the function. For example, $x = 1$ and $x = 3$ are each one unit from $x = 2$ and their functional values are

$$f(1) = 2(-1)^2 - 4 = -2$$
$$f(3) = 2(1)^2 - 4 = -2$$

Thus, the vertical line $x = 2$ is a line of symmetry—if the graph of equation (1) is drawn on a piece of paper and the paper folded along the line $x = 2$, then the two sides of the parabola will match exactly.

The above results are illustrated by graphing equation (1) or (2) and the line $x = 2$ in a suitable viewing window (Fig. 3).

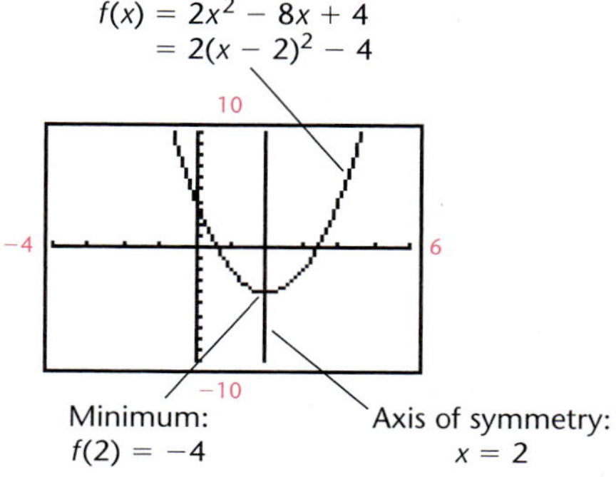

FIGURE 3 Graph of a quadratic function.

From the analysis of equation (2), illustrated by the graph in Figure 3, we conclude that $f(x)$ is decreasing on $(-\infty, 2]$ and increasing on $[2, \infty)$. Furthermore, $f(x)$ can assume any value greater than or equal to -4, but no values less than -4. Thus,

$$\text{Range of } f\text{: } y \geq -4 \qquad \text{or} \qquad [-4, \infty)$$

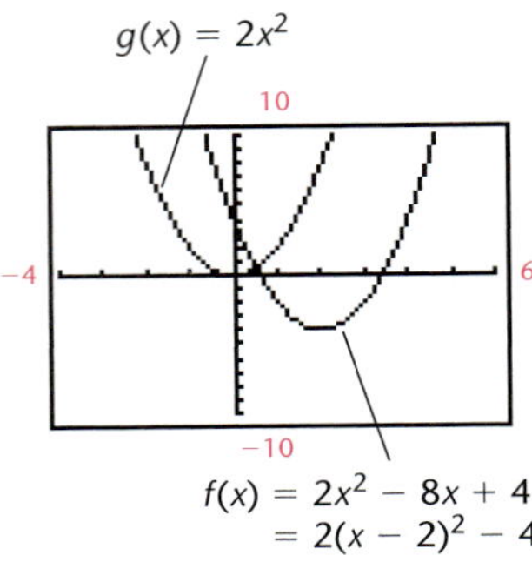

FIGURE 4 Graph of f is the graph of g translated.

In general, the graph of a quadratic function is a parabola with line of symmetry parallel to the vertical axis. The lowest or highest point on the parabola, whichever exists, is called the **vertex.** The maximum or minimum value of a quadratic function always occurs at the vertex of the graph. The vertical line of symmetry through the vertex is called the **axis** of the parabola. Thus, for the function $f(x) = 2x^2 - 8x + 4$, the vertical line $x = 2$ is the axis of the parabola and $(2, -4)$ is its vertex.

From equation (2), we can see that the graph of f is simply the graph of $g(x) = 2x^2$ translated to the right two units and down four units, as shown in Figure 4.

Notice the important results we have obtained from the vertex form of the quadratic function f:

→ The vertex of the parabola
→ The axis of the parabola
→ The minimum value of $f(x)$
→ The range of f
→ A relationship between the graph of f and the graph of g

EXPLORE/DISCUSS 3

Explore the effect of changing the constants a, h, and k on the graph of $f(x) = a(x - h)^2 + k$.

(A) Let $a = 1$ and $h = 5$. Graph function f for $k = -4, 0$, and 3 simultaneously in the same viewing window. Explain the effect of changing k on the graph of f.

(B) Let $a = 1$ and $k = 2$. Graph function f for $h = -4, 0$, and 5 simultaneously in the same viewing window. Explain the effect of changing h on the graph of f.

(C) Let $h = 5$ and $k = -2$. Graph function f for $a = 0.25, 1$, and 3 simultaneously in the same viewing window. Graph function f for $a = 1, -1$, and -0.25 simultaneously in the same viewing window. Explain the effect of changing a on the graph of f.

(D) Can all quadratic functions of the form $y = ax^2 + bx + c$ be rewritten as $a(x - h)^2 + k$?

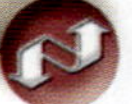

We generalize this discussion in the box:

Properties of a Quadratic Function and its Graph

Given a quadratic function and the vertex form obtained by completing the square

$$f(x) = ax^2 + bx + c = a(x - h)^2 + k \qquad a \neq 0$$

we summarize general properties as follows:

1. The graph of f is a parabola:

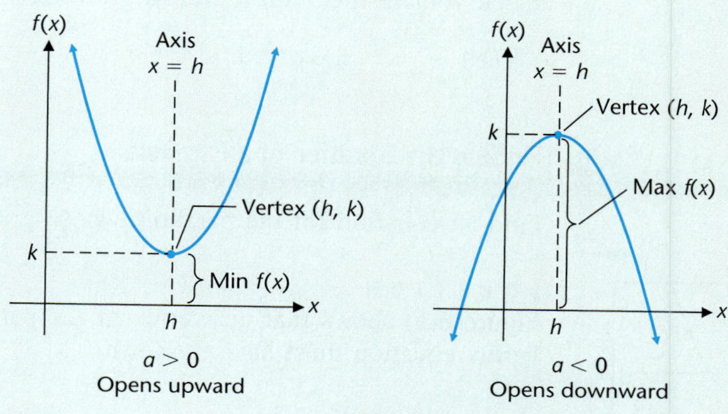

2. Vertex: (h, k) (parabola rises on one side of the vertex and falls on the other).
3. Axis (of symmetry): $x = h$ (parallel to y axis).
4. $f(h) = k$ is the minimum if $a > 0$ and the maximum if $a < 0$.
5. Domain: all real numbers; range: $(-\infty, k]$ if $a < 0$ or $[k, \infty)$ if $a > 0$.
6. The graph of f is the graph of $g(x) = ax^2$ translated horizontally h units and vertically k units.
7. $h = -\frac{b}{2a}$, $k = c - \frac{b^2}{4a}$

EXAMPLE 2 Analyzing a Quadratic Function

Find the vertex form for the following quadratic function, analyze the graph, and check your results with a graphing utility:

$$f(x) = -0.5x^2 - x + 5$$

SOLUTION

We complete the square to find the vertex form:

$$\begin{aligned} f(x) &= -0.5x^2 - x + 5 \\ &= -0.5(x^2 + 2x + ?) + 5 \\ &= -0.5(x^2 + 2x + 1) + 5 + 0.5 \\ &= -0.5(x + 1)^2 + 5.5 \end{aligned}$$

From the vertex form we see that $h = -1$ and $k = 5.5$. Therefore, the vertex is $(-1, 5.5)$, the axis of symmetry is $x = -1$, the maximum value is $f(-1) = 5.5$, and the range is $(-\infty, 5.5]$. The function f is increasing on $(-\infty, -1]$ and decreasing on $[-1, \infty)$. The graph of f is the graph of $g(x) = -0.5x^2$ shifted to the left one unit and upward five and one-half units. To check these results, we graph f and g simultaneously in the same viewing window, use the built-in maximum routine to locate the vertex, and add the graph of the axis of symmetry (Fig. 5).

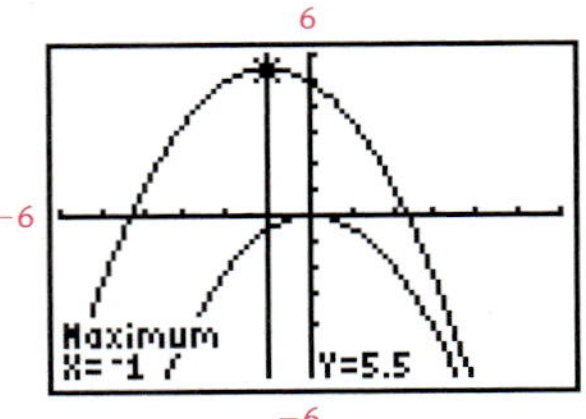

FIGURE 5

MATCHED 2 PROBLEM

Find the vertex form for the following quadratic function, analyze the graph, and check your results with a graphing utility:

$$f(x) = -x^2 + 3x - 1$$

EXAMPLE 3 Finding the Equation of a Parabola

Find an equation for the parabola whose graph is shown in Figure 6.

(a)

(b)

FIGURE 6

SOLUTION

Figure 6(a) shows that the vertex of the parabola is $(h, k) = (3, -2)$. Thus, the vertex equation must have the form

$$f(x) = a(x - 3)^2 - 2 \tag{3}$$

Figure 6(b) shows that $f(4) = 0$. Substituting in equation (3) and solving for a, we have

$$f(4) = a(4 - 3)^2 - 2 = 0$$
$$a = 2$$

Thus, the equation for the parabola is

$$f(x) = 2(x - 3)^2 - 2 = 2x^2 - 12x + 16$$

MATCHED 3 PROBLEM

Find the equation of the parabola with vertex (2, 4) and y intercept (0, 2).

EXAMPLE 4 Finding the Equation of a Parabola

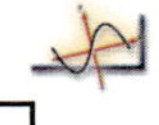

Find an equation for the parabola whose graph is shown in Figure 7.

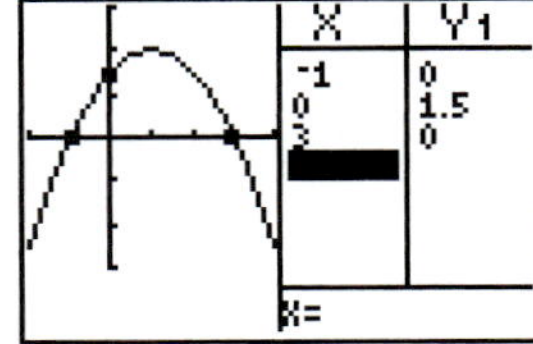

FIGURE 7

SOLUTION

Let

$$f(x) = a(x - h)^2 + k$$

Because $f(-1) = f(3)$, the axis of symmetry must be the midpoint of the interval $[-1, 3]$. Thus,

$$h = \frac{-1 + 3}{2} = 1 \qquad \text{and} \qquad f(x) = a(x - 1)^2 + k$$

Now we can use either x intercept to find a relationship between a and k. We choose $f(-1) = 0$.

$$\begin{aligned} f(-1) = a(-1 - 1)^2 + k &= 0 \\ 4a + k &= 0 \\ k &= -4a \end{aligned}$$

Now we can write

$$f(x) = a(x - 1)^2 - 4a$$

and use the y intercept to find a. (We can't use the other x intercept to find a. Try it to see why.)

$$\begin{aligned} f(0) = a(0 - 1)^2 - 4a &= 1.5 \\ a - 4a &= 1.5 \\ -3a &= 1.5 \\ a &= -0.5 \end{aligned}$$

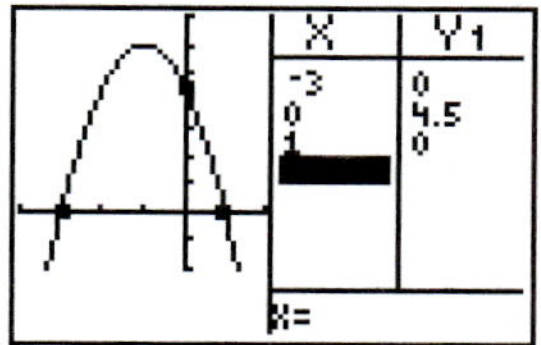

FIGURE 8

Thus,

$$f(x) = -0.5(x - 1)^2 + 2 = -0.5x^2 + x + 1.5$$

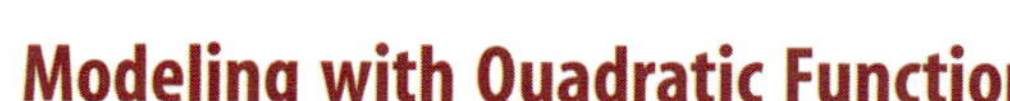

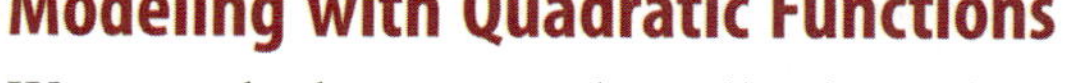

Find an equation for the parabola whose graph is shown in Figure 8.

Modeling with Quadratic Functions

We now look at several applications that can be modeled using quadratic functions.

EXAMPLE 5 Maximum Area

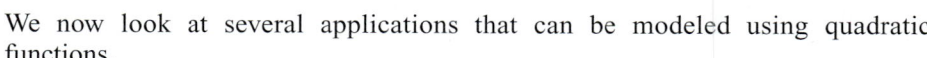

A dairy farm has a barn that is 150 feet long and 75 feet wide. The owner has 240 feet of fencing and wishes to use all of it in the construction of two identical adjacent outdoor pens with the long side of the barn as one side of the pens and a common fence between the two (Fig. 9). The owner wants the pens to be as large as possible.

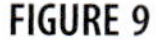

FIGURE 9

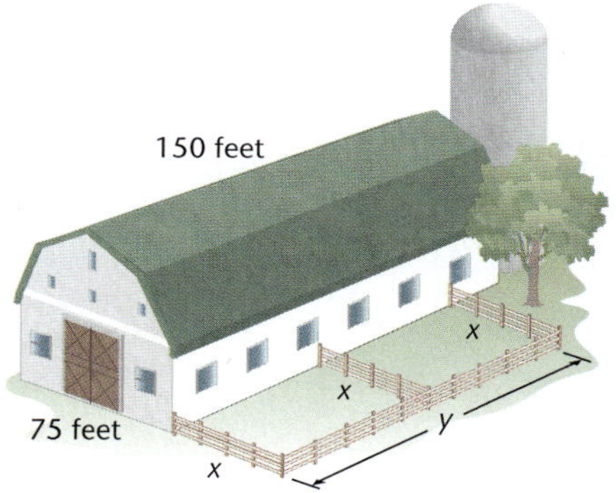

(A) Construct a mathematical model for the combined area of both pens in the form of a function $A(x)$ (see Fig. 9) and state the domain of A.

(B) Find the value of x that produces the maximum combined area.

(C) Find the dimensions and the area of each pen.

SOLUTION

(A) The combined area of the two pens is

$$A = xy$$

Building the pens will require $3x + y$ feet of fencing. Thus,

$$3x + y = 240$$
$$y = 240 - 3x$$

Because the distances x and y must be nonnegative, x and y must satisfy $x \geq 0$ and $y = 240 - 3x \geq 0$. It follows that $0 \leq x \leq 80$. Substituting for y in the combined area equation, we have the following model for this problem:

$$A(x) = x(240 - 3x) = 240x - 3x^2 \qquad 0 \leq x \leq 80$$

(B) Algebraic Solution

$$A(x) = 240x - 3x^2$$
$$= -3(x^2 - 80x + 1600) + 3 \cdot 1,600$$
$$= -3(x - 40)^2 + 4,800$$

Thus, the maximum combined area of 4,800 ft.^{2} occurs at $x = 40$ ft.

(B) Graphical Solution

Entering $y_1 = 240x - 3x^2$ and using the maximum command produces the graph in Figure 10. This shows that the maximum combined area of 4,800 ft.^{2} occurs at $x = 40$ ft.

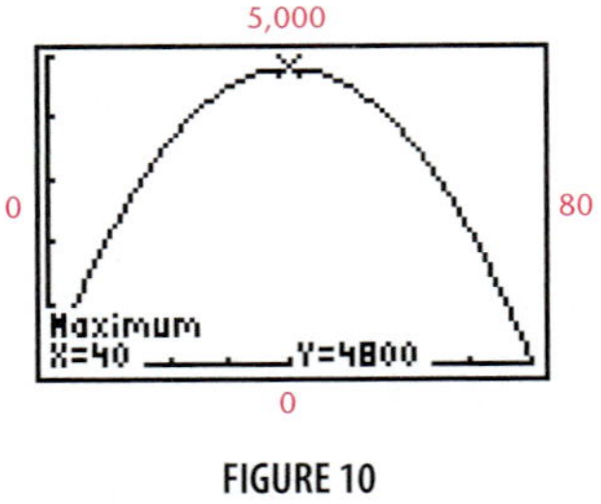

FIGURE 10

(C) Each pen is x by $y/2$ or 40 feet by 60 feet. The area of each pen is 40 feet × 60 feet = 2,400 ft^2.

MATCHED 5 PROBLEM

Repeat Example 5 with the owner constructing three identical adjacent pens instead of two.

Galileo was the first to discover that the distance an object falls is proportional to the square of the time it has been falling. Thus, a quadratic function is a good model for falling objects. Neglecting air resistance, the quadratic function

$$h(t) = h_0 - 16t^2$$

represents the **height of an object** t seconds after it is dropped from an initial height of h_0 feet. The constant -16 is related to the force of gravity and is dependent on the units used. That is, -16 only works for distances measured in feet and time measured in seconds. If the object is thrown either upward or downward, the quadratic model will also have a term involving t (see Problems 69 and 70 in Exercise 2.3).

EXAMPLE 6 Projectile Motion

A thermos bottle is dropped from a hot air balloon that is 200 feet in the air. When will the bottle hit the ground? Round answer to two decimal places.

SOLUTION

Because the initial height is 200 feet, the quadratic model for the height of the bottle is

$$h(t) = 200 - 16t^2$$

Because $h(t) = 0$ when the bottle hits the ground, we must solve this equation for t.

Algebraic Solution

$$\begin{aligned} h(t) = 200 - 16t^2 &= 0 \\ 16t^2 &= 200 \\ t^2 &= \frac{200}{16} = 12.5 \\ t &= \sqrt{12.5} \\ &= 3.54 \text{ seconds} \end{aligned}$$

Graphical Solution

Graphing $y_1 = 200 - 16x^2$ and using the zero command (Fig. 11) shows that $h = 0$ at $t = 3.54$ seconds.

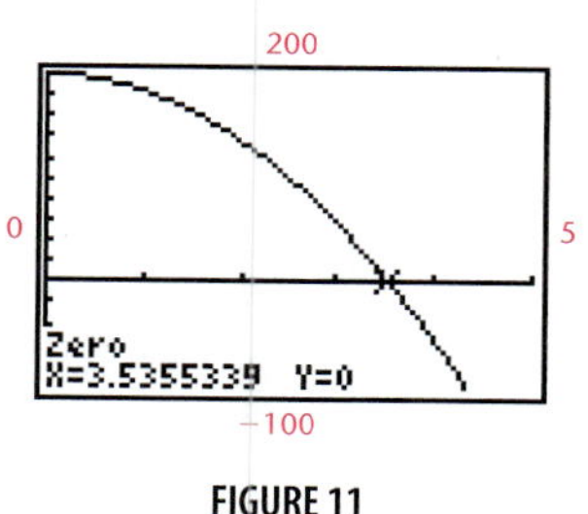

FIGURE 11

MATCHED 6 PROBLEM

A thermos bottle is dropped from a hot air balloon that is 300 feet in the air. When will the bottle hit the ground? Round answer to two decimal places.

ANSWERS MATCHED PROBLEMS

1. (A) $x^2 - 8x + 16 = (x - 4)^2$

   (B) $x^2 + \frac{7}{4}x + \frac{49}{64} = \left(x + \frac{7}{8}\right)^2$

   (C) $x^2 - \frac{2}{3}x + \frac{1}{9} = \left(x - \frac{1}{3}\right)^2$

2. Vertex form: $f(x) = -(x - 1.5)^2 + 1.25$. The vertex is (1.5, 1.25), the axis of symmetry is $x = 1.5$, the maximum value of $f(x)$ is 1.25, and the range of f is $(-\infty, 1.25]$. The function f is increasing on $(-\infty, 1.5]$ and decreasing on $[1.5, \infty)$. The graph of f is the graph of $g(x) = -x^2$ shifted one and a half units to the right and one and a quarter units upward.

3. $f(x) = -0.5(x - 2)^2 + 4$
   $= -0.5x^2 + 2x + 2$

4. $f(x) = -1.5(x + 1)^2 + 6$
   $= -1.5x^2 - 3x + 4.5$

5. (A) $A(x) = (240 - 4x)x$, $0 \le x \le 60$

   (B) The maximum combined area of 3,600 ft.[2] occurs at $x = 30$ feet.

   (C) Each pen is 30 feet by 40 feet with area 1,200 ft[2].

6. 4.33 seconds

EXERCISE 2.3

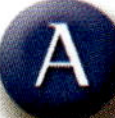

A

In Problems 1–6, complete the square and find the vertex form of each quadratic function.

1. $f(x) = x^2 - 4x + 5$
2. $g(x) = -x^2 - 2x - 3$
3. $h(x) = -x^2 - 2x - 1$
4. $k(x) = x^2 - 4x + 4$
5. $m(x) = x^2 - 4x + 1$
6. $n(x) = -x^2 - 2x + 3$

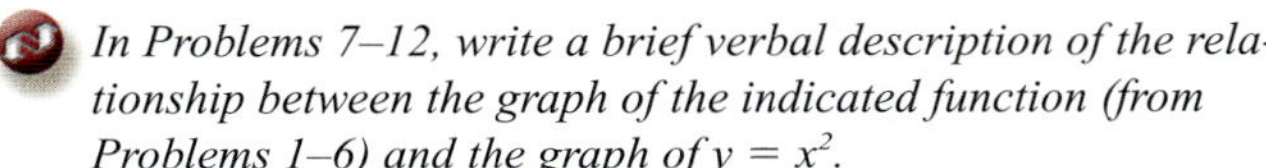

In Problems 7–12, write a brief verbal description of the relationship between the graph of the indicated function (from Problems 1–6) and the graph of $y = x^2$.

7. $f(x) = x^2 - 4x + 5$
8. $g(x) = -x^2 - 2x - 3$
9. $h(x) = -x^2 - 2x - 1$
10. $k(x) = x^2 - 4x + 4$
11. $m(x) = x^2 - 4x + 1$
12. $n(x) = -x^2 - 2x + 3$

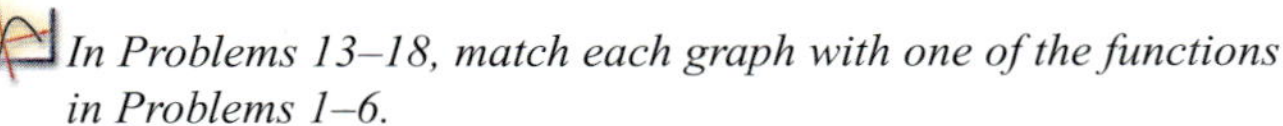

In Problems 13–18, match each graph with one of the functions in Problems 1–6.

13.

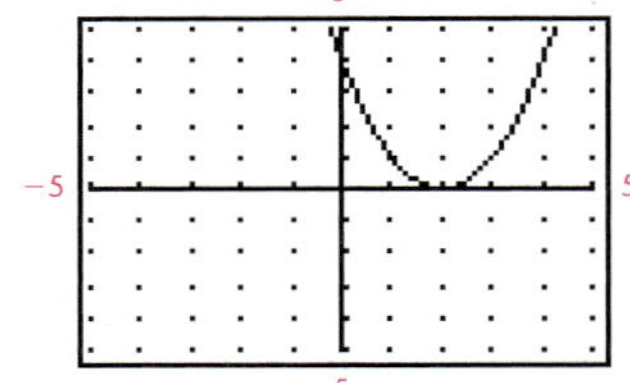

14.

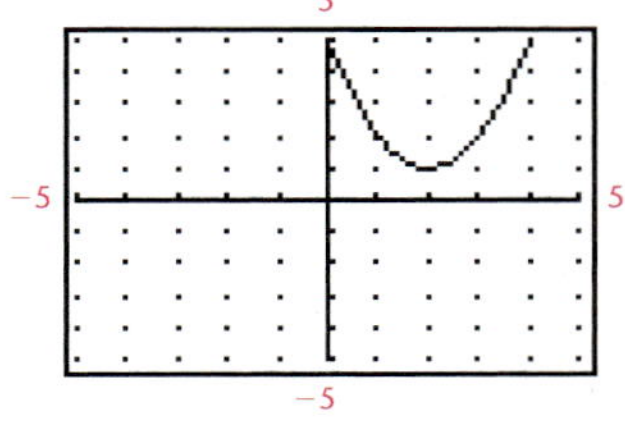

15.

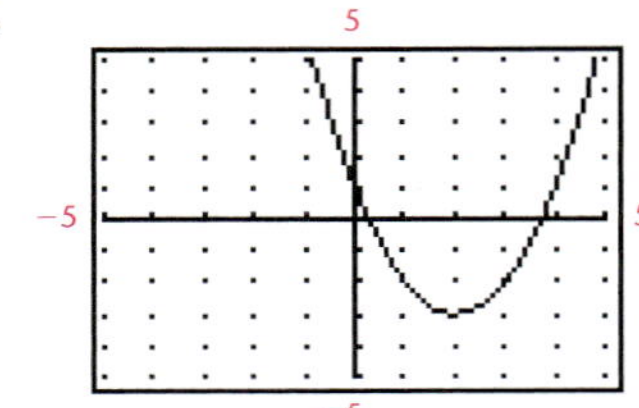

16.

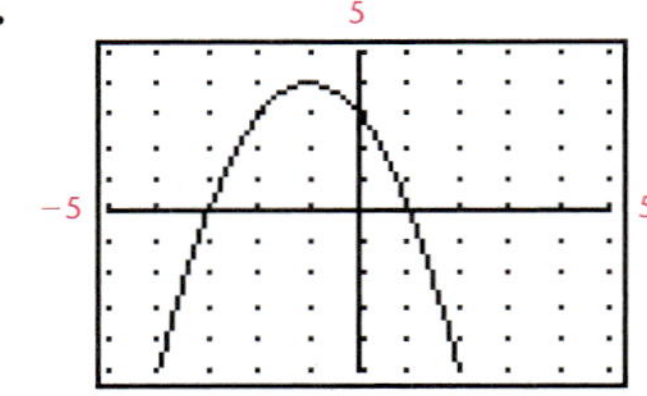

17.

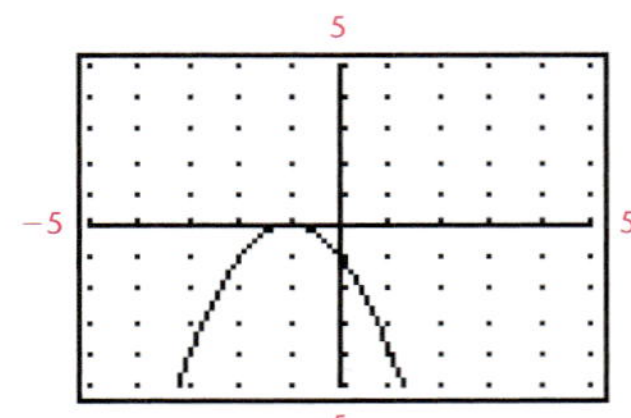

18.

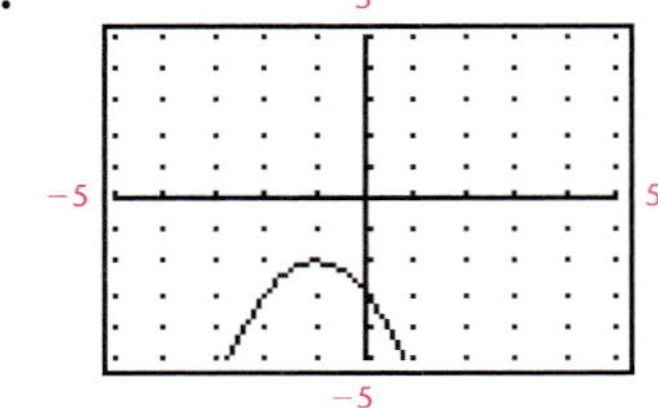

B

For each quadratic function in Problems 19–24, sketch a graph of the function and label the axis and the vertex.

19. $f(x) = 2x^2 - 24x + 90$

20. $f(x) = 3x^2 + 24x + 30$

21. $f(x) = -x^2 - 6x - 4$

22. $f(x) = -x^2 + 10x - 30$

23. $f(x) = 0.5x^2 - 2x - 7$

24. $f(x) = 0.4x^2 + 4x + 4$

In Problems 25–28, find the intervals where f is increasing, the intervals where f is decreasing, and the range. Express answers in interval notation.

25. $f(x) = 4x^2 - 18x + 25$

26. $f(x) = 5x^2 + 29x - 17$

27. $f(x) = -10x^2 + 44x + 12$

28. $f(x) = -8x^2 - 20x + 16$

In Problems 29–36, use the graph of the parabola to find the equation of the corresponding quadratic function.

29.

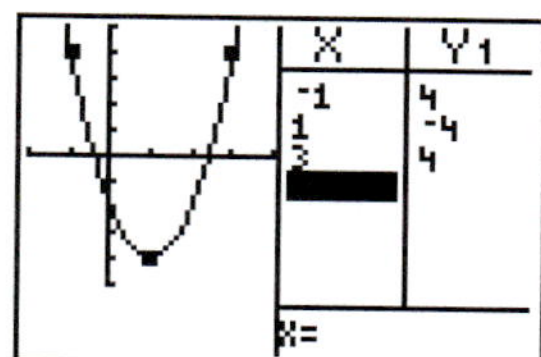

30.

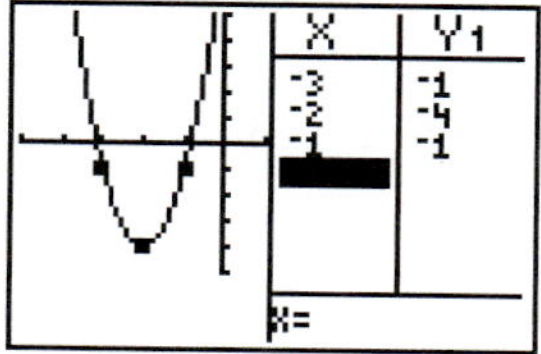

31.

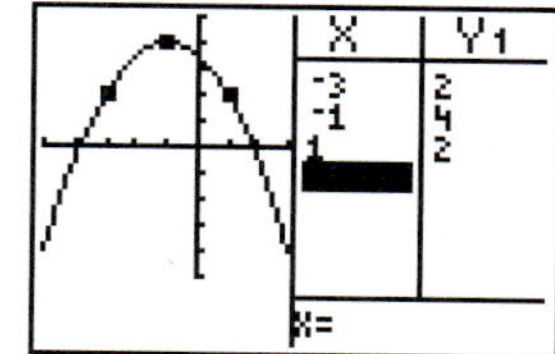

32.

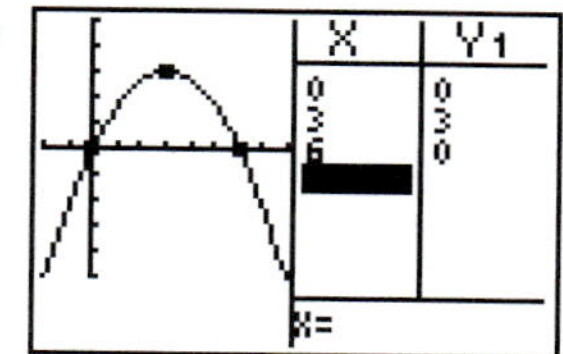

33.

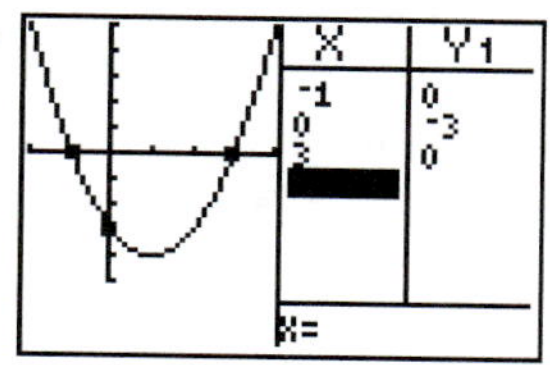

34.

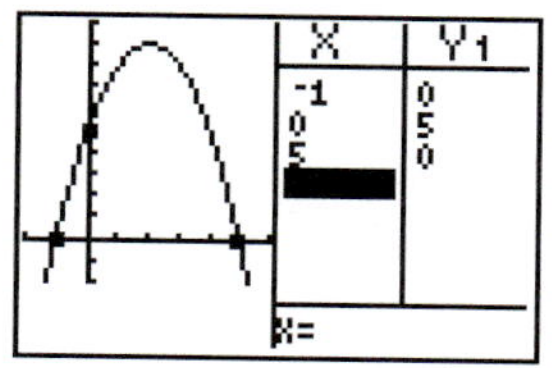

35.

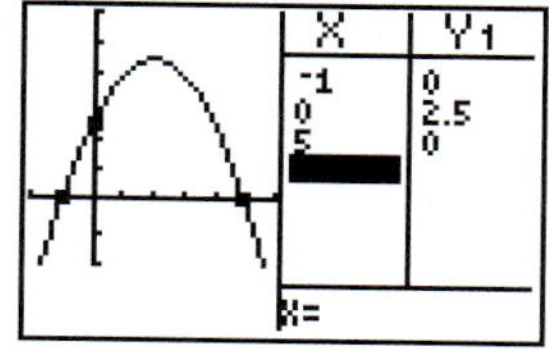

36.

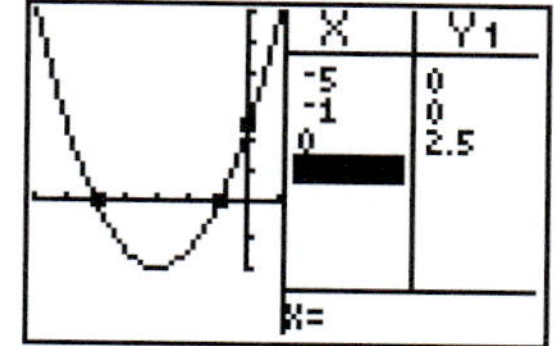

In problems 37–42, find the equation of a quadratic function whose graph satisfies the given conditions.

37. Vertex: (4, 8); x intercept: 6

38. Vertex: $(-2, -12)$; x intercept: -4

39. Vertex: $(-4, 12)$; y intercept: 4

40. Vertex: (5, 8); y intercept: -2

41. Vertex: $(-5, -25)$; additional point on graph: $(-2, 20)$

42. Vertex: $(6, -40)$; additional point on graph: (3, 50)

In Problems 43–46, find the inverse of the given function.

43. $g(x) = x^2 + 2x - 3, x \geq -1$

44. $h(x) = x^2 - 8x + 10, x \geq 4$

45. $p(x) = -x^2 + 2x + 3, x \leq 1$

46. $q(x) = -x^2 - 4x, x \leq -2$

47. Let $g(x) = x^2 + kx + 1$. Graph g for several different values of k and discuss the relationship among these graphs.

48. Confirm your conclusions in Problem 47 by finding the vertex form for g.

49. Let $f(x) = (x - 1)^2 + k$. Discuss the relationship between the values of k and the number of x intercepts for the graph of f. Generalize your comments to any function of the form

$$f(x) = a(x - h)^2 + k, a > 0$$

50. Let $f(x) = -(x - 2)^2 + k$. Discuss the relationship between the values of k and the number of x intercepts for the graph of f. Generalize your comments to any function of the form

$$f(x) = a(x - h)^2 + k, a < 0$$

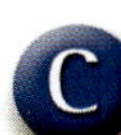

Recall that the standard equation of a circle with radius r and center (h, k) is

$$(x - h)^2 + (y - k)^2 = r^2$$

In Problems 51–54, use completing the square twice to find the center and radius of the circle with the given equation.

51. $x^2 + y^2 - 6x - 4y = 36$

52. $x^2 + y^2 - 2x - 10y = 55$

53. $x^2 + y^2 + 8x - 2y = 8$

54. $x^2 + y^2 - 4x + 12y = 24$

55. Let $f(x) = a(x - h)^2 + k$. Compare the values of $f(h + r)$ and $f(h - r)$ for any real number r. Interpret the results in terms of the graph of f.

56. Complete the square in $f(x) = ax^2 + bx + c, a \neq 0$, to show that $h = -\frac{b}{2a}$ and $k = c - \frac{b^2}{4a}$ (Property 7 on page 157).

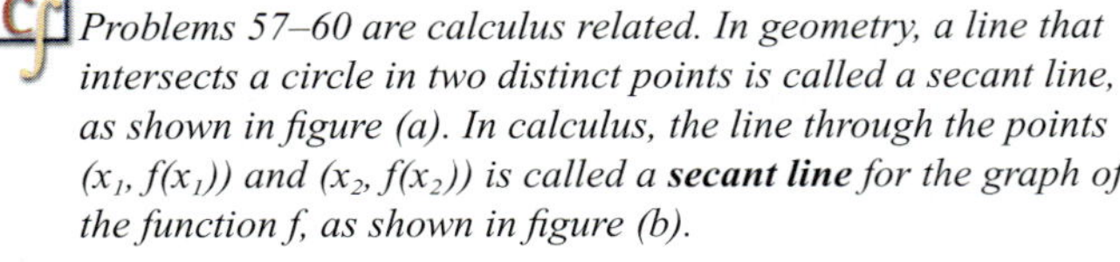

Problems 57–60 are calculus related. In geometry, a line that intersects a circle in two distinct points is called a secant line, as shown in figure (a). In calculus, the line through the points $(x_1, f(x_1))$ and $(x_2, f(x_2))$ is called a ***secant line*** *for the graph of the function f, as shown in figure (b).*

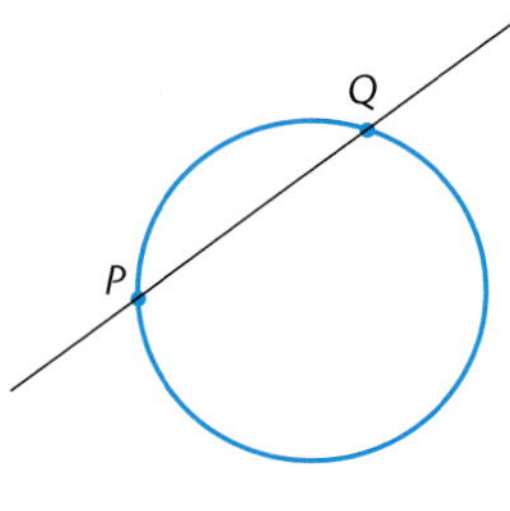

Secant line for a circle
(a)

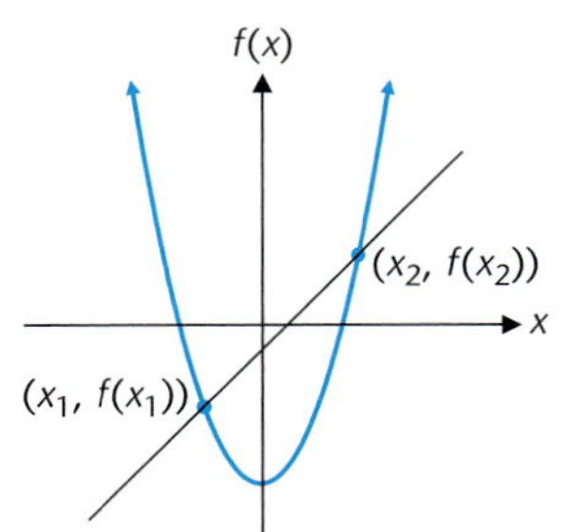

Secant line for the graph of a function
(b)

In Problems 57 and 58, find the equation of the secant line through the indicated points on the graph of f. Graph f and the secant line on the same coordinate system.

57. $f(x) = x^2 - 4$; $(-1, -3)$, $(3, 5)$

58. $f(x) = 9 - x^2$, $(-2, 5)$, $(4, -7)$

59. Let $f(x) = x^2 - 3x + 5$. If h is a nonzero real number, then $(2, f(2))$ and $(2 + h, f(2 + h))$ are two distinct points on the graph of f.

(A) Find the slope of the secant line through these two points.

(B) Evaluate the slope of the secant line for $h = 1$, $h = 0.1$, $h = 0.01$, and $h = 0.001$. What value does the slope seem to be approaching?

60. Repeat Problem 59 for $f(x) = x^2 + 2x - 6$.

61. Find the minimum product of two numbers whose difference is 30. Is there a maximum product? Explain.

62. Find the maximum product of two numbers whose sum is 60. Is there a minimum product? Explain.

APPLICATIONS

63. Construction. A horse breeder wants to construct a corral next to a horse barn that is 50 feet long, using all of the barn as one side of the corral (see the figure). He has 250 feet of fencing available and wants to use all of it.

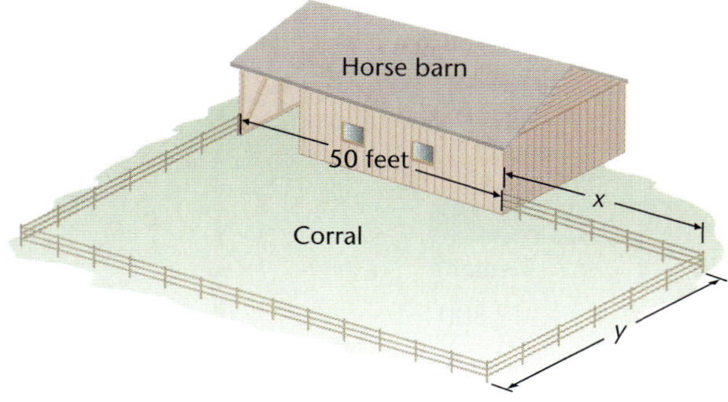

(A) Express the area $A(x)$ of the corral as a function of x and indicate its domain.

(B) Find the value of x that produces the maximum area.

(C) What are the dimensions of the corral with the maximum area?

64. Construction. Repeat Problem 63 if the horse breeder has only 140 feet of fencing available for the corral. Does the maximum value of the area function still occur at the vertex? Explain.

65. Falling Object. A sandbag is dropped off a high-altitude balloon at an altitude of 10,000 ft. When will the sandbag hit the ground?

66. Falling Object. A prankster drops a water balloon off the top of a 144-ft.-high building. When will the balloon hit the ground?

67. Falling Object. A cliff diver hits the water 2.5 seconds after diving off the cliff. How high is the cliff?

68. Falling Object. A forest ranger drops a coffee cup off a fire watchtower. If the cup hits the ground 1.5 seconds later, how high is the tower?

69. Projectile Flight. An arrow shot vertically into the air from a crossbow reaches a maximum height of 484 feet after 5.5 seconds of flight. Let the quadratic function $d(t)$ represent the distance above ground (in feet) t seconds after the arrow is released. (If air resistance is neglected, a quadratic model provides a good approximation for the flight of a projectile.)

(A) Find $d(t)$ and state its domain.

(B) At what times (to two decimal places) will the arrow be 250 feet above the ground?

70. Projectile Flight. Repeat Problem 69 if the arrow reaches a maximum height of 324 feet after 4.5 seconds of flight.

71. Engineering. The arch of a bridge is in the shape of a parabola 14 feet high at the center and 20 feet wide at the base (see the figure).

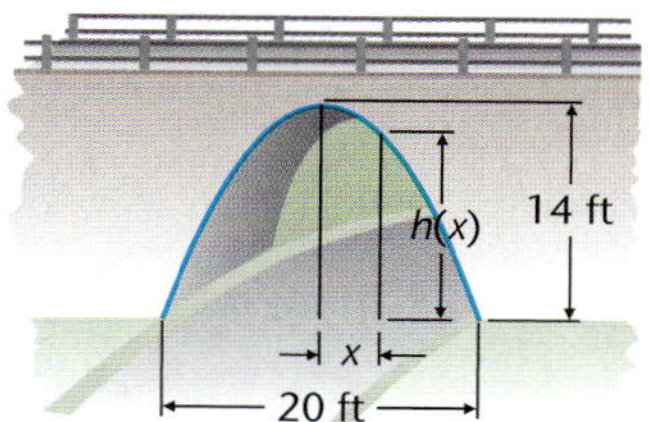

(A) Express the height of the arch $h(x)$ in terms of x and state its domain.

(B) Can a truck that is 8 feet wide and 12 feet high pass through the arch?

(C) What is the tallest 8-ft.-wide truck that can pass through the arch?

(D) What (to two decimal places) is the widest 12-ft.-high truck that can pass through the arch?

72. Engineering. The roadbed of one section of a suspension bridge is hanging from a large cable suspended between two towers that are 200 feet apart (see the figure). The cable forms a parabola that is 60 feet above the roadbed at the towers and 10 feet above the roadbed at the lowest point.

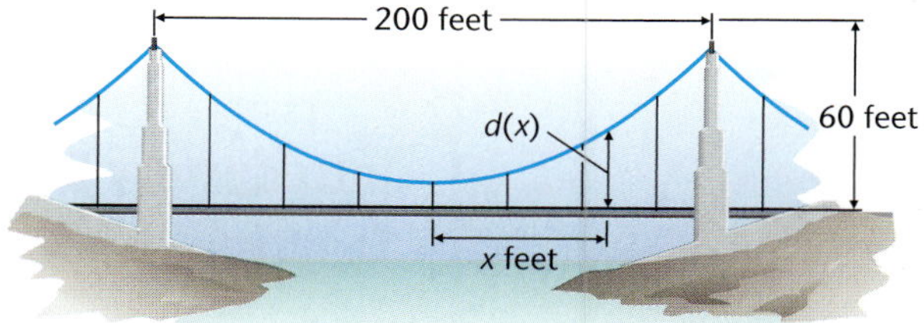

(A) Express the vertical distance $d(x)$ (in feet) from the roadbed to the suspension cable in terms of x and state the domain of d.

(B) The roadbed is supported by seven equally spaced vertical cables (see the figure). Find the combined total length of these supporting cables.

MODELING AND LINEAR REGRESSION

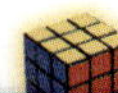

73. Maximizing Revenue. A company that manufactures flashlights has collected the price–demand data in Table 1. Round all numbers to three significant digits.

(A) Use the data in Table 1 and linear regression to find a linear price–demand function $p = d(x)$ where x is the number of flashlights (in thousands) that the company can sell at a price of p dollars.

(B) Find the price that maximizes the company's revenue from the sale of flashlights.

74. Maximizing Revenue. A company that manufactures pencil sharpeners has collected the price–demand data in Table 2. Round all numbers to three significant digits.

(A) Use the data in Table 2 and linear regression to find a linear price–demand function $p = d(x)$ where x is the number of pencil sharpeners (in thousands) that the company can sell at a price of p dollars.

(B) Find the price that maximizes the company's revenue from the sale of pencil sharpeners.

TABLE 1

Price	Demand
\$3.55	45,800
\$3.95	40,500
\$4.13	37,900
\$4.85	34,700
\$5.19	30,400
\$5.55	28,900
\$6.15	25,400

TABLE 2

Price	Demand
\$4.23	47,800
\$4.89	45,600
\$5.43	42,700
\$5.97	39,600
\$6.47	34,700
\$7.12	31,600
\$7.84	27,800

SECTION 2.4 Complex Numbers

Introductory Remarks • The Complex Number System • Complex Numbers and Radicals • Solving Equations Involving Complex Numbers

Introductory Remarks

The Pythagoreans (c. 500 B.C.) found that the simple quadratic equation

$$x^2 - 2 = 0 \tag{1}$$

had no rational number solution—it is not possible to find the ratio of two integers whose square is 2. If equation (1) were to have a solution, then a new kind of number had to be invented—an irrational number. The irrational numbers $\sqrt{2}$ and $-\sqrt{2}$ are both solutions to equation (1). Irrational numbers were not put on a firm mathematical foundation until the nineteenth century. The rational and irrational numbers together constitute the real number system. Is there any reason to invent any other kinds of numbers?

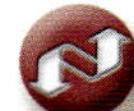

EXPLORE/DISCUSS 1

Graph $g(x) = x^2 - 1$ in a standard viewing window and discuss the relationship between the real zeros of the function and the x intercepts of its graph. Do the same for $f(x) = x^2 + 1$.

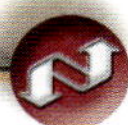

Does the simple quadratic equation

$$x^2 + 1 = 0 \tag{2}$$

have a solution? If equation (2) is to have a solution, x^2 must be negative. But the square of a real number is never negative. Thus, equation (2) cannot have any real number solutions. Once again, a new type of number must be invented—a number whose square can be negative. These new numbers are among the numbers called *complex numbers*. The complex numbers evolved over a long period, but, like the real numbers, it was not until the nineteenth century that they were given a firm mathematical foundation. Table 1 gives a brief history of the evolution of complex numbers.

TABLE 1 Brief History of Complex Numbers

Approximate Date A.D.	Person	Event
50	Heron of Alexandria	First recorded encounter of a square root of a negative number.
850	Mahavira of India	Said that a negative has no square root, because it is not a square.
1545	Cardano of Italy	Solutions to cubic equations involved square roots of negative numbers.
1637	Descartes of France	Introduced the terms *real* and *imaginary*.
1748	Euler of Switzerland	Used i for $\sqrt{-1}$.
1832	Gauss of Germany	Introduced the term *complex number*.

The Complex Number System

We start the development of the complex number system by defining a complex number and several special types of complex numbers. We then define equality, addition, and multiplication in this system, and from these definitions the important special properties and operational rules for addition, subtraction, multiplication, and division will follow.

DEFINITION 1

Complex Number

A **complex number** is a number of the form

$a + bi$ Standard Form

where a and b are real numbers and i is called the **imaginary unit.**

The imaginary unit i introduced in Definition 1 is not a real number. It is a special symbol used in the representation of the elements in this new complex number system.

Some examples of complex numbers are

$$3 - 2i \qquad \tfrac{1}{2} + 5i \qquad 2 - \tfrac{1}{3}i$$
$$0 + 3i \qquad 5 + 0i \qquad 0 + 0i$$

Particular kinds of complex numbers are given special names as follows:

DEFINITION 2

Names for Particular Kinds of Complex Numbers

Imaginary Unit:	i	
Complex Number:	$a + bi$	a and b real numbers
Imaginary Number:	$a + bi$	$b \neq 0$
Pure Imaginary Number:	$0 + bi = bi$	$b \neq 0$
Real Number:	$a + 0i = a$	
Zero:	$0 + 0i = 0$	
Conjugate of $a + bi$:	$a - bi$	

EXAMPLE 1 Special Types of Complex Numbers

Given the list of complex numbers:

$$3 - 2i \qquad \tfrac{1}{2} + 5i \qquad 2 - \tfrac{1}{3}i$$
$$0 + 3i = 3i \qquad 5 + 0i = 5 \qquad 0 + 0i = 0$$

(A) List all the imaginary numbers, pure imaginary numbers, real numbers, and zero.

(B) Write the conjugate of each.

SOLUTIONS

(A) Imaginary numbers: $3 - 2i, \frac{1}{2} + 5i, 2 - \frac{1}{3}i, 3i$
Pure imaginary numbers: $0 + 3i = 3i$
Real numbers: $5 + 0i = 5, 0 + 0i = 0$
Zero: $0 + 0i = 0$

(B) $3 + 2i \qquad \frac{1}{2} - 5i \qquad 2 + \frac{1}{3}i$
$0 - 3i = -3i \qquad 5 - 0i = 5 \qquad 0 - 0i = 0$

MATCHED PROBLEM 1

Given the list of complex numbers:

$$6 + 7i \qquad \sqrt{2} - \tfrac{1}{3}i \qquad 0 - i = -i$$
$$0 + \tfrac{2}{3}i = \tfrac{2}{3}i \qquad -\sqrt{3} + 0i = -\sqrt{3} \qquad 0 - 0i = 0$$

(A) List all the imaginary numbers, pure imaginary numbers, real numbers, and zero.

(B) Write the conjugate of each.

In Definition 2, notice that we identify a complex number of the form $a + 0i$ with the real number a, a complex number of the form $0 + bi$, $b \neq 0$, with the **pure imaginary number** bi, and the complex number $0 + 0i$ with the real number 0. Thus, a real number is also a complex number, just as a rational number is also a real number. Any complex number that is not a real number is called an **imaginary number.** If we combine the set of all real numbers with the set of all imaginary numbers, we obtain ***C*, the set of complex numbers.** The relationship of the complex number system to the other number systems we have studied is shown in Figure 1.

FIGURE 1

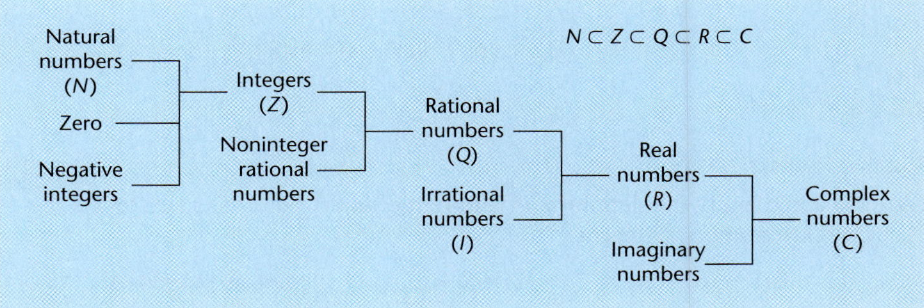

To use complex numbers, we must know how to add, subtract, multiply, and divide them. We start by defining equality, addition, and multiplication.

DEFINITION 3
Equality and Basic Operations

1. **Equality:** $a + bi = c + di$ if and only if $a = c$ and $b = d$
2. **Addition:** $(a + bi) + (c + di) = (a + c) + (b + d)i$
3. **Multiplication:** $(a + bi)(c + di) = (ac - bd) + (ad + bc)i$

We can show, using the definitions of addition and multiplication of complex numbers (Definition 3), that basic properties of the real number system* extend to the following basic properties of the complex number system.

Basic Properties of the Complex Number System

1. Addition and multiplication of complex numbers are commutative and associative operations.
2. There is an additive identity and a multiplicative identity for complex numbers.
3. Every complex number has an additive inverse or negative.
4. Every nonzero complex number has a multiplicative inverse or reciprocal.
5. Multiplication distributes over addition.

As a consequence of these properties you will not have to memorize the definitions of addition and multiplication of complex numbers in Definition 3.

*Basic properties of the real number system are discussed in the Basic Algebra Review, Section R.1, at www.mhhe.com/barnett

We can manipulate complex number symbols of the form $a + bi$ just like we manipulate real binomials of the form $a + bx$, as long as we remember that i is a special symbol for the imaginary unit, not for a real number.

The first arithmetic operation we consider is **addition.**

EXAMPLE 2 Addition of Complex Numbers

Carry out each operation and express the answer in standard form:

(A) $(2 - 3i) + (6 + 2i)$

(B) $(-5 + 4i) + (0 + 0i)$

SOLUTIONS

Algebraic Solutions

(A) We could apply the definition of addition directly, but it is easier to use complex number properties.

$$(2 - 3i) + (6 + 2i) = 2 - 3i + 6 + 2i \quad \text{Remove parentheses.}$$
$$= (2 + 6) + (-3 + 2)i \quad \text{Combine like terms.}$$
$$= 8 - i$$

(B) $(-5 + 4i) + (0 + 0i) = -5 + 4i + 0 + 0i$
$= -5 + 4i$

Graphical Solutions

```
(2-3i)+(6+2i)
            8-i
(-5+4i)+(0+0i)
          -5+4i
```

FIGURE 2

MATCHED 2 PROBLEM

Carry out each operation and express the answer in standard form:

(A) $(3 + 2i) + (6 - 4i)$ (B) $(0 + 0i) + (7 - 5i)$

Example 2, part B, and Matched Problem 2, part B, illustrate the following general result: For any complex number $a + bi$,

$$(a + bi) + (0 + 0i) = (0 + 0i) + (a + bi) = a + bi$$

Thus, $0 + 0i$ is the **additive identity** or **zero** for the complex numbers. We anticipated this result in Definition 2 when we identified the complex number $0 + 0i$ with the real number 0.

Not all graphing utilities make use of the $a + bi$ notation for complex numbers. For example, on the TI-86, the complex number $a + bi$ is entered as the ordered pair (a, b). Figure 3 shows the solution to Example 2 on a TI-86.

```
(2,-3)+(6,2)
              (8,-1)
(-5,4)+(0,0)
              (-5,4)
```

FIGURE 3 Complex number arithmetic on a TI-86.

We now turn to **negatives** and **subtraction,** which can be defined in terms of the additive inverse of a complex number. However, because of the already stated properties of complex numbers, we can manipulate $a + bi$ in the same way we manipulate the real binomial form $a + bx$.

EXAMPLE 3 Negation and Subtraction

Carry out each operation and express the answer in standard form:

(A) $-(4 - 5i)$ (B) $(7 - 3i) - (6 + 2i)$ (C) $(-2 + 7i) + (2 - 7i)$

SOLUTIONS

Algebraic Solutions

(A) $-(4 - 5i) = (-1)(4 - 5i) = -4 + 5i$

(B) $(7 - 3i) - (6 + 2i) = 7 - 3i - 6 - 2i$
$= 1 - 5i$

(C) $(-2 + 7i) + (2 - 7i) = -2 + 7i + 2 - 7i = 0$

Graphical Solutions

```
-(4-5i)
                -4+5i
(7-3i)-(6+2i)
                 1-5i
(-2+7i)+(2-7i)
                    0
```

FIGURE 4

MATCHED 3 PROBLEM

Carry out each operation and express the answer in standard form:

(A) $-(-3 + 2i)$ (B) $(3 - 5i) - (1 - 3i)$ (C) $(-4 + 9i) + (4 - 9i)$

In general, the **additive inverse** or **negative** of $a + bi$ is $-a - bi$ because

$$(a + bi) + (-a - bi) = (-a - bi) + (a + bi) = 0$$

(see Example 3, part C, and Matched Problem 3, part C).

Now we turn our attention to multiplication. First, we use the definition of multiplication to see what happens to the complex unit i when it is squared:

$$\begin{aligned} i^2 &= \overset{(a + bi)(c + di)}{(0 + 1i)(0 + 1i)} \\ &= \overset{(ac - bd) + (ad + bc)i}{(0 \cdot 0 - 1 \cdot 1) + (0 \cdot 1 + 1 \cdot 0)i} \\ &= -1 + 0i \\ &= -1 \end{aligned}$$

Thus, we have proved that

$$i^2 = -1$$

That is, the square of i is a negative real number and i is a solution to $x^2 + 1 = 0$. Because $i^2 = -1$, we define $\sqrt{-1}$ to be the imaginary unit i. Thus,

$$i = \sqrt{-1} \qquad \text{and} \qquad -i = -\sqrt{-1}$$

Just as in the case of addition and subtraction, **multiplication of complex numbers** can be carried out using the properties of complex numbers stated on p. 169. That is, we can manipulate $a + bi$ in the same way we manipulate the real binomial form $a + bx$. The key difference is that we replace i^2 with -1 each time it occurs.

EXAMPLE 4 Multiplying Complex Numbers

Carry out each operation and express the answer in standard form:

(A) $(2 - 3i)(6 + 2i)$ (B) $1(3 - 5i)$

(C) $i(1 + i)$ (D) $(3 + 4i)(3 - 4i)$

SOLUTIONS

Algebraic Solutions

(A) $(2 - 3i)(6 + 2i) = 2(6 + 2i) - 3i(6 + 2i)$

$= 12 + 4i - 18i + 6i^2$

$= 12 - 14i - 6(-1)$ Replace i^2 with -1.

$= 18 - 14i$

(B) $1(3 - 5i) = 1 \cdot 3 - 1 \cdot 5i = 3 - 5i$

(C) $i(1 + i) = i + i^2 = i - 1 = -1 + i$

(D) $(3 + 4i)(3 - 4i) = 9 - 12i + 12i - 16i^2$
$= 9 + 16 = 25$

Graphical Solutions

```
(2-3i)(6+2i)
            18-14i
1(3-5i)
              3-5i
i(1+i)
              -1+i
(3+4i)(3-4i)
                25
```

FIGURE 5

MATCHED 4 PROBLEM

Carry out each operation and express the answer in standard form:

(A) $(5 + 2i)(4 - 3i)$ (B) $3(-2 + 6i)$

(C) $i(2 - 3i)$ (D) $(2 + 3i)(2 - 3i)$

For any complex number $a + bi$,

$$1(a + bi) = (a + bi)1 = a + bi$$

(see Example 4, part B). Thus, 1 is the **multiplicative identity** for complex numbers, just as it is for real numbers.

Earlier we stated that every nonzero complex number has a multiplicative inverse or reciprocal. We will denote this as a fraction, just as we do with real numbers. Thus,

$$\frac{1}{a + bi} \quad \text{is the } \textbf{reciprocal} \text{ of} \quad a + bi \qquad a + bi \neq 0$$

The following important property of the conjugate of a complex number is used to express reciprocals and quotients in standard form.

THEOREM 1
Product of a Complex Number and Its Conjugate

$(a + bi)(a - bi) = a^2 + b^2$ A real number

We now turn to the fourth arithmetic operation, **division of complex numbers.** Division can be performed by making direct use of Theorem 1. As before, we can manipulate $a + bi$ in the same way we manipulate the real binomial form $a + bx$, except we replace i^2 with -1 each time it occurs. Example 5 should make the process clear.

EXAMPLE 5 Reciprocals and Quotients

Carry out each operation and express the answer in standard form:

SOLUTIONS

(A) $\dfrac{1}{2 + 3i}$ (B) $\dfrac{7 - 3i}{1 + i}$

Algebraic Solutions

(A) Multiply numerator and denominator by the conjugate of the denominator:

$$\frac{1}{2+3i} = \frac{1}{2+3i} \cdot \frac{2-3i}{2-3i} = \frac{2-3i}{4-9i^2} = \frac{2-3i}{4+9}$$

$$= \frac{2-3i}{13} = \frac{2}{13} - \frac{3}{13}i$$

(B)
$$\frac{7-3i}{1+i} = \frac{7-3i}{1+i} \cdot \frac{1-i}{1-i} = \frac{7-7i-3i+3i^2}{1-i^2}$$

$$= \frac{4-10i}{2} = 2 - 5i$$

Graphical Solutions

```
1/(2+3i)
.1538461538-.23...
Ans▸Frac
          2/13-3/13i
(7-3i)/(1+i)
                2-5i
```

FIGURE 6

In Figure 6, note that we used the Fraction command to convert the decimal form to the fraction form.

In real number arithmetic, we can verify a division with a multiplication. For example, $\frac{57}{3} = 19$ is correct because $3 \cdot 19 = 57$. The results in Example 5 are verified as follows:

$$(2 + 3i)\left(\frac{2}{13} - \frac{3}{13}i\right) = \frac{4}{13} - \frac{6}{13}i + \frac{6}{13}i - \frac{9}{13}i^2$$

$$= \frac{4}{13} + \frac{9}{13} = 1$$

$$(1 + i)(2 - 5i) = 2 - 5i + 2i - 5i^2 = 7 - 3i$$

MATCHED 5 PROBLEM

Carry out each operation and express the answer in standard form:

(A) $\dfrac{1}{4 + 2i}$ (B) $\dfrac{6 + 7i}{2 - i}$

EXAMPLE 6 Combined Operations

Carry out the indicated operations and write each answer in standard form:

SOLUTIONS

(A) $(3 - 2i)^2 - 6(3 - 2i) + 13$ (B) $\dfrac{2 - 3i}{2i}$

Algebraic Solutions

(A) $$\begin{aligned}(3 - 2i)^2 - 6(3 - 2i) + 13 &= 9 - 12i + 4i^2 - 18 + 12i + 13\\ &= 9 - 12i - 4 - 18 + 12i + 13\\ &= 0\end{aligned}$$

(B) If a complex number is divided by a pure imaginary number, we can make the denominator real by multiplying numerator and denominator by i.

$$\frac{2 - 3i}{2i} \cdot \frac{i}{i} = \frac{2i - 3i^2}{2i^2} = \frac{2i + 3}{-2} = -\frac{3}{2} - i$$

Graphical Solutions

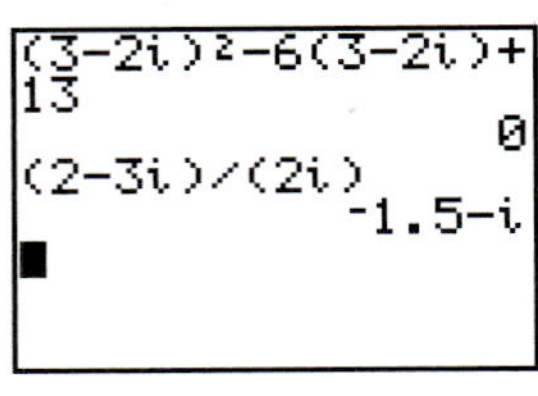

FIGURE 7

MATCHED 6 PROBLEM

Carry out the indicated operations and write each answer in standard form:

(A) $(3 + 2i)^2 - 6(3 + 2i) + 13$ (B) $\dfrac{4 - i}{3i}$

EXPLORE/DISCUSS 2

Natural number powers of i take on particularly simple forms:

i	$i^5 = i^4 \cdot i = (1)i = i$
$i^2 = -1$	$i^6 = i^4 \cdot i^2 = 1(-1) = -1$
$i^3 = i^2 \cdot i = (-1)i = -i$	$i^7 = i^4 \cdot i^3 = 1(-i) = -i$
$i^4 = i^2 \cdot i^2 = (-1)(-1) = 1$	$i^8 = i^4 \cdot i^4 = 1 \cdot 1 = 1$

In general, what are the possible values for i^n, n a natural number? Explain how you could easily evaluate i^n for any natural number n. Then evaluate each of the following:

(A) i^{17} (B) i^{24} (C) i^{38} (D) i^{47}

If your graphing utility can perform complex arithmetic, use it to check your calculations in parts A–D.

Complex Numbers and Radicals

Recall that we say that a is a square root of b if $a^2 = b$. If x is a positive real number, then x has two square roots, the principal square root, denoted by $\sqrt{x}$, and its negative, $-\sqrt{x}$. If x is a negative real number, then x still has two square roots, but now these square roots are imaginary numbers.

DEFINITION 4
Principal Square Root of a Negative Real Number

The **principal square root of a negative real number,** denoted by $\sqrt{-a}$, where a is positive, is defined by

$$\sqrt{-a} = i\sqrt{a} \qquad \sqrt{-3} = i\sqrt{3} \qquad \sqrt{-9} = i\sqrt{9} = 3i$$

The other square root of $-a$, $a > 0$, is $-\sqrt{-a} = -i\sqrt{a}$.

Note in Definition 4 that we wrote $i\sqrt{a}$ and $i\sqrt{3}$ in place of the standard forms $\sqrt{a}i$ and $\sqrt{3}i$. We follow this convention whenever it appears that i might accidentally slip under a radical sign ($\sqrt{a}i \neq \sqrt{ai}$, but $\sqrt{a}i = i\sqrt{a}$). Definition 4 is motivated by the fact that

$$(i\sqrt{a})^2 = i^2a = -a$$

EXAMPLE 7 Complex Numbers and Radicals

Write in standard form:

(A) $\sqrt{-4}$ (B) $4 + \sqrt{-5}$ (C) $\dfrac{-3 - \sqrt{-5}}{2}$ (D) $\dfrac{1}{1 - \sqrt{-9}}$

SOLUTIONS

Algebraic Solutions

(A) $\sqrt{-4} = i\sqrt{4} = 2i$

(B) $4 + \sqrt{-5} = 4 + i\sqrt{5}$

(C) $\dfrac{-3 - \sqrt{-5}}{2} = \dfrac{-3 - i\sqrt{5}}{2} = -\dfrac{3}{2} - \dfrac{\sqrt{5}}{2}i$

(D) $\dfrac{1}{1 - \sqrt{-9}} = \dfrac{1}{1 - 3i} = \dfrac{1 \cdot (1 + 3i)}{(1 - 3i) \cdot (1 + 3i)}$

$= \dfrac{1 + 3i}{1 - 9i^2} = \dfrac{1 + 3i}{10}$

$= \dfrac{1}{10} + \dfrac{3}{10}i$

Graphical Solutions

```
√(-4+0i)
                2i
4+√(-5+0i)
    4+2.236067977i
(-3-√(-5+0i))/2
-1.5-1.11803398...
1/(1-√(-9+0i))
            .1+.3i
```

FIGURE 8

Note that principal square roots like $\sqrt{-4}$ must be entered as $\sqrt{-4 + 0i}$ to indicate that we want to perform complex arithmetic rather than real arithmetic.

MATCHED 7 PROBLEM

Write in standard form:

(A) $\sqrt{-16}$ (B) $5 + \sqrt{-7}$ (C) $\dfrac{-5 - \sqrt{-2}}{2}$ (D) $\dfrac{1}{3 - \sqrt{-4}}$

EXPLORE/DISCUSS 3

From basic algebra, we know that if a and b are positive real numbers, then

$$\sqrt{a}\sqrt{b} = \sqrt{ab} \tag{3}$$

Thus, we can evaluate expressions like $\sqrt{9}\,\sqrt{4}$ two ways:

$$\sqrt{9}\,\sqrt{4} = \sqrt{(9)(4)} = \sqrt{36} = 6 \qquad \text{and} \qquad \sqrt{9}\,\sqrt{4} = (3)(2) = 6$$

Evaluate each of the following in two ways. Is equation (3) a valid property to use in all cases?

(A) $\sqrt{9}\,\sqrt{-4}$ (B) $\sqrt{-9}\,\sqrt{4}$ (C) $\sqrt{-9}\,\sqrt{-4}$

CAUTION

Note that in Example 7, part D, we wrote $1 - \sqrt{-9} = 1 - 3i$ before proceeding with the simplification. This is a necessary step because some of the properties of radicals that are true for real numbers turn out not to be true for complex numbers. In particular, for positive real numbers a and b,

$$\sqrt{a}\,\sqrt{b} = \sqrt{ab} \qquad \text{but} \qquad \sqrt{-a}\,\sqrt{-b} \neq \sqrt{(-a)(-b)}$$

(See Explore/Discuss 3.)

Solving Equations Involving Complex Numbers

EXAMPLE 8 Equations Involving Complex Numbers

(A) Solve for real numbers x and y:

$$(3x + 2) + (2y - 4)i = -4 + 6i$$

(B) Solve for complex number z:

$$(3 + 2i)z - 3 + 6i = 8 - 4i$$

SOLUTIONS

(A) Equate the real and imaginary parts of each side of the equation to form two equations:

$$\begin{aligned} 3x + 2 &= -4 \\ 3x &= -6 \\ x &= -2 \end{aligned} \qquad \begin{aligned} 2y - 4 &= 6 \\ 2y &= 10 \\ y &= 5 \end{aligned}$$

(B)
$$\begin{aligned} (3 + 2i)z - 3 + 6i &= 8 - 4i \\ (3 + 2i)z &= 11 - 10i \\ z &= \frac{11 - 10i}{3 + 2i} \\ &= \frac{(11 - 10i)(3 - 2i)}{(3 + 2i)(3 - 2i)} \\ &= \frac{13 - 52i}{13} \\ &= 1 - 4i \end{aligned}$$

A check is left to the reader.

MATCHED 8 PROBLEM

(A) Solve for real numbers x and y:

$$(2y - 7) + (3x + 4)i = 1 + i$$

(B) Solve for complex number z:

$$(1 + 3i)z + 4 - 5i = 3 + 2i$$

Early resistance to these new numbers is suggested by the words used to name them: *complex* and *imaginary*. In spite of this early resistance, complex numbers have come into widespread use in both pure and applied mathematics. They are used extensively, for example, in electrical engineering, physics, chemistry, statistics, and aeronautical engineering. Our first use of them will be in connection with solutions of second-degree equations in Section 2.5.

ANSWERS MATCHED PROBLEMS

1. (A) Imaginary numbers: $6 + 7i$, $\sqrt{2} - \frac{1}{3}i$, $0 - i = -i$, $0 + \frac{2}{3}i = \frac{2}{3}i$
   Pure imaginary numbers: $0 - i = -i$, $0 + \frac{2}{3}i = \frac{2}{3}i$
   Real numbers: $-\sqrt{3} + 0i = -\sqrt{3}$, $0 - 0i = 0$
   Zero: $0 - 0i = 0$
   (B) $6 - 7i$, $\sqrt{2} + \frac{1}{3}i$, $0 + i = i$, $0 - \frac{2}{3}i = -\frac{2}{3}i$, $-\sqrt{3} - 0i = -\sqrt{3}$, $0 + 0i = 0$
2. (A) $9 - 2i$ (B) $7 - 5i$
3. (A) $3 - 2i$ (B) $2 - 2i$ (C) 0
4. (A) $26 - 7i$ (B) $-6 + 18i$ (C) $3 + 2i$ (D) 13
5. (A) $\frac{1}{5} - \frac{1}{10}i$ (B) $1 + 4i$
6. (A) 0 (B) $-\frac{1}{3} - \frac{4}{3}i$
7. (A) $4i$ (B) $5 + i\sqrt{7}$ (C) $-\frac{5}{2} - (\sqrt{2}/2)i$ (D) $\frac{3}{13} + \frac{2}{13}i$
8. (A) $x = -1, y = 4$ (B) $z = 2 + i$

EXERCISE 2.4

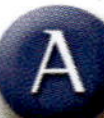

In Problems 1–26, perform the indicated operations and write each answer in standard form.

1. $(2 + 4i) + (5 + i)$

2. $(3 + i) + (4 + 2i)$

3. $(-2 + 6i) + (7 - 3i)$

4. $(6 - 2i) + (8 - 3i)$

5. $(6 + 7i) - (4 + 3i)$

6. $(9 + 8i) - (5 + 6i)$

7. $(3 + 5i) - (-2 - 4i)$

8. $(8 - 4i) - (11 - 2i)$

9. $(4 - 5i) + 2i$

10. $6 + (3 - 4i)$

11. $(4i)(6i)$

12. $(3i)(8i)$

13. $-3i(2 - 4i)$

14. $-2i(5 - 3i)$

15. $(3 + 3i)(2 - 3i)$

16. $(-2 - 3i)(3 - 5i)$

17. $(2 - 3i)(7 - 6i)$

18. $(3 + 2i)(2 - i)$

19. $(7 + 4i)(7 - 4i)$

20. $(5 + 3i)(5 - 3i)$

21. $\dfrac{1}{2 + i}$

22. $\dfrac{1}{3 - i}$

23. $\dfrac{3 + i}{2 - 3i}$

24. $\dfrac{2 - i}{3 + 2i}$

25. $\dfrac{13 + i}{2 - i}$

26. $\dfrac{15 - 3i}{2 - 3i}$

B

In Problems 27–34, evaluate and express results in standard form.

27. $\sqrt{2}\sqrt{8}$

28. $\sqrt{3}\sqrt{12}$

29. $\sqrt{2}\sqrt{-8}$

30. $\sqrt{-3}\sqrt{12}$

31. $\sqrt{-2}\sqrt{8}$

32. $\sqrt{3}\sqrt{-12}$

33. $\sqrt{-2}\sqrt{-8}$

34. $\sqrt{-3}\sqrt{-12}$

In Problems 35–44, convert imaginary numbers to standard form, perform the indicated operations, and express answers in standard form.

35. $(2 - \sqrt{-4}) + (5 - \sqrt{-9})$

36. $(3 - \sqrt{-4}) + (-8 + \sqrt{-25})$

37. $(9 - \sqrt{-9}) - (12 - \sqrt{-25})$

38. $(-2 - \sqrt{-36}) - (4 + \sqrt{-49})$

39. $(3 - \sqrt{-4})(-2 + \sqrt{-49})$

40. $(2 - \sqrt{-1})(5 + \sqrt{-9})$

41. $\dfrac{5 - \sqrt{-4}}{7}$

42. $\dfrac{6 - \sqrt{-64}}{2}$

43. $\dfrac{1}{2 - \sqrt{-9}}$

44. $\dfrac{1}{3 - \sqrt{-16}}$

Write Problems 45–50 in standard form.

45. $\dfrac{2}{5i}$

46. $\dfrac{1}{3i}$

47. $\dfrac{1 + 3i}{2i}$

48. $\dfrac{2 - i}{3i}$

49. $(2 - 3i)^2 - 2(2 - 3i) + 9$

50. $(2 - i)^2 + 3(2 - i) - 5$

51. Let $f(x) = x^2 - 2x + 2$.

(A) Show that the conjugate complex numbers $1 + i$ and $1 - i$ are both zeros of f.

(B) Does f have any real zeros? Any x intercepts? Explain.

52. Let $g(x) = -x^2 + 4x - 5$.

(A) Show that the conjugate complex numbers $2 + i$ and $2 - i$ are both zeros of g.

(B) Does g have any real zeros? Any x intercepts? Explain.

53. Simplify: i^{18}, i^{32}, and i^{67}.

54. Simplify: i^{21}, i^{43}, and i^{52}.

In Problems 55–58, solve for x and y.

55. $(2x - 1) + (3y + 2)i = 5 - 4i$

56. $3x + (y - 2)i = (5 - 2x) + (3y - 8)i$

57. $\dfrac{(1 + x) + (y - 2)i}{1 + i} = 2 - i$

58. $\dfrac{(2 + x) + (y + 3)i}{1 - i} = -3 + i$

In Problems 59–62, solve for z. Express answers in standard form.

59. $(2 + i)z + i = 4i$

60. $(3 - i)z + 2 = i$

61. $3iz + (2 - 4i) = (1 + 2i)z - 3i$

62. $(2 - i)z + (1 - 4i) = (-1 + 3i)z + (4 + 2i)$

63. Explain what is wrong with the following "proof" that $-1 = 1$:

$$-1 = i^2 = \sqrt{-1}\sqrt{-1} = \sqrt{(-1)(-1)} = \sqrt{1} = 1$$

64. Explain what is wrong with the following "proof" that $1/i = i$. What is the correct value of $1/i$?

$$\frac{1}{i} = \frac{1}{\sqrt{-1}} = \frac{\sqrt{1}}{\sqrt{-1}} = \sqrt{\frac{1}{-1}} = \sqrt{-1} = i$$

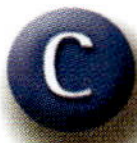

In Problems 65–70, perform the indicated operations, and write each answer in standard form.

65. $(a + bi) + (c + di)$

66. $(a + bi) - (c + di)$

67. $(a + bi)(a - bi)$

68. $(u - vi)(u + vi)$

69. $(a + bi)(c + di)$

70. $\dfrac{a + bi}{c + di}$

71. Show that $i^{4k} = 1$, k a natural number.

72. Show that $i^{4k+1} = i$, k a natural number.

73. Show that $2 - i$ and $-2 + i$ are square roots of $3 - 4i$.

74. Show that $-3 + 2i$ and $3 - 2i$ are square roots of $5 - 12i$.

75. Describe how you could find the square roots of $8 - 6i$ without using a graphing utility. What are the square roots of $8 - 6i$?

76. Describe how you could find the square roots of $2i$ without using a graphing utility. What are the square roots of $2i$?

77. Let $S_n = i + i^2 + i^3 + \cdots + i^n$, $n \geq 1$. Describe the possible values of S_n.

78. Let $T_n = i^2 + i^4 + i^6 + \cdots + i^{2n}$, $n \geq 1$. Describe the possible values of T_n.

Supply the reasons in the proofs for the theorems stated in Problems 79 and 80.

79. *Theorem*: The complex numbers are commutative under addition.

Proof: Let $a + bi$ and $c + di$ be two arbitrary complex numbers; then,

Statement

1. $(a + bi) + (c + di) = (a + c) + (b + d)i$

2. $= (c + a) + (d + b)i$

3. $= (c + di) + (a + bi)$

Reason

1.

2.

3.

80. *Theorem:* The complex numbers are commutative under multiplication.

Proof: Let $a + bi$ and $c + di$ be two arbitrary complex numbers; then,

Statement

1. $(a + bi) \cdot (c + di) = (ac - bd) + (ad + bc)i$

2. $= (ca - db) + (da + cb)i$

3. $= (c + di) + (a + bi)$

Reason

1.

2.

3.

Letters z and w are often used as complex variables, where $z = x + yi$, $w = u + vi$, and x, y, u, and v are real numbers. The conjugates of z and w, denoted by $\bar{z}$ and $\bar{w}$, respectively, are given by $\bar{z} = x - yi$ and $\bar{w} = u - vi$. In Problems 81–88, express each property of conjugates verbally and then prove the property.

81. $z\bar{z}$ is a real number.

82. $z + \bar{z}$ is a real number.

83. $\bar{z} = z$ if and only if z is real.

84. $\bar{\bar{z}} = z$

85. $\overline{z + w} = \bar{z} + \bar{w}$

86. $\overline{z - w} = \bar{z} - \bar{w}$

87. $\overline{zw} = \bar{z} \cdot \bar{w}$

88. $\overline{z/w} = \bar{z}/\bar{w}$

SECTION 2.5 Quadratic Equations and Models

Introduction • Solution by Factoring • Solution by Completing the Square • Solution by Quadratic Formula • Mathematical Modeling • Data Analysis and Regression

Introduction

In this book we are primarily interested in functions with real number domains and ranges. However, if we want to fully understand the nature of the zeros of a function or the roots of an equation, it is necessary to extend some of the definitions in Section 1.3 to include complex numbers. A complex number r is a **zero** of the function $f(x)$ and a **root** of the equation $f(x) = 0$ if $f(r) = 0$. As before, if r is a real number, then r is also an x intercept of the graph of f. An imaginary zero can never be an x intercept.

If a, b, and c are real numbers, $a \neq 0$, then associated with the quadratic function

$$f(x) = ax^2 + bx + c$$

is the **quadratic equation**

$$ax^2 + bx + c = 0$$

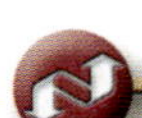

EXPLORE/DISCUSS 1

Match the zeros of each function on the left with one of the sets A, B, or C on the right:

Function	**Zeros**
$f(x) = x^2 - 1$	$A = \{1\}$
$g(x) = x^2 + 1$	$B = \{-1, 1\}$
$h(x) = (x - 1)^2$	$C = \{-i, i\}$

Which of these sets of zeros can be found using graphical approximation techniques? Which cannot?

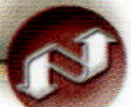

A graphing utility can be used to approximate the real roots of an equation, but not the imaginary roots. In this section we will develop algebraic techniques for finding the exact value of the roots of a quadratic equation, real or imaginary. In the process, we will derive the well-known *quadratic formula*, another important tool for our mathematical toolbox.

Solution by Factoring

If $ax^2 + bx + c$ can be written as the product of two first-degree factors, then the quadratic equation can be quickly and easily solved. The method of solution by factoring rests on the zero property of complex numbers, which generalizes the zero property of real numbers.

Zero Property

If m and n are complex numbers, then

$$m \cdot n = 0 \quad \text{if and only if} \quad m = 0 \text{ or } n = 0 \text{ (or both)}$$

EXAMPLE 1 Solving Quadratic Equations by Factoring

Solve by factoring:

(A) $6x^2 - 19x - 7 = 0$ (B) $x^2 - 6x + 5 = -4$ (C) $2x^2 = 3x$

SOLUTIONS

(A)
$$6x^2 - 19x - 7 = 0$$
$$(2x - 7)(3x + 1) = 0 \quad \text{Factor left side.}$$
$$2x - 7 = 0 \quad \text{or} \quad 3x + 1 = 0$$
$$x = \tfrac{7}{2} \qquad\qquad x = -\tfrac{1}{3}$$

The solution set is $\{-\frac{1}{3}, \frac{7}{2}\}$.

(B)
$$x^2 - 6x + 5 = -4$$
$$x^2 - 6x + 9 = 0 \quad \text{Add 4 to both sides.}$$
$$(x - 3)^2 = 0 \quad \text{Factor left side.}$$
$$x = 3$$

The solution set is $\{3\}$. The equation has one root, 3. But because it came from two factors, we call 3 a **double root** or a **root of multiplicity 2.**

(C)
$$2x^2 = 3x$$
$$2x^2 - 3x = 0$$
$$x(2x - 3) = 0$$
$$x = 0 \quad \text{or} \quad 2x - 3 = 0$$
$$x = \tfrac{3}{2}$$

Solution set: $\{0, \frac{3}{2}\}$

MATCHED PROBLEM 1

Solve by factoring:

(A) $3x^2 + 7x - 20 = 0$ (B) $4x^2 + 12x + 9 = 0$ (C) $4x^2 = 5x$

CAUTION

1. One side of an equation must be 0 before the zero property can be applied. Thus

$$x^2 - 6x + 5 = -4$$
$$(x - 1)(x - 5) = -4$$

does not imply that $x - 1 = -4$ or $x - 5 = -4$. See Example 1, part B, for the correct solution of this equation.

2. The equations

$$2x^2 = 3x \qquad \text{and} \qquad 2x = 3$$

are not equivalent. The first has solution set $\{0, \frac{3}{2}\}$, whereas the second has solution set $\{\frac{3}{2}\}$. The root $x = 0$ is lost when each member of the first equation is divided by the variable x. See Example 1, part C, for the correct solution of this equation.

Do not divide both members of an equation by an expression containing the variable for which you are solving. You may be dividing by 0.

REMARK It is common practice to represent solutions of quadratic equations informally by the last equation rather than by writing a solution set using set notation (see Example 1). From now on, we will follow this practice unless a particular emphasis is desired.

Solution by Completing the Square

Factoring is a specialized method that is very efficient if the factors can be quickly identified. However, not all quadratic equations are easy to factor. We now turn to a more general process that is guaranteed to work in all cases. This process is based on completing the square, discussed in Section 2.3, and the following square root property:

Square Root Property

If r is a complex number, s is a real number, and $r^2 = s$, then $r = \pm\sqrt{s}$.

EXAMPLE 2 Solution by Completing the Square

Use completing the square and the square root property to solve each of the following:

(A) $(x + \frac{1}{2})^2 - \frac{5}{4} = 0$ (B) $x^2 + 6x - 2 = 0$ (C) $2x^2 - 4x + 3 = 0$

SOLUTIONS

(A) This quadratic expression is already written in standard form. We solve for the squared term and then use the square root property:

$$\left(x + \tfrac{1}{2}\right)^2 - \tfrac{5}{4} = 0$$

$$\left(x + \tfrac{1}{2}\right)^2 = \tfrac{5}{4}$$ Apply the square root property.

$$x + \tfrac{1}{2} = \pm\sqrt{\tfrac{5}{4}}$$ Solve for x.

$$x = -\frac{1}{2} \pm \frac{\sqrt{5}}{2}$$

$$= \frac{-1 \pm \sqrt{5}}{2}$$

(B) We can speed up the process of completing the square by taking advantage of the fact that we are working with a quadratic equation, not a quadratic expression.

$$x^2 + 6x - 2 = 0$$

$$x^2 + 6x = 2$$

$$x^2 + 6x + 9 = 2 + 9$$ Complete the square on the left side, and add the same number to the right side.

$$(x + 3)^2 = 11$$

$$x + 3 = \pm\sqrt{11}$$

$$x = -3 \pm \sqrt{11}$$

(C) $2x^2 - 4x + 3 = 0$

$$x^2 - 2x + \tfrac{3}{2} = 0$$ Make the leading coefficient 1 by dividing by 2.

$$x^2 - 2x = -\tfrac{3}{2}$$

$$x^2 - 2x + 1 = -\tfrac{3}{2} + 1$$ Complete the square on the left side and add the same number to the right side.

$$(x - 1)^2 = -\tfrac{1}{2}$$ Factor the left side.

$$x - 1 = \pm\sqrt{-\tfrac{1}{2}}$$

$$x = 1 \pm i\sqrt{\tfrac{1}{2}}$$

$$= 1 \pm \frac{\sqrt{2}}{2}i$$ Answer in $a + bi$ form.

MATCHED 2 PROBLEM

Solve by completing the square:

(A) $\left(x + \tfrac{1}{3}\right)^2 - \tfrac{2}{9} = 0$ (B) $x^2 + 8x - 3 = 0$ (C) $3x^2 - 12x + 13 = 0$

CAUTION

Do not confuse completing the square in a quadratic function with completing the square in a quadratic equation. For functions, we add and subtract $\frac{b^2}{4}$. For equations, we add $\frac{b^2}{4}$ to both sides of the equation.

Function	**Equation**
$f(x) = x^2 - 6x - 4$	$x^2 - 6x - 4 = 0$
$= x^2 - 6x + 9 - 9 - 4$	$x^2 - 6x + 9 = 4 + 9$
$= (x - 3)^2 - 13$	$(x - 3)^2 = 13$

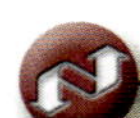

EXPLORE/DISCUSS 2

Graph the quadratic functions associated with the three quadratic equations in Example 2. Approximate the x intercepts of each function and compare with the roots found in Example 2. Which of these equations has roots that cannot be approximated graphically?

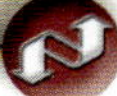

Solution by Quadratic Formula

Now consider the general quadratic equation with unspecified coefficients:

$$ax^2 + bx + c = 0 \qquad a \neq 0$$

We can solve it by completing the square exactly as we did in Example 2, part C. To make the leading coefficient 1, we must multiply both sides of the equation by $1/a$. Thus,

$$x^2 + \frac{b}{a}x + \frac{c}{a} = 0$$

Adding $-c/a$ to both sides of the equation and then completing the square of the left side, we have

$$x^2 + \frac{b}{a}x + \frac{b^2}{4a^2} = \frac{b^2}{4a^2} - \frac{c}{a}$$

We now factor the left side and solve using the square root property:

$$\left(x + \frac{b}{2a}\right)^2 = \frac{b^2 - 4ac}{4a^2}$$

$$x + \frac{b}{2a} = \pm\sqrt{\frac{b^2 - 4ac}{4a^2}}$$

$$x = -\frac{b}{2a} \pm \frac{\sqrt{b^2 - 4ac}}{2a} \qquad \text{See Problem 81.}$$

$$= \frac{-b \pm \sqrt{b^2 - 4ac}}{2a}$$

We have thus derived the well-known and widely used **quadratic formula:**

THEOREM 1
Quadratic Formula

If $ax^2 + bx + c = 0$, $a \neq 0$, then

$$x = \frac{-b \pm \sqrt{b^2 - 4ac}}{2a}$$

The quadratic formula and completing the square are equivalent methods. Either can be used to find the exact value of the roots of any quadratic equation.

EXAMPLE 3 Using the Quadratic Formula

Solve $2x + \frac{3}{2} = x^2$ by use of the quadratic formula. Leave the answer in simplest radical form.

SOLUTION

$$2x + \tfrac{3}{2} = x^2$$

$$4x + 3 = 2x^2 \qquad \text{Multiply both sides by 2.}$$

$$2x^2 - 4x - 3 = 0 \qquad \text{Write in standard form.}$$

$$x = \frac{-b \pm \sqrt{b^2 - 4ac}}{2a} \qquad a = 2, b = -4, c = -3$$

$$= \frac{-(-4) \pm \sqrt{(-4)^2 - 4(2)(-3)}}{2(2)}$$

$$= \frac{4 \pm \sqrt{40}}{4} = \frac{4 \pm 2\sqrt{10}}{4} = \frac{2 \pm \sqrt{10}}{2}$$

CAUTION

1. $-4^2 \neq (-4)^2 \qquad -4^2 = -16 \text{ and } (-4)^2 = 16$
2. $2 + \frac{\sqrt{10}}{2} \neq \frac{2 + \sqrt{10}}{2} \qquad 2 + \frac{\sqrt{10}}{2} = \frac{4 + \sqrt{10}}{2}$
3. $\frac{\cancel{4} \pm 2\sqrt{10}}{\cancel{4}} \neq \pm 2\sqrt{10} \qquad \frac{4 \pm 2\sqrt{10}}{4} = \frac{2(2 \pm \sqrt{10})}{4} = \frac{2 \pm \sqrt{10}}{2}$

MATCHED PROBLEM 3

Solve $x^2 - \frac{5}{2} = -3x$ using the quadratic formula. Leave the answer in simplest radical form.

EXPLORE/DISCUSS 3

Given the quadratic function $f(x) = ax^2 + bx + c$, let $D = b^2 - 4ac$. How many real zeros does f have if

(A) $D > 0$ (B) $D = 0$ (C) $D < 0$

In each of these three cases, what type of roots does the quadratic equation $f(x) = 0$ have?

Match each of the three cases with one of the following graphs.

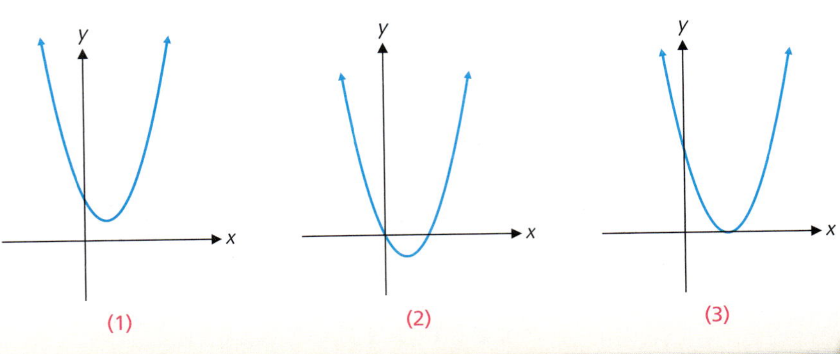

The quantity $b^2 - 4ac$ in the quadratic formula is called the **discriminant** and gives us information about the roots of the corresponding equation and the zeros of the associated quadratic function. This information is summarized in Table 1.

TABLE 1 Discriminants, Roots, and Zeros

Discriminant $b^2 - 4ac$	Roots of* $ax^2 + bx + c = 0$	Number of Real Zeros of* $f(x) = ax^2 + bx + c$
Positive	Two distinct real roots	2
0	One real root (a double root)	1
Negative	Two imaginary roots, one the conjugate of the other	0

*a, b, and c are real numbers with $a \neq 0$.

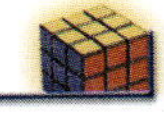

Mathematical Modeling

Now we want to consider some applications that involve quadratic equations.

EXAMPLE 4 Design

A rectangular picture frame of uniform width has outer dimensions of 12 inches by 18 inches. How wide (to the nearest tenth of an inch) must the frame be to display an area of 140 square inches?

SOLUTION

Constructing the Model

We begin by drawing and labeling a figure:

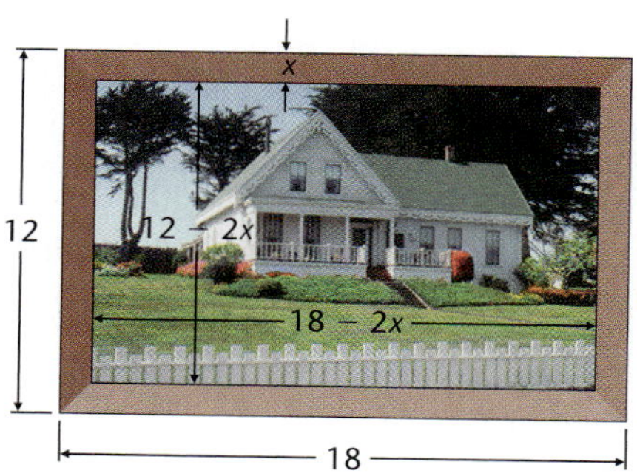

The width of the frame must be nonnegative, thus x must satisfy $x \geq 0$. Both the length and the width of the display area also must be nonnegative. This places additional restrictions on x.

$$\begin{aligned} 12 - 2x &\geq 0 \\ 12 &\geq 2x \\ 6 &\geq x \end{aligned} \qquad \begin{aligned} 18 - 2x &\geq 0 \\ 18 &\geq 2x \\ 9 &\geq x \end{aligned}$$

For both inequalities to be true, we must restrict x to satisfy $x \leq 6$. Because the area of the display is the product of the length and width, x must satisfy

$$(18 - 2x)(12 - 2x) = 140 \qquad 0 \leq x \leq 6 \tag{1}$$

Algebraic Solution

$$\begin{aligned} (18 - 2x)(12 - 2x) &= 140 \\ 216 - 36x - 24x + 4x^2 &= 140 \\ 4x^2 - 60x + 76 &= 0 \\ x^2 - 15x + 19 &= 0 \\ x &= \frac{15 \pm \sqrt{149}}{2} \end{aligned}$$

Thus, the quadratic equation has two solutions (rounded to one decimal place):

$$x = \frac{15 + \sqrt{149}}{2} = 13.6$$

and

$$x = \frac{15 - \sqrt{149}}{2} = 1.4$$

The first must be discarded because x must satisfy $x \leq 6$. So the width of the frame is 1.4 inches.

Graphical Solution

Entering $y_1 = (18 - 2x)(12 - 2x)$ and $y_2 = 140$ in the equation editor (Fig. 1) and using the zero command (Fig. 2) shows that the width of the frame is $x = 1.4$.

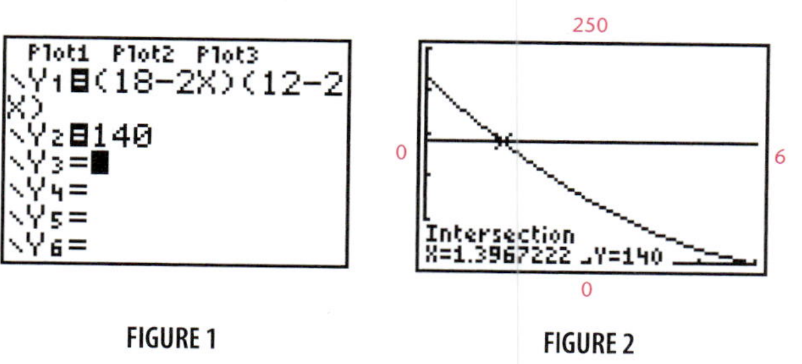

FIGURE 1

FIGURE 2

MATCHED 4 PROBLEM

A 1,200 square foot rectangular garden is enclosed with 150 feet of fencing. Find the dimensions of the garden to the nearest tenth of a foot.

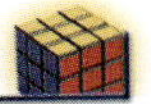

Data Analysis and Regression

Now that we have added quadratic functions to our mathematical toolbox, we can use this new tool in conjunction with another tool discussed previously—regression analysis. In Example 5, we use both of these tools to investigate the effect of recycling efforts on solid waste disposal.

EXAMPLE 5 Solid Waste Disposal

Franklin Associates, Ltd. of Prairie Village, Kansas, reported the data in Table 2 to the U.S. Environmental Protection Agency.

TABLE 2 Municipal Solid Waste Disposal

Year	Annual Landfill Disposal (Millions of Tons)	Per Person Per Day (Pounds)
1960	55.5	1.68
1970	88.2	2.37
1980	123.3	2.97
1985	136.4	3.13
1990	131.6	2.90
1995	118.4	2.50
2000	113.6	2.16

(A) Let x represent time in years with $x = 0$ corresponding to 1950, and let y represent the corresponding annual landfill disposal. Use regression analysis on a graphing utility to find a quadratic function $y = ax^2 + bx + c$ that models these data. (Round the constants a, b, and c to three significant digits.)

(B) If landfill disposal continues to follow the trend exhibited in Table 2, when (to the nearest year) will the annual landfill disposal return to the 1960 level?

(C) Is it reasonable to expect the annual landfill disposal to follow this trend indefinitely? Explain.

SOLUTIONS

(A) Because the values of y increase from 1960 to 1985 and then begin to decrease, a quadratic model seems a better choice than a linear one. Figure 3 shows the details of constructing the model on a graphing utility.

L1	L2	L3
10	55.5	
20	88.2	
30	123.3	
35	136.4	
40	131.6	
45	118.4	
50	113.6	

L3(1)=

(a) Data

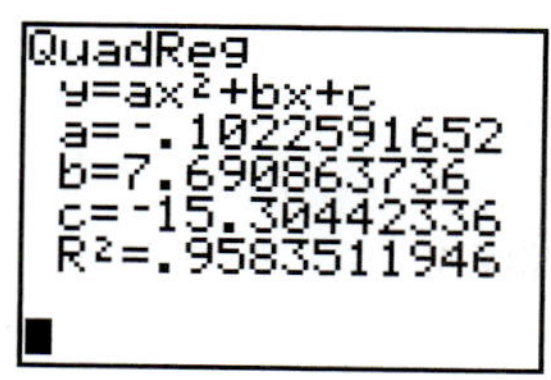

(b) Regression equation

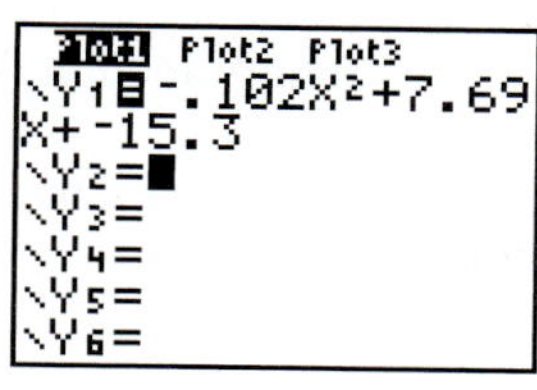

(c) Regression equation entered in equation editor

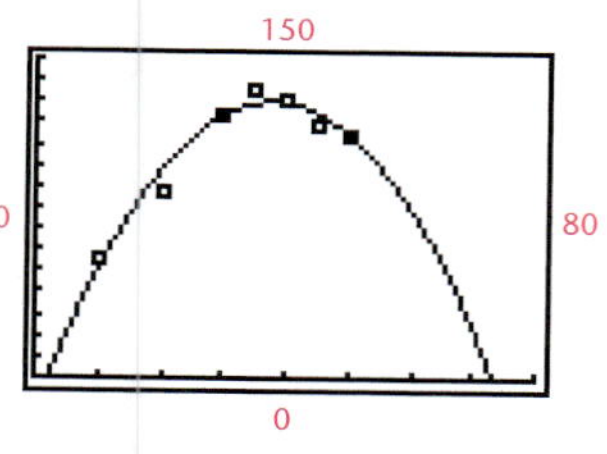

(d) Graph of data and regression equation

FIGURE 3

Rounding the constants to three significant digits, a quadratic regression equation for these data is

$$y_1 = -0.102x^2 + 7.69x - 15.3$$

The graph in Figure 3(d) indicates that this is a reasonable model for these data. It is, in fact, the "best" quadratic equation for these data.

(B) To determine when the annual landfill disposal returns to the 1960 level, we add the graph of $y_2 = 55.5$ to the graph (Fig. 4).

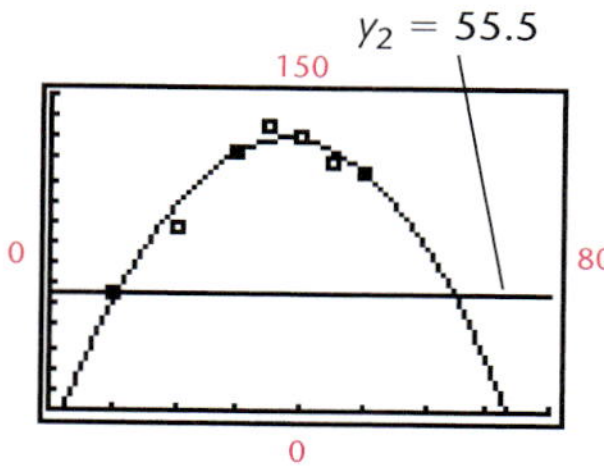

FIGURE 4

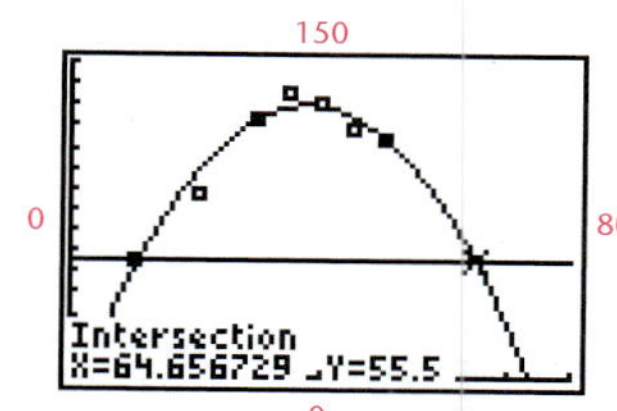

FIGURE 5

The graphs of y_1 and y_2 intersect twice, once at $x = 10$ (1960), and again at a later date. Using the intersect command (Fig. 5) shows that the x coordinate of the second intersection point (to the nearest integer) is 65. Thus, the annual landfill disposal returns to the 1960 level of 55.5 million tons in 2015.

(C) The graph of y_1 continues to decrease and reaches 0 somewhere between 2023 and 2024. It is highly unlikely that the annual landfall disposal will ever reach 0. As time goes by and more data become available, new models will have to be constructed to better predict future trends.

MATCHED 5 PROBLEM

Refer to Table 2.

(A) Let x represent time in years with $x = 0$ corresponding to 1950, and let y represent the corresponding landfill disposal per person per day. Use regression analysis on a graphing utility to find a quadratic function of the form $y = ax^2 + bx + c$ that models these data. (Round the constants a, b, and c to three significant digits.)

(B) If landfill disposal per person per day continues to follow the trend exhibited in Table 2, when (to the nearest year) will it fall below 1 pound per person per day?

(C) Is it reasonable to expect the landfill disposal per person per day to follow this trend indefinitely? Explain.

Most gasoline engines run more efficiently at a midrange speed than at either extremely high or extremely low speeds. Example 6 uses quadratic regression to determine the optimal speed for a speedboat.

EXAMPLE 6 Optimal Speed

TABLE 3

mph	mpg
4.6	3.07
7.3	3.17
21.0	6.77
29.8	6.62
40.2	2.77
44.6	2.37

Source: www.yamaha-motor.com

Table 3 contains performance data for a speedboat powered by a Yamaha outboard motor. In the work that follows, round all numbers to three significant digits.

(A) Let x be the speed of the boat in miles per hour (mph) and y the associated mileage in miles per gallon (mpg). Use the data in Table 3 to find a quadratic regression function $y = ax^2 + bx + c$ for this boat.

(B) A marina rents this boat for \$20 per hour plus the cost of the gasoline used. If gasoline costs \$1.50 per gallon and you take a 100-mile trip in this boat, construct a mathematical model and use it to answer the following questions:

What speed should you travel to minimize the rental charges?

What mileage will the boat get?

How long is the trip?

How much gasoline will you use?

How much will the trip cost you?

SOLUTIONS

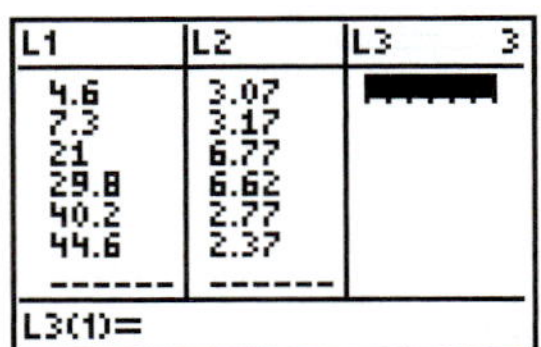

FIGURE 6

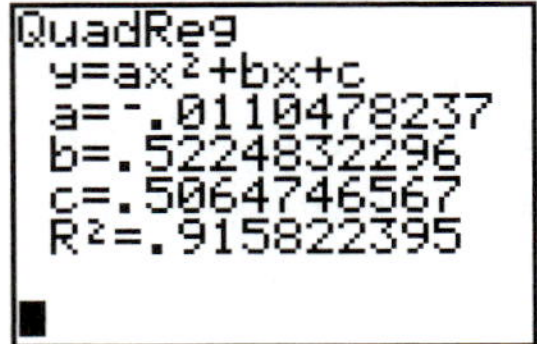

FIGURE 7

(A) Entering the data in the statistics editor (Fig. 6) and selecting the QuadReg option (Fig. 7) produces the following quadratic function relating speed and mileage:

$$y = -0.0110x^2 + 0.522x + 0.506$$

(B) If t is the number of hours the boat is rented and g is the number of gallons of gasoline used, then the cost of the rental (in dollars) is

$$C = 20t + 1.5g$$

If x is the speed of the boat and y is the associated mileage, then

$$xt = 100 \qquad \text{(miles/hour)(hours) = distance}$$

and

$$yg = 100 \qquad \text{(miles/gallon)(gallons) = distance}$$

Thus,

$$t = \frac{100}{x}$$

and, using the quadratic regression function from part A,

$$g = \frac{100}{y} = \frac{100}{-0.0110x^2 + 0.522x + 0.506}$$

Returning to the cost equation,

$$C = 20t + 1.5g$$

$$= 20\frac{100}{x} + 1.5\frac{100}{-0.0110x^2 + 0.522x + 0.506}$$

$$= \frac{2{,}000}{x} + \frac{150}{-0.0110x^2 + 0.522x + 0.506}$$

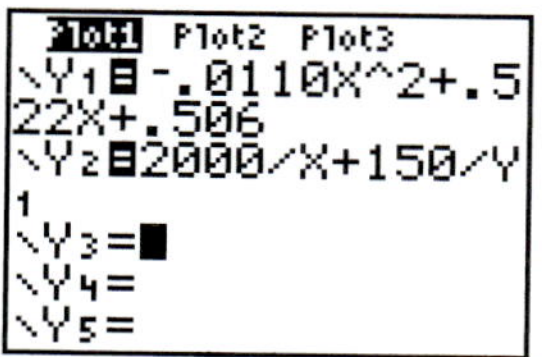

FIGURE 8

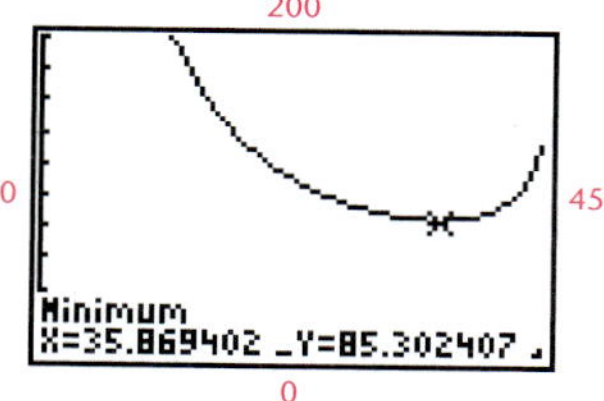

FIGURE 9

This is the mathematical model for the total cost of the trip. Our first objective is to find the speed x that produces the minimum cost. Entering the cost function in the equation editor (Fig. 8) and using the minimum command (Fig. 9), we see that the minimum cost occurs when the boat speed is 35.9 miles per hour. Evaluating the quadratic function in part A at 35.9, we find that the corresponding mileage is $y = 5.07$ miles per gallon. The trip will take $\frac{100}{35.9} = 2.79$ hours and consume $\frac{100}{5.07} = 19.7$ gallons of gas. We can see from Figure 9 that the trip will cost \$85.30. To check this, we can compute the cost directly

$$\begin{array}{ll} & \text{Rent} \quad \text{plus Gasoline} \\ C = & 20(2.79) \;+\; 1.5(19.7) = \$85.35 \end{array}$$

The small \$0.05 discrepancy between these two costs is caused by rounding all numbers to three significant digits.

MATCHED 6 PROBLEM

TABLE 4

mph	mpg
4.8	2.67
8.9	2.12
18.5	2.89
34.4	3.78
43.8	3.40
48.6	2.63

Source: www.yamaha-motor.com

Table 4 contains performance data for a speedboat powered by a Yamaha outboard motor. In the work that follows, round all numbers to three significant digits.

(A) Let x be the speed of the boat in miles per hour (mph) and y the associated mileage in miles per gallon (mpg). Use the data in Table 4 to find a quadratic regression function $y = ax^2 + bx + c$ for this boat.

(B) A marina rents this boat for \$15 per hour plus the cost of the gasoline used. If gasoline costs \$2 per gallon and you take a 200-mile trip in this boat, construct a mathematical model and use it to answer the following questions:

What speed should you travel to minimize the rental charges?

What mileage will the boat get?

How long does the trip take?

How much gasoline will you use?

How much will the trip cost you?

ANSWERS TO MATCHED PROBLEMS

1. (A) $\{-4, \frac{5}{3}\}$ (B) $\{-\frac{3}{2}\}$ (a double root) (C) $\{0, \frac{5}{4}\}$
2. (A) $x = (-1 \pm \sqrt{2})/3$ (B) $x = -4 \pm \sqrt{19}$ (C) $x = (6 \pm i\sqrt{3})/3$ or $2 \pm (\sqrt{3}/3)i$
3. $x = (-3 \pm \sqrt{19})/2$
4. 23.1 feet by 51.9 feet
5. (A) $y = -0.00272x^2 + 0.178x + 0.0896$ (B) 2010
6. (A) $y = -0.00176x^2 + 0.110x + 1.72$
   (B) The boat should travel at 43.0 miles per hour. The mileage is 3.20 miles per gallon. The trip will take 4.65 hours and will consume 62.5 gallons of gasoline. The trip will cost $195.

EXERCISE 2.5

A

In Problems 1–6, solve by factoring.

1. $4u^2 = 8u$
2. $3A^2 = -12A$
3. $9y^2 = 12y - 4$
4. $16x^2 + 8x = -1$
5. $11x = 2x^2 + 12$
6. $8 - 10x = 3x^2$

In Problems 7–18, solve by completing the square.

7. $x^2 - 6x - 3 = 0$
8. $y^2 - 10y - 3 = 0$
9. $t^2 - 4t + 8 = 0$
10. $w^2 - 6w + 25 = 0$
11. $m^2 + 2m + 9 = 0$
12. $n^2 + 8n + 34 = 0$
13. $2d^2 + 5d - 25 = 0$
14. $2u^2 + 7u + 3 = 0$
15. $2v^2 - 2v + 1 = 0$
16. $9x^2 - 12x + 5 = 0$
17. $4y^2 + 3y + 9 = 0$
18. $5t^2 + 2t + 5 = 0$

In Problems 19–26, solve using the quadratic formula.

19. $x^2 - 10x - 3 = 0$
20. $x^2 - 6x - 3 = 0$
21. $x^2 + 8 = 4x$
22. $y^2 + 3 = 2y$
23. $2x^2 + 1 = 4x$
24. $2m^2 + 3 = 6m$
25. $5x^2 + 2 = 2x$
26. $7x^2 + 6x + 4 = 0$

B

For each equation in Problems 27–32, use the discriminant to determine the number and type of zeros.

27. $2.4x^2 + 6.4x - 4.3 = 0$
28. $0.4x^2 - 3.2x + 6.4 = 0$
29. $6.5x^2 - 7.4x + 3.4 = 0$
30. $3.4x^2 - 2.5x - 1.5 = 0$
31. $0.3x^2 + 3.6x + 10.8 = 0$
32. $1.7x^2 + 2.4x + 1.4 = 0$

For each equation in Problems 33–38, use a graph to determine the number and type of zeros.

33. $0.2x^2 - 3.2x + 12.8 = 0$
34. $4.5x^2 + 1.7x - 0.4 = 0$
35. $3.4x^2 - 9.1x - 4.7 = 0$
36. $1.3x^2 - 1.5x + 0.8 = 0$
37. $2.4x^2 + 3.7x + 1.5 = 0$
38. $0.6x^2 + 6x + 15 = 0$

In Problems 39–48, solve algebraically and confirm with a graphing utility, if possible.

39. $x^2 - 6x - 3 = 0$
40. $y^2 - 10y - 3 = 0$
41. $2y^2 - 6y + 3 = 0$
42. $2d^2 - 4d + 1 = 0$
43. $3x^2 - 2x - 2 = 0$
44. $3x^2 + 5x - 4 = 0$
45. $12x^2 + 7x = 10$
46. $9x^2 + 9x = 4$
47. $x^2 = 3x + 1$
48. $x^2 + 2x = 2$

In Problems 49–52, solve for the indicated variable in terms of the other variables. Use positive square roots only.

49. $s = \frac{1}{2}gt^2$ for t
50. $a^2 + b^2 = c^2$ for a
51. $P = EI - RI^2$ for I
52. $A = P(1 + r)^2$ for r

In Problems 53–68, solve by any algebraic method and confirm graphically, if possible. Round any approximate solutions to three decimal places.

53. $x^2 - \sqrt{7}x + 2 = 0$
54. $x^2 + \sqrt{11}x + 3 = 0$
55. $x^2 - 2\sqrt{3}x + 3 = 0$
56. $x^2 - \sqrt{5}x - 5 = 0$
57. $x^2 + \sqrt{3}x - 4 = 0$
58. $x^2 + 2\sqrt{5}x + 5 = 0$

59. $1 + \frac{9}{x^2} = \frac{5}{x}$

60. $1 + \frac{25}{x^2} = \frac{9}{x}$

61. $1 + \frac{9}{x^2} = \frac{6}{x}$

62. $1 + \frac{25}{x^2} = \frac{10}{x}$

63. $1 + \frac{9}{x^2} = \frac{7}{x}$

64. $1 + \frac{25}{x^2} = \frac{11}{x}$

65. $3 + \frac{5}{x - 4} = \frac{7}{x + 4}$

66. $5 + \frac{6}{x - 2} = \frac{4}{x + 2}$

67. $\frac{8}{x - 5} = \frac{3}{x + 5} - 2$

68. $\frac{6}{x - 3} = \frac{4}{x + 3} - 3$

69. Consider the quadratic equation

$$x^2 + 4x + c = 0$$

where c is a real number. Discuss the relationship between the values of c and the three types of roots listed in Table 1 on page 186.

70. Consider the quadratic equation

$$x^2 - 2x + c = 0$$

where c is a real number. Discuss the relationship between the values of c and the three types of roots listed in Table 1 on page 186.

C

Solve Problems 71–74 and express answer in $a + bi$ form.

71. $x^2 + 3ix - 2 = 0$

72. $x^2 - 7ix - 10 = 0$

73. $x^2 + 2ix = 3$

74. $x^2 = 2ix - 3$

In Problems 75 and 76, find all solutions.

75. $x^3 - 1 = 0$

76. $x^4 - 1 = 0$

77. Can a quadratic equation with rational coefficients have one rational root and one irrational root? Explain.

78. Can a quadratic equation with real coefficients have one real root and one imaginary root? Explain.

79. Show that if r_1 and r_2 are the two roots of $ax^2 + bx + c = 0$, then $r_1 r_2 = c/a$.

80. For r_1 and r_2 in Problem 79, show that $r_1 + r_2 = -b/a$.

81. In one stage of the derivation of the quadratic formula, we replaced the expression

$$\pm\sqrt{(b^2 - 4ac)/4a^2}$$

with

$$\pm\sqrt{b^2 - 4ac}/2a$$

What justifies using $2a$ in place of $|2a|$?

82. Find the error in the following "proof" that two arbitrary numbers are equal to each other: Let a and b be arbitrary numbers such that $a \neq b$. Then

$$\begin{aligned}
(a - b)^2 &= a^2 - 2ab + b^2 = b^2 - 2ab + a^2 \\
(a - b)^2 &= (b - a)^2 \\
a - b &= b - a \\
2a &= 2b \\
a &= b
\end{aligned}$$

83. Find two numbers such that their sum is 21 and their product is 104.

84. Find all numbers with the property that when the number is added to itself the sum is the same as when the number is multiplied by itself.

85. Find two consecutive positive even integers whose product is 168.

86. Find two consecutive positive integers whose product is 600.

APPLICATIONS

87. Air Search. A search plane takes off from an airport at 6:00 A.M. and travels due north at 200 miles per hour. A second plane takes off at 6:30 A.M. and travels due east at 170 miles per hour. The planes carry radios with a maximum range of 500 miles. When (to the nearest minute) will these planes no longer be able to communicate with each other?

88. Navigation. A speedboat takes 1 hour longer to go 24 miles up a river than to return. If the boat cruises at 10 miles per hour in still water, what is the rate of the current?

89. Construction. A gardener has a 30 foot by 20 foot rectangular plot of ground. She wants to build a brick walkway of uniform width on the border of the plot (see the figure).

If the gardener wants to have 400 square feet of ground left for planting, how wide (to two decimal places) should she build the walkway?

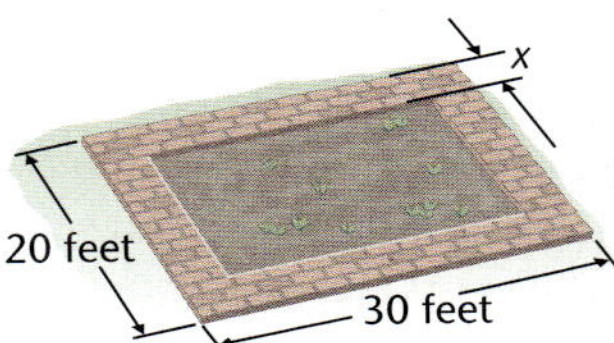

90. Construction. Refer to Problem 89. The gardener buys enough brick to build 160 square feet of walkway. Is this sufficient to build the walkway determined in Problem 89? If not, how wide (to two decimal places) can she build the walkway with these bricks?

91. Architecture. A developer wants to erect a rectangular building on a triangular-shaped piece of property that is 200 feet wide and 400 feet long (see the figure).

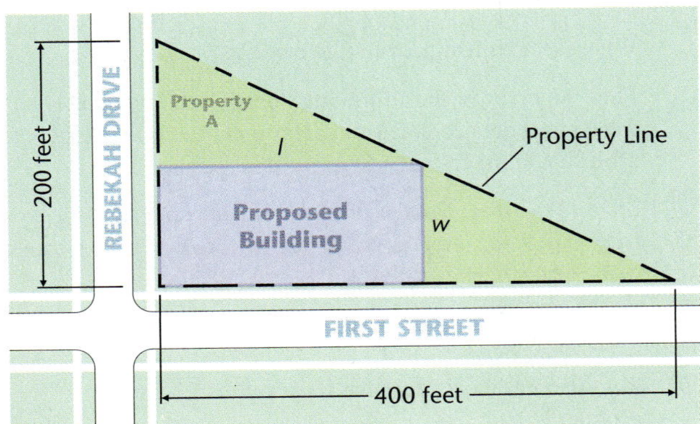

(A) Express the area $A(w)$ of the footprint of the building as a function of the width w and state the domain of this function. [*Hint:* Use Euclid's theorem* to find a relationship between the length l and width w.]

(B) Building codes require that this building have a footprint of at least 15,000 square feet. What are the widths of the building that will satisfy the building codes?

(C) Can the developer construct a building with a footprint of 25,000 square feet? What is the maximum area of the footprint of a building constructed in this manner?

**Euclid's theorem:* If two triangles are similar, their corresponding sides are proportional:

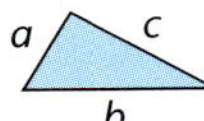

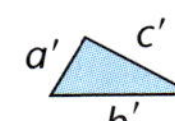

$$\frac{a}{a'} = \frac{b}{b'} = \frac{c}{c'}$$

★ **92. Architecture.** An architect is designing a small A-frame cottage for a resort area. A cross-section of the cottage is an isosceles triangle with a base of 5 meters and an altitude of 4 meters. The front wall of the cottage must accommodate a sliding door positioned as shown in the figure.

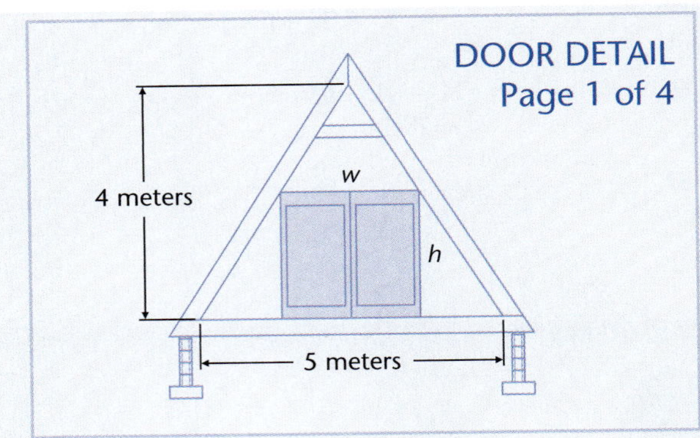

(A) Express the area $A(w)$ of the door as a function of the width w and state the domain of this function. [See the hint for Problem 91.]

(B) A provision of the building code requires that doorways must have an area of at least 4.2 square meters. Find the width of the doorways that satisfy this provision.

(C) A second provision of the building code requires all doorways to be at least 2 meters high. Discuss the effect of this requirement on the answer to part B.

93. Transportation. A delivery truck leaves a warehouse and travels north to factory A. From factory A the truck travels east to factory B and then returns directly to the warehouse (see the figure). The driver recorded the truck's odometer reading at the warehouse at both the beginning and the end of the trip and also at factory B, but forgot to record it at factory A (see the table). The driver does recall that it was further from the warehouse to factory A than it was from factory A to factory B. Because delivery charges are based on distance from the warehouse, the driver needs to know how far factory A is from the warehouse. Find this distance.

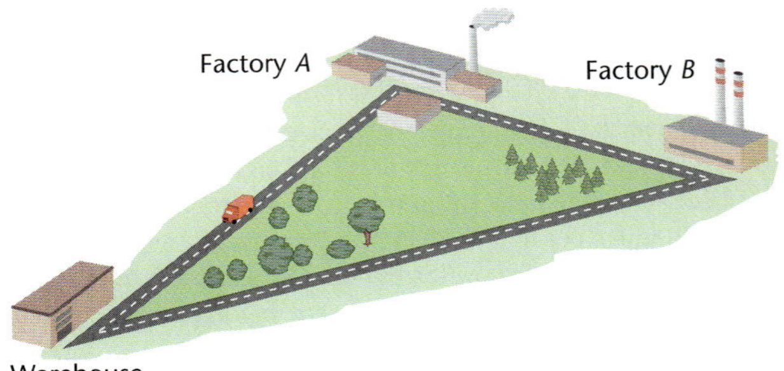

	Odometer Readings
Warehouse	52846
Factory A	52???
Factory B	52937
Warehouse	53002

★★ **94. Construction.** A $\frac{1}{4}$-mile track for racing stock cars consists of two semicircles connected by parallel straightaways (see the figure). To provide sufficient room for pit crews, emergency vehicles, and spectator parking, the track must enclose an area of 100,000 square feet. Find the length of the straightaways and the diameter of the semicircles to the nearest foot. [*Recall*: The area A and circumference C of a circle of diameter d are given by $A = \pi d^2/4$ and $C = \pi d$.]

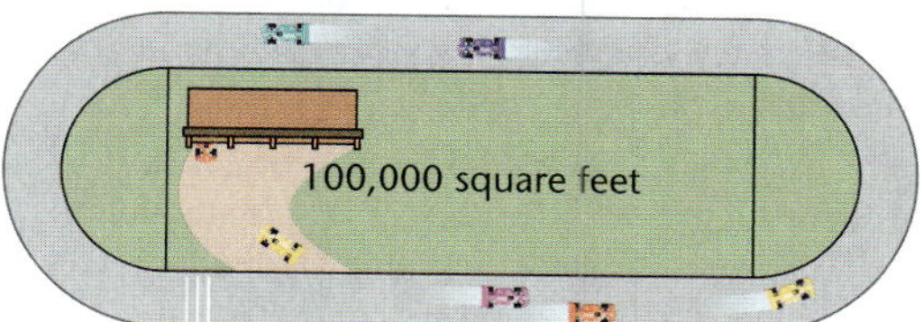

DATA ANALYSIS AND LINEAR REGRESSION 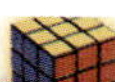

In Problems 95–102, unless directed otherwise, round all numbers to three significant digits,

95. Alcohol Consumption. Table 5 contains data related to the per capita ethanol consumption in the United States from 1960 to 1995.

(A) Let the independent variable x represent years since 1960. Find a quadratic regression model for the per capita beer consumption.

(B) If beer consumption continues to follow the trend exhibited in Table 5, when (to the nearest year) will the consumption return to the 1960 level?

(C) What does your model predict for beer consumption in the year 2000? Use the Internet or a library to compare your predicted results with the actual results.

TABLE 5 Per Capita Ethanol Consumption (in Gallons)

Year	Beer	Wine
1960	0.99	0.22
1965	1.04	0.24
1970	1.14	0.27
1975	1.26	0.32
1980	1.38	0.34
1985	1.33	0.38
1990	1.34	0.33
1995	1.25	0.29

Source: NIAAA

96. Alcohol Consumption. (See page 95.) Refer to Table 5.

(A) Let the independent variable x represent years since 1960. Find a quadratic regression model for the per capita wine consumption.

(B) If wine consumption continues to follow the trend exhibited in Table 5, when (to the nearest year) will the consumption return to the 1960 level?

(C) What does your model predict for wine consumption in the year 2000? Use the Internet or a library to compare your predicted results with the actual results.

97. Cigarette Production. Table 6 contains data related to the total production and per capita consumption of cigarettes in the United States from 1950 to 1995.

(A) Let the independent variable x represent years since 1950. Find a quadratic regression model for the total cigarette production.

(B) If cigarette production continues to follow the trend exhibited in Table 6, when (to the nearest year) will the production return to the 1950 level?

(C) What does your model predict for cigarette production in the year 2000? Use the Internet or a library to compare your predicted results with the actual results.

TABLE 6 Cigarette Consumption

Year	Production (Billions)	Per Capita Annual Consumption
1950	370	3,550
1955	396	3,600
1960	484	4,170
1965	529	4,260
1970	537	3,990
1975	607	4,120
1980	632	3,850
1985	594	3,370
1990	525	2,830
1995	487	2,520

Source: CDC

98. Cigarette Production. Refer to Table 6.

(A) Let the independent variable x represent years since 1950. Find a quadratic regression model for the per capita cigarette consumption.

(B) If per capita cigarette consumption continues to follow the trend exhibited in Table 6, when (to the nearest year) will the per capita consumption drop to 500 cigarettes?

(C) What does your model predict for per capita cigarette consumption in the year 2000? Use the Internet or a library to compare your predicted results with the actual results.

99. Stopping Distance. Table 7 contains data related to the length of the skid marks left by two different automobiles when making emergency stops.

(A) Let x be the speed of the vehicle in miles per hour. Find a quadratic regression model for the braking distance for auto A.

(B) An insurance investigator finds skid marks 200 feet long at the scene of an accident involving auto A. How fast (to the nearest mile) was auto A traveling when it made these skid marks?

TABLE 7 Skid Marks

Speed (mph)	Length of Skid Marks (in feet) Auto A	Auto B
20	21	29
30	44	53
40	76	86
50	114	124
60	182	193
70	238	263
80	305	332

100. Stopping Distance. Refer to Table 7.

(A) Let x be the speed of the vehicle in miles per hour. Find a quadratic regression model for the braking distance for auto B.

(B) An insurance investigator finds skid marks 165 feet long at the scene of an accident involving auto B. How fast (to the nearest mile per hour) was auto B traveling when it made these skid marks?

101. Optimal Speed. Table 8 contains performance data for two speedboats powered by Yamaha outboard motors.

(A) Let x be the speed of boat A in miles per hour (mph) and y the associated mileage in miles per gallon (mpg). Use the data in Table 8 to find a quadratic regression function $y = ax^2 + bx + c$ for this boat.

(B) A marina rents this boat for $10 per hour plus the cost of the gasoline used. If gasoline costs $1.50 per gallon and you take a 100-mile trip in this boat, construct a mathematical model and use it to answer the following questions:

What speed should you travel to minimize the rental charges?

What mileage will the boat get?

How long does the trip take?

How much gasoline will you use?

How much will the trip cost you?

TABLE 8 Performance Data

Boat A mph	mpg	Boat B mph	mpg
5.4	2.84	5.1	1.65
12.3	2.86	9.0	1.45
29.3	4.44	23.9	2.30
41.8	3.80	35.0	2.48
53.1	3.28	44.1	2.19
57.4	2.73	49.1	1.81

Source: www.yamaha-motor.com

102. Optimal Speed. Refer to Table 8.

(A) Let x be the speed of boat B in miles per hour (mph) and y the associated mileage in miles per gallon (mpg). Use the data in Table 8 to find a quadratic regression function $y = ax^2 + bx + c$ for this boat.

(B) A marina rents this boat for $15 per hour plus the cost of the gasoline used. If gasoline costs $2.00 per gallon and you take a 200-mile trip in this boat, construct a mathematical model and use it to answer the following questions:

What speed should you travel to minimize the rental charges?

What mileage will the boat get?

How long does the trip take?

How much gasoline will you use?

How much will the trip cost you?

SECTION 2.6 Additional Equation-Solving Techniques

Equations Involving Radicals • Equations of the Form $ax^{2p} + bx^p + c = 0$ • Mathematical Modeling

In this section we examine equations that can be transformed into quadratic equations by various algebraic manipulations. With proper interpretation, the solutions of the resulting quadratic equations will lead to the solutions of the original equations.

Equations Involving Radicals

Consider the equation

$$x = \sqrt{x + 2} \qquad (1)$$

Graphing both sides of the equation and using an intersection routine shows that $x = 2$ is a solution to the equation (Fig. 1). Is it the only solution?

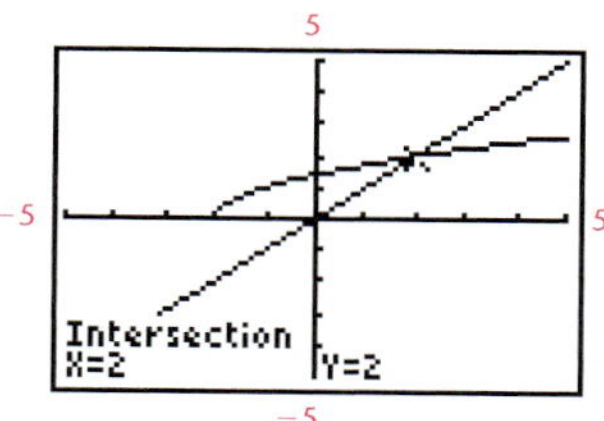

FIGURE 1
$y_1 = x,$
$y_2 = \sqrt{x + 2}.$

There may be other solutions not visible in this viewing window. Or there may be imaginary solutions (remember, graphical approximation applies only to real solutions). To solve this equation algebraically, we square each side of equation (1) and then proceed to solve the resulting quadratic equation. Thus,

$$\begin{aligned} x^2 &= (\sqrt{x + 2})^2 \qquad (2) \\ x^2 &= x + 2 \\ x^2 - x - 2 &= 0 \\ (x - 2)(x + 1) &= 0 \\ x &= 2, -1 \end{aligned}$$

These are the only solutions to the quadratic equation. We have already seen that $x = 2$ is a solution to the original equation. To check if -1 is a solution, we substitute in equation (1):

$$\begin{aligned} x &= \sqrt{x + 2} \\ -1 &\stackrel{?}{=} \sqrt{-1 + 2} \\ -1 &\stackrel{?}{=} \sqrt{1} \\ -1 &\neq 1 \end{aligned}$$

Thus, -1 is not a solution to equation (1). What have we gained by performing these algebraic manipulations? If we can be certain that all solutions of equation (1) must be among the solutions of equation (2), then we can rule out the possibility of any additional solutions to equation (1). Theorem 1 provides the necessary tool to do this.

THEOREM 1
Power Operation on Equations

If both sides of an equation are raised to the same natural number power, then the solution set of the original equation is a subset of the solution set of the new equation.

Equation	Solution Set
$x = 3$	$\{3\}$
$x^2 = 9$	$\{-3, 3\}$

Referring to equations (1) and (2) on page 197, we know that 2 and -1 are the only solutions to the quadratic equation (2). And we checked that -1 is not a solution to equation (1). Theorem 1 now implies that 2 must be the *only* solution to equation (1). We call -1 an *extraneous solution*. In general, an **extraneous solution** is a solution introduced during the solution process that does not satisfy the original equation.

Every solution of the new equation must be checked in the original equation to eliminate extraneous solutions.

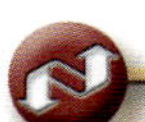

EXPLORE/DISCUSS 1

Figure 2 shows that $x = -1$ is a solution of the equation

$$\sqrt{x + 2} = 0.01x + 1.01$$

Are there any other solutions? Find any additional solutions both algebraically and graphically. What are some advantages and disadvantages of each of these solution methods?

FIGURE 2

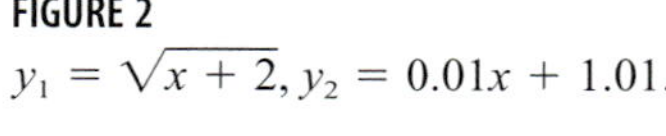

$y_1 = \sqrt{x + 2}, y_2 = 0.01x + 1.01.$

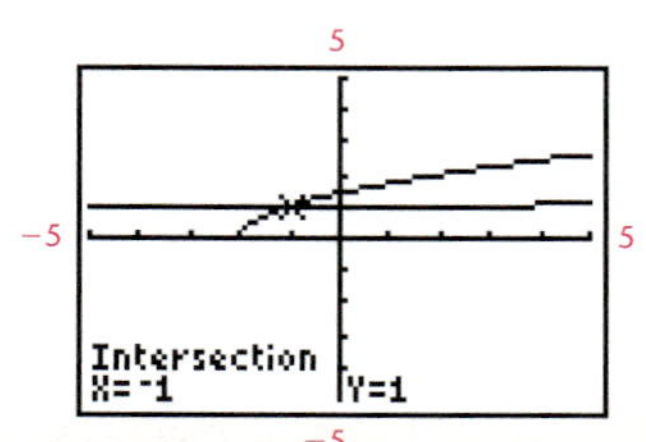

EXAMPLE 1 Solving Equations Involving Radicals

Solve algebraically $\sqrt{4x^2 + 8x + 7} - x = 1$.

SOLUTION

$$\sqrt{4x^2 + 8x + 7} - x = 1$$

$$\sqrt{4x^2 + 8x + 7} = x + 1 \quad \text{Isolate radical on one side.}$$

$$4x^2 + 8x + 7 = x^2 + 2x + 1 \quad \text{Square both sides.}$$

$$3x^2 + 6x + 6 = 0 \quad \text{Collect like terms.}$$

$$x^2 + 2x + 2 = 0 \quad \text{Use the quadratic formula.}$$

$$x = \frac{-2 \pm \sqrt{-4}}{2}$$

$$= -1 + i, -1 - i$$

CHECK

$x = -1 + i$

$$\sqrt{4x^2 + 8x + 7} - x = 1$$

$$\sqrt{4(-1 + i)^2 + 8(-1 + i) + 7} - (-1 + i) \stackrel{?}{=} 1$$

$$\sqrt{4 - 8i - 4 - 8 + 8i + 7} + 1 - i \stackrel{?}{=} 1$$

$$1 \stackrel{\checkmark}{=} 1$$

$x = -1 - i$

$$\sqrt{4x^2 + 8x + 7} - x = 1$$

$$\sqrt{4(-1 - i)^2 + 8(-1 - i) + 7} - (-1 - i) \stackrel{?}{=} 1$$

$$\sqrt{4 + 8i - 4 - 8 - 8i + 7} + 1 + i \stackrel{?}{=} 1$$

$$1 + 2i \neq 1$$

FIGURE 3
$y_1 = 1$,
$y_2 = \sqrt{4x^2 + 8x + 7} - x$.

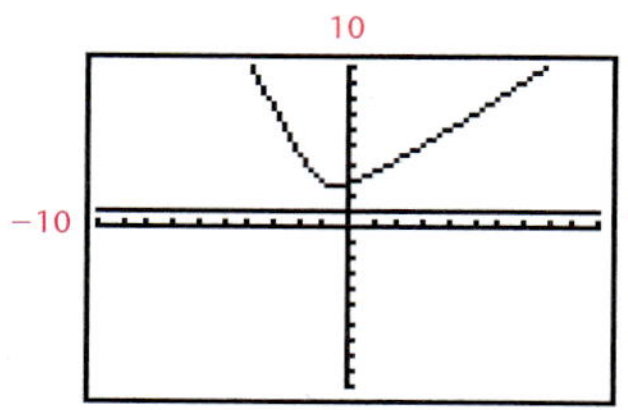

The check shows that $-1 + i$ is a solution to the original equation and $-1 - i$ is extraneous. Thus, the only solution is the imaginary number

$$x = -1 + i$$

Graphing both sides of the equation illustrates that there are no intersection points in a standard viewing window (Fig. 3). The algebraic solution shows that the equation has no real solutions, hence there cannot be any intersection points anywhere in the plane.

MATCHED 1 PROBLEM

Solve algebraically $\sqrt{x^2 - 2x - 2} + 2x = 2$.

EXAMPLE 2 Solving Equations Involving Two Radicals

Solve algebraically and graphically $\sqrt{2x + 3} - \sqrt{x - 2} = 2$.

SOLUTION

Algebraic Solution

$$\sqrt{2x+3} - \sqrt{x-2} = 2$$

$$\sqrt{2x+3} = \sqrt{x-2} + 2 \qquad \text{Isolate one of the radicals.}$$

$$2x + 3 = x - 2 + 4\sqrt{x-2} + 4 \qquad \text{Square both sides.}$$

$$x + 1 = 4\sqrt{x-2} \qquad \text{Isolate the remaining radical.}$$

$$x^2 + 2x + 1 = 16(x-2) \qquad \text{Square both sides.}$$

$$x^2 - 14x + 33 = 0$$

$$(x-3)(x-11) = 0$$

$$x = 3, 11$$

CHECK

$x = 3$

$$\sqrt{2x+3} - \sqrt{x-2} = 2$$

$$\sqrt{2(3)+3} - \sqrt{3-2} \stackrel{?}{=} 2$$

$$2 \stackrel{\checkmark}{=} 2$$

$x = 11$

$$\sqrt{2x+3} - \sqrt{x-2} = 2$$

$$\sqrt{2(11)+3} - \sqrt{11-2} \stackrel{?}{=} 2$$

$$2 \stackrel{\checkmark}{=} 2$$

Both solutions check. Thus,

$x = 3, 11$ Two solutions

Graphical Solution

Graphing $y_1 = \sqrt{2x+3} - \sqrt{x-2}$ and $y_2 = 2$ in a standard viewing window produces a graph that is not very useful (Fig. 4).

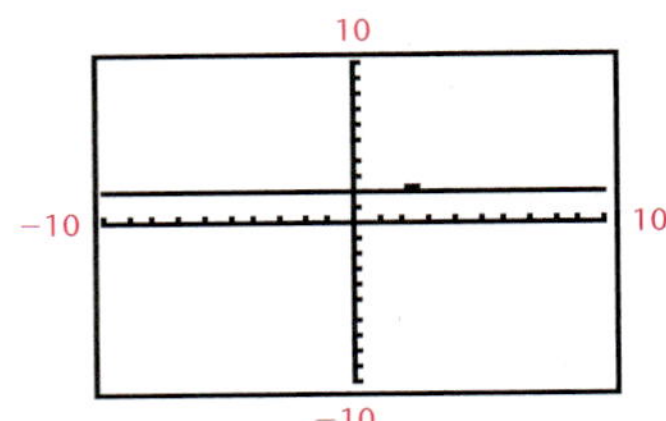

FIGURE 4

Examining a table of values (Fig. 5), suggests that choosing Xmin = 2, Xmax = 14, Ymin = 1.5, Ymax = 3 would produce a graph that clearly shows two intersection points.

X	Y1	Y2
2	2.6458	2
4	1.9024	2
6	1.873	2
8	1.9094	2
10	1.9674	2
12	2.0339	2
14	2.1037	2

Y1■√(2X+3)−√(X−...

FIGURE 5

Using the intersect command, the x coordinates of the intersection points are $x = 3$ (Fig. 6) and $x = 11$ (Fig. 7).

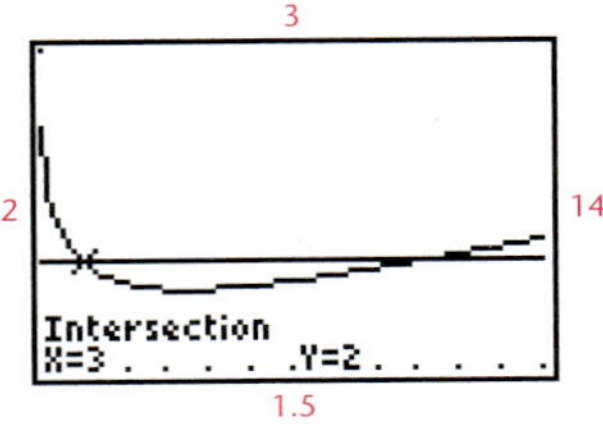

FIGURE 6

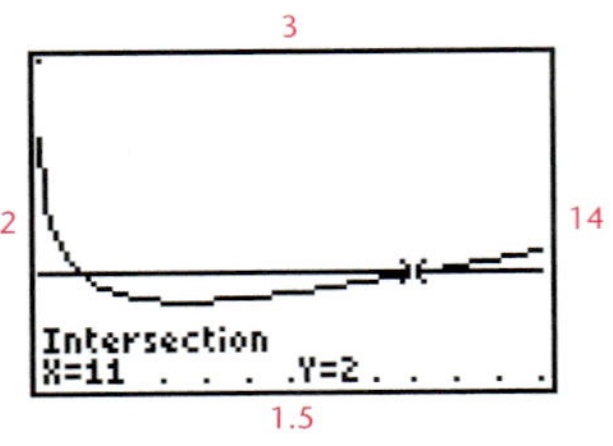

FIGURE 7

MATCHED 2 PROBLEM

Solve algebraically and graphically $\sqrt{2x + 5} + \sqrt{x + 2} = 5$.

How do you choose between algebraic and graphical solution methods? It depends on the type of solutions desired. If you want to find real and complex solutions, you must use algebraic methods, as we did in Example 1. If you are only interested in real solutions, then either method can be used, as in Example 2. As part of your learning experience, we recommend that you solve each equation algebraically and, when possible, confirm your solutions graphically.

Equations of the Form $ax^{2p} + bx^p + c = 0$

We introduce this topic with an example.

EXAMPLE 3 **Solving an Equation of the Form $ax^{2p} + bx^p + c = 0$**

Solve $x^{2/3} - x^{1/3} - 6 = 0$.

Method I. Direct solution:
Using the properties of exponents from basic algebra, we write $x^{2/3}$ as $(x^{1/3})^2$ and solve by factoring.

$$(x^{1/3})^2 - x^{1/3} - 6 = 0$$

$$(x^{1/3} - 3)(x^{1/3} + 2) = 0 \qquad \text{Factor left side.}$$

$$x^{1/3} = 3 \qquad \text{or} \qquad x^{1/3} = -2$$

$$(x^{1/3})^3 = 3^3 \qquad (x^{1/3})^3 = (-2)^3 \qquad \text{Cube both sides.}$$

$$x = 27 \qquad x = -8$$

Solution set: $\{-8, 27\}$

Method II. Using substitution:
Let $u = x^{1/3}$, solve for u, and then solve for x.

$$u^2 - u - 6 = 0$$

$$(u - 3)(u + 2) = 0$$

$$u = 3, -2$$

Replacing u with $x^{1/3}$, we obtain

$$x^{1/3} = 3 \qquad \text{or} \qquad x^{1/3} = -2$$

$$x = 27 \qquad x = -8$$

Solution set: $\{-8, 27\}$

FIGURE 8
$y_1 = x^{2/3} - x^{1/3} - 6$.

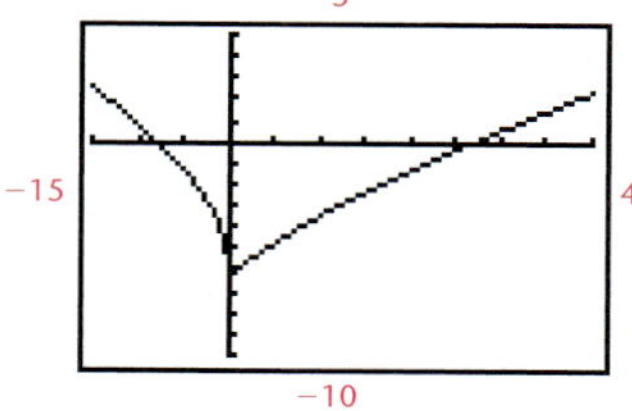

The graph in Figure 8 confirms these results. [*Note:* In some graphing utilities you may have to enter the left side of the equation in the form $y_1 = (x^2)^{1/3} - x^{1/3} - 6$ rather than $y_1 = x^{2/3} - x^{1/3} - 6$. Try both forms to see what happens.]

MATCHED 3 PROBLEM

Solve algebraically using both Method I and Method II and confirm graphically $x^{1/2} - 5x^{1/4} + 6 = 0$.

In general, if an equation that is not quadratic can be transformed to the form

$$au^2 + bu + c = 0$$

where u is an expression in some other variable, then the equation is called an **equation of quadratic type.** Equations of quadratic type often can be solved using quadratic methods.

EXPLORE/DISCUSS 2

Which of the following can be transformed into an equation of quadratic type by making a substitution of the form $u = x^n$? What is the resulting quadratic equation?

(A) $3x^{-4} + 2x^{-2} + 7 = 0$ (B) $7x^5 - 3x^2 + 3 = 0$

(C) $2x^5 + 4x^2\sqrt{x} - 6 = 0$ (D) $8x^{-2}\sqrt{x} - 5x^{-1}\sqrt{x} - 2 = 0$

In general, if a, b, c, m, and n are nonzero real numbers, when can an equation of the form $ax^m + bx^n + c = 0$ be transformed into an equation of quadratic type?

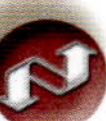

EXAMPLE 4 Solving Equations of Quadratic Type

Solve algebraically and confirm graphically, if possible $x^4 - 3x^2 - 4 = 0$.

SOLUTION

Algebraic Solution

The equation is quadratic in x^2. We solve for x^2 and then for x.

$$(x^2)^2 - 3x^2 - 4 = 0$$
$$(x^2 - 4)(x^2 + 1) = 0$$
$$x^2 = 4 \quad \text{or} \quad x^2 = -1$$
$$x = \pm 2 \quad \text{or} \quad x = \pm i$$

Because we did not raise each side of the equation to a natural number power, we do not have to check for extraneous solutions. (You should still check the accuracy of the solutions.)

Graphical Confirmation

Figures 9 and 10 show the two real solutions. The imaginary solutions cannot be confirmed graphically.

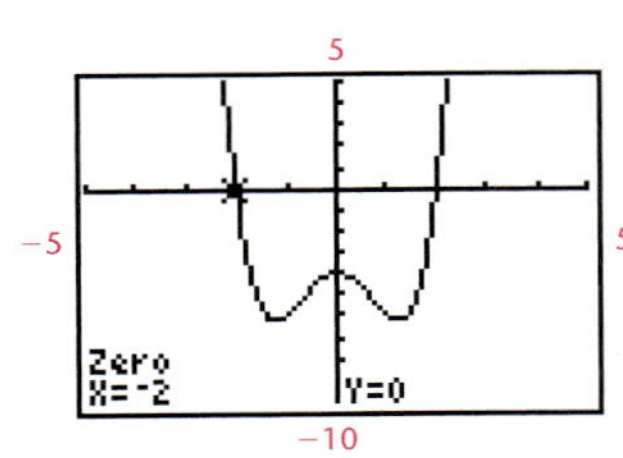

FIGURE 9

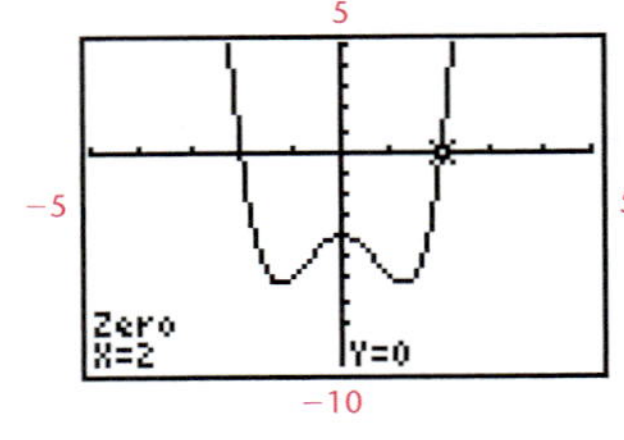

FIGURE 10

MATCHED 4 PROBLEM

Solve algebraically and confirm graphically, if possible, $x^4 + 3x^2 - 4 = 0$.

EXAMPLE 5 **Solving Equations of Quadratic Type**

Solve algebraically and graphically $3x^{-2/5} - 6x^{-1/5} + 2 = 0$.

SOLUTION

Algebraic Solution

The equation $3x^{-2/5} - 6x^{-1/5} + 2 = 0$ is quadratic in $x^{-1/5}$. We substitute $u = x^{-1/5}$ and solve for u:

$$3u^2 - 6u + 2 = 0$$

$$u = \frac{6 \pm \sqrt{12}}{6} \quad \text{Use the quadratic formula.}$$

$$= \frac{3 \pm \sqrt{3}}{3}$$

$$x = u^{-5}$$

$$= \left(\frac{3 \pm \sqrt{3}}{3}\right)^{-5}$$

Thus, the two solutions are

$$x = \left(\frac{3}{3 \pm \sqrt{3}}\right)^5 \quad \text{Use a calculator.}$$

$$\approx 0.102414, 74.147586$$

Graphical Solution

The graph of $y_1 = 3x^{-2/5} - 6x^{-1/5} + 2$ is the thick curve in Figure 11. The graph crosses the x axis near $x = 0$ and again near $x = 75$. The solution near $x = 75$ is easily approximated in this viewing window. The solution near the origin can be approximated in the same viewing window, but more insight is gained by changing the limits on the x axis (Fig. 12).

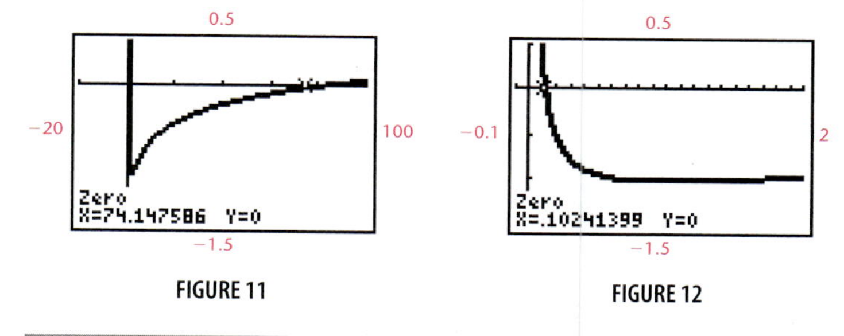

FIGURE 11

FIGURE 12

MATCHED 5 PROBLEM

Solve algebraically and graphically $3x^{-2/5} - x^{-1/5} - 2 = 0$.

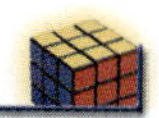

Mathematical Modeling

Examples 6 and 7 illustrate the use of radicals in constructing mathematical models.

EXAMPLE 6 **Depth of a Well**

The splash from a stone dropped into a deep well is heard 5 seconds after the stone is released (Fig. 13). How deep is the well? Round answer to the nearest foot.

SOLUTION

Constructing the Model

The time between the instant the stone is released and the instant the splash is heard can be broken down into two parts:

t_1 = Time stone is falling through the air
t_2 = Time sound travels back to surface

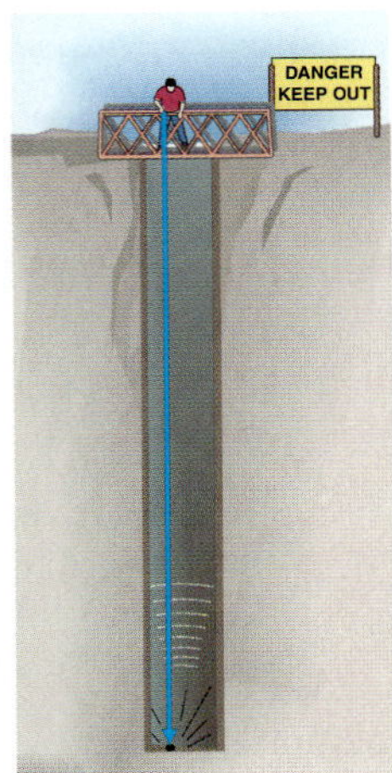

FIGURE 13

If x is the depth of the well in feet and both times are measured in seconds, then

$$x = 16t_1^2 \qquad \text{and} \qquad x = 1{,}100t_2$$

We have used the falling body formula for t_1 before. The formula for t_2 is based on the principle that sound travels through air at about 1,100 feet per second. Solving for t_1 and t_2, we have

$$16t_1^2 = x \qquad\qquad 1{,}100t_2 = x$$

$$t_1^2 = \frac{x}{16} \qquad\qquad t_2 = \frac{x}{1{,}100}$$

$$t_1 = \frac{\sqrt{x}}{4}$$

If we combine t_1 and t_2, we have a model for the total time t between releasing the stone and hearing the splash in terms of the depth of the well x:

$$t = t_1 + t_2 = \frac{\sqrt{x}}{4} + \frac{x}{1{,}100}$$

We are asked to find x when $t = 5$ seconds.

Algebraic Solution

$$\frac{\sqrt{x}}{4} + \frac{x}{1{,}100} = 5$$

$$275\sqrt{x} + x = 5{,}500$$

$$x + 275x^{1/2} - 5{,}500 = 0 \qquad \text{Let } u = x^{1/2}.$$

$$u^2 + 275u - 5{,}500 = 0$$

$$u = \frac{-275 \pm \sqrt{275^2 - 4(-5{,}500)}}{2} \qquad \text{Use the quadratic formula.}$$

$$= \frac{-275 \pm \sqrt{97{,}625}}{2}$$

$$= 18.724998 \quad \text{or} \quad -293.724998$$

Because $u = x^{1/2} > 0$, the second solution is discarded. Thus,

$$x = u^2$$

$$= 18.724998^2 \qquad \text{To the nearest foot}$$

$$= 351$$

The well is 351 feet deep.

Graphical Solution

Enter $y_1 = \frac{\sqrt{x}}{4} + \frac{x}{1{,}100}$ and $y_2 = 5$. To determine the window variables, examine a table of values (Fig. 14) with fairly large x values (remember, x is the depth of the well and wells can be thousands of feet deep).

X	Y1	Y2
0	0	5
100	2.5909	5
200	3.7174	5
300	4.6029	5
400	5.3636	5
500	6.0447	5
600	6.6692	5

Y1■√(X)/4+X/1100

FIGURE 14

Now graph y_1 and y_2 and use intersect (Fig. 15).

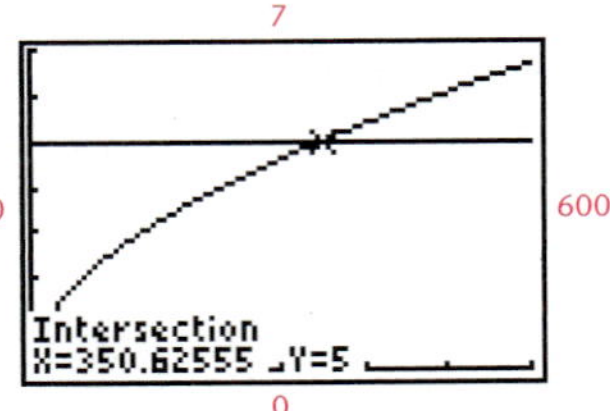

FIGURE 15

From Figure 15, we see that the well is 351 feet deep.

MATCHED 6 PROBLEM

The splash from a stone dropped into a deep well is heard 10 seconds after the stone is released. How deep is the well? Round answer to the nearest foot.

EXAMPLE 7 Design

FIGURE 16

A window in the shape of a semicircle with radius 20 inches contains a rectangular pane of glass as shown in Figure 16. Round all answers to three significant digits.

(A) Find a mathematical model for the area of the rectangle. Use one-half the length of the base of the rectangle for the independent variable in your model.

(B) Find the dimensions of the pane if the area of the pane is 320 square inches.

(C) Find the dimensions and the area of the largest possible rectangular pane of glass.

SOLUTIONS

(A) Place a rectangular coordinate system on the window (Fig. 17). Let x be one-half the base of the rectangle and y be the height of the rectangle.

FIGURE 17

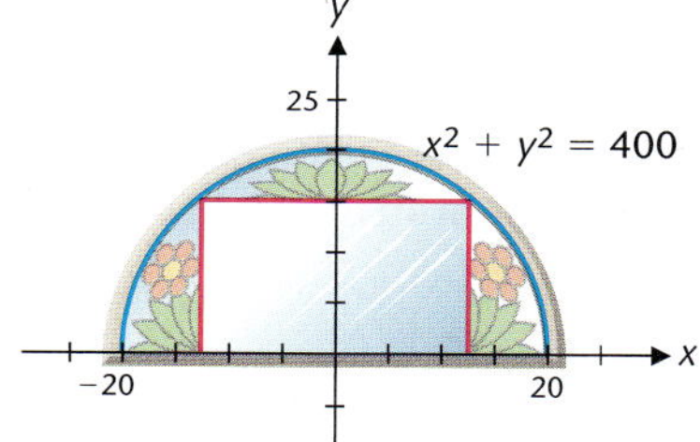

Because (x, y) are the coordinates of a point on the circle with radius 20 and center (0, 0), x and y must satisfy the equation of the circle.*

$$x^2 + y^2 = 400$$

$$y = \sqrt{400 - x^2} \qquad (3)$$

The area of the rectangle is

$$A = 2xy = 2x\sqrt{400 - x^2}$$

Thus, a model for the area is

$$A(x) = 2x\sqrt{400 - x^2} \qquad 0 \le x \le 20$$

(B) Solve the equation $A(x) = 320$.

*Circles are reviewed in Appendix A, Section A.3.

Algebraic Solution

$$2x\sqrt{400 - x^2} = 320$$
$$x\sqrt{400 - x^2} = 160$$
$$x^2(400 - x^2) = 25{,}600$$
$$400x^2 - x^4 = 25{,}600$$
$$400x^2 - x^4 - 25{,}600 = 0$$
$$x^4 - 400x^2 + 25{,}600 = 0$$

Use the quadratic formula to solve for x^2:

$$x^2 = \frac{400 \pm \sqrt{400^2 - 4 \cdot 25{,}600}}{2}$$
$$= \frac{400 \pm \sqrt{57{,}600}}{2}$$
$$= \frac{400 \pm 240}{2}$$
$$= 80, 320$$
$$x = \sqrt{80} = 8.94$$

or

$$x = \sqrt{320} = 17.9$$

A check shows that neither solution is extraneous.

Graphical Solution

To solve the equation $A(x) = 320$, enter both sides in the equation editor of a graphing utility (Fig. 18).

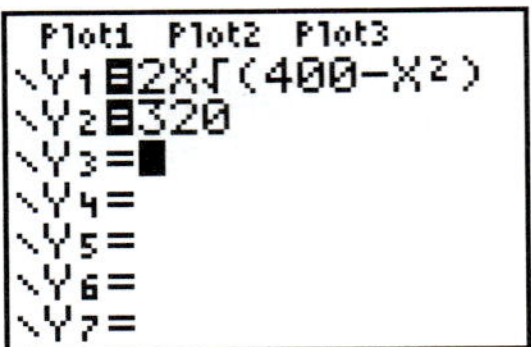

FIGURE 18

The values of x must satisfy $0 \le x \le 20$. Examining a table of values over this interval suggests that $0 \le y \le 500$ will produce a usable window (Fig. 19).

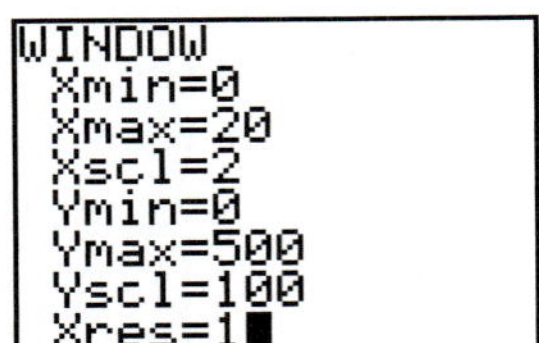

FIGURE 19

Graphing y_1 and y_2 and using the intersect command shows that the solutions are $x = 8.94$ (Fig. 20) and $x = 17.9$ (Fig. 21).

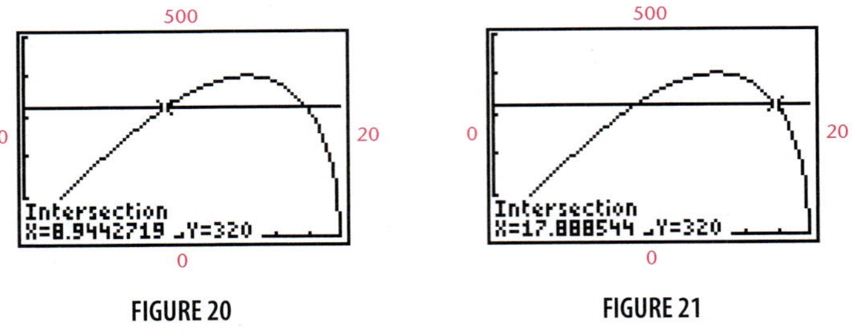

FIGURE 20 FIGURE 21

Now that we have determined the solutions to the equation $A(x) = 320$, we use equation (3) to find the dimensions of the two rectangles:

$$x = \sqrt{80} = 8.94 \quad \text{and} \quad y = \sqrt{400 - 80} = \sqrt{320} = 17.9$$
$$x = \sqrt{320} = 17.9 \quad \text{and} \quad y = \sqrt{400 - 320} = \sqrt{80} = 8.94$$

Recalling that x is one-half the base, the dimensions of the rectangles are

17.9 inches wide by 17.9 inches high or 35.8 inches wide by 8.94 inches high

Each solution is illustrated in Figure 22.

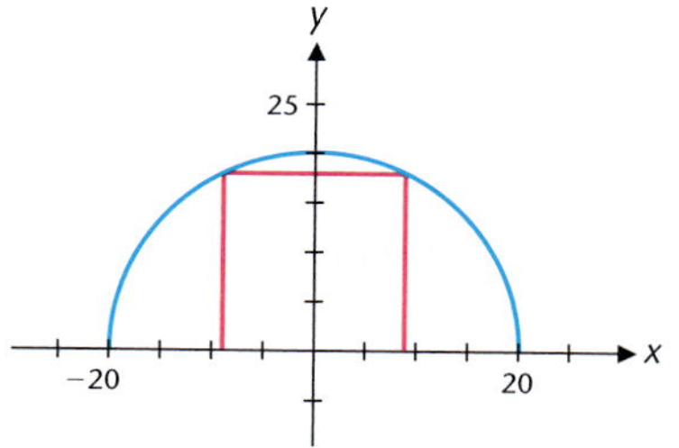

17.9 in. by 17.9 in.

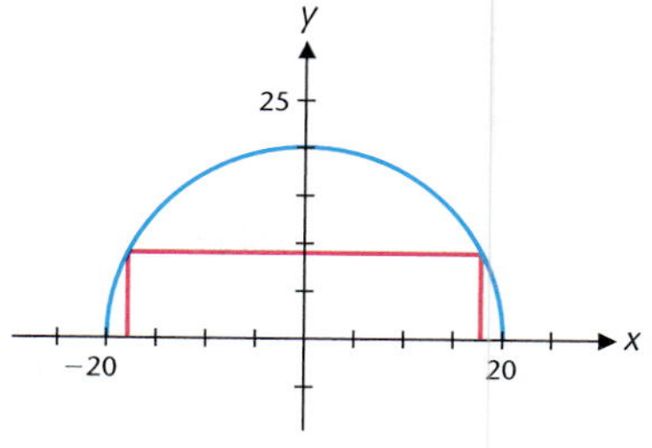

35.8 in. by 8.94 in.

FIGURE 22

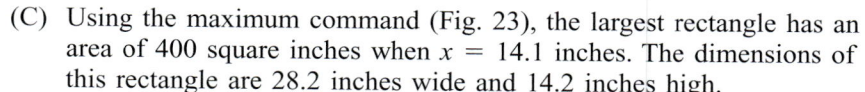

FIGURE 23

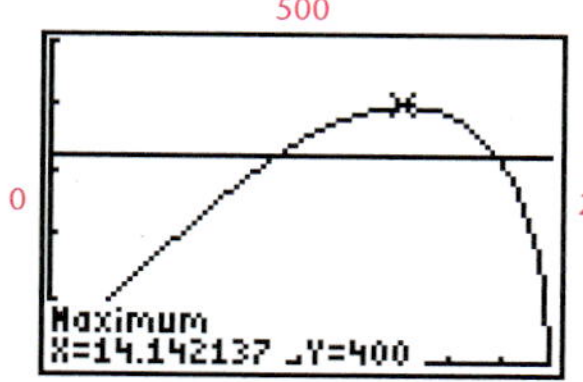

(C) Using the maximum command (Fig. 23), the largest rectangle has an area of 400 square inches when $x = 14.1$ inches. The dimensions of this rectangle are 28.2 inches wide and 14.2 inches high.

MATCHED 7 PROBLEM

A window in the shape of a semicircle with radius 25 inches contains a rectangular pane of glass as shown in Figure 16 in Example 7. Round all answers to three significant digits.

(A) Find a mathematical model for the area of the rectangle. Use one-half the length of the base of the rectangle for the independent variable in your model.

(B) Find the dimensions of the pane if the area of the pane is 500 square inches.

(C) Find the dimensions and the area of the largest possible rectangular pane of glass.

ANSWERS MATCHED PROBLEMS

1. $x = 1 - i$ **2.** $x = 2$ **3.** $x = 16, 81$

4. $x = \pm 1, \pm 2i$ **5.** $x = 1, -\frac{243}{32}$ **6.** 1,256 feet

7. (A) $A(x) = 2x\sqrt{625 - x^2}$, $0 \le x \le 25$

(B) 22.4 inches by 22.4 inches or 44.8 inches by 11.2 inches

(C) 35.4 inches by 17.7 inches, area = 625 square inches

EXERCISE 2.6

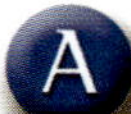

In Problems 1–6, determine the validity of each statement. If a statement is false, explain why.

1. If $x^2 = 5$, then $x = \pm\sqrt{5}$

2. $\sqrt{25} = \pm 5$

3. $(\sqrt{x - 1} + 1)^2 = x$

4. $(\sqrt{x} - 1)^2 + 1 = x$

5. If $x^3 = 2$, then $x = 8$

6. If $x^{1/3} = 2$, then $x = 8$

In Problems 7–12, transform each equation of quadratic type into a quadratic equation in u and state the substitution used in the transformation. If the equation is not an equation of quadratic type, say so.

7. $2x^{-6} - 4x^{-3} = 0$

8. $\frac{4}{7} - \frac{3}{x} + \frac{6}{x^2} = 0$

9. $3x^3 - 4x + 9 = 0$

10. $7x^{-1} + 3x^{-1/2} + 2 = 0$

11. $\frac{10}{9} + \frac{4}{x^2} - \frac{7}{x^4} = 0$

12. $3x^{3/2} - 5x^{1/2} + 12 = 0$

Solve Problems 13–28 algebraically and confirm graphically, if possible.

13. $\sqrt[3]{x + 5} = 3$

14. $\sqrt[4]{x - 3} = 2$

15. $\sqrt{5n + 9} = n - 1$

16. $m - 13 = \sqrt{m + 7}$

17. $\sqrt{x + 5} + 7 = 0$

18. $3 + \sqrt{2x - 1} = 0$

19. $\sqrt{3x + 4} = 2 + \sqrt{x}$

20. $\sqrt{3w - 2} - \sqrt{w} = 2$

21. $y^4 - 2y^2 - 8 = 0$

22. $x^4 - 7x^2 - 18 = 0$

23. $3x = \sqrt{x^2 - 2}$

24. $x = \sqrt{5x^2 + 9}$

25. $2x^{2/3} + 3x^{1/3} - 2 = 0$

26. $x^{2/3} - 3x^{1/3} - 10 = 0$

27. $(m^2 - m)^2 - 4(m^2 - m) = 12$

28. $(x^2 + 2x)^2 - (x^2 + 2x) = 6$

B

Solve Problems 29–44 algebraically and confirm graphically, if possible.

29. $\sqrt{u - 2} = 2 + \sqrt{2u + 3}$

30. $\sqrt{3t + 4} + \sqrt{t} = -3$

31. $\sqrt{3y - 2} = 3 - \sqrt{3y + 1}$

32. $\sqrt{2x - 1} - \sqrt{x - 4} = 2$

33. $\sqrt{7x - 2} - \sqrt{x + 1} = \sqrt{3}$

34. $\sqrt{3x + 6} - \sqrt{x + 4} = \sqrt{2}$

35. $\sqrt{4x^2 + 12x + 1} - 6x = 9$

36. $6x - \sqrt{4x^2 - 20x + 17} = 15$

37. $3n^{-2} - 11n^{-1} - 20 = 0$

38. $6x^{-2} - 5x^{-1} - 6 = 0$

39. $9y^{-4} - 10y^{-2} + 1 = 0$

40. $4x^{-4} - 17x^{-2} + 4 = 0$

41. $y^{1/2} - 3y^{1/4} + 2 = 0$

42. $4x^{-1} - 9x^{-1/2} + 2 = 0$

43. $(m - 5)^4 + 36 = 13(m - 5)^2$

44. $(x - 3)^4 + 3(x - 3)^2 = 4$

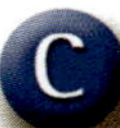

Solve Problems 45–48 algebraically and confirm graphically, if possible.

45. $\sqrt{5 - 2x} - \sqrt{x + 6} = \sqrt{x + 3}$

46. $\sqrt{2x + 3} - \sqrt{x - 2} = \sqrt{x + 1}$

47. $2 + 3y^{-4} = 6y^{-2}$

48. $4m^{-2} = 2 + m^{-4}$

Solve Problems 49–52 two ways: by squaring and by substitution. Confirm graphically, if possible.

49. $m - 7\sqrt{m} + 12 = 0$

50. $y - 6 + \sqrt{y} = 0$

51. $t - 11\sqrt{t} + 18 = 0$

52. $x = 15 - 2\sqrt{x}$

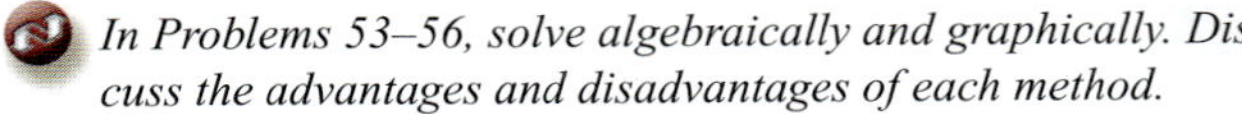

In Problems 53–56, solve algebraically and graphically. Discuss the advantages and disadvantages of each method.

53. $2\sqrt{x + 5} = 0.01x + 2.04$

54. $3\sqrt{x - 1} = 0.05x + 2.9$

55. $2x^{-2/5} - 5x^{-1/5} + 1 = 0$

56. $x^{-2/5} - 3x^{-1/5} + 1 = 0$

APPLICATIONS

57. Geometry. The diagonal of a rectangle is 10 inches and the area is 45 square inches. Find the dimensions of the rectangle, correct to one decimal place.

58. Geometry. The hypotenuse of a right triangle is 12 inches and the area is 24 square inches. Find the dimensions of the triangle, correct to one decimal place.

59. Physics–Well Depth. If the splash of a stone dropped into a well is heard 14 seconds after the stone is released, how deep (to the nearest foot) is the well?

60. Physics–Well Depth. If the splash of a stone dropped into a well is heard 2 seconds after the stone is released, how deep (to the nearest foot) is the well?

61. Manufacturing. A lumber mill cuts rectangular beams from circular logs that are 16 inches in diameter (see the figure).

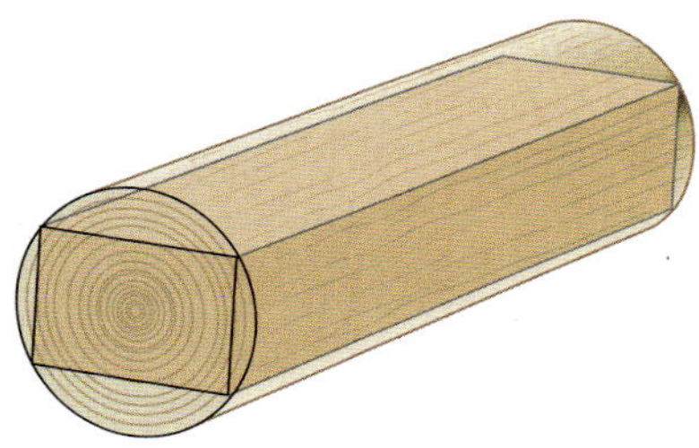

(A) Find a model for the cross-sectional area of the beam. Use the width of the beam as the independent variable.

(B) If the cross-sectional area of the beam is 120 square inches, find the dimensions correct to one decimal place.

(C) Find the dimensions of the beam that has the largest cross-sectional area and find this area. Round answers to one decimal place.

62. Design. A food-processing company packages an assortment of their products in circular metal tins 12 inches in diameter. Four identically sized rectangular boxes are used to divide the tin into eight compartments (see the figure).

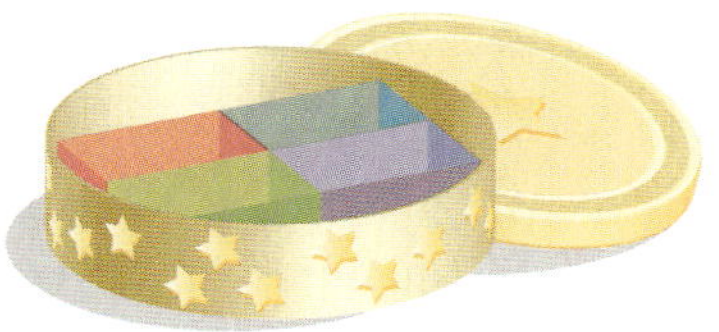

(A) Find a model for the cross-sectional area of one of these boxes. Use the width of the box as the independent variable.

(B) If the cross-sectional area of the box is 15 square inches, find the dimensions correct to one decimal place.

(C) Find the dimensions of the box that has the largest cross-sectional area and find this area. Round answers to one decimal place.

★ **63. Construction.** A water trough is constructed by bending a 4- by 6-foot rectangular sheet of metal down the middle and attaching triangular ends (see the figure). If the volume of the trough is 9 cubic feet, find the width correct to two decimal places.

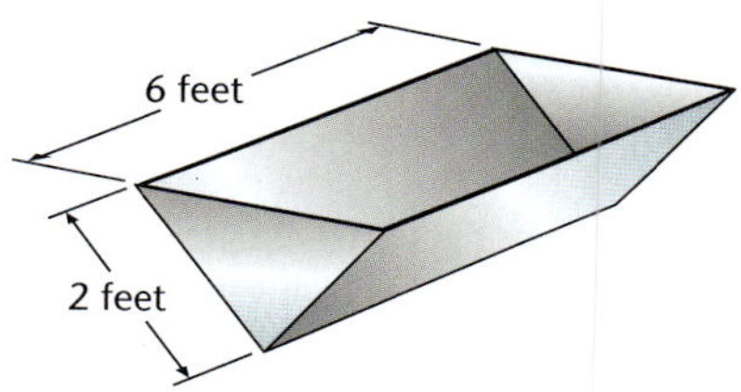

★ **64. Design.** A paper drinking cup in the shape of a right circular cone is constructed from 125 square centimeters of paper (see the figure). If the height of the cone is 10 centimeters, find the radius correct to two decimal places.

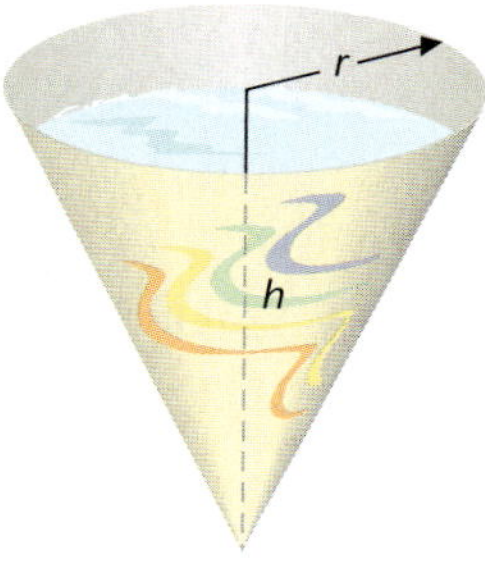

Lateral surface area:

$S = \pi r \sqrt{r^2 + h^2}$

SECTION 2.7 Solving Inequalities

Solving Linear Inequalities • Solving Inequalities Involving Absolute Value • Solving Quadratic Inequalities • Mathematical Modeling • Data Analysis and Regression

We now consider techniques for solving various types of inequalities and several applications that involve inequalities.

Solving Linear Inequalities

Any inequality that can be reduced to one of the four forms in (1) is called a **linear inequality in one variable.**

$$\left.\begin{aligned} mx + b &> 0 \\ mx + b &\geq 0 \\ mx + b &< 0 \\ mx + b &\leq 0 \end{aligned}\right\} \quad \text{Linear Inequalities} \qquad (1)$$

As was the case with equations, the **solution set of an inequality** is the set of all values of the variable that make the inequality a true statement. Each element of the solution set is called a **solution.** Two inequalities are said to be **equivalent** if they have the same solution set.

EXPLORE/DISCUSS 1

Associated with the linear equation and inequalities

$$3x - 12 = 0 \qquad 3x - 12 < 0 \qquad 3x - 12 > 0$$

is the linear function

$$f(x) = 3x - 12$$

(A) Graph the function f.

(B) From the graph of f describe verbally the values of x for which

$$f(x) = 0 \qquad f(x) < 0 \qquad f(x) > 0$$

(C) How are the answers to part (B) related to the solutions of

$$3x - 12 = 0 \qquad 3x - 12 < 0 \qquad 3x - 12 > 0$$

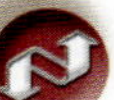

As you discovered in Explore/Discuss 1, solving inequalities graphically is both intuitive and efficient. Algebraic solution methods require an understanding of the algebraic operations that can be performed on an inequality to produce an equivalent inequality. The necessary facts are summarized in Theorem 1. (See Appendix A, Section A.1 for more information on inequalities and interval notation.)

THEOREM 1
Inequality Properties

An equivalent inequality will result and the **sense or direction will remain the same** if each side of the original inequality

1. Has the same real number added to or subtracted from it
2. Is multiplied or divided by the same positive number

An equivalent inequality will result and the **sense or direction will reverse** if each side of the original inequality

3. Is multiplied or divided by the same negative number

Note: Multiplication by 0 and division by 0 are not permitted.

Thus, we can perform essentially the same operations on inequalities that we perform on equations, with the exception that **the sense or direction of the inequality reverses if we multiply or divide both sides by a negative number.** Otherwise the sense or direction of the inequality does not change.

EXAMPLE 1 Solving a Linear Inequality

Solve $0.5x + 1 \le 0$.

SOLUTION

Algebraic Solution

$$0.5x + 1 \le 0$$

$$0.5x + 1 - 1 \le 0 - 1 \qquad \text{Add } -1 \text{ to both sides.}$$

$$0.5x \le -1$$

$$\frac{0.5x}{0.5} \le \frac{-1}{0.5} \qquad \text{Divide both sides by 0.5.}$$

$$x \le -2 \quad \text{or} \quad (-\infty, -2]$$

Graphical Solution

The graph of $f(x) = 0.5x + 1$ is shown in Figure 1. We see from the graph that $f(x)$ is negative to the left of -2 and positive to the right. Thus, the solution set of $0.5x + 1 \le 0$ is $x \le -2$ or $(-\infty, -2]$.

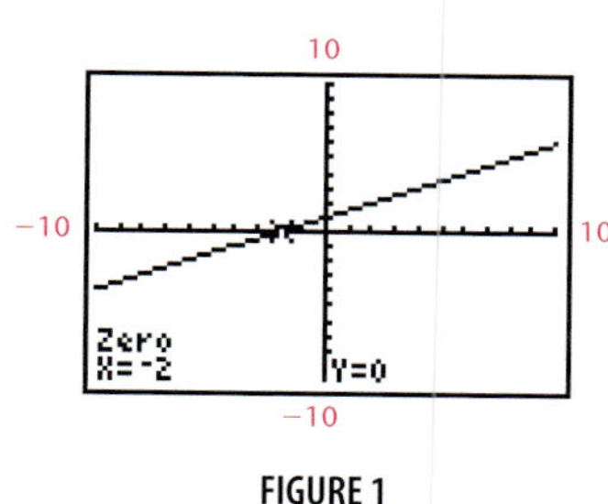

FIGURE 1

MATCHED PROBLEM 1

Solve $2x - 6 \ge 0$.

EXAMPLE 2 Solving Combined Inequalities

Solve $-3 \le 4 - 7x < 18$.

SOLUTION

Algebraic Solution

To solve algebraically, we perform operations on the combined inequality until we have isolated x in the middle with a coefficient of 1.

$$-3 \le 4 - 7x < 18$$

$$-3 - 4 \le 4 - 7x - 4 < 18 - 4$$ Subtract 4 from each member.

$$-7 \le -7x < 14$$

$$\frac{-7}{-7} \ge \frac{-7x}{-7} > \frac{14}{-7}$$ Divide each member by -7 and reverse each inequality.

$$1 \ge x > -2 \quad \text{or} \quad -2 < x \le 1 \quad \text{or} \quad (-2, 1] \qquad (2)$$

Graphical Solution

Enter $y_1 = -3$, $y_2 = 4 - 7x$, $y_3 = 18$, and find the intersection points (Fig. 2 and Fig. 3). It is clear from the graph that y_2 is between y_1 and y_3 for x between -2 and 1. Because $y_2 = -3$ at $x = 1$, we include 1 in the solution set, obtaining the same solution as shown in (2).

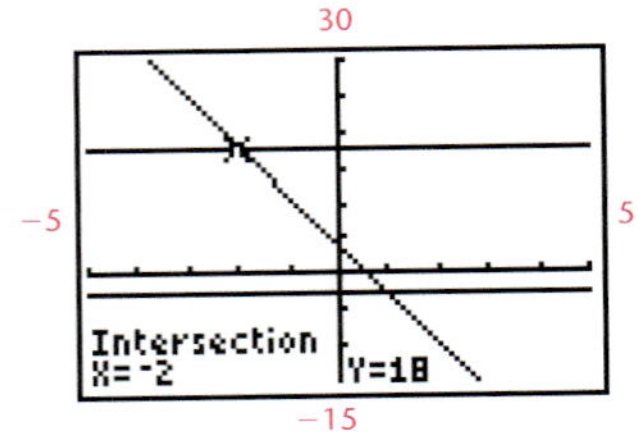

FIGURE 2

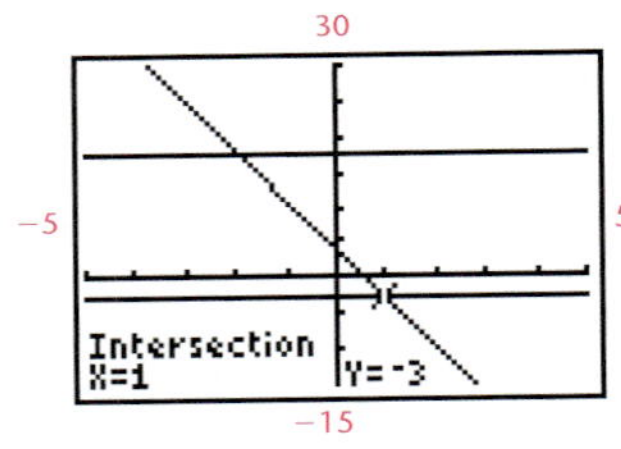

FIGURE 3

MATCHED PROBLEM 2

Solve $-3 < 7 - 2x \le 7$.

Solving Inequalities Involving Absolute Value

EXPLORE/DISCUSS 2

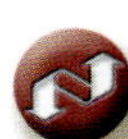

Recall the definition of the absolute value function (see Section 1.3)

$$f(x) = |x| = \begin{cases} x & \text{if } x \ge 0 \\ -x & \text{if } x < 0 \end{cases}$$

(A) Graph the absolute value function $f(x) = |x|$ and the constant function $g(x) = 3$ in the same viewing window.

(B) From the graph in part A, determine the values of x for which:

$|x| < 3$ $\quad |x| = 3 \quad$ $|x| > 3$

(C) Find all the points with coordinates $(x, 0)$ that are
Less than three units from the origin
Exactly three units from the origin
More than three units from the origin

(D) Compare the solutions found in parts (B) and (C).

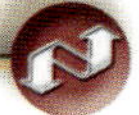

The absolute value function has a geometric interpretation that is useful for solving inequalities. The distance between the point $P = (x, 0)$ and the origin $O = (0, 0)$ is (see Appendix A, Section A.3)

$$\begin{aligned} d(P, O) &= \sqrt{(x - 0)^2 + (0 - 0)^2} \\ &= \sqrt{x^2} \\ &= |x| \end{aligned}$$

Thus, $|x|$ can be interpreted as the distance between the point with coordinates $(x, 0)$ and the origin. That is,

$|x| = p$ is equivalent to $x = p$ or $x = -p$

$|x| < p$ is equivalent to $-p < x < p$

$|x| > p$ is equivalent to $x < -p$ or $x > p$

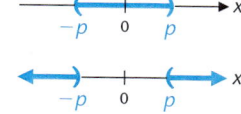

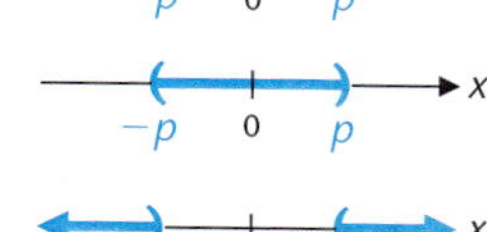

More generally, we have Theorem 2.

THEOREM 2
Geometric Interpretation of Absolute Value

For $p > 0$

1. $|ax + b| < p$ is equivalent to $-p < ax + b < p$.
2. $|ax + b| = p$ is equivalent to $ax + b = p$ or $ax + b = -p$.
3. $|ax + b| > p$ is equivalent to $ax + b < -p$ or $ax + b > p$.

EXAMPLE 3 Solving Inequalities Involving Absolute Value

Solve and write the solution in both inequality and interval notation for $|2x - 1| < 3$.

SOLUTION

Geometric Solution

The solution is the set of points x for which $2x - 1$ is less than three units from the origin. Thus,

$$-3 < 2x - 1 < 3$$

$$-3 + 1 < 2x - 1 + 1 < 3 + 1$$

$$-2 < 2x < 4$$

$$\frac{-2}{2} < \frac{2x}{2} < \frac{4}{2}$$

$$-1 < x < 2 \quad \text{or} \quad (-1, 2)$$

Graphical Solution

Enter $y_1 = |2x - 1|$ and $y_2 = 3$ and use intersect to find the intersection points (Fig. 4 and Fig. 5).

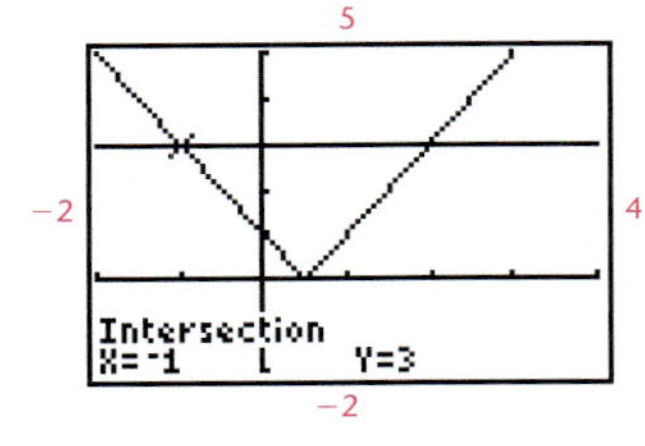

FIGURE 4

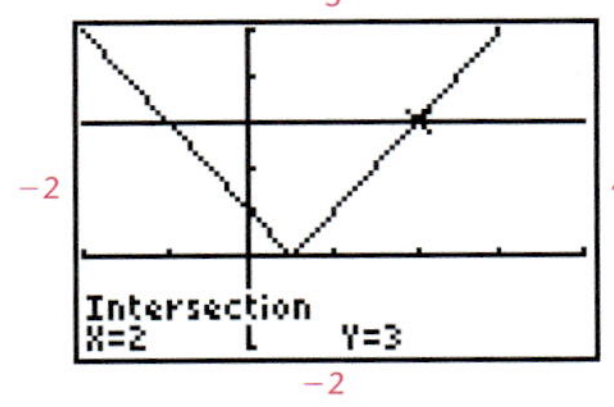

FIGURE 5

Examining these graphs, we see that the graph of y_1 is below the graph of y_2 for $-1 < x < 2$.

MATCHED 3 PROBLEM

Solve $|2x + 1| < 5$.

EXAMPLE 4 Solving an Inequality Involving Absolute Value

Solve and express the answer in both inequality and interval notation for $|4x - 5| \geq 2$.

SOLUTION

Geometric Solution

The solution is the set of points x for which $4x - 5$ is two or more units from the origin. Thus,

$$4x - 5 \leq -2 \quad \text{or} \quad 4x - 5 \geq 2$$

$$4x \leq 3 \quad \text{or} \quad 4x \geq 7$$

$$x \leq \frac{3}{4} = 0.75 \quad \text{or} \quad x \geq \frac{7}{4} = 1.75$$

The solution set is

$$\{x | x \leq 0.75 \text{ or } x \geq 1.75\}$$
$$= \{x | x \leq 0.75\} \cup \{x | x \geq 1.75\}$$
$$= (-\infty, 0.75] \cup [1.75, \infty)^* \qquad (3)$$

Graphical Solution

Enter $y_1 = |4x - 5|$ and $y_2 = 2$ and find the intersection points (Fig. 6 and Fig. 7).

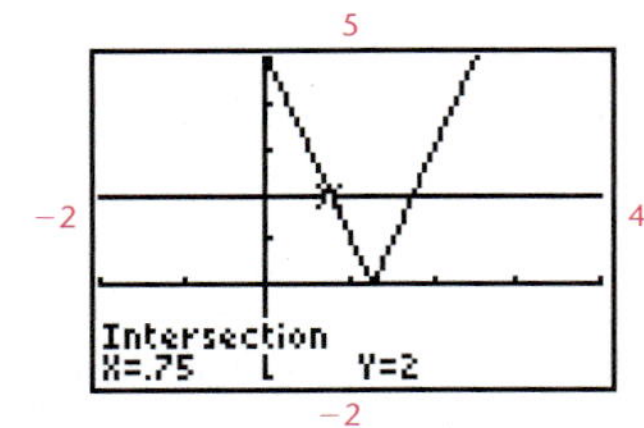

FIGURE 6

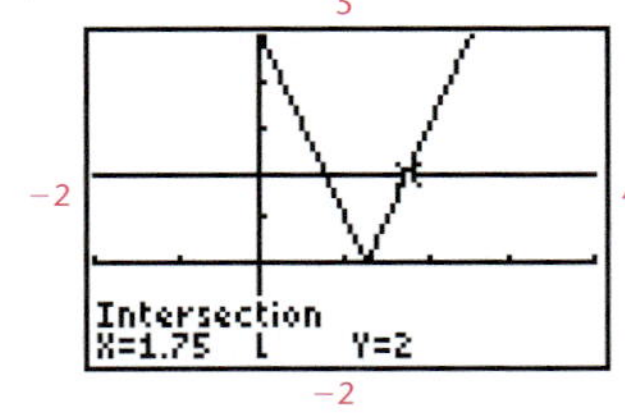

FIGURE 7

Examining these graphs, we see that if $x \leq 0.75$ or $x \geq 1.75$, then the graph of y_1 is on or above the graph of y_2. This is the same solution given in (3).

*The symbol ∪ denotes the union operation for sets. See Appendix A, Section A.1, for a discussion of interval notation and set operations.

MATCHED 4 PROBLEM

Solve and express the answer in both inequality and interval notation for $\left|\frac{2}{3}x + 1\right| \geq 2$.

Solving Quadratic Inequalities

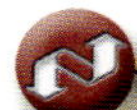

EXPLORE/DISCUSS 3

Graph $f(x) = (x + 2)(x - 3)$ and examine the graph to determine the solutions of the following inequalities:

(A) $f(x) > 0$ (B) $f(x) < 0$ (C) $f(x) \geq 0$ (D) $f(x) \leq 0$

Discuss algebraic methods that might be used to solve quadratic inequalities.

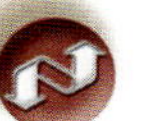

An inequality involving a quadratic function is called a **quadratic inequality.** One algebraic technique for solving quadratic inequalities involves factoring and the following theorem.

THEOREM 3
Products of Real Numbers

If a and b are real numbers, then
$ab > 0$ if and only if a and b have the same sign
$ab < 0$ if and only if a and b have opposite sign

EXAMPLE 5 Solving a Quadratic Inequality

Solve and express the answer in interval notation for $x^2 - x - 12 > 0$.

SOLUTION

Algebraic Solution

Factor the quadratic expression and consider two cases

$$x^2 - x - 12 = (x - 4)(x + 3) > 0$$

Graphical Solution

Enter $y_1 = x^2 - x - 12$. Using the zero command we see that $y_1 = 0$ at $x = -3$ (Fig. 8) and at $x = 4$ (Fig. 9).

Case 1: $x - 4 > 0$ and $x + 3 > 0$

$x > 4 \qquad x > -3$

Because both conditions must be satisfied, we conclude that the solution for this case is $(4, \infty)$.

Case 2: $x - 4 < 0$ and $x + 3 < 0$

$x < 4 \qquad x < -3$

In this case, we conclude that the solution is $(-\infty, -3)$. Combining the two cases, we have the solution:

$$(-\infty, -3) \cup (4, \infty)$$

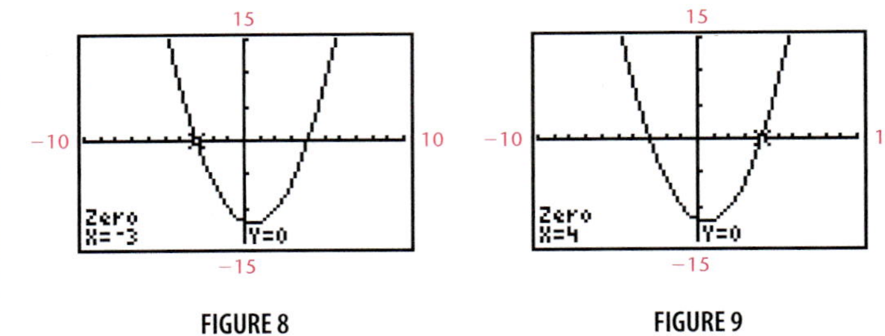

FIGURE 8

FIGURE 9

The graph of y_1 is above the x axis for $x < -3$ and also for $x > 4$. Thus, the solution to the inequality $y_1 > 0$ is

$$(-\infty, -3) \cup (4, \infty)$$

MATCHED 5 PROBLEM

Solve and express the answer in interval notation for $x^2 - x - 6 > 0$.

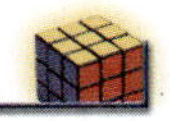

Mathematical Modeling

EXAMPLE 6 **Projectile Motion**

A projectile propelled straight upward from the ground reaches a maximum height of 576 feet above ground level after 6 seconds. Let the quadratic function $d(t)$ represent the distance above ground level (in feet) t seconds after the projectile is released.

(A) Find $d(t)$.

(B) At what times will the projectile be more than 320 feet above the ground? Express the answer in inequality notation.

SOLUTIONS

(A) Because the quadratic distance function d has a maximum value of 576 at $t = 6$, the vertex form for $d(t)$ is

$$d(t) = a(t - 6)^2 + 576$$

To determine a we use the fact that $d(0) = 0$.

$$\begin{aligned} d(0) = a(-6)^2 + 576 &= 0 \\ 36a &= -576 \\ a &= -16 \end{aligned}$$

Thus, the model for the flight of this projectile is

$$\begin{aligned} d(t) &= -16(t - 6)^2 + 576 \\ &= -16t^2 + 192t \end{aligned}$$

(B) To determine the times when the projectile is higher than 320 feet, we solve the inequality

$$d(t) = -16t^2 + 192t > 320$$

Algebraic Solution

$$-16t^2 + 192t > 320$$
$$-16t^2 + 192t - 320 > 0$$

$$\frac{-16t^2}{-16} + \frac{192t}{-16} - \frac{320}{-16} < 0$$

Divide each side by −16 and reverse the direction of the inequality.

$$t^2 - 12t + 20 < 0$$
$$(t - 2)(t - 10) < 0$$

Case 1: $t - 2 > 0$ and $t - 10 < 0$
$t > 2$ and $t < 10$
$2 < t < 10$

Case 2: $t - 2 < 0$ and $t - 10 > 0$
$t < 2$ and $t > 10$
No solution

Thus, the projectile is above 320 feet for $2 < t < 10$.

Graphical Solution

Graph $y_1 = -16x^2 + 192x$ and $y_2 = 320$ and find the intersection points (Fig. 10 and Fig. 11).

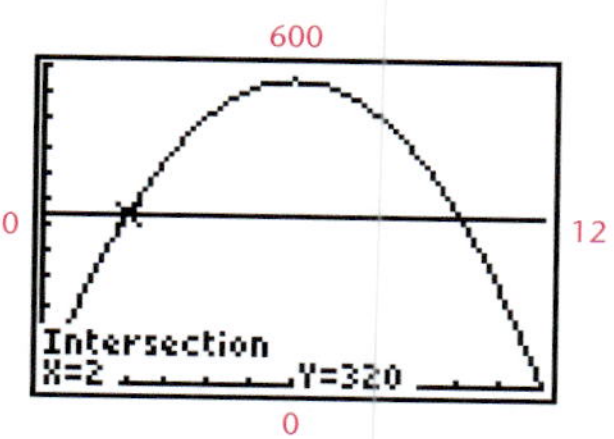

FIGURE 10

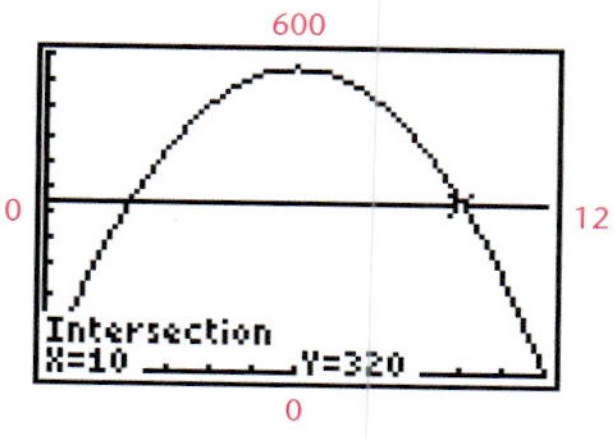

FIGURE 11

From these graphs we see that the projectile will be above 320 feet for $2 < t < 10$.

MATCHED 6 PROBLEM

Refer to the projectile equation in Example 6. At what times during its flight will the projectile be less than 432 feet above the ground? Express the answer in inequality notation.

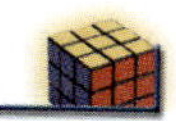

Data Analysis and Regression

EXAMPLE 7 **Break-Even, Profit, and Loss**

A paint manufacturer has weekly fixed costs of \$40,000 and variable costs of \$6.75 per gallon produced. Examining past records produces the price–demand data in Table 1 on page 218. Round all numbers to three significant digits.

(A) Use linear regression to find the price–demand equation $p = d(x)$ for the data in Table 1. What is the domain of $d(x)$?

TABLE 1 Price–Demand Data

Weekly Sales	Price per Gallon
5,610	\$20.50
5,810	\$18.70
5,990	\$17.90
6,180	\$16.20
6,460	\$15.40
6,730	\$13.80
6,940	\$12.90

(B) Find the revenue and cost functions as functions of the sales x. What is the domain of each function?

(C) Find the level of sales for which the company will break even. Describe verbally and graphically the sales levels that result in a profit and those that result in a loss.

(D) Find the sales and the price that will produce the maximum profit. Find the maximum profit.

SOLUTIONS

(A) Enter the data in Table 1 and select the LinReg($ax+b$) option (Fig. 12).

FIGURE 12

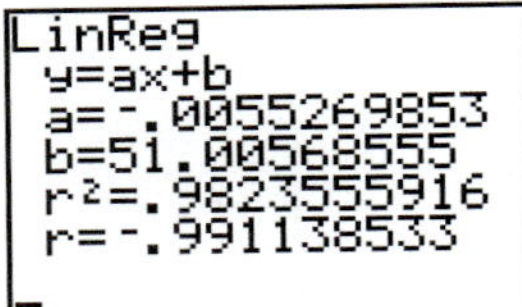

After rounding, the price–demand equation is

$$p = d(x) = 51.0 - 0.00553x$$

Because price is always nonnegative, x must satisfy

$$51.0 - 0.00553x \geq 0$$
$$51 \geq 0.00553x$$
$$x \leq \frac{51}{0.00553} = 9{,}220 \quad \text{To three significant digits}$$

Because sales are also nonnegative, the domain of the price–demand equation is $0 \leq x \leq 9{,}220$.

(B) The revenue function is

$$\begin{aligned} R(x) &= xp \\ &= x(51 - 0.00553x) \\ &= 51x - 0.00553x^2 \qquad 0 \leq x \leq 9{,}220 \end{aligned}$$

Note that the domain of R is the same as the domain of d. The cost function is

$$C(x) = 40{,}000 + 6.75x \qquad x \geq 0$$

FIGURE 13

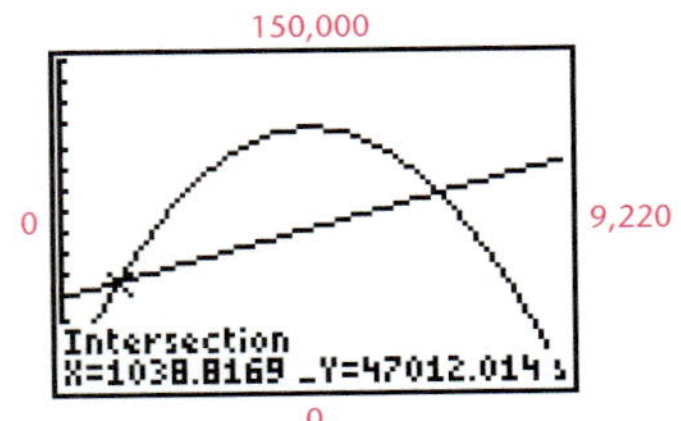

(C) The company will break even when revenue = cost, that is, when $R(x) = C(x)$. An intersection point on the graphs of R and C is often referred to as a **break-even point.** Graphs of both functions and their intersection points are shown in Figures 13 and 14.

Examining these graphs, we see that the company will break even if they sell 1,040 or 6,960 gallons of paint. If they sell between 1,040 and 6,960 gallons, then revenue is greater than cost and the company will make a profit. If they sell fewer than 1,040 or more than 6,960 gallons, then cost is greater than revenue and the company will lose money. These sales levels are illustrated in Figure 15.

FIGURE 14

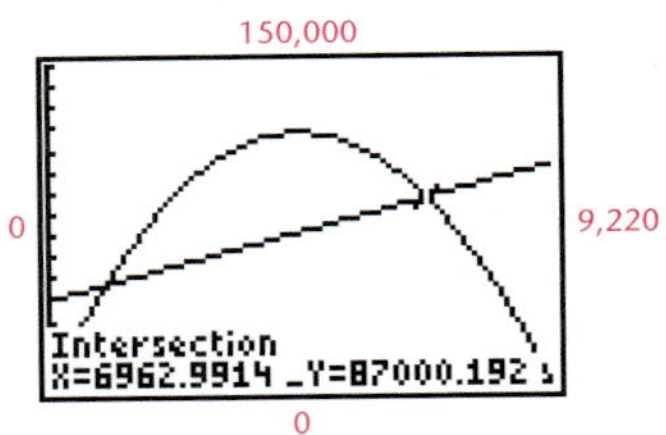

FIGURE 15

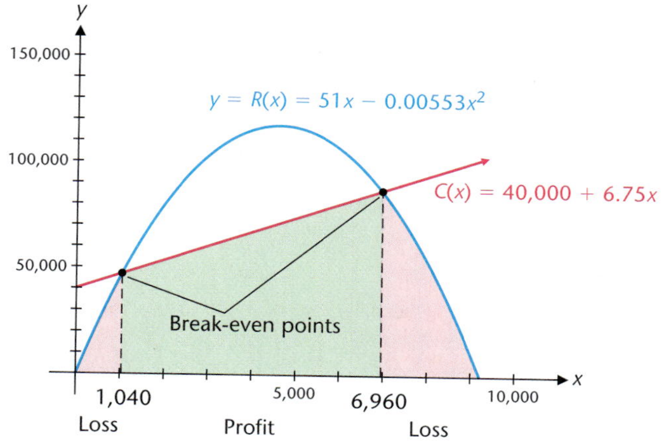

(D) The profit function for this manufacturer is

$$\begin{aligned} P(x) &= R(x) - C(x) \\ &= (51x - 0.00553x^2) - (40{,}000 + 6.75x) \\ &= 44.25x - 0.00553x^2 - 40{,}000 \end{aligned}$$

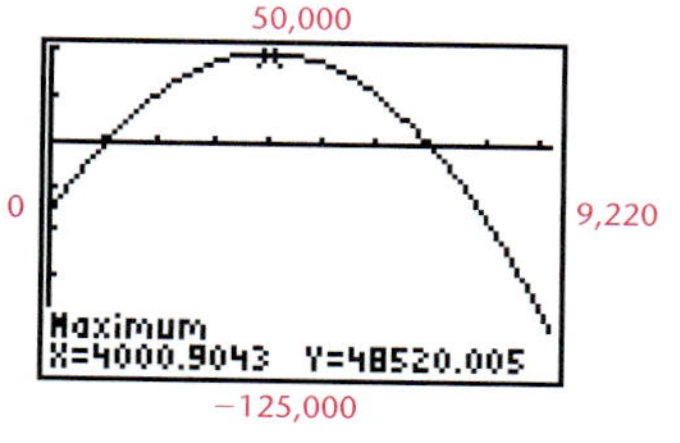

FIGURE 16

To find the largest profit, enter $y_1 = P(x)$ and use the maximum command (Fig. 16).

Thus, the maximum profit of \$48,500 occurs when 4,000 gallons of paint are sold at a price of

$$\begin{aligned} p = d(4{,}000) &= 51 - 0.00553(4{,}000) \\ &= \$28.90 \end{aligned}$$

MATCHED 7 PROBLEM

A paint manufacturer has weekly fixed costs of \$50,000 and variable costs of \$7.50 per gallon produced. Examining past records produces the price–demand data in Table 2. Round all numbers to three significant digits.

TABLE 2 Price–Demand Data

Weekly Sales	Price per Gallon
5,470	\$18.80
5,640	\$17.30
5,910	\$15.90
6,150	\$14.10
6,380	\$13.30
6,530	\$12.40
6,820	\$10.80

(A) Use linear regression to find the price–demand equation $p = d(x)$ for the data in Table 2. What is the domain of $d(x)$?

(B) Find the revenue and cost functions as functions of the sales x. What is the domain of each function?

(C) Find the level of sales for which the company will break even. Describe verbally and graphically the sales levels that result in a profit and those that result in a loss.

(D) Find the sales and the price that will produce the maximum profit. Find the maximum profit.

ANSWERS TO MATCHED PROBLEMS

1. $x \geq 3$ or $[3, \infty)$
2. $0 \leq x < 5$ or $[0, 5)$
3. $-3 < x < 2$ or $(-3, 2)$
4. $x \leq -4.5$ or $x \geq 1.5$; $(-\infty, -4.5] \cup [1.5, \infty)$
5. $(-\infty, -2) \cup (3, \infty)$
6. $0 \leq t < 3$ or $9 < t \leq 12$
7. (A) $p = d(x) = 50.0 - 0.00577x$
   $0 \leq x \leq 8{,}670$
   (B) $R(x) = 50x - 0.00577x^2$
   $0 \leq x \leq 8{,}670$
   $C(x) = 50{,}000 + 7.5x$
   $x \geq 0$
   (C) The company will break even if they sell 1,470 or 5,900 gallons of paint. If they sell between 1,470 and 5,900 gallons, then revenue is greater than cost and the company will make a profit. If they sell fewer than 1,470 or more than 5,900 gallons, then cost is greater than revenue and the company will lose money.

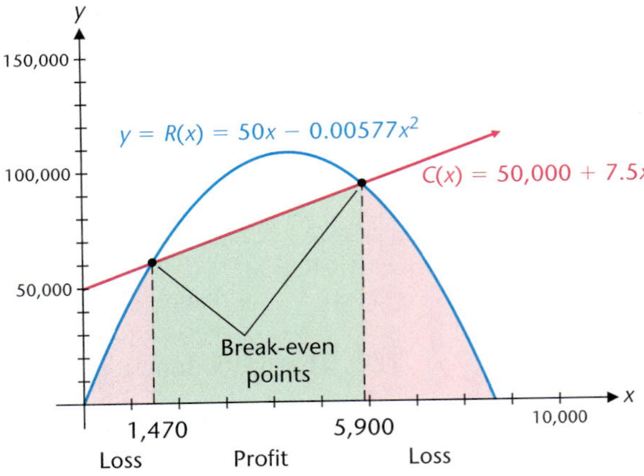

   (D) The company will make a maximum profit of $28,300 when they sell 3,680 gallons at $28.80 per gallon.

EXERCISE 2.7

A

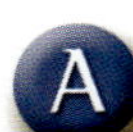

Use the graphs of functions u and v in the figure to solve the inequalities in Problems 1–8. (Assume the graphs continue as indicated beyond the portions shown here.) Express solutions in interval notation.

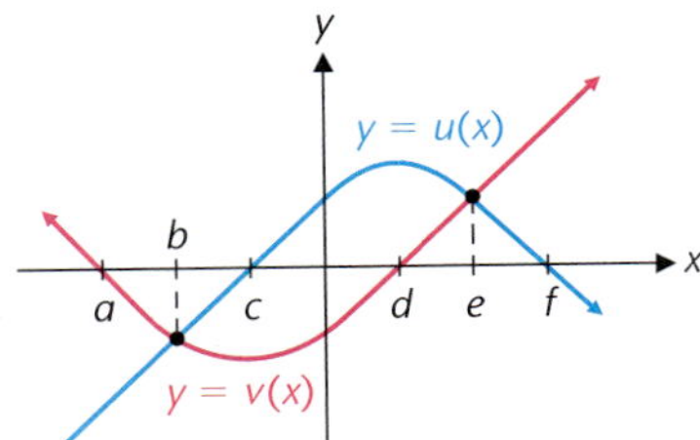

1. $u(x) > 0$
2. $v(x) \geq 0$
3. $v(x) \geq u(x)$
4. $v(x) < 0$
5. $u(x) \leq 0$
6. $u(x) - v(x) \geq 0$
7. $v(x) - u(x) > 0$
8. $v(x) < u(x)$

In Problems 9–16, write each statement as an absolute value inequality.

9. x is less than five units from 3.
10. w is more than four units from 2.
11. y is more than six units from -1.
12. z is less than eight units from -2.
13. a is no more than five units from 3.
14. c is no less than seven units from -4.
15. d is no less than four units from -2.
16. m is no more than six units from 1.

Solve Problems 17–30 and write answers in both interval and inequality notation.

17. $7x - 8 < 4x + 7$
18. $4x + 8 \geq x - 1$
19. $-5t < -10$
20. $-7n \geq 21$
21. $3 - m < 4(m - 3)$
22. $2(1 - u) \geq 5u$
23. $x^2 < 10 - 3x$
24. $x^2 + x < 12$
25. $x^2 + 21 > 10x$
26. $x^2 + 7x + 10 > 0$
27. $x^2 \leq 8x$
28. $x^2 + 6x \geq 0$
29. $x^2 + 5x \leq 0$
30. $x^2 \leq 4x$

B

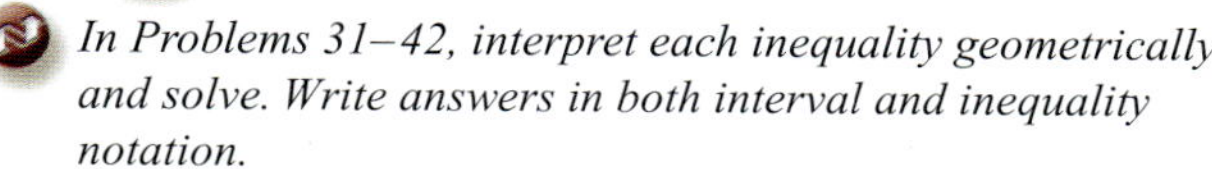

In Problems 31–42, interpret each inequality geometrically and solve. Write answers in both interval and inequality notation.

31. $|y| \le 7$ **32.** $|t| \le 5$ **33.** $|w| \ge 7$

34. $|r| \ge 5$ **35.** $|s - 5| < 3$ **36.** $|t - 3| < 4$

37. $|s - 5| > 3$ **38.** $|t - 3| > 4$ **39.** $|u + 8| \le 3$

40. $|x + 1| \le 5$ **41.** $|u + 8| \ge 3$ **42.** $|x + 1| \ge 5$

Solve Problems 43–60 and write answers in both interval and inequality notation.

43. $-4 < 5t + 6 \le 21$ **44.** $2 \le 3m - 7 < 14$

45. $-12 < \frac{3}{4}(2 - x) \le 24$ **46.** $24 \le \frac{2}{3}(x - 5) < 36$

47. $\frac{q}{7} - 3 > \frac{q - 4}{3} + 1$ **48.** $\frac{p}{3} - \frac{p - 2}{2} \le \frac{p}{4} - 4$

49. $x^2 + 1 < 2x$ **50.** $x^2 + 25 < 10x$

51. $x^2 < 4x - 3$ **52.** $x^2 - 8 > 2x$

53. $x^2 - 21 \ge 4x$ **54.** $x^2 + 13x + 40 \le 0$

55. $|3x - 7| \le 4$ **56.** $|5y + 2| \ge 8$

57. $|4 - 2t| > 6$ **58.** $|10 + 4s| < 6$

59. $|0.2u + 1.7| \ge 0.5$ **60.** $|0.5v - 2.5| > 1.6$

61. Discuss the possible signs of the numbers a and b given that

(A) $ab > 0$ **(B)** $ab < 0$

(C) $\frac{a}{b} > 0$ **(D)** $\frac{a}{b} < 0$

62. Discuss the possible signs of the numbers a, b, and c given that

(A) $abc > 0$ **(B)** $\frac{ab}{c} < 0$

(C) $\frac{a}{bc} > 0$ **(D)** $\frac{a^2}{bc} < 0$

In Problems 63–66, replace each question mark with < or > and explain why your choice makes the statement true.

63. If $a - b = 1$, then a ? b.

64. If $u - v = -2$, then u ? v.

65. If $a < 0$, $b < 0$, and $\frac{b}{a} > 1$, then a ? b.

66. If $a > 0$, $b > 0$, and $\frac{b}{a} > 1$, then a ? b.

C

Problems 67–70 are calculus related. Solve and write answers in interval notation.

67. $0 < |x - 3| < 0.1$ **68.** $0 < |x - 5| < 0.01$

69. $0 < |x - c| < 2c, c > 0$ **70.** $0 < |x - 2c| < c, c > 0$

In Problems 71–74, use the given information concerning the roots of the quadratic equation $ax^2 + bx + c = 0$ to describe the possible solution sets for the indicated inequality. Illustrate your conclusions with specific examples.

71. $ax^2 + bx + c > 0$, given distinct real roots r_1 and r_2 with $r_1 < r_2$.

72. $ax^2 + bx + c \le 0$, given distinct real roots r_1 and r_2 with $r_1 < r_2$.

73. $ax^2 + bx + c \ge 0$, given one (double) real root r.

74. $ax^2 + bx + c < 0$, given one (double) real root r.

75. Give an example of a quadratic inequality whose solution set is the entire real line.

76. Give an example of a quadratic inequality whose solution set is the empty set.

APPLICATIONS

77. Approximation. The area A of a region is approximately equal to 12.436. The error in this approximation is less than 0.001. Describe the possible values of this area both with an absolute value inequality and with interval notation.

78. Approximation. The volume V of a solid is approximately equal to 6.94. The error in this approximation is less than 0.02. Describe the possible values of this volume both with an absolute value inequality and with interval notation.

79. Break-Even Analysis. An electronics firm is planning to market a new graphing calculator. The fixed costs are \$650,000 and the variable costs are \$47 per calculator. The wholesale price of the calculator will be \$63. For the company to make a profit, revenues must be greater than costs.

(A) How many calculators must be sold for the company to make a profit?

(B) How many calculators must be sold for the company to break even?

(C) Discuss the relationship between the results in parts A and B.

80. Break-Even Analysis. A video game manufacturer is planning to market a 64-bit version of its game machine. The fixed costs are \$550,000 and the variable costs are \$120 per machine. The wholesale price of the machine will be \$140.

(A) How many game machines must be sold for the company to make a profit?

(B) How many game machines must be sold for the company to break even?

(C) Discuss the relationship between the results in parts A and B.

81. Break-Even Analysis. The electronics firm in Problem 79 finds that rising prices for parts increase the variable costs to \$50.50 per calculator.

(A) Discuss possible strategies the company might use to deal with this increase in costs.

(B) If the company continues to sell the calculators for \$63, how many must they sell now to make a profit?

(C) If the company wants to start making a profit at the same production level as before the cost increase, how much should they increase the wholesale price?

82. Break-Even Analysis. The video game manufacturer in Problem 80 finds that unexpected programming problems increase the fixed costs to \$660,000.

(A) Discuss possible strategies the company might use to deal with this increase in costs.

(B) If the company continues to sell the game machines for \$140, how many must they sell now to make a profit?

(C) If the company wants to start making a profit at the same production level as before the cost increase, how much should they increase the wholesale price?

★ **83. Significant Digits.** If $N = 2.37$ represents a measurement, then we assume an accuracy of 2.37 ± 0.005. Express the accuracy assumption using an absolute value inequality.

★ **84. Significant Digits.** If $N = 3.65 \times 10^{-3}$ is a number from a measurement, then we assume an accuracy of $3.65 \times 10^{-3} \pm 5 \times 10^{-6}$. Express the accuracy assumption using an absolute value inequality.

85. Profit Analysis. A screen printer produces custom silk-screen apparel. The cost $C(x)$ of printing x custom T-shirts and the revenue $R(x)$ from the sale of x T-shirts (both in dollars) are given by

$$C(x) = 200 + 2.25x$$
$$R(x) = 10x - 0.05x^2$$

Determine the production levels x (to the nearest integer) that will result in the printer showing a profit.

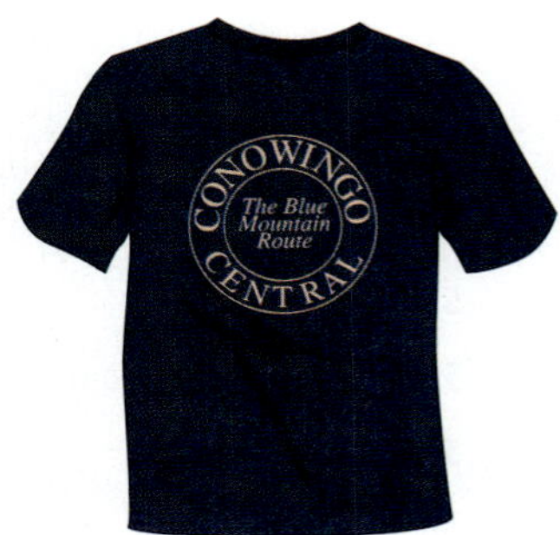

86. Profit Analysis. Refer to Problem 85. Determine the production levels x (to the nearest integer) that will result in the printer showing a profit of at least \$60.

87. Celsius/Fahrenheit. A formula for converting Celsius degrees to Fahrenheit degrees is given by the linear function

$$F = \frac{9}{5}C + 32$$

Determine to the nearest degree the Celsius range in temperature that corresponds to the Fahrenheit range of 60°F to 80°F.

88. Celsius/Fahrenheit. A formula for converting Fahrenheit degrees to Celsius degrees is given by the linear function

$$C = \frac{5}{9}(F - 32)$$

Determine to the nearest degree the Fahrenheit range in temperature that corresponds to a Celsius range of 20°C to 30°C.

89. Projectile Motion. A projectile propelled straight upward from the ground reaches a maximum height of 256 feet above ground level after 4 seconds. Let the quadratic function $d(t)$ represent the distance above ground level (in feet) t seconds after the projectile is released.

(A) Find $d(t)$.

(B) At what times will the projectile be more than 240 feet above the ground? Express the answer in inequality notation.

90. Projectile Motion. A projectile propelled straight upward from the ground reaches a maximum height of 784 feet above ground level after 7 seconds. Let the quadratic function $d(t)$ represent the distance above ground level (in feet) t seconds after the projectile is released.

(A) Find $d(t)$.

(B) At what times will the projectile be less than 640 feet above the ground? Express the answer in inequality notation.

91. Earth Science. Deeper and deeper holes are being bored into the Earth's surface every year in search of energy in the form of oil, gas, or heat. A bore at Windischeschenbach in the North German basin has reached a depth of more than 8 kilometers. The temperature in the bore is 30°C at a depth of 1 kilometer and increases 2.8°C for each additional 100 meters of depth. Find a mathematical model for the temperature T at a depth of x kilometers. At what interval of depths will the temperature be between 150°C and 200°C? Round answers to three decimal places.

92. Earth Science. A bore at Basel, Switzerland, has reached a depth of more than 5 kilometers. The temperature is 35°C at a depth of 1 kilometer and increases 3.6°C for each additional 100 meters of depth. Find a mathematical model for the temperature T at a depth of x kilometers. At what interval of depths will the temperature be between 100°C and 150°C? Round answers to three decimal places.

DATA ANALYSIS AND REGRESSION 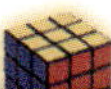

Twice each day 70 weather stations in the United States release high-altitude balloons containing instruments that send various data back to the station. Eventually, the balloons burst and the instruments parachute back to Earth to be reclaimed. The air pressure (in hectopascals), the altitude (in meters), and the temperature (in degrees Celsius) collected on the same day at two midwestern stations are given in Table 3. Round all numbers to three significant digits.*

TABLE 3 Upper-Air Weather Data

North Platte, NE			Minneapolis, MN		
PRES	**HGT**	**TEMP**	**PRES**	**HGT**	**TEMP**
745	2,574	8	756	2,438	2
700	3,087	5	728	2,743	0
627	3,962	−1	648	3,658	−4
559	4,877	−8	555	4,877	−10
551	4,992	−8	500	5,680	−17
476	6,096	−16	400	7,330	−28
404	7,315	−24	367	7,944	−32
387	7,620	−27	300	9,330	−45
300	9,410	−43	250	10,520	−55
259	10,363	−49	241	10,751	−57

Source: NOAA Air Resources Laboratory

*A unit of pressure equivalent to 1 millibar.

93. Weather. Let x be the altitude of the balloon released from North Platte and let y be the corresponding temperature. Use linear regression to find a linear function $y = ax + b$ that fits these data. For what altitudes will the temperature be between −10°C and −30°C?

94. Weather. Let x be the altitude of the balloon released from Minneapolis and let y be the corresponding temperature. Use linear regression to find a linear function $y = ax + b$ that fits these data. For what altitudes will the temperature be between −20°C and −40°C?

95. Weather. Let x be the altitude of the balloon released from North Platte and let y be the corresponding air pressure. Use linear regression to find a linear function $y = ax + b$ that fits these data. For what altitudes will the air pressure be between 350 hectopascals and 650 hectopascals?

96. Weather. Let x be the altitude of the balloon released from Minneapolis and let y be the corresponding air pressure. Use linear regression to find a linear function $y = ax + b$ that fits these data. For what altitudes will the air pressure be between 350 hectopascals and 650 hectopascals?

97. Break-Even Analysis. Table 4 on page 224 contains weekly price–demand data for orange juice and grapefruit juice for a fruit juice producer. The producer has weekly fixed cost of \$20,000 and variable cost of \$0.50 per gallon of orange juice produced.

(A) Use linear regression to find the price–demand equation $p = d(x)$ for the orange juice data in Table 4. What is the domain of $d(x)$?

(B) Find the revenue and cost functions as functions of the sales x. What is the domain of each function?

(C) Find the level of sales for which the company will break even. Describe verbally and graphically the sales levels that result in a profit and those that result in a loss.

(D) Find the sales and the price that will produce the maximum profit. Find the maximum profit.

TABLE 4 Fruit Juice Production

Orange Juice Demand (Gal.)	Price	Grapefruit Juice Demand (Gal.)	Price
21,800	\$1.95	2,130	\$2.32
24,300	\$1.81	2,480	\$2.21
26,700	\$1.43	2,610	\$2.07
28,900	\$1.37	2,890	\$1.87
29,700	\$1.28	3,170	\$1.81
33,700	\$1.14	3,640	\$1.68
34,800	\$0.96	4,350	\$1.56

98. Break-Even Analysis. The juice producer in Problem 97 has weekly fixed cost of \$3,000 and variable cost of \$0.40 per gallon of grapefruit juice produced.

(A) Use linear regression to find the price–demand equation $p = d(x)$ for the grapefruit juice data in Table 4. What is the domain of $d(x)$?

(B) Find the revenue and cost functions as functions of the sales x. What is the domain of each function?

(C) Find the level of sales for which the company will break even. Describe verbally and graphically the sales levels that result in a profit and those that result in a loss.

(D) Find the sales and the price that will produce the maximum profit. Find the maximum profit.

CHAPTER 2 REVIEW

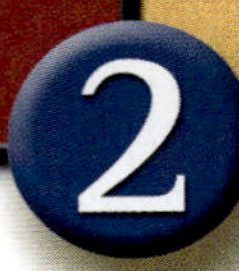

2.1 Linear Functions

A function f is a **linear function** if $f(x) = mx + b$, $m \neq 0$, where m and b are real numbers. The **domain** is the set of all real numbers and the **range** is the set of all real numbers. If $m = 0$, then f is called a **constant function,** $f(x) = b$, which has the set of all real numbers as its **domain** and the constant b as its **range.** The **standard form** for the equation of a line is $Ax + By = C$, where A, B, and C are real constants, and A and B are not both 0. Every straight line in a Cartesian coordinate system is the graph of an equation of this type. The **slope** of the line through the points (x_1, y_1) and (x_2, y_2) is

$$m = \frac{y_2 - y_1}{x_2 - x_1} \qquad x_1 \neq x_2$$

The slope is not defined for a vertical line where $x_1 = x_2$.

Equations of a Line

Standard Form	$Ax + By = C$	A and B not both 0
Slope–Intercept Form	$y = mx + b$	Slope: m; y intercept: b
Point–Slope Form	$y - y_1 = m(x - x_1)$	Slope: m; Point: (x_1, y_1)
Horizontal Line	$y = b$	Slope: 0
Vertical Line	$x = a$	Slope: Undefined

Two nonvertical lines with slopes m_1 and m_2 are **parallel** if and only if $m_1 = m_2$ and **perpendicular** if and only if $m_1m_2 = -1$. Slope can be interpreted as a **rate of change** or an **average rate of change.** The y intercept of a linear cost function is called the **fixed cost** and the slope is called the **variable cost.**

2.2 Linear Equations and Models

An equation that is true for all permissible values of the variable is called an **identity.** An equation that is true for some values of the variable and false for others is called a **conditional equation.** An equation that is false for all permissible values of the variable is called a **contradiction. Linear regression** is used to fit a curve to a data set. A **scatter diagram** is a graph of a data set. **Diagnostics** indicate how well a curve fits a data set. Supply and demand curves intersect at the **equilibrium point,** which consists of the **equilibrium price** and **equilibrium quantity.**

2.3 Quadratic Functions

If a, b, and c are real numbers with $a \neq 0$, then the function $f(x) = ax^2 + bx + c$ is a **quadratic function** and its graph is a **parabola. Completing the square** of the quadratic expression $x^2 + bx$ produces a perfect square:

$$x^2 + bx + \left(\frac{b}{2}\right)^2 = \left(x + \frac{b}{2}\right)^2$$

Completing the square for $f(x) = ax^2 + bx + c$ produces the **vertex form** $f(x) = a(x - h)^2 + k$ and the following properties:

1. The graph of f is a parabola:

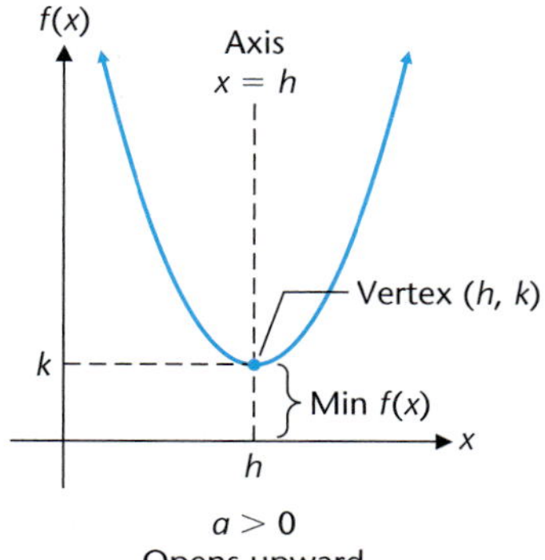

$a > 0$
Opens upward

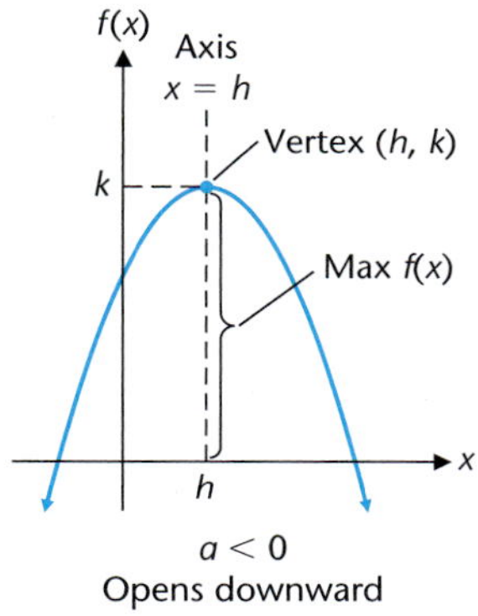

$a < 0$
Opens downward

2. Vertex: (h, k) (Parabola increases on one side of the vertex and decreases on the other.)
3. Axis (of symmetry): $x = h$ (parallel to y axis)
4. $f(h) = k$ is the minimum if $a > 0$ and the maximum if $a < 0$.
5. Domain: All real numbers
   Range: $(-\infty, k]$ if $a < 0$ or $[k, \infty)$ if $a > 0$
6. The graph of f is the graph of $g(x) = ax^2$ translated horizontally h units and vertically k units.
7. $h = -\dfrac{b}{2a}$, $k = c - \dfrac{b^2}{4a}$

2.4 Complex Numbers

A **complex number** in **standard form** is a number in the form $a + bi$ where a and b are real numbers and i is the **imaginary unit.** If $b \neq 0$ then $a + bi$ is also called an **imaginary number.** If $a = 0$ then $0 + bi = bi$ is also called a **pure imaginary number.** If $b = 0$ then $a + 0i = a$ is a **real number.** The complex **zero** is $0 + 0i = 0$. The **conjugate** of $a + bi$ is $a - bi$. **Equality, addition,** and **multiplication** are defined as follows:

1. $a + bi = c + di$ if and only if $a = c$ and $b = d$
2. $(a + bi) + (c + di) = (a + c) + (b + d)i$
3. $(a + bi)(c + di) = (ac - bd) + (ad + bc)i$

Because complex numbers obey the same commutative, associative, and distributive properties as real numbers, most operations with complex numbers are performed by using these properties and the fact that $\boldsymbol{i^2 = -1}$. The **property of conjugates,**

$$(a + bi)(a - bi) = a^2 + b^2$$

can be used to find **reciprocals** and **quotients.** If $a > 0$, then the **principal square root of the negative real number** $-a$ is $\sqrt{-a} = i\sqrt{a}$.

2.5 Quadratic Equations and Models

A **quadratic equation** is an equation that can be written in the form

$$ax^2 + bx + c = 0 \qquad a \neq 0$$

where x is a variable and a, b, and c are constants.
Algebraic methods of solution include:

1. **Factoring** and using the **zero property:**
   $m \cdot n = 0$ if and only if $m = 0$ or $n = 0$ (or both)
2. **Completing the square** and using the **square root property:**
   If A is a complex number, C is a real number, and $A^2 = C$, then $A = \pm\sqrt{C}$.
3. Using the **quadratic formula:**

$$x = \frac{-b \pm \sqrt{b^2 - 4ac}}{2a}$$

If the **discriminant** $\boldsymbol{b^2 - 4ac}$ is positive, the equation has two distinct **real roots;** if the discriminant is 0, the equation has one real **double root;** and if the discriminant is negative, the equation has two **imaginary roots,** each the conjugate of the other.

2.6 Additional Equation-Solving Techniques

A **radical** can be eliminated from an equation by isolating the radical on one side of the equation and raising both sides of the equation to the same natural number power to produce a new equation. The solution set of the original equation is a subset of the solution set of the new equation. The new equation may have **extraneous solutions** that are not solutions of the original equation. Consequently, **every solution of the new equation must be checked in the original equation to eliminate extraneous solutions.** If an equation contains more than one radical, then the process of isolating a radical and raising both sides to the same natural number power can be repeated until all radicals are eliminated. If a substitution transforms an equation into the form $au^2 + bu + c = 0$ where u is an expression in some other variable, then the equation is an **equation of quadratic type,** which can be solved by quadratic methods.

2.7 Solving Inequalities

Linear inequalities in one variable are expressed using the inequality symbols $<, >, \leq, \geq$. The **solution set of an inequality** is the set of all values of the variable that make the inequality a true statement. Each element of the solution set is called a **solution.** Two inequalities are **equivalent** if they have the same solution set. An equivalent inequality will result and the **sense or direction will remain the same** if each side of the original inequality:

1. Has the same real number added to or subtracted from it.
2. Is multiplied or divided by the same positive number.

An equivalent inequality will result and the **sense or direction will reverse** if each side of the original inequality:

3. Is multiplied or divided by the same negative number.

Note that multiplication by 0 and division by 0 are not permitted. The **absolute value function** $|x|$ can also be interpreted as the **distance** between x and the origin. More generally, for $p > 0$:

1. $|ax + b| < p$ is equivalent to $-p < ax + b < p$.
2. $|ax + b| = p$ is equivalent to $ax + b = p$ or $ax + b = -p$.
3. $|ax + b| > p$ is equivalent to $ax + b < -p$ or $ax + b > p$.

Quadratic inequalities are solved by factoring and considering cases based on the fact that $ab > 0$ if and only if a and b have the same sign and $ab < 0$ if and only if a and b have opposite signs. A **break-even point** is an intersection point for the graphs of a cost and a revenue equation.

CHAPTER 2 REVIEW EXERCISES

Work through all the problems in this chapter review and check answers in the back of the book. Answers to most review problems are there, and following each answer is a number in italics indicating the section in which that type of problem is discussed. Where weaknesses show up, review appropriate sections in the text.

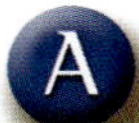

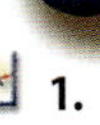

1. Use the graph of the linear function in the figure to find the rise, run, and slope. Write the equation of the line in the form $Ax + By = C$, where A, B, and C are integers with $A > 0$. (The horizontal and vertical line segments have integer lengths.)

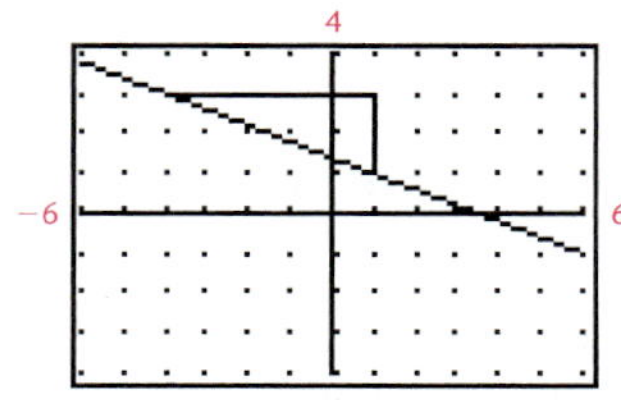

2. Graph $3x + 2y = 9$ and indicate its slope.

3. Write an equation of a line with x intercept 6 and y intercept 4. Write the final answer in the form $Ax + By = C$, where A, B, and C are integers with $A > 0$.

4. Write the slope intercept form of the equation of the line with slope $-\frac{2}{3}$ and y intercept 2.

5. Write the equations of the vertical and horizontal lines passing through the point $(-3, 4)$. What is the slope of each?

6. Solve algebraically and confirm graphically:
   (A) $0.05x + 0.25(30 - x) = 3.3$
   (B) $\frac{5x}{3} - \frac{4 + x}{2} = \frac{x - 2}{4} + 1$

In Problems 7 and 8,

(A) *Complete the square and find the vertex form of the function.*

(B) *Write a brief verbal description of the relationship between the graph of the function and the graph of $y = x^2$.*

(C) *Find the x intercepts algebraically and confirm graphically.*

7. $f(x) = -x^2 - 2x + 3$ **8.** $f(x) = x^2 - 3x - 2$

9. Perform the indicated operations and write the answers in standard form:

(A) $(-3 + 2i) + (6 - 8i)$ **(B)** $(3 - 3i)(2 + 3i)$

(C) $\dfrac{13 - i}{5 - 3i}$

In Problems 10–18, solve algebraically and confirm graphically, if possible.

10. $(x - 3)(x - 5) = (x - 2)(x - 6)$

11. $x^2 + x = (x - 4)(x + 5) + 20$

12. $x^2 - 2x + 3 = x^2 + 3x - 7$

13. $2x^2 - 7 = 0$ **14.** $2x^2 = 4x$

15. $2x^2 = 7x - 3$ **16.** $m^2 + m + 1 = 0$

17. $y^2 = \frac{3}{2}(y + 1)$ **18.** $\sqrt{5x - 6} - x = 0$

In Problems 19–21, solve and express answers in inequality and interval notation.

19. $3(2 - x) - 2 \le 2x - 1$

20. $x^2 + x < 20$ **21.** $x^2 > 4x + 12$

22. Discuss the use of the terms *rising, falling, increasing,* and *decreasing* as they apply to the descriptions of the following:

(A) A line with positive slope

(B) A line with negative slope

(C) A parabola that opens upward

(D) A parabola that opens downward

B

23. Find an equation of the line through the points $(-4, 3)$ and $(0, -3)$. Write the final answer in the form $Ax + By = C$, where A, B, and C are integers with $A > 0$.

24. Write the slope–intercept form of the equation of the line that passes through the point $(-2, 1)$ and is

(A) parallel to the line $6x + 3y = 5$

(B) perpendicular to the line $6x + 3y = 5$

In Problems 25–27, solve each inequality. Write answers in inequality notation.

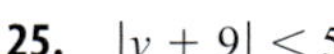

25. $|y + 9| < 5$ **26.** $|2x - 8| \ge 3$

27. $\sqrt{(1 - 2m)^2} \le 3$

In Problems 28 and 29, interpret each inequality geometrically and solve. Write answers in both interval and inequality notation.

28. $|y - 5| \le 2$ **29.** $|t + 6| > 9$

For each equation in Problems 30–32, use the discriminant to determine the number and type of zeros and confirm graphically.

30. $0.1x^2 + x + 1.5 = 0$ **31.** $0.1x^2 + x + 2.5 = 0$

32. $0.1x^2 + x + 3.5 = 0$

33. Let $f(x) = 0.5x^2 - 4x + 5$.

(A) Sketch the graph of f and label the axis and the vertex.

(B) Where is f increasing? Decreasing? What is the range? (Express answers in interval notation.)

34. Find the equations of the linear function g and the quadratic function f whose graphs are shown in the figure. This line is called the tangent line to the graph of f at the point $(-1, 0)$.

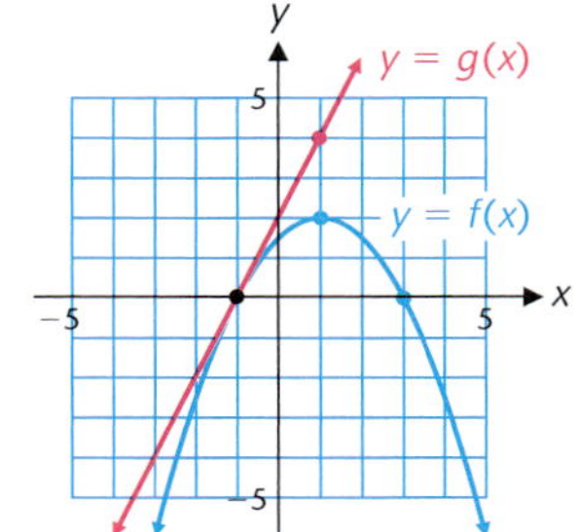

35. Perform the indicated operations and write the final answers in standard form:
(A) $(3 + i)^2 - 2(3 + i) + 3$ **(B)** i^{27}

36. Convert to $a + bi$ forms, perform the indicated operations, and write the final answers in standard form:
(A) $(2 - \sqrt{-4}) - (3 - \sqrt{-9})$
(B) $\frac{2 - \sqrt{-1}}{3 + \sqrt{-4}}$ **(C)** $\frac{4 + \sqrt{-25}}{\sqrt{-4}}$
(D) $\sqrt{-16}\sqrt{-25}$

Solve Problems 37–42 algebraically and confirm graphically, if possible.

37. $(x + \frac{5}{2})^2 = \frac{5}{4}$ 38. $1 + \frac{3}{u^2} = \frac{2}{u}$

39. $2x + 3\sqrt{4x^2 - 4x + 9} = 1$

40. $2x^{2/3} - 5x^{1/3} - 12 = 0$ 41. $m^4 + 5m^2 - 36 = 0$

42. $\sqrt{y - 2} - \sqrt{5y + 1} = -3$

43. Let $g(x) = x^2 - 4x + 5, x \geq 2$. Find g^{-1}.

44. Use linear regression to fit a line to each of the following data sets. How are the graphs of the two functions related? How are the two functions related?

(A)

x	y
3	1
4	3

(B)

x	y
1	3
3	4

45. Can a quadratic function have only imaginary zeros? If not, explain why. If so, give an example and discuss any special relationship between the zeros.

46. If a quadratic function has only imaginary zeros, can the function be graphed? If not, explain why. If so, what is the graph's relationship to the x axis?

47. Consider the quadratic equation

$$x^2 - 6x + c = 0$$

where c is a real number. Discuss the relationship between the values of c and the three types of roots listed in Table 1 in Section 2.5.

Solve Problems 48 and 49 for the indicated variable in terms of the other variables.

48. $P = M - Mdt$ for M (mathematics of finance)

49. $P = EI - RI^2$ for I (electrical engineering)

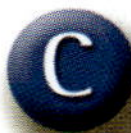

50. For what values of a and b is the following inequality true?

$$a + b < b - a$$

51. If a and b are negative numbers and $a > b$, then is a/b greater than 1 or less than 1?

52. Solve and graph. Write the answer using interval notation: $0 < |x - 6| < d$

53. Evaluate: $(a + bi)\left(\frac{a}{a^2 + b^2} - \frac{b}{a^2 + b^2}i\right)$; $a, b, \neq 0$

54. Are the graphs of $mx - y = b$ and $x + my = b$ parallel, perpendicular, or neither? Justify your answer.

55. Use completing the square to find the center and radius of the circle with equation:

$$x^2 - 4x + y^2 - 2y - 3 = 0$$

56. Refer to Problem 55. Find the equation of the line tangent to the circle at the point (4, 3). Graph the circle and the line on the same coordinate system.

57. Solve $3x^{-2/5} - 4x^{-1/5} + 1 = 0$ algebraically and graphically.

58. Find all solutions of $x^3 + 1 = 0$.

59. Find three consecutive integers whose sum is 144.

60. Find three consecutive even integers so that the first plus twice the second is twice the third.

Problems 61 and 62 refer to a triangle with base b and height h (see the figure). Write a mathematical expression in terms of b and h for each of the verbal statements in Problems 61 and 62.

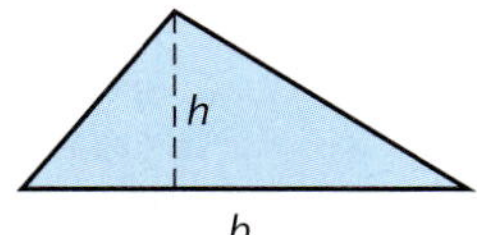

61. The base is five times the height.

62. The height is one-fourth of the base.

APPLICATIONS

63. Geometry. The diagonal of a rectangle is 32.5 inches and the area is 375 square inches. Find the dimensions of the rectangle, correct to one decimal place.

64. Falling Object. A worker at the top of a radio tower drops a hammer to the ground. If the hammer hits the ground 3.5 seconds after it is dropped, how high is the tower?

65. Cost Analysis. Cost equations for manufacturing companies are often quadratic—costs are high at low and high production levels. The weekly cost $C(x)$ (in dollars) for manufacturing x inexpensive calculators is

$$C(x) = 0.001x^2 - 9.5x + 30{,}000$$

Find the production level(s) (to the nearest integer) that

(A) Produces the minimum weekly cost. What is the minimum weekly cost (to the nearest cent)?

(B) Produces a weekly cost of \$12,000.

(C) Produces a weekly cost of \$6,000.

66. Break-Even Analysis. The manufacturing company in Problem 65 sells its calculators to wholesalers for \$3 each. How many calculators (to the nearest integer) must the company sell to break even?

67. Profit Analysis. Refer to Problems 65 and 66. Find the production levels that produce a profit. A loss. (Express answers in inequality notation.)

68. Linear Depreciation. A computer system was purchased by a small company for \$12,000 and is assumed to have a depreciated value of \$2,000 after 8 years. If the value is depreciated linearly from \$12,000 to \$2,000:

(A) Find the linear equation that relates value V (in dollars) to time t (in years).

(B) What would be the depreciated value of the system after 5 years?

69. Business–Pricing. A sporting goods store sells tennis shorts that cost \$30 for \$48 and sunglasses that cost \$20 for \$32.

(A) If the markup policy of the store for items that cost over \$10 is assumed to be linear and is reflected in the pricing of these two items, write an equation that expresses retail price R as a function of cost C.

(B) What should be the retail price of a pair of skis that cost \$105?

★ **70. Income.** A salesperson receives a base salary of \$200 per week and a commission of 10% on all sales over \$3,000 during the week. If x represents the salesperson's weekly sales, express the total weekly earnings $E(x)$ as a function of x. Find $E(2{,}000)$ and $E(5{,}000)$.

71. Construction. A farmer has 120 feet of fencing to be used in the construction of two identical rectangular pens sharing a common side (see the figure).

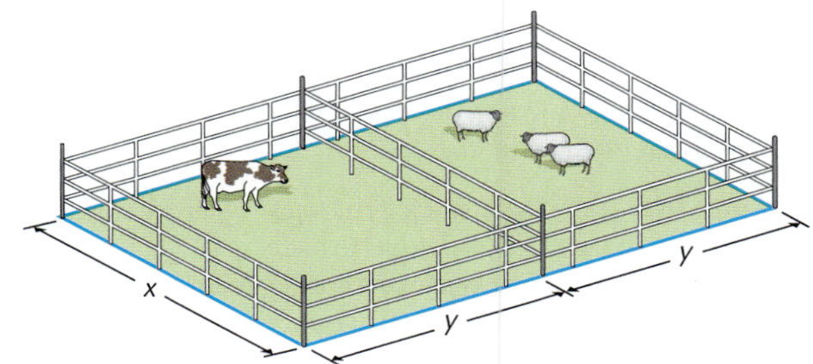

(A) Express the total area $A(x)$ enclosed by both pens as a function of the width x.

(B) From physical considerations, what is the domain of the function A?

(C) Find the dimensions of the pens that will make the total enclosed area maximum.

72. Sports Medicine. The following quotation was found in a sports medicine handout: "The idea is to raise and sustain your heart rate to 70% of its maximum safe rate for your age. One way to determine this is to subtract your age from 220 and multiply by 0.7."

(A) If H is the maximum safe sustained heart rate (in beats per minute) for a person of age A (in years), write a formula relating H and A.

(B) What is the maximum safe sustained heart rate for a 20-year-old?

(C) If the maximum safe sustained heart rate for a person is 126 beats per minute, how old is the person?

★73. Design. The pages of a textbook have uniform margins of 2 centimeters on all four sides (see the figure). If the area of the entire page is 480 square centimeters and the area of the printed portion is 320 square centimeters, find the dimensions of the page.

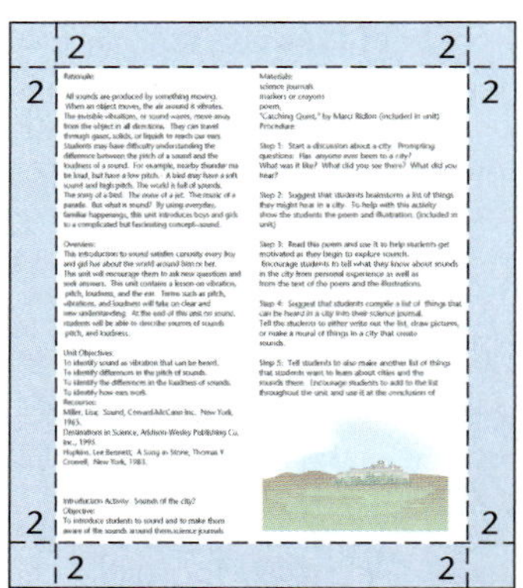

★74. Design. A landscape designer uses 8-foot timbers to form a pattern of isosceles triangles along the wall of a building (see the figure). If the area of each triangle is 24 square feet, find the base correct to two decimal places.

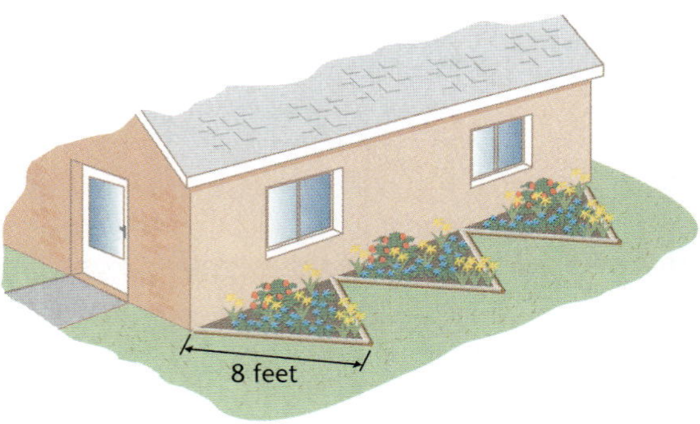

★75. Architecture. An entrance way in the shape of a parabola 12 feet wide and 12 feet high must enclose a rectangular door that is 8.4 feet high. What is the widest doorway (to the nearest tenth of a foot) that can be installed in the entrance way?

DATA ANALYSIS AND REGRESSION 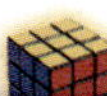

In Problems 76–80, unless directed otherwise, round all numbers to three significant digits,

76. Drug Use. The use of marijuana by teenagers declined throughout the 1980s, but began to increase during the 1990s. Table 1 gives the percentage of 12- to 17-year-olds who have ever used marijuana for selected years from 1979 to 1995.

TABLE 1 Marijuana Use: 12 to 17 Years Old

Year	Ever Used [%]
1979	26.7
1985	20.1
1990	12.7
1994	13.6
1995	16.2

Source: National Household Survey on Drug Abuse

(A) Find a quadratic regression model for the percentage of 12- to 17-year-olds who have ever used marijuana, using years since 1970 for the independent variable.

(B) Use your model to predict the year during which the percentage of marijuana users will return to the 1979 level.

77. Political Science. Association of economic class and party affiliation did not start with Roosevelt's New Deal; it goes back to the time of Andrew Jackson (1767–1845). Paul Lazarsfeld of Columbia University published an article in the November 1950 issue of *Scientific American* in which he discusses statistical investigations of the relationships between economic class and party affiliation. The data in Table 2 are taken from this article.

TABLE 2 Political Affiliations in 1836		
Ward	**Average Assessed Value per Person [in $100]**	**Democratic Votes [%]**
12	1.7	51
3	2.1	49
1	2.3	53
5	2.4	36
2	3.6	65
11	3.7	35
10	4.7	29
4	6.2	40
6	7.1	34
9	7.4	29
8	8.7	20
7	11.9	23

(A) Find a linear regression model for the data in the second and third columns of the table, using the average assessed value as the independent variable.

(B) Use the linear regression model to predict (to two decimal places) the percentage of votes for democrats in a ward with an average assessed value of $300.

78. Supply and Demand. Table 3 contains price–supply data and price–demand data for a broccoli grower. Find a linear model for the price–supply data where x is supply (in pounds) and y is price (in dollars). Do the same for the price–demand data. Find the equilibrium price for broccoli.

TABLE 3 Supply and Demand for Broccoli		
Price $/lb.	**Supply (lbs.)**	**Demand (lbs.)**
0.71	25,800	41,500
0.77	27,400	38,700
0.84	30,200	36,200
0.91	33,500	32,800
0.96	34,900	29,800
1.01	37,800	27,900
1.08	39,210	25,100

79. Break-Even Analysis. The broccoli grower in Problem 78 has fixed cost of $15,000 and variable cost of $0.20 per pound of broccoli produced.

(A) Find the revenue and cost functions as functions of the sales x. What is the domain of each function?

(B) Find the level of sales for which the company will break even. Describe verbally and graphically the sales levels that result in a profit and those that result in a loss.

(C) Find the sales and the price that will produce the maximum profit. Find the maximum profit.

80. Optimal Speed. Table 4 contains performance data for a speedboat powered by a Yamaha outboard motor.

(A) Let x be the speed of the boat in miles per hour (mph) and y the associated mileage in miles per gallon (mpg). Use the data in Table 4 to find a quadratic regression function $y = ax^2 + bx + c$ for this boat.

(B) A marina rents this boat for $15 per hour plus the cost of the gasoline used. If gasoline costs $1.60 per gallon and you take a 100-mile trip in this boat, construct a mathematical model and use it to answer the following questions:
What speed should you travel to minimize the rental charges?
What mileage will the boat get?
How long does the trip take?
How much gasoline will you use?
How much will the trip cost you?

TABLE 4 Performance Data	
mph	**mpg**
9.5	1.67
21.1	1.92
28.3	2.16
33.7	1.88
37.9	1.77
42.6	1.49

Source: www.yamaha-motor.com

CHAPTER 2 GROUP ACTIVITY

Mathematical Modeling in Population Studies

In a study on population growth in California, Tulane University demographer Leon Bouvier recorded the past population totals for every 10 years starting at 1900. Then, using sophisticated demographic techniques, he made high, low, and medium projections to the year 2040. Table 1 shows actual populations up to 1990 and medium projections (to the nearest million) to 2040.

TABLE 1 California Population 1900–2040

Years After 1900	Date	Population [Millions]	
0	1900	2	Actual
10	1910	3	
20	1920	4	
30	1930	5	
40	1940	5	
50	1950	10	
60	1960	15	
70	1970	20	
80	1980	23	
90	1990	30	
100	2000	35	Projected
110	2010	45	
120	2020	53	
130	2030	61	
140	2040	70	

1. Building a Mathematical Model.
   (A) Plot the first and last columns in Table 1 up to 1990 (actual populations). Would a linear or a quadratic function be the better model for these data? Why?
   (B) Use a graphing utility to compute a quadratic regression function to model the data you plotted in part A.
   (C) Graph this function and the data from part A for $0 \le x \le 150$. The results should look something like Figure 1.

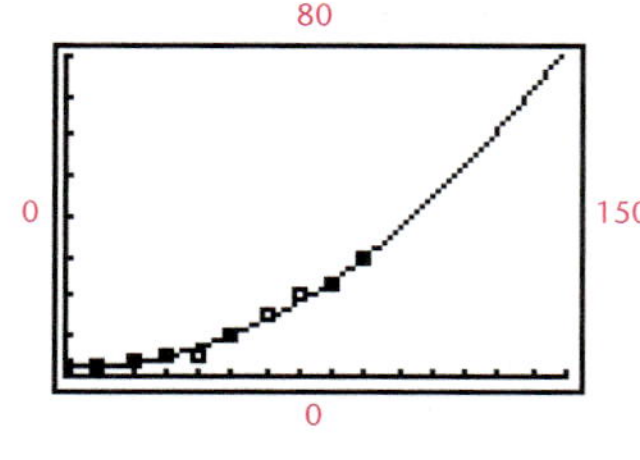

FIGURE 1

2. Using the Mathematical Model for Projections.

   Use the quadratic regression model to answer the following questions.
   (A) Calculate projected populations for California at 10-year intervals, starting at 2000 and ending at 2040. Compare your projections with Professor Bouvier's projections, both numerically and graphically.

(B) During what year would you project that the population will reach 40 million? 50 million?

(C) For what years would you project the population to be between 34 million and 68 million, inclusive?

CUMULATIVE REVIEW EXERCISES CHAPTERS 1 2

Work through all the problems in this chapter review and check answers in the back of the book. Answers to most review problems are there, and following each answer is a number in italics indicating the section in which that type of problem is discussed. Where weaknesses show up, review appropriate sections in the text.

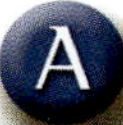

1. **(A)** Plot the points in the table in a rectangular coordinate system.

(B) Find the smallest viewing window that will contain all of these points. State your answer in terms of the window variables.

(C) Does this set of points define a function? Explain.

x	-3	1	-2	1	3
y	4	-2	4	-4	4

2. Given points $A = (3, 2)$ and $B = (5, 6)$, find

(A) Distance between A and B.

(B) Slope–intercept form of the equation of the line through A and B.

(C) Slope–intercept form of the equation of the line through B and perpendicular to the line through A and B.

(D) Standard form of the equation of the circle with center at A and passing through B.

(E) Graph the lines from parts B and C and the circle from part D on the same coordinate system.

3. Graph $2x - 3y = 6$ and indicate its slope and intercepts.

4. For $f(x) = x^2 - 2x + 5$ and $g(x) = 3x - 2$, find

(A) $f(-2) + g(-3)$

(B) $f(1) \cdot g(1)$

(C) $\dfrac{g(0)}{f(0)}$

5. How are the graphs of the following related to the graph of $y = |x|$?

(A) $y = 2|x|$

(B) $y = |x - 2|$

(C) $y = |x| - 2$

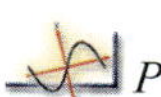 *Problems 6–8 refer to the function f given by the graph:*

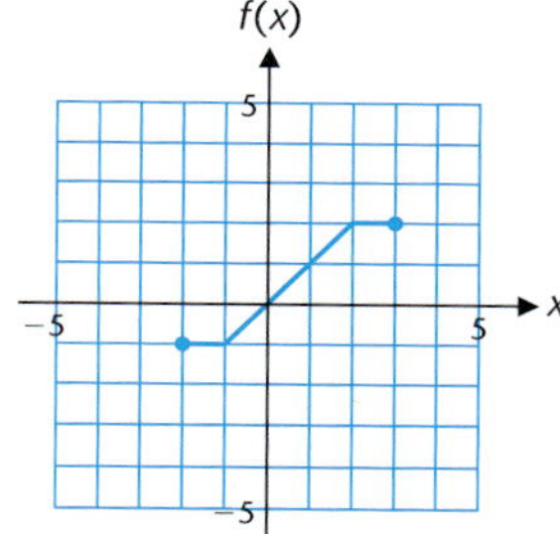

6. Find the domain and range of f. Express answers in interval notation.

7. Is f an even function, an odd function, or neither? Explain.

8. Use the graph of f to sketch a graph of the following:
(A) $y = -f(x + 1)$ **(B)** $y = 2f(x) - 2$

Solve Problems 9–13 algebraically and confirm graphically.

9. $\dfrac{7x}{5} - \dfrac{3 + 2x}{2} = \dfrac{x - 10}{3} + 2$

10. $3x^2 = -12x$

11. $4x^2 - 20 = 0$

12. $x^2 - 6x + 2 = 0$

13. $x - \sqrt{12 - x} = 0$

In Problems 14–16, solve and express answers in inequality and interval notation.

14. $2(3 - y) + 4 \le 5 - y$

15. $|x - 2| < 7$

16. $x^2 + 3x \ge 10$

17. Let $f(x) = x^2 - 4x - 1$.
(A) Find the vertex form of f.
(B) How is the graph of f related to the graph of $y = x^2$?
(C) Find the x intercepts algebraically and confirm graphically.

18. Perform the indicated operations and write the answer in standard form:
(A) $(2 - 3i) - (-5 + 7i)$
(B) $(1 + 4i)(3 - 5i)$
(C) $\dfrac{5 + i}{2 + 3i}$

B

19. Find each of the following for the function f given by the graph shown.

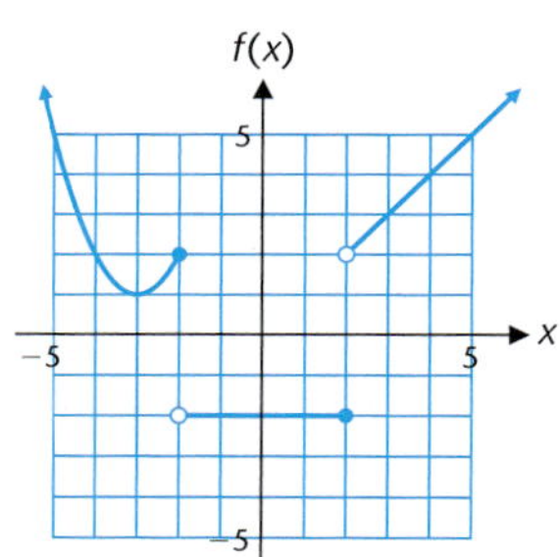

(A) The domain of f
(B) The range of f
(C) $f(-3) + f(-2) + f(2)$
(D) The intervals over which f is increasing.
(E) The x coordinates of any points of discontinuity.

20. Given $f(x) = 1/(x - 2)$ and $g(x) = (x + 3)/x$, find $f \circ g$. What is the domain of $f \circ g$?

21. Find $f^{-1}(x)$ for $f(x) = 2x + 5$.

22. Let $f(x) = \sqrt{x + 4}$
(A) Find $f^{-1}(x)$.
(B) Find the domain and range of f and f^{-1}.
(C) Graph f, f^{-1}, and $y = x$ on the same coordinate system and identify each graph.

23. Which of the following functions is one-to-one?
(A) $f(x) = x^3 + x$
(B) $g(x) = x^3 + x^2$

24. Write the slope–intercept form of the equation of the line passing through the point $(-6, 1)$ that is
(A) parallel to the line $3x + 2y = 12$.
(B) perpendicular to the line $3x + 2y = 12$.

25. Graph $f(x) = x^2 - 2x - 8$. Label the axis of symmetry and the coordinates of the vertex, and find the range, intercepts, and maximum or minimum value of $f(x)$.

In Problems 26 and 27, solve and express answers in inequality and interval notation.

26. $|4x - 9| > 3$

27. $\sqrt{(3m - 4)^2} \le 2$

28. Perform the indicated operations and write the final answers in standard form.
(A) $(2 - 3i)^2 - (4 - 5i)(2 - 3i) - (2 + 10i)$
(B) $\dfrac{3}{5} + \dfrac{4}{5}i + \dfrac{1}{\frac{3}{5} + \frac{4}{5}i}$
(C) i^{35}

29. Convert to $a + bi$ forms, perform the indicated operations, and write the final answers in standard form.
(A) $(5 + 2\sqrt{-9}) - (2 - 3\sqrt{-16})$
(B) $\dfrac{2 + 7\sqrt{-25}}{3 - \sqrt{-1}}$
(C) $\dfrac{12 - \sqrt{-64}}{\sqrt{-4}}$

30. Graph, finding the domain, range, and any points of discontinuity.

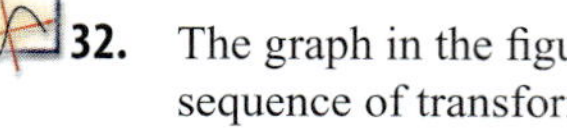

$$f(x) = \begin{cases} x - 1 & \text{if } x < 0 \\ x^2 + 1 & \text{if } x \ge 0 \end{cases}$$

31. Find the center and radius of the circle given by the equation $x^2 - 6x + y^2 + 2y = 0$. Graph the circle and show the center and the radius.

32. The graph in the figure is the result of applying a sequence of transformations to the graph of $y = |x|$. Describe the transformations verbally and write an equation for the graph in the figure.

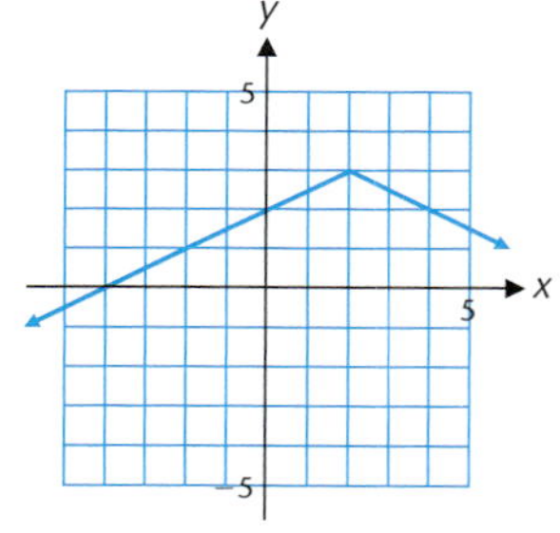

33. Find the standard form of the quadratic function whose graph is shown in the figure.

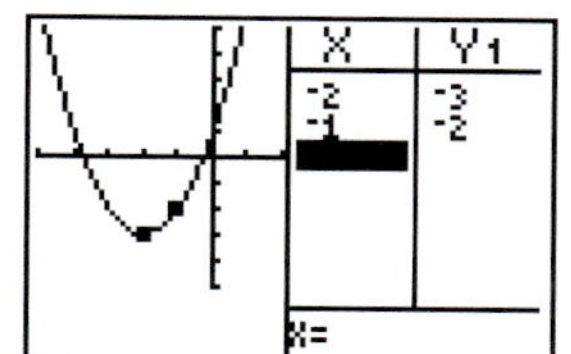

Solve Problems 34–38 algebraically and confirm graphically, if possible.

34. $1 + \frac{14}{y^2} = \frac{6}{y}$

35. $4x^{2/3} - 4x^{1/3} - 3 = 0$

36. $u^4 + u^2 - 12 = 0$

37. $\sqrt{8t - 2} - 2\sqrt{t} = 1$

38. $6x = \sqrt{9x^2 - 48}$

39. Consider the quadratic equation

$$x^2 + bx + 1 = 0$$

where b is a real number. Discuss the relationship between the values of b and the three types of roots listed in Table 1 in Section 2.5.

40. Give an example of an odd function. Of an even function. Can a function be both even and odd? Explain.

41. Can a quadratic equation with real coefficients have one imaginary root and one real root? One double imaginary root? Explain.

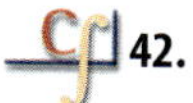

42. If $g(x) = -2x^2 + 3x - 1$, find $\frac{g(2 + h) - g(2)}{h}$.

43. The graph is the result of applying one or more transformations to the graph of one of the six basic functions in Figure 1, Section 1.4. Find an equation for the graph.

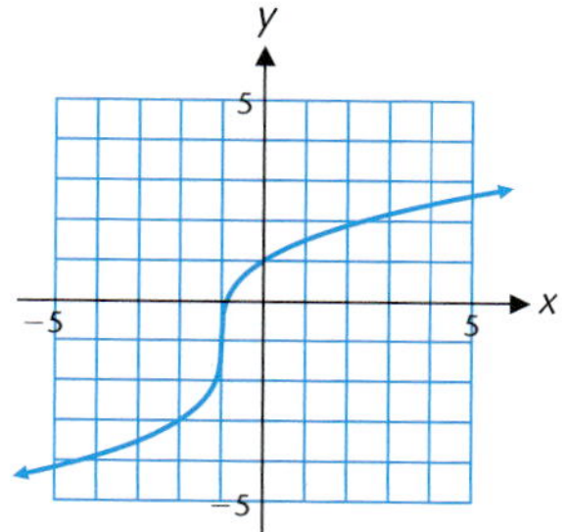

44. The total surface area of a right circular cylinder with radius r and height h is given by

$$A = 2\pi r(r + h) \qquad r > 0, h > 0$$

(A) Solve for h in terms of the other variables.
(B) Solve for r in terms of the other variables. Why is there only one solution?

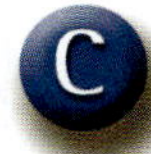

45. Given $f(x) = x^2$ and $g(x) = \sqrt{4 - x^2}$, find
(A) Domain of g
(B) f/g and its domain
(C) $f \circ g$ and its domain

46. Let $f(x) = x^2 - 2x - 3, x \geq 1$.
(A) Find $f^{-1}(x)$.
(B) Find the domain and range of f^{-1}.
(C) Graph f, f^{-1}, and $y = x$ on the same coordinate system.

47. Evaluate $x^2 - x + 2$ for $x = \frac{1}{2} - \frac{i}{2}\sqrt{7}$.

48. For what values of a and b is the inequality $a - b < b - a$ true?

49. Write in standard form: $\frac{a + bi}{a - bi}$; $a, b \neq 0$

Solve Problems 50–53 algebraically and confirm graphically, if possible.

50. $3x^2 = 2\sqrt{2}x - 1$

51. $1 + 13x^{-2} + 36x^{-4} = 0$

52. $\sqrt{16x^2 + 48x + 39} - 2x = 3$

53. $3x^{-2/5} - x^{-1/5} - 1 = 0$

54. Show that $5 + i$ and $-5 - i$ are the square roots of $24 + 10i$. Describe how you could find these square roots algebraically.

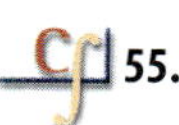

55. For $f(x) = 0.5x^2 - 3x - 7$, find
(A) $\frac{f(x + h) - f(x)}{h}$ **(B)** $\frac{f(x) - f(a)}{x - a}$

56. The function f is continuous for all real numbers and its graph passes through the points $(0, 4)$, $(5, -3)$, $(10, 2)$. Discuss the minimum and maximum number of x intercepts for f.

57. Let $f(x) = |x + 2| + |x - 2|$. Find a piecewise definition of f that does not involve the absolute value function. Graph f and find the domain and range.

58. Let $f(x) = 2x - [\![2x]\!]$. Write a piecewise definition for f and sketch the graph of f. Include sufficient intervals to clearly illustrate both the definition and the graph. Find the domain, range, and any point of discontinuity.

59. Find all solutions of $x^3 - 8 = 0$.

APPLICATIONS

60. **Break-Even Analysis.** The publisher's fixed costs for the production of a new cookbook are \$41,800. Variable costs are \$4.90 per book. If the book is sold to bookstores for \$9.65, how many must be sold for the publisher to break even?

61. **Finance.** An investor instructs a broker to purchase a certain stock whenever the price per share p of the stock is within \$10 of \$200. Express this instruction as an absolute value inequality.

62. **Profit and Loss Analysis.** At a price of \$$p$ per unit, the marketing department in a company estimates that the weekly cost C and the weekly revenue R, in thousands of dollars, will be given by the equations

$C = 88 - 12p$ **Cost equation**
$R = 15p - 2p^2$ **Revenue equation**

Find the prices for which the company has

(A) A profit **(B)** A loss

63. **Depreciation.** Office equipment was purchased for \$20,000 and is assumed to depreciate linearly to a scrap value of \$4,000 after 8 years.

(A) Find a linear function $v = d(t)$ that relates value v in dollars to time t in years.

(B) Find $t = d^{-1}(v)$.

★ **64.** **Shipping.** A ship leaves Port A, sails east to Port B, and then north to Port C, a total distance of 115 miles. The next day the ship sails directly from Port C back to Port A, a distance of 85 miles. Find the distance between Ports A and B and between Ports B and C.

65. **Price and Demand.** The weekly demand for mouthwash in a chain of drug stores is 1,160 bottles at a price of \$3.79 each. If the price is lowered to \$3.59, the weekly demand increases to 1,340 bottles. Assuming the relationship between the weekly demand x and the price per bottle p is linear, express x as a function of p. How many bottles would the store sell each week if the price were lowered to \$3.29?

66. **Business–Pricing.** A telephone company begins a new pricing plan that charges customers for local calls as follows: The first 60 calls each month are 6 cents each, the next 90 are 5 cents each, the next 150 are 4 cents each, and any additional calls are 3 cents each. If C is the cost, in dollars, of placing x calls per month, write a piecewise definition of C as a function of x and graph.

67. **Construction.** A home owner has 80 feet of chain-link fencing to be used to construct a dog pen adjacent to a house (see the figure).

(A) Express the area $A(x)$ enclosed by the pen as a function of the width x.

(B) From physical considerations, what is the domain of the function A?

(C) Graph A and determine the dimensions of the pen that will make the area maximum.

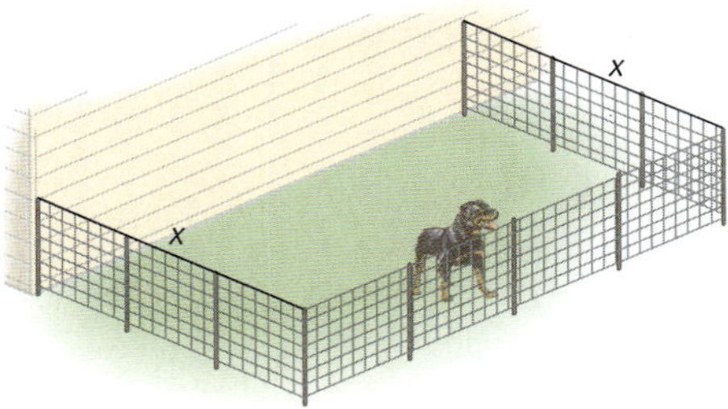

68. **Computer Science.** Let $f(x) = x - 2[\![x/2]\!]$. This function can be used to determine if an integer is odd or even.

(A) Find $f(1), f(2), f(3), f(4)$.

(B) Find $f(n)$ for any integer n. [*Hint:* Consider two cases, $n = 2k$ and $n = 2k + 1$, k is an integer.]

69. Price and Demand. The demand for barley q (in thousands of bushels) and the corresponding price p (in cents) at a midwestern grain exchange are shown in the figure.

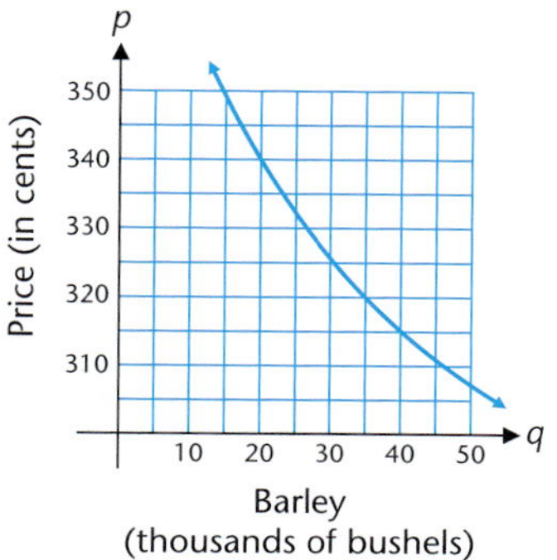

(A) What is the demand (to the nearest thousand bushels) when the price is 325 cents per bushel?

(B) Does the demand increase or decrease if the price is increased to 340 cents per bushel? By how much?

(C) Does the demand increase or decrease if the price is decreased to 315 cents per bushel? By how much?

(D) Write a brief description of the relationship between price and demand illustrated by this graph.

(E) Use the graph to estimate the price (to the nearest cent) when the demand is 20, 25, 30, 35, and 40 thousand bushels. Use these data to find a quadratic regression model for the price of barley using the demand as the independent variable.

DATA ANALYSIS AND REGRESSION 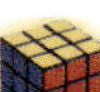

In Problems 70–72 round all values to three significant digits, unless directed otherwise.

70. Demand. Egg consumption per capita decreased from a high of about 400 per capita in 1945 to a low of about 230 in 1991, then it began to increase. Table 1 lists the annual per capita consumption of eggs in the United States since 1970.

TABLE 1 Per Capita Egg Consumption

1970	1975	1980	1985	1990	1995	2000
309	276	271	255	233	238	258

Source: Department of Agriculture.

(A) Find a quadratic regression equation $y = f(x)$ for the data in Table 1, where x is the number of years since 1970.

(B) Use the quadratic regression equation to project the year in which the per capita consumption will return to the 1970 level; to the 1945 level.

(C) Write a brief description of egg consumption from 1970 to 2000.

71. Stopping Distance. Table 2 contains data related to the length of the skid marks left by an automobile when making an emergency stop.

(A) Let x be the speed of the vehicle in miles per hour. Find a quadratic regression model for the braking distance.

(B) An insurance investigator finds skid marks 220 feet long at the scene of an accident involving this automobile. How fast (to the nearest mile per hour) was the automobile traveling when it made these skid marks?

TABLE 2 Skid Marks

Speed (mph)	Length of Skid Marks (in Feet)
20	24
30	48
40	81
50	118
60	187
70	246
80	312

72. Optimal Speed. Table 3 contains performance data for a speedboat powered by a Yamaha outboard motor.

(A) Let x be the speed of the boat in miles per hour (mph) and y the associated mileage in miles per gallon (mpg). Use the data in Table 3 to find a quadratic regression function $y = ax^2 + bx + c$ for this boat.

(B) A marina rents this boat for $10 per hour plus the cost of the gasoline used. If gasoline costs $1.40 per gallon and you take a 200-mile trip in this boat, construct a mathematical model and use it to answer the following questions:
What speed should you travel to minimize the rental charges?
What mileage will the boat get?
How long does the trip take?
How much gasoline will you use?
How much will the trip cost you?

TABLE 3 Performance Data

mph	mpg
4.8	1.41
8.6	1.65
11.9	1.85
26.7	1.78
35.9	1.51
44.5	1.08

Source: www.yamaha-motor.com

CHAPTER 3

Polynomial and Rational Functions

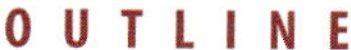

OUTLINE

RECALL THAT THE ZEROS OF A FUNCTION f ARE THE SOLUTIONS OR ROOTS of the equation $f(x) = 0$, if any exist. There are formulas that give the exact values of the zeros, real or imaginary, of any linear or quadratic function (Table 1).

TABLE 1 Zeros of Linear and Quadratic Functions

Function	Linear	Quadratic
Form	$f(x) = ax + b,\ a \neq 0$	$f(x) = ax^2 + bx + c,\ a \neq 0$
Equation	$ax + b = 0$	$ax^2 + bx + c = 0$
Zeros/Roots	$x = -\dfrac{b}{a}$	$x = \dfrac{-b \pm \sqrt{b^2 - 4ac}}{2a}$

Linear and quadratic functions are also called first- and second-degree *polynomial functions*, respectively. Thus, Table 1 contains formulas for the zeros of any first- or second-degree polynomial function. What about higher-degree polynomial functions such as

$$p(x) = 4x^3 - 2x^2 + 3x + 5 \qquad \text{Third degree (cubic)}$$

$$q(x) = -2x^4 + 5x^2 - 6 \qquad \text{Fourth degree (quartic)}$$

$$r(x) = x^5 - x^4 + x^3 - 10 \qquad \text{Fifth degree (quintic)}$$

Preparing for this chapter

Before getting started on this chapter, review the following concepts:

- Polynomials (Basic Algebra Review*, Sec. 2 and 3)
- Rational Expressions (Basic Algebra Review*, Section 4)
- Graphs of Functions (Chapter 1, Section 3)
- Linear Functions (Chapter 2, Section 1)
- Linear Regression (Chapter 2, Section 2)
- Quadratic Functions (Chapter 2, Section 3)
- Complex Numbers (Chapter 2, Section 4)
- Quadratic Formula (Chapter 2, Section 5)

*At www.mhhe.com/barnett

It turns out that there are direct, though complicated, methods for finding formulas for the zeros of any third- or fourth-degree polynomial function. However, the Frenchman Evariste Galois (1811–1832) proved at the age of 20 that for polynomial functions of degree greater than 4 there is no formula or finite step-by-step process that will always yield exact values for all zeros.* This does not mean that we give up looking for zeros of higher-degree polynomial functions. It just means that we will have to use a variety of specialized methods and sometimes we will have to approximate the zeros. The development of these methods is one of the primary objectives of Chapter 3. Throughout this chapter, we will always use the term *zero* to refer to an *exact value* and will indicate clearly when approximate values of the zero will suffice.

We begin in Section 3.1 by discussing the properties of graphs of polynomial functions. In Section 3.2 we introduce methods for locating the real zeros of polynomials with real coefficients. Once located, the real zeros are easily approximated with a graphing utility. In Section 3.3 we study the complex zeros of polynomials with complex coefficients (laying the foundation for the method of partial fraction decomposition discussed in Appendix B, Section B.2), and rational zeros of polynomials with rational coefficients. Section 3.4 provides a graphical approach to rational functions and rational inequalities that parallels the earlier development for polynomials.

SECTION 3.1 Polynomial Functions and Models

Graphs of Polynomial Functions • Polynomial Division • Remainder and Factor Theorems • Mathematical Modeling and Data Analysis

In this section we extend our work with linear and quadratic functions in Chapter 2 by defining polynomial functions and studying properties of their graphs. Algebraic tools (division and factorization of polynomials) are introduced to understand the general properties of graphs of polynomials, and for later use in analyzing the graphs of specific polynomials. Finally, we show how polynomials are applied in mathematical modeling and data analysis.

FIGURE 1 Graphs of linear and quadratic functions.

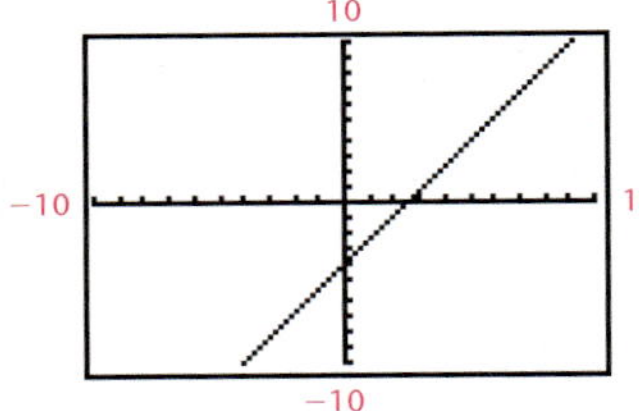

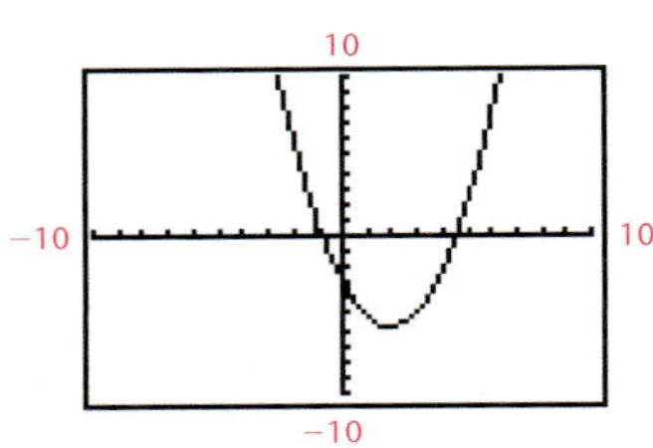

Graphs of Polynomial Functions

In Chapter 2 you were introduced to linear and quadratic functions and their graphs (Fig. 1):

$$f(x) = ax + b, \qquad a \neq 0 \qquad \text{Linear function}$$
$$f(x) = ax^2 + bx + c, \qquad a \neq 0 \qquad \text{Quadratic function}$$

A function such as

$$g(x) = 7x^4 - 5x^3 + (2 + 9i)x^2 + 3x - 1.95$$

*Galois's contribution, using the new concept of "group," was of the highest mathematical significance and originality. However, his contemporaries hardly read his papers, dismissing them as "almost unintelligible." At the age of 21, involved in political agitation, Galois met an untimely death in a duel. A short but fascinating account of Galois's tragic life can be found in E. T. Bell's *Men of Mathematics* (New York: Simon & Schuster, 1937), pp. 362–377.

which is the sum of a finite number of terms, each of the form ax^k, where a is a number and k is a nonnegative integer, is called a *polynomial function*. The polynomial function $g(x)$ is said to have *degree* 4 because x^4 is the highest power of x that appears among the terms of $g(x)$. Therefore, linear and quadratic functions are polynomial functions of degrees 1 and 2, respectively. The two functions $h(x) = x^{-1}$ and $k(x) = x^{1/2}$, however, are not polynomial functions (the exponents -1 and $\frac{1}{2}$ are not nonnegative integers).

DEFINITION 1
Polynomial Function

If n is a nonnegative integer, a function that can be written in the form

$$P(x) = a_n x^n + a_{n-1}x^{n-1} + \dots + a_1 x + a_0, \qquad a_n \neq 0$$

is called a **polynomial function of degree n.** The numbers $a_n, a_{n-1}, \dots, a_1, a_0$ are called the **coefficients** of $P(x)$.

We will assume that the coefficients of a polynomial function are complex numbers, or real numbers, or rational numbers, or integers, depending on our interest. Similarly, the domain of a polynomial function can be the set of complex numbers, the set of real numbers, or an appropriate subset of either, depending on our interest. In general, the context will dictate the choice of coefficients and domain. Definition 1 implies that a nonzero constant function has degree 0. The constant function with value 0 is considered to be a polynomial but is not assigned a degree.

DEFINITION 2
Zeros or Roots

A number r is said to be a **zero** or **root** of a function $P(x)$ if $P(r) = 0$.

The zeros of $P(x)$ are thus the solutions of the equation $P(x) = 0$. So if the coefficients of a polynomial $P(x)$ are real numbers, then the real zeros of $P(x)$ are just the x intercepts of the graph of $P(x)$. For example, the real zeros of the polynomial $P(x) = x^2 - 4$ are 2 and -2, the x intercepts of the graph of $P(x)$ [Fig. 2(a)]. However, a polynomial may have zeros that are not x intercepts. $Q(x) = x^2 + 4$, for example, has zeros $2i$ and $-2i$, but its graph has no x intercepts [Fig. 2(b)].

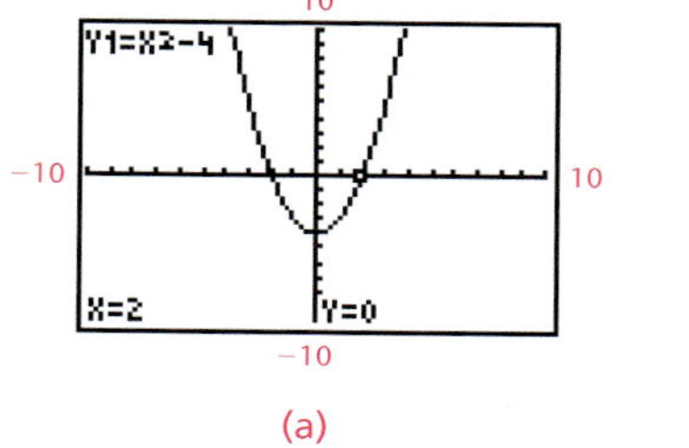

(a)

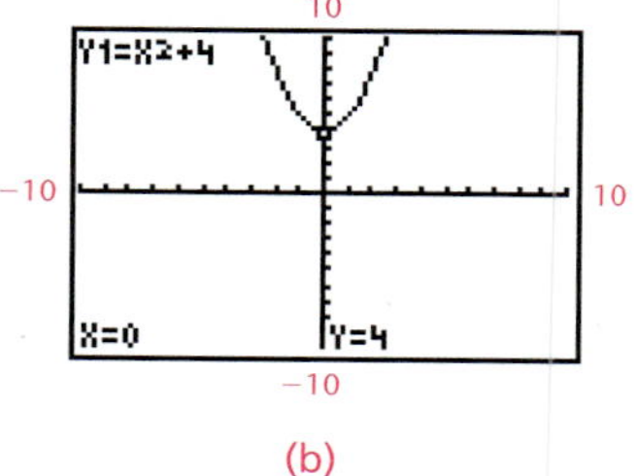

(b)

FIGURE 2 Real zeros are x intercepts.

EXAMPLE 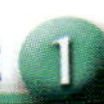 Zeros and x Intercepts

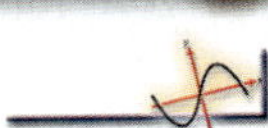
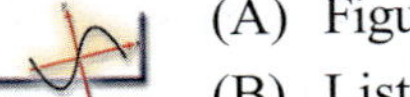

(A) Figure 3 shows the graph of a polynomial function of degree 5. List its real zeros.

(B) List all zeros of the polynomial function

$$P(x) = (x - 4)(x + 7)^3(x^2 + 9)(x^2 - 2x + 2)$$

Which zeros of $P(x)$ are x intercepts?

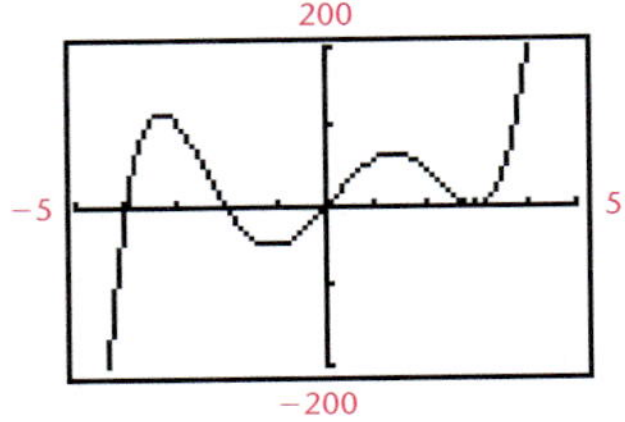

FIGURE 3

SOLUTION

(A) The real zeros are the x intercepts: -4, -2, 0, and 3.

(B) Note first that $P(x)$ is a polynomial because it can be written in the form of Definition 1 (it is not necessary to actually multiply out $P(x)$ to find that form). The zeros of $P(x)$ are the solutions to the equation $P(x) = 0$. Because a product equals 0 if and only if one of the factors equals 0, we solve each of the following equations (the last was solved using the quadratic formula):

$$x - 4 = 0 \qquad (x + 7)^3 = 0 \qquad x^2 + 9 = 0 \qquad x^2 - 2x + 2 = 0$$
$$x = 4 \qquad x = -7 \qquad x = \pm 3i \qquad x = 1 \pm i$$

Therefore, the zeros of $P(x)$, are 4, -7, $3i$, $-3i$, $1 + i$, and $1 - i$. Only two of the six zeros are real numbers and thus x intercepts: 4 and -7.

MATCHED 1 PROBLEM

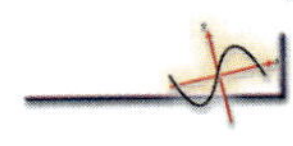

(A) Figure 4 shows the graph of a polynomial function of degree 4. List its real zeros.

(B) List all zeros of the polynomial function

$$P(x) = (x + 5)(x^2 - 4)(x^2 + 4)(x^2 + 2x + 5)$$

Which zeros of $P(x)$ are x intercepts?

FIGURE 4

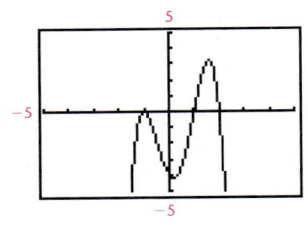

A point on a continuous graph that separates an increasing portion from a decreasing portion, or vice versa, is called a **turning point.** The vertex of a parabola, for example, is a turning point. Linear functions with real coefficients have exactly one real zero and no turning points; quadratic functions with real coefficients have at most two real zeros and exactly one turning point. (The y coordinate of any turning point is a local extremum. It is possible, however, that $f(c)$ is a local extremum of a continuous function f even though $(c, f(c))$ is not a turning point. See Problems 87 and 88 in Exercise 3.1.)

EXPLORE/DISCUSS 1

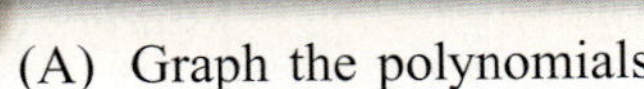

(A) Graph the polynomials

$$f(x) = 0.1x^4 - 0.5x^3 - 0.7x^2 + 4.5x - 3.1$$
$$g(x) = -1.1x^5 + 6.3x^3 - 8.2x - 0.1$$
$$h(x) = 0.01x^8 + x^6 - 7x^4 + 11x^2 - 3$$

in the standard viewing window.

(B) Assuming that all real zeros of $f(x)$, $g(x)$, and $h(x)$ appear in the standard viewing window, how is the number of real zeros of a polynomial related to its degree?

(C) Assuming that all turning points of $f(x)$, $g(x)$, and $h(x)$ appear in the standard viewing window, how is the number of turning points of a polynomial related to its degree?

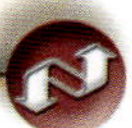

Explore/Discuss 1 suggests that the graphs of polynomial functions with real coefficients have the properties listed in Theorem 1, which we accept now without proof. Property 3 is proved later in Section 3.1. The other properties are established in calculus.

THEOREM 1

Properties of Graphs of Polynomial Functions

Let $P(x)$ be a polynomial of degree $n > 0$ with real coefficients. Then the graph of $P(x)$:

1. Is continuous for all real numbers
2. Has no sharp corners
3. Has at most n real zeros
4. Has at most $n - 1$ turning points
5. Increases or decreases without bound as $x \to \infty$ and as $x \to -\infty$*

Figure 5 shows graphs of representative polynomial functions of degrees 1 through 6, illustrating the five properties of Theorem 1.

FIGURE 5 Graphs of polynomial functions.

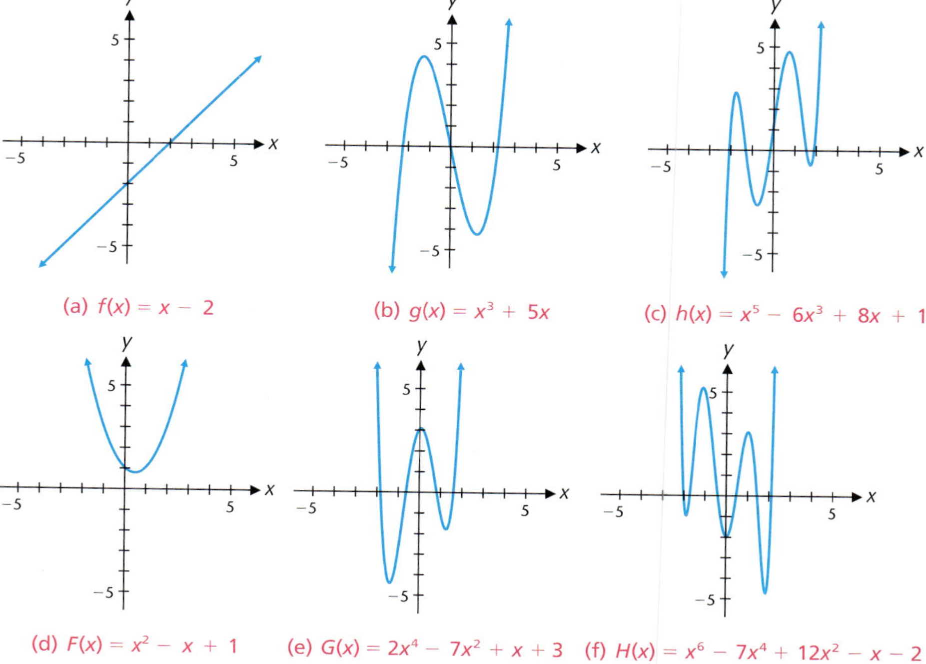

*Remember that ∞ and $-\infty$ are not real numbers. The statement *the graph of $P(x)$ increases without bound as $x \to -\infty$* means that for any horizontal line $y = b$ there is some interval $(-\infty, a] = \{x \mid x \le a\}$ on which the graph of $P(x)$ is above the horizontal line.

EXAMPLE 2 Properties of Graphs of Polynomials

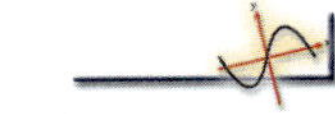

Explain why each graph is *not* the graph of a polynomial function by listing the properties of Theorem 1 that it *fails* to satisfy.

(A)

(B)

(C)

SOLUTION

(A) The graph has a sharp corner when $x = 0$. Property 2 fails.

(B) There are no points on the graph with x coordinate less than or equal to 0, so properties 1 and 5 fail.

(C) There are an infinite number of zeros and an infinite number of turning points, so properties 3 and 4 fail. Furthermore, the graph is bounded by the horizontal lines $y = \pm 1$, so property 5 fails.

MATCHED 2 PROBLEM

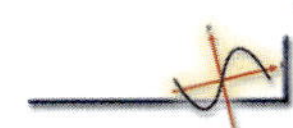

Explain why each graph is *not* the graph of a polynomial function by listing the properties of Theorem 1 that it *fails* to satisfy.

(A)

(B)

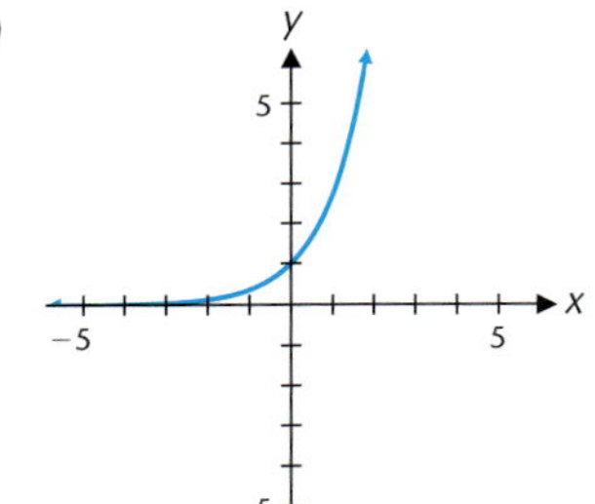

(C)

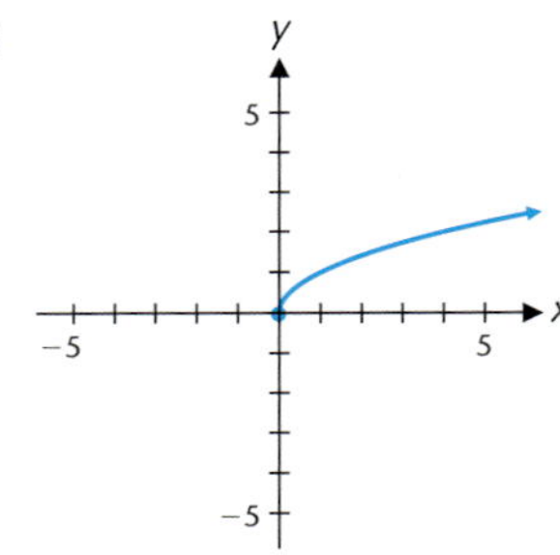

EXPLORE/DISCUSS 2

If n is a positive integer, then $y_1 = x^n$, $y_2 = x^{n-1}$, and $y_3 = x^n + x^{n-1}$ are all polynomial functions. Is the shape of y_3 more similar to the shape of y_1 or to the shape of y_2? Obtain evidence for your answer by graphing all three functions for several values of n.

Explore/Discuss 2 suggests that the shape of the graph of a polynomial function with real coefficients is similar to the shape of the graph of the **leading term,** that is, the term of highest degree. Figure 6 compares the graph of the polynomial $h(x) = x^5 - 6x^3 + 8x + 1$ from Figure 5 with the graph of its leading term $p(x) = x^5$. The graphs are dissimilar near the origin, but as we zoom out, the shapes of the two graphs become quite similar. The leading term in the polynomial dominates all other terms combined. Because the graph of $p(x)$ increases without bound as $x \rightarrow \infty$, the same is true of the graph of $h(x)$. And because the graph of $p(x)$ decreases without bound as $x \rightarrow -\infty$, the same is true of the graph of $h(x)$.

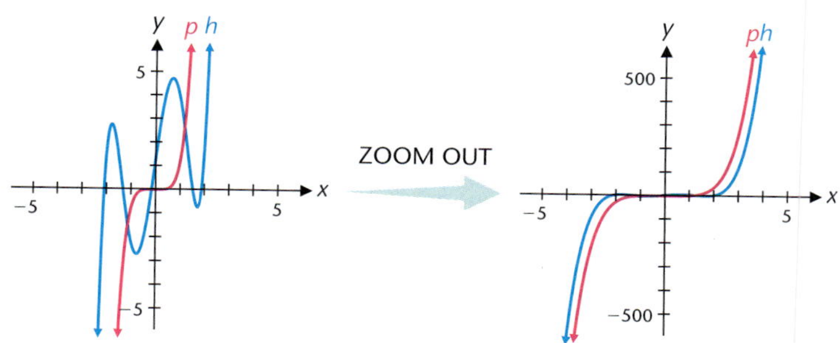

FIGURE 6 $p(x) = x^5$, $h(x) = x^5 - 6x^3 + 8x + 1$.

The left and right behavior of a polynomial function with real coefficients is determined by the left and right behavior of its leading term (see Fig. 6). Property 5 of Theorem 1 can therefore be refined. The various possibilities are summarized in Theorem 2.

THEOREM 2

Left and Right Behavior of Polynomial Functions

Let $P(x) = a_nx^n + a_{n-1}x^{n-1} + \ldots + a_1x + a_0$ be a polynomial function with real coefficients, $a_n \neq 0$, $n > 0$.

1. $a_n > 0$, n even: The graph of $P(x)$ increases without bound as $x \rightarrow \infty$ and increases without bound as $x \rightarrow -\infty$ (like the graphs of x^2, x^4, x^6, etc.).
2. $a_n > 0$, n odd: The graph of $P(x)$ increases without bound as $x \rightarrow \infty$ and decreases without bound as $x \rightarrow -\infty$ (like the graphs of x, x^3, x^5, etc.).
3. $a_n < 0$, n even: The graph of $P(x)$ decreases without bound as $x \rightarrow \infty$ and decreases without bound as $x \rightarrow -\infty$ (like the graphs of $-x^2$, $-x^4$, $-x^6$ etc.).
4. $a_n < 0$, n odd: The graph of $P(x)$ decreases without bound as $x \rightarrow \infty$ and increases without bound as $x \rightarrow -\infty$ (like the graphs of $-x$, $-x^3$, $-x^5$, etc.).

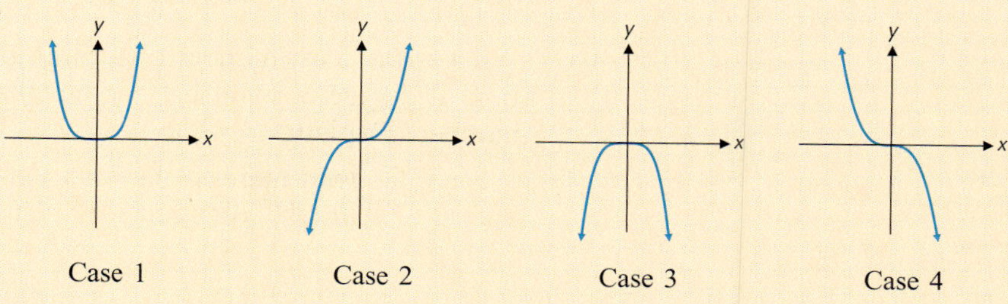

It is convenient to write $P(x) \to \infty$ as an abbreviation for the phrase *the graph of* $P(x)$ *increases without bound.* Using this notation, the left and right behavior in Case 4 of Theorem 2, for example, is: $P(x) \to -\infty$ as $x \to \infty$ and $P(x) \to \infty$ as $x \to -\infty$.

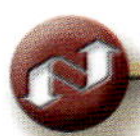

EXPLORE/DISCUSS 3

A student uses the terms *slide left, slide right, dome,* and *bowl* to describe the four cases of Theorem 2. Which term belongs to which case? Explain.

EXAMPLE 3 Left and Right Behavior of Polynomials

Determine the left and right behavior of each polynomial.

(A) $P(x) = 3 - x^2 + 4x^3 - x^4 - 2x^6$

(B) $Q(x) = 4x^5 + 8x^3 + 5x - 1$

SOLUTIONS

(A) The degree n of $P(x)$ is 6 (even) and the coefficient a_6 is -2 (negative), so the left and right behavior is the same as that of $-x^6$ (Case 3 of Theorem 2): $P(x) \to -\infty$ as $x \to \infty$ and $P(x) \to -\infty$ as $x \to -\infty$.

(B) The degree n of $Q(x)$ is 5 (odd) and the coefficient a_5 is 4 (positive), so the left and right behavior is the same as that of x^5 (Case 2 of Theorem 2): $P(x) \to \infty$ as $x \to \infty$ and $P(x) \to -\infty$ as $x \to -\infty$.

MATCHED PROBLEM 3

Determine the left and right behavior of each polynomial.

(A) $P(x) = 4x^9 - 3x^{11} + 5$

(B) $Q(x) = 1 - 2x^{50} + x^{100}$

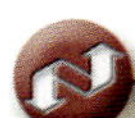

EXPLORE/DISCUSS 4

(A) What is the least number of turning points that a polynomial function of degree 5, with real coefficients, can have? The greatest number? Explain.

(B) What is the least number of x intercepts that a polynomial function of degree 5, with real coefficients, can have? The greatest number? Explain.

(C) What is the least number of turning points that a polynomial function of degree 6, with real coefficients, can have? The greatest number? Explain.

(D) What is the least number of x intercepts that a polynomial function of degree 6, with real coefficients, can have? The greatest number? Explain.

EXAMPLE 4 Analyzing the Graph of a Polynomial

Approximate to two decimal places the zeros and local extrema for

$$P(x) = x^3 - 14x^2 + 27x - 12$$

SOLUTION

Examining the graph of P in a standard viewing window [Fig. 7(a)], we see two zeros and a local maximum near $x = 1$. Zooming in shows these points more clearly [Fig. 7(b)]. Using familiar commands (details omitted), we find that $P(x) = 0$ for $x \approx 0.66$ and $x \approx 1.54$, and that $P(1.09) \approx 2.09$ is a local maximum value.

FIGURE 7
$P(x) = x^3 - 14x^2 + 27x - 12.$

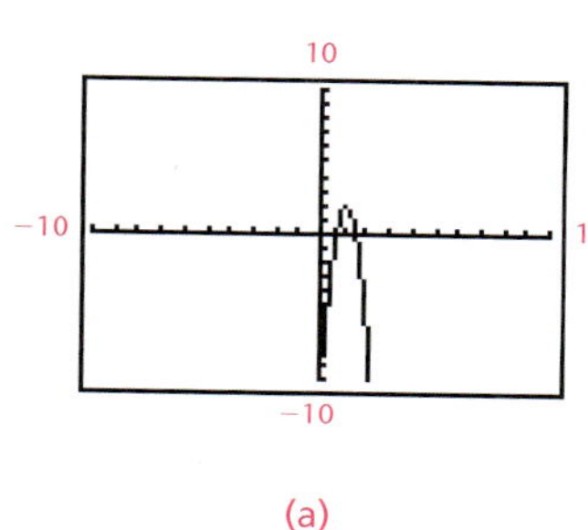

(a)

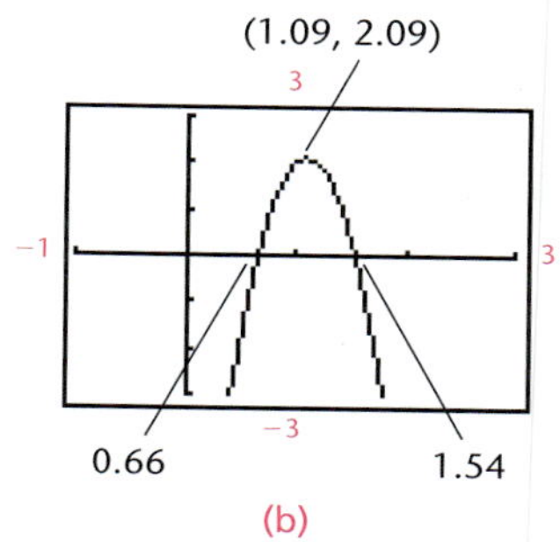

(b)

Have we found all the zeros and local extrema? The graph in Figure 7(a) seems to indicate that $P(x)$ is decreasing as x decreases to the left and as x increases to the right. However, the leading term for $P(x)$ is x^3. Because x^3 increases without bound as x increases to the right without bound, $P(x)$ must change direction at some point and become increasing. Thus, there must exist a local minimum and another zero that are not visible in this viewing window. Examining a table of values [Fig. 8(a)], we discover a local minimum near $x = 8$ and a zero near $x = 12$. Adjusting the window variables produces the graph in Figure 8(b). Using familiar commands (details omitted), we find that $P(x) = 0$ for $x \approx 11.80$ and that $P(8.24) \approx -180.61$ is a local minimum. Because a third-degree polynomial can have at most three zeros and two local extrema, we have found all the zeros and local extrema for this polynomial.

FIGURE 8
$P(x) = x^3 - 14x^2 + 27x - 12.$

X	Y1	
6	-138	
7	-166	
8	-180	
9	-174	
10	-142	
11	-78	
12	24	

Y1■X^3-14X²+27X...

(a)

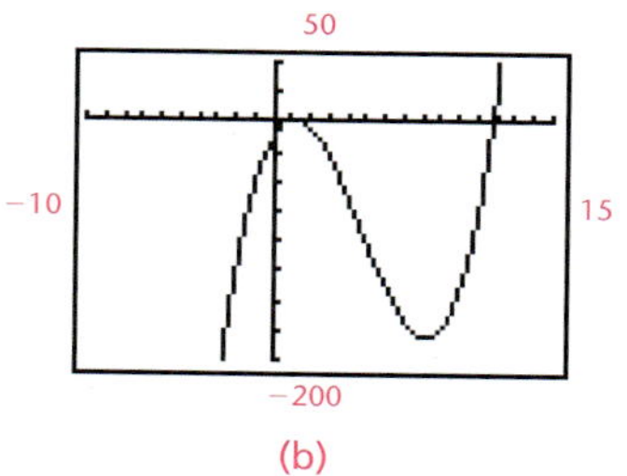

(b)

MATCHED 4 PROBLEM

Approximate to two decimal places the zeros and the coordinates of the local extrema for

$$P(x) = -x^3 - 14x^2 - 15x + 5$$

We have introduced the main properties of graphs of polynomial functions with real coefficients (Theorems 1 and 2) and have seen their usefulness in analyzing graphs (Example 4). But to understand why, for example, a polynomial function of degree n can have at most n real zeros, we must view polynomials from an algebraic perspective. Polynomials can be factored. We proceed to study the division and factorization of polynomials.

Polynomial Division

We can divide one polynomial (the dividend) by another (the divisor) by a long-division process similar to that used in arithmetic. Example 5 illustrates the process.

EXAMPLE 5 Algebraic Long Division

Divide $P(x) = 2x^4 + 3x^3 - x - 5$ by $x + 2$.

SOLUTION

$$\begin{array}{r l}
 & 2x^3 - x^2 + 2x - 5 \\
x + 2 \,\big) & 2x^4 + 3x^3 + 0x^2 - x - 5 \\
 & \underline{2x^4 + 4x^3} \\
 & \quad -x^3 + 0x^2 \\
 & \quad \underline{-x^3 - 2x^2} \\
 & \qquad 2x^2 - x \\
 & \qquad \underline{2x^2 + 4x} \\
 & \qquad\quad -5x - 5 \\
 & \qquad\quad \underline{-5x - 10} \\
 & \qquad\qquad\quad 5 = R
\end{array}$$

Remainder

Arrange the dividend and the divisor in descending powers of the variable. Insert, with 0 coefficients, any missing terms of degree less than 4. Divide the first term of the divisor into the first term of the dividend. Multiply the divisor by $2x^3$, line up like terms, subtract (change the sign and add) as in arithmetic, and bring down $0x^2$. Repeat the process until the degree of the remainder is less than that of the divisor.

Thus,

$$\frac{2x^4 + 3x^3 - x - 5}{x + 2} = 2x^3 - x^2 + 2x - 5 + \frac{5}{x + 2}$$

CHECK

$$\begin{aligned}
(x + 2)\left[2x^3 - x^2 + 2x - 5 + \frac{5}{x + 2}\right] \\
&= (x + 2)(2x^3 - x^2 + 2x - 5) + 5 \\
&= 2x^4 + 3x^3 - x - 5
\end{aligned}$$

MATCHED PROBLEM 5

Divide $6x^2 - 30 + 9x^3$ by $x - 2$.

The procedure illustrated in Example 5 is called the **division algorithm.** The concluding equation of Example 5 (before the check) may be multiplied by the divisor $x + 2$ to give the following form:

$$\underset{\text{Dividend}}{2x^4 + 3x^3 - x - 5} = \underset{\text{Divisor}\cdot\text{Quotient}}{(x + 2)(2x^3 - x^2 + 2x - 5)} + \underset{\text{Remainder}}{5}$$

This last equation is an *identity:* it is true for all replacements of x by real or complex numbers including $x = -2$. Theorem 3, which we state without proof, gives the general result of applying the division algorithm when the divisor has the form $x - r$.

THEOREM 3
Division Algorithm

For each polynomial $P(x)$ of degree greater than 0 and each number r, there exists a unique polynomial $Q(x)$ of degree 1 less than $P(x)$ and a unique number R such that

$$P(x) = (x - r)Q(x) + R$$

The polynomial $Q(x)$ is called the **quotient,** $x - r$ is the **divisor,** and R is the **remainder.** Note that R may be 0.

The long division of Example 5 can be carried out by a shortcut called *synthetic division.* The numerals that represent the essentials of the long-division process are indicated in color here.

$$\begin{array}{rrl}
 & 2x^3 - 1x^2 + 2x - 5 & \text{Quotient} \\
\text{Divisor} \quad x + 2 & \overline{)\,2x^4 + 3x^3 + 0x^2 - 1x - 5} & \text{Dividend} \\
 & \underline{2x^4 + 4x^3} \qquad\qquad\qquad\quad & \\
 & -1x^3 + 0x^2 \qquad\qquad & \\
 & \underline{-1x^3 - 2x^2} \qquad\qquad & \\
 & 2x^2 - 1x \qquad & \\
 & \underline{2x^2 + 4x} \qquad & \\
 & -5x - \ \ 5 & \\
 & \underline{-5x - 10} & \\
 & 5 & \text{Remainder}
\end{array}$$

The numerals printed in color can be arranged more conveniently as follows:

Dividend coefficients

$$\begin{array}{c|ccccc}
 & 2 & 3 & 0 & -1 & -5 \\
 & & 4 & -2 & 4 & -10 \\
\hline
2 & 2 & -1 & 2 & -5 & 5
\end{array}$$

Quotient coefficients — Remainder

Mechanically, we see that the second and third rows of numerals are generated as follows. The first coefficient, 2, of the dividend is brought down and multiplied by 2 from the divisor; and the product, 4, is placed under the second dividend coefficient, 3, and subtracted. The difference, -1, is again multiplied by the 2 from the divisor; and the product is placed under the third coefficient from the dividend and subtracted. This process is repeated until the remainder is reached. The process can be made a little faster, and less prone to sign errors, by changing $+2$ from the divisor to -2 and adding instead of subtracting. Thus

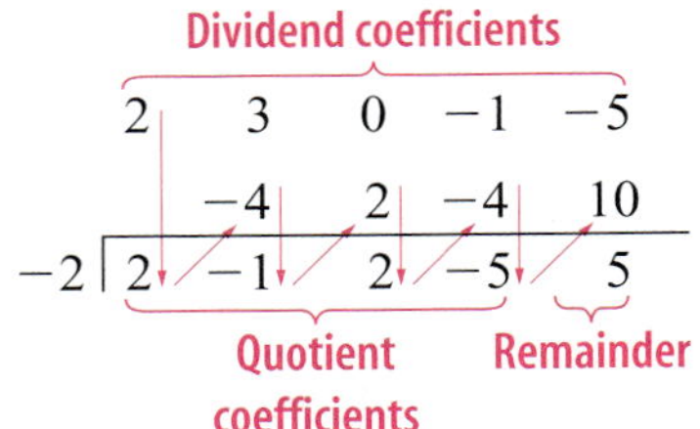

Key Steps in the Synthetic Division Process

To divide the polynomial $P(x)$ by $x - r$:

Step 1. Arrange the coefficients of $P(x)$ in order of descending powers of x. Write 0 as the coefficient for each missing power.

Step 2. After writing the divisor in the form $x - r$, use r to generate the second and third rows of numbers as follows. Bring down the first coefficient of the dividend and multiply it by r; then add the product to the second coefficient of the dividend. Multiply this sum by r, and add the product to the third coefficient of the dividend. Repeat the process until a product is added to the constant term of $P(x)$.

Step 3. The last number to the right in the third row of numbers is the remainder. The other numbers in the third row are the coefficients of the quotient, which is of degree 1 less than $P(x)$.

EXAMPLE 6 Synthetic Division

Use synthetic division to divide $P(x) = 4x^5 - 30x^3 - 50x - 2$ by $x + 3$. Find the quotient and remainder. Write the conclusion in the form $P(x) = (x - r)Q(x) + R$ of Theorem 3.

SOLUTION

Because $x + 3 = x - (-3)$, we have $r = -3$, and

$$\begin{array}{r|rrrrrr} & 4 & 0 & -30 & 0 & -50 & -2 \\ & & -12 & 36 & -18 & 54 & -12 \\ \hline -3 & 4 & -12 & 6 & -18 & 4 & -14 \end{array}$$

The quotient is $4x^4 - 12x^3 + 6x^2 - 18x + 4$ with a remainder of -14. Thus,

$$4x^5 - 30x^3 - 50x - 2 = (x + 3)(4x^4 - 12x^3 + 6x^2 - 18x + 4) - 14$$

MATCHED 6 PROBLEM

Repeat Example 6 with $P(x) = 3x^4 - 11x^3 - 18x + 8$ and divisor $x - 4$.

A calculator is a convenient tool for performing synthetic division. Any type of calculator can be used, although one with a memory will save some keystrokes. The flowchart in Figure 9 shows the repetitive steps in the synthetic division process, and Figure 10 illustrates the results of applying this process to Example 6 on a graphing calculator.

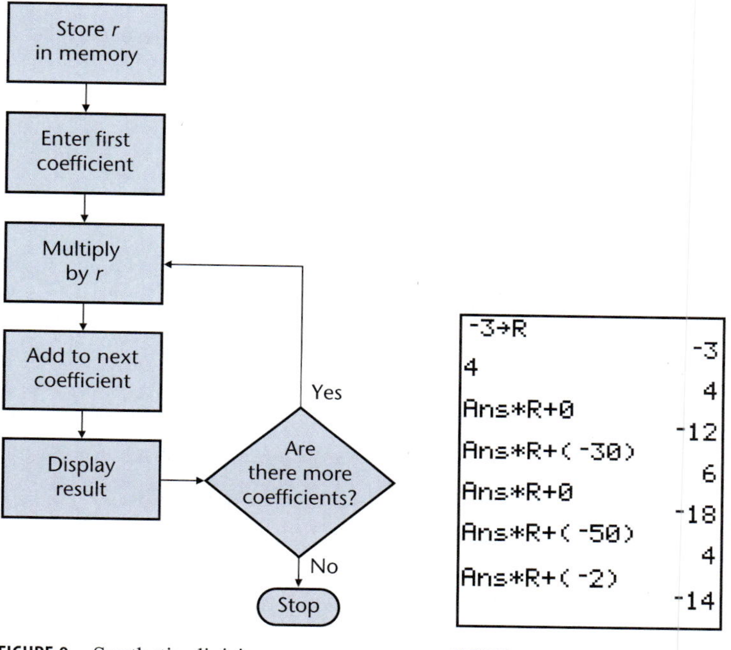

FIGURE 9 Synthetic division.

FIGURE 10

Remainder and Factor Theorems

EXPLORE/DISCUSS 5

Let $P(x) = x^3 - 3x^2 - 2x + 8$.

(A) Evaluate $P(x)$ for

(i) $x = -2$ (ii) $x = 1$ (iii) $x = 3$

(B) Use synthetic division to find the remainder when $P(x)$ is divided by

(i) $x + 2$ (ii) $x - 1$ (iii) $x - 3$

What conclusion does a comparison of the results in parts A and B suggest?

Explore/Discuss 5 suggests that when a polynomial $P(x)$ is divided by $x - r$, the remainder is equal to $P(r)$, the value of the polynomial $P(x)$ at $x = r$. This is true because the equation of the division algorithm, $P(x) = (x - r)Q(x) + R$, is an identity, valid for all replacements of the variable x by real or complex numbers; in particular, if $x = r$, then:

$$\begin{aligned} P(r) &= (r - r)Q(r) + R \\ &= 0 \cdot Q(r) + R \\ &= 0 + R \\ &= R \end{aligned}$$

We have proved the *remainder theorem.*

THEOREM 4

Remainder Theorem

If R is the remainder after dividing the polynomial $P(x)$ by $x - r$, then

$$P(r) = R$$

EXAMPLE 7 Two Methods for Evaluating Polynomials

If $P(x) = 4x^4 + 10x^3 + 19x + 5$, find $P(-3)$ by

(A) Using the remainder theorem and synthetic division

(B) Evaluating $P(-3)$ directly

SOLUTIONS

(A) Use synthetic division to divide $P(x)$ by $x - (-3)$.

$$\begin{array}{r|rrrrl} & 4 & 10 & 0 & 19 & 5 \\ & & -12 & 6 & -18 & -3 \\ \hline -3 & 4 & -2 & 6 & 1 & 2 = R = P(-3) \end{array}$$

(B) $P(-3) = 4(-3)^4 + 10(-3)^3 + 19(-3) + 5$
$= 2$

MATCHED 7 PROBLEM

Repeat Example 7 for $P(x) = 3x^4 - 16x^2 - 3x + 7$ and $x = -2$.

You might think the remainder theorem is not a very effective tool for evaluating polynomials. But let's consider the number of operations performed in parts A and B of Example 7. Synthetic division requires only four multiplications and four additions to find $P(-3)$, whereas the direct evaluation requires ten multiplications and four additions. [Note that evaluating $4(-3)^4$ actually requires five multiplications.] The difference becomes even larger as the degree of the polynomial increases. Computer programs that involve numerous polynomial evaluations often use synthetic division because of its efficiency. We will find synthetic division and the remainder theorem to be useful tools later in this chapter.

The remainder theorem implies that the division algorithm equation,

$$P(x) = (x - r)Q(x) + R$$

may be written in the form where R is replaced by $P(r)$:

$$P(x) = (x - r)Q(x) + P(r)$$

Therefore $x - r$ is a factor of $P(x)$ if and only if $P(r) = 0$, that is, if and only if r is a zero of the polynomial $P(x)$. This result is called the *factor theorem*.

THEOREM 5
Factor Theorem

If r is a zero of the polynomial $P(x)$, then $x - r$ is a factor of $P(x)$. Conversely, if $x - r$ is a factor of $P(x)$, then r is a zero of $P(x)$.

EXAMPLE 8 Factors of Polynomials

Use the factor theorem to show that $x + 1$ is a factor of $P(x) = x^{25} + 1$ but is not a factor of $Q(x) = x^{25} - 1$.

SOLUTION
Because

$$P(-1) = (-1)^{25} + 1 = -1 + 1 = 0$$

$x - (-1) = x + 1$ is a factor of $x^{25} + 1$. On the other hand,

$$Q(-1) = (-1)^{25} - 1 = -1 - 1 = -2$$

and $x + 1$ is not a factor of $x^{25} - 1$.

MATCHED PROBLEM 8

Use the factor theorem to show that $x - i$ is a factor of $P(x) = x^8 - 1$ but is not a factor of $Q(x) = x^8 + 1$.

If a polynomial $P(x)$ of degree n has n zeros, $r_1, r_2, \ldots, r_n$, then by the factor theorem,

$$P(x) = (x - r_1)Q_1(x), \text{ deg } Q_1(x) = n - 1$$

But $P(r_2) = 0$, so $Q_1(r_2) = 0$. Applying the factor theorem to $Q_1(x)$ we obtain

$$P(x) = (x - r_1)(x - r_2)Q_2(x), \text{ deg } Q_2(x) = n - 2$$

Continuing in this way,

$$P(x) = (x - r_1)(x - r_2) \ldots (x - r_n)Q_n(x), \text{ deg } Q_n(x) = 0$$

Because $Q_n(x)$ is a nonzero constant, the only roots of $P(x)$ are $r_1, r_2, \ldots, r_n$. We have proved Theorem 6.

THEOREM 6
Zeros of Polynomials

A polynomial of degree n has at most n zeros.

Theorem 6 implies that the graph of a polynomial of degree n with real coefficients has at most n real zeros (Property 3 of Theorem 1). The polynomial

$$H(x) = x^6 - 7x^4 + 12x^2 - x - 2$$

for example, has degree 6 and the maximum number of zeros [see Fig. 5(f), p. 243]. Of course polynomials of degree 6 may have fewer than six real zeros. In fact, $p(x) = x^6 + 1$ has no real zeros. However, it can be shown that the polynomial $p(x) = x^6 + 1$ has exactly six complex zeros.

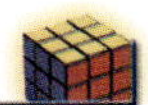

Mathematical Modeling and Data Analysis

In Chapter 2 we saw that regression techniques can be used to construct a linear or quadratic model for a set of data. Most graphing utilities have the ability to use a variety of functions for modeling data. We discuss polynomial regression models in this section and other types of regression models in later sections.

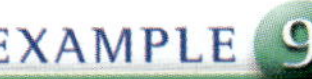

EXAMPLE 9 Estimating the Weight of Fish

Using the length of a fish to estimate its weight is of interest to both scientists and sport anglers. The data in Table 1 give the average weight of North American sturgeon for certain lengths. Use these data and regression techniques to find a cubic polynomial model that can be used to estimate the weight of a sturgeon for any length. Estimate (to the nearest ounce) the weights of sturgeon of lengths 45, 46, 47, 48, 49, and 50 inches, respectively.

TABLE 1 Sturgeon

Length (in.) x	Weight (oz.) y
18	13
22	26
26	46
30	75
34	115
38	166
44	282
52	492
60	796

Source: www.thefishernet.com

SOLUTION

The graph of the data in Table 1 [Fig. 11(a)] indicates that a linear regression model would not be appropriate for these data. And, in fact, we would not expect a linear relationship between length and weight. Instead, because weight is associated with volume, which involves three dimensions, it is more likely that the weight would be related to the cube of the length. We use a cubic regression polynomial to model these data [Fig. 11(b)]. Figure 11(c) adds the graph of the polynomial model to the graph of the data. The graph in Figure 11(c) shows that this cubic polynomial does provide a good fit for the data. (We will have more to say about the choice of functions and the accuracy of the fit provided by regression analysis later in the text.) Figure 11(d) shows the estimated weights for the requested lengths.

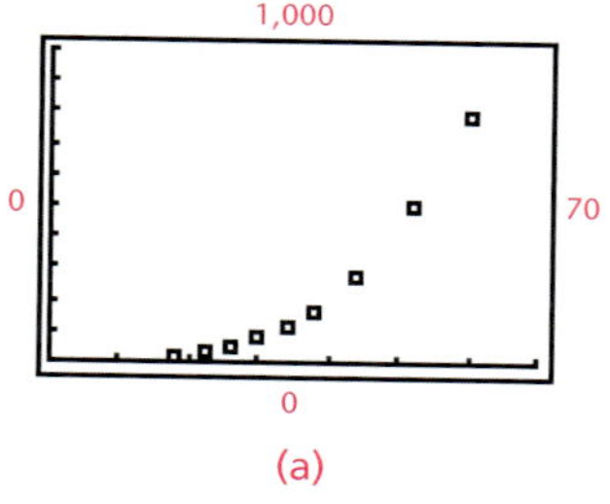

(a)

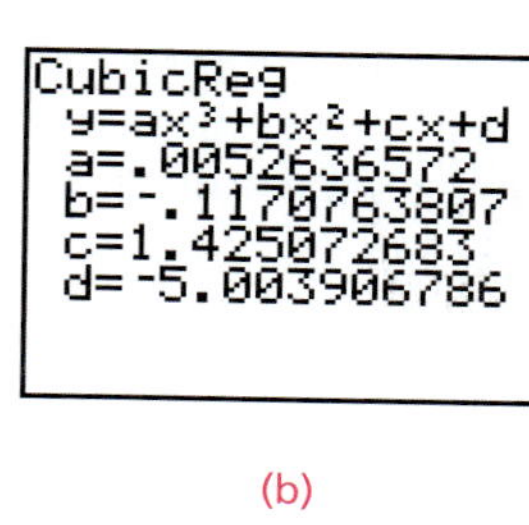

(b)

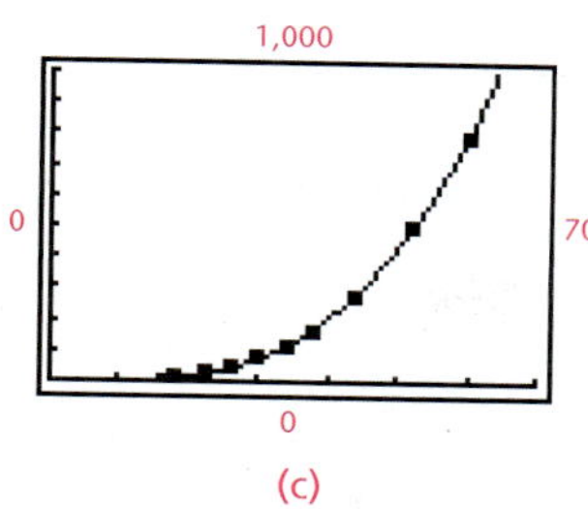

(c)

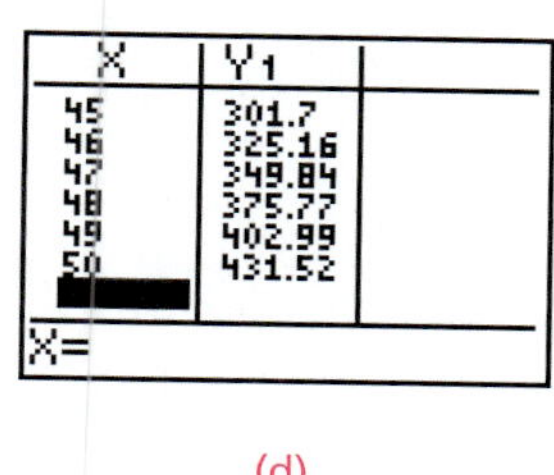

(d)

FIGURE 11

MATCHED 9 PROBLEM

Find a quadratic regression model for the data in Table 1 and compare it with the cubic regression model found in Example 9. Which model appears to provide a better fit for these data? Use numerical and/or graphical comparisons to support your choice.

EXAMPLE 10 Hydroelectric Power

TABLE 2

Year	U.S. Consumption of Hydroelectric Power (Quadrillion BTU)
1977	2.52
1979	3.14
1981	3.11
1983	3.90
1985	3.40
1987	3.12
1989	2.99
1991	3.14
1993	3.13
1995	3.48
1997	3.88
1999	3.47
2001	2.38

Source: U.S. Department of Energy

The data in Table 2 give the annual consumption of hydroelectric power (in quadrillion BTU) in the United States for selected years since 1977. Use regression techniques to find an appropriate polynomial model for the data. Discuss how well the model is expected to predict annual hydroelectric power consumption in the first decade of the twenty-first century.

SOLUTION

From Table 2 it appears that a polynomial model of the data would have three turning points—near 1983, 1989, and 1997. Because a polynomial with three turning points must have degree at least 4, we use quartic regression to find the polynomial of the form

$$y = ax^4 + bx^3 + cx^2 + dx + e$$

that best fits the data. Using $x = 0$ to represent the year 1970 and $x = 40$ to represent 2010, we enter the data [Fig. 12(a)], find the quartic regression model [Fig. 12(b)], and plot both the data points and the model [Fig. 12(c)]. The model is not expected to be a good predictor of consumption of hydroelectric power in the first decade of the twenty-first century. In fact, it predicts that consumption will be negative after the middle of 2003 [as indicated by the zero shown in Fig. 12(c)].

L1	L2	L3 3
7	2.52	
9	3.14	
11	3.11	
13	3.9	
15	3.4	
17	3.12	
19	2.99	

L3(1)=

(a)

QuarticReg
y=ax⁴+bx³+...+e
a=-1.591263E-4
b=.0120575381
c=-.3240175404
d=3.629543361
e=-10.90172993

(b)

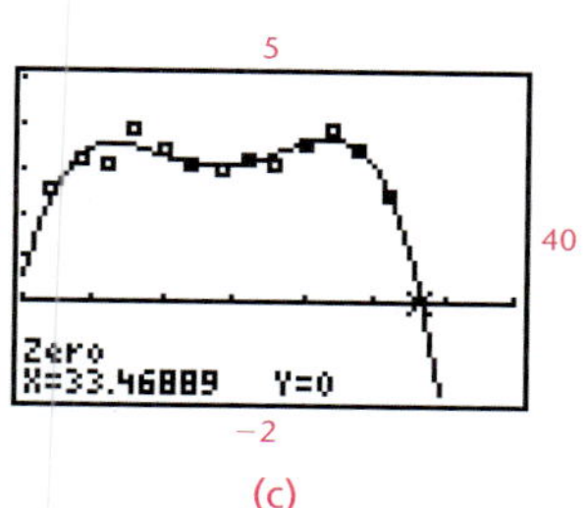

(c)

FIGURE 12

MATCHED 10 PROBLEM

Find the cubic regression model for the data of Table 2. Discuss which model is a better fit of the data—the cubic model or the quartic model of Example 10.

ANSWERS MATCHED PROBLEMS

1. (A) $-1, 1, 2$
   (B) The zeros are $-5, -2, 2, 2i, -2i, -1 + 2i$, and $-1 - 2i$; the x intercepts are $-5, -2$, and 2.
2. (A) Properties 1 and 5
   (B) Property 5
   (C) Properties 1 and 5
3. (A) $P(x) \to -\infty$ as $x \to \infty$ and $P(x) \to \infty$ as $x \to -\infty$.
   (B) $P(x) \to \infty$ as $x \to \infty$ and $P(x) \to \infty$ as $x \to -\infty$.
4. Zeros: $-12.80, -1.47, 0.27$; local maximum: $P(-0.57) \approx 9.19$; local minimum: $P(-8.76) \approx -265.71$
5. $9x^2 + 24x + 48 + \dfrac{66}{x - 2}$
6. $3x^4 - 11x^3 - 18x + 8 = (x - 4)(3x^3 + x^2 + 4x - 2)$
7. $P(-2) = -3$ for both parts, as it should
8. $P(i) = 0$, so $x - i$ is a factor of $x^8 - 1$; $Q(i) = 2$, so $x - i$ is not a factor of $x^8 + 1$
9. The cubic regression model provides a better model for these data, especially for $18 \le x \le 26$.

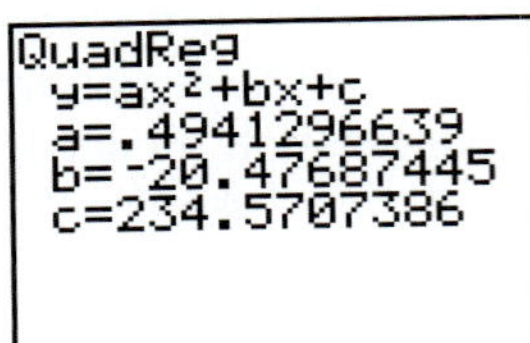

10.

CubicReg
y=ax³+bx²+cx+d
a=-3.605769E-5
b=-.0018095967
c=.1172176261
d=2.094498158

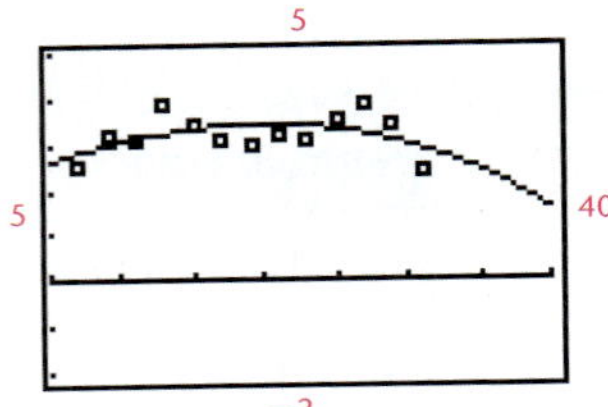

A visual inspection of the two graphs indicates that the quartic model is a better fit of the data than the cubic model.

EXERCISE 3.1

A

In Problems 1–4, a is a positive real number. Match each function with one of graphs (a)–(d).

1. $f(x) = -ax^3$
2. $g(x) = -ax^4$
3. $h(x) = ax^6$
4. $k(x) = -ax^5$

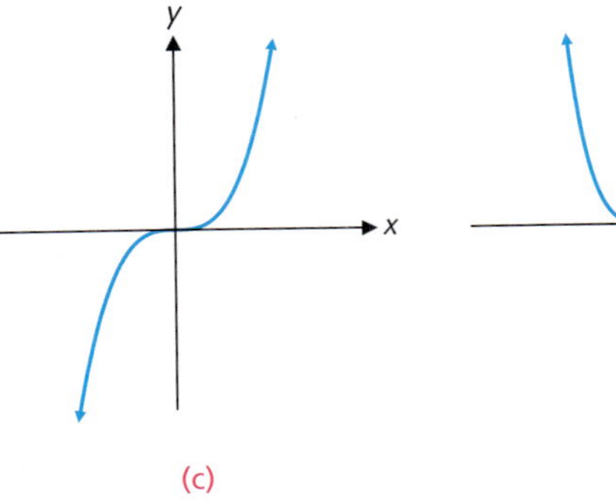

(c)

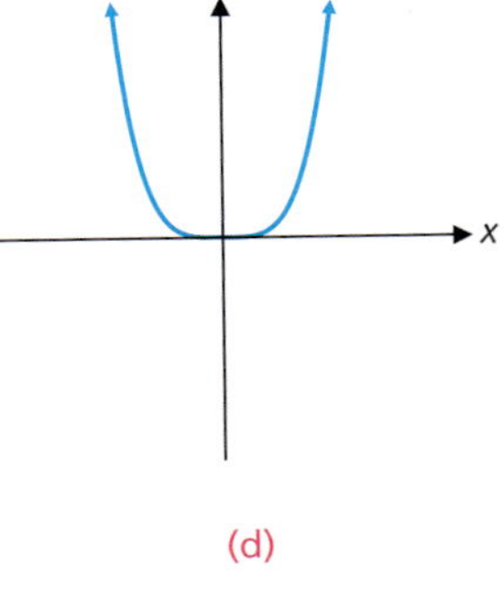

(d)

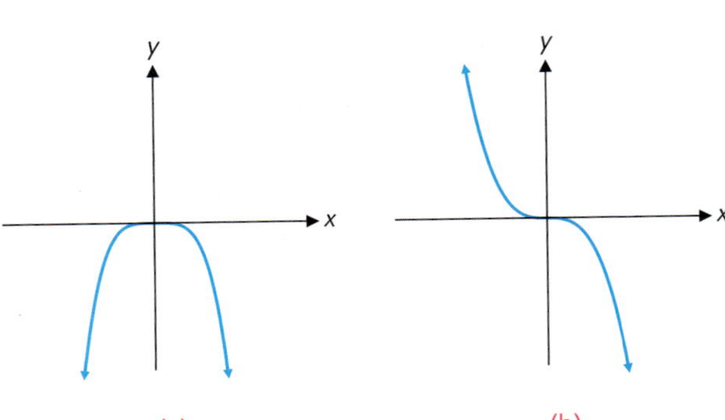

(a) (b)

In Problems 5–8, list the real zeros and turning points, and state the left and right behavior, of the polynomial function $P(x)$ that has the indicated graph.

5.

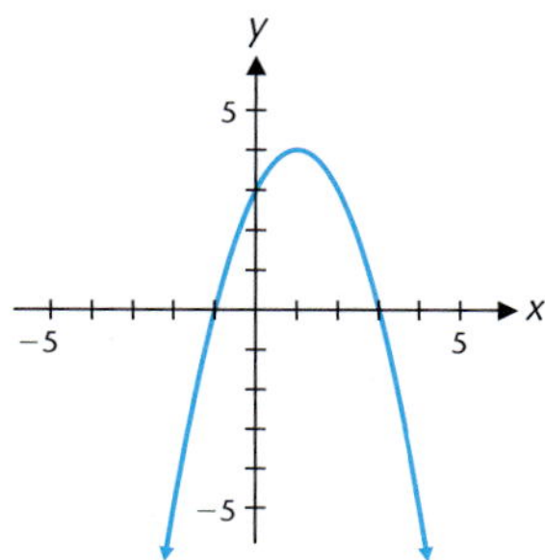

6.

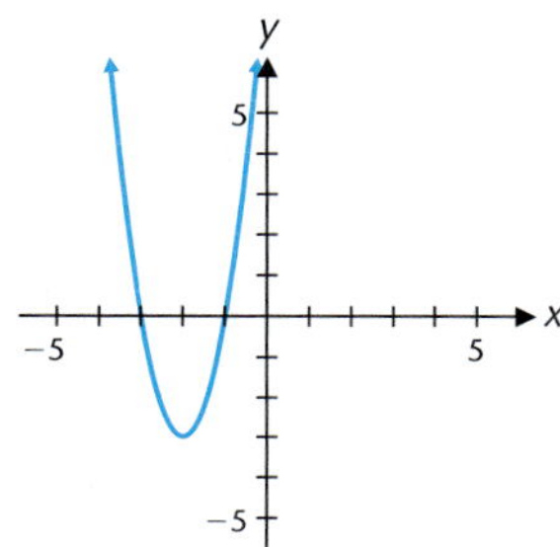

7.

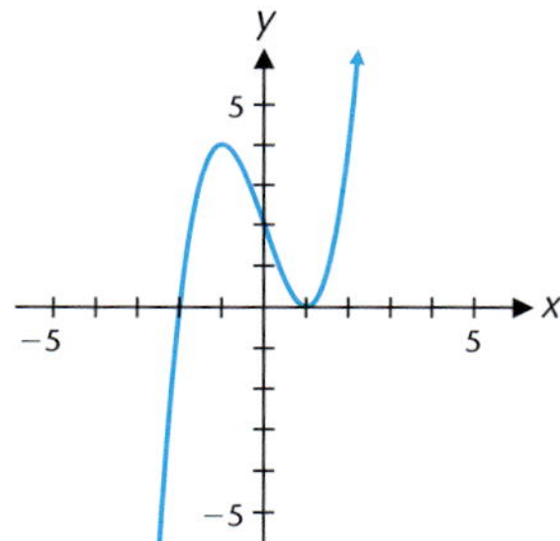

8.

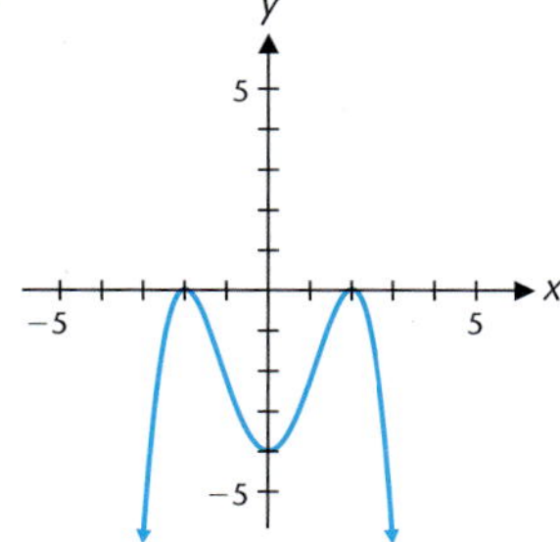

In Problems 9–12, explain why each graph is not the graph of a polynomial function.

9.

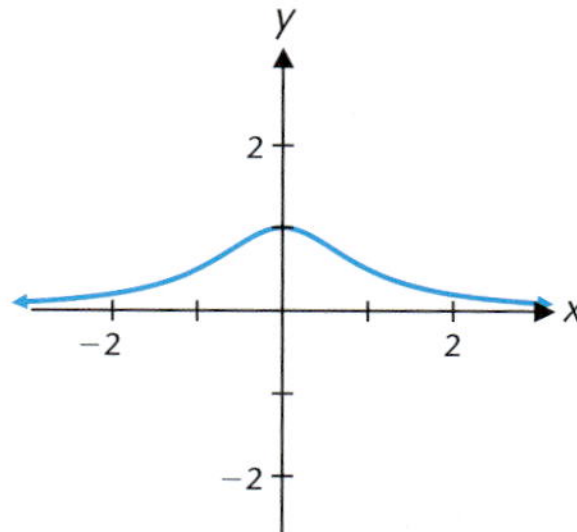

10.

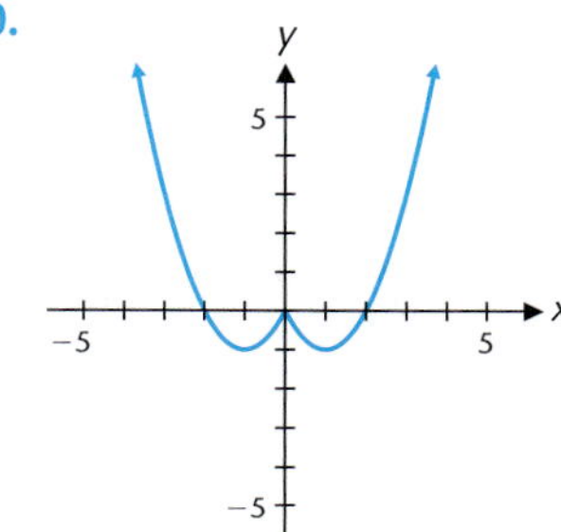

11.

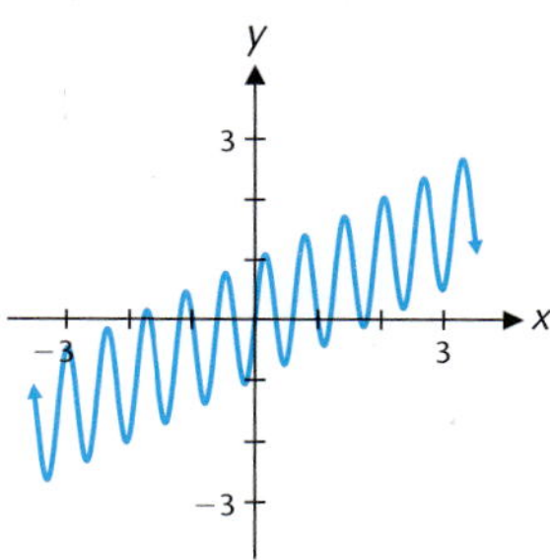

12.

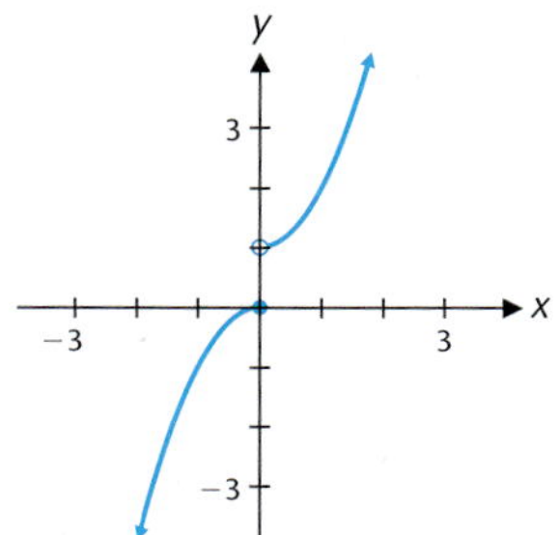

In Problems 13–16, list all zeros of each polynomial function, and specify those zeros that are x intercepts.

13. $P(x) = x(x^2 - 9)(x^2 + 4)$

14. $P(x) = (x^2 - 4)(x^4 - 1)$

15. $P(x) = (x + 5)(x^2 + 9)(x^2 + 16)$

16. $P(x) = (x^2 - 5x + 6)(x^2 - 5x + 7)$

In Problems 17–24, divide, using algebraic long division. Write the quotient, and indicate the remainder.

17. $(4m^2 - 1) \div (m - 1)$

18. $(y^2 - 9) \div (y + 3)$

19. $(6 - 6x + 8x^2) \div (x + 1)$

20. $(11x - 2 + 12x^2) \div (x + 2)$

21. $(x^3 - 1) \div (x - 1)$

22. $(a^3 + 27) \div (a + 3)$

23. $(3y - y^2 + 2y^3 - 1) \div (y + 2)$

24. $(3 + x^3 - x) \div (x - 3)$

In Problems 25–30, use synthetic division to write the quotient $P(x) \div (x - r)$ in the form $P(x)/(x - r) = Q(x) + R/(x - r)$, where R is a constant.

25. $(x^2 + 3x - 7) \div (x - 2)$

26. $(x^2 + 3x - 3) \div (x - 3)$

27. $(4x^2 + 10x - 9) \div (x + 3)$

28. $(2x^2 + 7x - 5) \div (x + 4)$

29. $(2x^3 - 3x + 1) \div (x - 2)$

30. $(x^3 + 2x^2 - 3x - 4) \div (x + 2)$

In Problems 31–34, determine whether the second polynomial is a factor of the first polynomial without dividing or using synthetic division. [Hint: Evaluate directly and use the factor theorem.]

31. $x^{18} - 1; x - 1$

32. $x^{18} - 1; x + 1$

33. $3x^3 - 7x^2 - 8x + 2; x + 1$

34. $3x^4 - 2x^3 + 5x - 6; x - 1$

B

Use synthetic division and the remainder theorem in Problems 35–40.

35. Find $P(-2)$, given $P(x) = 3x^2 - x - 10$.

36. Find $P(-3)$, given $P(x) = 4x^2 + 10x - 8$.

37. Find $P(2)$, given $P(x) = 2x^3 - 5x^2 + 7x - 7$.

38. Find $P(5)$, given $P(x) = 2x^3 - 12x^2 - x + 30$.

39. Find $P(-4)$, given $P(x) = x^4 - 10x^2 + 25x - 2$.

40. Find $P(-7)$, given $P(x) = x^4 + 5x^3 - 13x^2 - 30$.

In Problems 41–56, divide, using synthetic division. Write the quotient, and indicate the remainder. As coefficients get more involved, a calculator should prove helpful. Do not round off—all quantities are exact.

41. $(3x^4 - x - 4) \div (x + 1)$

42. $(5x^4 - 2x^2 - 3) \div (x - 1)$

43. $(x^5 + 1) \div (x + 1)$

44. $(x^4 - 16) \div (x - 2)$

45. $(3x^4 + 2x^3 - 4x - 1) \div (x + 3)$

46. $(x^4 - 3x^3 - 5x^2 + 6x - 3) \div (x - 4)$

47. $(2x^6 - 13x^5 + 75x^3 + 2x^2 - 50) \div (x - 5)$

48. $(4x^6 + 20x^5 - 24x^4 - 3x^2 - 13x + 30) \div (x + 6)$

49. $(4x^4 + 2x^3 - 6x^2 - 5x + 1) \div (x + \frac{1}{2})$

50. $(2x^3 - 5x^2 + 6x + 3) \div (x - \frac{1}{2})$

51. $(4x^3 + 4x^2 - 7x - 6) \div (x + \frac{3}{2})$

52. $(3x^3 - x^2 + x + 2) \div (x + \frac{2}{3})$

53. $(3x^4 - 2x^3 + 2x^2 - 3x + 1) \div (x - 0.4)$

54. $(4x^4 - 3x^3 + 5x^2 + 7x - 6) \div (x - 0.7)$

55. $(3x^5 + 2x^4 + 5x^3 - 7x - 3) \div (x + 0.8)$

56. $(7x^5 - x^4 + 3x^3 - 2x^2 - 5) \div (x + 0.9)$

For each polynomial function in Problems 57–62:
(A) State the left and right behavior, the maximum number of x intercepts, and the maximum number of local extrema.
(B) Approximate (to two decimal places) the x intercepts and the local extrema.

57. $P(x) = x^3 - 5x^2 + 2x + 6$

58. $P(x) = x^3 + 2x^2 - 5x - 3$

59. $P(x) = -x^3 + 4x^2 + x + 5$

60. $P(x) = -x^3 - 3x^2 + 4x - 4$

61. $P(x) = x^4 + x^3 - 5x^2 - 3x + 12$

62. $P(x) = -x^4 + 6x^2 - 3x - 16$

In Problems 63–66, either give an example of a polynomial with real coefficients that satisfies the given conditions or explain why such a polynomial cannot exist.

63. $P(x)$ is a third-degree polynomial with one x intercept.

64. $P(x)$ is a fourth-degree polynomial with no x intercepts.

65. $P(x)$ is a third-degree polynomial with no x intercepts.

66. $P(x)$ is a fourth-degree polynomial with no turning points.

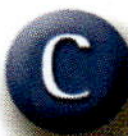

In Problems 67 and 68, divide, using synthetic division. Do not use a calculator.

67. $(x^3 - 3x^2 + x - 3) \div (x - i)$

68. $(x^3 - 2x^2 + x - 2) \div (x + i)$

69. Let $P(x) = x^2 + 2ix - 10$. Find

(A) $P(2 - i)$

(B) $P(5 - 5i)$

(C) $P(3 - i)$

(D) $P(-3 - i)$

70. Let $P(x) = x^2 - 4ix - 13$. Find

(A) $P(5 + 6i)$

(B) $P(1 + 2i)$

(C) $P(3 + 2i)$

(D) $P(-3 + 2i)$

In Problems 71–78, approximate (to two decimal places) the x intercepts and the local extrema.

71. $P(x) = 40 + 50x - 9x^2 - x^3$

72. $P(x) = 40 + 70x + 18x^2 + x^3$

73. $P(x) = 0.04x^3 - 10x + 5$

74. $P(x) = -0.01x^3 + 2.8x - 3$

75. $P(x) = 0.1x^4 + 0.3x^3 - 23x^2 - 23x + 90$

76. $P(x) = 0.1x^4 + 0.2x^3 - 19x^2 + 17x + 100$

77. $P(x) = x^4 - 24x^3 + 167x^2 - 275x + 131$

78. $P(x) = x^4 + 20x^3 + 118x^2 + 178x + 79$

79. **(A)** Divide $P(x) = a_2x^2 + a_1x + a_0$ by $x - r$, using both synthetic division and the long-division process, and compare the coefficients of the quotient and the remainder produced by each method.

(B) Expand the expression representing the remainder. What do you observe?

80. Repeat Problem 79 for

$$P(x) = a_3x^3 + a_2x^2 + a_1x + a_0$$

81. Polynomials also can be evaluated conveniently using a "nested factoring" scheme. For example, the polynomial $P(x) = 2x^4 - 3x^3 + 2x^2 - 5x + 7$ can be written in a nested factored form as follows:

$$\begin{aligned} P(x) &= 2x^4 - 3x^3 + 2x^2 - 5x + 7 \\ &= (2x - 3)x^3 + 2x^2 - 5x + 7 \\ &= [(2x - 3)x + 2]x^2 - 5x + 7 \\ &= \{[(2x - 3)x + 2]x - 5\}x + 7 \end{aligned}$$

Use the nested factored form to find $P(-2)$ and $P(1.7)$. [*Hint:* To evaluate $P(-2)$, store -2 in your calculator's memory and proceed from left to right recalling -2 as needed.]

82. Let $P(x) = 3x^4 + x^3 - 10x^2 + 5x - 2$. Find $P(-2)$ and $P(1.3)$ using the nested factoring scheme presented in Problem 81.

83. **(A)** What is the least number of turning points that a polynomial function of degree 4, with real coefficients, can have? The greatest number? Explain and give examples.

(B) What is the least number of x intercepts that a polynomial function of degree 4, with real coefficients, can have? The greatest number? Explain and give examples.

84. **(A)** What is the least number of turning points that a polynomial function of degree 3, with real coefficients, can have? The greatest number? Explain and give examples.

(B) What is the least number of x intercepts that a polynomial function of degree 3, with real coefficients, can have? The greatest number? Explain and give examples.

85. Is every polynomial of even degree an even function? Explain.

86. Is every polynomial of odd degree an odd function? Explain.

87. Let $f(x) = \begin{cases} x^2 & \text{if } |x| \geq 2 \\ 4 & \text{if } |x| < 2 \end{cases}$

(A) Graph f and observe that f is continuous.

(B) Find all numbers c such that $f(c)$ is a local extremum although $(c, f(c))$ is not a turning point.

88. Explain why the y coordinate of any turning point on the graph of a continuous function is a local extremum.

APPLICATIONS

89. Revenue. The price–demand equation for 8,000-BTU window air conditioners is given by

$$p = 0.0004x^2 - x + 569 \qquad 0 \le x \le 800$$

where x is the number of air conditioners that can be sold at a price of p dollars each.

(A) Find the revenue function.

(B) Find the number of air conditioners that must be sold to maximize the revenue, the corresponding price to the nearest dollar, and the maximum revenue to the nearest dollar.

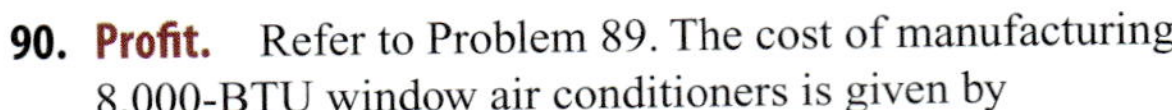

90. Profit. Refer to Problem 89. The cost of manufacturing 8,000-BTU window air conditioners is given by

$$C(x) = 10{,}000 + 90x$$

where $C(x)$ is the total cost in dollars of producing x air conditioners.

(A) Find the profit function.

(B) Find the number of air conditioners that must be sold to maximize the profit, the corresponding price to the nearest dollar, and the maximum profit to the nearest dollar.

91. Construction. A rectangular container measuring 1 foot by 2 feet by 4 feet is covered with a layer of lead shielding of uniform thickness (see the figure).

(A) Find the volume of lead shielding V as a function of the thickness x (in feet) of the shielding.

(B) Find the thickness of the lead shielding to three decimal places if the volume of the shielding is 3 cubic feet.

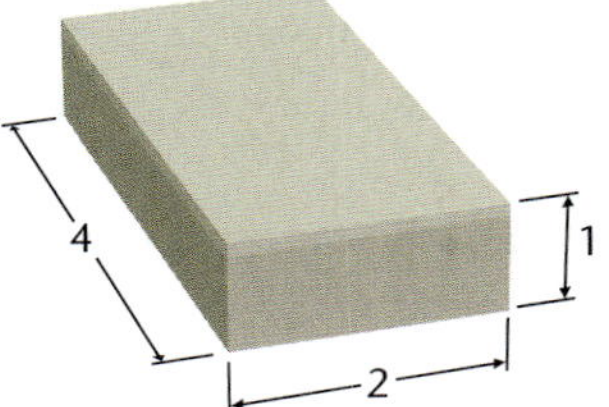

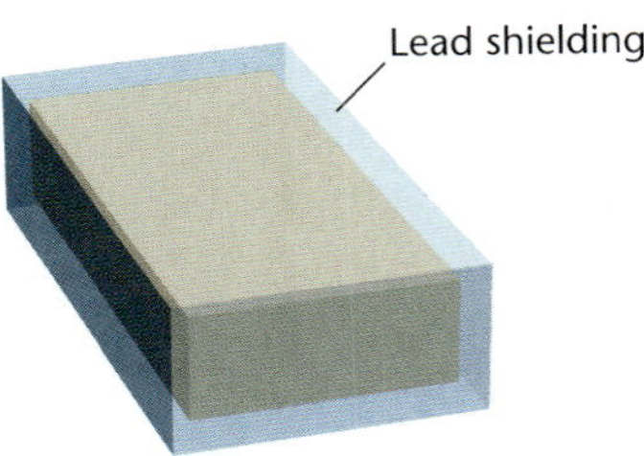

92. Manufacturing. A rectangular storage container measuring 2 feet by 2 feet by 3 feet is coated with a protective coating of plastic of uniform thickness.

(A) Find the volume of plastic V as a function of the thickness x (in feet) of the coating.

(B) Find the thickness of the plastic coating to four decimal places if the volume of the shielding is 0.1 cubic feet.

MODELING AND DATA ANALYSIS 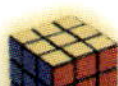

93. Health Care. Table 3 shows the total national expenditures (in billion dollars) and the per capita expenditures (in dollars) for selected years since 1960.

TABLE 3 National Health Expenditures

Year	Total Expenditures (Billion $)	Per Capita Expenditures ($)
1960	26.7	143
1970	73.1	348
1980	245.8	1,067
1990	695.6	2,737
1995	987.0	3,686
1997	1,093.9	4,011
1998	1,210.7	4,358

Source: U.S. Census Bureau.

(A) Let x represent the number of years since 1960 and find a cubic regression polynomial for the total national expenditures.

(B) Use the polynomial model from part A to estimate the total national expenditures (to the nearest tenth of a billion) for 2010.

94. Health Care. Refer to Table 3.

(A) Let x represent the number of years since 1960 and find a cubic regression polynomial for the per capita expenditures.

(B) Use the polynomial model from part A to estimate the per capita expenditures (to the nearest dollar) for 2010.

95. Marriage. Table 4 shows the marriage and divorce rates per 1,000 population for selected years since 1950.

TABLE 4 Marriages and Divorces (per 1,000 Population)

Year	Marriages	Divorces
1950	11.1	2.6
1960	8.5	2.2
1970	10.6	3.5
1980	10.6	5.2
1990	9.8	4.7
1999	8.6	4.1

Source: U.S. Census Bureau.

(A) Let x represent the number of years since 1950 and find a cubic regression polynomial for the marriage rate.

(B) Use the polynomial model from part A to estimate the marriage rate (to one decimal place) for 2008.

96. Divorce. Refer to Table 4.

(A) Let x represent the number of years since 1950 and find a cubic regression polynomial for the divorce rate.

(B) Use the polynomial model from part A to estimate the divorce rate (to one decimal place) for 2008.

SECTION 3.2 Real Zeros and Polynomial Inequalities

Upper and Lower Bounds for Real Zeros • Location Theorem and Bisection Method • Approximating Real Zeros at Turning Points • Polynomial Inequalities • Application

The real zeros of a polynomial $P(x)$ with real coefficients are just the x intercepts of the graph of $P(x)$. So an obvious strategy for finding the real zeros consists of two steps:

1. Graph $P(x)$ on a graphing utility.
2. Use the zero command to approximate each x intercept.

In this section we develop two important tools for carrying out this strategy: the *upper and lower bound theorem*, which determines appropriate window variables for step 1, and the *location theorem*, which leads to a simple approximation technique called the *bisection method* that underpins step 2. We also investigate some potential difficulties when the strategy is applied to polynomials that have a zero at a turning point, and we apply the strategy to solve polynomial inequalities.

In this section we restrict our attention to the real zeros of polynomials with real coefficients.

Upper and Lower Bounds for Real Zeros

A polynomial of degree n has at most n zeros. So if a calculator's viewing window displays three x intercepts of a cubic polynomial, then the zero command can be used to find all zeros of the polynomial.

EXAMPLE 1 Approximating Real Zeros

Approximate the zeros of $P(x) = x^3 - 6x^2 + 9x - 3$ to three decimal places.

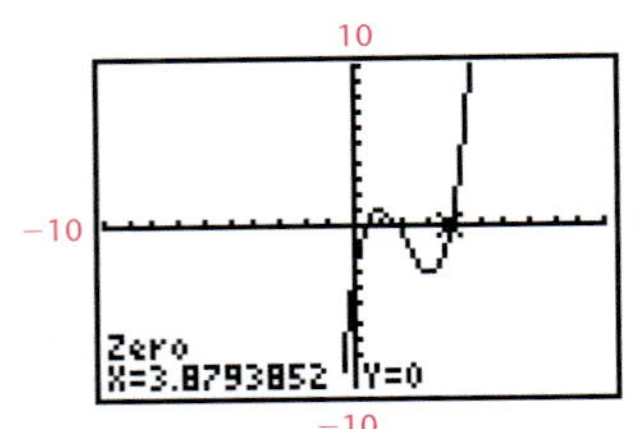

FIGURE 1 A zero of $P(x) = x^3 - 6x^2 + 9x - 3$.

SOLUTION

A graph of $P(x)$ in the standard viewing window shows three x intercepts (Fig. 1). We find each of them by applying the zero command: rounded to three decimal places they are 0.468, 1.653, and 3.879. Because a polynomial of degree 3 can have at most three zeros, we have found all of the zeros of $P(x)$.

MATCHED 1 PROBLEM

Approximate the zeros of $P(x) = 21 + 10x - 3x^2 - x^3$ to three decimal places.

A polynomial of degree 3 has at most three real zeros, but may have exactly one or exactly two (Fig. 2). If we cannot find a viewing window that displays more than one zero for a particular cubic polynomial, how long must we search before we can decide whether the polynomial has one, two, or three real zeros? Such a dilemma can be resolved by the *upper and lower bound theorem*. This theorem indicates how to find two numbers, a **lower bound** that is less than or equal to all real zeros of the polynomial, and an **upper bound** that is greater than or equal to all real zeros of the polynomial. Thus, all real zeros are guaranteed to lie between the lower bound and the upper bound.

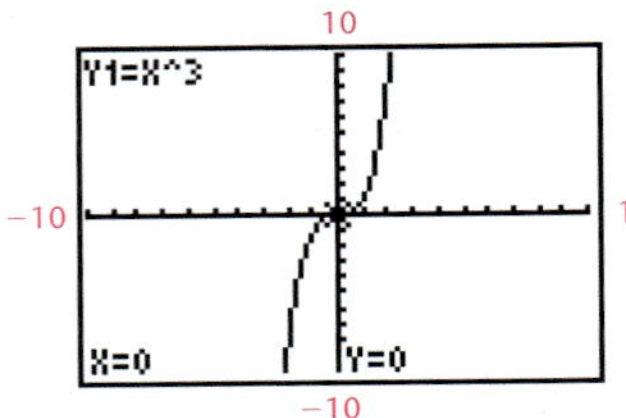

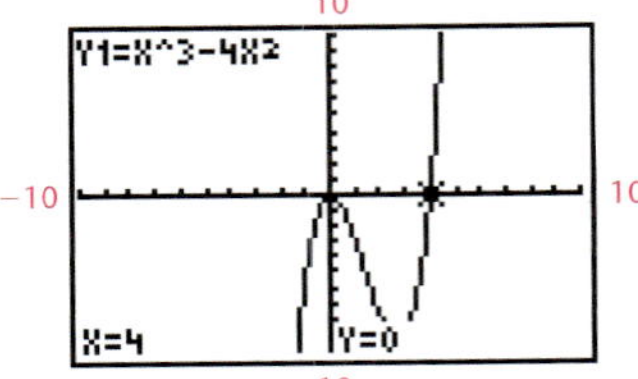

FIGURE 2 Cubic polynomials having one or two zeros.

THEOREM 1

Upper and Lower Bound Theorem

Let $P(x)$ be a polynomial of degree $n > 0$ with real coefficients, $a_n > 0$:

1. Upper bound: A number $r > 0$ is an upper bound for the real zeros of $P(x)$ if, when $P(x)$ is divided by $x - r$ by synthetic division, all numbers in the quotient row, including the remainder, are nonnegative.
2. Lower bound: A number $r < 0$ is a lower bound for the real zeros of $P(x)$ if, when $P(x)$ is divided by $x - r$ by synthetic division, all numbers in the quotient row, including the remainder, alternate in sign.

[*Note:* In the lower-bound test, if 0 appears in one or more places in the quotient row, including the remainder, the sign in front of it can be considered either positive or negative, but not both. For example, the numbers 1, 0, 1 can be considered to alternate in sign, whereas 1, 0, −1 cannot.]

We sketch a proof of part 1 of Theorem 1. The proof of part 2 is similar, only a little more difficult.

PROOF If all the numbers in the quotient row of the synthetic division are nonnegative after dividing $P(x)$ by $x - r$, then

$$P(x) = (x - r)Q(x) + R$$

where the coefficients of $Q(x)$ are nonnegative and R is nonnegative. If $x > r > 0$, then $x - r > 0$ and $Q(x) > 0$; hence,

$$P(x) = (x - r)Q(x) + R > 0$$

Thus, $P(x)$ cannot be 0 for any x greater than r; and r is an upper bound for the real zeros of $P(x)$.

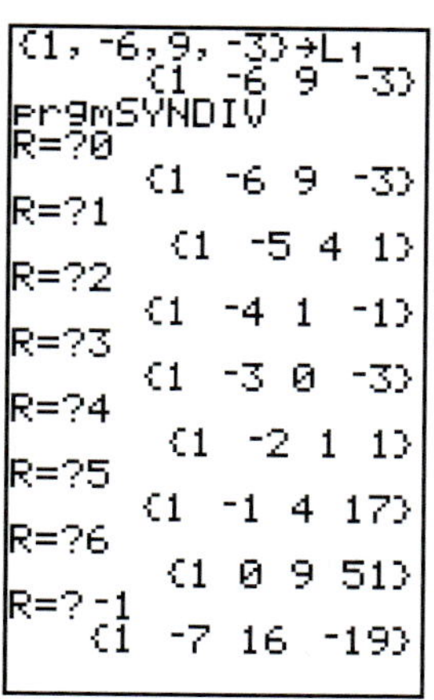

FIGURE 3 Synthetic division on a graphing utility.

Theorem 1 requires performing synthetic division repeatedly until the desired patterns occur in the quotient row. This is a simple, but tedious, operation to carry out by hand. SYNDIV* is a program that makes this process routine. Figure 3 shows the results of dividing the polynomial $P(x) = x^3 - 6x^2 + 9x - 3$ of Example 1 by $x - r$ for integer values of r from -1 through 6 using SYNDIV. (When $r = 2$, for example, the quotient is $x^2 - 4x + 1$ and the remainder is -1.)

From Figure 3 we see that the positive number 6 is an upper bound for the real zeros of $P(x)$ because all numbers in the quotient row, including the remainder, are nonnegative. Furthermore, the negative number -1 is a lower bound for the real zeros of $P(x)$ because all numbers in the quotient row, including the remainder, alternate in sign. We conclude that all zeros of $P(x)$ lie between -1 and 6.

EXAMPLE 2 Bounding Real Zeros

Let $P(x) = x^4 - 2x^3 - 10x^2 + 40x - 90$. Find the smallest positive integer and the largest negative integer that, by Theorem 1, are upper and lower bounds, respectively, for the real zeros of $P(x)$.

SOLUTION

We can use SYNDIV or hand calculations to perform synthetic division for $r = 1, 2, 3, \ldots$ until the quotient row turns nonnegative; then repeat this process for $r = -1, -2, -3, \ldots$ until the quotient row alternates in sign. We organize these results in the *synthetic division table* shown below. In a **synthetic division table** we dispense with writing the product of r with each coefficient in the quotient and simply list the results in the table.

		1	−2	−10	40	−90	
	1	1	−1	−11	29	−61	
	2	1	0	−10	20	−50	
	3	1	1	−7	19	−33	
	4	1	2	−2	32	38	
UB	5	1	3	5	65	235 ←	This quotient row is nonnegative; hence, 5 is an upper bound (UB).
	−1	1	−3	−7	47	−137	
	−2	1	−4	−2	44	−178	
	−3	1	−5	5	25	−165	
	−4	1	−6	14	−16	−26	
LB	−5	1	−7	25	−85	335 ←	This quotient row alternates in sign; hence, −5 is a lower bound (LB).

*Programs for TI-83 and TI-86 graphing calculators can be found at the website for this book (see Preface).

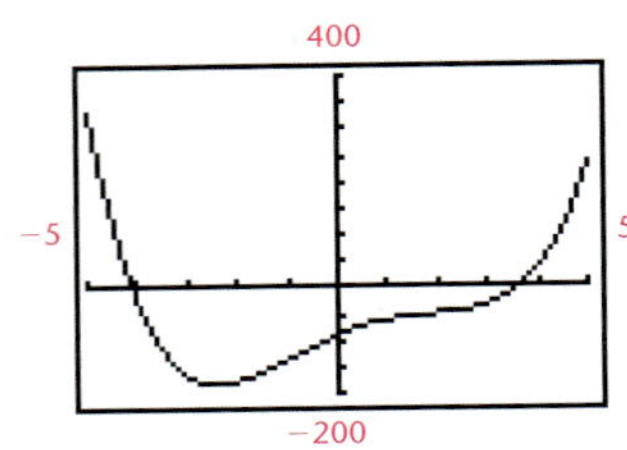

FIGURE 4 $P(x) = x^4 - 2x^3 - 10x^2 + 40x - 90$.

The graph of $P(x) = x^4 - 2x^3 - 10x^2 + 40x - 90$ for $-5 \le x \le 5$ is shown in Figure 4. Theorem 1 implies that all the real zeros of $P(x)$ are between -5 and 5. We can be certain that the graph does not change direction and cross the x axis somewhere outside the viewing window in Figure 4.

MATCHED 2 PROBLEM

Let $P(x) = x^4 - 5x^3 - x^2 + 40x - 70$. Find the smallest positive integer and the largest negative integer that, by Theorem 1, are upper and lower bounds, respectively, for the real zeros of $P(x)$.

EXAMPLE 3 Approximating Real Zeros

Let $P(x) = x^3 - 30x^2 + 275x - 720$.

(A) Find the smallest positive integer multiple of 10 and the largest negative integer multiple of 10 that, by Theorem 1, are upper and lower bounds, respectively, for the real zeros of $P(x)$.

(B) Approximate the real zeros of $P(x)$ to two decimal places.

SOLUTIONS

(A) We construct a synthetic division table to search for bounds for the zeros of $P(x)$. The size of the coefficients in $P(x)$ indicates that we can speed up this search by choosing larger increments between test values.

		1	−30	275	−720
	10	1	−20	75	30
	20	1	−10	75	780
UB	30	1	0	275	7,530
LB	−10	1	−40	675	−7,470

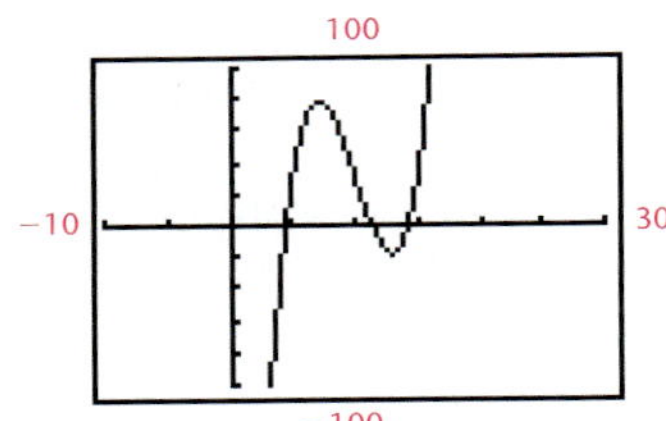

FIGURE 5 $P(x) = x^3 - 30x^2 + 275x - 720$.

Thus, all real zeros of $P(x) = x^3 - 30x^2 + 275x - 720$ must lie between -10 and 30.

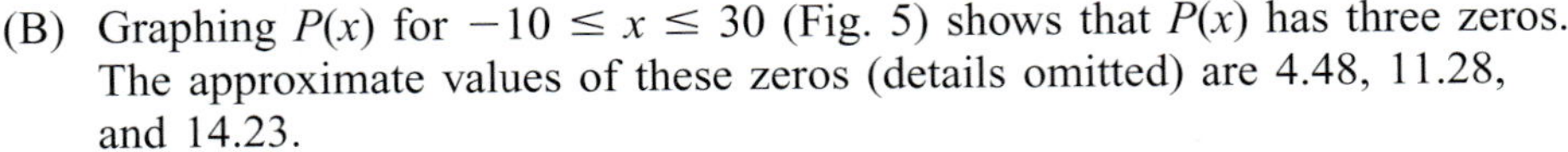

(B) Graphing $P(x)$ for $-10 \le x \le 30$ (Fig. 5) shows that $P(x)$ has three zeros. The approximate values of these zeros (details omitted) are 4.48, 11.28, and 14.23.

MATCHED 3 PROBLEM

Let $P(x) = x^3 - 25x^2 + 170x - 170$.

(A) Find the smallest positive integer multiple of 10 and the largest negative integer multiple of 10 that, by Theorem 1, are upper and lower bounds, respectively, for the real zeros of $P(x)$.

(B) Approximate the real zeros of $P(x)$ to two decimal places.

REMARK One of the most frequently asked questions concerning graphing utilities is how to determine the correct viewing window. The upper and lower bound theorem provides an answer to this question for polynomial functions. As

Example 3 illustrates, the upper and lower bound theorem and the zero approximation routine on a graphing utility are two important mathematical tools that work very well together.

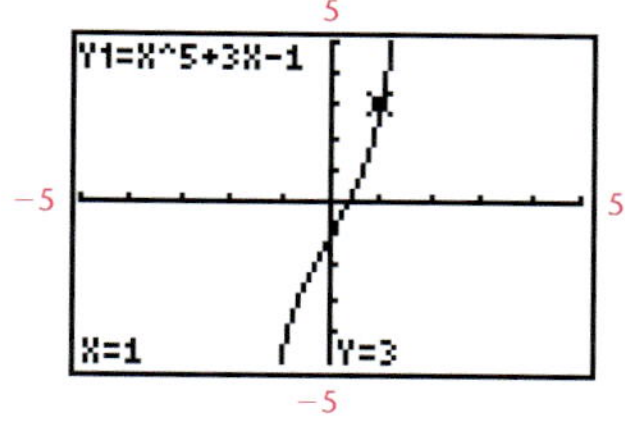

FIGURE 6 $P(x) = x^5 + 3x - 1$.

Location Theorem and Bisection Method

The graph of every polynomial function is continuous. Because the polynomial function $P(x) = x^5 + 3x - 1$ is negative when $x = 0$ $[P(0) = -1]$ and positive when $x = 1$ $[P(1) = 3]$, the graph of $P(x)$ must cross the x axis at least once between $x = 0$ and $x = 1$ (Fig. 6). This observation is the basis for Theorem 2 and leads to a simple method for approximating zeros.

THEOREM 2
Location Theorem*

Suppose that a function f is continuous on an interval I that contains numbers a and b. If $f(a)$ and $f(b)$ have opposite signs, then the graph of f has at least one x intercept between a and b.

The conclusion of Theorem 2 says that at least one zero of the function is "located" between a and b. There may be more than one zero between a and b: if $g(x) = x^3 + x^2 - 2x - 1$, then $g(-2)$ and $g(2)$ have opposite signs and there are three zeros between $x = -2$ and $x = 2$ [Fig. 7(a)]. The converse of Theorem 2 is false: $h(x) = x^2$ has an x intercept at $x = 0$ but does not change sign [Fig. 7(b)].

FIGURE 7 Polynomials may or may not change sign at a zero.

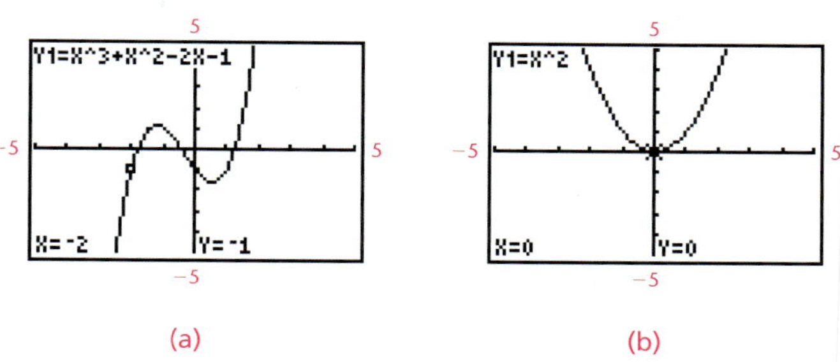

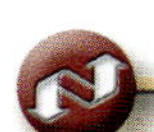

EXPLORE/DISCUSS 1

When synthetic division is used to divide a polynomial $P(x)$ by $x - 3$ the remainder is -33. When the same polynomial is divided by $x - 4$ the remainder is 38. Must $P(x)$ have a zero between 3 and 4? Explain.

*The location theorem is a formulation of the important intermediate value theorem of calculus.

Explore/Discuss 2 provides an introduction to the repeated systematic application of the location theorem (Theorem 2) called the *bisection method*. This method forms the basis for the zero approximation routines in many graphing utilities.

EXPLORE/DISCUSS 2

Let $P(x) = x^5 + 3x - 1$. Because $P(0)$ is negative and $P(1)$ is positive, the location theorem implies that $P(x)$ must have at least one zero in the interval $(0, 1)$.

(A) Is $P(0.5)$ positive or negative? Does the location theorem guarantee a zero of $P(x)$ in the interval $(0, 0.5)$ or in $(0.5, 1)$?

(B) Let m be the midpoint of the interval from part A that contains a zero of $P(x)$. Is $P(m)$ positive or negative? What does this tell you about the location of the zero?

(C) Explain how this process could be used repeatedly to approximate a zero to any desired accuracy.

(D) Check your answers to parts A and B by using the zero routine on a graphing utility.

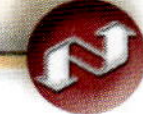

The **bisection method** is a systematic application of the procedure suggested in Explore/Discuss 2: Let $P(x)$ be a polynomial with real coefficients. If $P(x)$ has opposite signs at the endpoints of an interval (a, b), then by the location theorem $P(x)$ has a zero in (a, b). Bisect this interval (that is, find the midpoint $m = \frac{a + b}{2}$), check the sign of $P(m)$, and select the interval (a, m) or (m, b) that has opposite signs at the endpoints. We repeat this bisection procedure (producing a set of intervals, each contained in and half the length of the previous interval, and each containing the zero) until the desired accuracy is obtained. If at any point in the process $P(m) = 0$, we stop, because a real zero m has been found. An example will help clarify the process.

EXAMPLE 4 The Bisection Method

The polynomial $P(x) = x^4 - 2x^3 - 10x^2 + 40x - 90$ of Example 2 has a zero between 3 and 4. Use the bisection method to approximate it to one-decimal-place accuracy.

SOLUTION

We organize the results of our calculations in a table. Because the sign of $P(x)$ changes at the endpoints of the interval (3.5625, 3.625), we conclude that a real zero lies in this interval and is given by $r = 3.6$ to one-decimal-place accuracy (each endpoint rounds to 3.6).

TABLE 1 Bisection Approximation

Sign Change Interval (a, b)	Midpoint m	Sign of P: $P(a)$	$P(m)$	$P(b)$
(3, 4)	3.5	−	−	+
(3.5, 4)	3.75	−	+	+
(3.5, 3.75)	3.625	−	+	+
(3.5, 3.625)	3.5625	−	−	+
(3.5625, 3.625)	We stop here	−		+

Figure 8 illustrates the nested intervals produced by the bisection method in Table 1. Match each step in Table 1 with an interval in Figure 8. Note how each interval that contains a zero gets smaller and smaller and is contained in the preceding interval that contained the zero.

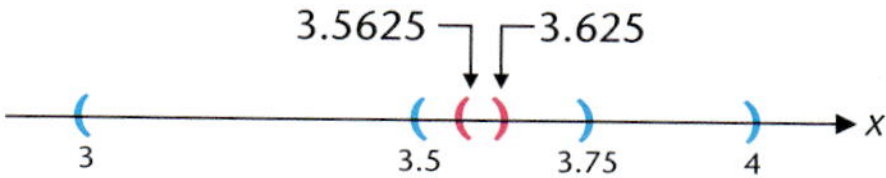

FIGURE 8 Nested intervals produced by the bisection method in Table 1.

If we had wanted two-decimal-place accuracy, we would have continued the process in Table 1 until the endpoints of a sign change interval rounded to the same two-decimal-place number.

MATCHED PROBLEM 4

The polynomial $P(x) = x^4 - 2x^3 - 10x^2 + 40x - 90$ of Example 2 has a zero between -5 and -4. Use the bisection method to approximate it to one-decimal-place accuracy.

Approximating Real Zeros at Turning Points

The bisection method for approximating zeros fails if a polynomial has a turning point at a zero, because the polynomial does not change sign at such a zero. Most graphing utilities use methods that are more sophisticated than the bisection method. Nevertheless, it is not unusual to get an error message when using the zero command to approximate a zero that is also a turning point. In this case, we can use the maximum or minimum command, as appropriate, to approximate the turning point, and thus the zero.

EXAMPLE 5 Approximating Zeros at Turning Points

Let $P(x) = x^5 + 6x^4 + 4x^3 - 24x^2 - 16x + 32$. Find the smallest positive integer and the largest negative integer that, by Theorem 1, are upper and lower bounds, respectively, for the real zeros of $P(x)$. Approximate the zeros to two decimal places, using maximum or minimum commands to approximate any zeros at turning points.

SOLUTION

The pertinent rows of a synthetic division table show that 2 is the upper bound and -6 is the lower bound:

	1	6	4	−24	−16	32
1	1	7	11	−13	−29	3
2	1	8	20	16	16	64
−5	1	1	−1	−19	79	−363
−6	1	0	4	−48	272	−1600

Examining the graph of $P(x)$ we find three zeros: the zero -3.24, found using the maximum command [Fig. 9(a)]; the zero -2, found using the zero command [Fig. 9(b)]; and the zero 1.24, found using the minimum command [Fig. 9(c)].

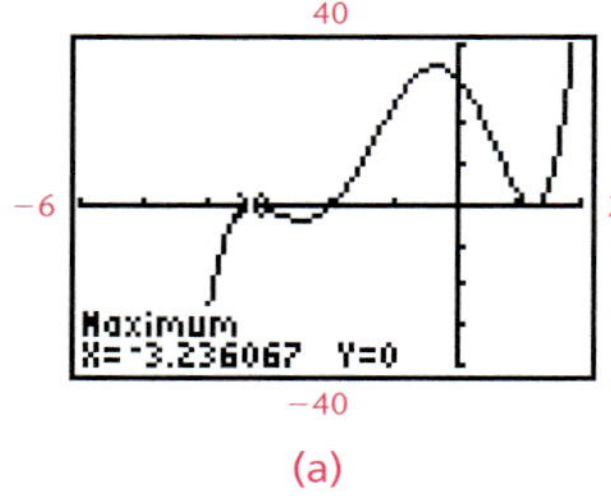

(a)

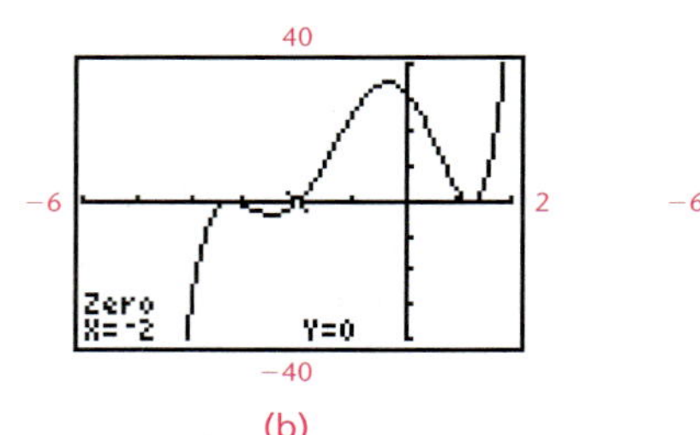

(b)

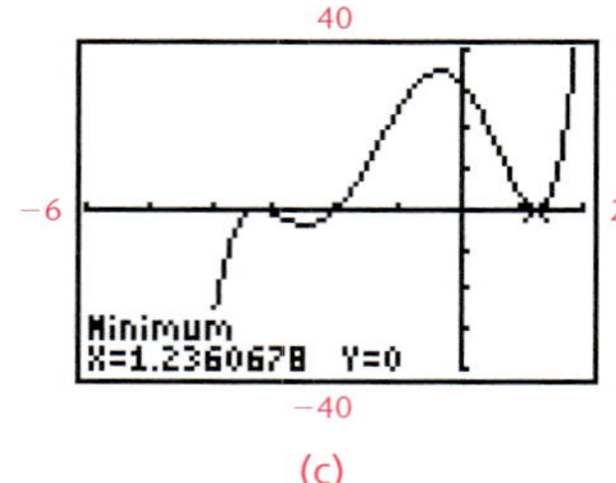

(c)

FIGURE 9 Zeros of $P(x) = x^5 + 6x^4 + 4x^3 - 24x^2 - 16x + 32$.

MATCHED 5 PROBLEM

Let $P(x) = x^5 - 6x^4 + 40x^2 - 12x - 72$. Find the smallest positive integer and the largest negative integer that, by Theorem 1, are upper and lower bounds, respectively, for the real zeros of $P(x)$. Approximate the zeros to two decimal places, using maximum or minimum commands to approximate any zeros at turning points.

Polynomial Inequalities

We can apply the techniques we have introduced for finding real zeros to solve polynomial inequalities. Consider, for example, the inequality

$$x^3 - 2x^2 - 5x + 6 > 0$$

The real zeros of $P(x) = x^3 - 2x^2 - 5x + 6$ are easily found to be -2, 1, and 3. They partition the x axis into four intervals

$$(-\infty, -2), (-2, 1), (1, 3), \text{ and } (3, \infty)$$

On any one of these intervals, the graph of P is either above the x axis or below the x axis, because, by the location theorem, **a continuous function can change sign only at a zero.**

One way to decide whether the graph of P is above or below the x axis on a given interval, say $(-2, 1)$, is to choose a "test number" that belongs to the interval, 0, for example, and evaluate P at the test number. Because $P(0) = 6 > 0$, the graph of P is above the x axis throughout the interval $(-2, 1)$. A second way to decide whether the graph of P is above or below the x axis on $(-2, 1)$ is to simply inspect the graph of P. Each technique has its advantages, and both are illustrated in the solutions to Example 6.

EXAMPLE 6 Solving Polynomial Inequalities

Solve the inequality $x^3 - 2x^2 - 5x + 6 > 0$.

SOLUTION

Algebraic Solution

Let $P(x) = x^3 - 2x^2 - 5x + 6$. Then

$$P(1) = 1^3 - 2(1^2) - 5 + 6 = 0$$

so 1 is a zero of P and $x - 1$ is a factor. Dividing $P(x)$ by $x - 1$ (details omitted) gives the quotient $x^2 - x - 6$. Therefore

$$P(x) = (x - 1)(x^2 - x - 6) = (x - 1)(x + 2)(x - 3)$$

The zeros of P are thus -2, 1, and 3. They partition the x axis into the four intervals shown in the table. A test number is chosen from each interval as indicated to determine whether $P(x)$ is positive (above the x axis) or negative (below the x axis) on that interval.

Interval	$(-\infty, -2)$	$(-2, 1)$	$(1, 3)$	$(3, \infty)$
Test number x	-3	0	2	4
$P(x)$	-24	6	-4	18
Sign of P	$-$	$+$	$-$	$+$

We conclude that the solution set of the inequality is

$$(-2, 1) \cup (3, \infty)$$

Graphical Solution

We graph P [Fig. 10(a)] and find that -2, 1, and 3 are the zeros of P. They partition the x axis into four intervals

$$(-\infty, -2), (-2, 1), (1, 3), \text{ and } (3, \infty)$$

By inspecting the graph of P we see that P is above the x axis on the intervals $(-2, 1)$ and $(3, \infty)$. Thus, the solution set of the inequality is

$$(-2, 1) \cup (3, \infty)$$

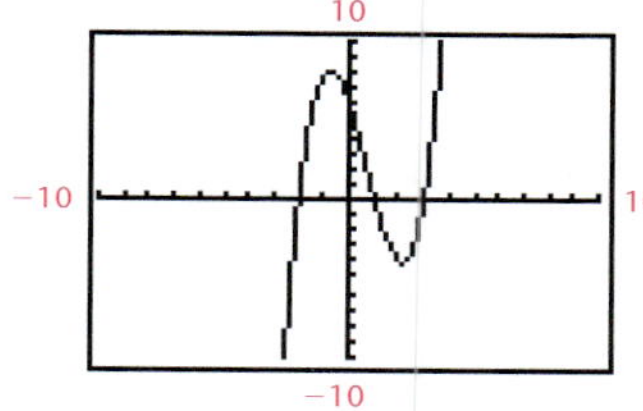

FIGURE 10 (a) $P(x) = x^3 - 2x^2 - 5x + 6$

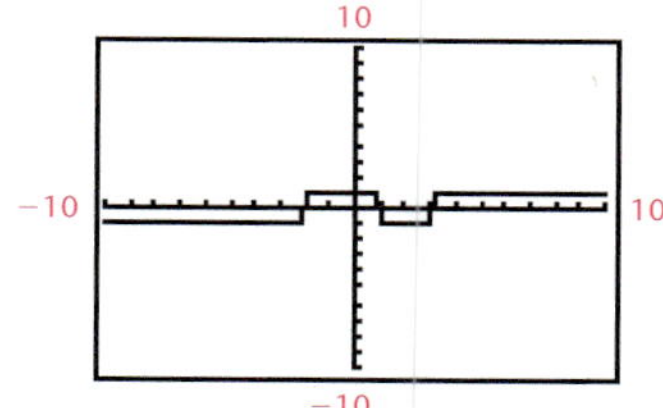

(b) $f(x) = \dfrac{P(x)}{|P(x)|}$

An alternative to inspecting the graph of P is to inspect the graph of

$$f(x) = \frac{P(x)}{|P(x)|}$$

The function $f(x)$ has the value 1 if $P(x)$ is positive, because then the absolute value of $P(x)$ is equal to $P(x)$. Similarly, $f(x)$ has the value -1 if $P(x)$ is negative. This technique makes it easy to identify the solution set of the original inequality [Fig. 10(b)] and often eliminates difficulties in choosing appropriate window variables.

MATCHED 6 PROBLEM

Solve the inequality $x^3 + x^2 - x - 1 < 0$.

EXPLORE/DISCUSS 3

Explain how Figure 10 may be used to write down the solution set for each inequality:

(A) $x^3 - 2x^2 - 5x + 6 < 0$

(B) $x^3 - 2x^2 - 5x + 6 \geq 0$

(C) $x^3 - 2x^2 - 5x + 6 \leq 0$

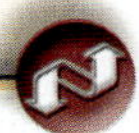

EXAMPLE 7 Solving Polynomial Inequalities

Solve $3x^2 + 12x - 4 \geq 2x^3 - 5x^2 + 7$ to three decimal places.

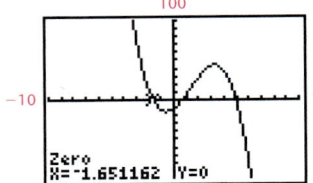

FIGURE 11

SOLUTION

Subtracting the right-hand side gives the equivalent inequality

$$P(x) = -2x^3 + 8x^2 + 12x - 11 \geq 0$$

The zeros of $P(x)$, to three decimal places, are -1.651, 0.669, and 4.983 (Fig. 11).

The graph of P is above the x axis on the intervals $(-\infty, -1.651)$ and $(0.669, 4.983)$. The solution set of the inequality is thus

$$(-\infty, -1.651] \cup [0.669, 4.983]$$

The square brackets indicate that the endpoints of the intervals—the zeros of the polynomial—also satisfy the inequality.

MATCHED 7 PROBLEM

Solve to three decimal places $5x^3 - 13x < 4x^2 + 10x - 5$.

Application

EXAMPLE 8 Construction

An oil tank is in the shape of a right circular cylinder with a hemisphere at each end (Fig. 12). The cylinder is 55 inches long, and the volume of the tank is $11{,}000\pi$ cubic inches (approximately 20 cubic feet). Let x denote the common radius of the hemispheres and the cylinder.

FIGURE 12

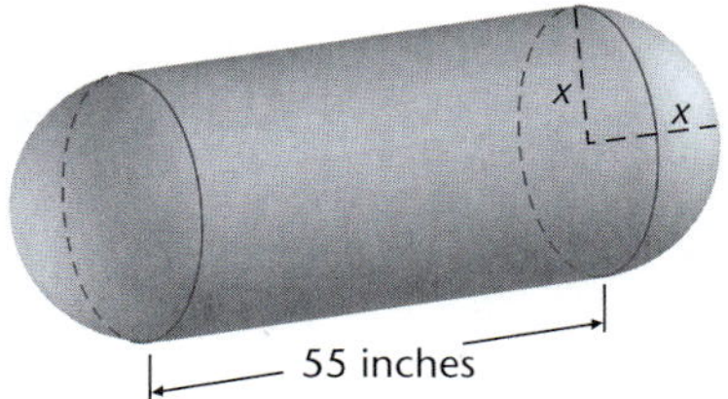

(A) Find a polynomial equation that x must satisfy.

(B) Approximate x to one decimal place.

SOLUTIONS

(A) If x is the common radius of the hemispheres and the cylinder in inches, then

$$\begin{pmatrix}\text{Volume}\\ \text{of}\\ \text{tank}\end{pmatrix} = \begin{pmatrix}\text{Volume}\\ \text{of two}\\ \text{hemispheres}\end{pmatrix} + \begin{pmatrix}\text{Volume}\\ \text{of}\\ \text{cylinder}\end{pmatrix}$$

$$11{,}000\pi = \tfrac{4}{3}\pi x^3 + 55\pi x^2 \qquad \text{Multiply by } 3/\pi.$$

$$33{,}000 = 4x^3 + 165x^2$$

$$0 = 4x^3 + 165x^2 - 33{,}000$$

Thus, x must be a positive zero of

$$P(x) = 4x^3 + 165x^2 - 33{,}000$$

(B) Because the coefficients of $P(x)$ are large, we use larger increments in the synthetic division table:

		4	165	0	−33,000
	10	4	205	2,050	−12,500
UB	20	4	245	4,900	65,000

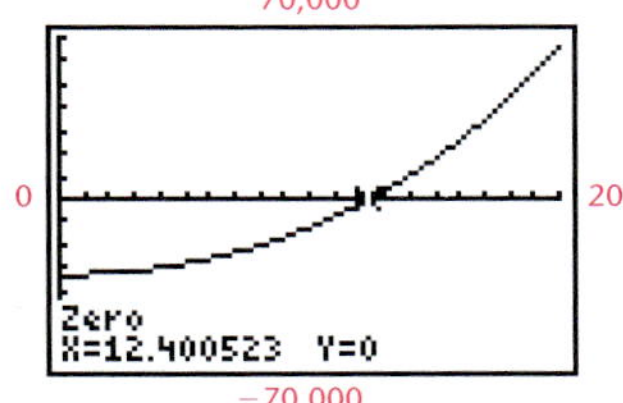

FIGURE 13
$P(x) = 4x^3 + 165x^2 - 33{,}000.$

Graphing $y = P(x)$ for $0 \le x \le 20$ (Fig. 13), we see that $x = 12.4$ inches (to one decimal place).

MATCHED 8 PROBLEM

Repeat Example 8 if the volume of the tank is $44{,}000\pi$ cubic inches.

ANSWERS MATCHED PROBLEMS

1. $-4.190, -1.721, 2.912$
2. Lower bound: -3; upper bound: 6
3. (A) Lower bound: -10; upper bound: 30 (B) Real zeros: 1.20, 11.46, 12.34
4. $x = -4.1$
5. Lower bound: -2; upper bound: 6; -1.65, 2, 3.65
6. $(-\infty, -1) \cup (-1, 1)$
7. $(-\infty, -1.899) \cup (0.212, 2.488)$
8. (A) $P(x) = 4x^3 + 165x^2 - 132{,}000 = 0$ (B) 22.7 inches

EXERCISE 3.2

A

In Problems 1–4, approximate the real zeros of each polynomial to three decimal places.

1. $P(x) = x^2 + 5x - 2$

2. $P(x) = 3x^2 - 7x + 1$

3. $P(x) = 2x^3 - 5x + 2$

4. $P(x) = x^3 - 4x^2 - 8x + 3$

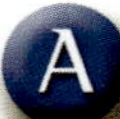

In Problems 5–8, use the graph of P(x) to write the solution set for each inequality.

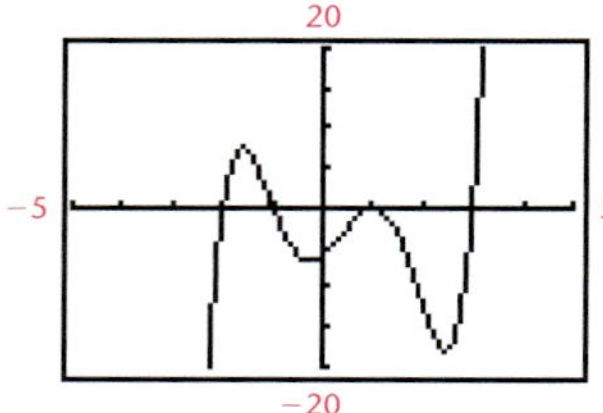

5. $P(x) \geq 0$

6. $P(x) < 0$

7. $P(x) > 0$

8. $P(x) \leq 0$

In Problems 9–12, solve each polynomial inequality to three decimal places (note the connection with Problems 1–4).

9. $x^2 + 5x - 2 > 0$

10. $3x^2 - 7x + 1 \geq 0$

11. $2x^3 - 5x + 2 \leq 0$

12. $x^3 - 4x^2 - 8x + 3 < 0$

Find the smallest positive integer and largest negative integer that, by Theorem 1, are upper and lower bounds, respectively, for the real zeros of each of the polynomials given in Problems 13–18.

13. $P(x) = x^3 - 3x + 1$

14. $P(x) = x^3 - 4x^2 + 4$

15. $P(x) = x^4 - 3x^3 + 4x^2 + 2x - 9$

16. $P(x) = x^4 - 4x^3 + 6x^2 - 4x - 7$

17. $P(x) = x^5 - 3x^3 + 3x^2 + 2x - 2$

18. $P(x) = x^5 - 3x^4 + 3x^2 + 2x - 1$

B

In Problems 19–26,
(A) Use the location theorem to explain why the polynomial function has a zero in the indicated interval.
(B) Determine the number of additional intervals required by the bisection method to obtain a one-decimal-place approximation to the zero and state the approximate value of the zero.

19. $P(x) = x^3 - 2x^2 - 5x + 4$; $(3, 4)$

20. $P(x) = x^3 + x^2 - 4x - 1$; $(1, 2)$

21. $P(x) = x^3 - 2x^2 - x + 5$; $(-2, -1)$

22. $P(x) = x^3 - 3x^2 - x - 2$; $(3, 4)$

23. $P(x) = x^4 - 2x^3 - 7x^2 + 9x + 7$; $(3, 4)$

24. $P(x) = x^4 - x^3 - 9x^2 + 9x + 4$; $(2, 3)$

25. $P(x) = x^4 - x^3 - 4x^2 + 4x + 3$; $(-1, 0)$

26. $P(x) = x^4 - 3x^3 - x^2 + 3x + 3$; $(2, 3)$

In Problems 27–34,
(A) Find the smallest positive integer and largest negative integer that, by Theorem 1, are upper and lower bounds, respectively, for the real zeros of P(x).
(B) Approximate the real zeros of each polynomial to two decimal places.

27. $P(x) = x^3 - 2x^2 + 3x - 8$

28. $P(x) = x^3 + 3x^2 + 4x + 5$

29. $P(x) = x^4 + x^3 - 5x^2 + 7x - 22$

30. $P(x) = x^4 - x^3 - 8x^2 - 12x - 25$

31. $P(x) = x^5 - 3x^3 - 4x + 4$

32. $P(x) = x^5 - x^4 - 2x^2 - 4x - 5$

33. $P(x) = x^5 + x^4 + 3x^3 + x^2 + 2x - 5$

34. $P(x) = x^5 - 2x^4 - 6x^2 - 9x + 10$

Problems 35–38 refer to the polynomial

$$P(x) = (x - 1)^2(x - 2)(x - 3)^4$$

35. Can the zero at $x = 1$ be approximated by the bisection method? Explain.

36. Can the zero at $x = 2$ be approximated by the bisection method? Explain.

37. Can the zero at $x = 3$ be approximated by the bisection method? Explain.

38. Which of the zeros can be approximated by a maximum approximation routine? By a minimum approximation routine? By the zero approximation routine on your graphing utility?

In Problems 39–44, approximate the zeros of each polynomial function to two decimal places, using maximum or minimum commands to approximate any zeros at turning points.

39. $P(x) = x^4 - 4x^3 - 10x^2 + 28x + 49$

40. $P(x) = x^4 + 4x^3 - 4x^2 - 16x + 16$

41. $P(x) = x^5 - 6x^4 + 4x^3 + 24x^2 - 16x - 32$

42. $P(x) = x^5 - 6x^4 + 2x^3 + 28x^2 - 15x + 2$

43. $P(x) = x^5 - 6x^4 + 11x^3 - 4x^2 - 3.75x - 0.5$

44. $P(x) = x^5 + 12x^4 + 47x^3 + 56x^2 - 15.75x + 1$

In Problems 45–54, solve each polynomial inequality to three decimal places.

45. $x^2 > 2$

46. $x^3 \le 10x$

47. $x^3 \ge 4x^2 + 7$

48. $3x^2 + 1 < 5x$

49. $x^2 + 7x - 3 \le x^3 + x + 4$

50. $x^4 + 1 > 3x^2$

51. $x^4 < 8x^3 - 17x^2 + 9x - 2$

52. $x^3 + 5x \ge 2x^3 - 4x^2 + 6$

53. $(x^2 + 2x - 2)^2 \ge 2$

54. $5 + 2x < (x^2 - 4)^2$

C

In Problems 55–64,
(A) Find the smallest positive integer multiple of 10 and largest negative integer multiple of 10 that, by Theorem 1, are upper and lower bounds, respectively, for the real zeros of each polynomial.
(B) Approximate the real zeros of each polynomial to two decimal places.

55. $P(x) = x^3 - 24x^2 - 25x + 10$

56. $P(x) = x^3 - 37x^2 + 70x - 20$

57. $P(x) = x^4 + 12x^3 - 900x^2 + 5{,}000$

58. $P(x) = x^4 - 12x^3 - 425x^2 + 7{,}000$

59. $P(x) = x^4 - 100x^2 - 1{,}000x - 5{,}000$

60. $P(x) = x^4 - 5x^3 - 50x^2 - 500x + 7{,}000$

61. $P(x) = 4x^4 - 40x^3 - 1{,}475x^2 + 7{,}875x - 10{,}000$

62. $P(x) = 9x^4 + 120x^3 - 3{,}083x^2 - 25{,}674x - 48{,}400$

63. $P(x) = 0.01x^5 - 0.1x^4 - 12x^3 + 9{,}000$

64. $P(x) = 0.1x^5 + 0.7x^4 - 18.775x^3 - 340x^2 - 1{,}645x - 2{,}450$

65. When synthetic division is used to divide a polynomial $P(x)$ by $x + 4$ the remainder is 10. When the same polynomial is divided by $x + 5$ the remainder is -8. Must $P(x)$ have a zero between -5 and -4? Explain.

66. When synthetic division is used to divide a polynomial $Q(x)$ by $x + 4$ the remainder is 10. When the same polynomial is divided by $x + 5$ the remainder is 8. Could $Q(x)$ have a zero between -5 and -4? Explain.

APPLICATIONS

Express the solutions to Problems 67–72 as the roots of a polynomial equation of the form $P(x) = 0$ and approximate these solutions to three decimal places.

★ **67. Geometry.** Find all points on the graph of $y = x^2$ that are one unit away from the point (1, 2). [*Hint:* Use the distance-between-two-points formula from Appendix A, Section A.3.]

★ **68. Geometry.** Find all points on the graph of $y = x^2$ that are one unit away from the point (2, 1).

★ **69. Manufacturing.** A box is to be made out of a piece of cardboard that measures 18 by 24 inches. Squares, x inches on a side, will be cut from each corner, and then the ends and sides will be folded up (see the figure). Find the value of x that would result in a box with a volume of 600 cubic inches.

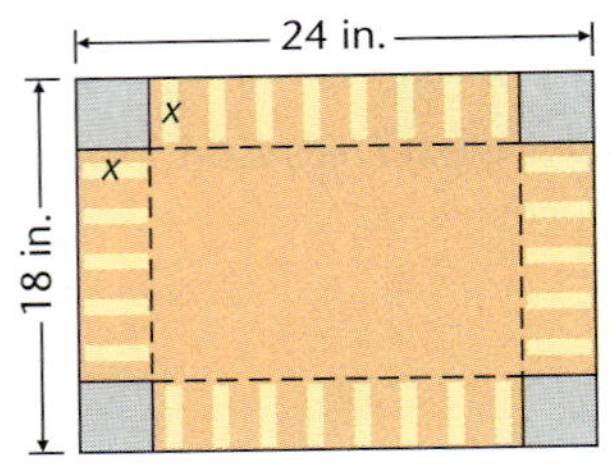

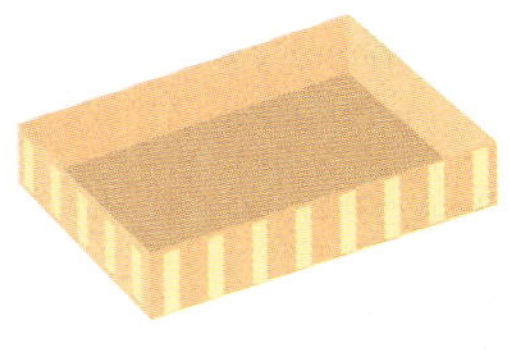

★ **70. Manufacturing.** A box with a hinged lid is to be made out of a piece of cardboard that measures 20 by 40 inches. Six squares, x inches on a side, will be cut from each corner and the middle, and then the ends and sides will be folded up to form the box and its lid (see the figure). Find the value of x that would result in a box with a volume of 500 cubic inches.

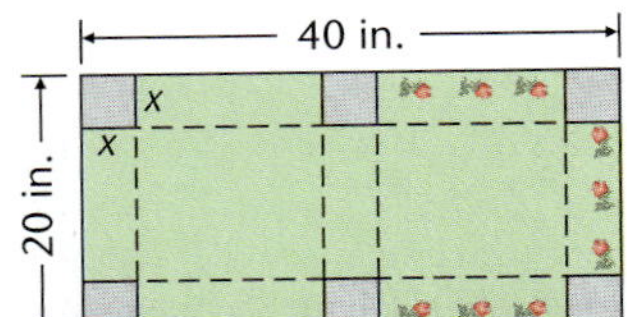

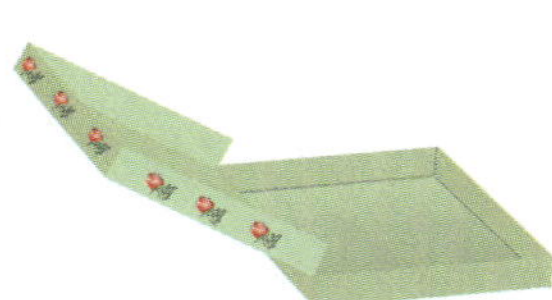

★ **71. Construction.** A propane gas tank is in the shape of a right circular cylinder with a hemisphere at each end (see the figure). If the overall length of the tank is 10 feet and the volume is 20π cubic feet, find the common radius of the hemispheres and the cylinder.

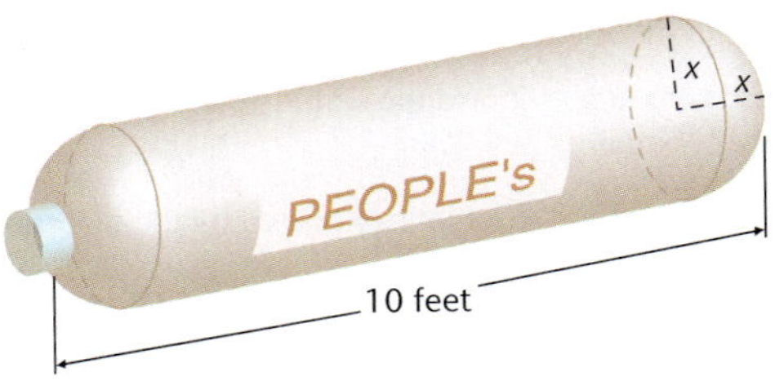

★ **72. Shipping.** A shipping box is reinforced with steel bands in all three directions (see the figure). A total of 20.5 feet of steel tape is to be used, with 6 inches of waste because of a 2-inch overlap in each direction. If the box has a square base and a volume of 2 cubic feet, find its dimensions.

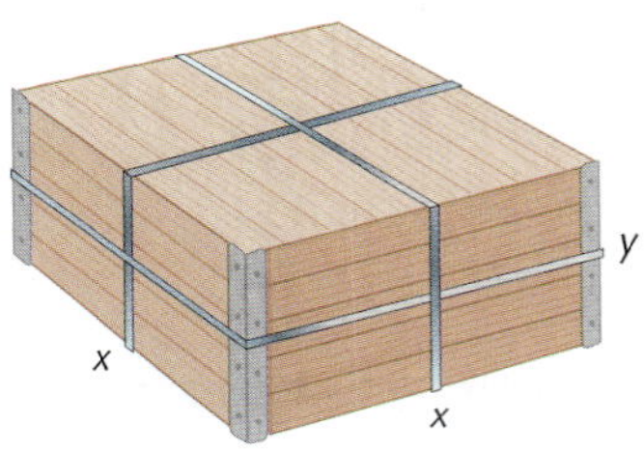

SECTION 3.3 Complex Zeros and Rational Zeros of Polynomials

Fundamental Theorem of Algebra • Factors of Polynomials with Real Coefficients • Graphs of Polynomials with Real Coefficients • Rational Zeros

The graph of the polynomial function $P(x) = x^2 + 4$ does not cross the x axis, so $P(x)$ has no real zeros. It does, however, have complex zeros, $2i$ and $-2i$; by the factor theorem, $x^2 + 4 = (x - 2i)(x + 2i)$. The *fundamental theorem of algebra* guarantees that *every* nonconstant polynomial with real or complex coefficients has a complex zero; it implies that such a polynomial can be factored as a product of linear factors. In Section 3.3 we study the fundamental theorem and its implications, including results on the graphs of polynomials with real coefficients. Finally, we consider a problem that has led to profound advances in mathematics: When can zeros of a polynomial be found *exactly*?

Fundamental Theorem of Algebra

The fundamental theorem of algebra was proved by Karl Friedrich Gauss (1777–1855), one of the greatest mathematicians of all time, in his doctoral thesis. A proof of the theorem is beyond the scope of this book, so we will state and use it without proof.

THEOREM 1
Fundamental Theorem of Algebra

Every polynomial of degree $n > 0$ with complex coefficients has a complex zero.

If $P(x)$ is a polynomial of degree $n > 0$ with complex coefficients, then by Theorem 1 it has a zero r_1. So $x - r_1$ is a factor of $P(x)$ by the factor theorem of Section 3.1. Thus,

$$P(x) = (x - r_1)Q(x), \text{ deg } Q(x) = n - 1$$

Now, applying the fundamental theorem to $Q(x)$, $Q(x)$ has a root r_2 and thus a factor $x - r_2$. (It is possible that r_2 is equal to r_1.) By continuing this reasoning we obtain a proof of Theorem 2.

THEOREM 2
***n* Linear Factors Theorem**

Every polynomial of degree $n > 0$ with complex coefficients can be factored as a product of n linear factors.

Suppose that a polynomial $P(x)$ is factored as a product of n linear factors. Any zero r of $P(x)$ must be a zero of one or more of the factors. The number of linear factors that have zero r is said to be the **multiplicity** of r. For example, the polynomial

$$P(x) = (x - 5)^3(x + 1)^2(x - 6i)(x + 2+3i) \quad (1)$$

has degree 7 and is written as a product of seven linear factors. $P(x)$ has just four zeros, namely 5, -1, $6i$, and $-2-3i$. Because the factor $x - 5$ appears to the power 3, we say that the zero 5 has *multiplicity* 3. Similarly, -1 has *multiplicity* 2, $6i$ has *multiplicity* 1, and $-2-3i$ has *multiplicity* 1. Note that the sum of the multiplicities is always equal to the degree of the polynomial: for $P(x)$ in (1), $3 + 2 + 1 + 1 = 7$.

EXAMPLE 1 Multiplicities of Zeros

Find the zeros and their multiplicities:

(A) $P(x) = (x + 2)^7(x - 4)^8(x^2 + 1)$

(B) $Q(x) = (x + 1)^3(x^2 - 1)(x + 1-i)$

SOLUTION

(A) Note that $x^2 + 1 = 0$ has the solutions i and $-i$. The zeros of $P(x)$ are -2 (multiplicity 7), 4 (multiplicity 8), i and $-i$ (each multiplicity 1).

(B) Note that $x^2 - 1 = (x - 1)(x + 1)$, so $x + 1$ appears four times as a factor of $Q(x)$. The zeros of $Q(x)$ are -1 (multiplicity 4), 1 (multiplicity 1), and $-1+i$ (multiplicity 1).

MATCHED PROBLEM 1

Find the zeros and their multiplicities:

(A) $P(x) = (x - 5)^3(x + 3)^2(x^2 + 16)$

(B) $Q(x) = (x^2 - 25)^3(x + 5)(x - i)$

Factors of Polynomials with Real Coefficients

If $p + qi$ is a zero of $P(x) = ax^2 + bx + c$, where a, b, c, p, and q are real numbers, then

$$P(p + qi) = 0$$

$$a(p + qi)^2 + b(p + qi) + c = 0 \quad \text{Take the conjugate of both sides.}$$

$$\overline{a(p + qi)^2 + b(p + qi) + c} = \overline{0} \quad \overline{z + w} = \overline{z} + \overline{w},\ \overline{zw} = \overline{z}\,\overline{w}$$

$$\overline{a}\,\overline{(p + qi)}^2 + \overline{b}\,\overline{(p + qi)} + \overline{c} = \overline{0} \quad \overline{z} = z \text{ if } z \text{ is real}, \overline{p + qi} = p - qi$$

$$a(p - qi)^2 + b(p - qi) + c = 0$$

$$P(p - qi) = 0$$

Therefore, $p - qi$ is also a zero of $P(x)$. This method of proof can be applied to any polynomial $P(x)$ of degree $n > 0$ with real coefficients, justifying Theorem 3.

THEOREM 3

Imaginary Zeros of Polynomials with Real Coefficients

Imaginary zeros of polynomials with real coefficients, if they exist, occur in conjugate pairs.

If a polynomial $P(x)$ of degree $n > 0$ has real coefficients and a linear factor of the form $x - (p + qi)$ where $q \neq 0$, then, by Theorem 3, $P(x)$ also has the linear factor $x - (p - qi)$. But

$$[x - (p + qi)][x - (p - qi)] = x^2 - 2px + p^2 - q^2$$

which is a quadratic factor of $P(x)$ with real coefficients and imaginary zeros. By this reasoning we can prove Theorem 4.

THEOREM 4
Linear and Quadratic Factors Theorem*

If $P(x)$ is a polynomial of degree $n > 0$ with real coefficients, then $P(x)$ can be factored as a product of linear factors (with real coefficients) and quadratic factors (with real coefficients and imaginary zeros).

EXAMPLE 2 Factors of Polynomials

Factor $P(x) = x^3 + x^2 + 4x + 4$ in two ways:

(A) As a product of linear factors (with real coefficients) and quadratic factors (with real coefficients and imaginary zeros)

(B) As a product of linear factors with complex coefficients

SOLUTION

(A) Note that $P(-1) = 0$, so -1 is a zero of $P(x)$ (or graph $P(x)$ and note that -1 is an x intercept). Therefore $x + 1$ is a factor of $P(x)$. Using synthetic division, the quotient is $x^2 + 4$, which has imaginary roots. Therefore

$$P(x) = (x + 1)(x^2 + 4)$$

An alternative solution is to factor by grouping:

$$\begin{aligned} x^3 + x^2 + 4x + 4 &= x^2(x + 1) + 4(x + 1) \\ &= (x^2 + 4)(x + 1) \end{aligned}$$

(B) Because $x^2 + 4$ has roots $2i$ and $-2i$,

$$P(x) = (x + 1)(x - 2i)(x + 2i)$$

MATCHED 2 PROBLEM

Factor $P(x) = x^5 - x^4 - x + 1$ in two ways:

(A) As a product of linear factors (with real coefficients) and quadratic factors (with real coefficients and imaginary zeros)

(B) As a product of linear factors with complex coefficients

---

*Theorem 4 underlies the technique of decomposing a rational function into partial fractions, which is useful in calculus. See Appendix B, Section B.2.

Graphs of Polynomials with Real Coefficients

The factorization described in Theorem 4 gives additional information about the graphs of polynomial functions with real coefficients. For certain polynomials the factorization of Theorem 4 will involve only linear factors; for others, only quadratic factors. Of course if only quadratic factors are present, then the degree of the polynomial $P(x)$ must be even. In other words, a polynomial $P(x)$ of odd degree with real coefficients must have a linear factor with real coefficients. This proves Theorem 5.

THEOREM 5
Real Zeros and Polynomials of Odd Degree

Every polynomial of odd degree with real coefficients has at least one real zero.

EXPLORE/DISCUSS 1

The graph of the polynomial $P(x) = x(x - 1)^2(x + 1)^4(x - 2)^3$ is shown in Figure 1. Find the real zeros of $P(x)$ and their multiplicities. How can a real zero of even multiplicity be distinguished from a real zero of odd multiplicity using only the graph?

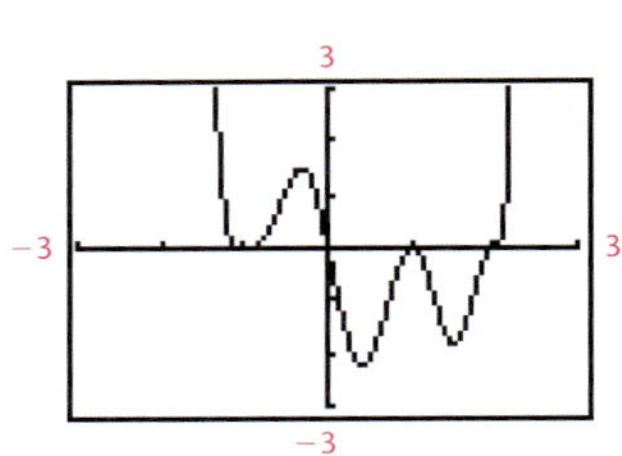

FIGURE 1 Graph of $P(x) = x(x - 1)^2(x + 1)^4(x - 2)^3$.

For polynomials with real coefficients, as suggested by Explore/Discuss 1, you can easily distinguish real zeros of even multiplicity from those of odd multiplicity using only the graph. Theorem 6, which we state without proof, gives distinguishing criteria.

THEOREM 6
Zeros of Even or Odd Multiplicity

Let $P(x)$ be a polynomial with real coefficients:

1. If r is a real zero of $P(x)$ of even multiplicity, then $P(x)$ has a turning point at r and does not change sign at r.
2. If r is a real zero of $P(x)$ of odd multiplicity, then $P(x)$ does not have a turning point at r and changes sign at r.

EXAMPLE 3 Multiplicities from Graphs

Figure 2 shows the graph of a polynomial function of degree 6. Find the real zeros and their multiplicities.

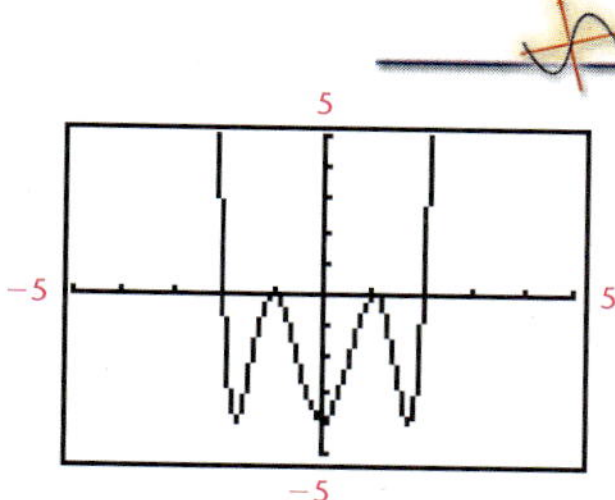

FIGURE 2

SOLUTION

The numbers -2, -1, 1, and 2 are real zeros (x intercepts). The graph has turning points at $x = \pm 1$ but not at $x = \pm 2$. Therefore, by Theorem 6, the zeros -1 and 1 have even multiplicity, and -2 and 2 have odd multiplicity. Because the sum of the multiplicities must equal 6 (the degree), the zeros -1 and 1 each have multiplicity 2, and the zeros -2 and 2 each have multiplicity 1.

MATCHED 3 PROBLEM

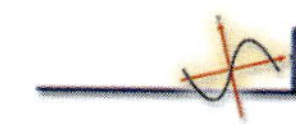

Figure 3 shows the graph of a polynomial function of degree 7. Find the real zeros and their multiplicities.

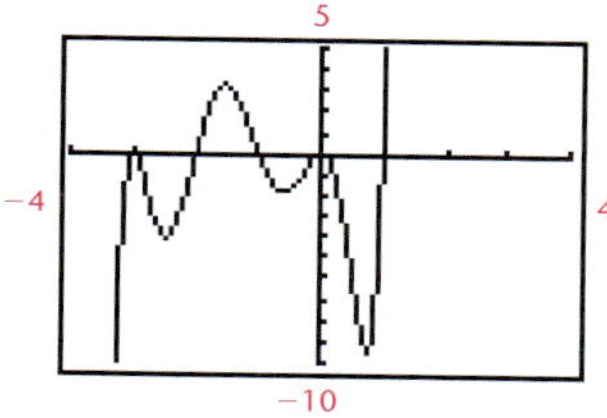

FIGURE 3

Rational Zeros

From a graphical perspective, finding a zero of a polynomial means finding a good approximation to an actual zero. A graphing calculator, for example, might give 2 as a zero of $P(x) = x^2 - (4 + 10^{-9})$ even though $P(2)$ is equal to -10^{-9}, not 0 (Fig. 4).

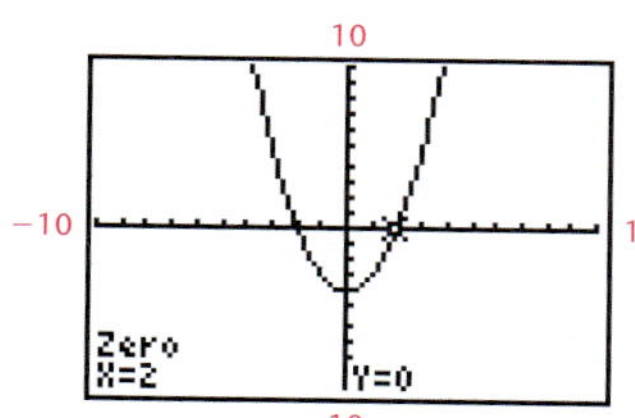

FIGURE 4 $P(x) = x^2 - (4 + 10^{-9})$.

It is natural, however, to want to find zeros *exactly*. Although this is impossible in general, we will adopt an algebraic strategy to find exact zeros in a special case, that of rational zeros of polynomials with rational coefficients. We will find a graphing utility to be helpful in carrying out the algebraic strategy.

First note that a polynomial with rational coefficients can always be written as a constant times a polynomial with integer coefficients. For example

$$P(x) = \frac{1}{2}x^3 - \frac{2}{3}x^2 + \frac{7}{4}x + 5$$

$$= \frac{1}{12}(6x^3 - 8x^2 + 21x + 60)$$

Because the zeros of $P(x)$ are the zeros of $6x^3 - 8x^2 + 21x + 60$, it is sufficient, for the purpose of finding rational zeros of polynomials with rational coefficients, to confine our attention to polynomials with integer coefficients.

We introduce the rational zero theorem by examining the following quadratic polynomial whose zeros can be found easily by factoring:

$$P(x) = 6x^2 - 13x - 5 = (2x - 5)(3x + 1)$$

$$\text{Zeros of } P(x)\text{:}\quad \frac{5}{2} \quad \text{and} \quad -\frac{1}{3} = \frac{-1}{3}$$

Notice that the numerators, 5 and -1, of the zeros are both integer factors of -5, the constant term in $P(x)$. The denominators 2 and 3 of the zeros are both integer factors of 6, the coefficient of the highest-degree term in $P(x)$. These observations are generalized in Theorem 7.

THEOREM 7
Rational Zero Theorem

If the rational number b/c, in lowest terms, is a zero of the polynomial

$$P(x) = a_nx^n + a_{n-1}x^{n-1} + \cdots + a_1x + a_0 \qquad a_n \neq 0$$

with integer coefficients, then b must be an integer factor of a_0 and c must be an integer factor of a_n.

$$P(x) = a_nx^n + a_{n-1}x^{n-1} + \cdots + a_1x + a_0$$

$\frac{b}{c}$ — b must be a factor of a_0; c must be a factor of a_n

The proof of Theorem 7 is not difficult and is instructive, so we sketch it here.

PROOF Because b/c is a zero of $P(x)$,

$$a_n\left(\frac{b}{c}\right)^n + a_{n-1}\left(\frac{b}{c}\right)^{n-1} + \cdots + a_1\left(\frac{b}{c}\right) + a_0 = 0 \tag{2}$$

If we multiply both sides of equation (2) by c^n, we obtain

$$a_nb^n + a_{n-1}b^{n-1}c + \cdots + a_1bc^{n-1} + a_0c^n = 0 \tag{3}$$

which can be written in the form

$$a_nb^n = c(-a_{n-1}b^{n-1} - \cdots - a_0c^{n-1}) \tag{4}$$

Because both sides of equation (4) are integers, c must be a factor of a_nb^n. And because the rational number b/c is given to be in lowest terms, b and c can have no common factors other than ±1. That is, b and c are **relatively prime.** This implies that b^n and c also are relatively prime. Hence, c must be a factor of a_n.

Now, if we solve equation (3) for a_0c^n and factor b out of the right side, we have

$$a_0c^n = b(-a_nb^{n-1} - \cdots - a_1c^{n-1})$$

We see that b is a factor of a_0c^n and, hence, a factor of a_0, because b and c are relatively prime.

EXPLORE/DISCUSS 2

Let $P(x) = a_3x^3 + a_2x^2 + a_1x + a_0$, where a_3, a_2, a_1, and a_0 are integers.

1. If $P(2) = 0$, there is one coefficient that must be an even integer. Identify this coefficient and explain why it must be even.

2. If $P(\frac{1}{2}) = 0$, there is one coefficient that must be an even integer. Identify this coefficient and explain why it must be even.
3. If $a_3 = a_0 = 1$, $P(-1) \neq 0$, and $P(1) \neq 0$, does $P(x)$ have any rational zeros? Support your conclusion with verbal arguments and/or examples.

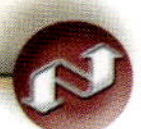

It is important to understand that Theorem 7 does not say that a polynomial *P*(*x*) with integer coefficients must have rational zeros.

It simply states that if $P(x)$ does have a rational zero, then the numerator of the zero must be an integer factor of a_0 and the denominator of the zero must be an integer factor of a_n. Because every integer has a finite number of integer factors, Theorem 7 enables us to construct a finite list of possible rational zeros. Finding any rational zeros then becomes a routine, although sometimes tedious, process of elimination.

As the next example illustrates, a graphing utility can greatly reduce the effort required to locate rational zeros.

EXAMPLE 4 Finding Rational Zeros

Find all the rational zeros for $P(x) = 2x^3 + 9x^2 + 7x - 6$.

SOLUTION

If b/c in lowest terms is a rational zero of $P(x)$, then b must be a factor of -6 and c must be a factor of 2.

Possible values of b are the integer factors of -6: $\pm 1, \pm 2, \pm 3, \pm 6$ (5)

Possible values of c are the integer factors of 2: $\pm 1, \pm 2$ (6)

Writing all possible fractions b/c where b is from (5) and c is from (6), we have

Possible rational zeros for $P(x)$: $\pm 1, \pm 2, \pm 3, \pm 6, \pm \frac{1}{2}, \pm \frac{3}{2}$ (7)

[Note that all fractions are in lowest terms and duplicates like $\pm 6/\pm 2 = \pm 3$ are not repeated.] If $P(x)$ has any rational zeros, they must be in list (7). We can test each number r in this list simply by evaluating $P(r)$. However, exploring the graph of $y = P(x)$ first will usually indicate which numbers in the list are the most likely candidates for zeros. Examining a graph of $P(x)$, we see that there are zeros near -3, near -2, and between 0 and 1, so we begin by evaluating $P(x)$ at -3, -2, and $\frac{1}{2}$ (Fig. 5).

FIGURE 5

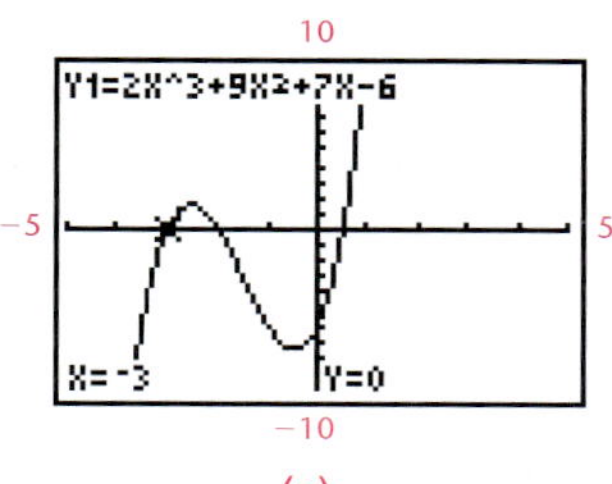

(a)

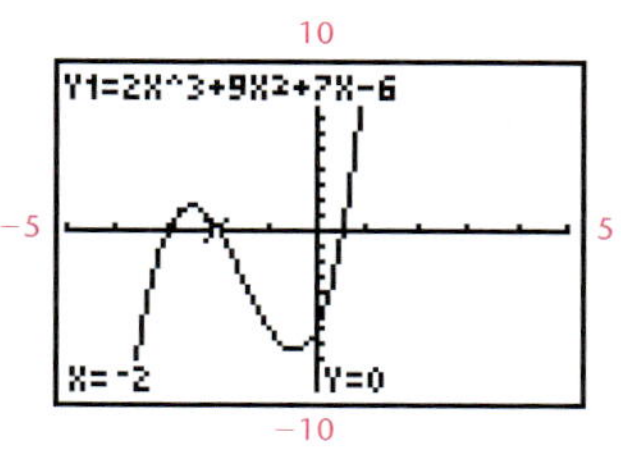

(b)

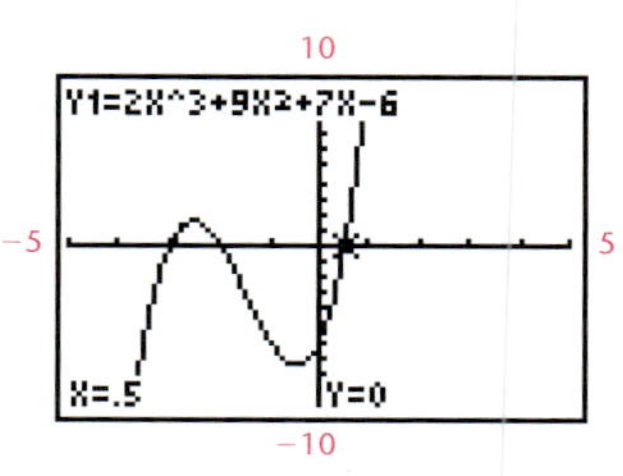

(c)

Thus -3, -2, and $\frac{1}{2}$ are rational zeros of $P(x)$. Because a third-degree polynomial can have at most three zeros, we have found all the rational zeros. There is no need to test the remaining candidates in list (7).

MATCHED 4 PROBLEM

Find all rational zeros for $P(x) = 2x^3 + x^2 - 11x - 10$.

As we saw in the solution of Example 4, rational zeros can be located by simply evaluating the polynomial. However, if we want to find multiple zeros, imaginary zeros, or exact values of irrational zeros, we need to consider *reduced polynomials*. If r is a zero of a polynomial $P(x)$, then we can write

$$P(x) = (x - r)Q(x)$$

where $Q(x)$ is a polynomial of degree one less than the degree of $P(x)$. The quotient polynomial $Q(x)$ is called the **reduced polynomial** for $P(x)$. In Example 4, after determining that -3 is a zero of $P(x)$, we can write

$$\begin{array}{r|rrrr} & 2 & 9 & 7 & -6 \\ & & -6 & -9 & 6 \\ \hline -3 & 2 & 3 & -2 & 0 \end{array}$$

$$\begin{aligned} P(x) &= 2x^3 + 9x^2 + 7x - 6 \\ &= (x + 3)(2x^2 + 3x - 2) \\ &= (x + 3)Q(x) \end{aligned}$$

Because the reduced polynomial $Q(x) = 2x^2 + 3x - 2$ is a quadratic, we can find its zeros by factoring or the quadratic formula. Thus,

$$P(x) = (x + 3)(2x^2 + 3x - 2) = (x + 3)(x + 2)(2x - 1)$$

and we see that the zeros of $P(x)$ are -3, -2, and $\frac{1}{2}$, as before.

EXAMPLE 5 Finding Rational and Irrational Zeros

Find all zeros exactly for $P(x) = 2x^3 - 7x^2 + 4x + 3$.

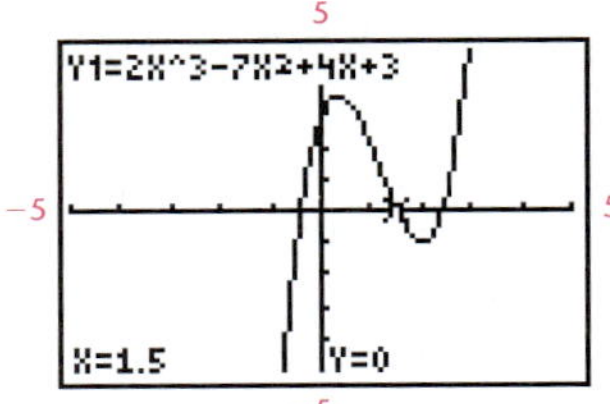

FIGURE 6

SOLUTION

First, list the possible rational zeros:

$$\pm 1, \pm 3, \pm \tfrac{1}{2}, \pm \tfrac{3}{2}$$

Examining the graph of $y = P(x)$ (Fig. 6), we see that there is a zero between -1 and 0, another between 1 and 2, and a third between 2 and 3. We test the only likely candidates, $-\frac{1}{2}$ and $\frac{3}{2}$:

$$P(-\tfrac{1}{2}) = -1 \quad \text{and} \quad P(\tfrac{3}{2}) = 0$$

Thus, $\frac{3}{2}$ is a zero, but $-\frac{1}{2}$ is not. Using synthetic division (details omitted), we can write

$$P(x) = (x - \tfrac{3}{2})(2x^2 - 4x - 2)$$

Because the reduced polynomial is quadratic, we can use the quadratic formula to find the exact values of the remaining zeros:

$$2x^2 - 4x - 2 = 0$$

$$x^2 - 2x - 1 = 0$$

$$x = \frac{2 \pm \sqrt{4 - 4(1)(-1)}}{2}$$

$$= \frac{2 \pm 2\sqrt{2}}{2} = 1 \pm \sqrt{2}$$

Thus, the exact zeros of $P(x)$ are $\frac{3}{2}$ and $1 \pm \sqrt{2}$.*

MATCHED 5 PROBLEM

Find all zeros exactly for $P(x) = 3x^3 - 10x^2 + 5x + 4$.

EXAMPLE 6 Finding Rational and Imaginary Zeros

Find all zeros exactly for $P(x) = x^4 - 6x^3 + 14x^2 - 14x + 5$.

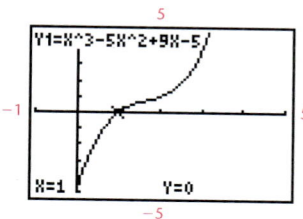

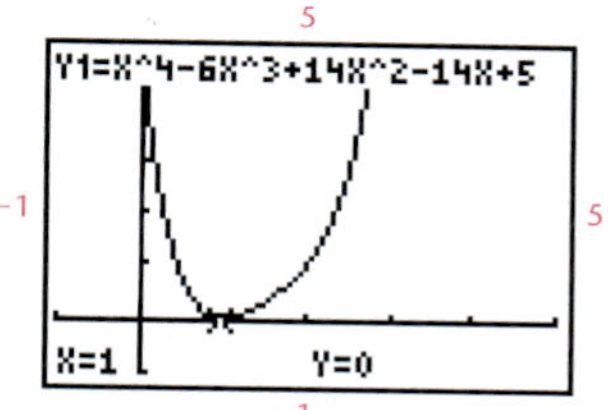

FIGURE 7

FIGURE 8

SOLUTION

The possible rational zeros are ± 1 and ± 5. Examining the graph of $P(x)$ (Fig. 7), we see that 1 is a zero. Because the graph of $P(x)$ does not appear to change sign at 1, this may be a multiple root. Using synthetic division (details omitted), we find that

$$P(x) = (x - 1)(x^3 - 5x^2 + 9x - 5)$$

The possible rational zeros of the reduced polynomial

$$Q(x) = x^3 - 5x^2 + 9x - 5$$

are ± 1 and ± 5. Examining the graph of $Q(x)$ (Fig. 8), we see that 1 is a rational zero. After a division, we have a quadratic reduced polynomial:

$$Q(x) = (x - 1)Q_1(x) = (x - 1)(x^2 - 4x + 5)$$

We use the quadratic formula to find the zeros of $Q_1(x)$:

$$x^2 - 4x + 5 = 0$$

$$x = \frac{4 \pm \sqrt{16 - 4(1)(5)}}{2}$$

$$= \frac{4 \pm \sqrt{-4}}{2} = 2 \pm i$$

Thus, the exact zeros of $P(x)$ are 1 (multiplicity 2), $2 - i$, and $2 + i$.

---

*By analogy with Theorem 3 (imaginary zeros of polynomials with real coefficients occur in conjugate pairs), it can be shown that if $r + s\sqrt{t}$ is a zero of a polynomial with rational coefficients, where r, s, and t are rational but t is not the square of a rational, then $r - s\sqrt{t}$ is also a zero.

MATCHED 6 PROBLEM

Find all zeros exactly for $P(x) = x^4 + 4x^3 + 10x^2 + 12x + 5$.

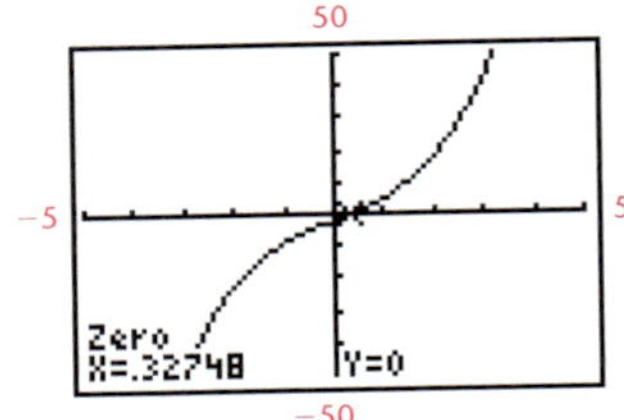

FIGURE 9 $P(x) = x^3 + 6x - 2$.

REMARK We were successful in finding all the zeros of the polynomials in Examples 5 and 6 because we could find sufficient rational zeros to reduce the original polynomial to a quadratic. This is not always possible. For example, the polynomial

$$P(x) = x^3 + 6x - 2$$

has no rational zeros, but does have an irrational zero at $x \approx 0.32748$ (Fig. 9). The other two zeros are imaginary. The techniques we have developed will not find the exact value of these roots.

EXPLORE/DISCUSS 3

There is a technique for finding the exact zeros of cubic polynomials, usually referred to as Cardano's formula.* This formula shows that the exact value of the irrational zero of $P(x) = x^3 + 6x - 2$ (see Fig. 9) is

$$x = \sqrt[3]{4} - \sqrt[3]{2}$$

(A) Verify that this is correct by expanding and simplifying

$$P(\sqrt[3]{4} - \sqrt[3]{2})$$

(B) Cardano's formula also shows that the two imaginary zeros are

$$-\tfrac{1}{2}(\sqrt[3]{4} - \sqrt[3]{2}) \pm \tfrac{1}{2}i\sqrt{3}(\sqrt[3]{4} - \sqrt[3]{2})$$

If you like algebraic manipulation, you can also verify that these are correct.

(C) Find a reference for Cardano's formula in a library or on the Internet. Use this formula to find the exact value of the irrational zero of

$$P(x) = x^3 + 9x - 6$$

Check your answer by comparing it with the approximate value obtained on a graphing utility.

ANSWERS MATCHED PROBLEMS

1. (A) 5 (multiplicity 3), -3 (multiplicity 2), $4i$ and $-4i$ (each multiplicity 1)
(B) -5 (multiplicity 4), 5 (multiplicity 3), i (multiplicity 1)

2. (A) $(x + 1)(x - 1)^2(x^2 + 1)$
(B) $(x + 1)(x - 1)^2(x + i)(x - i)$

3. -3 (multiplicity 2), -2 (multiplicity 1), -1 (multiplicity 1), 0 (multiplicity 2), 1 (multiplicity 1)

4. $-2, -1, \frac{5}{2}$

5. $\frac{4}{3}, 1 - \sqrt{2}, 1 + \sqrt{2}$

6. -1 (multiplicity 2), $-1 - 2i$, $-1 + 2i$

*Girolamo Cardano (1501–1576), an Italian mathematician and physician, was the first to publish a formula for the solution to cubic equations of the form $x^3 + ax + b = 0$ and the first to realize that this technique could be used to solve other cubic equations. Having predicted that he would live to the age of 75, Cardano committed suicide in 1576.

EXERCISE 3.3

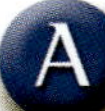

Write the zeros of each polynomial in Problems 1–8, and indicate the multiplicity of each if more than 1. What is the degree of each polynomial?

1. $P(x) = (x + 8)^3(x - 6)^2$

2. $P(x) = (x - 5)(x + 7)^2$

3. $P(x) = 3(x + 4)^3(x - 3)^2(x + 1)$

4. $P(x) = 5(x - 2)^3(x + 3)^2(x - 1)$

5. $P(x) = (x^2 + 4)^3(x^2 - 4)^5(x + 2i)$

6. $P(x) = (x^2 + 7x + 10)^2(x^2 + 6x + 10)^3$

7. $P(x) = (x^3 - 9x)(x^2 + 9)(x + 9)^2$

8. $P(x) = (x^3 - 3x^2 + 3x - 1)(x^2 - 1)(x - i)$

In Problems 9–14, find a polynomial P(x) of lowest degree, with leading coefficient 1, that has the indicated set of zeros. Leave the answer in a factored form. Indicate the degree of the polynomial.

9. 3 (multiplicity 2) and -4

10. -2 (multiplicity 3) and 1 (multiplicity 2)

11. -7 (multiplicity 3), $-3 + \sqrt{2}$, $-3 - \sqrt{2}$

12. $\frac{1}{3}$ (multiplicity 2), $5 + \sqrt{7}$, $5 - \sqrt{7}$

13. $(2 - 3i)$, $(2 + 3i)$, -4 (multiplicity 2)

14. $i\sqrt{3}$ (multiplicity 2), $-i\sqrt{3}$ (multiplicity 2), and 4 (multiplicity 3)

In Problems 15–20, find a polynomial of lowest degree, with leading coefficient 1, that has the indicated graph. Assume all zeros are integers. Leave the answer in a factored form. Indicate the degree of each polynomial.

15.

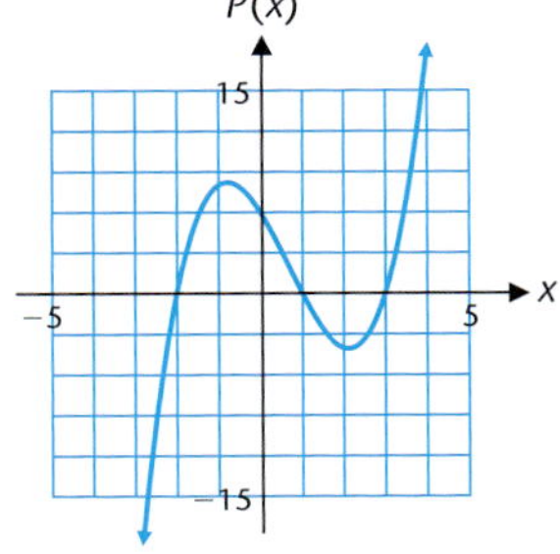

16.

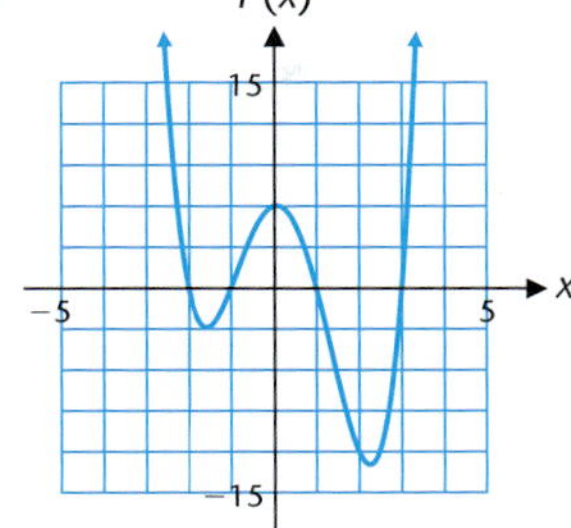

17.

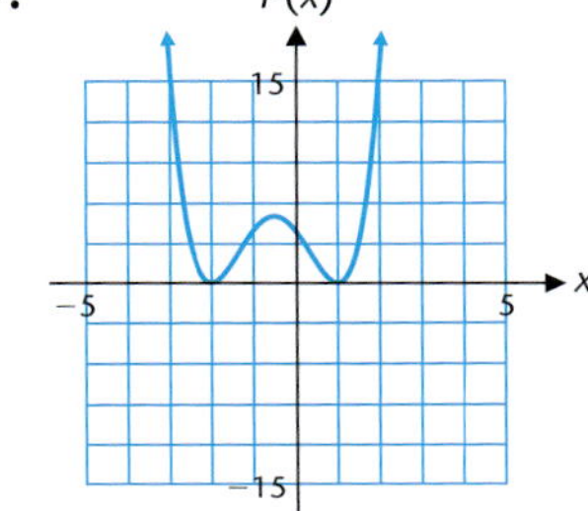

18.

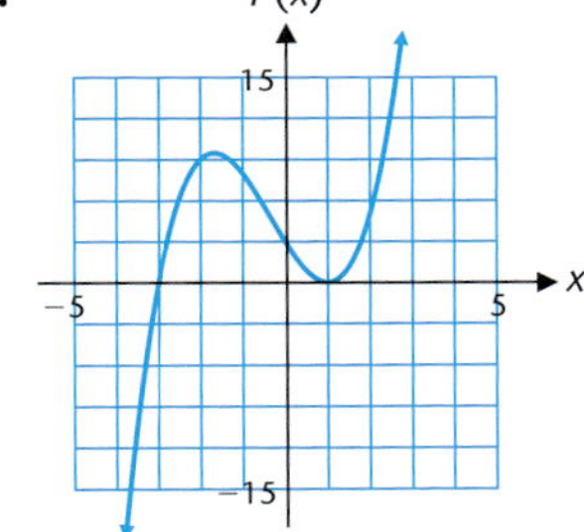

19.

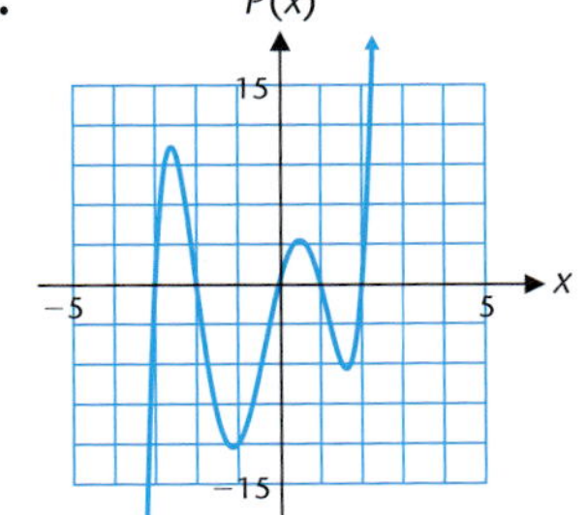

20.

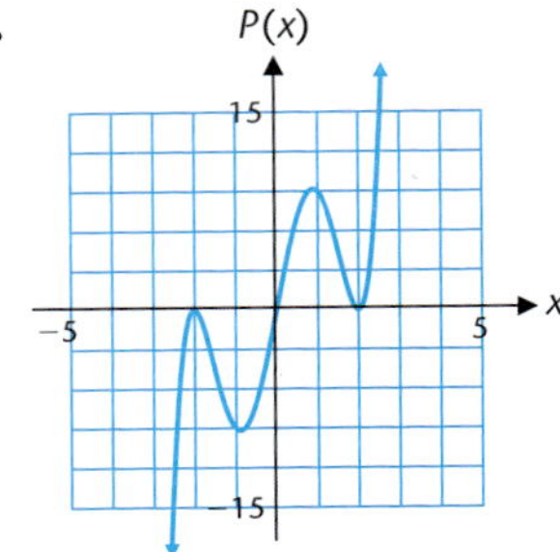

In Problems 21–24, factor each polynomial in two ways:
(A) As a product of linear factors (with real coefficients) and quadratic factors (with real coefficients and imaginary zeros)
(B) As a product of linear factors with complex coefficients

21. $P(x) = x^4 + 5x^2 + 4$

22. $P(x) = x^4 + 18x^2 + 81$

23. $P(x) = x^3 - x^2 + 25x - 25$

24. $P(x) = x^5 + x^4 - x - 1$

B

For each polynomial in problems 25–30, list all possible rational zeros (Theorem 7).

25. $P(x) = x^3 - 2x^2 - 5x + 6$

26. $P(x) = x^3 + 3x^2 - 6x - 8$

27. $P(x) = 3x^3 - 11x^2 + 8x + 4$

28. $P(x) = 2x^3 + x^2 - 4x - 3$

29. $P(x) = 12x^3 - 16x^2 - 5x + 3$

30. $P(x) = 2x^3 - 9x^2 + 14x - 5$

In Problems 31–36, write P(x) as a product of linear terms.

31. $P(x) = x^3 + 9x^2 + 24x + 16$; -1 is a zero

32. $P(x) = x^3 - 4x^2 - 3x + 18$; 3 is a double zero

33. $P(x) = x^4 - 1$; 1 and -1 are zeros

34. $P(x) = x^4 + 2x^2 + 1$; i is a double zero

35. $P(x) = 2x^3 - 17x^2 + 90x - 41$; $\frac{1}{2}$ is a zero

36. $P(x) = 3x^3 - 10x^2 + 31x + 26$; $-\frac{2}{3}$ is a zero

In Problems 37–46, find all roots exactly (rational, irrational, and imaginary) for each polynomial equation.

37. $2x^3 - 5x^2 + 1 = 0$

38. $2x^3 - 10x^2 + 12x - 4 = 0$

39. $x^4 + 4x^3 - x^2 - 20x - 20 = 0$

40. $x^4 - 4x^2 - 4x - 1 = 0$

41. $x^4 - 2x^3 - 5x^2 + 8x + 4 = 0$

42. $x^4 - 2x^2 - 16x - 15 = 0$

43. $x^4 + 10x^2 + 9 = 0$

44. $x^4 + 29x^2 + 100 = 0$

45. $2x^5 - 3x^4 - 2x + 3 = 0$

46. $2x^5 + x^4 - 6x^3 - 3x^2 - 8x - 4 = 0$

In Problems 47–56, find all zeros exactly (rational, irrational, and imaginary) for each polynomial.

47. $P(x) = x^3 - 19x + 30$

48. $P(x) = x^3 - 7x^2 + 36$

49. $P(x) = x^4 - \frac{21}{10}x^3 + \frac{2}{5}x$

50. $P(x) = x^4 + \frac{7}{6}x^3 - \frac{7}{3}x^2 - \frac{5}{2}x$

51. $P(x) = x^4 - 5x^3 + \frac{15}{2}x^2 - 2x - 2$

52. $P(x) = x^4 - \frac{13}{4}x^2 - \frac{5}{2}x - \frac{1}{4}$

53. $P(x) = x^4 + 11x^2 + 30$

54. $P(x) = x^4 + 5x^2 + 6$

55. $P(x) = 3x^5 - 5x^4 - 8x^3 + 16x^2 + 21x + 5$

56. $P(x) = 2x^5 - 3x^4 - 6x^3 + 23x^2 - 26x + 10$

In Problems 57–62, write each polynomial as a product of linear factors.

57. $P(x) = 6x^3 + 13x^2 - 4$

58. $P(x) = 6x^3 - 17x^2 - 4x + 3$

59. $P(x) = x^3 + 2x^2 - 9x - 4$

60. $P(x) = x^3 - 8x^2 + 17x - 4$

61. $P(x) = 4x^4 - 4x^3 - 9x^2 + x + 2$

62. $P(x) = 2x^4 + 3x^3 - 4x^2 - 3x + 2$

In Problems 63–68, multiply.

63. $[x - (4 - 5i)][x - (4 + 5i)]$

64. $[x - (2 - 3i)][x - (2 + 3i)]$

65. $[x - (3 + 4i)][x - (3 - 4i)]$

66. $[x - (5 + 2i)][x - (5 - 2i)]$

67. $[x - (a + bi)][x - (a - bi)]$

68. $(x - bi)(x + bi)$

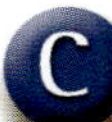

In Problems 69–74, find all other zeros of $P(x)$, given the indicated zero.

69. $P(x) = x^3 - 5x^2 + 4x + 10$; $3 - i$ is one zero

70. $P(x) = x^3 + x^2 - 4x + 6$; $1 + i$ is one zero

71. $P(x) = x^3 - 3x^2 + 25x - 75$; $-5i$ is one zero

72. $P(x) = x^3 + 2x^2 + 16x + 32$; $4i$ is one zero

73. $P(x) = x^4 - 4x^3 + 3x^2 + 8x - 10$; $2 + i$ is one zero

74. $P(x) = x^4 - 2x^3 + 7x^2 - 18x - 18$; $-3i$ is one zero

Prove that each of the real numbers in Problems 75–78 is not rational by writing an appropriate polynomial and making use of Theorem 7.

75. $\sqrt{6}$

76. $\sqrt{12}$

77. $\sqrt[3]{5}$

78. $\sqrt[5]{8}$

In Problems 79–82, determine the number of real zeros of each polynomial $P(x)$, and explain why $P(x)$ has no rational zeros.

79. $P(x) = x^4 - 5x^2 - 6$

80. $P(x) = 3x^4 + x^2 + 12$

81. $P(x) = x^3 - 3x - 1$

82. $P(x) = x^3 - 5x + 3$

In Problems 83–88, find all zeros (rational, irrational, and imaginary) exactly.

83. $P(x) = 3x^3 - 37x^2 + 84x - 24$

84. $P(x) = 2x^3 - 9x^2 - 2x + 30$

85. $P(x) = 4x^4 + 4x^3 + 49x^2 + 64x - 240$

86. $P(x) = 6x^4 + 35x^3 + 2x^2 - 233x - 360$

87. $P(x) = 4x^4 - 44x^3 + 145x^2 - 192x + 90$

88. $P(x) = x^5 - 6x^4 + 6x^3 + 28x^2 - 72x + 48$

89. The solutions to the equation $x^3 - 1 = 0$ are all the cube roots of 1.

(A) How many cube roots of 1 are there?

(B) 1 is obviously a cube root of 1; find all others.

90. The solutions to the equation $x^3 - 8 = 0$ are all the cube roots of 8.

(A) How many cube roots of 8 are there?

(B) 2 is obviously a cube root of 8; find all others.

91. If P is a polynomial function with real coefficients of degree n, with n odd, then what is the maximum number of times the graph of $y = P(x)$ can cross the x axis? What is the minimum number of times?

92. Answer the questions in Problem 91 for n even.

93. Given $P(x) = x^2 + 2ix - 5$ with $2 - i$ a zero, show that $2 + i$ is not a zero of $P(x)$. Does this contradict Theorem 3? Explain.

94. If $P(x)$ and $Q(x)$ are two polynomials of degree n, and if $P(x) = Q(x)$ for more than n values of x, then how are $P(x)$ and $Q(x)$ related?

APPLICATIONS

Find all rational solutions exactly, and find irrational solutions to two decimal places.

95. Storage. A rectangular storage unit has dimensions 1 by 2 by 3 feet. If each dimension is increased by the same amount, how much should this amount be to create a new storage unit with volume 10 times the old?

96. Construction. A rectangular box has dimensions 1 by 1 by 2 feet. If each dimension is increased by the same amount, how much should this amount be to create a new box with volume six times the old?

★ **97. Packaging.** An open box is to be made from a rectangular piece of cardboard that measures 8 by 5 inches, by cutting out squares of the same size from each corner and bending up the sides (see the figure). If the volume of the box is to be 14 cubic inches, how large a square should be cut from each corner? [*Hint:* Determine the domain of x from physical considerations before starting.]

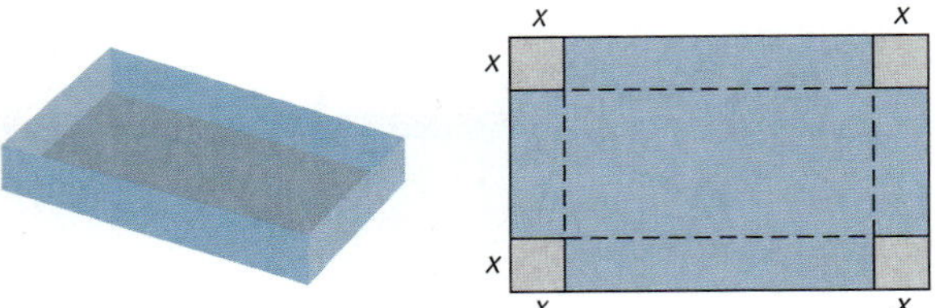

★ **98. Fabrication.** An open metal chemical tank is to be made from a rectangular piece of stainless steel that measures 10 by 8 feet, by cutting out squares of the same size from each corner and bending up the sides (see the figure for Problem 97). If the volume of the tank is to be 48 cubic feet, how large a square should be cut from each corner?

SECTION 3.4 Rational Functions and Inequalities

Rational Functions and Properties of Their Graphs • Vertical and Horizontal Asymptotes • Analyzing the Graph of a Rational Function • Rational Inequalities

In Section 3.4 we apply our knowledge of graphs of polynomial functions to study the graphs of *rational functions*, that is, functions that are quotients of polynomials. Although a graphing utility is an important aid in analyzing the graph of a rational function, a firm understanding of the properties of rational functions is necessary to correctly interpret graphing utility screens. The final goal is to produce a hand sketch showing all the important features of the graph.

Rational Functions and Properties of Their Graphs

The number $\frac{7}{13}$ is called a *rational number* because it is a quotient (or ratio) of integers. The function

$$f(x) = \frac{x + 1}{x^2 - x - 6}$$

is called a *rational function* because it is a quotient of polynomials.

DEFINITION 1
Rational Function

A function f is a **rational function** if it can be written in the form

$$f(x) = \frac{p(x)}{q(x)}$$

where $p(x)$ and $q(x)$ are polynomials of degrees m and n, respectively.

We will assume that the coefficients of $p(x)$ and $q(x)$ are real numbers, and that the domain of f is the set of all real numbers x such that $q(x) \neq 0$.

If a real number c is a zero of both $p(x)$ and $q(x)$, then, by the factor theorem, $x - c$ is a factor of both $p(x)$ and $q(x)$. The graphs of

$$f(x) = \frac{p(x)}{q(x)} = \frac{(x - c)\,\bar{p}(x)}{(x - c)\,\bar{q}(x)} \quad \text{and} \quad \bar{f}(x) = \frac{\bar{p}(x)}{\bar{q}(x)}$$

are then identical, except possibly for a "hole" at $x = c$ (Fig. 1).

FIGURE 1

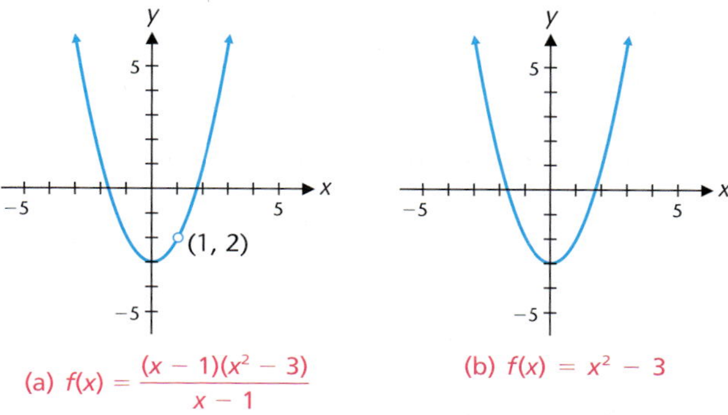

(a) $f(x) = \dfrac{(x - 1)(x^2 - 3)}{x - 1}$ (b) $f(x) = x^2 - 3$

Later in Section 3.4 we will explain how to handle the minor complication caused by common real zeros of $p(x)$ and $q(x)$. But to avoid that complication now, **unless stated to the contrary, we will assume that for any rational function f we consider, $p(x)$ and $q(x)$ have no real zero in common.**

Because the polynomial $q(x)$ of degree n has at most n real zeros, there are at most n real numbers that are not in the domain of f. Because a fraction equals 0 only if its numerator is 0, the x intercepts of the graph of f are the real zeros of the polynomial $p(x)$ of degree m. The number of x intercepts is thus at most m.

EXAMPLE 1 Domain and x Intercepts

Find the domain and x intercepts for $f(x) = \dfrac{2x^2 - 2x - 4}{x^2 - 9}$.

SOLUTION

$$f(x) = \frac{p(x)}{q(x)} = \frac{2x^2 - 2x - 4}{x^2 - 9} = \frac{2(x - 2)(x + 1)}{(x - 3)(x + 3)}$$

Because $q(x) = 0$ for $x = 3$ and $x = -3$, the domain of f is

$$x \neq \pm 3 \quad \text{or} \quad (-\infty, -3) \cup (-3, 3) \cup (3, \infty)$$

Because $p(x) = 0$ for $x = 2$ and $x = -1$, the zeros of f, and thus the x intercepts of f, are -1 and 2.

MATCHED 1 PROBLEM

Find the domain and x intercepts for $f(x) = \dfrac{3x^2 - 12}{x^2 + 2x - 3}$.

The graph of the rational function

$$f(x) = \frac{x^2 - 1.44}{x^3 - x}$$

is shown in Figure 2.

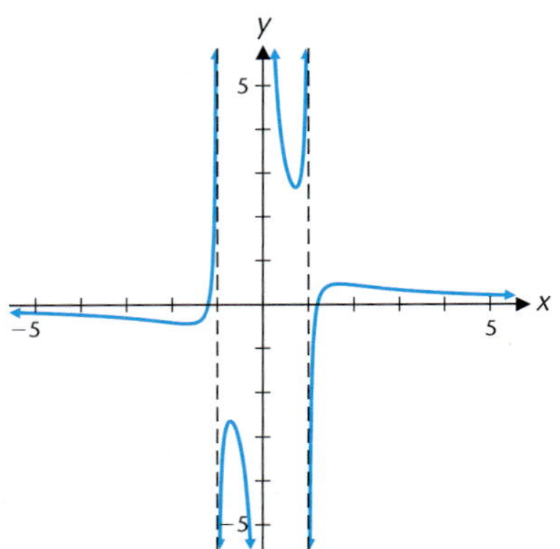

FIGURE 2 $f(x) = \dfrac{x^2 - 1.44}{x^3 - x}$.

The domain of f consists of all real numbers except $x = -1$, $x = 0$, and $x = 1$ (the zeros of the denominator $x^3 - x$). The dotted vertical lines at $x = \pm 1$ indicate that those values of x are excluded from the domain (a dotted vertical line at $x = 0$ would coincide with the y axis and is omitted). The graph is discontinuous at $x = -1$, $x = 0$, and $x = 1$, but is continuous elsewhere and has no sharp corners. The zeros of f are the zeros of the numerator $x^2 - 1.44$, namely $x = -1.2$ and $x = 1.2$. The graph of f has four turning points. Its left and right behavior is the same as that of the function $g(x) = \frac{1}{x}$ (the graph is close to the x axis for very large and very small values of x). The graph of f illustrates the general properties of rational functions that are listed in Theorem 1. We have already justified Property 3; the other properties are established in calculus.

THEOREM 1
Properties of Rational Functions

Let $f(x) = \frac{p(x)}{q(x)}$ be a rational function where $p(x)$ and $q(x)$ are polynomials of degrees m and n, respectively. Then the graph of $f(x)$:

1. Is continuous with the exception of at most n real numbers
2. Has no sharp corners
3. Has at most m real zeros
4. Has at most $m + n - 1$ turning points
5. Has the same left and right behavior as the quotient of the leading terms of $p(x)$ and $q(x)$

Figure 3 shows graphs of several rational functions, illustrating the properties of Theorem 1.

FIGURE 3 Graphs of rational functions.

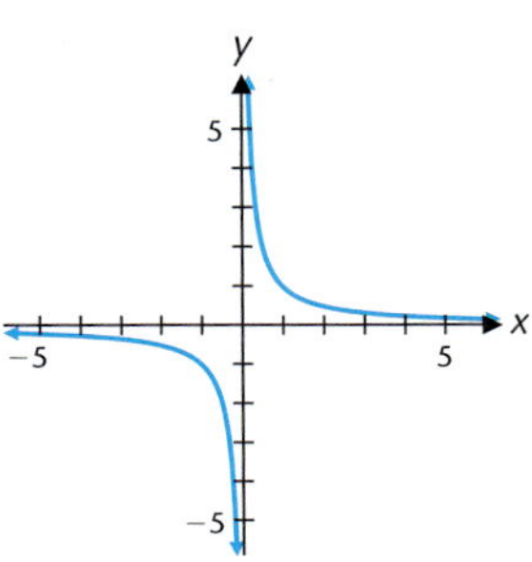

(a) $f(x) = \frac{1}{x}$

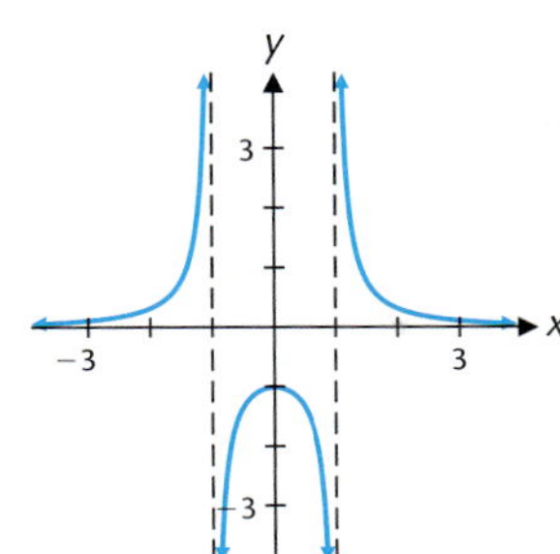

(b) $g(x) = \frac{1}{x^2 - 1}$

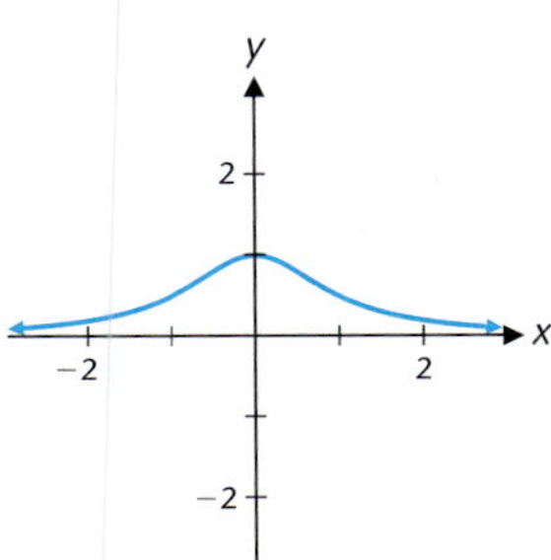

(c) $h(x) = \frac{1}{x^2 + 1}$

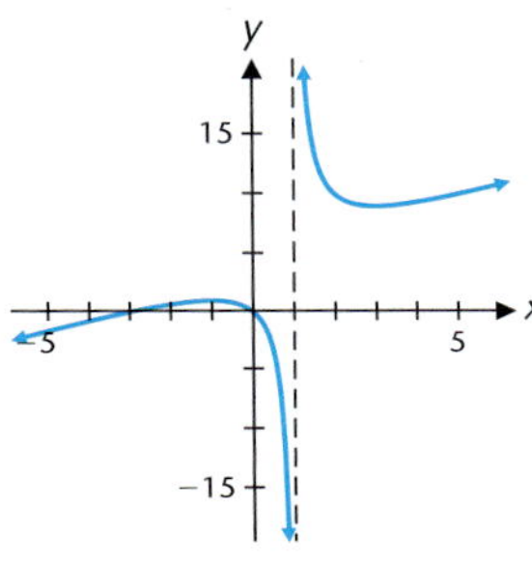

(d) $F(x) = \frac{x^2 + 3x}{x - 1}$

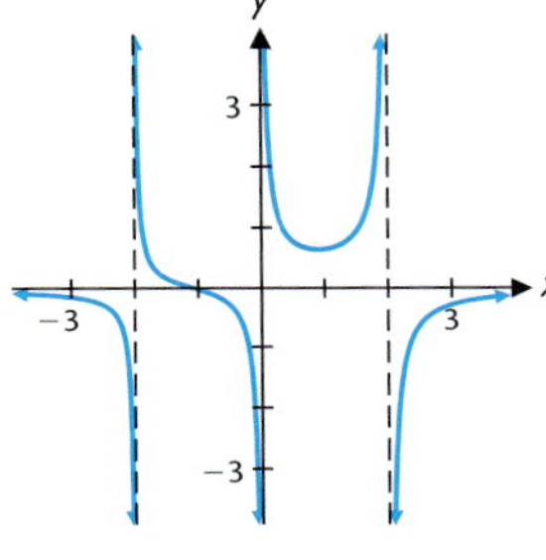

(e) $G(x) = \frac{-x - 1}{x^3 - 4x}$

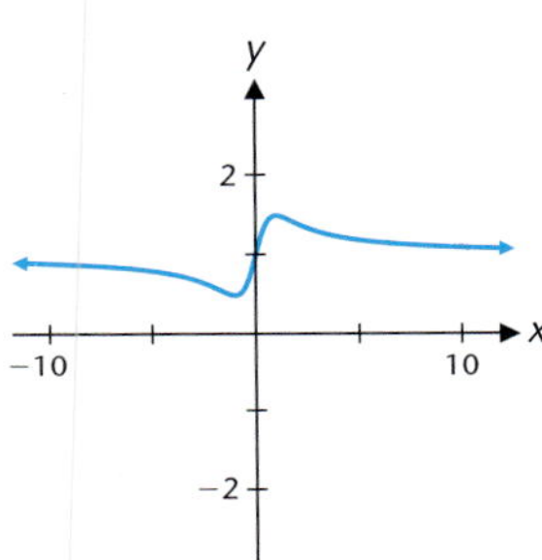

(f) $H(x) = \frac{x^2 + x + 1}{x^2 + 1}$

EXAMPLE 2 Properties of Graphs of Rational Functions

Use Theorem 1 to explain why each graph is not the graph of a rational function.

(A)

(B)

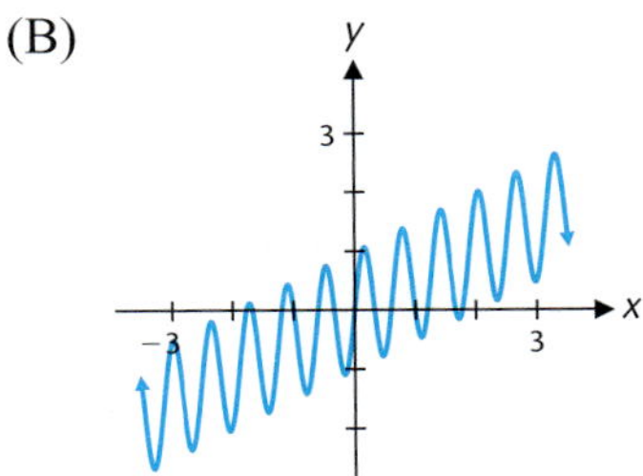

(C)

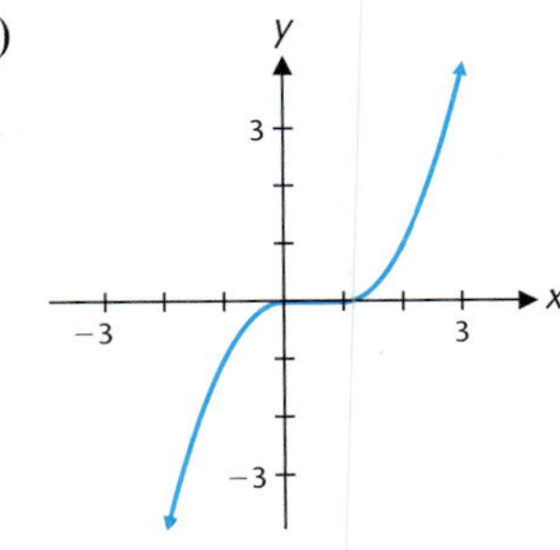

SOLUTION

(A) The graph has a sharp corner when $x = 0$, so Property 2 is not satisfied.

(B) The graph has an infinite number of turning points, so Property 4 is not satisfied.

(C) The graph has an infinite number of zeros (all values of x between 0 and 1, inclusive, are zeros), so Property 3 is not satisfied.

MATCHED 2 PROBLEM

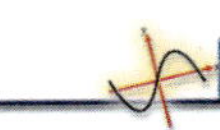

Use Theorem 1 to explain why each graph is not the graph of a rational function.

(A)

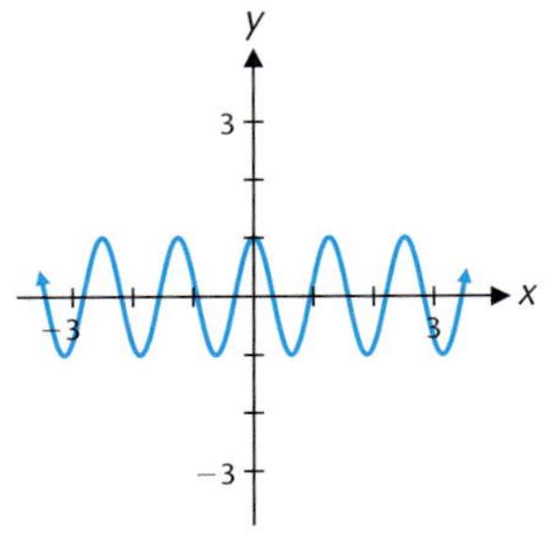

(B)

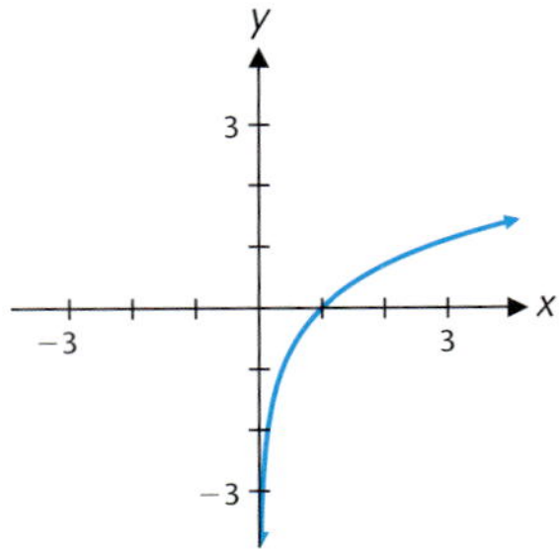

(C)

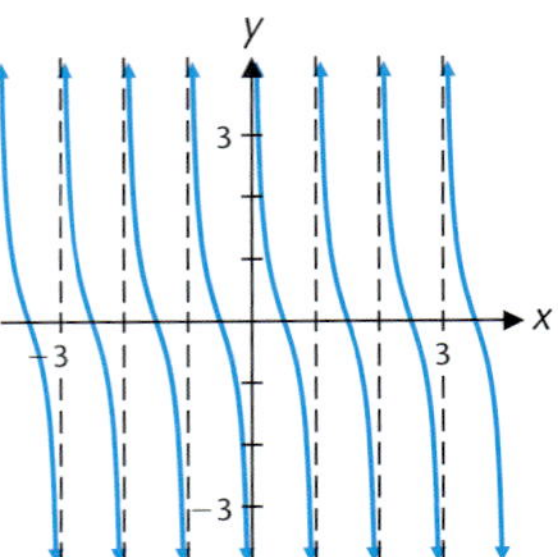

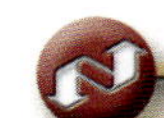

EXPLORE/DISCUSS 1

Give an example of a nonconstant rational function that has no zeros and no turning points. Could such a function be continuous? Explain.

Vertical and Horizontal Asymptotes

The graphs of Figure 3 exhibit similar behaviors near points of discontinuity that can be described using the concept of *vertical asymptote*. Consider, for example, the rational function $f(x) = \frac{1}{x}$ of Figure 3(a). As x approaches 0 from the right, the points $(x, \frac{1}{x})$ on the graph have larger and larger y coordinates—that is, $\frac{1}{x}$ increases without bound—as confirmed by Table 1. We write this symbolically as

$$\frac{1}{x} \to \infty \qquad \text{as} \qquad x \to 0^+$$

and say that the line $x = 0$ (the y axis) is a *vertical asymptote* for the graph of f.

TABLE 1 Behavior of $1/x$ as $x \to 0^+$

x	1	0.1	0.01	0.001	0.0001	0.000 01	0.000 001	. . .	x approaches 0 from the right ($x \to 0^+$)
$1/x$	1	10	100	1,000	10,000	100,000	1,000,000	. . .	$1/x$ increases without bound ($1/x \to \infty$)

If x approaches 0 from the left, the points $(x, \frac{1}{x})$ on the graph have smaller and smaller y coordinates—that is, $\frac{1}{x}$ decreases without bound—as confirmed by Table 2. We write this symbolically as

$$\frac{1}{x} \to -\infty \qquad \text{as} \qquad x \to 0^-$$

TABLE 2 Behavior of $1/x$ as $x \to 0^-$

x	−1	−0.1	−0.01	−0.001	−0.0001	−0.000 01	−0.000 001	. . .	x approaches 0 from the left ($x \to 0^-$)
$1/x$	−1	−10	−100	−1,000	−10,000	−100,000	−1,000,000	. . .	$1/x$ decreases without bound ($1/x \to -\infty$)

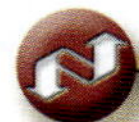

EXPLORE/DISCUSS 2

Construct tables similar to Tables 1 and 2 for $g(x) = \frac{1}{x^2}$ and discuss the behavior of the graph of $g(x)$ near $x = 0$.

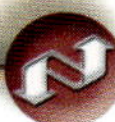

DEFINITION 2
Vertical Asymptote

The vertical line $x = a$ is a **vertical asymptote** for the graph of $y = f(x)$ if

$$f(x) \to \infty \quad \text{or} \quad f(x) \to -\infty \quad \text{as} \quad x \to a^+ \quad \text{or as} \quad x \to a^-$$

(that is, if $f(x)$ either increases or decreases without bound as x approaches a from the right or from the left).

THEOREM 2
Vertical Asymptotes of Rational Functions

Let $f(x) = \frac{p(x)}{q(x)}$ be a rational function. If a is a zero of $q(x)$, then the line $x = a$ is a vertical asymptote of the graph of f.

For example,

$$f(x) = \frac{x^2 - 1.44}{x^3 - x} = \frac{x^2 - 1.44}{x(x - 1)(x + 1)}$$

has three vertical asymptotes, $x = -1$, $x = 0$, and $x = 1$ (see Fig. 2 on p. 290).

The left and right behavior of some, but not all, rational functions can be described using the concept of *horizontal asymptote*. Consider $f(x) = \frac{1}{x}$. As values of x get larger and larger—that is, as x increases without bound—the points $(x, \frac{1}{x})$ have y coordinates that are positive and approach 0, as confirmed by Table 3. Similarly, as values of x get smaller and smaller—that is, as x decreases without bound—the points $(x, \frac{1}{x})$ have y coordinates that are negative and approach 0, as confirmed by Table 4. We write these facts symbolically as

$$\frac{1}{x} \to 0 \quad \text{as} \quad x \to \infty \quad \text{and as} \quad x \to -\infty$$

and say that the line $y = 0$ (the x axis) is a *horizontal asymptote* for the graph of f.

TABLE 3 Behavior of $1/x$ as $x \to \infty$

x	1	10	100	1,000	10,000	100,000	1,000,000	. . .	x increases without bound ($x \to \infty$)
$1/x$	1	0.1	0.01	0.001	0.0001	0.000 01	0.000 001	. . .	$1/x$ approaches 0 ($1/x \to 0$)

TABLE 4 Behavior of $1/x$ as $x \rightarrow -\infty$

x	-1	-10	-100	$-1,000$	$-10,000$	$-100,000$	$-1,000,000$	...	x decreases without bound ($x \rightarrow -\infty$)
$1/x$	-1	-0.1	-0.01	-0.001	-0.0001	$-0.000\,01$	$-0.000\,001$	...	$1/x$ approaches 0 ($1/x \rightarrow 0$)

EXPLORE/DISCUSS 3

Construct tables similar to Tables 3 and 4 for each of the following functions, and discuss the behavior of each as $x \rightarrow \infty$ and as $x \rightarrow -\infty$:

(A) $f(x) = \dfrac{3x}{x^2 + 1}$ (B) $g(x) = \dfrac{3x^2}{x^2 + 1}$ (C) $h(x) = \dfrac{3x^3}{x^2 + 1}$

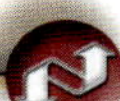

DEFINITION 3
Horizontal Asymptote

The horizontal line $y = b$ is a **horizontal asymptote** for the graph of $y = f(x)$ if

$$f(x) \rightarrow b \quad \text{as} \quad x \rightarrow -\infty \quad \text{or as} \quad x \rightarrow \infty$$

(that is, if $f(x)$ approaches b as x increases without bound or as x decreases without bound).

A rational function $f(x) = \frac{p(x)}{q(x)}$ has the same left and right behavior as the quotient of the leading terms of $p(x)$ and $q(x)$ (Property 5 of Theorem 1). Consequently, **a rational function has at most one horizontal asymptote.** Moreover, we can determine easily whether a rational function has a horizontal asymptote, and if it does, find its equation. Theorem 3 gives the details.

THEOREM 3
Horizontal Asymptotes of Rational Functions

Consider the rational function

$$f(x) = \frac{a_m x^m + \ldots + a_1 x + a_0}{b_n x^n + \ldots + b_1 x + b_0}$$

where $a_m \neq 0$, $b_n \neq 0$.

1. If $m < n$, the line $y = 0$ (the x axis) is a horizontal asymptote.
2. If $m = n$, the line $y = a_m/b_n$ is a horizontal asymptote.
3. If $m > n$, there is no horizontal asymptote.

In 1 and 2, the graph of f approaches the horizontal asymptote both as $x \rightarrow \infty$ and as $x \rightarrow -\infty$.

EXAMPLE 3 **Finding Vertical and Horizontal Asymptotes for a Rational Function**

Find all vertical and horizontal asymptotes for

$$f(x) = \frac{p(x)}{q(x)} = \frac{2x^2 - 2x - 4}{x^2 - 9}$$

SOLUTION

Because $q(x) = x^2 - 9 = (x - 3)(x + 3)$, the graph of $f(x)$ has vertical asymptotes at $x = 3$ and $x = -3$ (Theorem 1). Because $p(x)$ and $q(x)$ have the same degree, the line

$$y = \frac{a_2}{b_2} = \frac{2}{1} = 2 \qquad a_2 = 2, b_2 = 1$$

is a horizontal asymptote (Theorem 3, part 2).

MATCHED 3 PROBLEM

Find all vertical and horizontal asymptotes for

$$f(x) = \frac{3x^2 - 12}{x^2 + 2x - 3}$$

Analyzing the Graph of a Rational Function

We now use the techniques for locating asymptotes, along with other graphing aids discussed in the text, to graph several rational functions. First, we outline a systematic approach to the problem of graphing rational functions.

Analyzing and Sketching the Graph of a Rational Function: $f(x) = p(x)/q(x)$

Step 1. *Intercepts*. Find the real solutions of the equation $p(x) = 0$ and use these solutions to plot any x intercepts of the graph of f. Evaluate $f(0)$, if it exists, and plot the y intercept.

Step 2. *Vertical Asymptotes*. Find the real solutions of the equation $q(x) = 0$ and use these solutions to determine the domain of f, the points of discontinuity, and the vertical asymptotes. Sketch any vertical asymptotes as dashed lines.

Step 3. *Horizontal Asymptotes*. Determine whether there is a horizontal asymptote and if so, sketch it as a dashed line.

Step 4. *Complete the Sketch*. Using a graphing utility graph as an aid and the information determined in steps 1–3, sketch the graph.

EXAMPLE 4 **Graphing a Rational Function**

Graph $y = f(x) = \dfrac{2x}{x - 3}$.

SOLUTION

$$f(x) = \frac{2x}{x - 3} = \frac{p(x)}{q(x)}$$

Step 1. *Intercepts.* Find real zeros of $p(x) = 2x$ and find $f(0)$:

$$\begin{aligned} 2x &= 0 \\ x &= 0 \quad \text{x intercept} \\ f(0) &= 0 \quad \text{y intercept} \end{aligned}$$

The graph crosses the coordinate axes only at the origin. Plot this intercept, as shown in Figure 4.

FIGURE 4

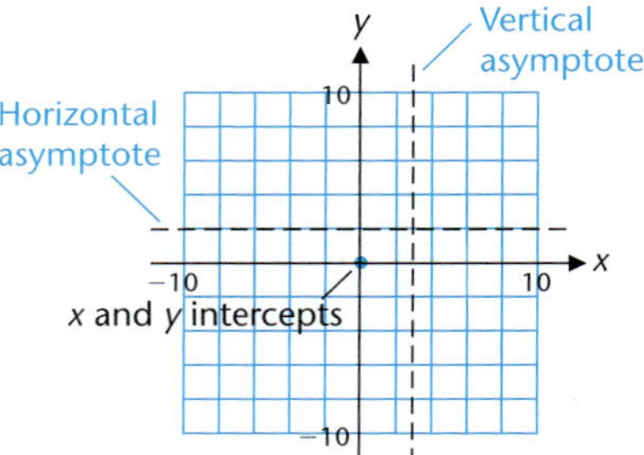

Intercepts and asymptotes

Step 2. *Vertical Asymptotes.* Find real zeros of $q(x) = x - 3$:

$$\begin{aligned} x - 3 &= 0 \\ x &= 3 \end{aligned}$$

The domain of f is $(-\infty, 3) \cup (3, \infty)$, f is discontinuous at $x = 3$, and the graph has a vertical asymptote at $x = 3$. Sketch this asymptote, as shown in Figure 4.

Step 3. *Horizontal Asymptote.* Because $p(x)$ and $q(x)$ have the same degree, the line $y = 2$ is a horizontal asymptote, as shown in Figure 4.

Step 4. *Complete the Sketch.* Using the graphing utility graph in Figure 5(a), we obtain the graph in Figure 5(b). Notice that the graph is a smooth continuous curve over the interval $(-\infty, 3)$ and over the interval $(3, \infty)$. As expected, there is a break in the graph at $x = 3$.

FIGURE 5

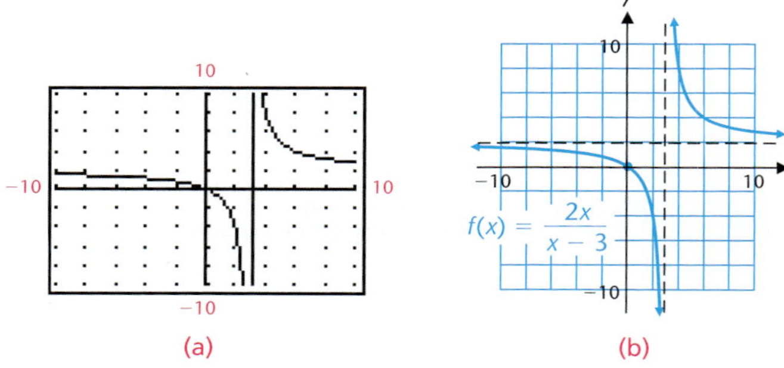

MATCHED 4 PROBLEM

Proceed as in Example 4 and sketch the graph of $y = f(x) = \dfrac{3x}{x + 2}$.

REMARK Refer to Example 4. When $f(x) = 2x/(x - 3)$ is graphed on a graphing utility [Fig. 4(a)], it appears that the graphing utility has also drawn the vertical asymptote at $x = 3$, but this is not the case. Most graphing utilities, when set in *connected mode,* calculate points on a graph and connect these points with line segments. The last point plotted to the left of the asymptote and the first plotted to the right of the asymptote will usually have very large y coordinates. If these y coordinates have opposite signs, then the graphing utility may connect the two points with a nearly vertical line segment, which gives the appearance of an asymptote. If you wish, you can set the calculator in *dot mode* to plot the points without the connecting line segments [Fig. 6(a)].

Depending on the scale, a graph may even appear to be continuous at a vertical asymptote [Fig. 6(b)]. It is important to always locate the vertical asymptotes as we did in step 2 before turning to the graphing utility graph to complete the sketch.

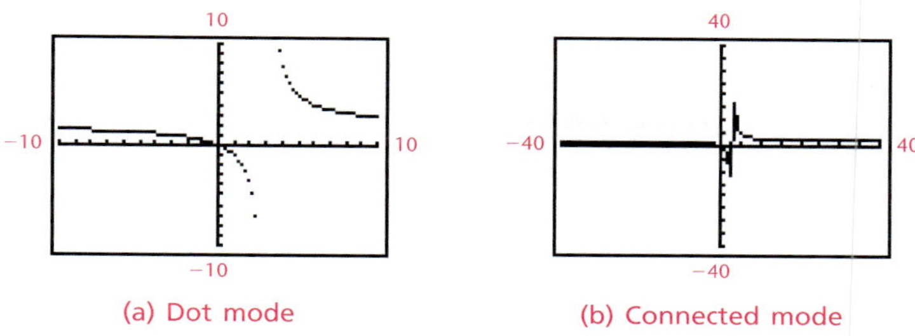

FIGURE 6 Graphing utility graphs of $f(x) = \dfrac{2x}{x - 3}$.

In the remaining examples we will just list the results of each step in the graphing strategy and omit the computational details.

EXAMPLE 5 Graphing a Rational Function

Graph $y = f(x) = \dfrac{x^2 - 6x + 9}{x^2 + x - 2}$.

SOLUTION

$$f(x) = \frac{x^2 - 6x + 9}{x^2 + x - 2} = \frac{(x - 3)^2}{(x + 2)(x - 1)}$$

x intercept: $x = 3$

y intercept: $y = f(0) = -\frac{9}{2} = -4.5$

Domain: $(-\infty, -2) \cup (-2, 1) \cup (1, \infty)$

Points of discontinuity: $x = -2$ and $x = 1$

Vertical asymptotes: $x = -2$ and $x = 1$

Horizontal asymptote: $y = 1$

Using the graphing utility graph in Figure 7 as an aid, sketch in the intercepts and asymptotes, then sketch the graph of f (Fig. 8).

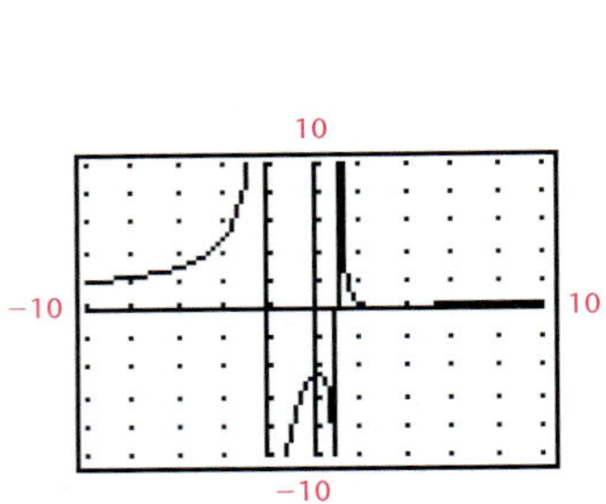

FIGURE 7

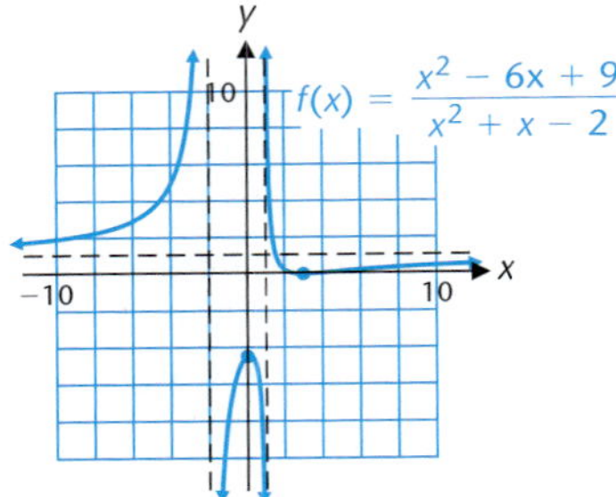

FIGURE 8

MATCHED 5 PROBLEM

Graph $y = f(x) = \dfrac{x^2}{x^2 - 7x + 10}$.

CAUTION

The graph of a function cannot cross a vertical asymptote, but the same statement is not true for horizontal asymptotes. The rational function

$$f(x) = \frac{2x^6 + x^5 - 5x^3 + 4x + 2}{x^6 + 1}$$

has the line $y = 2$ as a horizontal asymptote. The graph of f in Figure 9 clearly shows that **the graph of a function can cross a horizontal asymptote.** The definition of a horizontal asymptote requires $f(x)$ to approach b as x increases or decreases without bound, but it does not preclude the possibility that $f(x) = b$ for one or more values of x.

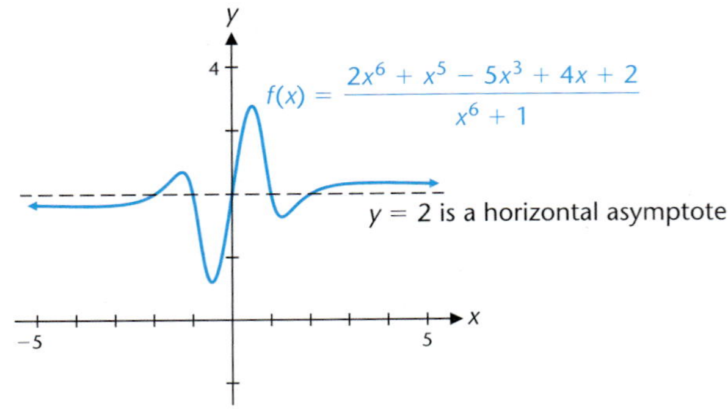

FIGURE 9 Multiple intersections of a graph and a horizontal asymptote.

EXAMPLE 6 Graphing a Rational Function

Graph $y = f(x) = \dfrac{x^2 - 3x - 4}{x - 2}$.

SOLUTION

$$f(x) = \frac{x^2 - 3x - 4}{x - 2} = \frac{(x + 1)(x - 4)}{x - 2}$$

x intercepts: $x = -1$ and $x = 4$
y intercept: $y = f(0) = 2$
Domain: $(-\infty, 2) \cup (2, \infty)$
Points of discontinuity: $x = 2$
Vertical asymptote: $x = 2$
No horizontal asymptote

Although the graph of f does not have a horizontal asymptote, we can still gain some useful information about the behavior of the graph as $x \to -\infty$ and as $x \to \infty$ if we first perform a long division:

$$\begin{array}{r} x - 1 \\ x - 2\overline{)x^2 - 3x - 4} \\ \underline{x^2 - 2x} \\ -x - 4 \\ \underline{-x + 2} \\ -6 \end{array}$$

Quotient: $x - 1$
Remainder: -6

Thus,

$$f(x) = \frac{x^2 - 3x - 4}{x - 2} = x - 1 - \frac{6}{x - 2}$$

As $x \to -\infty$ or $x \to \infty$, $6/(x - 2) \to 0$ and the graph of f approaches the line $y = x - 1$. This line is called an **oblique asymptote** for the graph of f. A graphing utility graph, including the oblique asymptote, is shown in Figure 10, and the graph of f is sketched in Figure 11.

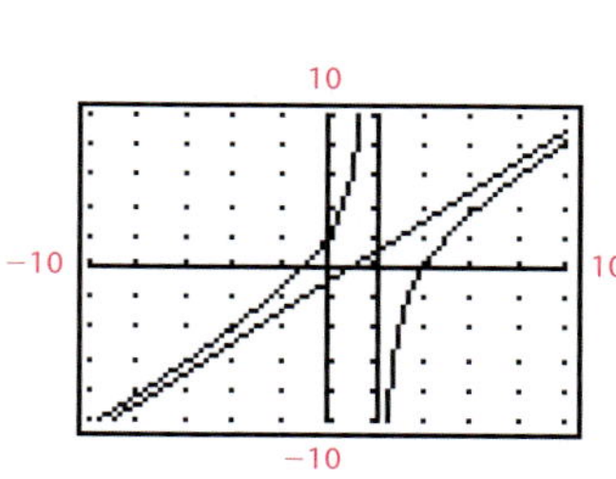

FIGURE 10

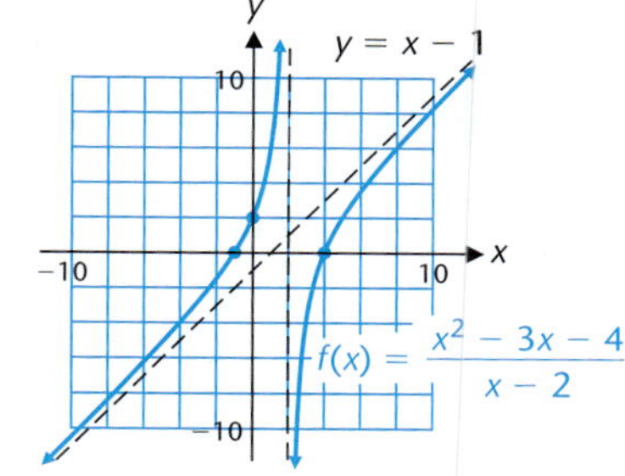

FIGURE 11

Generalizing the results of Example 6, we have Theorem 4.

THEOREM 4
Oblique Asymptotes and Rational Functions

If $f(x) = p(x)/q(x)$, where $p(x)$ and $q(x)$ are polynomials and the degree of $p(x)$ is 1 more than the degree of $q(x)$, then $f(x)$ can be expressed in the form

$$f(x) = mx + b + \frac{r(x)}{q(x)}$$

where the degree of $r(x)$ is less than the degree of $q(x)$. The line

$$y = mx + b$$

is an oblique asymptote for the graph of f. That is,

$$[f(x) - (mx + b)] \to 0 \qquad \text{as} \qquad x \to -\infty \qquad \text{or} \qquad x \to \infty$$

MATCHED 6 PROBLEM

Graph, including any oblique asymptotes, $y = f(x) = \dfrac{x^2 + 5}{x + 1}$.

At the beginning of Section 3.4 we made the assumption that for a rational function $f(x) = p(x)/q(x)$, the polynomials $p(x)$ and $q(x)$ have no common real zero. Now we abandon that assumption. Suppose that $p(x)$ and $q(x)$ have one or more real zeros in common. Then, by the factor theorem, $p(x)$ and $q(x)$ have one or more linear factors in common. We proceed to cancel common linear factors in

$$f(x) = \frac{p(x)}{q(x)}$$

until we obtain a rational function

$$\overline{f}(x) = \frac{\overline{p}(x)}{\overline{q}(x)}$$

in which $\overline{p}(x)$ and $\overline{q}(x)$ have no common real zero. We analyze and graph $\overline{f}(x)$, then insert "holes" as required in the graph of $\overline{f}$ to obtain the graph of f. Example 7 illustrates the details.

EXAMPLE 7 Graphing Arbitrary Rational Functions

Graph $f(x) = \dfrac{2x^5 - 4x^4 - 6x^3}{x^5 - 3x^4 - 3x^3 + 7x^2 + 6x}$.

SOLUTION

The real zeros of

$$p(x) = 2x^5 - 4x^4 - 6x^3$$

(obtained by graphing or factoring) are -1, 0, and 3.

The real zeros of

$$q(x) = x^5 - 3x^4 - 3x^3 + 7x^2 + 6x$$

are -1, 0, 2, and 3. The common zeros are -1, 0, and 3. Factoring and cancelling common linear factors gives

$$f(x) = \frac{2x^3(x+1)(x-3)}{x(x+1)^2(x-2)(x-3)} \quad \text{and} \quad \bar{f}(x) = \frac{2x^2}{(x+1)(x-2)}$$

We analyze $\bar{f}(x)$ as usual:

x intercept: $x = 0$

y intercept: $y = \bar{f}(0) = 0$

Domain: $(-\infty, -1) \cup (-1, 2) \cup (2, \infty)$

Points of discontinuity: $x = -1$, $x = 2$

Vertical asymptotes: $x = -1$, $x = 2$

Horizontal asymptote: $y = 2$

The graph of f is identical to the graph of $\bar{f}$ except possibly at the common real zeros -1, 0, and 3. We consider each common zero separately.

$x = -1$: Both f and $\bar{f}$ are undefined (no difference in their graphs).

$x = 0$: f is undefined but $\bar{f}(0) = 0$, so the graph of f has a hole at (0, 0).

$x = 3$: f is undefined but $\bar{f}(3) = 4.5$, so the graph of f has a hole at (3, 4.5).

Therefore, $f(x)$ has the following analysis:

x intercepts: none

y intercepts: none

Domain: $(-\infty, -1) \cup (-1, 0) \cup (0, 2) \cup (2, 3) \cup (3, \infty)$

Points of discontinuity: $x = -1$, $x = 0$, $x = 2$, $x = 3$

Vertical asymptotes: $x = -1$, $x = 2$

Horizontal asymptotes: $y = 2$

Holes: (0, 0), (3, 4.5)

Figure 12 shows the graphs of f and $\bar{f}$.

FIGURE 12

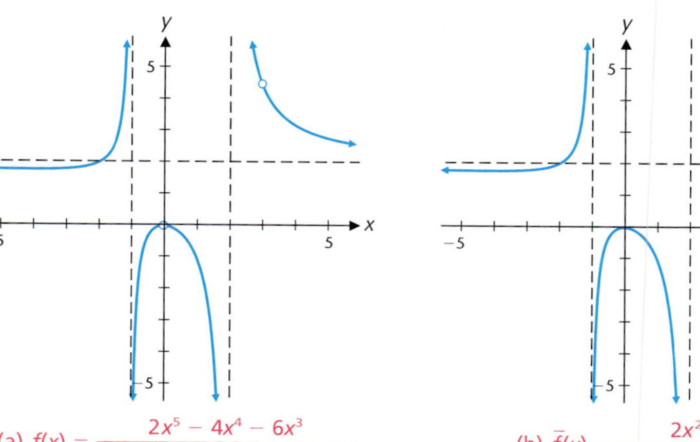

(a) $f(x) = \dfrac{2x^5 - 4x^4 - 6x^3}{x^5 - 3x^4 - 3x^3 + 7x^2 + 6x}$ (b) $\bar{f}(x) = \dfrac{2x^2}{(x+1)(x-2)}$

MATCHED 7 PROBLEM

Graph $f(x) = \dfrac{x^3 - x}{x^4 - x^2}$.

Rational Inequalities

A rational function $f(x) = p(x)/q(x)$ can change sign at a real zero of $p(x)$ (where f has an x intercept) or at a real zero of $q(x)$ (where f is discontinuous), but nowhere else (because f is continuous except where it is not defined). Rational inequalities can therefore be solved in the same way as polynomial inequalities, except that the partition of the x axis is determined by the zeros of $p(x)$ and the zeros of $q(x)$.

EXAMPLE 8 Solving Rational Inequalities

Solve $\dfrac{x^3 + 4x^2}{x^2 - 4} < 0$.

SOLUTION

Algebraic Solution

Let

$$f(x) = \frac{p(x)}{q(x)} = \frac{x^3 + 4x^2}{x^2 - 4}$$

The zeros of

$$p(x) = x^3 + 4x^2 = x^2(x + 4)$$

are 0 and -4. The zeros of

$$q(x) = x^2 - 4 = (x + 2)(x - 2)$$

are -2 and 2. These four zeros partition the x axis into the five intervals shown in the table. A test number is chosen from each interval as indicated to determine whether $f(x)$ is positive or negative.

Interval	Test number x	$f(x)$	Sign of f
$(-\infty, -4)$	-5	$-25/21$	$-$
$(-4, -2)$	-3	$9/5$	$+$
$(-2, 0)$	-1	-1	$-$
$(0, 2)$	1	$-5/3$	$-$
$(2, \infty)$	3	$63/5$	$+$

We conclude that the solution set of the inequality is

$$(-\infty, -4) \cup (-2, 0) \cup (0, 2)$$

Graphical Solution

Let

$$f(x) = \frac{p(x)}{q(x)} = \frac{x^3 + 4x^2}{x^2 - 4}$$

By graphing or factoring $p(x)$ we find that its zeros are 0 and -4. Similarly, the zeros of $q(x)$ are 2 and -2. These four zeros partition the x axis into five intervals:

$$(-\infty, -4), (-4, -2), (-2, 0), (0, 2), \text{ and } (2, \infty)$$

By inspecting the graph of f [Fig. 13(a)] or the graph of $f/|f|$ [Fig. 13(b)], we see that f is below the x axis on the intervals $(-\infty, -4)$, $(-2, 0)$, and $(0, 2)$.

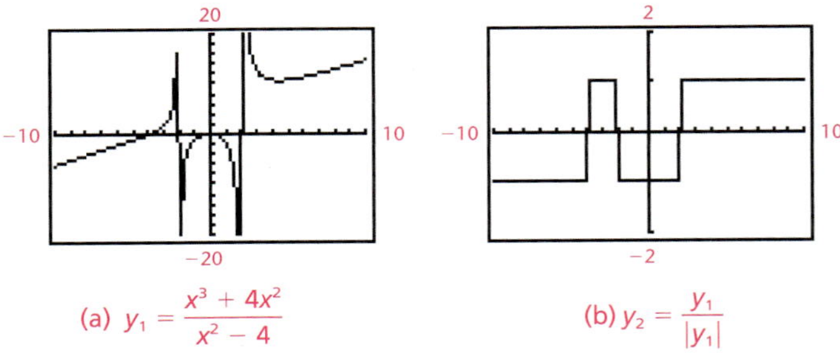

(a) $y_1 = \dfrac{x^3 + 4x^2}{x^2 - 4}$ (b) $y_2 = \dfrac{y_1}{|y_1|}$

FIGURE 13

We note that $f(0) = 0$, so $x = 0$ is not a solution to the inequality. We conclude that the solution set is

$$(-\infty, -4) \cup (-2, 0) \cup (0, 2)$$

MATCHED 8 PROBLEM

Solve $\dfrac{x^2 - 1}{x^2 - 9} \geq 0.$

EXAMPLE 9 Solving Rational Inequalities

Solve $1 \geq \dfrac{9x - 9}{x^2 + x - 3}$ to three decimal places.

SOLUTION

First we convert the inequality to an equivalent inequality in which one side is 0:

$$1 \geq \frac{9x - 9}{x^2 + x - 3} \qquad \text{Subtract } \frac{9x - 9}{x^2 + x - 3} \text{ from both sides.}$$

$$1 - \frac{9x - 9}{x^2 + x - 3} \geq 0 \qquad \text{Find a common denominator.}$$

$$\frac{x^2 + x - 3}{x^2 + x - 3} - \frac{9x - 9}{x^2 + x - 3} \geq 0 \qquad \text{Simplify.}$$

$$\frac{x^2 - 8x + 6}{x^2 + x - 3} \geq 0$$

The zeros of $x^2 - 8x + 6$, to three decimal places, are 0.838 and 7.162. The zeros of $x^2 + x - 3$ are -2.303 and 1.303. These four zeros partition the x axis into five intervals:

$(-\infty, -2.303)$, $(-2.303, 0.838)$, $(0.838, 1.303)$, $(1.303, 7.162)$, and $(7.162, \infty)$

We graph

$$f(x) = \frac{x^2 - 8x + 6}{x^2 + x - 3} \quad \text{and} \quad g(x) = \frac{f(x)}{|f(x)|}$$

(Fig. 14) and observe that the graph of f is above the x axis on the intervals $(-\infty, -2.303)$, $(0.838, 1.303)$, and $(7.162, \infty)$. The solution set of the inequality is thus

$(-\infty, -2.303) \cup [0.838, 1.303) \cup [7.162, \infty)$

Note that the endpoints that are zeros of f are included in the solution set of the inequality, but not the endpoints at which f is undefined.

FIGURE 14

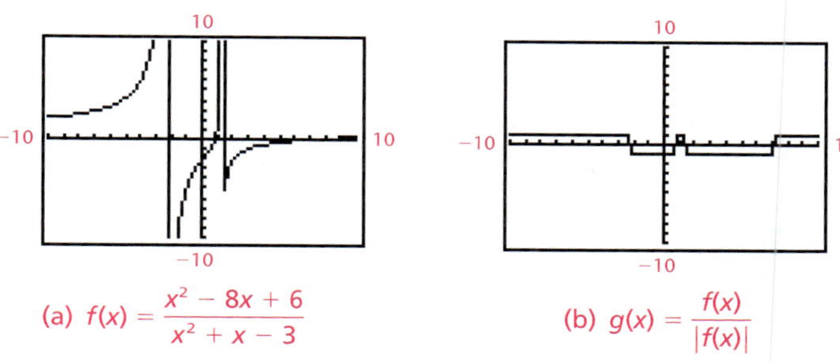

(a) $f(x) = \dfrac{x^2 - 8x + 6}{x^2 + x - 3}$ (b) $g(x) = \dfrac{f(x)}{|f(x)|}$

MATCHED 9 PROBLEM

Solve $\dfrac{x^3 + 4x^2 - 7}{x^2 - 5x - 1} \geq 0$ to three decimal places.

ANSWERS MATCHED PROBLEMS

1. Domain: $(-\infty, -3) \cup (-3, 1) \cup (1, \infty)$; x intercepts: $x = -2$, $x = 2$
2. (A) Properties 3 and 4 are not satisfied.
   (B) Property 1 is not satisfied.
   (C) Properties 1 and 3 are not satisfied.
3. Vertical asymptotes: $x = -3$, $x = 1$; horizontal asymptote: $y = 3$
4. 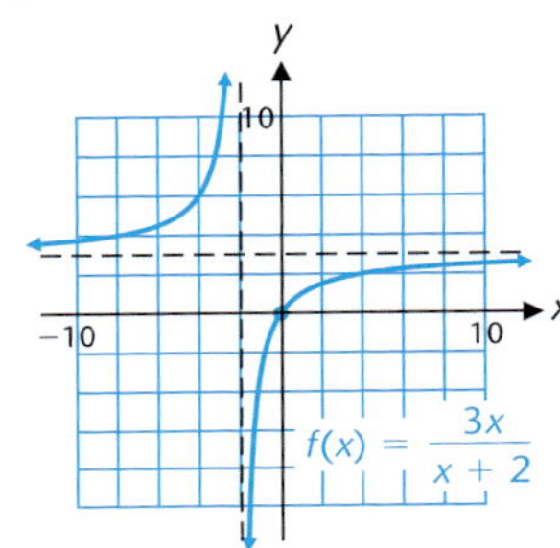

5. 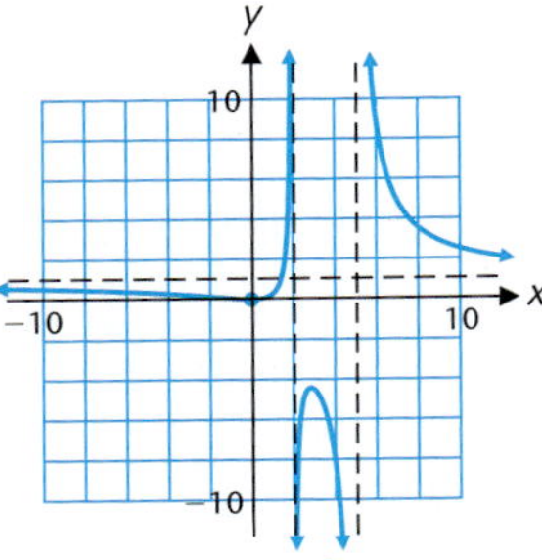

6. 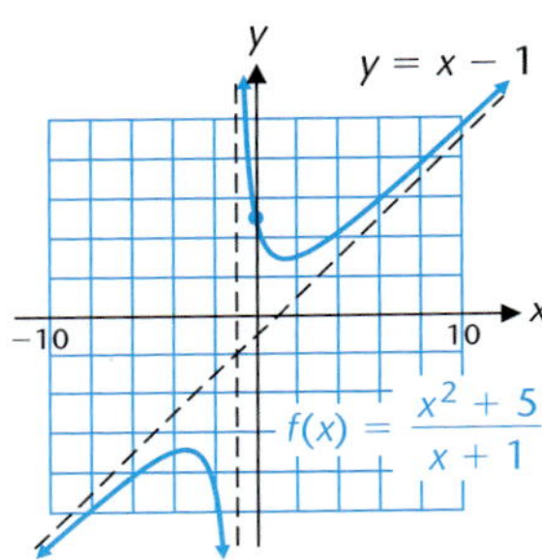

7. 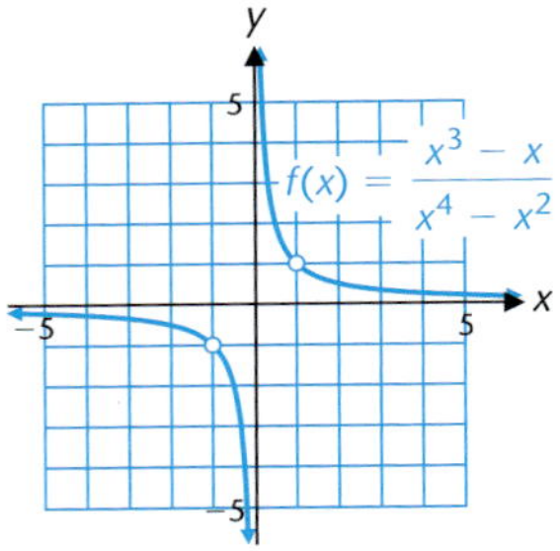

8. $(-\infty, -3) \cup [-1, 1] \cup (3, \infty)$
9. $[-3.391, -1.773] \cup (-0.193, 1.164] \cup (5.193, \infty)$

EXERCISE 3.4

A

In Problems 1–4, match each graph with one of the following functions:

$$f(x) = \frac{2x - 4}{x + 2} \qquad g(x) = \frac{2x + 4}{2 - x}$$

$$h(x) = \frac{2x + 4}{x - 2} \qquad k(x) = \frac{4 - 2x}{x + 2}$$

1. 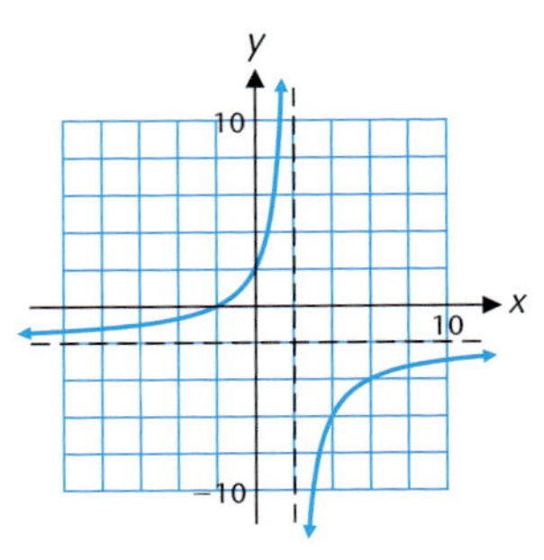

2. 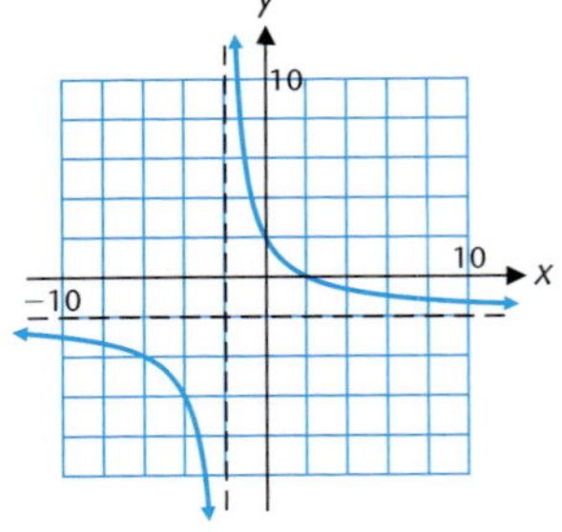

3.

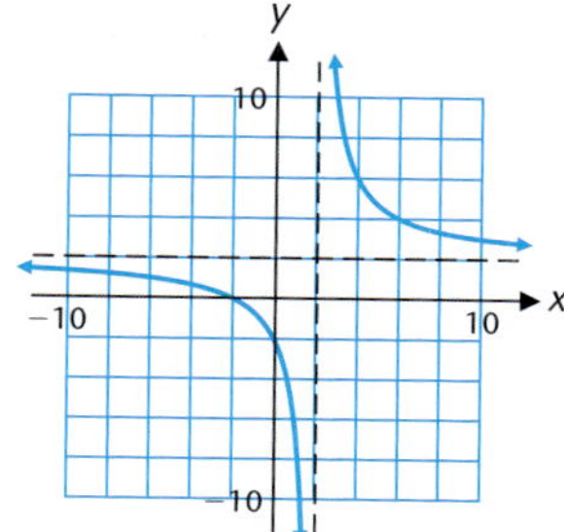

4.

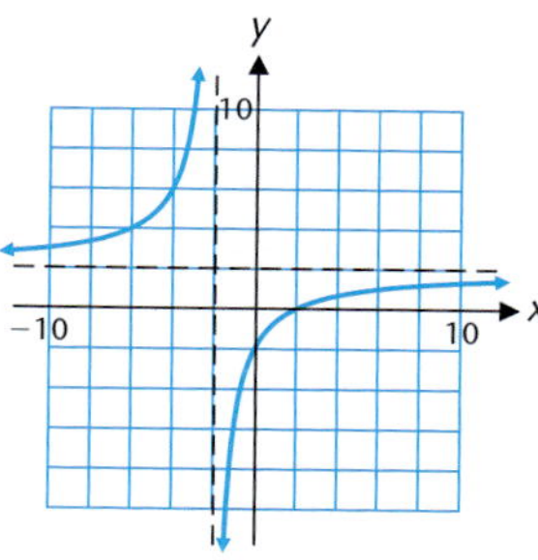

In Problems 5–12, find the domain and x intercepts.

5. $f(x) = \dfrac{2x - 4}{x + 1}$

6. $g(x) = \dfrac{3x + 6}{x - 1}$

7. $h(x) = \dfrac{x^2 - 1}{x^2 - 16}$

8. $k(x) = \dfrac{x^2 - 36}{x^2 - 25}$

9. $r(x) = \dfrac{x^2 - x - 6}{x^2 - x - 12}$

10. $s(x) = \dfrac{x^2 + x - 12}{x^2 + x - 6}$

11. $F(x) = \dfrac{x}{x^2 + 4}$

12. $G(x) = \dfrac{x^2}{x^2 + 16}$

In Problems 13–20, find all vertical and horizontal asymptotes.

13. $f(x) = \dfrac{2x}{x - 4}$

14. $h(x) = \dfrac{3x}{x + 5}$

15. $s(x) = \dfrac{2x^2 + 3x}{3x^2 - 48}$

16. $r(x) = \dfrac{5x^2 - 7x}{2x^2 - 50}$

17. $p(x) = \dfrac{2x}{x^4 + 1}$

18. $q(x) = \dfrac{5x^4}{2x^2 + 3x - 2}$

19. $t(x) = \dfrac{6x^4}{3x^2 - 2x - 5}$

20. $g(x) = \dfrac{3x}{x^4 + 2x^2 + 1}$

In Problems 21–24, explain why each graph is not the graph of a rational function.

21.

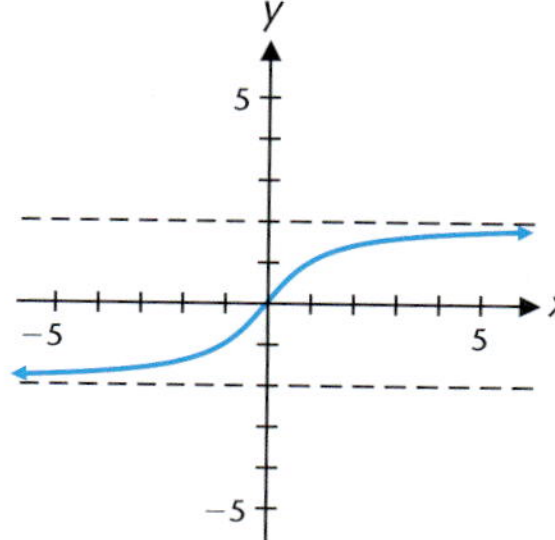

22.

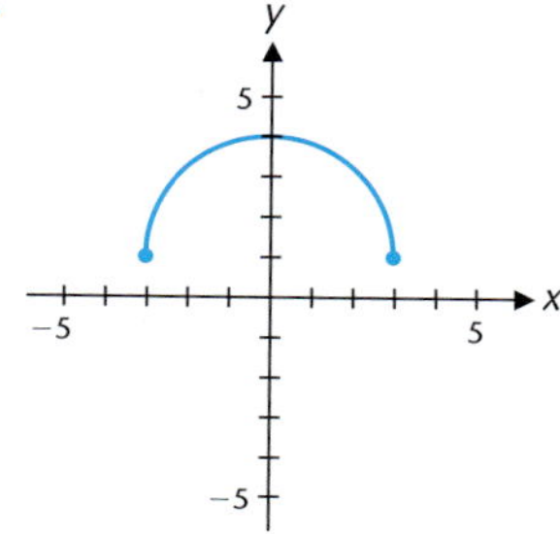

23.

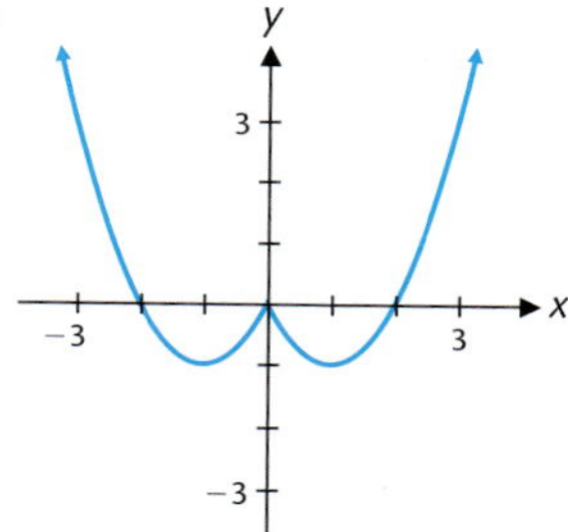

24.

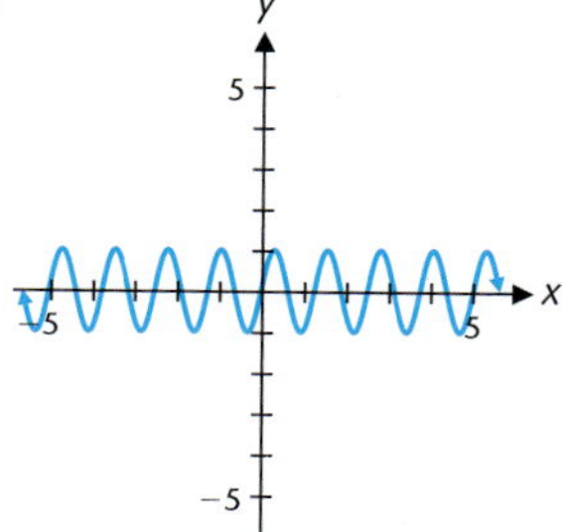

B

In Problems 25–44, use the graphing strategy outlined in the text to sketch the graph of each function.

25. $f(x) = \dfrac{1}{x - 4}$

26. $g(x) = \dfrac{1}{x + 3}$

27. $f(x) = \dfrac{x}{x + 1}$

28. $f(x) = \dfrac{3x}{x - 3}$

29. $h(x) = \dfrac{x}{2x - 2}$

30. $p(x) = \dfrac{3x}{4x + 4}$

31. $f(x) = \dfrac{2x - 4}{x + 3}$

32. $f(x) = \dfrac{3x + 3}{2 - x}$

33. $g(x) = \dfrac{1 - x^2}{x^2}$

34. $f(x) = \dfrac{x^2 + 1}{x^2}$

35. $f(x) = \dfrac{9}{x^2 - 9}$

36. $g(x) = \dfrac{6}{x^2 - x - 6}$

37. $f(x) = \dfrac{x}{x^2 - 1}$

38. $p(x) = \dfrac{x}{1 - x^2}$

39. $g(x) = \dfrac{2}{x^2 + 1}$

40. $f(x) = \dfrac{x}{x^2 + 1}$

41. $f(x) = \dfrac{12x^2}{(3x + 5)^2}$

42. $f(x) = \dfrac{7x^2}{(2x - 3)^2}$

43. $f(x) = \dfrac{x^2 - 1}{x^2 + 7x + 10}$

44. $f(x) = \dfrac{x^2 + 6x + 8}{x^2 - x - 2}$

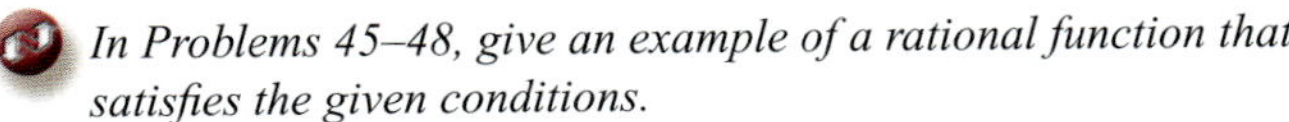

In Problems 45–48, give an example of a rational function that satisfies the given conditions.

45. Real zeros: $-2, -1, 1, 2$
vertical asymptotes: none
horizontal asymptote: $y = 3$

46. Real zeros: none
vertical asymptotes: $x = 4$
horizontal asymptote: $y = -2$

47. Real zeros: none
vertical asymptotes: $x = 10$
oblique asymptote: $y = 2x + 5$

48. Real zeros: 1, 2, 3
vertical asymptotes: none
oblique asymptote: $y = 2 - x$

In Problems 49–60, solve each rational inequality to three decimal places.

49. $\dfrac{x}{x - 4} \le 0$

50. $\dfrac{2x + 1}{x + 3} > 0$

51. $\dfrac{x^2 + 7x + 3}{x + 2} > 0$

52. $\dfrac{x^3 + 4}{x^2 + x - 3} \le 0$

53. $\dfrac{5}{x^2} - \dfrac{1}{x + 3} < 0$

54. $\dfrac{3x}{x + 4} - \dfrac{x + 1}{x^2} \ge 0$

55. $\dfrac{9}{x} - \dfrac{5}{x^2} \le 1$

56. $\dfrac{x + 4}{x^2 + 1} > 2$

57. $\dfrac{3x + 2}{x - 5} > 10$

58. $\dfrac{x}{x^2 + 5x - 6} \le 0.5$

59. $\dfrac{4}{x + 1} \ge \dfrac{7}{x}$

60. $\dfrac{1}{x^2 - 1} < \dfrac{x^2}{x^4 + 1}$

In Problems 61–66, find all vertical, horizontal, and oblique asymptotes.

61. $f(x) = \dfrac{2x^2}{x - 1}$

62. $g(x) = \dfrac{3x^2}{x + 2}$

63. $p(x) = \dfrac{x^3}{x^2 + 1}$

64. $q(x) = \dfrac{x^5}{x^3 - 8}$

65. $r(x) = \dfrac{2x^2 - 3x + 5}{x}$

66. $s(x) = \dfrac{-3x^2 + 5x + 9}{x}$

In Problems 67–70, investigate the behavior of each function as $x \to \infty$ and as $x \to -\infty$, and find any horizontal asymptotes (note that these functions are not rational).

67. $f(x) = \dfrac{5x}{\sqrt{x^2 + 1}}$

68. $f(x) = \dfrac{2x}{\sqrt{x^2 - 1}}$

69. $f(x) = \dfrac{4\sqrt{x^2 - 4}}{x}$

70. $f(x) = \dfrac{3\sqrt{x^2 + 1}}{x - 1}$

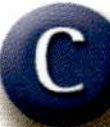

In Problems 71–76, use the graphing strategy outlined in the text to sketch the graph of each function. Include any oblique asymptotes.

71. $f(x) = \dfrac{x^2 + 1}{x}$

72. $g(x) = \dfrac{x^2 - 1}{x}$

73. $k(x) = \dfrac{x^2 - 4x + 3}{2x - 4}$

74. $h(x) = \dfrac{x^2 + x - 2}{2x - 4}$

75. $F(x) = \dfrac{8 - x^3}{4x^2}$

76. $G(x) = \dfrac{x^4 + 1}{x^3}$

If $f(x) = n(x)/d(x)$, where the degree of $n(x)$ is greater than the degree of $d(x)$, then long division can be used to write $f(x) = p(x) + q(x)/d(x)$, where $p(x)$ and $q(x)$ are polynomials with the degree of $q(x)$ less than the degree of $d(x)$. In Problems 77–80, perform the long division and discuss the relationship between the graphs of $f(x)$ and $p(x)$ as $x \to \infty$ and as $x \to -\infty$.

77. $f(x) = \dfrac{x^4}{x^2 + 1}$

78. $f(x) = \dfrac{x^5}{x^2 + 1}$

79. $f(x) = \dfrac{x^5}{x^2 - 1}$

80. $f(x) = \dfrac{x^5}{x^3 - 1}$

In calculus, it is often necessary to consider rational functions that are not in lowest terms, such as the functions given in Problems 81–84. For each function, state the domain, reduce the function to lowest terms, and sketch its graph. Remember to exclude from the graph any points with x values that are not in the domain.

81. $f(x) = \dfrac{x^2 - 4}{x - 2}$

82. $g(x) = \dfrac{x^2 - 1}{x + 1}$

83. $r(x) = \dfrac{x + 2}{x^2 - 4}$

84. $s(x) = \dfrac{x - 1}{x^2 - 1}$

APPLICATIONS

85. Employee Training. A company producing electronic components used in television sets has established that on the average, a new employee can assemble $N(t)$ components per day after t days of on-the-job training, as given by

$$N(t) = \frac{50t}{t + 4} \qquad t \geq 0$$

Sketch the graph of N, including any vertical or horizontal asymptotes. What does N approach as $t \to \infty$?

86. Physiology. In a study on the speed of muscle contraction in frogs under various loads, researchers W. O. Fems and J. Marsh found that the speed of contraction decreases with increasing loads. More precisely, they found that the relationship between speed of contraction S (in centimeters per second) and load w (in grams) is given approximately by

$$S(w) = \frac{26 + 0.06w}{w} \qquad w \geq 5$$

Sketch the graph of S, including any vertical or horizontal asymptotes. What does S approach as $w \to \infty$?

87. Retention. An experiment on retention is conducted in a psychology class. Each student in the class is given 1 day to memorize the same list of 40 special characters. The lists are turned in at the end of the day, and for each succeeding day for 20 days each student is asked to turn in a list of as many of the symbols as can be recalled. Averages are taken, and it is found that a good approximation of the average number of symbols, $N(t)$, retained after t days is given by

$$N(t) = \frac{5t + 30}{t} \qquad t \geq 1$$

Sketch the graph of N, including any vertical or horizontal asymptotes. What does N approach as $t \to \infty$?

88. Learning Theory. In 1917, L. L. Thurstone, a pioneer in quantitative learning theory, proposed the function

$$f(x) = \frac{a(x + c)}{(x + c) + b}$$

to describe the number of successful acts per unit time that a person could accomplish after x practice sessions. Suppose that for a particular person enrolling in a typing class,

$$f(x) = \frac{50(x + 1)}{x + 5} \qquad x \geq 0$$

where $f(x)$ is the number of words per minute the person is able to type after x weeks of lessons. Sketch the graph of f, including any vertical or horizontal asymptotes. What does f approach as $x \to \infty$?

Problems 89–92 are calculus related.

★ **89. Replacement Time.** A desktop office copier has an initial price of \$2,500. A maintenance/service contract costs \$200 for the first year and increases \$50 per year thereafter. It can be shown that the total cost of the copier after n years is given by

$$C(n) = 2{,}500 + 175n + 25n^2$$

The average cost per year for n years is $\overline{C}(n) = C(n)/n$.

(A) Find the rational function $\overline{C}$.

(B) When is the average cost per year minimum? (This is frequently referred to as the *replacement time* for this piece of equipment.)

(C) Sketch the graph of C, including any asymptotes.

★ **90. Average Cost.** The total cost of producing x units of a certain product is given by

$$C(x) = \tfrac{1}{5}x^2 + 2x + 2{,}000$$

The average cost per unit for producing x units is $\overline{C}(x) = C(x)/x$.

(A) Find the rational function $\overline{C}$.

(B) At what production level will the average cost per unit be minimal?

(C) Sketch the graph of $\overline{C}$, including any asymptotes.

★ **91. Construction.** A rectangular dog pen is to be made to enclose an area of 225 square feet.

(A) If x represents the width of the pen, express the total length $L(x)$ of the fencing material required for the pen in terms of x.

(B) Considering the physical limitations, what is the domain of the function L?

(C) Find the dimensions of the pen that will require the least amount of fencing material.

(D) Graph the function L, including any asymptotes.

★ **92. Construction.** Rework Problem 91 with the added assumption that the pen is to be divided into two sections, as shown in the figure. (Approximate dimensions to three decimal places.)

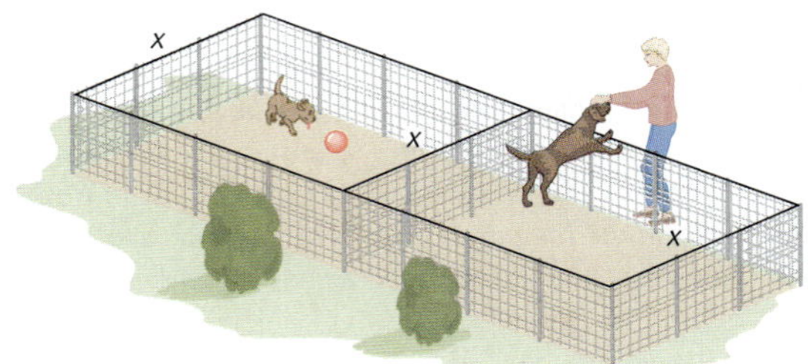

CHAPTER 3 REVIEW

3.1 Polynomial Functions and Models

A function that can be written in the form $P(x) = a_nx^n + a_{n-1}x^{n-1} + \ldots + a_1x + a_0$, $a_n \neq 0$, is a **polynomial function of degree *n*.** In this chapter, when not specified otherwise, the coefficients $a_n, a_{n-1}, \ldots, a_1, a_0$ are complex numbers and the domain of P is the set of complex numbers. A number r is said to be a **zero** (or **root**) of a function $P(x)$ if $P(r) = 0$. The zeros of $P(x)$ are thus the solutions of the equation $P(x) = 0$. The real zeros of $P(x)$ are just the x intercepts of the graph of $P(x)$. A point on a continuous graph that separates an increasing portion from a decreasing portion, or vice versa, is called a **turning point.** If $P(x)$ is a polynomial of degree $n > 0$ with real coefficients, then the graph of $P(x)$:

1. Is continuous for all real numbers
2. Has no sharp corners
3. Has at most n real zeros
4. Has at most $n - 1$ turning points
5. Increases or decreases without bound as $x \to \infty$ and as $x \to -\infty$

The left and right behavior of such a polynomial $P(x)$ is determined by its highest degree or **leading term:** As $x \to \pm\infty$, both a_nx^n and $P(x)$ approach $\pm\infty$, with the sign depending on n and the sign of a_n.

Let $P(x)$ be a polynomial of degree $n > 0$ and r a complex number. Then we have the following important results:

Division Algorithm

$P(x) = (x - r)Q(x) + R$ where the **quotient** $Q(x)$ and **remainder** R are unique.

Remainder Theorem

$P(r) = R$

Factor Theorem

$x - r$ is a factor of $P(x)$ if and only if $R = 0$.

Zeros Theorem

$P(x)$ has at most n zeros.

Synthetic division is an efficient method for dividing polynomials by linear terms of the form $x - r$ that is well suited to calculator use.

3.2 Real Zeros and Polynomial Inequalities

The following theorems are useful in locating and approximating all real zeros of a polynomial $P(x)$ of degree $n > 0$ with real coefficients, $a_n > 0$:

Upper and Lower Bound Theorem

1. Upper bound: A number $r > 0$ is an upper bound for the real zeros of $P(x)$ if, when $P(x)$ is divided by $x - r$ by synthetic division, all numbers in the quotient row, including the remainder, are nonnegative.
2. Lower bound: A number $r < 0$ is a lower bound for the real zeros of $P(x)$ if, when $P(x)$ is divided by $x - r$ by synthetic division, all numbers in the quotient row, including the remainder, alternate in sign.

Location Theorem

Suppose that a function f is continuous on an interval I that contains numbers a and b. If $f(a)$ and $f(b)$ have opposite signs, then the graph of f has at least one x intercept between a and b.

Polynomial inequalities can be solved by finding the zeros and inspecting the graph of an appropriate polynomial with real coefficients.

3.3 Complex Zeros and Rational Zeros of Polynomials

If $P(x)$ is a polynomial of degree $> n$ we have the following important theorems:

Fundamental Theorem of Algebra

$P(x)$ has at least one zero.

***n* Linear Factors Theorem**

$P(x)$ can be factored as a product of n linear factors.

If $P(x)$ is factored as a product of linear factors, the number of linear factors that have zero r is said to be the **multiplicity** of r.

Imaginary Zeros Theorem

Imaginary zeros of polynomials with real coefficients, if they exist, occur in conjugate pairs.

Linear and Quadratic Factors Theorem

If $P(x)$ has real coefficients, then $P(x)$ can be factored as a product of linear factors (with real coefficients) and quadratic factors (with real coefficients and imaginary zeros).

Real Zeros and Polynomials of Odd Degree

If $P(x)$ has odd degree and real coefficients, then the graph of P has at least one x intercept.

Zeros of Even or Odd Multiplicity

Let $P(x)$ have real coefficients:

1. If r is a real zero of $P(x)$ of even multiplicity, then $P(x)$ has a turning point at r and does not change sign at r.
2. If r is a real zero of $P(x)$ of odd multiplicity, then $P(x)$ does not have a turning point at r and changes sign at r.

Rational Zero Theorem

If the rational number b/c, in lowest terms, is a zero of the polynomial

$$P(x) = a_n x^n + a_{n-1}x^{n-1} + \cdots + a_1 x + a_0 \qquad a_n \neq 0$$

with integer coefficients, then b must be an integer factor of a_0 and c must be an integer factor of a_n.

If $P(x) = (x - r)Q(x)$, then $Q(x)$ is called a **reduced polynomial** for $P(x)$.

3.4 Rational Functions and Inequalities

A function f is a **rational function** if it can be written in the form

$$f(x) = \frac{p(x)}{q(x)}$$

where $p(x)$ and $q(x)$ are polynomials of degrees m and n, respectively (we assume $p(x)$ and $q(x)$ have no common factor). The graph of a rational function $f(x)$:

1. Is continuous with the exception of at most n real numbers
2. Has no sharp corners
3. Has at most m real zeros
4. Has at most $m + n - 1$ turning points
5. Has the same left and right behavior as the quotient of the leading terms of $p(x)$ and $q(x)$

The vertical line $x = a$ is a **vertical asymptote** for the graph of $y = f(x)$ if $f(x) \to \infty$ or $f(x) \to -\infty$ as $x \to a^+$ or as $x \to a^-$. If a is a zero of $q(x)$, then the line $x = a$ is a vertical asymptote of the graph of f. The horizontal line $y = b$ is a **horizontal asymptote** for the graph of $y = f(x)$ if $f(x) \to b$ as $x \to -\infty$ or as $x \to \infty$. The line $y = mx + b$ is an **oblique asymptote** if $[f(x) - (mx + b)] \to 0$ as $x \to -\infty$ or as $x \to \infty$.

Let $f(x) = \dfrac{a_m x^m + \ldots + a_1 x + a_0}{b_n x^n + \ldots + b_1 x + b_0}$, $a_m \neq 0$, $b_n \neq 0$.

1. If $m < n$, the line $y = 0$ (the x axis) is a horizontal asymptote.
2. If $m = n$, the line $y = a_m/b_n$ is a horizontal asymptote.
3. If $m > n$, there is no horizontal asymptote.

Analyzing and Sketching the Graph of a Rational Function: $f(x) = p(x)/q(x)$

Step 1. *Intercepts.* Find the real solutions of the equation $p(x) = 0$ and use these solutions to plot any x intercepts of the graph of f. Evaluate $f(0)$, if it exists, and plot the y intercept.

Step 2. *Vertical Asymptotes.* Find the real solutions of the equation $q(x) = 0$ and use these solutions to determine the domain of f, the points of discontinuity, and the vertical asymptotes. Sketch any vertical asymptotes as dashed lines.

Step 3. *Horizontal Asymptotes.* Determine whether there is a horizontal asymptote and if so, sketch it as a dashed line.

Step 4. *Complete the Sketch.* Using a graphing utility graph as an aid and the information determined in steps 1–3, sketch the graph.

Rational inequalities can be solved by finding the zeros of $p(x)$ and $q(x)$, for an appropriate rational function $f(x) = \frac{p(x)}{q(x)}$, and inspecting the graph of f.

CHAPTER 3 REVIEW EXERCISES

Work through all the problems in this chapter review, and check answers in the back of the book. Answers to all review problems are there, and following each answer is a number in italics indicating the section in which that type of problem is discussed. Where weaknesses show up, review appropriate sections in the text.

A

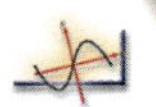

1. List the real zeros and turning points, and state the left and right behavior, of the polynomial function that has the indicated graph.

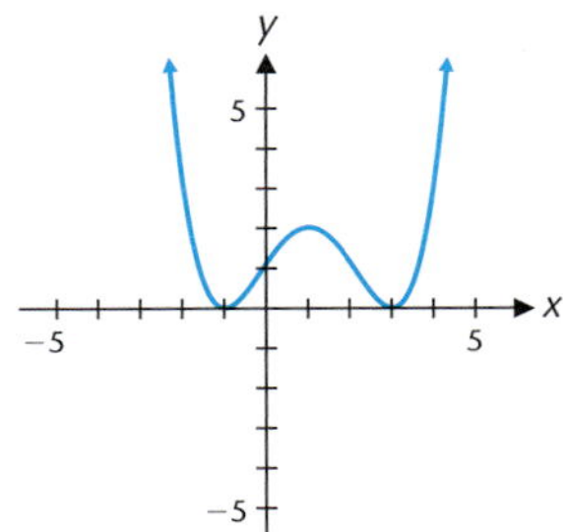

2. Use synthetic division to divide $P(x) = 2x^3 + 3x^2 - 1$ by $D(x) = x + 2$, and write the answer in the form $P(x) = D(x)Q(x) + R$.

3. If $P(x) = x^5 - 4x^4 + 9x^2 - 8$, find $P(3)$ using the remainder theorem and synthetic division.

4. What are the zeros of $P(x) = 3(x - 2)(x + 4)(x + 1)$?

5. If $P(x) = x^2 - 2x + 2$ and $P(1 + i) = 0$, find another zero of $P(x)$.

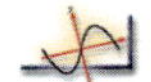

6. Let $P(x)$ be the polynomial whose graph is shown in the figure at the top of the next page.

(A) Assuming that $P(x)$ has integer zeros and leading coefficient 1, find the lowest-degree equation that could produce this graph.

(B) Describe the left and right behavior of $P(x)$.

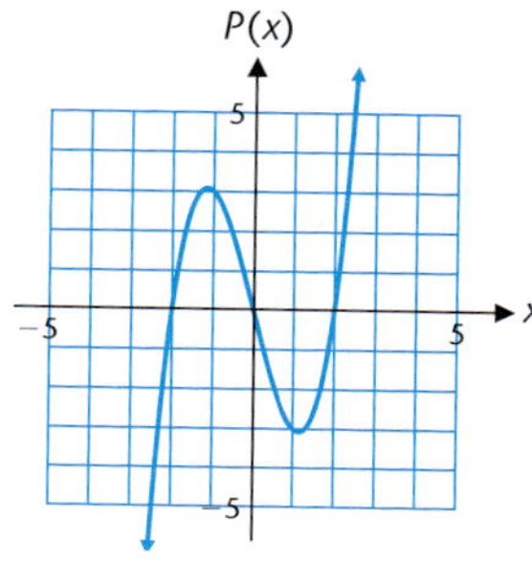

7. According to the upper and lower bound theorem, which of the following are upper or lower bounds of the zeros of $P(x) = x^3 - 4x^2 + 2$?

$-2, -1, 3, 4$

8. How do you know that $P(x) = 2x^3 - 3x^2 + x - 5$ has at least one real zero between 1 and 2?

9. Write the possible rational zeros for

$P(x) = x^3 - 4x^2 + x + 6.$

10. Find all rational zeros for $P(x) = x^3 - 4x^2 + x + 6$.

11. Find the domain and x intercept(s) for:

(A) $f(x) = \dfrac{2x - 3}{x + 4}$

(B) $g(x) = \dfrac{3x}{x^2 - x - 6}$

12. Find the horizontal and vertical asymptotes for the functions in Problem 11.

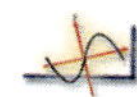

13. Explain why the graph is not the graph of a polynomial function.

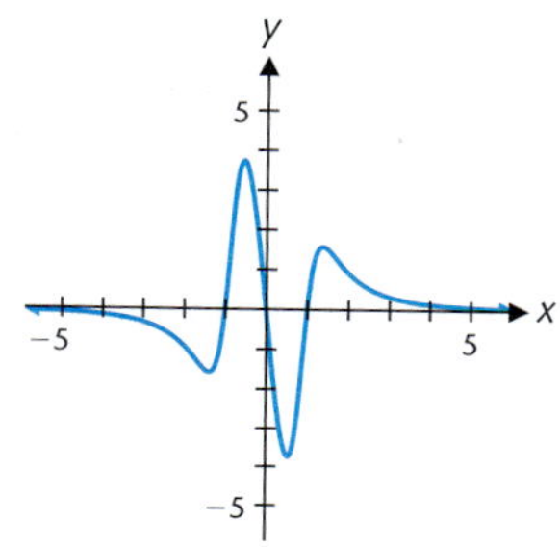

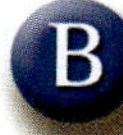

14. Let $P(x) = x^3 - 3x^2 - 3x + 4$.

(A) Graph $P(x)$ and describe the graph verbally, including the number of x intercepts, the number of turning points, and the left and right behavior.

(B) Approximate the largest x intercept to two decimal places.

15. If $P(x) = 8x^4 - 14x^3 - 13x^2 - 4x + 7$, find $Q(x)$ and R such that $P(x) = (x - \frac{1}{4})Q(x) + R$. What is $P(\frac{1}{4})$?

16. If $P(x) = 4x^3 - 8x^2 - 3x - 3$, find $P(-\frac{1}{2})$ using the remainder theorem and synthetic division.

17. Use the quadratic formula and the factor theorem to factor $P(x) = x^2 - 2x - 1$.

18. Is $x + 1$ a factor of $P(x) = 9x^{26} - 11x^{17} + 8x^{11} - 5x^4 - 7$? Explain, without dividing or using synthetic division.

19. Determine all rational zeros of $P(x) = 2x^3 - 3x^2 - 18x - 8$.

20. Factor the polynomial in Problem 19 into linear factors.

21. Find all rational zeros of $P(x) = x^3 - 3x^2 + 5$.

22. Find all zeros (rational, irrational, and imaginary) exactly for $P(x) = 2x^4 - x^3 + 2x - 1$.

23. Factor the polynomial in Problem 22 into linear factors.

24. If $P(x) = (x - 1)^2(x + 1)^3(x^2 - 1)(x^2 + 1)$, what is its degree? Write the zeros of $P(x)$, indicating the multiplicity of each if greater than 1.

25. Factor $P(x) = x^4 + 5x^2 - 36$ in two ways:

(A) As a product of linear factors (with real coefficients) and quadratic factors (with real coefficients and imaginary zeros)

(B) As a product of linear factors with complex coefficients

26. Let $P(x) = x^5 - 10x^4 + 30x^3 - 20x^2 - 15x - 2$.

(A) Approximate the zeros of $P(x)$ to two decimal places and state the multiplicity of each zero.

(B) Can any of these zeros be approximated with the bisection method? A maximum routine? A minimum routine? Explain.

27. Let $P(x) = x^4 - 2x^3 - 30x^2 - 25$.

(A) Find the smallest positive and largest negative integers that, by Theorem 1 in Section 3.2, are upper and lower bounds, respectively, for the real zeros of $P(x)$.

(B) If $(k, k + 1)$, k an integer, is the interval containing the largest real zero of $P(x)$, determine how many additional intervals are required in the bisection method to approximate this zero to one decimal place.

(C) Approximate the real zeros of $P(x)$ to two decimal places.

28. Let $f(x) = \dfrac{x - 1}{2x + 2}$.

(A) Find the domain and the intercepts for f.

(B) Find the vertical and horizontal asymptotes for f.

(C) Sketch a graph of f. Draw vertical and horizontal asymptotes with dashed lines.

29. Solve each polynomial inequality to three decimal places:

(A) $x^3 - 5x + 4 < 0$

(B) $x^3 - 5x + 4 < 2$

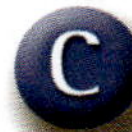

30. Explain why the graph is not the graph of a rational function.

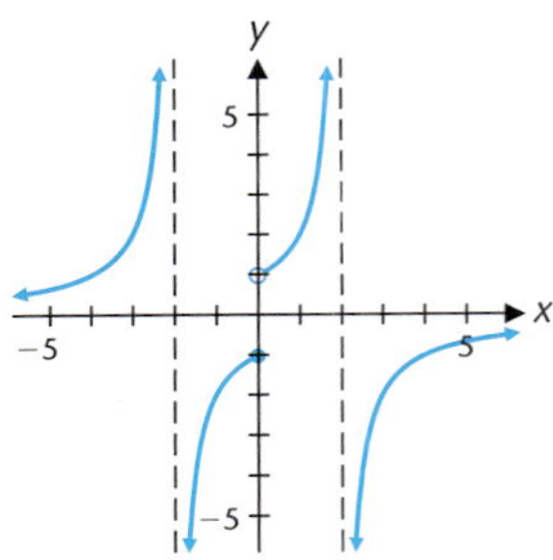

C

31. Use synthetic division to divide $P(x) = x^3 + 3x + 2$ by $[x - (1 + i)]$, and write the answer in the form $P(x) = D(x)Q(x) + R$.

32. Find a polynomial of lowest degree with leading coefficient 1 that has zeros $-\frac{1}{2}$ (multiplicity 2), -3, and 1 (multiplicity 3). (Leave the answer in factored form.) What is the degree of the polynomial?

33. Repeat Problem 32 for a polynomial $P(x)$ with zeros -5, $2 - 3i$, and $2 + 3i$.

34. Find all zeros (rational, irrational, and imaginary) exactly for $P(x) = 2x^5 - 5x^4 - 8x^3 + 21x^2 - 4$.

35. Factor the polynomial in Problem 34 into linear factors.

36. Let $P(x) = x^4 + 16x^3 + 47x^2 - 137x + 73$. Approximate (to two decimal places) the x intercepts and the local extrema.

37. What is the minimal degree of a polynomial $P(x)$, given that $P(-1) = -4$, $P(0) = 2$, $P(1) = -5$, and $P(2) = 3$? Justify your conclusion.

38. If $P(x)$ is a cubic polynomial with integer coefficients and if $1 + 2i$ is a zero of $P(x)$, can $P(x)$ have an irrational zero? Explain.

39. The solutions to the equation $x^3 - 27 = 0$ are the cube roots of 27.

(A) How many cube roots of 27 are there?

(B) 3 is obviously a cube root of 27; find all others.

40. Let $P(x) = x^4 + 2x^3 - 500x^2 - 4{,}000$.

(A) Find the smallest positive integer multiple of 10 and the largest negative integer multiple of 10 that, by Theorem 1 in Section 3.2, are upper and lower bounds, respectively, for the real zeros of $P(x)$.

(B) Approximate the real zeros of $P(x)$ to two decimal places.

41. Graph

$$f(x) = \frac{x^2 + 2x + 3}{x + 1}$$

Indicate any vertical, horizontal, or oblique asymptotes with dashed lines.

42. Use a graphing utility to find any horizontal asymptotes for

$$f(x) = \frac{2x}{\sqrt{x^2 + 3x + 4}}$$

43. Solve each rational inequality to three decimal places:

(A) $\dfrac{x^2 - 3}{x^3 - 3x + 1} \le 0$

(B) $\dfrac{x^2 - 3}{x^3 - 3x + 1} > \dfrac{5}{x^2}$

44. If $P(x) = x^3 - x^2 - 5x + 4$, determine the number of real zeros of $P(x)$ and explain why $P(x)$ has no rational zeros.

45. Give an example of a rational function $f(x)$ that satisfies the following conditions: the real zeros of f are -3, 0, and 2; the vertical asymptotes of f are the lines $x = -1$ and $x = 4$; and the line $y = 5$ is a horizontal asymptote.

APPLICATIONS

In Problems 46–49, express the solutions as the roots of a polynomial equation of the form $P(x) = 0$. Find rational solutions exactly and irrational solutions to three decimal places.

46. Architecture. An entryway is formed by placing a rectangular door inside an arch in the shape of the parabola with graph $y = 16 - x^2$, x and y in feet (see the figure). If the area of the door is 48 square feet, find the dimensions of the door.

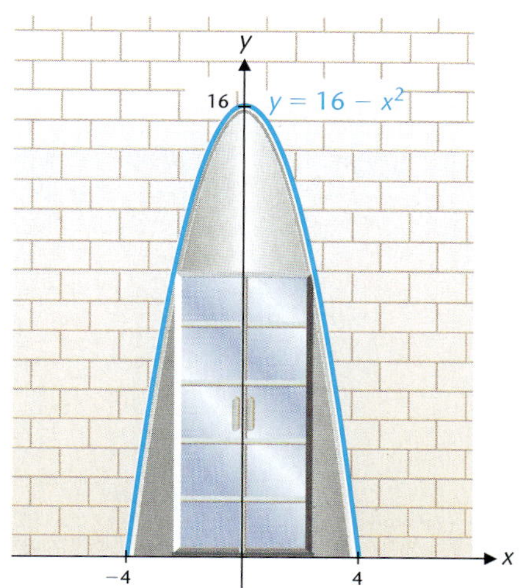

47. Construction. A grain silo is formed by attaching a hemisphere to the top of a right circular cylinder (see the figure). If the cylinder is 18 feet high and the volume of the silo is 486π cubic feet, find the common radius of the cylinder and the hemisphere.

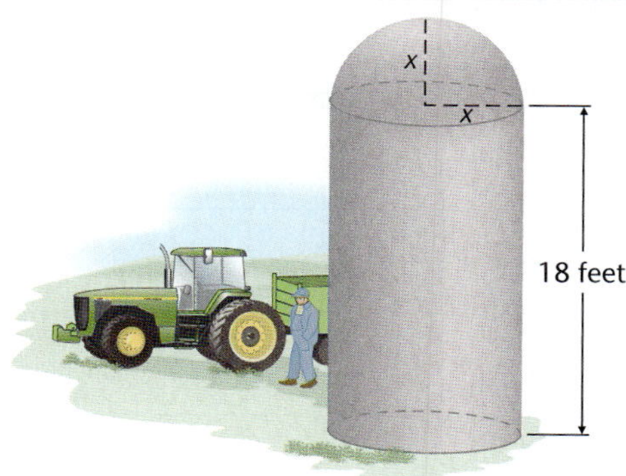

★ **48. Manufacturing.** A box is to be made out of a piece of cardboard that measures 15 by 20 inches. Squares, x inches on a side, will be cut from each corner, and then the ends and sides will be folded up (see the figure). Find the value of x that would result in a box with a volume of 300 cubic inches.

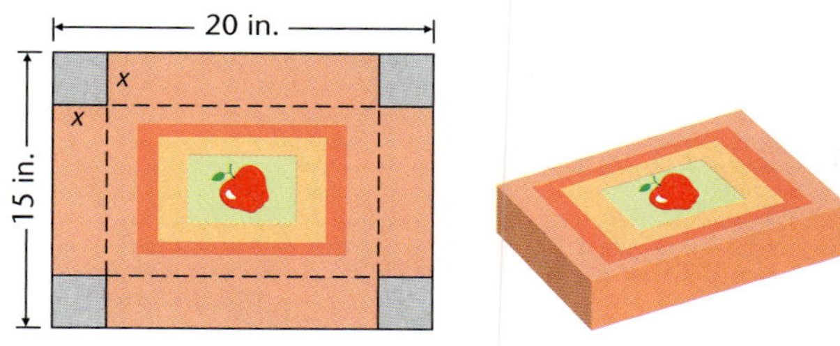

★ **49. Geometry.** Find all points on the graph of $y = x^2$ that are three units from the point (1, 4).

MODELING AND DATA ANALYSIS 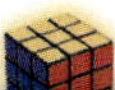

50. Advertising. A chain of appliance stores uses television ads to promote the sale of refrigerators. Analyzing past records produced the data in the table, where x is the number of ads placed monthly and y is the number of refrigerators sold that month.

(A) Find a cubic regression equation for these data using the number of ads as the independent variable.

(B) Estimate (to the nearest integer) the number of refrigerators that would be sold if 15 ads are placed monthly.

(C) Estimate (to the nearest integer) the number of ads that should be placed to sell 750 refrigerators monthly.

Number of Ads x	Number of Refrigerators y
10	270
20	430
25	525
30	630
45	890
48	915

51. Women in the Workforce. It is reasonable to conjecture from the data given in the table that many Japanese women tend to leave the workforce to marry and have children, but then reenter the workforce when the children are grown.

(A) Explain why you might expect cubic regression to provide a better fit to the data than linear or quadratic regression.

(B) Find a cubic regression model for these data using age as the independent variable.

(C) Use the regression equation to estimate (to the nearest year) the ages at which 65% of the women are in the workforce.

Women in the Workforce in Japan (1997)

Age	Percentage of Women Employed
22	73
27	65
32	56
37	63
42	71
47	72
52	68
57	59
62	42

CHAPTER 3 GROUP ACTIVITY

Interpolating Polynomials

Given two points in the plane, we can use the point–slope form of the equation of a line to find a polynomial whose graph passes through these two points. How can we proceed if we are given more than two points? For example, how can we find the equation of a polynomial $P(x)$ whose graph passes through the points listed in Table 1 and graphed in Figure 1?

TABLE 1

x	1	2	3	4
$P(x)$	1	3	−3	1

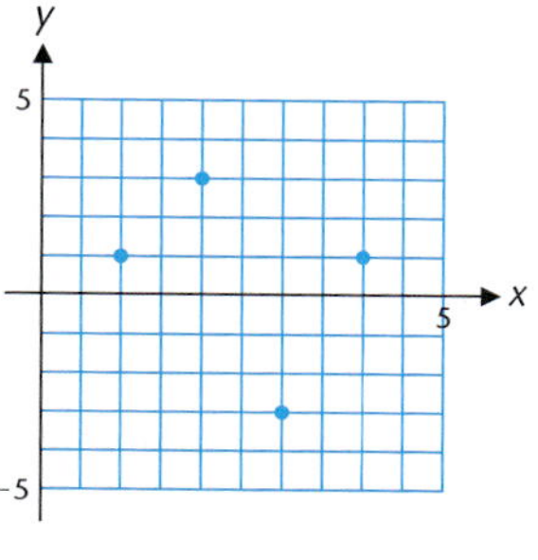

FIGURE 1

The key to solving this problem is to write the unknown polynomial $P(x)$ in the following special form:

$$P(x) = a_0 + a_1(x - 1) + a_2(x - 1)(x - 2) + a_3(x - 1)(x - 2)(x - 3) \qquad (1)$$

Because the graph of $P(x)$ is to pass through each point in Table 1, we can substitute each value of x in equation (1) to determine the coefficients a_0, a_1, a_2, and a_3. First we evaluate equation (1) at $x = 1$ to determine a_0:

$$\begin{aligned} 1 &= P(1) \\ &= a_0 \qquad \text{All other terms in equation (1) are 0 when } x = 1. \end{aligned}$$

Using this value for a_0 in equation (1) and evaluating at $x = 2$, we have

$$\begin{aligned} 3 &= P(2) = 1 + a_1(1) \qquad \text{All other terms are 0.} \\ 2 &= a_1 \end{aligned}$$

Continuing in this manner, we have

$$\begin{aligned} -3 &= P(3) = 1 + 2(2) + a_2(2)(1) \\ -8 &= 2a_2 \\ -4 &= a_2 \\ 1 &= P(4) = 1 + 2(3) - 4(3)(2) + a_3(3)(2)(1) \\ 18 &= 6a_3 \\ 3 &= a_3 \end{aligned}$$

We have now evaluated all the coefficients in equation (1) and can write

$$P(x) = 1 + 2(x - 1) - 4(x - 1)(x - 2) + 3(x - 1)(x - 2)(x - 3) \qquad (2)$$

If we expand the products in equation (2) and collect like terms, we can express $P(x)$ in the more conventional form (verify this):

$$P(x) = 3x^3 - 22x^2 + 47x - 27$$

(A) To check these calculations, evaluate $P(x)$ at $x = 1, 2, 3$, and 4 and compare the results with Table 1. Then add the graph of $P(x)$ to Figure 1.

(B) Write a verbal description of the special form of $P(x)$ in equation (1).

In general, given a set of $n + 1$ points:

x	x_0	x_1	$\cdots$	x_n
y	y_0	y_1	$\cdots$	y_n

the **interpolating polynomial** for these points is the polynomial $P(x)$ of degree less than or equal to n that satisfies $P(x_k) = y_k$ for $k = 0, 1, \ldots, n$. The **general form** of the interpolating polynomial is

$$P(x) = a_0 + a_1(x - x_0) + a_2(x - x_0)(x - x_1) + \cdots + a_n(x - x_0)(x - x_1) \cdot \cdots \cdot (x - x_{n-1})$$

(C) Summarize the procedure for using the points in the table to find the coefficients in the general form.

(D) Give an example to show that the interpolating polynomial can have degree strictly less than n.

(E) Could there be two different polynomials of degree less than or equal to n whose graph passes through the given $n + 1$ points? Justify your answer.

(F) Find the interpolating polynomial for each of Tables 2 and 3. Check your answers by evaluating the polynomial, and illustrate by graphing the points in the table and the polynomial in the same viewing window.

TABLE 2

x	−1	0	1	2
y	5	3	3	11

TABLE 3

x	−2	−1	0	1	2
y	−3	0	5	0	−3

A surprisingly short program on a graphing utility can be used to calculate the coefficients in the general form of an interpolating polynomial.* Figure 2 shows the output generated when we use the program to find the coefficients of the interpolating polynomial for Table 1.

FIGURE 2

```
{1,2,3,4}→L1
            {1 2 3 4}
{1,3,-3,1}→L2
           {1 3 -3 1}
INTERP
           {1 2 -4 3}
                 Done
```

*This program is available at the website for this book (see Preface).

CHAPTER 4

Exponential and Logarithmic Functions

OUTLINE

MOST OF THE FUNCTIONS WE HAVE CONSIDERED SO FAR have been polynomial or rational functions, with a few others involving roots of polynomial or rational functions. Functions that can be expressed in terms of addition, subtraction, multiplication, division, and the taking of roots of variables and constants are called *algebraic functions*.

In Chapter 4 we introduce and investigate the properties of *exponential functions* and *logarithmic functions*. These functions are not algebraic; they belong to the class of *transcendental functions*. Exponential and logarithmic functions are used to model a variety of real-world phenomena: growth of populations of people, animals, and bacteria; radioactive decay; epidemics; absorption of light as it passes through air, water, or glass; magnitudes of sounds and earthquakes. We consider applications in these areas plus many more in the sections that follow.

Preparing for this chapter

Before getting started on this chapter, review the following concepts:

- Exponents
  (Basic Algebra Review*, Sec. 5 and 6)
- Functions
  (Chapter 1, Section 2)
- Graphs of Functions
  (Chapter 1, Section 3)
- Quadratic Equations
  (Chapter 2, Section 5)
- Equation-Solving Techniques
  (Chapter 2, Section 6)

*At www.mhhe.com/barnett

SECTION 4.1 Exponential Functions

Exponential Functions • Graphs of Exponential Functions • Additional Exponential Properties • Base *e* Exponential Function • Compound Interest • Continuous Compound Interest

In Section 4.1 we introduce exponential functions and investigate their properties and graphs. We also study applications of exponential functions in the mathematics of finance.

Exponential Functions

Let's start by noting that the functions f and g given by

$$f(x) = 2^x \qquad \text{and} \qquad g(x) = x^2$$

are not the same function. Whether a variable appears as an exponent with a constant base or as a base with a constant exponent makes a big difference. The function g is a quadratic function, which we have already discussed. The function f is a new type of function called an *exponential function*.

The values of the exponential function $f(x) = 2^x$ for x an integer are easy to compute [Fig. 1(a)]. If $x = m/n$ is a rational number, then $f(m/n) = \sqrt[n]{2^m}$, which can be evaluated on almost any calculator [Fig. 1(b)]. Finally, a graphing utility can graph the function $f(x) = 2^x$ [Fig. 1(c)] for any given interval of x values.

FIGURE 1 $f(x) = 2^x$.

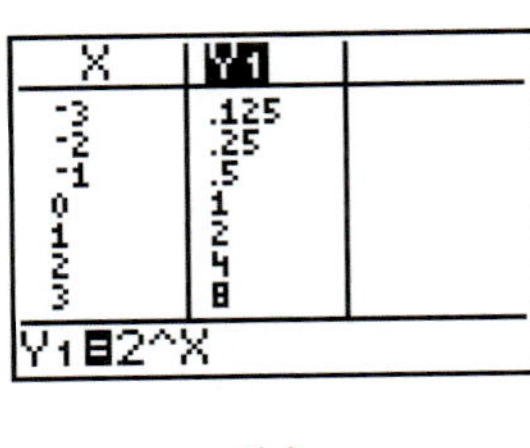

(a)

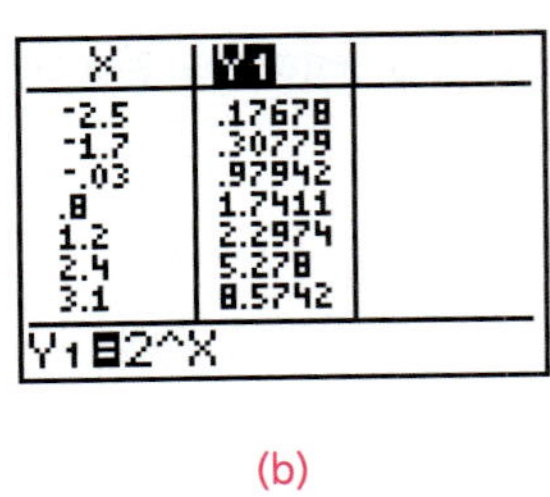

(b)

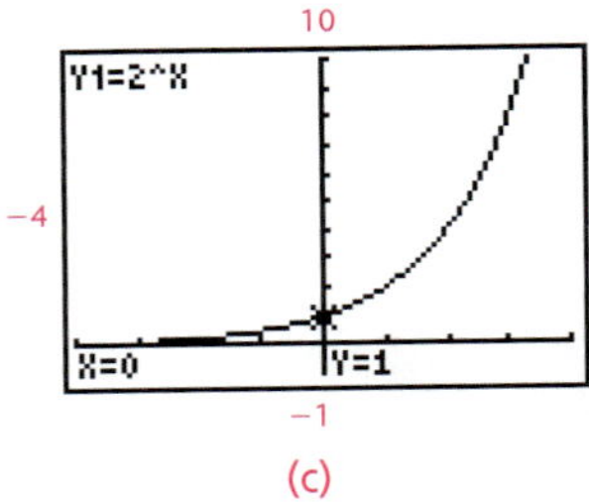

(c)

The only catch is that we have not yet defined 2^x for *all* real numbers. For example, what does

$$2^{\sqrt{2}}$$

mean? The question is not easy to answer at this time. In fact, a precise definition of $2^{\sqrt{2}}$ must wait for more advanced courses, where we can show that, if b is a positive real number and x is any real number, then

$$b^x$$

names a real number, and the graph of $f(x) = 2^x$ is as indicated in Figure 1. We also can show that for x irrational, b^x can be approximated as closely as we like by using rational number approximations for x. Because $\sqrt{2} = 1.414213 \ldots$, for example, the sequence

$$2^{1.4}, 2^{1.41}, 2^{1.414}, \ldots$$

approximates $2^{\sqrt{2}}$, and as we use more decimal places, the approximation improves.

DEFINITION 1
Exponential Function

The equation

$$f(x) = b^x \qquad b > 0,\ b \neq 1$$

defines an **exponential function** for each different constant b, called the **base.** The independent variable x may assume any real value.

Thus, the **domain of f** is the set of all real numbers, and it can be shown that the **range of f** is the set of all positive real numbers. We require the base b to be positive to avoid imaginary numbers such as $(-2)^{1/2}$.

Graphs of Exponential Functions

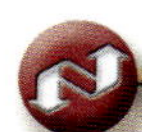

EXPLORE/DISCUSS 1

Compare the graphs of $f(x) = 3^x$ and $g(x) = 2^x$ by graphing both functions in the same viewing window. Find all points of intersection of the graphs. For which values of x is the graph of f above the graph of g? Below the graph of g? Are the graphs of f and g close together as $x \to \infty$? As $x \to -\infty$? Discuss.

It is useful to compare the graphs of $y = 2^x$ and $y = (\frac{1}{2})^x = 2^{-x}$ by plotting both on the same coordinate system, as shown in Figure 2(a). The graph of

$$f(x) = b^x \qquad b > 1 \qquad \text{Fig. 2(b)}$$

looks very much like the graph of the particular case $y = 2^x$, and the graph of

$$f(x) = b^x \qquad 0 < b < 1 \qquad \text{Fig. 2(b)}$$

looks very much like the graph of $y = (\frac{1}{2})^x$. Note in both cases that the x axis is a *horizontal asymptote* for the graph.

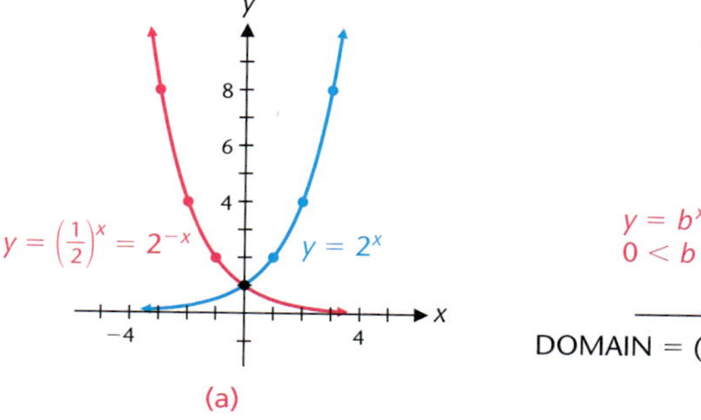

FIGURE 2
Basic exponential graphs.

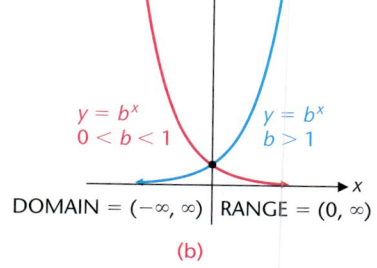

The graphs in Figure 2 suggest that the graphs of exponential functions have the properties listed in Theorem 1, which we state without proof.

THEOREM 1

Properties of Graphs of Exponential Functions

Let $f(x) = b^x$ be an exponential function, $b > 0$, $b \neq 1$. Then the graph of $f(x)$:

1. Is continuous for all real numbers
2. Has no sharp corners
3. Passes through the point (0, 1)
4. Lies above the x axis, which is a horizontal asymptote
5. Increases as x increases if $b > 1$; decreases as x increases if $0 < b < 1$
6. Intersects any horizontal line at most once (that is, f is one-to-one)

Property 4 of Theorem 1 implies that the graph of an exponential function cannot be the graph of a polynomial function. Properties 4 and 5 together imply that the graph of an exponential function cannot be the graph of a rational function. Property 6 implies that exponential functions have inverses; those inverses, called *logarithmic functions,* are discussed in Section 4.3.

Graphing exponential functions on a graphing utility is routine, but interpreting the results requires an understanding of the preceding properties.

EXAMPLE 1 Graphing Exponential Functions

Let $f(x) = \frac{1}{2}(4^x)$. Construct a table of values (rounded to two decimal places) for $f(x)$ using integer values from -3 to 3. Graph f on a graphing utility and then sketch a graph by hand.

SOLUTION

Set the graphing utility in two-decimal-place mode, construct the table [Fig. 3(a)], and graph the function [Fig. 3(b)]. The points on the graph of $f(x)$ for $x < 0$ are indistinguishable from the x axis in Figure 3(b). However, from the properties of an exponential function, we know that $f(x) > 0$ for all real numbers x and that $f(x) \to 0$ as $x \to -\infty$. The hand sketch in Figure 3(c) illustrates the behavior for $x < 0$ more clearly. Of course, zooming in on the graphing utility will also illustrate this behavior.

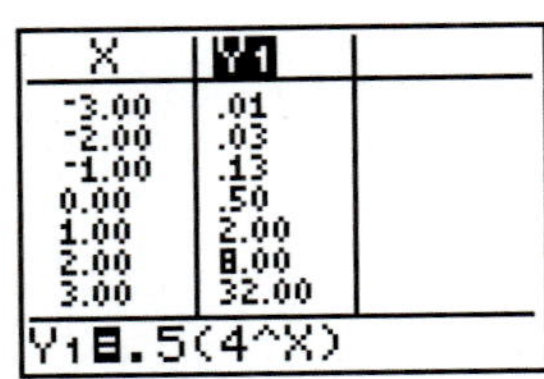

X	Y1
-3.00	.01
-2.00	.03
-1.00	.13
0.00	.50
1.00	2.00
2.00	8.00
3.00	32.00

Y1=.5(4^X)

(a)

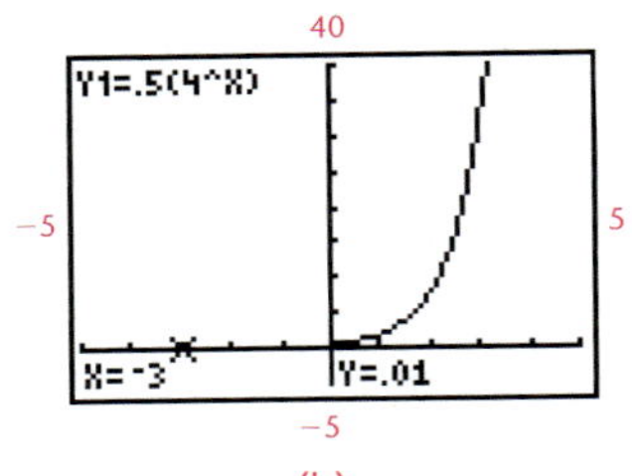

(b)

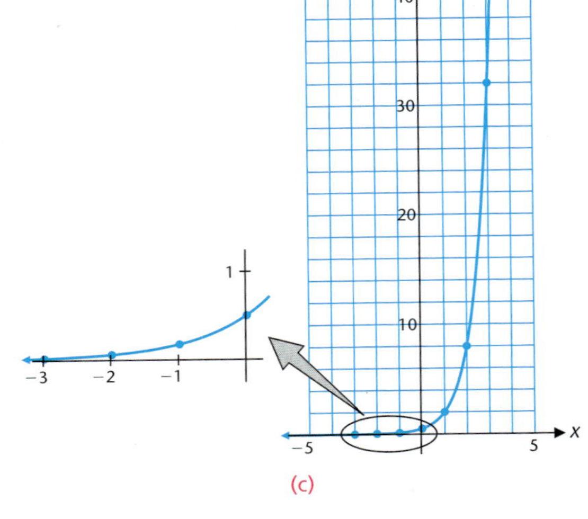

(c)

FIGURE 3

MATCHED 1 PROBLEM

Repeat Example 1 for $y = \frac{1}{2}\left(\frac{1}{4}\right)^x = \frac{1}{2}(4^{-x})$.

Additional Exponential Properties

Exponential functions whose domains include irrational numbers obey the familiar laws of exponents for rational exponents. We summarize these exponent laws here and add two other important and useful properties.

Exponential Function Properties

For a and b positive, $a \neq 1$, $b \neq 1$, and x and y real:

1. Exponent laws:

$$a^x a^y = a^{x+y} \qquad (a^x)^y = a^{xy} \qquad (ab)^x = a^x b^x$$

$$\left(\frac{a}{b}\right)^x = \frac{a^x}{b^x} \qquad \frac{a^x}{a^y} = a^{x-y} \qquad \frac{2^{5x}}{2^{7x}} = 2^{5x-7x} = 2^{-2x}$$

2. $a^x = a^y$ if and only if $x = y$. If $6^{4x} = 6^{2x+4}$, then $4x = 2x + 4$, and $x = 2$.
3. For $x \neq 0$, $a^x = b^x$ if and only if $a = b$. If $a^4 = 3^4$, then $a = 3$.

Property 2 is another way to express the fact that the exponential function $f(x) = a^x$ is one-to-one (see property 6 of Theorem 1). Because all exponential functions pass through the point (0, 1) (see property 3 of Theorem 1), property 3 indicates that the graphs of exponential functions with different bases do not intersect at any other points.

EXAMPLE 2 **Using Exponential Function Properties**

Solve $4^{x-3} = 8$ for x.

SOLUTION

Algebraic Solution

Express both sides in terms of the same base, and use property 2 to equate exponents.

$$4^{x-3} = 8$$

$$(2^2)^{x-3} = 2^3 \quad \text{Express 4 and 8 as powers of 2.}$$

$$2^{2x-6} = 2^3 \quad (a^x)^y = a^{xy}$$

$$2x - 6 = 3 \quad \text{Property 2}$$

$$2x = 9$$

$$x = \frac{9}{2}$$

CHECK

$$4^{(9/2)-3} = 4^{3/2} = (\sqrt{4})^3 = 2^3 \overset{\checkmark}{=} 8$$

Graphical Solution

Graph $y_1 = 4^{x-3}$ and $y_2 = 8$. Use the intersect command to obtain $x = 4.5$ (Fig. 4).

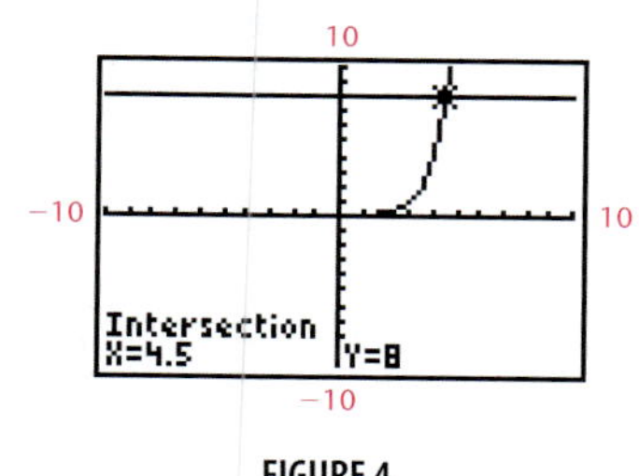

FIGURE 4

MATCHED 2 PROBLEM

Solve $27^{x+1} = 9$ for x.

Base e Exponential Function

Surprisingly, among the exponential functions it is not the function $g(x) = 2^x$ with base 2 or the function $h(x) = 10^x$ with base 10 that is used most frequently in mathematics. Instead, it is the function $f(x) = e^x$ with base e, where e is the limit of the expression

$$\left(1 + \frac{1}{x}\right)^x \tag{1}$$

as x gets larger and larger.

EXPLORE/DISCUSS 2

(A) Calculate the values of $[1 + (1/x)]^x$ for $x = 1, 2, 3, 4$, and 5. Are the values increasing or decreasing as x gets larger?

(B) Graph $y = [1 + (1/x)]^x$ and discuss the behavior of the graph as x increases without bound.

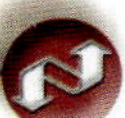

TABLE 1

x	$\left(1 + \frac{1}{x}\right)^x$
1	2
10	2.593 74 . . .
100	2.704 81 . . .
1,000	2.716 92 . . .
10,000	2.718 14 . . .
100,000	2.718 27 . . .
1,000,000	2.718 28 . . .

By calculating the value of expression (1) for larger and larger values of x (Table 1), it appears that $[1 + (1/x)]^x$ approaches a number close to 2.7183. In a calculus course we can show that as x increases without bound, the value of $[1 + (1/x)]^x$ approaches an irrational number that we call e. Just as irrational numbers such as π and $\sqrt{2}$ have unending, nonrepeating decimal representations, e also has an unending, nonrepeating decimal representation. To 12 decimal places (see Basic Algebra Review, Sec. 1, www.mhhe.com/barnett),

$$e = 2.718\ 281\ 828\ 459$$

Exactly who discovered e is still being debated. It is named after the great Swiss mathematician Leonhard Euler (1707–1783), who computed e to 23 decimal places using $[1 + (1/x)]^x$.

The constant e turns out to be an ideal base for an exponential function because in calculus and higher mathematics many operations take on their simplest form using this base. This is why you will see e used extensively in expressions and formulas that model real-world phenomena.

DEFINITION 2

Exponential Function with Base e

For x a real number, the equation

$$f(x) = e^x$$

defines the **exponential function with base e.**

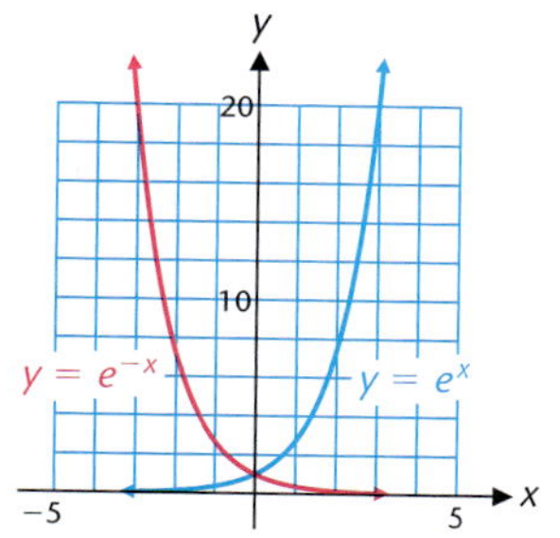

FIGURE 5 Exponential functions with base e.

The exponential function with base e is used so frequently that it is often referred to as *the* exponential function. The graphs of $y = e^x$ and $y = e^{-x}$ are shown in Figure 5.

EXPLORE/DISCUSS 3

(A) Graph $y_1 = e^x$, $y_2 = e^{0.5x}$, and $y_3 = e^{2x}$ in the same viewing window. How do these graphs compare with the graph of $y = b^x$ for $b > 1$?

(B) Graph $y_1 = e^{-x}$, $y_2 = e^{-0.5x}$, and $y_3 = e^{-2x}$ in the same viewing window. How do these graphs compare with the graph of $y = b^x$ for $0 < b < 1$?

(C) Use the properties of exponential functions to show that all of these functions are exponential functions.

EXAMPLE 3 Analyzing a Graph

Describe the graph of $f(x) = 4 - e^{x/2}$, including x and y intercepts, increasing and decreasing properties, and horizontal asymptotes. Round any approximate values to two decimal places.

SOLUTION

The graph of f is shown in Figure 6(a). The y intercept is $f(0) = 4 - 1 = 3$ and the x intercept is 2.77 (to two decimal places). The graph shows that f is decreasing for all x. Because the exponential function $e^{x/2} \to 0$ as $x \to -\infty$, it follows that $f(x) = 4 - e^{x/2} \to 4$ as $x \to -\infty$. The table in Figure 6(b) confirms this. Thus, the line $y = 4$ is a horizontal asymptote for the graph.

FIGURE 6 $f(x) = 4 - e^{x/2}$.

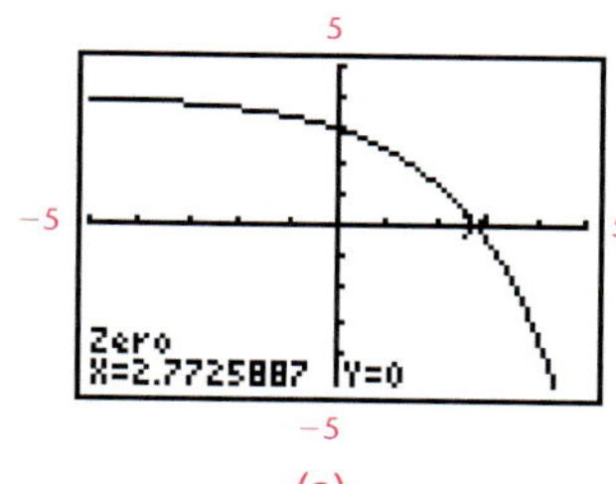

(a)

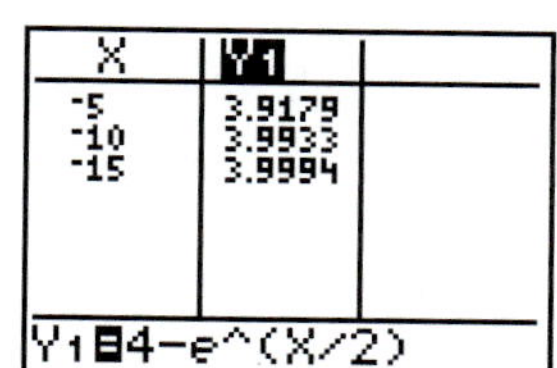

X	Y1
-5	3.9179
-10	3.9933
-15	3.9994

Y1=4−e^(X/2)

(b)

MATCHED 3 PROBLEM

Describe the graph of $f(x) = 2e^{x/2} - 5$, including x and y intercepts, increasing and decreasing properties, and horizontal asymptotes. Round any approximate values to two decimal places.

Compound Interest

The fee paid to use another's money is called **interest.** It is usually computed as a percentage, called the **interest rate,** of the principal over a given time. If, at the end of a payment period, the interest due is reinvested at the same rate, then the interest earned as well as the principal will earn interest during the next payment period. Interest paid on interest reinvested is called **compound interest.**

Suppose you deposit \$1,000 in a savings and loan that pays 8% compounded semiannually. How much will the savings and loan owe you at the end of 2 years? Compounded semiannually means that interest is paid to your account at the end of each 6-month period, and the interest will in turn earn interest. The **interest rate per period** is the annual rate, 8% = 0.08, divided by the number of compounding periods per year, 2. If we let A_1, A_2, A_3, and A_4 represent the new amounts due at the end of the first, second, third, and fourth periods, respectively, then

$$\begin{aligned}
A_1 &= \$1{,}000 + \$1{,}000\left(\frac{0.08}{2}\right) \\
&= \$1{,}000(1 + 0.04) && P\left(1 + \frac{r}{n}\right) \\
A_2 &= A_1(1 + 0.04) \\
&= [\$1{,}000(1 + 0.04)](1 + 0.04) \\
&= \$1{,}000(1 + 0.04)^2 && P\left(1 + \frac{r}{n}\right)^2 \\
A_3 &= A_2(1 + 0.04) \\
&= [\$1{,}000(1 + 0.04)^2](1 + 0.04) \\
&= \$1{,}000(1 + 0.04)^3 && P\left(1 + \frac{r}{n}\right)^3 \\
A_4 &= A_3(1 + 0.04) \\
&= [\$1{,}000(1 + 0.04)^3](1 + 0.04) \\
&= \$1{,}000(1 + 0.04)^4 && P\left(1 + \frac{r}{n}\right)^4
\end{aligned}$$

What do you think the savings and loan will owe you at the end of 6 years? If you guessed

$$A = \$1{,}000(1 + 0.04)^{12}$$

you have observed a pattern that is generalized in the following compound interest formula:

Compound Interest

If a **principal** P is invested at an annual **rate** r compounded m times a year, then the **amount** A in the account at the end of n compounding periods is given by

$$A = P\left(1 + \frac{r}{m}\right)^n$$

The annual rate r is expressed in decimal form.

EXAMPLE 4 Compound Interest

If you deposit \$5,000 in an account paying 9% compounded daily, how much will you have in the account in 5 years? Compute the answer to the nearest cent.

SOLUTION

Algebraic Solution

We use the compound interest formula with $P = 5{,}000$, $r = 0.09$, $m = 365$, and $n = 5(365) = 1{,}825$:

$$A = P\left(1 + \frac{r}{m}\right)^n$$

$$= 5{,}000\left(1 + \frac{0.09}{365}\right)^{1825}$$

$$= \$7{,}841.13$$

Graphical Solution

Graphing

$$A = 5{,}000\left(1 + \frac{0.09}{365}\right)^x$$

and using trace (Fig. 7) shows $A = \$7{,}841.13$.

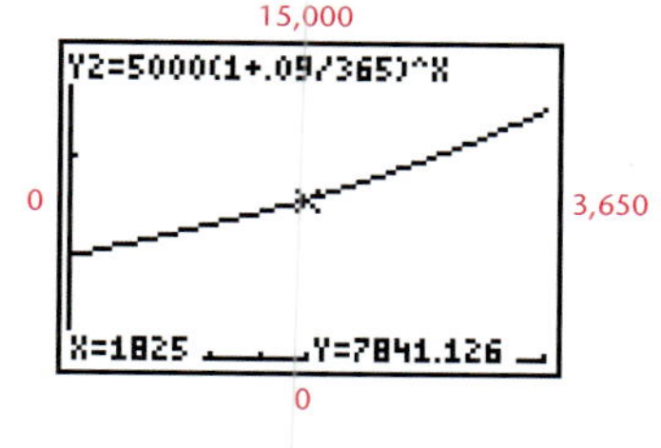

FIGURE 7

MATCHED PROBLEM 4

If \$1,000 is invested in an account paying 10% compounded monthly, how much will be in the account at the end of 10 years? Compute the answer to the nearest cent.

EXAMPLE 5 Comparing Investments

If \$1,000 is deposited into an account earning 10% compounded monthly and, at the same time, \$2,000 is deposited into an account earning 4% compounded monthly, will the first account ever be worth more than the second? If so, when?

SOLUTION

Let y_1 and y_2 represent the amounts in the first and second accounts, respectively, then

$$y_1 = 1{,}000(1 + 0.10/12)^x$$
$$y_2 = 2{,}000(1 + 0.04/12)^x$$

where x is the number of compounding periods (months). Examining the graphs of y_1 and y_2 [Fig. 8(a)], we see that the graphs intersect at $x \approx 139.438$ months. Because compound interest is paid at the end of each compounding period, we compare the amount in the accounts after 139 months and after 140 months [Fig. 8(b)]. Thus, the first account is worth more than the second for $x \geq 140$ months or 11 years and 8 months.

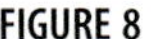

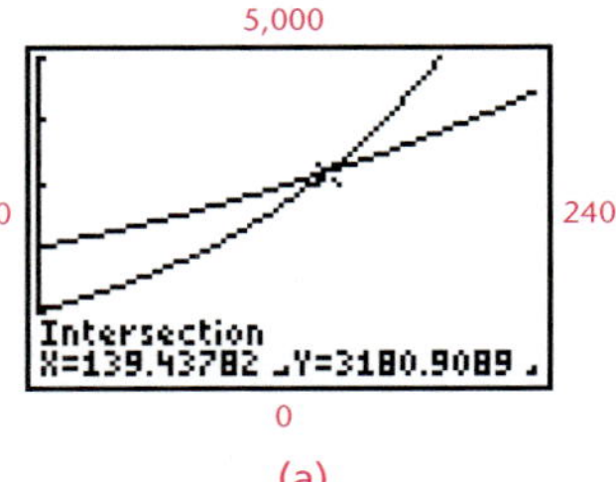

(a)

X	Y1	Y2
139	3169.4	3176.3
140	3195.8	3186.9

X=

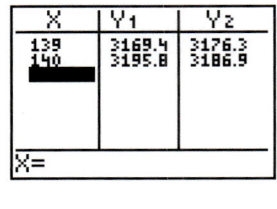

(b)

MATCHED 5 PROBLEM

If \$4,000 is deposited into an account earning 10% compounded quarterly and, at the same time, \$5,000 is deposited into an account earning 6% compounded quarterly, when will the first account be worth more than the second?

Continuous Compound Interest

If \$100 is deposited in an account that earns compound interest at an annual rate of 8% for 2 years, how will the amount A change if the number of compounding periods is increased? If m is the number of compounding periods per year, then

$$A = 100\left(1 + \frac{0.08}{m}\right)^{2m}$$

The amount A is computed for several values of m in Table 2. Notice that the largest gain appears in going from annually to semiannually. Then, the gains slow down as m increases. In fact, it appears that A might be tending to something close to \$117.35 as m gets larger and larger.

TABLE 2 Effect of Compounding Frequency

Compounding Frequency	m	$A = 100\left(1 + \frac{0.08}{m}\right)^{2m}$
Annually	1	$116.6400
Semiannually	2	116.9859
Quarterly	4	117.1659
Weekly	52	117.3367
Daily	365	117.3490
Hourly	8,760	117.3501

We now return to the general problem to see if we can determine what happens to $A = P[1 + (r/m)]^{mt}$ as m increases without bound. A little algebraic manipulation of the compound interest formula will lead to an answer and a significant result in the mathematics of finance:

$$A = P\left(1 + \frac{r}{m}\right)^{mt}$$

$$= P\left(1 + \frac{1}{m/r}\right)^{(m/r)rt} \qquad \text{Change algebraically.}$$

$$= \left[\left(1 + \frac{1}{x}\right)^{x}\right]^{rt} \qquad \text{Let } x = m/r.$$

The expression within the square brackets should look familiar. Recall from the first part of this section that

$$\left(1 + \frac{1}{x}\right)^{x} \to e \qquad \text{as} \qquad x \to \infty$$

Because r is fixed, $x = m/r \to \infty$ as $m \to \infty$. Thus,

$$P\left(1 + \frac{r}{m}\right)^{mt} \to Pe^{rt} \qquad \text{as} \qquad m \to \infty$$

and we have arrived at the **continuous compound interest formula,** a very important and widely used formula in business, banking, and economics.

Continuous Compound Interest Formula

If a principal P is invested at an annual rate r compounded continuously, then the amount A in the account at the end of t years is given by

$$A = Pe^{rt}$$

The annual rate r is expressed as a decimal.

EXAMPLE 6 Continuous Compound Interest

If \$100 is invested at an annual rate of 8% compounded continuously, what amount, to the nearest cent, will be in the account after 2 years?

SOLUTION

Algebraic Solution

Use the continuous compound interest formula to find A when $P = \$100$, $r = 0.08$, and $t = 2$:

$$A = Pe^{rt}$$
$$= \$100e^{(0.08)(2)} \quad \text{8\% is equivalent to } r = 0.08.$$
$$= \$117.35$$

Compare this result with the values calculated in Table 2.

Graphical Solution

Graphing

$$A = 100e^{0.08x}$$

and using trace (Fig. 9) shows $A = \$117.35$.

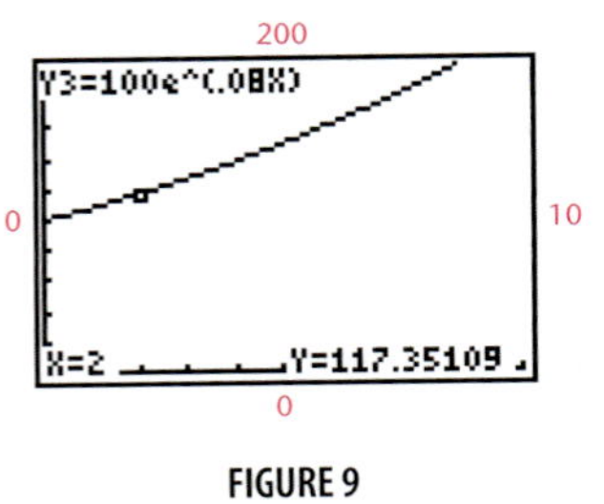

FIGURE 9

MATCHED PROBLEM 6

What amount will an account have after 5 years if \$100 is invested at an annual rate of 12% compounded annually? Quarterly? Continuously? Compute answers to the nearest cent.

ANSWERS MATCHED PROBLEMS

1. $y = \frac{1}{2}(4^{-x})$

x	y
−3	32.00
−2	8.00
−1	2.00
0	0.50
1	0.13
2	0.03
3	0.01

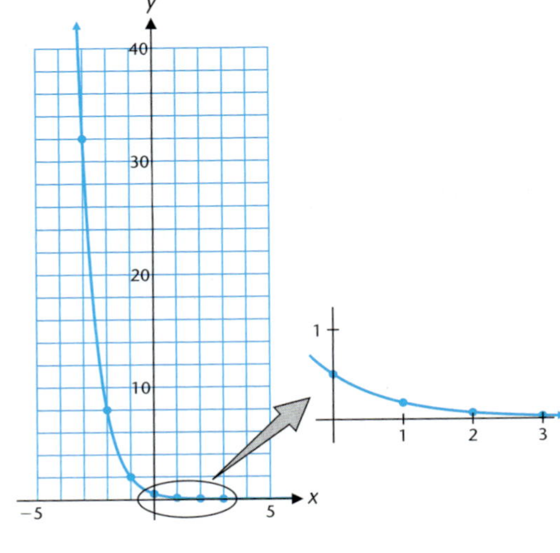

2. $x = -\frac{1}{3}$
3. y intercept: -3; x intercept: 1.83; increasing for all x; horizontal asymptote: $y = -5$
4. \$2,707.04
5. After 23 quarters
6. Annually: \$176.23; quarterly: \$180.61; continuously: \$182.21

EXERCISE 4.1

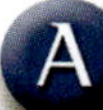

1. Match each equation with the graph of f, g, m, or n in the figure.

   (A) $y = (0.2)^x$ **(B)** $y = 2^x$

   (C) $y = (\frac{1}{3})^x$ **(D)** $y = 4^x$

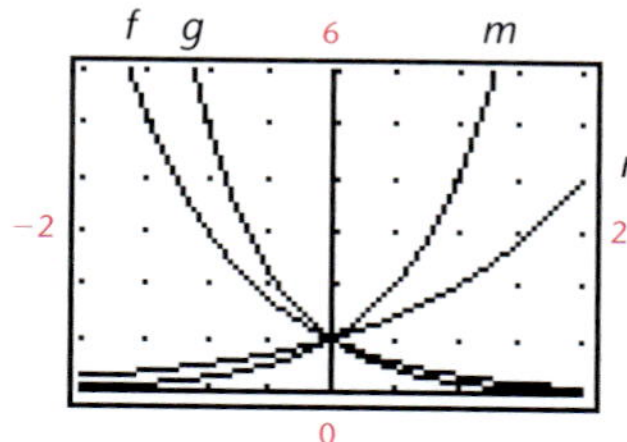

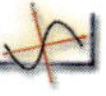

2. Match each equation with the graph of f, g, m, or n in the figure.

   (A) $y = e^{-1.2x}$ **(B)** $y = e^{0.7x}$

   (C) $y = e^{-0.4x}$ **(D)** $y = e^{1.3x}$

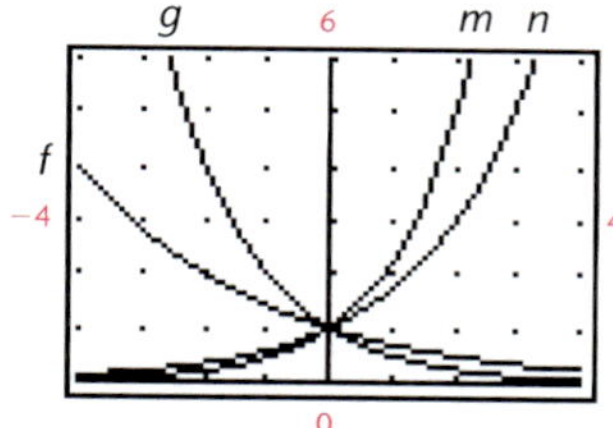

In Problems 3–10, compute answers to four significant digits.

3. $5^{\sqrt{3}}$ **4.** $3^{-\sqrt{2}}$

5. $e^2 + e^{-2}$ **6.** $e - e^{-1}$

7. $\sqrt{e}$ **8.** $e^{\sqrt{2}}$

9. $\dfrac{2^\pi + 2^{-\pi}}{2}$ **10.** $\dfrac{3^\pi - 3^{-\pi}}{2}$

In Problems 11–18, simplify.

11. $10^{3x-1}10^{4-x}$ **12.** $(4^{3x})^{2y}$ **13.** $\dfrac{3^x}{3^{1-x}}$

14. $\dfrac{5^{x-3}}{5^{x-4}}$ **15.** $\left(\dfrac{4^x}{5^y}\right)^{3z}$ **16.** $(2^x3^y)^z$

17. $\dfrac{e^{5x}}{e^{2x+1}}$ **18.** $\dfrac{e^{4-3x}}{e^{2-5x}}$

19. **(A)** Explain what is wrong with the following reasoning about the expression $[1 + (1/x)]^x$: As x gets large, $1 + (1/x)$ approaches 1 because $1/x$ approaches 0, and 1 raised to any power is 1, so $[1 + 1/x]^x$ approaches 1.

(B) Which number does $[1 + (1/x)]^x$ approach as x approaches ∞?

20. **(A)** Explain what is wrong with the following reasoning about the expression $[1 + (1/x)]^x$: If $b > 1$, then the exponential function b^x approaches ∞ as x approaches ∞, and $1 + (1/x)$ is greater than 1, so $[1 + (1/x)]^x$ approaches infinity as $x \to \infty$.

(B) Which number does $[1 + (1/x)]^x$ approach as x approaches ∞?

B

Before graphing the functions in Problems 21–30, classify each function as increasing or decreasing, find the x and y intercepts, and identify any asymptotes. Round any approximate values to two decimal places. Examine the graph to check your answers.

21. $y = 3^x$ **22.** $y = 5^x$

23. $y = (\frac{1}{3})^x = 3^{-x}$ **24.** $y = (\frac{1}{5})^x = 5^{-x}$

25. $g(x) = -3^{-x}$ **26.** $f(x) = -5^x$

27. $F(x) = 2 - e^{-x}$ **28.** $G(x) = e^{2x} - 3$

29. $m(t) = e^{3t} - 2$ **30.** $n(t) = 3 + e^{-2t}$

In Problems 31–42, solve for x.

31. $5^{3x} = 5^{4x-2}$ **32.** $10^{2-3x} = 10^{5x-6}$

33. $7^{x^2} = 7^{2x+3}$ **34.** $4^{5x-x^2} = 4^{-6}$

35. $(1 - x)^5 = (2x - 1)^5$ **36.** $5^3 = (x + 2)^3$

37. $2xe^{-x} = 0$

38. $(x - 3)e^x = 0$

39. $x^2e^x - 5xe^x = 0$

40. $3xe^{-x} + x^2e^{-x} = 0$

41. $9^{x^2} = 3^{3x-1}$

42. $4^{x^2} = 2^{x+3}$

43. Find all real numbers a such that $a^2 = a^{-2}$. Explain why this does not violate the second exponential function property in the box on page 321.

44. Find real numbers a and b such that $a \neq b$ but $a^4 = b^4$. Explain why this does not violate the third exponential function property in the box on page 321.

45. Examine the graph of $y = 1^x$ on a graphing utility and explain why 1 cannot be the base for an exponential function.

46. Examine the graph of $y = 0^x$ on a graphing utility and explain why 0 cannot be the base for an exponential function. [*Hint:* Turn the axes off before graphing.]

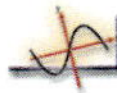

Graph each function in Problems 47–54 using the graph of f shown in the figure.

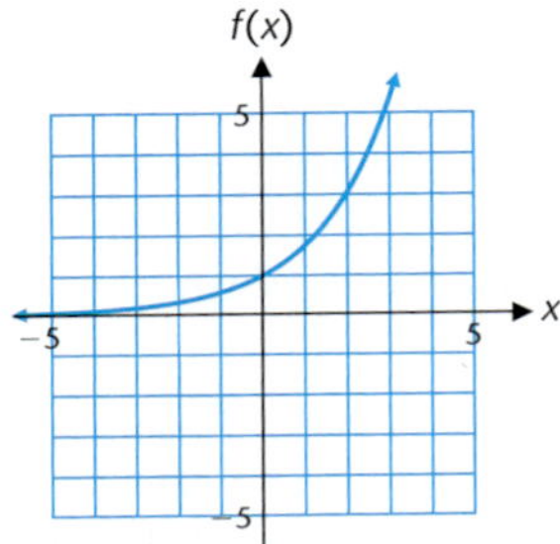

47. $y = f(x) - 2$

48. $y = f(x) + 1$

49. $y = f(x - 2)$

50. $y = f(x + 1)$

51. $y = 2f(x) - 4$

52. $y = 3 - 5f(x)$

53. $y = 2 - 3f(x - 4)$

54. $y = 2f(x + 1) - 1$

In Problems 55–60, describe the transformations that can be used to obtain the graph of g from the graph of $f(x) = e^x$ (see Section 1.4). Check your answers by graphing f and g in the same viewing window.

55. $g(x) = e^{x-2}$

56. $g(x) = e^{x+3}$

57. $g(x) = e^x + 2$

58. $g(x) = e^x - 1$

59. $g(x) = 2e^{-(x+2)}$

60. $g(x) = 0.5e^{-(x-1)}$

In Problems 61–64, simplify.

61. $\dfrac{-2x^3e^{-2x} - 3x^2e^{-2x}}{x^6}$

62. $\dfrac{5x^4e^{5x} - 4x^3e^{5x}}{x^8}$

63. $(e^x + e^{-x})^2 + (e^x - e^{-x})^2$

64. $e^x(e^{-x} + 1) - e^{-x}(e^x + 1)$

In Problems 65–76, use a graphing utility to find local extrema, y intercepts, and x intercepts. Investigate the behavior as $x \to \infty$ and as $x \to -\infty$ and identify any horizontal asymptotes. Round any approximate values to two decimal places.

65. $f(x) = 2 + e^{x-2}$

66. $g(x) = -3 + e^{1+x}$

67. $m(x) = e^{|x|}$

68. $n(x) = e^{-|x|}$

69. $s(x) = e^{-x^2}$

70. $r(x) = e^{x^2}$

71. $F(x) = \dfrac{200}{1 + 3e^{-x}}$

72. $G(x) = \dfrac{100}{1 + e^{-x}}$

73. $m(x) = 2x(3^{-x}) + 2$

74. $h(x) = 3x(2^{-x}) - 1$

75. $f(x) = \dfrac{2^x + 2^{-x}}{2}$

76. $g(x) = \dfrac{3^x + 3^{-x}}{2}$

C

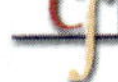

77. Use a graphing utility to investigate the behavior of $f(x) = (1 + x)^{1/x}$ as x approaches 0.

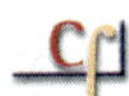

78. Use a graphing utility to investigate the behavior of $f(x) = (1 + x)^{1/x}$ as x approaches ∞.

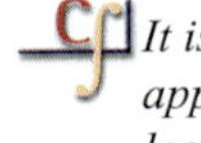

It is common practice in many applications of mathematics to approximate nonpolynomial functions with appropriately selected polynomials. For example, the polynomials in Problems 79–82, called ***Taylor polynomials,*** *can be used to approximate the exponential function $f(x) = e^x$. To illustrate this approximation graphically, in each problem graph $f(x) = e^x$ and the indicated polynomial in the same viewing window, $-4 \le x \le 4$ and $-5 \le y \le 50$.*

79. $P_1(x) = 1 + x + \frac{1}{2}x^2$

80. $P_2(x) = 1 + x + \frac{1}{2}x^2 + \frac{1}{6}x^3$

81. $P_3(x) = 1 + x + \frac{1}{2}x^2 + \frac{1}{6}x^3 + \frac{1}{24}x^4$

82. $P_4(x) = 1 + x + \frac{1}{2}x^2 + \frac{1}{6}x^3 + \frac{1}{24}x^4 + \frac{1}{120}x^5$

83. Investigate the behavior of the functions $f_1(x) = x/e^x$, $f_2(x) = x^2/e^x$, and $f_3(x) = x^3/e^x$ as $x \to \infty$ and as $x \to -\infty$, and find any horizontal asymptotes. Generalize to functions of the form $f_n(x) = x^n/e^x$, where n is any positive integer.

84. Investigate the behavior of the functions $g_1(x) = xe^x$, $g_2(x) = x^2e^x$, and $g_3(x) = x^3e^x$ as $x \to \infty$ and as $x \to -\infty$, and find any horizontal asymptotes. Generalize to functions of the form $g_n(x) = x^n e^x$, where n is any positive integer.

85. Explain why the graph of an exponential function cannot be the graph of a polynomial function.

86. Explain why the graph of an exponential function cannot be the graph of a rational function.

APPLICATIONS

★ **87. Finance.** A couple just had a new child. How much should they invest now at 8.25% compounded daily to have \$40,000 for the child's education 17 years from now? Compute the answer to the nearest dollar.

★ **88. Finance.** A person wishes to have \$15,000 cash for a new car 5 years from now. How much should be placed in an account now if the account pays 9.75% compounded weekly? Compute the answer to the nearest dollar.

89. Money Growth. If you invest \$5,250 in an account paying 11.38% compounded continuously, how much money will be in the account at the end of

(A) 6.25 years? **(B)** 17 years?

90. Money Growth. If you invest \$7,500 in an account paying 8.35% compounded continuously, how much money will be in the account at the end of

(A) 5.5 years? **(B)** 12 years?

★ **91. Finance.** If \$3,000 is deposited into an account earning 8% compounded daily and, at the same time, \$5,000 is deposited into an account earning 5% compounded daily, will the first account be worth more than the second? If so, when?

★ **92. Finance.** If \$4,000 is deposited into an account earning 9% compounded weekly and, at the same time, \$6,000 is deposited into an account earning 7% compounded weekly, will the first account be worth more than the second? If so, when?

★ **93. Finance.** Will an investment of \$10,000 at 8.9% compounded daily ever be worth more at the end of a quarter than an investment of \$10,000 at 9% compounded quarterly? Explain.

★ **94. Finance.** A sum of \$5,000 is invested at 13% compounded semiannually. Suppose that a second investment of \$5,000 is made at interest rate r compounded daily. For which values of r, to the nearest tenth of a percent, is the second investment better than the first? Discuss.

★ **95. Present Value.** A promissory note will pay \$30,000 at maturity 10 years from now. How much should you be willing to pay for the note now if the note gains value at a rate of 9% compounded continuously?

★ **96. Present Value.** A promissory note will pay \$50,000 at maturity $5\frac{1}{2}$ years from now. How much should you be willing to pay for the note now if the note gains value at a rate of 10% compounded continuously?

97. Money Growth. *Barron's,* a national business and financial weekly, published the following "Top Savings Deposit Yields" for $2\frac{1}{2}$-year certificate of deposit accounts:

Gill Savings	8.30% (CC)
Richardson Savings and Loan	8.40% (CQ)
USA Savings	8.25% (CD)

where CC represents compounded continuously, CQ compounded quarterly, and CD compounded daily. Compute the value of \$1,000 invested in each account at the end of $2\frac{1}{2}$ years.

98. Money Growth. Refer to Problem 97. In another issue of *Barron's,* 1-year certificate of deposit accounts included:

Alamo Savings	8.25% (CQ)
Lamar Savings	8.05% (CC)

Compute the value of \$10,000 invested in each account at the end of 1 year.

99. Finance. Suppose \$4,000 is invested at 11% compounded weekly. How much money will be in the account in

(A) $\frac{1}{2}$ year? **(B)** 10 years?

Compute answers to the nearest cent.

100. Finance. Suppose \$2,500 is invested at 7% compounded quarterly. How much money will be in the account in

(A) $\frac{3}{4}$ year? **(B)** 15 years?

Compute answers to the nearest cent.

SECTION 4.2 Exponential Models

Mathematical Modeling • Data Analysis and Regression • A Comparison of Exponential Growth Phenomena

In Section 4.2 we use exponential functions to model a wide variety of real-world phenomena, including growth of populations of people, animals, and bacteria; radioactive decay; spread of epidemics; propagation of rumors; light intensity; atmospheric pressure; and electric circuits. The regression techniques introduced in Chapter 2 to construct linear and quadratic models are extended to construct exponential models.

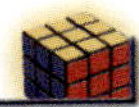

Mathematical Modeling

Populations tend to grow exponentially and at different rates. A convenient and easily understood measure of growth rate is the **doubling time**—that is, the time it takes for a population to double. Over short periods the **doubling time growth model** is often used to model population growth:

$$P = P_0 2^{t/d}$$

where P = Population at time t

P_0 = Population at time $t = 0$

d = Doubling time

Note that when $t = d$,

$$P = P_0 2^{d/d} = P_0 2$$

and the population is double the original, as it should be. We use this model to solve a population growth problem in Example 1.

EXAMPLE 1 Population Growth

Mexico has a population of around 100 million people, and it is estimated that the population will double in 21 years. If population growth continues at the same rate, what will be the population:

(A) 15 years from now? (B) 30 years from now?

Calculate answers to three significant digits.

SOLUTIONS

Algebraic Solutions

We use the doubling time growth model:

$$P = P_0 2^{t/d}$$

Graphical Solutions

We graph

$$P = 100(2^{x/21})$$

and construct a table of values (Fig. 2).

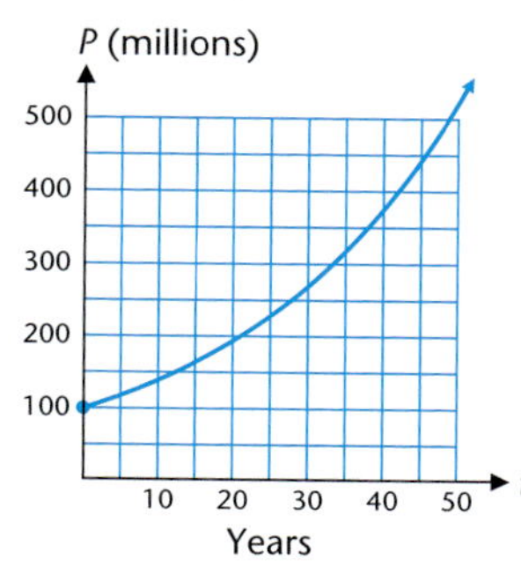

FIGURE 1 $P = 100(2^{t/21})$.

Substituting $P_0 = 100$ and $d = 21$, we obtain

$$P = 100(2^{t/21}) \quad \text{Figure 1.}$$

(A) Find P when $t = 15$ years:

$$P = 100(2^{15/21})$$
$$\approx 164 \text{ million people}$$

(B) Find P when $t = 30$ years:

$$P = 100(2^{30/21})$$
$$\approx 269 \text{ million people}$$

(A) When $x = 15$ years,
$P \approx 164$ million people.

(B) When $x = 30$ years,
$P \approx 269$ million people.

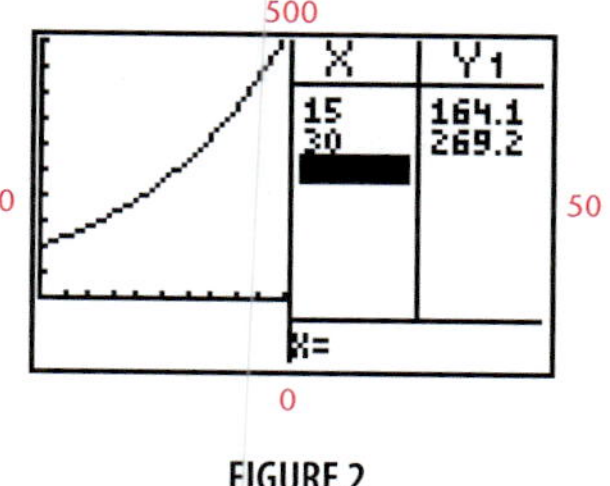

FIGURE 2

MATCHED 1 PROBLEM

The bacterium *Escherichia coli* (*E. coli*) is found naturally in the intestines of many mammals. In a particular laboratory experiment, the doubling time for *E. coli* is found to be 25 minutes. If the experiment starts with a population of 1,000 *E. coli* and there is no change in the doubling time, how many bacteria will be present:

(A) In 10 minutes? (B) In 5 hours?

Write answers to three significant digits.

EXPLORE/DISCUSS 1

The doubling time growth model would *not* be expected to give accurate results over long periods. According to the doubling time growth model of Example 1, what was the population of Mexico 500 years ago at the height of Aztec civilization? What will the population of Mexico be 200 years from now? Explain why these results are unrealistic. Discuss factors that affect human populations that are not taken into account by the doubling time growth model.

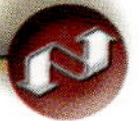

As an alternative to the doubling time growth model, we can use the equation $y = ce^{kt}$, where c and k are positive constants, to model population growth. Example 2 illustrates this approach.

EXAMPLE 2 Medicine—Bacteria Growth

Cholera, an intestinal disease, is caused by a cholera bacterium that multiplies exponentially by cell division as modeled by

$$N = N_0 e^{1.386t}$$

where N is the number of bacteria present after t hours and N_0 is the number of bacteria present at $t = 0$. If we start with 1 bacterium, how many bacteria will be present in

(A) 5 hours? (B) 12 hours?

Compute the answers to three significant digits.

SOLUTIONS

Algebraic Solutions

(A) Use $N_0 = 1$ and $t = 5$:

$$\begin{aligned} N &= N_0 e^{1.386t} \\ &= e^{1.386(5)} \\ &= 1{,}020 \end{aligned}$$

(B) Use $N_0 = 1$ and $t = 12$:

$$\begin{aligned} N &= N_0 e^{1.386t} \\ &= e^{1.386(12)} \\ &= 16{,}700{,}000 \end{aligned}$$

Graphical Solutions

We graph

$$N = e^{1.386x}$$

and construct a table of values (Fig. 3).

(A) When $x = 5$ hours, $N \approx 1{,}020$ bacteria.

(B) When $x = 12$ hours, $N \approx 16{,}700{,}000$ bacteria.

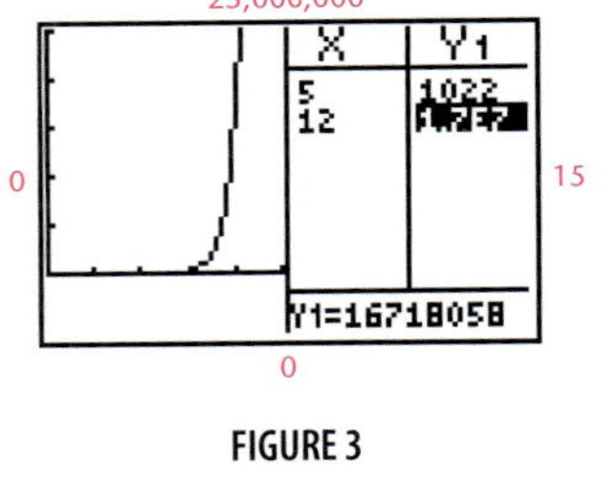

FIGURE 3

MATCHED 2 PROBLEM

Repeat Example 2 if $N = N_0 e^{0.783t}$ and all other information remains the same.

Exponential functions can also be used to model radioactive decay, which is sometimes referred to as *negative growth*. Radioactive materials are used extensively in medical diagnosis and therapy, as power sources in satellites, and as power sources in many countries. If we start with an amount A_0 of a particular radioactive isotope, the amount declines exponentially in time. The rate of decay varies from isotope to isotope. A convenient and easily understood measure of the rate of decay is the **half-life** of the isotope—that is, the time it takes for half of a particular material to decay. We use the following **half-life decay model:**

$$\begin{aligned} A &= A_0\left(\tfrac{1}{2}\right)^{t/h} \\ &= A_0 2^{-t/h} \end{aligned}$$

where A = Amount at time t

A_0 = Amount at time $t = 0$

h = Half-life

Note that when $t = h$,

$$A = A_0 2^{-h/h} = A_0 2^{-1} = \frac{A_0}{2}$$

and the amount of isotope is half the original amount, as it should be.

EXAMPLE 3 Radioactive Decay

The radioactive isotope gallium 67 (^{67}Ga), used in the diagnosis of malignant tumors, has a biological half-life of 46.5 hours. If we start with 100 milligrams of the isotope, how many milligrams will be left after

(A) 24 hours? (B) 1 week?

Compute answers to three significant digits.

SOLUTIONS

Algebraic Solutions

We use the half-life decay model:

$$A = A_0(\tfrac{1}{2})^{t/h} = A_0 2^{-t/h}$$

Using $A_0 = 100$ and $h = 46.5$, we obtain

$$A = 100(2^{-t/46.5}) \quad \text{Figure 4.}$$

(A) Find A when $t = 24$ hours:

$$A = 100(2^{-24/46.5})$$
$$= 69.9 \text{ milligrams}$$

(B) Find A when $t = 168$ hours (1 week = 168 hours):

$$A = 100(2^{-168/46.5})$$
$$= 8.17 \text{ milligrams}$$

FIGURE 4 $A = 100(2^{-t/46.5})$.

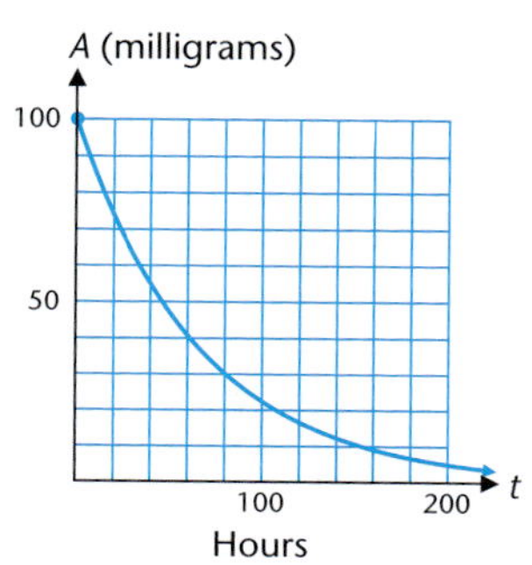

Graphical Solutions

We graph

$$A = 100(2^{-x/46.5})$$

and construct a table of values (Fig. 5).

(A) When $x = 24$ hours, $A \approx 69.9$ milligrams.

(B) When $x = 168$ hours, $A \approx 8.17$ milligrams.

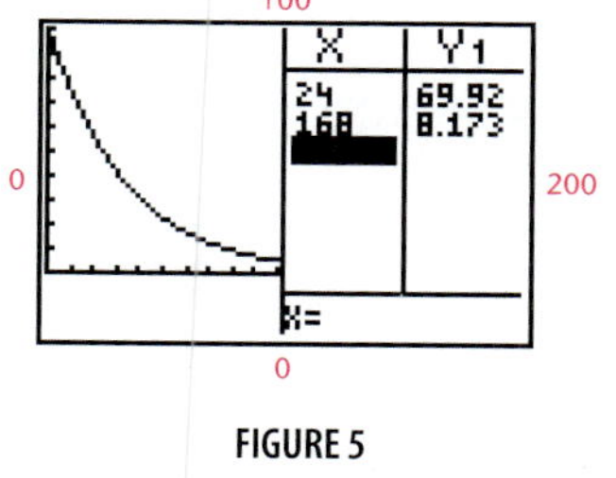

FIGURE 5

MATCHED 3 PROBLEM

Radioactive gold 198 (^{198}Au), used in imaging the structure of the liver, has a half-life of 2.67 days. If we start with 50 milligrams of the isotope, how many milligrams will be left after:

(A) $\frac{1}{2}$ day? (B) 1 week?

Compute answers to three significant digits.

As an alternative to the half-life decay model, we can use the equation $y = ce^{-kt}$, where c and k are positive constants, to model radioactive decay. Example 4 illustrates this approach.

EXAMPLE 4 Carbon-14 Dating

Cosmic-ray bombardment of the atmosphere produces neutrons, which in turn react with nitrogen to produce radioactive carbon-14. Radioactive carbon-14 enters all living tissues through carbon dioxide, which is first absorbed by plants. As long as a plant or animal is alive, carbon-14 is maintained in the living organism at a constant level. Once the organism dies, however, carbon-14 decays according to the equation

$$A = A_0e^{-0.000124t}$$

where A is the amount of carbon-14 present after t years and A_0 is the amount present at time $t = 0$. If 1,000 milligrams of carbon-14 are present at the start, how many milligrams will be present in

(A) 10,000 years? (B) 50,000 years?

Compute answers to three significant digits.

SOLUTIONS

Algebraic Solutions

Substituting $A_0 = 1{,}000$ in the decay equation, we have

$$A = 1{,}000e^{-0.000124t}$$ Figure 6.

(A) Solve for A when $t = 10{,}000$:

$$A = 1{,}000e^{-0.000124(10{,}000)} = 289 \text{ milligrams}$$

(B) Solve for A when $t = 50{,}000$:

$$A = 1{,}000e^{-0.000124(50{,}000)} = 2.03 \text{ milligrams}$$

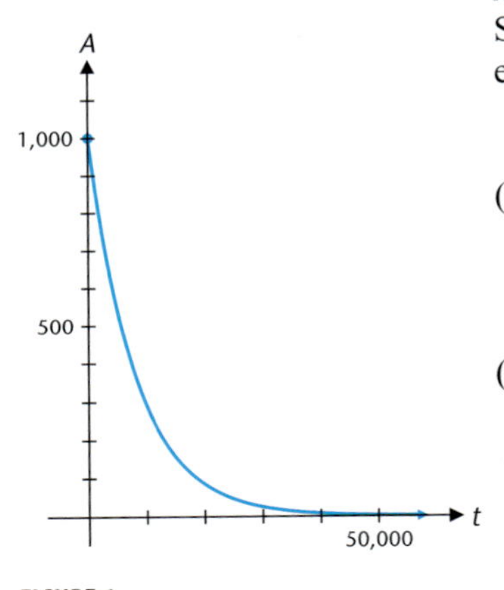

FIGURE 6

Graphical Solutions

We graph

$$A = 1{,}000e^{-0.000124x}$$

and construct a table of values (Fig. 7).

(A) When $x = 10{,}000$ years, $A \approx 289$ milligrams.

(B) When $x = 50{,}000$ years, $A \approx 2.03$ milligrams.

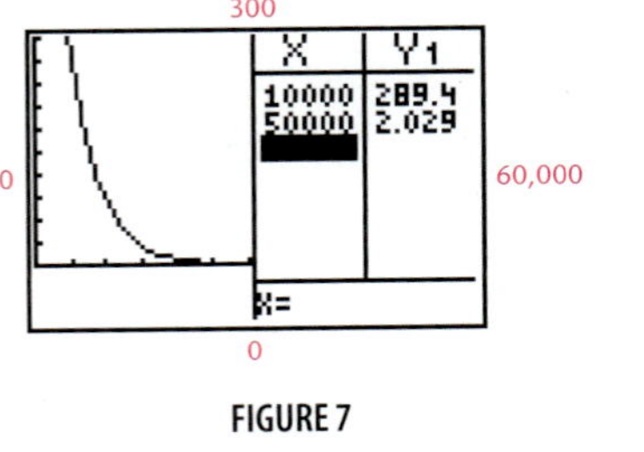

FIGURE 7

More will be said about carbon-14 dating in Exercise 4.5, where we will be interested in solving for t after being given information about A and A_0.

MATCHED 4 PROBLEM

Referring to Example 4, how many milligrams of carbon-14 would have to be present at the beginning to have 10 milligrams present after 20,000 years? Compute the answer to four significant digits.

We can model phenomena such as learning curves, for which growth has an upper bound, by the equation $y = c(1 - e^{-kt})$, where c and k are positive constants. Example 5 illustrates such limited growth.

EXAMPLE 5 Learning Curve

People assigned to assemble circuit boards for a computer manufacturing company undergo on-the-job training. From past experience, it was found that the learning curve for the average employee is given by

$$N = 40(1 - e^{-0.12t})$$

where N is the number of boards assembled per day after t days of training (Fig. 8).

(A) How many boards can an average employee produce after 3 days of training? After 5 days of training? Round answers to the nearest integer.

(B) How many days of training will it take until an average employee can assemble 25 boards a day? Round answers to the nearest integer.

(C) Does N approach a limiting value as t increases without bound? Explain.

FIGURE 8 Limited growth.

SOLUTION

(A) When $t = 3$,

$$N = 40(1 - e^{-0.12(3)}) = 12 \quad \text{Rounded to nearest integer}$$

so the average employee can produce 12 boards after 3 days of training. Similarly, when $t = 5$,

$$N = 40(1 - e^{-0.12(5)}) = 18 \quad \text{Rounded to nearest integer}$$

so the average employee can produce 18 boards after 5 days of training.

(B) Solve the equation $40(1 - e^{-0.12t}) = 25$ for t by graphing

$$y_1 = 40(1 - e^{-0.12t}) \qquad \text{and} \qquad y_2 = 25$$

and using the intersect command (Fig. 9). It will take 8 days of training.

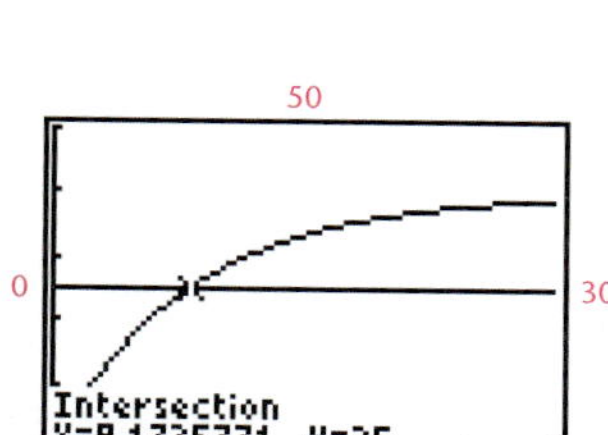

FIGURE 9 $y_1 = 40(1 - e^{-0.12t})$, $y_2 = 25$.

(C) Because $e^{-0.12t}$ approaches 0 as t increases without bound,

$$N = 40(1 - e^{-0.12t}) \to 40(1 - 0) = 40$$

So the limiting value of N is 40 boards per day.

MATCHED 5 PROBLEM

A company is trying to expose as many people as possible to a new product through television advertising in a large metropolitan area with 2 million potential viewers. A model for the number of people N, in millions, who are aware of the product after t days of advertising was found to be

$$N = 2(1 - e^{-0.037t})$$

(A) How many viewers are aware of the product after 2 days? After 10 days? Express answers as integers, rounded to three significant digits.

(B) How many days will it take until half of the potential viewers will become aware of the product? Round answer to the nearest integer.

(C) Does N approach a limiting value as t increases without bound? Explain.

We can model phenomena such as the spread of an epidemic or the propagation of a rumor by the *logistic equation,*

$$y = \frac{M}{(1 + ce^{-kt})}$$

where M, c, and k are positive constants. Logistic growth, illustrated in Example 6, approaches a limiting value as t increases without bound.

EXAMPLE 6 Logistic Growth in an Epidemic

A community of 1,000 individuals is assumed to be homogeneously mixed. One individual who has just returned from another community has influenza. Assume the community has not had influenza shots and all are susceptible. The spread of the disease in the community is predicted to be given by the logistic curve

$$N(t) = \frac{1,000}{1 + 999e^{-0.3t}}$$

where N is the number of people who have contracted influenza after t days.

(A) How many people have contracted influenza after 10 days? After 20 days? Round answers to the nearest integer.

(B) How many days will it take until half the community has contracted influenza? Round answer to the nearest integer.

(C) Does N approach a limiting value as t increases without bound? Explain.

SOLUTIONS

(A) The table in Figure 10(a) shows that $N(10) \approx 20$ individuals and $N(20) \approx 288$ individuals.

FIGURE 10
Logistic growth.

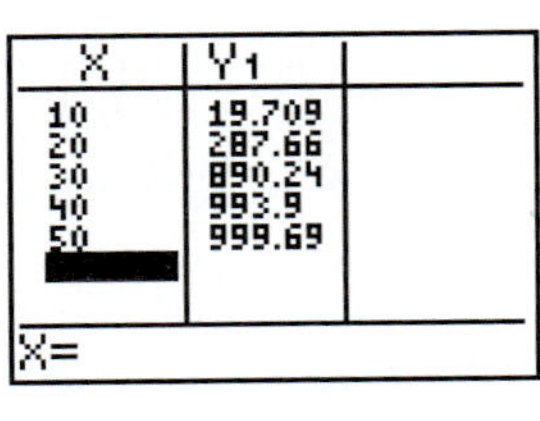

X	Y1
10	19.709
20	287.66
30	890.24
40	993.9
50	999.69

X=

(a)

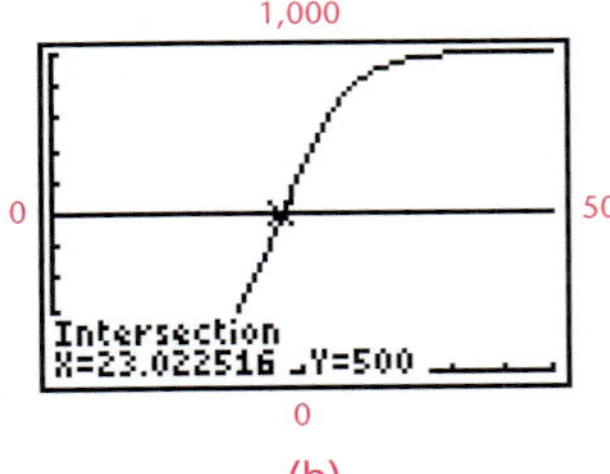

(b)

(B) Figure 10(b) shows that the graph of $N(t)$ intersects the line $y = 500$ after approximately 23 days.

(C) The values in Figure 10(a) and the graph in Figure 10(b) both indicate that N approaches 1,000 as t increases without bound. We can confirm this algebraically by noting that because $999e^{-0.3t} \to 0$ as t increases without bound,

$$N(t) = \frac{1{,}000}{1 + 999e^{-0.3t}} \to \frac{1{,}000}{1 + 0} = 1{,}000$$

Thus, the upper limit on the growth of N is 1,000, the total number of people in the community.

MATCHED 6 PROBLEM

A group of 400 parents, relatives, and friends are waiting anxiously at Kennedy Airport for a charter flight returning students after a year in Europe. It is stormy and the plane is late. A particular parent thought he had heard that the plane's radio had gone out and related this news to some friends, who in turn passed it on to others. The propagation of this rumor is predicted to be given by

$$N(t) = \frac{400}{1 + 399e^{-0.4t}}$$

where N is the number of people who have heard the rumor after t minutes.

(A) How many people have heard the rumor after 10 minutes? After 20 minutes? Round answers to the nearest integer.

(B) How many minutes will it take until half the group has heard the rumor? Round answer to the nearest integer.

(C) Does N approach a limiting value as t increases without bound? Explain.

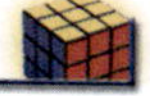

Data Analysis and Regression

We use exponential regression to fit a function of the form $y = ab^x$ to a set of data points, and logistic regression to fit a function of the form

$$y = \frac{c}{1 + ae^{-bx}}$$

to a set of data points. The techniques are similar to those introduced in Chapter 2 for linear and quadratic functions.

EXAMPLE 7 **Infectious Diseases**

The U.S. Department of Health and Human Services published the data in Table 1.

TABLE 1 Reported Cases of Infectious Diseases

Year	Mumps	Rubella
1970	104,953	56,552
1980	8,576	3,904
1990	5,292	1,125
1995	906	128
2000	323	152

(A) Let x represent time in years with $x = 0$ representing 1970, and let y represent the corresponding number of reported cases of mumps. Use regression analysis on a graphing utility to find an exponential function of the form $y = ab^x$ that models the data. (Round the constants a and b to three significant digits.)

(B) Use the exponential regression function to predict the number of reported cases of mumps in 2010.

SOLUTION

(A) Figure 11 shows the details of constructing the model on a graphing utility.

L1	L2	L3
0	104953	
10	8576	
20	5292	
25	906	
30	323	

L3(1)=

(a) Data

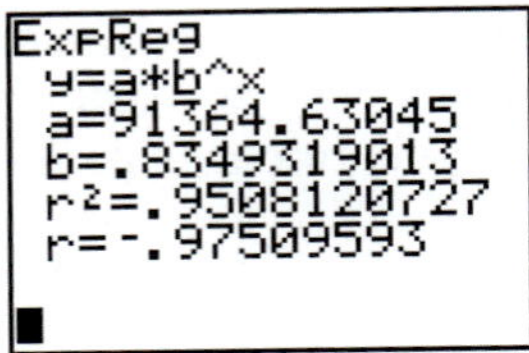

(b) Regression equation

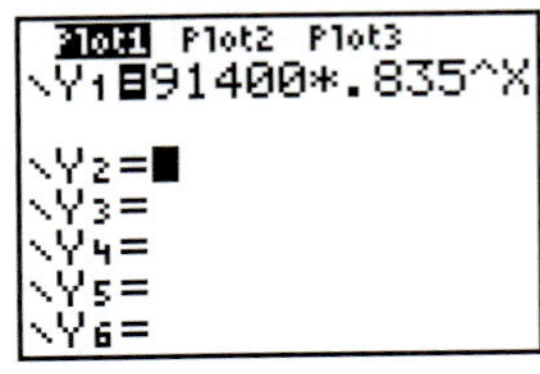

(c) Regression equation entered in equation editor

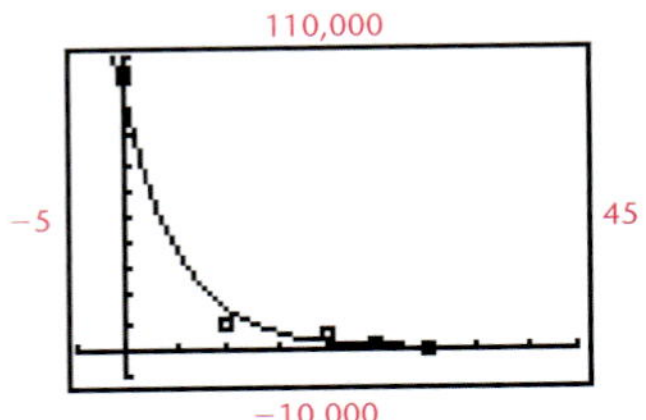

(d) Graph of data and regression equation

FIGURE 11

(B) Evaluating $y_1 = 91{,}400(0.835)^x$ at $x = 40$ gives a prediction of 67 cases of mumps in 2010.

MATCHED PROBLEM 7

Repeat Example 7 for reported cases of rubella.

EXAMPLE 8 AIDS Cases and Deaths

The U.S. Department of Health and Human Services published the data in Table 2.

TABLE 2 Acquired Immunodeficiency Syndrome (AIDS) Cases and Deaths in the United States

Year	Cases Diagnosed to Date	Known Deaths to Date
1985	23,185	12,648
1988	107,755	62,468
1991	261,259	159,294
1994	493,713	296,507
1997	672,970	406,179
2000	774,467	447,648

(A) Let x represent time in years with $x = 0$ representing 1985, and let y represent the corresponding number of AIDS cases diagnosed to date. Use regression analysis on a graphing utility to find a logistic function of the form

$$y = \frac{c}{1 + ae^{-bx}}$$

that models the data. (Round the constants a, b, and c to three significant digits.)

(B) Use the logistic regression function to predict the number of cases of AIDS diagnosed by 2010.

SOLUTION

(A) Figure 12 shows the details of constructing the model on a graphing utility.

L1	L2	L3
0	23185	
3	107755	
6	261259	
9	493713	
12	672970	
15	774467	

L3(1)=

(a) Data

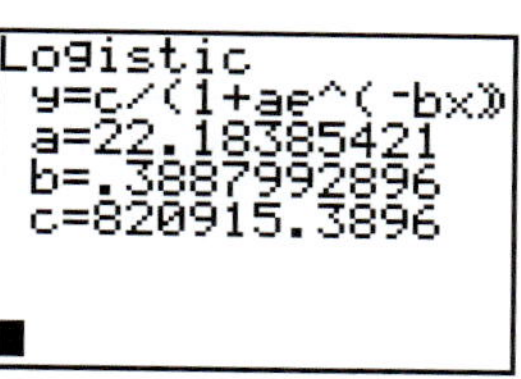

(b) Regression equation

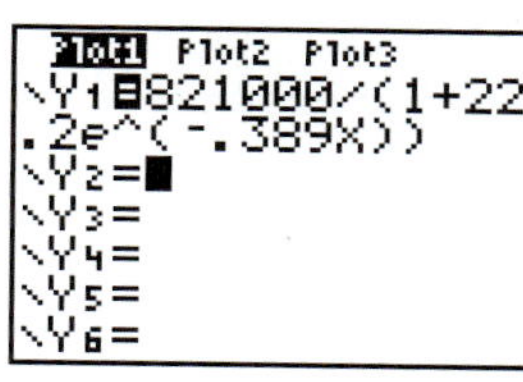

(c) Regression equation entered in equation editor

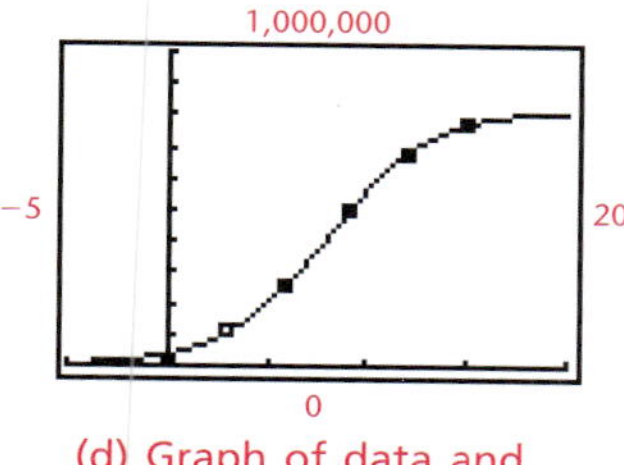

(d) Graph of data and regression equation

FIGURE 12

(B) Evaluating

$$y_1 = \frac{821{,}000}{1 + 22.2e^{-0.389x}}$$

at $x = 25$ gives a prediction of approximately 820,000 cases of AIDS diagnosed by 2010.

MATCHED 8 PROBLEM

Repeat Example 8 for known deaths from AIDS to date.

A Comparison of Exponential Growth Phenomena

The equations and graphs given in Table 3 compare several widely used growth models. These are divided basically into two groups: unlimited growth and limited growth. Following each equation and graph is a short, incomplete list of areas in which the models are used. We have only touched on a subject that has been extensively developed and that you are likely to study in greater depth in the future.

TABLE 3 Exponential Growth and Decay

Description	Equation	Graph	Uses
Unlimited growth	$y = ce^{kt}$ $c, k > 0$	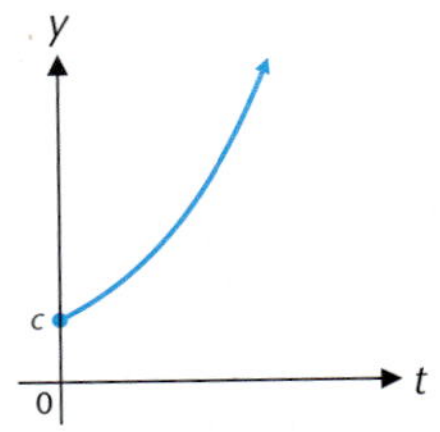	Short-term population growth (people, bacteria, etc.); growth of money at continuous compound interest
Exponential decay	$y = ce^{-kt}$ $c, k > 0$	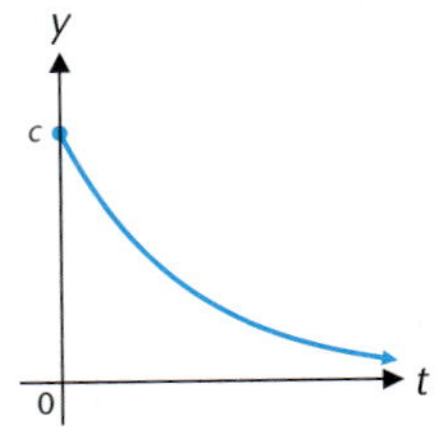	Radioactive decay; light absorption in water, glass, and the like; atmospheric pressure; electric circuits
Limited growth	$y = c(1 - e^{-kt})$ $c, k > 0$	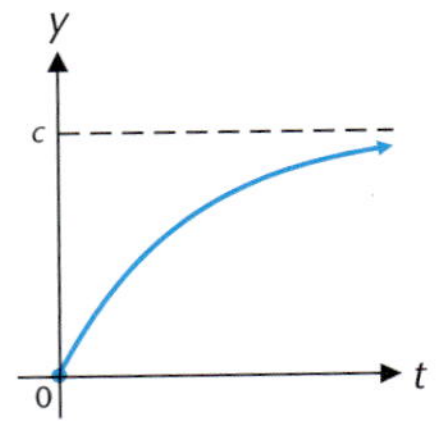	Learning skills; sales fads; company growth; electric circuits
Logistic growth	$y = \dfrac{M}{1 + ce^{-kt}}$ $c, k, M > 0$	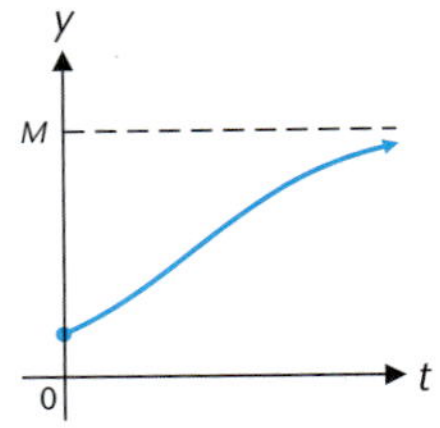	Long-term population growth; epidemics; sales of new products; company growth

ANSWERS MATCHED PROBLEMS

1. (A) 1,320 bacteria
(B) $4{,}100{,}000 = 4.10 \times 10^6$ bacteria

2. (A) 50 bacteria (B) 12,000 bacteria

3. (A) 43.9 milligrams (B) 8.12 milligrams

4. 119.4 milligrams

5. (A) 143,000 viewers; 619,000 viewers (B) 19 days
(C) N approaches an upper limit of 2 million, the number of potential viewers

6. (A) 48 individuals; 353 individuals (B) 15 minutes
(C) N approaches an upper limit of 400, the number of people in the entire group.

7. (A) $y = 44{,}500(0.815)^x$ (B) 12 cases

8. (A) $y = \dfrac{470{,}000}{1 + 23.9e^{-0.415x}}$
(B) 470,000 known deaths

EXERCISE 4.2

APPLICATIONS

1. **Gaming.** A person bets on red and black on a roulette wheel using a *Martingale strategy*. That is, a \$2 bet is placed on red, and the bet is doubled each time until a win occurs. The process is then repeated. If black occurs n times in a row, then $L = 2^n$ dollars is lost on the nth bet. Graph this function for $1 \le n \le 10$. Although the function is defined only for positive integers, points on this type of graph are usually joined with a smooth curve as a visual aid.

2. **Bacterial Growth.** If bacteria in a certain culture double every $\frac{1}{2}$ hour, write an equation that gives the number of bacteria N in the culture after t hours, assuming the culture has 100 bacteria at the start. Graph the equation for $0 \le t \le 5$.

3. **Population Growth.** Because of its short life span and frequent breeding, the fruit fly *Drosophila* is used in some genetic studies. Raymond Pearl of Johns Hopkins University, for example, studied 300 successive generations of descendants of a single pair of *Drosophila* flies. In a laboratory situation with ample food supply and space, the doubling time for a particular population is 2.4 days. If we start with 5 male and 5 female flies, how many flies should we expect to have in

   (A) 1 week? **(B)** 2 weeks?

4. **Population Growth.** If Kenya has a population of about 30,000,000 people and a doubling time of 19 years and if the growth continues at the same rate, find the population in

   (A) 10 years **(B)** 30 years

   Compute answers to two significant digits.

5. **Insecticides.** The use of the insecticide DDT is no longer allowed in many countries because of its long-term adverse effects. If a farmer uses 25 pounds of active DDT, assuming its half-life is 12 years, how much will still be active after

   (A) 5 years? **(B)** 20 years?

   Compute answers to two significant digits.

6. **Radioactive Tracers.** The radioactive isotope technetium-99m (^{99m}Tc) is used in imaging the brain. The isotope has a half-life of 6 hours. If 12 milligrams are used, how much will be present after

   (A) 3 hours? **(B)** 24 hours?

   Compute answers to three significant digits.

7. **Population Growth.** If the world population is about 6 billion people now and if the population grows continuously at an annual rate of 1.7%, what will the population be in 10 years? Compute the answer to two significant digits.

8. **Population Growth.** If the population in Mexico is around 100 million people now and if the population grows continuously at an annual rate of 2.3%, what will the population be in 8 years? Compute the answer to two significant digits.

9. **Population Growth.** In 1996 the population of Russia was 148 million and the population of Nigeria was 104 million. If the populations of Russia and Nigeria grow continuously at annual rates of -0.62% and 3.0%, respectively, when will Nigeria have a greater population than Russia?

10. **Population Growth.** In 1996 the population of Germany was 84 million and the population of Egypt was 64 million. If the populations of Germany and Egypt grow continuously at annual rates of -0.15% and 1.9%, respectively, when will Egypt have a greater population than Germany?

11. **Space Science.** Radioactive isotopes, as well as solar cells, are used to supply power to space vehicles. The isotopes gradually lose power because of radioactive decay. On a particular space vehicle the nuclear energy source has a power output of P watts after t days of use as given by

    $$P = 75e^{-0.0035t}$$

    Graph this function for $0 \le t \le 100$.

12. **Earth Science.** The atmospheric pressure P, in pounds per square inch, decreases exponentially with altitude h, in miles above sea level, as given by

    $$P = 14.7e^{-0.21h}$$

    Graph this function for $0 \le h \le 10$.

13. **Marine Biology.** Marine life is dependent upon the microscopic plant life that exists in the *photic zone,* a zone that goes to a depth where about 1% of the surface light still remains. Light intensity I relative to depth d, in feet, for one of the clearest bodies of water in the world, the Sargasso Sea in the West Indies, can be approximated by

    $$I = I_0e^{-0.00942d}$$

    where I_0 is the intensity of light at the surface. What percentage of the surface light will reach a depth of

    (A) 50 feet? **(B)** 100 feet?

14. **Marine Biology.** Refer to Problem 13. In some waters with a great deal of sediment, the photic zone may go down only 15 to 20 feet. In some murky harbors, the

intensity of light d feet below the surface is given approximately by

$$I = I_0 e^{-0.23d}$$

What percentage of the surface light will reach a depth of

(A) 10 feet? **(B)** 20 feet?

15. AIDS Epidemic. In June 1996 the World Health Organization estimated that 7.7 million cases of AIDS had occurred worldwide since the beginning of the epidemic. Assuming that the disease spreads continuously at an annual rate of 17%, estimate the total number of AIDS cases that will have occurred by June of the year

(A) 2005 **(B)** 2010

16. AIDS Epidemic. In June 1996 the World Health Organization estimated that 28 million people worldwide had been infected with human immunodeficiency virus (HIV) since the beginning of the AIDS epidemic. Assuming that HIV infection spreads continuously at an annual rate of 19%, estimate the total number of people who will have been infected with HIV by June of the year

(A) 2005 **(B)** 2010

17. Newton's Law of Cooling. This law states that the rate at which an object cools is proportional to the difference in temperature between the object and its surrounding medium. The temperature T of the object t hours later is given by

$$T = T_m + (T_0 - T_m)e^{-kt}$$

where T_m is the temperature of the surrounding medium and T_0 is the temperature of the object at $t = 0$. Suppose a bottle of wine at a room temperature of 72°F is placed in the refrigerator to cool before a dinner party. If the temperature in the refrigerator is kept at 40°F and $k = 0.4$, find the temperature of the wine, to the nearest degree, after 3 hours. (In Exercise 4.5 we will find out how to determine k.)

18. Newton's Law of Cooling. Refer to Problem 17. What is the temperature, to the nearest degree, of the wine after 5 hours in the refrigerator?

19. Photography. An electronic flash unit for a camera is activated when a capacitor is discharged through a filament of wire. After the flash is triggered, and the capacitor is discharged, the circuit (see the figure) is connected and the battery pack generates a current to recharge the capacitor. The time it takes for the capacitor to recharge is called the *recycle time*. For a particular flash unit using a 12-volt battery pack, the charge q, in coulombs, on the capacitor t seconds after recharging has started is given by

$$q = 0.0009(1 - e^{-0.2t})$$

Find the value that q approaches as t increases without bound and interpret.

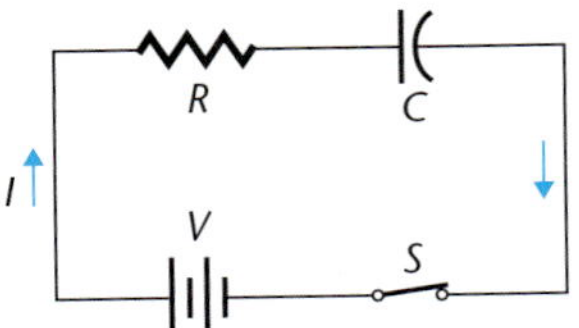

20. Medicine. An electronic heart pacemaker uses the same type of circuit as the flash unit in Problem 19, but it is designed so that the capacitor discharges 72 times a minute. For a particular pacemaker, the charge on the capacitor t seconds after it starts recharging is given by

$$q = 0.000\,008(1 - e^{-2t})$$

Find the value that q approaches as t increases without bound and interpret.

21. Wildlife Management. A herd of 20 white-tailed deer is introduced to a coastal island where there had been no deer before. Their population is predicted to increase according to the logistic curve

$$N = \frac{100}{1 + 4e^{-0.14t}}$$

where N is the number of deer expected in the herd after t years.

(A) How many deer will be present after 2 years? After 6 years? Round answers to the nearest integer.

(B) How many years will it take for the herd to grow to 50 deer? Round answer to the nearest integer.

(C) Does N approach a limiting value as t increases without bound? Explain.

22. Training. A trainee is hired by a computer manufacturing company to learn to test a particular model of a personal computer after it comes off the assembly line. The learning curve for an average trainee is given by

$$N = \frac{200}{4 + 21e^{-0.1t}}$$

(A) How many computers can an average trainee be expected to test after 3 days of training? After 6 days? Round answers to the nearest integer.

(B) How many days will it take until an average trainee can test 30 computers per day? Round answer to the nearest integer.

(C) Does N approach a limiting value as t increases without bound? Explain.

MODELING AND DATA ANALYSIS

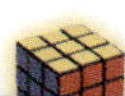

23. Depreciation. Table 4 gives the market value of a minivan (in dollars) x years after its purchase. Find an exponential regression model of the form $y = ab^x$ for this data set. Estimate the purchase price of the van. Estimate the value of the van 10 years after its purchase. Round answers to the nearest dollar.

TABLE 4

x	Value ($)
1	12,575
2	9,455
3	8,115
4	6,845
5	5,225
6	4,485

24. Depreciation. Table 5 gives the market value of a luxury sedan (in dollars) x years after its purchase. Find an exponential regression model of the form $y = ab^x$ for this data set. Estimate the purchase price of the sedan. Estimate the value of the sedan 10 years after its purchase. Round answers to the nearest dollar.

TABLE 5

x	Value ($)
1	23,125
2	19,050
3	15,625
4	11,875
5	9,450
6	7,125

25. Nuclear Power. Table 6 gives data on nuclear power generation by region for the years 1980–1999.

TABLE 6 Nuclear Power Generation

	(Billion Kilowatt-Hours)	
Year	**North America**	**Central and South America**
1980	287.0	2.2
1985	440.8	8.4
1990	649.0	9.0
1995	774.4	9.5
1998	750.2	10.3
1999	807.5	10.5

(A) Let x represent time in years with $x = 0$ representing 1980. Find a logistic regression model $\left(y = \frac{c}{1 + ae^{-bx}}\right)$ for the generation of nuclear power in North America. (Round the constants a, b, and c to three significant digits.)

(B) Use the logistic regression model to predict the generation of nuclear power in North America in 2010.

26. Nuclear Power. Refer to Table 6.

(A) Let x represent time in years with $x = 0$ representing 1980. Find a logistic regression model $\left(y = \frac{c}{1 + ae^{-bx}}\right)$ for the generation of nuclear power in Central and South America. (Round the constants a, b, and c to three significant digits.)

(B) Use the logistic regression model to predict the generation of nuclear power in Central and South America in 2010.

SECTION 4.3 Logarithmic Functions

Logarithmic Functions • From Logarithmic Form to Exponential Form, and Vice Versa • Properties of Logarithmic Functions • Common and Natural Logarithms • Change of Base

In Section 4.3 we introduce the inverses of the exponential functions—the logarithmic functions—and study their properties and graphs.

Logarithmic Functions

The exponential function $f(x) = b^x$, where $b > 0$, $b \neq 1$, is a one-to-one function, and therefore has an inverse. Its inverse, denoted $f^{-1}(x) = \log_b x$ (read "log to the base b of x"), is called the *logarithmic function with base b*. A point (x, y) lies on the graph of f^{-1} if and only if the point (y, x) lies on the graph of f; in other words,

$$y = \log_b x \text{ if and only if } x = b^y$$

We can use this fact to deduce information about the logarithmic functions from our knowledge of exponential functions. For example, the graph of f^{-1} is the graph of f reflected in the line $y = x$; and the domain and range of f^{-1} are, respectively, the range and domain of f.

Consider the exponential function $f(x) = 2^x$ and its inverse $f^{-1}(x) = \log_2 x$. Figure 1 shows the graphs of both functions and a table of selected points on those graphs. Because

$$y = \log_2 x \text{ if and only if } x = 2^y$$

$\log_2 x$ is the exponent to which 2 must be raised to obtain x: $2^{\log_2 x} = 2^y = x$.

FIGURE 1 Logarithmic function with base 2.

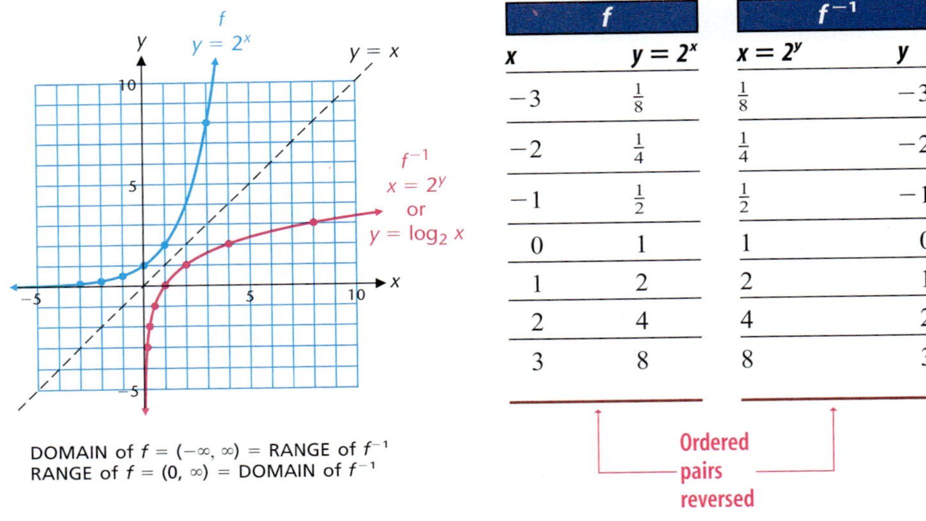

f: x	$y = 2^x$	f^{-1}: $x = 2^y$	y
−3	$\frac{1}{8}$	$\frac{1}{8}$	−3
−2	$\frac{1}{4}$	$\frac{1}{4}$	−2
−1	$\frac{1}{2}$	$\frac{1}{2}$	−1
0	1	1	0
1	2	2	1
2	4	4	2
3	8	8	3

DEFINITION 1
Logarithmic Function

For $b > 0$, $b \neq 1$, the inverse of $f(x) = b^x$, denoted $f^{-1}(x) = \log_b x$, is the **logarithmic function with base** b.

Logarithmic form		Exponential form
$y = \log_b x$	is equivalent to	$x = b^y$

The log to the base b of x is the exponent to which b must be raised to obtain x.

$y = \log_{10} x$	is equivalent to	$x = 10^y$
$y = \log_e x$	is equivalent to	$x = e^y$

Remember: A logarithm is an exponent.

It is very important to remember that $y = \log_b x$ and $x = b^y$ define the same function, and as such can be used interchangeably.

Because the domain of an exponential function includes all real numbers and its range is the set of positive real numbers, the **domain** of a logarithmic function is the set of all positive real numbers and its **range** is the set of all real numbers. Thus, $\log_{10} 3$ is defined, but $\log_{10} 0$ and $\log_{10} (-5)$ are not defined. That is, 3 is a logarithmic domain value, but 0 and -5 are not. Typical logarithmic curves are shown in Figure 2.

FIGURE 2 Typical logarithmic graphs.

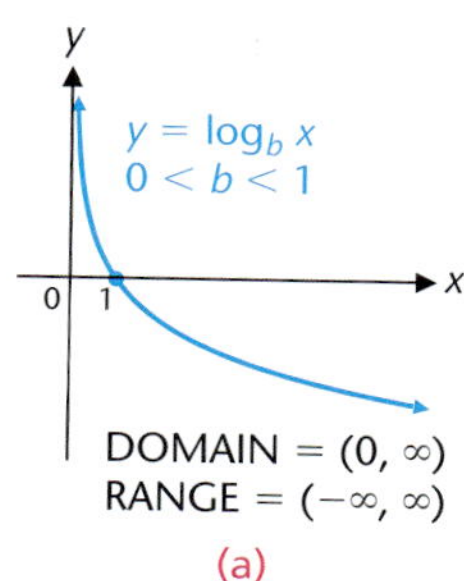

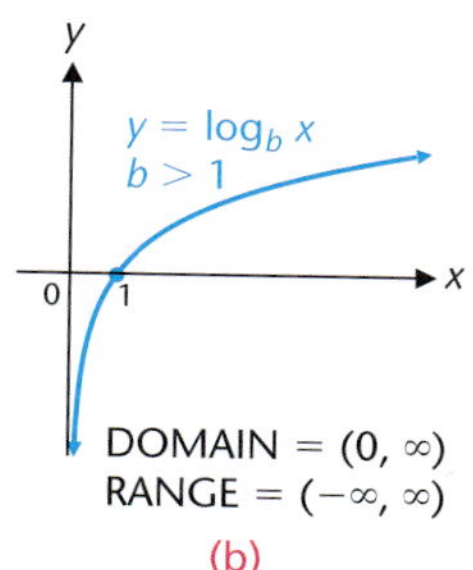

EXPLORE/DISCUSS 1

For the exponential function $f = \{(x, y) \mid y = (\frac{2}{3})^x\}$, graph f and $y = x$ on the same coordinate system. Then sketch the graph of f^{-1}. Use the Draw Inverse routine on a graphing utility to check your work. Discuss the domains and ranges of f and its inverse. By what other name is f^{-1} known?

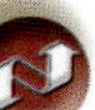

From Logarithmic Form to Exponential Form, and Vice Versa

We now look into the matter of converting logarithmic forms to equivalent exponential forms, and vice versa.

EXAMPLE 1 Logarithmic–Exponential Conversions

Change each logarithmic form to an equivalent exponential form.

(A) $\log_2 8 = 3$ (B) $\log_{25} 5 = \frac{1}{2}$ (C) $\log_2 (\frac{1}{4}) = -2$

SOLUTIONS

(A) $\log_2 8 = 3$ is equivalent to $8 = 2^3$.

(B) $\log_{25} 5 = \frac{1}{2}$ is equivalent to $5 = 25^{1/2}$.

(C) $\log_2 (\frac{1}{4}) = -2$ is equivalent to $\frac{1}{4} = 2^{-2}$.

MATCHED 1 PROBLEM

Change each logarithmic form to an equivalent exponential form.

(A) $\log_3 27 = 3$ (B) $\log_{36} 6 = \frac{1}{2}$ (C) $\log_3 (\frac{1}{9}) = -2$

EXAMPLE 2 Logarithmic–Exponential Conversions

Change each exponential form to an equivalent logarithmic form.

(A) $49 = 7^2$ (B) $3 = \sqrt{9}$ (C) $\frac{1}{5} = 5^{-1}$

SOLUTIONS

(A) $49 = 7^2$ is equivalent to $\log_7 49 = 2$.

(B) $3 = \sqrt{9}$ is equivalent to $\log_9 3 = \frac{1}{2}$.

(C) $\frac{1}{5} = 5^{-1}$ is equivalent to $\log_5 (\frac{1}{5}) = -1$.

MATCHED 2 PROBLEM

Change each exponential form to an equivalent logarithmic form.

(A) $64 = 4^3$ (B) $2 = \sqrt[3]{8}$ (C) $\frac{1}{16} = 4^{-2}$

To gain a little deeper understanding of logarithmic functions and their relationship to the exponential functions, we consider a few problems where we want to find x, b, or y in $y = \log_b x$, given the other two values. All values were chosen so that the problems can be solved without a calculator.

EXAMPLE 3 Solutions of the Equation $y = \log_b x$

Find x, b, or y as indicated.

(A) Find y: $y = \log_4 8$. (B) Find x: $\log_3 x = -2$.

(C) Find b: $\log_b 1{,}000 = 3$.

SOLUTIONS

(A) Write $y = \log_4 8$ in equivalent exponential form.

$$8 = 4^y$$

$$2^3 = 2^{2y} \quad \text{Write each number to the same base 2.}$$

$$2y = 3 \quad \text{Recall that } b^m = b^n \text{ if and only if } m = n.$$

$$y = \frac{3}{2}$$

Thus, $\frac{3}{2} = \log_4 8$.

(B) Write $\log_3 x = -2$ in equivalent exponential form.

$$x = 3^{-2}$$

$$= \frac{1}{3^2} = \frac{1}{9}$$

Thus, $\log_3 (\frac{1}{9}) = -2$.

(C) Write $\log_b 1{,}000 = 3$ in equivalent exponential form:

$$1{,}000 = b^3$$
$$10^3 = b^3 \quad \text{Write 1,000 as a third power.}$$
$$b = 10$$

Thus, $\log_{10} 1{,}000 = 3$.

MATCHED 3 PROBLEM

Find x, b, or y as indicated.

(A) Find y: $y = \log_9 27$. (B) Find x: $\log_2 x = -3$.

(C) Find b: $\log_b 100 = 2$.

Properties of Logarithmic Functions

The familiar properties of exponential functions imply corresponding properties of logarithmic functions.

EXPLORE/DISCUSS 2

Discuss the connection between the exponential equation and the logarithmic equation, and explain why each equation is valid.

(A) $2^4\, 2^7 = 2^{11}$; $\log_2 2^4 + \log_2 2^7 = \log_2 2^{11}$

(B) $2^{13}/2^5 = 2^8$; $\log_2 2^{13} - \log_2 2^5 = \log_2 2^8$

(C) $(2^6)^9 = 2^{54}$; $9 \log_2 2^6 = \log_2 2^{54}$

Several of the powerful and useful properties of logarithmic functions are listed in Theorem 1.

THEOREM 1
Properties of Logarithmic Functions

If b, M, and N are positive real numbers, $b \neq 1$, and p and x are real numbers, then

1. $\log_b 1 = 0$
2. $\log_b b = 1$
3. $\log_b b^x = x$
4. $b^{\log_b x} = x, x > 0$
5. $\log_b MN = \log_b M + \log_b N$
6. $\log_b \dfrac{M}{N} = \log_b M - \log_b N$
7. $\log_b M^p = p \log_b M$
8. $\log_b M = \log_b N$ if and only if $M = N$

The first two properties in Theorem 1 follow directly from the definition of a logarithmic function:

$$\log_b 1 = 0 \qquad \text{because} \qquad b^0 = 1$$
$$\log_b b = 1 \qquad \text{because} \qquad b^1 = b$$

The third and fourth properties are "inverse properties." They follow directly from the fact that exponential and logarithmic functions are inverses of each other. Recall from Section 1.6 that if f is one-to-one, then f^{-1} is a one-to-one function satisfying

$$f^{-1}(f(x)) = x \qquad \text{for all } x \text{ in the domain of } f$$
$$f(f^{-1}(x)) = x \qquad \text{for all } x \text{ in the domain of } f^{-1}$$

Applying these general properties to $f(x) = b^x$ and $f^{-1}(x) = \log_b x$, we see that

$$f^{-1}(f(x)) = x \qquad f(f^{-1}(x)) = x$$
$$\log_b (f(x)) = x \qquad b^{f^{-1}(x)} = x$$
$$\log_b b^x = x \qquad b^{\log_b x} = x$$

Properties 5 to 7 enable us to convert multiplication into addition, division into subtraction, and power and root problems into multiplication. The proofs of these properties are based on properties of exponents. A sketch of a proof of the fifth property follows: To bring exponents into the proof, we let

$$u = \log_b M \qquad \text{and} \qquad v = \log_b N$$

and convert these to the equivalent exponential forms

$$M = b^u \qquad \text{and} \qquad N = b^v$$

Now, see if you can provide the reasons for each of the following steps:

$$\log_b MN = \log_b b^u b^v = \log_b b^{u+v} = u + v = \log_b M + \log_b N$$

The other properties are established in a similar manner (see Problems 104 and 105 in Exercise 4.3.)

Finally, the eighth property follows from the fact that logarithmic functions are one-to-one.

EXAMPLE 4 Using Logarithmic Properties

Simplify, using the properties in Theorem 1.

(A) $\log_e 1$ (B) $\log_{10} 10$ (C) $\log_e e^{2x+1}$

(D) $\log_{10} 0.01$ (E) $10^{\log_{10} 7}$ (F) $e^{\log_e x^2}$

SOLUTIONS

(A) $\log_e 1 = 0$

(B) $\log_{10} 10 = 1$

(C) $\log_e e^{2x+1} = 2x + 1$

(D) $\log_{10} 0.01 = \log_{10} 10^{-2} = -2$

(E) $10^{\log_{10} 7} = 7$

(F) $e^{\log_e x^2} = x^2$

MATCHED 4 PROBLEM

Simplify, using the properties in Theorem 1.

(A) $\log_{10} 10^{-5}$ (B) $\log_5 25$ (C) $\log_{10} 1$

(D) $\log_e e^{m+n}$ (E) $10^{\log_{10} 4}$ (F) $e^{\log_e (x^4+1)}$

Common and Natural Logarithms

John Napier (1550–1617) is credited with the invention of logarithms, which evolved out of an interest in reducing the computational strain in research in astronomy. This new computational tool was immediately accepted by the scientific world. Now, with the availability of inexpensive calculators, logarithms have lost most of their importance as a computational device. However, the logarithmic concept has been greatly generalized since its conception, and logarithmic functions are used widely in both theoretical and applied sciences.

Of all possible logarithmic bases, the base e and the base 10 are used almost exclusively. To use logarithms in certain practical problems, we need to be able to approximate the logarithm of any positive number to either base 10 or base e. And conversely, if we are given the logarithm of a number to base 10 or base e, we need to be able to approximate the number. Historically, tables were used for this purpose, but now calculators are used because they are faster and can find far more values than any table can possibly include.

Common logarithms, also called **Briggsian logarithms,** are logarithms with base 10. **Natural logarithms,** also called **Napierian logarithms,** are logarithms with base e. Most calculators have a function key labeled "log" and a function key labeled "ln." The former represents the common logarithmic function and the latter the natural logarithmic function. In fact, "log" and "ln" are both used extensively in mathematical literature, and whenever you see either used in this book without a base indicated, they should be interpreted as in the box.

Logarithmic Functions

$y = \log x = \log_{10} x$ Common logarithmic function

$y = \ln x = \log_e x$ Natural logarithmic function

EXPLORE/DISCUSS 3

(A) Sketch the graph of $y = 10^x$, $y = \log x$, and $y = x$ in the same coordinate system and state the domain and range of the common logarithmic function.

(B) Sketch the graph of $y = e^x$, $y = \ln x$, and $y = x$ in the same coordinate system and state the domain and range of the natural logarithmic function.

EXAMPLE 5 Calculator Evaluation of Logarithms

Use a calculator to evaluate each to six decimal places.

(A) log 3,184 (B) ln 0.000 349 (C) log (−3.24)

SOLUTIONS

(A) log 3,184 = 3.502 973

(B) ln 0.000 349 = −7.960 439

(C) log (−3.24) = Error

Why is an error indicated in part C? Because −3.24 is not in the domain of the log function. [*Note:* Calculators display error messages in various ways. Some calculators use a more advanced definition of logarithmic functions that involves complex numbers. They will display an ordered pair, representing a complex number, as the value of log (−3.24), rather than an error message. You should interpret such a display as indicating that the number entered is not in the domain of the logarithmic function as we have defined it.]

MATCHED 5 PROBLEM

Use a calculator to evaluate each to six decimal places.

(A) log 0.013 529 (B) ln 28.693 28 (C) ln (−0.438)

When working with common and natural logarithms, we follow the common practice of using the equal sign "=" where it might be more appropriate to use the approximately equal sign "≈." No harm is done as long as we keep in mind that in a statement such as log 3.184 = 0.503, the number on the right is only assumed accurate to three decimal places and is not exact.

EXPLORE/DISCUSS 4

Graphs of the functions $f(x) = \log x$ and $g(x) = \ln x$ are shown in the graphing utility display of Figure 3. Which graph belongs to which function? It appears from the display that one of the functions may be a constant multiple of the other. Is that true? Find and discuss the evidence for your answer.

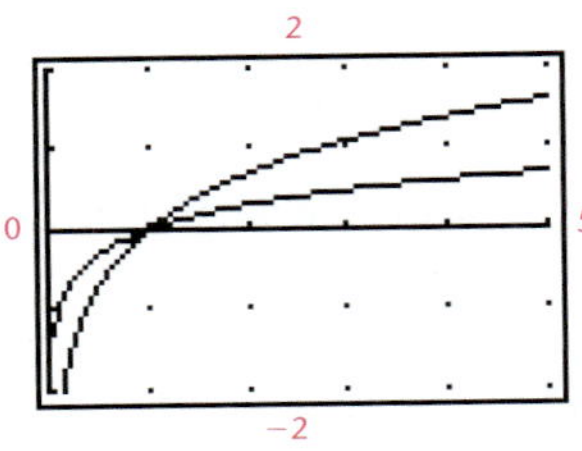

FIGURE 3

EXAMPLE 6 Calculator Evaluation of Logarithms

Use a calculator to evaluate each expression to three decimal places.

(A) $\dfrac{\log 2}{\log 1.1}$ (B) $\log \dfrac{2}{1.1}$ (C) $\log 2 - \log 1.1$

SOLUTIONS

(A) $\dfrac{\log 2}{\log 1.1} = 7.273$

(B) $\log \dfrac{2}{1.1} = 0.260$

(C) $\log 2 - \log 1.1 = 0.260$. Note that $\dfrac{\log 2}{\log 1.1} \neq \log 2 - \log 1.1$, but

$\log \dfrac{2}{1.1} = \log 2 - \log 1.1$ (see Theorem 1).

MATCHED 6 PROBLEM

Use a calculator to evaluate each to three decimal places.

(A) $\dfrac{\ln 3}{\ln 1.08}$ (B) $\ln \dfrac{3}{1.08}$ (C) $\ln 3 - \ln 1.08$

We now turn to the second problem: Given the logarithm of a number, find the number. To solve this problem, we make direct use of the logarithmic–exponential relationships.

Logarithmic–Exponential Relationships

$\log x = y$	is equivalent to	$x = 10^y$
$\ln x = y$	is equivalent to	$x = e^y$

EXAMPLE 7 Solving $\log_b x = y$ for x

Find x to three significant digits, given the indicated logarithms.

(A) $\log x = -9.315$ (B) $\ln x = 2.386$

SOLUTIONS

(A) $\log x = -9.315$

$x = 10^{-9.315}$ Change to exponential form (Definition 1)

$= 4.84 \times 10^{-10}$

Notice that the answer is displayed in scientific notation in the calculator.

(B) $\ln x = 2.386$

$x = e^{2.386}$ Change to exponential form (Definition 1)

$= 10.9$

MATCHED 7 PROBLEM

Find x to four significant digits, given the indicated logarithms.

(A) $\ln x = -5.062$ (B) $\log x = 12.0821$

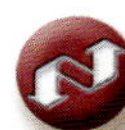

EXPLORE/DISCUSS 5

Example 7 was solved algebraically using the logarithmic–exponential relationships. Use the intersection routine on a graphing utility to solve this problem graphically. Discuss the relative merits of the two approaches.

Change of Base

How would you find the logarithm of a positive number to a base other than 10 or e? For example, how would you find $\log_3 5.2$? In Example 8 we evaluate this logarithm using a direct process. Then we develop a change-of-base formula to find such logarithms in general. You may find it easier to remember the process than the formula.

EXAMPLE 8 Evaluating a Base 3 Logarithm

Evaluate $\log_3 5.2$ to four decimal places.

SOLUTIONS

Let $y = \log_3 5.2$ and proceed as follows:

$$\log_3 5.2 = y$$

$$5.2 = 3^y \qquad \text{Change to exponential form.}$$

$$\ln 5.2 = \ln 3^y \qquad \text{Take the natural log (or common log) of each side.}$$

$$= y \ln 3 \qquad \log_b M^p = p \log_b M$$

$$y = \frac{\ln 5.2}{\ln 3} \qquad \text{Solve for } y.$$

Replace y with $\log_3 5.2$ from the first step, and use a calculator to evaluate the right side:

$$\log_3 5.2 = \frac{\ln 5.2}{\ln 3} = 1.5007$$

MATCHED 8 PROBLEM

Evaluate $\log_{0.5} 0.0372$ to four decimal places.

To develop a change-of-base formula for arbitrary positive bases, with neither base equal to 1, we proceed as above. Let $y = \log_b N$, where N and b are positive and $b \neq 1$. Then

$$\begin{aligned} \log_b N &= y \\ N &= b^y && \text{Write in exponential form.} \\ \log_a N &= \log_a b^y && \text{Take the log of each side to another positive base } a, a \neq 1. \\ &= y \log_a b && \log_b M^p = p \log_b M \\ y &= \frac{\log_a N}{\log_a b} && \text{Solve for } y. \end{aligned}$$

Replacing y with $\log_b N$ from the first step, we obtain the **change-of-base formula:**

$$\log_b N = \frac{\log_a N}{\log_a b}$$

In words, this formula states that the logarithm of a number to a given base is the logarithm of that number to a new base divided by the logarithm of the old base to the new base. In practice, we usually choose either e or 10 for the new base so that a calculator can be used to evaluate the necessary logarithms:

$$\log_b N = \frac{\ln N}{\ln b} \qquad \text{or} \qquad \log_b N = \frac{\log N}{\log b}$$

We used the first of these options in Example 8.

EXPLORE/DISCUSS 6

If b is any positive real number different from 1, the change-of-base formula implies that the function $y = \log_b x$ is a constant multiple of the natural logarithmic function; that is, $\log_b x = k \ln x$ for some k.

(A) Graph the functions $y = \ln x$, $y = 2 \ln x$, $y = 0.5 \ln x$, and $y = -3 \ln x$.

(B) Write each function of part A in the form $y = \log_b x$ by finding the base b to two decimal places.

(C) Is every exponential function $y = b^x$ a constant multiple of $y = e^x$? Explain.

CAUTION

We conclude this section by noting two common errors:

1. $\dfrac{\log_b M}{\log_b N} \neq \log_b M - \log_b N$ — $\log_b M - \log_b N = \log_b \dfrac{M}{N}$; $\dfrac{\log_b M}{\log_b N}$ cannot be simplified.

2. $\log_b (M + N) \neq \log_b M + \log_b N$ — $\log_b M + \log_b N = \log_b MN$; $\log_b (M + N)$ cannot be simplified.

ANSWERS TO MATCHED PROBLEMS

1. (A) $27 = 3^3$ (B) $6 = 36^{1/2}$ (C) $\frac{1}{9} = 3^{-2}$
2. (A) $\log_4 64 = 3$ (B) $\log_8 2 = \frac{1}{3}$ (C) $\log_4(\frac{1}{16}) = -2$
3. (A) $y = \frac{3}{2}$ (B) $x = \frac{1}{8}$ (C) $b = 10$
4. (A) -5 (B) 2 (C) 0 (D) $m + n$ (E) 4 (F) $x^4 + 1$
5. (A) $-1.868\ 734$ (B) 3.356 663 (C) Not possible
6. (A) 14.275 (B) 1.022 (C) 1.022
7. (A) $x = 0.006\ 333$ (B) $x = 1.208 \times 10^{12}$
8. 4.7486

EXERCISE 4.3

A

Rewrite Problems 1–8 in equivalent exponential form.

1. $\log_3 81 = 4$
2. $\log_5 125 = 3$
3. $\log_{10} 0.001 = -3$
4. $\log_{10} 1{,}000 = 3$
5. $\log_{81} 3 = \frac{1}{4}$
6. $\log_4 2 = \frac{1}{2}$
7. $\log_{1/2} 16 = -4$
8. $\log_{1/3} 27 = -3$

Rewrite Problems 9–16 in equivalent logarithmic form.

9. $0.0001 = 10^{-4}$
10. $10{,}000 = 10^4$
11. $8 = 4^{3/2}$
12. $9 = 27^{2/3}$
13. $\frac{1}{2} = 32^{-1/5}$
14. $\frac{1}{8} = 2^{-3}$
15. $7 = \sqrt{49}$
16. $4 = \sqrt[3]{64}$

In Problems 17–30, simplify each expression using Theorem 1.

17. $\log_{16} 1$
18. $\log_{25} 1$
19. $\log_{0.5} 0.5$
20. $\log_7 7$
21. $\log_e e^4$
22. $\log_{10} 10^5$
23. $\log_{10} 0.01$
24. $\log_{10} 100$
25. $\log_5 \sqrt[3]{5}$
26. $\log_2 \sqrt{8}$
27. $e^{\log_e \sqrt{x}}$
28. $e^{\log_e (x-1)}$
29. $e^{2 \log_e x}$
30. $10^{-3 \log_{10} u}$

In Problems 31–38, evaluate to four decimal places.

31. log 82,734
32. log 843,250
33. log 0.001 439
34. log 0.035 604
35. ln 43.046
36. ln 2,843,100
37. ln 0.081 043
38. ln 0.000 032 4

In Problems 39–46, evaluate x to four significant digits, given:

39. $\log x = 5.3027$
40. $\log x = 1.9168$
41. $\log x = -3.1773$
42. $\log x = -2.0411$
43. $\ln x = 3.8655$
44. $\ln x = 5.0884$
45. $\ln x = -0.3916$
46. $\ln x = -4.1083$

B

Find x, y, or b, as indicated in Problems 47–60.

47. $\log_2 x = 2$
48. $\log_3 x = 3$
49. $\log_4 16 = y$
50. $\log_8 64 = y$
51. $\log_b 16 = 2$
52. $\log_b 10^{-3} = -3$
53. $\log_b 1 = 0$
54. $\log_b b = 1$
55. $\log_4 x = \frac{1}{2}$
56. $\log_8 x = \frac{1}{3}$
57. $\log_{1/3} 9 = y$
58. $\log_{49}(\frac{1}{7}) = y$
59. $\log_b 1{,}000 = \frac{3}{2}$
60. $\log_b 4 = \frac{2}{3}$

In Problems 61–68, evaluate to three decimal places.

61. $n = \dfrac{\log 2}{\log 1.15}$
62. $n = \dfrac{\log 2}{\log 1.12}$
63. $n = \dfrac{\ln 3}{\ln 1.15}$
64. $n = \dfrac{\ln 4}{\ln 1.2}$
65. $x = \dfrac{\ln 0.5}{-0.21}$
66. $x = \dfrac{\ln 0.1}{-0.0025}$
67. $t = \dfrac{\ln 150}{\ln 3}$
68. $t = \dfrac{\log 200}{\log 2}$

In Problems 69–76, evaluate x to five significant digits.

69. $x = \log(5.3147 \times 10^{12})$

70. $x = \log(2.0991 \times 10^{17})$

71. $x = \ln(6.7917 \times 10^{-12})$

72. $x = \ln(4.0304 \times 10^{-8})$

73. $\log x = 32.068\,523$

74. $\log x = -12.731\,64$

75. $\ln x = -14.667\,13$

76. $\ln x = 18.891\,143$

In Problems 77–80, find f^{-1}. Check by graphing f, f^{-1}, and $y = x$ in the same viewing window on a graphing utility.

77. $f(x) = 2\ln(x + 2)$

78. $f(x) = 2\ln x + 2$

79. $f(x) = 4\ln x - 3$

80. $f(x) = 4\ln(x - 3)$

C

In Problems 81–84, find domain and range, x and y intercepts, and asymptotes. Round all approximate values to two decimal places.

81. $f(x) = -2 + \ln(1 + x^2)$

82. $f(x) = 2 - \ln(1 + |x|)$

83. $f(x) = 1 + \ln(1 - x^2)$

84. $f(x) = -1 + \ln(|1 - x^2|)$

85. Find the fallacy.

$$1 < 3$$

$$\frac{1}{27} < \frac{3}{27} \quad \text{Divide both sides by 27.}$$

$$\frac{1}{27} < \frac{1}{9}$$

$$(\tfrac{1}{3})^3 < (\tfrac{1}{3})^2$$

$$\log(\tfrac{1}{3})^3 < \log(\tfrac{1}{3})^2$$

$$3\log\tfrac{1}{3} < 2\log\tfrac{1}{3}$$

$$3 < 2 \quad \text{Divide both sides by } \log\tfrac{1}{3}.$$

86. Find the fallacy.

$$3 > 2$$

$$3\log\tfrac{1}{2} > 2\log\tfrac{1}{2} \quad \text{Multiply both sides by } \log\tfrac{1}{2}.$$

$$\log(\tfrac{1}{2})^3 > \log(\tfrac{1}{2})^2$$

$$(\tfrac{1}{2})^3 > (\tfrac{1}{2})^2$$

$$\tfrac{1}{8} > \tfrac{1}{4}$$

$$1 > 2 \quad \text{Multiply both sides by 8.}$$

87. The function $f(x) = \log x$ increases extremely slowly as $x \to \infty$, but the composite function $g(x) = \log(\log x)$ increases still more slowly.

(A) Illustrate this fact by computing the values of both functions for several large values of x.

(B) Determine the domain and range of the function g.

(C) Discuss the graphs of both functions.

88. The function $f(x) = \ln x$ increases extremely slowly as $x \to \infty$, but the composite function $g(x) = \ln(\ln x)$ increases still more slowly.

(A) Illustrate this fact by computing the values of both functions for several large values of x.

(B) Determine the domain and range of the function g.

(C) Discuss the graphs of both functions.

In Problems 89–92, use a graphing utility to find the coordinates of all points of intersection to two decimal places.

89. $f(x) = \ln x$, $g(x) = 0.1x - 0.2$

90. $f(x) = \log x$, $g(x) = 4 - x^2$

91. $f(x) = \ln x$, $g(x) = x^{1/3}$

92. $f(x) = 3\ln(x - 2)$, $g(x) = 4e^{-x}$

The polynomials in Problems 93–96, called ***Taylor polynomials,*** *can be used to approximate the function $g(x) = \ln(1 + x)$. To illustrate this approximation graphically, in each problem, graph $g(x) = \ln(1 + x)$ and the indicated polynomial in the same viewing window, $-1 \le x \le 3$ and $-2 \le y \le 2$.*

93. $P_1(x) = x - \frac{1}{2}x^2$

94. $P_2(x) = x - \frac{1}{2}x^2 + \frac{1}{3}x^3$

95. $P_3(x) = x - \frac{1}{2}x^2 + \frac{1}{3}x^3 - \frac{1}{4}x^4$

96. $P_4(x) = x - \frac{1}{2}x^2 + \frac{1}{3}x^3 - \frac{1}{4}x^4 + \frac{1}{5}x^5$

In Problems 97–100,

(A) Use the graph of $y = \log_2 x$ (Fig. 1) and graph transformations to sketch the graph of f.

(B) Find f^{-1} and use the Draw Inverse routine on a graphing utility to check the graph in part A.

97. $f(x) = \log_2(x - 2)$

98. $f(x) = \log_2(x + 3)$

99. $f(x) = \log_2 x - 2$

100. $f(x) = \log_2 x + 3$

101. **(A)** For $f = \{(x, y) \mid y = (\frac{1}{2})^x = 2^{-x}\}$, graph f, f^{-1}, and $y = x$ on the same coordinate system.

(B) Indicate the domain and range of f and f^{-1}.

(C) What other name can you use for the inverse of f?

102. Explain why the graph of the reflection of the function $y = 3^{x^2}$ in the line $y = x$ is not the graph of a function.

103. Explain why the graph of the reflection of the function $y = 2^{|x|}$ in the line $y = x$ is not the graph of a function.

104. Prove that $\log_b (M/N) = \log_b M - \log_b N$ under the hypotheses of Theorem 1.

105. Prove that $\log_b M^p = p \log_b M$ under the hypotheses of Theorem 1.

SECTION 4.4 Logarithmic Models

Logarithmic Scales • Rocket Flight • Data Analysis and Regression

In Section 4.4 we study the logarithmic scales that are used to compare intensities of sounds, magnitudes of earthquakes, and the brightness of stars. We construct logarithmic models using regression techniques.

Logarithmic Scales

SOUND INTENSITY The human ear is able to hear sound over an incredible range of intensities. The loudest sound a healthy person can hear without damage to the eardrum has an intensity 1 trillion (1,000,000,000,000) times that of the softest sound a person can hear. Working directly with numbers over such a wide range is very cumbersome. Because the logarithm, with base greater than 1, of a number increases much more slowly than the number itself, logarithms are often used to create more convenient compressed scales. The decibel scale for sound intensity is an example of such a scale. The **decibel,** named after the inventor of the telephone, Alexander Graham Bell (1847–1922), is defined as follows:

$$D = 10 \log \frac{I}{I_0} \qquad \text{Decibel scale} \qquad (1)$$

where D is the **decibel level** of the sound, I is the **intensity** of the sound measured in watts per square meter (W/m^2), and I_0 is the intensity of the least audible sound that an average healthy young person can hear. The latter is standardized to be $I_0 = 10^{-12}$ watts per square meter. Table 1 lists some typical sound intensities from familiar sources.

TABLE 1 Typical Sound Intensities

Sound Intensity (W/m^2)	Sound
1.0×10^{-12}	Threshold of hearing
5.2×10^{-10}	Whisper
3.2×10^{-6}	Normal conversation
8.5×10^{-4}	Heavy traffic
3.2×10^{-3}	Jackhammer
1.0×10^{0}	Threshold of pain
8.3×10^{2}	Jet plane with afterburner

EXAMPLE 1 Sound Intensity

Find the number of decibels from a whisper with sound intensity 5.20×10^{-10} watts per square meter. Compute the answer to two decimal places.

SOLUTION
We use the decibel formula (1):

$$\begin{aligned} D &= 10 \log \frac{I}{I_0} \\ &= 10 \log \frac{5.2 \times 10^{-10}}{10^{-12}} \\ &= 10 \log 520 \\ &= 27.16 \text{ decibels} \end{aligned}$$

MATCHED PROBLEM 1

Find the number of decibels from a jackhammer with sound intensity 3.2×10^{-3} watts per square meter. Compute the answer to two decimal places.

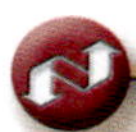

EXPLORE/DISCUSS 1

Imagine using a large sheet of graph paper, ruled with horizontal and vertical lines $\frac{1}{8}$-inch apart, to plot the sound intensities of Table 1 on the x axis and the corresponding decibel levels on the y axis. Suppose that each $\frac{1}{8}$-inch unit on the x axis represents the intensity of the least audible sound (10^{-12} W/m^2), and each $\frac{1}{8}$-inch unit on the y axis represents 1 decibel. If the point corresponding to a jet plane with afterburner is plotted on the graph paper, how far is it from the x axis? From the y axis? (Give the first answer in inches and the second in miles!) Discuss.

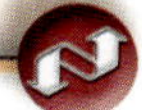

EARTHQUAKE INTENSITY The energy released by the largest earthquake recorded, measured in joules, is about 100 billion (100,000,000,000) times the energy released by a small earthquake that is barely felt. Over the past 150 years several people from various countries have devised different types of measures of earthquake magnitudes so that their severity could be easily compared. In 1935 the California seismologist Charles Richter devised a logarithmic scale that bears his name and is still widely used in the United States. The **magnitude** M on the **Richter scale*** is given as follows:

$$M = \frac{2}{3} \log \frac{E}{E_0} \qquad \text{Richter scale} \qquad (2)$$

*Originally, Richter defined the magnitude of an earthquake in terms of logarithms of the maximum seismic wave amplitude, in thousandths of a millimeter, measured on a standard seismograph. Formula (2) gives essentially the same magnitude that Richter obtained for a given earthquake but in terms of logarithms of the energy released by the earthquake.

where E is the energy released by the earthquake, measured in joules, and E_0 is the energy released by a very small reference earthquake, which has been standardized to be

$$E_0 = 10^{4.40} \text{ joules}$$

The destructive power of earthquakes relative to magnitudes on the Richter scale is indicated in Table 2.

TABLE 2 The Richter Scale

Magnitude on Richter Scale	Destructive Power
$M < 4.5$	Small
$4.5 < M < 5.5$	Moderate
$5.5 < M < 6.5$	Large
$6.5 < M < 7.5$	Major
$7.5 < M$	Greatest

EXAMPLE 2 Earthquake Intensity

The 1906 San Francisco earthquake released approximately 5.96×10^{16} joules of energy. What was its magnitude on the Richter scale? Compute the answer to two decimal places.

SOLUTION

We use the magnitude formula (2):

$$\begin{aligned} M &= \frac{2}{3} \log \frac{E}{E_0} \\ &= \frac{2}{3} \log \frac{5.96 \times 10^{16}}{10^{4.40}} \\ &= 8.25 \end{aligned}$$

MATCHED PROBLEM 2

The 1985 earthquake in central Chile released approximately 1.26×10^{16} joules of energy. What was its magnitude on the Richter scale? Compute the answer to two decimal places.

EXAMPLE 3 Earthquake Intensity

If the energy release of one earthquake is 1,000 times that of another, how much larger is the Richter scale reading of the larger than the smaller?

SOLUTION

Let

$$M_1 = \frac{2}{3}\log\frac{E_1}{E_0} \qquad \text{and} \qquad M_2 = \frac{2}{3}\log\frac{E_2}{E_0}$$

be the Richter equations for the smaller and larger earthquakes, respectively. Substituting $E_2 = 1{,}000E_1$ into the second equation, we obtain

$$\begin{aligned} M_2 &= \frac{2}{3}\log\frac{1{,}000E_1}{E_0} \\ &= \frac{2}{3}\left(\log 10^3 + \log\frac{E_1}{E_0}\right) && \log MN = \log M + \log N \\ &= \frac{2}{3}\left(3 + \log\frac{E_1}{E_0}\right) && \log 10^x = x \\ &= \frac{2}{3}(3) + \frac{2}{3}\log\frac{E_1}{E_0} && \text{Distributive property} \\ &= 2 + M_1 \end{aligned}$$

Thus, an earthquake with 1,000 times the energy of another has a Richter scale reading of 2 more than the other.

MATCHED PROBLEM 3

If the energy release of one earthquake is 10,000 times that of another, how much larger is the Richter scale reading of the larger than the smaller?

ROCKET FLIGHT The theory of rocket flight uses advanced mathematics and physics to show that the velocity v of a rocket at burnout (depletion of fuel supply) is given by

$$v = c\ln\frac{W_t}{W_b} \qquad \text{Rocket equation} \tag{3}$$

where c is the exhaust velocity of the rocket engine, W_t is the takeoff weight (fuel, structure, and payload), and W_b is the burnout weight (structure and payload).

Because of the Earth's atmospheric resistance, a launch vehicle velocity of at least 9.0 kilometers per second is required to achieve the minimum altitude needed for a stable orbit. It is clear that to increase velocity v, either the weight ratio W_t/W_b must be increased or the exhaust velocity c must be increased. The weight ratio can be increased by the use of solid fuels, and the exhaust velocity can be increased by improving the fuels, solid or liquid.

EXAMPLE 4 Rocket Flight Theory

A typical single-stage, solid-fuel rocket may have a weight ratio $W_t/W_b = 18.7$ and an exhaust velocity $c = 2.38$ kilometers per second. Would this rocket reach a launch velocity of 9.0 kilometers per second?

SOLUTION

We use the rocket equation (3):

$$\begin{aligned} v &= c \ln \frac{W_t}{W_b} \\ &= 2.38 \ln 18.7 \\ &= 6.97 \text{ kilometers per second} \end{aligned}$$

The velocity of the launch vehicle is far short of the 9.0 kilometers per second required to achieve orbit. This is why multiple-stage launchers are used—the deadweight from a preceding stage can be jettisoned into the ocean when the next stage takes over.

MATCHED PROBLEM 4

A launch vehicle using liquid fuel, such as a mixture of liquid hydrogen and liquid oxygen, can produce an exhaust velocity of $c = 4.7$ kilometers per second. However, the weight ratio W_t/W_b must be low—around 5.5 for some vehicles—because of the increased structural weight to accommodate the liquid fuel. How much more or less than the 9.0 kilometers per second required to reach orbit will be achieved by this vehicle?

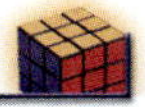

Data Analysis and Regression

We use logarithmic regression to fit a function of the form $y = a + b \ln x$ to a set of data points, making use of the techniques introduced earlier for linear, quadratic, exponential, and logistic functions.

EXAMPLE 5 Home Ownership Rates

The U.S. Census Bureau published the data in Table 3 on home ownership rates.

TABLE 3 Home Ownership Rates

Year	Home Ownership Rate (%)
1940	43.6
1950	55.0
1960	61.9
1970	62.9
1980	64.4
1990	64.2
2000	67.4

(A) Let x represent time in years with $x = 0$ representing 1900, and let y represent the corresponding home ownership rate. Use regression analysis on a graphing utility to find a logarithmic function of the form $y = a + b \ln x$ that models the data. (Round the constants a and b to three significant digits.)

(B) Use the logarithmic regression function to predict the home ownership rate in 2010.

SOLUTION

(A) Figure 1 shows the details of constructing the model on a graphing utility.

FIGURE 1

L1	L2	L3 1
40	43.6	------
50	55	
60	61.9	
70	62.9	
80	64.4	
90	64.2	
100	67.4	

L1(1)=40

(a) Data

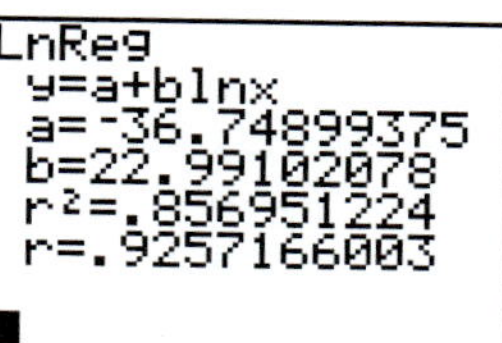

(b) Regression equation

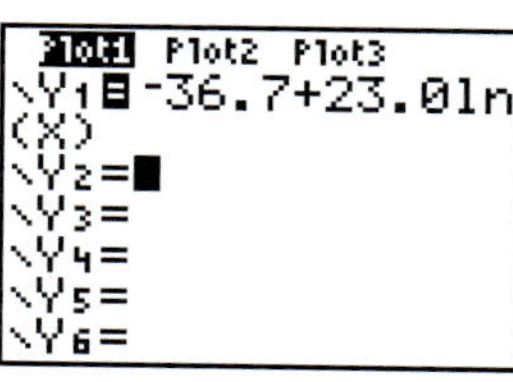

(c) Regression equation entered in equation editor

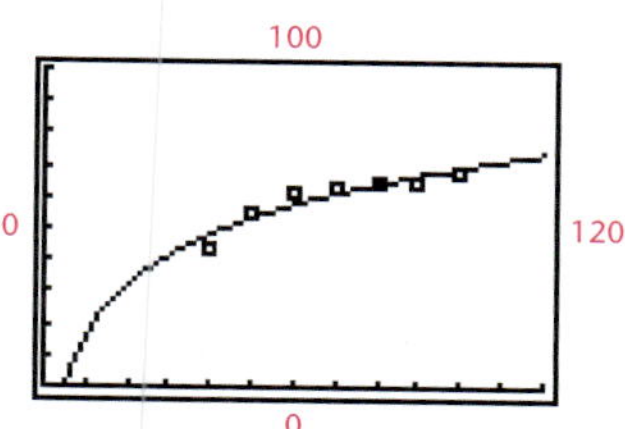

(d) Graph of data and regression equation

(B) Evaluating $y_1 = -36.7 + 23.0 \ln x$ at $x = 110$ predicts a home ownership rate of 71.4% in 2010.

MATCHED 5 PROBLEM

Refer to Example 5. The home ownership rate in 1995 was 64.7%.

(A) Find a logarithmic regression equation for the expanded data set.

(B) Predict the home ownership rate in 2010.

ANSWERS MATCHED PROBLEMS

1. 95.05 decibels
2. 7.80
3. 2.67
4. 1 kilometer per second less
5. (A) $-31.5 + 21.7 \ln x$
   (B) 70.5%

EXERCISE 4.4

APPLICATIONS

1. **Sound.** What is the decibel level of
   (A) The threshold of hearing, 1.0×10^{-12} watts per square meter?
   (B) The threshold of pain, 1.0 watt per square meter?
   Compute answers to two significant digits.

2. **Sound.** What is the decibel level of
   (A) A normal conversation, 3.2×10^{-6} watts per square meter?
   (B) A jet plane with an afterburner, 8.3×10^{2} watts per square meter?
   Compute answers to two significant digits.

3. Sound. If the intensity of a sound from one source is 1,000 times that of another, how much more is the decibel level of the louder sound than the quieter one?

4. Sound. If the intensity of a sound from one source is 10,000 times that of another, how much more is the decibel level of the louder sound than the quieter one?

5. Earthquakes. The strongest recorded earthquake to date was in Colombia in 1906, with an energy release of 1.99×10^{17} joules. What was its magnitude on the Richter scale? Compute the answer to one decimal place.

6. Earthquakes. Anchorage, Alaska, had a major earthquake in 1964 that released 7.08×10^{16} joules of energy. What was its magnitude on the Richter scale? Compute the answer to one decimal place.

★★ **7. Earthquakes.** The 1933 Long Beach, California, earthquake had a Richter scale reading of 6.3, and the 1964 Anchorage, Alaska, earthquake had a Richter scale reading of 8.3. How many times more powerful was the Anchorage earthquake than the Long Beach earthquake?

★★ **8. Earthquakes.** Generally, an earthquake requires a magnitude of over 5.6 on the Richter scale to inflict serious damage. How many times more powerful than this was the great 1906 Colombia earthquake, which registered a magnitude of 8.6 on the Richter scale?

9. Space Vehicles. A new solid-fuel rocket has a weight ratio $W_t/W_b = 19.8$ and an exhaust velocity $c = 2.57$ kilometers per second. What is its velocity at burnout? Compute the answer to two decimal places.

10. Space Vehicles. A liquid-fuel rocket has a weight ratio $W_t/W_b = 6.2$ and an exhaust velocity $c = 5.2$ kilometers per second. What is its velocity at burnout? Compute the answer to two decimal places.

11. Chemistry. The hydrogen ion concentration of a substance is related to its acidity and basicity. Because hydrogen ion concentrations vary over a very wide range, logarithms are used to create a compressed **pH scale,** which is defined as follows:

$$\text{pH} = -\log\,[\text{H}^+]$$

where $[\text{H}^+]$ is the hydrogen ion concentration, in moles per liter. Pure water has a pH of 7, which means it is neutral. Substances with a pH less than 7 are acidic, and those with a pH greater than 7 are basic. Compute the pH of each substance listed, given the indicated hydrogen ion concentration.

(A) Seawater, 4.63×10^{-9}

(B) Vinegar, 9.32×10^{-4}

Also, indicate whether each substance is acidic or basic. Compute answers to one decimal place.

12. Chemistry. Refer to Problem 11. Compute the pH of each substance below, given the indicated hydrogen ion concentration. Also, indicate whether it is acidic or basic. Compute answers to one decimal place.

(A) Milk, 2.83×10^{-7}

(B) Garden mulch, 3.78×10^{-6}

★ **13. Ecology.** Refer to Problem 11. Many lakes in Canada and the United States will no longer sustain some forms of wildlife because of the increase in acidity of the water from acid rain and snow caused by sulfur dioxide emissions from industry. If the pH of a sample of rainwater is 5.2, what is its hydrogen ion concentration in moles per liter? Compute the answer to two significant digits.

★ **14. Ecology.** Refer to Problem 11. If normal rainwater has a pH of 5.7, what is its hydrogen ion concentration in moles per liter? Compute the answer to two significant digits.

★★ **15. Astronomy.** The brightness of stars is expressed in terms of magnitudes on a numerical scale that increases as the brightness decreases. The magnitude m is given by the formula

$$m = 6 - 2.5 \log \frac{L}{L_0}$$

where L is the light flux of the star and L_0 is the light flux of the dimmest stars visible to the naked eye.

(A) What is the magnitude of the dimmest stars visible to the naked eye?

(B) How many times brighter is a star of magnitude 1 than a star of magnitude 6?

16. Astronomy. An optical instrument is required to observe stars beyond the sixth magnitude, the limit of ordinary vision. However, even optical instruments have their limitations. The limiting magnitude L of any optical telescope with lens diameter D, in inches, is given by

$$L = 8.8 + 5.1 \log D$$

(A) Find the limiting magnitude for a homemade 6-inch reflecting telescope.

(B) Find the diameter of a lens that would have a limiting magnitude of 20.6.

Compute answers to three significant digits.

MODELING AND DATA ANALYSIS

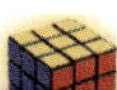

17. **Agriculture.** Table 4 shows the yield (bushels per acre) and the total production (millions of bushels) for corn in the United States for selected years since 1950. Let x represent years since 1900.

TABLE 4 United States Corn Production

Year	x	Yield (Bushels per Acre)	Total Production (Million Bushels)
1950	50	37.6	2,782
1960	60	55.6	3,479
1970	70	81.4	4,802
1980	80	97.7	6,867
1990	90	115.6	7,802

Source: U.S. Department of Agriculture.

(A) Find a logarithmic regression model ($y = a + b \ln x$) for the yield. Estimate (to one decimal place) the yield in 1996 and in 2010.

(B) The actual yield in 1996 was 127.1 bushels per acre. How does this compare with the estimated yield in part A? What effect will this additional 1996 information have on the estimate for 2010? Explain.

18. **Agriculture.** Refer to Table 4.

(A) Find a logarithmic regression model ($y = a + b \ln x$) for the total production. Estimate (to the nearest million) the production in 1996 and in 2010.

(B) The actual production in 1996 was 7,949 million bushels. How does this compare with the estimated production in part A? What effect will this 1996 production information have on the estimate for 2010? Explain.

SECTION 4.5 Exponential and Logarithmic Equations

Exponential Equations • Logarithmic Equations

Equations involving exponential and logarithmic functions, such as

$$2^{3x-2} = 5 \qquad \text{and} \qquad \log(x + 3) + \log x = 1$$

are called **exponential** and **logarithmic equations,** respectively. Logarithmic properties play a central role in their solution. Of course, a graphing utility can be used to find approximate solutions for many exponential and logarithmic equations. However, there are situations in which the algebraic solution is necessary. In Section 4.5, we emphasize algebraic solutions and use a graphing utility as a check, when appropriate.

Exponential Equations

The following examples illustrate the use of logarithmic properties in solving exponential equations.

EXAMPLE 1 Solving an Exponential Equation

Solve $2^{3x-2} = 5$ for x to four decimal places.

SOLUTION

Algebraic Solution

How can we get x out of the exponent? Use logs!

$$2^{3x-2} = 5$$

$$\log 2^{3x-2} = \log 5$$ Take the common or natural log of both sides.

$$(3x - 2)\log 2 = \log 5$$ Use $\log_b N^p = p \log_b N$ to get $3x - 2$ out of the exponent position.

$$3x - 2 = \frac{\log 5}{\log 2}$$

$$x = \frac{1}{3}\left(2 + \frac{\log 5}{\log 2}\right)$$ *Remember:* $\frac{\log 5}{\log 2} \neq \log 5 - \log 2.$

$$= 1.4406$$ To four decimal places.

Graphical Solution

Graph $y_1 = 2^{3x-2}$ and $y_2 = 5$ and use the intersect command (Fig. 1).

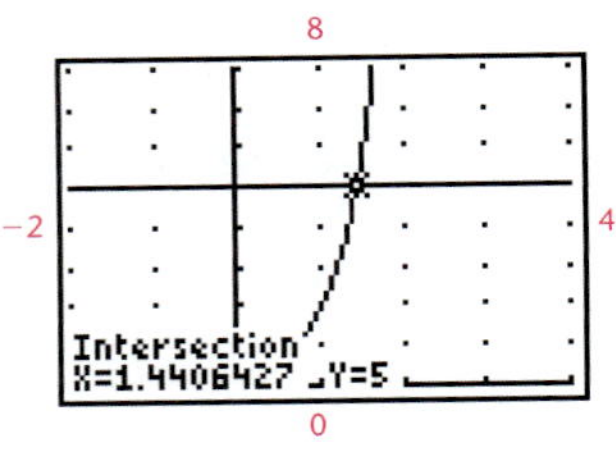

FIGURE 1 $y_1 = 2^{3x-2}, y_2 = 5.$

MATCHED 1 PROBLEM

Solve $35^{1-2x} = 7$ for x to four decimal places.

EXAMPLE 2 Compound Interest

A certain amount of money P (principal) is invested at an annual rate r compounded annually. The amount of money A in the account after t years, assuming no withdrawals, is given by

$$A = P\left(1 + \frac{r}{m}\right)^n = P(1 + r)^n$$ $m = 1$ for annual compounding.

How many years to the nearest year will it take the money to double if it is invested at 6% compounded annually?

SOLUTION

Algebraic Solution

To find the doubling time, we replace A in $A = P(1.06)^n$ with $2P$ and solve for n.

$$2P = P(1.06)^n$$

$$2 = 1.06^n$$ Divide both sides by P.

$$\log 2 = \log 1.06^n$$ Take the common or natural log of both sides.

$$= n \log 1.06$$ Note how log properties are used to get n out of the exponent position.

$$n = \frac{\log 2}{\log 1.06}$$

$$= 12 \text{ years}$$ To the nearest year.

Graphical Solution

Graph $y_1 = 1.06^x$ and $y_2 = 2$ and use the intersect command (Fig. 2).

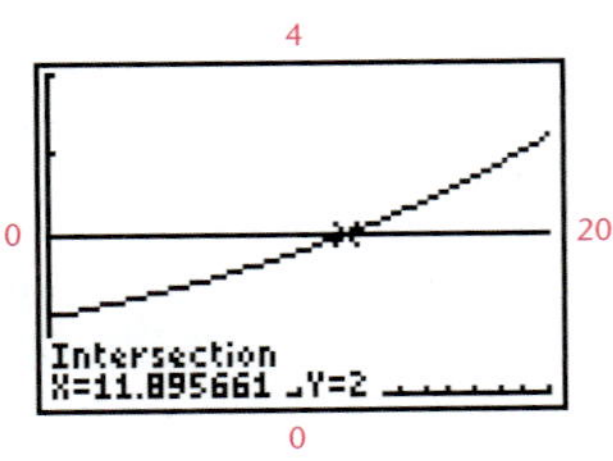

FIGURE 2 $y_1 = 1.06^x, y_2 = 2.$

MATCHED 2 PROBLEM

Repeat Example 2, changing the interest rate to 9% compounded annually.

EXAMPLE 3 Atmospheric Pressure

The atmospheric pressure P, in pounds per square inch, at x miles above sea level is given approximately by

$$P = 14.7e^{-0.21x}$$

At what height will the atmospheric pressure be half the sea-level pressure? Compute the answer to two significant digits.

SOLUTION

Algebraic Solution

Sea-level pressure is the pressure at $x = 0$. Thus,

$$P = 14.7e^0 = 14.7$$

One-half of sea-level pressure is $14.7/2 = 7.35$. Now our problem is to find x so that $P = 7.35$; that is, we solve $7.35 = 14.7e^{-0.21x}$ for x:

$$7.35 = 14.7e^{-0.21x}$$

$$0.5 = e^{-0.21x}$$ Divide both sides by 14.7 to simplify.

$$\ln 0.5 = \ln e^{-0.21x}$$ Because the base is e, take the natural log of both sides.

$$= -0.21x$$ $\ln e^a = a$

$$x = \frac{\ln 0.5}{-0.21}$$

$$= 3.3 \text{ miles}$$ To two significant digits.

Graphical Solution

Graph $y_1 = 14.7e^{-0.21x}$ and $y_2 = 7.35$ and use the intersect command (Fig. 3).

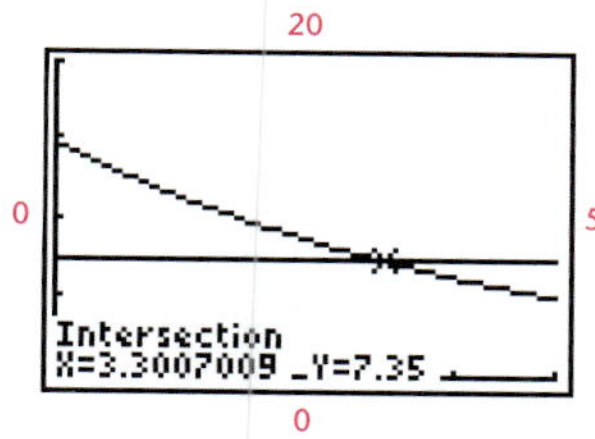

FIGURE 3 $y_1 = 14.7e^{-0.21x}$, $y_2 = 7.35$.

MATCHED 3 PROBLEM

Using the formula in Example 3, find the altitude in miles so that the atmospheric pressure will be one-eighth that at sea level. Compute the answer to two significant digits.

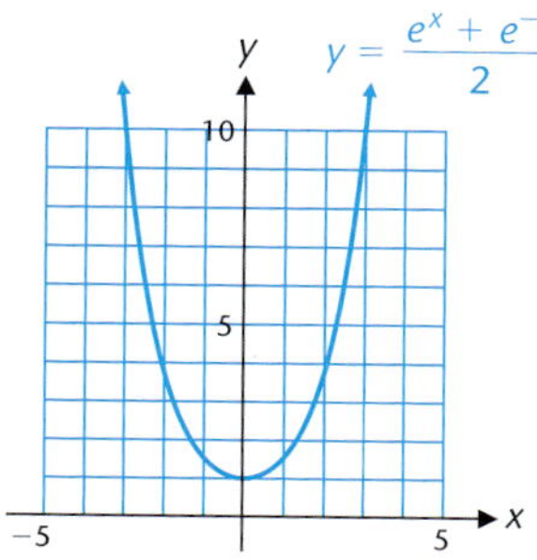

FIGURE 4 Catenary.

The graph of

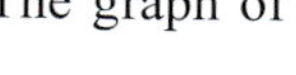

$$y = \frac{e^x + e^{-x}}{2} \tag{1}$$

is a curve called a **catenary** (Fig. 4). A uniform cable suspended between two fixed points is a physical example of such a curve.

EXAMPLE 4 Solving an Exponential Equation

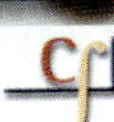

Given equation (1), find x for $y = 2.5$. Compute the answer to four decimal places.

SOLUTION

Algebraic Solution

$$y = \frac{e^x + e^{-x}}{2}$$

$$2.5 = \frac{e^x + e^{-x}}{2}$$

$$5 = e^x + e^{-x}$$

$$5e^x = e^{2x} + 1 \quad \text{Multiply both sides by } e^x.$$

$$e^{2x} - 5e^x + 1 = 0 \quad \text{This is a quadratic in } e^x.$$

Let $u = e^x$, then

$$u^2 - 5u + 1 = 0$$

$$u = \frac{5 \pm \sqrt{25 - 4(1)(1)}}{2}$$

$$= \frac{5 \pm \sqrt{21}}{2}$$

$$e^x = \frac{5 \pm \sqrt{21}}{2} \quad \text{Replace } u \text{ with } e^x \text{ and solve for } x.$$

$$\ln e^x = \ln \frac{5 \pm \sqrt{21}}{2} \quad \text{Take the natural log of both sides (both values on the right are positive).}$$

$$x = \ln \frac{5 \pm \sqrt{21}}{2} \quad \log_b b^x = x.$$

$$= -1.5668, 1.5668$$

Note that the algebraic method produces exact solutions, an important consideration in certain calculus applications (see Problems 51–54 of Exercise 4.5).

Graphical Solution

Graph $y_1 = (e^x + e^{-x})/2$ and $y_2 = 2.5$ and use the intersect command (Fig. 5).

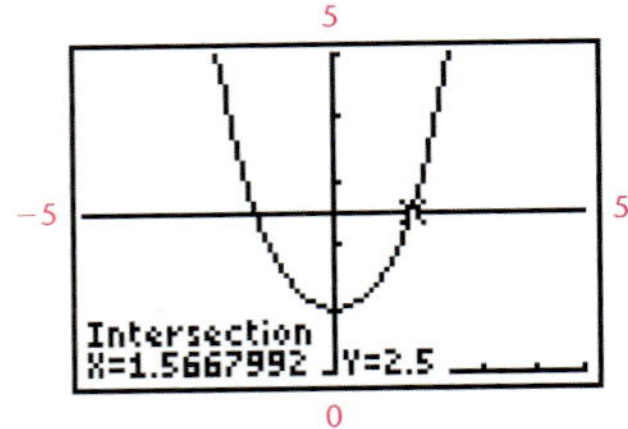

FIGURE 5 $y_1 = \dfrac{e^x + e^{-x}}{2}$, $y_2 = 2.5$.

Therefore $x = 1.5668$ is the positive solution. The negative solution can be found similarly using the intersect command. Alternatively, simply observe that the function

$$f(x) = \frac{e^x + e^{-x}}{2}$$

is an even function (that is, $f(x) = f(-x)$), so the graph of f is symmetrical with respect to the y axis (see Section 1.4). Because $(1.5668, 2.5)$ is a point on the graph of f, so is $(-1.5668, 2.5)$, and thus $x = -1.5668$ is the negative solution.

MATCHED 4 PROBLEM

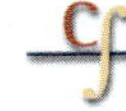

Given $y = (e^x - e^{-x})/2$, find x for $y = 1.5$. Compute the answer to three decimal places.

EXPLORE/DISCUSS 1

Let $y = e^{2x} + 3e^x + e^{-x}$

(A) Try to find x when $y = 7$ using the method of Example 4. Explain the difficulty that arises.

(B) Use a graphing utility to find x when $y = 7$.

Logarithmic Equations

We now illustrate the solution of several types of logarithmic equations.

EXAMPLE 5 Solving a Logarithmic Equation

Solve $\log(x + 3) + \log x = 1$, and check.

SOLUTION

Algebraic Solution

First use properties of logarithms to express the left side as a single logarithm, then convert to exponential form and solve for x.

$\log(x + 3) + \log x = 1$

$\log[x(x + 3)] = 1$ Combine left side using $\log M + \log N = \log MN$.

$x(x + 3) = 10^1$ Change to equivalent exponential form.

$x^2 + 3x - 10 = 0$ Write in $ax^2 + bx + c = 0$ form and solve.

$(x + 5)(x - 2) = 0$

$x = -5, 2$

CHECK

$x = -5$: $\log(-5 + 3) + \log(-5)$ is not defined because the domain of the log function is $(0, \infty)$.

$x = 2$: $\log(2 + 3) + \log 2 = \log 5 + \log 2$
$= \log(5 \cdot 2) = \log 10 \overset{\checkmark}{=} 1$

Thus, the only solution to the original equation is $x = 2$. Remember, answers should be checked in the original equation to see whether any should be discarded.

Graphical Solution

Graph $y_1 = \log(x + 3) + \log x$ and $y_2 = 1$ and use the intersect command. Figure 6 shows that $x = 2$ is a solution, and also shows that y_1 (the left side of the original equation) is not defined at $x = -5$, the extraneous solution produced by the algebraic method.

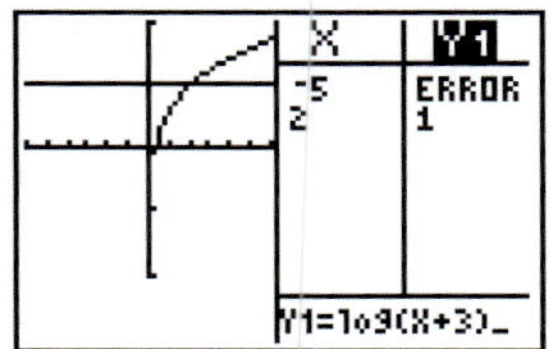

FIGURE 6 $y_1 = \log(x + 3) + \log x$, $y_2 = 1$.

MATCHED PROBLEM 5

Solve $\log(x - 15) = 2 - \log x$, and check.

EXAMPLE 6 Solving a Logarithmic Equation

Solve $(\ln x)^2 = \ln x^2$.

SOLUTION

Algebraic Solution

There are no logarithmic properties for simplifying $(\ln x)^2$. However, we can simplify $\ln x^2$, obtaining an equation involving $\ln x$ and $(\ln x)^2$.

$$(\ln x)^2 = \ln x^2$$
$$= 2 \ln x$$

This is a quadratic equation in ln *x*. Move all nonzero terms to the left and factor.

$$(\ln x)^2 - 2 \ln x = 0$$
$$(\ln x)(\ln x - 2) = 0$$
$$\ln x = 0 \quad \text{or} \quad \ln x - 2 = 0$$
$$x = e^0 \qquad \ln x = 2$$
$$= 1 \qquad x = e^2$$

Checking that both $x = 1$ and $x = e^2$ are solutions to the original equation is left to you.

Graphical Solution

Graph $y_1 = (\ln x)^2$ and $y_2 = \ln x^2$ and use the intersect command to obtain the solutions $x = 1$ and $x = 7.3890561$ (Fig. 7). The second solution is not exact; it is an approximation to e^2.

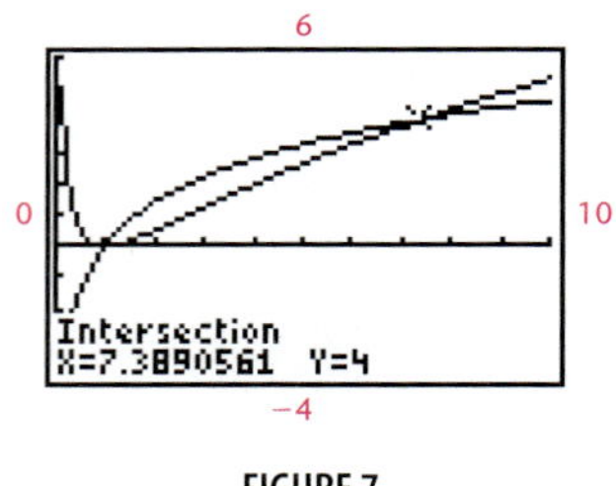

FIGURE 7

MATCHED 6 PROBLEM

Solve $\log x^2 = (\log x)^2$.

CAUTION

Note that

$$(\log_b x)^2 \neq \log_b x^2$$

$(\log_b x)^2 = (\log_b x)(\log_b x)$
$\log_b x^2 = 2 \log_b x$

EXAMPLE 7 Earthquake Intensity

Recall from Section 4.4 that the magnitude of an earthquake on the Richter scale is given by

$$M = \frac{2}{3} \log \frac{E}{E_0}$$

Solve for E in terms of the other symbols.

SOLUTION

$$M = \frac{2}{3} \log \frac{E}{E_0}$$

$$\log \frac{E}{E_0} = \frac{3M}{2}$$ Multiply both sides by $\frac{3}{2}$.

$$\frac{E}{E_0} = 10^{3M/2}$$ Change to exponential form.

$$E = E_0 10^{3M/2}$$

MATCHED 7 PROBLEM

Solve the rocket equation from Section 4.4 for W_b in terms of the other symbols:

$$v = c \ln \frac{W_t}{W_b}$$

ANSWERS MATCHED PROBLEMS

1. $x = 0.2263$
2. More than double in 9 years, but not quite double in 8 years
3. 9.9 miles
4. $x = 1.195$
5. $x = 20$
6. $x = 1{,}100$
7. $W_b = W_t e^{-v/c}$

EXERCISE 4.5

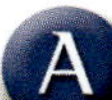

Solve Problems 1–12 algebraically and check graphically. Round answers to three significant digits.

1. $10^{-x} = 0.0347$ **2.** $10^{x} = 14.3$ **3.** $10^{3x+1} = 92$

4. $10^{5x-2} = 348$ **5.** $e^{x} = 3.65$ **6.** $e^{-x} = 0.0142$

7. $e^{2x-1} = 405$ **8.** $e^{3x+5} = 23.8$ **9.** $5^{x} = 18$

10. $3^{x} = 4$ **11.** $2^{-x} = 0.238$ **12.** $3^{-x} = 0.074$

Solve Problems 13–18 exactly.

13. $\log 5 + \log x = 2$ **14.** $\log x - \log 8 = 1$

15. $\log x + \log (x - 3) = 1$

16. $\log (x - 9) + \log 100x = 3$

17. $\log (x + 1) - \log (x - 1) = 1$

18. $\log (2x + 1) = 1 + \log (x - 2)$

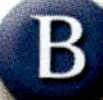

Solve Problems 19–26 algebraically and check graphically. Round answers to three significant digits.

19. $2 = 1.05^{x}$ **20.** $3 = 1.06^{x}$

21. $e^{-1.4x} = 13$ **22.** $e^{0.32x} = 632$

23. $123 = 500e^{-0.12x}$ **24.** $438 = 200e^{0.25x}$

25. $e^{-x^2} = 0.23$ **26.** $e^{x^2} = 125$

Solve Problems 27–38 exactly.

27. $\log x - \log 5 = \log 2 - \log (x - 3)$

28. $\log (6x + 5) - \log 3 = \log 2 - \log x$

29. $\ln x = \ln (2x - 1) - \ln (x - 2)$

30. $\ln (x + 1) = \ln (3x + 1) - \ln x$

31. $\log (2x + 1) = 1 - \log (x - 1)$

32. $1 - \log (x - 2) = \log (3x + 1)$

33. $(\ln x)^3 = \ln x^4$ **34.** $(\log x)^3 = \log x^4$

35. $\ln (\ln x) = 1$ **36.** $\log (\log x) = 1$

37. $x^{\log x} = 100x$ **38.** $3^{\log x} = 3x$

In Problems 39–40,
(A) Explain the difficulty in solving the equation exactly.
(B) Determine the number of solutions by graphing the functions on each side of the equation.

39. $e^{x/2} = 5 \ln x$ **40.** $\ln (\ln x) + \ln x = 2$

In Problems 41–42,
(A) Explain the difficulty in solving the equation exactly.
(B) Use a graphing utility to find all solutions to three decimal places.

41. $3^{x} + 2 = 7 + x - e^{-x}$ **42.** $e^{x/4} = 5 \log x + 4 \ln x$

C

Solve Problems 43–50 for the indicated variable in terms of the remaining symbols. Use the natural log for solving exponential equations.

43. $A = Pe^{rt}$ for r (finance)

44. $A = P\left(1 + \frac{r}{n}\right)^{nt}$ for t (finance)

45. $D = 10 \log \frac{I}{I_0}$ for I (sound)

46. $t = \frac{-1}{k}(\ln A - \ln A_0)$ for A (decay)

47. $M = 6 - 2.5 \log \frac{I}{I_0}$ for I (astronomy)

48. $L = 8.8 + 5.1 \log D$ for D (astronomy)

49. $I = \frac{E}{R}(1 - e^{-Rt/L})$ for t (circuitry)

50. $S = R\frac{(1 + i)^n - 1}{i}$ for n (annuity)

The following combinations of exponential functions define four of six ***hyperbolic functions,*** *an important class of functions in calculus and higher mathematics. Solve Problems 51–54 for x in terms of y. The results are used to define* ***inverse hyperbolic functions,*** *another important class of functions in calculus and higher mathematics.*

51. $y = \frac{e^x + e^{-x}}{2}$

52. $y = \frac{e^x - e^{-x}}{2}$

53. $y = \frac{e^x - e^{-x}}{e^x + e^{-x}}$

54. $y = \frac{e^x + e^{-x}}{e^x - e^{-x}}$

In Problems 55–66, use a graphing utility to approximate to two decimal places any solutions of the equation in the interval $0 \le x \le 1$. None of these equations can be solved exactly using any step-by-step algebraic process.

55. $2^{-x} - 2x = 0$

56. $3^{-x} - 3x = 0$

57. $x3^x - 1 = 0$

58. $x2^x - 1 = 0$

59. $e^{-x} - x = 0$

60. $xe^{2x} - 1 = 0$

61. $xe^x - 2 = 0$

62. $e^{-x} - 2x = 0$

63. $\ln x + 2x = 0$

64. $\ln x + x^2 = 0$

65. $\ln x + e^x = 0$

66. $\ln x + x = 0$

APPLICATIONS

Solve Problems 67–78 algebraically or graphically, whichever seems more appropriate.

67. Compound Interest. How many years, to the nearest year, will it take a sum of money to double if it is invested at 15% compounded annually?

68. Compound Interest. How many years, to the nearest year, will it take money to quadruple if it is invested at 20% compounded annually?

69. Compound Interest. At what annual rate compounded continuously will \$1,000 have to be invested to amount to \$2,500 in 10 years? Compute the answer to three significant digits.

70. Compound Interest. How many years will it take \$5,000 to amount to \$8,000 if it is invested at an annual rate of 9% compounded continuously? Compute the answer to three significant digits.

71. World Population. A mathematical model for world population growth over short periods is given by

$$P = P_0e^{rt}$$

where P is the population after t years, P_0 is the population at $t = 0$, and the population is assumed to grow continuously at the annual rate r. How many years, to the nearest year, will it take the world population to double if it grows continuously at an annual rate of 2%?

★ **72. World Population.** Refer to Problem 71. Starting with a world population of 4 billion people and assuming that the population grows continuously at an annual rate of 2%, how many years, to the nearest year, will it be before there is only 1 square yard of land per person? Earth contains approximately 1.7×10^{14} square yards of land.

★ **73. Archaeology—Carbon-14 Dating.** As long as a plant or animal is alive, carbon-14 is maintained in a constant amount in its tissues. Once dead, however, the plant or animal ceases taking in carbon, and carbon-14 diminishes by radioactive decay according to the equation

$$A = A_0 e^{-0.000124t}$$

where A is the amount after t years and A_0 is the amount when $t = 0$. Estimate the age of a skull uncovered in an archaeological site if 10% of the original amount of carbon-14 is still present. Compute the answer to three significant digits.

★ **74. Archaeology—Carbon-14 Dating.** Refer to Problem 73. What is the half-life of carbon-14? That is, how long will it take for half of a sample of carbon-14 to decay? Compute the answer to three significant digits.

★ **75. Photography.** An electronic flash unit for a camera is activated when a capacitor is discharged through a filament of wire. After the flash is triggered and the capacitor is discharged, the circuit (see the figure) is connected and the battery pack generates a current to recharge the capacitor. The time it takes for the capacitor to recharge is called the *recycle time.* For a particular flash unit using a 12-volt battery pack, the charge q, in coulombs, on the capacitor t seconds after recharging has started is given by

$$q = 0.0009(1 - e^{-0.2t})$$

How many seconds will it take the capacitor to reach a charge of 0.0007 coulomb? Compute the answer to three significant digits.

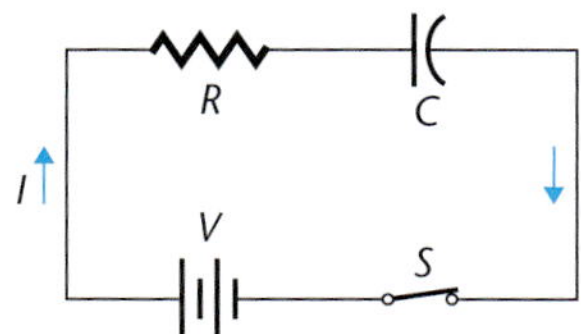

★ **76. Advertising.** A company is trying to expose as many people as possible to a new product through television advertising in a large metropolitan area with 2 million possible viewers. A model for the number of people N, in millions, who are aware of the product after t days of advertising was found to be

$$N = 2(1 - e^{-0.037t})$$

How many days, to the nearest day, will the advertising campaign have to last so that 80% of the possible viewers will be aware of the product?

★★ **77. Newton's Law of Cooling.** This law states that the rate at which an object cools is proportional to the difference in temperature between the object and its surrounding medium. The temperature T of the object t hours later is given by

$$T = T_m + (T_0 - T_m)e^{-kt}$$

where T_m is the temperature of the surrounding medium and T_0 is the temperature of the object at $t = 0$. Suppose a bottle of wine at a room temperature of 72°F is placed in a refrigerator at 40°F to cool before a dinner party. After an hour the temperature of the wine is found to be 61.5°F. Find the constant k, to two decimal places, and the time, to one decimal place, it will take the wine to cool from 72 to 50°F.

★ **78. Marine Biology.** Marine life is dependent upon the microscopic plant life that exists in the *photic zone,* a zone that goes to a depth where about 1% of the surface light still remains. Light intensity is reduced according to the exponential function

$$I = I_0 e^{-kd}$$

where I is the intensity d feet below the surface and I_0 is the intensity at the surface. The constant k is called the *coefficient of extinction*. At Crystal Lake in Wisconsin it was found that half the surface light remained at a depth of 14.3 feet. Find k, and find the depth of the photic zone. Compute answers to three significant digits.

CHAPTER 4 REVIEW

4.1 Exponential Functions

The equation $f(x) = b^x$, $b > 0$, $b \neq 1$, defines an **exponential function** with **base *b*.** The domain of f is $(-\infty, \infty)$ and the range is $(0, \infty)$. The **graph** of f is a continuous curve that has no sharp corners; passes through (0, 1); lies above the x axis, which is a horizontal asymptote; increases as x increases if $b > 1$; decreases as x increases if $b < 1$; and intersects any horizontal line at most once. The function f is one-to-one and has an inverse. We have the following **exponential function properties:**

1. $a^x a^y = a^{x+y}$ $\quad (a^x)^y = a^{xy}$ $\quad (ab)^x = a^x b^x$

$$\left(\frac{a}{b}\right)^x = \frac{a^x}{b^x} \qquad \frac{a^x}{a^y} = a^{x-y}$$

2. $a^x = a^y$ if and only if $x = y$.
3. For $x \neq 0$, $a^x = b^x$ if and only if $a = b$.

As x approaches ∞, the expression $[1 + (1/x)]^x$ approaches the irrational number $e \approx 2.718\ 281\ 828\ 459$. The function $f(x) = e^x$ is called the **exponential function with base *e*.** The growth of money in an account paying **compound interest** is described by $A = P(1 + r/m)^n$, where P is the **principal,** r is the annual **rate,** m is the number of compounding periods in 1 year, and A is the **amount** in the account after n compounding periods.

If the account pays **continuous compound interest,** the amount A in the account after t years is given by $A = Pe^{rt}$.

4.2 Exponential Models

Exponential functions are used to model various types of growth:

1. **Population growth** can be modeled by using the **doubling time growth model** $P = P_0 2^{t/d}$, where P is population at time t, P_0 is the population at time $t = 0$, and d is the **doubling time**—the time it takes for the population to double. Another model of population growth, $y = ce^{kt}$, where c and k are positive constants, uses the exponential function with base e.
2. **Radioactive decay** can be modeled by using the **half-life decay model** $A = A_0(\frac{1}{2})^{t/h} = A_0 2^{-t/h}$, where A is the amount at time t, A_0 is the amount at time $t = 0$, and h is the **half-life**—the time it takes for half the material to decay. Another model of radioactive decay, $y = ce^{-kt}$, where c and k are positive constants, uses the exponential function with base e.
3. **Limited growth—**the growth of a company or proficiency at learning a skill, for example—can often be modeled by the equation $y = c(1 - e^{-kt})$, where c and k are positive constants.
4. **Logistic growth**—the spread of an epidemic or sales of a new product, for example—can often be modeled by the equation $y = M/(1 + ce^{-kt})$ where c, k, and M are positive constants.

4.3 Logarithmic Functions

The **logarithmic function with base *b*** is defined to be the inverse of the exponential function with base b and is denoted by $y = \log_b x$. Thus, $y = \log_b x$ if and only if $x = b^y$, $b > 0$, $b \neq 1$.

The domain of a logarithmic function is $(0, \infty)$ and the range is $(-\infty, \infty)$. The graph of a logarithmic function is a continuous curve that always passes through the point $(1, 0)$ and has the y axis as a vertical asymptote. We have the following **properties of logarithmic functions:**

1. $\log_b 1 = 0$
2. $\log_b b = 1$
3. $\log_b b^x = x$
4. $b^{\log_b x} = x,\ x > 0$
5. $\log_b MN = \log_b M + \log_b N$
6. $\log_b \dfrac{M}{N} = \log_b M - \log_b N$
7. $\log_b M^p = p \log_b M$
8. $\log_b M = \log_b N$ if and only if $M = N$

Logarithms to the base 10 are called **common logarithms** and are denoted by $\log x$. Logarithms to the base e are called **natural logarithms** and are denoted by $\ln x$. Thus, $\log x = y$ is equivalent to $x = 10^y$, and $\ln x = y$ is equivalent to $x = e^y$.

The **change-of-base formula,** $\log_b N = (\log_a N)/(\log_a b)$, relates logarithms to two different bases and can be used, along with a calculator, to evaluate logarithms to bases other than e or 10.

4.4 Logarithmic Models

The following applications involve logarithmic functions:

1. The **decibel** is defined by $D = 10 \log (I/I_0)$, where D is the **decibel level** of the sound, I is the **intensity** of the sound, and $I_0 = 10^{-12}$ watts per square meter is a standardized sound level.
2. The **magnitude** M of an earthquake on the **Richter scale** is given by $M = \frac{2}{3} \log (E/E_0)$, where E is the energy released by the earthquake and $E_0 = 10^{4.40}$ joules is a standardized energy level.
3. The **velocity** v of a rocket at burnout is given by the **rocket equation** $v = c \ln (W_t/W_b)$, where c is the exhaust velocity, W_t is the takeoff weight, and W_b is the burnout weight.

Logarithmic regression is used to fit a function of the form $y = a + b \ln x$ to a set of data points.

4.5 Exponential and Logarithmic Equations

Various techniques for solving **exponential equations,** such as $2^{3x-2} = 5$, and **logarithmic equations,** such as $\log (x + 3) + \log x = 1$, are illustrated by examples.

CHAPTER 4 REVIEW EXERCISES

Work through all the problems in this chapter review and check answers in the back of the book. Answers to all review problems are there, and following each answer is a number in italics indicating the section in which that type of problem is discussed. Where weaknesses show up, review appropriate sections in the text.

A

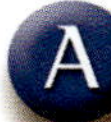

1. Match each equation with the graph of f, g, m, or n in the figure.

(A) $y = \log_2 x$ **(B)** $y = 0.5^x$
(C) $y = \log_{0.5} x$ **(D)** $y = 2^x$

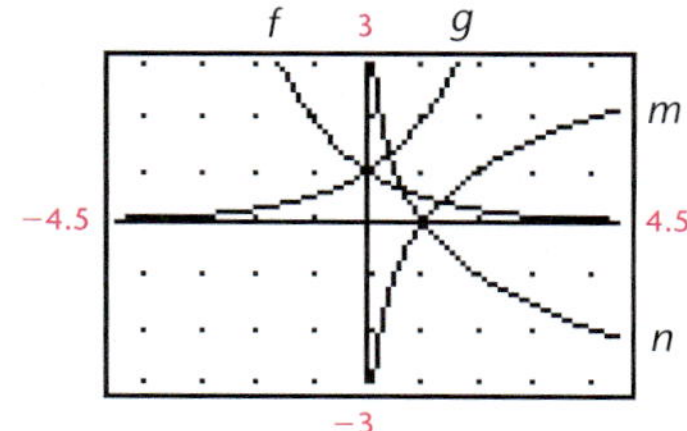

2. Write in logarithmic form using base 10: $m = 10^n$.

3. Write in logarithmic form using base e: $x = e^y$.

Write Problems 4 and 5 in exponential form.

4. $\log x = y$ **5.** $\ln y = x$

In Problems 6 and 7, simplify.

6. $\dfrac{7^{x+2}}{7^{2-x}}$ **7.** $\left(\dfrac{e^x}{e^{-x}}\right)^x$

Solve Problems 8–10 for x exactly. Do not use a calculator or table.

8. $\log_2 x = 3$ **9.** $\log_x 25 = 2$ **10.** $\log_3 27 = x$

Solve Problems 11–14 for x to three significant digits.

11. $10^x = 17.5$ **12.** $e^x = 143{,}000$

13. $\ln x = -0.015\,73$ **14.** $\log x = 2.013$

Evaluate Problems 15–18 to four significant digits using a calculator.

15. $\ln \pi$ **16.** $\log(-e)$

17. $\pi^{\ln 2}$ **18.** $\dfrac{e^{\pi} + e^{-\pi}}{2}$

B

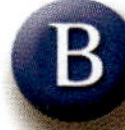

Solve Problems 19–29 for x exactly. Do not use a calculator or table.

19. $\ln(2x - 1) = \ln(x + 3)$

20. $\log(x^2 - 3) = 2\log(x - 1)$

21. $e^{x^2-3} = e^{2x}$ **22.** $4^{x-1} = 2^{1-x}$

23. $2x^2e^{-x} = 18e^{-x}$ **24.** $\log_{1/4} 16 = x$

25. $\log_x 9 = -2$ **26.** $\log_{16} x = \frac{3}{2}$

27. $\log_x e^5 = 5$ **28.** $10^{\log_{10} x} = 33$

29. $\ln x = 0$

Solve Problems 30–39 for x to three significant digits.

30. $x = 2(10^{1.32})$ **31.** $x = \log_5 23$

32. $\ln x = -3.218$ **33.** $x = \log(2.156 \times 10^{-7})$

34. $x = \dfrac{\ln 4}{\ln 2.31}$ **35.** $25 = 5(2^x)$

36. $4{,}000 = 2{,}500(e^{0.12x})$ **37.** $0.01 = e^{-0.05x}$

38. $5^{2x-3} = 7.08$ **39.** $\dfrac{e^x - e^{-x}}{2} = 1$

Solve Problems 40–45 for x exactly. Do not use a calculator or table.

40. $\log 3x^2 - \log 9x = 2$

41. $\log x - \log 3 = \log 4 - \log (x + 4)$

42. $\ln (x + 3) - \ln x = 2 \ln 2$

43. $\ln (2x + 1) - \ln (x - 1) = \ln x$

44. $(\log x)^3 = \log x^9$ **45.** $\ln (\log x) = 1$

In Problems 46 and 47, simplify.

46. $(e^x + 1)(e^{-x} - 1) - e^x(e^{-x} - 1)$

47. $(e^x + e^{-x})(e^x - e^{-x}) - (e^x - e^{-x})^2$

In Problems 48–51, find domain and range, intercepts, and asymptotes. Round all approximate values to two decimal places.

48. $y = 2^{x-1}$ **49.** $f(t) = 10e^{-0.08t}$

50. $y = \ln (x - 1)$ **51.** $N = \dfrac{100}{1 + 3e^{-t}}$

52. If the graph of $y = e^x$ is reflected in the line $y = x$, the graph of the function $y = \ln x$ is obtained. Discuss the functions that are obtained by reflecting the graph of $y = e^x$ in the x axis and the y axis.

53. **(A)** Explain why the equation $e^{-x/3} = 4 \ln (x + 1)$ has exactly one solution.
(B) Find the solution of the equation to three decimal places.

54. Approximate all real zeros of $f(x) = 4 - x^2 + \ln x$ to three decimal places.

55. Find the coordinates of the points of intersection of $f(x) = 10^{x-3}$ and $g(x) = 8 \log x$ to three decimal places.

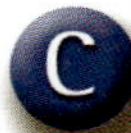

Solve Problems 56–59 for the indicated variable in terms of the remaining symbols.

56. $D = 10 \log \dfrac{I}{I_0}$ for I (sound intensity)

57. $y = \dfrac{1}{\sqrt{2\pi}} e^{-x^2/2}$ for x (probability)

58. $x = -\dfrac{1}{k} \ln \dfrac{I}{I_0}$ for I (X-ray intensity)

59. $r = P\dfrac{i}{1 - (1 + i)^{-n}}$ for n (finance)

60. Write $\ln y = -5t + \ln c$ in an exponential form free of logarithms; then solve for y in terms of the remaining symbols.

61. For $f = \{(x, y) \mid y = \log_2 x\}$, graph f and f^{-1} on the same coordinate system. What are the domains and ranges for f and f^{-1}?

62. Explain why 1 cannot be used as a logarithmic base.

63. Prove that $\log_b (M/N) = \log_b M - \log_b N$.

APPLICATIONS

Solve these application problems algebraically or graphically, whichever seems more appropriate.

64. Population Growth. Many countries have a population growth rate of 3% (or more) per year. At this rate, how many years will it take a population to double? Use the annual compounding growth model $P = P_0(1 + r)^t$. Compute the answer to three significant digits.

65. Population Growth. Repeat Problem 64 using the continuous compounding growth model $P = P_0e^{rt}$.

66. Carbon 14-Dating. How many years will it take for carbon-14 to diminish to 1% of the original amount after the death of a plant or animal? Use the formula $A = A_0e^{-0.000124t}$. Compute the answer to three significant digits.

★ **67. Medicine.** One leukemic cell injected into a healthy mouse will divide into two cells in about $\frac{1}{2}$ day. At the end of the day these two cells will divide into four. This doubling continues until 1 billion cells are formed; then the animal dies with leukemic cells in every part of the body.

(A) Write an equation that will give the number N of leukemic cells at the end of t days.

(B) When, to the nearest day, will the mouse die?

68. Money Growth. Assume $1 had been invested at an annual rate of 3% compounded continuously at the birth of Christ. What would be the value of the account in the year 2000? Compute the answer to two significant digits.

69. Present Value. Solving $A = Pe^{rt}$ for P, we obtain $P = Ae^{-rt}$, which is the **present value** of the amount A due in t years if money is invested at a rate r compounded continuously.

(A) Graph $P = 1{,}000(e^{-0.08t})$, $0 \le t \le 30$.

(B) What does it appear that P tends to as t tends to infinity? [*Conclusion:* The longer the time until the amount A is due, the smaller its present value, as we would expect.]

70. Earthquakes. The 1971 San Fernando, California, earthquake released 1.99×10^{14} joules of energy. Compute its magnitude on the Richter scale using the formula $M = \frac{2}{3} \log (E/E_0)$, where $E_0 = 10^{4.40}$ joules. Compute the answer to one decimal place.

71. Earthquakes. Refer to Problem 70. If the 1906 San Francisco earthquake had a magnitude of 8.3 on the Richter scale, how much energy was released? Compute the answer to three significant digits.

★ **72. Sound.** If the intensity of a sound from one source is 100,000 times that of another, how much more is the decibel level of the louder sound than the softer one? Use the formula $D = 10 \log (I/I_0)$.

★★ **73. Marine Biology.** The intensity of light entering water is reduced according to the exponential function

$$I = I_0e^{-kd}$$

where I is the intensity d feet below the surface, I_0 is the intensity at the surface, and k is the coefficient of extinction. Measurements in the Sargasso Sea in the West Indies have indicated that half the surface light reaches a depth of 73.6 feet. Find k, and find the depth at which 1% of the surface light remains. Compute answers to three significant digits.

★ **74. Wildlife Management.** A lake formed by a newly constructed dam is stocked with 1,000 fish. Their population is expected to increase according to the logistic curve

$$N = \frac{30}{1 + 29e^{-1.35t}}$$

where N is the number of fish, in thousands, expected after t years. The lake will be open to fishing when the number of fish reaches 20,000. How many years, to the nearest year, will this take?

MODELING AND DATA ANALYSIS 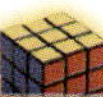

75. Medicare. The annual expenditures for Medicare (in billions of dollars) by the U.S. government for selected years since 1980 are shown in Table 1 (Bureau of the Census). Let x represent years since 1980.

(A) Find an exponential regression model of the form $y = ab^x$ for these data. Estimate (to the nearest billion) the total expenditures in 1996 and in 2010.

(B) When (to the nearest year) will the total expenditures reach $500 billion?

TABLE 1 Medicare Expenditures

Year	Billion $
1980	37
1985	72
1990	111
1995	181

Source: U.S. Bureau of the Census.

76. Agriculture. The total U.S. corn consumption (in millions of bushels) is shown in Table 2 for selected years since 1975. Let x represent years since 1900.

TABLE 2 Corn Consumption

Year	x	Total Consumption (Million Bushels)
1975	75	522
1980	80	659
1985	85	1,152
1990	90	1,373
1995	95	1,690

Source: U.S. Department of Agriculture.

(A) Find a logarithmic regression model of the form $y = a + b \ln x$ for these data. Estimate (to the nearest million) the total consumption in 1996 and in 2010.

(B) The actual consumption in 1996 was 1,583 million bushels. How does this compare with the estimated consumption in part A? What effect will this additional 1996 information have on the estimate for 2010? Explain.

CHAPTER 4 GROUP ACTIVITY

Comparing Regression Models

We have used polynomial, exponential, and logarithmic regression models to fit curves to data sets. And there are other equations that can be used for curve fitting (the TI-83 graphing calculator has 12 different equations on its STAT-CALC menu). How can we determine which equation provides the best fit for a given set of data? There are two principal ways to select models. The first is to use information about the type of data to help make a choice. For example, we expect the weight of a fish to be related to the cube of its length. And we expect most populations to grow exponentially, at least over the short term. The second method for choosing among equations involves developing a measure of how closely an equation fits a given data set. This is best introduced through an example. Consider the data set in Figure 1, where L1 represents the x coordinates and L2 represents the y coordinates. The graph of this data set is shown in Figure 2. Suppose we arbitrarily choose the equation $y_1 = 0.6x + 1$ to model these data (Fig. 3).

L1	L2	L3
2	4	
4	5	
6	3	
8	7	

L3(1)=

FIGURE 1

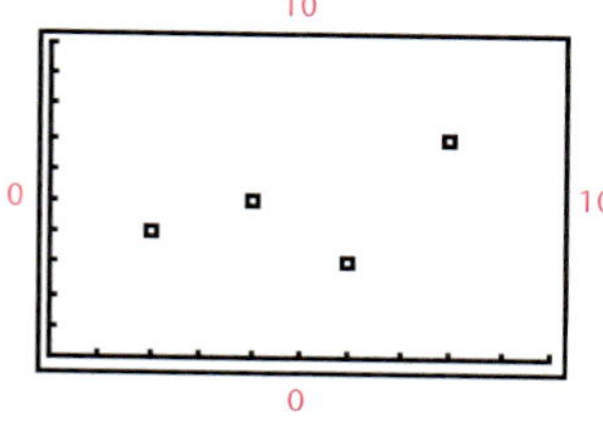

FIGURE 2

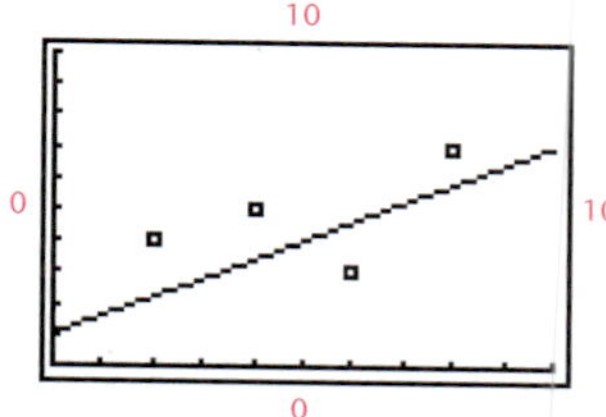

FIGURE 3 $y_1 = 0.6x + 1$.

To measure how well the graph of y_1 fits these data, we examine the difference between the y coordinates in the data set and the corresponding y coordinates on the graph of y_1 (L3 in Figs. 4 and 5). Each of these differences is called a **residual.** The most commonly accepted measure of the fit provided by a given model is the **sum of the squares of the residuals (SSR).** Computing this quantity is a simple matter on a graphing utility (Fig. 6).

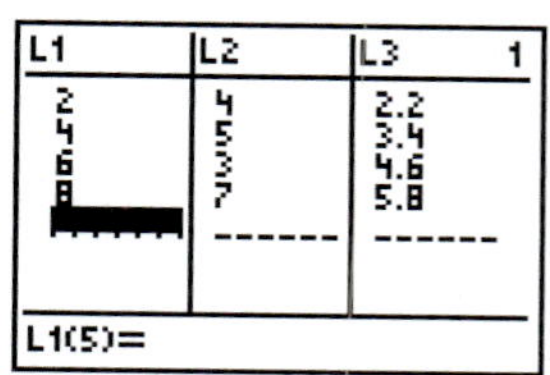

FIGURE 4

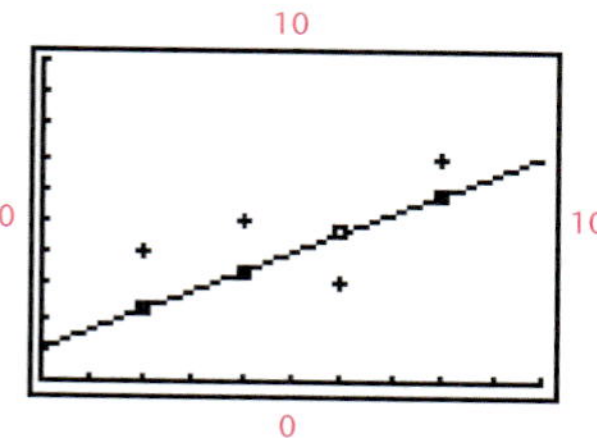

FIGURE 5 Here + is L2 and □ is L3.

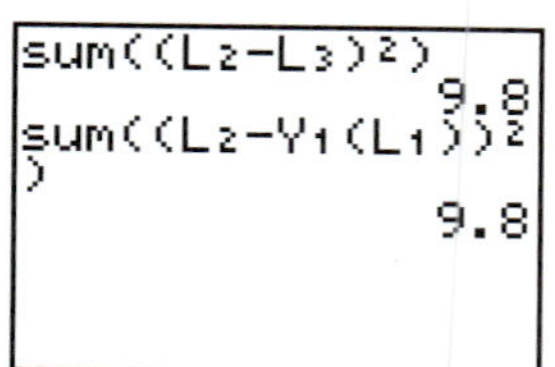

FIGURE 6 Two ways to calculate SSR.

(A) Find the linear regression model for the data in Figure 1, compute the SSR for this equation, and compare it with the one we computed for y_1.

It turns out that among all possible linear polynomials, **the linear regression model minimizes the sum of the squares of the residuals.** For this reason, the linear regression model is often called the **least-squares line.** A similar statement can be made for polynomials of any fixed degree. That is, the quadratic regression model minimizes the SSR over all quadratic polynomials, the cubic regression model minimizes the SSR over all cubic polynomials, and so on. The same statement cannot be made for exponential or logarithmic regression models. Nevertheless, the SSR can still be used to compare exponential, logarithmic, and polynomial models.

(B) Find the exponential and logarithmic regression models for the data in Figure 1, compute their SSRs, and compare with the linear model.

(C) National annual advertising expenditures for selected years since 1950 are shown in Table 1 where x is years since 1950 and y is total expenditures in billions of dollars. Which regression model would fit this data best: a quadratic model, a cubic model, or an exponential model? Use the SSRs to support your choice.

TABLE 1 Annual Advertising Expenditures, 1950–1995

x (years)	0	5	10	15	20	25	30	35	40	45
y (billion \$)	5.7	9.2	12.0	15.3	19.6	27.9	53.6	94.8	128.6	160.9

Source: U.S. Bureau of the Census.

CUMULATIVE REVIEW EXERCISES — CHAPTERS 3 4

Work through all the problems in this cumulative review and check answers in the back of the book. Answers to all review problems are there, and following each answer is a number in italics indicating the section in which that type of problem is discussed. Where weaknesses show up, review appropriate sections in the text.

A

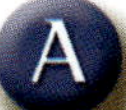
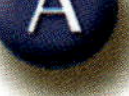

1. Let $P(x)$ be the polynomial whose graph is shown in the figure.

(A) Assuming that $P(x)$ has integer zeros and leading coefficient 1, find the lowest-degree equation that could produce this graph.

(B) Describe the left and right behavior of $P(x)$.

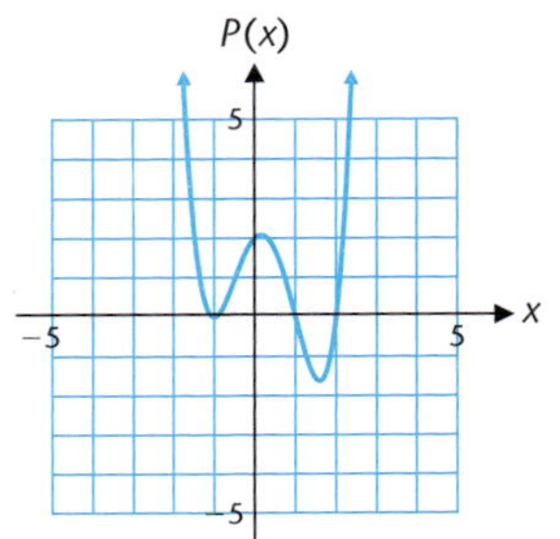

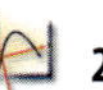

2. Match each equation with the graph of f, g, m, or n in the figure.

(A) $y = (\frac{3}{4})^x$ **(B)** $y = (\frac{4}{3})^x$

(C) $y = (\frac{3}{4})^x + (\frac{4}{3})^x$ **(D)** $y = (\frac{4}{3})^x - (\frac{3}{4})^x$

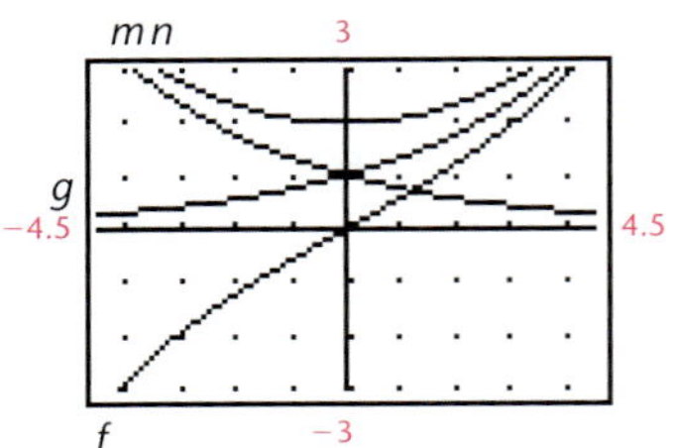

3. For $P(x) = 3x^3 + 5x^2 - 18x - 3$ and $D(x) = x + 3$, use synthetic division to divide $P(x)$ by $D(x)$, and write the answer in the form $P(x) = D(x)Q(x) + R$.

4. Let $P(x) = 2(x + 2)(x - 3)(x - 5)$. What are the zeros of $P(x)$?

5. Let $P(x) = 4x^3 - 5x^2 - 3x - 1$. How do you know that $P(x)$ has at least one real zero between 1 and 2?

6. Let $P(x) = x^3 + x^2 - 10x + 8$. Find all rational zeros for $P(x)$.

7. Solve for x.
(A) $y = 10^x$ **(B)** $y = \ln x$

8. Simplify.
(A) $(2e^x)^3$ **(B)** $\dfrac{e^{3x}}{e^{-2x}}$

9. Solve for x exactly. Do not use a calculator or a table.
(A) $\log_3 x = 2$
(B) $\log_3 81 = x$
(C) $\log_x 4 = -2$

10. Solve for x to three significant digits.
(A) $10^x = 2.35$ **(B)** $e^x = 87{,}500$
(C) $\log x = -1.25$ **(D)** $\ln x = 2.75$

B

11. Explain why the graph in the figure is not the graph of a polynomial function.

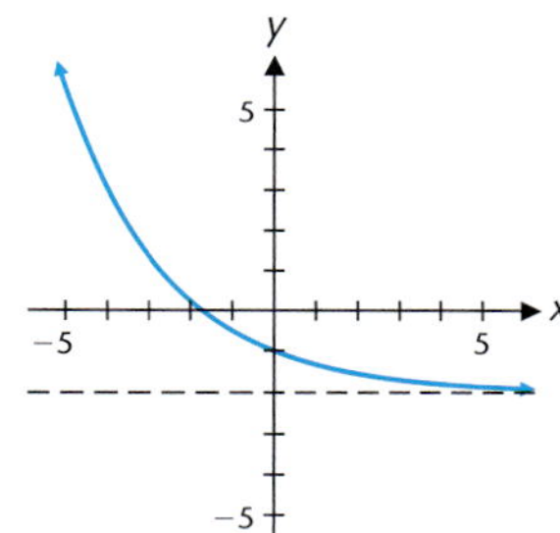

12. Explain why the graph in the figure is not the graph of a rational function.

13. The function f subtracts the square root of the domain element from three times the natural log of the domain element. Write an algebraic definition of f.

14. Write a verbal description of the function $f(x) = 100e^{0.5x} - 50$.

15. Let $f(x) = \dfrac{2x + 8}{x + 2}$.
(A) Find the domain and the intercepts for f.
(B) Find the vertical and horizontal asymptotes for f.
(C) Sketch the graph of f. Draw vertical and horizontal asymptotes with dashed lines.

16. Find all zeros of $P(x) = (x^3 + 4x)(x + 4)$, and specify those zeros that are x intercepts.

17. Solve $(x^3 + 4x)(x + 4) \leq 0$.

18. If $P(x) = 2x^3 - 5x^2 + 3x + 2$, find $P(\frac{1}{2})$ using the remainder theorem and synthetic division.

19. Which of the following is a factor of

$$P(x) = x^{25} - x^{20} + x^{15} + x^{10} - x^5 + 1$$

(A) $x - 1$ **(B)** $x + 1$

20. Let $P(x) = x^4 - 8x^2 + 3$.
(A) Graph $P(x)$ and describe the graph verbally, including the number of x intercepts, the number of turning points, and the left and right behavior.
(B) Approximate the largest x intercept to two decimal places.

21. Let $P(x) = x^5 - 8x^4 + 17x^3 + 2x^2 - 20x - 8$.
(A) Approximate the zeros of $P(x)$ to two decimal places and state the multiplicity of each zero.
(B) Can any of these zeros be approximated with the bisection method? A maximum routine? A minimum routine? Explain.

22. Let $P(x) = x^4 + 2x^3 - 20x^2 - 30$.
(A) Find the smallest positive and largest negative integers that, by Theorem 1 in Section 3.2, are upper and lower bounds, respectively, for the real zeros of $P(x)$.
(B) If $(k, k + 1)$, k an integer, is the interval containing the largest real zero of $P(x)$, determine how many additional intervals are required in the bisection method to approximate this zero to one decimal place.
(C) Approximate the real zeros of $P(x)$ to two decimal places.

23. Find all zeros (rational, irrational, and imaginary) exactly for $P(x) = 4x^3 - 20x^2 + 29x - 15$.

24. Find all zeros (rational, irrational, and imaginary) exactly for $P(x) = x^4 + 5x^3 + x^2 - 15x - 12$, and factor $P(x)$ into linear factors.

Solve Problems 25–34 for x exactly. Do not use a calculator or a table.

25. $2^{x^2} = 4^{x+4}$

26. $2x^2e^{-x} + xe^{-x} = e^{-x}$

27. $e^{\ln x} = 2.5$

28. $\log_x 10^4 = 4$

29. $\log_9 x = -\frac{3}{2}$

30. $\ln (x + 4) - \ln (x - 4) = 2 \ln 3$

31. $\ln (2x^2 + 2) = 2 \ln (2x - 4)$

32. $\log x + \log (x + 15) = 2$

33. $\log (\ln x) = -1$

34. $4 (\ln x)^2 = \ln x^2$

Solve Problems 35–39 for x to three significant digits.

35. $x = \log_3 41$

36. $\ln x = 1.45$

37. $4(2^x) = 20$

38. $10e^{-0.5x} = 1.6$

39. $\dfrac{e^x - e^{-x}}{e^x + e^{-x}} = \dfrac{1}{2}$

In Problems 40–44, find domain, range, intercepts, and asymptotes. Round all approximate values to two decimal places.

40. $f(x) = 3^{1-x}$

41. $g(x) = \ln (2 - x)$

42. $A(t) = 100e^{-0.3t}$

43. $h(x) = -2e^{-x} + 3$

44. $N(t) = \dfrac{6}{2 + e^{-0.1t}}$

45. If the graph of $y = \ln x$ is reflected in the line $y = x$, the graph of the function $y = e^x$ is obtained. Discuss the functions that are obtained by reflecting the graph of $y = \ln x$ in the x axis and in the y axis.

46. **(A)** Explain why the equation $e^{-x} = \ln x$ has exactly one solution.
(B) Approximate the solution of the equation to two decimal places.

In Problems 47 and 48, factor each polynomial in two ways:
(A) As a product of linear factors (with real coefficients) and quadratic factors (with real coefficients and imaginary zeros)
(B) As a product of linear factors with complex coefficients

47. $P(x) = x^4 + 9x^2 + 18$

48. $P(x) = x^4 - 23x^2 - 50$

C

49. Graph f and indicate any horizontal, vertical, or oblique asymptotes with dashed lines:

$$f(x) = \frac{x^2 + 4x + 8}{x + 2}$$

50. Let $P(x) = x^4 - 28x^3 + 262x^2 - 922x + 1{,}083$. Approximate (to two decimal places) the x intercepts and the local extrema.

51. Find a polynomial of lowest degree with leading coefficient 1 that has zeros -1 (multiplicity 2), 0 (multiplicity 3), $3 + 5i$, and $3 - 5i$. Leave the answer in factored form. What is the degree of the polynomial?

52. If $P(x)$ is a fourth-degree polynomial with integer coefficients and if i is a zero of $P(x)$, can $P(x)$ have any irrational zeros? Explain.

53. Let $P(x) = x^4 + 9x^3 - 500x^2 + 20{,}000$.
(A) Find the smallest positive integer multiple of 10 and the largest negative integer multiple of 10 that, by Theorem 1 in Section 3.2, are upper and lower bounds, respectively, for the real zeros of $P(x)$.
(B) Approximate the real zeros of $P(x)$ to two decimal places.

54. Find all zeros (rational, irrational, and imaginary) exactly for

$$P(x) = x^5 - 4x^4 + 3x^3 + 10x^2 - 10x - 12$$

and factor $P(x)$ into linear factors.

55. Find rational roots exactly and irrational roots to two decimal places for

$$P(x) = x^5 + 4x^4 + x^3 - 11x^2 - 8x + 4$$

56. Give an example of a rational function $f(x)$ that satisfies the following conditions: the real zeros of f are 5 and 8; $x = 1$ is the only vertical asymptote; and the line $y = 3$ is a horizontal asymptote.

57. Use natural logarithms to solve for n.

$$A = P\frac{(1 + i)^n - 1}{i}$$

58. Solve $\ln y = 5x + \ln A$ for y. Express the answer in a form that is free of logarithms.

59. Solve for x.

$$y = \frac{e^x - 2e^{-x}}{2}$$

60. Solve $\frac{x^3 - x}{x^3 - 8} \geq 0$.

61. Solve (to three decimal places)

$$\frac{4x}{x^2 - 1} < 3$$

APPLICATIONS

62. **Shipping.** A mailing service provides customers with rectangular shipping containers. The length plus the girth of one of these containers is 10 feet (see the figure). If the end of the container is square and the volume is 8 cubic feet, find the dimensions. Find rational solutions exactly and irrational solutions to two decimal places.

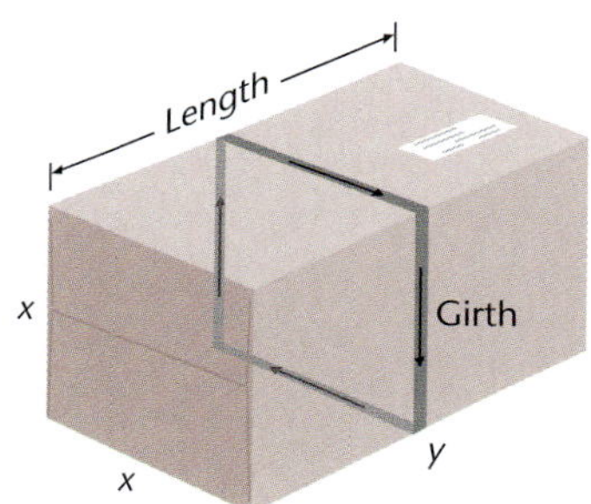

63. **Geometry.** The diagonal of a rectangle is 2 feet longer than one of the sides, and the area of the rectangle is 6 square feet. Find the dimensions of the rectangle. Find rational solutions exactly and irrational solutions to two decimal places.

64. **Population Growth.** If the Republic of the Congo has a population of about 40 million people and a doubling time of 22 years, find the population in
(A) 5 years **(B)** 30 years
Compute answers to three significant digits.

65. **Compound Interest.** How long will it take money invested in an account earning 7% compounded annually to double? Use the annual compounding growth model $P = P_0(1 + r)^t$, and compute the answer to three significant digits.

66. **Compound Interest.** Repeat Problem 65 using the continuous compound interest model $P = P_0e^{rt}$.

67. **Earthquakes.** If the 1906 and 1989 San Francisco earthquakes registered 8.3 and 7.1, respectively, on the Richter scale, how many times more powerful was the 1906 earthquake than the 1989 earthquake? Use the formula $M = \frac{2}{3} \log (E/E_0)$, where $E_0 = 10^{4.40}$ joules, and compute the answer to one decimal place.

68. **Sound.** If the decibel level at a rock concert is 88, find the intensity of the sound at the concert. Use the formula $D = 10 \log (I/I_0)$, where $I_0 = 10^{-12}$ watts per square meter, and compute the answer to two significant digits.

MODELING AND DATA ANALYSIS 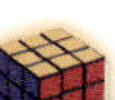

69. Table 1 shows the life expectancy (in years) at birth for residents of the United States from 1970 to 1995. Let x represent years since 1970. Use the indicated regression model to estimate the life expectancy (to the nearest tenth of a year) for a U.S. resident born in 2010.
(A) Linear regression
(B) Quadratic regression
(C) Cubic regression
(D) Exponential regression

TABLE 1

Year	Life Expectancy
1970	70.8
1975	72.6
1980	73.7
1985	74.7
1990	75.4
1995	75.9

Source: U.S. Census Bureau.

70. Refer to Problem 69. The Census Bureau projected the life expectancy for a U.S. resident born in 2010 to be 77.6 years. Which of the models in Problem 69 is closest to the Census Bureau projection?

CHAPTER 5

Trigonometric Functions

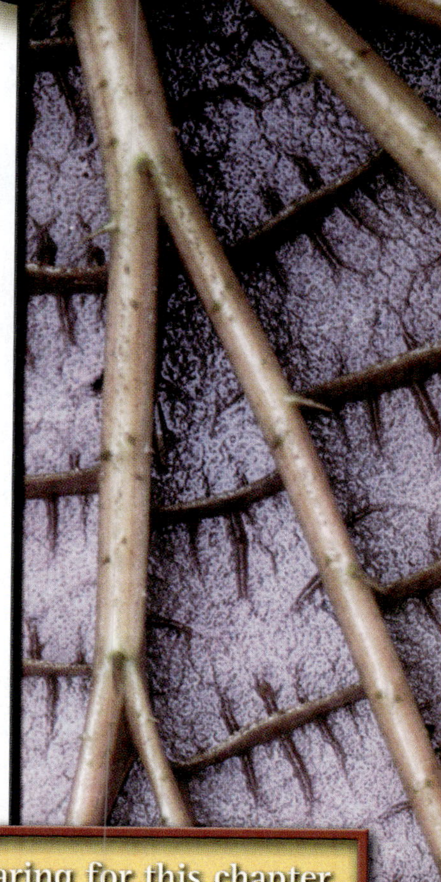

OUTLINE

TRIGONOMETRIC FUNCTIONS SEEM TO HAVE HAD THEIR ORIGINS WITH THE Greeks' investigation of the indirect measurement of distances and angles in the "celestial sphere." (The ancient Egyptians had used some elementary geometry to build the pyramids and remeasure lands flooded by the Nile, but neither they nor the ancient Babylonians had developed the concept of angle measure.) The word *trigonometry,* based on the Greek words for "triangle measurement," was first used as the title for a text by the German mathematician Pitiscus in A.D. 1600.

Originally the trigonometric functions were restricted to angles and their applications to the indirect measurement of angles and distances. These functions gradually broke free of these restrictions, and we now have trigonometric functions of real numbers. Modern applications range over many types of problems that have little or nothing to do with angles or triangles—applications involving periodic phenomena such as sound, light, and electrical waves; business cycles; and planetary motion.

In our approach to the subject we define the trigonometric functions in terms of coordinates of points on the unit circle.

Preparing for this chapter

Before getting started on this chapter, review the following concepts:

- Cartesian Coordinate System (Appendix A, Section A.1)
- Functions (Chapter 1, Section 2)
- Graphs of Functions (Chapter 1, Section 3)
- Transformations (Chapter 1, Section 4)
- Asymptotes (Chapter 3, Section 4)
- Operations on Functions (Chapter 1, Section 5)
- Inverse Functions (Chapter 1, Section 6)
- Significant Digits (Appendix B, Section B.1)
- Pythagorean Theorem (Appendix C)

SECTION 5.1 Angles and Their Measure

Angles • Degree and Radian Measure • Converting Degrees to Radians and Vice Versa • Linear and Angular Speed

In Section 5.1 we introduce the concept of angle and two measures of angles, *degree* and *radian*.

FIGURE 1 Angle θ or angle PVQ or $\angle V$.

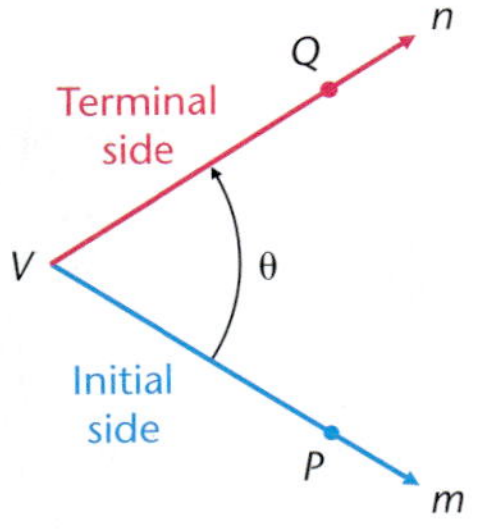

Angles

The study of trigonometry depends on the concept of angle. An **angle** is formed by rotating (in a plane) a ray m, called the **initial side** of the angle, around its endpoint until it coincides with a ray n, called the **terminal side** of the angle. The common endpoint V of m and n is called the **vertex** (Fig. 1).

A counterclockwise rotation produces a **positive angle,** and a clockwise rotation produces a **negative angle,** as shown in Figures 2(a) and 2(b). The amount of rotation in either direction is not restricted. Two different angles may have the same initial and terminal sides, as shown in Figure 2(c). Such angles are said to be **coterminal.**

FIGURE 2 Angles and rotation.

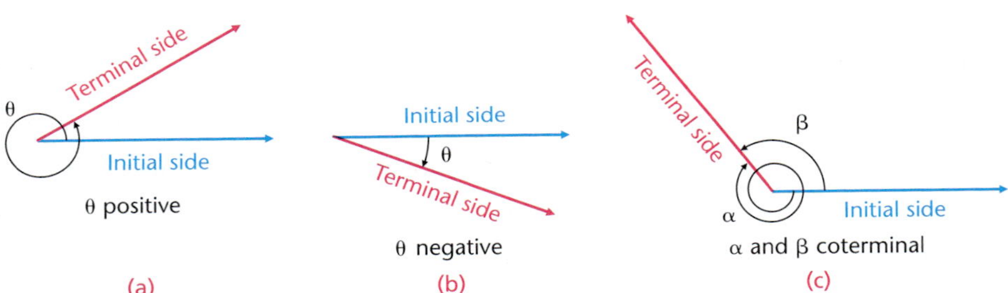

An angle in a rectangular coordinate system is said to be in **standard position** if its vertex is at the origin and the initial side is along the positive x axis. If the terminal side of an angle in standard position lies along a coordinate axis, the angle is said to be a **quadrantal angle.** If the terminal side does not lie along a coordinate axis, then the angle is often referred to in terms of the quadrant in which the terminal side lies (Fig. 3).

FIGURE 3 Angles in standard positions.

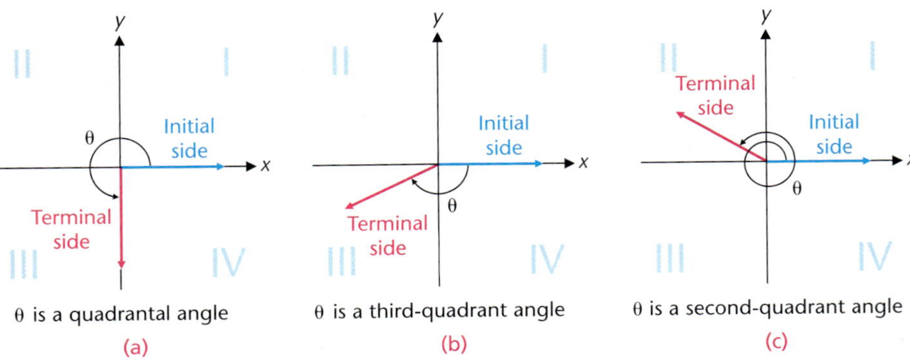

Degree and Radian Measure

Just as line segments are measured in centimeters, meters, inches, or miles, angles are measured in different units. The two most commonly used units for angle measure are *degree* and *radian*.

> **DEFINITION 1**
> **Degree Measure**
>
> A positive angle formed by one complete rotation is said to have a measure of *360 degrees* (360°). A positive angle formed by $\frac{1}{360}$ of a complete rotation is said to have a measure of *1 degree* (1°). The symbol ° denotes degrees.

Definition 1 is extended to all angles, not just the positive (counterclockwise) ones, in the obvious way. So, for example, a negative angle formed by $\frac{1}{4}$ of a complete clockwise rotation has a measure of $-90°$, and an angle for which the initial and terminal sides coincide, without rotation, has a measure of 0°.

Certain angles have special names that indicate their degree measure. Figure 4 shows a **straight angle,** a **right angle,** an **acute angle,** and an **obtuse angle.**

FIGURE 4 Types of angles.

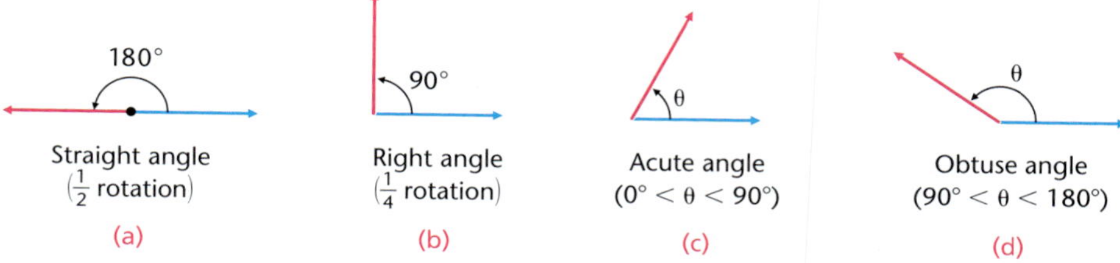

Two positive angles are **complementary** if their sum is 90°; they are **supplementary** if their sum is 180°.

A degree can be divided further using decimal notation. For example, 42.75° represents an angle of degree measure 42 plus three-quarters of 1 degree. A degree can also be divided further using minutes and seconds just as an hour is divided into minutes and seconds. Each degree is divided into 60 equal parts called **minutes,** and each minute is divided into 60 equal parts called **seconds.** Symbolically, minutes are represented by ′ and seconds by ″. Thus,

$$12°23'14''$$

is a concise way of writing 12 degrees, 23 minutes, and 14 seconds.

Decimal degrees (DD) are useful in some instances and degrees–minutes–seconds (DMS) are useful in others. You should be able to go from one form to the other as demonstrated in Example 1.

> **Conversion Accuracy**
>
> If an angle is measured to the nearest second, the converted decimal form should not go beyond three decimal places, and vice versa.

EXAMPLE 1 From DMS to DD and Back

(A) Convert 21°47′12″ to decimal degrees.

(B) Convert 105.183° to degree–minute–second form.

SOLUTIONS

(A) $21°47'12'' = \left(21 + \frac{47}{60} + \frac{12}{3{,}600}\right)^{\circ} = 21.787°$

(B) $$\begin{aligned} 105.183° &= 105°\ (0.183 \cdot 60)' \\ &= 105°\ 10.98' \\ &= 105°\ 10'\ (0.98 \cdot 60)'' \\ &= 105°\ 10'\ 59'' \end{aligned}$$

MATCHED 1 PROBLEM

(A) Convert 193°17′34″ to DD form.

(B) Convert 237.615° to DMS form.

Some scientific and some graphing calculators can convert the DD and DMS forms automatically, but the process differs significantly among the various types of calculators. Check your owner's manual for your particular calculator. The conversion methods outlined in Example 1 show you the reasoning behind the process, and are sometimes easier to use than the "automatic" methods for some calculators.

Degree measure of angles is used extensively in engineering, surveying, and navigation. Another unit of angle measure, called the *radian,* is better suited for certain mathematical developments, scientific work, and engineering applications.

DEFINITION 2

Radian Measure

If the vertex of a positive angle θ is placed at the center of a circle with radius $r > 0$, and the length of the arc opposite θ on the circumference is s, then the *radian measure* of θ is given by

$$\theta = \frac{s}{r} \text{ radians}$$

If $s = r$, then

$$\theta = \frac{r}{r} = 1 \text{ radian}$$

Thus, *1 radian* is the measure of the central angle of a circle that intercepts an arc that has the same length as the radius of the circle. [*Note:* s and r must be measured in the same units.]

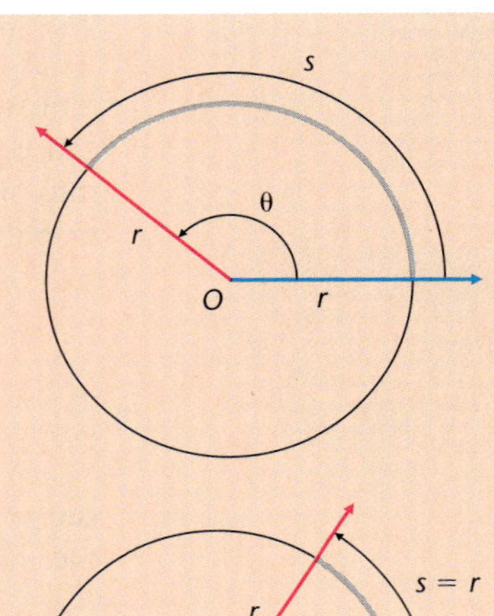

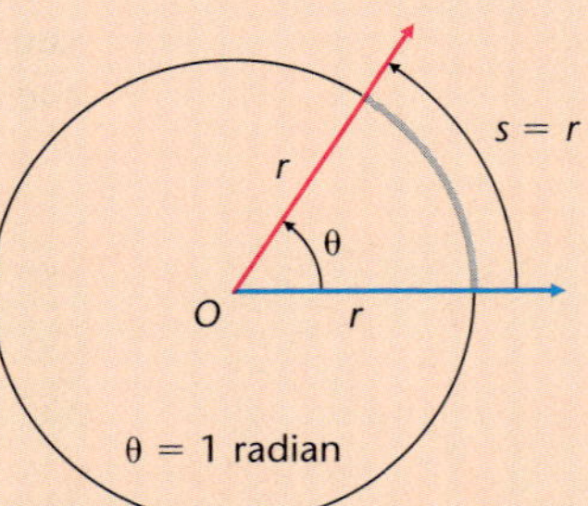

The circumference of a circle of radius r is $2\pi r$, so the radian measure of a positive angle formed by one complete rotation is

$$\theta = \frac{s}{r} = \frac{2\pi r}{r} = 2\pi \approx 6.283 \text{ radians}$$

Just as for degree measure, the definition is extended to apply to all angles; if θ is a negative angle, its radian measure is given by $\theta = -\frac{s}{r}$. Note that in the preceding sentence, as well as in Definition 2, the symbol θ is used in two ways: as the name of the angle and as the measure of the angle. The context indicates the meaning.

EXAMPLE 2 Computing Radian Measure

What is the radian measure of a central angle θ opposite an arc of 24 meters in a circle of radius 6 meters?

SOLUTION

$$\theta = \frac{s}{r} = \frac{24 \text{ meters}}{6 \text{ meters}} = 4 \text{ radians}$$

MATCHED PROBLEM 2

What is the radian measure of a central angle θ opposite an arc of 60 feet in a circle of radius 12 feet?

REMARK It is customary to omit the word *radians* when giving the radian measure of an angle. But if an angle is measured in any other units, the units must be stated explicitly. For example, if $\theta = 17$, then $\theta \approx 974°$.

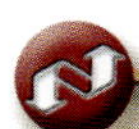

EXPLORE/DISCUSS 1

Discuss why the radian measure of an angle is independent of the size of the circle having the angle as a central angle.

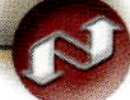

Converting Degrees to Radians and Vice Versa

What is the radian measure of an angle of 180°? Let θ be a central angle of 180° in a circle of radius r. Then the length s of the arc opposite θ is $\frac{1}{2}$ the circumference C of the circle. Therefore,

$$s = \frac{C}{2} = \frac{2\pi r}{2} = \pi r \qquad \text{and} \qquad \theta = \frac{s}{r} = \frac{\pi r}{r} = \pi \text{ radians}$$

Hence, 180° corresponds to π* radians. This is important to remember, because the radian measures of many special angles can be obtained from this correspondence. For example, 90° is 180°/2; therefore, 90° corresponds to $\pi/2$ radians.

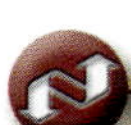

EXPLORE/DISCUSS 2

Write the radian measure of each of the following angles in the form $\frac{a}{b}\pi$, where a and b are positive integers and fraction $\frac{a}{b}$ is reduced to lowest terms: 15°, 30°, 45°, 60°, 75°, 90°, 105°, 120°, 135°, 150°, 165°, 180°.

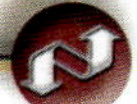

Some key results from Explore/Discuss 2 are summarized in Figure 5 for easy reference. These correspondences and multiples of them will be used extensively in work that follows.

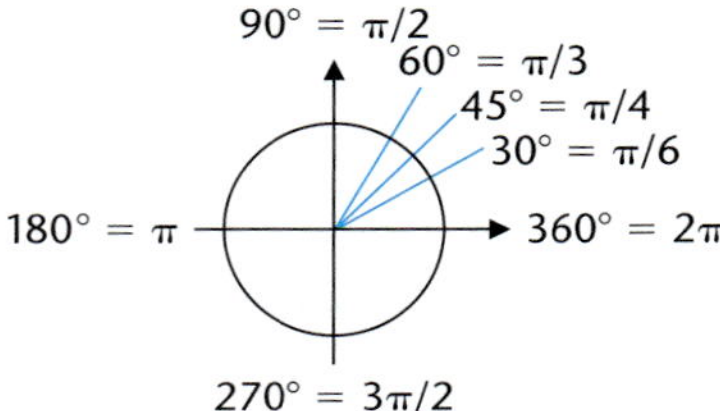

FIGURE 5 Radian–degree correspondences.

In general, the following proportion can be used to convert degree measure to radian measure and vice versa.

Radian–Degree Conversion Formulas

$$\frac{\theta_{\text{deg}}}{180°} = \frac{\theta_{\text{rad}}}{\pi \text{ radians}} \qquad \textbf{Basic proportion}$$

$$\theta_{\text{deg}} = \frac{180°}{\pi \text{ radians}}\,\theta_{\text{rad}} \qquad \textbf{Radians to degrees}$$

$$\theta_{\text{rad}} = \frac{\pi \text{ radians}}{180°}\,\theta_{\text{deg}} \qquad \textbf{Degrees to radians}$$

[*Note:* The basic proportion is usually easier to remember. Also we will omit units in calculations until the final answer. If your calculator does not have a key labeled π, use $\pi \approx 3.14159$.]

---

*The constant π has a long and interesting history; a few important dates are listed below:

1650 B.C.	Rhind Papyrus	$\pi \approx \frac{256}{81} = 3.16049\ldots$
240 B.C.	Archimedes	$3\frac{10}{71} < \pi < 3\frac{1}{7}$ $(3.1408\ldots < \pi < 3.1428\ldots)$
A.D. 264	Liu Hui	$\pi \approx 3.14159$
A.D. 470	Tsu Ch'ung-chih	$\pi \approx \frac{355}{113} = 3.1415929\ldots$
A.D. 1674	Leibniz	$\pi = 4(1 - \frac{1}{3} + \frac{1}{5} - \frac{1}{7} + \frac{1}{9} - \frac{1}{11} + \cdots)$
		$\approx 3.1415926535897932384626$
		(This and other series can be used to compute π to any decimal accuracy desired.)
A.D. 1761	Johann Lambert	Showed π to be irrational (π as a decimal is nonrepeating and nonterminating)

Some scientific and graphing calculators can automatically convert radian measure to degree measure, and vice versa. Check the owner's manual for your particular calculator.

EXAMPLE 3 Radian–Degree Conversions

(A) Find the radian measure, exact and to three significant digits, of an angle of 75°.

(B) Find the degree measure, exact and to four significant digits, of an angle of 5 radians.

(C) Find the radian measure to two decimal places of an angle of 41°12′.

SOLUTIONS

Exact — Three significant digits

$$\text{(A)}\quad \theta_{\text{rad}} = \frac{\pi \text{ radians}}{180°}\theta_{\text{deg}} = \frac{\pi}{180}(75) = \frac{5\pi}{12} = 1.31$$

Exact — Four significant digits

$$\text{(B)}\quad \theta_{\text{deg}} = \frac{180°}{\pi \text{ radians}}\theta_{\text{rad}} = \frac{180}{\pi}(5) = \frac{900}{\pi} = 286.5°$$

$$\text{(C)}\quad 41°12' = \left(41 + \frac{12}{60}\right)^{\circ} = 41.2°$$ Change 41°12′ to DD first.

$$\theta_{\text{rad}} = \frac{\pi \text{ radians}}{180°}\theta_{\text{deg}} = \frac{\pi}{180}(41.2) = 0.72$$ To two decimal places

FIGURE 6 Automatic conversion.

```
75°
                 1.31
5ʳ
                286.5
(41°12')°
                  .72
```

Figure 6 shows the three preceding conversions done automatically on a graphing calculator by selecting the appropriate angle mode.

MATCHED 3 PROBLEM

(A) Find the radian measure, exact and to three significant digits, of an angle of 240°.

(B) Find the degree measure, exact and to three significant digits, of an angle of 1 radian.

(C) Find the radian measure to three decimal places of an angle of 125°23′.

REMARK We will write θ in place of θ_{deg} and θ_{rad} when it is clear from the context whether we are dealing with degree or radian measure.

EXAMPLE 4 Engineering

A belt connects a pulley of 2-inch radius with a pulley of 5-inch radius. If the larger pulley turns through 10 radians, through how many radians will the smaller pulley turn?

SOLUTION

First we draw a sketch (Fig. 7).

FIGURE 7

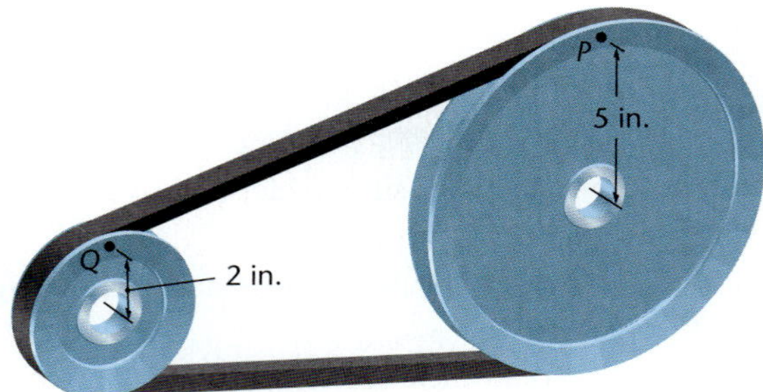

When the larger pulley turns through 10 radians, the point P on its circumference will travel the same distance (arc length) that point Q on the smaller circle travels. For the larger pulley,

$$\theta = \frac{s}{r}$$

$$s = r\theta = (5)(10) = 50 \text{ inches}$$

For the smaller pulley,

$$\theta = \frac{s}{r} = \frac{50}{2} = 25 \text{ radians}$$

MATCHED 4 PROBLEM

In Example 4, through how many radians will the larger pulley turn if the smaller pulley turns through 4 radians?

Linear and Angular Speed

FIGURE 8

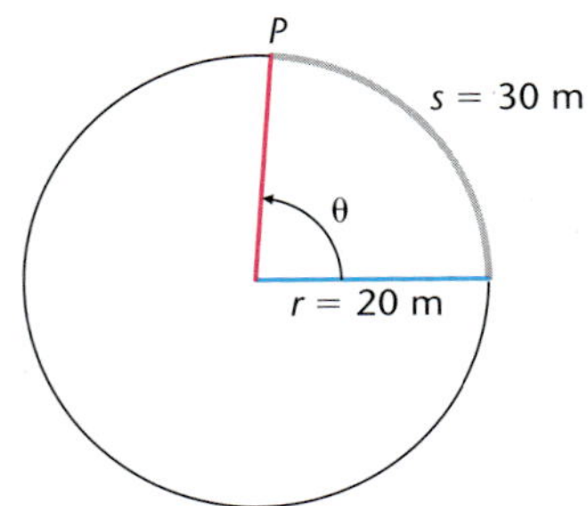

The **average speed** v of an object that travels a distance $d = 30$ meters in time $t = 3$ seconds is given by

$$v = \frac{d}{t} = \frac{30 \text{ meters}}{3 \text{ seconds}} = 10 \text{ meters per second}$$

Suppose that a point P moves an arc length of $s = 30$ meters in $t = 3$ seconds on the circumference of a circle of radius $r = 20$ meters (Fig. 8). Then, in those 3 seconds, the point P has moved through an angle of

$$\theta = \frac{s}{r} = \frac{30}{20} = 1.5 \text{ radians}$$

We call the average speed of point P, given by

$$v = \frac{s}{t} = 10 \text{ meters per second}$$

the (average) **linear speed** to distinguish it from the (average) **angular speed** that is given by

$$\omega = \frac{\theta}{t} = \frac{1.5}{3} = 0.5 \text{ radians per second}$$

Note that $v = r\omega$ (because $s = r\theta$). These concepts are summarized in the box.

Linear Speed and Angular Speed

Suppose a point P moves through an angle θ and arc length s, in time t, on the circumference of a circle of radius r. The (average) **linear speed** of P is

$$v = \frac{s}{t}$$

and the (average) **angular speed** is

$$\omega = \frac{\theta}{t}$$

Furthermore, $v = r\omega$.

EXAMPLE 5 Wind Power

A wind turbine of rotor diameter 15 meters makes 62 revolutions per minute. Find the angular speed (in radians per second) and the linear speed (in meters per second) of the rotor tip.

SOLUTION

The radius of the rotor is $15/2 = 7.5$ meters. In 1 minute the rotor moves through an angle of $62(2\pi) = 124\pi$ radians. Therefore, the angular speed is

$$\omega = \frac{\theta}{t} = \frac{124\pi \text{ radians}}{60 \text{ seconds}} \approx 6.49 \text{ radians per second}$$

and the linear speed of the rotor tip is

$$v = r\omega = 7.5\,\frac{124\pi}{60} \approx 48.69 \text{ meters per second}$$

MATCHED 5 PROBLEM

A wind turbine of rotor diameter 12 meters has a rotor tip speed of 34.2 meters per second. Find the angular speed of the rotor (in radians per second) and the number of revolutions per minute.

ANSWERS MATCHED PROBLEMS

1. (A) 193.293° (B) 237°36′54″
2. 5 radians
3. (A) $\frac{4\pi}{3} = 4.19$ (B) $\frac{180}{\pi} = 57.3°$ (C) 2.188
4. 1.6 radians
5. 5.7 radians per second; 54.43 revolutions per minute

EXERCISE 5.1

In all problems, if angle measure is expressed by a number that is not in degrees, it is assumed to be in radians.

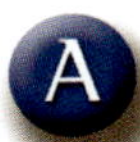

Find the degree measure of each of the angles in Problems 1–6, keeping in mind that an angle of one complete rotation corresponds to 360°.

1. $\frac{1}{9}$ rotation **2.** $\frac{1}{5}$ rotation **3.** $\frac{3}{4}$ rotation

4. $\frac{3}{8}$ rotation **5.** $\frac{9}{8}$ rotations **6.** $\frac{7}{6}$ rotations

Find the radian measure of a central angle θ opposite an arc s in a circle of radius r, where r and s are as given in Problems 7–10.

7. $r = 4$ centimeters, $s = 24$ centimeters

8. $r = 8$ inches, $s = 16$ inches

9. $r = 12$ feet, $s = 30$ feet

10. $r = 18$ meters, $s = 27$ meters

Find the radian measure of each angle in Problems 11–16, keeping in mind that an angle of one complete rotation corresponds to 2π radians.

11. $\frac{1}{8}$ rotation **12.** $\frac{1}{6}$ rotation **13.** $\frac{3}{4}$ rotation

14. $\frac{5}{12}$ rotation **15.** $\frac{13}{12}$ rotations **16.** $\frac{11}{8}$ rotations

B

Find the exact radian measure, in terms of π, of each angle in Problems 17–20.

17. 30°, 60°, 90°, 120°, 150°, 180°

18. 60°, 120°, 180°, 240°, 300°, 360°

19. −45°, −90°, −135°, −180°

20. −90°, −180°, −270°, −360°

Find the exact degree measure of each angle in Problems 21–24.

21. $\frac{\pi}{3}, \frac{2\pi}{3}, \pi, \frac{4\pi}{3}, \frac{5\pi}{3}, 2\pi$ **22.** $\frac{\pi}{6}, \frac{\pi}{3}, \frac{\pi}{2}, \frac{2\pi}{3}, \frac{5\pi}{6}, \pi$

23. $-\frac{\pi}{2}, -\pi, -\frac{3\pi}{2}, -2\pi$ **24.** $-\frac{\pi}{4}, -\frac{\pi}{2}, -\frac{3\pi}{4}, -\pi$

In Problems 25–30, determine whether the statement is true or false. If true, explain why. If false, give a counterexample.

25. If two angles in standard position have the same measure, then they are coterminal.

26. If two angles in standard position are coterminal, then they have the same measure.

27. If two positive angles are complementary, then both are acute.

28. If two positive angles are supplementary, then one is obtuse and the other is acute.

29. If the terminal side of an angle in standard position lies in quadrant I, then the angle is positive.

30. If the initial and terminal sides of an angle coincide, then the measure of the angle is zero.

Convert each angle in Problems 31–34 to decimal degrees to three decimal places.

31. 5°51′33″ **32.** 14°18′37″

33. 354°8′29″ **34.** 184°31′7″

Convert each angle in Problems 35–38 to degree–minute–second form.

35. 3.042° **36.** 49.715°

37. 403.223° **38.** 156.808°

Find the radian measure to three decimal places for each angle in Problems 39–44.

39. 64° **40.** 25° **41.** 108.413°

42. 203.097° **43.** 13°25′14″ **44.** 56°11′52″

Find the degree measure to two decimal places for each angle in Problems 45–50.

45. 0.93 **46.** 0.08 **47.** 1.13

48. 3.07 **49.** −2.35 **50.** −1.72

Indicate whether each angle in Problems 51–70 is a first-, second-, third-, or fourth-quadrant angle or a quadrantal angle. All angles are in standard position in a rectangular coordinate system. (A sketch may be of help in some problems.)

51. 187° **52.** 135° **53.** −200°

54. −60° **55.** 4 **56.** 3

57. 270° **58.** 360° **59.** −1

60. −6 **61.** $\frac{5\pi}{3}$ **62.** $\frac{2\pi}{3}$

63. $-\frac{7\pi}{6}$ **64.** $-\frac{3\pi}{4}$ **65.** $-\pi$

66. $-\frac{3\pi}{2}$ **67.** 820° **68.** −565°

69. $\frac{13\pi}{4}$ **70.** $\frac{23\pi}{3}$

71. Verbally describe the meaning of a central angle in a circle with radian measure 1.

72. Verbally describe the meaning of an angle with degree measure 1.

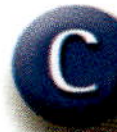

Which angles in Problems 73–78 are coterminal with 30° if all angles are placed in standard position in a rectangular coordinate system?

73. 390° **74.** 330° **75.** $\frac{\pi}{6}$

76. $-\frac{11\pi}{6}$ **77.** −690° **78.** 750°

Which angles in problems 79–84 are coterminal with $3\pi/4$ if all angles are placed in standard position in a rectangular coordinate system?

79. $-\frac{3\pi}{4}$ **80.** $-\frac{7\pi}{4}$ **81.** 135°

82. −225° **83.** $\frac{11\pi}{4}$ **84.** $-\frac{5\pi}{4}$

APPLICATIONS

85. Circumference of the Earth. The early Greeks used the proportion $s/C = \theta°/360°$, where s is an arc length on a circle, $\theta°$ is degree measure of the corresponding central angle, and C is the circumference of the circle ($C = 2\pi r$). Eratosthenes (240 B.C.), in his famous calculation of the circumference of the Earth, reasoned as follows: He knew at Syene (now Aswan) during the summer solstice the noon sun was directly overhead and shined on the water straight down a deep well. On the same day at the same time, 5,000 stadia (approx. 500 miles) due north in Alexandria, sun rays crossed a vertical pole at an angle of 7.5° as indicated in the figure. Carry out Eratosthenes' calculation for the circumference of the Earth to the nearest thousand miles. (The current calculation for the equatorial circumference is 24,902 miles.)

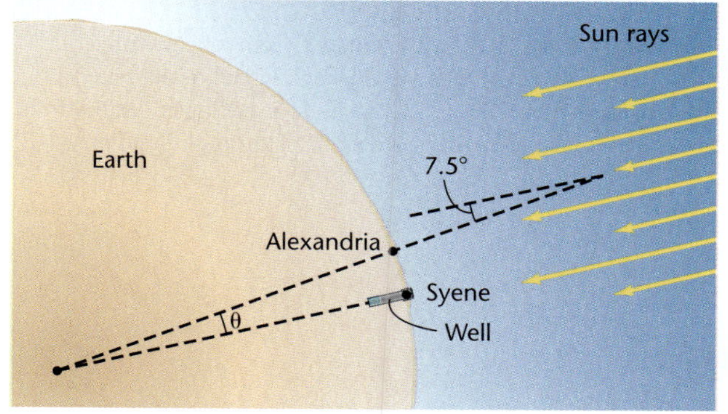

86. Circumference of the Earth. Repeat Problem 85 with the sun crossing the vertical pole in Alexandria at 7°12′.

87. Circumference of the Earth. In Problem 85, verbally explain how θ in the figure was determined.

88. Circumference of the Earth. Verbally explain how the radius, surface area, and volume of the Earth can be determined from the result of Problem 85.

89. Angular Speed. A wheel with diameter 6 feet makes 200 revolutions per minute. Find the angular speed (in radians per second) and the linear speed (in feet per second) of a point on the rim.

90. Angular Speed. A point on the rim of a wheel with diameter 6 feet has a linear speed of 100 feet per second. Find the angular speed (in radians per second) and the number of revolutions per minute.

91. Radian Measure. What is the radian measure of the larger angle made by the hands of a clock at 4:30? Express the answer exactly in terms of π.

92. Radian Measure. What is the radian measure of the smaller angle made by the hands of a clock at 1:30? Express the answer exactly in terms of π.

93. Engineering. Through how many radians does a pulley of 10-centimeter diameter turn when 10 meters of rope are pulled through it without slippage?

94. Engineering. Through how many radians does a pulley of 6-inch diameter turn when 4 feet of rope are pulled through it without slippage?

95. Astronomy. A line from the sun to the Earth sweeps out an angle of how many radians in 1 week? Assume the Earth's orbit is circular and there are 52 weeks in a year. Express the answer in terms of π and as a decimal to two decimal places.

96. Astronomy. A line from the center of the Earth to the equator sweeps out an angle of how many radians in 9 hours? Express the answer in terms of π and as a decimal to two decimal places.

★ **97. Engineering.** A trail bike has a front wheel with a diameter of 40 centimeters and a back wheel of diameter 60 centimeters. Through what angle in radians does the front wheel turn if the back wheel turns through 8 radians?

★ **98. Engineering.** In Problem 97, through what angle in radians will the back wheel turn if the front wheel turns through 15 radians?

99. Angular Speed. If the trail bike of Problem 97 travels at a speed of 10 kilometers per hour, find the angular speed (in radians per second) of each wheel.

100. Angular Speed. If a car travels at a speed of 60 miles per hour, find the angular speed (in radians per second) of a tire that has a diameter of 2 feet.

The arc length on a circle is easy to compute if the corresponding central angle is given in radians and the radius of the circle is known ($s = r\theta$). If the radius of a circle is large and a central angle is small, then an arc length is often used to approximate the length of the corresponding chord as shown in the figure. If an angle is given in degree measure, converting to radian measure first may be helpful in certain problems. This information will be useful in Problems 101–104.

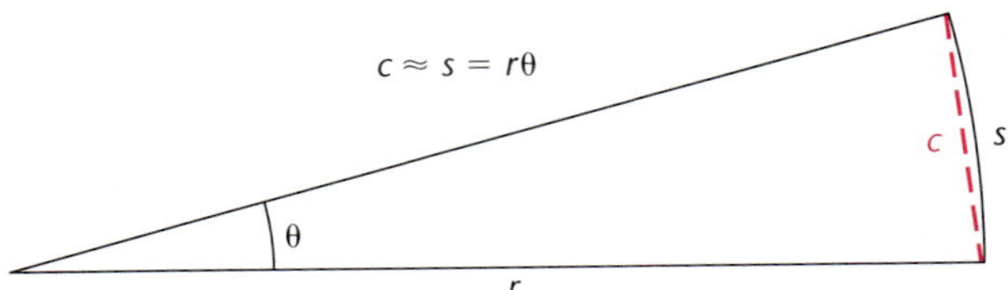

101. Astronomy. The sun is about 9.3×10^7 mi from the Earth. If the angle subtended by the diameter of the sun on the surface of the Earth is 9.3×10^{-3} rad, approximately what is the diameter of the sun to the nearest thousand miles in standard decimal notation?

102. Astronomy. The moon is about 381,000 kilometers from the Earth. If the angle subtended by the diameter of the moon on the surface of the Earth is 0.0092 radians, approximately what is the diameter of the moon to the nearest hundred kilometers?

103. Photography. The angle of view of a 1,000-millimeter telephoto lens is 2.5°. At 750 feet, what is the width of the field of view to the nearest foot?

104. Photography. The angle of view of a 300-millimeter lens is 8°. At 500 feet, what is the width of the field of view to the nearest foot?

SECTION 5.2 Trigonometric Functions: A Unit Circle Approach

The Wrapping Function • Definitions of the Trigonometric Functions • Graphs of the Trigonometric Functions

In Section 5.2 we introduce the six trigonometric functions in terms of the coordinates of points on the unit circle.

The Wrapping Function

FIGURE 1

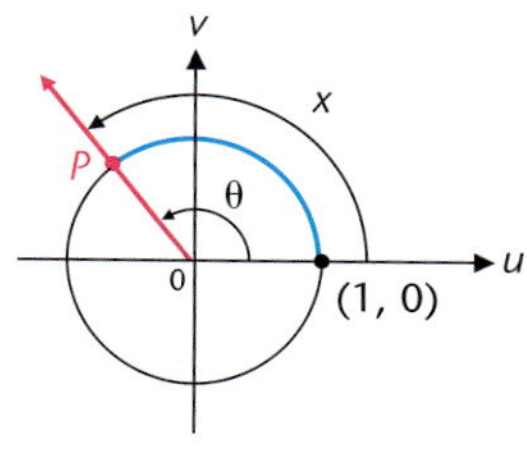

Consider a positive angle θ in standard position, and let P denote the point of intersection of the terminal side of θ with the unit circle $u^2 + v^2 = 1$ (Fig. 1).* Let x denote the length of the arc opposite θ on the unit circle. Because the unit circle has radius $r = 1$, the radian measure of θ is given by

$$\theta = \frac{x}{r} = \frac{x}{1} = x \text{ radians}$$

In other words, on the unit circle, the radian measure of a positive angle is equal to the length of the intercepted arc; similarly, on the unit circle, the radian measure of a negative angle is equal to the negative of the length of the intercepted arc. Because $\theta = x$, we may consider the real number x to be the name of the angle θ, when convenient. The function W that associates with each real number x the point $W(x) = P$ is called the *wrapping function*. The point P is called a *circular point.*

Consider, for example, the angle in standard position that has radian measure $\pi/2$. Its terminal side intersects the unit circle at the point (0, 1). Therefore, $W(\pi/2) = (0, 1)$. Similarly, we can find the circular point associated with any angle that is an integer multiple of $\pi/2$ (Fig. 2).

FIGURE 2 Circular points on the coordinate axes.

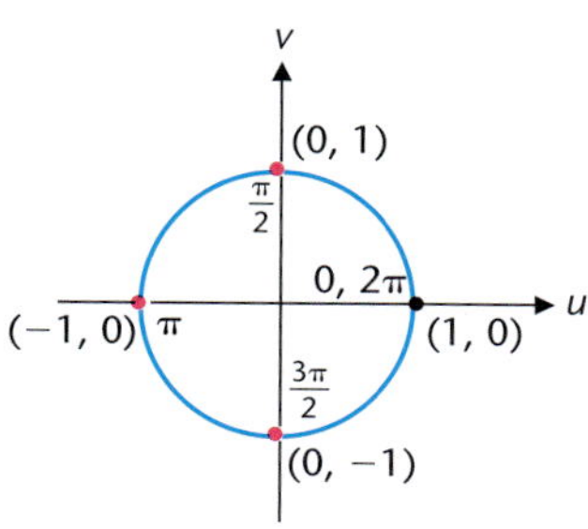

$$W(0) = (1, 0)$$

$$W\left(\frac{\pi}{2}\right) = (0, 1)$$

$$W(\pi) = (-1, 0)$$

$$W\left(\frac{3\pi}{2}\right) = (0, -1)$$

$$W(2\pi) = (1, 0)$$

*We use the variables u and v instead of x and y so that x can be used without ambiguity as an independent variable in defining the wrapping function and the trigonometric functions.

EXPLORE/DISCUSS 1

The name *wrapping function* stems from visualizing the correspondence as a wrapping of the real number line, with origin at (1, 0), around the unit circle—the positive real axis is wrapped counterclockwise, and the negative real axis is wrapped clockwise—so that each real number is paired with a unique circular point (Fig. 3).

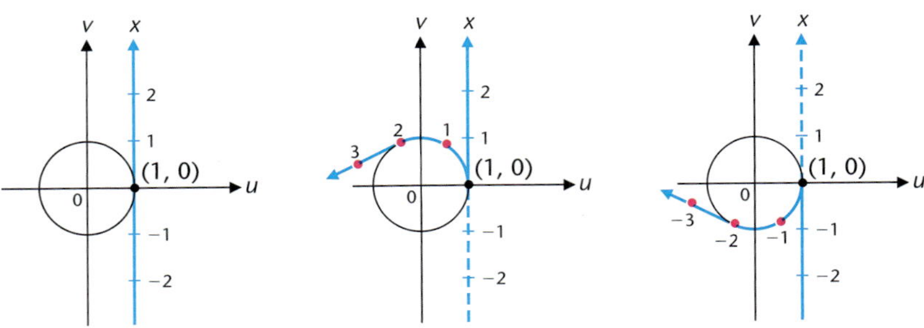

FIGURE 3 The wrapping function.

(A) Explain why the wrapping function is not one-to-one.

(B) In which quadrant is the circular point $W(1)$? $W(-10)$? $W(100)$?

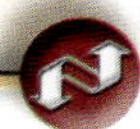

FIGURE 4

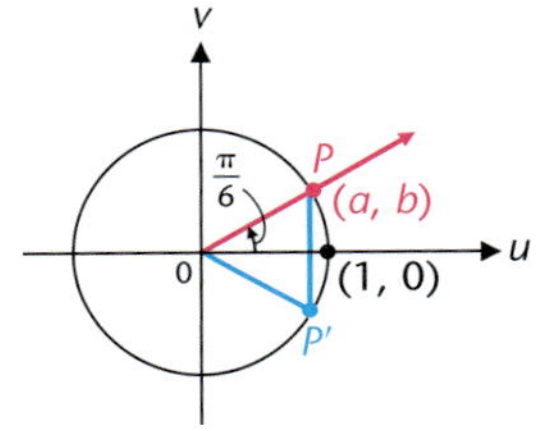

Given a real number x, it is difficult, in general, to find the coordinates (a, b) of the circular point $W(x)$ that is associated with x. (It is trigonometry that overcomes this difficulty.) For certain real numbers x, however, we can find the coordinates (a, b) of $W(x)$ by using simple geometric facts. For example, consider $x = \pi/6$ and let P denote the circular point $W(x) = (a, b)$ that is associated with x. Let P' be the reflection of P in the u axis (Fig. 4).

Then triangle $0PP'$ is equiangular (each angle has measure $\pi/3$ radians or 60°) and thus equilateral. Therefore $b = 1/2$. Because (a, b) lies on the unit circle, we solve for a:

$$a^2 + b^2 = 1$$

$$a^2 + \left(\frac{1}{2}\right)^2 = 1$$

$$a^2 = \frac{3}{4}$$

$$a = \pm\frac{\sqrt{3}}{2} \qquad a = -\frac{\sqrt{3}}{2} \text{ must be discarded (Why?)}$$

Thus,

$$W\left(\frac{\pi}{6}\right) = \left(\frac{\sqrt{3}}{2}, \frac{1}{2}\right)$$

EXAMPLE 1 Coordinates of Circular Points

Find the coordinates of the following circular points:

(A) $W(-\pi/2)$

(B) $W(5\pi/2)$

(C) $W(\pi/3)$

(D) $W(7\pi/6)$

(E) $W(\pi/4)$

FIGURE 5

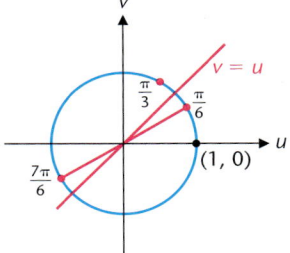

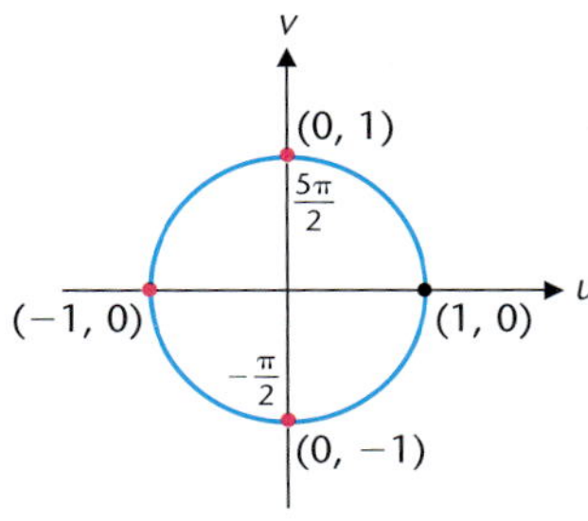

SOLUTIONS

(A) Because the circumference of the unit circle is 2π, $-\pi/2$ is the radian measure of a negative angle that is $\frac{1}{4}$ of a complete clockwise rotation. Thus, $W(-\pi/2) = (0, -1)$ (Fig. 5).

(B) Starting at (1, 0) and proceeding counterclockwise, we count quarter-circle steps, $\pi/2$, $2\pi/2$, $3\pi/2$, $4\pi/2$, and end at $5\pi/2$. Thus, the circular point is on the positive vertical axis, and $W(5\pi/2) = (0, 1)$ (see Fig. 5).

(C) The circular point $W(\pi/3)$ is the reflection of the point $W(\pi/6) = (\sqrt{3}/2, 1/2)$ in the line $u = v$. Thus, $W(\pi/3) = (1/2, \sqrt{3}/2)$ (Fig. 6).

FIGURE 6

(D) The circular point $W(7\pi/6)$ is the reflection of the point $W(\pi/6) = (\sqrt{3}/2, 1/2)$ in the origin. Thus, $W(7\pi/6) = (-\sqrt{3}/2, -1/2)$ (see Fig. 6).

(E) The circular point $W(\pi/4)$ lies on the line $u = v$, so $a = b$.

$$a^2 + b^2 = 1 \qquad \text{Substitute } b = a$$

$$2a^2 = 1$$

$$a^2 = \frac{1}{2}$$

$$a = \frac{1}{\sqrt{2}}$$

FIGURE 7

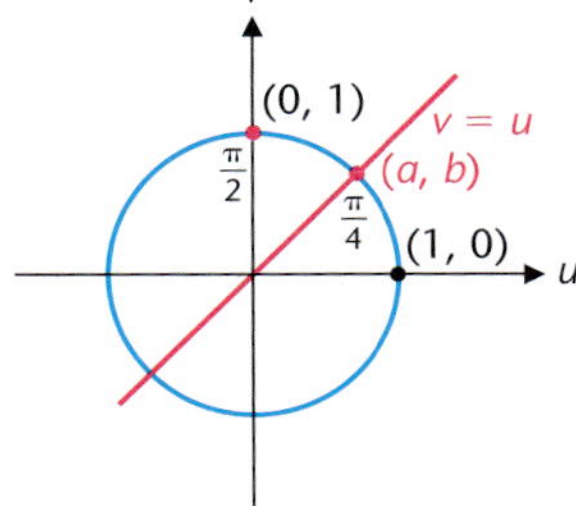

Therefore, $W(\pi/4) = (1/\sqrt{2}, 1/\sqrt{2})$ (Fig. 7).

MATCHED 1 PROBLEM

Find the coordinates of the following circular points:

(A) $W(3\pi)$

(B) $W(-7\pi/2)$

(C) $W(5\pi/6)$

(D) $W(-\pi/3)$

(E) $W(5\pi/4)$

Some key results from Example 1 are summarized in Figure 8. If x is any integer multiple of $\pi/6$ or $\pi/4$, then the coordinates of $W(x)$ can be determined easily from Figure 8 by using symmetry properties.

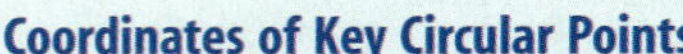

Coordinates of Key Circular Points

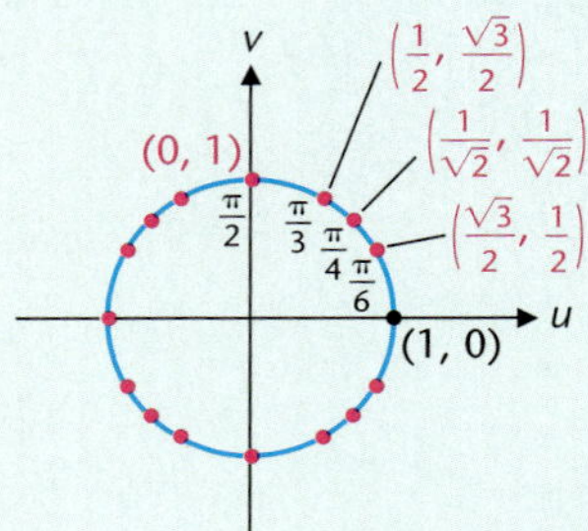

FIGURE 8

EXPLORE/DISCUSS 2

An effective **memory aid** for recalling the coordinates of the key circular points in Figure 8 can be created by writing the coordinates of the circular points $W(0)$, $W(\pi/6)$, $W(\pi/4)$, $W(\pi/3)$, and $W(\pi/2)$, keeping this order, in a form where each numerator is the square root of an appropriate number and each denominator is 2. For example, $W(0) = (1, 0) = (\sqrt{4}/2, \sqrt{0}/2)$. Describe the pattern that results.

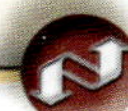

Definitions of the Trigonometric Functions

We use the correspondence between real numbers and circular points to define the six **trigonometric functions: sine, cosine, tangent, cotangent, secant,** and **cosecant.** The values of these functions at a real number x are denoted by **sin x, cos x, tan x, cot x, sec x,** and **csc x,** respectively.

DEFINITION 1
Trigonometric Functions

Let x be a real number and let (a, b) be the coordinates of the circular point $W(x)$ that lies on the terminal side of the angle with radian measure x. Then:

$$\sin x = b \qquad \csc x = \frac{1}{b} \quad b \neq 0$$

$$\cos x = a \qquad \sec x = \frac{1}{a} \quad a \neq 0$$

$$\tan x = \frac{b}{a} \quad a \neq 0 \qquad \cot x = \frac{a}{b} \quad b \neq 0$$

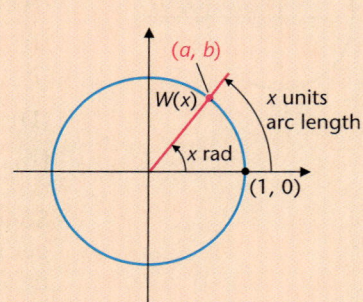

The domain of both the sine and cosine functions is the set of real numbers R. The range of both the sine and cosine functions is $[-1, 1]$. This is the set of numbers assumed by b, for sine, and a, for cosine, as the circular point (a, b) moves around the unit circle. The domain of cosecant is the set of real numbers x such that b in $W(x) = (a, b)$ is not 0. Similar restrictions are made on the domains of the other three trigonometric functions. We will have more to say about the domains and ranges of all six trigonometric functions in subsequent sections.

Note from Definition 1 that csc x is the reciprocal of sin x, provided that $\sin x \neq 0$. Therefore sin x is the reciprocal of csc x. Similarly, cos x and sec x are reciprocals of each other, as are tan x and cot x. We call these useful facts the *reciprocal identities*.

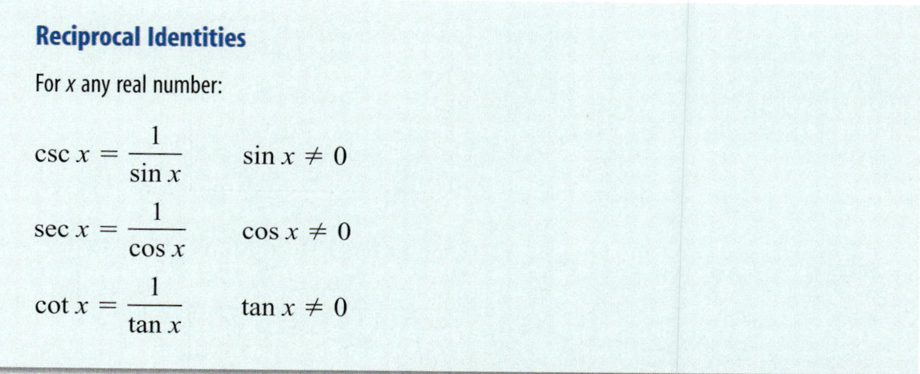

Reciprocal Identities

For x any real number:

$$\csc x = \frac{1}{\sin x} \qquad \sin x \neq 0$$

$$\sec x = \frac{1}{\cos x} \qquad \cos x \neq 0$$

$$\cot x = \frac{1}{\tan x} \qquad \tan x \neq 0$$

In Example 1 we were able to give a simple geometric argument to find, for example, that the coordinates of $W(7\pi/6)$ are $(-\sqrt{3}/2, -1/2)$. Therefore, $\sin(7\pi/6) = -1/2$ and $\cos(7\pi/6) = -\sqrt{3}/2$. These exact values correspond to the approximations given by a calculator [Fig. 9(a)]. For most values of x, however, simple geometric arguments fail to give the exact coordinates of $W(x)$. But a calculator, set in radian mode, can be used to give approximations. For example, if $x = \pi/7$, then $W(\pi/7) \approx (0.901, 0.434)$ [Fig. 9(b)].

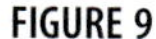
FIGURE 9

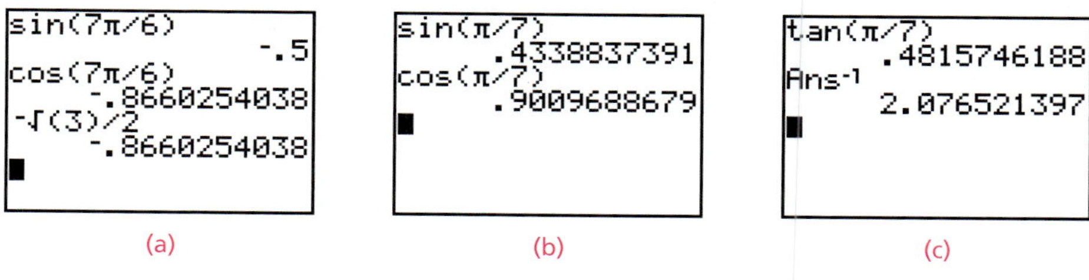

(a) (b) (c)

Most calculators have function keys for the sine, cosine, and tangent functions, but not for the cotangent, secant, and cosecant. Because the cotangent, secant, and cosecant are the reciprocals of the tangent, cosine, and sine, respectively, they can be evaluated easily. For example, $\cot(\pi/7) = 1/\tan(\pi/7) \approx 2.077$ [Fig. 9(c)]. Do not use the calculator function keys marked $\sin^{-1}$, $\cos^{-1}$, or $\tan^{-1}$ for this purpose—these keys are used to evaluate the inverse trigonometric functions of Section 5.6, not reciprocals.

EXAMPLE 2 Calculator Evaluation

Evaluate to four significant digits.

(A) tan 1.5

(B) csc (−6.27)

(C) sec $(11\pi/12)$

(D) The coordinates (a, b) of $W(1)$

SOLUTIONS

(A) tan 1.5 = 14.10

(B) csc (−6.27) = 1/sin (−6.27) = 75.84

```
sin(-6.27)
           .0131849251
Ans-1
           75.84419251
```

(C) sec $(11\pi/12)$ = 1/cos $(11\pi/12)$ = −1.035

```
cos(11π/12)
          -.9659258263
Ans-1
           -1.03527618
```

(D) $W(1)$ = (cos 1, sin 1) = (0.5403, 0.8415)

MATCHED 2 PROBLEM

Evaluate to four significant digits.

(A) cot (−8.25)

(B) sec $(7\pi/8)$

(C) csc (4.67)

(D) The coordinates (a, b) of $W(100)$

Graphs of the Trigonometric Functions

The graph of $y = \sin x$ is the set of all ordered pairs (x, y) of real numbers that satisfy the equation. Because $\sin x$, by Definition 1, is the second coordinate of the circular point $W(x)$, our knowledge of the coordinates of certain circular points (Table 1) gives the following solutions to $y = \sin x$: $(0, 0)$, $(\pi/2, 1)$, $(\pi, 0)$, and $(3\pi/2, -1)$.

TABLE 1

x	0	$\pi/2$	π	$3\pi/2$
$W(x)$	(0, **0**)	(0, **1**)	(−1, **0**)	(0, **−1**)
$\sin x$	**0**	**1**	**0**	**−1**

As x increases from 0 to $\pi/2$, the circular point $W(x)$ moves on the circumference of the unit circle from (0, 0) to (0, 1), and so $\sin x$ [the second coordinate of $W(x)$] increases from 0 to 1. Similarly, as x increases from $\pi/2$ to π, the circular point $W(x)$ moves on the circumference of the unit circle from (0, 1) to (−1, 0), and so $\sin x$ decreases from 1 to 0. These observations are in agreement with the graph of $y = \sin x$, obtained from a graphing calculator in radian mode [Fig. 10(a)].

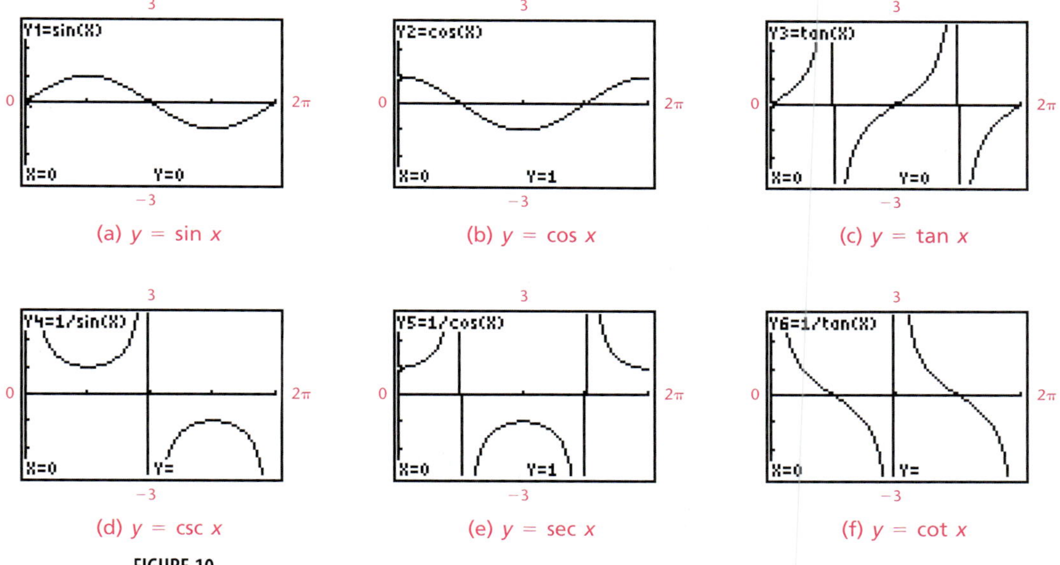

(a) $y = \sin x$ (b) $y = \cos x$ (c) $y = \tan x$

(d) $y = \csc x$ (e) $y = \sec x$ (f) $y = \cot x$

FIGURE 10

Figure 10 shows the graphs of all six trigonometric functions from $x = 0$ to $x = 2\pi$. The functions $y = \sin x$ and $y = \cos x$ are bounded; their maximum values are 1 and their minimum values are −1. The functions $y = \tan x$, $y = \cot x$, $y = \sec x$, and $y = \csc x$ are unbounded; they have vertical asymptotes at the values of x for which they are undefined. It is instructive to study and compare the graphs of reciprocal pairs, for example, $y = \cos x$ and $y = \sec x$. Note that $\sec x$ is undefined when $\cos x$ equals 0, and that because the maximum positive value of $\cos x$ is 1, the minimum positive value of $\sec x$ is 1. We will study the properties of trigonometric functions and their graphs in Section 5.4.

A graphing utility can be used to interactively explore the relationship between the unit circle definition of the sine function and the graph of the sine function. Explore/Discuss 3 provides the details.

EXPLORE/DISCUSS 3

Set your graphing utility in radian and parametric modes. Make the entries as indicated in Figure 11 to obtain the indicated graph (2π is entered as Tmax and Xmax, $\pi/2$ is entered as Xscl).

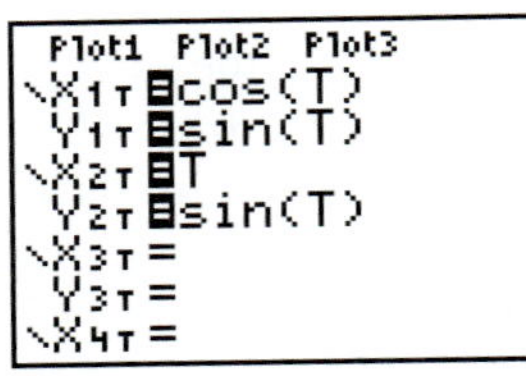

WINDOW
Tmin=0
Tmax=6.2831853…
Tstep=.1
Xmin=-1
Xmax=6.2831853…
Xscl=1.5707963…
Ymin=-2.4
Ymax=2.4
Yscl=1

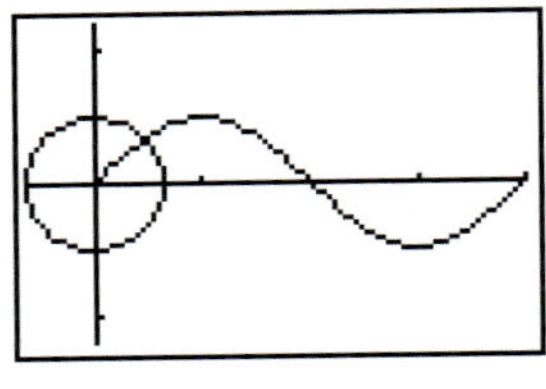

FIGURE 11

Use TRACE and move back and forth between the unit circle and the graph of the sine function for various values of T as T increases from 0 to 2π. Discuss what happens in each case. Figure 12 illustrates the case for T = 0.

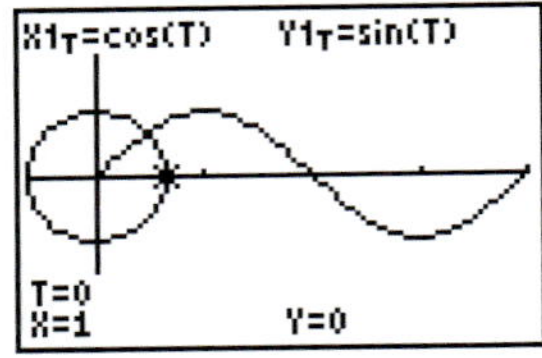

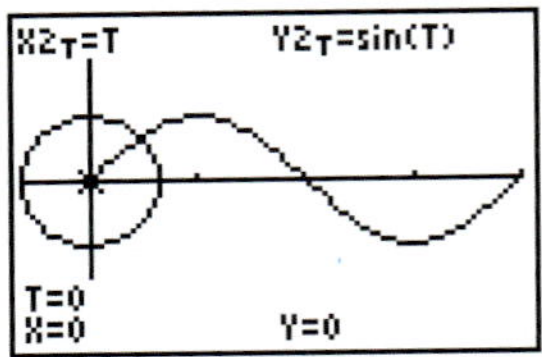

FIGURE 12

Repeat the exploration with $Y_{2T} = \cos(T)$

EXAMPLE 3 Zeros and Turning Points

Find the zeros and turning points of $y = \cos x$ on the interval $[-\pi/2, 3\pi/2]$.

SOLUTION

Recall that a *turning point* is a point on a graph that separates an increasing portion from a decreasing portion, or vice versa. A visual inspection of the graph of $y = \cos x$ [Fig. 13(a)] suggests that $(0, 1)$ and $(\pi, -1)$ are turning points, and that $-\pi/2$, $\pi/2$, and $3\pi/2$ are zeros. These observations are confirmed by noting that as x increases from $-\pi/2$ to $3\pi/2$, the first coordinate of the circular point $W(x)$ (that is, $\cos x$) has a maximum value of 1 (when $x = 0$), a minimum value of -1 (when $x = \pi$), and has the value 0 when $x = -\pi/2$, $\pi/2$, and $3\pi/2$ [Fig. 13(b)].

FIGURE 13

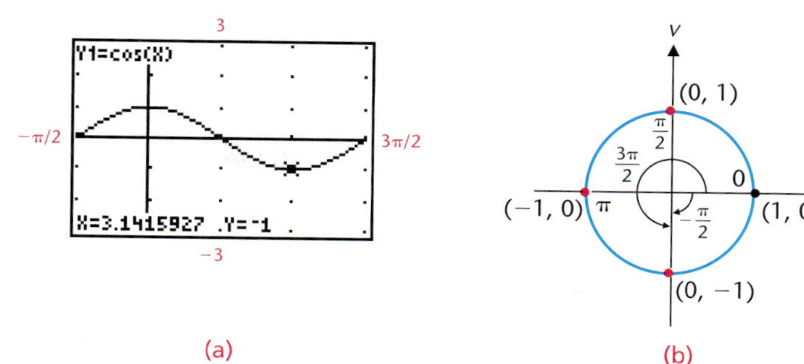

(a) (b)

MATCHED 3 PROBLEM

Find all zeros and turning points of $y = \csc x$ on the interval $(0, 4\pi)$.

EXAMPLE 4 Solving a Trigonometric Equation

Find all solutions of the equation $\sin x = 0.35x + 0.1$ to three decimal places.

SOLUTION

Graph $y_1 = \sin x$ and $y_2 = 0.35x + 0.1$ and use the intersect command (Fig. 14). The solutions are $x = -2.339$, 0.155, and 2.132.

FIGURE 14

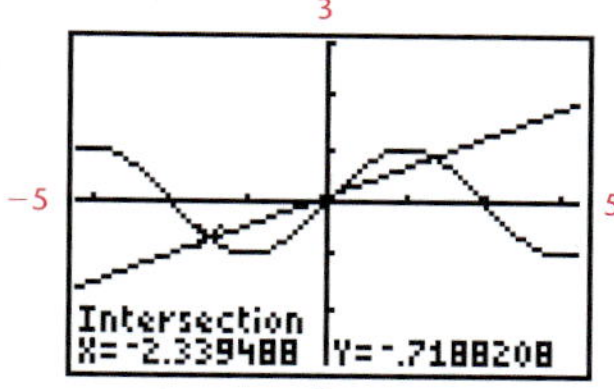

MATCHED 4 PROBLEM

Find all solutions of the equation $\cot x = x$ on the interval $(0, 2\pi)$ to three decimal places.

CAUTION

A common cause of error is to forget to set a calculator in the correct mode, degree or radian, before graphing or evaluating a function. In radian mode, a calculator will give 1 as the value of $\sin(\pi/2)$; in degree mode, it will give 0.0274 as the value of $\sin(\pi/2)$ [because $(\pi/2)° \approx 1.5708°$].

ANSWERS MATCHED PROBLEMS

1. (A) $(-1, 0)$
   (B) $(0, 1)$
   (C) $(-\sqrt{3}/2, 1/2)$
   (D) $(1/2, -\sqrt{3}/2)$
   (E) $(-1/\sqrt{2}, -1/\sqrt{2})$
2. (A) 0.4181
   (B) −1.082
   (C) −1.001
   (D) (0.8623, −0.5064)
3. Zeros: none; turning points: $(\pi/2, 1)$, $(3\pi/2, -1)$, $(5\pi/2, 1)$, $(7\pi/2, -1)$
4. 0.860, 3.426

EXERCISE 5.2

A

In Problems 1–16, find the coordinates of each circular point.

1. $W(3\pi/2)$
2. $W(-5\pi)$
3. $W(-6\pi)$
4. $W(-15\pi/2)$
5. $W(\pi/4)$
6. $W(\pi/3)$
7. $W(\pi/6)$
8. $W(-\pi/6)$
9. $W(-\pi/3)$
10. $W(-\pi/4)$
11. $W(2\pi/3)$
12. $W(11\pi/6)$
13. $W(-3\pi/4)$
14. $W(-7\pi/6)$
15. $W(13\pi/4)$
16. $W(-10\pi/3)$

In Problems 17–32, use your answers to Problems 1–16 to give the exact value of the expression (if it exists).

17. $\sin(3\pi/2)$
18. $\tan(-5\pi)$
19. $\cos(-6\pi)$
20. $\cot(-15\pi/2)$
21. $\sec(\pi/4)$
22. $\csc(\pi/3)$
23. $\tan(\pi/6)$
24. $\cos(-\pi/6)$
25. $\sin(-\pi/3)$
26. $\sec(-\pi/4)$
27. $\csc(2\pi/3)$
28. $\cot(11\pi/6)$
29. $\cos(-3\pi/4)$
30. $\tan(-7\pi/6)$
31. $\cot(13\pi/4)$
32. $\sin(-10\pi/3)$

In Problems 33–38, in which quadrants must W(x) lie so that:

33. $\cos x < 0$
34. $\tan x > 0$
35. $\sin x > 0$
36. $\sec x > 0$
37. $\cot x < 0$
38. $\csc x < 0$

Evaluate Problems 39–48 to four significant digits using a calculator set in radian mode.

39. $\cos 2.288$
40. $\sin 3.104$
41. $\tan(-4.644)$
42. $\sec(-1.555)$
43. $\csc 1.571$
44. $\cot 0.7854$
45. $\sin(\cos 0.3157)$
46. $\cos(\tan 5.183)$
47. $\cos[\csc(-1.408)]$
48. $\sec[\cot(-3.566)]$

Evaluate Problems 49–58 to four significant digits using a calculator. Make sure your calculator is in the correct mode (degree or radian) for each problem.

49. $\sin 25°$
50. $\tan 89°$
51. $\cot 12$
52. $\csc 13$
53. $\sin 2.137$
54. $\tan 4.327$
55. $\cot(-431.41°)$
56. $\sec(-247.39°)$
57. $\sin 113°27'13''$
58. $\cos 235°12'47''$

B

In Problems 59–62, find all zeros and turning points of each function on $[0, 4\pi]$.

59. $y = \sec x$
60. $y = \sin x$
61. $y = \tan x$
62. $y = \cot x$

In Problems 63–66, find all solutions of the trigonometric equation on the interval $[0, 2\pi]$ to three decimal places.

63. $\cos(2x) = x - 2.5$
64. $\tan x = 10 - x^2$
65. $\sec x = x^2 - 5$
66. $\sin(x/2) = \cos x$

Determine the signs of a and b for the coordinates (a, b) of each circular point indicated in Problems 67–76. First determine the quadrant in which each circular point lies. [Note: $\pi/2 \approx 1.57$, $\pi \approx 3.14$, $3\pi/2 \approx 4.71$, and $2\pi \approx 6.28$.]

67. $W(2)$
68. $W(1)$
69. $W(3)$
70. $W(4)$
71. $W(5)$
72. $W(7)$
73. $W(-2.5)$
74. $W(-4.5)$
75. $W(-6.1)$
76. $W(-1.8)$

In Problems 77–80, for each equation find all solutions for $0 \le x < 2\pi$, then write an expression that represents all solutions for the equation without any restrictions on x.

77. $W(x) = (1, 0)$ **78.** $W(x) = (-1, 0)$

79. $W(x) = (-1/\sqrt{2}, 1/\sqrt{2})$ **80.** $W(x) = (1/\sqrt{2}, -1/\sqrt{2})$

81. Describe in words why $W(x) = W(x + 4\pi)$ for every real number x.

82. Describe in words why $W(x) = W(x - 6\pi)$ for every real number x.

C

If $W(x) = (a, b)$, indicate whether the statements in Problems 83–88 are true (T) or false (F). Sketching figures should help you decide.

83. $W(x + \pi) = (-a, -b)$ **84.** $W(x + \pi) = (a, b)$

85. $W(-x) = (-a, b)$ **86.** $W(-x) = (a, -b)$

87. $W(x + 2\pi) = (a, b)$ **88.** $W(x + 2\pi) = (-a, -b)$

In Problems 89–92, find the value of each to one significant digit. Use only the accompanying figure, Definition 1, and a calculator as necessary for multiplication and division. Check your results by evaluating each directly on a calculator.

89. **(A)** $\sin 0.4$ **(B)** $\cos 0.4$ **(C)** $\tan 0.4$

90. **(A)** $\sin 0.8$ **(B)** $\cos 0.8$ **(C)** $\cot 0.8$

91. **(A)** $\sec 2.2$ **(B)** $\tan 5.9$ **(C)** $\cot 3.8$

92. **(A)** $\csc 2.5$ **(B)** $\cot 5.6$ **(C)** $\tan 4.3$

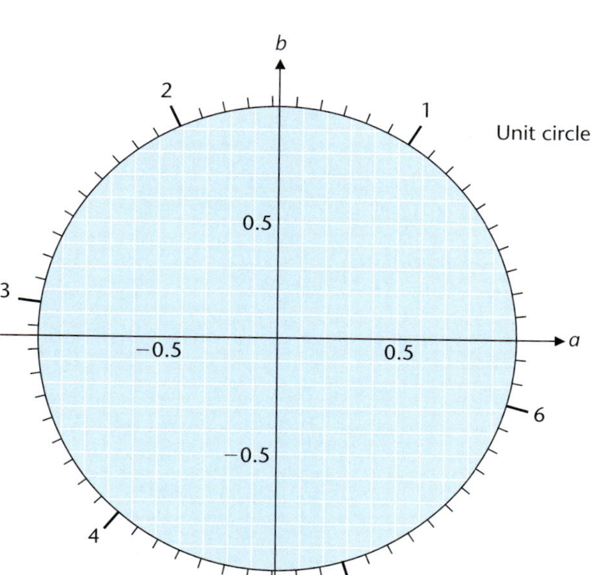

In Problems 93–96, in which quadrants are the statements true and why?

93. $\sin x < 0$ and $\cot x < 0$ **94.** $\cos x > 0$ and $\tan x < 0$

95. $\cos x < 0$ and $\sec x > 0$ **96.** $\sin x > 0$ and $\csc x < 0$

For which values of x, $0 \le x \le 2\pi$, is each of Problems 97–102 not defined?

97. $\cos x$ **98.** $\sin x$ **99.** $\tan x$

100. $\cot x$ **101.** $\sec x$ **102.** $\csc x$

APPLICATIONS

If an n-sided regular polygon is inscribed in a circle of radius r, then it can be shown that the area of the polygon is given by

$$A = \frac{1}{2}nr^2 \sin \frac{2\pi}{n}$$

Compute each area exactly and then to four significant digits using a calculator if the area is not an integer.

103. $n = 12, r = 5$ meters

104. $n = 4, r = 3$ inches

105. $n = 3, r = 4$ inches

106. $n = 8, r = 10$ centimeters

Approximating π. *Problems 107 and 108 refer to a sequence of numbers generated as follows:*

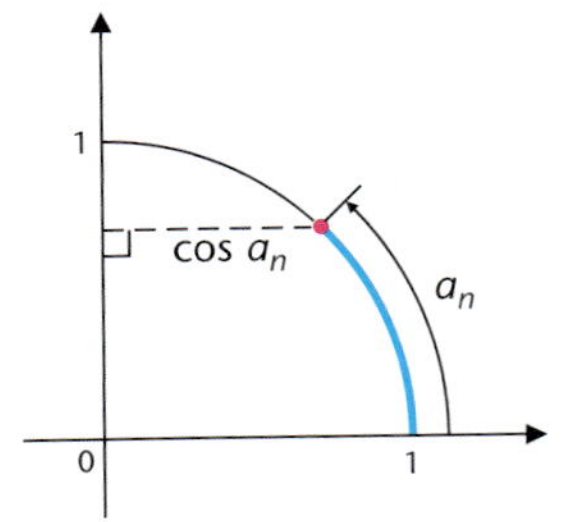

$$a_1$$
$$a_2 = a_1 + \cos a_1$$
$$a_3 = a_2 + \cos a_2$$
$$\vdots$$
$$a_{n+1} = a_n + \cos a_n$$

107. Let $a_1 = 0.5$, and compute the first five terms of the sequence to six decimal places and compare the fifth term with $\pi/2$ computed to six decimal places.

108. Repeat Problem 107, starting with $a_1 = 1$.

SECTION 5.3 Solving Right Triangles*

FIGURE 1

A **right triangle** is a triangle with one 90° angle (Fig. 1).

If only the angles of a right triangle are known, it is impossible to solve for the sides. (Why?) But if we are given two sides, or one acute angle and a side, then it is possible to solve for the remaining three quantities. This process is called **solving the right triangle.** We use the trigonometric functions to solve right triangles.

If a right triangle is located in the first quadrant as indicated by Figure 2, then, by similar triangles, the coordinates of the circular point Q are $(a/c, b/c)$.

FIGURE 2

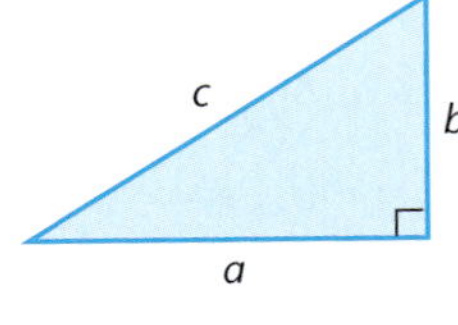

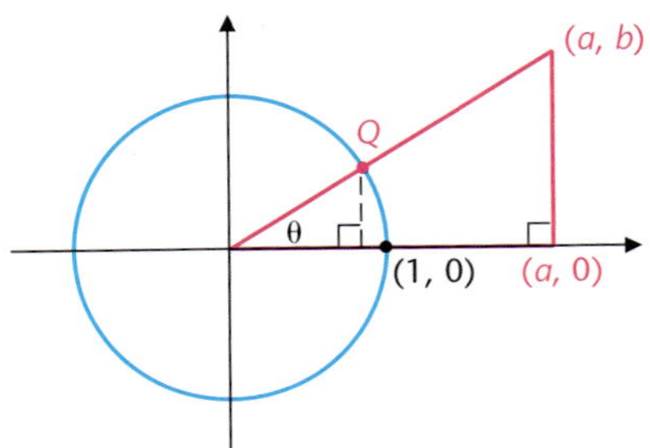

*This section provides a significant application of trigonometric functions to real-world problems. However, it may be postponed or omitted without loss of continuity, if desired. Some may want to cover the section just before Sections 7.1 and 7.2.

Therefore, using the definition of the trigonometric functions, $\sin\theta = b/c$ and $\cos\theta = a/c$. (Calculations using such trigonometric ratios are valid if θ is measured in either degrees or radians, provided your calculator is set in the correct mode—in this section we use degree measure.) All six trigonometric ratios are displayed in the box.

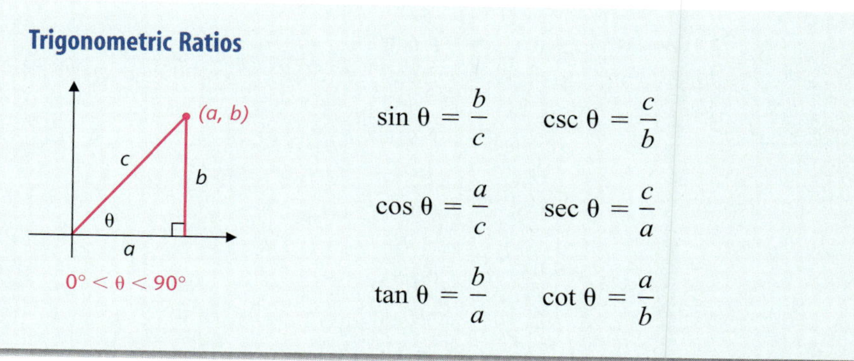

Side b is often referred to as the **side opposite** angle θ, a as the **side adjacent** to angle θ, and c as the **hypotenuse.** Using these designations for an arbitrary right triangle removed from a coordinate system, we have the following:

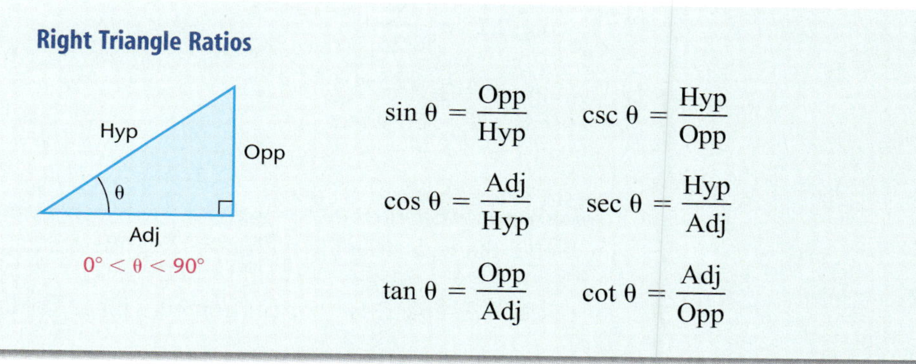

EXPLORE/DISCUSS 1

For a given value θ, $0 < \theta < 90°$, explain why the value of each of the six trigonometric functions is independent of the size of the right triangle that contains θ.

TABLE 1

Angle to Nearest	Significant Digits for Side Measure
1°	2
10′ or 0.1°	3
1′ or 0.01°	4
10″ or 0.001°	5

The use of the trigonometric ratios for right triangles is made clear in Examples 1 through 4. Regarding computational accuracy, we use Table 1 as a guide. (The table is also printed inside the front cover of this book for easy reference.) We will use = rather than ≈ in many places, realizing the accuracy indicated in Table 1 is all that is assumed. Another word of caution: when using your calculator be sure it is set in degree mode.

EXAMPLE 1 Right Triangle Solution

Solve the right triangle with $c = 6.25$ feet and $\beta = 32.2°$.

SOLUTION

First draw a figure and label the parts (Fig. 3):

FIGURE 3

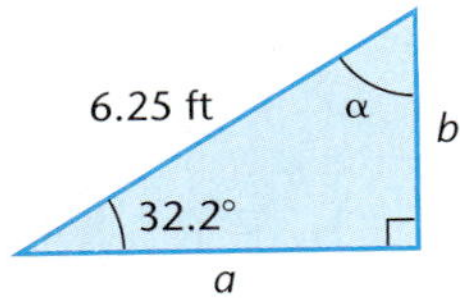

SOLVE FOR α

$\alpha = 90° - 32.2° = 57.8°$ α and β are complementary.

SOLVE FOR *b*

$$\sin \beta = \frac{b}{c}$$

Or use $\csc \beta = \frac{c}{b}$.

$$\sin 32.2° = \frac{b}{6.25}$$

$$b = 6.25 \sin 32.2°$$

$$= 3.33 \text{ feet}$$

SOLVE FOR *a*

$$\cos \beta = \frac{a}{c}$$

Or use $\sec \beta = \frac{c}{a}$.

$$\cos 32.2° = \frac{a}{6.25}$$

$$a = 6.25 \cos 32.2°$$

$$= 5.29 \text{ feet}$$

MATCHED PROBLEM 1

Solve the right triangle with $c = 27.3$ meters and $\alpha = 47.8°$.

In Example 2 we are confronted with a problem of the type: Find θ given

$$\sin \theta = 0.4196$$

We know how to find (or approximate) sin θ given θ, but how do we reverse the process? How do we find θ given sin θ? First, we note that the solution to the problem can be written symbolically as either

$$\theta = \arcsin 0.4196$$

or

"arcsin" and "$\sin^{-1}$" both represent the same thing.

$$\theta = \sin^{-1} 0.4196$$

Both expressions are read "θ is the angle whose sine is 0.4196."

CAUTION

It is important to note that $\sin^{-1} 0.4196$ does not mean $1/(\sin 0.4196)$. The superscript $^{-1}$ is part of a function symbol, and **$\sin^{-1}$ represents the inverse sine function.** Inverse trigonometric functions are developed in detail in Section 5.6.

Fortunately, we can find θ directly using a calculator. Most calculators of the type used in this book have the function keys $\boxed{\sin^{-1}}$, $\boxed{\cos^{-1}}$, and $\boxed{\tan^{-1}}$ or their equivalents (check your manual). These function keys take us from a trigonometric ratio back to the corresponding acute angle in degree measure when the calculator is in degree mode. Thus, if $\sin\theta = 0.4196$, then we can write $\theta = \arcsin 0.4196$ or $\theta = \sin^{-1} 0.4196$. We choose the latter and proceed as follows:

$$\theta = \sin^{-1} 0.4196$$
$$= 24.81° \quad \text{To the nearest hundredth degree}$$
$$\text{or } 24°49' \quad \text{To the nearest minute}$$

CHECK
$\sin 24.81° = 0.4196$

EXPLORE/DISCUSS 2

Solve each of the following for θ to the nearest hundredth of a degree using a calculator. Explain why an error message occurs in one of the problems.

(A) $\cos\theta = 0.2044$ (B) $\tan\theta = 1.4138$
(C) $\sin\theta = 1.4138$

EXAMPLE 2 Right Triangle Solution

Solve the right triangle with $a = 4.32$ centimeters and $b = 2.62$ centimeters. Compute the angle measures to the nearest 10′.

SOLUTION
Draw a figure and label the known parts (Fig. 4):

FIGURE 4

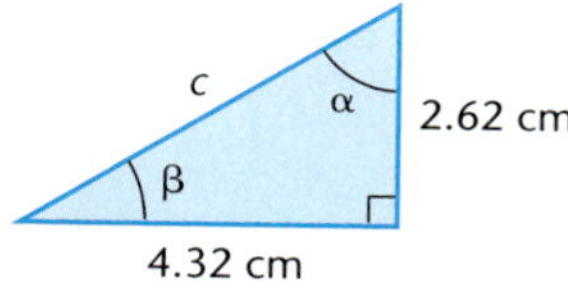

SOLVE FOR β

$$\tan\beta = \frac{2.62}{4.32}$$

$$\beta = \tan^{-1}\frac{2.62}{4.32}$$

$$= 31.2° \text{ or } 31°10' \quad 0.2° = [(0.2)(60)]' = 12' \approx 10' \text{ to nearest } 10'$$

SOLVE FOR α

$$\alpha = 90° - 31°10' = 89°60' - 31°10' = 58°50'$$

SOLVE FOR c

$$\sin\beta = \frac{2.62}{c}$$ Or use $\csc\beta = \frac{c}{2.62}$.

$$c = \frac{2.62}{\sin 31.2°} = 5.06 \text{ centimeters}$$

or, using the Pythagorean theorem,

$$c = \sqrt{4.32^2 + 2.62^2} = 5.05 \text{ centimeters}$$

Note the slight difference in the values obtained for c (5.05 versus 5.06). This was caused by rounding β to the nearest 10′ in the first calculation for c.

MATCHED 2 PROBLEM

Solve the right triangle with $a = 1.38$ kilometers and $b = 6.73$ kilometers.

EXAMPLE 3 Geometry

If a regular pentagon (a five-sided regular polygon) is inscribed in a circle of radius 5.35 centimeters, find the length of one side of the pentagon.

SOLUTION

Sketch a figure and insert triangle ACB with C at the center (Fig. 5). Add the auxiliary line CD as indicated. We will find AD and double it to find the length of the side wanted.

FIGURE 5

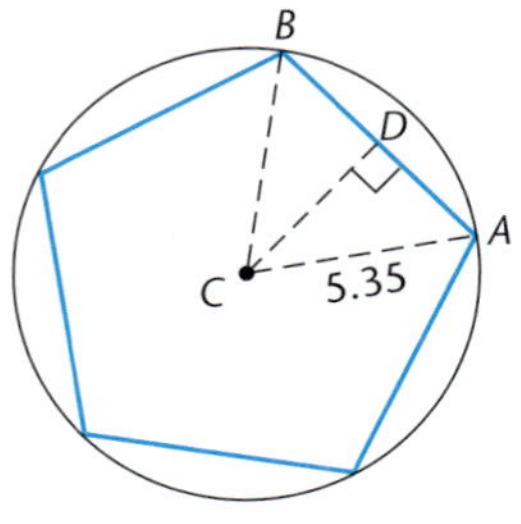

$$\text{Angle } ACB = \frac{360°}{5} = 72° \quad \text{Exact}$$

$$\text{Angle } ACD = \frac{72°}{2} = 36° \quad \text{Exact}$$

$$\sin(\text{angle } ACD) = \frac{AD}{AC}$$

$$\begin{aligned} AD &= AC \sin(\text{angle } ACD) \\ &= 5.35 \sin 36° \\ &= 3.14 \text{ centimeters} \\ AB &= 2AD = 6.28 \text{ centimeters} \end{aligned}$$

MATCHED 3 PROBLEM

If a square of side 43.6 meters is inscribed in a circle, what is the radius of the circle?

EXAMPLE 4 Architecture

In designing a house an architect wishes to determine the amount of overhang of a roof so that it shades the entire south wall at noon during the summer solstice (Fig. 6). Minimally, how much overhang should be provided for this purpose?

FIGURE 6

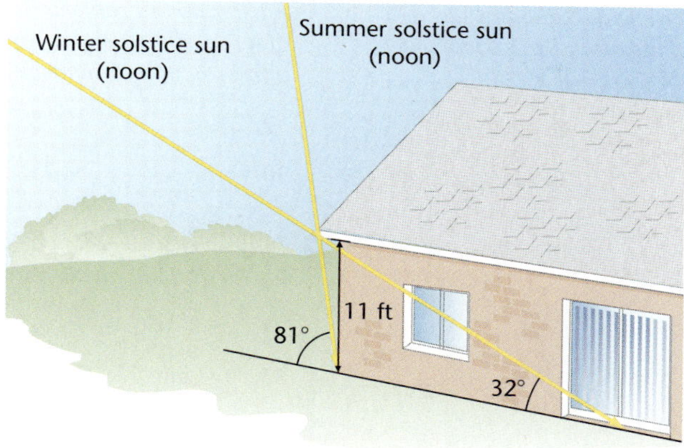

SOLUTION

From the figure we draw the following related right triangle (Fig. 7) and solve for x:

FIGURE 7

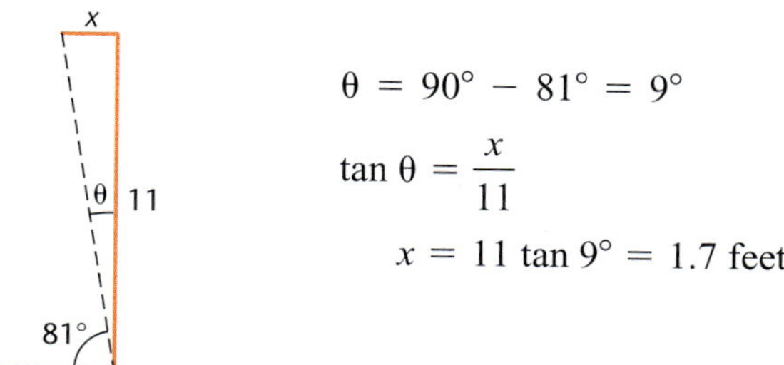

$$\theta = 90° - 81° = 9°$$

$$\tan \theta = \frac{x}{11}$$

$$x = 11 \tan 9° = 1.7 \text{ feet}$$

MATCHED 4 PROBLEM

With the overhang found in Example 4, how far will the shadow of the overhang come down the wall at noon during the winter solstice?

ANSWERS MATCHED PROBLEMS

1. $\beta = 42.2°$, $a = 20.2$ meters, $b = 18.3$ meters
2. $\alpha = 11°40'$, $\beta = 78°20'$, $c = 6.87$ kilometers
3. 30.8 meters
4. 1.1 feet

EXERCISE 5.3

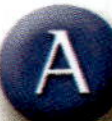

In Problems 1–6, use the figure to write the ratio of sides that corresponds to each trigonometric function. Do not look back at the definitions.

1. $\sin\theta$ **2.** $\cot\theta$ **3.** $\csc\theta$

4. $\cos\theta$ **5.** $\tan\theta$ **6.** $\sec\theta$

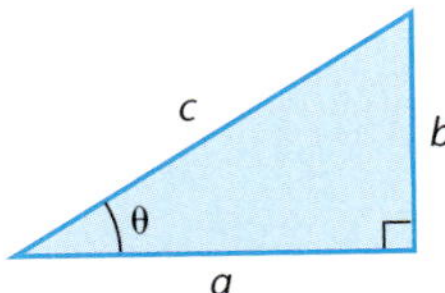

Figure for Problems 1–6

Each ratio in Problems 7–12 defines a trigonometric function of θ (refer to the figure for Problems 1–6). Indicate which function without looking back at the definitions.

7. a/c **8.** b/a **9.** c/a

10. b/c **11.** a/b **12.** c/b

In Problems 13–18, find each acute angle θ in degree measure to two decimal places using a calculator.

13. $\cos\theta = 0.4917$ **14.** $\sin\theta = 0.0859$

15. $\theta = \tan^{-1} 8.031$ **16.** $\theta = \cos^{-1} 0.5097$

17. $\sin\theta = 0.6031$ **18.** $\tan\theta = 1.993$

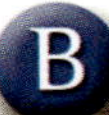

In Problems 19–30, use the figure and the given information to solve each triangle.

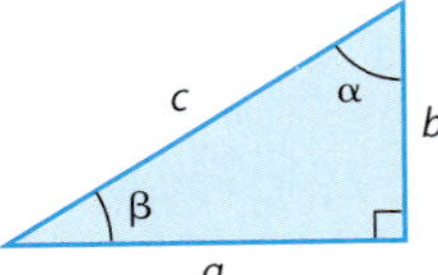

19. $\beta = 17.8°, c = 3.45$ **20.** $\beta = 33.7°, b = 22.4$

21. $\beta = 43°20', a = 123$ **22.** $\beta = 62°30', c = 42.5$

23. $\alpha = 23°0', a = 54.0$ **24.** $\alpha = 54°, c = 4.3$

25. $\alpha = 53.21°, b = 23.82$ **26.** $\alpha = 35.73°, b = 6.482$

27. $a = 6.00, b = 8.46$ **28.** $a = 22.0, b = 46.2$

29. $b = 10.0, c = 12.6$ **30.** $b = 50.0, c = 165$

In Problems 31–36, determine whether the statement is true or false. If true, explain why. If false, give a counterexample.

31. If any two angles of a right triangle are known, then it is possible to solve for the remaining angle and the three sides.

32. If any two sides of a right triangle are known, then it is possible to solve for the remaining side and the three angles.

33. If α and β are the acute angles of a right triangle, then $\sin\alpha = \cos\beta$.

34. If α and β are the acute angles of a right triangle, then $\tan\alpha = \cot\beta$.

35. If α and β are the acute angles of a right triangle, then $\sec\alpha = \cos\beta$.

36. If α and β are the acute angles of a right triangle, then $\csc\alpha = \sec\beta$.

Problems 37–42 give a ***geometric interpretation of the trigonometric ratios.*** *Refer to the figure, where O is the center of a circle of radius 1, θ is the acute angle AOD, D is the intersection point of the terminal side of angle θ with the circle, and EC is tangent to the circle at D.*

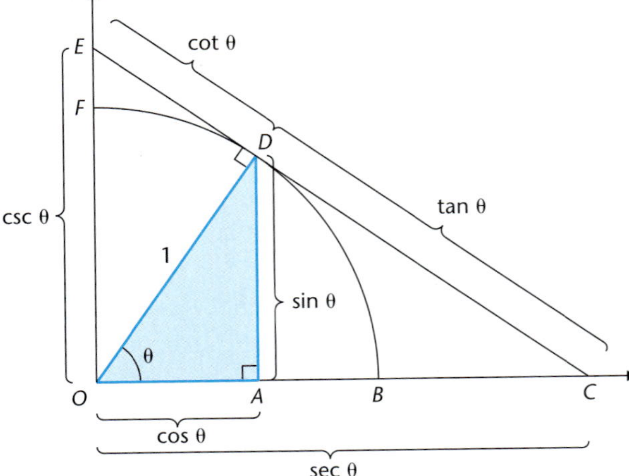

37. Explain why

(A) $\cos\theta = OA$ **(B)** $\cot\theta = DE$ **(C)** $\sec\theta = OC$

38. Explain why

(A) $\sin\theta = AD$ **(B)** $\tan\theta = DC$ **(C)** $\csc\theta = OE$

39. Explain what happens to each of the following as the acute angle θ approaches 90°.

(A) $\cos\theta$ **(B)** $\cot\theta$ **(C)** $\sec\theta$

40. Explain what happens to each of the following as the acute angle θ approaches 90°.

(A) $\sin\theta$ **(B)** $\tan\theta$ **(C)** $\csc\theta$

41. Explain what happens to each of the following as the acute angle θ approaches 0°.

(A) $\sin\theta$ **(B)** $\tan\theta$ **(C)** $\csc\theta$

42. Explain what happens to each of the following as the acute angle θ approaches 0°.

(A) $\cos\theta$ **(B)** $\cot\theta$ **(C)** $\sec\theta$

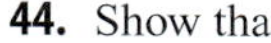

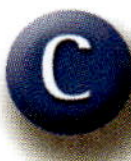

43. Show that

$$h = \frac{d}{\cot\alpha - \cot\beta}$$

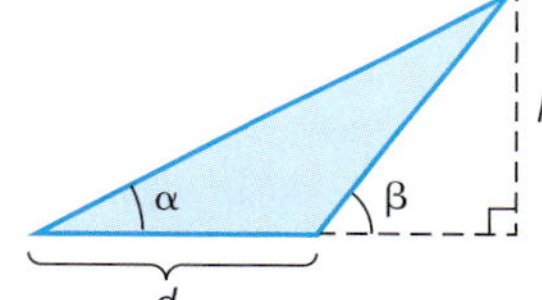

44. Show that

$$h = \frac{d}{\cot\alpha + \cot\beta}$$

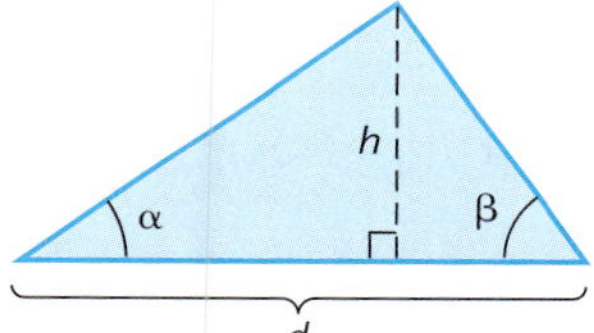

APPLICATIONS

45. Surveying. Find the height of a tree (growing on level ground) if at a point 105 feet from the base of the tree the angle to its top relative to the horizontal is found to be 65.3°.

46. Air Safety. To measure the height of a cloud ceiling over an airport, a searchlight is directed straight upward to produce a lighted spot on the clouds. Five hundred meters away an observer reports the angle of the spot relative to the horizontal to be 32.2°. How high (to the nearest meter) are the clouds above the airport?

47. Engineering. If a train climbs at a constant angle of 1°23′, how many vertical feet has it climbed after going 1 mile? (1 mile = 5,280 feet)

48. Air Safety. If a jet airliner climbs at an angle of 15°30′ with a constant speed of 315 miles per hour, how long will it take (to the nearest minute) to reach an altitude of 8.00 miles? Assume there is no wind.

★ **49. Astronomy.** Find the diameter of the moon (to the nearest mile) if at 239,000 miles from Earth it produces an angle of 32′ relative to an observer on Earth.

★ **50. Astronomy.** If the sun is 93,000,000 miles from Earth and its diameter is opposite an angle of 32′ relative to an observer on Earth, what is the diameter of the sun (to two significant digits)?

★ **51. Geometry.** If a circle of radius 4 centimeters has a chord of length 3 centimeters, find the central angle that is opposite this chord (to the nearest degree).

★ **52. Geometry.** Find the length of one side of a nine-sided regular polygon inscribed in a circle of radius 4.06 inches.

53. Physics. In a course in physics it is shown that the velocity v of a ball rolling down an inclined plane (neglecting air resistance and friction) is given by

$$v = gt\sin\theta$$

where g is a gravitational constant (acceleration due to gravity), t is time, and θ is the angle of inclination of the plane (see the figure on page 416). Galileo (1564–1642) used this equation in the form

$$g = \frac{v}{t\sin\theta}$$

to estimate g after measuring v experimentally. (At that time, no timing devices existed to measure the velocity of a free-falling body, so Galileo used the inclined plane to slow the motion down.) A steel ball is rolled down a glass plane inclined at 8.0°. Approximate g to one decimal place if at the end of 3.0 seconds the ball has a measured velocity of 4.2 meters per second.

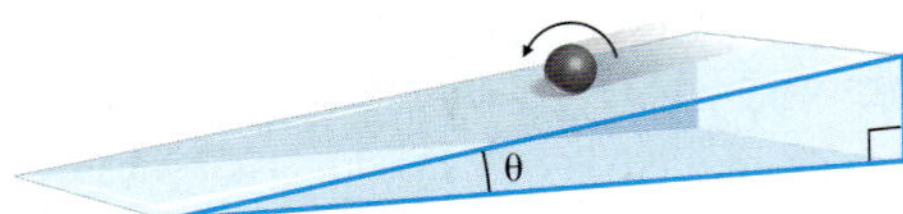

54. Physics. Refer to Problem 53. A steel ball is rolled down a glass plane inclined at 4.0°. Approximate g to one decimal place if at the end of 4.0 seconds the ball has a measured velocity of 9.0 feet per second.

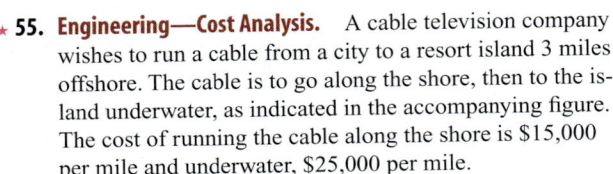

★ **55. Engineering—Cost Analysis.** A cable television company wishes to run a cable from a city to a resort island 3 miles offshore. The cable is to go along the shore, then to the island underwater, as indicated in the accompanying figure. The cost of running the cable along the shore is \$15,000 per mile and underwater, \$25,000 per mile.

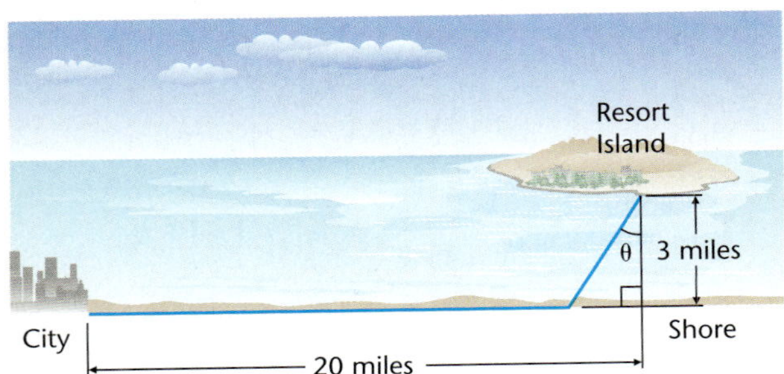

(A) Referring to the figure, show that the cost in terms of θ is given by

$$C(\theta) = 75{,}000 \sec\theta - 45{,}000 \tan\theta + 300{,}000$$

(B) Calculate a table of costs, each cost to the nearest dollar, for the following values of θ: 10°, 20°, 30°, 40°, and 50°. (Notice how the costs vary with θ. In a course in calculus, students are asked to find θ so that the cost is minimized.)

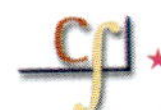

★ **56. Engineering—Cost Analysis.** Refer to Problem 55. Suppose the island is 4 miles offshore and the cost of running the cable along the shore is \$20,000 per mile and underwater, \$30,000 per mile.

(A) Referring to the figure for Problem 55 with appropriate changes, show that the cost in terms of θ is given by

$$C(\theta) = 120{,}000 \sec\theta - 80{,}000 \tan\theta + 400{,}000$$

(B) Calculate a table of costs, each cost to the nearest dollar, for the following values of θ: 10°, 20°, 30°, 40°, and 50°.

★★ **57. Geometry.** Find r in the accompanying figure (to two significant digits) so that the circle is tangent to all three sides of the isosceles triangle. [*Hint:* The radius of a circle is perpendicular to a tangent line at the point of tangency.]

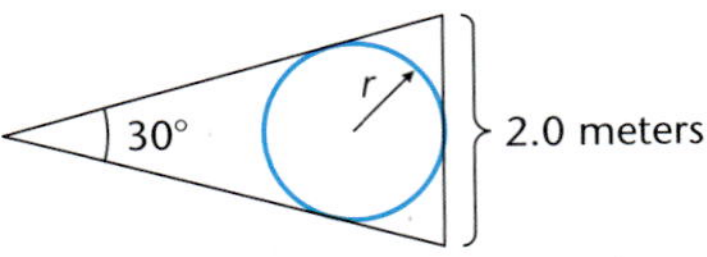

★★ **58. Geometry.** Find r in the accompanying figure (to two significant digits) so that the smaller circle is tangent to the larger circle and the two sides of the angle. [See the hint in Problem 57.]

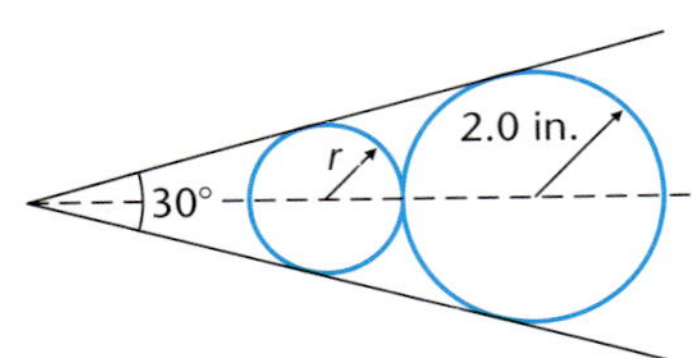

SECTION 5.4 Properties of Trigonometric Functions

Basic Identities • Sign Properties • Periodic Functions • Reference Triangles

In Section 5.4 we study properties of the trigonometric functions that distinguish them from the polynomial, rational, exponential, and logarithmic functions. The trigonometric functions are *periodic*, and as a consequence, have infinitely many zeros, or infinitely many turning points, or both.

Basic Identities

The definition of trigonometric functions provides several useful relationships among these functions. For convenience, we restate that definition.

DEFINITION 1

Trigonometric Functions

Let x be a real number and let (a, b) be the coordinates of the circular point $W(x)$ that lies on the terminal side of the angle with radian measure x. Then:

$$\sin x = b \qquad \csc x = \frac{1}{b} \quad b \neq 0$$

$$\cos x = a \qquad \sec x = \frac{1}{a} \quad a \neq 0$$

$$\tan x = \frac{b}{a} \quad a \neq 0 \qquad \cot x = \frac{a}{b} \quad b \neq 0$$

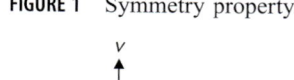

Because $\sin x = b$ and $\cos x = a$, we obtain the following equations:

$$\csc x = \frac{1}{b} = \frac{1}{\sin x} \tag{1}$$

$$\sec x = \frac{1}{a} = \frac{1}{\cos x} \tag{2}$$

$$\cot x = \frac{a}{b} = \frac{1}{b/a} = \frac{1}{\tan x} \tag{3}$$

$$\tan x = \frac{b}{a} = \frac{\sin x}{\cos x} \tag{4}$$

$$\cot x = \frac{a}{b} = \frac{\cos x}{\sin x} \tag{5}$$

Because the circular points $W(x)$ and $W(-x)$ are symmetrical with respect to the horizontal axis (Fig. 1), we have the following sign properties:

FIGURE 1 Symmetry property.

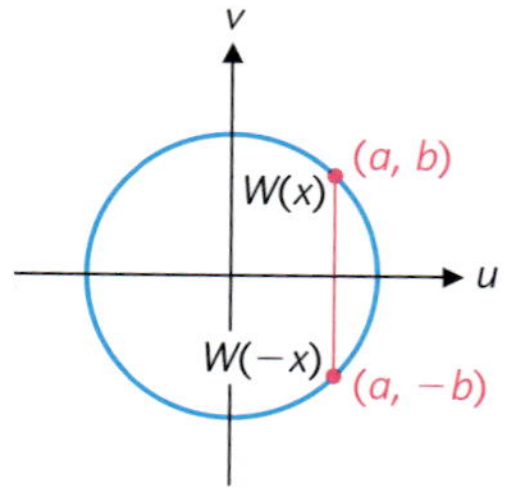

$$\sin(-x) = -b = -\sin x \tag{6}$$

$$\cos(-x) = a = \cos x \tag{7}$$

$$\tan(-x) = \frac{-b}{a} = -\frac{b}{a} = -\tan x \tag{8}$$

Finally, because $(a, b) = (\cos x, \sin x)$ is on the unit circle $u^2 + v^2 = 1$, it follows that

$$(\cos x)^2 + (\sin x)^2 = 1$$

which is usually written as

$$\sin^2 x + \cos^2 x = 1 \tag{9}$$

where $\sin^2 x$ and $\cos^2 x$ are concise ways of writing $(\sin x)^2$ and $(\cos x)^2$, respectively.

CAUTION

$(\sin x)^2 \neq \sin x^2$

$(\cos x)^2 \neq \cos x^2$

Equations (1)–(9) are called **basic identities.** They hold true for all replacements of x by real numbers for which both sides of an equation are defined. These basic identities must be memorized along with the definitions of the six trigonometric functions, because the material is used extensively in developments that follow. Note that most of Chapter 6 is devoted to trigonometric identities.

We summarize the basic identities for convenient reference in Theorem 1.

THEOREM 1

Basic Trigonometric Identities

For x any real number (in all cases restricted so that both sides of an equation are defined),

Reciprocal identities

(1) $\csc x = \dfrac{1}{\sin x}$ (2) $\sec x = \dfrac{1}{\cos x}$ (3) $\cot x = \dfrac{1}{\tan x}$

Quotient identities

(4) $\tan x = \dfrac{\sin x}{\cos x}$ (5) $\cot x = \dfrac{\cos x}{\sin x}$

Identities for negatives

(6) $\sin(-x) = -\sin x$ (7) $\cos(-x) = \cos x$ (8) $\tan(-x) = -\tan x$

Pythagorean identity

(9) $\sin^2 x + \cos^2 x = 1$

EXAMPLE 1 Using Basic Identities

Use the basic identities to find the values of the other five trigonometric functions given $\sin x = -\frac{1}{2}$ and $\tan x > 0$.

SOLUTION

We first note that the circular point $W(x)$ is in quadrant III, because that is the only quadrant in which $\sin x < 0$ and $\tan x > 0$. We next find $\cos x$ using identity (9):

$$\sin^2 x + \cos^2 x = 1 \qquad \text{Pythagorean identity (9)}$$

$$\left(-\tfrac{1}{2}\right)^2 + \cos^2 x = 1$$

$$\cos^2 x = \tfrac{3}{4}$$

$$\cos x = -\frac{\sqrt{3}}{2} \qquad \text{Because } W(x) \text{ is in quadrant III.}$$

Now, because we have values for sin x and cos x, we can find values for the other four trigonometric functions using identities (1), (2), (4), and (5):

$$\csc x = \frac{1}{\sin x} = \frac{1}{-\frac{1}{2}} = -2 \qquad \text{Reciprocal identity (1)}$$

$$\sec x = \frac{1}{\cos x} = \frac{1}{-\sqrt{3}/2} = -\frac{2}{\sqrt{3}} \qquad \text{Reciprocal identity (2)}$$

$$\tan x = \frac{\sin x}{\cos x} = \frac{-\frac{1}{2}}{-\sqrt{3}/2} = \frac{1}{\sqrt{3}} \qquad \text{Quotient identity (4)}$$

$$\cot x = \frac{\cos x}{\sin x} = \frac{-\sqrt{3}/2}{-\frac{1}{2}} = \sqrt{3} \qquad \text{Quotient identity (5)}$$

[*Note:* We could also use identity (3).]

It is important to note that we were able to find the values of the other five trigonometric functions without finding x.

MATCHED 1 PROBLEM

Use the basic identities to find the values of the other five trigonometric functions given $\cos x = 1/\sqrt{2}$ and $\cot x < 0$.

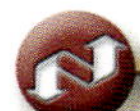

EXPLORE/DISCUSS 1

Suppose that $\sin x = -\frac{1}{2}$ and $\tan x > 0$, as in Example 1. Using basic identities and the results in Example 1, find each of the following:

(A) $\sin(-x)$ (B) $\sec(-x)$ (C) $\tan(-x)$

Verbally justify each step in your solution process.

Sign Properties

As a circular point $W(x)$ moves from quadrant to quadrant, its coordinates (a, b) undergo sign changes. Hence, the trigonometric functions also undergo sign changes. It is important to know the sign of each trigonometric function in each quadrant. Table 1 shows the sign behavior for each function. It is not necessary to memorize Table 1, because the sign of each function for each quadrant is easily determined from its definition (which *should* be memorized).

TABLE 1 Sign Properties

Trigonometric Function	Sign in Quadrant I	II	III	IV
$\sin x = b$	+	+	−	−
$\csc x = 1/b$	+	+	−	−
$\cos x = a$	+	−	−	+
$\sec x = 1/a$	+	−	−	+
$\tan x = b/a$	+	−	+	−
$\cot x = a/b$	+	−	+	−

Periodic Functions

Because the unit circle has a circumference of 2π, we find that for a given value of x (Fig. 2) we will return to the circular point $W(x) = (a, b)$ if we add any integer multiple of 2π to x. Think of a point P moving around the unit circle in either direction. Every time P covers a distance of 2π, the circumference of the circle, it is back at the point where it started. Thus, for x any real number,

$$\sin(x + 2k\pi) = \sin x \qquad k \text{ any integer}$$
$$\cos(x + 2k\pi) = \cos x \qquad k \text{ any integer}$$

Functions with this kind of repetitive behavior are called **periodic functions.** In general, we have Definition 2.

FIGURE 2

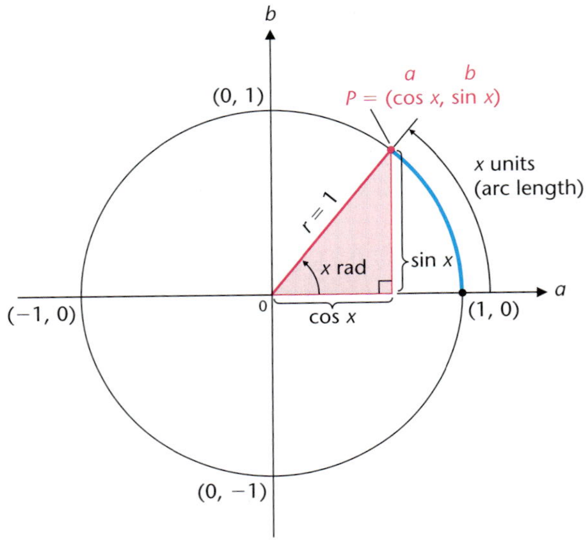

DEFINITION 2
Periodic Functions

A function f is **periodic** if there exists a positive real number p such that

$$f(x + p) = f(x)$$

for all x in the domain of f. The smallest such positive p, if it exists, is called the **fundamental period of f** (or often just the **period of f**).

Both the sine and cosine functions are periodic with period 2π. Once the graph for one period is known, the entire graph is obtained by repetition. The domain of both functions is the set of all real numbers, and the range of both is $[-1, 1]$. Because $b = 0$ at the circular points $(1, 0)$ and $(-1, 0)$, the zeros of the sine function are $k\pi$, k any integer. Because $a = 0$ at the circular points $(0, 1)$ and $(0, -1)$, the zeros of the cosine function are $\pi/2 + k\pi$, k any integer. By the basic identity $\sin(-x) = -\sin x$, the sine function is symmetrical with respect to the origin.

Because $\cos(-x) = \cos x$, the cosine function is symmetrical with respect to the y axis. Figures 3 and 4 summarize these properties and show the graphs of the sine and cosine functions, respectively.

Graph of $y = \sin x$

FIGURE 3

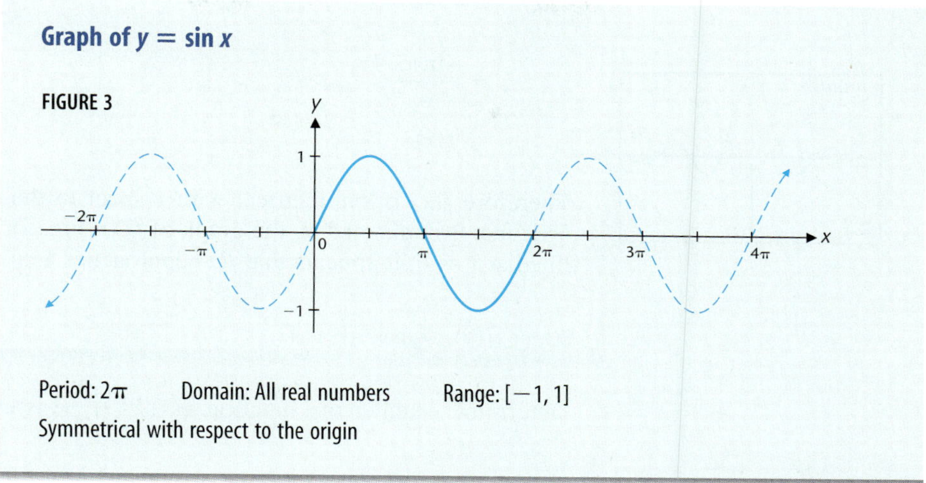

Period: 2π Domain: All real numbers Range: $[-1, 1]$

Symmetrical with respect to the origin

Graph of $y = \cos x$

FIGURE 4

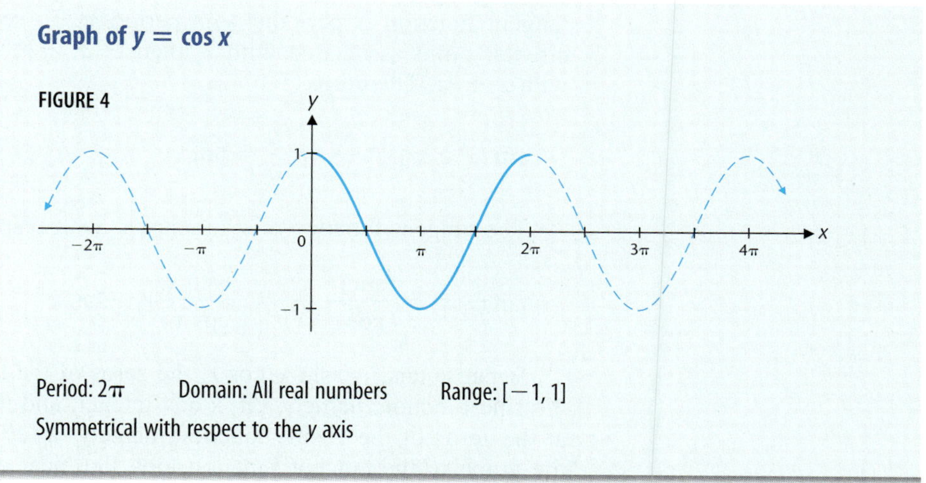

Period: 2π Domain: All real numbers Range: $[-1, 1]$

Symmetrical with respect to the y axis

In the terminology of Section 1.4, we say that the sine function is *odd* (because it is symmetrical with respect to the origin) and the cosine function is *even* (because it is symmetrical with respect to the y axis).

EXAMPLE 2 Symmetry

Determine whether the function $f(x) = \dfrac{\sin x}{x}$ is even, odd, or neither.

SOLUTION

$$f(-x) = \frac{\sin(-x)}{-x} \qquad \text{Sine function is odd}$$

$$= \frac{-\sin x}{-x}$$

$$= \frac{\sin x}{x}$$

$$= f(x)$$

FIGURE 5 $y_1 = \dfrac{\sin x}{x}$

Therefore $f(x)$ is symmetrical with respect to the y axis and is an even function. This fact is confirmed by the graph of $f(x)$ (Fig. 5). Note that although $f(x)$ is undefined at $x = 0$, it appears that $f(x)$ approaches 1 as x approaches 0 from either side.

MATCHED 2 PROBLEM

Determine whether the function $g(x) = \dfrac{\cos x}{x}$ is even, odd, or neither.

Because the tangent function is the quotient of the sine and cosine functions, you might expect that it would also be periodic with period 2π. Surprisingly, the tangent function is periodic with period π. To see this, note that if (a, b) is the circular point associated with x, then $(-a, -b)$ is the circular point associated with $x + \pi$. Therefore,

$$\tan(x + \pi) = \frac{-b}{-a} = \frac{b}{a} = \tan x$$

The tangent function is symmetrical with respect to the origin because

$$\tan(-x) = \frac{\sin(-x)}{\cos(-x)} = \frac{-\sin x}{\cos x} = -\tan x$$

Because $\tan x = \sin x/\cos x$, the zeros of the tangent function are the zeros of the sine function, namely, $k\pi$, k any integer, and the tangent function is undefined at the zeros of the cosine function, namely, $\pi/2 + k\pi$, k any integer. What does the graph of the tangent function look like near one of the values of x, say $\pi/2$, at which it is undefined? If $x < \pi/2$ but x is close to $\pi/2$, then b is close to 1 and a is positive and close to 0, so the ratio b/a is large and positive. Thus,

$$\tan x \to \infty \quad \text{as} \quad x \to (\pi/2)^-$$

Similarly, if $x > \pi/2$ but x is close to $\pi/2$, then b is close to 1 and a is negative and close to 0, so the ratio b/a is large in absolute value and negative. Thus,

$$\tan x \to -\infty \quad \text{as} \quad x \to (\pi/2)^+$$

Therefore the line $x = \pi/2$ is a vertical asymptote for the tangent function and, by periodicity, so are the vertical lines $x = \pi/2 + k\pi$, k any integer. Figure 6 summarizes these properties of the tangent function and shows its graph. The analogous properties of the cotangent function and its graph are shown in Figure 7.

Graph of $y = \tan x$

FIGURE 6

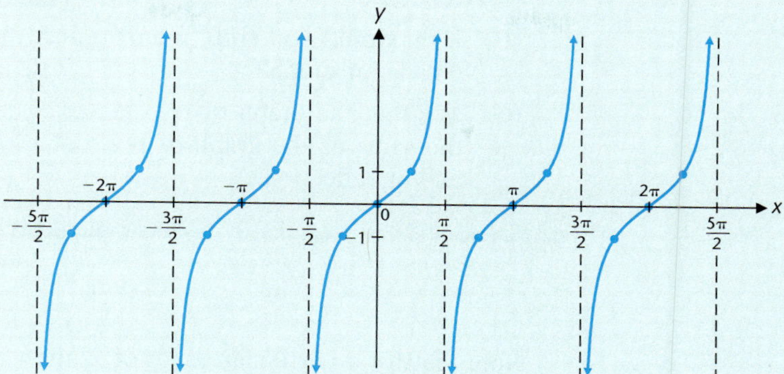

Period: π

Domain: All real numbers except $\pi/2 + k\pi$, k an integer

Range: All real numbers

Symmetrical with respect to the origin

Increasing function between consecutive asymptotes

Discontinuous at $x = \pi/2 + k\pi$, k an integer

Graph of $y = \cot x$

FIGURE 7

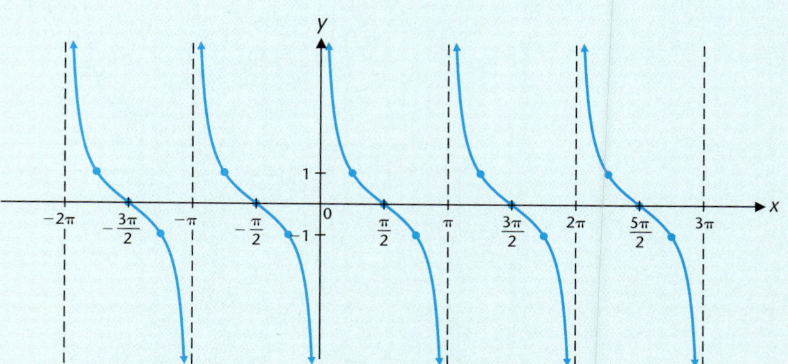

Period: π

Domain: All real numbers except $k\pi$, k an integer

Range: All real numbers

Symmetrical with respect to the origin

Decreasing function between consecutive asymptotes

Discontinuous at $x = k\pi$, k an integer

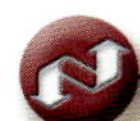

EXPLORE/DISCUSS 2

(A) Discuss how the graphs of the tangent and cotangent functions are related.

(B) How would you shift and/or reflect the tangent graph to obtain the cotangent graph?

(C) Is either the graph of $y = \tan (x - \pi/2)$ or $y = -\tan (x - \pi/2)$ the same as the graph of $y = \cot x$? Explain in terms of shifts and/or reflections.

Note that for a particular value of x, the y value on the graph of $y = \cot x$ is the reciprocal of the y value on the graph of $y = \tan x$. The vertical asymptotes of $y = \cot x$ occur at the zeros of $y = \tan x$, and vice versa.

The graphs of $y = \csc x$ and $y = \sec x$ can be obtained by taking the reciprocals of the y values of the graphs of $y = \sin x$ and $y = \cos x$, respectively. Vertical asymptotes occur at the zeros of $y = \sin x$ or $y = \cos x$. Figures 8 and 9 summarize the properties and show the graphs of $y = \csc x$ and $y = \sec x$. To emphasize the reciprocal relationships, the graphs of $y = \sin x$ and $y = \cos x$ are indicated in broken lines.

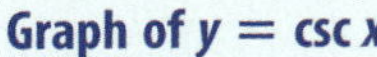

Graph of $y = \csc x$

FIGURE 8

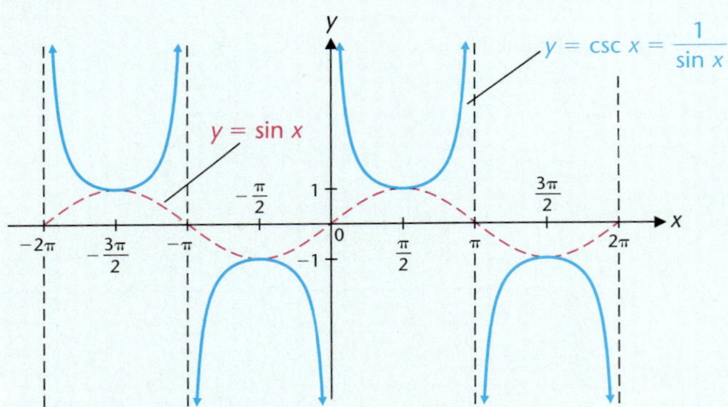

Period: 2π

Domain: All real numbers except $k\pi$, k an integer

Range: All real numbers y such that $y \leq -1$ or $y \geq 1$

Symmetrical with respect to the origin

Discontinuous at $x = k\pi$, k an integer

Graph of $y = \sec x$

FIGURE 9

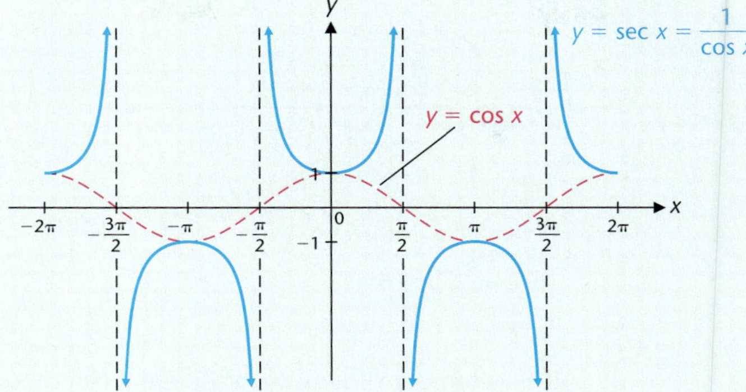

Period: 2π

Domain: All real numbers except $\pi/2 + k\pi$, k an integer

Range: All real numbers y such that $y \leq -1$ or $y \geq 1$

Symmetrical with respect to the y axis

Discontinuous at $x = \pi/2 + k\pi$, k an integer

Reference Triangles

Consider an angle θ in standard position. Let $P = (a, b)$ be the point of intersection of the terminal side of θ with a circle of radius $r > 0$. Then

$$a^2 + b^2 = r^2$$

so dividing both sides of the equation by r^2,

$$(a/r)^2 + (b/r)^2 = 1$$

Therefore the circular point Q on the terminal side of θ has coordinates $(a/r, b/r)$ (Fig. 10). By Definition 1,

$$\sin\theta = \frac{b}{r} \qquad \csc\theta = \frac{r}{b},\quad b \neq 0$$

$$\cos\theta = \frac{a}{r} \qquad \sec\theta = \frac{r}{a},\quad a \neq 0$$

$$\tan\theta = \frac{b}{a},\quad a \neq 0 \qquad \cot\theta = \frac{a}{b},\quad b \neq 0$$

FIGURE 10

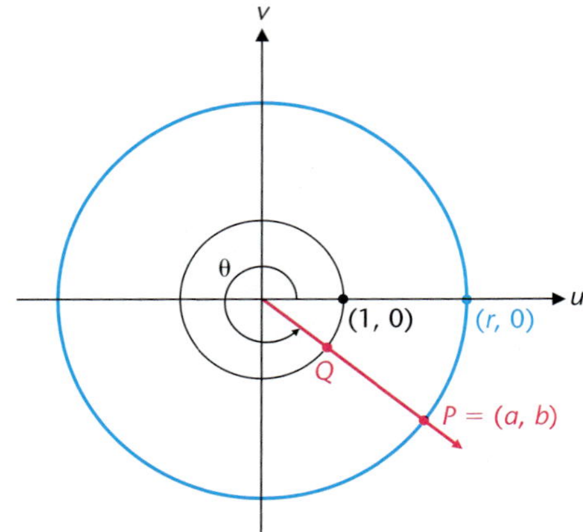

It is often convenient to associate a reference triangle and reference angle with θ, and to label the horizontal side, vertical side, and hypotenuse of the reference triangle with a, b, and r, respectively, to easily obtain the values of the trigonometric functions of θ.

Reference Triangle and Reference Angle

1. To form a **reference triangle** for θ, draw a perpendicular from a point $P = (a, b)$ on the terminal side of θ to the horizontal axis.
2. The **reference angle** α is the acute angle (always taken positive) between the terminal side of θ and the horizontal axis.

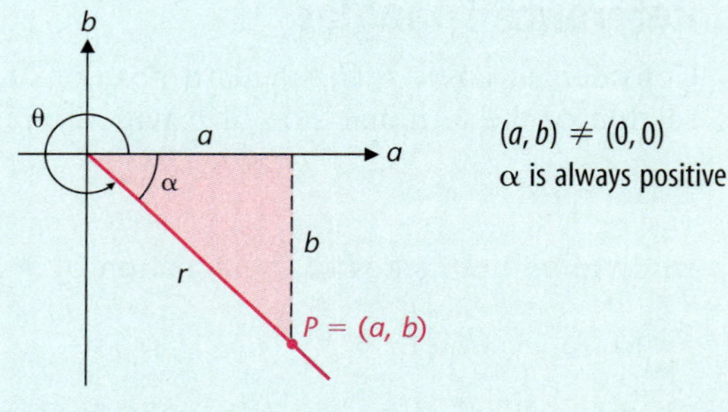

If Adj and Opp denote the labels a and b (possibly negative) on the horizontal and vertical sides of the reference triangle, and Hyp denotes the length r of the hypotenuse, then

$$\sin\theta = \frac{\text{Opp}}{\text{Hyp}} \qquad \csc\theta = \frac{\text{Hyp}}{\text{Opp}}$$

$$\cos\theta = \frac{\text{Adj}}{\text{Hyp}} \qquad \sec\theta = \frac{\text{Hyp}}{\text{Adj}}$$

$$\tan\theta = \frac{\text{Opp}}{\text{Adj}} \qquad \cot\theta = \frac{\text{Adj}}{\text{Opp}}$$

EXAMPLE 3 Values of the Trigonometric Functions

If $\sin\theta = 4/7$ and $\cos\theta < 0$, find the values of each of the other five trigonometric functions of θ.

SOLUTION

Because the sine of θ is positive and the cosine is negative, the angle θ is in quadrant II. We sketch a reference triangle (Fig. 11) and use the Pythagorean theorem to calculate the length of the horizontal side:

$$\sqrt{7^2 - 4^2} = \sqrt{33}$$

Therefore Adj $= -\sqrt{33}$, Opp $= 4$, Hyp $= 7$. The values of the other five trigonometric functions are:

FIGURE 11

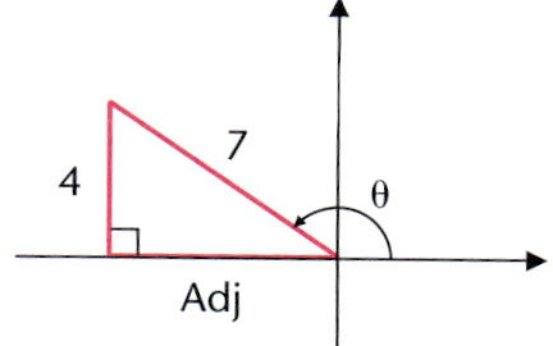

$$\cos\theta = \frac{-\sqrt{33}}{7}$$

$$\tan\theta = \frac{-4}{\sqrt{33}}$$

$$\csc\theta = \frac{7}{4}$$

$$\sec\theta = \frac{-7}{\sqrt{33}}$$

$$\cot\theta = \frac{-\sqrt{33}}{4}$$

MATCHED 3 PROBLEM

If $\tan\theta = 10$ and $\sin\theta < 0$, find the values of each of the other five trigonometric functions of θ.

ANSWERS MATCHED PROBLEMS

1. $\sin x = -1/\sqrt{2}$, $\tan x = -1$, $\csc x = -\sqrt{2}$, $\sec x = \sqrt{2}$, $\cot x = -1$ **2.** Odd
3. $\sin\theta = -10/\sqrt{101}$, $\cos\theta = -1/\sqrt{101}$, $\csc\theta = -\sqrt{101}/10$, $\sec\theta = -\sqrt{101}$, $\cot\theta = 1/10$

EXERCISE 5.4

The figure will be useful in many of the problems in this exercise.

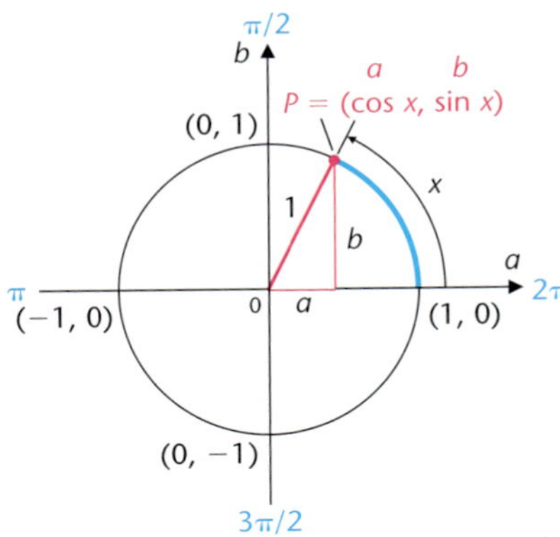

Figure for Problems 1–10.

A

Answer Problems 1–10 without looking back in the text or using a calculator. You can refer to the figure.

1. What are the periods of the sine, cotangent, and cosecant functions?

2. What are the periods of the cosine, tangent, and secant functions?

3. How far does the graph of each function deviate from the x axis?

(A) $y = \cos x$ **(B)** $y = \tan x$ **(C)** $y = \csc x$

4. How far does the graph of each function deviate from the x axis?

(A) $y = \sin x$ **(B)** $y = \cot x$ **(C)** $y = \sec x$

5. What are the x intercepts for the graph of each function over the interval $-2\pi \le x \le 2\pi$?

(A) $y = \sin x$ **(B)** $y = \cot x$ **(C)** $y = \csc x$

6. What are the x intercepts for the graph of each function over the interval $-2\pi \le x \le 2\pi$?

(A) $y = \cos x$ **(B)** $y = \tan x$ **(C)** $y = \sec x$

7. For what values of x, $-2\pi \le x \le 2\pi$, are the following functions not defined?

(A) $y = \cos x$ **(B)** $y = \tan x$ **(C)** $y = \csc x$

8. For what values of x, $-2\pi \le x \le 2\pi$, are the following functions not defined?

(A) $y = \sin x$ **(B)** $y = \cot x$ **(C)** $y = \sec x$

9. At what points, $-2\pi \le x \le 2\pi$, do the vertical asymptotes for the following functions cross the x axis?

(A) $y = \cos x$ **(B)** $y = \tan x$ **(C)** $y = \csc x$

10. At what points, $-2\pi \le x \le 2\pi$, do the vertical asymptotes for the following functions cross the x axis?

(A) $y = \sin x$ **(B)** $y = \cot x$ **(C)** $y = \sec x$

B

11. (A) Describe a shift and/or reflection that will transform the graph of $y = \csc x$ into the graph of $y = \sec x$.

(B) Is either the graph of $y = -\csc(x + \pi/2)$ or $y = -\csc(x - \pi/2)$ the same as the graph of $y = \sec x$? Explain in terms of shifts and/or reflections.

12. (A) Describe a shift and/or reflection that will transform the graph of $y = \sec x$ into the graph of $y = \csc x$.

(B) Is either the graph of $y = -\sec(x - \pi/2)$ or $y = -\sec(x + \pi/2)$ the same as the graph of $y = \csc x$? Explain in terms of shifts and/or reflections.

In Problems 13–20, determine whether each function is even, odd, or neither. Check your answer by graphing.

13. $y = \dfrac{\tan x}{x}$ **14.** $y = \dfrac{\sec x}{x}$

15. $y = \dfrac{\csc x}{x}$ **16.** $y = \dfrac{\cot x}{x}$

17. $y = \sin x \cos x$ **18.** $y = x \sin x \cos x$

19. $y = x^2 \sin x$ **20.** $y = x^3 \sin x$

Find the value of each of the six trigonometric functions for an angle θ that has a terminal side containing the point indicated in Problems 21–24.

21. $(6, 8)$ **22.** $(-3, 4)$

23. $(-1, \sqrt{3})$ **24.** $(\sqrt{3}, 1)$

Find the reference angle α for each angle θ in Problems 25–30.

25. $\theta = 300°$ **26.** $\theta = 135°$

27. $\theta = \dfrac{7\pi}{6}$ **28.** $\theta = \dfrac{\pi}{4}$

29. $\theta = -\dfrac{5\pi}{3}$ **30.** $\theta = -\dfrac{5\pi}{4}$

In Problems 31–36, find the smallest positive θ in degree and radian measure for which

31. $\cos\theta = \dfrac{-1}{2}$ **32.** $\sin\theta = \dfrac{-\sqrt{3}}{2}$

33. $\sin\theta = \dfrac{-1}{2}$ **34.** $\tan\theta = -\sqrt{3}$

35. $\csc\theta = \dfrac{-2}{\sqrt{3}}$ **36.** $\sec\theta = -\sqrt{2}$

Find the value of each of the other five trigonometric functions for an angle θ, without finding θ, given the information indicated in Problems 37–40. Sketching a reference triangle should be helpful.

37. $\sin\theta = \frac{3}{5}$ and $\cos\theta < 0$

38. $\tan\theta = -\frac{4}{3}$ and $\sin\theta < 0$

39. $\cos\theta = -\sqrt{5}/3$ and $\cot\theta > 0$

40. $\cos\theta = -\sqrt{5}/3$ and $\tan\theta > 0$

41. Which trigonometric functions are not defined when the terminal side of an angle lies along the vertical axis. Why?

42. Which trigonometric functions are not defined when the terminal side of an angle lies along the horizontal axis? Why?

43. Find exactly, all θ, $0° \le \theta < 360°$, for which $\cos\theta = -\sqrt{3}/2$.

44. Find exactly, all θ, $0° \le \theta < 360°$, for which $\cot\theta = -1/\sqrt{3}$.

45. Find exactly, all θ, $0 \le \theta < 2\pi$, for which $\tan\theta = 1$.

46. Find exactly, all θ, $0 \le \theta < 2\pi$, for which $\sec\theta = -\sqrt{2}$.

Problems 47–52 offer a preliminary investigation into the relationships of the graphs of $y = \sin x$ and $y = \cos x$ with the graphs of $y = A \sin x$, $y = A \cos x$, $y = \sin Bx$, $y = \cos Bx$, $y = \sin (x + C)$, and $y = \cos (x + C)$. This important topic is discussed in detail in Section 5.5.

47. **(A)** Graph $y = A \cos x$, $(-2\pi \le x \le 2\pi, -3 \le y \le 3)$, for $A = 1, 2$, and -3, all in the same viewing window.

(B) Do the x intercepts change? If so, where?

(C) How far does each graph deviate from the x axis? (Experiment with additional values of A.)

(D) Describe how the graph of $y = \cos x$ is changed by changing the values of A in $y = A \cos x$?

48. **(A)** Graph $y = A \sin x$, $(-2\pi \le x \le 2\pi, -3 \le y \le 3)$, for $A = 1, 3$, and -2, all in the same viewing window.

(B) Do the x intercepts change? If so, where?

(C) How far does each graph deviate from the x axis? (Experiment with additional values of A.)

(D) Describe how the graph of $y = \sin x$ is changed by changing the values of A in $y = A \sin x$?

49. **(A)** Graph $y = \sin Bx$ $(-\pi \le x \le \pi, -2 \le y \le 2)$, for $B = 1, 2$, and 3, all in the same viewing window.

(B) How many periods of each graph appear in this viewing rectangle? (Experiment with additional positive integer values of B.)

(C) Based on the observations in part B, how many periods of the graph of $y = \sin nx$, n a positive integer, would appear in this viewing window?

50. **(A)** Graph $y = \cos Bx$ $(-\pi \le x \le \pi, -2 \le y \le 2)$, for $B = 1, 2$, and 3, all in the same viewing window.

(B) How many periods of each graph appear in this viewing rectangle? (Experiment with additional positive integer values of B.)

(C) Based on the observations in part B, how many periods of the graph of $y = \cos nx$, n a positive integer, would appear in this viewing window?

51. **(A)** Graph $y = \cos (x + C)$, $-2\pi \le x \le 2\pi$, $-1.5 \le y \le 1.5$, for $C = 0, -\pi/2$, and $\pi/2$, all in the same viewing window. (Experiment with additional values of C.)

(B) Describe how the graph of $y = \cos x$ is changed by changing the values of C in $y = \cos (x + C)$?

52. **(A)** Graph $y = \sin(x + C)$, $-2\pi \le x \le 2\pi$, $-1.5 \le y \le 1.5$, for $C = 0$, $-\pi/2$, and $\pi/2$, all in the same viewing window. (Experiment with additional values of C.)

(B) Describe how the graph of $y = \sin x$ is changed by changing the values of C in $y = \sin(x + C)$?

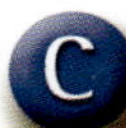

53. Try to calculate each of the following on your calculator. Explain the results.

(A) $\sec(\pi/2)$ **(B)** $\tan(-\pi/2)$ **(C)** $\cot(-\pi)$

54. Try to calculate each of the following on your calculator. Explain the results.

(A) $\csc \pi$ **(B)** $\tan(\pi/2)$ **(C)** $\cot 0$

55. Graph $f(x) = \sin x$ and $g(x) = x$ in the same viewing window $(-1 \le x \le 1, -1 \le y \le 1)$.

(A) What do you observe about the two graphs when x is close to 0, say $-0.5 \le x \le 0.5$?

(B) Complete the table to three decimal places (use the table feature on your graphing utility if it has one):

x	-0.3	-0.2	-0.1	0.0	0.1	0.2	0.3
$\sin x$							

(In applied mathematics certain derivations, formulas, and calculations are simplified by replacing $\sin x$ with x for small values of $|x|$.)

56. Graph $h(x) = \tan x$ and $g(x) = x$ in the same viewing window $(-1 \le x \le 1, -1 \le y \le 1)$.

(A) What do you observe about the two graphs when x is close to 0, say $-0.5 \le x \le 0.5$?

(B) Complete the table to three decimal places (use the table feature on your graphing utility if it has one):

x	-0.3	-0.2	-0.1	0.0	0.1	0.2	0.3
$\tan x$							

(In applied mathematics certain derivations, formulas, and calculations are simplified by replacing $\tan x$ with x for small values of $|x|$.)

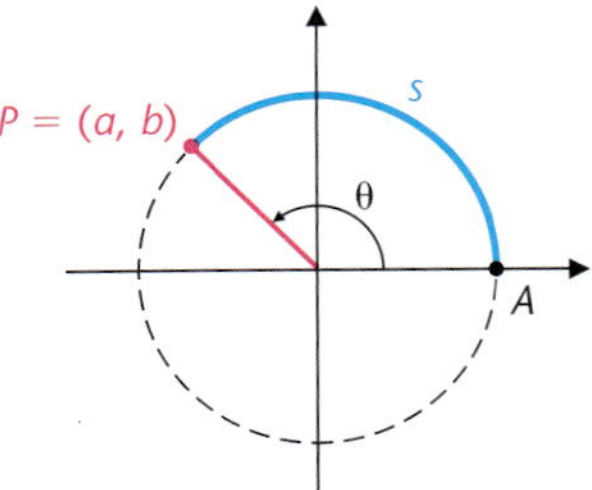

57. If the coordinates of A are (4, 0) and arc length s is 7 units, find

(A) The exact radian measure of θ

(B) The coordinates of P to three decimal places

58. If the coordinates of A are (2, 0) and arc length s is 8 units, find

(A) The exact radian measure of θ

(B) The coordinates of P to three decimal places

59. In a rectangular coordinate system, a circle with center at the origin passes through the point $(6\sqrt{3}, 6)$. What is the length of the arc on the circle in quadrant I between the positive horizontal axis and the point $(6\sqrt{3}, 6)$?

60. In a rectangular coordinate system, a circle with center at the origin passes through the point $(2, 2\sqrt{3})$. What is the length of the arc on the circle in quadrant I between the positive horizontal axis and the point $(2, 2\sqrt{3})$?

APPLICATIONS

61. Solar Energy. The intensity of light I on a solar cell changes with the angle of the sun and is given by the formula $I = k \cos \theta$, where k is a constant (see the figure). Find light intensity I in terms of k for $\theta = 0°$, $\theta = 30°$, and $\theta = 60°$.

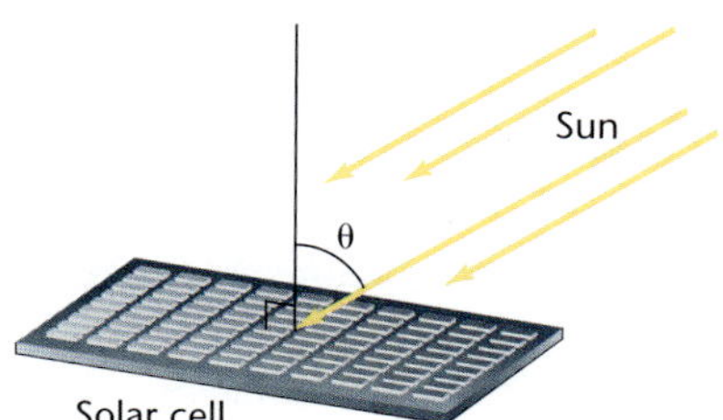

62. Solar Energy. Refer to Problem 61. Find light intensity I in terms of k for $\theta = 20°$, $\theta = 50°$, and $\theta = 90°$.

63. Physics—Engineering. The figure illustrates a piston connected to a wheel that turns 3 revolutions per second; hence, the angle θ is being generated at $3(2\pi) = 6\pi$ radians per second, or $\theta = 6\pi t$, where t is time in seconds. If P is at $(1, 0)$ when $t = 0$, show that

$$y = b + \sqrt{4^2 - a^2}$$
$$= \sin 6\pi t + \sqrt{16 - (\cos 6\pi t)^2}$$

for $t \geq 0$.

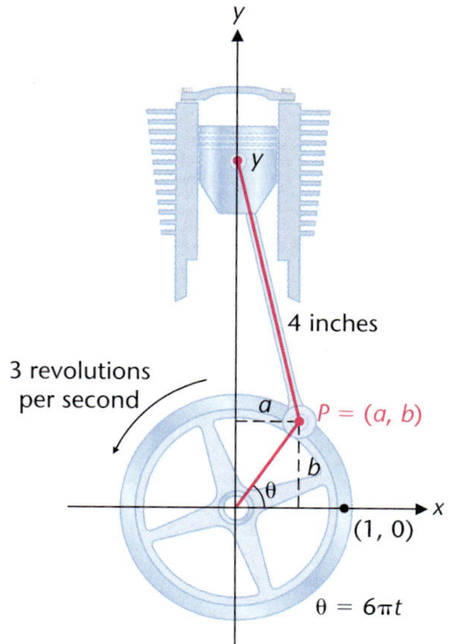

64. Physics—Engineering. In Problem 63, find the position of the piston y when $t = 0.2$ second (to three significant digits).

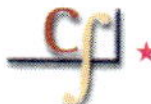

★ **65. Geometry.** The area of a regular n-sided polygon circumscribed about a circle of radius 1 is given by

$$A = n \tan \frac{180°}{n}$$

(A) Find A for $n = 8$, $n = 100$, $n = 1{,}000$, and $n = 10{,}000$. Compute each to five decimal places.

(B) What number does A seem to approach as $n \to \infty$? (What is the area of a circle with radius 1?)

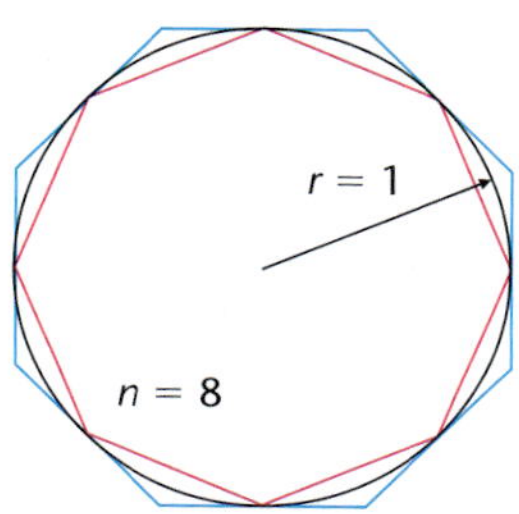

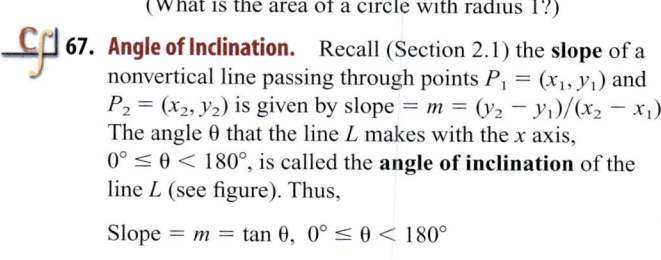

★ **66. Geometry.** The area of a regular n-sided polygon inscribed in a circle of radius 1 is given by

$$A = \frac{n}{2} \sin \frac{360°}{n}$$

(A) Find A for $n = 8$, $n = 100$, $n = 1{,}000$, and $n = 10{,}000$. Compute each to five decimal places.

(B) What number does A seem to approach as $n \to \infty$? (What is the area of a circle with radius 1?)

67. Angle of Inclination. Recall (Section 2.1) the **slope** of a nonvertical line passing through points $P_1 = (x_1, y_1)$ and $P_2 = (x_2, y_2)$ is given by slope $= m = (y_2 - y_1)/(x_2 - x_1)$. The angle θ that the line L makes with the x axis, $0° \leq \theta < 180°$, is called the **angle of inclination** of the line L (see figure). Thus,

$$\text{Slope} = m = \tan \theta, \quad 0° \leq \theta < 180°$$

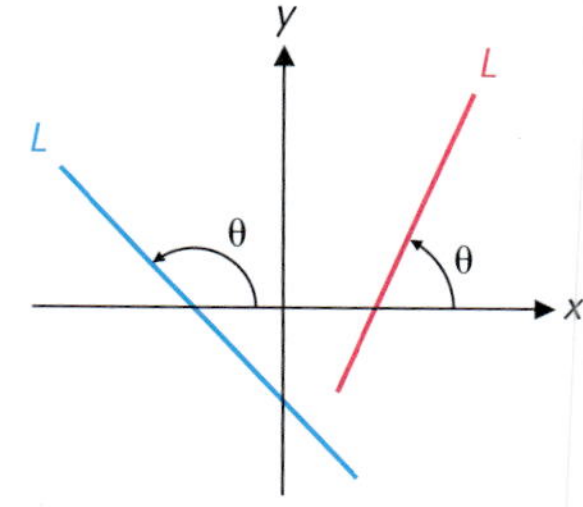

(A) Compute the slopes to two decimal places of the lines with angles of inclination 88.7° and 162.3°.

(B) Find the equation of a line passing through (−4, 5) with an angle of inclination 137°. Write the answer in the form $y = mx + b$, with m and b to two decimal places.

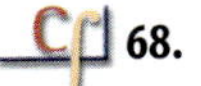

68. Angle of Inclination. Refer to Problem 67.

(A) Compute the slopes to two decimal places of the lines with angles of inclination 5.34° and 92.4°.

(B) Find the equation of a line passing through (6, −4) with an angle of inclination 106°. Write the answer in the form $y = mx + b$, with m and b to two decimal places.

SECTION 5.5 More General Trigonometric Functions and Models

Graphs of $y = A \sin Bx$ and $y = A \cos Bx$ • Graphs of $y = A \sin (Bx + C)$ and $y = A \cos (Bx + C)$ • Finding an Equation from the Graph of a Simple Harmonic • Modeling and Data Analysis

Imagine a weight suspended from the ceiling by a spring. If the weight were pulled downward and released, then, assuming no air resistance or friction, it would move up and down with the same frequency and amplitude forever. This idealized motion is an example of **simple harmonic motion.** Simple harmonic motion can be described by functions of the form $y = A \sin (Bx + C)$ or $y = A \cos (Bx + C)$, called **simple harmonics.**

Simple harmonics are extremely important in both pure and applied mathematics. In applied mathematics they are used in the analysis of sound waves, radio waves, X-rays, gamma rays, visible light, infrared radiation, ultraviolet radiation, seismic waves, ocean waves, electric circuits, electric generators, vibrations, bridge and building construction, spring–mass systems, bow waves of boats, sonic booms, and so on. Analysis involving simple harmonics is called *harmonic analysis*.

In Section 5.5 we study properties, graphs, and applications of simple harmonics. A brief review of graph transformations (Section 1.4) should prove helpful.

Graphs of $y = A \sin Bx$ and $y = A \cos Bx$

We visualize the graphs of functions of the form $y = A \sin Bx$ or $y = A \cos Bx$, and determine their zeros and turning points, by understanding how each of the constants A and B transforms the graph of $y = \sin x$ or $y = \cos x$.

EXAMPLE 1 Zeros and Turning Points

Find the zeros and turning points of each function on the interval $[0, 2\pi]$.

(A) $y = \dfrac{1}{2} \sin x$

(B) $y = -2 \sin x$

SOLUTIONS

(A) The function $y = \frac{1}{2} \sin x$ is the vertical contraction of $y = \sin x$ that is obtained by multiplying each ordinate value by $\frac{1}{2}$ (Fig. 1). Therefore its zeros on $[0, 2\pi]$ are identical to the zeros of $y = \sin x$, namely $x = 0$, π, and 2π. Because the turning points of $y = \sin x$ are $(\pi/2, 1)$ and $(3\pi/2, -1)$, the turning points of $y = \frac{1}{2} \sin x$ are $(\pi/2, 1/2)$ and $(3\pi/2, -1/2)$.

(B) The function $y = -2 \sin x$ is the vertical expansion of $y = \sin x$ that is obtained by multiplying each ordinate value by 2, followed by a reflection in the x axis (see Fig. 1). Therefore its zeros on $[0, 2\pi]$ are identical to the zeros of $y = \sin x$, namely $x = 0$, π, and 2π. Because the turning points of $y = \sin x$ are $(\pi/2, 1)$ and $(3\pi/2, -1)$, the turning points of $y = -2 \sin x$ are $(\pi/2, -2)$ and $(3\pi/2, 2)$.

FIGURE 1

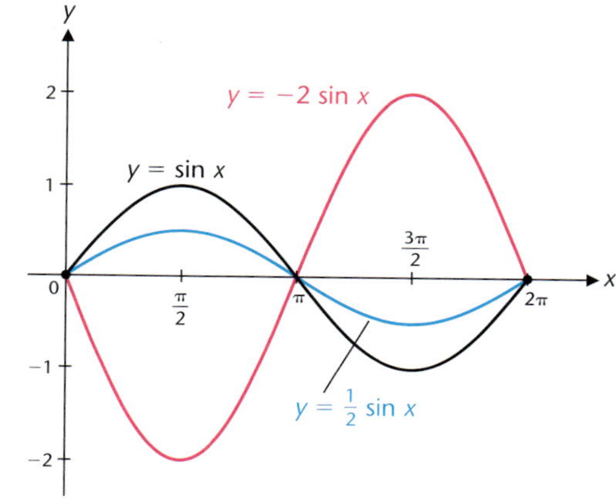

MATCHED 1 PROBLEM

Find the zeros and turning points of each function on the interval $[\pi/2, 5\pi/2]$.

(A) $y = -5 \cos x$

(B) $y = \dfrac{1}{3} \cos x$

As Example 1 illustrates, the graph of $y = A \sin x$ can be obtained from the graph of $y = \sin x$ by multiplying each y value of $y = \sin x$ by the constant A. The graph of $y = A \sin x$ still crosses the x axis where the graph of $y = \sin x$ crosses the x axis, because $A \cdot 0 = 0$. Because the maximum value of $\sin x$ is 1, the maximum value of $A \sin x$ is $|A| \cdot 1 = |A|$. The constant $|A|$ is called the **amplitude** of the graph of $y = A \sin x$ and indicates the maximum deviation of the graph of $y = A \sin x$ from the x axis.

The period of $y = A \sin x$ (assuming $A \neq 0$) is the same as the period of $y = \sin x$, namely 2π, because $A \sin (x + 2\pi) = A \sin x$.

EXAMPLE 2 Periods

Find the period of each function.

(A) $y = \sin 2x$

(B) $y = \sin (x/2)$

SOLUTIONS

(A) Because the function $y = \sin x$ has period 2π, the function $y = \sin 2x$ completes one cycle as $2x$ varies from

$$2x = 0 \qquad \text{to} \qquad 2x = 2\pi$$

or as x varies from

$$x = 0 \qquad \text{to} \qquad x = \pi \qquad \text{Half the period for } \sin x.$$

Therefore the period of $y = \sin 2x$ is π (Fig. 2).

(B) Because the function $y = \sin x$ has period 2π, the function $y = \sin (x/2)$ completes one cycle as $x/2$ varies from

$$\frac{x}{2} = 0 \qquad \text{to} \qquad \frac{x}{2} = 2\pi$$

or as x varies from

$$x = 0 \qquad \text{to} \qquad x = 4\pi \qquad \text{Double the period for } \sin x.$$

Therefore the period of $y = \sin (x/2)$ is 4π (see Fig. 2).

FIGURE 2

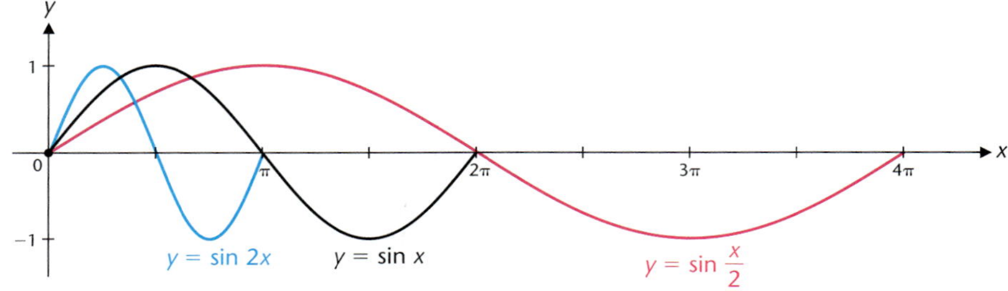

MATCHED 2 PROBLEM

Find the period of each function.

(A) $y = \cos (x/10)$

(B) $y = \cos (6\pi x)$

As Example 2 illustrates, the graph of $y = \sin Bx$, for a positive constant B, completes one cycle as Bx varies from

$$Bx = 0 \qquad \text{to} \qquad Bx = 2\pi$$

or as x varies from

$$x = 0 \qquad \text{to} \qquad x = \frac{2\pi}{B}$$

Therefore the period of $y = \sin Bx$ is $\frac{2\pi}{B}$. Note that the amplitude of $y = \sin Bx$ is 1, the same as the amplitude of $y = \sin x$. The effect of the constant B is to compress or stretch the basic sine curve by changing the period of the function, but not its amplitude. A similar analysis applies to $y = \cos Bx$, for $B > 0$, where it can be shown that the period is also $\frac{2\pi}{B}$. We combine and summarize our results on period and amplitude as follows:

Period and Amplitude

For $y = A \sin Bx$ or $y = A \cos Bx$, $A \neq 0$, $B > 0$:

$$\text{Amplitude} = |A| \qquad \text{Period} = \frac{2\pi}{B}$$

If $0 < B < 1$, the basic sine or cosine curve is stretched.
If $B > 1$, the basic sine or cosine curve is compressed.

You can either memorize the formula for the period, $\frac{2\pi}{B}$, or use the reasoning we used in deriving the formula. Recall, $\sin Bx$ or $\cos Bx$ completes one cycle as Bx varies from

$$Bx = 0 \qquad \text{to} \qquad Bx = 2\pi$$

that is, as x varies from

$$x = 0 \qquad \text{to} \qquad x = \frac{2\pi}{B}$$

Some prefer to memorize a formula, others a process.

EXAMPLE 3 Amplitude, Period, and Turning Points

Find the amplitude, period, and turning points of $y = -3 \cos (\pi x/2)$ on the interval $[-4, 4]$.

FIGURE 3

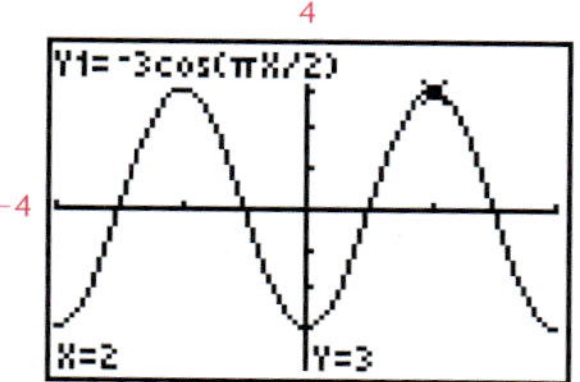

SOLUTION

$$\text{Amplitude} = |-3| = 3 \qquad \text{Period} = \frac{2\pi}{(\pi/2)} = 4$$

Because $y = \cos x$ has turning points at $x = 0$ and $x = \pm\pi$ (half of a complete cycle), $y = -3 \cos (\pi x/2)$ has turning points at $x = 0$ and $x = \pm 2$. The turning points on the interval $[-4, 4]$ are thus $(-2, 3)$, $(0, -3)$, and $(2, 3)$. These results are confirmed by graphing $y = -3 \cos (\pi x/2)$ (Fig. 3).

MATCHED 3 PROBLEM

Find the amplitude, period, and turning points of $y = \frac{1}{4} \sin (3\pi x)$ on the interval $[0, 1]$.

EXPLORE/DISCUSS 1

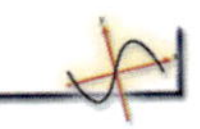

Find an equation of the form $y = A \cos Bx$ that produces the following graph. Check the result with a graphing utility.

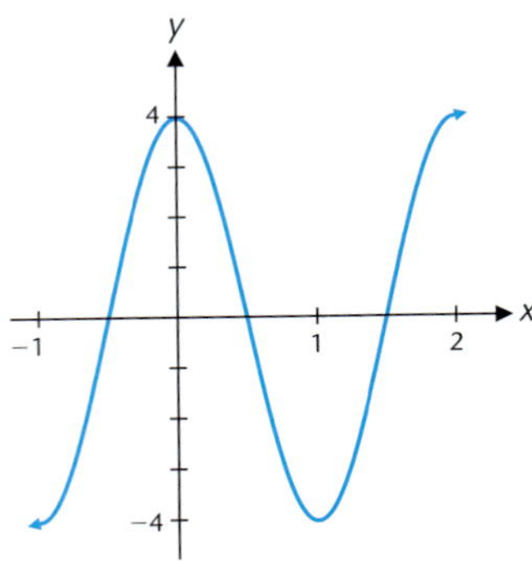

Is it possible for an equation of the form $y = A \sin Bx$ to produce the same graph? Explain.

Graphs of $y = A \sin (Bx + C)$ and $y = A \cos (Bx + C)$

The graph of $y = A \sin (Bx + C)$ is a horizontal translation of the graph of the function $y = A \sin Bx$. In fact, because the period of the sine function is 2π, $y = A \sin (Bx + C)$ completes one cycle as $Bx + C$ varies from

$$Bx + C = 0 \qquad \text{to} \qquad Bx + C = 2\pi$$

or (solving for x in each equation) as x varies from

Phase shift — Period

$$x = -\frac{C}{B} \qquad \text{to} \qquad x = -\frac{C}{B} + \frac{2\pi}{B}$$

We conclude that $y = A \sin (Bx + C)$ has a period of $2\pi/B$, and its graph is the graph of $y = A \sin Bx$ translated $|-C/B|$ units to the right if $-C/B$ is positive and $|-C/B|$ units to the left if $-C/B$ is negative. The number $-C/B$ is referred to as the **phase shift.**

EXAMPLE 4 Amplitude, Period, Phase Shift, and Zeros

Find the amplitude, period, phase shift, and zeros of $y = \frac{1}{2} \cos (4x - \pi/2)$.

SOLUTION

$$\text{Amplitude} = |A| = \left|\frac{1}{2}\right| = \frac{1}{2}$$

The graph completes one cycle as $4x - \pi/2$ varies from

$$4x - \frac{\pi}{2} = 0 \qquad \text{to} \qquad 4x - \frac{\pi}{2} = 2\pi$$

or as x varies from

$$x = \frac{\pi}{8} \qquad \text{to} \qquad x = \frac{\pi}{8} + \frac{\pi}{2}$$

$$\text{Phase shift} = \frac{\pi}{8} \qquad \text{Period} = \frac{\pi}{2}$$

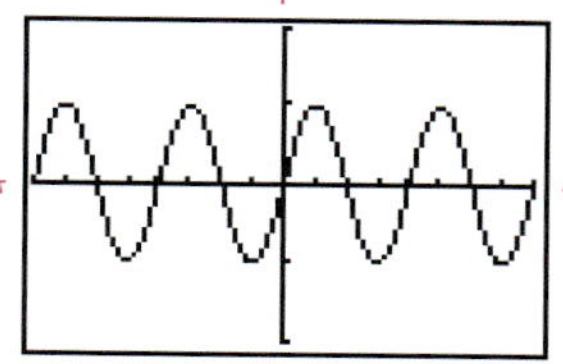

FIGURE 4

The zeros of $y = \frac{1}{2} \cos (4x - \pi/2)$ are obtained by shifting the zeros of $y = \frac{1}{2} \cos (4x)$ to the right by $\pi/8$ units. Because $x = \pi/8$ and $x = 3\pi/8$ are zeros of $y = \frac{1}{2} \cos (4x)$, $x = \pi/8 + \pi/8 = \pi/4$ and $x = 3\pi/8 + \pi/8 = \pi/2$ are zeros of $y = \frac{1}{2} \cos (4x - \pi/2)$. By periodicity the zeros of $y = \frac{1}{2} \cos (4x - \pi/2)$ are $x = k\pi/4$, k any integer. These results are confirmed by graphing $y = \frac{1}{2} \cos (4x - \pi/2)$ (Fig. 4).

MATCHED 4 PROBLEM

Find the amplitude, period, phase shift, and zeros of $y = \frac{3}{4} \sin (2x + \pi)$.

EXPLORE/DISCUSS 2

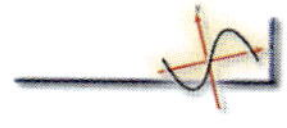

Find an equation of the form $y = A \sin (Bx + C)$ that produces the following graph. Check the results with a graphing utility.

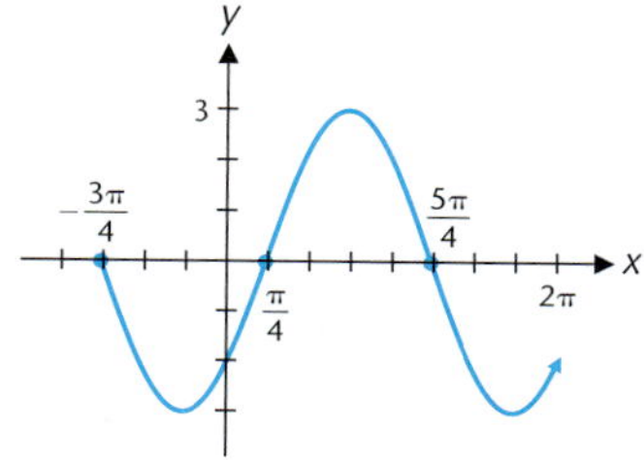

Is it possible for an equation of the form $y = A \cos (Bx + C)$ to produce the same graph? Explain.

The graphs of $y = A \sin (Bx + C) + k$ and $y = A \cos (Bx + C) + k$ are vertical translations (up k units if $k > 0$, down k units if $k < 0$) of the graphs of $y = A \sin (Bx + C)$ and $y = A \cos (Bx + C)$, respectively.

Because $y = \sec x$ and $y = \csc x$ are unbounded functions, amplitude is not defined for functions of the form $y = A \sec (Bx + C)$ and $y = A \csc (Bx + C)$. However, because both the secant and cosecant functions have period 2π, the functions $y = A \csc (Bx + C)$ and $y = A \sec (Bx + C)$ have period $2\pi/B$ and phase shift $-C/B$.

Because $y = \tan x$ and $y = \cot x$ are unbounded functions, amplitude is not defined for functions of the form $y = A \tan (Bx + C)$ or $y = A \cot (Bx + C)$. The tangent and cotangent functions both have period π, so the functions $y = A \tan (Bx + C)$ and $y = A \cot (Bx + C)$ have period π/B and phase shift $-C/B$.

Our results on amplitude, period, and phase shift are summarized in the box.

Amplitude, Period, and Phase Shift

Let A, B, C be constants such that $A \neq 0$ and $B > 0$.
For $y = A \sin (Bx + C)$ and $y = A \cos (Bx + C)$:

$$\text{Amplitude} = |A| \qquad \text{Period} = \frac{2\pi}{B} \qquad \text{Phase shift} = \frac{-C}{B}$$

For $y = A \sec (Bx + C)$ and $y = A \csc (Bx + C)$:

$$\text{Period} = \frac{2\pi}{B} \qquad \text{Phase shift} = \frac{-C}{B}$$

For $y = A \tan (Bx + C)$ and $y = A \cot (Bx + C)$:

$$\text{Period} = \frac{\pi}{B} \qquad \text{Phase shift} = \frac{-C}{B}$$

Note: Amplitude is not defined for the secant, cosecant, tangent, and cotangent functions, all of which are unbounded.

Finding an Equation from the Graph of a Simple Harmonic

Given the graph of a simple harmonic, we wish to find an equation of the form $y = A \sin (Bx + C)$ or $y = A \cos (Bx + C)$ that produces the graph. Example 5 illustrates the process.

EXAMPLE 5 Finding an Equation of a Simple Harmonic Graph

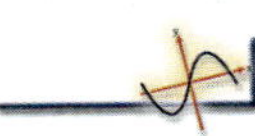

Graph $y_1 = 3 \sin x + 4 \cos x$ using a graphing utility, and find an equation of the form $y_2 = A \sin (Bx + C)$ that has the same graph as y_1. Find A and B exactly and C to three decimal places.

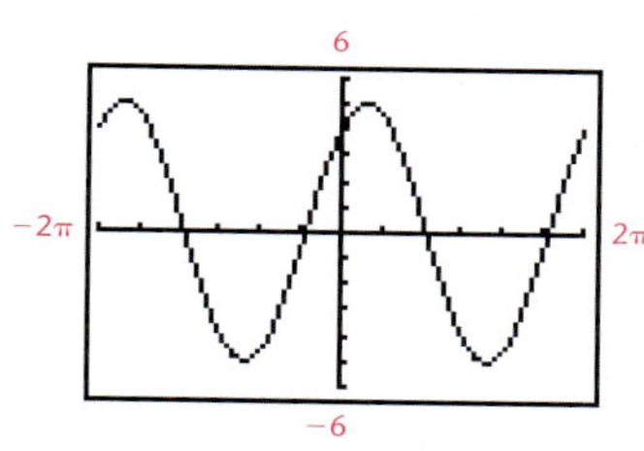

FIGURE 5 $y_1 = 3 \sin x + 4 \cos x$.

SOLUTION

The graph of y_1 is shown in Figure 5. The graph appears to be a sine curve shifted to the left. The amplitude and period appear to be 5 and 2π, respectively. (We will assume this for now and check it at the end.) Thus, $A = 5$, and because $P = 2\pi/B$, then $B = 2\pi/P = 2\pi/2\pi = 1$. Using a graphing utility, we find that the x intercept closest to the origin, to three decimal places, is -0.927. To find C, substitute $B = 1$ and $x = -0.927$ into the phase-shift formula $x = -C/B$ and solve for C:

$$x = -\frac{C}{B}$$

$$-0.927 = -\frac{C}{1}$$

$$C = 0.927$$

We now have the equation we are looking for:

$$y_2 = 5 \sin (x + 0.927)$$

CHECK

Graph y_1 and y_2 in the same viewing window. If the graphs are the same, it appears that only one graph is drawn—the second graph is drawn over the first. To check further that the graphs are the same, use TRACE and switch back and forth between y_1 and y_2 at different values of x. Figure 6 shows a comparison at $x = 0$ (both graphs appear in the same viewing window).

FIGURE 6

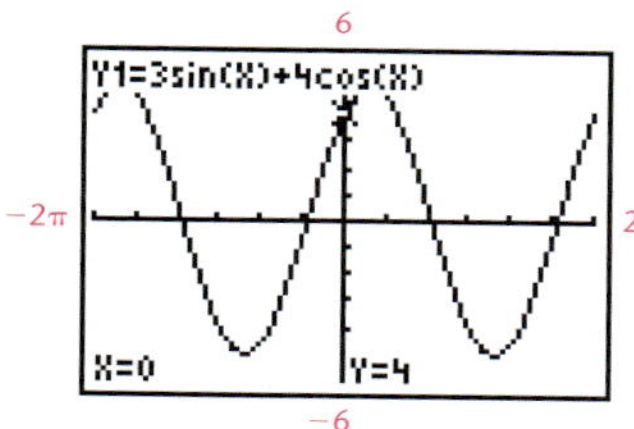

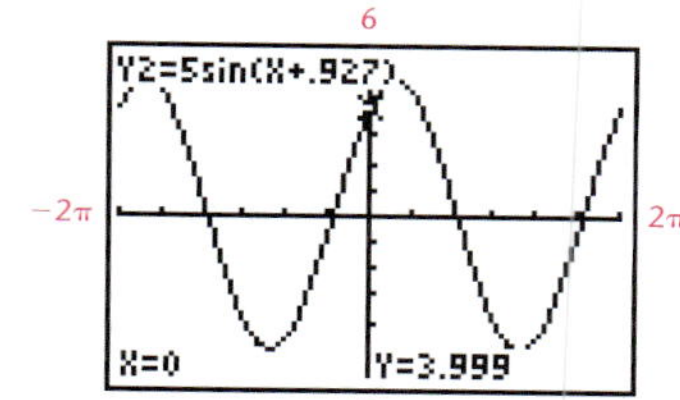

MATCHED 5 PROBLEM

Graph $y_1 = 4 \sin x - 3 \cos x$ using a graphing utility, and find an equation of the form $y_2 = A \sin (Bx + C)$ that has the same graph as y_1. (Find the x intercept closest to the origin to three decimal places.)

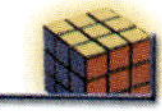

Modeling and Data Analysis

A graphing utility can be used to fit a function of the form $y = A \sin (Bx + C) + k$ to a set of data points. This tool, called **sinusoidal regression,** can be used to model periodic phenomena.

EXAMPLE 6 Temperature Variation

The monthly average high temperatures in Fairbanks, Alaska, are given in Table 1. Use sinusoidal regression to find the function $y = A \sin (Bx + C) + k$ that best fits the data. Round the constants A, B, C, and k to three significant digits and use the sinusoidal regression function to estimate the average high temperature on April 1.

TABLE 1 Temperatures in Fairbanks, Alaska

Month	1	2	3	4	5	6	7	8	9	10	11	12
Average High (°F)	0	8	25	44	61	71	73	66	54	31	11	3
Average Low (°F)	−19	−15	−3	20	37	49	52	46	35	16	−7	−15

SOLUTION

To observe the cyclical behavior of the data, we enter the average high temperatures for two consecutive years, from $x = 1$ to $x = 24$. The data, the sinusoidal regression function, and a plot of the data and graph of the regression function are shown in Figure 7. Rounding constants to three significant digits, the sinusoidal regression function is

$$y = 37.4 \sin (0.523x - 1.93) + 37.2$$

To estimate the average high temperature on April 1 we substitute $x = 3.5$, obtaining a temperature of 33.5°. (Note the slight discrepancy, due to rounding, from the estimate shown in Figure 7(c).)

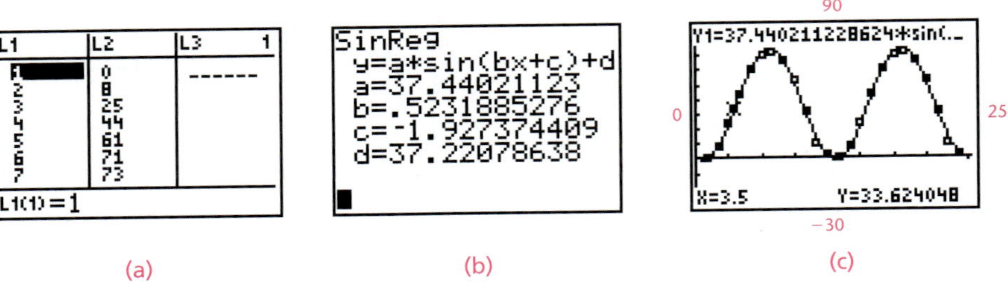

FIGURE 7 (a) (b) (c)

MATCHED PROBLEM 6

The monthly average low temperatures in Fairbanks, Alaska, are given in Table 1. Use sinusoidal regression to find the function $y = A \sin (Bx + C) + k$ that best fits the data. Round the constants A, B, C, and k to three significant digits and use the sinusoidal regression function to estimate the average low temperature on April 1.

ANSWERS MATCHED PROBLEMS

1. (A) Zeros: $\pi/2$, $3\pi/2$, $5\pi/2$; turning points: $(\pi, 5)$, $(2\pi, -5)$
   (B) Zeros: $\pi/2$, $3\pi/2$, $5\pi/2$; turning points: $(\pi, -1/3)$, $(2\pi, 1/3)$
2. (A) 20π (B) $1/3$
3. Amplitude: $1/4$; period: $2/3$; turning points: $(1/6, 1/4)$, $(1/2, -1/4)$, $(5/6, 1/4)$
4. Amplitude: $3/4$; period: π; phase shift: $-\pi/2$; zeros: $k\pi/2$, k any integer
5. $y_2 = 5 \sin (x - 0.644)$

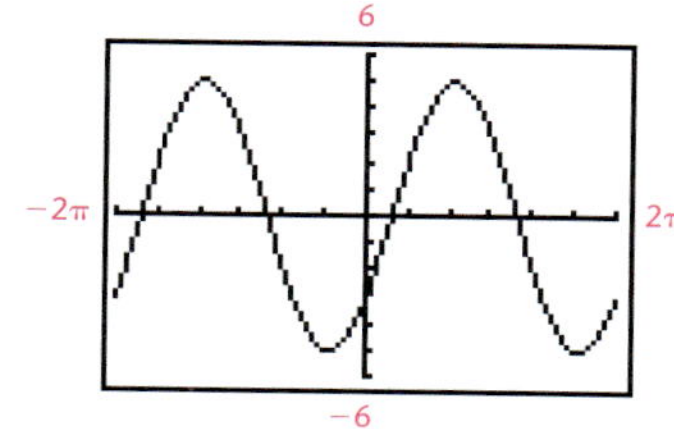

6. $y = 36.7 \sin (0.524x - 2.05) + 16.4$; 8.53°F

EXERCISE 5.5

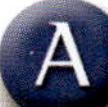

In Problems 1–12, find the amplitude (if applicable) and period. Check your answers by graphing.

1. $y = 3 \sin x, -2\pi \le x \le 2\pi$
2. $y = \frac{1}{4} \cos x, -2\pi \le x \le 2\pi$
3. $y = -\frac{1}{2} \cos x, -2\pi \le x \le 2\pi$
4. $y = -2 \sin x, -2\pi \le x \le 2\pi$
5. $y = \sin 3x, -\pi \le x \le \pi$
6. $y = \cos 2x, -\pi \le x \le \pi$
7. $y = 2 \cot 4x, 0 < x < \pi/2$
8. $y = 3 \tan 2x, -\pi \le x \le \pi$
9. $y = -\frac{1}{4} \tan 8\pi x, 0 < x < \frac{1}{2}$
10. $y = -\frac{1}{2} \cot 2\pi x, 0 < x < 1$
11. $y = \csc (x/2), -3\pi \le x \le 3\pi$
12. $y = \sec \pi x, -1.5 \le x \le 3.5$

In Problems 13–16, find the amplitude (if applicable), the period, and all zeros in the given interval. Check your answers by graphing.

13. $y = \sin \pi x, -2 \le x \le 2$
14. $y = \cos \pi x, -2 \le x \le 2$
15. $y = \frac{1}{2} \cot (x/2), 0 < x < 4\pi$
16. $y = \frac{1}{2} \tan (x/2), -\pi < x < 3\pi$

In Problems 17–20, find the amplitude (if applicable), the period, and all turning points in the given interval. Check your answers by graphing.

17. $y = 3 \cos 2x, -\pi \le x \le \pi$
18. $y = 2 \sin 4x, -\pi \le x \le \pi$
19. $y = 2 \sec \pi x, -1 \le x \le 3$
20. $y = 2 \csc (x/2), 0 < x < 8\pi$

B

In Problems 21–24, find the equation of the form $y = A \sin Bx$ that produces the graph shown. Check the results with a graphing utility.

21.

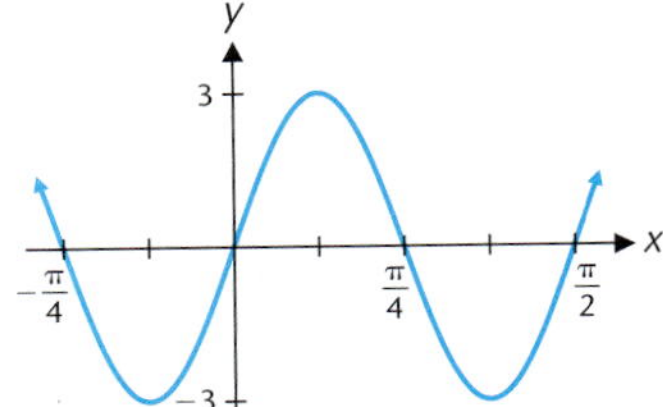

22.

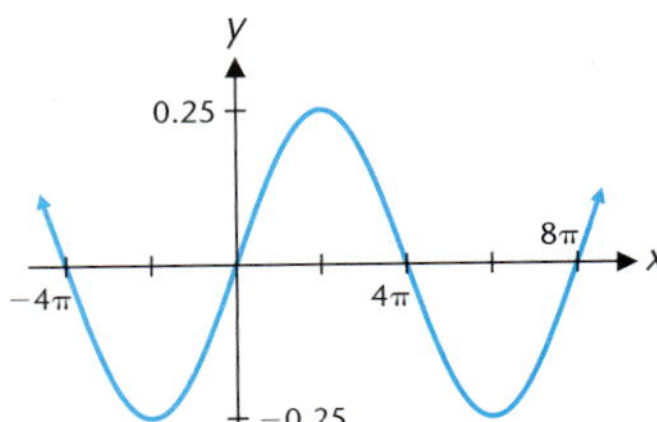

23.

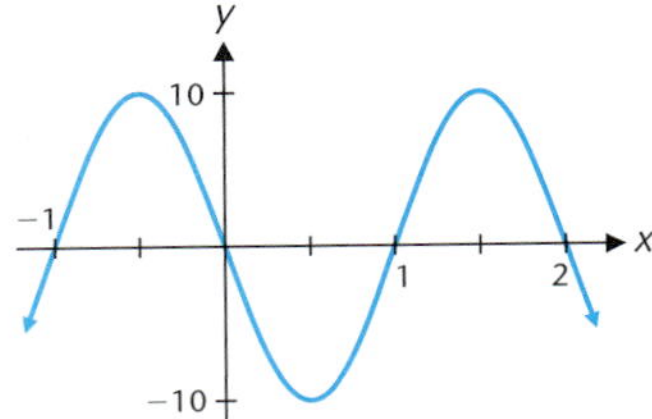

24.

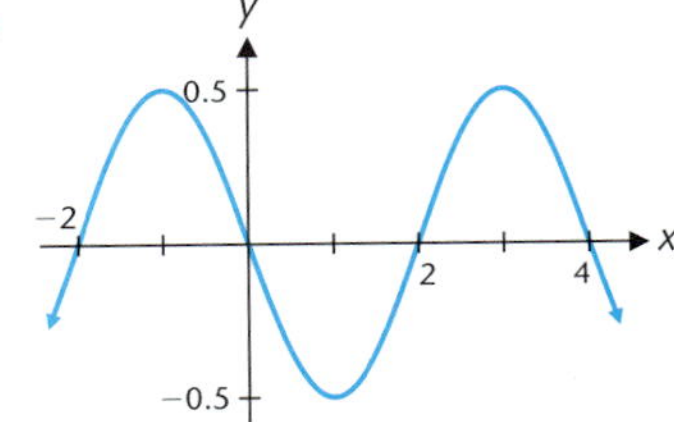

In Problems 25–28, find the equation of the form $y = A \cos Bx$ that produces the graph shown. Check the results with a graphing utility.

25.

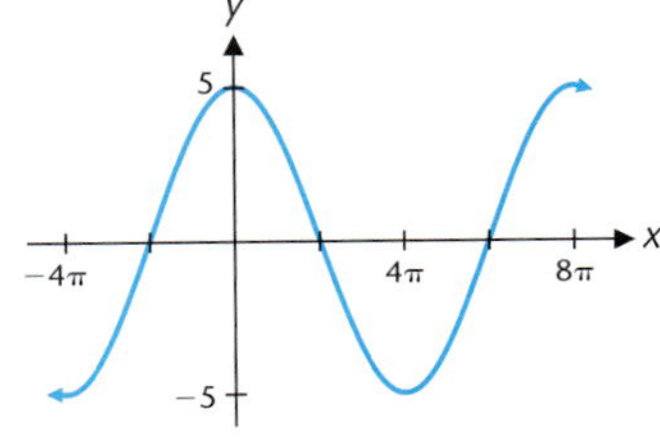

26.

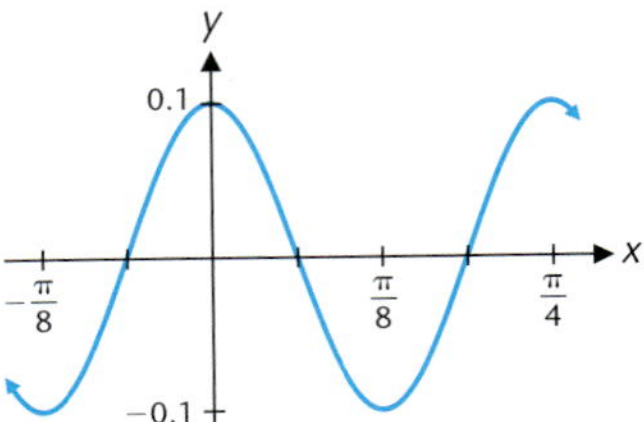

27.

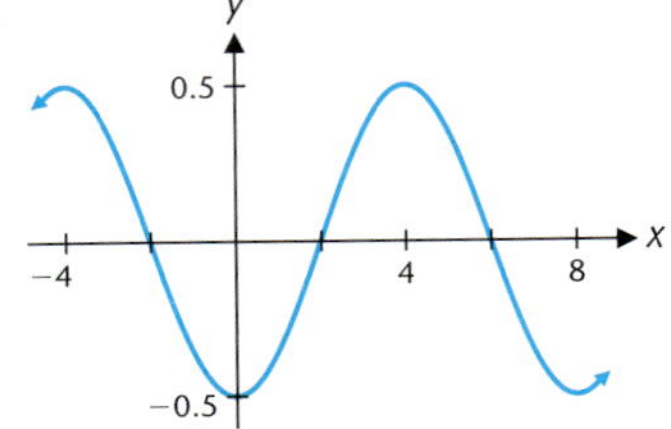

28.

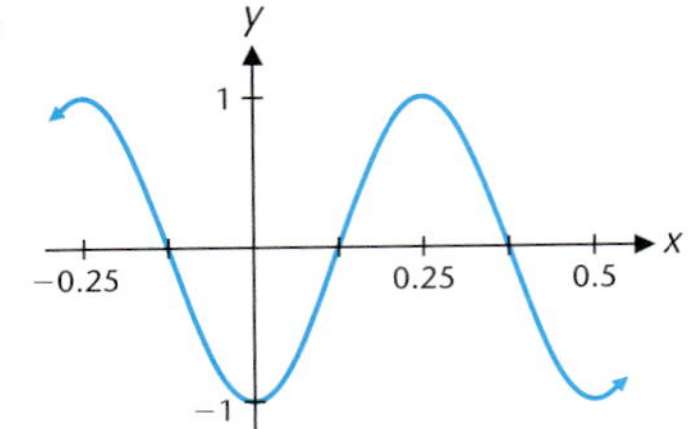

In Problems 29–44, find the amplitude (if applicable), period, and phase shift. Check your answers by graphing.

29. $y = \sin(x + \pi), -\pi \le x \le 3\pi$

30. $y = \cos(x - \pi), -\pi \le x \le 3\pi$

31. $y = \frac{1}{2}\cos(x - \pi/4), -\pi \le x \le 3\pi$

32. $y = 2\sin(x + \pi/4), -2\pi \le x \le 2\pi$

33. $y = \cot\left(x + \frac{\pi}{2}\right), -\frac{\pi}{2} < x < \frac{3\pi}{2}$

34. $y = \tan\left(x - \frac{\pi}{2}\right), -\pi < x < \pi$

35. $y = \sin[\pi(x - 1)], -2 \le x \le 3$

36. $y = \cos[2\pi(x - \frac{1}{2})], -1 \le x \le 2$

37. $y = 3\cos(\pi x + \pi/2), -2 \le x \le 2$

38. $y = 2\sin(\pi x - \pi/4), -1 \le x \le 3$

39. $y = \tan(2x + \pi), -\frac{3\pi}{4} < x < \frac{3\pi}{4}$

40. $y = \cot(2x - \pi), -\frac{\pi}{2} \le x \le \frac{\pi}{2}$

41. $y = -4 \cos (2x - \pi), -\pi \le x \le 3\pi$

42. $y = -2 \cos (4x + \pi), -\pi \le x \le \pi$

43. $y = \sec \left(\pi x + \dfrac{\pi}{2}\right), -1 < x < 1$

44. $y = \csc \left(\pi x - \dfrac{\pi}{2}\right), -1 < x < 1$

Graph each function in Problems 45–48. (Select the dimensions of each viewing window so that at least two periods are visible.) Find an equation of the form $y = k + A \sin Bx$ or $y = k + A \cos Bx$ that has the same graph as the given equation. (These problems suggest the existence of further identities in addition to the basic identities discussed in Section 5.4.)

45. $y = \cos^2 x - \sin^2 x$

46. $y = \sin x \cos x$

47. $y = 2 \sin^2 x$

48. $y = 2 \cos^2 x$

In Problems 49–56, graph at least two cycles of the given equation in a graphing utility, then find an equation of the form $y = A \tan Bx$, $y = A \cot Bx$, $y = A \sec Bx$, or $y = A \csc Bx$ that has the same graph. (These problems suggest additional identities beyond those discussed in Section 5.4. Additional identities are discussed in detail in Chapter 6.)

49. $y = \cot x - \tan x$

50. $y = \cot x + \tan x$

51. $y = \csc x + \cot x$

52. $y = \csc x - \cot x$

53. $y = \sin 3x + \cos 3x \cot 3x$

54. $y = \cos 2x + \sin 2x \tan 2x$

55. $y = \dfrac{\sin 4x}{1 + \cos 4x}$

56. $y = \dfrac{\sin 6x}{1 - \cos 6x}$

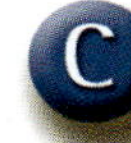

Problems 57 and 58 refer to the following graph:

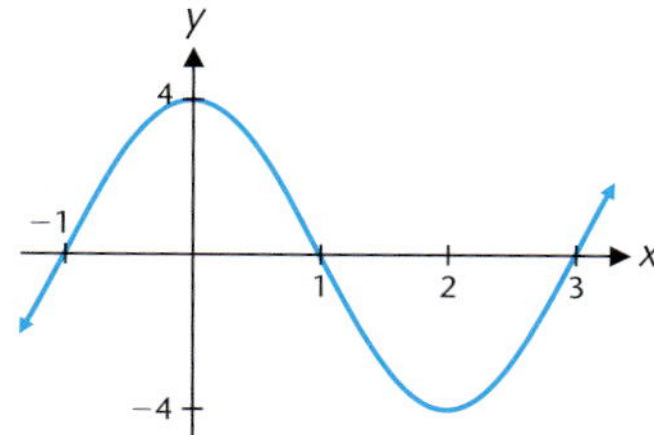

57. If the graph is a graph of an equation of the form $y = A \sin (Bx + C)$, $0 < -C/B < 2$, find the equation. Check the results with a graphing utility.

58. If the graph is a graph of an equation of the form $y = A \sin (Bx + C)$, $-2 < -C/B < 0$, find the equation. Check the results with a graphing utility.

Problems 59 and 60 refer to the following graph:

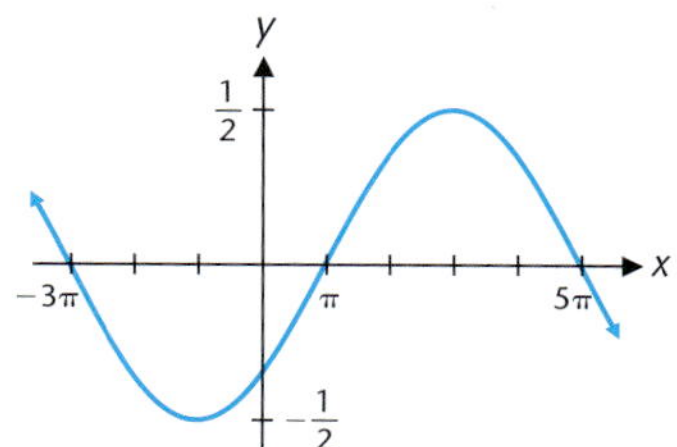

59. If the graph is a graph of an equation of the form $y = A \cos (Bx + C)$, $0 < -C/B < 4\pi$, find the equation. Check the results with a graphing utility.

60. If the graph is a graph of an equation of the form $y = A \cos (Bx + C)$, $-2\pi < -C/B < 0$, find the equation. Check the results with a graphing utility.

In Problems 61–64, state the amplitude, period, and phase shift of each function and sketch a graph of the function with the aid of a graphing utility.

61. $y = 3.5 \sin \left[\dfrac{\pi}{2}(t + 0.5)\right], 0 \le t \le 10$

62. $y = 5.4 \sin \left[\dfrac{\pi}{2.5}(t - 1)\right], 0 \le t \le 6$

63. $y = 50 \cos [2\pi(t - 0.25)], 0 \le t \le 2$

64. $y = 25 \cos [5\pi(t - 0.1)], 0 \le t \le 2$

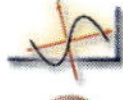

In Problems 65–70, graph each equation. (Select the dimensions of each viewing window so that at least two periods are visible.) Find an equation of the form $y = A \sin (Bx + C)$ that has the same graph as the given equation. Find A and B exactly and C to three decimal places. Use the x intercept closest to the origin as the phase shift.

65. $y = \sqrt{2} \sin x + \sqrt{2} \cos x$

66. $y = \sqrt{2} \sin x - \sqrt{2} \cos x$

67. $y = \sqrt{3}\sin x - \cos x$

68. $y = \sin x + \sqrt{3}\cos x$

69. $y = 4.8\sin 2x - 1.4\cos 2x$

70. $y = 1.4\sin 2x + 4.8\cos 2x$

Problems 71–76 illustrate combinations of functions that occur in harmonic analysis applications. Graph parts A, B, and C of each problem in the same viewing window. In Problems 71–74, what is happening to the amplitude of the function in part C? Give an example of a physical phenomenon that might be modeled by a similar function.

71. $0 \le x \le 16$

(A) $y = \dfrac{1}{x}$ **(B)** $y = -\dfrac{1}{x}$ **(C)** $y = \dfrac{1}{x}\sin\dfrac{\pi}{2}x$

72. $0 \le x \le 10$

(A) $y = \dfrac{2}{x}$ **(B)** $y = -\dfrac{2}{x}$ **(C)** $y = \dfrac{2}{x}\cos\pi x$

73. $0 \le x \le 10$

(A) $y = x$ **(B)** $y = -x$ **(C)** $y = x\sin\dfrac{\pi}{2}x$

74. $0 \le x \le 10$

(A) $y = \dfrac{x}{2}$ **(B)** $y = -\dfrac{x}{2}$ **(C)** $y = \dfrac{x}{2}\cos\pi x$

75. $0 \le x \le 2\pi$

(A) $y = \sin x$ **(B)** $y = \sin x + \dfrac{\sin 3x}{3}$

(C) $y = \sin x + \dfrac{\sin 3x}{3} + \dfrac{\sin 5x}{5}$

76. $0 \le x \le 4$

(A) $y = \sin\pi x$ **(B)** $y = \sin\pi x + \dfrac{\sin 2\pi x}{2}$

(C) $y = \sin\pi x + \dfrac{\sin 2\pi x}{2} + \dfrac{\sin 3\pi x}{3}$

APPLICATIONS

77. Spring-Mass System. A 6-pound weight hanging from the end of a spring is pulled $\frac{1}{3}$ foot below the equilibrium position and then released (see figure). If air resistance and friction are neglected, the distance x that the weight is from the equilibrium position relative to time t (in seconds) is given by

$$x = \tfrac{1}{3}\cos 8t$$

State the period P and amplitude A of this function, and graph it for $0 \le t \le \pi$.

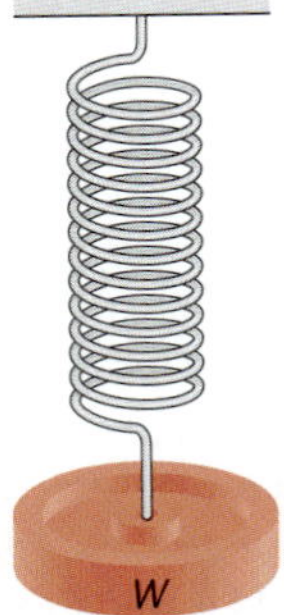

78. Electrical Circuit. An alternating current generator generates a current given by

$$I = 30\sin 120t$$

where t is time in seconds. What are the amplitude A and period P of this function? What is the frequency of the current; that is, how many cycles (periods) will be completed in 1 second?

★ **79. Spring-Mass System.** Assume the motion of the weight in Problem 77 has an amplitude of 8 inches and a period of 0.5 second, and that its position when $t = 0$ is 8 inches below its position at rest (displacement above rest position is positive and below is negative). Find an equation of the form $y = A\cos Bt$ that describes the motion at any time $t \ge 0$. (Neglect any damping forces—that is, friction and air resistance.)

★ **80. Electrical Circuit.** If the voltage E in an electrical circuit has an amplitude of 110 volts and a period of $\frac{1}{60}$ second, and if $E = 110$ volts when $t = 0$ seconds, find an equation of the form $E = A\cos Bt$ that gives the voltage at any time $t \ge 0$.

81. Pollution. The amount of sulfur dioxide pollutant from heating fuels released in the atmosphere in a city varies seasonally. Suppose the number of tons of pollutant released into the atmosphere during the nth week after January 1 for a particular city is given by

$$A(n) = 1.5 + \cos\frac{n\pi}{26} \qquad 0 \le n \le 104$$

Graph the function over the indicated interval and describe what the graph shows.

82. Medicine. A seated normal adult breathes in and exhales about 0.82 liter of air every 4.00 seconds. The volume of air in the lungs t seconds after exhaling is approximately

$$V(t) = 0.45 - 0.37\cos\frac{\pi t}{2} \qquad 0 \le t \le 8$$

Graph the function over the indicated interval and describe what the graph shows.

83. Electrical Circuit. The current in an electrical circuit is given by $I = 15 \cos (120\pi t + \pi/2)$, $0 \le t \le \frac{2}{60}$, where I is measured in amperes. State the amplitude A, period P, and phase shift. Graph the equation.

84. Electrical Circuit. The current in an electrical circuit is given by $I = 30 \cos (120\pi t - \pi)$, $0 \le t \le \frac{3}{60}$, where I is measured in amperes. State the amplitude A, period P, and phase shift. Graph the equation.

85. Physics—Engineering. The thin, plastic disk shown in the figure is rotated at 3 revolutions per second, starting at $\theta = 0$ (thus at the end of t seconds, $\theta = 6\pi t$—why?). If the disk has a radius of 3, show that the position of the shadow on the y scale from the small steel ball B is given by

$$y = 3 \sin 6\pi t$$

Graph this equation for $0 \le t \le 1$.

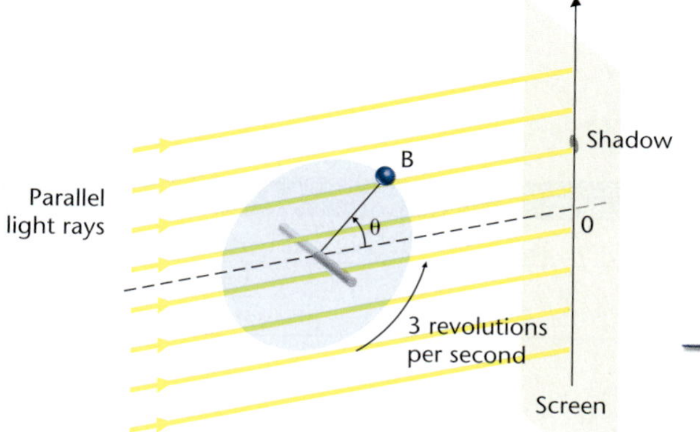

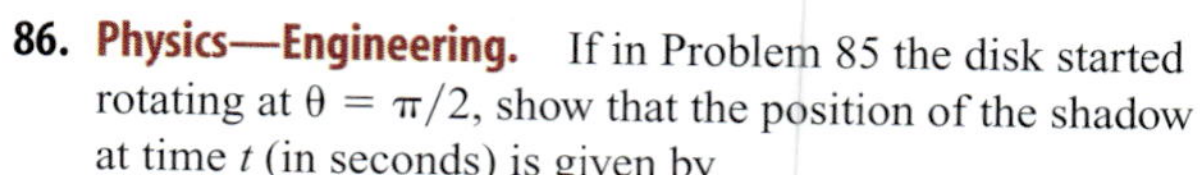

86. Physics—Engineering. If in Problem 85 the disk started rotating at $\theta = \pi/2$, show that the position of the shadow at time t (in seconds) is given by

$$y = 3 \sin \left(6\pi t + \frac{\pi}{2}\right)$$

Graph this equation for $0 \le t \le 1$.

★**87.** A beacon light 20 feet from a wall rotates clockwise at the rate of 1/4 revolutions per second (rps) (see the figure), thus, $\theta = \pi t/2$.

(A) Start counting time in seconds when the light spot is at N and write an equation for the length c of the light beam in terms of t.

(B) Graph the equation found in part A for the time interval [0, 1].

(C) Describe what happens to the length c of the light beam as t goes from 0 to 1.

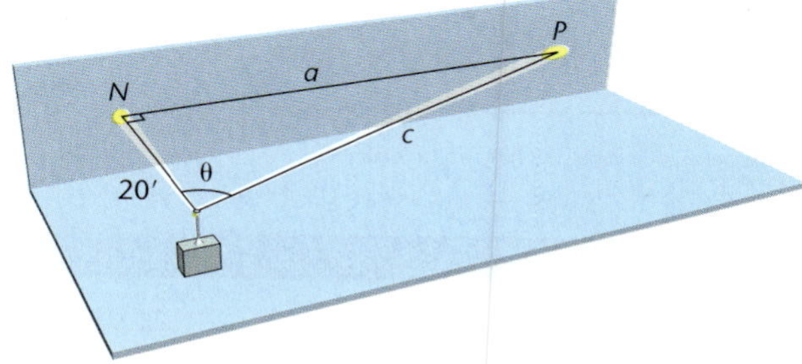

★**88.** Refer to Problem 87.

(A) Write an equation for the distance a the light spot travels along the wall in terms of time t.

(B) Graph the equation found in part A for the time interval [0, 1].

(C) Describe what happens to the distance a along the wall as t goes from 0 to 1.

MODELING AND DATA ANALYSIS 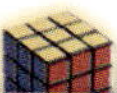

89. Modeling Sunset Times. Sunset times for the fifth of each month over a period of 1 year were taken from a tide booklet for the San Francisco Bay to form Table 2. Daylight savings time was ignored and the times are for a 24-hour clock starting at midnight.

(A) Using 1 month as the basic unit of time, enter the data for a 2-year period in your graphing utility and produce a scatter plot in the viewing window. Before entering Table 2 data into your graphing utility, convert sunset times from hours and minutes to decimal hours rounded to two decimal places. Choose $15 \le y \le 20$ for the viewing window.

(B) It appears that a sine curve of the form

$$y = k + A \sin (Bx + C)$$

will closely model these data. The constants k, A, and B are easily determined from Table 2 as follows: $A = (\max y - \min y)/2$, $B = 2\pi/\text{Period}$, and $k = \min y + A$. To estimate C, visually estimate to one decimal place the smallest positive phase shift from the plot in part A. After determining A, B, k, and C, write the resulting equation. (Your value of C may differ slightly from the answer in the back of the book.)

(C) Plot the results of parts A and B in the same viewing window. (An improved fit may result by adjusting your value of C slightly.)

(D) If your graphing utility has a sinusoidal regression feature, check your results from parts B and C by finding and plotting the regression equation.

90. Modeling Temperature Variation. The 30-year average monthly temperature, °F, for each month of the year for Washington, D.C., is given in Table 3 (*World Almanac*).

(A) Using 1 month as the basic unit of time, enter the data for a 2-year period in your graphing utility and produce a scatter plot in the viewing window. Choose $0 \le y \le 80$ for the viewing window.

(B) It appears that a sine curve of the form

$$y = k + A \sin (Bx + C)$$

will closely model these data. The constants k, A, and B are easily determined from Table 3 as follows: $A = (\max y - \min y)/2$, $B = 2\pi/\text{Period}$, and $k = \min y + A$. To estimate C, visually estimate to one decimal place the smallest positive phase shift from the plot in part A. After determining A, B, k, and C, write the resulting equation.

(C) Plot the results of parts A and B in the same viewing window. (An improved fit may result by adjusting your value of C slightly.)

(D) If your graphing utility has a sinusoidal regression feature, check your results from parts B and C by finding and plotting the regression equation.

TABLE 2

x (months)	1	2	3	4	5	6	7	8	9	10	11	12
y (sunset)*	17:05	17:38	18:07	18:36	19:04	19:29	19:35	19:15	18:34	17:47	17:07	16:51

*Time on a 24-hr clock, starting at midnight.

TABLE 3

x (months)	1	2	3	4	5	6	7	8	9	10	11	12
y (temp.)	31	34	43	53	62	71	76	74	67	55	45	35

SECTION 5.6 Inverse Trigonometric Functions

Inverse Sine Function • Inverse Cosine Function • Inverse Tangent Function • Summary • Inverse Cotangent, Secant, and Cosecant Functions (Optional)

A brief review of the general concept of inverse functions discussed in Section 1.6 should prove helpful before proceeding with Section 5.6. In the box we restate a few important facts about inverse functions from Section 1.6.

Facts About Inverse Functions

For a one-to-one function f and its inverse f^{-1}:

1. If (a, b) is an element of f, then (b, a) is an element of f^{-1}, and conversely.
2. Range of f = Domain of f^{-1}
   Domain of f = Range of f^{-1}

3. 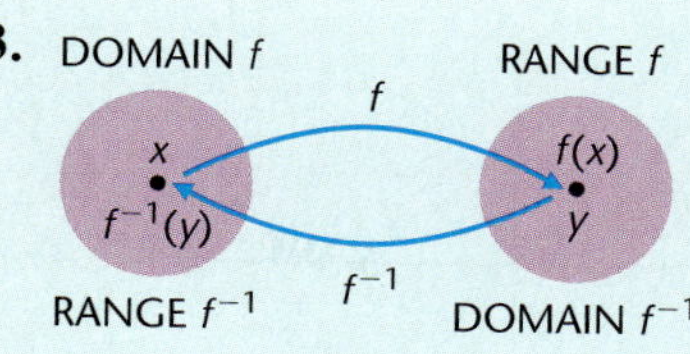

4. If $x = f^{-1}(y)$, then $y = f(x)$ for y in the domain of f^{-1} and x in the domain of f, and conversely.

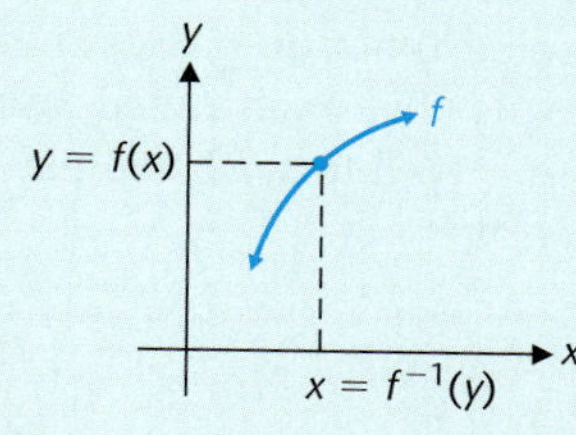

5. $f(f^{-1}(y)) = y$ for y in the domain of f^{-1}
$f^{-1}(f(x)) = x$ for x in the domain of f

All trigonometric functions are periodic; hence, each range value can be associated with infinitely many domain values (Fig. 1). As a result, no trigonometric function is one-to-one. Without restrictions, no trigonometric function has an inverse function. To resolve this problem, we restrict the domain of each function so that it is one-to-one over the restricted domain. Thus, for this restricted domain, an inverse function is guaranteed.

FIGURE 1 $y = \sin x$ is not one-to-one over $(-\infty, \infty)$.

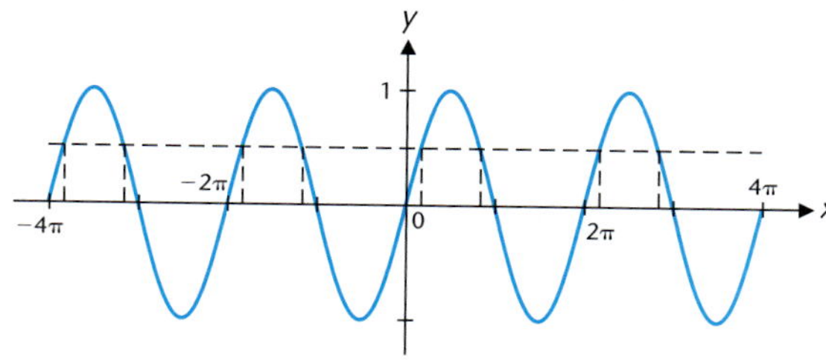

Inverse trigonometric functions represent another group of basic functions that are added to our library of elementary functions. These functions are used in many applications and mathematical developments, and will be particularly useful to us when we solve trigonometric equations in Section 6.5.

Inverse Sine Function

How can the domain of the sine function be restricted so that it is one-to-one? This can be done in infinitely many ways. A fairly natural and generally accepted way is illustrated in Figure 2.

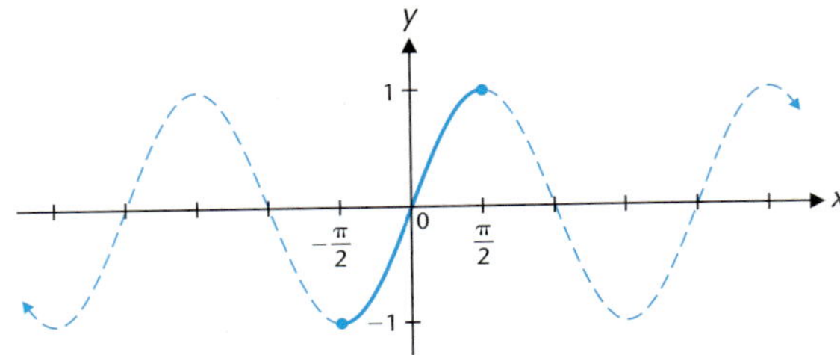

FIGURE 2 $y = \sin x$ is one-to-one over $[-\pi/2, \pi/2]$.

If the domain of the sine function is restricted to the interval $[-\pi/2, \pi/2]$, we see that the restricted function passes the horizontal line test (Section 1.6) and thus is one-to-one. Note that each range value from -1 to 1 is assumed exactly once as x moves from $-\pi/2$ to $\pi/2$. We use this restricted sine function to define the *inverse sine function.*

DEFINITION 1

Inverse Sine function

The **inverse sine function,** denoted by $\sin^{-1}$ or arcsin, is defined as the inverse of the restricted sine function $y = \sin x, -\pi/2 \leq x \leq \pi/2$. Thus,

$y = \sin^{-1} x$ and $y = \arcsin x$

are equivalent to

$\sin y = x \qquad \text{where} \qquad -\pi/2 \leq y \leq \pi/2, \ -1 \leq x \leq 1$

In words, the inverse sine of x, or the arcsine of x, is the number or angle y, $-\pi/2 \leq y \leq \pi/2$, whose sine is x.

To graph $y = \sin^{-1} x$, take each point on the graph of the restricted sine function and reverse the order of the coordinates. For example, because $(-\pi/2, -1)$, $(0, 0)$, and $(\pi/2, 1)$ are on the graph of the restricted sine function [Fig. 3(a)], then $(-1, -\pi/2)$, $(0, 0)$, and $(1, \pi/2)$ are on the graph of the inverse sine function, as shown in Figure 3(b). Using these three points provides us with a quick way of sketching the graph of the inverse sine function. A more accurate graph can be obtained by using a calculator.

FIGURE 3 Inverse sine function.

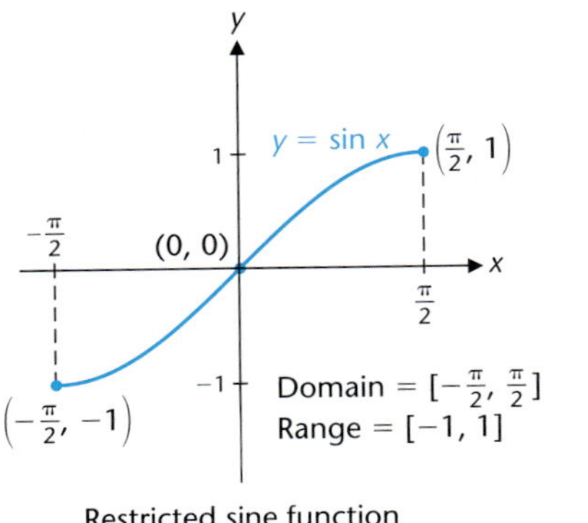

Restricted sine function

(a)

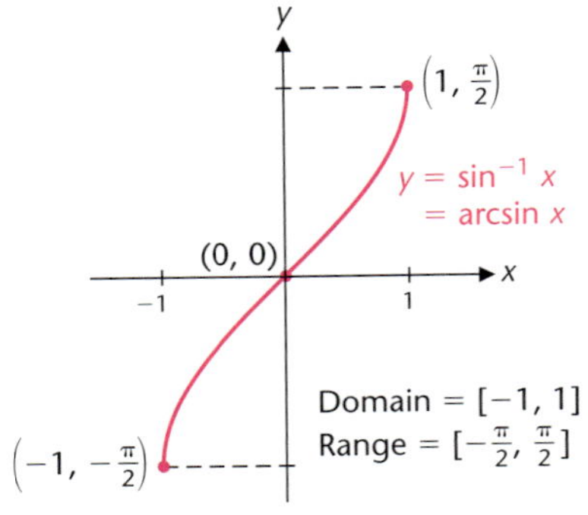

Inverse sine function

(b)

EXPLORE/DISCUSS 1

A graphing calculator produced the graph in Figure 4 for $y_1 = \sin^{-1} x$, $-2 \le x \le 2$, and $-2 \le y \le 2$. (Try this on your own graphing utility.) Explain why there are no parts of the graph on the intervals $[-2, -1)$ and $(1, 2]$.

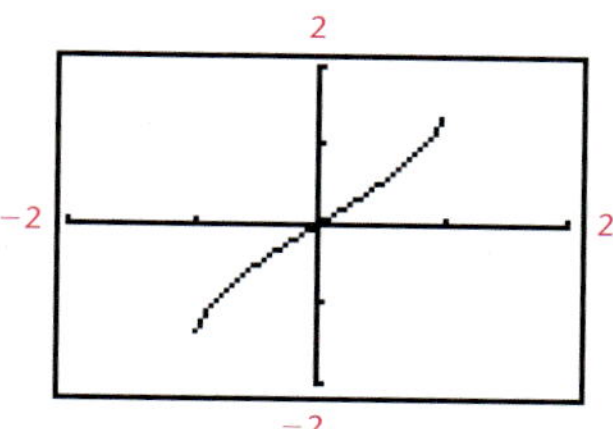

FIGURE 4

We state the important sine–inverse sine identities that follow from the general properties of inverse functions given in the box at the beginning of Section 5.6.

Sine–Inverse Sine Identities

$\sin(\sin^{-1} x) = x$	$-1 \le x \le 1$	$f(f^{-1}(x)) = x$
$\sin^{-1}(\sin x) = x$	$-\pi/2 \le x \le \pi/2$	$f^{-1}(f(x)) = x$

$\sin(\sin^{-1} 0.7) = 0.7$ $\quad$ $\sin(\sin^{-1} 1.3) \ne 1.3$

$\sin^{-1}[\sin(-1.2)] = -1.2$ $\quad$ $\sin^{-1}[\sin(-2)] \ne -2$

[*Note:* The number 1.3 is not in the domain of the inverse sine function, and -2 is not in the restricted domain of the sine function. Try calculating all these examples with your calculator and see what happens!]

EXAMPLE 1 Exact Values

Find exact values without using a calculator.

(A) $\arcsin(-\frac{1}{2})$ (B) $\sin^{-1}(\sin 1.2)$ (C) $\cos[\sin^{-1}(\frac{2}{3})]$

SOLUTIONS

(A) $y = \arcsin(-\frac{1}{2})$ is equivalent to

$$\sin y = -\tfrac{1}{2} \qquad -\frac{\pi}{2} \le y \le \frac{\pi}{2}$$

$$y = -\frac{\pi}{6} = \arcsin(-\tfrac{1}{2})$$

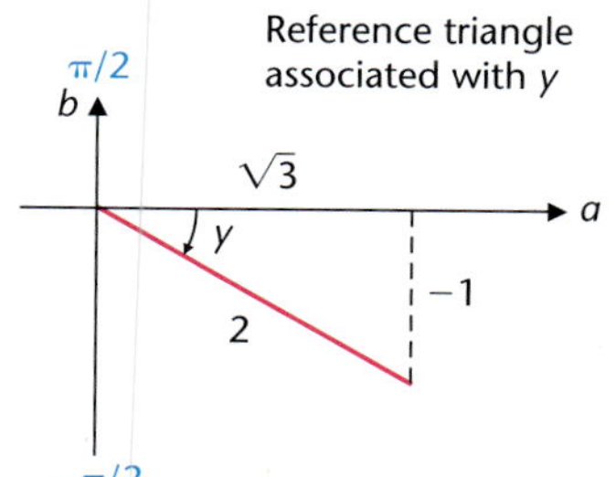

[*Note:* $y \neq 11\pi/6$, even though $\sin(11\pi/6) = -\frac{1}{2}$ because y must be between $-\pi/2$ and $\pi/2$, inclusive.]

(B) $\sin^{-1}(\sin 1.2) = 1.2$ Sine–inverse sine identity, because $-\pi/2 \leq 1.2 \leq \pi/2$

(C) Let $y = \sin^{-1}(\frac{2}{3})$; then $\sin y = (\frac{2}{3})$, $-\pi/2 \leq y \leq \pi/2$. Draw the reference triangle associated with y. Then $\cos y = \cos[\sin^{-1}(\frac{2}{3})]$ can be determined directly from the triangle (after finding the third side) without actually finding y.

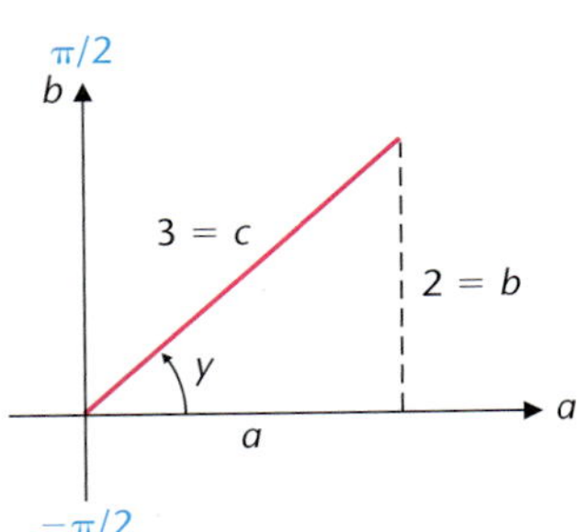

$$a^2 + b^2 = c^2$$
$$a = \sqrt{3^2 - 2^2} \quad \text{Because } a > 0 \text{ in quadrant I}$$
$$= \sqrt{5}$$

Thus, $\cos[\sin^{-1}(\frac{2}{3})] = \cos y = \sqrt{5}/3$.

MATCHED 1 PROBLEM

Find exact values without using a calculator.

(A) $\arcsin(\sqrt{2}/2)$ (B) $\sin[\sin^{-1}(-0.4)]$ (C) $\tan[\sin^{-1}(-1/\sqrt{5})]$

EXAMPLE 2 **Calculator Values**

Find to four significant digits using a calculator.

(A) $\arcsin(-0.3042)$ (B) $\sin^{-1} 1.357$ (C) $\cot[\sin^{-1}(-0.1087)]$

SOLUTIONS

The function keys used to represent inverse trigonometric functions vary among different brands of calculators, so read the user's manual for your calculator. Set your calculator in radian mode and follow your manual for key sequencing.

(A) $\arcsin(-0.3042) = -0.3091$

(B) $\sin^{-1} 1.357 = \text{Error}$ 1.357 is not in the domain of $\sin^{-1}$

(C) $\cot[\sin^{-1}(-0.1087)] = -9.145$

MATCHED 2 PROBLEM

Find to four significant digits using a calculator.

(A) $\sin^{-1} 0.2903$ (B) $\arcsin(-2.305)$ (C) $\cot[\sin^{-1}(-0.3446)]$

Inverse Cosine Function

To restrict the cosine function so that it becomes one-to-one, we choose the interval $[0, \pi]$. Over this interval the restricted function passes the horizontal line test, and each range value is assumed exactly once as x moves from 0 to π (Fig. 5). We use this restricted cosine function to define the *inverse cosine function.*

FIGURE 5 $y = \cos x$ is one-to-one over $[0, \pi]$.

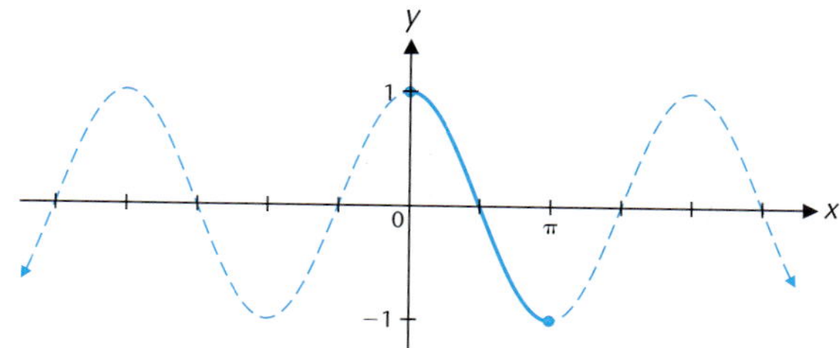

DEFINITION 2
Inverse Cosine Function

The **inverse cosine function,** denoted by $\cos^{-1}$ or arccos, is defined as the inverse of the restricted cosine function $y = \cos x, 0 \le x \le \pi$. Thus,

$$y = \cos^{-1} x \qquad \text{and} \qquad y = \arccos x$$

are equivalent to

$$\cos y = x \qquad \text{where} \qquad 0 \le y \le \pi, \ -1 \le x \le 1$$

In words, the inverse cosine of x, or the arccosine of x, is the number or angle y, $0 \le y \le \pi$, whose cosine is x.

Figure 6 compares the graphs of the restricted cosine function and its inverse. Notice that $(0, 1)$, $(\pi/2, 0)$, and $(\pi, -1)$ are on the restricted cosine graph. Reversing the coordinates gives us three points on the graph of the inverse cosine function.

FIGURE 6 Inverse cosine function.

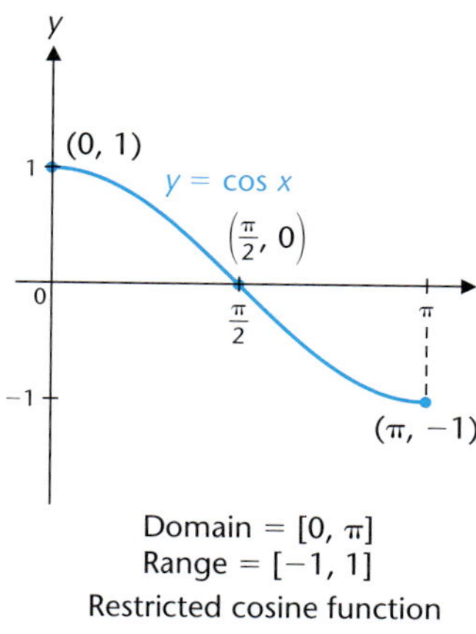

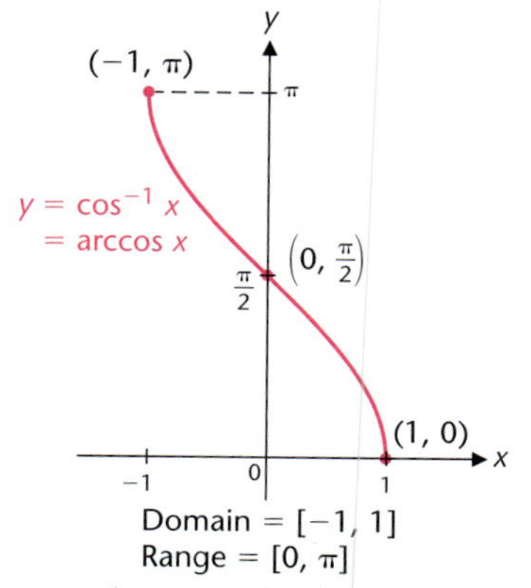

EXPLORE/DISCUSS 2

A graphing calculator produced the graph in Figure 7 for $y_1 = \cos^{-1} x$, $-2 \le x \le 2$, and $0 \le y \le 4$. (Try this on your own graphing utility.) Explain why there are no parts of the graph on the intervals $[-2, -1)$ and $(1, 2]$.

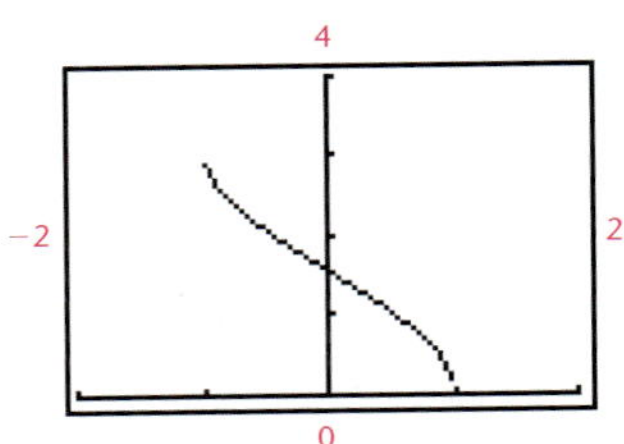

FIGURE 7

We complete the discussion by giving the cosine–inverse cosine identities:

Cosine–Inverse Cosine Identities

$\cos(\cos^{-1} x) = x$	$-1 \le x \le 1$	$f(f^{-1}(x)) = x$
$\cos^{-1}(\cos x) = x$	$0 \le x \le \pi$	$f^{-1}(f(x)) = x$

EXPLORE/DISCUSS 3

Evaluate each of the following with a calculator. Which illustrate a cosine–inverse cosine identity and which do not? Discuss why.

(A) $\cos(\cos^{-1} 0.2)$ (B) $\cos[\cos^{-1}(-2)]$

(C) $\cos^{-1}(\cos 2)$ (D) $\cos^{-1}[\cos(-3)]$

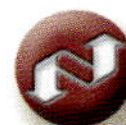

EXAMPLE 3 Exact Values

Find exact values without using a calculator.

(A) $\arccos(-\sqrt{3}/2)$ (B) $\cos(\cos^{-1} 0.7)$ (C) $\sin[\cos^{-1}(-\frac{1}{3})]$

SOLUTIONS

(A) $y = \arccos(-\sqrt{3}/2)$ is equivalent to

$$\cos y = -\frac{\sqrt{3}}{2} \qquad 0 \le y \le \pi$$

$$y = \frac{5\pi}{6} = \arccos\left(-\frac{\sqrt{3}}{2}\right)$$

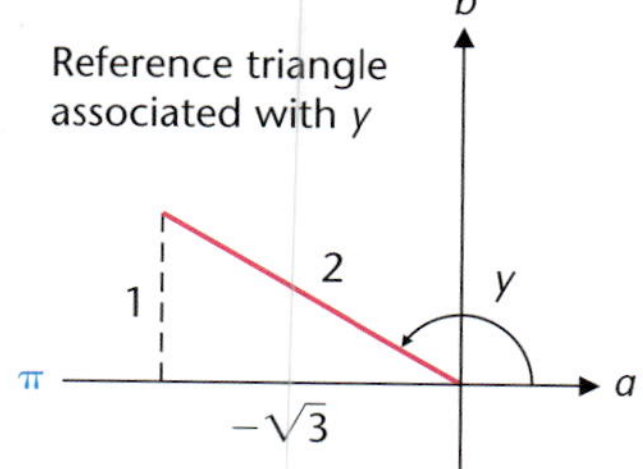

[*Note:* $y \neq -5\pi/6$, even though $\cos(-5\pi/6) = -\sqrt{3}/2$ because y must be between 0 and π, inclusive.]

(B) $\cos(\cos^{-1} 0.7) = 0.7$ Cosine–inverse cosine identity, because $-1 \le 0.7 \le 1$

(C) Let $y = \cos^{-1}(-\frac{1}{3})$; then $\cos y = -\frac{1}{3}$, $0 \le y \le \pi$. Draw a reference triangle associated with y. Then $\sin y = \sin[\cos^{-1}(-\frac{1}{3})]$ can be determined directly from the triangle (after finding the third side) without actually finding y.

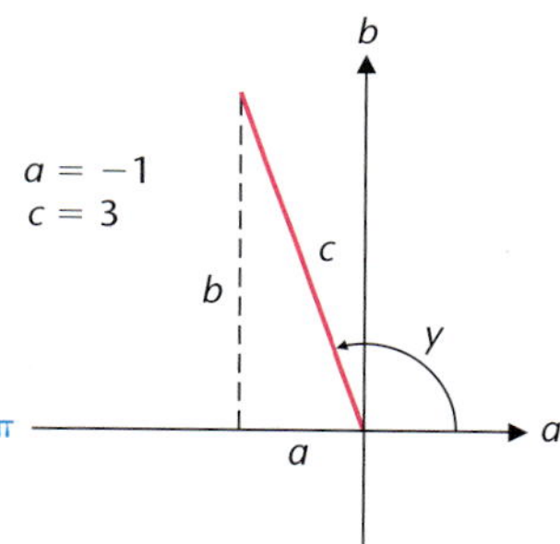

$$a^2 + b^2 = c^2$$

$$b = \sqrt{3^2 - (-1^2)} \qquad \text{Because } b > 0 \text{ in quadrant II}$$

$$= \sqrt{8} = 2\sqrt{2}$$

Thus, $\sin[\cos^{-1}(-\frac{1}{3})] = \sin y = 2\sqrt{2}/3$.

MATCHED 3 PROBLEM

Find exact values without using a calculator.

(A) $\arccos(\sqrt{2}/2)$ (B) $\cos^{-1}(\cos 3.05)$ (C) $\cot[\cos^{-1}(-1/\sqrt{5})]$

EXAMPLE 4 Calculator Values

Find to four significant digits using a calculator.

(A) arccos 0.4325 (B) $\cos^{-1} 2.137$ (C) $\csc[\cos^{-1}(-0.0349)]$

SOLUTIONS

Set your calculator in radian mode.

(A) arccos 0.4325 = 1.124

(B) $\cos^{-1} 2.137$ = Error 2.137 is not in the domain of $\cos^{-1}$

(C) $\csc[\cos^{-1}(-0.0349)] = 1.001$

MATCHED 4 PROBLEM

Find to four significant digits using a calculator.

(A) $\cos^{-1}(0.6773)$ (B) $\arccos(-1.003)$ (C) $\cot[\cos^{-1}(-0.5036)]$

Inverse Tangent Function

To restrict the tangent function so that it becomes one-to-one, we choose the interval $(-\pi/2, \pi/2)$. Over this interval the restricted function passes the horizontal line test, and each range value is assumed exactly once as x moves across this restricted domain (Fig. 8). We use this restricted tangent function to define the *inverse tangent function.*

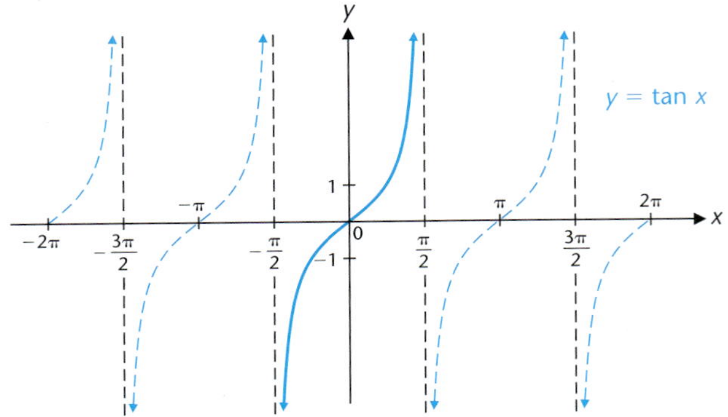

FIGURE 8 $y = \tan x$ is one-to-one over $(-\pi/2, \pi/2)$.

DEFINITION 3
Inverse Tangent Function

The **inverse tangent function,** denoted by $\tan^{-1}$ or arctan, is defined as the inverse of the restricted tangent function $y = \tan x$, $-\pi/2 < x < \pi/2$. Thus,

$$y = \tan^{-1} x \qquad \text{and} \qquad y = \arctan x$$

are equivalent to

$$\tan y = x \qquad \text{where} \qquad -\pi/2 < y < \pi/2 \text{ and } x \text{ is a real number}$$

In words, the inverse tangent of x, or the arctangent of x, is the number or angle y, $-\pi/2 < y < \pi/2$, whose tangent is x.

Figure 9 compares the graphs of the restricted tangent function and its inverse. Notice that $(-\pi/4, -1)$, $(0, 0)$, and $(\pi/4, 1)$ are on the restricted tangent graph. Reversing the coordinates gives us three points on the graph of the inverse tangent function. Also note that the vertical asymptotes become horizontal asymptotes for the inverse function.

FIGURE 9 Inverse tangent function.

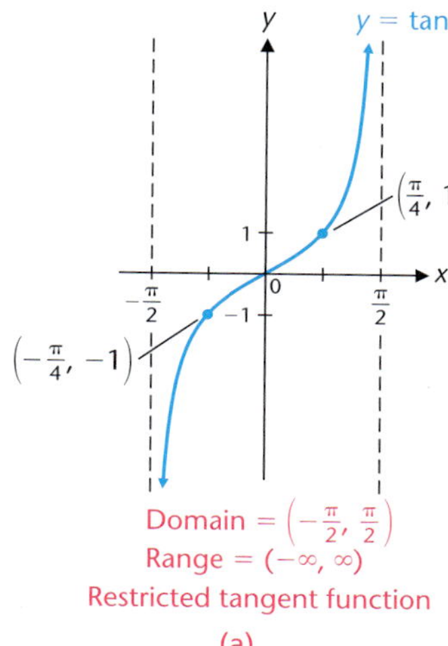

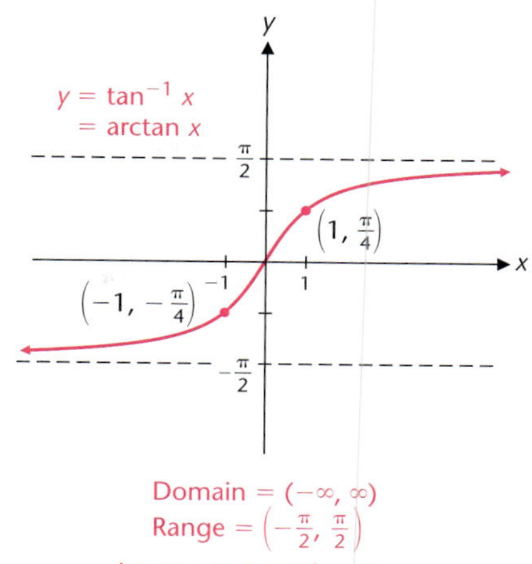

We now state the tangent–inverse tangent identities.

Tangent–Inverse Tangent Identities

$\tan(\tan^{-1} x) = x$	$-\infty < x < \infty$	$f(f^{-1}(x)) = x$
$\tan^{-1}(\tan x) = x$	$-\pi/2 < x < \pi/2$	$f^{-1}(f(x)) = x$

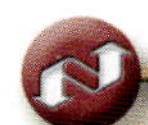

EXPLORE/DISCUSS 4

Evaluate each of the following with a calculator. Which illustrate a tangent–inverse tangent identity and which do not? Discuss why.

(A) $\tan(\tan^{-1} 30)$ (B) $\tan[\tan^{-1}(-455)]$

(C) $\tan^{-1}(\tan 1.4)$ (D) $\tan^{-1}[\tan(-3)]$

EXAMPLE 5 Exact Values

Find exact values without using a calculator.

(A) $\tan^{-1}(-1/\sqrt{3})$ (B) $\tan^{-1}(\tan 0.63)$

SOLUTIONS

(A) $y = \tan^{-1}(-1/\sqrt{3})$ is equivalent to

$$\tan y = -\frac{1}{\sqrt{3}} \qquad -\frac{\pi}{2} < y < \frac{\pi}{2}$$

$$y = -\frac{\pi}{6} = \tan^{-1}\left(-\frac{1}{\sqrt{3}}\right)$$

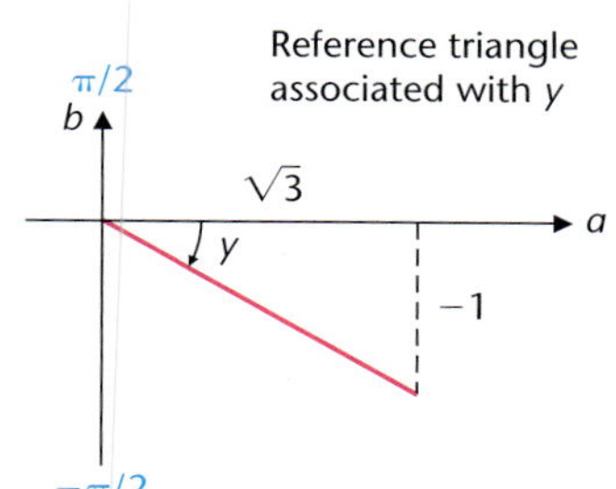

[*Note:* y cannot be $11\pi/6$ because y must be between $-\pi/2$ and $\pi/2$.]

(B) $\tan^{-1}(\tan 0.63) = 0.63$ Tangent–inverse tangent identity, because $-\pi/2 \le 0.63 \le \pi/2$

MATCHED 5 PROBLEM

Find exact values without using a calculator.

(A) $\arctan(-\sqrt{3})$ (B) $\tan(\tan^{-1} 43)$

Summary

We summarize the definitions and graphs of the inverse trigonometric functions discussed so far for convenient reference.

Summary of $\sin^{-1}$, $\cos^{-1}$, and $\tan^{-1}$

$y = \sin^{-1} x$ is equivalent to $x = \sin y$ where $-1 \le x \le 1$, $-\pi/2 \le y \le \pi/2$

$y = \cos^{-1} x$ is equivalent to $x = \cos y$ where $-1 \le x \le 1$, $0 \le y \le \pi$

$y = \tan^{-1} x$ is equivalent to $x = \tan y$ where $-\infty < x < \infty$, $-\pi/2 < y < \pi/2$

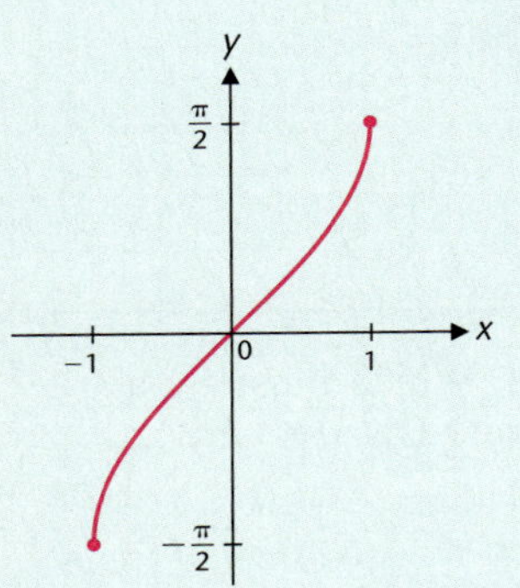

$y = \sin^{-1} x$
Domain = $[-1, 1]$
Range = $[-\frac{\pi}{2}, \frac{\pi}{2}]$

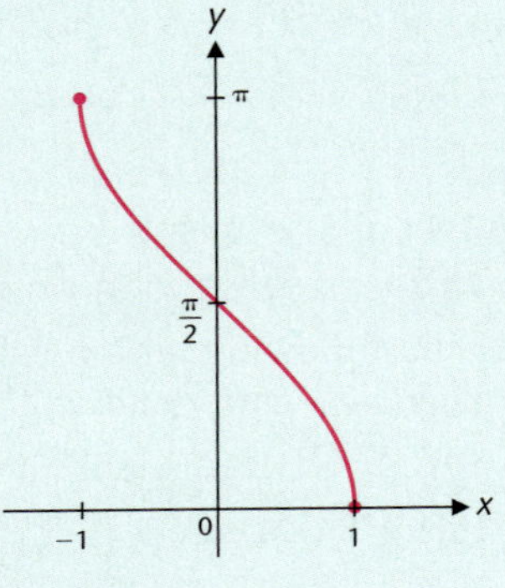

$y = \cos^{-1} x$
Domain = $[-1, 1]$
Range = $[0, \pi]$

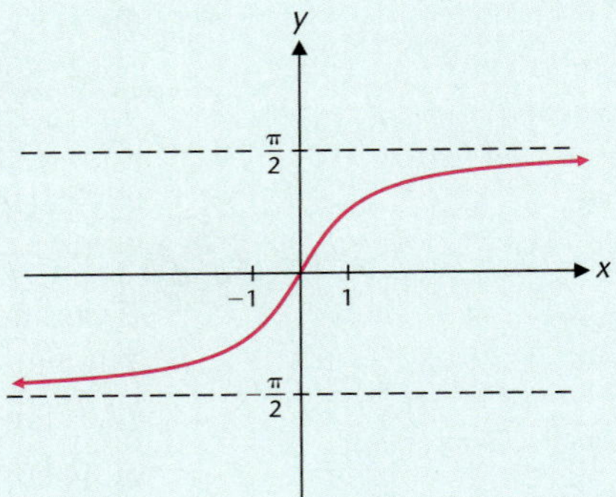

$y = \tan^{-1} x$
Domain = $(-\infty, \infty)$
Range = $\left(-\frac{\pi}{2}, \frac{\pi}{2}\right)$

Inverse Cotangent, Secant, and Cosecant Functions [Optional]

For completeness, we include the definitions and graphs of the inverse cotangent, secant, and cosecant functions.

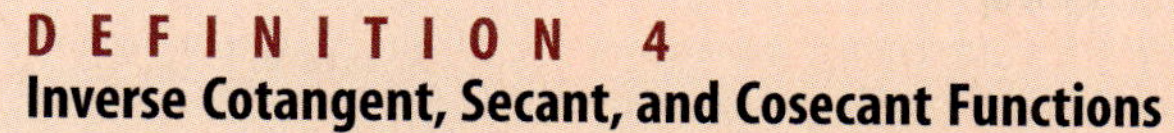

DEFINITION 4
Inverse Cotangent, Secant, and Cosecant Functions

$y = \cot^{-1} x$	is equivalent to $x = \cot y$	where $0 < y < \pi$, $-\infty < x < \infty$
$y = \sec^{-1} x$	is equivalent to $x = \sec y$	where $0 \le y \le \pi$, $y \ne \pi/2$, $\lvert x\rvert \ge 1$
$y = \csc^{-1} x$	is equivalent to $x = \csc y$	where $-\pi/2 \le y \le \pi/2$, $y \ne 0$, $\lvert x\rvert \ge 1$

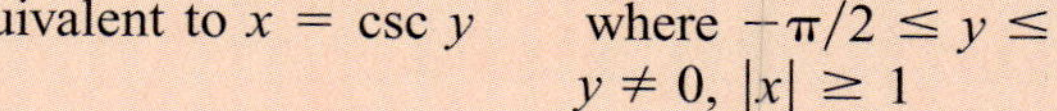

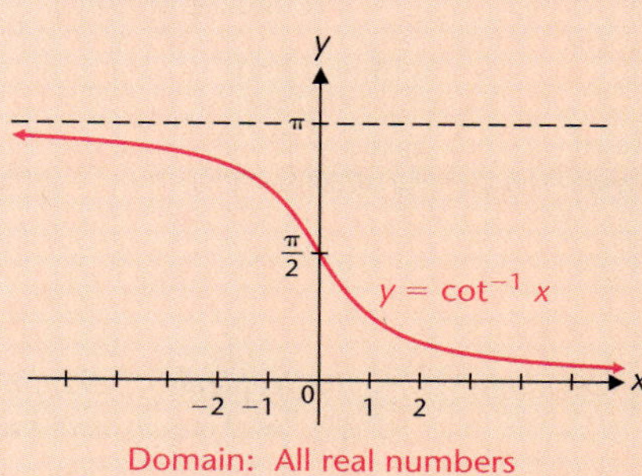

Domain: All real numbers
Range: $0 < y < \pi$

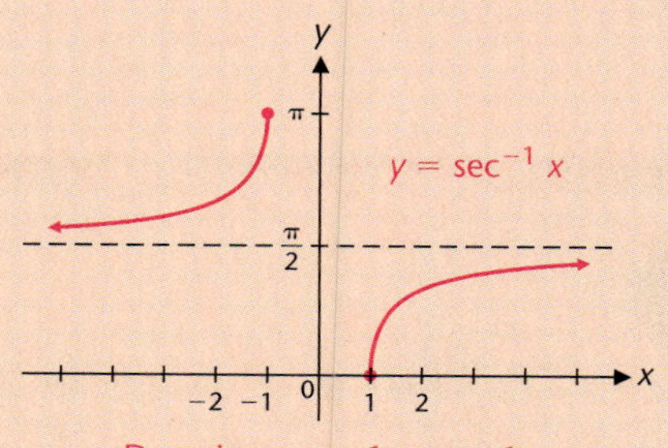

Domain: $x \le -1$ or $x \ge 1$
Range: $0 \le y \le \pi$, $y \ne \pi/2$

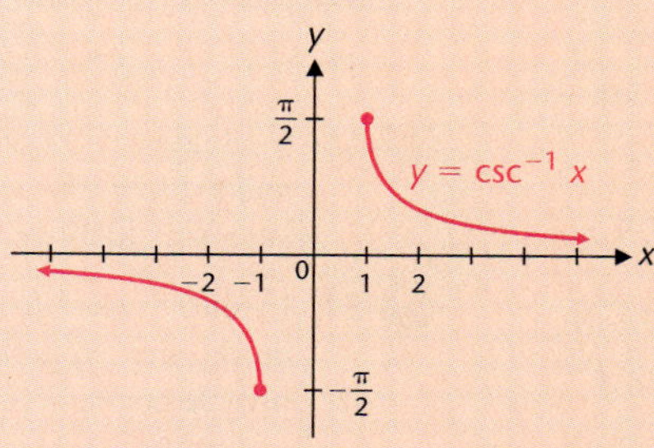

Domain: $x \le -1$ or $x \ge 1$
Range: $-\pi/2 \le y \le \pi/2$, $y \ne 0$

[*Note:* The definitions of $\sec^{-1}$ and $\csc^{-1}$ are not universally agreed upon.]

ANSWERS MATCHED PROBLEMS

1. (A) $\pi/4$ (B) -0.4 (C) $-1/2$
2. (A) 0.2945 (B) Not defined (C) -2.724
3. (A) $\pi/4$ (B) 3.05 (C) $-1/2$
4. (A) 0.8267 (B) Not defined (C) -0.5829
5. (A) $-\pi/3$ (B) 43

EXERCISE 5.6

Unless stated to the contrary, the inverse trigonometric functions are assumed to have real number ranges (use radian mode in calculator problems). A few problems involve ranges with angles in degree measure, and these are clearly indicated (use degree mode in calculator problems).

A

In Problems 1–12, find exact values without using a calculator.

1. $\cos^{-1} 0$

2. $\sin^{-1} 0$

3. $\arcsin(\sqrt{3}/2)$

4. $\arccos(\sqrt{3}/2)$

5. $\arctan \sqrt{3}$

6. $\tan^{-1} 1$

7. $\sin^{-1}(\sqrt{2}/2)$

8. $\cos^{-1}(\frac{1}{2})$

9. $\arccos 1$

10. $\arctan(1/\sqrt{3})$

11. $\sin^{-1}(\frac{1}{2})$

12. $\tan^{-1} 0$

In Problems 13–18, evaluate to four significant digits using a calculator.

13. $\sin^{-1} 0.9103$

14. $\cos^{-1} 0.4038$

15. $\arctan 103.7$

16. $\tan^{-1} 43.09$

17. $\arccos 3.051$

18. $\arcsin 1.131$

B

In Problems 19–34, find exact values without using a calculator.

19. $\arcsin(-\sqrt{2}/2)$

20. $\arccos(-\frac{1}{2})$

21. $\tan^{-1}(-\sqrt{3})$

22. $\tan^{-1}(-1)$

23. $\tan(\tan^{-1} 25)$

24. $\sin[\sin^{-1}(-0.6)]$

25. $\cos^{-1}(\cos 2.3)$

26. $\tan^{-1}[\tan(-1.5)]$

27. $\sin(\cos^{-1} \sqrt{3}/2)$

28. $\tan[\cos^{-1}(\frac{1}{2})]$

29. $\csc[\tan^{-1}(-1)]$

30. $\cos[\sin^{-1}(-\sqrt{2}/2)]$

31. $\sin^{-1}[\sin \pi]$

32. $\cos^{-1}[\cos(-\pi/2)]$

33. $\cos^{-1}[\cos(4\pi/3)]$

34. $\sin^{-1}[\sin(5\pi/4)]$

In Problems 35–40, evaluate to four significant digits using a calculator.

35. $\arctan(-10.04)$

36. $\tan^{-1}(-4.038)$

37. $\cot[\cos^{-1}(-0.7003)]$

38. $\sec[\sin^{-1}(-0.0399)]$

39. $\sqrt{5 + \cos^{-1}(1 - \sqrt{2})}$

40. $\sqrt{2} + \tan^{-1} \sqrt[3]{5}$

In Problems 41–46, find the exact degree measure of each without the use of a calculator.

41. $\sin^{-1}(-\sqrt{2}/2)$

42. $\cos^{-1}(-\frac{1}{2})$

43. $\arctan(-\sqrt{3})$

44. $\arctan(-1)$

45. $\cos^{-1}(-1)$

46. $\sin^{-1}(-1)$

In Problems 47–52, find the degree measure of each to two decimal places using a calculator set in degree mode.

47. $\cos^{-1} 0.7253$

48. $\tan^{-1} 12.4304$

49. $\arcsin(-0.3662)$

50. $\arccos(-0.9206)$

51. $\tan^{-1}(-837)$

52. $\sin^{-1}(-0.7071)$

53. Evaluate $\sin^{-1}(\sin 2)$ with a calculator set in radian mode, and explain why this does or does not illustrate the inverse sine–sine identity.

54. Evaluate $\cos^{-1}[\cos(-0.5)]$ with a calculator set in radian mode, and explain why this does or does not illustrate the inverse cosine–cosine identity.

In Problems 55–62, graph each function in a graphing utility over the indicated interval.

55. $y = \sin^{-1} x, -1 \le x \le 1$

56. $y = \cos^{-1} x, -1 \le x \le 1$

57. $y = \cos^{-1}(x/3), -3 \le x \le 3$

58. $y = \sin^{-1}(x/2), -2 \le x \le 2$

59. $y = \sin^{-1}(x - 2), 1 \le x \le 3$

60. $y = \cos^{-1}(x + 1), -2 \le x \le 0$

61. $y = \tan^{-1}(2x - 4), -2 \le x \le 6$

62. $y = \tan^{-1}(2x + 3), -5 \le x \le 2$

63. The identity $\cos(\cos^{-1} x) = x$ is valid for $-1 \le x \le 1$.

(A) Graph $y = \cos(\cos^{-1} x)$ for $-1 \le x \le 1$.

(B) What happens if you graph $y = \cos(\cos^{-1} x)$ over a larger interval, say $-2 \le x \le 2$? Explain.

64. The identity $\sin(\sin^{-1} x) = x$ is valid for $-1 \le x \le 1$.

(A) Graph $y = \sin(\sin^{-1} x)$ for $-1 \le x \le 1$.

(B) What happens if you graph $y = \sin(\sin^{-1} x)$ over a larger interval, say $-2 \le x \le 2$? Explain.

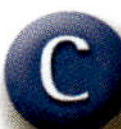

In Problems 65–68, write each expression as an algebraic expression in x free of trigonometric or inverse trigonometric functions.

65. $\cos(\sin^{-1} x)$

66. $\sin(\cos^{-1} x)$

67. $\cos(\arctan x)$

68. $\tan(\arcsin x)$

In Problems 69 and 70, find $f^{-1}(x)$. How must x be restricted in $f^{-1}(x)$?

69. $f(x) = 4 + 2\cos(x - 3),\ 3 \le x \le (3 + \pi)$

70. $f(x) = 3 + 5\sin(x - 1),\ (1 - \pi/2) \le x \le (1 + \pi/2)$

71. The identity $\cos^{-1}(\cos x) = x$ is valid for $0 \le x \le \pi$.

(A) Graph $y = \cos^{-1}(\cos x)$ for $0 \le x \le \pi$.

(B) What happens if you graph $y = \cos^{-1}(\cos x)$ over a larger interval, say $-2\pi \le x \le 2\pi$? Explain.

72. The identity $\sin^{-1}(\sin x) = x$ is valid for $-\pi/2 \le x \le \pi/2$.

(A) Graph $y = \sin^{-1}(\sin x)$ for $-\pi/2 \le x \le \pi/2$.

(B) What happens if you graph $y = \sin^{-1}(\sin x)$ over a larger interval, say $-2\pi \le x \le 2\pi$? Explain.

APPLICATIONS

73. Photography. The viewing angle changes with the focal length of a camera lens. A 28-millimeter wide-angle lens has a wide viewing angle and a 300-millimeter telephoto lens has a narrow viewing angle. For a 35-millimeter format camera the viewing angle θ, in degrees, is given by

$$\theta = 2\tan^{-1}\frac{21.634}{x}$$

where x is the focal length of the lens being used. What is the viewing angle (in decimal degrees to two decimal places) of a 28-millimeter lens? Of a 100-millimeter lens?

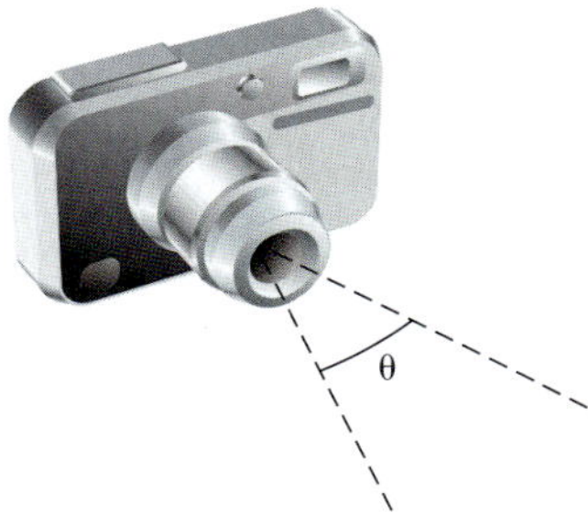

74. Photography. Referring to Problem 73, what is the viewing angle (in decimal degrees to two decimal places) of a 17-millimeter lens? Of a 70-millimeter lens?

75. (A) Graph the function in Problem 73 in a graphing utility using degree mode. The graph should cover lenses with focal lengths from 10 millimeters to 100 millimeters.

(B) What focal-length lens, to two decimal places, would have a viewing angle of 40°? Solve by graphing $\theta = 40$ and $\theta = 2\tan^{-1}(21.634/x)$ in the same viewing window and finding the point of intersection using an approximation routine.

76. (A) Graph the function in Problem 73 in a graphing utility, in degree mode, with the graph covering lenses with focal lengths from 100 millimeters to 1,000 millimeters.

(B) What focal length lens, to two decimal places, would have a viewing angle of 10°? Solve by graphing $\theta = 10$ and $\theta = \tan^{-1}(21.634/x)$ in the same viewing window and finding the point of intersection using an approximation routine.

★ **77. Engineering.** The length of the belt around the two pulleys in the figure is given by

$$L = \pi D + (d - D)\theta + 2C\sin\theta$$

where θ (in radians) is given by

$$\theta = \cos^{-1}\frac{D - d}{2C}$$

Verify these formulas, and find the length of the belt to two decimal places if $D = 4$ inches, $d = 2$ inches, and $C = 6$ inches.

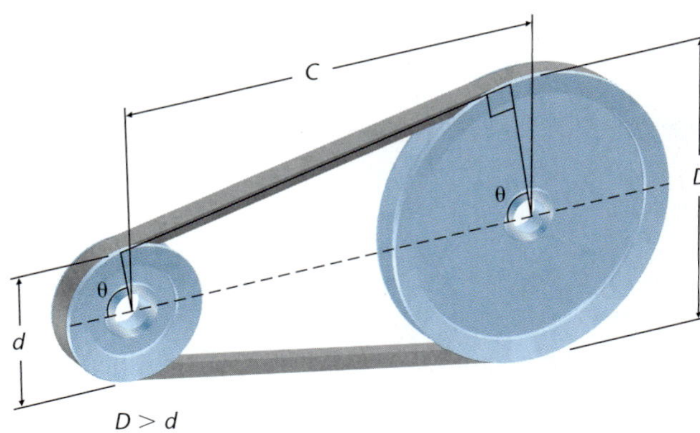

★ **78. Engineering.** For Problem 77, find the length of the belt if $D = 6$ inches, $d = 4$ inches, and $C = 10$ inches.

79. Engineering. The function

$$y_1 = 4\pi - 2\cos^{-1}\frac{1}{x} + 2x\sin\left(\cos^{-1}\frac{1}{x}\right)$$

represents the length of the belt around the two pulleys in Problem 77 when the centers of the pulleys are x inches apart.

(A) Graph y_1 in a graphing utility (in radian mode), with the graph covering pulleys with their centers from 3 to 10 inches apart.

(B) How far, to two decimal places, should the centers of the two pulleys be placed to use a belt 24 inches long? Solve by graphing y_1 and $y_2 = 24$ in the same viewing window and finding the point of intersection using an approximation routine.

80. Engineering. The function

$$y_1 = 6\pi - 2\cos^{-1}\frac{1}{x} + 2x\sin\left(\cos^{-1}\frac{1}{x}\right)$$

represents the length of the belt around the two pulleys in Problem 78 when the centers of the pulleys are x inches apart.

(A) Graph y_1 in a graphing utility (in radian mode), with the graph covering pulleys with their centers from 3 to 20 inches apart.

(B) How far, to two decimal places, should the centers of the two pulleys be placed to use a belt 36 inches long? Solve by graphing y_1 and $y_2 = 36$ in the same viewing window and finding the point of intersection using an approximation routine.

★ **81. Motion.** The figure represents a circular courtyard surrounded by a high stone wall. A floodlight located at E shines into the courtyard.

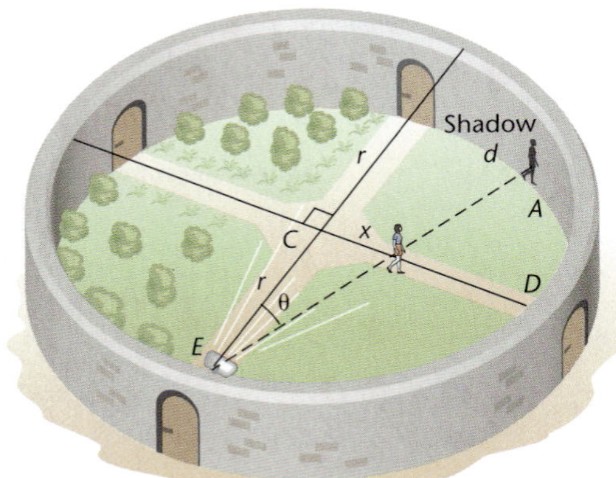

(A) If a person walks x feet away from the center along DC, show that the person's shadow will move a distance given by

$$d = 2r\theta = 2r\tan^{-1}\frac{x}{r}$$

where θ is in radians. [*Hint:* Draw a line from A to C.]

(B) Find d to two decimal places if $r = 100$ feet and $x = 40$ feet.

★ **82. Motion.** In Problem 81, find d for $r = 50$ feet and $x = 25$ feet.

CHAPTER 5 REVIEW

5.1 Angles and Their Measure

An **angle** is formed by rotating (in a plane) a ray m, called the **initial side** of the angle, around its endpoint until it coincides with a ray n, called the **terminal side** of the angle. The common endpoint of m and n is called the **vertex.** If the rotation is counterclockwise, the angle is **positive;** if clockwise, **negative.**

An angle is in **standard position** in a rectangular coordinate system if its vertex is at the origin and its initial side is along the positive x axis. **Quadrantal angles** have their terminal sides on a coordinate axis. An angle of **1 degree** is $\frac{1}{360}$ of a complete rotation. An angle of **1 radian** is a central angle of a circle subtended by an arc having the same length as the radius.

Radian measure: $\theta = \dfrac{s}{r}$

Radian–degree conversion: $\dfrac{\theta_{\text{deg}}}{180°} = \dfrac{\theta_{\text{rad}}}{\pi \text{ radians}}$

If a point P moves through an angle θ and arc length s, in time t, on the circumference of a circle of radius r, then the (average) **linear speed** of P is

$$v = \frac{s}{t}$$

and the (average) **angular speed** is

$$\omega = \frac{\theta}{t}$$

Because $s = r\theta$ it follows that $v = r\omega$.

5.2 Trigonometric Functions: A Unit Circle Approach

If θ is a positive angle in standard position, and P is the point of intersection of the terminal side of θ with the unit circle, then the radian measure of θ equals the length x of the arc opposite θ; and if θ is negative, the radian measure of θ equals the negative of the length of the intercepted arc. The function W that associates with each real number x the point $W(x) = P$ is called the **wrapping function,** and the point P is called a **circular point.** The function $W(x)$ can be visualized as a wrapping of the real number line, with origin at (1, 0), around the unit circle—the positive real axis is wrapped counterclockwise and the negative real axis is wrapped clockwise—so that each real number is paired with a unique circular point. The function $W(x)$ is not one-to-one: for example, each of the real numbers $2\pi k$, k any integer, corresponds to the circular point (1, 0).

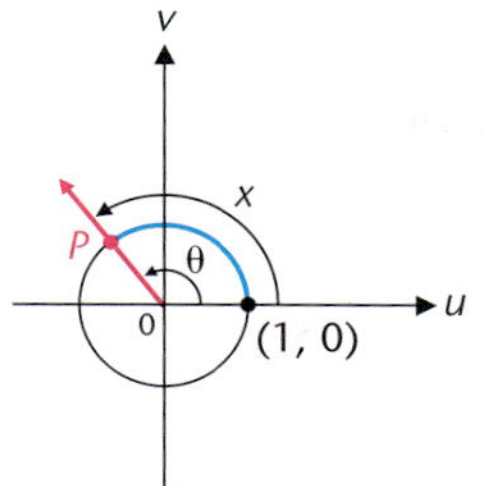

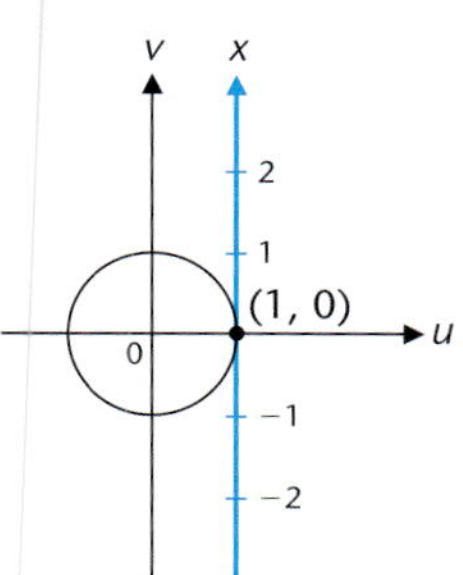

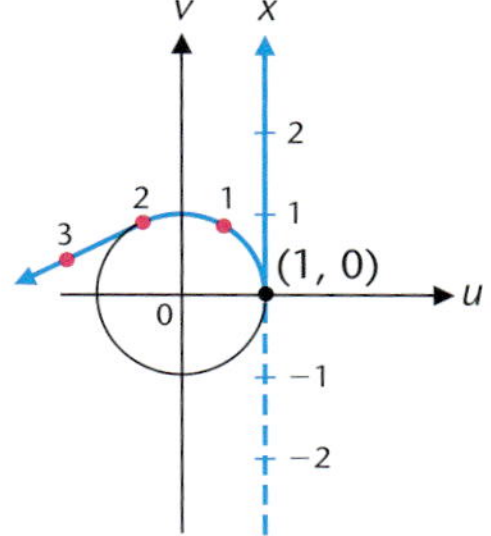

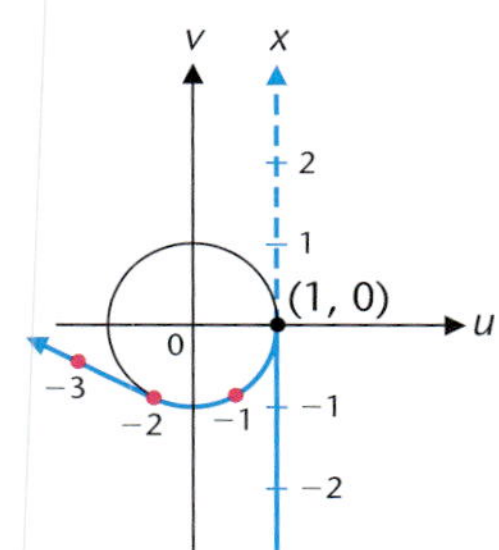

The coordinates of key circular points in the first quadrant can be found using simple geometric facts; the coordinates of the circular point associated with any multiple of $\pi/6$ or $\pi/4$ can then be determined using symmetry properties.

Coordinates of Key Circular Points

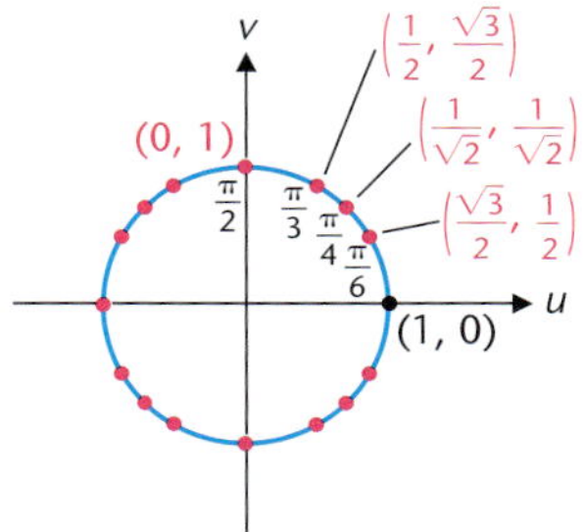

The six **trigonometric functions—sine, cosine, tangent, cotangent, secant, and cosecant**—are defined in terms of the coordinates (a, b) of the circular point $W(x)$ that lies on the terminal side of the angle with radian measure x:

$\sin x = b$	$\csc x = \dfrac{1}{b} \quad b \neq 0$
$\cos x = a$	$\sec x = \dfrac{1}{a} \quad a \neq 0$
$\tan x = \dfrac{b}{a} \quad a \neq 0$	$\cot x = \dfrac{a}{b} \quad b \neq 0$

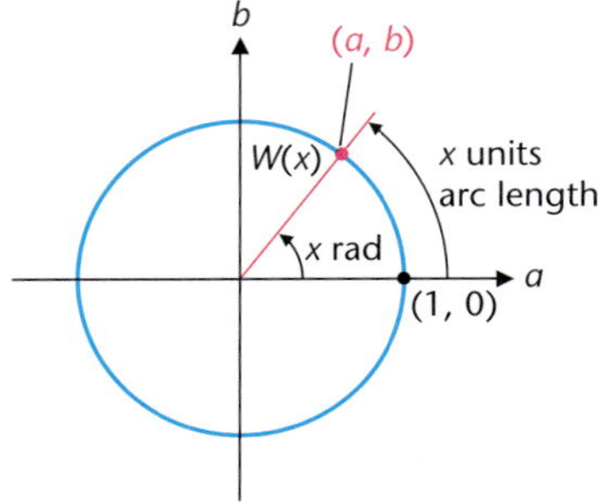

The trigonometric functions of any multiple of $\pi/6$ or $\pi/4$ can be determined exactly from the coordinates of the circular point. A graphing calculator can be used to graph the trigonometric functions and approximate their values at arbitrary angles.

5.3 Solving Right Triangles

A **right triangle** is a triangle with one 90° angle. To **solve a right triangle** is to find all unknown angles and sides, given the measures of two sides or the measures of one side and an acute angle.

Trigonometric Functions of Acute Angles

$$\sin\theta = \frac{\text{Opp}}{\text{Hyp}} \qquad \csc\theta = \frac{\text{Hyp}}{\text{Opp}}$$

$$\cos\theta = \frac{\text{Adj}}{\text{Hyp}} \qquad \sec\theta = \frac{\text{Hyp}}{\text{Adj}}$$

$$\tan\theta = \frac{\text{Opp}}{\text{Adj}} \qquad \cot\theta = \frac{\text{Adj}}{\text{Opp}}$$

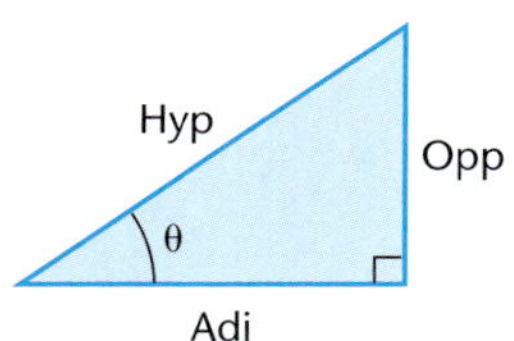

Computational Accuracy	
Angle to Nearest	**Significant Digits for Side Measure**
1°	2
10′ or 0.1°	3
1′ or 0.01°	4
10″ or 0.001°	5

5.4 Properties of Trigonometric Functions

The definition of the trigonometric functions implies that the following **basic identities** hold true for all replacements of x by real numbers for which both sides of an equation are defined:

Reciprocal identities

$$\csc x = \frac{1}{\sin x} \qquad \sec x = \frac{1}{\cos x} \qquad \cot x = \frac{1}{\tan x}$$

Quotient identities

$$\tan x = \frac{\sin x}{\cos x} \qquad \cot x = \frac{\cos x}{\sin x}$$

Identities for negatives

$$\sin(-x) = -\sin x \qquad \cos(-x) = \cos x$$

$$\tan(-x) = -\tan x$$

Pythagorean identity

$$\sin^2 x + \cos^2 x = 1$$

A function f is **periodic** if there exists a positive real number p such that

$$f(x + p) = f(x)$$

for all x in the domain of f. The smallest such positive p, if it exists, is called the **fundamental period of *f*,** or often just the **period of *f*.** All the trigonometric functions are periodic.

Graph of $y = \sin x$:

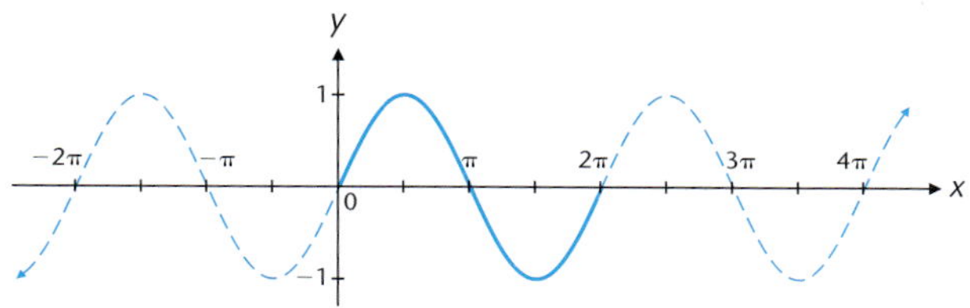

Period: 2π

Domain: All real numbers

Range: $[-1, 1]$

Graph of $y = \cos x$:

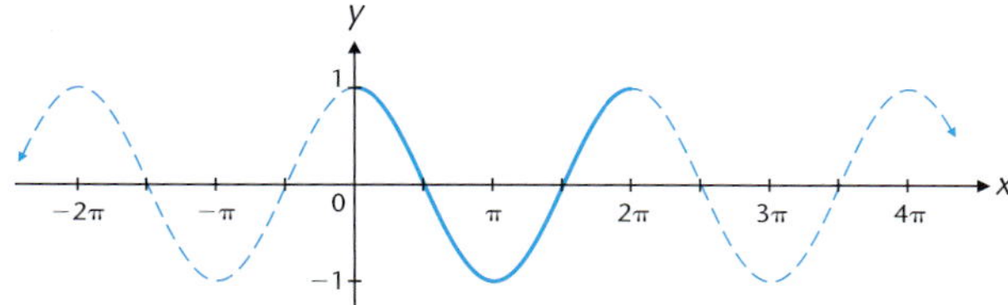

Period: 2π

Domain: All real numbers

Range: $[-1, 1]$

Graph of $y = \tan x$:

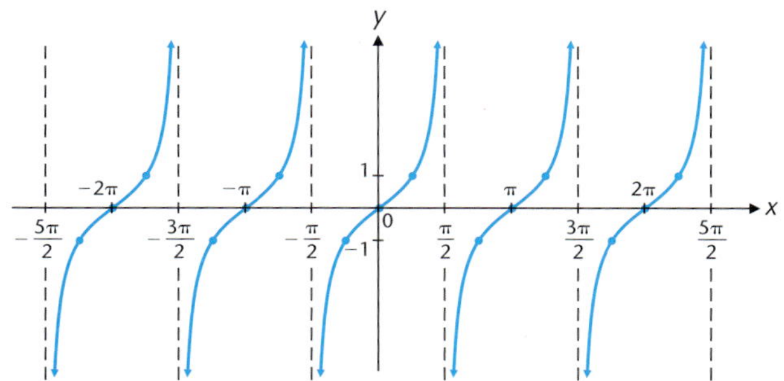

Period: π

Domain: All real numbers except $\pi/2 + k\pi$, k an integer

Range: All real numbers

Graph of $y = \cot x$:

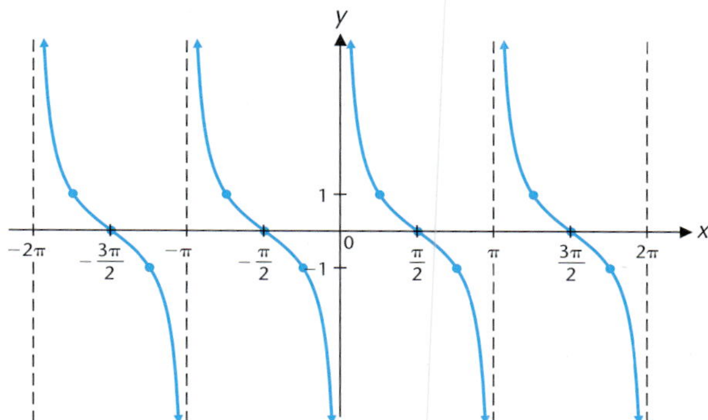

Period: π

Domain: All real numbers except $k\pi$, k an integer

Range: All real numbers

Graph of $y = \csc x$:

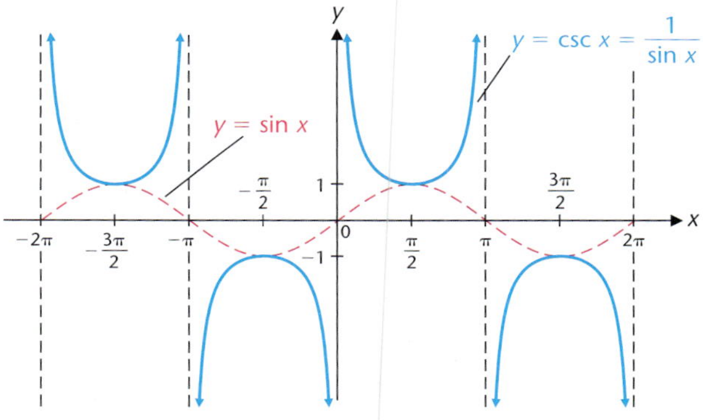

Period: 2π

Domain: All real numbers except $k\pi$, k an integer

Range: All real numbers y such that $y \leq -1$ or $y \geq 1$

Graph of $y = \sec x$:

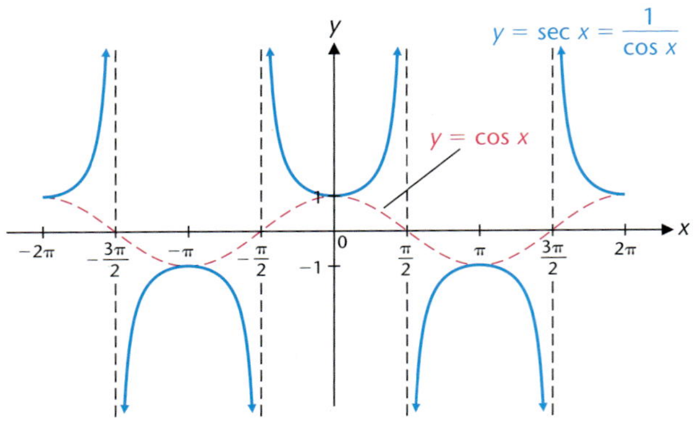

Period: 2π

Domain: All real numbers except $\pi/2 + k\pi$, k an integer

Range: All real numbers y such that $y \le -1$ or $y \ge 1$

Associated with each angle that does not terminate on a coordinate axis is a **reference triangle** for θ. The reference triangle is formed by drawing a perpendicular from point $P = (a, b)$ on the terminal side of θ to the horizontal axis. The **reference angle** α is the acute angle, always taken positive, between the terminal side of θ and the horizontal axis as indicated in the following figure.

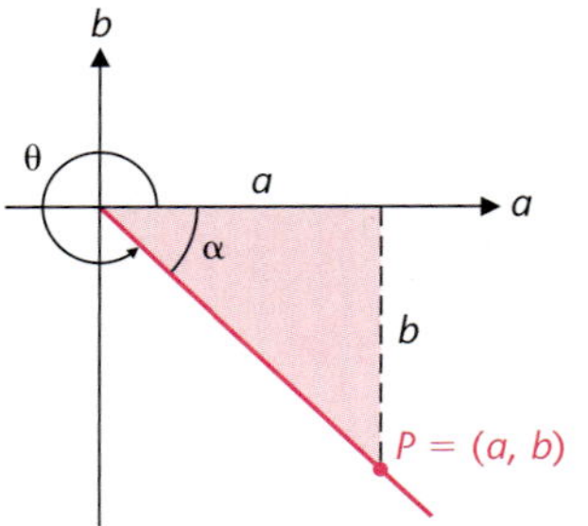

Reference Triangle
$(a, b) \ne (0, 0)$
α is always positive

5.5 More General Trigonometric Functions and Models

Let A, B, C be constants such that $A \ne 0$ and $B > 0$.
If $y = A \sin(Bx + C)$ or $y = A \cos(Bx + C)$:

$$\text{Amplitude} = |A| \qquad \text{Period} = \frac{2\pi}{B} \qquad \text{Phase shift} = \frac{-C}{B}$$

If $y = A \sec(Bx + C)$ or $y = \csc(Bx + C)$:

$$\text{Period} = \frac{2\pi}{B} \qquad \text{Phase shift} = \frac{-C}{B}$$

If $y = A \tan(Bx + C)$ or $y = A \cot(Bx + C)$:

$$\text{Period} = \frac{\pi}{B} \qquad \text{Phase shift} = \frac{-C}{B}$$

(Amplitude is not defined for the secant, cosecant, tangent, and cotangent functions, all of which are unbounded.)

Sinusoidal regression is used to find the function of the form $y = A \sin(Bx + C) + k$ that best fits a set of data points.

5.6 Inverse Trigonometric Functions

$y = \sin^{-1} x = \arcsin x$ if and only if $\sin y = x$, $-\pi/2 \le y \le \pi/2$ and $-1 \le x \le 1$.

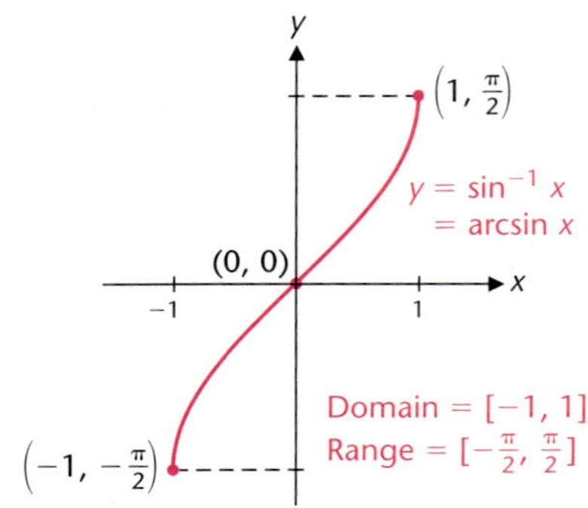

Inverse sine function

$y = \cos^{-1} x = \arccos x$ if and only if $\cos y = x$, $0 \le y \le \pi$ and $-1 \le x \le 1$.

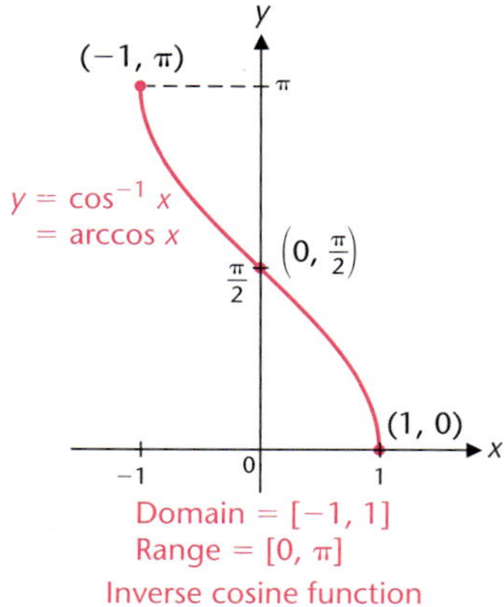

$y = \tan^{-1} x = \arctan x$ if and only if $\tan y = x$, $-\pi/2 < y < \pi/2$ and x is any real number.

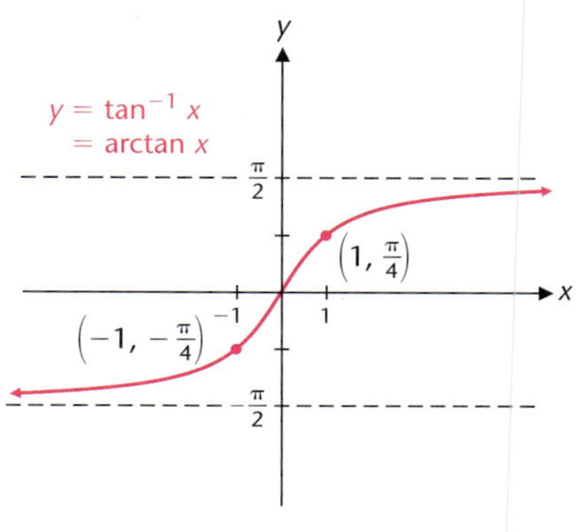

CHAPTER 5 REVIEW EXERCISES

Work through all the problems in this chapter review and check answers in the back of the book. Answers to all review problems are there, and following each answer is a number in italics indicating the section in which that type of problem is discussed. Where weaknesses show up, review appropriate sections in the text.

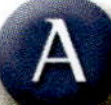

1. Find the radian measure of a central angle opposite an arc 15 centimeters long on a circle of radius 6 centimeters.

2. In a circle of radius 3 centimeters, find the length of an arc opposite an angle of 2.5 radians.

3. Solve the triangle:

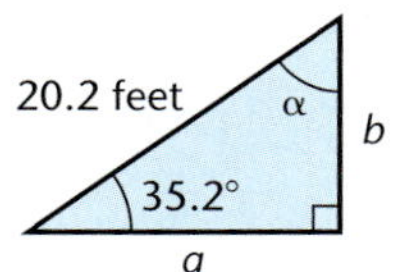

4. Find the reference angle associated with each angle θ.
(A) $\theta = \pi/3$ **(B)** $\theta = -120°$
(C) $\theta = -13\pi/6$ **(D)** $\theta = 210°$

5. In which quadrants is each negative?
(A) $\sin\theta$ **(B)** $\cos\theta$ **(C)** $\tan\theta$

6. If $(4, -3)$ is on the terminal side of angle θ, find
(A) $\sin\theta$ **(B)** $\sec\theta$ **(C)** $\cot\theta$

7. Complete Table 1 using exact values. Do not use a calculator.

TABLE 1

θ°	θ rad	sin θ	cos θ	tan θ	csc θ	sec θ	cot θ
0°					ND*		
30°							
45°	$\pi/4$		$1/\sqrt{2}$				
60°							
90°							
180°							
270°							
360°							

*ND = Not defined

8. What is the period of each of the following?
(A) $y = \cos x$ **(B)** $y = \csc x$ **(C)** $y = \tan x$

9. Indicate the domain and range of each.
(A) $y = \sin x$ **(B)** $y = \tan x$

10. Sketch a graph of $y = \sin x$, $-2\pi \le x \le 2\pi$.

11. Sketch a graph of $y = \cot x$, $-\pi < x < \pi$.

12. Verbally describe the meaning of a central angle in a circle with radian measure 0.5.

13. Describe the smallest shift of the graph of $y = \sin x$ that produces the graph of $y = \cos x$.

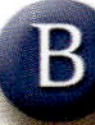

14. Change 1.37 radians to decimal degrees to two decimal places.

15. Solve the triangle:

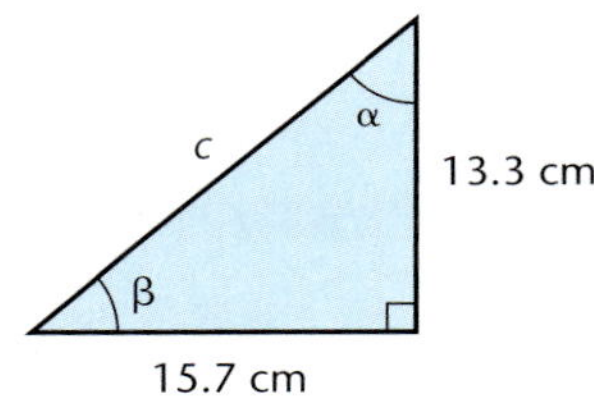

16. Indicate whether the angle is a quadrant I, II, III, or IV angle or a quadrantal angle.
(A) $-210°$ **(B)** $5\pi/2$ **(C)** 4.2 radians

17. Which of the following angles are coterminal with 120°?
(A) $-240°$ **(B)** $-7\pi/6$ **(C)** $840°$

18. Which of the following have the same value as cos 3?
(A) cos 3° **(B)** cos (3 radians) **(C)** $\cos(3 + 2\pi)$

19. For which values of x, $0 \le x < 2\pi$, is each of the following not defined?
(A) $\tan x$ **(B)** $\cot x$ **(C)** $\csc x$

20. A circular point $P = (a, b)$ moves clockwise around the circumference of a unit circle starting at (1, 0) and stops after covering a distance of 8.305 units. Explain how you would find the coordinates of point P at its final position and how you would determine which quadrant P is in. Find the coordinates of P to three decimal places and the quadrant for the final position of P.

In Problems 21–36, evaluate exactly without the use of a calculator.

21. $\tan 0$ **22.** $\sec 90°$

23. $\cos^{-1} 1$ **24.** $\cos\left(-\frac{3\pi}{4}\right)$

25. $\sin^{-1} \frac{\sqrt{2}}{2}$ **26.** $\csc 300°$

27. $\arctan \sqrt{3}$ **28.** $\sin 570°$

29. $\tan^{-1}(-1)$ **30.** $\cot\left(-\frac{4\pi}{3}\right)$

31. $\arcsin\left(-\frac{1}{2}\right)$ **32.** $\cos^{-1}\left(-\frac{\sqrt{3}}{2}\right)$

33. $\cos(\cos^{-1} 0.33)$ **34.** $\csc[\tan^{-1}(-1)]$

35. $\sin\left[\arccos\left(-\frac{1}{2}\right)\right]$ **36.** $\tan\left(\sin^{-1}\frac{-4}{5}\right)$

Evaluate Problems 37–44 to four significant digits using a calculator.

37. $\cos 423.7°$

38. $\tan 93°46'17''$

39. $\sec(-2.073)$

40. $\sin^{-1}(-0.8277)$

41. $\arccos(-1.3281)$

42. $\tan^{-1} 75.14$

43. $\csc[\cos^{-1}(-0.4081)]$

44. $\sin^{-1}(\tan 1.345)$

45. Find the exact degree measure of each without a calculator.
(A) $\theta = \sin^{-1}(-\frac{1}{2})$ **(B)** $\theta = \arccos(-\frac{1}{2})$

46. Find the degree measure of each to two decimal places using a calculator.
(A) $\theta = \cos^{-1}(-0.8763)$ **(B)** $\theta = \arctan 7.3771$

47. Evaluate $\cos^{-1}[\cos(-2)]$ with a calculator set in radian mode, and explain why this does or does not illustrate the inverse cosine–cosine identity.

48. Sketch a graph of $y = -2 \cos \pi x$, $-1 \le x \le 3$. Indicate amplitude A and period P.

49. Sketch a graph of $y = -2 + 3 \sin(x/2)$, $-4\pi \le x \le 4\pi$.

50. Find the equation of the form $y = A \cos Bx$ that has the graph shown below, then check the results with a graphing utility.

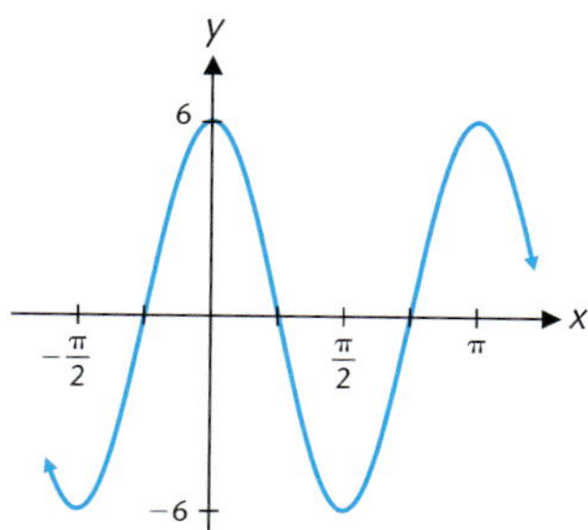

51. Find the equation of the form $y = A \sin Bx$ that has the graph shown below, then check the results with a graphing utility.

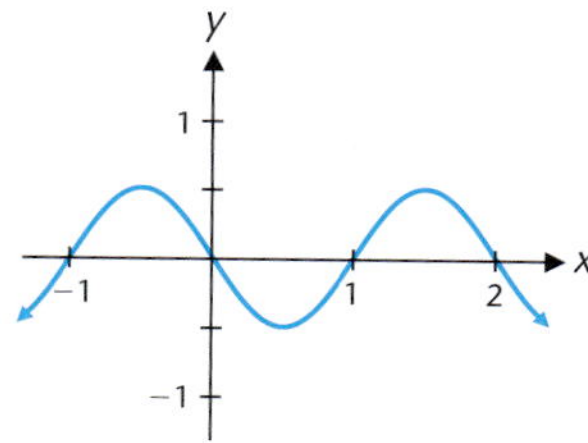

52. Describe the smallest shift and/or reflection that transforms the graph of $y = \tan x$ into the graph of $y = \cot x$.

53. Simplify each of the following using appropriate basic identities:
(A) $\sin(-x)\cot(-x)$ **(B)** $\dfrac{\sin^2 x}{1 - \sin^2 x}$

54. Sketch a graph of $y = 3 \sin[(x/2) + (\pi/2)]$ over the interval $-4\pi \le x \le 4\pi$.

55. Indicate the amplitude A, period P, and phase shift for the graph of $y = -2 \cos[(\pi/2)x - (\pi/4)]$. Do not graph.

56. Sketch a graph of $y = \cos^{-1} x$, and indicate the domain and range.

57. Graph $y = 1/(1 + \tan^2 x)$ in a graphing utility that displays at least two full periods of the graph. Find an equation of the form $y = k + A \sin Bx$ or $y = k + A \cos Bx$ that has the same graph.

58. Graph each equation in a graphing utility and find an equation of the form $y = A \tan Bx$ or $y = A \cot Bx$ that has the same graph as the given equation. Select the dimensions of the viewing window so that at least two periods are visible.
(A) $y = \dfrac{2 \sin^2 x}{\sin 2x}$ **(B)** $y = \dfrac{2 \cos^2 x}{\sin 2x}$

59. Determine whether each function is even, odd, or neither. Check your answer by graphing.
(A) $f(x) = \dfrac{1}{1 + \tan^2 x}$
(B) $g(x) = \dfrac{1}{1 + \tan x}$

In Problems 60 and 61, determine whether the statement is true or false. If true, explain why. If false, give a counterexample.

60. If α and β are the acute angles of a right triangle, then $\sin \alpha = \csc \beta$.

61. If α and β are the acute angles of a right triangle and $\alpha = \beta$, then all six trigonometric functions of α are greater than $\frac{1}{2}$ and less than $\frac{3}{2}$.

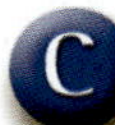

62. If in the figure the coordinates of A are $(8, 0)$ and arc length s is 20 units, find:

(A) The exact radian measure of θ

(B) The coordinates of P to three significant digits

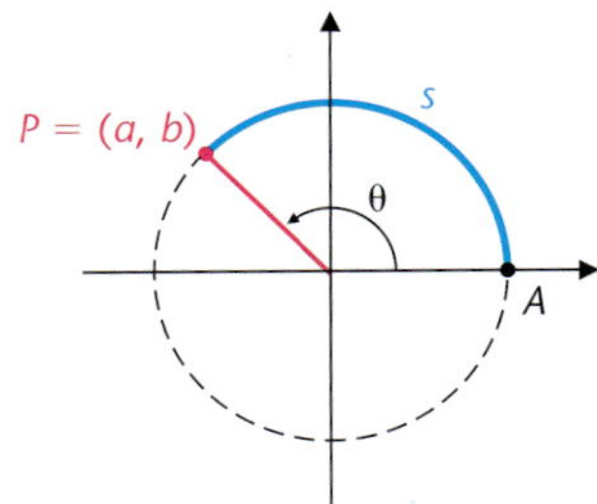

63. Find exactly the least positive real number for which

(A) $\cos x = -\frac{1}{2}$ **(B)** $\csc x = -\sqrt{2}$

64. Sketch a graph of $y = \sec x$, $-\pi/2 < x < 3\pi/2$.

65. Sketch a graph of $y = \tan^{-1} x$, and indicate the domain and range.

66. Indicate the period P and phase shift for the graph of $y = -5 \tan (\pi x + \pi/2)$. Do not graph.

67. Indicate the period and phase shift for the graph of $y = 3 \csc (x/2 - \pi/4)$. Do not graph.

68. Indicate whether each is symmetrical with respect to the x axis, y axis, or origin.

(A) Sine **(B)** Cosine **(C)** Tangent

69. Write as an algebraic expression in x free of trigonometric or inverse trigonometric functions:

$\sec (\sin^{-1} x)$

70. Try to calculate each of the following on your calculator. Explain the results.

(A) $\csc (-\pi)$ **(B)** $\tan (-3\pi/2)$ **(C)** $\sin^{-1} 2$

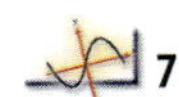

71. The accompanying graph is a graph of an equation of the form $y = A \sin (Bx + C)$, $-1 < -C/B < 0$. Find the equation, and check the results in a graphing utility.

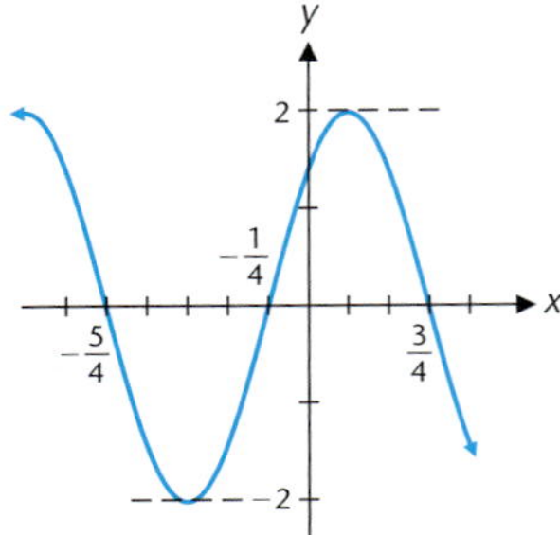

72. Graph $y = 1.2 \sin 2x + 1.6 \cos 2x$ in a graphing utility. (Select the dimensions of the viewing window so that at least two periods are visible.) Find an equation of the form $y = A \sin (Bx + C)$ that has the same graph as the given equation. Find A and B exactly and C to three decimal places. Use the x intercept closest to the origin as the phase shift.

73. A particular waveform is approximated by the first six terms of a Fourier series:

$$y = \frac{4}{\pi}\left(\sin x + \frac{\sin 3x}{3} + \frac{\sin 5x}{5} + \frac{\sin 7x}{7} + \frac{\sin 9x}{9} + \frac{\sin 11x}{11}\right)$$

(A) Graph this equation in a graphing utility for $-3\pi \le x \le 3\pi$ and $-2 \le y \le 2$.

(B) The graph in part A approximates a waveform that is made up entirely of straight line segments. Sketch by hand the waveform that the Fourier series approximates.

This waveform is called a **pulse wave** or a **square wave,** and is used, for example, to test distortion and to synchronize operations in computers.

APPLICATIONS

74. Astronomy. A line from the sun to the Earth sweeps out an angle of how many radians in 73 days? Express the answer in terms of π.

★ **75. Geometry.** Find the perimeter of a square inscribed in a circle of radius 5.00 centimeters.

76. Angular Speed. A wind turbine of rotor diameter 40 feet makes 80 revolutions per minute. Find the angular speed (in radians per second) and the linear speed (in feet per second) of the rotor tip.

★ **77. Alternating Current.** The current I in alternating electrical current has an amplitude of 30 amperes and a period of $\frac{1}{60}$ second. If $I = 30$ amperes when $t = 0$, find an equation of the form $I = A \cos Bt$ that gives the current at any time $t \geq 0$.

78. Restricted Access. A 10-foot-wide canal makes a right turn into a 15-foot-wide canal. Long narrow logs are to be floated through the canal around the right angle turn (see the figure). We are interested in finding the longest log that will go around the corner, ignoring the log's diameter.

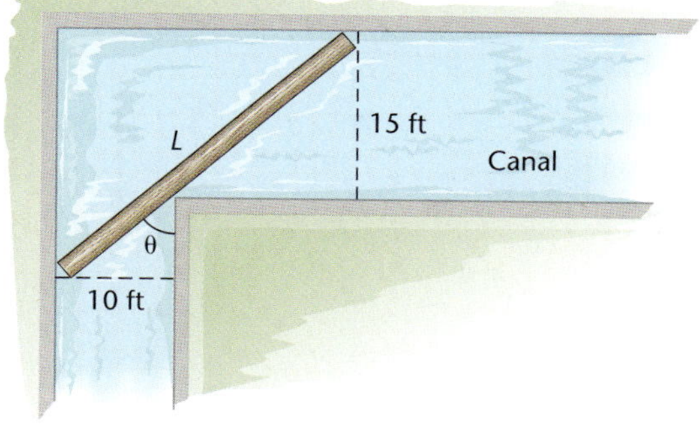

(A) Express the length L of the line that touches the two outer sides of the canal and the inside corner in terms of θ.

(B) Complete Table 2, each to one decimal place, and estimate from the table the longest log to the nearest foot that can make it around the corner. (The longest log is the shortest distance L.)

TABLE 2

θ (radians)	0.4	0.5	0.6	0.7	0.8	0.9	1.0
L (feet)	42.0						

(C) Graph the function in part A in a graphing utility and use an approximation method to find the shortest distance L to one decimal place; hence, the length of the longest log that can make it around the corner.

(D) Explain what happens to the length L as θ approaches 0 or $\pi/2$.

MODELING AND DATA ANALYSIS 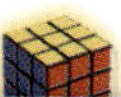

79. Modeling Seasonal Business Cycles. A soft drink company has revenues from sales over a 2-year period as shown by the accompanying graph, where $R(t)$ is revenue (in millions of dollars) for a month of sales t months after February 1.

(A) Find an equation of the form $R(t) = k + A \cos Bt$ that produces this graph, and check the result by graphing.

(B) Verbally interpret the graph

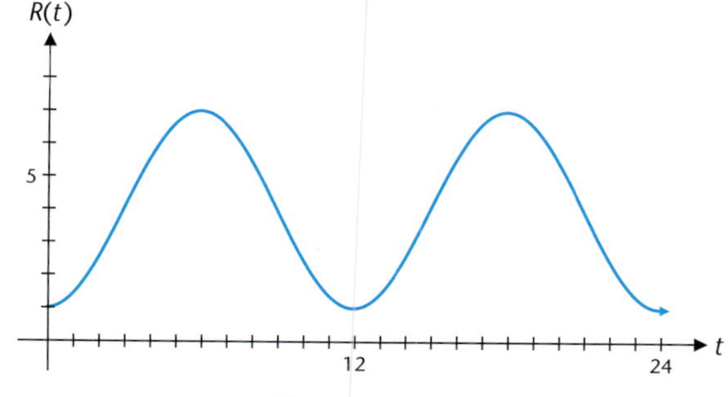

80. Modeling Temperature Variation. The 30-year average monthly temperature, °F, for each month of the year for Los Angeles is given in Table 3 (*World Almanac*).

(A) Using 1 month as the basic unit of time, enter the data for a 2-year period in your graphing utility and produce a scatter plot in the viewing window. Choose $40 \le y \le 90$ for the viewing window.

(B) It appears that a sine curve of the form

$$y = k + A \sin(Bx + C)$$

will closely model these data. The constants k, A, and B are easily determined from Table 3. To estimate C, visually estimate to one decimal place the smallest positive phase shift from the plot in part A. After determining A, B, k, and C, write the resulting equation. (Your value of C may differ slightly from the answer at the back of the book.)

(C) Plot the results of parts A and B in the same viewing window. (An improved fit may result by adjusting your value of C slightly.)

(D) If your graphing utility has a sinusoidal regression feature, check your results from parts B and C by finding and plotting the regression equation.

TABLE 3

x (months)	1	2	3	4	5	6	7	8	9	10	11	12
y (temperature)	58	60	61	63	66	70	74	75	74	70	63	58

CHAPTER 5 GROUP ACTIVITY

A Predator–Prey Analysis Involving Mountain Lions and Deer

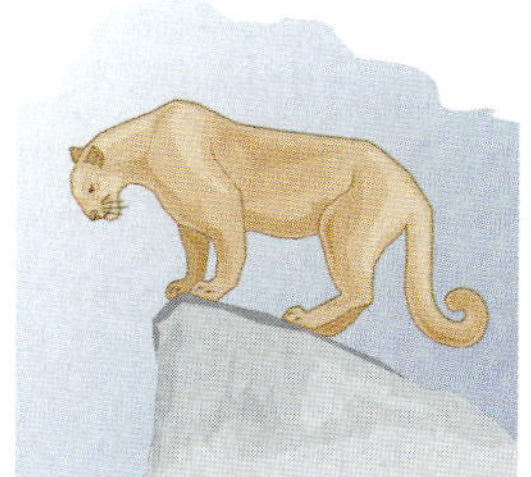

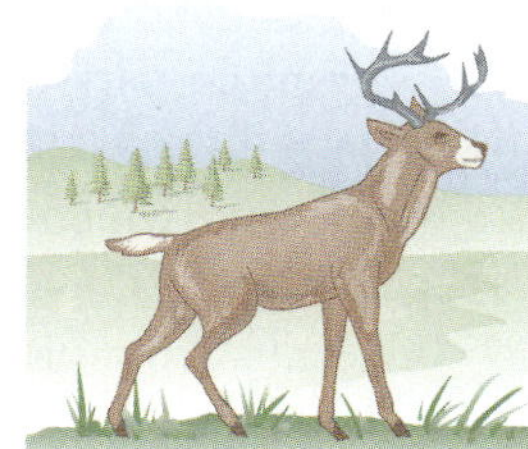

In some western state wilderness areas, deer and mountain lion populations are interrelated, because the mountain lions rely on the deer as a food source. The population of each species goes up and down in cycles, but out of phase with each other. A wildlife management research team estimated the respective populations in a particular region every 2 years over a 16-year period, with the results shown in Table 1.

TABLE 1 Mountain Lion/Deer Populations

Years	0	2	4	6	8	10	12	14	16
Deer	1,272	1,523	1,152	891	1,284	1,543	1,128	917	1,185
Mountain Lions	39	47	63	54	37	48	60	46	40

(A) Deer Population Analysis

1. Enter the data for the deer population for the time interval [0, 16] in a graphing utility and produce a scatter plot of the data.
2. A function of the form $y = k + A \sin(Bx + C)$ can be used to model these data. Use the data in Table 1 to determine k, A, and B. Use the graph in part 1 to visually estimate C to one decimal place.
3. Plot the data from part 1 and the equation from part 2 in the same viewing window. If necessary, adjust the value of C for a better fit.
4. If your graphing utility has a sinusoidal regression feature, check your results from parts 2 and 3 by finding and plotting the regression equation.
5. Write a summary of the results, describing fluctuations and cycles of the deer population.

(B) Mountain Lion Population Analysis

1. Enter the data for the mountain lion population for the time interval [0, 16] in a graphing utility and produce a scatter plot of the data.
2. A function of the form $y = k + A \sin (Bx + C)$ can be used to model these data. Use the data in Table 1 to determine k, A, and B. Use the graph in part 1 to visually estimate C to one decimal place.
3. Plot the data from part 1 and the equation from part 2 in the same viewing window. If necessary, adjust the value of C for a better fit.
4. If your graphing utility has a sinusoidal regression feature, check your results from parts 2 and 3 by finding and plotting the regression equation.
5. Write a summary of the results, describing fluctuations and cycles of the mountain lion population.

(C) Interrelationship of the Two Populations

1. Discuss the relationship of the maximum predator populations to the maximum prey populations relative to time.
2. Discuss the relationship of the minimum predator populations to the minimum prey populations relative to time.
3. Discuss the dynamics of the fluctuations of the two interdependent populations. What causes the two populations to rise and fall, and why are they out of phase with one another?

CHAPTER 6

Trigonometric Identities and Conditional Equations

OUTLINE

TRIGONOMETRIC FUNCTIONS ARE WIDELY USED IN SOLVING real-world problems and in the development of mathematics. Whatever their use, it is often of value to be able to change a trigonometric expression from one form to an equivalent more useful form. This involves the use of identities. Recall that an equation in one or more variables is said to be an *identity* if the left side is equal to the right side for all replacements of the variables for which both sides are defined.

For example, the equation

$$\sin^2 x + \cos^2 x = 1$$

is an identity, but the equation

$$\sin x + \cos x = 1$$

is not. The latter equation is called a **conditional equation,** because it holds for certain values of x (for example, $x = 0$ and $x = \pi/2$) but not for other values for which both sides are defined (for example,

Preparing for this chapter

Before getting started on this chapter, review the following concepts:

- Operations on Polynomials (Basic Algebra Review*, Section R.2)
- Factoring Polynomials (Basic Algebra Review*, Section R.3)
- Rational Expressions (Basic Algebra Review*, Section R.4)
- Cartesian Coordinate System (Appendix A, Section A.2)
- Quadratic Equations (Chapter 2, Section 5)
- Basic Identities (Chapter 5, Section 4)

*At www.mhhe.com/barnett

$x = \pi/4$). Sections 1 through 4 of Chapter 6 deal with trigonometric identities, and Section 6.5 with conditional trigonometric equations.

SECTION 6.1 Basic Identities and Their Use

Basic Identities • Establishing Other Identities

In Section 6.1 we review the basic identities introduced in Section 5.4 and show how they are used to establish other identities.

Basic Identities

In the box we list for convenient reference the basic identities introduced in Section 5.4. These identities will be used very frequently in the work that follows and should be memorized.

Basic Trigonometric Identities

Reciprocal Identities

$$\csc x = \frac{1}{\sin x} \qquad \sec x = \frac{1}{\cos x} \qquad \cot x = \frac{1}{\tan x}$$

Quotient Identities

$$\tan x = \frac{\sin x}{\cos x} \qquad \cot x = \frac{\cos x}{\sin x}$$

Identities for Negatives

$$\sin(-x) = -\sin x \qquad \cos(-x) = \cos x \qquad \tan(-x) = -\tan x$$

Pythagorean Identities

$$\sin^2 x + \cos^2 x = 1 \qquad \tan^2 x + 1 = \sec^2 x \qquad 1 + \cot^2 x = \csc^2 x$$

All these identities, with the exception of the second and third Pythagorean identities, were established in Section 5.4. The two exceptions can be derived from the first Pythagorean identity (see Explore/Discuss 1 and Problems 87 and 88 in Exercise 6.1).

EXPLORE/DISCUSS 1

Discuss an easy way to recall the second and third Pythagorean identities from the first. [*Hint:* Divide through the first Pythagorean identity by appropriate expressions.]

Establishing Other Identities

Identities are established to convert one form to an equivalent form that may be more useful. To *verify an identity* means to prove that both sides of an equation are equal for all replacements of the variables for which both sides are defined. Such a proof might use basic identities or other verified identities and algebraic operations such as multiplication, factoring, combining and reducing fractions, and so on. Examples 1 through 6 illustrate some of the techniques used to verify certain identities. The steps illustrated are not necessarily unique—often, there is more than one path to a desired goal. To become proficient in the use of identities, it is important that you work out many problems on your own.

EXAMPLE 1 Identity Verification

Verify the identity $\cos x \tan x = \sin x$.

VERIFICATION

Generally, we proceed by starting with the more complicated of the two sides, and transform that side into the other side in one or more steps using basic identities, algebra, or other established identities. Thus,

$$\cos x \tan x = \cos x \frac{\sin x}{\cos x} \qquad \text{Quotient identity}$$

$$= \sin x \qquad \text{Algebra}$$

MATCHED PROBLEM 1

Verify the identity $\sin x \cot x = \cos x$.

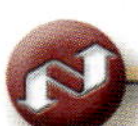

EXPLORE/DISCUSS 2

Graph the left and right sides of the identity in Example 1 in a graphing utility by letting $y_1 = \cos x \tan x$ and $y_2 = \sin x$. Use TRACE, moving back and forth between the graphs of y_1 and y_2, to compare values of y for given values of x. What does this investigation illustrate?

EXAMPLE 2 Identity Verification

Verify the identity $\sec(-x) = \sec x$.

VERIFICATION

$$\sec(-x) = \frac{1}{\cos(-x)} \qquad \text{Reciprocal identity}$$

$$= \frac{1}{\cos x} \qquad \text{Identity for negatives}$$

$$= \sec x \qquad \text{Reciprocal identity}$$

MATCHED 2 PROBLEM

Verify the identity $\csc(-x) = -\csc x$.

EXAMPLE 3 Identity Verification

Verify the identity $\cot x \cos x + \sin x = \csc x$.

VERIFICATION

$$\begin{aligned}
\cot x \cos x + \sin x &= \frac{\cos x}{\sin x}\cos x + \sin x && \text{Quotient identity}\\
&= \frac{\cos^2 x}{\sin x} + \sin x && \text{Algebra}\\
&= \frac{\cos^2 x + \sin^2 x}{\sin x} && \text{Algebra}\\
&= \frac{1}{\sin x} && \text{Pythagorean identity}\\
&= \csc x && \text{Reciprocal identity}
\end{aligned}$$

KEY ALGEBRAIC STEPS IN EXAMPLE 3

$$\frac{a}{b}a + b = \frac{a^2}{b} + b = \frac{a^2 + b^2}{b}$$

MATCHED 3 PROBLEM

Verify the identity $\tan x \sin x + \cos x = \sec x$.

To verify an identity, proceed from one side to the other, or from both sides to the middle, making sure all steps are reversible. Do not use properties of equality to perform the same operation on both sides of the equation. Although there is no fixed method of verification that works for all identities, there are certain steps that help in many cases.

Suggested Steps in Verifying Identities

1. Start with the more complicated side of the identity, and transform it into the simpler side.
2. Try algebraic operations such as multiplying, factoring, combining fractions, and splitting fractions.
3. If other steps fail, express each function in terms of sine and cosine functions, and then perform appropriate algebraic operations.
4. At each step, keep the other side of the identity in mind. This often reveals what you should do to get there.

EXAMPLE 4 Identity Verification

Verify the identity $\dfrac{1+\sin x}{\cos x}+\dfrac{\cos x}{1+\sin x}=2\sec x$.

VERIFICATION

$$\begin{aligned}
\frac{1+\sin x}{\cos x}+\frac{\cos x}{1+\sin x} &= \frac{(1+\sin x)^2+\cos^2 x}{\cos x\,(1+\sin x)} && \text{Algebra}\\
&= \frac{1+2\sin x+\sin^2 x+\cos^2 x}{\cos x\,(1+\sin x)} && \text{Algebra}\\
&= \frac{1+2\sin x+1}{\cos x\,(1+\sin x)} && \text{Pythagorean identity}\\
&= \frac{2+2\sin x}{\cos x\,(1+\sin x)} && \text{Algebra}\\
&= \frac{2(1+\sin x)}{\cos x\,(1+\sin x)} && \text{Algebra}\\
&= \frac{2}{\cos x} && \text{Algebra}\\
&= 2\sec x && \text{Reciprocal identity}
\end{aligned}$$

KEY ALGEBRAIC STEPS IN EXAMPLE 4

$\dfrac{a}{b}+\dfrac{b}{a}=\dfrac{a^2+b^2}{ba}$ $\quad(1+c)^2=1+2c+c^2$ $\quad\dfrac{m(a+b)}{n(a+b)}=\dfrac{m}{n}$

MATCHED PROBLEM 4

Verify the identity $\dfrac{1+\cos x}{\sin x}+\dfrac{\sin x}{1+\cos x}=2\csc x$.

EXAMPLE 5 Identity Verification

Verify the identity $\dfrac{\sin^2 x+2\sin x+1}{\cos^2 x}=\dfrac{1+\sin x}{1-\sin x}$.

VERIFICATION

$$\begin{aligned}
\frac{\sin^2 x+2\sin x+1}{\cos^2 x} &= \frac{(\sin x+1)^2}{\cos^2 x} && \text{Algebra}\\
&= \frac{(\sin x+1)^2}{1-\sin^2 x} && \text{Pythagorean identity}\\
&= \frac{(1+\sin x)^2}{(1-\sin x)(1+\sin x)} && \text{Algebra}\\
&= \frac{1+\sin x}{1-\sin x} && \text{Algebra}
\end{aligned}$$

KEY ALGEBRAIC STEPS IN EXAMPLE 5

$a^2+2a+1=(a+1)^2$ $\quad 1-b^2=(1-b)(1+b)$

MATCHED 5 PROBLEM

Verify the identity $\sec^4 x - 2\sec^2 x \tan^2 x + \tan^4 x = 1$.

EXAMPLE 6 Identity Verification

Verify the identity $\dfrac{\tan x - \cot x}{\tan x + \cot x} = 1 - 2\cos^2 x$.

VERIFICATION

$$\frac{\tan x - \cot x}{\tan x + \cot x} = \frac{\dfrac{\sin x}{\cos x} - \dfrac{\cos x}{\sin x}}{\dfrac{\sin x}{\cos x} + \dfrac{\cos x}{\sin x}}$$

Change to sines and cosines (quotient identities).

$$= \frac{(\sin x)(\cos x)\left(\dfrac{\sin x}{\cos x} - \dfrac{\cos x}{\sin x}\right)}{(\sin x)(\cos x)\left(\dfrac{\sin x}{\cos x} + \dfrac{\cos x}{\sin x}\right)}$$

Multiply numerator and denominator by $(\sin x)(\cos x)$, and use algebra to transform the compound fraction into a simple fraction.

$$= \frac{\sin^2 x - \cos^2 x}{\sin^2 x + \cos^2 x}$$

$$= \frac{1 - \cos^2 x - \cos^2 x}{1}$$

Pythagorean identity

$$= 1 - 2\cos^2 x$$

Algebra

KEY ALGEBRAIC STEPS IN EXAMPLE 6

$$\frac{\dfrac{a}{b} - \dfrac{b}{a}}{\dfrac{a}{b} + \dfrac{b}{a}} = \frac{ab\left(\dfrac{a}{b} - \dfrac{b}{a}\right)}{ab\left(\dfrac{a}{b} + \dfrac{b}{a}\right)} = \frac{a^2 - b^2}{a^2 + b^2}$$

MATCHED 6 PROBLEM

Verify the identity $\cot x - \tan x = \dfrac{2\cos^2 x - 1}{\sin x \cos x}$.

Just observing how others verify identities won't make *you* good at it. You must verify a large number on your own. With practice the process will seem less complicated.

EXAMPLE 7 Testing Identities Using a Graphing Utility

Use a graphing utility to test whether each of the following is an identity. If an equation appears to be an identity, verify it. If the equation does not appear to be an identity, find a value of x for which both sides are defined but are not equal.

(A) $\tan x + 1 = (\sec x)(\sin x - \cos x)$

(B) $\tan x - 1 = (\sec x)(\sin x - \cos x)$

SOLUTIONS

FIGURE 1

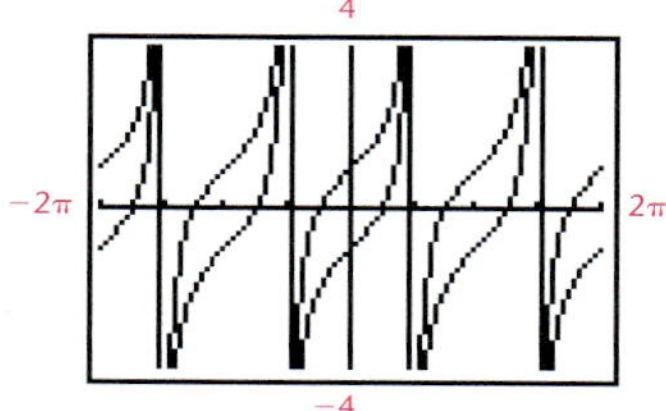

(A) Graph each side of the equation in the same viewing window (Fig. 1). The equation is not an identity, because the graphs do not match. Try $x = 0$.

Left side: $\tan 0 + 1 = 1$

Right side: $(\sec 0)(\sin 0 - \cos 0) = -1$

Finding one value of x for which both sides are defined, but are not equal, is enough to verify that the equation is not an identity.

FIGURE 2

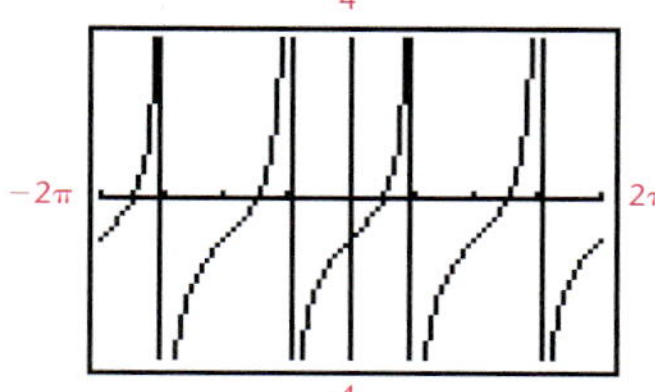

(B) Graph each side of the equation in the same viewing window (Fig. 2). The equation appears to be an identity, which we now verify:

$$
\begin{aligned}
&(\sec x)\,(\sin x - \cos x) \\
&\quad = \left(\frac{1}{\cos x}\right)(\sin x - \cos x) \\
&\quad = \left(\frac{\sin x}{\cos x} - \frac{\cos x}{\cos x}\right) \\
&\quad = \tan x - 1
\end{aligned}
$$

MATCHED 7 PROBLEM

Use a graphing utility to test whether each of the following is an identity. If an equation appears to be an identity, verify it. If the equation does not appear to be an identity, find a value of x for which both sides are defined but are not equal.

(A) $\dfrac{\sin x}{1 - \cos^2 x} = \csc x$ (B) $\dfrac{\sin x}{1 - \cos^2 x} = \sec x$

ANSWERS MATCHED PROBLEMS

In the following identity verifications, other correct sequences of steps are possible—the process is not unique.

1. $\sin x \cot x = \sin x \dfrac{\cos x}{\sin x} = \cos x$ **2.** $\csc(-x) = \dfrac{1}{\sin(-x)} = \dfrac{1}{-\sin x} = -\csc x$

3. $\tan x \sin x + \cos x = \dfrac{\sin^2 x}{\cos x} + \cos x = \dfrac{\sin^2 x + \cos^2 x}{\cos x} = \dfrac{1}{\cos x} = \sec x$

4. $\dfrac{1 + \cos x}{\sin x} + \dfrac{\sin x}{1 + \cos x} = \dfrac{(1 + \cos x)^2 + \sin^2 x}{\sin x\,(1 + \cos x)} = \dfrac{1 + 2\cos x + \cos^2 x + \sin^2 x}{\sin x\,(1 + \cos x)} = \dfrac{2(1 + \cos x)}{\sin x\,(1 + \cos x)} = 2\csc x$

5. $\sec^4 x - 2\sec^2 x \tan^2 x + \tan^4 x = (\sec^2 x - \tan^2 x)^2 = 1^2 = 1$

6. $\cot x - \tan x = \dfrac{\cos x}{\sin x} - \dfrac{\sin x}{\cos x} = \dfrac{\cos^2 x - \sin^2 x}{\sin x \cos x} = \dfrac{\cos^2 x - (1 - \cos^2 x)}{\sin x \cos x} = \dfrac{2\cos^2 x - 1}{\sin x \cos x}$

7. (A) An identity:

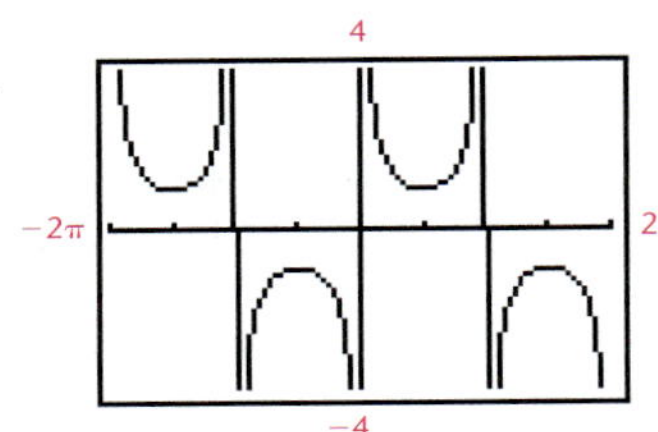

$$\frac{\sin x}{1 - \cos^2 x} = \frac{\sin x}{\sin^2 x} = \frac{1}{\sin x} = \csc x$$

(B) Not an identity: the left side is not equal to the right side for $x = 1$, for example.

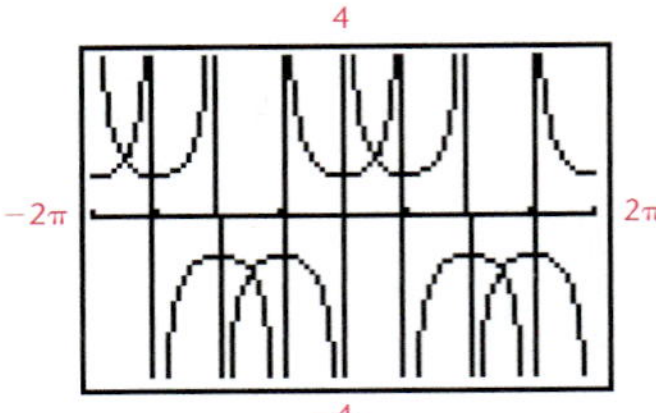

EXERCISE 6.1

A

Verify that Problems 1–26 are identities.

1. $\sin\theta \sec\theta = \tan\theta$
2. $\cos\theta \csc\theta = \cot\theta$
3. $\cot u \sec u \sin u = 1$
4. $\tan\theta \csc\theta \cos\theta = 1$
5. $\dfrac{\sin(-x)}{\cos(-x)} = -\tan x$
6. $\cot(-x)\tan x = -1$
7. $\sin\alpha = \dfrac{\tan\alpha \cot\alpha}{\csc\alpha}$
8. $\tan\alpha = \dfrac{\cos\alpha \sec\alpha}{\cot\alpha}$
9. $\cot u + 1 = (\csc u)(\cos u + \sin u)$
10. $\tan u + 1 = (\sec u)(\sin u + \cos u)$
11. $\dfrac{\cos x - \sin x}{\sin x \cos x} = \csc x - \sec x$
12. $\dfrac{\cos^2 x - \sin^2 x}{\sin x \cos x} = \cot x - \tan x$
13. $\dfrac{\sin^2 t}{\cos t} + \cos t = \sec t$
14. $\dfrac{\cos^2 t}{\sin t} + \sin t = \csc t$
15. $\dfrac{\cos x}{1 - \sin^2 x} = \sec x$
16. $\dfrac{\sin u}{1 - \cos^2 u} = \csc u$
17. $(1 - \cos u)(1 + \cos u) = \sin^2 u$
18. $(1 - \sin t)(1 + \sin t) = \cos^2 t$
19. $\cos^2 x - \sin^2 x = 1 - 2\sin^2 x$
20. $(\sin x + \cos x)^2 = 1 + 2\sin x \cos x$
21. $(\sec t + 1)(\sec t - 1) = \tan^2 t$
22. $(\csc t - 1)(\csc t + 1) = \cot^2 t$
23. $\csc^2 x - \cot^2 x = 1$
24. $\sec^2 u - \tan^2 u = 1$
25. $\cot x + \sec x = \dfrac{\cos x + \tan x}{\sin x}$
26. $\sin m(\csc m - \sin m) = \cos^2 m$

In Problems 27–30, graph all parts of each problem in the same viewing window in a graphing utility.

27. $-\pi \le x \le \pi$
    (A) $y = \sin^2 x$ **(B)** $y = \cos^2 x$
    (C) $y = \sin^2 x + \cos^2 x$

28. $-\pi \le x \le \pi$
    (A) $y = \sec^2 x$ **(B)** $y = \tan^2 x$
    (C) $y = \sec^2 x - \tan^2 x$

29. $-\pi \le x \le \pi$

(A) $y = \dfrac{\cos x}{\cot x \sin x}$ (B) $y = 1$

30. $-\pi \le x \le \pi$

(A) $y = \dfrac{\sin x}{\cos x \tan x}$ (B) $y = 1$

B

In Problems 31–38, is the equation an identity? Explain.

31. $\dfrac{x^2 - 9}{x + 3} = x - 3$

32. $\dfrac{5x}{|x|} = 5$

33. $\sqrt{x^2 + 4x + 4} = x + 2$

34. $\dfrac{1}{\sqrt{x}} = \dfrac{\sqrt{x}}{|x|}$

35. $\sin x - \cos x = 1$

36. $\sin x + \cos x = 1$

37. $\sin^2 x - \cos^2 x = 1$

38. $\sin^3 x + \cos^3 x = 1$

Verify that Problems 39–68 are identities.

39. $\dfrac{1 - (\sin x - \cos x)^2}{\sin x} = 2 \cos x$

40. $\dfrac{1 - \cos^2 y}{(1 - \sin y)(1 + \sin y)} = \tan^2 y$

41. $\cos \theta + \sin \theta = \dfrac{\cot \theta + 1}{\csc \theta}$

42. $\sin \theta + \cos \theta = \dfrac{\tan \theta + 1}{\sec \theta}$

43. $\dfrac{1 + \cos y}{1 - \cos y} = \dfrac{\sin^2 y}{(1 - \cos y)^2}$

44. $1 - \sin y = \dfrac{\cos^2 y}{1 + \sin y}$

45. $\tan^2 x - \sin^2 x = \tan^2 x \sin^2 x$

46. $\sec^2 x + \csc^2 x = \sec^2 x \csc^2 x$

47. $\dfrac{\csc \theta}{\cot \theta + \tan \theta} = \cos \theta$

48. $\dfrac{1 + \sec \theta}{\sin \theta + \tan \theta} = \csc \theta$

49. $\ln (\tan x) = \ln (\sin x) - \ln (\cos x)$

50. $\ln (\cot x) = \ln (\cos x) - \ln (\sin x)$

51. $\ln (\cot x) = -\ln (\tan x)$

52. $\ln (\csc x) = -\ln (\sin x)$

53. $\dfrac{1 - \cos A}{1 + \cos A} = \dfrac{\sec A - 1}{\sec A + 1}$

54. $\dfrac{1 - \csc y}{1 + \csc y} = \dfrac{\sin y - 1}{\sin y + 1}$

55. $\sin^4 w - \cos^4 w = 1 - 2 \cos^2 w$

56. $\sin^4 x + 2 \sin^2 x \cos^2 x + \cos^4 x = 1$

57. $\sec x - \dfrac{\cos x}{1 + \sin x} = \tan x$

58. $\csc n - \dfrac{\sin n}{1 + \cos n} = \cot n$

59. $\dfrac{\cos^2 z - 3 \cos z + 2}{\sin^2 z} = \dfrac{2 - \cos z}{1 + \cos z}$

60. $\dfrac{\sin^2 t + 4 \sin t + 3}{\cos^2 t} = \dfrac{3 + \sin t}{1 - \sin t}$

61. $\dfrac{\cos^3 \theta - \sin^3 \theta}{\cos \theta - \sin \theta} = 1 + \sin \theta \cos \theta$

62. $\dfrac{\cos^3 u + \sin^3 u}{\cos u + \sin u} = 1 - \sin u \cos u$

63. $(\sec x - \tan x)^2 = \dfrac{1 - \sin x}{1 + \sin x}$

64. $(\cot u - \csc u)^2 = \dfrac{1 - \cos u}{1 + \cos u}$

65. $\dfrac{\csc^4 x - 1}{\cot^2 x} = 2 + \cot^2 x$

66. $\dfrac{\sec^4 x - 1}{\tan^2 x} = 2 + \tan^2 x$

67. $\dfrac{1 + \sin v}{\cos v} = \dfrac{\cos v}{1 - \sin v}$

68. $\dfrac{\sin x}{1 - \cos x} = \dfrac{1 + \cos x}{\sin x}$

Use a graphing utility to test whether each of the following is an identity. If an equation appears to be an identity, verify it. If the equation does not appear to be an identity, find a value of x for which both sides are defined but are not equal.

69. $\dfrac{\sin (-x)}{\cos (-x) \tan (-x)} = -1$

70. $\dfrac{\cos (-x)}{\sin x \cot (-x)} = 1$

71. $\dfrac{\sin x}{\cos x \tan (-x)} = -1$

72. $\dfrac{\cos x}{\sin (-x) \cot (-x)} = 1$

73. $\sin x + \dfrac{\cos^2 x}{\sin x} = \sec x$

74. $\dfrac{1 - \tan^2 x}{1 - \cot^2 x} = \tan^2 x$

75. $\sin x + \dfrac{\cos^2 x}{\sin x} = \csc x$

76. $\dfrac{\tan^2 x - 1}{1 - \cot^2 x} = \tan^2 x$

77. $\dfrac{\tan x}{\sin x - 2 \tan x} = \dfrac{1}{\cos x - 2}$

78. $\dfrac{\cos x}{1 - \sin x} + \dfrac{\cos x}{1 + \sin x} = 2 \sec x$

79. $\dfrac{\tan x}{\sin x + 2 \tan x} = \dfrac{1}{\cos x - 2}$

80. $\dfrac{\cos x}{\sin x + 1} - \dfrac{\cos x}{\sin x - 1} = 2 \csc x$

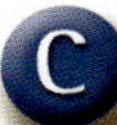

Verify that Problems 81–86 are identities.

81. $\dfrac{2 \sin^2 x + 3 \cos x - 3}{\sin^2 x} = \dfrac{2 \cos x - 1}{1 + \cos x}$

82. $\dfrac{3 \cos^2 z + 5 \sin z - 5}{\cos^2 z} = \dfrac{3 \sin z - 2}{1 + \sin z}$

83. $\dfrac{\tan u + \sin u}{\tan u - \sin u} - \dfrac{\sec u + 1}{\sec u - 1} = 0$

84. $\dfrac{\sin x \cos y + \cos x \sin y}{\cos x \cos y - \sin x \sin y} = \dfrac{\tan x + \tan y}{1 - \tan x \tan y}$

85. $\tan \alpha + \cot \beta = \dfrac{\tan \beta + \cot \alpha}{\tan \beta \cot \alpha}$

86. $\dfrac{\cot \alpha + \cot \beta}{\cot \alpha \cot \beta - 1} = \dfrac{\tan \alpha + \tan \beta}{1 - \tan \alpha \tan \beta}$

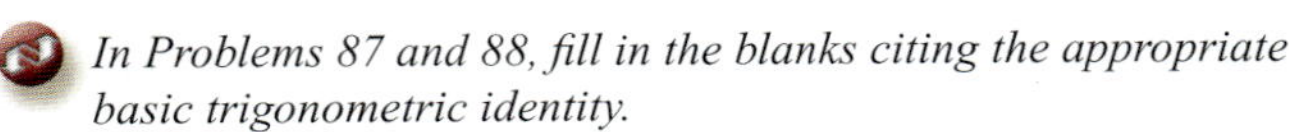

In Problems 87 and 88, fill in the blanks citing the appropriate basic trigonometric identity.

87.

Statement	Reason
$\cot^2 x + 1 = \left(\dfrac{\cos x}{\sin x}\right)^2 + 1$	(A) ______
$= \dfrac{\cos^2 x}{\sin^2 x} + 1$	Algebra
$= \dfrac{\cos^2 x + \sin^2 x}{\sin^2 x}$	Algebra
$= \dfrac{1}{\sin^2 x}$	(B) ______
$= \left(\dfrac{1}{\sin x}\right)^2$	Algebra
$= \csc^2 x$	(C) ______

88.

Statement	Reason
$\tan^2 x + 1 = \left(\dfrac{\sin x}{\cos x}\right)^2 + 1$	(A) ______
$= \dfrac{\sin^2 x}{\cos^2 x} + 1$	Algebra
$= \dfrac{\sin^2 x + \cos^2 x}{\cos^2 x}$	Algebra
$= \dfrac{1}{\cos^2 x}$	(B) ______
$= \left(\dfrac{1}{\cos x}\right)^2$	Algebra
$= \sec^2 x$	(C) ______

In Problems 89–94, examine the graph of $f(x)$ in a graphing utility to find a function of the form $g(x) = k + AT(x)$ that has the same graph as $f(x)$, where k and A are constants and $T(x)$ is one of the six trigonometric functions. Verify the identity $f(x) = g(x)$.

89. $f(x) = \dfrac{1 - \sin^2 x}{\tan x} + \sin x \cos x$

90. $f(x) = \dfrac{1 + \sin x}{2 \cos x} - \dfrac{\cos x}{2 + 2 \sin x}$

91. $f(x) = \dfrac{\cos^2 x}{1 + \sin x - \cos^2 x}$

92. $f(x) = \dfrac{\tan x \sin x}{1 - \cos x}$

93. $f(x) = \dfrac{1 + \cos x - 2 \cos^2 x}{1 - \cos x} - \dfrac{\sin^2 x}{1 + \cos x}$

94. $f(x) = \dfrac{3 \sin x - 2 \sin x \cos x}{1 - \cos x} - \dfrac{1 + \cos x}{\sin x}$

Each of the equations in Problems 95–102 is an identity in certain quadrants associated with x. Indicate which quadrants.

95. $\sqrt{1 - \cos^2 x} = -\sin x$ **96.** $\sqrt{1 - \sin^2 x} = \cos x$

97. $\sqrt{1 - \cos^2 x} = \sin x$ **98.** $\sqrt{1 - \sin^2 x} = -\cos x$

99. $\sqrt{1 - \sin^2 x} = |\cos x|$ **100.** $\sqrt{1 - \cos^2 x} = |\sin x|$

101. $\dfrac{\sin x}{\sqrt{1 - \sin^2 x}} = \tan x$ **102.** $\dfrac{\sin x}{\sqrt{1 - \sin^2 x}} = -\tan x$

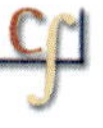

In calculus, trigonometric substitutions provide an effective way to rationalize the radical forms $\sqrt{a^2 - u^2}$ and $\sqrt{a^2 + u^2}$, which in turn leads to the solution to an important class of problems. Problems 103–106 involve such transformations. [Recall: $\sqrt{x^2} = |x|$ for all real numbers x.]

103. In the radical form $\sqrt{a^2 - u^2}$, $a > 0$, let $u = a \sin x$, $-\pi/2 < x < \pi/2$. Simplify, using a basic identity, and write the final form free of radicals.

104. In the radical form $\sqrt{a^2 - u^2}$, $a > 0$, let $u = a \cos x$, $0 < x < \pi$. Simplify, using a basic identity, and write the final form free of radicals.

105. In the radical form $\sqrt{a^2 + u^2}$, $a > 0$, let $u = a \tan x$, $0 < x < \pi/2$. Simplify, using a basic identity, and write the final form free of radicals.

106. In the radical form $\sqrt{a^2 + u^2}$, $a > 0$, let $u = a \cot x$, $0 < x < \pi/2$. Simplify, using a basic identity, and write the final form free of radicals.

SECTION 6.2 Sum, Difference, and Cofunction Identities

Sum and Difference Identities for Cosine • Cofunction Identities • Sum and Difference Identities for Sine and Tangent • Summary and Use

The basic identities discussed in Section 6.1 involved only one variable. In Section 6.2, we consider identities that involve two variables.

Sum and Difference Identities for Cosine

We start with the important **difference identity for cosine:**

$$\cos(x - y) = \cos x \cos y + \sin x \sin y \tag{1}$$

Many other useful identities can be readily verified from this particular one.

Here, we sketch a proof of equation (1) assuming x and y are in the interval $(0, 2\pi)$ and $x > y > 0$. It then follows easily, by periodicity and basic identities, that equation (1) holds for all real numbers x and y.

First, associate x and y with arcs and angles on the unit circle as indicated in Figure 1(a). Using the definitions of the circular functions given in Section 5.2, label the terminal points of x and y as shown in Figure 1(a).

FIGURE 1 Difference identity.

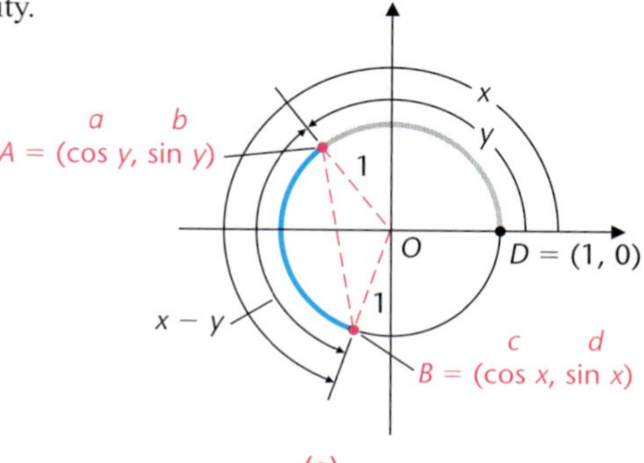

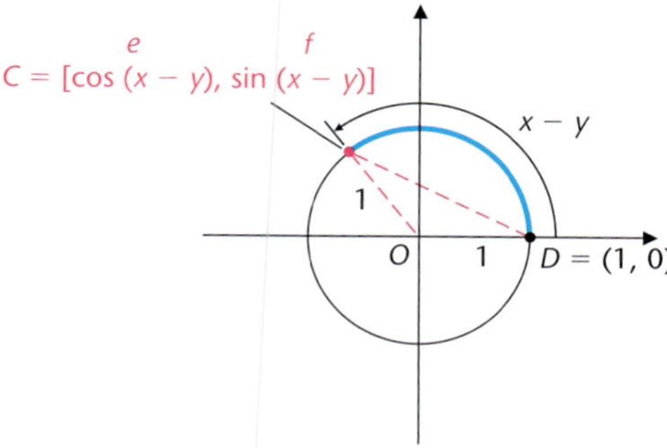

Now if you rotate the triangle AOB clockwise about the origin until the terminal point A coincides with $D = (1, 0)$, then terminal point B will be at C, as shown in Figure 1(b). Thus, because rotation preserves lengths,

$$\begin{aligned} d(A, B) &= d(C, D) \\ \sqrt{(c-a)^2 + (d-b)^2} &= \sqrt{(1-e)^2 + (0-f)^2} \\ (c-a)^2 + (d-b)^2 &= (1-e)^2 + f^2 \\ c^2 - 2ac + a^2 + d^2 - 2db + b^2 &= 1 - 2e + e^2 + f^2 \\ (c^2 + d^2) + (a^2 + b^2) - 2ac - 2db &= 1 - 2e + (e^2 + f^2) \end{aligned} \tag{2}$$

Because points A, B, and C are on unit circles, $c^2 + d^2 = 1$, $a^2 + b^2 = 1$, and $e^2 + f^2 = 1$, and equation (2) simplifies to

$$e = ac + bd \tag{3}$$

Replacing e, a, c, b, and d with $\cos(x - y)$, $\cos y$, $\cos x$, $\sin y$, and $\sin x$, respectively (see Fig. 1), we obtain

$$\begin{aligned} \cos(x - y) &= \cos y \cos x + \sin y \sin x \\ &= \cos x \cos y + \sin x \sin y \end{aligned} \tag{4}$$

We have thus established the difference identity for cosine.

If we replace y with $-y$ in equation (4) and use the identities for negatives (a good exercise for you), we obtain

$$\cos(x + y) = \cos x \cos y - \sin x \sin y \tag{5}$$

This is the **sum identity for cosine.**

EXPLORE/DISCUSS 1

Discuss how you would show that, in general,

$$\cos(x - y) \neq \cos x - \cos y$$

and

$$\cos(x + y) \neq \cos x + \cos y$$

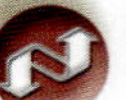

Cofunction Identities

To obtain sum and difference identities for the sine and tangent functions, we first derive *cofunction identities* directly from equation (1), the difference identity for cosine:

$$\begin{aligned} \cos(x - y) &= \cos x \cos y + \sin x \sin y \\ \cos\left(\frac{\pi}{2} - y\right) &= \cos\frac{\pi}{2}\cos y + \sin\frac{\pi}{2}\sin y \\ &= (0)(\cos y) + (1)(\sin y) \\ &= \sin y \end{aligned}$$

Thus, we have the **cofunction identity for cosine:**

$$\cos\left(\frac{\pi}{2} - y\right) = \sin y \tag{6}$$

for y any real number or angle in radian measure. If y is in degree measure, replace $\pi/2$ with 90°.

Now, if we let $y = \pi/2 - x$ in equation (6), we have

$$\cos\left[\frac{\pi}{2} - \left(\frac{\pi}{2} - x\right)\right] = \sin\left(\frac{\pi}{2} - x\right)$$
$$\cos x = \sin\left(\frac{\pi}{2} - x\right)$$

This is the **cofunction identity for sine;** that is,

$$\sin\left(\frac{\pi}{2} - x\right) = \cos x \tag{7}$$

where x is any real number or angle in radian measure. If x is in degree measure, replace $\pi/2$ with 90°.

Finally, we state the **cofunction identity for tangent** (and leave its derivation to Problem 10 in Exercise 6.2):

$$\tan\left(\frac{\pi}{2} - x\right) = \cot x \tag{8}$$

for x any real number or angle in radian measure. If x is in degree measure, replace $\pi/2$ with 90°.

REMARK If $0 < x < 90°$, then x and $90° - x$ are complementary angles. Originally, *cosine, cotangent,* and *cosecant* meant, respectively, "complements sine," "complements tangent," and "complements secant." Now we simply refer to cosine, cotangent, and cosecant as **cofunctions** of sine, tangent, and secant, respectively.

Sum and Difference Identities for Sine and Tangent

To derive a difference identity for sine, we use equations (1), (6), and (7) as follows:

$$\sin(x - y) = \cos\left[\frac{\pi}{2} - (x - y)\right] \qquad \text{Use equation (6).}$$

$$= \cos\left[\left(\frac{\pi}{2} - x\right) - (-y)\right] \qquad \text{Algebra}$$

$$= \cos\left(\frac{\pi}{2} - x\right)\cos(-y) + \sin\left(\frac{\pi}{2} - x\right)\sin(-y) \qquad \text{Use equation (1).}$$

$$= \sin x \cos y - \cos x \sin y \qquad \text{Use equations (6) and (7) and identities for negatives.}$$

The same result is obtained by replacing $\pi/2$ with 90°. Thus,

$$\sin(x - y) = \sin x \cos y - \cos x \sin y \tag{9}$$

is the **difference identity for sine.**

Now, if we replace y in equation (9) with $-y$ (a good exercise for you), we obtain

$$\sin(x + y) = \sin x \cos y + \cos x \sin y \tag{10}$$

the **sum identity for sine.**

It is not difficult to derive sum and difference identities for the tangent function. See if you can supply the reason for each step:

$$\tan(x - y) = \frac{\sin(x - y)}{\cos(x - y)}$$

$$= \frac{\sin x \cos y - \cos x \sin y}{\cos x \cos y + \sin x \sin y}$$

$$= \frac{\dfrac{\sin x \cos y}{\cos x \cos y} - \dfrac{\cos x \sin y}{\cos x \cos y}}{\dfrac{\cos x \cos y}{\cos x \cos y} + \dfrac{\sin x \sin y}{\cos x \cos y}}$$

Divide the numerator and denominator by $\cos x$ and $\cos y$.

$$= \frac{\dfrac{\sin x}{\cos x} - \dfrac{\sin y}{\cos y}}{1 + \dfrac{\sin x \sin y}{\cos x \cos y}}$$

$$= \frac{\tan x - \tan y}{1 + \tan x \tan y}$$

Thus,

$$\tan(x - y) = \frac{\tan x - \tan y}{1 + \tan x \tan y} \tag{11}$$

for all angles or real numbers x and y for which both sides are defined. This is the **difference identity for tangent.**

If we replace y in equation (11) with $-y$ (another good exercise for you), we obtain

$$\tan(x + y) = \frac{\tan x + \tan y}{1 - \tan x \tan y} \tag{12}$$

the **sum identity for tangent.**

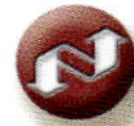

EXPLORE/DISCUSS 2

Discuss how you would show that, in general,

$$\tan(x - y) \neq \tan x - \tan y$$

and

$$\tan(x + y) \neq \tan x + \tan y$$

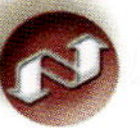

Summary and Use

Before proceeding with examples illustrating the use of these new identities, review the list given in the box.

Summary of Identities

Sum Identities

$$\sin(x + y) = \sin x \cos y + \cos x \sin y$$
$$\cos(x + y) = \cos x \cos y - \sin x \sin y$$
$$\tan(x + y) = \frac{\tan x + \tan y}{1 - \tan x \tan y}$$

Difference Identities

$$\sin(x - y) = \sin x \cos y - \cos x \sin y$$
$$\cos(x - y) = \cos x \cos y + \sin x \sin y$$
$$\tan(x - y) = \frac{\tan x - \tan y}{1 + \tan x \tan y}$$

Cofunction Identities

(Replace $\pi/2$ with $90°$ if x is in degrees.)

$$\cos\left(\frac{\pi}{2} - x\right) = \sin x \qquad \sin\left(\frac{\pi}{2} - x\right) = \cos x \qquad \tan\left(\frac{\pi}{2} - x\right) = \cot x$$

EXAMPLE 1 Using the Difference Identity

Simplify $\cos(x - \pi)$ using the difference identity.

SOLUTION

$$\begin{aligned}\cos(x - y) &= \cos x \cos y + \sin x \sin y \\ \cos(x - \pi) &= \cos x \cos \pi + \sin x \sin \pi \\ &= (\cos x)(-1) + (\sin x)(0) \\ &= -\cos x\end{aligned}$$

MATCHED PROBLEM 1

Simplify $\sin(x + 3\pi/2)$ using a sum identity.

EXAMPLE 2 Checking the Use of Sum and Difference Identities on a Graphing Utility

Simplify $\sin(x - \pi)$ using a difference identity. Enter the original form as y_1 and the converted form as y_2 in a graphing utility, then graph both in the same viewing window.

SOLUTION

$$\begin{aligned}\sin(x - y) &= \sin x \cos y - \cos x \sin y \\ \sin(x - \pi) &= \sin x \cos \pi - \cos x \sin \pi \\ &= (\sin x)(-1) - (\cos x)(0) \\ &= -\sin x\end{aligned}$$

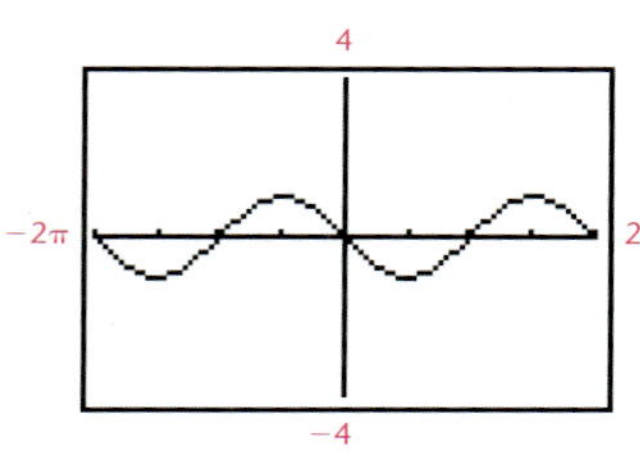

FIGURE 2

Graph $y_1 = \sin(x - \pi)$ and $y_2 = -\sin x$ in the same viewing window (Fig. 2). Use TRACE and move back and forth between y_1 and y_2 for different values of x to see that the corresponding y values are the same, or nearly the same.

MATCHED PROBLEM 2

Simplify $\cos(x + 3\pi/2)$ using a sum identity. Enter the original form as y_1 and the converted form as y_2 in a graphing utility, then graph both in the same viewing window.

EXAMPLE 3 Finding Exact Values

Find the exact value of $\tan 75°$ in radical form.

SOLUTION

Because we can write $75° = 45° + 30°$, the sum of two special angles, we can use the sum identity for tangents with $x = 45°$ and $y = 30°$:

$$\tan(x + y) = \frac{\tan x + \tan y}{1 - \tan x \tan y}$$

$$\tan(45° + 30°) = \frac{\tan 45° + \tan 30°}{1 - \tan 45° \tan 30°} \qquad \text{Sum identity}$$

$$= \frac{1 + (1/\sqrt{3})}{1 - 1(1/\sqrt{3})} \qquad \text{Evaluate functions exactly.}$$

$$= \frac{\sqrt{3} + 1}{\sqrt{3} - 1} \qquad \text{Multiply numerator and denominator by } \sqrt{3} \text{ and simplify.}$$

$$= 2 + \sqrt{3} \qquad \text{Rationalize denominator and simplify.}$$

MATCHED 3 PROBLEM

Find the exact value of cos 15° in radical form.

EXAMPLE 4 Finding Exact Values

Find the exact value of $\cos(x + y)$, given $\sin x = \frac{3}{5}$, $\cos y = \frac{4}{5}$, x is an angle in quadrant II, and y is an angle in quadrant I. Do not use a calculator.

SOLUTION

We start with the sum identity for cosine,

$$\cos(x + y) = \cos x \cos y - \sin x \sin y$$

We know $\sin x$ and $\cos y$ but not $\cos x$ and $\sin y$. We find the latter two using two different methods as follows (use the method that is easiest for you).

Given $\sin x = \frac{3}{5}$ and x is an angle in quadrant II, find $\cos x$:

Method I. Use a reference triangle:

$$\cos x = \frac{a}{5}$$

$$a^2 + 3^2 = 5^2$$

$$a^2 = 16$$

$$a = \pm 4$$

In quadrant II, $a = -4$

Therefore, $\cos x = -\frac{4}{5}$

Method II. Use a unit circle:

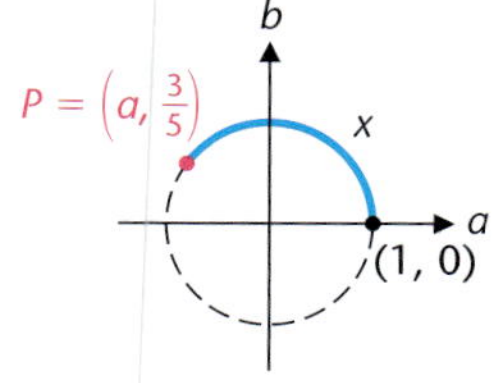

$$\cos x = a$$

$$a^2 + \left(\tfrac{3}{5}\right)^2 = 1$$

$$a^2 = \tfrac{16}{25}$$

$$a = \pm\tfrac{4}{5}$$

In quadrant II, $a = -\frac{4}{5}$

Therefore, $\cos x = -\frac{4}{5}$

Given $\cos y = \frac{4}{5}$ and y is an angle in quadrant I, find $\sin y$:

Method I. Use a reference triangle:

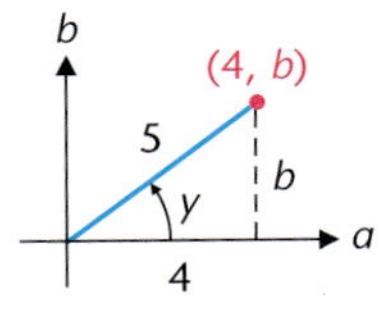

$$\sin y = \frac{b}{5}$$
$$4^2 + b^2 = 5^2$$
$$b^2 = 9$$
$$b = \pm 3$$

In quadrant I, $b = 3$

Therefore, $\sin y = \frac{3}{5}$

Method II. Use a unit circle:

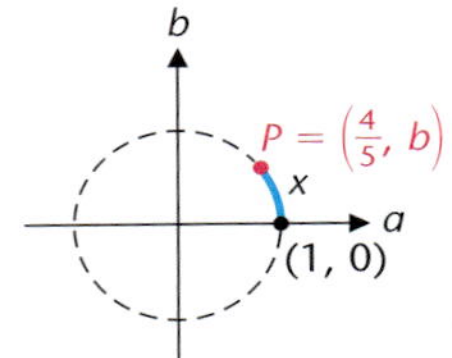

$$\sin y = b$$
$$\left(\tfrac{4}{5}\right)^2 + b^2 = 1$$
$$b^2 = \tfrac{9}{25}$$
$$b = \pm\tfrac{3}{5}$$

In quadrant I, $b = \frac{3}{5}$

Therefore, $\sin y = \frac{3}{5}$

We can now evaluate $\cos (x + y)$ without knowing x and y:

$$\begin{aligned}\cos (x + y) &= \cos x \cos y - \sin x \sin y \\ &= \left(-\tfrac{4}{5}\right)\left(\tfrac{4}{5}\right) - \left(\tfrac{3}{5}\right)\left(\tfrac{3}{5}\right) = -\tfrac{25}{25} = -1\end{aligned}$$

MATCHED 4 PROBLEM

Find the exact value of $\sin (x - y)$, given $\sin x = -\frac{2}{3}$, $\cos y = \sqrt{5}/3$, x is an angle in quadrant III, and y is an angle in quadrant IV. Do not use a calculator.

EXAMPLE 5 Identity Verification

Verify the identity $\tan x + \cot y = \dfrac{\cos (x - y)}{\cos x \sin y}$.

VERIFICATION

$$\begin{aligned}\frac{\cos (x - y)}{\cos x \sin y} &= \frac{\cos x \cos y + \sin x \sin y}{\cos x \sin y} && \text{Difference identity for cosine} \\ &= \frac{\cancel{\cos x} \cos y}{\cancel{\cos x} \sin y} + \frac{\sin x \cancel{\sin y}}{\cos x \cancel{\sin y}} && \text{Algebra} \\ &= \cot y + \tan x && \text{Quotient identities} \\ &= \tan x + \cot y\end{aligned}$$

MATCHED 5 PROBLEM

Verify the identity $\cot y - \cot x = \dfrac{\sin (x - y)}{\sin x \sin y}$.

ANSWERS TO MATCHED PROBLEMS

1. $-\cos x$ **2.** $y_1 = \cos(x + 3\pi/2)$, $y_2 = \sin x$ **3.** $(1 + \sqrt{3})/2\sqrt{2}$ or $(\sqrt{6} + \sqrt{2})/4$ **4.** $-4\sqrt{5}/9$

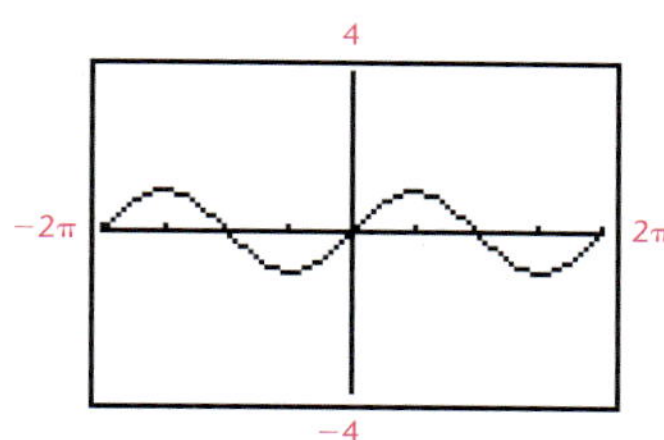

5. $\dfrac{\sin(x - y)}{\sin x \sin y} = \dfrac{\sin x \cos y - \cos x \sin y}{\sin x \sin y} = \dfrac{\cancel{\sin x} \cos y}{\cancel{\sin x} \sin y} - \dfrac{\cos x \cancel{\sin y}}{\sin x \cancel{\sin y}} = \cot y - \cot x$

EXERCISE 6.2

A

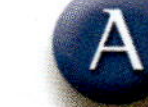

In Problems 1–8, is the equation an identity? Explain, making use of the sum or difference identities.

1. $\tan(x - \pi) = \tan x$

2. $\cos(x + \pi) = \cos x$

3. $\sin(x - \pi) = \sin x$

4. $\cot(x + \pi) = \cot x$

5. $\csc(2\pi - x) = \csc x$

6. $\sec(2\pi - x) = \sec x$

7. $\sin(x - \pi/2) = -\cos x$

8. $\cos(x - \pi/2) = -\sin x$

Verify each identity in Problems 9–12 using cofunction identities for sine and cosine and basic identities discussed in Section 6.1.

9. $\cot\left(\dfrac{\pi}{2} - x\right) = \tan x$

10. $\tan\left(\dfrac{\pi}{2} - x\right) = \cot x$

11. $\csc\left(\dfrac{\pi}{2} - x\right) = \sec x$

12. $\sec\left(\dfrac{\pi}{2} - x\right) = \csc x$

Convert Problems 13–18 to forms involving sin x, cos x, and/or tan x using sum or difference identities.

13. $\sin(30° - x)$

14. $\sin(x - 45°)$

15. $\sin(180° - x)$

16. $\cos(x + 180°)$

17. $\tan\left(x + \dfrac{\pi}{3}\right)$

18. $\tan\left(\dfrac{\pi}{4} - x\right)$

B

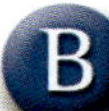

Use appropriate identities to find exact values for Problems 19–26. Do not use a calculator.

19. $\sec 75°$

20. $\sin 75°$

21. $\sin\dfrac{7\pi}{12}$ $\left[\textit{Hint}: \dfrac{7\pi}{12} = \dfrac{\pi}{3} + \dfrac{\pi}{4}\right]$

22. $\cos\dfrac{\pi}{12}$ $\left[\textit{Hint}: \dfrac{\pi}{12} = \dfrac{\pi}{4} - \dfrac{\pi}{6}\right]$

23. $\cos 74° \cos 44° + \sin 74° \sin 44°$

24. $\sin 22° \cos 38° + \cos 22° \sin 38°$

25. $\dfrac{\tan 27° + \tan 18°}{1 - \tan 27° \tan 18°}$

26. $\dfrac{\tan 110° - \tan 50°}{1 + \tan 110° \tan 50°}$

Find $\sin(x - y)$ and $\tan(x + y)$ exactly without a calculator using the information given in Problems 27–30.

27. $\sin x = -\frac{3}{5}$, $\sin y = \sqrt{8}/3$, x is a quadrant IV angle, y is a quadrant I angle.

28. $\sin x = \frac{2}{3}$, $\cos y = -\frac{1}{4}$, x is a quadrant II angle, y is a quadrant III angle.

29. $\tan x = \frac{3}{4}$, $\tan y = -\frac{1}{2}$, x is a quadrant III angle, y is a quadrant IV angle.

30. $\cos x = -\frac{1}{3}$, $\tan y = \frac{1}{2}$, x is a quadrant II angle, y is a quadrant III angle.

Verify each identity in Problems 31–44.

31. $\cos 2x = \cos^2 x - \sin^2 x$ **32.** $\sin 2x = 2 \sin x \cos x$

33. $\cot(x + y) = \dfrac{\cot x \cot y - 1}{\cot x + \cot y}$

34. $\cot(x - y) = \dfrac{\cot x \cot y + 1}{\cot y - \cot x}$

35. $\tan 2x = \dfrac{2 \tan x}{1 - \tan^2 x}$ **36.** $\cot 2x = \dfrac{\cot^2 x - 1}{2 \cot x}$

37. $\dfrac{\sin(v + u)}{\sin(v - u)} = \dfrac{\cot u + \cot v}{\cot u - \cot v}$

38. $\dfrac{\sin(u + v)}{\sin(u - v)} = \dfrac{\tan u + \tan v}{\tan u - \tan v}$

39. $\cot x - \tan y = \dfrac{\cos(x + y)}{\sin x \cos y}$

40. $\tan x - \tan y = \dfrac{\sin(x - y)}{\cos x \cos y}$

41. $\tan(x - y) = \dfrac{\cot y - \cot x}{\cot x \cot y + 1}$

42. $\tan(x + y) = \dfrac{\cot x + \cot y}{\cot x \cot y - 1}$

43. $\dfrac{\cos(x + h) - \cos x}{h} = \cos x\left(\dfrac{\cos h - 1}{h}\right) - \sin x\left(\dfrac{\sin h}{h}\right)$

44. $\dfrac{\sin(x + h) - \sin x}{h} = \sin x\left(\dfrac{\cos h - 1}{h}\right) + \cos x\left(\dfrac{\sin h}{h}\right)$

Evaluate both sides of the difference identity for sine and the sum identity for tangent for the values of x and y indicated in Problems 45–48. Evaluate to four significant digits using a calculator.

45. $x = 5.288$, $y = 1.769$ **46.** $x = 3.042$, $y = 2.384$

47. $x = 42.08°$, $y = 68.37°$ **48.** $x = 128.3°$, $y = 25.62°$

49. Explain how you would show that, in general, $\sec(x - y) \neq \sec x - \sec y$

50. Explain how you would show that, in general, $\csc(x + y) \neq \csc x + \csc y$

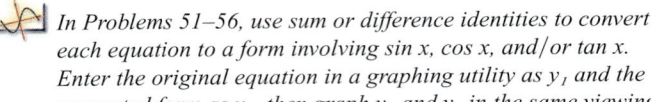

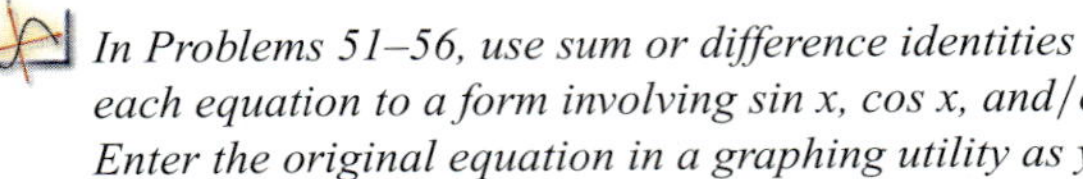

In Problems 51–56, use sum or difference identities to convert each equation to a form involving sin x, cos x, and/or tan x. Enter the original equation in a graphing utility as y_1 and the converted form as y_2, then graph y_1 and y_2 in the same viewing window. Use TRACE to compare the two graphs.

51. $y = \sin(x + \pi/6)$ **52.** $y = \sin(x - \pi/3)$

53. $y = \cos(x - 3\pi/4)$ **54.** $y = \cos(x + 5\pi/6)$

55. $y = \tan(x + 2\pi/3)$ **56.** $y = \tan(x - \pi/4)$

C

In Problems 57–60, evaluate exactly as real numbers without the use of a calculator.

57. $\sin[\cos^{-1}(-\frac{4}{5}) + \sin^{-1}(-\frac{3}{5})]$

58. $\cos[\sin^{-1}(-\frac{3}{5}) + \cos^{-1}(\frac{4}{5})]$

59. $\sin[\arccos \frac{1}{2} + \arcsin(-1)]$

60. $\cos[\arccos(-\sqrt{3}/2) - \arcsin(-\frac{1}{2})]$

61. Express $\sin(\sin^{-1} x + \cos^{-1} y)$ in an equivalent form free of trigonometric and inverse trigonometric functions.

62. Express $\cos(\sin^{-1} x - \cos^{-1} y)$ in an equivalent form free of trigonometric and inverse trigonometric functions.

Verify the identities in Problems 63 and 64.

63. $\cos(x + y + z) = \cos x \cos y \cos z - \sin x \sin y \cos z - \sin x \cos y \sin z - \cos x \sin y \sin z$

64. $\sin(x + y + z) = \sin x \cos y \cos z + \cos x \sin y \cos z + \cos x \cos y \sin z - \sin x \sin y \sin z$

In Problems 65 and 66, write each equation in terms of a single trigonometric function. Enter the original equation in a graphing utility as y_1 and the converted form as y_2, then graph y_1 and y_2 in the same viewing window. Use TRACE to compare the two graphs.

65. $y = \cos 1.2x \cos 0.8x - \sin 1.2x \sin 0.8x$

66. $y = \sin 0.8x \cos 0.3x - \cos 0.8x \sin 0.3x$

APPLICATIONS

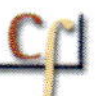

67. Analytic Geometry. Use the information in the figure to show that

$$\tan(\theta_2 - \theta_1) = \frac{m_2 - m_1}{1 + m_1 m_2}$$

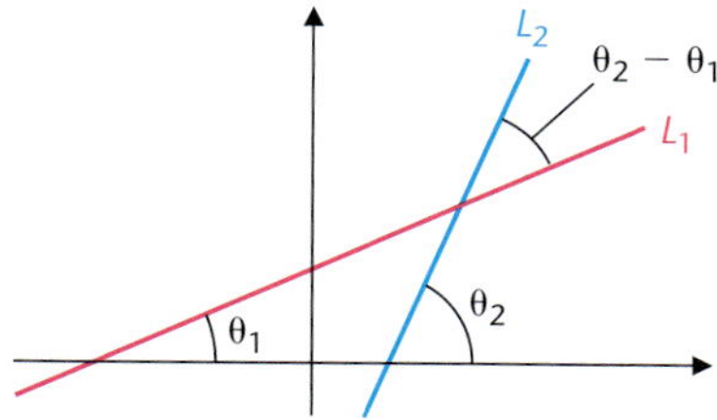

$\tan \theta_1 = \text{Slope of } L_1 = m_1$

$\tan \theta_2 = \text{Slope of } L_2 = m_2$

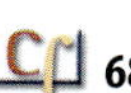

68. Analytic Geometry. Find the acute angle of intersection between the two lines $y = 3x + 1$ and $y = \frac{1}{2}x - 1$. (Use the results of Problem 67.)

★★ **69. Light Refraction.** Light rays passing through a plate glass window are refracted when they enter the glass and again when they leave to continue on a path parallel to the entering rays (see the figure). If the plate glass is M inches thick, the parallel displacement of the light rays is N inches, the angle of incidence is α, and the angle of refraction is β, show that

$$\tan \beta = \tan \alpha - \frac{N}{M} \sec \alpha$$

[*Hint:* First use geometric relationships to obtain

$$\frac{M}{\sec(90° - \beta)} = \frac{N}{\sin(\alpha - \beta)}$$

then use difference identities and fundamental identities to complete the derivation.]

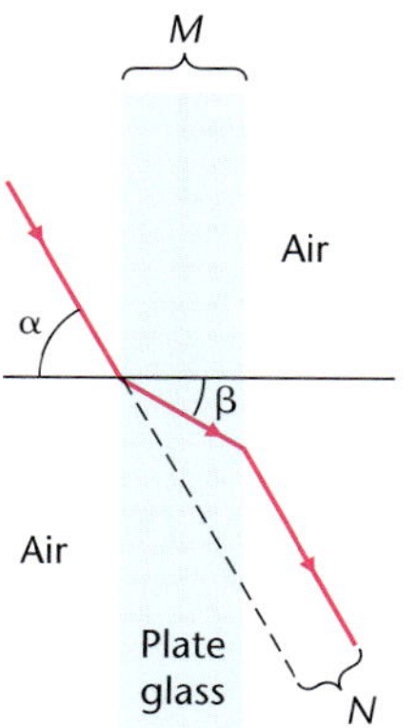

70. Light Refraction. Use the results of Problem 69 to find β to the nearest degree if $\alpha = 43°$, $M = 0.25$ inch, and $N = 0.11$ inch.

★**71. Surveying.** El Capitan is a large monolithic granite peak that rises straight up from the floor of Yosemite Valley in Yosemite National Park. It attracts rock climbers worldwide. At certain times, the reflection of the peak can be seen in the Merced River that runs along the valley floor. How can the height H of El Capitan above the river be determined by using only a sextant h feet high to measure the angle of elevation, β, to the top of the peak, and the angle of depression, α, of the reflected peak top in the river? (See accompanying figure, which is not to scale.)

(A) Using right triangle relationships, show that

$$H = h\left(\frac{1 + \tan \beta \cot \alpha}{1 - \tan \beta \cot \alpha}\right)$$

(B) Using sum or difference identities, show that the result in part A can be written in the form

$$H = h\left[\frac{\sin(\alpha + \beta)}{\sin(\alpha - \beta)}\right]$$

(C) If a sextant of height 4.90 feet measures α to be 46.23° and β to be 46.15°, compute the height H of El Capitan above the Merced River to three significant digits.

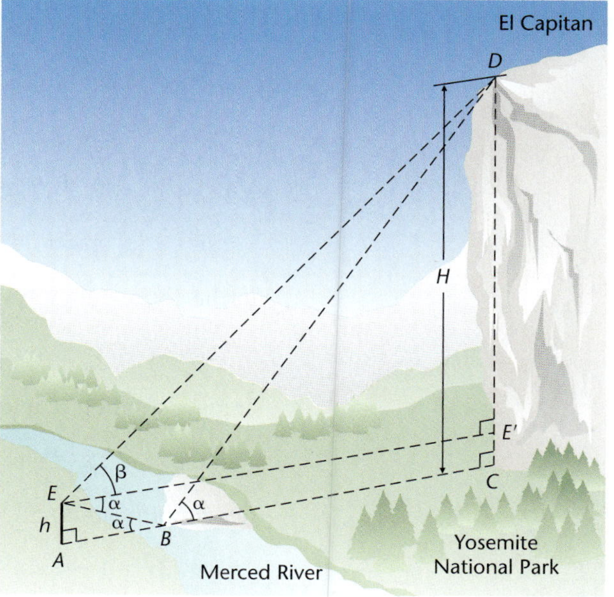

SECTION 6.3 Double-Angle and Half-Angle Identities

Double-Angle Identities • Half-Angle Identities

Section 6.3 develops another important set of identities called *double-angle* and *half-angle identities*. We can derive these identities directly from the sum and difference identities given in Section 6.2. Although the names use the word *angle,* the new identities hold for real numbers as well.

Double-Angle Identities

Start with the sum identity for sine,

$$\sin(x + y) = \sin x \cos y + \cos x \sin y$$

and replace y with x to obtain

$$\sin(x + x) = \sin x \cos x + \cos x \sin x$$

On simplification, this gives

$$\sin 2x = 2 \sin x \cos x \qquad \text{Double-angle identity for sine} \qquad (1)$$

If we start with the sum identity for cosine,

$$\cos(x + y) = \cos x \cos y - \sin x \sin y$$

and replace y with x, we obtain

$$\cos(x + x) = \cos x \cos x - \sin x \sin x$$

On simplification, this gives

$$\cos 2x = \cos^2 x - \sin^2 x \qquad \text{First double-angle identity for cosine} \qquad (2)$$

Now, using the Pythagorean identity

$$\sin^2 x + \cos^2 x = 1 \qquad (3)$$

in the form

$$\cos^2 x = 1 - \sin^2 x \qquad (4)$$

and substituting it into equation (2), we get

$$\cos 2x = 1 - \sin^2 x - \sin^2 x$$

On simplification, this gives

$$\cos 2x = 1 - 2\sin^2 x \qquad \text{Second double-angle identity for cosine} \qquad (5)$$

Or, if we use equation (3) in the form

$$\sin^2 x = 1 - \cos^2 x$$

and substitute it into equation (2), we get

$$\cos 2x = \cos^2 x - (1 - \cos^2 x)$$

On simplification, this gives

$$\cos 2x = 2\cos^2 x - 1 \qquad \text{Third double-angle identity for cosine} \tag{6}$$

Double-angle identities can be established for the tangent function in the same way by starting with the sum formula for tangent (a good exercise for you).

We list the double-angle identities below for convenient reference.

Double-Angle Identities

$$\sin 2x = 2 \sin x \cos x$$

$$\cos 2x = \cos^2 x - \sin^2 x = 1 - 2\sin^2 x = 2\cos^2 x - 1$$

$$\tan 2x = \frac{2\tan x}{1 - \tan^2 x} = \frac{2\cot x}{\cot^2 x - 1} = \frac{2}{\cot x - \tan x}$$

The identities in the second row can be solved for $\sin^2 x$ and $\cos^2 x$ to obtain the identities

$$\sin^2 x = \frac{1 - \cos 2x}{2} \qquad \cos^2 x = \frac{1 + \cos 2x}{2}$$

These are useful in calculus to transform a power form to a nonpower form.

EXPLORE/DISCUSS 1

(A) Discuss how you would show that, in general,

$$\sin 2x \neq 2 \sin x \qquad \cos 2x \neq 2\cos x \qquad \tan 2x \neq 2\tan x$$

(B) Graph $y_1 = \sin 2x$ and $y_2 = 2\sin x$ in the same viewing window. Conclusion? Repeat the process for the other two statements in part A.

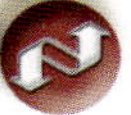

EXAMPLE 1 Identity Verification

Verify the identity $\cos 2x = \dfrac{1 - \tan^2 x}{1 + \tan^2 x}$.

VERIFICATION

We start with the right side:

$$\frac{1-\tan^2 x}{1+\tan^2 x} = \frac{1-\dfrac{\sin^2 x}{\cos^2 x}}{1+\dfrac{\sin^2 x}{\cos^2 x}} \qquad \text{Quotient identities}$$

$$= \frac{\cos^2 x\left(1-\dfrac{\sin^2 x}{\cos^2 x}\right)}{\cos^2 x\left(1+\dfrac{\sin^2 x}{\cos^2 x}\right)} \qquad \text{Algebra}$$

$$= \frac{\cos^2 x - \sin^2 x}{\cos^2 x + \sin^2 x} \qquad \text{Algebra}$$

$$= \cos^2 x - \sin^2 x \qquad \text{Pythagorean identity}$$

$$= \cos 2x \qquad \text{Double-angle identity}$$

KEY ALGEBRAIC STEPS IN EXAMPLE 1

$$\frac{1-\dfrac{a^2}{b^2}}{1+\dfrac{a^2}{b^2}} = \frac{b^2\left(1-\dfrac{a^2}{b^2}\right)}{b^2\left(1+\dfrac{a^2}{b^2}\right)} = \frac{b^2-a^2}{b^2+a^2}$$

MATCHED PROBLEM 1

Verify the identity $\sin 2x = \dfrac{2\tan x}{1+\tan^2 x}$.

EXAMPLE 2 Finding Exact Values

Find the exact values, without using a calculator, of $\sin 2x$ and $\cos 2x$ if $\tan x = -\frac{3}{4}$ and x is a quadrant IV angle.

SOLUTION

First draw the reference triangle for x and find any unknown sides:

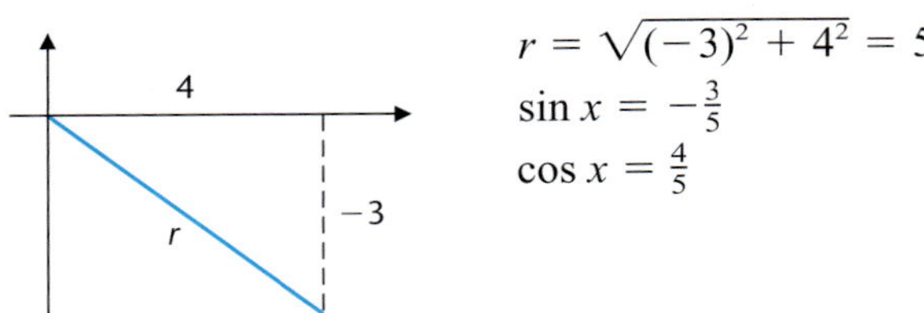

$$r = \sqrt{(-3)^2 + 4^2} = 5$$
$$\sin x = -\tfrac{3}{5}$$
$$\cos x = \tfrac{4}{5}$$

Now use double-angle identities for sine and cosine:

$$\sin 2x = 2\sin x\cos x = 2(-\tfrac{3}{5})(\tfrac{4}{5}) = -\tfrac{24}{25}$$
$$\cos 2x = 2\cos^2 x - 1 = 2(\tfrac{4}{5})^2 - 1 = \tfrac{7}{25}$$

MATCHED 2 PROBLEM

Find the exact values, without using a calculator, of $\cos 2x$ and $\tan 2x$ if $\sin x = \frac{4}{5}$ and x is a quadrant II angle.

Half-Angle Identities

Half-angle identities are simply double-angle identities stated in an alternate form. Let's start with the double-angle identity for cosine in the form

$$\cos 2m = 1 - 2\sin^2 m$$

Now replace m with $x/2$ and solve for $\sin(x/2)$ [if $2m$ is twice m, then m is half of $2m$—think about this]:

$$\cos x = 1 - 2\sin^2\frac{x}{2}$$

$$\sin^2\frac{x}{2} = \frac{1-\cos x}{2}$$

$$\sin\frac{x}{2} = \pm\sqrt{\frac{1-\cos x}{2}} \qquad \text{Half-angle identity for sine} \qquad (7)$$

where the choice of the sign is determined by the quadrant in which $x/2$ lies.

To obtain a half-angle identity for cosine, start with the double-angle identity for cosine in the form

$$\cos 2m = 2\cos^2 m - 1$$

and let $m = x/2$ to obtain

$$\cos\frac{x}{2} = \pm\sqrt{\frac{1+\cos x}{2}} \qquad \text{Half-angle identity for cosine} \qquad (8)$$

where the sign is determined by the quadrant in which $x/2$ lies.

To obtain a *half-angle identity for tangent,* use the quotient identity and the half-angle formulas for sine and cosine:

$$\tan\frac{x}{2} = \frac{\sin\frac{x}{2}}{\cos\frac{x}{2}} = \frac{\pm\sqrt{\frac{1-\cos x}{2}}}{\pm\sqrt{\frac{1+\cos x}{2}}} = \pm\sqrt{\frac{1-\cos x}{1+\cos x}}$$

Thus,

$$\tan\frac{x}{2} = \pm\sqrt{\frac{1-\cos x}{1+\cos x}} \qquad \text{Half-angle identity for tangent} \qquad (9)$$

where the sign is determined by the quadrant in which $x/2$ lies.

Simpler versions of equation (9) can be obtained as follows:

$$\left|\tan \frac{x}{2}\right| = \sqrt{\frac{1 - \cos x}{1 + \cos x}} \tag{10}$$

$$= \sqrt{\frac{1 - \cos x}{1 + \cos x} \cdot \frac{1 + \cos x}{1 + \cos x}}$$

$$= \sqrt{\frac{1 - \cos^2 x}{(1 + \cos x)^2}}$$

$$= \sqrt{\frac{\sin^2 x}{(1 + \cos x)^2}}$$

$$= \frac{\sqrt{\sin^2 x}}{\sqrt{(1 + \cos x)^2}}$$

$$= \frac{|\sin x|}{1 + \cos x}$$

$\sqrt{\sin^2 x} = |\sin x|$ and $\sqrt{(1 + \cos x)^2} = 1 + \cos x$, because $1 + \cos x$ is never negative.

All absolute value signs can be dropped, because it can be shown that tan $(x/2)$ and sin x always have the same sign (a good exercise for you). Thus,

$$\tan \frac{x}{2} = \frac{\sin x}{1 + \cos x} \tag{11}$$

Half-angle identity for tangent

By multiplying the numerator and the denominator in the radicand in equation (10) by $1 - \cos x$ and reasoning as before, we also can obtain

$$\tan \frac{x}{2} = \frac{1 - \cos x}{\sin x} \tag{12}$$

Half-angle identity for tangent

We now list all the half-angle identities for convenient reference.

Half-Angle Identities

$$\sin \frac{x}{2} = \pm\sqrt{\frac{1 - \cos x}{2}}$$

$$\cos \frac{x}{2} = \pm\sqrt{\frac{1 + \cos x}{2}}$$

$$\tan \frac{x}{2} = \pm\sqrt{\frac{1 - \cos x}{1 + \cos x}} = \frac{\sin x}{1 + \cos x} = \frac{1 - \cos x}{\sin x}$$

where the sign is determined by the quadrant in which $x/2$ lies.

EXPLORE/DISCUSS 2

(A) Discuss how you would show that, in general,

$$\sin\frac{x}{2} \neq \frac{1}{2}\sin x \qquad \cos\frac{x}{2} \neq \frac{1}{2}\cos x \qquad \tan\frac{x}{2} \neq \frac{1}{2}\tan x$$

(B) Graph $y_1 = \sin\frac{x}{2}$ and $y_2 = \frac{1}{2}\sin x$ in the same viewing window. Conclusion? Repeat the process for the other two statements in part A.

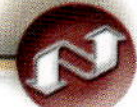

EXAMPLE 3 Finding Exact Values

Compute the exact value of sin 165° without a calculator using a half-angle identity.

SOLUTION

$$\sin 165° = \sin\frac{330°}{2}$$

$$= \sqrt{\frac{1-\cos 330°}{2}}$$ Use half-angle identity for sine with a positive radical, because sin 165° is positive.

$$= \sqrt{\frac{1-(\sqrt{3}/2)}{2}}$$

$$= \frac{\sqrt{2-\sqrt{3}}}{2}$$

MATCHED 3 PROBLEM

Compute the exact value of tan 105° without a calculator using a half-angle identity.

EXAMPLE 4 Finding Exact Values

Find the exact values of $\cos(x/2)$ and $\cot(x/2)$ without using a calculator if $\sin x = -\frac{3}{5}$, $\pi < x < 3\pi/2$.

SOLUTION

Draw a reference triangle in the third quadrant, and find cos x. Then use appropriate half-angle identities.

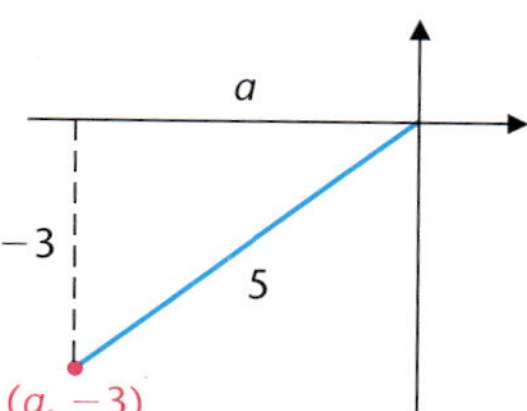

$$a = -\sqrt{5^2 - (-3)^2} = -4$$

$$\cos x = -\tfrac{4}{5}$$

If $\pi < x < 3\pi/2$, then

$$\frac{\pi}{2} < \frac{x}{2} < \frac{3\pi}{4}$$ Divide each member of $\pi < x < 3\pi/2$ by 2.

Thus, $x/2$ is an angle in the second quadrant where cosine and cotangent are negative, and

$$\cos\frac{x}{2} = -\sqrt{\frac{1+\cos x}{2}} = -\sqrt{\frac{1+(-\frac{4}{5})}{2}} = -\sqrt{\frac{1}{10}} \text{ or } \frac{-\sqrt{10}}{10}$$

$$\cot\frac{x}{2} = \frac{1}{\tan(x/2)} = \frac{\sin x}{1-\cos x} = \frac{-\frac{3}{5}}{1-(-\frac{4}{5})} = -\frac{1}{3}$$

MATCHED 4 PROBLEM

Find the exact values of $\sin(x/2)$ and $\tan(x/2)$ without using a calculator if $\cot x = -\frac{4}{3}$, $\pi/2 < x < \pi$.

EXAMPLE 5 Identity Verification

Verify the identity $\sin^2\frac{x}{2} = \frac{\tan x - \sin x}{2\tan x}$.

VERIFICATION

$$\sin\frac{x}{2} = \pm\sqrt{\frac{1-\cos x}{2}}$$ Half-angle identity for sine

$$\sin^2\frac{x}{2} = \frac{1-\cos x}{2}$$ Square both sides.

$$= \frac{\tan x}{\tan x}\cdot\frac{1-\cos x}{2}$$ Algebra

$$= \frac{\tan x - \tan x\cos x}{2\tan x}$$ Algebra

$$= \frac{\tan x - \sin x}{2\tan x}$$ Quotient identity

MATCHED 5 PROBLEM

Verify the identity $\cos^2\frac{x}{2} = \frac{\tan x + \sin x}{2\tan x}$.

ANSWERS MATCHED PROBLEMS

1. $\dfrac{2\tan x}{1+\tan^2 x} = \dfrac{2\left(\dfrac{\sin x}{\cos x}\right)}{1+\dfrac{\sin^2 x}{\cos^2 x}} = \dfrac{\cos^2 x\left[2\left(\dfrac{\sin x}{\cos x}\right)\right]}{\cos^2 x\left(1+\dfrac{\sin^2 x}{\cos^2 x}\right)} = \dfrac{2\sin x\cos x}{\cos^2 x+\sin^2 x} = 2\sin x\cos x = \sin 2x$

2. $\cos 2x = -\frac{7}{25}$, $\tan 2x = \frac{24}{7}$ 3. $-\sqrt{3}-2$ 4. $\sin(x/2) = 3\sqrt{10}/10$, $\tan(x/2) = 3$

5. $\cos^2\dfrac{x}{2} = \dfrac{1+\cos x}{2} = \dfrac{\tan x}{\tan x}\cdot\dfrac{1+\cos x}{2} = \dfrac{\tan x+\tan x\cos x}{2\tan x} = \dfrac{\tan x+\sin x}{2\tan x}$

EXERCISE 6.3

A

In Problems 1–6, verify each identity for the values indicated.

1. $\cos 2x = \cos^2 x - \sin^2 x$, $x = 30°$
2. $\sin 2x = 2\sin x\cos x$, $x = 45°$
3. $\tan 2x = \dfrac{2}{\cot x - \tan x}$, $x = \dfrac{\pi}{3}$
4. $\tan 2x = \dfrac{2\tan x}{1-\tan^2 x}$, $x = \dfrac{\pi}{6}$
5. $\sin\dfrac{x}{2} = \pm\sqrt{\dfrac{1-\cos x}{2}}$, $x = \pi$
   (Choose the correct sign.)
6. $\cos\dfrac{x}{2} = \pm\sqrt{\dfrac{1+\cos x}{2}}$, $x = \dfrac{\pi}{2}$
   (Choose the correct sign.)

In Problems 7–10, find the exact value without a calculator using double-angle and half-angle identities.

7. $\sin 22.5°$
8. $\tan 75°$
9. $\cos 67.5°$
10. $\tan 15°$

In Problems 11–14, graph y_1 and y_2 in the same viewing window for $-2\pi \le x \le 2\pi$. Use TRACE to compare the two graphs.

11. $y_1 = \cos 2x$, $y_2 = \cos^2 x - \sin^2 x$
12. $y_1 = \sin 2x$, $y_2 = 2\sin x\cos x$
13. $y_1 = \tan\dfrac{x}{2}$, $y_2 = \dfrac{\sin x}{1+\cos x}$
14. $y_1 = \tan 2x$, $y_2 = \dfrac{2\tan x}{1-\tan^2 x}$

B

Verify the identities in Problems 15–32.

15. $(\sin x + \cos x)^2 = 1 + \sin 2x$
16. $\sin 2x = (\tan x)(1 + \cos 2x)$
17. $\sin^2 x = \frac{1}{2}(1 - \cos 2x)$
18. $\cos^2 x = \frac{1}{2}(\cos 2x + 1)$
19. $1 - \cos 2x = \tan x\sin 2x$
20. $1 + \sin 2t = (\sin t + \cos t)^2$
21. $\sin^2\dfrac{x}{2} = \dfrac{1-\cos x}{2}$
22. $\cos^2\dfrac{x}{2} = \dfrac{1+\cos x}{2}$
23. $\cot 2x = \dfrac{1-\tan^2 x}{2\tan x}$
24. $\cot 2x = \dfrac{\cot x - \tan x}{2}$
25. $\cot\dfrac{\theta}{2} = \dfrac{\sin\theta}{1-\cos\theta}$
26. $\cot\dfrac{\theta}{2} = \dfrac{1+\cos\theta}{\sin\theta}$
27. $\cos 2u = \dfrac{1-\tan^2 u}{1+\tan^2 u}$
28. $\dfrac{\cos 2u}{1-\sin 2u} = \dfrac{1+\tan u}{1-\tan u}$
29. $2\csc 2x = \dfrac{1+\tan^2 x}{\tan x}$
30. $\sec 2x = \dfrac{\sec^2 x}{2-\sec^2 x}$
31. $\cos\alpha = \dfrac{1-\tan^2(\alpha/2)}{1+\tan^2(\alpha/2)}$
32. $\cos 2\alpha = \dfrac{\cot\alpha - \tan\alpha}{\cot\alpha + \tan\alpha}$

In Problems 33–38, is the equation an identity? Explain.

33. $\sin 4x = 4 \sin x \cos x$

34. $\csc 2x = 2 \csc x \sec x$

35. $\cot 2x = \dfrac{\tan x (\cot^2 x - 1)}{2}$

36. $\tan 4x = 4 \tan x$

37. $\cos 2x = 1 - 2 \cos^2 x$

38. $\tan 2x = \dfrac{2}{\tan x - \cot x}$

Compute the exact values of sin 2x, cos 2x, and tan 2x using the information given in Problems 39–42 and appropriate identities. Do not use a calculator.

39. $\sin x = \frac{3}{5}, \pi/2 < x < \pi$ **40.** $\cos x = -\frac{4}{5}, \pi/2 < x < \pi$

41. $\tan x = -\frac{5}{12}, -\pi/2 < x < 0$

42. $\cot x = -\frac{5}{12}, -\pi/2 < x < 0$

In Problems 43–46, compute the exact values of $\sin (x/2)$, $\cos (x/2)$, *and* $\tan (x/2)$ *using the information given and appropriate identities. Do not use a calculator.*

43. $\sin x = -\frac{1}{3}, \pi < x < 3\pi/2$

44. $\cos x = -\frac{1}{4}, \pi < x < 3\pi/2$

45. $\cot x = \frac{3}{4}, -\pi < x < -\pi/2$

46. $\tan x = \frac{3}{4}, -\pi < x < -\pi/2$

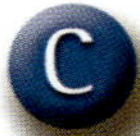

Suppose you are tutoring a student who is having difficulties in finding the exact values of sin θ and cos θ from the information given in Problems 47 and 48. Assuming you have worked through each problem and have identified the key steps in the solution process, proceed with your tutoring by guiding the student through the solution process using the following questions. Record the expected correct responses from the student.

(A) The angle 2θ is in what quadrant and how do you know?

(B) How can you find sin 2θ and cos 2θ? Find each.

(C) What identities relate sin θ and cos θ with either sin 2θ or cos 2θ?

(D) How would you use the identities in part C to find sin θ and cos θ exactly, including the correct sign?

(E) What are the exact values for sin θ and cos θ?

47. Find the exact values of $\sin \theta$ and $\cos \theta$, given $\tan 2\theta = -\frac{4}{3}, 0° < \theta < 90°$.

48. Find the exact values of $\sin \theta$ and $\cos \theta$, given $\sec 2\theta = -\frac{5}{4}, 0° < \theta < 90°$.

Verify each of the following identities for the value of x indicated in Problems 49–52. Compute values to five significant digits using a calculator.

(A) $\tan 2x = \dfrac{2 \tan x}{1 - \tan^2 x}$ *(B)* $\cos \dfrac{x}{2} = \pm\sqrt{\dfrac{1 + \cos x}{2}}$

(Choose the correct sign.)

49. $x = 252.06°$ **50.** $x = 72.358°$

51. $x = 0.934\,57$ **52.** $x = 4$

In Problems 53–56, graph y_1 *and* y_2 *in the same viewing window for* $-2\pi \le x \le 2\pi$, *and state the intervals for which the equation* $y_1 = y_2$ *is an identity.*

53. $y_1 = \cos (x/2), y_2 = \sqrt{\dfrac{1 + \cos x}{2}}$

54. $y_1 = \cos (x/2), y_2 = -\sqrt{\dfrac{1 + \cos x}{2}}$

55. $y_1 = \sin (x/2), y_2 = -\sqrt{\dfrac{1 - \cos x}{2}}$

56. $y_1 = \sin (x/2), y_2 = \sqrt{\dfrac{1 - \cos x}{2}}$

C

Verify the identities in Problems 57–60.

57. $\cos 3x = 4 \cos^3 x - 3 \cos x$

58. $\sin 3x = 3 \sin x - 4 \sin^3 x$

59. $\cos 4x = 8 \cos^4 x - 8 \cos^2 x + 1$

60. $\sin 4x = (\cos x)(4 \sin x - 8 \sin^3 x)$

In Problems 61–66, find the exact value of each without using a calculator.

61. $\cos [2 \cos^{-1} (\frac{3}{5})]$ **62.** $\sin [2 \cos^{-1} (\frac{3}{5})]$

63. $\tan [2 \cos^{-1} (-\frac{4}{5})]$ **64.** $\tan [2 \tan^{-1} (-\frac{3}{4})]$

65. $\cos [\frac{1}{2} \cos^{-1} (-\frac{3}{5})]$ **66.** $\sin [\frac{1}{2} \tan^{-1} (-\frac{4}{3})]$

In Problems 67–72, graph f(x) in a graphing utility, find a simpler function g(x) that has the same graph as f(x), and verify the identity f(x) = g(x). [Assume g(x) = k + A T(Bx) where k, A, and B are constants and T(x) is one of the six trigonometric functions.]

67. $f(x) = \csc x - \cot x$

68. $f(x) = \csc x + \cot x$

69. $f(x) = \dfrac{1 - 2\cos 2x}{2\sin x - 1}$

70. $f(x) = \dfrac{1 + 2\cos 2x}{1 + 2\cos x}$

71. $f(x) = \dfrac{1}{\cot x \sin 2x - 1}$

72. $f(x) = \dfrac{\cot x}{1 + \cos 2x}$

APPLICATIONS

★ **73. Indirect Measurement.** Find the exact value of x in the figure; then find x and θ to three decimal places. [*Hint:* Use $\cos 2\theta = 2\cos^2\theta - 1$.]

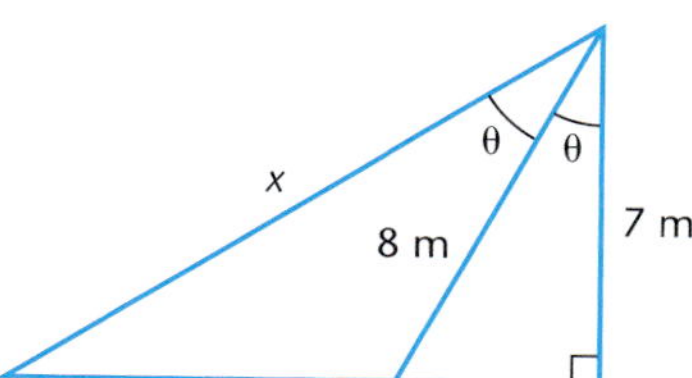

★ **74. Indirect Measurement.** Find the exact value of x in the figure; then find x and θ to three decimal places. [*Hint:* Use $\tan 2\theta = (2\tan\theta)/(1 - \tan^2\theta)$.]

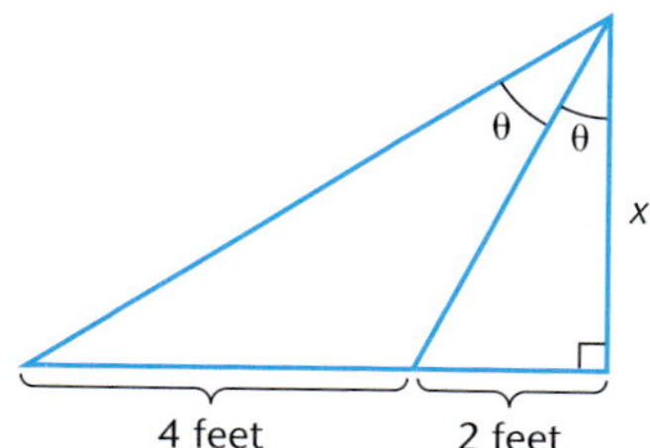

★ **75. Sports—Physics.** The theoretical distance d that a shot-putter, discus thrower, or javelin thrower can achieve on a given throw is found in physics to be given approximately by

$$d = \frac{2v_0^2 \sin\theta\cos\theta}{32 \text{ feet per second per second}}$$

where v_0 is the initial speed of the object thrown (in feet per second) and θ is the angle above the horizontal at which the object leaves the hand (see the figure).

(A) Write the formula in terms of $\sin 2\theta$ by using a suitable identity.

(B) Using the resulting equation in part A, determine the angle θ that will produce the maximum distance d for a given initial speed v_0. This result is an important consideration for shot-putters, javelin throwers, and discus throwers.

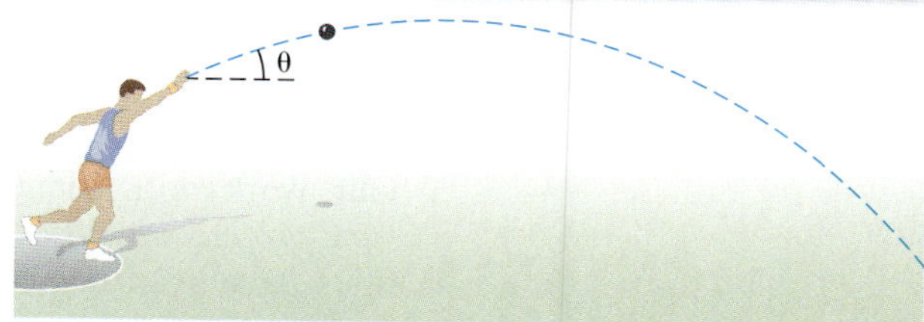

★★ **76. Geometry.** In part (a) of the figure, M and N are the midpoints of the sides of a square. Find the exact value of $\cos\theta$. [*Hint:* The solution uses the Pythagorean theorem, the definition of sine and cosine, a half-angle identity, and some auxiliary lines as drawn in part (b) of the figure.]

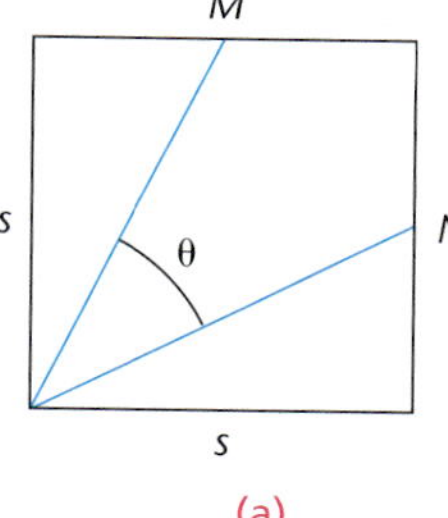

(a)

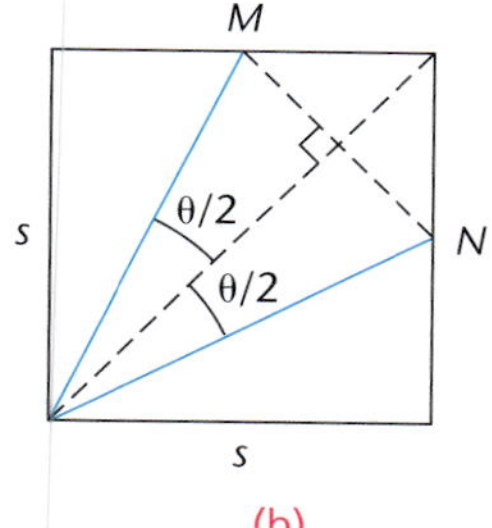

(b)

77. Area. An n-sided regular polygon is inscribed in a circle of radius R.

(A) Show that the area of the n-sided polygon is given by

$$A_n = \frac{1}{2} nR^2 \sin\frac{2\pi}{n}$$

[*Hint:* (Area of a triangle) = $(\frac{1}{2})$(base)(altitude). Also, a double-angle identity is useful.]

(B) For a circle of radius 1, complete Table 1, to five decimal places, using the formula in part A:

TABLE 1

n	10	100	1,000	10,000
A_n				

(C) What number does A_n seem to approach as n increases without bound? (What is the area of a circle of radius 1?)

(D) Will A_n exactly equal the area of the circumscribed circle for some sufficiently large n? How close can A_n be to the area of the circumscribed circle? [In calculus, the area of the circumscribed circle is called the *limit* of A_n as n increases without bound. In symbols, for a circle of radius 1, we would write $\lim_{n\to\infty} A_n = \pi$. The limit concept is the cornerstone on which calculus is constructed.]

SECTION 6.4 Product–Sum and Sum–Product Identities

Product–Sum Identities • Sum–Product Identities

Our work with identities is concluded by developing the *product–sum* and *sum–product identities,* which are easily derived from the sum and difference identities developed in Section 6.2. These identities are used in calculus to convert product forms to more convenient sum forms. They also are used in the study of sound waves in music to convert sum forms to more convenient product forms.

Product–Sum Identities

First, add left side to left side and right side to right side, the sum and difference identities for sine:

$$\begin{aligned} \sin(x+y) &= \sin x \cos y + \cos x \sin y \\ \sin(x-y) &= \sin x \cos y - \cos x \sin y \\ \hline \sin(x+y) + \sin(x-y) &= 2 \sin x \cos y \end{aligned}$$

or

$$\sin x \cos y = \tfrac{1}{2}[\sin(x+y) + \sin(x-y)]$$

Similarly, by adding or subtracting the appropriate sum and difference identities, we can obtain three other **product–sum identities.** These are listed below for convenient reference.

Product–Sum Identities

$$\sin x \cos y = \tfrac{1}{2}[\sin(x+y) + \sin(x-y)]$$

$$\cos x \sin y = \tfrac{1}{2}[\sin(x+y) - \sin(x-y)]$$

$$\sin x \sin y = \tfrac{1}{2}[\cos(x-y) - \cos(x+y)]$$

$$\cos x \cos y = \tfrac{1}{2}[\cos(x+y) + \cos(x-y)]$$

EXAMPLE 1 A Product as a Difference

Write the product $\cos 3t \sin t$ as a sum or difference.

SOLUTION

$$\cos x \sin y = \tfrac{1}{2}[\sin(x + y) - \sin(x - y)] \qquad \text{Let } x = 3t \text{ and } y = t.$$
$$\cos 3t \sin t = \tfrac{1}{2}[\sin(3t + t) - \sin(3t - t)]$$
$$= \tfrac{1}{2}\sin 4t - \tfrac{1}{2}\sin 2t$$

MATCHED 1 PROBLEM

Write the product $\cos 5\theta \cos 2\theta$ as a sum or difference.

EXAMPLE 2 Finding Exact Values

Evaluate $\sin 105° \sin 15°$ exactly using an appropriate product–sum identity.

SOLUTION

$$\sin x \sin y = \tfrac{1}{2}[\cos(x - y) - \cos(x + y)]$$
$$\sin 105° \sin 15° = \tfrac{1}{2}[\cos(105° - 15°) - \cos(105° + 15°)]$$
$$= \tfrac{1}{2}[\cos 90° - \cos 120°]$$
$$= \tfrac{1}{2}[0 - (-\tfrac{1}{2})] = \tfrac{1}{4} \text{ or } 0.25$$

MATCHED 2 PROBLEM

Evaluate $\cos 165° \sin 75°$ exactly using an appropriate product–sum identity.

Sum–Product Identities

The product–sum identities can be transformed into equivalent forms called **sum–product identities.** These identities are used to express sums and differences involving sines and cosines as products involving sines and cosines. We illustrate the transformation for one identity. The other three identities can be obtained by following similar procedures.

We start with a product–sum identity:

$$\sin \alpha \cos \beta = \tfrac{1}{2}[\sin(\alpha + \beta) + \sin(\alpha - \beta)] \tag{1}$$

We would like

$$\alpha + \beta = x$$
$$\alpha - \beta = y$$

Solving this system, we have

$$\alpha = \frac{x + y}{2} \qquad \beta = \frac{x - y}{2} \tag{2}$$

Substituting equation (2) into equation (1) and simplifying, we obtain

$$\sin x + \sin y = 2 \sin \frac{x + y}{2} \cos \frac{x - y}{2}$$

All four sum–product identities are listed next for convenient reference.

Sum–Product Identities

$$\sin x + \sin y = 2 \sin \frac{x + y}{2} \cos \frac{x - y}{2}$$

$$\sin x - \sin y = 2 \cos \frac{x + y}{2} \sin \frac{x - y}{2}$$

$$\cos x + \cos y = 2 \cos \frac{x + y}{2} \cos \frac{x - y}{2}$$

$$\cos x - \cos y = -2 \sin \frac{x + y}{2} \sin \frac{x - y}{2}$$

EXAMPLE 3 A Difference as a Product

Write the difference $\sin 7\theta - \sin 3\theta$ as a product.

SOLUTION

$$\sin x - \sin y = 2 \cos \frac{x + y}{2} \sin \frac{x - y}{2}$$

$$\sin 7\theta - \sin 3\theta = 2 \cos \frac{7\theta + 3\theta}{2} \sin \frac{7\theta - 3\theta}{2}$$

$$= 2 \cos 5\theta \sin 2\theta$$

MATCHED 3 PROBLEM

Write the sum $\cos 3t + \cos t$ as a product.

EXAMPLE 4 Finding Exact Values

Find the exact value of $\sin 105° - \sin 15°$ using an appropriate sum–product identity.

SOLUTION

$$\sin x - \sin y = 2 \cos \frac{x + y}{2} \sin \frac{x - y}{2}$$

$$\sin 105° - \sin 15° = 2 \cos \frac{105° + 15°}{2} \sin \frac{105° - 15°}{2}$$

$$= 2 \cos 60° \sin 45°$$

$$= 2\left(\frac{1}{2}\right)\left(\frac{\sqrt{2}}{2}\right) = \frac{\sqrt{2}}{2}$$

MATCHED 4 PROBLEM

Find the exact value of cos 165° − cos 75° using an appropriate sum–product identity.

EXPLORE/DISCUSS 1

The following "proof without words" of two of the sum–product identities is based on a similar "proof" by Sidney H. Kung, Jacksonville University, that was printed in the October 1996 issue of *Mathematics Magazine.* Discuss how the relationships below the figure are verified from the figure.

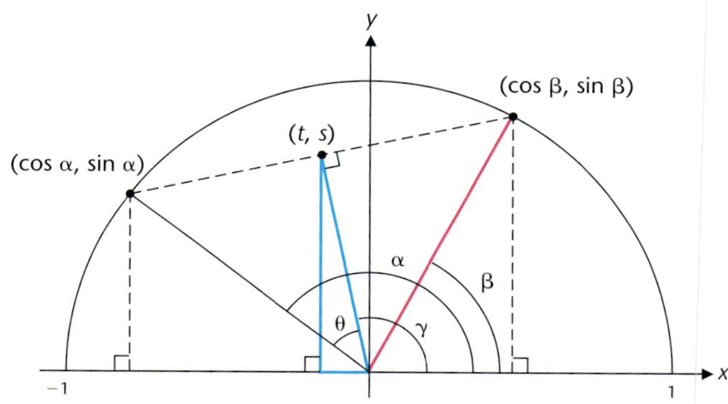

$$\theta = \frac{\alpha - \beta}{2} \qquad \gamma = \frac{\alpha + \beta}{2}$$

$$\frac{\sin \alpha + \sin \beta}{2} = s = \cos \frac{\alpha - \beta}{2} \sin \frac{\alpha + \beta}{2}$$

$$\frac{\cos \alpha + \cos \beta}{2} = t = \cos \frac{\alpha - \beta}{2} \cos \frac{\alpha + \beta}{2}$$

ANSWERS MATCHED PROBLEMS

1. $\frac{1}{2}\cos 7\theta + \frac{1}{2}\cos 3\theta$ **2.** $(-\sqrt{3} - 2)/4$ **3.** $2 \cos 2t \cos t$ **4.** $-\sqrt{6}/2$

EXERCISE 6.4

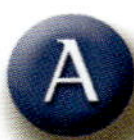

In Problems 1–4, write each product as a sum or difference involving sine and cosine.

1. $\sin 3m \cos m$

2. $\cos 7A \cos 5A$

3. $\sin u \sin 3u$

4. $\cos 2\theta \sin 3\theta$

In Problems 5–8, write each difference or sum as a product involving sines and cosines.

5. $\sin 3t + \sin t$

6. $\cos 7\theta + \cos 5\theta$

7. $\cos 5w - \cos 9w$

8. $\sin u - \sin 5u$

B

Evaluate Problems 9–12 exactly using an appropriate identity.

9. $\sin 195° \cos 75°$

10. $\cos 75° \sin 15°$

11. $\cos 15° \cos 75°$

12. $\sin 105° \sin 165°$

Evaluate Problems 13–16 exactly using an appropriate identity.

13. $\cos 285° + \cos 195°$

14. $\sin 195° + \sin 105°$

15. $\cos 15° - \cos 105°$

16. $\sin 75° - \sin 165°$

Use sum and difference identities to verify the identities in Problems 17 and 18.

17. $\cos x \cos y = \frac{1}{2}[\cos (x + y) + \cos (x - y)]$

18. $\sin x \sin y = \frac{1}{2}[\cos (x - y) - \cos (x + y)]$

19. Explain how you can transform the product–sum identity

$$\sin u \sin v = \tfrac{1}{2}[\cos (u - v) - \cos (u + v)]$$

into the sum–product identity

$$\cos x - \cos y = -2 \sin \frac{x + y}{2} \sin \frac{x - y}{2}$$

by a suitable substitution.

20. Explain how you can transform the product–sum identity

$$\cos u \cos v = \tfrac{1}{2}[\cos (u + v) + \cos (u - v)]$$

into the sum–product identity

$$\cos x + \cos y = 2 \cos \frac{x + y}{2} \cos \frac{x - y}{2}$$

by a suitable substitution.

Verify each identity in Problems 21–28.

21. $\dfrac{\sin 2t + \sin 4t}{\cos 2t - \cos 4t} = \cot t$

22. $\dfrac{\cos t - \cos 3t}{\sin t + \sin 3t} = \tan t$

23. $\dfrac{\sin x - \sin y}{\cos x - \cos y} = -\cot \dfrac{x + y}{2}$

24. $\dfrac{\sin x + \sin y}{\cos x + \cos y} = \tan \dfrac{x + y}{2}$

25. $\dfrac{\cos x + \cos y}{\sin x - \sin y} = \cot \dfrac{x - y}{2}$

26. $\dfrac{\cos x - \cos y}{\sin x + \sin y} = -\tan \dfrac{x - y}{2}$

27. $\dfrac{\cos x + \cos y}{\cos x - \cos y} = -\cot \dfrac{x + y}{2} \cot \dfrac{x - y}{2}$

28. $\dfrac{\sin x + \sin y}{\sin x - \sin y} = \dfrac{\tan [\frac{1}{2}(x + y)]}{\tan [\frac{1}{2}(x - y)]}$

In Problems 29–34, is the equation an identity? Explain.

29. $\sin 3x - \sin x = 2 \cos 2x \sin x$

30. $2 \sin x \cos 2x = \sin x + \sin 3x$

31. $\cos 3x - \cos x = 2 \sin 2x \sin x$

32. $2 \cos 3x \cos 5x = \cos 8x + \cos 2x$

33. $\cos x + \cos 5x = 2 \cos 2x \cos 3x$

34. $2 \sin 4x \cos 2x = \sin 8x + \sin 2x$

Verify each of the following identities for the values of x and y indicated in Problems 35–38. Evaluate each side to five significant digits.
(A) $\cos x \sin y = \frac{1}{2}[\sin (x + y) - \sin (x - y)]$
(B) $\cos x + \cos y = 2 \cos \dfrac{x + y}{2} \cos \dfrac{x - y}{2}$

35. $x = 172.63°, y = 20.177°$

36. $x = 50.137°, y = 18.044°$

37. $x = 1.1255, y = 3.6014$

38. $x = 0.039\,17, y = 0.610\,52$

In Problems 39–46, write each as a product if y is a sum or difference, or as a sum or difference if y is a product. Enter the original equation in a graphing utility as y_1, the converted form as y_2, and graph y_1 and y_2 in the same viewing window. Use TRACE to compare the two graphs.

39. $y = \sin 2x + \sin x$

40. $y = \cos 3x + \cos x$

41. $y = \cos 1.7x - \cos 0.3x$

42. $y = \sin 2.1x - \sin 0.5x$

43. $y = \sin 3x \cos x$

44. $y = \cos 5x \cos 3x$

45. $y = \sin 2.3x \sin 0.7x$

46. $y = \cos 1.9x \sin 0.5x$

C

Verify each identity in Problems 47 and 48.

47. $\cos x \cos y \cos z = \frac{1}{4}[\cos(x + y - z) + \cos(y + z - x) + \cos(z + x - y) + \cos(x + y + z)]$

48. $\sin x \sin y \sin z = \frac{1}{4}[\sin(x + y - z) + \sin(y + z - x) + \sin(z + x - y) - \sin(x + y + z)]$

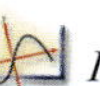

In Problems 49–52,

(A) Graph y_1, y_2, and y_3 in a graphing utility for $0 \le x \le 1$ and $-2 \le y \le 2$.

(B) Convert y_1 to a sum or difference and repeat part A.

49. $y_1 = 2\cos(28\pi x)\cos(2\pi x)$
$y_2 = 2\cos(2\pi x)$
$y_3 = -2\cos(2\pi x)$

50. $y_1 = 2\sin(24\pi x)\sin(2\pi x)$
$y_2 = 2\sin(2\pi x)$
$y_3 = -2\sin(2\pi x)$

51. $y_1 = 2\sin(20\pi x)\cos(2\pi x)$
$y_2 = 2\cos(2\pi x)$
$y_3 = -2\cos(2\pi x)$

52. $y_1 = 2\cos(16\pi x)\sin(2\pi x)$
$y_2 = 2\sin(2\pi x)$
$y_3 = -2\sin(2\pi x)$

APPLICATIONS

Problems 53 and 54 involve the phenomenon of sound called beats. If two tones having the same loudness and close together in pitch (frequency) are sounded, one following the other, most people have difficulty in differentiating the two tones. However, if the tones are sounded simultaneously, they will interact with each other, producing a low warbling sound called a **beat.** *Musicians, when tuning an instrument with other instruments or a tuning fork, listen for these lower beat frequencies and try to eliminate them by adjusting their instruments. Problems 53 and 54 provide a visual illustration of the beat phenomenon.*

53. Music—Beat Frequencies. The equations $y = 0.5\cos 128\pi t$ and $y = -0.5\cos 144\pi t$ model sound waves with frequencies 64 and 72 hertz, respectively. If both sounds are emitted simultaneously, a *beat* frequency results.

(A) Show that

$$0.5\cos 128\pi t - 0.5\cos 144\pi t = \sin 8\pi t \sin 136\pi t$$

(The product form is more useful to sound engineers.)

(B) Graph each equation in a different viewing window for $0 \le t \le 0.25$:

$y = 0.5\cos 128\pi t$
$y = -0.5\cos 144\pi t$
$y = 0.5\cos 128\pi t - 0.5\cos 144\pi t$
$y = \sin 8\pi t \sin 136\pi t$

54. Music—Beat Frequencies. The equations $y = 0.25\cos 256\pi t$ and $y = -0.25\cos 288\pi t$ model sound waves with frequencies 128 and 144 hertz, respectively. If both sounds are emitted simultaneously, a *beat* frequency results.

(A) Show that

$$0.25\cos 256\pi t - 0.25\cos 288\pi t = 0.5\sin 16\pi t \sin 272\pi t$$

(The product form is more useful to sound engineers.)

(B) Graph each equation in a different viewing window for $0 \le t \le 0.125$:

$y = 0.25\cos 256\pi t$
$y = -0.25\cos 288\pi t$
$y = 0.25\cos 256\pi t - 0.25\cos 288\pi t$
$y = 0.5\sin 16\pi t \sin 272\pi t$

SECTION 6.5 Trigonometric Equations

Solving Trigonometric Equations Using an Algebraic Approach • Solving Trigonometric Equations Using a Graphing Utility

Sections 6.1 through 6.4 of this chapter consider trigonometric equations called *identities*. These are equations that are true for all replacements of the variable(s) for which both sides are defined. We now consider another class of trigonometric equations, called **conditional equations,** which may be true for some replacements of the variable but false for others. For example,

$$\cos x = \sin x$$

is a conditional equation, because it is true for some values of x, for example, $x = \pi/4$, and false for others, such as $x = 0$. (Check both values.)

Section 6.5 considers two approaches for solving conditional trigonometric equations: an algebraic approach and a graphing utility approach. Solving trigonometric equations using an algebraic approach often requires the use of algebraic manipulation, identities, and ingenuity. In some cases algebraic methods lead to exact solutions, which are very useful in certain contexts. Graphing utility methods can be used to approximate solutions to a greater variety of trigonometric equations, but usually do not produce exact solutions. Each method has its strengths.

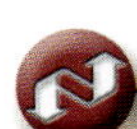

EXPLORE/DISCUSS 1

We are interested in solutions to the equation

$$\cos x = 0.5$$

The figure shows a partial graph of the left and right sides of the equation.

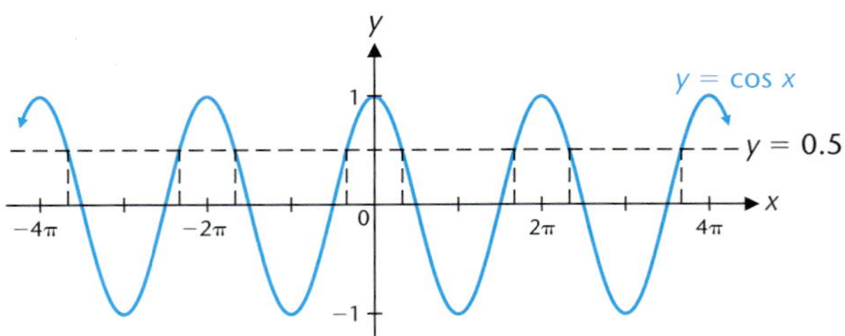

(A) How many solutions does the equation have on the interval $[0, 2\pi)$? What are they?

(B) How many solutions does the equation have on the interval $(-\infty, \infty)$? Discuss a method of writing all solutions to the equation.

Solving Trigonometric Equations Using an Algebraic Approach

You might find the following suggestions for solving trigonometric equations using an algebraic approach useful:

Suggestions For Solving Trigonometric Equations Algebraically

1. Regard one particular trigonometric function as a variable, and solve for it.
   (A) Consider using algebraic manipulation such as factoring, combining or separating fractions, and so on.
   (B) Consider using identities.
2. After solving for a trigonometric function, solve for the variable.

Examples 1–5 should help make the algebraic approach clear.

EXAMPLE 1 Exact Solutions Using Factoring

Find all solutions exactly for $2\cos^2 x - \cos x = 0$.

SOLUTION

Step 1. *Solve for cos x.*

$$2\cos^2 x - \cos x = 0 \qquad 2a^2 - a = a(2a - 1)$$

$$\cos x\,(2\cos x - 1) = 0 \qquad ab = 0 \text{ only if } a = 0 \text{ or } b = 0$$

$$\cos x = 0 \quad \text{or} \quad 2\cos x - 1 = 0$$

$$\cos x = \tfrac{1}{2}$$

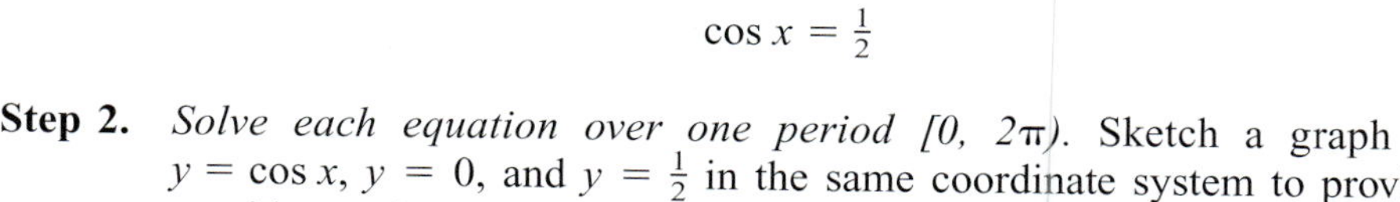

Step 2. *Solve each equation over one period [0, 2π).* Sketch a graph of $y = \cos x$, $y = 0$, and $y = \frac{1}{2}$ in the same coordinate system to provide an aid to writing all solutions over one period (Fig. 1).

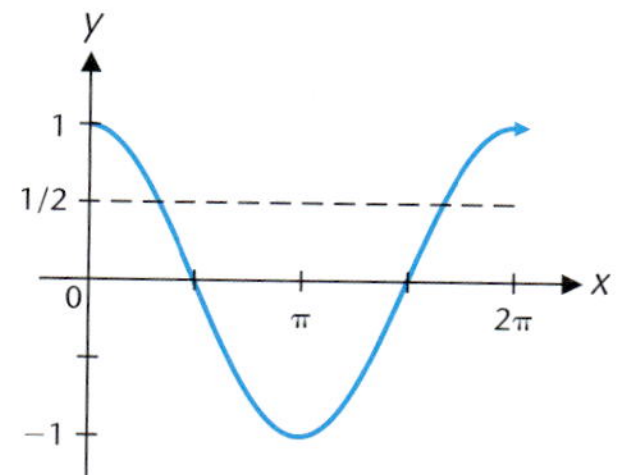

FIGURE 1

$$\cos x = 0 \qquad\qquad \cos x = \tfrac{1}{2}$$

$$x = \pi/2,\ 3\pi/2 \qquad\qquad x = \pi/3,\ 5\pi/3$$

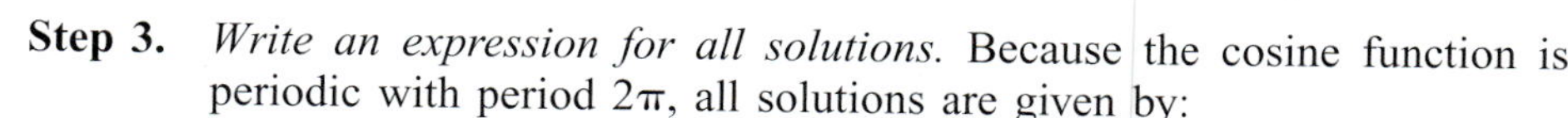

Step 3. *Write an expression for all solutions.* Because the cosine function is periodic with period 2π, all solutions are given by:

$$x = \begin{cases} \pi/3 + 2k\pi \\ \pi/2 + 2k\pi \\ 3\pi/2 + 2k\pi \\ 5\pi/3 + 2k\pi \end{cases} \quad k \text{ any integer}$$

MATCHED PROBLEM 1

Find all solutions exactly for $2\sin^2 x + \sin x = 0$.

EXAMPLE 2 Approximate Solutions Using Identities and Factoring

Find all real solutions for $3\cos^2 x + 8\sin x = 7$. Compute all inverse functions to four decimal places.

SOLUTION

Step 1. *Solve for sin x and/or cos x.* Move all nonzero terms to the left of the equal sign and express the left side in terms of $\sin x$:

$$
\begin{aligned}
3\cos^2 x + 8\sin x &= 7 \\
3\cos^2 x + 8\sin x - 7 &= 0 \qquad \cos^2 x = 1 - \sin^2 x \\
3(1 - \sin^2 x) + 8\sin x - 7 &= 0 \\
3\sin^2 x - 8\sin x + 4 &= 0 \qquad 3u^2 - 8u + 4 = (u-2)(3u-2) \\
(\sin x - 2)(3\sin x - 2) &= 0 \qquad ab = 0 \text{ only if } a = 0 \text{ or } b = 0 \\
\sin x - 2 &= 0 \qquad \text{or} \qquad 3\sin x - 2 = 0 \\
\sin x &= 2 \qquad\qquad\qquad \sin x = \tfrac{2}{3}
\end{aligned}
$$

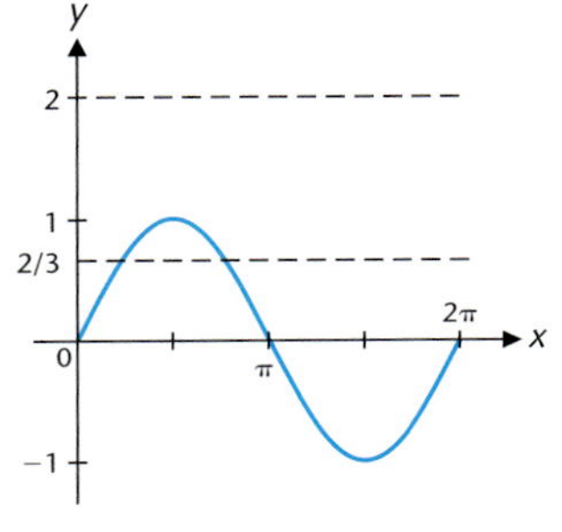

FIGURE 2

Step 2. *Solve each equation over one period [0, 2π):* Sketch a graph of $y = \sin x$, $y = 2$, and $y = \frac{2}{3}$ in the same coordinate system to provide an aid to writing all solutions over one period (Fig. 2).

Solve the first equation:

$\sin x = 2$ No solution, because $-1 \le \sin x \le 1$.

Solve the second equation:

$\sin x = \frac{2}{3}$ From the graph we see there are solutions in the first and second quadrants.

$x = \sin^{-1}\frac{2}{3} = 0.7297$ First quadrant solution

$x = \pi - 0.7297 = 2.4119$ Second quadrant solution

CHECK

$\sin 0.7297 = 0.6667$; $\sin 2.4119 = 0.6666$

(Checks may not be exact because of roundoff errors.)

Step 3. *Write an expression for all solutions.* Because the sine function is periodic with period 2π, all solutions are given by:

$$x = \begin{cases} 0.7297 + 2k\pi \\ 2.4119 + 2k\pi \end{cases} \quad k \text{ any integer}$$

MATCHED PROBLEM 2

Find all real solutions to $8\sin^2 x = 5 - 10\cos x$. Compute all inverse functions to four decimal places.

EXAMPLE 3 Approximate Solutions Using Substitution

Find θ in degree measure to three decimal places so that $5 \sin (2\theta - 5) = -3.045$, $0° \le 2\theta - 5 \le 360°$.

SOLUTION

Step 1. *Make a substitution.* Let $u = 2\theta - 5$ to obtain

$$5 \sin u = -3.045, \quad 0° \le u \le 360°$$

Step 2. *Solve for sin u.*

$$\sin u = \frac{-3.045}{5} = -0.609$$

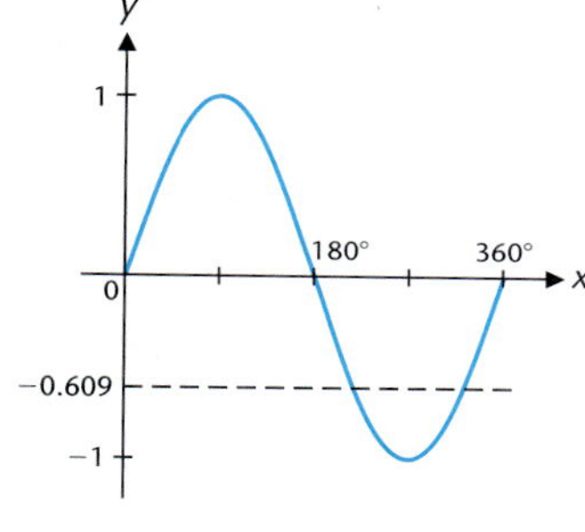

FIGURE 3

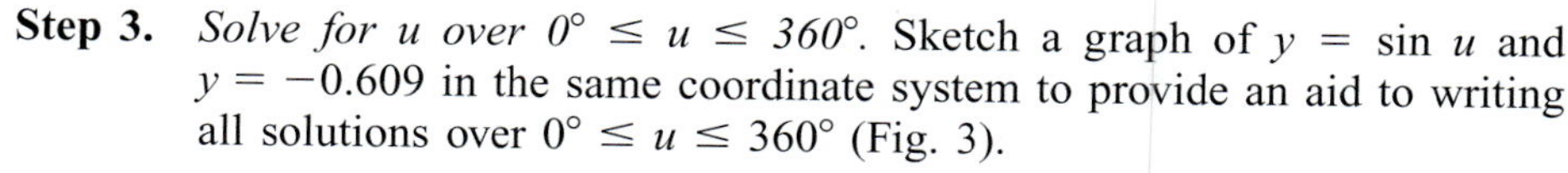

Step 3. *Solve for u over $0° \le u \le 360°$.* Sketch a graph of $y = \sin u$ and $y = -0.609$ in the same coordinate system to provide an aid to writing all solutions over $0° \le u \le 360°$ (Fig. 3).

Solutions are in the third and fourth quadrants. If the reference angle is α, then $u = 180° + \alpha$ or $u = 360° - \alpha$.

$$\alpha = \sin^{-1} 0.609 = 37.517° \qquad \text{Reference angle}$$
$$u = 180° + 37.517°$$
$$= 217.517° \qquad \text{Third quadrant solution}$$
$$u = 360° - 37.517°$$
$$= 322.483° \qquad \text{Fourth quadrant solution}$$

CHECK

$\sin 217.517° = -0.609$; $\sin 322.483° = -0.609$

Step 4. *Now solve for θ:*

$u = 217.517°$	$u = 322.483°$
$2\theta - 5 = 217.517°$	$2\theta - 5 = 322.483°$
$\theta = 111.259°$	$\theta = 163.742°$

A final check in the original equation is left to the reader.

MATCHED 3 PROBLEM

Find θ in degree measure to three decimal places so that $8 \tan (6\theta + 15) = -64.328$, $-90° < 6\theta + 15 < 90°$.

EXAMPLE 4 Exact Solutions Using Identities and Factoring

Find exact solutions for $\sin^2 x = \frac{1}{2} \sin 2x$, $0 \le x < 2\pi$.

SOLUTION

The following solution includes only the key steps. Sketch graphs as appropriate on scratch paper.

$$\sin^2 x = \tfrac{1}{2}\sin 2x \qquad \text{Use double-angle identity.}$$
$$= \tfrac{1}{2}(2\sin x\cos x)$$
$$\sin^2 x - \sin x\cos x = 0 \qquad a^2 - ab = a(a-b)$$
$$\sin x(\sin x - \cos x) = 0 \qquad a(a-b) = 0 \text{ only if } a = 0 \text{ or } a - b = 0$$
$$\sin x = 0 \quad \text{or} \quad \sin x - \cos x = 0$$
$$x = 0, \pi \qquad \sin x = \cos x$$
$$\frac{\sin x}{\cos x} = 1$$
$$\tan x = 1$$
$$x = \pi/4, 5\pi/4$$

Combining the solutions from both equations, we have the complete set of solutions:

$x = 0, \pi/4, \pi, 5\pi/4$

MATCHED 4 PROBLEM

Find exact solutions for $\sin 2x = \sin x$, $0 \le x < 2\pi$.

EXAMPLE 5 Approximate Solutions Using Identities and the Quadratic Formula

Solve $\cos 2x = 4\cos x - 2$ for all real x. Compute inverse functions to four decimal places.

SOLUTION

Step 1. *Solve for cos x.*

$$\cos 2x = 4\cos x - 2 \qquad \text{Use double-angle identity.}$$
$$2\cos^2 x - 1 = 4\cos x - 2$$
$$2\cos^2 x - 4\cos x + 1 = 0 \qquad \text{Quadratic in } \cos x\text{. Left side does not factor using integer coefficients. Solve using quadratic formula.}$$

$$\cos x = \frac{4 \pm \sqrt{16 - 4(2)(1)}}{2(2)}$$
$$= 1.707107 \text{ or } 0.292893$$

Step 2. *Solve each equation over one period [0, 2π):* Sketch a graph of $y = \cos x$, $y = 1.707107$, and $y = 0.292893$ in the same coordinate system to provide an aid to writing all solutions over one period (Fig. 4).

FIGURE 4

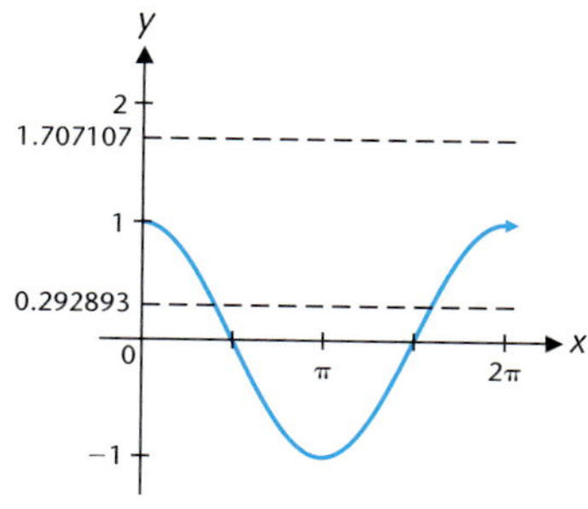

Solve the first equation:

$\cos x = 1.707107$ No solution, because $-1 \leq \cos x \leq 1$

Solve the second equation:

$\cos x = 0.292893$

Figure 4 indicates a first quadrant solution and a fourth quadrant solution. If the reference angle is α, then $x = \alpha$ or $x = 2\pi - \alpha$.

$$\alpha = \cos^{-1} 0.292893 = 1.2735$$
$$2\pi - \alpha = 2\pi - 1.2735 = 5.0096$$

CHECK

$\cos 1.2735 = 0.292936$; $\cos 5.0096 = 0.292854$

Step 3. *Write an expression for all solutions.* Because the cosine function is periodic with period 2π, all solutions are given by:

$$x = \begin{cases} 1.2735 + 2k\pi \\ 5.0096 + 2k\pi \end{cases} \quad k \text{ any integer}$$

MATCHED 5 PROBLEM

Solve $\cos 2x = 2(\sin x - 1)$ for all real x. Compute inverse functions to four decimal places.

Solving Trigonometric Equations Using a Graphing Utility

All the trigonometric equations that were solved earlier with algebraic methods can also be solved, though usually not exactly, with graphing utility methods. In addition, there are many trigonometric equations that can be solved (to any decimal accuracy desired) using graphing utility methods, but cannot be solved in a finite sequence of steps using algebraic methods. Examples 6–8 are examples of such equations.

EXAMPLE 6 Solution Using a Graphing Utility

Find all real solutions to four decimal places for $2 \cos 2x = 1.35x - 2$.

SOLUTION

This relatively simple trigonometric equation cannot be solved using a finite number of algebraic steps (try it!). However, it can be solved rather easily to the accuracy desired using a graphing utility. Graph $y_1 = 2 \cos 2x$ and $y_2 = 1.35x - 2$ in the same viewing window, and find any points of intersection using the intersect command. The first point of intersection is shown in Figure 5. It appears there may be more than one point of intersection, but zooming in on the portion of the graph in question shows that the two graphs do not intersect in that region (Fig. 6). The only solution is

$$x = 0.9639$$

CHECK

Left side: $2 \cos 2(0.9639) = -0.6989$
Right side: $1.35(0.9639) - 2 = -0.6987$

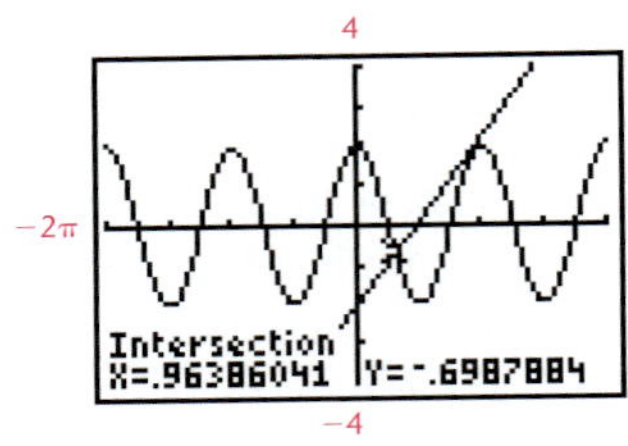

FIGURE 5

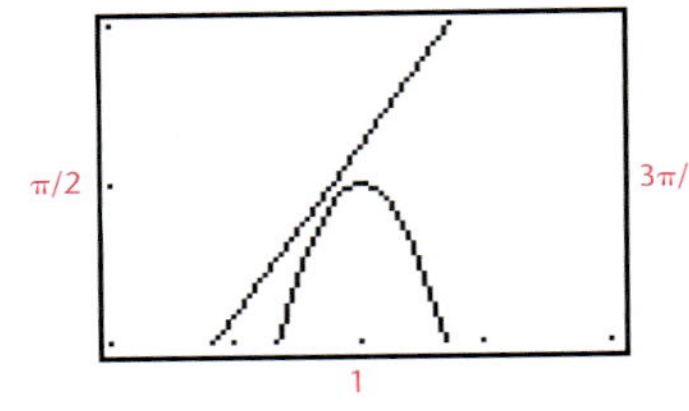

FIGURE 6

MATCHED 6 PROBLEM

Find all real solutions to four decimal places for $\sin x/2 = 0.2x - 0.5$.

EXAMPLE 7 Geometric Application

A 10-centimeter arc on a circle has an 8-centimeter chord. What is the radius of the circle to four decimal places? What is the radian measure of the central angle, to four decimal places, subtended by the arc?

SOLUTION

Sketch a figure with auxiliary lines (Fig. 7).
From the figure, θ in radians is

$$\theta = \frac{5}{R} \qquad \text{and} \qquad \sin \theta = \frac{4}{R}$$

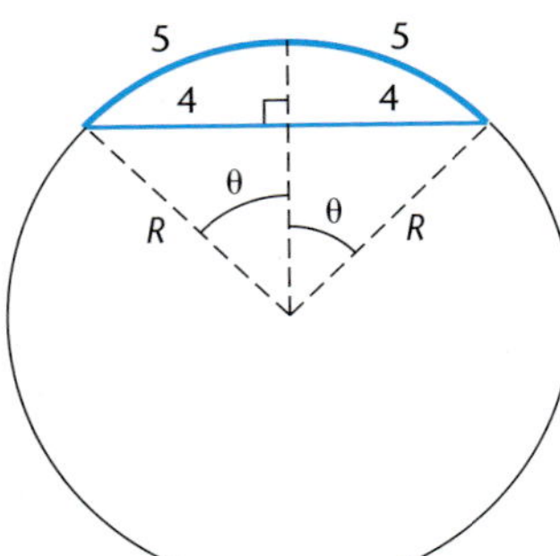

FIGURE 7

Thus,

$$\sin\frac{5}{R} = \frac{4}{R}$$

and our problem is to solve this trigonometric equation for R. Algebraic methods will not isolate R, so turn to the use of a graphing utility. Start by graphing $y_1 = \sin 5/x$ and $y_2 = 4/x$ in the same viewing window for $1 \le x \le 10$ and $-2 \le y \le 2$ (Fig. 8). It appears that the graphs intersect for x between 4 and 5. To get a clearer look at the intersection point we change the window dimensions to $4 \le x \le 5$ and $0.5 \le y \le 1.5$, and use the intersect command to find the point of intersection (Fig. 9).

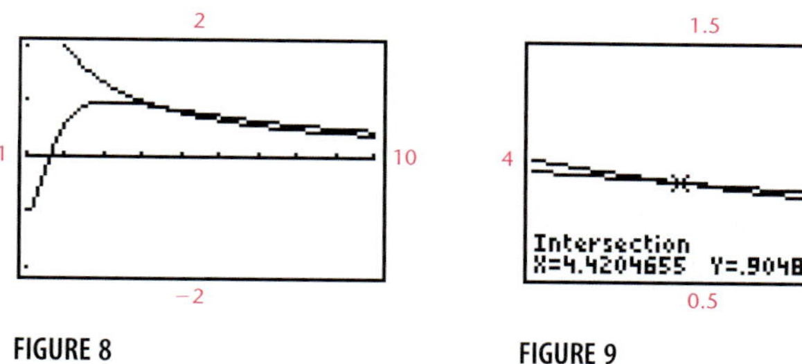

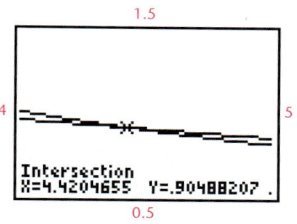

FIGURE 8

FIGURE 9

From Figure 9, we see that

$$R = 4.4205 \text{ centimeters}$$

CHECK

$\sin 5/R = \sin(5/4.4205) = 0.9049$; $4/R = 4/4.4205 = 0.9049$

Having R, we can compute the radian measure of the central angle subtended by the 10-centimeter arc:

$$\text{Central Angle} = \frac{10}{R} = \frac{10}{4.4205} = 2.2622 \text{ radians}$$

MATCHED 7 PROBLEM

An 8.2456-inch arc on a circle has a 6.0344-inch chord. What is the radius of the circle to four decimal places? What is the measure of the central angle, to four decimal places, subtended by the arc?

EXAMPLE 8 Solution Using a Graphing Utility

Find all real solutions, to four decimal places, for $\tan(x/2) = 1/x$, $-\pi < x \le 3\pi$.

SOLUTION

Graph $y_1 = \tan(x/2)$ and $y_2 = 1/x$ in the same viewing window for $-\pi < x < 3\pi$ (Fig. 10). Solutions are at points of intersection.

FIGURE 10

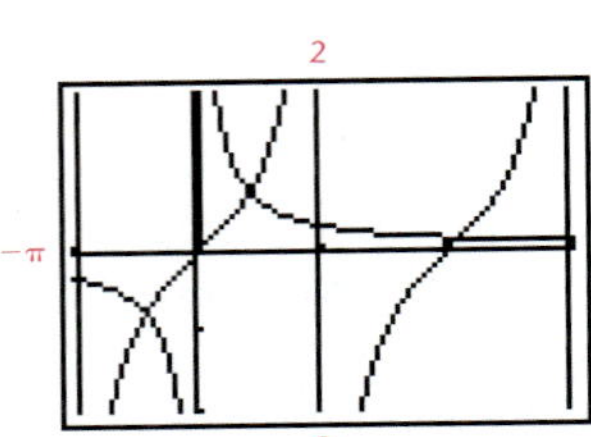

Using the intersect command, the three solutions are found to be

$$x = -1.3065, 1.3065, 6.5846$$

Checking these solutions is left to the reader.

MATCHED 8 PROBLEM

Find all real solutions, to four decimal places, for $0.25 \tan (x/2) = \ln x$, $0 < x < 4\pi$.

Solving trigonometric inequalities using a graphing utility is as easy as solving trigonometric equations using a graphing utility. Example 9 illustrates the process.

EXAMPLE 9 Solving a Trigonometric Inequality

Solve $\sin x - \cos x < 0.25x - 0.5$, using two-decimal-place accuracy.

FIGURE 11

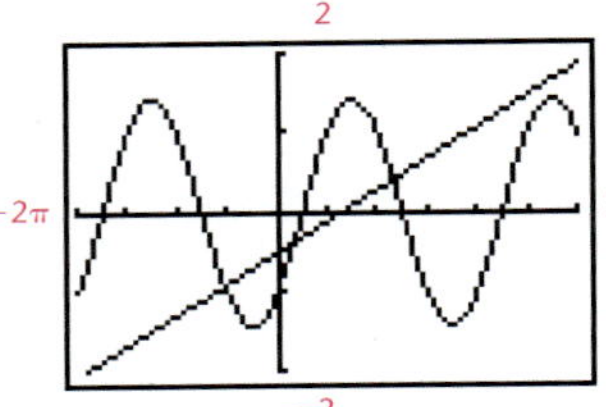

SOLUTION
Graph $y_1 = \sin x - \cos x$ and $y_2 = 0.25x - 0.5$ in the same viewing window (Fig. 11).

Finding the three points of intersection by the intersect command, we see that the graph of y_1 is below the graph of y_2 on the following two intervals: $(-1.65, 0.52)$ and $(3.63, \infty)$. Thus, the solution set to the inequality is $(-1.65, 0.52) \cup (3.63, \infty)$.

MATCHED 9 PROBLEM

Solve $\cos x - \sin x > 0.4 - 0.3x$, using two-decimal-place accuracy.

EXPLORE/DISCUSS 2

How many solutions does the following equation have?

$$\sin (1/x) = 0$$

Graph $y_1 = \sin (1/x)$ and $y_2 = 0$ for each of the indicated intervals in parts A–G. From each graph estimate the number of solutions that equation (1) appears to have. What final conjecture would you be willing to make regarding the number of solutions to equation (1)? Explain.

(A) $[-20, 20]$; Can 0 be a solution? Explain.

(B) $[-2, 2]$ (C) $[-1, 1]$ (D) $[-0.1, 0.1]$ (E) $[-0.01, 0.01]$

(F) $[-0.001, 0.001]$ (G) $[-0.0001, 0.0001]$

ANSWERS MATCHED PROBLEMS

1. $x = \begin{cases} 0 + 2k\pi \\ \pi + 2k\pi \\ 7\pi/6 + 2k\pi \\ 11\pi/6 + 2k\pi \end{cases}$ k any integer **2.** $x = \begin{cases} 1.8235 + 2k\pi \\ 4.4597 + 2k\pi \end{cases}$ k any integer **3.** $-16.318°$

4. $x = 0, \pi/3, \pi, 5\pi/3$ **5.** $x = \begin{cases} 0.9665 + 2k\pi \\ 2.1751 + 2k\pi \end{cases}$ k any integer **6.** $x = 5.1609$

7. $R = 3.1103$ inches; central angle $= 2.6511$ radians **8.** $x = 1.1828, 2.6369, 9.2004$ **9.** $(-1.67, 0.64) \cup (3.46, \infty)$

EXERCISE 6.5

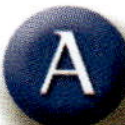

A

In Problems 1–12, find exact solutions over the indicated intervals (x a real number, θ in degrees).

1. $2 \sin x + 1 = 0, 0 \le x < 2\pi$

2. $2 \cos x + 1 = 0, 0 \le x < 2\pi$

3. $2 \sin x + 1 = 0$, all real x

4. $2 \cos x + 1 = 0$, all real x

5. $\tan x + \sqrt{3} = 0, 0 \le x < \pi$

6. $\sqrt{3} \tan x + 1 = 0, 0 \le x < \pi$

7. $\tan x + \sqrt{3} = 0$, all real x

8. $\sqrt{3} \tan x + 1 = 0$, all real x

9. $2 \cos \theta - \sqrt{3} = 0, 0° \le \theta < 360°$

10. $\sqrt{2} \sin \theta - 1 = 0, 0° \le \theta < 360°$

11. $2 \cos \theta - \sqrt{3} = 0$, all θ

12. $\sqrt{2} \sin \theta - 1 = 0$, all θ

Solve Problems 13–18 to four decimal places (θ in degrees, x real).

13. $7 \cos x - 3 = 0, 0 \le x < 2\pi$

14. $5 \cos x - 2 = 0, 0 \le x < 2\pi$

15. $2 \tan \theta - 7 = 0, 0° \le \theta < 180°$

16. $4 \tan \theta + 15 = 0, 0° \le \theta < 180°$

17. $1.3224 \sin x + 0.4732 = 0$, all real x

18. $5.0118 \sin x - 3.1105 = 0$, all real x

Solve Problems 19–22 to four decimal places using a graphing utility.

19. $1 - x = 2 \sin x$, all real x

20. $2x - \cos x = 0$, all real x

21. $\tan (x/2) = 8 - x, 0 \le x < \pi$

22. $\tan 2x = 1 + 3x, 0 \le x < \pi/4$

B

In Problems 23–34, find exact solutions for x real and θ in degrees.

23. $2 \sin^2 \theta + \sin 2\theta = 0$, all θ

24. $\cos^2 \theta = \frac{1}{2} \sin 2\theta$, all θ

25. $\tan x = -2 \sin x, 0 \le x < 2\pi$

26. $\cos x = \cot x, 0 \le x < 2\pi$

27. $\sec (x/2) + 2 = 0, 0 \le x < 2\pi$

28. $\tan (x/2) - 1 = 0, 0 \le x < 2\pi$

29. $2 \cos^2 \theta + 3 \sin \theta = 0, 0° \le \theta < 360°$

30. $\sin^2 \theta + 2 \cos \theta = -2, 0° \le \theta < 360°$

31. $\cos 2\theta + \cos \theta = 0, 0° \le \theta < 360°$

32. $\cos 2\theta + \sin^2 \theta = 0, 0° \le \theta < 360°$

33. $2 \sin^2 (x/2) - 3 \sin (x/2) + 1 = 0, 0 \le x \le 2\pi$

34. $4 \cos^2 2x - 4 \cos 2x + 1 = 0, 0 \le x \le 2\pi$

Solve Problems 35–40 (x real and θ in degrees). Compute inverse functions to four significant digits.

35. $6 \sin^2 \theta + 5 \sin \theta = 6, 0° \le \theta \le 90°$

36. $4 \cos^2 \theta = 7 \cos \theta + 2, 0° \le \theta \le 180°$

37. $3 \cos^2 x - 8 \cos x = 3, 0 \le x \le \pi$

38. $8 \sin^2 x + 10 \sin x = 3, 0 \le x \le \pi/2$

39. $2 \sin x = \cos 2x, 0 \le x < 2\pi$

40. $\cos 2x + 10 \cos x = 5, 0 \le x < 2\pi$

Solve Problems 41 and 42 for all real number solutions. Compute inverse functions to four significant digits.

41. $2 \sin^2 x = 1 - 2 \sin x$

42. $\cos^2 x = 3 - 5 \cos x$

Solve Problems 43–52 to four decimal places using a graphing utility.

43. $2 \sin x = \cos 2x, 0 \le x < 2\pi$

44. $\cos 2x + 10 \cos x = 5, 0 \le x < 2\pi$

45. $2 \sin^2 x = 1 - 2 \sin x$, all real x

46. $\cos^2 x = 3 - 5 \cos x$, all real x

47. $\cos 2x > x^2 - 2$, all real x

48. $2 \sin (x - 2) < 3 - x^2$, all real x

49. $\cos (2x + 1) \le 0.5x - 2$, all real x

50. $\sin (3 - 2x) \ge 1 - 0.4x$, all real x

51. $e^{\sin x} = 2x - 1$, all real x

52. $e^{-\sin x} = 3 - x$, all real x

53. Explain the difference between evaluating $\tan^{-1}(-5.377)$ and solving the equation $\tan x = -5.377$.

54. Explain the difference between evaluating $\cos^{-1}(-0.7334)$ and solving the equation $\cos x = -0.7334$.

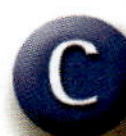

Find exact solutions to Problems 55–58. [Hint: Square both sides at an appropriate point, solve, then eliminate extraneous solutions at the end.]

55. $\cos x - \sin x = 1, 0 \le x < 2\pi$

56. $\sin x + \cos x = 1, 0 \le x < 2\pi$

57. $\tan x - \sec x = 1, 0 \le x < 2\pi$

58. $\sec x + \tan x = 1, 0 \le x < 2\pi$

Solve Problems 59–60 to four significant digits using a graphing utility.

59. $\sin (1/x) = 1.5 - 5x, 0.04 \le x \le 0.2$

60. $2 \cos (1/x) = 950x - 4, 0.006 < x < 0.007$

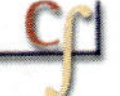

61. We are interested in the zeros of the function $f(x) = \sin (1/x)$ for $x > 0$.

(A) Explore the graph of f over different intervals $[0.1, b]$ for various values of b, $b > 0.1$. Does the function f have a largest zero? If so, what is it (to four decimal places)? Explain what happens to the graph of f as x increases without bound. Does the graph have an asymptote? If so, what is its equation?

(B) Explore the graph of f over different intervals $(0, b]$ for various values of b, $0 < b \le 0.1$. How many zeros exist between 0 and b, for any $b > 0$, however small? Explain why this happens. Does f have a smallest positive zero? Explain.

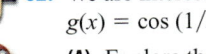

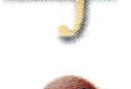

62. We are interested in the zeros of the function $g(x) = \cos (1/x)$ for $x > 0$.

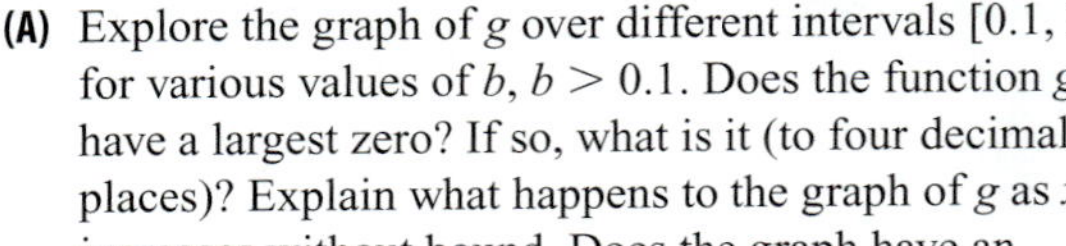

(A) Explore the graph of g over different intervals $[0.1, b]$ for various values of b, $b > 0.1$. Does the function g have a largest zero? If so, what is it (to four decimal places)? Explain what happens to the graph of g as x increases without bound. Does the graph have an asymptote? If so, what is its equation?

(B) Explore the graph of g over different intervals $(0, b]$ for various values of b, $0 < b \le 0.1$. How many zeros exist between 0 and b, for any $b > 0$, however small? Explain why this happens. Does g have a smallest positive zero? Explain.

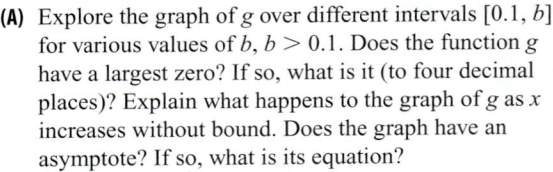

APPLICATIONS

63. Electric Current. An alternating current generator produces a current given by the equation

$I = 30 \sin 120\pi t$

where t is time in seconds and I is current in amperes. Find the smallest positive t (to four significant digits) such that $I = -10$ amperes.

64. Electric Current. Refer to Problem 63. Find the smallest positive t (to four significant digits) such that $I = 25$ amperes.

65. Optics. A polarizing filter for a camera contains two parallel plates of polarizing glass, one fixed and the other able to rotate. If θ is the angle of rotation from the position of maximum light transmission, then the intensity of light leaving the filter is $\cos^2 \theta$ times the intensity I of light entering the filter (see the figure).

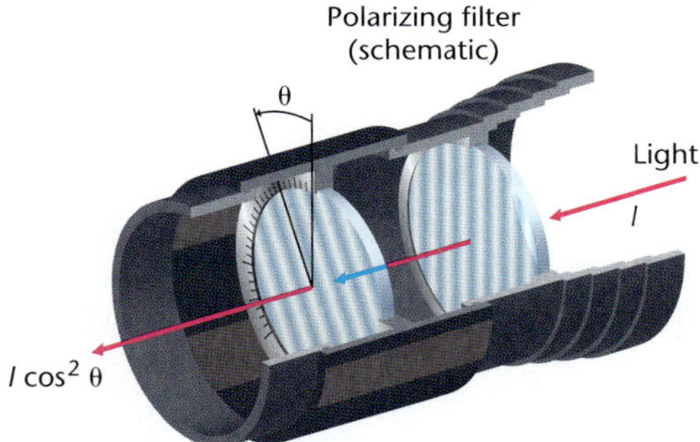

Find the smallest positive θ (in decimal degrees to two decimal places) so that the intensity of light leaving the filter is 40% of that entering.

66. Optics. Refer to Problem 65. Find the smallest positive θ so that the light leaving the filter is 70% of that entering.

67. Astronomy. The planet Mercury travels around the sun in an elliptical orbit given approximately by

$$r = \frac{3.44 \times 10^7}{1 - 0.206 \cos \theta}$$

(see the figure). Find the smallest positive θ (in decimal degrees to three significant digits) such that Mercury is 3.09×10^7 miles from the sun.

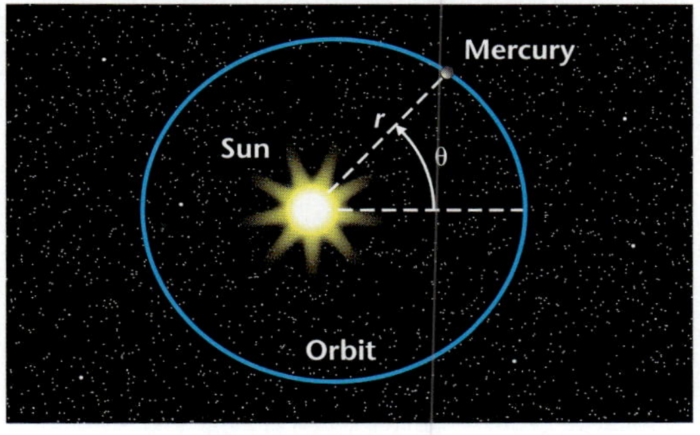

68. Astronomy. Refer to Problem 67. Find the smallest positive θ (in decimal degrees to three significant digits) such that Mercury is 3.78×10^7 miles from the sun.

69. Geometry. The area of the segment of a circle in the figure is given by

$A = \frac{1}{2} R^2 (\theta - \sin \theta)$

where θ is in radian measure. Use a graphing utility to find the radian measure, to three decimal places, of angle θ, if the radius is 8 inches and the area of the segment is 48 square inches.

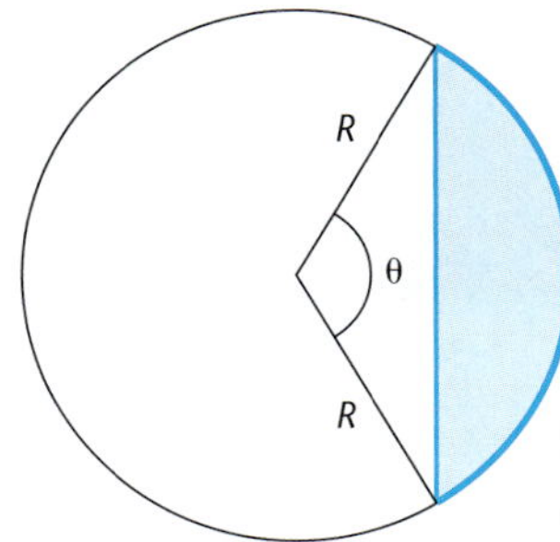

70. Geometry. Repeat Problem 69, if the radius is 10 centimeters and the area of the segment is 40 square centimeters.

71. Eye Surgery. A surgical technique for correcting an astigmatism involves removing small pieces of tissue to change the curvature of the cornea.* In the cross section of a cornea shown in the figure, the circular arc, with radius R and central angle 2θ, represents a cross section of the surface of the cornea.

(A) If $a = 5.5$ millimeters and $b = 2.5$ millimeters, find L correct to four decimal places.

*Based on the article "The Surgical Correction of Astigmatism" by Sheldon Rothman and Helen Strassberg in the *UMAP Journal,* Vol. v, no. 2, 1984.

(B) Reducing the chord length $2a$ without changing the length L of the arc has the effect of pushing the cornea outward and giving it a rounder, yet still a circular, shape. With the aid of a graphing utility in part of the solution, approximate b to four decimal places if a is reduced to 5.4 millimeters and L remains the same as it was in part A.

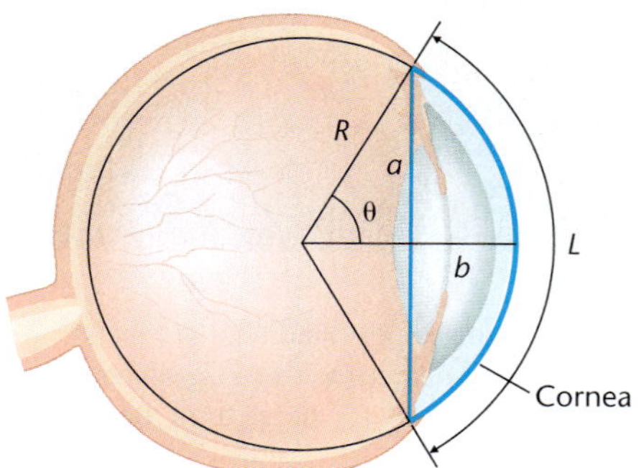

72. Eye Surgery. Refer to Problem 71.

(A) If in the figure $a = 5.4$ millimeters and $b = 2.4$ millimeters, find L correct to four decimal places.

(B) Increasing the chord length without changing the arc length L has the effect of pulling the cornea inward and giving it a flatter, yet still circular, shape. With the aid of a graphing utility in part of the solution, approximate b to four decimal places if a is increased to 5.5 millimeters and L remains the same as it was in part A.

Analytic Geometry. *Find simultaneous solutions for each system of equations in Problems 73 and 74 ($0° \le \theta \le 360°$). These are polar equations, which will be discussed in Chapter 7.*

★ **73.** $r = 2 \sin \theta$
$r = \sin 2\theta$

★ **74.** $r = 2 \sin \theta$
$r = 2(1 - \sin \theta)$

Problems 75 and 76 are related to rotation of axes in analytic geometry.

★★ **75. Analytic Geometry.** Given the equation $2xy = 1$, replace x and y with

$$x = u \cos \theta - v \sin \theta$$
$$y = u \sin \theta + v \cos \theta$$

and simplify the left side of the resulting equation. Find the smallest positive θ in degree measure so that the coefficient of the uv term is 0.

★★ **76. Analytic Geometry.** Repeat Problem 75 for $xy = -2$.

CHAPTER 6 REVIEW

6.1 Basic Identities and Their Use

The following 11 identities are basic to the process of changing trigonometric expressions to equivalent but more useful forms:

Reciprocal Identities

$$\csc x = \frac{1}{\sin x} \qquad \sec x = \frac{1}{\cos x} \qquad \cot x = \frac{1}{\tan x}$$

Quotient Identities

$$\tan x = \frac{\sin x}{\cos x} \qquad \cot x = \frac{\cos x}{\sin x}$$

Identities for Negatives

$$\sin(-x) = -\sin x \qquad \cos(-x) = \cos x$$
$$\tan(-x) = -\tan x$$

Pythagorean Identities

$$\sin^2 x + \cos^2 x = 1 \qquad \tan^2 x + 1 = \sec^2 x$$
$$1 + \cot^2 x = \csc^2 x$$

Although there is no fixed method of verification that works for all identities, the following suggested steps are helpful in many cases.

Suggested Steps in Verifying Identities

1. Start with the more complicated side of the identity, and transform it into the simpler side.
2. Try algebraic operations such as multiplying, factoring, combining fractions, and splitting fractions.
3. If other steps fail, express each function in terms of sine and cosine functions, and then perform appropriate algebraic operations.
4. At each step, keep the other side of the identity in mind. This often reveals what you should do to get there.

6.2 Sum, Difference, and Cofunction Identities

Sum Identities

$$\sin(x + y) = \sin x \cos y + \cos x \sin y$$
$$\cos(x + y) = \cos x \cos y - \sin x \sin y$$
$$\tan(x + y) = \frac{\tan x + \tan y}{1 - \tan x \tan y}$$

Difference Identities

$$\sin(x - y) = \sin x \cos y - \cos x \sin y$$
$$\cos(x - y) = \cos x \cos y + \sin x \sin y$$
$$\tan(x - y) = \frac{\tan x - \tan y}{1 + \tan x \tan y}$$

Cofunction Identities

(Replace $\pi/2$ with 90° if x is in degrees.)

$$\cos\left(\frac{\pi}{2} - x\right) = \sin x$$
$$\sin\left(\frac{\pi}{2} - x\right) = \cos x$$
$$\tan\left(\frac{\pi}{2} - x\right) = \cot x$$

6.3 Double-Angle and Half-Angle Identities

Double-Angle Identities

$$\sin 2x = 2 \sin x \cos x$$
$$\cos 2x = \cos^2 x - \sin^2 x = 1 - 2\sin^2 x = 2\cos^2 x - 1$$
$$\tan 2x = \frac{2\tan x}{1 - \tan^2 x} = \frac{2\cot x}{\cot^2 x - 1} = \frac{2}{\cot x - \tan x}$$

Half-Angle Identities

$$\sin\frac{x}{2} = \pm\sqrt{\frac{1 - \cos x}{2}}$$
$$\cos\frac{x}{2} = \pm\sqrt{\frac{1 + \cos x}{2}}$$
$$\tan\frac{x}{2} = \pm\sqrt{\frac{1 - \cos x}{1 + \cos x}} = \frac{\sin x}{1 + \cos x} = \frac{1 - \cos x}{\sin x}$$

6.4 Product–Sum and Sum–Product Identities

Product–Sum Identities

$$\sin x \cos y = \tfrac{1}{2}[\sin(x + y) + \sin(x - y)]$$
$$\cos x \sin y = \tfrac{1}{2}[\sin(x + y) - \sin(x - y)]$$
$$\sin x \sin y = \tfrac{1}{2}[\cos(x - y) - \cos(x + y)]$$
$$\cos x \cos y = \tfrac{1}{2}[\cos(x + y) + \cos(x - y)]$$

Sum–Product Identities

$$\sin x + \sin y = 2\sin\frac{x + y}{2}\cos\frac{x - y}{2}$$
$$\sin x - \sin y = 2\cos\frac{x + y}{2}\sin\frac{x - y}{2}$$
$$\cos x + \cos y = 2\cos\frac{x + y}{2}\cos\frac{x - y}{2}$$
$$\cos x - \cos y = -2\sin\frac{x + y}{2}\sin\frac{x - y}{2}$$

6.5 Trigonometric Equations

Sections 6.1 through 6.4 of the chapter considered trigonometric equations called **identities.** Identities are true for all replacements of the variable(s) for which both sides are defined. Section 6.5 considered **conditional equations.** Conditional equations may be true for some variable replacements, but are false for other variable replacements for which both sides are defined. The equation $\sin x = \cos x$ is a conditional equation.

In **solving a trigonometric equation using an algebraic approach,** no particular rule will always lead to all solutions of every trigonometric equation you are likely to encounter. Solving trigonometric equations algebraically often requires the use of algebraic manipulation, identities, and ingenuity.

Suggestions for Solving Trigonometric Equations Algebraically

1. Regard one particular trigonometric function as a variable, and solve for it.

(A) Consider using algebraic manipulation such as factoring, combining or separating fractions, and so on.
(B) Consider using identities.

2. After solving for a trigonometric function, solve for the variable.

In **solving a trigonometric equation using a graphing utility approach** you can solve a larger variety of problems than with the algebraic approach. The solutions are generally approximations (to whatever decimal accuracy desired).

CHAPTER 6 REVIEW EXERCISES

Work through all the problems in this chapter review and check answers in the back of the book. Answers to all review problems, except verifications, are there, and following each answer is a number in italics indicating the section in which that type of problem is discussed. Where weaknesses show up, review appropriate sections in the text.

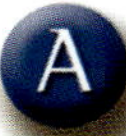

Verify each identity in Problems 1–4.

1. $\tan x + \cot x = \sec x \csc x$

2. $\sec^4 x - 2\sec^2 x \tan^2 x + \tan^4 x = 1$

3. $\dfrac{1}{1 - \sin x} + \dfrac{1}{1 + \sin x} = 2\sec^2 x$

4. $\cos\left(x - \dfrac{3\pi}{2}\right) = -\sin x$

5. Write as a sum: $\sin 5\alpha \cos 3\alpha$.

6. Write as a product: $\cos 7x - \cos 5x$.

7. Simplify: $\sin\left(x + \dfrac{9\pi}{2}\right)$.

Solve Problems 8 and 9 exactly (θ in degrees, x real).

8. $\sqrt{2}\cos\theta + 1 = 0$, all θ

9. $\sin x \tan x - \sin x = 0$, all real x

Solve Problems 10–13 to four decimal places (θ in degrees and x real).

10. $\sin x = 0.7088$, all real x

11. $\cos\theta = 0.2557$, all θ

12. $\cot x = -0.1692$, $-\pi/2 < x < \pi/2$

13. $3\tan(11 - 3x) = 23.46$, $-\pi/2 < 11 - 3x < \pi/2$

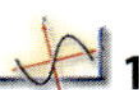

14. Use a graphing utility to test whether each of the following is an identity. If an equation appears to be an identity, verify it. If the equation does not appear to be an identity, find a value of x for which both sides are defined but are not equal.
(A) $(\sin x + \cos x)^2 = 1 - 2\sin x \cos x$
(B) $\cos^2 x - \sin^2 x = 1 - 2\sin^2 x$

B

Verify each identity in Problems 15–23.

15. $\dfrac{1 - 2\cos x - 3\cos^2 x}{\sin^2 x} = \dfrac{1 - 3\cos x}{1 - \cos x}$

16. $(1 - \cos x)(\csc x + \cot x) = \sin x$

17. $\dfrac{1 + \sin x}{\cos x} = \dfrac{\cos x}{1 - \sin x}$

18. $\cos 2x = \dfrac{1 - \tan^2 x}{1 + \tan^2 x}$

19. $\cot \frac{x}{2} = \frac{\sin x}{1 - \cos x}$

20. $\cot x - \tan x = \frac{4 \cos^2 x - 2}{\sin 2x}$

21. $\left(\frac{1 - \cot x}{\csc x}\right)^2 = 1 - \sin 2x$

22. $\tan m + \tan n = \frac{\sin (m + n)}{\cos m \cos n}$

23. $\tan (x + y) = \frac{\cot x + \cot y}{\cot x \cot y - 1}$

Evaluate Problems 24 and 25 exactly using appropriate sum–product or product–sum identities.

24. $\cos 195° \sin 75°$

25. $\cos 195° + \cos 105°$

In Problems 26–29, is the equation an identity? Explain.

26. $\cot^2 x = \csc^2 x + 1$

27. $\cos 3x = \cos x (\cos 2x - 2 \sin^2 x)$

28. $\sin (x + 3\pi/2) = \cos x$

29. $\cos (x - 3\pi/2) = \sin x$

Solve Problems 30–34 exactly (θ in degrees, x real).

30. $4 \sin^2 x - 3 = 0, 0 \le x < 2\pi$

31. $2 \sin^2 \theta + \cos \theta = 1, 0° \le \theta \le 180°$

32. $2 \sin^2 x - \sin x = 0$, all real x

33. $\sin 2x = \sqrt{3} \sin x$, all real x

34. $2 \sin^2 \theta + 5 \cos \theta + 1 = 0$, all θ

Solve Problems 35–37 to four significant digits (θ in degrees, x real).

35. $\tan \theta = 0.2557$, all θ

36. $\sin^2 x + 2 = 4 \sin x$, all real x

37. $\tan^2 x = 2 \tan x + 1, 0 \le x < \pi$

Solve Problems 38–41 to four decimal places.

38. $3 \sin 2x = 2x - 2.5$, all real x

39. $3 \sin 2x > 2x - 2.5$, all real x

40. $2 \sin^2 x - \cos 2x = 1 - x^2$, all real x

41. $2 \sin^2 x - \cos 2x \le 1 - x^2$, all real x

42. Given the equation $\tan (x + y) = \tan x + \tan y$:
(A) Is $x = 0$ and $y = \pi/4$ a solution?
(B) Is the equation an identity or a conditional equation? Explain.

43. Explain the difference in evaluating $\sin^{-1} 0.3351$ and solving the equation $\sin x = 0.3351$.

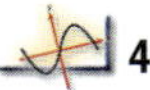

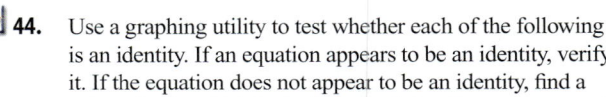

44. Use a graphing utility to test whether each of the following is an identity. If an equation appears to be an identity, verify it. If the equation does not appear to be an identity, find a value of x for which both sides are defined but are not equal.

(A) $\frac{\tan x}{\sin x + 2 \tan x} = \frac{1}{\cos x - 2}$

(B) $\frac{\tan x}{\sin x - 2 \tan x} = \frac{1}{\cos x - 2}$

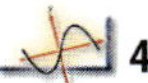

45. Use a sum or difference identity to convert $y = \cos (x - \pi/3)$ to a form involving $\sin x$ and/or $\cos x$. Enter the original equation in a graphing utility as y_1, the converted form as y_2, and graph y_1 and y_2 in the same viewing window. Use TRACE to compare the two graphs.

46. **(A)** Solve $\tan (x/2) = 2 \sin x$ exactly, $0 \le x < 2\pi$, using algebraic methods.
(B) Solve $\tan (x/2) = 2 \sin x$, $0 \le x < 2\pi$, to four decimal places using a graphing utility.

47. Solve $3 \cos (x - 1) = 2 - x^2$ for all real x, to three decimal places using a graphing utility.

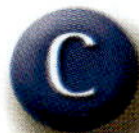

Solve Problems 48–50 exactly without the use of a calculator.

48. Given $\tan x = -\frac{3}{4}$, $\pi/2 \le x \le \pi$, find
(A) $\sin (x/2)$ **(B)** $\cos 2x$

49. $\sin [2 \tan^{-1} (-\frac{3}{4})]$

50. $\sin [\sin^{-1} (\frac{3}{5}) + \cos^{-1} (\frac{4}{5})]$

51. **(A)** Solve $\cos^2 2x = \cos 2x + \sin^2 2x$, $0 \le x < \pi$, exactly using algebraic methods.

(B) Solve $\cos^2 2x = \cos 2x + \sin^2 2x$, $0 \le x < \pi$, to four decimal places using a graphing utility.

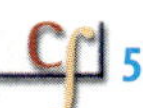

52. We are interested in the zeros of $f(x) = \sin \dfrac{1}{x - 1}$ for $x > 0$.

(A) Explore the graph of f over different intervals $[a, b]$ for various values of a and b, $0 < a < b$. Does the function f have a smallest zero? If so, what is it (to four decimal places)? Does the function have a largest zero? If so, what is it (to four decimal places)?

(B) Explain what happens to the graph as x increases without bound. Does the graph have an asymptote? If so, what is its equation?

(C) Explore the graph of f over smaller and smaller intervals containing $x = 1$. How many zeros exist on any interval containing $x = 1$? Is $x = 1$ a zero? Explain.

APPLICATIONS

53. Indirect Measurement. Find the exact value of x in the figure, then find x and θ to three decimal places. [*Hint:* Use a suitable identity involving $\tan 2\theta$.]

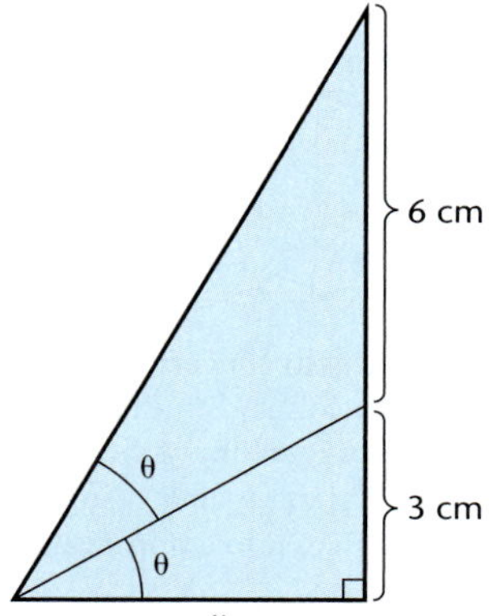

54. Electric Current. An alternating current generator produces a current given by the equation

$$I = 50 \sin 120\pi(t - 0.001)$$

where t is time in seconds and I is current in amperes. Find the smallest positive t, to three significant digits, such that $I = 40$ amperes.

55. Music—Beat Frequencies. The equations $y = 0.6 \cos 184\pi t$ and $y = -0.6 \cos 208\pi t$ model sound waves with frequencies 92 and 104 hertz, respectively. If both sounds are emitted simultaneously, a beat frequency results.

(A) Show that

$$0.6 \cos 184\pi t - 0.6 \cos 208\pi t = 1.2 \sin 12\pi t \sin 196\pi t$$

(B) Graph each of the following equations in a different viewing window for $0 \le t \le 0.2$.

$$y = 0.6 \cos 184\pi t$$
$$y = -0.6 \cos 208\pi t$$
$$y = 0.6 \cos 184\pi t - 0.6 \cos 208\pi t$$
$$y = 1.2 \sin 12\pi t \sin 196\pi t$$

56. Engineering. The circular arch of a bridge has an arc length of 36 feet and spans a 32-foot canal (see the figure). Determine the height of the circular arch above the water at the center of the bridge, and the radius of the circular arch, both to three decimal places. Start by drawing auxiliary lines in the figure, labeling appropriate parts, then explain how the trigonometric equation

$$\sin \theta = \frac{8}{9}\theta$$

is related to the problem. After solving the trigonometric equation for θ, the radius is easy to find and the height of the arch above the water can be found with a little ingenuity.

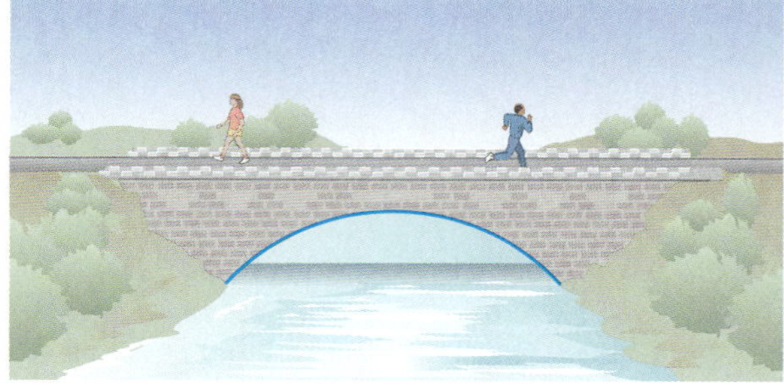

CHAPTER 6 GROUP ACTIVITY

From $M \sin Bt + N \cos Bt$ to $A \sin (Bt + C)$—A Harmonic Analysis Tool

In solving certain kinds of more advanced applied mathematical problems—problems dealing with electrical circuits, spring-mass systems, heat flow, and so on—the solution process leads naturally to a function of the form

$$y = M \sin Bt + N \cos Bt \tag{1}$$

(A) *Graphing Utility Exploration.* Use a graphing utility to explore the nature of the graph of equation (1) for various values of M, N, and B. Does the graph appear to be simple harmonic; that is, does it appear to be a graph of an equation of the form $y = A \sin (Bt + C)$?

The graph of $y = 2 \sin (\pi t) - 3 \cos (\pi t)$, which is typical of the various graphs from equation (1), is shown in Figure 1. It turns out that the graph in Figure 1 can also be obtained from an equation of the form

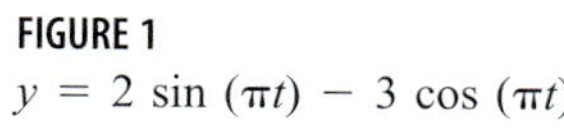

FIGURE 1
$y = 2 \sin (\pi t) - 3 \cos (\pi t)$.

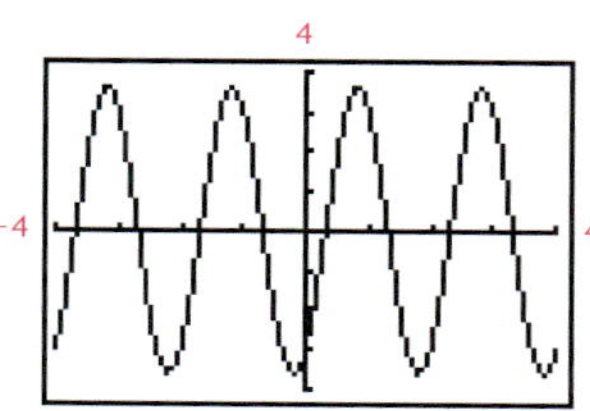

$$y = A \sin (Bt + C) \tag{2}$$

for suitable values of A, B, and C.

The problem now is: given M, N, and B in equation (1), find A, B, and C in equation (2) so that equation (2) produces the same graph as equation (1). The form of equation (2) is often preferred over (1), because from (2) you can easily read amplitude, period, and phase shift and recognize a phenomenon as simple harmonic.

The process of finding A, B, and C, given M, N, and B, requires a little ingenuity and the use of the sum identity

$$\sin (x + y) = \sin x \cos y + \cos x \sin y \tag{3}$$

How do we proceed? We start by trying to get the right side of equation (1) to look like the right side of identity (3). Then we use equation (3), from right to left, to obtain equation (2).

(B) *Establishing a Transformation Identity.* Show that

$$y = M \sin Bt + N \cos Bt = \sqrt{M^2 + N^2} \sin (Bt + C) \tag{4}$$

where C is any angle (in radians if t is real) having $P = (M, N)$ on its terminal side. [*Hint:* A first step is the following:

$$M \sin Bt + N \cos Bt = \frac{\sqrt{M^2 + N^2}}{\sqrt{M^2 + N^2}} (M \sin Bt + N \cos Bt)]$$

(C) *Use of Transformation Identity.* Use equation (4) to transform

$$y_1 = -4 \sin (t/2) + 3 \cos (t/2)$$

into the form $y_2 = A \sin (Bt + C)$, where C is chosen so that $|C|$ is minimum. Compute C to three decimal places. From the new equation, determine the amplitude, period, and phase shift.

(D) *Graphing Utility Visualization and Verification.* Graph y_1 and y_2 from part C in the same viewing window.

(E) *Physics Application.* A weight suspended from a spring, with spring constant 64, is pulled 4 centimeters below its equilibrium position and is then given a downward thrust to produce an initial downward velocity of 24 centimeters per second. In more advanced mathematics (differential equations) the equation of motion (neglecting air resistance and friction) is found to be given approximately by

$$y_1 = -3 \sin 8t - 4 \cos 8t$$

where y_1 is the coordinate of the bottom of the weight in Figure 2 at time t (y is in centimeters and t is in seconds). Transform the equation into the form

$$y_2 = A \sin (Bt + C)$$

and indicate the amplitude, period, and phase shift of the motion. Choose the least positive C and keep A positive.

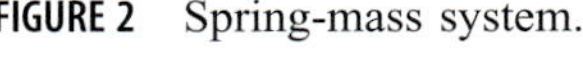

FIGURE 2 Spring-mass system.

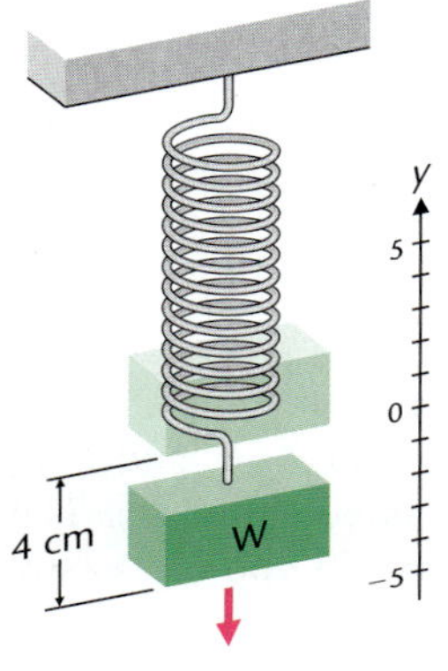

(F) *Graphing Utility Visualization and Verification.* Graph y_1 and y_2 from part E in the same viewing window of a graphing utility, $0 \le t \le 6$. How many times will the bottom of the weight pass $y = 2$ in the first 6 seconds?

(G) *Solving a Trigonometric Equation.* How long, to three decimal places, will it take the bottom of the weight to reach $y = 2$ for the first time?

CHAPTER 7

Additional Topics in Trigonometry

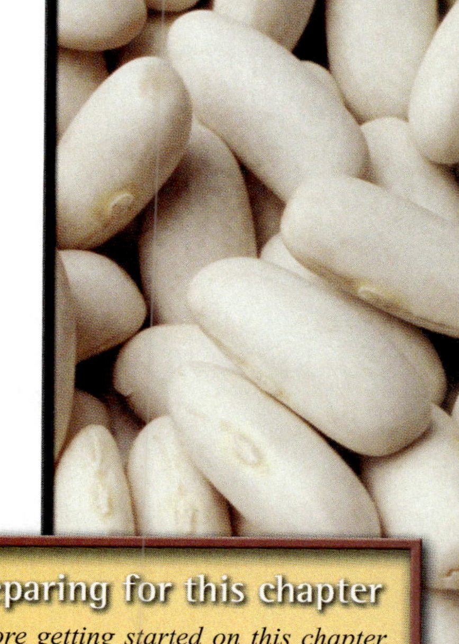

OUTLINE

IN CHAPTER 7 A NUMBER OF ADDITIONAL TOPICS INVOLVING trigonometry are considered. First, we return to the problem of solving triangles—not just right triangles, but any triangle. Then some of these ideas are used to develop the important concept of vector. With our knowledge of trigonometry, we introduce the *polar coordinate system,* probably the most important coordinate system after the rectangular coordinate system. After considering polar equations and their graphs, we represent complex numbers in *polar form*. Once a complex number is in polar form, it will be possible to find nth powers and nth roots of the number using an ingenious theorem established by De Moivre.

Preparing for this chapter

Before getting started on this chapter, review the following concepts:

- Rational Exponents (Basic Algebra Review*, Section R.6)
- Radicals (Basic Algebra Review*, Section R.7)
- Complex Numbers (Chapter 2, Section 4)
- Inverse Functions (Chapter 1, Section 6)
- Solving Right Triangles (Chapter 5, Section 3)
- Difference Identities (Chapter 6, Section 2)
- Significant Digits (Appendix B, Section B.1)

*At www.mhhe.com/barnett

SECTION 7.1 Law of Sines

Law of Sines Derivation • Solving the ASA and AAS Cases • Solving the SSA Case—Including the Ambiguous Case

In Chapter 5 we used trigonometric functions to solve problems concerning right triangles. We now consider analogous problems for *oblique triangles*—triangles without a right angle.

Every oblique triangle is either **acute,** all angles between 0° and 90°, or **obtuse,** one angle between 90° and 180°. Figure 1 illustrates both types of triangles.

FIGURE 1 Oblique triangles.

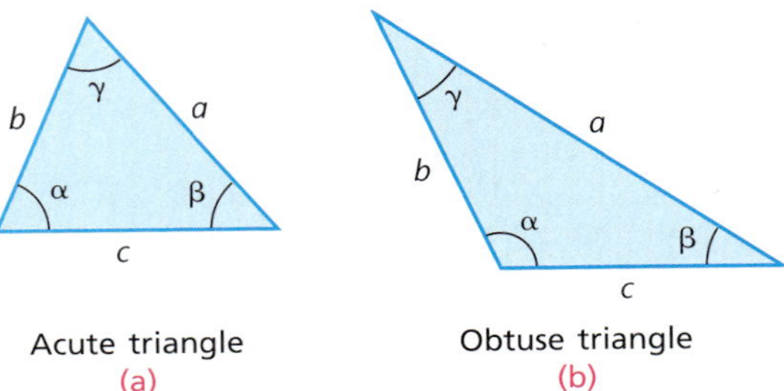

Note how the sides and angles of the oblique triangles in Figure 1 have been labeled: Side a is opposite angle α, side b is opposite angle β, and side c is opposite angle γ. Also note that the largest side of a triangle is opposite the largest angle. Given any three of the six quantities indicated in Figure 1, we are interested in finding the remaining three, if possible. This process is called *solving the triangle.*

If only the three angles of a triangle are known, it is impossible to solve for the sides. (Why?) But if we are given two angles and a side, or two sides and an angle, or all three sides, then it is possible to determine whether a triangle having the given quantities exists, and, if so, to solve for the remaining quantities.

The basic tools for solving oblique triangles are the *law of sines,* developed in Section 7.1, and the *law of cosines,* developed in Section 7.2.

Before proceeding with specific examples, it is important to recall the rules in Table 1 regarding accuracy of angle and side measure. Table 1 is repeated inside the front cover of the text for easy reference.

TABLE 1 Triangles and Significant Digits

Angle to Nearest	Significant Digits for Side Measure
1°	2
10′ or 0.1°	3
1′ or 0.01°	4
10″ or 0.001°	5

Calculator Calculations

When solving for a particular side or angle, carry out all operations within the calculator and then round to the appropriate number of significant digits (as specified in Table 1) at the end of the calculation. Your answers may still differ slightly from those in the book, depending on the order in which you solve for the sides and angles.

FIGURE 2

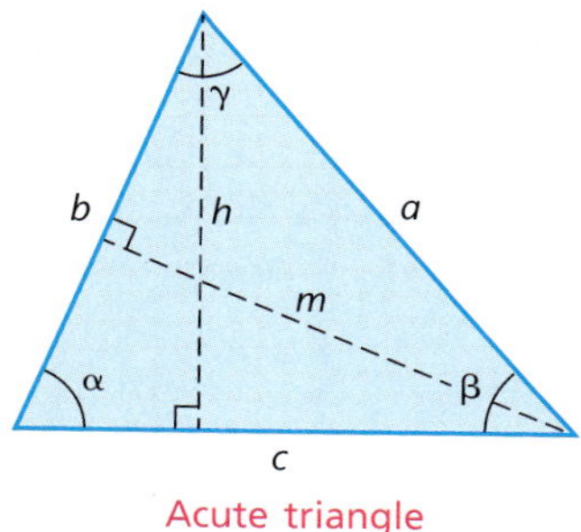

Acute triangle
(a)

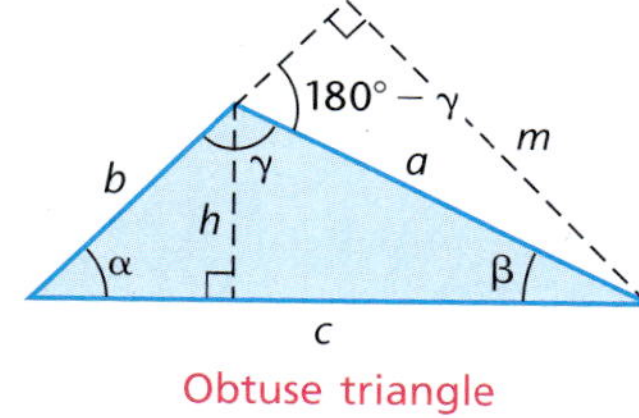

Obtuse triangle
(b)

Law of Sines Derivation

The law of sines is relatively easy to prove using the right triangle properties studied in Chapter 5. We will also use the fact that

$$\sin(180° - x) = \sin x$$

which is readily obtained using a difference identity (a good exercise for you). Referring to Figure 2, we proceed as follows: Angles α and β in Figure 2(a), and also in Figure 2(b), satisfy

$$\sin\alpha = \frac{h}{b} \qquad \text{and} \qquad \sin\beta = \frac{h}{a}$$

Solving each equation for h, we obtain

$$h = b\sin\alpha \qquad \text{and} \qquad h = a\sin\beta$$

Thus,

$$b\sin\alpha = a\sin\beta$$

$$\frac{\sin\alpha}{a} = \frac{\sin\beta}{b} \tag{1}$$

Similarly, angles α and γ in Figure 2(a), and also in Figure 2(b), satisfy

$$\sin\alpha = \frac{m}{c} \qquad \text{and} \quad \sin\gamma = \sin(180° - \gamma) = \frac{m}{a}$$

Solving each equation for m, we obtain

$$m = c\sin\alpha \qquad \text{and} \qquad m = a\sin\gamma$$

Thus,

$$c\sin\alpha = a\sin\gamma$$

$$\frac{\sin\alpha}{a} = \frac{\sin\gamma}{c} \tag{2}$$

If we combine equations (1) and (2), we obtain the **law of sines.**

THEOREM 1

Law of Sines

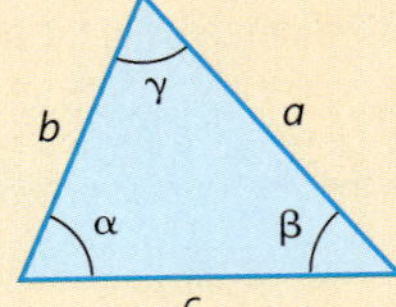

$$\frac{\sin \alpha}{a} = \frac{\sin \beta}{b} = \frac{\sin \gamma}{c}$$

In words, the ratio of the sine of an angle to its opposite side is the same as the ratio of the sine of either of the other angles to its opposite side.

Suppose that an angle of a triangle and its opposite side are known. Then the ratio of Theorem 1 can be calculated. So if one additional part of the triangle, either of the other angles or either of the other sides, is known, then the law of sines can be used to solve the triangle.

Thus, the law of sines is used to solve triangles, given:

1. Two sides and an angle opposite one of them (SSA), or
2. Two angles and any side (ASA or AAS)

If the given information for a triangle consists of two sides and the included angle (SAS) or three sides (SSS), then the law of sines cannot be applied. The key to handling these two cases, the law of cosines, is developed in Section 7.2.

We will apply the law of sines to the easier ASA and AAS cases first, and then will turn to the more challenging SSA case.

Solving the ASA and AAS Cases

EXAMPLE 1 Solving the ASA Case

Solve the triangle in Figure 3.

FIGURE 3

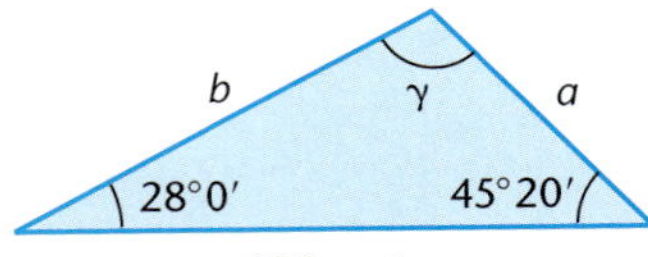

SOLUTION

We are given two angles and the included side, which is the ASA case. Find the third angle, then solve for the other two sides using the law of sines.

We solve for γ:

$$\begin{aligned} \alpha + \beta + \gamma &= 180° \\ \gamma &= 180° - (\alpha + \beta) \\ &= 180° - (28°0' + 45°20') \\ &= 106°40' \end{aligned}$$

We solve for a:

$$\frac{\sin \alpha}{a} = \frac{\sin \gamma}{c} \qquad \text{Law of sines}$$

$$\begin{aligned} a &= \frac{c \sin \alpha}{\sin \gamma} \\ &= \frac{120 \sin 28°0'}{\sin 106°40'} \\ &= 58.8 \text{ meters} \end{aligned}$$

We solve for b:

$$\frac{\sin \beta}{b} = \frac{\sin \gamma}{c} \qquad \text{Law of sines}$$

$$\begin{aligned} b &= \frac{c \sin \beta}{\sin \gamma} \\ &= \frac{120 \sin 45°20'}{\sin 106°40'} \\ &= 89.1 \text{ meters} \end{aligned}$$

MATCHED 1 PROBLEM

Solve the triangle in Figure 4.

FIGURE 4

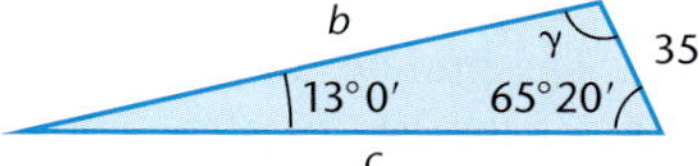

Note that the AAS case can always be converted to the ASA case by first solving for the third angle. For the ASA or AAS case to determine a unique triangle, the sum of the two angles must be between 0° and 180°, because the sum of all three angles in a triangle is 180° and no angle can be zero or negative.

Solving the SSA Case—Including the Ambiguous Case

We now look at the case where we are given two sides and an angle opposite one of the sides—the SSA case. This case has several possible outcomes, depending on the measures of the two sides and the angle. Table 2 illustrates the various possibilities.

TABLE 2 SSA Variations

α	a [$h = b \sin \alpha$]	Number of Triangles	Figure	Case
Acute	$0 < a < h$	0		(a)
Acute	$a = h$	1		(b)
Acute	$h < a < b$	2		(c)
Acute	$a \geq b$	1		(d)
Obtuse	$0 < a \leq b$	0		(e)
Obtuse	$a > b$	1		(f)

It is unnecessary to memorize Table 2 to solve triangles in the SSA case. Instead, given sides a, b, and angle α, we use the law of sines to solve for the angle β opposite side b. The number of triangles is equal to the number of solutions β, $0° < \beta < 180°$, of the law of sines equation

$$\frac{\sin \beta}{b} = \frac{\sin \alpha}{a} \tag{3}$$

that satisfy

$$\alpha + \beta < 180° \tag{4}$$

Thus, in practice, we check each solution of equation (3) to determine whether inequality (4) is satisfied. If it is, we can easily solve for the remaining parts of the triangle. Examples 2–4 will make the procedure clear.

EXAMPLE 2 The SSA Case: One Triangle

Solve the triangle(s) with $a = 47$ centimeters, $b = 23$ centimeters, and $\alpha = 123°$.

SOLUTION

We solve for β:

$$\frac{\sin \beta}{b} = \frac{\sin \alpha}{a} \qquad \text{Law of sines}$$

$$\sin \beta = \frac{b \sin \alpha}{a} = \frac{23 \sin 123°}{47}$$

This equation has two solutions between 0° and 180°:

$$\beta = \sin^{-1}\left(\frac{23 \sin 123°}{47}\right) = 24°$$

$$\beta' = 180° - 24° = 156°$$

Because

$$\alpha + \beta = 123° + 24° = 147° < 180°$$
$$\alpha + \beta' = 123° + 156° = 279° \geq 180°$$

there is only one triangle. [Note that this conclusion is consistent with Table 2. Because α is obtuse and $a > b$, we are in Case (f).]

We solve for γ:

$$\alpha + \beta + \gamma = 180°$$
$$\gamma = 180° - 123° - 24° = 33°$$

We solve for c:

$$\frac{\sin \alpha}{a} = \frac{\sin \gamma}{c}$$

$$c = \frac{a \sin \gamma}{\sin \alpha} = \frac{47 \sin 33°}{\sin 123°} = 31 \text{ centimeters}$$

MATCHED 2 PROBLEM

Solve the triangle(s) with $a = 88$ meters, $b = 62$ meters, and $\alpha = 81°$.

EXAMPLE 3 The SSA Case: No Triangle

Solve the triangle(s) with $a = 27$ inches, $b = 28$ inches, and $\alpha = 110°$.

SOLUTION

We solve for β:

$$\frac{\sin \beta}{b} = \frac{\sin \alpha}{a} \qquad \text{Law of sines}$$

$$\sin \beta = \frac{b \sin \alpha}{a} = \frac{28 \sin 110°}{27}$$

This equation has two solutions between 0° and 180°:

$$\beta = \sin^{-1}\left(\frac{28 \sin 110°}{27}\right) = 77°$$

$$\beta' = 180° - 77° = 103°$$

Because

$$\alpha + \beta = 110° + 77° = 187° \geq 180°$$
$$\alpha + \beta' = 110° + 103° = 213° \geq 180°$$

there is no triangle. [Note that this conclusion is consistent with Table 2. Because α is obtuse and $a \leq b$, we are in Case (e).]

MATCHED 3 PROBLEM

Solve the triangle(s) with $a = 64$ feet, $b = 79$ feet, and $\alpha = 57°$.

EXAMPLE 4 The SSA Case: Two Triangles

Solve the triangle(s) with $a = 1.0$ meters, $b = 1.8$ meters, and $\alpha = 26°$.

SOLUTION

We solve for β:

$$\frac{\sin \beta}{b} = \frac{\sin \alpha}{a} \qquad \text{Law of sines}$$

$$\sin \beta = \frac{b \sin \alpha}{a} = \frac{1.8 \sin 26°}{1.0}$$

This equation has two solutions between 0° and 180°:

$$\beta = \sin^{-1}\left(\frac{1.8 \sin 26°}{1.0}\right) = 52°$$
$$\beta' = 180° - 52° = 128°$$

Because

$$\alpha + \beta = 26° + 52° = 78° < 180°$$
$$\alpha + \beta' = 26° + 128° = 154° < 180°$$

there are two triangles. [Note that this conclusion is consistent with Table 2. Because α is acute and $h = b \sin \alpha < a < b$, we are in Case (c), the ambiguous case.]

We solve for γ and γ':

$$\gamma = 180° - 26° - 52° = 102°$$
$$\gamma' = 180° - 26° - 128° = 26°$$

We solve for c and c':

$$c = \frac{a \sin \gamma}{\sin \alpha} = \frac{1.0 \sin 102°}{\sin 26°} = 2.2 \text{ meters}$$

$$c' = \frac{a \sin \gamma'}{\sin \alpha} = \frac{1.0 \sin 26°}{\sin 26°} = 1.0 \text{ meters}$$

In summary:

Triangle I:	$\beta = 52°$	$\gamma = 102°$	$c = 2.2$ meters
Triangle II:	$\beta' = 128°$	$\gamma' = 26°$	$c' = 1.0$ meters

MATCHED 4 PROBLEM

Solve the triangle(s) with $a = 8$ kilometers, $b = 10$ kilometers, and $\alpha = 35°$.

EXPLORE/DISCUSS 1

Sides a and b and acute angle α of a triangle are given. Explain which case(s) of Table 2 could apply if, in solving the triangle, it is found that

(A) $\sin \beta > 1$

(B) $\sin \beta = 1$

(C) $\sin \beta < 1$

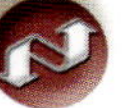

The law of sines is useful in many applications, as can be seen in Example 5 and the applications in Exercise 7.1.

EXAMPLE 5 Surveying

To measure the length d of a lake (Fig. 5), a base line AB is established and measured to be 125 meters. Angles A and B are measured to be 41.6° and 124.3°, respectively. How long is the lake?

FIGURE 5

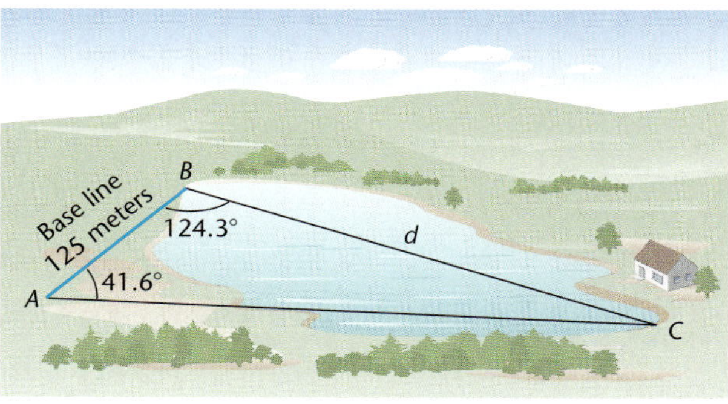

SOLUTION

Find angle C and use the law of sines.

$$\text{Angle } C = 180° - (124.3° + 41.6°) = 14.1°$$

$$\frac{\sin 14.1°}{125} = \frac{\sin 41.6°}{d}$$

$$d = 125\left(\frac{\sin 41.6°}{\sin 14.1°}\right) = 341 \text{ meters}$$

MATCHED 5 PROBLEM

In Example 5, find the distance AC.

ANSWERS MATCHED PROBLEMS

1. $\gamma = 101°40'$, $b = 141$, $c = 152$
2. $\beta = 44°$, $\gamma = 55°$, $c = 73$ meters
3. No solution
4. Triangle I: $\beta = 46°$, $\gamma = 99°$, $c = 14$ kilometers; Triangle II: $\beta' = 134°$, $\gamma' = 11°$, $c' = 2.7$ kilometers
5. 424 meters

EXERCISE 7.1

The labeling in the figure below is the convention we will follow in this exercise set. Your answers to some problems may differ slightly from those in the book, depending on the order in which you solve for the sides and angles of a given triangle.

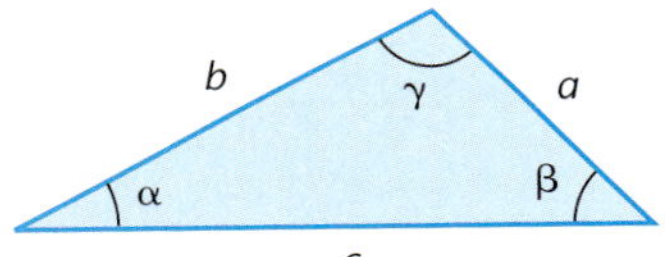

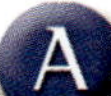

Solve each triangle in Problems 1–8.

1. $\alpha = 73°$, $\beta = 28°$, $c = 42$ feet
2. $\alpha = 41°$, $\beta = 33°$, $c = 21$ centimeters
3. $\alpha = 122°$, $\gamma = 18°$, $b = 12$ kilometers
4. $\beta = 43°$, $\gamma = 36°$, $a = 92$ millimeters
5. $\beta = 112°$, $\gamma = 19°$, $c = 23$ yards
6. $\alpha = 52°$, $\gamma = 105°$, $c = 47$ meters
7. $\alpha = 52°$, $\gamma = 47°$, $a = 13$ centimeters
8. $\beta = 83°$, $\gamma = 77°$, $c = 25$ miles

In Problems 9–16, determine whether the information in each problem allows you to construct zero, one, or two triangles. Do not solve the triangle. Explain which case in Table 2 applies.

9. $a = 2$ inches, $b = 4$ inches, $\alpha = 30°$
10. $a = 3$ feet, $b = 6$ feet, $\alpha = 30°$
11. $a = 6$ inches, $b = 4$ inches, $\alpha = 30°$
12. $a = 8$ feet, $b = 6$ feet, $\alpha = 30°$
13. $a = 1$ inch, $b = 4$ inches, $\alpha = 30°$
14. $a = 2$ feet, $b = 6$ feet, $\alpha = 30°$
15. $a = 3$ inches, $b = 4$ inches, $\alpha = 30°$
16. $a = 5$ feet, $b = 6$ feet, $\alpha = 30°$

B

Solve each triangle in Problems 17–32. If a problem has no solution, say so.

17. $\alpha = 118.3°$, $\gamma = 12.2°$, $b = 17.3$ feet

18. $\beta = 27.5°$, $\gamma = 54.5°$, $a = 9.27$ inches

19. $\alpha = 67.7°$, $\beta = 54.2°$, $b = 123$ meters

20. $\alpha = 122.7°$, $\beta = 34.4°$, $b = 18.3$ kilometers

21. $\alpha = 46.5°$, $a = 7.9$ millimeters, $b = 13.1$ millimeters

22. $\alpha = 26.3°$, $a = 14.7$ inches, $b = 35.2$ inches

23. $\alpha = 15.9°$, $a = 22.4$ inches, $b = 29.6$ inches

24. $\alpha = 43.5°$, $a = 138$ centimeters, $b = 172$ centimeters

25. $\beta = 38.9°$, $a = 42.7$ inches, $b = 30.0$ inches

26. $\beta = 27.3°$, $a = 244$ centimeters, $b = 135$ centimeters

27. $\alpha = 123.2°$, $a = 101$ yards, $b = 152$ yards

28. $\alpha = 137.3°$, $a = 13.9$ meters, $b = 19.1$ meters

29. $\beta = 29°30'$, $a = 43.2$ millimeters, $b = 56.5$ millimeters

30. $\beta = 33°50'$, $a = 673$ meters, $b = 1{,}240$ meters

31. $\alpha = 30°$, $a = 29$ feet, $b = 58$ feet

32. $\beta = 30°$, $a = 92$ inches, $b = 46$ inches

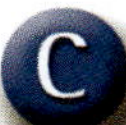

33. Let $\alpha = 42.3°$ and $b = 25.2$ centimeters. Determine a value k so that if $0 < a < k$, there is no solution; if $a = k$, there is one solution; and if $k < a < b$, there are two solutions.

34. Let $\alpha = 37.3°$ and $b = 42.8$ centimeters. Determine a value k so that if $0 < a < k$, there is no solution; if $a = k$, there is one solution; and if $k < a < b$, there are two solutions.

35. Mollweide's equation,

$$(a - b)\cos\frac{\gamma}{2} = c \sin\frac{\alpha - \beta}{2}$$

is often used to check the final solution of a triangle, because all six parts of a triangle are involved in the equation. If the left side does not equal the right side after substitution, then an error has been made in solving a triangle. Use this equation to check Problem 1. (Because of rounding errors, both sides may not be exactly the same.)

36. (A) Use the law of sines and suitable identities to show that for any triangle

$$\frac{a - b}{a + b} = \frac{\tan\dfrac{\alpha - \beta}{2}}{\tan\dfrac{\alpha + \beta}{2}}$$

(B) Verify the formula with values from Problem 1.

APPLICATIONS

37. Coast Guard. Two lookout posts, A and B (10.0 miles apart), are established along a coast to watch for illegal ships coming within the 3-mile limit. If post A reports a ship S at angle $BAS = 37°30'$ and post B reports the same ship at angle $ABS = 20°0'$, how far is the ship from post A? How far is the ship from the shore (assuming the shore is along the line joining the two observation posts)?

38. Fire Lookout. A fire at F is spotted from two fire lookout stations, A and B, which are 10.0 miles apart. If station B reports the fire at angle $ABF = 53°0'$ and station A reports the fire at angle $BAF = 28°30'$, how far is the fire from station A? From station B?

★ **39. Natural Science.** The tallest trees in the world grow in Redwood National Park in California; they are taller than a football field is long. Find the height of one of these trees, given the information in the figure on page 540. (The 100-foot measurement is accurate to three significant digits.)

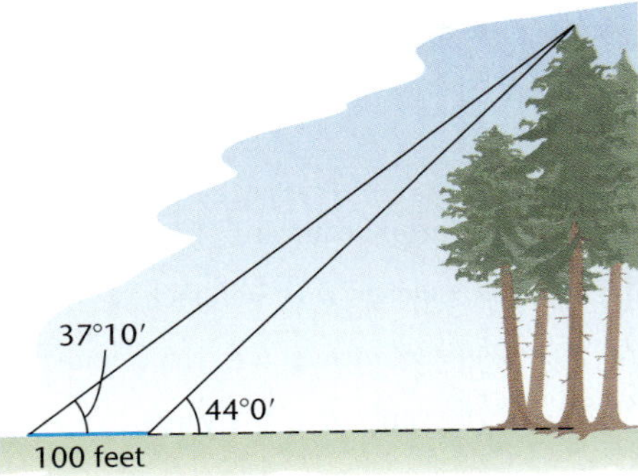

★ **40. Surveying.** To measure the height of Mt. Whitney in California, surveyors used a scheme like the one shown in the figure in Problem 39. They set up a horizontal base line 2,000 feet long at the foot of the mountain and found the angle nearest the mountain to be 43°5′; the angle farthest from the mountain was found to be 38°0′. If the base line was 5,000 feet above sea level, how high is Mt. Whitney above sea level?

41. Engineering. A 4.5-inch piston rod joins a piston to a 1.5-inch crankshaft (see the figure). How far is the base of the piston from the center of the crankshaft (distance d) when the rod makes an angle of 9° with the centerline? There are two answers to the problem.

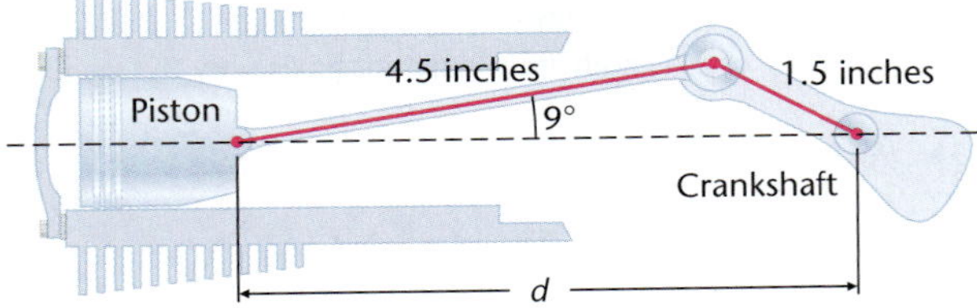

42. Engineering. Repeat Problem 41 if the piston rod is 6.3 inches, the crankshaft is 1.7 inches, and the angle is 11°.

43. Astronomy. The orbits of the Earth and Venus are approximately circular, with the sun at the center. A sighting of Venus is made from Earth, and the angle α is found to be 18°40′. If the radius of the orbit of the Earth is 1.495×10^8 kilometers and the radius of the orbit of Venus is 1.085×10^8 kilometers, what are the possible distances from the Earth to Venus? (See the figure.)

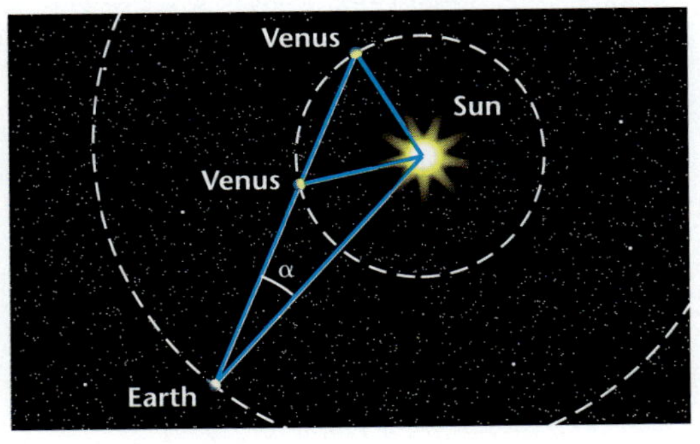

44. Astronomy. In Problem 43, find the maximum angle α. [*Hint:* The angle is maximum when a straight line joining the Earth and Venus is tangent to Venus's orbit.]

★ **45. Surveying.** A tree growing on a hillside casts a 102-foot shadow straight down the hill (see the figure). Find the vertical height of the tree if, relative to the horizontal, the hill slopes 15.0° and the angle of elevation of the sun is 62.0°.

★ **46. Surveying.** Find the height of the tree in Problem 45 if the shadow length is 157 feet and, relative to the horizontal, the hill slopes 11.0° and the angle of elevation of the sun is 42.0°.

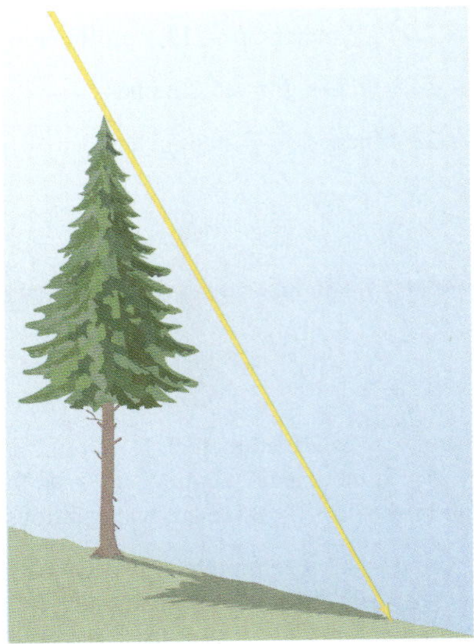

★ **47. Life Science.** A cross-section of the cornea of an eye, a circular arc, is shown in the figure. Find the arc radius R and the arc length s, given the chord length $C = 11.8$ millimeters and the central angle $\theta = 98.9°$.

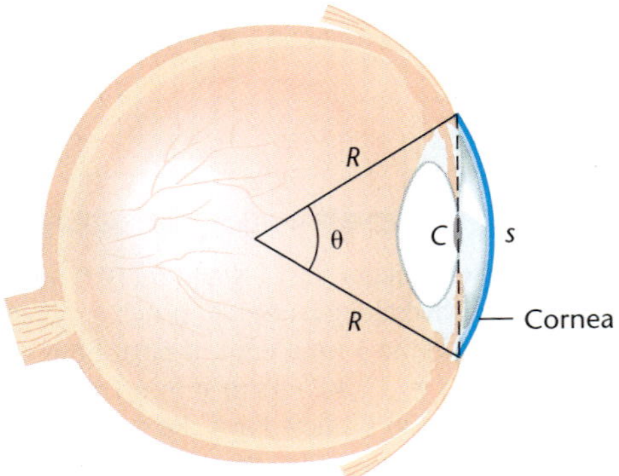

★ **48. Life Science.** Referring to the figure, find the arc radius R and the arc length s, given the chord length $C = 10.2$ millimeters and the central angle $\theta = 63.2°$.

★ **49. Surveying.** The procedure illustrated in Problems 39 and 40 is used to determine an inaccessible height h when a base line d on a line perpendicular to h can be established (see the figure) and the angles α and β can be measured. Show that

$$h = d\left[\frac{\sin \alpha \sin \beta}{\sin (\beta - \alpha)}\right]$$

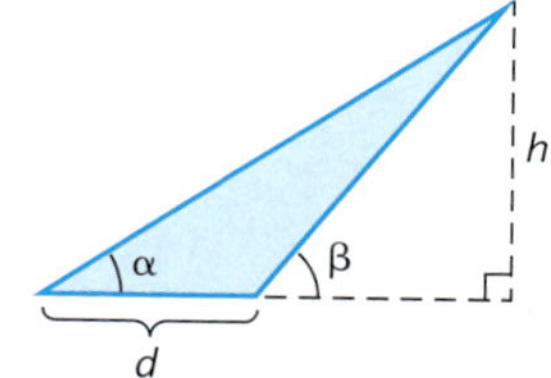

★★ **50. Surveying.** The layout in the figure is used to determine an inaccessible height h when a base line d in a plane perpendicular to h can be established and the angles α, β, and γ can be measured. Show that

$$h = d \sin \alpha \csc (\alpha + \beta) \tan \gamma$$

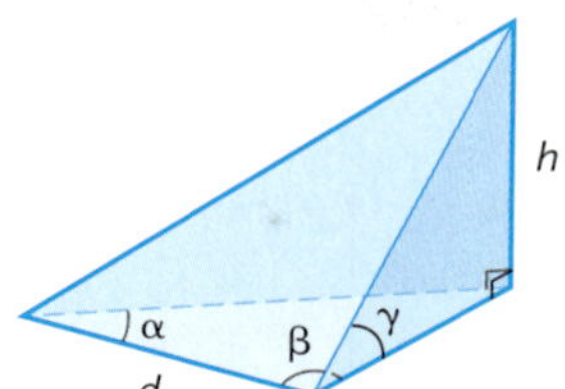

SECTION 7.2 Law of Cosines

Law of Cosines Derivation • Solving the SAS Case • Solving the SSS Case

If in a triangle two sides and the included angle are given (SAS), or three sides are given (SSS), the law of sines cannot be used to solve the triangle—neither case involves an angle and its opposite side (Fig. 1). Both cases can be solved starting with the *law of cosines,* which is the subject matter for Section 7.2.

FIGURE 1

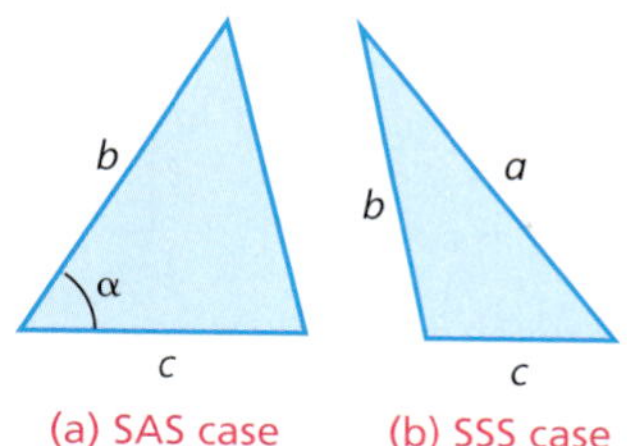

(a) SAS case (b) SSS case

Law of Cosines Derivation

Theorem 1 states the **law of cosines.**

THEOREM 1
Law of Cosines

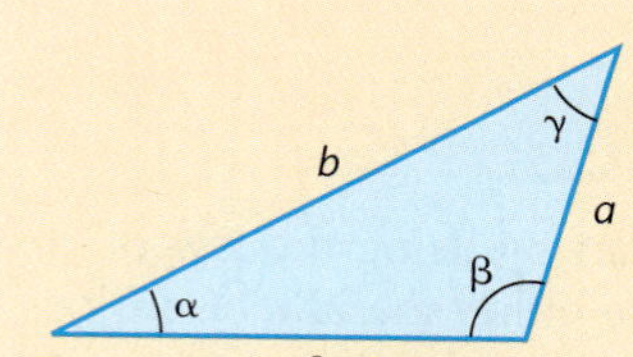

$$a^2 = b^2 + c^2 - 2bc \cos \alpha$$
$$b^2 = a^2 + c^2 - 2ac \cos \beta$$
$$c^2 = a^2 + b^2 - 2ab \cos \gamma$$

All three equations say essentially the same thing.

The law of cosines is used to solve triangles, given:

1. Two sides and the included angle (SAS), or
2. Three sides (SSS)

We will establish the first equation in Theorem 1. The other two equations then can be obtained from this one simply by relabeling the figure. We start by locating a triangle in a rectangular coordinate system. Figure 2 shows three typical triangles.

For an arbitrary triangle located as in Figure 2, the distance-between-two-points formula is used to obtain

$$a = \sqrt{(h - c)^2 + (k - 0)^2}$$

$$a^2 = (h - c)^2 + k^2 \qquad \text{Square both sides.} \tag{1}$$

$$= h^2 - 2hc + c^2 + k^2$$

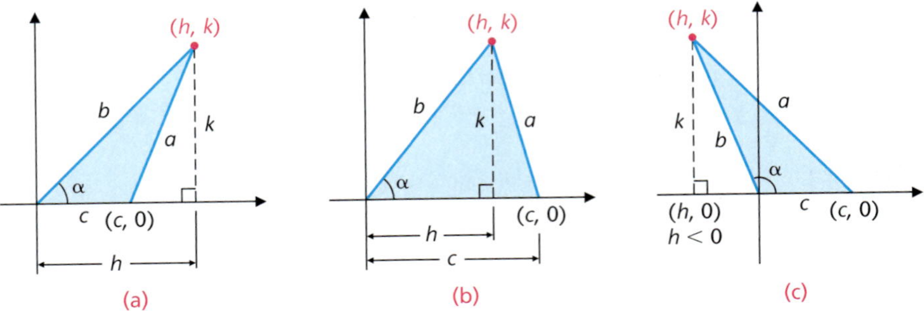

FIGURE 2 Three representative triangles.

From Figure 2, we note that

$$b^2 = h^2 + k^2$$

Substituting b^2 for $h^2 + k^2$ in equation (1), we obtain

$$a^2 = b^2 + c^2 - 2hc \tag{2}$$

But

$$\cos \alpha = \frac{h}{b}$$

$$h = b \cos \alpha$$

Thus, by replacing h in equation (2) with $b \cos \alpha$, we reach our objective:

$$a^2 = b^2 + c^2 - 2bc \cos \alpha$$

[*Note:* If α is acute, then $\cos \alpha$ is positive; if α is obtuse, then $\cos \alpha$ is negative.]

Solving the SAS Case

For the SAS case, start by using the law of cosines to find the side opposite the given angle. Then use either the law of cosines or the law of sines to find a second angle. Because of the simpler computations, the law of sines will generally be used to find the second angle.

EXPLORE/DISCUSS 1

After using the law of cosines to find the side opposite the angle for an SAS case, the law of sines is used to find a second angle. Figure 2 shows that there are two choices for a second angle.

(A) If the given angle is obtuse, can either of the remaining angles be obtuse? Explain.

(B) If the given angle is acute, then one of the remaining angles may or may not be obtuse. Explain why choosing the angle opposite the shorter side guarantees the selection of an acute angle.

(C) Starting with $(\sin \alpha)/a = (\sin \beta)/b$, show that

$$\alpha = \sin^{-1}\left(\frac{a \sin \beta}{b}\right) \tag{3}$$

(D) Explain why equation (3) gives us the correct angle α only if α is acute.

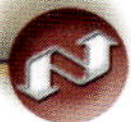

The preceding discussion leads to the following strategy for solving the SAS case:

Strategy for Solving the SAS Case

Step	Find	Method
1	Side opposite given angle	Law of cosines
2	Second angle (Find the angle opposite the shorter of the two given sides—this angle will always be acute.)	Law of sines
3	Third angle	Subtract the sum of the measures of the given angle and the angle found in step 2 from 180°.

EXAMPLE 1 Solving the SAS Case

FIGURE 3

Solve the triangle in Figure 3.

SOLUTION

We solve for b:

$$b^2 = a^2 + c^2 - 2ac \cos \beta$$ Law of cosines

$$b = \sqrt{a^2 + c^2 - 2ac \cos \beta}$$

$$= \sqrt{(10.3)^2 + (6.45)^2 - 2(10.3)(6.45) \cos 32.4°}$$

$$= 5.96 \text{ cm}$$

We solve for γ (the angle opposite the shorter side):

$$\frac{\sin \gamma}{c} = \frac{\sin \beta}{b}$$ Law of sines

$$\sin \gamma = \frac{c \sin \beta}{b}$$ Solve for γ.

$$\gamma = \sin^{-1}\left(\frac{c \sin \beta}{b}\right)$$ Because γ is acute, the inverse sine function gives us γ directly.

$$= \sin^{-1}\left(\frac{6.45 \sin 32.4°}{5.96}\right)$$

$$= 35.4°$$

We solve for α:

$$\alpha = 180° - (\beta + \gamma)$$
$$= 180° - (32.4° + 35.4°) = 112.2°$$

MATCHED 1 PROBLEM

Solve the triangle with $\alpha = 77.5°$, $b = 10.4$ feet, and $c = 17.7$ feet.

Solving the SSS Case

Starting with three sides of a triangle, the problem is to find the three angles. Subsequent calculations are simplified if we solve for the obtuse angle first, if present. The law of cosines is used for this purpose. A second angle, which must be acute, can be found using either law, although computations are usually simpler with the law of sines.

EXPLORE/DISCUSS 2

(A) Starting with $a^2 = b^2 + c^2 - 2bc \cos \alpha$, show that

$$\alpha = \cos^{-1}\left(\frac{a^2 - b^2 - c^2}{-2bc}\right) \tag{4}$$

(B) Does equation (4) give us the correct angle α irrespective of whether α is acute or obtuse? Explain.

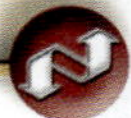

The preceding discussion leads to the following strategy for solving the SSS case:

Strategy for Solving the SSS Case

Step	Find	Method
1	Angle opposite longest side—this will take care of an obtuse angle, if present.	Law of cosines
2	Either of the remaining angles, which will be acute. (Why?)	Law of sines
3	Third angle	Subtract the sum of the measures of the angles found in steps 1 and 2 from 180°.

EXAMPLE 2 Solving the SSS Case

Solve the triangle with $a = 27.3$ meters, $b = 17.8$ meters, and $c = 35.2$ meters.

SOLUTION

Three sides of the triangle are given and we are to find the three angles. This is the SSS case.

Sketch the triangle (Fig. 4) and use the law of cosines to find the largest angle, then use the law of sines to find one of the two remaining acute angles.

FIGURE 4

We solve for γ:

$$c^2 = a^2 + b^2 - 2ab\cos\gamma \qquad \text{Law of cosines}$$

$$\cos\gamma = \frac{a^2 + b^2 - c^2}{2ab} \qquad \text{Solve for } \gamma.$$

$$\gamma = \cos^{-1}\left(\frac{a^2 + b^2 - c^2}{2ab}\right)$$

$$= \cos^{-1}\left[\frac{(27.3)^2 + (17.8)^2 - (35.2)^2}{2(27.3)(17.8)}\right]$$

$$= 100.5°$$

We solve for α:

$$\frac{\sin\alpha}{a} = \frac{\sin\gamma}{c} \qquad \text{Law of sines}$$

$$\sin\alpha = \frac{a\sin\gamma}{c} = \frac{27.3\sin 100.5°}{35.2} \qquad \text{Solve for } \alpha.$$

$$\alpha = \sin^{-1}\left(\frac{27.3\sin 100.5°}{35.2}\right)$$

$$= 49.7° \qquad \alpha \text{ is acute.}$$

We solve for β:

$$\begin{aligned} \alpha + \beta + \gamma &= 180^\circ \\ \beta &= 180^\circ - (\alpha + \gamma) \\ &= 180^\circ - (49.7^\circ + 100.5^\circ) \\ &= 29.8^\circ \end{aligned}$$

MATCHED 2 PROBLEM

Solve the triangle with $a = 1.25$ yards, $b = 2.05$ yards, and $c = 1.52$ yards.

EXAMPLE 3 Finding the Side of a Regular Polygon

If a seven-sided regular polygon is inscribed in a circle of radius 22.8 centimeters, find the length of one side of the polygon.

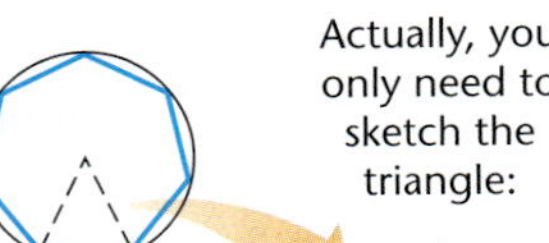

Actually, you only need to sketch the triangle:

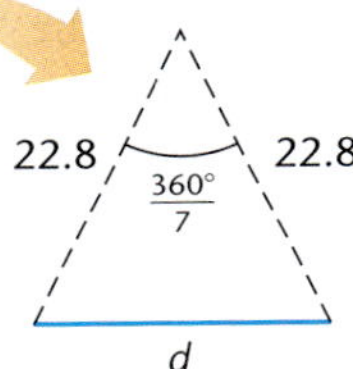

FIGURE 5

SOLUTION

Sketch a figure (Fig. 5) and use the law of cosines.

$$d^2 = 22.8^2 + 22.8^2 - 2(22.8)(22.8)\cos\frac{360^\circ}{7}$$

$$d = \sqrt{2(22.8)^2 - 2(22.8)^2\cos\frac{360^\circ}{7}}$$

$$= 19.8 \text{ centimeters}$$

MATCHED 3 PROBLEM

If an 11-sided regular polygon is inscribed in a circle with radius 4.63 inches, find the length of one side of the polygon.

ANSWERS MATCHED PROBLEMS

1. $a = 18.5$ feet, $\beta = 33.3^\circ$, $\gamma = 69.2^\circ$ **2.** $\alpha = 37.4^\circ$, $\beta = 95.0^\circ$, $\gamma = 47.6^\circ$ **3.** 2.61 inches

EXERCISE 7.2

The labeling in the figure below is the convention we will follow in this exercise set. Your answers to some problems may differ slightly from those in the book, depending on the order in which you solve for the sides and angles of a given triangle.

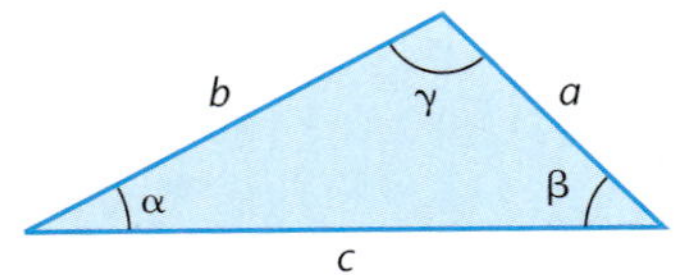

A

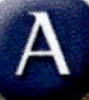

1. Referring to the figure, if $\alpha = 47.3^\circ$, $b = 11.7$ centimeters, and $c = 6.04$ centimeters, which of the two angles, β or γ, can you say for certain is acute and why?

2. Referring to the figure, if $\alpha = 93.5^\circ$, $b = 5.34$ inches, and $c = 8.77$ inches, which of the two angles, β or γ, can you say for certain is acute and why?

Solve each triangle in Problems 3–6.

3. $\alpha = 71.2°$, $b = 5.32$ yards, $c = 5.03$ yards

4. $\beta = 57.3°$, $a = 6.08$ centimeters, $c = 5.25$ centimeters

5. $\gamma = 120°20'$, $a = 5.73$ millimeters, $b = 10.2$ millimeters

6. $\alpha = 135°50'$, $b = 8.44$ inches, $c = 20.3$ inches

B

7. Referring to the figure at the beginning of the exercise set, if $a = 13.5$ feet, $b = 20.8$ feet, and $c = 8.09$ feet, then, if the triangle has an obtuse angle, which angle must it be and why?

8. Suppose you are told that a triangle has sides $a = 12.5$ centimeters, $b = 25.3$ centimeters, and $c = 10.7$ centimeters. Explain why the triangle has no solution.

Solve each triangle in Problems 9–12 if the triangle has a solution. Use decimal degrees for angle measure.

9. $a = 4.00$ meters, $b = 10.2$ meters, $c = 9.05$ meters

10. $a = 10.5$ miles, $b = 20.7$ miles, $c = 12.2$ miles

11. $a = 6.00$ kilometers, $b = 5.30$ kilometers, $c = 5.52$ kilometers

12. $a = 31.5$ meters, $b = 29.4$ meters, $c = 33.7$ meters

Problems 13–28 represent a variety of problems involving both the law of sines and the law of cosines. Solve each triangle. If a problem does not have a solution, say so.

13. $\alpha = 94.5°$, $\gamma = 88.3°$, $b = 23.7$ centimeters

14. $\beta = 85.6°$, $\gamma = 97.3°$, $a = 14.3$ millimeters

15. $\beta = 104.5°$, $a = 17.2$ inches, $c = 11.7$ inches

16. $\beta = 27.3°$, $a = 13.7$ yards, $c = 20.1$ yards

17. $\alpha = 57.2°$, $\gamma = 112.0°$, $c = 24.8$ meters

18. $\beta = 132.4°$, $\gamma = 17.3°$, $b = 67.6$ kilometers

19. $\beta = 38.4°$, $a = 11.5$ inches, $b = 14.0$ inches

20. $\gamma = 66.4°$, $b = 25.5$ meters, $c = 25.5$ meters

21. $a = 32.9$ meters, $b = 42.4$ meters, $c = 20.4$ meters

22. $a = 10.5$ centimeters, $b = 5.23$ centimeters, $c = 8.66$ centimeters

23. $\gamma = 58.4°$, $b = 7.23$ meters, $c = 6.54$ meters

24. $\alpha = 46.7°$, $a = 18.1$ meters, $b = 22.6$ meters

25. $\beta = 39.8°$, $a = 12.5$ inches, $b = 7.31$ inches

26. $\gamma = 47.9°$, $b = 35.2$ inches, $c = 25.5$ inches

27. $\beta = 13.6°$, $b = 21.6$ meters, $c = 58.4$ meters

28. $\beta = 25.1°$, $b = 53.7$ meters, $c = 98.5$ meters

C

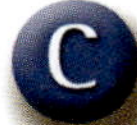

29. Show, using the law of cosines, that if $\gamma = 90°$, then $c^2 = a^2 + b^2$ (the Pythagorean theorem).

30. Show, using the law of cosines, that if $c^2 = a^2 + b^2$, then $\gamma = 90°$.

31. Show that for any triangle,

$$\frac{a^2 + b^2 + c^2}{2abc} = \frac{\cos\alpha}{a} + \frac{\cos\beta}{b} + \frac{\cos\gamma}{c}$$

32. Show that for any triangle,

$$a = b\cos\gamma + c\cos\beta$$

APPLICATIONS

33. Surveying. To find the length AB of a small lake, a surveyor measured angle ACB to be 96°, AC to be 91 yards, and BC to be 71 yards. What is the approximate length of the lake?

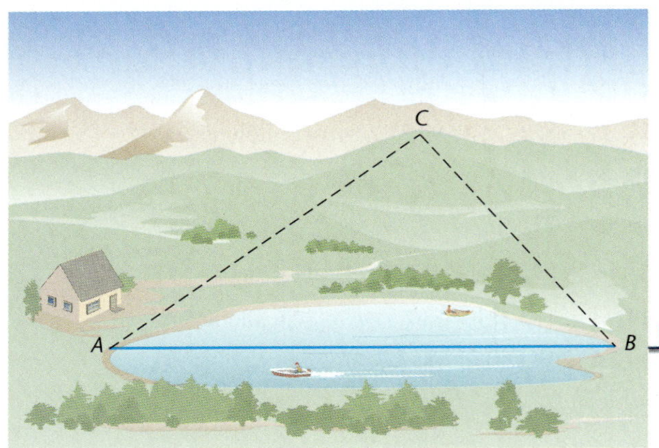

34. Surveying. Refer to problem 33. If a surveyor finds $\angle ACB = 110°$, $AC = 85$ meters, and $BC = 73$ meters, what is the approximate length of the lake?

35. Geometry. Find the measure in decimal degrees of a central angle subtended by a chord of length 112 millimeters in a circle of radius 72.8 millimeters.

36. Geometry. Find the measure in decimal degrees of a central angle subtended by a chord of length 13.8 feet in a circle of radius 8.26 feet.

37. Geometry. Two adjacent sides of a parallelogram meet at an angle of 35°10′ and have lengths of 3 and 8 feet. What is the length of the shorter diagonal of the parallelogram (to three significant digits)?

38. Geometry. What is the length of the longer diagonal of the parallelogram in problem 37 (to three significant digits)?

39. Navigation. Los Angeles and Las Vegas are approximately 200 miles apart. A pilot 80 miles from Los Angeles finds that she is 6°20′ off course relative to her start in Los Angeles. How far is she from Las Vegas at this time? (Compute the answer to three significant digits.)

40. Search and Rescue. At noon, two search planes set out from San Francisco to find a downed plane in the ocean. Plane A travels due west at 400 miles per hour, and plane B flies northwest at 500 miles per hour. At 2 P.M. plane A spots the survivors of the downed plane and radios plane B to come and assist in the rescue. How far is plane B from plane A at this time (to three significant digits)?

41. Geometry. Find the perimeter of a pentagon inscribed in a circle of radius 12.6 meters.

42. Geometry. Find the perimeter of a nine-sided regular polygon inscribed in a circle of radius 7.09 centimeters.

★**43. Analytic Geometry.** If point A in the figure has coordinates (3, 4) and point B has coordinates (4, 3), find the radian measure of angle θ to three decimal places.

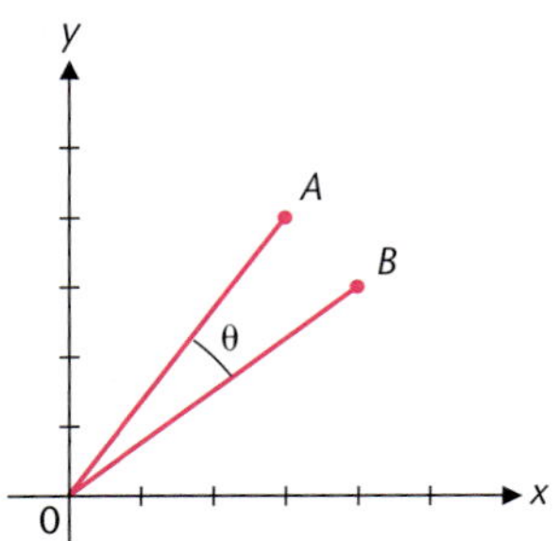

★**44. Analytic Geometry.** If point A has coordinates (4, 3) and point B has coordinates (5, 1), find the radian measure of $\angle AOB$ to three decimal places.

★**45. Engineering.** Three circles of radius 2.03, 5.00, and 8.20 centimeters are tangent to one another (see the figure). Find the three angles formed by the lines joining their centers (to the nearest 10′).

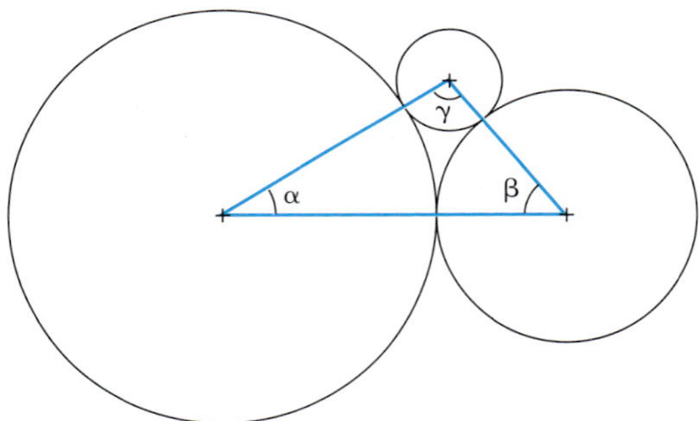

★**46. Engineering.** Three circles of radius 2.00, 5.00, and 8.00 inches are tangent to each other (see the figure). Find the three angles formed by the lines joining their centers (to the nearest 10′).

47. Geometry. A rectangular solid has sides as indicated in the figure. Find $\angle CAB$ to the nearest degree.

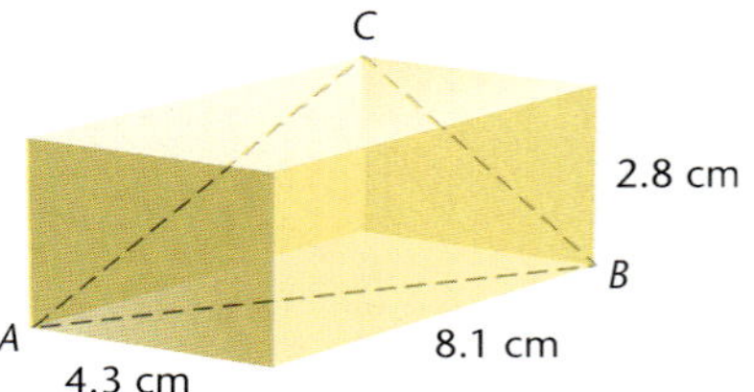

48. Geometry. Referring to problem 47, find $\angle ACB$ to the nearest degree.

★ **49. Space Science.** For communications between a space shuttle and the White Sands tracking station in southern New Mexico, two satellites are placed in geostationary orbit, 130° apart relative to the center of the Earth, and 22,300 miles above the surface of the Earth (see the figure). (A satellite in **geostationary** orbit remains stationary above a fixed point on the surface of the Earth.) Radio signals are sent from the tracking station by way of the satellites to the shuttle, and vice versa. How far to the nearest 100 miles is one of the geostationary satellites from the White Sands tracking station, W? The radius of the Earth is 3,964 miles.

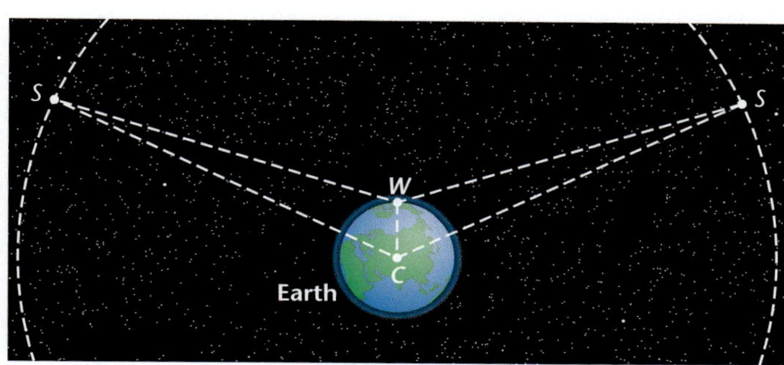

★ **50. Space Science.** A satellite S, in circular orbit around the Earth, is sighted by a tracking station T (see the figure). The distance TS is determined by radar to be 1,034 miles, and the angle of elevation above the horizon is 32.4°. How high is the satellite above the Earth at the time of the sighting? The radius of the Earth is 3,964 miles.

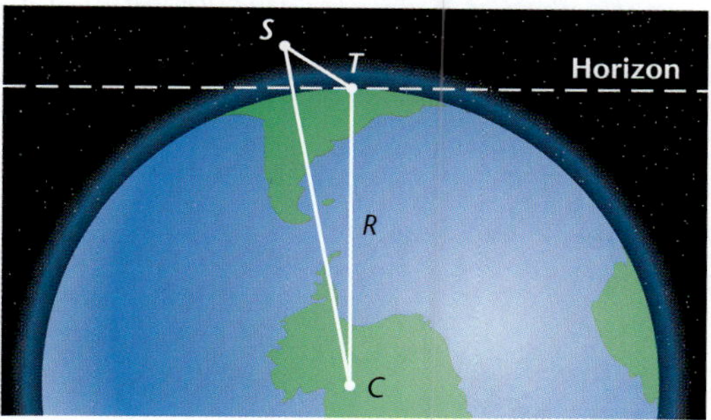

SECTION 7.3 Geometric Vectors

Geometric Vectors and Vector Addition • Velocity Vectors • Force Vectors • Resolution of Vectors into Vector Components

Many physical quantities, such as length, area, or volume, can be completely specified by a single real number. Other quantities, such as directed distances, velocities, and forces, require for their complete specification both a magnitude and a direction. The former are often called **scalar quantities,** and the latter are called **vector quantities.**

In Section 7.3 we limit our discussion to the intuitive idea of geometric vectors in a plane. In Section 7.4 we introduce algebraic vectors, a first step in the generalization of a concept that has far-reaching consequences. Vectors are widely used in many areas of science and engineering.

Geometric Vectors and Vector Addition

FIGURE 1 Vector $\overrightarrow{OP}$, or **v**.

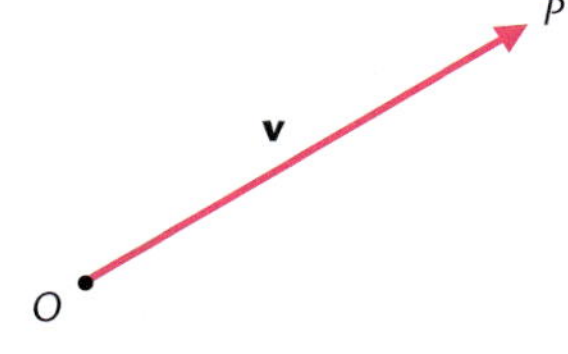

A line segment to which a direction has been assigned is called a **directed line segment.** A **geometric vector** is a directed line segment and is represented by an arrow (Fig. 1). A vector with an **initial point** O and a **terminal point** P (the end with the arrowhead) is denoted by $\overrightarrow{OP}$. Vectors are also denoted by a boldface letter, such as **v.** Because it is difficult to write boldface on paper, we suggest that you use an arrow over a single letter, such as $\vec{v}$, when you want the letter to denote a vector.

FIGURE 2 Vector addition: tail-to-tip rule.

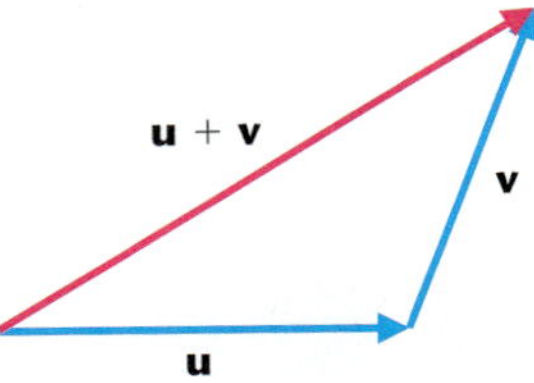

FIGURE 3 Vector addition: parallelogram rule.

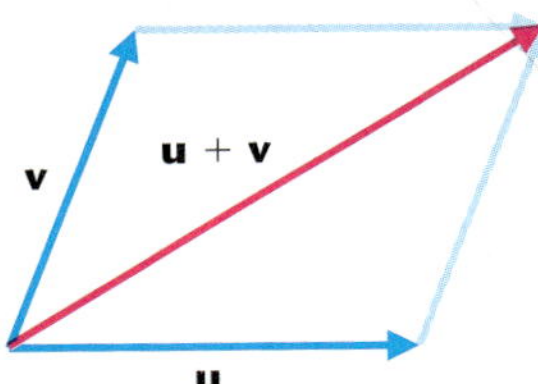

The **magnitude** of the vector $\overrightarrow{OP}$, denoted by $|\overrightarrow{OP}|$, $|\vec{v}|$, or $|\mathbf{v}|$, is the length of the directed line segment. Two vectors have the **same direction** if they are parallel and point in the same direction. Two vectors have **opposite direction** if they are parallel and point in opposite directions. The **zero vector,** denoted by $\vec{0}$ or **0,** has a magnitude of zero and an arbitrary direction. Two vectors are **equal** if they have the same magnitude and direction. Thus, a vector may be **translated** from one location to another as long as the magnitude and direction do not change.

The **sum of two vectors u and v** can be defined using the **tail-to-tip rule:** Translate **v** so that its tail end (initial point) is at the tip end (terminal point) of **u.** Then, the vector from the tail end of **u** to the tip end of **v** is the sum, denoted by $\mathbf{u} + \mathbf{v}$, of the vectors **u** and **v** (Fig. 2).

The sum of two nonparallel vectors also can be defined using the **parallelogram rule:** The **sum of two nonparallel vectors u and v** is the diagonal of the parallelogram formed using **u** and **v** as adjacent sides (Fig. 3). If **u** and **v** are parallel, use the tail-to-tip rule.

Both rules give the same sum. The choice of which rule to use depends on the situation and what seems most natural.

The vector $\mathbf{u} + \mathbf{v}$ is also called the **resultant** of the two vectors **u** and **v,** and **u** and **v** are called **vector components** of $\mathbf{u} + \mathbf{v}$. It is useful to observe that vector addition is **commutative** and **associative.** That is, $\mathbf{u} + \mathbf{v} = \mathbf{v} + \mathbf{u}$ and $\mathbf{u} + (\mathbf{v} + \mathbf{w}) = (\mathbf{u} + \mathbf{v}) + \mathbf{w}$.

EXPLORE/DISCUSS 1

If **a, b,** and **c** represent three arbitrary geometric vectors, illustrate using either definition of vector addition that

1. $\mathbf{a} + \mathbf{b} = \mathbf{b} + \mathbf{a}$
2. $\mathbf{a} + (\mathbf{b} + \mathbf{c}) = (\mathbf{a} + \mathbf{b}) + \mathbf{c}$

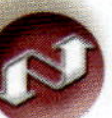

FIGURE 4

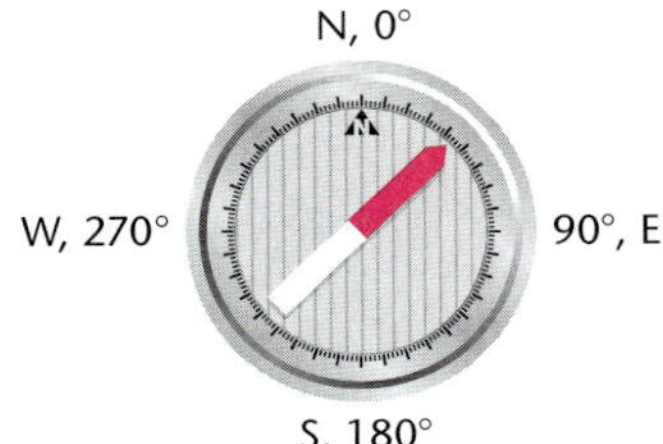

Navigational compass

Velocity Vectors

A vector that represents the direction and speed of an object in motion is called a **velocity vector.** Problems involving objects in motion often can be analyzed using vector methods. Many of these problems involve the use of a **navigational compass,** which is marked clockwise in degrees starting at north as indicated in Figure 4.

EXAMPLE 1 Apparent and Actual Velocity

An airplane has a compass heading (the direction the plane is pointing) of 85° and an airspeed (relative to the air) of 140 miles per hour. The wind is blowing from north to south at 66 miles per hour. The velocity of a plane relative to the air is called **apparent velocity,** and the velocity relative to the ground is called **resultant,** or **actual, velocity.** The resultant velocity is the vector sum of the apparent velocity and the wind velocity. Find the resultant velocity; that is, find the actual speed and direction of the airplane relative to the ground. Directions are given to the nearest degree and magnitudes to two significant digits.

SOLUTION

Geometric vectors [Fig. 5(a)] are used to represent the apparent velocity vector and the wind velocity vector. Add the two vectors using the tail-to-tip method of addition of vectors to obtain the resultant (actual) velocity vector [Fig. 5(b)]. From the vector diagram [Fig. 5(b)], we obtain the triangle in Figure 6 and solve for γ, c, and α.

FIGURE 5

(a) N, 85°, 180°, Apparent velocity, Wind velocity

(b) N, 85°, α, γ, Apparent velocity, Wind velocity, Actual velocity

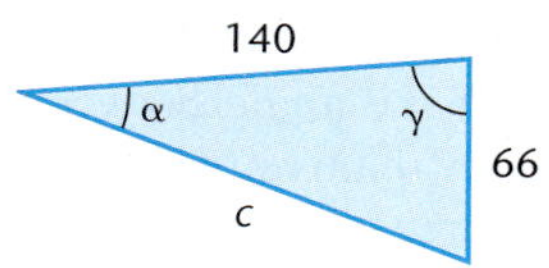

FIGURE 6

SOLVE FOR γ

Because the wind velocity vector is parallel to the north–south line, $\gamma = 85°$ [alternate interior angles of two parallel lines cut by a transversal are equal—see Fig. 5(b)].

SOLVE FOR c

Use the law of cosines:

$$c^2 = a^2 + b^2 - 2ab \cos \gamma$$

$$c = \sqrt{a^2 + b^2 - 2ab \cos \gamma}$$

$$= \sqrt{66^2 + 140^2 - 2(66)(140) \cos 85°}$$

$$= 150 \text{ miles per hour}$$

Speed relative to the ground

SOLVE FOR α

Use the law of sines:

$$\frac{\sin \alpha}{a} = \frac{\sin \gamma}{c}$$

$$\alpha = \sin^{-1}\left(\frac{a \sin \gamma}{c}\right)$$

$$= \sin^{-1}\left(\frac{66 \sin 85}{150}\right) = 26°$$

Actual heading $= 85° + \alpha = 85° + 26° = 111°$. Thus, the magnitude and direction of the resultant velocity vector are 150 miles per hour and 111°, respectively. That is, the plane, relative to the ground, is traveling at 150 miles per hour in a direction of 111°.

MATCHED 1 PROBLEM

A river is flowing southwest (225°) at 3.0 miles per hour. A boat crosses the river with a compass heading of 90°. If the speedometer on the boat reads 5.0 miles per hour (the boat's speed relative to the water), what is the resultant velocity? That is, what is the boat's actual speed and direction relative to the ground? Directions are to the nearest degree, and magnitudes are to two significant digits.

Force Vectors

A vector that represents the direction and magnitude of an applied force is called a **force vector.** If an object is subjected to two forces, then the sum of these two forces, the **resultant force,** is a single force. If the resultant force replaced the original two forces, it would act on the object in the same way as the two original forces taken together. In physics it is shown that the resultant force vector can be obtained using vector addition to add the two individual force vectors. It seems natural to use the parallelogram rule for adding force vectors, as is illustrated in Example 2.

EXAMPLE 2 Finding the Resultant Force

Two forces of 30 and 70 pounds act on a point in a plane. If the angle between the force vectors is 40°, what are the magnitude and direction (relative to the 70-pound force) of the resultant force? The magnitudes of the forces are to two significant digits and the angles to the nearest degree.

SOLUTION

We start with a diagram (Fig. 7), letting geometric vectors represent the various forces. Because adjacent angles in a parallelogram are supplementary, the measure of angle $OCB = 180° - 40° = 140°$. We can now find the magnitude of the resultant vector **R** using the law of cosines (Fig. 8).

FIGURE 7

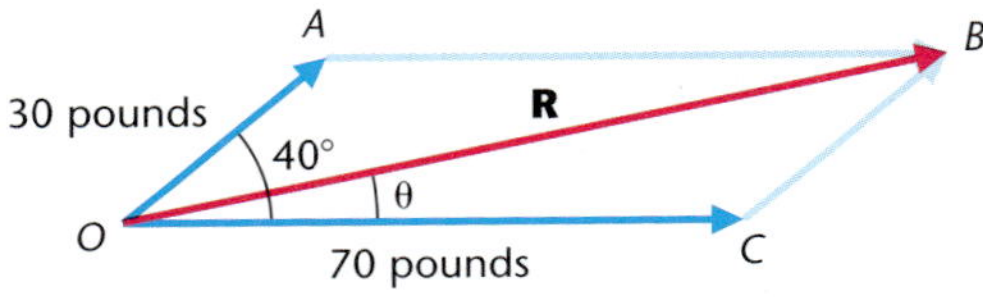

$$|\mathbf{R}|^2 = 30^2 + 70^2 - 2(30)(70)\cos 140°$$
$$|\mathbf{R}| = \sqrt{30^2 + 70^2 - 2(30)(70)\cos 140°}$$
$$= 95 \text{ pounds}$$

FIGURE 8

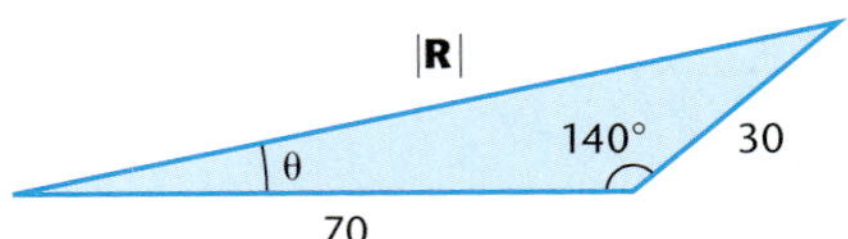

To find θ, the direction of **R,** we use the law of sines (Fig. 9).

$$\frac{\sin\theta}{30} = \frac{\sin 140°}{95}$$
$$\sin\theta = \frac{30\sin 140°}{95}$$
$$\theta = \sin^{-1}\left(\frac{30\sin 140°}{95}\right) = 12°$$

95
140°
30
θ
70

FIGURE 9

Thus, the two given forces are equivalent to a single force of 95 pounds in the direction of 12° (relative to the 70-pound force).

MATCHED 2 PROBLEM

Repeat Example 2 using an angle of 100° between the two forces.

Resolution of Vectors into Vector Components

Instead of adding vectors, many problems require the resolution of vectors into components. As we indicated earlier, whenever a vector is expressed as the sum or resultant of two vectors, the two vectors are called **vector components** of the given vector. Example 3 illustrates an application of the process of resolving a vector into vector components.

EXAMPLE 3 **Resolving a Force Vector into Components**

A car weighing 3,210 pounds is on a driveway inclined 20.2° to the horizontal. Neglecting friction, find the magnitude of the force parallel to the driveway that will keep the car from rolling down the hill.

SOLUTION

We start by drawing a vector diagram (Fig. 10).

FIGURE 10

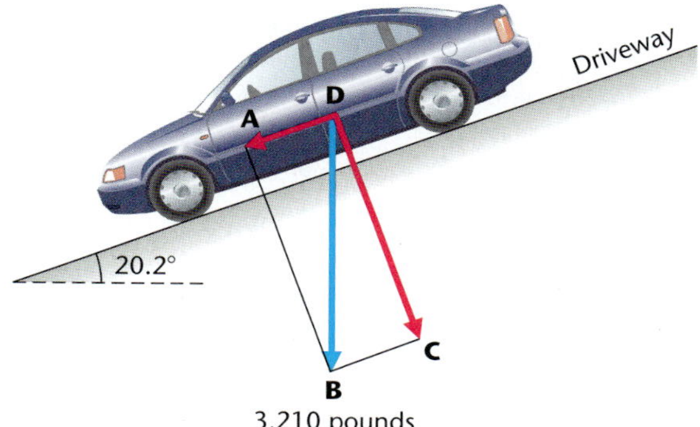

The force vector $\overrightarrow{DB}$ acts in a downward direction and represents the weight of the car. Note that $\overrightarrow{DB} = \overrightarrow{DC} + \overrightarrow{DA}$, where $\overrightarrow{DC}$ is the perpendicular component of $\overrightarrow{DB}$ relative to the driveway and $\overrightarrow{DA}$ is the parallel component of $\overrightarrow{DB}$ relative to the driveway.

To keep the car at D from rolling down the hill, we need a force with the magnitude of $\overrightarrow{DA}$ but oppositely directed. To find $|\overrightarrow{DA}|$, we note that $\angle ABD = 20.2°$. This is true because $\angle ABD$ and the driveway angle have the same complement, $\angle ADB$.

$$\sin 20.2° = \frac{|\overrightarrow{DA}|}{3{,}210}$$

$$|\overrightarrow{DA}| = 3{,}210 \sin 20.2°$$

$$= 1{,}110 \text{ pounds}$$

MATCHED 3 PROBLEM

Find the magnitude of the perpendicular component of $\overrightarrow{DB}$ in Example 3.

ANSWERS MATCHED PROBLEMS

1. Resultant velocity: magnitude = 3.6 miles per hour, direction = 126°
2. $|\mathbf{R}| = 71$ pounds, $\theta = 25°$
3. $|\overrightarrow{DC}| = 3{,}010$ pounds

EXERCISE 7.3

Express all angle measures in decimal degrees. In navigation problems, refer to the figure of a navigational compass.

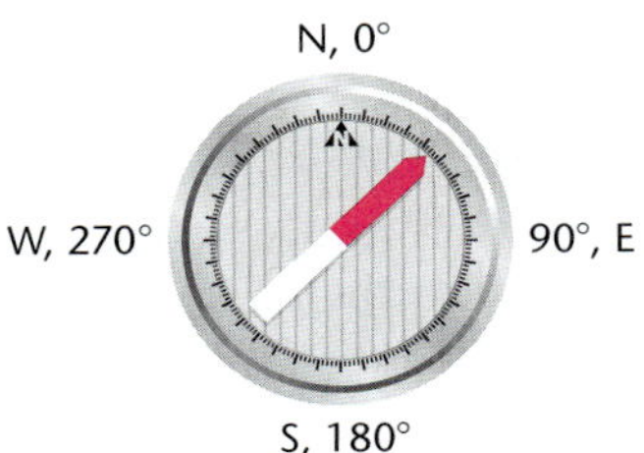

Navigational compass

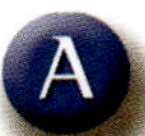

Problems 1–10 refer to figures (a) and (b) showing vector addition for vectors **u** *and* **v** *at right angles to each other.*

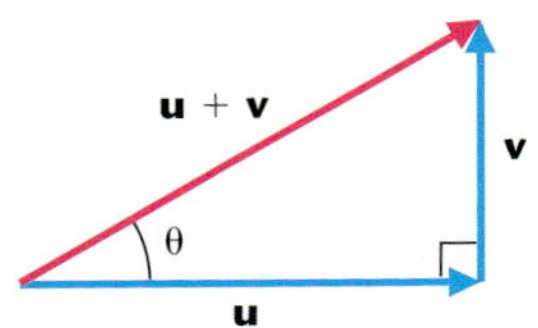

Tail-to-tip rule

(a)

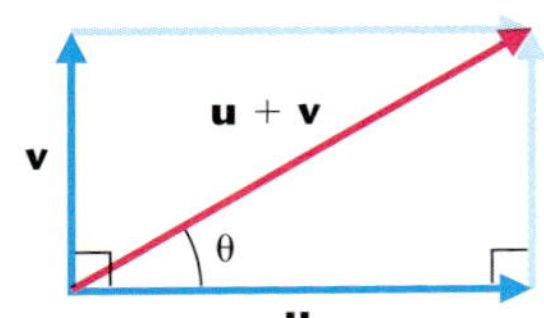

Parallelogram rule

(b)

In Problems 1–6, find $|\mathbf{u} + \mathbf{v}|$ and θ, given $|\mathbf{u}|$ and $|\mathbf{v}|$ in figures (a) and (b).

1. $|\mathbf{u}| = 37$ miles per hour, $|\mathbf{v}| = 45$ miles per hour

2. $|\mathbf{u}| = 62$ miles per hour, $|\mathbf{v}| = 34$ miles per hour

3. $|\mathbf{u}| = 38$ kilograms, $|\mathbf{v}| = 53$ kilograms

4. $|\mathbf{u}| = 48$ kilograms, $|\mathbf{v}| = 31$ kilograms

5. $|\mathbf{u}| = 434$ kilometers per hour, $|\mathbf{v}| = 105$ kilometers per hour

6. $|\mathbf{u}| = 143$ kilometers per hour, $|\mathbf{v}| = 57.4$ kilometers per hour

In Problems 7–10, find $|\mathbf{u}|$ and $|\mathbf{v}|$, the magnitudes of the horizontal and vertical components of **u** + **v**, *given $|\mathbf{u} + \mathbf{v}|$ and θ in figures (a) and (b).*

7. $|\mathbf{u} + \mathbf{v}| = 32$ pounds, $\theta = 22°$

8. $|\mathbf{u} + \mathbf{v}| = 250$ pounds, $\theta = 65°$

9. $|\mathbf{u} + \mathbf{v}| = 230$ miles per hour, $\theta = 72°$

10. $|\mathbf{u} + \mathbf{v}| = 28$ miles per hour, $\theta = 12°$

11. If two vectors have the same magnitude, are they equal? Explain why or why not.

12. Can the magnitude of a vector ever be negative? Explain why or why not.

B

Problems 13–20 refer to figures (a) and (b) showing vector addition for vectors **u** *and* **v**.

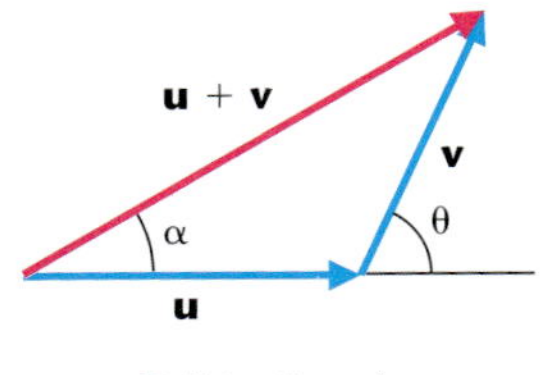

Tail-to-tip rule

(a)

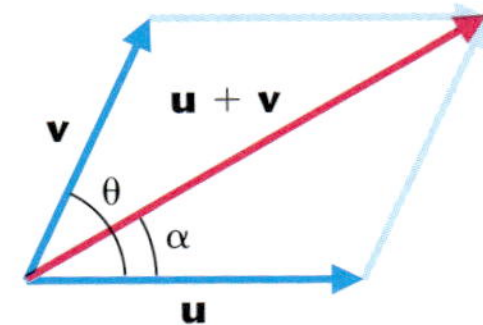

Parallelogram rule

(b)

In Problems 13–16, find $|\mathbf{u} + \mathbf{v}|$ and α given $|\mathbf{u}|$, $|\mathbf{v}|$, and θ in figures (a) and (b).

13. $|\mathbf{u}| = 66$ grams, $|\mathbf{v}| = 22$ grams, $\theta = 68°$

14. $|\mathbf{u}| = 120$ grams, $|\mathbf{v}| = 84$ grams, $\theta = 44°$

15. $|\mathbf{u}| = 21$ knots, $|\mathbf{v}| = 3.2$ knots, $\theta = 53°$

16. $|\mathbf{u}| = 8.0$ knots, $|\mathbf{v}| = 2.0$ knots, $\theta = 64°$

In Problems 17–20, find $|\mathbf{u}|$ *and* $|\mathbf{v}|$, *given* $|\mathbf{u} + \mathbf{v}|$, α *and* θ *in figures (a) and (b).*

17. $|\mathbf{u} + \mathbf{v}| = 14$ kilograms, $\alpha = 25°$, $\theta = 79°$

18. $|\mathbf{u} + \mathbf{v}| = 33$ kilograms, $\alpha = 17°$, $\theta = 43°$

19. $|\mathbf{u} + \mathbf{v}| = 223$ miles per hour, $\alpha = 42.3°$, $\theta = 69.4°$

20. $|\mathbf{u} + \mathbf{v}| = 437$ miles per hour, $\alpha = 17.8°$, $\theta = 50.5°$

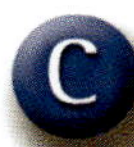

21. Is it correct to say that the zero vector is perpendicular to every vector? Explain.

22. Is it correct to say that the zero vector is parallel to every vector? Explain.

APPLICATIONS

In Problems 23–26, assume the north, east, south, and west directions are exact.

23. Navigation. An airplane is flying with a compass heading of 285° and an airspeed of 230 miles per hour. A steady wind of 35 miles per hour is blowing in the direction of 260°. What is the plane's actual velocity; that is, what is its speed and direction relative to the ground?

24. Navigation. A power boat crossing a wide river has a compass heading of 25° and speed relative to the water of 15 miles per hour. The river is flowing in the direction of 135° at 3.9 miles per hour. What is the boat's actual velocity; that is, what is its speed and direction relative to the ground?

★ **25. Navigation.** Two docks are directly opposite each other on a southward-flowing river. A boat pilot wishes to go in a straight line from the east dock to the west dock in a ferryboat with a cruising speed in still water of 8.0 knots. If the river's current is 2.5 knots, what compass heading should be maintained while crossing the river? What is the actual speed of the boat relative to the land?

★ **26. Navigation.** An airplane can cruise at 255 miles per hour in still air. If a steady wind of 46.0 miles per hour is blowing from the west, what compass heading should the pilot fly for the course of the plane relative to the ground to be north (0°)? Compute the ground speed for this course.

★ **27. Resultant Force.** A large ship has gone aground in a harbor and two tugs, with cables attached, attempt to pull it free. If one tug pulls with a compass course of 52° and a force of 2,300 pounds and a second tug pulls with a compass course of 97° and a force of 1,900 pounds, what is the compass direction and the magnitude of the resultant force?

★ **28. Resultant Force.** Repeat Problem 27 if one tug pulls with a compass direction of 161° and a force of 2,900 kilograms and a second tug pulls with a compass direction of 192° and a force of 3,600 kilograms.

29. Resolution of Forces. An automobile weighing 4,050 pounds is standing on a driveway inclined 5.5° with the horizontal.

(A) Find the magnitude of the force parallel to the driveway necessary to keep the car from rolling down the hill.

(B) Find the magnitude of the force perpendicular to the driveway.

30. Resolution of Forces. Repeat Problem 29 for a car weighing 2,500 pounds parked on a hill inclined at 15° to the horizontal.

★ **31. Resolution of Forces.** If two weights are fastened together and placed on inclined planes as shown in the figure, neglecting friction, which way will they slide?

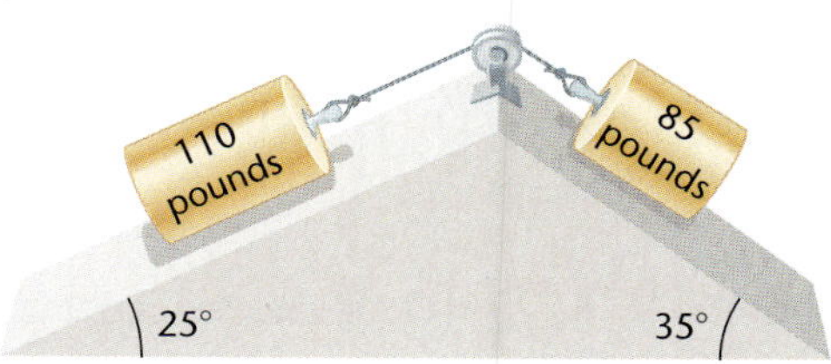

★ **32. Resolution of Forces.** If two weights are fastened together and placed on inclined planes as indicated in the figure, neglecting friction, which way will they slide?

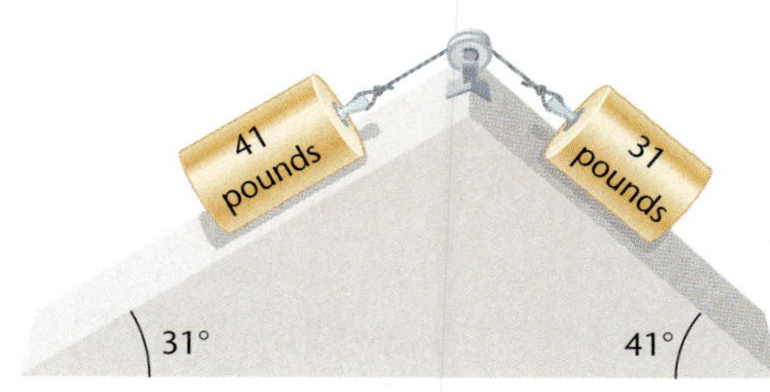

SECTION 7.4 Algebraic Vectors

From Geometric Vectors to Algebraic Vectors • Vector Addition and Scalar Multiplication • Unit Vectors • Algebraic Properties • Static Equilibrium

Geometric vectors in a plane are readily generalized to three-dimensional space. However, to generalize vectors further to higher-dimensional abstract spaces, it is essential to define the vector concept algebraically. This is done in such a way that the geometric vectors become special cases of the more general algebraic vectors. Algebraic vectors have many advantages over geometric vectors. One advantage will become apparent when we consider static equilibrium problems at the end of Section 7.4.

The development of algebraic vectors in this book is introductory and is restricted to the plane. Further study of vectors in three- and higher-dimensional spaces is reserved for more advanced mathematical courses.

FIGURE 1 $\overrightarrow{OP}$ is the standard vector for $\overrightarrow{AB}$.

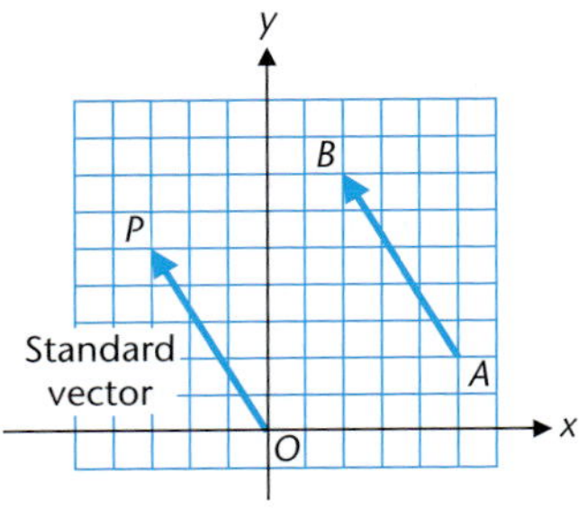

From Geometric Vectors to Algebraic Vectors

The transition from geometric vectors to algebraic vectors is begun by placing geometric vectors in a rectangular coordinate system. A geometric vector $\overrightarrow{AB}$ in a rectangular coordinate system translated so that its initial point is at the origin is said to be in **standard position.** The vector $\overrightarrow{OP}$ such that $\overrightarrow{OP} = \overrightarrow{AB}$ is said to be the **standard vector** for $\overrightarrow{AB}$ (Fig. 1).

Note that the vector $\overrightarrow{OP}$ in Figure 1 is the standard vector for infinitely many vectors—all vectors with the same magnitude and direction as $\overrightarrow{OP}$.

EXPLORE/DISCUSS 1

(A) In a copy of Figure 1, draw in three other vectors having $\overrightarrow{OP}$ as their standard vector.

(B) If the tail of a vector is at point $A = (-3, 2)$ and its tip is at $B = (6, 4)$, discuss how you would find the coordinates of P so that $\overrightarrow{OP}$ is the standard vector for $\overrightarrow{AB}$.

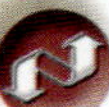

Given the coordinates of the endpoints of a geometric vector in a rectangular coordinate system, how do we find its corresponding standard vector? The process is not difficult. The coordinates of the initial point, O, of $\overrightarrow{OP}$ are always $(0, 0)$. Thus, we have only to find the coordinates of P, the terminal point of $\overrightarrow{OP}$. The coordinates of P are given by

$$(x_p, y_p) = (x_b - x_a, y_b - y_a) \tag{1}$$

where the coordinates of A are (x_a, y_a) and the coordinates of B are (x_b, y_b). Example 1 illustrates the use of equation (1).

EXAMPLE 1 Finding a Standard Vector for a Given Vector

Given the geometric vector $\overrightarrow{AB}$ with initial point $A = (3, 4)$ and terminal point $B = (7, -1)$, find the standard vector $\overrightarrow{OP}$ for $\overrightarrow{AB}$. That is, find the coordinates of the point P such that $\overrightarrow{OP} = \overrightarrow{AB}$.

FIGURE 2

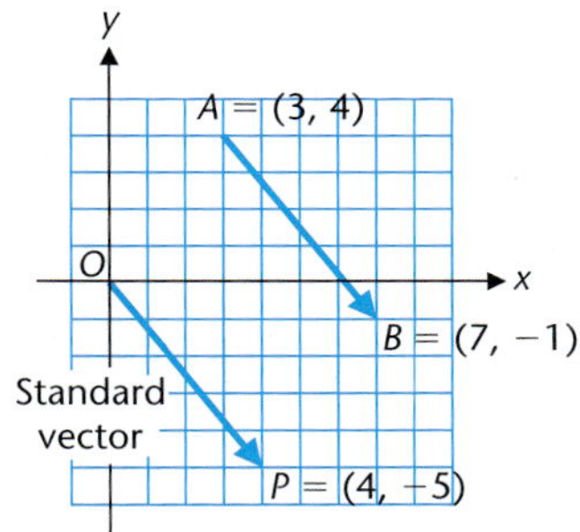

SOLUTION

The coordinates of P are given by

$$\begin{aligned}(x_p, y_p) &= (x_b - x_a, y_b - y_a)\\ &= (7 - 3, -1 - 4)\\ &= (4, -5)\end{aligned}$$

Note in Figure 2 that if we start at A, then move to the right four units and down five units, we will be at B. If we start at the origin, then move to the right four units and down five units, we will be at P.

MATCHED PROBLEM 1

Given the geometric vector $\overrightarrow{AB}$ with initial point $A = (8, -3)$ and terminal point $B = (4, 5)$, find the standard vector $\overrightarrow{OP}$ for $\overrightarrow{AB}$.

The preceding discussion suggests another way of looking at vectors. Because, given any geometric vector $\overrightarrow{AB}$ in a rectangular coordinate system, there always exists a point $P = (x_p, y_p)$ such that $\overrightarrow{OP} = \overrightarrow{AB}$, the point $P = (x_p, y_p)$ completely specifies the vector $\overrightarrow{AB}$, except for its position. And we are not concerned about its position because we are free to translate $\overrightarrow{AB}$ anywhere we please. Conversely, given any point $P = (x_p, y_p)$ in a rectangular coordinate system, the directed line segment joining O to P forms the geometric vector $\overrightarrow{OP}$.

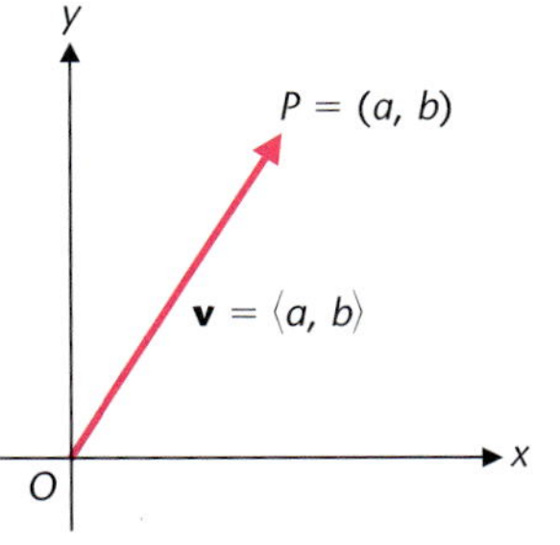

FIGURE 3 Algebraic vector $\langle a, b\rangle$ associated with a geometric vector $\overrightarrow{OP}$.

This leads us to define an **algebraic vector** as an ordered pair of real numbers. To avoid confusing a point (a, b) with a vector (a, b), **we use $\langle a, b\rangle$ to represent an algebraic vector.** Geometrically, the algebraic vector $\langle a, b\rangle$ corresponds to the standard (geometric) vector $\overrightarrow{OP}$ with terminal point $P = (a, b)$ and initial point $O = (0, 0)$, as illustrated in Figure 3.

The real numbers a and b are **scalar components** of the vector $\langle a, b\rangle$. The word **scalar** means real number and is often used in the context of vectors in reference to "scalar quantities" as opposed to "vector quantities." Thus, we talk about "scalar components" and "vector components" of a given vector. The words *scalar* and *vector* are often dropped if the meaning of component is clear from the context.

Two vectors $\mathbf{u} = \langle a, b\rangle$ and $\mathbf{v} = \langle c, d\rangle$ are said to be **equal** if their corresponding components are equal, that is, if $a = c$ and $b = d$. The **zero vector** is denoted by $\mathbf{0} = \langle 0, 0\rangle$.

Geometric vectors are limited to spaces we can visualize, that is, to two- and three-dimensional spaces. Algebraic vectors do not have these restrictions. The following are algebraic vectors from two-, three-, four-, and five-dimensional spaces:

$$\langle -2, 5\rangle \qquad \langle 3, 0, -8\rangle \qquad \langle 5, 1, 1, -2\rangle \qquad \langle -1, 0, 1, 3, 4\rangle$$

As we said earlier, the discussion in this book is limited to algebraic vectors in a two-dimensional space, which represents a plane.

We now define the *magnitude* of an algebraic vector.

FIGURE 4 Magnitude of vector $\langle a, b\rangle$ geometrically interpreted.

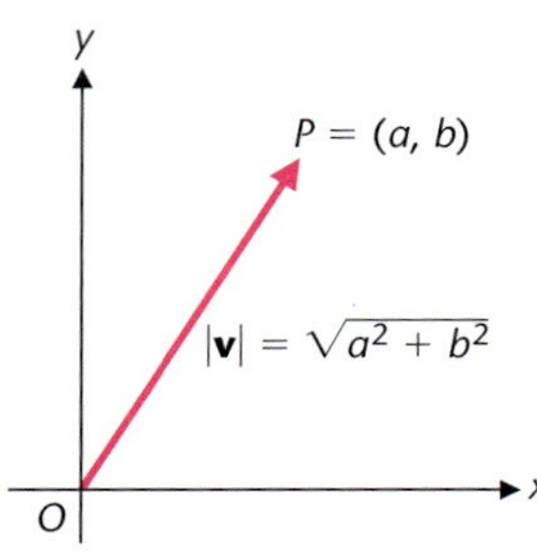

DEFINITION 1
Magnitude of v = ⟨a, b⟩

The **magnitude,** or **norm,** of a vector $\mathbf{v} = \langle a, b\rangle$ is denoted by $|\mathbf{v}|$ and is given by

$$|\mathbf{v}| = \sqrt{a^2 + b^2}$$

Geometrically, $\sqrt{a^2 + b^2}$ is the length of the standard geometric vector $\overrightarrow{OP}$ associated with the algebraic vector $\langle a, b\rangle$ (Fig. 4).

The definition of magnitude is readily generalized to higher-dimensional vector spaces. For example, if $|\mathbf{v}| = \langle a, b, c, d\rangle$, then the magnitude, or norm, is given by $\sqrt{a^2 + b^2 + c^2 + d^2}$. But now we are not able to interpret the result in terms of geometric vectors.

EXAMPLE 2 **Finding the Magnitude of a Vector**

Find the magnitude of the vector $\mathbf{v} = \langle 3, -5\rangle$.

SOLUTION

$$|\mathbf{v}| = \sqrt{3^2 + (-5)^2} = \sqrt{34}$$

MATCHED 2 PROBLEM

Find the magnitude of the vector $\mathbf{v} = \langle -2, 4\rangle$.

Vector Addition and Scalar Multiplication

To add two algebraic vectors, add the corresponding components as indicated in Definition 2:

DEFINITION 2
Vector Addition

If $\mathbf{u} = \langle a, b\rangle$ and $\mathbf{v} = \langle c, d\rangle$, then

$$\mathbf{u} + \mathbf{v} = \langle a + c, b + d\rangle$$

Addition of algebraic vectors is consistent with the parallelogram and tail-to-tip definitions for adding geometric vectors given in Section 7.3, as is investigated in Explore/Discuss 2 on page 559.

EXPLORE/DISCUSS 2

If $\mathbf{u} = \langle -3, 2\rangle$, $\mathbf{v} = \langle 7, 3\rangle$, then $\mathbf{u} + \mathbf{v} = \langle -3 + 7, 2 + 3\rangle = \langle 4, 5\rangle$. Locate **u**, **v**, and $\mathbf{u} + \mathbf{v}$ in a rectangular coordinate system and interpret geometrically in terms of the parallelogram and tail-to-tip rules discussed in Section 7.3.

To multiply a vector by a scalar (a real number) multiply each component by the scalar:

FIGURE 5 Scalar multiplication geometrically interpreted.

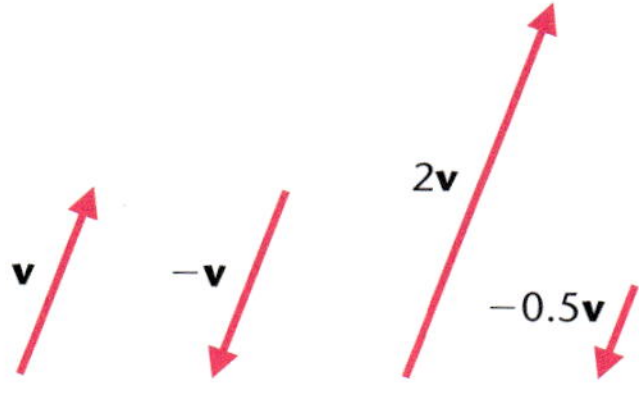

DEFINITION 3
Scalar Multiplication

If $\mathbf{u} = \langle a, b\rangle$ and k is a scalar, then

$$k\mathbf{u} = k\langle a, b\rangle = \langle ka, kb\rangle$$

Geometrically, if a vector **v** is multiplied by a scalar k, the magnitude of the vector **v** is multiplied by $|k|$. If k is positive, then $k\mathbf{v}$ has the same direction as **v.** If k is negative, then $k\mathbf{v}$ has the opposite direction as **v.** These relationships are illustrated in Figure 5.

EXAMPLE 3 Vector Addition and Scalar Multiplication

Let $\mathbf{u} = \langle 4, -3\rangle$, $\mathbf{v} = \langle 2, 3\rangle$, and $\mathbf{w} = \langle 0, -5\rangle$. Find

(A) $\mathbf{u} + \mathbf{v}$ (B) $-2\mathbf{u}$ (C) $2\mathbf{u} - 3\mathbf{v}$ (D) $3\mathbf{u} + 2\mathbf{v} - \mathbf{w}$

SOLUTIONS

(A) $\mathbf{u} + \mathbf{v} = \langle 4, -3\rangle + \langle 2, 3\rangle = \langle 6, 0\rangle$

(B) $-2\mathbf{u} = -2\langle 4, -3\rangle = \langle -8, 6\rangle$

(C) $2\mathbf{u} - 3\mathbf{v} = 2\langle 4, -3\rangle - 3\langle 2, 3\rangle$
$= \langle 8, -6\rangle + \langle -6, -9\rangle = \langle 2, -15\rangle$

(D) $3\mathbf{u} + 2\mathbf{v} - \mathbf{w} = 3\langle 4, -3\rangle + 2\langle 2, 3\rangle - \langle 0, -5\rangle$
$= \langle 12, -9\rangle + \langle 4, 6\rangle + \langle 0, 5\rangle$
$= \langle 16, 2\rangle$

MATCHED PROBLEM 3

Let $\mathbf{u} = \langle -5, 3\rangle$, $\mathbf{v} = \langle 4, -6\rangle$, and $\mathbf{w} = \langle -2, 0\rangle$. Find

(A) $\mathbf{u} + \mathbf{v}$ (B) $-3\mathbf{u}$ (C) $3\mathbf{u} - 2\mathbf{v}$ (D) $2\mathbf{u} - \mathbf{v} + 3\mathbf{w}$

Unit Vectors

If $|\mathbf{v}| = 1$, then $\mathbf{v}$ is called a **unit vector.** A unit vector can be formed from an arbitrary nonzero vector as follows:

A Unit Vector with the Same Direction as v

If $\mathbf{v}$ is a nonzero vector, then

$$\mathbf{u} = \frac{1}{|\mathbf{v}|}\mathbf{v}$$

is a unit vector with the same direction as **v.**

EXAMPLE 4 Finding a Unit Vector with the Same Direction as a Given Vector

Given a vector $\mathbf{v} = \langle 1, -2\rangle$, find a unit vector $\mathbf{u}$ with the same direction as $\mathbf{v}$.

SOLUTION

$$|\mathbf{v}| = \sqrt{1^2 + (-2)^2} = \sqrt{5}$$

$$\mathbf{u} = \frac{1}{|\mathbf{v}|}\mathbf{v} = \frac{1}{\sqrt{5}}\langle 1, -2\rangle$$

$$= \left\langle \frac{1}{\sqrt{5}}, \frac{-2}{\sqrt{5}} \right\rangle$$

CHECK

$$|\mathbf{u}| = \sqrt{\left(\frac{1}{\sqrt{5}}\right)^2 + \left(\frac{-2}{\sqrt{5}}\right)^2} = \sqrt{\frac{1}{5} + \frac{4}{5}} = \sqrt{1} = 1$$

And we see that $\mathbf{u}$ is a unit vector with the same direction as $\mathbf{v}$.

MATCHED PROBLEM 4

Given a vector $\mathbf{v} = \langle 3, 1\rangle$, find a unit vector $\mathbf{u}$ with the same direction as $\mathbf{v}$.

We now define two very important unit vectors, the **i** and **j** unit vectors.

The i and j Unit Vectors

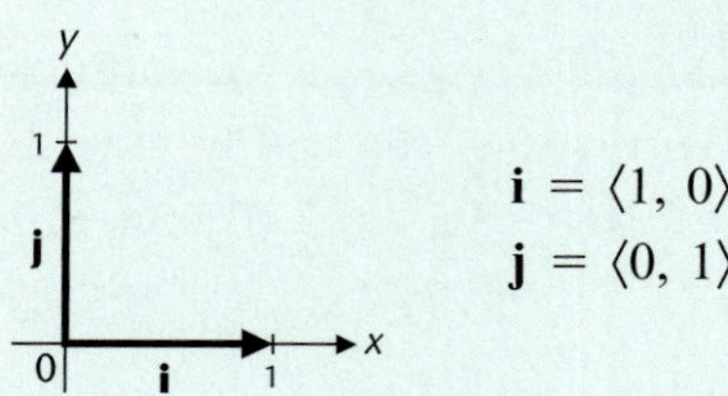

$$\mathbf{i} = \langle 1, 0\rangle$$
$$\mathbf{j} = \langle 0, 1\rangle$$

Why are the **i** and **j** unit vectors so important? One of the reasons is that any vector $\mathbf{v} = \langle a, b \rangle$ can be expressed as a linear combination of these two vectors; that is, as $a\mathbf{i} + b\mathbf{j}$.

$$\begin{aligned} \mathbf{v} = \langle a, b \rangle &= \langle a, 0 \rangle + \langle 0, b \rangle \\ &= a\langle 1, 0 \rangle + b\langle 0, 1 \rangle = a\mathbf{i} + b\mathbf{j} \end{aligned}$$

EXAMPLE 5 **Expressing a Vector in Terms of the i and j Vectors**

Express each vector as a linear combination of the **i** and **j** unit vectors.

(A) $\langle -2, 4 \rangle$ (B) $\langle 2, 0 \rangle$ (C) $\langle 0, -7 \rangle$

SOLUTIONS

(A) $\langle -2, 4 \rangle = -2\mathbf{i} + 4\mathbf{j}$

(B) $\langle 2, 0 \rangle = 2\mathbf{i} + 0\mathbf{j} = 2\mathbf{i}$

(C) $\langle 0, -7 \rangle = 0\mathbf{i} - 7\mathbf{j} = -7\mathbf{j}$

MATCHED 5 PROBLEM

Express each vector as a linear combination of the **i** and **j** unit vectors.

(A) $\langle 5, -3 \rangle$ (B) $\langle -9, 0 \rangle$ (C) $\langle 0, 6 \rangle$

Algebraic Properties

Vector addition and scalar multiplication possess algebraic properties similar to the real numbers. These properties enable us to manipulate symbols representing vectors and scalars in much the same way we manipulate symbols that represent real numbers in algebra. These properties are listed here for convenient reference.

Algebraic Properties of Vectors

A. Addition Properties. For all vectors **u**, **v**, and **w**,

1. $\mathbf{u} + \mathbf{v} = \mathbf{v} + \mathbf{u}$ **Commutative Property**
2. $\mathbf{u} + (\mathbf{v} + \mathbf{w}) = (\mathbf{u} + \mathbf{v}) + \mathbf{w}$ **Associative Property**
3. $\mathbf{u} + \mathbf{0} = \mathbf{0} + \mathbf{u} = \mathbf{u}$ **Additive Identity**
4. $\mathbf{u} + (-\mathbf{u}) = (-\mathbf{u}) + \mathbf{u} = \mathbf{0}$ **Additive Inverse**

B. Scalar Multiplication Properties. For all vectors **u** and **v** and all scalars m and n:

1. $m(n\mathbf{u}) = (mn)\mathbf{u}$ **Associative Property**
2. $m(\mathbf{u} + \mathbf{v}) = m\mathbf{u} + m\mathbf{v}$ **Distributive Property**
3. $(m + n)\mathbf{u} = m\mathbf{u} + n\mathbf{u}$ **Distributive Property**
4. $1\mathbf{u} = \mathbf{u}$ **Multiplicative Identity**

EXAMPLE 6 Algebraic Operations on Vectors Expressed in Terms of the i and j Vectors

For $\mathbf{u} = \mathbf{i} - 2\mathbf{j}$ and $\mathbf{v} = 5\mathbf{i} + 2\mathbf{j}$, compute each of the following:

(A) $\mathbf{u} + \mathbf{v}$ (B) $\mathbf{u} - \mathbf{v}$ (C) $2\mathbf{u} + 3\mathbf{v}$

SOLUTIONS

(A) $\mathbf{u} + \mathbf{v} = (\mathbf{i} - 2\mathbf{j}) + (5\mathbf{i} + 2\mathbf{j})$
$= \mathbf{i} - 2\mathbf{j} + 5\mathbf{i} + 2\mathbf{j} = 6\mathbf{i} + 0\mathbf{j} = 6\mathbf{i}$

(B) $\mathbf{u} - \mathbf{v} = (\mathbf{i} - 2\mathbf{j}) - (5\mathbf{i} + 2\mathbf{j})$
$= \mathbf{i} - 2\mathbf{j} - 5\mathbf{i} - 2\mathbf{j} = -4\mathbf{i} - 4\mathbf{j}$

(C) $2\mathbf{u} + 3\mathbf{v} = 2(\mathbf{i} - 2\mathbf{j}) + 3(5\mathbf{i} + 2\mathbf{j})$
$= 2\mathbf{i} - 4\mathbf{j} + 15\mathbf{i} + 6\mathbf{j} = 17\mathbf{i} + 2\mathbf{j}$

MATCHED 6 PROBLEM

For $\mathbf{u} = 2\mathbf{i} - \mathbf{j}$ and $\mathbf{v} = 4\mathbf{i} + 5\mathbf{j}$, compute each of the following:

(A) $\mathbf{u} + \mathbf{v}$ (B) $\mathbf{u} - \mathbf{v}$ (C) $3\mathbf{u} - 2\mathbf{v}$

Static Equilibrium

Algebraic vectors can be used to solve many types of problems in physics and engineering. We complete Section 7.4 by considering a few problems involving *static equilibrium*. Fundamental to our approach are two basic principles regarding forces and objects subject to these forces:

Conditions for Static Equilibrium

1. An object at rest is said to be in **static equilibrium.**
2. For an object located at the origin in a rectangular coordinate system to remain in static equilibrium, at rest, it is necessary that the sum of all the force vectors acting on the object be the zero vector.

Example 7 shows how some important physics and engineering problems can be solved using algebraic vectors and the conditions for static equilibrium. It is assumed that you know how to solve a system of two equations with two variables. In case you need a reminder, procedures are reviewed in Section 8.1.

EXAMPLE 7 Tension in Cables

A cable car, used to ferry people and supplies across a river, weighs 2,500 pounds fully loaded. The car stops when partway across and deflects the cable relative to the horizontal, as indicated in Figure 6. What is the tension in each part of the cable running to each tower?

FIGURE 6

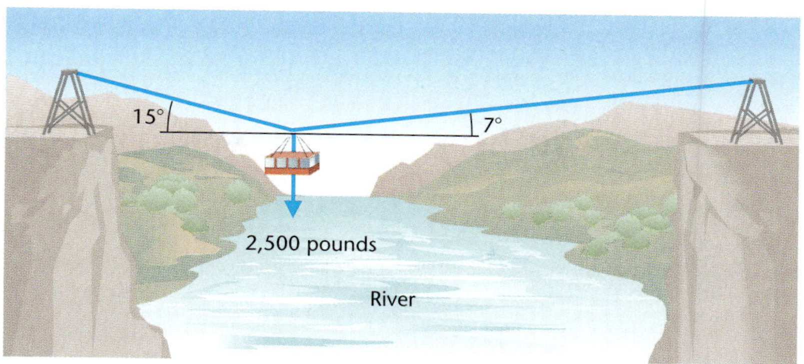

SOLUTION

Step 1. Draw a force diagram with all force vectors in standard position at the origin (Fig. 7). The objective is to find $|\mathbf{u}|$ and $|\mathbf{v}|$.

Step 2. Write each force vector in terms of the **i** and **j** unit vectors:

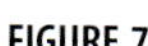

FIGURE 7

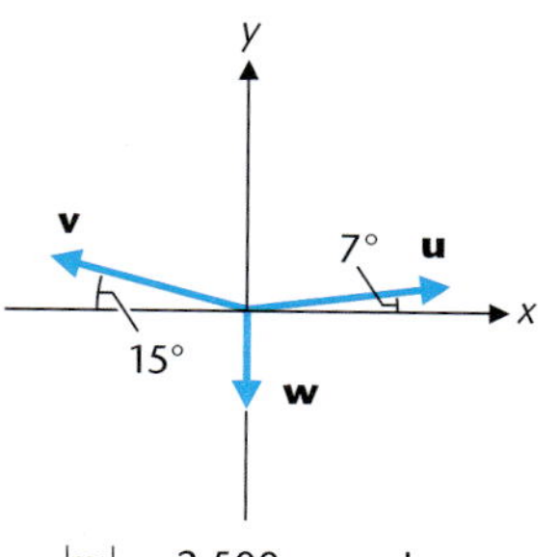

$|\mathbf{w}| = 2{,}500$ pounds

$$\mathbf{u} = |\mathbf{u}|(\cos 7°)\mathbf{i} + |\mathbf{u}|(\sin 7°)\mathbf{j}$$
$$\mathbf{v} = |\mathbf{v}|(-\cos 15°)\mathbf{i} + |\mathbf{v}|(\sin 15°)\mathbf{j}$$
$$\mathbf{w} = -2{,}500\mathbf{j}$$

Step 3. For the system to be in static equilibrium, the sum of the force vectors must be the zero vector. That is,

$$\mathbf{u} + \mathbf{v} + \mathbf{w} = \mathbf{0}$$

Replacing vectors **u**, **v**, and **w** from step 2, we obtain

$$[|\mathbf{u}|(\cos 7°)\mathbf{i} + |\mathbf{u}|(\sin 7°)\mathbf{j}] + [|\mathbf{v}|(-\cos 15°)\mathbf{i} + |\mathbf{v}|(\sin 15°)\mathbf{j}] - 2{,}500\mathbf{j} = 0\mathbf{i} + 0\mathbf{j}$$

which upon combining **i** and **j** vectors, becomes

$$[|\mathbf{u}|(\cos 7°) + |\mathbf{v}|(-\cos 15°)]\mathbf{i} + [|\mathbf{u}|(\sin 7°) + |\mathbf{v}|(\sin 15°) - 2{,}500]\mathbf{j} = 0\mathbf{i} + 0\mathbf{j}$$

Because two vectors are equal if and only if their corresponding components are equal, we are led to the following system of two equations in the two variables $|\mathbf{u}|$ and $|\mathbf{v}|$:

$$\begin{aligned}(\cos 7°)|\mathbf{u}| + (-\cos 15°)|\mathbf{v}| \qquad\quad &= 0\\ (\sin 7°)|\mathbf{u}| + \quad (\sin 15°)|\mathbf{v}| - 2{,}500 &= 0\end{aligned}$$

Solving this system by standard methods, we find that

$$|\mathbf{u}| = 6{,}400 \text{ pounds} \qquad \text{and} \qquad |\mathbf{v}| = 6{,}600 \text{ pounds}$$

Did you expect that the tension in each part of the cable is more than the weight hanging from the cable?

MATCHED 7 PROBLEM

Repeat Example 7 with 15° replaced with 13°, 7° replaced with 9°, and the 2,500 pounds replaced with 1,900 pounds.

ANSWERS MATCHED PROBLEMS

1. $P = (-4, 8)$ 2. $2\sqrt{5}$ 3. (A) $\langle -1, -3\rangle$ (B) $\langle 15, -9\rangle$ (C) $\langle -23, 21\rangle$ (D) $\langle -20, 12\rangle$
4. $\mathbf{u} = \langle 3/\sqrt{10}, 1/\sqrt{10}\rangle$ 5. (A) $5\mathbf{i} - 3\mathbf{j}$ (B) $-9\mathbf{i}$ (C) $6\mathbf{j}$
6. (A) $6\mathbf{i} + 4\mathbf{j}$ (B) $-2\mathbf{i} - 6\mathbf{j}$ (C) $-2\mathbf{i} - 13\mathbf{j}$ 7. $|\mathbf{u}| = 4{,}900$ pounds, $|\mathbf{v}| = 5{,}000$ pounds

EXERCISE 7.4

A

In Problems 1–6, represent each geometric vector $\overrightarrow{AB}$, with end-points as indicated, as an algebraic vector in the form $\langle a, b\rangle$.

1. $A = (3, -2), B = (0, -5)$
2. $A = (-1, 7), B = (1, -1)$
3. $A = (6, 0), B = (0, 7)$
4. $A = (0, -1), B = (-2, 0)$
5. $A = (0, 0), B = (3, 5)$
6. $A = (0, 0), B = (-2, -1)$

In Problems 7–12, find the magnitude of each vector.

7. $\langle 4, -3\rangle$
8. $\langle -3, 4\rangle$
9. $\langle 3, -5\rangle$
10. $\langle -5, -2\rangle$
11. $\langle -25, 0\rangle$
12. $\langle 0, -67\rangle$
13. Explain when two algebraic vectors are equal.
14. Explain when two geometric vectors are equal.

B

In Problems 15–18, find:

(A) $\mathbf{u} + \mathbf{v}$ **(B)** $\mathbf{u} - \mathbf{v}$ **(C)** $2\mathbf{u} - \mathbf{v} + 3\mathbf{w}$

15. $\mathbf{u} = \langle 2, 1\rangle, \mathbf{v} = \langle -1, 3\rangle, \mathbf{w} = \langle 3, 0\rangle$
16. $\mathbf{u} = \langle -1, 2\rangle, \mathbf{v} = \langle 3, -2\rangle, \mathbf{w} = \langle 0, -2\rangle$
17. $\mathbf{u} = \langle -4, -1\rangle, \mathbf{v} = \langle 2, 2\rangle, \mathbf{w} = \langle 0, 1\rangle$
18. $\mathbf{u} = \langle -3, 2\rangle, \mathbf{v} = \langle -2, 2\rangle, \mathbf{w} = \langle -3, 0\rangle$

In Problems 19–24, express $\mathbf{v}$ *in terms of the* $\mathbf{i}$ *and* $\mathbf{j}$ *unit vectors.*

19. $\mathbf{v} = \langle -3, 4\rangle$
20. $\mathbf{v} = \langle 2, -5\rangle$
21. $\mathbf{v} = \langle 3, 0\rangle$
22. $\mathbf{v} = \langle 0, -27\rangle$
23. $\mathbf{v} = \overrightarrow{AB}$, where $A = (2, 3)$ and $B = (-3, 1)$
24. $\mathbf{v} = \overrightarrow{AB}$, where $A = (-2, -1)$ and $B = (0, 2)$

In Problems 25–30, let $\mathbf{u} = 3\mathbf{i} - 2\mathbf{j}$, $\mathbf{v} = 2\mathbf{i} + 4\mathbf{j}$, *and* $\mathbf{w} = 2\mathbf{i}$, *and perform the indicated operations.*

25. $\mathbf{u} + \mathbf{v}$
26. $\mathbf{u} - \mathbf{v}$
27. $2\mathbf{u} - 3\mathbf{v}$
28. $3\mathbf{u} + 2\mathbf{v}$
29. $2\mathbf{u} - \mathbf{v} - 2\mathbf{w}$
30. $\mathbf{u} - 3\mathbf{v} + 2\mathbf{w}$

In Problems 31–34, find a unit vector $\mathbf{u}$ *with the same direction as* $\mathbf{v}$.

31. $\mathbf{v} = \langle -3, 4\rangle$
32. $\mathbf{v} = \langle 4, -3\rangle$
33. $\mathbf{v} = \langle -5, 3\rangle$
34. $\mathbf{v} = \langle 2, -3\rangle$
35. If exactly three nonzero force vectors with different directions are acting on an object at rest, how is any one of the force vectors related to the other two if the object is to remain at rest?
36. If exactly two nonzero force vectors are acting on an object at rest, what can you say about the vectors for the object to remain at rest?

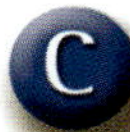

In Problems 37–44, let $\mathbf{u} = \langle a, b \rangle$, $\mathbf{v} = \langle c, d \rangle$, *and* $\mathbf{w} = \langle e, f \rangle$ *be vectors and m and n be scalars. Prove each of the following vector properties using appropriate properties of real numbers and the definitions of vector addition and scalar multiplication.*

37. $\mathbf{u} + (\mathbf{v} + \mathbf{w}) = (\mathbf{u} + \mathbf{v}) + \mathbf{w}$

38. $\mathbf{u} + \mathbf{v} = \mathbf{v} + \mathbf{u}$

39. $\mathbf{u} + \mathbf{0} = \mathbf{u}$

40. $\mathbf{u} + (-\mathbf{u}) = \mathbf{0}$

41. $(m + n)\mathbf{u} = m\mathbf{u} + n\mathbf{u}$

42. $m(\mathbf{u} + \mathbf{v}) = m\mathbf{u} + m\mathbf{v}$

43. $m(n\mathbf{u}) = (mn)\mathbf{u}$

44. $1\mathbf{u} = \mathbf{u}$

APPLICATIONS

In Problems 45–52, compute all answers to three significant digits.

45. Static Equilibrium. A unicyclist at a certain point on a tightrope deflects the rope as indicated in the figure. If the total weight of the cyclist and the unicycle is 155 pounds, how much tension is in each part of the cable?

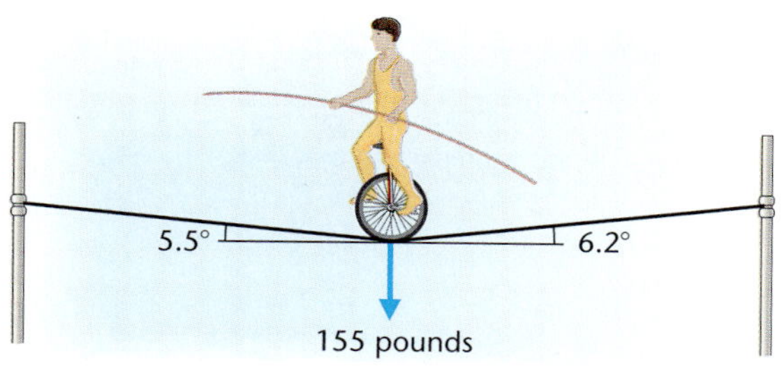

46. Static Equilibrium. Repeat Problem 45 with the left angle 4.2°, the right angle 5.3°, and the total weight 112 pounds.

47. Static Equilibrium. A weight of 1,000 pounds is suspended from two cables as shown in the figure. What is the tension in each cable?

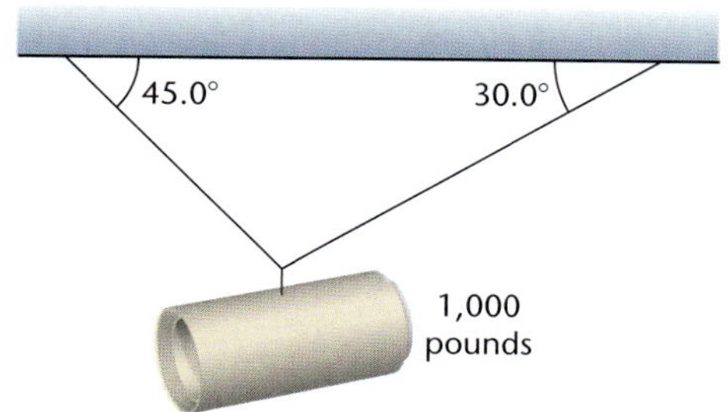

48. Static Equilibrium. A weight of 500 pounds is supported by two cables as illustrated. What is the tension in each cable?

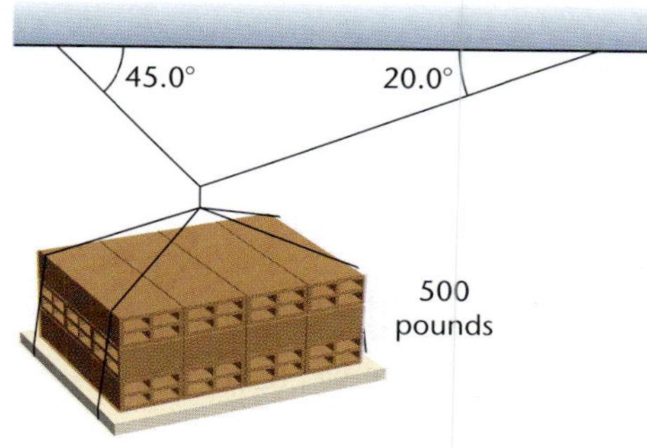

49. Static Equilibrium. A 400-pound sign is suspended as shown in figure (a) at the top of page 566. The corresponding force diagram (b) is formed by observing the following: Member AB is "pushing" at B and is under compression. This "pushing" force also can be thought of as the force vector $\mathbf{a}$ "pulling" to the right at B. The force vector $\mathbf{b}$ reflects the fact that member CB is under tension—that is, it is "pulling" at B. The force vector $\mathbf{c}$ corresponds to the weight of the sign "pulling" down at B. Find the magnitudes of the forces in the rigid supporting members; that is, find $|\mathbf{a}|$ and $|\mathbf{b}|$ in the force diagram (b).

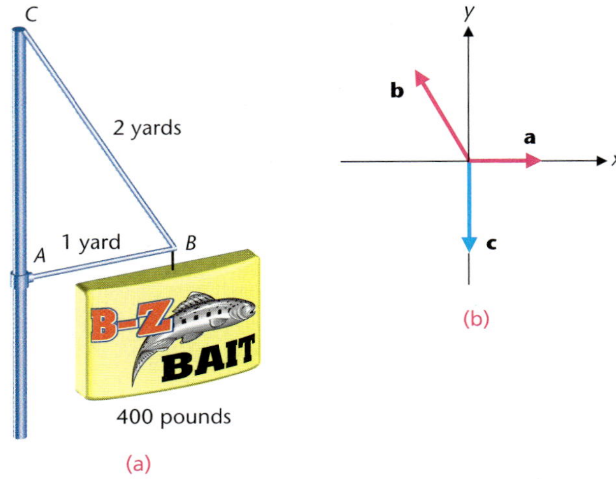

50. Static Equilibrium. A weight of 1,000 kilograms is supported as shown in the figure. What are the magnitudes of the forces on the members AB and BC?

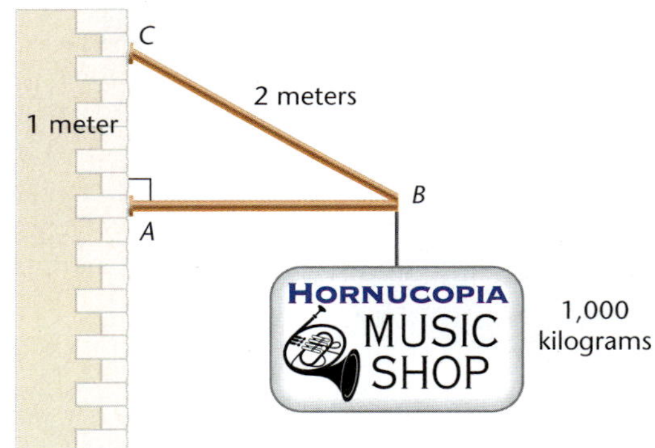

51. Static Equilibrium. A 1,250-pound weight is hanging from a hoist as indicated in the figure. What are the magnitudes of the forces on the members AB and BC?

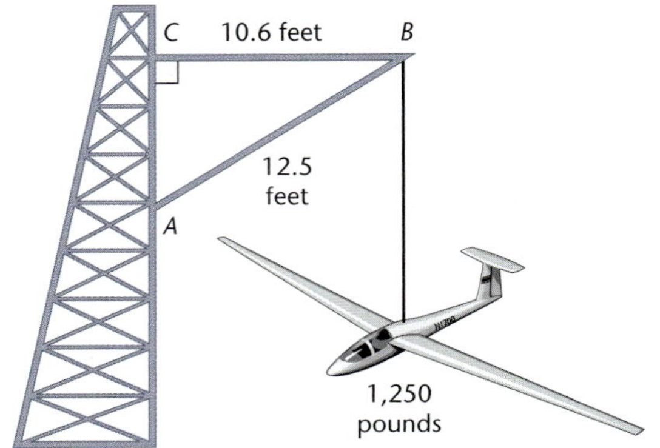

52. Static Equilibrium. A weight of 5,000 kilograms is supported as shown in the figure. What are the magnitudes of the forces on the members AB and BC?

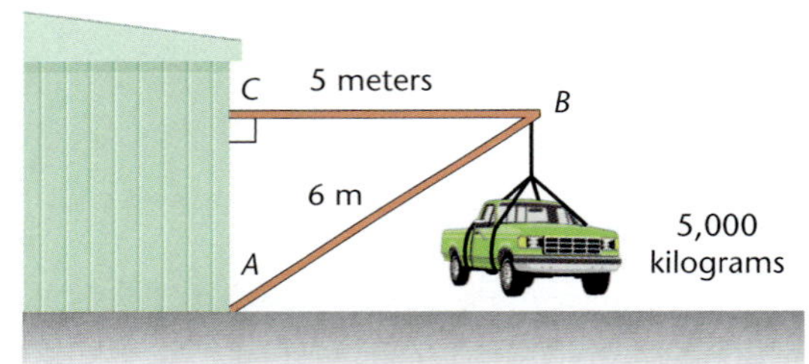

SECTION 7.5 Polar Coordinates and Graphs

Polar Coordinate System • Converting from Polar to Rectangular Form, and Vice Versa • Graphing Polar Equations • Some Standard Polar Curves • Application

Up until now we have used only the rectangular coordinate system. Other coordinate systems have particular advantages in certain situations. Of the many that are possible, the *polar coordinate system* ranks second in importance to the rectangular coordinate system and is the subject matter of Section 7.5.

FIGURE 1 Polar coordinate system.

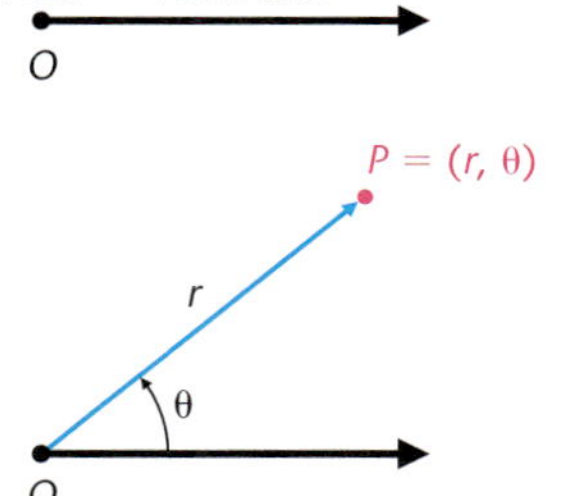

Polar Coordinate System

To form a **polar coordinate system** in a plane (Fig. 1), start with a fixed point O and call it the **pole,** or **origin.** From this point draw a half-line, or ray (usually horizontal and to the right), and call this line the **polar axis.**

If P is an arbitrary point in a plane, then associate polar coordinates (r, θ) with it as follows: Starting with the polar axis as the initial side of an angle, rotate the terminal side until it, or the extension of it through the pole, passes through the point. The θ coordinate in (r, θ) is this angle, in degree or radian measure. The angle θ is positive if the rotation is counterclockwise and negative if the rotation is clockwise. The r coordinate in (r, θ) is the directed distance from the pole to the point P. It is positive if measured from the pole along the terminal side of θ and negative if measured along the terminal side extended through the pole.

Figure 2 illustrates a point P with three different sets of polar coordinates. Study this figure carefully. The pole has polar coordinates $(0, \theta)$ for arbitrary θ. For example, $(0, 0°)$, $(0, \pi/3)$, and $(0, -371°)$ are all coordinates of the pole.

FIGURE 2 Polar coordinates of a point.

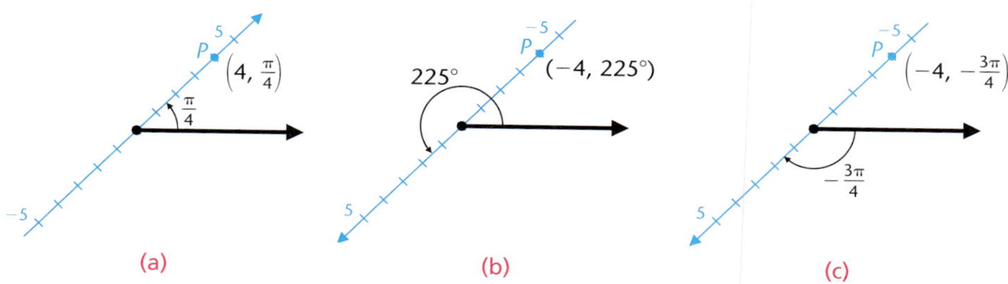

We now see a distinct difference between rectangular and polar coordinates for the given point. For a given point in a rectangular coordinate system, there exists exactly one set of rectangular coordinates. On the other hand, in a polar coordinate system, a point has infinitely many sets of polar coordinates.

Just as graph paper with a rectangular grid is readily available for plotting rectangular coordinates, polar graph paper is available for plotting polar coordinates.

EXAMPLE 1 Plotting Points in a Polar Coordinate System

Plot the following points in a polar coordinate system:

(A) $A = (3, 30°)$, $B = (-8, 180°)$, $C = (5, -135°)$, $D = (-10, -45°)$

(B) $A = (5, \pi/3)$, $B = (-6, 5\pi/6)$, $C = (7, -\pi/2)$, $D = (-4, -\pi/6)$

SOLUTIONS

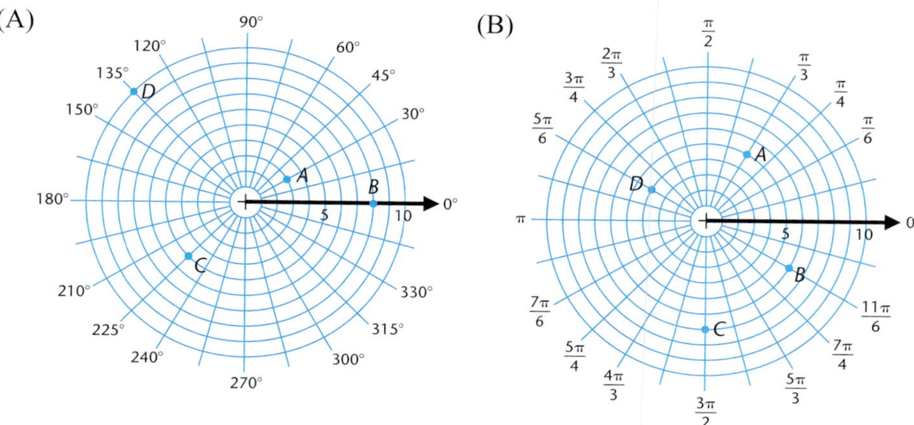

MATCHED 1 PROBLEM

Plot the following points in a polar coordinate system:

(A) $A = (8, 45°)$, $B = (-5, 150°)$, $C = (4, -210°)$, $D = (-6, -90°)$

(B) $A = (9, \pi/6)$, $B = (-3, -\pi)$, $C = (-7, 7\pi/4)$, $D = (5, -5\pi/6)$

EXPLORE/DISCUSS 1

A point in a polar coordinate system has coordinates $(5, 30°)$. How many other polar coordinates does the point have for θ restricted to $-360° \le \theta \le 360°$? Find the other coordinates of the point and explain how they are found.

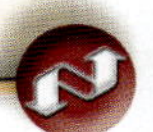

Converting from Polar to Rectangular Form, and Vice Versa

Often, it is necessary to transform coordinates or equations in rectangular form into polar form, or vice versa. The following polar–rectangular relationships are useful in this regard:

Polar–Rectangular Relationships

We have the following relationships between rectangular coordinates (x, y) and polar coordinates (r, θ):

$$r^2 = x^2 + y^2$$

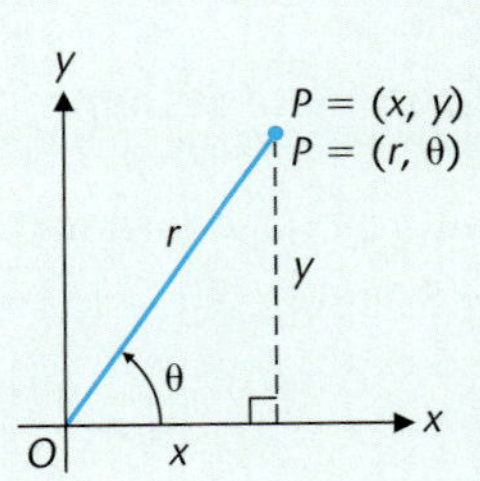

$$\sin\theta = \frac{y}{r} \quad \text{or} \quad y = r\sin\theta$$

$$\cos\theta = \frac{x}{r} \quad \text{or} \quad x = r\cos\theta$$

$$\tan\theta = \frac{y}{x}$$

[*Note:* The signs of x and y determine the quadrant for θ. The angle θ is chosen so that $-\pi < \theta \le \pi$ or $-180° < \theta \le 180°$, unless directed otherwise.]

Many calculators can automatically convert rectangular coordinates to polar form, and vice versa. (Read the manual for your particular calculator.) Example 2 illustrates calculator conversions in both directions.

EXAMPLE 2 **Converting from Polar to Rectangular Form, and Vice Versa**

(A) Convert the polar coordinates $(-4, 1.077)$ to rectangular coordinates to three decimal places.

(B) Convert the rectangular coordinates $(-3.207, -5.719)$ to polar coordinates with θ in degree measure, $-180° < \theta \leq 180°$ and $r \geq 0$.

SOLUTIONS

(A) Use a calculator set in radian mode.

$$(r, \theta) = (-4, 1.077)$$
$$x = r\cos\theta = (-4)\cos 1.077 = -1.896$$
$$y = r\sin\theta = (-4)\sin 1.077 = -3.522$$

Rectangular coordinates are $(-1.896, -3.522)$.

Figure 3 shows the same conversion done in a graphing calculator with a built-in conversion routine.

FIGURE 3

```
P▸Rx(-4,1.077)
          -1.895888451
P▸Ry(-4,1.077)
           -3.52215942
```

(B) Use a calculator set in degree mode.

$$(x, y) = (-3.207, -5.719)$$
$$r = \sqrt{x^2 + y^2} = \sqrt{(-3.207)^2 + (-5.719)^2} = 6.557$$
$$\tan\theta = \frac{y}{x} = \frac{-5.719}{-3.207}$$

θ is a third-quadrant angle and is to be chosen so that $-180° < \theta \leq 180°$.

$$\theta = -180° + \tan^{-1}\frac{-5.719}{-3.207} = -119.28°$$

Polar coordinates are $(6.557, -119.28°)$.

Figure 4 shows the same conversion done in a graphing calculator with a built-in conversion routine.

FIGURE 4

```
R▸Pr(-3.207,-5.7
19)
           6.556814013
R▸Pθ(-3.207,-5.7
19)
          -119.2820682
```

MATCHED 2 PROBLEM

(A) Convert the polar coordinates $(8.677, -1.385)$ to rectangular coordinates to three decimal places.

(B) Convert the rectangular coordinates $(-6.434, 4.023)$ to polar coordinates with θ in degree measure, $-180° < \theta \leq 180°$ and $r \geq 0$.

Generally, a more important use of the polar–rectangular relationships is in the conversion of equations in rectangular form to polar form, and vice versa.

EXAMPLE 3 Converting an Equation from Rectangular Form to Polar Form

Change $x^2 + y^2 - 4y = 0$ to polar form.

SOLUTION

Use $r^2 = x^2 + y^2$ and $y = r\sin\theta$.

$$x^2 + y^2 - 4y = 0$$
$$r^2 - 4r\sin\theta = 0$$
$$r(r - 4\sin\theta) = 0$$
$$r = 0 \quad \text{or} \quad r - 4\sin\theta = 0$$

The graph of $r = 0$ is the pole. Because the pole is included in the graph of $r - 4 \sin \theta = 0$ (let $\theta = 0$), we can discard $r = 0$ and keep only

$$r - 4 \sin \theta = 0$$

or

$$r = 4 \sin \theta \qquad \text{The polar form of } x^2 + y^2 - 4y = 0$$

MATCHED 3 PROBLEM

Change $x^2 + y^2 - 6x = 0$ to polar form.

EXAMPLE 4 Converting an Equation from Polar Form to Rectangular Form

Change $r = -3 \cos \theta$ to rectangular form.

SOLUTION

The transformation of this equation as it stands into rectangular form is fairly difficult. With a little trick, however, it becomes easy. We multiply both sides by r, which simply adds the pole to the graph. But the pole is already part of the graph of $r = -3 \cos \theta$ (let $\theta = \pi/2$), so we haven't actually changed anything.

$$\begin{aligned} r &= -3 \cos \theta \\ r^2 &= -3r \cos \theta && \text{Multiply both sides by } r. \\ x^2 + y^2 &= -3x && r^2 = x^2 + y^2 \text{ and } r \cos \theta = x \\ x^2 + y^2 + 3x &= 0 \end{aligned}$$

MATCHED 4 PROBLEM

Change $r + 2 \sin \theta = 0$ to rectangular form.

Graphing Polar Equations

We now turn to graphing polar equations. The **graph** of a polar equation, such as $r = 3\theta$ or $r = 6 \cos \theta$, in a polar coordinate system is the set of all points having coordinates that satisfy the polar equation. Certain curves have simpler representations in polar coordinates, and other curves have simpler representations in rectangular coordinates.

To establish fundamentals in graphing polar equations, we start with a point-by-point graph. We then consider a more rapid way of making rough sketches of certain polar curves. And, finally, we show how polar curves are graphed in a graphing utility.

To plot a polar equation using **point-by-point plotting,** just as in rectangular coordinates, make a table of values that satisfy the equation, plot these points, then join them with a smooth curve. Example 5 illustrates the process.

EXAMPLE 5 Point-by-Point Plotting

(A) Graph $r = 8 \cos \theta$ with θ in radians.

(B) Convert the polar equation in part A to rectangular form, and identify the graph.

SOLUTIONS

(A) We construct a table using multiples of $\pi/6$, plot these points, then join the points with a smooth curve (Fig. 5).

θ	r
0	8.0
$\pi/6$	6.9
$\pi/3$	4.0
$\pi/2$	0.0
$2\pi/3$	−4.0
$5\pi/6$	−6.9
π	8.0
Graph repeats	

FIGURE 5

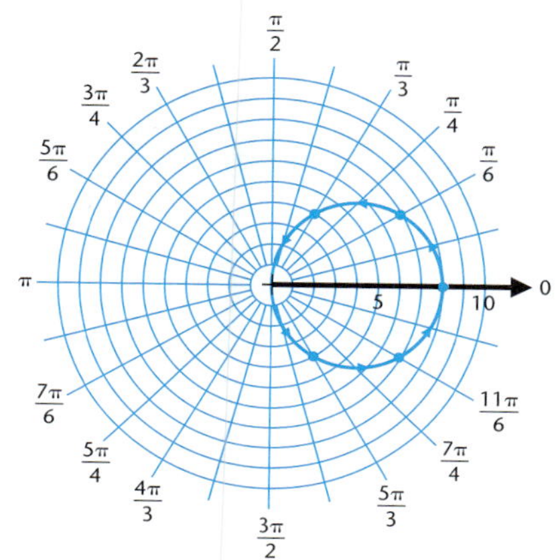

(B)

$$
\begin{aligned}
r &= 8 \cos \theta \\
r^2 &= 8r \cos \theta && \text{Multiply both sides by } r. \\
x^2 + y^2 &= 8x && \text{Change to rectangular form.} \\
x^2 - 8x + y^2 &= 0 \\
x^2 - 8x + 16 + y^2 &= 16 && \text{Complete the square on the left side.} \\
(x - 4)^2 + y^2 &= 4^2 && \text{Standard equation of a circle.}
\end{aligned}
$$

The graph in part A is a circle with center at (4, 0) and radius 4 (see Appendix A, Section A.3).

MATCHED 5 PROBLEM

(A) Graph $r = 8 \sin \theta$ with θ in degrees.

(B) Convert the polar equation in part A to rectangular form, and identify the graph.

If only a rough sketch of a polar equation involving $\sin \theta$ or $\cos \theta$ is desired, you can speed up the point-by-point graphing process by taking advantage of the uniform variation of $\sin \theta$ and $\cos \theta$ as θ moves around a unit circle. This process is referred to as **rapid polar sketching.** It is convenient to visualize Figure 6 in the process. With a little practice most of the table work in rapid sketching can be done mentally and a rough sketch can be made directly from the equation.

FIGURE 6

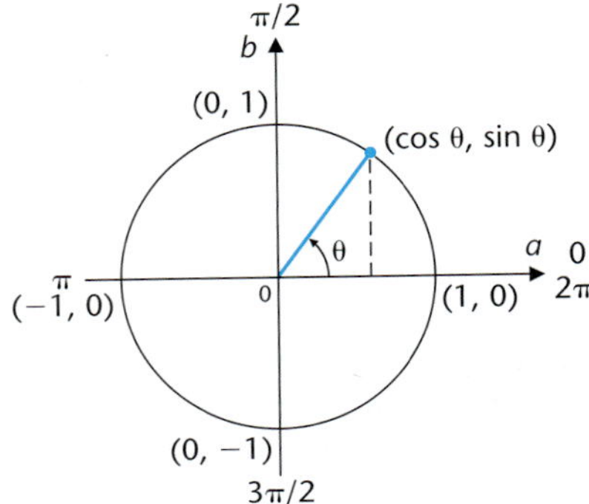

EXAMPLE 6 Rapid Polar Sketching

Sketch $r = 4 + 4 \cos \theta$ using rapid sketching techniques with θ in radians.

SOLUTION

We set up a table that indicates how r varies as we let θ vary through each set of quadrant values:

θ Varies from	$\cos \theta$ Varies from	$4 \cos \theta$ Varies from	$r = 4 + 4 \cos \theta$ Varies from
0 to $\pi/2$	1 to 0	4 to 0	8 to 4
$\pi/2$ to π	0 to -1	0 to -4	4 to 0
π to $3\pi/2$	-1 to 0	-4 to 0	0 to 4
$3\pi/2$ to 2π	0 to 1	0 to 4	4 to 8

Notice that as θ increases from 0 to $\pi/2$, $\cos \theta$ decreases from 1 to 0, $4 \cos \theta$ decreases from 4 to 0, and $r = 4 + 4 \cos \theta$ decreases from 8 to 4, and so on. Sketching these values, we obtain the graph in Figure 7, called a **cardioid.**

FIGURE 7

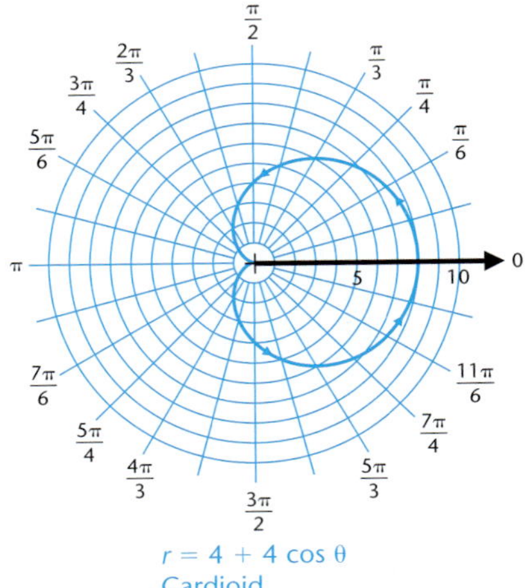

MATCHED 6 PROBLEM

Sketch $r = 5 + 5 \sin \theta$ using rapid sketching techniques with θ in radians.

EXAMPLE 7 Rapid Polar Sketching

Sketch $r = 8 \cos 2\theta$ with θ in radians.

SOLUTION

Start by letting 2θ (instead of θ) range through each set of quadrant values. That is, start with values for 2θ in the second column of the table, fill in the table to the right, and then fill in the first column for θ.

Start with the second column

θ Varies from	2θ Varies from	$\cos 2\theta$ Varies from	$r = 8 \cos 2\theta$ Varies from
0 to $\pi/4$	0 to $\pi/2$	1 to 0	8 to 0
$\pi/4$ to $\pi/2$	$\pi/2$ to π	0 to -1	0 to -8
$\pi/2$ to $3\pi/4$	π to $3\pi/2$	-1 to 0	-8 to 0
$3\pi/4$ to π	$3\pi/2$ to 2π	0 to 1	0 to 8
π to $5\pi/4$	2π to $5\pi/2$	1 to 0	8 to 0
$5\pi/4$ to $3\pi/2$	$5\pi/2$ to 3π	0 to -1	0 to -8
$3\pi/2$ to $7\pi/4$	3π to $7\pi/2$	-1 to 0	-8 to 0
$7\pi/4$ to 2π	$7\pi/2$ to 4π	0 to 1	0 to 8

As 2θ increases from 0 to $\pi/2$, θ increases from 0 to $\pi/4$, and r decreases from 8 to 0. As 2θ increases from $\pi/2$ to π, θ increases from $\pi/4$ to $\pi/2$, and r decreases from 0 to -8, and so on. Continue until the graph starts to repeat. Plotting the values, we obtain the graph in Figure 8, called a **four-leafed rose:**

FIGURE 8

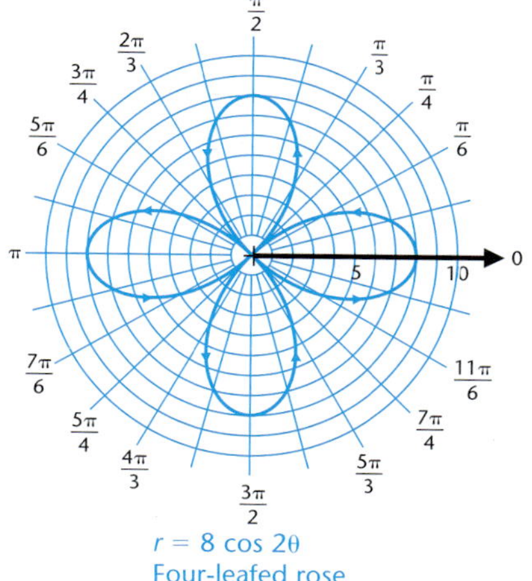

$r = 8 \cos 2\theta$
Four-leafed rose

MATCHED 7 PROBLEM

Sketch $r = 6 \sin 2\theta$ with θ in radians.

We now turn to **graphing polar equations in a graphing utility.** Example 8 illustrates the process.

EXAMPLE 8 Graphing in a Graphing Utility

Graph each of the following polar equations in a graphing utility (parts B and C are from Examples 6 and 7).

(A) $r = 3\theta,\ 0 \le \theta \le 3\pi/2$ (Archimedes' spiral)

(B) $r = 4 + 4 \cos \theta$ (cardioid)

(C) $r = 8 \cos 2\theta$ (four-leafed rose)

SOLUTION

Set the graphing utility in polar mode and select polar coordinates and radian measure. Adjust window values to accommodate the whole graph. A squared graph is often desirable in showing the true shape of the curve, and is used here. Many graphing utilities, including the one used here, do not show a polar grid. When using TRACE, many graphing utilities offer a choice between polar coordinates and rectangular coordinates for points on the polar curve. The graphs of the polar equations above are shown in Figure 9.

FIGURE 9

(A) $r = 3\theta,\ 0 \le \theta \le 3\pi/2$

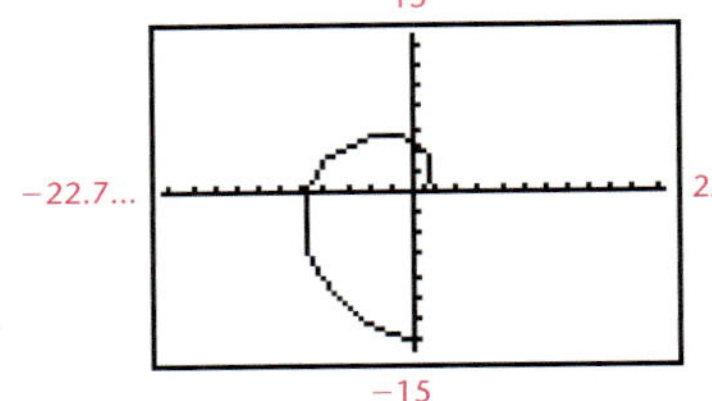

(B) $r = 4 + 4 \cos \theta$

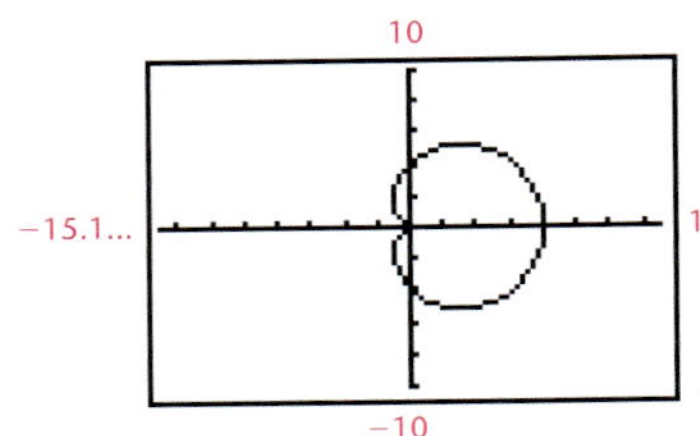

(C) $r = 8 \cos 2\theta$

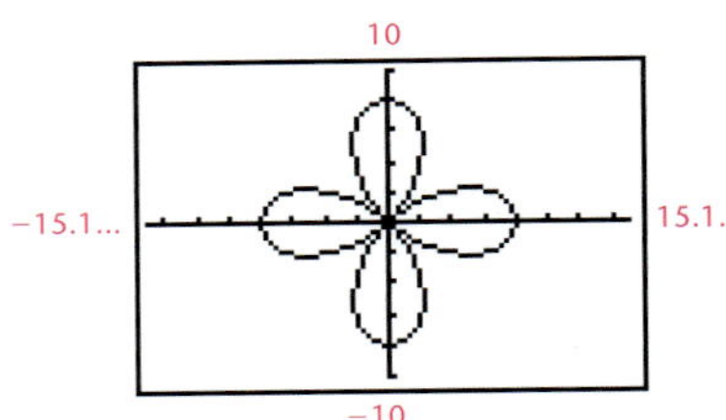

MATCHED 8 PROBLEM

Graph each of the following polar equations in a graphing utility.

(A) $r = 2\theta,\ 0 \le \theta \le 2\pi$

(B) $r = 5 + 5 \sin \theta$

(C) $r = 6 \sin 2\theta$

EXPLORE/DISCUSS 2

(A) Graph $r_1 = 10 \sin \theta$ and $r_2 = 10 \cos \theta$ in the same viewing window. Use TRACE on r_1 and estimate the polar coordinates where the two graphs intersect. Do the same thing for r_2. Which intersection point appears to have the same polar coordinates on each curve and consequently represents a simultaneous solution to both equations? Which intersection point appears to have different polar coordinates on each curve and consequently does not represent a simultaneous solution? Solve the system for r and θ.

(B) Explain how rectangular coordinate systems differ from polar coordinate systems relative to intersection points and simultaneous solutions of systems of equations in the respective systems.

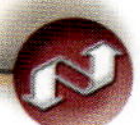

Some Standard Polar Curves

In a rectangular coordinate system the simplest types of equations to graph are found by setting the rectangular variables x and y equal to constants:

$$x = a \qquad \text{and} \qquad y = b$$

The graphs are straight lines: The graph of $x = a$ is a vertical line, and the graph of $y = b$ is a horizontal line. A glance at Table 1 shows that horizontal and vertical lines do not have simple equations in polar coordinates.

TABLE 1 Standard Polar Graphs

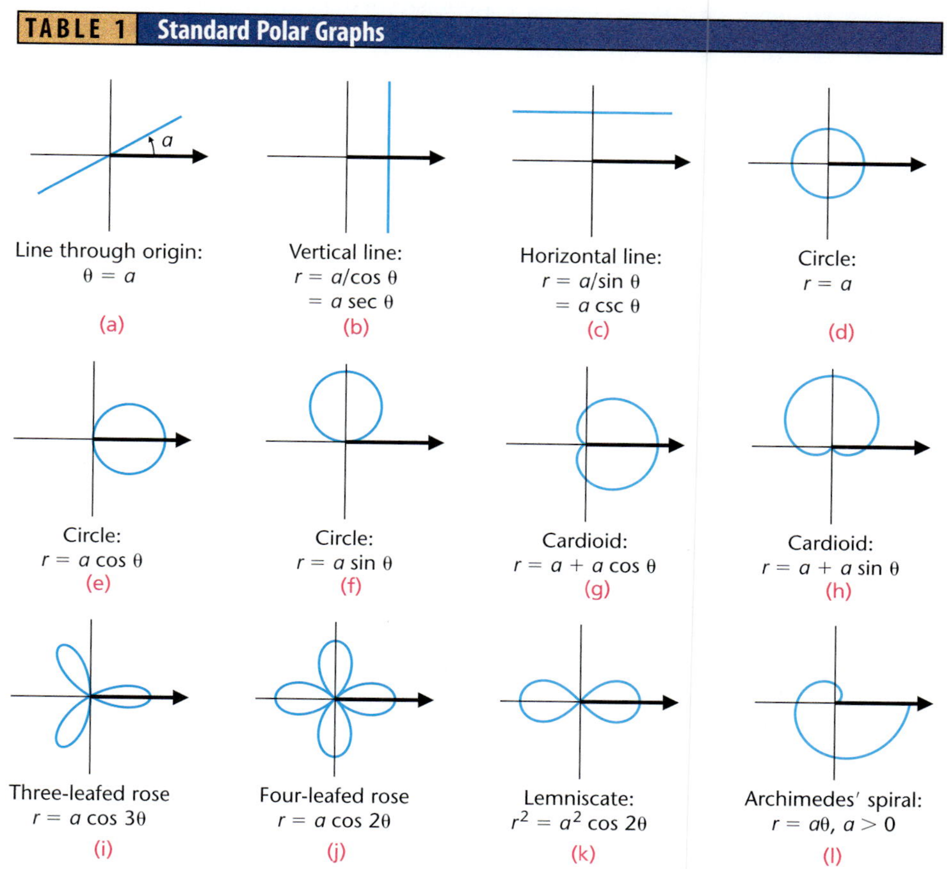

Two of the simplest types of polar equations to graph in a polar coordinate system are found by setting the polar variables r and θ equal to constants:

$$r = a \qquad \text{and} \qquad \theta = b$$

Figure 10 illustrates the graphs of $\theta = \pi/4$ and $r = 5$.

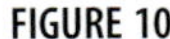
FIGURE 10

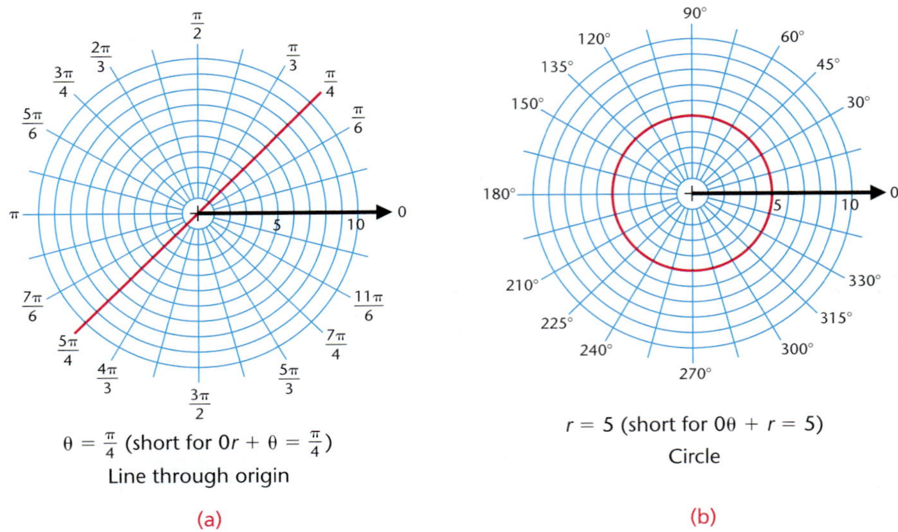

Table 1 illustrates a number of standard polar graphs and their equations. Polar graphing is often made easier if you have some idea of the final form.

Application

Serious sailboat racers make polar plots of boat speeds at various angles to the wind with various sail combinations at different wind speeds. With many polar plots for different sizes and types of sails at different wind speeds, they are able to accurately choose a sail for the optimum performance for different points of sail relative to any given wind strength. Figure 11 illustrates one such polar plot, where the maximum speed appears to be about 7.5 knots at 105° off the wind (with spinnaker sail set).

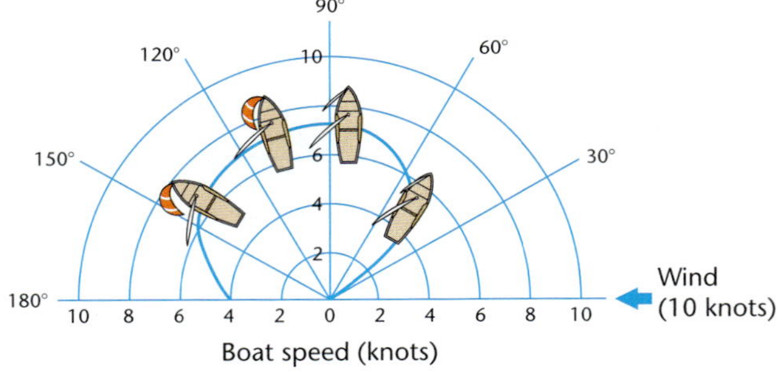

FIGURE 11 Polar diagram showing optimum sailing speed at different sailing angles to the wind.

ANSWERS MATCHED PROBLEMS

1. (A)

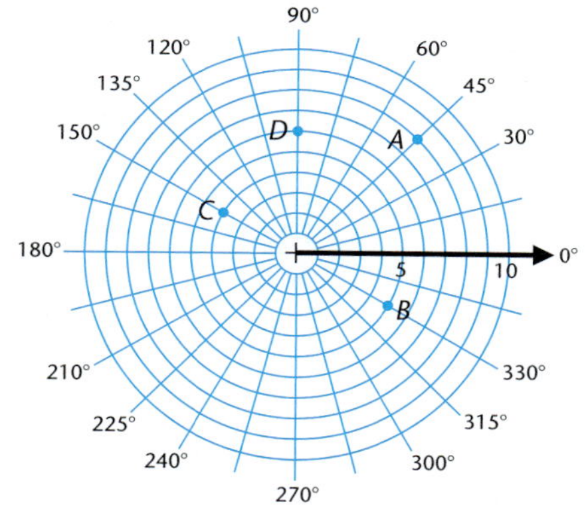

(B)

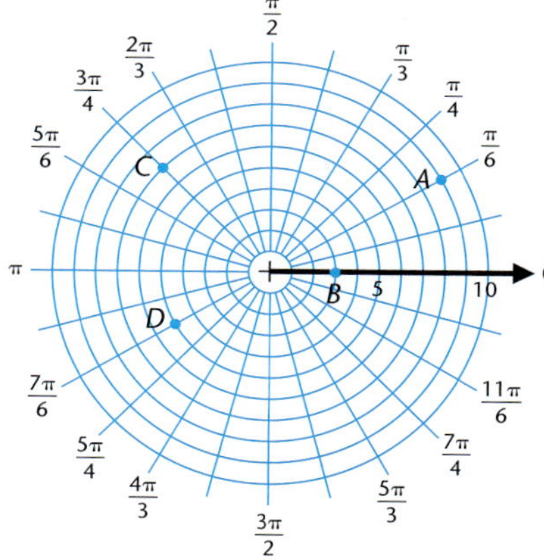

2. (A) (1.603, −8.528) (B) (7.588, 147.98°) **3.** 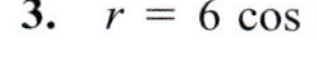 $r = 6 \cos \theta$ **4.** $x^2 + y^2 + 2y = 0$

5. (A)

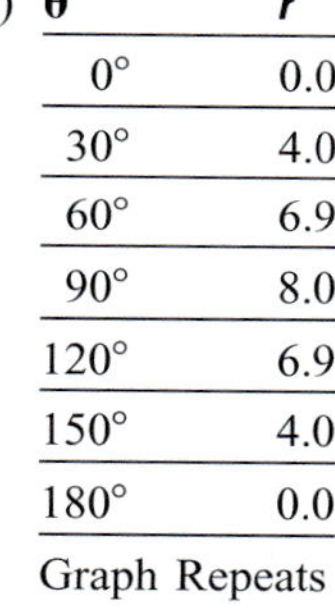

θ	r
0°	0.0
30°	4.0
60°	6.9
90°	8.0
120°	6.9
150°	4.0
180°	0.0
Graph Repeats	

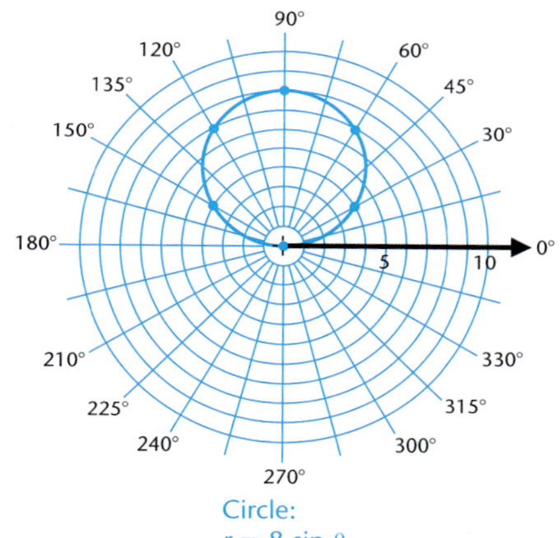

Circle:
$r = 8 \sin \theta$

(B) $x^2 + (y - 4)^2 = 4^2$, a circle with center at (0, 4) and radius 4

6. $r = 5 + 5 \sin \theta$, cardioid

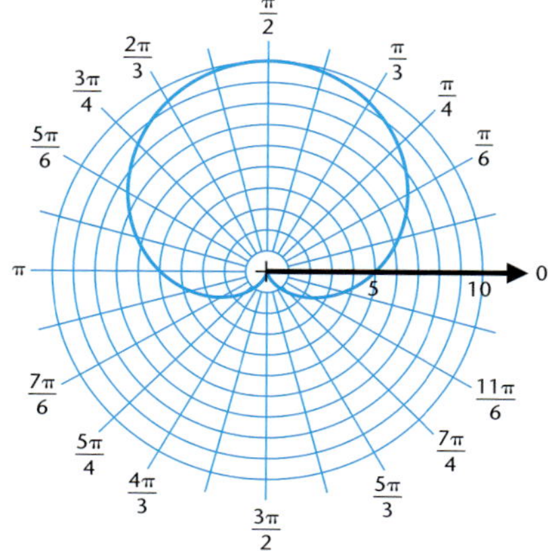

7. $r = 6 \sin 2\theta$, four-leafed rose

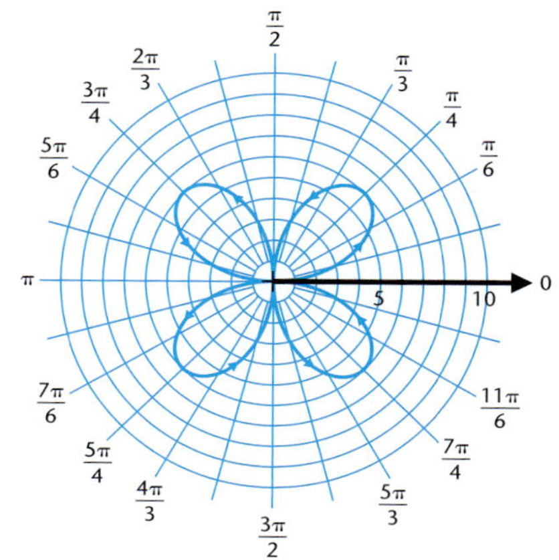

8. (A) $r = 2\theta$, $0 \le \theta \le 2\pi$

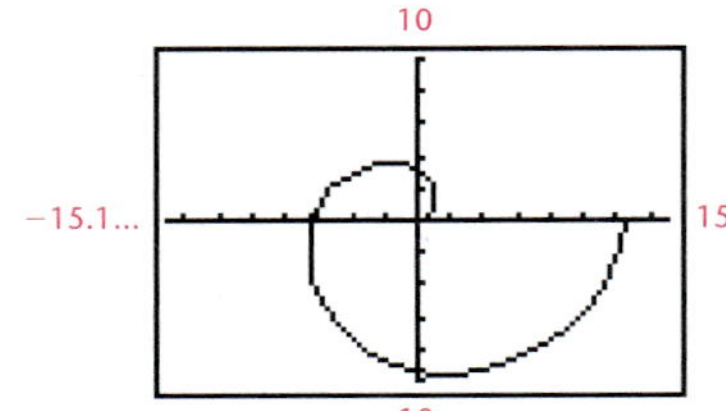

(B) $r = 5 + 5\sin\theta$

10
−15.1...
15.1...
−10

(C) $r = 6\sin 2\theta$

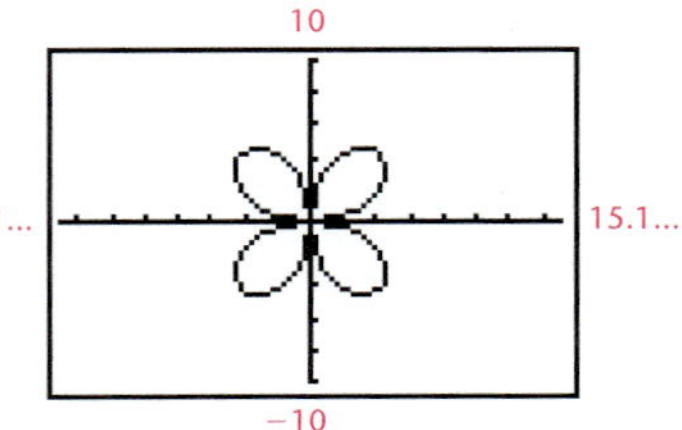

EXERCISE 7.5

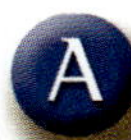

Plot A, B, and C in Problems 1–8 in a polar coordinate system.

1. $A = (4, 0°)$, $B = (7, 180°)$, $C = (9, 45°)$

2. $A = (8, 0°)$, $B = (5, 90°)$, $C = (6, 30°)$

3. $A = (-4, 0°)$, $B = (-7, 180°)$, $C = (-9, 45°)$

4. $A = (-8, 0°)$, $B = (-5, 90°)$, $C = (-6, 30°)$

5. $A = (8, -\pi/3)$, $B = (4, -\pi/4)$, $C = (10, -\pi/6)$

6. $A = (6, -\pi/6)$, $B = (5, -\pi/2)$, $C = (8, -\pi/4)$

7. $A = (-6, -\pi/6)$, $B = (-5, -\pi/2)$, $C = (-8, -\pi/4)$

8. $A = (-6, -\pi/2)$, $B = (-5, -\pi/3)$, $C = (-8, -\pi/4)$

9. A point in a polar coordinate system has coordinates $(-5, 3\pi/4)$. Find all other polar coordinates for the point, $-2\pi \le \theta \le 2\pi$, and verbally describe how the coordinates are associated with the point.

10. A point in a polar coordinate system has coordinates $(6, -30°)$. Find all other polar coordinates for the point, $-360° \le \theta \le 360°$, and verbally describe how the coordinates are associated with the point.

Graph Problems 11 and 12 in a polar coordinate system using point-by-point plotting and the special values 0, $\pi/6$, $\pi/4$, $\pi/3$, $\pi/2$, $2\pi/3$, $3\pi/4$, $5\pi/6$, and π for θ.

11. $r = 10\sin\theta$ **12.** $r = 10\cos\theta$

Verify the graphs of Problems 11 and 12 on a graphing utility.

Graph Problems 13–16 in a polar coordinate system.

13. $r = 8$ **14.** $r = 5$

15. $\theta = \pi/3$ **16.** $\theta = \pi/6$

In Problems 17–22, convert the polar coordinates to rectangular coordinates to three decimal places.

17. $(6, \pi/6)$ **18.** $(7, 2\pi/3)$

19. $(-2, 7\pi/8)$ **20.** $(3, -3\pi/7)$

21. $(-4.233, -2.084)$ **22.** $(-9.028, -0.663)$

B

In Problems 23–28, convert the rectangular coordinates to polar coordinates with θ in degree measure, $-180° < \theta \le 180°$, and $r \ge 0$.

23. $(3.5, 7.1)$ **24.** $(6.9, 4.7)$

25. $(22, -14)$ **26.** $(16, -27)$

27. $(-7.33, -2.04)$ **28.** $(-8.33, 4.29)$

In Problems 29–38, use rapid graphing techniques to sketch the graph of each polar equation. Check by graphing on a graphing utility.

29. $r = 4\sin\theta$ **30.** $r = 4\cos\theta$

31. $r = 10\sin 2\theta$ **32.** $r = 8\cos 2\theta$

33. $r = 5 \cos 3\theta$

34. $r = 6 \sin 3\theta$

35. $r = 2 + 2 \sin \theta$

36. $r = 3 + 3 \cos \theta$

37. $r = 2 + 4 \sin \theta$

38. $r = 2 + 4 \cos \theta$

Problems 39–44 are exploratory problems requiring the use of a graphing utility.

39. Graph each polar equation in its own viewing window: $r = 2 + 2 \sin \theta$, $r = 4 + 2 \sin \theta$, $r = 2 + 4 \sin \theta$.

40. Graph each polar equation in its own viewing window: $r = 2 + 2 \cos \theta$, $r = 4 + 2 \cos \theta$, $r = 2 + 4 \cos \theta$.

41. **(A)** Graph each polar equation in its own viewing window: $r = 4 \sin \theta$, $r = 4 \sin 3\theta$, $r = 4 \sin 5\theta$.

(B) What would you guess to be the number of leaves for $r = 4 \sin 7\theta$?

(C) What would you guess to be the number of leaves for $r = a \sin n\theta$, $a > 0$ and n odd?

42. **(A)** Graph each polar equation in its own viewing window: $r = 4 \cos \theta$, $r = 4 \cos 3\theta$, $r = 4 \cos 5\theta$.

(B) What would you guess to be the number of leaves for $r = 4 \cos 7\theta$?

(C) What would you guess to be the number of leaves for $r = a \cos n\theta$, $a > 0$ and n odd?

43. **(A)** Graph each polar equation in its own viewing window: $r = 4 \sin 2\theta$, $r = 4 \sin 4\theta$, $r = 4 \sin 6\theta$.

(B) What would you guess to be the number of leaves for $r = 4 \sin 8\theta$?

(C) What would you guess to be the number of leaves for $r = a \sin n\theta$, $a > 0$ and n even?

44. **(A)** Graph each polar equation in its own viewing window: $r = 4 \cos 2\theta$, $r = 4 \cos 4\theta$, $r = 4 \cos 6\theta$.

(B) What would you guess to be the number of leaves for $r = 4 \cos 8\theta$?

(C) What would you guess to be the number of leaves for $r = a \cos n\theta$, $a > 0$ and n even?

In Problems 45–50, change each rectangular equation to polar form.

45. $y^2 = 5y - x^2$

46. $6x - x^2 = y^2$

47. $y = x$

48. $x^2 + y^2 = 9$

49. $y^2 = 4x$

50. $2xy = 1$

In Problems 51–56, change each polar equation to rectangular form.

51. $r(3 \cos \theta - 4 \sin \theta) = -1$

52. $r(2 \cos \theta + \sin \theta) = 4$

53. $r = -2 \sin \theta$

54. $r = 8 \cos \theta$

55. $\theta = \pi/4$

56. $r = 4$

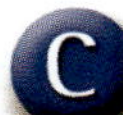

Problems 57 and 58 are exploratory problems requiring the use of a graphing utility.

57. Graph $r = 1 + 2 \sin (n\theta)$ for various values of n, n a natural number. Describe how n is related to the number of large petals and the number of small petals on the graph and how the large and small petals are related to each other relative to n.

58. Graph $r = 1 + 2 \cos (n\theta)$ for various values of n, n a natural number. Describe how n is related to the number of large petals and the number of small petals on the graph and how the large and small petals are related to each other relative to n.

In Problems 59–62 graph each system of equations on the same set of polar coordinate axes. Then solve the system simultaneously. [Note: Any solution (r_1, θ_1) to the system must satisfy each equation in the system and thus identifies a point of intersection of the two graphs. However, there may be other points of intersection of the two graphs that do not have any coordinates that satisfy both equations. This represents a major difference between the rectangular coordinate system and the polar coordinate system.]

59. $r = 4 \cos \theta$
$r = -4 \sin \theta$
$0 \le \theta \le \pi$

60. $r = 2 \cos \theta$
$r = 2 \sin \theta$
$0 \le \theta \le \pi$

61. $r = 6 \cos \theta$
$r = 6 \sin 2\theta$
$0° \le \theta \le 360°$

62. $r = 8 \sin \theta$
$r = 8 \cos 2\theta$
$0° \le \theta \le 360°$

APPLICATIONS

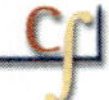

63. Analytic Geometry. A distance formula for the distance between two points in a polar coordinate system follows directly from the law of cosines:

$$d^2 = r_1^2 + r_2^2 - 2r_1r_2\cos(\theta_2 - \theta_1)$$
$$d = \sqrt{r_1^2 + r_2^2 - 2r_1r_2\cos(\theta_2 - \theta_1)}$$

Find the distance (to three decimal places) between the two points $P_1 = (4, \pi/4)$ and $P_2 = (1, \pi/2)$.

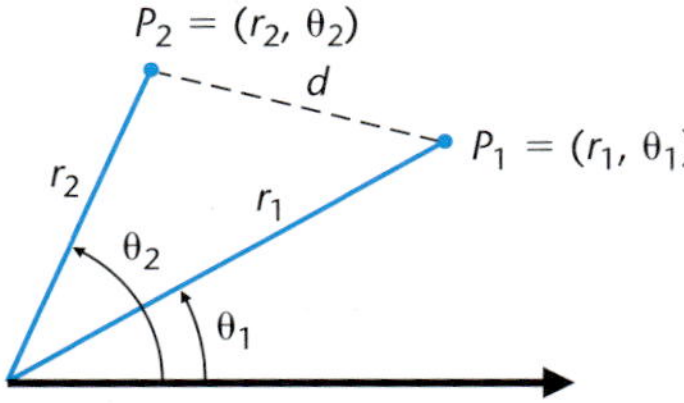

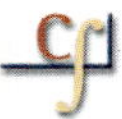

64. Analytic Geometry. Refer to Problem 63. Find the distance (to three decimal places) between the two points $P_1 = (2, 30°)$ and $P_2 = (3, 60°)$.

Problems 65–66 refer to the polar diagram in the figure. Polar diagrams of this type are used extensively by serious sailboat racers, and this polar diagram represents speeds in knots of a high-performance sailboat sailing at various angles to a wind blowing at 20 knots.

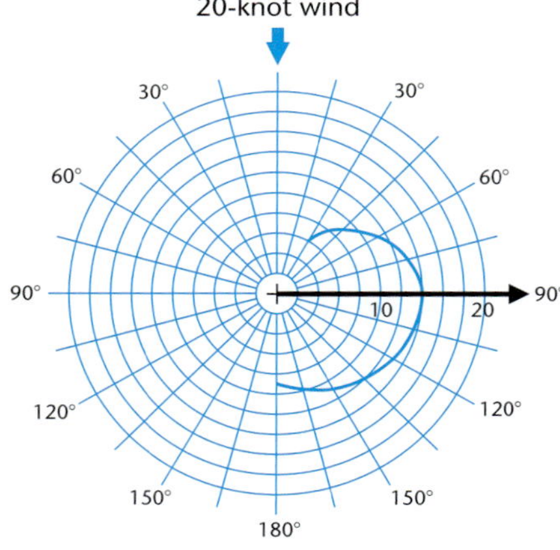

65. Sailboat Racing. Referring to the figure, estimate to the nearest knot the speed of the sailboat sailing at the following angles to the wind: 30°, 75°, 135°, and 180°.

66. Sailboat Racing. Referring to the figure, estimate to the nearest knot the speed of the sailboat sailing at the following angles to the wind: 45°, 90°, 120°, and 150°.

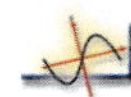

67. Conic Sections. Using a graphing utility, graph the equation

$$r = \frac{8}{1 - e\cos\theta}$$

for the following values of e (called the **eccentricity** of the conic) and identify each curve as a hyperbola, an ellipse, or a parabola.

(A) $e = 0.4$ **(B)** $e = 1$ **(C)** $e = 1.6$

(It is instructive to explore the graph for other positive values of e. See the Chapter 7 Group Activity.)

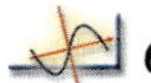

68. Conic Sections. Using a graphing utility, graph the equation

$$r = \frac{8}{1 - e\cos\theta}$$

for the following values of e and identify each curve as a hyperbola, an ellipse, or a parabola.

(A) $e = 0.6$ **(B)** $e = 1$ **(C)** $e = 2$

★ **69. Astronomy.**

(A) The planet Mercury travels around the sun in an elliptical orbit given approximately by

$$r = \frac{3.442 \times 10^7}{1 - 0.206\cos\theta}$$

where r is measured in miles and the sun is at the pole. Graph the orbit. Use TRACE to find the distance from Mercury to the sun at **aphelion** (greatest distance from the sun) and at **perihelion** (shortest distance from the sun).

(B) Johannes Kepler (1571–1630) showed that a line joining a planet to the sun sweeps out equal areas in space in equal intervals in time (see the figure). Use this information to determine whether a planet travels faster or slower at aphelion than at perihelion. Explain your answer.

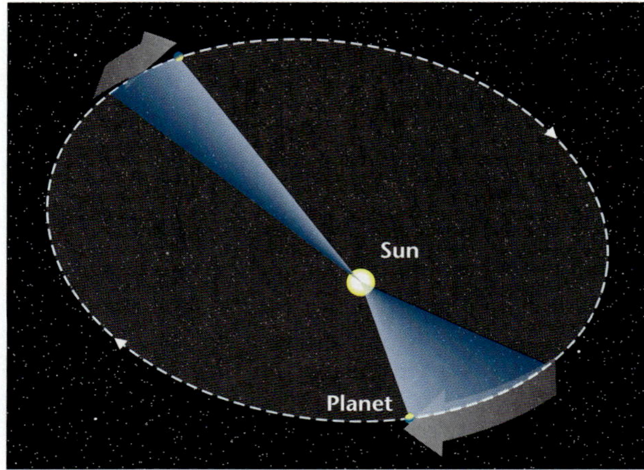

SECTION 7.6 Complex Numbers in Rectangular and Polar Forms

Rectangular Form • Polar Form • Multiplication and Division in Polar Form • Historical Note

Utilizing polar concepts studied in Sections 7.4 and 7.5, we now show how complex numbers can be written in polar form, which can be very useful in many applications. A brief review of Section 2.4 on complex numbers should prove helpful before proceeding further.

Rectangular Form

Recall from Section 2.4 that a complex number is any number that can be written in the form

$$a + bi$$

where a and b are real numbers and i is the imaginary unit. Thus, associated with each complex number $a + bi$ is a unique ordered pair of real numbers (a, b), and vice versa. For example,

$$3 - 5i \qquad \text{corresponds to} \qquad (3, -5)$$

FIGURE 1 Complex plane.

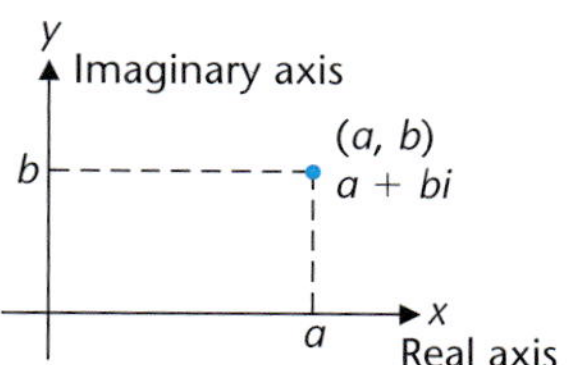

Associating these ordered pairs of real numbers with points in a rectangular coordinate system, we obtain a **complex plane** (Fig. 1). When complex numbers are associated with points in a rectangular coordinate system, we refer to the x axis as the **real axis** and the y axis as the **imaginary axis.** The complex number $a + bi$ is said to be in **rectangular form.**

EXAMPLE 1 Plotting in the Complex Plane

Plot the following complex numbers in a complex plane:

$$A = 2 + 3i \qquad B = -3 + 5i \qquad C = -4 \qquad D = -3i$$

SOLUTION

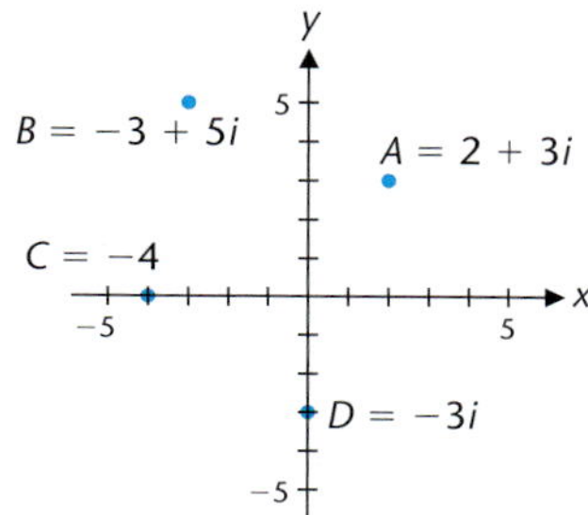

MATCHED 1 PROBLEM

Plot the following complex numbers in a complex plane:

$$A = 4 + 2i \qquad B = 2 - 3i \qquad C = -5 \qquad D = 4i$$

EXPLORE/DISCUSS 1

On a *real number line* there is a one-to-one correspondence between the set of real numbers and the set of points on the line: each real number is associated with exactly one point on the line and each point on the line is associated with exactly one real number. Does such a correspondence exist between the set of complex numbers and the set of points in an extended plane? Explain how a one-to-one correspondence can be established.

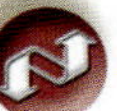

Polar Form

FIGURE 2 Rectangular–polar relationship.

Complex numbers also can be written in **polar form.** Using the polar–rectangular relationships from Section 7.5,

$$x = r\cos\theta \qquad \text{and} \qquad y = r\sin\theta$$

we can write the complex number $z = x + iy$ in polar form as follows:

$$z = x + iy = r\cos\theta + ir\sin\theta = r(\cos\theta + i\sin\theta) \tag{1}$$

This rectangular–polar relationship is illustrated in Figure 2. In a more advanced treatment of the subject, the following famous equation is established:

$$e^{i\theta} = \cos\theta + i\sin\theta \tag{2}$$

where $e^{i\theta}$ obeys all the basic laws of exponents. Thus, equation (1) takes on the form

$$z = x + yi = r(\cos\theta + i\sin\theta) = re^{i\theta} \tag{3}$$

FIGURE 3 $(1 + i) = 1.41e^{0.79i}$.

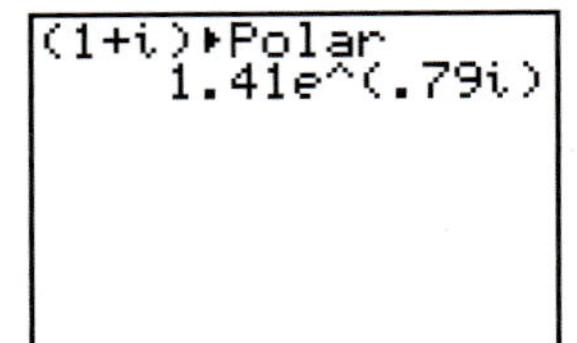

We will freely use $re^{i\theta}$ to denote the polar form of a complex number. In fact, some graphing calculators display the polar form of $x + iy$ this way (see Fig. 3 where θ is in radians and numbers are displayed to two decimal places).

Because $\cos\theta$ and $\sin\theta$ are both periodic with period 2π, we have

$$\begin{aligned} \cos(\theta + 2k\pi) &= \cos\theta \\ \sin(\theta + 2k\pi) &= \sin\theta \end{aligned} \qquad k \text{ any integer}$$

Thus, we can write a more general polar form for a complex number $z = x + iy$, as given in Definition 1, and observe that $re^{i\theta}$ is periodic (with respect to θ) with period $2k\pi$, k any integer.

DEFINITION 1

General Polar Form of a Complex Number

For k any integer

$$z = x + iy = r[\cos(\theta + 2k\pi) + i\sin(\theta + 2k\pi)]$$
$$z = re^{i(\theta + 2k\pi)}$$

The number r is called the **modulus,** or **absolute value,** of z and is denoted by **mod** z or $|z|$. The polar angle that the line joining z to the origin makes with the polar axis is called the **argument** of z and is denoted by **arg** z. From Figure 2 we see the following relationships:

DEFINITION 2

Modulus and Argument for $z = x + iy$

$\text{mod}\, z = r = \sqrt{x^2 + y^2}$ Never negative

$\arg z = \theta + 2k\pi$ k any integer

where $\sin\theta = y/r$ and $\cos\theta = x/r$. The argument θ is usually chosen so that $-180° < \theta \le 180°$ or $-\pi < \theta \le \pi$.

EXAMPLE 2 From Rectangular to Polar Form

Write parts A–C in polar form, θ in radians, $-\pi < \theta \le \pi$. Compute the modulus and arguments for parts A and B exactly; compute the modulus and argument for part C to two decimal places.

(A) $z_1 = 1 - i$ (B) $z_2 = -\sqrt{3} + i$ (C) $z = -5 - 2i$

SOLUTIONS

Locate in a complex plane first; then if x and y are associated with special angles, r and θ can often be determined by inspection.

FIGURE 4

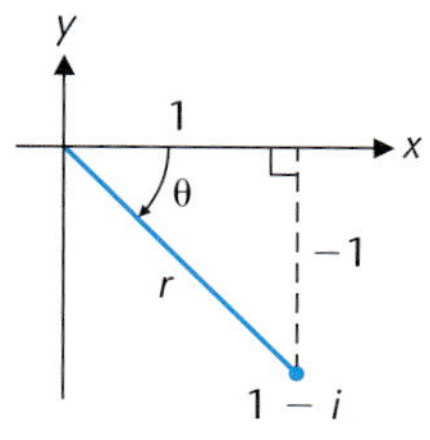

(A) A sketch shows that z_1 is associated with a special 45° triangle (Fig. 4). Thus, by inspection, $r = \sqrt{2}$, $\theta = -\pi/4$ (not $7\pi/4$), and

$$\begin{aligned} z_1 &= \sqrt{2}[\cos(-\pi/4) + i\sin(-\pi/4)] \\ &= \sqrt{2}\, e^{(-\pi/4)i} \end{aligned}$$

FIGURE 5

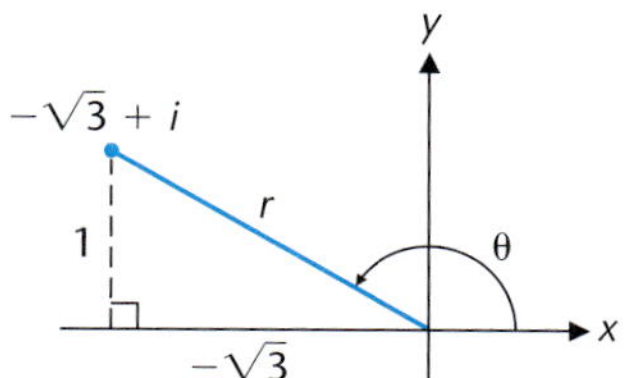

(B) A sketch shows that z_2 is associated with a special 30°– 60° triangle (Fig. 5). Thus by inspection, $r = 2$, $\theta = 5\pi/6$, and

$$\begin{aligned} z_2 &= 2(\cos 5\pi/6 + i\sin 5\pi/6) \\ &= 2e^{(5\pi/6)i} \end{aligned}$$

FIGURE 6

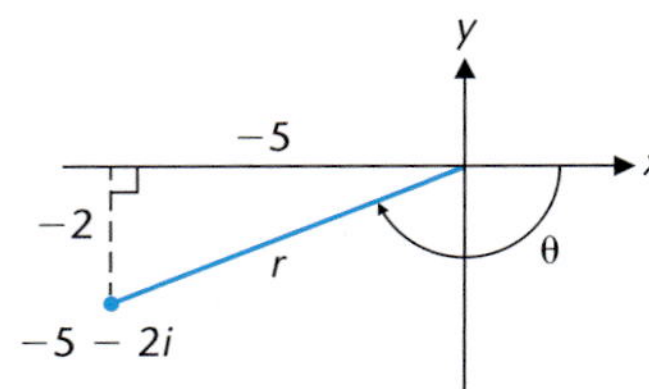

(C) A sketch shows that z_3 is not associated with a special triangle (Fig. 6). So, we proceed as follows:

$$r = \sqrt{(-5)^2 + (-2)^2} = 5.39 \qquad \text{To two decimal places}$$

$$\theta = -\pi + \tan^{-1}\left(\tfrac{2}{5}\right) = -2.76 \qquad \text{To two decimal places}$$

Thus,

$$z_3 = 5.39[\cos(-2.76) + i \sin(-2.76)]$$
$$= 5.39e^{(-2.76)i} \qquad \text{To two decimal places}$$

FIGURE 7
$(-5 - 2i) = 5.39e^{(-2.76)i}$.

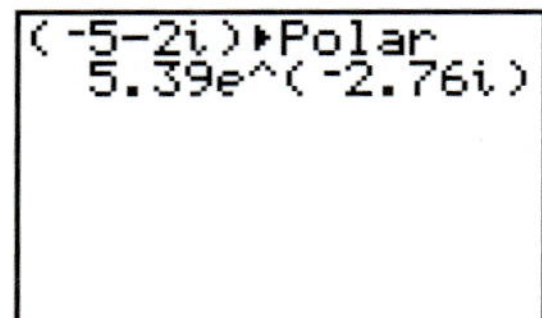

Figure 7 shows the same conversion done by a graphing calculator with a built-in conversion routine (with numbers displayed to two decimal places).

MATCHED 2 PROBLEM

Write parts A–C in polar form, θ in radians, $-\pi < \theta \le \pi$. Compute the modulus and arguments for parts A and B exactly; compute the modulus and argument for part C to two decimal places.

(A) $-1 + i$ (B) $1 + i\sqrt{3}$ (C) $-3 - 7i$

EXAMPLE 3 From Polar to Rectangular Form

Write parts A–C in rectangular form. Compute the exact values for parts A and B; for part C, compute a and b for $a + bi$ to two decimal places.

(A) $z_1 = 2e^{(5\pi/6)i}$ (B) $z_2 = 3e^{(-60°)i}$ (C) $z_3 = 7.19e^{(-2.13)i}$

SOLUTIONS

(A)
$$x + iy = 2e^{(5\pi/6)i}$$
$$= 2[\cos(5\pi/6) + i \sin(5\pi/6)]$$
$$= 2\left(\frac{-\sqrt{3}}{2}\right) + i2\left(\frac{1}{2}\right)$$
$$= -\sqrt{3} + i$$

(B)
$$x + iy = 3e^{(-60°)i}$$
$$= 3[\cos(-60°) + i \sin(-60°)]$$
$$= 3\left(\frac{1}{2}\right) + i3\left(\frac{-\sqrt{3}}{2}\right)$$
$$= \frac{3}{2} - \frac{3\sqrt{3}}{2}i$$

FIGURE 8
$7.19e^{(-2.13)i} = -3.81 - 6.09i$.

```
7.19e^(-2.13i)▸R
ect
        -3.81-6.09i
```

(C)
$$x + iy = 7.19e^{(-2.13)i}$$
$$= 7.19[\cos(-2.13) + i \sin(-2.13)]$$
$$= -3.81 - 6.09\,i$$

Figure 8 shows the same conversion done by a graphing calculator with a built-in conversion routine.

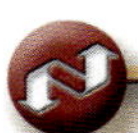

EXPLORE/DISCUSS 2

If your calculator has a built-in polar-to-rectangular conversion routine, try it on $\sqrt{2}e^{45°i}$ and $\sqrt{2}e^{(\pi/4)i}$, then reverse the process to see if you get back where you started. (For complex numbers in exponential polar form, some calculators require θ to be in radian mode for calculations. Check your user's manual.)

MATCHED 3 PROBLEM

Write parts A–C in rectangular form. Compute the exact values for parts A and B; for part C compute a and b for $a + bi$ to two decimal places.

(A) $z_1 = \sqrt{2}e^{(-\pi/2)i}$ (B) $z_2 = 3e^{120°i}$ (C) $z_3 = 6.49e^{(-2.08)i}$

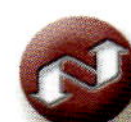

EXPLORE/DISCUSS 3

Let $z_1 = \sqrt{3} + i$ and $z_2 = 1 + i\sqrt{3}$.

(A) Find z_1z_2 and z_1/z_2 using the rectangular forms of z_1 and z_2.

(B) Find z_1z_2 and z_1/z_2 using the polar forms of z_1 and z_2, θ in degrees. (Assume the product and quotient exponent laws hold for $e^{i\theta}$.)

(C) Convert the results from part B back to rectangular form and compare with the results in part A.

Multiplication and Division in Polar Form

There is a particular advantage in representing complex numbers in polar form: multiplication and division become very easy. Theorem 1 provides the reason. (The polar form of a complex number obeys the product and quotient rules for exponents: $b^mb^n = b^{m+n}$ and $b^m/b^n = b^{m-n}$.)

THEOREM 1

Products and Quotients in Polar Form

If $z_1 = r_1e^{i\theta_1}$ and $z_2 = r_2e^{i\theta_2}$, then

1. $z_1z_2 = r_1e^{i\theta_1}r_2e^{i\theta_2} = r_1r_2e^{i(\theta_1+\theta_2)}$

2. $\dfrac{z_1}{z_2} = \dfrac{r_1e^{i\theta_1}}{r_2e^{i\theta_2}} = \dfrac{r_1}{r_2}e^{i(\theta_1-\theta_2)}$

We establish the multiplication property and leave the quotient property for Problem 32 in Exercise 7.6.

$$
\begin{aligned}
z_1z_2 &= r_1e^{i\theta_1}r_2e^{i\theta_2} \\
&= r_1r_2(\cos\theta_1 + i\sin\theta_1)(\cos\theta_2 + i\sin\theta_2) && \text{Multiply.} \\
&= r_1r_2(\cos\theta_1\cos\theta_2 + i\cos\theta_1\sin\theta_2 \\
&\quad + i\sin\theta_1\cos\theta_2 - \sin\theta_1\sin\theta_2) \\
&= r_1r_2[(\cos\theta_1\cos\theta_2 - \sin\theta_1\sin\theta_2) && \text{Use sum identities.} \\
&\quad + i(\cos\theta_1\sin\theta_2 + \sin\theta_1\cos\theta_2)] \\
&= r_1r_2[\cos(\theta_1+\theta_2) + i\sin(\theta_1+\theta_2)] \\
&= r_1r_2e^{i(\theta_1+\theta_2)}
\end{aligned}
$$

EXAMPLE 4 Products and Quotients

If $z_1 = 8e^{45°i}$ and $z_2 = 2e^{30°i}$, find

(A) z_1z_2 (B) z_1/z_2

SOLUTIONS

(A) $z_1z_2 = 8e^{45°i} \cdot 2e^{30°i}$

$$= 8 \cdot 2e^{i(45°+30°)} = 16e^{75°i}$$

(B) $\dfrac{z_1}{z_2} = \dfrac{8e^{45°i}}{2e^{30°i}}$

$$= \frac{8}{2}e^{i(45°-30°)} = 4e^{15°i}$$

MATCHED PROBLEM 4

If $z_1 = 9e^{165°i}$ and $z_2 = 3e^{55°i}$, find

(A) z_1z_2 (B) z_1/z_2

Historical Note

There is hardly an area in mathematics that does not have some imprint of the famous Swiss mathematician Leonhard Euler (1707–1783), who spent most of his productive life at the New St. Petersburg Academy in Russia and the Prussian Academy in Berlin. One of the most prolific writers in the history of the subject, he is credited with making the following familiar notations standard:

$f(x)$ function notation
e natural logarithmic base
i imaginary unit, $\sqrt{-1}$

For our immediate interest, he is also responsible for the extraordinary relationship

$$e^{i\theta} = \cos\theta + i\sin\theta$$

If we let $\theta = \pi$, we obtain an equation that relates five of the most important numbers in the history of mathematics:

$$e^{i\pi} + 1 = 0$$

ANSWERS MATCHED PROBLEMS

1.

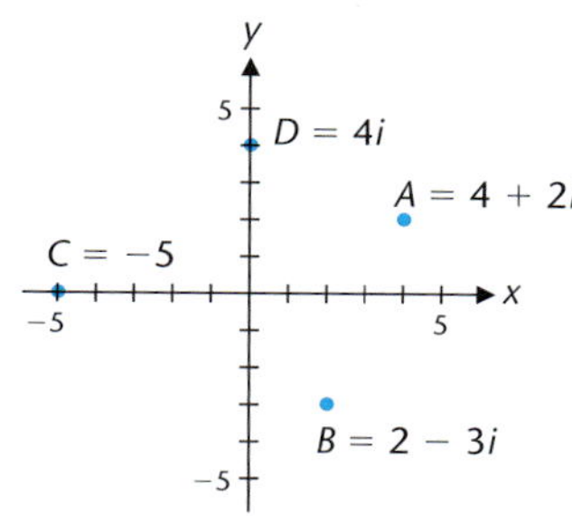

2. (A) $\sqrt{2}[\cos(3\pi/4) + i\sin(3\pi/4)] = \sqrt{2}e^{(3\pi/4)i}$
(B) $2[\cos(\pi/3) + i\sin(\pi/3)] = 2e^{(\pi/3)i}$
(C) $7.62[\cos(-1.98) + i\sin(-1.98)] = 7.62e^{(-1.98)i}$

3. (A) $-i\sqrt{2}$ (B) $-\frac{3}{2} + \frac{3\sqrt{3}}{2}i$ (C) $-3.16 - 5.67i$

4. (A) $z_1z_2 = 27e^{220°i}$ (B) $z_1/z_2 = 3e^{110°i}$

EXERCISE 7.6

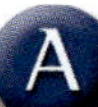

A

In Problems 1–8, plot each set of complex numbers in a complex plane.

1. $A = 3 + 4i, B = -2 - i, C = 2i$
2. $A = 4 + i, B = -3 + 2i, C = -3i$
3. $A = 3 - 3i, B = 4, C = -2 + 3i$
4. $A = -3, B = -2 - i, C = 4 + 4i$
5. $A = 2e^{(\pi/3)i}, B = \sqrt{2}e^{(\pi/4)i}, C = 4e^{(\pi/2)i}$
6. $A = 2e^{(\pi/6)i}, B = 4e^{\pi i}, C = \sqrt{2}e^{(3\pi/4)i}$
7. $A = 4e^{(-150°)i}, B = 3e^{20°i}, C = 5e^{(-90°)i}$
8. $A = 2e^{150°i}, B = 3e^{(-50°)i}, C = 4e^{75°i}$

B

In Problems 9–12, convert to the polar form $re^{i\theta}$. For Problems 9 and 10, choose θ in degrees, $-180° < \theta \le 180°$; for Problems 11 and 12 choose θ in radians, $-\pi < \theta \le \pi$. Compute the modulus and arguments for parts A and B exactly; compute the modulus and argument for part C to two decimal places.

9. **(A)** $\sqrt{3} + i$ **(B)** $-1 - i$ **(C)** $5 - 6i$
10. **(A)** $-1 + i\sqrt{3}$ **(B)** $-3i$ **(C)** $-7 - 4i$
11. **(A)** $-i\sqrt{3}$ **(B)** $-\sqrt{3} - i$ **(C)** $-8 + 5i$
12. **(A)** $\sqrt{3} - i$ **(B)** $-2 + 2i$ **(C)** $6 - 5i$

In Problems 13–16, change parts A–C to rectangular form. Compute the exact values for parts A and B; for part C compute a and b for a + bi to two decimal places.

13. **(A)** $2e^{(\pi/3)i}$ **(B)** $\sqrt{2}e^{(-45°)i}$ **(C)** $3.08e^{2.44i}$

14. **(A)** $2e^{30°i}$ **(B)** $\sqrt{2}e^{(-3\pi/4)i}$ **(C)** $5.71e^{(-0.48)i}$

15. **(A)** $6e^{(\pi/6)i}$ **(B)** $\sqrt{7}e^{(-90°)i}$ **(C)** $4.09e^{(-122.88°)i}$

16. **(A)** $\sqrt{3}e^{(-\pi/2)i}$ **(B)** $\sqrt{2}e^{135°i}$ **(C)** $6.83e^{(-108.82°)i}$

In Problems 17–22, find z_1z_2 and z_1/z_2 in the polar form $re^{i\theta}$.

17. $z_1 = 7e^{82°i}, z_2 = 2e^{31°i}$ **18.** $z_1 = 6e^{132°i}, z_2 = 3e^{93°i}$

19. $z_1 = 5e^{52°i}, z_2 = 2e^{83°i}$ **20.** $z_1 = 3e^{67°i}, z_2 = 2e^{97°i}$

21. $z_1 = 3.05e^{1.76i}, z_2 = 11.94e^{2.59i}$

22. $z_1 = 7.11e^{0.79i}, z_2 = 2.66e^{1.07i}$

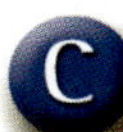

Simplify Problems 23–28 directly and by using polar forms. Write answers in both rectangular form and the polar form $re^{i\theta}$ (θ is in degrees).

23. $(-1 + i)^2$ **24.** $(1 + i)^2$

25. $(-1 + i)(1 + i)$ **26.** $(1 + i\sqrt{3})(\sqrt{3} + i)$

27. $(1 - i)^3$ **28.** $(1 + i)^3$

C

29. Show that $r^{1/3}e^{(\theta/3)i}$ is a cube root of $re^{i\theta}$.

30. Show that $r^{1/2}e^{(\theta/2)i}$ is a square root of $re^{i\theta}$.

31. If $z = re^{i\theta}$, show that $z^2 = r^2e^{2\theta i}$ and $z^3 = r^3e^{3\theta i}$. What do you think z^n will be for n a natural number?

32. Prove

$$\frac{z_1}{z_2} = \frac{r_1e^{i\theta_1}}{r_2e^{i\theta_2}} = \frac{r_1}{r_2}e^{i(\theta_1-\theta_2)}$$

APPLICATIONS

33. Forces and Complex Numbers. An object is located at the pole, and two forces **u** and **v** act on the object. Let the forces be vectors going from the pole to the complex numbers $20e^{0°i}$ and $10e^{60°i}$, respectively. Force **u** has a magnitude of 20 pounds in a direction of 0°. Force **v** has a magnitude of 10 pounds in a direction of 60°.

(A) Convert the polar forms of these complex numbers to rectangular form and add.

(B) Convert the sum from part A back to polar form.

(C) The vector going from the pole to the complex number in part B is the resultant of the two original forces. What is its magnitude and direction?

34. Forces and Complex Numbers. Repeat Problem 33 with forces **u** and **v** associated with the complex numbers $8e^{0°i}$ and $6e^{30°i}$, respectively.

SECTION 7.7 De Moivre's Theorem

De Moivre's Theorem, *n* a Natural Number • *n*th Roots of *z*

Abraham De Moivre (1667–1754), of French birth, spent most of his life in London doing private tutoring, writing, and publishing mathematics. He belonged to many prestigious professional societies in England, Germany, and France and was a close friend of Isaac Newton.

Using the polar form for a complex number, De Moivre established a theorem that still bears his name for raising complex numbers to natural number powers. More importantly, the theorem is the basis for the *nth root theorem,* which enables us to find *all n n*th roots of any complex number, real or imaginary.

De Moivre's Theorem, *n* a Natural Number

We start with Explore/Discuss 1 and generalize from this exploration.

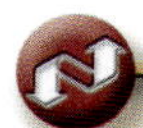

EXPLORE/DISCUSS 1

By repeated use of the product formula for the polar form $re^{i\theta}$, discussed in Section 7.6, establish the following:

1. $(x + iy)^2 = (re^{\theta i})^2 = r^2e^{2\theta i}$
2. $(x + iy)^3 = (re^{\theta i})^3 = r^3e^{3\theta i}$
3. $(x + iy)^4 = (re^{\theta i})^4 = r^4e^{4\theta i}$

Based on forms 1–3, and for n a natural number, what do you think the polar form of $(x + iy)^n$ would be?

If you guessed that the polar form of $(x + iy)^n$ is $r^ne^{n\theta i}$, you have arrived at De Moivre's theorem, which we now state without proof. A full proof of the theorem for all natural numbers n requires a method of proof, called *mathematical induction,* which is discussed in Section 10.2.

THEOREM 1

De Moivre's Theorem

If $z = x + iy = re^{i\theta}$, and n is a natural number, then

$$z^n = (x + iy)^n = (re^{i\theta})^n = r^ne^{n\theta i}$$

EXAMPLE 1 The Natural Number Power of a Complex Number

Use De Moivre's theorem to find $(1 + i)^{10}$. Write the answer in exact rectangular form.

SOLUTION

$(1 + i)^{10} = (\sqrt{2}e^{45°i})^{10}$	Convert $1 + i$ to polar form.
$= (\sqrt{2})^{10}e^{(10\cdot 45°)i}$	By De Moivre's theorem
$= 32e^{450°i}$	Change to rectangular form.
$= 32(\cos 450° + i \sin 450°)$	
$= 32(0 + i)$	
$= 32i$	Rectangular form

MATCHED 1 PROBLEM

Use De Moivre's theorem to find $(1 + i\sqrt{3})^5$. Write the answer in exact polar and rectangular forms.

EXAMPLE 2 The Natural Number Power of a Complex Number

Use De Moivre's theorem to find $(-\sqrt{3} + i)^6$. Write the answer in exact rectangular form.

SOLUTION

$$
\begin{aligned}
(-\sqrt{3} + i)^6 &= (2e^{150^\circ i})^6 && \text{Convert } -\sqrt{3} + i \text{ to polar form.}\\
&= 2^6 e^{(6\cdot 150^\circ)i} && \text{By De Moivre's theorem}\\
&= 64e^{900^\circ i} && \text{Change to rectangular form.}\\
&= 64\,(\cos 900^\circ + i \sin 900^\circ)\\
&= 64\,(-1 + i0)\\
&= -64 && \text{Rectangular form}
\end{aligned}
$$

[*Note:* $-\sqrt{3} + i$ must be a sixth root of -64, because $(-\sqrt{3} + i)^6 = -64$.]

MATCHED 2 PROBLEM

Use De Moivre's theorem to find $(1 - i\sqrt{3})^4$. Write the answer in exact polar and rectangular forms.

*n*th Roots of *z*

We now consider roots of complex numbers. We say ***w* is an *n*th root of *z***, n a natural number, if $w^n = z$. For example, if $w^2 = z$, then w is a square root of z. If $w^3 = z$, then w is a cube root of z. And so on.

EXPLORE/DISCUSS 2

If $z = re^{i\theta}$, then use De Moivre's theorem to show that $r^{1/2}e^{(\theta/2)i}$ is a square root of z and $r^{1/3}e^{(\theta/3)i}$ is a cube root of z.

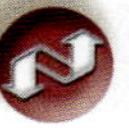

We can proceed in the same way as in Explore/Discuss 2 to show that $r^{1/n}e^{(\theta/n)i}$ is an nth root of $re^{i\theta}$, n a natural number:

$$
\begin{aligned}
[r^{1/n}e^{(\theta/n)i}]^n &= (r^{1/n})^n e^{n(\theta/n)i}\\
&= re^{\theta i}
\end{aligned}
$$

But we can do even better than this. The nth-root theorem (Theorem 2) shows us how to find *all* the nth roots of a complex number.

THEOREM 2
nth-Root Theorem

For n a positive integer greater than 1,

$$r^{1/n}e^{(\theta/n + k360°/n)i} \qquad k = 0, 1, \ldots, n - 1$$

are the n distinct nth roots of $re^{i\theta}$, and there are no others.

The proof of Theorem 2 is left to Problems 31 and 32 in Exercise 7.7.

EXAMPLE 3 Finding All Sixth Roots of a Complex Number

Find six distinct sixth roots of $-1 + i\sqrt{3}$, and plot them in a complex plane.

SOLUTION

First write $-1 + i\sqrt{3}$ in polar form:

$$-1 + i\sqrt{3} = 2e^{120°i}$$

Using the nth-root theorem, all six roots are given by

$$2^{1/6}e^{(120°/6 + k360°/6)i} = 2^{1/6}e^{(20° + k60°)i} \qquad k = 0, 1, 2, 3, 4, 5$$

Thus,

$$\begin{aligned}
w_1 &= 2^{1/6}e^{(20° + 0\cdot60°)i} = 2^{1/6}e^{20°i}\\
w_2 &= 2^{1/6}e^{(20° + 1\cdot60°)i} = 2^{1/6}e^{80°i}\\
w_3 &= 2^{1/6}e^{(20° + 2\cdot60°)i} = 2^{1/6}e^{140°i}\\
w_4 &= 2^{1/6}e^{(20° + 3\cdot60°)i} = 2^{1/6}e^{200°i}\\
w_5 &= 2^{1/6}e^{(20° + 4\cdot60°)i} = 2^{1/6}e^{260°i}\\
w_6 &= 2^{1/6}e^{(20° + 5\cdot60°)i} = 2^{1/6}e^{320°i}
\end{aligned}$$

FIGURE 1

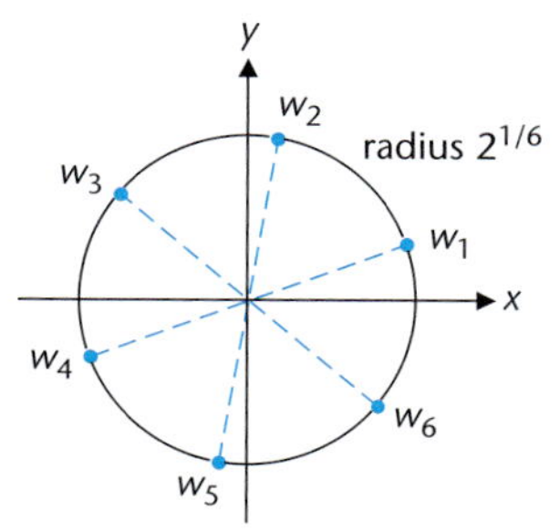

All roots are easily graphed in the complex plane after the first root is located. The root points are equally spaced around a circle of radius $2^{1/6}$ at an angular increment of 60° from one root to the next (Fig. 1).

MATCHED PROBLEM 3

Find five distinct fifth roots of $1 + i$. Leave the answers in polar form and plot them in a complex plane.

EXAMPLE 4 Solving a Cubic Equation

Solve $x^3 + 1 = 0$. Write final answers in rectangular form, and plot them in a complex plane.

SOLUTION

$$x^3 + 1 = 0$$
$$x^3 = -1$$

We see that x is a cube root of -1, and there are a total of three roots. To find the three roots, we first write -1 in polar form:

$$-1 = 1e^{180°i}$$

Using the nth-root theorem, all three cube roots of -1 are given by

$$1^{1/3}e^{(180°/3+k360°/3)i} = 1e^{(60°+k120°)i} \qquad k = 0, 1, 2$$

Thus,

$$w_1 = 1e^{60°i} = \cos 60° + i \sin 60° = \frac{1}{2} + i\frac{\sqrt{3}}{2}$$
$$w_2 = 1e^{180°i} = \cos 180° + i \sin 180° = -1$$
$$w_3 = 1e^{300°i} = \cos 300° + i \sin 300° = \frac{1}{2} - i\frac{\sqrt{3}}{2}$$

FIGURE 2

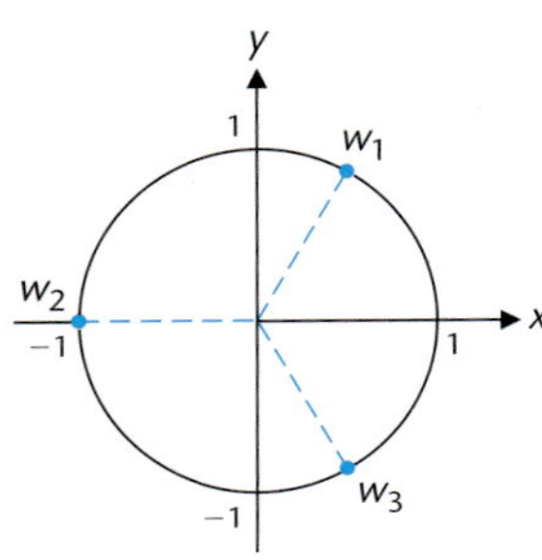

[*Note:* This problem can also be solved using factoring and the quadratic formula—try it.]

The three roots are graphed in Figure 2.

MATCHED 4 PROBLEM

Solve $x^3 - 1 = 0$. Write final answers in rectangular form, and plot them in a complex plane.

ANSWERS MATCHED PROBLEMS

1. $32e^{300°i} = 16 - i16\sqrt{3}$ 2. $16e^{(-240°)i} = -8 + i8\sqrt{3}$

3. $w_1 = 2^{1/10}e^{9°i}$, $w_2 = 2^{1/10}e^{81°i}$, $w_3 = 2^{1/10}e^{153°i}$, $w_4 = 2^{1/10}e^{225°i}$, $w_5 = 2^{1/10}e^{297°i}$

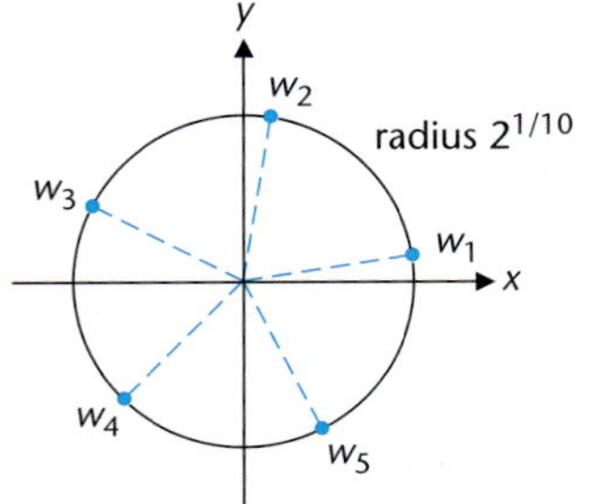

4. $1, -\frac{1}{2} + i\frac{\sqrt{3}}{2}, -\frac{1}{2} - i\frac{\sqrt{3}}{2}$

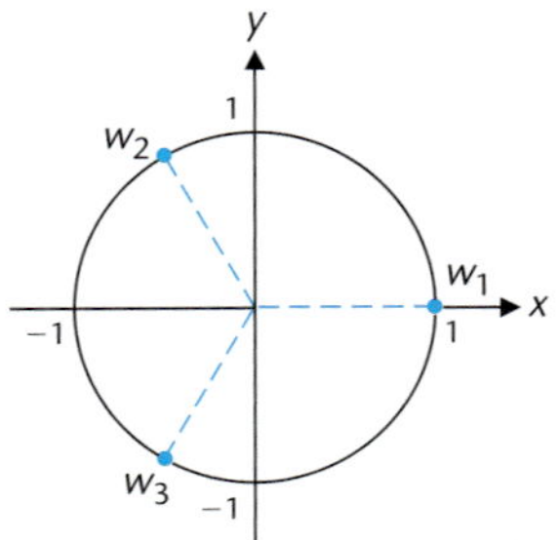

EXERCISE 7.7

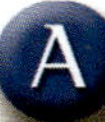

In Problems 1–6, use De Moivre's theorem to evaluate each. Leave answers in polar form.

1. $(2e^{30°i})^3$ **2.** $(5e^{15°i})^3$ **3.** $(\sqrt{2}e^{10°i})^6$

4. $(\sqrt{2}e^{15°i})^8$ **5.** $(1 + i\sqrt{3})^3$ **6.** $(\sqrt{3} + i)^8$

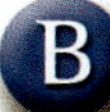

In Problems 7–12, find the value of each expression and write the final answer in exact rectangular form. (Verify the results in Problems 7–12 by evaluating each directly on a calculator.)

7. $(-\sqrt{3} - i)^4$ **8.** $(-1 + i)^4$ **9.** $(1 - i)^8$

10. $(-\sqrt{3} + i)^5$ **11.** $\left(-\frac{1}{2} + \frac{\sqrt{3}}{2}i\right)^3$ **12.** $\left(-\frac{1}{2} - \frac{\sqrt{3}}{2}i\right)^3$

For n and z as indicated in Problems 13–18, find all nth roots of z. Leave answers in the polar form $re^{i\theta}$.

13. $z = 8e^{30°i}, n = 3$ **14.** $z = 8e^{45°i}, n = 3$

15. $z = 81e^{60°i}, n = 4$ **16.** $z = 16e^{90°i}, n = 4$

17. $z = 1 - i, n = 5$ **18.** $z = -1 + i, n = 3$

For n and z as indicated in Problems 19–24, find all nth roots of z. Write answers in the polar form $re^{i\theta}$ and plot in a complex plane.

19. $z = 8, n = 3$ **20.** $z = 1, n = 4$

21. $z = -16, n = 4$ **22.** $z = -8, n = 3$

23. $z = i, n = 6$ **24.** $z = -i, n = 5$

25. **(A)** Show that $1 + i$ is a root of $x^4 + 4 = 0$. How many other roots does the equation have?

(B) The root $1 + i$ is located on a circle of radius $\sqrt{2}$ in the complex plane as indicated in the figure. Locate the other three roots of $x^4 + 4 = 0$ on the figure and explain geometrically how you found their location.

(C) Verify that each complex number found in part B is a root of $x^4 + 4 = 0$.

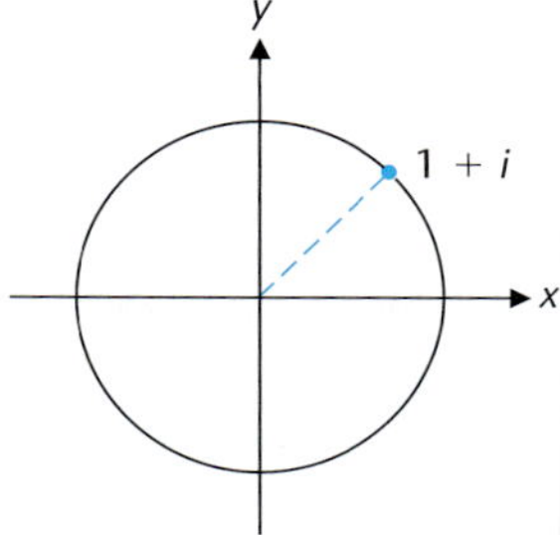

26. **(A)** Show that -2 is a root of $x^3 + 8 = 0$. How many other roots does the equation have?

(B) The root -2 is located on a circle of radius 2 in the complex plane as indicated in the figure. Locate the other two roots of $x^3 + 8 = 0$ on the figure and explain geometrically how you found their location.

(C) Verify that each complex number found in part B is a root of $x^3 + 8 = 0$.

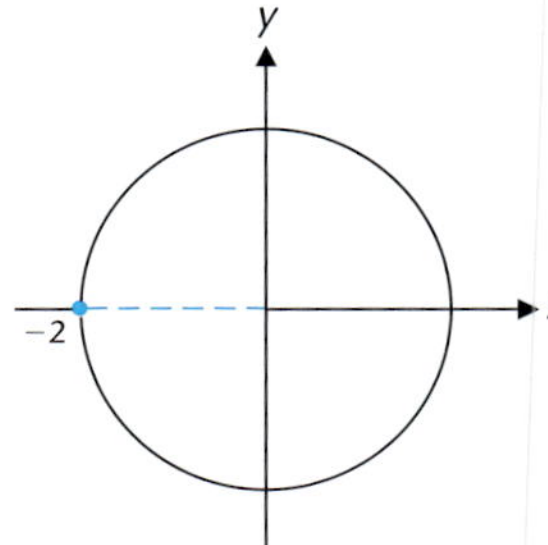

In Problems 27–30, solve each equation for all roots. Write final answers in the polar form $re^{i\theta}$ and exact rectangular form.

27. $x^3 + 64 = 0$ **28.** $x^3 - 64 = 0$

29. $x^3 - 27 = 0$ **30.** $x^3 + 27 = 0$

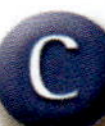

31. Show that

$$[r^{1/n}e^{(\theta/n+k360^\circ/n)i}]^n = re^{i\theta}$$

for any natural number n and any integer k.

32. Show that

$$r^{1/n}e^{(\theta/n+k360^\circ/n)i}$$

is the same number for $k = 0$ and $k = n$.

In Problems 33–36, write answers in the polar form $re^{i\theta}$.

33. Find all complex zeros for $P(x) = x^5 - 32$.

34. Find all complex zeros for $P(x) = x^6 + 1$.

35. Solve $x^5 + 1 = 0$ in the set of complex numbers.

36. Solve $x^3 - i = 0$ in the set of complex numbers.

In Problems 37 and 38, write answers using exact rectangular forms.

37. Write $P(x) = x^6 + 64$ as a product of linear factors.

38. Write $P(x) = x^6 - 1$ as a product of linear factors.

CHAPTER 7 REVIEW

7.1 Law of Sines

An **oblique triangle** is a triangle without a right angle. An oblique triangle is **acute** if all angles are between 0° and 90° and **obtuse** if one angle is between 90° and 180°. The labeling convention shown in these figures is followed in Chapter 7.

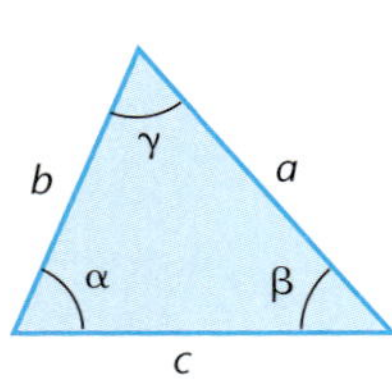

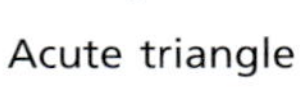
Acute triangle

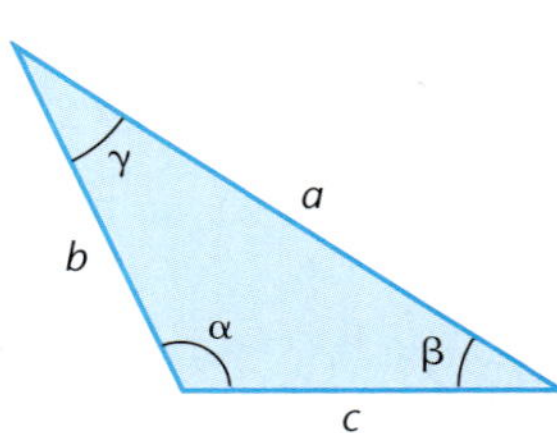

Obtuse triangle

TABLE 1 Triangles and Significant Digits

Angle to Nearest	Significant Digits for Side Measure
1°	2
10′ or 0.1°	3
1′ or 0.01°	4
10″ or 0.001°	5

The objective in Sections 7.1 and 7.2 is to solve an oblique triangle given any three of the six quantities indicated in either figure, if a solution exists. The law of sines, discussed in Section 7.1, and the law of cosines, discussed in Section 7.2, are used for this purpose. Accuracy in computation is governed by Table 1.

The **law of sines** is given as

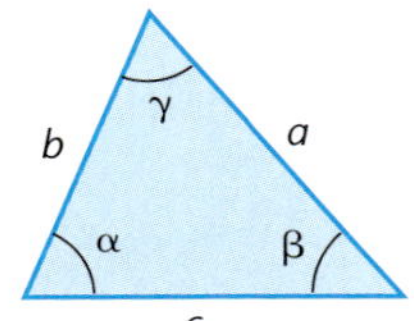

$$\frac{\sin\alpha}{a} = \frac{\sin\beta}{b} = \frac{\sin\gamma}{c}$$

and is generally used to solve the ASA, AAS, and SSA cases for oblique triangles. The AAS case is easily reduced to the ASA case by solving for the third angle first. The SSA case has a number of variations, including the ambiguous case. These variations are summarized in Table 2. Note that the ambiguous case always results in two triangles, one obtuse and one acute.

TABLE 2 SSA Variations

α	a [$h = b \sin \alpha$]	Number of Triangles	Figure
Acute	$0 < a < h$	0	
Acute	$a = h$	1	
Acute	$h < a < b$	2	Ambiguous case
Acute	$a \geq b$	1	
Obtuse	$0 < a \leq b$	0	
Obtuse	$a > b$	1	

7.2 Law of Cosines

The **law of cosines** is given as

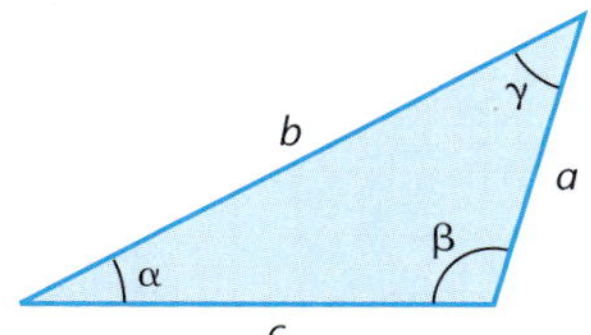

$$a^2 = b^2 + c^2 - 2bc \cos \alpha$$
$$b^2 = a^2 + c^2 - 2ac \cos \beta$$
$$c^2 = a^2 + b^2 - 2ab \cos \gamma$$

and is generally used as the first step in solving the SAS and SSS cases for oblique triangles. After a side or angle is found using the law of cosines, it is usually easier to continue the solving process with the law of sines.

7.3 Geometric Vectors

A **scalar** is a real number. A **geometric vector** in a plane is a directed line segment and is represented by an arrow as indicated in the figure.

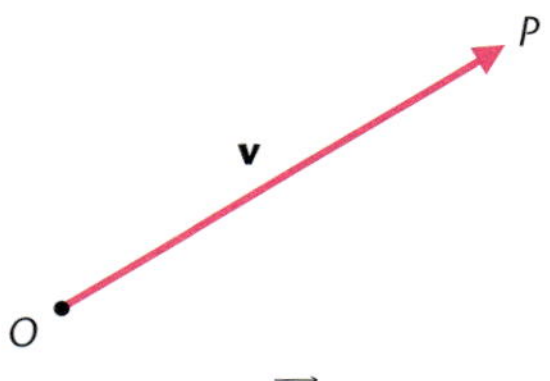

Vector $\overrightarrow{OP}$, or **v**

The point O is called the **initial point,** and the point P is called the **terminal point.**

The **magnitude** of the vector $\overrightarrow{AB}$, denoted by $|\overrightarrow{AB}|$, $|\vec{v}|$, or $|\mathbf{v}|$, is the length of the directed line segment. Two vectors have the **same direction** if they are parallel and point in the same direction. Two vectors have **opposite direction** if they are parallel and point in opposite directions. The **zero vector,** denoted by $\vec{0}$ or **0,** has a magnitude of zero and an arbitrary direction. Two vectors are **equal** if they have the same magnitude and direction. Thus, a vector may be **translated** from one location to another as long as the magnitude and direction do not change.

The **sum of two vectors u and v** can be defined using the **tail-to-tip rule.** The sum of two nonparallel vectors also can be defined using the **parallelogram rule.** Both forms are shown in the following figure:

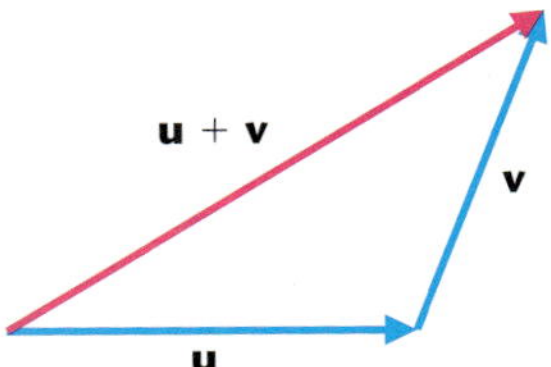

Vector addition: tail-to-tip rule

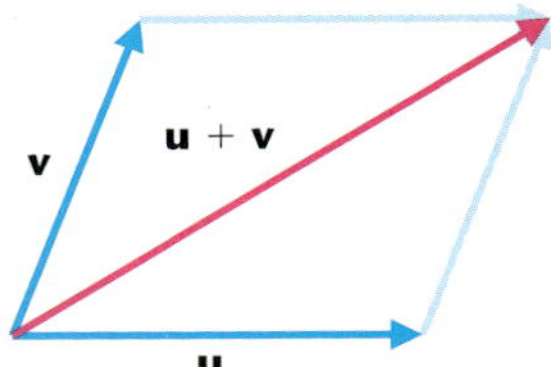

Vector addition: parallelogram rule

The vector **u** + **v** is also called the **resultant** of the two vectors **u** and **v**, and **u** and **v** are called **vector components** of **u** + **v**. Vector addition is **commutative;** that is, **u** + **v** = **v** + **u**.

A vector that represents the direction and speed of an object in motion is called a **velocity vector.** The velocity of an airplane relative to the air is called the **apparent velocity,** and the velocity relative to the ground is called the **resultant,** or **actual, velocity.** The resultant velocity is the vector sum of the apparent velocity and wind velocity. Similar statements apply to objects in water subject to currents.

A vector that represents the direction and magnitude of an applied force is called a **force vector.** If an object is subjected to two forces, then the sum of these two forces, the **resultant force,** is a single force acting on the object in the same way as the two original forces taken together.

7.4 Algebraic Vectors

A geometric vector $\overrightarrow{AB}$ in a rectangular coordinate system translated so that its initial point is at the origin is said to be in **standard position.** The vector $\overrightarrow{OP}$ such that $\overrightarrow{OP} = \overrightarrow{AB}$ is said to be the **standard vector** for $\overrightarrow{AB}$. This is shown in the following figure.

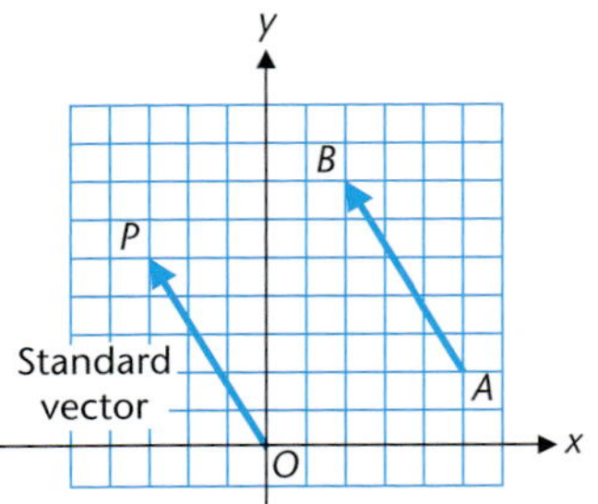

$\overrightarrow{OP}$ is the standard vector for $\overrightarrow{AB}$

Note that the vector $\overrightarrow{OP}$ in the figure is the standard vector for infinitely many vectors—all vectors with the same magnitude and direction as $\overrightarrow{OP}$.

Referring to the figure, if the coordinates of A are (x_a, y_a) and the coordinates of B are (x_b, y_b), then the coordinates of P are given by

$$(x_p, y_p) = (x_b - x_a, y_b - y_a)$$

Each geometric vector in a coordinate system can be associated with an ordered pair of real numbers, the coordinates of the terminal point of its standard vector. Conversely, every ordered pair of real numbers can be associated with a unique geometric standard vector. This leads to the definition of an **algebraic vector** as an ordered pair of real numbers, denoted by $\langle a, b\rangle$. The real numbers a and b are **scalar components** of the vector $\langle a, b\rangle$.

Two vectors $\mathbf{u} = \langle a, b\rangle$ and $\mathbf{v} = \langle c, d\rangle$ are said to be **equal** if their corresponding components are equal, that is, if $a = c$ and $b = d$. The **zero vector** is denoted by $\mathbf{0} = \langle 0, 0\rangle$ and has arbitrary direction.

The **magnitude,** or **norm,** of a vector $\mathbf{v} = \langle a, b\rangle$ is denoted by $|\mathbf{v}|$ and is given by

$$|\mathbf{v}| = \sqrt{a^2 + b^2}$$

Geometrically, $\sqrt{a^2 + b^2}$ is the length of the standard geometric vector $\overrightarrow{OP}$ associated with the algebraic vector $\langle a, b\rangle$.

If $\mathbf{u} = \langle a, b\rangle$, $\mathbf{v} = \langle c, d\rangle$, and k is a scalar, then the **sum** of **u** and **v** is given by

$$\mathbf{u} + \mathbf{v} = \langle a + c, b + d\rangle$$

and **scalar multiplication** of **u** by k is given by

$$k\mathbf{u} = k\langle a, b\rangle = \langle ka, kb\rangle$$

If **v** is a nonzero vector, then

$$\mathbf{u} = \frac{1}{|\mathbf{v}|}\mathbf{v}$$

is **a unit vector with the same direction as v.** The **i** and **j** unit vectors are defined as follows:

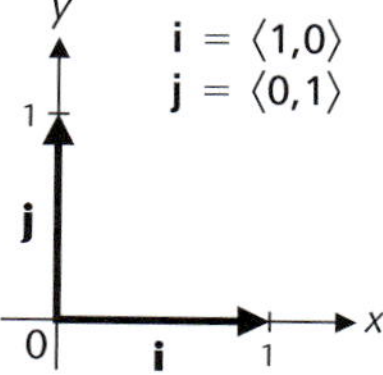

Every algebraic vector can be expressed in terms of the **i** and **j** unit vectors:

$$\mathbf{v} = \langle a, b\rangle = a\mathbf{i} + b\mathbf{j}$$

The following algebraic properties of vector addition and scalar multiplication enable us to manipulate symbols representing vectors and scalars in much the same way we manipulate symbols that represent real numbers in algebra.

Algebraic Properties of Vectors

A. Addition Properties. For all vectors **u, v,** and **w:**

1. $\mathbf{u} + \mathbf{v} = \mathbf{v} + \mathbf{u}$	**Commutative Property**
2. $\mathbf{u} + (\mathbf{v} + \mathbf{w}) = (\mathbf{u} + \mathbf{v}) + \mathbf{w}$	**Associative Property**
3. $\mathbf{u} + \mathbf{0} = \mathbf{0} + \mathbf{u} = \mathbf{u}$	**Additive Identity**
4. $\mathbf{u} + (-\mathbf{u}) = (-\mathbf{u}) + \mathbf{u} = \mathbf{0}$	**Additive Inverse**

B. Scalar Multiplication Properties. For all vectors **u** and **v** and all scalars *m* and *n*:

1. $m(n\mathbf{u}) = (mn)\mathbf{u}$	**Associative Property**
2. $m(\mathbf{u} + \mathbf{v}) = m\mathbf{u} + m\mathbf{v}$	**Distributive Property**
3. $(m + n)\mathbf{u} = m\mathbf{u} + n\mathbf{u}$	**Distributive Property**
4. $1\mathbf{u} = \mathbf{u}$	**Multiplicative Identity**

Certain *static equilibrium* problems can be solved using the material developed in Section 7.4. The conditions for static equilibrium are

1. An object at rest is said to be in **static equilibrium.**
2. For an object located at the origin in a rectangular coordinate system to remain in static equilibrium, at rest, it is necessary that the sum of all the force vectors acting on the object be the zero vector.

7.5 Polar Coordinates and Graphs

The figure illustrates a **polar coordinate system.** The fixed point O is called the **pole** or **origin,** and the horizontal arrow is called the **polar axis.** We have the following **relationships between rectangular coordinates (x, y) and polar coordinates (r, θ):**

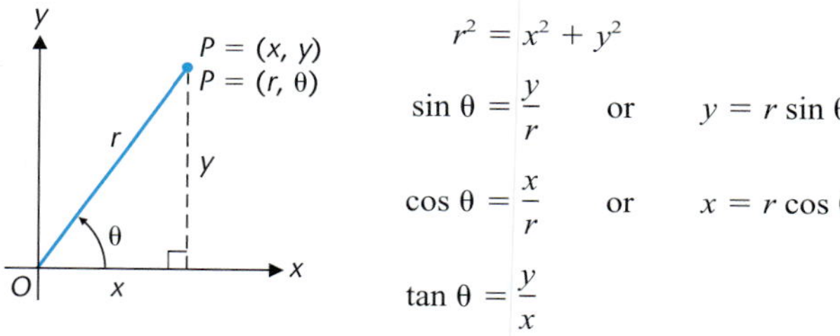

$$r^2 = x^2 + y^2$$

$$\sin\theta = \frac{y}{r} \quad \text{or} \quad y = r\sin\theta$$

$$\cos\theta = \frac{x}{r} \quad \text{or} \quad x = r\cos\theta$$

$$\tan\theta = \frac{y}{x}$$

[*Note:* The signs of x and y determine the quadrant for θ. The angle θ is chosen so that $-\pi < \theta \le \pi$ or $-180° < \theta \le 180°$, unless directed otherwise.]

Polar graphs can be obtained by **point-by-point plotting** much in the same ways graphs in rectangular coordinates are formed. Make a table of values that satisfy the polar equation, plot these points, then join them with a smooth curve.

Graphs can also be obtained by **rapid graphing techniques.** If only a rough sketch of a polar equation involving $\sin\theta$ or $\cos\theta$ is desired, we can speed up the point-by-point graphing process by taking advantage of the uniform variation of $\sin\theta$ and $\cos\theta$ as θ moves through each set of quadrant values. **Graphing utilities** can produce polar graphs almost instantly.

The table shows some standard polar curves with their equations:

Standard Polar Graphs

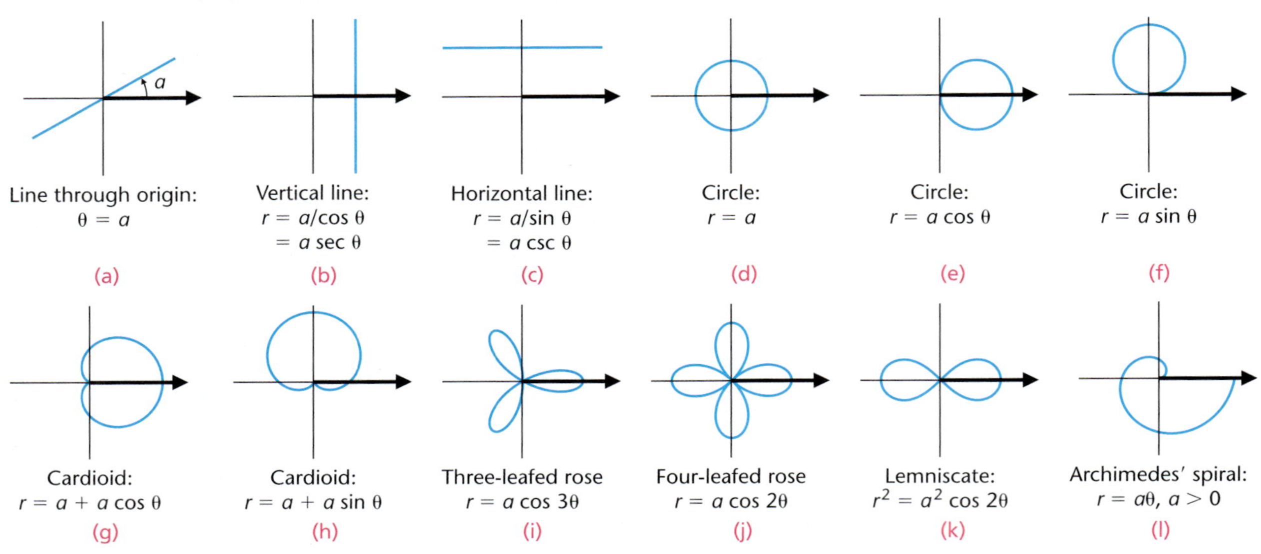

7.6 Complex Numbers in Rectangular and Polar Forms

A **complex number** is a number of the form

$$a + bi$$

where a and b are real numbers and i is the **imaginary unit.** The figure shows a complex number $a + bi$ plotted in a **complex plane.**

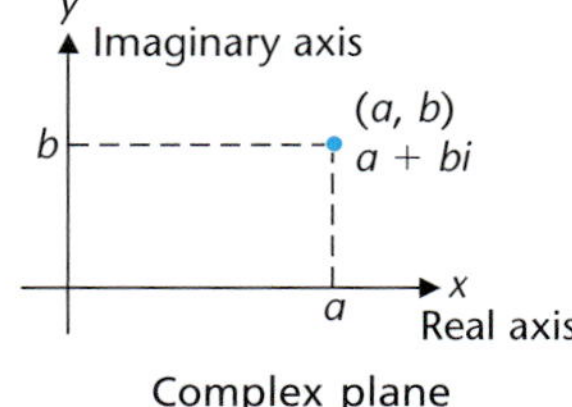

Complex plane

When complex numbers are associated with points in a rectangular coordinate system, we refer to the x axis as the **real axis** and the y axis as the **imaginary axis.** The complex number $a + bi$ is said to be in **rectangular form.**

Complex numbers can also be written in **polar form** using $x = r\cos\theta$ and $y = r\sin\theta$ as shown in the figure:

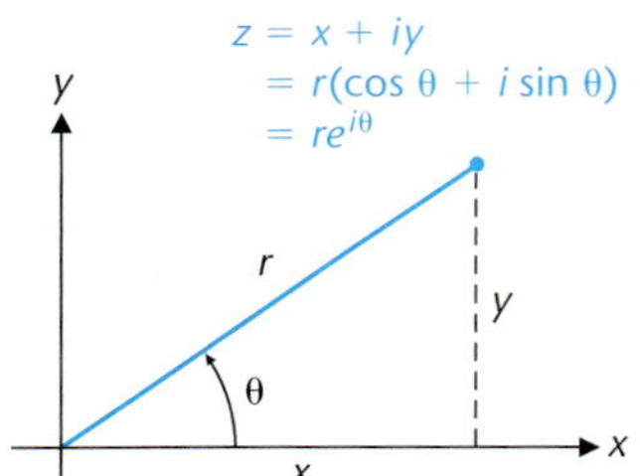

Rectangular–polar relationship

Because of the periodic nature of sine and cosine functions, we have the more **general polar form** for a complex number $z = x + iy$:

$$z = x + iy = r\,[\cos(\theta + 2k\pi) + i\sin(\theta + 2k\pi)]$$
$$= re^{(\theta + 2k\pi)i}$$

where

$$e^{i\theta} = \cos\theta + i\sin\theta$$

and the quadrant for θ is determined by x and y.

The number r is called the **modulus,** or **absolute value,** of z and is denoted by **mod z or $|z|$**. The polar angle that the line joining z to the origin makes with the polar axis is called the

argument of z and is denoted by **arg** z. From the figure illustrating the rectangular–polar relationships we have the following representations of the modulus and argument for $z = x + iy$:

$$\text{mod } z = r = \sqrt{x^2 + y^2} \qquad \text{Never negative}$$
$$\arg z = \theta + 2k\pi \qquad k \text{ any integer}$$

where $\sin\theta = y/r$ and $\cos\theta = x/r$, and θ is usually chosen so that $-\pi < \theta \le \pi$ or $-180° < \theta \le 180°$.

Products and **quotients** of complex numbers in polar form are found as follows: If

$$z_1 = r_1 e^{i\theta_1} \qquad \text{and} \qquad z_2 = r_2 e^{i\theta_2}$$

then

1. $z_1 z_2 = r_1 e^{i\theta_1} r_2 e^{i\theta_2} = r_1 r_2 e^{i(\theta_1+\theta_2)}$
2. $\dfrac{z_1}{z_2} = \dfrac{r_1 e^{i\theta_1}}{r_2 e^{i\theta_2}} = \dfrac{r_1}{r_2} e^{i(\theta_1-\theta_2)}$

7.7 De Moivre's Theorem

Section 7.7 discusses the famous De Moivre theorem and the related nth-root theorem. These theorems make the process of finding natural number powers and all the nth roots of a complex number relatively easy. **De Moivre's theorem** is stated as follows: If

$$z = x + iy = re^{i\theta}$$

and n is a natural number, then

$$z^n = (x + iy)^n = (re^{i\theta})^n = r^n e^{n\theta i}$$

From De Moivre's theorem, we can derive the ***n*th-root theorem:** For n a positive integer greater than 1,

$$r^{1/n} e^{(\theta/n + k\,360°/n)i} \qquad k = 0, 1, \ldots, n-1$$

are the n distinct nth roots of $re^{i\theta}$, and there are no others.

CHAPTER 7 REVIEW EXERCISES

Work through all the problems in this chapter review and check answers in the back of the book. Answers to all review problems are there, and following each answer is a number in italics indicating the section in which that type of problem is discussed. Where weaknesses show up, review appropriate sections in the text.

Problems in this exercise use the following labeling of sides and angles:

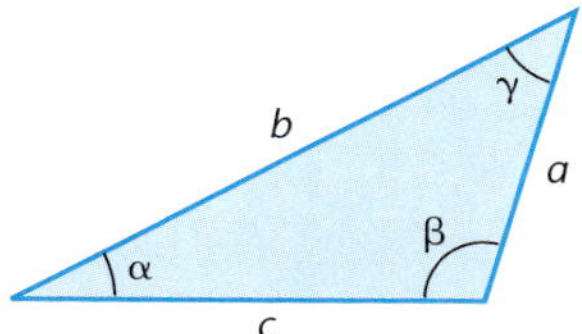

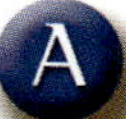

In Problems 1–3, determine whether the information in each problem allows you to construct 0, 1, or 2 triangles. Do not solve the triangle.

1. $a = 11$ meters, $b = 3.7$ meters, $\alpha = 67°$
2. $c = 15$ centimeters, $\alpha = 97°$, $\beta = 84°$
3. $a = 18$ feet, $b = 22$ feet, $\alpha = 54°$
4. Referring to the figure at the beginning of the exercise, if $\alpha = 52.6°$, $b = 57.1$ centimeters, and $c = 79.5$ centimeters, which of the two angles, β or γ, can you say for certain is acute and why?

In Problems 5–7, solve each triangle, given the indicated information.

5. $\alpha = 67°$, $\beta = 38°$, and $c = 49$ meters
6. $\alpha = 15°$, $b = 9.1$ feet, and $c = 12$ feet
7. $\gamma = 121°$, $c = 11$ centimeters, and $b = 4.2$ centimeters

8. Given geometric vectors **u** and **v** as indicated in the figure, find $|\mathbf{u} + \mathbf{v}|$ and θ, given $|\mathbf{u}| = 160$ miles per hour and $|\mathbf{v}| = 55$ miles per hour.

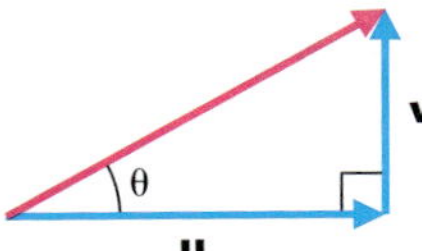

9. Write the algebraic vector $\langle a, b\rangle$ corresponding to the geometric vector $\overrightarrow{AB}$ with endpoints $A = (2, 6)$ and $B = (5, -1)$.

10. Find the magnitude of the vector $\langle -3, -5\rangle$.

11. Sketch a graph of $\theta = \pi/6$ in a polar coordinate system.

12. Sketch a graph of $r = 6$ in a polar coordinate system.

13. Plot in a complex plane: $A = 3 + 5i$, $B = -1 - i$, $C = -3i$.

14. A point in a polar coordinate system has coordinates $(10, -30°)$. Find all other polar coordinates for the point, $-360° \le \theta \le 360°$, and verbally describe how the coordinates are associated with the point.

15. Plot in a complex plane: $A = 5e^{30°i}$, $B = 10e^{(-\pi/2)i}$, $C = 7e^{(3\pi/4)i}$.

16. **(A)** Change $1 - i\sqrt{3}$ to the polar form $re^{i\theta}$, $r \ge 0$, $-180° < \theta \le 180°$.
(B) Change $4e^{(-30°)i}$ to exact rectangular form.

17. **(A)** Find $[(-1/2) - (\sqrt{3}/2)i]^3$ using De Moivre's theorem. Write the final answer in exact rectangular form.
(B) Verify the results in part A with a calculator.

18. Find $(2e^{15°i})^4$ using De Moivre's theorem, and write the final answer in exact rectangular form.

B

19. Referring to the figure at the beginning of the exercise, if $a = 434$ meters, $b = 302$ meters, and $c = 197$ meters, then if the triangle has an obtuse angle which angle must it be and why?

In Problems 20–23, solve each triangle. If a problem does not have a solution, say so. If a triangle has two solutions, say so, and solve the obtuse case.

20. $\beta = 115.4°$, $a = 5.32$ centimeters, $c = 7.05$ centimeters

21. $\alpha = 63.2°$, $a = 179$ millimeters, $b = 205$ millimeters

22. $\alpha = 26.4°$, $a = 52.2$ kilometers, $b = 84.6$ kilometers

23. $a = 19.0$ inches, $b = 27.8$ inches, $c = 26.1$ inches

24. If four nonzero force vectors with different magnitudes and directions are acting on an object at rest, what must the sum of all four vectors be for the object to remain at rest?

25. Given geometric vectors **u** and **v** as indicated in the figure, find $|\mathbf{u} + \mathbf{v}|$ and α, given $|\mathbf{u}| = 75.2$ kilograms, $|\mathbf{v}| = 34.2$ kilograms, and $\theta = 57.2°$.

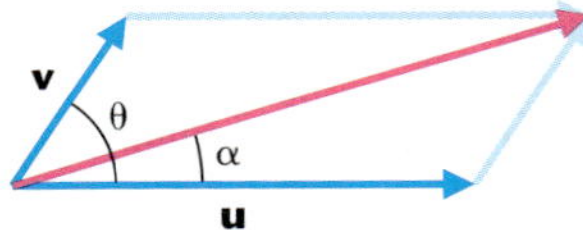

26. Express each vector in terms of **i** and **j** unit vectors:
(A) $\mathbf{u} = \langle -3, 9\rangle$ **(B)** $\mathbf{v} = \langle 0, -2\rangle$

For the indicated vectors in Problems 27 and 28, find
(A) $\mathbf{u} - \mathbf{v}$ **(B)** $3\mathbf{u} - \mathbf{v} + 2\mathbf{w}$

27. $\mathbf{u} = \langle -2, 3\rangle$, $\mathbf{v} = \langle 2, -4\rangle$, $\mathbf{w} = \langle -3, 0\rangle$

28. $\mathbf{u} = \mathbf{i} - 2\mathbf{j}$, $\mathbf{v} = 3\mathbf{i} + 2\mathbf{j}$, $\mathbf{w} = -\mathbf{j}$

29. Find a unit vector **u** with the same direction as $\mathbf{v} = \langle -1, -3\rangle$.

In Problems 30–33, use rapid sketching techniques to sketch each graph in a polar coordinate system. Check by graphing on a graphing utility.

30. $r = 6 + 4\cos\theta$ **31.** $r = 8 + 8\sin\theta$

32. $r = 10\cos 2\theta$ **33.** $r = 8\sin 3\theta$

34. Graph $r = 6\cos\frac{\theta}{7}$ for $0 \le \theta \le 7\pi$.

35. Graph $r = 6\cos\frac{\theta}{9}$ for $0 \le \theta \le 9\pi$.

36. Graph $r = 8(\sin\theta)^{2n}$, for $n = 1, 2$, and 3. How many leaves do you expect the graph will have for arbitrary n?

37. Graph $r = 3/(1 - e\cos\theta)$ for the following values of e and identify each curve as an ellipse, a parabola, or a hyperbola:
(A) $e = 0.55$ **(B)** $e = 1$ **(C)** $e = 1.7$

38. Convert $x^2 + y^2 = 6x$ to polar form.

39. Convert $r = 5\cos\theta$ to rectangular form.

40. Change the following complex numbers to the polar form $re^{i\theta}$, $r \ge 0$, $-180° < \theta \le 180°$: $z_1 = -1 + i$, $z_2 = -1 - i\sqrt{3}$, $z_3 = 5$.

41. Change the following complex numbers to exact rectangular form: $z_1 = \sqrt{2}e^{(\pi/4)i}$, $z_2 = 3e^{210°i}$, $z_3 = 2e^{(-2\pi/3)i}$.

42. If $z_1 = 8e^{25°i}$ and $z_2 = 4e^{19°i}$, find
(A) z_1z_2 **(B)** z_1/z_2

Leave answers in the polar form $re^{i\theta}$.

43. **(A)** Write $(1 + i\sqrt{3})^4$ in exact rectangular form. Use De Moivre's theorem.
(B) Verify part A by evaluating $(1 + i\sqrt{3})^4$ directly on a calculator.

44. Find all cube roots of i. Write final answers in exact rectangular form, and locate the roots on a circle in the complex plane.

45. Find all cube roots of $-4\sqrt{3} + 4i$ exactly. Leave answers in the polar form $re^{i\theta}$.

46. Show that $4e^{15°i}$ is a square root of $8\sqrt{3} + 8i$.

47. Change the rectangular coordinates $(5.17, -2.53)$ to polar coordinates to two decimal places, $r \ge 0$, $-180° < \theta \le 180°$.

48. Change the polar coordinates $(5.81, -2.72)$ to rectangular coordinates to two decimal places.

49. Change the complex number $-3.18 + 4.19i$ to the polar form $re^{i\theta}$ to two decimal places, $r \ge 0$, $-180° < \theta \le 180°$.

50. Change the complex number $7.63e^{(-162.27°)i}$ to rectangular form $a + bi$, where a and b are computed to two decimal places.

51. **(A)** The cube root of a complex number is shown in the figure. Geometrically locate all other cube roots of the number on the figure, and explain how they were located.
(B) Determine geometrically the other cube roots of the number in exact rectangular form.
(C) Cube each cube root from parts A and B.

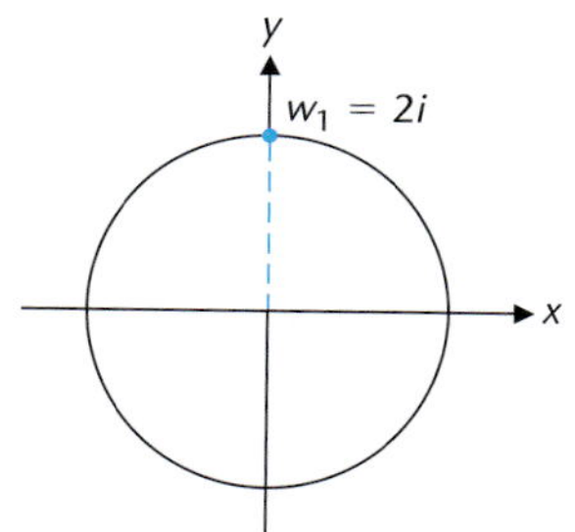

C

52. For an oblique triangle with $\alpha = 23.4°$, $b = 44.6$ millimeters, and a the side opposite angle α, determine a value k so that if $0 < a < k$, there is no solution; if $a = k$, there is one solution; and if $k < a < b$, there are two solutions.

53. Show that for any triangle

$$\frac{a^2 + b^2 + c^2}{2abc} = \frac{\cos\alpha}{a} + \frac{\cos\beta}{b} + \frac{\cos\gamma}{c}$$

54. Let $\mathbf{u} = \langle a, b\rangle$ and $\mathbf{v} = \langle c, d\rangle$ be vectors and m a scalar; prove
(A) $(\mathbf{u} + \mathbf{v}) = (\mathbf{v} + \mathbf{u})$
(B) $m(\mathbf{u} + \mathbf{v}) = m\mathbf{u} + m\mathbf{v}$

55. Given the polar equation $r = 4 + 4\cos(\theta/2)$.
(A) Sketch a graph of the equation using rapid graphing techniques.
(B) Verify the graph in part A on a graphing utility.

56. **(A)** Graph $r = -8 \sin \theta$ and $r = 8 \cos \theta$, $0 \le \theta \le \pi$, in the same viewing window. Use TRACE to determine which intersection point has coordinates that satisfy both equations simultaneously.

(B) Solve the equations simultaneously to verify the results in part A.

(C) Explain why the pole is not a simultaneous solution, even though the two curves intersect at the pole.

57. Find all solutions, real and imaginary, for $x^8 - 1 = 0$. Write roots in exact rectangular form.

58. Write $P(x) = x^3 - 8i$ as a product of linear factors.

APPLICATIONS

For Problems 59–61, use the navigational compass shown. Assume directions given in terms of north, east, south, and west are exact.

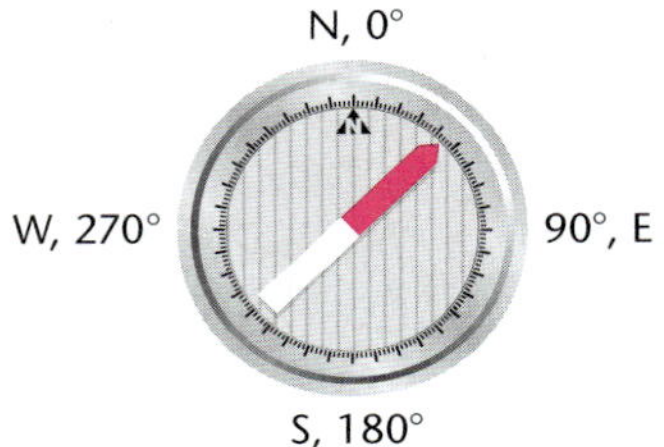

Navigational compass

59. Navigation. An airplane flies east at 256 miles per hour, and another airplane flies southeast at 304 miles per hour. After 2 hours, how far apart are the two planes?

60. Navigation. An airplane flies with an airspeed of 450 miles per hour and a compass heading of 75°. If the wind is blowing at 65 miles per hour out of the north (from north to south), what is the plane's actual direction and speed relative to the ground? Compute direction to the nearest degree and speed to the nearest mile per hour.

★ **61. Navigation.** An airplane that can cruise at 500 miles per hour in still air is to fly due east. If the wind is blowing from the northeast at 50 miles per hour, what compass heading should the pilot choose? What will be the actual speed of the plane relative to the ground? Compute direction to the nearest degree and speed to the nearest mile per hour.

★ **62. Coastal Navigation.** The owner of a pleasure boat cruising along a coast wants to pass a rocky point at a safe distance (see the figure). Sightings of the rocky point are made at A and at B, 1.0 mile apart. If the boat continues on the same course, how close will it come to the point? That is, find d in the figure to the nearest tenth of a mile.

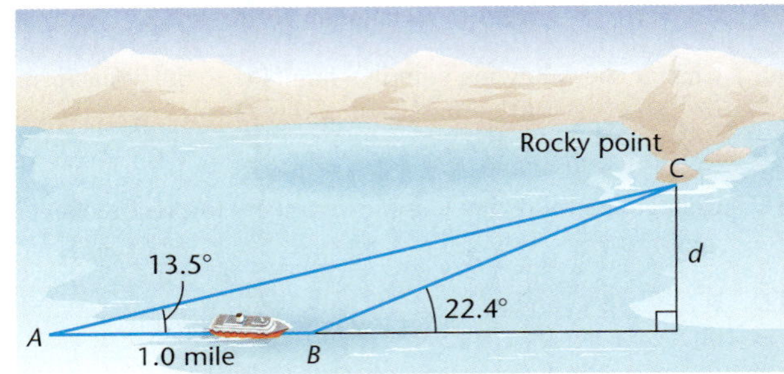

63. Forces. Two forces **u** and **v** are acting on an object as indicated in the figure. Find the direction and magnitude of the resultant force **u** + **v** relative to force **v**.

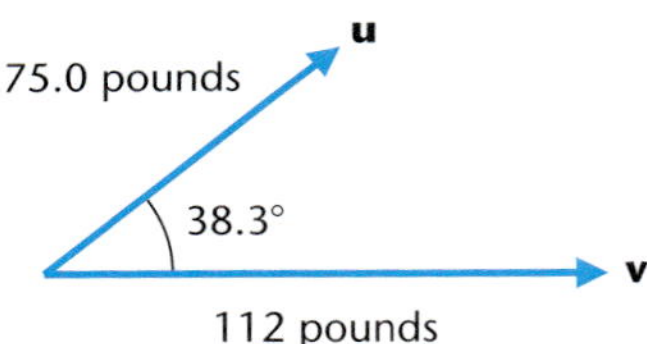

★ **64. Static Equilibrium.** Two forces **u** and **v** are acting on an object as indicated in the figure. What third force **w** must be added to achieve static equilibrium? Give direction relative to **u**.

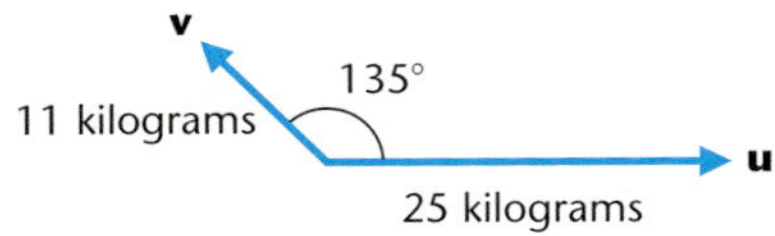

★ **65. Engineering.** A cable car weighing 1,000 pounds is used to cross a river (see the figure). What is the tension in each half of the cable when the car is located as indicated? Compute the answer to three significant digits.

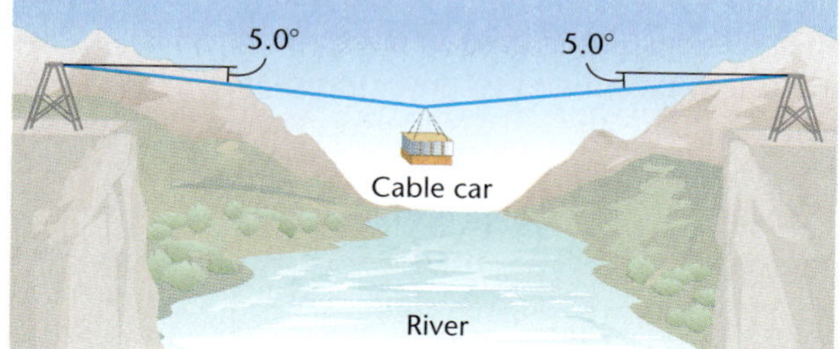

66. Astronomy.

(A) The planet Mars travels around the sun in an elliptical orbit given approximately by

$$r = \frac{1.41 \times 10^8}{1 - 0.0934 \cos \theta} \qquad (1)$$

where r is measured in miles and the sun is at the pole. Graph the orbit. Use TRACE to find the distance (to three significant digits) from Mars to the sun at **aphelion** (greatest distance from the sun) and at **perihelion** (shortest distance from the sun).

(B) Referring to equation (1), r is maximum when the denominator is minimum, and r is minimum when the denominator is maximum. Use this information to find the distance from Mars to the sun at aphelion and at perihelion.

Conic Sections and Planetary Orbits

I Conic Sections in Polar Form

(A) Introduction to Conics. To understand orbits of planets, comets and other celestial bodies, you must know something of the nature and properties of conic sections. (Conic sections are treated in detail in Chapter 11. Here our treatment will be brief and limited to polar representations.) **Conic sections** get their name because the curves are formed by cutting a complete right circular cone of two nappes with a plane (Fig. 1). Any plane perpendicular to the axis of the cone cuts a section that is a **circle.** Tilt the plane slightly and the section becomes an **ellipse.** If the plane is parallel to one edge of the cone, it will cut only one nappe and the section will be a **parabola.** Tilt the plane further to the vertical, then it will cut both nappes of the cone and produce a **hyperbola** with two branches. Closed orbits are ellipses or circles. Open (or escape) orbits are parabolas or hyperbolas.

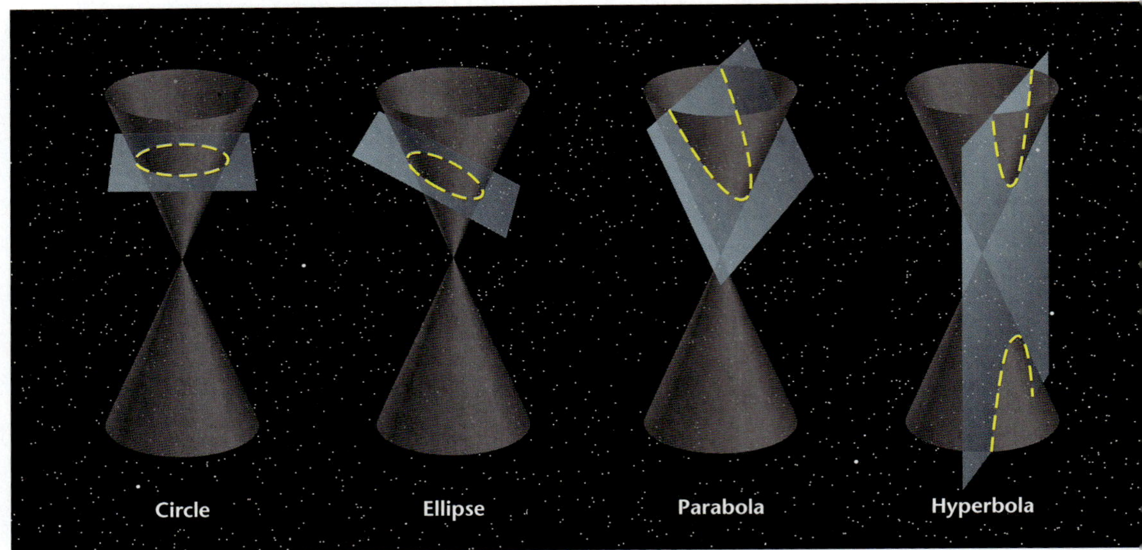

FIGURE 1 Conic sections.

(B) Conics and Eccentricity. Another way of defining conic sections is in terms of their eccentricity. Let F be a fixed point, called the **focus,** and let d be a fixed line, called the **directrix** (Fig. 2). For positive values of **eccentricity** e, a conic section can be defined as the set of points $\{P\}$ having the property that the ratio of the distance from P to the focus F to the distance from P to the directrix d is the constant e. As we will see, an ellipse, a parabola, or a hyperbola can be obtained by choosing e appropriately.

(C) Polar Representation of Conics. A unified treatment of conic sections can be obtained by use of the polar coordinate system. Polar equations of conics are used extensively in celestial mechanics to describe and analyze orbits of planets, comets, satellites, and other celestial bodies.

Problem 1: Polar Equation of a Conic. Use the eccentricity definition of a conic section given in part B to show that the polar equation of a conic is given by

$$r = \frac{ep}{1 - e\cos\theta} \tag{1}$$

FIGURE 2 Conic sections.

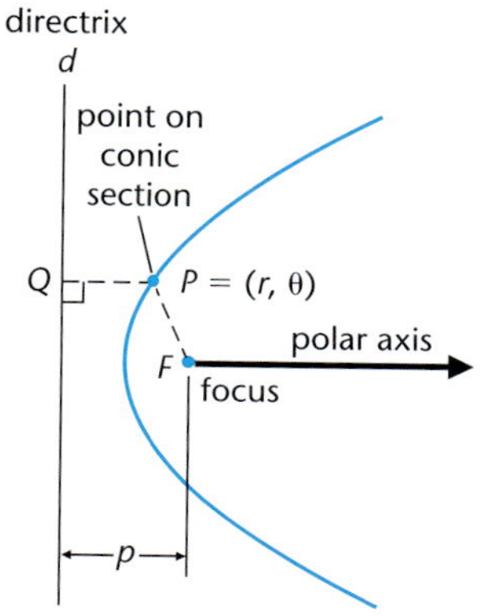

where p is the distance between the focus F and the directrix d, the pole of the polar axis is at F, and the polar axis is perpendicular to d and is pointing away from d (see Fig. 2).

Problem 2: Graphing Utility Exploration, $0 < e < 1$. For $0 < e < 1$, use a graphing utility to systematically explore the nature of the changes in the graph of equation (1) as you change the eccentricity e and the distance p. Summarize the results of holding e fixed and changing p, and the results of holding p fixed and changing e. For $0 < e < 1$, which conic section is produced?

Problem 3: Graphing Utility Exploration, $e = 1$. For $e = 1$, use a graphing utility to systematically explore the nature of the changes in the graph of equation (1) as you change the distance p. Summarize the results of holding e to 1 and changing p. For $e = 1$, which conic section is produced?

Problem 4: Graphing Utility Exploration, $e > 1$. For $e > 1$, use a graphing utility to systematically explore the nature of the changes in the graph of equation (1) as you change the eccentricity e and the distance p. Summarize the results of holding e fixed and changing p, and the results of holding p fixed and changing e. For $e > 1$, which conic section is produced?

II Planetary Orbits

We are now interested in finding polar equations for the orbits of specific planets where the sun is at the pole. Then these equations can be graphed in a graphing utility and further questions about the orbits can be answered. The material in Table 1, found in the readily available *World Almanac* and rounded to three significant digits, gives us more than enough information to find the polar equation for any planet's orbit.

TABLE 1 The Planets

Planet	Eccentricity	Maximum Distance from Sun (Millions of Miles)	Minimum Distance from Sun (Millions of Miles)
Mercury	0.206	43.4	28.6
Venus	0.00677	67.7	66.8
Earth	0.0167	94.6	91.4
Mars	0.0934	155	129
Jupiter	0.0485	507	461
Saturn	0.0555	938	838
Uranus	0.0463	1,860	1,670
Neptune	0.00899	2,820	2,760
Pluto	0.249	4,550	2,760

Problem 5: Polar Equations for the Orbits of Mercury, Earth, and Mars. In all cases the polar axis intersects the planet's orbit at aphelion (the greatest distance from the sun).

(A) Show that Mercury's orbit is given approximately by

$$r = \frac{3.44 \times 10^7}{1 - 0.206 \cos \theta}$$

(B) Show that Earth's orbit is given approximately by

$$r = \frac{9.30 \times 10^7}{1 - 0.0167 \cos \theta}$$

(C) Show that Mars' orbit is given approximately by

$$r = \frac{1.41 \times 10^8}{1 - 0.0934 \cos \theta}$$

Problem 6: Plotting the Orbits for Mercury, Earth, and Mars. Plot all three orbits (Mercury, Earth, and Mars) from the equations in parts A, B, and C in the same viewing window of a graphing utility. Choose the window dimensions so that Mars' orbit fills up most of the window.

Problem 7: Finding Distances and Angles Related to Orbits. Figure 3 represents a schematic drawing showing Earth at two locations during its orbit. Find the straight-line distance between the position at A and the position at B to three significant digits. Find the measures of the angles BAO and ABO in degree measure to one decimal place. The Earth's orbit crosses the polar axis at aphelion (the greatest distance from the sun).

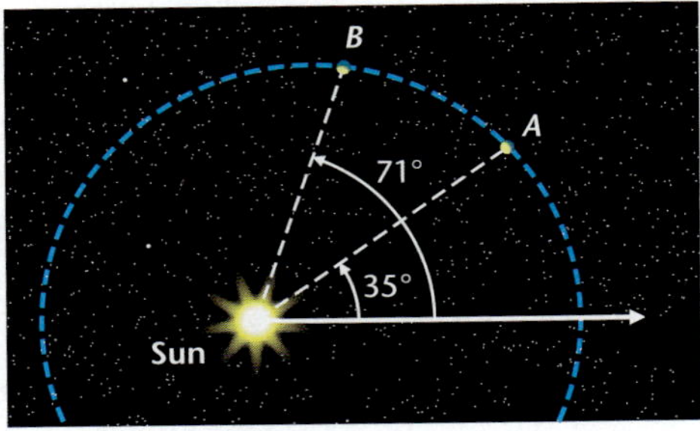

FIGURE 3 Earth's orbit.

CUMULATIVE REVIEW EXERCISES

CHAPTERS 5 6 7

Work through all the problems in this cumulative review and check answers in the back of the book. Answers to all review problems, except verifications, are there, and following each answer is a number in italics indicating the section in which that type of problem is discussed. Where weaknesses show up, review appropriate sections in the text.

A

1. In a circle of radius 6 meters, find the length of an arc opposite an angle of 0.31 radians.

2. Solve the triangle.

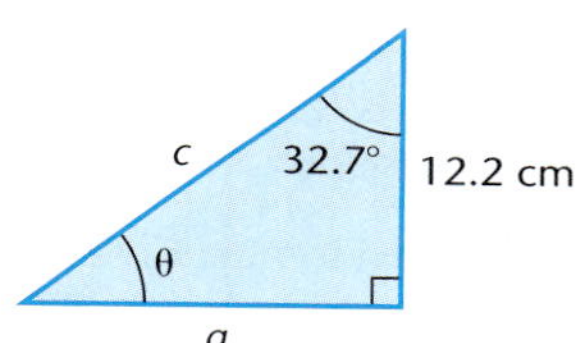

3. In which quadrants is each positive?
(A) $\sin\theta$ **(B)** $\cos\theta$ **(C)** $\tan\theta$

4. If $(-3, 4)$ is on the terminal side of an angle θ, find
(A) $\cos\theta$ **(B)** $\csc\theta$ **(C)** $\tan\theta$

5. Find the reference angle associated with each angle θ:
(A) $-3\pi/4$ **(B)** $245°$ **(C)** $-30°$

6. Indicate the domain, range, and period of each.
(A) $y = \sin x$ **(B)** $y = \cos x$ **(C)** $y = \tan x$

7. Sketch a graph of $y = \cos x$, $-\pi/2 \le x \le 5\pi/2$.

8. Sketch a graph of $y = \tan x$, $-\pi/2 < x < 3\pi/2$.

9. Describe the meaning of a central angle in a circle with radian measure 2.

10. Describe the smallest shift of the graph of $y = \cos x$ to produce the graph of $y = \sin x$.

Verify each identity in Problems 11–14.

11. $\cot\theta \sec\theta = \csc\theta$

12. $\sec x - \cos x = \tan x \sin x$

13. $\sin(x - \pi/2) = -\cos x$

14. $\csc 2x = \frac{1}{2}\csc x \sec x$

15. Use a graphing utility to test whether each of the following is an identity. If an equation appears to be an identity, verify it. If the equation does not appear to be an identity, find a value of x for which both sides are defined but are not equal.

(A) $\dfrac{\sin^2 x}{\cos x} + \cos x = \csc x$

(B) $\dfrac{\sin^2 x}{\cos x} + \cos x = \sec x$

16. If in a triangle, $a = 32.5$ feet, $c = 77.2$ feet, and $\beta = 61.3°$, without solving the triangle or drawing any pictures, which of the two angles, α or γ, can you say for certain is acute and why?

Solve Problems 17 and 18 to four decimal places.

17. $\sin x = 0.3188$, $0 \le x \le 2\pi$

18. $\tan\theta = -4.076$, $-90° < \theta < 90°$

19. Solve the triangle.

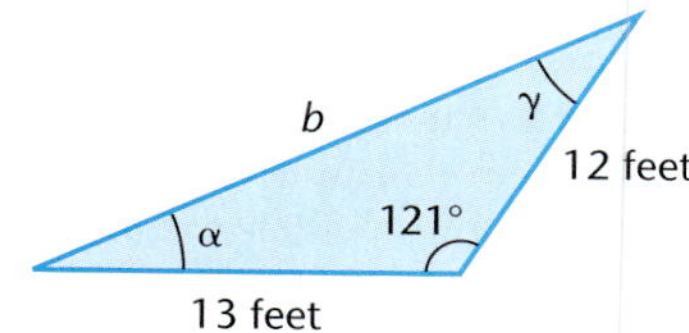

20. Write the algebraic vector $\langle a, b\rangle$ corresponding to the geometric vector $\overrightarrow{AB}$ with endpoints $A = (-3, 2)$ and $B = (3, -1)$.

21. A point in a polar coordinate system has coordinates $(-5, 150°)$. Find all other polar coordinates for the point, $-360° \le \theta \le 360°$, and verbally describe how the coordinates are associated with the point.

22. Sketch a graph of $r = 6\cos\theta$ in a polar coordinate system.

23. Plot in a complex plane: $A = -3 + 4i$ and $B = 4e^{60°i}$.

24. Find $(2e^{10°i})^3$. Write the final answer in exact rectangular form.

B

25. Which of the following angles are coterminal with 150°: 30°, $-7\pi/6$, 870°?

26. Change 1.31 radians to decimal degrees to two decimal places.

27. Which of the following have the same value as cos 8?
(A) $\cos(8 \text{ rad})$ **(B)** $\cos 8°$ **(C)** $\cos(8 - 4\pi)$

Evaluate Problems 28–37 exactly without a calculator. If the function is not defined at the value, say so.

28. $\sin(-5\pi/6)$ **29.** $\tan(\pi/2)$ **30.** $\cot(7\pi/4)$

31. $\sec 330°$ **32.** $\cos^{-1}(-1)$ **33.** $\sin^{-1} 1.5$

34. $\arccos(-\frac{1}{2})$ **35.** $\sin(\sin^{-1} 0.55)$ **36.** $\cos[\sin^{-1}(-\frac{4}{5})]$

37. $\cos[\tan^{-1}(-2)]$

38. Evaluate to four significant digits using a calculator. If a function is not defined, say so.
(A) $\tan 84°12'55''$ **(B)** $\sec(-1.8409)$
(C) $\tan^{-1}(-84.32)$ **(D)** $\cos^{-1}(\tan 2.314)$

39. Sketch a graph of $y = 2 - 2\cos(\pi x/2)$, $-1 \le x \le 5$.

40. **(A)** Find the exact degree measure of $\theta = \cos^{-1}(-\sqrt{3}/2)$ without a calculator.
(B) Find the degree measure of $\theta = \sin^{-1}(-0.338)$ to three decimal places using a calculator.

41. Evaluate $\sin^{-1}(\sin 3)$ with a calculator set in radian mode, and explain why this does or does not illustrate a sine–inverse sine identity.

42. A circular point $P = (a, b)$ moves counterclockwise around the circumference of a unit circle starting at (1, 0) and stops after covering a distance of 11.205 units. Explain how you would find the coordinates of point P at its final position and how you would determine which quadrant P is in. Find the coordinates of P to three decimal places and the quadrant for the final position of P.

43. Explain the difference in solving the equation $\tan x = -24.5$ and evaluating $\tan^{-1}(-24.5)$.

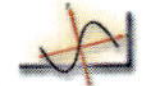

44. Find an equation of the form $y = k + a \sin Bx$ that produces the graph shown at the top of the next column.

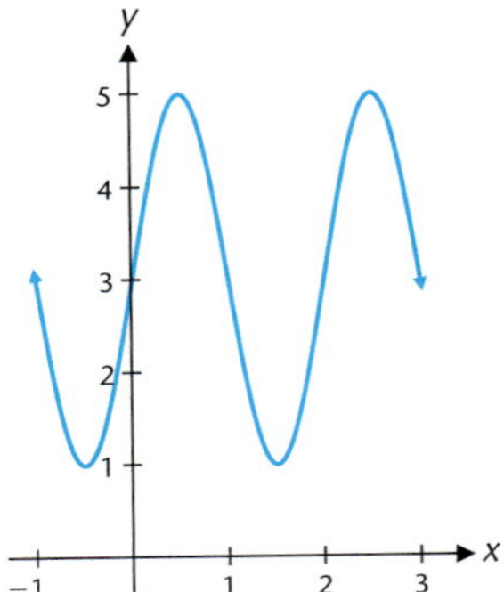

45. Sketch a graph of $y = 3\sin(2x - \pi)$, $-\pi \le x \le 2\pi$. Indicate amplitude A, period P, and phase shift $P.S.$

46. Sketch a graph of $y = 2\tan(\pi x/2 - \pi/2)$, $0 < x < 4$. Indicate the period P and phase shift $P.S.$

47. Sketch a graph of $y = \sin x$ and $y = \csc x$ in the same coordinate system.

48. Describe the smallest left shift and/or reflection that transforms the graph of $y = \cot x$ into the graph of $y = \tan x$.

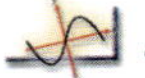

49. Graph $y = 1/(\cot^2 x + 1)$ in a graphing utility that displays at least two full periods of the graph. Find an equation of the form $y = k + A\sin Bx$ or $y = k + A\cos Bx$ that has the same graph. Graph both equations in the same viewing window and use TRACE to verify that both graphs are the same.

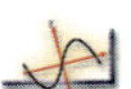

50. Graph $y = (2 - 2\sin^2 x)/(\sin 2x)$ in a graphing utility that displays at least two full periods of the graph. Find an equation of the form $y = A\tan Bx$ or $y = A\cot Bx$ that has the same graph. Graph both equations in the same viewing window and use TRACE to verify that both graphs are the same.

51. Given the equation $\sin 2x = 2\sin x$,
(A) Are $x = 0$ and $x = \pi$ solutions?
(B) Is the equation an identity or a conditional equation? Explain.

Verify each identity in Problems 52–57.

52. $\dfrac{\sin u}{1 + \cos u} + \cot u = \csc u$

53. $\sec x + \tan x = \dfrac{\cos x}{1 - \sin x}$

54. $\tan \dfrac{x}{2} = \csc x - \cot x$

55. $\csc^2 \dfrac{x}{2} = 2 \csc x (\csc x + \cot x)$

56. $\dfrac{2}{1 + \cos 2x} = \sec^2 x$

57. $\dfrac{\cos x + \cos y}{\sin x - \sin y} = \cot \dfrac{x - y}{2}$

[*Hint:* Use sum–product identities.]

58. Use a graphing utility to test whether each of the following is an identity. If an equation appears to be an identity, verify it. If the equation does not appear to be an identity, find a value of x for which both sides are defined but are not equal.

(A) $\dfrac{\tan x}{2 \tan x - \sin x} = \dfrac{1}{2 + \sin x}$

(B) $\dfrac{\tan x}{2 \tan x - \sin x} = \dfrac{1}{2 - \cos x}$

59. Find $\cos (x - y)$ exactly without a calculator given $\sin x = (-2/\sqrt{5})$, $\cos y = (-2/\sqrt{5})$, x a quadrant IV angle, and y a quadrant III angle.

60. Compute the exact value of $\sin 2x$ and $\cos (x/2)$ without a calculator, given $\sin x = \frac{3}{5}$, $\pi/2 \le x \le \pi$.

Solve Problems 61 and 62 exactly without a calculator, θ in degrees and x real.

61. $2 \sin^2 \theta + \sin \theta = 1, 0 \le \theta < 360°$

62. $\sin 2x = \sin x$, all real solutions

63. **(A)** Solve $\cot x = -2 \cos x$ exactly, $0 \le x \le 2\pi$.
(B) Solve $\cot x = -2 \cos x$ to three decimal places using a graphing utility, $0 \le x \le 2\pi$.

64. Solve $2 \cos x = x - \cos 2x$ to three decimal places for all real solutions using a graphing utility.

In Problems 65–67, solve each triangle labeled as in the figure at the top of the next column. If a problem does not have a solution, say so. If a triangle has two solutions, solve the obtuse case.

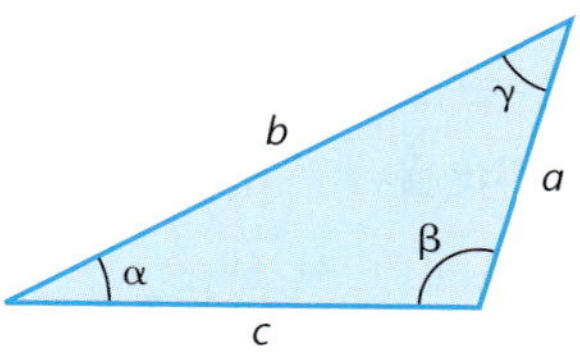

65. $a = 21.3$ meters, $b = 37.4$ meters, $c = 48.2$ meters

66. $\alpha = 125.4°$, $b = 25.4$ millimeters, $a = 20.3$ millimeters

67. $\alpha = 52.9°$, $b = 37.1$ inches, $a = 34.4$ inches

68. Assume in a triangle that γ is acute, $a = 92.5$ centimeters, and $b = 43.4$ centimeters. Which of the angles, α or β, can you say for certain is acute and why?

69. Given geometric vectors as indicated in the figures, find $|\mathbf{u} + \mathbf{v}|$ and α, given $|\mathbf{u}| = 25.3$ pounds, $|\mathbf{v}| = 13.4$ pounds, and $\theta = 48.3°$.

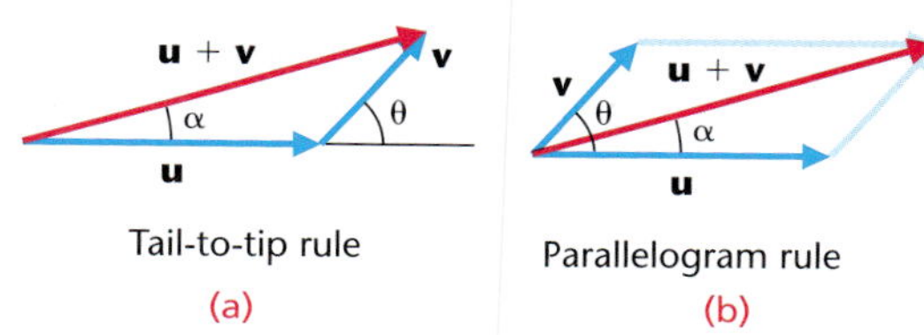

Tail-to-tip rule
(a)

Parallelogram rule
(b)

70. Find $2\mathbf{u} - \mathbf{v} + 3\mathbf{w}$ for,
(A) $\mathbf{u} = \langle -1, 2\rangle, \mathbf{v} = \langle 0, -2\rangle, \mathbf{w} = \langle 1, -1\rangle$
(B) $\mathbf{u} = 2\mathbf{i} - \mathbf{j}, \mathbf{v} = \mathbf{i} + 3\mathbf{j}, \mathbf{w} = 2\mathbf{j}$

71. Convert to polar form: $x^2 + y^2 = 8y$.

72. Convert $r = -4 \cos \theta$ to rectangular form.

Use rapid sketching techniques to graph Problems 73 and 74 in a polar coordinate system. Check by graphing on a graphing utility.

73. $r = 4 + 4 \cos \theta$

74. $r = 6 \sin 3\theta$

75. Graph $r = 5(\cos 2\theta)^{2n}$, for $n = 1, 2,$ and 3. How many leaves do you expect the graph will have for arbitrary n?

76. Graph $r = e^{(\cos \theta)} - 2 \cos (4\theta)$ using a squared window and 0.05 for a step size for θ. The resulting curve is often referred to as *a butterfly curve.*

77. Change the rectangular coordinates $(-2.78, -3.19)$ to polar coordinates to two decimal places, $r \ge 0$, $-180° < \theta \le 180°$.

78. Change the polar coordinates $(6.22, -4.08)$ to rectangular coordinates to two decimal places.

79. Change $2e^{(-\pi/6)i}$ to exact rectangular form.

80. Change $z = -1 + i\sqrt{3}$ to the polar form $re^{i\theta}$, θ in degrees.

81. Compute $(1 - i\sqrt{3})^6$ using De Moivre's theorem and write the final answer in $a + bi$ form.

82. Find all cube roots of $-i$ exactly. Write final answers in the form $a + bi$, and locate the roots on a circle in the complex plane.

83. Change the complex number $-4.88 - 3.17i$ to the polar form $re^{i\theta}$ to two decimal places, $r \geq 0$, $-180° < \theta \leq 180°$.

84. Change the complex number $6.97e^{163.87°i}$ to rectangular form $a + bi$, where a and b are computed to two decimal places.

85. **(A)** The fourth root of a complex number is shown in the figure. Geometrically locate all other fourth roots of the number on the figure, and explain how they were located.
(B) Determine geometrically the other fourth roots of the number in exact rectangular form.
(C) Raise each fourth root from parts A and B to the fourth power.

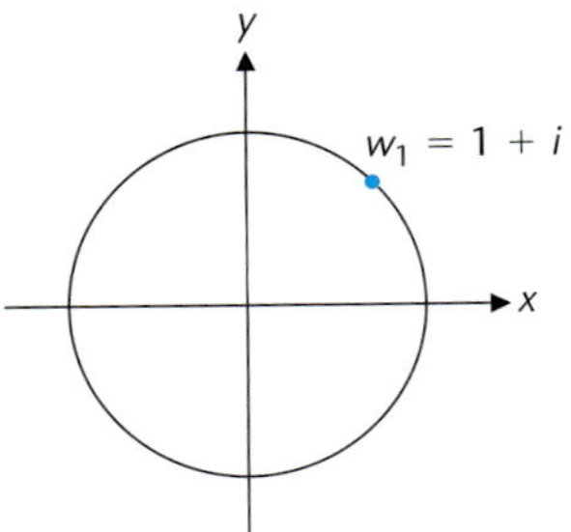

C

86. If, in the figure, the coordinates of A are $(1, 0)$ and arc length s is 1.2 units, find the coordinates of P to three significant digits.

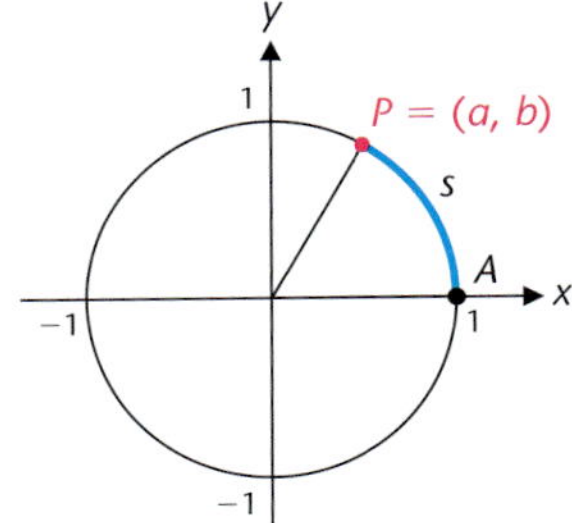

87. Sketch a graph of $y = 1 + \sec x$, $-3\pi/2 < x < 3\pi/2$.

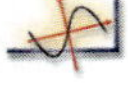

88. The accompanying graph is a graph of an equation of the form $y = A \cos (Bx + C)$, $0 < -B/C < 1$. Find the equation by finding A, B, and C exactly. What are the period, amplitude, and phase shift?

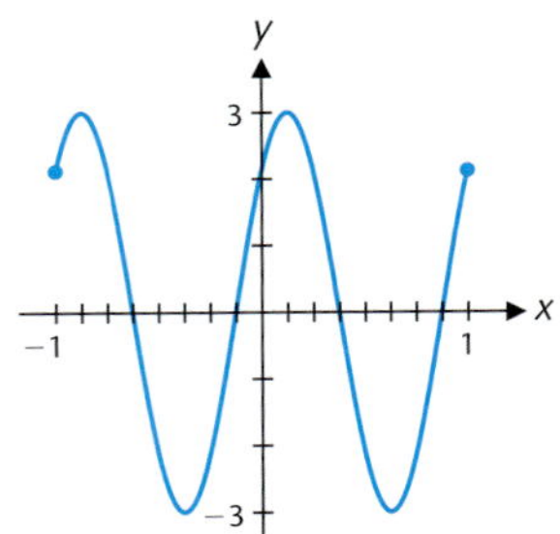

89. Graph $1.6 \sin 2x - 1.2 \cos 2x$ in a graphing utility. (Select the dimensions of a viewing window so that at least two periods are visible.) Find an equation of the form $y = A \sin (Bx + C)$ that has the same graph as the given equation. Find A and B exactly and C to three decimal places. Use the x intercept closest to the origin as the phase shift. To check your results graph both equations in the same viewing window and use TRACE while shifting back and forth between the two graphs.

90. Write $\csc (\cos^{-1} x)$ as an algebraic expression in x free of trigonometric or inverse trigonometric functions.

Solve Problems 91 and 92 without a calculator.

91. $\sin [2 \cot^{-1} (\frac{3}{4})] = ?$

92. Given $\sec x = -5/3$, $\pi/2 \leq x \leq \pi$, find
(A) $\sin (x/2)$ **(B)** $\cos 2x$

93. **(A)** Solve $2 \sin^2 x = 3 \cos x$ exactly for all real solutions, $0 \leq x \leq 2\pi$.
(B) Solve $2 \sin^2 x = 3 \cos x$ to four decimal places using a graphing utility, $0 \leq x \leq 2\pi$.

94. **(A)** Use rapid sketching techniques to sketch a graph of the polar equation $r^2 = 36 \cos 2\theta$.
(B) Verify the graph in part A using a graphing utility.

95. **(A)** Graph $r_1 = 2 + 2 \cos \theta$ and $r_2 = 6 \cos \theta$ in the same viewing window, $0 \leq \theta \leq 2\pi$.

(B) Use TRACE to determine how many times the graph of r_2 crosses the graph of r_1 as θ goes from 0 to 2π.

(C) Solve the two equations simultaneously to find the exact solutions for $0 \leq \theta \leq 2\pi$.

(D) Explain why the number of solutions found in part C does not agree with the number of times r_1 crosses r_2, $0 \leq \theta \leq 2\pi$.

96. Write $P(x) = x^3 + i$ as a product of linear factors.

APPLICATIONS

97. Astronomy. A line from the sun to the Earth sweeps out an angle of how many radians in 5 days?

98. Meteorology. A weather balloon is released and rises vertically. Two weather stations C and D in the same vertical plane as the balloon and 1,000 meters apart sight the balloon at the same time and record the information given in the figure. At the time of sighting, how high was the balloon to the nearest meter?

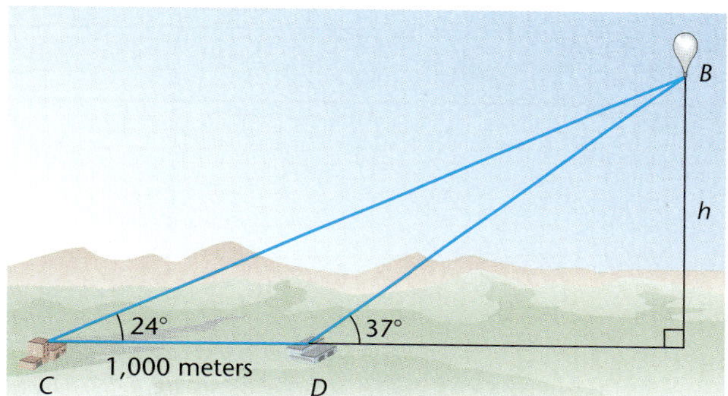

99. Geometry. Find the length to two decimal places of one side of a regular pentagon inscribed in a circle with radius 5 inches.

100. Geometry. Find $\angle ABC$ to the nearest degree in the rectangular solid shown in the figure.

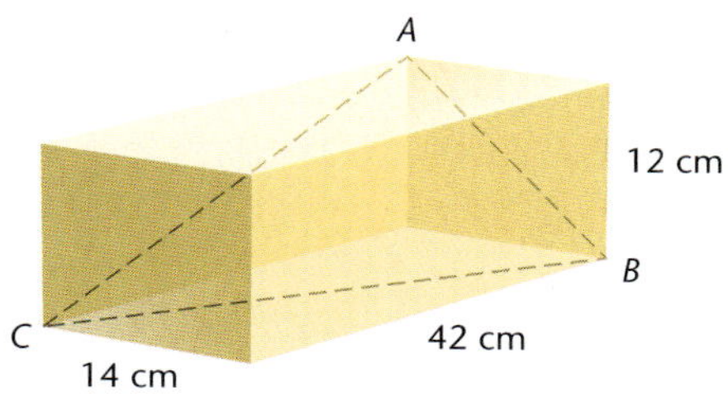

101. Electrical Circuit. The current I in an alternating electrical circuit has an amplitude of 50 amperes and a period of $\frac{1}{110}$ second. If $I = 50$ amperes when $t = 0$, find an equation of the form $I = A \cos Bt$ that gives the current at time $t \geq 0$.

102. Navigation. An airplane flies with an airspeed of 260 miles per hour and a compass heading of 110°. If a 36 mile per hour wind is blowing out of the north, what is the plane's actual heading and ground speed? Compute direction to the nearest degree and ground speed to the nearest mile per hour.

103. Engineering. A 65-pound child glides across a small river on a homemade cable trolley (see the figure). What is the tension on each half of the support cable when the child is in the center? Compute answer to nearest pound.

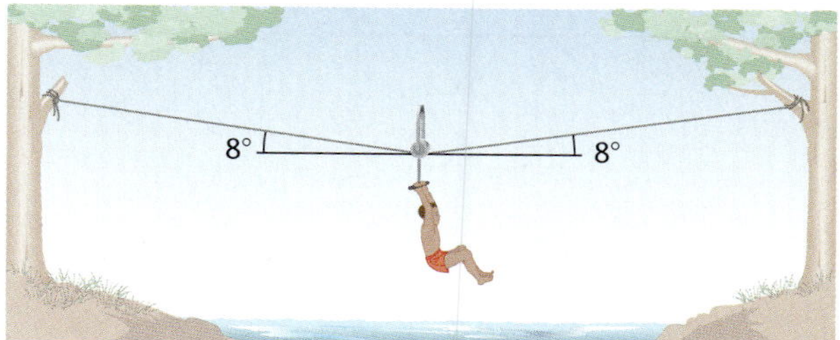

104. Geometry. A circular arc of 10 centimeters has a chord of 8 centimeters as shown in the figure.

(A) Explain how the radius is given by the equation

$$\sin \frac{5}{R} = \frac{4}{R}$$

(B) What difficulties do you encounter in trying to solve the equation in part A exactly using algebraic and trigonometric methods?

(C) Show on a graphing utility how to approximate the radius of the circle R, and find R to three decimal places.

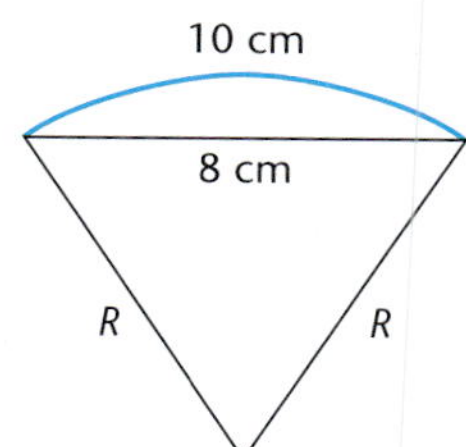

MODELING AND DATA ANALYSIS 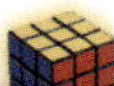

105. Modeling Temperature Variation. The 30-year average monthly temperature, in degrees Fahrenheit, for each month of the year for Washington, D.C., is given in Table 1 (from the *World Almanac*).

(A) Using 1 month as the basic unit of time, enter the data for a 2-year period in your graphing utility and produce a scatter plot in the viewing window. Choose $25 \leq y \leq 80$ for the viewing window.

(B) It appears that a sine curve of the form

$$y = k + A \sin (Bx + C)$$

will closely model these data. The constants k, A, and B are easily determined from Table 1. To estimate C, visually estimate to one decimal place the smallest positive phase shift from the plot in part A. After determining A, B, k, and C, write the resulting equation. (Your value of C may differ slightly from the answer in the book.)

(C) Plot the results of parts A and B in the same viewing window. (An improved fit may result by adjusting your value of C a little.)

(D) If your graphing utility has a sinusoidal regression feature, check your results from parts B and C by finding and plotting the regression equation.

TABLE 1 Monthly Average Temperatures, Washington, D.C.

x (months)	1	2	3	4	5	6	7	8	9	10	11	12
y (temperature)	31	34	43	53	62	71	76	74	67	55	45	35

CHAPTER 8

Modeling with Linear Systems

OUTLINE

IN CHAPTER 8 WE FIRST DISCUSS HOW SYSTEMS OF LINEAR EQUATIONS involving two variables are solved graphically and algebraically. Because these techniques are not suitable for linear systems involving larger numbers of equations and variables, we then turn to a different method of solution involving the concept of an *augmented matrix,* which arises quite naturally when dealing with larger linear systems. Finally, we discuss systems of linear inequalities and linear programming, a relatively new and powerful mathematical tool that will be used to solve a variety of interesting practical problems.

Preparing for this chapter

Before getting started on this chapter, review the following concepts:

- Properties of Real Numbers (Basic Algebra Review*, Section R.1)
- Linear Equations and Inequalities (Appendix A, Section A.1)
- Linear Functions (Chapter 2, Sections 1 and 2)

*At www.mhhe.com/barnett

SECTION 8.1 Systems of Linear Equations in Two Variables

Systems of Equations • Graphing • Substitution • Modeling with Systems of Linear Equations

In Section 8.1 we discuss both graphical and algebraic methods for solving systems of linear equations in two variables. Then we use systems of this type to construct and solve mathematical models for several applications.

Systems of Equations

To establish basic concepts, consider the following example. At a computer fair, student tickets cost \$2 and general admission tickets cost \$3. If a total of 7 tickets are purchased for a total cost of \$18, how many of each type were purchased?

Let

x = Number of student tickets

y = Number of general admission tickets

Then

$$x + y = 7 \quad \text{Total number of tickets purchased}$$
$$2x + 3y = 18 \quad \text{Total purchase cost}$$

We now have a system of two linear equations in two variables. Thus, we can solve this problem by finding all pairs of numbers x and y that satisfy both equations.

In general, we are interested in solving linear systems of the type

$$ax + by = h \quad \text{System of two linear equations in two variables}$$
$$cx + dy = k$$

where x and y are variables, a, b, c, and d are real numbers called the **coefficients** of x and y, and h and k are real numbers called the **constant terms** in the equations. A pair of numbers $x = x_0$ and $y = y_0$ is a **solution** of this system if each equation is satisfied by the pair. The set of all such pairs of numbers is called the **solution set** for the system. To **solve** a system is to find its solution set.

Graphing

Recall that the graph of a linear equation is the line consisting of all ordered pairs that satisfy the equation. To solve the ticket problem by graphing, we graph both equations in the same coordinate system. The coordinates of any points that the lines have in common must be solutions to the system, because they must satisfy both equations.

EXAMPLE 1 Solving a System by Graphing

Solve the ticket problem by graphing:

$$\begin{aligned} x + y &= 7 \\ 2x + 3y &= 18 \end{aligned}$$

SOLUTIONS

Hand-Drawn Solution

Find the x and y intercepts for each line.

$x + y = 7$

x	y
0	7
7	0

$2x + 3y = 18$

x	y
0	6
9	0

Plot these points, graph the two lines, estimate the intersection point visually (Fig. 1), and check the estimate.

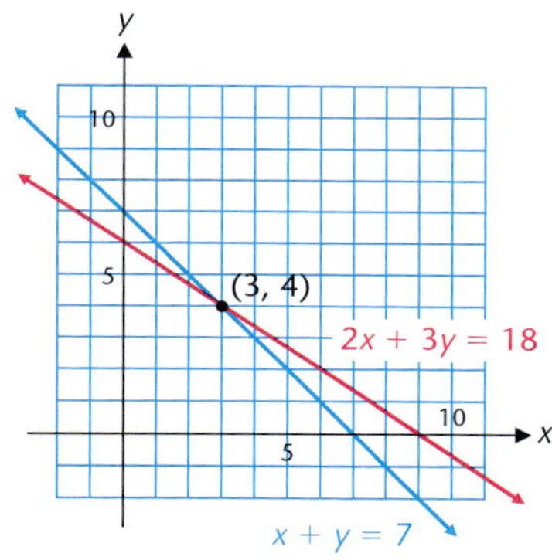

FIGURE 1

$x = 3$ Student tickets

$y = 4$ General admission tickets

CHECK

$$\begin{aligned} x + y &= 7 \\ 3 + 4 &\stackrel{?}{=} 7 \\ 7 &\stackrel{\checkmark}{=} 7 \end{aligned} \qquad \begin{aligned} 2x + 3y &= 18 \\ 2(3) + 3(4) &\stackrel{?}{=} 18 \\ 18 &\stackrel{\checkmark}{=} 18 \end{aligned}$$

Graphing Utility Solution

First, solve each equation for y:

$$\begin{aligned} x + y &= 7 \\ y &= 7 - x \end{aligned} \qquad \begin{aligned} 2x + 3y &= 18 \\ 3y &= 18 - 2x \\ y &= 6 - \tfrac{2}{3}x \end{aligned}$$

Next, enter these functions in the equation editor of a graphing utility (Fig. 2) and use the intersect command to find the intersection point (Fig. 3).

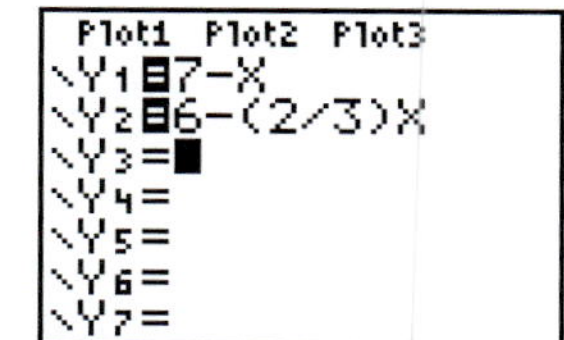

FIGURE 2

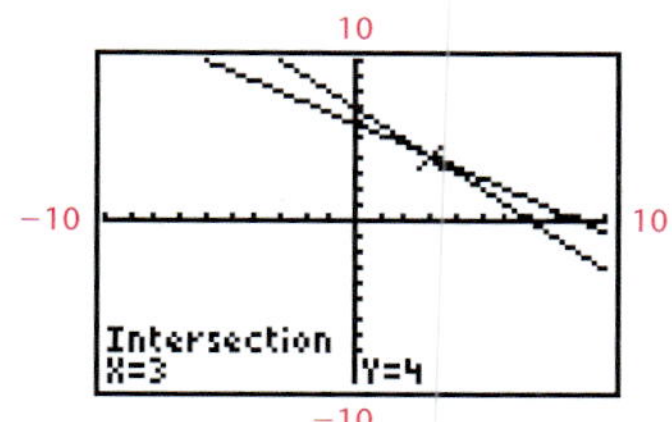

FIGURE 3

From Figure 3, we see that the solution is

$x = 3$ Student tickets

$y = 4$ General admission tickets

MATCHED PROBLEM 1

Solve two ways as in Example 1:

$$\begin{aligned} x - y &= 3 \\ x + 2y &= -3 \end{aligned}$$

It is clear that Example 1 has exactly one solution, because the lines have exactly one point of intersection. In general, lines in a rectangular coordinate system are related to each other in one of three ways, as illustrated in Example 2.

EXAMPLE 2 Determining the Nature of Solutions

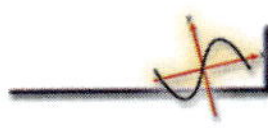

Match each of the following systems with one of the graphs in Figure 4 and discuss the nature of the solutions:

(A) $2x - 3y = 2$
$x + 2y = 8$

(B) $4x + 6y = 12$
$2x + 3y = -6$

(C) $2x - 3y = -6$
$-x + \frac{3}{2}y = 3$

FIGURE 4

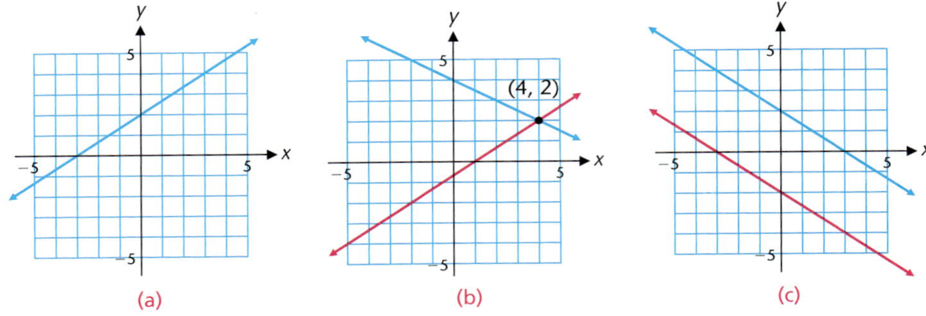

SOLUTIONS

(A) Graph (b). The lines intersect in a single point. There is exactly one solution: $x = 4$, $y = 2$.

(B) Graph (c). The lines are parallel and never intersect. There are no solutions.

(C) Graph (a). The lines coincide. Every point on the line is a solution. There are an infinite number of solutions.

MATCHED PROBLEM 2

Solve each of the following systems by graphing:

(A) $2x + 3y = 12$
$x - 3y = -3$

(B) $x - 3y = -3$
$-2x + 6y = 12$

(C) $2x - 3y = 12$
$-x + \frac{3}{2}y = -6$

We now define some terms that can be used to describe the different types of solutions to systems of equations illustrated in Example 2.

Systems of Linear Equations: Basic Terms

A system of linear equations is **consistent** if it has one or more solutions and **inconsistent** if no solutions exist. Furthermore, a consistent system is said to be **independent** if it has exactly one solution (often referred to as the **unique solution**) and **dependent** if it has more than one solution.

Referring to the three systems in Example 2, the system in part A is consistent and independent, with the unique solution $x = 4$ and $y = 2$. The system in part B is inconsistent, with no solution. And the system in part C is consistent and dependent, with an infinite number of solutions: all the points on the two coinciding lines.

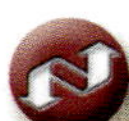

EXPLORE/DISCUSS 1

Can a consistent and dependent system have exactly two solutions? Exactly three solutions? Explain.

By geometrically interpreting a system of two linear equations in two variables, we gain useful information about what to expect in the way of solutions to the system. In general, any two lines in a rectangular coordinate plane must intersect in exactly one point, be parallel, or coincide (have identical graphs). Thus, the systems in Example 2 illustrate the only three possible types of solutions for systems of two linear equations in two variables. These ideas are summarized in Theorem 1.

THEOREM 1
Possible Solutions to a Linear System

The linear system

$$ax + by = h$$
$$cx + dy = k$$

must have

1. Exactly one solution — Consistent and independent

   or

2. No solution — Inconsistent

   or

3. Infinitely many solutions — Consistent and dependent

There are no other possibilities.

One drawback of finding a solution by graphing is the inaccuracy of hand-drawn graphs. Graphic solutions performed on a graphing utility, however, provide both a useful geometric interpretation and an accurate approximation of the solution to a system of linear equations in two variables.

EXAMPLE 3 Solving a System Using a Graphing Utility

Solve to two decimal places using a graphing utility:
$$5x - 3y = 13$$
$$2x + 4y = 15$$

SOLUTION

First solve each equation for y:

$5x - 3y = 13$	$2x + 4y = 15$
$-3y = -5x + 13$	$4y = -2x + 15$
$y = \frac{5}{3}x - \frac{13}{3}$	$y = -0.5x + 3.75$

Next, enter each equation in a graphing utility [Fig. 5(a)], graph in an appropriate viewing window, and approximate the intersection point [Fig. 5(b)].

FIGURE 5

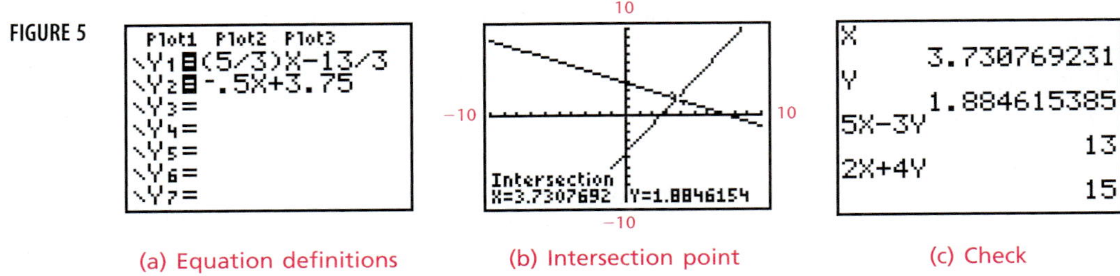

(a) Equation definitions (b) Intersection point (c) Check

Rounding the values in Figure 5(b) to two decimal places, we see that the solution is

$$x = 3.73 \text{ and } y = 1.88 \qquad \text{or} \qquad (3.73, 1.88)$$

Figure 5(c) shows a check of this solution.

MATCHED 3 PROBLEM

Solve to two decimal places using a graphing utility:
$$\begin{aligned} 2x - 5y &= -25 \\ 4x + 3y &= 5 \end{aligned}$$

REMARK In the solution to Example 3, you might wonder why we checked a solution produced by a graphing utility. After all, we don't expect a graphing utility to make an error. But the equations in the original system and the equations entered in Figure 5(a) are not identical. We might have made an error when solving the original equations for y. The check in Figure 5(c) eliminates this possibility.

Graphic methods help us visualize a system and its solutions, frequently reveal relationships that might otherwise be hidden, and, with the assistance of a graphing utility, provide very accurate approximations to solutions.

Substitution

There are a number of different algebraic techniques that can also be used to solve systems of linear equations in two variables. One of the simplest is the *substitution method.* To solve a system by **substitution,** we first choose one of the two equations in a system and solve for one variable in terms of the other. (We make a choice that avoids fractions, if possible.) Then we substitute the result in the other equation and solve the resulting linear equation in one variable. Finally, we substitute this result back into the expression obtained in the first step to find the second variable. We return to the ticket problem stated on p. 614 to illustrate this process.

EXAMPLE 4 Solving a System by Substitution

Use substitution to solve the ticket problem:
$$\begin{aligned} x + y &= 7 \\ 2x + 3y &= 18 \end{aligned}$$

SOLUTION

Solve either equation for one variable and substitute into the remaining equation. We choose to solve the first equation for y in terms of x:

$$x + y = 7 \qquad \text{Solve the first equation for } y \text{ in terms of } x.$$
$$y = 7 - x \qquad \text{Substitute into the second equation.}$$

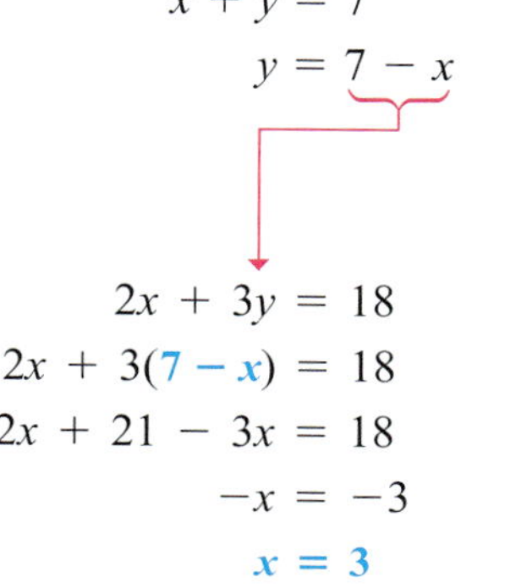

$$\begin{aligned} 2x + 3y &= 18 \\ 2x + 3(7 - x) &= 18 \\ 2x + 21 - 3x &= 18 \\ -x &= -3 \\ x &= 3 \end{aligned}$$

Now, replace x with 3 in $y = 7 - x$:

$$\begin{aligned} y &= 7 - x \\ y &= 7 - 3 \\ y &= 4 \end{aligned}$$

Thus the solution is 3 student tickets and 4 general admission tickets.

CHECK

$x + y = 7$	$2x + 3y = 18$
$3 + 4 \stackrel{?}{=} 7$	$2(3) + 3(4) \stackrel{?}{=} 18$
$7 \stackrel{\checkmark}{=} 7$	$18 \stackrel{\checkmark}{=} 18$

MATCHED 4 PROBLEM

Solve by substitution and check:
$$\begin{aligned} x - y &= 3 \\ x + 2y &= -3 \end{aligned}$$

EXAMPLE 5 Solving a System by Substitution

Solve by substitution and check:
$$\begin{aligned} 2x - 3y &= 7 \\ 3x - y &= 7 \end{aligned}$$

SOLUTION

To avoid fractions, we choose to solve the second equation for y:

$$\begin{aligned} 3x - y &= 7 && \text{Solve for } y \text{ in terms of } x. \\ -y &= -3x + 7 \\ y &= 3x - 7 && \text{Substitute into first equation.} \\ 2x - 3y &= 7 && \text{First equation} \\ 2x - 3(3x - 7) &= 7 && \text{Solve for } x. \\ 2x - 9x + 21 &= 7 \\ -7x &= -14 \\ x &= 2 && \text{Substitute } x = 2 \text{ in } y = 3x - 7. \\ y &= 3x - 7 \\ y &= 3(2) - 7 \\ y &= -1 \end{aligned}$$

Thus, the solution is $x = 2$ and $y = -1$.

CHECK

$$\begin{aligned} 2x - 3y &= 7 \\ 2(2) - 3(-1) &\stackrel{?}{=} 7 \\ 7 &\stackrel{\checkmark}{=} 7 \end{aligned} \qquad \begin{aligned} 3x - y &= 7 \\ 3(2) - (-1) &\stackrel{?}{=} 7 \\ 7 &\stackrel{\checkmark}{=} 7 \end{aligned}$$

MATCHED 5 PROBLEM

Solve by substitution and check: $\begin{aligned} 3x - 4y &= 18 \\ 2x + y &= 1 \end{aligned}$

EXPLORE/DISCUSS 2

Use substitution to solve each of the following systems. Discuss the nature of the solution sets you obtain.

$$\begin{aligned} x + 3y &= 4 \\ 2x + 6y &= 7 \end{aligned} \qquad \begin{aligned} x + 3y &= 4 \\ 2x + 6y &= 8 \end{aligned}$$

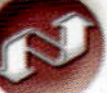

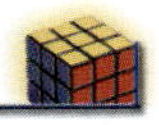

Modeling with Systems of Linear Equations

Examples 6, 7, and 8 illustrate the use of systems of linear equations to construct models for applied problems. Each model can be solved by either graphing or substitution—the choice is really a matter of personal preference.

EXAMPLE 6 Food Processing

A food manufacturer produces regular and lite smoked sausages. A regular sausage is 72% pork and 28% turkey and a lite sausage is 22% pork and 78% turkey. The company has just received a shipment of 2,000 pounds of pork and 2,000 pounds of turkey. How many pounds of each type of sausage should be produced to use all the meat in this shipment?

SOLUTION

First we define the relevant variables:

x = Pounds of regular sausage

y = Pounds of lite sausage

Next we summarize the given information in Table 1. It is convenient to organize the table so that the quantities represented by variables correspond to columns in the table (rather than to rows), as shown.

TABLE 1

	Regular Sausage	Lite Sausage	Total
Pork	72%	22%	2,000
Turkey	28%	78%	2,000

Now we use the information in the table to form equations involving x and y:

$$\begin{pmatrix}\text{Pork in } x \text{ pounds} \\ \text{of regular sausage}\end{pmatrix} + \begin{pmatrix}\text{Pork in } y \text{ pounds} \\ \text{of lite sausage}\end{pmatrix} = \begin{pmatrix}\text{Total} \\ \text{pork}\end{pmatrix}$$

$$0.72x \quad + \quad 0.22y \quad = 2{,}000$$

$$\begin{pmatrix}\text{Turkey in } x \text{ pounds} \\ \text{of regular sausage}\end{pmatrix} + \begin{pmatrix}\text{Turkey in } y \text{ pounds} \\ \text{of lite sausage}\end{pmatrix} = \begin{pmatrix}\text{Total} \\ \text{turkey}\end{pmatrix}$$

$$0.28x \quad + \quad 0.78y \quad = 2{,}000$$

We will solve this system graphically. Figure 6(a) shows the equations after they have been solved for y and entered in the equation editor of a graphing utility. From Figure 6(b), we conclude that producing 2,240 pounds of regular sausage and 1,760 pounds of lite sausage will use all the available pork and turkey.

FIGURE 6

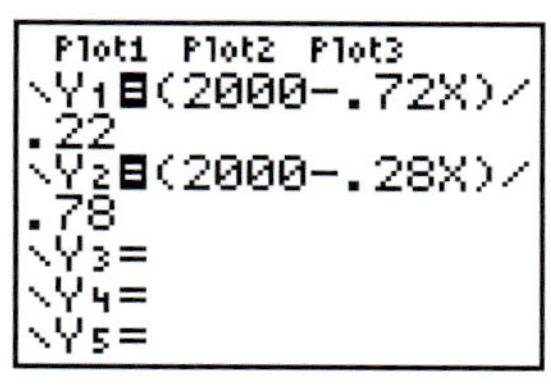

(a)

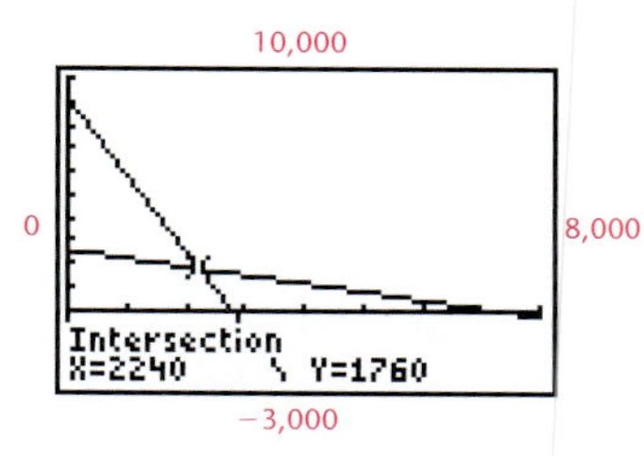

(b)

MATCHED 6 PROBLEM

A food manufacturer produces regular and deluxe rice mixtures by mixing wild rice with long-grain rice. The regular rice mixture is 5% wild rice and 95% long-grain rice and the deluxe rice mixture is 10% wild rice and 90% long-grain rice. The company has just received a shipment of 120 pounds of wild rice and 1,500 pounds of long-grain rice. How many pounds of each type of rice mixture should be produced to use all the rice in this shipment?

EXAMPLE 7 Airspeed

An airplane makes the 2,400-mile trip from Washington, D.C. to San Francisco in 7.5 hours and makes the return trip in 6 hours. Assuming that the plane travels at a constant airspeed and that the wind blows at a constant rate from west to east, find the plane's airspeed and the wind rate.

SOLUTION

Let x represent the airspeed of the plane and let y represent the rate at which the wind is blowing (both in miles per hour). The ground speed of the plane is determined by combining these two rates; that is,

$$x - y = \text{Ground speed flying east to west (headwind)}$$
$$x + y = \text{Ground speed flying west to east (tailwind)}$$

Applying the familiar formula $D = RT$ to each leg of the trip leads to the following system of equations:

$$2{,}400 = 7.5(x - y) \quad \text{From Washington to San Francisco}$$
$$2{,}400 = 6(x + y) \quad \text{From San Francisco to Washington}$$

After simplification, we have

$$x - y = 320$$
$$x + y = 400$$

Solve using substitution:

$$x = y + 320 \quad \text{Solve first equation for } x.$$
$$y + 320 + y = 400 \quad \text{Substitute in second equation.}$$
$$2y = 80$$
$$y = 40 \text{ miles per hour} \quad \text{Wind rate}$$
$$x = 40 + 320$$
$$x = 360 \text{ miles per hour} \quad \text{Airspeed}$$

CHECK

$$2{,}400 = 7.5(x - y) \qquad 2{,}400 = 6(x + y)$$
$$2{,}400 \stackrel{?}{=} 7.5(360 - 40) \qquad 2{,}400 \stackrel{?}{=} 6(360 + 40)$$
$$2{,}400 \stackrel{\checkmark}{=} 2{,}400 \qquad 2{,}400 \stackrel{\checkmark}{=} 2{,}400$$

MATCHED 7 PROBLEM

A boat takes 8 hours to travel 80 miles upstream and 5 hours to return to its starting point. Find the speed of the boat in still water and the speed of the current.

EXAMPLE 8 Supply and Demand

The price–demand and price–supply equations for the sale of cherries each day in a particular city are

$$p = -0.3q + 5 \qquad \text{Demand equation (consumer)}$$
$$p = 0.06q + 0.68 \qquad \text{Supply equation (supplier)}$$

where q represents the quantity in thousands of pounds and p represents the price per pound in dollars.

(A) Discuss the relationship between supply and demand when cherries are selling for \$1.70 per pound.

(B) Discuss the relationship between supply and demand when cherries are selling for \$1.10 per pound.

(C) Find the equilibrium quantity and the equilibrium price.

SOLUTIONS

(A) Enter $y_1 = -0.3x + 5$, $y_2 = 0.06x + 0.68$, and $y_3 = 1.70$. Use the intersect command to find the supply (Fig. 7) and the demand (Fig. 8) when the price is \$1.70.

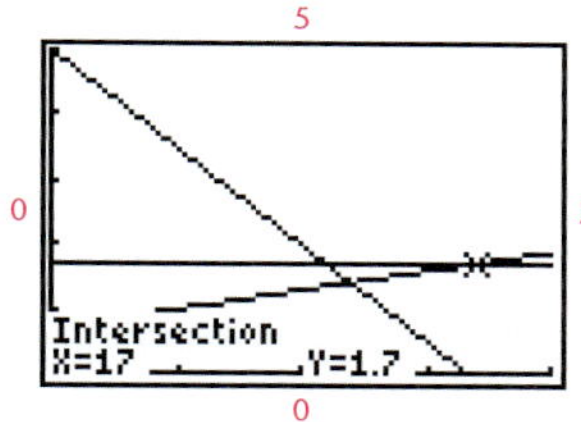

FIGURE 7

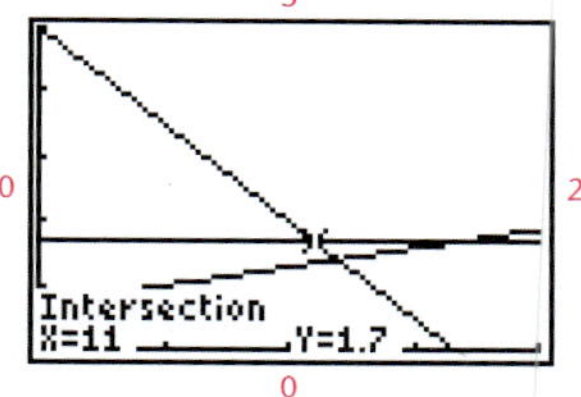

FIGURE 8

From Figures 7 and 8, we find that at a price of \$1.70 per pound, suppliers are willing to supply 17,000 pounds of cherries, but consumers will purchase only 11,000 pounds. The supply exceeds the demand at this price, and the price will come down.

(B) Changing y_3 to $y_3 = 1.10$ and proceeding as before (details omitted), we find that at this price consumers will purchase 13,000 pounds of cherries, but suppliers will supply only 7,000 pounds. Thus, at \$1.10 per pound the demand exceeds the supply and the price will go up.

FIGURE 9

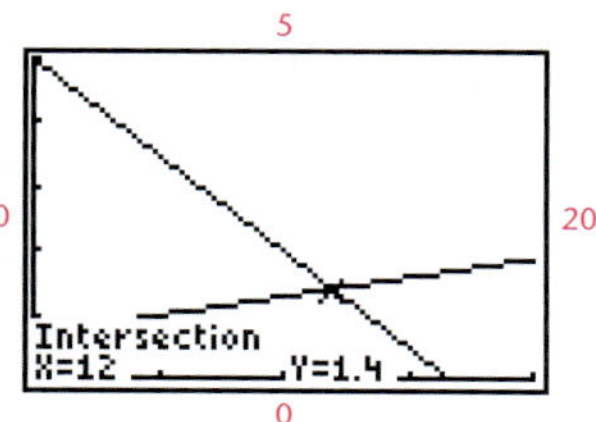

(C) The equilibrium price is the price at which supply will equal demand (see Section 2.2) and the equilibrium quantity is the common value of supply and demand. Using the intersect command (Fig. 9), we see that the equilibrium quantity is 12,000 pounds and the equilibrium price is \$1.40 per pound.

MATCHED PROBLEM 8

The price–demand and price–supply equations for strawberries each day in a certain city are

$$p = -0.2q + 4 \qquad \text{Demand equation}$$
$$p = 0.04q + 1.84 \qquad \text{Supply equation}$$

where q represents the quantity in thousands of pounds and p represents the price per pound in dollars.

Find the equilibrium quantity and the equilibrium price.

ANSWERS MATCHED PROBLEMS

1.

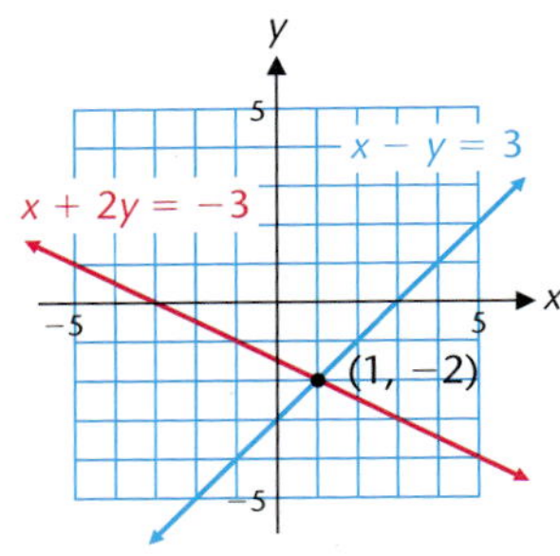

$x = 1, y = -2$

Check:

$$x - y = 3$$
$$1 - (-2) \stackrel{?}{=} 3$$
$$3 \stackrel{\checkmark}{=} 3$$
$$x + 2y = -3$$
$$1 + 2(-2) \stackrel{?}{=} -3$$
$$-3 \stackrel{\checkmark}{=} -3$$

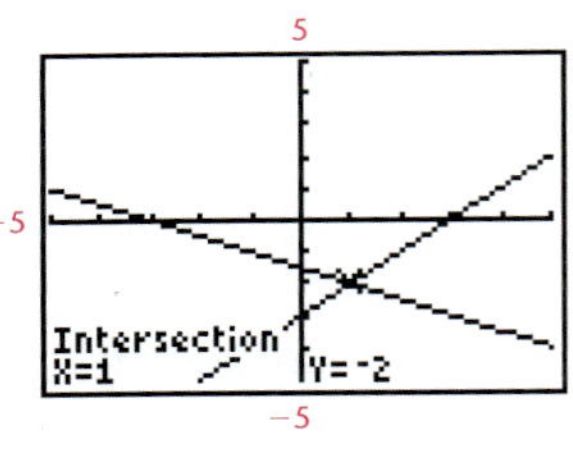

2. (A) (3, 2) or $x = 3$ and $y = 2$ (B) No solutions (C) Infinite number of solutions

3. $(-1.92, 4.23)$ or $x = -1.92$ and $y = 4.23$

4. $x = 1, y = -2$

5. $x = 2, y = -3$

6. 840 pounds of regular mix, 780 pounds of deluxe mix

7. Boat: 13 miles per hour; current: 3 miles per hour

8. Equilibrium quantity = 9,000 pounds; equilibrium price = \$2.20 per pound

EXERCISE 8.1

A

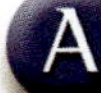
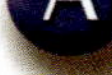

Match each system in Problems 1–4 with one of the following graphs, and use the graph to solve the system.

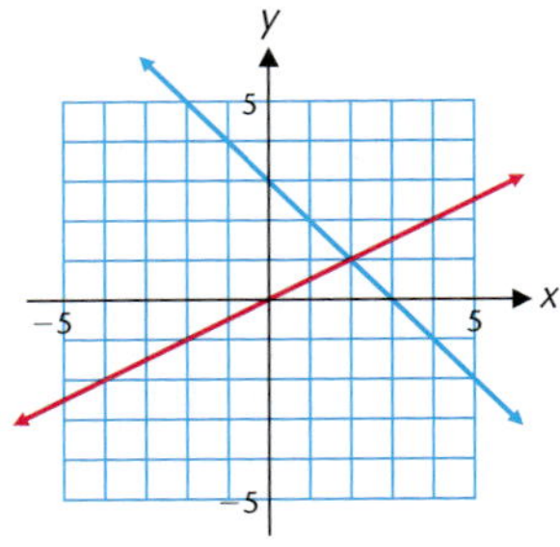

(a)

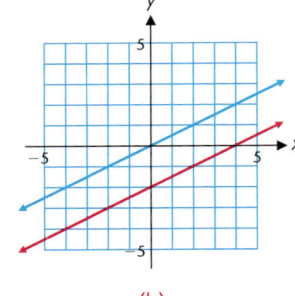

(b)

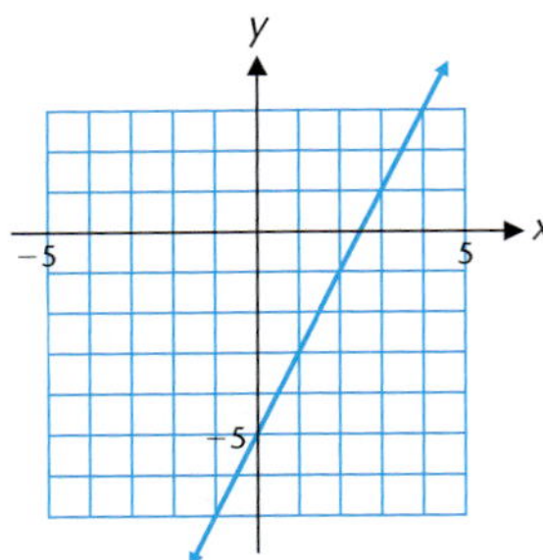

(c)

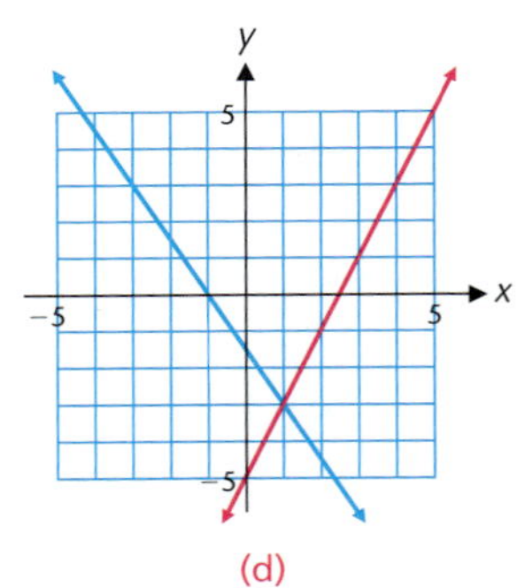

(d)

1. $\begin{aligned} 2x - 4y &= 8 \\ x - 2y &= 0 \end{aligned}$

2. $\begin{aligned} x + y &= 3 \\ x - 2y &= 0 \end{aligned}$

3. $\begin{aligned} 2x - y &= 5 \\ 3x + 2y &= -3 \end{aligned}$

4. $\begin{aligned} 4x - 2y &= 10 \\ 2x - y &= 5 \end{aligned}$

Solve Problems 5–10 by graphing.

5. $\begin{aligned} x + y &= 7 \\ x - y &= 3 \end{aligned}$

6. $\begin{aligned} x - y &= 2 \\ x + y &= 4 \end{aligned}$

7. $\begin{aligned} 3x - 2y &= 12 \\ 7x + 2y &= 8 \end{aligned}$

8. $\begin{aligned} 3x - y &= 2 \\ x + 2y &= 10 \end{aligned}$

9. $\begin{aligned} 3u + 5v &= 15 \\ 6u + 10v &= -30 \end{aligned}$

10. $\begin{aligned} m + 2n &= 4 \\ 2m + 4n &= -8 \end{aligned}$

Solve Problems 11–16 by substitution.

11. $\begin{aligned} y &= 2x + 3 \\ y &= 3x - 5 \end{aligned}$

12. $\begin{aligned} y &= x + 4 \\ y &= 5x - 8 \end{aligned}$

13. $\begin{aligned} x - y &= 4 \\ x + 3y &= 12 \end{aligned}$

14. $\begin{aligned} 2x - y &= 3 \\ x + 2y &= 14 \end{aligned}$

15. $\begin{aligned} 3x - y &= 7 \\ 2x + 3y &= 1 \end{aligned}$

16. $\begin{aligned} 2x + y &= 6 \\ x - y &= -3 \end{aligned}$

B

Solve Problems 17–30 by either method. Round any approximate values to two decimal places.

17. $\begin{aligned} 4x + 3y &= 26 \\ 3x - 11y &= -7 \end{aligned}$

18. $\begin{aligned} 9x - 3y &= 24 \\ 11x + 2y &= 1 \end{aligned}$

19. $\begin{aligned} 7m + 12n &= -1 \\ 5m - 3n &= 7 \end{aligned}$

20. $\begin{aligned} 3p + 8q &= 4 \\ 15p + 10q &= -10 \end{aligned}$

21. $\begin{aligned} y &= 0.08x \\ y &= 100 + 0.04x \end{aligned}$

22. $\begin{aligned} y &= 0.07x \\ y &= 80 + 0.05x \end{aligned}$

23. $\begin{aligned} 0.2u - 0.5v &= 0.07 \\ 0.8u - 0.3v &= 0.79 \end{aligned}$

24. $\begin{aligned} 0.3s - 0.6t &= 0.18 \\ 0.5s - 0.2t &= 0.54 \end{aligned}$

25. $\begin{aligned} \tfrac{2}{5}x + \tfrac{3}{2}y &= 2 \\ \tfrac{7}{3}x - \tfrac{5}{4}y &= -5 \end{aligned}$

26. $\begin{aligned} \tfrac{7}{2}x - \tfrac{5}{6}y &= 10 \\ \tfrac{2}{5}x + \tfrac{4}{3}y &= 6 \end{aligned}$

27. $\begin{aligned} 2x - 3y &= -5 \\ 3x + 4y &= 13 \end{aligned}$

28. $\begin{aligned} 7x - 3y &= 20 \\ 5x + 2y &= 8 \end{aligned}$

29. $\begin{aligned} 3.5x - 2.4y &= 0.1 \\ 2.6x - 1.7y &= -0.2 \end{aligned}$

30. $\begin{aligned} 5.4x + 4.2y &= -12.9 \\ 3.7x + 6.4y &= -4.5 \end{aligned}$

31. In the process of solving a system by substitution, suppose you encounter a contradiction, such as $0 = 1$. How would you describe the solutions to such a system? Illustrate your ideas with the system

$$\begin{aligned} x - 2y &= -3 \\ -2x + 4y &= 7 \end{aligned}$$

32. Repeat Problem 31 for the following system:

$$\begin{aligned} 2x - y &= 4 \\ -4x + 2y &= -7 \end{aligned}$$

33. In the process of solving a system by substitution, suppose you encounter an identity, such as $0 = 0$. How would you describe the solutions to such a system? Illustrate your ideas with the system

$$\begin{aligned} x - 2y &= -3 \\ -2x + 4y &= 6 \end{aligned}$$

34. Repeat Problem 33 for the following system:

$$\begin{aligned} 2x - y &= 4 \\ -4x + 2y &= -8 \end{aligned}$$

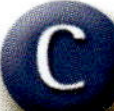

C

In Problems 35 and 36, solve each system for p and q in terms of x and y. Explain how you could check your solution and then perform the check.

35. $\begin{aligned} x &= 2 + p - 2q \\ y &= 3 - p + 3q \end{aligned}$

36. $\begin{aligned} x &= -1 + 2p - q \\ y &= 4 - p + q \end{aligned}$

Problems 37 and 38 refer to the system

$$\begin{aligned} ax + by &= h \\ cx + dy &= k \end{aligned}$$

where x and y are variables and a, b, c, d, h, and k are real constants.

37. Solve the system for x and y in terms of the constants a, b, c, d, h, and k. Clearly state any assumptions you must make about the constants during the solution process.

38. Discuss the nature of solutions to systems that do not satisfy the assumptions you made in Problem 37.

APPLICATIONS

39. Airspeed. It takes a private airplane 8.75 hours to make the 2,100-mile flight from Atlanta to Los Angeles and 5 hours to make the return trip. Assuming that the wind blows at a constant rate from Los Angeles to Atlanta, find the airspeed of the plane and the wind rate.

40. Airspeed. A plane carries enough fuel for 20 hours of flight at an airspeed of 150 miles per hour. How far can it fly into a 30 mile per hour headwind and still have enough fuel to return to its starting point? (This distance is called the *point of no return.*)

41. Rate–Time. A crew of eight can row 20 kilometers per hour in still water. The crew rows upstream and then returns to its starting point in 15 minutes. If the river is flowing at 2 kilometers per hour, how far upstream did the crew row?

42. Rate–Time. It takes a boat 2 hours to travel 20 miles down a river and 3 hours to return upstream to its starting point. What is the rate of the current in the river?

43. Chemistry. A chemist has two solutions of hydrochloric acid in stock: a 50% solution and an 80% solution. How much of each should be used to obtain 100 milliliters of a 68% solution?

44. Business. A jeweler has two bars of gold alloy in stock, one of 12 carats and the other of 18 carats (24-carat gold is pure gold, 12-carat is $\frac{12}{24}$ pure, 18-carat gold is $\frac{18}{24}$ pure, and so on). How many grams of each alloy must be mixed to obtain 10 grams of 14-carat gold?

45. Finance. Suppose you have \$12,000 to invest. If part is invested at 10% and the rest at 15%, how much should be invested at each rate to yield 12% on the total amount invested?

46. Finance. An investor has \$20,000 to invest. If part is invested at 8% and the rest at 12%, how much should be invested at each rate to yield 11% on the total amount invested?

47. Production. A supplier for the electronics industry manufactures keyboards and screens for graphing calculators at plants in Mexico and Taiwan. The hourly production rates at each plant are given in the table. How many hours should each plant be operated to exactly fill an order for 4,000 keyboards and screens?

Plant	Keyboards	Screens
Mexico	40	32
Taiwan	20	32

48. Production. A company produces Italian sausages and bratwursts at plants in Green Bay and Sheboygan. The hourly production rates at each plant are given in the table. How many hours should each plant be operated to exactly fill an order for 62,250 Italian sausages and 76,500 bratwursts?

Plant	Italian Sausage	Bratwurst
Green Bay	800	800
Sheboygan	500	1,000

49. Nutrition. Animals in an experiment are to be kept on a strict diet. Each animal is to receive, among other things, 20 grams of protein and 6 grams of fat. The laboratory technician is able to purchase two food mixes of the following compositions: Mix A has 10% protein and 6% fat; mix B has 20% protein and 2% fat. How many grams of each mix should be used to obtain the right diet for a single animal?

50. Nutrition. A fruit grower can use two types of fertilizer in an orange grove, brand A and brand B. Each bag of brand A contains 8 pounds of nitrogen and 4 pounds of phosphoric acid. Each bag of brand B contains 7 pounds of nitrogen and 7 pounds of phosphoric acid. Tests indicate that the grove needs 720 pounds of nitrogen and 500 pounds of phosphoric acid. How many bags of each brand should be used to provide the required amounts of nitrogen and phosphoric acid?

51. Supply and Demand. Suppose the supply and demand equations for printed T-shirts in a resort town for a particular week are

$$p = 0.007q + 3 \quad \text{Supply equation}$$
$$p = -0.018q + 15 \quad \text{Demand equation}$$

where p is the price in dollars and q is the quantity.

(A) Find the supply and the demand (to the nearest unit) if T-shirts are priced at \$4 each. Discuss the stability of the T-shirt market at this price level.

(B) Find the supply and the demand (to the nearest unit) if T-shirts are priced at \$8 each. Discuss the stability of the T-shirt market at this price level.

(C) Find the equilibrium price and quantity.

(D) Graph the two equations in the same coordinate system and identify the equilibrium point, supply curve, and demand curve.

52. Supply and Demand. Suppose the supply and demand equations for printed baseball caps in a resort town for a particular week are

$$p = 0.006q + 2 \qquad \text{Supply equation}$$
$$p = -0.014q + 13 \qquad \text{Demand equation}$$

where p is the price in dollars and q is the quantity in hundreds.

(A) Find the supply and the demand (to the nearest unit) if baseball caps are priced at \$4 each. Discuss the stability of the baseball cap market at this price level.

(B) Find the supply and the demand (to the nearest unit) if baseball caps are priced at \$8 each. Discuss the stability of the baseball cap market at this price level.

(C) Find the equilibrium price and quantity.

(D) Graph the two equations in the same coordinate system and identify the equilibrium point, supply curve, and demand curve.

★ **53. Supply and Demand.** At \$0.60 per bushel, the daily supply for wheat is 450 bushels and the daily demand is 645 bushels. When the price is raised to \$0.90 per bushel, the daily supply increases to 750 bushels and the daily demand decreases to 495 bushels. Assume that the supply and demand equations are linear.

(A) Find the supply equation. [*Hint:* Write the supply equation in the form $p = aq + b$ and solve for a and b.]

(B) Find the demand equation.

(C) Find the equilibrium price and quantity.

★ **54. Supply and Demand.** At \$1.40 per bushel, the daily supply for soybeans is 1,075 bushels and the daily demand is 580 bushels. When the price falls to \$1.20 per bushel, the daily supply decreases to 575 bushels and the daily demand increases to 980 bushels. Assume that the supply and demand equations are linear.

(A) Find the supply equation. [See the hint in Problem 53.]

(B) Find the demand equation.

(C) Find the equilibrium price and quantity.

★ **55. Physics.** An object dropped off the top of a tall building falls vertically with constant acceleration. If s is the distance of the object above the ground (in feet) t seconds after its release, then s and t are related by an equation of the form

$$s = a + bt^2$$

where a and b are constants. Suppose the object is 180 feet above the ground 1 second after its release and 132 feet above the ground 2 seconds after its release.

(A) Find the constants a and b.

(B) How high is the building?

(C) How long does the object fall?

★ **56. Physics.** Repeat Problem 55 if the object is 240 feet above the ground after 1 second and 192 feet above the ground after 2 seconds.

★ **57. Earth Science.** An earthquake emits a primary wave and a secondary wave. Near the surface of the Earth the primary wave travels at about 5 miles per second and the secondary wave at about 3 miles per second. From the time lag between the two waves arriving at a given receiving station, it is possible to estimate the distance to the quake. (The *epicenter* can be located by obtaining distance bearings at three or more stations.) Suppose a station measured a time difference of 16 seconds between the arrival of the two waves. How long did each wave travel, and how far was the earthquake from the station?

★ **58. Earth Science.** A ship using sound-sensing devices above and below water recorded a surface explosion 6 seconds sooner by its underwater device than its above-water device. Sound travels in air at about 1,100 feet per second and in seawater at about 5,000 feet per second.

(A) How long did it take each sound wave to reach the ship?

(B) How far was the explosion from the ship?

SECTION 8.2 Systems of Linear Equations and Augmented Matrices

Elimination by Addition • Matrices • Solving Linear Systems Using Augmented Matrices

Most real-world applications of linear systems involve a large number of variables and equations. Computers are usually used to solve these larger systems. Although very effective for systems involving two variables, the graphing and substitution methods discussed in Section 8.1 are not well-suited for computer use in the solution of larger systems. In Section 8.2, we begin the development of a method that will work for systems of any size and that lends itself to computer implementation.

Elimination by Addition

We begin with an algebraic solution method called **elimination by addition.** As we will see, this method is readily generalized to larger systems. The method involves the replacement of systems of equations with simpler *equivalent systems,* by performing appropriate operations, until we obtain a system with an obvious solution. **Equivalent systems** of equations are systems that have the same solution set. Theorem 1 lists operations that produce equivalent systems.

THEOREM 1
Equation Operations Producing Equivalent Systems

A system of linear equations is transformed into an equivalent system if

1. Two equations are interchanged.
2. An equation is multiplied by a nonzero constant.
3. A constant multiple of one equation is added to another equation.

Any one of the three operations in Theorem 1 can be used to produce an equivalent system, but operations 2 and 3 will be of most use to us now. Operation 1 becomes more important later in the section. The use of Theorem 1 is best illustrated by examples.

EXAMPLE 1 Solving a System Using Elimination by Addition

Solve using elimination by addition:
$$\begin{aligned} 3x - 2y &= 8 \\ 2x + 5y &= -1 \end{aligned}$$

SOLUTION

We use Theorem 1 to eliminate one of the variables and thus obtain a system with an obvious solution.

$$\begin{aligned} 3x - 2y &= 8 \\ 2x + 5y &= -1 \end{aligned}$$

If we multiply the top equation by 5, the bottom by 2, and then add, we can eliminate y.

$$\begin{aligned} 15x - 10y &= 40 \\ 4x + 10y &= -2 \\ 19x \qquad &= 38 \\ x &= 2 \end{aligned}$$

The equation $x = 2$ paired with either of the two original equations produces an equivalent system. Thus, we can substitute $x = 2$ back into either of the two original equations to solve for y. We choose the second equation.

$$
\begin{aligned}
2(2) + 5y &= -1 \\
5y &= -5 \\
y &= -1
\end{aligned}
$$

Solution: $x = 2$, $y = -1$ or $(2, -1)$.

CHECK

$$
\begin{aligned}
3x - 2y &= 8 \\
3(2) - 2(-1) &\stackrel{?}{=} 8 \\
8 &\stackrel{\checkmark}{=} 8
\end{aligned}
\qquad
\begin{aligned}
2x + 5y &= -1 \\
2(2) + 5(-1) &\stackrel{?}{=} -1 \\
-1 &\stackrel{\checkmark}{=} -1
\end{aligned}
$$

MATCHED 1 PROBLEM

Solve using elimination by addition:

$$
\begin{aligned}
6x + 3y &= 3 \\
5x + 4y &= 7
\end{aligned}
$$

EXPLORE/DISCUSS 1

In each of the following systems, compare the results of applying elimination by addition with the graphical solution and discuss the nature of the solution sets.

(A) $\begin{aligned} 3x + 2y &= 6 \\ 6x + 4y &= 12 \end{aligned}$ (B) $\begin{aligned} 3x + 2y &= 6 \\ 6x + 4y &= 13 \end{aligned}$

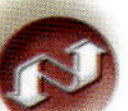

Let's see what happens in the elimination process when a system either has no solution or has infinitely many solutions. Consider the following system:

$$
\begin{aligned}
2x + 6y &= -3 \\
x + 3y &= 2
\end{aligned}
$$

Multiplying the second equation by -2 and adding, we obtain

$$
\begin{aligned}
2x + 6y &= -3 \\
-2x - 6y &= -4 \\
\hline
0 &= -7
\end{aligned}
$$

We have obtained a contradiction. An assumption that the original system has solutions must be false, otherwise, we have proved that $0 = -7$! Thus, the system has no solution. The graphs of the equations are parallel and the system is inconsistent.

Now consider the system

$$\begin{aligned} x - \tfrac{1}{2}y &= 4 \\ -2x + y &= -8 \end{aligned}$$

If we multiply the top equation by 2 and add the result to the bottom equation, we get

$$\begin{aligned} 2x - y &= 8 \\ -2x + y &= -8 \\ \hline 0 &= 0 \end{aligned}$$

Obtaining $0 = 0$ by addition implies that the two original equations are equivalent. That is, their graphs coincide and the system is dependent. If we let $x = t$, where t is any real number, and solve either equation for y, we obtain $y = 2t - 8$. Thus,

$$(t, 2t - 8) \qquad t \text{ a real number}$$

describes the solution set for the system. The variable t is called a **parameter,** and replacing t with a real number produces a **particular solution** to the system. For example, some particular solutions to this system are

$t = -1$	$t = 2$	$t = 5$	$t = 9.4$
$(-1, -10)$	$(2, -4)$	$(5, 2)$	$(9.4, 10.8)$

Example 2 illustrates that elimination by addition provides an efficient method for solving applied problems.

EXAMPLE 2 Diet

A woman wants to use milk and orange juice to increase the amount of calcium and vitamin A in her daily diet. An ounce of milk contains 41 milligrams of calcium and 59 micrograms* of vitamin A. An ounce of orange juice contains 5 milligrams of calcium and 75 micrograms of vitamin A. How many ounces of milk and orange juice should she drink each day to provide exactly 550 milligrams of calcium and 1,300 micrograms of vitamin A?

SOLUTION

First we define the relevant variables:

x = Number of ounces of milk

y = Number of ounces of orange juice

Next we summarize the given information in Table 1.

TABLE 1

	Milk	Orange Juice	Total Needed
Calcium (mg)	41	5	550
Vitamin A (μg)	59	75	1,300

*A microgram (μg) is one-millionth (10^{-6}) of a gram.

Now we use the information in Table 1 to form equations involving x and y:

$$\begin{pmatrix}\text{Calcium in } x \text{ ounces}\\ \text{of milk}\end{pmatrix} + \begin{pmatrix}\text{Calcium in } y \text{ ounces}\\ \text{of orange juice}\end{pmatrix} = \begin{pmatrix}\text{Total calcium}\\ \text{needed (milligrams)}\end{pmatrix}$$

$$41x + 5y = 550$$

$$\begin{pmatrix}\text{Vitamin A in } x \text{ ounces}\\ \text{of milk}\end{pmatrix} + \begin{pmatrix}\text{Vitamin A in } y \text{ ounces}\\ \text{of orange juice}\end{pmatrix} = \begin{pmatrix}\text{Total vitamin A}\\ \text{needed (micrograms)}\end{pmatrix}$$

$$59x + 75y = 1{,}300$$

Solve using elimination by addition:

$$\begin{aligned} -615x - 75y &= -8{,}250 \\ 59x + 75y &= 1{,}300 \\ \hline -556x &= -6{,}950 \\ x &= 12.5 \end{aligned} \qquad \begin{aligned} 41(12.5) + 5y &= 550 \\ 5y &= 37.5 \\ y &= 7.5 \end{aligned}$$

Drinking 12.5 ounces of milk and 7.5 ounces of orange juice each day will provide the required amounts of calcium and vitamin A.

CHECK

$$\begin{aligned} 41x + 5y &= 550 \\ 41(12.5) + 5(7.5) &\overset{?}{=} 550 \\ 550 &\overset{\checkmark}{=} 550 \end{aligned} \qquad \begin{aligned} 59x + 75y &= 1{,}300 \\ 59(12.5) + 75(7.5) &\overset{?}{=} 1{,}300 \\ 1{,}300 &\overset{\checkmark}{=} 1{,}300 \end{aligned}$$

MATCHED 2 PROBLEM

A man wants to use cottage cheese and yogurt to increase the amount of protein and calcium in his daily diet. An ounce of cottage cheese contains 3 grams of protein and 15 milligrams of calcium. An ounce of yogurt contains 1 gram of protein and 41 milligrams of calcium. How many ounces of cottage cheese and yogurt should he eat each day to provide exactly 62 grams of protein and 760 milligrams of calcium?

Matrices

In solving systems of equations using elimination by addition, the coefficients of the variables and the constant terms play a central role. The process can be made more efficient for generalization and computer work by the introduction of a mathematical form called a *matrix*. A **matrix** is a rectangular array of numbers written within brackets. Two examples are

$$A = \begin{bmatrix} 1 & -3 & 7 \\ 5 & 0 & -4 \end{bmatrix} \qquad B = \begin{bmatrix} -5 & 4 & 11 \\ 0 & 1 & 6 \\ -2 & 12 & 8 \\ -3 & 0 & -1 \end{bmatrix}$$

Each number in a matrix is called an **element** of the matrix. Matrix A has six elements arranged in two rows and three columns. Matrix B has 12 elements arranged in four rows and three columns. If a matrix has m rows and n columns, it is called an $\boldsymbol{m \times n}$ **matrix** (read "m by n matrix"). The expression $m \times n$ is called the **size** of the matrix, and the numbers m and n are called the **dimensions** of the matrix. It is important to note that the number of rows is always given first. Referring to matrices A and B on page 631, A is a 2×3 matrix and B is a 4×3 matrix. A matrix with n rows and n columns is called a **square matrix of order *n*.** A matrix with only one column is called a **column matrix,** and a matrix with only one row is called a **row matrix.** These definitions are illustrated by the following:

3×3	4×1	1×4
$\begin{bmatrix} 0.5 & 0.2 & 1.0 \\ 0.0 & 0.3 & 0.5 \\ 0.7 & 0.0 & 0.2 \end{bmatrix}$	$\begin{bmatrix} 3 \\ -2 \\ 1 \\ 0 \end{bmatrix}$	$\begin{bmatrix} 2 & \frac{1}{2} & 0 & -\frac{2}{3} \end{bmatrix}$
Square matrix of order 3	Column matrix	Row matrix

The **position** of an element in a matrix is the row and column containing the element. This is usually denoted using **double subscript notation** a_{ij}, where i is the row and j is the column containing the element a_{ij}, as illustrated:

$$A = \begin{bmatrix} 1 & 5 & -3 \\ 6 & 0 & -4 \end{bmatrix} \qquad \begin{array}{l} a_{11} = 1, a_{12} = 5, a_{13} = -3 \\ a_{21} = 6, a_{22} = 0, a_{23} = -4 \end{array}$$

Note that a_{12} is read "a one two," not "a twelve." The elements $a_{11} = 1$ and $a_{22} = 0$ make up the *principal diagonal* of A. In general, the **principal diagonal** of a matrix A consists of the elements $a_{11}, a_{22}, a_{33}, \ldots$.

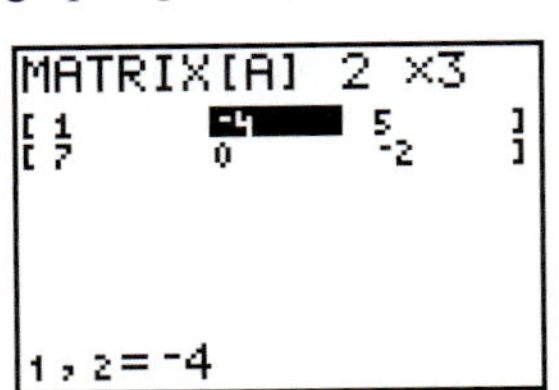

FIGURE 1 Matrix notation on a graphing utility.

REMARK Most graphing utilities are capable of storing and manipulating matrices. Figure 1 shows matrix A displayed in the editing screen of a particular graphing calculator. The size of the matrix is given at the top of the screen. The position and the value of the currently selected element is given at the bottom. Notice that a comma is used in the notation of the position. This is common practice on graphing utilities but not in mathematical literature.

The coefficients and constant terms in a system of linear equations can be used to form several matrices of interest to our work. Related to the system

$$\begin{aligned} 2x - 3y &= 5 \\ x + 2y &= -3 \end{aligned} \tag{1}$$

are the following matrices:

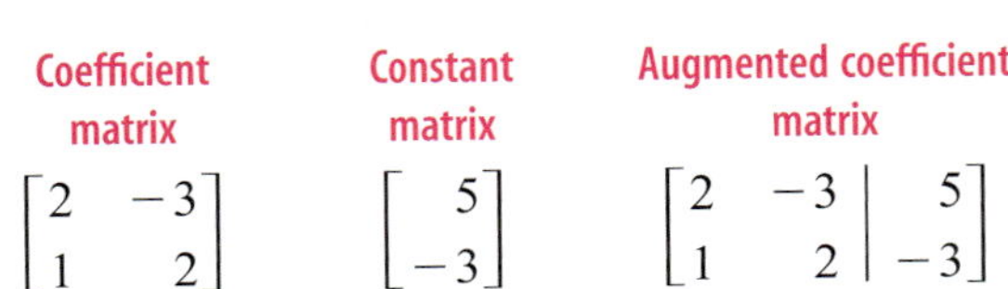

Coefficient matrix	Constant matrix	Augmented coefficient matrix
$\begin{bmatrix} 2 & -3 \\ 1 & 2 \end{bmatrix}$	$\begin{bmatrix} 5 \\ -3 \end{bmatrix}$	$\left[\begin{array}{cc\|c} 2 & -3 & 5 \\ 1 & 2 & -3 \end{array}\right]$

The augmented coefficient matrix will be used in this section. The other matrices will be used in later sections. The augmented coefficient matrix contains the essential parts of the system—both the coefficients and the constants. The vertical bar is included only as a visual aid to help us separate the coefficients from the constant terms. (Matrices entered and displayed on a graphing utility will not display this line.)

For ease of generalization to the larger systems in the following sections, we are now going to change the notation for the variables in system (1) to a subscript form (we would soon run out of letters, but we will not run out of subscripts). That is, in place of x and y, we will use x_1 and x_2, respectively, and (1) will be written as

$$\begin{aligned} 2x_1 - 3x_2 &= 5 \\ x_1 + 2x_2 &= -3 \end{aligned}$$

In general, associated with each linear system of the form

$$\begin{aligned} a_{11}x_1 + a_{12}x_2 &= k_1 \\ a_{21}x_1 + a_{22}x_2 &= k_2 \end{aligned} \qquad (2)$$

where x_1 and x_2 are variables, is the **augmented matrix** of the system:

$$\left[\begin{array}{cc|c} a_{11} & a_{12} & k_1 \\ a_{21} & a_{22} & k_2 \end{array}\right]$$

This matrix contains the essential parts of system (2). Our objective is to learn how to manipulate augmented matrices in such a way that a solution to system (2) will result, if a solution exists.

In our earlier discussion of using elimination by addition, we said that two systems were equivalent if they had the same solution. And we used the operations in Theorem 1 to transform a system into an equivalent system. Paralleling this approach, we now say that two augmented matrices are **row-equivalent,** denoted by the symbol $\sim$ between the two matrices, if they are augmented matrices of equivalent systems of equations. And we use the operations listed in Theorem 2 to transform augmented matrices into row-equivalent matrices. Note that Theorem 2 is a direct consequence of Theorem 1.

THEOREM 2
Row Operations Producing Row-Equivalent Matrices

An augmented matrix is transformed into a row-equivalent matrix if any of the following **row operations** is performed:

1. Two rows are interchanged ($R_i \leftrightarrow R_j$).
2. A row is multiplied by a nonzero constant ($kR_i \rightarrow R_i$).
3. A constant multiple of one row is added to another row ($kR_j + R_i \rightarrow R_i$).

[*Note:* The arrow means "replaces."]

Solving Linear Systems Using Augmented Matrices

The use of Theorem 2 in solving systems in the form of (2) is best illustrated by examples.

EXAMPLE 3 Solving a System Using Augmented Matrix Methods

Solve, using augmented matrix methods:

$$\begin{aligned} 3x_1 + 4x_2 &= 1 \\ x_1 - 2x_2 &= 7 \end{aligned} \tag{3}$$

SOLUTION

We start by writing the augmented matrix corresponding to system (3):

$$\left[\begin{array}{cc|c} 3 & 4 & 1 \\ 1 & -2 & 7 \end{array}\right] \tag{4}$$

Our objective is to use row operations from Theorem 2 to try to transform matrix (4) into the form

$$\left[\begin{array}{cc|c} 1 & 0 & m \\ 0 & 1 & n \end{array}\right] \tag{5}$$

where m and n are real numbers. The solution to system (3) will then be obvious, because matrix (5) will be the augmented matrix of the following system:

$$\begin{aligned} x_1 &= m \qquad x_1 + 0x_2 = m \\ x_2 &= n \qquad 0x_1 + x_2 = n \end{aligned}$$

We now proceed to use row operations to transform (4) into form (5).

Step 1. To get a 1 in the upper left corner, we interchange rows 1 and 2—Theorem 2, part 1:

$$\left[\begin{array}{cc|c} 3 & 4 & 1 \\ 1 & -2 & 7 \end{array}\right] \begin{array}{c} R_1 \leftrightarrow R_2 \\ \sim \end{array} \left[\begin{array}{cc|c} 1 & -2 & 7 \\ 3 & 4 & 1 \end{array}\right]$$

Now you see why we wanted Theorem 2, part 1.

Step 2. To get a 0 in the lower left corner, we multiply R_1 by -3 and add to R_2—Theorem 2, part 3. This changes R_2 but not R_1. Some people find it useful to write $(-3)R_1$ outside the matrix to help reduce errors in arithmetic:

$$\left[\begin{array}{cc|c} 1 & -2 & 7 \\ 3 & 4 & 1 \end{array}\right] \begin{array}{c} (-3)R_1 + R_2 \rightarrow R_2 \\ \sim \end{array} \left[\begin{array}{cc|c} 1 & -2 & 7 \\ 0 & 10 & -20 \end{array}\right]$$

$-3 \quad 6 \quad -21$

Step 3. To get a 1 in the second row, second column, we multiply R_2 by $\frac{1}{10}$—Theorem 2, part 2:

$$\left[\begin{array}{cc|c} 1 & -2 & 7 \\ 0 & 10 & -20 \end{array}\right] \begin{array}{c} \frac{1}{10}R_2 \rightarrow R_2 \\ \sim \end{array} \left[\begin{array}{cc|c} 1 & -2 & 7 \\ 0 & 1 & -2 \end{array}\right]$$

Step 4. To get a 0 in the first row, second column, we multiply R_2 by 2 and add the result to R_1—Theorem 2, part 3. This changes R_1 but not R_2.

$$\begin{matrix} 0 & 2 & -4 \end{matrix}$$
$$\left[\begin{array}{cc|c} 1 & -2 & 7 \\ 0 & 1 & -2 \end{array}\right] \underset{\sim}{2R_2 + R_1 \to R_1} \left[\begin{array}{cc|c} 1 & 0 & 3 \\ 0 & 1 & -2 \end{array}\right]$$

We have accomplished our objective! The last matrix is the augmented matrix for the system

$$\begin{aligned} x_1 &= 3 \\ x_2 &= -2 \end{aligned} \qquad (6)$$

Because system (6) is equivalent to the original system (3), we have solved system (3). That is, $x_1 = 3$ and $x_2 = -2$.

CHECK

$$\begin{aligned} 3x_1 + 4x_2 &= 1 \\ 3(3) + 4(-2) &\overset{?}{=} 1 \\ 1 &\overset{\checkmark}{=} 1 \end{aligned} \qquad \begin{aligned} x_1 - 2x_2 &= 7 \\ 3 - 2(-2) &\overset{?}{=} 7 \\ 7 &\overset{\checkmark}{=} 7 \end{aligned}$$

The above process is written more compactly as follows:

Step 1. Need a 1 here
$$\left[\begin{array}{cc|c} 3 & 4 & 1 \\ 1 & -2 & 7 \end{array}\right] \qquad R_1 \leftrightarrow R_2$$

Step 2. Need a 0 here
$$\sim \left[\begin{array}{cc|c} 1 & -2 & 7 \\ 3 & 4 & 1 \end{array}\right] \qquad (-3)R_1 + R_2 \to R_2$$
$$\begin{matrix} -3 & 6 & -21 \end{matrix}$$

Step 3. Need a 1 here
$$\sim \left[\begin{array}{cc|c} 1 & -2 & 7 \\ 0 & 10 & -20 \end{array}\right] \qquad \tfrac{1}{10}R_2 \to R_2$$

Step 4. Need a 0 here
$$\begin{matrix} 0 & 2 & -4 \end{matrix}$$
$$\sim \left[\begin{array}{cc|c} 1 & -2 & 7 \\ 0 & 1 & -2 \end{array}\right] \qquad 2R_2 + R_1 \to R_1$$

$$\sim \left[\begin{array}{cc|c} 1 & 0 & 3 \\ 0 & 1 & -2 \end{array}\right]$$

Therefore, $x_1 = 3$ and $x_2 = -2$.

MATCHED 3 PROBLEM

Solve, using augmented matrix methods:
$$\begin{aligned} 2x_1 - x_2 &= -7 \\ x_1 + 2x_2 &= 4 \end{aligned}$$

EXPLORE/DISCUSS 2

The summary at the end of Example 3 shows five augmented coefficient matrices. Write the linear system that each matrix represents, solve each system graphically, and discuss the relationship between these solutions.

EXAMPLE 4 Solving a System Using Augmented Matrix Methods

Solve, using augmented matrix methods:

$$\begin{aligned} 2x_1 - 3x_2 &= 7 \\ 3x_1 + 4x_2 &= 2 \end{aligned}$$

SOLUTION

Step 1. Need a 1 here

$$\left[\begin{array}{rr|r} 2 & -3 & 7 \\ 3 & 4 & 2 \end{array}\right] \qquad \tfrac{1}{2}R_1 \to R_1$$

Step 2. Need a 0 here

$$\sim \left[\begin{array}{rr|r} 1 & -\frac{3}{2} & \frac{7}{2} \\ 3 & 4 & 2 \end{array}\right] \qquad (-3)R_1 + R_2 \to R_2$$

$$-3 \qquad \tfrac{9}{2} \qquad -\tfrac{21}{2}$$

Step 3. Need a 1 here

$$\sim \left[\begin{array}{rr|r} 1 & -\frac{3}{2} & \frac{7}{2} \\ 0 & \frac{17}{2} & -\frac{17}{2} \end{array}\right] \qquad \tfrac{2}{17}R_2 \to R_2$$

$$0 \qquad \tfrac{3}{2} \qquad -\tfrac{3}{2}$$

Step 4. Need a 0 here

$$\sim \left[\begin{array}{rr|r} 1 & -\frac{3}{2} & \frac{7}{2} \\ 0 & 1 & -1 \end{array}\right] \qquad \tfrac{3}{2}R_2 + R_1 \to R_1$$

$$\sim \left[\begin{array}{rr|r} 1 & 0 & 2 \\ 0 & 1 & -1 \end{array}\right]$$

Thus, $x_1 = 2$ and $x_2 = -1$. You should check this solution in the original system.

MATCHED PROBLEM 4

Solve, using augmented matrix methods:

$$\begin{aligned} 5x_1 - 2x_2 &= 12 \\ 2x_1 + 3x_2 &= 1 \end{aligned}$$

EXAMPLE 5 Solving a System Using Augmented Matrix Methods

Solve, using augmented matrix methods:

$$\begin{aligned} 2x_1 - x_2 &= 4 \\ -6x_1 + 3x_2 &= -12 \end{aligned} \qquad (7)$$

SOLUTION

$$\begin{bmatrix} 2 & -1 & | & 4 \\ -6 & 3 & | & -12 \end{bmatrix} \quad \begin{array}{l} \tfrac{1}{2}R_1 \to R_1 \text{ (This produces a 1 in the upper left corner.)} \\ \tfrac{1}{3}R_2 \to R_2 \text{ (This simplifies } R_2\text{.)} \end{array}$$

$$\sim \begin{bmatrix} 1 & -\tfrac{1}{2} & | & 2 \\ -2 & 1 & | & -4 \end{bmatrix} \quad 2R_1 + R_2 \to R_2 \text{ (This produces a 0 in the lower left corner.)}$$

$$\begin{matrix} 2 & -1 & & 4 \end{matrix}$$

$$\sim \begin{bmatrix} 1 & -\tfrac{1}{2} & | & 2 \\ 0 & 0 & | & 0 \end{bmatrix}$$

The last matrix corresponds to the system

$$\begin{aligned} x_1 - \tfrac{1}{2}x_2 &= 2 & \qquad x_1 - \tfrac{1}{2}x_2 &= 2 \\ 0 &= 0 & \qquad 0x_1 + 0x_2 &= 0 \end{aligned}$$

Thus, $x_1 = \frac{1}{2}x_2 + 2$. Hence, for any real number t, if $x_2 = t$, then $x_1 = \frac{1}{2}t + 2$. That is, the solution set is described by

$$(\tfrac{1}{2}t + 2,\ t) \qquad t \text{ a real number} \tag{8}$$

For example, if $t = 6$, then $(5, 6)$ is a particular solution; if $t = -2$, then $(1, -2)$ is another particular solution; and so on. Geometrically, the graphs of the two original equations coincide and there are infinitely many solutions.

> **In general, if we end up with a row of 0s in an augmented matrix for a two-equation–two-variable system, the system is dependent and there are infinitely many solutions.**

CHECK

The following is a check that (8) provides a solution for system (7) for any real number t:

$$\begin{aligned} 2x_1 - x_2 &= 4 \\ 2(\tfrac{1}{2}t + 2) - t &\overset{?}{=} 4 \\ t + 4 - t &\overset{?}{=} 4 \\ 4 &\overset{\checkmark}{=} 4 \end{aligned} \qquad \begin{aligned} -6x_1 + 3x_2 &= -12 \\ -6(\tfrac{1}{2}t + 2) + 3t &\overset{?}{=} -12 \\ -3t - 12 + 3t &\overset{?}{=} -12 \\ -12 &\overset{\checkmark}{=} -12 \end{aligned}$$

MATCHED 5 PROBLEM

Solve, using augmented matrix methods:

$$\begin{aligned} -2x_1 + 6x_2 &= 6 \\ 3x_1 - 9x_2 &= -9 \end{aligned}$$

EXPLORE/DISCUSS 3

Most graphing utilities can perform row operations. Figure 2 shows the solution to Example 5 on a particular graphing calculator. Consult your manual to see how to perform row operations, and solve Matched Problem 5 on your graphing utility.

```
[A]
   [[2   -1 4   ]
    [-6 3   -12]]
*row(1/2,[A],1)→
[A]
   [[1  -.5 2  ]
    [-6 3   -12]]
```

```
*row(1/3,[A],2)→
[A]
   [[1  -.5 2 ]
    [-2 1   -4]]
*row+(2,[A],1,2)
→[A]
     [[1 -.5 2]
      [0 0   0]]
```

FIGURE 2 Performing row operations on a graphing utility.

EXAMPLE 6 Solving a System Using Augmented Matrix Methods

Solve, using augmented matrix methods:

$$\begin{aligned} 2x_1 + 6x_2 &= -3 \\ x_1 + 3x_2 &= 2 \end{aligned}$$

SOLUTION

$$\left[\begin{array}{cc|c} 2 & 6 & -3 \\ 1 & 3 & 2 \end{array}\right] \quad R_1 \leftrightarrow R_2$$

$$\sim \left[\begin{array}{cc|c} 1 & 3 & 2 \\ 2 & 6 & -3 \end{array}\right] \quad (-2)R_1 + R_2 \rightarrow R_2$$

$$\begin{array}{ccc} -2 & -6 & -4 \end{array}$$

$$\sim \left[\begin{array}{cc|c} 1 & 3 & 2 \\ 0 & 0 & -7 \end{array}\right] \quad R_2 \text{ implies the contradiction: } 0 = -7$$

This is the augmented matrix of the system

$$\begin{aligned} x_1 + 3x_2 &= 2 & \qquad x_1 + 3x_2 &= 2 \\ 0 &= -7 & \qquad 0x_1 + 0x_2 &= -7 \end{aligned}$$

The second equation is not satisfied by any ordered pair of real numbers. Hence, the original system is inconsistent and has no solution. Otherwise, we have proved that $0 = -7$!

Thus, if we obtain all 0s to the left of the vertical bar and a nonzero number to the right of the bar in a row of an augmented matrix, then the system is inconsistent and there are no solutions.

MATCHED PROBLEM 6

Solve, using augmented matrix methods:

$$\begin{aligned} 2x_1 - x_2 &= 3 \\ 4x_1 - 2x_2 &= -1 \end{aligned}$$

> **Summary**
>
> The augmented matrix of any system of two linear equations in two variables can be transformed by row operations into a row-equivalent matrix having one of the following forms, where m, n, and p are real numbers and $p \neq 0$:
>
> | Form 1: A Unique Solution (Consistent and Independent) | Form 2: Infinitely Many Solutions (Consistent and Dependent) | Form 3: No Solution (Inconsistent) |
> | --- | --- | --- |
> | $\left[\begin{array}{cc\|c} 1 & 0 & m \\ 0 & 1 & n \end{array}\right]$ | $\left[\begin{array}{cc\|c} 1 & m & n \\ 0 & 0 & 0 \end{array}\right]$ | $\left[\begin{array}{cc\|c} 1 & m & n \\ 0 & 0 & p \end{array}\right]$ |

The process of solving systems of equations described in Section 8.2 is referred to as *Gauss–Jordan elimination*. We will use this method to solve larger-scale systems in Section 8.3, including systems where the number of equations and the number of variables are not the same.

ANSWERS MATCHED PROBLEMS

1. $(-1, 3)$, or $x = -1$ and $y = 3$
2. 16.5 ounces of cottage cheese, 12.5 ounces of yogurt
3. $x_1 = -2, x_2 = 3$
4. $x_1 = 2, x_2 = -1$
5. The system is dependent. For t any real number, $x_2 = t$, $x_1 = 3t - 3$ is a solution.
6. Inconsistent—no solution

EXERCISE 8.2

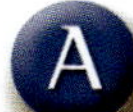

Solve Problems 1–4 using elimination by addition.

1. $\begin{aligned} 2x + 3y &= 1 \\ 3x - y &= 7 \end{aligned}$

2. $\begin{aligned} 2m - n &= 10 \\ m - 2n &= -4 \end{aligned}$

3. $\begin{aligned} 4x - 3y &= 15 \\ 3x + 4y &= 5 \end{aligned}$

4. $\begin{aligned} 5x + 2y &= 1 \\ 2x - 3y &= -11 \end{aligned}$

Problems 5–14 refer to the following matrices:

$$A = \begin{bmatrix} 3 & -2 & 0 \\ 4 & 1 & -6 \end{bmatrix} \qquad B = \begin{bmatrix} -2 & 8 & 0 \\ -3 & 6 & 9 \\ 4 & 2 & 0 \end{bmatrix}$$

$$C = \begin{bmatrix} 3 & -2 & 0 \end{bmatrix} \qquad D = \begin{bmatrix} -4 \\ 7 \end{bmatrix}$$

5. What is the size of A? Of C?

6. What is the size of B? Of D?

7. Identify all row matrices.

8. Identify all column matrices.

9. Identify all square matrices.

10. How many additional rows would matrix A need to be a square matrix?

11. For matrix A, find a_{12} and a_{23}.

12. For matrix A, find a_{21} and a_{13}.

13. Find the elements on the principal diagonal of matrix B.

14. Find the elements on the principal diagonal of matrix A.

Perform each of the row operations indicated in Problems 15–26 on the following matrix:

$$\left[\begin{array}{cc|c} 1 & -3 & 2 \\ 4 & -6 & -8 \end{array}\right]$$

15. $R_1 \leftrightarrow R_2$

16. $\frac{1}{2}R_2 \rightarrow R_2$

17. $-4R_1 \to R_1$

18. $-2R_1 \to R_1$

19. $2R_2 \to R_2$

20. $-1R_2 \to R_2$

21. $(-4)R_1 + R_2 \to R_2$

22. $(-\frac{1}{2})R_2 + R_1 \to R_1$

23. $(-2)R_1 + R_2 \to R_2$

24. $(-3)R_1 + R_2 \to R_2$

25. $(-1)R_1 + R_2 \to R_2$

26. $1R_1 + R_2 \to R_2$

B

Each of the matrices in Problems 27–32 is the result of performing a single row operation on the matrix A shown below. Identify the row operation. Check your work by performing the row operation you identified on a graphing utility.

$$A = \left[\begin{array}{cc|c} -1 & 2 & -3 \\ 6 & -3 & 12 \end{array}\right]$$

27. $\left[\begin{array}{cc|c} -1 & 2 & -3 \\ 2 & -1 & 4 \end{array}\right]$

28. $\left[\begin{array}{cc|c} -2 & 4 & -6 \\ 6 & -3 & 12 \end{array}\right]$

29. $\left[\begin{array}{cc|c} -1 & 2 & -3 \\ 0 & 9 & -6 \end{array}\right]$

30. $\left[\begin{array}{cc|c} 3 & 0 & 5 \\ 6 & -3 & 12 \end{array}\right]$

31. $\left[\begin{array}{cc|c} 1 & 1 & 1 \\ 6 & -3 & 12 \end{array}\right]$

32. $\left[\begin{array}{cc|c} -1 & 2 & -3 \\ 2 & 5 & 0 \end{array}\right]$

Solve Problems 33–38 using augmented matrix methods. Write the linear system represented by each augmented matrix in your solution, and solve each of these systems graphically. Discuss the relationship between the solutions of these systems.

33. $\begin{aligned} x_1 + x_2 &= 7 \\ x_1 - x_2 &= 1 \end{aligned}$

34. $\begin{aligned} x_1 + x_2 &= 5 \\ x_1 - x_2 &= -3 \end{aligned}$

35. $\begin{aligned} 0.5x_1 - x_2 &= -2 \\ -x_1 + 2x_2 &= 4 \end{aligned}$

36. $\begin{aligned} -x_1 + 0.5x_2 &= 1 \\ 2x_1 - x_2 &= -2 \end{aligned}$

37. $\begin{aligned} 0.5x_1 - x_2 &= -2 \\ -x_1 + 2x_2 &= 8 \end{aligned}$

38. $\begin{aligned} -x_1 + 0.5x_2 &= 1 \\ 2x_1 - x_2 &= -4 \end{aligned}$

Solve Problems 39–50 using augmented matrix methods.

39. $\begin{aligned} x_1 - 4x_2 &= -2 \\ -2x_1 + x_2 &= -3 \end{aligned}$

40. $\begin{aligned} x_1 - 3x_2 &= -5 \\ -3x_1 - x_2 &= 5 \end{aligned}$

41. $\begin{aligned} 3x_1 - x_2 &= 2 \\ x_1 + 2x_2 &= 10 \end{aligned}$

42. $\begin{aligned} 2x_1 + x_2 &= 0 \\ x_1 - 2x_2 &= -5 \end{aligned}$

43. $\begin{aligned} x_1 + 2x_2 &= 4 \\ 2x_1 + 4x_2 &= -8 \end{aligned}$

44. $\begin{aligned} 2x_1 - 3x_2 &= -2 \\ -4x_1 + 6x_2 &= 7 \end{aligned}$

45. $\begin{aligned} 2x_1 + x_2 &= 6 \\ x_1 - x_2 &= -3 \end{aligned}$

46. $\begin{aligned} 3x_1 - x_2 &= -5 \\ x_1 + 3x_2 &= 5 \end{aligned}$

47. $\begin{aligned} 3x_1 - 6x_2 &= -9 \\ -2x_1 + 4x_2 &= 6 \end{aligned}$

48. $\begin{aligned} 2x_1 - 4x_2 &= -2 \\ -3x_1 + 6x_2 &= 3 \end{aligned}$

49. $\begin{aligned} 4x_1 - 2x_2 &= 2 \\ -6x_1 + 3x_2 &= -3 \end{aligned}$

50. $\begin{aligned} -6x_1 + 2x_2 &= 4 \\ 3x_1 - x_2 &= -2 \end{aligned}$

C

51. The coefficients of the three systems below are very similar. You might guess that the solution sets to the three systems would also be nearly identical. Develop evidence for or against this guess by considering graphs of the systems and solutions obtained using elimination by addition.

(A) $\begin{aligned} 4x + 5y &= 4 \\ 9x + 11y &= 4 \end{aligned}$

(B) $\begin{aligned} 4x + 5y &= 4 \\ 8x + 11y &= 4 \end{aligned}$

(C) $\begin{aligned} 4x + 5y &= 4 \\ 8x + 10y &= 4 \end{aligned}$

52. Repeat Problem 51 for the following systems.

(A) $\begin{aligned} 5x - 6y &= -10 \\ 11x - 13y &= -20 \end{aligned}$

(B) $\begin{aligned} 5x - 6y &= -10 \\ 10x - 13y &= -20 \end{aligned}$

(C) $\begin{aligned} 5x - 6y &= -10 \\ 10x - 12y &= -20 \end{aligned}$

Solve Problems 53–56 using augmented matrix methods. Use a graphing utility to perform the row operations.

53. $\begin{aligned} 0.8x_1 + 2.88x_2 &= 4 \\ 1.25x_1 + 4.34x_2 &= 5 \end{aligned}$

54. $\begin{aligned} 2.7x_1 - 15.12x_2 &= 27 \\ 3.25x_1 - 18.52x_2 &= 33 \end{aligned}$

55. $\begin{aligned} 4.8x_1 - 40.32x_2 &= 295.2 \\ -3.75x_1 + 28.7x_2 &= -211.2 \end{aligned}$

56. $\begin{aligned} 5.7x_1 - 8.55x_2 &= -35.91 \\ 4.5x_1 + 5.73x_2 &= 76.17 \end{aligned}$

APPLICATIONS

57. Puzzle. A friend of yours came out of the post office having spent \$19.50 on 32¢ and 23¢ stamps. If she bought 75 stamps in all, how many of each type did she buy?

58. Puzzle. A parking meter contains only nickels and dimes worth \$6.05. If there are 89 coins in all, how many of each type are there?

59. Investments. Bond A pays 6% compounded annually and bond B pays 9% compounded annually. If a \$200,000 investment in a combination of the two bonds returns \$14,775 annually, how much is invested in each bond?

60. Investments. Past history indicates that mutual fund A will earn 14.6% annually and mutual fund B will earn 9.8% annually. How should an investment be divided between the two funds to produce an expected return of 11%?

61. Chemistry. A chemist has two solutions of sulfuric acid: a 20% solution and an 80% solution. How much of each should be used to obtain 100 liters of a 62% solution?

62. Chemistry. A chemist has two solutions: one containing 40% alcohol and another containing 70% alcohol. How much of each should be used to obtain 80 liters of a 49% solution?

63. Nutrition. Animals in an experiment are to be kept on a strict diet. Each animal is to receive, among other things, 54 grams of protein and 24 grams of fat. The laboratory technician is able to purchase two food mixes of the following compositions: Mix A has 15% protein and 10% fat; mix B has 30% protein and 5% fat. How many grams of each mix should be used to obtain the right diet for a single animal?

64. Nutrition—Plants. A fruit grower can use two types of fertilizer in his orange grove, brand A and brand B. Each bag of brand A contains 9 pounds of nitrogen and 5 pounds of phosphoric acid. Each bag of brand B contains 8 pounds of nitrogen and 6 pounds of phosphoric acid. Tests indicate that the grove needs 770 pounds of nitrogen and 490 pounds of phosphoric acid. How many bags of each brand should be used to provide the required amounts of nitrogen and phosphoric acid?

65. Delivery Charges. United Express, a nationwide package delivery service, charges a base price for overnight delivery of packages weighing 1 pound or less and a surcharge for each additional pound (or fraction thereof). A customer is billed \$27.75 for shipping a 5-pound package and \$64.50 for shipping a 20-pound package. Find the base price and the surcharge for each additional pound.

66. Delivery Charges. Refer to Problem 65. Federated Shipping, a competing overnight delivery service, informs the customer in Problem 65 that it would ship the 5-pound package for \$29.95 and the 20-pound package for \$59.20.

(A) If Federated Shipping computes its cost in the same manner as United Express, find the base price and the surcharge for Federated Shipping.

(B) Devise a simple rule that the customer can use to choose the cheaper of the two services for each package shipped. Justify your answer.

67. Resource Allocation. A coffee manufacturer uses Colombian and Brazilian coffee beans to produce two blends, robust and mild. A pound of the robust blend requires 12 ounces of Colombian beans and 4 ounces of Brazilian beans. A pound of the mild blend requires 6 ounces of Colombian beans and 10 ounces of Brazilian beans. Coffee is shipped in 132-pound burlap bags. The company has 50 bags of Colombian beans and 40 bags of Brazilian beans on hand. How many pounds of each blend should it produce to use all the available beans?

68. Resource Allocation. Refer to Problem 67.

(A) If the company decides to discontinue production of the robust blend and only produce the mild blend, how many pounds of the mild blend can it produce and how many beans of each type will it use? Are there any beans that are not used?

(B) Repeat part A if the company decides to discontinue production of the mild blend and only produce the robust blend.

SECTION 8.3 Gauss–Jordan Elimination

Reduced Matrices • Solving Systems by Gauss–Jordan Elimination • Mathematical Modeling

Now that you have had some experience with row operations on simple augmented matrices, we will consider systems involving more than two variables. In addition, we will not require that a system have the same number of equations as variables. It turns out that the results for two-variable–two-equation linear systems, stated in Theorem 1 in Section 8.1, actually hold for linear systems of any size.

Possible Solutions to a Linear System

It can be shown that any linear system must have exactly one solution, no solution, or an infinite number of solutions, regardless of the number of equations or the number of variables in the system. The terms *unique, consistent, inconsistent, dependent,* and *independent* are used to describe these solutions, just as they are for systems with two variables.

Reduced Matrices

In Section 8.2 we used row operations to transform the augmented coefficient matrix for a system of two equations in two variables

$$\left[\begin{array}{cc|c} a_{11} & a_{12} & k_1 \\ a_{21} & a_{22} & k_2 \end{array}\right] \qquad \begin{aligned} a_{11}x_1 + a_{12}x_2 &= k_1 \\ a_{21}x_1 + a_{22}x_2 &= k_2 \end{aligned}$$

into one of the following simplified forms:

$$\underset{\text{Form 1}}{\left[\begin{array}{cc|c} 1 & 0 & m \\ 0 & 1 & n \end{array}\right]} \qquad \underset{\text{Form 2}}{\left[\begin{array}{cc|c} 1 & m & n \\ 0 & 0 & 0 \end{array}\right]} \qquad \underset{\text{Form 3}}{\left[\begin{array}{cc|c} 1 & m & n \\ 0 & 0 & p \end{array}\right]} \tag{1}$$

where m, n, and p are real numbers, $p \neq 0$. Each of these reduced forms represents a system that has a different type of solution set, and no two of these forms are row-equivalent. Thus, we consider each of these to be a different simplified form. Now we want to consider larger systems with more variables and more equations.

EXPLORE/DISCUSS 1

Forms 1, 2, and 3 above represent systems that have, respectively, a unique solution, an infinite number of solutions, and no solution. Discuss the number of solutions for the systems of three equations in three variables represented by the following augmented coefficient matrices.

$$\text{(A)} \left[\begin{array}{ccc|c} 1 & 1 & 1 & 2 \\ 0 & 0 & 0 & 3 \\ 0 & 0 & 0 & 0 \end{array}\right] \qquad \text{(B)} \left[\begin{array}{ccc|c} 1 & 1 & 1 & 2 \\ 0 & 0 & 0 & 0 \\ 0 & 0 & 0 & 0 \end{array}\right] \qquad \text{(C)} \left[\begin{array}{ccc|c} 1 & 0 & 0 & 2 \\ 0 & 1 & 0 & 3 \\ 0 & 0 & 1 & 4 \end{array}\right]$$

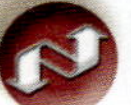

Because there is no upper limit on the number of variables or the number of equations in a linear system, it is not feasible to explicitly list all possible "simplified forms" for larger systems, as we did for systems of two equations in two variables. Instead, we state a general definition of a simplified form called a *reduced matrix* that can be applied to all matrices and systems, regardless of size.

DEFINITION 1
Reduced Matrix

A matrix is in **reduced form*** if

1. Each row consisting entirely of 0s is below any row having at least one nonzero element.
2. The leftmost or leading nonzero element in each row is 1.
3. The column containing the leading 1 of a given row has 0s above and below the 1.
4. The leading 1 in any row is to the right of the leading 1 in the preceding row.

EXAMPLE 1 Reduced Forms

The matrices below are not in reduced form. Indicate which condition in Definition 1 is violated for each matrix. State the row operation(s) required to transform the matrix to reduced form, and find the reduced form.

(A) $\left[\begin{array}{cc|c} 0 & 1 & -2 \\ 1 & 0 & 3 \end{array}\right]$ (B) $\left[\begin{array}{ccc|c} 1 & 2 & -2 & 3 \\ 0 & 0 & 1 & -1 \end{array}\right]$

(C) $\left[\begin{array}{cc|c} 1 & 0 & -3 \\ 0 & 0 & 0 \\ 0 & 1 & -2 \end{array}\right]$ (D) $\left[\begin{array}{ccc|c} 1 & 0 & 0 & -1 \\ 0 & 2 & 0 & 3 \\ 0 & 0 & 1 & -5 \end{array}\right]$

SOLUTIONS

(A) Condition 4 is violated: The leading 1 in row 2 is not to the right of the leading 1 in row 1. Perform the row operation $R_1 \leftrightarrow R_2$ to obtain the reduced form:

$$\left[\begin{array}{cc|c} 1 & 0 & 3 \\ 0 & 1 & -2 \end{array}\right]$$

(B) Condition 3 is violated: The column containing the leading 1 in row 2 does not have a 0 above the 1. To obtain the reduced form, perform the row operation $2R_2 + R_1 \rightarrow R_1$

$$\left[\begin{array}{ccc|c} 1 & 2 & 0 & 1 \\ 0 & 0 & 1 & -1 \end{array}\right]$$

(C) Condition 1 is violated: The second row contains all 0s, and it is above a row having at least one nonzero element. Perform the row operation $R_2 \leftrightarrow R_3$ to obtain the reduced form:

$$\left[\begin{array}{cc|c} 1 & 0 & -3 \\ 0 & 1 & -2 \\ 0 & 0 & 0 \end{array}\right]$$

---

*The reduced form we have defined here is often referred to as the **reduced row-echelon form** to distinguish it from other reduced forms (see Problems 57–62 in Exercise 8.3). Most graphing utilities use the abbreviation **rref** to refer to this reduced form.

(D) Condition 2 is violated: The leading nonzero element in row 2 is not a 1. Perform the row operation $\frac{1}{2}R_2 \to R_2$ to obtain the reduced form:

$$\left[\begin{array}{ccc|c} 1 & 0 & 0 & -1 \\ 0 & 1 & 0 & \frac{3}{2} \\ 0 & 0 & 1 & -5 \end{array}\right]$$

MATCHED PROBLEM 1

The matrices below are not in reduced form. Indicate which condition in Definition 1 is violated for each matrix. State the row operation(s) required to transform the matrix to reduced form and find the reduced form.

(A) $\left[\begin{array}{cc|c} 1 & 0 & 2 \\ 0 & 3 & -6 \end{array}\right]$ (B) $\left[\begin{array}{ccc|c} 1 & 5 & 4 & 3 \\ 0 & 1 & 2 & -1 \\ 0 & 0 & 0 & 0 \end{array}\right]$

(C) $\left[\begin{array}{ccc|c} 0 & 1 & 0 & -3 \\ 1 & 0 & 0 & 0 \\ 0 & 0 & 1 & 2 \end{array}\right]$ (D) $\left[\begin{array}{ccc|c} 1 & 2 & 0 & 3 \\ 0 & 0 & 0 & 0 \\ 0 & 0 & 1 & 4 \end{array}\right]$

Solving Systems by Gauss–Jordan Elimination

We are now ready to outline the Gauss–Jordan elimination method for solving systems of linear equations. The method systematically transforms an augmented matrix into a reduced form. The system corresponding to a reduced augmented coefficient matrix is called a **reduced system.** As we will see, reduced systems are easy to solve.

The Gauss–Jordan elimination method is named after the German mathematician Karl Friedrich Gauss (1777–1855) and the German geodesist Wilhelm Jordan (1842–1899). Gauss, one of the greatest mathematicians of all time, used a method of solving systems of equations that was later generalized by Jordan to solve problems in large-scale surveying.

EXAMPLE 2 Solving a System Using Gauss–Jordan Elimination

Solve by Gauss–Jordan elimination:

$$\begin{aligned} 2x_1 - 2x_2 + x_3 &= 3 \\ 3x_1 + x_2 - x_3 &= 7 \\ x_1 - 3x_2 + 2x_3 &= 0 \end{aligned}$$

SOLUTION

Write the augmented matrix and follow the steps indicated at the right to produce a reduced form.

Need a 1 here.

$$\left[\begin{array}{ccc|c} 2 & -2 & 1 & 3 \\ 3 & 1 & -1 & 7 \\ 1 & -3 & 2 & 0 \end{array}\right] \quad R_1 \leftrightarrow R_3$$

Step 1. Choose the leading nonzero column and get a 1 at the top.

Need 0s here.
$$\sim\left[\begin{array}{rrr|r} 1 & -3 & 2 & 0 \\ 3 & 1 & -1 & 7 \\ 2 & -2 & 1 & 3 \end{array}\right] \begin{array}{l} \\ (-3)R_1 + R_2 \to R_2 \\ (-2)R_1 + R_3 \to R_3 \end{array}$$

Step 2. Use multiples of the row containing the 1 from step 1 to get zeros in all remaining places in the column containing this 1.

Need a 1 here.
$$\sim\left[\begin{array}{rrr|r} 1 & -3 & 2 & 0 \\ 0 & 10 & -7 & 7 \\ 0 & 4 & -3 & 3 \end{array}\right] \quad 0.1R_2 \to R_2$$

Step 3. Repeat step 1 with the *submatrix* formed by (mentally) deleting the top row.

Need 0s here.
$$\sim\left[\begin{array}{rrr|r} 1 & -3 & 2 & 0 \\ 0 & 1 & -0.7 & 0.7 \\ 0 & 4 & -3 & 3 \end{array}\right] \begin{array}{l} 3R_2 + R_1 \to R_1 \\ \\ (-4)R_2 + R_3 \to R_3 \end{array}$$

Step 4. Repeat step 2 with the *entire matrix*.

Need a 1 here.
$$\sim\left[\begin{array}{rrr|r} 1 & 0 & -0.1 & 2.1 \\ 0 & 1 & -0.7 & 0.7 \\ 0 & 0 & -0.2 & 0.2 \end{array}\right] \quad (-5)R_3 \to R_3$$

Step 5. Repeat step 1 with the *submatrix* formed by (mentally) deleting the top two rows.

Need 0s here.
$$\sim\left[\begin{array}{rrr|r} 1 & 0 & -0.1 & 2.1 \\ 0 & 1 & -0.7 & 0.7 \\ 0 & 0 & 1 & -1 \end{array}\right] \begin{array}{l} 0.1R_3 + R_1 \to R_1 \\ 0.7R_3 + R_2 \to R_2 \\ \\ \end{array}$$

Step 6. Repeat step 2 with the *entire matrix*.

$$\sim\left[\begin{array}{rrr|r} 1 & 0 & 0 & 2 \\ 0 & 1 & 0 & 0 \\ 0 & 0 & 1 & -1 \end{array}\right]$$

The matrix is now in reduced form, and we can proceed to solve the corresponding reduced system.

$$\begin{aligned} x_1 \quad\quad &= 2 \\ x_2 \quad &= 0 \\ x_3 &= -1 \end{aligned}$$

The solution to this system is $x_1 = 2$, $x_2 = 0$, $x_3 = -1$. You should check this solution in the original system.

Gauss–Jordan Elimination

Step 1. Choose the leading nonzero column of an augmented matrix and use appropriate row operations to get a 1 at the top.

Step 2. Use multiples of the row containing the 1 from step 1 to get zeros in all remaining places in the column containing this 1.

Step 3. Repeat step 1 with the **submatrix** formed by (mentally) deleting the row used in step 2 and all rows above this row.

Step 4. Repeat step 2 with the **entire matrix,** including the mentally deleted rows. Continue this process until the entire matrix is in reduced form.

[*Note:* If at any point in this process we obtain a row with all zeros to the left of the vertical line and a nonzero number to the right, we can stop, because we will have a contradiction: $0 = n, n \neq 0$. We can then conclude that the system has no solution.]

REMARKS

1. Although each matrix has a unique reduced form, the sequence of steps (algorithm) presented here for transforming a matrix into a reduced form is not unique. That is, other sequences of steps (using row operations) can produce a reduced matrix. (For example, it is possible to use row operations in such a way that computations involving fractions are minimized.) But we emphasize again that we are not interested in the most efficient hand methods for transforming small matrices into reduced forms. Our main interest is in giving you a little experience with a method that is suitable for solving large-scale systems on a computer or graphing utility.
2. Most graphing utilities have the ability to find reduced forms, either directly or with some programming. Figure 1 illustrates the solution of Example 2 on a graphing utility that has a built-in command for finding reduced forms. Notice that in row 2 and column 4 of the reduced form the graphing utility has displayed the very small number −3.5E-13 instead of the exact value 0. This is a common occurrence on a graphing utility and causes no problems. Just replace any very small numbers displayed in scientific notation with 0.

FIGURE 1 Gauss–Jordan elimination on a graphing calculator.

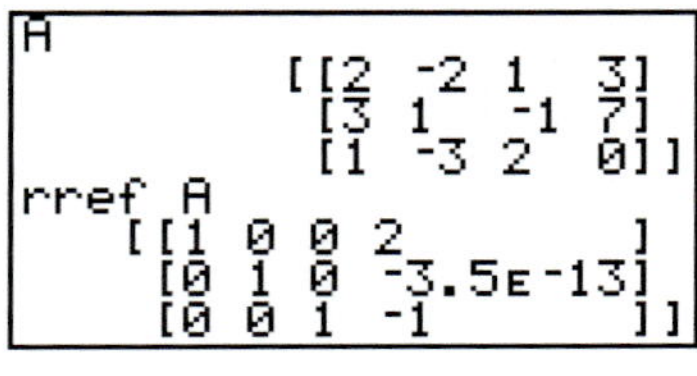

A
[[2 -2 1 3]
[3 1 -1 7]
[1 -3 2 0]]
rref A
[[1 0 0 2]
[0 1 0 -3.5E-13]
[0 0 1 -1]]

(a) TI-85/86

[A]
[[2 -2 1 3]
[3 1 -1 7]
[1 -3 2 0]]
rref([A])
[[1 0 0 2 ...
[0 1 0 -3.5E-1...
[0 0 1 -1 ...

(b) TI-83

REMARK

Most of the graphing calculator screens displayed in the text were produced on a Texas Instruments TI-83. However, in this chapter there are a few screens that were produced on a TI-85/86 to display more of the matrix (see Fig. 1).

MATCHED PROBLEM 2

Solve by Gauss–Jordan elimination:

$$\begin{aligned} 3x_1 + x_2 - 2x_3 &= 2 \\ x_1 - 2x_2 + x_3 &= 3 \\ 2x_1 - x_2 - 3x_3 &= 3 \end{aligned}$$

EXAMPLE 3 Solving a System Using Gauss–Jordan Elimination

Solve by Gauss–Jordan elimination:

$$\begin{aligned} 2x_1 - 4x_2 + x_3 &= -4 \\ 4x_1 - 8x_2 + 7x_3 &= 2 \\ -2x_1 + 4x_2 - 3x_3 &= 5 \end{aligned}$$

SOLUTION

Algebraic Solution

$$\left[\begin{array}{rrr|r} 2 & -4 & 1 & -4 \\ 4 & -8 & 7 & 2 \\ -2 & 4 & -3 & 5 \end{array}\right] \quad 0.5R_1 \to R_1$$

$$\sim \left[\begin{array}{rrr|r} 1 & -2 & 0.5 & -2 \\ 4 & -8 & 7 & 2 \\ -2 & 4 & -3 & 5 \end{array}\right] \quad \begin{array}{l} (-4)R_1 + R_2 \to R_2 \\ 2R_1 + R_3 \to R_3 \end{array}$$

$$\sim \left[\begin{array}{rrr|r} 1 & -2 & 0.5 & -2 \\ 0 & 0 & 5 & 10 \\ 0 & 0 & -2 & 1 \end{array}\right] \quad 0.2R_2 \to R_2$$

Note that column 3 is the leading nonzero column in this submatrix.

$$\sim \left[\begin{array}{rrr|r} 1 & -2 & 0.5 & -2 \\ 0 & 0 & 1 & 2 \\ 0 & 0 & -2 & 1 \end{array}\right] \quad \begin{array}{l} (-0.5)R_2 + R_1 \to R_1 \\ 2R_2 + R_3 \to R_3 \end{array}$$

$$\sim \left[\begin{array}{rrr|r} 1 & -2 & 0 & -3 \\ 0 & 0 & 1 & 2 \\ 0 & 0 & 0 & 5 \end{array}\right]$$

We stop the Gauss–Jordan elimination, although the matrix is not in reduced form, because the last row produces a contradiction: $0x_1 + 0x_2 + 0x_3 = 5$

The system is inconsistent and has no solution.

Graphing Utility Solution

Enter the augmented coefficient matrix and use the rref command to find the reduced form (Fig. 2).

```
A
        [[2   -4  1   -4]
         [4   -8  7   2 ]
         [-2  4   -3  5 ]]
rref A
         [[1  -2  0  0]
          [0  0   1  0]
          [0  0   0  1]]
```

FIGURE 2

The last row of the reduced form in Figure 2 indicates an inconsistent system with no solution. Notice that the graphing utility program does not stop when a contradiction first occurs, as we did in the algebraic solution, but continues on to find the reduced form.

MATCHED PROBLEM 3

Solve by Gauss–Jordan elimination:

$$\begin{aligned} 2x_1 - 4x_2 - x_3 &= -8 \\ 4x_1 - 8x_2 + 3x_3 &= 4 \\ -2x_1 + 4x_2 + x_3 &= 11 \end{aligned}$$

EXAMPLE 4 Solving a System Using Gauss–Jordan Elimination

Solve by Gauss–Jordan elimination:

$$\begin{aligned} 3x_1 + 6x_2 - 9x_3 &= 15 \\ 2x_1 + 4x_2 - 6x_3 &= 10 \\ -2x_1 - 3x_2 + 4x_3 &= -6 \end{aligned}$$

SOLUTION

First we find the reduced form of the augmented coefficient matrix.

Algebraic Method

$$\begin{bmatrix} 3 & 6 & -9 & | & 15 \\ 2 & 4 & -6 & | & 10 \\ -2 & -3 & 4 & | & -6 \end{bmatrix} \quad \tfrac{1}{3}R_1 \rightarrow R_1$$

$$\sim \begin{bmatrix} 1 & 2 & -3 & | & 5 \\ 2 & 4 & -6 & | & 10 \\ -2 & -3 & 4 & | & -6 \end{bmatrix} \quad \begin{matrix} (-2)R_1 + R_2 \rightarrow R_2 \\ 2R_1 + R_3 \rightarrow R_3 \end{matrix}$$

$$\sim \begin{bmatrix} 1 & 2 & -3 & | & 5 \\ 0 & 0 & 0 & | & 0 \\ 0 & 1 & -2 & | & 4 \end{bmatrix} \quad R_2 \leftrightarrow R_3$$

Note that we must interchange rows 2 and 3 to obtain a nonzero entry at the top of the second column of this submatrix.

$$\sim \begin{bmatrix} 1 & 2 & -3 & | & 5 \\ 0 & 1 & -2 & | & 4 \\ 0 & 0 & 0 & | & 0 \end{bmatrix} \quad (-2)R_2 + R_1 \rightarrow R_1$$

$$\sim \begin{bmatrix} 1 & 0 & 1 & | & -3 \\ 0 & 1 & -2 & | & 4 \\ 0 & 0 & 0 & | & 0 \end{bmatrix}$$

This matrix is now in reduced form.

Graphing Utility Method

Enter the augmented coefficient matrix and use rref (Fig. 3).

```
A
      [[3  6  -9 15]
       [2  4  -6 10]
       [-2 -3 4  -6]]
rref A
      [[1 0 1   -3]
       [0 1 -2  4 ]
       [0 0 0   0 ]]
```

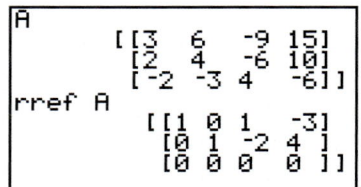

FIGURE 3

Using the reduced form from either of the above methods, we write the corresponding system and solve.

$$\begin{aligned} x_1 \quad + x_3 &= -3 \\ x_2 - 2x_3 &= 4 \end{aligned}$$

We discard the equation corresponding to the third (all 0) row in the reduced form, because it is satisfied by all values of x_1, x_2, and x_3.

Note that the leading variable in each equation appears in one and only one equation. We solve for the leading variables x_1 and x_2 in terms of the remaining variable x_3:

$$\begin{aligned} x_1 &= -x_3 - 3 \\ x_2 &= 2x_3 + 4 \end{aligned}$$

This dependent system has an infinite number of solutions. We will use a parameter to represent all the solutions. If we let $x_3 = t$, then for any real number t,

$$\begin{aligned} x_1 &= -t - 3 \\ x_2 &= 2t + 4 \\ x_3 &= t \end{aligned}$$

You should check that $(-t - 3, 2t + 4, t)$ is a solution of the original system for any real number t. Some particular solutions are

$t = 0$	$t = -2$	$t = 3.5$
$(-3, 4, 0)$	$(-1, 0, -2)$	$(-6.5, 11, 3.5)$

MATCHED 4 PROBLEM

Solve by Gauss–Jordan elimination:

$$\begin{aligned} 2x_1 - 2x_2 - 4x_3 &= -2 \\ 3x_1 - 3x_2 - 6x_3 &= -3 \\ -2x_1 + 3x_2 + x_3 &= 7 \end{aligned}$$

In general,

If the number of leading 1s in a reduced augmented coefficient matrix is less than the number of variables in the system and there are no contradictions, then the system is dependent and has infinitely many solutions.

There are many different ways to use the reduced augmented coefficient matrix to describe the infinite number of solutions of a dependent system. We will always proceed as follows: Solve each equation in a reduced system for its leading variable and then introduce a different parameter for each remaining variable. As the solution to Example 4 illustrates, this method produces a concise and useful representation of the solutions to a dependent system. Example 5 illustrates a dependent system where two parameters are required to describe the solution.

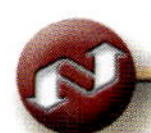

EXPLORE/DISCUSS 2

Explain why the definition of reduced form ensures that each leading variable in a reduced system appears in one and only one equation and no equation contains more than one leading variable. Discuss methods for determining if a consistent system is independent or dependent by examining the reduced form.

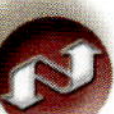

EXAMPLE 5 Solving a System Using Gauss–Jordan Elimination

Solve by using rref on a graphing utility:

$$\begin{aligned} x_1 + 2x_2 + 4x_3 + x_4 - x_5 &= 1 \\ 2x_1 + 4x_2 + 8x_3 + 3x_4 - 4x_5 &= 2 \\ x_1 + 3x_2 + 7x_3 + \phantom{x_4 - {}} 3x_5 &= -2 \end{aligned}$$

SOLUTION

The augmented coefficient matrix and its reduced form are shown in Figure 4. Write the corresponding reduced system and solve.

FIGURE 4

```
A
    [[1 2 4 1 -1 1 ]
     [2 4 8 3 -4 2 ]
     [1 3 7 0 3  -2]]
rref A
   [[1 0 -2 0 -3 7 ]
    [0 1 3  0 2  -3]
    [0 0 0  1 -2 0 ]]
```

$$\begin{aligned} x_1 \quad - 2x_3 \quad\quad - 3x_5 &= 7 \\ x_2 + 3x_3 \quad\quad + 2x_5 &= -3 \\ x_4 - 2x_5 &= 0 \end{aligned}$$

Solve for the leading variables x_1, x_2, and x_4 in terms of the remaining variables x_3 and x_5.

$$\begin{aligned} x_1 &= 2x_3 + 3x_5 + 7 \\ x_2 &= -3x_3 - 2x_5 - 3 \\ x_4 &= 2x_5 \end{aligned}$$

If we let $x_3 = s$ and $x_5 = t$, then for any real numbers s and t,

$$\begin{aligned} x_1 &= 2s + 3t + 7 \\ x_2 &= -3s - 2t - 3 \\ x_3 &= s \\ x_4 &= 2t \\ x_5 &= t \end{aligned}$$

is a solution. The check is left for you to perform.

MATCHED 5 PROBLEM

Solve by using rref on a graphing utility:

$$\begin{aligned} x_1 - x_2 + 2x_3 \quad\quad - 2x_5 &= 3 \\ -2x_1 + 2x_2 - 4x_3 - x_4 + x_5 &= -5 \\ 3x_1 - 3x_2 + 7x_3 + x_4 - 4x_5 &= 6 \end{aligned}$$

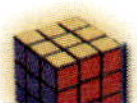

Mathematical Modeling

Dependent systems of linear equations provide an excellent opportunity to discuss mathematical modeling in a little more detail. The process of using mathematics to solve real-world problems can be broken down into three steps (Fig. 5):

Step 1. *Construct* a mathematical model whose solution will provide information about the real-world problem.
Step 2. *Solve* the mathematical model.
Step 3. *Interpret* the solution to the mathematical model in terms of the original real-world problem.

FIGURE 5

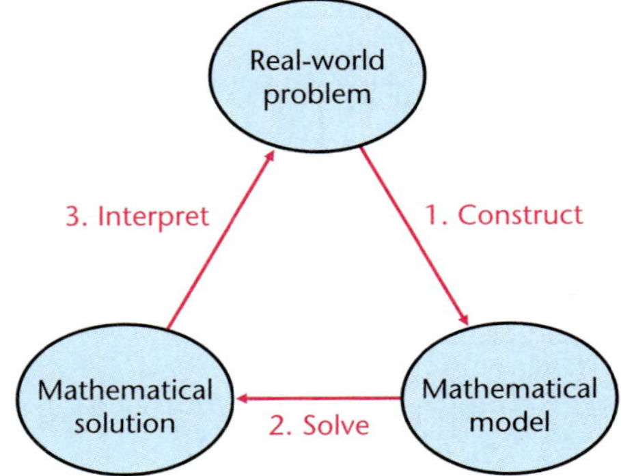

In more complex problems, this cycle may have to be repeated several times to obtain the required information about the real-world problem.

EXAMPLE 6 Purchasing

A chemical manufacturer wants to purchase a fleet of 24 railroad tank cars with a combined carrying capacity of 250,000 gallons. Tank cars with three different carrying capacities are available: 6,000 gallons, 8,000 gallons, and 18,000 gallons. How many of each type of tank car should be purchased?

SOLUTION

Let

x_1 = Number of 6,000-gallon tank cars

x_2 = Number of 8,000-gallon tank cars

x_3 = Number of 18,000-gallon tank cars

Then

$$x_1 + x_2 + x_3 = 24 \quad \text{Total number of tank cars}$$
$$6{,}000x_1 + 8{,}000x_2 + 18{,}000x_3 = 250{,}000 \quad \text{Total carrying capacity}$$

The augmented coefficient matrix and its reduced form are shown in Figure 6. Write the corresponding reduced system and solve.

FIGURE 6

```
A
[[1     1     1     24...
 [6000  8000  18000 25...
rref A
        [[1 0 -5 -29]
         [0 1 6   53 ]]
```

$$x_1 - 5x_3 = -29 \quad \text{or} \quad x_1 = 5x_3 - 29$$
$$x_2 + 6x_3 = 53 \quad \text{or} \quad x_2 = -6x_3 + 53$$

Let $x_3 = t$. Then for t any real number,

$$x_1 = 5t - 29$$
$$x_2 = -6t + 53$$
$$x_3 = t$$

is a solution—or is it? Because the variables in this system represent the number of tank cars purchased, the values of x_1, x_2, and x_3 must be nonnegative integers. Thus, the third equation requires that t must be a nonnegative integer. The first equation requires that $5t - 29 \geq 0$, so t must be at least 6. The middle equation requires that $-6t + 53 \geq 0$, so t can be no larger than 8. Thus, 6, 7, and 8 are the only possible values for t. There are only three possible combinations that meet the company's specifications of 24 tank cars with a total carrying capacity of 250,000 gallons, as shown in Table 1.

TABLE 1

t	6,000-Gallon Tank Cars x_1	8,000-Gallon Tank Cars x_2	18,000-Gallon Tank Cars x_3
6	1	17	6
7	6	11	7
8	11	5	8

The final choice would probably be influenced by other factors. For example, the company might want to minimize the cost of the 24 tank cars.

MATCHED 6 PROBLEM

A commuter airline wants to purchase a fleet of 30 airplanes with a combined carrying capacity of 960 passengers. The three available types of planes carry 18, 24, and 42 passengers, respectively. How many of each type of plane should be purchased?

ANSWERS MATCHED PROBLEMS

1. (A) Condition 2 is violated: the 3 in row 2 and column 2 should be a 1. Perform the operation $\frac{1}{3}R_2 \to R_2$ to obtain

$$\left[\begin{array}{cc|c} 1 & 0 & 2 \\ 0 & 1 & -2 \end{array}\right]$$

(B) Condition 3 is violated: the 5 in row 1 and column 2 should be a 0. Perform the operation $(-5)R_2 + R_1 \to R_1$ to obtain

$$\left[\begin{array}{ccc|c} 1 & 0 & -6 & 8 \\ 0 & 1 & 2 & -1 \\ 0 & 0 & 0 & 0 \end{array}\right]$$

(C) Condition 4 is violated: the leading 1 in the second row is not to the right of the leading 1 in the first row. Perform the operation $R_1 \leftrightarrow R_2$ to obtain

$$\left[\begin{array}{ccc|c} 1 & 0 & 0 & 0 \\ 0 & 1 & 0 & -3 \\ 0 & 0 & 1 & 2 \end{array}\right]$$

(D) Condition 1 is violated: the all-zero second row should be at the bottom. Perform the operation $R_2 \leftrightarrow R_3$ to obtain

$$\left[\begin{array}{ccc|c} 1 & 2 & 0 & 3 \\ 0 & 0 & 1 & 4 \\ 0 & 0 & 0 & 0 \end{array}\right]$$

2. $x_1 = 1, x_2 = -1, x_3 = 0$

3. Inconsistent; no solution

4. $x_1 = 5t + 4, x_2 = 3t + 5, x_3 = t$, t any real number

5. $x_1 = s + 7, x_2 = s, x_3 = t - 2, x_4 = -3t - 1, x_5 = t$, s and t any real numbers

6.

t	18-Passenger Planes x_1	24-Passenger Planes x_2	42-Passenger Planes x_3
14	2	14	14
15	5	10	15
16	8	6	16
17	11	2	17

EXERCISE 8.3

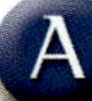

A

In Problems 1–10, if a matrix is in reduced form, say so. If not, explain why and indicate the row operation(s) necessary to transform the matrix into reduced form.

1. $\left[\begin{array}{cc|c} 1 & 0 & 3 \\ 0 & 1 & -2 \end{array}\right]$

2. $\left[\begin{array}{cc|c} 0 & 1 & 3 \\ 1 & 0 & -2 \end{array}\right]$

3. $\left[\begin{array}{ccc|c} 1 & 0 & 3 & 2 \\ 0 & 0 & 0 & 0 \\ 0 & 1 & -1 & 5 \end{array}\right]$

4. $\left[\begin{array}{ccc|c} 1 & 0 & 0 & -4 \\ 0 & 1 & 0 & 0 \\ 0 & 0 & 1 & 1 \end{array}\right]$

5. $\left[\begin{array}{ccc|c} 0 & 1 & 0 & 4 \\ 0 & 0 & 3 & -2 \\ 0 & 0 & 0 & 0 \end{array}\right]$

6. $\left[\begin{array}{ccc|c} 1 & 1 & -3 & 2 \\ 0 & 0 & 1 & 5 \\ 0 & 0 & 0 & 0 \end{array}\right]$

7. $\left[\begin{array}{ccc|c} 1 & 1 & 0 & -1 \\ 0 & 0 & 1 & 1 \\ 0 & 0 & 0 & 0 \end{array}\right]$

8. $\left[\begin{array}{ccc|c} 1 & 0 & -1 & 4 \\ 0 & 2 & 1 & 3 \\ 0 & 0 & 0 & 0 \end{array}\right]$

9. $\left[\begin{array}{cccc|c} 1 & 0 & -3 & 0 & 1 \\ 0 & 0 & 1 & 1 & 0 \end{array}\right]$

10. $\left[\begin{array}{cccc|c} 1 & -3 & 0 & 0 & 1 \\ 0 & 0 & 1 & 1 & 0 \end{array}\right]$

In Problems 11–18, classify the system corresponding to each reduced matrix as consistent and independent, consistent and dependent, or inconsistent. Write the corresponding system and solve, if possible.

11. $\left[\begin{array}{ccc|c} 1 & 0 & 0 & -2 \\ 0 & 1 & 0 & 3 \\ 0 & 0 & 1 & 0 \end{array}\right]$

12. $\left[\begin{array}{cccc|c} 1 & 0 & 0 & 0 & -2 \\ 0 & 1 & 0 & 0 & 0 \\ 0 & 0 & 1 & 0 & 1 \\ 0 & 0 & 0 & 1 & 3 \end{array}\right]$

13. $\left[\begin{array}{ccc|c} 1 & 0 & -2 & 3 \\ 0 & 1 & 1 & -5 \\ 0 & 0 & 0 & 0 \end{array}\right]$

14. $\left[\begin{array}{ccc|c} 1 & -2 & 0 & -3 \\ 0 & 0 & 1 & 5 \\ 0 & 0 & 0 & 0 \end{array}\right]$

15. $\left[\begin{array}{cc|c} 1 & 0 & 0 \\ 0 & 1 & 0 \\ 0 & 0 & 1 \end{array}\right]$

16. $\left[\begin{array}{cc|c} 1 & 0 & 5 \\ 0 & 1 & -3 \\ 0 & 0 & 1 \end{array}\right]$

17. $\left[\begin{array}{cccc|c} 1 & -2 & 0 & -3 & -5 \\ 0 & 0 & 1 & 3 & 2 \end{array}\right]$

18. $\left[\begin{array}{cccc|c} 1 & 0 & -2 & 3 & 4 \\ 0 & 1 & -1 & 2 & -1 \end{array}\right]$

B

Use row operations to change each matrix in Problems 19–24 to reduced form.

19. $\left[\begin{array}{cc|c} 1 & 2 & -1 \\ 0 & 1 & 3 \end{array}\right]$

20. $\left[\begin{array}{cc|c} 1 & 3 & 1 \\ 0 & 2 & -4 \end{array}\right]$

21. $\left[\begin{array}{ccc|c} 1 & 0 & -3 & 1 \\ 0 & 1 & 2 & 0 \\ 0 & 0 & 3 & -6 \end{array}\right]$

22. $\left[\begin{array}{ccc|c} 1 & 0 & 4 & 0 \\ 0 & 1 & -3 & -1 \\ 0 & 0 & -2 & 2 \end{array}\right]$

23. $\left[\begin{array}{ccc|c} 1 & 2 & -2 & -1 \\ 0 & 3 & -6 & 1 \\ 0 & -1 & 2 & -\frac{1}{3} \end{array}\right]$

24. $\left[\begin{array}{ccc|c} 0 & -2 & 8 & 1 \\ 2 & -2 & 6 & -4 \\ 0 & -1 & 4 & \frac{1}{2} \end{array}\right]$

Solve Problems 25–44 using Gauss–Jordan elimination.

25. $\begin{aligned} 2x_1 + 4x_2 - 10x_3 &= -2 \\ 3x_1 + 9x_2 - 21x_3 &= 0 \\ x_1 + 5x_2 - 12x_3 &= 1 \end{aligned}$

26. $\begin{aligned} 3x_1 + 5x_2 - x_3 &= -7 \\ x_1 + x_2 + x_3 &= -1 \\ 2x_1 + 11x_3 &= 7 \end{aligned}$

27. $\begin{aligned} 3x_1 + 8x_2 - x_3 &= -18 \\ 2x_1 + x_2 + 5x_3 &= 8 \\ 2x_1 + 4x_2 + 2x_3 &= -4 \end{aligned}$

28. $\begin{aligned} 2x_1 + 7x_2 + 15x_3 &= -12 \\ 4x_1 + 7x_2 + 13x_3 &= -10 \\ 3x_1 + 6x_2 + 12x_3 &= -9 \end{aligned}$

29. $\begin{aligned} 2x_1 - x_2 - 3x_3 &= 8 \\ x_1 - 2x_2 &= 7 \end{aligned}$

30. $\begin{aligned} 2x_1 + 4x_2 - 6x_3 &= 10 \\ 3x_1 + 3x_2 - 3x_3 &= 6 \end{aligned}$

31. $\begin{aligned} 2x_1 - x_2 &= 0 \\ 3x_1 + 2x_2 &= 7 \\ x_1 - x_2 &= -1 \end{aligned}$

32. $\begin{aligned} 2x_1 - x_2 &= 0 \\ 3x_1 + 2x_2 &= 7 \\ x_1 - x_2 &= -2 \end{aligned}$

33. $\begin{aligned} 3x_1 - 4x_2 - x_3 &= 1 \\ 2x_1 - 3x_2 + x_3 &= 1 \\ x_1 - 2x_2 + 3x_3 &= 2 \end{aligned}$

34. $\begin{aligned} 3x_1 + 7x_2 - x_3 &= 11 \\ x_1 + 2x_2 - x_3 &= 3 \\ 2x_1 + 4x_2 - 2x_3 &= 10 \end{aligned}$

35. $\begin{aligned} -2x_1 + x_2 + 3x_3 &= -7 \\ x_1 - 4x_2 + 2x_3 &= 0 \\ x_1 - 3x_2 + x_3 &= 1 \end{aligned}$

36. $\begin{aligned} 2x_1 + 5x_2 + 4x_3 &= -7 \\ -4x_1 - 5x_2 + 2x_3 &= 9 \\ -2x_1 - x_2 + 4x_3 &= 3 \end{aligned}$

37. $\begin{aligned} 2x_1 - 2x_2 - 4x_3 &= -2 \\ -3x_1 + 3x_2 + 6x_3 &= 3 \end{aligned}$

38. $\begin{aligned} 2x_1 + 8x_2 - 6x_3 &= 4 \\ -3x_1 - 12x_2 + 9x_3 &= -6 \end{aligned}$

39. $\begin{aligned} 4x_1 - x_2 + 2x_3 &= 3 \\ -4x_1 + x_2 - 3x_3 &= -10 \\ 8x_1 - 2x_2 + 9x_3 &= -1 \end{aligned}$

40. $\begin{aligned} 4x_1 - 2x_2 + 2x_3 &= 5 \\ -6x_1 + 3x_2 - 3x_3 &= -2 \\ 10x_1 - 5x_2 + 9x_3 &= 4 \end{aligned}$

41. $\begin{aligned} 2x_1 - 5x_2 - 3x_3 &= 7 \\ -4x_1 + 10x_2 + 2x_3 &= 6 \\ 6x_1 - 15x_2 - x_3 &= -19 \end{aligned}$

42. $\begin{aligned} -4x_1 + 8x_2 + 10x_3 &= -6 \\ 6x_1 - 12x_2 - 15x_3 &= 9 \\ -8x_1 + 14x_2 + 19x_3 &= -8 \end{aligned}$

43. $\begin{aligned} 5x_1 - 3x_2 + 2x_3 &= 13 \\ 2x_1 - x_2 - 3x_3 &= 1 \\ 4x_1 - 2x_2 + 4x_3 &= 12 \end{aligned}$

44. $\begin{aligned} 4x_1 - 2x_2 + 3x_3 &= 3 \\ 3x_1 - x_2 - 2x_3 &= -10 \\ 2x_1 + 4x_2 - x_3 &= -1 \end{aligned}$

45. Consider a consistent system of three linear equations in three variables. Discuss the nature of the solution set for the system if the reduced form of the augmented coefficient matrix has

(A) One leading 1 **(B)** Two leading 1s

(C) Three leading 1s **(D)** Four leading 1s

46. Consider a system of three linear equations in three variables. Give examples of two reduced forms that are not row equivalent if the system is

(A) Consistent and dependent

(B) Inconsistent

In Problems 47–50, discuss the relationship between the number of solutions of the system and the constant k.

47. $\begin{aligned} x_1 - x_2 &= 4 \\ 3x_1 + kx_2 &= 7 \end{aligned}$

48. $\begin{aligned} x_1 + 2x_2 &= 4 \\ -2x_1 + kx_2 &= -8 \end{aligned}$

49. $\begin{aligned} x_1 + kx_2 &= 3 \\ 2x_1 + 6x_2 &= 6 \end{aligned}$

50. $\begin{aligned} x_1 + kx_2 &= 3 \\ 2x_1 + 4x_2 &= 8 \end{aligned}$

C

In Problems 51–56, solve using Gauss–Jordan elimination. Use rref on a graphing utility to find the reduced forms.

51. $\begin{aligned} x_1 + 2x_2 - 4x_3 - x_4 &= 7 \\ 2x_1 + 5x_2 - 9x_3 - 4x_4 &= 16 \\ x_1 + 5x_2 - 7x_3 - 7x_4 &= 13 \end{aligned}$

52. $\begin{aligned} 2x_1 + 4x_2 + 5x_3 + 4x_4 &= 8 \\ x_1 + 2x_2 + 2x_3 + x_4 &= 3 \end{aligned}$

53. $\begin{aligned} x_1 - x_2 + 3x_3 - 2x_4 &= 1 \\ -2x_1 + 4x_2 - 3x_3 + x_4 &= 0.5 \\ 3x_1 - x_2 + 10x_3 - 4x_4 &= 2.9 \\ 4x_1 - 3x_2 + 8x_3 - 2x_4 &= 0.6 \end{aligned}$

54. $\begin{aligned} x_1 + x_2 + 4x_3 + x_4 &= 1.3 \\ -x_1 + x_2 - x_3 &= 1.1 \\ 2x_1 + x_3 + 3x_4 &= -4.4 \\ 2x_1 + 5x_2 + 11x_3 + 3x_4 &= 5.6 \end{aligned}$

55. $\begin{aligned} x_1 - 2x_2 + x_3 + x_4 + 2x_5 &= 2 \\ -2x_1 + 4x_2 + 2x_3 + 2x_4 - 2x_5 &= 0 \\ 3x_1 - 6x_2 + x_3 + x_4 + 5x_5 &= 4 \\ -x_1 + 2x_2 + 3x_3 + x_4 + x_5 &= 3 \end{aligned}$

56. $\begin{aligned} x_1 - 3x_2 + x_3 + x_4 + 2x_5 &= 2 \\ -x_1 + 5x_2 + 2x_3 + 2x_4 - 2x_5 &= 0 \\ 2x_1 - 6x_2 + 2x_3 + 2x_4 + 4x_5 &= 4 \\ -x_1 + 3x_2 - x_3 - x_5 &= -3 \end{aligned}$

Most graphing utilities also have a routine that produces the row-echelon form (ref) of a matrix. Problems 57–62 require a graphing utility with this routine. In Problems 57–60 use the row-echelon form to solve the indicated system.

57. $\begin{aligned} x_1 + 2x_2 &= 6 \\ 2x_1 - 5x_2 &= 3 \end{aligned}$

58. $\begin{aligned} 3x_1 - 4x_2 &= 10 \\ 4x_1 + 2x_2 &= 6 \end{aligned}$

59. $\begin{aligned} x_1 - 2x_2 + x_3 &= -7 \\ 2x_1 + x_2 - 2x_3 &= -6 \\ -2x_1 - x_2 + 4x_3 &= 14 \end{aligned}$

60. $\begin{aligned} 2x_1 - x_2 + 3x_3 &= 17 \\ -x_1 + 2x_2 - x_3 &= -12 \\ 4x_1 + x_2 - 5x_3 &= 3 \end{aligned}$

61. Based on the results in Problems 57–60, discuss the differences between the row-echelon form and the reduced row-echelon form of a matrix.

62. Describe a general procedure for using the row-echelon form to find the solution of a linear system.

APPLICATIONS

Solve Problems 63–78 using Gauss–Jordan elimination.

★ **63. Puzzle.** A friend of yours came out of the post office after spending $14.00 on 15¢, 20¢, and 35¢ stamps. If she bought 45 stamps in all, how many of each type did she buy?

★ **64. Puzzle.** A parking meter accepts only nickels, dimes, and quarters. If the meter contains 32 coins with a total value of $6.80, how many of each type are there?

★★ **65. Chemistry.** A chemist can purchase a 10% saline solution in 500 cubic centimeter containers, a 20% saline solution in 500 cubic centimeter containers, and a 50% saline solution in 1,000 cubic centimeter containers. He needs 12,000 cubic centimeters of 30% saline solution. How many containers of each type of solution should he purchase to form this solution?

★★ **66. Chemistry.** Repeat Problem 65 if the 50% saline solution is available only in 1,500 cubic centimeter containers.

67. Geometry. Find a, b, and c so that the graph of the parabola with equation $y = a + bx + cx^2$ passes through the points $(-2, 3)$, $(-1, 2)$, and $(1, 6)$.

68. Geometry. Find a, b, and c so that the graph of the parabola with equation $y = a + bx + cx^2$ passes through the points $(1, 3)$, $(2, 2)$, and $(3, 5)$.

69. Geometry. Find a, b, and c so that the graph of the circle with equation $x^2 + y^2 + ax + by + c = 0$ passes through the points $(6, 2)$, $(4, 6)$, and $(-3, -1)$.

70. Geometry. Find a, b, and c so that the graph of the circle with equation $x^2 + y^2 + ax + by + c = 0$ passes through the points $(-4, 1)$, $(-1, 2)$, and $(3, -6)$.

71. Production Scheduling. A small manufacturing plant makes three types of inflatable boats: one-person, two-person, and four-person models. Each boat requires the services of three departments, as listed in the table. The cutting, assembly, and packaging departments have available a maximum of 380, 330, and 120 labor-hours per week, respectively.

	One-Person Boat	Two-Person Boat	Four-Person Boat
Cutting department	0.5 h	1.0 h	1.5 h
Assembly department	0.6 h	0.9 h	1.2 h
Packaging department	0.2 h	0.3 h	0.5 h

(A) How many boats of each type must be produced each week for the plant to operate at full capacity?

(B) How is the production schedule in part A affected if the packaging department is no longer used?

(C) How is the production schedule in part A affected if the four-person boat is no longer produced?

72. Production Scheduling. Repeat Problem 71 assuming the cutting, assembly, and packaging departments have available a maximum of 350, 330, and 115 labor-hours per week, respectively.

73. Nutrition. A dietitian in a hospital is to arrange a special diet using three basic foods. The diet is to include exactly 340 units of calcium, 180 units of iron, and 220 units of vitamin A. The number of units per ounce of each nutrient for each of the foods is indicated in the table.

	Units Per Ounce		
	Food *A*	Food *B*	Food *C*
Calcium	30	10	20
Iron	10	10	20
Vitamin A	10	30	20

(A) How many ounces of each food must be used to meet the diet requirements?

(B) How is the diet in part A affected if food C is not used?

(C) How is the diet in part A affected if the vitamin A requirement is dropped?

74. Nutrition. Repeat Problem 73 if the diet is to include exactly 400 units of calcium, 160 units of iron, and 240 units of vitamin A.

75. Agriculture. A farmer can buy four types of fertilizer. Each barrel of mix A contains 30 pounds of phosphoric acid, 50 pounds of nitrogen, and 30 pounds of potash; each barrel of mix B contains 30 pounds of phosphoric acid, 75 pounds of nitrogen, and 20 pounds of potash; each barrel of mix C contains 30 pounds of phosphoric acid, 25 pounds of nitrogen, and 20 pounds of potash; and each barrel of mix D contains 60 pounds of phosphoric acid, 25 pounds of nitrogen, and 50 pounds of potash. Soil tests indicate that a particular field needs 900 pounds of phosphoric acid, 750 pounds of nitrogen, and 700 pounds of potash. How many barrels of each type of fertilizer should the farmer mix together to supply the necessary nutrients for the field?

76. Animal Nutrition. In a laboratory experiment, rats are to be fed five packets of food containing a total of 80 units of vitamin E. There are four different brands of food packets that can be used. A packet of brand A contains 5 units of vitamin E, a packet of brand B contains 10 units of vitamin E, a packet of brand C contains 15 units of vitamin E, and a packet of brand D contains 20 units of vitamin E. How many packets of each brand should be mixed and fed to the rats?

77. Sociology. Two sociologists have grant money to study school busing in a particular city. They wish to conduct an opinion survey using 600 telephone contacts and 400 house contacts. Survey company A has personnel to do 30 telephone and 10 house contacts per hour; survey company B can handle 20 telephone and 20 house contacts per hour. How many hours should be scheduled for each firm to produce exactly the number of contacts needed?

78. Sociology. Repeat Problem 77 if 650 telephone contacts and 350 house contacts are needed.

SECTION 8.4 Systems of Linear Inequalities

Graphing Linear Inequalities in Two Variables • Solving Systems of Linear Inequalities • Mathematical Modeling with Systems of Linear Inequalities

Many applications of mathematics involve systems of inequalities rather than systems of equations. A graph is often the most convenient way to represent the solutions of a system of inequalities in two variables. In Section 8.4 we discuss techniques for graphing both a single linear inequality in two variables and a system of linear inequalities in two variables.

Graphing Linear Inequalities in Two Variables

We know how to graph first-degree equations such as

$$y = 2x - 3 \quad \text{and} \quad 2x - 3y = 5$$

but how do we graph first-degree inequalities such as

$$y \leq 2x - 3 \quad \text{and} \quad 2x - 3y > 5$$

Actually, graphing these inequalities is almost as easy as graphing the equations. But before we begin, we must discuss some important subsets of a plane in a rectangular coordinate system.

A line divides a plane into two halves called **half-planes.** A vertical line divides a plane into **left** and **right half-planes** [Fig. 1(a)]; a nonvertical line divides a plane into **upper** and **lower half-planes** [Fig. 1(b)].

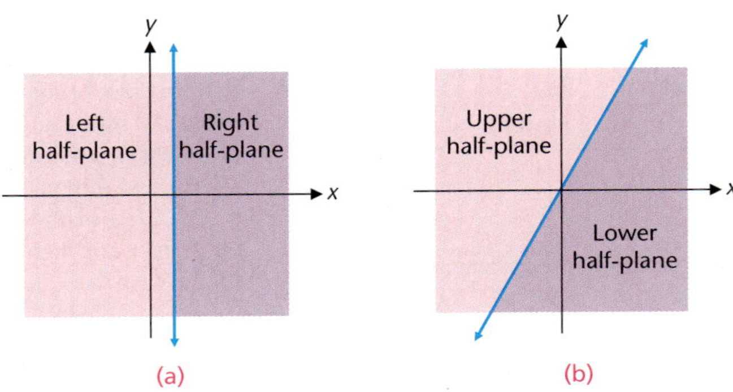

FIGURE 1 Half-planes.

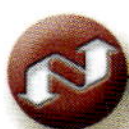

EXPLORE/DISCUSS 1

Consider the following linear equation and related linear inequalities:

(1) $2x - 3y = 12$ (2) $2x - 3y < 12$ (3) $2x - 3y > 12$

(A) Graph the line with equation (1).

(B) Find the point on this line with x coordinate 3 and draw a vertical line through this point. Discuss the relationship between the y coordinates of the points on this line and statements (1), (2), and (3).

(C) Repeat part B for $x = -3$. For $x = 9$.

(D) Based on your observations in parts B and C, write a verbal description of all the points in the plane that satisfy equation (1), those that satisfy inequality (2), and those that satisfy inequality (3).

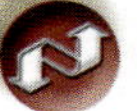

Now let's investigate the half-planes determined by the linear equation $y = 2x - 3$. We start by graphing $y = 2x - 3$ (Fig. 2). For any given value of x, there is exactly one value for y such that (x, y) lies on the line. For the same x, if the point (x, y) is below the line, then $y < 2x - 3$. Thus, the lower half-plane corresponds to the solution of the inequality $y < 2x - 3$. Similarly, the upper half-plane corresponds to the solution of the inequality $y > 2x - 3$, as shown in Figure 2.

FIGURE 2

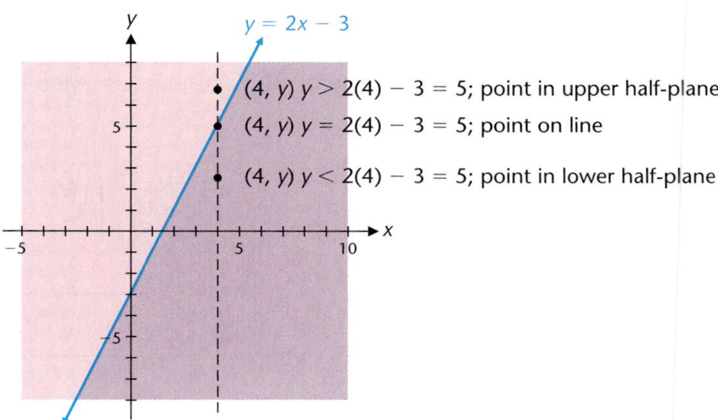

The four inequalities formed from the equation $y = 2x - 3$ by replacing the $=$ sign by $\geq$, $>$, $\leq$, and $<$, respectively, are

$$y \geq 2x - 3 \qquad y > 2x - 3 \qquad y \leq 2x - 3 \qquad y < 2x - 3$$

The graph of each is a half-plane. The line $y = 2x - 3$, called the **boundary line** for the half-plane, is included for $\geq$ and $\leq$ and excluded for $>$ and $<$. In Figure 3, the half-planes are indicated with small arrows on the graph of $y = 2x - 3$ and then graphed as shaded regions. Included boundary lines are shown as solid lines, and excluded boundary lines are shown as dashed lines.

FIGURE 3

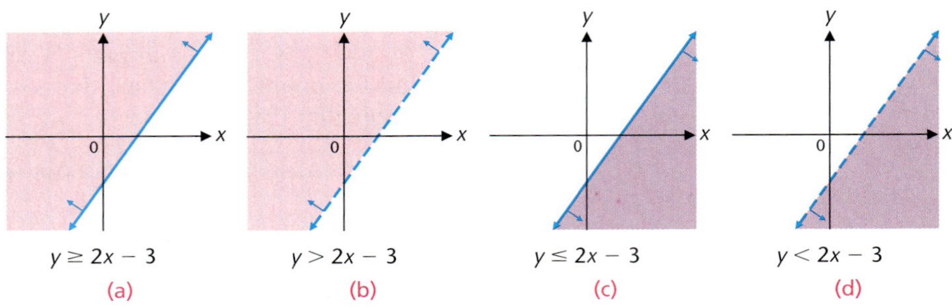

Most graphing utilities give the user the option to shade the region above or below an equation by changing the icon to the left of the equation* [Fig. 4(a)]. Graphing y_1 with the shade-above option selected produces the graph in Figure 4(b) and graphing the same function in y_2 with the shade-below option produces the graph in Figure 4(c). Note that although it is possible to graph a dashed line on a graphing utility, it is not possible to distinguish between a dashed line and a solid line when using the shading options. We will indicate in words if the boundary line is not part of a solution graphed on a graphing utility.

FIGURE 4

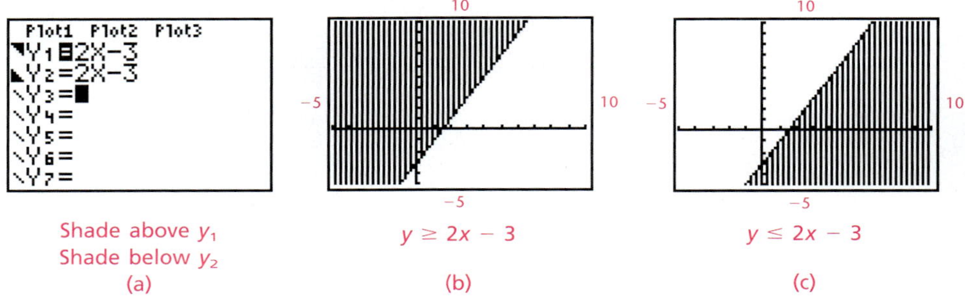

*On most graphing calculators, you can select a shading option by moving the cursor to the icon at the left side of the screen and pressing ENTER repeatedly to toggle through the various choices.

THEOREM 1

Graphs of Linear Inequalities in Two Variables

The graph of a linear inequality

$$Ax + By < C \quad \text{or} \quad Ax + By > C$$

with $B \neq 0$, is either the upper half-plane or the lower half-plane (but not both) determined by the line $Ax + By = C$.

If $B = 0$, then the graph of

$$Ax < C \quad \text{or} \quad Ax > C$$

is either the left half-plane or the right half-plane (but not both) determined by the line $Ax = C$.

As a consequence of Theorem 1, we state simple and quick procedures for graphing a linear inequality by hand and on a graphing utility.

Algebraic Procedure for Graphing Linear Inequalities in Two Variables

Step 1. Graph $Ax + By = C$ as a dashed line if equality is not included in the original statement or as a solid line if equality is included.

Step 2. Choose a test point anywhere in the plane not on the line and substitute the coordinates into the inequality. The origin (0, 0) often requires the least computation.

Step 3. The graph of the original inequality includes the half-plane containing the test point if the inequality is satisfied by that point, or the half-plane not containing that point if the inequality is not satisfied by that point.

Graphing Utility Procedure for Graphing Linear Inequalities in Two Variables

Step 1. Solve the inequality for y.

Step 2. Enter the equation of the boundary line and select a shading option as follows:

$$\left.\begin{matrix} y > mx + b \\ \text{or} \\ y \geq mx + b \end{matrix}\right\} \quad \text{Select shade above.}$$

$$\left.\begin{matrix} y < mx + b \\ \text{or} \\ y \leq mx + b \end{matrix}\right\} \quad \text{Select shade below.}$$

Step 3. Graph the solution.

EXAMPLE 1 Graphing a Linear Inequality

Graph: $3x - 4y \leq 12$

SOLUTION

Algebraic Solution

Step 1. Graph $3x - 4y = 12$ as a solid line, because equality is included in the original statement (Fig. 5).

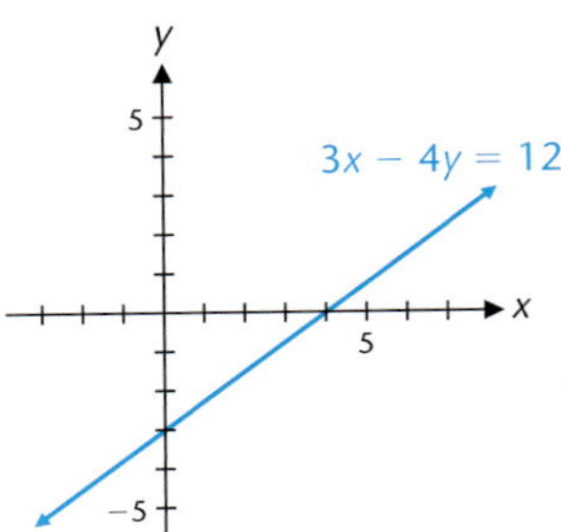

FIGURE 5

Step 2. Pick a convenient test point above or below the line. The origin (0, 0) requires the least computation. Substituting (0, 0) into the inequality

$$3x - 4y \le 12$$

$$3(0) - 4(0) = 0 \le 12$$

produces a true statement; therefore, (0, 0) is in the solution set.

Step 3. The line $3x - 4y = 12$ and the half-plane containing the origin form the graph of $3x - 4y \le 12$ (Fig. 6).

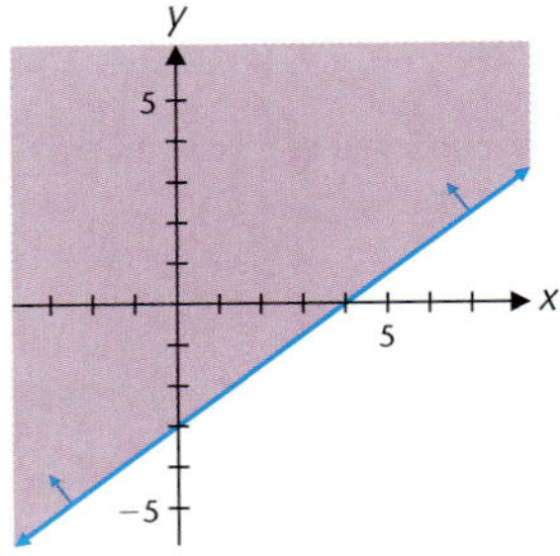

FIGURE 6

Graphing Utility Solution

Step 1. Solve the inequality for y.

$$3x - 4y \le 12$$

$$-4y \le -3x + 12$$

$$\frac{-4y}{-4} \ge \frac{-3x}{-4} + \frac{12}{-4}$$

$$y \ge 0.75x - 3$$

Step 2. Enter $y_1 = 0.75x - 3$ and select the shade-above option (Fig. 7).

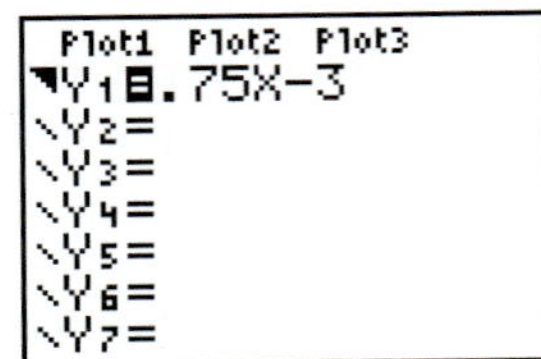

FIGURE 7

Step 3. Graph the solution (Fig. 8).

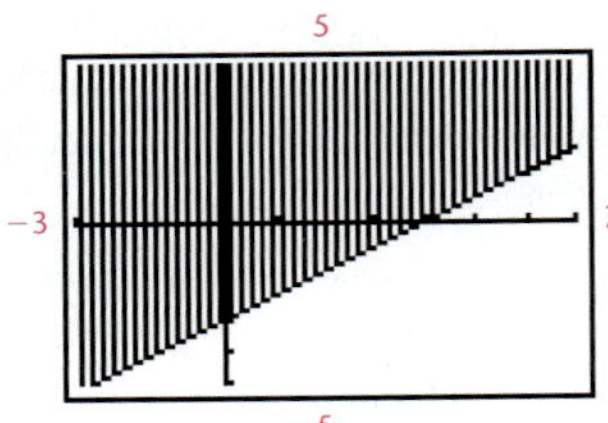

FIGURE 8

The boundary line is part of the solution.

MATCHED PROBLEM 1

Graph by hand and on a graphing utility: $2x + 3y < 6$

EXAMPLE 2 Graphing a Linear Inequality

Graph: (A) $y > -3$ (B) $2x \le 5$

SOLUTIONS

(A) The graph of $y > -3$ is shown in Figure 9.

(B) The graph of $2x \le 5$ is shown in Figure 10.

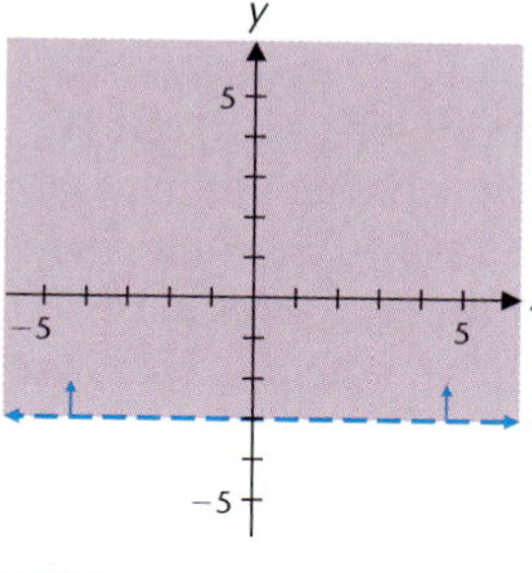

FIGURE 9

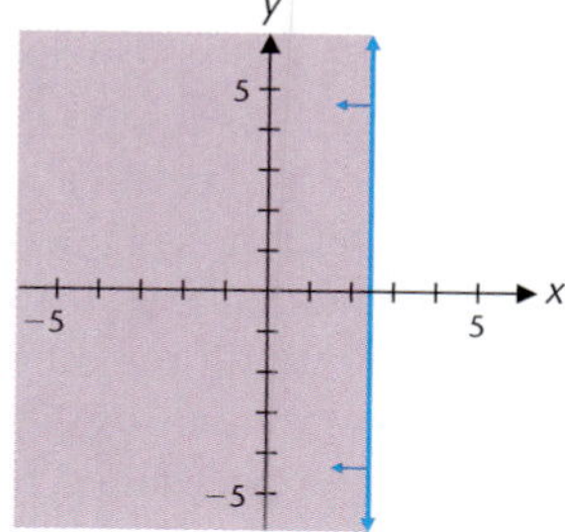

FIGURE 10

MATCHED 2 PROBLEM

Graph: (A) $y \le 2$ (B) $3x > -8$

Solving Systems of Linear Inequalities

We now consider systems of linear inequalities such as

$$\begin{aligned} x + y &\ge 6 \\ 2x - y &\ge 0 \end{aligned} \qquad \text{and} \qquad \begin{aligned} 2x + y &\le 22 \\ x + y &\le 13 \\ 2x + 5y &\le 50 \\ x &\ge 0 \\ y &\ge 0 \end{aligned}$$

We wish to solve such systems graphically—that is, to find the graph of all ordered pairs of real numbers (x, y) that simultaneously satisfy all the inequalities in the system. The graph is called the **solution region** for the system. To find the solution region, we graph each inequality in the system and then take the intersection of all the graphs. To simplify the discussion that follows, **we will consider only systems of linear inequalities where equality is included in each statement in the system.**

EXAMPLE 3 Solving a System of Linear Inequalities

Graph the solution region for the following system of linear inequalities:

$$x + y \geq 6$$
$$2x - y \geq 0$$

SOLUTION

Algebraic Solution

First, graph the line $x + y = 6$ and shade the region that satisfies the inequality $x + y \geq 6$. This region is shaded in blue in Figure 11(a). Next, graph the line $2x - y = 0$ and shade the region that satisfies the inequality $2x - y \geq 0$. This region is shaded in red in Figure 11(a). The solution region for the system of inequalities is the intersection of these two regions. This is the region shaded in both red and blue in Figure 11(a), which is redrawn in Figure 11(b) with only the solution region shaded for clarity. The coordinates of any point in the shaded region of Figure 11(b) specify a solution to the system. For example, the points (2, 4), (6, 3), and (7.43, 8.56) are three of infinitely many solutions, as can be easily checked. The intersection point (2, 4) can be obtained by solving the equations $x + y = 6$ and $2x - y = 0$ simultaneously.

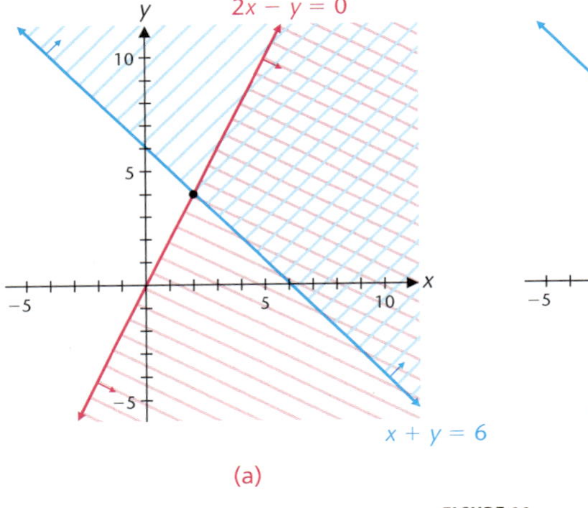

(a)

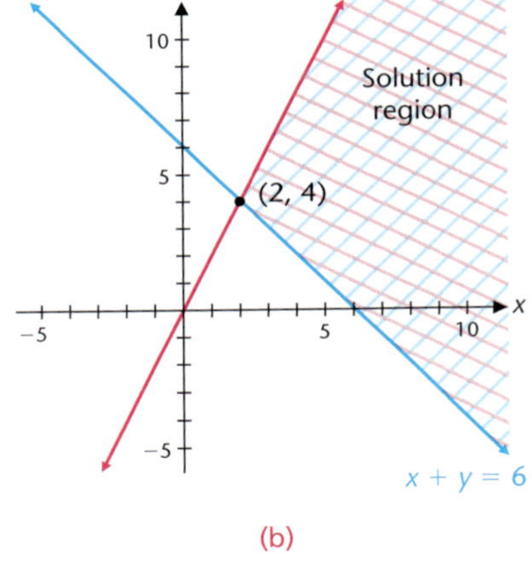

(b)

FIGURE 11

Graphing Utility Solution

Solve each inequality for y:

$$x + y \geq 6$$
$$y \geq 6 - x$$
$$2x - y \geq 0$$
$$-y \geq -2x$$
$$y \leq 2x$$

Enter $y_1 = 6 - x$ and select the shade-above option (Fig. 12). Enter $y_2 = 2x$ and select the shade-below option (Fig. 12).

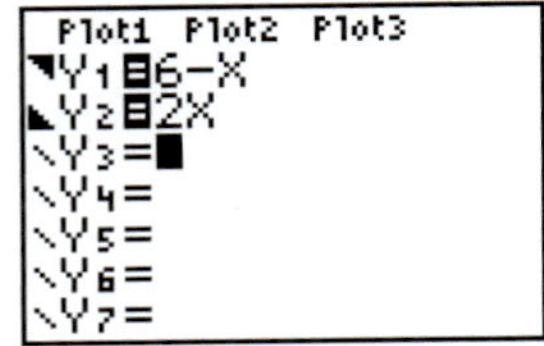

FIGURE 12

Graph the inequalities (Fig. 13). The region shaded with both horizontal and vertical lines is the solution region.

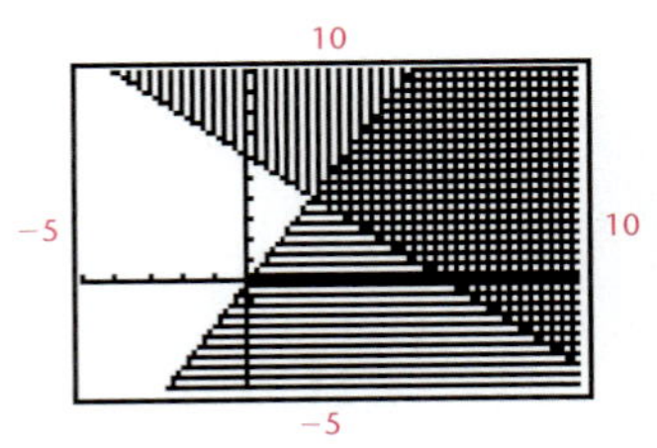

FIGURE 13

MATCHED PROBLEM 3

Graph the solution region for the following system of linear inequalities two ways, as in Example 3:

$$3x + y \leq 21$$
$$x - 2y \leq 0$$

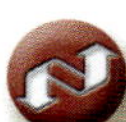

EXPLORE/DISCUSS 2

Refer to Example 3. Graph each boundary line and shade the regions obtained by reversing each inequality. That is, shade the region of the plane that corresponds to the inequality $x + y < 6$ and then shade the region that corresponds to the inequality $2x - y < 0$. What portion of the plane is left unshaded? Compare this method with the one used in the solution to Example 3.

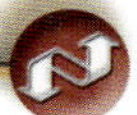

The method of solving inequalities investigated in Explore/Discuss 2 works very well on a graphing utility that allows the user to shade above and below a graph. Referring to Example 3, the unshaded region in Figure 14(b) corresponds to the solution region in Figure 13.

FIGURE 14

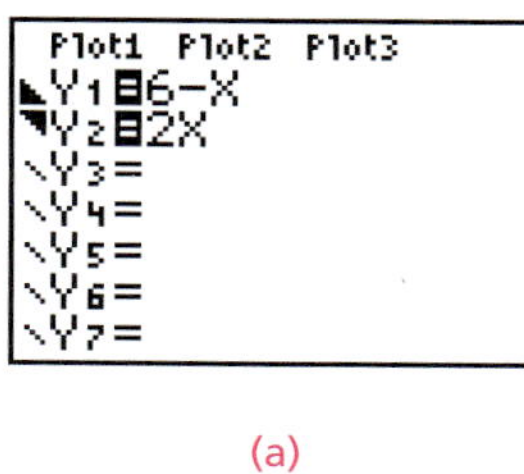

(a)

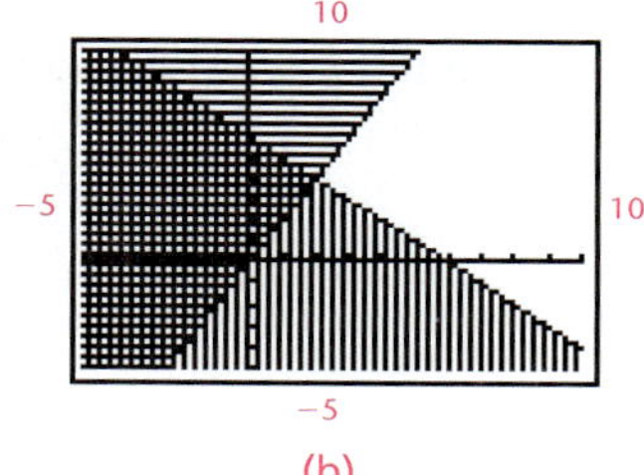

(b)

The points of intersection of the lines that form the boundary of a solution region play a fundamental role in the solution of linear programming problems, which are discussed in Section 8.5.

DEFINITION 1
Corner Point

A **corner point** of a solution region is a point in the solution region that is the intersection of two boundary lines.

The point (2, 4) is the only corner point of the solution region in Example 3; see Figure 11(b).

EXAMPLE 4 Solving a System of Linear Inequalities

Graph the solution region for the following system of linear inequalities, and find the corner points.

$$\begin{aligned} 2x + y &\le 22 \\ x + y &\le 13 \\ 2x + 5y &\le 50 \\ x &\ge 0 \\ y &\ge 0 \end{aligned}$$

SOLUTIONS

Algebraic Solution

The inequalities $x \geq 0$ and $y \geq 0$, called **nonnegative restrictions,** occur frequently in applications involving systems of inequalities because x and y often represent quantities that can't be negative—number of units produced, number of hours worked, and the like. The solution region lies in the first quadrant, and we can restrict our attention to that portion of the plane. First we graph the lines.

$$2x + y = 22$$
$$x + y = 13$$
$$2x + 5y = 50$$

Find the x and y intercepts of each line; then sketch the line through these points, as shown in Figure 15.

Next, choosing (0, 0) as a test point, we see that the graph of each of the first three inequalities in the system consists of its corresponding line and the half-plane lying below it. Thus, the solution region of the system consists of the points in the first quadrant that simultaneously lie below all three of these lines (Fig. 15).

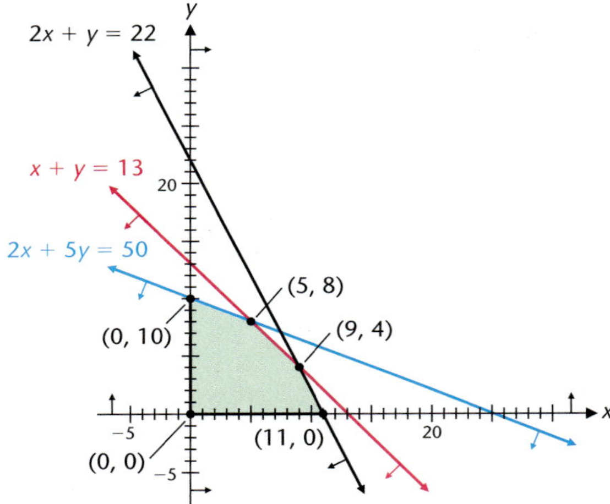

FIGURE 15

The corner points (0, 0), (0, 10), and (11, 0) can be determined from the graph. The other two corner points are determined as follows:

Solve the system

$$2x + 5y = 50$$
$$x + y = 13$$

to obtain (5, 8).

Solve the system

$$2x + y = 22$$
$$x + y = 13$$

to obtain (9, 4).

Note that the lines $2x + 5y = 50$ and $2x + y = 22$ also intersect, but the intersection point is not part of the solution region, and hence, is not a corner point.

Graphing Utility Solution

Solve each inequality for y.

$$2x + y \leq 22$$
$$y \leq 22 - 2x$$
$$x + y \leq 13$$
$$y \leq 13 - x$$
$$2x + 5y \leq 50$$
$$5y \leq 50 - 2x$$
$$y \leq 10 - 0.4x$$

Enter the equation of each boundary line and select the desired shading option (Fig. 16).

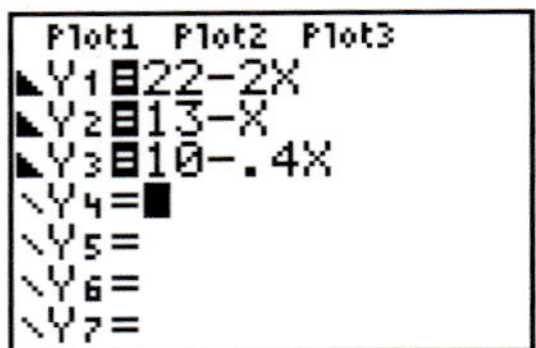

FIGURE 16

The nonnegative restrictions, $x \geq 0$ and $y \geq 0$, indicate that the solution region lies in the first quadrant. To restrict the graph on a graphing utility to the first quadrant, simply choose Xmin = 0 and Ymin = 0. The solution region is the five-sided polygon in the lower left corner of the screen that is shaded horizontally, vertically, and diagonally (Fig. 17).

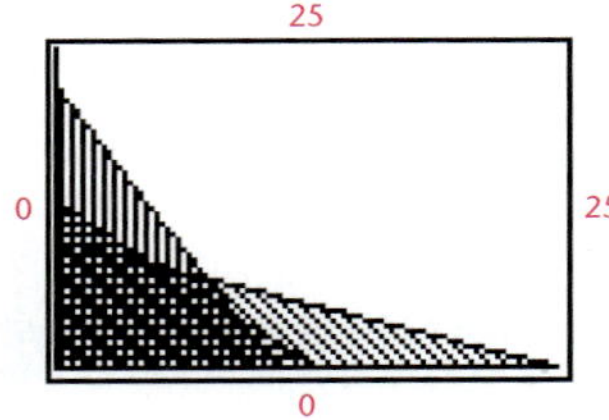

FIGURE 17

The corner points are (0, 0), (0, 10), (11, 0), (9, 4), and (5, 8). The last two were found by using the intersect command (details omitted).

MATCHED 4 PROBLEM

Graph the solution region of the following system of linear inequalities two ways, as in Example 4.

$$\begin{aligned} 5x + y &\ge 20 \\ x + y &\ge 12 \\ x + 3y &\ge 18 \\ x &\ge 0 \\ y &\ge 0 \end{aligned}$$

If you find it difficult to recognize the solution region in Figure 17, you might want to consider the technique of shading the complement discussed in Explore/Discuss 2 and illustrated in Figure 18.

FIGURE 18

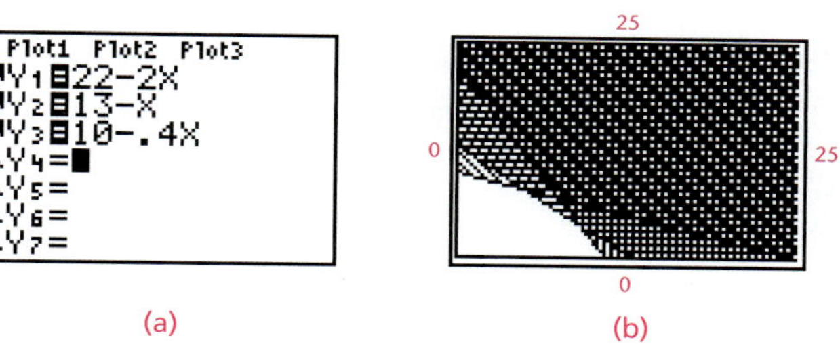

(a) (b)

If we compare the solution regions of Examples 3 and 4, we see that there is a fundamental difference between these two regions. We can draw a circle around the solution region in Example 4. However, it is impossible to include all the points in the solution region in Example 3 in any circle, no matter how large we draw it. This leads to the following definition.

DEFINITION 2

Bounded and Unbounded Solution Regions

A solution region of a system of linear inequalities is **bounded** if it can be enclosed within a circle. If it cannot be enclosed within a circle, then it is **unbounded.**

Thus, the solution region for Example 4 is bounded and the solution region for Example 3 is unbounded. This definition will be important in Section 8.5.

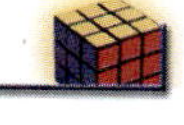

Mathematical Modeling with Systems of Linear Inequalities

EXAMPLE 5 **Production Scheduling**

A manufacturer of surfboards makes a standard model and a competition model. Each standard board requires 6 labor-hours for fabricating and 1 labor-hour for finishing. Each competition board requires 8 labor-hours for fabricating and 3 labor-hours for finishing. The maximum labor-hours available per week in the fabricating and finishing departments are 120 and 30, respectively. What combinations of boards can be produced each week so as not to exceed the number of labor-hours available in each department per week?

SOLUTION

To clarify relationships, we summarize the information in the following table:

	Standard Model (Labor-Hours per Board)	Competition Model (Labor-Hours per Board)	Maximum Labor-Hours Available per Week
Fabricating	6	8	120
Finishing	1	3	30

Let

x = Number of standard boards produced per week

y = Number of competition boards produced per week

These variables are restricted as follows:

Fabricating department restriction:

$$\begin{pmatrix}\text{Weekly fabricating}\\ \text{time for } x\\ \text{standard boards}\end{pmatrix} + \begin{pmatrix}\text{Weekly fabricating}\\ \text{time for } y\\ \text{competition boards}\end{pmatrix} \le \begin{pmatrix}\text{Maximum labor-hours}\\ \text{available per week}\end{pmatrix}$$

$$6x \quad + \quad 8y \quad \le \quad 120$$

Finishing department restriction:

$$\begin{pmatrix}\text{Weekly finishing}\\ \text{time for } x\\ \text{standard boards}\end{pmatrix} + \begin{pmatrix}\text{Weekly finishing}\\ \text{time for } y\\ \text{competition boards}\end{pmatrix} \le \begin{pmatrix}\text{Maximum labor-hours}\\ \text{available per week}\end{pmatrix}$$

$$1x \quad + \quad 3y \quad \le \quad 30$$

Because it is not possible to manufacture a negative number of boards, x and y also must satisfy the nonnegative restrictions

$$x \ge 0$$
$$y \ge 0$$

Thus, x and y must satisfy the following system of linear inequalities:

$$\begin{aligned} 6x + 8y &\le 120 && \text{Fabricating department restriction}\\ x + 3y &\le 30 && \text{Finishing department restriction}\\ x &\ge 0 && \text{Nonnegative restriction}\\ y &\ge 0 && \text{Nonnegative restriction} \end{aligned}$$

Graphing this system of linear inequalities, we obtain the set of **feasible solutions,** or the **feasible region,** as shown in Figure 19. For problems of this type and for the linear programming problems we consider in Section 8.5, solution regions are often referred to as feasible regions. Any point within the shaded area, including the boundary lines, represents a possible production schedule. Any point outside the shaded area represents an impossible schedule. For example, it would be possible to produce 12 standard boards and 5 competition boards per week, but it would not be possible to produce 12 standard boards and 7 competition boards per week (see Fig. 19).

FIGURE 19

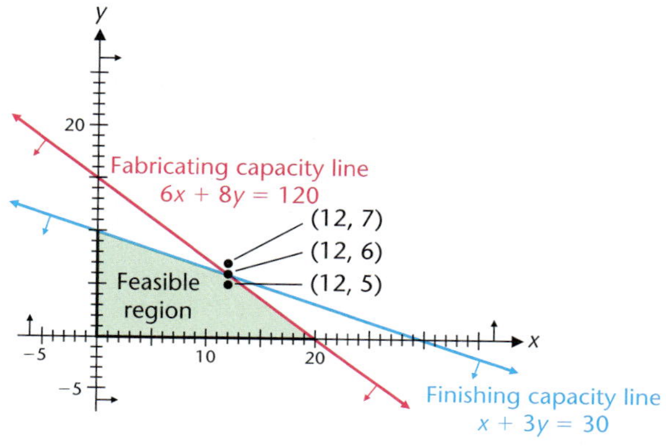

MATCHED 5 PROBLEM

Repeat Example 5 using 5 hours for fabricating a standard board and a maximum of 27 labor-hours for the finishing department.

REMARK Refer to Example 5. How do we interpret a production schedule of 10.5 standard boards and 4.3 competition boards? It is not possible to manufacture a fraction of a board. But it is possible to *average* 10.5 standard and 4.3 competition boards per week. In general, we will assume that all points in the feasible region represent acceptable solutions, although noninteger solutions might require special interpretation.

ANSWERS MATCHED PROBLEMS

1.

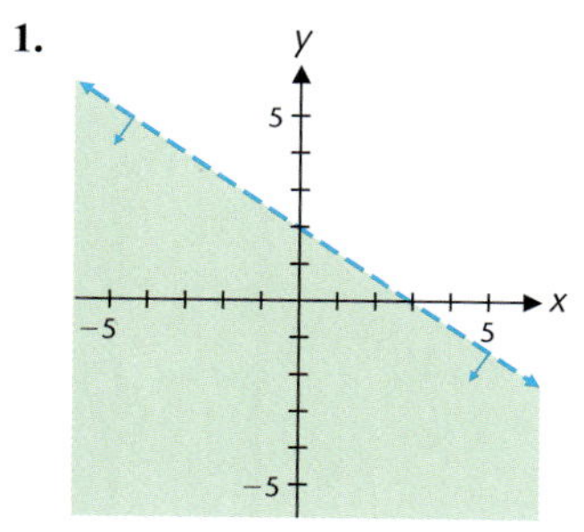

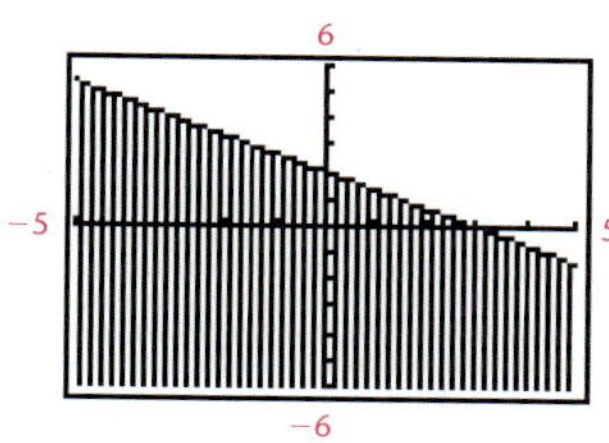

The boundary line is not part of the solution.

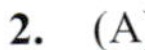

2. (A)

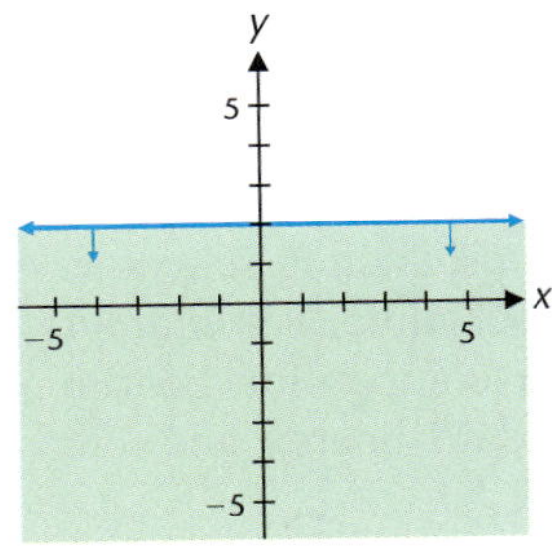

(B)

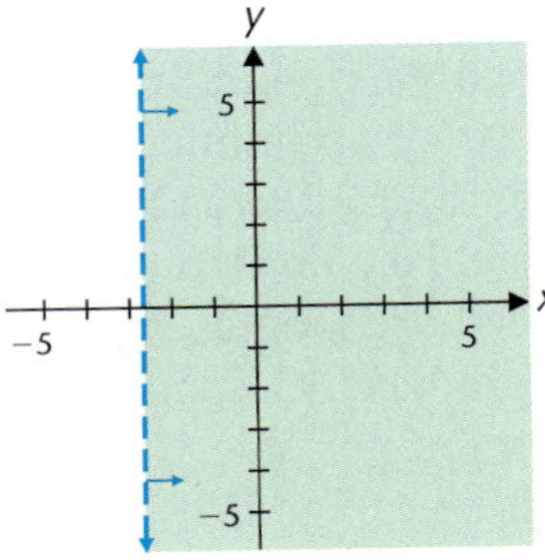

3.

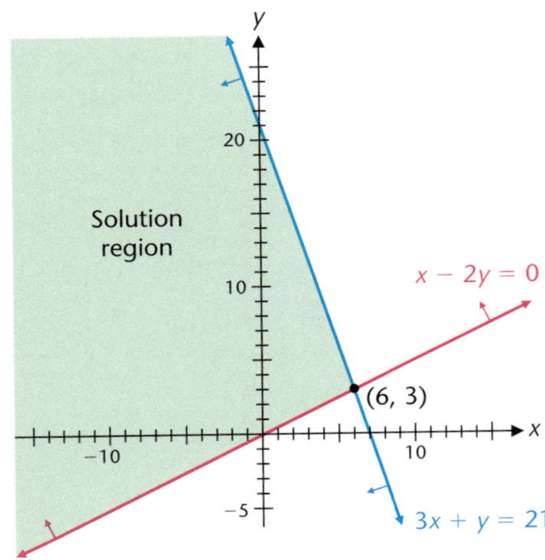

30

−15 15

−10

The boundary of the solution region is part of the solution region.

4.

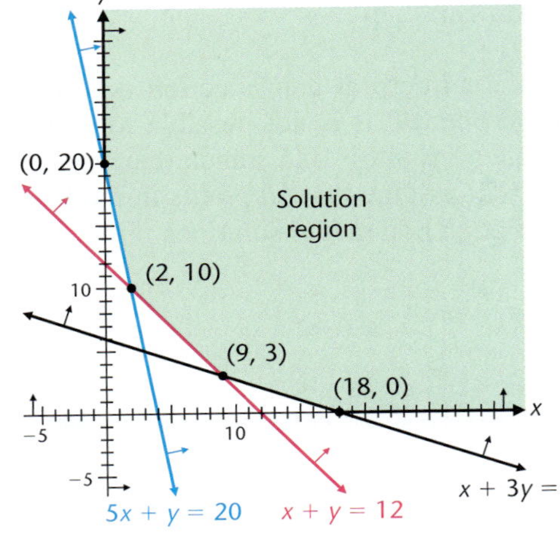

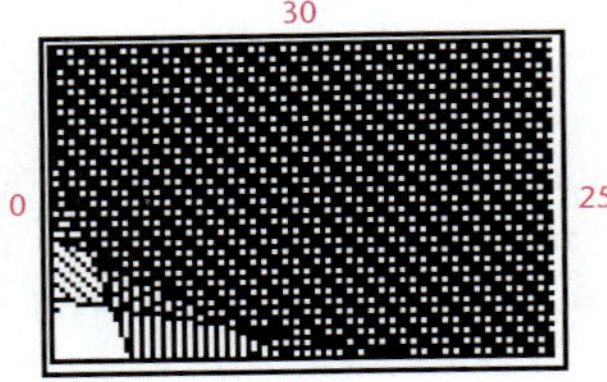

The boundary of the solution region is part of the solution region.

5.

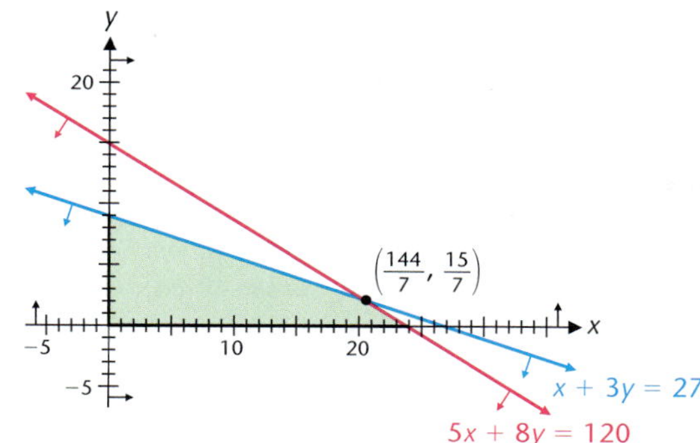

EXERCISE 8.4

A

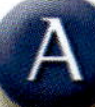

In Problems 1–10, graph the solution region for each inequality and write a verbal description of the solution region.

1. $2x - 3y < 6$
2. $3x + 4y < 12$
3. $3x + 2y \geq 18$
4. $3y - 2x \geq 24$
5. $y \leq \frac{2}{3}x + 5$
6. $y \geq \frac{1}{3}x - 2$
7. $y < 8$
8. $x > -5$
9. $-3 \leq y < 2$
10. $-1 < x \leq 3$

In Problems 11–14, match the solution region of each system of linear inequalities with one of the four regions shown in the figure.

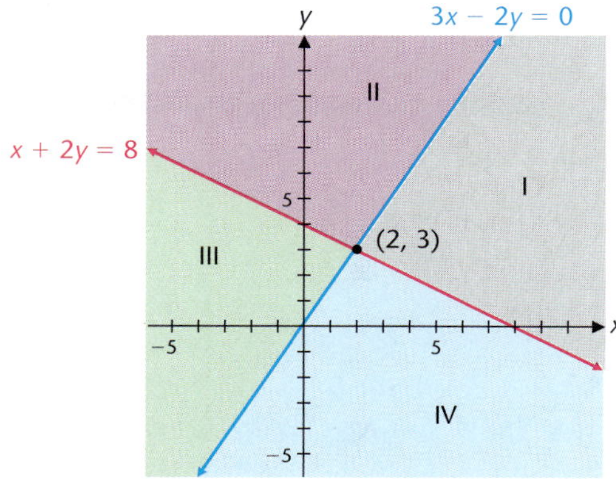

11. $x + 2y \leq 8$
    $3x - 2y \geq 0$
12. $x + 2y \geq 8$
    $3x - 2y \leq 0$
13. $x + 2y \geq 8$
    $3x - 2y \geq 0$
14. $x + 2y \leq 8$
    $3x - 2y \leq 0$

In Problems 15–20, graph the solution region for each system of linear inequalities.

15. $x \geq 5$
    $y \leq 6$
16. $x \leq 4$
    $y \geq 2$
17. $3x + y \geq 6$
    $x \leq 4$
18. $3x + 4y \leq 12$
    $y \geq -3$
19. $x - 2y \leq 12$
    $2x + y \geq 4$
20. $2x + 5y \leq 20$
    $x - 5y \leq -5$

Problems 21–24 require a graphing utility that gives the user the option of shading above or below a graph.

(A) Graph the boundary lines in a standard viewing window and shade the region that contains the points that satisfy each inequality.

(B) Repeat part A, but this time shade the region that contains the points that do not satisfy each inequality (see Explore/Discuss 2)

Explain how you can recognize the solution region in each graph.

21. $x + y \leq 5$
    $2x - y \leq 1$
22. $x - 2y \leq 1$
    $x + 3y \geq 12$
23. $2x + y \geq 4$
    $3x - y \leq 7$
24. $3x + y \geq -2$
    $x - 2y \geq -6$

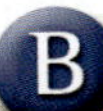

In Problems 25–28, match the solution region of each system of linear inequalities with one of the four regions shown in the figure below. Identify the corner points of each solution region.

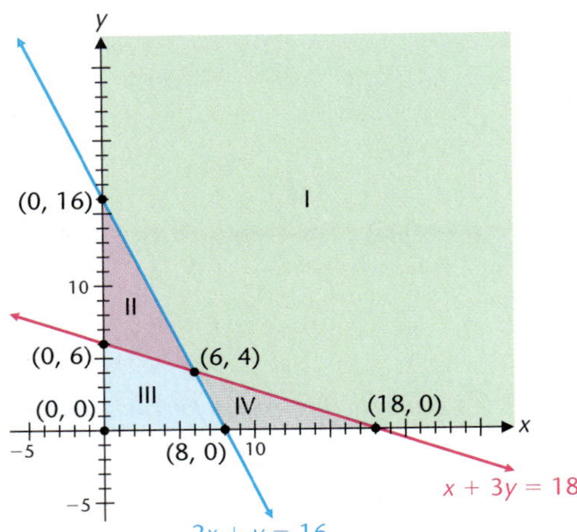

25. $\begin{aligned} x + 3y &\le 18 \\ 2x + y &\ge 16 \\ x &\ge 0 \\ y &\ge 0 \end{aligned}$

26. $\begin{aligned} x + 3y &\le 18 \\ 2x + y &\le 16 \\ x &\ge 0 \\ y &\ge 0 \end{aligned}$

27. $\begin{aligned} x + 3y &\ge 18 \\ 2x + y &\ge 16 \\ x &\ge 0 \\ y &\ge 0 \end{aligned}$

28. $\begin{aligned} x + 3y &\ge 18 \\ 2x + y &\le 16 \\ x &\ge 0 \\ y &\ge 0 \end{aligned}$

In Problems 29–40, graph the solution region for each system, and indicate whether each solution region is bounded or unbounded. Find the coordinates of each corner point.

29. $\begin{aligned} 2x + 3y &\le 6 \\ x &\ge 0 \\ y &\ge 0 \end{aligned}$

30. $\begin{aligned} 4x + 3y &\le 12 \\ x &\ge 0 \\ y &\ge 0 \end{aligned}$

31. $\begin{aligned} 4x + 5y &\ge 20 \\ x &\ge 0 \\ y &\ge 0 \end{aligned}$

32. $\begin{aligned} 5x + 6y &\ge 30 \\ x &\ge 0 \\ y &\ge 0 \end{aligned}$

33. $\begin{aligned} 2x + y &\le 8 \\ x + 3y &\le 12 \\ x &\ge 0 \\ y &\ge 0 \end{aligned}$

34. $\begin{aligned} x + 2y &\le 10 \\ 3x + y &\le 15 \\ x &\ge 0 \\ y &\ge 0 \end{aligned}$

35. $\begin{aligned} 4x + 3y &\ge 24 \\ 2x + 3y &\ge 18 \\ x &\ge 0 \\ y &\ge 0 \end{aligned}$

36. $\begin{aligned} x + 2y &\ge 8 \\ 2x + y &\ge 10 \\ x &\ge 0 \\ y &\ge 0 \end{aligned}$

37. $\begin{aligned} 2x + y &\le 12 \\ x + y &\le 7 \\ x + 2y &\le 10 \\ x &\ge 0 \\ y &\ge 0 \end{aligned}$

38. $\begin{aligned} 3x + y &\le 21 \\ x + y &\le 9 \\ x + 3y &\le 21 \\ x &\ge 0 \\ y &\ge 0 \end{aligned}$

39. $\begin{aligned} x + 2y &\ge 16 \\ x + y &\ge 12 \\ 2x + y &\ge 14 \\ x &\ge 0 \\ y &\ge 0 \end{aligned}$

40. $\begin{aligned} 3x + y &\ge 30 \\ x + y &\ge 16 \\ x + 3y &\ge 24 \\ x &\ge 0 \\ y &\ge 0 \end{aligned}$

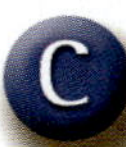

In Problems 41–48, graph the solution region for each system, and indicate whether each solution region is bounded or unbounded. Find the coordinates of each corner point.

41. $\begin{aligned} x + y &\le 11 \\ 5x + y &\ge 15 \\ x + 2y &\ge 12 \end{aligned}$

42. $\begin{aligned} 4x + y &\le 32 \\ x + 3y &\le 30 \\ 5x + 4y &\ge 51 \end{aligned}$

43. $\begin{aligned} 3x + 2y &\ge 24 \\ 3x + y &\le 15 \\ x &\ge 4 \end{aligned}$

44. $\begin{aligned} 3x + 4y &\le 48 \\ x + 2y &\ge 24 \\ y &\le 9 \end{aligned}$

45. $\begin{aligned} x + y &\le 10 \\ 3x + 5y &\ge 15 \\ 3x - 2y &\le 15 \\ -5x + 2y &\le 6 \end{aligned}$

46. $\begin{aligned} 3x - y &\ge 1 \\ -x + 5y &\ge 9 \\ x + y &\le 9 \\ y &\le 5 \end{aligned}$

47. $\begin{aligned} 16x + 13y &\le 119 \\ 12x + 16y &\ge 101 \\ -4x + 3y &\le 11 \end{aligned}$

48. $\begin{aligned} 8x + 4y &\le 41 \\ -15x + 5y &\le 19 \\ 2x + 6y &\ge 37 \end{aligned}$

APPLICATIONS

49. Manufacturing—Resource Allocation. A manufacturing company makes two types of water skis: a trick ski and a slalom ski. The trick ski requires 6 labor-hours for fabricating and 1 labor-hour for finishing. The slalom ski requires 4 labor-hours for fabricating and 1 labor-hour for finishing. The maximum labor-hours available per day for fabricating and finishing are 108 and 24, respectively. If x is the number of trick skis and y is the number of slalom skis produced per day, write a system of inequalities that indicates appropriate restraints on x and y. Graph the set of feasible solutions for the number of each type of ski that can be produced.

50. Manufacturing—Resource Allocation. A furniture manufacturing company manufactures dining room tables and chairs. A table requires 8 labor-hours for assembling and 2 labor-hours for finishing. A chair requires 2 labor-hours for assembling and 1 labor-hour for finishing. The maximum labor-hours available per day for assembly and finishing are 400 and 120, respectively. If x is the number of tables and y is the number of chairs produced per day, write a system of inequalities that indicates appropriate restraints on x and y. Graph the set of feasible solutions for the number of tables and chairs that can be produced.

★ **51. Manufacturing—Resource Allocation.** Refer to Problem 49. The company makes a profit of \$50 on each trick ski and a profit of \$60 on each slalom ski.

(A) If the company makes 10 trick and 10 slalom skis per day, the daily profit will be \$1,100. Are there other feasible production schedules that will result in a daily profit of \$1,100? How are these schedules related to the graph of the line $50x + 60y = 1{,}100$?

(B) Find a feasible production schedule that will produce a daily profit greater than \$1,100 and repeat part A for this schedule.

(C) Discuss methods for using lines like those in parts A and B to find the largest possible daily profit.

★ **52. Manufacturing—Resource Allocation.** Refer to Problem 50. The company makes a profit of \$50 on each table and a profit of \$15 on each chair.

(A) If the company makes 20 tables and 20 chairs per day, the daily profit will be \$1,300. Are there other feasible production schedules that will result in a daily profit of \$1,300? How are these schedules related to the graph of the line $50x + 15y = 1{,}300$?

(B) Find a feasible production schedule that will produce a daily profit greater than \$1,300 and repeat part A for this schedule.

(C) Discuss methods for using lines like those in parts A and B to find the largest possible daily profit.

53. Nutrition—Plants. A farmer can buy two types of fertilizer, mix A and mix B. Each cubic yard of mix A contains 20 pounds of phosphoric acid, 30 pounds of nitrogen, and 5 pounds of potash. Each cubic yard of mix B contains 10 pounds of phosphoric acid, 30 pounds of nitrogen, and 10 pounds of potash. The minimum requirements are 460 pounds of phosphoric acid, 960 pounds of nitrogen, and 220 pounds of potash. If x is the number of cubic yards of mix A used and y is the number of cubic yards of mix B used, write a system of inequalities that indicates appropriate restraints on x and y. Graph the set of feasible solutions for the amount of mix A and mix B that can be used.

54. Nutrition. A dietitian in a hospital is to arrange a special diet using two foods. Each ounce of food M contains 30 units of calcium, 10 units of iron, and 10 units of vitamin A. Each ounce of food N contains 10 units of calcium, 10 units of iron, and 30 units of vitamin A. The minimum requirements in the diet are 360 units of calcium, 160 units of iron, and 240 units of vitamin A. If x is the number of ounces of food M used and y is the number of ounces of food N used, write a system of linear inequalities that reflects the conditions indicated. Graph the set of feasible solutions for the amount of each kind of food that can be used.

55. Sociology. A city council voted to conduct a study on inner-city community problems. A nearby university was contacted to provide sociologists and research assistants. Each sociologist will spend 10 hours per week collecting data in the field and 30 hours per week analyzing data in the research center. Each research assistant will spend 30 hours per week in the field and 10 hours per week in the research center. The minimum weekly labor-hour requirements are 280 hours in the field and 360 hours in the research center. If x is the number of sociologists hired for the study and y is the number of research assistants hired for the study, write a system of linear inequalities that indicates appropriate restrictions on x and y. Graph the set of feasible solutions.

56. Psychology. In an experiment on conditioning, a psychologist uses two types of Skinner (conditioning) boxes with mice and rats. Each mouse spends 10 minutes per day in box A and 20 minutes per day in box B. Each rat spends 20 minutes per day in box A and 10 minutes per day in box B. The total maximum time available per day is 800 minutes for box A and 640 minutes for box B. We are interested in the various numbers of mice and rats that can be used in the experiment under the conditions stated. If x is the number of mice used and y is the number of rats used, write a system of linear inequalities that indicates appropriate restrictions on x and y. Graph the set of feasible solutions.

SECTION 8.5 Linear Programming

A Linear Programming Problem • Linear Programming—A General Description • Mathematical Modeling and Linear Programming

Several problems in Section 8.4 are related to the general type of problems called *linear programming problems.* Linear programming is a mathematical process that has been developed to help management in decision making, and it has become one of the most widely used and best known tools of management science and industrial engineering. We will use an intuitive graphical approach based on the techniques discussed in Section 8.4 to illustrate this process for problems involving two variables.

The American mathematician George B. Dantzig (1914–) formulated the first linear programming problem in 1947 and introduced a solution technique, called the *simplex method,* that does not rely on graphing and is readily adaptable to computer solutions. Today, it is quite common to use a computer to solve applied linear programming problems involving thousands of variables and thousands of inequalities.

A Linear Programming Problem

We begin our discussion with an example that will lead to a general procedure for solving linear programming problems in two variables.

EXAMPLE 1 Production Scheduling

A manufacturer of fiberglass camper tops for pickup trucks makes a compact model and a regular model. Each compact top requires 5 hours from the fabricating department and 2 hours from the finishing department. Each regular top requires 4 hours from the fabricating department and 3 hours from the finishing department. The maximum labor-hours available per week in the fabricating department and the finishing department are 200 and 108, respectively. If the company makes a profit of \$40 on each compact top and \$50 on each regular top, how many tops of each type should be manufactured each week to maximize the total weekly profit, assuming all tops can be sold? What is the maximum profit?

SOLUTION

This is an example of a linear programming problem. To see relationships more clearly, we summarize the manufacturing requirements, objectives, and restrictions in the table:

	Compact Model (Labor-Hours per Top)	Regular Model (Labor-Hours per Top)	Maximum Labor-Hours Available per Week
Fabricating	5	4	200
Finishing	2	3	108
Profit per top	\$40	\$50	

We now proceed to formulate a *mathematical model* for the problem and then to solve it using graphical methods.

OBJECTIVE FUNCTION The *objective* of management is to *decide* how many of each camper top model should be produced each week to *maximize* profit. Let

x = Number of compact tops produced per week
y = Number of regular tops produced per week
} Decision variables

The following function gives the total profit P for x compact tops and y regular tops manufactured each week:

$P = 40x + 50y$ Objective function

Mathematically, management needs to decide on values for the **decision variables** (x and y) that achieve its objective, that is, maximizing the **objective function** (profit) $P = 40x + 50y$. It appears that the profit can be made as large as we like by manufacturing more and more tops—or can it?

CONSTRAINTS Any manufacturing company, no matter how large or small, has manufacturing limits imposed by available resources, plant capacity, demand, and so forth. These limits are referred to as **problem constraints.**

Fabricating department constraint:

(Weekly fabricating time for x compact tops) + (Weekly fabricating time for y regular tops) ≤ (Maximum labor-hours available per week)

$$5x + 4y \le 200$$

Finishing department constraint:

(Weekly finishing time for x compact tops) + (Weekly finishing time for y regular tops) ≤ (Maximum labor-hours available per week)

$$2x + 3y \le 108$$

Nonnegative constraints: It is not possible to manufacture a negative number of tops; thus, we have the **nonnegative constraints**

$x \ge 0$
$y \ge 0$

which we usually write in the form

$x, y \ge 0$

MATHEMATICAL MODEL We now have a **mathematical model** for the problem under consideration:

$$\begin{aligned} &\text{Maximize} && P = 40x + 50y && \text{Objective function} \\ &\text{Subject to} && \left.\begin{aligned} 5x + 4y &\le 200 \\ 2x + 3y &\le 108 \end{aligned}\right\} && \text{Problem constraints} \\ &&& x, y \ge 0 && \text{Nonnegative constraints} \end{aligned}$$

GRAPHICAL SOLUTION Graphing the solution region for the system of linear inequality constraints, as in Section 8.4, we obtain the feasible region for production schedules, as shown in Figure 1.

FIGURE 1

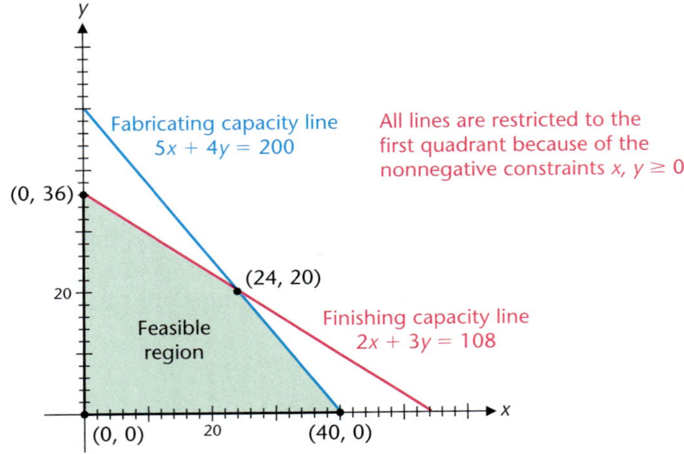

By choosing a production schedule (x, y) from the feasible region, a profit can be determined using the objective function $P = 40x + 50y$. For example, if $x = 24$ and $y = 10$, then the profit for the week is

$$P = 40(24) + 50(10) = \$1,460$$

Or if $x = 15$ and $y = 20$, then the profit for the week is

$$P = 40(15) + 50(20) = \$1,600$$

The question is, out of all possible production schedules (x, y) from the feasible region, which schedule(s) produces the maximum profit? Such a schedule, if it exists, is called an **optimal solution** to the problem because it produces the maximum value of the objective function and is in the feasible region. It is not practical to use point-by-point checking to find the optimal solution. Even if we consider only points with integer coordinates, there are over 800 such points in the feasible region for this problem. Instead, we use the theory that has been developed to solve linear programming problems. Using advanced techniques, it can be shown that:

If the feasible region is bounded, then one or more of the corner points of the feasible region is an optimal solution to the problem.

The maximum value of the objective function is unique; however, there can be more than one feasible production schedule that will produce this unique value. We will have more to say about this later in this section.

Corner Point (x, y)	Objective Function $P = 40x + 50y$	
(0, 0)	0	
(0, 36)	1,800	
(24, 20)	1,960	Maximum value of P
(40, 0)	1,600	

Because the feasible region for this problem is bounded, at least one of the corner points, (0, 0), (0, 36), (24, 20), or (40, 0), is an optimal solution. To find which one, we evaluate $P = 40x + 50y$ at each corner point and choose the corner point that produces the largest value of P. It is convenient to organize these calculations in a table, as shown in the margin.

Examining the values in the table, we see that the maximum value of P at a corner point is $P = 1{,}960$ at $x = 24$ and $y = 20$. Because the maximum value of P over the entire feasible region must always occur at a corner point, we conclude that the maximum profit is \$1,960 when 24 compact tops and 20 regular tops are produced each week.

MATCHED 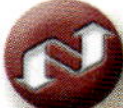 PROBLEM

We now convert the surfboard problem discussed in Section 8.4 into a linear programming problem. A manufacturer of surfboards makes a standard model and a competition model. Each standard board requires 6 labor-hours for fabricating and 1 labor-hour for finishing. Each competition board requires 8 labor-hours for fabricating and 3 labor-hours for finishing. The maximum labor-hours available per week in the fabricating and finishing departments are 120 and 30, respectively. If the company makes a profit of \$40 on each standard board and \$75 on each competition board, how many boards of each type should be manufactured each week to maximize the total weekly profit?

(A) Identify the decision variables.

(B) Write the objective function P.

(C) Write the problem constraints and the nonnegative constraints.

(D) Graph the feasible region, identify the corner points, and evaluate P at each corner point.

(E) How many boards of each type should be manufactured each week to maximize the profit? What is the maximum profit?

EXPLORE/DISCUSS 1

Refer to Example 1. If we assign the profit P in $P = 40x + 50y$ a particular value and plot the resulting equation in the coordinate system shown in Figure 1, we obtain a **constant-profit line (isoprofit line).** Every point in the feasible region on this line represents a production schedule that will produce the same profit. Figure 2 shows the constant-profit lines for $P = \$1{,}000$ and $P = \$1{,}500$.

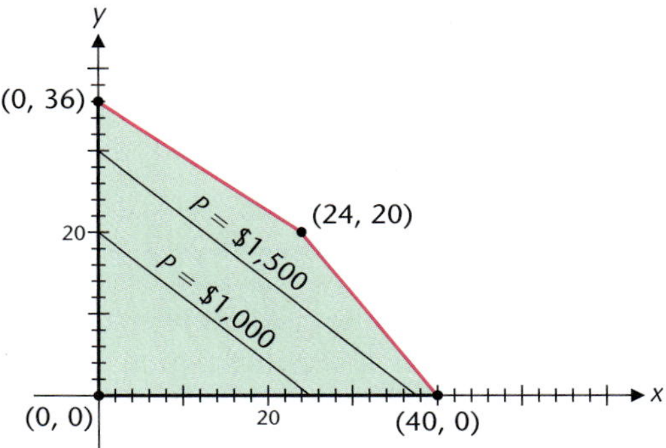

FIGURE 2

(A) How are all the constant-profit lines related?

(B) Place a straightedge along the constant-profit line for $P = \$1{,}000$ and slide it as far as possible in the direction of increasing profit without changing its slope and without leaving the feasible region. Explain how this process can be used to identify the optimal solution to a linear programming problem.

(C) If P is changed to $P = 25x + 75y$, graph the constant-profit lines for $P = \$1{,}000$ and $P = \$1{,}500$, and use a straightedge to identify the optimal solution. Check your answer by evaluating P at each corner point.

(D) Repeat part C for $P = 75x + 25y$.

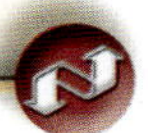

Linear Programming—A General Description

The linear programming problems considered in Example 1 and Matched Problem 1 were *maximization problems,* where we wanted to maximize profits. The same technique can be used to solve *minimization problems,* where, for example, we may want to minimize costs. Before considering additional examples, we state a few general definitions.

A **linear programming problem** is one that is concerned with finding the **optimal value** (maximum or minimum value) of a linear *objective function* of the form

$$z = ax + by$$

where the *decision variables* x and y are subject to *problem constraints* in the form of linear inequalities and to *nonnegative constraints* $x, y \geq 0$. The set of points satisfying both the problem constraints and the nonnegative constraints is called the *feasible region* for the problem. Any point in the feasible region that produces the optimal value of the objective function over the feasible region is called an *optimal solution*.

Theorem 1 is fundamental to the solving of linear programming problems.

THEOREM 1

Fundamental Theorem of Linear Programming

Let S be the feasible region for a linear programming problem, and let $z = ax + by$ be the objective function. If S is bounded, then z has both a maximum and a minimum value on S and each of these occurs at a corner point of S. If S is unbounded, then a maximum or minimum value of z on S may not exist. However, if either does exist, then it must occur at a corner point of S.

We will not consider any problems with unbounded feasible regions in this brief introduction. If a feasible region is bounded, then Theorem 1 provides the basis for the following simple procedure for solving the associated linear programming problem:

Solution of Linear Programming Problems

Step 1. Form a mathematical model for the problem:

(A) Introduce decision variables and write a linear objective function.

(B) Write problem constraints in the form of linear inequalities.

(C) Write nonnegative constraints.

Step 2. Graph the feasible region and find the corner points.

Step 3. Evaluate the objective function at each corner point to determine the optimal solution.

Before considering additional applications, we use this procedure to solve a linear programming problem where the model has already been determined.

EXAMPLE 2 Solving a Linear Programming Problem

Minimize and maximize $z = 5x + 15y$

Subject to

$$\begin{aligned} x + 3y &\le 60 \\ x + y &\ge 10 \\ x - y &\le 0 \\ x, y &\ge 0 \end{aligned}$$

SOLUTION

This problem is a combination of two linear programming problems—a minimization problem and a maximization problem. Because the feasible region is the same for both problems, we can solve these problems together. To begin, we graph the feasible region S, as shown in Figure 3, and find the coordinates of each corner point.

FIGURE 3

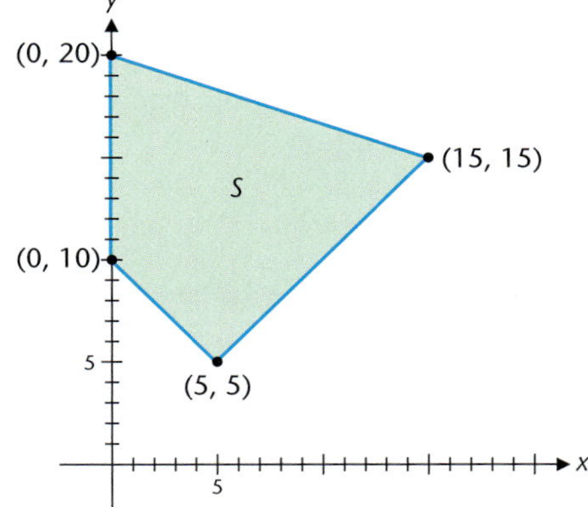

Next, we evaluate the objective function at each corner point, with the results given in the table:

Corner Point (x, y)	Objective Function $z = 5x + 15y$		
(0, 10)	150		
(0, 20)	300	Maximum value	Multiple optimal solutions
(15, 15)	300	Maximum value	
(5, 5)	100	Minimum value	

Examining the values in the table, we see that the minimum value of z on the feasible region S is 100 at (5, 5). Thus, (5, 5) is the optimal solution to the minimization problem. The maximum value of z on the feasible region S is 300, which occurs at (0, 20) and at (15, 15). Thus, the maximization problem has **multiple optimal solutions.** In general,

> **If two corner points are both optimal solutions of the same type (both produce the same maximum value or both produce the same minimum value) to a linear programming problem, then any point on the line segment joining the two corner points is also an optimal solution of that type.**

It can be shown that this is the only way that an optimal value occurs at more than one point.

MATCHED 2 PROBLEM

Minimize and maximize $z = 10x + 5y$

$$\begin{aligned} \text{Subject to} \quad 2x + y &\geq 40 \\ 3x + y &\leq 150 \\ 2x - y &\geq 0 \\ x, y &\geq 0 \end{aligned}$$

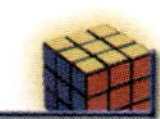

Mathematical Modeling and Linear Programming

Now we consider another application where we must first find the mathematical model and then find its solution.

EXAMPLE 3 **Agriculture**

	Pounds per Cubic Yard	
	Mix *A*	**Mix *B***
Nitrogen	10	5
Potash	8	24
Phosphoric acid	9	6

A farmer can use two types of fertilizer, mix A and mix B. The amounts (in pounds) of nitrogen, phosphoric acid, and potash in a cubic yard of each mix are given in the table. Tests performed on the soil in a large field indicate that the field needs at least 840 pounds of potash and at least 350 pounds of nitrogen. The tests also indicate that no more than 630 pounds of phosphoric acid should be added to the field. A cubic yard of mix A costs \$7, and a cubic yard of mix B costs \$9. How many cubic yards of each mix should the farmer add to the field to supply the necessary nutrients at minimal cost?

SOLUTION

Let

x = Number of cubic yards of mix A added to the field
y = Number of cubic yards of mix B added to the field } Decision variables

We form the linear objective function

$$C = 7x + 9y$$

which gives the cost of adding x cubic yards of mix A and y cubic yards of mix B to the field. Using the data in the table and proceeding as in Example 1, we formulate the mathematical model for the problem:

Minimize	$C = 7x + 9y$	Objective function
Subject to	$10x + 5y \geq 350$	Nitrogen constraint
	$8x + 24y \geq 840$	Potash constraint
	$9x + 6y \leq 630$	Phosphoric acid constraint
	$x, y \geq 0$	Nonnegative constraints

Solving the system of constraint inequalities graphically, we obtain the feasible region S shown in Figure 4, and then we find the coordinates of each corner point.

FIGURE 4

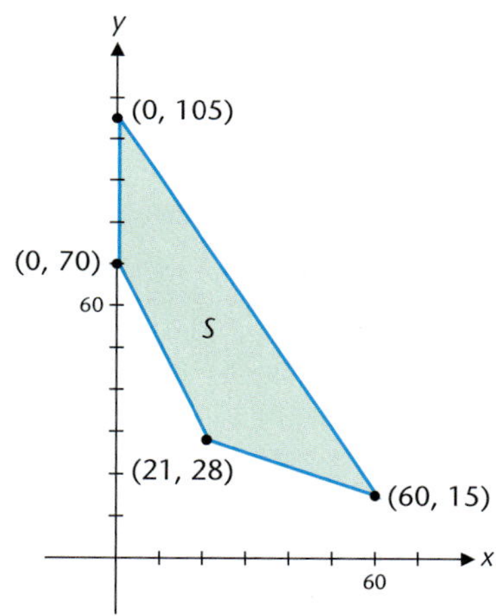

Corner Point (x, y)	Objective Function $C = 7x + 9y$	
(0, 105)	945	
(0, 70)	630	
(21, 28)	399	Minimum value of C
(60, 15)	555	

Next, we evaluate the objective function at each corner point, as shown in the table.

The optimal value is $C = 399$ at the corner point (21, 28). Thus, the farmer should add 21 cubic yards of mix A and 28 cubic yards of mix B at a cost of \$399. This will result in adding the following nutrients to the field:

Nitrogen:	$10(21) + 5(28)$	$= 350$ pounds
Potash:	$8(21) + 24(28)$	$= 840$ pounds
Phosphoric acid:	$9(21) + 6(28)$	$= 357$ pounds

All the nutritional requirements are satisfied.

MATCHED 3 PROBLEM

Repeat Example 3 if the tests indicate that the field needs at least 400 pounds of nitrogen with all other conditions remaining the same.

ANSWERS MATCHED PROBLEMS

1. (A) x = Number of standard boards manufactured each week
   y = Number of competition boards manufactured each week

   (B) $P = 40x + 75y$ (C) $6x + 8y \le 120$ Fabricating constraint
   $x + 3y \le 30$ Finishing constraint
   $x, y \ge 0$ Nonnegative constraints

   (D)

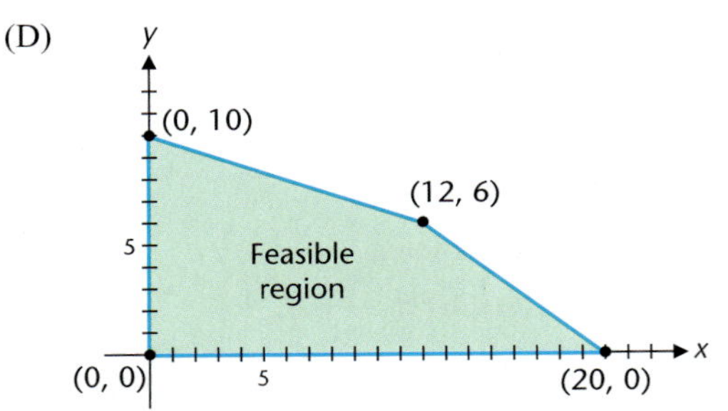

Corner Point (x, y)	Objective Function $P = 40x + 75y$
(0, 0)	0
(0, 10)	750
(12, 6)	930
(20, 0)	800

   (E) 12 standard boards and 6 competition boards for a maximum profit of $930

2. Max $z = 600$ at (30, 60); min $z = 200$ at (10, 20) and (20, 0) (multiple optimal solutions)
3. 27 cubic yards of mix A, 26 cubic yards of mix B; min $C = \$423$

EXERCISE 8.5

A

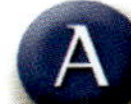
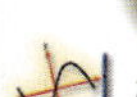

In Problems 1–4, find the maximum value of each objective function over the feasible region S shown in the figure.

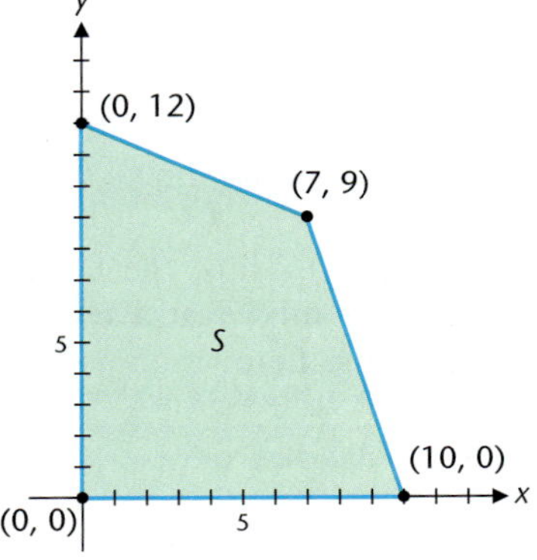

1. $z = x + y$
2. $z = 4x + y$
3. $z = 3x + 7y$
4. $z = 9x + 3y$

Problems 5–8 refer to the feasible region S shown and the constant-value lines discussed in Explore/Discuss 1. For each objective function, draw the line that passes through the feasible point (5, 5) and use the straightedge method from Explore/Discuss 1 to find the maximum value. Check your answer by evaluating the objective function at each corner point.

5. $z = x + 2y$
6. $z = 3x + y$
7. $z = 7x + 2y$
8. $z = 2x + 8y$

In Problems 9–12, find the minimum value of each objective function over the feasible region T shown in the figure.

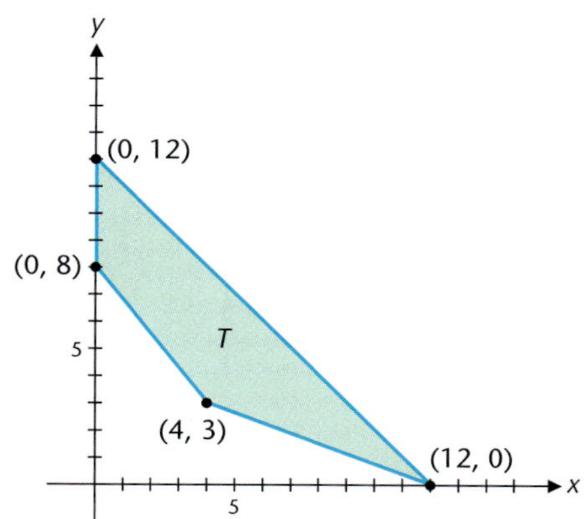

9. $z = 7x + 4y$

10. $z = 7x + 9y$

11. $z = 3x + 8y$

12. $z = 5x + 4y$

Problems 13–16 refer to the feasible region T shown. For each objective function, draw the constant-value line that passes through the feasible point (5, 5) and use the straightedge method from Explore/Discuss 1 to find the minimum value. Check your answer by evaluating the objective function at each corner point.

13. $z = x + 2y$

14. $z = 2x + y$

15. $z = 5x + 4y$

16. $z = 2x + 8y$

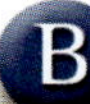

B

In Problems 17–30, solve the linear programming problems.

17. Maximize $z = 3x + 2y$
Subject to
$$\begin{aligned} x + 2y &\le 10 \\ 3x + y &\le 15 \\ x, y &\ge 0 \end{aligned}$$

18. Maximize $z = 4x + 5y$
Subject to
$$\begin{aligned} 2x + y &\le 12 \\ x + 3y &\le 21 \\ x, y &\ge 0 \end{aligned}$$

19. Minimize $z = 3x + 4y$
Subject to
$$\begin{aligned} 2x + y &\ge 8 \\ x + 2y &\le 10 \\ x, y &\ge 0 \end{aligned}$$

20. Minimize $z = 2x + y$
Subject to
$$\begin{aligned} 4x + 3y &\ge 24 \\ 4x + y &\le 16 \\ x, y &\ge 0 \end{aligned}$$

21. Maximize $z = 3x + 4y$
Subject to
$$\begin{aligned} x + 2y &\le 24 \\ x + y &\le 14 \\ 2x + y &\le 24 \\ x, y &\ge 0 \end{aligned}$$

22. Maximize $z = 5x + 3y$
Subject to
$$\begin{aligned} 3x + y &\le 24 \\ x + y &\le 10 \\ x + 3y &\le 24 \\ x, y &\ge 0 \end{aligned}$$

23. Minimize $z = 5x + 6y$
Subject to
$$\begin{aligned} x + 4y &\ge 20 \\ 4x + y &\ge 20 \\ x + y &\le 20 \\ x, y &\ge 0 \end{aligned}$$

24. Minimize $z = x + 2y$
Subject to
$$\begin{aligned} 2x + 3y &\ge 30 \\ 3x + 2y &\ge 30 \\ x + y &\le 15 \\ x, y &\ge 0 \end{aligned}$$

25. Minimize and maximize $z = 25x + 50y$
Subject to
$$\begin{aligned} x + 2y &\le 120 \\ x + y &\ge 60 \\ x - 2y &\ge 0 \\ x, y &\ge 0 \end{aligned}$$

26. Minimize and maximize $z = 15x + 30y$
Subject to
$$\begin{aligned} x + 2y &\ge 100 \\ 2x - y &\le 0 \\ 2x + y &\le 200 \\ x, y &\ge 0 \end{aligned}$$

27. Minimize and maximize $z = 25x + 15y$

Subject to
$$\begin{aligned} 4x + 5y &\geq 100 \\ 3x + 4y &\leq 240 \\ x &\leq 60 \\ y &\leq 45 \\ x, y &\geq 0 \end{aligned}$$

28. Minimize and maximize $z = 25x + 30y$

Subject to
$$\begin{aligned} 2x + 3y &\geq 120 \\ 3x + 2y &\leq 360 \\ x &\leq 80 \\ y &\leq 120 \\ x, y &\geq 0 \end{aligned}$$

29. Maximize $P = 525x_1 + 478x_2$

Subject to
$$\begin{aligned} 275x_1 + 322x_2 &\leq 3{,}381 \\ 350x_1 + 340x_2 &\leq 3{,}762 \\ 425x_1 + 306x_2 &\leq 4{,}114 \\ x_1, x_2 &\geq 0 \end{aligned}$$

30. Maximize $P = 300x_1 + 460x_2$

Subject to
$$\begin{aligned} 245x_1 + 452x_2 &\leq 4{,}181 \\ 290x_1 + 379x_2 &\leq 3{,}888 \\ 390x_1 + 299x_2 &\leq 4{,}407 \\ x_1, x_2 &\geq 0 \end{aligned}$$

C

31. The corner points for the feasible region determined by the problem constraints

$$\begin{aligned} 2x + y &\leq 10 \\ x + 3y &\leq 15 \\ x, y &\geq 0 \end{aligned}$$

are $O = (0, 0)$, $A = (5, 0)$, $B = (3, 4)$, and $C = (0, 5)$. If $z = ax + by$ and $a, b > 0$, determine conditions on a and b that ensure that the maximum value of z occurs

(A) Only at A

(B) Only at B

(C) Only at C

(D) At both A and B

(E) At both B and C

32. The corner points for the feasible region determined by the problem constraints

$$\begin{aligned} x + y &\geq 4 \\ x + 2y &\geq 6 \\ 2x + 3y &\leq 12 \\ x, y &\geq 0 \end{aligned}$$

are $A = (6, 0)$, $B = (2, 2)$, and $C = (0, 4)$. If $z = ax + by$ and $a, b > 0$, determine conditions on a and b that ensure that the minimum value of z occurs

(A) Only at A

(B) Only at B

(C) Only at C

(D) At both A and B

(E) At both B and C

APPLICATIONS

33. Resource Allocation. A manufacturing company makes two types of water skis, a trick ski and a slalom ski. The relevant manufacturing data are given in the table.

(A) If the profit on a trick ski is \$40 and the profit on a slalom ski is \$30, how many of each type of ski should be manufactured each day to realize a maximum profit? What is the maximum profit?

(B) Discuss the effect on the production schedule and the maximum profit if the profit on a slalom ski decreases to \$25 and all other data remain the same.

(C) Discuss the effect on the production schedule and the maximum profit if the profit on a slalom ski increases to \$45 and all other data remain the same.

	Trick Ski (Labor-Hours per Ski)	Slalom Ski (Labor-Hours per Ski)	Maximum Labor-Hours Available per Day
Fabricating department	6	4	108
Finishing department	1	1	24

34. Psychology. In an experiment on conditioning, a psychologist uses two types of Skinner boxes with mice and rats.

The amount of time (in minutes) each mouse and each rat spends in each box per day is given in the table. What is the maximum total number of mice and rats that can be used in this experiment? How many mice and how many rats produce this maximum?

	Mice (Minutes)	Rats (Minutes)	Max. Time Available per Day (Minutes)
Skinner box *A*	10	20	800
Skinner box *B*	20	10	640

35. Purchasing. A trucking firm wants to purchase a maximum of 15 new trucks that will provide at least 36 tons of additional shipping capacity. A model *A* truck holds 2 tons and costs $15,000. A model *B* truck holds 3 tons and costs $24,000. How many trucks of each model should the company purchase to provide the additional shipping capacity at minimal cost? What is the minimal cost?

36. Transportation. The officers of a high school senior class are planning to rent buses and vans for a class trip. Each bus can transport 40 students, requires 3 chaperones, and costs $1,200 to rent. Each van can transport 8 students, requires 1 chaperone, and costs $100 to rent. The officers want to be able to accommodate at least 400 students with no more than 36 chaperones. How many vehicles of each type should they rent to minimize the transportation costs? What are the minimal transportation costs?

★ **37. Resource Allocation.** A furniture company manufactures dining room tables and chairs. Each table requires 8 hours from the assembly department and 2 hours from the finishing department and contributes a profit of $90. Each chair requires 2 hours from the assembly department and 1 hour from the finishing department and contributes a profit of $25. The maximum labor-hours available each day in the assembly and finishing departments are 400 and 120, respectively.

(A) How many tables and how many chairs should be manufactured each day to maximize the daily profit? What is the maximum daily profit?

(B) Discuss the effect on the production schedule and the maximum profit if the marketing department of the company decides that the number of chairs produced should be at least four times the number of tables produced.

★ **38. Resource Allocation.** An electronics firm manufactures two types of personal computers, a desktop model and a portable model. The production of a desktop computer requires a capital expenditure of $400 and 40 hours of labor. The production of a portable computer requires a capital expenditure of $250 and 30 hours of labor. The firm has $20,000 capital and 2,160 labor-hours available for production of desktop and portable computers.

(A) What is the maximum number of computers the company is capable of producing?

(B) If each desktop computer contributes a profit of $320 and each portable contributes a profit of $220, how much profit will the company make by producing the maximum number of computers determined in part A? Is this the maximum profit? If not, what is the maximum profit?

39. Pollution Control. Because of new federal regulations on pollution, a chemical plant introduced a new process to supplement or replace an older process used in the production of a particular chemical. The older process emitted 20 grams of sulfur dioxide and 40 grams of particulate matter into the atmosphere for each gallon of chemical produced. The new process emits 5 grams of sulfur dioxide and 20 grams of particulate matter for each gallon produced. The company makes a profit of 60¢ per gallon and 20¢ per gallon on the old and new processes, respectively.

(A) If the regulations allow the plant to emit no more than 16,000 grams of sulfur dioxide and 30,000 grams of particulate matter daily, how many gallons of the chemical should be produced by each process to maximize daily profit? What is the maximum daily profit?

(B) Discuss the effect on the production schedule and the maximum profit if the regulations restrict emissions of sulfur dioxide to 11,500 grams daily and all other data remain unchanged.

(C) Discuss the effect on the production schedule and the maximum profit if the regulations restrict emissions of sulfur dioxide to 7,200 grams daily and all other data remain unchanged.

★★ **40. Sociology.** A city council voted to conduct a study on inner-city community problems. A nearby university was contacted to provide a maximum of 40 sociologists and research assistants. Allocation of time and cost per week are given in the table.

(A) How many sociologists and research assistants should be hired to meet the weekly labor-hour requirements and minimize the weekly cost? What is the weekly cost?

(B) Discuss the effect on the solution in part A if the council decides that they should not hire more sociologists than research assistants and all other data remain unchanged.

	Sociologist (Labor-Hours)	Research Assistant (Labor-Hours)	Minimum Labor-Hours Needed per Week
Fieldwork	10	30	280
Research center	30	10	360
Cost per week	$500	$300	

★★ **41. Plant Nutrition.** A fruit grower can use two types of fertilizer in her orange grove, brand A and brand B. The amounts (in pounds) of nitrogen, phosphoric acid, potash, and chloride in a bag of each mix are given in the table. Tests indicate that the grove needs at least 480 pounds of phosphoric acid, at least 540 pounds of potash, and at most 620 pounds of chloride. If the grower always uses a combination of bags of brand A and brand B that will satisfy the constraints of phosphoric acid, potash, and chloride, discuss the effect that this will have on the amount of nitrogen added to the field.

	Pounds per Bag	
	Brand *A*	**Brand *B***
Nitrogen	6	7
Phosphoric acid	2	4
Potash	6	3
Chloride	3	4

★★ **42. Diet.** A dietitian in a hospital is to arrange a special diet composed of two foods, M and N. Each ounce of food M contains 16 units of calcium, 5 units of iron, 6 units of cholesterol, and 8 units of vitamin A. Each ounce of food N contains 4 units of calcium, 25 units of iron, 4 units of cholesterol, and 4 units of vitamin A. The diet requires at least 320 units of calcium, at least 575 units of iron, and at most 300 units of cholesterol. If the dietitian always selects a combination of foods M and N that will satisfy the constraints for calcium, iron, and cholesterol, discuss the effects that this will have on the amount of vitamin A in the diet.

CHAPTER 8 REVIEW

8.1 Systems of Linear Equations in Two Variables

A system of two linear equations with two variables is a system of the form

$$\begin{aligned} ax + by &= h \\ cx + dy &= k \end{aligned} \qquad (1)$$

where x and y are variables, a, b, c, and d are real numbers called the **coefficients** of x and y, and h and k are real numbers called the **constant terms** in the equations. The ordered pair of numbers (x_0, y_0) is a **solution** to system (1) if each equation is satisfied by the pair. The set of all such ordered pairs of numbers is called the **solution set** for the system. To **solve** a system is to find its solution set.

In general, a system of linear equations has exactly one solution, no solution, or infinitely many solutions. A system of linear equations is **consistent** if it has one or more solutions and **inconsistent** if no solutions exist. A consistent system is said to be **independent** if it has exactly one solution and **dependent** if it has more than one solution.

Two standard methods for solving system (1) were discussed: **graphing** and **substitution.**

8.2 Systems of Linear Equations and Augmented Matrices

Two systems of equations are **equivalent** if both have the same solution set. A system of linear equations is transformed into an equivalent system if

1. Two equations are interchanged.
2. An equation is multiplied by a nonzero constant.
3. A constant multiple of one equation is added to another equation.

These operations form the basis of solution using **elimination by addition.**

The method of solution using elimination by addition is transformed into a more efficient method for larger-scale systems by the introduction of an *augmented matrix*. A **matrix** is a rectangular array of numbers written within brackets. Each number in a matrix is called an **element** of the matrix. If a matrix has m rows and n columns, it is called an $\boldsymbol{m \times n}$ **matrix** (read "m by n matrix"). The expression $m \times n$ is called the **size** of the matrix, and the numbers m and n are called the **dimensions** of the matrix. A matrix with n rows and n columns is called a **square matrix of order *n*.** A matrix with only one column is called a **column matrix,** and a matrix with only one row is called a **row matrix.** The **position** of an element in a matrix is the row and column containing the element. This is usually denoted using **double subscript notation** a_{ij}, where i is the row and j is the column containing the element a_{ij}.

For ease of generalization to larger systems, we change the notation for variables and constants in system (1) to a subscript form:

$$\begin{aligned} a_{11}x_1 + a_{12}x_2 &= k_1 \\ a_{21}x_1 + a_{22}x_2 &= k_2 \end{aligned} \qquad (2)$$

Associated with each linear system of the form (2), where x_1 and x_2 are variables, is the **augmented matrix** of the system:

Column 1 (C_1)
Column 2 (C_2)
Column 3 (C_3)

$$\left[\begin{array}{cc|c} a_{11} & a_{12} & k_1 \\ a_{21} & a_{22} & k_2 \end{array}\right] \begin{array}{l} \leftarrow \text{Row 1 } (R_1) \\ \leftarrow \text{Row 2 } (R_2) \end{array} \qquad (3)$$

Two augmented matrices are **row-equivalent,** denoted by the symbol ~ between the two matrices, if they are augmented matrices of equivalent systems of equations. An augmented matrix is transformed into a row-equivalent matrix if any of the following **row operations** is performed:

1. Two rows are interchanged.
2. A row is multiplied by a nonzero constant.
3. A constant multiple of one row is added to another row.

The following symbols are used to describe these row operations:

1. $R_i \leftrightarrow R_j$ means "interchange row i with row j."
2. $kR_i \to R_i$ means "multiply row i by the constant k."
3. $kR_j + R_i \to R_i$ means "multiply row j by the constant k and add to R_i."

In solving system (2) using row operations, the objective is to transform the augmented matrix (3) into the form

$$\left[\begin{array}{cc|c} 1 & 0 & m \\ 0 & 1 & n \end{array}\right]$$

If this can be done, then (m, n) is the unique solution of system (2). If (3) is transformed into the form

$$\left[\begin{array}{cc|c} 1 & m & n \\ 0 & 0 & 0 \end{array}\right]$$

then system (2) has infinitely many solutions. If (3) is transformed into the form

$$\left[\begin{array}{cc|c} 1 & m & n \\ 0 & 0 & p \end{array}\right] \qquad p \neq 0$$

then system (2) does not have a solution.

8.3 Gauss–Jordan Elimination

In Section 8.2 we were actually using *Gauss–Jordan elimination* to solve a system of two equations with two variables. The method generalizes completely for systems with more than two variables, and the number of variables does not have to be the same as the number of equations.

As before, our objective is to start with the augmented matrix of a linear system and transform it using row operations into a simple form where the solution can be read by inspection. The simple form, called the **reduced form,** is achieved if:

1. Each row consisting entirely of 0s is below any row having at least one nonzero element.
2. The leftmost or leading nonzero element in each row is 1.
3. The column containing the leading 1 of a given row has 0s above and below the 1.
4. The leading 1 in any row is to the right of the leading 1 in the preceding row.

A **reduced system** is a system of linear equations that corresponds to a reduced augmented matrix. When a reduced system has more variables than equations and contains no contradictions, the system is dependent and has infinitely many solutions.

The **Gauss–Jordan elimination** procedure for solving a system of linear equations is given in step-by-step form as follows:

Step 1. Choose the leading nonzero column of an augmented matrix, and use appropriate row operations to get a 1 at the top.

Step 2. Use multiples of the row containing the 1 from step 1 to get zeros in all remaining places in the column containing this 1.

Step 3. Repeat step 1 with the **submatrix** formed by (mentally) deleting the row used in step 2 and all rows above this row.

Step 4. Repeat step 2 with the **entire matrix,** including the mentally deleted rows. Continue this process until the entire matrix is in reduced form.

If at any point in the above process we obtain a row with all 0s to the left of the vertical line and a nonzero number n to the right, we can stop, because we have a contradiction: $0 = n$, $n \neq 0$. We can then conclude that the system has no solution. If this does not happen and we obtain an augmented matrix in reduced form without any contradictions, the solution can be read by inspection.

8.4 Systems of Linear Inequalities

A graph is often the most convenient way to represent the solution of a linear inequality in two variables or of a system of linear inequalities in two variables.

A vertical line divides a plane into **left** and **right half-planes.** A nonvertical line divides a plane into **upper** and **lower half-planes.** If A, B, and C are real numbers with A and B not both zero, then the **graph of the linear inequality**

$$Ax + By < C \quad \text{or} \quad Ax + By > C$$

with $B \neq 0$, is either the upper half-plane or the lower half-plane (but not both) determined by the line $Ax + By = C$. If $B = 0$, then the graph of

$$Ax < C \quad \text{or} \quad Ax > C$$

is either the left half-plane or the right half-plane (but not both) determined by the line $Ax = C$. There are two **step-by-step procedures for graphing a linear inequality in two variables:**

Algebraic Procedure

Step 1. Graph $Ax + By = C$ as a broken line if equality is not included in the original statement or as a solid line if equality is included.

Step 2. Choose a test point anywhere in the plane not on the line and substitute the coordinates into the inequality. The origin (0, 0) often requires the least computation.

Step 3. The graph of the original inequality includes the half-plane containing the test point if the inequality is satisfied by that point, or the half-plane not containing that point if the inequality is not satisfied by that point.

Graphing Utility Procedure

Step 1. Solve the inequality for y.

Step 2. Enter the equation of the boundary line and select a shading option as follows:

$$\left.\begin{array}{c} y > mx + b \\ \text{or} \\ y \geq mx + b \end{array}\right\} \quad \text{Select shade above.}$$

$$\left.\begin{array}{c} y < mx + b \\ \text{or} \\ y \leq mx + b \end{array}\right\} \quad \text{Select shade below.}$$

Step 3. Graph the solution.

We now turn to systems of linear inequalities in two variables. The **solution to a system of linear inequalities in two variables** is the set of all ordered pairs of real numbers that simultaneously satisfy all the inequalities in the system. The graph is called the **solution region.** In many applications the solution region is also referred to as the **feasible region.** To **find the solution region,** we graph each inequality in the system and then take the intersection of all the graphs. A **corner point** of a solution region is a point in the solution region that is the intersection of two boundary lines. A solution region is **bounded** if it can be enclosed within a circle. If it cannot be enclosed within a circle, then it is **unbounded.**

8.5 Linear Programming

Linear programming is a mathematical process that has been developed to help management in decision-making, and it has become one of the most widely used and best-known tools of management science and industrial engineering.

A **linear programming problem** is one that is concerned with finding the **optimal value** (maximum or minimum value) of a linear **objective function** of the form $z = ax + by$, where the **decision variables** x and y are subject to **problem constraints** in the form of linear inequalities and **nonnegative constraints** $x, y \geq 0$. The set of points satisfying both the problem constraints and the nonnegative constraints is called the *feasible region* for the problem. Any point in the feasible region that produces the optimal value of the objective function over the feasible region is called an **optimal solution.** The **fundamental theorem of linear programming** is basic to the solving of linear programming problems: Let S be the feasible region for a linear programming problem, and let $z = ax + by$ be the objective function. If S is bounded, then z has both a maximum and a minimum value on S and each of these occurs at a corner point of S. If S is unbounded, then a maximum or minimum value of z on S may not exist. However, if either does exist, then it must occur at a corner point of S.

Problems with unbounded feasible regions are not considered in this brief introduction. The fundamental theorem leads to a simple **step-by-step solution to linear programming problems with a bounded feasible region:**

Step 1. Form a mathematical model for the problem:
(A) Introduce decision variables and write a linear objective function.
(B) Write problem constraints in the form of linear inequalities.
(C) Write nonnegative constraints.

Step 2. Graph the feasible region and find the corner points.

Step 3. Evaluate the objective function at each corner point to determine the optimal solution.

If two corner points are both optimal solutions of the same type (both produce the same maximum value or both produce the same minimum value) to a linear programming problem, then any point on the line segment joining the two corner points is also an optimal solution of that type.

CHAPTER 8 REVIEW EXERCISES

Work through all the problems in this chapter review and check answers in the back of the book. Answers to all review problems are there, and following each answer is a number in italics indicating the section in which that type of problem is discussed. Where weaknesses show up, review appropriate sections in the text.

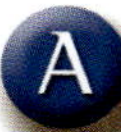

Solve Problems 1 and 2 by substitution.

1. $y = 4x - 9$
$y = -x + 6$

2. $3x + 2y = 5$
$4x - y = 14$

Solve Problems 3–6 by graphing.

3. $3x - 2y = 8$
$x + 3y = -1$

4. $2x + y = 4$
$-2x + 7y = 9$

5. $3x - 4y \geq 24$

6. $2x + y \leq 2$
$x + 2y \geq -2$

Perform each of the row operations indicated in Problems 7–9 on the following augmented matrix:

$$\left[\begin{array}{cc|c} 1 & -4 & 5 \\ 3 & -6 & 12 \end{array}\right]$$

7. $R_1 \leftrightarrow R_2$

8. $\frac{1}{3}R_2 \rightarrow R_2$

9. $(-3)R_1 + R_2 \rightarrow R_2$

In Problems 10–12, write the linear system corresponding to each reduced augmented matrix and solve.

10. $\left[\begin{array}{cc|c} 1 & 0 & 4 \\ 0 & 1 & -7 \end{array}\right]$

11. $\left[\begin{array}{cc|c} 1 & -1 & 4 \\ 0 & 0 & 1 \end{array}\right]$

12. $\left[\begin{array}{cc|c} 1 & -1 & 4 \\ 0 & 0 & 0 \end{array}\right]$

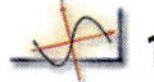

13. Find the maximum and minimum values of $z = 5x + 3y$ over the feasible region S shown in the figure.

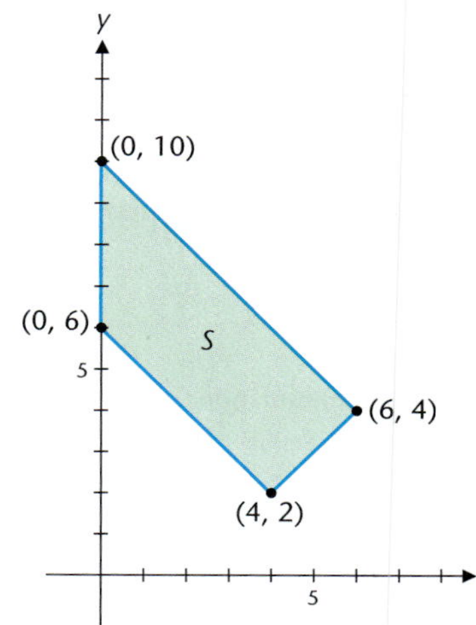

14. Use Gauss–Jordan elimination to solve the system

$$\begin{aligned} x_1 - x_2 &= 4 \\ 2x_1 + x_2 &= 2 \end{aligned}$$

Then write the linear system represented by each augmented matrix in your solution, and solve each of these systems graphically. Discuss the relationship between the solutions of these systems.

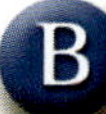

Solve Problems 15–20 using Gauss–Jordan elimination.

15. $$\begin{aligned} 3x_1 + 2x_2 &= 3 \\ x_1 + 3x_2 &= 8 \end{aligned}$$

16. $$\begin{aligned} x_1 + x_2 &= 1 \\ x_1 - x_3 &= -2 \\ x_2 + 2x_3 &= 4 \end{aligned}$$

17. $$\begin{aligned} x_1 + 2x_2 + 3x_3 &= 1 \\ 2x_1 + 3x_2 + 4x_3 &= 3 \\ x_1 + 2x_2 + x_3 &= 3 \end{aligned}$$

18. $$\begin{aligned} x_1 + 2x_2 - x_3 &= 2 \\ 2x_1 + 3x_2 + x_3 &= -3 \\ 3x_1 + 5x_2 &= -1 \end{aligned}$$

19. $$\begin{aligned} x_1 - 2x_2 &= 1 \\ 2x_1 - x_2 &= 0 \\ x_1 - 3x_2 &= -2 \end{aligned}$$

20. $$\begin{aligned} x_1 + 2x_2 - x_3 &= 2 \\ 3x_1 - x_2 + 2x_3 &= -3 \end{aligned}$$

Solve the systems in Problems 21–23 graphically, and indicate whether each solution region is bounded or unbounded. Find the coordinates of each corner point.

21. $$\begin{aligned} 2x + y &\le 8 \\ 2x + 3y &\le 12 \\ x, y &\ge 0 \end{aligned}$$

22. $$\begin{aligned} 2x + y &\ge 8 \\ x + 3y &\ge 12 \\ x, y &\ge 0 \end{aligned}$$

23. $$\begin{aligned} x + y &\le 20 \\ x + 4y &\ge 20 \\ x - y &\ge 0 \end{aligned}$$

Solve the linear programming problems in Problems 24–26.

24. Maximize $z = 7x + 9y$
Subject to
$$\begin{aligned} x + 2y &\le 8 \\ 2x + y &\le 10 \\ x, y &\ge 0 \end{aligned}$$

25. Minimize $z = 5x + 10y$
Subject to
$$\begin{aligned} x + y &\le 20 \\ 3x + y &\ge 15 \\ x + 2y &\ge 15 \\ x, y &\ge 0 \end{aligned}$$

26. Minimize and maximize $z = 5x + 8y$
Subject to
$$\begin{aligned} x + 2y &\le 20 \\ 3x + y &\le 15 \\ x + y &\ge 7 \\ x, y &\ge 0 \end{aligned}$$

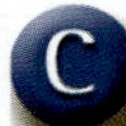

27. Solve using Gauss–Jordan elimination:

$$\begin{aligned} x_1 + x_2 + x_3 &= 7{,}000 \\ 0.04x_1 + 0.05x_2 + 0.06x_3 &= 360 \\ 0.04x_1 + 0.05x_2 - 0.06x_3 &= 120 \end{aligned}$$

28. Maximize $z = 30x + 20y$
Subject to
$$\begin{aligned} 1.2x + 0.6y &\le 960 \\ 0.04x + 0.03y &\le 36 \\ 0.2x + 0.3y &\le 270 \\ x, y &\ge 0 \end{aligned}$$

29. Discuss the number of solutions for the system corresponding to the reduced form shown below if
(A) $m \ne 0$ **(B)** $m = 0$ and $n \ne 0$
(C) $m = 0$ and $n = 0$

$$\left[\begin{array}{ccc|c} 1 & 0 & -3 & 4 \\ 0 & 1 & 2 & 5 \\ 0 & 0 & m & n \end{array}\right]$$

APPLICATIONS

30. Business. A container holds 120 packages. Some of the packages weigh $\frac{1}{2}$ pound each, and the rest weigh $\frac{1}{3}$ pound each. If the total contents of the container weigh 48 pounds, how many are there of each type of package? Solve using two-equation–two-variable methods.

★ **31. Geometry.** Find the dimensions of a rectangle with an area of 48 square meters and a perimeter of 28 meters. Solve using two-equation–two-variable methods.

★ **32. Diet.** A laboratory assistant wishes to obtain a food mix that contains, among other things, 27 grams of protein, 5.4 grams of fat, and 19 grams of moisture. He has available mixes A, B, and C with the compositions listed in the table. How many grams of each mix should be used to get the desired diet mix? Set up a system of equations and solve using Gauss–Jordan elimination.

Mix	Protein (%)	Fat (%)	Moisture (%)
A	30	3	10
B	20	5	20
C	10	4	10

★★ **33. Puzzle.** A piggy bank contains 30 coins worth \$1.90.

(A) If the bank contains only nickels and dimes, how many coins of each type does it contain?

(B) If the bank contains nickels, dimes, and quarters, how many coins of each type does it contain?

★ **34. Resource Allocation.** North Star Sail Loft manufactures regular and competition sails. Each regular sail takes 1 labor-hour to cut and 3 labor-hours to sew. Each competition sail takes 2 labor-hours to cut and 4 labor-hours to sew. There are 140 labor-hours available in the cutting department and 360 labor-hours available in the sewing department.

(A) If the loft makes a profit of \$60 on each regular sail and \$100 on each competition sail, how many sails of each type should the company manufacture to maximize its profit? What is the maximum profit?

(B) An increase in the demand for competition sails causes the profit on a competition sail to rise to \$125. Discuss the effect of this change on the number of sails manufactured and on the maximum profit.

(C) A decrease in the demand for competition sails causes the profit on a competition sail to drop to \$75. Discuss the effect of this change on the number of sails manufactured and on the maximum profit.

★ **35. Nutrition—Animals.** A special diet for laboratory animals is to contain at least 800 units of vitamins, at least 800 units of minerals, and at most 1,300 calories. There are two feed mixes available, mix A and mix B. A gram of mix A contains 5 units of vitamins, 2 units of minerals, and 4 calories. A gram of mix B contains 2 units of vitamins, 4 units of minerals, and 4 calories.

(A) If mix A costs \$0.07 per gram and mix B costs \$0.04 per gram, how many grams of each mix should be used to satisfy the requirements of the diet at minimal cost? What is the minimal cost?

(B) If the price of mix B decreases to \$0.02 per gram, discuss the effect of this change on the solution in part A.

(C) If the price of mix B increases to \$0.15 per gram, discuss the effect of this change on the solution in part A.

Modeling with Systems of Linear Equations

In this group activity we consider two real-world problems that can be solved using systems of linear equations: heat conduction and traffic flow. Both problems involve using a grid and a basic assumption to construct the model (the system of equations). Gauss–Jordan elimination is then used to solve the model. In the heat conduction problem, the solution of the model is easily interpreted in terms of the original problem. The system in the second problem is dependent, and the solution requires a more careful interpretation.

I Heat Conduction

A metal grid consists of four thin metal bars. The end of each bar of the grid is kept at a constant temperature, as shown in Figure 1. We assume that the temperature at each intersection point in the grid is the average of the temperatures at the four adjacent points in the grid (adjacent points are either other intersection points or ends of bars). Thus, the temperature x_1 at the intersection point in the upper left-hand corner of the grid must satisfy

$$x_1 = \tfrac{1}{4}(\underset{\text{Left}}{40} + \underset{\text{Above}}{0} + \underset{\text{Right}}{x_2} + \underset{\text{Below}}{x_3})$$

Find equations for the temperature at the other three intersection points, and solve the resulting system to find the temperature at each intersection point in the grid.

FIGURE 1

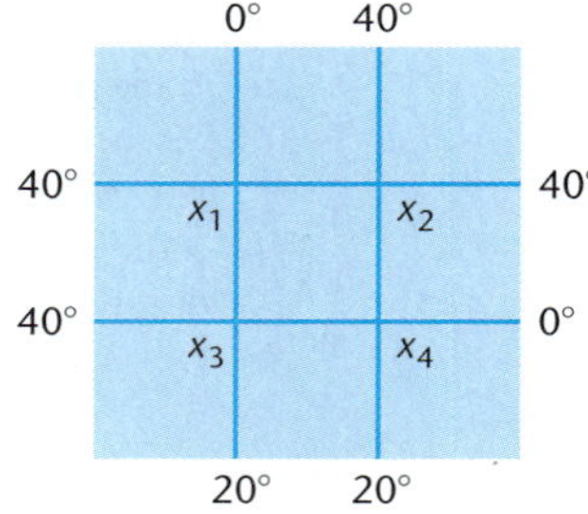

II Traffic Flow

The rush-hour traffic flow for a network of four one-way streets in a city is shown in Figure 2. The numbers next to each street indicate the number of vehicles per hour that enter and leave the network on that street. The variables x_1, x_2, x_3, and x_4 represent the flow of traffic between the four intersections in the network. For a smooth flow of traffic, we assume that the number of vehicles entering each intersection should always equal the number leaving. For example, because 1,500 vehicles enter the intersection of 5th Street and Washington Avenue each hour and $x_1 + x_4$ vehicles leave this intersection, we see that $x_1 + x_4 = 1{,}500$.

(A) Find the equations determined by the traffic flow at each of the other three intersections.

(B) Find the solution to the system in part A.

(C) What is the maximum number of vehicles that can travel from Washington Avenue to Lincoln Avenue on 5th Street? What is the minimum number?

(D) If traffic lights are adjusted so that 1,000 vehicles per hour travel from Washington Avenue to Lincoln Avenue on 5th Street, determine the flow around the rest of the network.

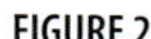
FIGURE 2

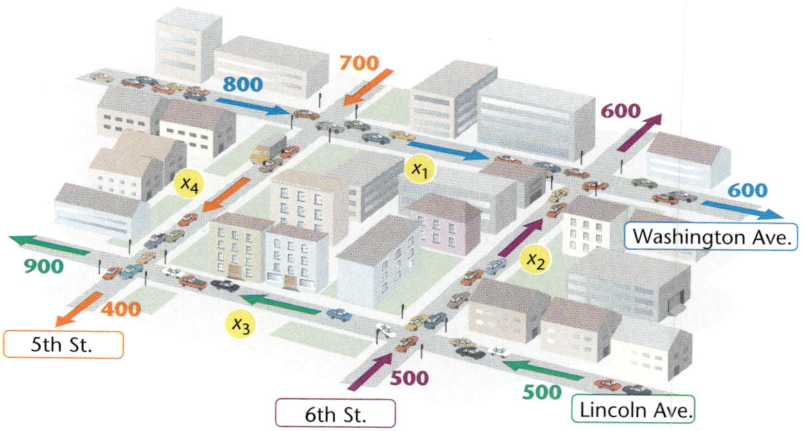

CHAPTER 9

Matrices and Determinants

OUTLINE

In Chapter 9 we discuss matrices in more detail. In the first three sections we define and study some algebraic operations on matrices, including addition, multiplication, and inversion. The next three sections deal with the determinant of a matrix.

In Chapter 8 we used row operations and Gauss–Jordan elimination to solve systems of linear equations. Row operations play a prominent role in the development of several topics in Chapter 9. One consequence of our discussion will be the development of two additional methods for solving systems of linear equations: one method involves inverse matrices and the other determinants.

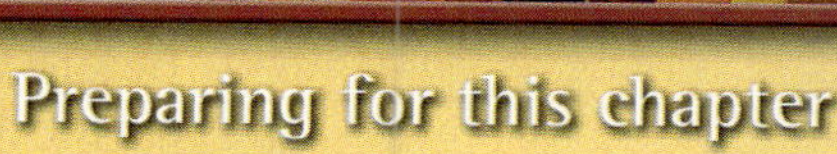

Preparing for this chapter

Before getting started on this chapter, review the following concepts:

- Properties of Real Numbers (Basic Algebra Review*, Section R.1)
- Linear Equations (Appendix A, Section A.1)
- Matrices and Linear Systems (Chapter 8, Sections 2 and 3)

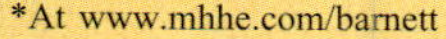

*At www.mhhe.com/barnett

SECTION 9.1 Matrix Operations

Addition and Subtraction • Multiplication of a Matrix by a Number • Matrix Product

Matrices are both a very ancient and a very current mathematical concept. References to matrices and systems of equations can be found in Chinese manuscripts dating back to around 200 B.C. Over the years, mathematicians and scientists have found many applications of matrices. More recently, the advent of personal and large-scale computers has increased the use of matrices in a wide variety of applications. In 1979 Dan Bricklin and Robert Frankston introduced VisiCalc, the first electronic spreadsheet program for personal computers. Simply put, a *spreadsheet* is a computer program that allows the user to enter and manipulate numbers, often using matrix notation and operations. Spreadsheets were initially used by businesses in areas such as budgeting, sales projections, and cost estimation. However, many other applications have begun to appear. For example, a scientist can use a spreadsheet to analyze the results of an experiment, or a teacher can use one to record and average grades. There are even spreadsheets that can be used to help compute an individual's income tax.

In Section 8.2 we introduced basic matrix terminology and solved systems of equations by performing row operations on augmented coefficient matrices. Matrices have many other useful applications and possess an interesting mathematical structure in their own right. As we will see, matrix addition and multiplication are similar to real number addition and multiplication in many respects, but there are some important differences.

Addition and Subtraction

Before we can discuss arithmetic operations for matrices, we have to define equality for matrices. Two matrices are **equal** if they have the same size and their corresponding elements are equal. For example,

$$\underset{2 \times 3}{\begin{bmatrix} a & b & c \\ d & e & f \end{bmatrix}} = \underset{2 \times 3}{\begin{bmatrix} u & v & w \\ x & y & z \end{bmatrix}} \quad \text{if and only if} \quad \begin{matrix} a = u & b = v & c = w \\ d = x & e = y & f = z \end{matrix}$$

The **sum of two matrices** of the same size is a matrix with elements that are the sums of the corresponding elements of the two given matrices.

Addition is not defined for matrices of different sizes.

EXAMPLE 1 Matrix Addition

Add:

(A) $\begin{bmatrix} a & b \\ c & d \end{bmatrix} + \begin{bmatrix} w & x \\ y & z \end{bmatrix}$

(B) $\begin{bmatrix} 2 & -3 & 0 \\ 1 & 2 & -5 \end{bmatrix} + \begin{bmatrix} 3 & 1 & 2 \\ -3 & 2 & 5 \end{bmatrix}$

(C) $\begin{bmatrix} 2 & 1 & 4 \\ 3 & 2 & -3 \end{bmatrix} + \begin{bmatrix} 0 & 2 \\ -3 & 5 \\ -1 & 4 \end{bmatrix}$

SOLUTIONS

(A) $\begin{bmatrix} a & b \\ c & d \end{bmatrix} + \begin{bmatrix} w & x \\ y & z \end{bmatrix} = \begin{bmatrix} (a+w) & (b+x) \\ (c+y) & (d+z) \end{bmatrix}$

(B) Algebraic Solution

$$\begin{bmatrix} 2 & -3 & 0 \\ 1 & 2 & -5 \end{bmatrix} + \begin{bmatrix} 3 & 1 & 2 \\ -3 & 2 & 5 \end{bmatrix} = \begin{bmatrix} (2+3) & (-3+1) & (0+2) \\ (1-3) & (2+2) & (-5+5) \end{bmatrix}$$

$$= \begin{bmatrix} 5 & -2 & 2 \\ -2 & 4 & 0 \end{bmatrix}$$

(B) Graphing Utility Solution

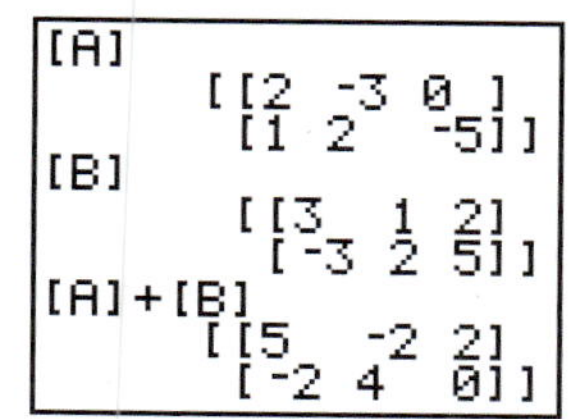

FIGURE 1

(C) Algebraic Solution

$$\begin{bmatrix} 2 & 1 & 4 \\ 3 & 2 & -3 \end{bmatrix} + \begin{bmatrix} 0 & 2 \\ -3 & 5 \\ -1 & 4 \end{bmatrix}$$

Because the first matrix is 2×3 and the second is 3×2, this sum is not defined.

(C) Graphing Utility Solution

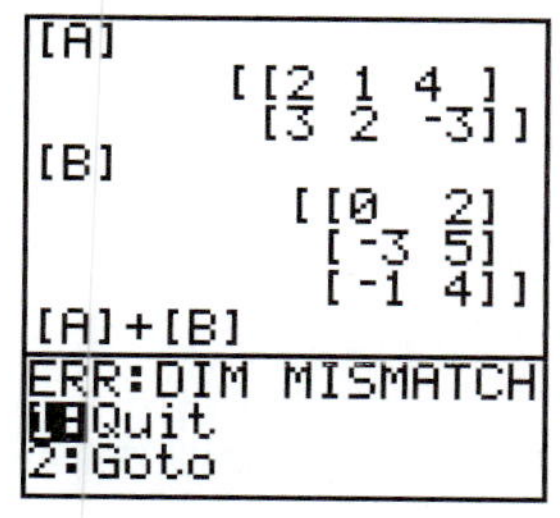

FIGURE 2

MATCHED 1 PROBLEM

Add: $\begin{bmatrix} 3 & 2 \\ -1 & -1 \\ 0 & 3 \end{bmatrix} + \begin{bmatrix} -2 & 3 \\ 1 & -1 \\ 2 & -2 \end{bmatrix}$

Because we add two matrices by adding their corresponding elements, it follows from the properties of real numbers that matrices of the same size are commutative and associative relative to addition. That is, if A, B, and C are matrices of the same size, then

$$A + B = B + A \qquad \text{Commutative}$$
$$(A + B) + C = A + (B + C) \qquad \text{Associative}$$

A matrix with elements that are all 0s is called a **zero matrix.** For example, the following are zero matrices of different sizes:

$$[0 \quad 0 \quad 0] \qquad \begin{bmatrix} 0 & 0 \\ 0 & 0 \end{bmatrix} \qquad \begin{bmatrix} 0 \\ 0 \\ 0 \\ 0 \end{bmatrix} \qquad \begin{bmatrix} 0 & 0 & 0 & 0 \\ 0 & 0 & 0 & 0 \\ 0 & 0 & 0 & 0 \end{bmatrix}$$

[*Note:* "0" can be used to denote the zero matrix of any size.]

The **negative of a matrix *M*,** denoted by $-M$, is a matrix with elements that are the negatives of the elements in M. Thus, if

$$M = \begin{bmatrix} a & b \\ c & d \end{bmatrix}$$

then

$$-M = \begin{bmatrix} -a & -b \\ -c & -d \end{bmatrix}$$

Note that $M + (-M) = 0$ (a zero matrix).

If A and B are matrices of the same size, then we define **subtraction** as follows:

$$A - B = A + (-B)$$

Thus, to subtract matrix B from matrix A, we simply subtract corresponding elements.

EXAMPLE 2 Matrix Subtraction

Subtract: $\begin{bmatrix} 3 & -2 \\ 5 & 0 \end{bmatrix} - \begin{bmatrix} -2 & 2 \\ 3 & 4 \end{bmatrix}$

SOLUTION

Algebraic Solution

$$\begin{bmatrix} 3 & -2 \\ 5 & 0 \end{bmatrix} - \begin{bmatrix} -2 & 2 \\ 3 & 4 \end{bmatrix} = \begin{bmatrix} 3 & -2 \\ 5 & 0 \end{bmatrix} + \begin{bmatrix} 2 & -2 \\ -3 & -4 \end{bmatrix}$$

$$= \begin{bmatrix} 5 & -4 \\ 2 & -4 \end{bmatrix}$$

Graphing Utility Solution

```
[A]
                [[3  -2]
                 [5  0 ]]
[B]
                [[-2  2]
                 [3   4]]
[A]-[B]
                [[5  -4]
                 [2  -4]]
```

FIGURE 3

MATCHED 2 PROBLEM

Subtract: $[2 \quad -3 \quad 5] - [3 \quad -2 \quad 1]$

Multiplication of a Matrix by a Number

The **product of a number k and a matrix M,** denoted by **kM,** is a matrix formed by multiplying each element of M by k.

EXAMPLE 3 Multiplication of a Matrix by a Number

Multiply: $-2\begin{bmatrix} 3 & -1 & 0 \\ -2 & 1 & 3 \\ 0 & -1 & -2 \end{bmatrix}$

SOLUTION

Algebraic Solution

$$-2\begin{bmatrix} 3 & -1 & 0 \\ -2 & 1 & 3 \\ 0 & -1 & -2 \end{bmatrix} = \begin{bmatrix} -2(3) & -2(-1) & -2(0) \\ -2(-2) & -2(1) & -2(3) \\ -2(0) & -2(-1) & -2(-2) \end{bmatrix}$$

$$= \begin{bmatrix} -6 & 2 & 0 \\ 4 & -2 & -6 \\ 0 & 2 & 4 \end{bmatrix}$$

Graphing Utility Solution

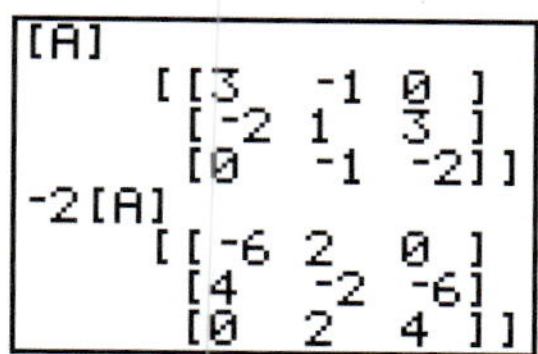

FIGURE 4

MATCHED 3 PROBLEM

Find: $10\begin{bmatrix} 1.3 \\ 0.2 \\ 3.5 \end{bmatrix}$

EXPLORE/DISCUSS 1

Multiplication of two numbers can be interpreted as repeated addition if one of the numbers is a positive integer. That is,

$$2a = a + a \qquad 3a = a + a + a \qquad 4a = a + a + a + a$$

and so on. Discuss this interpretation for the product of an integer k and a matrix M. Use specific examples to illustrate your remarks.

We now consider an application that uses various matrix operations.

EXAMPLE 4 Sales and Commissions

Ms. Fong and Mr. Petris are salespeople for a new car agency that sells only two models. August was the last month for this year's models, and next year's models were introduced in September. Gross dollar sales for each month are given in the following matrices:

$$\begin{array}{c} \\ \text{Fong} \\ \text{Petris} \end{array} \begin{array}{c} \text{AUGUST SALES} \\ \begin{array}{cc} \text{Compact} & \text{Luxury} \end{array} \\ \begin{bmatrix} \$36{,}000 & \$72{,}000 \\ \$72{,}000 & \$0 \end{bmatrix} = A \end{array} \qquad \begin{array}{c} \text{SEPTEMBER SALES} \\ \begin{array}{cc} \text{Compact} & \text{Luxury} \end{array} \\ \begin{bmatrix} \$144{,}000 & \$288{,}000 \\ \$180{,}000 & \$216{,}000 \end{bmatrix} = B \end{array}$$

For example, Ms. Fong had $36,000 in compact sales in August and Mr. Petris had $216,000 in luxury car sales in September.

(A) What are the combined dollar sales in August and September for each salesperson and each model?

(B) What was the increase in dollar sales from August to September?

(C) If both salespeople receive a 3% commission on gross dollar sales, compute the commission for each salesperson for each model sold in September.

SOLUTIONS

(A) Algebraic Solution

$$A + B = \begin{bmatrix} 36{,}000 & 72{,}000 \\ 72{,}000 & 0 \end{bmatrix} + \begin{bmatrix} 144{,}000 & 288{,}000 \\ 180{,}000 & 216{,}000 \end{bmatrix}$$

$$= \begin{array}{c} \begin{array}{cc} \text{Compact} & \text{Luxury} \end{array} \\ \begin{bmatrix} \$180{,}000 & \$360{,}000 \\ \$252{,}000 & \$216{,}000 \end{bmatrix} \end{array} \begin{array}{c} \\ \text{Fong} \\ \text{Petris} \end{array}$$

(A) Graphing Utility Solution

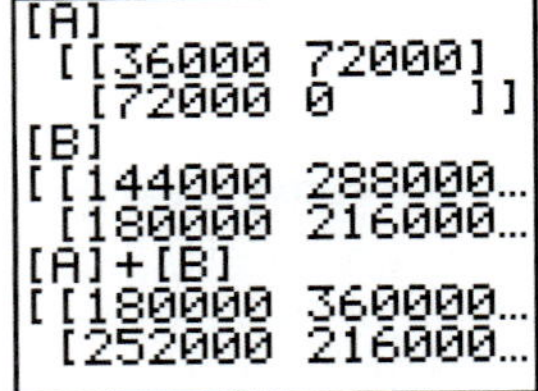

FIGURE 5

(B) Algebraic Solution

$$B - A = \begin{bmatrix} 144{,}000 & 288{,}000 \\ 180{,}000 & 216{,}000 \end{bmatrix} - \begin{bmatrix} 36{,}000 & 72{,}000 \\ 72{,}000 & 0 \end{bmatrix}$$

$$= \begin{array}{c} \begin{array}{cc} \text{Compact} & \text{Luxury} \end{array} \\ \begin{bmatrix} \$108{,}000 & \$216{,}000 \\ \$108{,}000 & \$216{,}000 \end{bmatrix} \end{array} \begin{array}{c} \\ \text{Fong} \\ \text{Petris} \end{array}$$

(B) Graphing Utility Solution

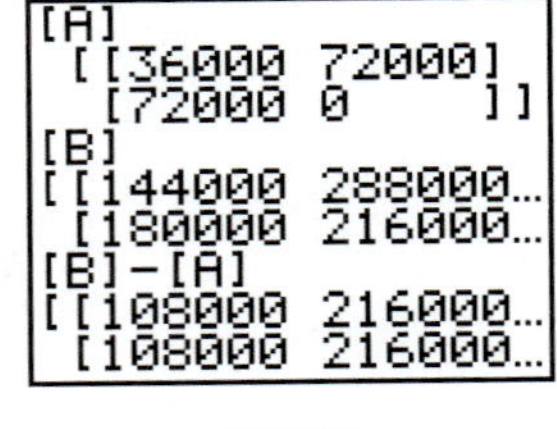

FIGURE 6

(C) Algebraic Solution

$$0.03B = \begin{bmatrix} (0.03)(144{,}000) & (0.03)(288{,}000) \\ (0.03)(180{,}000) & (0.03)(216{,}000) \end{bmatrix}$$

$$\begin{matrix} \text{Compact} & \text{Luxury} & \\ \end{matrix}$$

$$= \begin{bmatrix} \$4{,}320 & \$8{,}640 \\ \$5{,}400 & \$6{,}480 \end{bmatrix} \begin{matrix} \text{Fong} \\ \text{Petris} \end{matrix}$$

(C) Graphing Utility Solution

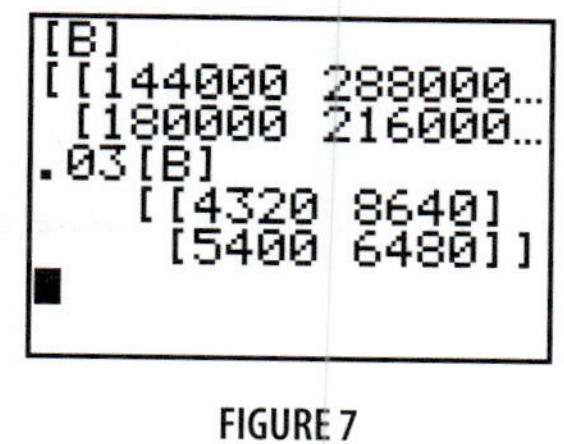

FIGURE 7

MATCHED 4 PROBLEM

Repeat Example 4 with

$$A = \begin{bmatrix} \$72{,}000 & \$72{,}000 \\ \$36{,}000 & \$72{,}000 \end{bmatrix} \quad \text{and} \quad B = \begin{bmatrix} \$180{,}000 & \$216{,}000 \\ \$144{,}000 & \$216{,}000 \end{bmatrix}$$

Example 4 involved an agency with only two salespeople and two models. A more realistic problem might involve 20 salespeople and 15 models. Problems of this size are often solved with the aid of a spreadsheet on a personal computer. Figure 8 illustrates a computer spreadsheet solution for Example 4.

FIGURE 8

	A	B	C	D	E	F	G
1		Compact	Luxury	Compact	Luxury	Compact	Luxury
2		August Sales		September Sales		September Commissions	
3	Fong	$36,000	$72,000	$144,000	$288,000	$4,320	$8,640
4	Petris	$72,000	$0	$180,000	$216,000	$5,400	$6,480
5		Combined Sales		Sales Increases			
6	Fong	$180,000	$360,000	$108,000	$216,000		
7	Petris	$252,000	$216,000	$108,000	$216,000		

Matrix Product

Now we are going to introduce a matrix multiplication that may at first seem rather strange. In spite of its apparent strangeness, this operation is well-founded in the general theory of matrices and, as we will see, is extremely useful in many practical problems.

Historically, matrix multiplication was introduced by the English mathematician Arthur Cayley (1821–1895) in studies of linear equations and linear transformations. In Section 9.3 you will see how matrix multiplication is central to the process of expressing systems of equations as matrix equations and to the process of solving matrix equations. Matrix equations and their solutions provide us with an alternate method of solving linear systems with the same number of variables as equations.

We start by defining the product of two special matrices, a row matrix and a column matrix.

DEFINITION 1

Product of a Row Matrix and a Column Matrix

The **product** of a $1 \times n$ row matrix and an $n \times 1$ column matrix is a 1×1 matrix given by

$$\overset{1 \times n}{[a_1 \; a_2 \; \cdots \; a_n]} \overset{n \times 1}{\begin{bmatrix} b_1 \\ b_2 \\ \vdots \\ b_n \end{bmatrix}} = [a_1b_1 + a_2b_2 + \cdots + a_nb_n]$$

Note that the number of elements in the row matrix and in the column matrix must be the same for the product to be defined.

EXAMPLE 5 Product of a Row Matrix and a Column Matrix

Multiply: $[2 \quad -3 \quad 0]\begin{bmatrix} -5 \\ 2 \\ -2 \end{bmatrix}$

SOLUTION

Algebraic Solution

$$[2 \quad -3 \quad 0]\begin{bmatrix} -5 \\ 2 \\ -2 \end{bmatrix} = [(2)(-5) + (-3)(2) + (0)(-2)]$$

$$= [-10 + (-6) + 0] = [-16]$$

Graphing Utility Solution

```
[A]
        [[2  -3  0]]
[B]
             [[-5]
              [2 ]
              [-2]]
[A][B]
             [[-16]]
```

FIGURE 9

MATCHED PROBLEM 5

$$[-1 \quad 0 \quad 3 \quad 2]\begin{bmatrix} 2 \\ 3 \\ 4 \\ -1 \end{bmatrix} = ?$$

Refer to Example 5. The distinction between the real number -16 and the 1×1 matrix $[-16]$ is a technical one, and it is common to see 1×1 matrices written as real numbers without brackets. In the work that follows, we will frequently refer to 1×1 matrices as real numbers and omit the brackets whenever it is convenient to do so.

EXAMPLE 6 Production Scheduling

A factory produces a slalom water ski that requires 4 labor-hours in the fabricating department and 1 labor-hour in the finishing department. Fabricating personnel receive \$10 per hour, and finishing personnel receive \$8 per hour. Total labor cost per ski is given by the product

$$[4 \quad 1]\begin{bmatrix} 10 \\ 8 \end{bmatrix} = [(4)(10) + (1)(8)] = [40 + 8] = [48] \text{ or } \$48 \text{ per ski}$$

MATCHED 6 PROBLEM

If the factory in Example 6 also produces a trick water ski that requires 6 labor-hours in the fabricating department and 1.5 labor-hours in the finishing department, write a product between appropriate row and column matrices that gives the total labor cost for this ski. Compute the cost.

We now use the product of a $1 \times n$ row matrix and an $n \times 1$ column matrix to extend the definition of matrix product to more general matrices.

DEFINITION 2
Matrix Product

If A is an $m \times p$ matrix and B is a $p \times n$ matrix, then the **matrix product** of A and B, denoted AB, is an $m \times n$ matrix whose element in the ith row and jth column is the real number obtained from the product of the ith row of A and the jth column of B. If the number of columns in A does not equal the number of rows in B, then the matrix product AB is **not defined.**

It is important to check sizes before starting the multiplication process. If A is an $a \times b$ matrix and B is a $c \times d$ matrix, then if $b = c$, the product AB will exist and will be an $a \times d$ matrix (Fig. 10). If $b \neq c$, then the product AB does not exist.

FIGURE 10

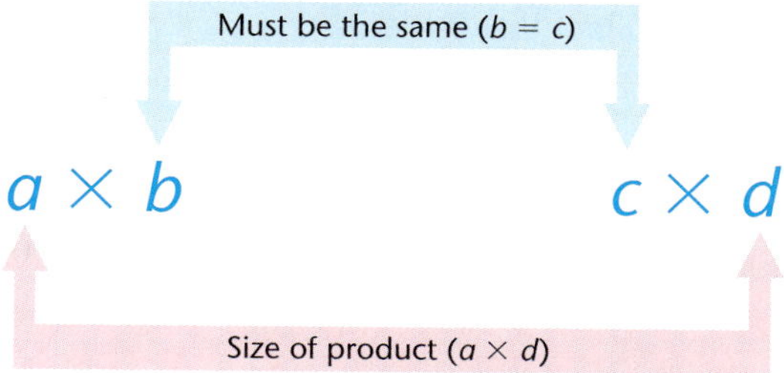

The definition is not as complicated as it might first seem. An example should help clarify the process. For

$$A = \begin{bmatrix} 2 & 3 & -1 \\ -2 & 1 & 2 \end{bmatrix} \quad \text{and} \quad B = \begin{bmatrix} 1 & 3 \\ 2 & 0 \\ -1 & 2 \end{bmatrix}$$

A is 2×3, B is 3×2, and so AB is 2×2. To find the first row of AB, we take the product of the first row of A with every column of B and write each result as a real number, not a 1×1 matrix. The second row of AB is computed in the same manner. The four products of row and column matrices used to produce the four elements in AB are shown in the dashed box. These products are usually calculated mentally, or with the aid of a calculator, and need not be written out. The shaded portions highlight the steps involved in computing the element in the first row and second column of AB.

$$\underset{2 \times 3}{\begin{bmatrix} 2 & 3 & -1 \\ -2 & 1 & 2 \end{bmatrix}} \underset{3 \times 2}{\begin{bmatrix} 1 & 3 \\ 2 & 0 \\ -1 & 2 \end{bmatrix}} = \begin{bmatrix} [2 \quad 3 \quad -1]\begin{bmatrix} 1 \\ 2 \\ -1 \end{bmatrix} & [2 \quad 3 \quad -1]\begin{bmatrix} 3 \\ 0 \\ 2 \end{bmatrix} \\ [-2 \quad 1 \quad 2]\begin{bmatrix} 1 \\ 2 \\ -1 \end{bmatrix} & [-2 \quad 1 \quad 2]\begin{bmatrix} 3 \\ 0 \\ 2 \end{bmatrix} \end{bmatrix}$$

$$= \underset{2 \times 2}{\begin{bmatrix} 9 & 4 \\ -2 & -2 \end{bmatrix}}$$

EXAMPLE 7 Matrix Product

Given

$$A = \begin{bmatrix} 2 & 1 \\ 1 & 0 \\ -1 & 2 \end{bmatrix} \qquad B = \begin{bmatrix} 1 & -1 & 0 & 1 \\ 2 & 1 & 2 & 0 \end{bmatrix}$$

$$C = \begin{bmatrix} 2 & 6 \\ -1 & -3 \end{bmatrix} \qquad D = \begin{bmatrix} 1 & 2 \\ 3 & 6 \end{bmatrix}$$

Find each product that is defined:

(A) AB (B) BA (C) CD (D) DC

SOLUTIONS

(A) Algebraic Solution

$$AB = \underset{3 \times 2}{\begin{bmatrix} 2 & 1 \\ 1 & 0 \\ -1 & 2 \end{bmatrix}} \underset{2 \times 4}{\begin{bmatrix} 1 & -1 & 0 & 1 \\ 2 & 1 & 2 & 0 \end{bmatrix}}$$

$$= \begin{bmatrix} 4 & -1 & 2 & 2 \\ 1 & -1 & 0 & 1 \\ 3 & 3 & 4 & -1 \end{bmatrix}$$

(A) Graphing Utility Solution

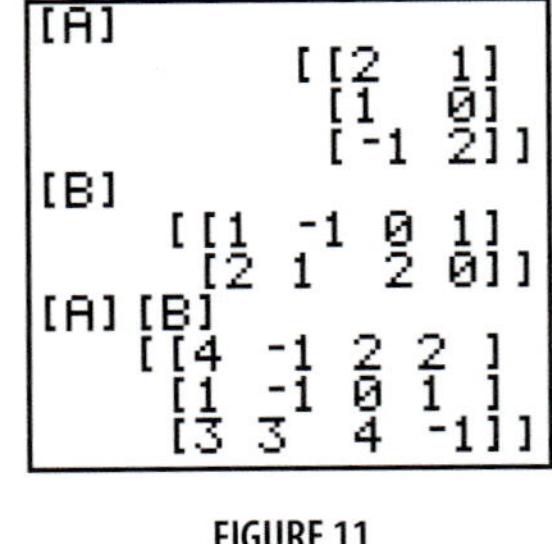

FIGURE 11

(B) Algebraic Solution

$$BA = \overset{2 \times 4}{\begin{bmatrix} 1 & -1 & 0 & 1 \\ 2 & 1 & 2 & 0 \end{bmatrix}} \overset{3 \times 2}{\begin{bmatrix} 2 & 1 \\ 1 & 0 \\ -1 & 2 \end{bmatrix}}$$

Because B has four columns and A has three rows, the product BA is not defined.

(B) Graphing Utility Solution

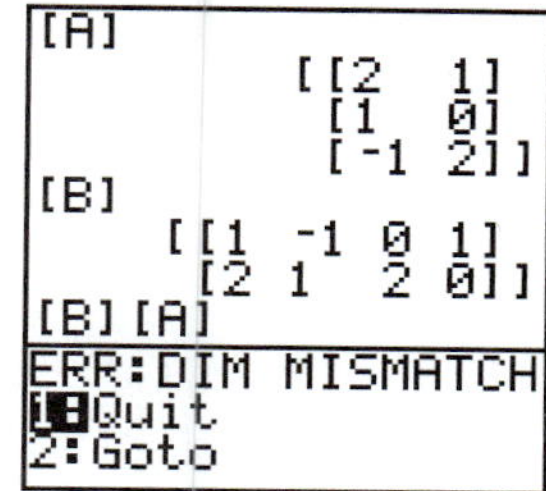

FIGURE 12

(C) Algebraic Solution

$$CD = \begin{bmatrix} 2 & 6 \\ -1 & -3 \end{bmatrix} \begin{bmatrix} 1 & 2 \\ 3 & 6 \end{bmatrix}$$

$$= \begin{bmatrix} 20 & 40 \\ -10 & -20 \end{bmatrix}$$

(C) Graphing Utility Solution

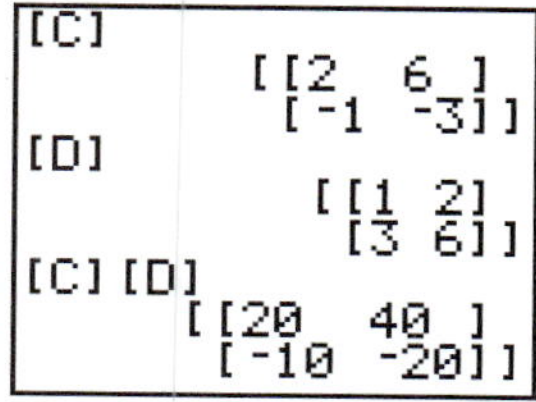

FIGURE 13

(D) Algebraic Solution

$$DC = \begin{bmatrix} 1 & 2 \\ 3 & 6 \end{bmatrix} \begin{bmatrix} 2 & 6 \\ -1 & -3 \end{bmatrix}$$

$$= \begin{bmatrix} 0 & 0 \\ 0 & 0 \end{bmatrix}$$

(D) Graphing Utility Solution

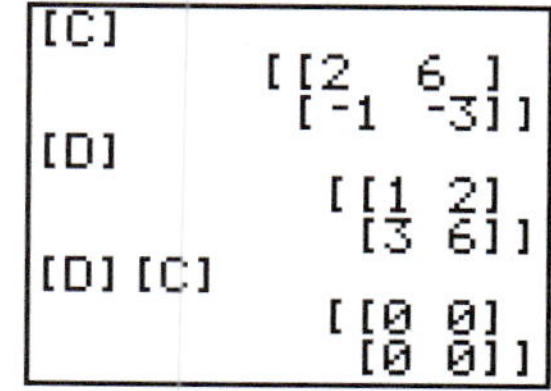

FIGURE 14

MATCHED 7 PROBLEM

Find each product, if it is defined:

(A) $\begin{bmatrix} -1 & 0 & 3 & -2 \\ 1 & 2 & 2 & 0 \end{bmatrix} \begin{bmatrix} -1 & 1 \\ 2 & 3 \\ 1 & 0 \end{bmatrix}$ (B) $\begin{bmatrix} -1 & 1 \\ 2 & 3 \\ 1 & 0 \end{bmatrix} \begin{bmatrix} -1 & 0 & 3 & -2 \\ 1 & 2 & 2 & 0 \end{bmatrix}$

(C) $\begin{bmatrix} 1 & 2 \\ -1 & -2 \end{bmatrix} \begin{bmatrix} -2 & 4 \\ 1 & -2 \end{bmatrix}$ (D) $\begin{bmatrix} -2 & 4 \\ 1 & -2 \end{bmatrix} \begin{bmatrix} 1 & 2 \\ -1 & -2 \end{bmatrix}$

(E) $\begin{bmatrix} 3 & -2 & 1 \end{bmatrix} \begin{bmatrix} 4 \\ 2 \\ 3 \end{bmatrix}$ (F) $\begin{bmatrix} 4 \\ 2 \\ 3 \end{bmatrix} \begin{bmatrix} 3 & -2 & 1 \end{bmatrix}$

In the arithmetic of real numbers it does not matter in which order we multiply; for example, $5 \times 7 = 7 \times 5$. In matrix multiplication, however, it does make a difference. That is, AB does not always equal BA, even if both multiplications are defined and both products are the same size (see Example 7, parts C and D).

Thus,

Matrix multiplication is not commutative.

Also, AB may be zero with neither A nor B equal to zero (see Example 7, part D). Thus,

The zero property does not hold for matrix multiplication.

(The zero property, namely, $ab = 0$ if and only if $a = 0$ or $b = 0$, does hold for all real numbers a and b.)

Just as we used the familiar algebraic notation AB to represent the product of matrices A and B, we use the notation A^2 for AA, the product of A with itself, A^3 for AAA, and so on.

EXPLORE/DISCUSS 2

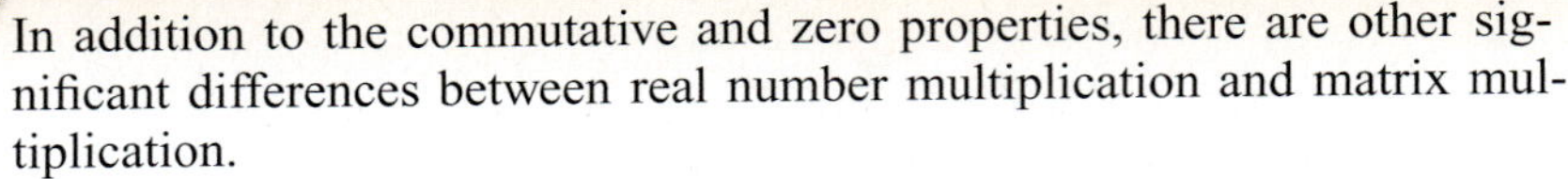

In addition to the commutative and zero properties, there are other significant differences between real number multiplication and matrix multiplication.

(A) In real number multiplication, the only real number whose square is 0 is the real number 0 ($0^2 = 0$). Find at least one 2×2 matrix A with all elements nonzero such that $A^2 = 0$, where 0 is the 2×2 zero matrix.

(B) In real number multiplication, the only nonzero real number that is equal to its square is the real number 1 ($1^2 = 1$). Find at least one 2×2 matrix A with all elements nonzero such that $A^2 = A$.

We will continue our discussion of properties of matrix multiplication later in Chapter 9. Now we consider an application of matrix multiplication.

EXAMPLE 8 Labor Costs

Let us combine the time requirements for slalom and trick water skis discussed in Example 6 and Matched Problem 6 into one matrix:

Labor-hours per ski

$$\begin{array}{l} \\ \text{Trick ski} \\ \text{Slalom ski} \end{array} \begin{array}{c} \begin{array}{cc} \text{Assembly department} & \text{Finishing department} \end{array} \\ \begin{bmatrix} 6\text{ h} & 1.5\text{ h} \\ 4\text{ h} & 1\text{ h} \end{bmatrix} \end{array} = L$$

Now suppose that the company has two manufacturing plants, X and Y, in different parts of the country and that the hourly wages for each department are given in the following matrix:

$$\begin{array}{l} \\ \\ \text{Assembly department} \\ \text{Finishing department} \end{array} \begin{array}{c} \text{Hourly wages} \\ \begin{array}{cc} \text{Plant} & \text{Plant} \\ X & Y \end{array} \\ \begin{bmatrix} \$10 & \$12 \\ \$\ 8 & \$10 \end{bmatrix} = H \end{array}$$

Because H and L are both 2×2 matrices, we can take the product of H and L in either order and the result will be a 2×2 matrix:

$$HL = \begin{bmatrix} 10 & 12 \\ 8 & 10 \end{bmatrix} \begin{bmatrix} 6 & 1.5 \\ 4 & 1 \end{bmatrix} = \begin{bmatrix} 108 & 27 \\ 88 & 22 \end{bmatrix}$$

$$LH = \begin{bmatrix} 6 & 1.5 \\ 4 & 1 \end{bmatrix} \begin{bmatrix} 10 & 12 \\ 8 & 10 \end{bmatrix} = \begin{bmatrix} 72 & 87 \\ 48 & 58 \end{bmatrix}$$

How can we interpret the elements in these products? Let's begin with the product HL. The element 108 in the first row and first column of HL is the product of the first row matrix of H and the first column matrix of L:

$$\begin{array}{cc} \text{Plant} & \text{Plant} \\ X & Y \end{array}$$
$$[10 \quad 12] \begin{bmatrix} 6 \\ 4 \end{bmatrix} \begin{array}{l} \text{Trick} \\ \text{Slalom} \end{array} = 10(6) + 12(4) = 60 + 48 = 108$$

Notice that \$60 is the labor cost for assembling a trick ski at plant X and \$48 is the labor cost for assembling a slalom ski at plant Y. Although both numbers represent labor costs, it makes no sense to add them together. They do not pertain to the same type of ski or to the same plant. Thus, although the product HL happens to be defined mathematically, it has no useful interpretation in this problem.

Now let's consider the product LH. The element 72 in the first row and first column of LH is given by the following product:

$$\begin{array}{cc} \text{Assembly} & \text{Finishing} \end{array}$$
$$[6 \quad 1.5] \begin{bmatrix} 10 \\ 8 \end{bmatrix} \begin{array}{l} \text{Assembly} \\ \text{Finishing} \end{array} = 6(10) + 1.5(8)$$
$$= 60 + 12 = 72$$

where \$60 is the labor cost for assembling a trick ski at plant X and \$12 is the labor cost for finishing a trick ski at plant X. Thus, the sum is the total labor cost for producing a trick ski at plant X. The other elements in LH also represent total labor costs, as indicated by the row and column labels shown below:

$$\begin{array}{c} \text{Labor costs per ski} \\ \begin{array}{cc} \text{Plant} & \text{Plant} \\ X & Y \end{array} \end{array}$$
$$LH = \begin{bmatrix} \$72 & \$87 \\ \$48 & \$58 \end{bmatrix} \begin{array}{l} \text{Trick ski} \\ \text{Slalom ski} \end{array}$$

MATCHED 8 PROBLEM

Refer to Example 8. The company wants to know how many hours to schedule in each department to produce 1,000 trick skis and 2,000 slalom skis. These production requirements can be represented by either of the following matrices:

$$P = \overset{\text{Trick skis} \quad \text{Slalom skis}}{[1{,}000 \quad 2{,}000]} \qquad Q = \begin{bmatrix} 1{,}000 \\ 2{,}000 \end{bmatrix} \begin{matrix} \text{Trick skis} \\ \text{Slalom skis} \end{matrix}$$

Using the labor-hour matrix L from Example 8, find PL or LQ, whichever has a meaningful interpretation for this problem, and label the rows and columns accordingly.

Figure 15 shows a solution to Example 8 on a spreadsheet.

FIGURE 15 Matrix multiplication in a spreadsheet.

	A	B	C	D	E	F
1		Labor-hours per ski			Hourly wages	
2		Fabricating	Finishing		Plant X	Plant Y
3	Trick ski	6	1.5	Fabricating	$10	$12
4	Slalom ski	4	1	Finishing	$8	$10
5		Labor costs per ski				
6		Plant X	Plant Y			
7	Trick ski	$72	$87			
8	Slalom ski	$48	$58			

CAUTION

Example 8 and Matched Problem 8 illustrate an important point about matrix multiplication. Even if you are using a graphing utility to perform the calculations in a matrix product, it is still necessary for you to know the definition of matrix multiplication so that you can interpret the results correctly.

ANSWERS MATCHED PROBLEMS

1. $\begin{bmatrix} 1 & 5 \\ 0 & -2 \\ 2 & 1 \end{bmatrix}$ **2.** $[-1 \quad -1 \quad 4]$ **3.** $\begin{bmatrix} 13 \\ 2 \\ 35 \end{bmatrix}$

4. (A) $\begin{bmatrix} \$252{,}000 & \$288{,}000 \\ \$180{,}000 & \$288{,}000 \end{bmatrix}$ (B) $\begin{bmatrix} \$108{,}000 & \$144{,}000 \\ \$108{,}000 & \$144{,}000 \end{bmatrix}$ (C) $\begin{bmatrix} \$5{,}400 & \$6{,}480 \\ \$4{,}320 & \$6{,}480 \end{bmatrix}$

5. $[8]$ **6.** $[6 \quad 1.5]\begin{bmatrix} 10 \\ 8 \end{bmatrix} = [72]$ or \$72

7. (A) Not defined (B) $\begin{bmatrix} 2 & 2 & -1 & 2 \\ 1 & 6 & 12 & -4 \\ -1 & 0 & 3 & -2 \end{bmatrix}$ (C) $\begin{bmatrix} 0 & 0 \\ 0 & 0 \end{bmatrix}$ (D) $\begin{bmatrix} -6 & -12 \\ 3 & 6 \end{bmatrix}$ (E) $[11]$ (F) $\begin{bmatrix} 12 & -8 & 4 \\ 6 & -4 & 2 \\ 9 & -6 & 3 \end{bmatrix}$

8. $PL = \overset{\text{Assembly} \quad \text{Finishing}}{[14{,}000 \quad 3{,}500]}$ Labor hours

EXERCISE 9.1

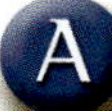

Perform the indicated operations in Problems 1–18, if possible.

1. $\begin{bmatrix} -1 & 4 \\ 2 & -6 \end{bmatrix} + \begin{bmatrix} 1 & -2 \\ 0 & 5 \end{bmatrix}$

2. $\begin{bmatrix} 2 & -1 \\ 3 & 0 \end{bmatrix} + \begin{bmatrix} -3 & 1 \\ 2 & -3 \end{bmatrix}$

3. $\begin{bmatrix} -3 & 5 \\ 2 & 0 \\ 1 & 4 \end{bmatrix} + \begin{bmatrix} 2 & 1 \\ -6 & 3 \\ 0 & -5 \end{bmatrix}$

4. $\begin{bmatrix} 4 & -1 & 0 \\ 2 & 1 & 3 \end{bmatrix} + \begin{bmatrix} -2 & 1 & 3 \\ 5 & 6 & -8 \end{bmatrix}$

5. $\begin{bmatrix} -3 & 5 \\ 2 & 0 \\ 1 & 4 \end{bmatrix} + \begin{bmatrix} -2 & 1 & 3 \\ 5 & 6 & -8 \end{bmatrix}$

6. $\begin{bmatrix} 4 & -1 & 0 \\ 2 & 1 & 3 \end{bmatrix} + \begin{bmatrix} 2 & 1 \\ -6 & 3 \\ 0 & -5 \end{bmatrix}$

7. $\begin{bmatrix} 6 & 2 & -3 \\ 0 & -4 & 5 \end{bmatrix} - \begin{bmatrix} 4 & -1 & 2 \\ -5 & 1 & -2 \end{bmatrix}$

8. $\begin{bmatrix} 4 & -5 \\ 1 & 0 \\ 1 & -3 \end{bmatrix} - \begin{bmatrix} -1 & 2 \\ 6 & -2 \\ 1 & -7 \end{bmatrix}$

9. $10\begin{bmatrix} 2 & -1 & 3 \\ 0 & -4 & 5 \end{bmatrix}$

10. $5\begin{bmatrix} 1 & -2 & 0 & 4 \\ -3 & 2 & -1 & 6 \end{bmatrix}$

11. $[2 \quad 4]\begin{bmatrix} 3 \\ 1 \end{bmatrix}$

12. $[1 \quad 5]\begin{bmatrix} 6 \\ 2 \end{bmatrix}$

13. $\begin{bmatrix} 3 & 4 \\ -1 & -2 \end{bmatrix}\begin{bmatrix} -1 \\ 2 \end{bmatrix}$

14. $\begin{bmatrix} -1 & 1 \\ 2 & -3 \end{bmatrix}\begin{bmatrix} 4 \\ -2 \end{bmatrix}$

15. $\begin{bmatrix} 2 & -3 \\ 1 & 2 \end{bmatrix}\begin{bmatrix} 1 & -1 \\ 0 & -2 \end{bmatrix}$

16. $\begin{bmatrix} -3 & 2 \\ 4 & -1 \end{bmatrix}\begin{bmatrix} -2 & 5 \\ -1 & 3 \end{bmatrix}$

17. $\begin{bmatrix} 1 & -1 \\ 0 & -2 \end{bmatrix}\begin{bmatrix} 2 & -3 \\ 1 & 2 \end{bmatrix}$

18. $\begin{bmatrix} -2 & 5 \\ -1 & 3 \end{bmatrix}\begin{bmatrix} -3 & 2 \\ 4 & -1 \end{bmatrix}$

B

Find the products in Problems 19–26.

19. $[4 \quad -2]\begin{bmatrix} -5 \\ -3 \end{bmatrix}$

20. $[2 \quad -1]\begin{bmatrix} 3 \\ -4 \end{bmatrix}$

21. $\begin{bmatrix} -5 \\ -3 \end{bmatrix}[4 \quad -2]$

22. $\begin{bmatrix} 3 \\ -4 \end{bmatrix}[2 \quad -1]$

23. $[3 \quad -2 \quad -4]\begin{bmatrix} 1 \\ 2 \\ -3 \end{bmatrix}$

24. $[1 \quad -2 \quad 2]\begin{bmatrix} 2 \\ -1 \\ 1 \end{bmatrix}$

25. $\begin{bmatrix} 1 \\ 2 \\ -3 \end{bmatrix}[3 \quad -2 \quad -4]$

26. $\begin{bmatrix} 2 \\ -1 \\ 1 \end{bmatrix}[1 \quad -2 \quad 2]$

Problems 27–44 refer to the following matrices.

$$A = \begin{bmatrix} 2 & -1 & 3 \\ 0 & 4 & -2 \end{bmatrix} \qquad B = \begin{bmatrix} -3 & 1 \\ 2 & 5 \end{bmatrix}$$

$$C = \begin{bmatrix} -1 & 0 & 2 \\ 4 & -3 & 1 \\ -2 & 3 & 5 \end{bmatrix} \qquad D = \begin{bmatrix} 3 & -2 \\ 0 & -1 \\ 1 & 2 \end{bmatrix}$$

Perform the indicated operations, if possible.

27. CA **28.** AC **29.** BA

30. AB **31.** C^2 **32.** B^2

33. $C + DA$ **34.** $B + AD$ **35.** $0.2CD$

36. $0.1DB$ **37.** $2DB + 5CD$ **38.** $3BA + 4AC$

39. $(-1)AC + 3DB$ **40.** $(-2)BA + 6CD$

41. CDA **42.** ACD

43. DBA **44.** BAD

In Problems 45 and 46, calculate $B, B^2, B^3, \ldots,$ and $AB, AB^2, AB^3, \ldots.$ Describe any patterns you observe in each sequence of matrices.

45. $A = [0.3 \quad 0.7]$ and $B = \begin{bmatrix} 0.4 & 0.6 \\ 0.2 & 0.8 \end{bmatrix}$

46. $A = [0.4 \quad 0.6]$ and $B = \begin{bmatrix} 0.9 & 0.1 \\ 0.3 & 0.7 \end{bmatrix}$

47. Find a, b, c, and d so that

$$\begin{bmatrix} a & b \\ c & d \end{bmatrix} + \begin{bmatrix} 2 & -3 \\ 0 & 1 \end{bmatrix} = \begin{bmatrix} 1 & -2 \\ 3 & -4 \end{bmatrix}$$

48. Find w, x, y, and z so that

$$\begin{bmatrix} 4 & -2 \\ -3 & 0 \end{bmatrix} + \begin{bmatrix} w & x \\ y & z \end{bmatrix} = \begin{bmatrix} 2 & -3 \\ 0 & 5 \end{bmatrix}$$

49. Find x and y so that

$$\begin{bmatrix} 3x & 5 \\ -1 & 4x \end{bmatrix} + \begin{bmatrix} 2y & -3 \\ -6 & -y \end{bmatrix} = \begin{bmatrix} 7 & 2 \\ -7 & 2 \end{bmatrix}$$

50. Find x and y so that

$$\begin{bmatrix} 4 & 2x \\ -4x & -3 \end{bmatrix} + \begin{bmatrix} -5 & -3y \\ 5y & 3 \end{bmatrix} = \begin{bmatrix} -1 & 1 \\ 1 & 0 \end{bmatrix}$$

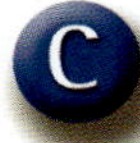

In Problems 51 and 52, let a, b, and c be any nonzero real numbers, let

$$A = \begin{bmatrix} a & b \\ c & -a \end{bmatrix} \quad \text{and} \quad I = \begin{bmatrix} 1 & 0 \\ 0 & 1 \end{bmatrix}$$

51. If $A^2 = 0$, how are a, b, and c related? Use this relationship to provide several examples of 2×2 matrices with no zero entries whose square is the zero matrix.

52. If $A^2 = I$, how are a, b, and c related? Use this relationship to provide several examples of 2×2 matrices with no zero entries whose square is the matrix I.

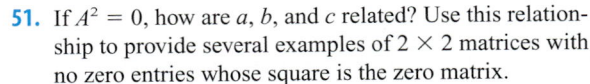

Problems 53 and 54 refer to the matrices

$$A = \begin{bmatrix} a & b \\ c & d \end{bmatrix} \quad \text{and} \quad B = \begin{bmatrix} 1 & 1 \\ 1 & 1 \end{bmatrix}$$

53. If $AB = 0$, how are a, b, c, and d related? Use this relationship to provide several examples of 2×2 matrices A with no zero entries that satisfy $AB = 0$.

54. If $BA = 0$, how are a, b, c, and d related? Use this relationship to provide several examples of 2×2 matrices A with no zero entries that satisfy $BA = 0$.

C

55. Find x and y so that

$$\begin{bmatrix} 1 & 3 \\ -2 & -2 \end{bmatrix} \begin{bmatrix} x & 1 \\ 3 & 2 \end{bmatrix} = \begin{bmatrix} y & 7 \\ y & -6 \end{bmatrix}$$

56. Find x and y so that

$$\begin{bmatrix} x & -1 \\ 1 & 0 \end{bmatrix} \begin{bmatrix} 2 & 1 \\ 4 & 1 \end{bmatrix} = \begin{bmatrix} y & y \\ 2 & 1 \end{bmatrix}$$

57. Find a, b, c, and d so that

$$\begin{bmatrix} 1 & 3 \\ 1 & 4 \end{bmatrix} \begin{bmatrix} a & b \\ c & d \end{bmatrix} = \begin{bmatrix} 6 & -5 \\ 7 & -7 \end{bmatrix}$$

58. Find a, b, c, and d so that

$$\begin{bmatrix} 1 & -2 \\ 2 & -3 \end{bmatrix} \begin{bmatrix} a & b \\ c & d \end{bmatrix} = \begin{bmatrix} 1 & 0 \\ 3 & 2 \end{bmatrix}$$

59. A square matrix is a **diagonal matrix** if all elements not on the principal diagonal are zero. Thus, a 2×2 diagonal matrix has the form

$$A = \begin{bmatrix} a & 0 \\ 0 & d \end{bmatrix}$$

where a and d are any real numbers. Discuss the validity of each of the following statements. If the statement is always true, explain why. If not, give examples.

(A) If A and B are 2×2 diagonal matrices, then $A + B$ is a 2×2 diagonal matrix.

(B) If A and B are 2×2 diagonal matrices, then $A + B = B + A$.

(C) If A and B are 2×2 diagonal matrices, then AB is a 2×2 diagonal matrix.

(D) If A and B are 2×2 diagonal matrices, then $AB = BA$.

60. A square matrix is an **upper triangular matrix** if all elements below the principal diagonal are zero. Thus, a 2×2 upper triangular matrix has the form

$$A = \begin{bmatrix} a & b \\ 0 & d \end{bmatrix}$$

where a, b, and d are any real numbers. Discuss the validity of each of the following statements. If the statement is always true, explain why. If not, give examples.

(A) If A and B are 2×2 upper triangular matrices, then $A + B$ is a 2×2 upper triangular matrix.

(B) If A and B are 2×2 upper triangular matrices, then $A + B = B + A$.

(C) If A and B are 2×2 upper triangular matrices, then AB is a 2×2 upper triangular matrix.

(D) If A and B are 2×2 upper triangular matrices, then $AB = BA$.

APPLICATIONS

61. Cost Analysis. A company with two different plants manufactures guitars and banjos. Its production costs for each instrument are given in the following matrices:

Plant X (Guitar, Banjo); Plant Y (Guitar, Banjo); rows: Materials, Labor

$$\begin{bmatrix} \$30 & \$25 \\ \$60 & \$80 \end{bmatrix} = A \qquad \begin{bmatrix} \$36 & \$27 \\ \$54 & \$74 \end{bmatrix} = B$$

Find $\frac{1}{2}(A + B)$, the average cost of production for the two plants.

62. Cost Analysis. If both labor and materials at plant X in Problem 61 are increased 20%, find $\frac{1}{2}(1.2A + B)$, the new average cost of production for the two plants.

63. Markup. An import car dealer sells three models of a car. Current dealer invoice price (cost) and the retail price for the basic models and the indicated options are given in the following two matrices (where "Air" means air conditioning):

Dealer invoice price

	Basic car	Air	AM/FM radio	Cruise control
Model A	\$10,400	\$682	\$215	\$182
Model B	\$12,500	\$721	\$295	\$182
Model C	\$16,400	\$827	\$443	\$192

$= M$

Retail price

	Basic car	Air	AM/FM radio	Cruise control
Model A	\$13,900	\$783	\$263	\$215
Model B	\$15,000	\$838	\$395	\$236
Model C	\$18,300	\$967	\$573	\$248

$= N$

We define the markup matrix to be $N - M$ (**markup** is the difference between the retail price and the dealer invoice price). Suppose the value of the dollar has had a sharp decline and the dealer invoice price is to have an across-the-board 15% increase next year. To stay competitive with domestic cars, the dealer increases the retail prices only 10%. Calculate a markup matrix for next year's models and the indicated options. (Compute results to the nearest dollar.)

64. Markup. Referring to Problem 63, what is the markup matrix resulting from a 20% increase in dealer invoice prices and an increase in retail prices of 15%? (Compute results to the nearest dollar.)

65. Labor Costs. A company with manufacturing plants located in different parts of the country has labor-hour and wage requirements for the manufacturing of three types of inflatable boats as given in the following two matrices:

Labor-hours per boat

	Cutting department	Assembly department	Packaging department
One-person boat	0.6 h	0.6 h	0.2 h
Two-person boat	1.0 h	0.9 h	0.3 h
Four-person boat	1.5 h	1.2 h	0.4 h

$= M$

Hourly wages

	Plant I	Plant II
Cutting department	\$8	\$9
Assembly department	\$10	\$12
Packaging department	\$5	\$6

$= N$

(A) Find the labor costs for a one-person boat manufactured at plant I.

(B) Find the labor costs for a four-person boat manufactured at plant II.

(C) Discuss possible interpretations of the elements in the matrix products MN and NM.

(D) If either of the products MN or NM has a meaningful interpretation, find the product and label its rows and columns.

66. Inventory Value. A personal computer retail company sells five different computer models through three stores located in a large metropolitan area. The inventory of each model on hand in each store is summarized in matrix M. Wholesale (W) and retail (R) values of each model computer are summarized in matrix N.

Model

	A	B	C	D	E
Store 1	4	2	3	7	1
Store 2	2	3	5	0	6
Store 3	10	4	3	4	3

$= M$

	W	R
A	\$700	\$840
B	\$1,400	\$1,800
C	\$1,800	\$2,400
D	\$2,700	\$3,300
E	\$3,500	\$4,900

$= N$

(A) What is the retail value of the inventory at store 2?

(B) What is the wholesale value of the inventory at store 3?

(C) Discuss possible interpretations of the elements in the matrix products MN and NM.

(D) If either of the products MN or NM has a meaningful interpretation, find the product and label its rows and columns.

(E) Discuss methods of matrix multiplication that can be used to find the total inventory of each model on hand at all three stores. State the matrices that can be used, and perform the necessary operations.

(F) Discuss methods of matrix multiplication that can be used to find the total inventory of all five models at each store. State the matrices that can be used, and perform the necessary operations.

67. Airfreight. A nationwide airfreight service has connecting flights between five cities, as illustrated in the figure. To represent this schedule in matrix form, we construct a 5×5 **incidence matrix** A, where the rows represent the origins of each flight and the columns represent the destinations. We place a 1 in the ith row and jth column of this matrix if there is a connecting flight from the ith city to the jth city and a 0 otherwise. We also place 0s on the principal diagonal, because a connecting flight with the same origin and destination does not make sense.

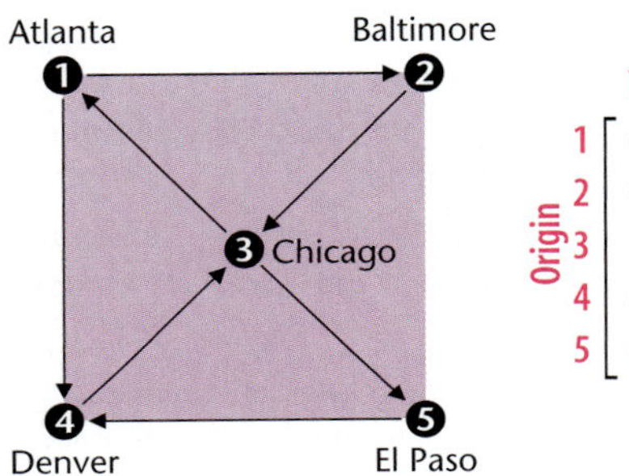

Destination

$$\text{Origin}\ \begin{matrix} & 1 & 2 & 3 & 4 & 5 \\ 1 & 0 & 1 & 0 & 1 & 0 \\ 2 & 0 & 0 & 1 & 0 & 0 \\ 3 & 1 & 0 & 0 & 0 & 1 \\ 4 & 0 & 0 & 1 & 0 & 0 \\ 5 & 0 & 0 & 0 & 1 & 0 \end{matrix} = A$$

Now that the schedule has been represented in the mathematical form of a matrix, we can perform operations on this matrix to obtain information about the schedule.

(A) Find A^2. What does the 1 in row 2 and column 1 of A^2 indicate about the schedule? What does the 2 in row 1 and column 3 indicate about the schedule? In general, how would you interpret each element off the principal diagonal of A^2? [*Hint:* Examine the diagram for possible connections between the ith city and the jth city.]

(B) Find A^3. What does the 1 in row 4 and column 2 of A^3 indicate about the schedule? What does the 2 in row 1 and column 5 indicate about the schedule? In general, how would you interpret each element off the principal diagonal of A^3?

(C) Compute $A, A + A^2, A + A^2 + A^3, \ldots$, until you obtain a matrix with no zero elements (except possibly on the principal diagonal), and interpret.

68. Airfreight. Find the incidence matrix A for the flight schedule illustrated in the figure. Compute $A, A + A^2, A + A^2 + A^3, \ldots$, until you obtain a matrix with no zero elements (except possibly on the principal diagonal), and interpret.

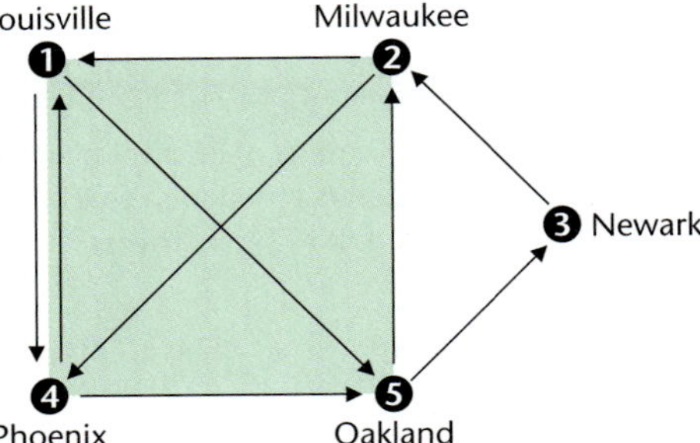

69. Politics. In a local election, a group hired a public relations firm to promote its candidate in three ways: telephone, house calls, and letters. The cost per contact is given in matrix M:

$$M = \begin{bmatrix} \$0.80 \\ \$1.50 \\ \$0.40 \end{bmatrix} \begin{matrix} \text{Telephone} \\ \text{House call} \\ \text{Letter} \end{matrix} \quad (\text{Cost per contact})$$

The number of contacts of each type made in two adjacent cities is given in matrix N:

$$\begin{matrix} & \text{Telephone} & \text{House call} & \text{Letter} & \\ N = & \begin{bmatrix} 1{,}000 & 500 & 5{,}000 \\ 2{,}000 & 800 & 8{,}000 \end{bmatrix} & & & \begin{matrix} \text{Berkeley} \\ \text{Oakland} \end{matrix} \end{matrix}$$

(A) Find the total amount spent in Berkeley.

(B) Find the total amount spent in Oakland.

(C) Discuss possible interpretations of the elements in the matrix products MN and NM.

(D) If either of the products MN or NM has a meaningful interpretation, find the product and label its rows and columns.

(E) Discuss methods of matrix multiplication that can be used to find the total number of telephone calls, house calls, and letters. State the matrices that can be used, and perform the necessary operations.

(F) Discuss methods of matrix multiplication that can be used to find the total number of contacts in Berkeley and in Oakland. State the matrices that can be used, and perform the necessary operations.

70. Nutrition. A nutritionist for a cereal company blends two cereals in different mixes. The amounts of protein, carbohydrate, and fat (in grams per ounce) in each cereal are given by matrix M. The amounts of each cereal used in the three mixes are given by matrix N.

$$M = \begin{bmatrix} 4 \text{ g/oz} & 2 \text{ g/oz} \\ 20 \text{ g/oz} & 16 \text{ g/oz} \\ 3 \text{ g/oz} & 1 \text{ g/oz} \end{bmatrix} \begin{matrix} \text{Protein} \\ \text{Carbohydrate} \\ \text{Fat} \end{matrix}$$

(columns: Cereal A, Cereal B)

$$N = \begin{bmatrix} 15 \text{ oz} & 10 \text{ oz} & 5 \text{ oz} \\ 5 \text{ oz} & 10 \text{ oz} & 15 \text{ oz} \end{bmatrix} \begin{matrix} \text{Cereal } A \\ \text{Cereal } B \end{matrix}$$

(columns: Mix X, Mix Y, Mix Z)

(A) Find the amount of protein in mix X.

(B) Find the amount of fat in mix Z.

(C) Discuss possible interpretations of the elements in the matrix products MN and NM.

(D) If either of the products MN or NM has a meaningful interpretation, find the product and label its rows and columns.

71. Dominance Relation. To rank players for an upcoming tennis tournament, a club decides to have each player play one set with every other player. The results are given in the table.

Player	Defeated
1. Aaron	Charles, Dan, Elvis
2. Bart	Aaron, Dan, Elvis
3. Charles	Bart, Dan
4. Dan	Frank
5. Elvis	Charles, Dan, Frank
6. Frank	Aaron, Bart, Charles

(A) Express the outcomes as an incidence matrix A by placing a 1 in the ith row and jth column of A if player i defeated player j and a 0 otherwise (see Problem 67).

(B) Compute the matrix $B = A + A^2$.

(C) Discuss matrix multiplication methods that can be used to find the sum of each of the rows in B. State the matrices that can be used and perform the necessary operations.

(D) Rank the players from strongest to weakest. Explain the reasoning behind your ranking.

72. Dominance Relation. Each member of a chess team plays one match with every other player. The results are given in the table.

Player	Defeated
1. Anne	Diane
2. Bridget	Anne, Carol, Diane
3. Carol	Anne
4. Diane	Carol, Erlene
5. Erlene	Anne, Bridget, Carol

(A) Express the outcomes as an incidence matrix A by placing a 1 in the ith row and jth column of A if player i defeated player j and a 0 otherwise (see Problem 67).

(B) Compute the matrix $B = A + A^2$.

(C) Discuss matrix multiplication methods that can be used to find the sum of each of the rows in B. State the matrices that can be used and perform the necessary operations.

(D) Rank the players from strongest to weakest. Explain the reasoning behind your ranking.

SECTION 9.2 Inverse of a Square Matrix

Identity Matrix for Multiplication • Inverse of a Square Matrix • Application: Cryptography

In Section 9.2 we introduce the identity matrix and the inverse of a square matrix. These matrix forms, along with matrix multiplication, are then used to solve some systems of equations written in matrix form in Section 9.3.

Identity Matrix for Multiplication

We know that for any real number a

$$(1)a = a(1) = a$$

The number 1 is called the *identity* for real number multiplication. Does the set of all matrices of a given dimension have an identity element for multiplication? That is, if M is an arbitrary $m \times n$ matrix, does M have an identity element I such that $IM = MI = M$? The answer in general is no. However, the set of all **square matrices of order n** (matrices with n rows and n columns) does have an identity.

DEFINITION 1
Identity Matrix

The **identity matrix for multiplication** for the set of all square matrices of order n is the square matrix of order n, denoted by I, with 1s along the principal diagonal (from upper left corner to lower right corner) and 0s elsewhere.

For example,

$$\begin{bmatrix} 1 & 0 \\ 0 & 1 \end{bmatrix} \quad \text{and} \quad \begin{bmatrix} 1 & 0 & 0 \\ 0 & 1 & 0 \\ 0 & 0 & 1 \end{bmatrix}$$

are the identity matrices for all square matrices of order 2 and 3, respectively.

Most graphing utilities have a built-in command for generating the identity matrix of a given order (Fig. 1).

FIGURE 1 Identity matrices.

```
identity(2)
         [[1 0]
          [0 1]]
identity(3)
       [[1 0 0]
        [0 1 0]
        [0 0 1]]
```

EXAMPLE 1 Identity Matrix Multiplication

(A) $\begin{bmatrix} 1 & 0 & 0 \\ 0 & 1 & 0 \\ 0 & 0 & 1 \end{bmatrix} \begin{bmatrix} a & b & c \\ d & e & f \\ g & h & i \end{bmatrix} = \begin{bmatrix} a & b & c \\ d & e & f \\ g & h & i \end{bmatrix}$

(B) $\begin{bmatrix} a & b & c \\ d & e & f \\ g & h & i \end{bmatrix} \begin{bmatrix} 1 & 0 & 0 \\ 0 & 1 & 0 \\ 0 & 0 & 1 \end{bmatrix} = \begin{bmatrix} a & b & c \\ d & e & f \\ g & h & i \end{bmatrix}$

(C) $\begin{bmatrix} 1 & 0 \\ 0 & 1 \end{bmatrix} \begin{bmatrix} a & b & c \\ d & e & f \end{bmatrix} = \begin{bmatrix} a & b & c \\ d & e & f \end{bmatrix}$

(D) $\begin{bmatrix} a & b & c \\ d & e & f \end{bmatrix} \begin{bmatrix} 1 & 0 & 0 \\ 0 & 1 & 0 \\ 0 & 0 & 1 \end{bmatrix} = \begin{bmatrix} a & b & c \\ d & e & f \end{bmatrix}$

MATCHED 1 PROBLEM

Multiply:

(A) $\begin{bmatrix} 1 & 0 \\ 0 & 1 \end{bmatrix}\begin{bmatrix} 3 & -5 \\ 4 & 6 \end{bmatrix}$ and $\begin{bmatrix} 3 & -5 \\ 4 & 6 \end{bmatrix}\begin{bmatrix} 1 & 0 \\ 0 & 1 \end{bmatrix}$

(B) $\begin{bmatrix} 1 & 0 & 0 \\ 0 & 1 & 0 \\ 0 & 0 & 1 \end{bmatrix}\begin{bmatrix} 5 & -7 \\ 2 & 4 \\ 6 & -8 \end{bmatrix}$ and $\begin{bmatrix} 5 & -7 \\ 2 & 4 \\ 6 & -8 \end{bmatrix}\begin{bmatrix} 1 & 0 \\ 0 & 1 \end{bmatrix}$

In general, we can show that if M is a square matrix of order n and I is the identity matrix of order n, then

$$IM = MI = M$$

If M is an $m \times n$ matrix that is not square ($m \neq n$), then it is still possible to multiply M on the left and on the right by an identity matrix, but not with the same-size identity matrix (see Example 1, parts C and D). To avoid the complications involved with associating two different identity matrices with each nonsquare matrix, we restrict our attention in Section 9.2 to square matrices.

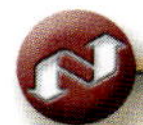

EXPLORE/DISCUSS 1

The only real number solutions to the equation $x^2 = 1$ are $x = 1$ and $x = -1$.

(A) Show that $A = \begin{bmatrix} 0 & 1 \\ 1 & 0 \end{bmatrix}$ satisfies $A^2 = I$, where I is the 2×2 identity matrix.

(B) Show that $B = \begin{bmatrix} 0 & -1 \\ -1 & 0 \end{bmatrix}$ satisfies $B^2 = I$.

(C) Find a 2×2 matrix with all elements nonzero whose square is the 2×2 identity matrix.

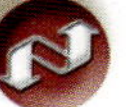

Inverse of a Square Matrix

In the set of real numbers, we know that for each real number a, except 0, there exists a real number a^{-1} such that

$$a^{-1}a = 1$$

The number a^{-1} is called the *inverse* of the number a relative to multiplication, or the *multiplicative inverse* of a. For example, 2^{-1} is the multiplicative inverse of 2, because $2^{-1}(2) = 1$. We use this idea to define the *inverse of a square matrix.*

DEFINITION 2

Inverse of a Square Matrix

If A is a square matrix of order n and if there exists a matrix A^{-1} (read "A inverse") such that

$$A^{-1}A = AA^{-1} = I$$

then A^{-1} is called the **multiplicative inverse of *A*** or, more simply, the **inverse of *A*.** If no such matrix exists, then A is said to be a **singular matrix.**

EXPLORE/DISCUSS 2

Let $A = \begin{bmatrix} 4 & 2 \\ 2 & 2 \end{bmatrix} \quad B = \begin{bmatrix} 0.25 & 0.5 \\ 0.5 & 0.5 \end{bmatrix} \quad C = \begin{bmatrix} 0.5 & -0.5 \\ -0.5 & 1 \end{bmatrix}$

(A) How are the entries in A and B related?

(B) Find AB. Is B the inverse of A?

(C) Find AC. Is C the inverse of A?

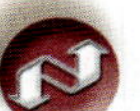

The multiplicative inverse of a nonzero real number a also can be written as $1/a$. This notation is not used for matrix inverses.

Let's use Definition 2 to find A^{-1}, if it exists, for

$$A = \begin{bmatrix} 2 & 3 \\ 1 & 2 \end{bmatrix}$$

We are looking for

$$A^{-1} = \begin{bmatrix} a & c \\ b & d \end{bmatrix}$$

such that

$$AA^{-1} = A^{-1}A = I$$

Thus, we write

$$\overset{A}{\begin{bmatrix} 2 & 3 \\ 1 & 2 \end{bmatrix}} \overset{A^{-1}}{\begin{bmatrix} a & c \\ b & d \end{bmatrix}} = \overset{I}{\begin{bmatrix} 1 & 0 \\ 0 & 1 \end{bmatrix}}$$

and try to find a, b, c, and d so that the product of A and A^{-1} is the identity matrix I. Multiplying A and A^{-1} on the left side, we obtain

$$\begin{bmatrix} (2a + 3b) & (2c + 3d) \\ (a + 2b) & (c + 2d) \end{bmatrix} = \begin{bmatrix} 1 & 0 \\ 0 & 1 \end{bmatrix}$$

which is true only if

$$\begin{aligned} 2a + 3b &= 1 \qquad & 2c + 3d &= 0 \\ a + 2b &= 0 \qquad & c + 2d &= 1 \end{aligned}$$

Solving these two systems, we find that $a = 2$, $b = -1$, $c = -3$, and $d = 2$. Thus,

$$A^{-1} = \begin{bmatrix} 2 & -3 \\ -1 & 2 \end{bmatrix}$$

as is easily checked:

$$\overset{A}{\begin{bmatrix} 2 & 3 \\ 1 & 2 \end{bmatrix}} \overset{A^{-1}}{\begin{bmatrix} 2 & -3 \\ -1 & 2 \end{bmatrix}} = \overset{I}{\begin{bmatrix} 1 & 0 \\ 0 & 1 \end{bmatrix}} = \overset{A^{-1}}{\begin{bmatrix} 2 & -3 \\ -1 & 2 \end{bmatrix}} \overset{A}{\begin{bmatrix} 2 & 3 \\ 1 & 2 \end{bmatrix}}$$

Unlike nonzero real numbers, inverses do not always exist for nonzero square matrices. For example, if

$$B = \begin{bmatrix} 2 & 1 \\ 4 & 2 \end{bmatrix}$$

then, proceeding as before, we are led to the systems

$$\begin{aligned} 2a + b &= 1 \qquad & 2c + d &= 0 \\ 4a + 2b &= 0 \qquad & 4c + 2d &= 1 \end{aligned}$$

These systems are both inconsistent and have no solution. Hence, B^{-1} does not exist, and B is a singular matrix.

Most graphing utilities can find matrix inverses and can identify singular matrices. Figure 2 shows the calculation of A^{-1} for the matrix A discussed earlier. Figure 3 shows the error message that results when the inverse operation is applied to the singular matrix B discussed earlier.

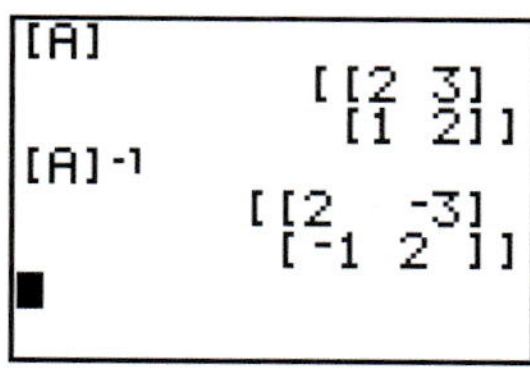

FIGURE 2

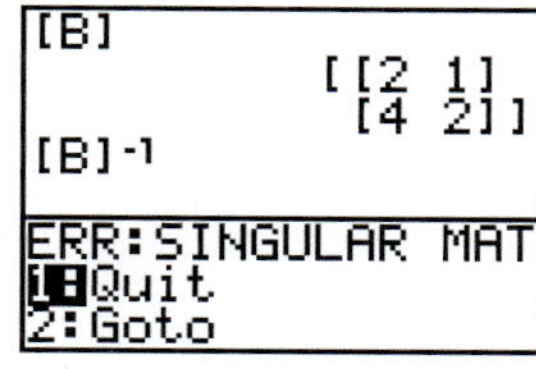

FIGURE 3

[A]^(-1)

ERR:DOMAIN
1:Quit
2:Goto

FIGURE 4

Note that the inverse operation is performed by pressing the x^{-1} key. Entering [A] ^ (−1) results in an error message (Fig. 4).

Being able to find inverses, when they exist, leads to direct and simple solutions to many practical problems. In Section 9.3, for example, we will show how inverses can be used to solve systems of linear equations.

The algebraic method outlined for finding the inverse, if it exists, gets very involved for matrices of order larger than 2. Now that we know what we are looking for, we can use augmented matrices, as in Section 8.3, to make the process more efficient. Details are illustrated in Example 2.

EXAMPLE 2 Finding an Inverse

Find the inverse, if it exists, of

$$A = \begin{bmatrix} 1 & -1 & 1 \\ 0 & 2 & -1 \\ 2 & 3 & 0 \end{bmatrix}$$

SOLUTION

We start as before and write

$$\overset{A}{\begin{bmatrix} 1 & -1 & 1 \\ 0 & 2 & -1 \\ 2 & 3 & 0 \end{bmatrix}} \overset{A^{-1}}{\begin{bmatrix} a & d & g \\ b & e & h \\ c & f & i \end{bmatrix}} = \overset{I}{\begin{bmatrix} 1 & 0 & 0 \\ 0 & 1 & 0 \\ 0 & 0 & 1 \end{bmatrix}}$$

This is true only if

$$\begin{aligned} a - b + c &= 1 \\ 2b - c &= 0 \\ 2a + 3b &= 0 \end{aligned} \qquad \begin{aligned} d - e + f &= 0 \\ 2e - f &= 1 \\ 2d + 3e &= 0 \end{aligned} \qquad \begin{aligned} g - h + i &= 0 \\ 2h - i &= 0 \\ 2g + 3h &= 1 \end{aligned}$$

Now we write augmented matrices for each of the three systems:

$$\overset{\text{First}}{\left[\begin{array}{rrr|r} 1 & -1 & 1 & 1 \\ 0 & 2 & -1 & 0 \\ 2 & 3 & 0 & 0 \end{array}\right]} \quad \overset{\text{Second}}{\left[\begin{array}{rrr|r} 1 & -1 & 1 & 0 \\ 0 & 2 & -1 & 1 \\ 2 & 3 & 0 & 0 \end{array}\right]} \quad \overset{\text{Third}}{\left[\begin{array}{rrr|r} 1 & -1 & 1 & 0 \\ 0 & 2 & -1 & 0 \\ 2 & 3 & 0 & 1 \end{array}\right]}$$

Because each matrix to the left of the vertical bar is the same, exactly the same row operations can be used on each augmented matrix to transform it into a reduced form. We can speed up the process substantially by combining all three augmented matrices into the single augmented matrix form

$$\left[\begin{array}{rrr|rrr} 1 & -1 & 1 & 1 & 0 & 0 \\ 0 & 2 & -1 & 0 & 1 & 0 \\ 2 & 3 & 0 & 0 & 0 & 1 \end{array}\right] = [A \mid I] \tag{1}$$

We now try to perform row operations on matrix (1) until we obtain a row-equivalent matrix that looks like matrix (2):

$$\left[\begin{array}{ccc|ccc} 1 & 0 & 0 & a & d & g \\ 0 & 1 & 0 & b & e & h \\ 0 & 0 & 1 & c & f & i \end{array}\right] = [I \mid B] \tag{2}$$

(with I labeling the left block and B the right block)

If this can be done, then the new matrix to the right of the vertical bar is A^{-1}! Now let's try to transform matrix (1) into a form like that of matrix (2). We follow the same sequence of steps as in the solution of linear systems by Gauss–Jordan elimination (see Section 8.3):

$$\begin{array}{c} \quad A \qquad\qquad\qquad I \\ \left[\begin{array}{rrr|rrr} 1 & -1 & 1 & 1 & 0 & 0 \\ 0 & 2 & -1 & 0 & 1 & 0 \\ 2 & 3 & 0 & 0 & 0 & 1 \end{array}\right] \end{array} \quad (-2)R_1 + R_3 \rightarrow R_3$$

$$\sim \left[\begin{array}{rrr|rrr} 1 & -1 & 1 & 1 & 0 & 0 \\ 0 & 2 & -1 & 0 & 1 & 0 \\ 0 & 5 & -2 & -2 & 0 & 1 \end{array}\right] \quad \tfrac{1}{2}R_2 \rightarrow R_2$$

$$\sim \left[\begin{array}{rrr|rrr} 1 & -1 & 1 & 1 & 0 & 0 \\ 0 & 1 & -\frac{1}{2} & 0 & \frac{1}{2} & 0 \\ 0 & 5 & -2 & -2 & 0 & 1 \end{array}\right] \quad \begin{array}{l} R_2 + R_1 \rightarrow R_1 \\ \\ (-5)R_2 + R_3 \rightarrow R_3 \end{array}$$

$$\sim \left[\begin{array}{rrr|rrr} 1 & 0 & \frac{1}{2} & 1 & \frac{1}{2} & 0 \\ 0 & 1 & -\frac{1}{2} & 0 & \frac{1}{2} & 0 \\ 0 & 0 & \frac{1}{2} & -2 & -\frac{5}{2} & 1 \end{array}\right] \quad 2R_3 \rightarrow R_3$$

$$\sim \left[\begin{array}{rrr|rrr} 1 & 0 & \frac{1}{2} & 1 & \frac{1}{2} & 0 \\ 0 & 1 & -\frac{1}{2} & 0 & \frac{1}{2} & 0 \\ 0 & 0 & 1 & -4 & -5 & 2 \end{array}\right] \quad \begin{array}{l} (-\frac{1}{2})R_3 + R_1 \rightarrow R_1 \\ \frac{1}{2}R_3 + R_2 \rightarrow R_2 \end{array}$$

$$\sim \left[\begin{array}{rrr|rrr} 1 & 0 & 0 & 3 & 3 & -1 \\ 0 & 1 & 0 & -2 & -2 & 1 \\ 0 & 0 & 1 & -4 & -5 & 2 \end{array}\right] = [I|B]$$

Converting back to systems of equations equivalent to our three original systems (we won't have to do this step in practice), we have

$$\begin{array}{lll} a = 3 & d = 3 & g = -1 \\ b = -2 & e = -2 & h = 1 \\ c = -4 & f = -5 & i = 2 \end{array}$$

And these are just the elements of A^{-1} that we are looking for! Hence,

$$A^{-1} = \begin{bmatrix} 3 & 3 & -1 \\ -2 & -2 & 1 \\ -4 & -5 & 2 \end{bmatrix}$$

Note that this is the matrix to the right of the vertical line in the last augmented matrix.

CHECK

Because the definition of matrix inverse requires that

$$A^{-1}A = I \qquad \text{and} \qquad AA^{-1} = I \tag{3}$$

it appears that we must compute both $A^{-1}A$ and AA^{-1} to check our work. However, it can be shown that if one of the equations in (3) is satisfied, then the other is also satisfied. Thus, for checking purposes it is sufficient to compute either $A^{-1}A$ or AA^{-1}—we don't need to do both.

$$A^{-1}A = \begin{bmatrix} 3 & 3 & -1 \\ -2 & -2 & 1 \\ -4 & -5 & 2 \end{bmatrix} \begin{bmatrix} 1 & -1 & 1 \\ 0 & 2 & -1 \\ 2 & 3 & 0 \end{bmatrix} = \begin{bmatrix} 1 & 0 & 0 \\ 0 & 1 & 0 \\ 0 & 0 & 1 \end{bmatrix} = I$$

MATCHED 2 PROBLEM

Let $A = \begin{bmatrix} 3 & -1 & 1 \\ -1 & 1 & 0 \\ 1 & 0 & 1 \end{bmatrix}$

(A) Form the augmented matrix $[A \mid I]$.

(B) Use row operations to transform $[A \mid I]$ into $[I \mid B]$.

(C) Verify by multiplication that $B = A^{-1}$.

The procedure used in Example 2 can be used to find the inverse of any square matrix, if the inverse exists, and will also indicate when the inverse does not exist. These ideas are summarized in Theorem 1.

THEOREM 1

Inverse of a Square Matrix *A*

If $[A \mid I]$ is transformed by row operations into $[I \mid B]$, then the resulting matrix B is A^{-1}. If, however, we obtain all 0s in one or more rows to the left of the vertical line, then A^{-1} does not exist.

EXPLORE/DISCUSS 3

(A) Suppose that the square matrix A has a row of all zeros. Explain why A has no inverse.

(B) Suppose that the square matrix A has a column of all zeros. Explain why A has no inverse.

EXAMPLE 3 **Finding a Matrix Inverse**

Find A^{-1}, given $A = \begin{bmatrix} 4 & -1 \\ -6 & 2 \end{bmatrix}$

SOLUTION

Algebraic Solution

$$\left[\begin{array}{rr|rr} 4 & -1 & 1 & 0 \\ -6 & 2 & 0 & 1 \end{array}\right] \quad \tfrac{1}{4}R_1 \to R_1$$

$$\sim \left[\begin{array}{rr|rr} 1 & -\frac{1}{4} & \frac{1}{4} & 0 \\ -6 & 2 & 0 & 1 \end{array}\right] \quad 6R_1 + R_2 \to R_2$$

$$\sim \left[\begin{array}{rr|rr} 1 & -\frac{1}{4} & \frac{1}{4} & 0 \\ 0 & \frac{1}{2} & \frac{3}{2} & 1 \end{array}\right] \quad 2R_2 \to R_2$$

$$\sim \left[\begin{array}{rr|rr} 1 & -\frac{1}{4} & \frac{1}{4} & 0 \\ 0 & 1 & 3 & 2 \end{array}\right] \quad \tfrac{1}{4}R_2 + R_1 \to R_1$$

$$\sim \left[\begin{array}{rr|rr} 1 & 0 & 1 & \frac{1}{2} \\ 0 & 1 & 3 & 2 \end{array}\right]$$

Thus,

$$A^{-1} = \begin{bmatrix} 1 & \frac{1}{2} \\ 3 & 2 \end{bmatrix}$$

Check by showing $A^{-1}A = I$.

Graphing Utility Solution

Enter A and use the inverse key (Fig. 5).

[A]
[[4 -1]
[-6 2]]
[A]-1
[[1 .5]
[3 2]]

FIGURE 5

From Figure 5, we see that

$$A^{-1} = \begin{bmatrix} 1 & 0.5 \\ 3 & 2 \end{bmatrix}$$

MATCHED 3 PROBLEM

Find A^{-1}, given $A = \begin{bmatrix} 2 & -6 \\ 1 & -2 \end{bmatrix}$

EXAMPLE 4 Finding an Inverse

Find B^{-1}, if it exists, given $B = \begin{bmatrix} 10 & -2 \\ -5 & 1 \end{bmatrix}$

SOLUTION

Algebraic Solution

$$\left[\begin{array}{rr|rr} 10 & -2 & 1 & 0 \\ -5 & 1 & 0 & 1 \end{array}\right] \sim \left[\begin{array}{rr|rr} 1 & -\frac{1}{5} & \frac{1}{10} & 0 \\ -5 & 1 & 0 & 1 \end{array}\right]$$

$$\sim \left[\begin{array}{rr|rr} 1 & -\frac{1}{5} & \frac{1}{10} & 0 \\ 0 & 0 & \frac{1}{2} & 1 \end{array}\right]$$

We have all 0s in the second row to the left of the vertical line. Therefore, B^{-1} does not exist.

Graphing Utility Solution

Enter B and use the inverse key (Fig. 6).

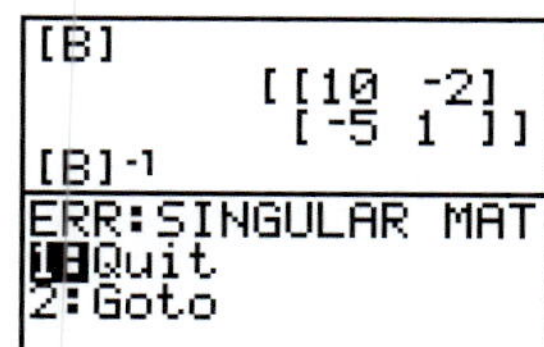

FIGURE 6

From Figure 6, we see that B is a singular matrix and B^{-1} does not exist.

MATCHED 4 PROBLEM

Find B^{-1}, if it exists, given $B = \begin{bmatrix} 6 & -3 \\ -2 & 1 \end{bmatrix}$

Application: Cryptography

Matrix inverses can be used to provide a simple and effective procedure for encoding and decoding messages. To begin, we assign the numbers 1 to 26 to the letters in the alphabet, as shown below. We also assign the number 27 to a blank to provide for space between words. (A more sophisticated code could include both uppercase and lowercase letters and punctuation symbols.)

A	B	C	D	E	F	G	H	I	J	K	L	M	N
1	2	3	4	5	6	7	8	9	10	11	12	13	14
O	P	Q	R	S	T	U	V	W	X	Y	Z	Blank	
15	16	17	18	19	20	21	22	23	24	25	26	27	

Thus, the message I LOVE MATH corresponds to the sequence

9 27 12 15 22 5 27 13 1 20 8

Any matrix whose elements are positive integers and whose inverse exists can be used as an **encoding matrix.** For example, to use the 2×2 matrix

$$A = \begin{bmatrix} 4 & 3 \\ 5 & 4 \end{bmatrix}$$

to encode the above message, first we divide the numbers in the sequence into groups of 2 and use these groups as the columns of a matrix B with two rows:

$$B = \begin{bmatrix} 9 & 12 & 22 & 27 & 1 & 8 \\ 27 & 15 & 5 & 13 & 20 & 27 \end{bmatrix}$$

Proceed down the columns, not across the rows.

(Notice that we added an extra blank at the end of the message to make the columns come out even.) Then we multiply this matrix on the left by A:

$$AB = \begin{bmatrix} 4 & 3 \\ 5 & 4 \end{bmatrix} \begin{bmatrix} 9 & 12 & 22 & 27 & 1 & 8 \\ 27 & 15 & 5 & 13 & 20 & 27 \end{bmatrix}$$

$$= \begin{bmatrix} 117 & 93 & 103 & 147 & 64 & 113 \\ 153 & 120 & 130 & 187 & 85 & 148 \end{bmatrix}$$

The coded message is

117 153 93 120 103 130 147 187 64 85 113 148

This message can be decoded simply by putting it back into matrix form and multiplying on the left by the **decoding matrix** A^{-1}. Because A^{-1} is easily determined if A is known, the encoding matrix A is the only key needed to decode messages encoded in this manner. Although simple in concept, codes of this type can be very difficult to crack.

EXAMPLE 5 Cryptography

The message

31 54 69 37 64 82 23 50 66 51 69 75 23 30 36 65 84 84

was encoded with the matrix A shown below. Decode this message.

$$A = \begin{bmatrix} 0 & 2 & 1 \\ 1 & 2 & 1 \\ 2 & 1 & 1 \end{bmatrix}$$

SOLUTION

We begin by entering the 3×3 encoding matrix A (Fig. 7). Then we enter the coded message in the columns of a matrix C with three rows (Fig. 7). If B is the matrix containing the uncoded message, then B and C are related by $C = AB$. To find B, we multiply both sides of the equation $C = AB$ by A^{-1} (Fig. 8).

```
A
            [[0 2 1]
             [1 2 1]
             [2 1 1]]
C
[[31 37 23 51 23 65]
 [54 64 50 69 30 84]
 [69 82 66 75 36 84]]
```

FIGURE 7

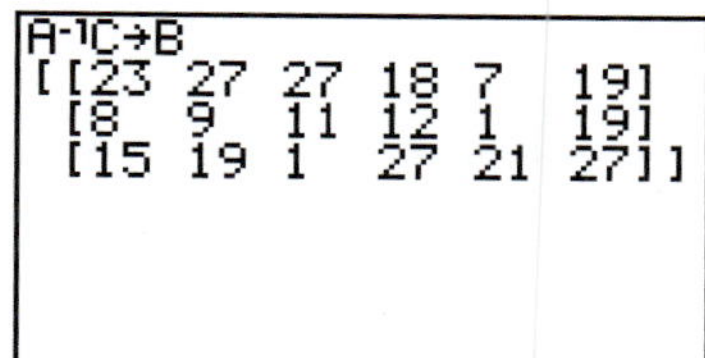

A⁻¹C→B
[[23 27 27 18 7 19]
[8 9 11 12 1 19]
[15 19 1 27 21 27]]

FIGURE 8

Writing the numbers in the columns of this matrix in sequence and using the correspondence between numbers and letters noted earlier produces the decoded message:

23	8	15	27	9	19	27	11	1	18	12	27	7	1	21	19	19	27
W	H	O		I	S		K	A	R	L		G	A	U	S	S	

The answer to this question can be found in Chapter 8.

MATCHED 5 PROBLEM

The message

46 84 85 55 101 100 59 95 132 25 42 53 52 91 90 43 71 83 19 37 25

was encoded with the matrix A shown below. Decode this message.

$$A = \begin{bmatrix} 1 & 1 & 1 \\ 2 & 1 & 2 \\ 2 & 3 & 1 \end{bmatrix}$$

ANSWERS MATCHED PROBLEMS

1. (A) $\begin{bmatrix} 3 & -5 \\ 4 & 6 \end{bmatrix}$ (B) $\begin{bmatrix} 5 & -7 \\ 2 & 4 \\ 6 & -8 \end{bmatrix}$

2. (A) $\left[\begin{array}{rrr|rrr} 3 & -1 & 1 & 1 & 0 & 0 \\ -1 & 1 & 0 & 0 & 1 & 0 \\ 1 & 0 & 1 & 0 & 0 & 1 \end{array}\right]$ (B) $\left[\begin{array}{rrr|rrr} 1 & 0 & 0 & 1 & 1 & -1 \\ 0 & 1 & 0 & 1 & 2 & -1 \\ 0 & 0 & 1 & -1 & -1 & 2 \end{array}\right]$ (C) $\begin{bmatrix} 1 & 1 & -1 \\ 1 & 2 & -1 \\ -1 & -1 & 2 \end{bmatrix}\begin{bmatrix} 3 & -1 & 1 \\ -1 & 1 & 0 \\ 1 & 0 & 1 \end{bmatrix} = \begin{bmatrix} 1 & 0 & 0 \\ 0 & 1 & 0 \\ 0 & 0 & 1 \end{bmatrix}$

3. $\begin{bmatrix} -1 & 3 \\ -\frac{1}{2} & 1 \end{bmatrix}$ **4.** Does not exist **5.** WHO IS WILHELM JORDAN

EXERCISE 9.2

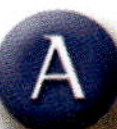

Perform the indicated operations in Problems 1–8.

1. $\begin{bmatrix} 1 & 0 \\ 0 & 1 \end{bmatrix}\begin{bmatrix} 2 & -3 \\ 4 & 5 \end{bmatrix}$

2. $\begin{bmatrix} 1 & 0 \\ 0 & 1 \end{bmatrix}\begin{bmatrix} 4 & -3 \\ 0 & 2 \end{bmatrix}$

3. $\begin{bmatrix} 2 & -3 \\ 4 & 5 \end{bmatrix}\begin{bmatrix} 1 & 0 \\ 0 & 1 \end{bmatrix}$

4. $\begin{bmatrix} 4 & -3 \\ 0 & 2 \end{bmatrix}\begin{bmatrix} 1 & 0 \\ 0 & 1 \end{bmatrix}$

5. $\begin{bmatrix} 1 & 0 & 0 \\ 0 & 1 & 0 \\ 0 & 0 & 1 \end{bmatrix}\begin{bmatrix} -2 & 1 & 3 \\ 2 & 4 & -2 \\ 5 & 1 & 0 \end{bmatrix}$

6. $\begin{bmatrix} 1 & 0 & 0 \\ 0 & 1 & 0 \\ 0 & 0 & 1 \end{bmatrix}\begin{bmatrix} -3 & 0 & 2 \\ 1 & 1 & 5 \\ 2 & -1 & 7 \end{bmatrix}$

7. $\begin{bmatrix} -2 & 1 & 3 \\ 2 & 4 & -2 \\ 5 & 1 & 0 \end{bmatrix}\begin{bmatrix} 1 & 0 & 0 \\ 0 & 1 & 0 \\ 0 & 0 & 1 \end{bmatrix}$

8. $\begin{bmatrix} -3 & 0 & 2 \\ 1 & 1 & 5 \\ 2 & -1 & 7 \end{bmatrix}\begin{bmatrix} 1 & 0 & 0 \\ 0 & 1 & 0 \\ 0 & 0 & 1 \end{bmatrix}$

In Problems 9–18, examine the product of the two matrices to determine if each is the inverse of the other.

9. $\begin{bmatrix} 3 & -4 \\ -2 & 3 \end{bmatrix}; \begin{bmatrix} 3 & 4 \\ 2 & 3 \end{bmatrix}$

10. $\begin{bmatrix} -2 & -1 \\ -4 & 2 \end{bmatrix}; \begin{bmatrix} 1 & -1 \\ 2 & -2 \end{bmatrix}$

11. $\begin{bmatrix} 2 & 2 \\ -1 & -1 \end{bmatrix}; \begin{bmatrix} 1 & 1 \\ -1 & -1 \end{bmatrix}$

12. $\begin{bmatrix} 5 & -7 \\ -2 & 3 \end{bmatrix}; \begin{bmatrix} 3 & 7 \\ 2 & 5 \end{bmatrix}$

13. $\begin{bmatrix} -5 & 2 \\ -8 & 3 \end{bmatrix}; \begin{bmatrix} 3 & -2 \\ 8 & -5 \end{bmatrix}$

14. $\begin{bmatrix} 7 & 4 \\ -5 & -3 \end{bmatrix}; \begin{bmatrix} 3 & 4 \\ -5 & -7 \end{bmatrix}$

15. $\begin{bmatrix} 1 & 2 & 0 \\ 0 & 1 & 0 \\ -1 & -1 & 1 \end{bmatrix}; \begin{bmatrix} 1 & -2 & 0 \\ 0 & 1 & 0 \\ 1 & -1 & 0 \end{bmatrix}$

16. $\begin{bmatrix} 1 & 0 & 1 \\ -3 & 1 & -2 \\ 0 & 0 & 1 \end{bmatrix}; \begin{bmatrix} 1 & 0 & -1 \\ 3 & 1 & -1 \\ 0 & 0 & 1 \end{bmatrix}$

17. $\begin{bmatrix} 1 & -1 & 1 \\ 0 & 2 & -1 \\ 2 & 3 & 0 \end{bmatrix}; \begin{bmatrix} 3 & 3 & -1 \\ -2 & -2 & 1 \\ -4 & -5 & 2 \end{bmatrix}$

18. $\begin{bmatrix} 1 & 0 & -1 \\ 3 & 1 & -1 \\ 0 & 0 & 0 \end{bmatrix}; \begin{bmatrix} 1 & 0 & -1 \\ -3 & 1 & -2 \\ 0 & 0 & 1 \end{bmatrix}$

B

Given A in Problems 19–28, find A^{-1}, and show that $A^{-1}A = I$.

19. $\begin{bmatrix} 0 & -1 \\ 1 & 4 \end{bmatrix}$ **20.** $\begin{bmatrix} -1 & 5 \\ 0 & -1 \end{bmatrix}$ **21.** $\begin{bmatrix} 1 & 2 \\ 1 & 3 \end{bmatrix}$

22. $\begin{bmatrix} 2 & 1 \\ 5 & 3 \end{bmatrix}$ **23.** $\begin{bmatrix} 1 & 3 \\ 2 & 7 \end{bmatrix}$ **24.** $\begin{bmatrix} 2 & 1 \\ 1 & 1 \end{bmatrix}$

25. $\begin{bmatrix} 1 & -2 & 0 \\ 0 & 1 & 1 \\ 2 & -1 & 2 \end{bmatrix}$

26. $\begin{bmatrix} 1 & 3 & 0 \\ 1 & 2 & 3 \\ 0 & -1 & 2 \end{bmatrix}$

27. $\begin{bmatrix} 1 & 1 & 0 \\ 0 & 2 & -1 \\ 1 & 0 & 1 \end{bmatrix}$

28. $\begin{bmatrix} 1 & 0 & -1 \\ 2 & -1 & 0 \\ 1 & 1 & -4 \end{bmatrix}$

Find the inverse of each matrix in Problems 29–32, if it exists.

29. $\begin{bmatrix} 3 & 9 \\ 2 & 6 \end{bmatrix}$

30. $\begin{bmatrix} 2 & -4 \\ -3 & 6 \end{bmatrix}$

31. $\begin{bmatrix} 2 & 3 \\ 3 & 5 \end{bmatrix}$

32. $\begin{bmatrix} -5 & 4 \\ 4 & -3 \end{bmatrix}$

In Problems 33–38, explain how you can determine that each matrix is singular without performing any calculations.

33. $A = \begin{bmatrix} 1 & 2 & -1 \\ 1 & -1 & 3 \\ 1 & 2 & -1 \end{bmatrix}$

34. $B = \begin{bmatrix} 2 & 3 & -2 \\ 0 & 0 & 0 \\ 1 & -1 & 4 \end{bmatrix}$

35. $A = \begin{bmatrix} 1 & 0 & -1 \\ 1 & 0 & 3 \\ 5 & 0 & -2 \end{bmatrix}$

36. $B = \begin{bmatrix} 0 & 3 & -2 \\ 0 & 1 & -3 \\ 0 & -1 & 4 \end{bmatrix}$

37. $A = \begin{bmatrix} 1 & 2 & -1 \\ 1 & -1 & 3 \\ 1 & -1 & 3 \end{bmatrix}$

38. $B = \begin{bmatrix} 2 & 3 & -2 \\ 2 & 3 & -2 \\ 1 & -1 & 4 \end{bmatrix}$

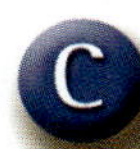

Find the inverse of each matrix in Problems 39–44, if it exists.

39. $\begin{bmatrix} 2 & 2 & -1 \\ 0 & 4 & -1 \\ -1 & -2 & 1 \end{bmatrix}$

40. $\begin{bmatrix} 4 & 2 & -1 \\ 1 & 1 & -1 \\ -3 & -1 & 1 \end{bmatrix}$

41. $\begin{bmatrix} 2 & 1 & 1 \\ 1 & 1 & 0 \\ -1 & -1 & 0 \end{bmatrix}$

42. $\begin{bmatrix} 1 & -1 & 0 \\ 2 & -1 & 1 \\ 0 & 1 & 1 \end{bmatrix}$

43. $\begin{bmatrix} 1 & 5 & 10 \\ 0 & 1 & 4 \\ 1 & 6 & 15 \end{bmatrix}$

44. $\begin{bmatrix} 1 & -5 & -10 \\ 0 & 1 & 6 \\ 1 & -4 & -3 \end{bmatrix}$

45. Discuss the existence of A^{-1} for 2×2 diagonal matrices of the form

$$A = \begin{bmatrix} a & 0 \\ 0 & d \end{bmatrix}$$

46. Discuss the existence of A^{-1} for 2×2 upper triangular matrices of the form

$$A = \begin{bmatrix} a & b \\ 0 & d \end{bmatrix}$$

47. Find A^{-1} and A^2 for each of the following matrices.

(A) $A = \begin{bmatrix} 3 & 2 \\ -4 & -3 \end{bmatrix}$ **(B)** $A = \begin{bmatrix} -2 & -1 \\ 3 & 2 \end{bmatrix}$

48. Based on your observations in Problem 47, if $A = A^{-1}$ for a square matrix A, what is A^2? Give a mathematical argument to support your conclusion.

49. Find $(A^{-1})^{-1}$ for each of the following matrices.

(A) $A = \begin{bmatrix} 4 & 2 \\ 1 & 3 \end{bmatrix}$ **(B)** $A = \begin{bmatrix} 5 & 5 \\ -1 & 3 \end{bmatrix}$

50. Based on your observations in Problem 49, if A^{-1} exists for a square matrix A, what is $(A^{-1})^{-1}$? Give a mathematical argument to support your conclusion.

51. Find $(AB)^{-1}$, $A^{-1}B^{-1}$, and $B^{-1}A^{-1}$ for each of the following pairs of matrices.

(A) $A = \begin{bmatrix} 3 & 4 \\ 2 & 3 \end{bmatrix}$ and $B = \begin{bmatrix} 3 & 7 \\ 2 & 5 \end{bmatrix}$

(B) $A = \begin{bmatrix} 1 & -1 \\ 2 & 3 \end{bmatrix}$ and $B = \begin{bmatrix} 6 & 2 \\ 2 & 1 \end{bmatrix}$

52. Based on your observations in Problem 51, which of the following is a true statement? Give a mathematical argument to support your conclusion.

(A) $(AB)^{-1} = A^{-1}B^{-1}$

(B) $(AB)^{-1} = B^{-1}A^{-1}$

APPLICATIONS

Problems 53–56 refer to the encoding matrix $A = \begin{bmatrix} 3 & 5 \\ 1 & 2 \end{bmatrix}$

53. Cryptography. Encode the message CAT IN THE HAT with the matrix A given.

54. Cryptography. Encode the message FOX IN SOCKS with the matrix A given.

55. Cryptography. The following message was encoded with the matrix A given. Decode this message.

111 43 40 15 177 68 50 19 116 45 86
29 62 22 121 43 68 27

56. Cryptography. The following message was encoded with the matrix A given. Decode this message.

99 38 154 58 115 43 121 43 20 7 149
56 86 29 196 73 99 38

Problems 57–60 refer to the encoding matrix

$$B = \begin{bmatrix} 1 & 0 & 1 & 0 & 1 \\ 0 & 1 & 1 & 0 & 3 \\ 2 & 1 & 1 & 1 & 1 \\ 0 & 0 & 1 & 0 & 2 \\ 1 & 1 & 1 & 2 & 1 \end{bmatrix}$$

57. Cryptography. Encode the message DWIGHT DAVID EISENHOWER with the matrix B given.

58. Cryptography. Encode the message JOHN FITZGERALD KENNEDY with the matrix B given.

59. Cryptography. The following message was encoded with the matrix B given. Decode this message.

41 84 82 44 74 25 56 67 20 54 43
54 89 39 102 44 67 86 44 90 68 135
136 81 149

60. Cryptography. The following message was encoded with the matrix B given. Decode this message.

22 15 57 5 47 54 58 89 45 84 46
80 87 53 96 51 68 116 39 113 68 135
136 81 149

SECTION 9.3 Matrix Equations and Systems of Linear Equations

Matrix Equations • Matrix Equations and Systems of Linear Equations • Application

The identity matrix and inverse matrix discussed in Section 9.2 can be put to immediate use in the solving of certain simple matrix equations. Being able to solve a matrix equation gives us another important method of solving a system of equations having the same number of variables as equations. If the system either has fewer variables than equations or more variables than equations, then we must return to the Gauss–Jordan method of elimination.

Matrix Equations

Before we discuss the solution of matrix equations, you will probably find it helpful to briefly review the basic properties of linear equations discussed in Appendix A, Section A.1.

EXPLORE/DISCUSS 1

Let a, b, and c be real numbers, with $a \neq 0$. Solve each equation for x.

(A) $ax = b$ (B) $ax + b = c$

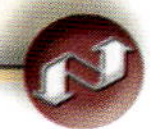

Solving simple matrix equations follows very much the same procedures used in solving real number equations. We have, however, less freedom with matrix equations, because matrix multiplication is not commutative. In solving matrix equations, we will be guided by the properties of matrices summarized in Theorem 1.

THEOREM 1
Basic Properties of Matrices

Assuming all products and sums are defined for the indicated matrices A, B, C, I, and 0, then

Addition Properties

Associative:	$(A + B) + C = A + (B + C)$
Commutative:	$A + B = B + A$
Additive Identity:	$A + 0 = 0 + A = A$
Additive Inverse:	$A + (-A) = (-A) + A = 0$

Multiplication Properties

Associative Property:	$A(BC) = (AB)C$
Multiplicative Identity:	$AI = IA = A$
Multiplicative Inverse:	If A is a square matrix and A^{-1} exists, then $AA^{-1} = A^{-1}A = I$.

Combined Properties

Left Distributive:	$A(B + C) = AB + AC$
Right Distributive:	$(B + C)A = BA + CA$

Equality

Addition:	If $A = B$, then $A + C = B + C$.
Left Multiplication:	If $A = B$, then $CA = CB$.
Right Multiplication:	If $A = B$, then $AC = BC$.

The process of solving certain types of simple matrix equations is best illustrated by an example.

EXAMPLE 1 Solving a Matrix Equation

Given an $n \times n$ matrix A and $n \times 1$ column matrices B and X, solve $AX = B$ for X. Assume all necessary inverses exist.

SOLUTION

We are interested in finding a column matrix X that satisfies the matrix equation $AX = B$. To solve this equation, we multiply both sides, on the left, by A^{-1}, assuming it exists, to isolate X on the left side.

$$
\begin{aligned}
AX &= B \\
A^{-1}(AX) &= A^{-1}B && \text{Use the left multiplication property.} \\
(A^{-1}A)X &= A^{-1}B && \text{Associative property} \\
IX &= A^{-1}B && A^{-1}A = I \\
X &= A^{-1}B && IX = X
\end{aligned}
$$

CAUTION

1. Do not mix the left multiplication property and the right multiplication property. If $AX = B$, then

$$A^{-1}(AX) \neq BA^{-1}$$

2. Matrix division is not defined. If a, b, and x are real numbers, then the solution of $ax = b$ can be written either as $x = a^{-1}b$ or as $x = \frac{b}{a}$. But if A, B, and X are matrices, the solution of $AX = B$ must be written as $X = A^{-1}B$. The expression $\frac{B}{A}$ is not defined for matrices.

MATCHED 1 PROBLEM

Given an $n \times n$ matrix A and $n \times 1$ column matrices B, C, and X, solve $AX + C = B$ for X. Assume all necessary inverses exist.

Matrix Equations and Systems of Linear Equations

We now show how independent systems of linear equations with the same number of variables as equations can be solved by first converting the system into a matrix equation of the form $AX = B$ and using $X = A^{-1}B$ as obtained in Example 1.

EXAMPLE 2 Using Inverses to Solve Systems of Equations

Use matrix inverse methods to solve the system

$$
\begin{aligned}
x_1 - x_2 + x_3 &= 1 \\
2x_2 - x_3 &= 1 \\
2x_1 + 3x_2 &= 1
\end{aligned}
\tag{1}
$$

SOLUTION

The inverse of the coefficient matrix

$$A = \begin{bmatrix} 1 & -1 & 1 \\ 0 & 2 & -1 \\ 2 & 3 & 0 \end{bmatrix}$$

provides an efficient method for solving this system. To see how, we convert system (1) into a matrix equation:

$$\overset{A}{\begin{bmatrix} 1 & -1 & 1 \\ 0 & 2 & -1 \\ 2 & 3 & 0 \end{bmatrix}} \overset{X}{\begin{bmatrix} x_1 \\ x_2 \\ x_3 \end{bmatrix}} = \overset{B}{\begin{bmatrix} 1 \\ 1 \\ 1 \end{bmatrix}} \qquad (2)$$

Check that matrix equation (2) is equivalent to system (1) by finding the product of the left side and then equating corresponding elements on the left with those on the right. Now you see another important reason for defining matrix multiplication as we did.

We are interested in finding a column matrix X that satisfies the matrix equation $AX = B$. In Example 1 we found that if $AX = B$ and if A^{-1} exists, then

$$X = A^{-1}B$$

Algebraic Solution

The inverse of A was found in Example 2 in Section 9.2 to be

$$A^{-1} = \begin{bmatrix} 3 & 3 & -1 \\ -2 & -2 & 1 \\ -4 & -5 & 2 \end{bmatrix}$$

Thus,

$$\overset{X}{\begin{bmatrix} x_1 \\ x_2 \\ x_3 \end{bmatrix}} = \overset{A^{-1}}{\begin{bmatrix} 3 & 3 & -1 \\ -2 & -2 & 1 \\ -4 & -5 & 2 \end{bmatrix}} \overset{B}{\begin{bmatrix} 1 \\ 1 \\ 1 \end{bmatrix}} = \begin{bmatrix} 5 \\ -3 \\ -7 \end{bmatrix}$$

and we can conclude that $x_1 = 5$, $x_2 = -3$, and $x_3 = -7$. Check this result in system (1).

Graphing Utility Solution

To solve this problem on a graphing utility, enter A and B (Fig. 1) and simply type $A^{-1}B$ (Fig. 2).

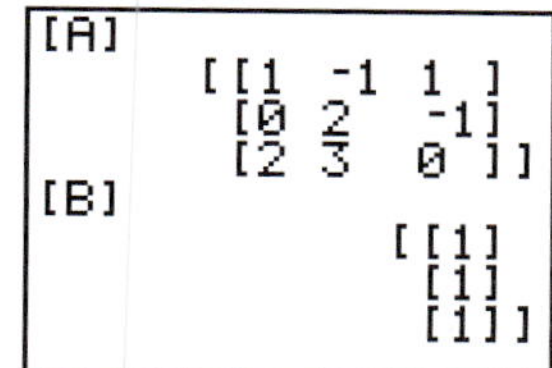

FIGURE 1

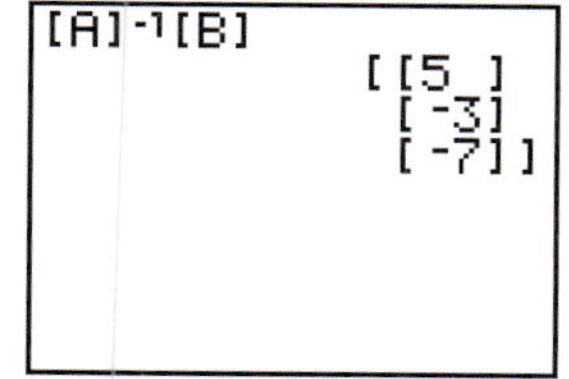

FIGURE 2

MATCHED 2 PROBLEM

Use matrix inverse methods to solve the system (see Matched Problem 2 in Section 9.2 for the inverse of the coefficient matrix):

$$\begin{aligned} 3x_1 - x_2 + x_3 &= 1 \\ -x_1 + x_2 \qquad &= 3 \\ x_1 \qquad + x_3 &= 2 \end{aligned}$$

At first glance, using matrix inverse methods seems to require the same amount of effort as using Gauss–Jordan elimination. In either case, row operations must be applied to an augmented matrix involving the coefficients of the system. The advantage of the inverse matrix method becomes readily apparent when solving a number of systems with a common coefficient matrix and different constant terms.

EXAMPLE 3 Using Inverses to Solve Systems of Equations

Use matrix inverse methods to solve each of the following systems:

(A)
$$\begin{aligned} x_1 - x_2 + x_3 &= 3 \\ 2x_2 - x_3 &= 1 \\ 2x_1 + 3x_2 \quad &= 4 \end{aligned}$$

(B)
$$\begin{aligned} x_1 - x_2 + x_3 &= -5 \\ 2x_2 - x_3 &= 2 \\ 2x_1 + 3x_2 \quad &= -3 \end{aligned}$$

SOLUTIONS

Notice that both systems have the same coefficient matrix A as system (1) in Example 2. Only the constant terms have been changed. Thus, we can use A^{-1} to solve these systems just as we did in Example 2.

(A) Algebraic Solution

$$\overset{X}{\begin{bmatrix} x_1 \\ x_2 \\ x_3 \end{bmatrix}} = \overset{A^{-1}}{\begin{bmatrix} 3 & 3 & -1 \\ -2 & -2 & 1 \\ -4 & -5 & 2 \end{bmatrix}} \overset{B}{\begin{bmatrix} 3 \\ 1 \\ 4 \end{bmatrix}} = \begin{bmatrix} 8 \\ -4 \\ -9 \end{bmatrix}$$

Thus, $x_1 = 8$, $x_2 = -4$, and $x_3 = -9$

(A) Graphing Utility Solution

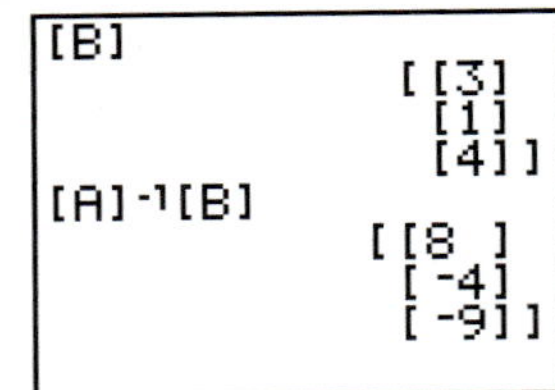

FIGURE 3

(B) Algebraic Solution

$$\overset{X}{\begin{bmatrix} x_1 \\ x_2 \\ x_3 \end{bmatrix}} = \overset{A^{-1}}{\begin{bmatrix} 3 & 3 & -1 \\ -2 & -2 & 1 \\ -4 & -5 & 2 \end{bmatrix}} \overset{B}{\begin{bmatrix} -5 \\ 2 \\ -3 \end{bmatrix}} = \begin{bmatrix} -6 \\ 3 \\ 4 \end{bmatrix}$$

Thus, $x_1 = -6$, $x_2 = 3$, and $x_3 = 4$

(B) Graphing Utility Solution

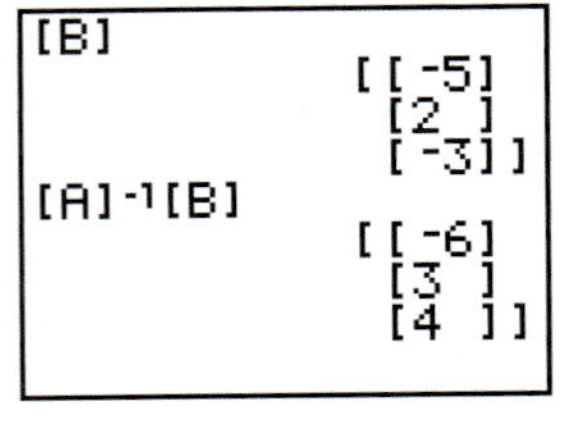

FIGURE 4

MATCHED 3 PROBLEM

Use matrix inverse methods to solve each of the following systems (see Matched Problem 2):

(A)
$$\begin{aligned} 3x_1 - x_2 + x_3 &= 3 \\ -x_1 + x_2 \quad &= -3 \\ x_1 \quad + x_3 &= 2 \end{aligned}$$

(B)
$$\begin{aligned} 3x_1 - x_2 + x_3 &= -5 \\ -x_1 + x_2 \quad &= 1 \\ x_1 \quad + x_3 &= -4 \end{aligned}$$

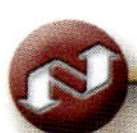

EXPLORE/DISCUSS 2

Use matrix inverse methods to solve each of the following systems, if possible, otherwise use Gauss–Jordan elimination. Describe the types of systems that can be solved by inverse methods and those that cannot. Are there any systems that cannot be solved by Gauss–Jordan elimination?

(A)
$$\begin{aligned} x_1 - x_2 &= 1 \\ x_1 + x_2 &= 7 \\ 3x_1 - x_2 &= 9 \end{aligned}$$

(B)
$$\begin{aligned} x_1 - x_2 + x_3 &= 1 \\ x_1 + x_2 - x_3 &= 7 \\ 3x_1 - x_2 + x_3 &= 9 \end{aligned}$$

(C)
$$\begin{aligned} x_1 - x_2 + x_3 &= 1 \\ x_1 + x_2 - x_3 &= 7 \\ 3x_1 - x_2 + x_3 &= 8 \end{aligned}$$

(D)
$$\begin{aligned} x_1 - x_2 + x_3 &= 1 \\ x_1 + x_2 - x_3 &= 7 \\ 3x_1 - x_2 + 2x_3 &= 8 \end{aligned}$$

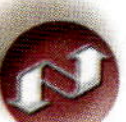

Using Inverse Methods to Solve Systems of Equations

If the number of equations in a system equals the number of variables and the coefficient matrix has an inverse, then the system will always have a unique solution that can be found by using the inverse of the coefficient matrix to solve the corresponding matrix equation.

Matrix equation	Solution
$AX = B$	$X = A^{-1}B$

REMARK What happens if the coefficient matrix does not have an inverse? In this case, it can be shown that the system does not have a unique solution and is either dependent or inconsistent. Gauss–Jordan elimination must be used to determine which is the case. Also, as we mentioned earlier, Gauss–Jordan elimination must always be used if the number of variables is not the same as the number of equations.

Application

The application in Example 4 illustrates the usefulness of the inverse method.

EXAMPLE 4 Investment Allocation

An investment adviser currently has two types of investments available for clients: an investment M that pays 10% per year and an investment N of higher risk that pays 20% per year. Clients may divide their investments between the two to achieve any total return desired between 10% and 20%. However, the higher the desired return, the higher the risk. How should each client listed in the table invest to achieve the indicated return?

	Client			
	1	2	3	k
Total investment	\$20,000	\$50,000	\$10,000	k_1
Annual return desired	\$2,400	\$7,500	\$1,300	k_2
	(12%)	(15%)	(13%)	

SOLUTION

We first solve the problem for an arbitrary client k using inverses, and then apply the result to the three specific clients.

Let

x_1 = Amount invested in M

x_2 = Amount invested in N

Then

$$x_1 + x_2 = k_1 \qquad \text{Total invested}$$
$$0.1x_1 + 0.2x_2 = k_2 \qquad \text{Total annual return}$$

Write as a matrix equation:

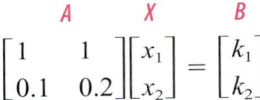

$$\overset{A}{\begin{bmatrix} 1 & 1 \\ 0.1 & 0.2 \end{bmatrix}} \overset{X}{\begin{bmatrix} x_1 \\ x_2 \end{bmatrix}} = \overset{B}{\begin{bmatrix} k_1 \\ k_2 \end{bmatrix}}$$

FIGURE 5

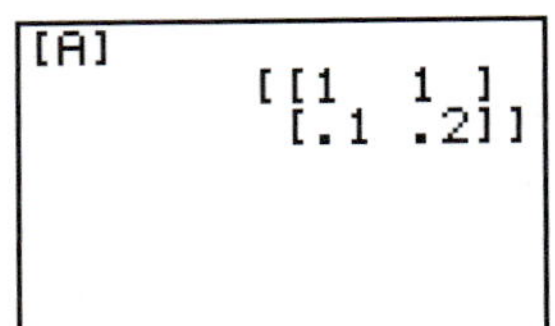

If A^{-1} exists, then

$$X = A^{-1}B$$

To solve each client's investment problem on a graphing utility, first we enter A (Fig. 5), then we enter the appropriate values for B and compute $A^{-1}B$ (Fig. 6).

FIGURE 6

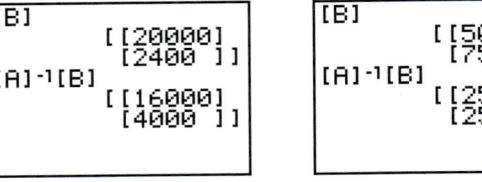

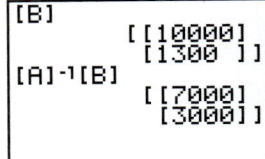

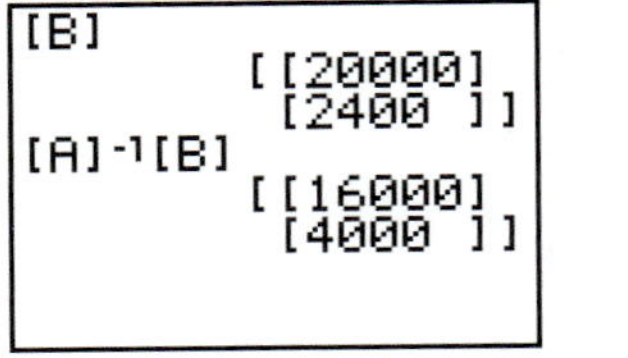

(a) Client 1

[B]
[[50000]
[7500]]
[A]⁻¹[B]
[[25000]
[25000]]

(b) Client 2

[B]
[[10000]
[1300]]
[A]⁻¹[B]
[[7000]
[3000]]

(c) Client 3

From Figure 6, we see that Client 1 should invest \$16,000 in investment M and \$4,000 in investment N, Client 2 should invest \$25,000 in investment M and \$25,000 in investment N, and Client 3 should invest \$7,000 in investment M and \$3,000 in investment N.

MATCHED 4 PROBLEM

Repeat Example 4 with investment M paying 8% and investment N paying 24%.

ANSWERS TO MATCHED PROBLEMS

1.
$$AX + C = B$$
$$\begin{aligned} (AX + C) - C &= B - C \\ AX + (C - C) &= B - C \\ AX + 0 &= B - C \end{aligned}$$
$$AX = B - C$$
$$\begin{aligned} A^{-1}(AX) &= A^{-1}(B - C) \\ (A^{-1}A)X &= A^{-1}(B - C) \\ IX &= A^{-1}(B - C) \end{aligned}$$
$$X = A^{-1}(B - C)$$

2. $x_1 = 2, x_2 = 5, x_3 = 0$ **3.** (A) $x_1 = -2, x_2 = -5, x_3 = 4$ (B) $x_1 = 0, x_2 = 1, x_3 = -4$

4. $A^{-1} = \begin{bmatrix} 1.5 & -6.25 \\ -0.5 & 6.25 \end{bmatrix}$; Client 1: \$15,000 in M and \$5,000 in N; Client 2: \$28,125 in M and \$21,875 in N; Client 3: \$6,875 in M and \$3,125 in N

EXERCISE 9.3

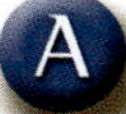

Write Problems 1–4 as systems of linear equations without matrices.

1. $\begin{bmatrix} 2 & -1 \\ 1 & 3 \end{bmatrix}\begin{bmatrix} x_1 \\ x_2 \end{bmatrix} = \begin{bmatrix} 3 \\ -2 \end{bmatrix}$ **2.** $\begin{bmatrix} -3 & 1 \\ -1 & 2 \end{bmatrix}\begin{bmatrix} x_1 \\ x_2 \end{bmatrix} = \begin{bmatrix} -2 \\ 5 \end{bmatrix}$

3. $\begin{bmatrix} -2 & 0 & 1 \\ 1 & 2 & 1 \\ 0 & 1 & -1 \end{bmatrix}\begin{bmatrix} x_1 \\ x_2 \\ x_3 \end{bmatrix} = \begin{bmatrix} 3 \\ -4 \\ 2 \end{bmatrix}$

4. $\begin{bmatrix} 1 & -2 & 0 \\ -3 & 1 & -1 \\ 2 & 0 & 4 \end{bmatrix}\begin{bmatrix} x_1 \\ x_2 \\ x_3 \end{bmatrix} = \begin{bmatrix} 3 \\ -2 \\ 5 \end{bmatrix}$

Write each system in Problems 5–8 as a matrix equation of the form $AX = B$.

5. $\begin{aligned} 4x_1 - 3x_2 &= 2 \\ x_1 + 2x_2 &= 1 \end{aligned}$ **6.** $\begin{aligned} x_1 - 2x_2 &= 7 \\ -3x_1 + x_2 &= -3 \end{aligned}$

7. $\begin{aligned} x_1 - 2x_2 + x_3 &= -1 \\ -x_1 + x_2 &= 2 \\ 2x_1 + 3x_2 + x_3 &= -3 \end{aligned}$ **8.** $\begin{aligned} 2x_1 + 3x_3 &= 5 \\ x_1 - 2x_2 + x_3 &= -4 \\ -x_1 + 3x_2 &= 2 \end{aligned}$

In Problems 9–12, find x_1 and x_2.

9. $\begin{bmatrix} x_1 \\ x_2 \end{bmatrix} = \begin{bmatrix} 3 & -2 \\ 1 & 4 \end{bmatrix}\begin{bmatrix} -2 \\ 1 \end{bmatrix}$ **10.** $\begin{bmatrix} x_1 \\ x_2 \end{bmatrix} = \begin{bmatrix} -2 & 1 \\ -1 & 2 \end{bmatrix}\begin{bmatrix} 3 \\ -2 \end{bmatrix}$

11. $\begin{bmatrix} x_1 \\ x_2 \end{bmatrix} = \begin{bmatrix} -2 & 3 \\ 2 & -1 \end{bmatrix}\begin{bmatrix} 3 \\ 2 \end{bmatrix}$ **12.** $\begin{bmatrix} x_1 \\ x_2 \end{bmatrix} = \begin{bmatrix} 3 & -1 \\ 0 & 2 \end{bmatrix}\begin{bmatrix} -2 \\ 1 \end{bmatrix}$

In Problems 13–16, find x_1 and x_2.

13. $\begin{bmatrix} 1 & -1 \\ 1 & -2 \end{bmatrix}\begin{bmatrix} x_1 \\ x_2 \end{bmatrix} = \begin{bmatrix} 5 \\ 7 \end{bmatrix}$ **14.** $\begin{bmatrix} 1 & 3 \\ 1 & 4 \end{bmatrix}\begin{bmatrix} x_1 \\ x_2 \end{bmatrix} = \begin{bmatrix} 9 \\ 6 \end{bmatrix}$

15. $\begin{bmatrix} 1 & 1 \\ 2 & -3 \end{bmatrix}\begin{bmatrix} x_1 \\ x_2 \end{bmatrix} = \begin{bmatrix} 15 \\ 10 \end{bmatrix}$ **16.** $\begin{bmatrix} 1 & 1 \\ 3 & -2 \end{bmatrix}\begin{bmatrix} x_1 \\ x_2 \end{bmatrix} = \begin{bmatrix} 10 \\ 20 \end{bmatrix}$

B

Write each system in Problems 17–24 as a matrix equation and solve using inverses.

17. $\begin{aligned} x_1 + 2x_2 &= k_1 \\ x_1 + 3x_2 &= k_2 \end{aligned}$

(A) $k_1 = 1, k_2 = 3$
(B) $k_1 = 3, k_2 = 5$
(C) $k_1 = -2, k_2 = 1$

18. $\begin{aligned} 2x_1 + x_2 &= k_1 \\ 5x_1 + 3x_2 &= k_2 \end{aligned}$

(A) $k_1 = 2, k_2 = 13$
(B) $k_1 = -2, k_2 = 4$
(C) $k_1 = 1, k_2 = -3$

19. $\begin{aligned} x_1 + 3x_2 &= k_1 \\ 2x_1 + 7x_2 &= k_2 \end{aligned}$

(A) $k_1 = 2, k_2 = -1$
(B) $k_1 = 1, k_2 = 0$
(C) $k_1 = 3, k_2 = -1$

20. $\begin{aligned} 2x_1 + x_2 &= k_1 \\ x_1 + x_2 &= k_2 \end{aligned}$

(A) $k_1 = -1, k_2 = -2$
(B) $k_1 = 2, k_2 = 3$
(C) $k_1 = 2, k_2 = 0$

21. $\begin{aligned} x_1 - 2x_2 &= k_1 \\ x_2 + x_3 &= k_2 \\ 2x_1 - x_2 + 2x_3 &= k_3 \end{aligned}$

(A) $k_1 = 1, k_2 = 0, k_3 = 2$
(B) $k_1 = -1, k_2 = 1, k_3 = 0$
(C) $k_1 = 2, k_2 = -2, k_3 = 1$

22. $\begin{aligned} x_1 + 3x_2 &= k_1 \\ x_1 + 2x_2 + 3x_3 &= k_2 \\ -x_2 + 2x_3 &= k_3 \end{aligned}$

(A) $k_1 = 0, k_2 = 2, k_3 = 1$
(B) $k_1 = -2, k_2 = 0, k_3 = 1$
(C) $k_1 = 3, k_2 = 1, k_3 = 0$

23. $\begin{aligned} x_1 + x_2 &= k_1 \\ 2x_2 - x_3 &= k_2 \\ x_1 + x_3 &= k_3 \end{aligned}$

(A) $k_1 = 2, k_2 = 0, k_3 = 4$
(B) $k_1 = 0, k_2 = 4, k_3 = -2$
(C) $k_1 = 4, k_2 = 2, k_3 = 0$

24. $\begin{aligned} x_1 - x_3 &= k_1 \\ 2x_1 - x_2 &= k_2 \\ x_1 + x_2 - 4x_3 &= k_3 \end{aligned}$

(A) $k_1 = 4, k_2 = 8, k_3 = 0$
(B) $k_1 = 4, k_2 = 0, k_3 = -4$
(C) $k_1 = 0, k_2 = 8, k_3 = -8$

In Problems 25–30, explain why the system cannot be solved by matrix inverse methods. Discuss methods that could be used and then solve the system.

25. $\begin{aligned} -2x_1 + 4x_2 &= -5 \\ 6x_1 - 12x_2 &= 15 \end{aligned}$

26. $\begin{aligned} -2x_1 + 4x_2 &= 5 \\ 6x_1 - 12x_2 &= 15 \end{aligned}$

27. $\begin{aligned} x_1 - 3x_2 - 2x_3 &= -1 \\ -2x_1 + 6x_2 + 4x_3 &= 3 \end{aligned}$

28. $\begin{aligned} x_1 - 3x_2 - 2x_3 &= -1 \\ -2x_1 + 7x_2 + 3x_3 &= 3 \end{aligned}$

29. $\begin{aligned} x_1 - 2x_2 + 3x_3 &= 1 \\ 2x_1 - 3x_2 - 2x_3 &= 3 \\ x_1 - x_2 - 5x_3 &= 2 \end{aligned}$

30. $\begin{aligned} x_1 - 2x_2 + 3x_3 &= 1 \\ 2x_1 - 3x_2 - 2x_3 &= 3 \\ x_1 - x_2 - 5x_3 &= 4 \end{aligned}$

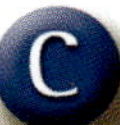

C

For $n \times n$ matrices A and B and $n \times 1$ matrices C, D, and X, solve each matrix equation in Problems 31–36 for X. Assume all necessary inverses exist.

31. $AX - BX = C$

32. $AX + BX = C$

33. $AX + X = C$

34. $AX - X = C$

35. $AX - C = D - BX$

36. $AX + C = BX + D$

37. Use matrix inverse methods to solve the following system for the indicated values of k_1 and k_2.

$$\begin{aligned} x_1 + 2.001x_2 &= k_1 \\ x_1 + 2x_2 &= k_2 \end{aligned}$$

(A) $k_1 = 1, k_2 = 1$
(B) $k_1 = 1, k_2 = 0$
(C) $k_1 = 0, k_2 = 1$

Discuss the effect of small changes in the constant terms on the solution set of this system.

38. Repeat Problem 37 for the following system:

$$\begin{aligned} x_1 - 3.001x_2 &= k_1 \\ x_1 - 3x_2 &= k_2 \end{aligned}$$

APPLICATIONS

Solve using systems of equations and matrix inverse methods.

39. Resource Allocation. A concert hall has 10,000 seats and two categories of ticket prices, \$4 and \$8. Assume all seats in each category can be sold.

	Concert		
	1	2	3
Tickets sold	10,000	10,000	10,000
Return required	\$56,000	\$60,000	\$68,000

(A) How many tickets of each category should be sold to bring in each of the returns indicated in the table?

(B) Is it possible to bring in a return of \$9,000? Of \$3,000? Explain.

(C) Describe all the possible returns.

40. Production Scheduling. Labor and material costs for manufacturing two guitar models are given in the following table:

Guitar Model	Labor Cost	Material Cost
A	\$30	\$20
B	\$40	\$30

(A) If a total of \$3,000 a week is allowed for labor and material, how many of each model should be produced each week to exactly use each of the allocations of the \$3,000 indicated in the following table?

	Weekly Allocation		
	1	2	3
Labor	\$1,800	\$1,750	\$1,720
Material	\$1,200	\$1,250	\$1,280

(B) Is it possible to use an allocation of \$1,600 for labor and \$1,400 for material? Of \$2,000 for labor and \$1,000 for material? Explain.

★ **41. Circuit Analysis.** A direct current electric circuit consisting of conductors (wires), resistors, and batteries is diagrammed in the figure.

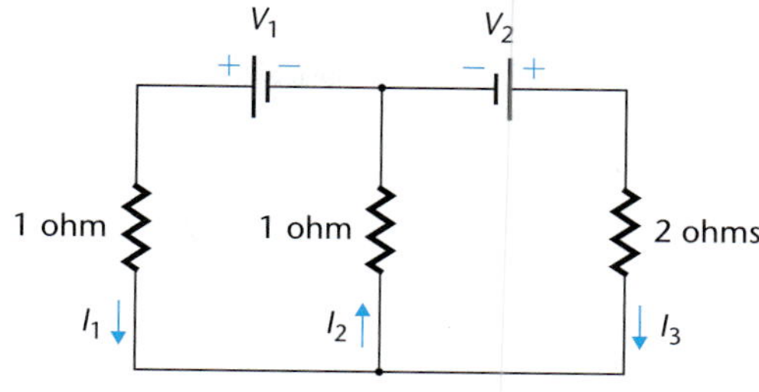

If I_1, I_2, and I_3 are the currents (in amperes) in the three branches of the circuit and V_1 and V_2 are the voltages (in volts) of the two batteries, then *Kirchhoff's* laws* can be used to show that the currents satisfy the following system of equations:

$$\begin{aligned} I_1 - I_2 + I_3 &= 0 \\ I_1 + I_2 \quad\quad &= V_1 \\ I_2 + 2I_3 &= V_2 \end{aligned}$$

Solve this system for

(A) $V_1 = 10$ volts, $V_2 = 10$ volts

(B) $V_1 = 10$ volts, $V_2 = 15$ volts

(C) $V_1 = 15$ volts, $V_2 = 10$ volts

★ **42. Circuit Analysis.** Repeat Problem 41 for the electric circuit shown in the figure.

$$\begin{aligned} I_1 - I_2 + I_3 &= 0 \\ I_1 + 2I_2 \quad\quad &= V_1 \\ 2I_2 + 2I_3 &= V_2 \end{aligned}$$

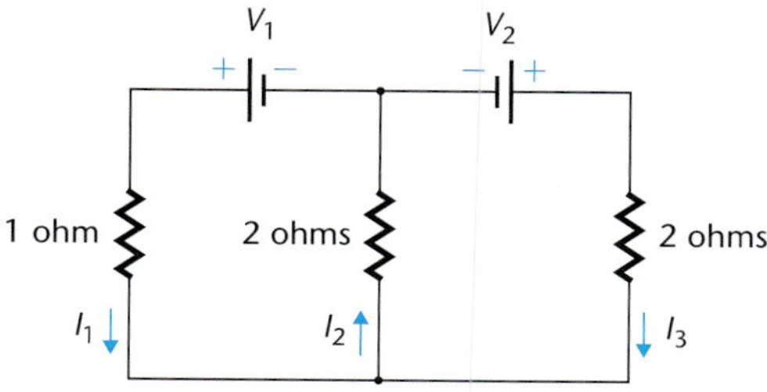

★★ **43. Geometry.** The graph of $f(x) = ax^2 + bx + c$ passes through the points $(1, k_1)$, $(2, k_2)$, and $(3, k_3)$. Determine a, b, and c for

(A) $k_1 = -2, k_2 = 1, k_3 = 6$

(B) $k_1 = 4, k_2 = 3, k_3 = -2$

(C) $k_1 = 8, k_2 = -5, k_3 = 4$

*Gustav Kirchhoff (1824–1887), a German physicist, was among the first to apply theoretical mathematics to physics. He is best known for his development of certain properties of electric circuits, which are now known as **Kirchhoff's laws.**

★★ **44. Geometry.** Repeat Problem 43 if the graph passes through the points $(-1, k_1)$, $(0, k_2)$, and $(1, k_3)$.

Check your answers in Problems 43 and 44 by graphing $y = f(x)$ on a graphing utility and verifying that the graph passes through the indicated points.

45. Diets. A biologist has available two commercial food mixes with the following percentages of protein and fat:

Mix	Protein (%)	Fat (%)
A	20	2
B	10	6

(A) How many ounces of each mix should be used to prepare each of the diets listed in the following table?

	Diet		
	1	2	3
Protein	20 oz	10 oz	10 oz
Fat	6 oz	4 oz	6 oz

(B) Is it possible to prepare a diet consisting of 20 ounces of protein and 14 ounces of fat? Of 20 ounces of protein and 1 ounce of fat? Explain.

SECTION 9.4 Determinants

Determinants • Second-Order Determinants • Third-Order Determinants • Higher-Order Determinants

Determinants

In Section 9.4 we are going to associate with each square matrix a real number, called the **determinant** of the matrix. If A is a square matrix, then the determinant of A is denoted by **det *A*,** or simply by writing the array of elements in A using vertical lines in place of square brackets. For example,

$$\det\begin{bmatrix} 2 & -3 \\ 5 & 1 \end{bmatrix} = \begin{vmatrix} 2 & -3 \\ 5 & 1 \end{vmatrix} \qquad \det\begin{bmatrix} 1 & -2 & 3 \\ 0 & 5 & -7 \\ -2 & 1 & 6 \end{bmatrix} = \begin{vmatrix} 1 & -2 & 3 \\ 0 & 5 & -7 \\ -2 & 1 & 6 \end{vmatrix}$$

A determinant of **order *n*** is a determinant with n rows and n columns. In this section we concentrate most of our attention on determining the values of determinants of orders 2 and 3. But many of the results and procedures discussed can be generalized completely to determinants of order n.

Second-Order Determinants

In general, a **second-order determinant** is written as

$$\begin{vmatrix} a_{11} & a_{12} \\ a_{21} & a_{22} \end{vmatrix}$$

and represents a real number as given in Definition 1.

DEFINITION 1

Value of a Second-Order Determinant

$$\begin{vmatrix} a_{11} & a_{12} \\ a_{21} & a_{22} \end{vmatrix} = a_{11}a_{22} - a_{21}a_{12} \qquad (1)$$

Formula (1) is easily remembered if you notice that the expression on the right is the product of the **principal diagonal,** from upper left to lower right, minus the product of the **secondary diagonal,** from lower left to upper right.

EXAMPLE 1 Evaluating a Second-Order Determinant

Find $\begin{vmatrix} -1 & 2 \\ -3 & -4 \end{vmatrix}$

SOLUTION

Algebraic Solution

$$\begin{vmatrix} -1 & 2 \\ -3 & -4 \end{vmatrix} = (-1)(-4) - (-3)(2) = 4 - (-6) = 10$$

Graphing Utility Solution

Enter the given matrix and use the det command (Fig. 1).

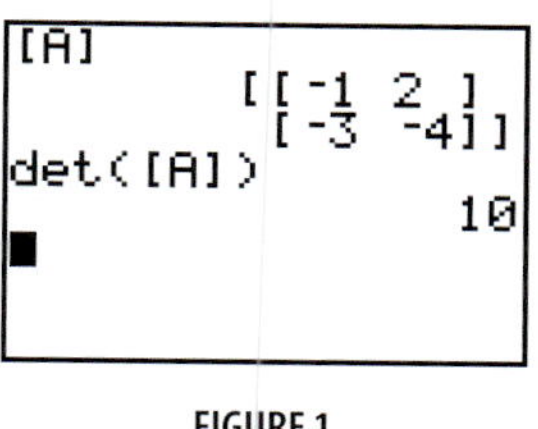

FIGURE 1

MATCHED 1 PROBLEM

Find: $\begin{vmatrix} 3 & -5 \\ 4 & -2 \end{vmatrix}$

Third-Order Determinants

A determinant of order 3 is a square array of nine elements and represents a real number given by Definition 2, which is a special case of the general definition of the value of an nth-order determinant. Note that each term in the expansion on the right of equation (2) contains exactly one element from each row and each column.

DEFINITION 2
Value of a Third-Order Determinant

$$\begin{vmatrix} a_{11} & a_{12} & a_{13} \\ a_{21} & a_{22} & a_{23} \\ a_{31} & a_{32} & a_{33} \end{vmatrix} = a_{11}a_{22}a_{33} - a_{11}a_{32}a_{23} + a_{21}a_{32}a_{13} - a_{21}a_{12}a_{33} + a_{31}a_{12}a_{23} - a_{31}a_{22}a_{13} \tag{2}$$

Don't panic! You don't need to memorize formula (2). After we introduce the ideas of *minor* and *cofactor* below, we will state a theorem that can be used to obtain the same result with much less trouble.

The **minor of an element** in a third-order determinant is a second-order determinant obtained by deleting the row and column that contains the element. For example, in the determinant in formula (2),

$$\text{Minor of } a_{23} = \begin{vmatrix} a_{11} & a_{12} \\ a_{31} & a_{32} \end{vmatrix} \qquad \begin{vmatrix} a_{11} & a_{12} & a_{13} \\ a_{21} & a_{22} & a_{23} \\ a_{31} & a_{32} & a_{33} \end{vmatrix}$$

Deletions are usually done mentally.

$$\text{Minor of } a_{32} = \begin{vmatrix} a_{11} & a_{13} \\ a_{21} & a_{23} \end{vmatrix} \qquad \begin{vmatrix} a_{11} & a_{12} & a_{13} \\ a_{21} & a_{22} & a_{23} \\ a_{31} & a_{32} & a_{33} \end{vmatrix}$$

EXPLORE/DISCUSS 1

Write the minors of the other seven elements in the determinant in formula (2).

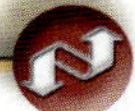

A quantity closely associated with the minor of an element is the **cofactor of an element** a_{ij} (from the ith row and jth column), which is the product of the minor of a_{ij} and $(-1)^{i+j}$.

DEFINITION 3
Cofactor

Cofactor of $a_{ij} = (-1)^{i+j}(\text{Minor of } a_{ij})$

Thus, a cofactor of an element is nothing more than a signed minor. The sign is determined by raising -1 to a power that is the sum of the numbers indicating the row and column in which the element appears. Note that $(-1)^{i+j}$ is 1 if $i + j$ is even and -1 if $i + j$ is odd. Thus, if we are given the determinant

$$\begin{vmatrix} a_{11} & a_{12} & a_{13} \\ a_{21} & a_{22} & a_{23} \\ a_{31} & a_{32} & a_{33} \end{vmatrix}$$

then

$$\text{Cofactor of } a_{23} = (-1)^{2+3}\begin{vmatrix} a_{11} & a_{12} \\ a_{31} & a_{32} \end{vmatrix} = -\begin{vmatrix} a_{11} & a_{12} \\ a_{31} & a_{32} \end{vmatrix}$$

$$\text{Cofactor of } a_{11} = (-1)^{1+1}\begin{vmatrix} a_{22} & a_{23} \\ a_{32} & a_{33} \end{vmatrix} = \begin{vmatrix} a_{22} & a_{23} \\ a_{32} & a_{33} \end{vmatrix}$$

EXAMPLE 2 Finding Cofactors

Find the cofactors of -2 and 5 in the determinant

$$\begin{vmatrix} -2 & 0 & 3 \\ 1 & -6 & 5 \\ -1 & 2 & 0 \end{vmatrix}$$

SOLUTION

$$\text{Cofactor of } -2 = (-1)^{1+1}\begin{vmatrix} -6 & 5 \\ 2 & 0 \end{vmatrix} = \begin{vmatrix} -6 & 5 \\ 2 & 0 \end{vmatrix}$$

$$= (-6)(0) - (2)(5) = -10$$

$$\text{Cofactor of } 5 = (-1)^{2+3}\begin{vmatrix} -2 & 0 \\ -1 & 2 \end{vmatrix} = -\begin{vmatrix} -2 & 0 \\ -1 & 2 \end{vmatrix}$$

$$= -[(-2)(2) - (-1)(0)] = 4$$

MATCHED 2 PROBLEM

Find the cofactors of 2 and 3 in the determinant in Example 2.

[*Note:* The sign in front of the minor, $(-1)^{i+j}$, can be determined rather mechanically by using a checkerboard pattern of + and − signs over the determinant, starting with + in the upper left-hand corner:

$$\begin{matrix} + & - & + \\ - & + & - \\ + & - & + \end{matrix}$$

Use either the checkerboard or the exponent method—whichever is easier for you—to determine the sign in front of the minor.]

Now we are ready for the key theorem of this section, Theorem 1. This theorem provides us with an efficient step-by-step procedure, called an *algorithm,* for evaluating third-order determinants.

> **THEOREM 1**
> **Value of a Third-Order Determinant**
>
> The value of a determinant of order 3 is the sum of three products obtained by multiplying each element of any one row (or each element of any one column) by its cofactor.

To prove this theorem we must show that the expansions indicated by the theorem for any row or any column (six cases) produce the expression on the right of formula (2). Proofs of special cases of this theorem are left to the C problems in Exercise 9.4.

EXAMPLE 3 Evaluating a Third-Order Determinant

$$\text{Evaluate } \begin{vmatrix} 2 & -2 & 0 \\ -3 & 1 & 2 \\ 1 & -3 & -1 \end{vmatrix}$$

SOLUTION

Algebraic Solution

Expanding the first row, we have

$$\begin{vmatrix} 2 & -2 & 0 \\ -3 & 1 & 2 \\ 1 & -3 & -1 \end{vmatrix}$$

$$= a_{11}\begin{pmatrix}\text{Cofactor}\\ \text{of } a_{11}\end{pmatrix} + a_{12}\begin{pmatrix}\text{Cofactor}\\ \text{of } a_{12}\end{pmatrix} + a_{13}\begin{pmatrix}\text{Cofactor}\\ \text{of } a_{13}\end{pmatrix}$$

$$= 2\left[(-1)^{1+1}\begin{vmatrix} 1 & 2 \\ -3 & -1 \end{vmatrix}\right] + (-2)\left[(-1)^{1+2}\begin{vmatrix} -3 & 2 \\ 1 & -1 \end{vmatrix}\right] + 0$$

$$= (2)(1)[(1)(-1) - (-3)(2)] + (-2)(-1)[(-3)(-1) - (1)(2)]$$

$$= (2)(5) + (2)(1) = 12$$

Graphing Utility Solution

Enter the matrix and use the det command (Fig. 2).

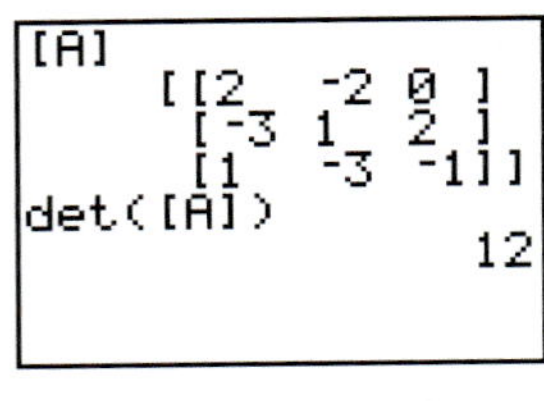

FIGURE 2

MATCHED PROBLEM 3

$$\text{Evaluate } \begin{vmatrix} 2 & 1 & -1 \\ -2 & -3 & 0 \\ -1 & 2 & 1 \end{vmatrix}$$

Refer to Example 3. According to Theorem 1, we should get the same value if we expand any other row or column. Example 4 verifies this for the second column.

EXAMPLE 4 Expanding a Different Row or Column

$$\text{Expand the second column to evaluate } \begin{vmatrix} 2 & -2 & 0 \\ -3 & 1 & 2 \\ 1 & -3 & -1 \end{vmatrix}$$

SOLUTION

$$\begin{vmatrix} 2 & -2 & 0 \\ -3 & 1 & 2 \\ 1 & -3 & -1 \end{vmatrix} = a_{12}\begin{pmatrix}\text{Cofactor} \\ \text{of } a_{12}\end{pmatrix} + a_{22}\begin{pmatrix}\text{Cofactor} \\ \text{of } a_{22}\end{pmatrix} + a_{32}\begin{pmatrix}\text{Cofactor} \\ \text{of } a_{32}\end{pmatrix}$$

$$= (-2)\left[(-1)^{1+2}\begin{vmatrix} -3 & 2 \\ 1 & -1 \end{vmatrix}\right] + (1)\left[(-1)^{2+2}\begin{vmatrix} 2 & 0 \\ 1 & -1 \end{vmatrix}\right]$$
$$+ (-3)\left[(-1)^{3+2}\begin{vmatrix} 2 & 0 \\ -3 & 2 \end{vmatrix}\right]$$

$$= (-2)(-1)[(-3)(-1) - (1)(2)] + (1)(1)[(2)(-1) - (1)(0)]$$
$$+ (-3)(-1)[(2)(2) - (-3)(0)]$$

$$= (2)(1) + (1)(-2) + (3)(4) = 12$$

MATCHED 4 PROBLEM

Expand a row or column different from the one you used to solve Matched Problem 3 to evaluate

$$\begin{vmatrix} 2 & 1 & -1 \\ -2 & -3 & 0 \\ -1 & 2 & 1 \end{vmatrix}$$

Higher-Order Determinants

Theorem 1 and the definitions of minor and cofactor generalize completely for determinants of order higher than 3. These concepts are illustrated for a fourth-order determinant in Example 5.

EXAMPLE 5 Evaluating a Fourth-Order Determinant

Given the fourth-order determinant

$$\begin{vmatrix} 0 & -1 & 0 & 2 \\ -5 & -6 & 0 & -3 \\ 4 & 5 & -2 & 6 \\ 0 & 3 & 0 & -4 \end{vmatrix}$$

(A) Find the minor in determinant form of the element 3.

(B) Find the cofactor in determinant form of the element -5.

(C) Find the value of the fourth-order determinant.

SOLUTIONS

(A) Minor of 3 $= \begin{vmatrix} 0 & 0 & 2 \\ -5 & 0 & -3 \\ 4 & -2 & 6 \end{vmatrix}$

(B) Cofactor of $-5 = (-1)^{2+1}\begin{vmatrix} -1 & 0 & 2 \\ 5 & -2 & 6 \\ 3 & 0 & -4 \end{vmatrix} = -\begin{vmatrix} -1 & 0 & 2 \\ 5 & -2 & 6 \\ 3 & 0 & -4 \end{vmatrix}$

(C) Algebraic Solution

Generalizing Theorem 1, the value of this fourth-order determinant is the sum of four products obtained by multiplying each element of any one row (or each element of any one column) by its cofactor. The work involved in this evaluation is greatly reduced if we choose the row or column with the greatest number of zeros. Because column 3 has three zeros, we expand along this column:

$$\begin{vmatrix} 0 & -1 & 0 & 2 \\ -5 & -6 & 0 & -3 \\ 4 & 5 & -2 & 6 \\ 0 & 3 & 0 & -4 \end{vmatrix} = 0 + 0 + (-2)(-1)^{3+3}\begin{vmatrix} 0 & -1 & 2 \\ -5 & -6 & -3 \\ 0 & 3 & -4 \end{vmatrix} + 0$$

$$= (-2)\begin{vmatrix} 0 & -1 & 2 \\ -5 & -6 & -3 \\ 0 & 3 & -4 \end{vmatrix}$$ Expand this determinant along the first column.

$$= (-2)\left[0 + (-5)(-1)^{2+1}\begin{vmatrix} -1 & 2 \\ 3 & -4 \end{vmatrix} + 0\right]$$

$$= (-2)(-5)(-1)(-2) = 20$$

Graphing Utility Solution

Enter the matrix and use the det command (Fig. 3).

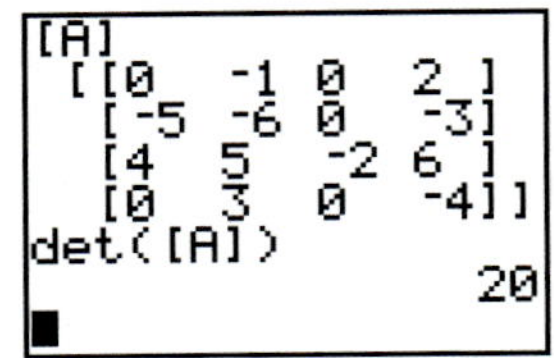

FIGURE 3

MATCHED 5 PROBLEM

Repeat Example 5 for the following fourth-order determinant:

$$\begin{vmatrix} 0 & 4 & -2 & 0 \\ -3 & 3 & -1 & 2 \\ 0 & 6 & 0 & 0 \\ 5 & -6 & -5 & -4 \end{vmatrix}$$

EXPLORE/DISCUSS 2

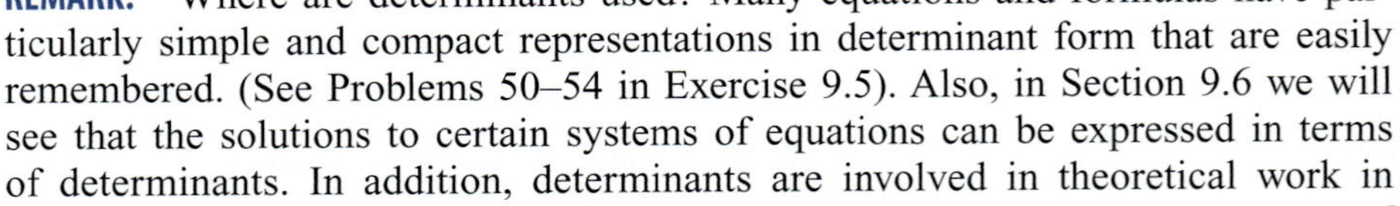

Write a checkerboard pattern of + and − signs for a fourth-order determinant, and use it to determine the signs of the minors in Example 5.

REMARK. Where are determinants used? Many equations and formulas have particularly simple and compact representations in determinant form that are easily remembered. (See Problems 50–54 in Exercise 9.5). Also, in Section 9.6 we will see that the solutions to certain systems of equations can be expressed in terms of determinants. In addition, determinants are involved in theoretical work in advanced mathematics courses. For example, it can be shown that the inverse of a square matrix exists if and only if its determinant is not 0.

ANSWERS MATCHED PROBLEMS

1. 14
2. Cofactor of 2 = 13; cofactor of 3 = −4
3. 3
4. 3
5. (A) $\begin{vmatrix} 0 & -2 & 0 \\ 0 & 0 & 0 \\ 5 & -5 & -4 \end{vmatrix}$ (B) $-\begin{vmatrix} 0 & 4 & 0 \\ -3 & 3 & 2 \\ 0 & 6 & 0 \end{vmatrix}$ (C) −24

EXERCISE 9.4

A

Evaluate each second-order determinant in Problems 1–6.

1. $\begin{vmatrix} 5 & 4 \\ 2 & 3 \end{vmatrix}$ **2.** $\begin{vmatrix} 8 & -3 \\ 4 & 1 \end{vmatrix}$

3. $\begin{vmatrix} 3 & -7 \\ -5 & 6 \end{vmatrix}$ **4.** $\begin{vmatrix} 9 & -2 \\ 4 & 0 \end{vmatrix}$

5. $\begin{vmatrix} 4.3 & -1.2 \\ -5.1 & 3.7 \end{vmatrix}$ **6.** $\begin{vmatrix} -0.7 & -2.3 \\ 1.9 & -4.8 \end{vmatrix}$

Problems 7–14 pertain to the determinant below:

$$\begin{vmatrix} 5 & -1 & -3 \\ 3 & 4 & 6 \\ 0 & -2 & 8 \end{vmatrix}$$

Write the minor of each element given in Problems 7–10. Leave the answer in determinant form.

7. a_{11} **8.** a_{33}

9. a_{23} **10.** a_{12}

Write the cofactor of each element given in Problems 11–14, and evaluate each.

11. a_{11} **12.** a_{33}

13. a_{23} **14.** a_{12}

Evaluate Problems 15–20 using cofactors and using a graphing utility.

15. $\begin{vmatrix} 1 & 0 & 0 \\ -2 & 4 & 3 \\ 5 & -2 & 1 \end{vmatrix}$ **16.** $\begin{vmatrix} 2 & -3 & 5 \\ 0 & -3 & 1 \\ 0 & 6 & 2 \end{vmatrix}$

17. $\begin{vmatrix} 0 & 1 & 5 \\ 3 & -7 & 6 \\ 0 & -2 & -3 \end{vmatrix}$ **18.** $\begin{vmatrix} 4 & -2 & 0 \\ 9 & 5 & 4 \\ 1 & 2 & 0 \end{vmatrix}$

19. $\begin{vmatrix} -1 & 2 & -3 \\ -2 & 0 & -6 \\ 4 & -3 & 2 \end{vmatrix}$ **20.** $\begin{vmatrix} 0 & 2 & -1 \\ -6 & 3 & 1 \\ 7 & -9 & -2 \end{vmatrix}$

B

Given the determinant

$$\begin{vmatrix} a_{11} & a_{12} & a_{13} & a_{14} \\ a_{21} & a_{22} & a_{23} & a_{24} \\ a_{31} & a_{32} & a_{33} & a_{34} \\ a_{41} & a_{42} & a_{43} & a_{44} \end{vmatrix}$$

write the cofactor in determinant form of each element in Problems 21–24.

21. a_{11} **22.** a_{44}

23. a_{43} **24.** a_{23}

Evaluate each determinant in Problems 25–34 using cofactors and using a graphing utility.

25. $\begin{vmatrix} 3 & -2 & -8 \\ -2 & 0 & -3 \\ 1 & 0 & -4 \end{vmatrix}$ **26.** $\begin{vmatrix} 4 & -4 & 6 \\ 2 & 8 & -3 \\ 0 & -5 & 0 \end{vmatrix}$

27. $\begin{vmatrix} 1 & 4 & 1 \\ 1 & 1 & -2 \\ 2 & 1 & -1 \end{vmatrix}$ **28.** $\begin{vmatrix} 3 & 2 & 1 \\ -1 & 5 & 1 \\ 2 & 3 & 1 \end{vmatrix}$

29. $\begin{vmatrix} 1 & 4 & 3 \\ 2 & 1 & 6 \\ 3 & -2 & 9 \end{vmatrix}$

30. $\begin{vmatrix} 4 & -6 & 3 \\ -1 & 4 & 1 \\ 5 & -6 & 3 \end{vmatrix}$

31. $\begin{vmatrix} 2 & 6 & 1 & 7 \\ 0 & 3 & 0 & 0 \\ 3 & 4 & 2 & 5 \\ 0 & 9 & 0 & 2 \end{vmatrix}$

32. $\begin{vmatrix} 0 & 1 & 0 & 1 \\ 2 & 4 & 7 & 6 \\ 0 & 3 & 0 & 1 \\ 0 & 6 & 2 & 5 \end{vmatrix}$

33. $\begin{vmatrix} -2 & 0 & 0 & 0 & 0 \\ 9 & -1 & 0 & 0 & 0 \\ 2 & 1 & 3 & 0 & 0 \\ -1 & 4 & 2 & 2 & 0 \\ 7 & -2 & 3 & 5 & 5 \end{vmatrix}$

34. $\begin{vmatrix} 2 & 0 & 0 & 0 & 0 \\ 0 & 3 & 0 & 0 & 0 \\ 0 & 0 & 2 & 0 & 0 \\ 0 & 0 & 0 & 1 & 0 \\ 0 & 0 & 0 & 0 & 4 \end{vmatrix}$

If A is a 3 × 3 matrix, det A can be evaluated by the following ***diagonal expansion.*** *Form a 3 × 5 matrix by augmenting A on the right with its first two columns, and compute the diagonal products $p_1, p_2, \ldots, p_6$ indicated by the arrows:*

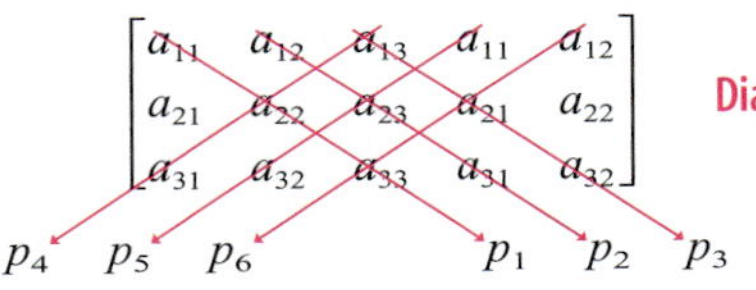

Diagonal expansion formula

The determinant of A is given by [compare with formula (2)]

$$\begin{aligned} \det A &= p_1 + p_2 + p_3 - p_4 - p_5 - p_6 \\ &= a_{11}a_{22}a_{33} + a_{12}a_{23}a_{31} + a_{13}a_{21}a_{32} - a_{13}a_{22}a_{31} \\ &\quad - a_{11}a_{23}a_{32} - a_{12}a_{21}a_{33} \end{aligned}$$

[Caution: The diagonal expansion procedure works only for 3 × 3 matrices. Do not apply it to matrices of any other size.]

Use the diagonal expansion formula to evaluate the determinants in Problems 35 and 36.

35. $\begin{vmatrix} 2 & 6 & -1 \\ 5 & 3 & -7 \\ -4 & -2 & 1 \end{vmatrix}$

36. $\begin{vmatrix} 4 & 1 & -5 \\ 1 & 2 & -6 \\ -3 & -1 & 7 \end{vmatrix}$

A square matrix is called an ***upper triangular matrix*** *if all elements below the principal diagonal are zero. In Problems 37–40, determine whether the statement is true or false. If true, explain why. If false, give a counterexample.*

37. If the determinant of an upper triangular matrix is 0, then the elements on the principal diagonal are all 0.

38. If A and B are upper triangular matrices, then $\det(A + B) = \det A + \det B$.

39. The determinant of an upper triangular matrix is the product of the elements on the principal diagonal.

40. If A and B are upper triangular matrices, then $\det(AB) = (\det A)(\det B)$.

C

In Problems 41–46, all the letters represent real numbers. Find an equation that each pair of determinants satisfies, and describe the relationship between the two determinants verbally.

41. $\begin{vmatrix} a & b \\ c & d \end{vmatrix}, \begin{vmatrix} c & d \\ a & b \end{vmatrix}$

42. $\begin{vmatrix} a & b \\ c & d \end{vmatrix}, \begin{vmatrix} b & a \\ d & c \end{vmatrix}$

43. $\begin{vmatrix} a & b \\ c & d \end{vmatrix}, \begin{vmatrix} ka & b \\ kc & d \end{vmatrix}$

44. $\begin{vmatrix} a & b \\ c & d \end{vmatrix}, \begin{vmatrix} a & b \\ kc & kd \end{vmatrix}$

45. $\begin{vmatrix} a & b \\ c & d \end{vmatrix}, \begin{vmatrix} kc + a & kd + b \\ c & d \end{vmatrix}$

46. $\begin{vmatrix} a & b \\ c & d \end{vmatrix}, \begin{vmatrix} a & ka + b \\ c & kc + d \end{vmatrix}$

47. Show that the expansion of the determinant

$$\begin{vmatrix} a_{11} & a_{12} & a_{13} \\ a_{21} & a_{22} & a_{23} \\ a_{31} & a_{32} & a_{33} \end{vmatrix}$$

by the first column is the same as its expansion by the third row.

48. Repeat Problem 47, using the second row and the third column.

49. If

$$A = \begin{bmatrix} 2 & 3 \\ 1 & -2 \end{bmatrix} \quad \text{and} \quad B = \begin{bmatrix} -1 & 3 \\ 2 & 1 \end{bmatrix}$$

show that $\det(AB) = (\det A)(\det B)$.

50. If

$$A = \begin{bmatrix} a & b \\ c & d \end{bmatrix} \quad \text{and} \quad B = \begin{bmatrix} w & x \\ y & z \end{bmatrix}$$

show that $\det(AB) = (\det A)(\det B)$.

If A is an $n \times n$ matrix and I is the $n \times n$ identity matrix, then the function $f(x) = |xI - A|$ is called the ***characteristic polynomial*** *of A, and the zeros of $f(x)$ are called the* ***eigenvalues*** *of A. Characteristic polynomials and eigenvalues have many important applications that are discussed in more advanced treatments of matrices. In Problems 51–54, find the characteristic polynomial and the eigenvalues of each matrix.*

51. $\begin{bmatrix} 5 & -4 \\ 2 & -1 \end{bmatrix}$

52. $\begin{bmatrix} 8 & -6 \\ 3 & -1 \end{bmatrix}$

53. $\begin{bmatrix} 4 & -4 & 0 \\ 2 & -2 & 0 \\ 4 & -8 & -4 \end{bmatrix}$

54. $\begin{bmatrix} -2 & 2 & 0 \\ -1 & 1 & 0 \\ -2 & 4 & 2 \end{bmatrix}$

SECTION 9.5 Properties of Determinants

Discussion of Determinant Properties • Summary of Determinant Properties

Determinants have useful properties that can greatly reduce the labor in evaluating determinants of order 3 or greater. These properties and their use are the subject matter for Section 9.5.

Discussion of Determinant Properties

We now state and discuss five general determinant properties in the form of theorems. Because the proofs for the general cases of these theorems are involved and notationally difficult, we will sketch only informal proofs for determinants of order 3. The theorems, however, apply to determinants of any order.

THEOREM 1
Multiplying a Row or Column by a Constant

If each element of any row (or column) of a determinant is multiplied by a constant k, the new determinant is k times the original.

PARTIAL PROOF Let C_{ij} be the cofactor of a_{ij}. Then expanding by the first row, we have

$$\begin{vmatrix} ka_{11} & ka_{12} & ka_{13} \\ a_{21} & a_{22} & a_{23} \\ a_{31} & a_{32} & a_{33} \end{vmatrix} = ka_{11}C_{11} + ka_{12}C_{12} + ka_{13}C_{13}$$

$$= k(a_{11}C_{11} + a_{12}C_{12} + a_{13}C_{13})$$

$$= k\begin{vmatrix} a_{11} & a_{12} & a_{13} \\ a_{21} & a_{22} & a_{23} \\ a_{31} & a_{32} & a_{33} \end{vmatrix}$$

Theorem 1 also states that a factor common to all elements of a row (or column) can be taken out as a factor of the determinant.

EXAMPLE 1 Taking Out a Common Factor of a Column

$$\begin{vmatrix} 6 & 1 & 3 \\ -2 & 7 & -2 \\ 4 & 5 & 0 \end{vmatrix} = 2\begin{vmatrix} 3 & 1 & 3 \\ -1 & 7 & -2 \\ 2 & 5 & 0 \end{vmatrix}$$

where 2 is a common factor of the first column.

MATCHED 1 PROBLEM

Take out factors common to any row or any column:

$$\begin{vmatrix} 3 & 2 & 1 \\ 6 & 3 & -9 \\ 1 & 0 & -5 \end{vmatrix}$$

EXPLORE/DISCUSS 1

(A) How are $\begin{vmatrix} a & b \\ c & d \end{vmatrix}$ and $\begin{vmatrix} ka & kb \\ kc & kd \end{vmatrix}$ related?

(B) How are $\begin{vmatrix} a & b & c \\ d & e & f \\ g & h & i \end{vmatrix}$ and $\begin{vmatrix} ka & kb & kc \\ kd & ke & kf \\ kg & kh & ki \end{vmatrix}$ related?

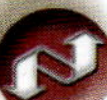

THEOREM 2
Row or Column of Zeros

If every element in a row (or column) is 0, the value of the determinant is 0.

Theorem 2 is an immediate consequence of Theorem 1, and its proof is left as an exercise. It is illustrated in the following example:

$$\begin{vmatrix} 3 & -2 & 5 \\ 0 & 0 & 0 \\ -1 & 4 & 9 \end{vmatrix} = 0$$

THEOREM 3
Interchanging Rows or Columns

If two rows (or two columns) of a determinant are interchanged, the new determinant is the negative of the original.

A proof of Theorem 3 even for a determinant of order 3 is notationally involved. We suggest that you partially prove the theorem by direct expansion of the determinants before and after the interchange of two rows (or columns). The theorem is illustrated by the following example, where the second and third columns are interchanged:

$$\begin{vmatrix} 1 & 0 & 9 \\ -2 & 1 & 5 \\ 3 & 0 & 7 \end{vmatrix} = -\begin{vmatrix} 1 & 9 & 0 \\ -2 & 5 & 1 \\ 3 & 7 & 0 \end{vmatrix}$$

EXPLORE/DISCUSS 2

(A) What are the cofactors of each element in the first row of the following determinant? What is the value of the determinant?

$$\begin{vmatrix} a & b & c \\ d & e & f \\ d & e & f \end{vmatrix}$$

(B) What are the cofactors of each element in the second column of the following determinant? What is the value of the determinant?

$$\begin{vmatrix} a & b & a \\ d & e & d \\ g & h & g \end{vmatrix}$$

THEOREM 4
Equal Rows or Columns

If the corresponding elements are equal in two rows (or columns), the value of the determinant is 0.

PROOF The general proof of Theorem 4 follows directly from Theorem 3. If we start with a determinant D that has two rows (or columns) equal and we interchange the equal rows (or columns), the new determinant will be the same as the original. But by Theorem 3,

$$D = -D$$

hence,

$$2D = 0$$
$$D = 0$$

THEOREM 5
Addition of Rows or Columns

If a multiple of any row (or column) of a determinant is added to any other row (or column), the value of the determinant is not changed.

PARTIAL PROOF If, in a general third-order determinant, we add a k multiple of the second column to the first and then expand by the first column, we obtain (where C_{ij} is the cofactor of a_{ij} in the original determinant)

$$\begin{vmatrix} a_{11} + ka_{12} & a_{12} & a_{13} \\ a_{21} + ka_{22} & a_{22} & a_{23} \\ a_{31} + ka_{32} & a_{32} & a_{33} \end{vmatrix} = (a_{11} + ka_{12})C_{11} + (a_{21} + ka_{22})C_{21} + (a_{31} + ka_{32})C_{31}$$

$$= (a_{11}C_{11} + a_{21}C_{21} + a_{31}C_{31}) + k(a_{12}C_{11} + a_{22}C_{21} + a_{32}C_{31})$$

$$= \begin{vmatrix} a_{11} & a_{12} & a_{13} \\ a_{21} & a_{22} & a_{23} \\ a_{31} & a_{32} & a_{33} \end{vmatrix} + k\begin{vmatrix} a_{12} & a_{12} & a_{13} \\ a_{22} & a_{22} & a_{23} \\ a_{32} & a_{32} & a_{33} \end{vmatrix} = \begin{vmatrix} a_{11} & a_{12} & a_{13} \\ a_{21} & a_{22} & a_{23} \\ a_{31} & a_{32} & a_{33} \end{vmatrix}$$

The determinant following k is 0 because the first and second columns are equal.

Note the similarity in the process described in Theorem 5 to that used to obtain row-equivalent matrices. We use this theorem to transform a determinant without 0 elements into one that contains a row or column with all elements 0 but one. The transformed determinant can then be easily expanded by this row (or column). An example best illustrates the process.

EXAMPLE 2 Evaluating a Determinant

Evaluate the determinant $\begin{vmatrix} 3 & -1 & 2 \\ -2 & 4 & -3 \\ 4 & -2 & 5 \end{vmatrix}$

SOLUTION

Algebraic Solution

We use Theorem 5 to obtain two zeros in the first row, and then expand the determinant by this row. To start, we replace the third column with the sum of it and 2 times the second column to obtain a 0 in the a_{13} position:

$$\begin{vmatrix} 3 & -1 & 2 \\ -2 & 4 & -3 \\ 4 & -2 & 5 \end{vmatrix} = \begin{vmatrix} 3 & -1 & 0 \\ -2 & 4 & 5 \\ 4 & -2 & 1 \end{vmatrix} \quad 2C_2 + C_3 \to C_3{}^*$$

Next, to obtain a 0 in the a_{11} position, we replace the first column with the sum of it and 3 times the second column:

$$\begin{vmatrix} 3 & -1 & 0 \\ -2 & 4 & 5 \\ 4 & -2 & 1 \end{vmatrix} = \begin{vmatrix} 0 & -1 & 0 \\ 10 & 4 & 5 \\ -2 & -2 & 1 \end{vmatrix} \quad 3C_2 + C_1 \to C_1$$

Now it is an easy matter to expand this last determinant by the first row to obtain

$$\begin{vmatrix} 0 & -1 & 0 \\ 10 & 4 & 5 \\ -2 & -2 & 1 \end{vmatrix} = 0 + (-1)\left[(-1)^{1+2}\begin{vmatrix} 10 & 5 \\ -2 & 1 \end{vmatrix}\right] + 0 = 20$$

Graphing Utility Solution

Enter the matrix and use the det command (Fig. 1).

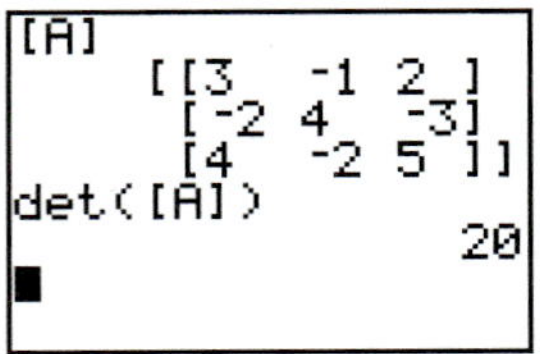

FIGURE 1

*C_1, C_2, and C_3 represent columns 1, 2, and 3, respectively.

MATCHED 2 PROBLEM

Evaluate the following determinant by first using Theorem 5 to obtain zeros in the a_{11} and a_{31} positions, and then expand by the first column.

$$\begin{vmatrix} 3 & 10 & -5 \\ 1 & 6 & -3 \\ 2 & 3 & 4 \end{vmatrix}$$

Summary of Determinant Properties

We now summarize the five determinant properties discussed earlier in Table 1 for convenient reference. Although these properties hold for determinants of any order, for simplicity, we illustrate each property in terms of second-order determinants.

TABLE 1 Summary of Determinant Properties

Property	Examples
1. If each element of any row (or column) of a determinant is multiplied by a constant k, the new determinant is k times the original.	$\begin{vmatrix} 2a & 2b \\ c & d \end{vmatrix} = 2\begin{vmatrix} a & b \\ c & d \end{vmatrix}$ $3\begin{vmatrix} a & b \\ c & d \end{vmatrix} = \begin{vmatrix} 3a & b \\ 3c & d \end{vmatrix}$
2. If every element in a row (or column) is 0, the value of the determinant is 0.	$\begin{vmatrix} a & b \\ 0 & 0 \end{vmatrix} = 0$ $\begin{vmatrix} 0 & b \\ 0 & d \end{vmatrix} = 0$
3. If two rows (or two columns) of a determinant are interchanged, the new determinant is the negative of the original.	$\begin{vmatrix} a & b \\ c & d \end{vmatrix} = -\begin{vmatrix} c & d \\ a & b \end{vmatrix}$ $\begin{vmatrix} a & b \\ c & d \end{vmatrix} = -\begin{vmatrix} b & a \\ d & c \end{vmatrix}$
4. If the corresponding elements are equal in two rows (or columns), the value of the determinant is 0.	$\begin{vmatrix} a & b \\ a & b \end{vmatrix} = 0$ $\begin{vmatrix} a & a \\ c & c \end{vmatrix} = 0$
5. If a multiple of any row (or column) of a determinant is added to any other row (or column), the value of the determinant is not changed.	$\begin{vmatrix} a & b \\ c & d \end{vmatrix} = \begin{vmatrix} a & b \\ c+ka & d+kb \end{vmatrix}$ $\begin{vmatrix} a & b \\ c & d \end{vmatrix} = \begin{vmatrix} a+kb & b \\ c+kd & d \end{vmatrix}$

ANSWERS MATCHED PROBLEMS

1. $3\begin{vmatrix} 3 & 2 & 1 \\ 2 & 1 & -3 \\ 1 & 0 & -5 \end{vmatrix}$ **2.** 44

EXERCISE 9.5

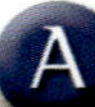

A

For each statement in Problems 1–10, identify the theorem from Section 9.5 that justifies it. Do not evaluate.

1. $\begin{vmatrix} 16 & 8 \\ 0 & -1 \end{vmatrix} = 8\begin{vmatrix} 2 & 1 \\ 0 & -1 \end{vmatrix}$

2. $\begin{vmatrix} 1 & -9 \\ 0 & -6 \end{vmatrix} = -3\begin{vmatrix} 1 & 3 \\ 0 & 2 \end{vmatrix}$

3. $-2\begin{vmatrix} 2 & 1 \\ -3 & 4 \end{vmatrix} = \begin{vmatrix} -4 & 1 \\ 6 & 4 \end{vmatrix}$

4. $4\begin{vmatrix} -1 & 3 \\ 2 & 1 \end{vmatrix} = \begin{vmatrix} -4 & 12 \\ 2 & 1 \end{vmatrix}$

5. $\begin{vmatrix} 3 & 0 \\ -2 & 0 \end{vmatrix} = 0$

6. $\begin{vmatrix} 5 & -7 \\ 0 & 0 \end{vmatrix} = 0$

7. $\begin{vmatrix} 5 & -1 \\ 8 & 0 \end{vmatrix} = -\begin{vmatrix} -1 & 5 \\ 0 & 8 \end{vmatrix}$

8. $\begin{vmatrix} 6 & 9 \\ 0 & 1 \end{vmatrix} = -\begin{vmatrix} 0 & 1 \\ 6 & 9 \end{vmatrix}$

9. $\begin{vmatrix} 4 & 3 \\ 1 & 2 \end{vmatrix} = \begin{vmatrix} 4-4 & 3-8 \\ 1 & 2 \end{vmatrix}$

10. $\begin{vmatrix} 3 & 2 \\ 5 & 1 \end{vmatrix} = \begin{vmatrix} 3+4 & 2 \\ 5+2 & 1 \end{vmatrix}$

In Problems 11–14, Theorem 5 was used to transform the determinant on the left to that on the right. Replace each letter x with an appropriate numeral to complete the transformation.

11. $\begin{vmatrix} -1 & 3 \\ 2 & -4 \end{vmatrix} = \begin{vmatrix} -1 & x \\ 2 & 2 \end{vmatrix}$

12. $\begin{vmatrix} -1 & 3 \\ 5 & -2 \end{vmatrix} = \begin{vmatrix} -1 & 3 \\ x & 13 \end{vmatrix}$

13. $\begin{vmatrix} -1 & 2 & 3 \\ 2 & 1 & 4 \\ 1 & 3 & 2 \end{vmatrix} = \begin{vmatrix} -1 & 2 & 0 \\ 2 & 1 & 10 \\ 1 & 3 & x \end{vmatrix}$

14. $\begin{vmatrix} -1 & 2 & 3 \\ 2 & 1 & 4 \\ 1 & 3 & 2 \end{vmatrix} = \begin{vmatrix} -1 & 0 & 3 \\ 2 & x & 4 \\ 1 & 5 & 2 \end{vmatrix}$

Given that

$$\begin{vmatrix} a & b \\ c & d \end{vmatrix} = 10$$

use the properties of determinants discussed in Section 9.5 to evaluate each determinant in Problems 15–20.

15. $\begin{vmatrix} c & d \\ a & b \end{vmatrix}$

16. $\begin{vmatrix} 2a & 2b \\ c & d \end{vmatrix}$

17. $\begin{vmatrix} a+c & b+d \\ c & d \end{vmatrix}$

18. $\begin{vmatrix} a+b & b \\ c+d & d \end{vmatrix}$

19. $\begin{vmatrix} a & a-b \\ c & c-d \end{vmatrix}$

20. $\begin{vmatrix} a+c & b+d \\ -a & -b \end{vmatrix}$

In Problems 21–24, transform each determinant into one that contains a row (or column) with all elements 0 but one, if possible. Then expand the transformed determinant by this row (or column).

21. $\begin{vmatrix} -1 & 0 & 3 \\ 2 & 5 & 4 \\ 1 & 5 & 2 \end{vmatrix}$

22. $\begin{vmatrix} -1 & 2 & 0 \\ 2 & 1 & 10 \\ 1 & 3 & 5 \end{vmatrix}$

23. $\begin{vmatrix} 3 & 5 & 0 \\ 1 & 1 & -2 \\ 2 & 1 & -1 \end{vmatrix}$

24. $\begin{vmatrix} 2 & 0 & 1 \\ -1 & -3 & 4 \\ 1 & 2 & 3 \end{vmatrix}$

B

For each statement in Problems 25–30, identify the theorem from Section 9.5 that justifies it.

25. $-2\begin{vmatrix} 1 & 0 & 2 \\ 3 & -2 & 4 \\ 0 & 1 & 1 \end{vmatrix} = \begin{vmatrix} 1 & 0 & 2 \\ -6 & 4 & -8 \\ 0 & 1 & 1 \end{vmatrix}$

26. $\begin{vmatrix} 8 & 0 & 1 \\ 12 & -1 & 0 \\ 4 & 3 & 2 \end{vmatrix} = 4\begin{vmatrix} 2 & 0 & 1 \\ 3 & -1 & 0 \\ 1 & 3 & 2 \end{vmatrix}$

27. $\begin{vmatrix} 1 & 2 & 0 \\ -1 & 3 & 0 \\ 0 & 1 & 0 \end{vmatrix} = 0$

28. $\begin{vmatrix} -2 & 5 & 13 \\ 1 & 7 & 12 \\ 0 & 8 & 15 \end{vmatrix} = -\begin{vmatrix} 5 & -2 & 13 \\ 7 & 1 & 12 \\ 8 & 0 & 15 \end{vmatrix}$

29. $\begin{vmatrix} 4 & 2 & -1 \\ 2 & 0 & 2 \\ -3 & 5 & -2 \end{vmatrix} = \begin{vmatrix} 4-4 & 2 & -1 \\ 2+8 & 0 & 2 \\ -3-8 & 5 & -2 \end{vmatrix}$

30. $\begin{vmatrix} 7 & 7 & 1 \\ -3 & -3 & 11 \\ 2 & 2 & 0 \end{vmatrix} = 0$

In Problems 31–34, Theorem 5 was used to transform the determinant on the left to that on the right. Replace each letter x and y with an appropriate numeral to complete the transformation.

31. $\begin{vmatrix} 2 & 1 & -1 \\ 3 & 4 & 1 \\ 1 & 2 & -2 \end{vmatrix} = \begin{vmatrix} 0 & 0 & -1 \\ x & 5 & 1 \\ -3 & y & -2 \end{vmatrix}$

32. $\begin{vmatrix} 3 & -1 & 1 \\ -2 & 4 & 3 \\ 1 & 5 & 2 \end{vmatrix} = \begin{vmatrix} 0 & -1 & 0 \\ 10 & 4 & 7 \\ x & 5 & y \end{vmatrix}$

33. $\begin{vmatrix} 7 & 9 & 4 \\ 2 & 3 & 1 \\ 3 & 4 & -2 \end{vmatrix} = \begin{vmatrix} -1 & x & 0 \\ 2 & 3 & 1 \\ 7 & y & 0 \end{vmatrix}$

34. $\begin{vmatrix} 5 & 2 & 3 \\ 3 & 1 & 2 \\ -4 & -3 & 5 \end{vmatrix} = \begin{vmatrix} x & 0 & -1 \\ 3 & 1 & 2 \\ 5 & 0 & y \end{vmatrix}$

In Problems 35–42, transform each determinant into one that contains a row (or column) with all elements 0 but one, if possible. Then expand the transformed determinant by this row (or column).

35. $\begin{vmatrix} 1 & 5 & 3 \\ 4 & 2 & 1 \\ 3 & 1 & 2 \end{vmatrix}$

36. $\begin{vmatrix} -1 & 5 & 1 \\ 2 & 3 & 1 \\ 3 & 2 & 1 \end{vmatrix}$

37. $\begin{vmatrix} 5 & 2 & -3 \\ -2 & 4 & 4 \\ 1 & -1 & 3 \end{vmatrix}$

38. $\begin{vmatrix} 5 & 3 & -6 \\ -1 & 1 & 4 \\ 4 & 3 & -6 \end{vmatrix}$

39. $\begin{vmatrix} 3 & -4 & 1 \\ 6 & -1 & 2 \\ 9 & 2 & 3 \end{vmatrix}$

40. $\begin{vmatrix} 2 & 3 & -1 \\ 5 & 4 & 7 \\ -4 & -6 & 2 \end{vmatrix}$

41. $\begin{vmatrix} 0 & 1 & 0 & 1 \\ 1 & -2 & 4 & 3 \\ 2 & 1 & 5 & 4 \\ 1 & 2 & 1 & 2 \end{vmatrix}$

42. $\begin{vmatrix} 2 & 3 & 1 & -1 \\ 3 & 1 & 2 & 1 \\ 0 & 5 & 4 & 0 \\ -1 & 2 & 3 & 0 \end{vmatrix}$

C

Transform each determinant in Problems 43 and 44 into one that contains a row (or column) with all elements 0 but one, if possible. Then expand the transformed determinant by this row (or column).

43. $\begin{vmatrix} 3 & 2 & 3 & 1 \\ 3 & -2 & 8 & 5 \\ 2 & 1 & 3 & 1 \\ 4 & 5 & 4 & -3 \end{vmatrix}$

44. $\begin{vmatrix} -1 & 4 & 2 & 1 \\ 5 & -1 & -3 & -1 \\ 2 & -1 & -2 & 3 \\ -3 & 3 & 3 & 3 \end{vmatrix}$

Problems 45–48 are representative cases of theorems discussed in Section 9.5. Use cofactor expansions to verify each statement directly, without reference to the theorem it represents.

45. $\begin{vmatrix} a & b & a \\ d & e & d \\ g & h & g \end{vmatrix} = 0$

46. $\begin{vmatrix} a & b & c \\ kd & ke & kf \\ g & h & i \end{vmatrix} = k\begin{vmatrix} a & b & c \\ d & e & f \\ g & h & i \end{vmatrix}$

47. $\begin{vmatrix} a_1 & b_1 & c_1 \\ a_2 & b_2 & c_2 \\ a_3 & b_3 & c_3 \end{vmatrix} = -\begin{vmatrix} b_1 & a_1 & c_1 \\ b_2 & a_2 & c_2 \\ b_3 & a_3 & c_3 \end{vmatrix}$

48. $\begin{vmatrix} a_1 & b_1 & c_1 \\ a_2 & b_2 & c_2 \\ a_3 & b_3 & c_3 \end{vmatrix} = \begin{vmatrix} a_1 + kc_1 & b_1 & c_1 \\ a_2 + kc_2 & b_2 & c_2 \\ a_3 + kc_3 & b_3 & c_3 \end{vmatrix}$

49. Without expanding, explain why (2, 5) and (−3, 4) satisfy the equation

$$\begin{vmatrix} x & y & 1 \\ 2 & 5 & 1 \\ -3 & 4 & 1 \end{vmatrix} = 0$$

50. Show that

$$\begin{vmatrix} x & y & 1 \\ 2 & 3 & 1 \\ -1 & 2 & 1 \end{vmatrix} = 0$$

is the equation of a line that passes through (2, 3) and (−1, 2).

51. Show that

$$\begin{vmatrix} x & y & 1 \\ x_1 & y_1 & 1 \\ x_2 & y_2 & 1 \end{vmatrix} = 0$$

is the equation of a line that passes through (x_1, y_1) and (x_2, y_2).

52. In analytic geometry it is shown that the area of a triangle with vertices (x_1, y_1), (x_2, y_2), and (x_3, y_3) is the absolute value of

$$\frac{1}{2}\begin{vmatrix} x_1 & y_1 & 1 \\ x_2 & y_2 & 1 \\ x_3 & y_3 & 1 \end{vmatrix}$$

Use this result to find the area of a triangle with vertices $(-1, 4)$, $(4, 8)$, and $(1, 1)$.

53. What can we say about the three points (x_1, y_1), (x_2, y_2), and (x_3, y_3) if the following equation is true?

$$\begin{vmatrix} x_1 & y_1 & 1 \\ x_2 & y_2 & 1 \\ x_3 & y_3 & 1 \end{vmatrix} = 0$$

[*Hint:* See Problem 52.]

54. If the three points (x_1, y_1), (x_2, y_2), and (x_3, y_3) are all on the same line, what can we say about the value of the determinant below?

$$\begin{vmatrix} x_1 & y_1 & 1 \\ x_2 & y_2 & 1 \\ x_3 & y_3 & 1 \end{vmatrix}$$

SECTION 9.6 Determinants and Cramer's Rule

Two Equations–Two Variables • Three Equations–Three Variables

Now let's see how determinants arise rather naturally in the process of solving systems of linear equations. We start by investigating two equations and two variables, and then extend our results to three equations and three variables.

Two Equations–Two Variables

Instead of thinking of each system of linear equations in two variables as a different problem, let's see what happens when we attempt to solve the general system

$$a_{11}x + a_{12}y = k_1 \qquad (1A)$$

$$a_{21}x + a_{22}y = k_2 \qquad (1B)$$

once and for all, in terms of the unspecified real constants a_{11}, a_{12}, a_{21}, a_{22}, k_1, and k_2.

We proceed by multiplying equations (1A) and (1B) by suitable constants so that when the resulting equations are added, left side to left side and right side to right side, one of the variables drops out. Suppose we choose to eliminate y. What constant should we use to make the coefficients of y the same except for the signs? Multiply equation (1A) by a_{22} and (1B) by $-a_{12}$; then add:

$$\begin{array}{lrl} a_{22}(1A): & a_{11}a_{22}x + a_{12}a_{22}y = & k_1a_{22} \\ -a_{12}(1B): & -a_{21}a_{12}x - a_{12}a_{22}y = & -k_2a_{12} \\ \hline & a_{11}a_{22}x - a_{21}a_{12}x + 0y = & k_1a_{22} - k_2a_{12} \\ & (a_{11}a_{22} - a_{21}a_{12})x = & k_1a_{22} - k_2a_{12} \end{array}$$

$$x = \frac{k_1a_{22} - k_2a_{12}}{a_{11}a_{22} - a_{21}a_{12}} \qquad a_{11}a_{22} - a_{21}a_{12} \neq 0$$

What do the numerator and denominator remind you of? From your experience with determinants in Sections 9.4 and 9.5, you should recognize these expressions as

$$x = \frac{\begin{vmatrix} k_1 & a_{12} \\ k_2 & a_{22} \end{vmatrix}}{\begin{vmatrix} a_{11} & a_{12} \\ a_{21} & a_{22} \end{vmatrix}}$$

Similarly, starting with system (1A) and (1B) and eliminating x (this is left as an exercise), we obtain

$$y = \frac{\begin{vmatrix} a_{11} & k_1 \\ a_{21} & k_2 \end{vmatrix}}{\begin{vmatrix} a_{11} & a_{12} \\ a_{21} & a_{22} \end{vmatrix}}$$

These results are summarized in Theorem 1, **Cramer's rule,** which is named after the Swiss mathematician Gabriel Cramer (1704–1752).

THEOREM 1

Cramer's Rule for Two Equations and Two Variables

Given the system

$$\begin{aligned} a_{11}x + a_{12}y &= k_1 \\ a_{21}x + a_{22}y &= k_2 \end{aligned} \quad \text{with} \quad D = \begin{vmatrix} a_{11} & a_{12} \\ a_{21} & a_{22} \end{vmatrix} \neq 0$$

then

$$x = \frac{\begin{vmatrix} k_1 & a_{12} \\ k_2 & a_{22} \end{vmatrix}}{D} \quad \text{and} \quad y = \frac{\begin{vmatrix} a_{11} & k_1 \\ a_{21} & k_2 \end{vmatrix}}{D}$$

The determinant D is called the **coefficient determinant.** If $D \neq 0$, then the system has exactly one solution, which is given by Cramer's rule. If, on the other hand, $D = 0$, then it can be shown that the system is either inconsistent and has no solutions or is dependent and has an infinite number of solutions. We must use other methods, such as those discussed in Chapter 8, to determine the exact nature of the solutions when $D = 0$.

EXAMPLE 1 Solving a System with Cramer's Rule

Solve using Cramer's rule:

$$\begin{aligned} 3x - 5y &= 2 \\ -4x + 3y &= -1 \end{aligned}$$

SOLUTION

Algebraic Solution

$$D = \begin{vmatrix} 3 & -5 \\ -4 & 3 \end{vmatrix} = -11$$

$$x = \frac{\begin{vmatrix} 2 & -5 \\ -1 & 3 \end{vmatrix}}{-11} = -\frac{1}{11}$$

$$y = \frac{\begin{vmatrix} 3 & 2 \\ -4 & -1 \end{vmatrix}}{-11} = -\frac{5}{11}$$

Graphing Utility Solution

Store the coefficient matrix in A, the matrix with the constants in the first column in B, and the matrix with the constants in the second column in C. Then use the det command to apply Cramer's rule (Fig. 1).

```
[A]
          [[3  -5]
           [-4 3 ]]
[B]
          [[2  -5]
           [-1 3 ]]
det([B])/det([A]
)
      -.0909090909
Ans▸Frac
             -1/11
[C]
          [[3  2 ]
           [-4 -1]]
det([C])/det([A]
)
      -.4545454545
Ans▸Frac
             -5/11
```

FIGURE 1

MATCHED 1 PROBLEM

Solve using Cramer's rule:

$$\begin{aligned} 3x + 2y &= -4 \\ -4x + 3y &= -10 \end{aligned}$$

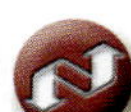

EXPLORE/DISCUSS 1

Recall that a system of linear equations must have zero, one, or an infinite number of solutions. Discuss the number of solutions for the system

$$\begin{aligned} ax + 3y &= b \\ 4x + 2y &= 8 \end{aligned}$$

where a and b are real numbers. Use Cramer's rule where appropriate and Gauss–Jordan elimination otherwise.

Three Equations–Three Variables

Cramer's rule can be generalized completely for any size linear system that has the *same number of variables as equations*. However, it cannot be used to solve systems where the number of variables is not equal to the number of equations. In Theorem 2 we state without proof Cramer's rule for three equations and three variables.

THEOREM 2

Cramer's Rule for Three Equations and Three Variables

Given the system

$$\begin{aligned} a_{11}x + a_{12}y + a_{13}z &= k_1 \\ a_{21}x + a_{22}y + a_{23}z &= k_2 \\ a_{31}x + a_{32}y + a_{33}z &= k_3 \end{aligned} \qquad \text{with} \qquad D = \begin{vmatrix} a_{11} & a_{12} & a_{13} \\ a_{21} & a_{22} & a_{23} \\ a_{31} & a_{32} & a_{33} \end{vmatrix} \neq 0$$

then

$$x = \frac{\begin{vmatrix} k_1 & a_{12} & a_{13} \\ k_2 & a_{22} & a_{23} \\ k_3 & a_{32} & a_{33} \end{vmatrix}}{D} \qquad y = \frac{\begin{vmatrix} a_{11} & k_1 & a_{13} \\ a_{21} & k_2 & a_{23} \\ a_{31} & k_3 & a_{33} \end{vmatrix}}{D} \qquad z = \frac{\begin{vmatrix} a_{11} & a_{12} & k_1 \\ a_{21} & a_{22} & k_2 \\ a_{31} & a_{32} & k_3 \end{vmatrix}}{D}$$

You can easily remember these determinant formulas for x, y, and z if you observe the following:

1. Determinant D is formed from the coefficients of x, y, and z, keeping the same relative position in the determinant as found in the system of equations.
2. Determinant D appears in the denominators for x, y, and z.
3. The numerator for x can be obtained from D by replacing the coefficients of x (a_{11}, a_{21}, and a_{31}) with the constants k_1, k_2, and k_3, respectively. Similar statements can be made for the numerators for y and z.

EXAMPLE 2 Solving a System with Cramer's Rule

Solve using Cramer's rule:

$$\begin{aligned} x + y \quad &= 2 \\ 3y - z &= -4 \\ x \quad + z &= 3 \end{aligned}$$

SOLUTION

Algebraic Solution

$$D = \begin{vmatrix} 1 & 1 & 0 \\ 0 & 3 & -1 \\ 1 & 0 & 1 \end{vmatrix} = 2$$

$$x = \frac{\begin{vmatrix} 2 & 1 & 0 \\ -4 & 3 & -1 \\ 3 & 0 & 1 \end{vmatrix}}{2} = \frac{7}{2}$$

$$y = \frac{\begin{vmatrix} 1 & 2 & 0 \\ 0 & -4 & -1 \\ 1 & 3 & 1 \end{vmatrix}}{2} = -\frac{3}{2}$$

$$z = \frac{\begin{vmatrix} 1 & 1 & 2 \\ 0 & 3 & -4 \\ 1 & 0 & 3 \end{vmatrix}}{2} = -\frac{1}{2}$$

Graphing Utility Solution

Store the coefficient matrix in A, and the matrix with the constants replacing the coefficients of x, y, and z in B, C, and D, respectively. Then use the det command to apply Cramer's rule (Fig. 2).

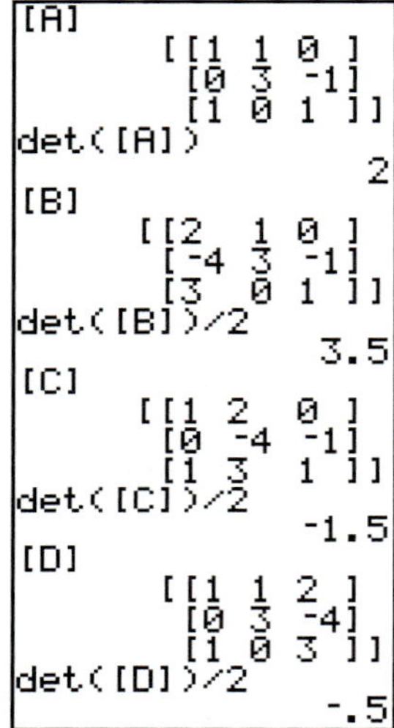

FIGURE 2

MATCHED 2 PROBLEM

Solve using Cramer's rule:

$$\begin{aligned} 3x \quad - z &= 5 \\ x - y + z &= 0 \\ x + y \quad &= 1 \end{aligned}$$

In practice, Cramer's rule is rarely used to solve systems of order higher than 2 or 3 by hand, because more efficient methods are available using computer methods. However, Cramer's rule is a valuable tool in more advanced theoretical and applied mathematics.

ANSWERS MATCHED PROBLEMS

1. $x = \frac{8}{17}$, $y = -\frac{46}{17}$ 2. $x = \frac{6}{5}$, $y = -\frac{1}{5}$, $z = -\frac{7}{5}$

EXERCISE 9.6

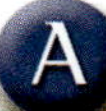

A

Solve Problems 1–8 using Cramer's rule.

1. $\begin{aligned} x + 2y &= 1 \\ x + 3y &= -1 \end{aligned}$

2. $\begin{aligned} x + 2y &= 3 \\ x + 3y &= 5 \end{aligned}$

3. $\begin{aligned} 2x + y &= 1 \\ 5x + 3y &= 2 \end{aligned}$

4. $\begin{aligned} x + 3y &= 1 \\ 2x + 8y &= 0 \end{aligned}$

5. $\begin{aligned} 2x - y &= -3 \\ -x + 3y &= 3 \end{aligned}$

6. $\begin{aligned} -3x + 2y &= 1 \\ 2x - 3y &= -3 \end{aligned}$

7. $\begin{aligned} 4x - 3y &= 4 \\ 3x + 2y &= -2 \end{aligned}$

8. $\begin{aligned} 5x + 2y &= -1 \\ 2x - 3y &= 2 \end{aligned}$

B

Solve Problems 9–12 to two significant digits using Cramer's rule.

9. $\begin{aligned} 0.9925x - 0.9659y &= 0 \\ 0.1219x + 0.2588y &= 2{,}500 \end{aligned}$

10. $\begin{aligned} 0.9877x - 0.9744y &= 0 \\ 0.1564x + 0.2250y &= 1{,}900 \end{aligned}$

11. $\begin{aligned} 0.9954x - 0.9942y &= 0 \\ 0.0958x + 0.1080y &= 155 \end{aligned}$

12. $\begin{aligned} 0.9973x - 0.9957y &= 0 \\ 0.0732x + 0.0924y &= 112 \end{aligned}$

Solve Problems 13–20 using Cramer's rule:

13. $\begin{aligned} x + y &= 0 \\ 2y + z &= -5 \\ -x + z &= -3 \end{aligned}$

14. $\begin{aligned} x + y &= -4 \\ 2y + z &= 0 \\ -x + z &= 5 \end{aligned}$

15. $\begin{aligned} x + y &= 1 \\ 2y + z &= 0 \\ -y + z &= 1 \end{aligned}$

16. $\begin{aligned} x + 3y &= -3 \\ 2y + z &= 3 \\ -x + 3z &= 7 \end{aligned}$

17. $\begin{aligned} 3y + z &= -1 \\ x + 2z &= 3 \\ x - 3y &= -2 \end{aligned}$

18. $\begin{aligned} x - z &= 3 \\ 2x - y &= -3 \\ x + y + z &= 1 \end{aligned}$

19. $\begin{aligned} 2y - z &= -3 \\ x - y - z &= 2 \\ x - y + 2z &= 4 \end{aligned}$

20. $\begin{aligned} 2x + y &= 2 \\ x - y + z &= -1 \\ x + y + z &= 2 \end{aligned}$

Discuss the number of solutions for the systems in Problems 21 and 22 where a and b are real numbers. Use Cramer's rule where appropriate and Gauss–Jordan elimination otherwise.

21. $\begin{aligned} ax + 3y &= b \\ 2x + 4y &= 5 \end{aligned}$

22. $\begin{aligned} 2x + ay &= b \\ 3x + 4y &= 7 \end{aligned}$

C

In Problems 23 and 24, use Cramer's rule to solve for x only.

23. $\begin{aligned} 2x - 3y + z &= -3 \\ -4x + 3y + 2z &= -11 \\ x - y - z &= 3 \end{aligned}$

24. $\begin{aligned} x + 4y - 3z &= 25 \\ 3x + y - z &= 2 \\ -4x + y + 2z &= 1 \end{aligned}$

In Problems 25 and 26, use Cramer's rule to solve for y only.

25. $\begin{aligned} 12x - 14y + 11z &= 5 \\ 15x + 7y - 9z &= -13 \\ 5x - 3y + 2z &= 0 \end{aligned}$

26. $\begin{aligned} 2x - y + 4z &= 15 \\ -x + y + 2z &= 5 \\ 3x + 4y - 2z &= 4 \end{aligned}$

In Problems 27 and 28, use Cramer's rule to solve for z only.

27. $\begin{aligned} 3x - 4y + 5z &= 18 \\ -9x + 8y + 7z &= -13 \\ 5x - 7y + 10z &= 33 \end{aligned}$

28. $$\begin{aligned} 13x + 11y + 10z &= 2 \\ 10x + 8y + 7z &= 1 \\ 8x + 5y + 4z &= 4 \end{aligned}$$

It is clear that $x = 0$, $y = 0$, $z = 0$ is a solution to each of the systems given in Problems 29 and 30. Use Cramer's rule to determine whether this solution is unique. [Hint: If $D \neq 0$, what can you conclude? If $D = 0$, what can you conclude?]

29. $$\begin{aligned} x - 4y + 9z &= 0 \\ 4x - y + 6z &= 0 \\ x - y + 3z &= 0 \end{aligned}$$

30. $$\begin{aligned} 3x - y + 3z &= 0 \\ 5x + 5y - 9z &= 0 \\ -2x + y - 3z &= 0 \end{aligned}$$

31. Prove Theorem 1 for y.

32. (Omit this problem if you have not studied trigonometry.) The angles α, β, and γ and the sides a, b, and c of a triangle (see the figure) satisfy

$$\begin{aligned} c &= b\cos\alpha + a\cos\beta \\ b &= c\cos\alpha \qquad\qquad + a\cos\gamma \\ a &= \qquad\quad c\cos\beta + b\cos\gamma \end{aligned}$$

Use Cramer's rule to express $\cos\alpha$ in terms of a, b, and c, thereby deriving the familiar law of cosines from trigonometry:

$$\cos\alpha = \frac{b^2 + c^2 - a^2}{2bc}$$

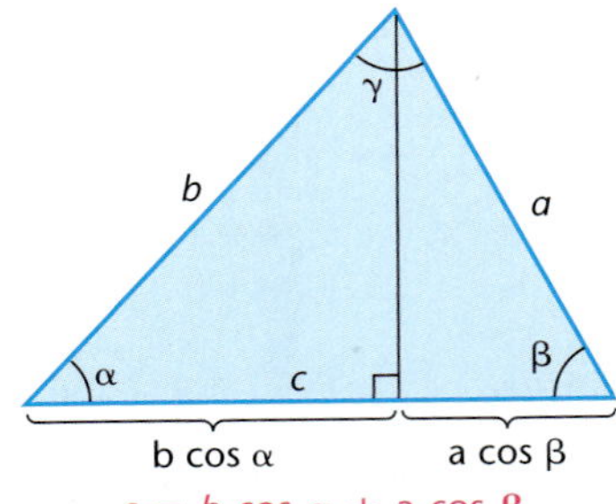

$c = b\cos\alpha + a\cos\beta$

APPLICATIONS

33. Revenue Analysis. A supermarket sells two brands of coffee: brand A at \$$p$ per pound and brand B at \$$q$ per pound. The daily demand equations for brands A and B are, respectively,

$$\begin{aligned} x &= 200 - 6p + 4q \\ y &= 300 + 2p - 3q \end{aligned} \tag{1}$$

(both in pounds). The daily revenue R is given by

$$R = xp + yq$$

(A) To analyze the effect of price changes on the daily revenue, an economist wants to express the daily revenue R in terms of p and q only. Use (1) to eliminate x and y in the equation for R, thus expressing the daily revenue in terms of p and q.

(B) To analyze the effect of changes in demand on the daily revenue, the economist now wants to express the daily revenue in terms of x and y only. Use Cramer's rule to solve system (1) for p and q in terms of x and y and then express the daily revenue R in terms of x and y.

34. Revenue Analysis. A company manufactures ten-speed and three-speed bicycles. The weekly demand equations are

$$\begin{aligned} p &= 230 - 10x + 5y \\ q &= 130 + 4x - 4y \end{aligned} \tag{2}$$

where \$$p$ is the price of a ten-speed bicycle, \$$q$ is the price of a three-speed bicycle, x is the weekly demand for ten-speed bicycles, and y is the weekly demand for three-speed bicycles. The weekly revenue R is given by

$$R = xp + yq$$

(A) Use (2) to express the daily revenue in terms of x and y only.

(B) Use Cramer's rule to solve system (2) for x and y in terms of p and q, and then express the daily revenue R in terms of p and q only.

9.1 Matrix Operations

Two matrices are **equal** if they are the same size and their corresponding elements are equal. The **sum of two matrices** of the same size is a matrix with elements that are the sums of the corresponding elements of the two given matrices. Matrix addition is **commutative** and **associative.** A matrix with all zero elements is called the **zero matrix.** The **negative of a matrix M,** denoted $-M$, is a matrix with elements that are the negatives of the elements in M. If A and B are matrices of the same size, then we define **subtraction** as follows: $A - B = A + (-B)$. The **product of a number k and a matrix M,** denoted by kM, is a matrix formed by multiplying each element of M by k. The **product** of a $1 \times n$ row matrix and an $n \times 1$ column matrix is a 1×1 matrix given by

$$\underset{1 \times n}{[a_1\ a_2 \cdots a_n]} \underset{n \times 1}{\begin{bmatrix} b_1 \\ b_2 \\ \vdots \\ b_n \end{bmatrix}} = \underset{1 \times 1}{[a_1b_1 + a_2b_2 + \cdots + a_nb_n]}$$

If A is an $m \times p$ matrix and B is a $p \times n$ matrix, then the **matrix product** of A and B, denoted AB, is an $m \times n$ matrix whose element in the ith row and jth column is the real number obtained from the product of the ith row of A and the jth column of B. If the number of columns in A does not equal the number of rows in B, then the matrix product AB is **not defined. Matrix multiplication is not commutative,** and the **zero property does not hold for matrix multiplication.** That is, for matrices A and B, the matrix product AB can be zero without either A or B being the zero matrix.

9.2 Inverse of a Square Matrix

The **identity matrix for multiplication** for the set of all square matrices of order n is the square matrix of order n, denoted by I, with ones along the **principal diagonal** (from upper left corner to lower right corner) and zeros elsewhere. If M is a square matrix of order n and I is the identity matrix of order n, then

$$IM = MI = M$$

If M is a square matrix of order n and if there exists a matrix M^{-1} (read "M inverse") such that

$$M^{-1}M = MM^{-1} = I$$

then M^{-1} is called the **multiplicative inverse of M** or, more simply, the **inverse of M.** If the augmented matrix $[M \mid I]$ is transformed by row operations into $[I \mid B]$, then the resulting matrix B is M^{-1}. If, however, we obtain all zeros in one or more rows to the left of the vertical line, then M^{-1} does not exist and M is called a **singular matrix.**

9.3 Matrix Equations and Systems of Linear Equations

The following properties of matrices are fundamental to the process of solving matrix equations. Assuming all products and sums are defined for the indicated matrices A, B, C, I, and 0, then:

Addition Properties

Associative:	$(A + B) + C = A + (B + C)$
Commutative:	$A + B = B + A$
Additive Identity:	$A + 0 = 0 + A = A$
Additive Inverse:	$A + (-A) = (-A) + A = 0$

Multiplication Properties

Associative Property:	$A(BC) = (AB)C$
Multiplicative Identity:	$AI = IA = A$
Multiplicative Inverse:	If A is a square matrix and A^{-1} exists, then $AA^{-1} = A^{-1}A = I$.

Combined Properties

Left Distributive:	$A(B + C) = AB + AC$
Right Distributive:	$(B + C)A = BA + CA$

Equality

Addition:	If $A = B$, then $A + C = B + C$.
Left Multiplication:	If $A = B$, then $CA = CB$.
Right Multiplication:	If $A = B$, then $AC = BC$.

A system of linear equations with the same number of variables as equations such as

$$\begin{aligned} a_{11}x_1 + a_{12}x_2 + a_{13}x_3 &= k_1 \\ a_{21}x_1 + a_{22}x_2 + a_{23}x_3 &= k_2 \\ a_{31}x_1 + a_{32}x_2 + a_{33}x_3 &= k_3 \end{aligned}$$

can be written as the matrix equation

$$\underset{A}{\begin{bmatrix} a_{11} & a_{12} & a_{13} \\ a_{21} & a_{22} & a_{23} \\ a_{31} & a_{32} & a_{33} \end{bmatrix}} \underset{X}{\begin{bmatrix} x_1 \\ x_2 \\ x_3 \end{bmatrix}} = \underset{B}{\begin{bmatrix} k_1 \\ k_2 \\ k_3 \end{bmatrix}}$$

If the inverse of A exists, then the matrix equation has a unique solution given by

$$X = A^{-1}B$$

After multiplying B by A^{-1} from the left, it is easy to read the solution to the original system of equations.

9.4 Determinants

Associated with each square matrix A is a real number called the **determinant** of the matrix. The determinant of A is denoted by

det A, or simply by writing the array of elements in A using vertical lines in place of square brackets. For example,

$$\det\begin{bmatrix} a_{11} & a_{12} \\ a_{21} & a_{22} \end{bmatrix} = \begin{vmatrix} a_{11} & a_{12} \\ a_{21} & a_{22} \end{vmatrix}$$

A determinant of **order n** is a determinant with n rows and n columns.

The **value of a second-order determinant** is the real number given by

$$\begin{vmatrix} a_{11} & a_{12} \\ a_{21} & a_{22} \end{vmatrix} = a_{11}a_{22} - a_{21}a_{12}$$

The **value of a third-order determinant** is the sum of three products obtained by multiplying each element of any one row (or each element of any one column) by its cofactor. The **cofactor of an element** a_{ij} (from the ith row and jth column) is the product of the minor of a_{ij} and $(-1)^{i+j}$. The **minor of an element** a_{ij} is the determinant remaining after deleting the ith row and jth column. A similar process can be used to evaluate determinants of order higher than 3.

9.5 Properties of Determinants

The use of the following five determinant properties can greatly reduce the effort in evaluating determinants of order 3 or greater:

1. If each element of any row (or column) of a determinant is multiplied by a constant k, the new determinant is k times the original.
$$\begin{vmatrix} 2a & 2b \\ c & d \end{vmatrix} = 2\begin{vmatrix} a & b \\ c & d \end{vmatrix} \qquad 3\begin{vmatrix} a & b \\ c & d \end{vmatrix} = \begin{vmatrix} 3a & b \\ 3c & d \end{vmatrix}$$
2. If every element in a row (or column) is 0, the value of the determinant is 0.
$$\begin{vmatrix} a & b \\ 0 & 0 \end{vmatrix} = 0 \qquad \begin{vmatrix} 0 & b \\ 0 & d \end{vmatrix} = 0$$
3. If two rows (or two columns) of a determinant are interchanged, the new determinant is the negative of the original.
$$\begin{vmatrix} a & b \\ c & d \end{vmatrix} = -\begin{vmatrix} c & d \\ a & b \end{vmatrix} \qquad \begin{vmatrix} a & b \\ c & d \end{vmatrix} = -\begin{vmatrix} b & a \\ d & c \end{vmatrix}$$
4. If the corresponding elements are equal in two rows (or columns), the value of the determinant is 0.
$$\begin{vmatrix} a & b \\ a & b \end{vmatrix} = 0 \qquad \begin{vmatrix} a & a \\ c & c \end{vmatrix} = 0$$
5. If a multiple of any row (or column) of a determinant is added to any other row (or column), the value of the determinant is not changed.
$$\begin{vmatrix} a & b \\ c & d \end{vmatrix} = \begin{vmatrix} a & b \\ c+ka & d+kb \end{vmatrix} \qquad \begin{vmatrix} a & b \\ c & d \end{vmatrix} = \begin{vmatrix} a+kb & b \\ c+kd & d \end{vmatrix}$$

9.6 Determinants and Cramer's Rule

Systems of equations having the same number of variables as equations can also be solved using determinants and Cramer's rule. **Cramer's rule for three equations and three variables** is as follows: Given the system

$$\begin{aligned} a_{11}x + a_{12}y + a_{13}z &= k_1 \\ a_{21}x + a_{22}y + a_{23}z &= k_2 \\ a_{31}x + a_{32}y + a_{33}z &= k_3 \end{aligned} \qquad \text{with} \qquad D = \begin{vmatrix} a_{11} & a_{12} & a_{13} \\ a_{21} & a_{22} & a_{23} \\ a_{31} & a_{32} & a_{33} \end{vmatrix} \neq 0$$

then

$$x = \frac{\begin{vmatrix} k_1 & a_{12} & a_{13} \\ k_2 & a_{22} & a_{23} \\ k_3 & a_{32} & a_{33} \end{vmatrix}}{D} \qquad y = \frac{\begin{vmatrix} a_{11} & k_1 & a_{13} \\ a_{21} & k_2 & a_{23} \\ a_{31} & k_3 & a_{33} \end{vmatrix}}{D} \qquad z = \frac{\begin{vmatrix} a_{11} & a_{12} & k_1 \\ a_{21} & a_{22} & k_2 \\ a_{31} & a_{32} & k_3 \end{vmatrix}}{D}$$

Cramer's rule can be generalized completely for any size linear system that has the same number of variables as equations. The formulas are easily remembered if you observe the following:

1. Determinant D is formed from the coefficients of x, y, and z, keeping the same relative position in the determinant as found in the system of equations.
2. Determinant D appears in the denominators for x, y, and z.
3. The numerator for x can be obtained from D by replacing the coefficients of x (a_{11}, a_{21}, and a_{31}) with the constants k_1, k_2, and k_3, respectively. Similar statements can be made for the numerators for y and z.

Cramer's rule is rarely used to solve systems of order higher than 3 by hand, because more efficient methods are available. Cramer's rule, however, is a valuable tool in more advanced theoretical and applied mathematics.

CHAPTER 9 REVIEW EXERCISES

Work through all the problems in this chapter review and check answers in the back of the book. Answers to all review problems are there, and following each answer is a number in italics indicating the section in which that type of problem is discussed. Where weaknesses show up, review appropriate sections in the text.

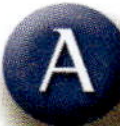

In Problems 1–9, perform the operations that are defined, given the following matrices:

$$A = \begin{bmatrix} 4 & -2 \\ 0 & 3 \end{bmatrix} \quad B = \begin{bmatrix} -1 & 5 \\ -4 & 6 \end{bmatrix} \quad C = \begin{bmatrix} -1 & 4 \end{bmatrix} \quad D = \begin{bmatrix} 3 \\ -2 \end{bmatrix}$$

1. AB **2.** CD **3.** CB

4. AD **5.** $A + B$ **6.** $C + D$

7. $A + C$ **8.** $2A - 5B$ **9.** $CA + C$

10. Find the inverse of

$$A = \begin{bmatrix} 4 & 7 \\ -1 & -2 \end{bmatrix}$$

Show that $A^{-1}A = I$.

11. Write the system

$$3x_1 + 2x_2 = k_1$$
$$4x_1 + 3x_2 = k_2$$

as a matrix equation, and solve using matrix inverse methods for:

(A) $k_1 = 3, k_2 = 5$ **(B)** $k_1 = 7, k_2 = 10$
(C) $k_1 = 4, k_2 = 2$

Evaluate the determinants in Problems 12 and 13.

12. $\begin{vmatrix} 2 & -3 \\ -5 & -1 \end{vmatrix}$ **13.** $\begin{vmatrix} 2 & 3 & -4 \\ 0 & 5 & 0 \\ 1 & -4 & -2 \end{vmatrix}$

14. Solve the system using Cramer's rule:

$$3x - 2y = 8$$
$$x + 3y = -1$$

15. Use properties of determinants to find each of the following, given that

$$\begin{vmatrix} a & b & c \\ d & e & f \\ g & h & i \end{vmatrix} = 2$$

(A) $\begin{vmatrix} g & h & i \\ d & e & f \\ a & b & c \end{vmatrix}$ **(B)** $\begin{vmatrix} a & 3b & c \\ d & 3e & f \\ g & 3h & i \end{vmatrix}$

(C) $\begin{vmatrix} a & b & a+b+c \\ d & e & d+e+f \\ g & h & g+h+i \end{vmatrix}$

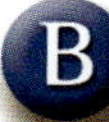

In Problems 16–21, perform the operations that are defined, given the following matrices:

$$A = \begin{bmatrix} 1 & 2 \\ 4 & 5 \\ -3 & -1 \end{bmatrix} \quad B = \begin{bmatrix} 6 \\ 0 \\ -4 \end{bmatrix} \quad C = \begin{bmatrix} 2 & 4 & -1 \end{bmatrix}$$

$$D = \begin{bmatrix} 7 & 0 & -5 \\ 0 & 8 & -2 \end{bmatrix} \quad E = \begin{bmatrix} 9 & -3 \\ -6 & 2 \end{bmatrix}$$

16. AD **17.** DA **18.** BC

19. CB **20.** DE **21.** ED

22. Find the inverse of

$$A = \begin{bmatrix} 1 & 0 & 4 \\ -2 & 1 & 0 \\ 4 & -1 & 4 \end{bmatrix}$$

Show that $AA^{-1} = I$.

23. Write the system

$$\begin{aligned} x_1 + 2x_2 + 3x_3 &= k_1 \\ 2x_1 + 3x_2 + 4x_3 &= k_2 \\ x_1 + 2x_2 + x_3 &= k_3 \end{aligned}$$

as a matrix equation, and solve using matrix inverse methods for:

(A) $k_1 = 1, k_2 = 3, k_3 = 3$
(B) $k_1 = 0, k_2 = 0, k_3 = -2$
(C) $k_1 = -3, k_2 = -4, k_3 = 1$

Evaluate the determinants in Problems 24 and 25.

24. $\begin{vmatrix} -\frac{1}{4} & \frac{3}{2} \\ \frac{1}{2} & \frac{2}{3} \end{vmatrix}$

25. $\begin{vmatrix} 2 & -1 & 1 \\ -3 & 5 & 2 \\ 1 & -2 & 4 \end{vmatrix}$

26. Solve for y only using Cramer's rule:

$$\begin{aligned} x - 2y + z &= -6 \\ y - z &= 4 \\ 2x + 2y + z &= 2 \end{aligned}$$

(Find the numerator and denominator first; then reduce.)

27. Discuss the number of solutions for a system of n equations in n variables if the coefficient matrix:

(A) Has an inverse **(B)** Does not have an inverse

28. If A is a nonzero square matrix of order n satisfying $A^2 = 0$, can A^{-1} exist? Explain.

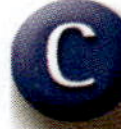

C

29. For $n \times n$ matrices A and C and $n \times 1$ column matrices B and X, solve for X assuming all necessary inverses exist:

$$AX - B = CX$$

30. Find the inverse of

$$A = \begin{bmatrix} 4 & 5 & 6 \\ 4 & 5 & -6 \\ 1 & 1 & 1 \end{bmatrix}$$

Show that $A^{-1}A = I$.

31. Clear the decimals in the system

$$\begin{aligned} 0.04x_1 + 0.05x_2 + 0.06x_3 &= 360 \\ 0.04x_1 + 0.05x_2 - 0.06x_3 &= 120 \\ x_1 + x_2 + x_3 &= 7{,}000 \end{aligned}$$

by multiplying the first two equations by 100. Then write the resulting system as a matrix equation and solve using the inverse found in Problem 30.

32. $\begin{vmatrix} -1 & 4 & 1 & 1 \\ 5 & -1 & 2 & -1 \\ 2 & -1 & 0 & 3 \\ -3 & 3 & 0 & 3 \end{vmatrix} = ?$

33. Show that

$$\begin{vmatrix} u & v \\ w & x \end{vmatrix} = \begin{vmatrix} u + kv & v \\ w + kx & x \end{vmatrix}$$

34. Explain why the points $(1, 2)$ and $(-1, 5)$ must satisfy the equation

$$\begin{vmatrix} x & y & 1 \\ 1 & 2 & 1 \\ -1 & 5 & 1 \end{vmatrix} = 0$$

Describe the set of all points that satisfy this equation.

APPLICATIONS

35. Resource Allocation. A Colorado mining company operates mines at Big Bend and Saw Pit. The Big Bend mine produces ore that is 5% nickel and 7% copper. The Saw Pit mine produces ore that is 3% nickel and 4% copper. How many tons of ore should be produced at each mine to obtain the amounts of nickel and copper listed in the table? Set up a matrix equation and solve using matrix inverses.

	Nickel	Copper
(A)	3.6 tons	5 tons
(B)	3 tons	4.1 tons
(C)	3.2 tons	4.4 tons

36. Labor Costs. A company with manufacturing plants in North Carolina and South Carolina has labor-hour and wage requirements for the manufacturing of computer desks and printer stands as given in matrices L and H:

Labor-hour requirements (columns: Fabricating department, Assembly department, Packaging department; rows: Desk, Stand)

$$L = \begin{bmatrix} 1.7\text{ h} & 2.4\text{ h} & 0.8\text{ h} \\ 0.9\text{ h} & 1.8\text{ h} & 0.6\text{ h} \end{bmatrix} \begin{matrix} \text{Desk} \\ \text{Stand} \end{matrix}$$

Hourly wages (columns: North Carolina plant, South Carolina plant; rows: Fabricating department, Assembly department, Packaging department)

$$H = \begin{bmatrix} \$11.50 & \$10.00 \\ \$9.50 & \$8.50 \\ \$5.00 & \$4.50 \end{bmatrix} \begin{matrix} \text{Fabricating department} \\ \text{Assembly department} \\ \text{Packaging department} \end{matrix}$$

(A) Find the labor cost for producing one printer stand at the South Carolina plant.

(B) Discuss possible interpretations of the elements in the matrix products HL and LH.

(C) If either of the products HL or LH has a meaningful interpretation, find the product and label its rows and columns.

37. Labor Costs. The monthly production of computer desks and printer stands for the company in Problem 36 for the months of January and February is given in matrices J and F:

January production (columns: North Carolina plant, South Carolina plant)

$$J = \begin{bmatrix} 1{,}500 & 1{,}650 \\ 850 & 700 \end{bmatrix} \begin{matrix} \text{Desks} \\ \text{Stands} \end{matrix}$$

February production (columns: North Carolina plant, South Carolina plant)

$$F = \begin{bmatrix} 1{,}700 & 1{,}810 \\ 930 & 740 \end{bmatrix} \begin{matrix} \text{Desks} \\ \text{Stands} \end{matrix}$$

(A) Find the average monthly production for the months of January and February.

(B) Find the increase in production from January to February.

(C) Find $J\begin{bmatrix} 1 \\ 1 \end{bmatrix}$ and interpret.

38. Cryptography. The following message was encoded with the matrix B shown. Decode the message:

25 8 26 24 25 33 21 41 48 41 30 50
21 32 41 52 52 79

$$B = \begin{bmatrix} 1 & 1 & 0 \\ 1 & 0 & 1 \\ 1 & 1 & 1 \end{bmatrix}$$

CHAPTER 9 GROUP ACTIVITY

Using Matrices to Find Cost, Revenue, and Profit

A toy distributor purchases model train components from various suppliers and packages these components in three different ready-to-run train sets: the Limited, the Empire, and the Comet. The components used in each set are listed in Table 1. For convenience, the total labor time (in minutes) required to prepare a set for shipping is included as a component.

TABLE 1 Product Components

	Train Sets		
Components	**Limited**	**Empire**	**Comet**
Locomotives	1	1	2
Cars	5	6	8
Track pieces	20	24	32
Track switches	1	2	4
Power pack	1	1	1
Labor (min)	15	18	24

The current costs of the components are given in Table 2, and the distributor's selling prices for the sets are given in Table 3.

TABLE 2 Component Costs

Components	Cost per Unit ($)
Locomotive	12.52
Car	1.43
Track piece	0.25
Track switch	2.29
Power pack	12.54
Labor (per min)	0.15

TABLE 3 Selling Prices

Set	Price
Limited	$54.60
Empire	$62.28
Comet	$81.15

The distributor has just received the order shown in Table 4 from a retail toy store.

TABLE 4 Customer Order

Set	Quantity
Limited	48
Empire	24
Comet	12

The distributor wants to store the information in each table in a matrix and use matrix operations to find the following information:

1. The inventory (parts and labor) required to fill the order
2. The cost (parts and labor) of filling the order
3. The revenue (sales) received from the customer
4. The profit realized on the order

(A) Use a single letter to designate the matrix representing each table, and write matrix expressions in terms of these letters that will provide the required information. Discuss the size of the matrix you must use to represent each table so that all the pertinent matrix operations are defined.

(B) Evaluate the matrix expressions in part A.

Shortly after filling the order in Table 4, a supplier informs the distributor that the cars and locomotives used in these train sets are no longer available. The distributor currently has 30 locomotives and 134 cars in stock.

(C) How many train sets of each type can the distributor produce using all the available locomotives and cars? Assume that the distributor has unlimited quantities of the other components used in these sets.

(D) How much profit will the distributor make if all these sets are sold? If there is more than one way to use all the available locomotives and cars, which one will produce the largest profit?

CUMULATIVE REVIEW EXERCISES — CHAPTERS 8 9

Work through all the problems in this cumulative review and check answers in the back of the book. Answers to all review problems are there, and following each answer is a number in italics indicating the section in which that type of problem is discussed. Where weaknesses show up, review appropriate sections in the text.

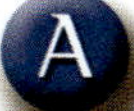

1. Solve using substitution or elimination by addition:

$$\begin{aligned} 3x - 5y &= 11 \\ 2x + 3y &= 1 \end{aligned}$$

2. Solve by graphing: $\begin{aligned} 2x - y &= -4 \\ 3x + y &= -1 \end{aligned}$

3. Solve by substitution or elimination by addition:

$$\begin{aligned} -6x + 3y &= 2 \\ 2x - y &= 1 \end{aligned}$$

4. Solve by graphing: $\begin{aligned} 3x + 5y &\le 15 \\ x, y &\ge 0 \end{aligned}$

5. Find the maximum and minimum value of $z = 2x + 3y$ over the feasible region S:

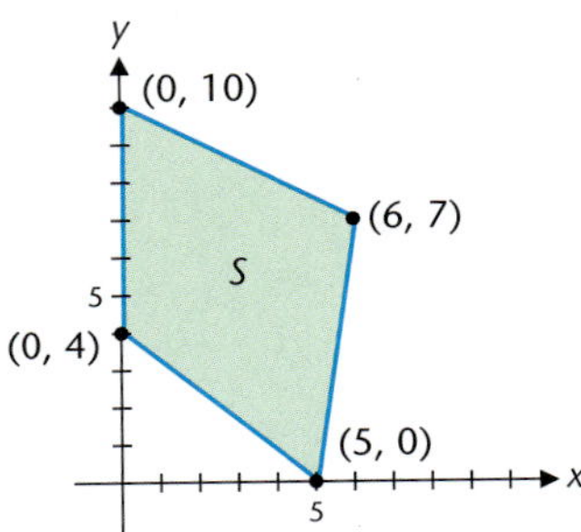

6. Perform the operations that are defined, given the following matrices:

$$M = \begin{bmatrix} 2 & 1 \\ 1 & -3 \end{bmatrix} \qquad N = \begin{bmatrix} 1 & 2 \\ -1 & 3 \end{bmatrix}$$

$$P = \begin{bmatrix} 1 & 2 \end{bmatrix} \qquad Q = \begin{bmatrix} -1 \\ 2 \end{bmatrix}$$

(A) $M - 2N$ **(B)** $P + Q$ **(C)** PQ
(D) MN **(E)** PN **(F)** QM

7. Evaluate: $\begin{vmatrix} 0 & 2 & 0 \\ 1 & 3 & 2 \\ -1 & 4 & 3 \end{vmatrix}$

8. Write the linear system corresponding to each augmented matrix and solve:

(A) $\left[\begin{array}{cc|c} 1 & 0 & 3 \\ 0 & 1 & -4 \end{array}\right]$ **(B)** $\left[\begin{array}{cc|c} 1 & -2 & 3 \\ 0 & 0 & 0 \end{array}\right]$

(C) $\left[\begin{array}{cc|c} 1 & -2 & 3 \\ 0 & 0 & 1 \end{array}\right]$

9. Given the system: $\begin{aligned} x_1 + x_2 &= 3 \\ -x_1 + x_2 &= 5 \end{aligned}$

(A) Write the augmented matrix for the system.
(B) Transform the augmented matrix into reduced form.
(C) Write the solution to the system.

10. Given the system: $\begin{aligned} x_1 - 3x_2 &= k_1 \\ 2x_1 - 5x_2 &= k_2 \end{aligned}$

(A) Write the system as a matrix equation of the form $AX = B$.
(B) Find the inverse of the coefficient matrix A.
(C) Use A^{-1} to find the solution for $k_1 = -2$ and $k_2 = 1$.
(D) Use A^{-1} to find the solution for $k_1 = 1$ and $k_2 = -2$.

11. Given the system: $\begin{aligned} 2x - 3y &= 1 \\ 4x - 5y &= 2 \end{aligned}$

(A) Find the determinant of the coefficient matrix.
(B) Solve the system using Cramer's rule.

12. Use Gauss–Jordan elimination to solve the system

$$\begin{aligned} x_1 + 3x_2 &= 10 \\ 2x_1 - x_2 &= -1 \end{aligned}$$

Then write the linear system represented by each augmented matrix in your solution, and solve each of these systems graphically. Discuss the relationship between the solutions of these systems.

13. Solve graphically to two decimal places:

$$\begin{aligned} -2x + 3y &= 7 \\ 3x + 4y &= 18 \end{aligned}$$

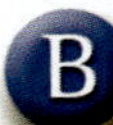

Solve Problems 14–16 using Gauss–Jordan elimination.

14. $\begin{aligned} x_1 + 2x_2 - x_3 &= 3 \\ x_2 + x_3 &= -2 \\ 2x_1 + 3x_2 + x_3 &= 0 \end{aligned}$

15. $\begin{aligned} x_1 + x_2 - x_3 &= 2 \\ 4x_2 + 6x_3 &= -1 \\ 6x_2 + 9x_3 &= 0 \end{aligned}$

16. $\begin{aligned} x_1 - 2x_2 + x_3 &= 1 \\ 3x_1 - 2x_2 - x_3 &= -5 \end{aligned}$

17. Given $M = \begin{bmatrix} 1 & 2 & -1 \end{bmatrix}$ and $N = \begin{bmatrix} 1 \\ -1 \\ 2 \end{bmatrix}$. Find:

(A) MN **(B)** NM

18. Given

$$L = \begin{bmatrix} 2 & -1 & 0 \\ 1 & 2 & 1 \end{bmatrix} \qquad M = \begin{bmatrix} 1 & 2 \\ -1 & 0 \\ 1 & 1 \end{bmatrix} \qquad N = \begin{bmatrix} 2 & 1 \\ -1 & 0 \end{bmatrix}$$

Find, if defined: **(A)** $LM - 2N$ **(B)** $ML + N$

19. Graph the solution region and indicate whether the solution region is bounded or unbounded. Find the coordinates of each corner point.

$$\begin{aligned} 3x + 2y &\geq 12 \\ x + 2y &\geq 8 \\ x, y &\geq 0 \end{aligned}$$

20. Solve the linear programming problem:

Maximize $z = 4x + 9y$

Subject to
$$\begin{aligned} x + 2y &\le 14 \\ 2x + y &\le 16 \\ x, y &\ge 0 \end{aligned}$$

21. Given the system:
$$\begin{aligned} x_1 + 4x_2 + 2x_3 &= k_1 \\ 2x_1 + 6x_2 + 3x_3 &= k_2 \\ 2x_1 + 5x_2 + 2x_3 &= k_3 \end{aligned}$$

(A) Write the system as a matrix equation in the form $AX = B$.

(B) Find the inverse of the coefficient matrix A.

(C) Use A^{-1} to solve the system when $k_1 = -1$, $k_2 = 2$, and $k_3 = 1$.

(D) Use A^{-1} to solve the system when $k_1 = 2$, $k_2 = 0$, and $k_3 = -1$.

22. Given the system:
$$\begin{aligned} x + 2y - z &= 1 \\ 2x + 8y + z &= -2 \\ -x + 3y + 5z &= 2 \end{aligned}$$

(A) Evaluate the coefficient determinant D.

(B) Solve for z using Cramer's rule.

C

23. Discuss the number of solutions for the system corresponding to the reduced form shown below if

(A) $m = 0$ and $n = 0$ **(B)** $m = 0$ and $n \ne 0$

(C) $m \ne 0$

$$\left[\begin{array}{ccc|c} 1 & 0 & -5 & 2 \\ 0 & 1 & 3 & 6 \\ 0 & 0 & m & n \end{array}\right]$$

24. If a square matrix A satisfies the equation $A^2 = A$, find A. Assume that A^{-1} exists.

25. Which of the following augmented matrices are in reduced form?

$$L = \left[\begin{array}{ccc|c} 1 & 0 & 0 & 2 \\ 0 & 1 & 0 & 0 \\ 0 & 0 & 1 & -1 \end{array}\right] \quad M = \left[\begin{array}{ccc|c} 1 & 0 & 3 & 3 \\ 0 & 1 & -2 & 2 \\ 0 & 0 & 0 & 0 \end{array}\right]$$

$$N = \left[\begin{array}{cc|c} 0 & 0 & 0 \\ 1 & 0 & 2 \\ 0 & 1 & -3 \end{array}\right] \quad P = \left[\begin{array}{cccc|c} 1 & 2 & 0 & 2 & -2 \\ 0 & 0 & 1 & 3 & 1 \end{array}\right]$$

26. Show that

$$k\begin{vmatrix} a & b \\ c & d \end{vmatrix} = \begin{vmatrix} ka & b \\ kc & d \end{vmatrix}$$

27. Show that

$$\begin{vmatrix} a & b \\ c & d \end{vmatrix} = \begin{vmatrix} a & b \\ c + ka & d + kb \end{vmatrix}$$

28. If $M = \begin{vmatrix} a & b \\ c & d \end{vmatrix}$ and $\det M \ne 0$, show that

$$M^{-1} = \frac{1}{\det M}\begin{bmatrix} d & -b \\ -c & a \end{bmatrix}$$

Recall that a square matrix is called ***upper triangular*** *if all elements below the principal diagonal are zero, and it is called* ***diagonal*** *if all elements not on the principal diagonal are zero. A square matrix is called* ***lower triangular*** *if all elements above the principal diagonal are zero. In Problems 29–36, determine whether the statement is true or false. If true, explain why. If false, give a counterexample.*

29. The sum of two upper triangular matrices is upper triangular.

30. The product of two lower triangular matrices is lower triangular.

31. The sum of an upper triangular matrix and a lower triangular matrix is a diagonal matrix.

32. The product of an upper triangular matrix and a lower triangular matrix is a diagonal matrix.

33. A matrix that is both upper triangular and lower triangular is a diagonal matrix.

34. If a diagonal matrix has no zero elements on the principal diagonal, then it has an inverse.

35. The determinant of a diagonal matrix is the product of the elements on the principal diagonal.

36. The determinant of a lower triangular matrix is the product of the elements on the principal diagonal.

APPLICATIONS

37. Finance. An investor has \$12,000 to invest. If part is invested at 8% and the rest in a higher-risk investment at 14%, how much should be invested at each rate to produce the same yield as if all had been invested at 10%?

38. Diet. In an experiment involving mice, a zoologist needs a food mix that contains, among other things, 23 grams of protein, 6.2 grams of fat, and 16 grams of moisture. She has on hand mixes of the following compositions: Mix A contains 20% protein, 2% fat, and 15% moisture; mix B contains 10% protein, 6% fat, and 10% moisture; and mix C contains 15% protein, 5% fat, and 5% moisture. How many grams of each mix should be used to get the desired diet mix?

39. Purchasing. A soft-drink distributor has budgeted \$300,000 for the purchase of 12 new delivery trucks. If a model A truck costs \$18,000, a model B truck costs \$22,000, and a model C truck costs \$30,000, how many trucks of each model should the distributor purchase to use exactly all the budgeted funds?

40. Manufacturing. A manufacturer makes two types of day packs, a standard model and a deluxe model. Each standard model requires 0.5 labor-hour from the fabricating department and 0.3 labor-hour from the sewing department. Each deluxe model requires 0.5 labor-hour from the fabricating department and 0.6 labor-hour from the sewing department. The maximum number of labor-hours available per week in the fabricating department and the sewing department are 300 and 240, respectively.

(A) If the profit on a standard day pack is \$8 and the profit on a deluxe day pack is \$12, how many of each type of pack should be manufactured each day to realize a maximum profit? What is the maximum profit?

(B) Discuss the effect on the production schedule and the maximum profit if the profit on a standard day pack decreases by \$3 and the profit on a deluxe day pack increases by \$3.

(C) Discuss the effect on the production schedule and the maximum profit if the profit on a standard day pack increases by \$3 and the profit on a deluxe day pack decreases by \$3.

41. Averaging Tests. A teacher has given four tests to a class of five students and stored the results in the following matrix:

$$\begin{array}{l} \\ \\ \text{Ann} \\ \text{Bob} \\ \text{Carol} \\ \text{Dan} \\ \text{Eric} \end{array} \begin{array}{c} \text{Tests} \\ \begin{array}{cccc} 1 & 2 & 3 & 4 \end{array} \\ \begin{bmatrix} 78 & 84 & 81 & 86 \\ 91 & 65 & 84 & 92 \\ 95 & 90 & 92 & 91 \\ 75 & 82 & 87 & 91 \\ 83 & 88 & 81 & 76 \end{bmatrix} \end{array} = M$$

Discuss methods of matrix multiplication that the teacher can use to obtain the indicated information in parts A–C below. In each case, state the matrices to be used and then perform the necessary multiplications.

(A) The average on all four tests for each student, assuming that all four tests are given equal weight

(B) The average on all four tests for each student, assuming that the first three tests are given equal weight and the fourth is given twice this weight

(C) The class average on each of the four tests

CHAPTER 10

Sequences, Induction, and Probability

OUTLINE

THE LISTS

1, 4, 9, 16, 25, 36, 49, 64, . . .

and

3, 6, 3, 1, 4, 2, 1, 4, . . .

are examples of sequences. The first sequence exhibits a great deal of regularity. You no doubt recognize it as the sequence of perfect squares. Its terms are increasing, and the differences between terms form a striking pattern. You probably do not recognize the second sequence, whose terms do not suggest an obvious pattern. In fact, the second sequence records the results of repeatedly tossing a single die. Sequences, and the related concept of series, are useful tools in almost all areas of mathematics. In Chapter 10 they play roles in the development of several topics: a method of proof called *mathematical induction,* techniques for counting, and probability.

Preparing for this chapter

Before getting started on this chapter, review the following concepts:

- Set Notation (Basic Algebra Review*, Section R.1)
- Operations on Polynomials (Basic Algebra Review*, Section R.2)
- Integer Exponents (Basic Algebra Review*, Section R.5)
- Functions (Chapter 1, Section 2)
- Set Operations (Appendix A, Section A.1)

*At www.mhhe.com/barnett

SECTION 10.1 Sequences and Series

Sequences • Series

In Section 10.1 we introduce special notation and formulas for representing and generating sequences and sums of sequences.

Sequences

Consider the function f given by

$$f(n) = 2n - 1 \tag{1}$$

where the domain of f is the set of natural numbers N. Note that

$$f(1) = 1, f(2) = 3, f(3) = 5, \ldots$$

The function f is an example of a sequence. A **sequence** is a function with domain a set of successive integers. However, a sequence is hardly ever represented in the form of equation (1). A special notation for sequences has evolved, which we describe here.

To start, the range value $f(n)$ is usually symbolized more compactly with a symbol such as a_n. Thus, in place of equation (1) we write

$$a_n = 2n - 1$$

The domain is understood to be the set of natural numbers N unless stated to the contrary or the context indicates otherwise. The elements in the range are called **terms of the sequence:** a_1 is the first term, a_2 the second term, and a_n the nth term, or the **general term:**

$$\begin{aligned} a_1 &= 2(1) - 1 = 1 && \text{First term} \\ a_2 &= 2(2) - 1 = 3 && \text{Second term} \\ a_3 &= 2(3) - 1 = 5 && \text{Third term} \\ &\vdots \qquad\qquad\quad \vdots \end{aligned}$$

The ordered list of elements

$$1, 3, 5, \ldots, 2n - 1, \ldots$$

in which the terms of a sequence are written in their natural order with respect to the domain values, is often informally referred to as a sequence. A sequence is also represented in the abbreviated form $\{a_n\}$, where a symbol for the nth term is placed between braces. For example, we can refer to the sequence

$$1, 3, 5, \ldots, 2n - 1, \ldots$$

as the sequence $\{2n - 1\}$.

If the domain of a function is a finite set of successive integers, then the sequence is called a **finite sequence.** If the domain is an infinite set of successive integers, then the sequence is called an **infinite sequence.** The sequence $\{2n - 1\}$ above is an example of an infinite sequence.

EXPLORE/DISCUSS 1

The sequence $\{2n - 1\}$ is a function whose domain is the set of natural numbers, and so it may be graphed in the same way as any function whose domain and range are sets of real numbers (Fig. 1).

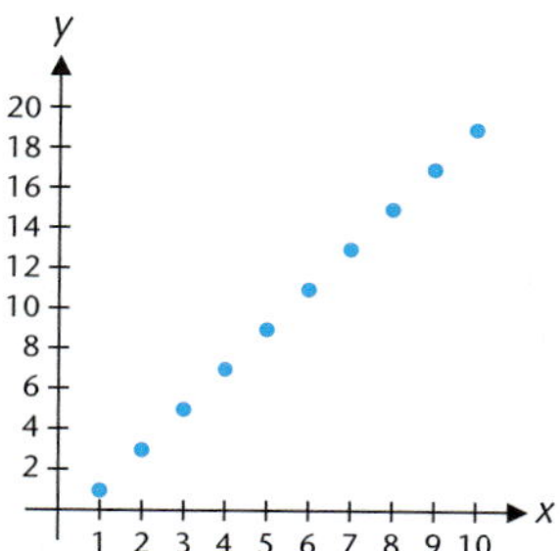

FIGURE 1 Graph of $\{2n - 1\}$.

(A) Explain why the graph of the sequence $\{2n - 1\}$ is not continuous.

(B) Explain why the points on the graph of $\{2n - 1\}$ lie on a line. Find an equation for that line.

(C) Graph the sequence $\left\{\dfrac{2n^2 - n + 1}{n}\right\}$. How are the graphs of $\{2n - 1\}$ and $\left\{\dfrac{2n^2 - n + 1}{n}\right\}$ related?

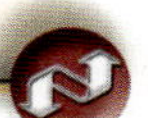

There are several different ways a graphing utility can be used in the study of sequences. Refer to Explore/Discuss 1. Figure 2(a) shows the sequence $\{2n - 1\}$ entered as a function in an equation editor. This produces a continuous graph [Fig. 2(b)] that contains the points in the graph of the sequence (Fig. 1). Figure 2(c) shows the points on the graph of the sequence displayed in a table.

FIGURE 2

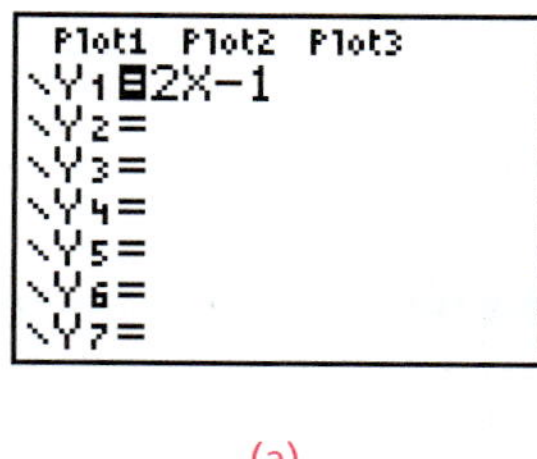

(a)

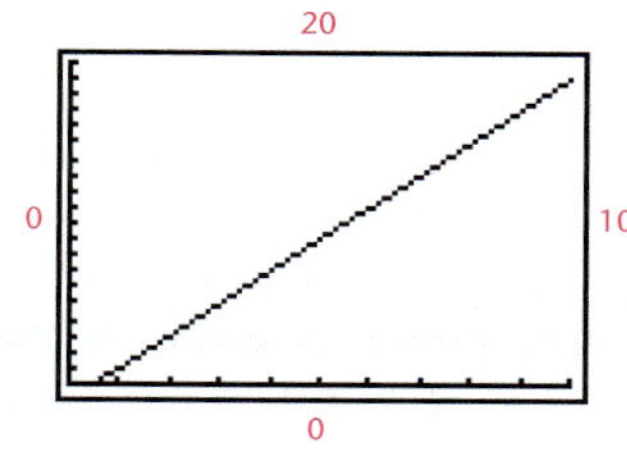

(b)

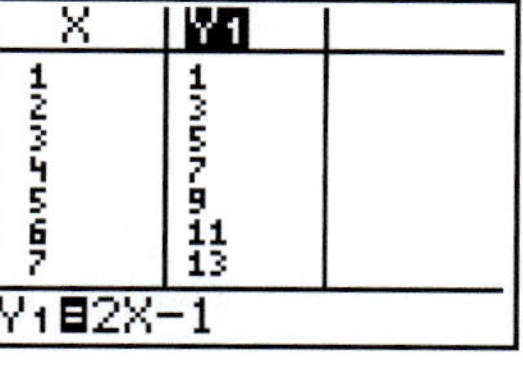

(c)

In Figure 3(a), sequence commands are used to store the first and second coordinates of the first 10 points on the graph of the sequence $\{2n - 1\}$ in lists L_1 and L_2, respectively. A statistical plot routine is used to graph these points [Fig. 3(b)], and a statistical editor is used to display the points on the graph [Fig. 3(c)].

FIGURE 3

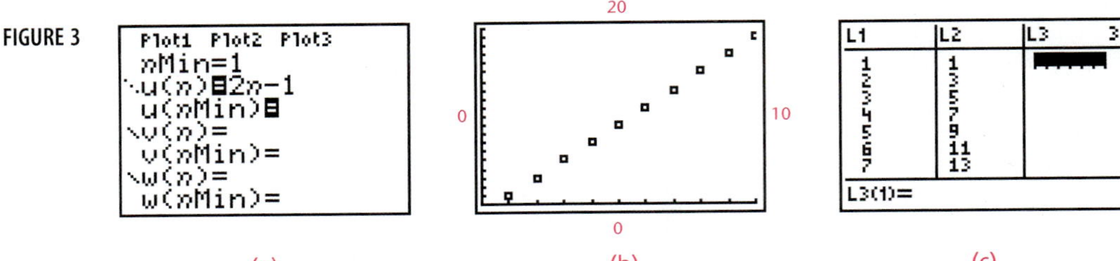

(a) (b) (c)

Most graphing utilities can produce the results shown in Figures 2 and 3. The Texas Instruments TI-83 has a special sequence mode that is very useful for studying sequences. Figure 4(a) shows the sequence $\{2n - 1\}$ entered in the sequence editor, Figure 4(b) shows the graph of this sequence, and Figure 4(c) displays the points on the graph in a table.

FIGURE 4

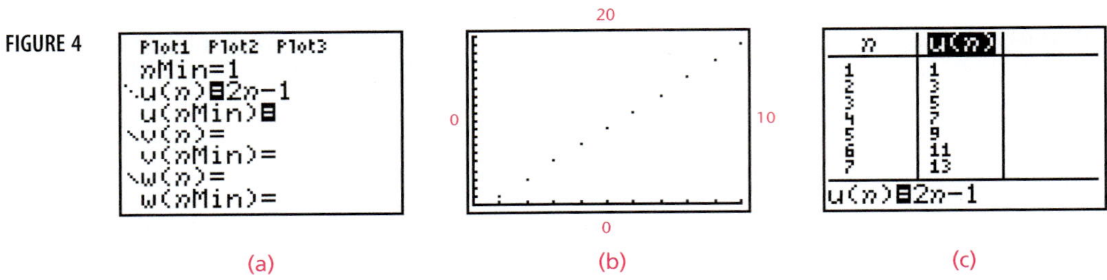

(a) (b) (c)

Examining graphs and displaying values are very helpful activities when working with sequences. Consult your manual to see which of the methods illustrated in Figures 2–4 works on your graphing utility.

Some sequences are specified by a **recursion formula**—that is, a formula that defines each term in terms of one or more preceding terms. The sequence we have chosen to illustrate a recursion formula is a very famous sequence in the history of mathematics called the **Fibonacci sequence.** It is named after the most celebrated mathematician of the thirteenth century, Leonardo Fibonacci from Italy (1180?–1250?).

EXAMPLE 1 Fibonacci Sequence

List the first seven terms of the sequence specified by

$$a_1 = 1$$

$$a_2 = 1$$

$$a_n = a_{n-2} + a_{n-1} \qquad n \geq 3$$

SOLUTION

Algebraic Solution

$$\begin{aligned} a_1 &= 1 \\ a_2 &= 1 \\ a_3 &= a_1 + a_2 = 1 + 1 = 2 \\ a_4 &= a_2 + a_3 = 1 + 2 = 3 \\ a_5 &= a_3 + a_4 = 2 + 3 = 5 \\ a_6 &= a_4 + a_5 = 3 + 5 = 8 \\ a_7 &= a_5 + a_6 = 5 + 8 = 13 \end{aligned}$$

Graphing Utility Solution

RECUR* is a program that will compute the terms in any recursion formula of the form

$$a_n = Aa_{n-2} + Ba_{n-1}$$

where A is the coefficient of the first term and B is the coefficient of the second term. Entering 1 for a_{n-2}, A, a_{n-1}, and B, and requesting seven terms produces the first seven terms in the Fibonacci sequence (Fig. 5).

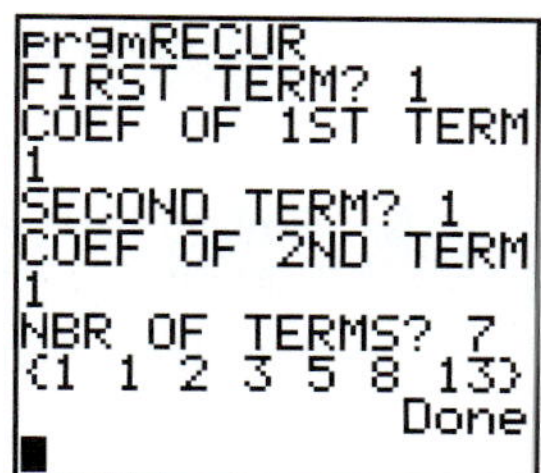

FIGURE 5

MATCHED 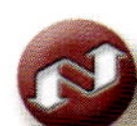 PROBLEM

List the first seven terms of the sequence specified by

$$a_1 = 1$$
$$a_2 = 1$$
$$a_n = a_{n-2} - a_{n-1} \qquad n \geq 3$$

EXPLORE/DISCUSS 2

A multiple-choice test question asked for the next term in the sequence:

1, 3, 9, . . .

and gave the following choices:

(A) 16 (B) 19 (C) 27

Which is the correct answer?

Compare the first four terms of the following sequences:

(A) $a_n = 3^{n-1}$ (B) $b_n = 1 + 2(n - 1)^2$ (C) $c_n = 8n + \dfrac{12}{n} - 19$

Now which of the choices appears to be correct?

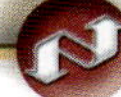

*Programs for TI-83 and TI-86 graphing calculators can be found at the website for this book.

Now we consider the reverse problem. That is, can a sequence be defined just by listing the first three or four terms of the sequence? And can we then use these initial terms to find a formula for the nth term? In general, without other information, the answer to the first question is no. As Explore/Discuss 2 illustrates, many different sequences may start off with the same terms. Simply listing the first three terms, or any other finite number of terms, does not specify a particular sequence. In fact, it can be shown that given any list of m numbers, there are an infinite number of sequences whose first m terms agree with these given numbers.

What about the second question? That is, given a few terms, can we find the general formula for at least one sequence whose first few terms agree with the given terms? The answer to this question is a qualified yes. If we can observe a simple pattern in the given terms, then we may be able to construct a general term that will produce the pattern. The next example illustrates this approach.

EXAMPLE 2 Finding the General Term of a Sequence

Find the general term of a sequence whose first four terms are

(A) 5, 6, 7, 8, . . . (B) 2, −4, 8, −16, . . .

SOLUTIONS

(A) Because these terms are consecutive integers, one solution is $a_n = n$, $n \geq 5$. If we want the domain of the sequence to be all natural numbers, then another solution is $b_n = n + 4$.

(B) Each of these terms can be written as the product of a power of 2 and a power of -1:

$$\begin{aligned} 2 &= (-1)^0 2^1 \\ -4 &= (-1)^1 2^2 \\ 8 &= (-1)^2 2^3 \\ -16 &= (-1)^3 2^4 \end{aligned}$$

If we choose the domain to be all natural numbers, then a solution is

$$a_n = (-1)^{n-1} 2^n$$

MATCHED 2 PROBLEM

Find the general term of a sequence whose first four terms are

(A) 2, 4, 6, 8, . . . (B) $1, -\frac{1}{2}, \frac{1}{4}, -\frac{1}{8}, \ldots$

In general, there is usually more than one way of representing the nth term of a given sequence. This was seen in the solution of Example 2, part A. However, unless stated to the contrary, we assume the domain of the sequence is the set of natural numbers N.

EXPLORE/DISCUSS 3

The sequence with general term $b_n = \frac{\sqrt{5}}{5}\left(\frac{1 + \sqrt{5}}{2}\right)^n$ is closely related to the Fibonacci sequence. Compute the first 20 terms of both sequences and discuss the relationship. [The first seven values of b_n are shown in Fig. 6(b)].

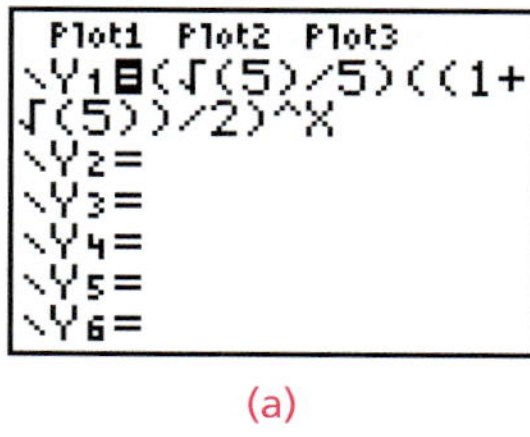

(a)

X	Y1
1	.72361
2	1.1708
3	1.8944
4	3.0652
5	4.9597
6	8.0249
7	12.985

Y1■(√(5)/5)((1+...

(b)

FIGURE 6

Series

If $a_1, a_2, a_3, \ldots, a_n, \ldots$ is a sequence, then the expression

$$a_1 + a_2 + a_3 + \cdots + a_n + \cdots$$

is called a **series.** If the sequence is finite, the corresponding series is a **finite series.** If the sequence is infinite, the corresponding series is an **infinite series.** For example,

$1, 2, 4, 8, 16$ — Finite sequence

$1 + 2 + 4 + 8 + 16$ — Finite series

We restrict our discussion to finite series in Section 10.1.

Series are often represented in a compact form called **summation notation** using the symbol $\sum$, which is a stylized version of the Greek letter sigma. Consider the following examples:

$$\sum_{k=1}^{4} a_k = a_1 + a_2 + a_3 + a_4$$

$$\sum_{k=3}^{7} b_k = b_3 + b_4 + b_5 + b_6 + b_7$$

$$\sum_{k=0}^{n} c_k = c_0 + c_1 + c_2 + \cdots + c_n$$

Domain is the set of integers k satisfying $0 \le k \le n$.

The terms on the right are obtained from the expression on the left by successively replacing the **summing index** k with integers, starting with the first number indicated below $\sum$ and ending with the number that appears above $\sum$. Thus, for example, if we are given the sequence

$$\frac{1}{2}, \frac{1}{4}, \frac{1}{8}, \ldots, \frac{1}{2^n}$$

the corresponding series is

$$\sum_{k=1}^{n} \frac{1}{2^k} = \frac{1}{2} + \frac{1}{4} + \frac{1}{8} + \cdots + \frac{1}{2^n}$$

EXAMPLE 3 Writing the Terms of a Series

Write without summation notation: $\displaystyle\sum_{k=1}^{5} \frac{k-1}{k}$

SOLUTION

$$\sum_{k=1}^{5} \frac{k-1}{k} = \frac{1-1}{1} + \frac{2-1}{2} + \frac{3-1}{3} + \frac{4-1}{4} + \frac{5-1}{5}$$
$$= 0 + \frac{1}{2} + \frac{2}{3} + \frac{3}{4} + \frac{4}{5}$$

MATCHED 3 PROBLEM

Write without summation notation: $\displaystyle\sum_{k=0}^{5} \frac{(-1)^k}{2k+1}$

If the terms of a series are alternately positive and negative, it is called an **alternating series.** Example 4 deals with the representation of such a series.

EXAMPLE 4 Writing a Series in Summation Notation

Write the following series using summation notation:

$$1 - \frac{1}{2} + \frac{1}{3} - \frac{1}{4} + \frac{1}{5} - \frac{1}{6}$$

(A) Start the summing index at $k = 1$.

(B) Start the summing index at $k = 0$.

SOLUTIONS

(A) $(-1)^{k-1}$ provides the alternation of sign, and $1/k$ provides the other part of each term. Thus, we can write

$$\sum_{k=1}^{6} \frac{(-1)^{k-1}}{k}$$

as can be easily checked.

(B) $(-1)^k$ provides the alternation of sign, and $1/(k+1)$ provides the other part of each term. Thus, we write

$$\sum_{k=0}^{5} \frac{(-1)^k}{k+1}$$

as can be checked.

MATCHED 4 PROBLEM

Write the following series using summation notation:

$$1 - \frac{2}{3} + \frac{4}{9} - \frac{8}{27} + \frac{16}{81}$$

(A) Start with $k = 1$. (B) Start with $k = 0$.

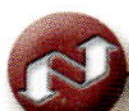

EXPLORE/DISCUSS 4

(A) Find the smallest number of terms of the infinite series

$$1 + \frac{1}{2} + \frac{1}{3} + \cdots + \frac{1}{n} + \cdots$$

that, when added together, give a number greater than 3.

(B) Enter the function $y_1 = \text{sum}(\text{seq}(1/N,N,1,X))$ in the equation editor of a graphing utility and examine a table (Fig. 7) to find the smallest number of terms of the infinite series in part A that, when added together, give a number greater than 4.

X	Y1	
1	1	
2	1.5	
3	1.8333	
4	2.0833	
5	2.2833	
6	2.45	
7	2.5929	

Y1■sum(seq(1/N,...

FIGURE 7

(C) Find the smallest number of terms of the infinite series

$$\frac{1}{2} + \frac{1}{4} + \cdots + \frac{1}{2^n} + \cdots$$

that, when added together, give a number greater than 0.99. Greater than 0.999. Can the sum ever exceed 1? Explain.

ANSWERS MATCHED PROBLEMS

1. 1, 1, 0, 1, −1, 2, −3
2. (A) $a_n = 2n$

   (B) $a_n = (-1)^{n-1}\left(\frac{1}{2}\right)^{n-1}$
3. $1 - \frac{1}{3} + \frac{1}{5} - \frac{1}{7} + \frac{1}{9} - \frac{1}{11}$
4. (A) $\sum_{k=1}^{5} (-1)^{k-1}\left(\frac{2}{3}\right)^{k-1}$

   (B) $\sum_{k=0}^{4} (-1)^{k}\left(\frac{2}{3}\right)^{k}$

EXERCISE 10.1

Write the first four terms for each sequence in Problems 1–6.

1. $a_n = n - 2$
2. $a_n = n + 3$
3. $a_n = \dfrac{n-1}{n+1}$
4. $a_n = \left(1 + \dfrac{1}{n}\right)^n$
5. $a_n = (-2)^{n+1}$
6. $a_n = \dfrac{(-1)^{n+1}}{n^2}$
7. Write the eighth term in the sequence in Problem 1.
8. Write the tenth term in the sequence in Problem 2.
9. Write the one-hundredth term in the sequence in Problem 3.
10. Write the two-hundredth term in the sequence in Problem 4.

In Problems 11–16, write each series in expanded form without summation notation.

11. $\sum_{k=1}^{5} k$
12. $\sum_{k=1}^{4} k^2$
13. $\sum_{k=1}^{3} \dfrac{1}{10^k}$
14. $\sum_{k=1}^{5} \left(\dfrac{1}{3}\right)^k$
15. $\sum_{k=1}^{4} (-1)^k$
16. $\sum_{k=1}^{6} (-1)^{k+1}k$

B

Write the first five terms of each sequence in Problems 17–26.

17. $a_n = (-1)^{n+1}n^2$
18. $a_n = (-1)^{n+1}\left(\dfrac{1}{2^n}\right)$
19. $a_n = \dfrac{1}{3}\left(1 - \dfrac{1}{10^n}\right)$
20. $a_n = n[1 - (-1)^n]$
21. $a_n = (-\frac{1}{2})^{n-1}$
22. $a_n = (-\frac{3}{2})^{n-1}$
23. $a_1 = 7;\ a_n = a_{n-1} - 4,\ n \geq 2$
24. $a_1 = 3;\ a_n = a_{n-1} + 5,\ n \geq 2$
25. $a_1 = 4;\ a_n = \frac{1}{4}a_{n-1},\ n \geq 2$
26. $a_1 = 2;\ a_n = 2a_{n-1},\ n \geq 2$

In Problems 27–30, write the first seven terms of each sequence. Use the program RECUR to check your answers.*

27. $a_1 = 1, a_2 = 2, a_n = a_{n-2} + 2a_{n-1}, n \geq 3$
28. $a_1 = 1, a_2 = -1, a_n = a_{n-2} - a_{n-1}, n \geq 3$
29. $a_1 = -1, a_2 = 2, a_n = 2a_{n-2} + a_{n-1}, n \geq 3$
30. $a_1 = 2, a_2 = 1, a_n = -a_{n-2} + a_{n-1}, n \geq 3$

In Problems 31–42, find the general term of a sequence whose first four terms are given.

31. $4, 5, 6, 7, \ldots$
32. $-2, -1, 0, 1, \ldots$
33. $3, 6, 9, 12, \ldots$
34. $-2, -4, -6, -8, \ldots$
35. $\frac{1}{2}, \frac{2}{3}, \frac{3}{4}, \frac{4}{5}, \ldots$
36. $\frac{1}{2}, \frac{3}{4}, \frac{5}{6}, \frac{7}{8}, \ldots$
37. $1, -1, 1, -1, \ldots$
38. $1, -2, 3, -4, \ldots$
39. $-2, 4, -8, 16, \ldots$
40. $1, -3, 5, -7, \ldots$
41. $x, \dfrac{x^2}{2}, \dfrac{x^3}{3}, \dfrac{x^4}{4}, \ldots$
42. $x, -x^3, x^5, -x^7, \ldots$

In Problems 43–46, find nth-term formulas for two different sequences with the property that both sequences have identical terms for $n = 1, 2,$ and 3. [Hint: Graph the points (n, a_n) and apply quadratic regression to find one formula.]

43. $1, 2, 4, \ldots$
44. $1, 4, 16, \ldots$
45. $1, 8, 27, \ldots$
46. $1, 16, 81, \ldots$

In Problems 47–50, use a graphing utility to graph the first 20 terms of each sequence.

47. $a_n = 1/n$
48. $a_n = 2 + \pi n$
49. $a_n = (-0.9)^n$
50. $a_1 = -1, a_n = \frac{2}{3}a_{n-1} + \frac{1}{2}$

*Programs for TI-83 and TI-86 graphing calculators can be found at the website for this book.

In Problems 51–56, write each series in expanded form without summation notation.

51. $\sum_{k=1}^{4} \frac{(-2)^{k+1}}{k}$

52. $\sum_{k=1}^{5} (-1)^{k+1}(2k-1)^2$

53. $\sum_{k=1}^{3} \frac{1}{k} x^{k+1}$

54. $\sum_{k=1}^{5} x^{k-1}$

55. $\sum_{k=1}^{5} \frac{(-1)^{k+1}}{k} x^k$

56. $\sum_{k=0}^{4} \frac{(-1)^k x^{2k+1}}{2k+1}$

In Problems 57–64, write each series using summation notation with the summing index k starting at k = 1.

57. $1^2 + 2^2 + 3^2 + 4^2$

58. $2 + 3 + 4 + 5 + 6$

59. $\frac{1}{2} + \frac{1}{2^2} + \frac{1}{2^3} + \frac{1}{2^4} + \frac{1}{2^5}$

60. $1 - \frac{1}{2} + \frac{1}{3} - \frac{1}{4}$

61. $1 + \frac{1}{2^2} + \frac{1}{3^2} + \cdots + \frac{1}{n^2}$

62. $2 + \frac{3}{2} + \frac{4}{3} + \cdots + \frac{n+1}{n}$

63. $1 - 4 + 9 - \cdots + (-1)^{n+1}n^2$

64. $\frac{1}{2} - \frac{1}{4} + \frac{1}{8} - \cdots + \frac{(-1)^{n+1}}{2^n}$

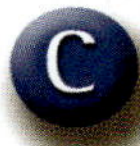

C

The sequence

$$a_n = \frac{a_{n-1}^2 + M}{2a_{n-1}} \qquad n \ge 2, M \text{ a positive real number}$$

can be used to find $\sqrt{M}$ to any decimal-place accuracy desired. To start the sequence, choose a_1 arbitrarily from the positive real numbers. Problems 65 and 66 are related to this sequence. SQRT is a program that can be used to investigate this sequence.*

65. **(A)** Find the first four terms of the sequence

$$a_1 = 3 \qquad a_n = \frac{a_{n-1}^2 + 2}{2a_{n-1}} \qquad n \ge 2$$

(B) Compare the terms with $\sqrt{2}$ from a calculator.

(C) Repeat parts A and B letting a_1 be any other positive number, say 1.

66. **(A)** Find the first four terms of the sequence

$$a_1 = 2 \qquad a_n = \frac{a_{n-1}^2 + 5}{2a_{n-1}} \qquad n \ge 2$$

(B) Find $\sqrt{5}$ with a calculator, and compare with the results of part A.

(C) Repeat parts A and B letting a_1 be any other positive number, say 3.

*Programs for TI-83 and TI-86 graphing calculators can be found at the website for this book.

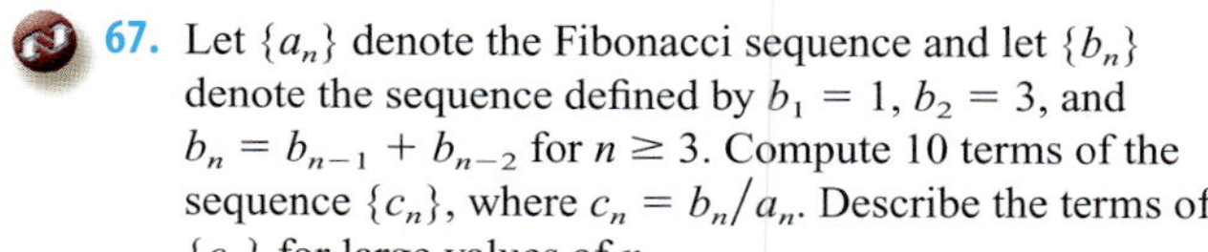

67. Let $\{a_n\}$ denote the Fibonacci sequence and let $\{b_n\}$ denote the sequence defined by $b_1 = 1$, $b_2 = 3$, and $b_n = b_{n-1} + b_{n-2}$ for $n \ge 3$. Compute 10 terms of the sequence $\{c_n\}$, where $c_n = b_n/a_n$. Describe the terms of $\{c_n\}$ for large values of n.

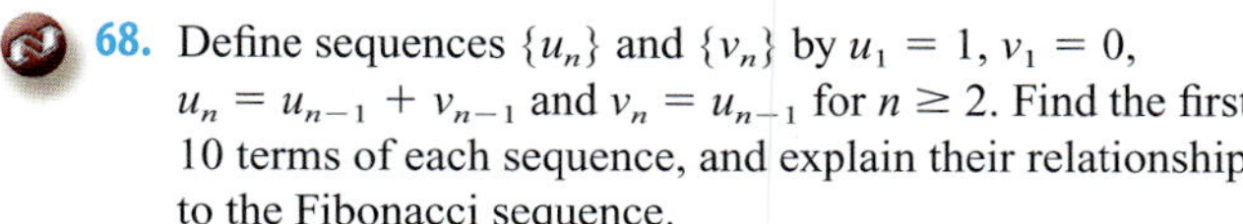

68. Define sequences $\{u_n\}$ and $\{v_n\}$ by $u_1 = 1$, $v_1 = 0$, $u_n = u_{n-1} + v_{n-1}$ and $v_n = u_{n-1}$ for $n \ge 2$. Find the first 10 terms of each sequence, and explain their relationship to the Fibonacci sequence.

In calculus, it can be shown that

$$e^x = \sum_{k=0}^{\infty} \frac{x^k}{k!} \approx 1 + \frac{x}{1!} + \frac{x^2}{2!} + \frac{x^3}{3!} + \cdots + \frac{x^n}{n!}$$

where the larger n is, the better the approximation. Problems 69 and 70 refer to this series. Note that n!, read "n factorial," is defined by $0! = 1$ and $n! = 1 \cdot 2 \cdot 3 \cdot \cdots \cdot n$ for $n \in N$.

69. Approximate $e^{0.2}$ using the first five terms of the series. Compare this approximation with your calculator evaluation of $e^{0.2}$.

70. Approximate $e^{-0.5}$ using the first five terms of the series. Compare this approximation with your calculator evaluation of $e^{-0.5}$.

71. Show that $\sum_{k=1}^{n} ca_k = c\sum_{k=1}^{n} a_k$

72. Show that $\sum_{k=1}^{n} (a_k + b_k) = \sum_{k=1}^{n} a_k + \sum_{k=1}^{n} b_k$

SECTION 10.2 Mathematical Induction

Introduction

In common usage, the word *induction* means the generalization from particular cases or facts. The ability to formulate general hypotheses from a limited number of facts is a distinguishing characteristic of a creative mathematician. The creative process does not stop here, however. These hypotheses must then be proved or disproved. In mathematics, a special method of proof called **mathematical induction** ranks among the most important basic tools in a mathematician's toolbox. In Section 10.2, mathematical induction will be used to prove a variety of mathematical statements, some new and some that up to now we have just assumed to be true.

We illustrate the process of formulating hypotheses by an example. Suppose we are interested in the sum of the first n consecutive odd integers, where n is a positive integer. We begin by writing the sums for the first few values of n to see if we can observe a pattern:

$$
\begin{aligned}
1 &= 1 && n = 1\\
1 + 3 &= 4 && n = 2\\
1 + 3 + 5 &= 9 && n = 3\\
1 + 3 + 5 + 7 &= 16 && n = 4\\
1 + 3 + 5 + 7 + 9 &= 25 && n = 5
\end{aligned}
$$

Is there any pattern to the sums 1, 4, 9, 16, and 25? You no doubt observed that each is a perfect square and, in fact, each is the square of the number of terms in the sum. Thus, the following conjecture seems reasonable:

Conjecture P_n: For each positive integer n,

$$1 + 3 + 5 + \cdots + (2n - 1) = n^2$$

That is, the sum of the first n odd integers is n^2 for each positive integer n.

So far ordinary induction has been used to generalize the pattern observed in the first few cases listed. But at this point conjecture P_n is simply that—a conjecture. How do we prove that P_n is a true statement? Continuing to list specific cases will never provide a general proof—not in your lifetime or all your descendants' lifetimes! Mathematical induction is the tool we will use to establish the validity of conjecture P_n.

Before discussing this method of proof, let's consider another conjecture:

Conjecture Q_n: For each positive integer n, the number $n^2 - n + 41$ is a prime number.

TABLE 1

n	$n^2 - n + 41$	Prime?
1	41	Yes
2	43	Yes
3	47	Yes
4	53	Yes
5	61	Yes

It is important to recognize that a conjecture can be proved false if it fails for only one case. A single case or example for which a conjecture fails is called a **counterexample.** We check the conjecture for a few particular cases in Table 1. From the table, it certainly appears that conjecture Q_n has a good chance of being true. You may want to check a few more cases. If you persist, you will find that conjecture Q_n is true for n up to 41. What happens at $n = 41$?

$$41^2 - 41 + 41 = 41^2$$

which is not prime. Thus, because $n = 41$ provides a counterexample, conjecture Q_n is false. Here we see the danger of generalizing without proof from a few special cases. This example was discovered by Euler (1707–1783).

EXPLORE/DISCUSS 1

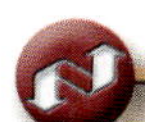

Prove that the following statement is false by finding a counterexample: If $n \geq 2$, then at least one-third of the positive integers less than or equal to n are prime.

Mathematical Induction

We begin by stating the *principle of mathematical induction,* which forms the basis for all our work in Section 10.2.

THEOREM 1
Principle of Mathematical Induction

Let P_n be a statement associated with each positive integer n, and suppose the following conditions are satisfied:

1. P_1 is true.
2. For any positive integer k, if P_k is true, then P_{k+1} is also true.

Then the statement P_n is true for all positive integers n.

Theorem 1 must be read very carefully. At first glance, it seems to say that if we assume a statement is true, then it is true. But that is not the case at all. If the two conditions in Theorem 1 are satisfied, then we can reason as follows:

P_1 is true.	Condition 1
P_2 is true, because P_1 is true.	Condition 2
P_3 is true, because P_2 is true.	Condition 2
P_4 is true, because P_3 is true.	Condition 2
⋮	⋮

Because this chain of implications never ends, we will eventually reach P_n for any positive integer n.

FIGURE 1 Interpreting mathematical induction.

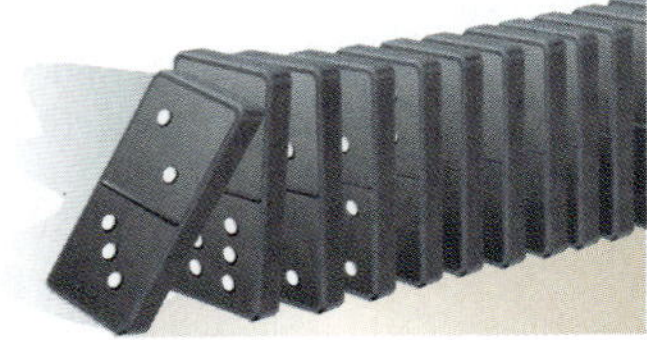

Condition 1: The first domino can be pushed over.
(a)

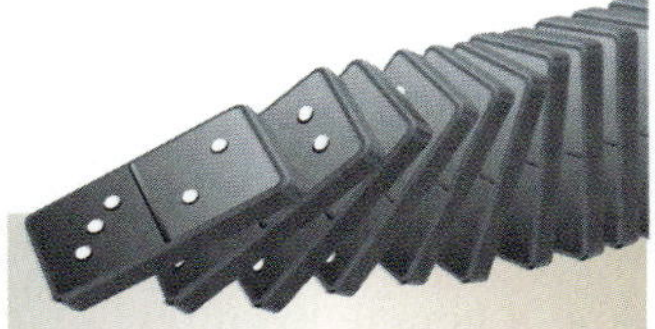

Condition 2: If the kth domino falls, then so does the $(k + 1)$st.
(b)

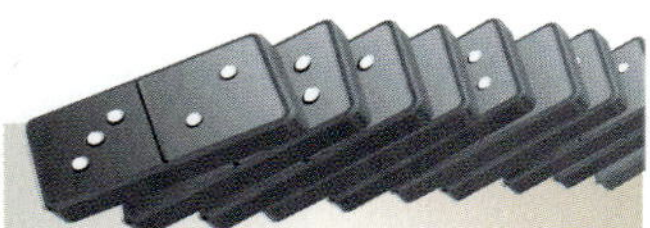

Conclusion: All the dominoes will fall.
(c)

To help visualize this process, picture a row of dominoes that goes on forever (Fig. 1) and interpret the conditions in Theorem 1 as follows: Condition 1 says that the first domino can be pushed over. Condition 2 says that if the kth domino falls, then so does the $(k + 1)$st domino. Together, these two conditions imply that all the dominoes must fall.

Now, to illustrate the process of proof by mathematical induction, we return to the conjecture P_n discussed earlier, which we restate below:

$$P_n\text{: } 1 + 3 + 5 + \cdots + (2n - 1) = n^2 \qquad n \text{ any positive integer}$$

We already know that P_1 is a true statement. In fact, we demonstrated that P_1 through P_5 are all true by direct calculation. Thus, condition 1 in Theorem 1 is satisfied. To show that condition 2 is satisfied, we assume that P_k is a true statement:

$$P_k\text{: } 1 + 3 + 5 + \cdots + (2k - 1) = k^2$$

Now we must show that this assumption implies that P_{k+1} is also a true statement:

$$P_{k+1}\text{: } 1 + 3 + 5 + \cdots + (2k - 1) + (2k + 1) = (k + 1)^2$$

Because we have assumed that P_k is true, we can perform the operations on this equation. Note that the left side of P_{k+1} is the left side of P_k plus $(2k + 1)$. So we start by adding $(2k + 1)$ to both sides of P_k:

$$1 + 3 + 5 + \cdots + (2k - 1) = k^2 \qquad P_k$$

$$1 + 3 + 5 + \cdots + (2k - 1) + (2k + 1) = k^2 + (2k + 1) \qquad \text{Add } 2k + 1 \text{ to both sides.}$$

Factoring the right side of this equation, we have

$$1 + 3 + 5 + \cdots + (2k - 1) + (2k + 1) = (k + 1)^2 \qquad P_{k+1}$$

But this last equation is P_{k+1}. Thus, we have started with P_k, the statement we assumed true, and performed valid operations to produce P_{k+1}, the statement we want to be true. In other words, we have shown that if P_k is true, then P_{k+1} is also true. Because both conditions in Theorem 1 are satisfied, P_n is true for all positive integers n.

Additional Examples of Mathematical Induction

Now we will consider some additional examples of proof by induction. The first is another summation formula. Mathematical induction is the primary tool for proving that formulas of this type are true.

EXAMPLE 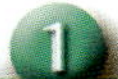 1 Proving a Summation Formula

Prove that for all positive integers n

$$\frac{1}{2} + \frac{1}{4} + \frac{1}{8} + \cdots + \frac{1}{2^n} = \frac{2^n - 1}{2^n}$$

PROOF

State the conjecture:

$$P_n\colon\ \frac{1}{2}+\frac{1}{4}+\frac{1}{8}+\cdots+\frac{1}{2^n}=\frac{2^n-1}{2^n}$$

PART 1 Show that P_1 is true.

$$P_1\colon\ \frac{1}{2}=\frac{2^1-1}{2^1}$$
$$=\frac{1}{2}$$

Thus, P_1 is true.

PART 2 Show that if P_k is true, then P_{k+1} is true. It is a good practice to always write out both P_k and P_{k+1} at the beginning of any induction proof to see what is assumed and what must be proved:

$$P_k\colon\ \frac{1}{2}+\frac{1}{4}+\frac{1}{8}+\cdots+\frac{1}{2^k}=\frac{2^k-1}{2^k}$$ We assume P_k is true.

$$P_{k+1}\colon\ \frac{1}{2}+\frac{1}{4}+\frac{1}{8}+\cdots+\frac{1}{2^k}+\frac{1}{2^{k+1}}=\frac{2^{k+1}-1}{2^{k+1}}$$ We must show that P_{k+1} follows from P_k.

We start with the true statement P_k, add $1/2^{k+1}$ to both sides, and simplify the right side:

$$\frac{1}{2}+\frac{1}{4}+\frac{1}{8}+\cdots+\frac{1}{2^k}=\frac{2^k-1}{2^k}$$ P_k

$$\frac{1}{2}+\frac{1}{4}+\frac{1}{8}+\cdots+\frac{1}{2^k}+\frac{1}{2^{k+1}}=\frac{2^k-1}{2^k}+\frac{1}{2^{k+1}}$$
$$=\frac{2^k-1}{2^k}\cdot\frac{2}{2}+\frac{1}{2^{k+1}}$$
$$=\frac{2^{k+1}-2+1}{2^{k+1}}$$
$$=\frac{2^{k+1}-1}{2^{k+1}}$$

Thus,

$$\frac{1}{2}+\frac{1}{4}+\frac{1}{8}+\cdots+\frac{1}{2^k}+\frac{1}{2^{k+1}}=\frac{2^{k+1}-1}{2^{k+1}}$$ P_{k+1}

and we have shown that if P_k is true, then P_{k+1} is true.

CONCLUSION Both conditions in Theorem 1 are satisfied. Thus, P_n is true for all positive integers n.

MATCHED 1 PROBLEM

Prove that for all positive integers n

$$1 + 2 + 3 + \cdots + n = \frac{n(n + 1)}{2}$$

Example 2 provides a proof of a law of exponents that previously we had to assume was true. First we redefine a^n for n a positive integer, using a recursion formula:

DEFINITION 1
Recursive Definition of a^n

For n a positive integer

$$a^1 = a$$
$$a^{n+1} = a^n a \qquad n > 1$$

EXAMPLE 2 Proving a Law of Exponents

Prove that $(xy)^n = x^n y^n$ for all positive integers n.

PROOF
State the conjecture:

P_n: $(xy)^n = x^n y^n$

PART 1 Show that P_1 is true.

$(xy)^1 = xy$ Definition 1
$= x^1 y^1$ Definition 1

Thus, P_1 is true.

PART 2 Show that if P_k is true, then P_{k+1} is true.

P_k: $(xy)^k = x^k y^k$ Assume P_k is true.
P_{k+1}: $(xy)^{k+1} = x^{k+1} y^{k+1}$ Show that P_{k+1} follows from P_k.

Here we start with the left side of P_{k+1} and use P_k to find the right side of P_{k+1}:

$(xy)^{k+1} = (xy)^k (xy)^1$ Definition 1
$= x^k y^k xy$ Use P_k: $(xy)^k = x^k y^k$.
$= (x^k x)(y^k y)$ Property of real numbers
$= x^{k+1} y^{k+1}$ Definition 1

Thus, $(xy)^{k+1} = x^{k+1} y^{k+1}$, and we have shown that if P_k is true, then P_{k+1} is true.

CONCLUSION Both conditions in Theorem 1 are satisfied. Thus, P_n is true for all positive integers n.

MATCHED 2 PROBLEM

Prove that $(x/y)^n = x^n/y^n$ for all positive integers n.

Example 3 deals with factors of integers. Before we start, recall that an integer p is *divisible* by an integer q if $p = qr$ for some integer r.

EXAMPLE 3 Proving a Divisibility Property

Prove that $4^{2n} - 1$ is divisible by 5 for all positive integers n.

PROOF

Use the definition of divisibility to state the conjecture as follows:

P_n: $4^{2n} - 1 = 5r$ for some integer r

PART 1 Show that P_1 is true.

P_1: $4^2 - 1 = 15 = 5 \cdot 3$

Thus, P_1 is true.

PART 2 Show that if P_k is true, then P_{k+1} is true.

P_k: $4^{2k} - 1 = 5r$ for some integer r **Assume P_k is true.**

P_{k+1}: $4^{2(k+1)} - 1 = 5s$ for some integer s **Show that P_{k+1} must follow.**

As before, we start with the true statement P_k:

$$
\begin{aligned}
4^{2k} - 1 &= 5r && P_k \\
4^2(4^{2k} - 1) &= 4^2(5r) && \text{Multiply both sides by } 4^2. \\
4^{2k+2} - 16 &= 80r && \text{Simplify.} \\
4^{2(k+1)} - 1 &= 80r + 15 && \text{Add 15 to both sides.} \\
&= 5(16r + 3) && \text{Factor out 5.}
\end{aligned}
$$

Thus,

$4^{2(k+1)} - 1 = 5s$ P_{k+1}

where $s = 16r + 3$ is an integer, and we have shown that if P_k is true, then P_{k+1} is true.

CONCLUSION Both conditions in Theorem 1 are satisfied. Thus, P_n is true for all positive integers n.

MATCHED 3 PROBLEM

Prove that $8^n - 1$ is divisible by 7 for all positive integers n.

In some cases, a conjecture may be true only for $n \geq m$, where m is a positive integer, rather than for all $n \geq 0$. For example, see Problems 49 and 50 in Exercise 10.2. The principle of mathematical induction can be extended to cover cases like this as follows:

THEOREM 2

Extended Principle of Mathematical Induction

Let m be a positive integer, let P_n be a statement associated with each integer $n \geq m$, and suppose the following conditions are satisfied:

1. P_m is true.
2. For any integer $k \geq m$, if P_k is true, then P_{k+1} is also true.

Then the statement P_n is true for all integers $n \geq m$.

Three Famous Problems

The problem of determining whether a certain statement about the positive integers is true may be extremely difficult. Proofs may require remarkable insight and ingenuity and the development of techniques far more advanced than mathematical induction. Consider, for example, the famous problems of proving the following statements:

1. **Lagrange's Four Square Theorem, 1772:** Each positive integer can be expressed as the sum of four or fewer squares of positive integers.
2. **Fermat's Last Theorem, 1637:** For $n > 2$, $x^n + y^n = z^n$ does not have solutions in the natural numbers.
3. **Goldbach's Conjecture, 1742:** Every positive even integer greater than 2 is the sum of two prime numbers.

The first statement was considered by the early Greeks and finally proved in 1772 by Lagrange. Fermat's last theorem, defying the best mathematical minds for over 350 years, finally succumbed to a 200-page proof by Professor Andrew Wiles of Princeton University in 1993. To this date no one has been able to prove or disprove Goldbach's conjecture.

EXPLORE/DISCUSS 2

(A) Explain the difference between a theorem and a conjecture.

(B) Why is "Fermat's last theorem" a misnomer? Suggest more accurate names for the result.

ANSWERS MATCHED PROBLEMS

1. Sketch of proof. State the conjecture:

P_n: $1 + 2 + 3 + \cdots + n = \frac{n(n+1)}{2}$

Part 1. $1 = \frac{1(1+1)}{2}$. P_1 is true.

Part 2. Show that if P_k is true, then P_{k+1} is true.

$$1 + 2 + 3 + \cdots + k = \frac{k(k+1)}{2} \qquad P_k$$

$$1 + 2 + 3 + \cdots + k + (k+1) = \frac{k(k+1)}{2} + (k+1)$$

$$= \frac{(k+1)(k+2)}{2} \qquad P_{k+1}$$

Conclusion: P_n is true.

2. Sketch of proof. State the conjecture: P_n: $\left(\frac{x}{y}\right)^n = \frac{x^n}{y^n}$

Part 1. $\left(\frac{x}{y}\right)^1 = \frac{x}{y} = \frac{x^1}{y^1}$. P_1 is true.

Part 2. Show that if P_k is true, then P_{k+1} is true.

$$\left(\frac{x}{y}\right)^{k+1} = \left(\frac{x}{y}\right)^k\left(\frac{x}{y}\right) = \frac{x^k}{y^k}\left(\frac{x}{y}\right) = \frac{x^k x}{y^k y} = \frac{x^{k+1}}{y^{k+1}}$$

Conclusion: P_n is true.

3. Sketch of proof. State the conjecture: P_n: $8^n - 1 = 7r$ for some integer r

Part 1. $8^1 - 1 = 7 = 7 \cdot 1$. P_1 is true.

Part 2. Show that if P_k is true, then P_{k+1} is true.

$$8^k - 1 = 7r \qquad P_k$$

$$8(8^k - 1) = 8(7r)$$

$$8^{k+1} - 1 = 56r + 7 = 7(8r + 1) = 7s \qquad P_{k+1}$$

Conclusion: P_n is true.

EXERCISE 10.2

A

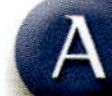

In Problems 1–4, find the first positive integer n that causes the statement to fail.

1. $(3 + 5)^n = 3^n + 5^n$
2. $n < 10$
3. $n^2 = 3n - 2$
4. $n^3 + 11n = 6n^2 + 6$

Verify each statement P_n in Problems 5–10 for n = 1, 2, and 3.

5. P_n: $2 + 6 + 10 + \cdots + (4n - 2) = 2n^2$
6. P_n: $4 + 8 + 12 + \cdots + 4n = 2n(n + 1)$
7. P_n: $a^5 a^n = a^{5+n}$
8. P_n: $(a^5)^n = a^{5n}$
9. P_n: $9^n - 1$ is divisible by 4
10. P_n: $4^n - 1$ is divisible by 3

Write P_k and P_{k+1} for P_n as indicated in Problems 11–16.

11. P_n in Problem 5
12. P_n in Problem 6
13. P_n in Problem 7
14. P_n in Problem 8
15. P_n in Problem 9
16. P_n in Problem 10

In Problems 17–22, use mathematical induction to prove that each P_n holds for all positive integers n.

17. P_n in Problem 5
18. P_n in Problem 6
19. P_n in Problem 7
20. P_n in Problem 8
21. P_n in Problem 9
22. P_n in Problem 10

B

In Problems 23–26, prove the statement is false by finding a counterexample.

23. If $n > 2$, then any polynomial of degree n has at least one real zero.
24. Any positive integer $n > 7$ can be written as the sum of three or fewer squares of positive integers.
25. If n is a positive integer, then there is at least one prime number p such that $n < p < n + 6$.

26. If a, b, c, and d are positive integers such that $a^2 + b^2 = c^2 + d^2$, then $a = c$ or $a = d$.

In Problems 27–42, use mathematical induction to prove each proposition for all positive integers n, unless restricted otherwise.

27. $2 + 2^2 + 2^3 + \cdots + 2^n = 2^{n+1} - 2$

28. $\frac{1}{2} + \frac{1}{4} + \frac{1}{8} + \cdots + \frac{1}{2^n} = 1 - \left(\frac{1}{2}\right)^n$

29. $1^2 + 3^2 + 5^2 + \cdots + (2n - 1)^2 = \frac{1}{3}(4n^3 - n)$

30. $1 + 8 + 16 + \cdots + 8(n - 1) = (2n - 1)^2;\ n > 1$

31. $1^2 + 2^2 + 3^2 + \cdots + n^2 = \frac{n(n + 1)(2n + 1)}{6}$

32. $1 \cdot 2 + 2 \cdot 3 + 3 \cdot 4 + \cdots + n(n + 1) = \frac{n(n + 1)(n + 2)}{3}$

33. $\frac{a^n}{a^3} = a^{n-3};\ n > 3$

34. $\frac{a^5}{a^n} = \frac{1}{a^{n-5}};\ n > 5$

35. $a^m a^n = a^{m+n};\ m, n \in N$ [*Hint:* Choose m as an arbitrary element of N, and then use induction on n.]

36. $(a^n)^m = a^{mn};\ m, n \in N$

37. $x^n - 1$ is divisible by $x - 1$; $x \neq 1$ [*Hint:* Divisible means that $x^n - 1 = (x - 1)Q(x)$ for some polynomial $Q(x)$.]

38. $x^n - y^n$ is divisible by $x - y$; $x \neq y$

39. $x^{2n} - 1$ is divisible by $x - 1$; $x \neq 1$

40. $x^{2n} - 1$ is divisible by $x + 1$; $x \neq -1$

41. $1^3 + 2^3 + 3^3 + \cdots + n^3 = (1 + 2 + 3 + \cdots + n)^2$ [*Hint:* See Matched Problem 1 following Example 1.]

42. $\frac{1}{1 \cdot 2 \cdot 3} + \frac{1}{2 \cdot 3 \cdot 4} + \frac{1}{3 \cdot 4 \cdot 5} + \cdots + \frac{1}{n(n + 1)(n + 2)} = \frac{n(n + 3)}{4(n + 1)(n + 2)}$

C

In Problems 43–46, suggest a formula for each expression, and prove your hypothesis using mathematical induction, $n \in N$.

43. $2 + 4 + 6 + \cdots + 2n$

44. $\frac{1}{1 \cdot 2} + \frac{1}{2 \cdot 3} + \frac{1}{3 \cdot 4} + \cdots + \frac{1}{n(n + 1)}$

45. The number of lines determined by n points in a plane, no three of which are collinear

46. The number of diagonals in a polygon with n sides

Prove Problems 47–50 true for all integers n as specified.

47. $a > 1 \Rightarrow a^n > 1;\ n \in N$

48. $0 < a < 1 \Rightarrow 0 < a^n < 1;\ n \in N$

49. $n^2 > 2n;\ n \geq 3$

50. $2^n > n^2;\ n \geq 5$

51. Prove or disprove the generalization of the following two facts:

$$3^2 + 4^2 = 5^2$$
$$3^3 + 4^3 + 5^3 = 6^3$$

52. Prove or disprove: $n^2 + 21n + 1$ is a prime number for all natural numbers n.

If $\{a_n\}$ and $\{b_n\}$ are two sequences, we write $\{a_n\} = \{b_n\}$ if and only if $a_n = b_n$, $n \in N$. In Problems 53–56, use mathematical induction to show that $\{a_n\} = \{b_n\}$.

53. $a_1 = 1, a_n = a_{n-1} + 2;\ b_n = 2n - 1$

54. $a_1 = 2, a_n = a_{n-1} + 2;\ b_n = 2n$

55. $a_1 = 2, a_n = 2^2 a_{n-1};\ b_n = 2^{2n-1}$

56. $a_1 = 2, a_n = 3a_{n-1};\ b_n = 2 \cdot 3^{n-1}$

SECTION 10.3 Arithmetic and Geometric Sequences

Arithmetic and Geometric Sequences • *n*th-Term Formulas • Sum Formulas for Finite Arithmetic Series • Sum Formulas for Finite Geometric Series • Sum Formula for Infinite Geometric Series

For most sequences it is difficult to sum an arbitrary number of terms of the sequence without adding term by term. But particular types of sequences, *arithmetic sequences* and *geometric sequences,* have certain properties that lead to convenient and useful formulas for the sums of the corresponding *arithmetic series* and *geometric series.*

Arithmetic and Geometric Sequences

The sequence 5, 7, 9, 11, 13, . . . , $5 + 2(n - 1)$, . . . , where each term after the first is obtained by adding 2 to the preceding term, is an example of an arithmetic sequence. The sequence 5, 10, 20, 40, 80, . . . , $5(2)^{n-1}$, . . . , where each term after the first is obtained by multiplying the preceding term by 2, is an example of a geometric sequence.

DEFINITION 1
Arithmetic Sequence

A sequence

$$a_1, a_2, a_3, \ldots, a_n, \ldots$$

is called an **arithmetic sequence,** or **arithmetic progression,** if there exists a constant *d*, called the **common difference,** such that

$$a_n - a_{n-1} = d$$

That is,

$$a_n = a_{n-1} + d \qquad \text{for every } n > 1$$

DEFINITION 2
Geometric Sequence

A sequence

$$a_1, a_2, a_3, \ldots, a_n, \ldots$$

is called a **geometric sequence,** or **geometric progression,** if there exists a nonzero constant *r*, called the **common ratio,** such that

$$\frac{a_n}{a_{n-1}} = r$$

That is,

$$a_n = ra_{n-1} \qquad \text{for every } n > 1$$

EXPLORE/DISCUSS 1

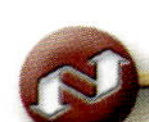

(A) Graph the arithmetic sequence 5, 7, 9, ...
Describe the graphs of all arithmetic sequences with common difference 2.

(B) Graph the geometric sequence 5, 10, 20, ...
Describe the graphs of all geometric sequences with common ratio 2.

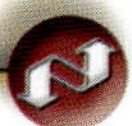

EXAMPLE 1 Recognizing Arithmetic and Geometric Sequences

Which of the following can be the first four terms of an arithmetic sequence? Of a geometric sequence?

(A) 1, 2, 3, 5, . . . (B) $-1, 3, -9, 27, \ldots$

(C) 3, 3, 3, 3, . . . (D) 10, 8.5, 7, 5.5, . . .

SOLUTIONS

(A) Because $2 - 1 \neq 5 - 3$, there is no common difference, so the sequence is not an arithmetic sequence. Because $\frac{2}{1} \neq \frac{3}{2}$, there is no common ratio, so the sequence is not geometric either.

(B) The sequence is geometric with common ratio -3, but it is not arithmetic.

(C) The sequence is arithmetic with common difference 0 and it is also geometric with common ratio 1.

(D) The sequence is arithmetic with common difference -1.5, but it is not geometric.

MATCHED 1 PROBLEM

Which of the following can be the first four terms of an arithmetic sequence? Of a geometric sequence?

(A) 8, 2, 0.5, 0.125, . . . (B) $-7, -2, 3, 8, \ldots$ (C) 1, 5, 25, 100, . . .

*n*th-Term Formulas

If $\{a_n\}$ is an arithmetic sequence with common difference d, then

$$a_2 = a_1 + d$$
$$a_3 = a_2 + d = a_1 + 2d$$
$$a_4 = a_3 + d = a_1 + 3d$$

This suggests Theorem 1, which can be proved by mathematical induction (see Problem 63 in Exercise 10.3).

THEOREM 1
The *n*th Term of an Arithmetic Sequence

$a_n = a_1 + (n - 1)d \qquad$ for every $n > 1$

Similarly, if $\{a_n\}$ is a geometric sequence with common ratio r, then

$$a_2 = a_1 r$$
$$a_3 = a_2 r = a_1 r^2$$
$$a_4 = a_3 r = a_1 r^3$$

This suggests Theorem 2, which can also be proved by mathematical induction (see Problem 69 in Exercise 10.3).

THEOREM 2
The *n*th Term of a Geometric Sequence

$a_n = a_1 r^{n-1} \qquad$ for every $n > 1$

EXAMPLE 2 Finding Terms in Arithmetic and Geometric Sequences

(A) If the first and tenth terms of an arithmetic sequence are 3 and 30, respectively, find the fiftieth term of the sequence.

(B) If the first and tenth terms of a geometric sequence are 1 and 4, find the seventeenth term to three decimal places.

SOLUTIONS

(A) First use Theorem 1 with $a_1 = 3$ and $a_{10} = 30$ to find d:

$$a_n = a_1 + (n - 1)d$$
$$a_{10} = a_1 + (10 - 1)d$$
$$30 = 3 + 9d$$
$$d = 3$$

Now find a_{50}:

$$a_{50} = a_1 + (50 - 1)3$$
$$= 3 + 49 \cdot 3$$
$$= 150$$

(B) First let $n = 10$, $a_1 = 1$, $a_{10} = 4$ and use Theorem 2 to find r.

$$a_n = a_1 r^{n-1}$$
$$4 = 1r^{10-1}$$
$$r = 4^{1/9}$$

Now use Theorem 2 again, this time with $n = 17$.

$$a_{17} = a_1 r^{16} = 1(4^{1/9})^{16} = 4^{16/9} \approx 11.758$$

MATCHED 2 PROBLEM

(A) If the first and fifteenth terms of an arithmetic sequence are -5 and 23, respectively, find the seventy-third term of the sequence.

(B) Find the eighth term of the geometric sequence $\frac{1}{64}, -\frac{1}{32}, \frac{1}{16}, \ldots$.

Sum Formulas for Finite Arithmetic Series

If $a_1, a_2, a_3, \ldots, a_n$ is a finite arithmetic sequence, then the corresponding series $a_1 + a_2 + a_3 + \cdots + a_n$ is called an *arithmetic series*. We will derive two simple and very useful formulas for the sum of an arithmetic series. Let d be the common difference of the arithmetic sequence $a_1, a_2, a_3, \ldots, a_n$ and let S_n denote the sum of the series $a_1 + a_2 + a_3 + \cdots + a_n$.

Then

$$S_n = a_1 + (a_1 + d) + \cdots + [a_1 + (n - 2)d] + [a_1 + (n - 1)d]$$

Reversing the order of the sum, we obtain

$$S_n = [a_1 + (n - 1)d] + [a_1 + (n - 2)d] + \cdots + (a_1 + d) + a_1$$

Adding the left sides of these two equations and corresponding elements of the right sides, we see that

$$\begin{aligned} 2S_n &= [2a_1 + (n - 1)d] + [2a_1 + (n - 1)d] + \cdots + [2a_1 + (n - 1)d] \\ &= n[2a_1 + (n - 1)d] \end{aligned}$$

This can be restated as in Theorem 3.

THEOREM 3
Sum of an Arithmetic Series—First Form

$$S_n = \frac{n}{2}[2a_1 + (n - 1)d]$$

By replacing $a_1 + (n - 1)d$ with a_n, we obtain a second useful formula for the sum.

THEOREM 4
Sum of an Arithmetic Series—Second Form

$$S_n = \frac{n}{2}(a_1 + a_n)$$

The proof of the first sum formula by mathematical induction is left as an exercise (see Problem 64 in Exercise 10.3).

EXAMPLE 3 Finding the Sum of an Arithmetic Series

Find the sum of the first 26 terms of an arithmetic series if the first term is -7 and $d = 3$.

SOLUTION

Algebraic Solution

Let $n = 26$, $a_1 = -7$, $d = 3$, and use Theorem 3.

$$S_n = \frac{n}{2}[2a_1 + (n - 1)d]$$

$$S_{26} = \tfrac{26}{2}[2(-7) + (26 - 1)3]$$

$$= 793$$

Graphing Utility Solution

Use the seq command to generate the sequence of 26 terms, then use the sum command to compute the sum of these terms (Fig. 1).

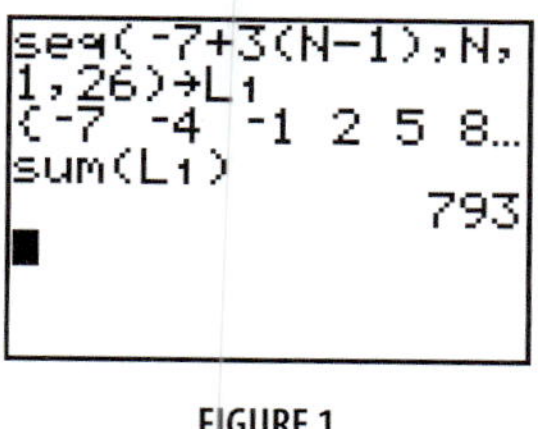

FIGURE 1

MATCHED 3 PROBLEM

Find the sum of the first 52 terms of an arithmetic series if the first term is 23 and $d = -2$.

EXAMPLE 4 Finding the Sum of an Arithmetic Series

Find the sum of all the odd numbers between 51 and 99, inclusive.

SOLUTION

Algebraic Solution

First, use $a_1 = 51$, $a_n = 99$, and Theorem 1 to find n:

$$a_n = a_1 + (n - 1)d$$

$$99 = 51 + (n - 1)2$$

$$n = 25$$

Now use Theorem 4 to find S_{25}:

$$S_n = \frac{n}{2}(a_1 + a_n)$$

$$S_{25} = \tfrac{25}{2}(51 + 99)$$

$$= 1{,}875$$

Graphing Utility Solution

Use the seq command to generate the required sequence of odd numbers, then use the sum command to compute the sum of these terms (Fig. 2).

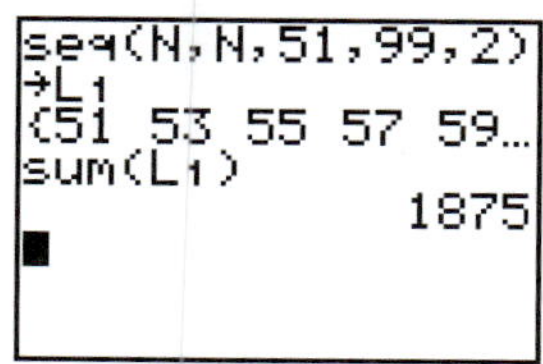

FIGURE 2

MATCHED 4 PROBLEM

Find the sum of all the even numbers between -22 and 52, inclusive.

EXAMPLE 5 **Prize Money**

A 16-team bowling league has \$8,000 to be awarded as prize money. If the last-place team is awarded \$275 in prize money and the award increases by the same amount for each successive finishing place, how much will the first-place team receive?

SOLUTION

If a_1 is the award for the first-place team, a_2 is the award for the second-place team, and so on, then the prize money awards form an arithmetic sequence with $n = 16$, $a_{16} = 275$, and $S_{16} = 8{,}000$. Use Theorem 4 to find a_1.

$$S_n = \frac{n}{2}(a_1 + a_n)$$

$$8{,}000 = \tfrac{16}{2}(a_1 + 275)$$

$$a_1 = 725$$

Thus, the first-place team receives \$725.

MATCHED 5 PROBLEM

Refer to Example 5. How much prize money is awarded to the second-place team?

Sum Formulas for Finite Geometric Series

If $a_1, a_2, a_3, \ldots, a_n$ is a finite geometric sequence, then the corresponding series $a_1 + a_2 + a_3 + \cdots + a_n$ is called a *geometric series*. As with arithmetic series, we can derive two simple and very useful formulas for the sum of a geometric series. Let r be the common ratio of the geometric sequence $a_1, a_2, a_3, \ldots, a_n$ and let S_n denote the sum of the series $a_1 + a_2 + a_3 + \cdots + a_n$. Then

$$S_n = a_1 + a_1r + a_1r^2 + a_1r^3 + \cdots + a_1r^{n-2} + a_1r^{n-1}$$

Multiply both sides of this equation by r to obtain

$$rS_n = a_1r + a_1r^2 + a_1r^3 + \cdots + a_1r^{n-1} + a_1r^n$$

Now subtract the left side of the second equation from the left side of the first, and the right side of the second equation from the right side of the first to obtain

$$S_n - rS_n = a_1 - a_1r^n$$

$$S_n(1 - r) = a_1 - a_1r^n$$

Thus, solving for S_n, we obtain the following formula for the sum of a geometric series:

THEOREM 5
Sum of a Geometric Series—First Form

$$S_n = \frac{a_1 - a_1 r^n}{1 - r} \qquad r \neq 1$$

Because $a_n = a_1 r^{n-1}$, or $ra_n = a_1 r^n$, the sum formula also can be written in the following form:

THEOREM 6
Sum of a Geometric Series—Second Form

$$S_n = \frac{a_1 - ra_n}{1 - r} \qquad r \neq 1$$

The proof of the first sum formula (Theorem 5) by mathematical induction is left as an exercise (see Problem 70, Exercise 10.3).

If $r = 1$, then

$$S_n = a_1 + a_1(1) + a_1(1^2) + \cdots + a_1(1^{n-1}) = na_1$$

EXAMPLE 6 Finding the Sum of a Geometric Series

Find the sum of the first 20 terms of a geometric series if the first term is 1 and $r = 2$.

SOLUTION

Algebraic Solution

Let $n = 20$, $a_1 = 1$, $r = 2$, and use Theorem 5.

$$S_n = \frac{a_1 - a_1 r^n}{1 - r}$$

$$= \frac{1 - 1 \cdot 2^{20}}{1 - 2} = 1{,}048{,}575$$

Graphing Utility Solution

Use the seq command to generate the required sequence of powers of 2, then use the sum command to compute the sum of these terms (Fig. 3).

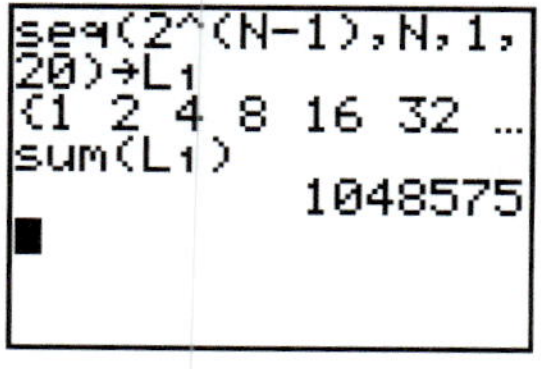

FIGURE 3

MATCHED 6 PROBLEM

Find the sum, to two decimal places, of the first 14 terms of a geometric series if the first term is $\frac{1}{64}$ and $r = -2$.

Sum Formula for Infinite Geometric Series

Consider a geometric series with $a_1 = 5$ and $r = \frac{1}{2}$. What happens to the sum S_n as n increases? To answer this question, we first write the sum formula in the more convenient form

$$S_n = \frac{a_1 - a_1 r^n}{1 - r} = \frac{a_1}{1 - r} - \frac{a_1 r^n}{1 - r} \tag{1}$$

For $a_1 = 5$ and $r = \frac{1}{2}$,

$$S_n = 10 - 10\left(\frac{1}{2}\right)^n$$

Thus,

$$S_2 = 10 - 10\left(\frac{1}{4}\right)$$

$$S_4 = 10 - 10\left(\frac{1}{16}\right)$$

$$S_{10} = 10 - 10\left(\frac{1}{1{,}024}\right)$$

$$S_{20} = 10 - 10\left(\frac{1}{1{,}048{,}576}\right)$$

It appears that $(\frac{1}{2})^n$ becomes smaller and smaller as n increases and that the sum gets closer and closer to 10.

In general, it is possible to show that, if $|r| < 1$, then r^n will get closer and closer to 0 as n increases. Symbolically, $r^n \to 0$ as $n \to \infty$. Thus, the term

$$\frac{a_1 r^n}{1 - r}$$

in equation (1) will tend to 0 as n increases, and S_n will tend to

$$\frac{a_1}{1 - r}$$

In other words, if $|r| < 1$, then S_n can be made as close to

$$\frac{a_1}{1 - r}$$

as we wish by taking n sufficiently large. Thus, we define the **sum of an infinite geometric series** by the following formula:

DEFINITION 3
Sum of an Infinite Geometric Series

$$S_\infty = \frac{a_1}{1 - r} \qquad |r| < 1$$

If $|r| \ge 1$, an infinite geometric series has no sum.

EXAMPLE 7 Expressing a Repeating Decimal as a Fraction

Represent the repeating decimal $0.454\,545 \cdots = 0.\overline{45}$ as the quotient of two integers. Recall that a repeating decimal names a rational number and that any rational number can be represented as the quotient of two integers.

SOLUTION

$0.\overline{45} = 0.45 + 0.0045 + 0.000\,045 + \cdots$

The right side of the equation is an infinite geometric series with $a_1 = 0.45$ and $r = 0.01$. Thus,

$$S_\infty = \frac{a_1}{1 - r} = \frac{0.45}{1 - 0.01} = \frac{0.45}{0.99} = \frac{5}{11}$$

Hence, $0.\overline{45}$ and $\frac{5}{11}$ name the same rational number. Check the result by dividing 5 by 11.

MATCHED 7 PROBLEM

Repeat Example 7 for $0.818\,181 \cdots = 0.\overline{81}$.

EXAMPLE 8 Economy Stimulation

A state government uses proceeds from a lottery to provide a tax rebate for property owners. Suppose an individual receives a \$500 rebate and spends 80% of this, and each of the recipients of the money spent by this individual also spends 80% of what he or she receives, and this process continues without end. According to the **multiplier doctrine** in economics, the effect of the original \$500 tax rebate on the economy is multiplied many times. What is the total amount spent if the process continues as indicated?

SOLUTION

The individual receives \$500 and spends 0.8(500) = \$400. The recipients of this \$400 spend 0.8(400) = \$320, the recipients of this \$320 spend 0.8(320) = \$256, and so on. Thus, the total spending generated by the \$500 rebate is

$$400 + 320 + 256 + \cdots = 400 + 0.8(400) + (0.8)^2(400) + \cdots$$

which we recognize as an infinite geometric series with $a_1 = 400$ and $r = 0.8$. Thus, the total amount spent is

$$S_\infty = \frac{a_1}{1 - r} = \frac{400}{1 - 0.8} = \frac{400}{0.2} = \$2{,}000$$

MATCHED 8 PROBLEM

Repeat Example 8 if the tax rebate is \$1,000 and the percentage spent by all recipients is 90%.

EXPLORE/DISCUSS 2

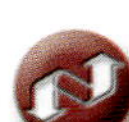

(A) Find an infinite geometric series with $a_1 = 10$ whose sum is 1,000.

(B) Find an infinite geometric series with $a_1 = 10$ whose sum is 6.

(C) Suppose that an infinite geometric series with $a_1 = 10$ has a sum. Explain why that sum must be greater than 5.

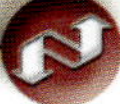

ANSWERS MATCHED PROBLEMS

1. (A) The sequence is geometric with $r = \frac{1}{4}$, but not arithmetic.
   (B) The sequence is arithmetic with $d = 5$, but not geometric.
   (C) The sequence is neither arithmetic nor geometric.
2. (A) 139 (B) −2 **3.** −1,456 **4.** 570 **5.** \$695 **6.** −85.33 **7.** $\frac{9}{11}$ **8.** \$9,000

EXERCISE 10.3

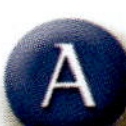

In Problems 1 and 2, determine whether the following can be the first three terms of an arithmetic or geometric sequence, and, if so, find the common difference or common ratio and the next two terms of the sequence.

1. (A) $-11, -16, -21, \ldots$ **(B)** $2, -4, 8, \ldots$
   (C) $1, 4, 9, \ldots$ **(D)** $\frac{1}{2}, \frac{1}{6}, \frac{1}{18}, \ldots$

2. (A) $5, 20, 100, \ldots$ **(B)** $-5, -5, -5, \ldots$
   (C) $7, 6.5, 6, \ldots$ **(D)** $512, 256, 128, \ldots$

Let $a_1, a_2, a_3, \ldots, a_n, \ldots$ be an arithmetic sequence. In Problems 3–10, find the indicated quantities.

3. $a_1 = -5, d = 4; a_2 = ?, a_3 = ?, a_4 = ?$

4. $a_1 = -18, d = 3; a_2 = ?, a_3 = ?, a_4 = ?$

5. $a_1 = -3, d = 5; a_{15} = ?, S_{11} = ?$

6. $a_1 = 3, d = 4; a_{22} = ?, S_{21} = ?$

7. $a_1 = 1, a_2 = 5; S_{21} = ?$

8. $a_1 = 5, a_2 = 11; S_{11} = ?$

9. $a_1 = 7, a_2 = 5; a_{15} = ?$

10. $a_1 = -3, d = -4; a_{10} = ?$

Let $a_1, a_2, a_3, \ldots, a_n, \ldots$ be a geometric sequence. In Problems 11–16, find each of the indicated quantities.

11. $a_1 = -6, r = -\frac{1}{2}; a_2 = ?, a_3 = ?, a_4 = ?$

12. $a_1 = 12, r = \frac{2}{3}; a_2 = ?, a_3 = ?, a_4 = ?$

13. $a_1 = 81, r = \frac{1}{3}; a_{10} = ?$

14. $a_1 = 64, r = \frac{1}{2}; a_{13} = ?$

15. $a_1 = 3, a_7 = 2{,}187, r = 3; S_7 = ?$

16. $a_1 = 1, a_7 = 729, r = -3; S_7 = ?$

B

Let $a_1, a_2, a_3, \ldots, a_n, \ldots$ be an arithmetic sequence. In Problems 17–24, find the indicated quantities.

17. $a_1 = 3, a_{20} = 117; d = ?, a_{101} = ?$

18. $a_1 = 7, a_8 = 28; d = ?, a_{25} = ?$

19. $a_1 = -12, a_{40} = 22; S_{40} = ?$

20. $a_1 = 24, a_{24} = -28; S_{24} = ?$

21. $a_1 = \frac{1}{3}, a_2 = \frac{1}{2}; a_{11} = ?, S_{11} = ?$

22. $a_1 = \frac{1}{6}, a_2 = \frac{1}{4}; a_{19} = ?, S_{19} = ?$

23. $a_3 = 13, a_{10} = 55; a_1 = ?$

24. $a_9 = -12, a_{13} = 3; a_1 = ?$

Let $a_1, a_2, a_3, \ldots, a_n, \ldots$ be a geometric sequence. Find each of the indicated quantities in Problems 25–30.

25. $a_1 = 100, a_6 = 1; r = ?$

26. $a_1 = 10, a_{10} = 30; r = ?$

27. $a_1 = 5, r = -2; S_{10} = ?$

28. $a_1 = 3, r = 2; S_{10} = ?$

29. $a_1 = 9, a_4 = \frac{8}{3}; a_2 = ?, a_3 = ?$

30. $a_1 = 12, a_4 = -\frac{4}{9}; a_2 = ?, a_3 = ?$

31. $S_{51} = \sum_{k=1}^{51} (3k + 3) = ?$

32. $S_{40} = \sum_{k=1}^{40} (2k - 3) = ?$

33. $S_7 = \sum_{k=1}^{7} (-3)^{k-1} = ?$

34. $S_7 = \sum_{k=1}^{7} 3^k = ?$

35. Find $g(1) + g(2) + g(3) + \cdots + g(51)$ if $g(t) = 5 - t$.

36. Find $f(1) + f(2) + f(3) + \cdots + f(20)$ if $f(x) = 2x - 5$.

37. Find $g(1) + g(2) + \cdots + g(10)$ if $g(x) = (\frac{1}{2})^x$.

38. Find $f(1) + f(2) + \cdots + f(10)$ if $f(x) = 2^x$.

39. Find the sum of all the even integers between 21 and 135.

40. Find the sum of all the odd integers between 100 and 500.

41. Show that the sum of the first n odd natural numbers is n^2, using appropriate formulas from Section 10.3.

42. Show that the sum of the first n even natural numbers is $n + n^2$, using appropriate formulas from Section 10.3.

43. Find a positive number x so that $-2 + x - 6$ is a three-term geometric series.

44. Find a positive number x so that $6 + x + 8$ is a three-term geometric series.

45. For a given sequence in which $a_1 = -3$ and $a_n = a_{n-1} + 3$, $n > 1$, find a_n in terms of n.

46. For the sequence in Problem 45, find $S_n = \sum_{k=1}^{n} a_k$ in terms of n.

In Problems 47–50, find the least positive integer n such that $a_n < b_n$ by graphing the sequences $\{a_n\}$ and $\{b_n\}$ with a graphing utility. Check your answer by using a graphing utility to display both sequences in table form.

47. $a_n = 5 + 8n, b_n = 1.1^n$

48. $a_n = 96 + 47n, b_n = 8(1.5)^n$

49. $a_n = 1{,}000(0.99)^n, b_n = 2n + 1$

50. $a_n = 500 - n, b_n = 1.05^n$

In Problems 51–56, find the sum of each infinite geometric series that has a sum.

51. $3 + 1 + \frac{1}{3} + \cdots$

52. $16 + 4 + 1 + \cdots$

53. $2 + 4 + 8 + \cdots$

54. $4 + 6 + 9 + \cdots$

55. $2 - \frac{1}{2} + \frac{1}{8} - \cdots$

56. $21 - 3 + \frac{3}{7} - \cdots$

In Problems 57–62, represent each repeating decimal as the quotient of two integers.

57. $0.\overline{7} = 0.7777\cdots$

58. $0.\overline{5} = 0.5555\cdots$

59. $0.\overline{54} = 0.545\,454\cdots$

60. $0.\overline{27} = 0.272\,727\cdots$

61. $3.\overline{216} = 3.216\,216\,216\cdots$

62. $5.\overline{63} = 5.636\,363\cdots$

C

63. Prove, using mathematical induction, that if $\{a_n\}$ is an arithmetic sequence, then

$$a_n = a_1 + (n - 1)d \qquad \text{for every } n > 1$$

64. Prove, using mathematical induction, that if $\{a_n\}$ is an arithmetic sequence, then

$$S_n = \frac{n}{2}[2a_1 + (n - 1)d]$$

65. If in a given sequence, $a_1 = -2$ and $a_n = -3a_{n-1}$, $n > 1$, find a_n in terms of n.

66. For the sequence in Problem 65, find $S_n = \sum_{k=1}^{n} a_k$ in terms of n.

67. Show that $(x^2 + xy + y^2)$, $(z^2 + xz + x^2)$, and $(y^2 + yz + z^2)$ are consecutive terms of an arithmetic progression if x, y, and z form an arithmetic progression. (From U.S.S.R. Mathematical Olympiads, 1955–1956, Grade 9.)

68. Take 121 terms of each arithmetic progression 2, 7, 12, . . . and 2, 5, 8, ... How many numbers will there be in common? (From U.S.S.R. Mathematical Olympiads, 1955–1956, Grade 9.)

69. Prove, using mathematical induction, that if $\{a_n\}$ is a geometric sequence, then

$$a_n = a_1 r^{n-1} \qquad n \in N$$

70. Prove, using mathematical induction, that if $\{a_n\}$ is a geometric sequence, then

$$S_n = \frac{a_1 - a_1 r^n}{1 - r} \qquad n \in N, r \neq 1$$

71. Given the system of equations

$$ax + by = c$$
$$dx + ey = f$$

where a, b, c, d, e, f is any arithmetic progression with a nonzero constant difference, show that the system has a unique solution.

72. The sum of the first and fourth terms of an arithmetic sequence is 2, and the sum of their squares is 20. Find the sum of the first eight terms of the sequence.

APPLICATIONS

73. Business. In investigating different job opportunities, you find that firm A will start you at \$25,000 per year and guarantee you a raise of \$1,200 each year whereas firm B will start you at \$28,000 per year but will guarantee you a raise of only \$800 each year. Over a period of 15 years, how much would you receive from each firm?

74. Business. In Problem 73, what would be your annual salary at each firm for the tenth year?

75. Economics. The government, through a subsidy program, distributes \$1,000,000. If we assume that each individual or agency spends 0.8 of what is received, and 0.8 of this is spent, and so on, how much total increase in spending results from this government action?

76. Economics. Because of reduced taxes, an individual has an extra \$600 in spendable income. If we assume that the individual spends 70% of this on consumer goods, that the producers of these goods in turn spend 70% of what they receive on consumer goods, and that this process continues indefinitely, what is the total amount spent on consumer goods?

★ **77. Business.** If \$$P$ is invested at r% compounded annually, the amount A present after n years forms a geometric progression with a common ratio $1 + r$. Write a formula for the amount present after n years. How long will it take a sum of money P to double if invested at 6% interest compounded annually?

★ **78. Population Growth.** If a population of A_0 people grows at the constant rate of r% per year, the population after t years forms a geometric progression with a common ratio $1 + r$. Write a formula for the total population after t years. If the world's population is increasing at the rate of 2% per year, how long will it take to double?

79. Finance. Eleven years ago an investment earned \$7,000 for the year. Last year the investment earned \$14,000. If the earnings from the investment have increased the same amount each year, what is the yearly increase and how much income has accrued from the investment over the past 11 years?

80. Air Temperature. As dry air moves upward, it expands. In so doing, it cools at the rate of about 5°F for each 1,000-foot rise. This is known as the **adiabatic process.**

(A) Temperatures at altitudes that are multiples of 1,000 feet form what kind of a sequence?

(B) If the ground temperature is 80°F, write a formula for the temperature T_n in terms of n, if n is in thousands of feet.

81. Engineering. A rotating flywheel coming to rest rotates 300 revolutions the first minute (see the figure). If in each subsequent minute it rotates two-thirds as many times as in the preceding minute, how many revolutions will the wheel make before coming to rest?

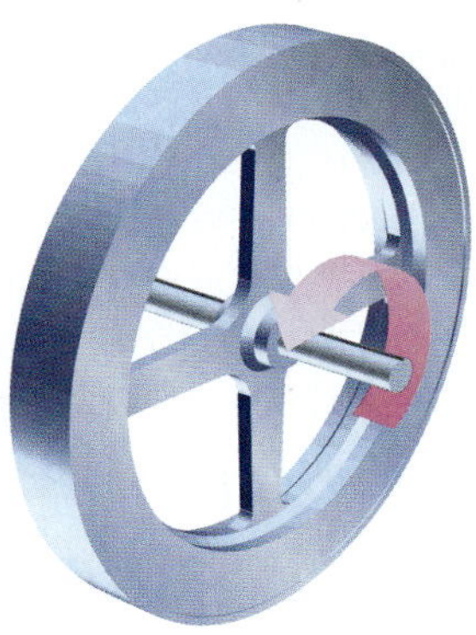

82. Physics. The first swing of a bob on a pendulum is 10 inches. If on each subsequent swing it travels 0.9 as far as on the preceding swing, how far will the bob travel before coming to rest?

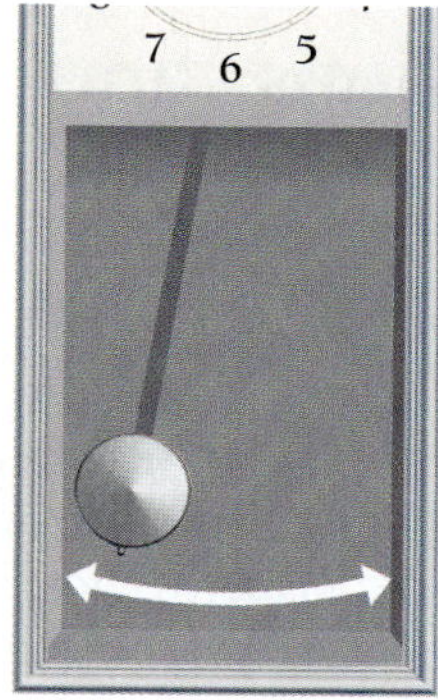

83. Food Chain. A plant is eaten by an insect, an insect by a trout, a trout by a salmon, a salmon by a bear, and the bear is eaten by you. If only 20% of the energy is transformed from one stage to the next, how many calories must be supplied by plant food to provide you with 2,000 calories from the bear meat?

★ **84. Genealogy.** If there are 30 years in a generation, how many direct ancestors did each of us have 600 years ago? By *direct* ancestors we mean parents, grandparents, great-grandparents, and so on.

★ **85. Physics.** An object falling from rest in a vacuum near the surface of the Earth falls 16 feet during the first second, 48 feet during the second second, 80 feet during the third second, and so on.

(A) How far will the object fall during the eleventh second?

(B) How far will the object fall in 11 seconds?

(C) How far will the object fall in t seconds?

★ **86. Physics.** In Problem 85, how far will the object fall during:

(A) The twentieth second? **(B)** The tth second?

★ **87. Bacteria Growth.** A single cholera bacterium divides every $\frac{1}{2}$ hour to produce two complete cholera bacteria. If we start with a colony of A_0 bacteria, how many bacteria will we have in t hours, assuming adequate food supply?

★ **88. Cell Division.** One leukemic cell injected into a healthy mouse will divide into two cells in about $\frac{1}{2}$ day. At the end of the day these two cells will divide again, with the doubling process continuing each $\frac{1}{2}$ day until there are 1 billion cells, at which time the mouse dies. On which day after the experiment is started does this happen?

★★ **89. Astronomy.** Ever since the time of the Greek astronomer Hipparchus, second century B.C., the brightness of stars has been measured in terms of magnitude. The brightest stars, excluding the sun, are classed as magnitude 1, and the dimmest visible to the eye are classed as magnitude 6. In 1856, the English astronomer N. R. Pogson showed that first-magnitude stars are 100 times brighter than sixth-magnitude stars. If the ratio of brightness between consecutive magnitudes is constant, find this ratio. [*Hint:* If b_n is the brightness of an nth-magnitude star, find r for the geometric progression $b_1, b_2, b_3, \ldots$, given $b_1 = 100b_6$.]

★ **90. Music.** The notes on a piano, as measured in cycles per second, form a geometric progression.

(A) If A is 400 cycles per second and A′, 12 notes higher, is 800 cycles per second, find the constant ratio r.

(B) Find the cycles per second for C, three notes higher than A.

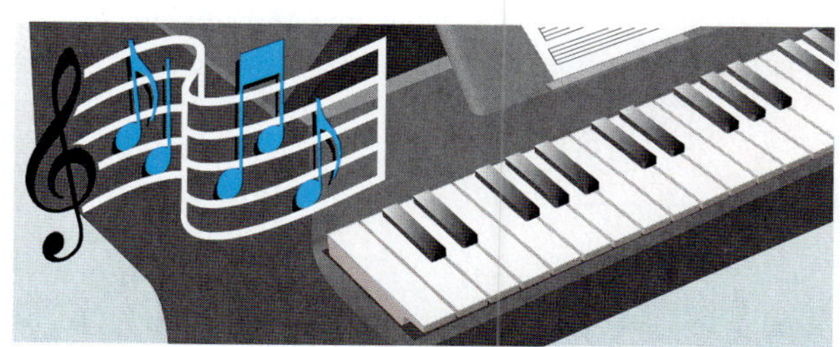

91. Puzzle. If you place 1¢ on the first square of a chessboard, 2¢ on the second square, 4¢ on the third, and so on, continuing to double the amount until all 64 squares are covered, how much money will be on the sixty-fourth square? How much money will there be on the whole board?

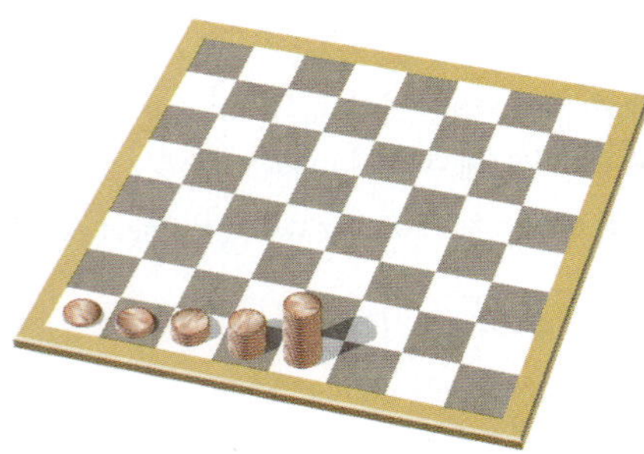

★ **92. Puzzle.** If a sheet of very thin paper 0.001-inch thick is torn in half, and each half is again torn in half, and this process is repeated for a total of 32 times, how high will the stack of paper be if the pieces are placed one on top of the other? Give the answer to the nearest mile.

★ **93. Atmospheric Pressure.** If atmospheric pressure decreases roughly by a factor of 10 for each 10-mile increase in altitude up to 60 miles, and if the pressure is 15 pounds per square inch at sea level, what will the pressure be 40 miles up?

94. Zeno's Paradox. Visualize a hypothetical 440-yard oval racetrack that has tapes stretched across the track at the halfway point and at each point that marks the halfway point of each remaining distance thereafter. A runner running around the track has to break the first tape before the second, the second before the third, and so on. From this point of view it appears that he will never finish the race. This famous paradox is attributed to the Greek philosopher Zeno (495–435 B.C.). If we assume the runner runs at 440 yards per minute, the times between tape breakings form an infinite geometric progression. What is the sum of this progression?

95. Geometry. If the midpoints of the sides of an equilateral triangle are joined by straight lines, the new figure will be an equilateral triangle with a perimeter equal to half the original. If we start with an equilateral triangle with perimeter 1 and form a sequence of "nested" equilateral triangles proceeding as described, what will be the total perimeter of all the triangles that can be formed in this way?

96. Photography. The shutter speeds and f-stops on a camera are given as follows:

Shutter speeds: $1, \frac{1}{2}, \frac{1}{4}, \frac{1}{8}, \frac{1}{15}, \frac{1}{30}, \frac{1}{60}, \frac{1}{125}, \frac{1}{250}, \frac{1}{500}$

f-stops: 1.4, 2, 2.8, 4, 5.6, 8, 11, 16, 22

These are very close to being geometric progressions. Estimate their common ratios.

★★ **97. Geometry.** We know that the sum of the interior angles of a triangle is 180°. Show that the sums of the interior angles of polygons with 3, 4, 5, 6, . . . sides form an arithmetic sequence. Find the sum of the interior angles for a 21-sided polygon.

SECTION 10.4 Multiplication Principle, Permutations, and Combinations

Multiplication Principle • Factorial • Permutations • Combinations

FIGURE 1

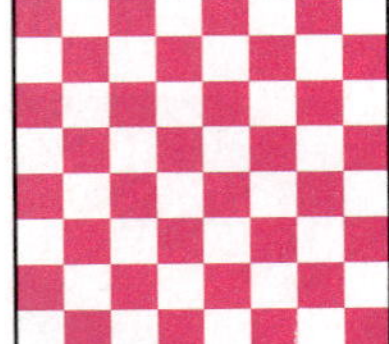

Section 10.4 introduces some new mathematical tools that are usually referred to as *counting techniques*. In general, a **counting technique** is a mathematical method of determining the number of objects in a set without actually enumerating the objects in the set as 1, 2, 3, ... For example, we can count the number of squares in a checkerboard (Fig. 1) by counting 1, 2, 3, . . . , 64. This is enumeration. Or we can note that there are eight rows with eight squares in each row. Thus, the total number of squares must be $8 \times 8 = 64$. This is a very simple counting technique.

Now consider the problem of assigning telephone numbers. How many different seven-digit telephone numbers can be formed? As we will soon see, the answer is $10^7 = 10{,}000{,}000$, a number that is much too large to obtain by enumeration. Thus, counting techniques are essential tools if the number of elements in a set is very large. The techniques developed in Section 10.4 will be applied to a brief introduction to probability theory in Section 10.5, and to a famous algebraic formula in Section 10.6.

Multiplication Principle

We start with an example.

EXAMPLE 1 Combined Outcomes

Suppose we flip a coin and then throw a single die (Fig. 2). What are the possible combined outcomes?

SOLUTION

To solve this problem, we use a **tree diagram:**

FIGURE 2 Coin and die outcomes.

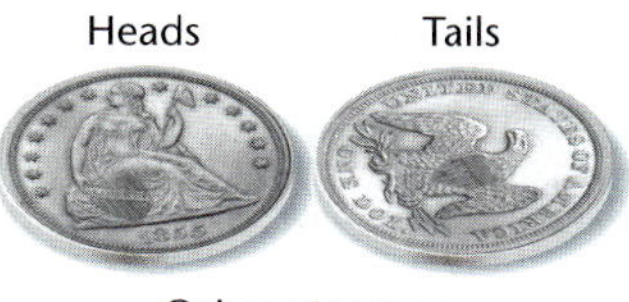

Coin outcomes

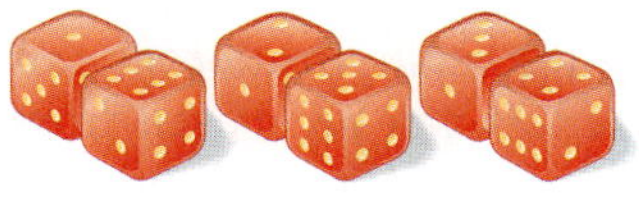

Die outcomes

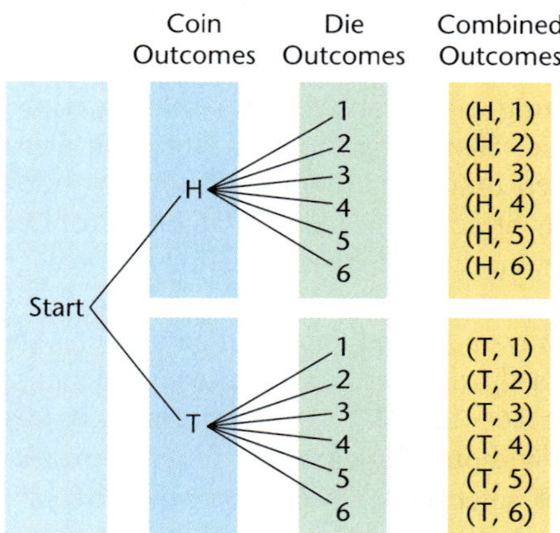

Thus, there are 12 possible combined outcomes—two ways in which the coin can come up followed by six ways in which the die can come up.

MATCHED PROBLEM 1

Use a tree diagram to determine the number of possible outcomes of throwing a single die followed by flipping a coin.

Now suppose you are asked, "From the 26 letters in the alphabet, how many ways can 3 letters appear in a row on a license plate if no letter is repeated?" To try to count the possibilities using a tree diagram would be extremely tedious, to say the least. The following **multiplication principle,** also called the **fundamental counting principle,** enables us to solve this problem easily. In addition, it forms the basis for several other counting techniques developed later in Section 10.4.

Multiplication Principle

1. If two operations O_1 and O_2 are performed in order with N_1 possible outcomes for the first operation and N_2 possible outcomes for the second operation, then there are

$$N_1 \cdot N_2$$

possible combined outcomes of the first operation followed by the second.

2. In general, if n operations $O_1, O_2, \ldots, O_n$ are performed in order, with possible number of outcomes $N_1, N_2, \ldots, N_n$, respectively, then there are

$$N_1 \cdot N_2 \cdot \cdots \cdot N_n$$

possible combined outcomes of the operations performed in the given order.

In Example 1, we see that there are two possible outcomes from the first operation of flipping a coin and six possible outcomes from the second operation of throwing a die. Hence, by the multiplication principle, there are $2 \cdot 6 = 12$ possible combined outcomes of flipping a coin followed by throwing a die. Use the multiplication principle to solve Matched Problem 1.

To answer the license plate question, we reason as follows: There are 26 ways the first letter can be chosen. After a first letter is chosen, 25 letters remain; hence there are 25 ways a second letter can be chosen. And after 2 letters are chosen, there are 24 ways a third letter can be chosen. Hence, using the multiplication principle, there are $26 \cdot 25 \cdot 24 = 15{,}600$ possible ways 3 letters can be chosen from the alphabet without allowing any letter to repeat. By not allowing any letter to repeat, earlier selections affect the choice of subsequent selections. If we allow letters to repeat, then earlier selections do not affect the choice in subsequent selections, and there are 26 possible choices for each of the 3 letters. Thus, if we allow letters to repeat, there are $26 \cdot 26 \cdot 26 = 26^3 = 17{,}576$ possible ways the 3 letters can be chosen from the alphabet.

EXAMPLE 2 Computer-Generated Tests

Many universities and colleges are now using computer-assisted testing procedures. Suppose a screening test is to consist of five questions, and a computer stores five equivalent questions for the first test question, eight equivalent questions for the second, six for the third, five for the fourth, and ten for the fifth. How many different five-question tests can the computer select? Two tests are considered different if they differ in one or more questions.

SOLUTION

O_1:	Select the first question	N_1:	five ways
O_2:	Select the second question	N_2:	eight ways
O_3:	Select the third question	N_3:	six ways
O_4:	Select the fourth question	N_4:	five ways
O_5:	Select the fifth question	N_5:	ten ways

Thus, the computer can generate

$5 \cdot 8 \cdot 6 \cdot 5 \cdot 10 = 12{,}000$ different tests

MATCHED 2 PROBLEM

Each question on a multiple-choice test has five choices. If there are five such questions on a test, how many different response sheets are possible if only one choice is marked for each question?

EXAMPLE 3 Counting Code Words

How many three-letter code words are possible using the first eight letters of the alphabet if:

(A) No letter can be repeated? (B) Letters can be repeated?

(C) Adjacent letters cannot be alike?

SOLUTIONS

(A) No letter can be repeated.

O_1: Select first letter	N_1: eight ways	
O_2: Select second letter	N_2: seven ways	Because one letter has been used
O_3: Select third letter	N_3: six ways	Because two letters have been used

Thus, there are

$8 \cdot 7 \cdot 6 = 336$ possible code words

(B) Letters can be repeated.

O_1: Select first letter	N_1: eight ways	
O_2: Select second letter	N_2: eight ways	Repeats are allowed.
O_3: Select third letter	N_3: eight ways	Repeats are allowed.

Thus, there are

$8 \cdot 8 \cdot 8 = 8^3 = 512$ possible code words

(C) Adjacent letters cannot be alike.

O_1: Select first letter	N_1: eight ways	
O_2: Select second letter	N_2: seven ways	Cannot be the same as the first
O_3: Select third letter	N_3: seven ways	Cannot be the same as the second, but can be the same as the first

Thus, there are

$8 \cdot 7 \cdot 7 = 392$ possible code words

MATCHED 3 PROBLEM

How many four-letter code words are possible using the first ten letters of the alphabet under the three conditions stated in Example 3?

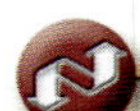

EXPLORE/DISCUSS 1

The postal service of a developing country is choosing a five-character postal code consisting of letters (of the English alphabet) and digits. At least half a million postal codes must be accommodated. Which format would you recommend to make the codes easy to remember?

The multiplication principle can be used to develop two additional counting techniques that are extremely useful in more complicated counting problems. Both of these methods use the factorial function, which we introduce next.

Factorial

For n a natural number, ***n* factorial**—denoted by $n!$—is the product of the first n natural numbers. **Zero factorial** is defined to be 1.

DEFINITION 1

***n* Factorial**

For n a natural number

$n! = n(n - 1) \cdot \cdots \cdot 2 \cdot 1$

$1! = 1$

$0! = 1$

It is also useful to note that

THEOREM 1

Recursion Formula for *n* Factorial

$n! = n \cdot (n - 1)!$

EXAMPLE 4 **Evaluating Factorials**

(A) $4! = 4 \cdot 3! = 4 \cdot 3 \cdot 2! = 4 \cdot 3 \cdot 2 \cdot 1! = 4 \cdot 3 \cdot 2 \cdot 1 = 24$

(B) $5! = 5 \cdot 4 \cdot 3 \cdot 2 \cdot 1 = 120$

(C) $\frac{7!}{6!} = \frac{7 \cdot \cancel{6!}}{\cancel{6!}} = 7$

(D) $\frac{8!}{5!} = \frac{8 \cdot 7 \cdot 6 \cdot \cancel{5!}}{\cancel{5!}} = 336$

(E) $\frac{9!}{6!3!} = \frac{\overset{3}{\cancel{9}} \cdot \overset{4}{\cancel{8}} \cdot 7 \cdot \cancel{6!}}{\cancel{6!}\,\cancel{3} \cdot \cancel{2} \cdot 1} = 84$

MATCHED 4 PROBLEM

Find (A) $6!$ (B) $\frac{6!}{5!}$ (C) $\frac{9!}{6!}$ (D) $\frac{10!}{7!3!}$

CAUTION

When reducing fractions involving factorials, don't confuse the single integer n with the symbol $n!$, which represents the product of n consecutive integers.

$$\frac{6!}{3!} \neq 2! \qquad \frac{6!}{3!} = \frac{6 \cdot 5 \cdot 4 \cdot 3!}{3!} = 6 \cdot 5 \cdot 4 = 120$$

EXPLORE/DISCUSS 2

A student used a calculator* to solve Matched Problem 4, as shown in Figure 3. Check these answers. If any are incorrect, explain why and find a correct calculator solution.

6!	
	720
6!/5!	
	6
9!/6!	
	504
10!/7!3!	
	4320

FIGURE 3

It is interesting and useful to note that $n!$ grows very rapidly. Compare the following:

$$5! = 120 \qquad 10! = 3{,}628{,}800 \qquad 15! = 1{,}307{,}674{,}368{,}000$$

*The factorial symbol ! and related symbols can be found under the MATH-PROB menus on a TI-83 or TI-86.

If $n!$ is too large for a calculator to store and display, an error message is displayed. Find the value of n such that your calculator will evaluate $n!$, but not $(n + 1)!$.

Permutations

Suppose four pictures are to be arranged from left to right on one wall of an art gallery. How many arrangements are possible? Using the multiplication principle, there are four ways of selecting the first picture. After the first picture is selected, there are three ways of selecting the second picture. After the first two pictures are selected, there are two ways of selecting the third picture. And after the first three pictures are selected, there is only one way to select the fourth. Thus, the number of arrangements possible for the four pictures is

$$4 \cdot 3 \cdot 2 \cdot 1 = 4! \qquad \text{or} \qquad 24$$

In general, we refer to a particular arrangement, or **ordering,** of n objects without repetition as a **permutation** of the n objects. How many permutations of n objects are there? From the reasoning above, there are n ways in which the first object can be chosen, there are $n - 1$ ways in which the second object can be chosen, and so on. Applying the multiplication principle, we have Theorem 2.

THEOREM 2
Permutations of *n* Objects

The number of permutations of n objects, denoted by $P_{n,n}$, is given by

$$P_{n,n} = n \cdot (n - 1) \cdot \cdots \cdot 1 = n!$$

Now suppose the director of the art gallery decides to use only two of the four available pictures on the wall, arranged from left to right. How many arrangements of two pictures can be formed from the four? There are four ways the first picture can be selected. After selecting the first picture, there are three ways the second picture can be selected. Thus, the number of arrangements of two pictures from four pictures, denoted by $P_{4,2}$, is given by

$$P_{4,2} = 4 \cdot 3 = 12$$

Or, in terms of factorials, multiplying $4 \cdot 3$ by 1 in the form $2!/2!$, we have

$$P_{4,2} = 4 \cdot 3 = \frac{4 \cdot 3 \cdot 2!}{2!} = \frac{4!}{2!}$$

This last form gives $P_{4,2}$ in terms of factorials, which is useful in some cases.

A **permutation of a set of *n* objects taken *r* at a time** is an arrangement of the r objects in a specific order. Thus, reasoning in the same way as in the example above, we find that the number of permutations of n objects taken r at a time, $0 \le r \le n$, denoted by $P_{n,r}$, is given by

$$P_{n,r} = n(n - 1)(n - 2) \cdot \cdots \cdot (n - r + 1)$$

Multiplying the right side of this equation by 1 in the form $(n - r)!/(n - r)!$, we obtain a factorial form for $P_{n,r}$:

$$P_{n,r} = n(n - 1)(n - 2) \cdot \cdots \cdot (n - r + 1)\,\frac{(n - r)!}{(n - r)!}$$

But

$$n(n - 1)(n - 2) \cdot \cdots \cdot (n - r + 1)(n - r)! = n!$$

Hence, we have Theorem 3.

THEOREM 3

Permutation of *n* Objects Taken *r* at a Time

The number of permutations of n objects taken r at a time is given by

$$P_{n,r} = \underbrace{n(n - 1)(n - 2) \cdot \cdots \cdot (n - r + 1)}_{r \text{ factors}}$$

or

$$P_{n,r} = \frac{n!}{(n - r)!} \qquad 0 \le r \le n$$

Note that if $r = n$, then the number of permutations of n objects taken n at a time is

$$P_{n,n} = \frac{n!}{(n - n)!} = \frac{n!}{0!} = n! \qquad \text{Recall, } 0! = 1.$$

which agrees with Theorem 2, as it should.

The permutation symbol $P_{n,r}$ *also can be denoted by* P^n_r, ${}_nP_r$, or $P(n, r)$. Many calculators use ${}_nP_r$ to denote the function that evaluates the permutation symbol.

EXAMPLE 5 Selecting Officers

From a committee of eight people, in how many ways can we choose a chair and a vice-chair, assuming one person cannot hold more than one position?

SOLUTION

We are actually asking for the number of permutations of eight objects taken two at a time—that is, $P_{8,2}$:

$$P_{8,2} = \frac{8!}{(8 - 2)!} = \frac{8!}{6!} = \frac{8 \cdot 7 \cdot 6!}{6!} = 56$$

MATCHED PROBLEM 5

From a committee of ten people, in how many ways can we choose a chair, vice-chair, and secretary, assuming one person cannot hold more than one position?

EXAMPLE 6 **Evaluating $P_{n,r}$**

Find the number of permutations of 25 objects taken

(A) Two at a time (B) Four at a time (C) Eight at a time

SOLUTION

Figure 4 shows the solution on a graphing utility.

FIGURE 4

```
25 nPr 2
                600
25 nPr 4
             303600
25 nPr 8
       4.3609104E10
```

MATCHED 6 PROBLEM

Find the number of permutations of 30 objects taken

(A) Two at a time (B) Four at a time (C) Six at a time

Combinations

Now suppose that an art museum owns eight paintings by a given artist and another art museum wishes to borrow three of these paintings for a special show. How many ways can three paintings be selected for shipment out of the eight available? Here, the order of the items selected doesn't matter. What we are actually interested in is how many subsets of three objects can be formed from a set of eight objects. We call such a subset a **combination** of eight objects taken three at a time. The total number of combinations is denoted by the symbol

$$C_{8,3} \quad \text{or} \quad \binom{8}{3}$$

To find the number of combinations of eight objects taken three at a time, $C_{8,3}$, we make use of the formula for $P_{n,r}$ and the multiplication principle. We know that the number of permutations of eight objects taken three at a time is given by $P_{8,3}$, and we have a formula for computing this quantity. Now suppose we think of $P_{8,3}$ in terms of two operations:

O_1: Select a subset of three objects (paintings)

N_1: $C_{8,3}$ ways

O_2: Arrange the subset in a given order

N_2: $3!$ ways

The combined operation, O_1 followed by O_2, produces a permutation of eight objects taken three at a time. Thus,

$$P_{8,3} = C_{8,3} \cdot 3!$$

To find $C_{8,3}$, we replace $P_{8,3}$ in the preceding equation with $8!/(8 - 3)!$ and solve for $C_{8,3}$:

$$\frac{8!}{(8-3)!} = C_{8,3} \cdot 3!$$

$$C_{8,3} = \frac{8!}{3!(8-3)!} = \frac{8 \cdot 7 \cdot 6 \cdot 5!}{3 \cdot 2 \cdot 1 \cdot 5!} = 56$$

Thus, the museum can make 56 different selections of three paintings from the eight available.

A **combination of a set of n objects taken r at a time** is an r-element subset of the n objects. Reasoning in the same way as in the example, the number of combinations of n objects taken r at a time, $0 \le r \le n$, denoted by $C_{n,r}$, can be obtained by solving for $C_{n,r}$ in the relationship

$$P_{n,r} = C_{n,r} \cdot r!$$

$$C_{n,r} = \frac{P_{n,r}}{r!}$$

$$= \frac{n!}{r!(n-r)!} \qquad P_{n,r} = \frac{n!}{(n-r)!}$$

THEOREM 4

Combination of n Objects Taken r at a Time

The number of combinations of n objects taken r at a time is given by

$$C_{n,r} = \binom{n}{r} = \frac{P_{n,r}}{r!} = \frac{n!}{r!(n-r)!} \qquad 0 \le r \le n$$

The combination symbols $C_{n,r}$ and $\binom{n}{r}$ also can be denoted by C_r^n, ${}_nC_r$, or $C(n, r)$.

EXAMPLE 7 Selecting Subcommittees

From a committee of eight people, in how many ways can we choose a subcommittee of two people?

SOLUTION

Notice how this example differs from Example 5, where we wanted to know how many ways a chair and a vice-chair can be chosen from a committee of eight people. In Example 5, ordering matters. In choosing a subcommittee of two people, the ordering does not matter. Thus, we are actually asking for the number of combinations of eight objects taken two at a time. The number is given by

$$C_{8,2} = \binom{8}{2} = \frac{8!}{2!(8-2)!} = \frac{8 \cdot 7 \cdot 6!}{2 \cdot 1 \cdot 6!} = 28$$

MATCHED PROBLEM 7

How many subcommittees of three people can be chosen from a committee of eight people?

EXAMPLE 8 Evaluating $C_{n,r}$

Find the number of combinations of 25 objects taken

(A) Two at a time (B) Four at a time (C) Eight at a time

FIGURE 5

```
25 nCr 2
             300
25 nCr 4
           12650
25 nCr 8
         1081575
```

SOLUTION

Figure 5 shows the solution on a graphing utility. Compare these results with Example 6.

MATCHED 8 PROBLEM

Find the number of combinations of 30 objects taken

(A) Two at a time (B) Four at a time (C) Six at a time

Remember: In a permutation, order counts. In a combination, order does not count.

To determine whether a permutation or combination is needed, decide whether rearranging the collection or listing makes a difference. If so, use permutations. If not, use combinations.

EXPLORE/DISCUSS 3

Each of the following is a selection without repetition. Would you consider the selection to be a combination? A permutation? Discuss your reasoning.

(A) A student checks out three books from the library.

(B) A baseball manager names his starting lineup.

(C) The newly elected president names his cabinet members.

(D) The president selects a delegation of three cabinet members to attend the funeral of a head of state.

(E) An orchestra conductor chooses three pieces of music for a symphony program.

FIGURE 6 A standard deck of cards.

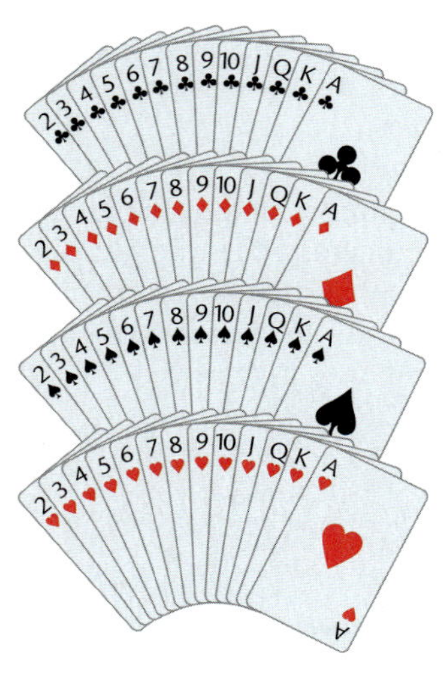

A **standard deck** of 52 cards (Fig. 6) has four 13-card suits: diamonds, hearts, clubs, and spades. Each 13-card suit contains cards numbered from 2 to 10, a jack, a queen, a king, and an ace. The jack, queen, and king are called **face cards.** Depending on the game, the ace may be counted as the lowest and/or the highest card in the suit. Example 9, as well as other examples and exercises in Chapter 10, refer to this standard deck.

EXAMPLE 9 Counting Card Hands

Out of a standard 52-card deck, how many 5-card hands will have three aces and two kings?

SOLUTION

O_1: Choose three aces out of four possible — **Order is not important.**

N_1: $C_{4,3}$

O_2: Choose two kings out of four possible — **Order is not important.**

N_2: $C_{4,2}$

Using the multiplication principle, we have

$$\text{Number of hands} = C_{4,3} \cdot C_{4,2} = 4 \cdot 6 = 24$$

MATCHED 9 PROBLEM

From a standard 52-card deck, how many 5-card hands will have three hearts and two spades?

EXAMPLE 10 Counting Serial Numbers

Serial numbers for a product are to be made using two letters followed by three numbers. If the letters are to be taken from the first eight letters of the alphabet with no repeats and the numbers from the 10 digits 0 through 9 with no repeats, how many serial numbers are possible?

SOLUTION

O_1: Choose two letters out of eight available — **Order is important.**

N_1: $P_{8,2}$

O_2: Choose three numbers out of ten available — **Order is important.**

N_2: $P_{10,3}$

Using the multiplication principle, we have

$$\text{Number of serial numbers} = P_{8,2} \cdot P_{10,3} = 40{,}320$$

MATCHED 10 PROBLEM

Repeat Example 10 under the same conditions, except the serial numbers are now to have three letters followed by two digits with no repeats.

ANSWERS MATCHED PROBLEMS

1.

HT	HT	HT	HT	HT	HT
1	2	3	4	5	6

Start

2. 5^5, or 3,125

3. (A) $10 \cdot 9 \cdot 8 \cdot 7 = 5{,}040$ (B) $10 \cdot 10 \cdot 10 \cdot 10 = 10{,}000$ (C) $10 \cdot 9 \cdot 9 \cdot 9 = 7{,}290$

4. (A) 720 (B) 6 (C) 504 (D) 120

5. $P_{10,3} = \dfrac{10!}{(10-3)!} = 720$

6. (A) 870 (B) 657,720 (C) 427,518,000 **7.** $C_{8,3} = \dfrac{8!}{3!(8-3)!} = 56$

8. (A) 435 (B) 27,405 (C) 593,775 **9.** $C_{13,3} \cdot C_{13,2} = 22,308$ **10.** $P_{8,3} \cdot P_{10,2} = 30,240$

EXERCISE 10.4

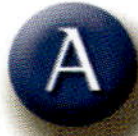

Evaluate Problems 1–12.

1. $9!$ **2.** $10!$ **3.** $11!$

4. $12!$ **5.** $\dfrac{11!}{8!}$ **6.** $\dfrac{14!}{12!}$

7. $\dfrac{5!}{2!3!}$ **8.** $\dfrac{6!}{4!2!}$ **9.** $\dfrac{7!}{4!(7-4)!}$

10. $\dfrac{8!}{3!(8-3)!}$ **11.** $\dfrac{7!}{7!(7-7)!}$ **12.** $\dfrac{8!}{0!(8-0)!}$

13. The figure shows calculator solutions to Problems 5, 7, and 9. Check these answers. If any are incorrect, explain why and find a correct calculator solution.

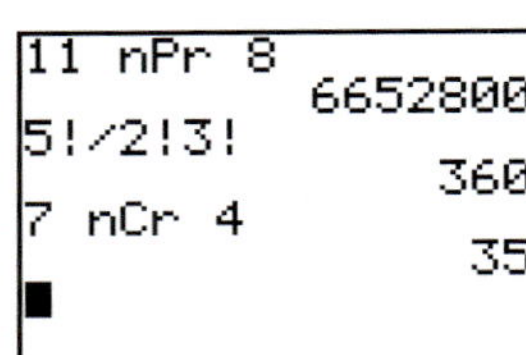

14. The figure shows calculator solutions to Problems 6, 8, and 10. Check these answers. If any are incorrect, explain why and find a correct calculator solution.

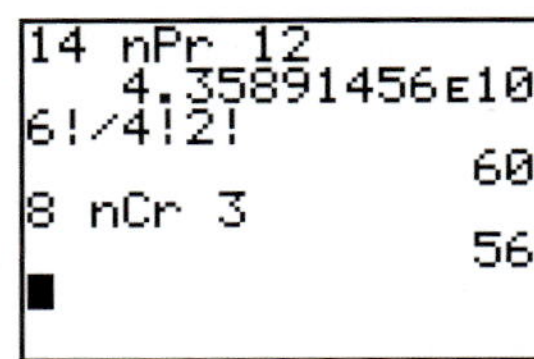

Evaluate Problems 15–22.

15. $P_{5,3}$ **16.** $P_{4,2}$ **17.** $P_{52,4}$

18. $P_{52,2}$ **19.** $C_{5,3}$ **20.** $C_{4,2}$

21. $C_{52,4}$ **22.** $C_{52,2}$

In Problems 23 and 24, would you consider the selection to be a combination or a permutation? Explain your reasoning.

23. **(A)** The recently elected chief executive officer (CEO) of a company named three new vice-presidents, of marketing, research, and manufacturing.

(B) The CEO selected three of her vice-presidents to attend the dedication ceremony of a new plant.

24. **(A)** An individual rented four videos from a rental store to watch over a weekend.

(B) The same individual did some holiday shopping by buying four videos, one for his father, one for his mother, one for his younger sister, and one for his older brother.

25. A particular new car model is available with five choices of color, three choices of transmission, four types of interior, and two types of engine. How many different variations of this model car are possible?

26. A deli serves sandwiches with the following options: three kinds of bread, five kinds of meat, and lettuce or sprouts. How many different sandwiches are possible, assuming one item is used out of each category?

27. In a horse race, how many different finishes among the first three places are possible for a 10-horse race? Exclude ties.

28. In a long-distance foot race, how many different finishes among the first five places are possible for a 50-person race? Exclude ties.

29. How many ways can a subcommittee of three people be selected from a committee of seven people? How many ways can a president, vice-president, and secretary be chosen from a committee of seven people?

30. Suppose nine cards are numbered with the nine digits from 1 to 9. A three-card hand is dealt, one card at a time. How many hands are possible where:

(A) Order is taken into consideration?

(B) Order is not taken into consideration?

31. There are 10 teams in a league. If each team is to play every other team exactly once, how many games must be scheduled?

32. Given seven points, no three of which are on a straight line, how many lines can be drawn joining two points at a time?

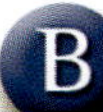

33. How many four-letter code words are possible from the first six letters of the alphabet, with no letter repeated? Allowing letters to repeat?

34. How many five-letter code words are possible from the first seven letters of the alphabet, with no letter repeated? Allowing letters to repeat?

35. A combination lock has five wheels, each labeled with the 10 digits from 0 to 9. How many opening combinations of five numbers are possible, assuming no digit is repeated? Assuming digits can be repeated?

36. A small combination lock on a suitcase has three wheels, each labeled with digits from 0 to 9. How many opening combinations of three numbers are possible, assuming no digit is repeated? Assuming digits can be repeated?

37. From a standard 52-card deck, how many 5-card hands will have all hearts?

38. From a standard 52-card deck, how many 5-card hands will have all face cards? All face cards, but no kings? Consider only jacks, queens, and kings to be face cards.

39. How many different license plates are possible if each contains three letters followed by three digits? How many of these license plates contain no repeated letters and no repeated digits?

40. How may five-digit zip codes are possible? How many of these codes contain no repeated digits?

41. From a standard 52-card deck, how many 7-card hands have exactly five spades and two hearts?

42. From a standard 52-card deck, how many 5-card hands will have two clubs and three hearts?

43. A catering service offers eight appetizers, ten main courses, and seven desserts. A banquet chairperson is to select three appetizers, four main courses, and two desserts for a banquet. How many ways can this be done?

44. Three research departments have 12, 15, and 18 members, respectively. If each department is to select a delegate and an alternate to represent the department at a conference, how many ways can this be done?

45. **(A)** Use a graphing utility to display the sequences $P_{10,0}$, $P_{10,1}, \ldots, P_{10,10}$ and $0!, 1!, \ldots, 10!$ in table form, and show that $P_{10,r} \geq r!$ for $r = 0, 1, \ldots, 10$.
(B) Find all values of r such that $P_{10,r} = r!$
(C) Explain why $P_{n,r} \geq r!$ whenever $0 \leq r \leq n$.

46. **(A)** How are the sequences $\frac{P_{10,0}}{0!}, \frac{P_{10,1}}{1!}, \ldots, \frac{P_{10,10}}{10!}$ and $C_{10,0}$, $C_{10,1}, \ldots, C_{10,10}$ related?
(B) Use a graphing utility to graph each sequence and confirm the relationship of part A.

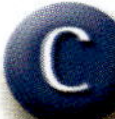

47. A sporting goods store has 12 pairs of ski gloves of 12 different brands thrown loosely in a bin. The gloves are all the same size. In how many ways can a left-hand glove and a right-hand glove be selected that do not match relative to brand?

48. A sporting goods store has six pairs of running shoes of six different styles thrown loosely in a basket. The shoes are all the same size. In how many ways can a left shoe and a right shoe be selected that do not match?

49. Eight distinct points are selected on the circumference of a circle.
(A) How many chords can be drawn by joining the points in all possible ways?
(B) How many triangles can be drawn using these eight points as vertices?
(C) How many quadrilaterals can be drawn using these eight points as vertices?

50. Five distinct points are selected on the circumference of a circle.

(A) How many chords can be drawn by joining the points in all possible ways?

(B) How many triangles can be drawn using these five points as vertices?

51. How many ways can two people be seated in a row of five chairs? Three people? Four people? Five people?

52. Each of two countries sends five delegates to a negotiating conference. A rectangular table is used with five chairs on each long side. If each country is assigned a long side of the table, how many seating arrangements are possible? [*Hint:* Operation 1 is assigning a long side of the table to each country.]

53. A basketball team has five distinct positions. Out of eight players, how many starting teams are possible if

(A) The distinct positions are taken into consideration?

(B) The distinct positions are not taken into consideration?

(C) The distinct positions are not taken into consideration, but either Mike or Ken, but not both, must start?

54. How many committees of four people are possible from a group of nine people if

(A) There are no restrictions?

(B) Both Juan and Mary must be on the committee?

(C) Either Juan or Mary, but not both, must be on the committee?

55. A 5-card hand is dealt from a standard 52-card deck. Which is more likely: the hand contains exactly one king or the hand contains no hearts?

56. A 10-card hand is dealt from a standard 52-card deck. Which is more likely: all cards in the hand are red or the hand contains all four aces?

SECTION 10.5 Sample Spaces and Probability

Experiments • Sample Spaces and Events • Probability of an Event • Equally Likely Assumption • Empirical Probability

Section 10.5 provides an introduction to probability, a topic to which whole books and courses are devoted. Probability involves many subtle notions, and care must be taken at the beginning to understand the fundamental concepts on which the subject is based. First, we develop a mathematical model for probability studies. Our development, because of space, must be somewhat informal. More formal and precise treatments can be found in books on probability.

Experiments

Our first step in constructing a mathematical model for probability studies is to describe the type of experiments on which probability studies are based. Some types of experiments do not yield the same results, no matter how carefully they are repeated under the same conditions. These experiments are called **random experiments.** Familiar examples of random experiments are flipping coins, rolling dice, observing the frequency of defective items from an assembly line, or observing the frequency of deaths in a certain age group.

Probability theory is a branch of mathematics that has been developed to deal with outcomes of random experiments, both real and conceptual. In the work that follows, the word **experiment** will be used to mean a random experiment.

Sample Spaces and Events

Associated with outcomes of experiments are *sample spaces* and *events.* Our second step in constructing a mathematical model for probability studies is to define these two terms. Set concepts will be useful in this regard.

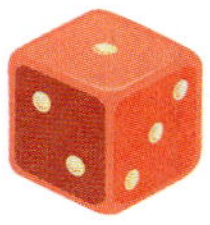

Consider the experiment, "A single six-sided die is rolled." What outcomes might we observe? We might be interested in the number of dots facing up, or whether the number of dots facing up is an even number, or whether the number of dots facing up is divisible by 3, and so on. The list of possible outcomes appears endless. In general, there is no unique method of analyzing all possible outcomes of an experiment. Therefore, before conducting an experiment, it is important to decide just what outcomes are of interest.

In the die experiment, suppose we limit our interest to the number of dots facing up when the die comes to rest. Having decided what to observe, we make a list of outcomes of the experiment, called *simple events,* such that in each trial of the experiment, one and only one of the results on the list will occur. The set of simple events for the experiment is called a **sample space** for the experiment. The sample space S we have chosen for the die-rolling experiment is

$$S = \{1, 2, 3, 4, 5, 6\}$$

Now consider the outcome, "The number of dots facing up is an even number." This outcome is not a simple event, because it will occur whenever 2, 4, or 6 dots appear, that is, whenever an element in the subset

$$E = \{2, 4, 6\}$$

occurs. Subset E is called a *compound event.* In general, we have the following definition:

DEFINITION 1

Event

Given a sample space S for an experiment, we define an **event *E*** to be any subset of S. If an event E has only one element in it, it is called a **simple event.** If event E has more than one element, it is called a **compound event.** We say that **an event *E* occurs** if any of the simple events in E occurs.

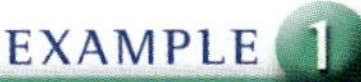

EXAMPLE 1 Choosing a Sample Space

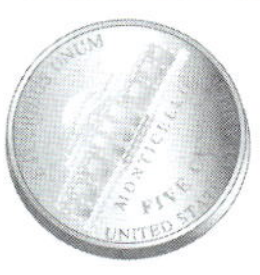

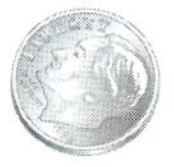

A nickel and a dime are tossed. How will we identify a sample space for this experiment?

SOLUTIONS

There are a number of possibilities, depending on our interest. We will consider three.

(A) If we are interested in whether each coin falls heads (H) or tails (T), then, using a tree diagram, we can easily determine an appropriate sample space for the experiment:

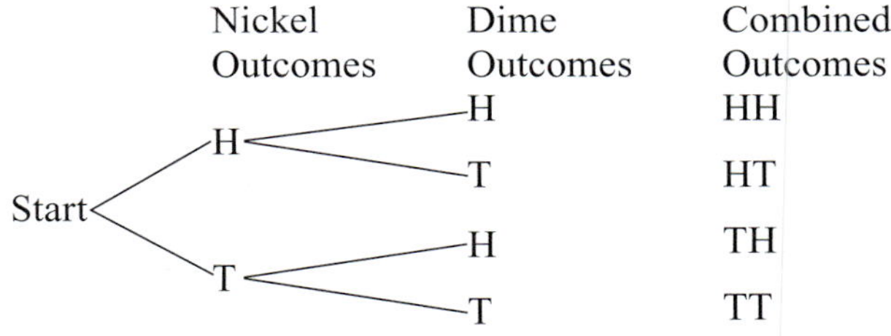

Thus,

$$S_1 = \{\text{HH, HT, TH, TT}\}$$

and there are four simple events in the sample space.

(B) If we are interested only in the number of heads that appear on a single toss of the two coins, then we can let

$$S_2 = \{0, 1, 2\}$$

and there are three simple events in the sample space.

(C) If we are interested in whether the coins match (M) or don't match (D), then we can let

$$S_3 = \{M, D\}$$

and there are only two simple events in the sample space.

MATCHED 1 PROBLEM

An experiment consists of recording the boy–girl composition of families with two children.

(A) What is an appropriate sample space if we are interested in the sex of each child in the order of their births? Draw a tree diagram.

(B) What is an appropriate sample space if we are interested only in the number of girls in a family?

(C) What is an appropriate sample space if we are interested only in whether the sexes are alike (A) or different (D)?

(D) What is an appropriate sample space for all three interests expressed above?

In Example 1, sample space S_1 contains more information than either S_2 or S_3. If we know which outcome has occurred in S_1, then we know which outcome has occurred in S_2 and S_3. However, the reverse is not true. In this sense, we say that S_1 is a more **fundamental sample space** than either S_2 or S_3.

Important Remark: There is no one correct sample space for a given experiment. When specifying a sample space for an experiment, we include as much detail as necessary to answer *all* questions of interest regarding the outcomes of the experiment. If in doubt, include more elements in the sample space rather than fewer.

Now let's return to the two-coin problem in Example 1 and the sample space

$$S_1 = \{\text{HH, HT, TH, TT}\}$$

Suppose we are interested in the outcome, "Exactly 1 head is up." Looking at S_1, we find that it occurs if either of the two simple events HT or TH occurs.* Thus,

*Technically, we should write {HT} and {TH}, because there is a logical distinction between an element of a set and a subset consisting of only that element. But we will just keep this in mind and drop the braces for simple events to simplify the notation.

to say that the event, "Exactly 1 head is up" occurs is the same as saying the experiment has an outcome in the set

$$E = \{\text{HT, TH}\}$$

This is a subset of the sample space S_1. The event E is a compound event.

EXAMPLE 2 Rolling Two Dice

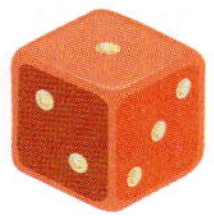
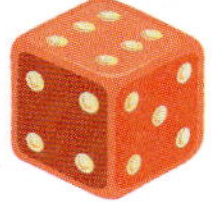

Consider an experiment of rolling two dice. A convenient sample space that will enable us to answer many questions about interesting events is shown in Figure 1. Let S be the set of all ordered pairs listed in the figure. Note that the simple event (3, 2) is to be distinguished from the simple event (2, 3). The former indicates a 3 turned up on the first die and a 2 on the second, whereas the latter indicates a 2 turned up on the first die and a 3 on the second. What is the event that corresponds to each of the following outcomes?

(A) A sum of 7 turns up. (B) A sum of 11 turns up.

(C) A sum less than 4 turns up. (D) A sum of 12 turns up.

FIGURE 1 A sample space for rolling two dice.

SECOND DIE (columns: 1–6); FIRST DIE (rows: 1–6)

FIRST DIE \ SECOND DIE	⚀	⚁	⚂	⚃	⚄	⚅
⚀	(1, 1)	(1, 2)	(1, 3)	(1, 4)	(1, 5)	(1, 6)
⚁	(2, 1)	(2, 2)	(2, 3)	(2, 4)	(2, 5)	(2, 6)
⚂	(3, 1)	(3, 2)	(3, 3)	(3, 4)	(3, 5)	(3, 6)
⚃	(4, 1)	(4, 2)	(4, 3)	(4, 4)	(4, 5)	(4, 6)
⚄	(5, 1)	(5, 2)	(5, 3)	(5, 4)	(5, 5)	(5, 6)
⚅	(6, 1)	(6, 2)	(6, 3)	(6, 4)	(6, 5)	(6, 6)

SOLUTIONS

(A) By "A sum of 7 turns up," we mean that the sum of all dots on both turned-up faces is 7. This outcome corresponds to the event

$$\{(6, 1), (5, 2), (4, 3), (3, 4), (2, 5), (1, 6)\}$$

(B) "A sum of 11 turns up" corresponds to the event

$$\{(6, 5), (5, 6)\}$$

(C) "A sum less than 4 turns up" corresponds to the event

$$\{(1, 1), (2, 1), (1, 2)\}$$

(D) "A sum of 12 turns up" corresponds to the event

$$\{(6, 6)\}$$

MATCHED 2 PROBLEM

Refer to the sample space in Example 2 (Fig. 1). What is the event that corresponds to each of the following outcomes?

(A) A sum of 5 turns up.

(B) A sum that is a prime number greater than 7 turns up.

Informally, to facilitate discussion, we often use the terms *event* and *outcome of an experiment* interchangeably. Thus, in Example 2 we might say "the event 'A sum of 11 turns up'" in place of "the outcome 'A sum of 11 turns up,'" or even write

$$E = \text{A sum of 11 turns up} = \{(6, 5), (5, 6)\}$$

Technically speaking, an event is the mathematical counterpart of an outcome of an experiment.

Probability of an Event

The next step in developing our mathematical model for probability studies is the introduction of a *probability function.* This is a function that assigns to an arbitrary event associated with a sample space a real number between 0 and 1, inclusive. We start by discussing ways in which probabilities are assigned to simple events in S.

DEFINITION 2
Probabilities for Simple Events

Given a sample space

$$S = \{e_1, e_2, \ldots, e_n\}$$

with n simple events, to each simple event e_i we assign a real number, denoted by $P(e_i)$, that is called the **probability of the event e_i.** These numbers may be assigned in an arbitrary manner as long as the following two conditions are satisfied:

1. $0 \le P(e_i) \le 1$
2. $P(e_1) + P(e_2) + \cdots + P(e_n) = 1$ The sum of the probabilities of all simple events in the sample space is 1.

Any probability assignment that meets conditions 1 and 2 is said to be an **acceptable probability assignment.**

Our mathematical theory does not explain how acceptable probabilities are assigned to simple events. These assignments are generally based on the expected or actual percentage of times a simple event occurs when an experiment is repeated a large number of times. Assignments based on this principle are called *reasonable.*

Let an experiment be the flipping of a single coin, and let us choose a sample space S to be

$$S = \{\text{H}, \text{T}\}$$

If a coin appears to be fair, we are inclined to assign probabilities to the simple events in S as follows:

$$P(\mathrm{H}) = \tfrac{1}{2} \qquad \text{and} \qquad P(\mathrm{T}) = \tfrac{1}{2}$$

These assignments are based on reasoning that, because there are two ways a coin can land, in the long run a head will turn up half the time and a tail will turn up half the time. These probability assignments are acceptable, because both of the conditions for acceptable probability assignments in Definition 2 are satisfied:

1. $0 \le P(\mathrm{H}) \le 1,\ 0 \le P(\mathrm{T}) \le 1$
2. $P(\mathrm{H}) + P(\mathrm{T}) = \frac{1}{2} + \frac{1}{2} = 1$

But there are other acceptable assignments. Maybe after flipping a coin 1,000 times we find that the head turns up 376 times and the tail turns up 624 times. With this result, we might suspect that the coin is not fair and assign the simple events in the sample space S the probabilities

$$P(\mathrm{H}) = .376 \qquad \text{and} \qquad P(\mathrm{T}) = .624$$

This is also an acceptable assignment. But the probability assignment

$$P(\mathrm{H}) = 1 \qquad \text{and} \qquad P(\mathrm{T}) = 0$$

though acceptable, is not reasonable, unless the coin has two heads. The assignment

$$P(\mathrm{H}) = .6 \qquad \text{and} \qquad P(\mathrm{T}) = .8$$

is not acceptable, because $.6 + .8 = 1.4$, which violates condition 2 in Definition 2.

In probability studies, the 0 to the left of the decimal is usually omitted. Thus, we write .8 and not 0.8.

It is important to keep in mind that out of the infinitely many possible acceptable probability assignments to simple events in a sample space, we are generally inclined to choose one assignment over another based on reasoning or experimental results.

Given an acceptable probability assignment for simple events in a sample space S, how do we define the probability of an arbitrary event E associated with S?

DEFINITION 3
Probability of an Event *E*

Given an acceptable probability assignment for the simple events in a sample space S, we define the **probability of an arbitrary event *E*,** denoted by $P(E)$, as follows:

1. If E is the empty set, then $P(E) = 0$.
2. If E is a simple event, then $P(E)$ has already been assigned.
3. If E is a compound event, then $P(E)$ is the sum of the probabilities of all the simple events in E.
4. If E is the sample space S, then $P(E) = P(S) = 1$. This is a special case of 3.

EXAMPLE 3 Finding Probabilities of Events

Let's return to Example 1, the tossing of a nickel and dime, and the sample space

$$S = \{\text{HH, HT, TH, TT}\}$$

Because there are four simple outcomes and the coins are assumed to be fair, it appears that each outcome should occur in the long run 25% of the time. Let's assign the same probability of $\frac{1}{4}$ to each simple event in S:

Simple event, e_i	HH	HT	TH	TT
$P(e_i)$	$\frac{1}{4}$	$\frac{1}{4}$	$\frac{1}{4}$	$\frac{1}{4}$

This is an acceptable assignment according to Definition 2 and a reasonable assignment for ideal coins that are perfectly balanced or coins close to ideal.

(A) What is the probability of getting exactly one head?

(B) What is the probability of getting at least one head?

(C) What is the probability of getting a head or a tail?

(D) What is the probability of getting three heads?

SOLUTIONS

(A) E_1 = Getting one head = {HT, TH}

Because E_1 is a compound event, we use item 3 in Definition 3 and find $P(E_1)$ by adding the probabilities of the simple events in E_1. Thus,

$$P(E_1) = P(\text{HT}) + P(\text{TH}) = \tfrac{1}{4} + \tfrac{1}{4} = \tfrac{1}{2}$$

(B)
$$\begin{aligned} E_2 &= \text{Getting at least 1 head} = \{\text{HH, HT, TH}\} \\ P(E_2) &= P(\text{HH}) + P(\text{HT}) + P(\text{TH}) \\ &= \tfrac{1}{4} + \tfrac{1}{4} + \tfrac{1}{4} = \tfrac{3}{4} \end{aligned}$$

(C)
$$\begin{aligned} E_3 &= \{\text{HH, HT, TH, TT}\} = S \\ P(E_3) &= P(S) = 1 \end{aligned}$$
$\frac{1}{4} + \frac{1}{4} + \frac{1}{4} + \frac{1}{4} = 1$

(D)
$$\begin{aligned} E_3 &= \text{Getting three heads} = \varnothing \\ P(\varnothing) &= 0 \end{aligned}$$
Empty set

Steps for Finding Probabilities of Events

Step 1. Set up an appropriate sample space S for the experiment.

Step 2. Assign acceptable probabilities to the simple events in S.

Step 3. To obtain the probability of an arbitrary event E, add the probabilities of the simple events in E.

The function P defined in steps 2 and 3 is called a **probability function.** The domain of this function is all possible events in the sample space S, and the range is a set of real numbers between 0 and 1, inclusive.

MATCHED 3 PROBLEM

Return to Matched Problem 1, recording the boy–girl composition of families with two children and the sample space

$$S = \{\text{BB, BG, GB, GG}\}$$

Statistics from the U.S. Census Bureau indicate that an acceptable and reasonable probability for this sample space is

Simple event, e_i	BB	BG	GB	GG
$P(e_i)$	.26	.25	.25	.24

Find the probabilities of the following events:

(A) E_1 = Having at least one girl in the family

(B) E_2 = Having at most one girl in the family

(C) E_3 = Having two children of the same sex in the family

Equally Likely Assumption

In tossing a nickel and dime (Example 3), we assigned the same probability, $\frac{1}{4}$, to each simple event in the sample space $S = \{\text{HH, HT, TH, TT}\}$. By assigning the same probability to each simple event in S, we are actually making the assumption that each simple event is as likely to occur as any other. We refer to this as an **equally likely assumption.** In general, we have Definition 4.

DEFINITION 4
Probability of a Simple Event Under an Equally Likely Assumption

If, in a sample space

$$S = \{e_1, e_2, \ldots, e_n\}$$

with n elements, we assume each simple event e_i is as likely to occur as any other, then we assign the probability $1/n$ to each. That is,

$$P(e_i) = \frac{1}{n}$$

Under an equally likely assumption, we can develop a very useful formula for finding probabilities of arbitrary events associated with a sample space S. Consider the following example.

If a single die is rolled and we assume each face is as likely to come up as any other, then for the sample space

$$S = \{1, 2, 3, 4, 5, 6\}$$

we assign a probability of $\frac{1}{6}$ to each simple event, because there are six simple events. Then the probability of

$$E = \text{Rolling a prime number} = \{2, 3, 5\}$$

is

$$P(E) = P(2) + P(3) + P(5) = \tfrac{1}{6} + \tfrac{1}{6} + \tfrac{1}{6} = \tfrac{3}{6} = \tfrac{1}{2}$$

Thus, under the assumption that each simple event is as likely to occur as any other, the computation of the probability of the occurrence of any event E in a sample space S is the number of elements in E divided by the number of elements in S.

THEOREM 1

Probability of an Arbitrary Event Under an Equally Likely Assumption

If we assume each simple event in sample space S is as likely to occur as any other, then the probability of an arbitrary event E in S is given by

$$P(E) = \frac{\text{Number of elements in } E}{\text{Number of elements in } S} = \frac{n(E)}{n(S)}$$

EXAMPLE 4 Finding Probabilities of Events

If in rolling two dice we assume each simple event in the sample space shown in Figure 1 on p. 817 is as likely as any other, find the probabilities of the following events:

(A) E_1 = A sum of 7 turns up

(B) E_2 = A sum of 11 turns up

(C) E_3 = A sum less than 4 turns up

(D) E_4 = A sum of 12 turns up

SOLUTIONS

Referring to Figure 1, we see that:

(A) $P(E_1) = \dfrac{n(E_1)}{n(S)} = \dfrac{6}{36} = \dfrac{1}{6}$

(B) $P(E_2) = \dfrac{n(E_2)}{n(S)} = \dfrac{2}{36} = \dfrac{1}{18}$

(C) $P(E_3) = \dfrac{n(E_3)}{n(S)} = \dfrac{3}{36} = \dfrac{1}{12}$

(D) $P(E_4) = \dfrac{n(E_4)}{n(S)} = \dfrac{1}{36}$

MATCHED 4 PROBLEM

Under the conditions in Example 4, find the probabilities of the following events:

(A) E_5 = A sum of 5 turns up

(B) E_6 = A sum that is a prime number greater than 7 turns up

EXPLORE/DISCUSS 1

A box contains four red balls and seven green balls. A ball is drawn at random and then, without replacing the first ball, a second ball is drawn. Discuss whether the equally likely assumption would be appropriate for the sample space $S = \{\text{RR}, \text{RG}, \text{GR}, \text{GG}\}$.

We now turn to some examples that make use of the counting techniques developed in Section 10.4.

EXAMPLE 5 Drawing Cards

In drawing 5 cards from a 52-card deck without replacement, what is the probability of getting five spades?

SOLUTION

Let the sample space S be the set of all 5-card hands from a 52-card deck. Because the order in a hand does not matter, $n(S) = C_{52,5}$. The event we seek is

E = Set of all 5-card hands from 13 spades

Again, the order does not matter and $n(E) = C_{13,5}$. Thus, assuming each 5-card hand is as likely as any other,

$$P(E) = \frac{n(E)}{n(S)} = \frac{C_{13,5}}{C_{52,5}} = \frac{13!/5!8!}{52!/5!47!} = \frac{13!}{5!8!} \cdot \frac{5!47!}{52!} \approx .0005$$

MATCHED 5 PROBLEM

In drawing 7 cards from a 52-card deck without replacement, what is the probability of getting seven hearts?

EXAMPLE 6 Selecting Committees

The board of regents of a university is made up of 12 men and 16 women. If a committee of six is chosen at random, what is the probability that it will contain three men and three women?

SOLUTION

Let S = Set of all 6-person committees out of 28 people:

$$n(S) = C_{28,6}$$

Let E = Set of all 6-person committees with 3 men and 3 women. To find $n(E)$, we use the multiplication principle and the following two operations:

O_1: Select 3 men out of the 12 available $\quad N_1$: $C_{12,3}$

O_2: Select 3 women out of the 16 available $\quad N_2$: $C_{16,3}$

Thus,

$$n(E) = C_{12,3} \cdot C_{16,3}$$

and

$$P(E) = \frac{n(E)}{n(S)} = \frac{C_{12,3} \cdot C_{16,3}}{C_{28,6}} \approx .327$$

MATCHED 6 PROBLEM

What is the probability that the committee in Example 6 will have four men and two women?

Empirical Probability

In the earlier examples in Section 10.5, we made a reasonable assumption about an experiment and used deductive reasoning to assign probabilities. For example, it is reasonable to assume that an ordinary coin will come up heads about as often as it will come up tails. Probabilities determined in this manner are called **theoretical probabilities.** No experiments are ever conducted. But what if the theoretical probabilities are not obvious? Then we assign probabilities to simple events based on the results of actual experiments. Probabilities determined from the results of actually performing an experiment are called **empirical probabilities.** As an experiment is repeated over and over, the percentage of times an event occurs may get closer and closer to a single fixed number. If so, this single fixed number is generally called the **actual probability** of the event.

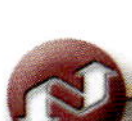

EXPLORE/DISCUSS 2

Like a coin, a thumbtack tossed into the air will land in one of two positions, point up or point down [Fig. 2(a)]. Unlike a coin, we would not expect both events to occur with the same frequency. Indeed, the frequencies of landing point up and point down may well vary from one thumbtack to another [Fig. 2(b)]. Find two thumbtacks of different sizes and guess which one is likely to land point up more frequently. Then toss each tack 100 times and record the number of times each lands point up. Did the experiment confirm your initial guess?

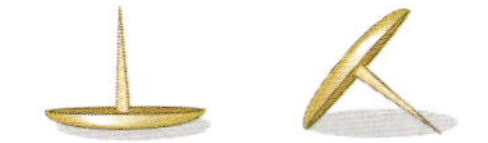

(a) Point up or point down

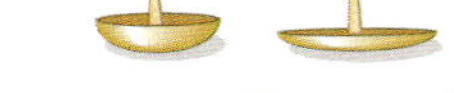

(b) Two different tacks

FIGURE 2

Suppose when tossing one of the thumbtacks in Explore/Discuss 2, we observe that the tack lands point up 43 times and point down 57 times. Based on this experiment, it seems reasonable to say that for this particular thumbtack

$$P(\text{Point up}) = \frac{43}{100} = .43$$

$$P(\text{Point down}) = \frac{57}{100} = .57$$

Probability assignments based on the results of repeated trials of an experiment are called **approximate empirical probabilities.**

In general, if we conduct an experiment n times and an event E occurs with **frequency** $f(E)$, then the ratio $f(E)/n$ is called the **relative frequency** of the occurrence of event E in n trials. We define the **empirical probability** of E, denoted by $P(E)$, by the number, if it exists, that the relative frequency $f(E)/n$ approaches as n gets larger and larger. Of course, for any particular n, the relative frequency $f(E)/n$ is generally only approximately equal to $P(E)$. However, as n increases, we expect the approximation to improve.

DEFINITION 5
Empirical Probability

If $f(E)$ is the frequency of event E in n trials, then

$$P(E) \approx \frac{\text{Frequency of occurrence of } E}{\text{Total number of trials}} = \frac{f(E)}{n}$$

If we can also deduce theoretical probabilities for an experiment, then we expect the approximate empirical probabilities to approach the theoretical probabilities. If this does not happen, then we should begin to suspect the manner in which the theoretical probabilities were computed. If $P(E)$ is the theoretical probability of an event E and the experiment is performed n times, then the **expected frequency** of the occurrence of E is $n \cdot P(E)$.

EXAMPLE 7 Finding Approximate Empirical and Theoretical Probabilities

Two coins are tossed 500 times with the following frequencies of outcomes:

Two heads: 121

One head: 262

Zero heads: 117

(A) Compute the approximate empirical probability for each outcome.

(B) Compute the theoretical probability for each outcome.

(C) Compute the expected frequency for each outcome.

SOLUTIONS

(A)

$$P(\text{two heads}) \approx \frac{121}{500} = .242$$

$$P(\text{one head}) \approx \frac{262}{500} = .524$$

$$P(\text{zero heads}) \approx \frac{117}{500} = .234$$

(B) A sample space of equally likely simple events is $S = \{\text{HH, HT, TH, TT}\}$. Let

$$E_1 = \text{two heads} = \{\text{HH}\}$$
$$E_2 = \text{one head} = \{\text{HT, TH}\}$$
$$E_3 = \text{zero heads} = \{\text{TT}\}$$

Then

$$P(E_1) = \frac{n(E_1)}{n(S)} = \frac{1}{4} = .25$$

$$P(E_2) = \frac{n(E_2)}{n(S)} = \frac{2}{4} = .50$$

$$P(E_3) = \frac{n(E_3)}{n(S)} = \frac{1}{4} = .25$$

(C) The expected frequencies are

E_1: 500(.25) = 125
E_2: 500(.5) = 250
E_3: 500(.25) = 125

The actual frequencies obtained from performing the experiment are reasonably close to the expected frequencies. Increasing the number of trials of the experiment would produce even better approximations.

MATCHED 7 PROBLEM

One die is rolled 500 times with the following frequencies of outcomes:

Outcome	1	2	3	4	5	6
Frequency	89	83	77	91	72	88

(A) Compute the approximate empirical probability for each outcome.
(B) Compute the theoretical probability for each outcome.
(C) Compute the expected frequency for each outcome.

FIGURE 3 Using a random number generator.

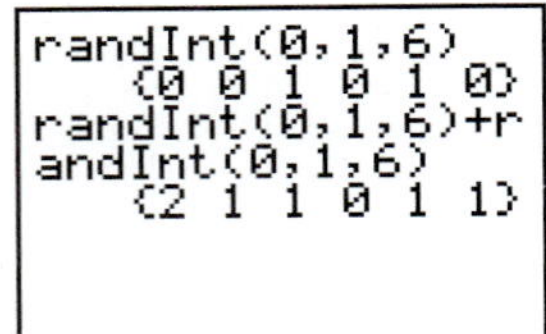

Tossing two coins 500 times is certainly a tedious task and we did not do this to generate the data in Example 7. Instead, we used a random number generator on a graphing utility to simulate this experiment. Specifically, we used the command **randInt(*i*,*k*,*n*)** on a Texas Instruments TI-83, which generates a random sequence of n integers between i and k, inclusively. If we let 0 represent tails and 1 represent heads, then a random sequence of 0s and 1s can be used to represent the outcomes of repeated tosses of one coin (see the first two lines of Fig. 3). Thus, in six tosses, we obtained two heads and four tails. To simulate tossing two coins, we simply add together two similar statements, as shown in lines three through five of Figure 3. We see that in these six tosses zero heads occurred once, one head occurred four times, and two heads occurred once. Of course, to obtain meaningful results, we need to toss the coins many more times. Figure 4(a) shows a command that will simulate 500 tosses of two coins. To determine the frequency of each outcome, we construct a histogram [Figs. 4(b) and 4(c)] and use the TRACE command to determine the following frequencies [Figs. 5(a), 5(b), and 5(c)].

FIGURE 4 Simulating 500 tosses of two coins.

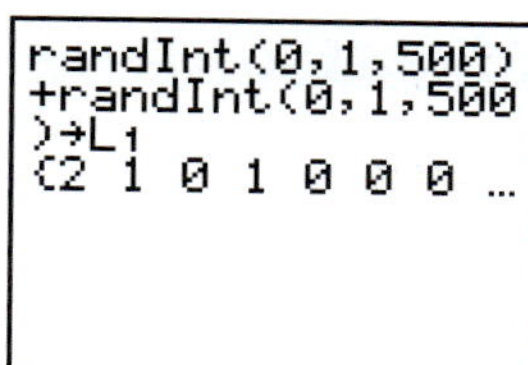

(a) Generating the random numbers

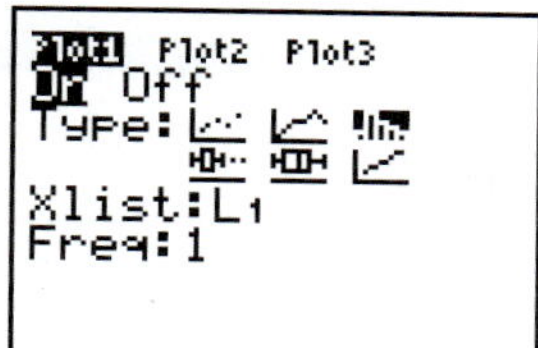

(b) Setting up the histogram

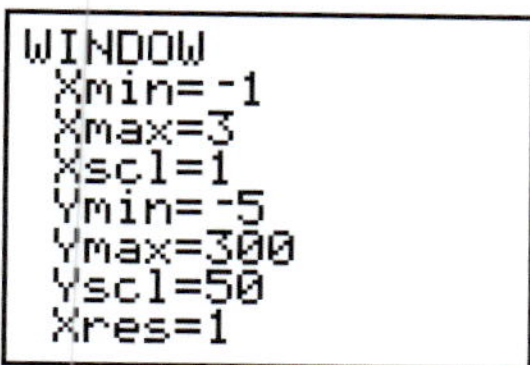

(c) Selecting the window variables

FIGURE 5 Results of the simulation.

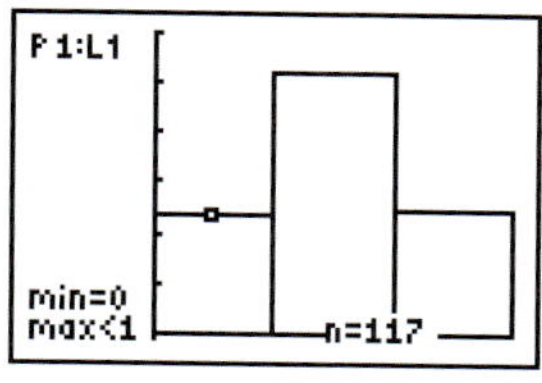

(a) 0 heads: 117

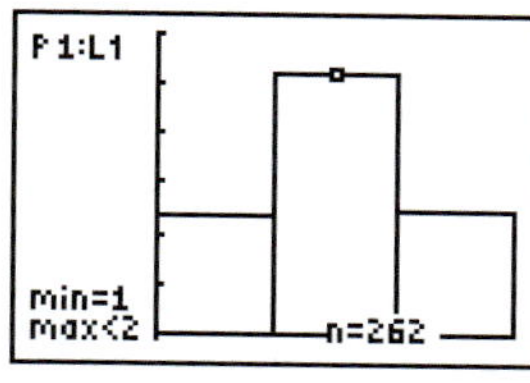

(b) 1 head: 262

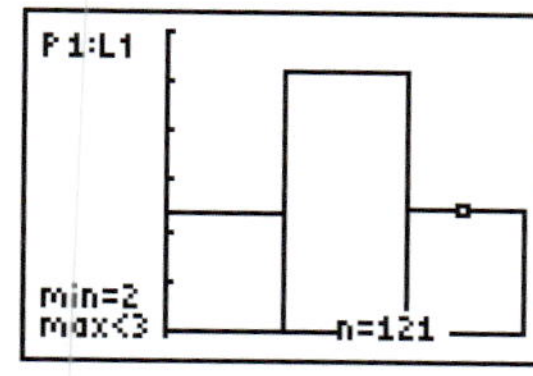

(c) 2 heads: 121

If you perform the same simulation on your graphing utility, you are not likely to get exactly the same results. But the approximate empirical probabilities you obtain will be close to the theoretical probabilities.

EXPLORE/DISCUSS 3

This discussion assumes that your graphing utility has the ability to generate and manipulate sequences of random integers.

(A) As an alternative to using the histogram in Figure 5 to count the outcomes of the sequence of random integers in Figure 4(a), enter the following function and evaluate it for $x = 0$, 1, and 2:

$$y_1 = \text{sum}(\text{seq}(L_1(I)=X,I,1,\text{dim}(L_1)))$$

(B) Simulate the experiment of rolling a single die and compare your empirical results with the results in Matched Problem 7.

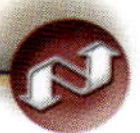

EXAMPLE 8 Empirical Probabilities for an Insurance Company

An insurance company selected 1,000 drivers at random in a particular city to determine a relationship between age and accidents. The data obtained are listed in Table 1. Compute the approximate empirical probabilities of the following events for a driver chosen at random in the city:

(A) E_1: being under 20 years old *and* having exactly three accidents in 1 year

(B) E_2: being 30–39 years old *and* having one or more accidents in 1 year

(C) E_3: having no accidents in 1 year

(D) E_4: being under 20 years old *or* having exactly three accidents in 1 year

TABLE 1

	Accidents in 1 Year				
Age	**0**	**1**	**2**	**3**	**Over 3**
Under 20	50	62	53	35	20
20–29	64	93	67	40	36
30–39	82	68	32	14	4
40–49	38	32	20	7	3
Over 49	43	50	35	28	24

SOLUTIONS

(A) $P(E_1) \approx \dfrac{35}{1,000} = .035$

(B) $P(E_2) \approx \dfrac{68 + 32 + 14 + 4}{1,000} = .118$

(C) $P(E_3) \approx \dfrac{50 + 64 + 82 + 38 + 43}{1,000} = .277$

(D) $P(E_4) \approx \dfrac{50 + 62 + 53 + 35 + 20 + 40 + 14 + 7 + 28}{1,000} = .309$

Notice that in this type of problem, which is typical of many realistic problems, approximate empirical probabilities are the only type we can compute.

MATCHED 8 PROBLEM

Referring to Table 1 in Example 8, compute the approximate empirical probabilities of the following events for a driver chosen at random in the city:

(A) E_1: being under 20 years old with no accidents in 1 year

(B) E_2: being 20–29 years old and having fewer than two accidents in 1 year

(C) E_3: not being over 49 years old

Approximate empirical probabilities are often used to test theoretical probabilities. Equally likely assumptions may not be justified in reality. In addition to this use, there are many situations in which it is either very difficult or impossible to compute the theoretical probabilities for given events. For example, insurance companies use past experience to establish approximate empirical probabilities to predict future accident rates; baseball teams use batting averages, which are approximate empirical probabilities based on past experience, to predict the future performance of a player; and pollsters use approximate empirical probabilities to predict outcomes of elections.

ANSWERS MATCHED PROBLEMS

1. (A) $S_1 = \{BB, BG, GB, GG\}$;

Sex of First Child	Sex of Second Child	Combined Outcomes
B	B	BB
	G	BG
G	B	GB
	G	GG

(B) $S_2 = \{0, 1, 2\}$ (C) $S_3 = \{A, D\}$ (D) The sample space in part A.

2. (A) $\{(4, 1), (3, 2), (2, 3), (1, 4)\}$ (B) $\{(6, 5), (5, 6)\}$ 3. (A) .74 (B) .76 (C) .5

4. (A) $P(E_5) = \frac{1}{9}$ (B) $P(E_6) = \frac{1}{18}$ 5. $C_{13,7}/C_{52,7} \approx .000013$ 6. $C_{12,4} \cdot C_{16,2}/C_{28,6} \approx .158$

7. (A) $P(E_1) \approx .178$, $P(E_2) \approx .166$, $P(E_3) \approx .154$, $P(E_4) \approx .182$, $P(E_5) \approx .144$, $P(E_6) \approx .176$
(B) $\frac{1}{6} \approx .167$ for each (C) 83.3 for each 8. (A) $P(E_1) \approx .05$ (B) $P(E_2) \approx .157$ (C) $P(E_3) \approx .82$

EXERCISE 10.5

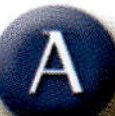

1. How would you interpret $P(E) = 1$?

2. How would you interpret $P(E) = 0$?

3. A spinner can land on four different colors: red (R), green (G), yellow (Y), and blue (B). If we do not assume each color is as likely to turn up as any other, which of the following probability assignments have to be rejected, and why?
   (A) $P(R) = .15$, $P(G) = -.35$, $P(Y) = .50$, $P(B) = .70$
   (B) $P(R) = .32$, $P(G) = .28$, $P(Y) = .24$, $P(B) = .30$
   (C) $P(R) = .26$, $P(G) = .14$, $P(Y) = .30$, $P(B) = .30$

4. Under the probability assignments in Problem 3, part C, what is the probability that the spinner will not land on blue?

5. Under the probability assignments in Problem 3, part C, what is the probability that the spinner will land on red or yellow?

6. Under the probability assignments in Problem 3, part C, what is the probability that the spinner will not land on red or yellow?

7. A ski jumper has jumped over 300 feet in 25 out of 250 jumps. What is the approximate empirical probability of the next jump being over 300 feet?

8. In a certain city there are 4,000 youths between 16 and 20 years old who drive cars. If 560 of them were involved in accidents last year, what is the approximate empirical probability of a youth in this age group being involved in an accident this year?

9. Out of 420 times at bat, a baseball player gets 189 hits. What is the approximate empirical probability that the player will get a hit next time at bat?

10. In a medical experiment, a new drug is found to help 2,400 out of 3,000 people. If a doctor prescribes the drug for a particular patient, what is the approximate empirical probability that the patient will be helped?

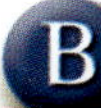

11. A small combination lock on a suitcase has three wheels, each labeled with the 10 digits from 0 to 9. If an opening combination is a particular sequence of three digits with no repeats, what is the probability of a person guessing the right combination?

12. A combination lock has five wheels, each labeled with the 10 digits from 0 to 9. If an opening combination is a particular sequence of five digits with no repeats, what is the probability of a person guessing the right combination?

Problems 13–18 involve an experiment consisting of dealing 5 cards from a standard 52-card deck. In Problems 13–16, what is the probability of being dealt:

13. Five black cards

14. Five hearts

15. Five face cards if an ace is considered to be a face card.

16. Five nonface cards if an ace is considered to be a one and not a face.

17. If we are interested in the number of aces in a 5-card hand, would $S = \{0, 1, 2, 3, 4\}$ be an acceptable sample space? Would it be an equally-likely sample space? Explain.

18. If we are interested in the number of black cards in a 5-card hand, would $S = \{0, 1, 2, 3, 4, 5\}$ be an acceptable sample space? Would it be an equally-likely sample space? Explain.

19. If four-digit numbers less than 5,000 are randomly formed from the digits 1, 3, 5, 7, and 9, what is the probability of forming a number divisible by 5? Digits may be repeated; for example, 1,355 is acceptable.

20. If code words of four letters are generated at random using the letters A, B, C, D, E, and F, what is the probability of forming a word without a vowel in it? Letters may be repeated.

21. Suppose five thank-you notes are written and five envelopes are addressed. Accidentally, the notes are randomly inserted into the envelopes and mailed without checking the addresses. What is the probability that all five notes will be inserted into the correct envelopes?

22. Suppose six people check their coats in a checkroom. If all claim checks are lost and the six coats are randomly returned, what is the probability that all six people will get their own coats back?

An experiment consists of rolling two fair dice and adding the dots on the two sides facing up. Using the sample space shown in Figure 1 (p. 817) and assuming each simple event is as likely as any other, find the probabilities of the sums of dots indicated in Problems 23–38.

23. Sum is 2.

24. Sum is 10.

25. Sum is 6.

26. Sum is 8.

27. Sum is less than 5.

28. Sum is greater than 8.

29. Sum is not 7 or 11.

30. Sum is not 2, 4, or 6.

31. Sum is 1.

32. Sum is not 13.

33. Sum is divisible by 3.

34. Sum is divisible by 4.

35. Sum is 7 or 11 (a "natural").

36. Sum is 2, 3, or 12 ("craps").

37. Sum is divisible by 2 or 3.

38. Sum is divisible by 2 and 3.

39. Five thousand people work in a large auto plant. An individual is selected at random and his or her birthday (month and day, not year) is recorded. Set up an appropriate sample space for this experiment and assign acceptable probabilities to the simple events.

40. In a hotly contested three-way race for governor of Minnesota, the leading candidates are running neck-and-neck while the third candidate is receiving half the support of either of the others. Registered voters are chosen at random and are asked for which of the three they are most likely to vote. Set up an appropriate sample space for the random survey experiment and assign acceptable probabilities to the simple events.

41. A pair of dice is rolled 500 times with the following frequencies:

Sum	2	3	4	5	6	7	8	9	10	11	12
Frequency	11	35	44	50	71	89	72	52	36	26	14

(A) Compute the approximate empirical probability for each outcome.

(B) Compute the theoretical probability for each outcome, assuming fair dice.

(C) Compute the expected frequency of each outcome.

(D) Describe how a random number generator could be used to simulate this experiment. If your graphing utility has a random number generator, use it to simulate 500 tosses of a pair of dice and compare your results with part C.

42. Three coins are flipped 500 times with the following frequencies of outcomes:

Three heads: 58 Two heads: 198

One head: 190 Zero heads: 54

(A) Compute the approximate empirical probability for each outcome.

(B) Compute the theoretical probability for each outcome, assuming fair coins.

(C) Compute the expected frequency of each outcome.

(D) Describe how a random number generator could be used to simulate this experiment. If your graphing utility has a random number generator, use it to simulate 500 tosses of three coins and compare your results with part C.

43. (A) Is it possible to get 29 heads in 30 flips of a fair coin? Explain.

(B) If you flip a coin 50 times and get 42 heads, would you suspect that the coin was unfair? Why or why not? If you suspect an unfair coin, what empirical probabilities would you assign to the simple events of the sample space?

44. **(A)** Is it possible to get nine double sixes in 12 rolls of a pair of fair dice? Explain.

(B) If you roll a pair of dice 40 times and get 14 double sixes, would you suspect that the dice were unfair? Why or why not? If you suspect loaded dice, what empirical probability would you assign to the event of rolling a double six?

An experiment consists of tossing three fair coins, but one of the three coins has a head on both sides. Compute the probabilities of obtaining the indicated results in Problems 45–50.

45. One head

46. Two heads

47. Three heads

48. Zero heads

49. More than one head

50. More than one tail

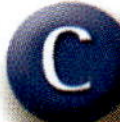

An experiment consists of rolling two fair dice and adding the dots on the two sides facing up. Each die has one dot on two opposite faces, two dots on two opposite faces, and three dots on two opposite faces. Compute the probabilities of obtaining the indicated sums in Problems 51–58.

51. 2

52. 3

53. 4

54. 5

55. 6

56. 7

57. An odd sum

58. An even sum

An experiment consists of dealing 5 cards from a standard 52-card deck. In Problems 59–66, what is the probability of being dealt the following cards?

59. Five cards, jacks through aces

60. Five cards, 2 through 10

61. Four aces

62. Four of a kind

63. Straight flush, ace high; that is, 10, jack, queen, king, ace in one suit

64. Straight flush, starting with 2; that is, 2, 3, 4, 5, 6 in one suit

65. Two aces and three queens

66. Two kings and three aces

APPLICATIONS

67. Market Analysis. A company selected 1,000 households at random and surveyed them to determine a relationship between income level and the number of television sets in a home. The information gathered is listed in the table:

	Televisions per Household				
Yearly Income ($)	**0**	**1**	**2**	**3**	**Above 3**
Less than 12,000	0	40	51	11	0
12,000–19,999	0	70	80	15	1
20,000–39,999	2	112	130	80	12
40,000–59,999	10	90	80	60	21
60,000 or more	30	32	28	25	20

Compute the approximate empirical probabilities:

(A) Of a household earning \$12,000–\$19,999 per year *and* owning exactly three television sets

(B) Of a household earning \$20,000–\$39,999 per year *and* owning more than one television set

(C) Of a household earning \$60,000 or more per year *or* owning more than three television sets

(D) Of a household not owning zero television sets

68. Market Analysis. Use the sample results in Problem 67 to compute the approximate empirical probabilities:

(A) Of a household earning \$40,000–\$59,999 per year *and* owning zero television sets

(B) Of a household earning \$12,000–\$39,999 per year *and* owning more than two television sets

(C) Of a household earning less than \$20,000 per year *or* owning exactly two television sets

(D) Of a household not owning more than three television sets

SECTION 10.6 Binomial Formula

Pascal's Triangle • The Binomial Formula • Proof of the Binomial Formula

The binomial form

$$(a + b)^n$$

where n is a natural number, appears more frequently than you might expect. It turns out that the coefficients in the expansion are related to probability concepts that we have already discussed.

Pascal's Triangle

Let's begin by expanding $(a + b)^n$ for the first few values of n. We include $n = 0$, which is not a natural number, for reasons of completeness that will become apparent later.

$$\begin{aligned}(a + b)^0 &= 1\\ (a + b)^1 &= a + b\\ (a + b)^2 &= a^2 + 2ab + b^2\\ (a + b)^3 &= a^3 + 3a^2b + 3ab^2 + b^3\end{aligned} \tag{1}$$

EXPLORE/DISCUSS 1

Based on the expansions in equations (1), how many terms would you expect $(a + b)^n$ to have? What is the first term? What is the last term?

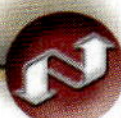

FIGURE 1 Pascal's triangle.

1

1 1

1 2 1

1 3 3 1

Now let's examine just the coefficients of the expansions in equations (1) arranged in a form that is usually referred to as **Pascal's triangle** (Fig. 1).

EXPLORE/DISCUSS 2

Refer to Figure 1.

(A) How is the middle element in the third row related to the elements in the row above it?

(B) How are the two inner elements in the fourth row related to the elements in the row above them?

(C) Based on your observations in parts A and B, make a conjecture about the fifth and sixth rows. Check your conjecture by expanding $(a + b)^4$ and $(a + b)^5$.

FIGURE 2

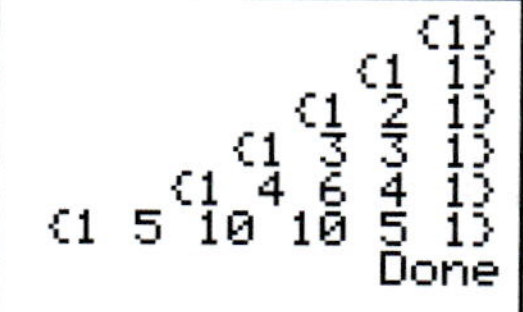

Many students find Pascal's triangle a useful tool for determining the coefficients in the expansion of $(a + b)^n$, especially for small values of n. Figure 2 contains output from a program called PASCAL.* You should recognize the output in the table—it is the first six lines of Pascal's triangle. The major drawback of using this triangle, whether it is generated by hand or on a graphing utility, is that to find the elements in a given row, you must write out all the preceding rows. It would be useful to find a formula that gives the coefficients for a binomial expansion directly. Fortunately, such a formula exists—the combination formula $C_{n,r}$ introduced in Section 10.4.

The Binomial Formula

When working with binomial expansions, it is customary to use another notation for the combination formula.

DEFINITION 1
Combination Formula

For nonnegative integers r and n, $0 \leq r \leq n$,

$$\binom{n}{r} = C_{n,r} = \frac{n!}{r!(n-r)!}$$

Theorem 1 establishes that the coefficients in a binomial expansion can always be expressed in terms of the combination formula. This is a very important theoretical result and a very practical tool. As we shall see, using this theorem in conjunction with a graphing utility provides a very efficient method for expanding binomials.

THEOREM 1
Binomial Formula

For n a positive integer

$$(a + b)^n = \sum_{k=0}^{n} \binom{n}{k} a^{n-k} b^k$$

We defer the proof of Theorem 1 until the end of Section 10.6. Because the values of the combination formula are the coefficients in a binomial expansion, it is natural to call them **binomial coefficients.**

EXAMPLE 1 Using the Binomial Formula

Use the binomial formula to expand $(x + y)^6$.

*Programs for TI-83 and TI-86 graphing calculators can be found at the website for this book.

SOLUTION

Compute the Coefficients

X	Y1
0	1
1	6
2	15
3	20
4	15
5	6
6	1

Y1=6 nCr X

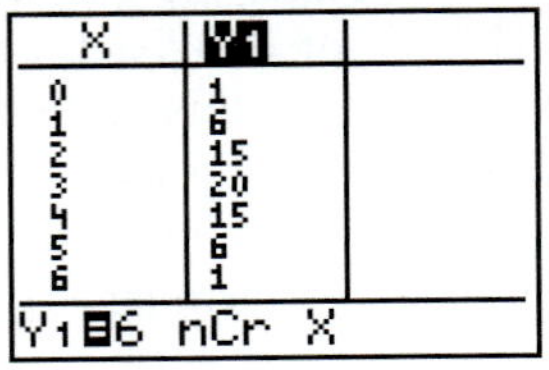

FIGURE 3

Write the Expansion

$$(x + y)^6 = \sum_{k=0}^{6} \binom{6}{k} x^{6-k} y^k$$

$$= \binom{6}{0}x^6 + \binom{6}{1}x^5y + \binom{6}{2}x^4y^2 + \binom{6}{3}x^3y^3 + \binom{6}{4}x^2y^4 + \binom{6}{5}xy^5 + \binom{6}{6}y^6$$

$$= x^6 + 6x^5y + 15x^4y^2 + 20x^3y^3 + 15x^2y^4 + 6xy^5 + y^6$$

See Figure 3.

MATCHED 1 PROBLEM

Use the binomial formula to expand $(x + 1)^5$.

EXAMPLE 2 Using the Binomial Formula

Use the binomial formula to expand $(3p - 2q)^4$.

SOLUTION

Compute the Coefficients

X	Y1
0	81
1	-216
2	216
3	-96
4	16
5	0
6	0

Y1=(4 nCr X)3^(...

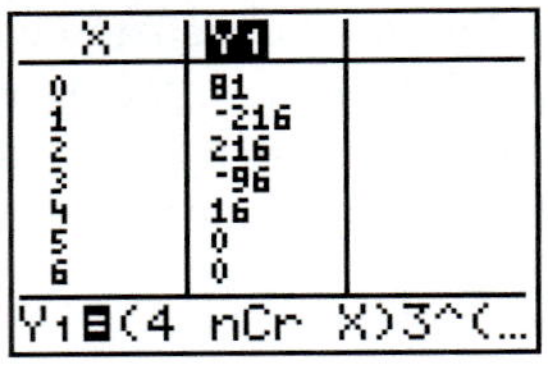

FIGURE 4 $y_1 = C_{4,x}3^{4-x}(-2)^x$.

Write the Expansion

$$(3p - 2q)^4 = [(3p) + (-2q)]^4$$

$$= \sum_{k=0}^{4} \binom{4}{k} (3p)^{4-k}(-2q)^k$$

$$= \sum_{k=0}^{4} \binom{4}{k} 3^{4-k}(-2)^k p^{4-k} q^k$$

$$= 81p^4 - 216p^3q + 216p^2q^2 - 96pq^3 + 16q^4$$

See Figure 4.

MATCHED 2 PROBLEM

Use the binomial formula to expand $(2m - 5n)^3$.

EXPLORE/DISCUSS 3

(A) Compute each term and also the sum of the alternating series

$$\binom{6}{0} - \binom{6}{1} + \binom{6}{2} - \cdots + \binom{6}{6}.$$

(B) What result about an alternating series can be deduced by letting $a = 1$ and $b = -1$ in the binomial formula?

EXAMPLE 3 Using the Binomial Formula

Find the term containing x^9 in the expansion of $(x + 3)^{14}$.

SOLUTION
In the expansion

$$(x+3)^{14} = \sum_{k=0}^{14} \binom{14}{k} x^{14-k} 3^k$$

the exponent of x is 9 when $k = 5$. Thus, the term containing x^9 is

$$\binom{14}{5} x^9 3^5 = (2{,}002)(243)x^9 = 486{,}486x^9$$

MATCHED 3 PROBLEM

Find the term containing y^8 in the expansion of $(2 + y)^{14}$.

EXAMPLE 4 Using the Binomial Formula

If the terms in the expansion of $(x - 2)^{20}$ are arranged in decreasing powers of x, find the fourth term and the sixteenth term.

SOLUTION
In the expansion of $(a + b)^n$, the exponent of b in the rth term is $r - 1$ and the exponent of a is $n - (r - 1)$. Thus,

Fourth term:

$$\binom{20}{3} x^{17}(-2)^3 = \frac{20 \cdot 19 \cdot 18}{3 \cdot 2 \cdot 1} x^{17}(-8) = -9{,}120x^{17}$$

Sixteenth term:

$$\binom{20}{15} x^5(-2)^{15} = \frac{20 \cdot 19 \cdot 18 \cdot 17 \cdot 16}{5 \cdot 4 \cdot 3 \cdot 2 \cdot 1} x^5(-32{,}768) = -508{,}035{,}072x^5$$

MATCHED 4 PROBLEM

If the terms in the expansion of $(u - 1)^{18}$ are arranged in decreasing powers of u, find the fifth term and the twelfth term.

Proof of the Binomial Formula

We now proceed to prove that the binomial formula holds for all natural numbers n using mathematical induction.

PROOF State the conjecture.

$$P_n: \quad (a+b)^n = \sum_{j=0}^{n} \binom{n}{j} a^{n-j} b^j$$

PART 1 Show that P_1 is true.

$$\sum_{j=0}^{1}\binom{1}{j}a^{1-j}b^{j} = \binom{1}{0}a + \binom{1}{1}b = a + b = (a + b)^1$$

Thus, P_1 is true.

PART 2 Show that if P_k is true, then P_{k+1} is true.

$$P_k\text{: } (a + b)^k = \sum_{j=0}^{k}\binom{k}{j}a^{k-j}b^{j} \qquad \textbf{Assume } P_k \textbf{ is true.}$$

$$P_{k+1}\text{: } (a + b)^{k+1} = \sum_{j=0}^{k+1}\binom{k+1}{j}a^{k+1-j}b^{j} \qquad \textbf{Show } P_{k+1} \textbf{ is true.}$$

We begin by multiplying both sides of P_k by $(a + b)$:

$$(a + b)^k(a + b) = \left[\sum_{j=0}^{k}\binom{k}{j}a^{k-j}b^{j}\right](a + b)$$

The left side of this equation is the left side of P_{k+1}. Now we multiply out the right side of the equation and try to obtain the right side of P_{k+1}:

$$\begin{aligned}
(a + b)^{k+1} &= \left[\binom{k}{0}a^k + \binom{k}{1}a^{k-1}b + \binom{k}{2}a^{k-2}b^2 + \cdots + \binom{k}{k}b^k\right](a + b)\\
&= \left[\binom{k}{0}a^{k+1} + \binom{k}{1}a^{k}b + \binom{k}{2}a^{k-1}b^2 + \cdots + \binom{k}{k}ab^k\right]\\
&\quad + \left[\binom{k}{0}a^{k}b + \binom{k}{1}a^{k-1}b^2 + \cdots + \binom{k}{k-1}ab^k + \binom{k}{k}b^{k+1}\right]\\
&= \binom{k}{0}a^{k+1} + \left[\binom{k}{0} + \binom{k}{1}\right]a^{k}b + \left[\binom{k}{1} + \binom{k}{2}\right]a^{k-1}b^2 + \cdots\\
&\quad + \left[\binom{k}{k-1} + \binom{k}{k}\right]ab^k + \binom{k}{k}b^{k+1}
\end{aligned}$$

We now use the following facts (the proofs are left as exercises; see Problems 59–61, Exercise 10.6).

$$\binom{k}{r-1} + \binom{k}{r} = \binom{k+1}{r} \qquad \binom{k}{0} = \binom{k+1}{0} \qquad \binom{k}{k} = \binom{k+1}{k+1}$$

to rewrite the right side as

$$\begin{aligned}
&\binom{k+1}{0}a^{k+1} + \binom{k+1}{1}a^{k}b + \binom{k+1}{2}a^{k-1}b^2 + \cdots\\
&\qquad + \binom{k+1}{k}ab^k + \binom{k+1}{k+1}b^{k+1} = \sum_{j=0}^{k+1}\binom{k+1}{j}a^{k+1-j}b^{j}
\end{aligned}$$

Because the right side of the last equation is the right side of P_{k+1}, we have shown that P_{k+1} follows from P_k.

CONCLUSION P_n is true. That is, the binomial formula holds for all positive integers n.

ANSWERS MATCHED PROBLEMS

1. $x^5 + 5x^4 + 10x^3 + 10x^2 + 5x + 1$ **2.** $8m^3 - 60m^2n + 150mn^2 - 125n^3$ **3.** $192{,}192y^8$ **4.** $3{,}060u^{14}$; $-31{,}824u^7$

EXERCISE 10.6

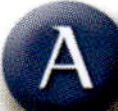

In Problems 1–8, use Pascal's triangle to evaluate each expression.

1. $\binom{5}{3}$

2. $\binom{6}{4}$

3. $\binom{4}{2}$

4. $\binom{7}{5}$

5. $C_{6,3}$

6. $C_{5,3}$

7. $C_{7,4}$

8. $C_{4,3}$

In Problems 9–16, use a graphing utility to evaluate each expression.

9. $\binom{9}{3}$

10. $\binom{10}{6}$

11. $\binom{12}{10}$

12. $\binom{13}{8}$

13. $\binom{17}{13}$

14. $\binom{20}{16}$

15. $\binom{50}{4}$

16. $\binom{50}{45}$

Expand Problems 17–28 using the binomial formula.

17. $(m + n)^3$ **18.** $(x + 2)^3$ **19.** $(2x - 3y)^3$

20. $(3u + 2v)^3$ **21.** $(x - 2)^4$ **22.** $(x - y)^4$

23. $(m + 3n)^4$ **24.** $(3p - q)^4$ **25.** $(2x - y)^5$

26. $(2x - 1)^5$ **27.** $(m + 2n)^6$ **28.** $(2x - y)^6$

B

In Problems 29–38, find the term of the binomial expansion containing the given power of x.

29. $(x + 1)^7$; x^4 **30.** $(x + 1)^8$; x^5 **31.** $(2x - 1)^{11}$; x^6

32. $(3x + 1)^{12}$; x^7 **33.** $(2x + 3)^{18}$; x^{14} **34.** $(3x - 2)^{17}$; x^5

35. $(x^2 - 1)^6$; x^8 **36.** $(x^2 - 1)^9$; x^7 **37.** $(x^2 + 1)^9$; x^{11}

38. $(x^2 + 1)^{10}$; x^{14}

In Problems 39–46, find the indicated term in each expansion if the terms of the expansion are arranged in decreasing powers of the first term in the binomial.

39. $(u + v)^{15}$; seventh term

40. $(a + b)^{12}$; fifth term

41. $(2m + n)^{12}$; eleventh term

42. $(x + 2y)^{20}$; third term

43. $[(w/2) - 2]^{12}$; seventh term

44. $(x - 3)^{10}$; fourth term

45. $(3x - 2y)^8$; sixth term

46. $(2p - 3q)^7$; fourth term

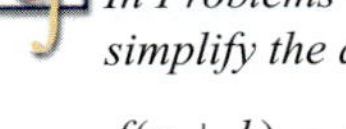

In Problems 47–50, use the binomial formula to expand and simplify the difference quotient

$$\frac{f(x + h) - f(x)}{h}$$

for the indicated function f. Discuss the behavior of the simplified form as h approaches 0.

47. $f(x) = x^3$

48. $f(x) = x^4$

49. $f(x) = x^5$

50. $f(x) = x^6$

In Problems 51–54, use a graphing utility to graph each sequence and to display it in table form.

51. Find the number of terms of the sequence

$$\binom{20}{0}, \binom{20}{1}, \binom{20}{2}, \ldots, \binom{20}{20}$$

that are greater than one-half of the largest term.

52. Find the number of terms of the sequence

$$\binom{40}{0}, \binom{40}{1}, \binom{40}{2}, \ldots, \binom{40}{40}$$

that are greater than one-half of the largest term.

53. **(A)** Find the largest term of the sequence $a_0, a_1, a_2, \ldots, a_{10}$ to three decimal places, where

$$a_k = \binom{10}{k}(0.6)^{10-k}(0.4)^k$$

(B) According to the binomial formula, what is the sum of the series $a_0 + a_1 + a_2 + \cdots + a_{10}$?

54. **(A)** Find the largest term of the sequence $a_0, a_1, a_2, \ldots, a_{10}$ to three decimal places, where

$$a_k = \binom{10}{k}(0.3)^{10-k}(0.7)^k$$

(B) According to the binomial formula, what is the sum of the series $a_0 + a_1 + a_2 + \cdots + a_{10}$?

C

55. Evaluate $(1.01)^{10}$ to four decimal places, using the binomial formula. [*Hint:* Let $1.01 = 1 + 0.01$.]

56. Evaluate $(0.99)^6$ to four decimal places, using the binomial formula.

57. Show that: $\binom{n}{r} = \binom{n}{n - r}$

58. Show that: $\binom{n}{0} = \binom{n}{n}$

59. Show that: $\binom{k}{r - 1} + \binom{k}{r} = \binom{k + 1}{r}$

60. Show that: $\binom{k}{0} = \binom{k + 1}{0}$

61. Show that: $\binom{k}{k} = \binom{k + 1}{k + 1}$

62. Show that: $\binom{n}{r}$ is given by the recursion formula

$$\binom{n}{r} = \frac{n - r + 1}{r}\binom{n}{r - 1}$$

where $\binom{n}{0} = 1$.

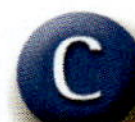

63. Write $2^n = (1 + 1)^n$ and expand, using the binomial formula to obtain

$$2^n = \binom{n}{0} + \binom{n}{1} + \binom{n}{2} + \cdots + \binom{n}{n}$$

64. Write $0 = (1 - 1)^n$ and expand, using the binomial formula, to obtain

$$0 = \binom{n}{0} - \binom{n}{1} + \binom{n}{2} - \cdots + (-1)^n\binom{n}{n}$$

10.1 Sequences and Series

A **sequence** is a function with the domain a set of successive integers. The symbol a_n, called the ***n*th term,** or **general term,** represents the range value associated with the domain value n. Unless specified otherwise, the domain is understood to be the set of natural numbers. A **finite sequence** has a finite domain, and an **infinite sequence** has an infinite domain. A **recursion formula** defines each term of a sequence in terms of one or more of the preceding terms. For example, the **Fibonacci sequence** is defined by $a_n = a_{n-1} + a_{n-2}$ for $n \geq 3$, where $a_1 = a_2 = 1$. If $a_1, a_2, \ldots, a_n, \ldots$ is a sequence, then the expression $a_1 + a_2 + \cdots + a_n + \cdots$ is called a **series.** A finite sequence produces a **finite series,** and an infinite sequence produces an **infinite series.** Series can be represented using the summation notation:

$$\sum_{k=m}^{n} a_k = a_m + a_{m+1} + \cdots + a_n$$

where k is called the **summing index.** If the terms in the series are alternately positive and negative, the series is called an **alternating series.**

10.2 Mathematical Induction

A wide variety of statements can be proven using the **principle of mathematical induction:** Let P_n be a statement associated with each positive integer n and suppose the following conditions are satisfied:

1. P_1 is true.
2. For any positive integer k, if P_k is true, then P_{k+1} is also true.

Then the statement P_n is true for all positive integers n.

To use mathematical induction to prove statements involving laws of exponents, it is convenient to state a **recursive definition of a^n:**

$$a^1 = a \qquad \text{and} \qquad a^{n+1} = a^n a \quad \text{for any integer } n > 1$$

To deal with conjectures that may be true only for $n \geq m$, where m is a positive integer, we use the **extended principle of mathematical induction:** Let m be a positive integer, let P_n be a statement associated with each integer $n \geq m$, and suppose the following conditions are satisfied:

1. P_m is true.
2. For any integer $k \geq m$, if P_k is true, then P_{k+1} is also true.

Then the statement P_n is true for all integers $n \geq m$.

10.3 Arithmetic and Geometric Sequences

A sequence is called an **arithmetic sequence,** or **arithmetic progression,** if there exists a constant d, called the **common difference,** such that

$$a_n - a_{n-1} = d \qquad \text{or} \qquad a_n = a_{n-1} + d$$
$$\text{for every } n > 1$$

The following formulas are useful when working with arithmetic sequences and their corresponding series:

$a_n = a_1 + (n - 1)d$ — ***n*th-Term Formula**

$S_n = \frac{n}{2}[2a_1 + (n - 1)d]$ — **Sum Formula—First Form**

$S_n = \frac{n}{2}(a_1 + a_n)$ — **Sum Formula—Second Form**

A sequence is called a **geometric sequence,** or a **geometric progression,** if there exists a nonzero constant r, called the **common ratio,** such that

$$\frac{a_n}{a_{n-1}} = r \qquad \text{or} \qquad a_n = ra_{n-1} \qquad \text{for every } n > 1$$

The following formulas are useful when working with geometric sequences and their corresponding series:

$a_n = a_1 r^{n-1}$ — ***n*th-Term Formula**

$S_n = \frac{a_1 - a_1 r^n}{1 - r} \quad r \neq 1$ — **Sum Formula—First Form**

$S_n = \frac{a_1 - ra_n}{1 - r} \quad r \neq 1$ — **Sum Formula—Second Form**

$S_\infty = \frac{a_1}{1 - r} \quad |r| < 1$ — **Sum of an Infinite Geometric Series**

10.4 Multiplication Principle, Permutations, and Combinations

Given a sequence of operations, **tree diagrams** are often used to list all the possible combined outcomes. To count the number of combined outcomes without actually listing them, we use the **multiplication principle:**

1. If operations O_1 and O_2 are performed in order with N_1 possible outcomes for the first operation and N_2 possible outcomes for the second operation, then there are

$$N_1 \cdot N_2$$

possible outcomes of the first operation followed by the second.
2. In general, if n operations $O_1, O_2, \ldots, O_n$ are performed in order, with possible number of outcomes $N_1, N_2, \ldots, N_n$, respectively, then there are

$$N_1 \cdot N_2 \cdot \cdots \cdot N_n$$

possible combined outcomes of the operations performed in the given order.

A particular arrangement or ordering of n objects without repetition is called a **permutation.** The number of permutations of n objects is given by

$$P_{n,n} = n \cdot (n-1) \cdot \cdots \cdot 1 = n!$$

and the number of permutations of n objects taken r at a time is given by

$$P_{n,r} = \frac{n!}{(n-r)!} \qquad 0 \le r \le n$$

A **combination of a set of n elements taken r at a time** is an r-element subset of the n objects. The number of combinations of n objects taken r at a time is given by

$$C_{n,r} = \binom{n}{r} = \frac{P_{n,r}}{r!} = \frac{n!}{r!(n-r)!} \qquad 0 \le r \le n$$

In a permutation, order is important. In a combination, order is not important.

10.5 Sample Spaces and Probability

The outcomes of an experiment are called **simple events** if one and only one of these results will occur in each trial of the experiment. The set of all simple events is called the **sample space.** Any subset of the sample space is called an **event.** An event is a **simple event** if it has only one element in it and a **compound event** if it has more than one element in it. We say that **an event E occurs** if any of the simple events in E occurs. A sample space S_1 is **more fundamental** than a second sample space S_2 if knowledge of which event occurs in S_1 tells us which event in S_2 occurs, but not conversely.

Given a sample space $S = \{e_1, e_2, \ldots, e_n\}$ with n simple events, to each simple event e_i we assign a real number, denoted by $P(e_i)$, that is called the **probability of the event e_i** and satisfies:

1. $0 \le P(e_i) \le 1$
2. $P(e_1) + P(e_2) + \cdots + P(e_n) = 1$

Any probability assignment that meets conditions 1 and 2 is said to be an **acceptable probability assignment.**

Given an acceptable probability assignment for the simple events in a sample space S, the **probability of an arbitrary event E** is defined as follows:

1. If E is the empty set, then $P(E) = 0$.
2. If E is a simple event, then $P(E)$ has already been assigned.
3. If E is a compound event, then $P(E)$ is the sum of the probabilities of all the simple events in E.
4. If E is the sample space S, then $P(E) = P(S) = 1$.

If each of the simple events in a sample space $S = \{e_1, e_2, \ldots, e_n\}$ with n simple events is **equally likely** to occur, then we assign the probability $1/n$ to each. If E is an arbitrary event in S, then

$$P(E) = \frac{\text{Number of elements in } E}{\text{Number of elements in } S} = \frac{n(E)}{n(S)}$$

If we conduct an experiment n times and event E occurs with **frequency** $f(E)$, then the ratio $f(E)/n$ is called the **relative frequency** of the occurrence of event E in n trials. As n increases, $f(E)/n$ usually approaches a number that is called the **empirical probability** $P(E)$. Thus, $f(E)/n$ is used as an **approximate empirical probability** for $P(E)$.

If $P(E)$ is the theoretical probability of an event E and the experiment is performed n times, then the **expected frequency** of the occurrence of E is $n \cdot P(E)$.

The command **randInt(*i*,*k*,*n*)** on a Texas Instruments TI-83 generates a random sequence of n integers between i and k, inclusively, that can be used to simulate repeated trials of experiments.

10.6 Binomial Formula

Pascal's triangle is a triangular array of coefficients for the expansion of the binomial $(a+b)^n$, where n is a positive integer. New notation for the combination formula is

$$\binom{n}{r} = C_{n,r} = \frac{n!}{r!(n-r)!}$$

For n a positive integer, the **binomial formula** is

$$(a+b)^n = \sum_{k=0}^{n} \binom{n}{k} a^{n-k} b^k$$

The numbers $\binom{n}{k}$, $0 \le k \le n$, are called **binomial coefficients.**

CHAPTER 10 REVIEW EXERCISES

Work through all the problems in this chapter review and check answers in the back of the book. Answers to all review problems are there, and following each answer is a number in italics indicating the section in which that type of problem is discussed. Where weaknesses show up, review appropriate sections in the text.

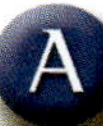

1. Determine whether each of the following can be the first three terms of a geometric sequence, an arithmetic sequence, or neither.
(A) $16, -8, 4, \ldots$ **(B)** $5, 7, 9, \ldots$
(C) $-8, -5, -2, \ldots$ **(D)** $2, 3, 5, \ldots$
(E) $-1, 2, -4, \ldots$

In Problems 2–5:
(A) Write the first four terms of each sequence.
(B) Find a_{10}. *(C) Find S_{10}.*

2. $a_n = 2n + 3$ **3.** $a_n = 32(\frac{1}{2})^n$

4. $a_1 = -8; a_n = a_{n-1} + 3, n \geq 2$

5. $a_1 = -1; a_n = (-2)a_{n-1}, n \geq 2$

6. Find S_∞ in Problem 3.

Evaluate Problems 7–10.

7. $6!$ **8.** $\dfrac{22!}{19!}$

9. $\dfrac{7!}{2!(7-2)!}$ **10.** $C_{6,2}$ and $P_{6,2}$

11. A single die is rolled and a coin is flipped. How many combined outcomes are possible? Solve
(A) By using a tree diagram
(B) By using the multiplication principle

12. How many seating arrangements are possible with six people and six chairs in a row? Solve by using the multiplication principle.

13. Solve Problem 12 using permutations or combinations, whichever is applicable.

14. In a single deal of 5 cards from a standard 52-card deck, what is the probability of being dealt five clubs?

15. Betty and Bill are members of a 15-person ski club. If the president and treasurer are selected by lottery, what is the probability that Betty will be president and Bill will be treasurer? A person cannot hold more than one office.

16. A drug has side effects for 50 out of 1,000 people in a test. What is the approximate empirical probability that a person using the drug will have side effects?

Verify Problems 17–19 for n = 1, 2, and 3.

17. P_n: $5 + 7 + 9 + \cdots + (2n + 3) = n^2 + 4n$

18. P_n: $2 + 4 + 8 + \cdots + 2^n = 2^{n+1} - 2$

19. P_n: $49^n - 1$ is divisible by 6

In Problems 20–22, write P_k and P_{k+1}.

20. For P_n in Problem 17 **21.** For P_n in Problem 18

22. For P_n in Problem 19

23. Either prove the statement is true or prove it is false by finding a counterexample: If n is a positive integer, then the sum of the series $1 + \frac{1}{2} + \frac{1}{3} + \cdots + \frac{1}{n}$ is less than 4.

B

Write Problems 24 and 25 without summation notation, and find the sum.

24. $S_{10} = \sum_{k=1}^{10} (2k - 8)$

25. $S_7 = \sum_{k=1}^{7} \frac{16}{2^k}$

26. $S_\infty = 27 - 18 + 12 + \cdots = ?$

27. Write

$$S_n = \frac{1}{3} - \frac{1}{9} + \frac{1}{27} + \cdots + \frac{(-1)^{n+1}}{3^n}$$

using summation notation, and find S_∞.

28. Someone tells you that the following approximate empirical probabilities apply to the sample space $\{e_1, e_2, e_3, e_4\}$: $P(e_1) \approx .1$, $P(e_2) \approx -.2$, $P(e_3) \approx .6$, $P(e_4) \approx 2$. There are three reasons why P cannot be a probability function. Name them.

29. Six distinct points are selected on the circumference of a circle. How many triangles can be formed using these points as vertices?

30. In an arithmetic sequence, $a_1 = 13$ and $a_7 = 31$. Find the common difference d and the fifth term a_5.

31. How many three-letter code words are possible using the first eight letters of the alphabet if no letter can be repeated? If letters can be repeated? If adjacent letters cannot be alike?

32. Two coins are flipped 1,000 times with the following frequencies:

Two heads:	210
One head:	480
Zero heads:	310

(A) Compute the empirical probability for each outcome.
(B) Compute the theoretical probability for each outcome.
(C) Using the theoretical probabilities computed in part B, compute the expected frequency of each outcome, assuming fair coins.

33. From a standard deck of 52 cards, what is the probability of obtaining a 5-card hand:
(A) Of all diamonds?
(B) Of three diamonds and two spades?
Write answers in terms of $C_{n,r}$ or $P_{n,r}$, as appropriate. Do not evaluate.

34. A group of 10 people includes one married couple. If four people are selected at random, what is the probability that the married couple is selected?

35. A spinning device has three numbers, 1, 2, 3, each as likely to turn up as the other. If the device is spun twice, what is the probability that:
(A) The same number turns up both times?
(B) The sum of the numbers turning up is 5?

36. Use the formula for the sum of an infinite geometric series to write $0.727\,272 \cdots = 0.\overline{72}$ as the quotient of two integers.

37. Solve the following problems using $P_{n,r}$ or $C_{n,r}$, as appropriate:
(A) How many three-digit opening combinations are possible on a combination lock with six digits if the digits cannot be repeated?
(B) Suppose five tennis players have made the finals. If each of the five players is to play every other player exactly once, how many games must be scheduled?

Evaluate Problems 38–40.

38. $\frac{20!}{18!(20-18)!}$ **39.** $\binom{16}{12}$ **40.** $\binom{11}{11}$

41. Expand $(x - y)^5$ using the binomial formula.

42. Find the term containing x^6 in the expansion of $(x + 2)^9$.

43. If the terms in the expansion of $(2x - y)^{12}$ are arranged in descending powers of x, find the tenth term.

Establish each statement in Problems 44–46 for all natural numbers, using mathematical induction.

44. P_n in Problem 17

45. P_n in Problem 18

46. P_n in Problem 19

In Problems 47 and 48, find the smallest positive integer n such that $a_n < b_n$ by graphing the sequences $\{a_n\}$ and $\{b_n\}$ with a graphing utility. Check your answer by using a graphing utility to display both sequences in table form.

47. $a_n = C_{50,n}, b_n = 3^n$

48. $a_1 = 100, a_n = 0.99a_{n-1} + 5, b_n = 9 + 7n$

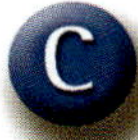

49. How many different families with five children are possible, excluding multiple births, where the sex of each child in the order of their birth is taken into consideration? How many families are possible if the order pattern is not taken into account?

50. A free-falling body travels $g/2$ feet in the first second, $3g/2$ feet during the next second, $5g/2$ feet the next, and so on. Find the distance fallen during the twenty-fifth second and the total distance fallen from the start to the end of the twenty-fifth second.

51. How many ways can two people be seated in a row of four chairs?

52. Expand $(x + i)^6$, where i is the imaginary unit, using the binomial formula.

53. If three people are selected from a group of seven men and three women, what is the probability that at least one woman is selected?

54. Three fair coins are tossed 1,000 times with the following frequencies of outcomes:

Number of heads	0	1	2	3
Frequency	120	360	350	170

(A) What is the approximate empirical probability of obtaining two heads?

(B) What is the theoretical probability of obtaining two heads?

(C) What is the expected frequency of obtaining two heads?

Prove that each statement in Problems 55–59 holds for all positive integers, using mathematical induction.

55. $\sum_{k=1}^{n} k^3 = \left(\sum_{k=1}^{n} k\right)^2$

56. $x^{2n} - y^{2n}$ is divisible by $x - y, x \neq y$

57. $\dfrac{a^n}{a^m} = a^{n-m}$; $n > m$, n, m positive integers

58. $\{a_n\} = \{b_n\}$, where $a_n = a_{n-1} + 2, a_1 = -3, b_n = -5 + 2n$

59. $(1!)1 + (2!)2 + (3!)3 + \cdots + (n!)n = (n + 1)! - 1$
(From U.S.S.R. Mathematical Olympiads, 1955–1956, Grade 10.)

APPLICATIONS

60. Loan Repayment. You borrow \$7,200 and agree to pay 1% of the unpaid balance each month for interest. If you decide to pay an additional \$300 each month to reduce the unpaid balance, how much interest will you pay over the 24 months it will take to repay this loan?

61. Economics. Due to reduced taxes, an individual has an extra \$2,400 in spendable income. If we assume that the individual spends 75% of this on consumer goods, and the producers of those consumer goods in turn spend 75% on consumer goods, and that this process continues indefinitely, what is the total amount (to the nearest dollar) spent on consumer goods?

62. Compound Interest. If \$500 is invested at 6% compounded annually, the amount A present after n years forms a geometric sequence with common ratio $1 + 0.06 = 1.06$. Use a geometric sequence formula to find the amount A in the account (to the nearest cent) after 10 years. After 20 years.

63. Transportation. A distribution center A wishes to distribute its products to five different retail stores, B, C, D, E, and F, in a city. How many different route plans can be constructed so that a single truck can start from A, deliver to each store exactly once, and then return to the center?

64. Market Analysis. A videocassette company selected 1,000 persons at random and surveyed them to determine a relationship between age of purchaser and annual videocassette purchases. The results are given in the table.

	Cassettes Purchased Annually				
Age	**0**	**1**	**2**	**Above 2**	**Totals**
Under 12	60	70	30	10	170
12–18	30	100	100	60	290
19–25	70	110	120	30	330
Over 25	100	50	40	20	210
Totals	260	330	290	120	1,000

Find the empirical probability that a person selected at random

(A) Is over 25 *and* buys exactly two cassettes annually.
(B) Is 12–18 years old *and* buys more than one cassette annually.
(C) Is 12–18 years old *or* buys more than one cassette annually.

★ **65. Quality Control.** Twelve precision parts, including two that are substandard, are sent to an assembly plant. The plant manager selects four at random and will return the whole shipment if one or more of the sample are found to be substandard. What is the probability that the shipment will be returned?

Sequences Specified by Recursion Formulas

The recursion formula* $a_n = 5a_{n-1} - 6a_{n-2}$, together with the initial values $a_1 = 4$, $a_2 = 14$, specifies the sequence $\{a_n\}$ whose first several terms are 4, 14, 46, 146, 454, 1394, ... The sequence $\{a_n\}$ is neither arithmetic nor geometric. Nevertheless, because it satisfies a simple recursion formula, it is possible to obtain an nth-term formula for $\{a_n\}$ that is analogous to the nth-term formulas for arithmetic and geometric sequences. Such an nth-term formula is valuable because it allows us to estimate a term of a sequence without computing all the preceding terms.

If the geometric sequence $\{r^n\}$ satisfies the recursion formula above, then $r^n = 5r^{n-1} - 6r^{n-2}$. Dividing both sides by r^{n-2} leads to the quadratic equation $r^2 - 5r + 6 = 0$, whose solutions are $r = 2$ and $r = 3$. Now it is easy to check that the geometric sequences $\{2^n\} = 2, 4, 8, 16, \dots$ and $\{3^n\} = 3, 9, 27, 81, \dots$ satisfy the recursion formula. Therefore, any sequence of the form $\{u2^n + v3^n\}$, where u and v are constants, will satisfy the same recursion formula.

We now find u and v so that the first two terms of $\{u2^n + v3^n\}$ are $a_1 = 4$, $a_2 = 14$. Letting $n = 1$ and $n = 2$ we see that u and v must satisfy the following linear system:

$$\begin{aligned} 2u + 3v &= 4 \\ 4u + 9v &= 14 \end{aligned}$$

Solving the system gives $u = -1$, $v = 2$. Therefore, an nth-term formula for the original sequence is $a_n = (-1)2^n + (2)3^n$.

Note that the nth-term formula was obtained by solving a quadratic equation and a system of two linear equations in two variables.

(A) Compute $(-1)2^n + (2)3^n$ for $n = 1, 2, \dots, 6$, and compare with the terms of $\{a_n\}$.

(B) Estimate the one-hundredth term of $\{a_n\}$.

(C) Show that any sequence of the form $\{u2^n + v3^n\}$, where u and v are constants, satisfies the recursion formula $a_n = 5a_{n-1} - 6a_{n-2}$.

(D) Find an nth-term formula for the sequence $\{b_n\}$ that is specified by $b_1 = 5$, $b_2 = 55$, $b_n = 3b_{n-1} + 4b_{n-2}$.

(E) Find an nth-term formula for the Fibonacci sequence.

(F) Find an nth-term formula for the sequence $\{c_n\}$ that is specified by $c_1 = -3$, $c_2 = 15$, $c_3 = 99$, $c_n = 6c_{n-1} - 3c_{n-2} - 10c_{n-3}$. (Because the recursion formula involves the three terms that precede c_n, our method will involve the solution of a cubic equation and a system of three linear equations in three variables.)

---

*The program RECUR, found at the website for this book, evaluates the terms in any sequence defined by this type of recursion formula.

CHAPTER 11

Additional Topics in Analytic Geometry

OUTLINE

ANALYTIC GEOMETRY, A UNION OF GEOMETRY AND ALGEBRA, enables us to analyze certain geometric concepts algebraically and to interpret certain algebraic relationships geometrically. Our two main concerns center on graphing algebraic equations and finding equations of useful geometric figures. We have discussed a number of topics in analytic geometry, such as straight lines and circles, in earlier chapters. In Chapter 11 we discuss additional analytic geometry topics: conic sections, translation of axes, and systems of quadratic equations.

René Descartes (1596–1650), the French philosopher–mathematician, is generally recognized as the founder of analytic geometry.

Preparing for this chapter

Before getting started on this chapter, review the following concepts:

- Graphs and Transformations (Chapter 1, Section 4)
- Cartesian Coordinate System (Appendix A, Section A.2)
- Basic Formulas in Analytic Geometry (Appendix A, Section A.3)
- Linear Functions (Chapter 2, Section 1)
- Quadratic Functions (Chapter 2, Section 3)
- Quadratic Equations (Chapter 2, Section 5)
- Equation-Solving Techniques (Chapter 2, Section 6)
- Asymptotes (Chapter 3, Section 4)

SECTION 11.1 Conic Sections; Parabola

Conic Sections • Definition of a Parabola • Drawing a Parabola • Standard Equations and Their Graphs • Applications

In Section 11.1 we introduce the general concept of a conic section and then discuss the particular conic section called a *parabola*. In Sections 11.2 and 11.3 we will discuss two other conic sections called *ellipses* and *hyperbolas*.

Conic Sections

In Section 2.1 we found that the graph of a first-degree equation in two variables,

$$Ax + By = C \tag{1}$$

where A and B are not both 0, is a straight line, and every straight line in a rectangular coordinate system has an equation of this form. What kind of graph will a second-degree equation in two variables,

$$Ax^2 + Bxy + Cy^2 + Dx + Ey + F = 0 \tag{2}$$

where A, B, and C are not all 0, yield for different sets of values of the coefficients? The graphs of equation (2) for various choices of the coefficients are plane curves obtainable by intersecting a cone* with a plane, as shown in Figure 1. These curves are called **conic sections.**

FIGURE 1 Conic sections.

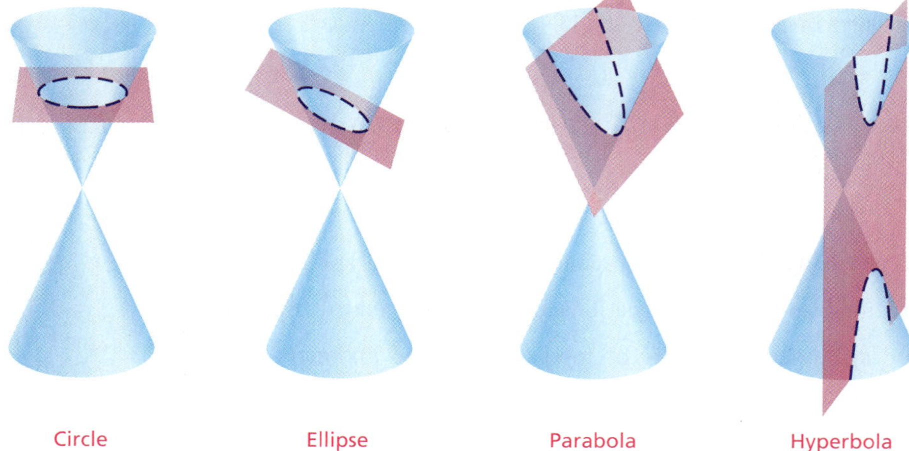

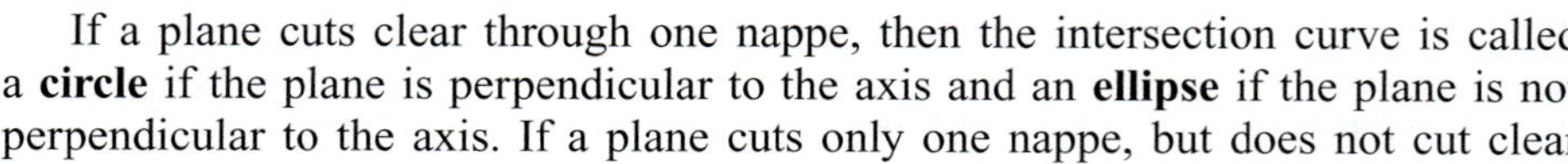

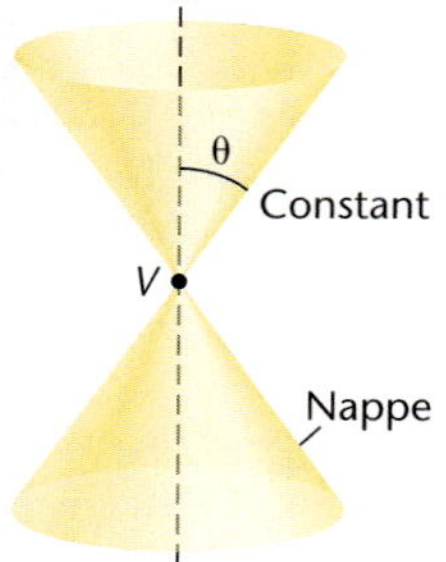

If a plane cuts clear through one nappe, then the intersection curve is called a **circle** if the plane is perpendicular to the axis and an **ellipse** if the plane is not perpendicular to the axis. If a plane cuts only one nappe, but does not cut clear

*Starting with a fixed line L and a fixed point V on L, the surface formed by all straight lines through V making a constant angle θ with L is called a **right circular cone.** The fixed line L is called the **axis** of the cone, and V is its **vertex.** The two parts of the cone separated by the vertex are called **nappes.**

through, then the intersection curve is called a **parabola.** Finally, if a plane cuts through both nappes, but not through the vertex, the resulting intersection curve is called a **hyperbola.** A plane passing through the vertex of the cone produces a **degenerate conic**—a point, a line, or a pair of lines.

Conic sections are very useful and are readily observed in your immediate surroundings: wheels (circle), the path of water from a garden hose (parabola), some serving platters (ellipses), and the shadow on a wall from a light surrounded by a cylindrical or conical lamp shade (hyperbola) are some examples (Fig. 2). We will discuss many applications of conics throughout the remainder of this chapter.

FIGURE 2 Examples of conics.

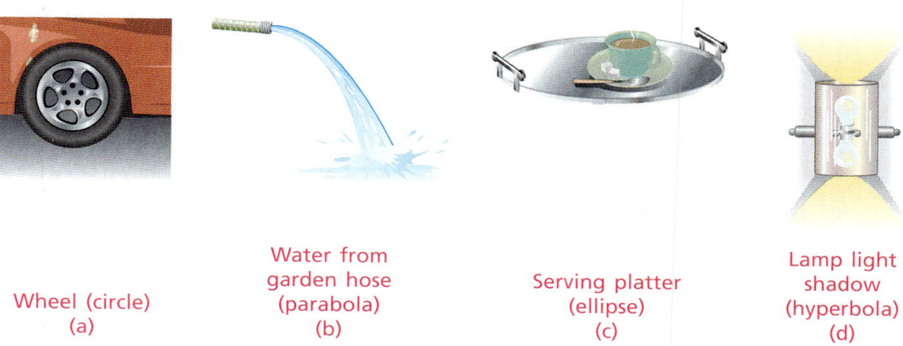

A definition of a conic section that does not depend on the coordinates of points in any coordinate system is called a **coordinate-free definition.** In Appendix A, Section A.3 we gave a coordinate-free definition of a circle and developed its standard equation in a rectangular coordinate system. In Sections 11.1, 11.2, and 11.3 we will give coordinate-free definitions of a parabola, ellipse, and hyperbola, and we will develop standard equations for each of these conics in a rectangular coordinate system.

Definition of a Parabola

The following definition of a parabola does not depend on the coordinates of points in any coordinate system:

DEFINITION 1
Parabola

A **parabola** is the set of all points in a plane equidistant from a fixed point F and a fixed line L in the plane. The fixed point F is called the **focus,** and the fixed line L is called the **directrix.** A line through the focus perpendicular to the directrix is called the **axis,** and the point on the axis halfway between the directrix and focus is called the **vertex.**

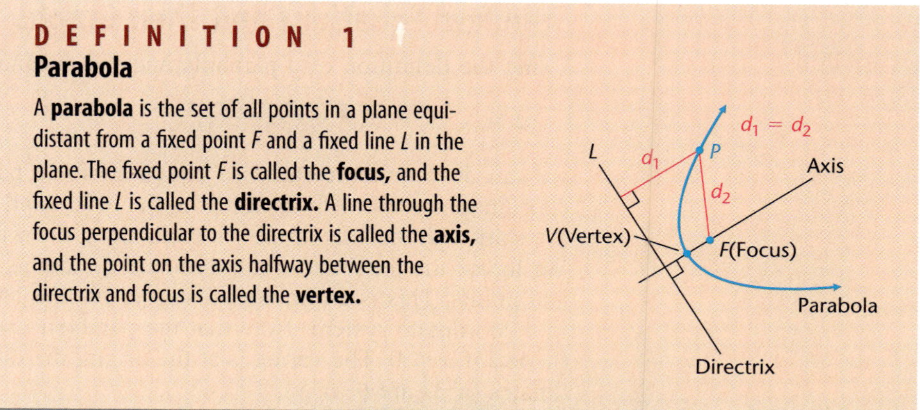

Drawing a Parabola

Using Definition 1, we can draw a parabola with fairly simple equipment—a straightedge, a right-angle drawing triangle, a piece of string, a thumbtack, and a pencil. Referring to Figure 3, tape the straightedge along the line AB and place the thumbtack above the line AB. Place one leg of the triangle along the straightedge as indicated, then take a piece of string the same length as the other leg, tie one end to the thumbtack, and fasten the other end with tape at C on the triangle. Now press the string to the edge of the triangle, and keeping the string taut, slide the triangle along the straightedge. Because DE will always equal DF, the resulting curve will be part of a parabola with directrix AB lying along the straightedge and focus F at the thumbtack.

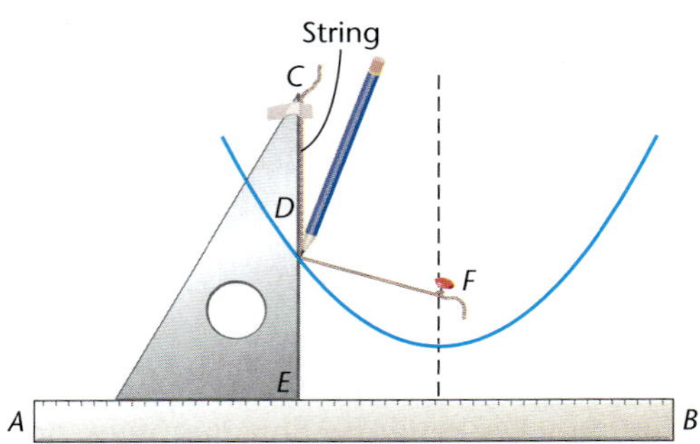

FIGURE 3 Drawing a parabola.

EXPLORE/DISCUSS 1

The line through the focus F that is perpendicular to the axis of a parabola intersects the parabola in two points G and H. Explain why the distance from G to H is twice the distance from F to the directrix of the parabola.

Standard Equations and Their Graphs

Using the definition of a parabola and the distance-between-two-points formula

$$d = \sqrt{(x_2 - x_1)^2 + (y_2 - y_1)^2} \tag{3}$$

we can derive simple standard equations for a parabola located in a rectangular coordinate system with its vertex at the origin and its axis along a coordinate axis. We start with the axis of the parabola along the x axis and the focus at $F = (a, 0)$. We locate the parabola in a coordinate system as in Figure 4 and label key lines and points. This is an important step in finding an equation of a geometric figure in a coordinate system. Note that the parabola opens to the right if $a > 0$ and to the left if $a < 0$. The vertex is at the origin, the directrix is $x = -a$, and the coordinates of M are $(-a, y)$.

FIGURE 4 Parabola with center at the origin and axis the x axis.

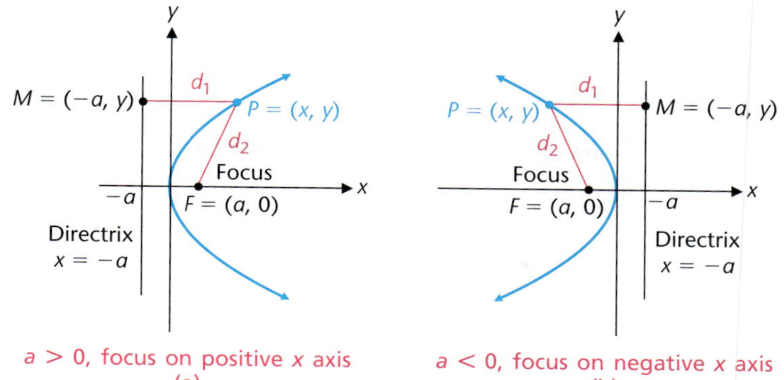

The point $P = (x, y)$ is a point on the parabola if and only if

$$\begin{aligned} d_1 &= d_2 \\ d(P, M) &= d(P, F) \\ \sqrt{(x + a)^2 + (y - y)^2} &= \sqrt{(x - a)^2 + (y - 0)^2} && \text{Use equation (3).} \\ (x + a)^2 &= (x - a)^2 + y^2 && \text{Square both sides.} \\ x^2 + 2ax + a^2 &= x^2 - 2ax + a^2 + y^2 && \text{Simplify.} \\ y^2 &= 4ax \end{aligned} \tag{4}$$

Equation (4) is the standard equation of a parabola with vertex at the origin, axis the x axis, and focus at $(a, 0)$.

Now we locate the vertex at the origin and focus on the y axis at $(0, a)$. Looking at Figure 5, we note that the parabola opens upward if $a > 0$ and downward if $a < 0$. The directrix is $y = -a$, and the coordinates of N are $(x, -a)$. The point $P = (x, y)$ is a point on the parabola if and only if

$$\begin{aligned} d_1 &= d_2 \\ d(P, N) &= d(P, F) \\ \sqrt{(x - x)^2 + (y + a)^2} &= \sqrt{(x - 0)^2 + (y - a)^2} && \text{Use equation (3).} \\ (y + a)^2 &= x^2 + (y - a)^2 && \text{Square both sides.} \\ y^2 + 2ay + a^2 &= x^2 + y^2 - 2ay + a^2 && \text{Simplify.} \\ x^2 &= 4ay \end{aligned} \tag{5}$$

Equation (5) is the standard equation of a parabola with vertex at the origin, axis the y axis, and focus at $(0, a)$.

FIGURE 5 Parabola with center at the origin and axis the y axis.

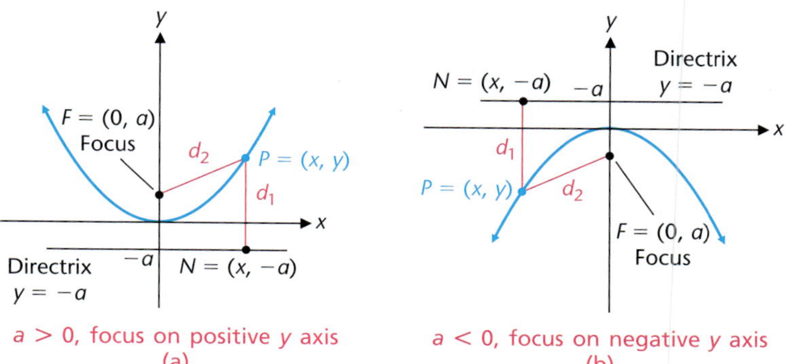

We summarize these results for easy reference in Theorem 1.

THEOREM 1

Standard Equations of a Parabola with Vertex at (0, 0)

1. $y^2 = 4ax$
   Vertex: (0, 0)
   Focus: $(a, 0)$
   Directrix: $x = -a$
   Symmetric with respect to the x axis
   Axis the x axis

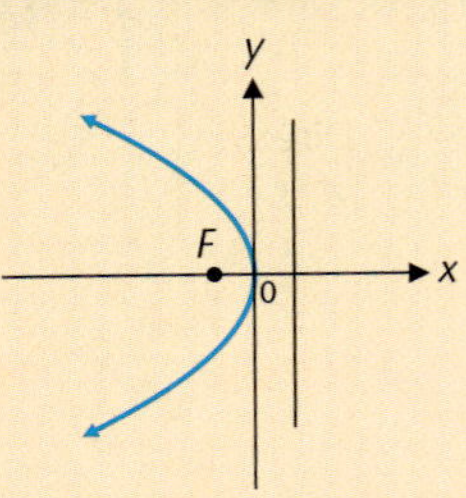

$a < 0$ (opens left)

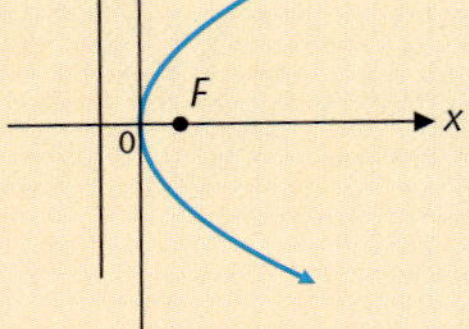

$a > 0$ (opens right)

2. $x^2 = 4ay$
   Vertex: (0, 0)
   Focus: $(0, a)$
   Directrix: $y = -a$
   Symmetric with respect to the y axis
   Axis the y axis

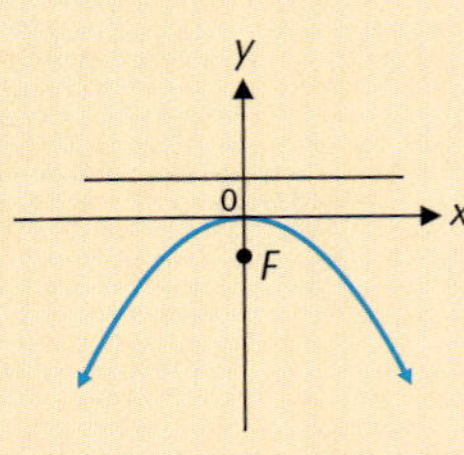

$a < 0$ (opens down)

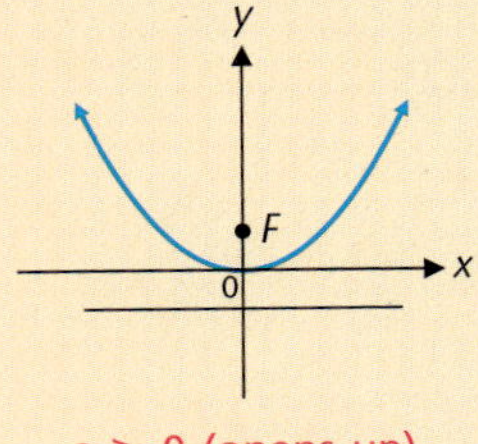

$a > 0$ (opens up)

EXAMPLE 1 Graphing $y^2 = 4ax$

Locate the focus and directrix and draw the graph of $y^2 = 16x$.

SOLUTION

Focus: $y^2 = 16x = 4(\overset{a}{4})x$ $\qquad$ Directrix: $x = -a$

$F = (a, 0) = (4, 0)$ $\qquad\qquad$ $= -4$

Graphing by Hand

To graph $y^2 = 16x$, it is convenient to assign x values that make the right side a perfect square, and solve for y. Note that x must be greater than or equal to 0 for y to be a real number. Because $a > 0$, the parabola opens to the right (Fig. 6).

x	0	1	4
y	0	± 4	± 8

Graphing Utility Graph

To graph $y^2 = 16x$ on a graphing utility, we solve this equation for y.

$$y^2 = 16x$$
$$y = \pm 4\sqrt{x}$$

This results in two functions, $y = 4\sqrt{x}$ and $y = -4\sqrt{x}$. Entering these functions in a graphing utility (Fig. 7) and graphing in a standard viewing window produces the graph of the parabola (Fig. 8).

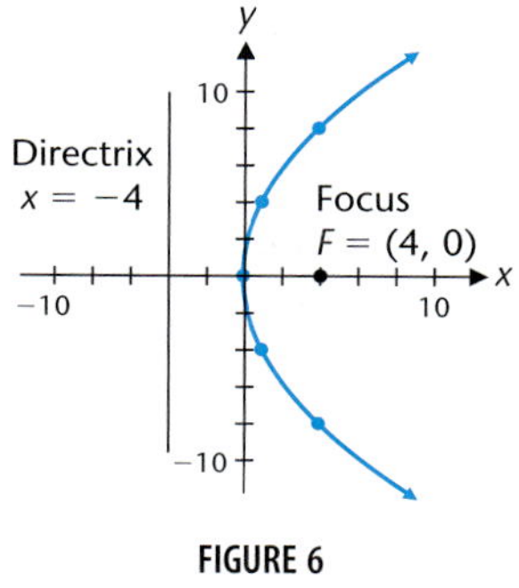

FIGURE 6

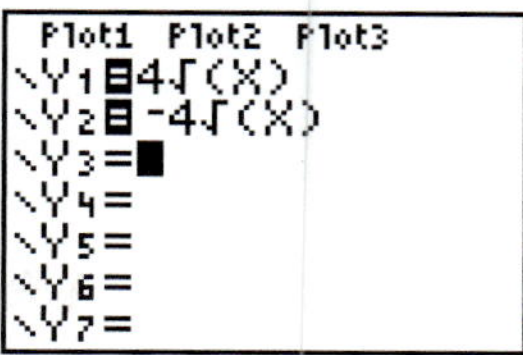

FIGURE 7

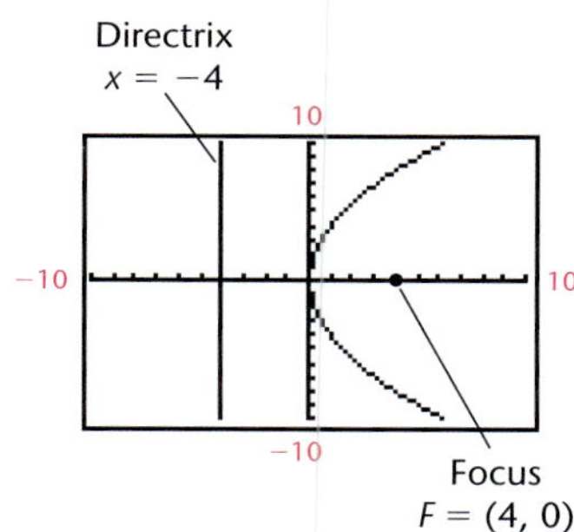

FIGURE 8

MATCHED 1 PROBLEM

Graph $y^2 = -8x$, and locate the focus and directrix.

CAUTION

A common error in making a quick sketch of $y^2 = 4ax$ or $x^2 = 4ay$ is to sketch the first with the y axis as its axis and the second with the x axis as its axis. The graph of $y^2 = 4ax$ is symmetric with respect to the x axis, and the graph of $x^2 = 4ay$ is symmetric with respect to the y axis, as a quick symmetry check will reveal.

EXAMPLE 2 Finding the Equation of a Parabola

(A) Find the equation of a parabola having the origin as its vertex, the y axis as its axis, and $(-10, -5)$ on its graph.

(B) Find the coordinates of its focus and the equation of its directrix.

SOLUTIONS

(A) The parabola is opening down and has an equation of the form $x^2 = 4ay$. Because $(-10, -5)$ is on the graph, we have

$$\begin{aligned} x^2 &= 4ay \\ (-10)^2 &= 4a(-5) \\ 100 &= -20a \\ a &= -5 \end{aligned}$$

Thus, the equation of the parabola is

$$\begin{aligned} x^2 &= 4(-5)y \\ &= -20y \end{aligned}$$

(B) Focus: $x^2 = -20y = 4(\overset{a}{-5})y$

$$F = (0, a) = (0, -5)$$

Directrix: $$\begin{aligned} y &= -a \\ &= -(-5) \\ &= 5 \end{aligned}$$

MATCHED 2 PROBLEM

(A) Find the equation of a parabola having the origin as its vertex, the x axis as its axis, and $(4, -8)$ on its graph.

(B) Find the coordinates of its focus and the equation of its directrix.

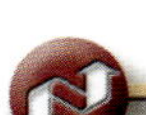

EXPLORE/DISCUSS 2

Consider the graph of an equation in the variables x and y. The equation of its magnification by a factor $k > 0$ is obtained by replacing x and y in the equation by x/k and y/k, respectively. (Of course, a magnification by a factor k between 0 and 1 means an actual reduction in size.)

(A) Show that the magnification by a factor 3 of the circle with equation $x^2 + y^2 = 1$ has equation $x^2 + y^2 = 9$.

(B) Explain why every circle with center at $(0, 0)$ is a magnification of the circle with equation $x^2 + y^2 = 1$.

(C) Find the equation of the magnification by a factor 3 of the parabola with equation $x^2 = y$. Graph both equations.

(D) Explain why every parabola with vertex $(0, 0)$ that opens upward is a magnification of the parabola with equation $x^2 = y$.

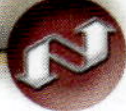

Applications

Parabolic forms are frequently encountered in the physical world. Suspension bridges, arch bridges, microphones, symphony shells, satellite antennas, radio and optical telescopes, radar equipment, solar furnaces, and searchlights are only a few of many items that use parabolic forms in their design.

Figure 9(a) illustrates a parabolic reflector used in all reflecting telescopes—from 3- to 6-inch home types to the 200-inch research instrument on Mount Palomar in California. Parallel light rays from distant celestial bodies are reflected to the focus off a parabolic mirror. If the light source is the sun, then the parallel rays are focused at F and we have a solar furnace. Temperatures of over 6,000°C have been achieved by such furnaces. If we locate a light source at F, then the rays in Figure 9(a) reverse, and we have a spotlight or a searchlight. Automobile headlights can use parabolic reflectors with special lenses over the light to diffuse the rays into useful patterns.

Figure 9(b) shows a suspension bridge, such as the Golden Gate Bridge in San Francisco. The suspension cable is a parabola. It is interesting to note that a free-hanging cable, such as a telephone line, does not form a parabola. It forms another curve called a *catenary.*

Figure 9(c) shows a concrete arch bridge. If all the loads on the arch are to be compression loads (concrete works very well under compression), then using physics and advanced mathematics, it can be shown that the arch must be parabolic.

FIGURE 9 Uses of parabolic forms.

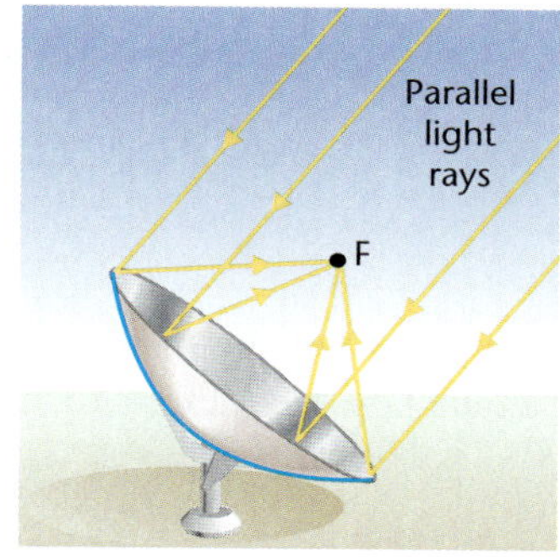

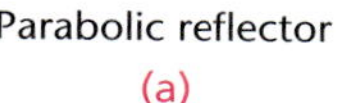
Parabolic reflector
(a)

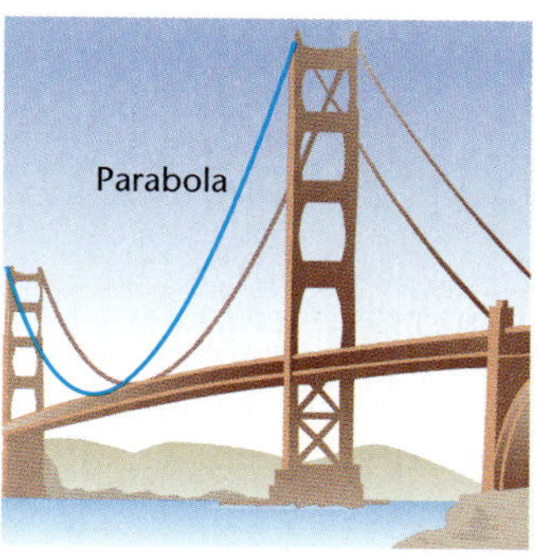

Suspension bridge
(b)

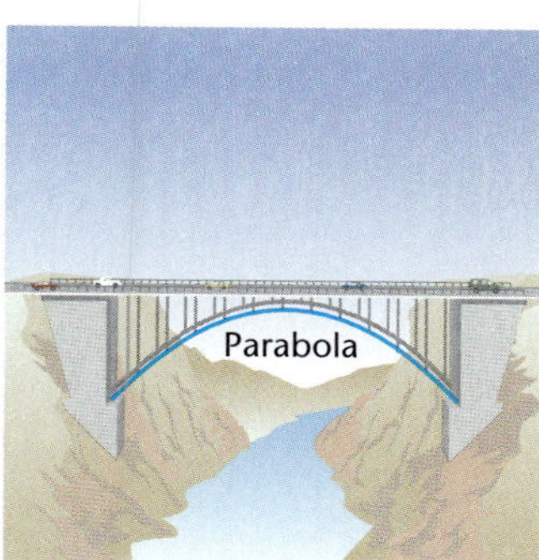

Arch bridge
(c)

EXAMPLE 3 Parabolic Reflector

A **paraboloid** is formed by revolving a parabola about its axis. A spotlight in the form of a paraboloid 5 inches deep has its focus 2 inches from the vertex. Find, to one decimal place, the radius R of the opening of the spotlight.

SOLUTION

Step 1. Locate a parabolic cross section containing the axis in a rectangular coordinate system, and label all known parts and parts to be found. This is a very important step and can be done in infinitely many ways. Because we are in charge, we can make things simpler for ourselves by locating the vertex at the origin and choosing a coordinate axis as the axis. We choose the y axis as the axis of the parabola with the parabola opening upward (Fig. 10).

FIGURE 10

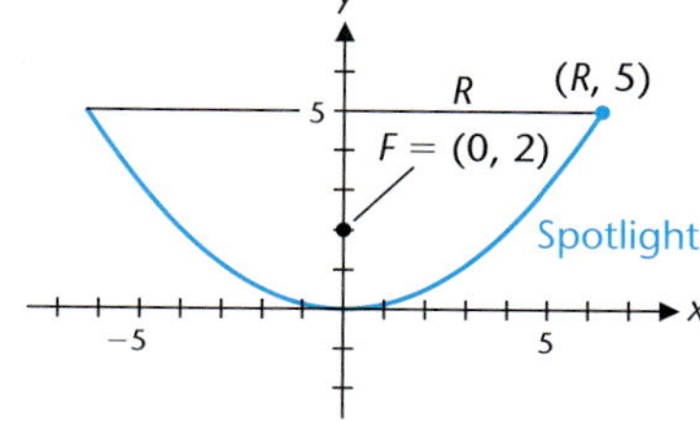

Step 2. Find the equation of the parabola in the figure. Because the parabola has the y axis as its axis and the vertex at the origin, the equation is of the form

$$x^2 = 4ay$$

We are given $F = (0, a) = (0, 2)$; thus, $a = 2$, and the equation of the parabola is

$$x^2 = 8y$$

Step 3. Use the equation found in step 2 to find the radius R of the opening. Because $(R, 5)$ is on the parabola, we have

$$R^2 = 8(5)$$
$$R = \sqrt{40} \approx 6.3 \text{ inches}$$

MATCHED 3 PROBLEM

Repeat Example 3 with a paraboloid 12 inches deep and a focus 9 inches from the vertex.

ANSWERS MATCHED PROBLEMS

1. Focus: $(-2, 0)$
Directrix: $x = 2$

x	0	-2
y	0	± 4

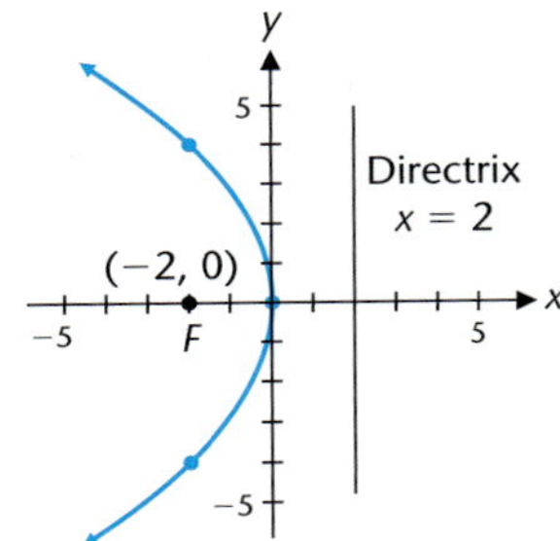

2. (A) $y^2 = 16x$ (B) Focus: $(4, 0)$; Directrix: $x = -4$ **3.** $R = 20.8$ inches

EXERCISE 11.1

In Problems 1–12, graph each equation, and locate the focus and directrix. Check by graphing on a graphing utility.

1. $y^2 = 4x$ **2.** $y^2 = 8x$

3. $x^2 = 8y$ **4.** $x^2 = 4y$

5. $y^2 = -12x$ **6.** $y^2 = -4x$

7. $x^2 = -4y$ **8.** $x^2 = -8y$

9. $y^2 = -20x$ **10.** $x^2 = -24y$

11. $x^2 = 10y$ **12.** $y^2 = 6x$

Find the coordinates to two decimal places of the focus for each parabola in Problems 13–18.

13. $y^2 = 39x$

14. $x^2 = 58y$

15. $x^2 = -105y$

16. $y^2 = -93x$

17. $y^2 = -77x$

18. $x^2 = -205y$

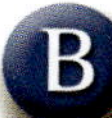

In Problems 19–26, find the equation of a parabola with vertex at the origin, axis the x or y axis, and

19. Directrix $y = -3$

20. Directrix $y = 4$

21. Focus $(0, -7)$

22. Focus $(0, 5)$

23. Directrix $x = 6$

24. Directrix $x = -9$

25. Focus $(2, 0)$

26. Focus $(-4, 0)$

In Problems 27–32, find the equation of the parabola having its vertex at the origin, its axis as indicated, and passing through the indicated point.

27. y axis; $(4, 2)$

28. x axis; $(4, 8)$

29. x axis; $(-3, 6)$

30. y axis; $(-5, 10)$

31. y axis; $(-6, -9)$

32. x axis; $(-6, -12)$

In Problems 33–36, find the first-quadrant points of intersection for each system of equations to three decimal places.

33. $x^2 = 4y$
$y^2 = 4x$

34. $y^2 = 3x$
$x^2 = 3y$

35. $y^2 = 6x$
$x^2 = 5y$

36. $x^2 = 7y$
$y^2 = 2x$

37. Consider the parabola with equation $x^2 = 4ay$.

(A) How many lines through $(0, 0)$ intersect the parabola in exactly one point? Find their equations.

(B) Find the coordinates of all points of intersection of the parabola with the line through $(0, 0)$ having slope $m \neq 0$.

38. Find the coordinates of all points of intersection of the parabola with equation $x^2 = 4ay$ and the parabola with equation $y^2 = 4bx$.

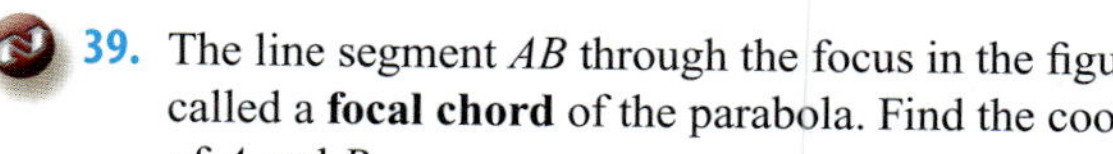

39. The line segment AB through the focus in the figure is called a **focal chord** of the parabola. Find the coordinates of A and B.

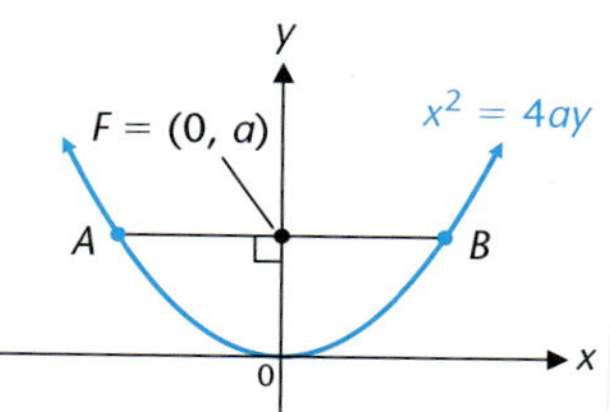

40. The line segment AB through the focus in the figure is called a **focal chord** of the parabola. Find the coordinates of A and B.

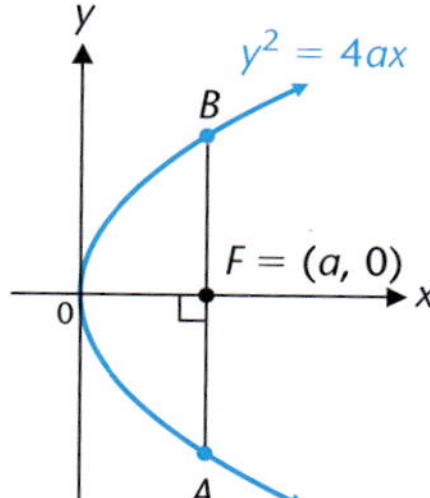

Refer to Explore/Discuss 2. In Problems 41–44, find the magnification of the given equation by the given factor. Graph both equations.

41. $y^2 = 4x$, factor 2

42. $y^2 = -4x$, factor 0.25

43. $x^2 = -4y$, factor 0.5

44. $x^2 = 4y$, factor 4

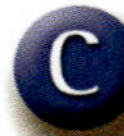

In Problems 45–48, use the definition of a parabola and the distance formula to find the equation of a parabola with

45. Directrix $y = -4$ and focus $(2, 2)$

46. Directrix $y = 2$ and focus $(-3, 6)$

47. Directrix $x = 2$ and focus $(6, -4)$

48. Directrix $x = -3$ and focus $(1, 4)$

APPLICATIONS

49. Engineering. The parabolic arch in the concrete bridge in the figure must have a clearance of 50 feet above the water and span a distance of 200 feet. Find the equation of the parabola after inserting a coordinate system with the origin at the vertex of the parabola and the vertical y axis (pointing upward) along the axis of the parabola.

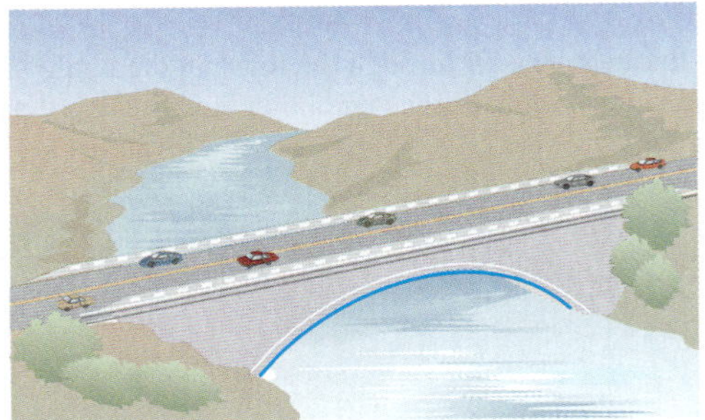

50. Astronomy. The cross section of a parabolic reflector with 6-inch diameter is ground so that its vertex is 0.15 inch below the rim (see the figure).

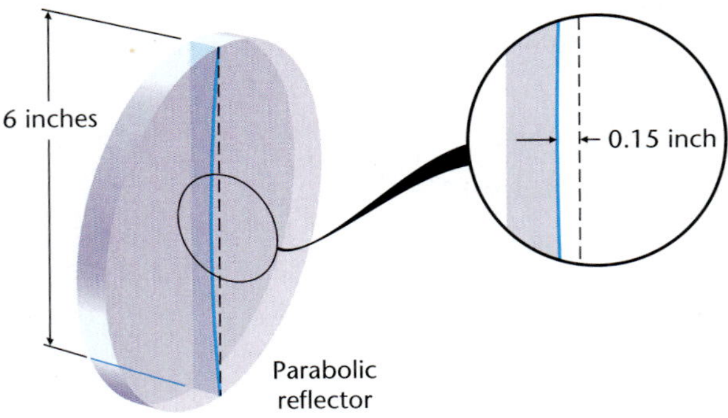

(A) Find the equation of the parabola after inserting an xy coordinate system with the vertex at the origin and the y axis (pointing upward) the axis of the parabola.

(B) How far is the focus from the vertex?

51. Space Science. A designer of a 200-foot-diameter parabolic electromagnetic antenna for tracking space probes wants to place the focus 100 feet above the vertex (see the figure).

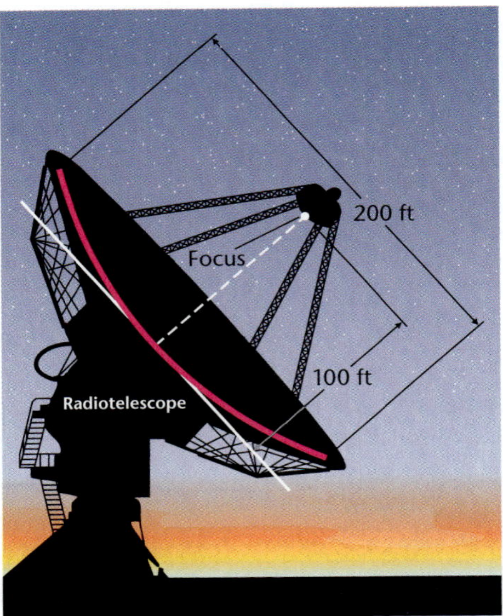

(A) Find the equation of the parabola using the axis of the parabola as the y axis (up positive) and vertex at the origin.

(B) Determine the depth of the parabolic reflector.

52. Signal Light. A signal light on a ship is a spotlight with parallel reflected light rays (see the figure). Suppose the parabolic reflector is 12 inches in diameter and the light source is located at the focus, which is 1.5 inches from the vertex.

(A) Find the equation of the parabola using the axis of the parabola as the x axis (right positive) and vertex at the origin.

(B) Determine the depth of the parabolic reflector.

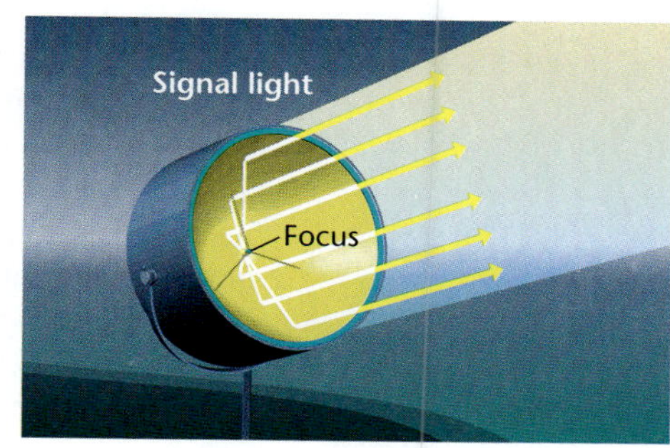

SECTION 11.2 Ellipse

Definition of an Ellipse • Drawing an Ellipse • Standard Equations and Their Graphs • Applications

We start our discussion of the ellipse with a coordinate-free definition. Using this definition, we show how an ellipse can be drawn and we derive standard equations for ellipses specially located in a rectangular coordinate system.

Definition of an Ellipse

The following is a coordinate-free definition of an ellipse:

DEFINITION 1
Ellipse

An **ellipse** is the set of all points P in a plane such that the sum of the distances of P from two fixed points in the plane is constant. Each of the fixed points, F' and F, is called a **focus,** and together they are called **foci.** Referring to the figure, the line segment $V'V$ through the foci is the **major axis.** The perpendicular bisector $B'B$ of the major axis is the **minor axis.** Each end of the major axis, V' and V, is called a **vertex.** The midpoint of the line segment $F'F$ is called the **center** of the ellipse.

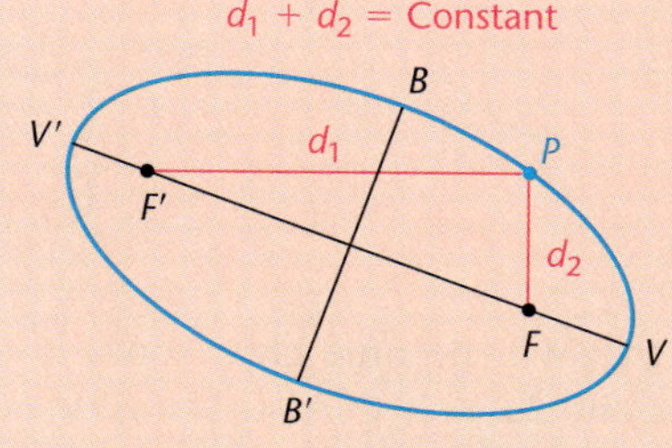

Drawing an Ellipse

An ellipse is easy to draw. All you need is a piece of string, two thumbtacks, and a pencil or pen (Fig. 1). Place the two thumbtacks in a piece of cardboard. These form the foci of the ellipse. Take a piece of string longer than the distance between the two thumbtacks—this represents the constant in the definition—and tie each end to a thumbtack. Finally, catch the tip of a pencil under the string and move it while keeping the string taut. The resulting figure is by definition an ellipse. Ellipses of different shapes result, depending on the placement of thumbtacks and the length of the string joining them.

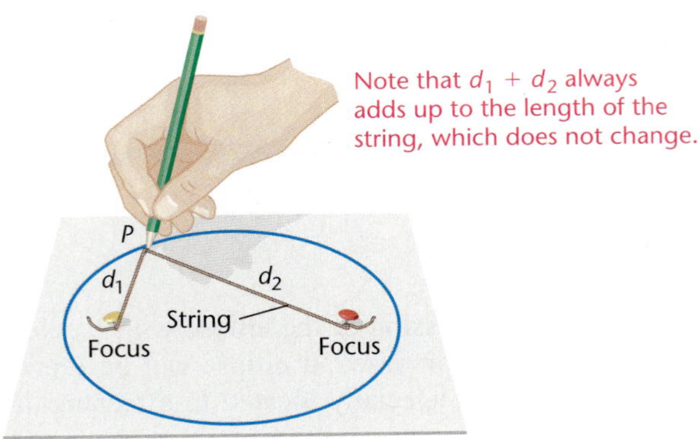

FIGURE 1 Drawing an ellipse.

Standard Equations and Their Graphs

Using the definition of an ellipse and the distance-between-two-points formula, we can derive standard equations for an ellipse located in a rectangular coordinate system. We start by placing an ellipse in the coordinate system with the foci on the x axis equidistant from the origin at $F' = (-c, 0)$ and $F = (c, 0)$, as in Figure 2.

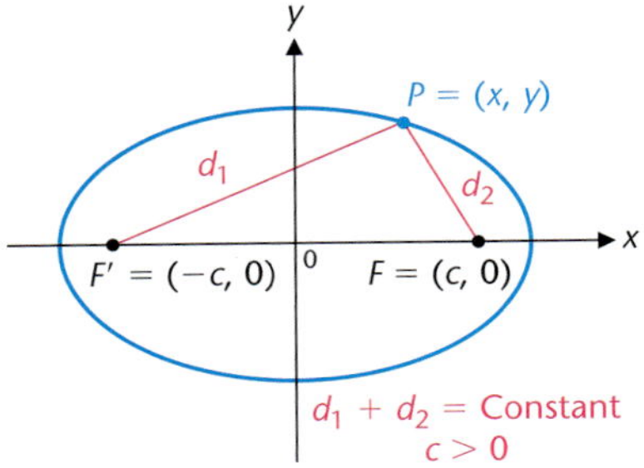

FIGURE 2 Ellipse with foci on x axis.

For reasons that will become clear soon, it is convenient to represent the constant sum $d_1 + d_2$ by $2a$, $a > 0$. Also, the geometric fact that the sum of the lengths of any two sides of a triangle must be greater than the third side can be applied to Figure 2 to derive the following useful result:

$$\begin{aligned} d(F', P) + d(P, F) &> d(F', F) \\ d_1 + d_2 &> 2c \\ 2a &> 2c \\ a &> c \end{aligned} \tag{1}$$

We will use this result in the derivation of the equation of an ellipse, which we now begin.

Referring to Figure 2, the point $P = (x, y)$ is on the ellipse if and only if

$$\begin{aligned} d_1 + d_2 &= 2a \\ d(P, F') + d(P, F) &= 2a \\ \sqrt{(x + c)^2 + (y - 0)^2} + \sqrt{(x - c)^2 + (y - 0)^2} &= 2a \end{aligned}$$

After eliminating radicals and simplifying, a good exercise for you, we obtain

$$(a^2 - c^2)x^2 + a^2y^2 = a^2(a^2 - c^2) \tag{2}$$

$$\frac{x^2}{a^2} + \frac{y^2}{a^2 - c^2} = 1 \tag{3}$$

Dividing both sides of equation (2) by $a^2(a^2 - c^2)$ is permitted, because neither a^2 nor $a^2 - c^2$ is 0. From equation (1), $a > c$; thus $a^2 > c^2$ and $a^2 - c^2 > 0$. The constant a was chosen positive at the beginning.

To simplify equation (3) further, we let

$$b^2 = a^2 - c^2 \qquad b > 0 \tag{4}$$

to obtain

$$\frac{x^2}{a^2} + \frac{y^2}{b^2} = 1 \tag{5}$$

From equation (5) we see that the x intercepts are $x = \pm a$ and the y intercepts are $y = \pm b$. The x intercepts are also the vertices. Thus,

Major axis length $= 2a$
Minor axis length $= 2b$

To see that the major axis is longer than the minor axis, we show that $2a > 2b$. Returning to equation (4),

$$\begin{aligned} b^2 &= a^2 - c^2 && a, b, c > 0 \\ b^2 + c^2 &= a^2 \\ b^2 &< a^2 && \text{Definition of } < \\ b^2 - a^2 &< 0 \\ (b - a)(b + a) &< 0 \\ b - a &< 0 && \text{Because } b + a \text{ is positive, } b - a \text{ must be negative.} \\ b &< a \\ 2b &< 2a \\ 2a &> 2b \end{aligned}$$

$$\begin{pmatrix}\text{Length of}\\\text{major axis}\end{pmatrix} > \begin{pmatrix}\text{Length of}\\\text{minor axis}\end{pmatrix}$$

If we start with the foci on the y axis at $F = (0, c)$ and $F' = (0, -c)$ as in Figure 3, instead of on the x axis as in Figure 2, then, following arguments similar to those used for the first derivation, we obtain

$$\frac{x^2}{b^2} + \frac{y^2}{a^2} = 1 \qquad a > b \tag{6}$$

where the relationship among a, b, and c remains the same as before:

$$b^2 = a^2 - c^2 \tag{7}$$

The center is still at the origin, but the major axis is now along the y axis and the minor axis is along the x axis.

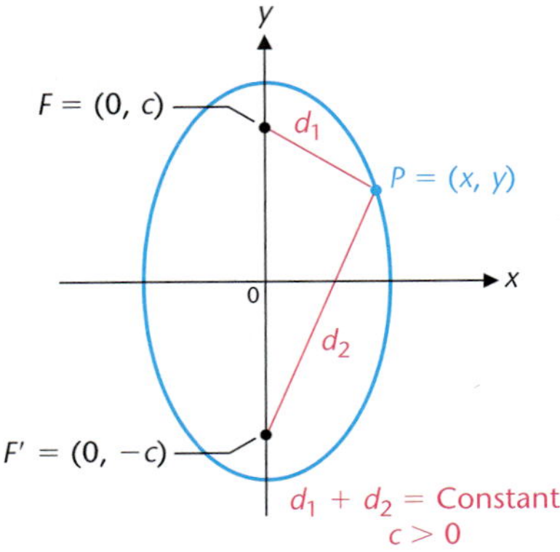

FIGURE 3 Ellipse with foci on y axis.

To sketch graphs of equations of the form of equations (5) or (6) is an easy matter. We find the x and y intercepts and sketch in an appropriate ellipse. Because replacing x with $-x$ or y with $-y$ produces an equivalent equation, we conclude that the graphs are symmetric with respect to the x axis, y axis, and origin. If further accuracy is required, additional points can be found with the aid of a calculator and the use of symmetry properties.

Given an equation of the form of equations (5) or (6), how can we find the coordinates of the foci without memorizing or looking up the relation $b^2 = a^2 - c^2$? There is a simple geometric relationship in an ellipse that enables us to get the same result using the Pythagorean theorem. To see this relationship, refer to Figure 4(a). Then, using the definition of an ellipse and $2a$ for the constant sum, as we did in deriving the standard equations, we see that

$$\begin{aligned} d + d &= 2a \\ 2d &= 2a \\ d &= a \end{aligned}$$

Thus,

The length of the line segment from the end of a minor axis to a focus is the same as half the length of a major axis.

This geometric relationship is illustrated in Figure 4(b). Using the Pythagorean theorem for the triangle in Figure 4(b), we have

$$b^2 + c^2 = a^2$$

or

$$b^2 = a^2 - c^2 \quad \text{Equations (4) and (7)}$$

or

$$c^2 = a^2 - b^2 \quad \text{Useful for finding the foci, given } a \text{ and } b$$

Thus, we can find the foci of an ellipse given the intercepts a and b simply by using the triangle in Figure 4(b) and the Pythagorean theorem.

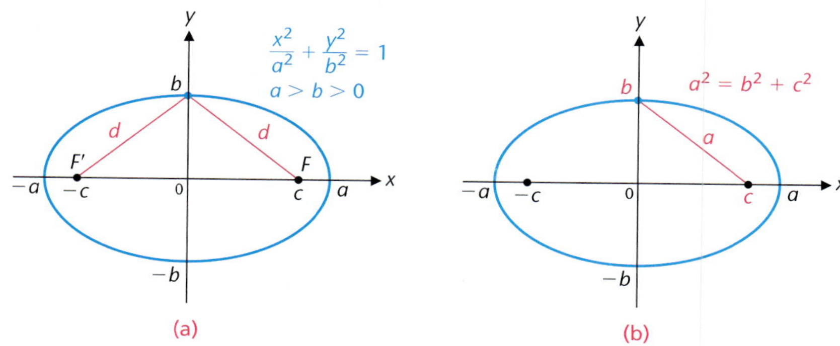

FIGURE 4 Geometric relationships.

We summarize all of these results for convenient reference in Theorem 1.

THEOREM 1

Standard Equations of an Ellipse with Center at (0, 0)

1. $\frac{x^2}{a^2} + \frac{y^2}{b^2} = 1 \qquad a > b > 0$

   x intercepts: $\pm a$ (vertices)
   y intercepts: $\pm b$
   Foci: $F' = (-c, 0)$, $F = (c, 0)$

   $$c^2 = a^2 - b^2$$

   Major axis length $= 2a$
   Minor axis length $= 2b$

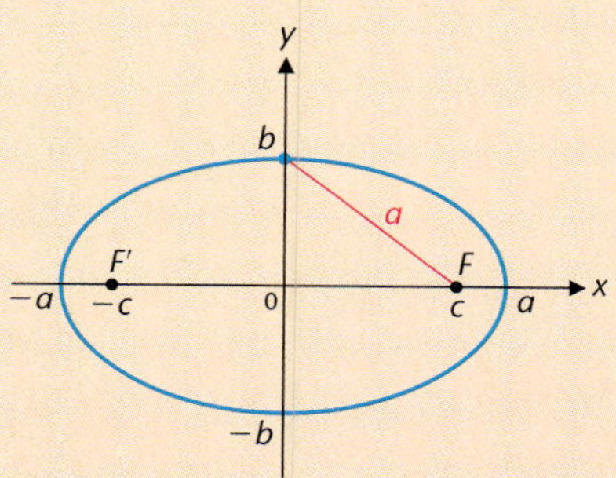

2. $\frac{x^2}{b^2} + \frac{y^2}{a^2} = 1 \qquad a > b > 0$

   x intercepts: $\pm b$
   y intercepts: $\pm a$ (vertices)
   Foci: $F' = (0, -c)$, $F = (0, c)$

   $$c^2 = a^2 - b^2$$

   Major axis length $= 2a$
   Minor axis length $= 2b$

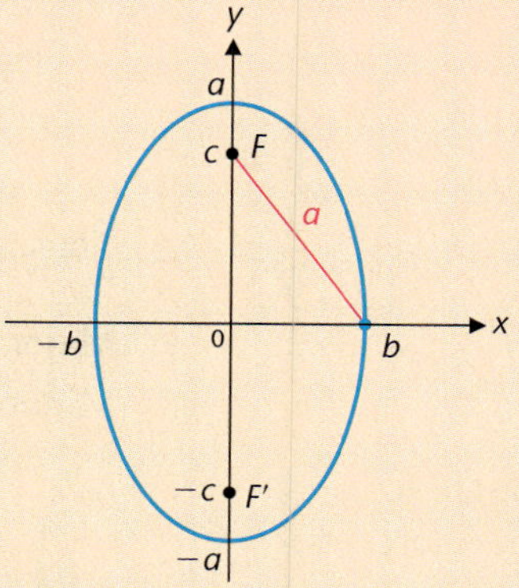

   [*Note:* Both graphs are symmetric with respect to the x axis, y axis, and origin. Also, the major axis is always longer than the minor axis.]

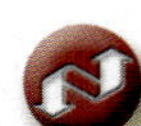

EXPLORE/DISCUSS 1

The line through a focus F of an ellipse that is perpendicular to the major axis intersects the ellipse in two points G and H. For each of the two standard equations of an ellipse with center (0, 0), find an expression in terms of a and b for the distance from G to H.

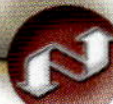

EXAMPLE 1 Graphing Ellipses

Find the coordinates of the foci, find the lengths of the major and minor axes, and graph the following equation:

$$9x^2 + 16y^2 = 144$$

SOLUTION

First, write the equation in standard form by dividing both sides by 144 and determine a and b:

$$9x^2 + 16y^2 = 144$$

$$\frac{9x^2}{144} + \frac{16y^2}{144} = \frac{144}{144}$$

$$\frac{x^2}{16} + \frac{y^2}{9} = 1$$

$a = 4 \qquad$ and $\qquad b = 3$

Foci: $c^2 = a^2 - b^2$

$= 16 - 9$

$= 7$

$c = \sqrt{7}$ c must be positive

Thus, the foci are $F' = (-\sqrt{7}, 0)$ and $F = (\sqrt{7}, 0)$.

Major axis length $2(4) = 8$

Minor axis length $2(3) = 6$

Hand-Drawn Graph

Locate the intercepts:

x intercepts: ± 4

y intercepts: ± 3

Plot the intercepts and sketch in the ellipse (Fig. 5).

Graphing Utility Solution

Solve the original equation for y:

$$9x^2 + 16y^2 = 144$$

$$y^2 = (144 - 9x^2)/16$$

$$y = \pm\sqrt{(144 - 9x^2)/16}$$

This produces the two functions whose graphs are shown in Figure 6. Notice that we used a squared viewing window to avoid distorting the

FIGURE 5

shape of the ellipse. Also note the gaps in the graph at ± 4. This is due to the relatively low resolution of a graphing utility screen.

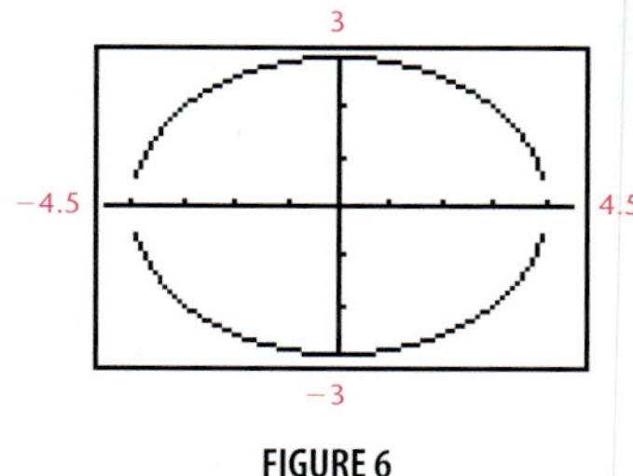

FIGURE 6

MATCHED 1 PROBLEM

Find the coordinates of the foci, find the lengths of the major and minor axes, and graph the following equation:

$$x^2 + 4y^2 = 4$$

EXAMPLE 2 Graphing Ellipses

Find the coordinates of the foci, find the lengths of the major and minor axes, and graph the following equation:

$$2x^2 + y^2 = 10$$

SOLUTION

First, write the equation in standard form by dividing both sides by 10 and determine a and b:

$$2x^2 + y^2 = 10$$

$$\frac{2x^2}{10} + \frac{y^2}{10} = \frac{10}{10}$$

$$\frac{x^2}{5} + \frac{y^2}{10} = 1$$

$a = \sqrt{10}$ and $b = \sqrt{5}$

Foci: $c^2 = a^2 - b^2$

$= 10 - 5$

$= 5$

$c = \sqrt{5}$ *c must be positive*

Thus, the foci are $F' = (0, -\sqrt{5})$ and $F = (0, \sqrt{5})$.

Major axis length $2\sqrt{10} \approx 6.32$

Minor axis length $2\sqrt{5} \approx 4.47$

Hand-Drawn Graph

Locate the intercepts:

x intercepts: $\pm\sqrt{5} \approx \pm 2.24$

y intercepts: $\pm\sqrt{10} \approx \pm 3.16$

Plot the intercepts and sketch in the ellipse (Fig. 7).

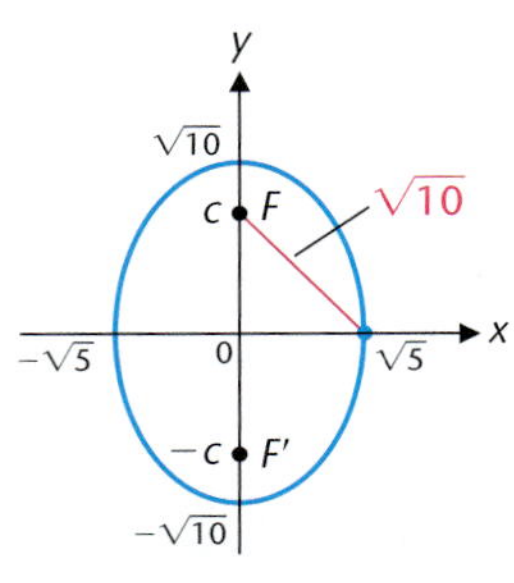

FIGURE 7

Graphing Utility Solution

Solve for y:

$$2x^2 + y^2 = 10$$
$$y^2 = 10 - 2x^2$$
$$y = \pm\sqrt{10 - 2x^2}$$

Graph $y_1 = \sqrt{10 - 2x^2}$ and $y_2 = -\sqrt{10 - 2x^2}$ in a squared viewing window (Fig. 8).

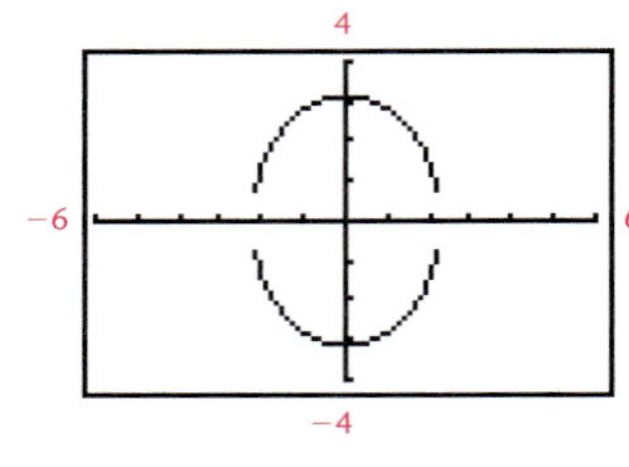

FIGURE 8

MATCHED 2 PROBLEM

Find the coordinates of the foci, find the lengths of the major and minor axes, and graph the following equation:

$$3x^2 + y^2 = 18$$

EXAMPLE 3 Finding the Equation of an Ellipse

Find an equation of an ellipse in the form

$$\frac{x^2}{M} + \frac{y^2}{N} = 1 \qquad M, N > 0$$

if the center is at the origin, the major axis is along the y axis, and

(A) Length of major axis = 20
Length of minor axis = 12

(B) Length of major axis = 10
Distance of foci from center = 4

SOLUTIONS

(A) Compute x and y intercepts and make a rough sketch of the ellipse, as shown in Figure 9.

FIGURE 9

$$\frac{x^2}{b^2} + \frac{y^2}{a^2} = 1$$

$$a = \frac{20}{2} = 10 \qquad b = \frac{12}{2} = 6$$

$$\frac{x^2}{36} + \frac{y^2}{100} = 1$$

FIGURE 10

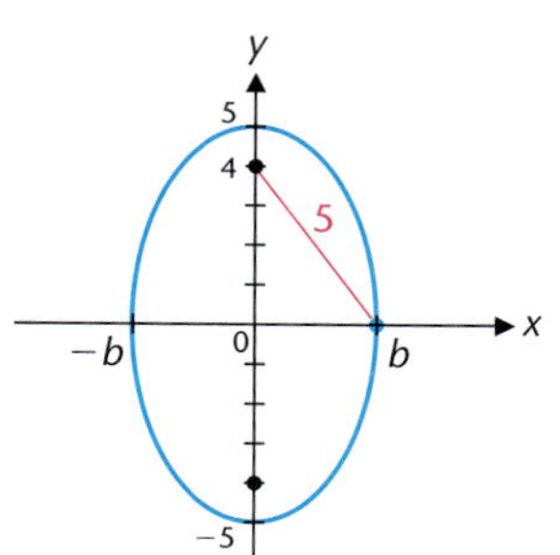

(B) Make a rough sketch of the ellipse, as shown in Figure 10; locate the foci and y intercepts, then determine the x intercepts using the special triangle relationship discussed earlier.

$$\frac{x^2}{b^2} + \frac{y^2}{a^2} = 1$$

$$a = \frac{10}{2} = 5 \qquad b^2 = 5^2 - 4^2 = 25 - 16 = 9$$

$$b = 3$$

$$\frac{x^2}{9} + \frac{y^2}{25} = 1$$

MATCHED 3 PROBLEM

Find an equation of an ellipse in the form

$$\frac{x^2}{M} + \frac{y^2}{N} = 1 \qquad M, N > 0$$

if the center is at the origin, the major axis is along the x axis, and

(A) Length of major axis = 50
Length of minor axis = 30

(B) Length of minor axis = 16
Distance of foci from center = 6

EXPLORE/DISCUSS 2

Consider the graph of an equation in the variables x and y. The equation of its magnification by a factor $k > 0$ is obtained by replacing x and y in the equation by x/k and y/k, respectively.

(A) Find the equation of the magnification by a factor 3 of the ellipse with equation $(x^2/4) + y^2 = 1$. Graph both equations.

(B) Give an example of an ellipse with center (0, 0) with $a > b$ that is not a magnification of $(x^2/4) + y^2 = 1$.

(C) Find the equations of all ellipses that are magnifications of $(x^2/4) + y^2 = 1$.

Applications

You are no doubt aware of many occurrences and uses of elliptical forms: orbits of satellites, planets, and comets; shapes of galaxies; gears and cams; some airplane wings, boat keels, and rudders; tabletops; public fountains; and domes in buildings are a few examples (Fig. 11).

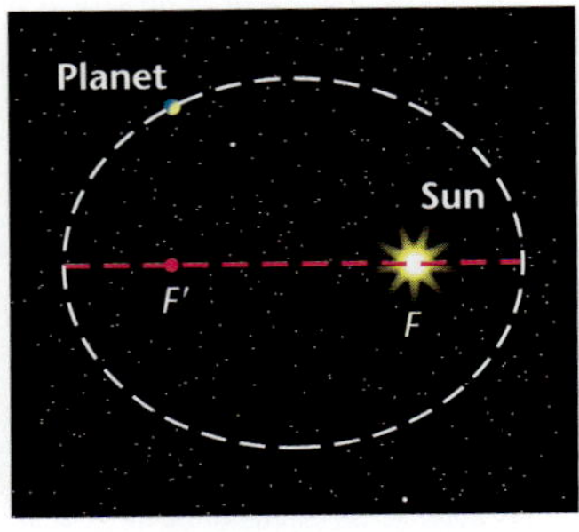

Planetary motion

(a)

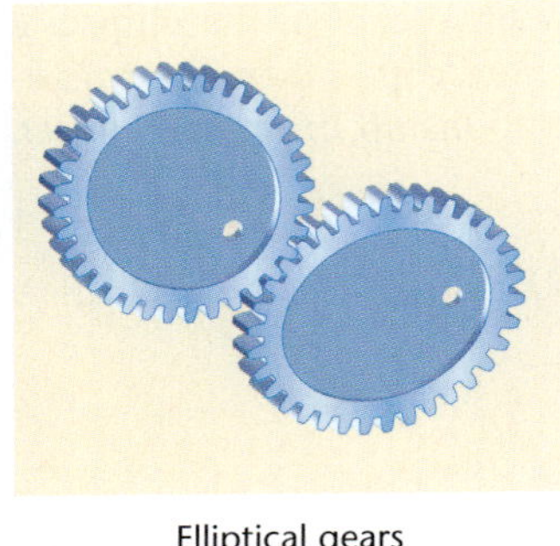

Elliptical gears

(b)

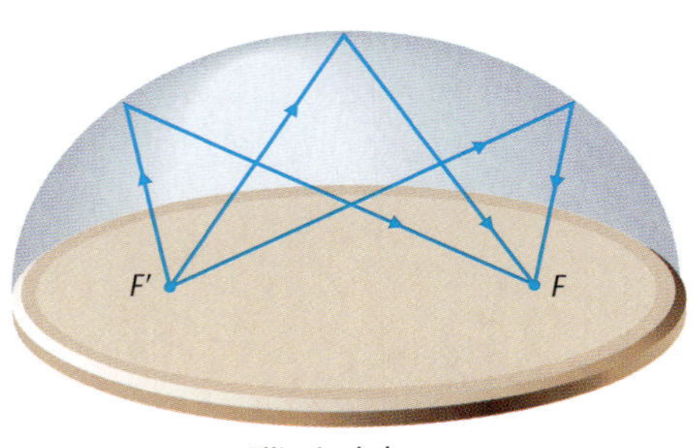

Elliptical dome

(c)

FIGURE 11 Uses of elliptical forms.

Johannes Kepler (1571–1630), a German astronomer, discovered that planets move in elliptical orbits, with the sun at a focus, and not in circular orbits as had been thought before [Fig. 11(a)]. Figure 11(b) shows a pair of elliptical gears with pivot points at foci. Such gears transfer constant rotational speed to variable rotational speed, and vice versa. Figure 11(c) shows an elliptical dome. An interesting property of such a dome is that a sound or light source at one focus will reflect off the dome and pass through the other focus. One of the chambers in the Capitol Building in Washington, D.C., has such a dome, and is referred to as a whispering room because a whispered sound at one focus can be easily heard at the other focus.

A fairly recent application in medicine is the use of elliptical reflectors and ultrasound to break up kidney stones. A device called a lithotripter is used to generate intense sound waves that break up the stone from outside the body, thus avoiding surgery. To be certain that the waves do not damage other parts of the body, the reflecting property of the ellipse is used to design and correctly position the lithotripter.

EXAMPLE 4 Medicinal Lithotripsy

A lithotripter is formed by rotating the portion of an ellipse below the minor axis around the major axis (Fig. 12). The lithotripter is 20 centimeters wide and 16 centimeters deep. If the ultrasound source is positioned at one focus of the ellipse and the kidney stone at the other, then all the sound waves will pass through the kidney stone. How far from the kidney stone should the point V on the base of the lithotripter be positioned to focus the sound waves on the kidney stone? Round the answer to one decimal place.

FIGURE 12 Lithotripter.

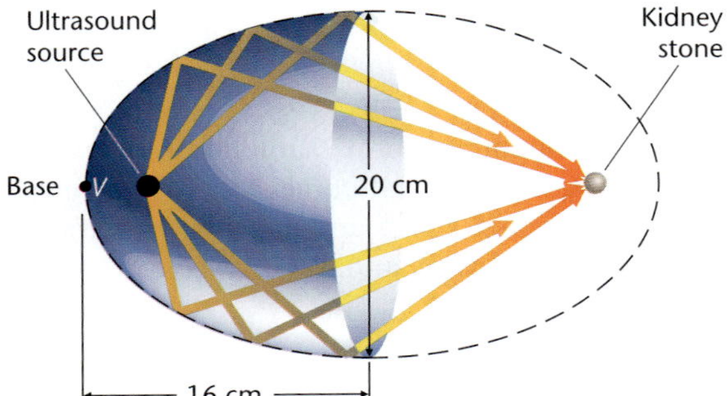

SOLUTION

From Figure 12 we see that $a = 16$ and $b = 10$ for the ellipse used to form the lithotripter. Thus, the distance c from the center to either the kidney stone or the ultrasound source is given by

$$c = \sqrt{a^2 - b^2} = \sqrt{16^2 - 10^2} = \sqrt{156} \approx 12.5$$

and the distance from the base of the lithotripter to the kidney stone is $16 + 12.5 = 28.5$ centimeters.

MATCHED 4 PROBLEM

Because lithotripsy is an external procedure, the lithotripter described in Example 4 can be used only on stones within 12.5 centimeters of the surface of the body. Suppose a kidney stone is located 14 centimeters from the surface. If the diameter is kept fixed at 20 centimeters, how deep must a lithotripter be to focus on this kidney stone? Round answer to one decimal place.

ANSWERS MATCHED PROBLEMS

1.

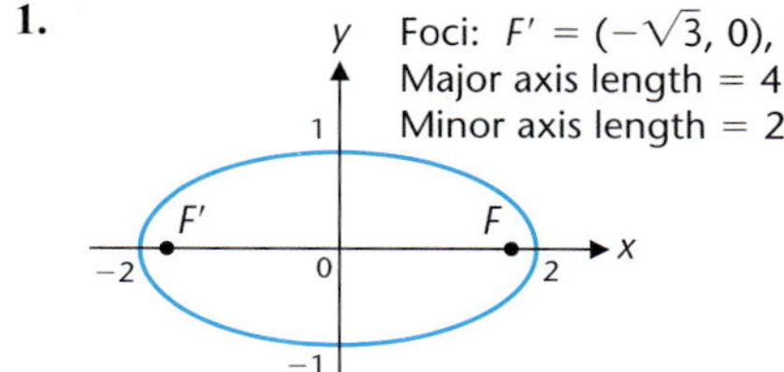

Foci: $F' = (-\sqrt{3}, 0)$, $F = (\sqrt{3}, 0)$
Major axis length = 4
Minor axis length = 2

2.

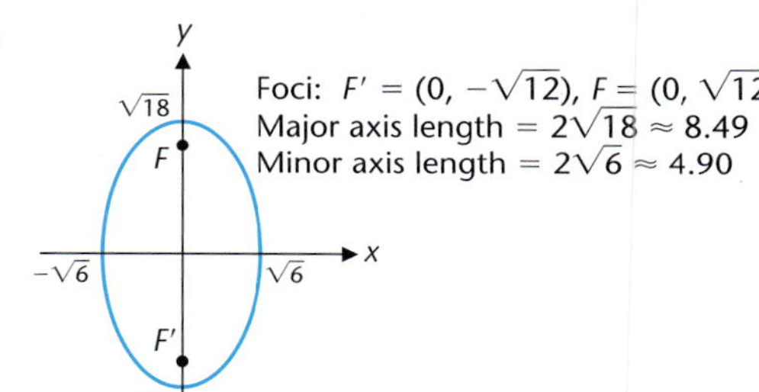

Foci: $F' = (0, -\sqrt{12})$, $F = (0, \sqrt{12})$
Major axis length = $2\sqrt{18} \approx 8.49$
Minor axis length = $2\sqrt{6} \approx 4.90$

3. (A) $\frac{x^2}{625} + \frac{y^2}{225} = 1$ (B) $\frac{x^2}{100} + \frac{y^2}{64} = 1$ **4.** 17.2 centimeters

EXERCISE 11.2

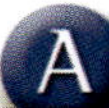

In Problems 1–6, sketch a graph of each equation, find the coordinates of the foci, and find the lengths of the major and minor axes. Check by graphing on a graphing utility.

1. $\frac{x^2}{25} + \frac{y^2}{4} = 1$

2. $\frac{x^2}{9} + \frac{y^2}{4} = 1$

3. $\frac{x^2}{4} + \frac{y^2}{25} = 1$

4. $\frac{x^2}{4} + \frac{y^2}{9} = 1$

5. $x^2 + 9y^2 = 9$

6. $4x^2 + y^2 = 4$

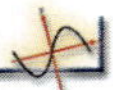

In Problems 7–10, match each equation with one of graphs (a)–(d).

7. $9x^2 + 16y^2 = 144$ **8.** $16x^2 + 9y^2 = 144$ **9.** $4x^2 + y^2 = 16$ **10.** $x^2 + 4y^2 = 16$

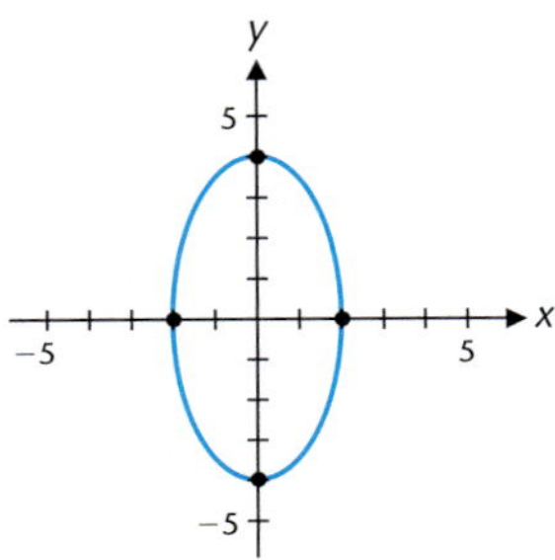

(a)

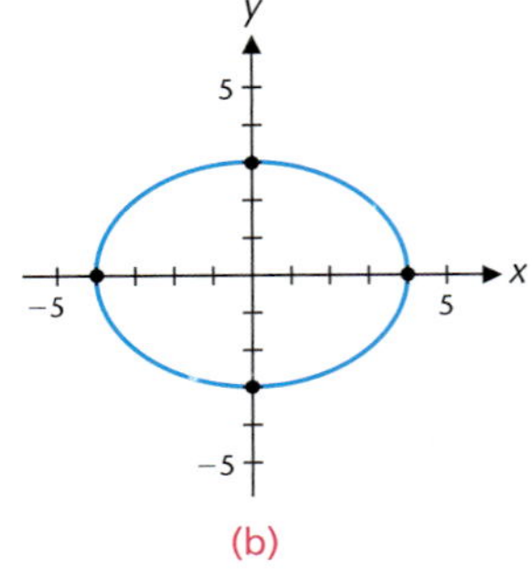

(b)

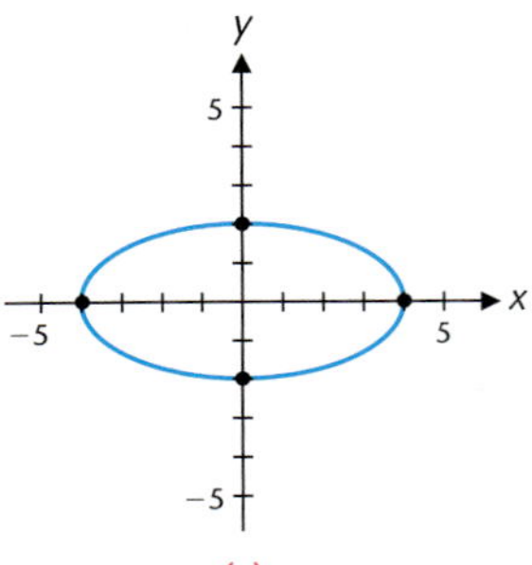

(c)

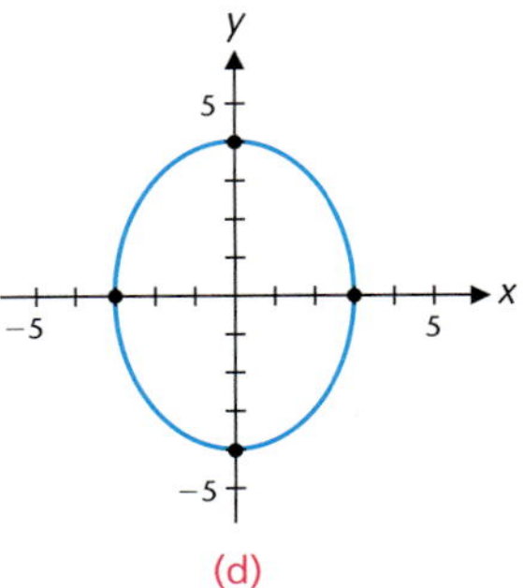

(d)

B

In Problems 11–16, sketch a graph of each equation, find the coordinates of the foci, and find the lengths of the major and minor axes. Check by graphing on a graphing utility.

11. $25x^2 + 9y^2 = 225$

12. $16x^2 + 25y^2 = 400$

13. $2x^2 + y^2 = 12$

14. $4x^2 + 3y^2 = 24$

15. $4x^2 + 7y^2 = 28$

16. $3x^2 + 2y^2 = 24$

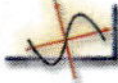

In Problems 17–28, find an equation of an ellipse in the form

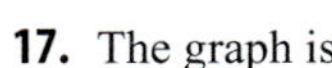

$$\frac{x^2}{M} + \frac{y^2}{N} = 1 \qquad M, N > 0$$

if the center is at the origin, and

17. The graph is

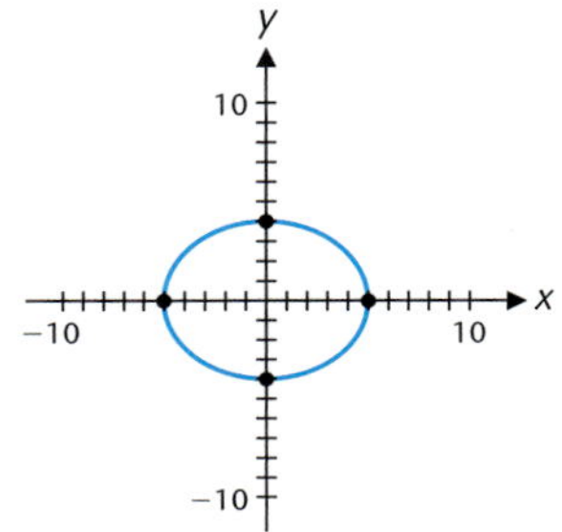

18. The graph is

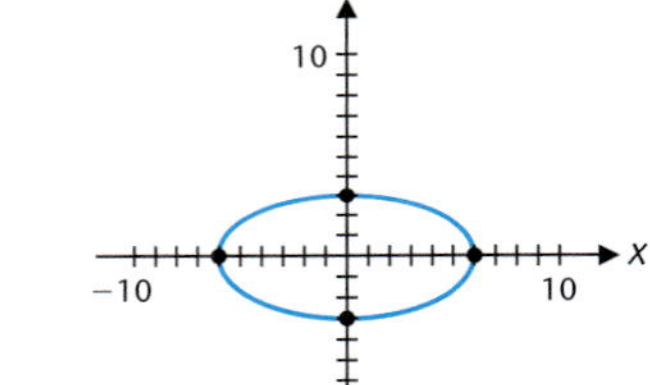

19. The graph is

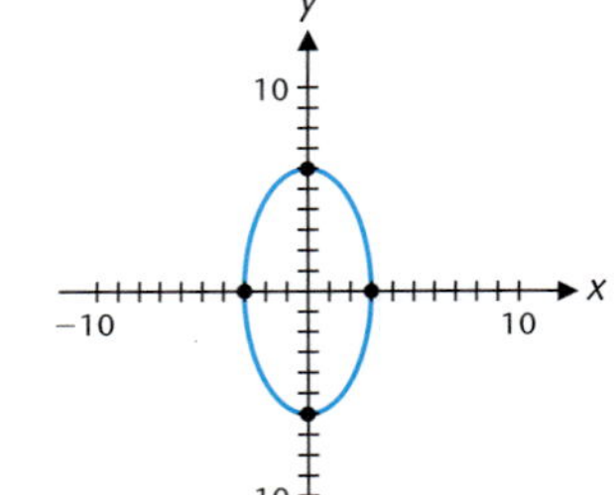

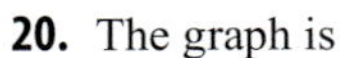

20. The graph is

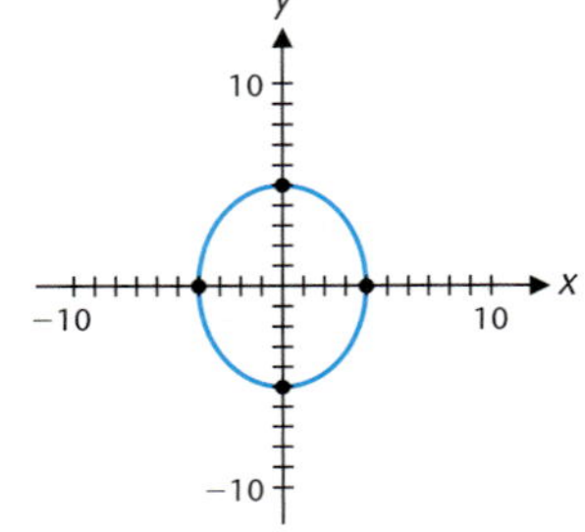

21. Major axis on x axis
Major axis length $= 10$
Minor axis length $= 6$

22. Major axis on x axis
Major axis length $= 14$
Minor axis length $= 10$

23. Major axis on y axis
Major axis length $= 22$
Minor axis length $= 16$

24. Major axis on y axis
Major axis length $= 24$
Minor axis length $= 18$

25. Major axis on x axis
Major axis length $= 16$
Distance of foci from center $= 6$

26. Major axis on y axis
Major axis length $= 24$
Distance of foci from center $= 10$

27. Major axis on y axis
Minor axis length $= 20$
Distance of foci from center $= \sqrt{70}$

28. Major axis on x axis
Minor axis length $= 14$
Distance of foci from center $= \sqrt{200}$

29. Explain why an equation whose graph is an ellipse does not define a function.

30. Consider all ellipses having $(0, \pm 1)$ as the ends of the minor axis. Describe the connection between the elongation of the ellipse and the distance from a focus to the origin.

Refer to Explore/Discuss 2. In Problems 31–34, find the magnification of the given equation by the given factor. Graph both equations.

31. $16x^2 + 9y^2 = 144$, factor 2

32. $9x^2 + 16y^2 = 144$, factor 0.25

33. $x^2 + 4y^2 = 16$, factor 0.5

34. $4x^2 + y^2 = 16$, factor 4

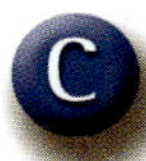

35. Find an equation of the set of points in a plane, each of whose distance from $(2, 0)$ is one-half its distance from the line $x = 8$. Identify the geometric figure.

36. Find an equation of the set of points in a plane, each of whose distance from $(0, 9)$ is three-fourths its distance from the line $y = 16$. Identify the geometric figure.

APPLICATIONS

37. Engineering. The semielliptical arch in the concrete bridge in the figure must have a clearance of 12 feet above the water and span a distance of 40 feet. Find the equation of the ellipse after inserting a coordinate system with the center of the ellipse at the origin and the major axis on the x axis. The y axis points up, and the x axis points to the right. How much clearance above the water is there 5 feet from the bank?

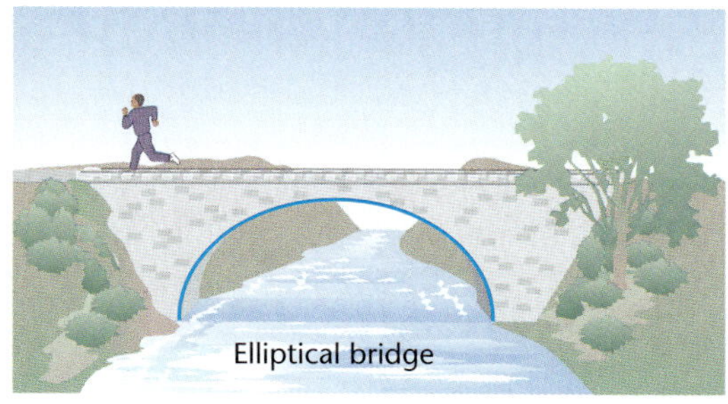

Elliptical bridge

38. Design. A 4×8 foot elliptical tabletop is to be cut out of a 4×8 foot rectangular sheet of teak plywood (see the figure). To draw the ellipse on the plywood, how far should the foci be located from each edge and how long a piece of string must be fastened to each focus to produce the ellipse (see Fig. 1 in the text)? Compute the answer to two decimal places.

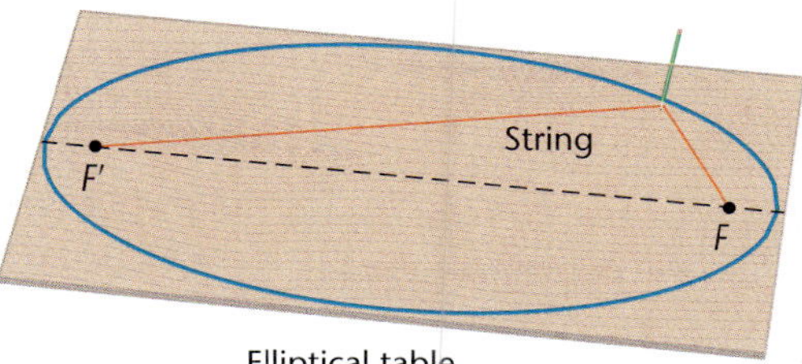

Elliptical table

★ **39. Aeronautical Engineering.** Of all possible wing shapes, it has been determined that the one with the least drag along the trailing edge is an ellipse. The leading edge may be a straight line, as shown in the figure. One of the most famous planes with this design was the World War II British Spitfire. The plane in the figure has a wingspan of 48.0 feet.

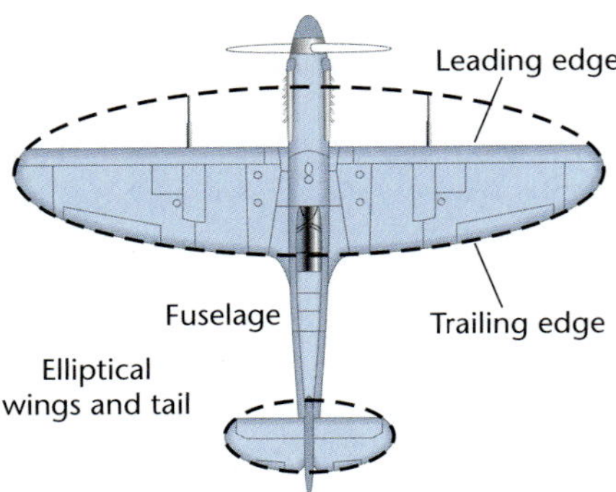

(A) If the straight-line leading edge is parallel to the major axis of the ellipse and is 1.14 feet in front of it, and if the leading edge is 46.0 feet long (including the width of the fuselage), find the equation of the ellipse. Let the x axis lie along the major axis (positive right), and let the y axis lie along the minor axis (positive forward).

(B) How wide is the wing in the center of the fuselage (assuming the wing passes through the fuselage)?

Compute quantities to three significant digits.

★ **40. Naval Architecture.** Currently, many high-performance racing sailboats use elliptical keels, rudders, and main sails for the reasons stated in Problem 39—less drag along the trailing edge. In the accompanying figure, the ellipse containing the keel has a 12.0-foot major axis. The straight-line leading edge is parallel to the major axis of the ellipse and 1.00 foot in front of it. The chord is 1.00 foot shorter than the major axis.

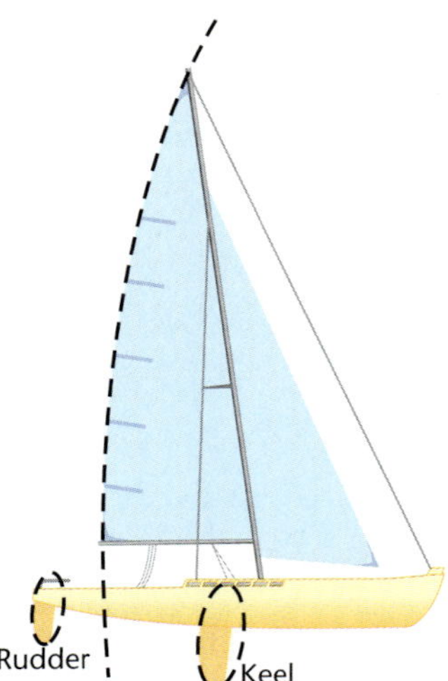

(A) Find the equation of the ellipse. Let the y axis lie along the minor axis of the ellipse, and let the x axis lie along the major axis, both with positive direction upward.

(B) What is the width of the keel, measured perpendicular to the major axis, 1 foot up the major axis from the bottom end of the keel?

Compute quantities to three significant digits.

SECTION 11.3 Hyperbola

Definition of a Hyperbola • Drawing a Hyperbola • Standard Equations and Their Graphs • Applications

As before, we start with a coordinate-free definition of a hyperbola. Using this definition, we show how a hyperbola can be drawn and we derive standard equations for hyperbolas specially located in a rectangular coordinate system.

Definition of a Hyperbola

The following is a coordinate-free definition of a hyperbola:

DEFINITION 1
Hyperbola

A **hyperbola** is the set of all points P in a plane such that the absolute value of the difference of the distances of P to two fixed points in the plane is a positive constant. Each of the fixed points, F' and F, is called a **focus.** The intersection points V' and V of the line through the foci and the two branches of the hyperbola are called **vertices,** and each is called a **vertex.** The line segment $V'V$ is called the **transverse axis.** The midpoint of the transverse axis is the **center** of the hyperbola.

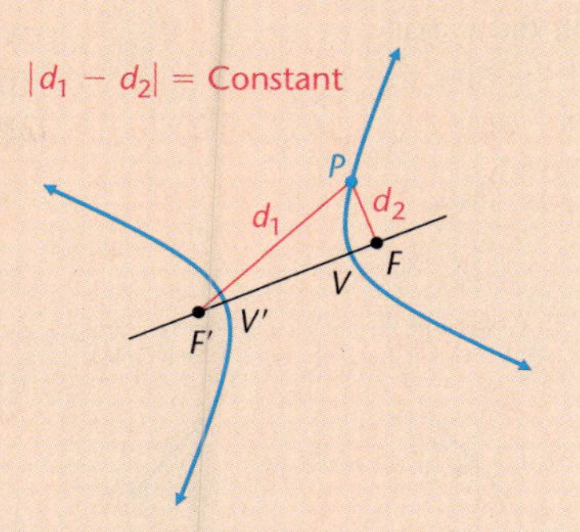

Drawing a Hyperbola

Thumbtacks, a straightedge, string, and a pencil are all that are needed to draw a hyperbola (Fig. 1). Place two thumbtacks in a piece of cardboard—these form the foci of the hyperbola. Rest one corner of the straightedge at the focus F' so that it is free to rotate about this point. Cut a piece of string shorter than the length of the straightedge, and fasten one end to the straightedge corner A and the other end to the thumbtack at F. Now push the string with a pencil up against the straightedge at B. Keeping the string taut, rotate the straightedge about F', keeping the corner at F'. The resulting curve will be part of a hyperbola. Other parts of the hyperbola can be drawn by changing the position of the straightedge and string. To see that the resulting curve meets the conditions of the definition, note that the difference of the distances BF' and BF is

$$\begin{aligned} BF' - BF &= BF' + BA - BF - BA \\ &= AF' - (BF + BA) \\ &= \begin{pmatrix}\text{Straightedge}\\ \text{length}\end{pmatrix} - \begin{pmatrix}\text{String}\\ \text{length}\end{pmatrix} \\ &= \text{Constant} \end{aligned}$$

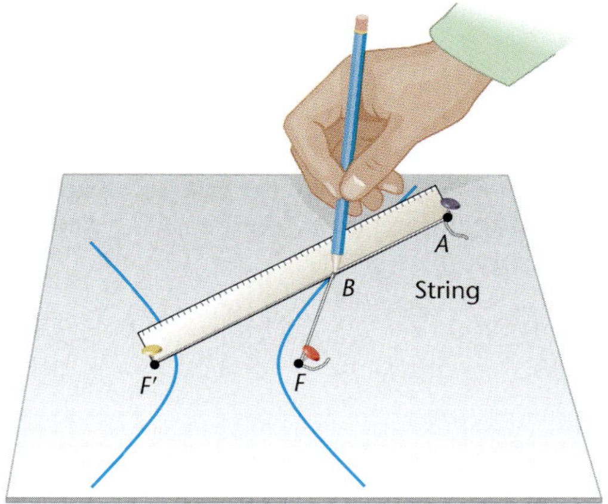

FIGURE 1 Drawing a hyperbola.

Standard Equations and Their Graphs

FIGURE 2 Hyperbola with foci on the x axis.

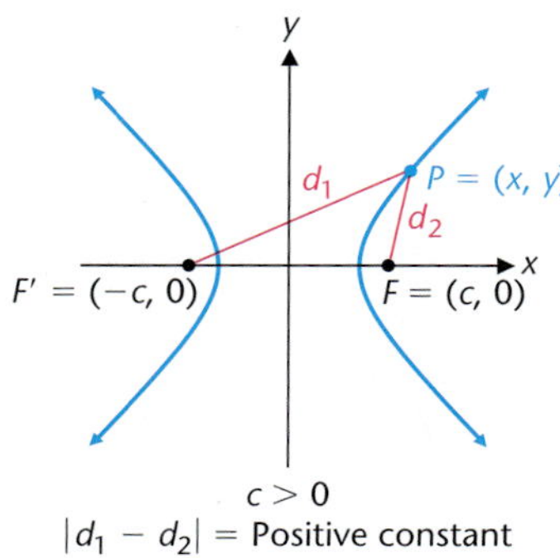

Using the definition of a hyperbola and the distance-between-two-points formula, we can derive the standard equations for a hyperbola located in a rectangular coordinate system. We start by placing a hyperbola in the coordinate system with the foci on the x axis equidistant from the origin at $F' = (-c, 0)$ and $F = (c, 0)$, $c > 0$, as in Figure 2.

Just as for the ellipse, it is convenient to represent the constant difference by $2a$, $a > 0$. Also, the geometric fact that the difference of two sides of a triangle is always less than the third side can be applied to Figure 2 to derive the following useful result:

$$\begin{aligned} |d_1 - d_2| &< 2c \\ 2a &< 2c \\ a &< c \end{aligned} \tag{1}$$

We will use this result in the derivation of the equation of a hyperbola, which we now begin.

Referring to Figure 2, the point $P = (x, y)$ is on the hyperbola if and only if

$$\begin{aligned} |d_1 - d_2| &= 2a \\ |d(P, F') - d(P, F)| &= 2a \\ \left|\sqrt{(x + c)^2 + y^2} - \sqrt{(x - c)^2 + y^2}\right| &= 2a \end{aligned}$$

After eliminating radicals and absolute value signs by appropriate use of squaring and simplifying, another good exercise for you, we have

$$(c^2 - a^2)x^2 - a^2y^2 = a^2(c^2 - a^2) \tag{2}$$

$$\frac{x^2}{a^2} - \frac{y^2}{c^2 - a^2} = 1 \tag{3}$$

Dividing both sides of equation (2) by $a^2(c^2 - a^2)$ is permitted, because neither a^2 nor $c^2 - a^2$ is 0. From equation (1), $a < c$; thus, $a^2 < c^2$ and $c^2 - a^2 > 0$. The constant a was chosen positive at the beginning.

To simplify equation (3) further, we let

$$b^2 = c^2 - a^2 \qquad b > 0 \tag{4}$$

to obtain

$$\frac{x^2}{a^2} - \frac{y^2}{b^2} = 1 \tag{5}$$

From equation (5) we see that the x intercepts, which are also the vertices, are $x = \pm a$ and there are no y intercepts. To see why there are no y intercepts, let $x = 0$ and solve for y:

$$\begin{aligned} \frac{0^2}{a^2} - \frac{y^2}{b^2} &= 1 \\ y^2 &= -b^2 \\ y &= \pm\sqrt{-b^2} \qquad \text{An imaginary number} \end{aligned}$$

FIGURE 3 Hyperbola with foci on the y axis.

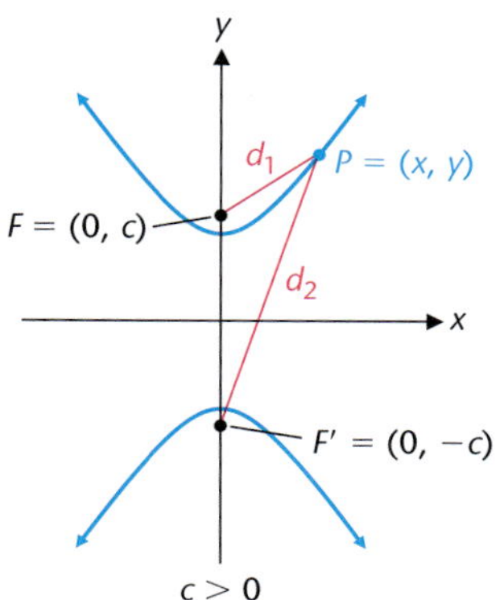

If we start with the foci on the y axis at $F' = (0, -c)$ and $F = (0, c)$ as in Figure 3, instead of on the x axis as in Figure 2, then, following arguments similar to those used for the first derivation, we obtain

$$\frac{y^2}{a^2} - \frac{x^2}{b^2} = 1 \tag{6}$$

where the relationship among a, b, and c remains the same as before:

$$b^2 = c^2 - a^2 \tag{7}$$

The center is still at the origin, but the transverse axis is now on the y axis.

As an aid to graphing equation (5), we solve the equation for y in terms of x, another good exercise for you, to obtain

$$y = \pm\frac{b}{a}x\sqrt{1 - \frac{a^2}{x^2}} \tag{8}$$

As x changes so that $|x|$ becomes larger, the expression $1 - (a^2/x^2)$ within the radical approaches 1. Hence, for large values of $|x|$, equation (5) behaves very much like the lines

$$y = \pm\frac{b}{a}x \tag{9}$$

These lines are **asymptotes** for the graph of equation (5). The hyperbola approaches these lines as a point $P = (x, y)$ on the hyperbola moves away from the origin (Fig. 4). An easy way to draw the asymptotes is to first draw the rectangle as in Figure 4, then extend the diagonals. We refer to this rectangle as the **asymptote rectangle.**

FIGURE 4 Asymptotes.

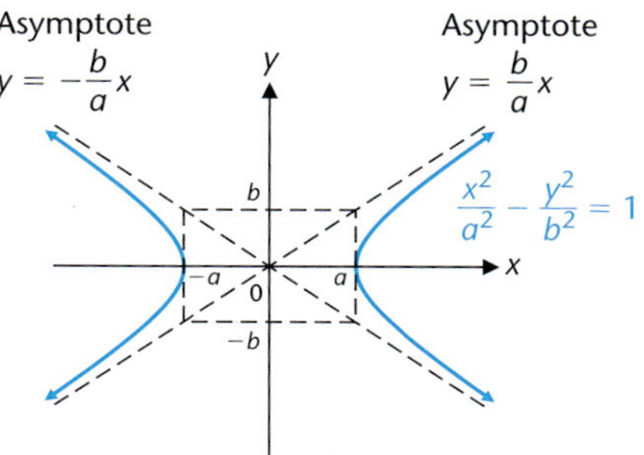

Starting with equation (6) and proceeding as we did for equation (5), we obtain the asymptotes for the graph of equation (6):

$$y = \pm\frac{a}{b}x \tag{10}$$

The perpendicular bisector of the transverse axis, extending from one side of the asymptote rectangle to the other, is called the **conjugate axis** of the hyperbola.

Given an equation of the form of equations (5) or (6), how can we find the coordinates of the foci without memorizing or looking up the relation $b^2 = c^2 - a^2$? Just as with the ellipse, there is a simple geometric relationship in a hyperbola that enables us to get the same result using the Pythagorean theorem. To see this relationship, we rewrite $b^2 = c^2 - a^2$ in the form

$$c^2 = a^2 + b^2 \tag{11}$$

Note in the figures in Theorem 1 below that the distance from the center to a focus is the same as the distance from the center to a corner of the asymptote rectangle. Stated in another way:

A circle, with center at the origin, that passes through all four corners of the asymptote rectangle also passes through all foci of hyperbolas with asymptotes determined by the diagonals of the rectangle.

We summarize all the preceding results in Theorem 1 for convenient reference.

THEOREM 1

Standard Equations of a Hyperbola with Center at (0, 0)

1. $\dfrac{x^2}{a^2} - \dfrac{y^2}{b^2} = 1$

x intercepts: $\pm a$ (vertices)

y intercepts: none

Foci: $F' = (-c, 0)$, $F = (c, 0)$

$$c^2 = a^2 + b^2$$

Transverse axis length $= 2a$

Conjugate axis length $= 2b$

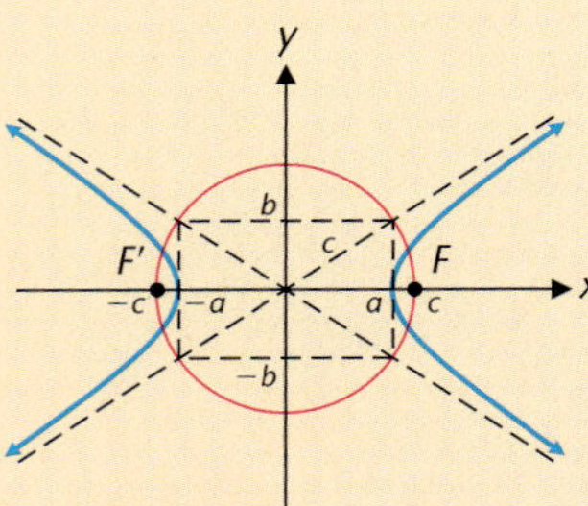

2. $\dfrac{y^2}{a^2} - \dfrac{x^2}{b^2} = 1$

x intercepts: none

y intercepts: $\pm a$ (vertices)

Foci: $F' = (0, -c)$, $F = (0, c)$

$$c^2 = a^2 + b^2$$

Transverse axis length $= 2a$

Conjugate axis length $= 2b$

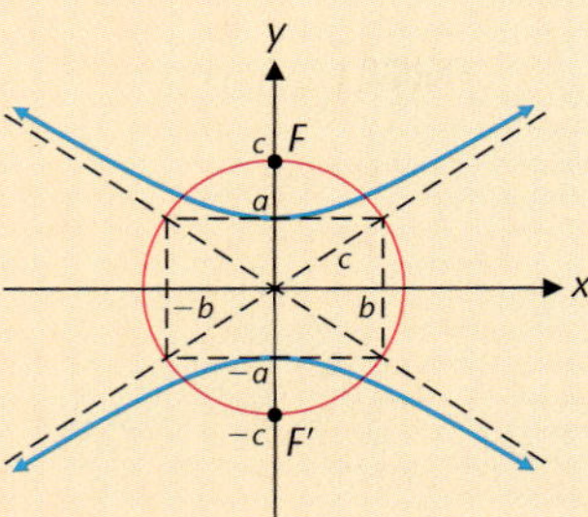

[*Note:* Both graphs are symmetric with respect to the x axis, y axis, and origin.]

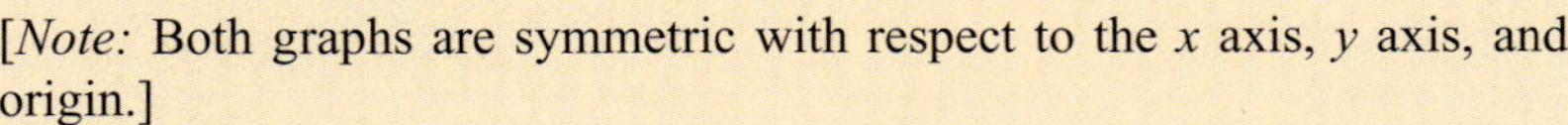

EXPLORE/DISCUSS 1

The line through a focus F of a hyperbola that is perpendicular to the transverse axis intersects the hyperbola in two points G and H. For each of the two standard equations of a hyperbola with center (0, 0), find an expression in terms of a and b for the distance from G to H.

EXAMPLE 1 Graphing Hyperbolas

Find the coordinates of the foci, find the lengths of the transverse and conjugate axes, and graph the following equation:

$$9x^2 - 16y^2 = 144$$

SOLUTION

First, write the equation in standard form by dividing both sides by 144 and determine a and b:

$$9x^2 - 16y^2 = 144$$

$$\frac{9x^2}{144} - \frac{16y^2}{144} = \frac{144}{144}$$

$$\frac{x^2}{16} - \frac{y^2}{9} = 1$$

$a = 4$ and $b = 3$

Foci: $c^2 = a^2 + b^2$

$= 16 + 9$

$= 25$

$c = 5$

Thus, the foci are $F' = (-5, 0)$ and $F = (5, 0)$.

Transverse axis length $= 2(4) = 8$

Conjugate axis length $= 2(3) = 6$

Hand-Drawn Solution

Locate the intercepts:

x intercepts: ± 4

y intercepts: none

Sketch the intercepts, and the asymptotes using the asymptote rectangle, then sketch the hyperbola (Fig. 5).

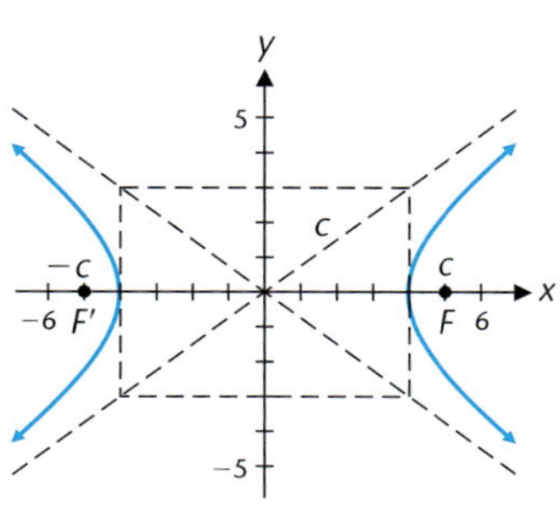

FIGURE 5

Graphing Utility Solution

Solve for y:

$$9x^2 - 16y^2 = 144$$

$$-16y^2 = 144 - 9x^2$$

$$y^2 = (9x^2 - 144)/16$$

$$y = \pm\sqrt{(9x^2 - 144)/16}$$

This produces the functions $y_1 = \sqrt{(9x^2 - 144)/16}$ and $y_2 = -\sqrt{(9x^2 - 144)/16}$ whose graphs are shown in Figure 6.

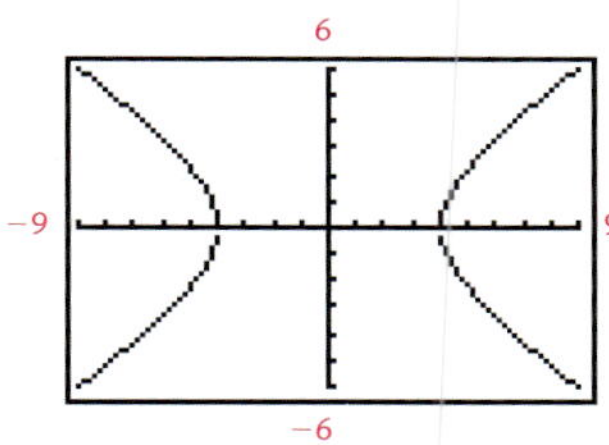

FIGURE 6

MATCHED 1 PROBLEM

Find the coordinates of the foci, find the lengths of the transverse and conjugate axes, and graph the following equation:

$$16x^2 - 25y^2 = 400$$

EXAMPLE 2 Graphing Hyperbolas

Find the coordinates of the foci, find the lengths of the transverse and conjugate axes, and graph the following equation:

$$16y^2 - 9x^2 = 144$$

SOLUTION

Write the equation in standard form:

$$16y^2 - 9x^2 = 144$$

$$\frac{y^2}{9} - \frac{x^2}{16} = 1$$

$$a = 3 \qquad \text{and} \qquad b = 4$$

$$\text{Foci: } c^2 = a^2 + b^2$$
$$= 9 + 16$$
$$= 25$$
$$c = 5$$

Thus, the foci are $F' = (0, -5)$ and $F = (0, 5)$.

$$\text{Transverse axis length} = 2(3) = 6$$
$$\text{Conjugate axis length} = 2(4) = 8$$

Hand-Drawn Solution

Locate the intercepts:

x intercepts: none

y intercepts: ± 3

Sketch the intercepts, and the asymptotes using the asymptote rectangle, then sketch the hyperbola (Fig. 7).

Graphing Utility Solution

Solve for y:

$$16y^2 - 9x^2 = 144$$
$$y^2 = (9x^2 + 144)/16$$
$$y = \pm\sqrt{(9x^2 + 144)/16}$$

This produces the functions $y_1 = \sqrt{(9x^2 + 144)/16}$ and $y_2 = -\sqrt{(9x^2 + 144)/16}$ whose graphs are shown in Figure 8.

FIGURE 7

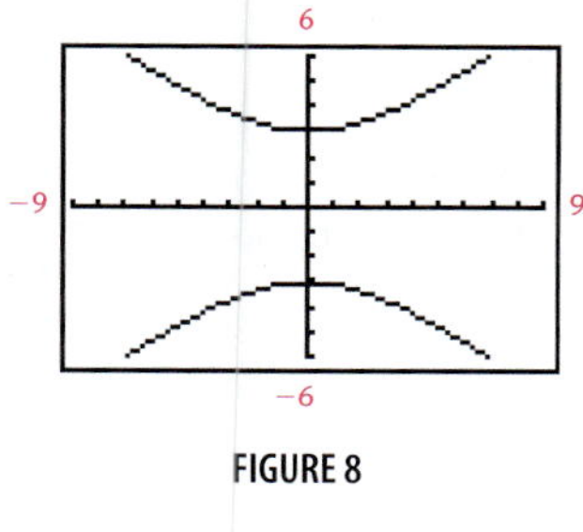

FIGURE 8

MATCHED 2 PROBLEM

Find the coordinates of the foci, find the lengths of the transverse and conjugate axes, and graph the following equation:

$$25y^2 - 16x^2 = 400$$

EXAMPLE 3 Graphing Hyperbolas

Find the coordinates of the foci, find the lengths of the transverse and conjugate axes, and graph the following equation:

$$2x^2 - y^2 = 10$$

SOLUTION

$$2x^2 - y^2 = 10$$

$$\frac{x^2}{5} - \frac{y^2}{10} = 1$$

$$a = \sqrt{5} \qquad \text{and} \qquad b = \sqrt{10}$$

$$\begin{aligned} \text{Foci: } c^2 &= a^2 + b^2 \\ &= 5 + 10 \\ &= 15 \\ c &= \sqrt{15} \end{aligned}$$

Thus, the foci are $F' = (-\sqrt{15}, 0)$ and $F = (\sqrt{15}, 0)$.

Transverse axis length $= 2\sqrt{5} \approx 4.47$
Conjugate axis length $= 2\sqrt{10} \approx 6.32$

Hand-Drawn Solution

Locate the intercepts:

x intercepts: $\pm\sqrt{5}$

y intercepts: none

Sketch the intercepts, and the asymptotes using the asymptote rectangle, then sketch the hyperbola (Fig. 9).

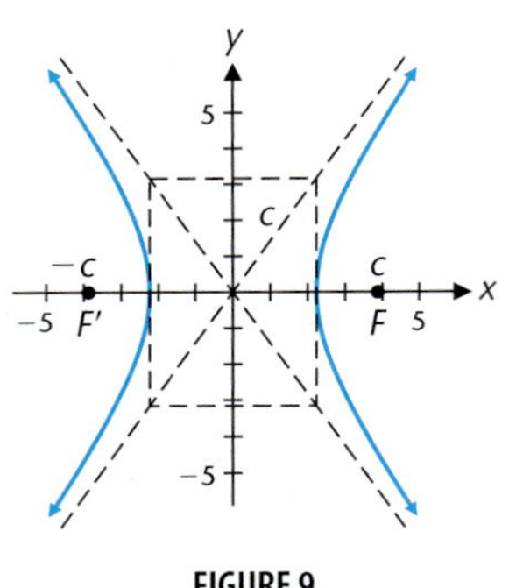

FIGURE 9

Graphing Utility Solution

Solve for y:

$$2x^2 - y^2 = 10$$
$$y^2 = 2x^2 - 10$$
$$y = \pm\sqrt{2x^2 - 10}$$

This produces the functions $y_1 = \sqrt{2x^2 - 10}$ and $y_2 = -\sqrt{2x^2 - 10}$ whose graphs are shown in Figure 10.

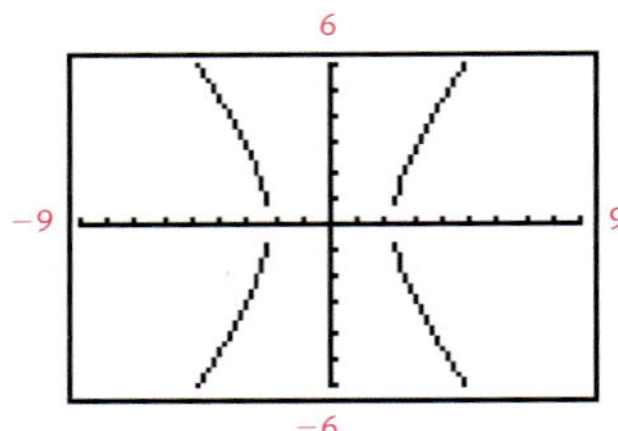

FIGURE 10

MATCHED 3 PROBLEM

Find the coordinates of the foci, find the lengths of the transverse and conjugate axes, and graph the following equation:

$$y^2 - 3x^2 = 12$$

Hyperbolas of the form

$$\frac{x^2}{M} - \frac{y^2}{N} = 1 \quad \text{and} \quad \frac{y^2}{N} - \frac{x^2}{M} = 1 \qquad M, N > 0$$

are called **conjugate hyperbolas.** In Examples 1 and 2 and in Matched Problems 1 and 2, the hyperbolas are conjugate hyperbolas—they share the same asymptotes.

CAUTION

When making a quick sketch of a hyperbola, it is a common error to have the hyperbola opening up and down when it should open left and right, or vice versa. The mistake can be avoided if you first locate the intercepts accurately.

EXAMPLE 4 Finding the Equation of a Hyperbola

Find an equation of a hyperbola in the form

$$\frac{y^2}{M} - \frac{x^2}{N} = 1 \qquad M, N > 0$$

if the center is at the origin, and:

(A) Length of transverse axis is 12
Length of conjugate axis is 20

(B) Length of transverse axis is 6
Distance of foci from center is 5

SOLUTIONS

(A) Start with

$$\frac{y^2}{a^2} - \frac{x^2}{b^2} = 1$$

and find a and b:

$$a = \frac{12}{2} = 6 \qquad \text{and} \qquad b = \frac{20}{2} = 10$$

Thus, the equation is

$$\frac{y^2}{36} - \frac{x^2}{100} = 1$$

(B) Start with

$$\frac{y^2}{a^2} - \frac{x^2}{b^2} = 1$$

and find a and b:

$$a = \frac{6}{2} = 3$$

FIGURE 11 Asymptote rectangle.

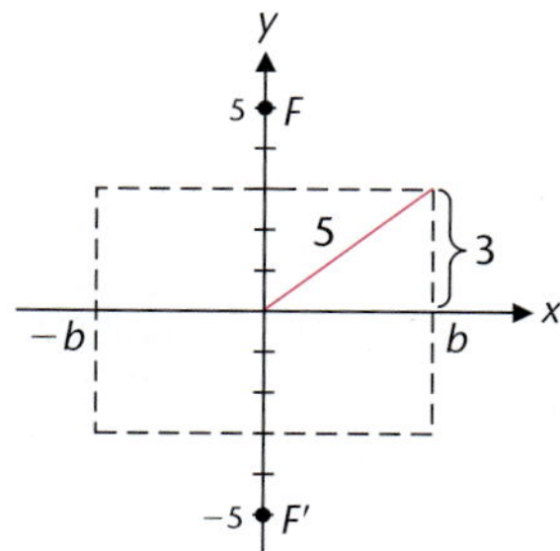

To find b, sketch the asymptote rectangle (Fig. 11), label known parts, and use the Pythagorean theorem:

$$\begin{aligned} b^2 &= 5^2 - 3^2 \\ &= 16 \\ b &= 4 \end{aligned}$$

Thus, the equation is

$$\frac{y^2}{9} - \frac{x^2}{16} = 1$$

MATCHED 4 PROBLEM

Find an equation of a hyperbola in the form

$$\frac{x^2}{M} - \frac{y^2}{N} = 1 \qquad M, N > 0$$

if the center is at the origin, and:

(A) Length of transverse axis is 50
Length of conjugate axis is 30

(B) Length of conjugate axis is 12
Distance of foci from center is 9

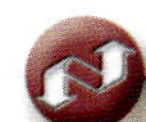

EXPLORE/DISCUSS 2

(A) Does the line with equation $y = x$ intersect the hyperbola with equation $x^2 - (y^2/4) = 1$? If so, find the coordinates of all intersection points.

(B) Does the line with equation $y = 3x$ intersect the hyperbola with equation $x^2 - (y^2/4) = 1$? If so, find the coordinates of all intersection points.

(C) For which values of m does the line with equation $y = mx$ intersect the hyperbola $\frac{x^2}{a^2} - \frac{y^2}{b^2} = 1$? Find the coordinates of all intersection points.

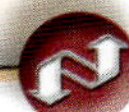

Applications

You may not be aware of the many important uses of hyperbolic forms. They are encountered in the study of comets; the loran system of navigation for pleasure boats, ships, and aircraft; sundials; capillary action; nuclear reactor cooling towers; optical and radiotelescopes; and contemporary architectural structures. The TWA building at Kennedy Airport is a *hyperbolic paraboloid,* and the St. Louis Science Center Planetarium is a *hyperboloid.* With such structures, thin concrete shells can span large spaces [Fig. 12(a)]. Some comets from outer space occasionally enter the sun's gravitational field, follow a hyperbolic path around the sun (with the sun as a focus), and then leave, never to be seen again [Fig. 12(b)]. Example 5 illustrates the use of hyperbolas in navigation.

FIGURE 12 Uses of hyperbolic forms.

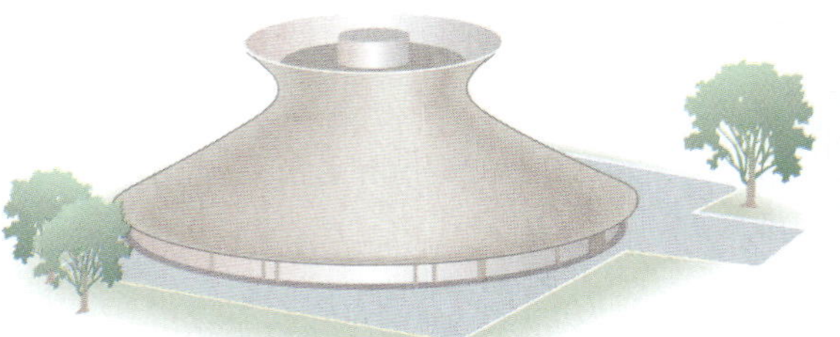

St. Louis Planetarium
(a)

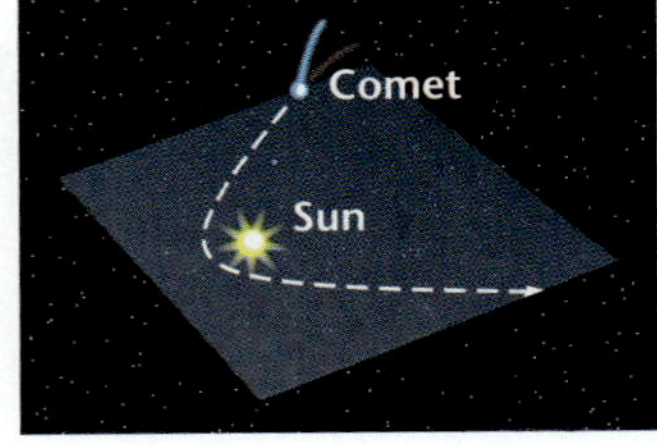

Comet around sun
(b)

EXAMPLE 5 Navigation

A ship is traveling on a course parallel to and 60 miles from a straight shoreline. Two transmitting stations, S_1 and S_2, are located 200 miles apart on the shoreline (Fig. 13). By timing radio signals from the stations, the ship's navigator determines that the ship is between the two stations and 50 miles closer to S_2 than to S_1. Find the distance from the ship to each station. Round answers to one decimal place.

FIGURE 13 $d_1 - d_2 = 50.$

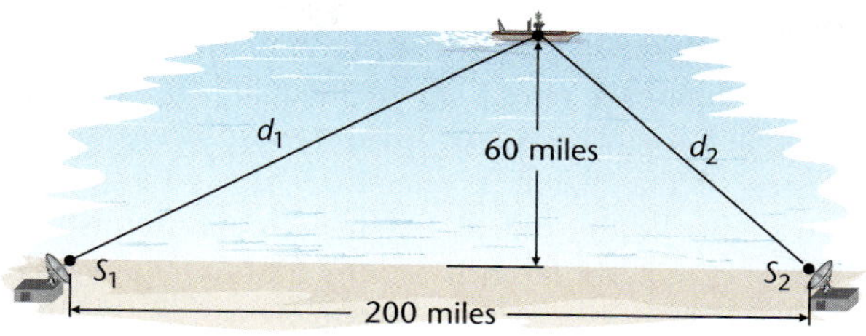

SOLUTION

If d_1 and d_2 are the distances from the ship to S_1 and S_2, respectively, then $d_1 - d_2 = 50$ and the ship must be on the hyperbola with foci at S_1 and S_2 and fixed difference 50, as illustrated in Figure 14. In the derivation of the equation of a hyperbola, we represented the fixed difference as $2a$. Thus, for the hyperbola in Figure 14 we have:

FIGURE 14

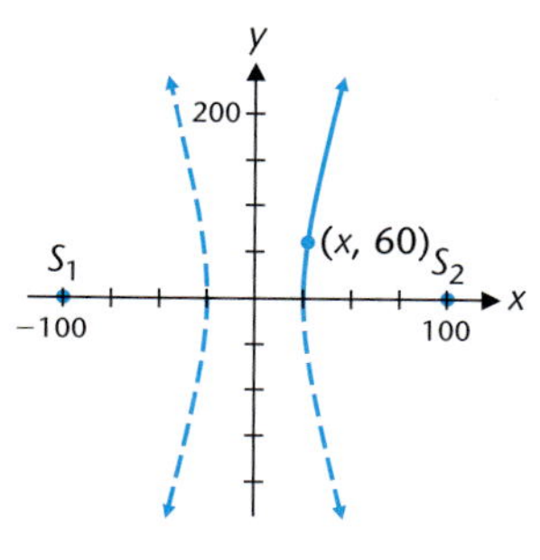

$$c = 100$$
$$a = \tfrac{1}{2}(50) = 25$$
$$b = \sqrt{100^2 - 25^2} = \sqrt{9{,}375}$$

The equation for this hyperbola is

$$\frac{x^2}{625} - \frac{y^2}{9{,}375} = 1$$

Substitute $y = 60$ and solve for x (see Fig. 14):

$$\frac{x^2}{625} - \frac{60^2}{9{,}375} = 1$$
$$\frac{x^2}{625} = \frac{3{,}600}{9{,}375} + 1$$
$$x^2 = 625\,\frac{3{,}600 + 9{,}375}{9{,}375}$$
$$= 865$$

Thus, $x = \sqrt{865} \approx 29.41$. (The negative square root is discarded, because the ship is closer to S_2 than to S_1.)

Distance from ship to S_1	Distance from ship to S_2
$d_1 = \sqrt{(29.41 + 100)^2 + 60^2}$	$d_2 = \sqrt{(29.41 - 100)^2 + 60^2}$
$= \sqrt{20{,}346.9841}$	$= \sqrt{8{,}582.9841}$
≈ 142.6 miles	≈ 92.6 miles

Notice that the difference between these two distances is 50, as it should be.

MATCHED 5 PROBLEM

Repeat Example 5 if the ship is 80 miles closer to S_2 than to S_1.

FIGURE 15 Loran navigation.

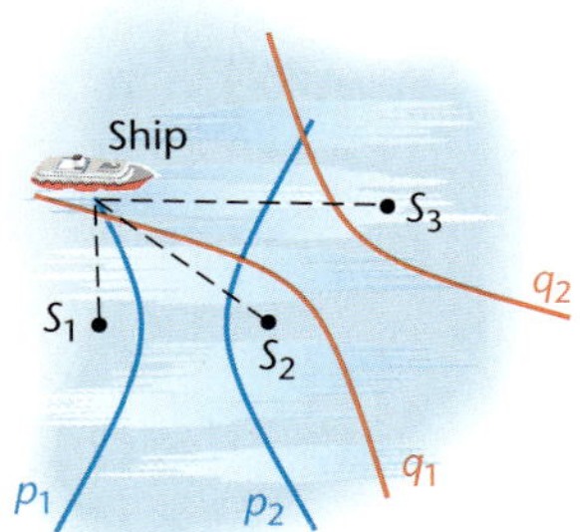

Example 5 illustrates a simplified form of the loran (LOng RAnge Navigation) system. In practice, three transmitting stations are used to send out signals simultaneously (Fig. 15), instead of the two used in Example 5. A computer onboard a ship will record these signals and use them to determine the differences of the distances that the ship is to S_1 and S_2, and to S_2 and S_3. Plotting all points so that these distances remain constant produces two branches, p_1 and p_2, of a hyperbola with foci S_1 and S_2, and two branches, q_1 and q_2, of a hyperbola with foci S_2 and S_3. It is easy to tell which branches the ship is on by comparing the signals from each station. The intersection of a branch of each hyperbola locates the ship and the computer expresses this in terms of longitude and latitude.

ANSWERS MATCHED PROBLEMS

1.

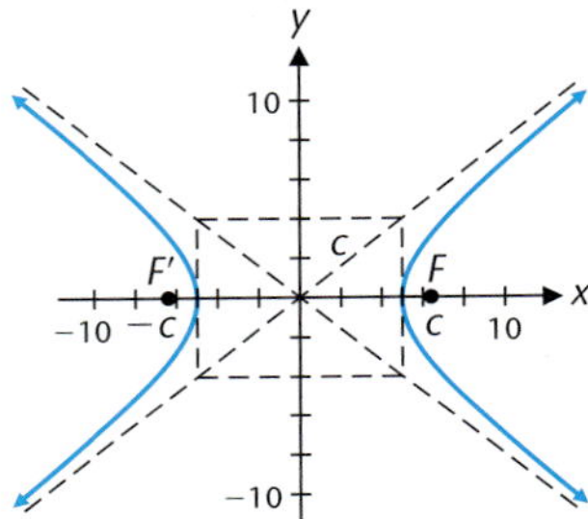

$$\frac{x^2}{25} - \frac{y^2}{16} = 1$$

Foci: $F' = (-\sqrt{41}, 0)$, $F = (\sqrt{41}, 0)$
Transverse axis length = 10
Conjugate axis length = 8

2.

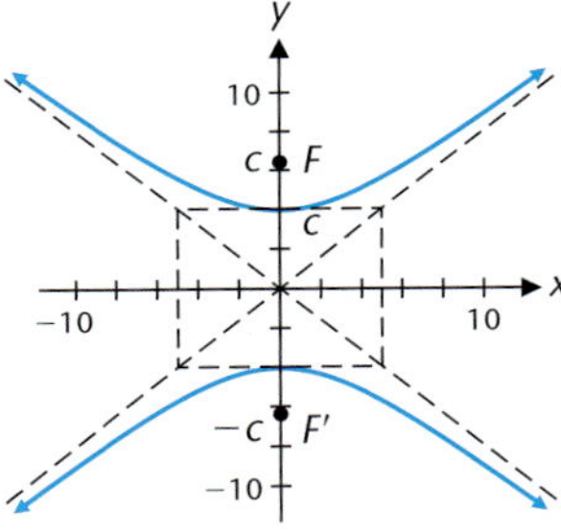

$$\frac{y^2}{16} - \frac{x^2}{25} = 1$$

Foci: $F' = (0, -\sqrt{41})$, $F = (0, \sqrt{41})$
Transverse axis length = 8
Conjugate axis length = 10

3.

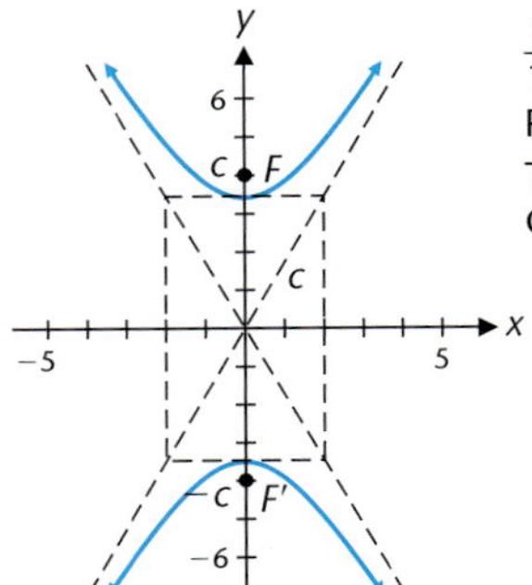

$$\frac{y^2}{12} - \frac{x^2}{4} = 1$$

Foci: $F' = (0, -4)$, $F = (0, 4)$
Transverse axis length = $2\sqrt{12} \approx 6.93$
Conjugate axis length = 4

4. (A) $\dfrac{x^2}{625} - \dfrac{y^2}{225} = 1$ (B) $\dfrac{x^2}{45} - \dfrac{y^2}{36} = 1$ **5.** $d_1 = 159.5$ miles, $d_2 = 79.5$ miles

EXERCISE 11.3

A

In Problems 1–4, match each equation with one of graphs (a)–(d).

1. $x^2 - y^2 = 1$ **2.** $y^2 - x^2 = 1$

3. $y^2 - x^2 = 4$ **4.** $x^2 - y^2 = 4$

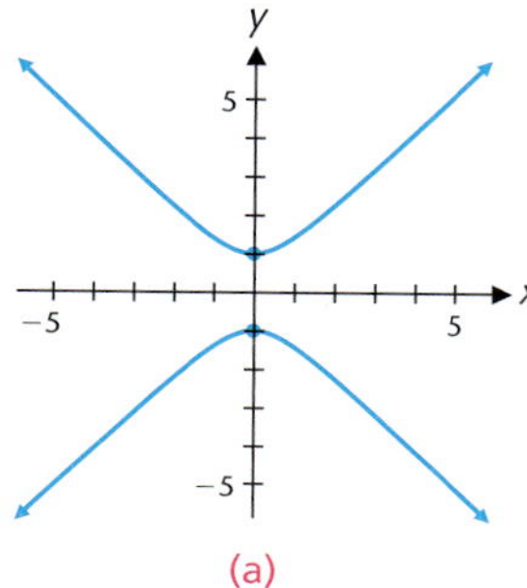
(a)

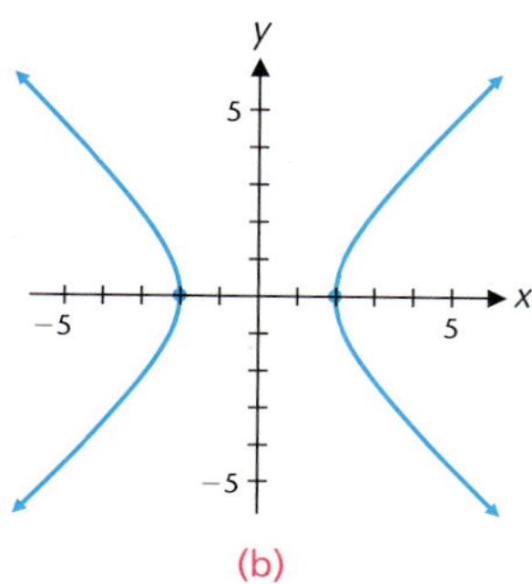
(b)

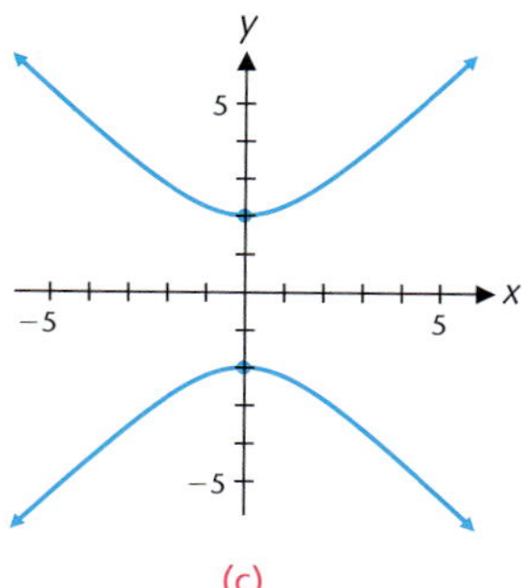
(c)

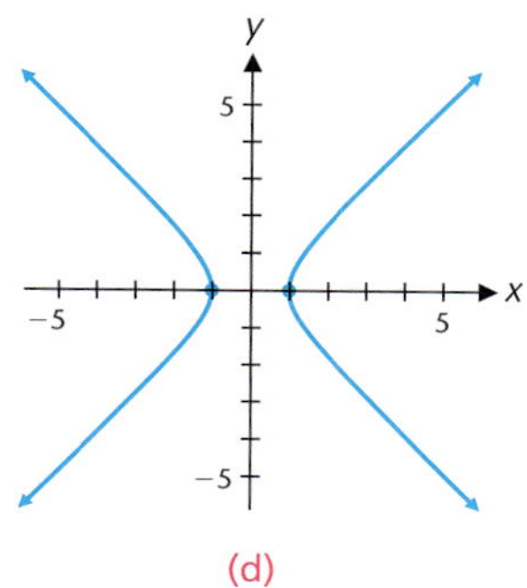
(d)

Sketch a graph of each equation in Problems 5–12, find the coordinates of the foci, and find the lengths of the transverse and conjugate axes.

5. $\frac{x^2}{9} - \frac{y^2}{4} = 1$ **6.** $\frac{x^2}{9} - \frac{y^2}{25} = 1$

7. $\frac{y^2}{4} - \frac{x^2}{9} = 1$ **8.** $\frac{y^2}{25} - \frac{x^2}{9} = 1$

9. $4x^2 - y^2 = 16$ **10.** $x^2 - 9y^2 = 9$

11. $9y^2 - 16x^2 = 144$ **12.** $4y^2 - 25x^2 = 100$

B

Sketch a graph of each equation in Problems 13–16, find the coordinates of the foci, and find the lengths of the transverse and conjugate axes.

13. $3x^2 - 2y^2 = 12$ **14.** $3x^2 - 4y^2 = 24$

15. $7y^2 - 4x^2 = 28$ **16.** $3y^2 - 2x^2 = 24$

In Problems 17–28, find an equation of a hyperbola in the form

$$\frac{x^2}{M} - \frac{y^2}{N} = 1 \quad \text{or} \quad \frac{y^2}{N} - \frac{x^2}{M} = 1 \quad M, N > 0$$

if the center is at the origin, and

17. The graph is

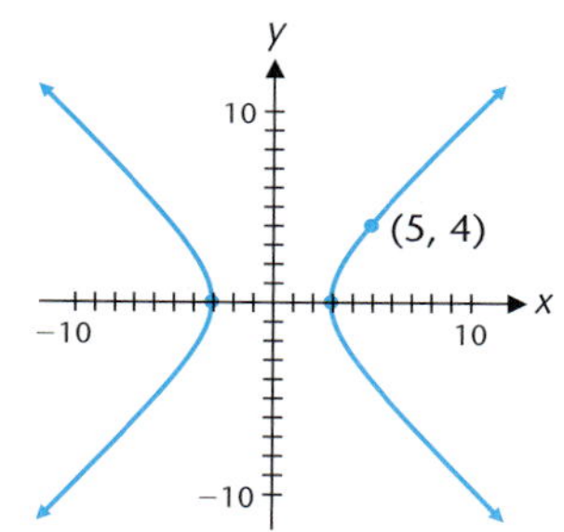

18. The graph is

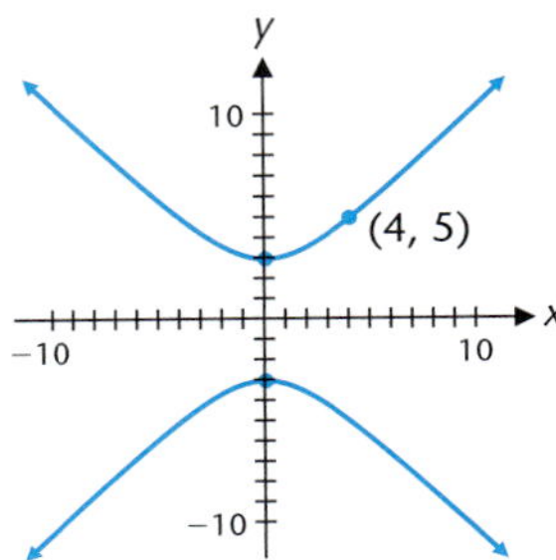

19. The graph is

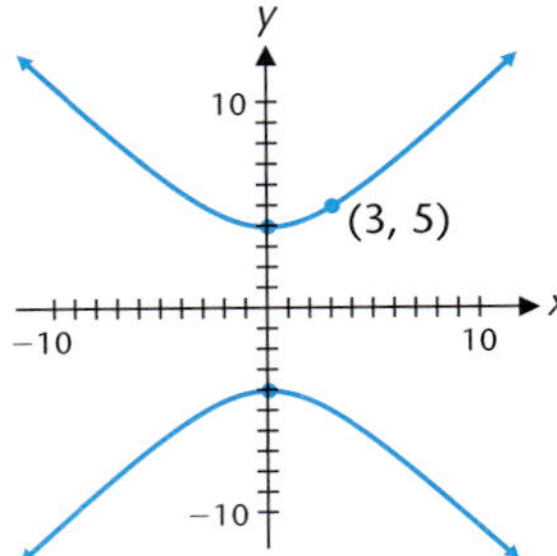

20. The graph is

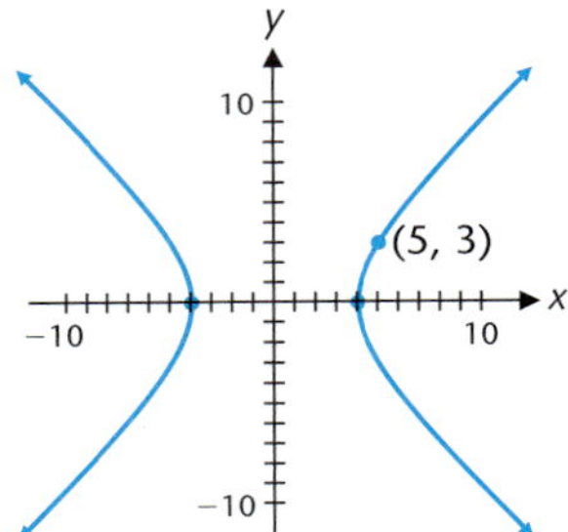

21. Transverse axis on x axis
Transverse axis length $= 14$
Conjugate axis length $= 10$

22. Transverse axis on x axis
Transverse axis length $= 8$
Conjugate axis length $= 6$

23. Transverse axis on y axis
Transverse axis length $= 24$
Conjugate axis length $= 18$

24. Transverse axis on y axis
Transverse axis length $= 16$
Conjugate axis length $= 22$

25. Transverse axis on x axis
Transverse axis length $= 18$
Distance of foci from center $= 11$

26. Transverse axis on x axis
Transverse axis length $= 16$
Distance of foci from center $= 10$

27. Conjugate axis on x axis
Conjugate axis length $= 14$
Distance of foci from center $= \sqrt{200}$

28. Conjugate axis on x axis
Conjugate axis length $= 10$
Distance of foci from center $= \sqrt{70}$

29. **(A)** How many hyperbolas have center at (0, 0) and a focus at (1, 0)? Find their equations.

(B) How many ellipses have center at (0, 0) and a focus at (1, 0)? Find their equations.

(C) How many parabolas have center at (0, 0) and focus at (1, 0)? Find their equations.

30. How many hyperbolas have the lines $y = \pm 2x$ as asymptotes? Find their equations.

31. Find all intersection points of the graph of the hyperbola $x^2 - y^2 = 1$ with the graph of each of the following lines:

(A) $y = 0.5x$ **(B)** $y = 2x$

For what values of m will the graph of the hyperbola and the graph of the line $y = mx$ intersect? Find the coordinates of these intersection points.

32. Find all intersection points of the graph of the hyperbola $y^2 - x^2 = 1$ with the graph of each of the following lines:

(A) $y = 0.5x$ **(B)** $y = 2x$

For what values of m will the graph of the hyperbola and the graph of the line $y = mx$ intersect? Find the coordinates of these intersection points.

33. Find all intersection points of the graph of the hyperbola $y^2 - 4x^2 = 1$ with the graph of each of the following lines:

(A) $y = x$ **(B)** $y = 3x$

For what values of m will the graph of the hyperbola and the graph of the line $y = mx$ intersect? Find the coordinates of these intersection points.

34. Find all intersection points of the graph of the hyperbola $4x^2 - y^2 = 1$ with the graph of each of the following lines:

(A) $y = x$ **(B)** $y = 3x$

For what values of m will the graph of the hyperbola and the graph of the line $y = mx$ intersect? Find the coordinates of these intersection points.

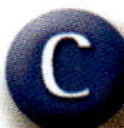

Eccentricity. *Problems 35 and 36 and Problems 35 and 36 in Exercise 11.2 are related to a property of conics called* ***eccentricity,*** *which is denoted by a positive real number E. Parabolas, ellipses, and hyperbolas all can be defined in terms of E, a fixed point called a focus, and a fixed line not containing the focus called a directrix as follows: The set of points in a plane each of whose distance from a fixed point is E times its distance from a fixed line is an ellipse if* $0 < E < 1$*, a parabola if* $E = 1$*, and a hyperbola if* $E > 1$.

35. Find an equation of the set of points in a plane each of whose distance from (3, 0) is three-halves its distance from the line $x = \frac{4}{3}$. Identify the geometric figure.

36. Find an equation of the set of points in a plane each of whose distance from (0, 4) is four-thirds its distance from the line $y = \frac{9}{4}$. Identify the geometric figure.

APPLICATIONS

37. Architecture. An architect is interested in designing a thin-shelled dome in the shape of a hyperbolic paraboloid, as shown in Figure (a). Find the equation of the hyperbola located in a coordinate system [Fig. (b)] satisfying the indicated conditions. How far is the hyperbola above the vertex 6 feet to the right of the vertex? Compute the answer to two decimal places.

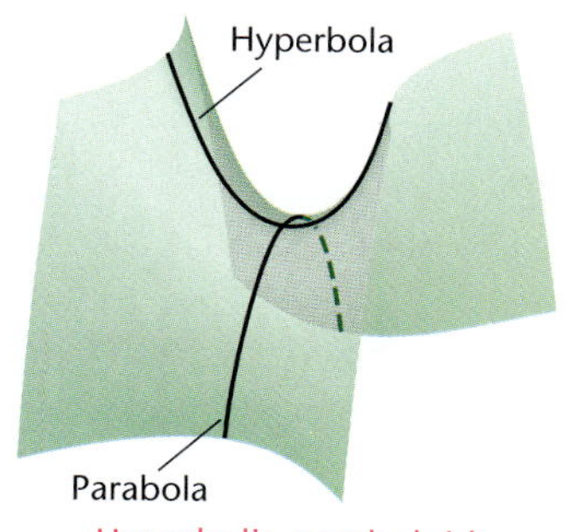

Hyperbolic paraboloid
(a)

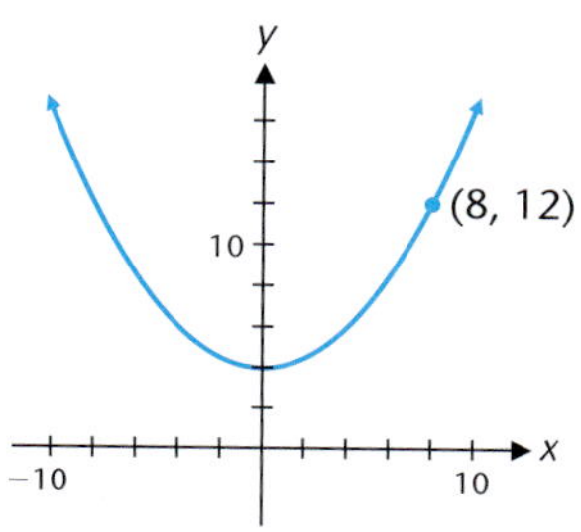

Hyperbola part of dome
(b)

38. Nuclear Power. A nuclear reactor cooling tower is a **hyperboloid,** that is, a hyperbola rotated around its conjugate axis, as shown in Figure (a). The equation of the hyperbola in Figure (b) used to generate the hyperboloid is

$$\frac{x^2}{100^2} - \frac{y^2}{150^2} = 1$$

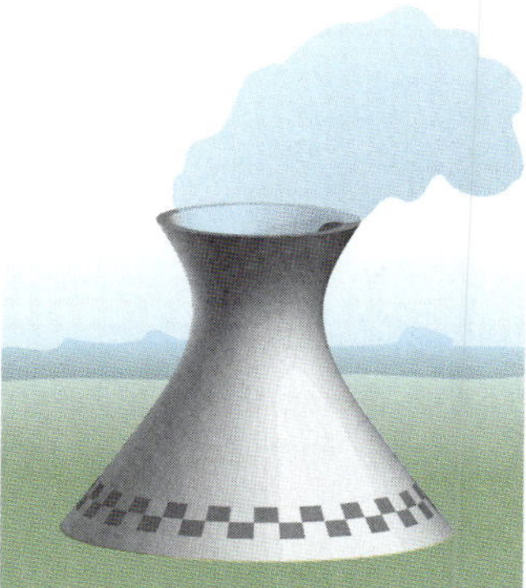

Nuclear reactor cooling tower
(a)

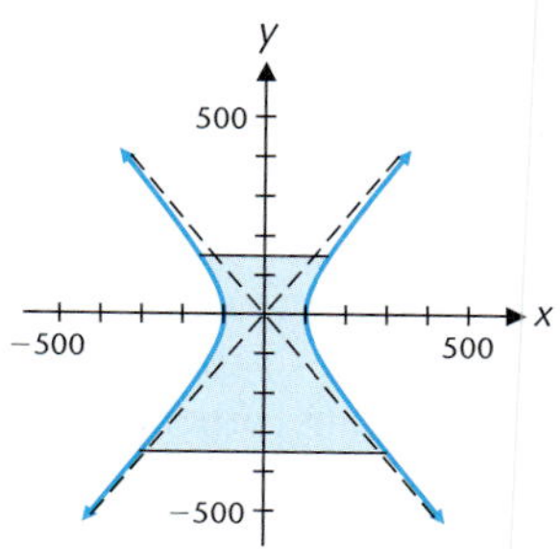

Hyperbola part of dome
(b)

If the tower is 500 feet tall, the top is 150 feet above the center of the hyperbola, and the base is 350 feet below the center, what is the radius of the top and the base? What is the radius of the smallest circular cross section in the tower? Compute answers to three significant digits.

39. Space Science. In tracking space probes to the outer planets, NASA uses large parabolic reflectors with diameters equal to two-thirds the length of a football field. Needless to say, many design problems are created by the weight of these reflectors. One weight problem is solved by using a hyperbolic reflector sharing the parabola's focus to reflect the incoming electromagnetic waves to the other focus of the hyperbola where receiving equipment is installed (see the figure).

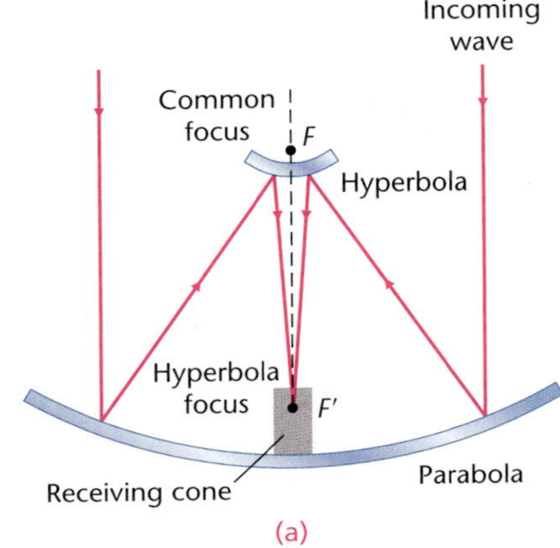

(a)

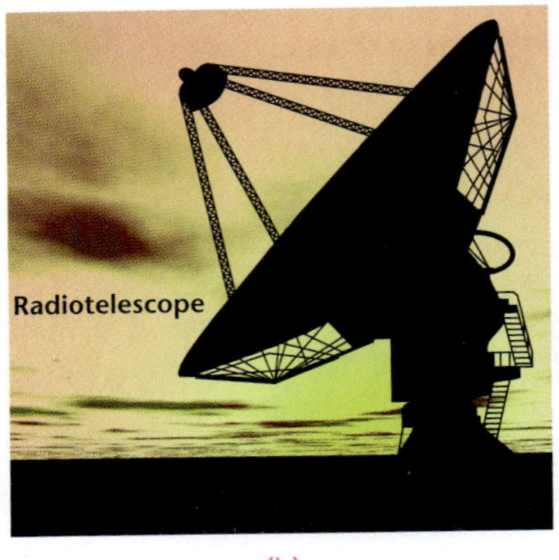

(b)

For the receiving antenna shown in the figure, the common focus F is located 120 feet above the vertex of the parabola, and focus F' (for the hyperbola) is 20 feet above the vertex. The vertex of the reflecting hyperbola is 110 feet above the vertex for the parabola. Introduce a coordinate system by using the axis of the parabola as the y axis (up positive), and let the x axis pass through the center of the hyperbola (right positive). What is the equation of the reflecting hyperbola? Write y in terms of x.

SECTION 11.4 Translation of Axes

Translation of Axes • Standard Equations of Translated Conics • Graphing Equations of the Form $Ax^2 + Cy^2 + Dx + Ey + F = 0$ • Finding Equations of Conics

In Sections 11.1, 11.2, and 11.3 we found standard equations for parabolas, ellipses, and hyperbolas located with their axes on the coordinate axes and centered relative to the origin. What happens if we move conics away from the origin while keeping their axes parallel to the coordinate axes? We will show that we can obtain new standard equations that are special cases of the equation

$$Ax^2 + Cy^2 + Dx + Ey + F = 0$$

where A and C are not both zero. The basic mathematical tool used in this endeavor is *translation of axes*. The usefulness of translation of axes is not limited to graphing conics, however. Translation of axes can be put to good use in many other graphing situations.

FIGURE 1 Translation of coordinates.

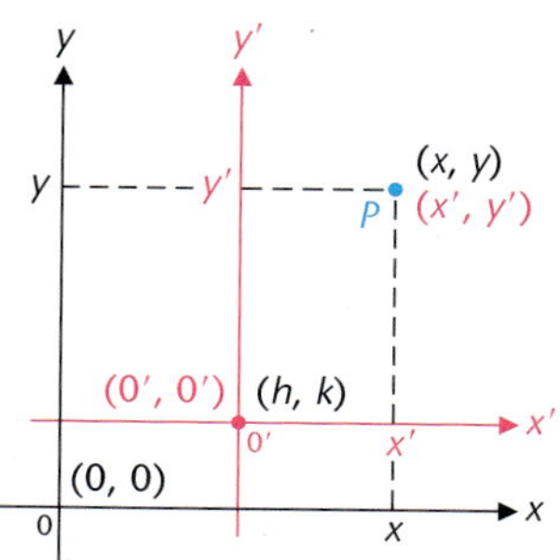

Translation of Axes

A **translation of coordinate axes** occurs when the new coordinate axes have the same direction as and are parallel to the original coordinate axes. To see how coordinates in the original system are changed when moving to the translated system, and vice versa, refer to Figure 1.

A point P in the plane has two sets of coordinates: (x, y) in the original system and (x', y') in the translated system. If the coordinates of the origin of the translated system are (h, k) relative to the original system, then the old and new coordinates are related as given in Theorem 1.

THEOREM 1

Translation Formulas

1. $x = x' + h$, $y = y' + k$

2. $x' = x - h$, $y' = y - k$

It can be shown that these formulas hold for (h, k) located anywhere in the original coordinate system.

EXAMPLE 1 Equation of a Curve in a Translated System

A curve has the equation

$$(x - 4)^2 + (y + 1)^2 = 36$$

If the origin is translated to $(4, -1)$, find the equation of the curve in the translated system and identify the curve.

SOLUTION

Because $(h, k) = (4, -1)$, use translation formulas

$$x' = x - h = x - 4$$
$$y' = y - k = y + 1$$

to obtain, after substitution,

$$x'^2 + y'^2 = 36$$

This is the equation of a circle of radius 6 with center at the new origin. The coordinates of the new origin in the original coordinate system are $(4, -1)$ (Fig. 2). Note that this result agrees with our general treatment of the circle in Appendix A, Section A.3.

FIGURE 2 $(x - 4)^2 + (y + 1)^2 = 36.$

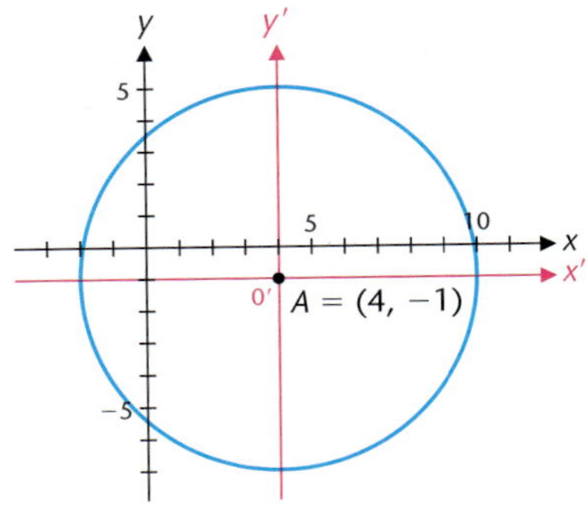

MATCHED 1 PROBLEM

A curve has the equation $(y + 2)^2 = 8(x - 3)$. If the origin is translated to $(3, -2)$, find an equation of the curve in the translated system and identify the curve.

Standard Equations of Translated Conics

We now proceed to find standard equations of conics translated away from the origin. We do this by first writing the standard equations found in earlier sections in the $x'y'$ coordinate system with $0'$ at (h, k). We then use translation equations to find the standard forms relative to the original xy coordinate system. The equations of translation in all cases are

$$x' = x - h$$
$$y' = y - k$$

For parabolas we have

$$x'^2 = 4ay' \qquad (x - h)^2 = 4a(y - k)$$
$$y'^2 = 4ax' \qquad (y - k)^2 = 4a(x - h)$$

For circles we have

$$x'^2 + y'^2 = r^2 \qquad (x - h)^2 + (y - k)^2 = r^2$$

For ellipses we have for $a > b > 0$

$$\frac{x'^2}{a^2} + \frac{y'^2}{b^2} = 1 \qquad \frac{(x - h)^2}{a^2} + \frac{(y - k)^2}{b^2} = 1$$

$$\frac{x'^2}{b^2} + \frac{y'^2}{a^2} = 1 \qquad \frac{(x - h)^2}{b^2} + \frac{(y - k)^2}{a^2} = 1$$

For hyperbolas we have

$$\frac{x'^2}{a^2} - \frac{y'^2}{b^2} = 1 \qquad \frac{(x - h)^2}{a^2} - \frac{(y - k)^2}{b^2} = 1$$

$$\frac{y'^2}{a^2} - \frac{x'^2}{b^2} = 1 \qquad \frac{(y - k)^2}{a^2} - \frac{(x - h)^2}{b^2} = 1$$

Table 1 summarizes these results with appropriate figures and some properties discussed earlier.

TABLE 1 Standard Equations for Translated Conics

Parabolas	
$(x - h)^2 = 4a(y - k)$	$(y - k)^2 = 4a(x - h)$
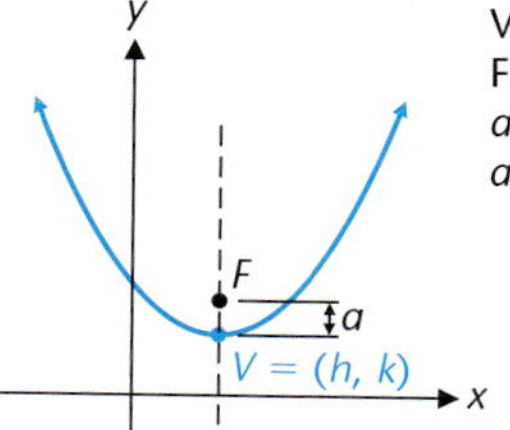 Vertex (h, k) Focus $(h, k + a)$ $a > 0$ opens up $a < 0$ opens down	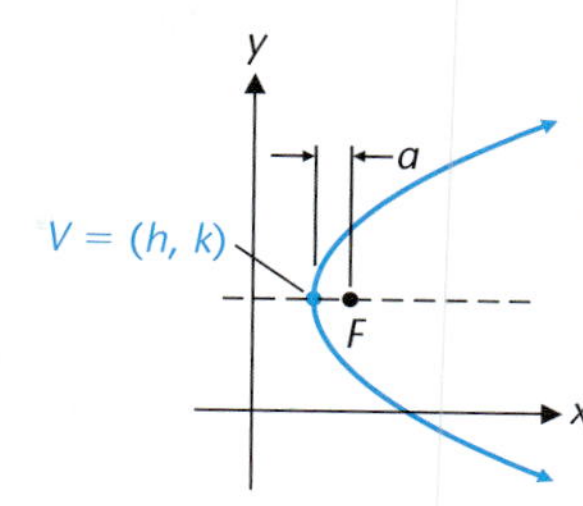 Vertex (h, k) Focus $(h + a, k)$ $a < 0$ opens left $a > 0$ opens right

Circles

$$(x - h)^2 + (y - k)^2 = r^2$$

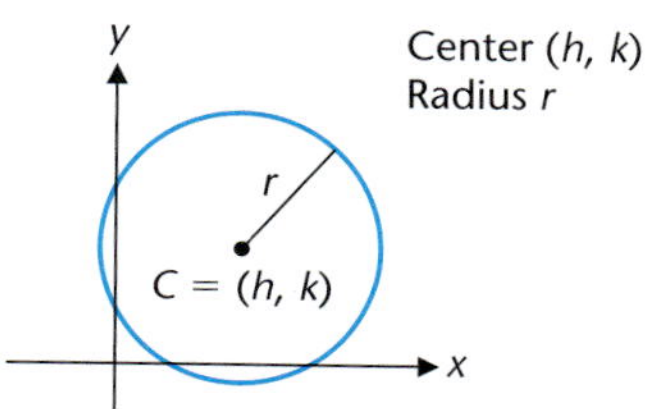

Center (h, k)
Radius r

Ellipses

$\dfrac{(x-h)^2}{a^2} + \dfrac{(y-k)^2}{b^2} = 1$	$a > b > 0$	$\dfrac{(x-h)^2}{b^2} + \dfrac{(y-k)^2}{a^2} = 1$
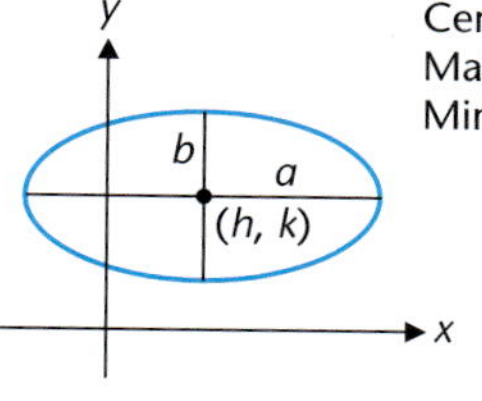 Center (h, k) Major axis $2a$ Minor axis $2b$		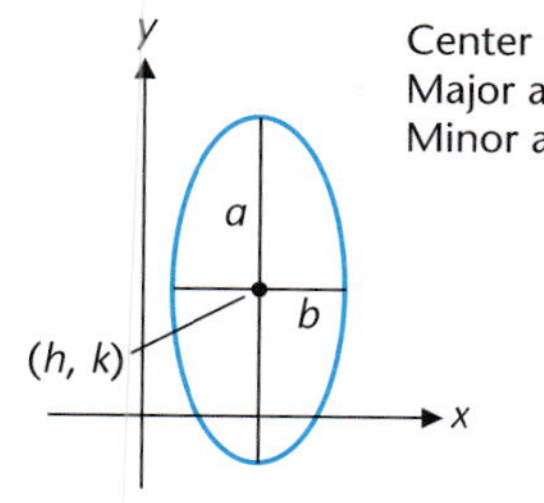 Center (h, k) Major axis $2a$ Minor axis $2b$

Hyperbolas

$\dfrac{(x-h)^2}{a^2} - \dfrac{(y-k)^2}{b^2} = 1$	$\dfrac{(y-k)^2}{a^2} - \dfrac{(x-h)^2}{b^2} = 1$
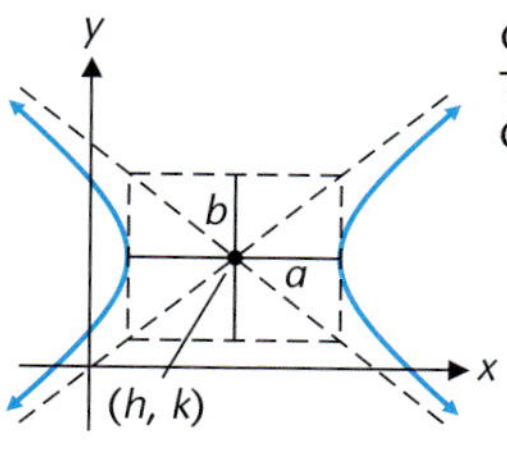 Center (h, k) Transverse axis $2a$ Conjugate axis $2b$	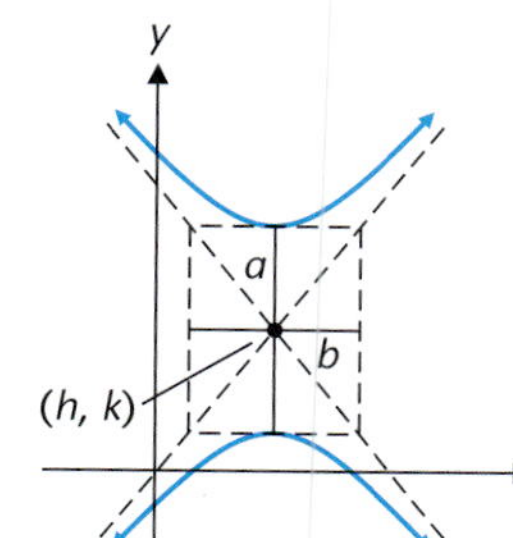 Center (h, k) Transverse axis $2a$ Conjugate axis $2b$

Graphing Equations of the Form $Ax^2 + Cy^2 + Dx + Ey + F = 0$

It can be shown that the graph of

$$Ax^2 + Cy^2 + Dx + Ey + F = 0 \tag{1}$$

where A and C are not both zero, is a conic or a degenerate conic or that there is no graph. If we can transform equation (1) into one of the standard forms in Table 1, then we will be able to identify its graph and sketch it rather quickly. The process of completing the square discussed in Section 2.3 will be our primary tool in accomplishing this transformation. A couple of examples should help make the process clear.

EXAMPLE 2 Graphing a Translated Conic

Given the equation

$$y^2 - 6y - 4x + 1 = 0 \tag{2}$$

(A) Transform the equation into one of the standard forms in Table 1 and identify the conic.

(B) Find the equation in the translated system.

(C) Graph the conic.

SOLUTIONS

(A) Complete the square in equation (2) relative to each variable that is squared—in this case y:

$$y^2 - 6y - 4x + 1 = 0$$
$$y^2 - 6y \quad = 4x - 1$$
$$y^2 - 6y \quad + 9 = 4x + 8$$

Add 9 to both sides to complete the square on the left side.

$$(y - 3)^2 = 4(x + 2) \tag{3}$$

From Table 1 we recognize equation (3) as an equation of a parabola opening to the right with vertex at $(h, k) = (-2, 3)$.

(B) Find the equation of the parabola in the translated system with origin $0'$ at $(h, k) = (-2, 3)$. The equations of translation are read directly from equation (3):

$$x' = x + 2$$
$$y' = y - 3$$

Making these substitutions in equation (3) we obtain

$$y'^2 = 4x' \tag{4}$$

the equation of the parabola in the $x'y'$ system.

(C) Hand-Drawn Solution
Graph equation (4) in the $x'y'$ system following the process discussed in Section 11.1. The resulting graph is the graph of the original equation relative to the original xy coordinate system (Fig. 3).

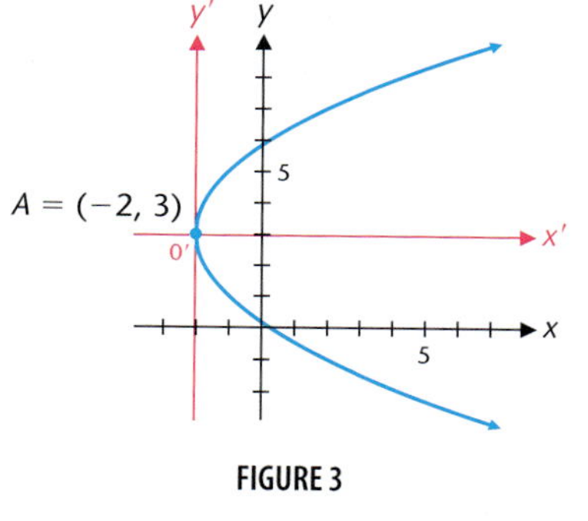

FIGURE 3

Graphing Utility Solution
To graph on a graphing utility, we can solve either equation (2) or equation (3) for y. Choosing equation (2) has the added benefit of providing a check of the derivation of equation (3).

$$y^2 - 6y - 4x + 1 = 0 \quad \text{Quadratic equation with } a = 1, b = -6, \text{ and } c = -4x + 1$$

$$y = \frac{6 \pm \sqrt{36 - 4(1)(-4x + 1)}}{2(1)}$$

$$= \frac{6 \pm \sqrt{32 + 16x}}{2}$$

$$= 3 \pm 2\sqrt{2 + x} \qquad (5)$$

Figure 4 shows the graph of the two functions determined by equation (5) and the vertex of the parabola.

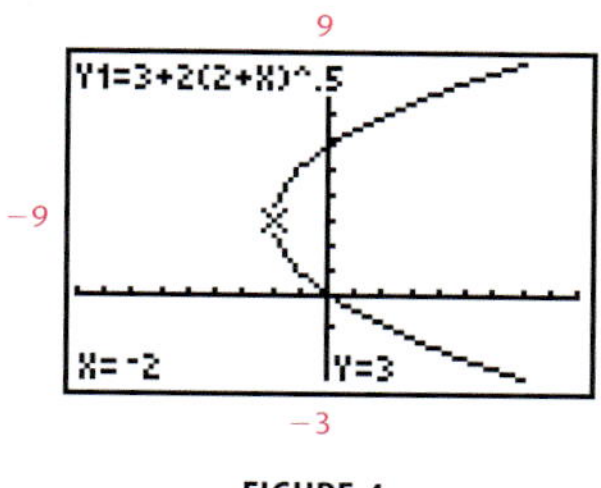

FIGURE 4

MATCHED 2 PROBLEM

Repeat Example 2 for the equation $x^2 + 4x + 4y - 12 = 0$.

EXAMPLE 3 Graphing a Translated Conic

Given the equation

$$9x^2 - 4y^2 - 36x - 24y - 36 = 0$$

(A) Transform the equation into one of the standard forms in Table 1 and identify the conic.

(B) Find the equation in the translated system.

(C) Graph the conic.

(D) Find the coordinates of any foci relative to the original system.

SOLUTIONS

(A) Complete the square relative to both x and y.

$$\begin{aligned} 9x^2 - 4y^2 - 36x - 24y - 36 &= 0 \\ 9x^2 - 36x \quad - 4y^2 - 24y &= 36 \\ 9(x^2 - 4x \quad) - 4(y^2 + 6y \quad) &= 36 \\ 9(x^2 - 4x + 4) - 4(y^2 + 6y + 9) &= 36 + 36 - 36 \\ 9(x - 2)^2 - 4(y + 3)^2 &= 36 \\ \frac{(x - 2)^2}{4} - \frac{(y + 3)^2}{9} &= 1 \end{aligned}$$

From Table 1 we recognize the last equation as an equation of a hyperbola opening left and right with center at $(h, k) = (2, -3)$.

(B) Find the equation of the hyperbola in the translated system with origin $0'$ at $(h, k) = (2, -3)$. The equations of translation are read directly from the last equation in part A:

$$\begin{aligned} x' &= x - 2 \\ y' &= y + 3 \end{aligned}$$

Making these substitutions, we obtain

$$\frac{x'^2}{4} - \frac{y'^2}{9} = 1$$

the equation of the hyperbola in the $x'y'$ system.

(C) Hand-Drawn Solution

Graph the equation obtained in part B in the $x'y'$ system following the process discussed in Section 11.3. The resulting graph is the graph of the original equation relative to the original xy coordinate system (Fig. 5).

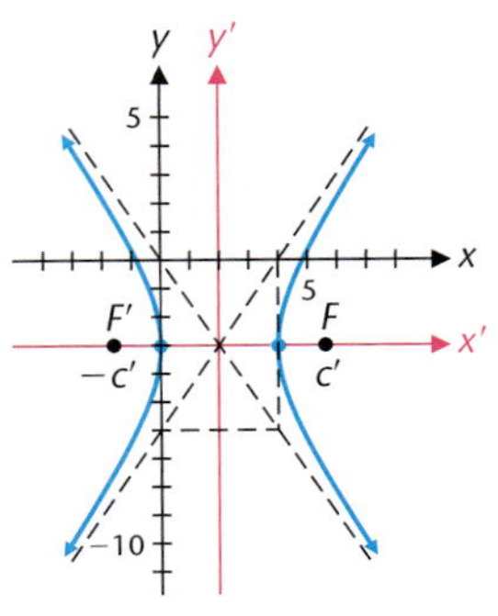

FIGURE 5

Graphing Utility Solution

We return to the original equation and use the quadratic formula to solve for y:

$$9x^2 - 4y^2 - 36x - 24y - 36 = 0 \qquad \text{Write in the form } ay^2 + by + c = 0.$$

$$4y^2 + 24y + (-9x^2 + 36x + 36) = 0$$

$$y = \frac{-24 \pm \sqrt{24^2 - 4(4)(-9x^2 + 36x + 36)}}{8}$$

$$= -3 \pm 1.5\sqrt{x^2 - 4x} \qquad (6)$$

The two functions determined by equation (6) are graphed in Figure 6.

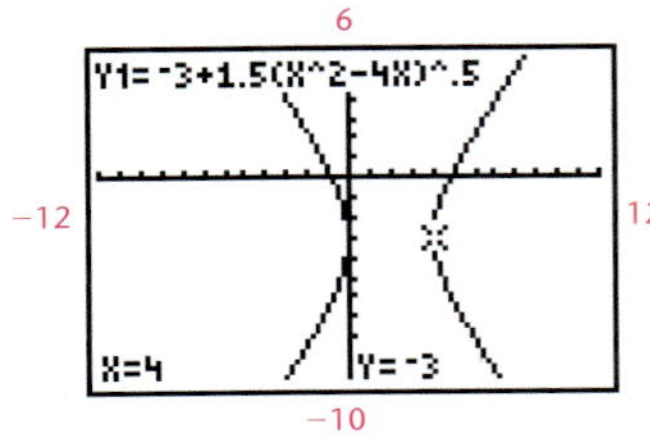

FIGURE 6

(D) Find the coordinates of the foci. To find the coordinates of the foci in the original system, first find the coordinates in the translated system:

$$c'^2 = 2^2 + 3^2 = 13$$
$$c' = \sqrt{13}$$
$$-c' = -\sqrt{13}$$

Thus, the coordinates in the translated system are

$$F' = (-\sqrt{13}, 0) \qquad \text{and} \qquad F = (\sqrt{13}, 0)$$

Now, use

$$x = x' + h = x' + 2$$
$$y = y' + k = y' - 3$$

to obtain

$$F' = (-\sqrt{13} + 2, -3) \qquad \text{and} \qquad F = (\sqrt{13} + 2, -3)$$

as the coordinates of the foci in the original system.

MATCHED 3 PROBLEM

Repeat Example 3 for the equation

$$9x^2 + 16y^2 + 36x - 32y - 92 = 0$$

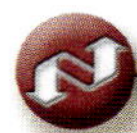

EXPLORE/DISCUSS 1

If $A \neq 0$ and $C \neq 0$, show that the translation of axes $x' = x + \dfrac{D}{2A}$, $y' = y + \dfrac{E}{2C}$ transforms the equation $Ax^2 + Cy^2 + Dx + Ey + F = 0$ into an equation of the form $Ax'^2 + Cy'^2 = K$.

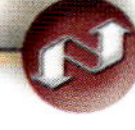

Finding Equations of Conics

We now reverse the problem: Given certain information about a conic in a rectangular coordinate system, find its equation.

EXAMPLE 4 **Finding the Equation of a Translated Conic**

Find the equation of a hyperbola with vertices on the line $x = -4$, conjugate axis on the line $y = 3$, length of the transverse axis $= 4$, and length of the conjugate axis $= 6$.

SOLUTION

Locate the vertices, asymptote rectangle, and asymptotes in the original coordinate system [Fig. 7(a)], then sketch the hyperbola and translate the origin to the center of the hyperbola [Fig. 7(b)].

FIGURE 7

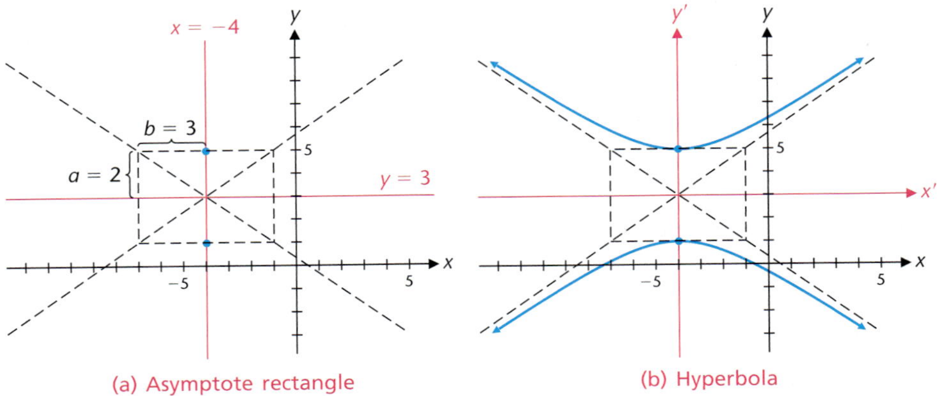

(a) Asymptote rectangle (b) Hyperbola

Next write the equation of the hyperbola in the translated system:

$$\frac{y'^2}{4} - \frac{x'^2}{9} = 1$$

The origin in the translated system is at $(h, k) = (-4, 3)$, and the translation formulas are

$$x' = x - h = x - (-4) = x + 4$$
$$y' = y - k = y - 3$$

Thus, the equation of the hyperbola in the original system is

$$\frac{(y - 3)^2}{4} - \frac{(x + 4)^2}{9} = 1$$

or, after simplifying and writing in the form of equation (1),

$$4x^2 - 9y^2 + 32x + 54y + 19 = 0$$

MATCHED 4 PROBLEM

Find the equation of an ellipse with foci on the line $x = 4$, minor axis on the line $y = -3$, length of the major axis $= 8$, and length of the minor axis $= 4$.

EXPLORE/DISCUSS 2

Use the strategy of completing the square to transform each equation to an equation in an $x'y'$ coordinate system. Note that the equation you obtain is not one of the standard forms in Table 1; instead, it is either the equation of a degenerate conic or the equation has no solution. If the solution set of the equation is not empty, graph it and identify the graph (a point, a line, two parallel lines, or two intersecting lines).

(A) $x^2 + 2y^2 - 2x + 16y + 33 = 0$

(B) $4x^2 - y^2 - 24x - 2y + 35 = 0$

(C) $y^2 - 2y - 15 = 0$

(D) $5x^2 + y^2 + 12y + 40 = 0$

(E) $x^2 - 18x + 81 = 0$

ANSWERS TO MATCHED PROBLEMS

1. $y'^2 = 8x'$; a parabola
2. (A) $(x + 2)^2 = -4(y - 4)$; a parabola (B) $x'^2 = -4y'$ (C)

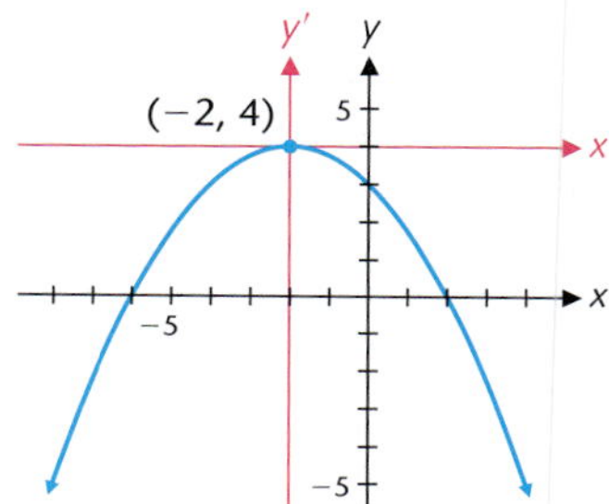

3. (A) $\dfrac{(x + 2)^2}{16} + \dfrac{(y - 1)^2}{9} = 1$; ellipse (B) $\dfrac{x'^2}{16} + \dfrac{y'^2}{9} = 1$ (C)

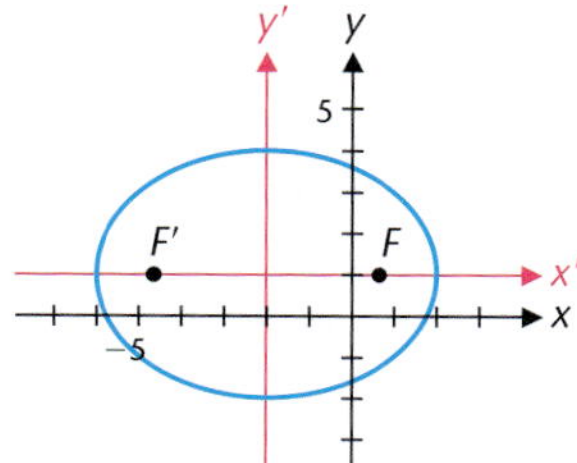

(D) Foci: $F' = (-\sqrt{7} - 2, 1)$, $F = (\sqrt{7} - 2, 1)$

4. $\dfrac{(x - 4)^2}{4} + \dfrac{(y + 3)^2}{16} = 1$, or $4x^2 + y^2 - 32x + 6y + 57 = 0$

EXERCISE 11.4

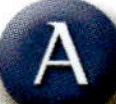

In Problems 1–8:

(A) Find translation formulas that translate the origin to the indicated point (h, k).

(B) Write the equation of the curve for the translated system.

(C) Identify the curve.

1. $(x - 3)^2 + (y - 5)^2 = 81$; $(3, 5)$

2. $(x - 3)^2 = 8(y + 2)$; $(3, -2)$

3. $\dfrac{(x+7)^2}{9} + \dfrac{(y-4)^2}{16} = 1; (-7, 4)$

4. $(x+2)^2 + (y+6)^2 = 36; (-2, -6)$

5. $(y+9)^2 = 16(x-4); (4, -9)$

6. $\dfrac{(y-9)^2}{10} - \dfrac{(x+5)^2}{6} = 1; (-5, 9)$

7. $\dfrac{(x+8)^2}{12} + \dfrac{(y+3)^2}{8} = 1; (-8, -3)$

8. $\dfrac{(x+7)^2}{25} - \dfrac{(y-8)^2}{50} = 1; (-7, 8)$

In Problems 9–14:
(A) Write each equation in one of the standard forms listed in Table 1.
(B) Identify the curve.

9. $16(x-3)^2 - 9(y+2)^2 = 144$

10. $(y+2)^2 - 12(x-3) = 0$

11. $6(x+5)^2 + 5(y+7)^2 = 30$

12. $12(y-5)^2 - 8(x-3)^2 = 24$

13. $(x+6)^2 + 24(y-4) = 0$

14. $4(x-7)^2 + 7(y-3)^2 = 28$

B

In Problems 15–22, transform each equation into one of the standard forms in Table 1. Identify the curve and graph it.

15. $4x^2 + 9y^2 - 16x - 36y + 16 = 0$

16. $16x^2 + 9y^2 + 64x + 54y + 1 = 0$

17. $x^2 + 8x + 8y = 0$

18. $y^2 + 12x + 4y - 32 = 0$

19. $x^2 + y^2 + 12x + 10y + 45 = 0$

20. $x^2 + y^2 - 8x - 6y = 0$

21. $-9x^2 + 16y^2 - 72x - 96y - 144 = 0$

22. $16x^2 - 25y^2 - 160x = 0$

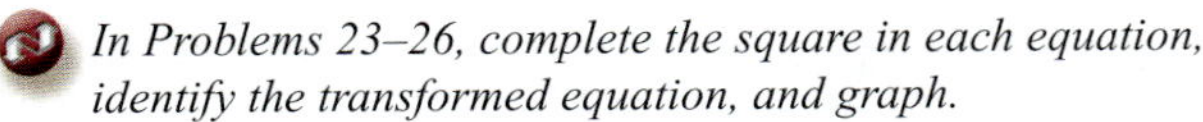

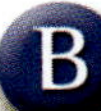

In Problems 23–26, complete the square in each equation, identify the transformed equation, and graph.

23. $x^2 - 2x + y^2 + 4y + 5 = 0$

24. $x^2 - 6x + 2y^2 + 4y + 11 = 0$

25. $x^2 + 8x - 4y^2 + 8y + 12 = 0$

26. $x^2 + 4x - y^2 + 6y - 5 = 0$

27. If $A \neq 0$, $C = 0$, and $E \neq 0$, find h and k so that the translation of axes $x = x' + h$, $y = y' + k$ transforms the equation $Ax^2 + Cy^2 + Dx + Ey + F = 0$ into one of the standard forms of Table 1.

28. If $A = 0$, $C \neq 0$, and $D \neq 0$, find h and k so that the translation of axes $x = x' + h$, $y = y' + k$ transforms the equation $Ax^2 + Cy^2 + Dx + Ey + F = 0$ into one of the standard forms of Table 1.

In Problems 29–40, use the given information to find the equation of each conic. Express the answer in the form $Ax^2 + Cy^2 + Dx + Ey + F = 0$ with integer coefficients and $A > 0$.

29. A parabola with vertex at (2, 5), axis the line $x = 2$, and passing through the point $(-2, 1)$.

30. A parabola with vertex at $(4, -1)$, axis the line $y = -1$, and passing through the point (2, 3).

31. An ellipse with major axis on the line $y = -3$, minor axis on the line $x = -2$, length of major axis = 8, and length of minor axis = 4.

32. An ellipse with major axis on the line $x = -4$, minor axis on the line $y = 1$, length of major axis = 4, and length of minor axis = 2.

33. An ellipse with vertices $(4, -7)$ and $(4, 3)$ and foci $(4, -6)$ and $(4, 2)$.

34. An ellipse with vertices $(-3, 1)$ and $(7, 1)$ and foci $(-1, 1)$ and $(5, 1)$.

35. A hyperbola with transverse axis on the line $x = 2$, length of transverse axis = 4, conjugate axis on the line $y = 3$, and length of conjugate axis = 2.

36. A hyperbola with transverse axis on the line $y = -5$, length of transverse axis = 6, conjugate axis on the line $x = 2$, and length of conjugate axis = 6.

37. An ellipse with the following graph:

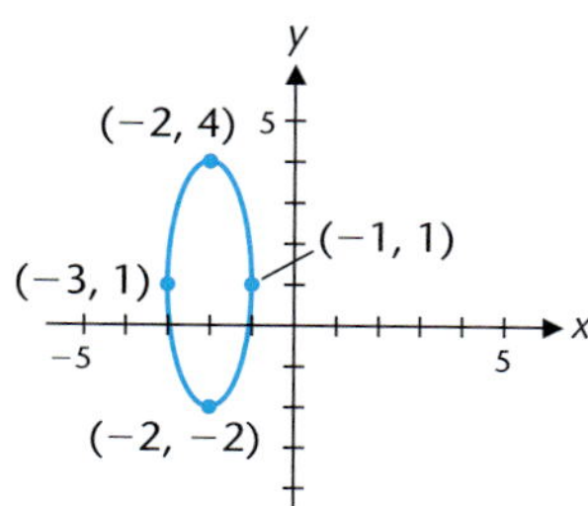

38. An ellipse with the following graph:

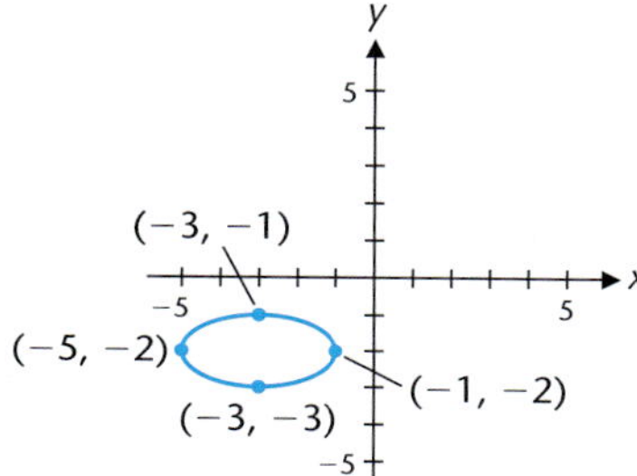

39. A hyperbola with the following graph:

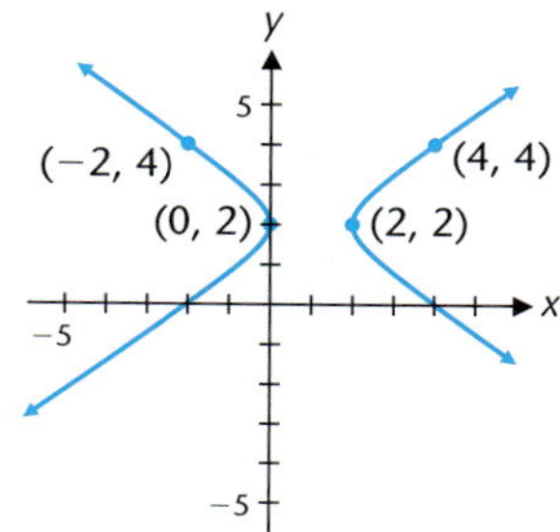

40. A hyperbola with the following graph:

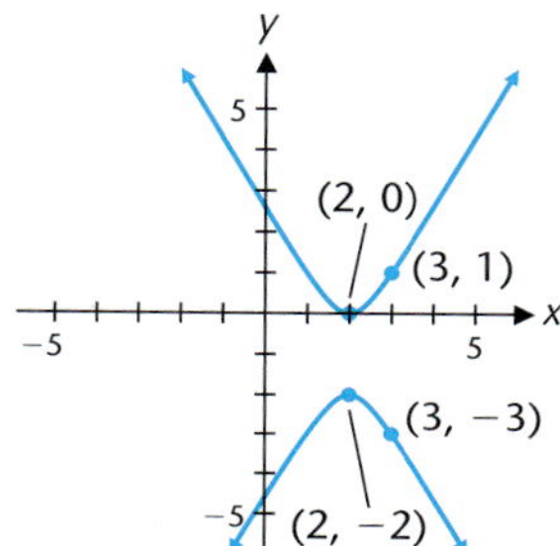

C

In Problems 41–46, find the coordinates of any foci relative to the original coordinate system.

41. Problem 15

42. Problem 16

43. Problem 17

44. Problem 18

45. Problem 21

46. Problem 22

SECTION 11.5 Rotation of Axes

Rotation of Axes • Rotation Used in Graphing • Identifying Conics

In Section 11.4 we found that when $B = 0$, the graph of

$$Ax^2 + Bxy + Cy^2 + Dx + Ey + F = 0 \qquad A, B, C \text{ not all } 0 \qquad (1)$$

is a conic or a degenerate conic or there is no graph. What happens if $B \neq 0$ in equation (1)? In this case we will show that a suitable rotation of axes can be used to transform equation (1) into a new equation in x' and y' with no $x'y'$ term.

Then we can proceed as before to find the standard form and the graph for this new equation. Thus, except for degenerate cases, the graph of a general second degree equation in two variables [equation (1)] is always one of the conics: a circle, a parabola, an ellipse, or a hyperbola.

Rotation of Axes

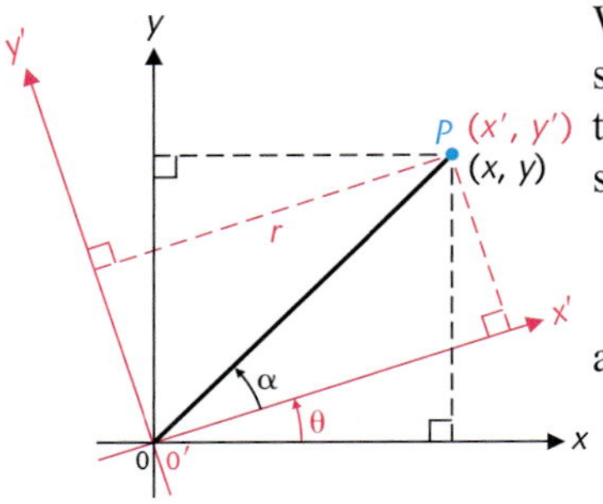

We now introduce a transformation of coordinates from an xy system to an $x'y'$ system that is accomplished by a rotation of axes. The origin is kept fixed and the x' and y' axes are obtained by rotating the x and y axes counterclockwise, as shown in Figure 1. Referring to Figure 1 and using trigonometry, we have

FIGURE 1

$$x' = r\cos\alpha \qquad y' = r\sin\alpha \tag{2}$$

and

$$x = r\cos(\theta + \alpha) \qquad y = r\sin(\theta + \alpha) \tag{3}$$

Using addition identities from trigonometry for the equations in (3), we obtain

$$\begin{aligned} x &= r\cos(\theta + \alpha) \\ &= r(\cos\theta\cos\alpha - \sin\theta\sin\alpha) \\ &= r\cos\theta\cos\alpha - r\sin\theta\sin\alpha \\ &= (r\cos\alpha)\cos\theta - (r\sin\alpha)\sin\theta \\ &= x'\cos\theta - y'\sin\theta \qquad \text{Substitute } x' = r\cos\alpha \text{ and } y' = r\sin\alpha \end{aligned} \tag{4}$$

$$\begin{aligned} y &= r\sin(\theta + \alpha) \\ &= r(\sin\theta\cos\alpha + \cos\theta\sin\alpha) \\ &= r\sin\theta\cos\alpha + r\cos\theta\sin\alpha \\ &= (r\cos\alpha)\sin\theta + (r\sin\alpha)\cos\theta \\ &= x'\sin\theta + y'\cos\theta \qquad \text{Substitute } x' = r\cos\alpha \text{ and } y' = r\sin\alpha \end{aligned} \tag{5}$$

Thus, equations (4) and (5) together transform the xy coordinate system into the $x'y'$ coordinate system.

Equations (4) and (5) can be solved for x' and y' in terms of x and y to produce formulas that transform the $x'y'$ coordinate system back into the xy coordinate system. Omitting the details, the formulas for the transformation in the reverse direction are

$$x' = x\cos\theta + y\sin\theta \qquad y' = -x\sin\theta + y\cos\theta \tag{6}$$

These results are summarized in Theorem 1.

THEOREM 1
Rotation Formulas

If the xy coordinate axes are rotated counterclockwise through an angle of θ then the xy and $x'y'$ coordinates of a point P are related by:

1. $x = x' \cos \theta - y' \sin \theta$
   $y = x' \sin \theta + y' \cos \theta$
2. $x' = x \cos \theta + y \sin \theta$
   $y' = -x \sin \theta + y \cos \theta$

These formulas hold for P any point in the original coordinate system and θ any counterclockwise rotation.

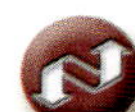

EXPLORE/DISCUSS 1

Let θ be the first quadrant angle satisfying $\sin \theta = \frac{3}{5}$ and $\cos \theta = \frac{4}{5}$ and let an xy coordinate system be transformed into an $x'y'$ coordinate system by a counterclockwise rotation through the angle θ.

(A) Sketch the $x'y'$ coordinate system in the xy coordinate system.

(B) Express x' and y' in terms of x and y.

(C) Solve $x' = 0$ to find the equation of the y' axis in the xy coordinate system.

(D) Solve $y' = 0$ to find the equation of the x' axis in the xy coordinate system.

(E) Use the results found in parts C and D to graph the $x'y'$ coordinate system in the xy coordinate system on a graphing utility, using a squared viewing window.

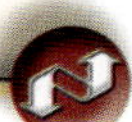

Rotation Used in Graphing

We now investigate how rotation formulas are used in graphing.

EXAMPLE 1 Using the Rotation of Axes Formulas

Transform the equation $xy = -2$ using a rotation of axes through 45°. Graph the new equation and identify the curve. Check by graphing on a graphing utility.

SOLUTION

Use the rotation formulas:

$$x = x'\cos 45° - y'\sin 45° = \frac{\sqrt{2}}{2}(x' - y')$$

$$y = x'\sin 45° + y'\cos 45° = \frac{\sqrt{2}}{2}(x' + y')$$

$$xy = -2$$

$$\frac{\sqrt{2}}{2}(x' - y')\frac{\sqrt{2}}{2}(x' + y') = -2$$

$$\frac{1}{2}(x'^2 - y'^2) = -2$$

$$\frac{x'^2}{2} - \frac{y'^2}{2} = -2$$

$$\frac{y'^2}{4} - \frac{x'^2}{4} = 1$$

This is a standard equation for a hyperbola. Summarizing, the graph of $xy = -2$ in the $x'y'$ coordinate system is a hyperbola with equation

$$\frac{y'^2}{4} - \frac{x'^2}{4} = 1$$

and the graph shown in Figure 2.

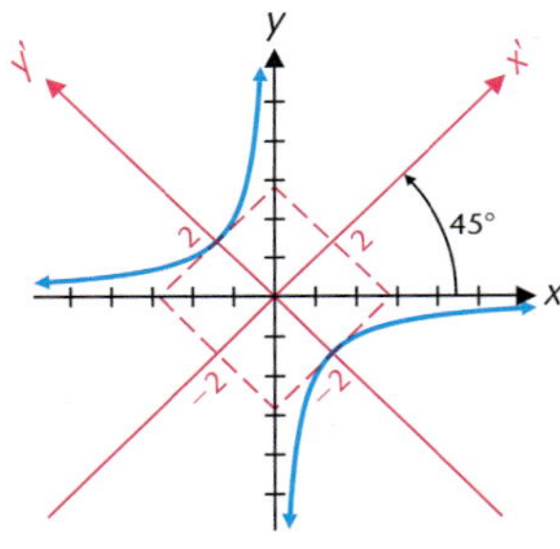

FIGURE 2

Notice that the asymptotes in the rotated system are the x and y axes in the original system.

Graphing Utility Check

To graph $xy = -2$ on a graphing utility, we must solve for y:

$$xy = -2$$

$$y = -\frac{2}{x}$$

To find equations for the x' and y' axes in the xy coordinate system, we use part 2 of Theorem 1 to write

$$x' = x\cos 45° + y\sin 45° = \frac{\sqrt{2}}{2}(x + y)$$

$$y' = -x\sin 45° + y\cos 45° = \frac{\sqrt{2}}{2}(-x + y)$$

Equations for the y' and x' axes in the xy coordinate system can be found by solving the equations $x' = 0$ and $y' = 0$ respectively.

$$x' = \frac{\sqrt{2}}{2}(x + y) = 0$$

$$x + y = 0$$

$$y = -x \qquad y' \text{ axis}$$

$$y' = \frac{\sqrt{2}}{2}(-x + y) = 0$$

$$-x + y = 0$$

$$y = x \qquad x' \text{ axis}$$

Entering $y_1 = -2/x$, $y_2 = -x$, and $y_3 = x$ and graphing in a squared viewing window produces the graph of the rotated hyperbola (Fig. 3).

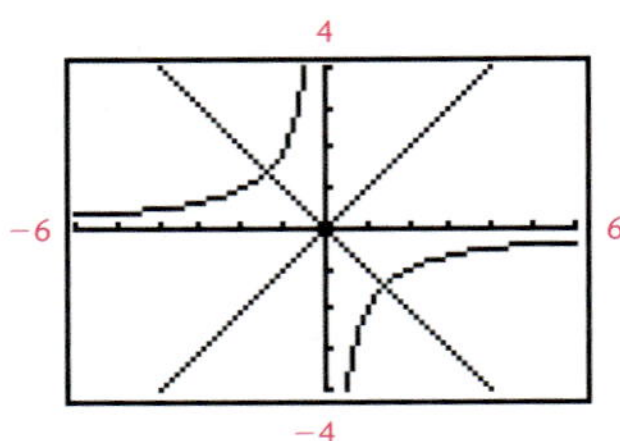

FIGURE 3

MATCHED 1 PROBLEM

Transform the equation $2xy = 1$ using a rotation of axes through 45°. Graph the new equation and identify the curve. Check by graphing on a graphing utility.

In Example 1, a 45° rotation transformed the original equation into one with no $x'y'$ term. This made it easy to recognize that the graph of the transformed equation was a hyperbola. In general, how do we determine the angle of rotation that will transform an equation with an xy term into one with no $x'y'$ term? To find out, we substitute

$$x = x'\cos\theta - y'\sin\theta \qquad \text{and} \qquad y = x'\sin\theta + y'\cos\theta$$

into equation (1) to obtain

$$\begin{aligned} &A(x'\cos\theta - y'\sin\theta)^2 + B(x'\cos\theta - y'\sin\theta)(x'\sin\theta + y'\cos\theta) \\ &\quad + C(x'\sin\theta + y'\cos\theta)^2 + D(x'\cos\theta - y'\sin\theta) \\ &\quad + E(x'\sin\theta + y'\cos\theta) + F = 0 \end{aligned}$$

After multiplying and collecting terms, we have

$$A'x'^2 + B'x'y' + C'y'^2 + D'x' + E'y' + F = 0 \tag{7}$$

where

$$B' = 2(C - A)\sin\theta\cos\theta + B(\cos^2\theta - \sin^2\theta) \tag{8}$$

For the $x'y'$ term in equation (7) to drop out, B' must be 0. We won't worry about A', C', D', and E' at this point; they will automatically be determined once we find θ so that $B' = 0$. We set the right side of equation (8) equal to 0 and solve for θ:

$$2(C - A)\sin\theta\cos\theta + B(\cos^2\theta - \sin^2\theta) = 0$$

Using the double-angle identities from trigonometry, $\sin 2\theta = 2\sin\theta\cos\theta$ and $\cos 2\theta = \cos^2\theta - \sin^2\theta$, we obtain

$$\begin{aligned} (C - A)\sin 2\theta + B\cos 2\theta &= 0 \\ B\cos 2\theta &= (A - C)\sin 2\theta \\ \frac{\cos 2\theta}{\sin 2\theta} &= \frac{A - C}{B} \\ \cot 2\theta &= \frac{A - C}{B} \end{aligned} \tag{9}$$

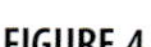
FIGURE 4

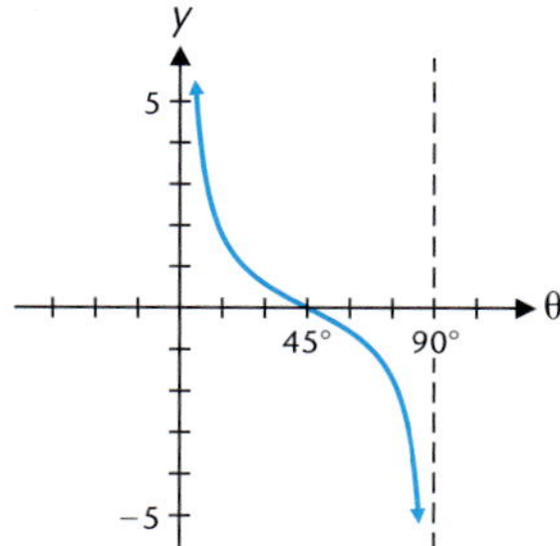

Thus, if we choose θ so that $\cot 2\theta = (A - C)/B$, then $B' = 0$ and the $x'y'$ term in equation (7) will drop out. There is always an angle θ between 0° and 90° that solves equation (9), because the range of $y = \cot 2\theta$ for $0° < \theta < 90°$ is the set of all real numbers (Fig. 4).

THEOREM 2

Angle of Rotation to Eliminate the $x'y'$ Term

To transform the equation

$$Ax^2 + Bxy + Cy^2 + Dx + Ey + F = 0$$

into an equation in x' and y' with no $x'y'$ term, find θ so that

$$\cot 2\theta = \frac{A - C}{B} \qquad \text{and} \qquad 0° < \theta < 90°$$

and use the rotation formulas in Theorem 1.

EXPLORE/DISCUSS 2

Find a rotation angle θ that will eliminate the xy term in each of the following equations.

(A) $x^2 + 3xy + y^2 = 15$

(B) $5x^2 + 2xy + 3y^2 = 20$

(C) $4x^2 + \sqrt{3}xy + y^2 = 10$

EXAMPLE 2 Identifying and Graphing an Equation with an xy Term

Given the equation $17x^2 - 6xy + 9y^2 = 72$, find the angle of rotation so that the transformed equation will have no $x'y'$ term. Sketch and identify the graph. Check by graphing on a graphing utility.

SOLUTION

$$17x^2 - 6xy + 9y^2 = 72 \qquad (10)$$

$$\cot 2\theta = \frac{A - C}{B} = \frac{17 - 9}{-6} = -\frac{4}{3}$$

Thus, 2θ is a quadrant II angle, and using the reference triangle in the figure, we can see that $\cos 2\theta = -\frac{4}{5}$. We can find the rotation formulas exactly by the use of the half-angle identities

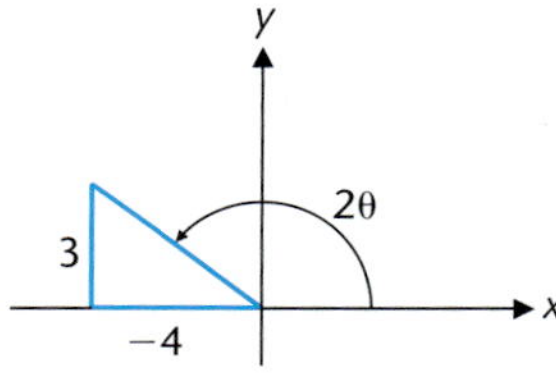

$$\sin \theta = \sqrt{\frac{1 - \cos 2\theta}{2}}$$

and

$$\cos \theta = \sqrt{\frac{1 + \cos 2\theta}{2}}$$

Using these identities, we obtain

$$\sin \theta = \sqrt{\frac{1 - (-\frac{4}{5})}{2}} = \frac{3}{\sqrt{10}}$$

Graphing Utility Check

First use the quadratic formula to solve equation (10) for y:

$$17x^2 - 6xy + 9y^2 = 72$$

$$9y^2 - 6xy + 17x^2 - 72 = 0$$

Use $a = 9, b = -6x$ and $c = 17x^2 - 72$ in the quadratic formula.

$$y = \frac{6x \pm \sqrt{(-6x)^2 - 4(9)(17x^2 - 72)}}{2(9)}$$

$$= \frac{6x \pm \sqrt{2{,}592 - 576x^2}}{18}$$

This produces two functions,

$$y_1 = \frac{6x + \sqrt{2{,}592 - 576x^2}}{18}$$

$$y_2 = \frac{6x - \sqrt{2{,}592 - 576x^2}}{18}$$

To find equations for the x' and y' axes in the xy coordinate system, we use part 2 of Theorem 1 to write

$$x' = \frac{1}{\sqrt{10}}x + \frac{3}{\sqrt{10}}y$$

and

$$\cos\theta = \sqrt{\frac{1 + (-\frac{4}{5})}{2}} = \frac{1}{\sqrt{10}}$$

Hence, the rotation formulas are

$$x = \frac{1}{\sqrt{10}}x' - \frac{3}{\sqrt{10}}y'$$

and (11)

$$y = \frac{3}{\sqrt{10}}x' + \frac{1}{\sqrt{10}}y'$$

Substituting equations (11) into equation (10), we have

$$17\left(\frac{1}{\sqrt{10}}x' - \frac{3}{\sqrt{10}}y'\right)^2 - 6\left(\frac{1}{\sqrt{10}}x' - \frac{3}{\sqrt{10}}y'\right)\left(\frac{3}{\sqrt{10}}x' + \frac{1}{\sqrt{10}}y'\right) + 9\left(\frac{3}{\sqrt{10}}x' + \frac{1}{\sqrt{10}}y'\right)^2 = 72$$

$$\frac{17}{10}(x' - 3y')^2 - \frac{6}{10}(x' - 3y')(3x' + y') + \frac{9}{10}(3x' + y')^2 = 72$$

Further simplification leads to

$$\frac{x'^2}{9} + \frac{y'^2}{4} = 1$$

which is a standard equation for an ellipse. To graph, we rotate the original axes through an angle θ determined as follows:

$$\cot 2\theta = -\frac{4}{3}$$
$$2\theta \approx 143.1301°$$
$$\theta \approx 71.57°$$

We could also use either

$$\sin\theta = \frac{3}{\sqrt{10}} \quad \text{or} \quad \cos\theta = \frac{1}{\sqrt{10}}$$

to determine the angle of rotation. Summarizing these results, the graph of $17x^2 - 6xy + 9y^2 = 72$ in the $x'y'$ coordinate system formed by a rotation of 71.57° is an ellipse with equation

$$\frac{x'^2}{9} + \frac{y'^2}{4} = 1$$

and

$$y' = -\frac{3}{\sqrt{10}}x + \frac{1}{\sqrt{10}}y$$

As before, equations for the y' and x' axes in the xy coordinate system can be found by solving the equations $x' = 0$ and $y' = 0$ respectively.

$$x' = \frac{1}{\sqrt{10}}x + \frac{3}{\sqrt{10}}y = 0$$
$$\frac{3}{\sqrt{10}}y = -\frac{1}{\sqrt{10}}x$$
$$y = -\frac{x}{3} \qquad y' \text{ axis}$$

$$y' = -\frac{3}{\sqrt{10}}x + \frac{1}{\sqrt{10}}y = 0$$
$$\frac{1}{\sqrt{10}}y = \frac{3}{\sqrt{10}}x$$
$$y = -3x \qquad x' \text{ axis}$$

Entering y_1, y_2, and the equations for the x' and y' axes (Fig. 6) and graphing in a squared viewing window produces the graph of a rotated ellipse (Fig. 7).

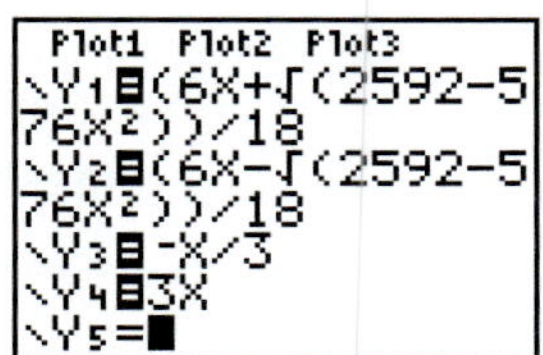

FIGURE 6

and the graph shown in Figure 5.

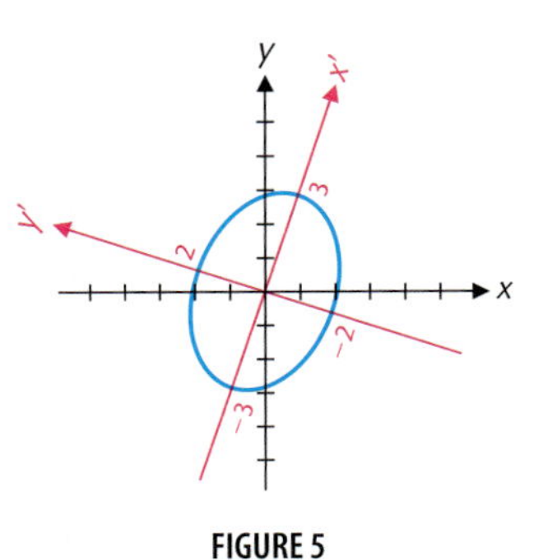

FIGURE 5

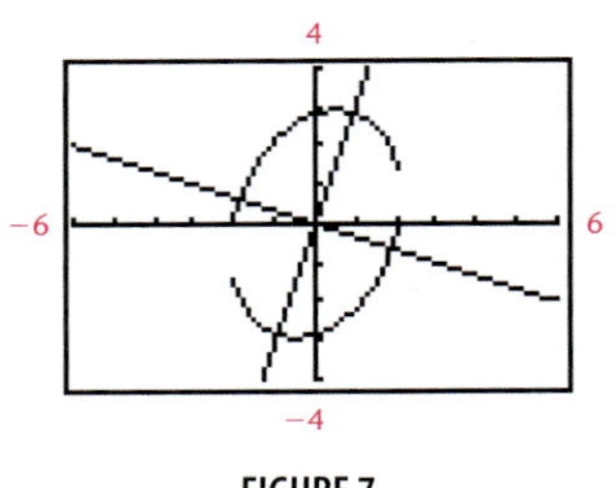

FIGURE 7

MATCHED 2 PROBLEM

Given the equation $3x^2 + 26\sqrt{3}xy - 23y^2 = 144$, find the angle of rotation so that the transformed equation will have no $x'y'$ term. Sketch and identify the graph. Check by graphing on a graphing utility.

Identifying Conics

The **discriminant** of the general second-degree equation in two variables [equation (1)] is $B^2 - 4AC$. It can be shown that the value of this expression does not change when the axes are rotated. This forms the basis for Theorem 3.

THEOREM 3
Identifying Conics

The graph of the equation

$$Ax^2 + Bxy + Cy^2 + Dx + Ey + F = 0$$

is, excluding degenerate cases,

1. A hyperbola if $B^2 - 4AC > 0$
2. A parabola if $B^2 - 4AC = 0$
3. An ellipse if $B^2 - 4AC < 0$

The proof of Theorem 3 is beyond the scope of this book. Its use is best illustrated by example.

EXAMPLE 3 **Identifying and Graphing Conics**

Identify the following conics and graph on a graphing utility.

(A) $x^2 - xy + y^2 = 5$

(B) $x^2 - xy - y^2 = 5$

(C) $x^2 - 4xy + 4y^2 + x = 5$

SOLUTIONS

(A) The discriminant is

$$B^2 - 4AC = (-1)^2 - 4(1)(1) = 1 - 4 = -3 < 0$$

Thus, this conic is an ellipse. Now we use the quadratic formula to solve for y.

$$x^2 - xy + y^2 = 5$$

$$y^2 - xy + x^2 - 5 = 0 \quad a = 1, b = -x, \text{ and } c = x^2 - 5$$

$$y = \frac{x \pm \sqrt{(-x)^2 - 4(1)(x^2 - 5)}}{2}$$

$$= \frac{x \pm \sqrt{20 - 3x^2}}{2}$$

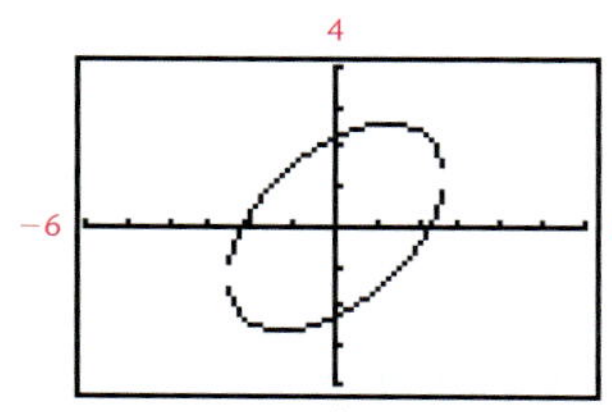

FIGURE 8

Graphing

$$y_1 = \frac{x + \sqrt{20 - 3x^2}}{2} \quad \text{and} \quad y_2 = \frac{x - \sqrt{20 - 3x^2}}{2}$$

produces the rotated ellipse in Figure 8.

(B) $B^2 - 4AC = (-1)^2 - 4(1)(-1) = 1 + 4 = 5 > 0$
This conic is a hyperbola.

$$x^2 - xy + y^2 = 5$$

$$y^2 + xy + 5 - x^2 = 0 \quad a = 1, b = x, \text{ and } c = 5 - x^2$$

$$y = \frac{-x \pm \sqrt{x^2 - 4(1)(5 - x^2)}}{2}$$

$$= \frac{-x \pm \sqrt{5x^2 - 20}}{2}$$

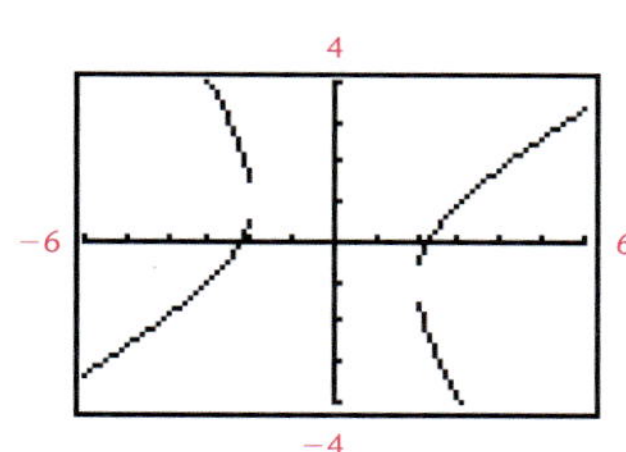

FIGURE 9

Graphing

$$y_1 = \frac{-x + \sqrt{5x^2 - 20}}{2} \quad \text{and} \quad y_2 = \frac{-x - \sqrt{5x^2 - 20}}{2}$$

produces the rotated hyperbola in Figure 9.

(C) $B^2 - 4AC = (-4)^2 - 4(1)(4) = 16 - 16 = 0$
This conic is a parabola.

$$x^2 - 4xy + y^2 + x = 5$$

$$y^2 - 4xy + 4x^2 + x - 5 = 0 \quad a = 1, b = -4x, \text{ and } c = 4x^2 + x - 5$$

$$y = \frac{4x \pm \sqrt{(-4x)^2 - 4(1)(4x^2 + x - 5)}}{2}$$

$$= \frac{4x \pm \sqrt{20 - 4x}}{2}$$

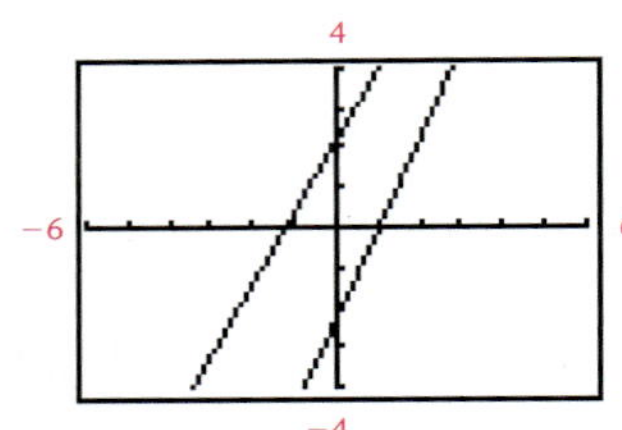

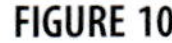
FIGURE 10

The graphs of

$$y_1 = \frac{4x + \sqrt{20 - 4x}}{2} \quad \text{and} \quad y_2 = \frac{4x - \sqrt{20 - 4x}}{2}$$

are shown in Figure 10. Because we know the graph is a parabola, we must enlarge the viewing window to find the vertex. Adjusting the viewing window produces the graph of the rotated parabola in Figure 11.

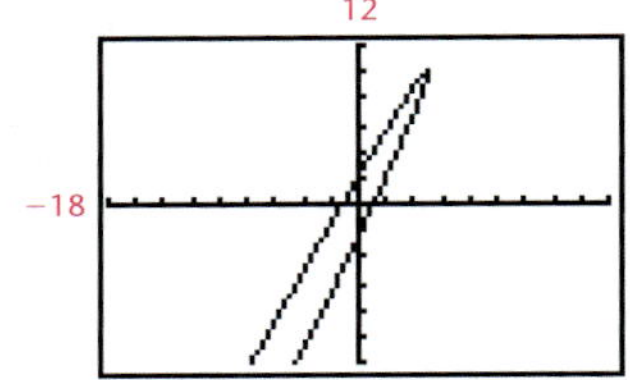

FIGURE 11

MATCHED 3 PROBLEM

Identify the following conics and graph on a graphing utility.

(A) $x^2 + xy + 2y^2 = 10$

(B) $x^2 + xy - 2y^2 = 10$

(C) $x^2 - 2xy + y^2 - x = 10$

ANSWERS MATCHED PROBLEMS

1. $x'^2 - y'^2 = 1$; hyperbola

2. $\frac{x'^2}{9} - \frac{y'^2}{4} = 1$; $\theta = 30°$; hyperbola

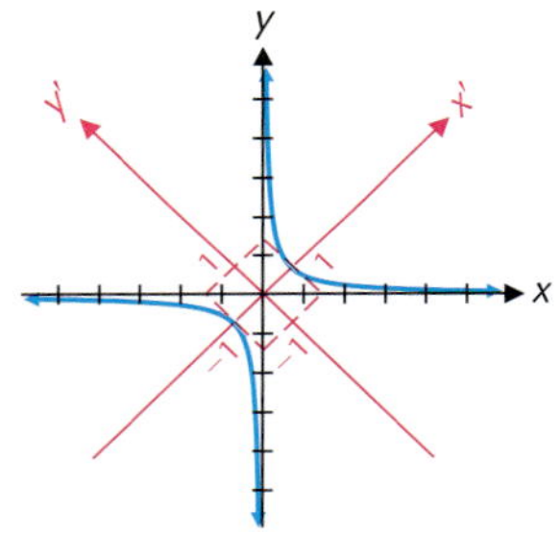

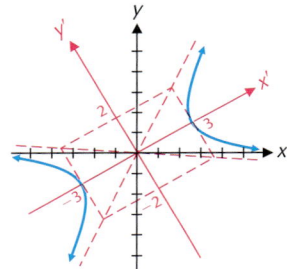

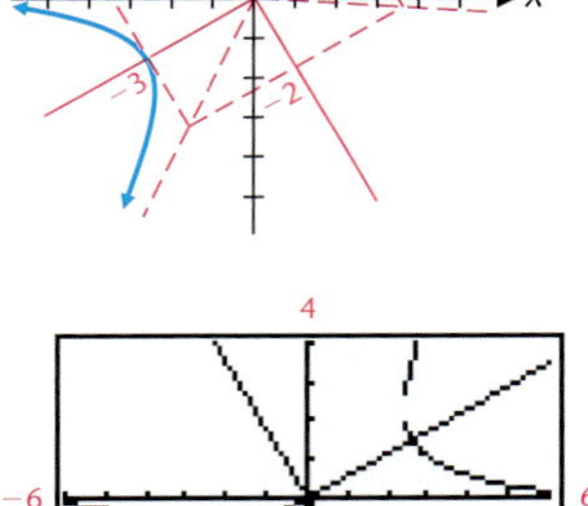

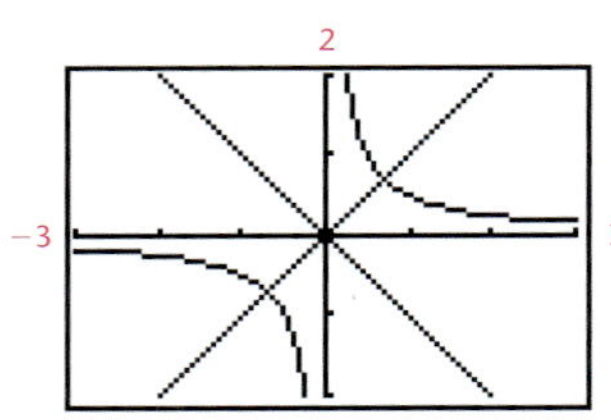

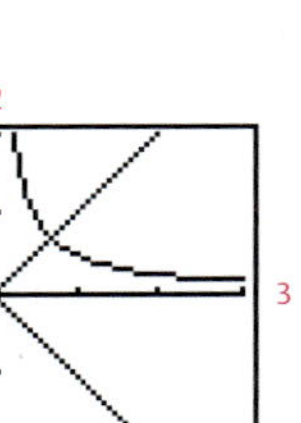

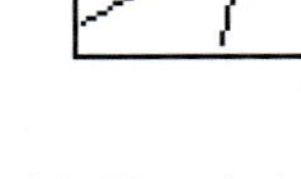

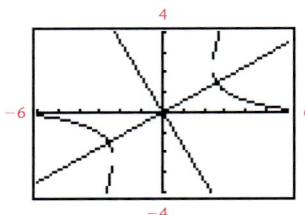

3. (A) Ellipse (B) Hyperbola (C) Parabola

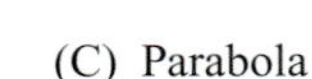

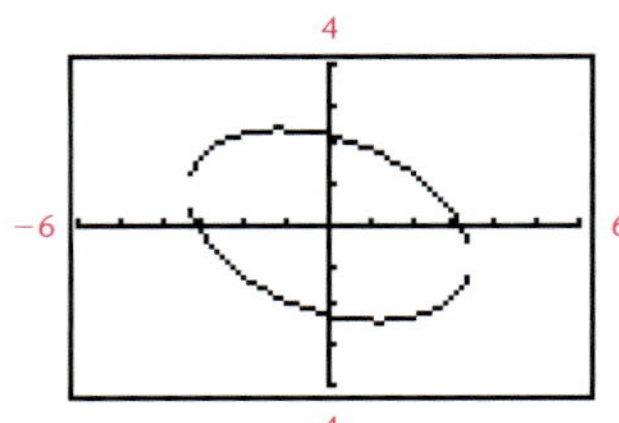

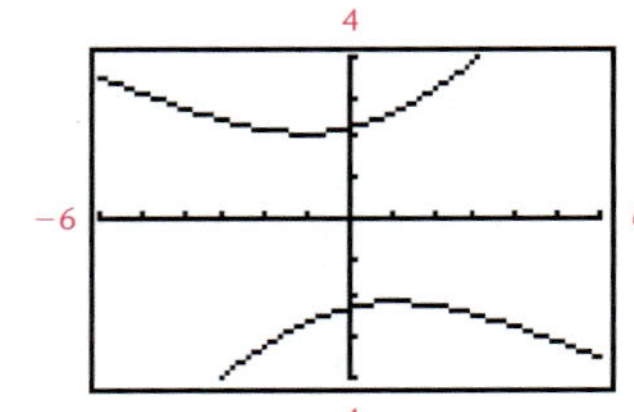

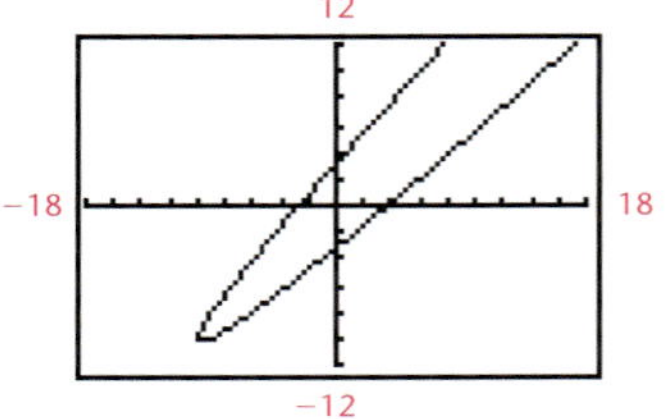

EXERCISE 11.5

A

In Problems 1–4, find the $x'y'$ coordinates of the given points if the coordinate axes are rotated through the indicated angle.

1. $(1, 0), (0, 1), (1, -1), (-3, 4), \theta = 30°$

2. $(1, 0), (0, 1), (-1, 2), (-2, 5), \theta = 60°$

3. $(1, 0), (0, 1), (-1, -2), (1, -3), \theta = 45°$

4. $(1, 1), (-1, -1), (1, -1), (-1, 1), \theta = 90°$

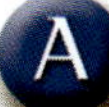
In Problems 5–8, find the equations of the x' and y' axes in terms of x and y if the xy coordinate axes are rotated through the indicated angle.

5. $\theta = 30°$ **6.** $\theta = 60°$

7. $\theta = 45°$ **8.** $\theta = 90°$

B

In Problems 9–12, find the transformed equation when the axes are rotated through the indicated angle. Sketch and identify the graph.

9. $x^2 + y^2 = 49, \theta = 45°$ **10.** $x^2 + y^2 = 25, \theta = 60°$

11. $2x^2 + \sqrt{3}xy + y^2 - 10 = 0, \theta = 30°$

12. $x^2 + 8xy + y^2 - 75 = 0, \theta = 45°$

In Problems 13–18, find the angle of rotation so that the transformed equation will have no $x'y'$ term. Sketch and identify the graph.

13. $x^2 - 4xy + y^2 = 12$ **14.** $x^2 + xy + y^2 = 6$

15. $8x^2 - 4xy + 5y^2 = 36$ **16.** $5x^2 - 4xy + 8y^2 = 36$

17. $x^2 - 2\sqrt{3}xy + 3y^2 - 16\sqrt{3}x - 16y = 0$

18. $x^2 - 2\sqrt{3}xy + 3y^2 + 8\sqrt{3}x - 8y = 0$

In Problems 19–24, use the discriminant to identify each graph. Graph on a graphing utility.

19. $13x^2 + 10xy + 13y^2 - 72 = 0$

20. $3x^2 - 10xy + 3y^2 + 8 = 0$

21. $x^2 - 6\sqrt{3}xy - 5y^2 - 8 = 0$

22. $16x^2 + 24xy + 9y^2 + 15x - 20y = 0$

23. $16x^2 - 24xy + 9y^2 + 60x + 80y = 0$

24. $7x^2 - 6\sqrt{3}xy + 13y^2 - 16 = 0$

C

In Problems 25 and 26, use a rotation followed by a translation to transform each equation into a standard form. Sketch and identify the curve.

25. $x^2 + 2\sqrt{3}xy + 3y^2 - 8\sqrt{3}x - 8y - 4 = 0$

26. $73x^2 + 72xy + 52y^2 - 260x - 320y + 400 = 0$

SECTION 11.6 Nonlinear Systems

Solution by Substitution • Solution by Elimination • Modeling with Nonlinear Systems

In Chapters 8 and 9 we considered systems of equations in which each equation in the system was linear. If a system of equations contains any equations that are not linear, then the system is called a **nonlinear system.** In Section 11.6 we investigate nonlinear systems that involve at least one second-degree quadratic equation, such as

$$\begin{aligned} x^2 + y^2 &= 5 \\ 3x + y &= 1 \end{aligned} \qquad \begin{aligned} x^2 - 2y^2 &= 2 \\ xy &= 2 \end{aligned} \qquad \begin{aligned} x^2 - y^2 &= 5 \\ x^2 + 2y^2 &= 17 \end{aligned}$$

It can be shown that **such systems have at most four solutions,** some of which may be imaginary. Because we are interested in finding both real and imaginary solutions to the systems we consider, we consider the replacement set for each variable to be the set of complex numbers.

Solution by Substitution

The substitution method used to solve linear systems of two equations in two variables is also an effective method for solving nonlinear systems. This process is best illustrated by examples.

Solving a Nonlinear System by Substitution

Solve the system $x^2 + y^2 = 5$
$3x + y = 1$

SOLUTION

Algebraic Solution

Solve the second equation for y in terms of x; then substitute for y in the first equation to obtain an equation that involves x alone.

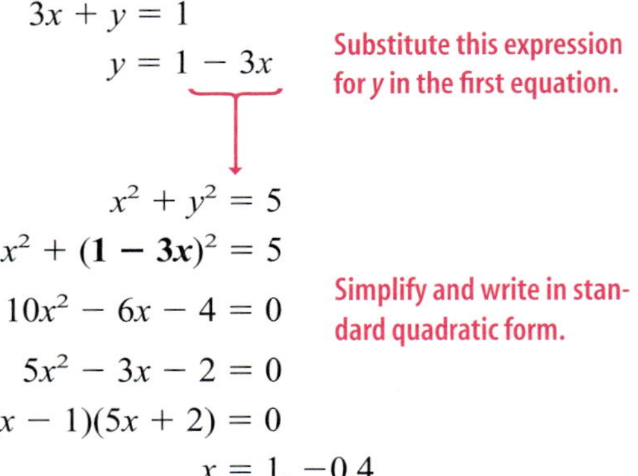

$$3x + y = 1$$
$$y = 1 - 3x$$

Substitute this expression for y in the first equation.

$$x^2 + y^2 = 5$$
$$x^2 + (\mathbf{1 - 3x})^2 = 5$$
$$10x^2 - 6x - 4 = 0$$

Simplify and write in standard quadratic form.

$$5x^2 - 3x - 2 = 0$$
$$(x - 1)(5x + 2) = 0$$
$$x = 1, -0.4$$

If we substitute these values back into the equation $y = 1 - 3x$, we obtain two solutions to the system:

$x = 1$ $\qquad$ $x = -0.4$

$y = 1 - 3(1) = -2$ $\qquad$ $y = 1 - 3(-0.4) = 2.2$

A check, which you should provide, verifies that $(1, -2)$ and $(-0.4, 2.2)$ are both solutions to the system.

Graphical Solution

We enter two equations to graph the circle and one to graph the line (Fig. 1). Using the intersect command, we find two solutions, $(1, -2)$ (Fig. 2) and $(-0.4, 2.2)$ (Fig. 3).

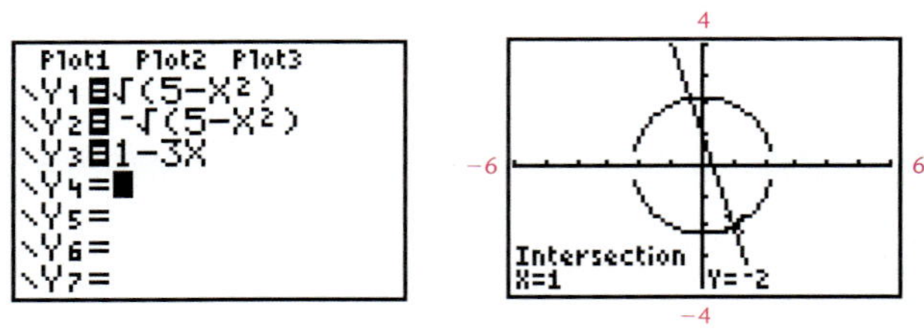

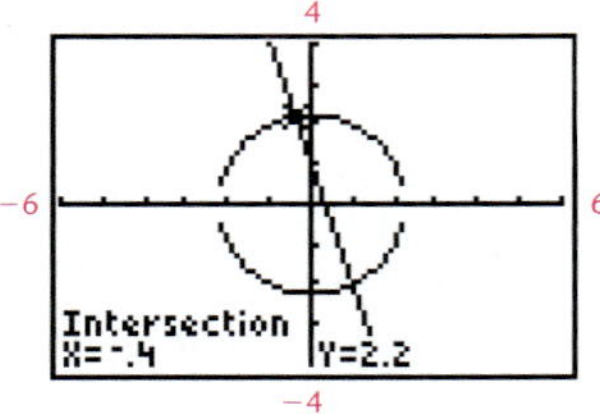

FIGURE 3

MATCHED 1 PROBLEM

Solve the system
$$\begin{aligned} x^2 + y^2 &= 10 \\ 2x + y &= 1 \end{aligned}$$

Refer to the algebraic solution of Example 1. If we substitute $x = 1$ and $x = -0.4$ back into the equation $x^2 + y^2 = 5$, we obtain

$$\begin{aligned} x &= 1 \\ 1^2 + y^2 &= 5 \\ y^2 &= 4 \\ y &= \pm 2 \end{aligned} \qquad\qquad \begin{aligned} x &= -0.4 \\ (-0.4)^2 + y^2 &= 5 \\ y^2 &= 4.84 \\ y &= \pm 2.2 \end{aligned}$$

It appears that we have found two additional solutions, (1, 2) and (−0.4, −2.2). But neither of these solutions satisfies the equation $3x + y = 1$, which you should verify. So, neither is a solution to the original system. We have produced two **extraneous roots,** apparent solutions that do not actually satisfy both equations in the system. This is a common occurrence when solving nonlinear systems.

It is always important to check the solutions of any nonlinear systems to ensure that extraneous roots have not been introduced.

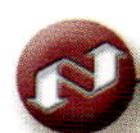

EXPLORE/DISCUSS 1

In Example 1, we saw that the line $3x + y = 1$ intersects the circle $x^2 + y^2 = 5$ in two points.

(A) Consider the system
$$\begin{aligned} x^2 + y^2 &= 5 \\ 3x + y &= 10 \end{aligned}$$

Are there any real solutions to this system? Are there any complex solutions? Find any real or complex solutions.

(B) Consider the family of lines given by

$$3x + y = b \qquad b \text{ any real number}$$

What do all these lines have in common? Illustrate graphically the lines in this family that intersect the circle $x^2 + y^2 = 5$ in exactly one point. How many such lines are there? What are the corresponding value(s) of b? What are the intersection points? How are these lines related to the circle?

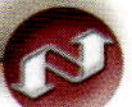

EXAMPLE 2 Solving a Nonlinear System by Substitution

Solve
$$\begin{aligned} x^2 - 2y^2 &= 2 \\ xy &= 2 \end{aligned}$$

SOLUTION

Algebraic Solution

Solve the second equation for y, substitute in the first equation, and proceed as before.

$$xy = 2$$

$$y = \frac{2}{x}$$

$$x^2 - 2\left(\frac{2}{x}\right)^2 = 2$$

$$x^2 - \frac{8}{x^2} = 2$$

$$x^4 - 2x^2 - 8 = 0 \quad \text{Multiply both sides by } x^2 \text{ and simplify.}$$

$$u^2 - 2u - 8 = 0 \quad \text{Substitute } u = x^2 \text{ to transform to quadratic form and solve.}$$

$$(u - 4)(u + 2) = 0$$

$$u = 4, -2$$

Thus,

$$x^2 = 4 \qquad \text{or} \qquad x^2 = -2$$

$$x = \pm 2 \qquad\qquad x = \pm\sqrt{-2} = \pm i\sqrt{2}$$

For $x = 2$, $y = \frac{2}{2} = 1$

For $x = -2$, $y = \frac{2}{-2} = -1$

For $x = i\sqrt{2}$, $y = \frac{2}{i\sqrt{2}} = -i\sqrt{2}$

For $x = -i\sqrt{2}$, $y = \frac{2}{-i\sqrt{2}} = i\sqrt{2}$

Thus, the four solutions to this system are (2, 1), (−2, −1), $(i\sqrt{2}, -i\sqrt{2})$, and $(-i\sqrt{2}, i\sqrt{2})$. You should check that each of these satisfies both equations in the system.

Graphical Solution

Solving the first equation for y, we have

$$x^2 - 2y^2 = 2$$

$$-2y^2 = -x^2 + 2$$

$$y^2 = 0.5x^2 - 1 \quad \text{Multiply both sides by } \tfrac{1}{-2} = -0.5.$$

$$y = \pm\sqrt{0.5x^2 - 1}$$

We enter these two equations and $y = 2/x$ (Fig. 4) in a graphing utility. Using the intersect command, we find the two real solutions, (2, 1) (Fig. 5) and (−2, −1) (Fig. 6). But we cannot find the two complex solutions.

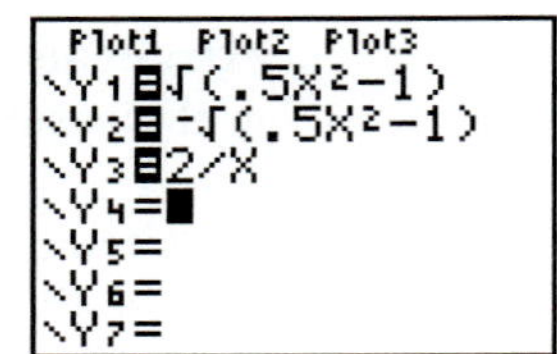

FIGURE 4

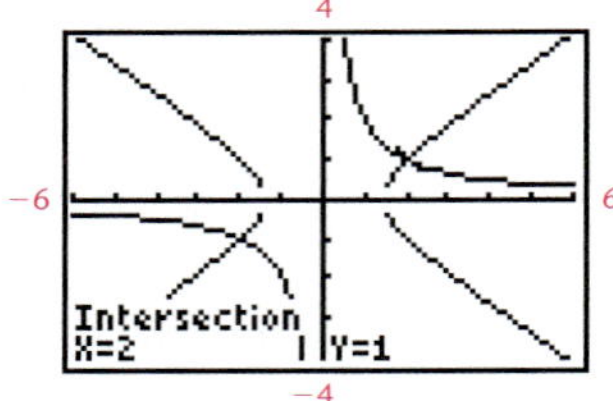

FIGURE 5

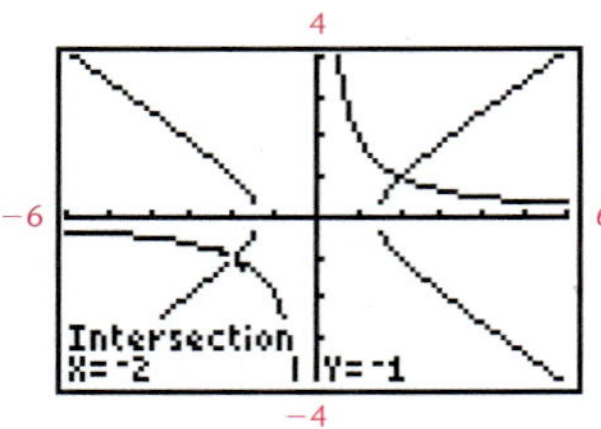

FIGURE 6

MATCHED 2 PROBLEM

Solve $3x^2 - y^2 = 6$

$xy = 3$

Compare the two solutions in Example 2. The algebraic solution found all four solution points, two real and two imaginary. But the graphical solution only found the two real solution points. This is true in general: **Graphical methods can never be used to find imaginary solutions.**

Solution by Elimination

Just as with systems of linear equations, we can add a multiple of one equation to another equation to produce an equivalent system—a system with the same solution set. Certain nonlinear systems can be solved in this manner.

EXAMPLE 3 Solving a Nonlinear System by Elimination

Solve
$$\begin{aligned} x^2 - y^2 &= 5 \\ x^2 + 2y^2 &= 17 \end{aligned}$$

SOLUTION

Algebraic Solution

Multiplying the second equation by -1 and adding it to the first equation will eliminate x^2:

$$\begin{aligned} x^2 - y^2 &= 5 \\ -x^2 - 2y^2 &= -17 \\ \hline -3y^2 &= -12 \\ y^2 &= 4 \\ y &= \pm 2 \end{aligned}$$

Now substitute $y^2 = 4$ back into either original equation to find x.

$$\begin{aligned} x^2 - y^2 &= 5 \\ x^2 - 4 &= 5 \\ x^2 &= 9 \\ x &= \pm 3 \end{aligned}$$

Thus, $(3, -2)$, $(3, 2)$, $(-3, -2)$, and $(-3, 2)$ are the four solutions to the system. The check of the solution is left to you.

Graphical Solution

Solving each equation in the original system for y produces the four functions shown in Figure 7. The graph is shown in Figure 8. Using intersect four times (details omitted) shows that the solutions are $(3, -2)$, $(3, 2)$, $(-3, -2)$, and $(-3, 2)$.

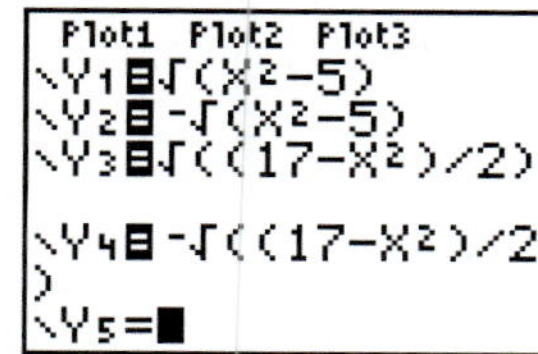

FIGURE 7

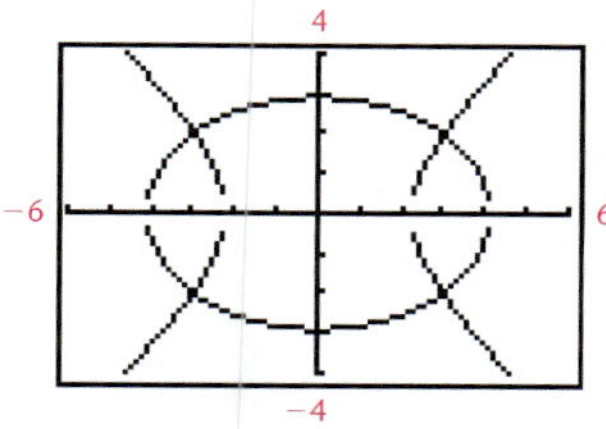

FIGURE 8

MATCHED PROBLEM 3

Solve
$$\begin{aligned} 2x^2 - 3y^2 &= 5 \\ 3x^2 + 4y^2 &= 16 \end{aligned}$$

EXAMPLE 4 Solving a Nonlinear System by Elimination

Solve $x^2 - y^2 = 25$
$x^2 + y^2 = 7$

SOLUTION

Algebraic Solution

Adding the second equation to the first eliminates y^2.

$$\begin{aligned} x^2 - y^2 &= 25 \\ x^2 + y^2 &= 7 \\ \hline 2x^2 &= 32 \\ x^2 &= 16 \\ x &= \pm 4 \end{aligned}$$

Now substitute $x^2 = 16$ back into either original equation to find y.

$$\begin{aligned} x^2 + y^2 &= 7 \\ 16 + y^2 &= 7 \\ y^2 &= -9 \\ y &= \pm\sqrt{-9} = \pm 3i \end{aligned}$$

Thus, $(4, -3i)$, $(4, 3i)$, $(-4, 3i)$, and $(-4, -3i)$ are the four solutions to the system. The check of the solutions is left to you.

Graphical Solution

Solving each equation in the original system for y produces the four functions shown in Figure 9. The graph (Fig. 10) shows that there are no real solutions. As before, the imaginary solutions cannot be found graphically.

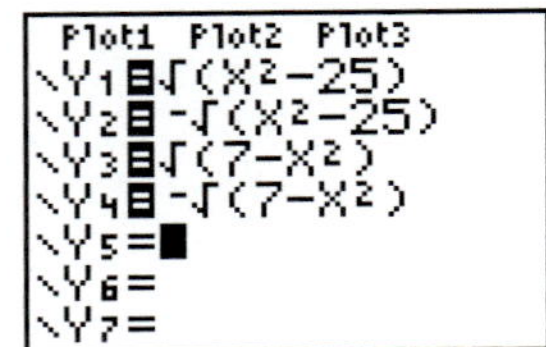

FIGURE 9

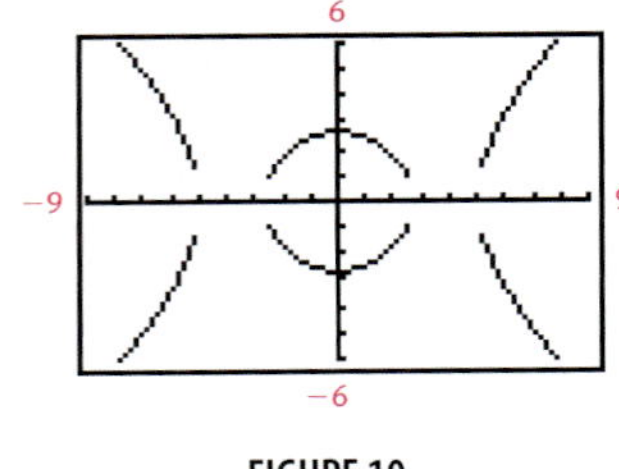

FIGURE 10

MATCHED PROBLEM 4

Solve $y^2 - 2x^2 = 25$
$y^2 + x^2 = 1$

Modeling with Nonlinear Systems

EXAMPLE 5 Design

An engineer is to design a rectangular computer screen with a 19-inch diagonal and a 175-square-inch area. Find the dimensions of the screen to the nearest tenth of an inch.

SOLUTION

Algebraic Solution

Sketch a rectangle letting x be the width and y the height (Fig. 11). We obtain the following system using the Pythagorean theorem and the formula for the area of a rectangle:

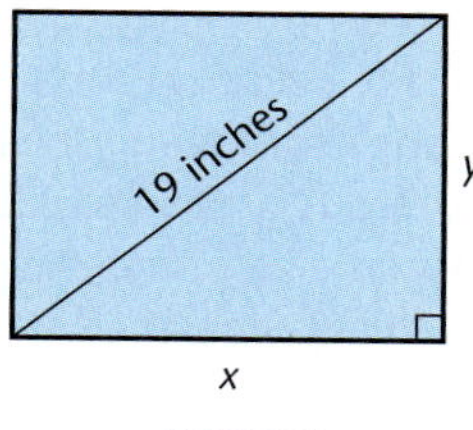

FIGURE 11

$$x^2 + y^2 = 19^2$$
$$xy = 175$$

This system is solved by substitution. We are only interested in real solutions. We start by solving the second equation for y in terms of x and substituting the result into the first equation.

$$y = \frac{175}{x}$$

$$x^2 + \frac{175^2}{x^2} = 19^2$$

$$x^4 + 30{,}625 = 361x^2 \quad \text{Multiply both sides by } x^2 \text{ and simplify.}$$

$$x^4 - 361x^2 + 30{,}625 = 0 \quad \text{Simplify quadratic in } x^2.$$

Solve the last equation for x^2 using the quadratic formula, then solve for x:

$$x = \sqrt{\frac{361 \pm \sqrt{361^2 - 4(1)(30{,}625)}}{2}}$$
$$= 15.0 \text{ inches or } 11.7 \text{ inches}$$

Substitute each choice of x into $y = 175/x$ to find the corresponding y values:

For $x = 15.0$ inches,

$$y = \frac{175}{15} = 11.7 \text{ inches}$$

For $x = 11.7$ inches,

$$y = \frac{175}{11.7} = 15.0 \text{ inches}$$

Assuming the screen is wider than it is high, the dimensions are 15.0 by 11.7 inches.

Graphical Solution

Figure 12 shows the three functions required to graph this system. The graph is shown in Figure 13. We are only interested in the solutions in the first quadrant. Zooming in and using intersect produces the results in Figures 14 and 15. Assuming that the screen is wider than it is high, its dimensions are 15.0 by 11.7 inches.

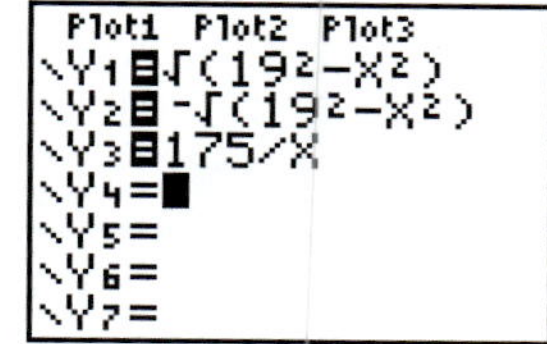

FIGURE 12

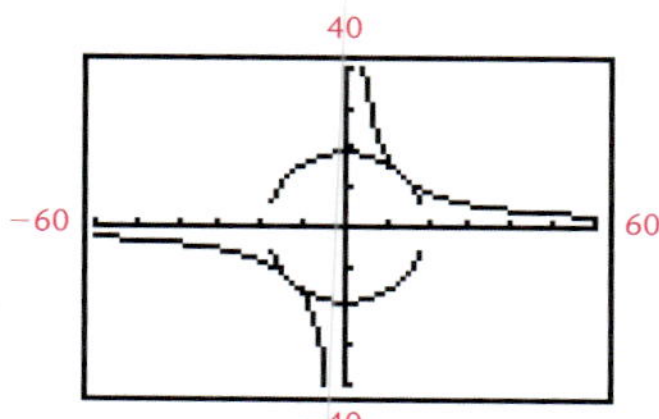

FIGURE 13

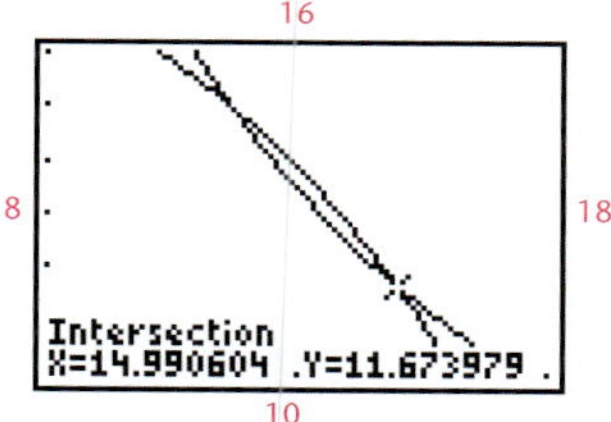

FIGURE 14

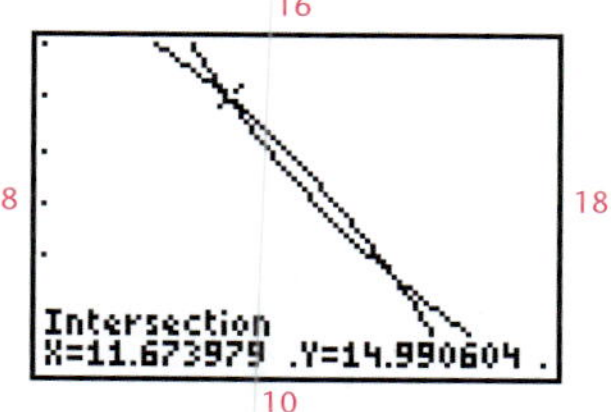

FIGURE 15

MATCHED PROBLEM 5

An engineer is to design a rectangular television screen with a 21-inch diagonal and a 209-square-inch area. Find the dimensions of the screen to the nearest tenth of an inch.

A **seismograph** is a machine that records the waves generated by an earthquake. It can determine the distance from the recording machine to the epicenter of the quake, but not the direction. A reading from a single station places the epicenter somewhere on a circle. A reading from a second station places the epicenter at one of the two intersection points of two circles. A reading from a third station determines the precise location of the epicenter. This procedure for locating the epicenter is called **triangulation.** Example 6 illustrates this process.

EXAMPLE 6 Triangulation

Seismographs are located at three stations. Station B is 500 miles east and 100 miles north of station A. Station C is 100 miles east and 600 miles north of station A. Readings from the three stations indicate that the epicenter of an earthquake is 300 miles from station A, 288 miles from station B, and 425 miles from station C. Use triangulation and a graphing utility to locate the epicenter of the earthquake relative to station A. Round answers to the nearest mile.

SOLUTION

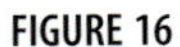
FIGURE 16

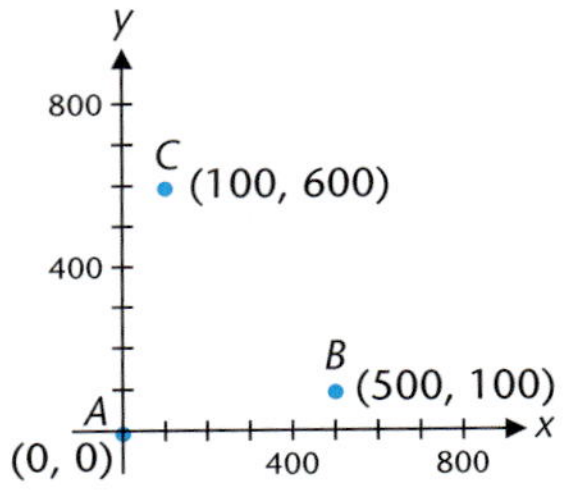

Because the locations of stations B and C are given in terms of station A, we introduce a coordinate system with station A at the origin (Fig. 16). Then we add station B at (500, 100) and station C at (100, 600). The epicenter is located at the intersection of the following three circles:

$$x^2 + y^2 = 300^2 \qquad \text{Station A}$$
$$(x - 500)^2 + (y - 100)^2 = 288^2 \qquad \text{Station B}$$
$$(x - 100)^2 + (y - 600)^2 = 425^2 \qquad \text{Station C}$$

FIGURE 17

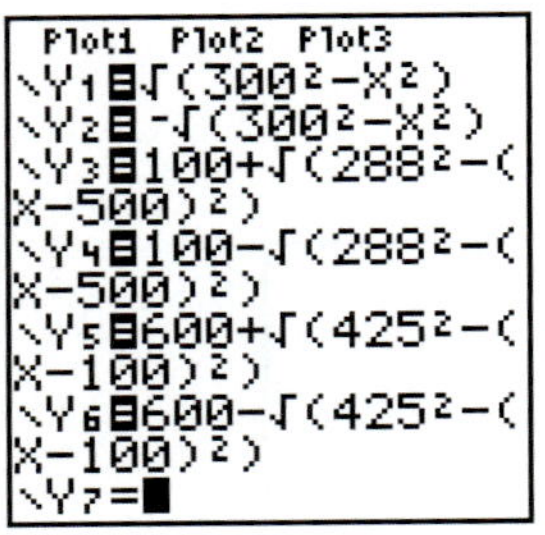

Solving these equations for y produces six functions (Fig. 17). Graphing the first four functions in a square viewing window and using intersect shows that the first two circles intersect at (228, 195) (Fig. 18) and (285, −92) (Fig. 19).

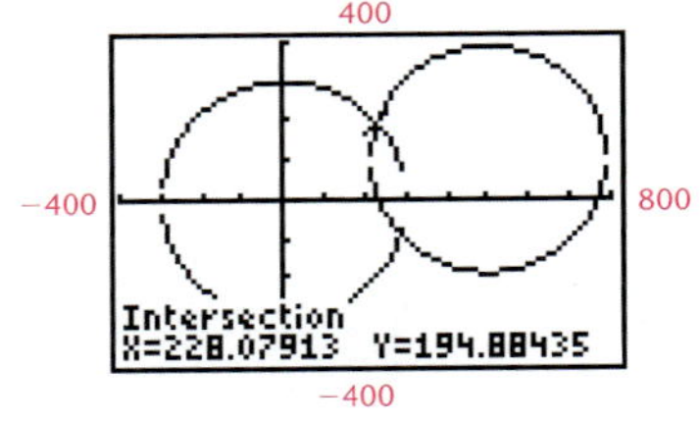

FIGURE 18

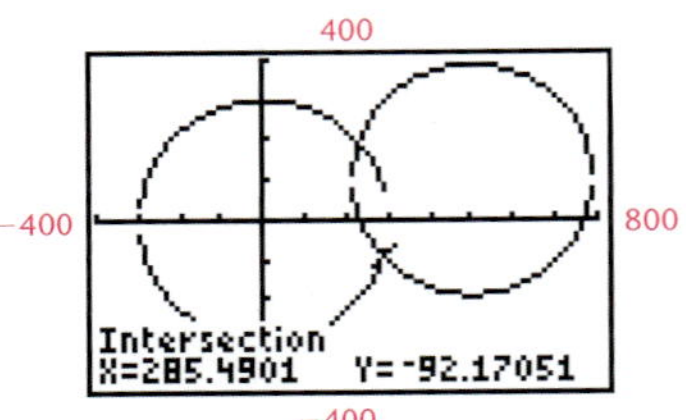

FIGURE 19

Graphing all six functions in a larger window and using trace shows that the third circle passes through the point (228, 195) (Fig. 20). Thus, the epicenter of the earthquake is 228 miles east and 195 miles north of station A.

FIGURE 20

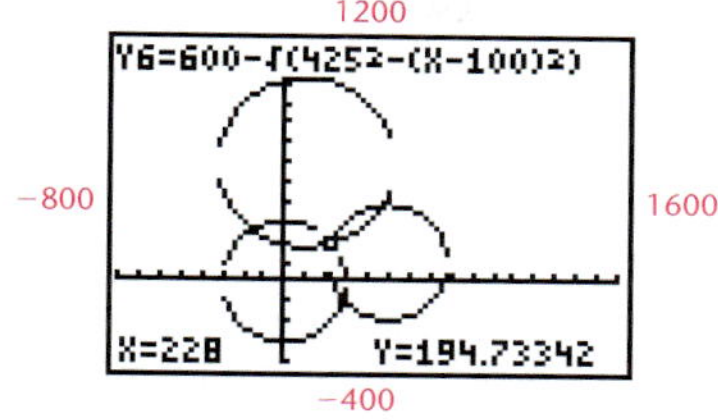

MATCHED 6 PROBLEM

Repeat Example 6 if the epicenter of an earthquake is 410 miles from station A, 310 miles from station B, and 338 miles from station C.

ANSWERS MATCHED PROBLEMS

1. $(-1, 3), \left(\frac{9}{5}, -\frac{13}{5}\right)$ **2.** $(\sqrt{3}, \sqrt{3}), (-\sqrt{3}, -\sqrt{3}), (i, -3i), (-i, 3i)$ **3.** $(2, 1), (2, -1), (-2, 1), (-2, -1)$
4. $(2i\sqrt{2}, 3), (2i\sqrt{2}, -3), (-2i\sqrt{2}, 3), (-2i\sqrt{2}, -3)$ **5.** 17.1 by 12.2 inches
6. The epicenter is 270 miles east and 308 miles north of station A.

EXERCISE 11.6

A

Solve each system in Problems 1–12.

1. $x^2 + y^2 = 169$
$x = -12$

2. $x^2 + y^2 = 25$
$y = -4$

3. $8x^2 - y^2 = 16$
$y = 2x$

4. $y^2 = 2x$
$x = y - \frac{1}{2}$

5. $3x^2 - 2y^2 = 25$
$x + y = 0$

6. $x^2 + 4y^2 = 32$
$x + 2y = 0$

7. $y^2 = x$
$x - 2y = 2$

8. $x^2 = 2y$
$3x = y + 2$

9. $2x^2 + y^2 = 24$
$x^2 - y^2 = -12$

10. $x^2 - y^2 = 3$
$x^2 + y^2 = 5$

11. $x^2 + y^2 = 10$
$16x^2 + y^2 = 25$

12. $x^2 - 2y^2 = 1$
$x^2 + 4y^2 = 25$

B

Solve each system in Problems 13–24.

13. $xy - 4 = 0$
$x - y = 2$

14. $xy - 6 = 0$
$x - y = 4$

15. $x^2 + 2y^2 = 6$
$xy = 2$

16. $2x^2 + y^2 = 18$
$xy = 4$

17. $2x^2 + 3y^2 = -4$
$4x^2 + 2y^2 = 8$

18. $2x^2 - 3y^2 = 10$
$x^2 + 4y^2 = -17$

19. $x^2 - y^2 = 2$
$y^2 = x$

20. $x^2 + y^2 = 20$
$x^2 = y$

21. $x^2 + y^2 = 9$
$x^2 = 9 - 2y$

22. $x^2 + y^2 = 16$
$y^2 = 4 - x$

23. $x^2 - y^2 = 3$
$xy = 2$

24. $y^2 = 5x^2 + 1$
$xy = 2$

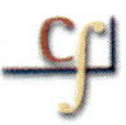

An important type of calculus problem is to find the area between the graphs of two functions. To solve some of these problems it is necessary to find the coordinates of the points of intersection of the two graphs. In Problems 25–32, find the coordinates of the points of intersection of the two given equations.

25. $y = 5 - x^2, y = 2 - 2x$

26. $y = 5x - x^2, y = x + 3$

27. $y = x^2 - x, y = 2x$

28. $y = x^2 + 2x, y = 3x$

29. $y = x^2 - 6x + 9, y = 5 - x$

30. $y = x^2 + 2x + 3, y = 2x + 4$

31. $y = 8 + 4x - x^2, y = x^2 - 2x$

32. $y = x^2 - 4x - 10, y = 14 - 2x - x^2$

33. Consider the circle with equation $x^2 + y^2 = 5$ and the family of lines given by $2x - y = b$, where b is any real number.

(A) Illustrate graphically the lines in this family that intersect the circle in exactly one point, and describe the relationship between the circle and these lines.

(B) Find the values of b corresponding to the lines in part A, and find the intersection points of the lines and the circle.

(C) How is the line with equation $x + 2y = 0$ related to this family of lines? How could this line be used to find the intersection points in part B?

34. Consider the circle with equation $x^2 + y^2 = 25$ and the family of lines given by $3x + 4y = b$, where b is any real number.

(A) Illustrate graphically the lines in this family that intersect the circle in exactly one point, and describe the relationship between the circle and these lines.

(B) Find the values of b corresponding to the lines in part A, and find the intersection points of the lines and the circle.

(C) How is the line with equation $4x - 3y = 0$ related to this family of lines? How could this line be used to find the intersection points and the values of b in part B?

35. Consider the system of equations

$$x^2 - y = 0$$
$$2x - y = b$$

where b is any real number. Determine the permissible values of b and describe verbally the graph of the system if:

(A) The system has two distinct real solutions.

(B) The system has one distinct real solution.

(C) The system has two distinct imaginary solutions.

36. Consider the system of equations

$$x^2 - y = 0$$
$$4x + y = b$$

where b is any real number. Determine the permissible values of b and describe verbally the graph of the system if:

(A) The system has two distinct real solutions.

(B) The system has one distinct real solution.

(C) The system has two distinct imaginary solutions.

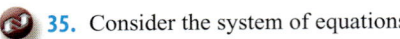

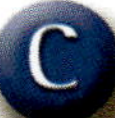

Solve each system in Problems 37–44.

37. $2x + 5y + 7xy = 8$
$xy - 3 = 0$

38. $2x + 3y + xy = 16$
$xy - 5 = 0$

39. $x^2 - 2xy + y^2 = 1$
$x - 2y = 2$

40. $x^2 + xy - y^2 = -5$
$y - x = 3$

41. $2x^2 - xy + y^2 = 8$
$x^2 - y^2 = 0$

42. $x^2 + 2xy + y^2 = 36$
$x^2 - xy = 0$

43. $x^2 + xy - 3y^2 = 3$
$x^2 + 4xy + 3y^2 = 0$

44. $x^2 - 2xy + 2y^2 = 16$
$x^2 - y^2 = 0$

In Problems 45–50, use a graphing utility to find the real solutions of each system to two decimal places.

45. $-x^2 + 2xy + y^2 = 1$
$3x^2 - 4xy + y^2 = 2$

46. $-x^2 + 4xy + y^2 = 2$
$8x^2 - 2xy + y^2 = 9$

47. $3x^2 - 4xy - y^2 = 2$
$2x^2 + 2xy + y^2 = 9$

48. $5x^2 + 4xy + y^2 = 4$
$4x^2 - 2xy + y^2 = 16$

49. $2x^2 - 2xy + y^2 = 9$
$4x^2 - 4xy + y^2 + x = 3$

50. $2x^2 + 2xy + y^2 = 12$
$4x^2 - 4xy + y^2 + x + 2y = 9$

APPLICATIONS

51. Numbers. Find two numbers such that their sum is 3 and their product is 1.

52. Numbers. Find two numbers such that their difference is 1 and their product is 1. (Let x be the larger number and y the smaller number).

53. Geometry. Find the lengths of the legs of a right triangle with an area of 30 square inches if its hypotenuse is 13 inches long.

54. Geometry. Find the dimensions of a rectangle with an area of 32 square meters if its perimeter is 36 meters long.

55. Design. An engineer is designing a small portable television set. According to the design specifications, the set must have a rectangular screen with a 7.5-inch diagonal and an area of 27 square inches. Find the dimensions of the screen.

56. Design. An artist is designing a logo for a business in the shape of a circle with an inscribed rectangle. The diameter of the circle is 6.5 inches, and the area of the rectangle is 15 square inches. Find the dimensions of the rectangle.

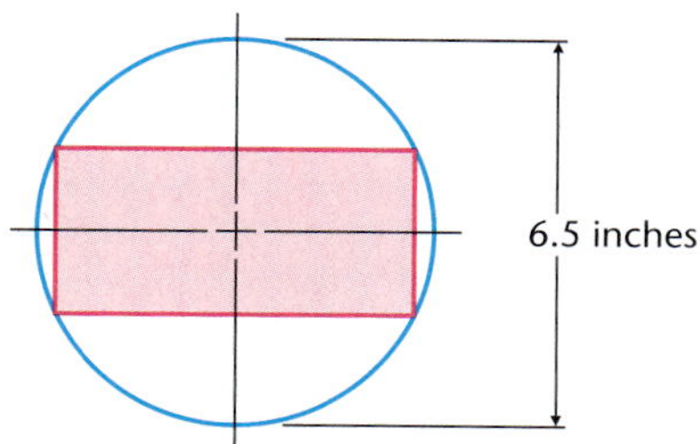

57. Television. High-definition television sets have screens with a width-to-height ratio of 16 to 9. Find the dimensions of a screen with a 60-inch diagonal. Round dimensions to three significant digits.

58. Television. Traditional television sets have screens with a width-to-height ratio of 4 to 3. Find the dimensions of a screen with a 32-inch diagonal. Round dimensions to three significant digits.

59. Triangulation. Seismographs are located at three stations. Station B is 400 miles due east of station A. Station C is 500 miles due north of station A. Readings from the three stations indicate that the epicenter of an earthquake is 200 miles from station A, 320 miles from station B, and 363 miles from station C. Use triangulation and a graphing utility to locate the epicenter of the earthquake. Express the answer relative to station A and round values to the nearest mile.

60. Triangulation. Seismographs are located at three stations. Station B is 600 miles due east of station A. Station C is 500 miles due north of station A. Readings from the three stations indicate that the epicenter of an earthquake is 400 miles from station A, 550 miles from station B, and 231 miles from station C. Use triangulation and a graphing utility to locate the epicenter of the earthquake. Express the answer relative to station A and round values to the nearest mile.

★ **61. Construction.** A rectangular swimming pool with a deck 5 feet wide is enclosed by a fence as shown in the figure. The surface area of the pool is 572 square feet, and the total area enclosed by the fence (including the pool and the deck) is 1,152 square feet. Find the dimensions of the pool.

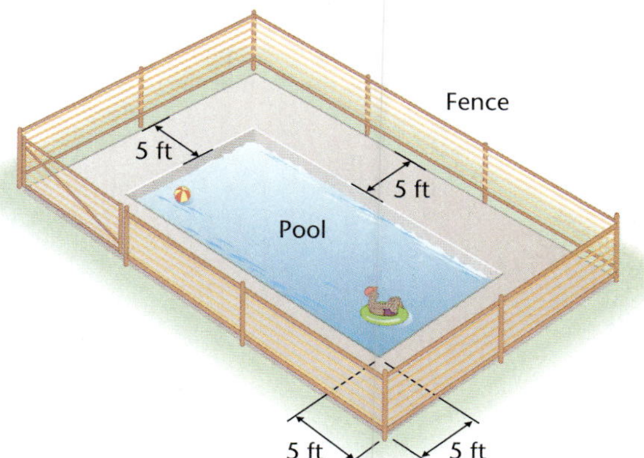

★ **62. Construction.** An open-topped rectangular box is formed by cutting a 6-inch square from each corner of a rectangular piece of cardboard and bending up the ends and sides. The area of the cardboard before the corners are removed is 768 square inches, and the volume of the box is 1,440 cubic inches. Find the dimensions of the original piece of cardboard.

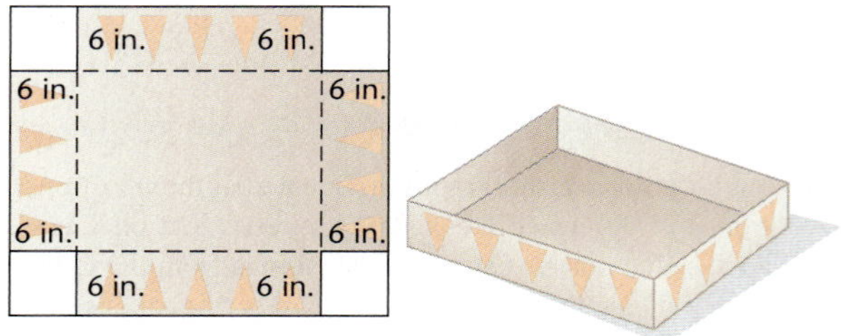

★★ **63. Transportation.** Two boats leave Bournemouth, England, at the same time and follow the same route on the 75-mile trip across the English Channel to Cherbourg, France. The average speed of boat A is 5 miles per hour greater than the average speed of boat B. Consequently, boat A arrives at Cherbourg 30 minutes before boat B. Find the average speed of each boat.

★★ **64. Transportation.** Bus A leaves Milwaukee at noon and travels west on Interstate 94. Bus B leaves Milwaukee 30 minutes later, travels the same route, and overtakes bus A at a point 210 miles west of Milwaukee. If the average speed of bus B is 10 miles per hour greater than the average speed of bus A, at what time did bus B overtake bus A?

CHAPTER 11 REVIEW

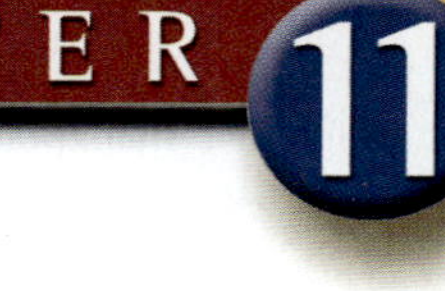

11.1 Conic Sections; Parabola

The plane curves obtained by intersecting a right circular cone with a plane are called **conic sections.** If the plane cuts clear through one nappe, then the intersection curve is called a **circle** if the plane is perpendicular to the axis and an **ellipse** if the plane is not perpendicular to the axis. If a plane cuts only one nappe, but does not cut clear through, then the intersection curve is called a **parabola.** If a plane cuts through both nappes, but not through the vertex, the resulting intersection curve is called a **hyperbola.** A plane passing through the vertex of the cone produces a **degenerate conic**—a point, a line, or a pair of lines. The figure illustrates the four nondegenerate conics.

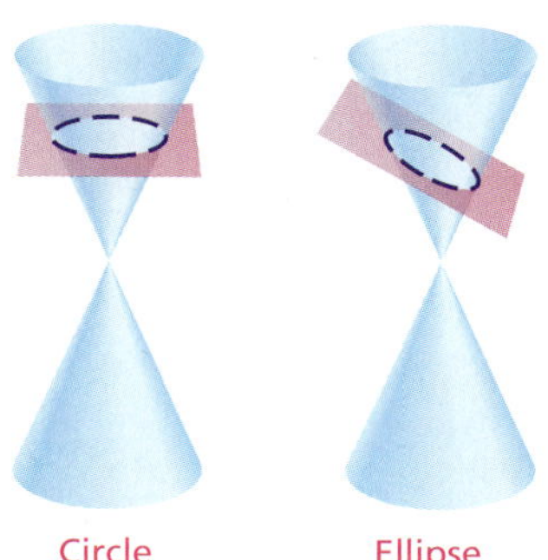

Circle Ellipse

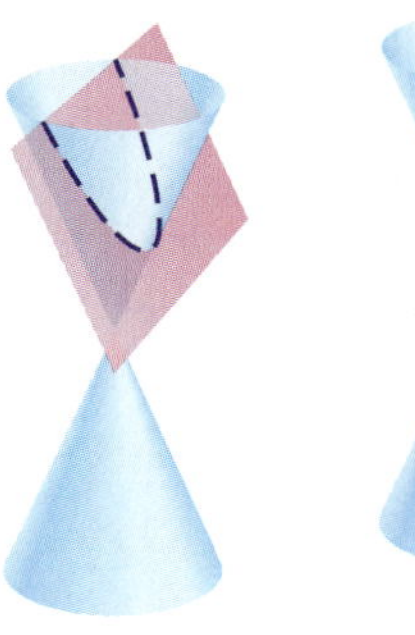

Parabola Hyperbola

The graph of

$$Ax^2 + Bxy + Cy^2 + Dx + Ey + F = 0$$

where A, B, and C are not all 0, is a conic.

The following is a coordinate-free definition of a parabola:

Parabola

A *parabola* is the set of all points in a plane equidistant from a fixed point F and a fixed line L in the plane. The fixed point F is called the **focus,** and the fixed line L is called the **directrix.** A line through the focus perpendicular to the directrix is called the

axis, and the point on the axis halfway between the directrix and focus is called the **vertex.**

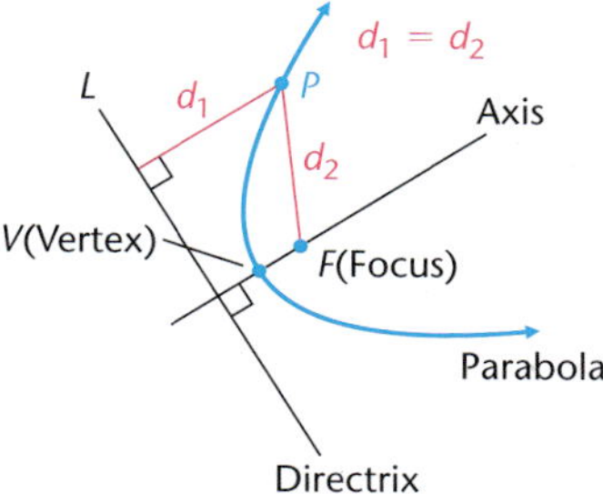

From the definition of a parabola, we can obtain the following standard equations:

Standard Equations of a Parabola with Vertex at (0, 0)

1. $y^2 = 4ax$
Vertex: (0, 0)
Focus: $(a, 0)$
Directrix: $x = -a$
Symmetric with respect to the x axis
Axis the x axis

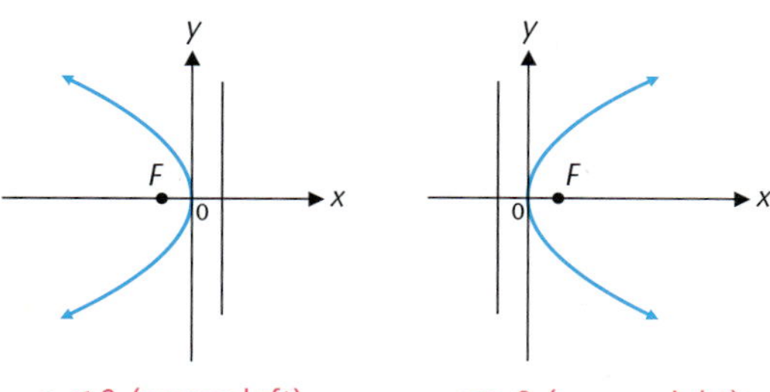

2. $x^2 = 4ay$
Vertex: (0, 0)
Focus: $(0, a)$
Directrix: $y = -a$
Symmetric with respect to the y axis
Axis the y axis

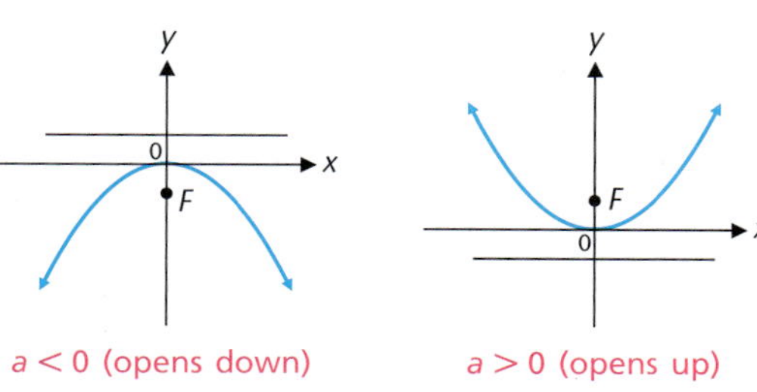

11.2 Ellipse

The following is a coordinate-free definition of an ellipse:

Ellipse

An *ellipse* is the set of all points P in a plane such that the sum of the distances of P from two fixed points in the plane is constant. Each of the fixed points, F' and F, is called a *focus,* and together they are called **foci.** Referring to the figure, the line segment $V'V$ through the foci is the **major axis.** The perpendicular bisector $B'B$ of the major axis is the **minor axis.** Each end of the major axis, V' and V, is called a *vertex*. The midpoint of the line segment $F'F$ is called the **center** of the ellipse.

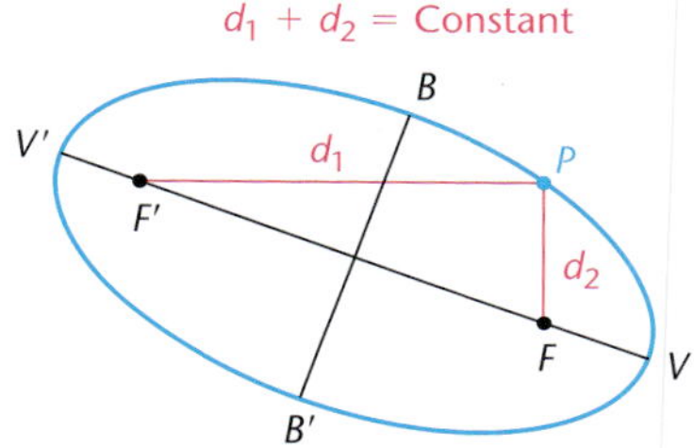

From the definition of an ellipse, we can obtain the following standard equations:

Standard Equations of an Ellipse with Center at (0, 0)

1. $\dfrac{x^2}{a^2} + \dfrac{y^2}{b^2} = 1 \qquad a > b > 0$
x intercepts: $\pm a$ (vertices)
y intercepts: $\pm b$
Foci: $F' = (-c, 0)$, $F = (c, 0)$
$c^2 = a^2 - b^2$
Major axis length $= 2a$
Minor axis length $= 2b$

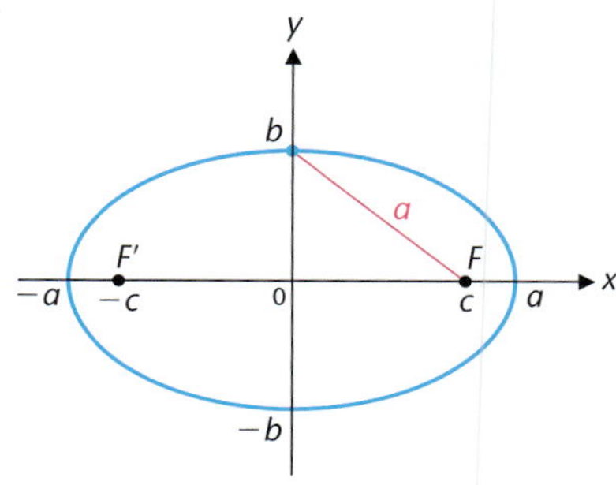

2. $\dfrac{x^2}{b^2} + \dfrac{y^2}{a^2} = 1 \qquad a > b > 0$

x intercepts: $\pm b$
y intercepts: $\pm a$ (vertices)
Foci: $F' = (0, -c)$, $F = (0, c)$

$$c^2 = a^2 - b^2$$

Major axis length $= 2a$
Minor axis length $= 2b$

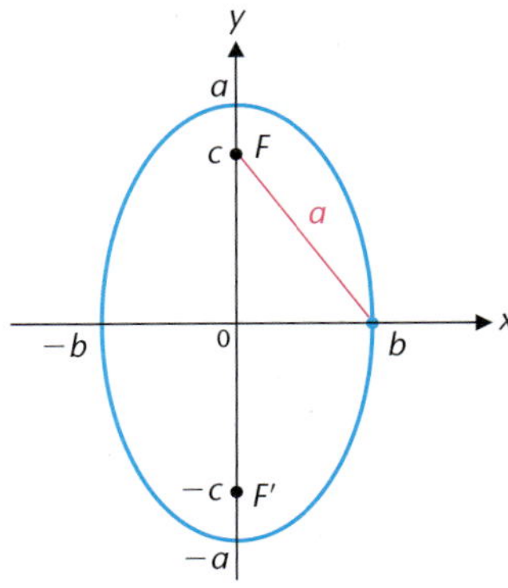

[*Note:* Both graphs are symmetric with respect to the x axis, y axis, and origin. Also, the major axis is always longer than the minor axis.]

11.3 Hyperbola

The following is a coordinate-free definition of a hyperbola:

Hyperbola

A *hyperbola* is the set of all points P in a plane such that the absolute value of the difference of the distances of P to two fixed points in the plane is a positive constant. Each of the fixed points, F' and F, is called a *focus.* The intersection points V' and V of the line through the foci and the two branches of the hyperbola are called **vertices,** and each is called a *vertex.* The line segment $V'V$ is called the **transverse axis.** The midpoint of the transverse axis is the *center* of the hyperbola.

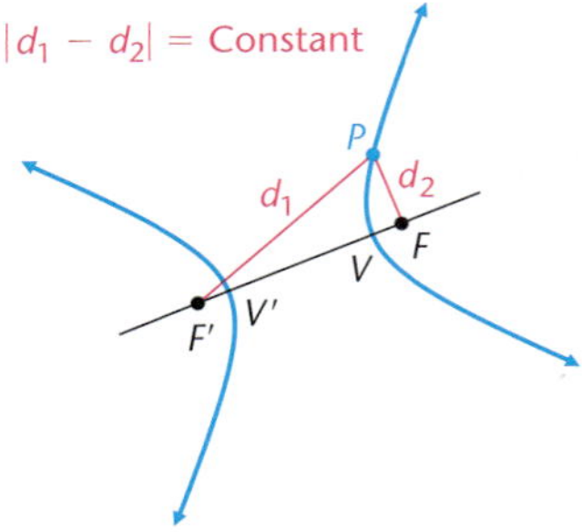

From the definition of a hyperbola, we can obtain the following standard equations:

Standard Equations of a Hyperbola with Center at (0, 0)

1. $\dfrac{x^2}{a^2} - \dfrac{y^2}{b^2} = 1$

x intercepts: $\pm a$ (vertices)
y intercepts: none
Foci: $F' = (-c, 0)$, $F = (c, 0)$

$$c^2 = a^2 + b^2$$

Transverse axis length $= 2a$
Conjugate axis length $= 2b$

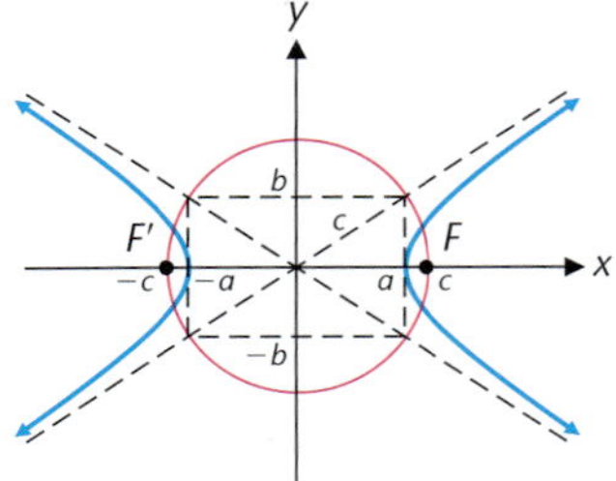

2. $\dfrac{y^2}{a^2} - \dfrac{x^2}{b^2} = 1$

x intercepts: none
y intercepts; $\pm a$ (vertices)
Foci: $F' = (0, -c)$, $F = (0, c)$

$$c^2 = a^2 + b^2$$

Transverse axis length $= 2a$
Conjugate axis length $= 2b$

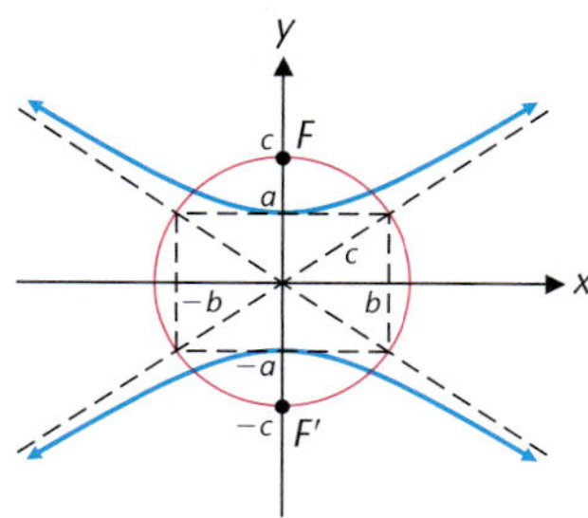

[*Note:* Both graphs are symmetric with respect to the x axis, y axis, and origin.]

11.4 Translation of Axes

In Sections 11.1, 11.2, and 11.3 we found standard equations for parabolas, ellipses, and hyperbolas located with their axes on the coordinate axes and centered relative to the origin. We now move the conics away from the origin while keeping their axes parallel to the coordinate axes. In this process we obtain new standard equations that are special cases of the equation $Ax^2 + Cy^2 + Dx + Ey + F = 0$, where A and C are not both zero. The basic mathematical tool used is *translation of axes.*

A **translation of coordinate axes** occurs when the new coordinate axes have the same direction as and are parallel to the original coordinate axes. **Translation formulas** are as follows:

1. $x = x' + h$ **2.** $x' = x - h$
$y = y' + k$ $y' = y - k$

where (h, k) are the coordinates of the origin $0'$ relative to the original system.

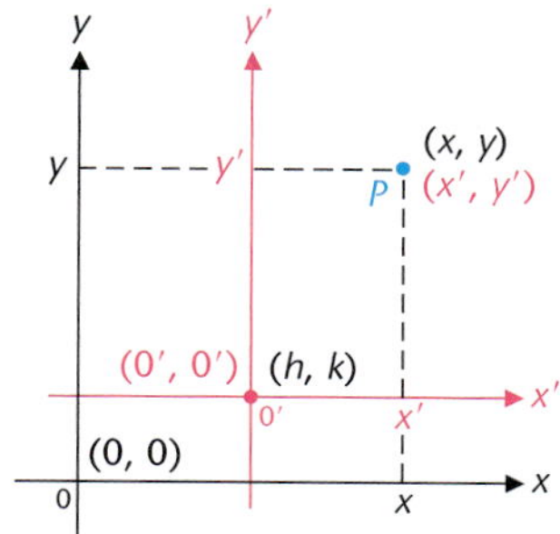

Table 1 on page 924 lists the standard equations for translated conics.

11.5 Rotation of Axes

If the xy coordinate axes are rotated counterclockwise through an angle θ into the $x'y'$ coordinate axes, then the xy and $x'y'$ coordinate systems are related by the **rotation formulas:**

1. $x = x' \cos\theta - y' \sin\theta$ **2.** $x' = x \cos\theta + y \sin\theta$
$y = x' \sin\theta + y' \cos\theta$ $y' = -x \sin\theta + y \cos\theta$

To transform the general quadratic equation

$$Ax^2 + Bxy + Cy^2 + Dx + Ey + F = 0$$

into an equation in x' and y' with no $x'y'$ term, choose the **angle of rotation** θ to satisfy $\cot 2\theta = (A - C)/B$ and $0° < \theta < 90°$. The **discriminant** of the general second-degree equation in two variables is $B^2 - 4AC$ and the graph is

1. A hyperbola if $B^2 - 4AC > 0$
2. A parabola if $B^2 - 4AC = 0$
3. An ellipse if $B^2 - 4AC < 0$

11.6 Nonlinear Systems

If a system contains any equations that are not linear, then the system is called a **nonlinear system.** Section 11.6 deals with systems involving second-degree terms. These systems have at most four solutions, some of which may be imaginary. These systems can be solved algebraically using **substitution** or **elimination.** Real solutions can also be found using a graphing utility, but imaginary solutions cannot.

TABLE 1 Standard Equations for Translated Conics

Parabolas

$$(x - h)^2 = 4a(y - k)$$

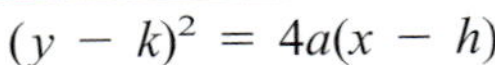

$$(y - k)^2 = 4a(x - h)$$

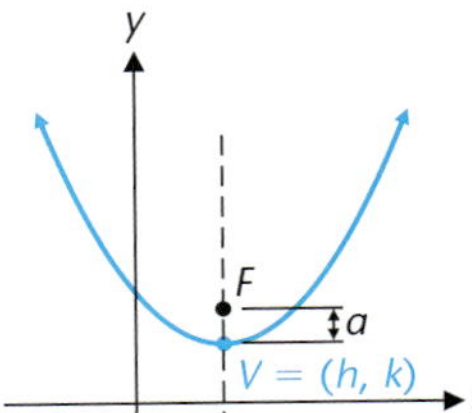

Vertex (h, k)
Focus $(h, k + a)$
$a > 0$ opens up
$a < 0$ opens down

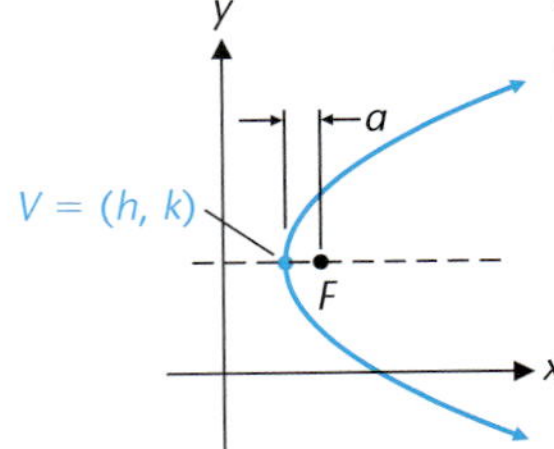

Vertex (h, k)
Focus $(h + a, k)$
$a < 0$ opens left
$a > 0$ opens right

Circles

$$(x - h)^2 + (y - k)^2 = r^2$$

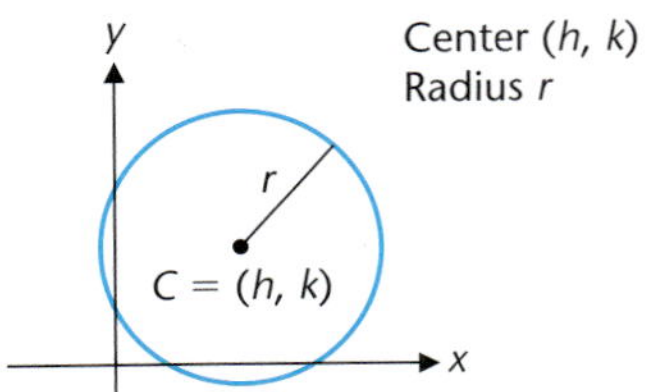

Center (h, k)
Radius r

Ellipses

$$\frac{(x - h)^2}{a^2} + \frac{(y - k)^2}{b^2} = 1 \qquad a > b > 0 \qquad \frac{(x - h)^2}{b^2} + \frac{(y - k)^2}{a^2} = 1$$

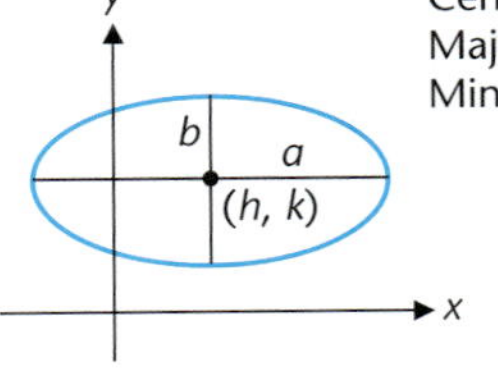

Center (h, k)
Major axis $2a$
Minor axis $2b$

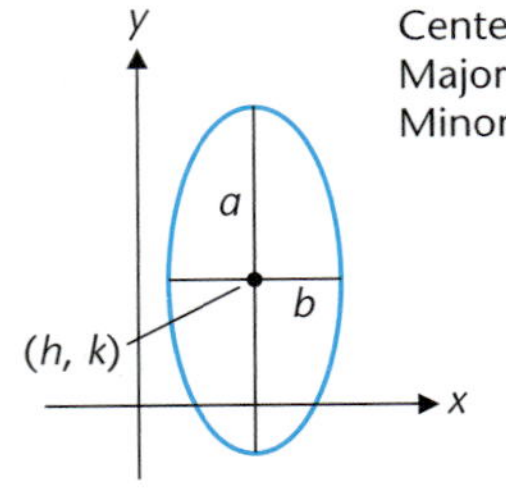

Center (h, k)
Major axis $2a$
Minor axis $2b$

Hyperbolas

$$\frac{(x - h)^2}{a^2} - \frac{(y - k)^2}{b^2} = 1$$

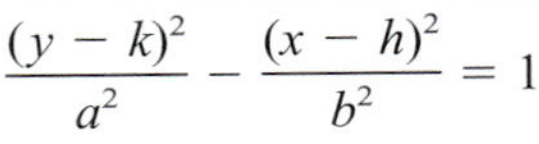

$$\frac{(y - k)^2}{a^2} - \frac{(x - h)^2}{b^2} = 1$$

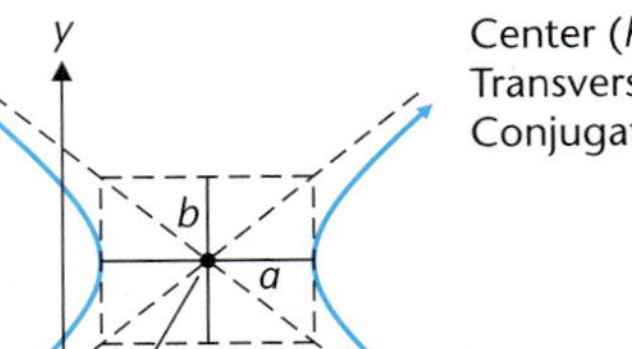

Center (h, k)
Transverse axis $2a$
Conjugate axis $2b$

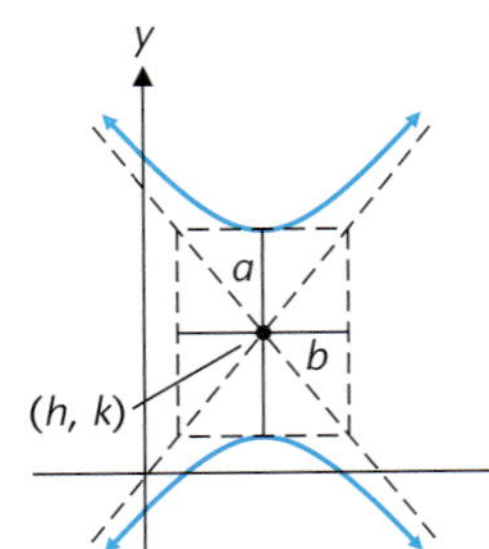

Center (h, k)
Transverse axis $2a$
Conjugate axis $2b$

CHAPTER 11 REVIEW EXERCISES

Work through all the problems in this chapter review and check answers in the back of the book. Answers to all review problems are there, and following each answer is a number in italics indicating the section in which that type of problem is discussed. Where weaknesses show up, review appropriate sections in the text.

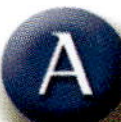

A

In Problems 1–3, graph each equation and locate foci. Locate the directrix for any parabolas. Find the lengths of major, minor, transverse, and conjugate axes where applicable.

1. $9x^2 + 25y^2 = 225$ **2.** $x^2 = -12y$

3. $25y^2 - 9x^2 = 225$

In Problems 4–6:
(A) Write each equation in one of the standard forms listed in Table 1 of the review.
(B) Identify the curve.

4. $4(y + 2)^2 - 25(x - 4)^2 = 100$

5. $(x + 5)^2 + 12(y + 4) = 0$

6. $16(x - 6)^2 + 9(y - 4)^2 = 144$

7. Find the $x'y'$ coordinates of the point (3, 4) when the axes are rotated through

(A) 30° **(B)** 45° **(C)** 60°

8. Find the equations of the x' and y' axes in terms of x and y if the axes are rotated through an angle of 75°.

In Problems 9–11, solve the system.

9. $y = x^2 - 5x - 3$
$y = -x + 2$

10. $x^2 + y^2 = 2$
$2x - y = 3$

11. $3x^2 - y^2 = -6$
$2x^2 + 3y^2 = 29$

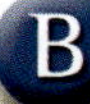

B

12. Find the equation of the parabola having its vertex at the origin, its axis the x axis, and $(-4, -2)$ on its graph.

13. Find an equation of an ellipse in the form

$$\frac{x^2}{M} + \frac{y^2}{N} = 1 \qquad M, N > 0$$

if the center is at the origin, the major axis is on the y axis, the minor axis length is 6, and the distance of the foci from the center is 4.

14. Find an equation of a hyperbola in the form

$$\frac{y^2}{M} - \frac{x^2}{N} = 1 \qquad M, N > 0$$

if the center is at the origin, the conjugate axis length is 8, and the foci are 5 units from the center.

In Problems 15–20, solve the system.

15. $x^2 + 4y^2 = 32$
$x + 2y = 0$

16. $16x^2 + 25y^2 = 400$
$16x^2 - 45y = 0$

17. $x^2 + y^2 = 10$
$16x^2 + y^2 = 25$

18. $x^2 - y^2 = 2$
$y^2 = x$

19. $x^2 + 2xy + y^2 = 1$
$xy = -2$

20. $2x^2 + xy + y^2 = 8$
$x^2 - y^2 = 0$

In Problems 21–23, transform each equation into one of the standard forms in Table 1 in the review. Identify the curve and graph it.

21. $16x^2 + 4y^2 + 96x - 16y + 96 = 0$

22. $x^2 - 4x - 8y - 20 = 0$

23. $4x^2 - 9y^2 + 24x - 36y - 36 = 0$

24. Given the equation $x^2 - \sqrt{3}xy + 2y^2 - 10 = 0$, find the transformed equation when the axes are rotated through 30°. Sketch and identify the graph.

25. Given the equation $5x^2 + 26xy + 5y^2 + 72 = 0$, find the angle of rotation so that the transformed equation will have no $x'y'$ term. Sketch and identify the graph.

26. Given the equation $3x^2 + 4xy + 2y^2 - 20 = 0$, identify the curve and graph on a graphing utility.

27. Use a graphing utility to graph $x^2 = y$ and $x^2 = 50y$ in the viewing window $-10 \le x, y \le 10$. Find m so that the graph of $x^2 = y$ in the viewing window $-m \le x, y \le m$, has the same appearance as the graph of $x^2 = 50y$ in $-10 \le x, y \le 10$. Explain.

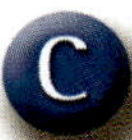

28. Use the definition of a parabola and the distance formula to find the equation of a parabola with directrix $x = 6$ and focus at (2, 4).

29. Find an equation of the set of points in a plane each of whose distance from (4, 0) is twice its distance from the line $x = 1$. Identify the geometric figure.

30. Find an equation of the set of points in a plane each of whose distance from (4, 0) is two-thirds its distance from the line $x = 9$. Identify the geometric figure.

In Problems 31–33, find the coordinates of any foci relative to the original coordinate system.

31. Problem 19 **32.** Problem 20 **33.** Problem 21

34. Solve:
$$x^2 - xy + y^2 = 4$$
$$x^2 + xy - 2y^2 = 0$$

In Problems 35 and 36, find all real solutions to two decimal places.

35.
$$x^2 - 3y^2 + 9x + 7y - 22 = 0$$
$$4x^2 + 5x + 10y - 53 = 0$$

36.
$$x^2 + 4xy + y^2 = 8$$
$$5x^2 + 2xy + y^2 = 25$$

APPLICATIONS

37. Communications. A parabolic satellite television antenna has a diameter of 8 feet and is 1 foot deep. How far is the focus from the vertex?

38. Engineering. An elliptical gear is to have foci 8 centimeters apart and a major axis 10 centimeters long. Letting the x axis lie along the major axis (right positive) and the y axis lie along the minor axis (up positive), write the equation of the ellipse in the standard form

$$\frac{x^2}{a^2} + \frac{y^2}{b^2} = 1$$

39. Space Science. A hyperbolic reflector for a radiotelescope (such as that illustrated in Problem 39, Exercise 11.3) has the equation

$$\frac{y^2}{40^2} - \frac{x^2}{30^2} = 1$$

If the reflector has a diameter of 30 feet, how deep is it? Compute the answer to three significant digits.

40. Triangulation. Seismographs are located at three stations. Station B is 200 miles due east of station A. Station C is 300 miles due north of station A. Readings from the three stations indicate that the epicenter of an earthquake is 179 miles from station A, 167 miles from station B, and 194 miles from station C. Use triangulation and a graphing utility to locate the epicenter of the earthquake. Express the answer relative to station A and round values to the nearest mile.

CHAPTER 11 GROUP ACTIVITY

Focal Chords

Many of the applications of the conic sections are based on their reflective or focal properties. One of the interesting algebraic properties of the conic sections concerns their focal chords.

If a line through a focus F contains two points G and H of a conic section, then the line segment GH is called a **focal chord.** Let $G = (x_1, y_1)$ and $H = (x_2, y_2)$ be points on the graph of $x^2 = 4ay$ such that GH is a focal chord. Let u denote the length of GF and v the length of FH (Fig. 1).

FIGURE 1 Focal chord GH of the parabola $x^2 = 4ay$.

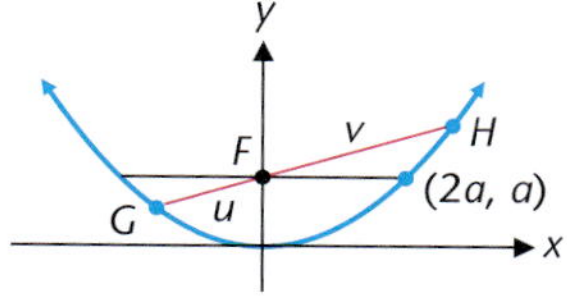

(A) Use the distance formula to show that $u = y_1 + a$.

(B) Show that G and H lie on the line $y - a = mx$, where $m = (y_2 - y_1)/(x_2 - x_1)$.

(C) Solve $y - a = mx$ for x and substitute in $x^2 = 4ay$, obtaining a quadratic equation in y. Explain why $y_1y_2 = a^2$.

(D) Show that $\frac{1}{u} + \frac{1}{v} = \frac{1}{a}$.

(E) Show that $u + v - 4a = \frac{(u - 2a)^2}{u - a}$. Explain why this implies that $u + v \geq 4a$, with equality if and only if $u = v = 2a$.

(F) Which focal chord is the shortest? Is there a longest focal chord?

(G) Is $\frac{1}{u} + \frac{1}{v}$ a constant for focal chords of the ellipse? For focal chords of the hyperbola? Obtain evidence for your answers by considering specific examples.

(H) The conic section with focus at the origin, directrix the line $x = D > 0$, and eccentricity $E > 0$ has the polar equation $r = \frac{DE}{1 + E\cos\theta}$. Explain how this polar equation makes it easy to show that $\frac{1}{u} + \frac{1}{v} = \frac{1}{a}$ for a parabola. Use the polar equation to determine the sum $\frac{1}{u} + \frac{1}{v}$ for a focal chord of an ellipse or hyperbola.

CUMULATIVE REVIEW EXERCISES CHAPTERS 10 11

Work through all the problems in this cumulative review and check answers in the back of the book. Answers to all review problems are there, and following each answer is a number in italics indicating the section in which that type of problem is discussed. Where weaknesses show up, review appropriate sections in the text.

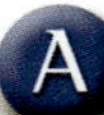

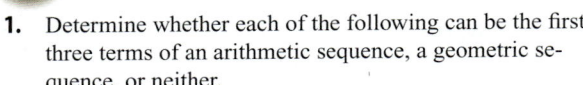

1. Determine whether each of the following can be the first three terms of an arithmetic sequence, a geometric sequence, or neither.
   (A) 20, 15, 10, . . . **(B)** 5, 25, 125, . . .
   (C) 5, 25, 50, . . . **(D)** 27, −9, 3, . . .
   (E) −9, −6, −3, . . .

In Problems 2–4:
(A) Write the first four terms of each sequence.
(B) Find a_8. (C) Find S_8.

2. $a_n = 2 \cdot 5^n$ 3. $a_n = 3n - 1$

4. $a_1 = 100;\ a_n = a_{n-1} - 6,\ n \ge 2$

5. Evaluate each of the following:
   (A) $8!$ **(B)** $\frac{32!}{30!}$ **(C)** $\frac{9!}{3!(9-3)!}$

6. Evaluate each of the following:
   (A) $\binom{7}{2}$ **(B)** $C_{7,2}$ **(C)** $P_{7,2}$

In Problems 7–9, graph each equation and locate foci. Locate the directrix for any parabolas. Find the lengths of major, minor, transverse, and conjugate axes where applicable.

7. $25x^2 - 36y^2 = 900$ 8. $25x^2 + 36y^2 = 900$

9. $25x^2 - 36y = 0$

10. What type of curve is the graph of

$$3x^2 - 4xy + 2y^2 - 7 = 0$$

11. Solve $x^2 + y^2 = 2$

$$2x - y = 1$$

12. A coin is flipped three times. How many combined outcomes are possible? Solve
   (A) By using a tree diagram
   (B) By using the multiplication principle

13. How many ways can four distinct books be arranged on a shelf? Solve
   (A) By using the multiplication principle
   (B) By using permutations or combinations, whichever is applicable

14. In a single deal of 3 cards from a standard 52-card deck, what is the probability of being dealt three diamonds?

15. Each of the 10 digits 0 through 9 is printed on 1 of 10 different cards. Four of these cards are drawn in succession without replacement. What is the probability of drawing the digits 4, 5, 6, and 7 by drawing 4 on the first draw, 5 on the second draw, 6 on the third draw, and 7 on the fourth draw? What is the probability of drawing the digits 4, 5, 6, and 7 in any order?

16. A thumbtack lands point down in 38 out of 100 tosses. What is the approximate empirical probability of the tack landing point up?

Verify Problems 17 and 18 for n = 1, 2, and 3.

17. P_n: $1 + 5 + 9 + \cdots + (4n - 3) = n(2n - 1)$

18. P_n: $n^2 + n + 2$ is divisible by 2

In Problems 19 and 20, write P_k and P_{k+1}.

19. For P_n in Problem 15 20. For P_n in Problem 16

B

21. Find the equation of the parabola having its vertex at the origin, its axis the y axis, and $(2, -8)$ on its graph.

22. Find an equation of an ellipse in the form

$$\frac{x^2}{M} + \frac{y^2}{N} = 1 \qquad M, N > 0$$

if the center is at the origin, the major axis is the x axis, the major axis length is 10, and the distance of the foci from the center is 3.

23. Find an equation of a hyperbola in the form

$$\frac{x^2}{M} - \frac{y^2}{N} = 1 \qquad M, N > 0$$

if the center is at the origin, the transverse axis length is 16, and the distance of the foci from the center is $\sqrt{89}$.

In Problems 24 and 25, solve the system.

24. $x^2 - 3xy + 3y^2 = 1$

$$xy = 1$$

25. $x^2 - 3xy + y^2 = -1$
$x^2 - xy = 0$

In Problems 26 and 27, find the angle of rotation so that the transformed equation will have no x'y' term. Identify the curve and graph it.

26. $2\sqrt{3}xy + 2y^2 + 3 = 0$

27. $x^2 + 2xy + y^2 + 4\sqrt{2}x - 4\sqrt{2}y = 0$

28. Find all real solutions to two decimal places

$x^2 + 2xy - y^2 = 1$
$9x^2 + 4xy + y^2 = 15$

29. Write $\sum_{k=1}^{5} k^k$ without summation notation and find the sum.

30. Write the series $\frac{2}{2!} - \frac{2^2}{3!} + \frac{2^3}{4!} - \frac{2^4}{5!} + \frac{2^5}{6!} - \frac{2^6}{7!}$ using summation notation with the summation index k starting at $k = 1$.

31. Find S_∞ for the geometric series $108 - 36 + 12 - 4 + \cdots$.

32. How many four-letter code words are possible using the first six letters of the alphabet if no letter can be repeated? If letters can be repeated? If adjacent letters cannot be alike?

33. A basketball team with 12 members has two centers. If 5 players are selected at random, what is the probability that both centers are selected? Express the answer in terms of $C_{n,r}$ or $P_{n,r}$, as appropriate, and evaluate.

34. A single die is rolled 1,000 times with the frequencies of outcomes shown in the table.

(A) What is the approximate empirical probability that the number of dots showing is divisible by 3?

(B) What is the theoretical probability that the number of dots showing is divisible by 3?

Number of dots facing up	1	2	3	4	5	6
Frequency	160	155	195	180	140	170

35. Let $a_n = 100(0.9)^n$ and $b_n = 10 + 0.03n$. Find the least positive integer n such that $a_n < b_n$ by graphing the sequences $\{a_n\}$ and $\{b_n\}$ with a graphing utility. Check your answer by using a graphing utility to display both sequences in table form.

36. Evaluate each of the following:

(A) $P_{25,5}$ **(B)** $C(25, 5)$ **(C)** $\binom{25}{20}$

37. Expand $(a + \frac{1}{2}b)^6$ using the binomial formula.

38. Find the fifth and the eighth terms in the expansion of $(3x - y)^{10}$.

Establish each statement in Problems 39 and 40 for all positive integers using mathematical induction.

39. P_n in Problem 15 **40.** P_n in Problem 16

41. Find the sum of all the odd integers between 50 and 500.

42. Use the formula for the sum of an infinite geometric series to write $2.\overline{45} = 2.454\ 545 \cdots$ as the quotient of two integers.

43. Let $a_k = \binom{30}{k}(0.1)^{30-k}(0.9)^k$ for $k = 0, 1, \ldots, 30$. Use a graphing utility to find the largest term of the sequence $\{a_k\}$ and the number of terms that are greater than 0.01.

In Problems 44–46, use a translation of coordinates to transform each equation into a standard equation for a nondegenerate conic. Identify the curve and graph it.

44. $4x + 4y - y^2 + 8 = 0$

45. $x^2 + 2x - 4y^2 - 16y + 1 = 0$

46. $4x^2 - 16x + 9y^2 + 54y + 61 = 0$

47. How many nine-digit zip codes are possible? How many of these have no repeated digits?

48. Use mathematical induction to prove that the following statement holds for all positive integers:

$$P_n\text{: } \frac{1}{1 \cdot 3} + \frac{1}{3 \cdot 5} + \frac{1}{5 \cdot 7} + \cdots + \frac{1}{(2n-1)(2n+1)} = \frac{n}{2n+1}$$

49. Three-digit numbers are randomly formed from the digits 1, 2, 3, 4, and 5. What is the probability of forming an even number if digits cannot be repeated? If digits can be repeated?

C

50. Use the binomial formula to expand $(x - 2i)^6$, where i is the imaginary unit.

51. Use the definition of a parabola and the distance formula to find the equation of a parabola with directrix $y = 3$ and focus $(6, 1)$.

52. An ellipse has vertices $(\pm 4, 0)$ and foci $(\pm 2, 0)$. Find the y intercepts.

53. A hyperbola has vertices $(2, \pm 3)$ and foci $(2, \pm 5)$. Find the length of the conjugate axis.

54. Seven distinct points are selected on the circumference of a circle. How many triangles can be formed using these seven points as vertices?

55. Use mathematical induction to prove that $2^n < n!$ for all integers $n > 3$.

56. Use mathematical induction to show that $\{a_n\} = \{b_n\}$, where $a_1 = 3$, $a_n = 2a_{n-1} - 1$ for $n > 1$, and $b_n = 2^n + 1$, $n \geq 1$.

57. Find an equation of the set of points in the plane each of whose distance from $(1, 4)$ is three times its distance from the x axis. Write the equation in the form $Ax^2 + Cy^2 + Dx + Ey + F = 0$, and identify the curve.

58. A box of 12 lightbulbs contains 4 defective bulbs. If three bulbs are selected at random, what is the probability of selecting at least one defective bulb?

APPLICATIONS

59. **Economics.** The government, through a subsidy program, distributes \$2,000,000. If we assume that each individual or agency spends 75% of what it receives, and 75% of this is spent, and so on, how much total increase in spending results from this government action?

60. **Geometry.** Find the dimensions of a rectangle with perimeter 24 meters and area 32 square meters.

61. **Engineering.** An automobile headlight contains a parabolic reflector with a diameter of 8 inches. If the light source is located at the focus, which is 1 inch from the vertex, how deep is the reflector?

62. **Architecture.** A sound whispered at one focus of a whispering chamber can be easily heard at the other focus. Suppose that a cross section of this chamber is a semielliptical arch which is 80 feet wide and 24 feet high (see the figure). How far is each focus from the center of the arch? How high is the arch above each focus?

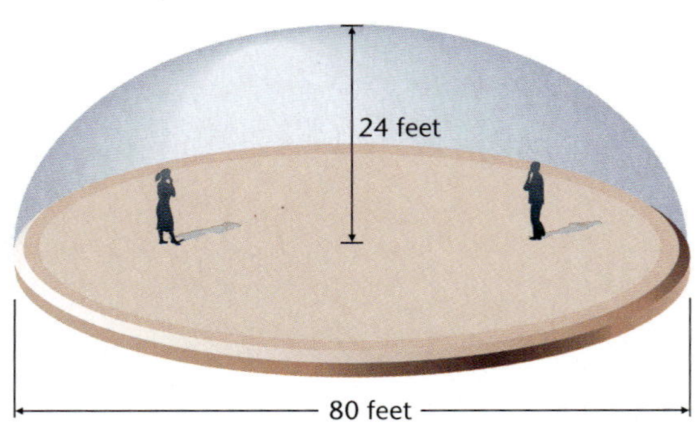

63. **Political Science.** A random survey of 1,000 residents in a state produced the following results:

	Party Affiliation			
Age	**Democrat**	**Republican**	**Independent**	**Totals**
Under 30	130	80	40	250
30–39	120	90	20	230
40–49	70	80	20	170
50–59	50	60	10	120
Over 59	90	110	30	230
Totals	460	420	120	1,000

Find the empirical probability that a person selected at random:

(A) Is under 30 *and* a Democrat
(B) Is under 40 *and* a Republican
(C) Is over 59 *or* is an Independent

APPENDIX A

Review of Equations and Graphing

OUTLINE

In this appendix we provide review and reference material especially important for a course in mathematics that emphasizes the use of graphing technology. Further review of basic algebra operations and concepts usually studied in earlier courses is available at the Online Learning Center for *College Algebra: Graphs and Models,* Second Edition. To access the Online Learning Center, visit www.mhhe.com/barnett and click on the image of this book's cover.

SECTION A.1 Linear Equations and Inequalities

Equations • Solving Linear Equations • Inequality Relations and Interval Notation • Solving Linear Inequalities

Equations

An **algebraic equation** is a mathematical statement that relates two algebraic expressions involving at least one variable. Some examples of equations with x as a variable are

$$3x - 2 = 7 \qquad \frac{1}{1+x} = \frac{x}{x-2}$$

$$2x^2 - 3x + 5 = 0 \qquad \sqrt{x+4} = x - 1$$

The **replacement set,** or **domain,** for a variable is defined to be the set of numbers that are permitted to replace the variable.

> **ASSUMPTION**
> **On Domains of Variables**
>
> Unless stated to the contrary, we assume that the domain for a variable is the set of those real numbers for which the algebraic expressions involving the variable are real numbers.

For example, the domain for the variable x in the expression

$$2x - 4$$

is R, the set of all real numbers, because $2x - 4$ represents a real number for all replacements of x by real numbers. The domain of x in the equation

$$\frac{1}{x} = \frac{2}{x-3}$$

is the set of all real numbers except 0 and 3. These values are excluded because the left member is not defined for $x = 0$ and the right member is not defined for $x = 3$. The left and right members represent real numbers for all other replacements of x by real numbers.

The **solution set** for an equation is defined to be the set of elements in the domain of the variable that make the equation true. Each element of the solution set is called a **solution,** or **root,** of the equation. To **solve an equation** is to find the solution set for the equation.

Knowing what we mean by the solution set of an equation is one thing; finding it is another. To this end we introduce the idea of equivalent equations. Two equations are said to be **equivalent** if they both have the same solution set for a given replacement set. A basic technique for solving equations is to perform operations on equations that produce simpler equivalent equations, and to continue the process until an equation is reached whose solution is obvious.

Application of any of the properties of equality given in Theorem 1 will produce equivalent equations.

THEOREM 1
Properties of Equality

For a, b, and c any real numbers,

1. If $a = b$, then $a + c = b + c$.	**Addition Property**
2. If $a = b$, then $a - c = b - c$.	**Subtraction Property**
3. If $a = b$, then $ca = cb$, $c \neq 0$.	**Multiplication Property**
4. If $a = b$, then $\frac{a}{c} = \frac{b}{c}$, $c \neq 0$.	**Division Property**
5. If $a = b$, then either may replace the other in any statement without changing the truth or falsity of the statement.	**Substitution Property**

Solving Linear Equations

We now turn our attention to methods of solving *first-degree,* or *linear, equations* in one variable.

DEFINITION 1
Linear Equation in One Variable

Any equation that can be written in the form

$ax + b = 0 \qquad a \neq 0$ **Standard Form**

where a and b are real constants and x is a variable, is called a **linear,** or **first-degree, equation** in one variable.

$5x - 1 = 2(x + 3)$ is a linear equation, because it can be written in the standard form $3x - 7 = 0$.

EXAMPLE 1 Solving a Linear Equation

Solve $5x - 9 = 3x + 7$ and check.

SOLUTION

We use the properties of equality to transform the given equation into an equivalent equation whose solution is obvious.

$5x - 9 = 3x + 7$	Original equation
$5x - 9 + 9 = 3x + 7 + 9$	Add 9 to both sides.
$5x = 3x + 16$	Combine like terms.
$5x - 3x = 3x + 16 - 3x$	Subtract $3x$ from both sides.
$2x = 16$	Combine like terms.
$\frac{2x}{2} = \frac{16}{2}$	Divide both sides by 2.
$x = 8$	Simplify.

The solution set for this last equation is obvious:

Solution set: $\{8\}$

And because the equation $x = 8$ is equivalent to all the preceding equations in our solution, $\{8\}$ is also the solution set for all these equations, including the original equation. [*Note:* If an equation has only one element in its solution set, we generally use the last equation (in this case, $x = 8$) rather than set notation to represent the solution.]

CHECK

$$
\begin{aligned}
5x - 9 &= 3x + 7 && \text{Original equation} \\
5(8) - 9 &\stackrel{?}{=} 3(8) + 7 && \text{Substitute } x = 8. \\
40 - 9 &\stackrel{?}{=} 24 + 7 && \text{Simplify each side.} \\
31 &\stackrel{\checkmark}{=} 31 && \text{A true statement}
\end{aligned}
$$

MATCHED PROBLEM 1

Solve $7x - 10 = 4x + 5$ and check.

EXAMPLE 2 Solving a Linear Equation

Solve $3x - 2(2x - 5) = 2(x + 3) - 8$ and check.

SOLUTION

$$
\begin{aligned}
3x - 2(2x - 5) &= 2(x + 3) - 8 && \text{Original equation} \\
3x - 4x + 10 &= 2x + 6 - 8 && \text{Clear parentheses.} \\
-x + 10 &= 2x - 2 && \text{Combine like terms.} \\
-3x &= -12 && \text{Subtract } 2x \text{ and 10 from both sides.} \\
x &= 4 && \text{Divide both sides by } -3.
\end{aligned}
$$

CHECK

$$
\begin{aligned}
3x - 2(2x - 5) &= 2(x + 3) - 8 \\
3(4) - 2[2(4) - 5] &\stackrel{?}{=} 2[(4) + 3] - 8 \\
6 &\stackrel{\checkmark}{=} 6
\end{aligned}
$$

MATCHED PROBLEM 2

Solve $2(3 - x) - (3x + 1) = 8 - 2(x + 2)$ and check.

Inequality Relations and Interval Notation

Just as we use $=$ to replace the words *is equal to,* we use the **inequality symbols** $<$ and $>$ to represent *is less than* and *is greater than,* respectively.

Although it probably seems obvious to you that

$$2 < 4 \qquad 5 > 0 \qquad 25{,}000 > 1$$

are true, it may not seem as obvious that

$$-4 < -2 \qquad 0 > -5 \qquad -25{,}000 < -1$$

To make the inequality relation precise so that we can interpret it relative to all real numbers, we need a precise definition of the concept.

DEFINITION 2

a* < *b* and *b* > *a

For a and b real numbers, we say that ***a* is less than *b*** or ***b* is greater than *a*** and write

$$a < b \qquad \text{or} \qquad b > a$$

if there exists a positive real number p such that $a + p = b$ (or equivalently, $b - a = p$).

We certainly expect that if a positive number is added to *any* real number, the sum is larger than the original. That is essentially what the definition states.

When we write

$$a \leq b$$

we mean $a < b$ or $a = b$ and say ***a* is less than or equal to *b*.** When we write

$$a \geq b$$

we mean $a > b$ or $a = b$ and say ***a* is greater than or equal to *b*.**

The inequality symbols $<$ and $>$ have a very clear geometric interpretation on the real number line. If $a < b$, then a is to the left of b; if $c > d$, then c is to the right of d (Fig. 1).

FIGURE 1 $a < b$, $c > d$.

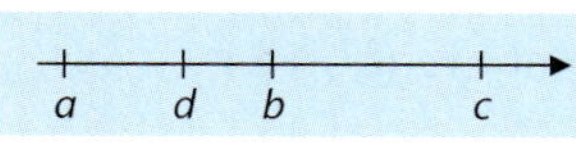

It is an interesting and useful fact that for any two real numbers a and b, either $a < b$, or $a > b$, or $a = b$. This is called the **trichotomy property** of real numbers.

The double inequality $a < x \leq b$ means that $x > a$ and $x \leq b$; that is, x is between a and b, including b but not including a. The set of all real numbers x satisfying the inequality $a < x \leq b$ is called an **interval** and is represented by $(a, b]$. Thus,

$$(a, b] = \{x \mid a < x \leq b\}^*$$

The number a is called the **left endpoint** of the interval, and the symbol (indicates that a is not included in the interval. The number b is called the **right endpoint** of the interval, and the symbol] indicates that b is included in the interval. Other types of intervals of real numbers are shown in Table 1.

---

*In general, $\{x \mid P(x)\}$ represents the set of all x such that statement $P(x)$ is true. To express this set verbally, just read the vertical bar as "such that."

TABLE 1 Interval Notation

Interval Notation	Inequality Notation	Line Graph	Type
$[a, b]$	$a \le x \le b$	a b x	Closed
$[a, b)$	$a \le x < b$	a b x	Half-open
$(a, b]$	$a < x \le b$	a b x	Half-open
(a, b)	$a < x < b$	a b x	Open
$[b, \infty)$	$x \ge b$	b x	Closed
(b, ∞)	$x > b$	b x	Open
$(-\infty, a]$	$x \le a$	a x	Closed
$(-\infty, a)$	$x < a$	a x	Open

Note that the symbol ∞, read *infinity,* used in Table 1 is not a numeral. When we write $[b, \infty)$, we are simply referring to the interval starting at b and continuing indefinitely to the right. We would never write $[b, \infty]$ or $b \le x \le \infty$, because ∞ cannot be used as an endpoint of an interval. The interval $(-\infty, \infty)$ represents the set of real numbers R, because its graph is the entire real number line.

CAUTION

It is important to note that

$$5 > x \ge -3 \qquad \text{is equivalent to } [-3, 5) \text{ and not to } (5, -3]$$

In interval notation, the smaller number is always written to the left. Thus, it may be useful to rewrite the inequality as $-3 \le x < 5$ before rewriting it in interval notation.

EXAMPLE 3 Graphing Intervals and Inequalities

Write each of the following in inequality notation and graph on a real number line:

(A) $[-2, 3)$ (B) $(-4, 2)$ (C) $[-2, \infty)$ (D) $(-\infty, 3)$

SOLUTIONS

(A) $-2 \le x < 3$

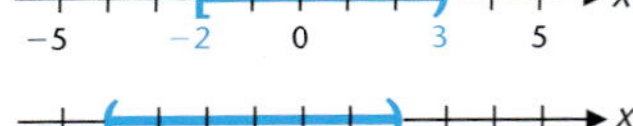

(B) $-4 < x < 2$

-5 -4 0 2 5 x

(C) $x \geq -2$

(D) $x < 3$

MATCHED 3 PROBLEM

Write each of the following in interval notation and graph on a real number line:

(A) $-3 < x \leq 3$ (B) $2 \geq x \geq -1$ (C) $x > 1$ (D) $x \leq 2$

EXPLORE/DISCUSS 1

Example 3, part C, shows the graph of the inequality $x \geq -2$. What is the graph of $x < -2$? What is the corresponding interval? Describe the relationship between these sets.

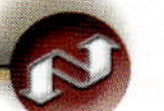

Because intervals are sets of real numbers, the set operations of *union* and *intersection* are often useful when working with intervals. The **union** of sets A and B, denoted by $A \cup B$, is the set formed by combining all the elements of A and all the elements of B. The **intersection** of sets A and B, denoted by $A \cap B$, is the set of elements of A that are also in B. Symbolically:

DEFINITION 3
Union and Intersection

Union: $A \cup B = \{x \mid x \text{ is in } A \text{ or } x \text{ is in } B\}$

$\{1, 2, 3\} \cup \{2, 3, 4, 5\} = \{1, 2, 3, 4, 5\}$

Intersection: $A \cap B = \{x \mid x \text{ is in } A \text{ and } x \text{ is in } B\}$

$\{1, 2, 3\} \cap \{2, 3, 4, 5\} = \{2, 3\}$

EXAMPLE 4 Graphing Unions and Intersections of Intervals

If $A = [-2, 3]$, $B = (1, 6)$, and $C = (4, \infty)$, graph the indicated sets and write as a single interval, if possible.

(A) $A \cup B$ and $A \cap B$ (B) $A \cup C$ and $A \cap C$

SOLUTIONS

(A)

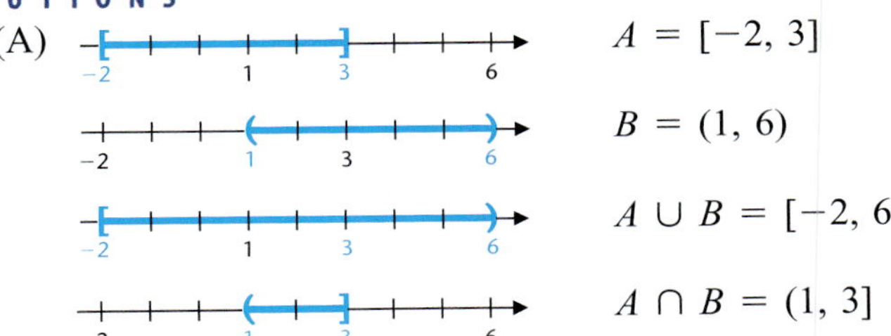

$A = [-2, 3]$

$B = (1, 6)$

$A \cup B = [-2, 6)$

$A \cap B = (1, 3]$

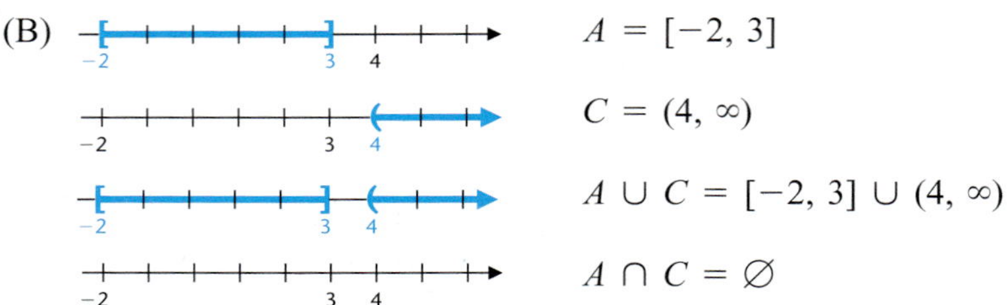

MATCHED 4 PROBLEM

If $D = [-4, 1)$, $E = (-1, 3]$, and $F = [2, \infty)$, graph the indicated sets and write as a single interval, if possible.

(A) $D \cup E$ (B) $D \cap E$ (C) $E \cup F$ (D) $E \cap F$

EXPLORE/DISCUSS 2

Replace ? with < or > in each of the following:

(A) $-1 \; ? \; 3$ and $2(-1) \; ? \; 2(3)$

(B) $-1 \; ? \; 3$ and $-2(-1) \; ? \; -2(3)$

(C) $12 \; ? \; -8$ and $\frac{12}{4} \; ? \; \frac{-8}{4}$

(D) $12 \; ? \; -8$ and $\frac{12}{-4} \; ? \; \frac{-8}{-4}$

Based on these examples, describe verbally the effect of multiplying both sides of an inequality by a number.

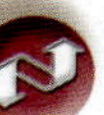

Solving Linear Inequalities

We now turn to the problem of solving linear inequalities in one variable, such as

$$2(2x + 3) < 6(x - 2) + 10 \qquad \text{and} \qquad -3 < 2x + 3 \le 9$$

The **solution set** for an inequality is the set of all values of the variable that make the inequality a true statement. Each element of the solution set is called a **solution** of the inequality. To **solve an inequality** is to find its solution set. Two inequalities are **equivalent** if they have the same solution set for a given replacement set. Just as with equations, we perform operations on inequalities that produce simpler equivalent inequalities, and continue the process until an inequality is reached whose solution is obvious. The properties of inequalities given in Theorem 2 can be used to produce equivalent inequalities.

THEOREM 2
Inequality Properties

For a, b, and c any real numbers,

1. If $a < b$ and $b < c$, then $a < c$. — **Transitive Property**
2. If $a < b$, then $a + c < b + c$. — **Addition Property**
   $-2 < 4$ $\quad -2 + 3 < 4 + 3$
3. If $a < b$, then $a - c < b - c$. — **Subtraction Property**
   $-2 < 4$ $\quad -2 - 3 < 4 - 3$
4. If $a < b$ and c is positive, then $ca < cb$.
   $-2 < 4$ $\quad 3(-2) < 3(4)$
5. If $a < b$ and c is negative, then $ca > cb$.
   $-2 < 4$ $\quad (-3)(-2) > (-3)(4)$

   Multiplication Property (Note difference between 4 and 5.)
6. If $a < b$ and c is positive, then $\frac{a}{c} < \frac{b}{c}$.
   $-2 < 4$ $\quad \frac{-2}{2} < \frac{4}{2}$
7. If $a < b$ and c is negative, then $\frac{a}{c} > \frac{b}{c}$.
   $-2 < 4$ $\quad \frac{-2}{-2} > \frac{4}{-2}$

   Division Property (Note difference between 6 and 7.)

Similar properties hold if each inequality sign is reversed, or if $<$ is replaced with $\leq$ and $>$ is replaced with $\geq$. Thus, we find that we can perform essentially the same operations on inequalities that we perform on equations. When working with inequalities, however, we have to be particularly careful of the use of the multiplication and division properties.

The order of the inequality reverses if we multiply or divide both sides of an inequality statement by a negative number.

EXPLORE/DISCUSS 3

Properties of equality are easily summarized. We can add, subtract, multiply, or divide both sides of an equation by any nonzero real number to produce an equivalent equation. Write a similar summary for the properties of inequalities.

Now let's see how the inequality properties are used to solve linear inequalities. Several examples will illustrate the process.

EXAMPLE 5 Solving a Linear Inequality

Solve $2(2x + 3) - 10 < 6(x - 2)$ and graph.

SOLUTION

$$2(2x + 3) - 10 < 6(x - 2)$$

$$4x + 6 - 10 < 6x - 12 \quad \text{Simplify left and right sides.}$$

$$4x - 4 < 6x - 12$$

$$4x - 4 + 4 < 6x - 12 + 4 \quad \text{Addition property}$$

$$4x < 6x - 8$$

$$4x - 6x < 6x - 8 - 6x \quad \text{Subtraction property}$$

$$-2x < -8$$

$$\frac{-2x}{-2} > \frac{-8}{-2} \quad \text{Division property—note that order reverses because } -2 \text{ is negative.}$$

$$x > 4 \quad \text{or} \quad (4, \infty) \quad \text{Solution set}$$

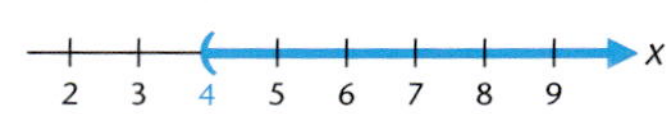

Graph of solution set

MATCHED PROBLEM 5

Solve $3(x - 1) \geq 5(x + 2) - 5$ and graph.

EXAMPLE 6 Solving a Double Inequality

Solve $-1 \leq 3 - 4x < 11$ and graph.

SOLUTION

We proceed as before, except we try to isolate x in the middle with a coefficient of 1.

$$-1 \leq 3 - 4x < 11$$

$$-1 - 3 \leq 3 - 4x - 3 < 11 - 3 \quad \text{Subtract 3 from each member.}$$

$$-4 \leq -4x < 8$$

$$\frac{-4}{-4} \geq \frac{-4x}{-4} > \frac{8}{-4} \quad \text{Divide each number by } -4 \text{ and reverse each inequality.}$$

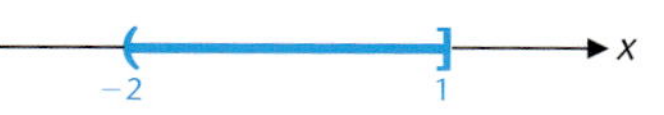

$$1 \geq x > -2 \quad \text{or} \quad -2 < x \leq 1 \quad \text{or} \quad (-2, 1]$$

MATCHED PROBLEM 6

Solve $-5 < 10 - 3x \leq 10$ and graph.

ANSWERS MATCHED PROBLEMS

1. $x = 5$

2. $x = \frac{1}{3}$

3. (A) $(-3, 3]$

(B) $[-1, 2]$

(C) $(1, \infty)$

(D) $(-\infty, 2]$

4. (A) $D \cup E = [-4, 3]$

(B) $D \cap E = (-1, 1)$

(C) $E \cup F = (-1, \infty)$

(D) $E \cap F = [2, 3]$

5. $x \leq -4$ or $(-\infty, -4]$

6. $5 > x \geq 0$
or
$0 \leq x < 5$ or $[0, 5)$

EXERCISE A.1

A

Solve Problems 1–6.

1. $x + 5 = 12$
2. $x - 9 = -2$
3. $2s - 7 = -2$
4. $7 - 3t = 1$
5. $2m + 8 = 5m - 7$
6. $3y + 5 = 6y - 10$

In Problems 7–12, rewrite in inequality notation and graph on a real number line.

7. $[-8, 7]$
8. $(-4, 8)$
9. $[-6, 6)$
10. $(-3, 3]$
11. $[-6, \infty)$
12. $(-\infty, 7)$

In Problems 13–18, rewrite in interval notation and graph on a real number line.

13. $-2 < x \leq 6$
14. $-5 \leq x \leq 5$
15. $-7 < x < 8$
16. $-4 \leq x < 5$
17. $x \leq -2$
18. $x > 3$

In Problems 19–22, write in interval and inequality notation.

19.

20.

21.

22.

In Problems 23–30, replace each ? with > or < to make the resulting statement true.

23. $12 ? 6$ and $12 + 5 ? 6 + 5$
24. $-4 ? -2$ and $-4 - 7 ? -2 - 7$
25. $-6 ? -8$ and $-6 - 3 ? -8 - 3$
26. $4 ? 9$ and $4 + 2 ? 9 + 2$
27. $2 ? -1$ and $-2(2) ? -2(-1)$
28. $-3 ? 2$ and $4(-3) ? 4(2)$
29. $2 ? 6$ and $\frac{2}{2} ? \frac{6}{2}$
30. $-10 ? -15$ and $\frac{-10}{5} ? \frac{-15}{5}$

In Problems 31–40, solve and graph.

31. $7x - 8 < 4x + 7$
32. $4x + 8 \geq x - 1$
33. $3 - x \geq 5(3 - x)$
34. $2(x - 3) + 5 < 5 - x$
35. $\frac{N}{-2} > 4$
36. $\frac{M}{-3} \leq -2$
37. $3 - m < 4(m - 3)$
38. $2(1 - u) \geq 5u$
39. $-2 - \frac{B}{4} \leq \frac{1 + B}{3}$
40. $\frac{y - 3}{4} - 1 > \frac{y}{2}$

B

Solve Problems 41–46.

41. $3 - \frac{2x - 3}{3} = \frac{5 - x}{2}$

42. $\frac{x - 2}{3} + 1 = \frac{x}{7}$

43. $0.1(x - 7) + 0.05x = 0.8$

44. $0.4(x + 5) - 0.3x = 17$

45. $0.3x - 0.04(x + 1) = 2.04$

46. $0.02x - 0.5(x - 2) = 5.32$

In Problems 47–58, graph the indicated set and write as a single interval, if possible.

47. $(-5, 5) \cup [4, 7]$

48. $(-5, 5) \cap [4, 7]$

49. $[-1, 4) \cap (2, 6]$

50. $[-1, 4) \cup (2, 6]$

51. $(-\infty, 1) \cup (-2, \infty)$

52. $(-\infty, 1) \cap (2, \infty)$

53. $(-\infty, -1) \cup [3, 7)$

54. $(1, 6] \cup [9, \infty)$

55. $[2, 3] \cup (1, 5)$

56. $[2, 3] \cap (1, 5)$

57. $(-\infty, 4) \cup (-1, 6]$

58. $(-3, 2) \cup [0, \infty)$

In Problems 59–68, solve and graph.

59. $\frac{q}{7} - 3 > \frac{q - 4}{3} + 1$

60. $\frac{p}{3} - \frac{p - 2}{2} \le \frac{p}{4} - 4$

61. $\frac{2x}{5} - \frac{1}{2}(x - 3) \le \frac{2x}{3} - \frac{3}{10}(x + 2)$

62. $\frac{2}{3}(x + 7) - \frac{x}{4} > \frac{1}{2}(3 - x) + \frac{x}{6}$

63. $-4 \le \frac{9}{5}x + 32 \le 68$

64. $-1 \le \frac{2}{3}A + 5 \le 11$

65. $16 < 7 - 3x \le 31$

66. $-1 \le 9 - 2x < 5$

67. $-6 < -\frac{2}{5}(1 - x) \le 4$

68. $15 \le 7 - \frac{2}{5}x \le 21$

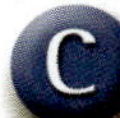

69. Indicate true (T) or false (F):

(A) If $p > q$ and $m > 0$, then $mp < mq$.

(B) If $p < q$ and $m < 0$, then $mp > mq$.

(C) If $p > 0$ and $q < 0$, then $p + q > q$.

70. Assume that $m > n > 0$; then

$$mn > n^2$$
$$mn - m^2 > n^2 - m^2$$
$$m(n - m) > (n + m)(n - m)$$
$$m > n + m$$
$$0 > n$$

But it was assumed that $n > 0$. Find the error.

Prove each inequality property in Problems 71–74, given a, b, and c are arbitrary real numbers.

71. If $a < b$, then $a + c < b + c$.

72. If $a < b$, then $a - c < b - c$.

73. (A) If $a < b$ and c is positive, then $ca < cb$.

(B) If $a < b$ and c is negative, then $ca > cb$.

74. (A) If $a < b$ and c is positive, then $\frac{a}{c} < \frac{b}{c}$.

(B) If $a < b$ and c is negative, then $\frac{a}{c} > \frac{b}{c}$.

SECTION A.2 Cartesian Coordinate System

Cartesian Coordinate System • Graphing: Point by Point • Data Analysis

Analytic geometry is the study of the relationship between geometric forms, such as circles and lines, and algebraic forms, such as equations and inequalities. The key to this relationship is the Cartesian coordinate system, named after the French mathematician and philosopher René Descartes (1596–1650) who was the first to combine the study of algebra and geometry into a single discipline. In Section A.2 we develop some of the basic tools used to graph equations.

Cartesian Coordinate System

Just as a real number line establishes a one-to-one correspondence between the points on a line and the elements in the set of real numbers, we can form a **real plane** by establishing a one-to-one correspondence between the points in a plane and elements in the set of all ordered pairs of real numbers. This can be done by means of a Cartesian coordinate system.

FIGURE 1 Cartesian coordinate system.

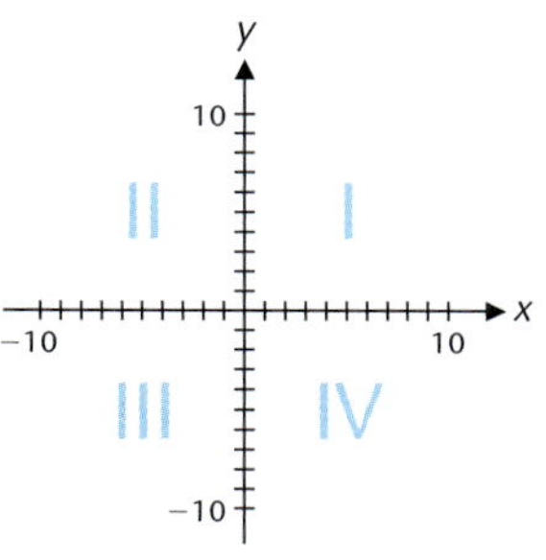

Recall that to form a **Cartesian** or **rectangular coordinate system,** we select two real number lines, one horizontal and one vertical, and let them cross through their origins as indicated in Figure 1. Up and to the right are the usual choices for the positive directions. These two number lines are called the **horizontal axis** and the **vertical axis,** or together, the **coordinate axes.** The horizontal axis is usually referred to as the ***x* axis** and the vertical axis as the ***y* axis,** and each is labeled accordingly. Other labels may be used in certain situations. The coordinate axes divide the plane into four parts called **quadrants,** which are numbered counterclockwise from I to IV (see Fig. 1).

Now we want to assign *coordinates* to each point in the plane. Given an arbitrary point P in the plane, pass horizontal and vertical lines through the point (Fig. 2). The vertical line will intersect the horizontal axis at a point with coordinate a, and the horizontal line will intersect the vertical axis at a point with coordinate b. These two numbers written as the ordered pair (a, b) form the **coordinates** of the point P. The first coordinate a is called the **abscissa** of P; the second coordinate b is called the **ordinate** of P. The abscissa of Q in Figure 2 is -10, and the ordinate of Q is 5. The coordinates of a point can also be referenced in terms of the axis labels. The ***x* coordinate** of R in Figure 2 is 5, and the ***y* coordinate** of R is 10. The point with coordinates (0, 0) is called the **origin.**

FIGURE 2 Coordinates in a plane.

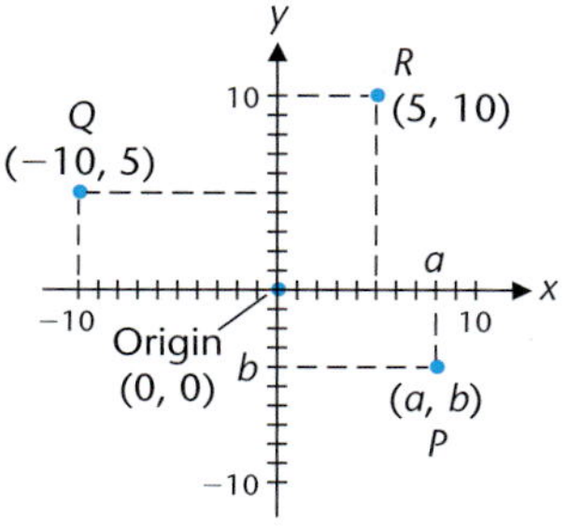

The procedure we have just described assigns to each point P in the plane a unique pair of real numbers (a, b). Conversely, if we are given an ordered pair of real numbers (a, b), then, reversing this procedure, we can determine a unique point P in the plane. Thus:

There is a one-to-one correspondence between the points in a plane and the elements in the set of all ordered pairs of real numbers.

This is often referred to as the **fundamental theorem of analytic geometry.** Because of this correspondence, we often speak of the point (a, b) when we are referring to the point with coordinates (a, b). We also write $P = (a, b)$ to identify the coordinates of the point P. Thus, in Figure 2, referring to Q as the point $(-10, 5)$ and writing $R = (5, 10)$ are both acceptable statements.

Given any set of ordered pairs S, the **graph** of S is the set of points in the plane corresponding to the ordered pairs in S.

Graphing: Point by Point

The fundamental theorem of analytic geometry enables us to look at algebraic forms geometrically and to look at geometric forms algebraically. We begin by considering an algebraic form, an equation in two variables:

$$y = x^2 - 4 \qquad (1)$$

A **solution** to equation (1) is an ordered pair of real numbers (a, b) such that

$$b = a^2 - 4$$

The **solution set** of equation (1) is the set of all these ordered pairs. More formally,

$$\text{Solution set of equation (1): } \{(x, y) \mid y = x^2 - 4\} \qquad (2)$$

To find a solution for equation (1) we simply replace x with a number and calculate the value of y. For example, if $x = 2$, then $y = 2^2 - 4 = 0$, and the ordered pair (2, 0) is a solution. Similarly, if $x = -3$, then $y = (-3)^2 - 4 = 5$, and the ordered pair $(-3, 5)$ is also a solution of equation (1). In fact, any real number substituted for x in equation (1) will produce a solution to the equation. Thus, the solution set shown in (2) must have an infinite number of elements. We now use a rectangular coordinate system to provide a geometric representation of this set.

The **graph of an equation** is the graph of its solution set. To *sketch the graph of an equation,* we plot enough points from its solution set so that the total graph is apparent and then connect these points with a smooth curve, proceeding from left to right. This process is called **point-by-point plotting.**

EXAMPLE 1 Graphing an Equation Using Point-by-Point Plotting

Sketch a graph of $y = x^2 - 4$.

SOLUTION

We make up a table of solutions—ordered pairs of real numbers that satisfy the given equation.

x	−4	−3	−2	−1	0	1	2	3	4
y	12	5	0	−3	−4	−3	0	5	12

FIGURE 3

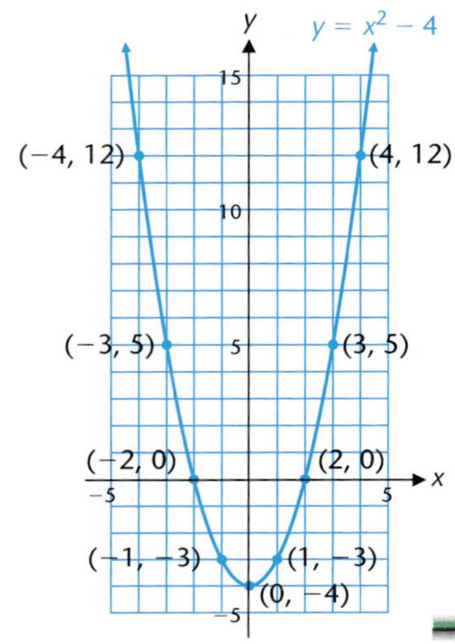

After plotting these solutions, if there are any portions of the graph that are unclear, we plot additional points until the shape of the graph is apparent. Then we join all these plotted points with a smooth curve, as shown in Figure 3. Arrowheads are used to indicate that the graph continues beyond the portion shown here with no significant changes in shape.

The resulting figure is called a *parabola.* Notice that if we fold the paper along the y axis, the right side will match the left side. We say that the graph is *symmetric with respect to the y axis* and call the y axis the *axis of the parabola.* More is said about parabolas elsewhere in the text.

MATCHED PROBLEM 1

Sketch a graph of $y = 8 - x^2$ using point-by-point plotting.

EXAMPLE 2 Graphing an Equation Using Point-by-Point Plotting

Sketch a graph of $y = \sqrt{x}$.

SOLUTION

Proceeding as before, we make up a table of solutions:

x	0	1	2	3	4	5	6	7	8	9
y	0	1	$\sqrt{2} \approx 1.4$	$\sqrt{3} \approx 1.7$	2	$\sqrt{5} \approx 2.2$	$\sqrt{6} \approx 2.4$	$\sqrt{7} \approx 2.6$	$\sqrt{8} \approx 2.8$	3

FIGURE 4

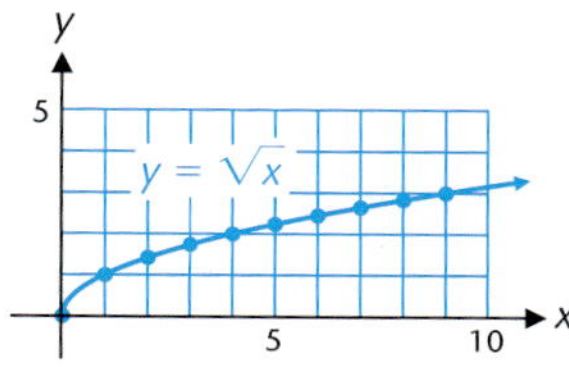

For graphing purposes, the irrational numbers in the table were evaluated on a calculator and rounded to one decimal place. Plotting these points and connecting them with a smooth curve produces the graph in Figure 4.

Notice that we did not include any negative values of x in the table. If x is a negative real number, then $\sqrt{x}$ is not a real number. Because the coordinates of a point in a rectangular coordinate system must be real numbers, *when graphing an equation, we consider only those values of the variables that produce real solutions to the equation.* We have more to say about numbers of the form $\sqrt{x}$, where x is negative, elsewhere in this book.

MATCHED PROBLEM 2

Sketch a graph of $y = 4 - \sqrt{x}$.

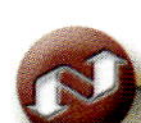

EXPLORE/DISCUSS 1

To graph the equation $y = -x^3 + 2x$, we use point-by-point plotting to obtain the graph in Figure 5.

x	y
−1	−1
0	0
1	1

FIGURE 5

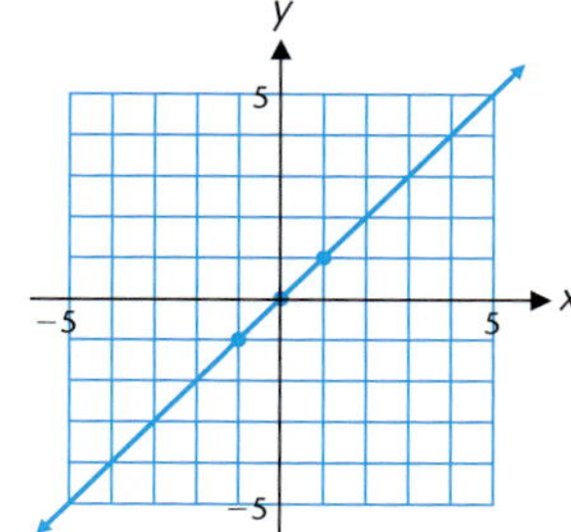

(A) Do you think this is the correct graph of the equation? If so, why? If not, why?

(B) Add points on the graph for $x = -2, -0.5, 0.5$, and 2.

(C) Now, what do you think the graph looks like? Sketch your version of the graph, adding more points as necessary.

(D) Write a short statement explaining any conclusions you might draw from parts (A), (B), and (C).

As Explore/Discuss 1 illustrates, sometimes it can be difficult to determine the apparent shape of a graph by simply plotting a few points. One of the major objectives of this book is to develop mathematical tools that will help us analyze graphs.

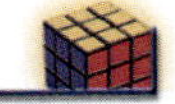

Data Analysis

In applications, numeric data are often collected and presented in graphical form. A graph illustrates the relationship between two quantities, and estimating coordinates of points on the graph provides specific examples of this relationship, even if no equation for the graph is available. Example 3 illustrates the analysis of data presented in graphical form.

EXAMPLE 3 Ozone Levels

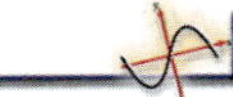

The ozone level during a 12-hour period in a suburb of Milwaukee, Wisconsin on a particular summer day is given in Figure 6, where L is ozone in parts per billion and t is time in hours. Use this graph to estimate the following ozone levels to the nearest integer and times to the nearest quarter hour:

(A) The ozone level at 6 P.M.

(B) The highest ozone level and the time when it occurs

(C) The time(s) when the ozone level is 90 ppb

FIGURE 6

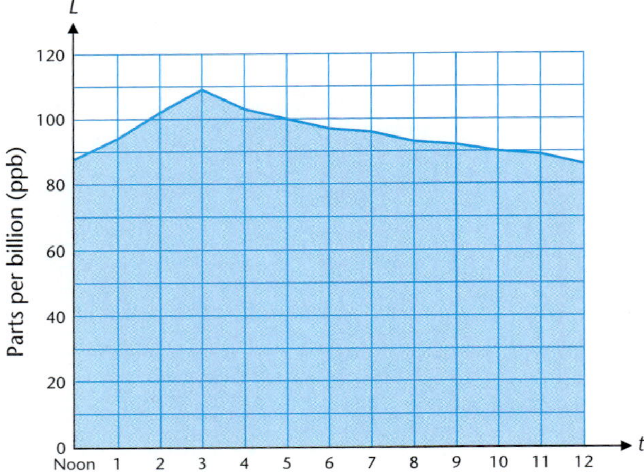

Source: Wisconsin Department of Natural Resources.

SOLUTION

(A) The L coordinate of the point on the graph with t coordinate 6 is approximately 97 parts per billion (Fig. 7).

(B) The highest ozone level is approximately 109 parts per billion at 3 P.M.

(C) The ozone level is 90 parts per billion at about 12:30 P.M. and again at 10 P.M.

FIGURE 7

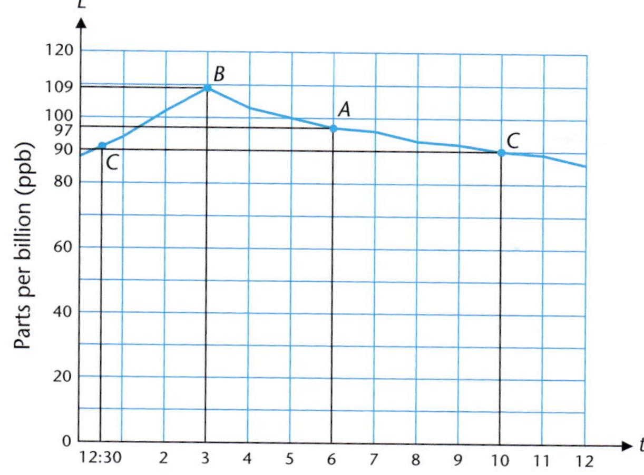

MATCHED 3 PROBLEM

Use the graph in Figure 6 to estimate the following ozone level to the nearest integer and time(s) to the nearest quarter hour.

(A) The ozone level at 7 P.M.

(B) The time(s) when the ozone level is 100 parts per billion.

ANSWERS MATCHED PROBLEMS

1.

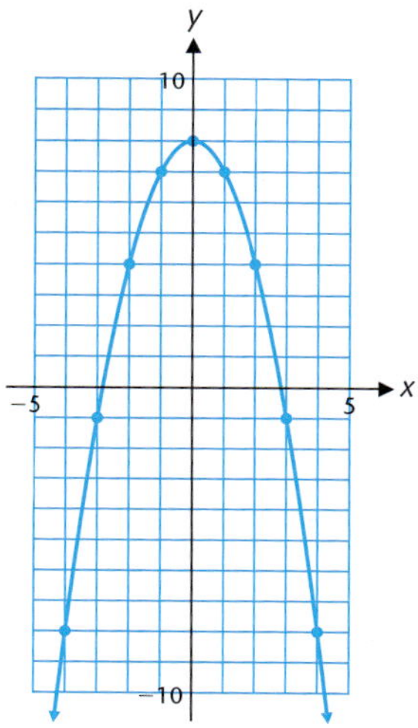

2.

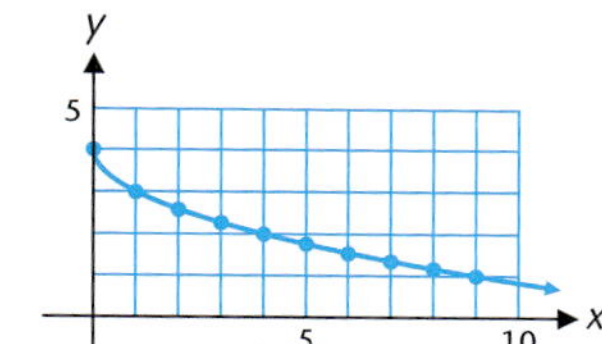

3. (A) 96 parts per billion
(B) 1:45 P.M. and 5 P.M.

EXERCISE A.2

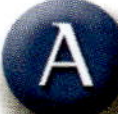

In Problems 1–4, plot the given points in a rectangular coordinate system.

1. (5, 0), (3, −2), (−4, 2), (4, 4) **2.** (0, 4), (−3, 2), (5, −1), (−2, −4) **3.** (0, −2), (−1, −3), (4, −5), (−2, 1)

4. (−2, 0), (3, 2), (1, −4), (−3, 5)

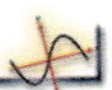

In Problems 5–8, find the coordinates of points A, B, C, and D.

5.

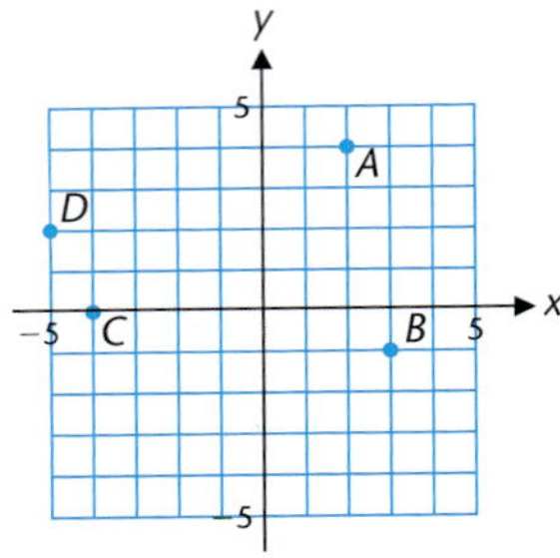

6.

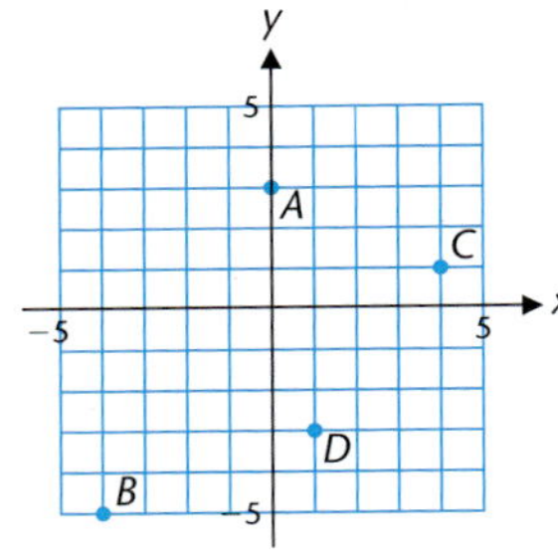

7.

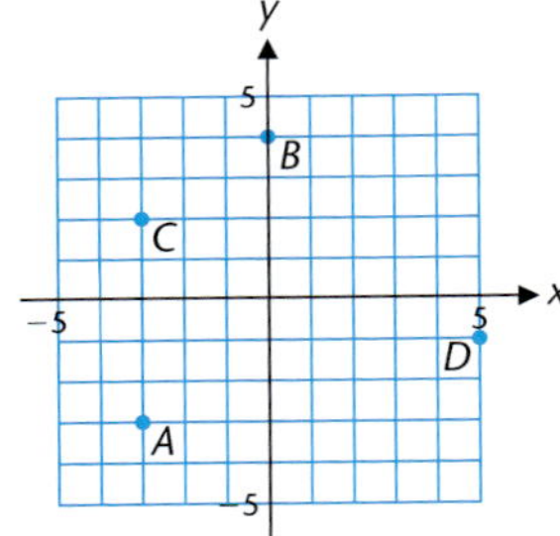

8.

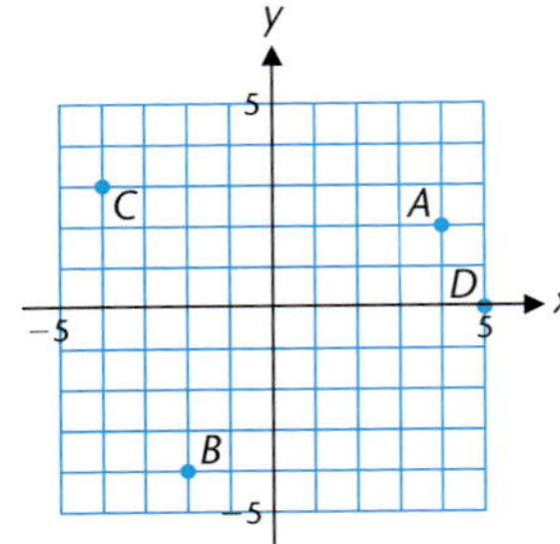

B

For each equation in Problems 9–14, make up a table of solutions using $x = -3, -2, -1, 0, 1, 2,$ and 3. Plot these solutions and graph the equation.

9. $y = x + 1$

10. $y = 2 - x$

11. $y = x^2 - 5$

12. $y = 4 - x^2$

13. $y = 3 + x - 0.5x^2$

14. $y = 4 - x - 0.5x^2$

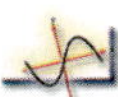

In Problems 15–18, use the graph to estimate to the nearest integer the missing coordinates of the indicated points. (Be sure you find all possible answers.)

15. **(A)** (8, ?) **(B)** (−5, ?) **(C)** (0, ?)
(D) (?, 6) **(E)** (?, −5) **(F)** (?, 0)

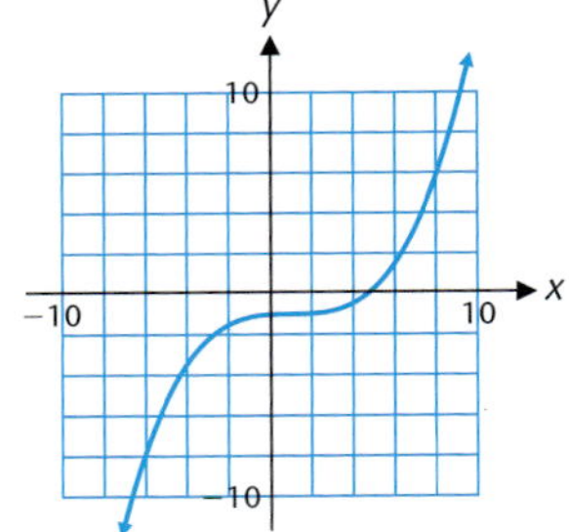

16. **(A)** (3, ?) **(B)** (−5, ?) **(C)** (0, ?)
(D) (?, 3) **(E)** (?, −4) **(F)** (?, 0)

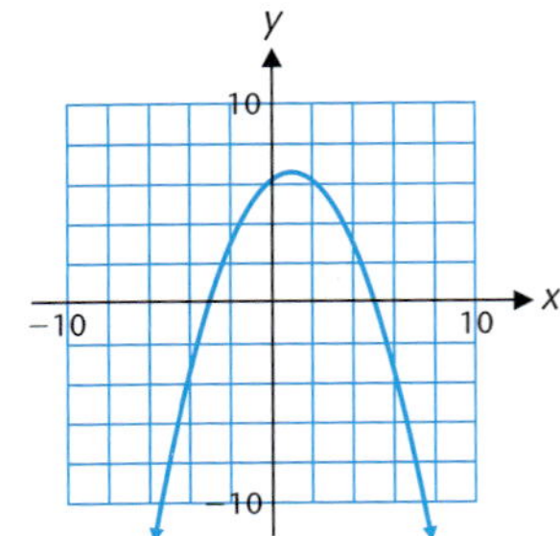

17. **(A)** (1, ?) **(B)** (−8, ?) **(C)** (0, ?)
(D) (?, −6) **(E)** (?, 4) **(F)** (?, 0)

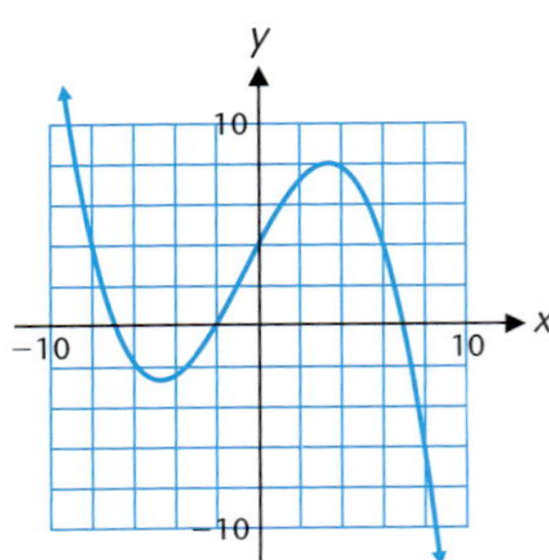

18. **(A)** (6, ?) **(B)** (−6, ?) **(C)** (0, ?)
(D) (?, −2) **(E)** (?, 1) **(F)** (?, 0)

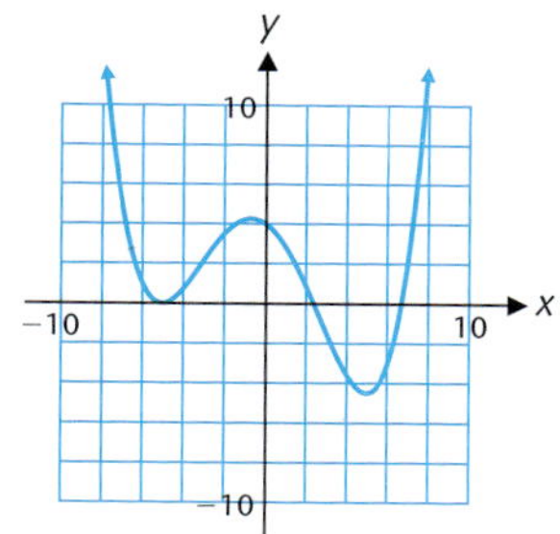

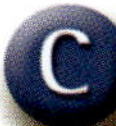

19. **(A)** Sketch a graph based on the solutions in the following table.

x	-2	0	2
y	-2	0	2

(B) Sketch a graph based on the solutions in the following table.

x	-1	0	1
y	2	0	-2

(C) Complete the following table for $y = x^3 - 3x$ and sketch a graph of the equation.

x	-2	-1	0	1	2
y					

(D) Write a short statement explaining any conclusions you might draw from parts (A), (B), and (C).

20. **(A)** Sketch a graph based on the solutions in the following table.

x	-1	1	3
y	4	2	0

(B) Sketch a graph based on the solutions in the following table.

x	0	1	2
y	0	2	4

(C) Complete the following table for $y = 3x^2 - x^3$ and sketch a graph of the equation.

x	-1	0	1	2	3
y					

(D) Write a short statement explaining any conclusions you might draw from parts (A), (B), and (C).

In Problems 21–28, graph each equation using point-by-point plotting.

21. $y = x^{1/3}$

22. $y = x^{2/3}$

23. $y = x^3$

24. $y = x^4$

25. $y = \sqrt{x - 1}$

26. $y = \sqrt{5 - x}$

27. $y = \sqrt{1 + x^2}$

28. $y = x\sqrt{1 + x^2}$

C

29. **(A)** Graph the triangle with vertices $A = (1, 1)$, $B = (7, 2)$, and $C = (4, 6)$.

(B) Now graph the triangle with vertices $A' = (1, -1)$, $B' = (7, -2)$, and $C' = (4, -6)$ in the same coordinate system.

(C) How are these two triangles related? How would you describe the effect of changing the sign of the y coordinate of all the points on a graph?

30. **(A)** Graph the triangle with vertices $A = (1, 1)$, $B = (7, 2)$, and $C = (4, 6)$.

(B) Now graph the triangle with vertices $A' = (-1, 1)$, $B' = (-7, 2)$, and $C' = (-4, 6)$ in the same coordinate system.

(C) How are these two triangles related? How would you describe the effect of changing the sign of the x coordinate of all the points on a graph?

31. **(A)** Graph the triangle with vertices $A = (1, 1)$, $B = (7, 2)$, and $C = (4, 6)$.

(B) Now graph the triangle with vertices $A' = (-1, -1)$, $B' = (-7, -2)$, and $C' = (-4, -6)$ in the same coordinate system.

(C) How are these two triangles related? How would you describe the effect of changing the signs of the x and y coordinates of all the points on a graph?

32. **(A)** Graph the triangle with vertices $A = (1, 2)$, $B = (1, 4)$, and $C = (3, 4)$.

(B) Now graph $y = x$ and the triangle obtained by reversing the coordinates for each vertex of the original triangle: $A' = (2, 1)$, $B' = (4, 1)$, $C' = (4, 3)$.

(C) How are these two triangles related? How would you describe the effect of reversing the coordinates of each point on a graph?

APPLICATIONS

33. Business. After extensive surveys, the marketing research department of a producer of popular cassette tapes developed the demand equation

$$n = 10 - p \qquad 5 \le p \le 10$$

where n is the number of units (in thousands) retailers are willing to buy per day at \$$p$ per tape. The company's daily revenue R (in thousands of dollars) is given by

$$R = np = (10 - p)p \qquad 5 \le p \le 10$$

Graph the revenue equation for the indicated values of p.

34. Business. Repeat Problem 33 for the demand equation

$$n = 8 - p \qquad 4 \le p \le 8$$

35. Physics. The speed (in meters per second) of a ball swinging at the end of a pendulum is given by

$$v = 0.5\sqrt{2 - x}$$

where x is the vertical displacement (in centimeters) of the ball from its position at rest (see the figure).

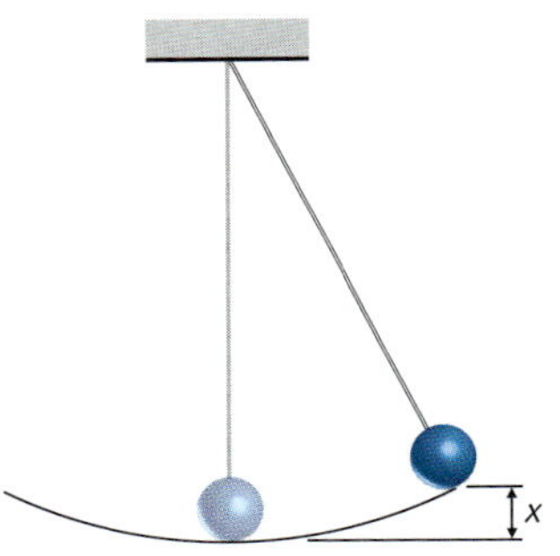

(A) Graph $v = 0.5\sqrt{2 - x}$ for $0 \le x \le 2$.

(B) Describe the relationship between this graph and the physical behavior of the ball as it swings back and forth.

36. Physics. The speed (in meters per second) of a ball oscillating at the end of a spring is given by

$$v = 4\sqrt{25 - x^2}$$

where x is the vertical displacement (in centimeters) of the ball from its position at rest (positive displacement measured downward—see the figure).

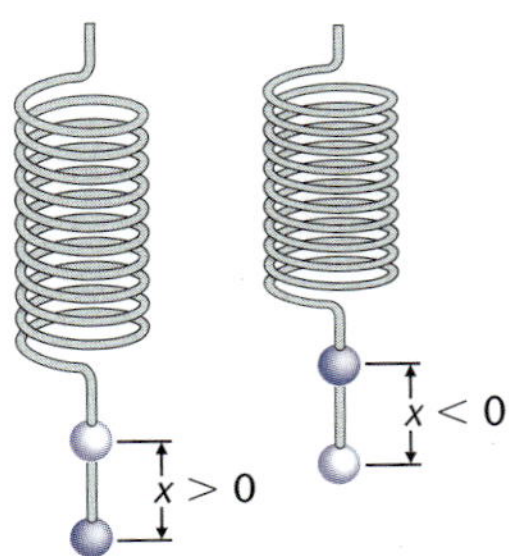

(A) Graph $v = 4\sqrt{25 - x^2}$ for $-5 \le x \le 5$.

(B) Describe the relationship between this graph and the physical behavior of the ball as it oscillates up and down.

DATA ANALYSIS 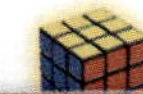

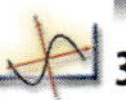

37. Price and Demand. The quantity of a product that consumers are willing to buy during some period depends on its price. The price p and corresponding weekly demand q for a particular brand of diet soda in a city are shown in the figure. Use this graph to estimate the following demands to the nearest 100 cases.

(A) What is the demand when the price is \$6.00 per case?

(B) Does the demand increase or decrease if the price is increased to \$6.30 per case? By how much?

(C) Does the demand increase or decrease if the price is decreased to \$5.70? By how much?

(D) Write a brief description of the relationship between price and demand illustrated by this graph.

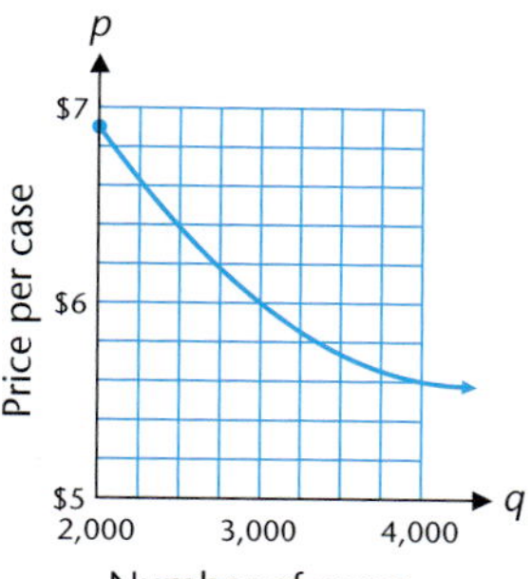

38. Price and Supply. The quantity of a product that suppliers are willing to sell during some period depends on its price. The price p and corresponding weekly supply q for a particular brand of diet soda in a city are shown in the figure. Use this graph to estimate the following supplies to the nearest 100 cases.

(A) What is the supply when the price is \$5.60 per case?

(B) Does the supply increase or decrease if the price is increased to \$5.80 per case? By how much?

(C) Does the supply increase or decrease if the price is decreased to \$5.40 per case? By how much?

(D) Write a brief description of the relationship between price and supply illustrated by this graph.

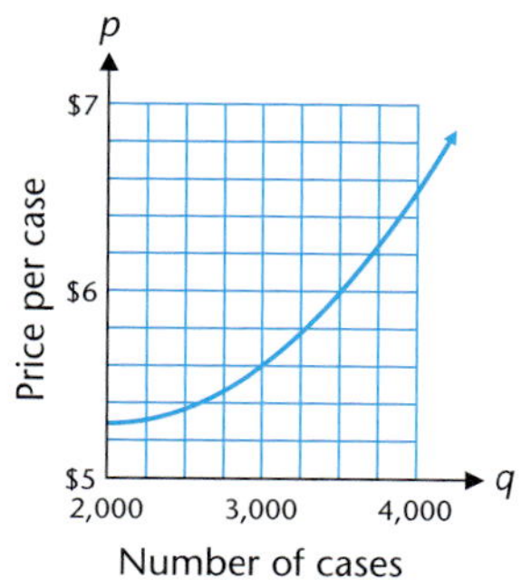

39. Temperature. The temperature (in degrees Fahrenheit) during a spring day in the Midwest is given in the figure. Use this graph to estimate the following temperatures to the nearest degree and times to the nearest hour:

(A) The temperature at 9:00 A.M.

(B) The highest temperature and the time when it occurs

(C) The time(s) when the temperature is 49°

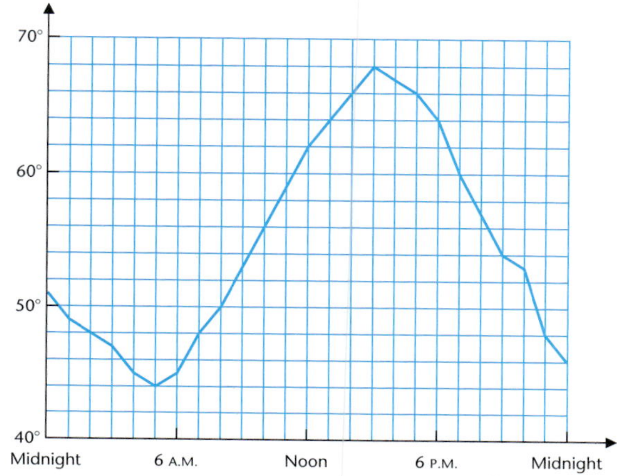

40. Temperature. Use the figure for Problem 39 to estimate the following temperatures to the nearest degree and times to the nearest half hour:

(A) The temperature at 7:00 P.M.

(B) The lowest temperature and the time when it occurs

(C) The time(s) when the temperature is 52°

SECTION A.3 Basic Formulas in Analytic Geometry

Distance between Two Points • Midpoint of a Line Segment • Circles

Two basic problems studied in analytic geometry are

1. Given an equation, find its graph.
2. Given a figure (line, circle, parabola, ellipse, etc.) in a coordinate system, find its equation.

The first problem was discussed in Section A.2. In Section A.3 we introduce some tools that are useful when studying the second problem.

Distance between Two Points

Given two points P_1 and P_2 in a rectangular coordinate system, we denote the **distance between P_1 and P_2 by $d(P_1, P_2)$.** We begin with an example.

EXAMPLE 1 Distance between Two Points

Find the distance between the points $P_1 = (1, 2)$ and $P_2 = (4, 6)$.

SOLUTION

Connecting the points P_1, P_2, and $P_3 = (4, 2)$ with straight line segments forms a right triangle (Fig. 1).

FIGURE 1

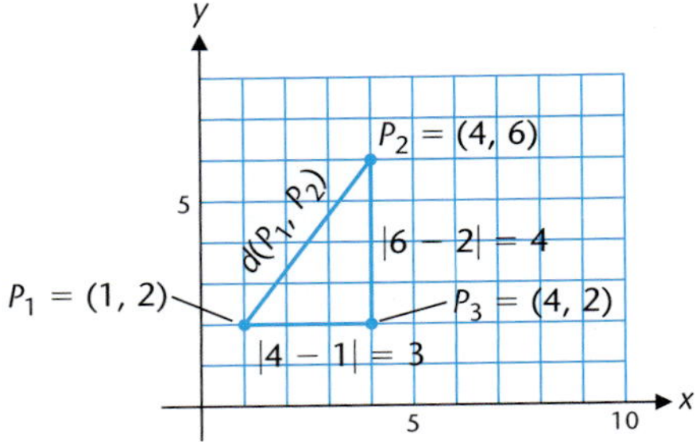

From the figure, we see that lengths of the legs of the triangle are

$$d(P_1, P_3) = |4 - 1| = 3$$

and

$$d(P_3, P_2) = |6 - 2| = 4$$

The length of the hypotenuse is $d(P_1, P_2)$, the distance we are seeking. Applying the Pythagorean theorem (see Appendix C) we have

$$\begin{aligned}[d(P_1, P_2)]^2 &= [d(P_1, P_3)]^2 + [d(P_3, P_2)]^2 \\ &= 3^2 + 4^2 \\ &= 9 + 16 \\ &= 25\end{aligned}$$

Thus,

$$d(P_1, P_2) = \sqrt{25} = 5$$

MATCHED PROBLEM 1

Find the distance between the points $P_1 = (1, 2)$ and $P_2 = (13, 7)$.

The ideas used in Example 1 can be generalized to any two distinct points in the plane. If $P_1 = (x_1, y_1)$ and $P_2 = (x_2, y_2)$ are two points in a rectangular coordinate system (Fig. 2), then

$$[d(P_1, P_2)]^2 = |x_2 - x_1|^2 + |y_2 - y_1|^2$$
$$= (x_2 - x_1)^2 + (y_2 - y_1)^2 \quad \text{Because } |N|^2 = N^2.$$

FIGURE 2 Distance between two points.

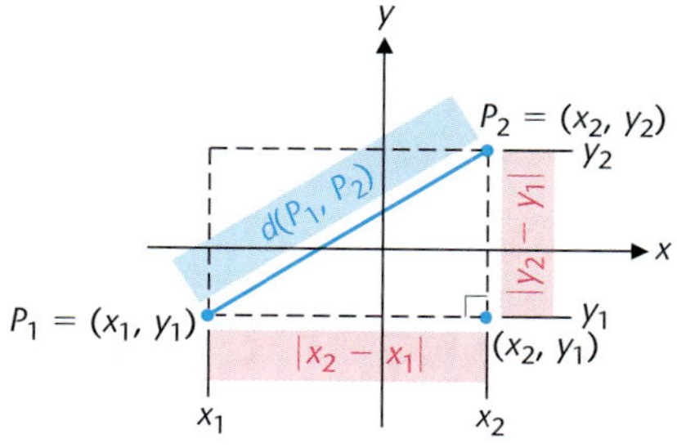

Thus,

THEOREM 1
Distance between $P_1 = (x_1, y_1)$ and $P_2 = (x_2, y_2)$

$$d(P_1, P_2) = \sqrt{(x_2 - x_1)^2 + (y_2 - y_1)^2}$$

EXAMPLE 2 Using the Distance-between-Two-Points Formula

Find the distance between the points $(-3, 5)$ and $(-2, -8)$.

SOLUTION
Let $(x_1, y_1) = (-3, 5)$ and $(x_2, y_2) = (-2, -8)$. Then,

$$d = \sqrt{[(-2) - (-3)]^2 + [(-8) - 5]^2}$$
$$= \sqrt{(-2 + 3)^2 + (-8 - 5)^2} = \sqrt{1^2 + (-13)^2} = \sqrt{1 + 169} = \sqrt{170}$$

Notice that if we choose $(x_1, y_1) = (-2, -8)$ and $(x_2, y_2) = (-3, 5)$, then

$$d = \sqrt{[(-3) - (-2)]^2 + [5 - (-8)]^2} = \sqrt{1 + 169} = \sqrt{170}$$

so it doesn't matter which point we designate as P_1 or P_2.

MATCHED 2 PROBLEM

Find the distance between the points (6, −3) and (−7, −5).

Midpoint of a Line Segment

EXPLORE/DISCUSS 1

(A) Graph the line segment L joining the points (1, 2) and (7, 10).

(B) Find the average a of the x coordinates of these two points.

(C) Find the average b of the y coordinates of these two points.

(D) Plot the point (a, b). Is it on the line segment L?

(E) Find the distance between (1, 2) and (a, b) and the distance between (a, b) and (7, 10). How would you describe the point (a, b)?

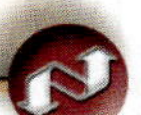

The **midpoint** of the line segment between two points is the point on the line segment that is equidistant from each of the points. A formula for finding the midpoint is given in Theorem 2. The proof is discussed in the exercises.

THEOREM 2

Midpoint Formula

The midpoint of the line segment joining $P_1 = (x_1, y_1)$ and $P_2 = (x_2, y_2)$ is

$$M = \left(\frac{x_1 + x_2}{2}, \frac{y_1 + y_2}{2}\right)$$

The point M is the unique point satisfying

$$d(P_1, M) = d(M, P_2) = \frac{1}{2}d(P_1, P_2)$$

Note that the coordinates of the midpoint are simply the averages of the respective coordinates of the two given points.

EXAMPLE 3 Using the Midpoint Formula

Find the midpoint M of the line segment joining $A = (-3, 2)$ and $B = (4, -5)$. Plot A, B, and M and verify that $d(A, M) = d(M, B) = \frac{1}{2}d(A, B)$.

SOLUTION

Finding the Midpoint

We use the midpoint formula with $(x_1, y_1) = (-3, 2)$ and $(x_2, y_2) = (4, -5)$ to obtain the coordinates of the midpoint M.

$$\begin{aligned} M &= \left(\frac{x_1 + x_2}{2}, \frac{y_1 + y_2}{2}\right) \\ &= \left(\frac{-3 + 4}{2}, \frac{2 + (-5)}{2}\right) \\ &= \left(\frac{1}{2}, \frac{-3}{2}\right) \\ &= (0.5, -1.5) \end{aligned}$$

The fraction form of M is probably more convenient for plotting the point. The decimal form is more convenient for computing distances.

Plotting and Verifying

Figure 3 shows the three points.

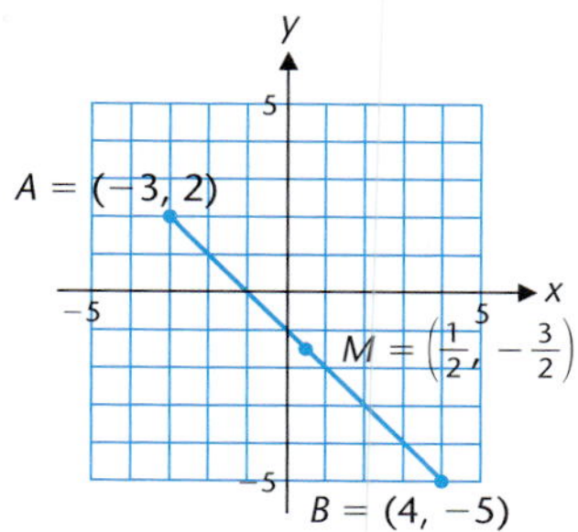

FIGURE 3

Find the distances $d(A, M)$, $d(M, B)$, and $\frac{1}{2}d(A, B)$

$$\begin{aligned} d(A, M) &= \sqrt{(-3 - 0.5)^2 + [2 - (-1.5)]^2} \\ &= \sqrt{12.25 + 12.25} = \sqrt{24.5} \\ d(M, B) &= \sqrt{(0.5 - 4)^2 + [-1.5 - (-5)]^2} \\ &= \sqrt{12.25 + 12.25} = \sqrt{24.5} \\ d(A, B) &= \sqrt{(-3 - 4)^2 + [2 - (-5)]^2} \\ &= \sqrt{49 + 49} = \sqrt{98} \\ \frac{1}{2}d(A, B) &= \frac{1}{2}\sqrt{98} = \sqrt{\frac{98}{4}} = \sqrt{24.5} \end{aligned}$$

Because $d(A, M) = d(M, B) = \frac{1}{2}d(A, B)$, M is the midpoint of the line segment joining A and B.

MATCHED 3 PROBLEM

Find the midpoint M of the line segment joining $A = (4, 1)$ and $B = (-3, -5)$. Plot A, B, and M and verify that $d(A, M) = d(M, B) = \frac{1}{2}d(A, B)$.

EXAMPLE 4 Using the Midpoint Formula

If $M = (1, 1)$ is the midpoint of the line segment joining $A = (-3, -1)$ and $B = (x, y)$, find the coordinates of B.

SOLUTION

Algebraic Solution

From the midpoint formula, we have

$$M = (1, 1) = \left(\frac{-3 + x}{2}, \frac{-1 + y}{2}\right)$$

We equate the corresponding coordinates and solve the resulting equations for x and y

$$1 = \frac{-3 + x}{2} \qquad 1 = \frac{-1 + y}{2}$$

$$2 = -3 + x \qquad 2 = -1 + y$$

$$2 + 3 = -3 + x + 3 \qquad 2 + 1 = -1 + y + 1$$

$$5 = x \qquad 3 = y$$

Thus,

$$B = (5, 3)$$

Graphical Solution

Plot A and M and draw right triangle ACM as shown in Figure 4.

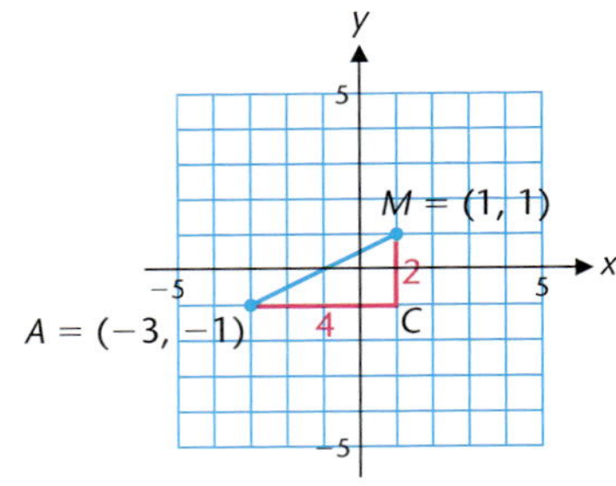

FIGURE 4

Now form triangle MDB by shifting triangle ACM four units to the right and two units up (Fig. 5).

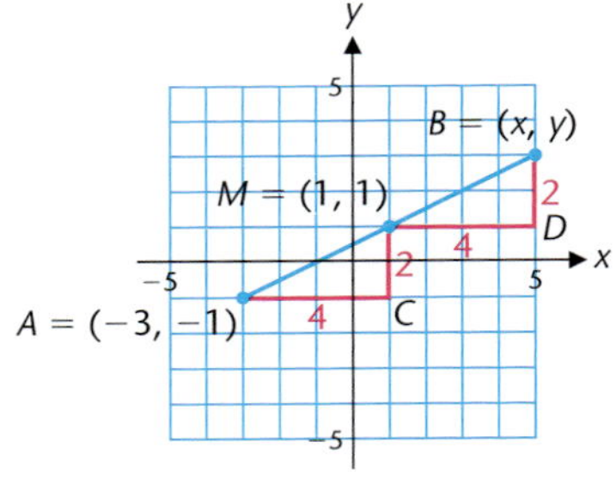

FIGURE 5

Because triangles ACM and MDB are **congruent** (see Appendix C), $d(A, M) = d(M, B)$ and M must be the midpoint of the segment joining A and B. From the graph we can see that

$$x = 1 + 4 = 5 \quad \text{and} \quad y = 1 + 2 = 3$$

Thus,

$$B = (5, 3)$$

MATCHED 4 PROBLEM

If $M = (1, -1)$ is the midpoint of the line segment joining $A = (-1, -5)$ and $B = (x, y)$, find the coordinates of B.

Circles

The distance-between-two-points formula would be helpful if its only use were to find actual distances between points, such as in Example 2. However, its more important use is in finding equations of figures in a rectangular coordinate system. We start with an example.

EXAMPLE 5 Equations and Graphs of Circles

Write an equation for the set of all points that are five units from the origin. Graph your equation.

SOLUTION

Writing the Equation

The distance between a point (x, y) and the origin is

$$d = \sqrt{(x - 0)^2 + (y - 0)^2} = \sqrt{x^2 + y^2}$$

Thus, an equation for the set of points that are five units from the origin is

$$\sqrt{x^2 + y^2} = 5$$

We square both sides of this equation to obtain an equation that does not contain any radicals.

$$x^2 + y^2 = 25$$

Graphing the Equation

We make up a table of solutions and plot the points in the table. Joining these points produces a familiar geometric object—a circle (Fig. 6).

x	y
-5	0
-4	± 3
-3	± 4
0	± 5
3	± 4
4	± 3
5	0

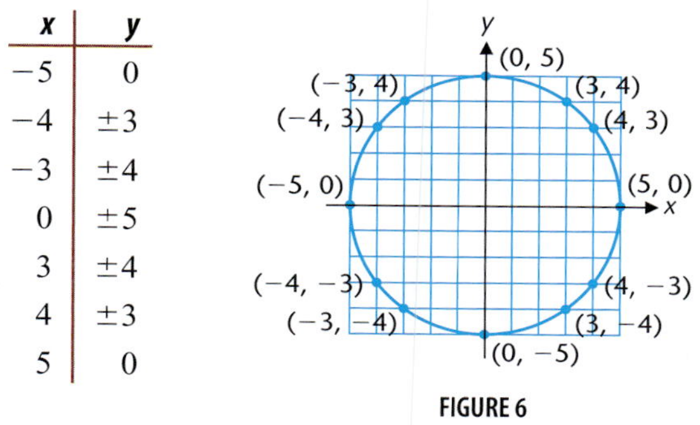

FIGURE 6

MATCHED 5 PROBLEM

Write an equation for the set of all points that are three units from the origin. Graph your equation.

In Example 5, we began with a verbal description of a set of points, produced an algebraic equation that these points must satisfy, constructed a numerical table listing some of these points, and then drew a graphical representation of this set of points. The interplay between verbal, algebraic, numerical, and graphical concepts is one of the central themes of this book.

Now we generalize the ideas introduced in Example 5.

DEFINITION 1

Circle

A **circle** is the set of all points in a plane equidistant from a fixed point. The fixed distance is called the **radius,** and the fixed point is called the **center.**

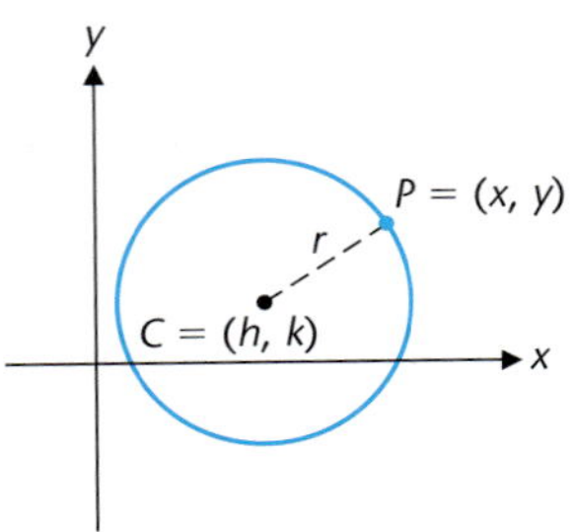

FIGURE 7 Circle.

Let's find the equation of a circle with radius r ($r > 0$) and center C at (h, k) in a rectangular coordinate system (Fig. 7). The circle consists of all points $P = (x, y)$ satisfying $d(P, C) = r$; that is, all points satisfying

$$\sqrt{(x - h)^2 + (y - k)^2} = r \quad r > 0$$

or, equivalently,

$$(x - h)^2 + (y - k)^2 = r^2 \quad r > 0$$

THEOREM 3

Standard Equation of a Circle

Circle with radius r and center at (h, k)

$$(x - h)^2 + (y - k)^2 = r^2 \quad r > 0$$

EXAMPLE 6 Equations and Graphs of Circles

Find the equation of a circle with radius 4 and center at $C = (-3, 6)$. Graph the equation.

SOLUTION

Writing the Equation

$$C = (h, k) = (-3, 6) \text{ and } r = 4$$
$$(x - h)^2 + (y - k)^2 = r^2$$
$$[x - (-3)]^2 + (y - 6)^2 = 4^2$$
$$(x + 3)^2 + (y - 6)^2 = 16$$

Graphing the Equation

To graph the equation, plot the center and a few points on the circle, then draw a circle of radius 4 (Fig. 8).

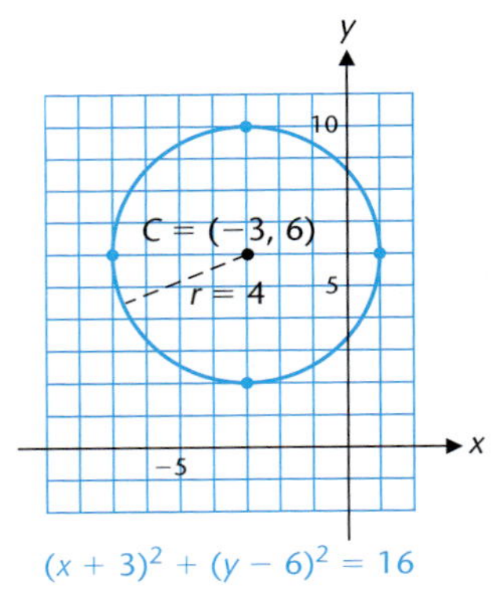

FIGURE 8

MATCHED PROBLEM 6

Find the equation of a circle with radius 3 and center at $C = (3, -2)$. Graph the equation.

Refer to the solution of Example 6. If we square both binomials and collect like terms we obtain the following:

$$\begin{aligned} (x + 3)^2 + (y - 6)^2 &= 16 \\ x^2 + 6x + 9 + y^2 - 12y + 36 &= 16 \\ x^2 + 6x + y^2 - 12y + 29 &= 0 \end{aligned}$$

With the exception of some degenerate cases, an equation of the form

$$x^2 + Ax + y^2 + By + C = 0$$

has a circle as its graph. To find the center and the radius, we use a technique called **completing the square,** which is based on the following equation:

$$\left(x + \frac{b}{2}\right)^2 = x^2 + bx + \left(\frac{b}{2}\right)^2$$

In the expanded form on the right, notice that the constant term $(b/2)^2$ is the square of one-half of the coefficient of the x term.

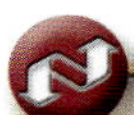

EXPLORE/DISCUSS 2

Replace ❒ and ❍ with numbers that make each equation true for all x.

(A) $(x + ❒)^2 = x^2 + 4x + ❍$

(B) $(x + ❒)^2 = x^2 - 6x + ❍$

(C) $(x + ❒)^2 = x^2 + 5x + ❍$

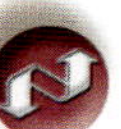

Example 7 illustrates the use of the completing-the-square technique.

EXAMPLE 7 Finding the Center and Radius of a Circle

Find the center and radius and sketch the graph of the circle with equation

$$x^2 - 10x + y^2 + 8y = 40$$

SOLUTION

We apply the completing-the-square technique to the first two terms and then to the second two terms. Because we are dealing with an equation, we must be certain to add the required numbers to both sides of the equation.

$$x^2 - 10x + \qquad y^2 + 8y \qquad = 40$$

$$x^2 - 10x + \left(\frac{10}{2}\right)^2 + y^2 + 8y + \left(\frac{8}{2}\right)^2 = 40 + \left(\frac{10}{2}\right)^2 + \left(\frac{8}{2}\right)^2$$

$$x^2 - 10x + 25 + y^2 + 8y + 16 = 40 + 25 + 16$$

$$(x - 5)^2 + (y + 4)^2 = 81$$

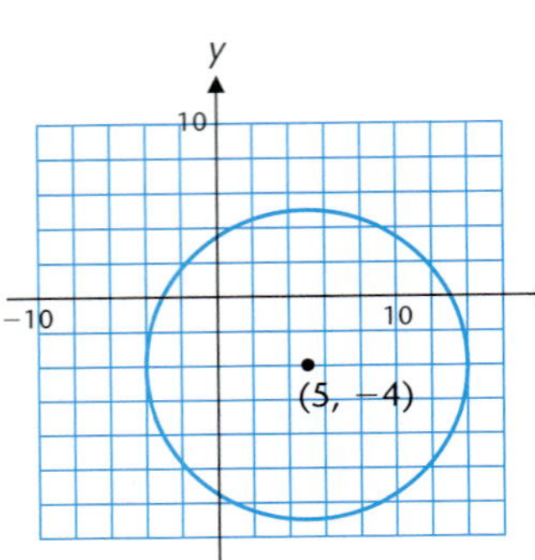

FIGURE 9

We now have the standard equation of a circle with radius 9 and center (5, −4). The graph is shown in Figure 9.

MATCHED 7 PROBLEM

Find the center and radius and sketch the graph of the circle with equation

$$x^2 + 4x + y^2 - 6y = 12$$

ANSWERS MATCHED PROBLEMS

1. 13
2. $\sqrt{173}$
3. $M = \left(\frac{1}{2}, -2\right) = (0.5, -2)$; $d(A, M) = \sqrt{21.25} = d(M, B)$

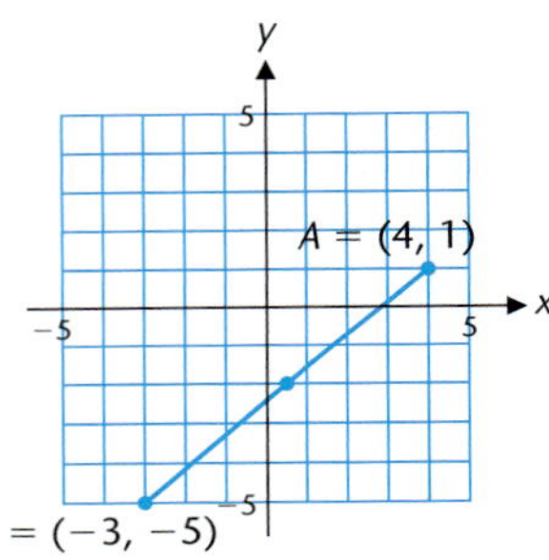

4. $B = (3, 3)$
5. $x^2 + y^2 = 9$

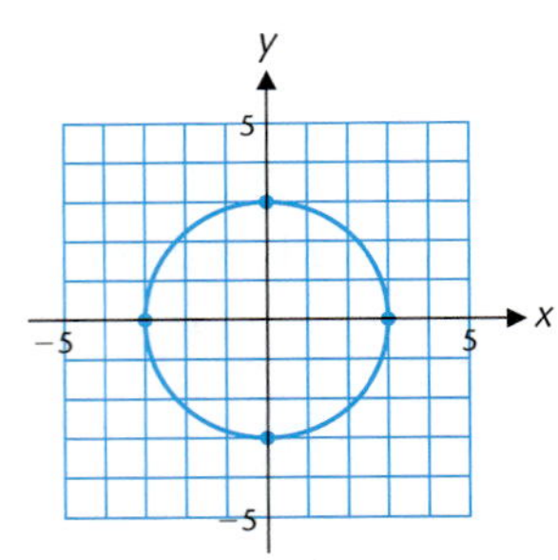

6. $(x - 3)^2 + (y + 2)^2 = 9$

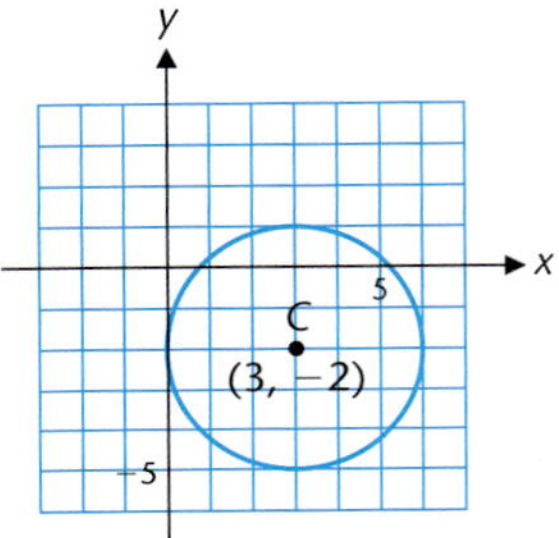

7. Circle with radius 5 and center (−2, 3)

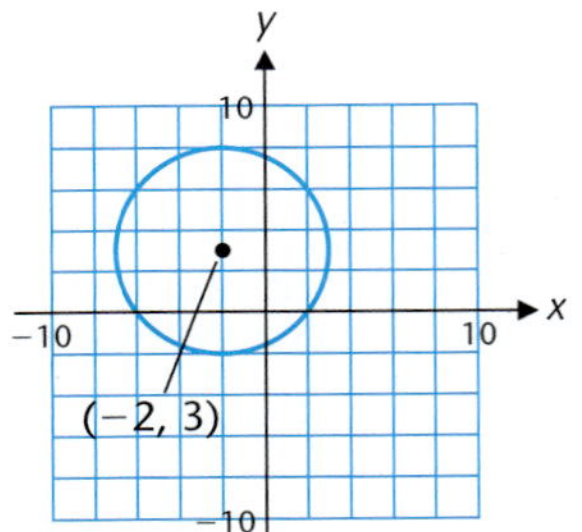

EXERCISE A.3

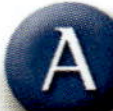

In Problems 1–8, find the distance between each pair of points and the midpoint of the line segment joining the points. Leave distance in radical form.

1. (1, 0), (4, 4)
2. (0, 1), (3, 5)
3. (0, −2), (5, 10)
4. (3, 0), (−2, −3)

5. $(-6, -4), (3, 4)$

6. $(-5, 4), (6, -1)$

7. $(6, 6), (4, -2)$

8. $(5, -3), (-1, 4)$

In Problems 9–16, write the equation of a circle with the indicated center and radius.

9. $C = (0, 0), r = 7$

10. $C = (0, 0), r = 5$

11. $C = (2, 3), r = 6$

12. $C = (5, 6), r = 2$

13. $C = (-4, 1), r = \sqrt{7}$

14. $C = (-5, 6), r = \sqrt{11}$

15. $C = (-3, -4), r = \sqrt{2}$

16. $C = (4, -1), r = \sqrt{5}$

In Problems 17–20, replace ❑ and ❍ with numbers that make the equation true for all x.

17. $(x + ❑)^2 = x^2 + 8x + ❍$

18. $(x + ❑)^2 = x^2 - 10x + ❍$

19. $(x + ❑)^2 = x^2 - 3x + ❍$

20. $(x + ❑)^2 = x^2 + x + ❍$

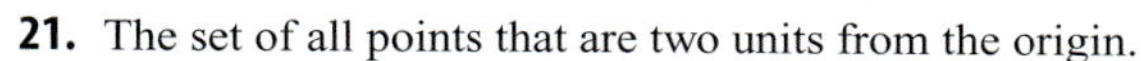

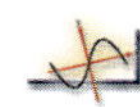

B

In Problems 21–26, write an equation for the given set of points. Graph your equation.

21. The set of all points that are two units from the origin.

22. The set of all points that are four units from the origin.

23. The set of all points that are one unit from (1, 0).

24. The set of all points that are one unit from $(0, -1)$.

25. The set of all points that are three units from $(-2, 1)$.

26. The set of all points that are two units from $(3, -2)$.

27. Let M be the midpoint of A and B, where

$A = (a_1, a_2)$, $B = (1, 3)$, and $M = (-2, 6)$.

(A) Use the fact that -2 is the average of a_1 and 1 to find a_1.

(B) Use the fact that 6 is the average of a_2 and 3 to find a_2.

(C) Find $d(A, M)$ and $d(M, B)$.

28. Let M be the midpoint of A and B, where

$A = (-3, 5)$, $B = (b_1, b_2)$, and $M = (4, -2)$.

(A) Use the fact that 4 is the average of -3 and b_1 to find b_1.

(B) Use the fact that -2 is the average of 5 and b_2 to find b_2.

(C) Find $d(A, M)$ and $d(M, B)$.

In Problems 29–32, write a verbal description of the graph and then write an equation that would produce the graph.

29.

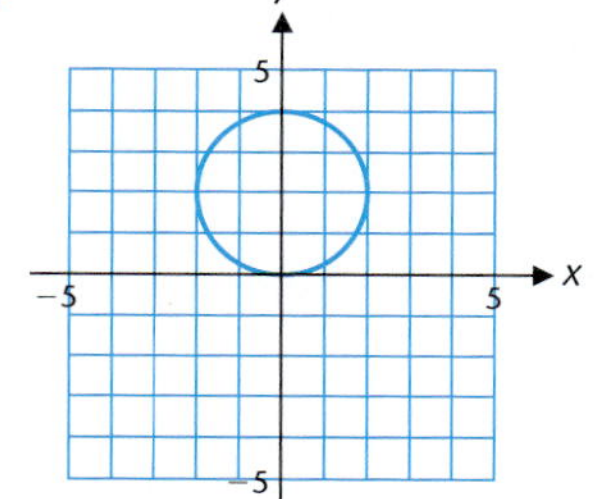

30.

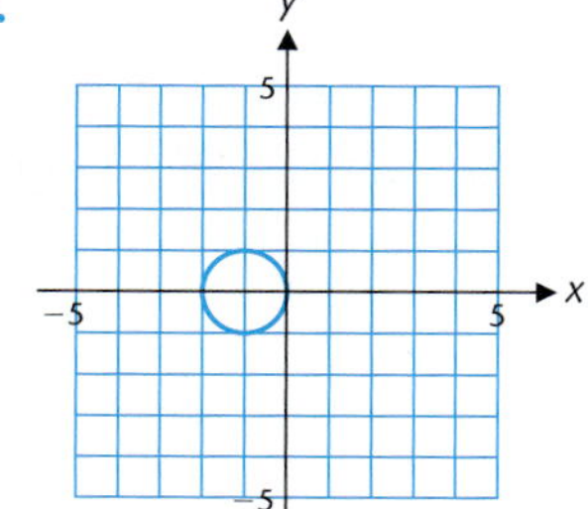

31.

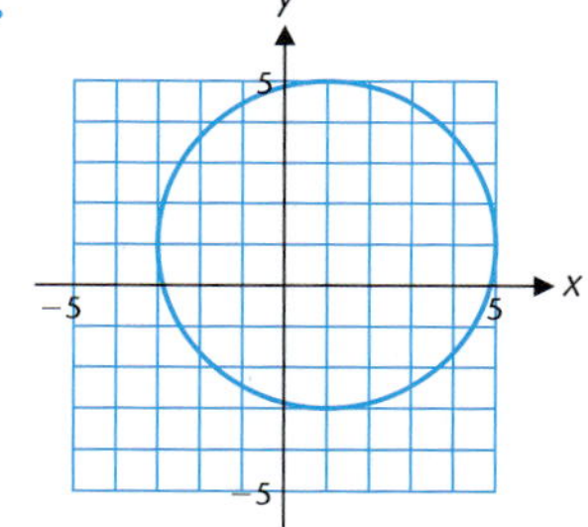

32.

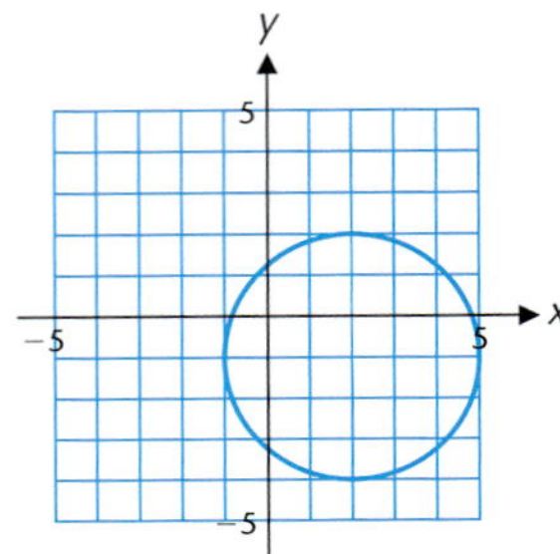

In Problems 33–38, M is the midpoint of A and B. Find the indicated point. Verify that $d(A, M) = d(M, B) = \frac{1}{2}d(A, B)$.

33. $A = (-4.3, 5.2)$, $B = (9.6, -1.7)$, $M = ?$

34. $A = (2.8, -3.5)$, $B = (-4.1, 7.6)$, $M = ?$

35. $A = (25, 10)$, $M = (-5, -2)$, $B = ?$

36. $M = (2.5, 3.5)$, $B = (12, 10)$, $A = ?$

37. $M = (-8, -6)$, $B = (2, 4)$, $A = ?$

38. $A = (-4, -2)$, $M = (-1.5, -4.5)$, $B = ?$

In Problems 39–44, find the center and radius of the circle with the given equation.

39. $x^2 + y^2 - 2y = 0$

40. $x^2 + 2x + y^2 = 0$

41. $x^2 - 2x + y^2 + 6y - 6 = 0$

42. $x^2 + 4x + y^2 - 8y + 16 = 0$

43. $x^2 + x + y^2 + 3y + 2 = 0$

44. $x^2 - 5x + y^2 - 3y + 5 = 0$

In Problems 45 and 46, show that the given points are the vertices of a right triangle (see the Pythagorean theorem in Appendix C). Find the length of the line segment from the midpoint of the hypotenuse to the opposite vertex.

45. $(-3, 2), (1, -2), (8, 5)$

46. $(-1, 3), (3, 5), (5, 1)$

Find the perimeter (to two decimal places) of the triangle with the vertices indicated in Problems 47 and 48.

47. $(-3, 1), (1, -2), (4, 3)$

48. $(-2, 4), (3, 1), (-3, -2)$

C

49. If $P_1 = (x_1, y_1)$, $P_2 = (x_2, y_2)$ and $M = \left(\frac{x_1 + x_2}{2}, \frac{y_1 + y_2}{2}\right)$, show that $d(P_1, M) = d(M, P_2) = \frac{1}{2}d(P_1, P_2)$. (This is one step in the proof of Theorem 2).

50. A parallelogram $ABCD$ is shown in the figure.

(A) Find the midpoint of the line segment joining A and C.

(B) Find the midpoint of the line segment joining B and D.

(C) What can you conclude about the diagonals of the parallelogram?

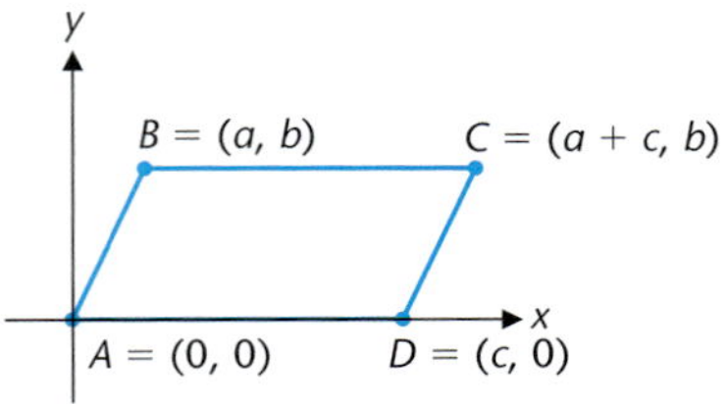

Find the equation of a circle that has a diameter with the end points given in Problems 51 and 52.

51. $(7, -3), (1, 7)$

52. $(-3, 2), (7, -4)$

53. Find the equation of a circle with center $(2, 2)$ whose graph passes through the point $(3, -5)$.

54. Find the equation of a circle with center $(-5, 4)$ whose graph passes through the point $(2, -3)$.

APPLICATIONS

55. Sports. A singles court for lawn tennis is a rectangle 27 feet wide and 78 feet long (see the figure). Points B and F are the midpoints of the end lines of the court.

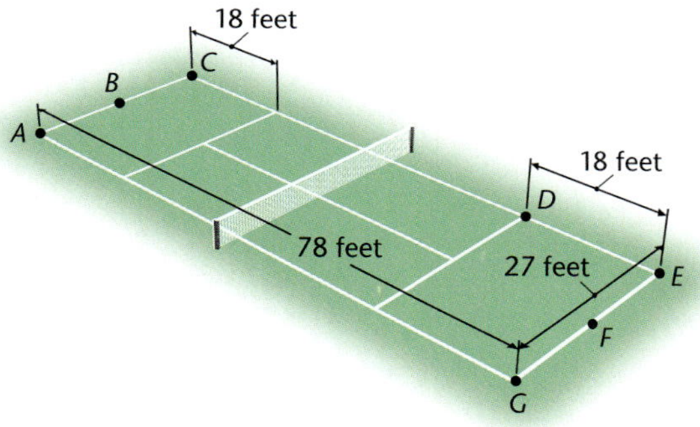

(A) Sketch a graph of the court with A at the origin of your coordinate system, C on the positive y axis, and G on the positive x axis. Find the coordinates of points A through G.

(B) Find $d(B, D)$ and $d(F, C)$ to the nearest foot.

56. Sports. Refer to Exercise 55. Find $d(A, D)$ and $d(C, G)$ to the nearest foot.

57. Architecture. An arched doorway is formed by placing a circular arc on top of a rectangle (see the figure). If the doorway is 4 feet wide and the height of the arc above its ends is 1 foot, what is the radius of the circle containing the arc? [*Hint:* Note that $(2, r - 1)$ must satisfy $x^2 + y^2 = r^2$.]

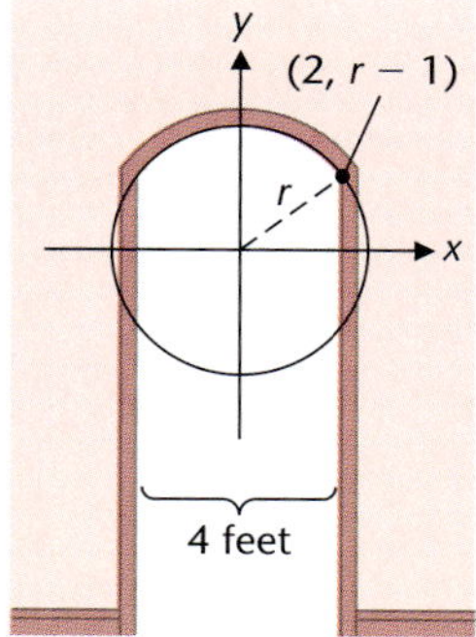

Arched doorway

58. Engineering. The cross-section of a rivet has a top that is an arc of a circle (see the figure). If the ends of the arc are 12 millimeters apart and the top is 4 millimeters above the ends, what is the radius of the circle containing the arc?

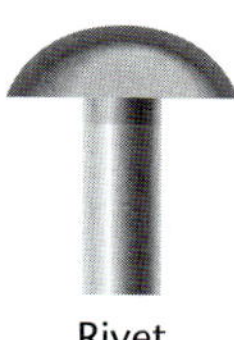

Rivet

★ **59. Construction.** Town B is located 36 miles east and 15 miles north of town A (see the figure). A local telephone company wants to position a relay tower so that the distance from the tower to town B is twice the distance from the tower to town A.

(A) Show that the tower must lie on a circle, find the center and radius of this circle, and graph.

(B) If the company decides to position the tower on this circle at a point directly east of town A, how far from town A should they place the tower? Compute answer to one decimal place.

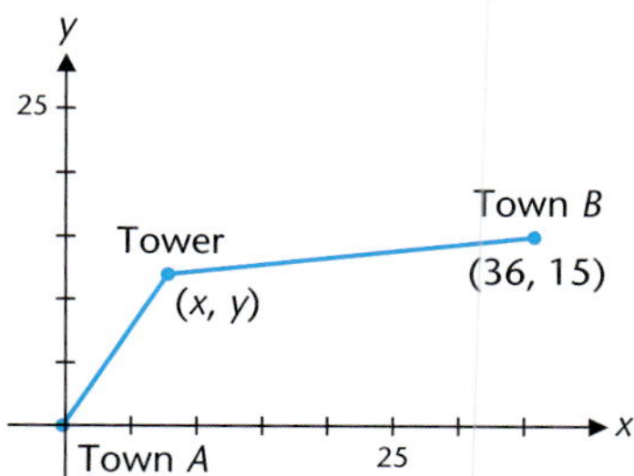

★ **60. Construction.** Repeat Problem 59 if the distance from the tower to town A is twice the distance from the tower to town B.

APPENDIX B

Special Topics

OUTLINE

SECTION B.1 Significant Digits

Significant Digits • Rounding Convention

Significant Digits

Most calculations involving problems of the real world deal with numbers that are only approximate. It therefore seems reasonable to assume that a final answer should not be any more accurate than the least accurate number used in the calculation. This is an important point, because calculators tend to give the impression that greater accuracy is achieved than is warranted.

Suppose we wish to compute the length of the diagonal of a rectangular field from measurements of its sides of 237.8 meters and 61.3 meters. Using the Pythagorean theorem and a calculator, we find

$$\begin{aligned} d &= \sqrt{237.8^2 + 61.3^2} \\ &= 245.573\ 878\ldots \end{aligned}$$

d

61.3 meters

237.8 meters

The calculator answer suggests an accuracy that is not justified. What accuracy is justified? To answer this question, we introduce the idea of *significant digits.*

Whenever we write a measurement such as 61.3 meters, we assume that the measurement is accurate to the last digit written. Thus, the measurement 61.3 meters indicates that the measurement was made to the nearest tenth of a meter. That is, the actual width is between 61.25 meters and 61.35 meters. In general, the digits in a number that indicate the accuracy of the number are called **significant digits.** If all the digits in a number are nonzero, then they are all significant. Thus, the measurement 61.3 meters has three significant digits, and the measurement 237.8 meters has four significant digits.

What are the significant digits in the number 7,800? The accuracy of this number is not clear. It could represent a measurement with any of the following accuracies:

Between 7,750 and 7,850	Correct to the hundreds place
Between 7,795 and 7,805	Correct to the tens place
Between 7,799.5 and 7,800.5	Correct to the units place

To give a precise definition of significant digits that resolves this ambiguity, we use scientific notation.

DEFINITION 1
Significant Digits

If a number x is written in scientific notation as

$$x = a \times 10^n \qquad 1 \le a < 10,\ n \text{ an integer}$$

then the number of significant digits in x is the number of digits in a.

Thus,

7.8×10^3	has two significant digits
7.80×10^3	has three significant digits
7.800×10^3	has four significant digits

All three of these measurements have the same decimal representation (7,800), but each represents a different accuracy.

Definition 1 tells us how to write a number so that the number of significant digits is clear, but it does not tell us how to interpret the accuracy of a number that is not written in scientific notation. We will use the following convention for numbers that are written as decimal fractions:

Significant Digits in Decimal Fractions

The number of significant digits in a number with no decimal point is found by counting the digits from left to right, starting with the first digit and ending with the last *nonzero* digit.

The number of significant digits in a number containing a decimal point is found by counting the digits from left to right, starting with the first *nonzero* digit and ending with the last digit.

Applying this rule to the number 7,800, we conclude that this number has two significant digits. If we want to indicate that it has three or four significant digits, we must use scientific notation.

EXAMPLE 1 Significant Digits in Decimal Fractions

Underline the significant digits in the following numbers:

(A) 70,007 (B) 82,000 (C) 5.600 (D) 0.0008 (E) 0.000 830

SOLUTIONS

(A) $\underline{70{,}007}$ (B) $\underline{82}{,}000$ (C) $\underline{5.600}$ (D) $0.000\underline{8}$ (E) $0.000\ \underline{830}$

MATCHED PROBLEM 1

Underline the significant digits in the following numbers:

(A) 5,009 (B) 12,300 (C) 23.4000 (D) 0.00050 (E) 0.0012

Rounding Convention

In calculations involving multiplication, division, powers, and roots, we adopt the following convention:

Rounding Calculated Values

The result of a calculation is rounded to the same number of significant digits as the number used in the calculation that has the least number of significant digits.

Thus, in computing the length of the diagonal of the rectangular field shown earlier, we write the answer rounded to three significant digits because the width has three significant digits and the length has four significant digits:

$d = 246$ meters   Three significant digits

One Final Note: In rounding a number that is exactly halfway between a larger and a smaller number, we use the convention of making the final result even.

EXAMPLE 2 Rounding Numbers

Round each number to three significant digits.

(A) 43.0690 (B) 48.05 (C) 48.15 (D) $8.017\ 632 \times 10^{-3}$

SOLUTIONS

(A) 43.1

(B) 48.0
(C) 48.2
} Use the convention of making the digit before the 5 even if it is odd, or leaving it alone if it is even.

(D) 8.02×10^{-3}

MATCHED 2 PROBLEM

Round each number to three significant digits.

(A) 3.1495 (B) 0.004 135 (C) 32,450 (D) $4.314\ 764\ 09 \times 10^{12}$

ANSWERS MATCHED PROBLEMS

1. (A) 5,009 (B) 12,300 (C) 23.4000
(D) 0.00050 (E) 0.0012

2. (A) 3.15 (B) 0.004 14 (C) 32,400
(D) 4.31×10^{12}

EXERCISE B.1

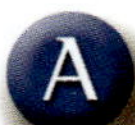

In Problems 1–12, underline the significant digits in each number.

1. 123,005

2. 3,400,002

3. 20,040

4. 300,600

5. 6.0

6. 7.00

7. 80.000

8. 900.0000

9. 0.012

10. 0.0015

11. 0.000 960

12. 0.000 700

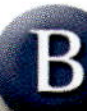

In Problems 13–22, round each number to three significant digits.

13. 3.0780

14. 4.0240

15. 924,300

16. 643,820

17. 23.65

18. 23.75

19. $2.816\,743 \times 10^3$

20. 56.114×10^4

21. $6.782\,045 \times 10^{-4}$

22. $5.248\,102 \times 10^{-3}$

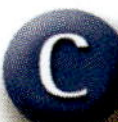

In Problems 23 and 24, find the diagonal of the rectangle with the indicated side measurements. Round answers to the number of significant digits appropriate for the given measurements.

23. 25 feet by 20 feet

24. 2,900 yards by 1,570 yards

SECTION B.2 Partial Fractions

Basic Theorems • Partial Fraction Decomposition

You have now had considerable experience combining two or more rational expressions into a single rational expression. For example, problems such as

$$\frac{2}{x+5} + \frac{3}{x-4} = \frac{2(x-4) + 3(x+5)}{(x+5)(x-4)} = \frac{5x+7}{(x+5)(x-4)}$$

should seem routine. Frequently in more advanced courses, particularly in calculus, it is advantageous to be able to reverse this process—that is, to be able to express a rational expression as the sum of two or more simpler rational expressions called **partial fractions.** As is often the case with reverse processes, the process of decomposing a rational expression into partial fractions is more difficult than combining rational expressions. Basic to the process is the factoring of polynomials, so many of the topics discussed in Chapter 3 can be put to effective use. Partial fraction decomposition is usually accomplished by solving a related system of linear equations. If you are familiar with basic techniques for solving linear systems discussed earlier in this book, such as Gauss–Jordan elimination, inverse matrix solutions, or Cramer's rule, you may use these as you see fit. However, all of the linear systems encountered in Section B.2 can also be solved by some special techniques developed here. Mathematically equivalent to the techniques mentioned, these special techniques are generally easier to use in partial fraction decomposition problems.

We confine our attention to rational expressions of the form $P(x)/D(x)$, where $P(x)$ and $D(x)$ are polynomials with real coefficients. In addition, we assume that the degree of $P(x)$ is less than the degree of $D(x)$. If the degree of $P(x)$ is greater than or equal to that of $D(x)$, we have only to divide $P(x)$ by $D(x)$ to obtain

$$\frac{P(x)}{D(x)} = Q(x) + \frac{R(x)}{D(x)}$$

where the degree of $R(x)$ is less than that of $D(x)$. For example,

$$\frac{x^4 - 3x^3 + 2x^2 - 5x + 1}{x^2 - 2x + 1} = x^2 - x - 1 + \frac{-6x + 2}{x^2 - 2x + 1}$$

If the degree of $P(x)$ is less than that of $D(x)$, then $P(x)/D(x)$ is called a **proper fraction.**

Basic Theorems

Our task now is to establish a systematic way to decompose a proper fraction into the sum of two or more partial fractions. Theorems 1, 2, and 3 take care of the problem completely.

THEOREM 1
Equal Polynomials

Two polynomials are equal to each other if and only if the coefficients of terms of like degree are equal.

For example, if

Equate the constant terms.

$$(A + 2B)x + B = 5x - 3$$

Equate the coefficients of x.

then

$$\begin{aligned} B &= -3 \\ A + 2B &= 5 \\ A + 2(-3) &= 5 \\ A &= 11 \end{aligned}$$

Substitute $B = -3$ into the second equation to solve for A.

EXPLORE/DISCUSS 1

If

$$x + 5 = A(x + 1) + B(x - 3) \qquad (1)$$

is a polynomial identity (that is, both sides represent the same polynomial), then equating coefficients produces the system

$$1 = A + B \qquad \text{Equating coefficients of } x$$
$$5 = A - 3B \qquad \text{Equating constant terms}$$

(A) Solve this system graphically.

(B) For an alternate method of solution, substitute $x = 3$ in equation (1) to find A and then substitute $x = -1$ in equation (1) to find B. Explain why this method is valid.

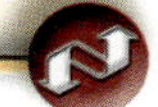

The Linear and Quadratic Factors Theorem from Chapter 3 (page 277) underlies the technique of decomposing a rational function into partial fractions. We restate the theorem here.

THEOREM 2
Linear and Quadratic Factors Theorem

For a polynomial of degree $n > 0$ with real coefficients, there always exists a factorization involving only linear and/or quadratic factors with real coefficients in which the quadratic factors have imaginary zeros.

The quadratic formula can be used to determine easily whether a given quadratic factor $ax^2 + bx + c$, with real coefficients, has imaginary zeros. If $b^2 - 4ac < 0$, then $ax^2 + bx + c$ has imaginary zeros. Otherwise its zeros are real. Therefore, $ax^2 + bx + c$ has imaginary zeros if and only if it cannot be factored as a product of linear factors with real coefficients.

Partial Fraction Decomposition

We are now ready to state Theorem 3, which forms the basis for partial fraction decomposition.

THEOREM 3
Partial Fraction Decomposition

Any proper fraction $P(x)/D(x)$ reduced to lowest terms can be decomposed into the sum of partial fractions as follows:

1. If $D(x)$ has a nonrepeating linear factor of the form $ax + b$, then the partial fraction decomposition of $P(x)/D(x)$ contains a term of the form

$$\frac{A}{ax+b} \qquad A \text{ a constant}$$

2. If $D(x)$ has a k-repeating linear factor of the form $(ax + b)^k$, then the partial fraction decomposition of $P(x)/D(x)$ contains k terms of the form

$$\frac{A_1}{ax+b} + \frac{A_2}{(ax+b)^2} + \cdots + \frac{A_k}{(ax+b)^k} \qquad A_1, A_2, \ldots, A_k \text{ constants}$$

3. If $D(x)$ has a nonrepeating quadratic factor of the form $ax^2 + bx + c$ that has imaginary zeros, then the partial fraction decomposition of $P(x)/D(x)$ contains a term of the form

$$\frac{Ax+B}{ax^2+bx+c} \qquad A, B \text{ constants}$$

4. If $D(x)$ has a k-repeating quadratic factor of the form $(ax^2 + bx + c)^k$, where $ax^2 + bx + c$ has imaginary zeros, then the partial fraction decomposition of $P(x)/D(x)$ contains k terms of the form

$$\frac{A_1x+B_1}{ax^2+bx+c} + \frac{A_2x+B_2}{(ax^2+bx+c)^2} + \cdots + \frac{A_kx+B_k}{(ax^2+bx+c)^k}$$

$$A_1, \ldots, A_k, B_1, \ldots, B_k \text{ constants}$$

Let's see how the theorem is used to obtain partial fraction decompositions in several examples.

EXAMPLE 1 Nonrepeating Linear Factors

Decompose into partial fractions $\dfrac{5x+7}{x^2+2x-3}$

SOLUTION

We first try to factor the denominator. If it can't be factored in the real numbers, then we can't go any further. In this example, the denominator factors, so we apply part 1 from Theorem 3:

$$\frac{5x+7}{(x-1)(x+3)} = \frac{A}{x-1} + \frac{B}{x+3} \tag{2}$$

To find the constants A and B, we combine the fractions on the right side of equation (2) to obtain

$$\frac{5x + 7}{(x - 1)(x + 3)} = \frac{A(x + 3) + B(x - 1)}{(x - 1)(x + 3)}$$

Because these fractions have the same denominator, their numerators must be equal. Thus

$$5x + 7 = A(x + 3) + B(x - 1) \tag{3}$$

We could multiply the right side and find A and B by using Theorem 1, but in this case it is easier to take advantage of the fact that equation (3) is an identity—that is, it must hold for all values of x. In particular, we note that if we let $x = 1$, then the second term of the right side drops out and we can solve for A:

$$\begin{aligned} 5 \cdot 1 + 7 &= A(1 + 3) + B(1 - 1) \\ 12 &= 4A \\ A &= 3 \end{aligned}$$

Similarly, if we let $x = -3$, the first term drops out and we find

$$\begin{aligned} -8 &= -4B \\ B &= 2 \end{aligned}$$

Hence,

$$\frac{5x + 7}{x^2 + 2x - 3} = \frac{3}{x - 1} + \frac{2}{x + 3} \tag{4}$$

as can easily be checked by adding the two fractions on the right.

MATCHED 1 PROBLEM

Decompose into partial fractions $\dfrac{7x + 6}{x^2 + x - 6}$

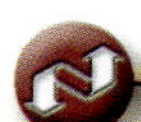

EXPLORE/DISCUSS 2

A graphing utility can also be used to check a partial fraction decomposition. To check Example 1, we graph the left and right sides of equation (4) in a graphing utility (Fig. 1). Discuss how the TRACE command on the graphing utility can be used to check that the graphing utility is displaying two identical graphs.

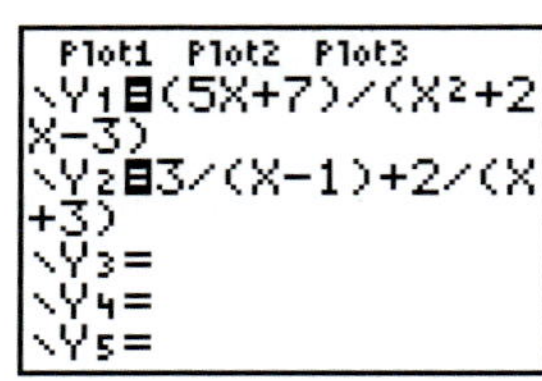

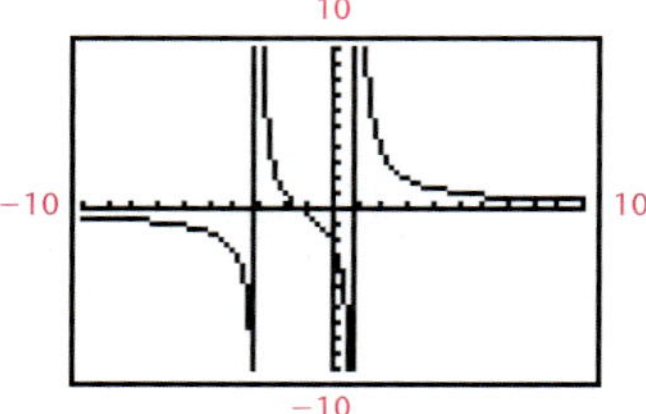

FIGURE 1

EXAMPLE 2 Repeating Linear Factors

Decompose into partial fractions $\dfrac{6x^2 - 14x - 27}{(x + 2)(x - 3)^2}$

SOLUTION

Using parts 1 and 2 from Theorem 3, we write

$$\frac{6x^2 - 14x - 27}{(x + 2)(x - 3)^2} = \frac{A}{x + 2} + \frac{B}{x - 3} + \frac{C}{(x - 3)^2}$$

$$= \frac{A(x - 3)^2 + B(x + 2)(x - 3) + C(x + 2)}{(x + 2)(x - 3)^2}$$

Thus, for all x,

$$6x^2 - 14x - 27 = A(x - 3)^2 + B(x + 2)(x - 3) + C(x + 2)$$

If $x = 3$, then $\qquad$ If $x = -2$, then

$$-15 = 5C \qquad 25 = 25A$$
$$C = -3 \qquad A = 1$$

There are no other values of x that will cause terms on the right to drop out. Because any value of x can be substituted to produce an equation relating A, B, and C, we let $x = 0$ and obtain

$$-27 = 9A - 6B + 2C \qquad \text{Substitute } A = 1 \text{ and } C = -3.$$
$$-27 = 9 - 6B - 6$$
$$B = 5$$

Thus,

$$\frac{6x^2 - 14x - 27}{(x + 2)(x - 3)^2} = \frac{1}{x + 2} + \frac{5}{x - 3} - \frac{3}{(x - 3)^2}$$

MATCHED 2 PROBLEM

Decompose into partial fractions $\dfrac{x^2 + 11x + 15}{(x - 1)(x + 2)^2}$

EXAMPLE 3 Nonrepeating Linear and Quadratic Factors

Decompose into partial fractions $\dfrac{5x^2 - 8x + 5}{(x - 2)(x^2 - x + 1)}$

SOLUTION

First, we see that the quadratic in the denominator can't be factored further in the real numbers. Then, we use parts 1 and 3 from Theorem 3 to write

$$\frac{5x^2 - 8x + 5}{(x - 2)(x^2 - x + 1)} = \frac{A}{x - 2} + \frac{Bx + C}{x^2 - x + 1}$$

$$= \frac{A(x^2 - x + 1) + (Bx + C)(x - 2)}{(x - 2)(x^2 - x + 1)}$$

Thus, for all x,

$$5x^2 - 8x + 5 = A(x^2 - x + 1) + (Bx + C)(x - 2)$$

If $x = 2$, then

$$9 = 3A$$
$$A = 3$$

If $x = 0$, then, using $A = 3$, we have

$$5 = 3 - 2C$$
$$C = -1$$

If $x = 1$, then, using $A = 3$ and $C = -1$, we have

$$2 = 3 + (B - 1)(-1)$$
$$B = 2$$

Hence,

$$\frac{5x^2 - 8x + 5}{(x - 2)(x^2 - x + 1)} = \frac{3}{x - 2} + \frac{2x - 1}{x^2 - x + 1}$$

MATCHED 3 PROBLEM

Decompose into partial fractions $\dfrac{7x^2 - 11x + 6}{(x - 1)(2x^2 - 3x + 2)}$

EXAMPLE 4 Repeating Quadratic Factors

Decompose into partial fractions $\dfrac{x^3 - 4x^2 + 9x - 5}{(x^2 - 2x + 3)^2}$

SOLUTION

Because $x^2 - 2x + 3$ can't be factored further in the real numbers, we proceed to use part 4 from Theorem 3 to write

$$\frac{x^3 - 4x^2 + 9x - 5}{(x^2 - 2x + 3)^2} = \frac{Ax + B}{x^2 - 2x + 3} + \frac{Cx + D}{(x^2 - 2x + 3)^2}$$

$$= \frac{(Ax + B)(x^2 - 2x + 3) + Cx + D}{(x^2 - 2x + 3)^2}$$

Thus, for all x,

$$x^3 - 4x^2 + 9x - 5 = (Ax + B)(x^2 - 2x + 3) + Cx + D$$

Because the substitution of carefully chosen values of x doesn't lead to the immediate determination of A, B, C, or D, we multiply and rearrange the right side to obtain

$$x^3 - 4x^2 + 9x - 5 = Ax^3 + (B - 2A)x^2 + (3A - 2B + C)x + (3B + D)$$

Now we use Theorem 1 to equate coefficients of terms of like degree:

$$\begin{aligned} A &= 1 \\ B - 2A &= -4 \\ 3A - 2B + C &= 9 \\ 3B + D &= -5 \end{aligned}$$

$1x^3 \quad -4x^2 \quad +9x \quad -5$

$Ax^3 + (B - 2A)x^2 + (3A - 2B + C)x + (3B + D)$

From these equations we easily find that $A = 1$, $B = -2$, $C = 2$, and $D = 1$. Now we can write

$$\frac{x^3 - 4x^2 + 9x - 5}{(x^2 - 2x + 3)^2} = \frac{x - 2}{x^2 - 2x + 3} + \frac{2x + 1}{(x^2 - 2x + 3)^2}$$

MATCHED PROBLEM 4

Decompose into partial fractions $\dfrac{3x^3 - 6x^2 + 7x - 2}{(x^2 - 2x + 2)^2}$

ANSWERS MATCHED PROBLEMS

1. $\dfrac{4}{x - 2} + \dfrac{3}{x + 3}$ **2.** $\dfrac{3}{x - 1} - \dfrac{2}{x + 2} + \dfrac{1}{(x + 2)^2}$ **3.** $\dfrac{2}{x - 1} + \dfrac{3x - 2}{2x^2 - 3x + 2}$ **4.** $\dfrac{3x}{x^2 - 2x + 2} + \dfrac{x - 2}{(x^2 - 2x + 2)^2}$

EXERCISE B.2

A

In Problems 1–4, find A and B so that the right side is equal to the left. After cross-multiplying to produce a polynomial equation, solve each problem two ways (see Explore/Discuss 1). First, equate the coefficients of both sides to determine a linear system for A and B and solve this system graphically. Second, solve for A and B by evaluating both sides for selected values of x.

1. $\dfrac{7x - 14}{(x - 4)(x + 3)} = \dfrac{A}{x - 4} + \dfrac{B}{x + 3}$

2. $\dfrac{9x + 21}{(x + 5)(x - 3)} = \dfrac{A}{x + 5} + \dfrac{B}{x - 3}$

3. $\dfrac{17x - 1}{(2x - 3)(3x - 1)} = \dfrac{A}{2x - 3} + \dfrac{B}{3x - 1}$

4. $\dfrac{x - 11}{(3x + 2)(2x - 1)} = \dfrac{A}{3x + 2} + \dfrac{B}{2x - 1}$

In Problems 5–10, find A, B, C, and D, so that the right side is equal to the left. Check each problem two ways. First, check by adding the fractions in your decomposition. Second, graph the original function and your equation in the same window of a graphing utility and use TRACE or TABLE to verify that the graphing utility is displaying two identical graphs.

5. $\dfrac{3x^2 + 7x + 1}{x(x + 1)^2} = \dfrac{A}{x} + \dfrac{B}{x + 1} + \dfrac{C}{(x + 1)^2}$

6. $\dfrac{x^2 - 6x + 11}{(x + 1)(x - 2)^2} = \dfrac{A}{x + 1} + \dfrac{B}{x - 2} + \dfrac{C}{(x - 2)^2}$

7. $\dfrac{3x^2 + x}{(x - 2)(x^2 + 3)} = \dfrac{A}{x - 2} + \dfrac{Bx + C}{x^2 + 3}$

8. $\dfrac{5x^2 - 9x + 19}{(x - 4)(x^2 + 5)} = \dfrac{A}{x - 4} + \dfrac{Bx + C}{x^2 + 5}$

9. $\dfrac{2x^2 + 4x - 1}{(x^2 + x + 1)^2} = \dfrac{Ax + B}{x^2 + x + 1} + \dfrac{Cx + D}{(x^2 + x + 1)^2}$

10. $\dfrac{3x^3 - 3x^2 + 10x - 4}{(x^2 - x + 3)^2} = \dfrac{Ax + B}{x^2 - x + 3} + \dfrac{Cx + D}{(x^2 - x + 3)^2}$

B

In Problems 11–22, decompose into partial fractions.

11. $\dfrac{-x + 22}{x^2 - 2x - 8}$

12. $\dfrac{-x - 21}{x^2 + 2x - 15}$

13. $\dfrac{3x - 13}{6x^2 - x - 12}$

14. $\dfrac{11x - 11}{6x^2 + 7x - 3}$

15. $\dfrac{x^2 - 12x + 18}{x^3 - 6x^2 + 9x}$

16. $\dfrac{5x^2 - 36x + 48}{x(x - 4)^2}$

17. $\dfrac{5x^2 + 3x + 6}{x^3 + 2x^2 + 3x}$

18. $\dfrac{6x^2 - 15x + 16}{x^3 - 3x^2 + 4x}$

19. $\dfrac{2x^3 + 7x + 5}{x^4 + 4x^2 + 4}$

20. $\dfrac{-5x^2 + 7x - 18}{x^4 + 6x^2 + 9}$

21. $\dfrac{x^3 - 7x^2 + 17x - 17}{x^2 - 5x + 6}$

22. $\dfrac{x^3 + x^2 - 13x + 11}{x^2 + 2x - 15}$

C

In Problems 23–30, decompose into partial fractions.

23. $\dfrac{4x^2 + 5x - 9}{x^3 - 6x - 9}$

24. $\dfrac{4x^2 - 8x + 1}{x^3 - x + 6}$

25. $\dfrac{x^2 + 16x + 18}{x^3 + 2x^2 - 15x - 36}$

26. $\dfrac{5x^2 - 18x + 1}{x^3 - x^2 - 8x + 12}$

27. $\dfrac{-x^2 + x - 7}{x^4 - 5x^3 + 9x^2 - 8x + 4}$

28. $\dfrac{-2x^3 + 12x^2 - 20x - 10}{x^4 - 7x^3 + 17x^2 - 21x + 18}$

29. $\dfrac{4x^5 + 12x^4 - x^3 + 7x^2 - 4x + 2}{4x^4 + 4x^3 - 5x^2 + 5x - 2}$

30. $\dfrac{6x^5 - 13x^4 + x^3 - 8x^2 + 2x}{6x^4 - 7x^3 + x^2 + x - 1}$

SECTION B.3 Descartes' Rule of Signs

Variation in Sign • Descartes' Rule of Signs

In Section B.3 we discuss *Descartes' rule of signs*, a theorem that gives information about the number of real zeros of a polynomial. When used in conjunction with the theorems discussed in Sections 3.2 and 3.3, Descartes' rule of signs can simplify the search for zeros.

Variation in Sign

When the terms of a polynomial with real coefficients are arranged in order of descending powers, we say that a **variation in sign** occurs if two successive terms have opposite signs. Missing terms (terms with 0 coefficients) are ignored. For a given polynomial, we are interested in the total number of variations in sign in both $P(x)$ and $P(-x)$.

EXAMPLE 1 Variations in Sign

If $P(x) = 3x^4 - 2x^3 + 3x - 5$, find the number of variations in sign in $P(x)$ and in $P(-x)$.

SOLUTION

The signs of the coefficients in

$$P(x) = 3x^4 - 2x^3 + 3x - 5$$

are

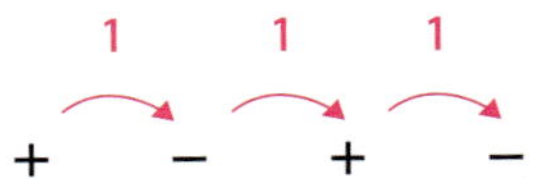

Thus, $P(x)$ has three variations in sign. The signs of the coefficients in

$$P(-x) = 3x^4 + 2x^3 - 3x - 5$$

are

$$\begin{matrix} & & \overset{1}{\curvearrowright} & \\ + & + & - & - \end{matrix}$$

$P(-x)$ has one variation in sign.

MATCHED 1 PROBLEM

If $P(x) = 2x^5 - x^4 - x^3 + x + 5$, find the number of variations in sign in $P(x)$ and in $P(-x)$.

Descartes' Rule of Signs

The number of variations in sign for $P(x)$ and $P(-x)$ gives us useful information about the number of real zeros of a polynomial with real coefficients. In 1636 René Descartes (1596–1650), a French philosopher and mathematician, gave the first proof of a simplified version of a theorem that now bears his name. We state Theorem 1 without proof, because a proof is beyond the scope of this book.

THEOREM 1
Descartes' Rule of Signs

Given a polynomial $P(x)$ with real coefficients and nonzero constant term

1. **Positive Zeros.** The number of positive zeros of $P(x)$ is never greater than the number of variations in sign in $P(x)$ and, if less, then always by an even number.
2. **Negative Zeros.** The number of negative zeros of $P(x)$ is never greater than the number of variations in sign in $P(-x)$ and, if less, then always by an even number.

It is important to understand that when we refer to positive zeros and negative zeros we are referring to real zeros. There are no positive or negative imaginary numbers.

EXAMPLE 2 Combinations of Zeros

Let $P(x) = x^3 + x - 2$. How many real zeros does $P(x)$ have? How many imaginary zeros does $P(x)$ have?

SOLUTION

We apply Descartes' rule of signs

$$P(x) = x^3 + x - 2 \quad \text{One variation in sign}$$
$$P(-x) = -x^3 - x - 2 \quad \text{No variations in sign}$$

According to Descartes' rule of signs, $P(x)$ must have at most one positive zero and the number of positive zeros must differ from one by an even number.

Because 0 does not differ from 1 by an even integer, 0 positive zeros is not a possibility, and $P(x)$ must have exactly one positive zero. Because $P(-x)$ has no variations in sign, $P(x)$ has no negative zeros. From the fundamental theorem of algebra (Section 3.3), $P(x)$ must have a total of three zeros. Thus, the two remaining zeros must be imaginary.

MATCHED 2 PROBLEM

Let $P(x) = x^4 + x^3 - 2$. How many real zeros does $P(x)$ have? How many imaginary zeros does $P(x)$ have?

EXPLORE/DISCUSS 1

Apply Descartes' rule of signs to each of the following polynomials to find the maximum possible number of real zeros. Then use a graphing utility to determine the exact number of real zeros and the exact number of imaginary zeros for each polynomial.

(A) $P(x) = x^4 - 4x^2 + 3$ (B) $Q(x) = x^4 - 4x^2 + 5$

Descartes' rule of signs enabled us to determine the exact number of real and imaginary zeros in Example 2. If there is more than zero or one variation in sign, then there can be more than one possible outcome. A table is a convenient way to summarize the various possibilities.

EXAMPLE 3 Possible Combinations of Zeros

Construct a table showing the possible combinations of positive, negative, and imaginary zeros of

(A) $P(x) = 3x^4 - 2x^3 + 3x - 5$

(B) $Q(x) = x^5 - 2x^4 + 5x^3 - 7x - 9$

SOLUTION

(A) $P(x)$ has three variations in sign and $P(-x)$ has one (see Example 1). Thus, $P(x)$ has either three positive zeros or one positive zero and exactly one negative zero. Because $P(x)$ must have a total of four zeros, any remaining zeros must be imaginary. The table summarizes the possible combinations of zeros, where $+$ = positive, $-$ = negative, I = imaginary. Note that the sum of each row is four, the degree of $P(x)$ and the total number of zeros.

+	−	*I*
3	1	0
1	1	2

(B) Apply Descartes' rule of signs,

$$Q(x) = x^5 - 2x^4 + 5x^3 - 7x - 9 \quad \text{Three variations in sign}$$
$$Q(-x) = -x^5 - 2x^4 - 5x^3 + 7x - 9 \quad \text{Two variations in sign}$$

Possible combinations of zeros are given in the table.

+	−	I
3	2	0
3	0	2
1	2	2
1	0	4

MATCHED 3 PROBLEM

Construct a table showing the possible combinations of positive, negative, and imaginary zeros of

(A) $P(x) = x^4 - 3x + 1$
(B) $Q(x) = 4x^5 + 2x^4 - x^3 + x + 5$

ANSWERS MATCHED PROBLEMS

1. $P(x)$ has two variations in sign and $P(-x)$ has three.
2. $P(x)$ has two real zeros and two imaginary zeros.
3. (A)

+	−	I
2	0	2
0	0	4

(B)

+	−	I
2	3	0
2	1	2
0	3	2
0	1	4

EXERCISE B.3

In Problems 1–6, find the number of variations in sign in $P(x)$ and $P(-x)$.

1. $P(x) = x^3 + 2x + 7$
2. $P(x) = x^3 + 4x - 5$
3. $P(x) = x^3 - 3x^2 - 9$
4. $P(x) = x^3 + 6x^2 + 8$
5. $P(x) = x^4 + 2x^3 + 3x - 5$
6. $P(x) = x^4 - 4x^3 - 2x - 3$

In Problems 7–12, use Descartes' rule of signs to find the number of real zeros and the number of imaginary zeros for each polynomial. Check your answer by graphing the polynomial on a graphing utility.

7. $P(x) = x^3 + 2x - 4$

8. $Q(x) = x^3 + 3x + 6$

9. $f(x) = x^3 + 2x^2 + 1$

10. $g(x) = x^3 - 5x^2 - 7$

11. $s(x) = x^4 - 2x^3 - 7x - 8$

12. $r(x) = x^4 + 3x^3 + 4x - 9$

B

In Problems 13–22, construct a table showing the possible combinations of positive, negative, and imaginary zeros of each polynomial.

13. $P(x) = x^3 - 3x^2 - 2x + 4$

14. $Q(x) = x^3 - 4x^2 + 3x + 5$

15. $t(x) = x^4 - 2x^3 - 4x^2 - 2x + 3$

16. $s(x) = x^4 + 2x^3 - 5x^2 + 3x - 6$

17. $f(x) = x^5 - x^4 + 3x^3 + 9x^2 - x + 5$

18. $g(x) = x^5 + x^4 - 4x^3 + 3x^2 - x - 1$

19. $P(x) = x^6 - 12$

20. $Q(x) = x^8 - 24$

21. $r(x) = x^7 + 32$

22. $w(x) = x^5 + 25$

In Problems 23–26, discuss the possible combinations of positive, negative, and imaginary zeros of $P(x) = x^2 + ax + b$ for the indicated values of a and b.

23. $a > 0, b > 0$

24. $a < 0, b > 0$

25. $a > 0, b < 0$

26. $a < 0, b < 0$

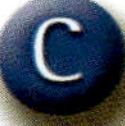

In Problems 27–32, construct a table showing the possible combinations of positive, negative, and imaginary zeros of each polynomial.

27. $f(x) = x^6 - 3x^5 + 4x^4 + 3x^3 - 2x + 5$

28. $g(x) = x^6 + 4x^5 - 2x^4 - 6x^3 - 5x^2 + 7$

29. $s(x) = x^7 + 3x^5 + 4x^2 - 3x + 5$

30. $w(x) = x^7 + 2x^5 - 3x^2 + 2x - 7$

31. $P(x) = x^8 - x - 1$

32. $Q(x) = x^9 + x - 1$

In Problems 33–36, discuss the possible combinations of positive, negative, and imaginary zeros of $P(x) = x^3 + ax + b$ for the indicated values of a and b.

33. $a > 0, b > 0$

34. $a > 0, b < 0$

35. $a < 0, b > 0$

36. $a < 0, b < 0$

SECTION B.4 Parametric Equations

Parametric Equations and Plane Curves • Parametric Equations and Conic Sections • Projectile Motion • Cycloid

Parametric Equations and Plane Curves

Consider the two equations

$$\begin{aligned} x &= t + 1 \\ y &= t^2 - 2t \end{aligned} \qquad -\infty < t < \infty \tag{1}$$

FIGURE 1 Graph of $x = t + 1$, $y = t^2 - 2t$, $-\infty < t < \infty$.

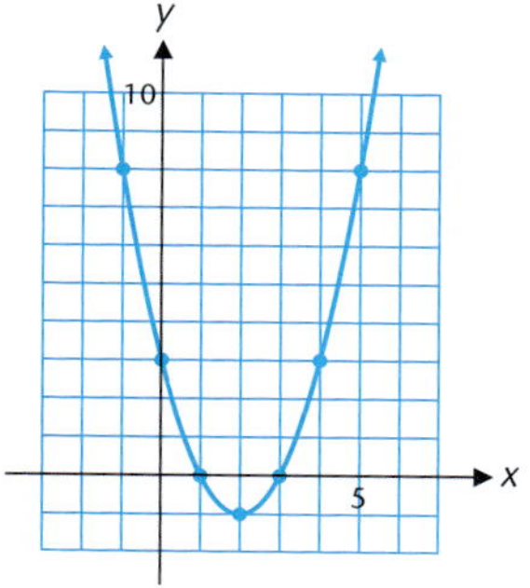

Each value of t determines a value of x, a value of y, and hence, an ordered pair (x, y). To graph the set of ordered pairs (x, y) determined by letting t assume all real values, we construct Table 1 listing selected values of t and the corresponding values of x and y. Then we plot the ordered pairs (x, y) and connect them with a continuous curve, as shown in Figure 1. The variable t is called a *parameter* and does not appear on the graph. Equations (1) are called *parametric equations* because both x and y are expressed in terms of the parameter t. The graph of the ordered pairs (x, y) is called a *plane curve*.

TABLE 1

t	**0**	**1**	**2**	**3**	**4**	**−1**	**−2**
x	1	2	3	4	5	0	−1
y	0	−1	0	3	8	3	8

Parametric equations can also be graphed on a graphing utility. Figure 2(a) shows the Parametric mode selected on a Texas Instruments TI-83 calculator. Figure 2(b) shows the equation editor with the parametric equations in (1) entered as x_{1T} and y_{1T}. In Figure 2(c), notice that there are three new window variables, Tmin, Tmax, and Tstep, that must be entered by the user.

FIGURE 2 Graphing parametric equations on a graphing utility.

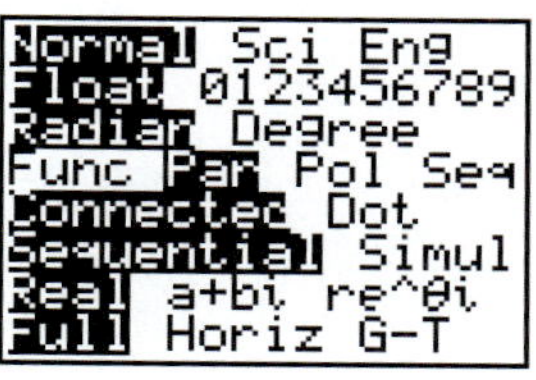

(a)

Plot1 Plot2 Plot3
\X1T=T+1
Y1T=T²−2T
\X2T=
Y2T=
\X3T=
Y3T=
\X4T=

(b)

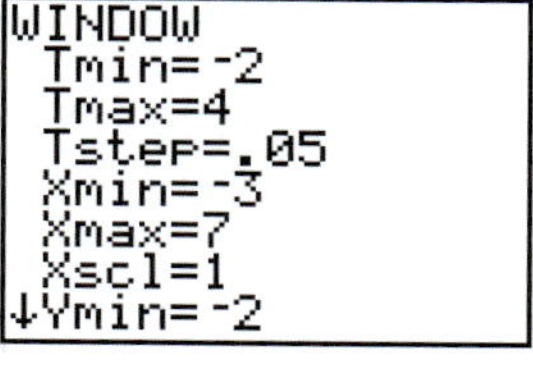

(c)

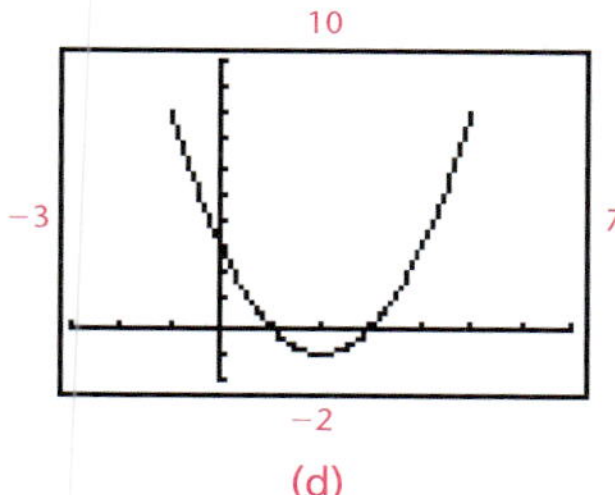

(d)

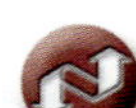

EXPLORE/DISCUSS 1

(A) Consult the manual for your graphing utility and reproduce Figure 2(a).

(B) Discuss the effect of using different values for Tmin and Tmax. Try Tmin $= -1$ and -3. Try Tmax $= 3$ and 5.

(C) Discuss the effect of using different values for Tstep. Try Tstep $= 1$, 0.1, and 0.01.

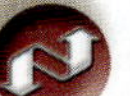

In some cases it is possible to eliminate the parameter by solving one of the equations for t and substituting into the other. In the example just considered, solving the first equation for t in terms of x, we have

$$t = x - 1$$

Then, substituting the result into the second equation, we obtain

$$\begin{aligned} y &= (x - 1)^2 - 2(x - 1) \\ &= x^2 - 4x + 3 \end{aligned}$$

We recognize this as the equation of a parabola, as we would guess from Figure 1.

In other cases, it may not be easy or possible to eliminate the parameter to obtain an equation in just x and y. For example, for

$$\begin{aligned} x &= t + \log t \\ y &= t - e^t \end{aligned} \qquad t > 0$$

you will not find it possible to solve either equation for t in terms of functions we have considered.

Is there more than one parametric representation for a plane curve? The answer is yes. In fact, there are an unlimited number of parametric representations for the same plane curve. The following are two additional representations of the parabola in Figure 1.

$$\begin{aligned} x &= t + 3 \\ y &= t^2 + 2t \end{aligned} \qquad -\infty < t < \infty \tag{2}$$

$$\begin{aligned} x &= t \\ y &= t^2 - 4t + 3 \end{aligned} \qquad -\infty < t < \infty \tag{3}$$

The concepts introduced in the preceding discussion are summarized in Definition 1.

DEFINITION 1

Parametric Equations and Plane Curves

A **plane curve** is the set of points (x, y) determined by the **parametric equations**

$$x = f(t)$$
$$y = g(t)$$

where the **parameter** t varies over an interval I and the functions f and g are both defined on the interval I.

Why are we interested in parametric representations of plane curves? It turns out that this approach is more general than using equations with two variables as we have been doing. In addition, the approach generalizes to curves in three- and higher-dimensional spaces. Other important reasons for using parametric representations of plane curves will be brought out in the discussion and examples that follow.

EXAMPLE 1 Eliminating the Parameter

Eliminate the parameter and identify the plane curve given parametrically by

$$x = \sqrt{t} \qquad 0 \le t \le 9 \qquad (4)$$
$$y = \sqrt{9 - t}$$

SOLUTION

To eliminate the parameter t, we solve each equation (4) for t:

$$x = \sqrt{t} \qquad\qquad y = \sqrt{9 - t}$$
$$x^2 = t \qquad\qquad y^2 = 9 - t$$
$$t = 9 - y^2$$

Equating the last two equations, we have

$$x^2 = 9 - y^2$$
$$x^2 + y^2 = 9 \qquad \text{A circle of radius 3 centered at (0, 0)}$$

Thus, the graph of the parametric equations in (4) is the quarter of the circle of radius 3 centered at the origin that lies in the first quadrant (Fig. 3).

FIGURE 3

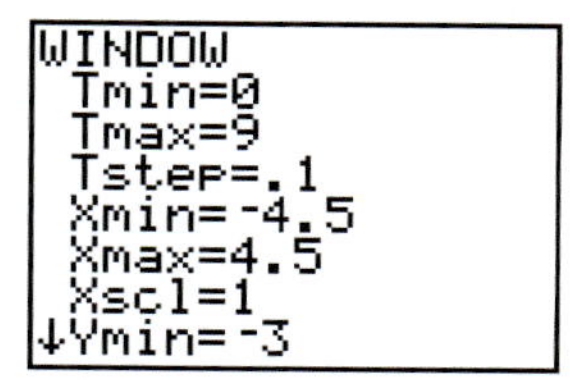

(a)

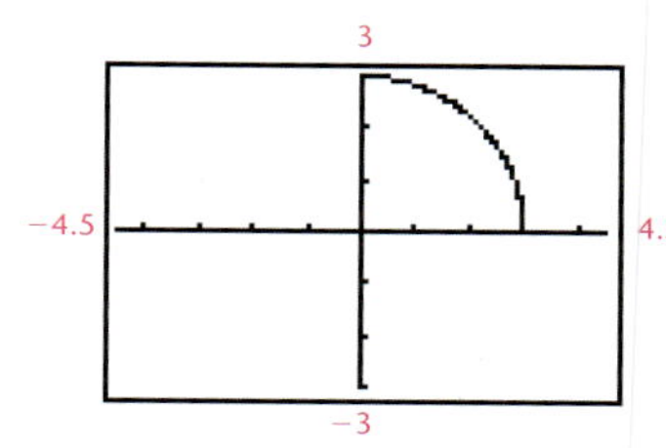

(b)

MATCHED 1 PROBLEM

Eliminate the parameter and identify the plane curve given parametrically by $x = \sqrt{4 - t}$, $y = -\sqrt{t}$, $0 \leq t \leq 4$.

Parametric Equations and Conic Sections

Trigonometric functions provide very effective representations for many conic sections. Examples 2 and 3 illustrate the basic concepts.

EXAMPLE 2 **Identifying a Conic Section in Parametric Form**

Eliminate the parameter θ and identify the plane curve given by

$$\begin{aligned} x &= 8 \cos \theta \\ y &= 4 \sin \theta \end{aligned} \qquad 0 \leq \theta \leq 2\pi \tag{5}$$

SOLUTION

To eliminate the parameter θ, we solve the first equation in (5) for $\cos\theta$, the second for $\sin\theta$, and substitute into the Pythagorean identity $\cos^2\theta + \sin^2\theta = 1$:

$$\cos\theta = \frac{x}{8} \qquad \text{and} \qquad \sin\theta = \frac{y}{4}$$

$$\cos^2\theta + \sin^2\theta = 1$$

$$\left(\frac{x}{8}\right)^2 + \left(\frac{y}{4}\right)^2 = 1$$

$$\frac{x^2}{64} + \frac{y^2}{16} = 1$$

The graph is an ellipse (Fig. 4).

FIGURE 4 Graph of $x = 8\cos\theta$, $y = 4\sin\theta$, $0 \leq \theta \leq 2\pi$.

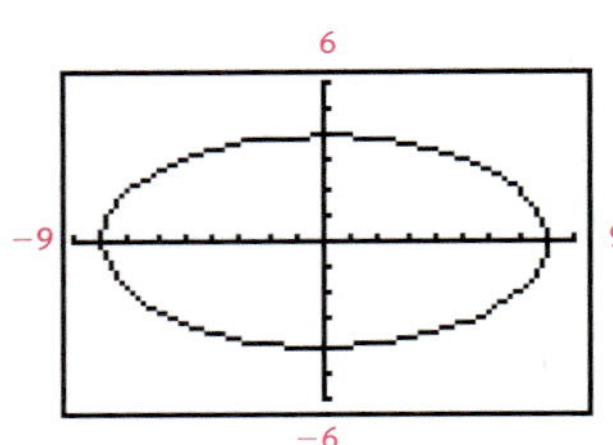

MATCHED 2 PROBLEM

Eliminate the parameter θ and identify the plane curve given by $x = 4\cos\theta$, $y = 4\sin\theta$, $0 \leq \theta \leq 2\pi$.

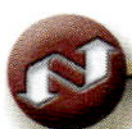

EXPLORE/DISCUSS 2

Graph one period ($0 \leq \theta \leq 2\pi$) of each of the three plane curves given parametrically by

$$\begin{aligned} x_1 &= 5 \cos \theta & x_2 &= 2 \cos \theta & x_3 &= 5 \cos \theta \\ y_1 &= 5 \sin \theta & y_2 &= 2 \sin \theta & y_3 &= 2 \sin \theta \end{aligned}$$

Identify the curves by eliminating the parameter. What happens if you graph less than one period? More than one period?

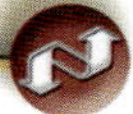

EXAMPLE 3 Parametric Equations for Conic Sections

Find parametric equations for the conic section with the given equation:

(A) $25x^2 + 9y^2 - 100x + 54y - 44 = 0$

(B) $x^2 - 16y^2 - 10x + 32y - 7 = 0$

SOLUTIONS

(A) By completing the square in x and y we obtain the standard form $\frac{(x-2)^2}{9} + \frac{(y+3)^2}{25} = 1$. So the graph is an ellipse with center $(2, -3)$ and major axis on the line $x = 2$. Because $\cos^2 \theta + \sin^2 \theta = 1$, a parametric representation with parameter θ is obtained by letting $\frac{x-2}{3} = \cos \theta$, $\frac{y+3}{5} = \sin \theta$:

$$x = 2 + 3 \cos \theta$$
$$y = -3 + 5 \sin \theta$$

Because $\sin \theta$ and $\cos \theta$ have period 2π, graphing these equations for $0 \leq \theta \leq 2\pi$ will produce a complete graph of the ellipse (Fig. 5).

FIGURE 5 $x = 2 + 3 \cos \theta$, $y = -3 + 5 \sin \theta$, $0 \leq \theta \leq 2\pi$.

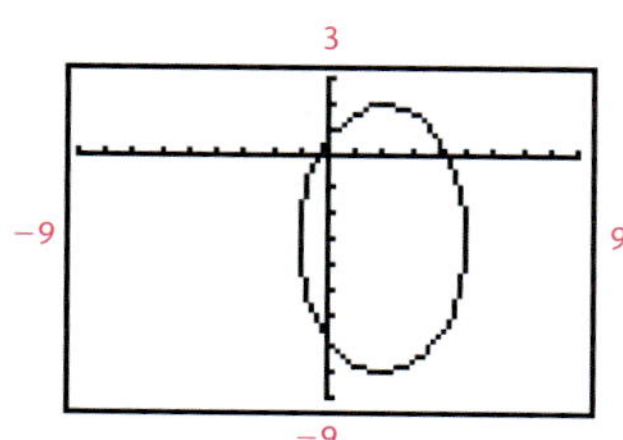

(B) By completing the square in x and y we obtain the standard form $\frac{(x-5)^2}{16} - (y-1)^2 = 1$. So the graph is a hyperbola with center $(5, 1)$ and transverse axis on the line $y = 1$. Because $\sec^2 \theta - \tan^2 \theta = 1$, a parametric representation with parameter θ is obtained by letting $\frac{x-5}{4} = \sec \theta$, $y - 1 = \tan \theta$:

$$x = 5 + 4 \sec \theta$$
$$y = 1 + \tan \theta$$

FIGURE 6 $x = 5 + 4 \sec \theta$, $y = 1 + \tan \theta$, $0 \leq \theta \leq 2\pi$, $\theta \neq \frac{\pi}{2}, \frac{3\pi}{2}$.

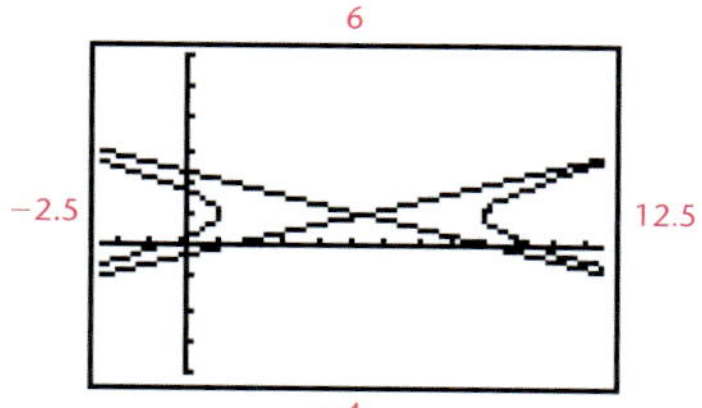

The period of $\tan \theta$ is π, but the period of $\sec \theta$ is 2π, so we have to use $0 \leq \theta \leq 2\pi$ to produce a complete graph of the hyperbola (Fig. 6). To be precise, we should exclude $\theta = \pi/2$ and $3\pi/2$, because the tangent function is not defined at these values. Including them does not affect the graph, because most graphing utilities ignore points where functions are undefined. Note that when the parametric equations are graphed in the connected mode, the graph appears to show the asymptotes of the hyperbola (see Fig. 6).

MATCHED 3 PROBLEM

Find parametric equations for the conic section with the given equation:

(A) $36x^2 + 16y^2 + 504x - 96y + 1{,}332 = 0$

(B) $16y^2 - 9x^2 - 36x + 128y + 76 = 0$

REMARK Refer to Example 3, part A. Any interval of the form $a \leq \theta \leq a + b$, where $b \geq 2\pi$, will produce a graph containing all the points on this ellipse. We will follow the practice of always choosing the shortest interval starting at 0 that will generate all the points on a conic section. For this ellipse, that interval is $[0, 2\pi]$.

Projectile Motion

Newton's laws and advanced mathematics can be used to determine the *path of a projectile.* If v_0 is the initial speed of the projectile at an angle α with the horizontal and a_0 is the initial altitude of the projectile (Fig. 7), then, neglecting air resistance, the **path of the projectile** is given by

$$\begin{aligned} x &= (v_0 \cos \alpha)t \\ y &= a_0 + (v_0 \sin \alpha)t - 4.9t^2 \end{aligned} \qquad 0 \leq t \leq b \tag{6}$$

The parameter t represents time in seconds, and x and y are distances measured in meters. Solving the first equation in equations (6) for t in terms of x, substituting into the second equation, and simplifying, produces the following equation:

$$y = a_0 + (\tan \alpha)x - \frac{4.9}{v_0^2 \cos^2 \alpha}x^2 \tag{7}$$

You should verify this by supplying the omitted details.

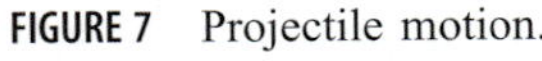

FIGURE 7 Projectile motion.

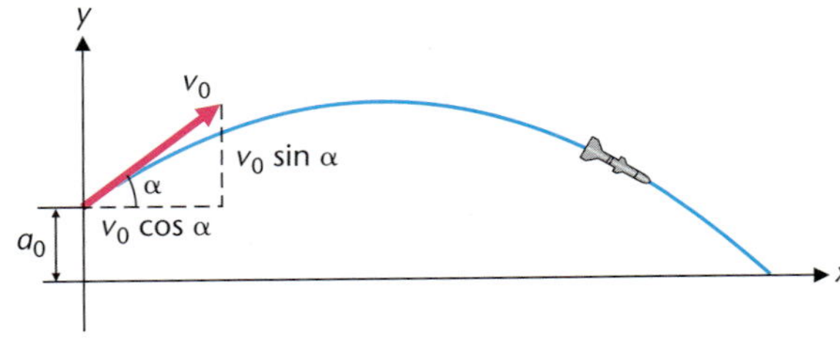

We recognize equation (7) as a parabola. This equation in x and y describes the path the projectile follows but tells us little else about its flight. On the other hand, the parametric equations (6) not only determine the path of the projectile but also tell us where it is at any time t. Furthermore, using concepts from physics and calculus, the parametric equations can be used to determine the velocity and acceleration of the projectile at any time t. This illustrates another advantage of using parametric representations of plane curves.

EXAMPLE 4 Projectile Motion

An automobile drives off a 50-meter cliff traveling at 25 meters per second (Fig. 8). When (to the nearest tenth of a second) will the automobile strike the ground? How far (to the nearest meter) from the base of the cliff is the point of impact?

FIGURE 8

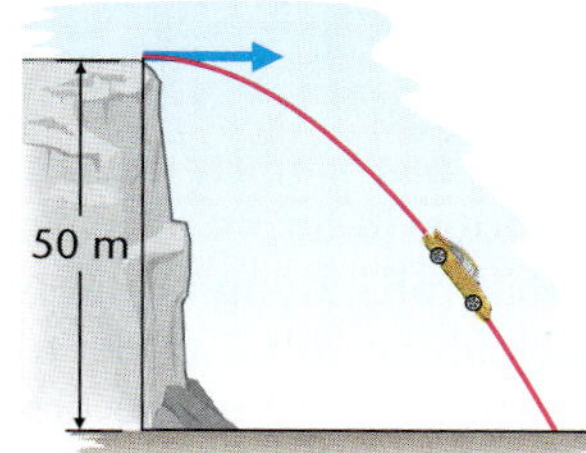

SOLUTION

At the instant the automobile leaves the cliff, the velocity is 25 meters per second, the angle with the horizontal is 0, and the altitude is 50 meters. Substituting these values in equations (6), the parametric equations for the path of the automobile are

$$x = 25t$$
$$y = 50 - 4.9t^2$$

The automobile strikes the ground when $y = 0$. Using the parametric equation for y, we have

$$y = 50 - 4.9t^2 = 0$$
$$-4.9t^2 = -50$$
$$t = \sqrt{\frac{-50}{-4.9}} \approx 3.2 \text{ seconds}$$

The distance from the base of the cliff is the same as the value of x. Substituting $t = 3.2$ in the first parametric equation, the distance from the base of the cliff at the point of impact is $x = 25(3.2) = 80$ meters.

MATCHED PROBLEM 4

A gardener is holding a hose in a horizontal position 1.5 meters above the ground. Water is leaving the hose at a speed of 5 meters per second. What is the distance (to the nearest tenth of a meter) from the gardener's feet to the point where the water hits the ground?

The **range of a projectile** at an altitude $a_0 = 0$ is the distance from the point of firing to the point of impact. If we keep the initial speed v_0 of the projectile constant and vary the angle α in Figure 7, we obtain different parabolic paths followed by the projectile and different ranges. The maximum range is obtained when $\alpha = 45°$. Furthermore, assuming that the projectile always stays in the same vertical plane, then there are points in the air and on the ground that the projectile cannot reach, irrespective of the angle α used, $0° \leq \alpha \leq 180°$. Using more advanced mathematics, it can be shown that the reachable region is separated from the nonreachable region by a parabola called an **envelope** of the other parabolas (Fig. 9).

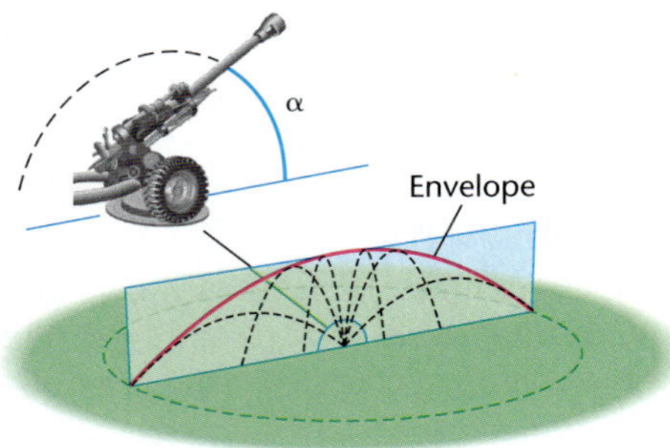

FIGURE 9 Reachable region of a projectile.

Cycloid

We now consider an unusual curve called a *cycloid,* which has a fairly simple parametric representation and a very complicated representation in terms of x and y only. The path traced by a point on the rim of a circle that rolls along a line is called a **cycloid.** To derive parametric equations for a cycloid we roll a circle of radius a along the x axis with the tracing point P on the rim starting at the origin (Fig. 10).

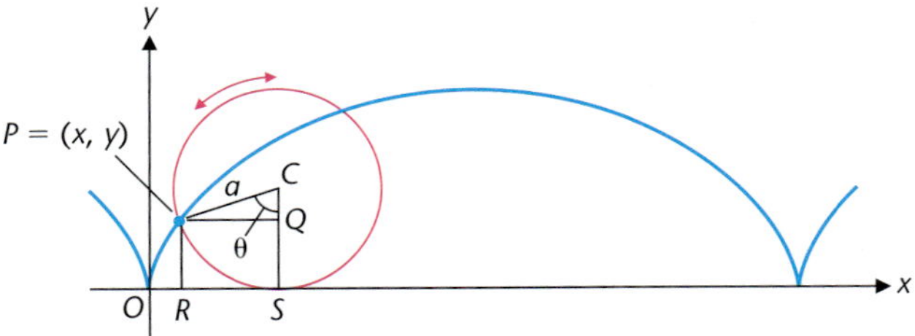

FIGURE 10 Cycloid.

Because the circle rolls along the x axis without slipping (see Fig. 10), we see that

$$\begin{aligned} d(O, S) &= \text{arc } PS \\ &= a\theta \qquad \theta \text{ in radians} \end{aligned} \tag{8}$$

where S is the point of contact between the circle and the x axis. Referring to triangle CPQ, we see that

$$d(P, Q) = a \sin \theta \qquad 0 \leq \theta \leq \pi/2 \tag{9}$$
$$d(Q, C) = a \cos \theta \qquad 0 \leq \theta \leq \pi/2 \tag{10}$$

Using these results, we have

$$\begin{aligned} x &= d(O, R) \\ &= d(O, S) - d(R, S) \\ &= (\text{arc } PS) - d(P, Q) \\ &= a\theta - a \sin \theta \end{aligned}$$

Use equations (8) and (9).

$$\begin{aligned} y &= d(R, P) \\ &= d(S, C) - d(Q, C) \\ &= a - a \cos \theta \end{aligned}$$

Use equation (10) and the fact that $d(S, C) = a$.

Although θ in equations (9) and (10) was restricted so that $0 \le \theta \le \pi/2$, it can be shown that the derived parametric equations generate the whole cycloid for $-\infty < \theta < \infty$. The graph specifies a periodic function with period $2\pi a$. Thus, in general, we have Theorem 1.

THEOREM 1

Parametric Equations for a Cycloid

For a circle of radius a rolled along the x axis, the resulting cycloid generated by a point on the rim starting at the origin is given by

$$x = a\theta - a \sin \theta \qquad -\infty < \theta < \infty$$
$$y = a - a \cos \theta$$

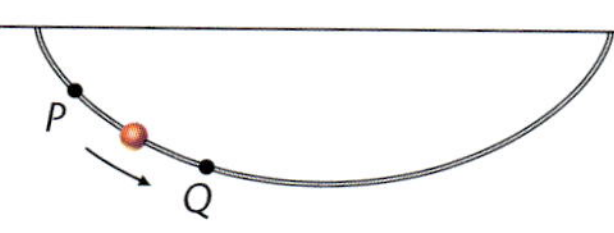

FIGURE 11 Cycloid path.

The cycloid is a good example of a curve that is very difficult to represent without the use of a parameter. A cycloid has a very interesting physical property. An object sliding without friction from a point P to a point Q lower than P, but not on the same vertical line as P, will arrive at Q in a shorter time traveling along a cycloid than on any other path (Fig. 11).

EXPLORE/DISCUSS 3

(A) Let Q be a point b units from the center of a wheel of radius a, where $0 < b < a$. If the wheel rolls along the x axis with the tracing point Q starting at $(0, a - b)$, explain why parametric equations for the path of Q are given by

$$x = a\theta - b \sin \theta$$
$$y = a - b \cos \theta$$

(B) Use a graphing utility to graph the paths of a point on the rim of a wheel of radius 1, and a point halfway between the rim and center, as the wheel makes two complete revolutions rolling along the x axis.

ANSWERS MATCHED PROBLEMS

1. The quarter of the circle of radius 2 centered at the origin that lies in the fourth quadrant.
2. $x^2 + y^2 = 16$, circle of radius 4 centered at $(0, 0)$
3. (A) Ellipse: $x = -7 + 4 \cos \theta$, $y = 3 + 6 \sin \theta$, $0 \le \theta \le 2\pi$
   (B) Hyperbola: $x = -2 + 4 \tan \theta$, $y = -4 + 3 \sec \theta$, $0 \le \theta \le 2\pi$, $\theta \ne \frac{\pi}{2}, \frac{3\pi}{2}$
4. 2.8 meters

EXERCISE B.4

A

1. If $x = t^2$ and $y = t^2 - 2$, then $y = x - 2$. Discuss the differences between the graph of the parametric equations and the graph of the line $y = x - 2$.

2. If $x = t^2$ and $y = t^4 - 2$, then $y = x^2 - 2$. Discuss the differences between the graph of the parametric equations and the graph of the parabola $y = x^2 - 2$.

In Problems 3 and 4, configure your graphing utility to produce Figure 2, as in Explore/Discuss 1.

3. Use TRACE to reproduce Table 1 on page A-53.
4. Use TABLE to reproduce Table 1 on page A-53.
5. Under the ZOOM menu, does ZDecimal affect the values of t or the values of x and y?
6. Under the ZOOM menu, does ZStandard affect the values of t or the values of x and y?
7. Under the ZOOM menu, does Zoom In affect the values of t or the values of x and y?
8. Under the ZOOM menu, does ZSquare affect the values of t or the values of x and y?

In Problems 9–18, the interval for the parameter is the whole real line. For each pair of parametric equations, eliminate the parameter t and find an equation for the curve in terms of x and y. Identify and graph the curve.

9. $x = -t$, $y = 2t - 2$
10. $x = t$, $y = t + 1$
11. $x = -t^2$, $y = 2t^2 - 2$
12. $x = t^2$, $y = t^2 + 1$
13. $x = 3t$, $y = -2t$
14. $x = 2t$, $y = t$
15. $x = \frac{1}{4}t^2$, $y = t$
16. $x = 2t$, $y = t^2$
17. $x = \frac{1}{4}t^4$, $y = t^2$
18. $x = 2t^2$, $y = t^4$

B

In Problems 19–30, obtain an equation in x and y by eliminating the parameter. Identify the curve.

19. $x = t - 2$, $y = 4 - 2t$
20. $x = t - 1$, $y = 2t + 2$
21. $x = t - 1$, $y = \sqrt{t}$, $t \ge 0$
22. $x = \sqrt{t}$, $y = t + 1$, $t \ge 0$
23. $x = \sqrt{t}$, $y = 2\sqrt{16 - t}$, $0 \le t \le 16$
24. $x = -3\sqrt{t}$, $y = \sqrt{25 - t}$, $0 \le t \le 25$
25. $x = -\sqrt{t + 1}$, $y = -\sqrt{t - 1}$, $t \ge 1$
26. $x = \sqrt{2 - t}$, $y = -\sqrt{4 - t}$, $t \le 2$
27. $x = 3 \sin \theta$, $y = 4 \cos \theta$, $0 \le \theta \le 2\pi$
28. $x = 3 \sin \theta$, $y = 3 \cos \theta$, $0 \le \theta \le 2\pi$
29. $x = 2 + 2 \sin \theta$, $y = 3 + 2 \cos \theta$, $0 \le \theta \le 2\pi$
30. $x = 3 + 4 \sin \theta$, $y = 2 + 2 \cos \theta$, $0 \le \theta \le 2\pi$
31. If $A \ne 0$, $C = 0$, and $E \ne 0$, find parametric equations for $Ax^2 + Cy^2 + Dx + Ey + F = 0$. Identify the curve.
32. If $A = 0$, $C \ne 0$, and $D \ne 0$, find parametric equations for $Ax^2 + Cy^2 + Dx + Ey + F = 0$. Identify the curve.

In Problems 33–36, eliminate the parameter and find the standard equation for the curve. Name the curve and find its center.

33. $x = 3 + 6\cos t,\ y = 2 + 4\sin t,\ 0 \le t \le 2\pi$

34. $x = 1 + 3\sec t,\ y = -2 + 2\tan t,\ 0 \le t \le 2\pi,\ t \ne \frac{\pi}{2}, \frac{3\pi}{2}$

35. $x = -3 + 2\tan t,\ y = -1 + 5\sec t,\ 0 \le t \le 2\pi,\ t \ne \frac{\pi}{2}, \frac{3\pi}{2}$

36. $x = -4 + 5\cos t,\ y = 1 + 8\sin t,\ 0 \le t \le 2\pi$

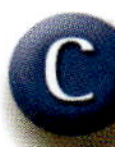

In Problems 37–42, the interval for the parameter is the entire real line. Obtain an equation in x and y by eliminating the parameter and identify the curve.

37. $x = \sqrt{t^2 + 1},\ y = \sqrt{t^2 + 9}$

38. $x = \sqrt{t^2 + 4},\ y = \sqrt{t^2 + 1}$

39. $x = \dfrac{2}{\sqrt{t^2 + 1}},\ y = \dfrac{2t}{\sqrt{t^2 + 1}}$

40. $x = \dfrac{3t}{\sqrt{t^2 + 1}},\ y = \dfrac{3}{\sqrt{t^2 + 1}}$

41. $x = \dfrac{8}{t^2 + 4},\ y = \dfrac{4t}{t^2 + 4}$

42. $x = \dfrac{4t}{t^2 + 1},\ y = \dfrac{4t^2}{t^2 + 1}$

In Problems 43–46, find the standard form of each equation. Name the curve and find its center. Then use trigonometric functions to find parametric equations for the curve.

43. $25x^2 - 200x - 9y^2 - 18y + 616 = 0$

44. $36x^2 + 360x + 4y^2 - 8y + 760 = 0$

45. $4x^2 - 24x + 49y^2 + 392y + 624 = 0$

46. $16x^2 + 32x - 9y^2 - 36y - 164 = 0$

47. Consider the following two pairs of parametric equations:

$x_1 = t,\ y_1 = e^t,\ -\infty < t < \infty$

$x_2 = e^t,\ y_2 = t,\ -\infty < t < \infty$

(A) Graph both pairs of parametric equations in a squared viewing window and discuss the relationship between the graphs.

(B) Eliminate the parameter and express each equation as a function of x. How are these functions related?

48. Consider the following two pairs of parametric equations:

$x_1 = t,\ y_1 = \log t,\ t > 0$

$x_2 = \log t,\ y_2 = t,\ t > 0$

(A) Graph both pairs of parametric equations in a squared viewing window and discuss the relationship between the graphs.

(B) Eliminate the parameter and express each equation as a function of x. How are these functions related?

APPLICATIONS

49. Projectile Motion. An airplane flying at an altitude of 1,000 meters is dropping medical supplies to hurricane victims on an island. The path of the plane is horizontal, the speed is 125 meters per second, and the supplies are dropped at the instant the plane crosses the shoreline. How far inland (to the nearest meter) will the supplies land?

50. Projectile Motion. One stone is dropped vertically from the top of a tower 40 meters high. A second stone is thrown horizontally from the top of the tower with a speed of 30 meters per second. How far apart (to the nearest tenth of a meter) are the stones when they land?

51. Projectile Motion. A projectile is fired with an initial speed of 300 meters per second at an angle of 45° to the horizontal. Neglecting air resistance, find

(A) The time of impact

(B) The horizontal distance covered (range) in meters and kilometers at time of impact

(C) The maximum height in meters of the projectile

Compute all answers to three decimal places.

52. Projectile Motion. Repeat Problem 51 if the same projectile is fired at 40° to the horizontal instead of 45°.

APPENDIX C

Geometric Formulas

Similar Triangles

(A) Two triangles are similar if two angles of one triangle have the same measure as two angles of the other.

(B) If two triangles are similar, their corresponding sides are proportional:

$$\frac{a}{a'} = \frac{b}{b'} = \frac{c}{c'}$$

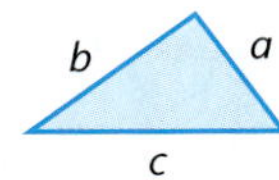

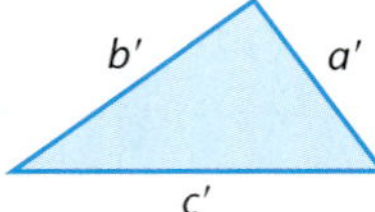

Pythagorean Theorem

$$c^2 = a^2 + b^2$$

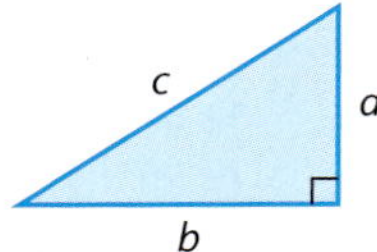

Rectangle

$A = ab$ Area

$P = 2a + 2b$ Perimeter

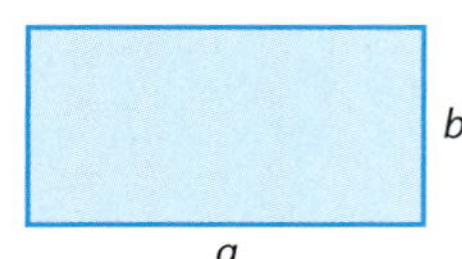

Parallelogram

$h =$ Height

$A = ah = ab \sin \theta$ Area

$P = 2a + 2b$ Perimeter

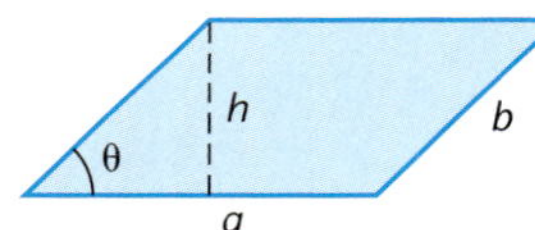

Triangle

$h =$ Height

$A = \frac{1}{2}hc$ Area

$P = a + b + c$ Perimeter

$s = \frac{1}{2}(a + b + c)$ Semiperimeter

$A = \sqrt{s(s - a)(s - b)(s - c)}$ Area—Heron's formula

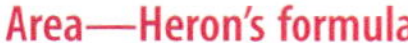

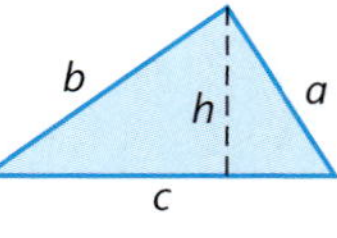

Trapezoid

Base a is parallel to base b.

$h =$ Height

$A = \frac{1}{2}(a + b)h$ Area

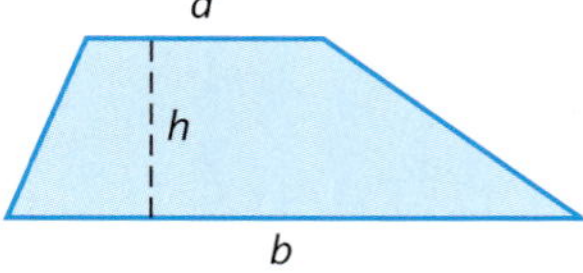

Circle

R = Radius
D = Diameter
$D = 2R$
$A = \pi R^2 = \frac{1}{4}\pi D^2$ Area
$C = 2\pi R = \pi D$ Circumference
$\frac{C}{D} = \pi$ For all circles
$\pi \approx 3.141\ 59$

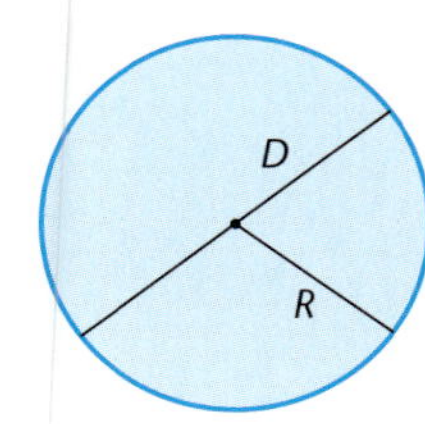

Rectangular Solid

$V = abc$ Volume
$T = 2ab + 2ac + 2bc$ Total surface area

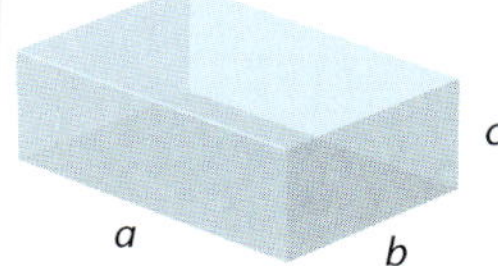

Right Circular Cylinder

R = Radius of base
h = Height
$V = \pi R^2 h$ Volume
$S = 2\pi Rh$ Lateral surface area
$T = 2\pi R(R + h)$ Total surface area

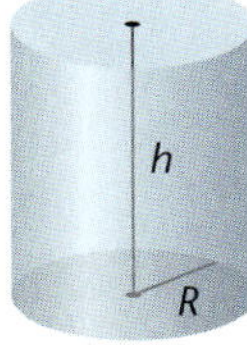

Right Circular Cone

R = Radius of base
h = Height
s = Slant height
$V = \frac{1}{3}\pi R^2 h$ Volume
$S = \pi Rs = \pi R\sqrt{R^2 + h^2}$ Lateral surface area
$T = \pi R(R + s) = \pi R(R + \sqrt{R^2 + h^2})$ Total surface area

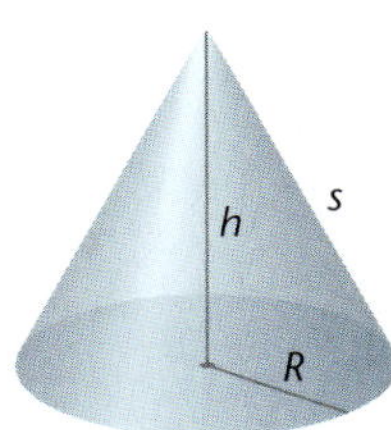

Sphere

R = Radius
D = Diameter
$D = 2R$
$V = \frac{4}{3}\pi R^3 = \frac{1}{6}\pi D^3$ Volume
$S = 4\pi R^2 = \pi D^2$ Surface area

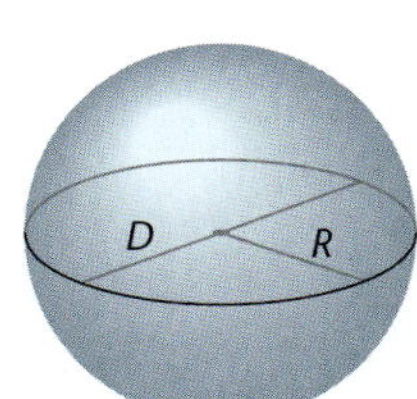

CHAPTER 1

Exercise 1.1

1. Yes **3.** No **5.** Yes

7. (A) Xmin = −7, Xmax = 6, Ymin = −9, Ymax = 14

9.

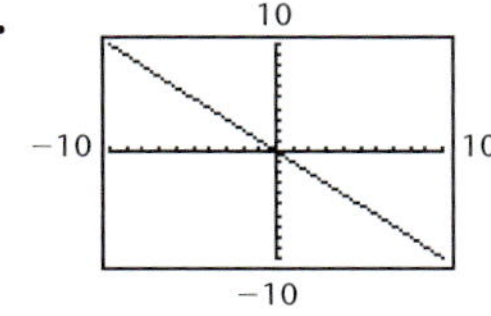

11.

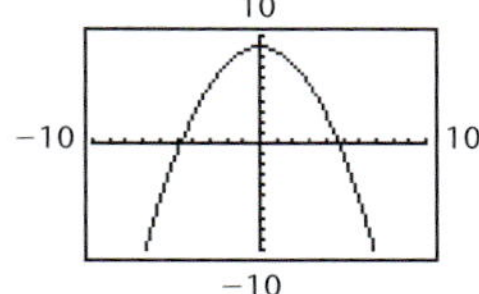

13.

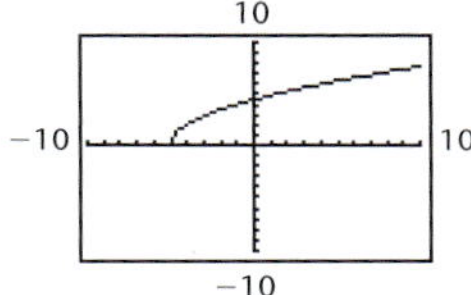

15.

x	−2	0	2	4	6
y	−8	4	8	4	−8

17.

x	−5	−3	−1	1	3	5
y	0	4	5.7	6.9	8	8.9

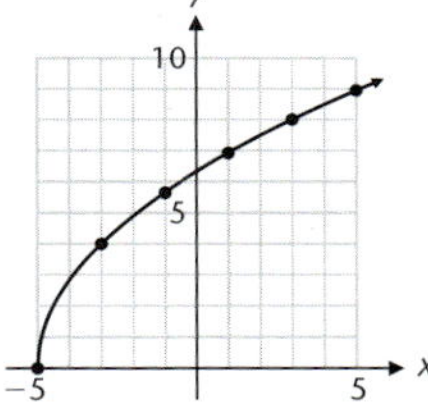

19.

x	−3	−1	1	3	5
y	10.5	−2.5	4.5	7.5	−17.5

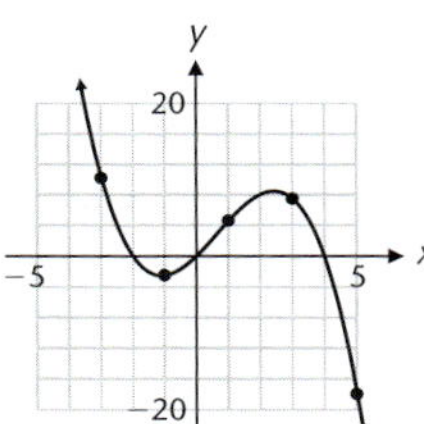

21. (A) −6.37 (B) 0.63

23. (A) 0.92 (B) −3.93

25. (6.77, −21.86), (−1.77, 20.86)

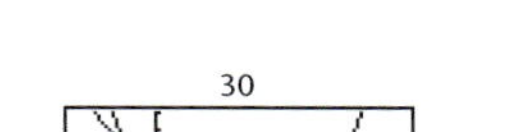
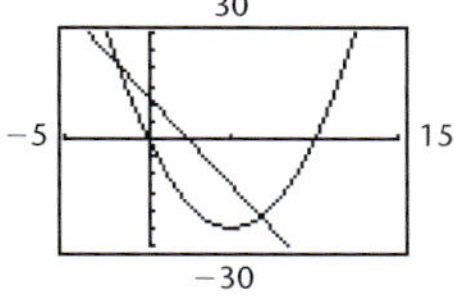

27. (0.54, 7.83), (18.46, −63.83)

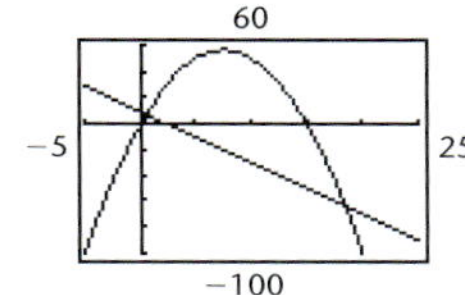

29. (0.36, 8.23), (19.64, 66.05)

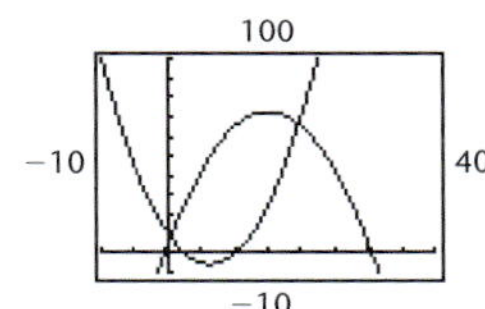

31. (18.84, 13.77), (106.16, 31.23)

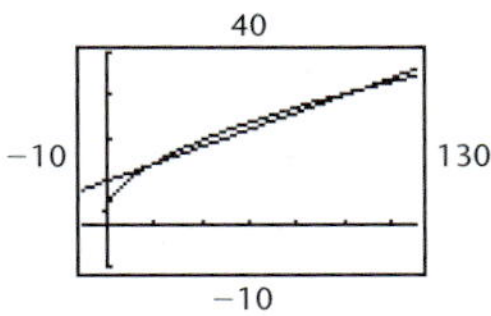

33. (−2.43, 2.75), (53.37, 7.96)

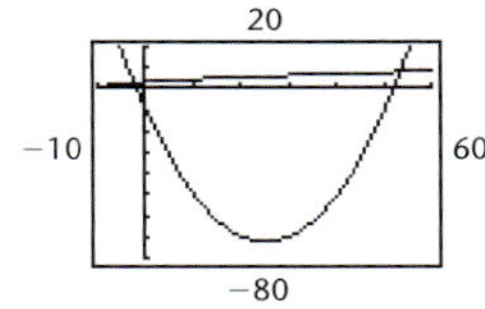

35. (A) The curve is a circle of radius 3 centered at the origin.

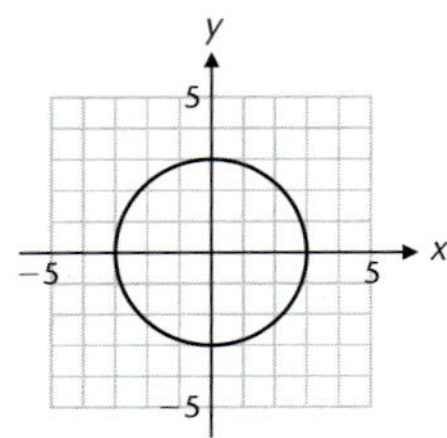

(B) This circle is distorted.

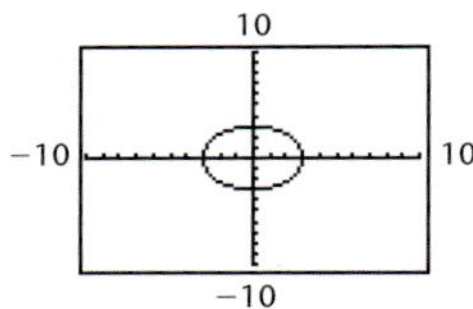

(C) ZDecimal produces a circle that fills the screen. ZSquare produces a small circle. ZoomFit produces another distorted circle.

37. (A) −7.99 (B) −5.85, 3.44, 12.41 (C) 14.60

39. (A) No solution (B) −8.81, 4.86

41. (A) −38.84, 11.16, 27.69 (B) −40, 20 (C) −41.07

43. (A) −17.84, 6.93 (B) −10 (C) No solution

45. 1.4142

49. There are two solutions: radius = 20.93 and height = 29.07 or radius = 43.17 and height = 6.83.

51. A 0.89-inch square or a 2.40-inch square can be cut out. Dimensions for smaller square: 0.89 inches × 9.23 inches × 6.73 inches; dimensions for larger square: 2.40 inches × 6.20 inches × 3.70 inches.

53. A 0.93-inch square or a 3.92-inch square can be cut out. Dimensions for smaller square: 0.93 inches × 10.14 inches × 10.61 inches; dimensions for larger square: 3.92 inches × 4.16 inches × 6.12 inches.

55. (A)

x	17,800	15,600	13,600
y	20	25	30

(B) Demand decreases 2,000 cases

(C) Demand increases 2,200 cases

57. (A)

y	20	25	30
R	356,000	390,000	408,000

(B) Revenue increases $18,000

(C) Revenue decreases $34,000

(D) The company should raise the price $5 to increase the revenue.

Exercise 1.2

1. A function **3.** Not a function **5.** A function

7. A function; domain = {2, 3, 4, 5}; range = {4, 6, 8, 10}

9. Not a function

11. A function; domain = {0, 1, 2, 3, 4, 5}; range = {1, 2}

13. A function **15.** Not a function **17.** Not a function

19. (A) A function (B) Not a function

21. A function **23.** Not a function

25. −8 **27.** −6 **29.** 1 **31.** 10 **33.** $-\frac{30}{17}$ **35.** 3

37. All real numbers; $(-\infty, \infty)$

39. $x \neq 4$; $(-\infty, 4) \cup (4, \infty)$

41. $t \geq 4$; $[4, \infty)$

43. $w \geq -\frac{7}{3}$; $[-\frac{7}{3}, \infty)$ **45.** All real numbers; $(-\infty, \infty)$

47. $v \neq 4, -4$; $(-\infty, -4) \cup (-4, 4) \cup (4, \infty)$

49. $x \geq -4, x \neq 1$; $[-4, 1) \cup (1, \infty)$

51. $t \geq 0, t \neq 9$; $[0, 9) \cup (9, \infty)$

53. $f(x) = 2x - 3$ **55.** $f(x) = 4x^2 - 2x + 9$ **57.** 3

59. $-6 - h$ **61.** $11 - 2h$ **63.** $g(x) = 3x + 1$

65. $F(x) = \dfrac{x}{8 + \sqrt{x}}$

67. Function f multiplies the domain element by 2 and then subtracts the product of 3 and the square of the domain element.

69. Function F takes the square root of the sum of the fourth power of the domain element and 9.

71. Function f multiplies the square of the input by 2, subtracts 4 times the input, and adds 6. $f(x) = 2x^2 - 4x + 6$

73. Function f multiplies the input by 4, subtracts 3 times the square root of the input, and adds 9. $f(x) = 4x - 3\sqrt{x} + 9$

75. (A) 3 (B) 3 **77.** (A) $2x + h$ (B) $x + a$

79. (A) $-6x - 3h + 9$ (B) $-3x - 3a + 9$

81. (A) $3x^2 + 3xh + h^2$ (B) $x^2 + ax + a^2$

83. The values of f are very close to 2 when x is close to 1.

85. (A) $s(0) = 0$, $s(1) = 16$, $s(2) = 64$, $s(3) = 144$

(B) $64 + 16h$

(C) Let $q(h) = [s(2 + h) - s(2)]/h$

h	−1	−0.1	−0.01	−0.001	0.001	0.01	0.1	1
$q(h)$	48	62.4	63.84	63.984	64.016	64.16	65.6	80

(D) $q(h)$, the average velocity from 2 to $2 + h$ seconds, approaches 64 feet per second

87. (A)

x	0	5,000	10,000	15,000	20,000	25,000	30,000
$B(x)$	212	203	194	185	176	167	158

(B) The boiling point drops 9°F for each 5,000-foot increase in altitude.

89. The rental charges are \$20 per day plus \$0.25 per mile driven.

91. (A)

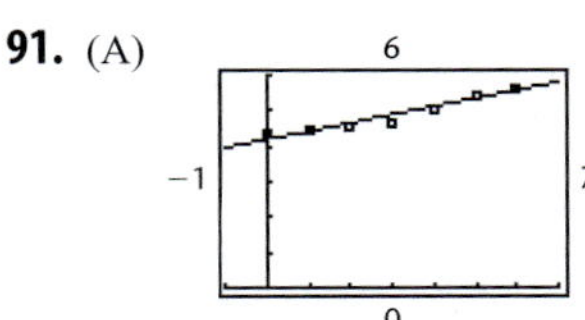

t	0	1	2	3	4	5	6
data	4.35	4.42	4.59	4.69	5.08	5.39	5.66
$A(t)$	4.2	4.43	4.66	4.89	5.12	5.35	5.58

(B) The estimated price of admission is \$5.81 in 2002 and \$6.04 in 2003.

93. (A)

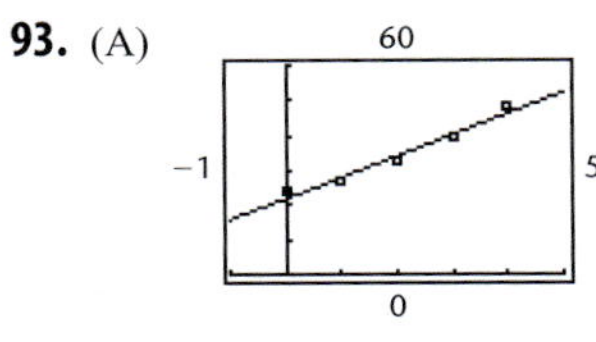

t	0	1	2	3	4
data	24	27	33	40	48
$A(t)$	22	28.1	34.2	40.3	46.4

(B) Merck's estimated sales are \$53 billion in 2002 and \$65 billion in 2004.

95. (A)

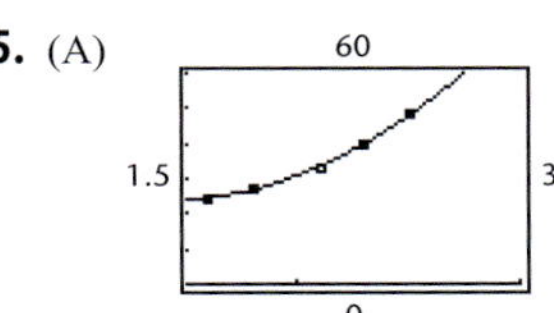

r	1.6	1.8	2.1	2.3	2.5
data	24	27	33	40	48
$S(r)$	24.6	27	33.6	40	48

(B) Merck's estimated sales are \$58 billion if they spend \$2.7 billion on R & D and \$75 billion if they spend \$3 billion on R & D.

Exercise 1.3

1. (A) $[-4, 4)$ (B) $[-3, 3)$ (C) 0 (D) 0 (E) $[-4, 4)$ (F) None (G) None (H) None

3. (A) $(-\infty, \infty)$ (B) $[-4, \infty)$ (C) $-3, 1$ (D) -3 (E) $[-1, \infty)$ (F) $(-\infty, -1]$ (G) None (H) None

5. (A) $(-\infty, 2) \cup (2, \infty)$ (The function is not defined at $x = 2$.) (B) $(-\infty, -1) \cup [1, \infty)$ (C) None (D) 1 (E) None (F) $(-\infty, -2] \cup (2, \infty)$ (G) $[-2, 2)$ (H) $x = 2$

7. Increasing: $[-2, 10]$; decreasing: $[-10, -2]$

9. Decreasing: $[-4, 3]$; constant: $[-10, -4]$, $[3, 10]$

11. Increasing: $[-4, 0]$, $[4, 10]$; decreasing: $[-10, -4]$, $[0, 4]$

13. x intercepts: -1.405, 6.405; y intercept: -9; local minimum: $f(2.5) = -15.25$

15. x intercept: 3.377; y intercept: 25; local minimum: $h(-1.155) = 21.921$; local maximum: $h(1.155) = 28.079$

17. x intercepts: $-\sqrt{12} \approx -3.464$, $\sqrt{12} \approx 3.464$; y intercept: $\sqrt{12} \approx 3.464$; local minima: $m(-\sqrt{12}) = 0$, $m(\sqrt{12}) = 0$

19. One possible answer:

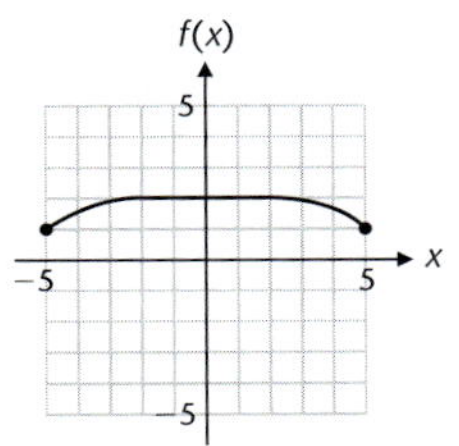

21. One possible answer:

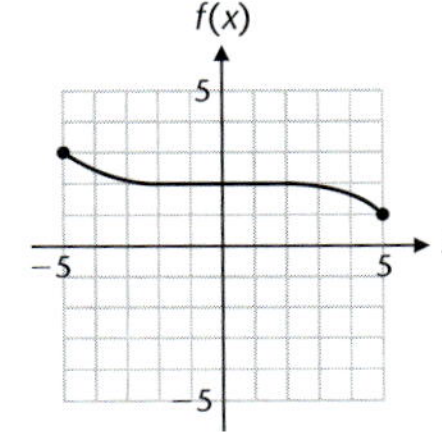

23. One possible answer:

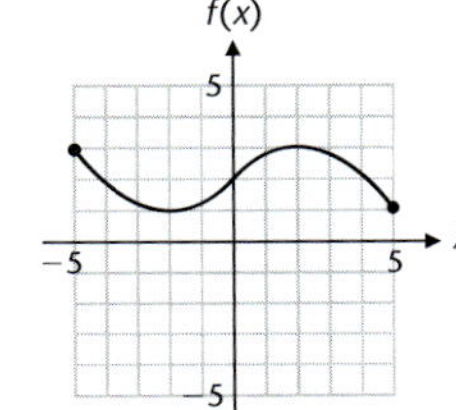

25. $f(-2) = -1$, $f(-1) = 0$, $f(1) = 0$, $f(2) = -1$

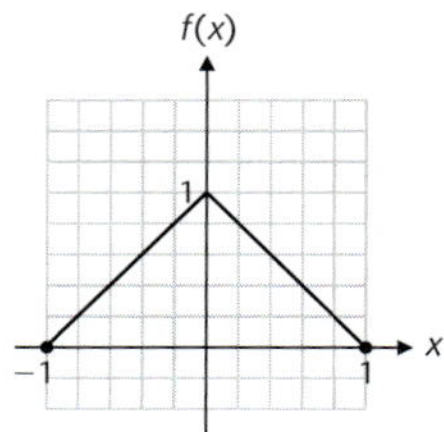

27. $f(-2) = 0$, $f(-1) = -3$, $f(1) = -1$, $f(2) = 0$

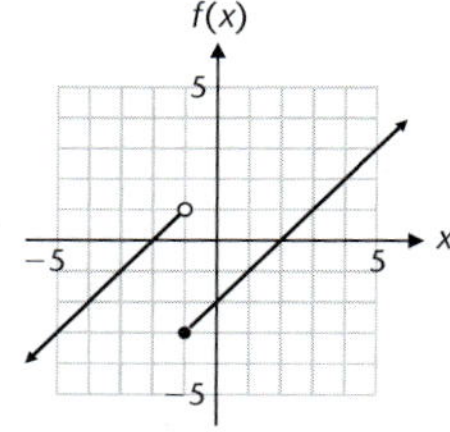

29. $f(-2) = 5$, $f(-1) = 2$, $f(1) = -2$, $f(2) = -5$

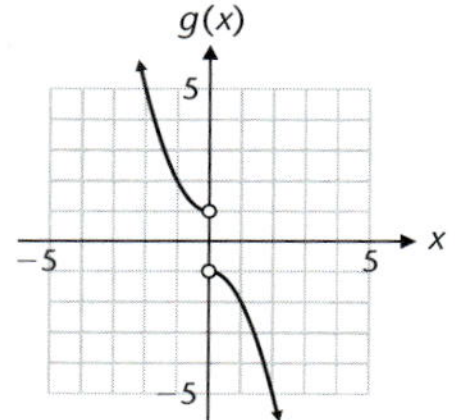

31. Domain: $(-\infty, \infty)$; range: $(-\infty, \infty)$; y intercept: -30; x intercepts: -44.99, -0.82, 0.81

33. Domain: $(-\infty, \infty)$; range: $(-\infty, 10{,}200]$; y intercept: 200; x intercepts: -14.18, 14.18

35. Domain: $[0, \infty)$; range: $(-\infty, 16]$; y intercept: 0; x intercepts: 0, 64

37. Domain: $[-5, \infty)$; range: $[-134.02, \infty)$; y intercept: -111.80; x intercepts: -4.79, 14.94

39. The graph of f is rising and f is increasing on $(-\infty, 0.13]$ and $[7.87, \infty)$. The graph of f is falling and f is decreasing on $[0.13, 7.87]$. $f(0.13) = -9.81$ is a local maximum and $f(7.87) = -242.19$ is a local minimum.

41. The graph of m is rising and m is increasing on $(-\infty, -12]$ and $[0, 12]$. The graph of m is falling and m is decreasing on $[-12, 0]$ and $[12, \infty)$. $m(-12) = 144$ and $m(12) = 144$ are local maxima and $m(0) = 0$ is a local minimum.

43. The graph of g is rising and g is increasing on $[-15, 2.5]$ and $[20, \infty)$. The graph of g is falling and g is decreasing on $(-\infty, -15]$ and $[2.5, 20]$. $g(2.5) = 306.25$ is a local maximum and $g(-15) = 0$ and $g(20) = 0$ are local minima.

45. The graph of f decreases on $[-10, -2.15]$ to a local minimum value, $f(-2.15) \approx -36.62$, and then increases on $[-2.15, 10]$.

47. The graph of h increases on $[-10, -4.64]$ to a local maximum value, $h(-4.64) \approx 281.93$, decreases on $[-4.64, 5.31]$ to a local minimum value, $h(5.31) \approx -211.41$, and then increases on $[5.31, 10]$.

49. The graph of p decreases on $[-10, -3.77]$ to a local minimum value, $p(-3.77) \approx 0$, increases on $[-3.77, 0.50]$ to a local maximum value, $p(0.50) = 18.25$, decreases on $[0.50, 4.77]$ to a local minimum value, $p(4.77) \approx 0$, and then increases on $[4.77, 10]$.

51. One possible answer:

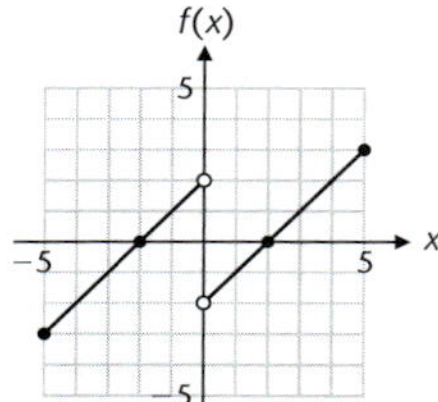

53. One possible answer:

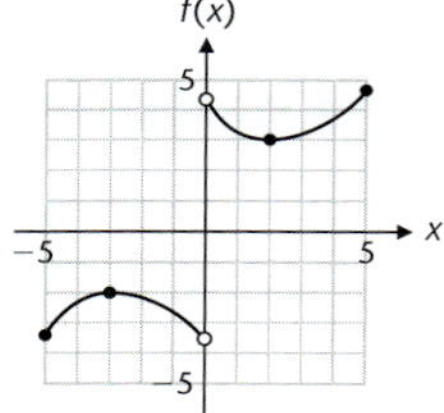

55. One possible answer:

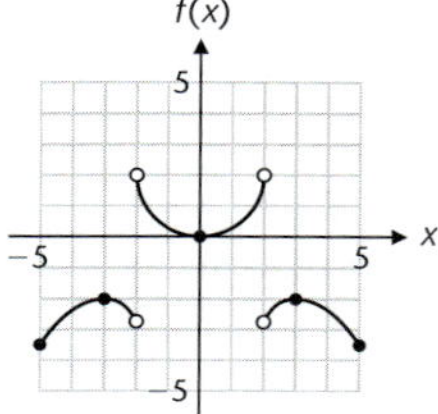

57. Domain: all real numbers except $x = 2$; range: $\{-5, 5\}$ (a set, not an interval); discontinuous at $x = 2$

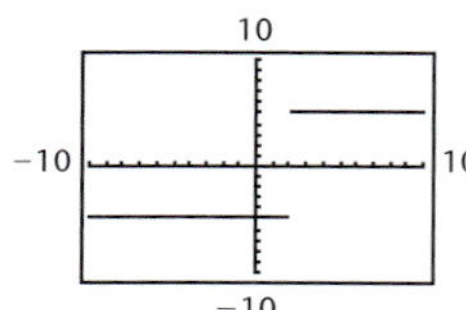

59. Domain: all real numbers except $x = 1$; range: $(-\infty, -3) \cup (5, \infty)$; discontinuous at $x = 1$

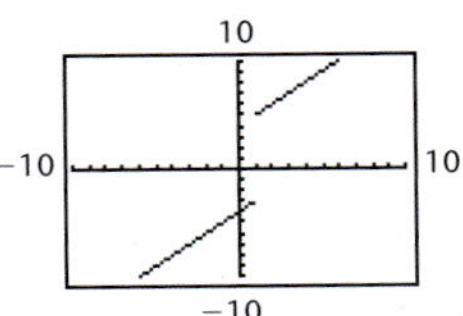

61. Domain: all real numbers except $x = 3$; range: $(0, \infty)$; discontinuous at $x = 3$

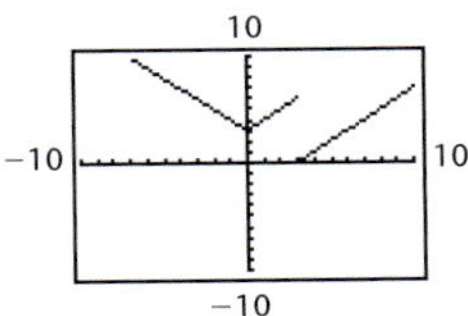

63. Domain: all real numbers; range: all integers; discontinuous at the even integers

$$f(x) = \begin{cases} \vdots & & \vdots \\ -2 & \text{if} & -4 \le x < -2 \\ -1 & \text{if} & -2 \le x < 0 \\ 0 & \text{if} & 0 \le x < 2 \\ 1 & \text{if} & 2 \le x < 4 \\ 2 & \text{if} & 4 \le x < 6 \\ \vdots & & \vdots \end{cases}$$

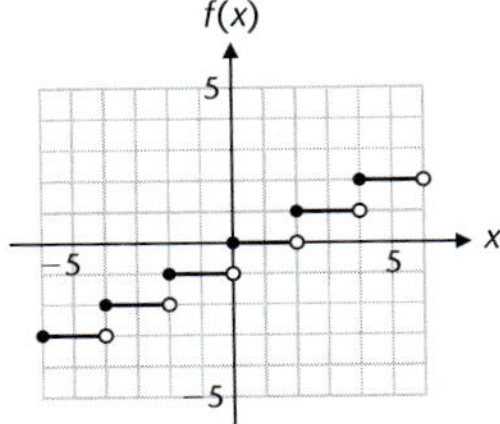

65. Domain: all real numbers; range: all integers; discontinuous at rational numbers of the form $\frac{k}{3}$, where k is an integer.

$$f(x) = \begin{cases} \vdots & & \vdots \\ -2 & \text{if} & -\frac{2}{3} \le x < -\frac{1}{3} \\ -1 & \text{if} & -\frac{1}{3} \le x < 0 \\ 0 & \text{if} & 0 \le x < \frac{1}{3} \\ 1 & \text{if} & \frac{1}{3} \le x < \frac{2}{3} \\ 2 & \text{if} & \frac{2}{3} \le x < 1 \\ \vdots & & \vdots \end{cases}$$

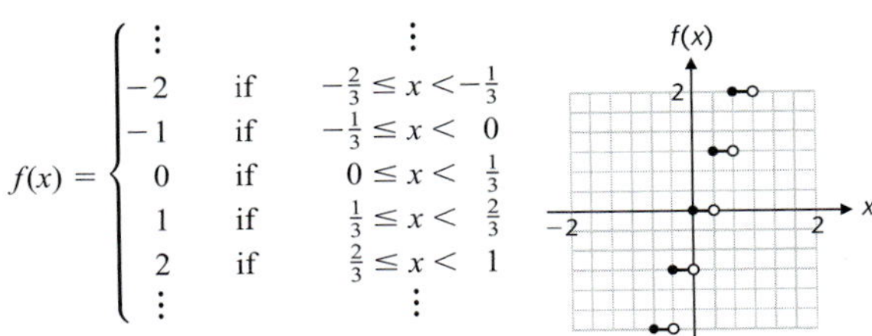

67. Domain: all real numbers; range: $[0, 1)$; discontinuous at all integers

$$f(x) = \begin{cases} \vdots & & \vdots \\ x + 2 & \text{if} & -2 \le x < -1 \\ x + 1 & \text{if} & -1 \le x < 0 \\ x & \text{if} & 0 \le x < 1 \\ x - 1 & \text{if} & 1 \le x < 2 \\ x - 2 & \text{if} & 2 \le x < 3 \\ \vdots & & \vdots \end{cases}$$

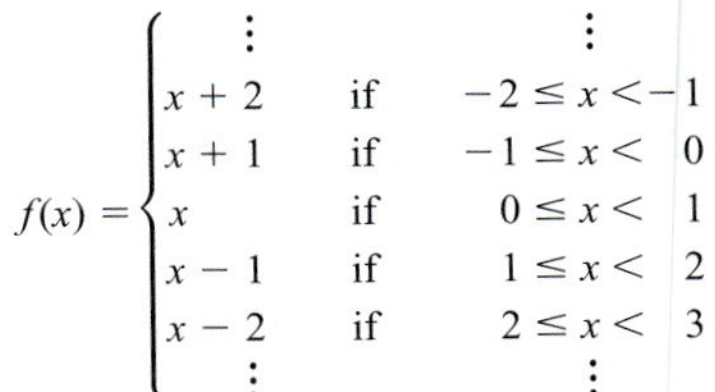

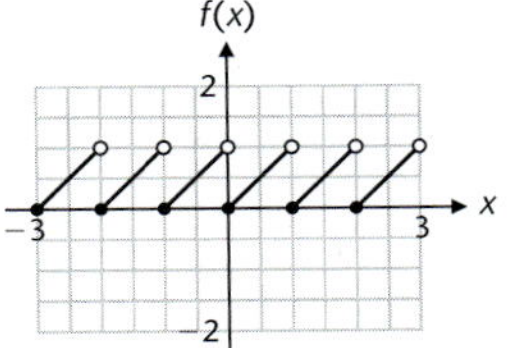

69. (A) One possible answer:

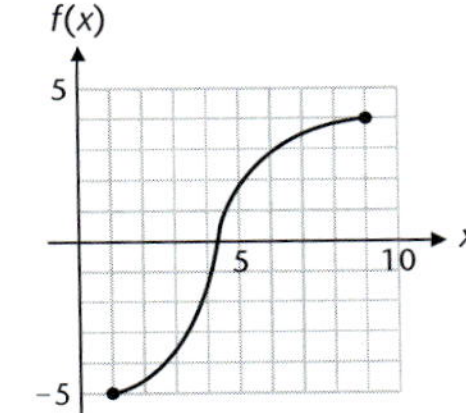

(B) The graph must cross the x axis exactly once.

71. (A) One possible answer:

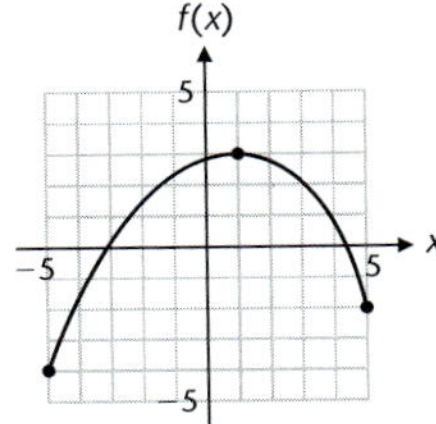

(B) The graph must cross the x axis at least twice. There is no upper limit on the number of times it can cross the x axis.

73. (A) One possible answer:

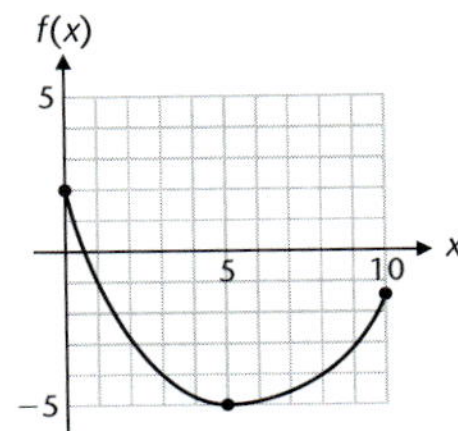

(B) The graph can cross the x axis zero, one, or two times.

75.

$$\left.\begin{aligned} f(4) &= 10[\![0.5 + 0.4]\!] &&= 10(0) &&= 0 \\ f(-4) &= 10[\![0.5 - 0.4]\!] &&= 10(0) &&= 0 \\ f(6) &= 10[\![0.5 + 0.6]\!] &&= 10(1) &&= 10 \\ f(-6) &= 10[\![0.5 - 0.6]\!] &&= 10(-1) &&= -10 \\ f(24) &= 10[\![0.5 + 2.4]\!] &&= 10(2) &&= 20 \\ f(25) &= 10[\![0.5 + 2.5]\!] &&= 10(3) &&= 30 \\ f(247) &= 10[\![0.5 + 24.7]\!] &&= 10(25) &&= 250 \\ f(-243) &= 10[\![0.5 - 24.3]\!] &&= 10(-24) &&= -240 \\ f(-245) &= 10[\![0.5 - 24.5]\!] &&= 10(-24) &&= -240 \\ f(-246) &= 10[\![0.5 - 24.6]\!] &&= 10(-25) &&= -250 \end{aligned}\right\} \text{ f rounds numbers to the tens place}$$

77. $f(x) = [\![100x + 0.5]\!]/100$

79. The maximum revenue is \$25,714 when 857 car seats are sold.

81. The maximum volume is 654.98 cubic inches when the side of each square is 3.39 inches.

83. The minimum cost is \$300,000 when the land portion of the pipe is 13.59 miles.

85. $R(x) = \begin{cases} 32 & \text{if } 0 \le x \le 100 \\ 16 + 0.16x & \text{if } x > 100 \end{cases}$

87. (A) $C(x) = \begin{cases} 15 & 0 < x \le 1 \\ 18 & 1 < x \le 2 \\ 21 & 2 < x \le 3 \\ 24 & 3 < x \le 4 \\ 27 & 4 < x \le 5 \\ 30 & 5 < x \le 6 \end{cases}$

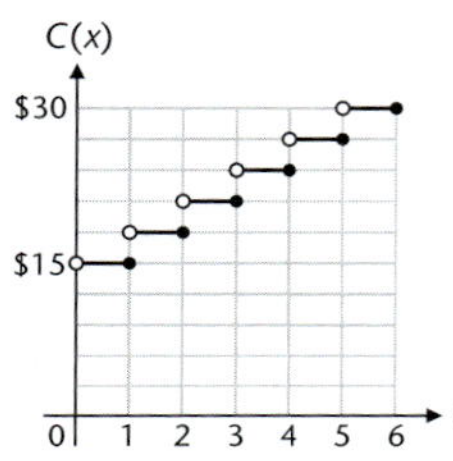

(B) No, because $f(x) \neq C(x)$ at $x = 1, 2, 3, 4, 5$, or 6

89. $E(x) = \begin{cases} 200 & \text{if} & 0 \le x \le 3{,}000 \\ 80 + 0.04x & \text{if} & 3{,}000 < x < 8{,}000 \\ 180 + 0.04x & \text{if} & 8{,}000 \le x \end{cases}$

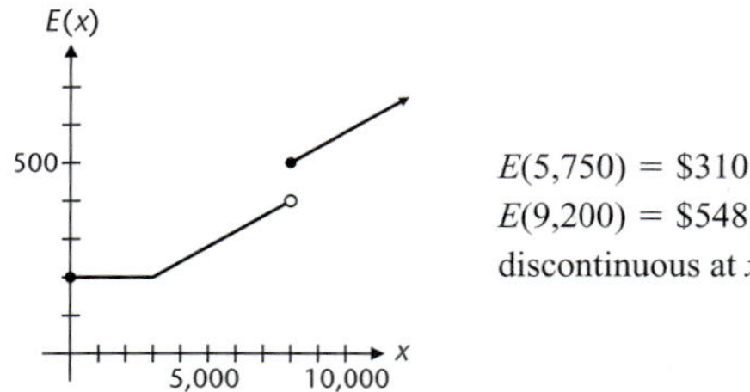

$E(5{,}750) = \$310$; $E(9{,}200) = \$548$; discontinuous at $x = 8{,}000$

91. $T(x) = \begin{cases} 0.0535x & \text{if} & 0 \le x \le 18{,}120 \\ 0.0705x - 308.46 & \text{if} & 18{,}120 < x \le 59{,}500 \\ 0.0785x - 783.75 & \text{if} & x > 59{,}500 \end{cases}$

$T(10{,}000) = \$535$; $T(30{,}000) = \$1{,}806.54$; $T(100{,}000) = \$7{,}066.25$

93. (A)

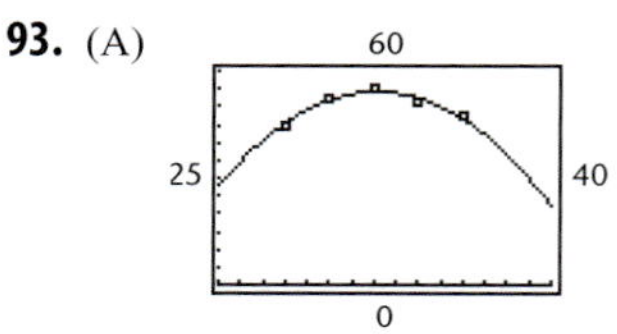

(B) 53.50 thousand miles, 49.95 thousand miles

(C) The mileage increases to a maximum value of 54.18 thousand miles at a pressure of 32.14 pounds per square inch and then begins to decrease.

Exercise 1.4

1. Odd **3.** Even **5.** Neither **7.** Even **9.** Neither

11.

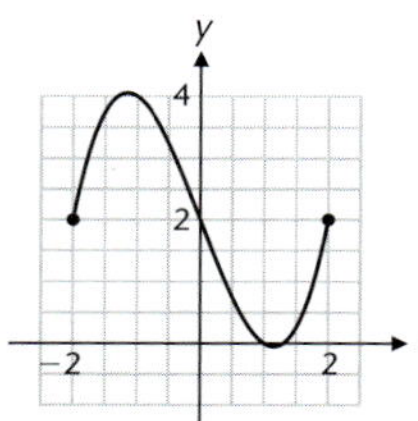

13.

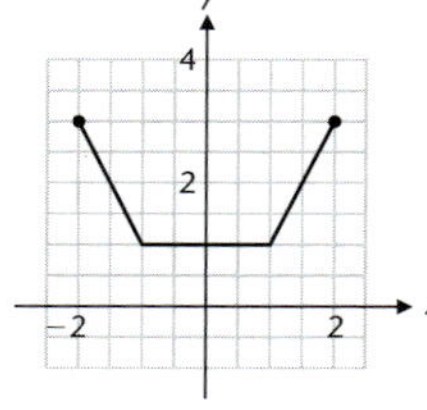

15.

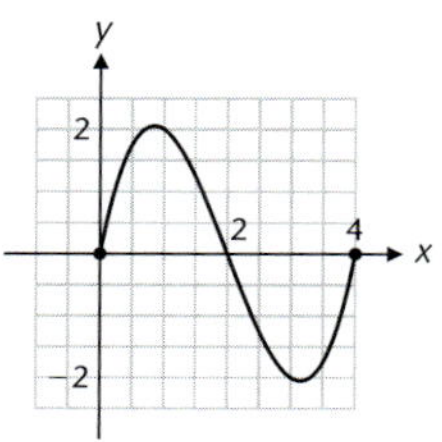

17.

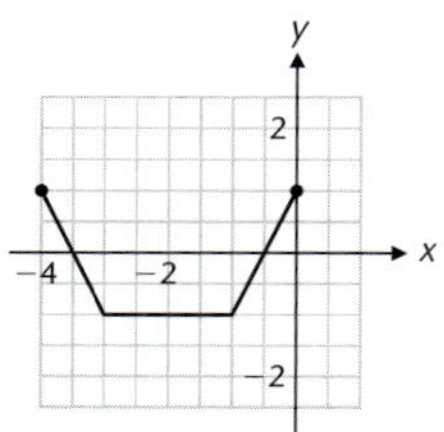

19.

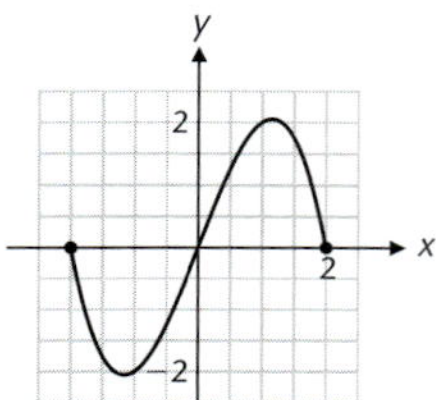

21.

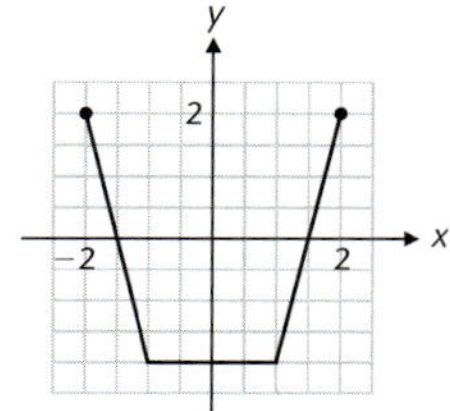

23.

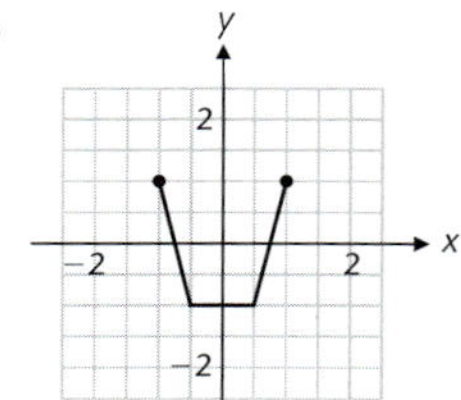

25.

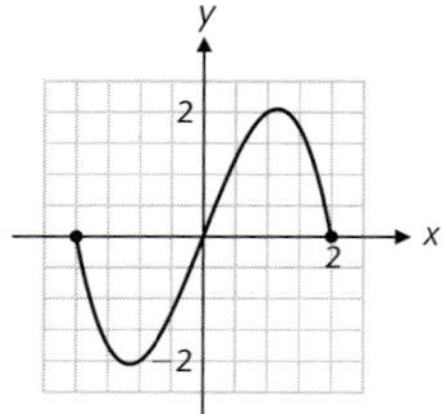

27. The graph of $y = x^2$ is shifted two units to the right; $y = (x - 2)^2$.
29. The graph of $y = x^3$ is shifted down two units; $y = x^3 - 2$.
31. The graph of $y = |x|$ is vertically contracted by a factor of 0.25; $y = 0.25|x|$.
33. The graph of $y = x^3$ is reflected in the x axis (or the y axis); $y = -x^3$.
35. $g(x) = \sqrt[3]{x + 4} - 5$ **37.** $g(x) = -0.5(6 + \sqrt{x})$
39. $g(x) = -2(x + 4)^2 - 2$ **41.** $g(x) = \sqrt{-0.5(x + 2)}$
43. The graph of $y = x^2$ is shifted seven units left and nine units up.
45. The graph of $y = |x|$ is shifted eight units right and reflected in the x axis.
47. The graph of $y = \sqrt{x}$ is reflected in the x axis and shifted three units up.
49. The graph of $y = x^2$ is vertically expanded by a factor of four and reflected in the x axis.
51. $y = |x + 2| + 2$ **53.** $y = 4 - \sqrt{x}$ **55.** $y = 4 - (x - 1)^2$
57. $y = 0.5(x - 3)^3 + 1$
59. Reversing the order does not change the result.
61. Reversing the order can change the result.
63. Reversing the order does not change the result.

65.

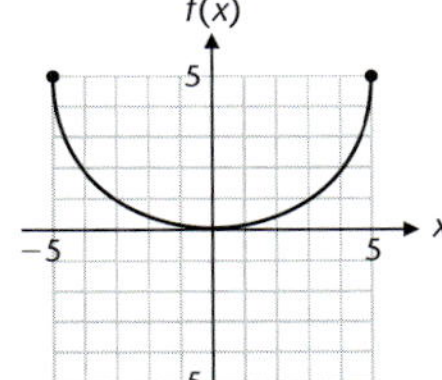

67.

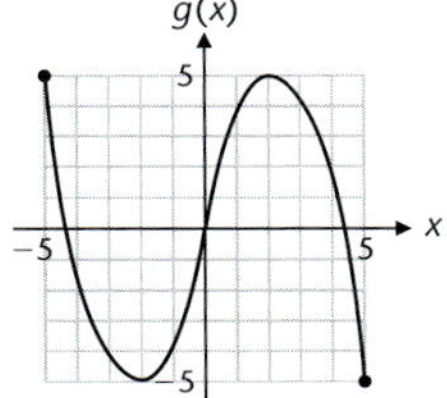

69. Conclusion: any function can be written as the sum of two other functions, one even and the other odd.
71. Graph of $f(x)$ Graph of $|f(x)|$

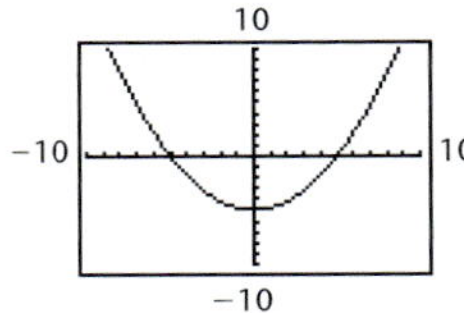

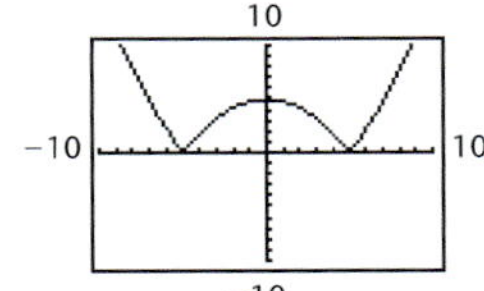

Graph of $-|f(x)|$

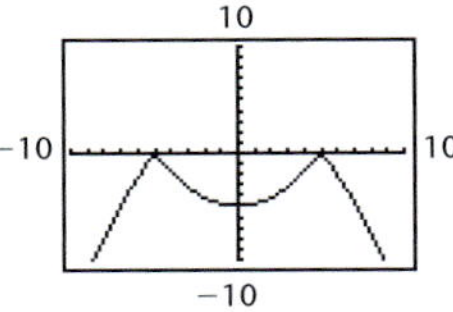

73. Graph of $f(x)$ 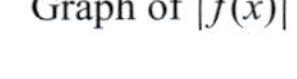
Graph of $|f(x)|$

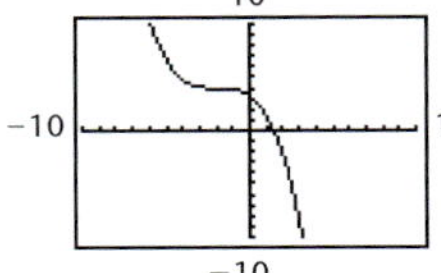

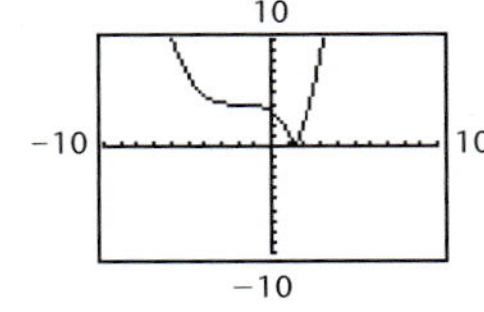

Graph of $-|f(x)|$

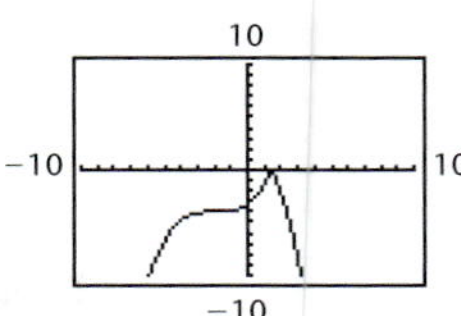

75. The graph of $y = |f(x)|$ is the same as the graph of $y = f(x)$ whenever $f(x) \geq 0$ and is the reflection of the graph of $y = f(x)$ with respect to the x axis whenever $f(x) < 0$.

77.

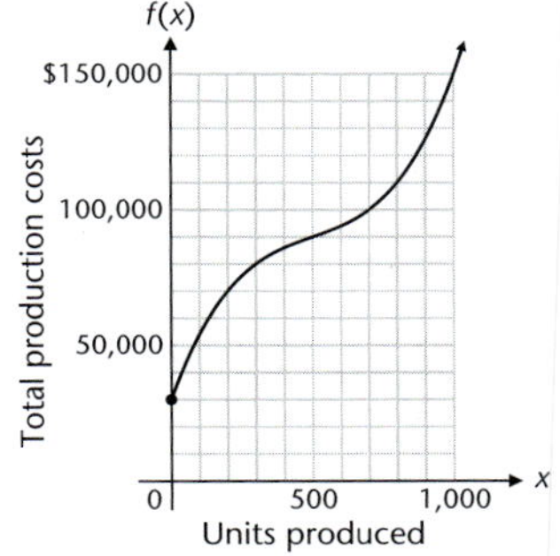

79. Each graph is a vertical translation of the graph of $y = 0.004(x - 10)^3$.

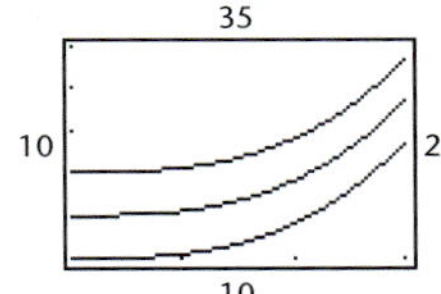

81. Each graph is a vertical contraction followed by a vertical translation of the graph of $y = x^2$.

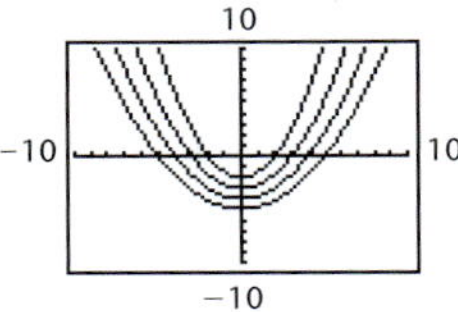

83. Each graph is a portion of the graph of a horizontal translation followed by a vertical expansion (except for $C = 8$) of the graph of $y = t^2$.

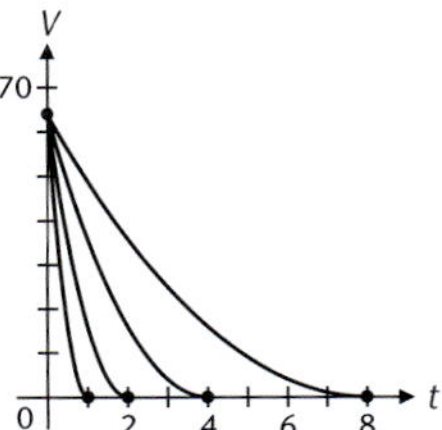

Exercise 1.5

1.

x	-3	-2	-1	0	1	2	3
$(f+g)(x)$	3	3	1	-1	-3	-3	-3

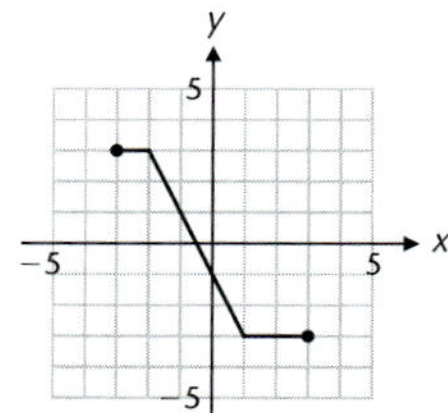

3.

x	-3	-2	-1	0	1	2	3
$(fg)(x)$	2	0	-2	-2	0	2	2

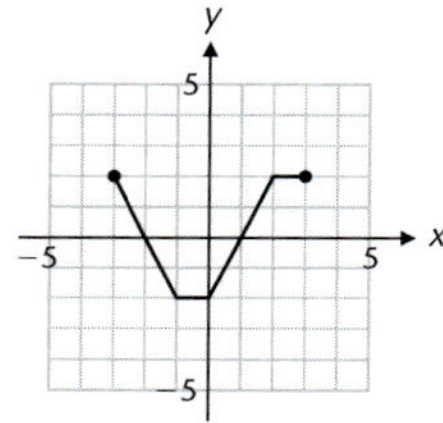

5. -2 **7.** 1 **9.** -2 **11.** 3

13. $(f+g)(x) = 5x + 1$; $(f-g)(x) = 3x - 1$; $(fg)(x) = 4x^2 + 4x$; $\left(\frac{f}{g}\right)(x) = \frac{4x}{x+1}$; domain $f+g, f-g, fg = (-\infty, \infty)$; domain of $f/g = (-\infty, -1) \cup (-1, \infty)$

15. $(f+g)(x) = 3x^2 + 1$; $(f-g)(x) = x^2 - 1$; $(fg)(x) = 2x^4 + 2x^2$; $\left(\frac{f}{g}\right)(x) = \frac{2x^2}{x^2+1}$; domain of each function: $(-\infty, \infty)$

17. $(f+g)(x) = x^2 + 3x + 4$; $(f-g)(x) = -x^2 + 3x + 6$; $(fg)(x) = 3x^3 + 5x^2 - 3x - 5$; $\left(\frac{f}{g}\right)(x) = \frac{3x+5}{x^2-1}$; domain $f+g, f-g, fg$: $(-\infty, \infty)$; domain of f/g: $(-\infty, -1) \cup (-1, 1) \cup (1, \infty)$

19. $(f \circ g)(x) = (x^2 - x + 1)^3$; domain: $(-\infty, \infty)$; $(g \circ f)(x) = x^6 - x^3 + 1$; domain: $(-\infty, \infty)$

21. $(f \circ g)(x) = |2x + 4|$; domain: $(-\infty, \infty)$; $(g \circ f)(x) = 2|x + 1| + 3$; domain: $(-\infty, \infty)$

23. $(f \circ g)(x) = (2x^3 + 4)^{1/3}$; domain: $(-\infty, \infty)$; $(g \circ f)(x) = 2x + 4$; domain: $(-\infty, \infty)$

25. $(f \circ g)(x) = (g \circ f)(x) = x$; symmetric with respect to the line $y = x$

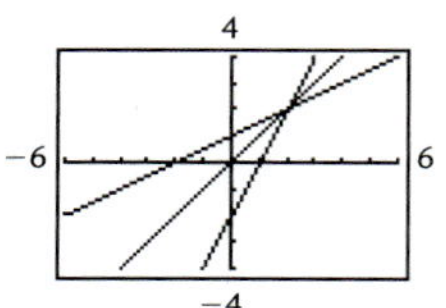

27. $(f \circ g)(x) = (g \circ f)(x) = x$; symmetric with respect to the line $y = x$

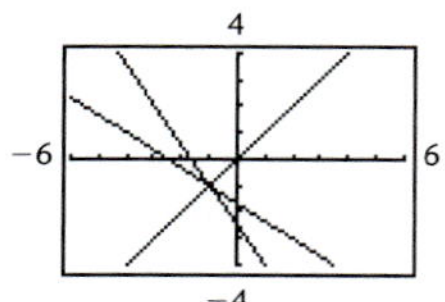

29. $(f+g)(x) = \sqrt{2-x} + \sqrt{x+3}$; $(f-g)(x) = \sqrt{2-x} - \sqrt{x+3}$; $(fg)(x) = \sqrt{6 - x - x^2}$; $\left(\frac{f}{g}\right)(x) = \sqrt{\frac{2-x}{x+3}}$. The domain of the functions $f+g, f-g$, and fg is $[-3, 2]$. The domain of $\frac{f}{g}$ is $(-3, 2]$.

31. $(f+g)(x) = 2\sqrt{x} - 2$; $(f-g)(x) = 6$; $(fg)(x) = x - 2\sqrt{x} - 8$; $\left(\frac{f}{g}\right)(x) = \frac{\sqrt{x}+2}{\sqrt{x}-4}$. The domain of $f+g, f-g$, and fg is $[0, \infty)$. Domain of $\frac{f}{g} = [0, 16) \cup (16, \infty)$.

33. $(f+g)(x) = \sqrt{x^2 + x - 6} + \sqrt{7 + 6x - x^2}$; $(f-g)(x) = \sqrt{x^2 + x - 6} - \sqrt{7 + 6x - x^2}$; $(fg)(x) = \sqrt{-x^4 + 5x^3 + 19x^2 - 29x - 42}$; $\left(\frac{f}{g}\right)(x) = \sqrt{\frac{x^2 + x - 6}{7 + 6x - x^2}}$. The domain of the functions $f+g, f-g$, and fg is $[2, 7]$. The domain of $\frac{f}{g}$ is $[2, 7)$.

35. $(f \circ g)(x) = \sqrt{x - 4}$; domain: $[4, \infty)$; $(g \circ f)(x) = \sqrt{x} - 4$; domain: $[0, \infty)$

37. $(f \circ g)(x) = \frac{1}{x} + 2$; domain: $(-\infty, 0) \cup (0, \infty)$; $(g \circ f)(x) = \frac{1}{x+2}$; domain: $(-\infty, -2) \cup (-2, \infty)$

39. $(f \circ g)(x) = \frac{1}{|x-1|}$; domain: $(-\infty, 1) \cup (1, \infty)$; $(g \circ f)(x) = \frac{1}{|x| - 1}$; domain: $(-\infty, -1) \cup (-1, 1) \cup (1, \infty)$

41. (e) **43.** (a) **45.** (c)

47. $g(x) = 2x - 7; f(x) = x^4; h(x) = (f \circ g)(x)$

49. $g(x) = 4 + 2x; f(x) = x^{1/2}; h(x) = (f \circ g)(x)$

51. $f(x) = x^7; g(x) = 3x - 5; h(x) = (g \circ f)(x)$

53. $f(x) = x^{-1/2}; g(x) = 4x + 3; h(x) = (g \circ f)(x)$

59. $(f + g)(x) = 2x; (f - g)(x) = \frac{2}{x}; (fg)(x) = x^2 - \frac{1}{x^2}; \left(\frac{f}{g}\right)(x) = \frac{x^2 + 1}{x^2 - 1}.$
The domain of $f + g, f - g$, and fg is $(-\infty, 0) \cup (0, \infty)$. The domain of $\frac{f}{g}$ is $(-\infty, -1) \cup (-1, 0) \cup (0, 1) \cup (1, \infty)$.

61. $(f + g)(x) = 2; (f - g)(x) = \frac{-2x}{|x|}; (fg)(x) = 0; \left(\frac{f}{g}\right)(x) = 0.$
The domain of $f + g, f - g$, and fg is $(-\infty, 0) \cup (0, \infty)$; domain of $\frac{f}{g}$ is $(0, \infty)$.

63. $(f \circ g)(x) = \sqrt{4 - x^2}$; domain of $f \circ g$ is $[-2, 2]$;
$(g \circ f)(x) = 4 - x$; domain of $g \circ f$ is $(-\infty, 4]$.

65. $(f \circ g)(x) = \frac{6x - 10}{x}$; domain of $f \circ g$ is $(-\infty, 0) \cup (0, 2) \cup (2, \infty)$;
$(g \circ f)(x) = \frac{x + 5}{5 - x}$; domain of $g \circ f$ is $(-\infty, 0) \cup (0, 5) \cup (5, \infty)$.

67. $(f \circ g)(x) = \sqrt{16 - x^2}$; domain of $f \circ g$ is $[-4, 4]$;
$(g \circ f)(x) = \sqrt{34 - x^2}$; domain of $g \circ f$ is $[-5, 5]$.

69. $(f \circ g)(x) = \sqrt{2 + x}$; domain of $f \circ g$ is $[-2, 3]$; the first graph is correct.

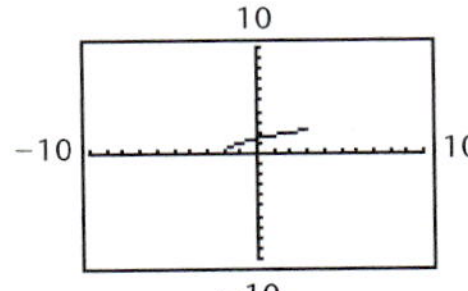

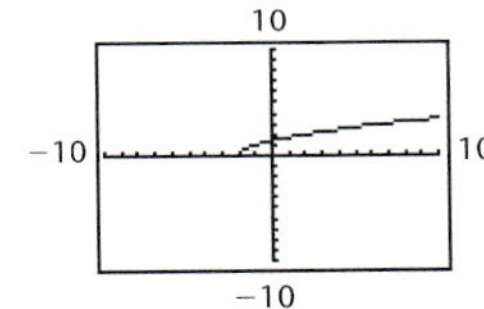

71. $(f \circ g)(x) = \sqrt{x^2 + 1}$; domain of $f \circ g$ is $(-\infty, -2] \cup [2, \infty)$; the first graph is correct.

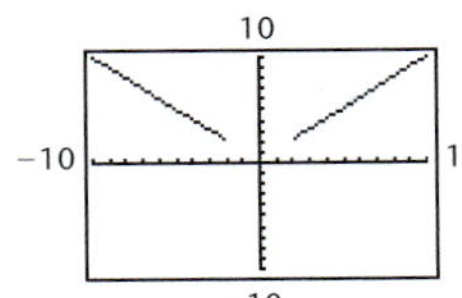

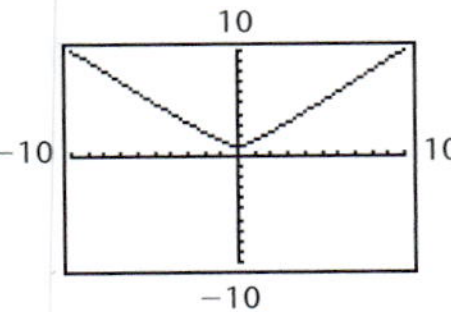

73. $(f \circ g)(x) = \sqrt{16 - x^2}$; domain of $f \circ g$ is $[-3, 3]$; the first graph is correct.

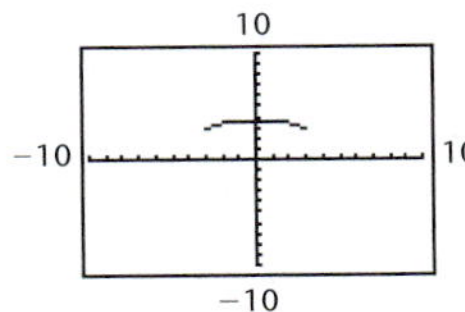

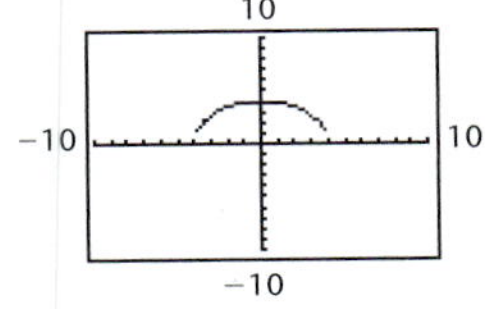

75. $P(p) = 4{,}400p - 200p^2 - 16{,}000$; maximum profit occurs when $p = \$11$.

77. $V(t) = 0.016\pi t^{2/3}$

79. (A) $r(h) = \frac{1}{2}h$ (B) $V(h) = \frac{1}{12}\pi h^3$ (C) $V(t) = \frac{0.125}{12}\pi t^{3/2}$

Exercise 1.6

1. The original set and the reversed set are both one-to-one functions.

3. The original set is a function. The reversed set is not a function.

5. Neither set is a function. **7.** One-to-one **9.** Not one-to-one

11. One-to-one **13.** Not one-to-one **15.** One-to-one

17. One-to-one **19.** One-to-one **21.** Not one-to-one

23. One-to-one

25. One-to-one **27.** Not one-to-one

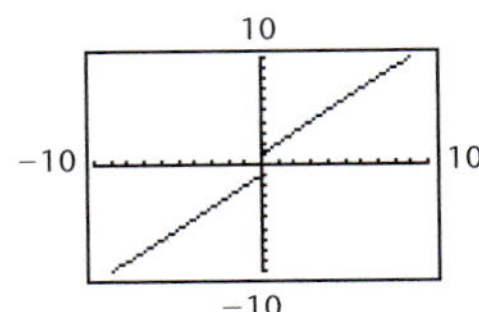

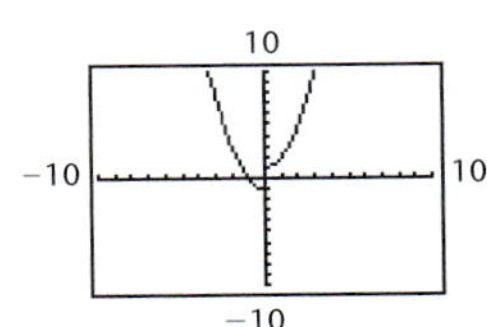

29. Not one-to-one **31.** One-to-one

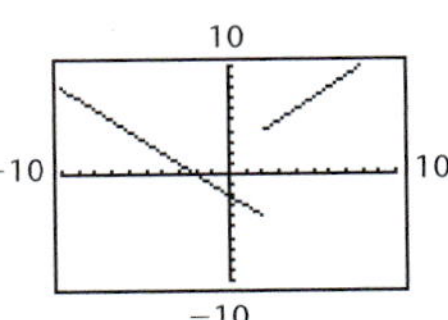

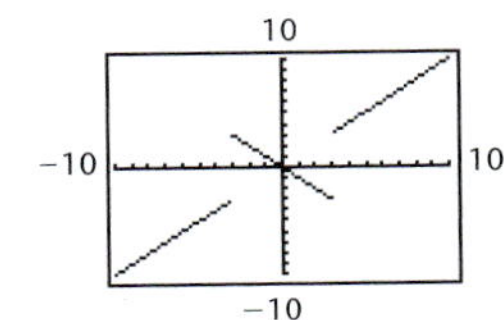

33. Yes **35.** No **37.** Yes

39. The function h multiplies the input by 3 and then subtracts 7. The inverse function adds 7 to the input and then divides by 3. The algebraic equation is $h^{-1}(x) = (x + 7)/3$.

41. The function m adds 11 to the input and then takes the cube root. The inverse function cubes the input and then subtracts 11. The algebraic equation is $m^{-1}(x) = x^3 - 11$.

43. The function s multiplies the input by 3, adds 17, and then raises this expression to the fifth power. The inverse function takes the fifth root of the input, subtracts 17, and then divides by 3. The algebraic equation is $s^{-1}(x) = (x^{1/5} - 17)/3$.

45. Range of $f^{-1} = [-4, 4]$; domain of $f^{-1} = [1, 5]$

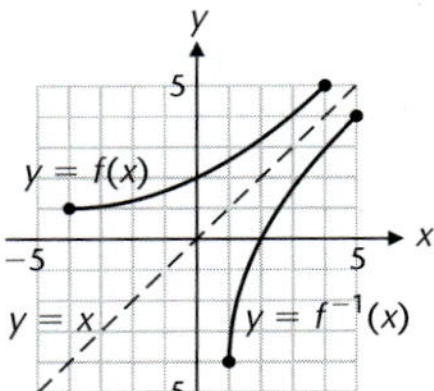

47. Range of $f^{-1} = [-5, 3]$; domain of $f^{-1} = [-3, 5]$

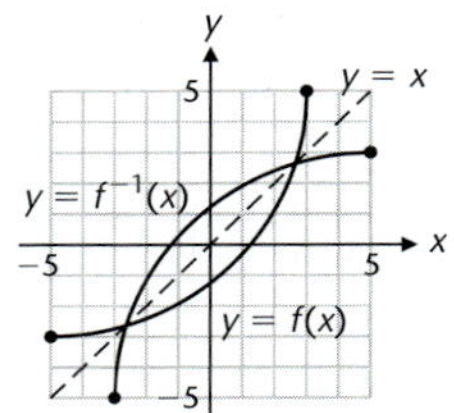

49.

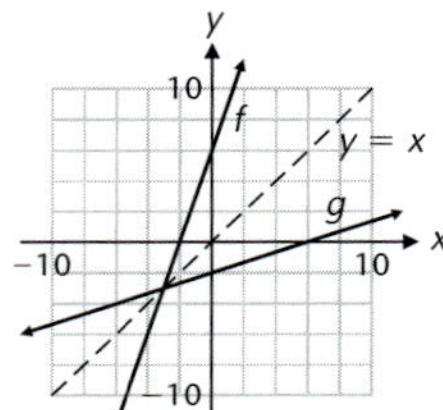

51.

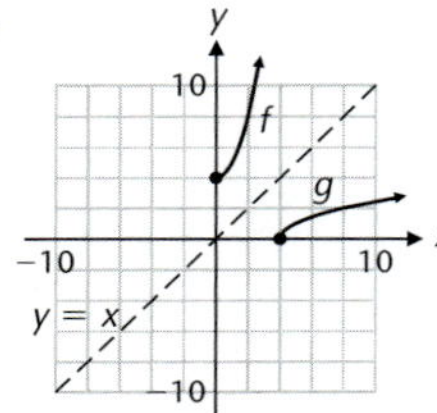

53.

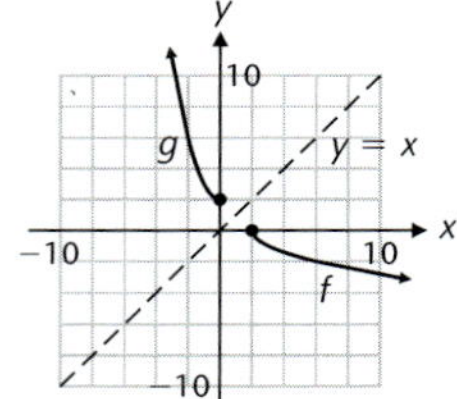

55. $f^{-1}(x) = \frac{1}{3}x$ **57.** $f^{-1}(x) = \dfrac{x+3}{4}$ **59.** $f^{-1}(x) = 10x - 6$

61. $f^{-1}(x) = \dfrac{x+2}{x}$ **63.** $f^{-1}(x) = \dfrac{2x}{1-x}$ **65.** $f^{-1}(x) = \dfrac{4x+5}{3x-2}$

67. $f^{-1}(x) = \sqrt[3]{x-1}$ **69.** $f^{-1}(x) = (4-x)^5 - 2$

71. $f^{-1}(x) = 16 - 4x^2, x \geq 0$ **73.** $f^{-1}(x) = (3-x)^2 + 2, x \leq 3$

75. The x intercept of f is the y intercept of f^{-1} and the y intercept of f is the x intercept of f^{-1}.

77. $f^{-1}(x) = 1 + \sqrt{x-2}$ **79.** $f^{-1}(x) = -1 - \sqrt{x+3}$

81. $f^{-1}(x) = \sqrt{9 - x^2}$; domain of $f^{-1} = [-3, 0]$; range of $f^{-1} = [0, 3]$

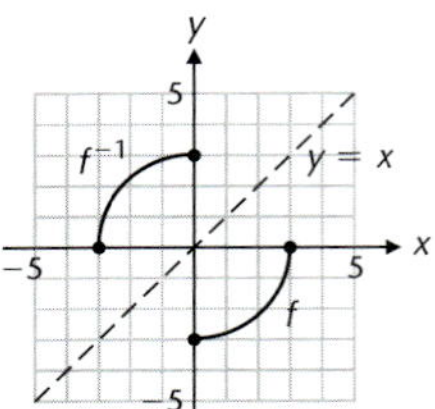

83. $f^{-1}(x) = -\sqrt{9 - x^2}$; domain: $f^{-1} = [0, 3]$; range of $f^{-1} = [-3, 0]$

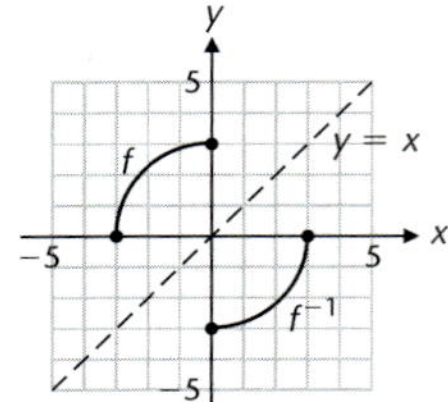

85. $f^{-1}(x) = \sqrt{2x - x^2}$; domain of $f^{-1} = [1, 2]$; range of $f^{-1} = [0, 1]$

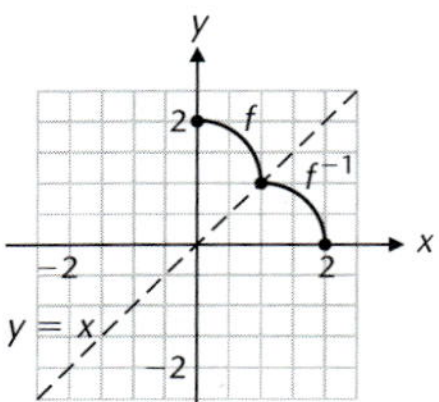

87. $f^{-1}(x) = -\sqrt{2x - x^2}$; domain of $f^{-1} = [0, 1]$; range of $f^{-1} = [-1, 0]$

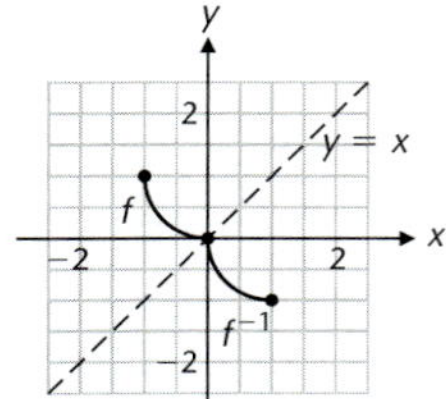

89. $f^{-1}(x) = \dfrac{x-b}{a}$

91. $a = 1$ and $b = 0$ or $a = -1$ and b arbitrary.

95. (A) $f^{-1}(x) = 2 - \sqrt{x}$ (B) $f^{-1}(x) = 2 + \sqrt{x}$

97. (A) $f^{-1}(x) = 2 - \sqrt{4 - x^2}, 0 \le x \le 2$
(B) $f^{-1}(x) = 2 + \sqrt{4 - x^2}, 0 \le x \le 2$

99. (A) $200 \le q \le 1{,}000$ (B) $p = \dfrac{15{,}000}{q} - 5$; domain: $200 \le q \le 1{,}000$; range: $10 \le p \le 70$

101. $R(x) = 50x - 0.025x^2$

Chapter 1 Review

1. Xmin = −4, Xmax = 9, Ymin = −6, Ymax = 7 *(1.1)*

2. (A) A one-to-one function with domain {1, 2, 3} and range {1, 4, 9}. The inverse function is {(1, 1), (4, 2), (9, 3)} with domain {1, 4, 9} and range {1, 2, 3}.
(B) Not a function
(C) A function that is not one-to-one. The domain is {−2, −1, 0, 1, 2} and the range is {2}.
(D) A one-to-one function with domain {−2, −1, 0, 1, 2} and range {−2, −1, 1, 2, 3}. The inverse function is {(2, −2), (3, −1), (−1, 0), (−2, 1), (1, 2)} with domain {−2, −1, 1, 2, 3} and range {−2, −1, 0, 1, 2}. *(1.2)*

3. (A) Not a function (B) A function (C) A function (D) Not a function *(1.2)*

4. (A) −1 (B) 24 (C) 0 (D) 0 *(1.2)*

5.

x	−3	−2	−1	0	1	2	3
$(f - g)(x)$	7	5	3	1	1	1	1

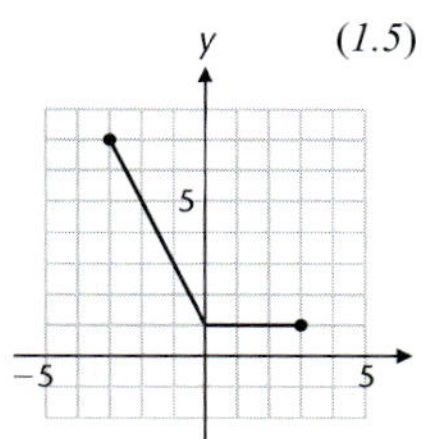

(1.5)

6.

x	−3	−2	−1	0	1	2	3
$(fg)(x)$	−12	−6	−2	0	0	2	6

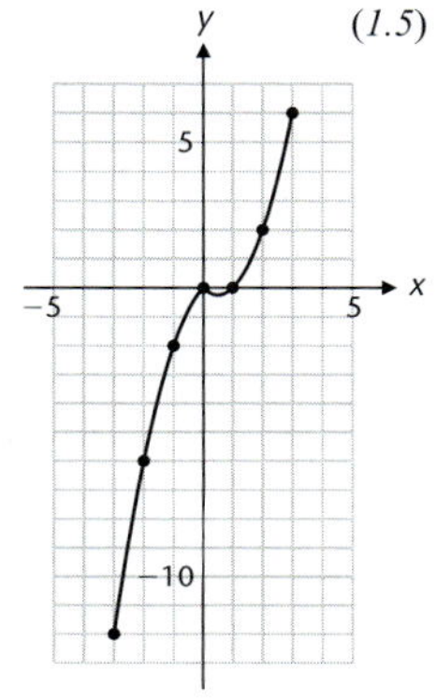

(1.5)

7. 2 *(1.5)* **8.** 1 *(1.5)* **9.** 0 *(1.5)* **10.** 2 *(1.5)*

11. No *(1.6)* **12.** Yes *(1.6)*

13. (A) Odd (B) Even (C) Neither *(1.4)*

14. $f(-4) = 4, f(0) = -4, f(3) = 0, f(5)$ is not defined *(1.2, 1.3)*

15. $x = -2, x = 1$ *(1.2, 1.3)*

16. Domain: [−4, 5); range: [−4, 4] *(1.3)*

17. Increasing: [0, 5); decreasing: [−4, 0] *(1.3)*

18. $x = 0$ *(1.3)*

19. *(1.4)*

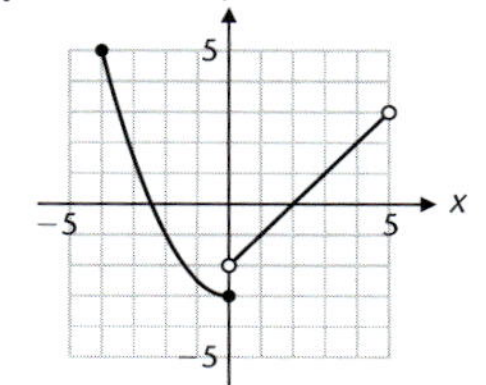

20. *(1.4)*

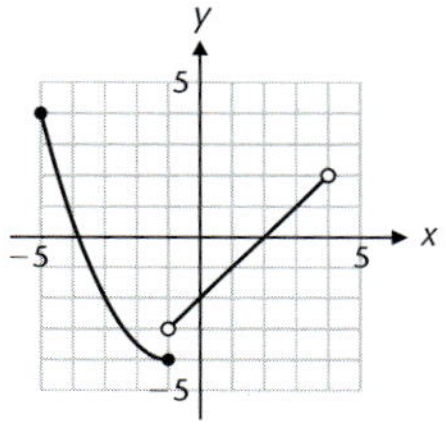

21. *(1.4)*

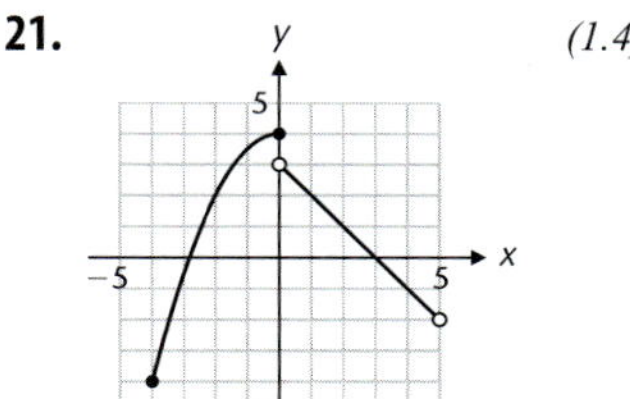

22. *(1.4)*

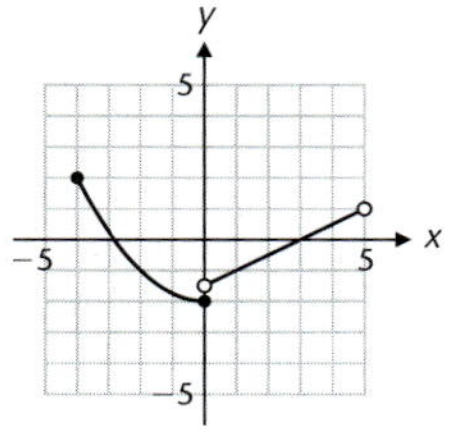

23. *(1.4)*

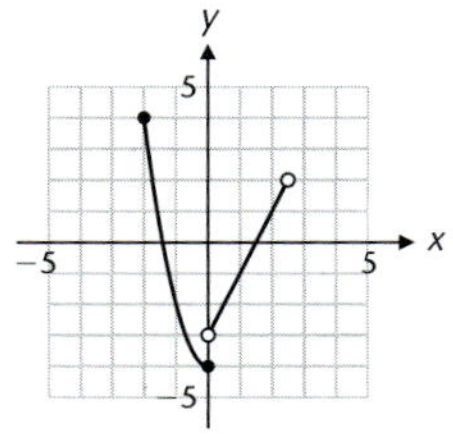

24. (1.4)

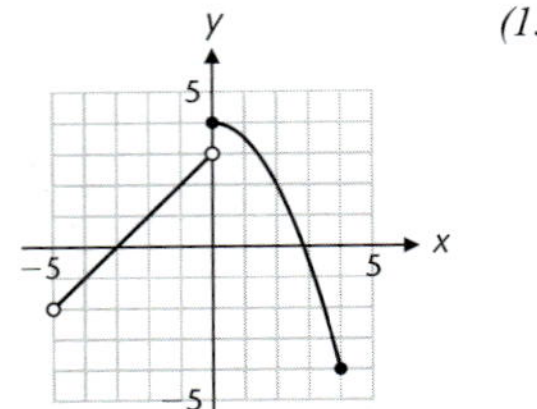

25. (A) g (B) m (C) n (D) f (1.4)

26. (A) $(f/g)(x) = (x^2 - 4)/(x + 3)$; domain of $f/g = (-\infty, -3) \cup (-3, \infty)$
(B) $(g/f)(x) = (x + 3)/(x^2 - 4)$; domain of $g/f = (-\infty, -2) \cup (-2, 2) \cup (2, \infty)$
(C) $(f \circ g)(x) = x^2 + 6x + 5$; domain of $f \circ g = (-\infty, \infty)$
(D) $(g \circ f)(x) = x^2 - 1$; domain of $g \circ f = (-\infty, \infty)$ (1.5)

27. (A) 0 (B) 1 (C) 2 (D) 0 (1.3)

28. (A) $-2, 0$ (B) $-1, 1$ (C) No solution
(D) $x = 3$ and $x < -2$ (1.3)

29. Domain $= (-\infty, \infty)$; range $= (-3, \infty)$ (1.3)

30. $[-2, -1], [1, \infty)$ (1.3) **31.** $[-1, 1)$ (1.3)

32. $(-\infty, -2)$ (1.3) **33.** $x = -2, x = 1$ (1.3)

34. (0.64, −12.43), (23.36, 78.43) (1.1)

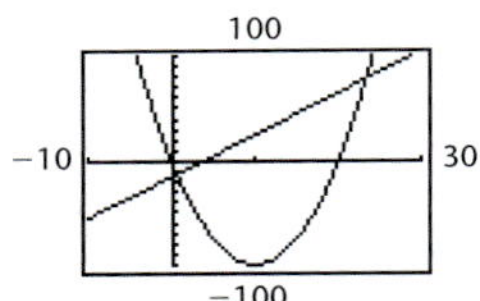

35. (−4.26, 2.72), (88.70, 30.61) (1.1)

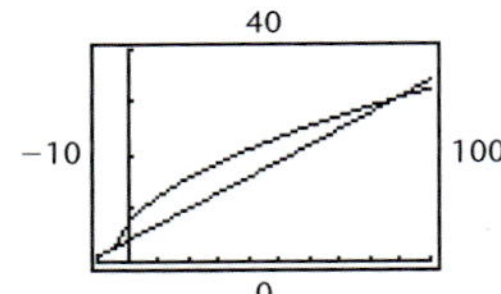

36. (A) −6.71, 5.67, 21.04 (B) 22.99 (1.1, 1.3)
(C) −7.97, 8.28, 19.70 (D) −9.84

37. This equation defines a function. For any real number x, the number $y = 5 - 0.5x$ is the only number that corresponds to x. (1.2)

38. This equation does not define a function. For example, the ordered pairs (2, 2) and (2, −2) both satisfy the equation. (1.2)

39. (A) All real numbers (B) All real numbers except $t = 5$
(C) $w \geq 0$ or $[0, \infty)$ (1.2)

40. $5 + 2h$ (1.2) **41.** $f(x) = 4x^3 - \sqrt{x}$ (1.2)

42. The function f multiplies the square of the domain element by 3, adds 4 times the domain element, and then subtracts 6. (1.2)

43. x intercepts: 0, 3.30; y intercept: 0; local maximum: $x = 1.31$, $y = 5.15$; domain: $[0, \infty)$; range: $(-\infty, 5.15]$ (1.3)

44. x intercepts: −26.58, −3.58, 3.15; y intercept: −300; local maximum: $x = -18.00$, $y = 2{,}616.00$; local minimum: $x = 0.00$, $y = -300$; domain: $(-\infty, \infty)$; range: $(-\infty, \infty)$ (1.3)

45. (A)

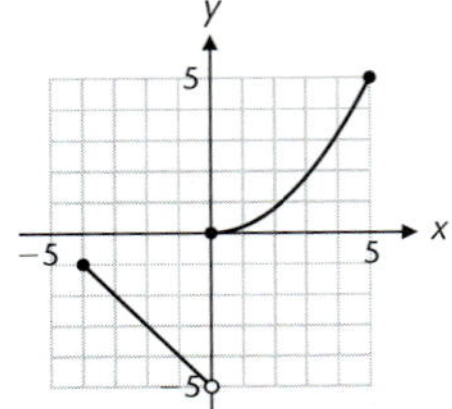

(B) Domain: $[-4, 5]$; range: $(-5, -1] \cup [0, 5]$ (C) $x = 0$
(D) Decreasing on $[-4, 0)$; increasing on $[0, 5]$ (1.3)

46. The graph of f increases on $(-\infty, -4.47]$ to a local maximum value, $f(-4.47) \approx 22.89$, decreases on $[-4.47, 4.47]$ to a local minimum value, $f(4.47) \approx -12.89$, and then increases on $[4.47, \infty)$. (1.3)

47. (A) Reflected in x axis (B) Shifted down three units
(C) Shifted left three units
(D) Contracted horizontally by $\frac{1}{2}$ (1.4)

48. (A) $-(x - 2)^2 + 4$ (B) $4 - 4\sqrt{x}$ (1.4)

49. $g(x) = 8 - 3|x - 4|$ (1.4)

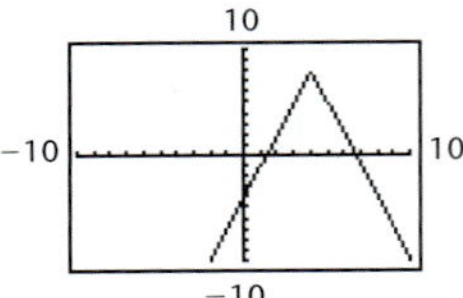

50. $t(x) = 0.25x^2 + x - 3$ (1.4)

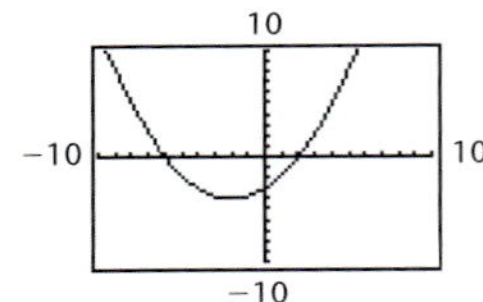

51. Yes (1.6)

52. The function k cubes the input and then adds 5. The inverse function subtracts 5 from the input and then takes the cube root of the result. $k^{-1}(x) = \sqrt[3]{x - 5}$ (1.6)

53. Domain: $x \geq 0, x \neq 9$ or $[0, 9) \cup (9, \infty)$ (1.2)

54. (A) $(f \circ g)(x) = \sqrt{|x|} - 8$; $(g \circ f)(x) = |\sqrt{x} - 8|$
(B) Domain of $f \circ g$ is the set of all real numbers. The domain of $(g \circ f)$ is $[0, \infty)$. (1.5)

55. Functions f, h, and F are one-to-one (1.6)

56. $f^{-1}(x) = (x + 7)/3$;
domain of f^{-1} = range of $f^{-1} = (-\infty, \infty)$ (1.6)

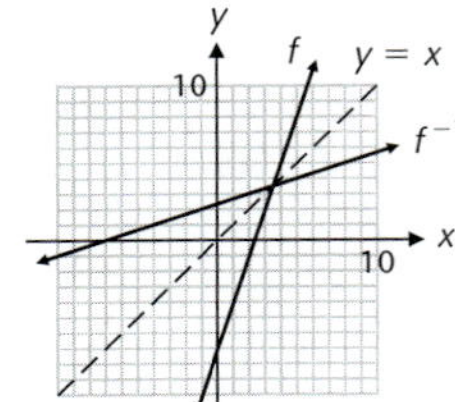

57. $f^{-1}(x) = x^2 + 1$;
domain of $f^{-1} = [0, \infty)$; range of $f^{-1} = [1, \infty)$ *(1.6)*

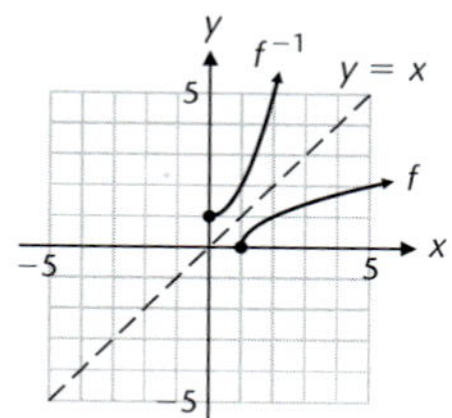

58. $f^{-1}(x) = \sqrt{x + 1}$;
domain of $f^{-1} = [-1, \infty)$; range of $f^{-1} = [0, \infty)$ *(1.6)*

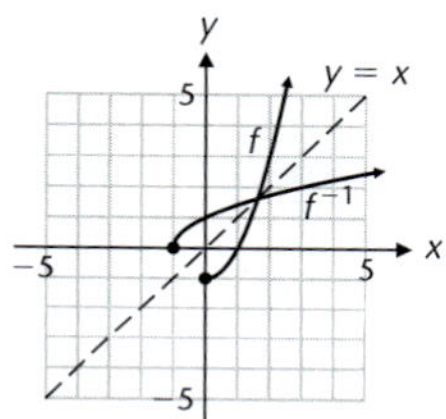

59. (A) One possible answer:

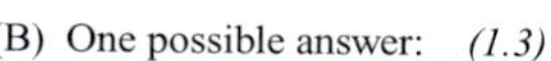
(B) One possible answer: *(1.3)*

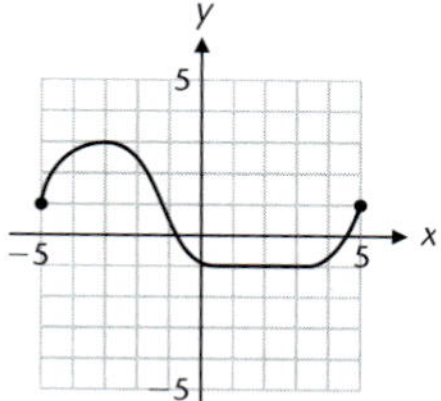

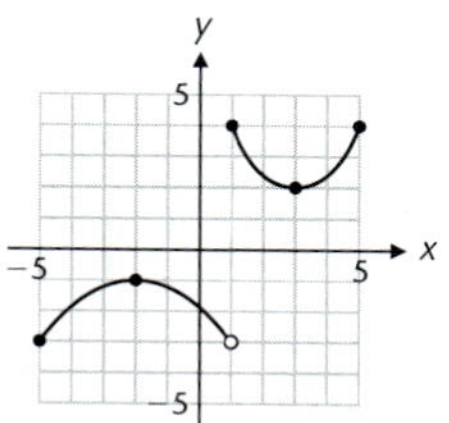

60. The function squares the input, multiplies the result by 2, then subtracts 4 times the input and adds 5. $g(t) = 2t^2 - 4t + 5$ *(1.2)*

61. Domain: all real numbers except $x = 2$; range: $y > -3$ or $(-3, \infty)$; discontinuous at $x = 2$ *(1.3)*

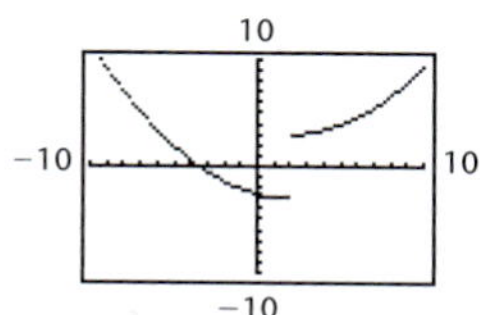

62. (A)

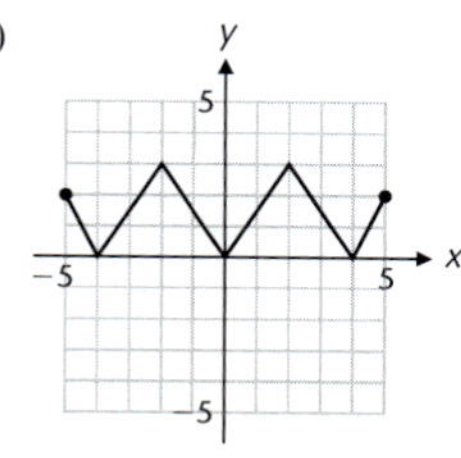

(B) *(1.4)*

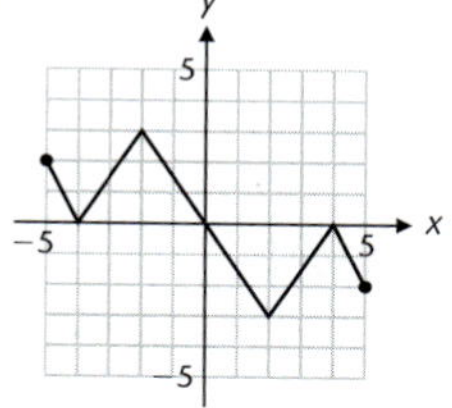

63. (A) $6x - 5 + 3h$ (B) $3x + 3a - 5$ *(1.2)*

64. (A) The graph must cross the x axis exactly once.
(B) The graph can cross the x axis at most once. *(1.3)*

65. (A) $f(x) = \begin{cases} 2 & \text{for } -3 < x \le -2 \\ 1 & \text{for } -2 < x \le -1 \\ 0 & \text{for } -1 < x < 1 \\ 1 & \text{for } 1 \le x < 2 \\ 2 & \text{for } 2 \le x < 3 \end{cases}$

(B)

(C) Range: nonnegative integers

(D) Discontinuous at all integers except 0 (E) Even *(1.3, 1.4)*

66. (A) [1, 3]
(B) $q = g^{-1}(p) = \dfrac{4{,}500}{p} - 500$; domain: [1, 3];
range: [1,000, 4,000]
(C) $R(p) = 4{,}500 - 500p$ (D) $R(q) = 9q/(1 + 0.002q)$ *(1.6)*

67. $P(p) = -6{,}500 + 600p - 10p^2$; the maximum profit is \$2,500 when the price is \$30. *(1.5)*

68. (A) 11.3 seconds (B) 155 meters *(1.3)*

69. The maximum value is approximately 10,480 cubic inches when the flap is 6.8 inches wide. *(1.3)*

70. (A)

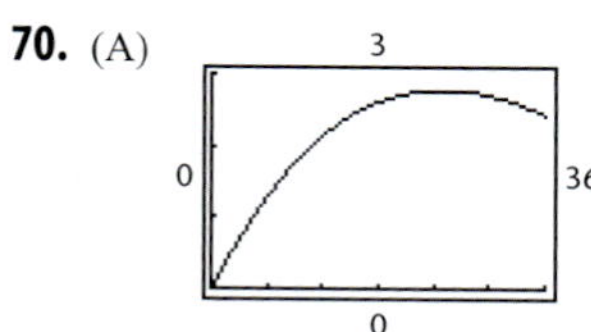

(B) The function increases on [0, 24.8] to a local maximum of 2.8 cubic centimeters per second, and then decreases on [24.8, 36]. *(1.3)*

71. (A)

x	1	2	3	4	5	6	7	8	9	10	11	12	13	14	15	16
$f(x)$	0	1	2	0	1	2	3	4	0	1	2	3	4	5	6	0

(B) $f(n^2) = 0$
(C) If $f(x) = 0$, then x is a perfect square integer. *(1.3)*

72. (A)

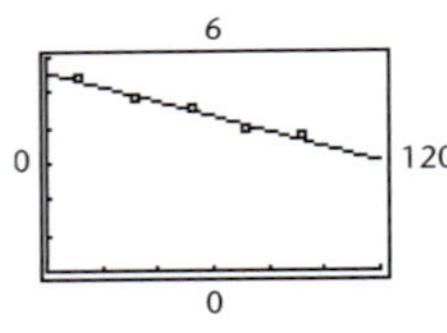

x	12	32	52	72	92
Data	5.41	4.81	4.51	4.00	3.75
$f(x)$	5.32	4.90	4.48	4.06	3.64

(B) 3.302 minutes *(1.2)*

73. $T(x) = \begin{cases} 0.02x & \text{if} \quad 0 \le x \le 3{,}000 \\ 0.03x - 30 & \text{if} \quad 3{,}000 < x \le 5{,}000 \\ 0.05x - 130 & \text{if} \quad 5{,}000 < x \le 17{,}000 \\ 0.0575x - 257.5 & \text{if} \quad 17{,}000 \le x \end{cases}$

x	\$2,000	\$4,000	\$10,000	\$30,000
$T(x)$	\$40	\$90.00	\$370	\$1,467.50

(1.3)

CHAPTER 2

Exercise 2.1

1. Rise = 3, run = 5, slope = $\frac{3}{5} = 0.6$, $3x - 5y = -4$
3. Rise = 2, run = 8, slope = $\frac{2}{8} = 0.25$, $x - 4y = -8$
5. Rise = -3, run = 5, slope = $-\frac{3}{5} = -0.6$, $3x + 5y = -2$
7. x intercept: $x = -2$, y intercept: $y = 2$, slope = 1, $y = x + 2$
9. x intercept: $x = -2$, y intercept: $y = -4$, slope = -2, $y = -2x - 4$
11. x intercept: $x = 3$, y intercept: $y = -1$, slope = $\frac{1}{3}$, $y = \frac{1}{3}x - 1$
13. The graph of f is vertically expanded by 3 and shifted down seven units.
15. The graph of f is vertically contracted by $\frac{1}{2}$, reflected in the x axis, and shifted down four units.
17. Not linear **19.** Linear **21.** Linear **23.** Linear
25. Not linear
27. x intercept: $\frac{20}{3}$; y intercept: 4; slope: $-\frac{3}{5}$

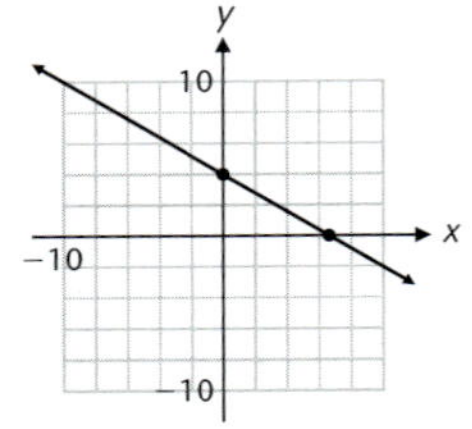

29. x intercept: 0; y intercept: 0; slope: $-\frac{3}{4}$

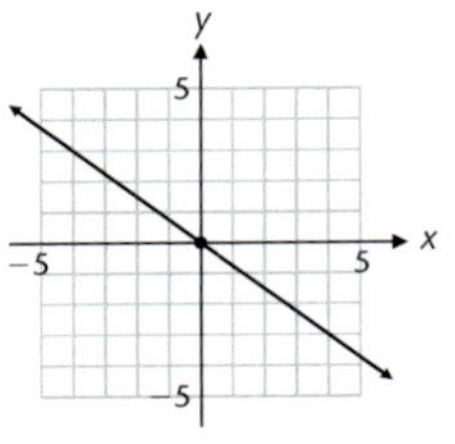

31. x intercept: $\frac{15}{2}$; y intercept: -5; slope: $\frac{2}{3}$

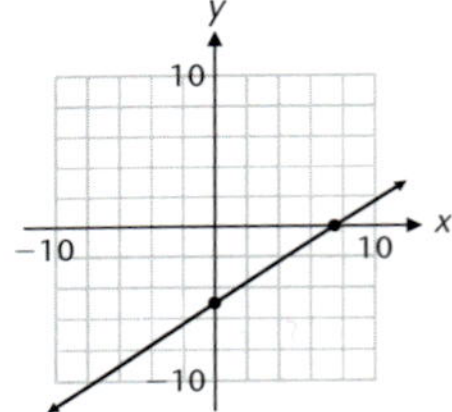

33. x intercept: -4; y intercept: 8; slope: 2

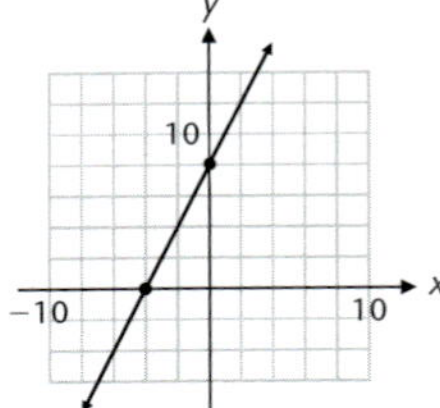

35. x intercept: -3; y intercept: none; slope is not defined

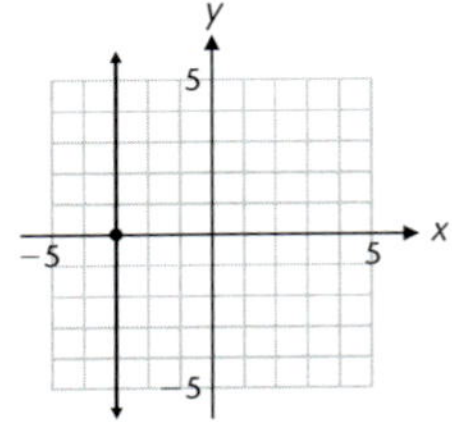

37. x intercept: none; y intercept: 3.5; slope: 0

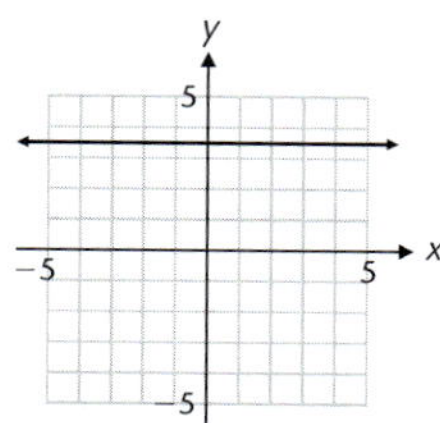

39. $y = x$ **41.** $y = -\frac{2}{3}x - 4$ **43.** $y = -3x + 4$
45. $y = -\frac{2}{5}x + 2$ **47.** $y = -2x + 8$ **49.** $y = -\frac{4}{3}x + \frac{8}{3}$
51. $y = 4$ **53.** $x = 4$ **55.** $y = \frac{3}{4}x + 3$ **57.** $3x - y = -13$
59. $3x - y = 9$ **61.** $x = 2$ **63.** $x = 3$ **65.** $3x - 2y = 15$
67. $3x - y = 4$
71. (A) $f(x) = \frac{5}{8}x - \frac{19}{8}$ (B) $g(x) = \frac{8}{5}x + \frac{19}{5}$
(C) f and g are inverse functions

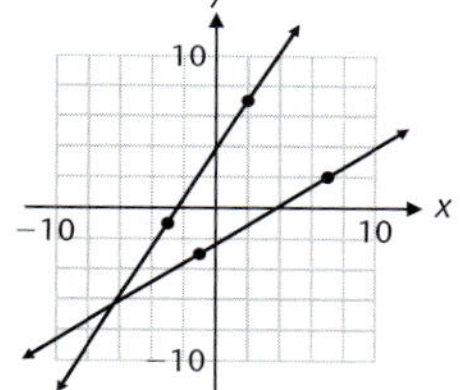

73. slope $AB = -\frac{3}{4} =$ slope DC
75. (slope AB) (slope BC) $= (-\frac{3}{4})(\frac{4}{3}) = -1$
77. $6x + 8y = -9$
79. $3x + 4y = 25$ **81.** $x - y = 10$

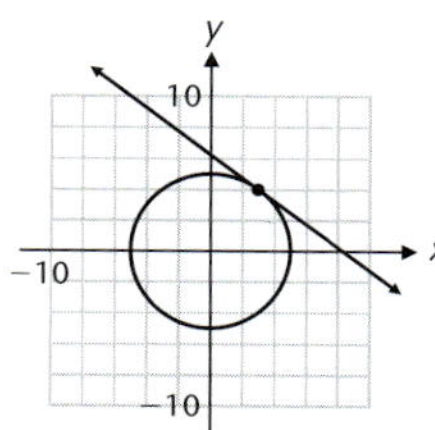

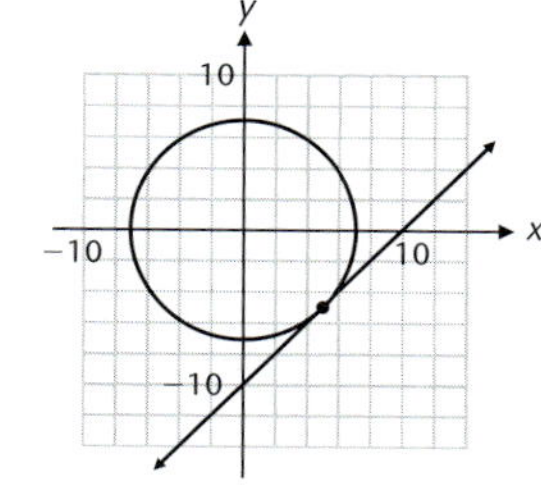

83. $232 = 5x - 12y$

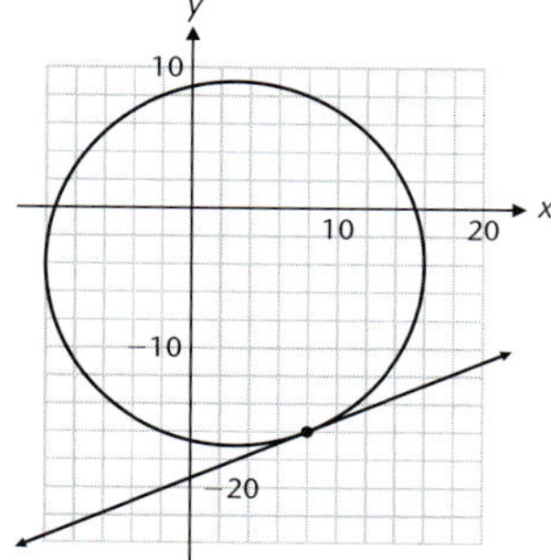

85. (A)

(B) Varying C produces a family of parallel lines.
87. The function g is never linear.
91. (A) $F = \frac{9}{5}C + 32$ (B) 68°F, 30°C (C) $\frac{9}{5}$
93. (A) $V = -1{,}600t + 8{,}000$, $0 \le t \le 5$ (B) $V = \$3{,}200$
(C) $-1{,}600$
95. $C(x) = 124 + 0.12x$, 1,050 doughnuts
97. (A) $C(x) = 2{,}147 + 75x$
(B) The daily fixed costs are \$2,147 and the variable cost per club is \$75.
99. (A) $R = 0.00152C - 0.159$, $C \ge 210$ (B) $R = 0.236$
(C) Slope $= 0.00152$; coronary risk increases 0.00152 per unit increase in cholesterol above the 210 cholesterol level.
101. (A) $T = -5A + 70$, $A \ge 0$ (B) $A = 14{,}000$ feet
(C) Slope $= -5$; the temperature changes -5°F for each 1,000-foot rise in altitude.
103. (A) $h = 1.13t + 12.8$ (B) $t = 32.9$ hours

Exercise 2.2

1. $x = c, x = f$ **3.** $x = b, x = e$ **5.** Identity
7. Conditional equation, $x = 0$ **9.** Contradiction
11. Contradiction **13.** Identity as long as $x \ne 1$
15. Conditional equation, $x = 0$ **17.** Contradiction
19. 18 **21.** 9 **23.** $\frac{11}{2}$ or 5.5 **25.** $t = -\frac{5}{4}$
27. 3 **29.** No solution **31.** $x = 7$ **33.** $x = \frac{6}{5}$
35. Identity **37.** $x = 3.4$ **39.** No solution

41. $d = \frac{a_n - a_1}{n - 1}$ **43.** $f = \frac{d_1 d_2}{d_2 + d_1}$ **45.** $a = \frac{A - 2bc}{2b + 2c}$
47. $x = \frac{5y + 3}{2 - 3y}$
49. The graphs are identical for $x \ge 0$. For $x < 0$, each is the reflection of the other in the x axis.
51. $l = 2w$ **53.** $w = \frac{1}{2}l$ **55.** $l = w + 3$ **57.** $l = w - 4$
59. (A) $y = 0.5x - 1.5$ (B) $y = 2x + 3$
The graphs are symmetric with respect to the line $y = x$. Each function is the inverse of the other.

61. $x = 2$ **63.** Identity, $x \neq 0, 1$
65. 27, 28, 29 **67.** 8, 10, 12, 14 **69.** 10 inches by 20 inches
71. $19,750 **73.** 5,000 trout **75.** 10 gallons
77. 11.25 liters **79.** (A) 216 miles (B) 225 miles
81. 90 miles
83. (A) $T = 30 + 25(x - 3)$ (B) 330°C (C) 13 kilometers
85. Model: $y = 0.227x + 4.20$, ticket price: $6.47
87. Men: $y = -0.111x + 51.6$, women: $y = -0.162x + 58.5$, in 2104 the times for men and women are equal.
89. Supply: $y = 4.95x - 4.22$, demand: $y = -7.65x + 25.7$, equilibrium price: $2.37 per bushel

Exercise 2.3

1. $f(x) = (x - 2)^2 + 1$ **3.** $h(x) = -(x + 1)^2$
5. $m(x) = (x - 2)^2 - 3$
7. The graph of f is the graph of $y = x^2$ shifted to the right two units and up one unit.
9. The graph of h is the graph of $y = x^2$ reflected in the x axis and shifted to the left one unit.
11. The graph of m is the graph of $y = x^2$ shifted to the right two units and down three units.
13. k **15.** m **17.** h

19.

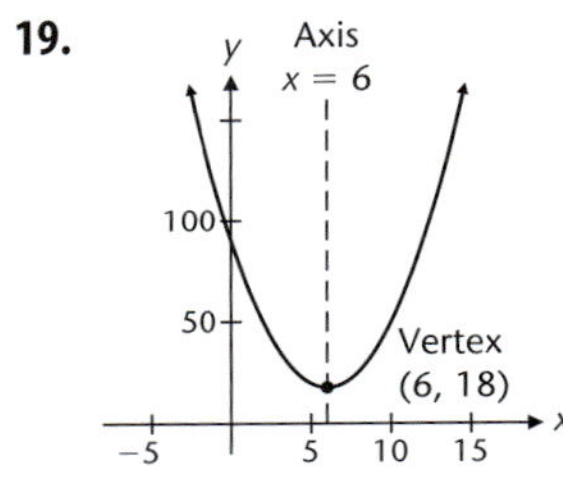

21.

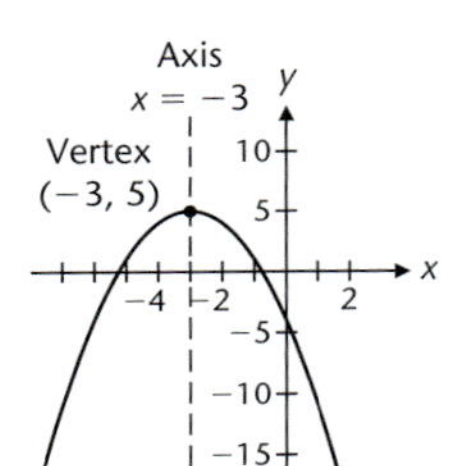

23.

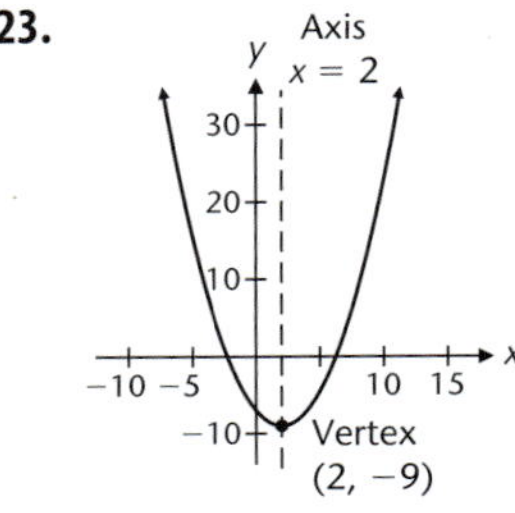

25. Increasing: $[2.25, \infty)$; decreasing: $(-\infty, 2.25]$; range: $[4.75, \infty)$
27. Increasing: $(-\infty, 2.2]$; decreasing: $[2.2, \infty)$; range: $(-\infty, 60.4]$
29. $y = 2x^2 - 4x - 2$ **31.** $y = -0.5x^2 - x + 3.5$
33. $y = x^2 - 2x - 3$ **35.** $y = -0.5x^2 + 2x + 2.5$
37. $y = -2x^2 + 16x - 24$ **39.** $y = -0.5x^2 - 4x + 4$
41. $y = 5x^2 + 50x + 100$ **43.** $g^{-1}(x) = -1 + \sqrt{x + 4}$
45. $p^{-1}(x) = 1 - \sqrt{4 - x}$
47. All the graphs are translations of the graph of $y = x^2$.
51. Center: (3, 2); radius: 7 **53.** Center: $(-4, 1)$; radius: 5

57. $y = 2x - 1$

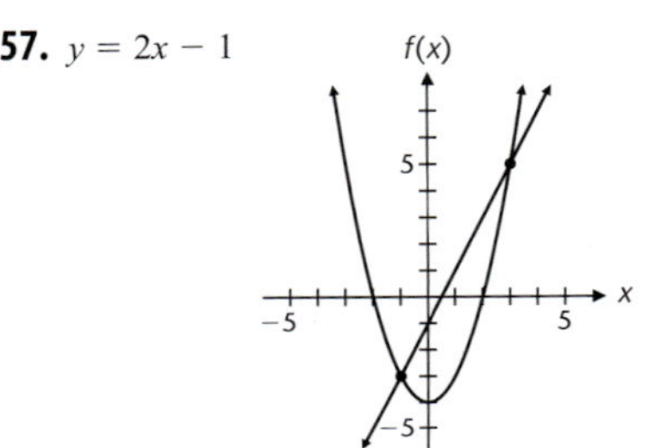

59. (A) $1 + h$ (B)

h	$slope = 1 + h$
1	2
0.1	1.1
0.01	1.01
0.001	1.001

; The slope seems to be approaching 1.

61. The minimum product is -225 for the numbers 15 and -15. There is no maximum product.
63. (A) $A(x) = 5{,}000 + 50x - x^2, 0 \le x \le 100$ (B) $x = 25$ (C) 75 feet × 75 feet
65. After 25 seconds **67.** 100 feet
69. (A) $d(t) = 176t - 16t^2, 0 \le t \le 11$ (B) 1.68 seconds; 9.32 seconds
71. (A) $h(x) = -0.14x^2 + 14, 0 \le x \le 10$ (B) No (C) 11.76 feet (D) 7.56 feet
73. (A) $p = d(x) = -0.129x + 9.27$ (B) $4.71

Exercise 2.4

1. $7 + 5i$ **3.** $5 + 3i$ **5.** $2 + 4i$ **7.** $5 + 9i$ **9.** $4 - 3i$
11. -24 or $-24 + 0i$ **13.** $-12 - 6i$ **15.** $15 - 3i$
17. $-4 - 33i$ **19.** 65 or $65 + 0i$ **21.** $\frac{2}{5} - \frac{1}{5}i$
23. $\frac{3}{13} + \frac{11}{13}i$ **25.** $5 + 3i$ **27.** 4 **29.** $4i$ **31.** $4i$
33. -4 **35.** $7 - 5i$ **37.** $-3 + 2i$ **39.** $8 + 25i$

41. $\frac{5}{7} - \frac{2}{7}i$ **43.** $\frac{2}{13} + \frac{3}{13}i$ **45.** $-\frac{2}{5}i$ or $0 - \frac{2}{5}i$ **47.** $\frac{3}{2} - \frac{1}{2}i$
49. $-6i$ or $0 - 6i$ **53.** $i^{18} = -1, i^{32} = 1, i^{67} = -i$
55. $x = 3, y = -2$ **57.** $x = 2, y = 3$ **59.** $0.6 + 1.2i$
61. $1.5 + 0.5i$ **65.** $(a + c) + (b + d)i$
67. $a^2 + b^2$ or $(a^2 + b^2) + 0i$ **69.** $(ac - bd) + (ad + bc)i$
71. $i^{4k} = (i^4)^k = (i^2 \cdot i^2)^k = [(-1)(-1)]^k = 1^k = 1$
75. $3 - i, -3 + i$
79. (1) Definition of addition; (2) Commutative (+) property for R; (3) Definition of addition

Exercise 2.5

1. $u = 0, 2$ **3.** $y = \frac{2}{3}$ (double root) **5.** $x = \frac{3}{2}, 4$
7. $x = 3 \pm 2\sqrt{3}$ **9.** $t = 2 \pm 2i$
11. $m = -1 \pm 2i\sqrt{2}$ **13.** $d = -5, 2.5$
15. $v = \frac{1}{2} \pm \frac{1}{2}i$ **17.** $y = -\frac{3}{8} \pm \frac{3i\sqrt{15}}{8}$
19. $x = 5 \pm 2\sqrt{7}$ **21.** $x = 2 \pm 2i$ **23.** $x = \frac{2 \pm \sqrt{2}}{2}$
25. $x = \frac{1}{5} \pm \frac{3}{5}i$
27. Two real zeros **29.** Two imaginary zeros
31. One real zero **33.** One real zero
35. Two real zeros **37.** Two imaginary zeros
39. $x = 3 \pm 2\sqrt{3}$ **41.** $y = \frac{3 \pm \sqrt{3}}{2}$ **43.** $x = \frac{1 \pm \sqrt{7}}{3}$
45. $x = -\frac{5}{4}, \frac{2}{3}$ **47.** $x = \frac{3 \pm \sqrt{13}}{2}$ **49.** $t = \sqrt{\frac{2s}{g}}$
51. $I = \frac{E + \sqrt{E^2 - 4RP}}{2R}$ **53.** $x = \frac{\sqrt{7} \pm i}{2}$ **55.** $x = \sqrt{3}$
57. $x = \frac{-\sqrt{3} \pm \sqrt{19}}{2}$ **59.** $x = \frac{5}{2} \pm \frac{i\sqrt{11}}{2}$ **61.** $x = 3$
63. $x = \frac{7}{2} \pm \frac{\sqrt{13}}{2}$ **65.** $x = 0, \frac{2}{3}$ **67.** $x = -\frac{5}{4} \pm \frac{i\sqrt{15}}{4}$
71. $x = -i, -2i$ **73.** $x = \sqrt{2} - i, -\sqrt{2} - i$
75. $x = 1, -\frac{1}{2} \pm \frac{1}{2}i\sqrt{3}$
81. The $\pm$ in front still yields the same two numbers even if a is negative.
83. 8, 13 **85.** 12, 14 **87.** At 8:06 A.M. **89.** 2.19 feet
91. (A) $A(w) = 400w - 2w^2, 0 \le w \le 200$ (B) $50 \le w \le 150$
(C) No, the maximum cross-sectional area is 20,000 square feet when $w = 100$ feet.
93. 52 miles
95. (A) $y = -0.000607x^2 + 0.0308x + 0.943$
(B) 2009 (C) 1.20 gallons
97. (A) $y = -0.326x^2 + 18.2x + 340$
(B) 2004 (C) 435 billion
99. (A) $y = 0.0476x^2 + 0.0452x - 0.357$ (B) 64 miles per hour
101. (A) $y = -0.00217x^2 + 0.140x + 1.90$
(B) 44.0 miles per hour, 3.86 miles per gallon, 2.27 hours, 25.9 gallons, $61.60

Exercise 2.6

1. T **3.** F **5.** F **7.** $2u^2 - 4u = 0, u = x^{-3}$
9. Not of quadratic type **11.** $\frac{10}{9} + 4u - 7u^2 = 0, u = \frac{1}{x^2}$
13. $x = 22$ **15.** $n = 8$ **17.** No solution **19.** $x = 0, 4$
21. $y = \pm 2, \pm i\sqrt{2}$ **23.** $x = \frac{1}{2}i$ **25.** $x = \frac{1}{8}, -8$
27. $m = 3, -2, \frac{1}{2} \pm \frac{\sqrt{7}}{2}i$ **29.** No solution **31.** $y = 1$
33. $x = 2$ **35.** $x = -\frac{3}{2} + \frac{1}{2}i$ **37.** $n = -\frac{3}{4}, \frac{1}{5}$ **39.** $y = \pm 3, \pm 1$
41. $y = 1, 16$ **43.** $m = 3, 7, 2, 8$ **45.** $x = -2$
47. $y = \pm\sqrt{\frac{3 \pm \sqrt{3}}{2}}$ (four roots) **49.** $m = 9, 16$ **51.** $t = 4, 81$
53. $x = -4, 39{,}596$ **55.** $x = \left(\frac{4}{5 \pm \sqrt{17}}\right)^5 \approx 0.016203, 1974.98$
57. 5.3 inches by 8.5 inches **59.** 2, 277 feet
61. (A) $A = w\sqrt{256 - w^2}, 0 < w < 16$
(B) 13.1 inches by 9.1 inches
(C) The maximum area is 128 square inches when the beam is a square with sides of 11.3 inches.
63. 1.65 feet or 3.65 feet

Exercise 2.7

1. (c, f) **3.** $(-\infty, b] \cup [e, \infty)$ **5.** $(-\infty, c] \cup [f, \infty)$
7. $(-\infty, b) \cup (e, \infty)$ **9.** $|x - 3| < 5$ **11.** $|y + 1| > 6$
13. $|a - 3| \le 5$ **15.** $|d + 2| \ge 4$ **17.** $(-\infty, 5); x < 5$
19. $(2, \infty); t > 2$ **21.** $(3, \infty); m > 3$ **23.** $(-5, 2); -5 < x < 2$
25. $(-\infty, 3) \cup (7, \infty); x < 3$ or $x > 7$ **27.** $[0, 8]; 0 \le x \le 8$
29. $[-5, 0]; -5 \le x \le 0$

31. y is no more than seven units from the origin: $[-7, 7]$; $-7 \le x \le 7$

33. w is at least seven units from the origin; $(-\infty, -7] \cup [7, \infty)$; $x \le -7$ or $x \ge 7$

35. s is less than three units from 5; $(2, 8)$; $2 < s < 8$

37. s is more than three units from 5; $(-\infty, 2) \cup (8, \infty)$; $s < 2$ or $s > 8$

39. u is no more than three units from -8; $[-11, -5]$; $-11 \le u \le -5$

41. u is at least three units from -8; $(-\infty, -11] \cup [-5, \infty)$; $u \le -11$ or $u \ge -5$

43. $(-2, 3]$; $-2 < t \le 3$ **45.** $[-30, 18)$; $-30 \le x < 18$

47. $(-\infty, -14)$; $q < -14$ **49.** No solution, $\varnothing$

51. $(1, 3)$; $1 < x < 3$ **53.** $(-\infty, -3] \cup [7, \infty)$; $x \le -3$ or $x \ge 7$

55. $[1, \frac{11}{3}]$; $1 \le x \le \frac{11}{3}$ **57.** $(-\infty, -1) \cup (5, \infty)$; $t < -1$ or $t > 5$

59. $(-\infty, -11] \cup [-6, \infty)$; $u \le -11$ or $u \ge -6$

61. (A) and (C): a and b have the same sign
(B) and (D): a and b have opposite signs

63. $>$ **65.** $>$ **67.** $(2.9, 3) \cup (3, 3.1)$ **69.** $(-c, c) \cup (c, 3c)$

71. If $a > 0$, the solution set is $(-\infty, r_1) \cup (r_2, \infty)$. If $a < 0$, the solution set is (r_1, r_2).

73. If $a > 0$, the solution set is R, the set of real numbers. If $a < 0$, the solution set is $\{r\}$.

75. $x^2 \ge 0$ **77.** $|A - 12.436| < 0.001$; $(12.435, 12.437)$

79. (A) $x > 40{,}625$ (B) $x = 40{,}625$

81. (B) $x > 52{,}000$ (C) Raise wholesale price \$3.50 to \$66.50

83. $|N - 2.37| \le 0.005$ **85.** $33 < x < 122$

87. 16°C to 27°C **89.** (A) $d(t) = -16t^2 + 128t$ (B) $3 < t < 5$

91. $T(x) = 28x + 2$; $5.286 < x < 7.071$

93. $y = -0.00738x + 28.2$; $5{,}180 < x < 7{,}890$

95. $y = -0.0623x + 877$; $3{,}640 < x < 8{,}460$

97. (A) $p = d(x) = -0.0000727x + 3.50$, $0 \le x \le 48{,}100$
(B) $R(x) = -0.0000727x^2 + 3.50x$, $0 \le x \le 48{,}100$
$C(x) = 20{,}000 + 0.5x$, $x \ge 0$
(C) The company will break even at the sales levels of 8,360 gallons or 32,900 gallons. Any sales between these two levels will produce a profit. Any sales less than 8,360 or greater than 32,900 will result in a loss.
(D) The maximum profit is \$10,900 when the sales are 20,600 gallons.

Chapter 2 Review

1. Rise $= -2$, run $= 5$, slope $= -\frac{2}{5} = -0.4$, $2x + 5y = 7$ *(2.1)*

2. Slope: $-\frac{3}{2}$ *(2.1)*

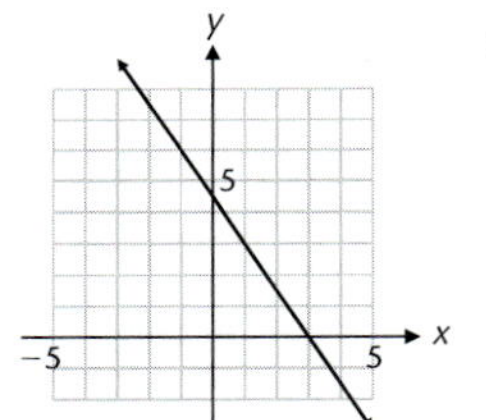

3. $2x + 3y = 12$ *(2.1)* **4.** $y = -\frac{2}{3}x + 2$ *(2.1)*

5. Vertical: $x = -3$, slope not defined; horizontal: $y = 4$, slope $= 0$ *(2.1)*

6. (A) $x = 21$ (B) $x = \frac{30}{11}$ *(2.2)*

7. (A) $f(x) = -(x + 1)^2 + 4$
(B) It is the same as the graph of $y = x^2$ reflected in the x axis, shifted left one unit, and up four units.
(C) $x = -3, 1$ *(2.3, 2.5)*

8. (A) $f(x) = (x - \frac{3}{2})^2 - \frac{17}{4}$
(B) It is the same as the graph of $y = x^2$ shifted right $\frac{3}{2}$ units, and down $\frac{17}{4}$ units.
(C) $x = \dfrac{3 \pm \sqrt{17}}{2}$ *(2.3, 2.5)*

9. (A) $3 - 6i$ (B) $15 + 3i$ (C) $2 + i$ *(2.4)*

10. No solution *(2.2)* **11.** All real numbers *(2.2)*

12. $x = 2$ *(2.1, 2.5)* **13.** $x = \pm\dfrac{\sqrt{14}}{2}$ *(2.6)*

14. $x = 0, 2$ *(2.5)* **15.** $x = \frac{1}{2}, 3$ *(2.5)*

16. $m = -\dfrac{1}{2} \pm \dfrac{\sqrt{3}}{2}i$ *(2.5)* **17.** $y = \dfrac{3 \pm \sqrt{33}}{4}$ *(2.5)*

18. $x = 2, 3$ *(2.6)* **19.** $x \ge 1$; $[1, \infty)$ *(2.2)*

20. $(-5, 4)$; $-5 < x < 4$; *(2.2)*

21. $x < -2$ or $x > 6$; $(-\infty, -2) \cup (6, \infty)$ *(2.5)*

23. $3x + 2y = -6$ *(2.1)*

24. (A) $y = -2x - 3$ (B) $y = \frac{1}{2}x + 2$ *(2.1)*

25. $-14 < y < -4$ *(2.2)* **26.** $x \le 2.5$ or $5.5 \le x$ *(2.2)*

27. $-1 \le m \le 2$ *(2.2)* **28.** $3 \le y \le 7$; $[3, 7]$ *(2.7)*

29. $t < -15$ or $t > 3$; $(-\infty, -15) \cup (3, \infty)$ *(2.7)*

30. Two real roots *(2.5)* **31.** One real root *(2.5)*

32. Two imaginary roots *(2.5)*

33. (A)

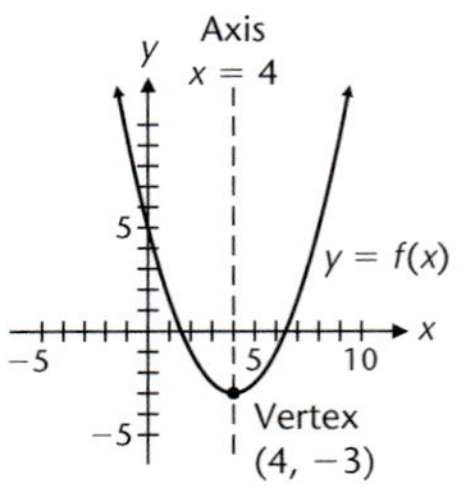

Axis $x = 4$; $y = f(x)$; Vertex $(4, -3)$

(B) Increasing: $[4, \infty)$; decreasing: $(-\infty, 4]$; range: $[-3, \infty)$ *(2.3)*

34. $g(x) = 2x + 2$; $f(x) = -0.5x^2 + x + 1.5$ *(2.1, 2.3)*

35. (A) $5 + 4i$ (B) $-i$ *(2.4)*

36. (A) $-1 + i$ (B) $\frac{4}{13} - \frac{7}{13}i$ (C) $\frac{5}{2} - 2i$ (D) -20 *(2.4)*

37. $x = \dfrac{-5 \pm \sqrt{5}}{2}$ *(2.5)* **38.** $u = 1 \pm i\sqrt{2}$ *(2.6)*

39. $\frac{1}{2} - \frac{3}{2}i$ *(2.6)* **40.** $x = -\frac{27}{8}, 64$ *(2.6)*

41. $m = \pm 3i, \pm 2$ *(2.6)* **42.** $y = \frac{9}{4}, 3$ *(2.6)*

43. $g^{-1}(x) = 2 + \sqrt{x - 1}$ *(2.3)*

44. (A) $y = 2x - 5$ (B) $y = 0.5x + 2.5$ *(2.2)*
The graphs are symmetric with respect to the line $y = x$.
Each function is the inverse of the other.

47. If $c < 9$ there are two distinct real roots, if $c = 9$ there is one real double root, and if $c > 9$ there are two distinct imaginary roots. *(2.5)*

48. $M = \dfrac{P}{1 - dt}$ *(2.2)* **49.** $I = \dfrac{E \pm \sqrt{E^2 - 4PR}}{2R}$ *(2.5)*

50. True for all real b and all negative a *(2.2)*

51. $\dfrac{a}{b}$ is less than 1. *(2.2)*

52. $6 - d < x < 6 + d, x \neq 6, (6 - d, 6) \cup (6, 6 + d)$ *(2.2)*

53. 1 *(2.4)* **54.** Perpendicular *(2.1)*

55. Center: (2, 1); radius: $2\sqrt{2}$ *(2.3)*

56. $y = -x + 7$ *(1.1, 2.1)*

57. $x = 1, 243$ *(2.6)* **58.** $x = -1; \dfrac{1 \pm i\sqrt{3}}{2}$ *(2.5)*

59. 47, 48, 49 *(2.2)* **60.** 4, 6, 8 *(2.2)* **61.** $b = 5h$ *(2.2)*

62. $h = 0.25b$ *(2.2)*

63. 12.5 inches by 30.0 inches *(2.5)* **64.** 196 feet *(2.6)*

65. (A) 4,750 calculators; $7,437.50 (B) 2,614 or 6,886 calculators
(C) None *(2.3)*

66. 3,240 or 9,260 calculators *(2.3)*

67. Profit: $3{,}240 < x < 9{,}260$; loss: $0 \le x < 3{,}240$ or $x > 9{,}260$ *(2.3)*

68. (A) $V = -1{,}250t + 12{,}000$ (B) $V = \$5{,}750$ *(2.1)*

69. (A) $R = 1.6C$ (B) $R = \$168$ *(2.1)*

70. $E(x) = \begin{cases} 200 & \text{if } 0 \le x \le 3{,}000 \\ 0.1x - 100 & \text{if } x > 3{,}000 \end{cases}$; $E(2{,}000) = 200$, $E(5{,}000) = 400$ *(2.1)*

71. (A) $A(x) = 60x - \frac{3}{2}x^2$ (B) $0 < x < 40$
(C) $x = 20, y = 15$ *(2.3)*

72. (A) $H = 0.7(220 - A)$ (B) $H = 140$ beats per minute
(C) $A = 40$ years old *(2.1)*

73. 20 centimeters by 24 centimeters *(2.5)*

74. $B = 14.58$ feet or 6.58 feet *(2.6)* **75.** 6.6 feet *(2.3)*

76. (A)

```
QuadReg
 y=ax²+bx+c
 a=.074496337
 b=-3.312225275
 c=51.05448718
```

(B) 2005 *(2.3)*

77. (A)

```
LinReg
 y=ax+b
 a=-3.100337191
 b=54.6334032
```

(B) 45.33% *(2.1)*

78. Supply: $y = 0.0000258x + 0.0533$; *(2.2)*
demand: $y = -0.0000221x + 1.63$; equilibrium price: $0.90

79. (A) $R(x) = -0.0000221x^2 + 1.63x, 0 \le x \le 73{,}800$ *(2.5)*
$C(x) = 15{,}000 + 0.2x, x \ge 0$
(B) The company will break even at sales levels of 13,200 pounds and 51,500 pounds. It will make a profit for any sales level between these two break-even levels and a loss for any sales level less than 13,200 pounds or greater than 51,500 pounds.
(C) The maximum profit is $8,130 when 32,400 pounds of broccoli are sold.

80. $y = -0.00174x^2 + 0.0865x + 0.980$, 33.6 miles per hour, 1.92 miles per gallon, 2.98 hours, 52.1 gallons, $128 *(2.5)*

Cumulative Review for Chapters 1 and 2

1. (A)

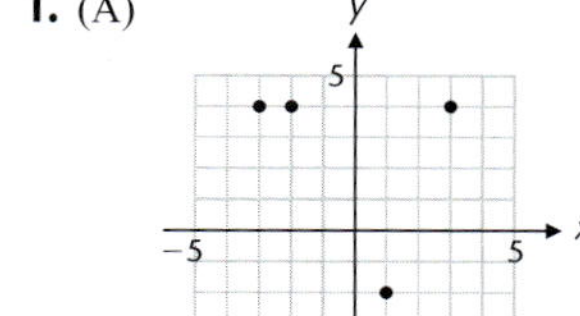

(B) Xmin = −3, Xmax = 3, Ymin = −4, Ymax = 4
(C) No *(1.1, 1.2)*

2. (A) $2\sqrt{5}$ (B) $y = 2x - 4$ (C) $y = -\frac{1}{2}x + \frac{17}{2}$
(D) $(x - 3)^2 + (y - 2)^2 = 20$

(E)

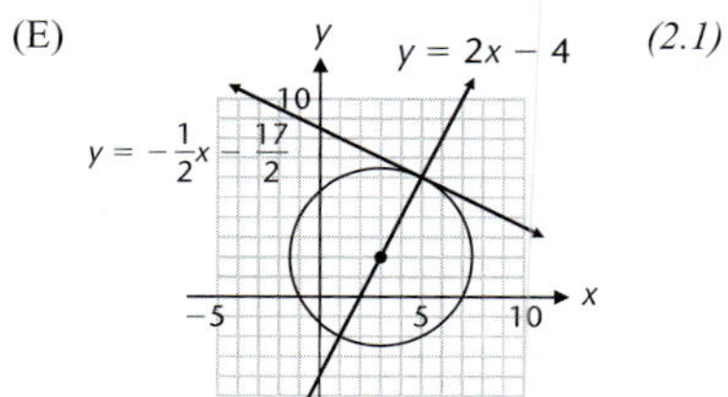

(2.1)

3. Slope: $\frac{2}{3}$; y intercept: -2; x intercept: 3 *(2.1)*

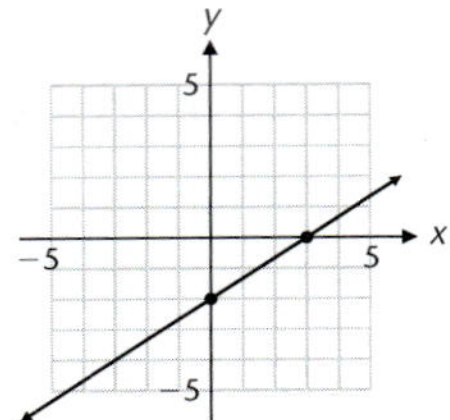

4. (A) 2 (B) 4 (C) $-\frac{2}{5}$ *(1.2)*
5. (A) Expanded by a factor of 2 (B) Shifted right two units (C) Shifted down two units *(1.4)*
6. Domain: $[-2, 3]$; range: $[-1, 2]$ *(1.2)* **7.** Neither *(1.4)*

8. (A)

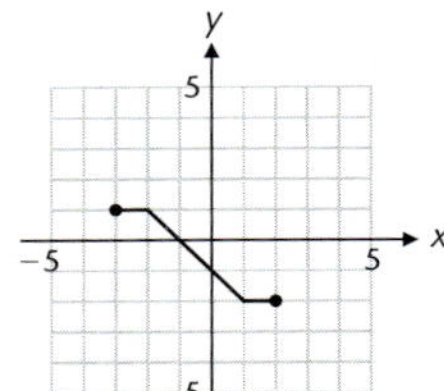

(B) *(1.4)*

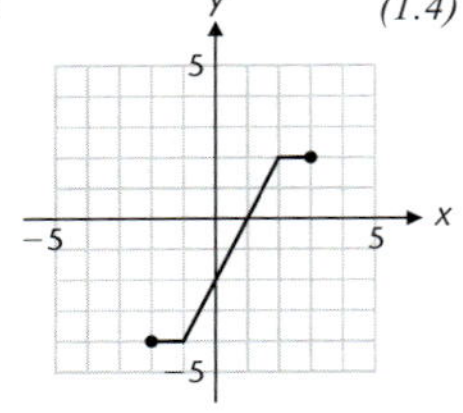

9. $x = \frac{5}{2}$ *(2.2)* **10.** $x = 0, -4$ *(2.5)* **11.** $x = \pm\sqrt{5}$ *(2.5)*
12. $x = 3 \pm \sqrt{7}$ *(2.5)* **13.** $x = 3$ *(2.6)* **14.** $y \geq 5$; $[5, \infty)$ *(2.7)*
15. $-5 < x < 9$; $(-5, 9)$ *(2.7)*
16. $x \leq -5$ or $x \geq 2$; $(-\infty, -5] \cup [2, \infty)$ *(2.7)*
17. (A) $f(x) = (x - 2)^2 - 5$
(B) It is the same as the graph of $y = x^2$ shifted to the right two units and down five units.
(C) $x = 2 \pm \sqrt{5}$ *(2.3, 2.5)*
18. (A) $7 - 10i$ (B) $23 + 7i$ (C) $1 - i$ *(2.4)*
19. (A) All real numbers $(-\infty, \infty)$ (B) $\{-2\} \cup [1, \infty)$
(C) 1 (D) $[-3, -2]$ and $[2, \infty)$ (E) $-2, 2$ *(1.4)*
20. $(f \circ g)(x) = \dfrac{x}{3 - x}$; domain: $(-\infty, 0) \cup (0, 3) \cup (3, \infty)$ *(1.5)*
21. $f^{-1}(x) = \dfrac{x - 5}{2}$ or $\dfrac{1}{2}x - \dfrac{5}{2}$ *(1.6)*
22. (A) $f^{-1}(x) = x^2 - 4$; domain: $x \geq 0$
(B) Domain of $f = [-4, \infty) =$ Range of f^{-1}
Range of $f = [0, \infty) =$ Domain of f^{-1}

(C) *(1.6)*

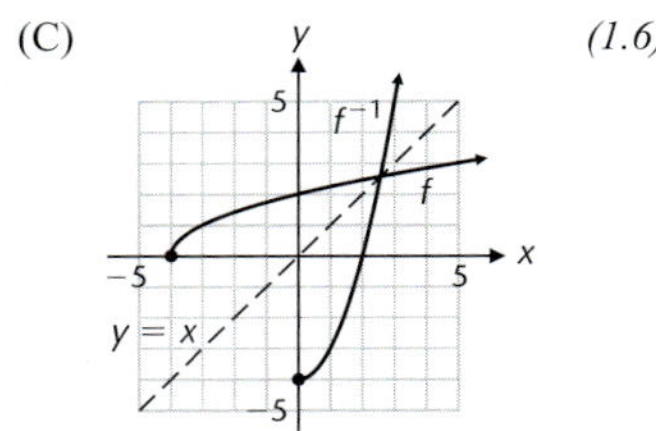

23. (A) *(1.6)*
24. (A) $y = -\frac{3}{2}x - 8$ (B) $y = \frac{2}{3}x + 5$ *(2.1)*

25. Range: $[-9, \infty)$; $\min f(x) = f\left(-\dfrac{b}{2a}\right) = -9$;
y intercept: $f(0) = -8$; x intercepts: $x = 4$ and $x = -2$ *(2.3)*

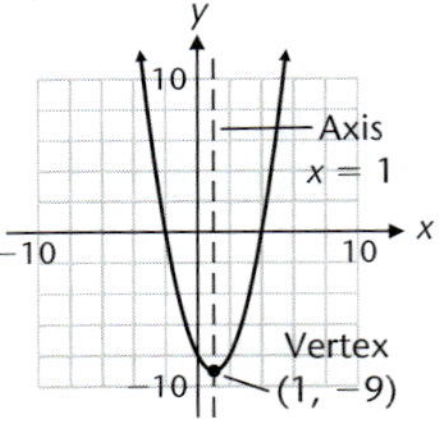

26. $x < \frac{3}{2}$ or $x > 3$; $(-\infty, \frac{3}{2}) \cup (3, \infty)$ *(2.7)*
27. $\frac{2}{3} \leq m \leq 2$; $[\frac{2}{3}, 2]$ *(2.7)*
28. (A) $0 + 0i$ or 0 (B) $\frac{6}{5}$
(C) $i^{35} = i^{32}i^3 = (i^4)^8(-i) = 1^8(-i) = -i$ *(2.4)*
29. (A) $3 + 18i$ (B) $-2.9 + 10.7i$ (C) $-4 - 6i$ *(2.4)*
30. Domain: all real numbers; range: $(-\infty, -1) \cup [1, \infty)$; discontinuous at: $x = 0$ *(1.4)*

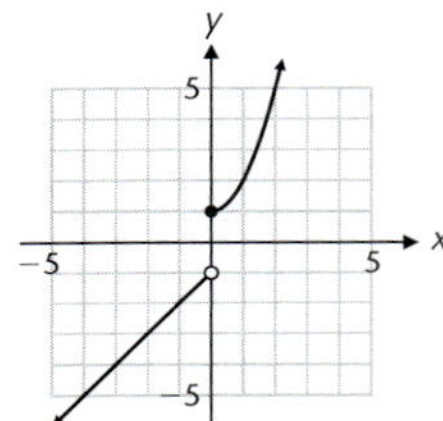

31. Center: $(3, -1)$; radius: $\sqrt{10}$ *(2.3)*

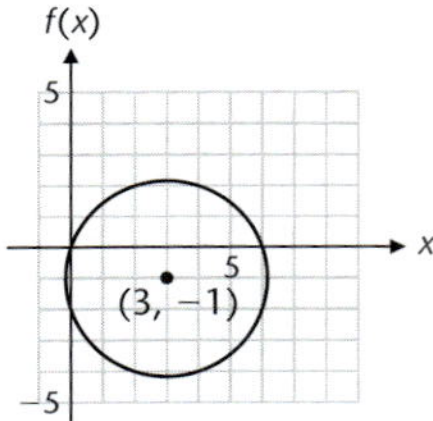

32. The graph of $y = |x|$ is contracted by $\frac{1}{2}$, reflected in the x axis, shifted two units to the right and three units up;
$y = -\frac{1}{2}|x - 2| + 3$. *(1.4)*
33. $y = (x + 2)^2 - 3$ *(2.3)* **34.** $y = 3 \pm i\sqrt{5}$ *(2.6)*
35. $x = \frac{27}{8}, -\frac{1}{8}$ *(2.6)* **36.** $u = \pm 2i, \pm\sqrt{3}$ *(2.6)*
37. $t = \frac{9}{4}$ *(2.6)* **38.** $x = \frac{4}{3}i$ *(2.4)*
39. If $b < -2$ or $b > 2$ there are two distinct real roots; if $b = -2$ or $b = 2$, there is one real double root; and if $-2 < b < 2$ there are two distinct imaginary roots. *(2.5)*
42. $-5 - 2h$ *(1.2)* **43.** $y = 2\sqrt[3]{x + 1} - 1$ *(1.4)*
44. (A) $h = \dfrac{A - 2\pi r^2}{2\pi r}$
(B) $r = -\dfrac{h}{2} + \sqrt{\dfrac{h^2}{4} + \dfrac{A}{2\pi}}$ The negative root is discarded because r must be positive. *(2.2, 2.5)*

45. (A) Domain g: $[-2, 2]$ (B) $\left(\frac{f}{g}\right)(x) = \frac{x^2}{\sqrt{4 - x^2}}$; domain of f/g is $(-2, 2)$ (C) $(f \circ g)(x) = 4 - x^2$; domain of $f \circ g$ is $[-2, 2]$ *(1.5)*

46. (A) $f^{-1}(x) = 1 + \sqrt{x + 4}$

(B) Domain of f^{-1} is $[-4, \infty)$. Range of f^{-1} = Domain of f is $[1, \infty)$

(C) *(1.6)*

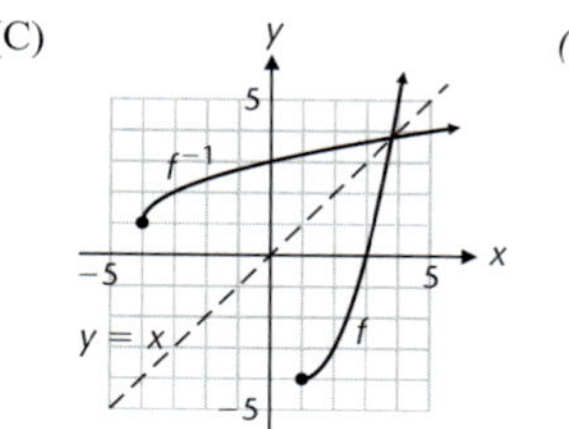

47. 0 *(2.4)*

48. All a and b such that $a < b$. *(2.2)*

49. $\frac{a^2 - b^2}{a^2 + b^2} + \frac{2ab}{a^2 + b^2}i$ *(2.4)* **50.** $x = \frac{\sqrt{2} \pm i}{3}$ *(2.5)*

51. $x = \pm 2i, \pm 3i$ *(2.6)* **52.** $x = -1.5 + 0.5i$ *(2.6)*

53. $x = \left(\frac{6}{1 \pm \sqrt{13}}\right)^5$ *(2.6)*

55. (A) $x - 3 + 0.5h$ (B) $0.5x + 0.5a - 3$ *(1.2)*

57. $f(x) = \begin{cases} -2x & \text{if } x < -2 \\ 4 & \text{if } -2 \le x \le 2 \\ 2x & \text{if } x > 2 \end{cases}$ Domain: all real numbers; range: $[4, \infty)$ *(1.3)*

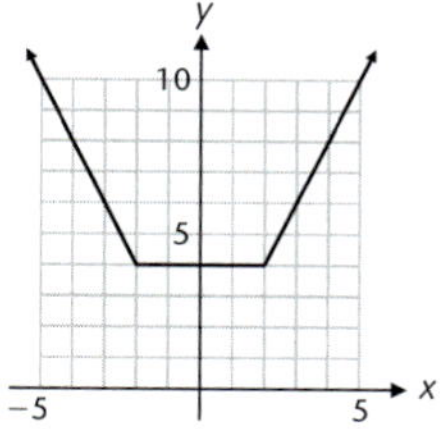

58. $f(x) = \begin{cases} \vdots & & \vdots \\ 2x + 2 & \text{if} & -1 \le x < -\frac{1}{2} \\ 2x + 1 & \text{if} & -\frac{1}{2} \le x < 0 \\ 2x & \text{if} & 0 \le x < \frac{1}{2} \\ 2x - 1 & \text{if} & \frac{1}{2} \le x < 1 \\ 2x - 2 & \text{if} & 1 \le x < \frac{3}{2} \\ 2x - 3 & \text{if} & \frac{3}{2} \le x < 2 \\ \vdots & & \vdots \end{cases}$ Domain: all real numbers; range: $[0, 1)$; discontinuous at $x = k/2$, k an integer *(1.3)*

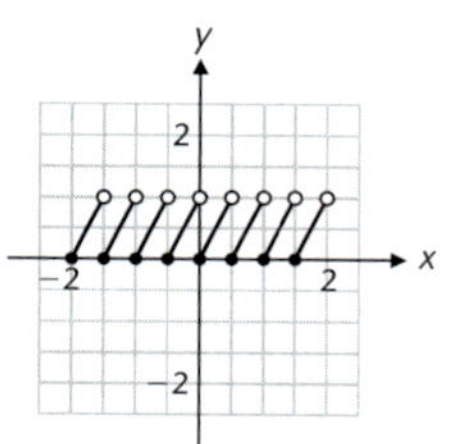

59. $x = 2, -1 \pm i\sqrt{3}$ *(2.5)* **60.** $x = 8{,}800$ books *(2.2)*

61. $|p - 200| \le 10$ *(2.2)*

62. (A) Profit: $\$5.5 < p < \$8$ or $(\$5.5, \$8)$

(B) Loss: $\$0 \le p < \$5.5$ or $p > \$8$. $[\$0, \$5.5) \cup (\$8, \infty)$ *(2.3)*

63. (A) $v = -2{,}000t + 20{,}000$ (B) $t = -0.0005v + 10$ *(1.6)*

64. 40 miles from A to B and 75 miles from B to C or 75 miles from A to B and 40 miles from B to C *(2.5)*

65. $x = -900(3.29) + 4{,}571$; 1,610 bottles *(2.2)*

66. $C(x) = \begin{cases} 0.06x & \text{if } 0 \le x \le 60 \\ 0.05x + 0.6 & \text{if } 60 < x \le 150 \\ 0.04x + 2.1 & \text{if } 150 < x \le 300 \\ 0.03x + 5.1 & \text{if } 300 < x \end{cases}$ *(1.3)*

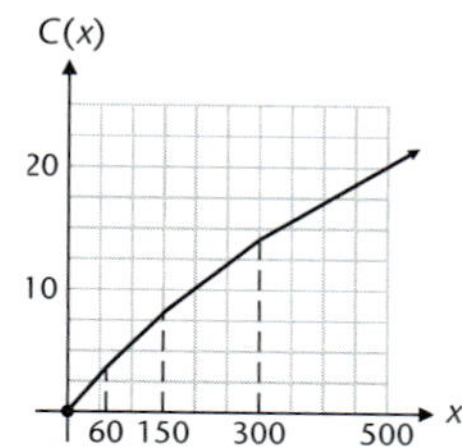

67. (A) $A(x) = 80x - 2x^2$

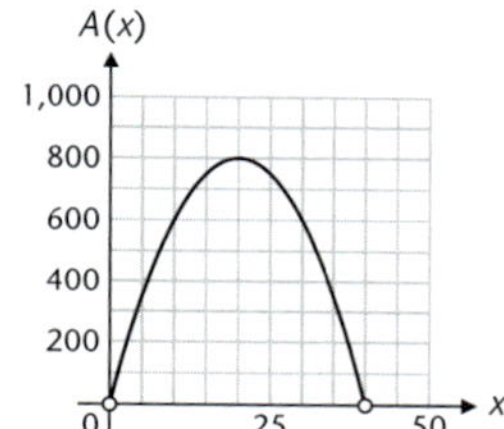

(B) $0 < x < 40$ (C) 20 feet by 40 feet *(1.3, 2.3)*

68. (A) $f(1) = 1; f(2) = 0; f(3) = 1; f(4) = 0$

(B) $f(n) = \begin{cases} 1 & \text{if } n \text{ is an odd integer} \\ 0 & \text{if } n \text{ is an even integer} \end{cases}$ *(1.3)*

69. (A) 30,000 bushels

(B) The demand decreases to 20,000 bushels.

(C) The demand increases to 40,000 bushels.

(E) *(1.1, 2.1)*

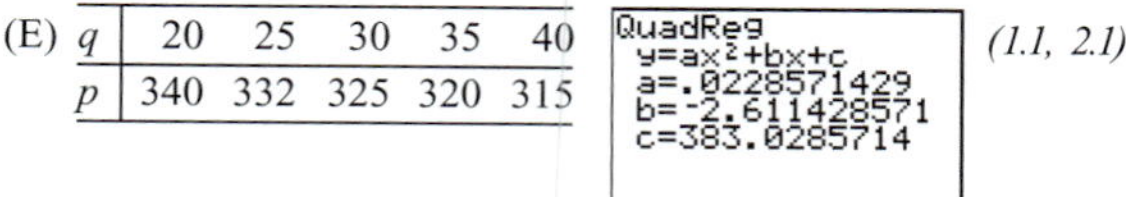

q	20	25	30	35	40
p	340	332	325	320	315

70. (A) $y = 0.144x^2 - 6.24x + 310$ *(2.5)*

(B) The per capita consumption will return to the 1970 level in 2013 and to the 1945 level in 2025.

71. (A) $y = 0.0481x^2 + 0.0690x + 2.21$ *(2.5)*

(B) 66.6 miles per hour

72. $y = -0.00149x^2 + 0.0627x + 1.21$, 29.6 miles per hour, 1.76 miles per gallon, 6.76 hours, 114 gallons, \$227 *(2.5)*

CHAPTER 3

Exercise 3.1

1. c **3.** d
5. Zeros: $-1, 3$; turning point: $(1, 4)$; $P(x) \to -\infty$ as $x \to \infty$ and $P(x) \to -\infty$ as $x \to -\infty$
7. Zeros: $-2, 1$; turning points: $(-1, 4), (1, 0)$; $P(x) \to \infty$ as $x \to \infty$ and $P(x) \to -\infty$ as $x \to -\infty$
9. The graph does not increase or decrease without bound as $x \to \infty$ and as $x \to -\infty$.
11. The graph has an infinite number of turning points.
13. Zeros: $-3, 0, 3, 2i, -2i$; x intercepts: $-3, 0, 3$
15. Zeros: $-5, 3i, -3i, 4i, -4i$; x intercepts: -5
17. $4m + 4, R = 3$ **19.** $8x - 14, R = 20$
21. $x^2 + x + 1, R = 0$ **23.** $2y^2 - 5y + 13, R = -27$
25. $\dfrac{x^2 + 3x - 7}{x - 2} = x + 5 + \dfrac{3}{x - 2}$
27. $\dfrac{4x^2 + 10x - 9}{x + 3} = 4x - 2 - \dfrac{3}{x + 3}$
29. $\dfrac{2x^3 - 3x + 1}{x - 2} = 2x^2 + 4x + 5 + \dfrac{11}{x - 2}$
31. Yes **33.** Yes **35.** 4 **37.** 3 **39.** -6
41. $3x^3 - 3x^2 + 3x - 4, R = 0$ **43.** $x^4 - x^3 + x^2 - x + 1, R = 0$
45. $3x^3 - 7x^2 + 21x - 67, R = 200$
47. $2x^5 - 3x^4 - 15x^3 + 2x + 10, R = 0$
49. $4x^3 - 6x - 2, R = 2$ **51.** $4x^2 - 2x - 4, R = 0$
53. $3x^3 - 0.8x^2 + 1.68x - 2.328, R = 0.0688$
55. $3x^4 - 0.4x^3 + 5.32x^2 - 4.256x - 3.5952, R = -0.12384$
57. (A) $P(x) \to \infty$ as $x \to \infty$ and $P(x) \to -\infty$ as $x \to -\infty$; three intercepts and two local extrema
(B) x intercepts: $-0.86, 1.68, 4.18$; local maximum: $P(0.21) \approx 6.21$; local minimum: $P(3.12) \approx -6.06$
59. (A) $P(x) \to -\infty$ as $x \to \infty$ and $P(x) \to \infty$ as $x \to -\infty$; three intercepts and two local extrema
(B) x intercept: 4.47; local minimum: $P(-0.12) \approx 4.94$; local maximum: $P(2.79) \approx 17.21$
61. (A) $P(x) \to \infty$ as $x \to \infty$ and as $x \to -\infty$; four intercepts and three local extrema
(B) x intercepts: none; local minimum: $P(-1.87) \approx 5.81$; local maximum: $P(-0.28) \approx 12.43$; local minimum: $P(1.41) \approx 4.59$
63. $P(x) = x^3$ **65.** No such polynomial exists.
67. $x^2 + (-3 + i)x - 3i, R = 0$
69. (A) -5 (B) $-40i$ (C) 0 (D) 0
71. x intercepts: $-12.69, -0.72, 4.41$; local maximum: $P(2.07) \approx 96.07$; local minimum: $P(-8.07) \approx -424.07$
73. x intercepts: $-16.06, 0.50, 15.56$; local maximum: $P(-9.13) \approx 65.86$; local minimum: $P(9.13) \approx -55.86$
75. x intercepts: $-16.15, -2.53, 1.56, 14.12$; local minimum: $P(-11.68) \approx -1{,}395.99$; local maximum: $P(-0.50) \approx 95.72$; local minimum: $P(9.92) \approx -1{,}140.27$
77. x intercepts: 1, 1.09; local minimum: $P(1.05) \approx -0.20$; local maximum: $P(6.01) \approx 605.01$; local minimum: $P(10.94) \approx 9.70$
79. (A) In both cases the coefficient of x is a_2, the constant term $a_2r + a_1$, and the remainder is $(a_2r + a_1)r + a_0$.
(B) The remainder expanded is $a_2r^2 + a_1r + a_0 = P(r)$.
81. $P(-2) = 81$; $P(1.7) = 6.2452$ or 6.2 **83.** (A) 1; 3 (B) 0; 4
85. No; $f(x) = x^2 + x$ has even degree, but $f(1) \neq f(-1)$.
87. (A)

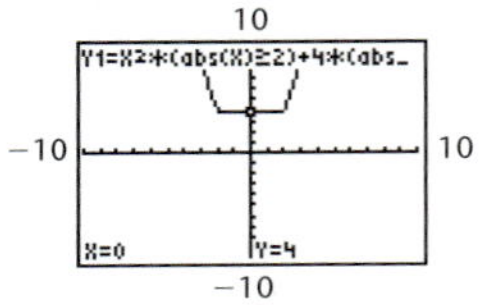

10
Y1=X²*(abs(X)≥2)+4*(abs_
−10 10
X=0 Y=4
−10

(B) $\{c \mid -2 \le c \le 2\}$
89. (A) $R(x) = 0.0004x^3 - x^2 + 569x$
(B) 364 air conditioners; price: \$258; max revenue: \$93,911
91. (A) $V(x) = (1 + 2x)(2 + 2x)(4 + 2x) - 8$ (B) 0.097 feet
93. (A)

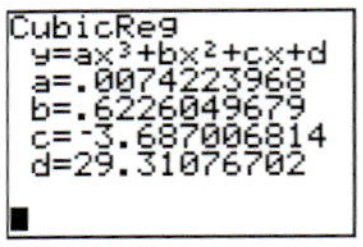

CubicReg
y=ax³+bx²+cx+d
a=.0074223968
b=.6226049679
c=-3.687006814
d=29.31076702

(B) \$2, 329.3 billion
95. (A)

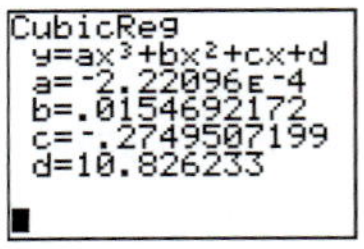

CubicReg
y=ax³+bx²+cx+d
a=-2.22096E-4
b=.0154692172
c=-.2749507199
d=10.826233

(B) 3.6 marriages per 1,000 population

Exercise 3.2

1. $-5.372, 0.372$ **3.** $-1.752, 0.432, 1.320$
5. $[-2, -1] \cup \{1\} \cup [3, \infty)$ **7.** $(-2, -1) \cup (3, \infty)$
9. $(-\infty, -5.372) \cup (0.372, \infty)$ **11.** $(-\infty, -1.752] \cup [0.432, 1.320]$
13. Upper bound: 2; lower bound: -2
15. Upper bound: 3; lower bound: -2
17. Upper bound: 2; lower bound: -3
19. (A) $P(3) < 0$ and $P(4) > 0$ (B) Five intervals; 3.2
21. (A) $P(-2) < 0$ and $P(-1) > 0$ (B) Six intervals; -1.4
23. (A) $P(3) < 0$ and $P(4) > 0$ (B) Four intervals; 3.1
25. (A) $P(-1) < 0$ and $P(0) > 0$ (B) Five intervals; -0.5
27. (A) Upper bound: 3; lower bound: -1 (B) 2.25
29. (A) Upper bound: 3; lower bound: -4 (B) $-3.51, 2.12$
31. (A) Upper bound: 2; lower bound: -3 (B) $-2.09, 0.75, 1.88$
33. (A) Upper bound: 1; lower bound: -1 (B) 0.83
39. $-1.83, 3.83$ **41.** $-1.24, 2, 3.24$ **43.** $-0.22, 2, 2.22$

45. $(-\infty, -1.414) \cup (1.414, \infty)$ **47.** $[4.367, \infty)$
49. $[-2.507, 1.222] \cup [2.285, \infty)$ **51.** $(2.484, 4.873)$
53. $(-\infty, -3.101] \cup [-2.259, 0.259] \cup [1.101, \infty)$
55. (A) Upper bound: 30; lower bound: -10
(B) $-1.29, 0.31, 24.98$
57. (A) Upper bound: 30; lower bound: -40
(B) $-36.53, -2.33, 2.40, 24.46$
59. (A) Upper bound: 20; lower bound: -10 (B) $-7.47, 14.03$
61. (A) Upper bound: 30; lower bound: -20
(B) -17.66, 2.5 (double zero), 22.66
63. (A) Upper bound: 40; lower bound: -40
(B) $-30.45, 9.06, 39.80$
65. Yes **67.** $x^4 - 3x^2 - 2x + 4 = 0$; (1, 1) and (1.659, 2.752)
69. $4x^3 - 84x^2 + 432x - 600 = 0$; 2.319 inches or 4.590 inches
71. $x^3 - 15x^2 + 30 = 0$; 1.490 feet

Exercise 3.3

1. -8 (multiplicity 3), 6 (multiplicity 2); degree of $P(x)$ is 5
3. -4 (multiplicity 3), 3 (multiplicity 2); -1; degree of $P(x)$ is 6
5. $2i$ (multiplicity 3), $-2i$ (multiplicity 4), -2 (multiplicity 5), 2 (multiplicity 5); degree of $P(x)$ is 17
7. -9 (multiplicity 2), $-3, 0, 3, 3i, -3i$; degree of $P(x)$ is 7
9. $P(x) = (x-3)^2(x+4)$; degree 3
11. $P(x) = (x+7)^3[x - (-3+\sqrt{2})][x - (-3-\sqrt{2})]$; degree 5
13. $P(x) = [x - (2-3i)][x - (2+3i)](x+4)^2$; degree 4
15. $(x+2)(x-1)(x-3)$; degree 3 **17.** $(x+2)^2(x-1)^2$; degree 4
19. $(x+3)(x+2)\,x(x-1)(x-2)$; degree 5
21. (A) $(x^2+1)(x^2+4)$ (B) $(x+i)(x-i)(x+2i)(x-2i)$
23. (A) $(x-1)(x^2+25)$ (B) $(x-1)(x+5i)(x-5i)$
25. $\pm1, \pm2, \pm3, \pm6$ **27.** $\pm1, \pm2, \pm4, \pm\frac{1}{3}, \pm\frac{2}{3}, \pm\frac{4}{3}$
29. $\pm1, \pm3, \pm\frac{1}{2}, \pm\frac{3}{2}, \pm\frac{1}{3}, \pm\frac{1}{4}, \pm\frac{3}{4}, \pm\frac{1}{6}, \pm\frac{1}{12}$
31. $P(x) = (x+4)^2(x+1)$ **33.** $P(x) = (x-1)(x+1)(x-i)(x+i)$
35. $P(x) = (2x-1)[x-(4+5i)][x-(4-5i)]$ **37.** $\frac{1}{2}, 1 \pm \sqrt{2}$
39. -2 (double), $\pm\sqrt{5}$ **41.** $\pm2, 1 \pm \sqrt{2}$ **43.** $\pm i, \pm 3i$
45. $\pm1, \frac{3}{2}, \pm i$ **47.** $2, 3, -5$ **49.** $0, 2, -\frac{2}{5}, \frac{1}{2}$
51. 2 (double), $\frac{1}{2} \pm \frac{1}{2}\sqrt{3}$ **53.** $\pm i\sqrt{5}, \pm i\sqrt{6}$,
55. -1 (double), $-\frac{1}{3}, 2 \pm i$ **57.** $P(x) = (x+2)(3x+2)(2x-1)$
59. $P(x) = (x+4)[x-(1+\sqrt{2})][x-(1-\sqrt{2})]$
61. $P(x) = (x-2)(x+1)(2x+1)(2x-1)$ **63.** $x^2 - 8x + 41$
65. $x^2 - 6x + 25$ **67.** $x^2 - 2ax + a^2 + b^2$ **69.** -1 and $3 + i$
71. $5i$ and 3 **73.** $2+i, 2-i, \sqrt{2}, -\sqrt{2}$
75. $\sqrt{6}$ is a zero of $P(x) = x^2 - 6$, but $P(x)$ has no rational zeros.
77. $\sqrt[3]{5}$ is a zero of $P(x) = x^3 - 5$, but $P(x)$ has no rational zeros.
79. 2 **81.** 3 **83.** $\frac{1}{3}, 6 \pm 2\sqrt{3}$ **85.** $\frac{3}{2}, -\frac{5}{2}, \pm 4i$
87. $\frac{3}{2}$ (double), $4 \pm \sqrt{6}$
89. (A) 3 (B) $-\frac{1}{2} + \frac{\sqrt{3}}{2}i$ and $-\frac{1}{2} - \frac{\sqrt{3}}{2}i$
91. Maximum of n; minimum of 1
93. No, because $P(x)$ is not a polynomial with real coefficients (the coefficient of x is the imaginary number $2i$).
95. 2 feet **97.** 0.5×0.5 inches or 1.59×1.59 inches

Exercise 3.4

1. $g(x)$ **3.** $h(x)$
5. Domain: $(-\infty, -1) \cup (-1, \infty)$; x intercept: 2
7. Domain: $(-\infty, -4) \cup (-4, 4) \cup (4, \infty)$; x intercepts: $-1, 1$
9. Domain: $(-\infty, -3) \cup (-3, 4) \cup (4, \infty)$; x intercepts: $-2, 3$
11. Domain: all real numbers; x intercept: 0
13. Vertical asymptote: $x = 4$; horizontal asymptote: $y = 2$
15. Vertical asymptotes: $x = -4, x = 4$; horizontal asymptote: $y = \frac{2}{3}$
17. No vertical asymptotes; horizontal asymptote: $y = 0$
19. Vertical asymptotes: $x = -1, x = \frac{5}{3}$; no horizontal asymptote
21. The graph has more than one horizontal asymptote.
23. The graph has a sharp corner at (0, 0).

25.
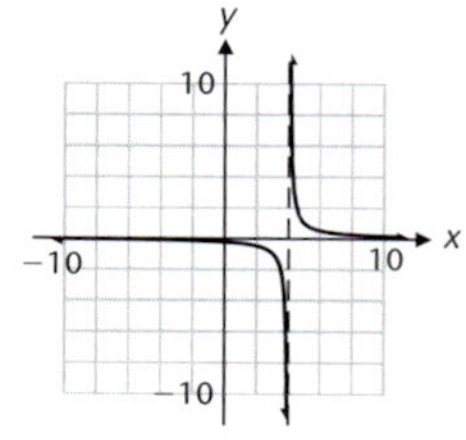

27.
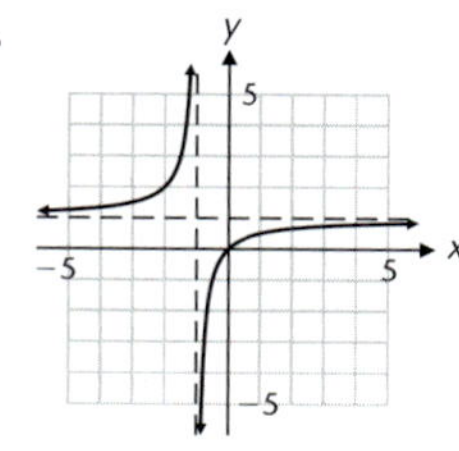

29.
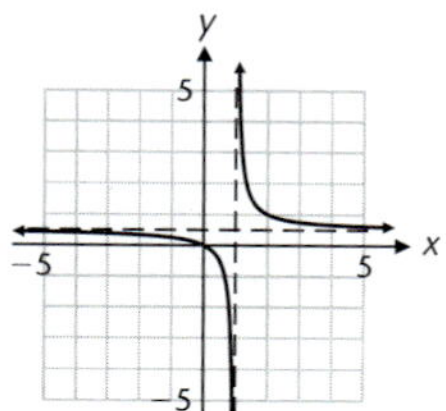

31.
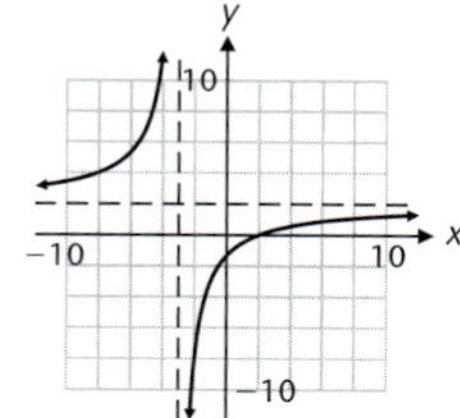

33.
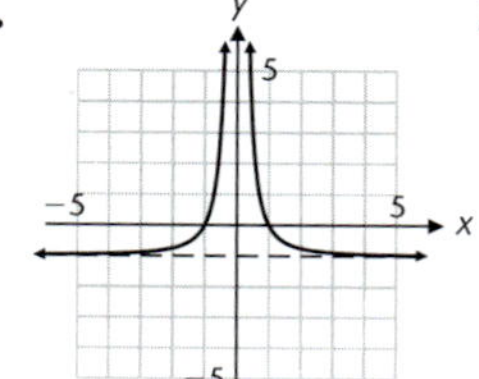

35.
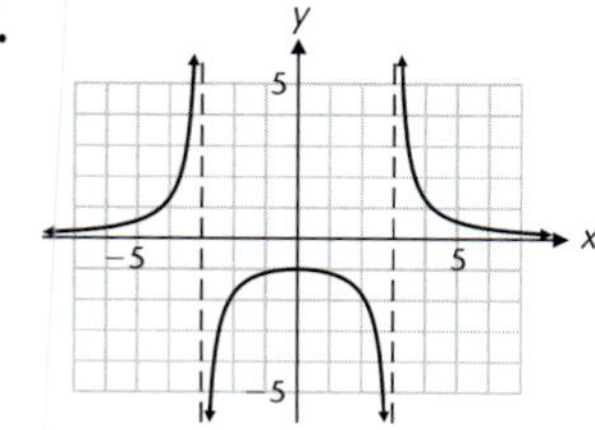

37.

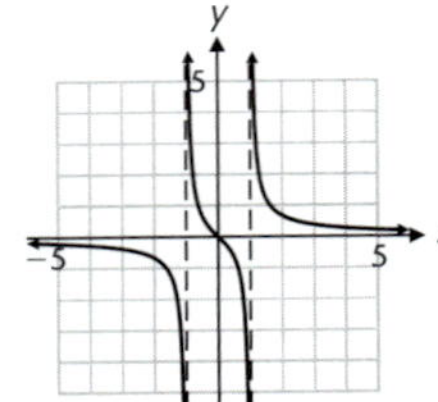

39.

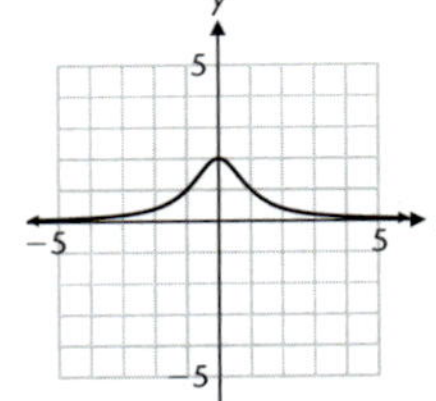

41.

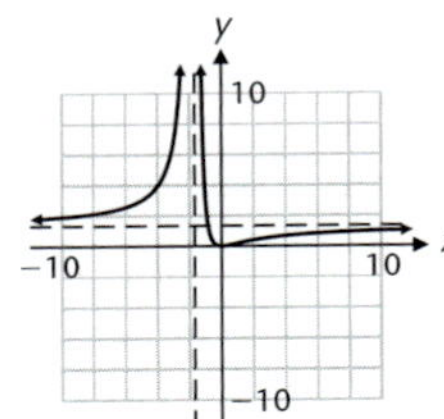

43.

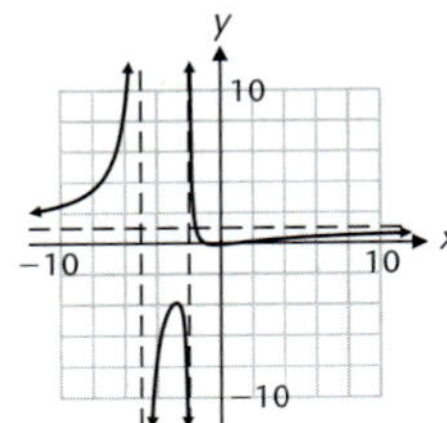

45. $f(x) = \dfrac{3(x^2 - 1)(x^2 - 4)}{x^4 + 1}$

47. $f(x) = \dfrac{(2x + 5)(x - 10) + 100}{x - 10}$

49. $[0, 4)$ **51.** $(-6.541, -2) \cup (-0.459, \infty)$

53. $(-3, -2.110) \cup (7.110, \infty)$

55. $(-\infty, 0) \cup (0, 0.595) \cup [8.405, \infty)$

57. $(5, 7.429)$ **59.** $(-\infty, -2.333] \cup (-1, 0)$

61. Vertical asymptote: $x = 1$; oblique asymptote: $y = 2x + 2$

63. Oblique asymptote: $y = x$

65. Vertical asymptote: $x = 0$; oblique asymptote: $y = 2x - 3$

67. $f(x) \to 5$ as $x \to \infty$ and $f(x) \to -5$ as $x \to -\infty$; the lines $y = 5$ and $y = -5$ are horizontal asymptotes.

69. $f(x) \to 4$ as $x \to \infty$ and $f(x) \to -4$ as $x \to -\infty$; the lines $y = 4$ and $y = -4$ are horizontal asymptotes.

71.

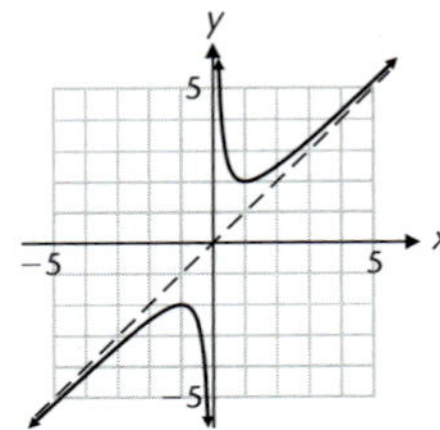

73.

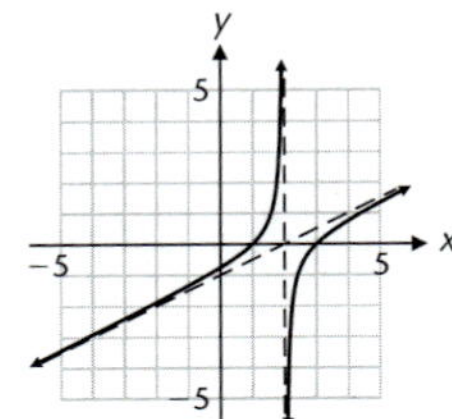

75.

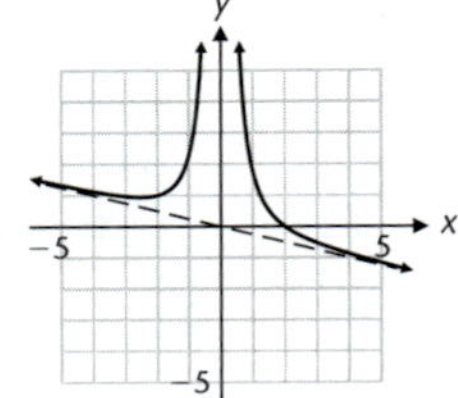

77. Let $p(x) = x^2 - 1$; $[f(x) - p(x)] \to 0$ as $x \to \infty$ and as $x \to -\infty$

79. $p(x) = x^3 + x$; $[f(x) - p(x)] \to 0$ as $x \to \infty$ and as $x \to -\infty$

81. Domain: $x \neq 2$, or $(-\infty, 2) \cup (2, \infty)$; $f(x) = x + 2$

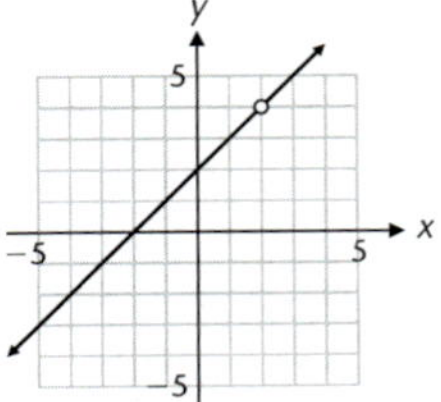

83. Domain: $x \neq 2, -2$ or $(-\infty, -2) \cup (-2, 2) \cup (2, \infty)$; $r(x) = \dfrac{1}{x - 2}$

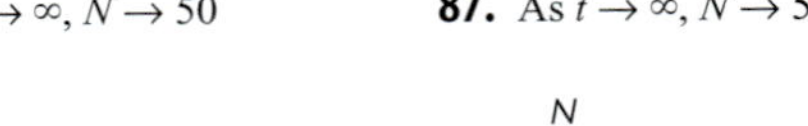

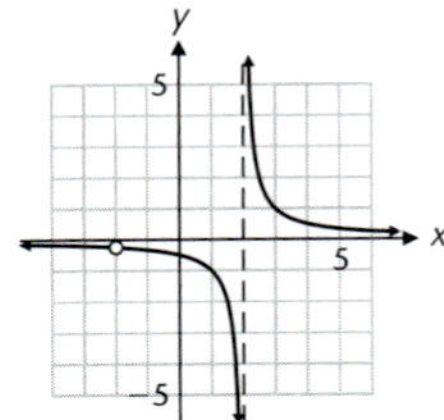

85. As $t \to \infty$, $N \to 50$

87. As $t \to \infty$, $N \to 5$

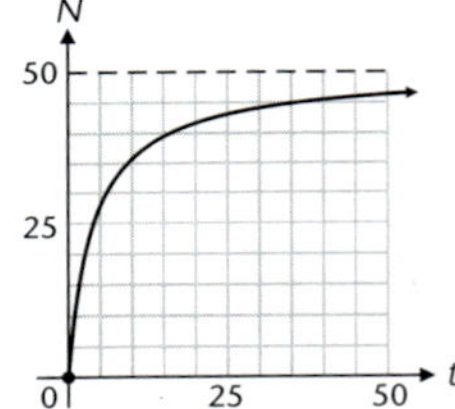

89. (A) $\overline{C}(n) = 25n + 175 + \dfrac{2{,}500}{n}$ (B) 10 years

(C)

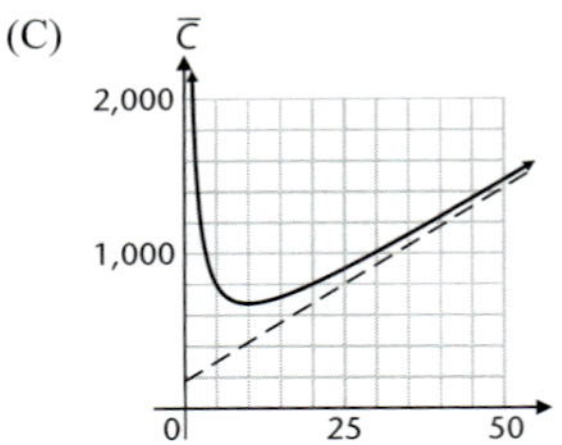

91. (A) $L(x) = 2x + \dfrac{450}{x} = \dfrac{2x^2 + 450}{x}$ (B) $(0, \infty)$

(C) 15 feet by 15 feet (D)

Chapter 3 Review

1. Zeros: $-1, 3$; turning points: $(-1, 0), (1, 2), (3, 0)$; $P(x) \to \infty$ as $x \to \infty$ and $P(x) \to \infty$ as $x \to -\infty$ *(3.1)*
2. $2x^3 + 3x^2 - 1 = (x + 2)(2x^2 - x + 2) - 5$ *(3.1)*
3. $P(3) = -8$ *(3.1)* **4.** $2, -4, -1$ *(3.1)*
5. $1 - i$ is a zero. *(3.3)*
6. (A) $P(x) = (x + 2)x(x - 2) = x^3 - 4x$
(B) $P(x) \to \infty$ as $x \to \infty$ and $P(x) \to -\infty$ as $x \to -\infty$ *(3.1)*
7. Lower bounds: $-2, -1$; upper bound: 4 *(3.2)*
8. $P(1) = -5$ and $P(2) = 1$ are of opposite sign. *(3.2)*
9. $\pm 1, \pm 2, \pm 3, \pm 6$ *(3.3)* **10.** $-1, 2, 3$ *(3.3)*
11. (A) Domain is $(-\infty, -4) \cup (-4, \infty)$; x intercept is $\frac{3}{2}$.
(B) Domain is $(-\infty, -2) \cup (-2, 3) \cup (3, \infty)$; x intercept is 0. *(3.4)*
12. (A) Horizontal asymptote: $y = 2$; vertical asymptote: $x = -4$
(B) Horizontal asymptote: $y = 0$; vertical asymptotes: $x = -2, x = 3$ *(3.4)*
13. The graph does not increase or decrease without bound as $x \to \infty$ and as $x \to -\infty$. *(3.1)*
14. (A) The graph of $P(x)$ has three x intercepts and two turning points; $P(x) \to \infty$ as $x \to \infty$ and $P(x) \to -\infty$ as $x \to -\infty$ (B) 3.53 *(3.1)*

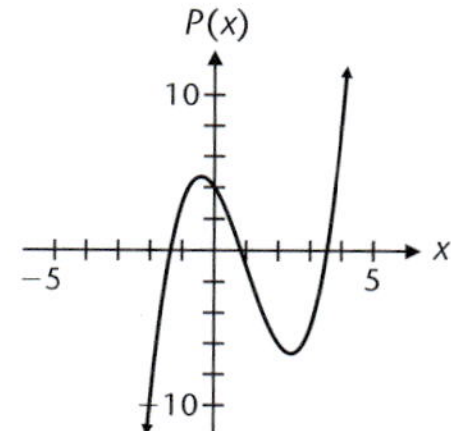

15. $Q(x) = 8x^3 - 12x^2 - 16x - 8, R = 5; P(\frac{1}{4}) = 5$ *(3.1)*
16. -4 *(3.1)* **17.** $P(x) = [x - (1 + \sqrt{2})][x - (1 - \sqrt{2})]$ *(3.1)*
18. Yes, because $P(-1) = 0, x - (-1) = x + 1$ must be a factor. *(3.1)*
19. $4, -\frac{1}{2}, -2$ *(3.3)* **20.** $(x - 4)(2x + 1)(x + 2)$ *(3.3)*
21. No rational zeros *(3.3)* **22.** $-1, \frac{1}{2}$, and $\frac{1 \pm i\sqrt{3}}{2}$ *(3.3)*
23. $(x + 1)(2x - 1)\left(x - \frac{1 + i\sqrt{3}}{2}\right)\left(x - \frac{1 - i\sqrt{3}}{2}\right)$ *(3.3)*
24. deg $P(x) = 9$; 1 (multiplicity 3), -1 (multiplicity 4), $i, -i$ *(3.3)*
25. (A) $(x - 2)(x + 2)(x^2 + 9)$
(B) $(x - 2)(x + 2)(x - 3i)(x + 3i)$ *(3.3)*
26. (A) -0.24 (double zero); 2 (simple zero); 4.24 (double zero)
(B) -0.24 can be approximated with a maximum routine; 2 can be approximated with the bisection; 4.24 can be approximated with a minimum routine. *(3.2)*
27. (A) Upper bound: 7; lower bound: -5 (B) Four intervals
(C) $-4.67, 6.62$ *(3.2)*
28. (A) Domain is $(-\infty, -1) \cup (-1, \infty)$; x intercept: $x = 1$; y intercept: $y = -\frac{1}{2}$
(B) Vertical asymptote: $x = -1$; horizontal asymptote: $y = \frac{1}{2}$
(C) *(3.4)*

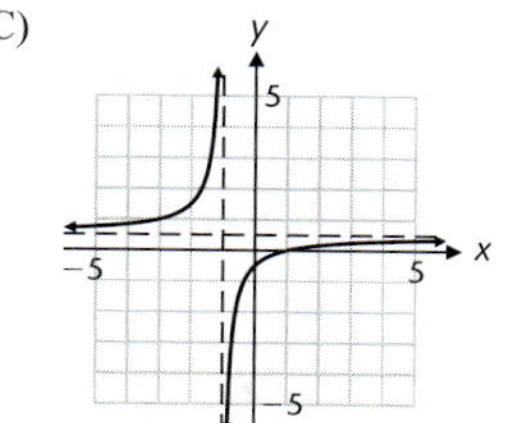

29. (A) $(-\infty, -2.562) \cup (1, 1.562)$
(B) $(-\infty, -2.414) \cup (0.414, 2)$ *(3.2)*
30. The graph is discontinuous at $x = 0$, but $x = 0$ is not a vertical asymptote. *(3.4)*
31. $P(x) = [x^2 + (1 + i)x + (3 + 2i)][x - (1 + i)] + 3 + 5i$ *(3.1)*
32. $P(x) = (x + \frac{1}{2})^2 (x + 3)(x - 1)^3$. The degree is 6. *(3.3)*
33. $P(x) = (x + 5)[x - (2 - 3i)][x - (2 + 3i)]$. The degree is 3. *(3.3)*
34. $\frac{1}{2}, \pm 2, 1 \pm \sqrt{2}$ *(3.3)*
35. $(x - 2)(x + 2)(2x - 1)[x - (1 - \sqrt{2})][x - (1 + \sqrt{2})]$ *(3.3)*
36. Zeros: 0.91, 1; local minimum: $P(-8.94) \approx 9.70$; local maximum: $P(-4.01) \approx 605.01$; local minimum: $P(0.95) \approx -0.20$ *(3.2)*
37. Because $P(x)$ changes sign three times, the minimal degree is 3. *(3.1)*
38. $P(x) = a(x - r)(x^2 - 2x + 5)$ and because the constant term, $-5ar$, must be an integer, r must be a rational number. *(3.3)*
39. (A) 3 (B) $-\frac{3}{2} \pm \frac{3i\sqrt{3}}{2}$ *(3.3)*
40. (A) Upper bound: 30; lower bound: -30
(B) $-23.54, 21.57$ *(3.2)*
41. *(3.4)*

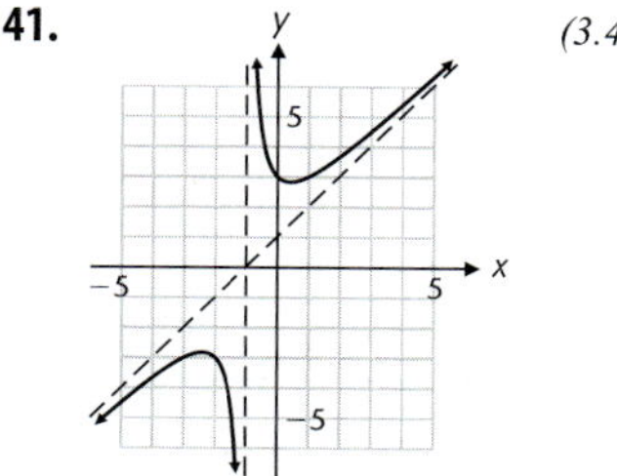

42. $y = 2$ and $y = -2$ *(3.4)*
43. (A) $(-\infty, -1.879) \cup [-1.732, 0.347) \cup (1.532, 1.732]$
(B) $(-1.879, -1.843) \cup (1.420, 1.532)$ *(3.4)*
44. 3; None of the candidates for rational zeros ($\pm 1, \pm 2$, and ± 4) are actually zeros. *(3.3)*
45. $f(x) = \frac{5x(x + 3)(x - 2)}{(x + 1)^2(x - 4)}$ *(3.4)*
46. $2x^3 - 32x + 48 = 0$, 4×12 feet or 5.211×9.211 feet *(3.2)*
47. $x^3 + 27x^2 - 729 = 0$, 4.789 feet *(3.2)*
48. $4x^3 - 70x^2 + 300x - 300 = 0$, 1.450 inches or 4.465 inches *(3.2)*
49. $x^4 - 7x^2 - 2x + 8 = 0$, $(-2, 4), (-1.562, 2.440), (1, 1), (2.562, 6.564)$ *(3.2)*

50. (A)

```
CubicReg
 y=ax³+bx²+cx+d
 a=-.0102791694
 b=.8486894107
 c=-2.575453575
 d=221.667258
```

(B) 339 refrigerators
(C) 36 ads *(3.1)*

51. (B) 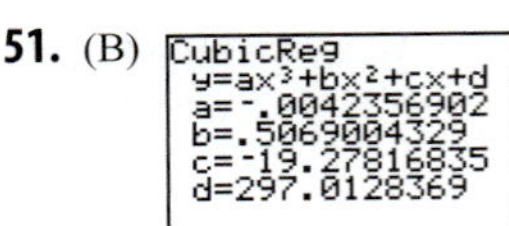

(C) 25 years, 39 years, 55 years *(3.1)*

CHAPTER 4

Exercise 4.1

1. (A) g (B) n (C) f (D) m **3.** 16.24 **5.** 7.524
7. 1.649 **9.** 4.469 **11.** 10^{2x+3}
13. 3^{2x-1} **15.** $\frac{4^{3xz}}{5^{3yz}}$ **17.** e^{3x-1} **19.** (B) e
21. Increasing; y intercept: 1; horizontal asymptote: $y = 0$
23. Decreasing; y intercept: 1; horizontal asymptote: $y = 0$
25. Increasing; y intercept: -1; horizontal asymptote: $y = 0$
27. Increasing; x intercept: -0.69; y intercept: 1; horizontal asymptote: $y = 2$
29. Increasing; x intercept: 0.23; y intercept: -1; horizontal asymptote: $y = -2$
31. $x = 2$ **33.** $x = -1, 3$ **35.** $x = \frac{2}{3}$ **37.** $x = 0$
39. $x = 0, 5$ **41.** $x = \frac{1}{2}, 1$ **43.** $a = 1$ or $a = -1$

47.

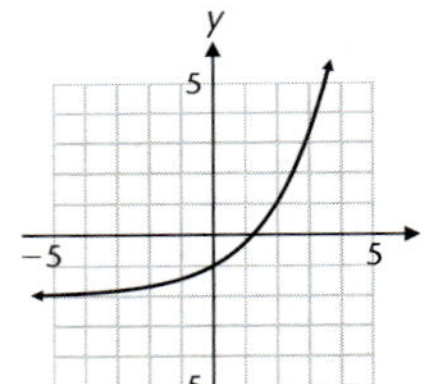

49.

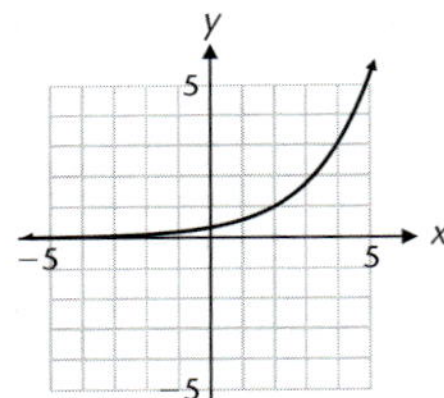

51.

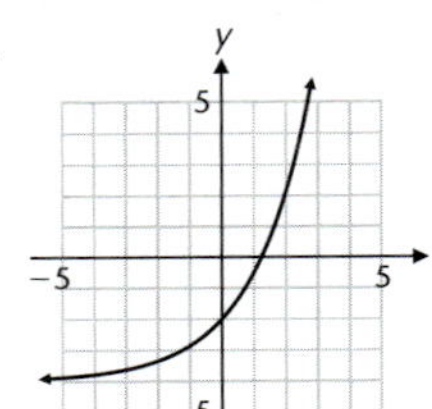

53.

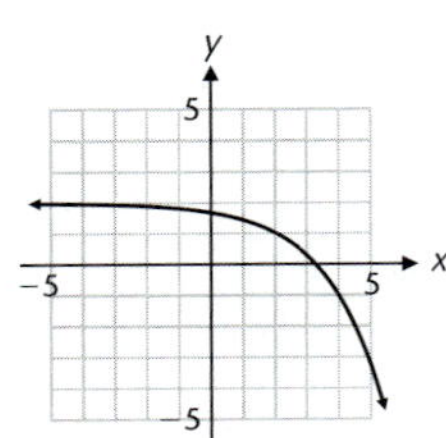

55. The graph of g is the same as the graph of f shifted to the right two units.
57. The graph of g is the same as the graph of f shifted upward two units.
59. The graph of g is the same as the graph of f reflected in the y axis, shifted to the left two units, and expanded vertically.
61. $\frac{e^{-2x}(-2x - 3)}{x^4}$ **63.** $2e^{2x} + 2e^{-2x}$
65. No local extrema; no x intercept; y intercept: 2.14; horizontal asymptote: $y = 2$
67. Local minimum: $m(0) = 1$; no x intercepts; y intercept: 1; no horizontal asymptotes
69. Local maximum: $s(0) = 1$; no x intercepts; y intercept: 1; horizontal asymptote: x axis
71. No local extrema; no x intercept; y intercept: 50; horizontal asymptotes: x axis and $y = 200$
73. Local maximum: $m(0.91) \approx 2.67$; x intercept: -0.55; horizontal asymptote: $y = 2$
75. Local minimum: $f(0) = 1$; no x intercepts; no horizontal asymptotes
77. $f(x) \to 2.7183 \approx e$ as $x \to 0$

79.

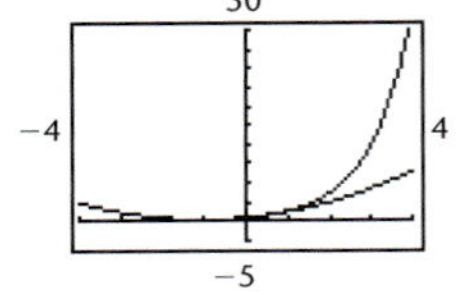

81.

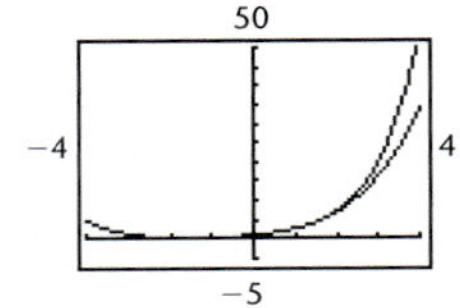

83. As $x \to \infty$, $f_n(x) \to 0$; the line $y = 0$ is a horizontal asymptote. As $x \to -\infty$, $f_1(x) \to -\infty$ and $f_3(x) \to -\infty$, while $f_2(x) \to \infty$. As $x \to -\infty$, $f_n(x) \to \infty$ if n is even and $f_n(x) \to -\infty$ if n is odd.
85. The graph of a nonconstant polynomial has no horizontal asymptote.
87. \$9,841 **89.** (A) \$10,691.81 (B) \$36,336.69
91. Yes, after 6,217 days **93.** No **95.** \$12,197.09
97. Gill Savings: \$1,230.60; Richardson S & L: \$1,231.00; U.S.A. Savings: \$1,229.03
99. (A) \$4,225.92 (B) \$12,002.75

Exercise 4.2

1.

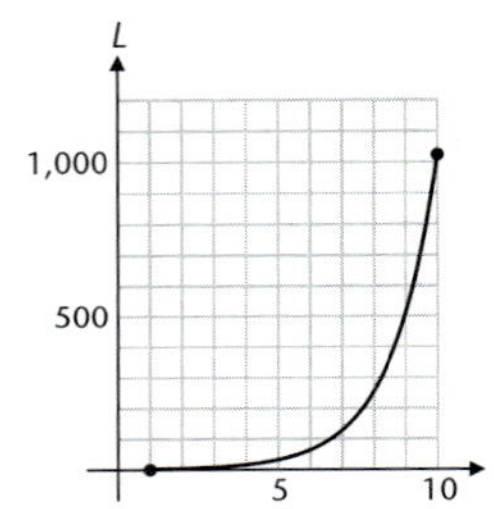

3. (A) 76 flies
(B) 570 flies

5. (A) 19 pounds (B) 7.9 pounds **7.** 7.1 billion **9.** 2006

11.

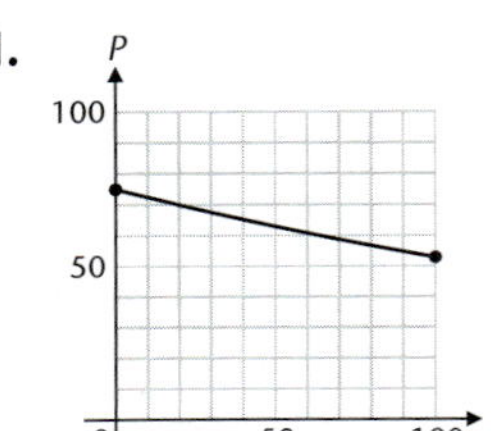

13. (A) 62% (B) 39%

15. (A) 36 million (B) 83 million **17.** $T = 50°F$

19. q approaches 0.0009 coulombs, the upper limit for the charge on the capacitor

21. (A) 25 deer; 37 deer (B) 10 years
(C) N approaches 100 deer, the upper limit for the number of deer the island can support

23.

ExpReg
y=a*b^x
a=14910.20311
b=.8162940177

Estimated purchase price: \$14,910; estimated value after 10 years: \$1,959

25. (A)

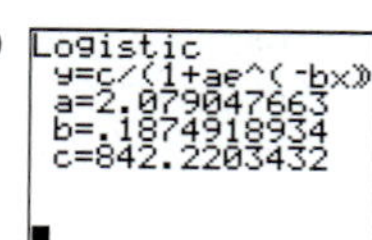

(B) 836.0 billion kilowatt-hours

Exercise 4.3

1. $81 = 3^4$ **3.** $0.001 = 10^{-3}$ **5.** $3 = 81^{1/4}$
7. $16 = (\frac{1}{2})^{-4}$ **9.** $\log_{10} 0.0001 = -4$ **11.** $\log_4 8 = \frac{3}{2}$
13. $\log_{32} \frac{1}{2} = -\frac{1}{5}$ **15.** $\log_{49} 7 = \frac{1}{2}$ **17.** 0 **19.** 1
21. 4 **23.** $\log_{10} 10^{-2} = -2$ **25.** $\frac{1}{3}$ **27.** $\sqrt{x}$ **29.** x^2
31. 4.9177 **33.** −2.8419 **35.** 3.7623 **37.** −2.5128
39. 200,800 **41.** 0.0006648 **43.** 47.73 **45.** 0.6760
47. $x = 2^2 = 4$ **49.** $y = 2$ **51.** $b = 4$
53. b is any positive real number except 1 **55.** $x = 2$
57. $y = -2$ **59.** $b = 100$ **61.** 4.959 **63.** 7.861
65. 3.301 **67.** 4.561 **69.** $x = 12.725$ **71.** −25.715
73. $x = 1.1709 \times 10^{32}$ **75.** 4.2672×10^{-7}
77. $f^{-1}(x) = e^{x/2} - 2$ **79.** $f^{-1}(x) = e^{(x+3)/4}$
81. Domain: $(-\infty, \infty)$; range: $[-2, \infty)$; x intercepts: ± 2.53; y intercept: -2; no asymptotes
83. Domain: $(-1, 1)$; range: $(-\infty, 1]$; x intercepts: ± 0.80; y intercept: 1; vertical asymptotes: $x = \pm 1$
85. The inequality sign in the last step reverses because $\log \frac{1}{3}$ is negative.
87. (B) Domain = $(1, \infty)$; range = $(-\infty, \infty)$
89. (0.90, −0.11), (38.51, 3.65) **91.** (6.41, 1.86), (93.35, 4.54)

93.

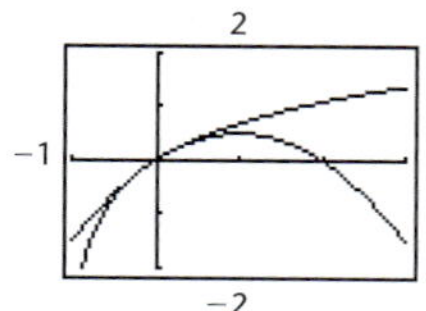

95.

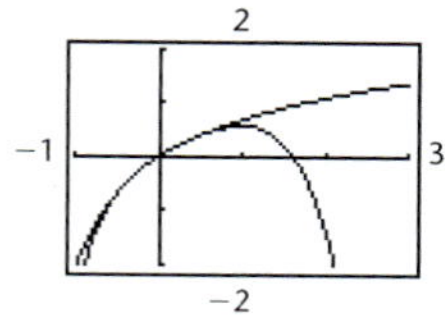

97. (A)

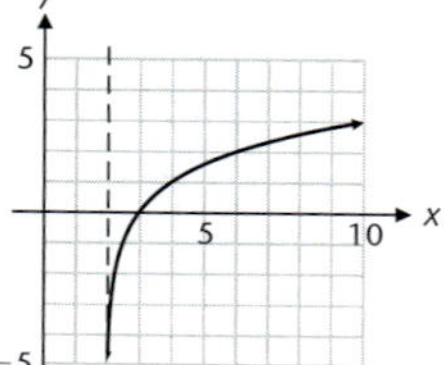

(B) $f^{-1}(x) = 2^x + 2$

99. (A)

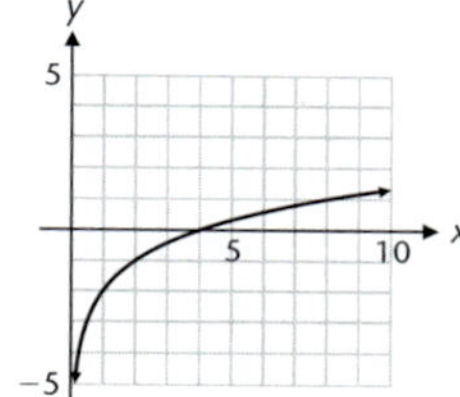

(B) $f^{-1}(x) = 2^{x+2}$

101. (A)

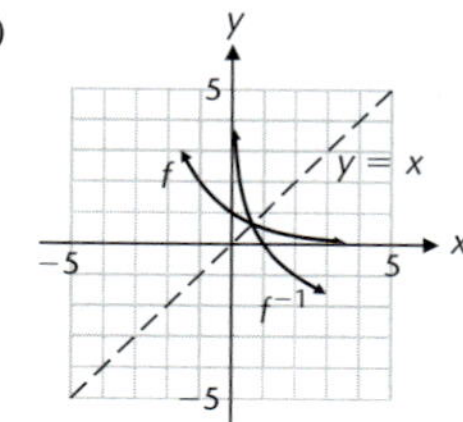

(B) Domain $f = (-\infty, \infty) =$ range f^{-1};
range $f = (0, \infty) =$ domain f^{-1}
(C) $f^{-1}(x) = \log_{1/2} x = -\log_2 x$

Exercise 4.4

1. (A) 0 decibels (B) 120 decibels **3.** 30 decibels
5. 8.6 **7.** 1,000 times as powerful
9. 7.67 kilometers per second
11. (A) 8.3, basic (B) 3.0, acidic
13. 6.3×10^{-6} moles per liter

15. (A) $m = 6$ (B) 100 times brighter
17. (A) 1996: 123.0 bushels per acre; 2010: 141.4 bushels per acre

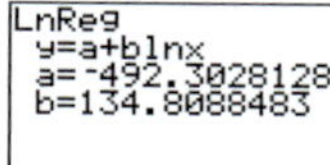

Exercise 4.5

1. 1.46 **3.** 0.321 **5.** 1.29 **7.** 3.50 **9.** 1.80
11. 2.07 **13.** 20 **15.** $x = 5$ **17.** $x = \frac{11}{9}$ **19.** 14.2
21. -1.83 **23.** 11.7 **25.** ± 1.21 **27.** $x = 5$
29. $2 + \sqrt{3}$ **31.** $\dfrac{1 + \sqrt{89}}{4}$ **33.** $1, e^2, e^{-2}$ **35.** $x = e^e$
37. $x = 100, 0.1$ **39.** (B) 2 **41.** (B) $-1.252, 1.707$
43. $r = \dfrac{1}{t} \ln \dfrac{A}{P}$ **45.** $I = I_0(10^{D/10})$ **47.** $I = I_0[10^{(6-M)/2.5}]$
49. $t = -\dfrac{L}{R} \ln\left(1 - \dfrac{RI}{E}\right)$ **51.** $x = \ln(y \pm \sqrt{y^2 - 1})$
53. $x = \dfrac{1}{2} \ln \dfrac{1 + y}{1 - y}$ **55.** 0.38 **57.** 0.55 **59.** 0.57
61. 0.85 **63.** 0.43 **65.** 0.27
67. $n = 5$ years to the nearest year **69.** $r = 0.0916$ or 9.16%
71. $t = 35$ years to the nearest year **73.** $t = 18{,}600$ years old
75. $t = 7.52$ seconds **77.** $k = 0.40$, $t = 2.9$ hours

Chapter 4 Review

1. (A) m (B) f (C) n (D) g *(4.1, 4.3)*
2. $\log m = n$ *(4.3)* **3.** $\ln x = y$ *(4.3)* **4.** $x = 10^y$ *(4.3)*
5. $y = e^x$ *(4.3)* **6.** 7^{2x} *(4.1)* **7.** e^{2x^2} *(4.1)*
8. $x = 8$ *(4.3)* **9.** $x = 5$ *(4.3)* **10.** $x = 3$ *(4.3)*
11. $x = 1.24$ *(4.5)* **12.** $x = 11.9$ *(4.5)* **13.** $x = 0.984$ *(4.3)*
14. $x = 103$ *(4.3)* **15.** 1.145 *(4.3)* **16.** Not defined *(4.3)*
17. 2.211 *(4.3)* **18.** 11.59 *(4.1)* **19.** $x = 4$ *(4.5)*
20. $x = 2$ *(4.5)* **21.** $x = 3, -1$ *(4.5)* **22.** $x = 1$ *(4.5)*
23. $x = 3, -3$ *(4.5)* **24.** $x = -2$ *(4.5)* **25.** $x = \frac{1}{3}$ *(4.5)*
26. $x = 64$ *(4.5)* **27.** $x = e$ *(4.5)* **28.** $x = 33$ *(4.5)*
29. $x = 1$ *(4.5)* **30.** $x = 41.8$ *(4.1)* **31.** $x = 1.95$ *(4.3)*
32. $x = 0.0400$ *(4.3)* **33.** $x = -6.67$ *(4.3)*
34. $x = 1.66$ *(4.3)* **35.** $x = 2.32$ *(4.5)* **36.** $x = 3.92$ *(4.5)*
37. $x = 92.1$ *(4.5)* **38.** $x = 2.11$ *(4.5)*
39. $x = 0.881$ *(4.5)* **40.** $x = 300$ *(4.5)* **41.** $x = 2$ *(4.5)*
42. $x = 1$ *(4.5)* **43.** $x = \dfrac{3 + \sqrt{13}}{2}$ *(4.5)*
44. $x = 1, 10^3, 10^{-3}$ *(4.5)* **45.** $x = 10^e$ *(4.5)*
46. $e^{-x} - 1$ *(4.1)* **47.** $2 - 2e^{-2x}$ *(4.1)*
48. Domain $= (-\infty, \infty)$; range $= (0, \infty)$; y intercept: 0.5; horizontal asymptote: $y = 0$ *(4.1)*
49. Domain $= (-\infty, \infty)$; range $= (0, \infty)$; y intercept: 10; horizontal asymptote: $y = 0$ *(4.1)*
50. Domain $= (1, \infty)$; range $= (-\infty, \infty)$; x intercept: 2; vertical asymptote: $x = 1$ *(4.3)*
51. Domain $= (-\infty, \infty)$; range $= (0, 100)$; y intercept: 25; horizontal asymptotes: $y = 0$ and $y = 100$ *(4.1)*
52. $y = -e^x$; $y = e^{-x}$ or $y = \dfrac{1}{e^x}$ or $y = \left(\dfrac{1}{e}\right)^x$ *(4.1)*
53. (A) $y = e^{-x/3}$ is decreasing while $y = 4 \ln(x + 1)$ is increasing without bound.
(B) 0.258 *(4.5)*
54. 0.018, 2.187 *(4.5)* **55.** (1.003, 0.010), (3.653, 4.502) *(4.5)*
56. $I = I_0(10^{D/10})$ *(4.5)* **57.** $x = \pm\sqrt{-2 \ln(\sqrt{2\pi}y)}$ *(4.5)*
58. $I = I_0(e^{-kx})$ *(4.5)* **59.** $n = -\dfrac{\ln(1 - \frac{Pi}{r})}{\ln(1 + i)}$ *(4.5)*
60. $y = ce^{-5t}$ *(4.5)*
61. Domain $f = (0, \infty) =$ Range f^{-1}
Range $f = (-\infty, \infty) =$ Domain f^{-1} *(4.3)*

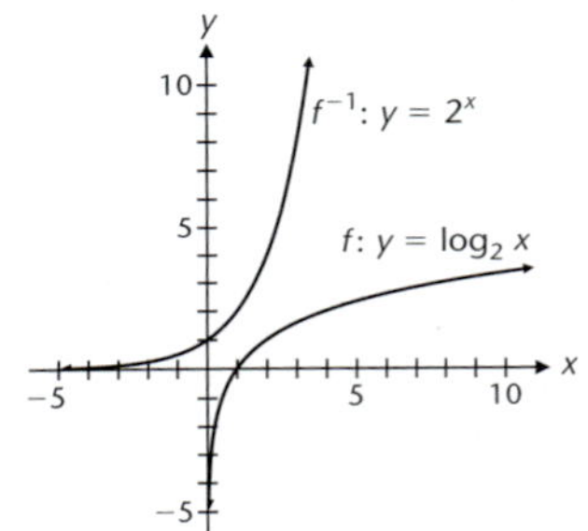

62. If $\log_1 x = y$, then we would have to have $1^y = x$; that is, $1 = x$ for arbitrary positive x, which is impossible. *(4.3)*
64. $t = 23.4$ years *(4.2)* **65.** $t = 23.1$ years *(4.2)*
66. $t = 37{,}100$ years *(4.2)*
67. (A) $N = 2^{2t}$ (or $N = 4^t$) (B) $t = 15$ days *(4.2)*
68. $A = 1.1 \times 10^{26}$ dollars *(4.1)*
69. (A) (B) 0 *(4.1)*
70. $M = 6.6$ *(4.4)* **71.** $E = 10^{16.85}$ or 7.08×10^{16} joules *(4.4)*
72. The level of the louder sound is 50 decibels more. *(4.4)*
73. $k = 0.00942$, $d = 489$ feet *(4.2)* **74.** $t = 3$ years *(4.2)*

75. (A) 1996: \$207 billion; 2010: \$886 billion

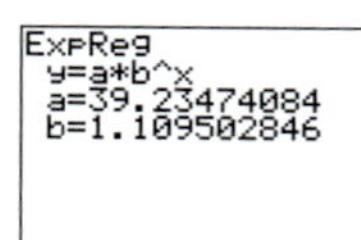

(B) Midway through 2004 *(4.2)*

76. (A) 1996: 1,724 million bushels; 2010: 2,426 million bushels *(4.4)*

LnReg
y=a+blnx
a=-21796.9294
b=5153.244133

Cumulative Review for Chapters 3 and 4

1. (A) $P(x) = (x + 1)^2(x - 1)(x - 2)$
(B) $P(x) \to \infty$ as $x \to \infty$ and as $x \to -\infty$ *(3.1)*
2. (A) m (B) g (C) n (D) f *(4.1)*
3. $3x^3 + 5x^2 - 18x - 3 = (x + 3)(3x^2 - 4x - 6) + 15$ *(3.1)*
4. $-2, 3, 5$ *(3.1)*
5. $P(1) = -5$ and $P(2) = 5$ are of opposite sign. *(3.2)*
6. $1, 2, -4$ *(3.3)* **7.** (A) $x = \log y$ (B) $x = e^y$ *(4.3)*
8. (A) $8e^{3x}$ (B) e^{5x} *(4.1)* **9.** (A) 9 (B) 4 (C) $\frac{1}{2}$ *(4.3)*
10. (A) 0.371 (B) 11.4 (C) 0.0562 (D) 15.6 *(4.3)*
11. The graph of a nonconstant polynomial has no horizontal asymptote. *(3.1)*
12. The graph does not approach the horizontal asymptote as $x \to -\infty$. *(3.4)*
13. $f(x) = 3 \ln x - \sqrt{x}$ *(4.3)*
14. The function f multiplies the base e raised to power of one-half the domain element by 100 and then subtracts 50. *(4.1)*
15. (A) Domain: $x \neq -2$; x intercept: $x = -4$; y intercept: $y = 4$
(B) Vertical asymptote: $x = -2$; horizontal asymptote: $y = 2$
(C) *(3.4)*

16. $0, -4, \pm 2i$; 0 and -4 are x intercepts *(3.1)*
17. $[-4,0]$ *(3.2)* **18.** $P(\frac{1}{2}) = \frac{5}{2}$ *(3.1)* **19.** (B) *(3.1)*
20. (A) The graph of $P(x)$ has four x intercepts and three turning points; $P(x) \to \infty$ as $x \to \infty$ and as $x \to -\infty$

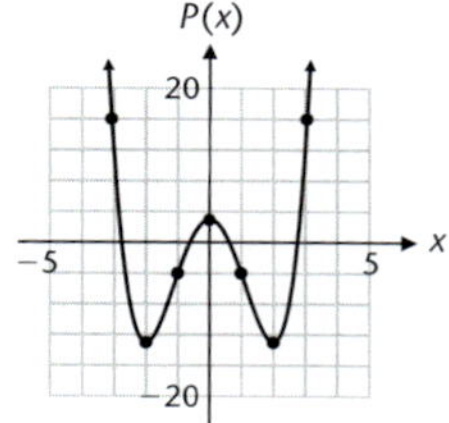

(B) 2.76 *(3.1)*

21. (A) -0.56 (double zero); 2 (simple zero); 3.56 (double zero)
(B) -0.56 can be approximated with a maximum routine; 2 can be approximated with the bisection; 3.56 can be approximated with a minimum routine *(3.2)*
22. (A) Upper bound: 4; lower bound: -6 (B) four intervals
(C) $-5.68, 3.80$ *(3.2)*
23. $3, 1 \pm \frac{1}{2}i$ *(3.3)*
24. $P(x) = (x + 1)(x + 4)(x^2 - 3) = (x + 1)(x + 4)(x - \sqrt{3})(x + \sqrt{3})$; the four zeros are $-1, -4, \pm\sqrt{3}$. *(3.3)*
25. $x = 4, -2$ *(4.5)* **26.** $\frac{1}{2}, -1$ *(4.5)* **27.** $x = 2.5$ *(4.5)*
28. $x = 10$ *(4.5)* **29.** $x = \frac{1}{27}$ *(4.5)* **30.** $x = 5$ *(4.5)*
31. $x = 7$ *(4.5)* **32.** $x = 5$ *(4.5)* **33.** $x = e^{0.1}$ *(4.5)*
34. $x = 1, e^{0.5}$ *(4.5)* **35.** $x = 3.38$ *(4.5)* **36.** $x = 4.26$ *(4.5)*
37. $x = 2.32$ *(4.5)* **38.** $x = 3.67$ *(4.5)* **39.** $x = 0.549$ *(4.5)*
40. Domain: $(-\infty, \infty)$; range: $(0, \infty)$; y intercept: 3; horizontal asymptote: $y = 0$ *(4.1)*
41. Domain: $(-\infty, 2)$; range: $(-\infty, \infty)$; x intercept: 1; y intercept: ln 2; vertical asymptote: $x = 2$ *(4.3)*
42. Domain: $(-\infty, \infty)$; range: $(0, \infty)$; y intercept: 100; horizontal asymptote: $y = 0$ *(4.1)*
43. Domain: $(-\infty, \infty)$; range: $(-\infty, 3)$; x intercept: -0.41; y intercept: 1; horizontal asymptote: $y = 3$ *(4.1)*
44. Domain: $(-\infty, \infty)$; range: $(0, 3)$; y intercept: 2; horizontal asymptotes: $y = 0$ and $y = 3$ *(4.1)*
45. A reflection in the x axis transforms the graph of $y = \ln x$ into the graph of $y = -\ln x$. A reflection in the y axis transforms the graph of $y = \ln x$ into the graph of $y = \ln(-x)$. *(4.3)*
46. (A) For $x > 0$, $y = e^{-x}$ decreases from 1 to 0 while ln x increases from $-\infty$ to ∞. Consequently, the graphs can intersect at exactly one point.
(B) 1.31 *(4.3)*
47. (A) $(x^2 + 3)(x^2 + 6)$
(B) $(x + i\sqrt{3})(x - i\sqrt{3})(x + i\sqrt{6})(x - i\sqrt{6})$ *(3.3)*
48. (A) $(x^2 - 25)(x^2 + 2)$
(B) $(x + 5)(x - 5)(x + i\sqrt{2})(x - i\sqrt{2})$ *(3.3)*
49. Vertical asymptote: $x = -2$;
oblique asymptote: $y = x + 2$ *(3.4)*

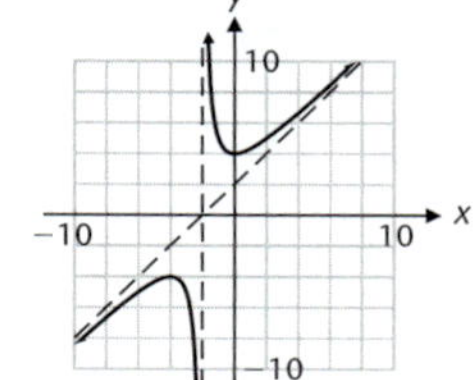

50. Zeros: 2.97, 3; local minimum: $P(2.98) \approx -0.02$; local maximum: $P(7.03) \approx 264.03$; local minimum: $P(10.98) \approx 15.98$ *(3.2)*
51. $P(x) = (x+1)^2x^3(x-3-5i)(x-3+5i)$; degree 7 *(3.3)*
52. Yes, for example, $P(x) = (x+i)(x-i)(x+\sqrt{2})(x-\sqrt{2}) = x^4 - x^2 - 2$ *(3.3)*
53. (A) Upper bound: 20; lower bound: -30
(B) $-26.68, -6.22, 7.23, 16.67$ *(3.2)*
54. 2, -1 (double), and $2 \pm i\sqrt{2}$;
$P(x) = (x-2)(x+1)^2(x-2-i\sqrt{2})(x-2+i\sqrt{2})$ *(3.3)*
55. -2 (double), $-1.88, 0.35, 1.53$ *(3.3)*
56. $f(x) = \dfrac{3(x-5)(x-8)}{(x-1)^2}$ *(3.4)*
57. $n = \dfrac{\ln(1 + \frac{Ai}{P})}{\ln(1+i)}$ *(4.5)* **58.** $y = Ae^{5x}$ *(4.5)*
59. $x = \ln(y + \sqrt{y^2+2})$ *(4.5)*
60. $(-\infty, -1] \cup [0, 1] \cup (2, \infty)$ *(3.4)*
61. $(-\infty, -1) \cup (-0.535, 1) \cup (1.869, \infty)$ *(3.4)*
62. $x = 2$ feet and $y = 2$ feet, or $x = 1.28$ feet and $y = 4.88$ feet *(3.3)*
63. 1.79 feet by 3.35 feet *(3.3)*
64. (A) 46.8 million (B) 103 million *(4.2)*
65. $t = 10.2$ years *(4.1)* **66.** $t = 9.90$ years *(4.1)*
67. 63.1 times as powerful *(4.4)*
68. $I = 6.31 \times 10^{-4}$ watts per square meter *(4.4)*
69. (A) 79.3 (B) 75.4 (C) 77.8 (D) 79.5 *(3.1, 4.2)*
70. Cubic regression *(3.1, 4.2)*

CHAPTER 5

Exercise 5.1

1. 40° **3.** 270° **5.** 405° **7.** 6 **9.** 2.5
11. $\frac{\pi}{4}$ **13.** $\frac{3\pi}{2}$ **15.** $\frac{13\pi}{6}$ **17.** $\frac{\pi}{6}, \frac{\pi}{3}, \frac{\pi}{2}, \frac{2\pi}{3}, \frac{5\pi}{6}, \pi$
19. $-\frac{\pi}{4}, -\frac{\pi}{2}, -\frac{3\pi}{4}, -\pi$ **21.** 60°, 120°, 180°, 240°, 300°, 360°
23. $-90°, -180°, -270°, -360°$ **25.** True **27.** True
29. False **31.** 5.859° **33.** 354.141° **35.** 3°2′31″
37. 403°13′23″ **39.** 1.117 **41.** 1.892 **43.** 0.234
45. 53.29° **47.** 64.74° **49.** $-134.65°$ **51.** Quadrant III
53. Quadrant II **55.** Quadrant III **57.** Quadrantal angle
59. Quadrant IV **61.** Quadrant IV **63.** Quadrant II
65. Quadrantal angle **67.** Quadrant II **69.** Quadrant III
71. A central angle of radian measure 1 is an angle subtended by an arc of the same length as the radius of the circle.
73. Coterminal **75.** Coterminal **77.** Coterminal
79. Not coterminal **81.** Coterminal **83.** Coterminal
85. 24,000 miles **87.** The 7.5° angle and θ have a common side. (An extended vertical pole in Alexandria will pass through the center of the Earth.) The sun's rays are essentially parallel when they arrive at the Earth. Thus, the other two sides of the angles are parallel, because a sun ray to the bottom of the well, when extended, will pass through the center of the Earth. From geometry we know that the alternate interior angles made by a line intersecting two parallel lines are equal. Therefore, $\theta = 7.5°$.
89. 20.94 radians/second; 62.83 feet/second **91.** $\frac{7\pi}{4}$ radians
93. 200 radians **95.** $\frac{\pi}{26} \approx 0.12$ radians **97.** 12
99. Front wheel: 13.89 radians/second; back wheel: 9.26 radians/second
101. 865,000 miles **103.** 33 feet

Exercise 5.2

1. $(0, -1)$ **3.** $(1, 0)$ **5.** $(1/\sqrt{2}, 1/\sqrt{2})$
7. $(\sqrt{3}/2, 1/2)$ **9.** $(1/2, -\sqrt{3}/2)$ **11.** $(-1/2, \sqrt{3}/2)$
13. $(-1/\sqrt{2}, -1/\sqrt{2})$ **15.** $(-1/\sqrt{2}, -1/\sqrt{2})$
17. -1 **19.** 1 **21.** $\sqrt{2}$ **23.** $1/\sqrt{3}$ **25.** $-\sqrt{3}/2$
27. $2\sqrt{3}$ **29.** $-1/\sqrt{2}$ **31.** 1 **33.** Quadrants II and III
35. Quadrants I and II **37.** Quadrants II and IV
39. -0.6573 **41.** -14.60 **43.** 1.000 **45.** 0.8138
47. 0.5290 **49.** 0.4226 **51.** -1.573 **53.** 0.8439
55. -0.3363 **57.** 0.9174
59. Zeros: none; turning points: $(\pi, -1), (2\pi, 1), (3\pi, -1)$
61. Zeros: $0, \pi, 2\pi, 3\pi, 4\pi$; turning points: none
63. 1.508, 2.208, 3.384 **65.** 4.769 **67.** a, $-$; b, $+$
69. a, $-$; b, $+$ **71.** a, $+$; b, $-$ **73.** a, $-$; b, $-$
75. a, $+$; b, $+$ **77.** 0; $2k\pi$, k any integer
79. $3\pi/4$; $3\pi/4 + 2k\pi$, k any integer
81. $W(x)$ is the coordinates of a point on a unit circle that is $|x|$ units from (1, 0), in a counterclockwise direction if x is positive and in a clockwise direction if x is negative. $W(x + 4\pi)$ has the same coordinates as $W(x)$, because we return to the same point every time we go around the unit circle any integer multiple of 2π units (the circumference of the circle) in either direction.
83. True **85.** False **87.** True
89. (A) $\sin 0.4 = 0.4$ (B) $\cos 0.4 = 0.9$ (C) $\tan 0.4 = 0.4$
91. (A) $\sec 2.2 = -2$ (B) $\tan 5.9 = -0.4$ (C) $\cot 3.8 = 1$
93. $\sin x < 0$ in quadrants III and IV; $\cot x < 0$ in quadrants II and IV; therefore, both are true in quadrant IV.
95. $\cos x < 0$ in quadrants II and III; $\sec x > 0$ in quadrants I and IV; therefore, it is not possible to have both true for the same value of x.

97. None **99.** $\frac{\pi}{2}$ and $\frac{3\pi}{2}$ **101.** $\frac{\pi}{2}$ and $\frac{3\pi}{2}$
103. 75 square meters **105.** $12\sqrt{3} \approx 20.78$ square inches

107. $a_1 = 0.5, a_2 = 1.377583, a_3 = 1.569596, a_4 = 1.570796, a_5 = 1.570796; \frac{\pi}{2} = 1.570796$

Exercise 5.3

1. b/c **3.** c/b **5.** b/a **7.** $\cos\theta$ **9.** $\sec\theta$
11. $\cot\theta$ **13.** 60.55° **15.** 82.90° **17.** 37.09°
19. $\alpha = 72.2°, a = 3.28, b = 1.05$
21. $\alpha = 46°40', b = 116, c = 169$
23. $\beta = 67°0', b = 127, c = 138$
25. $\beta = 36.79°, a = 31.85, c = 39.77$
27. $\beta = 54.6°$ or $54°40', \alpha = 35°20', c = 10.4$
29. $\beta = 52.5°$ or $52°30', \alpha = 37°30', a = 7.67$ **31.** False
33. True **35.** False
37. (A) $\cos\theta = OA/1 = OA$
(B) Angle $OED = \theta$; $\cot\theta = DE/1 = DE$
(C) $\sec\theta = OC/1 = OC$

39. (A) As θ approaches 90°, $OA = \cos\theta$ approaches 0.
(B) As θ approaches 90°, $DE = \cot\theta$ approaches 0.
(C) As θ approaches 90°, $OC = \sec\theta$ increases without bound.
41. (A) As θ approaches 0°, $AD = \sin\theta$ approaches 0.
(B) As θ approaches 0°, $CD = \tan\theta$ approaches 0.
(C) As θ approaches 0°, $OE = \csc\theta$ increases without bound.
45. 228 feet **47.** 127.5 feet **49.** 2,225 miles
51. 44° **53.** 9.8 meters/second^2
55. (B)

θ	$C(\theta)$
10°	$368,222
20°	$363,435
30°	$360,622
40°	$360,146
50°	$363,050

57. 0.77 meters

Exercise 5.4

1. $2\pi, \pi, 2\pi$ **3.** (A) 1 unit (B) Indefinitely far
(C) Indefinitely far
5. (A) $-2\pi, -\pi, 0, \pi, 2\pi$ (B) $-3\pi/2, -\pi/2, \pi/2, 3\pi/2$
(C) No x intercepts
7. (A) None (B) $-3\pi/2, -\pi/2, \pi/2, 3\pi/2$
(C) $-2\pi, -\pi, 0, \pi, 2\pi$
9. (A) No vertical asymptotes (B) $-3\pi/2, -\pi/2, \pi/2, 3\pi/2$
(C) $-2\pi, -\pi, 0, \pi, 2\pi$
11. (A) A shift of $\pi/2$ to the left will transform the cosecant graph into the secant graph. [The answer is not unique—see part B.]
(B) The graph of $y = -\csc(x - \pi/2)$ is a $\pi/2$ shift to the right and a reflection in the x axis of the graph of $y = \csc x$. The result is the graph of $y = \sec x$.
13. Even **15.** Even **17.** Odd **19.** Odd
21. $\sin\theta = \frac{4}{5}, \csc\theta = \frac{5}{4}, \cos\theta = \frac{3}{5}, \sec\theta = \frac{5}{3}, \tan\theta = \frac{4}{3}, \cot\theta = \frac{3}{4}$
23. $\sin\theta = \frac{\sqrt{3}}{2}, \csc\theta = \frac{2}{\sqrt{3}}, \cos\theta = -\frac{1}{2}, \sec\theta = -2,$
$\tan\theta = -\sqrt{3}, \cot\theta = -\frac{1}{\sqrt{3}}$ **25.** 60° **27.** $\frac{\pi}{6}$
29. $\frac{\pi}{3}$ **31.** 120° or $\frac{2\pi}{3}$ radians **33.** 210° or $\frac{7\pi}{6}$ radians
35. 240° or $\frac{4\pi}{3}$ radians
37. $\cos\theta = -\frac{4}{5}, \tan\theta = -\frac{3}{4}, \sec\theta = -\frac{5}{4}, \cot\theta = -\frac{4}{3}, \csc\theta = \frac{5}{3}$
39. $\sin\theta = -\frac{2}{3}, \sec\theta = -\frac{3}{\sqrt{5}}, \tan\theta = \frac{2}{\sqrt{5}}, \cot\theta = \frac{\sqrt{5}}{2},$
$\csc\theta = -\frac{3}{2}$

41. Tangent and secant, because $\tan\theta = b/a$ and $\sec\theta = r/a$ and $a = 0$ if $P = (a, b)$ is on the vertical axis (division by zero is not defined).
43. 150°, 210° **45.** $\frac{\pi}{4}, \frac{5\pi}{4}$
47. (A)

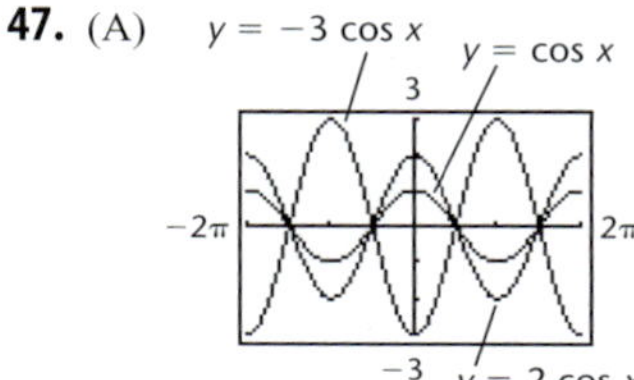

(B) No
(C) 1 unit; 2 units; 3 units
(D) The deviation of the graph from the x axis is changed by changing A. The deviation appears to be $|A|$.

49. (A)

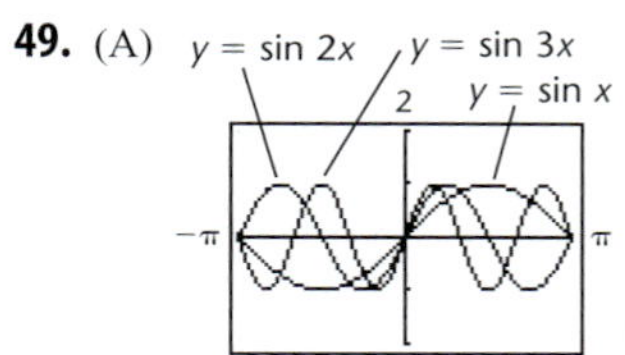

(B) 1; 2; 3 (C) n

51. (A)

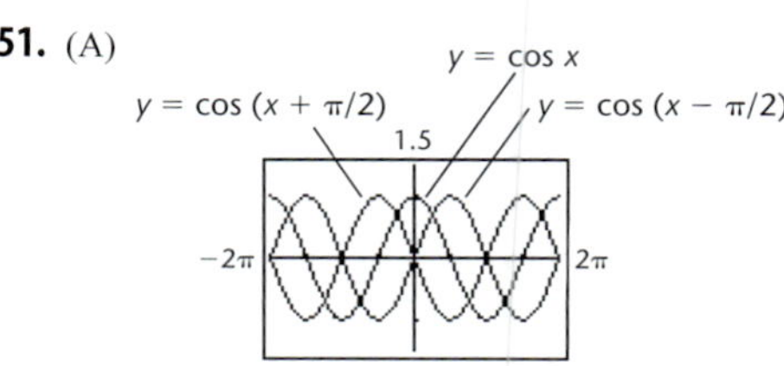

(B) The graph of $y = \cos x$ is shifted $|C|$ units to the right if $C < 0$ and $|C|$ units to the left if $C > 0$.

53. For each case, the number is not in the domain of the function and an error message of some type will appear.

55. (A) Both graphs are almost indistinguishable the closer the x is to the origin.

(B)

x	-0.3	-0.2	-0.1	0.0	0.1	0.2	0.3
$\sin x$	-0.296	-0.199	-0.100	0.000	0.100	0.199	0.296

57. (A) 1.75 radians (B) $(-0.713, 3.936)$

59. 2π units

61. k, $0.866k$, $0.5k$

65. (A) 3.31371, 3.14263, 3.14160, 3.14159
(B) $\pi = 3.1415926\ldots$

67. (A) 44.07; -0.32 (B) $y = -0.93x + 1.28$

Exercise 5.5

1. $A = 3$, $P = 2\pi$

3. $A = \frac{1}{2}$, $P = 2\pi$

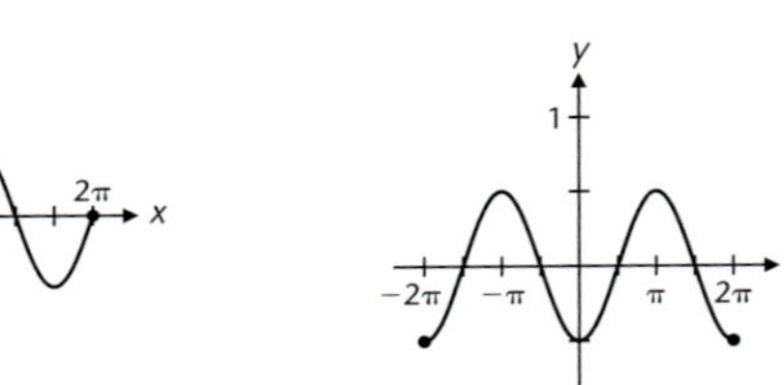

5. $A = 1$, $P = \dfrac{2\pi}{3}$

7. Period $= \dfrac{\pi}{4}$

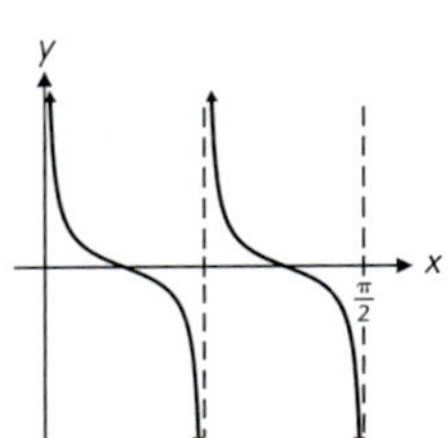

9. Period $= \dfrac{1}{8}$

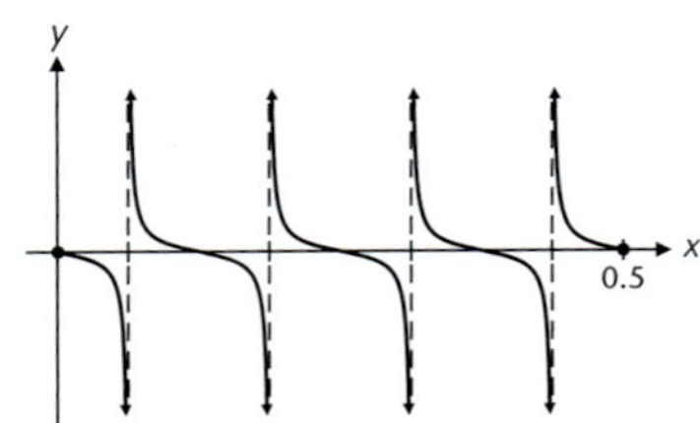

11. Period $= 4\pi$

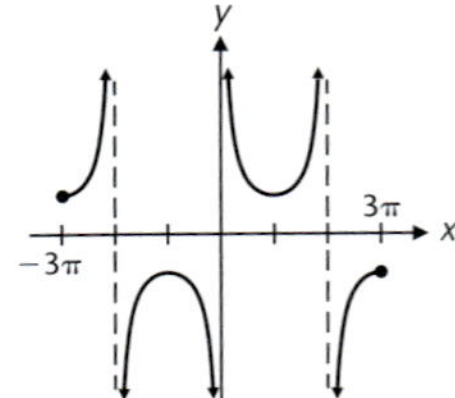

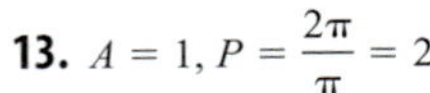

13. $A = 1$, $P = \dfrac{2\pi}{\pi} = 2$

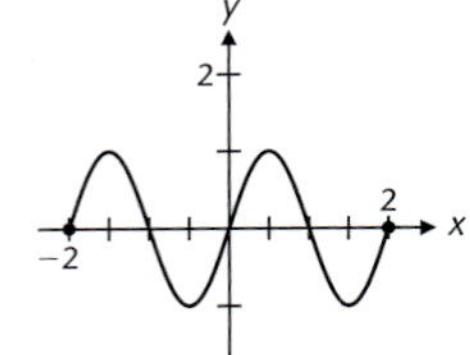

15. Period $= 2\pi$

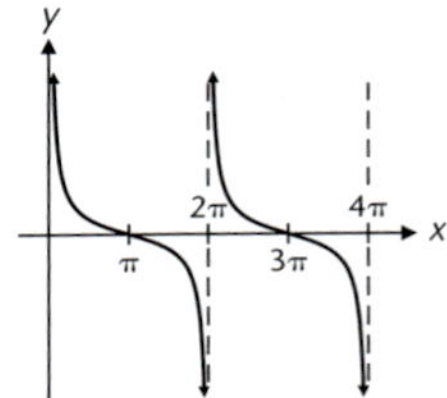

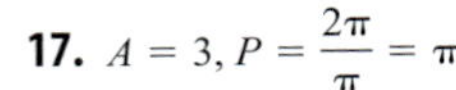

17. $A = 3$, $P = \dfrac{2\pi}{\pi} = \pi$

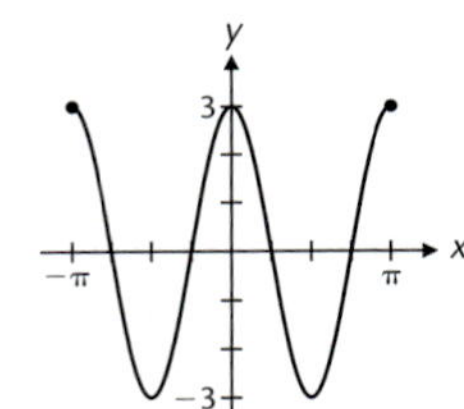

19. Period $= 2$

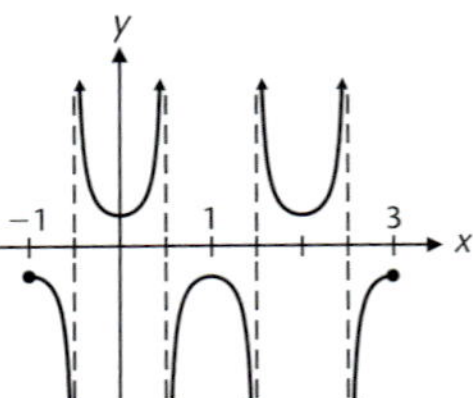

21. $A = 3$, $P = \dfrac{\pi}{2} = \dfrac{2\pi}{4}$; $B = 4$, $y = 3 \sin 4x$, $-\dfrac{\pi}{4} \le x \le \dfrac{\pi}{2}$

23. $A = 10, P = 2 = \dfrac{2\pi}{\pi}; B = \pi, y = -10 \sin \pi x, -1 \leq x \leq 2$

25. $A = 5, P = 8\pi = 2\pi \cdot 4 = 2\pi \div \frac{1}{4}; B = \frac{1}{4}, y = 5 \cos \frac{1}{4}x,$ $-4\pi \leq x \leq 8\pi$

27. $A = 0.5, P = 8 = 2\pi \cdot \dfrac{4}{\pi} = 2\pi \div \dfrac{\pi}{4}; B = \dfrac{\pi}{4}, y = -0.5 \cos \dfrac{\pi x}{4},$ $-4 \leq x \leq 8$

29. $A = 1, P = 2\pi$, phase shift $= -\pi$

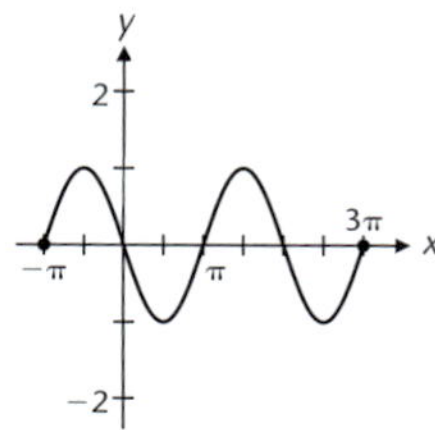

31. $A = \dfrac{1}{2}, P = 2\pi$, phase shift $= \dfrac{\pi}{4}$

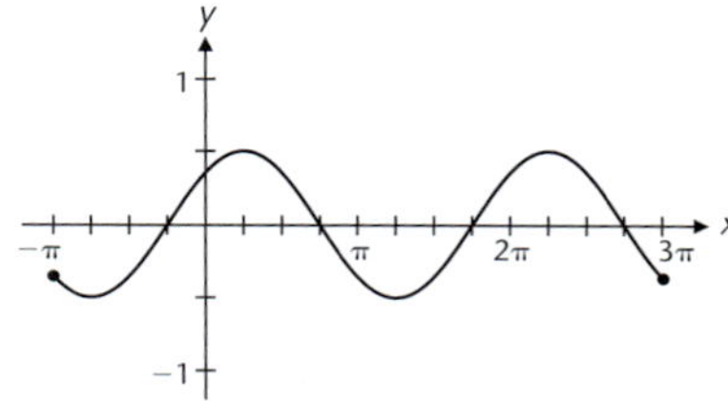

33. Period $= \pi$, phase shift $= -\dfrac{\pi}{2}$

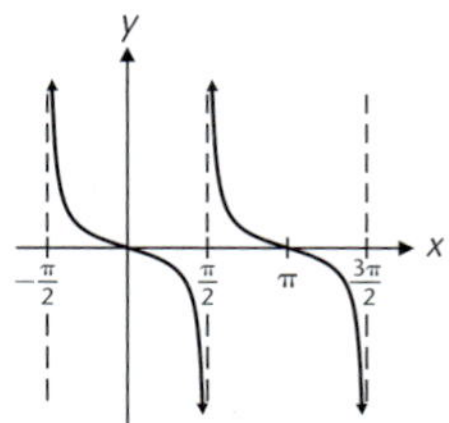

35. $A = 1, P = 2$, phase shift $= 1$

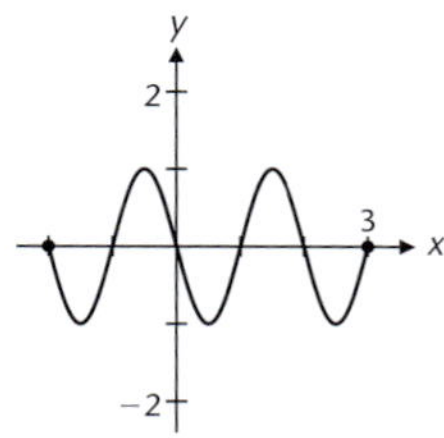

37. $A = 3, P = 2$, phase shift $= -\frac{1}{2}$

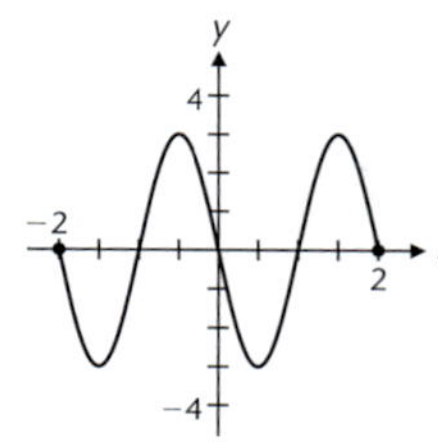

39. Period $= \dfrac{\pi}{2}$, phase shift $= -\dfrac{\pi}{2}$

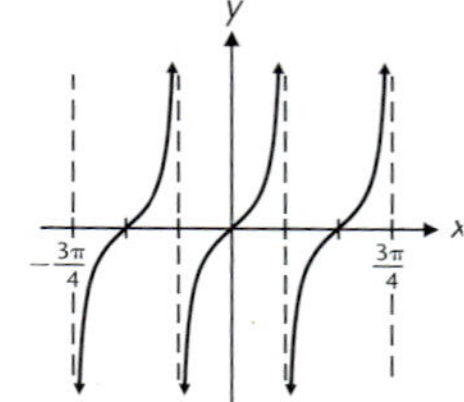

41. $A = 4, P = \pi$, phase shift $= \dfrac{\pi}{2}$

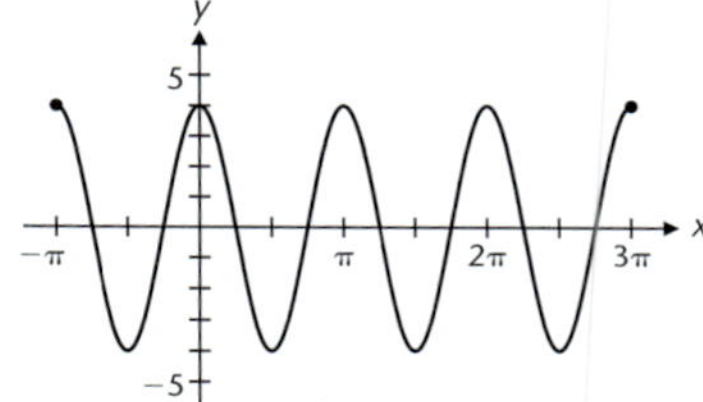

43. Period $= 2$, phase shift $= -\frac{1}{2}$

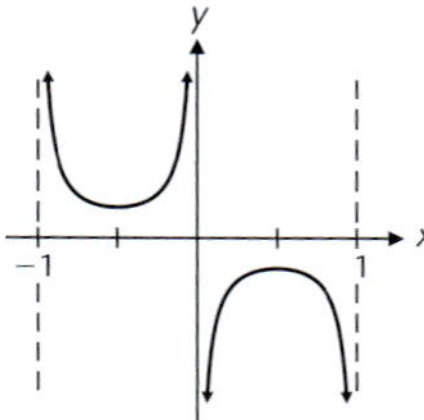

45. $y = \cos 2x$ **47.** $y = 1 - \cos 2x$

49. $y = 2 \cot 2x$

51. $y = \cot (x/2)$

53. $y = \csc 3x$

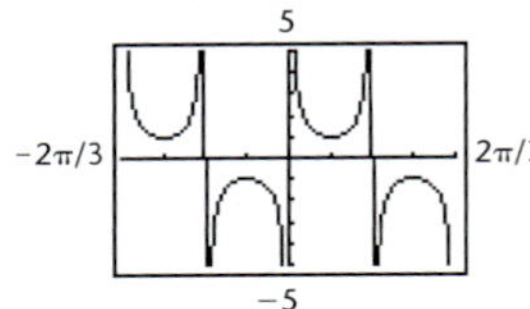

55. $y = \tan 2x$

57. $y = -4 \sin \left(\frac{\pi}{2}x - \frac{\pi}{2} \right)$

59. $y = \frac{1}{2} \cos \left(\frac{1}{4}x - \frac{3\pi}{4} \right)$

61. $A = 3.5, P = 4$, phase shift $= -0.5$

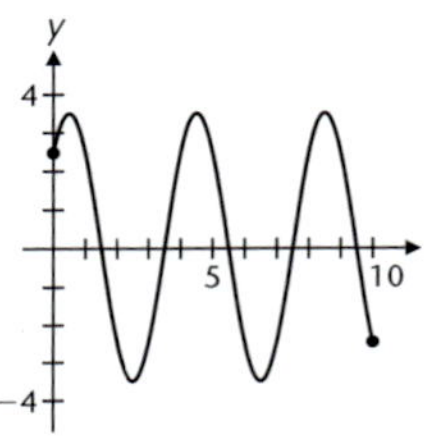

63. $A = 50, P = 1$, phase shift $= 0.25$

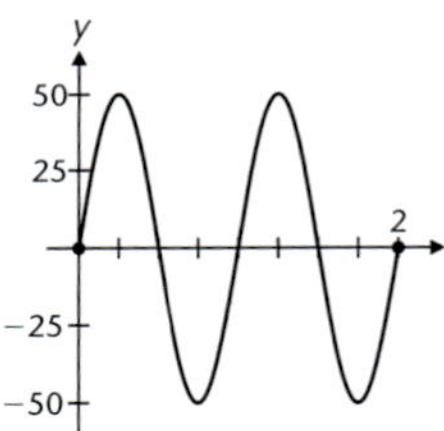

65. $y = 2 \sin (x + 0.785)$

67. $y = 2 \sin (x - 0.524)$

69. $y = 5 \sin (2x - 0.284)$

71. The amplitude is decreasing with time. This is often referred to as a *damped sine wave.* Examples are a car's vertical motion, which is damped by the suspension system after the car goes over a bump, and the slowing down of a pendulum that is released away from the vertical line of suspension (air resistance and friction).

73.

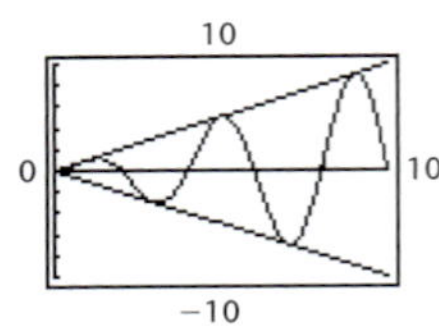

The amplitude is increasing with time. In physical and electrical systems this is referred to as *resonance.* Some examples are the swinging of a bridge during high winds and the movement of tall buildings during an earthquake. Some bridges and buildings are destroyed when the resonance reaches the elastic limits of the structure.

75.

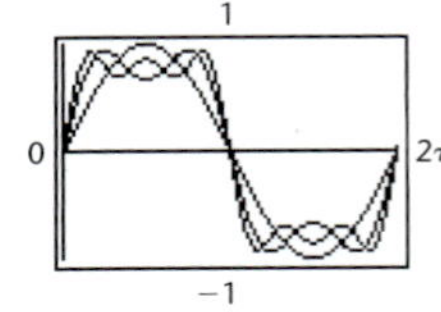

77. $A = \frac{1}{3}, P = \frac{2\pi}{8} = \frac{\pi}{4}$

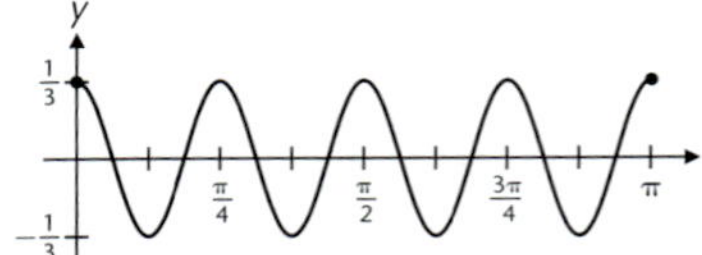

79. $y = -8 \cos 4\pi t$

81. The graph shows the seasonal changes of sulfur dioxide pollutant in the atmosphere; more is produced during winter months because of increased heating.

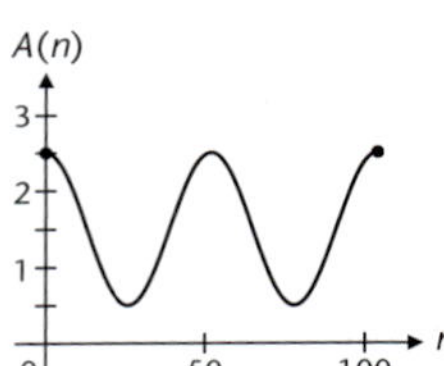

83. $A = 15, P = \frac{1}{60}$, phase shift $= -\frac{1}{240}$

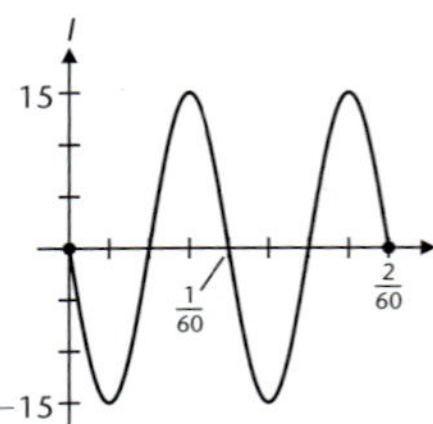

85. $A = 3, P = \frac{1}{3}$

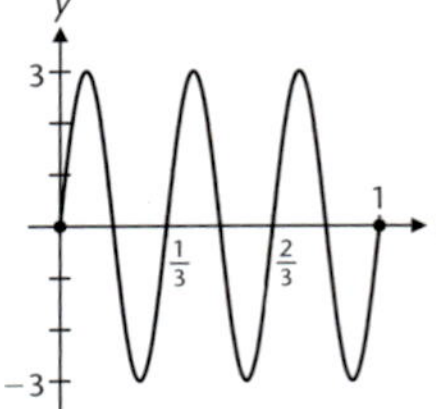

87. (A) $c = 20 \sec(\pi t/2)$, $[0, 1)$

(B)

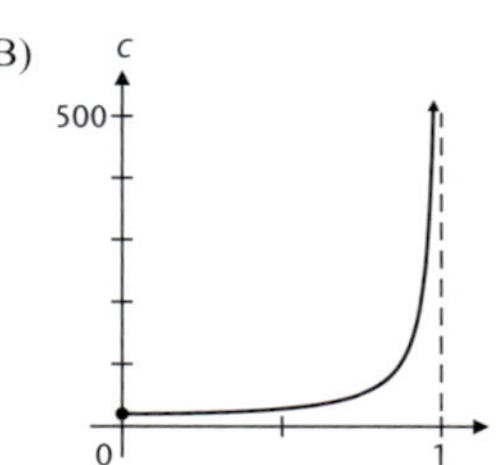

(C) The length of the light beam starts at 20 feet and increases slowly at first, then increases rapidly without end.

89. (A)

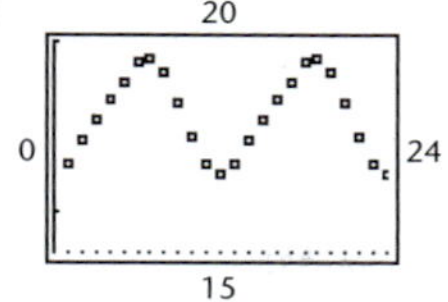

(B) $y = 18.22 + 1.37 \sin\left(\frac{\pi x}{6} - 1.75\right)$

(C)

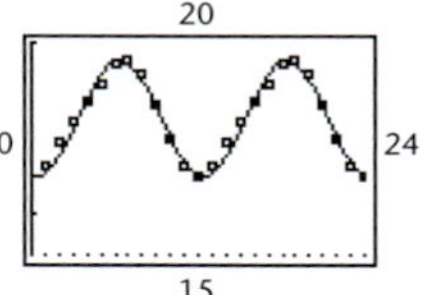

Exercise 5.6

1. $\frac{\pi}{2}$ **3.** $\frac{\pi}{3}$ **5.** $\frac{\pi}{3}$ **7.** $\frac{\pi}{4}$ **9.** 0 **11.** $\frac{\pi}{6}$
13. 1.144 **15.** 1.561 **17.** Not defined **19.** $-\frac{\pi}{4}$
21. $-\frac{\pi}{3}$ **23.** 25 **25.** 2.3 **27.** $\frac{1}{2}$ **29.** $-\sqrt{2}$
31. 0 **33.** $2\pi/3$ **35.** -1.472 **37.** -0.9810
39. 2.645 **41.** $-45°$ **43.** $-60°$ **45.** $180°$
47. $43.51°$ **49.** $-21.48°$ **51.** $-89.93°$
53. $\sin^{-1}(\sin 2) = 1.1416 \neq 2$. For the identity $\sin^{-1}(\sin x) = x$ to hold, x must be in the restricted domain of the sine function; that is, $-\frac{\pi}{2} \le x \le \frac{\pi}{2}$. The number 2 is not in the restricted domain.

55.

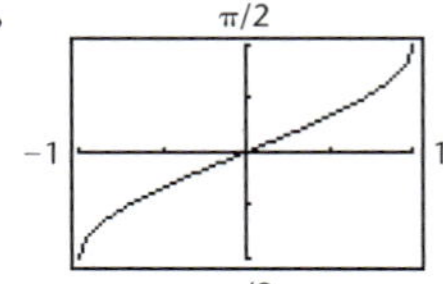

57.

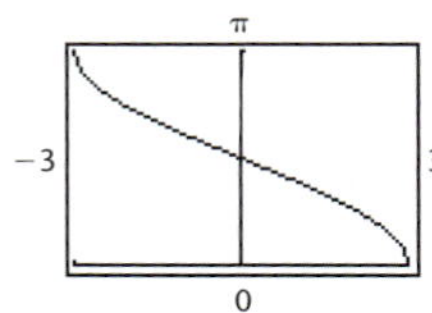

59.

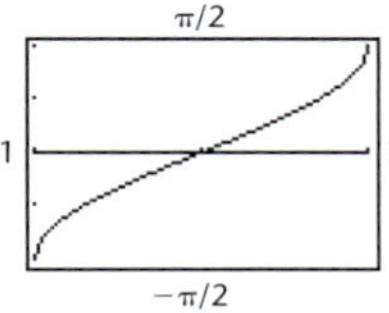

61.

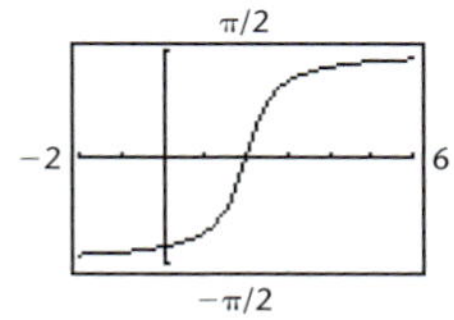

63. (A)

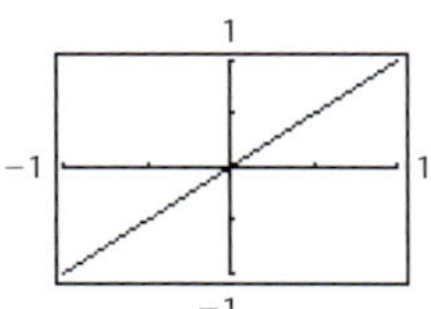

(B) The domain of $\cos^{-1}$ is restricted to $-1 \le x \le 1$; hence no graph will appear for other x.

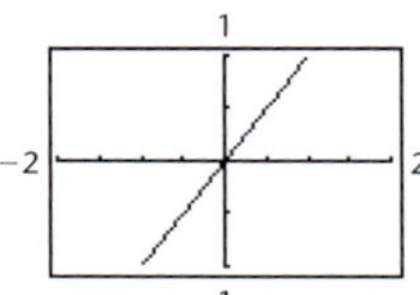

65. $\sqrt{1 - x^2}$ **67.** $\frac{1}{\sqrt{1 + x^2}}$

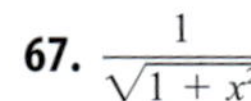

69. $f^{-1}(x) = 3 + \cos^{-1}\frac{x - 4}{2}$; $2 \le x \le 6$

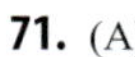

71. (A)

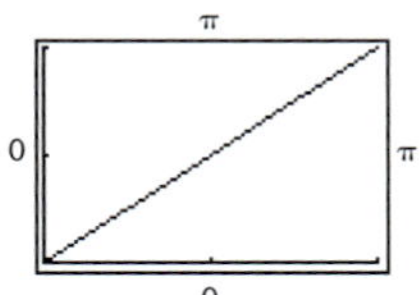

(B) The domain for $\cos x$ is $(-\infty, \infty)$ and the range is $[-1, 1]$, which is the domain for $\cos^{-1} x$. Thus, $y = \cos^{-1}(\cos x)$ has a graph over the interval $(-\infty, \infty)$, but $\cos^{-1}(\cos x) = x$ only on the restricted domain of $\cos x$, $[0, \pi]$.

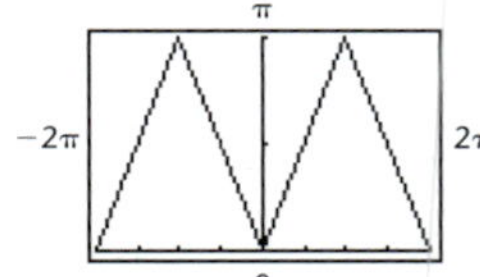

73. 75.38°; 24.41°

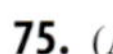
75. (A)

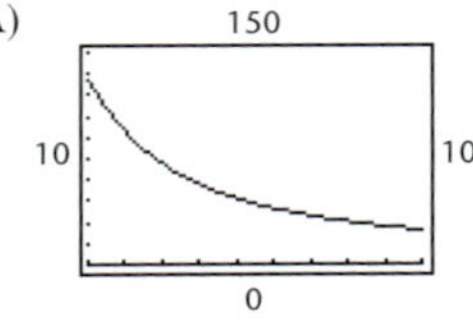

(B) 59.44 mm

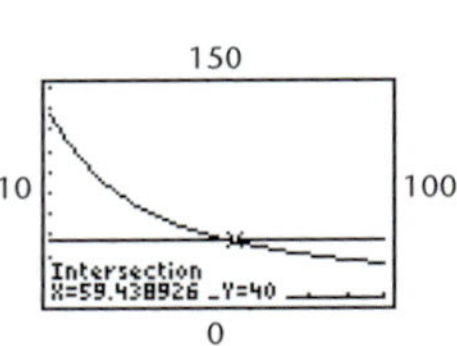

77. 21.59 inches

79. (A)

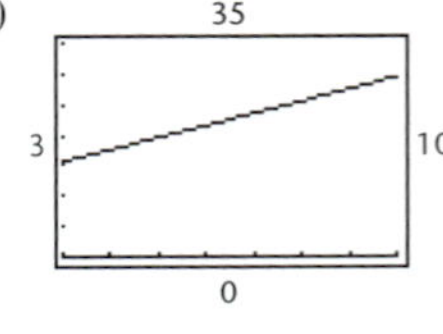

(B) 7.22 inches

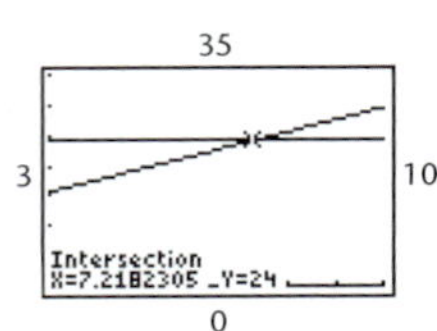

81. (B) 76.10 feet

Chapter 5 Review

1. 2.5 radians *(5.1)* **2.** 7.5 centimeters *(5.1)*
3. $\alpha = 54.8°$, $a = 16.5$ feet, $b = 11.6$ feet *(5.3)*
4. (A) $\frac{\pi}{3}$ (B) 60° (C) $\frac{\pi}{6}$ (D) 30° *(5.4)*
5. (A) III, IV (B) II, III (C) II, IV *(5.2)*
6. (A) $-\frac{3}{5}$ (B) $\frac{5}{4}$ (C) $-\frac{4}{3}$ *(5.4)*
7.

θ°	θ rad	sin θ	cos θ	tan θ	csc θ	sec θ	cot θ
0°	0	0	1	0	ND*	1	ND
30°	$\pi/6$	$1/2$	$\sqrt{3}/2$	$1/\sqrt{3}$	2	$2/\sqrt{3}$	$\sqrt{3}$
45°	$\pi/4$	$1/\sqrt{2}$	$1/\sqrt{2}$	1	$\sqrt{2}$	$\sqrt{2}$	1
60°	$\pi/3$	$\sqrt{3}/2$	$1/2$	$\sqrt{3}$	$2/\sqrt{3}$	2	$1/\sqrt{3}$
90°	$\pi/2$	1	0	ND	1	ND	0
180°	π	0	−1	0	ND	−1	ND
270°	$3\pi/2$	−1	0	ND	−1	ND	0
360°	2π	0	1	0	ND	1	ND

(5.1, 5.2)

*ND = not defined

8. (A) 2π (B) 2π (C) π *(5.4)*
9. (A) Domain = $(-\infty, \infty)$, range = $[-1, 1]$

(B) Domain is set of all real numbers except $x = \frac{2k+1}{2}\pi$, k an integer, range = all real numbers *(5.4)*

10. *(5.4)*

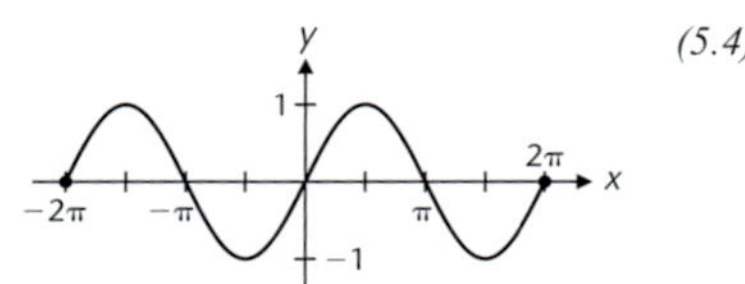

11. *(5.4)*

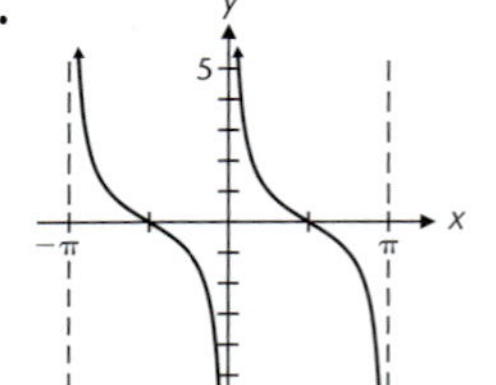

12. The central angle in a circle subtended by an arc of half the length of the radius. *(5.1)*

13. If the graph of $y = \sin x$ is shifted $\frac{\pi}{2}$ units to the left, the result will be the graph of $y = \cos x$. *(5.4)*

14. 78.50° *(5.1)* **15.** $\alpha = 49.7°$; $\beta = 40.3°$; $c = 20.6$ centimeters *(5.3)*
16. (A) II (B) Quadrantal (C) III *(5.1)*
17. (A) and (C) *(5.1)* **18.** (B) and (C) *(5.2)*
19. (A) $\frac{\pi}{2}, \frac{3\pi}{2}$ (B) $0, \pi$ (C) $0, \pi$ *(5.4)*

20. Because the coordinates of a point on a unit circle are given by $P = (a, b) = P = (\cos x, \sin x)$, we evaluate $P = (\cos(-8.305), \sin(-8.305))$—using a calculator set in radian mode—to obtain $P = (-0.436, -0.900)$. Note that $x = -8.305$, because P is moving clockwise. The quadrant in which $P = (a, b)$ lies can be determined by the signs of a and b. In this case, P is in the third quadrant, because a is negative and b is negative. *(5.1, 5.2)*

21. 0 *(5.2)* **22.** Not defined *(5.2)* **23.** 0 *(5.6)*
24. $-\frac{1}{\sqrt{2}}$ or $-\frac{\sqrt{2}}{2}$ *(5.2)* **25.** $\frac{\pi}{4}$ *(5.6)*
26. $-\frac{2}{\sqrt{3}}$ or $\frac{-2\sqrt{3}}{3}$ *(5.2)* **27.** $\frac{\pi}{3}$ *(5.6)* **28.** $-\frac{1}{2}$ *(5.2)*

29. $-\frac{\pi}{4}$ *(5.6)* **30.** $-\frac{1}{\sqrt{3}}$ or $-\frac{\sqrt{3}}{3}$ *(5.2)*

31. $-\frac{\pi}{6}$ *(5.6)* **32.** $\frac{5\pi}{6}$ *(5.6)* **33.** 0.33 *(5.6)*

34. $-\sqrt{2}$ *(5.6)* **35.** $\frac{\sqrt{3}}{2}$ *(5.6)* **36.** $-\frac{4}{3}$ *(5.6)*

37. 0.4431 *(5.2)* **38.** -15.17 *(5.2)* **39.** -2.077 *(5.2)*

40. -0.9750 *(5.6)* **41.** Not defined *(5.6)*

42. 1.557 *(5.6)* **43.** 1.095 *(5.6)*

44. Not defined *(5.6)*

45. (A) $\theta = -30°$ (B) $\theta = 120°$ *(5.6)*

46. (A) $\theta = 151.20°$ (B) $\theta = 82.28°$ *(5.6)*

47. $\cos^{-1}[\cos(-2)] = 2$. For the identity $\cos^{-1}(\cos x) = x$ to hold, x must be in the restricted domain of the cosine function; that is, $0 \le x \le \pi$. The number -2 is not in the restricted domain. *(5.6)*

48. $A = 2, P = 2$ *(5.5)*

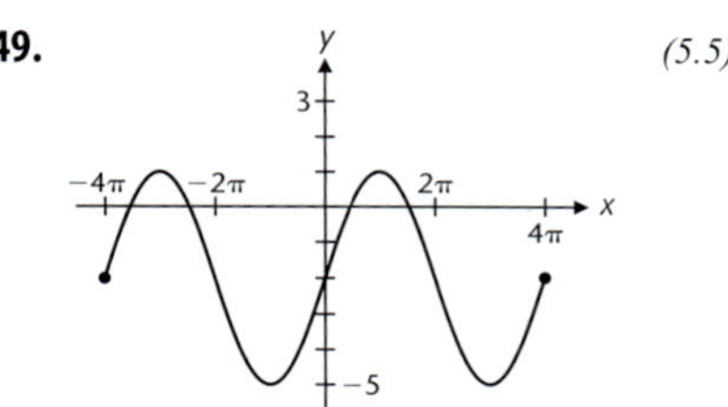

49. *(5.5)*

50. $y = 6\cos 2x;\ -\frac{\pi}{2} \le x \le \pi$ *(5.5)*

51. $y = -0.5\sin \pi x;\ -1 \le x \le 2$ *(5.5)*

52. If the graph of $y = \tan x$ is shifted $\frac{\pi}{2}$ units to the right and reflected in the x axis, the result will be the graph of $y = \cot x$. *(5.4)*

53. (A) $\cos x$ (B) $\tan^2 x$ *(5.4)*

54. *(5.5)*

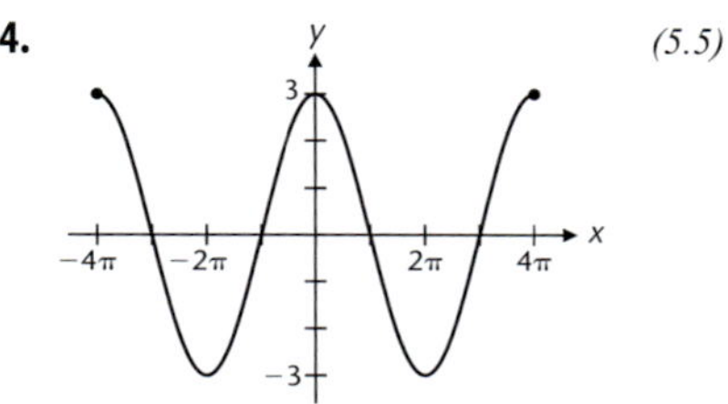

55. $A = 2;\ P = 4;$ phase shift $= \frac{1}{2}$ *(5.5)*

56. Domain $= [-1, 1]$; range $= [0, \pi]$ *(5.6)*

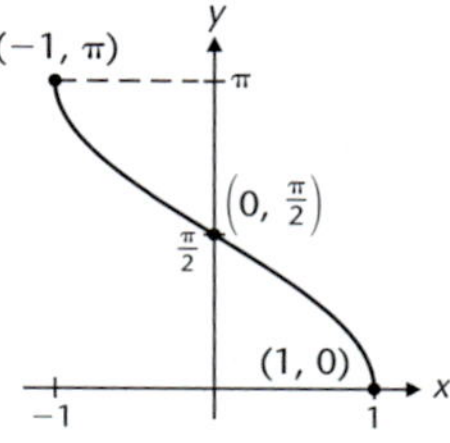

57. $y = \frac{1}{2}\cos 2x + \frac{1}{2}$ *(5.5)*

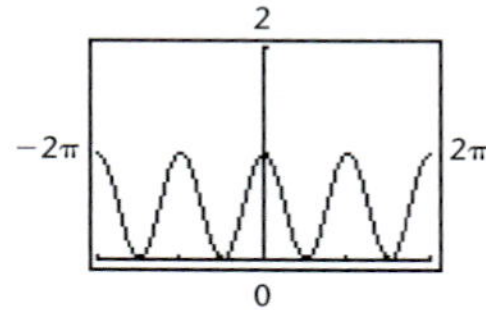

58. (A) $y = \tan x$ (B) $y = \cot x$ *(5.5)*

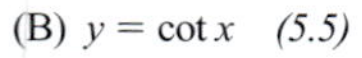

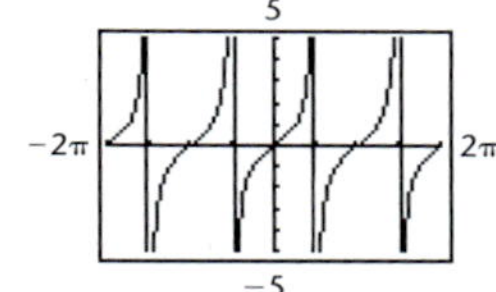

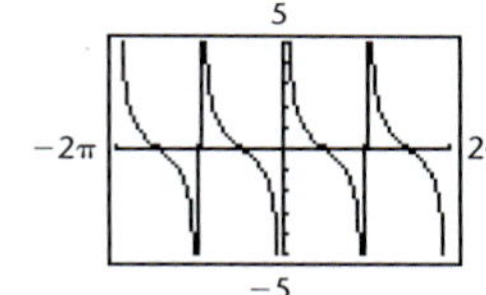

59. (A) Even (B) Neither *(5.4)*

60. False *(5.3)* **61.** True *(5.3)*

62. (A) 2.5 radians (B) $(-6.41, 4.79)$ *(5.1, 5.2)*

63. (A) $\frac{2\pi}{3}$ (B) $\frac{5\pi}{4}$ *(5.2)*

64. *(5.4)*

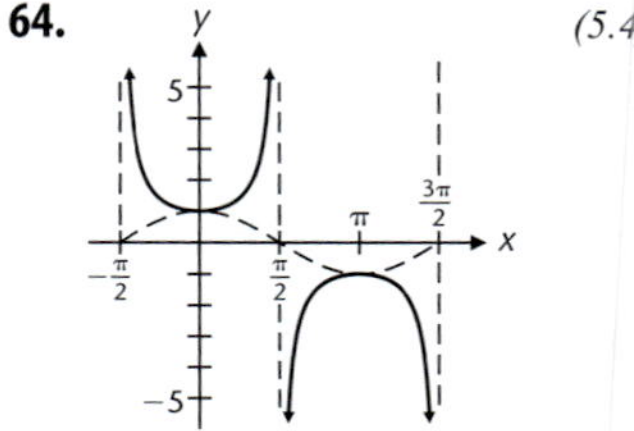

65.

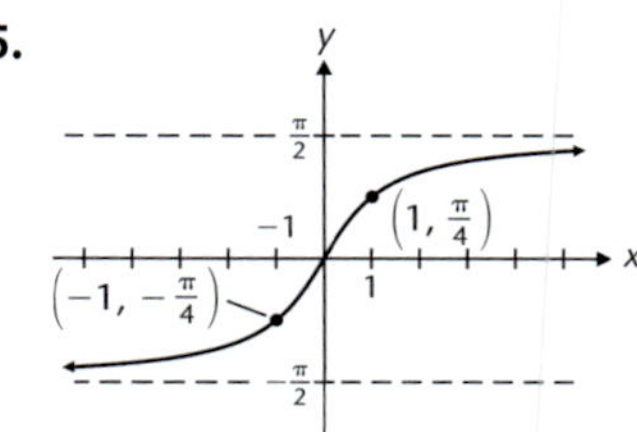

Domain = all real numbers; range $= \left(-\frac{\pi}{2}, \frac{\pi}{2}\right)$ *(5.6)*

66. Phase shift $= -\frac{1}{2}$; period $P = 1$ *(5.5)*

67. Phase shift $= \frac{\pi}{2}$; period $= 4\pi$ *(5.5)*

68. (A) Sine has origin symmetry.
(B) Cosine has y axis symmetry.
(C) Tangent has origin symmetry. *(5.4)*

69. $\frac{1}{\sqrt{1 - x^2}}$ *(5.6)*

70. For each case, the number is not in the domain of the function and an error message of some type will appear. *(5.2, 5.6)*

71. $y = 2 \sin\left(\pi x + \frac{\pi}{4}\right)$ *(5.5)*

72. $y = 2 \sin(2x + 0.928)$ *(5.5)*

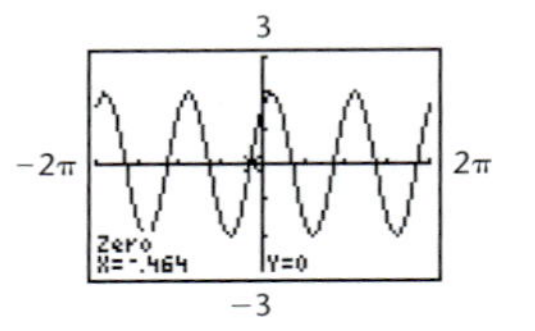

73. (A)

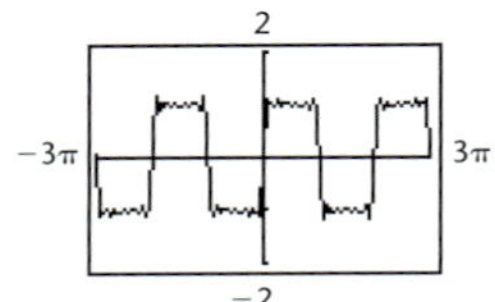

(B) *(5.5)*

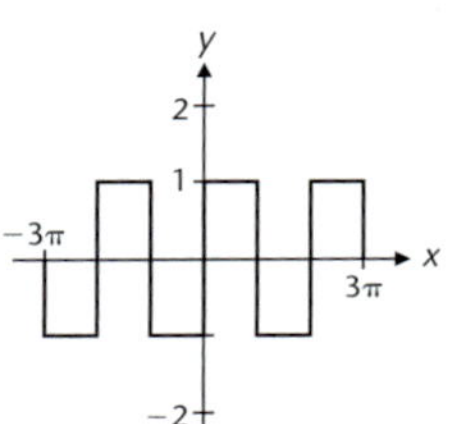

74. $\frac{2\pi}{5}$ radians *(5.1)*

75. 28.3 centimeters *(5.2)*

76. 8.38 radians/second; 167.55 feet/second *(5.1)*

77. $I = 30 \cos 120\pi t$ *(5.5)*

78. (A) $L = 10 \csc \theta + 15 \sec \theta$; $0 < \theta < \frac{\pi}{2}$

(B)

θ (radians)	0.4	0.5	0.6	0.7	0.8	0.9	1.0
L (feet)	42.0	38.0	35.9	35.1	35.5	36.9	39.6

35 feet is the length of the longest log that can make the corner.

(C) Length of longest log that can make the corner is 35.1 feet.

(D) Length L increases without bound. *(5.2, 5.3)*

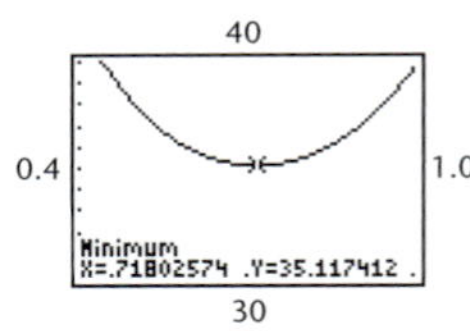

79. (A) $R(t) = 4 - 3 \cos \frac{\pi}{6} t$.

(B) The graph shows the seasonal changes in soft drink consumption. Most is consumed in August and the least in February. *(5.5)*

80. (A)

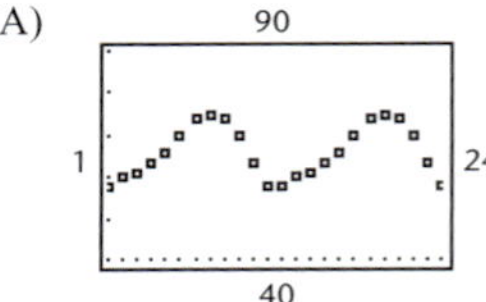

(B) $y = 66.5 + 8.5 \sin\left(\frac{\pi}{6}x - 2.4\right)$

(C) *(5.5)*

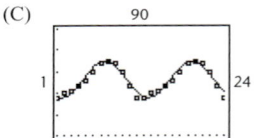

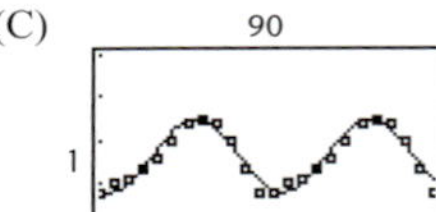

CHAPTER 6

Exercise 6.1

27.

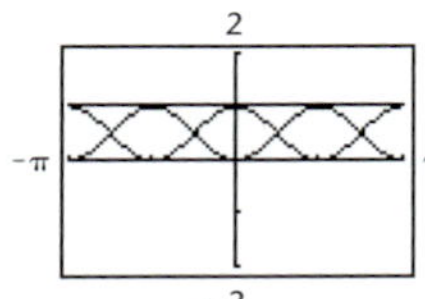

29.

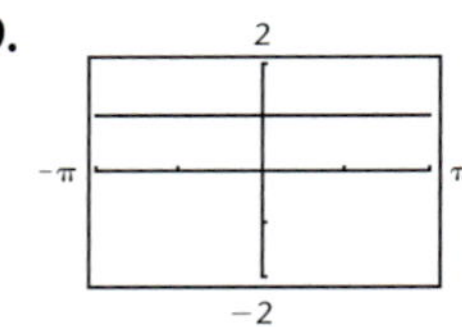

31. Yes **33.** No **35.** No **37.** No **69.** Not an identity

71. An identity **73.** Not an identity **75.** An identity
77. An identity **79.** Not an identity

87. (A) $\cot x = \frac{\cos x}{\sin x}$ (B) $\sin^2 x + \cos^2 x = 1$

(C) $\csc x = \frac{1}{\sin x}$

89. $g(x) = \cot x$

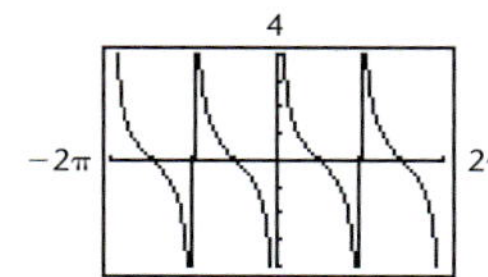

91. $g(x) = -1 + \csc x$

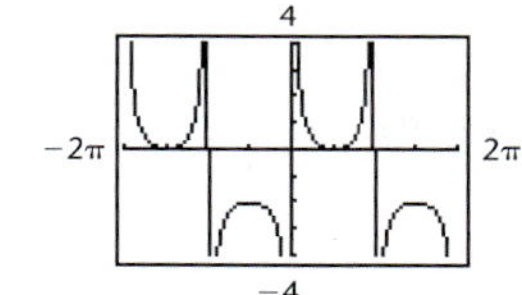

93. $g(x) = 3 \cos x$

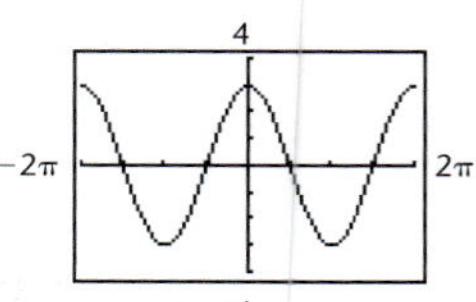

95. III, IV **97.** I, II **99.** All quadrants **101.** I, IV **103.** $a \cos x$ **105.** $a \sec x$

Exercise 6.2

1. Yes **3.** No **5.** No **7.** Yes **13.** $\frac{1}{2}(\cos x - \sqrt{3} \sin x)$ **15.** $\sin x$ **17.** $\dfrac{\tan x + \sqrt{3}}{1 - \sqrt{3} \tan x}$ **19.** $\dfrac{2\sqrt{2}}{\sqrt{3} - 1}$ **21.** $\dfrac{\sqrt{3} + 1}{2\sqrt{2}}$ **23.** $\dfrac{\sqrt{3}}{2}$ **25.** 1

27. $\sin(x - y) = \dfrac{-3 - 4\sqrt{8}}{15}$; $\tan(x + y) = \dfrac{4\sqrt{8} - 3}{4 + 3\sqrt{8}}$

29. $\sin(x - y) = \dfrac{-2}{\sqrt{5}}$; $\tan(x + y) = \frac{2}{11}$

45. $-0.3685, -0.3685; 0.9771, 0.9771$

47. $-0.4429, -0.4429; -2.682, -2.682$

49. Evaluate each side for a particular set of values of x and y for which each side is defined. If the left side is not equal to the right side, then the equation is not an identity. For example, for $x = 2$ and $y = 1$, both sides are defined, but are not equal.

51. $y_1 = \sin(x + \pi/6)$; $y_2 = \dfrac{\sqrt{3}}{2} \sin x + \dfrac{1}{2} \cos x$

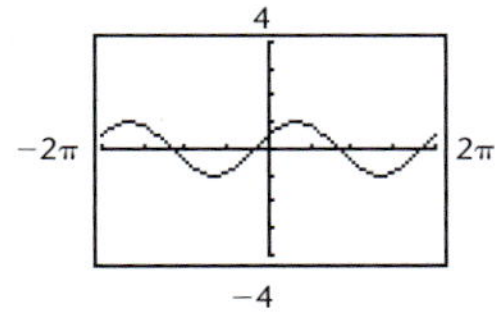

53. $y_1 = \cos(x - 3\pi/4)$; $y_2 = -\dfrac{\sqrt{2}}{2} \cos x + \dfrac{\sqrt{2}}{2} \sin x$

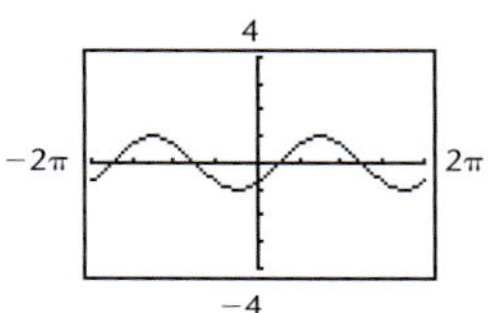

55. $y_1 = \tan(x + 2\pi/3)$; $y_2 = \dfrac{\tan x - \sqrt{3}}{1 + \sqrt{3} \tan x}$

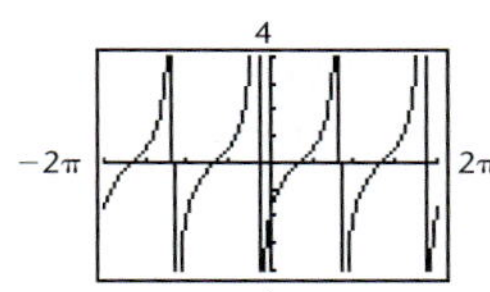

57. $\frac{24}{25}$ **59.** $-\frac{1}{2}$ **61.** $xy + \sqrt{1 - x^2}\sqrt{1 - y^2}$

65. $y_1 = \cos 1.2x \cos 0.8x - \sin 1.2x \sin 0.8x$; $y_2 = \cos 2x$

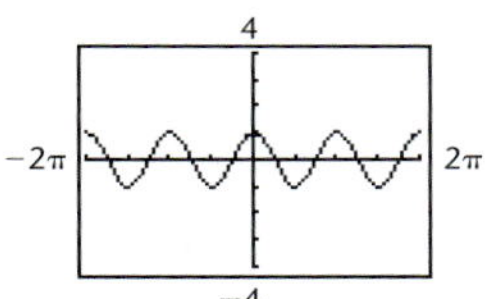

71. (C) 3,510 feet

Exercise 6.3

1. $\frac{1}{2} = \frac{1}{2}$ **3.** $-\sqrt{3} = -\sqrt{3}$ **5.** $1 = 1$ **7.** $\dfrac{\sqrt{2 - \sqrt{2}}}{2}$ **9.** $\dfrac{\sqrt{2 - \sqrt{2}}}{2}$

11.

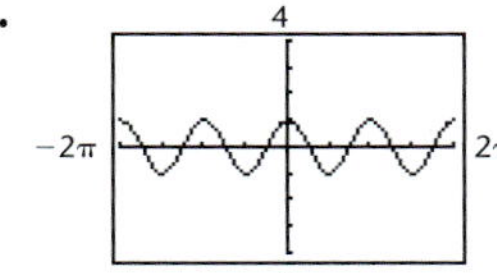

13.

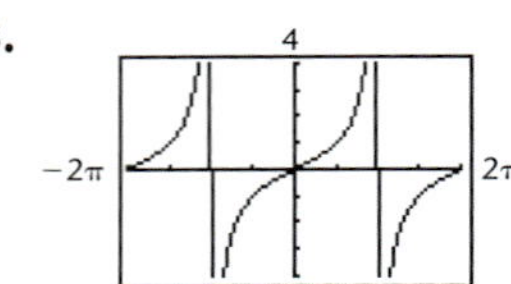

33. No **35.** Yes **37.** No

39. $\sin 2x = -\frac{24}{25}$, $\cos 2x = \frac{7}{25}$, $\tan 2x = -\frac{24}{7}$

41. $\sin 2x = -\frac{120}{169}$, $\cos 2x = \frac{119}{169}$, $\tan 2x = -\frac{120}{119}$

43. $\sin \frac{x}{2} = \sqrt{\frac{3 + 2\sqrt{2}}{6}}$, $\cos \frac{x}{2} = -\sqrt{\frac{3 - 2\sqrt{2}}{6}}$,

$\tan \frac{x}{2} = -3 - 2\sqrt{2}$

45. $\sin \frac{x}{2} = -\frac{2\sqrt{5}}{5}$, $\cos \frac{x}{2} = \frac{\sqrt{5}}{5}$, $\tan \frac{x}{2} = -2$

47. (A) 2θ is a second quadrant angle, because θ is a first quadrant angle and $\tan 2\theta$ is negative for 2θ in the second quadrant and not for 2θ in the first.

(B) Construct a reference triangle for 2θ in the second quadrant with $(a, b) = (-3, 4)$. Use the Pythagorean theorem to find $r = 5$. Thus, $\sin 2\theta = 4/5$ and $\cos 2\theta = -3/5$.

(C) The double-angle identities $\cos 2\theta = 1 - 2\sin^2 \theta$ and $\cos 2\theta = 2\cos^2 \theta - 1$.

(D) Use the identities in part C in the form

$\sin \theta = \sqrt{\frac{1 - \cos 2\theta}{2}}$ and $\cos \theta = \sqrt{\frac{1 + \cos 2\theta}{2}}$.

The positive radicals are used because θ is in quadrant one.

(E) $\sin \theta = 2\sqrt{5}/5$; $\cos \theta = \sqrt{5}/5$

49. (A) $-0.72335 = -0.72335$ (B) $-0.58821 = -0.58821$

51. (A) $-3.2518 = -3.2518$ (B) $0.89279 = 0.89279$

53. $y_1 = y_2$ for $[-\pi, \pi]$ **55.** $y_1 = y_2$ for $[-2\pi, 0]$

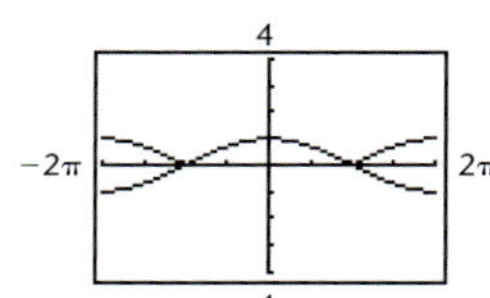

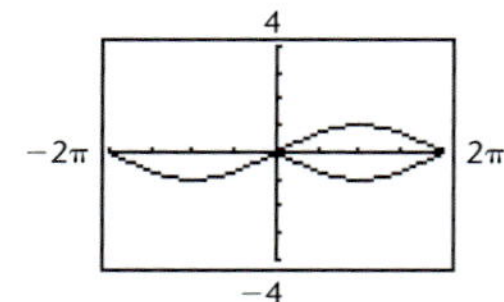

61. $-\frac{7}{25}$ **63.** $-\frac{24}{7}$ **65.** $\frac{\sqrt{5}}{5}$

67. $\tan \frac{x}{2}$ **69.** $1 + 2\sin x$

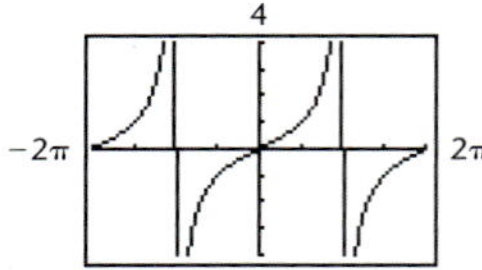

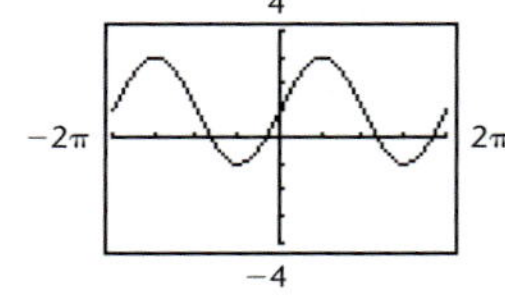

71. $\sec 2x$

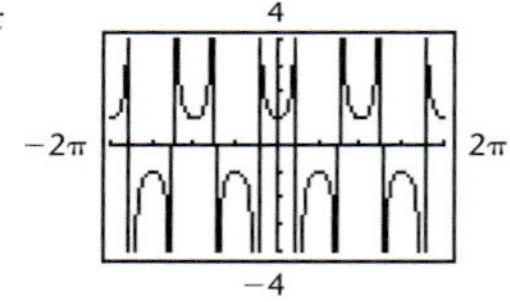

73. $x = \frac{224}{17} \approx 13.176$ m; $\theta = 28.955°$

75. (A) $d = \frac{v_0^2 \sin 2\theta}{32 \text{ ft/sec}^2}$ (B) $\theta = 45°$

77. (B) **TABLE 1**

n	10	100	1,000	10,000
A_n	2.93893	3.13953	3.14157	3.14159

(C) A_n appears to approach π, the area of the circle with radius 1.

(D) A_n will not exactly equal the area of the circumscribing circle for any n no matter how large n is chosen; however, A_n can be made as close to the area of the circumscribing circle as we like by making n sufficiently large.

Exercise 6.4

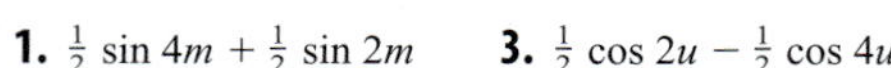

1. $\frac{1}{2}\sin 4m + \frac{1}{2}\sin 2m$ **3.** $\frac{1}{2}\cos 2u - \frac{1}{2}\cos 4u$

5. $2\sin 2t \cos t$ **7.** $2\sin 7w \sin 2w$

9. $\frac{\sqrt{3} - 2}{4}$ **11.** $\frac{1}{4}$ **13.** $-\frac{\sqrt{2}}{2}$ **15.** $\frac{\sqrt{6}}{2}$

19. Let $x = u + v$ and $y = u - v$ and solve the resulting system for u and v in terms of x and y, then substitute the results into the first identity. The second identity will result after a small amount of algebraic manipulation.

29. Yes **31.** No **33.** Yes

35. (A) $-0.34207 = -0.34207$ (B) $-0.05311 = -0.05311$

37. (A) $-0.19115 = -0.19115$ (B) $-0.46541 = -0.46541$

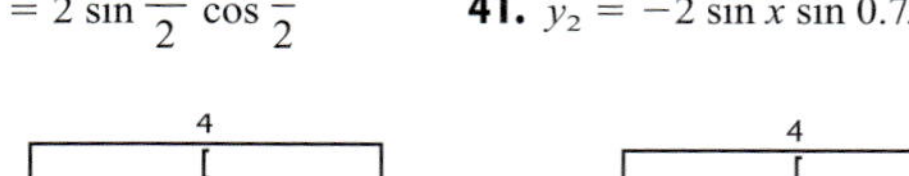

39. $y_2 = 2\sin \frac{3x}{2} \cos \frac{x}{2}$ **41.** $y_2 = -2\sin x \sin 0.7x$

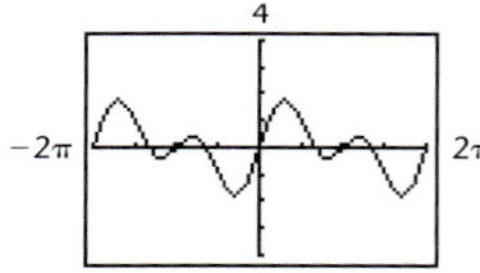

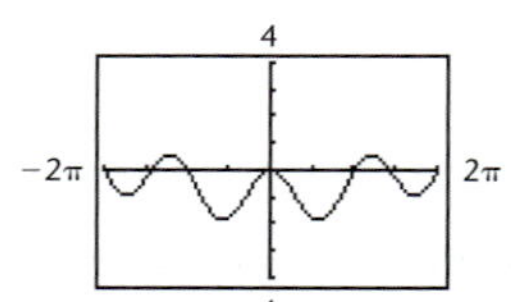

43. $y_2 = \frac{1}{2}(\sin 4x + \sin 2x)$

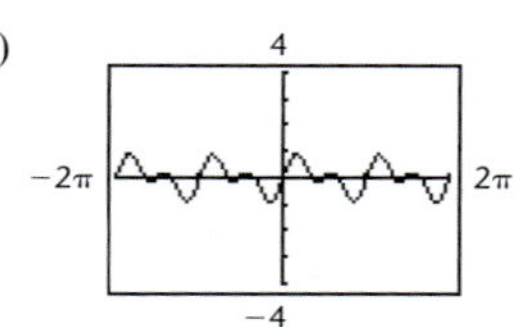

45. $y_2 = \frac{1}{2}(\cos 1.6x - \cos 3x)$

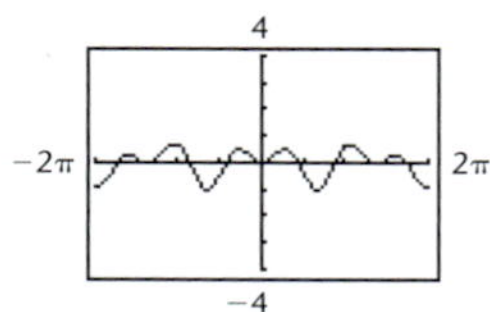

49. (A)

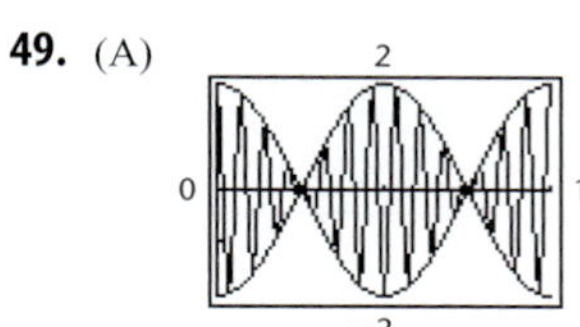

(B) $y_1 = \cos(30\pi x) + \cos(26\pi x)$; Graph same as part A

51. (A)

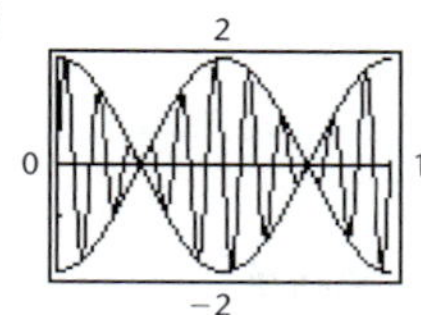

(B) $y_1 = \sin(22\pi x) + \sin(18\pi x)$; Graph same as part A

53. (B)

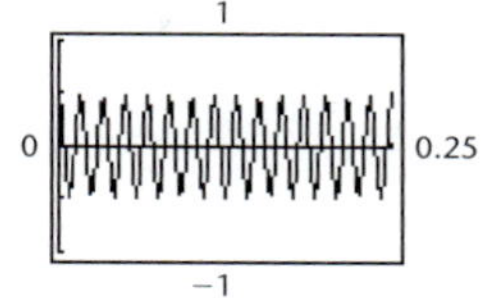

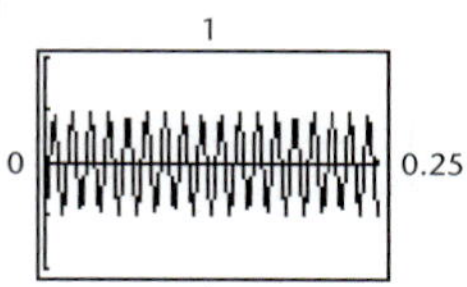

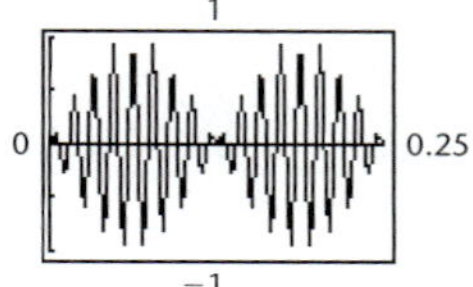

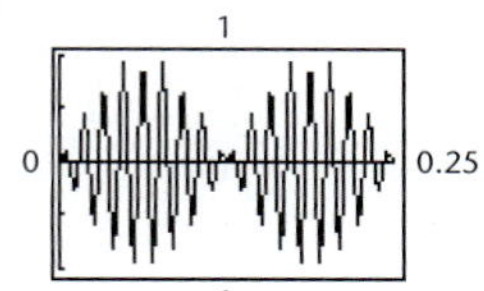

Exercise 6.5

1. $7\pi/6, 11\pi/6$ **3.** $7\pi/6 + 2k\pi, 11\pi/6 + 2k\pi$, k any integer
5. $2\pi/3$ **7.** $2\pi/3 + k\pi$, k any integer **9.** $30°, 330°$
11. $30° + k(360°), 330° + k(360°)$, k any integer **13.** 1.1279, 5.1553
15. 74.0546° **17.** $3.5075 + 2k\pi, 5.9172 + 2k\pi$, k any integer
19. 0.3376 **21.** 2.7642
23. $k(180°), 135° + k(180°)$, k any integer **25.** $0, 2\pi/3, \pi, 4\pi/3$
27. $4\pi/3$ **29.** 210°, 330° **31.** 60°, 180°, 300°
33. $\pi/3, \pi, 5\pi/3$ **35.** 41.81° **37.** 1.911
39. 0.3747, 2.767 **41.** $0.3747 + 2k\pi, 2.767 + 2k\pi$, k any integer
43. 0.3747, 2.7669 **45.** $0.3747 + 2k\pi, 2.7669 + 2k\pi$, k any integer
47. $(-1.1530, 1.1530)$ **49.** $[3.5424, 5.3778], [5.9227, \infty)$
51. 1.8183
53. $\tan^{-1}(-5.377)$ has exactly one value, -1.387; the equation $\tan x = -5.377$ has infinitely many solutions, which are found by adding $k\pi$, k any integer, to each solution in one period of $\tan x$.

55. $0, 3\pi/2$ **57.** π **59.** 0.1204, 0.1384
61. (A) The largest zero for f is 0.3183. As x increases without bound, $1/x$ tends to 0 through positive numbers, and $\sin(1/x)$ tends to 0 through positive numbers. $y = 0$ is a horizontal asymptote for the graph of f.
(B) Infinitely many zeros exist between 0 and b, for any b, however small. The exploration graphs suggest this conclusion, which is reinforced by the following reasoning. Note that for each interval $(0, b]$, however small, as x tends to zero through positive numbers, $1/x$ increases without bound, and as $1/x$ increases without bound, $\sin(1/x)$ will cross the x axis an unlimited number of times. The function f does not have a smallest zero, because, between 0 and b, no matter how small b is, there is always an unlimited number of zeros.
63. 0.009235 seconds **65.** 50.77° **67.** 123°
69. 2.267 radians
71. (A) 12.4575 millimeters (B) 2.6496 millimeters
73. $(r, \theta) = (0, 0°), (0, 180°), (0, 360°)$ **75.** $\theta = 45°$

Chapter 6 Review

5. $\frac{1}{2}\sin 8\alpha + \frac{1}{2}\sin 2\alpha$ *(6.4)* **6.** $-2\sin 6x \sin x$ *(6.4)*
7. $\cos x$ *(6.2)*
8. $135° + k360°, 225° + k360°$, k any integer *(6.5)*
9. $k\pi$ or $\frac{\pi}{4} + k\pi$, k any integer *(6.5)*
10. $x = \begin{cases} 0.7878 + 2k\pi \\ 2.3538 + 2k\pi \end{cases}$ k any integer *(6.5)*
11. $x = \begin{cases} 75.1849° + k360° \\ 284.8151° + k360° \end{cases}$ k any integer *(6.5)*

12. -1.4032 *(6.5)* **13.** 3.1855 *(6.5)*
14. (A) Not an identity (B) An identity *(6.1)*
24. $\frac{-2 - \sqrt{3}}{4}$ *(5.2, 6.4)* **25.** $-\frac{\sqrt{6}}{2}$ *(5.2, 6.4)*
26. No *(6.1)* **27.** Yes *(6.2)* **28.** No *(6.2)*
29. No *(6.2)* **30.** $\frac{\pi}{3}, \frac{2\pi}{3}, \frac{4\pi}{3}, \frac{5\pi}{3}$ *(6.5)*
31. 0°, 120° *(6.5)*
32. $x = 0 + 2k\pi, x = \pi + 2k\pi, x = \frac{\pi}{6} + 2k\pi, x = \frac{5\pi}{6} + 2k\pi$, k any integer. The first two can also be written together as $x = k\pi$, k any integer. *(6.5)*

33. $x = 0 + 2k\pi, x = \pi + 2k\pi, x = \frac{\pi}{6} + 2k\pi, x = \frac{11\pi}{6} + 2k\pi$, k any integer. The first two can also be written together as $x = k\pi$, k any integer. *(6.5)*
34. $120° + k360°, 240° + k360°$, k any integer *(6.5)*
35. $14.34° + k180°$ *(6.5)*
36. $x = \begin{cases} 0.6259 + 2k\pi \\ 2.516 + 2k\pi \end{cases}$ k any integer *(6.5)*
37. 1.178, 2.749 *(6.5)* **38.** 1.4903 *(6.5)*
39. $(-\infty, 1.4903)$ *(6.5)* **40.** $-0.6716, 0.6716$ *(6.5)*
41. $[-0.6716, 0.6716]$ *(6.5)*
42. (A) Yes
(B) Conditional equation, because the equation is false for $x = 1$ and $y = 1$, for example, and both sides are defined at $x = 1$ and $y = 1$. *(6.1)*
43. $\sin^{-1} 0.3351$ has exactly one value, whereas the equation $\sin x = 0.3351$ has infinitely many solutions. *(5.6, 6.5)*

44. (A) Not an identity (B) An identity *(6.1)*
45. $y_2 = \frac{1}{2}\cos x + \frac{\sqrt{3}}{2}\sin x$ *(6.2)*

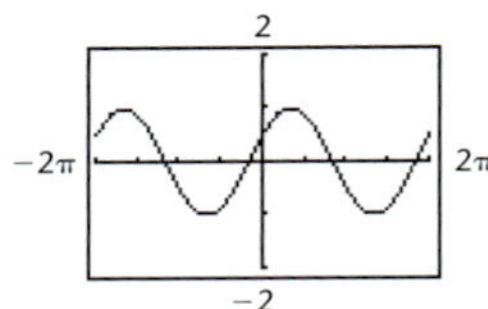

46. (A) $0, \frac{2\pi}{3}, \frac{4\pi}{3}$ (B) 0, 2.0944 and 4.1888 *(6.5)*
47. 0.149 and -2.233 *(6.5)*
48. (A) $\frac{3}{\sqrt{10}}$ or $\frac{3\sqrt{10}}{10}$ (B) $\frac{7}{25}$ *(6.3)*
49. $-\frac{24}{25}$ *(6.3)* **50.** $\frac{24}{25}$ *(6.2)*
51. (A) $0, \frac{\pi}{3}, \frac{2\pi}{3}$ (B) 0, 1.0472, 2.0944 *(6.5)*
52. (A) 0.6817, 1.3183
(B) As x increases without bound, $\frac{1}{x-1}$ tends to 0 through positive numbers and $\sin \frac{1}{x-1}$ tends to 0 through positive numbers. $y = 0$ is a horizontal asymptote for the graph of f.
(C) The exploratory graphs are left to the student. There are infinitely many zeros in any interval containing $x = 1$. The number $x = 1$ is not a zero because $\sin \frac{1}{x-1}$ is not defined at $x = 1$. *(6.5)*
53. $x = \sqrt{27}$; $x = 5.196$ centimeters, $\theta = 30.000°$ *(6.3)*
54. 0.00346 seconds *(6.5)*
55. (B) $y = 0.6 \cos 184\pi t$ $y = -0.6 \cos 208\pi t$

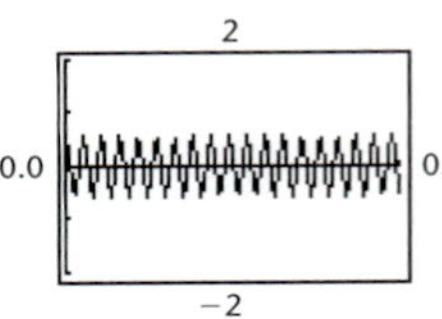

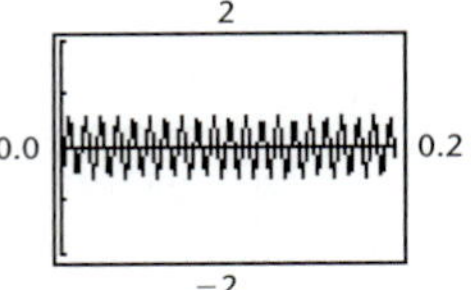

$y = 0.6 \cos 184\pi t - 0.6 \cos 208\pi t$

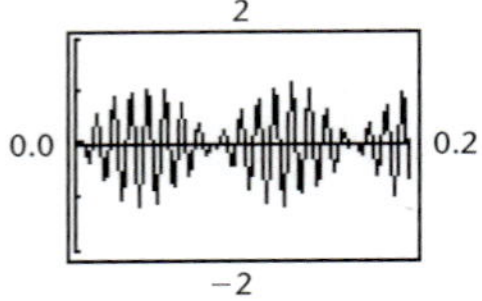

$y = 1.2 \sin 12\pi t \sin 196\pi t$ *(6.4)*

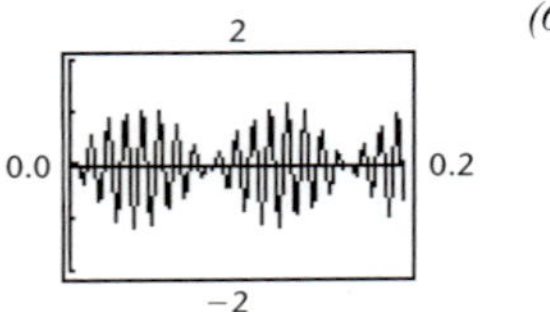

56. Height = 7.057 feet, radius = 21.668 feet

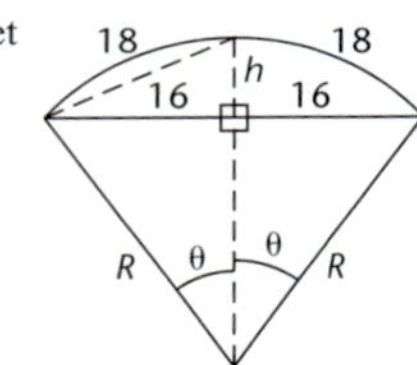

From the figure, $R\theta = 18$ and $\sin\theta = \frac{16}{R}$. From these two equations, solving each for R in terms of θ and setting the results equal to each other, we obtain the desired trigonometric equation. *(6.5)*

CHAPTER 7

Exercise 7.1

1. $\gamma = 79°, a = 41$ feet, $b = 20$ feet **3.** $\beta = 40°, a = 16$ kilometers, $c = 5.8$ kilometers **5.** $\alpha = 49°, a = 53$ yards, $b = 66$ yards
7. $\beta = 81°, b = 16$ centimeters, $c = 12$ centimeters
9. 1 triangle; the case where α is acute and $a = 2 = h$
11. 1 triangle; the case where α is acute and $a \geq b$ $(a = 6, b = 4)$
13. 0 triangles; the case where α is acute and $0 < a < h$ $(a = 1, h = 2)$
15. 2 triangles; the case where α is acute and $h < a < b$ $(h = 2, a = 3, b = 4)$
17. $\beta = 49.5°, a = 20.0$ feet, $c = 4.81$ feet
19. $\gamma = 58.1°, a = 140$ meters, $c = 129$ meters **21.** No solution
23. Triangle I: $\beta = 158.8°, \gamma = 5.3°, c = 7.55$ inches; triangle II: $\beta = 21.2°, \gamma = 142.9°, c = 49.3$ inches

25. Triangle I: $\alpha = 116.6°$, $\gamma = 24.5°$, $c = 19.8$ inches; triangle II: $\alpha = 63.4°$, $\gamma = 77.7°$, $c = 46.7$ inches
27. No solution **29.** $\alpha = 22°10'$, $\gamma = 128°20'$, $c = 89.9$ millimeters
31. $\beta = 90°$, $\gamma = 60°$, $c = 50$ feet **33.** $k = 25.2 \sin 42.3° = 17.0$

35. Left side: 16.204; right side: 16.073 **37.** 4.06 miles, 2.47 miles
39. 353 feet **41.** 5.8 inches, 3.1 inches
43. 4.42×10^7 kilometers, 2.39×10^8 kilometers **45.** 159 feet
47. $R = 7.76$ millimeters, $s = 13.4$ millimeters

Exercise 7.2

1. Angle γ is acute. A triangle can have at most one obtuse angle. Because α is acute, then, if the triangle has an obtuse angle it must be the angle opposite the longer of the two sides, b and c. Thus, γ, the angle opposite the shorter of the two sides, c, must be acute.
3. $a = 6.03$ yards, $\beta = 56.6°$, $\gamma = 52.2°$
5. $c = 14.0$ millimeters, $\alpha = 20°40'$, $\beta = 39°0'$
7. If the triangle has an obtuse angle, then it must be the angle opposite the longest side; in this case, β.
9. $\alpha = 23.0°$, $\beta = 94.9°$, $\gamma = 62.1°$
11. $\alpha = 67.3°$, $\beta = 54.6°$, $\gamma = 58.1°$
13. No solution, because $\alpha + \gamma > 180°$
15. $b = 23.1$ inches, $\alpha = 46.1°$, $\gamma = 29.4°$
17. $\beta = 10.8°$, $a = 22.5$ meters, $b = 5.01$ meters
19. $\alpha = 30.7°$, $\gamma = 110.9°$, $c = 21.0$ inches
21. $\alpha = 49.1°$, $\beta = 102.9°$, $\gamma = 28.0°$
23. Triangle I: $\beta = 109.7°$, $\alpha = 11.9°$, $a = 1.58$ meters; triangle II: $\beta = 70.3°$, $\alpha = 51.3°$, $a = 5.99$ meters
25. No solution
27. Triangle I: $\gamma = 140.5°$, $\alpha = 25.9°$, $a = 40.1$ meters; triangle II: $\gamma = 39.5°$, $\alpha = 126.9°$, $a = 73.5$ meters
33. 120 yards **35.** 100.6° **37.** 5.81 feet **39.** 121 miles
41. 74.1 meters **43.** 0.284 radians
45. $\alpha = 31°50'$, $\beta = 50°10'$, $\gamma = 98°0'$
47. $\angle CAB = 33°$ **49.** 24,800 miles

Exercise 7.3

1. $|\mathbf{u} + \mathbf{v}| = 58$ miles per hour, $\theta = 51°$
3. $|\mathbf{u} + \mathbf{v}| = 65$ kilograms, $\theta = 54°$
5. $|\mathbf{u} + \mathbf{v}| = 447$ kilometers per hour, $\theta = 13.6°$
7. $|\mathbf{u}| = 30$ pounds, $|\mathbf{v}| = 12$ pounds
9. $|\mathbf{u}| = 71$ miles per hour, $|\mathbf{v}| = 220$ miles per hour
11. No. Two vectors are equal if and only if they have the same magnitude and direction.
13. $|\mathbf{u} + \mathbf{v}| = 77$ grams, $\alpha = 15°$ **15.** $|\mathbf{u} + \mathbf{v}| = 23$ knots, $\alpha = 6°$
17. $|\mathbf{u}| = 12$ kilograms, $|\mathbf{v}| = 6.0$ kilograms
19. $|\mathbf{u}| = 109$ miles per hour, $|\mathbf{v}| = 160$ miles per hour
21. Because the zero vector has an arbitrary direction, it can be perpendicular to any vector.
23. 260 miles per hour at 282° **25.** 288°, 7.6 knots
27. 3,900 pounds at 72° **29.** (A) 388 pounds (B) 4,030 pounds
31. To the right

Exercise 7.4

1. $\langle -3, -3\rangle$ **3.** $\langle -6, 7\rangle$ **5.** $\langle 3, 5\rangle$ **7.** 5 **9.** $\sqrt{34}$
11. 25
13. Two algebraic vectors, $\langle a, b\rangle$ and $\langle c, d\rangle$, are equal if and only if $a = c$ and $b = d$.
15. (A) $\langle 1, 4\rangle$ (B) $\langle 3, -2\rangle$ (C) $\langle 14, -1\rangle$
17. (A) $\langle -2, 1\rangle$ (B) $\langle -6, -3\rangle$ (C) $\langle -10, -1\rangle$
19. $\mathbf{v} = -3\mathbf{i} + 4\mathbf{j}$ **21.** $\mathbf{v} = 3\mathbf{i}$ **23.** $\mathbf{v} = -5\mathbf{i} - 2\mathbf{j}$
25. $5\mathbf{i} + 2\mathbf{j}$ **27.** $-16\mathbf{j}$ **29.** $-8\mathbf{j}$ **31.** $\mathbf{u} = \langle -\frac{3}{5}, \frac{4}{5}\rangle$
33. $\mathbf{u} = \left\langle \frac{-5}{\sqrt{34}}, \frac{3}{\sqrt{34}} \right\rangle$
35. Any one of the force vectors must have the same magnitude as the sum vector of the other two and be oppositely directed as the sum vector.
45. 760 pounds to the left; 761 pounds to the right
47. 897 pounds to the left; 732 pounds to the right
49. This corresponds to a tension force of 462 pounds in member CB. This corresponds to a compression force of 231 pounds in member AB.
51. AB = a compression of 2,360 pounds; BC = a tension of 2,000 pounds

Exercise 7.5

1.

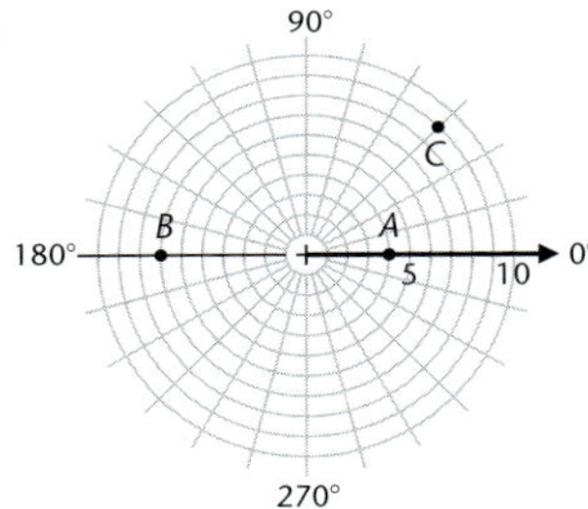

3.

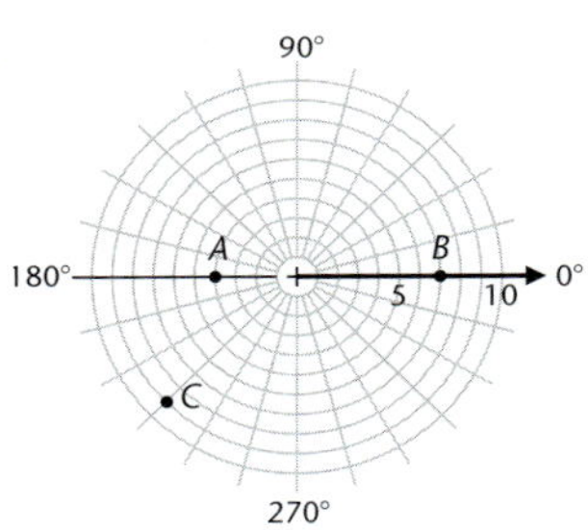

5.

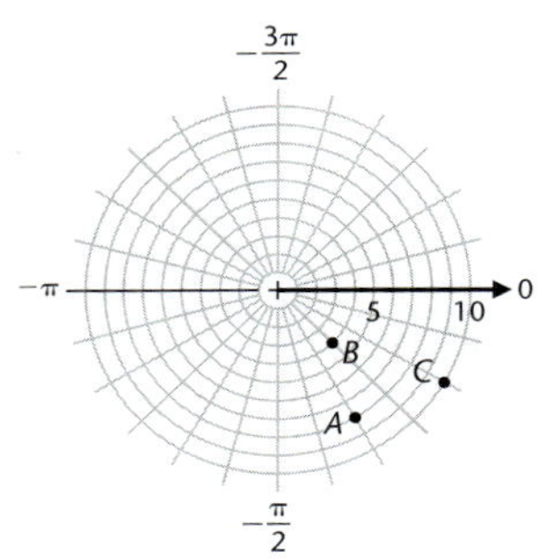

7.

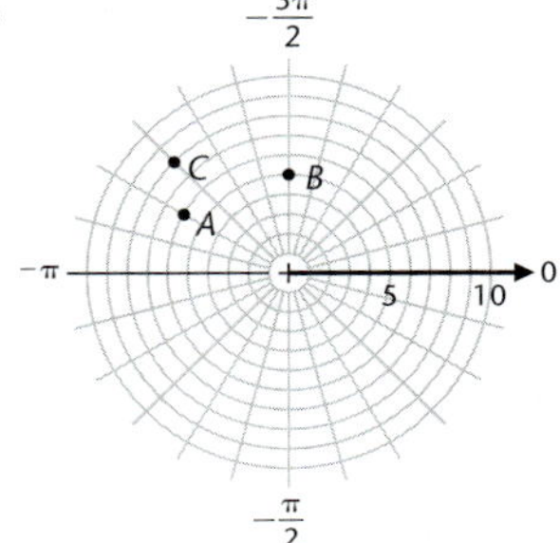

9. $(5, -\pi/4)$: The polar axis is rotated $\pi/4$ radians clockwise (negative direction) and the point is located five units from the pole along the positive polar axis. $(5, 7\pi/4)$: The polar axis is rotated $7\pi/4$ radians counterclockwise (positive direction) and the point is located five units from the pole along the positive polar axis. $(-5, -5\pi/4)$: The polar axis is rotated $5\pi/4$ radians clockwise (negative direction) and the point is located five units from the pole along the negative polar axis.

11.

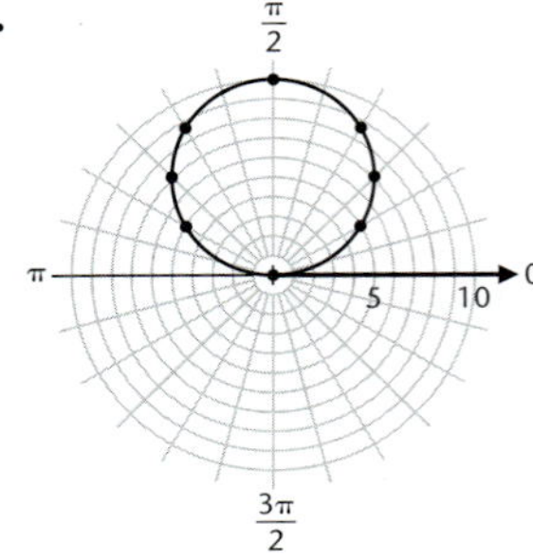

13.

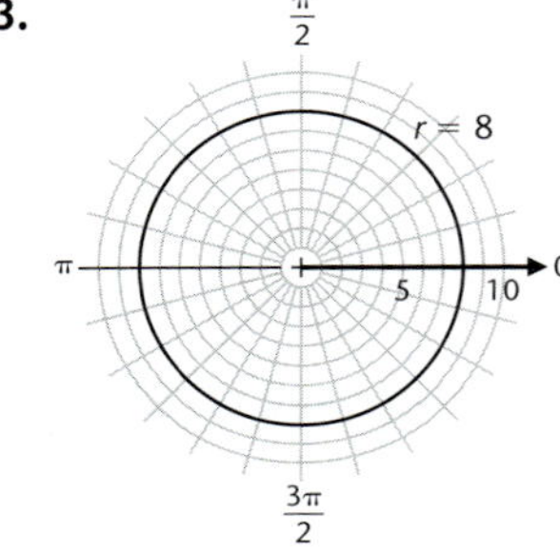

15.

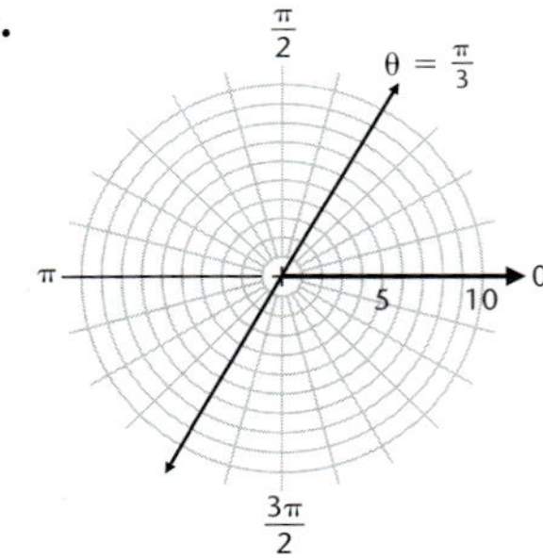

17. (5.196, 3.000) **19.** (1.848, −0.765) **21.** (2.078, 3.688)
23. (7.9, 64°) **25.** (26, −32°) **27.** (7.61, −164.4°)

29.

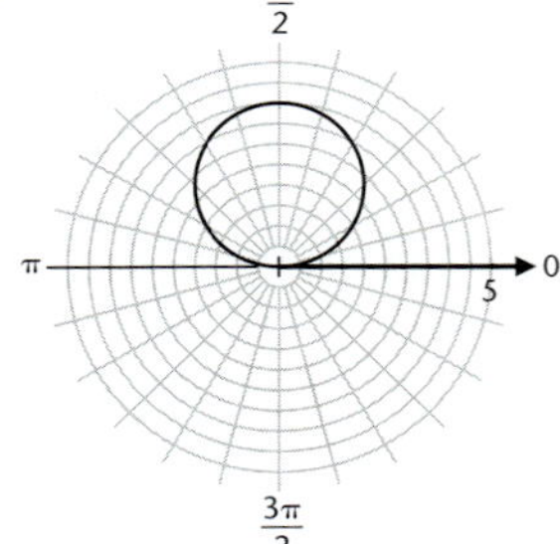

31.

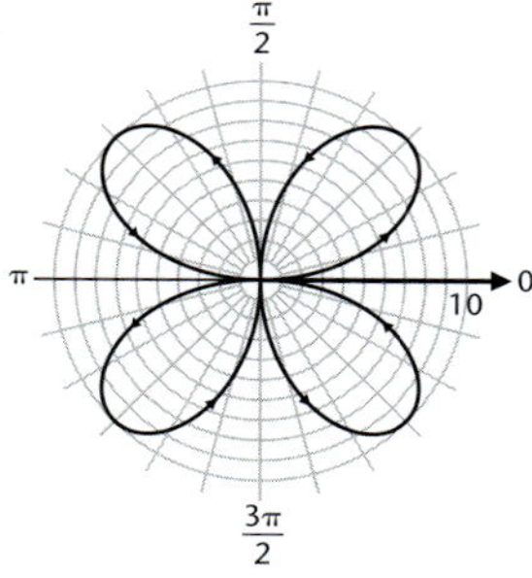

33.

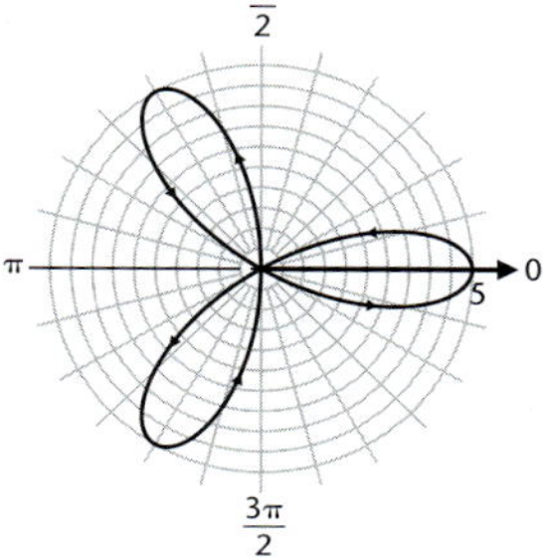

35.

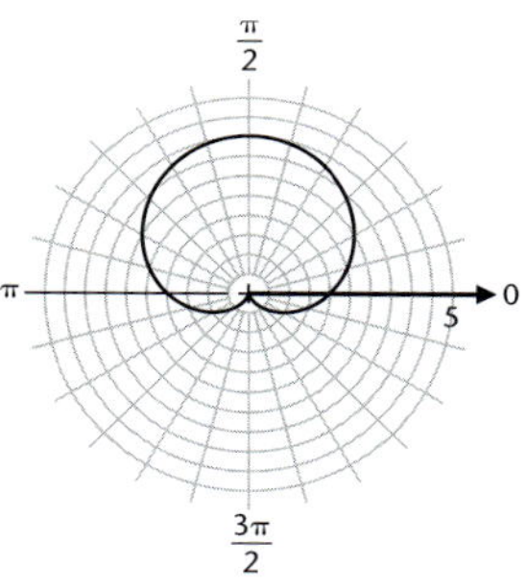

37.

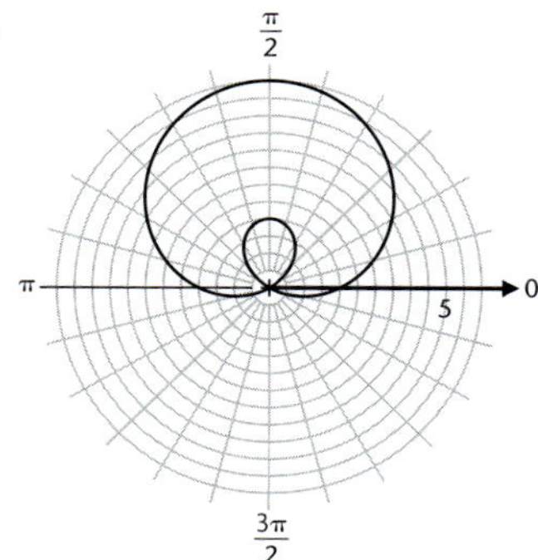

39.

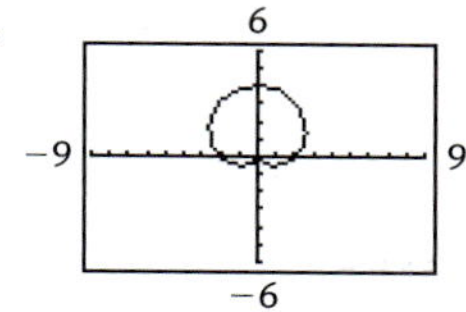

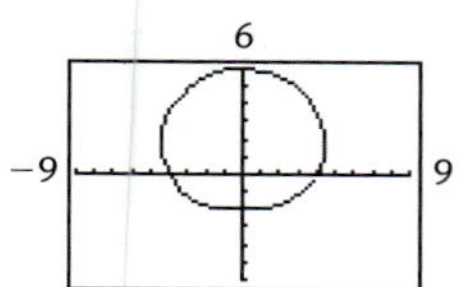

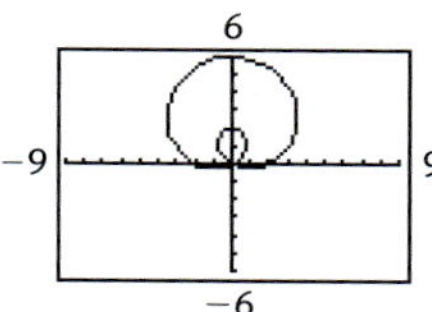

41. (A)

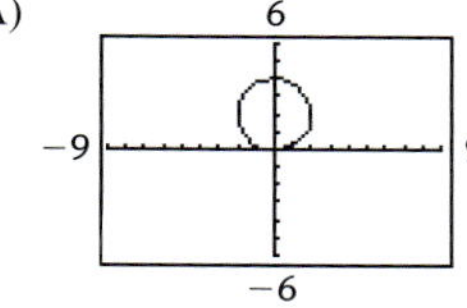

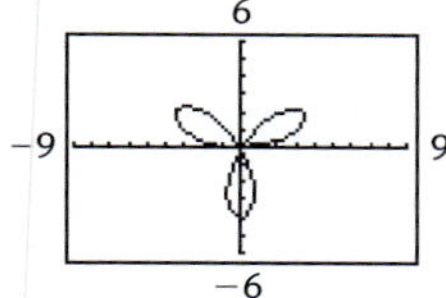

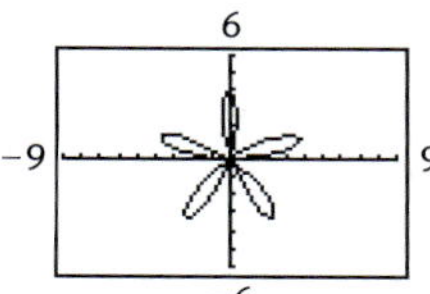

(B) 7 (C) n

43. (A)

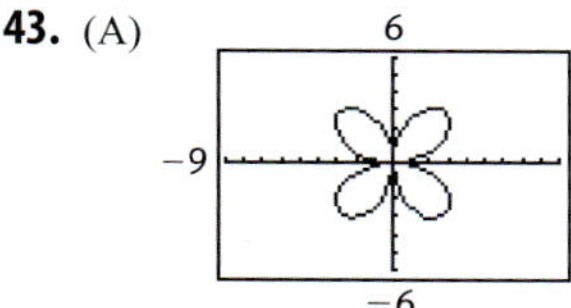

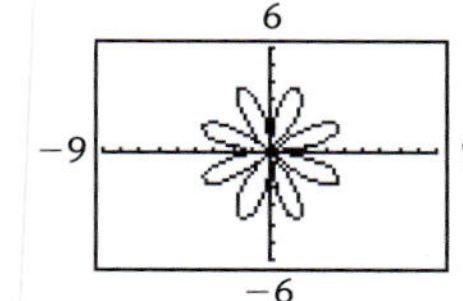

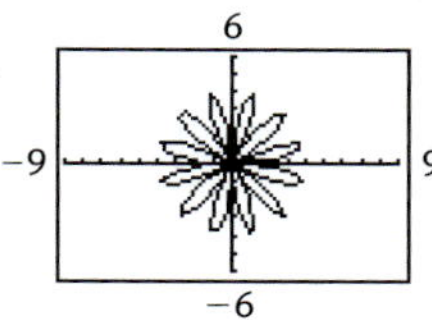

(B) 16 (C) $2n$

45. $r = 5 \sin \theta$ **47.** $\tan \theta = 1$ or $\theta = \frac{\pi}{4}$

49. $r = \frac{4 \cos \theta}{\sin^2 \theta} = 4 \cot \theta \csc \theta$ **51.** $3x - 4y = -1$

53. $x^2 + y^2 = -2y$ **55.** $y = x$

57. For each n, there are n large petals and n small petals. For n odd, the small petals are within the large petals; for n even, the small petals are between the large petals.

59.

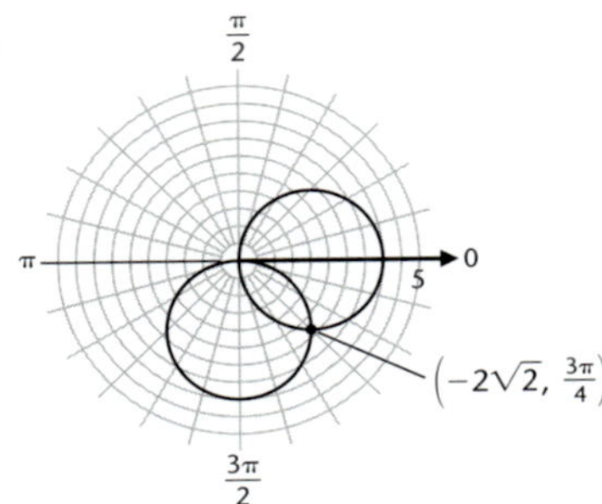

$(r, \theta) = \left(-2\sqrt{2}, \frac{3\pi}{4}\right)$ [*Note:* (0, 0) is not a solution of the system even though the graphs cross at the origin.]

61.

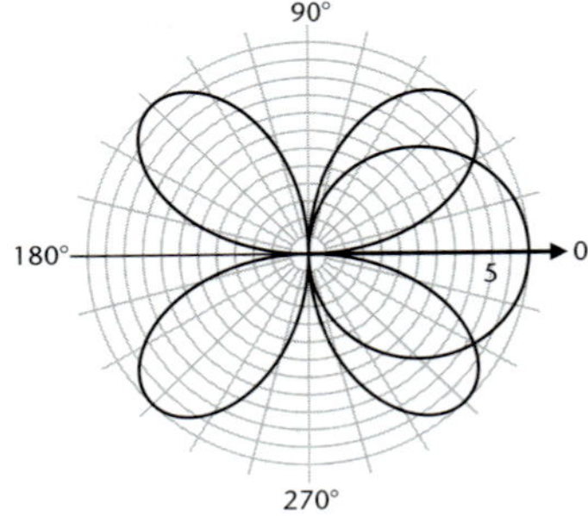

$(r, \theta) = (0, 90°), (0, 270°), (3\sqrt{3}, 30°), (-3\sqrt{3}, 150°)$ [*Note:* (0, 0) is not a solution of the system even though the graphs cross at the origin.]

63. 3.368 units **65.** 6 knots, 13 knots, 12 knots, 9 knots

67. (A) Ellipse (B) Parabola

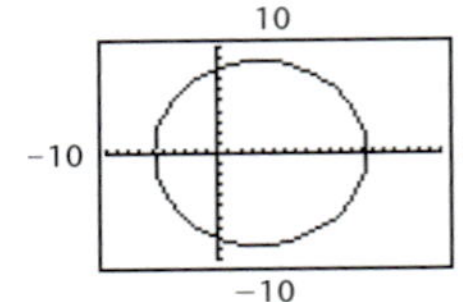

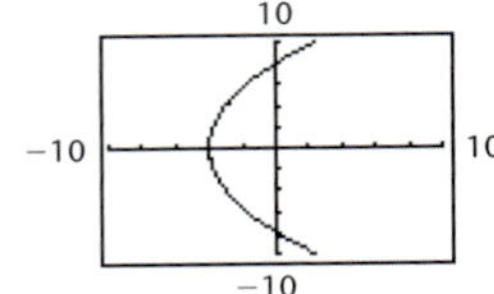

(C) Hyperbola

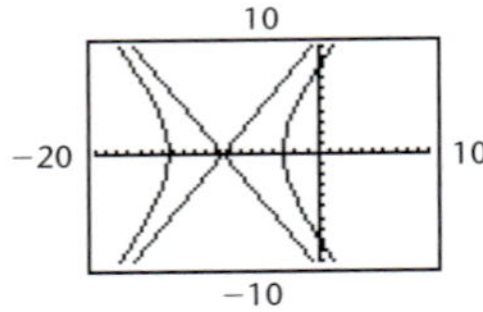

69. (A) Aphelion: 4.34×10^7 miles; perihelion: 2.85×10^7 miles

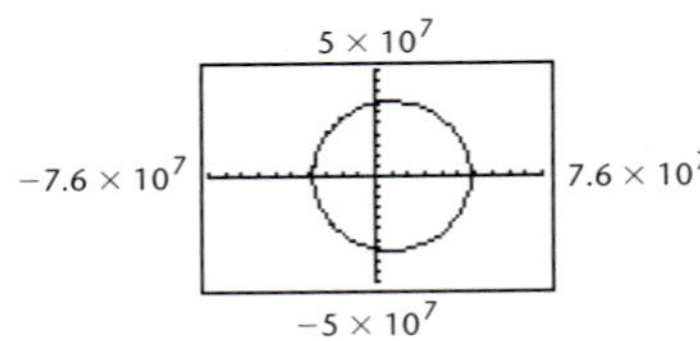

(B) Faster at perihelion. Because the distance from the sun to Mercury is less at perihelion than at aphelion, the planet must move faster near perihelion for the line joining Mercury to the sun to sweep out equal areas in equal intervals of time.

Exercise 7.6

1.

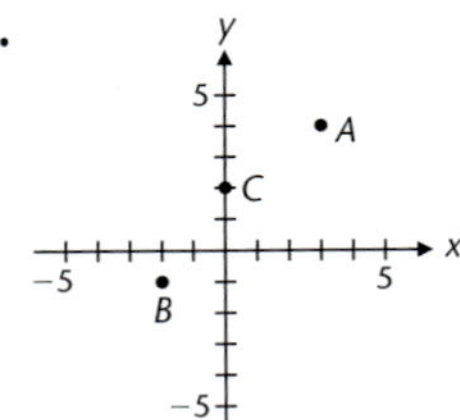

3.

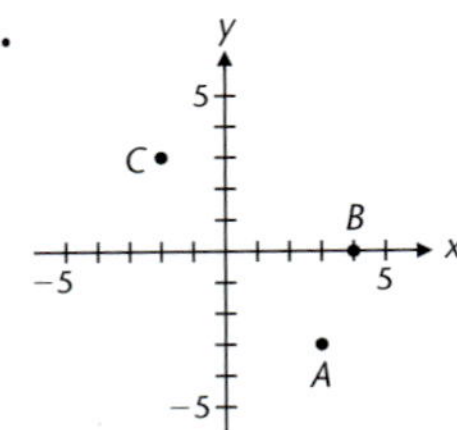

5.

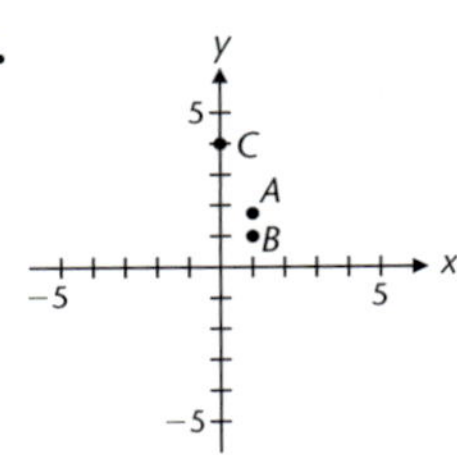

7.

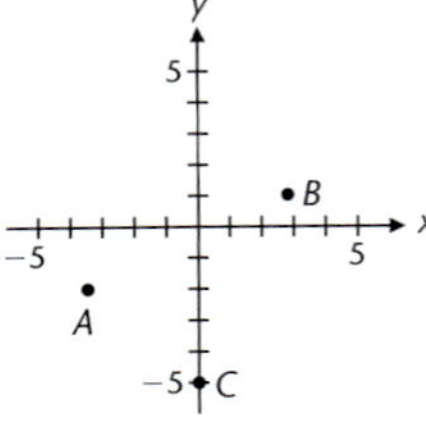

9. (A) $2e^{30°i}$ (B) $\sqrt{2}\,e^{(-135°)i}$ (C) $7.81e^{(-50.19°)i}$

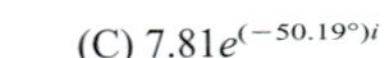

11. (A) $\sqrt{3}\,e^{(-\pi/2)i}$ (B) $2e^{(-5\pi/6)i}$ (C) $9.43e^{2.58i}$

13. (A) $1 + i\sqrt{3}$ (B) $1 - i$ (C) $-2.35 + 1.99i$

15. (A) $3\sqrt{3} + 3i$ (B) $-i\sqrt{7}$ (C) $-2.22 - 3.43i$

17. $14e^{113°i}$; $3.5e^{51°i}$ **19.** $10e^{135°i}$; $2.5e^{(-31°)i}$

21. $36.42e^{4.35i}$; $0.26e^{(-0.83i)}$ **23.** $-2i$; $2e^{(-90°)i}$

25. -2; $2e^{180°i}$ **27.** $-2 - 2i$, $2\sqrt{2}e^{(-135°)i}$ **31.** $z^n = r^n e^{n\theta i}$

33. (A) $(20 + 0i) + (5 + 5i\sqrt{3}) = 25 + 5i\sqrt{3}$ (B) $26.5e^{19.1°i}$

(C) 26.5 pounds at an angle of 19.1°

Exercise 7.7

1. $8e^{90°i}$ **3.** $8e^{60°i}$ **5.** $8e^{180°i}$ **7.** $-8 + 8\sqrt{3}i$
9. 16 **11.** 1 **13.** $w_1 = 2e^{10°i}$, $w_2 = 2e^{130°i}$, $w_3 = 2e^{250°i}$
15. $w_1 = 3e^{15°i}$, $w_2 = 3e^{105°i}$, $w_3 = 3e^{195°i}$, $w_4 = 3e^{285°i}$
17. $w_1 = 2^{1/10}e^{(-9°)i}$, $w_2 = 2^{1/10}e^{63°i}$, $w_3 = 2^{1/10}e^{135°i}$, $w_4 = 2^{1/10}e^{207°i}$, $w_5 = 2^{1/10}e^{279°i}$
19. $w_1 = 2e^{0°i}$, $w_2 = 2e^{120°i}$, $w_3 = 2e^{240°i}$

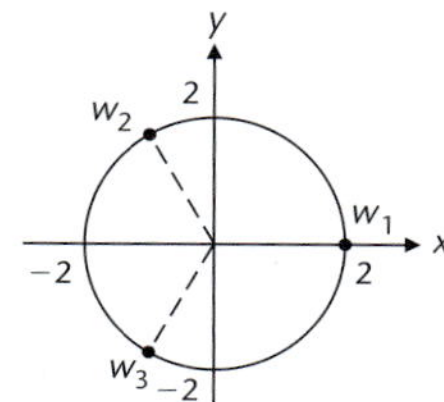

21. $w_1 = 2e^{45°i}$, $w_2 = 2e^{135°i}$, $w_3 = 2e^{225°i}$, $w_4 = 2e^{315°i}$

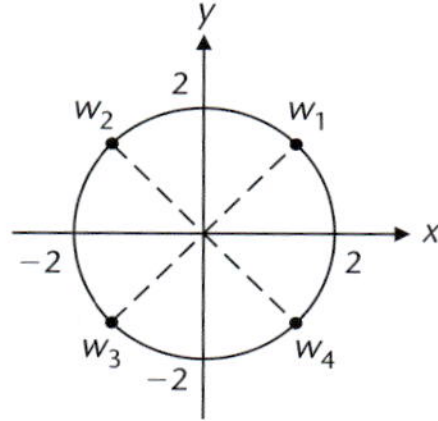

23. $w_1 = 1e^{15°i}$, $w_2 = 1e^{75°i}$, $w_3 = 1e^{135°i}$, $w_4 = 1e^{195°i}$, $w_5 = 1e^{255°i}$, $w_6 = 1e^{315°i}$

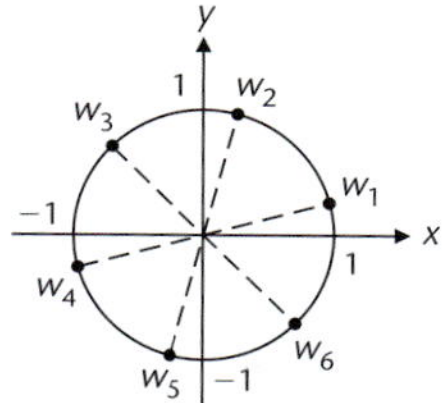

25. (A) $(1 + i)^4 + 4 = -4 + 4 = 0$. There are three other roots.
(B) The four roots are equally spaced around the circle. Because there are four roots, the angle between successive roots on the circle is $360°/4 = 90°$.

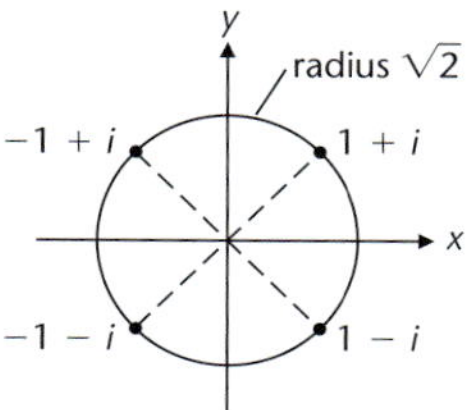

(C) $(-1 + i)^4 + 4 = -4 + 4 = 0$;
$(-1 - i)^4 + 4 = -4 + 4 = 0$; $(1 - i)^4 + 4 = -4 + 4 = 0$

27. $x_1 = 4e^{60°i} = 2 + 2\sqrt{3}i$, $x_2 = 4e^{180°i} = -4$, $x_3 = 4e^{300°i} = 2 - 2\sqrt{3}i$

29. $x_1 = 3e^{0°i} = 3$, $x_2 = 3e^{120°i} = -\frac{3}{2} + \frac{3\sqrt{3}}{2}i$, $x_3 = 3e^{240°i} = -\frac{3}{2} - \frac{3\sqrt{3}}{2}i$

33. $x_1 = 2e^{0°i}$, $x_2 = 2e^{72°i}$, $x_3 = 2e^{144°i}$, $x_4 = 2e^{216°i}$, $x_5 = 2e^{288°i}$
35. $w_1 = e^{36°i}$, $w_2 = e^{108°i}$, $w_3 = e^{180°i}$, $w_4 = e^{252°i}$, $w_5 = e^{324°i}$
37. $P(x) = (x - 2i)(x + 2i)[x - (-\sqrt{3} + i)][x - (-\sqrt{3} - i)][x - (\sqrt{3} + i)][x - (\sqrt{3} - i)]$

Chapter 7 Review

1. 1 *(7.1)* **2.** 0 *(7.1)* **3.** 2 *(7.1)*
4. Angle β is acute. A triangle can have at most one obtuse angle. Because α is acute, then, if the triangle has an obtuse angle it must be the angle opposite the longer of the two sides, *b* and *c*. Thus, β, the angle opposite the shorter of the two sides, *b*, must be acute. *(7.2)*
5. $\gamma = 75°$, $a = 47$ meters, $b = 31$ meters *(7.1)*
6. $a = 4.00$ feet, $\beta = 36°$, $\gamma = 129°$ *(7.1, 7.2)*
7. $\beta = 19°$, $\alpha = 40°$, $a = 8.2$ centimeters *(7.1)*
8. $|\mathbf{u} + \mathbf{v}| = 170$ miles per hour, $\theta = 19°$ *(7.3)*
9. $\langle 3, -7 \rangle$ *(7.4)* **10.** $\sqrt{34}$ *(7.4)*

11. *(7.5)*

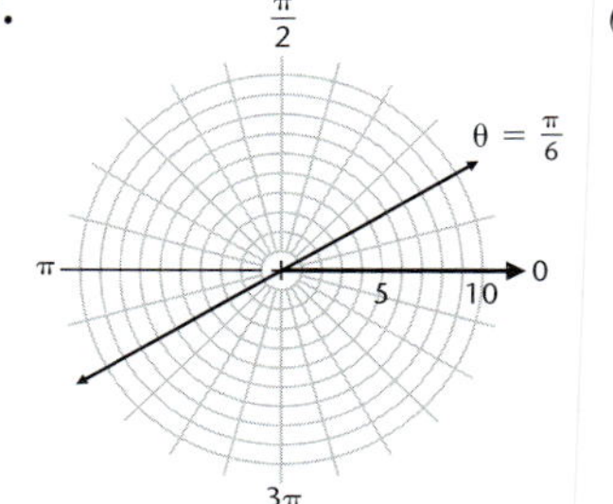

12. (7.5)

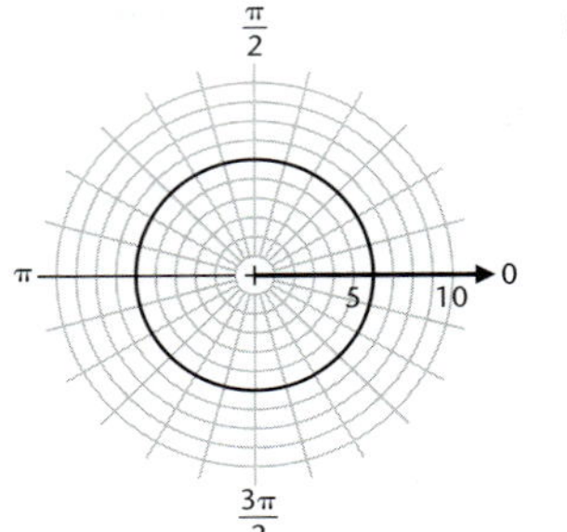

13. (7.6)

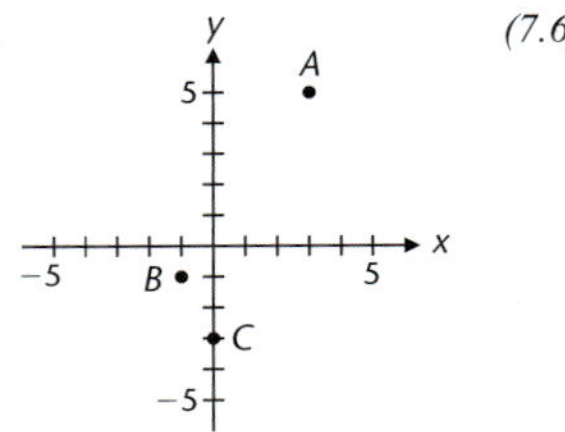

14. (−10, −210°): The polar axis is rotated 210° clockwise (negative direction) and the point is located 10 units from the pole along the negative polar axis. (−10, 150°): The polar axis is rotated 150° counterclockwise (positive direction) and the point is located 10 units from the pole along the negative polar axis. (10, 330°): The polar axis is rotated 330° counterclockwise and the point is located 10 units from the pole along the positive polar axis. *(7.5)*

15. (7.6)

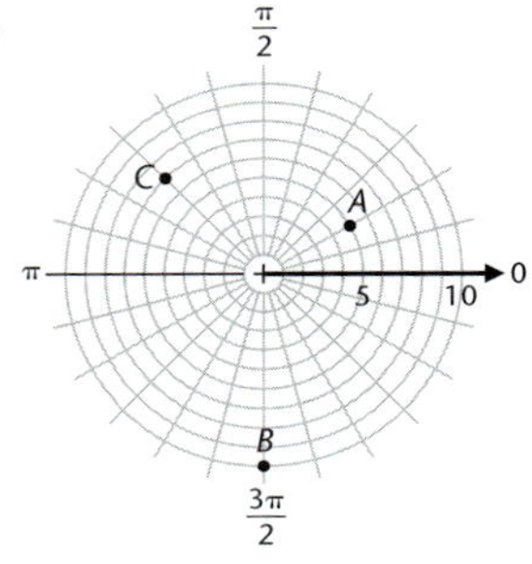

16. (A) $2e^{(-60°)i}$ (B) $2\sqrt{3} - 2i$ *(7.6)* **17.** (A) 1 *(7.7)*
18. $8 + 8i\sqrt{3}$ *(7.7)*
19. If the triangle has an obtuse angle, then it must be the angle opposite the longest side; in this case, α. *(7.2)*
20. $b = 10.5$ centimeters, $\alpha = 27.2°$, $\gamma = 37.4°$ *(7.2)*
21. No solution *(7.1)*
22. Two solutions. Obtuse case: $\beta = 133.9°$, $\gamma = 19.7°$, $c = 39.6$ kilometers *(7.1)*
23. $\alpha = 41.1°$, $\beta = 74.2°$, $\gamma = 64.7°$ *(7.1, 7.2)*
24. The sum of all of the force vectors must be the zero vector for the object to remain at rest. *(7.4)*
25. $|\mathbf{u} + \mathbf{v}| = 98.0$ kilograms, $\alpha = 17.1°$ *(7.3)*
26. (A) $\mathbf{u} = -3\mathbf{i} + 9\mathbf{j}$ (B) $\mathbf{v} = -2\mathbf{j}$ *(7.4)*
27. (A) $\langle -4, 7\rangle$ (B) $\langle -14, 13\rangle$ *(7.4)*
28. (A) $-2\mathbf{i} - 4\mathbf{j}$ (B) $-10\mathbf{j}$ *(7.4)*
29. $\mathbf{u} = \left\langle \frac{-1}{\sqrt{10}}, \frac{-3}{\sqrt{10}} \right\rangle$ *(7.4)*

30. (7.5)

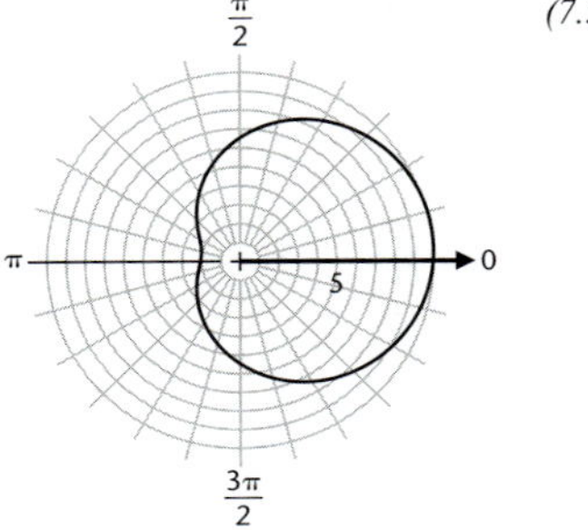

31. (7.5)

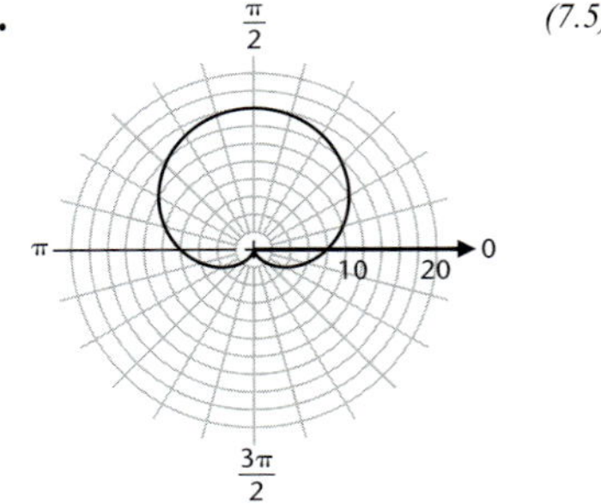

32. (7.5)

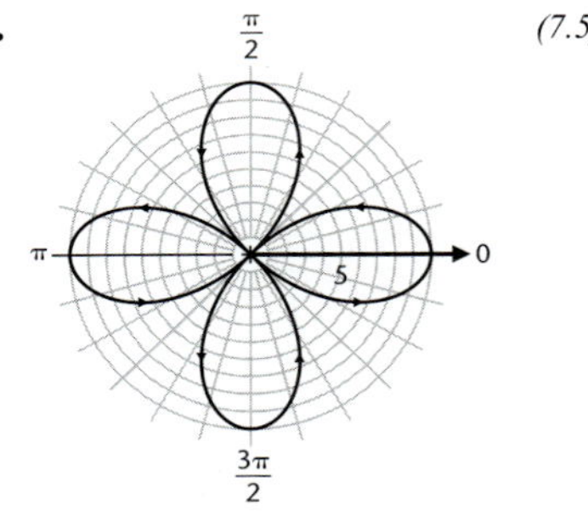

33. (7.5)

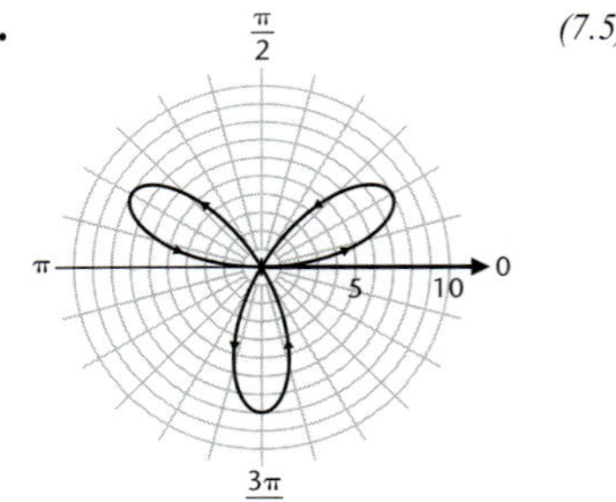

34. (7.5)

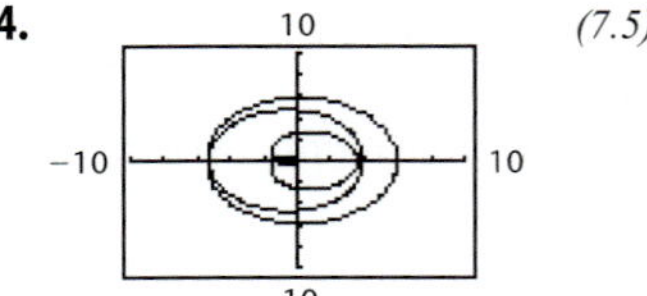

35. (7.5)

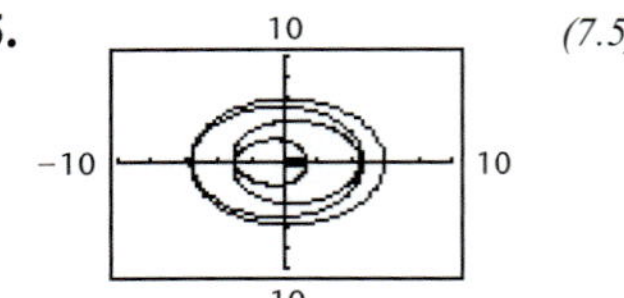

36. $n = 1$

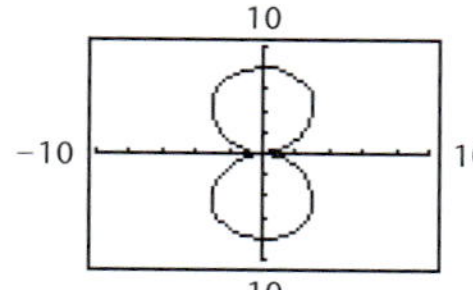

$n = 2$

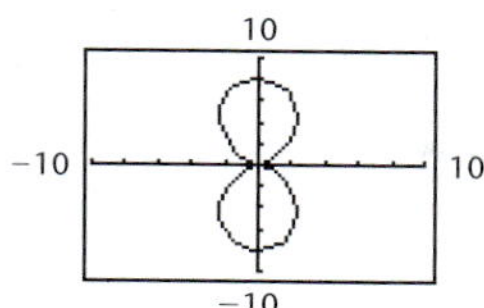

$n = 3$

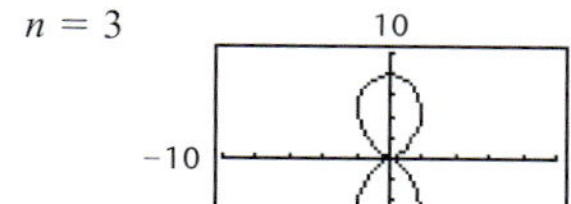

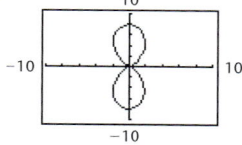

2 leaves for all n *(7.5)*

37. (A) Ellipse

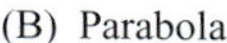
(B) Parabola

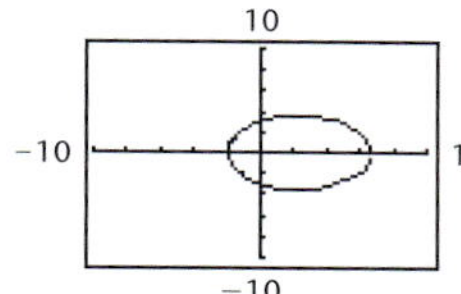

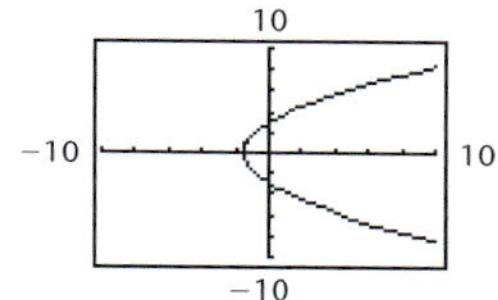

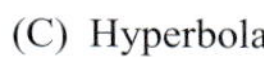
(C) Hyperbola

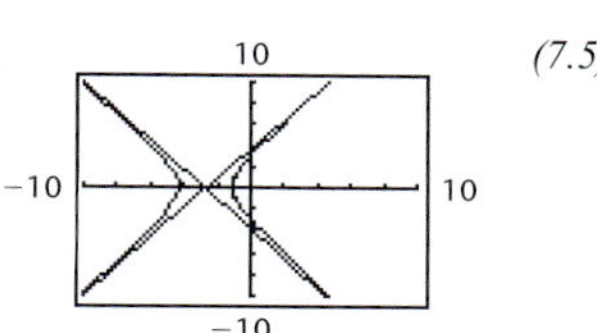

(7.5)

38. $r^2 = 6r \cos \theta$ or $r = 6 \cos \theta$ *(7.5)*

39. $x^2 + y^2 = 5x$ *(7.6)*

40. $z_1 = \sqrt{2}\, e^{135°i}$, $z_2 = 2e^{(-120°)i}$, $z_3 = 5e^{0°i}$ *(7.6)*

41. $z_1 = 1 + i$, $z_2 = (-3\sqrt{3}/2) - (3/2)i$, $z_3 = -1 - i\sqrt{3}$ *(7.6)*

42. (A) $32e^{44°i}$ (B) $2e^{6°i}$ *(7.6)*

43. (A) $-8 - 8\sqrt{3}i$ (B) $-8 - 13.86i$ *(7.7)*

44. $w_1 = (\sqrt{3}/2) + (1/2)i$, $w_2 = (-\sqrt{3}/2) + (1/2)i$, $w_3 = -i$ *(7.7)*

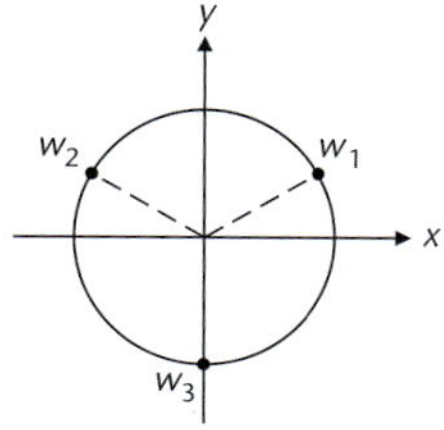

45. $2e^{50°i}, 2e^{170°i}, 2e^{290°i}$ *(7.7)*

46. $(4e^{15°i})^2 = 16e^{30°i} = 8\sqrt{3} + 8i$ *(7.7)*

47. $(5.76, -26.08°)$ *(7.5)* **48.** $(-5.30, -2.38)$ *(7.5)*

49. $5.26e^{127.20°i}$ *(7.6)* **50.** $-7.27 - 2.32i$ *(7.6)*

51. (A) There are a total of three cube roots and they are spaced equally around a circle of radius 2.

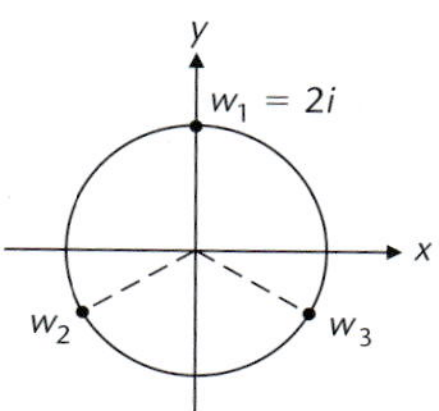

(B) $w_2 = -\sqrt{3} - i$, $w_3 = \sqrt{3} - i$

(C) The cube of each cube root is $-8i$. *(7.7)*

52. $k = 44.6 \sin 23.4°$ *(7.1)*

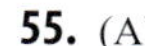
55. (A)

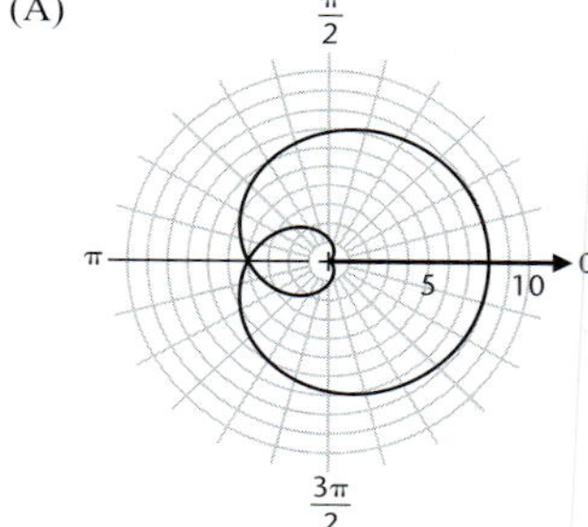

(B)

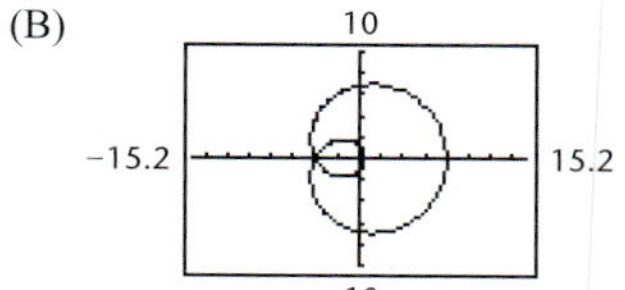

(7.5)

56. (A) The coordinates of P represent a simultaneous solution.

(B) $r = -4\sqrt{2}$, $\theta = 3\pi/4$

(C) The two graphs go through the pole at different values of θ. *(7.5)*

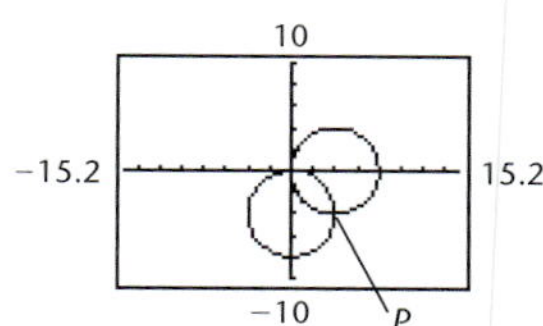

57. $1, -1, i, -i,$
$\frac{\sqrt{2}}{2} + i\frac{\sqrt{2}}{2}, \frac{\sqrt{2}}{2} - i\frac{\sqrt{2}}{2}, -\frac{\sqrt{2}}{2} + i\frac{\sqrt{2}}{2}, -\frac{\sqrt{2}}{2} - i\frac{\sqrt{2}}{2}$ *(7.7)*

58. $P(x) = (x + 2i)[x - (-\sqrt{3} + i)][x - (\sqrt{3} + i)]$ *(7.7)*

59. 438 miles *(7.3)*

60. 438 miles per hour at 83° *(7.3)*

61. 86°, 464 miles per hour *(7.3)*

62. 0.6 miles *(7.1)*

63. 177 pounds at 15.2° relative to **v** *(7.3)*

64. 19 kilograms at 204° relative to **u** *(7.4)*

65. 5,740 pounds *(7.4)*

66. (A) Distance at aphelion: 1.56×10^8 miles; distance at perihelion: 1.29×10^8 miles

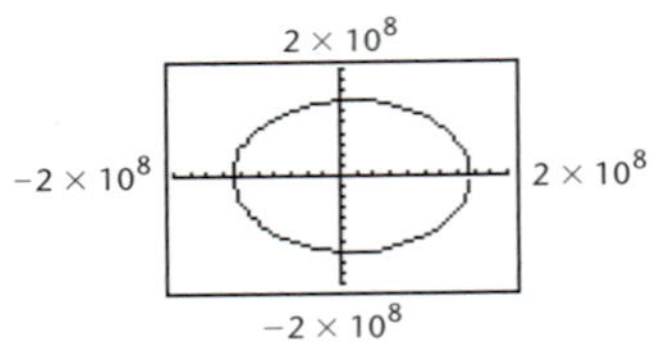

(B) Distance at aphelion: 1.56×10^8 miles; distance at perihelion: 1.29×10^8 miles *(7.5)*

Cumulative Review Exercise for Chapter 5, 6, and 7

1. 1.86 meters *(5.1)*

2. $\theta = 57.3°$, 14.5 centimeters, 7.83 centimeters *(5.3)*

3. (A) I, II (B) I, IV (C) I, III *(5.2)*

4. (A) $-\frac{3}{5}$ (B) $\frac{5}{4}$ (C) $-\frac{4}{3}$ *(5.4)* **5.** (A) $\frac{\pi}{4}$ (B) 65° (C) 30° *(5.4)*

6. (A) Domain: all real numbers; range: $-1 \le y \le 1$; period: 2π
(B) Domain: all real numbers; range: $-1 \le y \le 1$; period: 2π
(C) Domain: all real numbers except $x = \frac{\pi}{2} + k\pi$, k an integer; range: all real numbers; period: π *(5.4)*

7. *(5.4)*

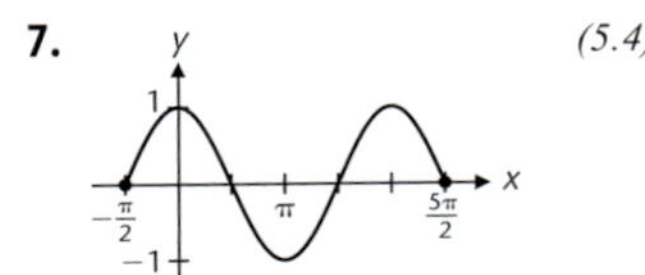

8. *(5.4)*

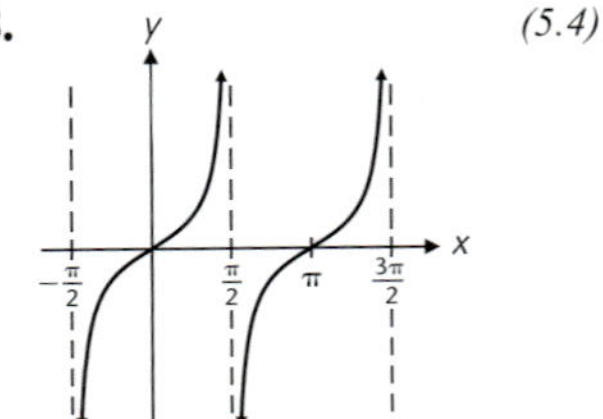

9. The central angle of a circle subtended by an arc of twice the length of the radius. *(5.1)*

10. If the graph of $y = \cos x$ is shifted $\pi/2$ units to the right, the result will be the graph of $y = \sin x$. *(5.4)*

15. (A) Not an identity (B) An identity *(6.1)*

16. Angle α is acute. A triangle can have at most one obtuse angle. Because β is acute, then, if the triangle has an obtuse angle it must be the angle opposite the longer of the two sides, a and c. Thus, α, the angle opposite the shorter of the two sides, a, must be acute. *(7.2)*

17. 0.3245, 2.8171 *(6.5)* **18.** $-76.2154°$ *(6.5)*

19. $b = 22$ feet, $\alpha = 28°$, $\gamma = 31°$ *(7.1, 7.2)* **20.** $\langle 6, -3\rangle$ *(7.4)*

21. $(5, -30°)$: The polar axis is rotated 30° clockwise (negative direction) and the point is located five units from the pole along the positive polar axis. $(-5, -210°)$: The polar axis is rotated 210° clockwise (negative direction) and the point is located five units from the pole along the negative polar axis. $(5, 330°)$: The polar axis is rotated 330° counterclockwise (positive direction) and the point is located five units from the pole along the positive polar axis. *(7.5)*

22. *(7.5)*

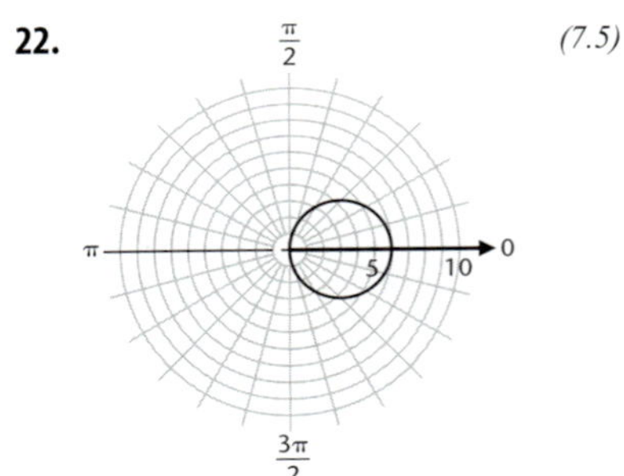

23. *(7.6)*

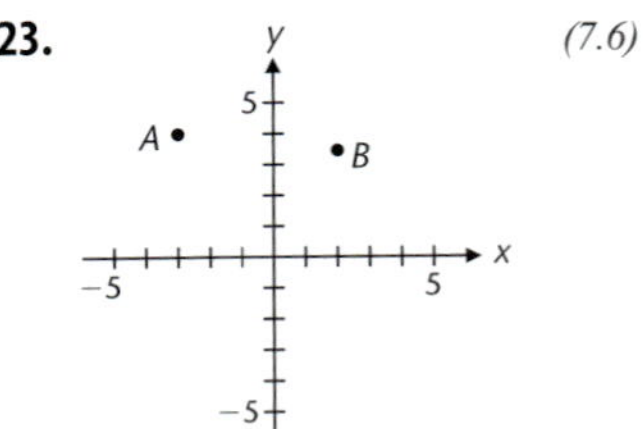

24. $4\sqrt{3} + 4i$ *(7.7)* **25.** $-\frac{7\pi}{6}$, 870° *(5.2)*

26. 75.06° *(5.1)* **27.** (A) and (C) *(5.2)*

28. $-\frac{1}{2}$ *(5.2)* **29.** Not defined *(5.2)* **30.** -1 *(5.2)*

31. $\frac{2}{\sqrt{3}}$ *(5.2)* **32.** π *(5.6)* **33.** Not defined *(5.6)*

34. $\frac{2\pi}{3}$ *(5.6)* **35.** 0.55 *(5.6)* **36.** $\frac{3}{5}$ *(5.6)*

37. $\frac{1}{\sqrt{5}}$ *(5.6)*

38. (A) 9.871 (B) -3.748
(C) -1.559 (D) Not defined *(5.2, 5.6)*

39. *(5.5)*

40. (A) 150° (B) $-19.755°$ *(5.6)*

41. $\sin^{-1}(\sin 3) = 0.142$. For the identity $\sin^{-1}(\sin x) = x$ to hold, x must be in the restricted domain of the sine function; that is, $-\pi/2 \le x \le \pi/2$. The number 3 is not in the restricted domain. *(5.6)*

42. Because the coordinates of a point on a unit circle are given by $P = (a, b) = (\cos x, \sin x)$, we evaluate $P = (\cos(11.205), \sin(11.205))$—using a calculator set in radian mode—to obtain $P = (0.208, -0.978)$. The quadrant in which $P = (a, b)$ lies can be determined by the signs of a and b. In this case P is in the fourth quadrant, because a is positive and b is negative. *(5.1, 5.2)*

43. The equation has infinitely many solutions $[x = \tan^{-1}(-24.5) + k\pi$, k any integer$]$; $\tan^{-1}(-24.5)$ has a unique value (-1.530 to three decimal places). *(5.6)*

44. $y = 3 + 2\sin \pi x$ *(5.5)*

45. $A = 3$, $P = \pi$, $P.S. = \frac{\pi}{2}$ *(5.5)*

46. Period $= 2$, $P.S. = 1$ *(5.5)*

47. *(5.4)*

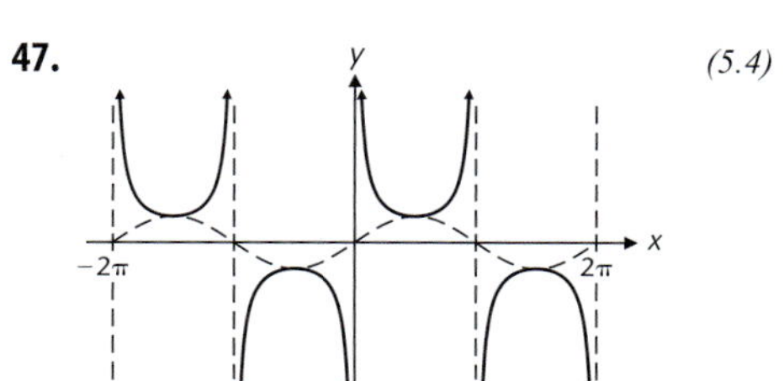

48. If the graph of $y = \cot x$ is shifted to the left $\pi/2$ units and reflected in the x axis, the result will be the graph of $y = \tan x$. *(5.4)*

49. $y = \frac{1}{2} - \frac{1}{2}\cos 2x$ *(5.5)* **50.** $y = \cot x$ *(5.5)*

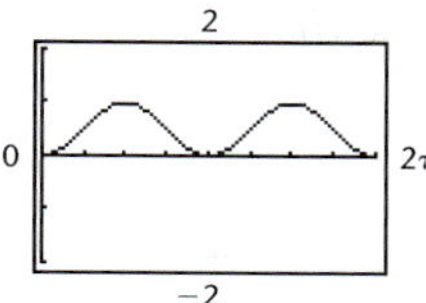

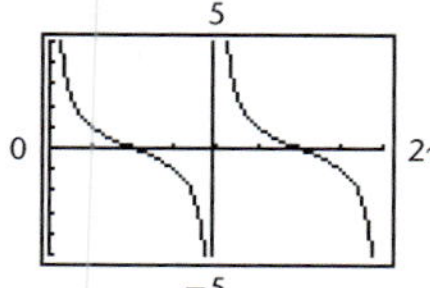

51. (A) Yes
(B) Conditional, because both sides are defined at $x = \pi/2$, for example, but $\pi/2$ is not a solution. *(6.1)*

58. (A) Not an identity (B) An identity *(6.1)* **59.** 0 *(6.2)*

60. $\sin 2x = -\frac{24}{25}$, $\cos\frac{x}{2} = \sqrt{\frac{1}{10}}$ or $\frac{\sqrt{10}}{10}$ *(6.3)*

61. 30°, 150°, 270° *(6.5)*

62. $x = k\pi, \frac{\pi}{3} + 2k\pi, -\frac{\pi}{3} + 2k\pi$, k any integer *(6.5)*

63. (A) $\pi/2, 3\pi/2, 7\pi/6, 11\pi/6$
(B) 1.571, 3.665, 4.712, 5.760 *(6.5)*

64. $x = 0.926$ *(6.5)*

65. $\gamma = 107.2°$, $\alpha = 25.0°$, $\beta = 47.8°$ *(7.1, 7.2)*

66. No solution *(7.1)*

67. $\beta = 120.7°$, $\gamma = 6.4°$, $c = 4.81$ inches *(7.1)*

68. β must be acute. A triangle can have at most one obtuse angle, and because γ is acute, the obtuse angle, if present, must be opposite the longer of the two sides a and b. *(7.2)*

69. $|\mathbf{u} + \mathbf{v}| = 35.6$ pounds, $\alpha = 16.3°$ *(7.1, 7.2, 7.3)*

70. (A) $\langle 1, 3 \rangle$ (B) $3\mathbf{i} + \mathbf{j}$ *(7.4)* **71.** $r = 8\sin\theta$ *(7.5)*

72. $x^2 + y^2 = -4x$ *(7.5)*

73. *(7.5)*

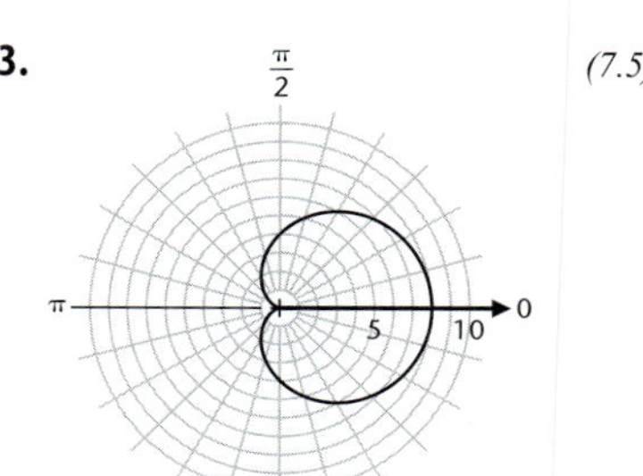

74. *(7.5)*

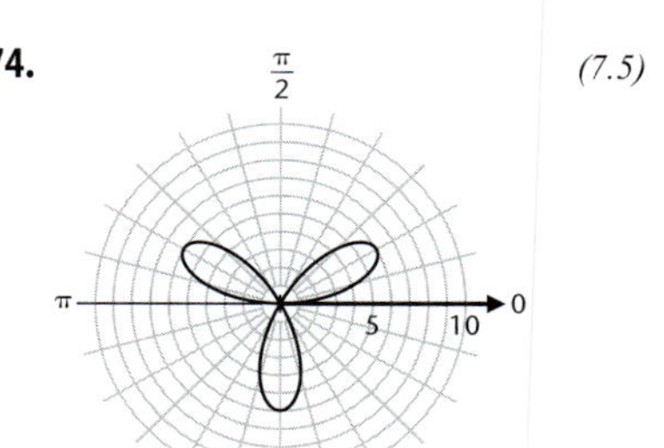

75. $n = 1$

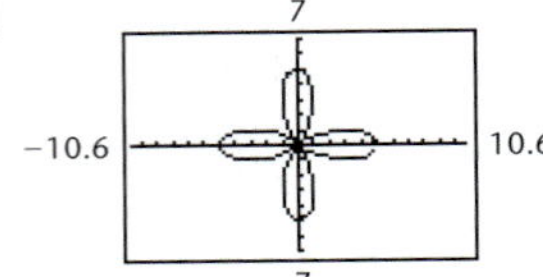

$n = 2$

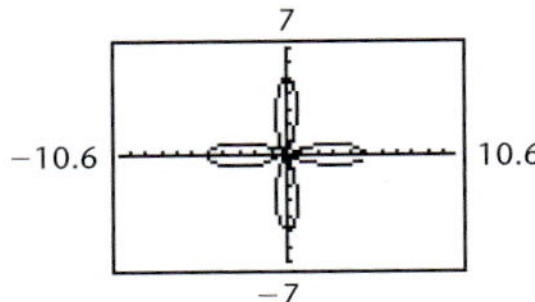

$n = 3$

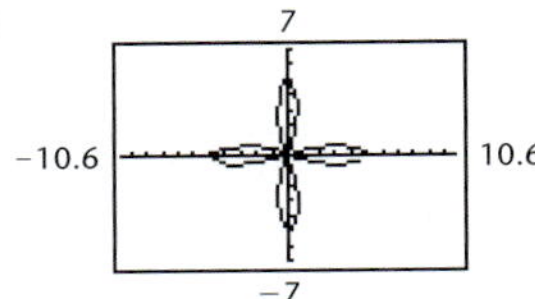

4 leaves for all n *(7.5)*

76.

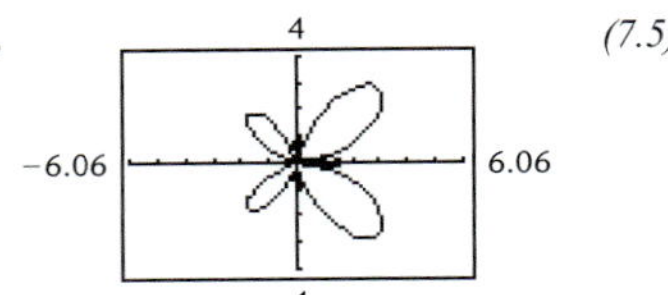

(7.5)

77. $(4.23, -131.07°)$ *(7.5)* **78.** $(-3.68, 5.02)$ *(7.5)*
79. $\sqrt{3} - i$ *(7.6)* **80.** $z = 2e^{120°i}$ *(7.6)*
81. $64 + 0i = 64$ *(7.7)*
82. $w_1 = \frac{\sqrt{3}}{2} - \frac{1}{2}i, w_2 = i, w_3 = -\frac{\sqrt{3}}{2} - \frac{1}{2}i$ *(7.8)*

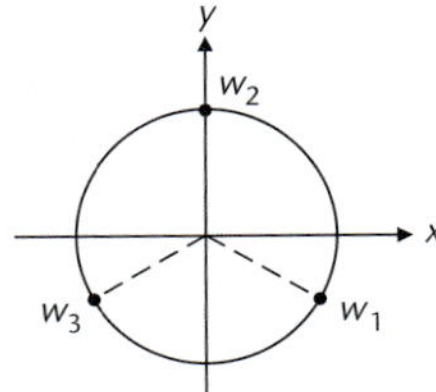

83. $5.82e^{(-146.99°)i}$ *(7.6)* **84.** $-6.70 + 1.94i$ *(7.6)*
85. (A) There are a total of four fourth roots and they are spaced equally around a circle of radius $\sqrt{2}$.

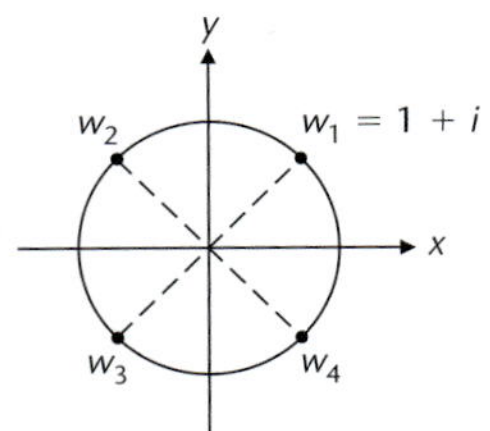

(B) $w_2 = -1 + i, w_3 = -1 - i, w_4 = 1 - i$
(C) The fourth power of each fourth root is -4. *(7.7)*

86. $a = \cos 1.2 = 0.362, b = \sin 1.2 = 0.932$ *(5.2)*

87. *(5.4)*

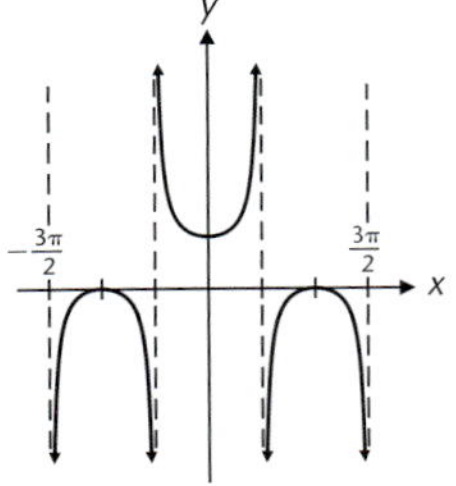

88. $y = 3 \cos (2\pi x - \pi/4)$; amplitude = 3, period = 1, *P.S.* = 1/8 *(5.5)*
89. $y = 2 \sin (2x - 0.644)$ *(5.5)*

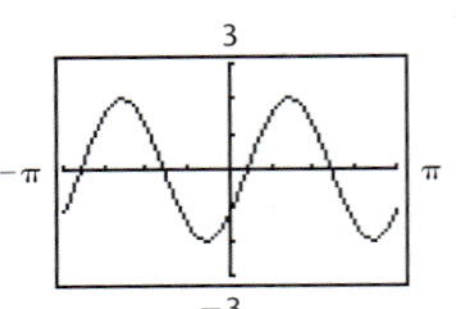

90. $\frac{1}{\sqrt{1 - x^2}}$ *(5.6)*

91. $\frac{24}{25}$ *(5.6, 6.3)*

92. (A) $\frac{2}{\sqrt{5}}$ or $\frac{2\sqrt{5}}{5}$ (B) $-\frac{7}{25}$ *(6.3)*

93. (A) $\pi/3, 5\pi/3$ (B) 1.0472, 5.2360 *(6.5)*

94. (A)

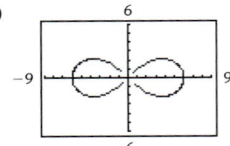

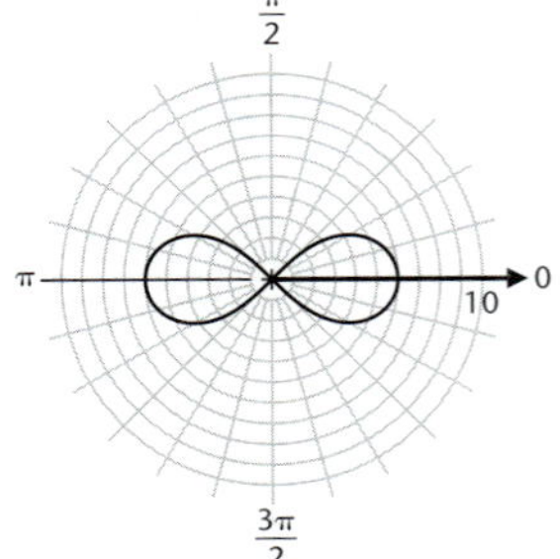

(B) *(7.5)*

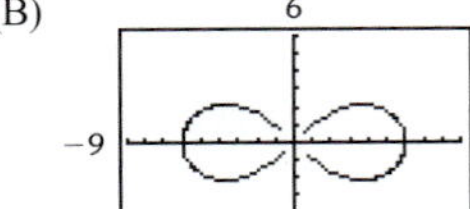

95. (A)

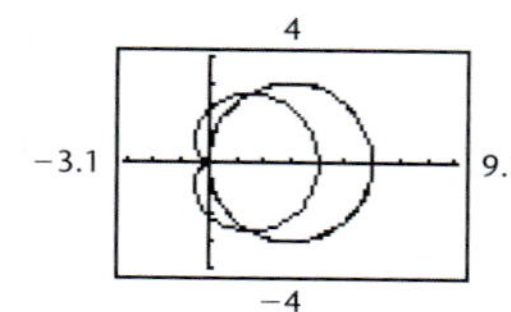

(B) 6
(C) $(3, \pi/3), (3, 5\pi/3)$
(D) The points on r_2 and r_1 arrive at the intersection points for different values of θ, except for the two found in part C. *(7.5)*

96. $P(x) = (x - i)[x - (\sqrt{3}/2 - i/2)][x - (-\sqrt{3}/2 - i/2)]$ *(7.7)*

97. $\frac{2\pi}{73}$ radians *(5.1)* **98.** 1,088 meters *(5.3)*

99. 5.88 inches *(5.3, 7.2)* **100.** 76° *(7.2)*

101. $I = 50 \cos 220\,\pi t$ *(5.5)*

102. 274 miles per hour at 117° *(7.3)*

103. Both have a tension of 234 pounds *(7.4)*

104. (A) Add the perpendicular bisector of the chord as shown in the figure. Then, $\sin \theta = 4/R$ and $\theta = 5/R$. Substituting the second into the first, we obtain $\sin 5/R = 4/R$.

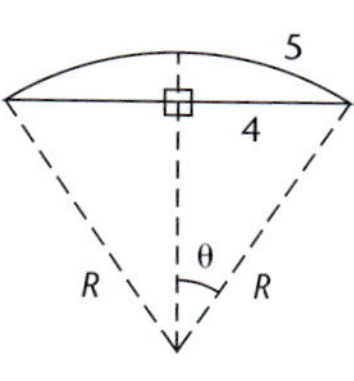

(B) R cannot be isolated on one side of the equation.
(C) Plot $y_1 = \sin 5/R$ and $y_2 = 4/R$ in the same viewing window and solve for R at the point of intersection using the intersect command (see figure). $R = 4.420$ centimeters. *(6.5)*

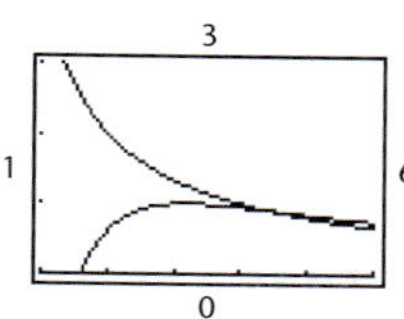

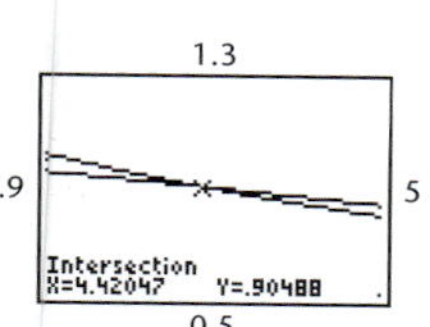

105. (A)

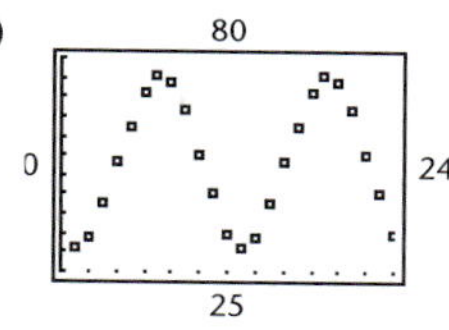

(B) $y = 53.5 + 22.5 \sin(\pi x/6 - 2.1)$

(C)

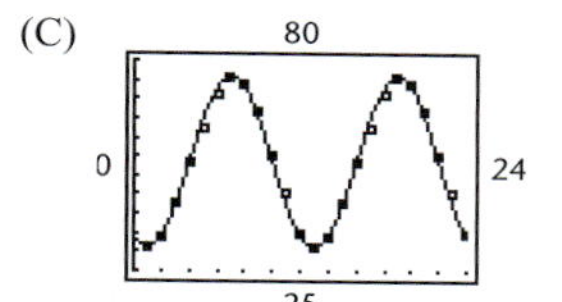

(5.7)

CHAPTER 8

Exercise 8.1

1. b, no solution **3.** d, $(1, -3)$ **5.** $(5, 2)$ **7.** $(2, -3)$

9. No solution (parallel lines) **11.** $x = 8, y = 19$

13. $x = 6, y = 2$ **15.** $x = 2, y = -1$ **17.** $x = 5, y = 2$

19. $m = 1, n = -2/3$ **21.** $x = 2{,}500, y = 200$

23. $u = 1.1, v = 0.3$ **25.** $x = -5/4, y = 5/3$

27. $(1.12, 2.41)$ **29.** $(-2.24, -3.31)$

31. The system has no solution.

33. The system has an infinite number of solutions.

35. $q = x + y - 5, p = 3x + 2y - 12$

37. $x = \dfrac{dh - bk}{ad - bc}, y = \dfrac{ak - ch}{ad - bc}, ad - bc \neq 0$

39. Airspeed = 330 miles per hour; wind rate = 90 miles per hour

41. 2.475 kilometers

43. 40 milliliters of 50% solution and 60 milliliters of 80% solution

45. \$7,200 invested at 10% and \$4,800 invested at 15%

47. Mexico plant: 75 hours; Taiwan plant: 50 hours

49. Mix A: 80 grams; mix B: 60 grams

51. (A) Supply: 143 T-shirts; demand: 611 T-shirts
(B) Supply: 714 T-shirts; demand: 389 T-shirts
(C) Equilibrium price: \$6.36; equilibrium quantity: 480 T-shirts

(D)

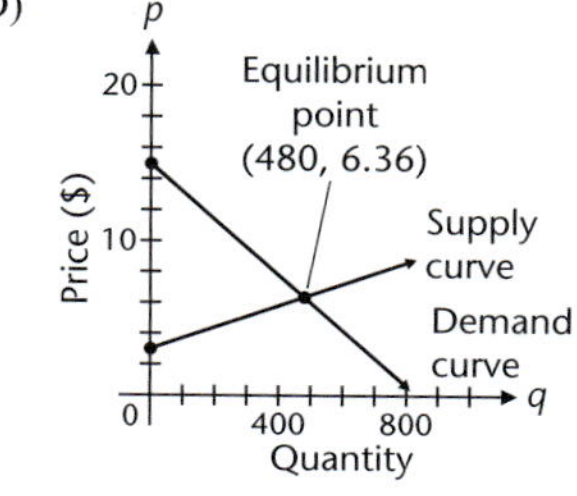

53. (A) $p = 0.001q + 0.15$ (B) $p = -0.002q + 1.89$
(C) Equilibrium price = \$0.73; equilibrium quantity = 580 bushels

55. (A) $a = 196, b = -16$ (B) 196 feet (C) 3.5 second

57. 40 seconds, 24 seconds, 120 miles

Exercise 8.2

1. (2, −1) **3.** (3, −1) **5.** 2 × 3, 1 × 3 **7.** *C* **9.** *B*

11. −2, −6 **13.** −2, 6, 0 **15.** $\left[\begin{array}{rr|r} 4 & -6 & -8 \\ 1 & -3 & 2 \end{array}\right]$

17. $\left[\begin{array}{rr|r} -4 & 12 & -8 \\ 4 & -6 & -8 \end{array}\right]$ **19.** $\left[\begin{array}{rr|r} 1 & -3 & 2 \\ 8 & -12 & -16 \end{array}\right]$

21. $\left[\begin{array}{rr|r} 1 & -3 & 2 \\ 0 & 6 & -16 \end{array}\right]$ **23.** $\left[\begin{array}{rr|r} 1 & -3 & 2 \\ 2 & 0 & -12 \end{array}\right]$

25. $\left[\begin{array}{rr|r} 1 & -3 & 2 \\ 3 & -3 & -10 \end{array}\right]$ **27.** $\frac{1}{3}R_2 \to R_2$ **29.** $6R_1 + R_2 \to R_2$

31. $\frac{1}{3}R_2 + R_1 \to R_1$

33. $x_1 = 4, x_2 = 3$; each pair of lines has the same intersection point.

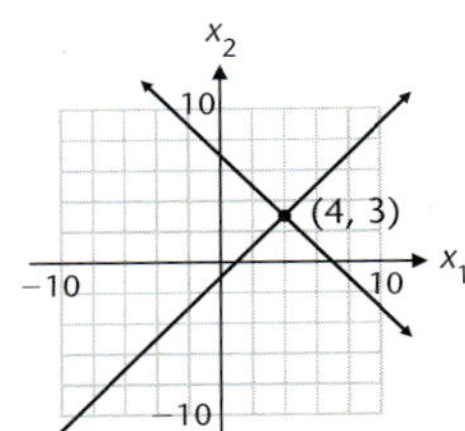

$x_1 + x_2 = 7$
$x_1 - x_2 = 1$

$x_1 + x_2 = 7$
$-2x_2 = -6$

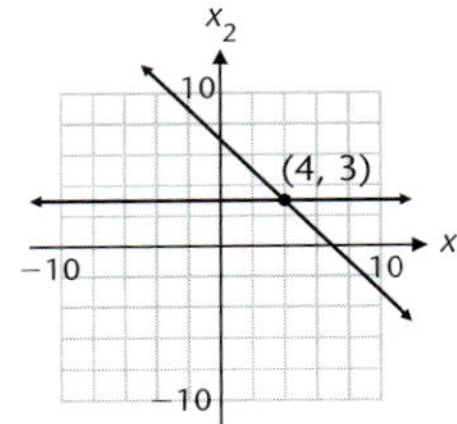

$x_1 + x_2 = 7$
$x_2 = 3$

$x_1 = 4$
$x_2 = 3$

35. $\{(2t - 4, t) \mid t \text{ any real number}\}$; the graph of each system is the same.

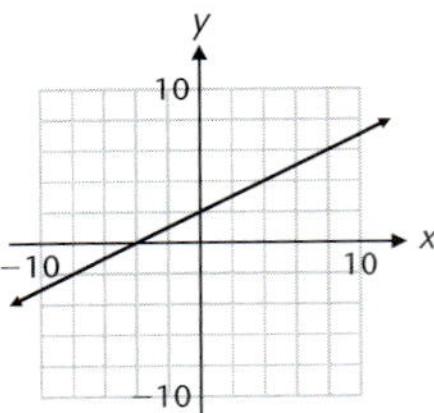

37. No solution; the graph of each system is two parallel lines until a contradiction is reached.

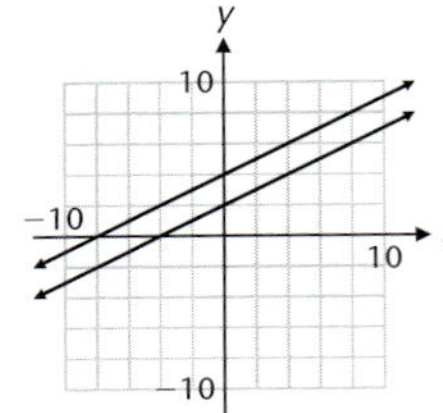

39. $x_1 = 2$ and $x_2 = 1$ **41.** $x_1 = 2$ and $x_2 = 4$

43. No solution **45.** $x_1 = 1$ and $x_2 = 4$

47. Infinitely many solutions: for any real number s, $x_2 = s$, $x_1 = 2s - 3$

49. Infinitely many solutions: for any real number s, $x_2 = s$, $x_1 = \frac{1}{2}s + \frac{1}{2}$

51. (A) (−24, 20) (B) (6, −4) (C) No solution

53. (−23.125, 7.8125) **55.** (3.225, −6.9375)

57. 25 32¢ stamps, 50 23¢ stamps

59. \$107,500 in bond *A* and \$92,500 in bond *B*

61. 30 liters of 20% solution and 70 liters of 80% solution

63. 200 grams of mix *A* and 80 grams of mix *B*

65. Base price = \$17.95, surcharge = \$2.45 per pound

67. 5,720 pounds of the robust blend and 6,160 pounds of the mild blend

Exercise 8.3

1. Reduced form **3.** Not reduced form; $R_2 \leftrightarrow R_3$

5. Not reduced form; $\frac{1}{3}R_2 \to R_2$ **7.** Reduced form

9. Not reduced form; $3R_2 + R_1 \to R_1$

11. Consistent and independent; $x_1 = -2, x_2 = 3, x_3 = 0$

13. Consistent and dependent; $x_1 = 2t + 3, x_2 = -t - 5, x_3 = t$ is the solution for t any real number.

15. Inconsistent; no solution

17. Consistent and dependent; $x_1 = 2s + 3t - 5, x_2 = s, x_3 = -3t + 2, x_4 = t$ is the solution for s and t any real numbers.

19. $\left[\begin{array}{rr|r} 1 & 0 & -7 \\ 0 & 1 & 3 \end{array}\right]$ **21.** $\left[\begin{array}{rrr|r} 1 & 0 & 0 & -5 \\ 0 & 1 & 0 & 4 \\ 0 & 0 & 1 & -2 \end{array}\right]$ **23.** $\left[\begin{array}{rrr|r} 1 & 0 & 2 & -\frac{5}{3} \\ 0 & 1 & -2 & \frac{1}{3} \\ 0 & 0 & 0 & 0 \end{array}\right]$

25. $x_1 = -2, x_2 = 3$, and $x_3 = 1$ **27.** $x_1 = 0, x_2 = -2$, and $x_3 = 2$

29. $x_1 = 2t + 3, x_2 = t - 2, x_3 = t$, t any real number

31. $x_1 = 1, x_2 = 2$ **33.** No solution

35. $x_1 = 2t + 4, x_2 = t + 1, x_3 = t$, t any real number

37. $x_1 = s + 2t - 1, x_2 = s, x_3 = t$, s and t any real numbers

39. No solution

41. $x_1 = 2.5t - 4, x_2 = t, x_3 = -5$ for t any real number

43. $x_1 = 1, x_2 = -2, x_3 = 1$

45. (A) Dependent with two parameters

(B) Dependent with one parameter

(C) Independent (D) Impossible

51. $x_1 = 2s - 3t + 3, x_2 = s + 2t + 2, x_3 = s, x_4 = t$, s and t any real numbers

53. $x_1 = -0.5, x_2 = 0.2, x_3 = 0.3, x_4 = -0.4$

55. $x_1 = 2s - 1.5t + 1, x_2 = s, x_3 = -t + 1.5, x_4 = 0.5t - 0.5$, $x_5 = t$ for s and t any real numbers

57. $x_1 = 4, x_2 = 1$ **59.** $x_1 = -1.4, x_2 = 4.8, x_3 = 4$

63. $x_1 = (3t - 100)$ 15¢ stamps, $x_2 = (145 - 4t)$ 20¢ stamps, $x_3 = t$ 35¢ stamps, where $t = 34, 35$, or 36

65. $x_1 = (6t - 24)$ 500-cubic centimeter containers of 10% solution, $x_2 = (48 - 8t)$ 500-cubic centimeter containers of 20% solution, $x_3 = t$ 1,000-cubic centimeter containers of 50% solution where $t = 4, 5$, or 6

67. $a = 3, b = 2, c = 1$ **69.** $a = -2, b = -4$, and $c = -20$

71. (A) $x_1 = 20$ one-person boats, $x_2 = 220$ two-person boats, $x_3 = 100$ four-person boats

(B) $x_1 = (t - 80)$ one-person boats, $x_2 = (-2t + 420)$ two-person boats, $x_3 = t$ four-person boats, $80 \le t \le 210$, t an integer

(C) No solution; no production schedule will use all the labor-hours in all departments.

73. (A) $x_1 = 8$ ounces food A, $x_2 = 2$ ounces food B, $x_3 = 4$ ounces food C

(B) No solution

(C) $x_1 = 8$ ounces food A, $x_2 = -2t + 10$ ounces food B, $x_3 = t$ ounces food C, $0 \le t \le 5$

75. $10 - t$ barrels of mix A, $t - 5$ barrels of mix B, $25 - 2t$ barrels of mix C, and t barrels of mix D, where t is an integer satisfying $5 \le t \le 10$

77. $x_1 = 10$ hours company A, $x_2 = 15$ hours company B

Exercise 8.4

1.

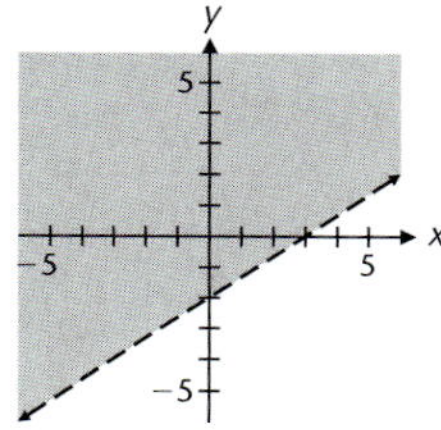

The solution region is the set of points (x, y) that are above the graph of the line $2x - 3y = 6$.

3.

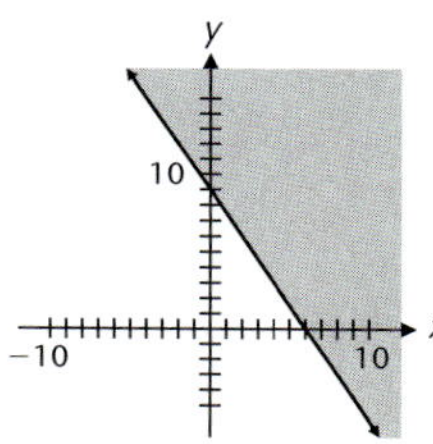

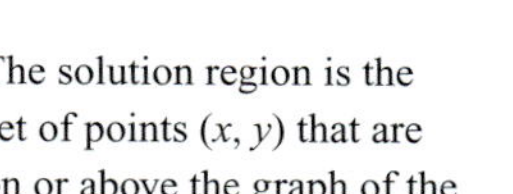

The solution region is the set of points (x, y) that are on or above the graph of the line $3x + 2y = 18$.

5.

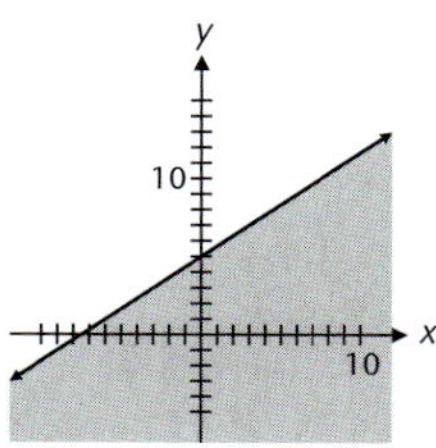

The solution region is the set of points (x, y) that are on or below the graph of the line $y = \frac{2}{3}x + 5$.

7.

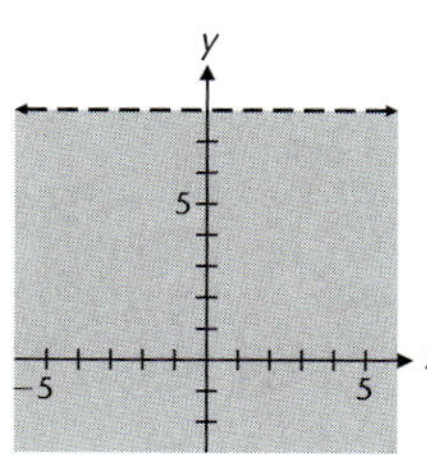

The solution region is the set of points (x, y) that are below the graph of the horizontal line $y = 8$.

9.

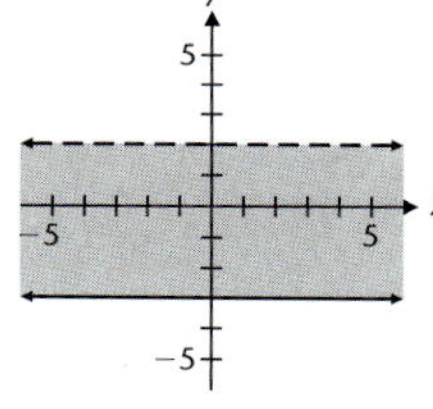

The solution region is the set of points (x, y) that are below the graph of the horizontal line $y = 2$ and on or above the graph of the horizontal line $y = -3$.

11. Region IV **13.** Region I **15.**

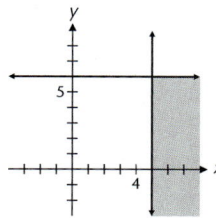

17.

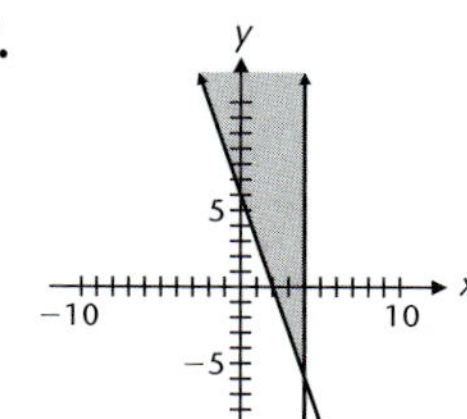

19.

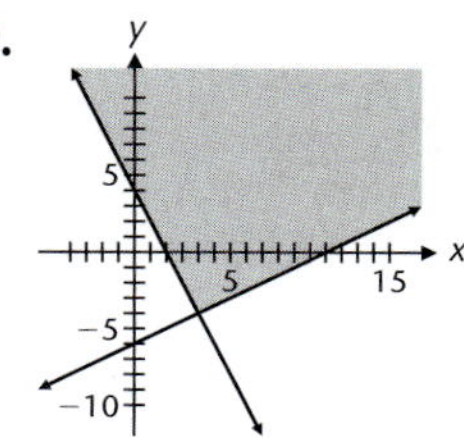

21. (A) Solution region is the double-shaded region.

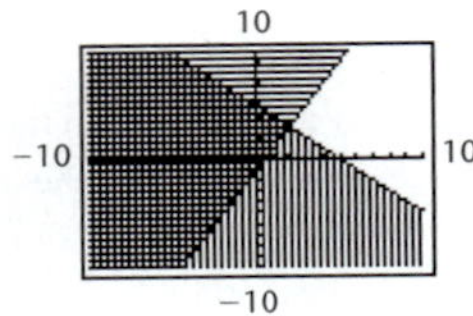

(B) Solution region is the unshaded region.

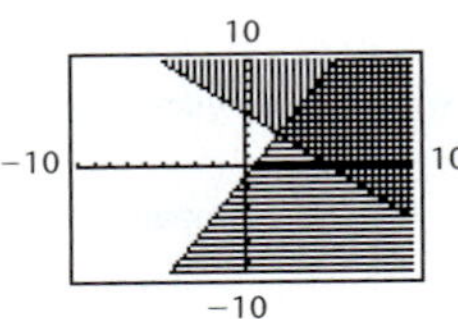

23. (A) Solution region is the double-shaded region.

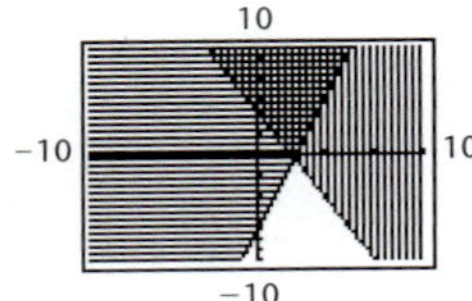

(B) Solution region is the unshaded region.

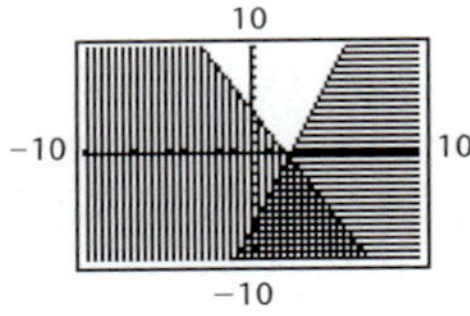

25. Region IV; corner points are (6, 4), (8, 0), and (18, 0)

27. Region I; corner points are (0, 16), (6, 4), and (18, 0)

29.

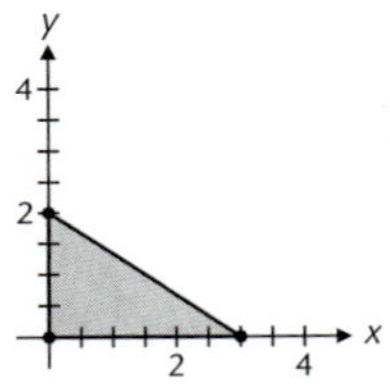

Corner points: (0, 0), (0, 2), (3, 0); bounded

31.

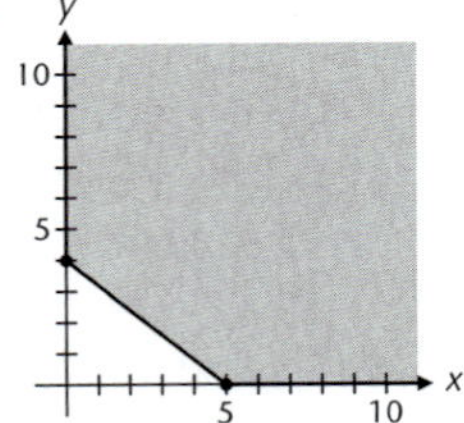

Corner points: (0, 4) and (5, 0); unbounded

33.

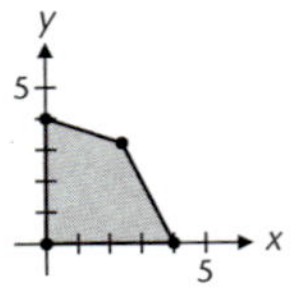

Corner points: (0, 4), (0, 0), $(\frac{12}{5}, \frac{16}{5})$, (4, 0); bounded

35.

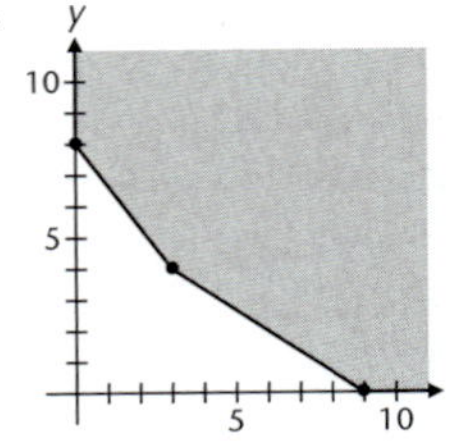

Corner points: (9, 0), (0, 8), and (3, 4); unbounded

37.

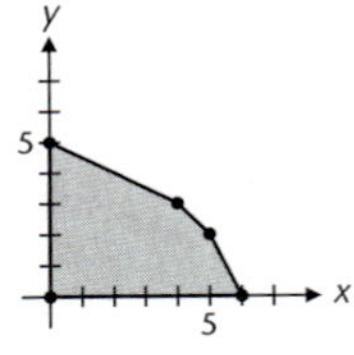

Corner points: (6, 0), (4, 3), (5, 2), (0, 0), and (0, 5); bounded

39.

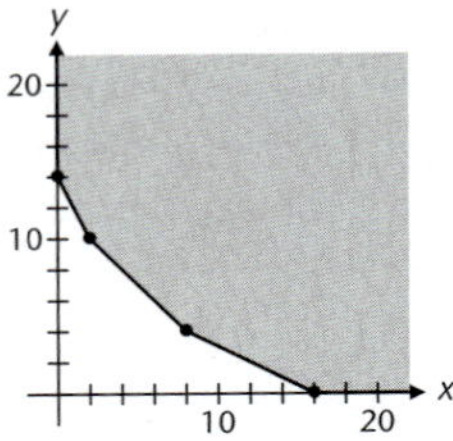

Corner points: (0, 14), (2, 10), (8, 4), (16, 0); unbounded

41.

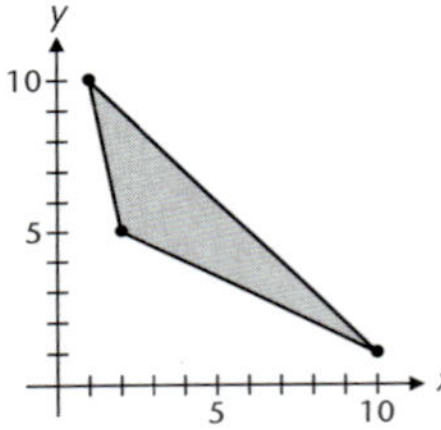

Corner points: (2, 5), (10, 1), (1, 10); bounded

43. The feasible region is empty.

45.

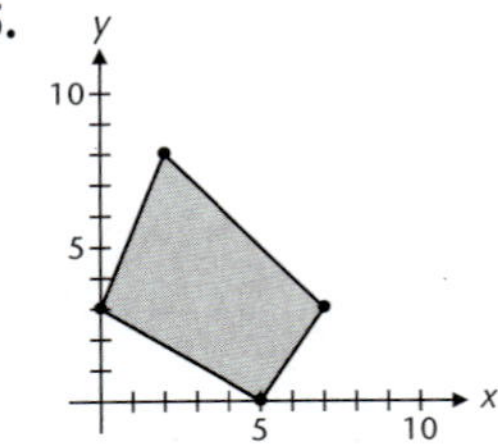

Corner points: (0, 3), (5, 0), (7, 3), (2, 8); bounded

47.

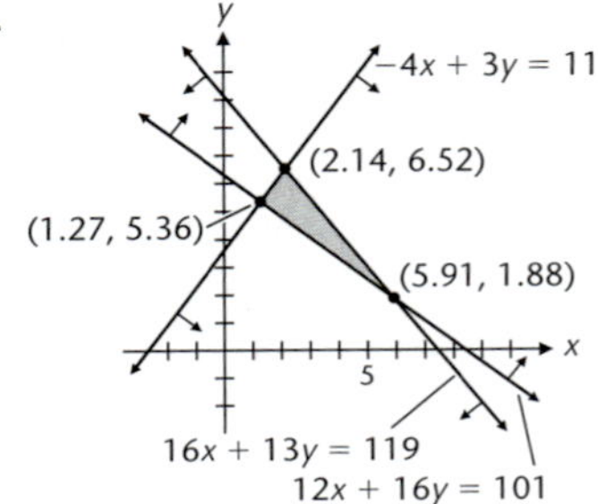

Corner points: (1.27, 5.36), (2.14, 6.52), (5.91, 1.88); bounded

49. 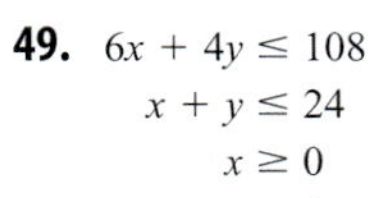
$6x + 4y \le 108$
$x + y \le 24$
$x \ge 0$
$y \ge 0$

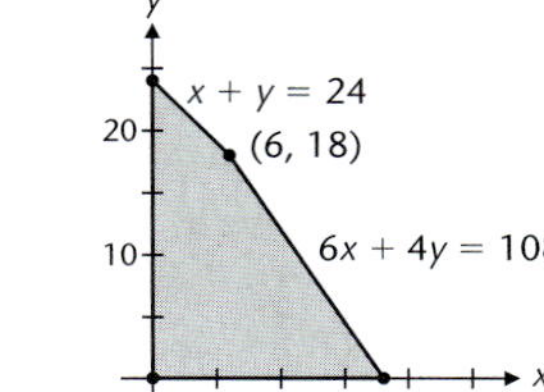

51. (A) All production schedules in the feasible region that are on the graph of $50x + 60y = 1{,}100$ will result in a profit of $1,100.
(B) There are many possible choices. For example, producing 5 trick and 15 slalom skis will produce a profit of $1,150. The graph of the line $50x + 60y = 1{,}150$ includes all the production schedules in the feasible region that result in a profit of $1,150.

53. 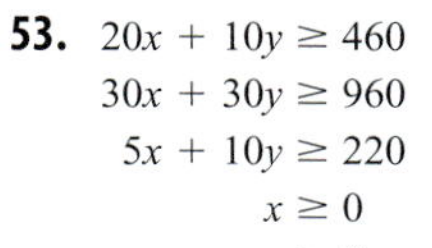
$20x + 10y \ge 460$
$30x + 30y \ge 960$
$5x + 10y \ge 220$
$x \ge 0$
$y \ge 0$

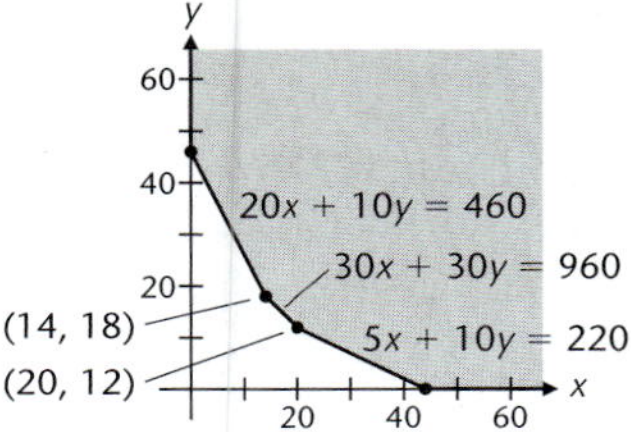

55. $10x + 30y \ge 280$
$30x + 10y \ge 360$
$x \ge 0$
$y \ge 0$

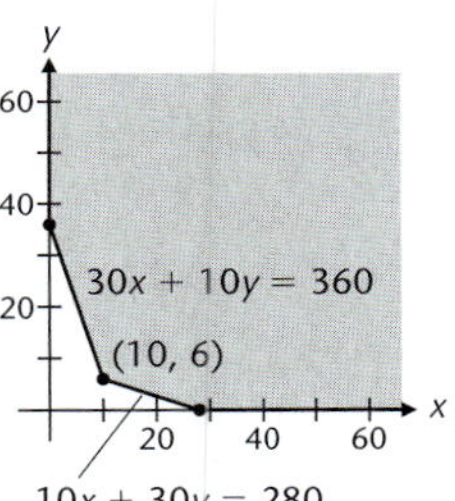

Exercise 8.5

1. Maximum value of z on S is 16 at (7, 9).
3. Maximum value of z on S is 84 at both (0, 12) and (7, 9).
5. Maximum value of z on S is 25 at (7, 9).
7. Maximum value of z on S is 70 at (10, 0).
9. Minimum value of z on T is 32 at (0, 8).
11. Minimum value of z on T is 36 at both (12, 0) and (4, 3).
13. Minimum value of z on T is 10 at (4, 3).
15. Minimum value of z on T is 32 at both (0, 8) and (4, 3).
17. Maximum value of z on S is 18 at (4, 3).
19. Minimum value of z on S is 12 at (4, 0).
21. Maximum value of z on S is 52 at (4, 10).
23. Minimum value of z on S is 44 at (4, 4).
25. The minimum value of z on S is 1,500 at (60, 0). The maximum value of z on S is 3,000 at (60, 30) and (120, 0) (multiple optimal solutions).
27. The minimum value of z on S is 300 at (0, 20). The maximum value of z on S is 1,725 at (60, 15).
29. Max $P = 5{,}507$ at $x_1 = 6.62$ and $x_2 = 4.25$
31. (A) $a > 2b$ (B) $\frac{1}{3}b < a < 2b$ (C) $a < \frac{1}{3}b$ or $b > 3a$
(D) $a = 2b$ (E) $b = 3a$

33. (A) 6 trick skis, 18 slalom skis; $780
(B) The maximum profit decreases to $720 when 18 trick and no slalom skis are produced.
(C) The maximum profit increases to $1,080 when no trick and 24 slalom skis are produced.
35. 9 model A trucks and 6 model B trucks to realize the minimum cost of $279,000
37. (A) 40 tables, 40 chairs; $4,600
(B) The maximum profit decreases to $3,800 when 20 tables and 80 chairs are produced.
39. (A) Max $P = \$450$ when 750 gallons are produced using the old process exclusively.
(B) The maximum profit decreases to $380 when 400 gallons are produced using the old process and 700 gallons using the new process.
(C) The maximum profit decreases to $288 when 1,440 gallons are produced using the new process exclusively.
41. The nitrogen will range from a minimum of 940 pounds when 40 bags of brand A and 100 bags of brand B are used to a maximum of 1,190 pounds when 140 bags of brand A and 50 bags of brand B are used.

Chapter 8 Review

1. $x = 3, y = 3$ *(8.1)* **2.** $x = 3, y = -2$ *(8.1)*
3. $x = 2, y = -1$ *(8.1)* **4.** $x = 1.1875, y = 1.625$ *(8.1)*

5. *(8.4)*

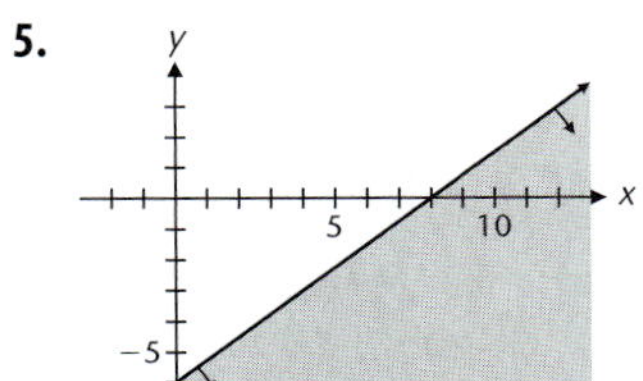

6. (8.4)

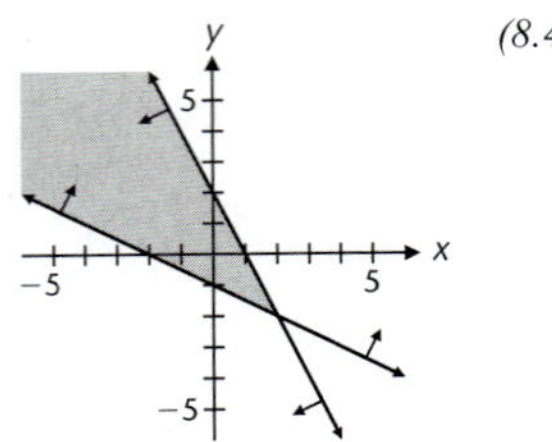

7. $\left[\begin{array}{rr|r} 3 & -6 & 12 \\ 1 & -4 & 5 \end{array}\right]$ (8.2) **8.** $\left[\begin{array}{rr|r} 1 & -4 & 5 \\ 1 & -2 & 4 \end{array}\right]$ (8.2)

9. $\left[\begin{array}{rr|r} 1 & -4 & 5 \\ 0 & 6 & -3 \end{array}\right]$ (8.2)

10. $x_1 = 4$
$x_2 = -7$
The solution is $(4, -7)$ (8.3)

11. $x_1 - x_2 = 4$
$0 = 1$
No solution (8.3)

12. $x_1 - x_2 = 4$
$x_1 = t + 4$, $x_2 = t$ is the solution, for t any real number (8.3)

13. The maximum value of z on S is 42 at (6, 4). The minimum value of z on S is 18 at (0, 6). (8.5)

14. $x_1 = 2$, $x_2 = -2$; each pair of lines has the same intersection point. (8.3)

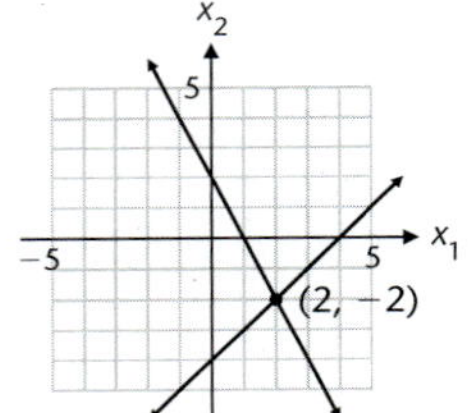

$x_1 - x_2 = 4$
$2x_1 + x_2 = 2$

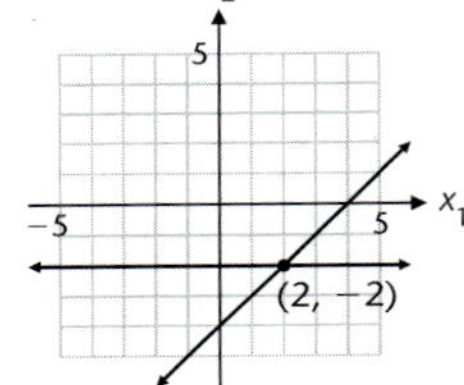

$x_1 - x_2 = 4$
$3x_2 = -6$

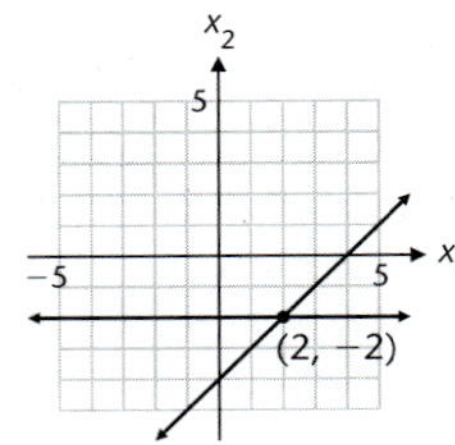

$x_1 - x_2 = 4$
$x_2 = -2$

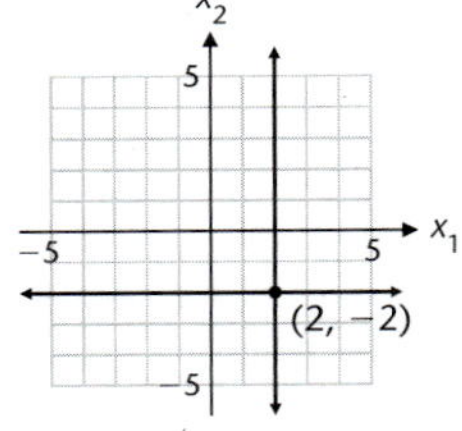

$x_1 = 2$
$x_2 = -2$

15. $x_1 = -1, x_2 = 3$ (8.3) **16.** $x_1 = -1, x_2 = 2, x_3 = 1$ (8.3)

17. $x_1 = 2, x_2 = 1, x_3 = -1$ (8.3)

18. $x_1 = -5t - 12$, $x_2 = 3t + 7$, $x_3 = t$ is a solution for every real number t. There are infinitely many solutions. (8.3)

19. No solution (8.3)

20. $x_1 = -\frac{3}{7}t - \frac{4}{7}$, $x_2 = \frac{5}{7}t + \frac{9}{7}$, $x_3 = t$ is a solution for every real number t. There are infinitely many solutions. (8.3)

21. Corners: (0, 4), (0, 0), (4, 0), and (3, 2); bounded (8.4)

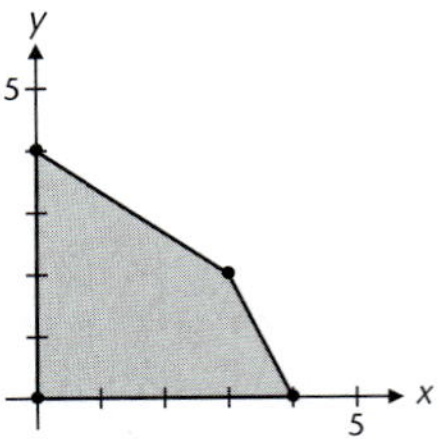

22. Corners: (0, 8), (12, 0), $(\frac{12}{5}, \frac{16}{5})$; unbounded (8.4)

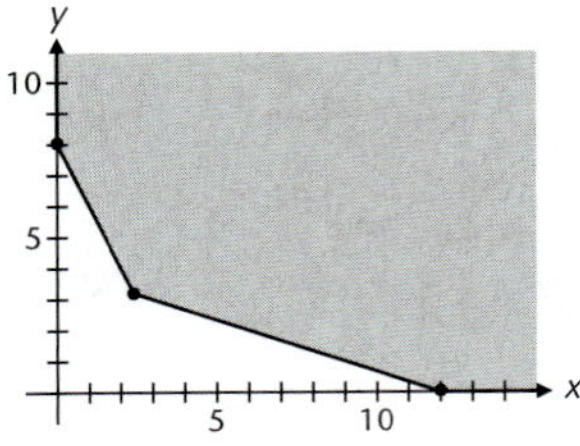

23. Corners: (4, 4), (10, 10), (20, 0); bounded (8.4)

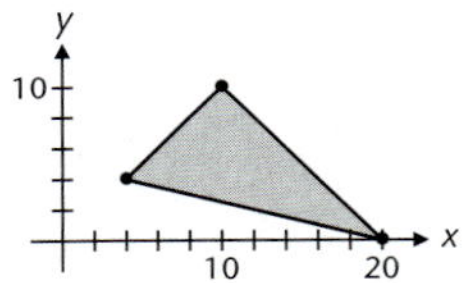

24. The maximum value of z on S is 46 at (4, 2). (8.5)

25. The minimum value of z on S is 75 at (3, 6) and (15, 0) (multiple optimal solutions). (8.5)

26. The minimum value of z on S is 44 at (4, 3). The maximum value of z on S is 82 at (2, 9). (8.5)

27. $x_1 = 1{,}000$, $x_2 = 4{,}000$, $x_3 = 2{,}000$ (8.3)

28. The maximum value of z on S is 26,000 at (600, 400). (8.5)

29. (A) A unique solution (B) No solution
(C) An infinite number of solutions (8.3)

30. 48 $\frac{1}{2}$-lb packages and 72 $\frac{1}{3}$-lb packages (8.1)

31. 6 meters by 8 meters (8.1)

32. $x_1 = 40$ grams mix A, $x_2 = 60$ grams mix B, $x_3 = 30$ grams mix C (8.3)

33. (A) $x_1 = 22$ nickels, $x_2 = 8$ dimes
(B) $x_1 = 3t + 22$ nickels, $x_2 = 8 - 4t$ dimes, $x_3 = t$ quarters, $t = 0, 1,$ or 2 (8.3)

34. (A) Maximum profit is $P = \$7{,}800$ when 80 regular and 30 competition sails are produced.

(B) The maximum profit increases to $8,750 when 70 competition and no regular sails are produced.
(C) The maximum profit decreases to $7,200 when no competition and 120 regular sales are produced. *(8.5)*

35. (A) The minimum cost is $C =$ $13 when 100 grams of mix A and 150 grams of mix B are used.
(B) The minimum cost decreases to $9 when 50 grams of mix A and 275 grams of mix B are used.
(C) The minimum cost increases to $28.75 when 250 grams of mix A and 75 grams of mix B are used. *(8.5)*

CHAPTER 9

Exercise 9.1

1. $\begin{bmatrix} 0 & 2 \\ 2 & -1 \end{bmatrix}$ **3.** $\begin{bmatrix} -1 & 6 \\ -4 & 3 \\ 1 & -1 \end{bmatrix}$ **5.** Not defined

7. $\begin{bmatrix} 2 & 3 & -5 \\ 5 & -5 & 7 \end{bmatrix}$ **9.** $\begin{bmatrix} 20 & -10 & 30 \\ 0 & -40 & 50 \end{bmatrix}$ **11.** $[10]$

13. $\begin{bmatrix} 5 \\ -3 \end{bmatrix}$ **15.** $\begin{bmatrix} 2 & 4 \\ 1 & -5 \end{bmatrix}$ **17.** $\begin{bmatrix} 1 & -5 \\ -2 & -4 \end{bmatrix}$ **19.** $[-14]$

21. $\begin{bmatrix} -20 & 10 \\ -12 & 6 \end{bmatrix}$ **23.** $[11]$ **25.** $\begin{bmatrix} 3 & -2 & -4 \\ 6 & -4 & -8 \\ -9 & 6 & 12 \end{bmatrix}$

27. Not defined **29.** $\begin{bmatrix} -6 & 7 & -11 \\ 4 & 18 & -4 \end{bmatrix}$ **31.** $\begin{bmatrix} -3 & 6 & 8 \\ -18 & 12 & 10 \\ 4 & 6 & 24 \end{bmatrix}$

33. $\begin{bmatrix} 5 & -11 & 15 \\ 4 & -7 & 3 \\ 0 & 10 & 4 \end{bmatrix}$ **35.** $\begin{bmatrix} -0.2 & 1.2 \\ 2.6 & -0.6 \\ -0.2 & 2.2 \end{bmatrix}$ **37.** $\begin{bmatrix} -31 & 16 \\ 61 & -25 \\ -3 & 77 \end{bmatrix}$

39. Not defined **41.** $\begin{bmatrix} -2 & 25 & -15 \\ 26 & -25 & 45 \\ -2 & 45 & -25 \end{bmatrix}$

43. $\begin{bmatrix} -26 & -15 & -25 \\ -4 & -18 & 4 \\ 2 & 43 & -19 \end{bmatrix}$ **45.** $B^n \to \begin{bmatrix} 0.25 & 0.75 \\ 0.25 & 0.75 \end{bmatrix}$, $AB^n \to [0.25 \;\; 0.75]$

47. $a = -1, b = 1, c = 3, d = -5$ **49.** $x = 1, y = 2$

51. $a^2 + bc = 0$

53. $a = -b$ and $c = -d$ **55.** $x = -5, y = 4$

57. $a = 3, b = 1, c = 1, d = -2$ **59.** All are true

61.

	Guitar	Banjo
Materials	$33	$26
Labor	$57	$77

63.

	Basic car	Air	AM/FM radio	Cruise control
			Markup	
Model A	$3,330	$77	$42	$27
Model B	$2,125	$93	$95	$50
Model C	$1,270	$113	$121	$52

65. (A) $11.80 (B) $30.30
(C) MN gives the labor costs per boat at each plant.
(D)

$MN =$

	Plant I	Plant II
One-person boat	$11.80	$13.80
Two-person boat	$18.50	$21.60
Four-person boat	$26.00	$30.30

67. (A) $A^2 = \begin{bmatrix} 0 & 0 & 2 & 0 & 0 \\ 1 & 0 & 0 & 0 & 1 \\ 0 & 1 & 0 & 2 & 0 \\ 1 & 0 & 0 & 0 & 1 \\ 0 & 0 & 1 & 0 & 0 \end{bmatrix}$

There is one way to travel from Baltimore to Atlanta with one intermediate connection; there are two ways to travel from Atlanta to Chicago with one intermediate connection. In general, the elements in A^2 indicate the number of different ways to travel from the ith city to the jth city with one intermediate connection.

(B) $A^3 = \begin{bmatrix} 2 & 0 & 0 & 0 & 2 \\ 0 & 1 & 0 & 2 & 0 \\ 0 & 0 & 3 & 0 & 0 \\ 0 & 1 & 0 & 2 & 0 \\ 1 & 0 & 0 & 0 & 1 \end{bmatrix}$

There is one way to travel from Denver to Baltimore with two intermediate connections; there are two ways to travel from Atlanta to El Paso with two intermediate connections. In general, the elements in A^3 indicate the number of different ways to travel from the ith city to the jth city with two intermediate connections.

(C) $A + A^2 + A^3 + A^4 = \begin{bmatrix} 2 & 3 & 2 & 5 & 2 \\ 1 & 1 & 4 & 2 & 1 \\ 4 & 1 & 3 & 2 & 4 \\ 1 & 1 & 4 & 2 & 1 \\ 1 & 1 & 1 & 3 & 1 \end{bmatrix}$

It is possible to travel from any origin to any destination with at most three intermediate connections.

69. (A) $3,550 (B) $6,000
(C) NM gives the total cost per town.
(D) $NM =$

	Cost/town
Berkeley	$3,550
Oakland	$6,000

(E) $[1 \;\; 1]N =$

Telephone call	House call	Letter
3,000	1,300	13,000

(F) $N\begin{bmatrix} 1 \\ 1 \\ 1 \end{bmatrix} =$

	Total contacts
Berkeley	6,500
Oakland	10,800

71. (A) $\begin{bmatrix} 0 & 0 & 1 & 1 & 1 & 0 \\ 1 & 0 & 0 & 1 & 1 & 0 \\ 0 & 1 & 0 & 1 & 0 & 0 \\ 0 & 0 & 0 & 0 & 0 & 1 \\ 0 & 0 & 1 & 1 & 0 & 1 \\ 1 & 1 & 1 & 0 & 0 & 0 \end{bmatrix}$ (B) $\begin{bmatrix} 0 & 1 & 2 & 3 & 1 & 2 \\ 1 & 0 & 2 & 3 & 2 & 2 \\ 1 & 1 & 0 & 2 & 1 & 1 \\ 1 & 1 & 1 & 0 & 0 & 1 \\ 1 & 2 & 2 & 2 & 0 & 2 \\ 2 & 2 & 2 & 3 & 2 & 0 \end{bmatrix}$

(C) $BC = \begin{bmatrix} 9 \\ 10 \\ 6 \\ 4 \\ 9 \\ 11 \end{bmatrix}$ where $C = \begin{bmatrix} 1 \\ 1 \\ 1 \\ 1 \\ 1 \\ 1 \end{bmatrix}$

(D) Frank, Bart, Aaron and Elvis (tie), Charles, Dan

Exercise 9.2

1. $\begin{bmatrix} 2 & -3 \\ 4 & 5 \end{bmatrix}$ **3.** $\begin{bmatrix} 2 & -3 \\ 4 & 5 \end{bmatrix}$ **5.** $\begin{bmatrix} -2 & 1 & 3 \\ 2 & 4 & -2 \\ 5 & 1 & 0 \end{bmatrix}$

7. $\begin{bmatrix} -2 & 1 & 3 \\ 2 & 4 & -2 \\ 5 & 1 & 0 \end{bmatrix}$ **9.** Yes **11.** No **13.** Yes **15.** No

17. Yes **19.** $\begin{bmatrix} 4 & 1 \\ -1 & 0 \end{bmatrix}$ **21.** $\begin{bmatrix} 3 & -2 \\ -1 & 1 \end{bmatrix}$

23. $\begin{bmatrix} 7 & -3 \\ -2 & 1 \end{bmatrix}$ **25.** $\begin{bmatrix} -3 & -4 & 2 \\ -2 & -2 & 1 \\ 2 & 3 & -1 \end{bmatrix}$ **27.** $\begin{bmatrix} 2 & -1 & -1 \\ -1 & 1 & 1 \\ -2 & 1 & 2 \end{bmatrix}$

29. Does not exist **31.** $\begin{bmatrix} 5 & -3 \\ -3 & 2 \end{bmatrix}$ **39.** $\begin{bmatrix} 1 & 0 & 1 \\ \frac{1}{2} & \frac{1}{2} & 1 \\ 2 & 1 & 4 \end{bmatrix}$

41. Does not exist **43.** $\begin{bmatrix} -9 & -15 & 10 \\ 4 & 5 & -4 \\ -1 & -1 & 1 \end{bmatrix}$

45. A^{-1} exists if and only if all the elements on the main diagonal are nonzero.

47. In both parts, $A^{-1} = A$ and $A^2 = I$

49. In both parts, $(A^{-1})^{-1} = A$

51. (A) $(AB)^{-1} = \begin{bmatrix} 29 & -41 \\ -12 & 17 \end{bmatrix}$, $A^{-1}B^{-1} = \begin{bmatrix} 23 & -33 \\ -16 & 23 \end{bmatrix}$,

$B^{-1}A^{-1} = \begin{bmatrix} 29 & -41 \\ -12 & 17 \end{bmatrix}$

(B) $(AB)^{-1} = \begin{bmatrix} 0.7 & -0.1 \\ -1.8 & 0.4 \end{bmatrix}$, $A^{-1}B^{-1} = \begin{bmatrix} 0.1 & 0 \\ -0.4 & 1 \end{bmatrix}$,

$B^{-1}A^{-1} = \begin{bmatrix} 0.7 & -0.1 \\ -1.8 & 0.4 \end{bmatrix}$

53. 14 5 195 74 97 37 181 67 49 18 121 43 103 41

55. GREEN EGGS AND HAM

57. 21 56 55 25 58 46 97 94 48 75 45 58 63 45 59 48 64 80 44 69 68 104 123 72 127

59. LYNDON BAINES JOHNSON

Exercise 9.3

1. $\begin{aligned} 2x_1 - x_2 &= 3 \\ x_1 + 3x_2 &= -2 \end{aligned}$ **3.** $\begin{aligned} -2x_1 \qquad + x_3 &= 3 \\ x_1 + 2x_2 + x_3 &= -4 \\ x_2 - x_3 &= 2 \end{aligned}$

5. $\begin{bmatrix} 4 & -3 \\ 1 & 2 \end{bmatrix}\begin{bmatrix} x_1 \\ x_2 \end{bmatrix} = \begin{bmatrix} 2 \\ 1 \end{bmatrix}$ **7.** $\begin{bmatrix} 1 & -2 & 1 \\ -1 & 1 & 0 \\ 2 & 3 & 1 \end{bmatrix}\begin{bmatrix} x_1 \\ x_2 \\ x_3 \end{bmatrix} = \begin{bmatrix} -1 \\ 2 \\ -3 \end{bmatrix}$

9. $x_1 = -8$ and $x_2 = 2$ **11.** $x_1 = 0$ and $x_2 = 4$

13. $x_1 = 3, x_2 = -2$ **15.** $x_1 = 11, x_2 = 4$

17. (A) $x_1 = -3, x_2 = 2$ (B) $x_1 = -1, x_2 = 2$
(C) $x_1 = -8, x_2 = 3$

19. (A) $x_1 = 17, x_2 = -5$ (B) $x_1 = 7, x_2 = -2$
(C) $x_1 = 24, x_2 = -7$

21. (A) $x_1 = 1, x_2 = 0, x_3 = 0$ (B) $x_1 = -1, x_2 = 0, x_3 = 1$
(C) $x_1 = 4, x_2 = 1, x_3 = -3$

23. (A) $x_1 = 0, x_2 = 2, x_3 = 4$ (B) $x_1 = -2, x_2 = 2, x_3 = 0$
(C) $x_1 = 6, x_2 = -2, x_3 = -6$

25. $x_1 = 2t + 2.5, x_2 = t$, t any real number **27.** No solution

29. $x_1 = 13t + 3, x_2 = 8t + 1, x_3 = t$, t any real number

31. $X = (A - B)^{-1}C$ $[X \neq C(A - B)^{-1}]$ **33.** $X = (A + I)^{-1}C$

35. $X = (A + B)^{-1}(C + D)$

37. (A) $x_1 = 1, x_2 = 0$ (B) $x_1 = -2{,}000, x_2 = 1{,}000$
(C) $x_1 = 2{,}001, x_2 = -1{,}000$

39. (A) Concert 1: 6,000 \$4 tickets and 4,000 \$8 tickets; concert 2: 5,000 \$4 tickets and 5,000 \$8 tickets; concert 3: 3,000 \$4 tickets and 7,000 \$8 tickets
(B) No (C) Between \$40,000 and \$80,000

41. (A) $I_1 = 4, I_2 = 6, I_3 = 2$ (B) $I_1 = 3, I_2 = 7, I_3 = 4$
(C) $I_1 = 7, I_2 = 8, I_3 = 1$

43. (A) $a = 1, b = 0, c = -3$ (B) $a = -2, b = 5, c = 1$
(C) $a = 11, b = -46, c = 43$

45. (A) Diet 1: 60 ounces mix A and 80 ounces mix B; diet 2: 20 ounces mix A and 60 ounces mix B; diet 3: 0 ounces mix A and 100 ounces mix B
(B) No

Exercise 9.4

1. 7 **3.** -17 **5.** 9.79 **7.** $\begin{vmatrix} 4 & 6 \\ -2 & 8 \end{vmatrix}$ **9.** $\begin{vmatrix} 5 & -1 \\ 0 & -2 \end{vmatrix}$
11. $(-1)^{1+1}\begin{vmatrix} 4 & 6 \\ -2 & 8 \end{vmatrix} = 44$ **13.** $(-1)^{2+3}\begin{vmatrix} 5 & -1 \\ 0 & -2 \end{vmatrix} = 10$
15. 10 **17.** -21 **19.** -40 **21.** $(-1)^{1+1}\begin{vmatrix} a_{22} & a_{23} & a_{24} \\ a_{32} & a_{33} & a_{34} \\ a_{42} & a_{43} & a_{44} \end{vmatrix}$
23. $(-1)^{4+3}\begin{vmatrix} a_{11} & a_{12} & a_{14} \\ a_{21} & a_{22} & a_{24} \\ a_{31} & a_{32} & a_{34} \end{vmatrix}$ **25.** 22 **27.** -12 **29.** 0
31. 6 **33.** 60 **35.** 114 **37.** False **39.** True
41. $\begin{vmatrix} a & b \\ c & d \end{vmatrix} = -\begin{vmatrix} c & d \\ a & b \end{vmatrix}$; interchanging the rows of this determinant changes the sign.
43. $\begin{vmatrix} ka & b \\ kc & d \end{vmatrix} = k\begin{vmatrix} a & b \\ c & d \end{vmatrix}$; multiplying a column of this determinant by a number k multiplies the value of the determinant by k.
45. $\begin{vmatrix} kc + a & kd + b \\ c & d \end{vmatrix} = \begin{vmatrix} a & b \\ c & d \end{vmatrix}$; adding a multiple of one row to the other row does not change the value of the determinant.
49. $49 = (-7)(-7)$ **51.** $f(x) = x^2 - 4x + 3$; 1, 3
53. $f(x) = x^3 + 2x^2 - 8x$; $-4, 0, 2$

Exercise 9.5

1. Theorem 1 **3.** Theorem 1 **5.** Theorem 2
7. Theorem 3 **9.** Theorem 5 **11.** $x = 0$ **13.** $x = 5$
15. -10 **17.** 10 **19.** -10 **21.** 25 **23.** -12
25. Theorem 1 **27.** Theorem 2 **29.** Theorem 5
31. $x = 5, y = 0$ **33.** $x = -3, y = 10$ **35.** -28
37. 106 **39.** 0 **41.** 6 **43.** 14
45. Expand the left side of the equation using minors.
47. Expand both sides of the equation and compare.
49. This follows from Theorem 4.
51. Expand the determinant about the first row to obtain $(y_1 - y_2)x - (x_1 - x_2)y + (x_1y_2 - x_2y_1) = 0$. Then show that the two points satisfy this linear equation.
53. If the determinant is 0, then the area of the triangle formed by the three points is 0. The only way this can happen is if the three points are on the same line—that is, the points are collinear.

Exercise 9.6

1. $x = 5, y = -2$ **3.** $x = 1, y = -1$ **5.** $x = -\frac{6}{5}, y = \frac{3}{5}$
7. $x = \frac{2}{17}, y = -\frac{20}{17}$ **9.** $x = 6{,}400, y = 6{,}600$
11. $x = 760, y = 760$ **13.** $x = 2, y = -2, z = -1$
15. $x = \frac{4}{3}, y = -\frac{1}{3}, z = \frac{2}{3}$ **17.** $x = -9, y = -\frac{7}{3}, z = 6$
19. $x = \frac{3}{2}, y = -\frac{7}{6}, z = \frac{2}{3}$
21. If $a = \frac{3}{2}$ and $b = \frac{15}{4}$, there are an infinite number of solutions. If $a = \frac{3}{2}$ and $b \neq \frac{15}{4}$, there are no solutions. If $a \neq \frac{3}{2}$, there is one solution.
23. $x = 4$ **25.** $y = 2$ **27.** $z = \frac{5}{2}$
29. Because $D = 0$, the system has either no solution or infinitely many. Because $x = 0, y = 0, z = 0$ is a solution, the second case must hold.
33. (A) $R = 200p + 300q - 6p^2 + 6pq - 3q^2$
(B) $p = -0.3x - 0.4y + 180$, $q = -0.2x - 0.6y + 220$, $R = 180x + 220y - 0.3x^2 - 0.6xy - 0.6y^2$

Chapter 9 Review

1. $\begin{bmatrix} 4 & 8 \\ -12 & 18 \end{bmatrix}$ *(9.1)* **2.** $[-11]$ *(9.1)*
3. $[-15 \quad 19]$ *(9.1)* **4.** $\begin{bmatrix} 16 \\ -6 \end{bmatrix}$ *(9.1)* **5.** $\begin{bmatrix} 3 & 3 \\ -4 & 9 \end{bmatrix}$ *(9.1)*
6. Not defined *(9.1)* **7.** Not defined *(9.1)*
8. $\begin{bmatrix} 13 & -29 \\ 20 & -24 \end{bmatrix}$ *(9.1)* **9.** $[-5 \quad 18]$ *(9.1)*
10. $\begin{bmatrix} 2 & 7 \\ -1 & -4 \end{bmatrix}$ *(9.2)*
11. (A) $x_1 = -1, x_2 = 3$ (B) $x_1 = 1, x_2 = 2$
(C) $x_1 = 8, x_2 = -10$ *(9.3)*
12. -17 *(9.4)* **13.** 0 *(9.4, 9.5)* **14.** $x = 2, y = -1$ *(9.6)*
15. (A) -2 (B) 6 (C) 2 *(9.5)*
16. $\begin{bmatrix} 7 & 16 & -9 \\ 28 & 40 & -30 \\ -21 & -8 & 17 \end{bmatrix}$ *(9.1)* **17.** $\begin{bmatrix} 22 & 19 \\ 38 & 42 \end{bmatrix}$ *(9.1)*

18. $\begin{bmatrix} 12 & 24 & -6 \\ 0 & 0 & 0 \\ -8 & -16 & 4 \end{bmatrix}$ *(9.1)* **19.** [16] *(9.1)*

20. Not defined *(9.1)*

21. $\begin{bmatrix} 63 & -24 & -39 \\ -42 & 16 & 26 \end{bmatrix}$ *(9.1)* **22.** $\begin{bmatrix} -1 & 1 & 1 \\ -2 & 3 & 2 \\ \frac{1}{2} & -\frac{1}{4} & -\frac{1}{4} \end{bmatrix}$ *(9.2)*

23. (A) $x_1 = 2, x_2 = 1, x_3 = -1$ (B) $x_1 = 1, x_2 = -2, x_3 = 1$
(C) $x_1 = -1, x_2 = 2, x_3 = -2$ *(9.3)*

24. $-\frac{11}{12}$ *(9.4)* **25.** 35 *(9.4, 9.5)* **26.** $y = \frac{10}{5} = 2$ *(9.6)*

27. (A) A unique solution
(B) Either no solution or an infinite number *(9.3)*

28. No *(9.3)* **29.** $X = (A - C)^{-1}B$ *(9.3)*

30. $\begin{bmatrix} -\frac{11}{12} & -\frac{1}{12} & 5 \\ \frac{10}{12} & \frac{2}{12} & -4 \\ \frac{1}{12} & -\frac{1}{12} & 0 \end{bmatrix}$ or $\frac{1}{12}\begin{bmatrix} -11 & -1 & 60 \\ 10 & 2 & -48 \\ 1 & -1 & 0 \end{bmatrix}$ *(9.2)*

31. $x_1 = 1{,}000, x_2 = 4{,}000, x_3 = 2{,}000$ *(9.3)* **32.** 42 *(9.5)*

33. $\begin{vmatrix} u + kv & v \\ w + kx & x \end{vmatrix} = (u + kv)x - (w + kx)v$
$= ux + kvx - wv - kvx = ux - wv = \begin{vmatrix} u & v \\ w & x \end{vmatrix}$ *(9.5)*

34. Theorem 4 in Section 9.5 implies that both points satisfy the equation. All other points on the line through the given points will also satisfy the equation. *(9.5)*

35. (A) 60 tons at Big Bend, 20 tons at Saw Pit
(B) 30 tons at Big Bend, 50 tons at Saw Pit
(C) 40 tons at Big Bend, 40 tons at Saw Pit *(9.3)*

36. (A) \$27
(B) Elements in LH give the total labor cost of manufacturing each product at each plant.
(C) *(9.1)*

	North Carolina	South Carolina	
$LH =$	\$46.35	\$41.00	Desks
	\$30.45	\$27.00	Stands

37. (A) $\begin{bmatrix} 1{,}600 & 1{,}730 \\ 890 & 720 \end{bmatrix}$ (B) $\begin{bmatrix} 200 & 160 \\ 80 & 40 \end{bmatrix}$
(C) $\begin{bmatrix} 3{,}150 \\ 1{,}550 \end{bmatrix}$ Desks / Stands
Total production of each item in January *(9.1)*

38. GRAPHING UTILITY *(9.2)*

Cumulative Review for Chapters 8 and 9

1. $x = 2, y = -1$ *(8.1, 8.2)* **2.** $(-1, 2)$ *(8.1)*

3. No solution *(8.1, 8.2)* **4.** *(8.4)*

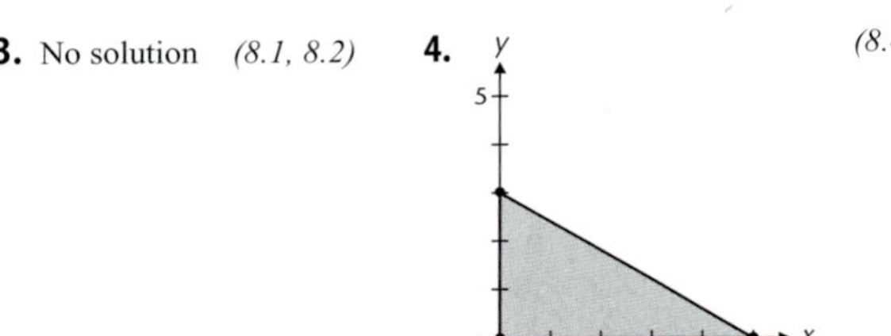

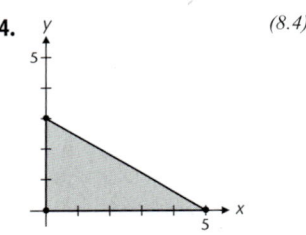

5. Maximum: 33; Minimum: 10 *(8.5)*

6. (A) $\begin{bmatrix} 0 & -3 \\ 3 & -9 \end{bmatrix}$ (B) Not defined (C) [3]
(D) $\begin{bmatrix} 1 & 7 \\ 4 & -7 \end{bmatrix}$ (E) $[-1, 8]$ (F) Not defined *(9.1)*

7. -10 *(9.4)*

8. (A) $x_1 = 3, x_2 = -4$
(B) $x_1 = 2t + 3, x_2 = t$, t any real number.
(C) No solution *(8.1)*

9. (A) $\left[\begin{array}{cc|c} 1 & 1 & 3 \\ -1 & 1 & 5 \end{array}\right]$ (B) $\left[\begin{array}{cc|c} 1 & 0 & -1 \\ 0 & 1 & 4 \end{array}\right]$
(C) $x_1 = -1, x_2 = 4$ *(8.1, 8.2)*

10. (A) $\begin{bmatrix} 1 & -3 \\ 2 & -5 \end{bmatrix}\begin{bmatrix} x_1 \\ x_2 \end{bmatrix} = \begin{bmatrix} k_1 \\ k_2 \end{bmatrix}$ (B) $A^{-1} = \begin{bmatrix} -5 & 3 \\ -2 & 1 \end{bmatrix}$
(C) $x_1 = 13, x_2 = 5$ (D) $x_1 = -11, x_2 = -4$ *(9.3)*

11. (A) 2 (B) $x = \frac{1}{2}, y = 0$ *(9.6)*

12. $x_1 = 1, x_2 = 3$; each pair of lines has the same intersection point. *(8.1)*

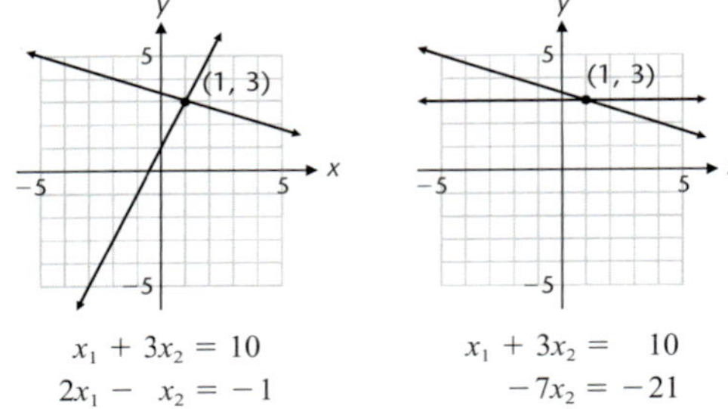

$x_1 + 3x_2 = 10$
$2x_1 - x_2 = -1$

$x_1 + 3x_2 = 10$
$-7x_2 = -21$

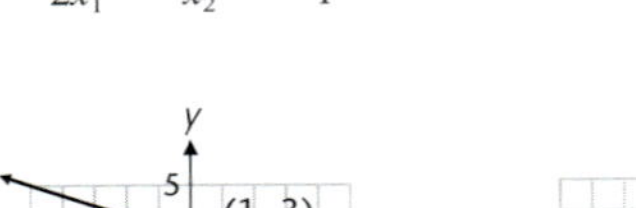

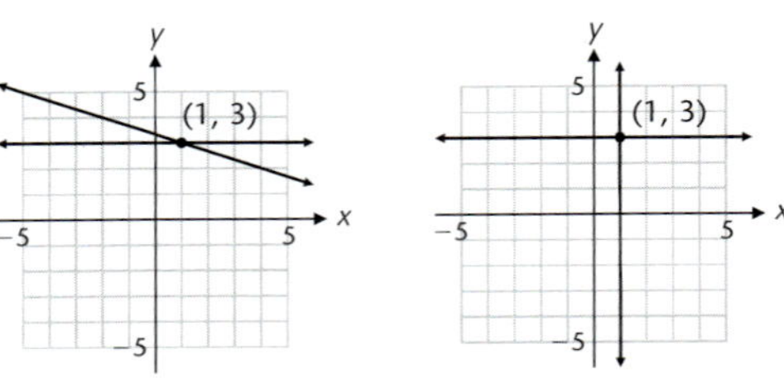

$x_1 + 3x_2 = 10$
$x_2 = 3$

$x_1 = 1$
$x_2 = 3$

13. (1.53, 3.35) *(8.1)* **14.** $(1, 0, -2)$ *(8.2)*

15. No solution *(8.2)*

16. $(t - 3, t - 2, t)$ t any real number *(8.2)*

17. (A) $[-3]$ (B) $\begin{bmatrix} 1 & 2 & -1 \\ -1 & -2 & 1 \\ 2 & 4 & -2 \end{bmatrix}$ *(9.1)*

18. (A) $\begin{bmatrix} -1 & 2 \\ 2 & 3 \end{bmatrix}$ (B) Not defined *(9.1)*

19. 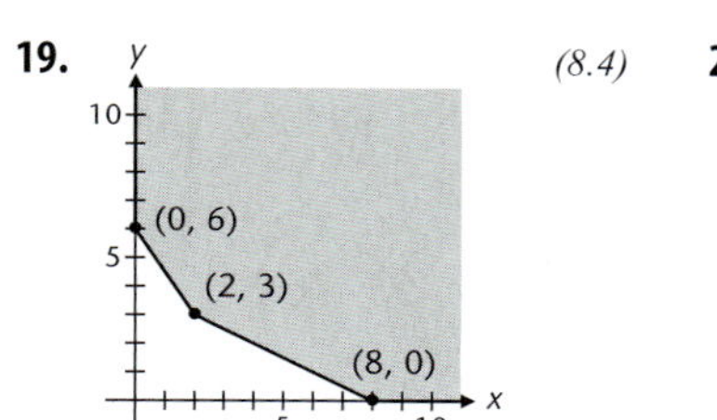

(8.4) **20.** 63 *(8.5)*

21. (A) $\begin{bmatrix}1 & 4 & 2\\2 & 6 & 3\\2 & 5 & 2\end{bmatrix}\begin{bmatrix}x_1\\x_2\\x_3\end{bmatrix}=\begin{bmatrix}k_1\\k_2\\k_3\end{bmatrix}$ (B) $A^{-1}=\begin{bmatrix}-3 & 2 & 0\\2 & -2 & 1\\-2 & 3 & -2\end{bmatrix}$
(C) $(7, -5, 6)$ (D) $(-6, 3, -2)$ *(9.3)*
22. (A) $D = 1$ (B) $z = 32$ *(9.5, 9.6)*
23. (A) Infinite number of solutions (B) No solution
(C) Unique solution *(8.2)*
24. $A = I$, the $n \times n$ identity *(9.3)* **25.** L, M, and P *(8.2)*
29. True *(9.1)* **30.** True *(9.1)* **31.** False *(9.1)*
32. False *(9.1)* **33.** True *(9.2)* **34.** True *(9.2)*
35. True *(9.4)* **36.** True *(9.4)*
37. \$8,000 at 8% and \$4,000 at 14% *(8.1, 8.2)*
38. 60 grams of mix A, 50 grams of mix B, 40 grams of mix C *(8.2)*

39. 1 model A truck, 6 model B trucks, and 5 model C trucks; or 3 model A trucks, 3 model B trucks, and 6 model C trucks; or 5 model A trucks and 7 model C trucks. *(8.2)*
40. (A) Manufacturing 400 standard and 200 deluxe day packs produces a maximum weekly profit of \$5,600.
(B) The maximum weekly profit increases to \$6,000 when 0 standard and 400 deluxe day packs are manufactured.
(C) The maximum weekly profit increases to \$6,000 when 600 standard and 0 deluxe day packs are manufactured. *(8.5)*
41. (A) $M\begin{bmatrix}0.25\\0.25\\0.25\\0.25\end{bmatrix}=\begin{bmatrix}82.25\\83\\92\\83.75\\82\end{bmatrix}$ Ann, Bob, Carol, Dan, Eric
(B) $M\begin{bmatrix}0.2\\0.2\\0.2\\0.4\end{bmatrix}=\begin{bmatrix}83\\84.8\\91.8\\85.2\\80.8\end{bmatrix}$ Ann, Bob, Carol, Dan, Eric
(C) Class averages: Test 1, Test 2, Test 3, Test 4
$[0.2\ \ 0.2\ \ 0.2\ \ 0.2\ \ 0.2]\,M = [84.4\ \ 81.8\ \ 85\ \ 87.2]$
(9.1)

CHAPTER 10

Exercise 10.1

1. $-1, 0, 1, 2$ **3.** $0, \frac{1}{3}, \frac{1}{2}, \frac{3}{5}$ **5.** $4, -8, 16, -32$ **7.** 6
9. $\frac{99}{101}$ **11.** $1 + 2 + 3 + 4 + 5$ **13.** $\frac{1}{10} + \frac{1}{100} + \frac{1}{1{,}000}$
15. $-1 + 1 - 1 + 1$ **17.** $1, -4, 9, -16, 25$
19. 0.3, 0.33, 0.333, 0.3333, 0.33333 **21.** $1, -\frac{1}{2}, \frac{1}{4}, -\frac{1}{8}, \frac{1}{16}$
23. $7, 3, -1, -5, -9$ **25.** $4, 1, \frac{1}{4}, \frac{1}{16}, \frac{1}{64}$
27. 1, 2, 5, 12, 29, 70, 169 **29.** $-1, 2, 0, 4, 4, 12, 20$
31. $a_n = n + 3$ **33.** $a_n = 3n$ **35.** $a_n = \frac{n}{n+1}$
37. $a_n = (-1)^{n+1}$ **39.** $a_n = (-2)^n$ **41.** $a_n = \frac{x^n}{n}$
43. $a_n = 2^{n-1}$ and $b_n = 0.5n^2 - 0.5n + 1$ are two of many correct answers.
45. $a_n = n^3$ and $b_n = 6n^2 - 11n + 6$ are two of many correct answers.

47. 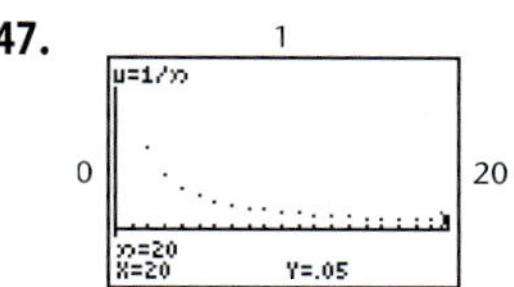

49. 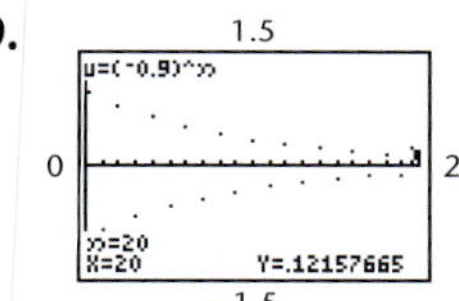

51. $\frac{4}{1} - \frac{8}{2} + \frac{16}{3} - \frac{32}{4}$ **53.** $x^2 + \frac{x^3}{2} + \frac{x^4}{3}$
55. $x - \frac{x^2}{2} + \frac{x^3}{3} - \frac{x^4}{4} + \frac{x^5}{5}$ **57.** $\sum_{k=1}^{4} k^2$ **59.** $\sum_{k=1}^{5} \frac{1}{2^k}$
61. $\sum_{k=1}^{n} \frac{1}{k^2}$ **63.** $\sum_{k=1}^{n} (-1)^{k+1}k^2$
65. (A) 3, 1.83, 1.46, 1.415 (B) Calculator $\sqrt{2} = 1.4142135\ldots$
(C) $a_1 = 1$; 1, 1.5, 1.417, 1.414
67. The values of c_n are approximately 2.236 (i.e., $\sqrt{5}$) for large values of n.
69. $e^{0.2} = 1.2214000$; $e^{0.2} = 1.2214028$ (calculator—direct evaluation)

Exercise 10.2

1. Fails at $n = 2$ **3.** Fails at $n = 3$
5. P_1: $2 = 2 \cdot 1^2$; P_2: $2 + 6 = 2 \cdot 2^2$; P_3: $2 + 6 + 10 = 2 \cdot 3^2$
7. P_1: $a^5a^1 = a^{5+1}$; P_2: $a^5a^2 = a^5(a^1a) = (a^5a)a = a^6a = a^7 = a^{5+2}$; P_3: $a^5a^3 = a^5(a^2a) = a^5(a^1a)a = [(a^5a)a]a = a^8 = a^{5+3}$
9. P_1: $9^1 - 1 = 8$ is divisible by 4; P_2: $9^2 - 1 = 80$ is divisible by 4; P_3: $9^3 - 1 = 728$ is divisible by 4

11. P_k: $2 + 6 + 10 + \cdots + (4k - 2) = 2k^2$;
P_{k+1}: $2 + 6 + 10 + \cdots + (4k - 2) + (4k + 2) = 2(k + 1)^2$
13. P_k: $a^5a^k = a^{5+k}$; P_{k+1}: $a^5a^{k+1} = a^{5+k+1}$
15. P_k: $9^k - 1 = 4r$ for some integer r;
P_{k+1}: $9^{k+1} - 1 = 4s$ for some integer s
23. $n = 4$, $p(x) = x^4 + 1$ **25.** $n = 23$
43. P_n: $2 + 4 + 6 + \cdots + 2n = n(n + 1)$
45. $1 + 2 + 3 + \cdots + (n - 1) = \dfrac{n(n - 1)}{2}, n \geq 2$
51. $3^4 + 4^4 + 5^4 + 6^4 \neq 7^4$

Exercise 10.3

1. (A) Arithmetic with $d = -5$; $-26, -31$
(B) Geometric with $r = -2$; $-16, 32$ (C) Neither
(D) Geometric with $r = \frac{1}{3}$; $\frac{1}{54}, \frac{1}{162}$
3. $a_2 = -1$; $a_3 = 3$; $a_4 = 7$
5. $a_{15} = 67$; $S_{11} = 242$ **7.** $S_{21} = 861$ **9.** $a_{15} = -21$
11. $a_2 = 3$; $a_3 = -\frac{3}{2}$; $a_4 = \frac{3}{4}$
13. $a_{10} = \frac{1}{243}$ **15.** $S_7 = 3{,}279$ **17.** $d = 6$; $a_{101} = 603$
19. $S_{40} = 200$ **21.** $a_{11} = 2$; $S_{11} = \frac{77}{6}$ **23.** $a_1 = 1$
25. $r = 0.398$ **27.** $S_{10} = -1{,}705$ **29.** $a_2 = 6$; $a_3 = 4$
31. $S_{51} = 4{,}131$ **33.** $S_7 = 547$ **35.** $-1{,}071$ **37.** $\dfrac{1{,}023}{1{,}024}$
39. 4,446 **43.** $x = 2\sqrt{3}$ **45.** $a_n = -3 + (n - 1)3$ or $3n - 6$
47. 66 **49.** 133 **51.** $S_\infty = \frac{9}{2}$ **53.** no sum
55. $S_\infty = \frac{8}{5}$ **57.** $\frac{7}{9}$ **59.** $\frac{6}{11}$ **61.** $3\frac{8}{37}$ or $\frac{119}{37}$
65. $a_n = (-2)(-3)^{n-1}$ **67.** *Hint:* $y = x + d$, $z = x + 2d$
71. $x = -1$, $y = 2$ **73.** Firm A: \$501,000; firm B: \$504,000
75. \$4,000,000 **77.** $P(1 + r)^n$; approximately 12 years
79. \$700 per year; \$115,500 **81.** 900 **83.** 1,250,000
85. (A) 336 feet (B) 1,936 feet (C) $16t^2$ feet
87. $A = A_0 2^{2t}$ **89.** $r = 10^{-0.4} = 0.398$
91. 9.22×10^{16} dollars; 1.845×10^{17} dollars
93. 0.0015 pounds per square inch **95.** 2 **97.** 3,420°

Exercise 10.4

1. 362,880 **3.** 39,916,800 **5.** 990 **7.** 10 **9.** 35
11. 1 **15.** 60 **17.** 6,497,400 **19.** 10 **21.** 270,725
25. $5 \cdot 3 \cdot 4 \cdot 2 = 120$ **27.** $P_{10,3} = 10 \cdot 9 \cdot 8 = 720$
29. $C_{7,3} = 35$ subcommittees; $P_{7,3} = 210$ **31.** $C_{10,2} = 45$
33. No repeats: $6 \cdot 5 \cdot 4 \cdot 3 = 360$; with repeats: $6 \cdot 6 \cdot 6 \cdot 6 = 1{,}296$
35. No repeats: $10 \cdot 9 \cdot 8 \cdot 7 \cdot 6 = 30{,}240$;
with repeats: $10 \cdot 10 \cdot 10 \cdot 10 \cdot 10 = 100{,}000$
37. $C_{13,5} = 1{,}287$
39. $26 \cdot 26 \cdot 26 \cdot 10 \cdot 10 \cdot 10 = 17{,}576{,}000$ possible license plates;
no repeats: $26 \cdot 25 \cdot 24 \cdot 10 \cdot 9 \cdot 8 = 11{,}232{,}000$
41. $C_{13,5} \cdot C_{13,2} = 100{,}386$ **43.** $C_{8,3} \cdot C_{10,4} \cdot C_{7,2} = 246{,}960$
45. (B) $r = 0, 10$
(C) Each is the product of r consecutive integers, the largest of which is n for $P_{n,r}$ and r for $r!$.
47. $12 \cdot 11 = 132$
49. (A) $C_{8,2} = 28$ (B) $C_{8,3} = 56$ (C) $C_{8,4} = 70$
51. Two people: $5 \cdot 4 = 20$; three people: $5 \cdot 4 \cdot 3 = 60$;
four people: $5 \cdot 4 \cdot 3 \cdot 2 = 120$;
five people: $5 \cdot 4 \cdot 3 \cdot 2 \cdot 1 = 120$
53. (A) $P_{8,5} = 6{,}720$ (B) $C_{8,5} = 56$ (C) $C_{2,1} \cdot C_{6,4} = 30$
55. There are $C_{4,1} \cdot C_{48,4} = 778{,}320$ hands that contain exactly one king, and $C_{39,5} = 575{,}757$ hands containing no hearts, so the former is more likely.

Exercise 10.5

1. Occurrence of E is certain
3. (A) No probability can be negative
(B) $P(R) + P(G) + P(Y) + P(B) \neq 1$
(C) Is an acceptable probability assignment.
5. $P(R) + P(Y) = .56$ **7.** .1 **9.** .45
11. $P(E) = \dfrac{n(E)}{n(S)} = \dfrac{1}{720} \approx .0014$ **13.** $\dfrac{C_{26,5}}{C_{52,5}} \approx .025$
15. $\dfrac{C_{16,5}}{C_{52,5}} \approx .0017$ **17.** Yes; no **19.** $P(E) = \dfrac{n(E)}{n(S)} = \dfrac{50}{250} = .2$
21. $P(E) = \dfrac{n(E)}{n(S)} = \dfrac{1}{120} \approx .008$ **23.** $\frac{1}{36}$ **25.** $\frac{5}{36}$ **27.** $\frac{1}{6}$
29. $\frac{7}{9}$ **31.** 0 **33.** $\frac{1}{3}$ **35.** $\frac{2}{9}$ **37.** $\frac{2}{3}$
39. $S = \{1, 2, 3, \ldots, 365\}$; $P(e_i) = 1/365$
41. (A) $P(2) \approx .022$, $P(3) \approx .07$, $P(4) \approx .088$, $P(5) \approx .1$,
$P(6) \approx .142$, $P(7) \approx .178$, $P(8) \approx .144$, $P(9) \approx .104$,
$P(10) \approx .072$, $P(11) \approx .052$, $P(12) \approx .028$
(B) $P(2) = \frac{1}{36}$, $P(3) = \frac{2}{36}$, $P(4) = \frac{3}{36}$, $P(5) = \frac{4}{36}$, $P(6) = \frac{5}{36}$,
$P(7) = \frac{6}{36}$, $P(8) = \frac{5}{36}$, $P(9) = \frac{4}{36}$, $P(10) = \frac{3}{36}$, $P(11) = \frac{2}{36}$,
$P(12) = \frac{1}{36}$

(C)

Sum	Expected frequency	Sum	Expected frequency
2	13.9	8	69.4
3	27.8	9	55.6
4	41.7	10	41.7
5	55.6	11	27.8
6	69.4	12	13.9
7	83.3		

45. $\frac{1}{4}$ **47.** $\frac{1}{4}$ **49.** $\frac{3}{4}$ **51.** $\frac{1}{9}$ **53.** $\frac{1}{3}$ **55.** $\frac{1}{9}$ **57.** $\frac{4}{9}$
59. $\frac{C_{16,5}}{C_{52,5}} \approx .00168$ **61.** $\frac{48}{C_{52,5}} \approx .000\,0185$
63. $\frac{4}{C_{52,5}} \approx .000\,0015$ **65.** $\frac{C_{4,2} \cdot C_{4,3}}{C_{52,5}} \approx .000\,009$
67. (A) .015 (B) .222 (C) .169 (D) .958

Exercise 10.6

1. 10 **3.** 6 **5.** 20 **7.** 35 **9.** 84 **11.** 66
13. 2,380 **15.** 230,300 **17.** $m^3 + 3m^2n + 3mn^2 + n^3$
19. $8x^3 - 36x^2y + 54xy^2 - 27y^3$ **21.** $x^4 - 8x^3 + 24x^2 - 32x + 16$
23. $m^4 + 12m^3n + 54m^2n^2 + 108mn^3 + 81n^4$
25. $32x^5 - 80x^4y + 80x^3y^2 - 40x^2y^3 + 10xy^4 - y^5$
27. $m^6 + 12m^5n + 60m^4n^2 + 160m^3n^3 + 240m^2n^4 + 192mn^5 + 64n^6$
29. $35x^4$ **31.** $-29{,}586x^6$ **33.** $4{,}060{,}938{,}240x^{14}$
35. $15x^8$ **37.** Does not exist **39.** $5{,}005u^9v^6$ **41.** $264m^2n^{10}$
43. $924w^6$ **45.** $-48{,}384x^3y^5$ **47.** $3x^2 + 3xh + h^2$; approaches $3x^2$
49. $5x^4 + 10x^3h + 10x^2h^2 + 5xh^3 + h^4$; approaches $5x^4$
51. 5 **53.** (A) $a_4 = 0.251$ (B) 1 **55.** 1.1046

Chapter 10 Review

1. (A) Geometric (B) Arithmetic (C) Arithmetic
(D) Neither (E) Geometric *(10.1, 10.3)*
2. (A) 5, 7, 9, 11 (B) $a_{10} = 23$ (C) $S_{10} = 140$ *(10.1, 10.3)*
3. (A) 16, 8, 4, 2 (B) $a_{10} = \frac{1}{32}$ (C) $S_{10} = 31\frac{31}{32}$ *(10.1, 10.3)*
4. (A) $-8, -5, -2, 1$ (B) $a_{10} = 19$ (C) $S_{10} = 55$ *(10.1, 10.3)*
5. (A) $-1, 2, -4, 8$ (B) $a_{10} = 512$ (C) $S_{10} = 341$ *(10.1, 10.3)*
6. $S_\infty = 32$ *(10.3)* **7.** 720 *(10.4)*
8. $\frac{22 \cdot 21 \cdot 20 \cdot 19!}{19!} = 9{,}240$ *(10.4)* **9.** 21 *(10.4)*
10. $C_{6,2} = 15$; $P_{6,2} = 30$ *(10.5)*
11. (A) 12 combined outcomes: (B) $6 \cdot 2 = 12$ *(10.5)*

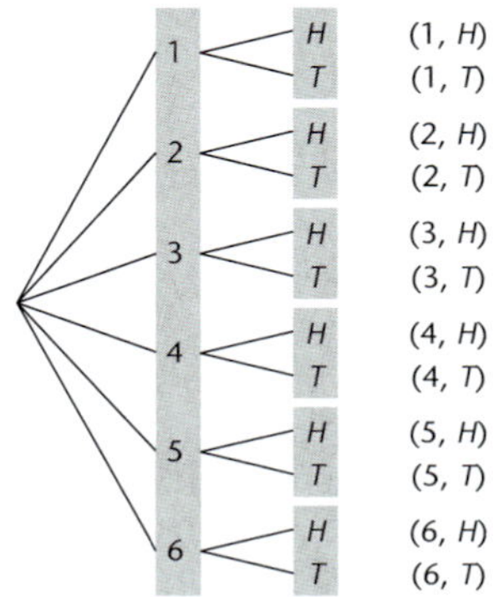

12. $6 \cdot 5 \cdot 4 \cdot 3 \cdot 2 \cdot 1 = 720$ *(10.5)* **13.** $P_{6,6} = 6! = 720$ *(10.5)*
14. $\frac{C_{13,5}}{C_{52,5}} \approx .0005$ *(10.5)* **15.** $\frac{1}{P_{15,2}} \approx .0048$ *(10.5)* **16.** .05 *(10.5)*
17. P_1: $5 = 1^2 + 4 \cdot 1 = 5$; P_2: $5 + 7 = 2^2 + 4 \cdot 2$;
P_3: $5 + 7 + 9 = 3^2 + 4 \cdot 3$ *(10.2)*
18. P_1: $2 = 2^{1+1} - 2$; P_2: $2 + 4 = 2^{2+1} - 2$;
P_3: $2 + 4 + 8 = 2^{3+1} - 2$ *(10.2)*
19. P_1: $49^1 - 1 = 48$ is divisible by 6; P_2: $49^2 - 1 = 2{,}400$ is divisible by 6; P_3: $49^3 - 1 = 117{,}648$ is divisible by 6 *(10.2)*
20. P_k: $5 + 7 + 9 + \cdots + (2k + 3) = k^2 + 4k$;
P_{k+1}: $5 + 7 + 9 + \cdots + (2k + 3) + (2k + 5) = (k + 1)^2 + 4(k + 1)$ *(10.2)*
21. P_k: $2 + 4 + 8 + \cdots + 2^k = 2^{k+1} - 2$;
P_{k+1}: $2 + 4 + 8 + \cdots + 2^k + 2^{k+1} = 2^{k+2} - 2$ *(10.2)*
22. P_k: $49^k - 1 = 6r$ for some integer r;
P_{k+1}: $49^{k+1} - 1 = 6s$ for some integer s *(10.2)*
23. $n = 31$ is a counterexample *(10.2)*
24. $S_{10} = (-6) + (-4) + (-2) + 0 + 2 + 4 + 6 + 8 + 10 + 12 = 30$ *(10.3)*
25. $S_7 = 8 + 4 + 2 + 1 + \frac{1}{2} + \frac{1}{4} + \frac{1}{8} = 15\frac{7}{8}$ *(10.3)*
26. $S_\infty = \frac{81}{5}$ *(10.3)* **27.** $S_n = \sum_{k=1}^{n} \frac{(-1)^{k+1}}{3^k}$; $S_\infty = \frac{1}{4}$ *(10.3)*
28. The probability of an event cannot be negative, but $P(e_2)$ is given as negative. The sum of the probabilities of the simple events must be 1, but it is given as 2.5. The probability of an event cannot be greater than 1, but $P(e_4)$ is given as 2. *(10.5)*
29. $C_{6,3} = 20$ *(10.4)* **30.** $d = 3$, $a_5 = 25$ *(10.3)*
31. 336; 512; 392 *(10.4)*
32. (A) P(2 heads) = .21; P(1 head) = .48; P(0 heads) = .31
(B) $P(E_1) = .25$; $P(E_2) = .5$; $P(E_3) = .25$
(C) 2 heads = 250; 1 head = 500; 0 heads = 250 *(10.5)*

33. (A) $\frac{C_{13,5}}{C_{52,5}}$ (B) $\frac{C_{13,3} \cdot C_{13,2}}{C_{52,5}}$ *(10.5)* **34.** $\frac{C_{8,2}}{C_{10,4}} = \frac{2}{15}$ *(10.5)*
35. (A) $\frac{1}{3}$ (B) $\frac{2}{9}$ *(10.5)* **36.** $\frac{8}{11}$ *(10.3)*
37. (A) $P_{6,3} = 120$ (B) $C_{5,2} = 10$ *(10.4)* **38.** 190 *(10.6)*
39. 1,820 *(10.6)* **40.** 1 *(10.6)*
41. $x^5 - 5x^4y + 10x^3y^2 - 10x^2y^3 + 5xy^4 - y^5$ *(10.6)*
42. $672x^6$ *(10.6)* **43.** $-1{,}760x^3y^9$ *(10.6)* **47.** 29 *(10.6)*
48. 26 *(10.1)* **49.** $2 \cdot 2 \cdot 2 \cdot 2 \cdot 2 = 32$; 6 *(10.4)*
50. $\frac{49g}{2}$ feet; $\frac{625g}{2}$ feet *(10.3)* **51.** 12 *(10.4)*
52. $x^6 + 6ix^5 - 15x^4 - 20ix^3 + 15x^2 + 6ix - 1$ *(10.6)*
53. $1 - \frac{C_{7,3}}{C_{10,3}} = \frac{17}{24}$ *(10.5)*
54. (A) .350 (B) $\frac{3}{8} = .375$ (C) 375 *(10.5)*
60. $900 *(10.3)* **61.** $7,200 *(10.3)*
62. $895.42; $1,603.57 *(10.3)* **63.** $P_{5,5} = 120$ *(10.4)*
64. (A) .04 (B) .16 (C) .54 *(10.5)*
65. $1 - \frac{C_{10,4}}{C_{12,4}} \approx .576$ *(10.5)*

CHAPTER 11

Exercise 11.1

1.

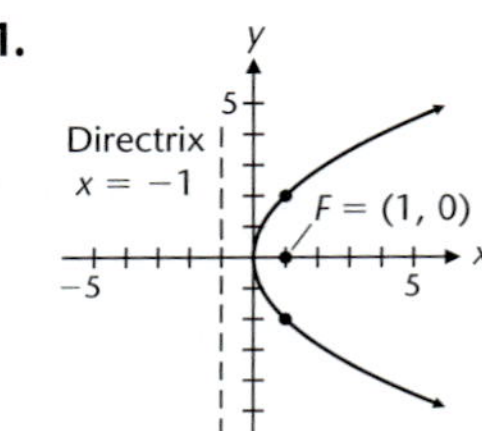

3.

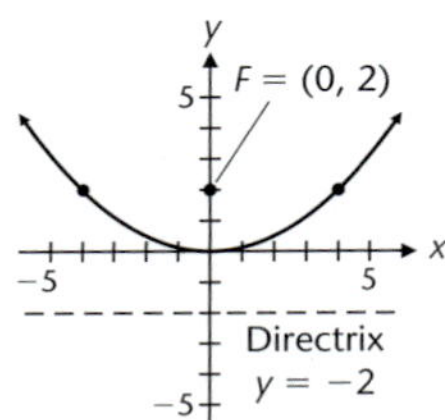

5.

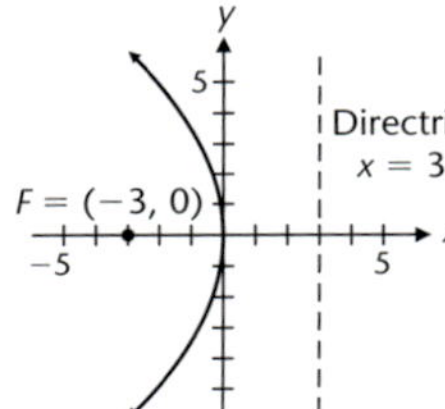

7.

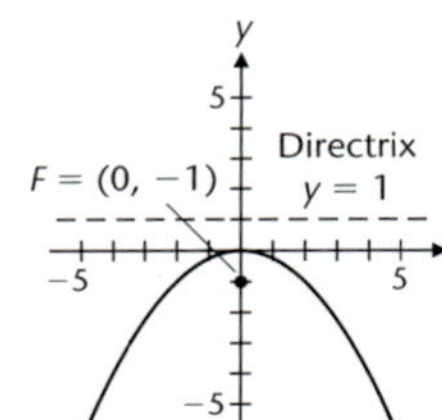

9.

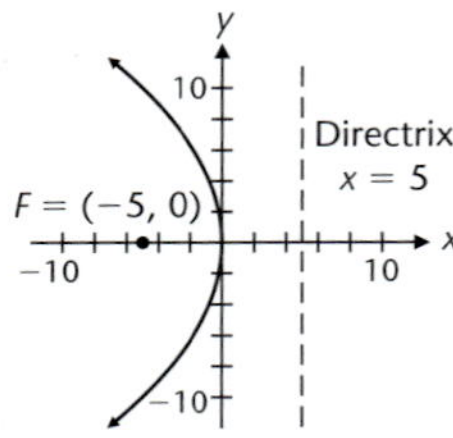

11.

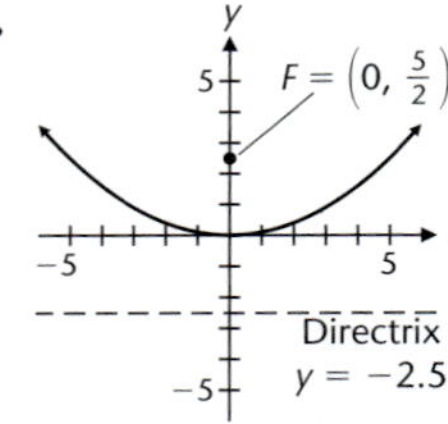

13. (9.75, 0) **15.** (0, −26.25) **17.** (−19.25, 0)
19. $x^2 = 12y$ **21.** $x^2 = -28y$ **23.** $y^2 = -24x$
25. $y^2 = 8x$ **27.** $x^2 = 8y$ **29.** $y^2 = -12x$ **31.** $x^2 = -4y$

33.

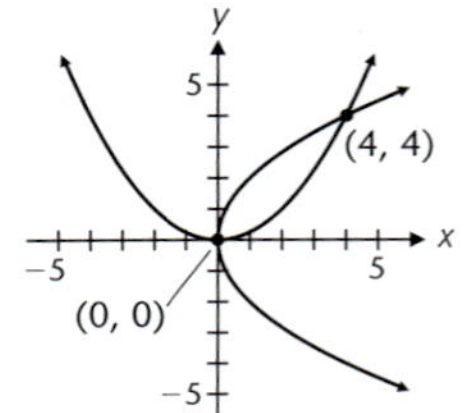

35.

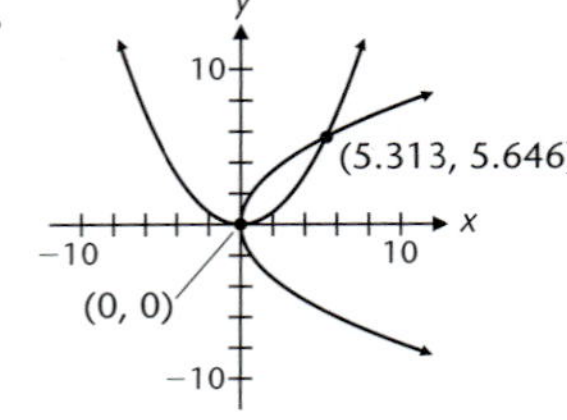

37. (A) 2; $x = 0$ and $y = 0$ (B) (0, 0), $(4am, 4am^2)$
39. $A = (-2a, a)$, $B = (2a, a)$
41. $y^2 = 8x$

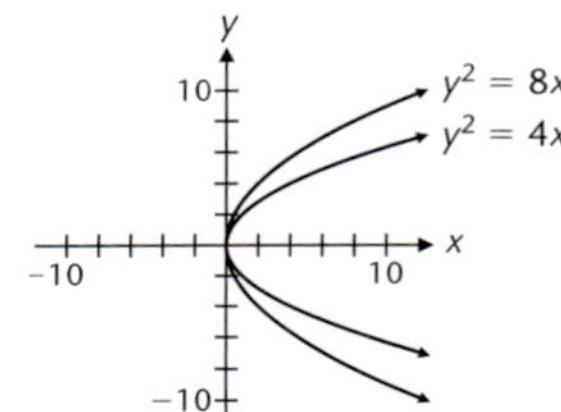

43. $x^2 = -2y$

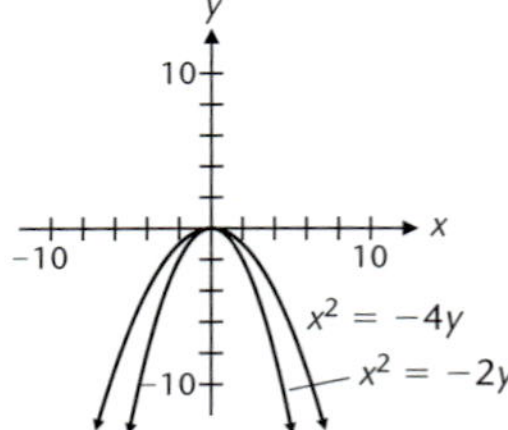

45. $x^2 - 4x - 12y - 8 = 0$ **47.** $y^2 + 8y - 8x + 48 = 0$
49. $x^2 = -200y$
51. (A) $y = 0.0025x^2$, $-100 \le x \le 100$ (B) 25 feet

Exercise 11.2

1. Foci: $F' = (-\sqrt{21}, 0)$, $F = (\sqrt{21}, 0)$; major axis length $= 10$; minor axis length $= 4$

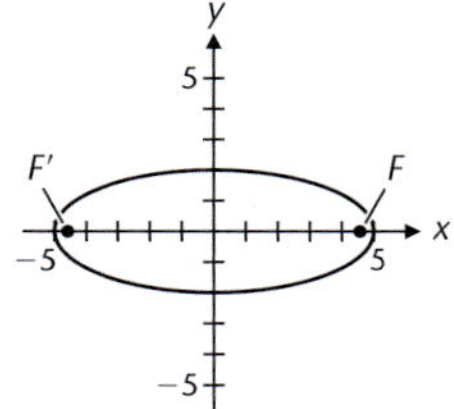

3. Foci: $F' = (0, -\sqrt{21})$, $F = (0, \sqrt{21})$; major axis length $= 10$; minor axis length $= 4$

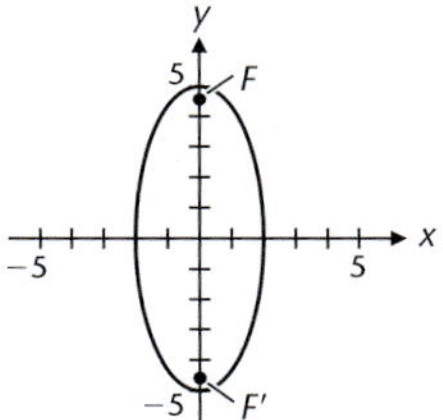

5. Foci: $F' = (-\sqrt{8}, 0)$, $F = (\sqrt{8}, 0)$; major axis length $= 6$; minor axis length $= 2$

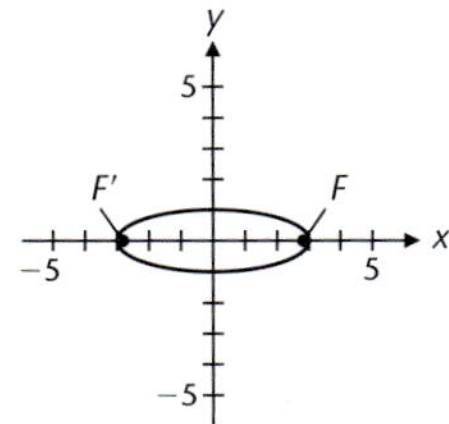

7. (b) **9.** (a)

11. Foci: $F' = (0, -4)$, $F = (0, 4)$; major axis length $= 10$; minor axis length $= 6$

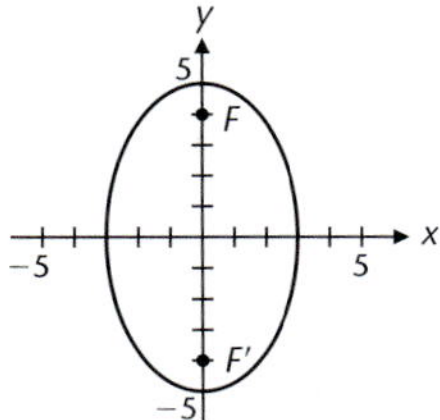

13. Foci: $F' = (0, -\sqrt{6})$, $F = (0, \sqrt{6})$; major axis length $= 2\sqrt{12} \approx 6.93$; minor axis length $= 2\sqrt{6} \approx 4.90$

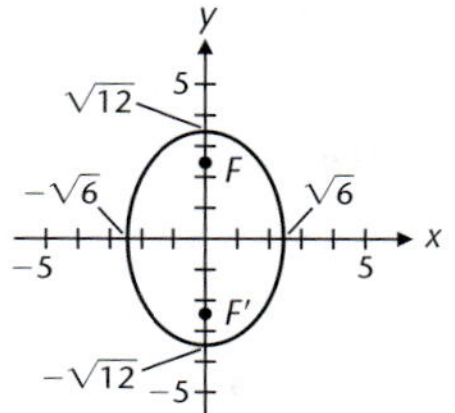

15. Foci: $F' = (-\sqrt{3}, 0)$, $F = (\sqrt{3}, 0)$; major axis length $= 2\sqrt{7} \approx 5.29$; minor axis length $= 4$

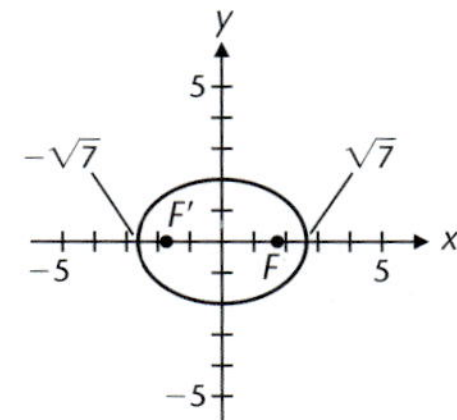

17. $\frac{x^2}{25} + \frac{y^2}{16} = 1$ **19.** $\frac{x^2}{9} + \frac{y^2}{36} = 1$ **21.** $\frac{x^2}{25} + \frac{y^2}{9} = 1$

23. $\frac{x^2}{64} + \frac{y^2}{121} = 1$ **25.** $\frac{x^2}{64} + \frac{y^2}{28} = 1$ **27.** $\frac{x^2}{100} + \frac{y^2}{170} = 1$

29. It does not pass the vertical line test.

31. $16x^2 + 9y^2 = 576$

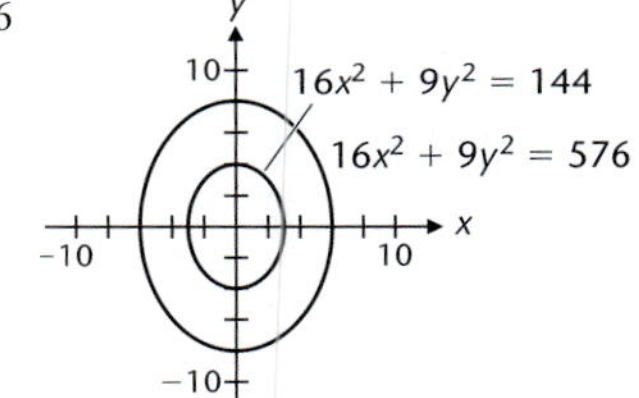

33. $x^2 + 4y^2 = 4$

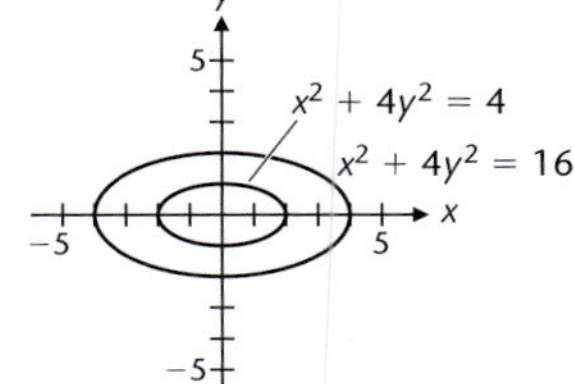

35. $\frac{x^2}{16} + \frac{y^2}{12} = 1$: ellipse

37. $\frac{x^2}{400} + \frac{y^2}{144} = 1$; 7.94 feet approximately

39. (A) $\frac{x^2}{576} + \frac{y^2}{15.9} = 1$ (B) 5.13 feet

Exercise 11.3

1. (d) **3.** (c)

5. Foci: $F' = (-\sqrt{13}, 0)$, $F = (\sqrt{13}, 0)$; transverse axis length = 6; conjugate axis length = 4

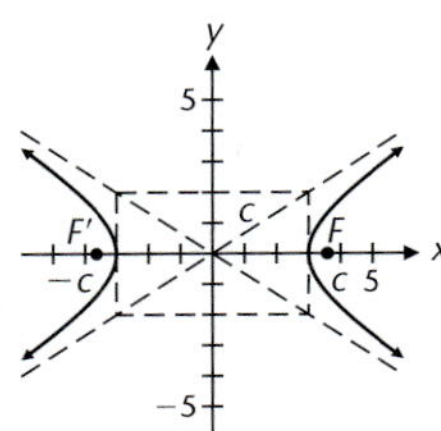

7. Foci: $F' = (0, -\sqrt{13})$, $F = (0, \sqrt{13})$; transverse axis length = 4; conjugate axis length = 6

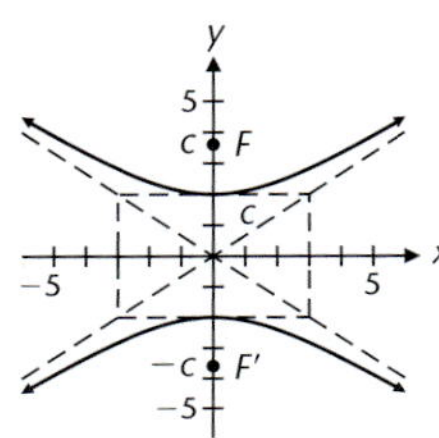

9. Foci: $F' = (-\sqrt{20}, 0)$, $F = (\sqrt{20}, 0)$; transverse axis length = 4; conjugate axis length = 8

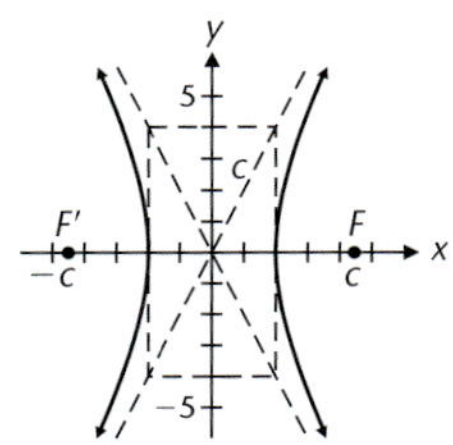

11. Foci: $F' = (0, -5)$, $F = (0, 5)$; transverse axis length = 8; conjugate axis length = 6

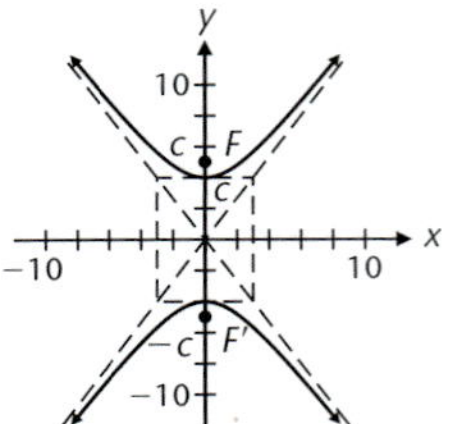

13. Foci: $F' = (-\sqrt{10}, 0)$, $F = (\sqrt{10}, 0)$; transverse axis length = 4; conjugate axis length = $2\sqrt{6} \approx 4.90$

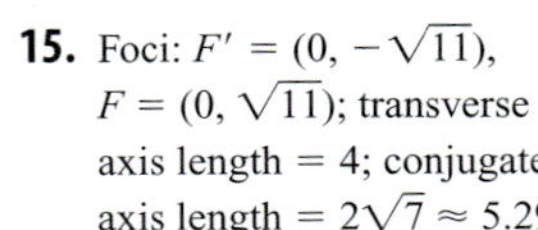
15. Foci: $F' = (0, -\sqrt{11})$, $F = (0, \sqrt{11})$; transverse axis length = 4; conjugate axis length = $2\sqrt{7} \approx 5.29$

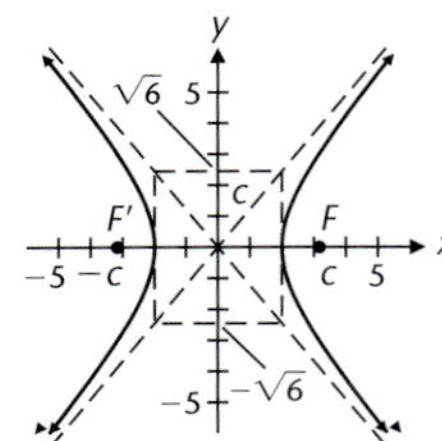

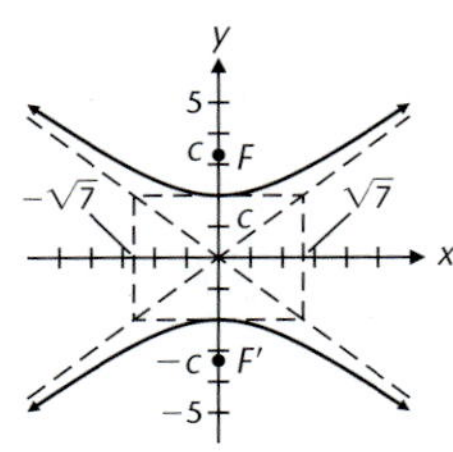

17. $\frac{x^2}{9} - \frac{y^2}{9} = 1$ **19.** $\frac{y^2}{16} - \frac{x^2}{16} = 1$ **21.** $\frac{x^2}{49} - \frac{y^2}{25} = 1$

23. $\frac{y^2}{144} - \frac{x^2}{81} = 1$ **25.** $\frac{x^2}{81} - \frac{y^2}{40} = 1$ **27.** $\frac{y^2}{151} - \frac{x^2}{49} = 1$

29. (A) Infinitely many; $\frac{x^2}{a^2} - \frac{y^2}{1 - a^2} = 1$ $(0 < a < 1)$

(B) Infinitely many; $\frac{x^2}{a^2} + \frac{y^2}{a^2 - 1} = 1$ $(a > 1)$

(C) One; $y^2 = 4x$

31. (A) $(2/\sqrt{3}, 1/\sqrt{3})$, $(-2/\sqrt{3}, -1/\sqrt{3})$

(B) No intersection points

The graphs intersect at $x = \pm 1/(\sqrt{1 - m^2})$ and $y = \pm m/(\sqrt{1 - m^2})$ for $-1 < m < 1$.

33. (A) No intersection points

(B) $(1/\sqrt{5}, 3/\sqrt{5})$, $(-1/\sqrt{5}, -3/\sqrt{5})$

The graphs intersect at $x = \pm 1/(\sqrt{m^2 - 4})$ and $y = \pm m/(\sqrt{m^2 - 4})$ for $m < -2$ or $m > 2$.

35. $\frac{x^2}{4} - \frac{y^2}{5} = 1$; hyperbola

37. $\frac{y^2}{16} - \frac{x^2}{8} = 1$; 5.38 feet above vertex **39.** $y = \frac{4}{3}\sqrt{x^2 + 30^2}$

Exercise 11.4

1. (A) $x' = x - 3; y' = y - 5$ (B) $x'^2 + y'^2 = 81$ (C) Circle

3. (A) $x' = x + 7, y' = y - 4$ (B) $\frac{x'^2}{9} + \frac{y'^2}{16} = 1$ (C) Ellipse

5. (A) $x' = x - 4, y' = y + 9$ (B) $y'^2 = 16x'$ (C) Parabola

7. (A) $x' = x + 8, y' = y + 3$ (B) $\frac{x'^2}{12} + \frac{y'^2}{8} = 1$ (C) Ellipse

9. (A) $\frac{(x - 3)^2}{9} - \frac{(y + 2)^2}{16} = 1$ (B) Hyperbola

11. (A) $\frac{(x + 5)^2}{5} + \frac{(y + 7)^2}{6} = 1$ (B) Ellipse

13. (A) $(x + 6)^2 = -24(y - 4)$ (B) Parabola

15. $\frac{(x - 2)^2}{9} + \frac{(y - 2)^2}{4} = 1$; ellipse

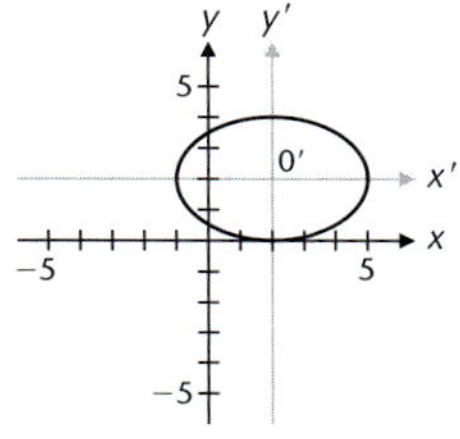

17. $(x + 4)^2 = -8(y - 2)$; parabola

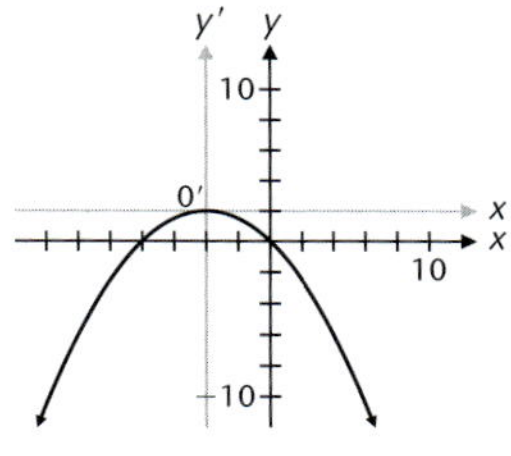

19. $(x + 6)^2 + (y + 5)^2 = 16$; circle

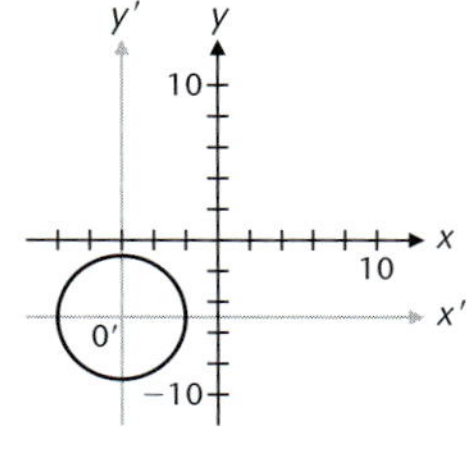

21. $\frac{(y - 3)^2}{9} - \frac{(x + 4)^2}{16} = 1$; hyperbola

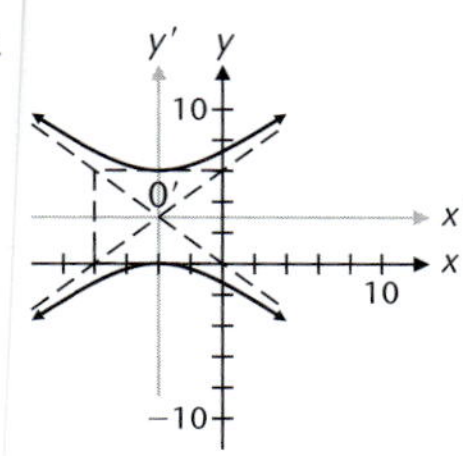

23. $(x - 1)^2 + (y + 2)^2 = 0$; the point $(1, -2)$ (a degenerate circle)

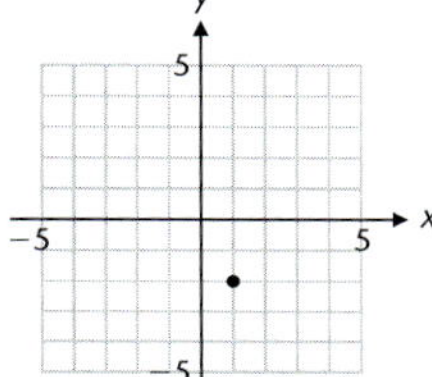

25. $(x + 4)^2 - 4(y - 1)^2 = 0$; the lines $y = 0.5x + 3$ and $y = -0.5x - 1$, intersecting at $(-4, 1)$ (a degenerate hyperbola)

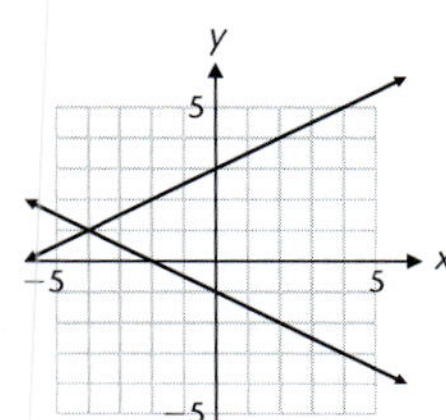

27. $h = \frac{-D}{2A}, k = \frac{D^2 - 4AF}{4AE}$

29. $x^2 - 4x + 4y - 16 = 0$

31. $x^2 + 4y^2 + 4x + 24y + 24 = 0$

33. $25x^2 + 9y^2 - 200x + 36y + 211 = 0$

35. $4x^2 - y^2 - 16x + 6y + 11 = 0$

37. $9x^2 + y^2 + 36x - 2y + 28 = 0$

39. $x^2 - 2y^2 - 2x + 8y - 8 = 0$

41. $F' = (-\sqrt{5} + 2, 2)$ and $F = (\sqrt{5} + 2, 2)$

43. $F = (-4, 0)$

45. $F' = (-4, -2), F = (-4, 8)$

Exercise 11.5

1. $(\sqrt{3}/2, -1/2), (1/2, \sqrt{3}/2), ((\sqrt{3} - 1)/2, (-1 - \sqrt{3})/2), ((-3\sqrt{3} + 4)/2, (3 + 4\sqrt{3})/2)$

3. $(\sqrt{2}/2, -\sqrt{2}/2), (\sqrt{2}/2, \sqrt{2}/2), (-3\sqrt{2}/2, -\sqrt{2}/2), (-\sqrt{2}, -2\sqrt{2})$

5. y' axis: $y = -\sqrt{3}x$; x' axis: $y = \frac{1}{\sqrt{3}}x$

7. y' axis: $y = -x$; x' axis: $y = x$

9. $x'^2 + y'^2 = 49$; circle

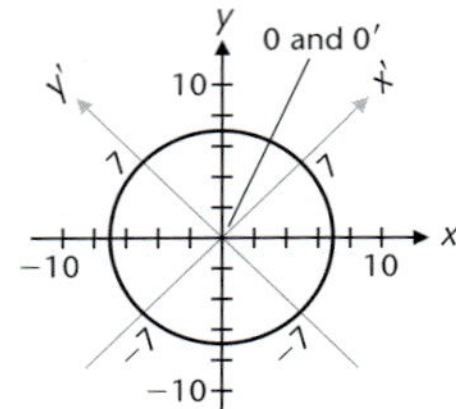

11. $\frac{x'^2}{4} + \frac{y'^2}{20} = 1$; ellipse

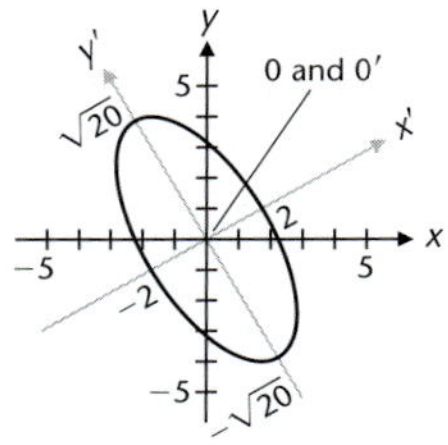

13. $\frac{y'^2}{4} - \frac{x'^2}{12} = 1$; hyperbola
$\theta = 45°$

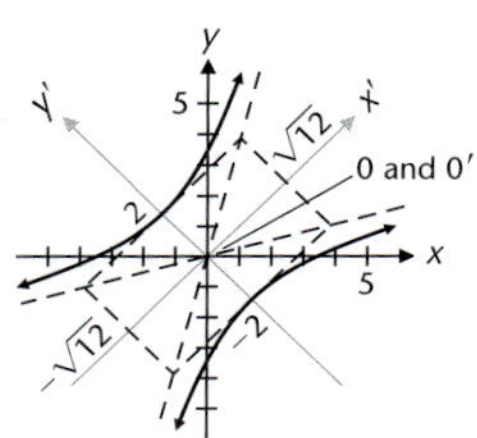

15. $\frac{x'^2}{9} + \frac{y'^2}{4} = 1$; ellipse,
$\theta = 63.43°$

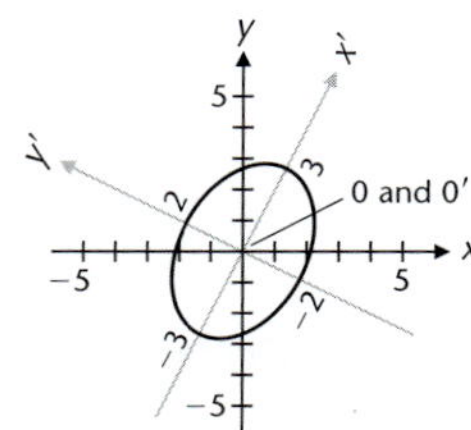

17. $y'^2 = 8x'$; parabola
$\theta = 30°$

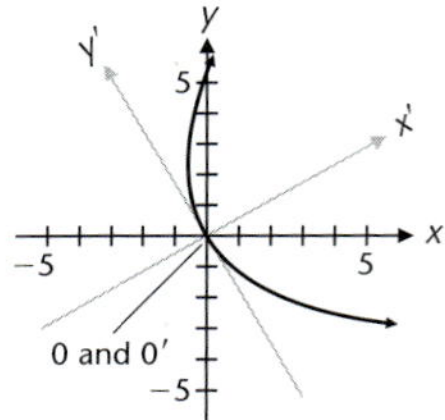

19. Ellipse

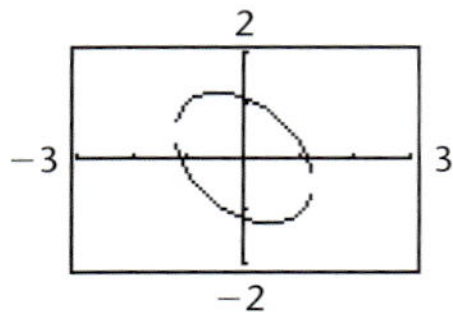

21. Hyperbola

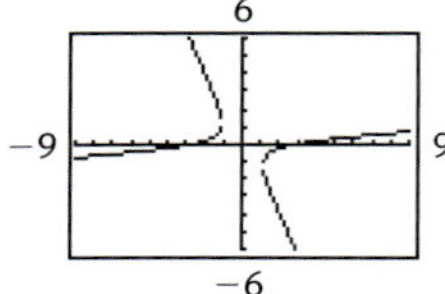

23. Parabola

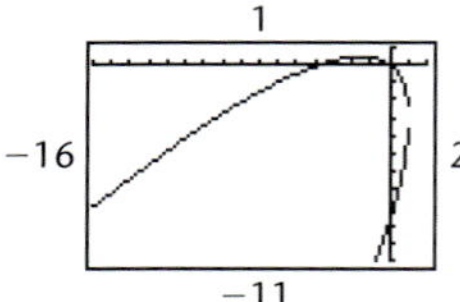

25. $\theta = 60°$; $x'^2 - 2\sqrt{3}x' + 2y' - 1 = 0$;
translate $0'$ to $(\sqrt{3}, 2)$; $x''^2 = -2y''$; parabola

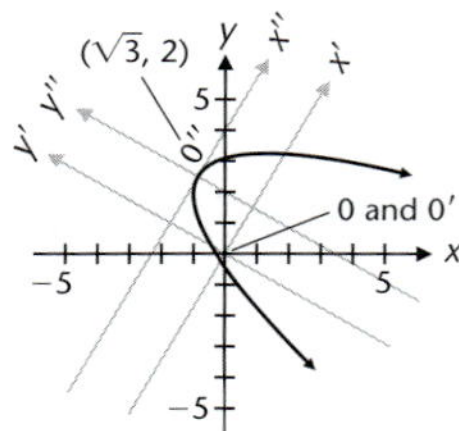

Exercise 11.6

1. $(-12, 5), (-12, -5)$ **3.** $(2, 4), (-2, -4)$
5. $(5, -5), (-5, 5)$ **7.** $(4 + 2\sqrt{3}, 1 + \sqrt{3}), (4 - 2\sqrt{3}, 1 - \sqrt{3})$
9. $(2, 4), (2, -4), (-2, 4), (-2, -4)$
11. $(1, 3), (1, -3), (-1, 3), (-1, -3)$
13. $(1 + \sqrt{5}, -1 + \sqrt{5}), (1 - \sqrt{5}, -1 - \sqrt{5})$
15. $(\sqrt{2}, \sqrt{2}), (-\sqrt{2}, -\sqrt{2}), (2, 1), (-2, -1)$
17. $(2, 2i), (2, -2i), (-2, 2i), (-2, -2i)$
19. $(2, \sqrt{2}), (2, -\sqrt{2}), (-1, i), (-1, -i)$
21. $(3, 0), (-3, 0), (\sqrt{5}, 2), (-\sqrt{5}, 2)$
23. $(2, 1), (-2, -1), (i, -2i), (-i, 2i)$ **25.** $(-1, 4), (3, -4)$
27. $(0, 0), (3, 6)$ **29.** $(1, 4), (4, 1)$ **31.** $(-1, 3), (4, 8)$
33. (A) The lines are tangent to the circle.

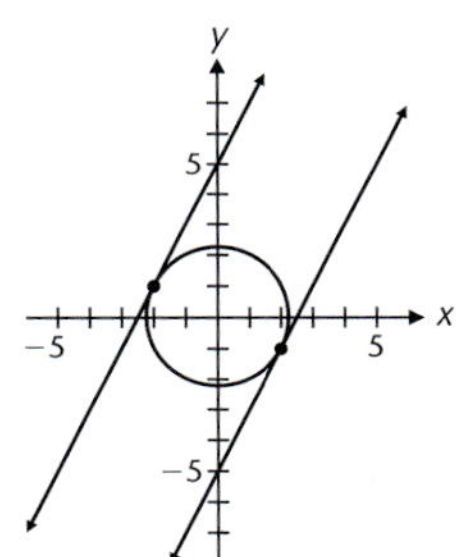

(B) $b = 5$, intersection point is $(2, -1)$; $b = -5$, intersection point is $(-2, 1)$
(C) The line $x + 2y = 0$ is perpendicular to all the lines in the family and intersects the circle at the intersection points found in part B. Solving the system $x^2 + y^2 = 5, x + 2y = 0$ would determine the intersection points.
35. (A) $b < 1$ (B) $b = 1$ (C) $b > 1$
37. $(-5, -\frac{3}{5}), (-\frac{3}{2}, -2)$ **39.** $(0, -1), (-4, -3)$
41. $(2, 2), (-2, -2), (\sqrt{2}, -\sqrt{2}), (-\sqrt{2}, \sqrt{2})$
43. $(-3, 1), (3, -1), (-i, i), (i, -i)$
45. $(-1.41, -0.82), (-0.13, 1.15), (0.13, -1.15), (1.41, 0.82)$
47. $(-1.66, -0.84), (-0.91, 3.77), (0.91, -3.77), (1.66, 0.84)$
49. $(-2.96, -3.47), (-0.89, -3.76), (1.39, 4.05), (2.46, 4.18)$
51. $\frac{1}{2}(3 - \sqrt{5}), \frac{1}{2}(3 + \sqrt{5})$ **53.** 5 inches and 12 inches
55. 6 by 4.5 inches **57.** 52.3 inches by 29.4 inches
59. 122 miles east and 158 miles north of station A
61. 22 by 26 feet
63. Boat A: 30 miles per hour; boat B: 25 miles per hour

Chapter 11 Review

1. Foci: $F' = (-4, 0), F = (4, 0)$; major axis length $= 10$; minor axis length $= 6$ *(11.2)*

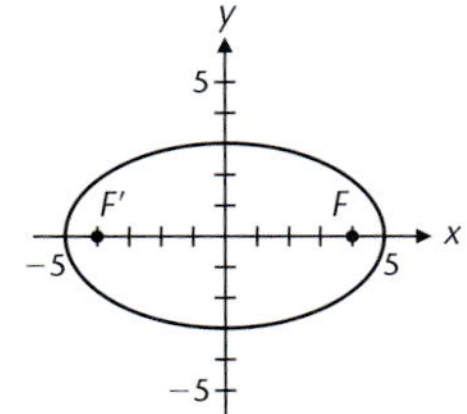

2. *(11.1)*

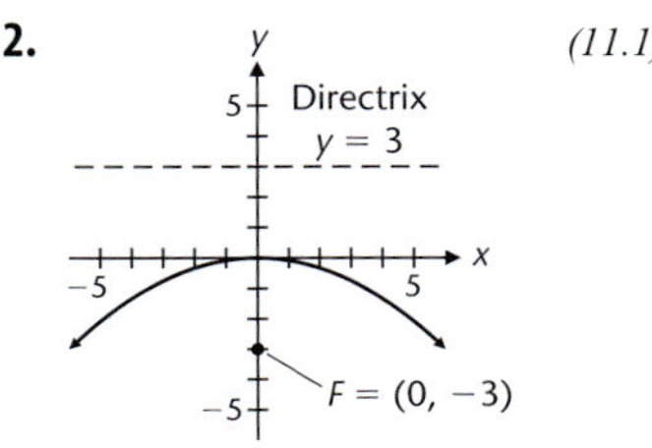

3. Foci: $F' = (0, -\sqrt{34}), F = (0, \sqrt{34})$; transverse axis length $= 6$; conjugate axis length $= 10$ *(11.3)*

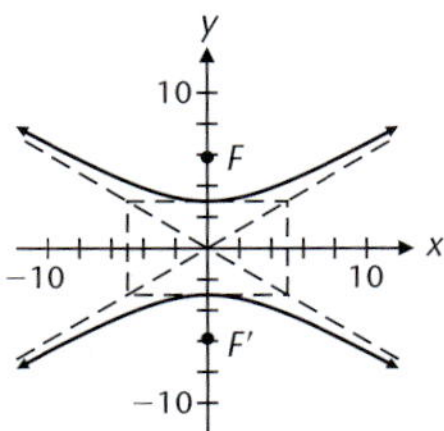

4. (A) $\frac{(y + 2)^2}{25} - \frac{(x - 4)^2}{4} = 1$ (B) Hyperbola *(11.4)*
5. (A) $(x + 5)^2 = -12(y + 4)$ (B) Parabola *(11.4)*
6. (A) $\frac{(x - 6)^2}{9} + \frac{(y - 4)^2}{16} = 1$ (B) Ellipse *(11.4)*
7. (A) $((3\sqrt{3} + 4)/2, (-3 + 4\sqrt{3})/2)$
(B) $(7\sqrt{2}/2, \sqrt{2}/2)$
(C) $((3 + 4\sqrt{3})/2, (-3\sqrt{3} + 4)/2)$ *(11.5)*
8. y' axis: $y = (2 - \sqrt{3})x$;
x' axis: $y = -(2 + \sqrt{3})x$ *(11.5)*

9. (5, −3), (−1, 3) *(11.5)* **10.** (1, −1), (1.4, −0.2) *(11.6)*
11. (1, 3), (1, −3), (−1, 3), (−1, −3) *(11.6)*
12. $y^2 = -x$ *(11.1)* **13.** $\frac{x^2}{9} + \frac{y^2}{25} = 1$ *(11.2)*
14. $\frac{y^2}{9} - \frac{x^2}{16} = 1$ *(11.3)*
15. (−4, 2), (4, −2) *(11.6)* **16.** (−3, 3.2), (3, 3.2) *(11.6)*
17. (1, 3), (1, −3), (−1, 3), (−1, −3) *(11.6)*
18. $(2, \sqrt{2}), (2, -\sqrt{2}), (-1, i), (-1, -i)$ *(11.6)*
19. (1, −2), (−1, 2), (2, −1), (−2, 1) *(11.6)*
20. $(2, -2), (-2, 2), (\sqrt{2}, \sqrt{2}), (-\sqrt{2}, -\sqrt{2})$ *(11.6)*
21. $\frac{(x+3)^2}{4} + \frac{(y-2)^2}{16} = 1$; ellipse *(11.4)*

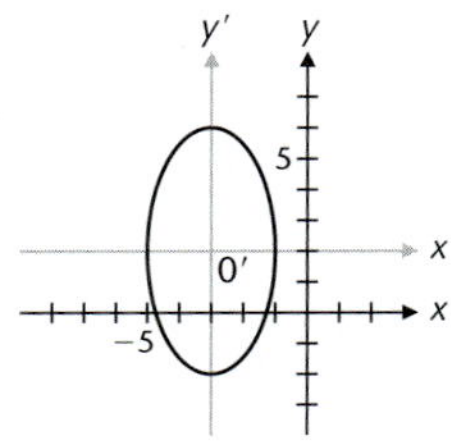

22. $(x-2)^2 = 4(2)(y+3)$; parabola *(11.4)*

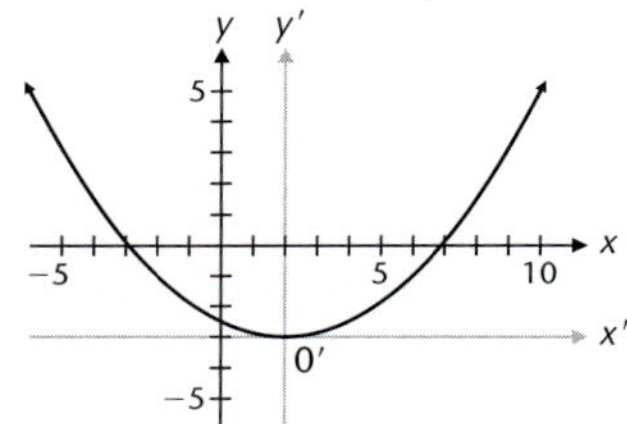

23. $\frac{(x+3)^2}{9} - \frac{(y+2)^2}{4} = 1$; hyperbola *(11.4)*

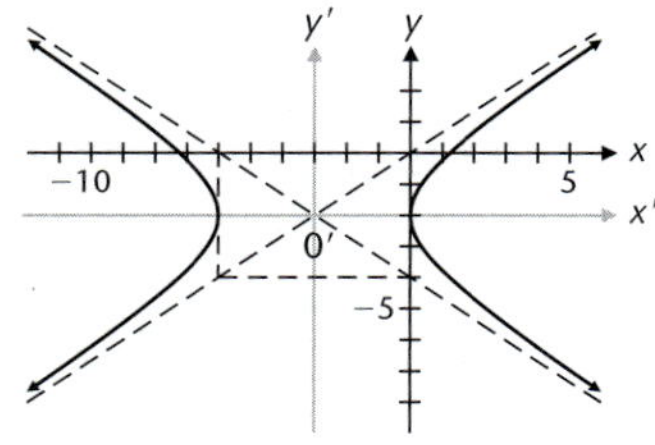

24. $\frac{x'^2}{20} + \frac{y'^2}{4} = 1$; ellipse *(11.5)*

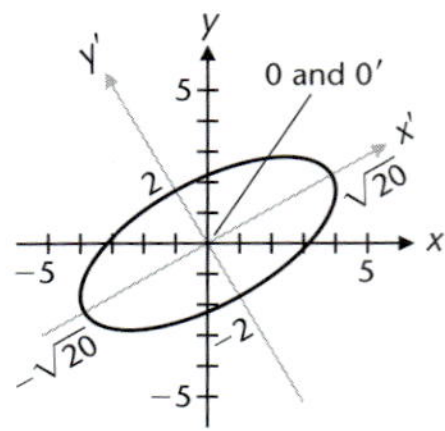

25. $\frac{y'^2}{9} - \frac{x'^2}{4} = 1$; hyperbola; $\theta = 45°$ *(11.5)*

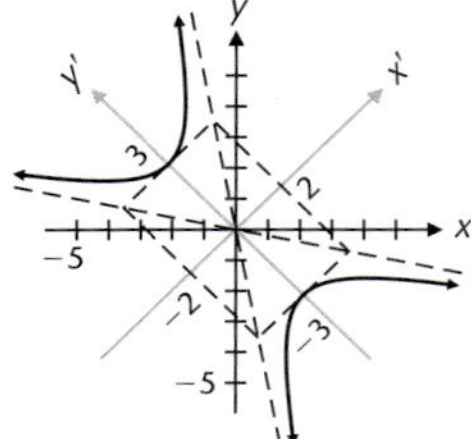

26. Ellipse *(11.5)*

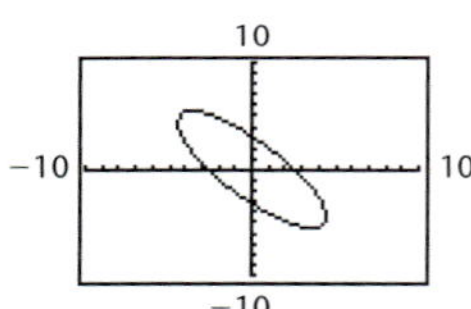

27. $m = 0.2$; $x^2 = 50y$ is a magnification by a factor 50 of $x^2 = y$ *(11.1)*
28. $(y-4)^2 = -8(x-4)$ or $y^2 - 8y + 8x - 16 = 0$ *(11.1)*
29. $\frac{x^2}{4} - \frac{y^2}{12} = 1$; hyperbola *(11.3)*
30. $\frac{x^2}{36} + \frac{y^2}{20} = 1$; ellipse *(11.2)*
31. $F' = (-3, -\sqrt{12} + 2)$ and $F = (-3, \sqrt{12} + 2)$ *(11.4)*
32. $F = (2, -1)$ *(11.4)*
33. $F' = (-\sqrt{13} - 3, -2)$ and $F = (\sqrt{13} - 3, -2)$ *(11.4)*
34. $(2, 2), (-2, -2), (\frac{4}{7}\sqrt{7}, -\frac{2}{7}\sqrt{7}), (-\frac{4}{7}\sqrt{7}, \frac{2}{7}\sqrt{7})$ *(11.6)*
35. (2.09, 2.50), (3.67, −1.92) *(11.4)*
36. (−2.16, −0.37), (−1.09, 5.59), (1.09, −5.59), (2.16, 0.37) *(11.6)*
37. 4 feet *(11.1)* **38.** $\frac{x^2}{5^2} + \frac{y^2}{3^2} = 1$ *(11.2)*
39. 4.72 feet deep *(11.3)*
40. 110 miles east and 141 miles north of station A *(11.6)*

Cumulative Review for Chapters 10 and 11

1. (A) Arithmetic (B) Geometric (C) Neither (D) Geometric (E) Arithmetic *(10.3)*

2. (A) 10, 50, 250, 1,250 (B) $a_8 = 781{,}250$ (C) $S_8 = 976{,}560$ *(10.3)*

3. (A) 2, 5, 8, 11 (B) $a_8 = 23$ (C) $S_8 = 100$ *(10.3)*

4. (A) 100, 94, 88, 82 (B) $a_8 = 58$ (C) $S_8 = 632$ *(10.3)*

5. (A) 40,320 (B) 992 (C) 84 *(10.4)*

6. (A) 21 (B) 21 (C) 42 *(10.4, 10.5)*

7. Foci: $F' = (-\sqrt{61}, 0)$, $F = (\sqrt{61}, 0)$; transverse axis length = 12; conjugate axis length = 10 *(11.3)*

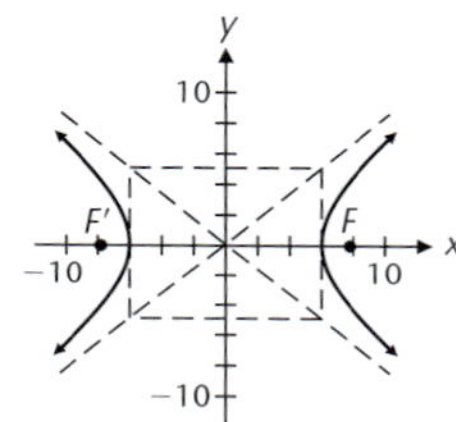

8. Foci: $F' = (-\sqrt{11}, 0)$, $F = (\sqrt{11}, 0)$; major axis length = 12; minor axis length = 10 *(11.2)*

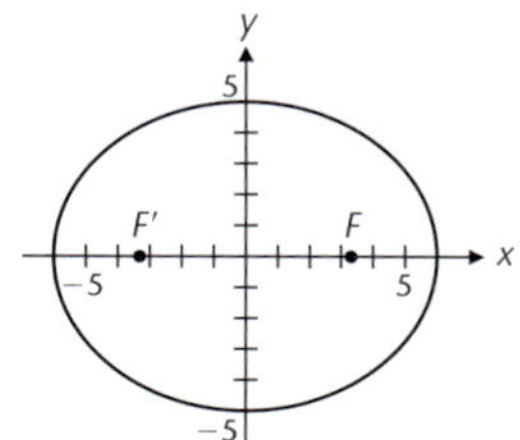

9. *(11.1)*

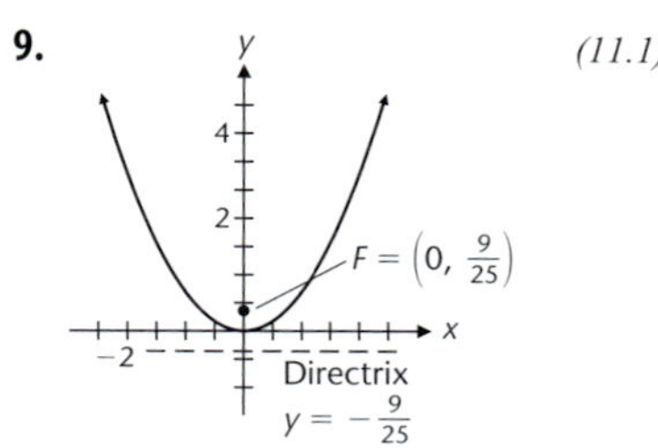

10. Ellipse *(11.5)*

11. (1, 1), (−1/5, −7/5) *(11.6)*

12. (A) 8 combined outcomes: (B) $2 \cdot 2 \cdot 2 = 8$ *(10.5)*

HHH
HHT
HTH
HTT
THH
THT
TTH
TTT

13. (A) $4 \cdot 3 \cdot 2 \cdot 1 = 24$ (B) $P_{4,4} = 4! = 24$ *(10.5)*

14. $\dfrac{C_{13,3}}{C_{52,3}} \approx .0129$ *(10.5)* **15.** $\dfrac{1}{P_{10,4}} \approx .0002$; $\dfrac{1}{C_{10,4}} \approx .0048$ *(10.5)*

16. .62 *(10.5)*

17. P_1: $1 = 1(2 \cdot 1 - 1)$; P_2: $1 + 5 = 2(2 \cdot 2 - 1)$; P_3: $1 + 5 + 9 = 3(2 \cdot 3 - 1)$ *(10.2)*

18. P_1: $1^2 + 1 + 2 = 4$ is divisible by 2; P_2: $2^2 + 2 + 2 = 8$ is divisible by 2; P_3: $3^2 + 3 + 2 = 14$ is divisible by 2 *(10.2)*

19. P_k: $1 + 5 + 9 + \cdots + (4k - 3) = k(2k - 1)$; P_{k+1}: $1 + 5 + 9 + \cdots + (4k - 3) + (4k + 1) = (k + 1)(2k + 1)$ *(10.2)*

20. P_k: $k^2 + k + 2 = 2r$ for some integer r; P_{k+1}: $(k + 1)^2 + (k + 1) + 2 = 2s$ for some integer s *(10.2)*

21. $y = -2x^2$ *(11.1)* **22.** $\dfrac{x^2}{25} + \dfrac{y^2}{16} = 1$ *(11.2)*

23. $\dfrac{x^2}{64} - \dfrac{y^2}{25} = 1$ *(11.3)*

24. (1, 1), (−1, −1), $(\sqrt{3}, \sqrt{3}/3)$, $(-\sqrt{3}, -\sqrt{3}/3)$ *(11.6)*

25. $(0, i)$, $(0, -i)$, (1, 1), (−1, −1) *(11.6)*

26. $\dfrac{y'^2}{3} - x'^2 = 1$; hyperbola; $\theta = 60°$ *(11.5)*

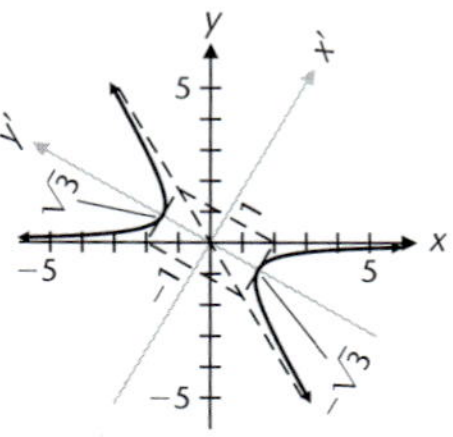

27. $x'^2 = 4y'$; parabola; $\theta = 45°$ *(11.5)*

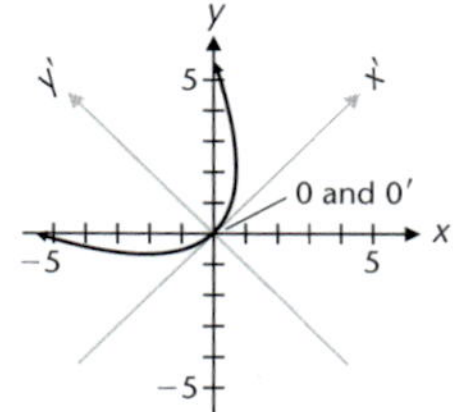

28. $(-1.35, 0.28), (-0.87, -1.60), (0.87, 1.60), (1.35, -0.28)$ *(11.6)*
29. $1 + 4 + 27 + 256 + 3{,}125 = 3{,}413$ *(10.1)*
30. $\sum_{k=1}^{6} (-1)^{k+1} \frac{2^k}{(k+1)!}$ *(10.1)* **31.** 81 *(10.3)*
32. 360; 1,296; 750 *(10.4)* **33.** $\frac{C_{10,3}}{C_{12,5}} = \frac{5}{33} = .\overline{15}$ *(10.5)*
34. (A) .365 (B) $\frac{1}{3}$ *(10.5)* **35.** $n = 22$ *(10.3)*
36. (A) 6,375,600 (B) 53,130 (C) 53,130 *(10.4, 10.6)*
37. $a^6 + 3a^5b + \frac{15}{4}a^4b^2 + \frac{5}{2}a^3b^3 + \frac{15}{16}a^2b^4 + \frac{3}{16}ab^5 + \frac{1}{64}b^6$ *(10.6)*
38. $153{,}090x^6y^4$; $-3{,}240x^3y^7$ *(10.6)* **41.** 61,875 *(10.3)*
42. $\frac{27}{11}$ *(10.3)* **43.** $a_{27} = 0.236$; 8 terms *(10.6)*
44. $4(x + 3) = (y - 2)^2$; parabola *(11.1)*

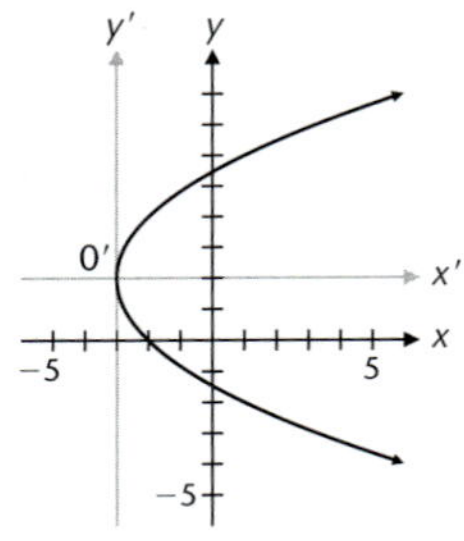

45. $\frac{(y + 2)^2}{4} - \frac{(x + 1)^2}{16} = 1$; hyperbola *(11.3)*

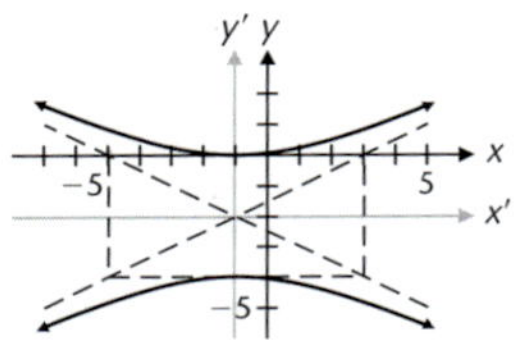

46. $\frac{(x - 2)^2}{9} + \frac{(y + 3)^2}{4} = 1$; ellipse *(11.2)*

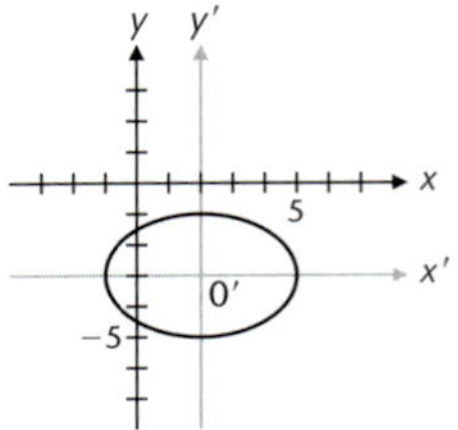

47. 10^9; 3,628,800 zip codes *(10.1)* **49.** $\frac{2}{5}$; $\frac{2}{5}$ *(10.5)*
50. $x^6 - 12ix^5 - 60x^4 + 160ix^3 + 240x^2 - 192ix - 64$ *(10.6)*
51. $x^2 - 12x + 4y + 28 = 0$ *(11.1)*
52. $\pm 2\sqrt{3}$ *(11.2)* **53.** 8 *(11.3)* **54.** $C_{7,3} = 35$ *(10.4)*
57. $x^2 - 8y^2 - 2x - 8y + 17 = 0$; hyperbola *(11.3)*
58. $1 - \frac{C_{8,3}}{C_{12,3}} = \frac{41}{55} = .7\overline{45}$ *(10.5)*
59. \$6,000,000 *(10.3)* **60.** 4 meters by 8 meters *(11.5)*
61. 4 inches *(11.1)* **62.** 32 feet, 14.4 feet *(11.2)*
63. (A) .13 (B) .17 (C) .32 *(10.5)*

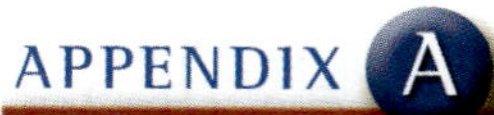

Exercise A.1

1. $x = 7$ **3.** $s = 2.5$ **5.** $m = 5$

7. $-8 \le x \le 7$;

9. $-6 \le x < 6$;

11. $x \ge -6$;

13. $(-2, 6]$;

15. $(-7, 8)$;

17. $(-\infty, -2]$;

19. $[-7, 2)$; $-7 \le x < 2$ **21.** $(-\infty, 0]$; $x \le 0$ **23.** $>, >$

25. $>, >$ **27.** $>, <$ **29.** $<, <$

31. $x < 5$; $(-\infty, 5)$;

33. $x \ge 3$; $[3, \infty)$;

35. $N < -8$; $(-\infty, -8)$;

37. $m > 3$; $(3, \infty)$;

39. $B \ge -4$; $[-4, \infty)$; **41.** 9 **43.** 10

45. 8 **47.** $(-5, 7]$; $-5 < x \le 7$;

49. $(2, 4)$; $2 < x < 4$;

51. $(-\infty, \infty)$; $-\infty < x < \infty$;

53. $(-\infty, -1) \cup [3, 7)$; $x < -1$ or $3 \le x < 7$;

55. $(1, 5)$; $1 < x < 5$;

57. $(-\infty, 6]$; $x \le 6$;

59. $q < -14$; $(-\infty, -14)$;

61. $x \ge 4.5$; $[4.5, \infty)$;

63. $-20 \le x \le 20$; $[-20, 20]$;

65. $-8 \le x < -3$; $[-8, -3)$;

67. $-14 < x \le 11$; $(-14, 11]$;

69. (A) F (B) T (C) T

Exercise A.2

1.

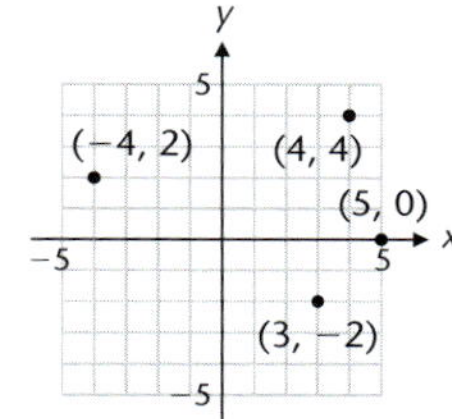

3.

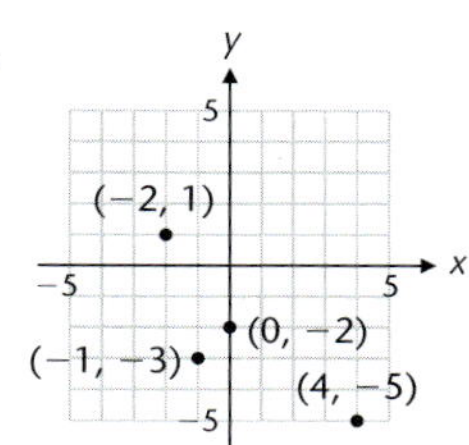

5. $A = (2, 4)$, $B = (3, -1)$, $C = (-4, 0)$, $D = (-5, 2)$

7. $A = (-3, -3)$, $B = (0, 4)$, $C = (-3, 2)$, $D = (5, -1)$

9.

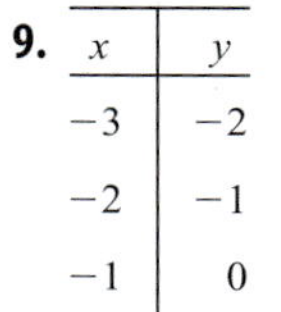
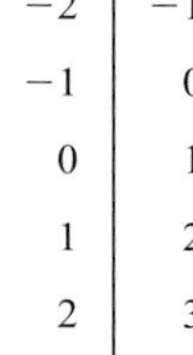

x	y
−3	−2
−2	−1
−1	0
0	1
1	2
2	3
3	4

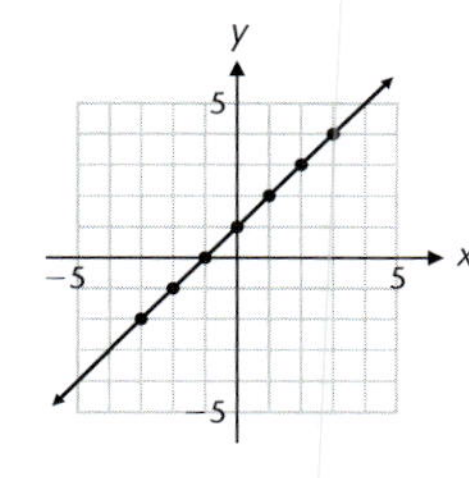

11.

x	y
−3	4
−2	−1
−1	−4
0	−5
1	−4
2	−1
3	4

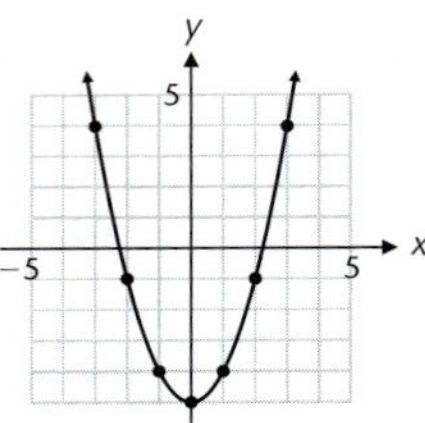

13.

x	y
−3	−4.5
−2	−1
−1	1.5
0	3
1	3.5
2	3
3	1.5

15. (A) 6 (B) −5 (C) −1 (D) 8 (E) −5 (F) 5

17. (A) 6 (B) 4 (C) 4 (D) 8 (E) −8, 0, 6 (F) −7, −2, 7

19. (A)

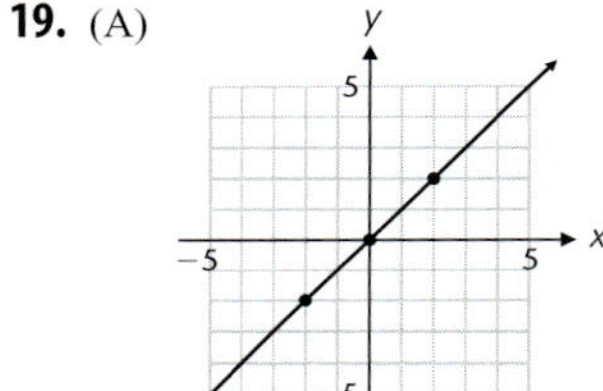

(B)

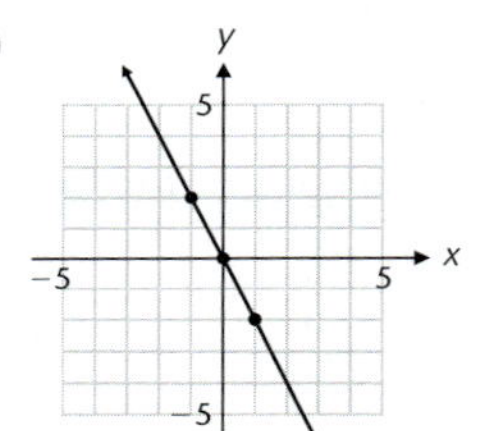

(C)

x	−2	−1	0	1	2
y	−2	2	0	−2	2

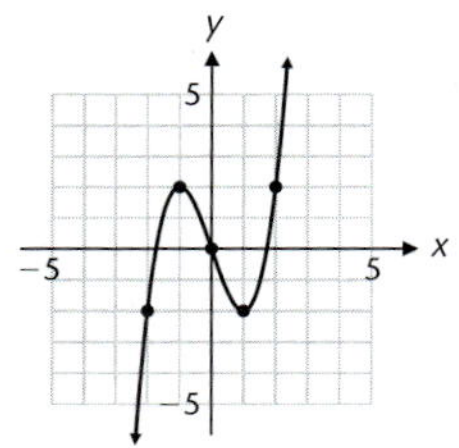

21.

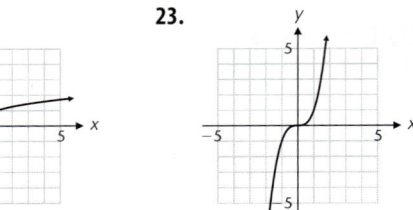

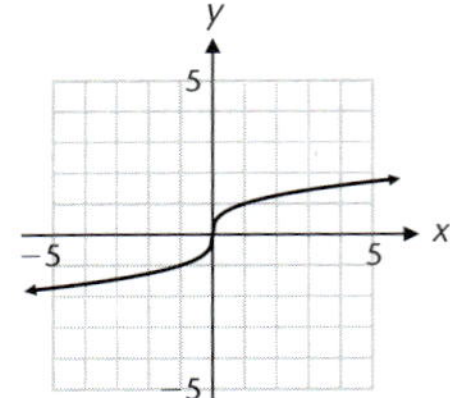

23.

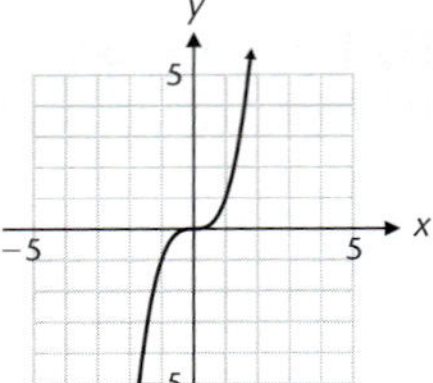

25.

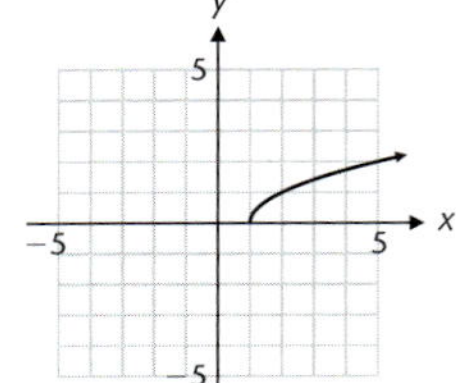

27.

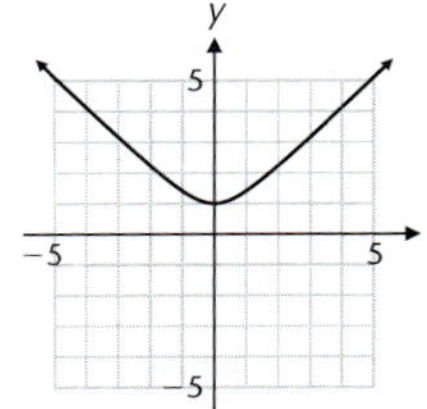

29.

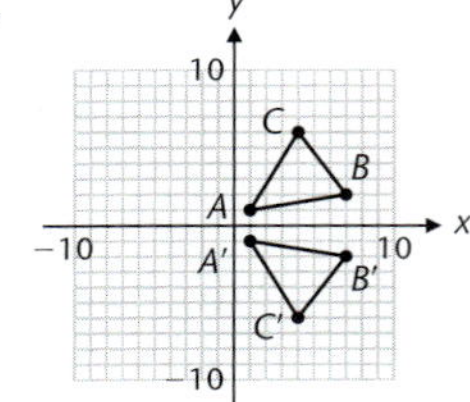

31.

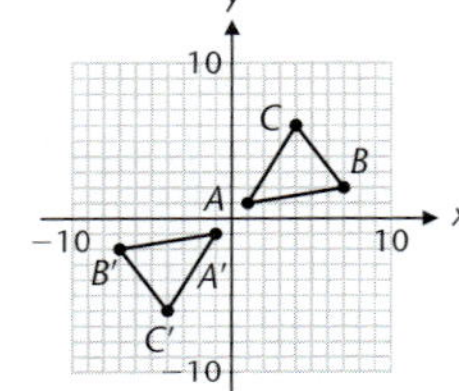

33.

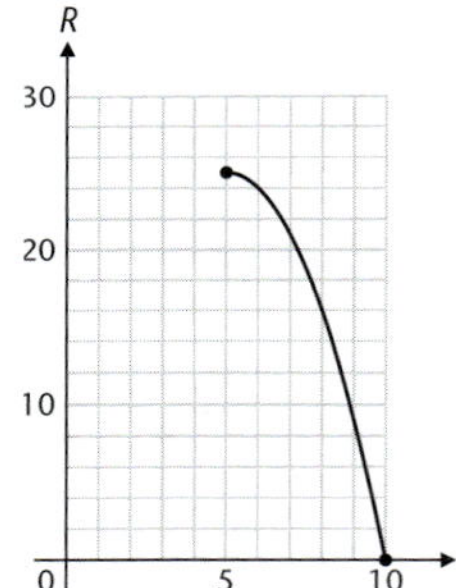

35.

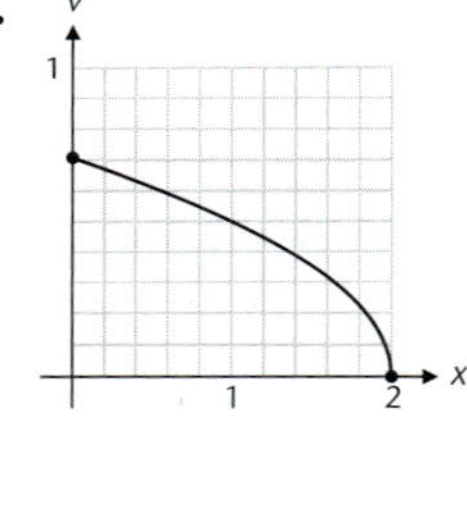

37. (A) 3,000 cases (B) Demand decreases by 400 cases (C) Demand increases by 600 cases

39. (A) 53°F (B) 68°F at 3 P.M. (C) 1 A.M., 7 A.M., 11 P.M.

Exercise A.3

1. $d = 5, M = (2.5, 2)$ **3.** $d = 13, M = (2.5, 4)$
5. $d = \sqrt{145}, M = (-1.5, 0)$ **7.** $d = \sqrt{68}, M = (5, 2)$
9. $x^2 + y^2 = 49$ **11.** $(x - 2)^2 + (y - 3)^2 = 36$
13. $(x + 4)^2 + (y - 1)^2 = 7$ **15.** $(x + 3)^2 + (y + 4)^2 = 2$
17. $(x + 4)^2 = x^2 + 8x + 16$ **19.** $(x - \frac{3}{2})^2 = x^2 - 3x + \frac{9}{4}$
21. $x^2 + y^2 = 4$ **23.** $(x - 1)^2 + y^2 = 1$

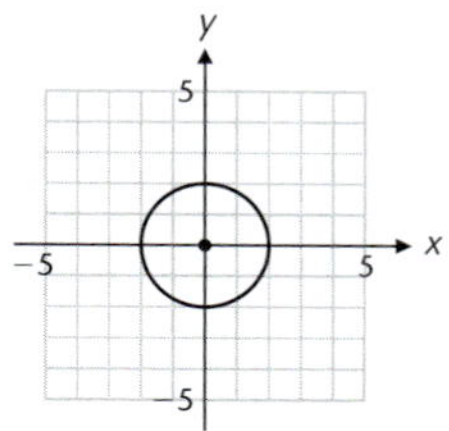

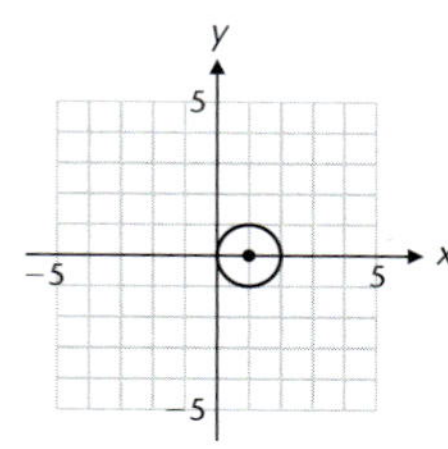

25. $(x + 2)^2 + (y - 1)^2 = 9$

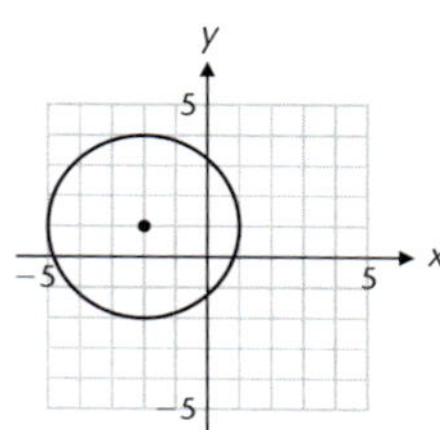

27. (A) -5 (B) 9 (C) $\sqrt{18}$
29. The set of all points that are two units from the point (0, 2). $x^2 + (y - 2)^2 = 4$
31. The set of all points that are four units from the point (1, 1). $(x - 1)^2 + (y - 1)^2 = 16$
33. (2.65, 1.75) **35.** $(-35, -14)$ **37.** $(-18, -16)$
39. Center: (0, 1); radius: 1 **41.** Center: $(1, -3)$; radius: 4
43. Center: $(-0.5, -1.5)$; radius: $\sqrt{0.5}$
45. $\sqrt{32.5}$ **47.** 18.11 **51.** $(x - 4)^2 + (y - 2)^2 = 34$
53. $(x - 2)^2 + (y - 2)^2 = 50$
55. (A) $A = (0, 0), B = (0, 13.5), C = (0, 27), D = (60, 27),$ $E = (78, 27), F = (78, 13.5), G = (78, 0)$
(B) 62 feet, 79 feet
57. 2.5 feet
59. (A) $(x + 12)^2 + (y + 5)^2 = 26^2$; center: $(-12, -5)$; radius: 26

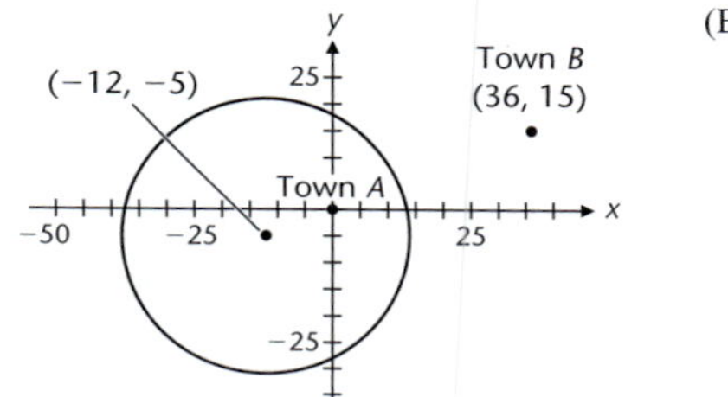

(B) 13.5 miles

APPENDIX B

Exercise B.1

1. <u>123, 005</u> **3.** <u>20,04</u>0 **5.** <u>6.0</u> **7.** <u>80.000</u>

9. 0.0<u>12</u> **11.** 0.000 <u>960</u> **13.** 3.08 **15.** 924, 000
17. 23.6 **19.** 2.82×10^3 **21.** 6.78×10^{-4} **23.** 30 feet

Exercise B.2

1. $A = 2, B = 5$ **3.** $A = 7, B = -2$
5. $A = 1, B = 2, C = 3$ **7.** $A = 2, B = 1, C = 3$
9. $A = 0, B = 2, C = 2, D = -3$ **11.** $\frac{-4}{x + 2} + \frac{3}{x - 4}$
13. $\frac{3}{3x + 4} - \frac{1}{2x - 3}$ **15.** $\frac{2}{x} - \frac{1}{x - 3} - \frac{3}{(x - 3)^2}$
17. $\frac{2}{x} + \frac{3x - 1}{x^2 + 2x + 3}$ **19.** $\frac{2x}{x^2 + 2} + \frac{3x + 5}{(x^2 + 2)^2}$

21. $x - 2 + \frac{3}{x - 2} - \frac{2}{x - 3}$ **23.** $\frac{2}{x - 3} + \frac{2x + 5}{x^2 + 3x + 3}$
25. $\frac{2}{x - 4} - \frac{1}{x + 3} + \frac{3}{(x + 3)^2}$
27. $\frac{2}{x - 2} - \frac{3}{(x - 2)^2} - \frac{2x}{x^2 - x + 1}$
29. $x + 2 - \frac{2}{x + 2} + \frac{1}{2x - 1} + \frac{x - 1}{2x^2 - x + 1}$

Exercise B.3

1. Zero variations in $P(x)$ and one in $P(-x)$

3. One variation in $P(x)$ and zero in $P(-x)$

5. One variation in $P(x)$ and one in $P(-x)$

7. One real zero and two imaginary zeros

9. One real zero and two imaginary zeros

11. Two real zeros and two imaginary zeros

13.

+	−	I
2	1	0
0	1	2

15.

+	−	I
2	2	0
2	0	2
0	2	2
0	0	4

17.

+	−	I
4	1	0
2	1	2
0	1	4

19.

+	−	I
1	1	4

21.

+	−	I
0	1	6

23. There are no positive zeros. There are either two negative zeros or two imaginary zeros.

25. There is one negative zero and one positive zero.

27.

+	−	I
4	2	0
4	0	2
2	2	2
2	0	4
0	2	4
0	0	6

29.

+	−	I
2	1	4
0	1	6

31.

+	−	I
1	1	6

33. There is one negative zero and there are two imaginary zeros.

35. There is one negative zero. There are either two positive or two imaginary zeros.

Exercise B.4

5. The values of x and y

7. The values of x and y

9. $y = -2x - 2$; straight line

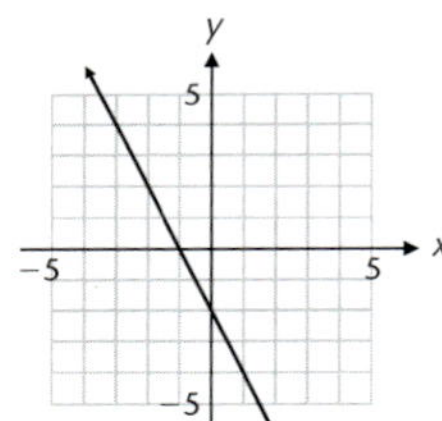

11. $y = -2x - 2, x \le 0$; a ray (part of a straight line)

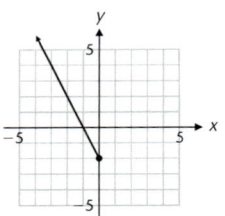

13. $y = -\frac{2}{3}x$; straight line

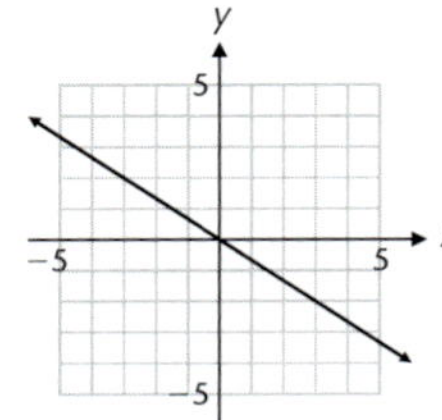

15. $y^2 = 4x$; parabola

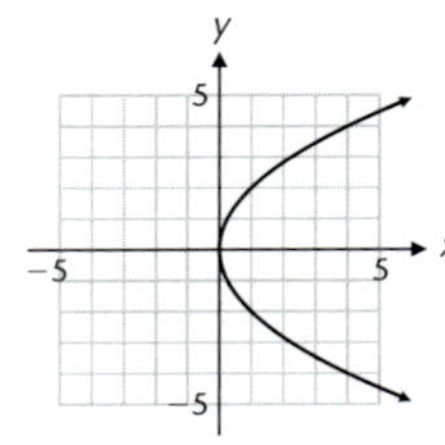

17. $y^2 = 4x, y \ge 0$; parabola (upper half)

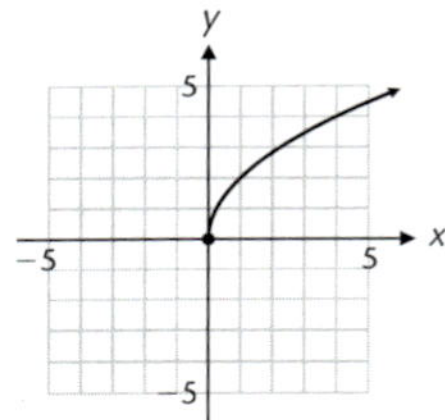

19. $y = -2x$; line

21. $y^2 = x + 1, y \ge 0, x \ge -1$; parabola (upper half)

23. $4x^2 + y^2 = 64, 0 \le x \le 4, 0 \le y \le 8$; ellipse (first quadrant portion)

25. $x^2 - y^2 = 2, x \le -\sqrt{2}, y \le 0$; hyperbola (third quadrant portion)

27. $\frac{x^2}{9} + \frac{y^2}{16} = 1$; ellipse

29. $(x - 2)^2 + (y - 3)^2 = 4$; circle

31. $x = t, y = \frac{At^2 + Dt + F}{-E}, -\infty < t < \infty$; parabola

33. $\frac{(x-3)^2}{36} + \frac{(y-2)^2}{16} = 1$; ellipse with center (3, 2)

35. $\frac{(y+1)^2}{25} - \frac{(x+3)^2}{4} = 1$; hyperbola with center $(-3, -1)$

37. $y^2 - x^2 = 8, x \geq 1, y \geq 3$; part of a hyperbola

39. $x^2 + y^2 = 4, 0 < x \leq 2, -2 < y < 2$; semicircle (excluding the end points)

41. $x^2 + y^2 = 2x, x \neq 0$ or $(x-1)^2 + y^2 = 1, x \neq 0$; circle (note hole at origin)

43. $\frac{(y+1)^2}{25} - \frac{(x-4)^2}{9} = 1$; hyperbola with center $(4, -1)$;
$x = 4 + 3 \tan t, y = -1 + 5 \sec t, 0 \leq t \leq 2\pi,$
$t \neq \frac{\pi}{2}, \frac{3\pi}{2}$

45. $\frac{(x-3)^2}{49} + \frac{(y+4)^2}{4} = 1$; ellipse with center $(3, -4)$;
$x = 3 + 7 \cos t, y = -4 + 2 \sin t, 0 \leq t \leq 2\pi$

49. 1,786 meters

51. (A) 43.292 seconds
(B) 9,183.620 meters, 9.184 kilometers
(C) 2,295.918 meters

APPLICATIONS INDEX